Book of Abstracts of the 74th A

he

European Federation of Animal Sc

The European Federation of Animal Science wishes to express its appreciation to the
Ministero delle Politiche Agricole Alimentari e Forestali (Italy) and the
Associazione Italiana Allevatori (Italy)
for their valuable support of its activities.

Book of Abstracts of the 74th Annual Meeting of the European Federation of Animal Science

Lyon, France, 26 August – 1 September, 2023

EAAP Scientific Committee:

EAN: 9789086863846
e-EAN: 9789086869367
ISBN: 978-90-8686-384-6
e-ISBN: 978-90-8686-936-7
DOI: 10.3920/978-90-8686-936-7

ISSN 1382-6077

First published, 2023

Welcome to the EAAP + WAAP + INTERBULL 2023 in Lyon

On behalf of the French Organising Committee, we are very delighted to welcome you to attend the Joint International Congress on Animal Science co-organised by the European Federation of Animal Science (EAAP), the World Association for Animal production (WAAP) and Interbull to be held in Lyon, the capital of the Auvergne-Rhône-Alpes French region, from August 26th to September 1st, 2023.

The general topic of this Congress is 'Climate change, biodiversity and global sustainability of animal production'. Different sessions are jointly organised between the EAAP, WAAP, Interbull and also other partners to cover various areas of knowledge related to animal science, such as genetics, nutrition, physiology, animal health and welfare.

The Auvergne-Rhône-Alpes region covers an area of 69,711 km^2 in the central-east part of the south of France. It is characterised by diverse natural resources and topographies (including high mountains to the east and volcanos to the west), climates, cultures, folklore, architecture, and languages. The region accounts for 11.9% of the French economic output. It hosts 48,500 farms (with an average of 59 ha), including 51.9% of them for the production of animal products and 10.4% with both plant and animal production. About 20% and 13% of the farms concern the production of beef and milk respectively, and 11% are with sheep, goats or other herbivores. This places the Auvergne-Rhône-Alpes region 3rd in place amongst the other French regions in agriculture. The region had 13% of organic farms, 30% with other labels or quality signs and 30% with short supply chains.

Apart from the outstanding scientific programme, social activities is organised in Lyon, which was the former capital of the Gauls at the time of the Roman empire, a major economic hub during the Renaissance and today the third-largest city and second-largest metropolitan area of France. The city is known as 'the gastronomic capital of the world' and is recognised for its historical and architectural landmarks and for its universities.

The city is also known for its light festival and is a major centre for chemical, pharmaceutical and biotech industries. Several exceptional sites of the region are inscribed on the UNESCO World Heritage List including the historical district of Lyon.

We welcome you and wish you an exceptional stay in Lyon!

Jean-François Hocquette
Senior scientist of INRAE
President of the French Association for Animal Production

The Joint International Congress on Animal Science (EAAP annual congress, WAAP congress and Interbull meeting) is locally co-organised by INRAE (The French National Research Institute for Agriculture, Food, and Environment) and the French Association for Animal Production (AFZ), on behalf of the European Federation of Animal Science (EAAP) and the World Association for Animal production (WAAP) and by "France Génétique Elevage" on behalf of Interbull.

National Organisers of the 74th EAAP Annual Meeting

Lyon, France

26 August – 1 September, 2023

French Steering Committees
Jean-François Hocquette (INRAE, AFZ)

Local organising Committee
Jean-François Hocquette (INRAE, AFZ)
Adeline Dubost (INRAE)
Sabrina Gasser (INRAE)
Valérie Heuzé (AFZ)
Jérôme Normand (IDELE)
Emmanuelle Caramelle-Holtz (IDELE)
Charlotte Chêne (VetAgro Sup)
Karima Latti (ISARA)
Denis Bastianelli (CIRAD)
Philippe Chemineau (WAAP)

French Scientific Committees

Global committee
Patrick Chapoutot
Jean-Louis Peyraud
Rene Baumont
Sabrina Gasser
Philippe Chemineau
Emmanuelle Gilot-Fromont
Fabienne Blanc
Karine Chalvet-Monfray
Marie-Pierre Ellies
Latifa Najar
Denis Bastianelli

Precision Livestock Farming:
Nathalie Hostiou
Amélie Fischer

Genetics:
Laurent Journaux
Jean-Pierre Bidanel

Nutrition:
Latifa Najar
Jaap Van Milgen

Health and Welfare:
Christian Ducrot
Alice de Boyer des Roches

Animal Physiology :
Isabelle Louveau
Xavier Druart

Livestock Farming Systems:
Vincent Thenard
Marie-Odile Nozieres-Petit

Pig Production:
Ludovic Brossard
Christine Roguet

Horse Production:
Marion Cressent
Léa Lansade

Insects:
Thomas Lefevre
Anna Zaidman-Rémy

Cattle Production:
René Baumont
Christophe Denoyelle

Sheep and Goat Production:
Jérémie Jost

Conference website: www.eaap2023.org

European Federation of Animal Science (EAAP)

President:	Isabel Casasús
Secretary General:	Andrea Rosati
Address:	Via G. Tomassetti 3, A/I I-00161 Rome, Italy
Phone/Fax:	+39 06 4420 2639
E-mail:	eaap@eaap.org
Web:	www.eaap.org

Council Members

President	Isabel Casasús (Spain)
Vice-Presidents	Peter Sanftleben (Germany) Hans Spoolder (the Netherlands)
Members	Ilan Halachmi (Israel) Gunnfríður Elín Hreiðarsdóttir (Iceland) Stéphane Ingrand (France) Martin Heinrich Lidauer (Finland) Denis Kučevič (Serbia) Nicolò Maciotta (Italy) Olga Moreira (Spain) Klemen Potocnik (Slovenia)
Auditors	Zdravko Barac (Croatia) Georgia Hadjipavlou (Cyprus)
Alternate Auditor	Jeanne Bormann (Luxembourg)
FAO Representative	Badi Besbes

The Organization

EAAP (The European Federation of Animal Science) organises every year an international meeting which attracts between 900 and 2,000 people. The main aims of EAAP are to promote, by means of active co-operation between its members and other relevant international and national organisations, the advancement of scientific research, sustainable development and systems of production; experimentation, application and extension; to improve the technical and economic conditions of the livestock sector; to promote the welfare of farm animals and the conservation of the rural environment; to control and optimise the use of natural resources in general and animal genetic resources in particular; to encourage the involvement of young scientists and technicians. More information on the organisation and its activities can be found at www.eaap.org

YoungEAAP

What is the YoungEAAP?

YoungEAAP is a group of young scientists organized under the EAAP umbrella. It aims to create a platform where scientists during their early career get the opportunity to meet and share their experiences, expectations and aspirations. This is done through activities at the Annual EAAP Meetings and social media. The large constituency and diversity of the EAAP member countries, commissions and delegates create a very important platform to stay up-to-date, close the gap between our training and the future employer expectations, while fine-tuning our skills and providing young scientists applied and industry-relevant research ideas.

Committee Members at a glance

- Ines Adriaens (president)
- Jana Obsteter (vice president)
- Giulia Gislon (secretary)

YoungEAAP promotes Young and Early Career Scientists to:

- Stay up-to-date (i.e. EAAP activities, social media);
- Close the gap between our training and the future employer expectations;
- Fine-tune our skills through EAAP meetings, expand the special young scientists' sessions, and/or start online webinars/trainings with industry and academic leaders;
- Meet to network and share our graduate school or early employment experiences;
- Develop research ideas, projects and proposals.

Who can be a Member of YoungEAAP?

All individual members of EAAP can join the YoungEAAP if they meet one of the following criteria: Researchers under 35 years of age OR within 10 years after PhD-graduation

Just request your membership form (ines.adriaens@kuleuven.be) and become member of this network!

Industry Members Club

EAAP started in 2023 a new initiative to create closer connections between European livestock industries and the animal science network. Therefore, the "EAAP Industry Club" was shaped with the specific aim of bringing together the important industries of the livestock sector with our European Federation of Animal Sciences. All companies dealing with animal production (nutrition, genetic, applied technologies, etc.) are invited to join the "EAAP Industry Club" because industries will have opportunity to increase their visibility, to be actively involved in European animal science activities, and to receive news and services necessary to industries. In addition, through the Club, industries will enlarge their scientific network and will receive specific discounts on sponsoring activities.

The Industries that already joined the "EAAP Industry Club" are:

The Club gives:

Visibility • Company name and logo at EAAP website and all relevant documents • Slides with name and logo at Official Events • Priority links with EAAP Socials • Invite, through EAAP dissemination tools and socials, people to events organized by your company

• Information disseminated through a brand new Industry Newsletter

Networking • Joining the Study Commissions and Working Groups • Suggest topics to be considered for Annual Meetings Scientific Sessions • Organize Professional Panel through the EAAP platforms

Economic Benefits • One free registration to each Annual Meeting and at every meeting organized by EAAP • Five individual memberships at no cost • Many possible discounts (-30%) to increase company visibility through: EAAP Newsletter, EAAP website, EAAP Annual Meetings and workshops

• Support young scientist by sponsoring scholarships named by the company • Co-Organize and sponsor webinars

Make yourself more visible within the livestock industry via the animal science network!

For more information please contact eaap@eaap.org

75th Annual Meeting of the European Federation of Animal Science

Florence, Italy, September 1st-5th, 2024

Organizing Committees

The 75th EAAP annual meeting is organized by the Italian Association of Animal Science and Production (ASPA).

Italian Steering Committee

Presidents
Nicolò Pietro Paolo Macciotta (ASPA President – University of Sassari), Marcello Mele (University of Pisa)

Committee
Antonella Baldi (University of Milano)
Giovanni Bittante (University of Padova)
Riccardo Bozzi (University of Florence)
Arianna Buccioni (University of Florence)
Giuseppe Campanile (University of Napoli)
Salvatore Claps (Research Centre for Animal Production and Aquaculture- CREA ZA)
Pasquale De Palo (University of Bari)
Andrea Formigoni (University of Bologna)
Riccardo Negrini (Italian Association of Animal Breeders AIA – University of Piacenza)
Giuliana Parisi (University of Florence)
Fabio Pilla (University of Campobasso)
Baldassare Portolano (University of Palermo)
Carolina Pugliese (University of Florence)
Giuseppe Pulina (University of Sassari)
Bruno Ronchi (University of Viterbo)
Agostino Sevi (University of Foggia)
Bruno Stefanon (University of Udine)
Francesco Tiezzi Mazzoni Della Stella Maestri (University of Florence)
Paolo Trevisi (University of Bologna)

Conference website: www.eaap2024.org

Commission on Animal Genetics

Filippo Miglior	President Canada	Guelph University fmiglior@uoguelph.ca
Marcin Pszczola	Vice-President Poland	Poznan University of Life Sciences marcin.pszczola@gmail.com
Morten Kargo	Vice-President Denmark	Aarhus University morten.kargo@qgg.au.dk
Ewa Sell-Kubiak	Vice-President Poland	Poznan University of Life Sciences ewa.sell-kubiak@puls.edu.pl
Francesco Tiezzi	Secretary Italy	University of Florence francesco.tiezzi2@unifi.it
Christa Egger-Danner	Industry rep. Austria	Zuchtdata egger-danner@zuchtdata.at
Guilherme Neumann	Young Club Germany	Humboldt University guilherme.neumann@hu-berlin.de
Ivan Pocrnic	Young Club UK	Roslin Institute ivan.pocrnic@roslin.ed.ac.uk

Commission on Animal Nutrition

Luciano Pinotti	President Italy	University of Milan luciano.pinotti@unimi.it
Sam De Campeneere	Vice-President Belgium	ILVO sam.decampeneere@ilvo.vlaanderen.be
Latifa Abdenneby -Najar	Vice-President France	IDELE latifa.najar@idele.fr
Maria José Ranilla García	Secretary Spain	Universidad de León mjrang@unileon.es
Sokratis Stergiadis	Secretary UK	University of Reading s.stergiadis@reading.ac.uk
Susanne Kreuzer-Redmer Vetmed Vienna	Austria	Secretary susanne.kreuzer-redmer@vetmeduni.ac.at
Javier Alvarez Rodriguez	Secretary Spain	University of Lleida javier.alvarez@udl.cat
Daniele Bonvicini	Industry Rep Italy	Prosol S.p.a d.bonvicini@prosol-spa.it
Geert Bruggeman	Industry Rep Belgium	Nusciencegroup geert.bruggeman@nusciencegroup.com
Eric Newton	Young Club UK	University of Reading Newtoeri000@gmail.com

Commission on Health and Welfare

Laura Boyle	President Ireland	Teagasc laura.boyle@teagasc.ie
Flaviana Gottardo	Vice-President Italy	University of Padova flaviana.gottardo@unipd.it
Giulietta Minozzi	Vice-President Italy	University of Milan giulietta.minozzi@unimi.it
Isabel Blanco Penedo	Secretary Spain	University of Lleida isabel.blancopenedo@udl.cat
Holinger Mirjam	Secretary Switzerland	Research Institute of Organic Agriculture FiBL mirjam.holinger@fibl.org
Angela Trocino	Secretary Italy	University of Padova angela.trocino@unipd.it
Julia Adriana Calderon Diaz	Industry Rep Spain	Genusplc juliaadriana.calderondiaz@genusplc.com
Mariana Dantas de Brito Almeida	Young Club Portugal	University of Tras-os-Montes and Alto Douro mdantas@utad.pt

Commission on Animal Physiology

Name	Position / Country	Institution / Email
David Kenny	President Ireland	Teagasc David.Kenny@teagasc.ie
Kate Keogh	Vice-President Ireland	Teagasc kate.keogh@teagasc.ie
Alan Kelly	Secretary Ireland	University College Dublin alan.kelly@ucd.ie
Yuri Montanholi	Secretary USA	North Dakota State University yuri.montanholi@ndsu.edu
Maya Zachut	Secretary Israel	Volcani Institute mayak@volcani.agri.gov.il
Federico Randi	Industry rep. France	Ceva Sante Animale federico.randi@ceva.com
Olaia Urrutia	Young Club Spain	Public University of Navarre olaia.urrutia@unavarra.es

Commission on Livestock Farming Systems

Name	Position / Country	Institution / Email
Michael Lee	President United Kingdom	Harper Adams University MRFLee@harper-adams.ac.uk
Enrico Sturaro	Vice President Italy	University of Padova enrico.sturaro@unipd.it
Tommy Boland	Secretary Ireland	University College Dublin tommy.boland@ucd.ie
Ioanna Poulopoulou	Secretary Italy	Free University of Bozen Ioanna.Poulopoulou@unibz.it
Vincent Thenard	Secretary France	INRAE vincent.thenard@inrae.fr
Saheed Salami	Industry rep. UK	Alltech saheed.salami@alltech.com
Tiago T. da Silva Siqueira	Young Club France	INRA tiago.teixeira.dasilva.siqueira@gmail.com

Commission on Cattle Production

Name	Position / Country	Institution / Email
Massimo De Marchi	President Italy	Padova University massimo.demarchi@unipd.it
Paul Galama	Vice-President Netherlands	Wageningen Livestock Research paul.galama@wur.nl
Joel Berard	Vice-President Switzerland	Agroscope joel.berard@agroscope.admin.ch
Jean François Hocquette	Secretary France	INRAE jean-francois.hocquette@inrae.fr
Poulad Pourazad	Industry rep Austria	Delacon Biotechnik GmbH poulad.pourazad@delacon.com
Angela Costa	Young Club Italy	University of Bologna angela.costa2@unibo.it

Commission on Sheep and Goat Production

Vasco Augusto Pilão Cadavez	President Portugal	CIMO - Mountain Research Centre vcadavez@ipb.pt
Lorenzo E. Hernandez Castellano	Vice-President Spain	Universidad de Las Palmas de Gran Canaria lorenzo.hernandez@ulpgc.es
Antonello Cannas	Secretary Italy	University of Sassari cannas@uniss.it
Georgia Hadjipavlou	Secretary Cyprus	Agricultural Research Institute georgiah@ari.gov.cy
Neil Keane	Industry rep. Ireland	Alltech nkeane@alltech.com
Christos Dadousis	Young Club Italy	University of Parma christos.dadousis@unipr.it

Commission on Pig Production

Sam Millet	President Belgium	ILVO sam.millet@ilvo.vlaanderen.be
Paolo Trevisi	Vice president Italy	Bologna University paolo.trevisi@unibo.it
Giuseppe Bee	Vice president Switzerland	Agroscope Liebefeld-Posieux ALP giuseppe.bee@agroscope.admin.ch
Katja Nilsson	Secretary Sweden	Swedish University of Agricultural Science katja.nilsson@slu.se
Katarzyna Stadnicka	Secretary Poland	Collegium Medicum Nicolaus Copernicus University katarzyna.stadnicka@cm.umk.pl
Grzegorz Brodziak	Industry rep. Poland	Goodvalley Agro S.A. grzegorz.brodziak@goodvalley.com
Tristan Chalvon-demersay	Industry rep. France	Metex-noovistago tristan.chalvon-demersay@metex-noovistago.com
Stafford Vigors	Young Club Ireland	University College Dublin staffordvigors1@ucd.ie

Commission on Horse Production

Rhys Evans	President Norway	Norwegian University College of Green Development rhys@hgut.no
Klemen Potočnik	Vice president Slovenija	University of Ljubljana klemen.potocnik@bf.uni-lj.si
Roberto Mantovani	Vice president Italy	University of Padua- DAFNAE roberto.mantovani@unipd.it
Isabel Cervantes Navarro	Vice president Spain	Complutense University of Madrid icervantes@vet.ucm.es
Pasquale De Palo	Secretary Italy	University of Bari pasquale.depalo@uniba.it
Jackie Tapprest	Secretary France	Animal health laboratory (ANSES) jackie.tapprest@anses.fr
Melissa Cox	Industry rep. Germany	Generatio GmbH – Center for Animal Genetics melissa.cox@centerforanimalgenetics.de
Juliette Auclair-Ranzaud	Young Club France	Institut français du cheval et de l'équitation juliette.auclair-ronzaud@ifce.fr
Kirsty Tan	Young Club Germany	Christian-Albrechts-Universität zu Kiel kirsty.tan89@gmail.com

Commission on Insects

Laura Gasco	President Italy	University of Turin laura.gasco@unito.it
Anton Gligorescu	Vice president Denmark	Aarhus University angl@bio.au.dk
Christoph Sandrock	Secretary Switzerland	Research Institute of Organic Agriculture FiBL christoph.sandrock@fibl.org
David Deruytter	Secretary Belgium	INAGRO david.deruytter@inagro.be
Maria Martinez Castillero	Industry rep. UK	Betabugs maria@betabugs.uk
Daniel Murta	Industry rep. Portugal	Ingredient Odyssey – EntoGreen daniel.murta@entogreen.com
Thomas Lefebvre	Industry rep. France	Ynsect thomas.lefebvre@ynsect.com
Matteo Ottoboni	Young Club Italy	Univerity of Milan matteo.ottoboni@unimi.it
Cassandra Maya	Young Club Denmark	Copenhagen University casma@nexs.ku.dk
Ilaria Biasato	Young Club Italy	University of Turin ilaria.biasato@unito.it

Commission on Precision Livestock Farming

Matti Pastell	President Finland	Natural Resources Institute Finland (Luke) matti.pastell@luke.fi
Jarissa Maselyne	Vice president Belgium	ILVO jarissa.maselyne@ilvo.vlaanderen.be
Francisco Maroto Molina	Vice president Spain	University of Cordoba g02mamof@uco.es
Claire Morgan-Davies	Vice president United Kingdom	Scotland's Rural College (SRUC) claire.morgan-davies@sruc.ac.uk
Ines Adriaens	Vice president Belgium	KU Leuven ines.adriaens@kuleuven.be
Jean-Marc Gautier	Vice president France	IDELE jean-marc.gautier@idele.fr
Shelly Druyan	Secretary Israel	ARO, The Volcani Center shelly.druyan@mail.huji.ac.il
Radovan Kasarda	Secretary Slovakia	Slovak University of Agriculture in Nitra radovan.kasarda@uniag.sk
Michael Odintsov	Secretary Italy	Regrowth s.r.l.s m.odintsov.vaintrub@gmail.com
Hiemke Knijn	Industry rep. The Netherlands	CRV hiemke.knijn@crv4all.com
Daniel Foy	Industry rep USA	AgriGates d.foy@agrigates.io
Victor Bloch	Young Club Finland	Luke victor.bloch@luke.fi

Founded on 9 October 1979 at the initiative of representative organisations in the cattle sector, INTERBEV was recognised on november 18th, 1980 as an interbranch organisation for meat and livestock by ministerial decree.

Its role today is to defend and promote the common interests of both breeding and small-scale, industrial and commercial activities within the meat sector, represented by 22 National Organisations of all industry sections. INTERBEV does this by way of five species specific sections and this ensures the development and promotion of each sector: bovines, calves, sheeps, horses and goats. INTERBEV is represented in the various regions of France by its 12 Regional Committees, which are responsible for putting into place the inter-professional strategies as well as relaying and adapting any national communications on a local level.

The inter-professional organisation's strategic base set up by INTERBEV in 2017, is a collective approach to social responsibility, integrating the industry's plans set out at the French National Food Conference.
This strategy, called PACTE, is based on 5 core values: Progress , Future, Dialogue, Transparency and Expertise .

Its objective is to provide a global response to the expectations of society and those involved in the industry in terms of good production and consumption practices, with evidence and guarantees in support of this. It aims to highlight the strengths of the sector, to identify areas for improvement by working particularly with environmental and animal welfare NGOs and to use collective tools, in order to provide sustainable food.

This approach is under the guidance of the internationally recognised ISO 26000 standard, entitled Social Responsibility. In 2018 INTERBEV was the first food industry inter-professional organisation to be awarded a level 3 out of 4 AFNOR certificate "CSR commitment confirmed", later confirmed in 2021.

ThermoFisher
SCIENTIFIC
Total support from the first step
to the next advancement
Count on Thermo Fisher Scientific as a partner that will
provide the right genotyping solutions at the right time
Learn more at thermofisher.com/agrigenomics
applied biosystems
For Research Use Only. Not for use in diagnostic procedures. © 2023 Thermo Fisher Scientific Inc. All rights reserved.
All trademarks are the property of Thermo Fisher Scientific and its subsidiaries unless otherwise specified. COL024278 0623

ADM
Building World-Class Nutrition With You
A globla leader in animal nutrition, ADM partners with you to provide high-quality, sustainable nutrition that meets all of your needs. With our experienced species specialists, a continuously growing product portfolio and access to global learnings, we work with you to advance nutritional performance and develop solutions to drive success today and tomorrow.
Your Edge. Our Expertise.
800-775-3295 I animalnutrition@adm.com I adm.com/animalnutrition

Acknowledgements

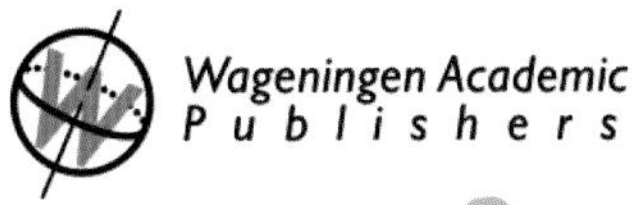

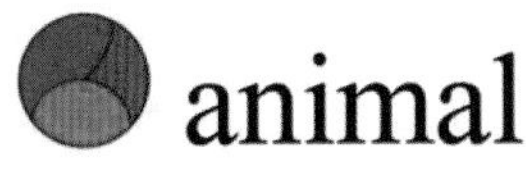

HoloRuminant
Understanding microbiomes of the ruminant holobiont

mEAT
quality

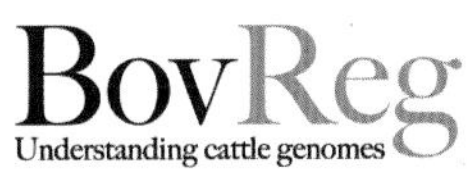

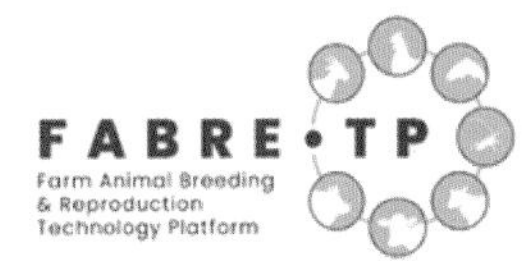

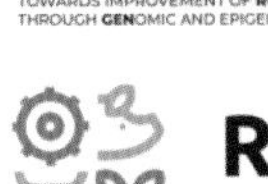

Thank you

to the 74th EAAP Annual Conference Sponsors and Friends

Diamond sponsors

illumina®

Platinum sponsor

FranceAgriMer

ÉTABLISSEMENT NATIONAL
DES PRODUITS DE L'AGRICULTURE ET DE LA MER

Gold sponsors

AMERICAN SOCIETY OF
ANIMAL SCIENCE

Silver sponsors

Bronze sponsors

Other sponsors

Scientific programme EAAP 2023

Sunday 27 August 2023 9.00 – 18.00	Monday 28 August 2023 8.30 – 13.00
Session 01 WAAP Plenary Session Chair: Chemineau 9:00 – 12:30 **Control of green house gas emissions in ruminants farming** Chair: Peyraud / Mottet 14:00 – 18:00 **Biodiversity as a lever for sustainable animal production** Chair: Rieu / Gauly	**Session 02** Sustainable livestock farming – defining metrics and rationalising trade-offs? – Part 1 Chair: Lee / O'Mara **Session 03** Up side and down side of 'genomic selection' Chair: Miglior / Jornaux **Session 04** Adaptation of breeding strategies and genetics to climate change, right animal for right environment Chair: Kargo / Sturaro **Session 05** Climate care dairy farming: follow up – Part 1 Chair: Galama / Rees **Session 06** Production of high quality animal products and implementation of new technologies in animal husbandry Chair: Georgieva / Hocquette **Session 07** Challenges of fish nutrition for sustainable aquaculture Chair: Medale / Skiba **Session 08** Various topics in pig production Chair: Trevisi / Vigors **Session 09** Insect applications Chair: Sanchez / Gligorescu **Session 10** Impact of epigenetics and genetics in determining animal physiology Chair: Jammes **Session 11** Improving pig welfare in conventional production systems Chair: Gottardo / Blanco Penedo **Session 12** Technologies for outdoor systems: which application and opportunities? Chair: Maroto-Molina / Gautier **Session 13** Innovations and technological advancements in small ruminant production with particular emphasis in Mediterranean semi-extensive or extensive systems Chair: Hadjipavlou / Lauvie

Monday 28 August 2023 14.00 – 18.00	Tuesday 29 August 2023 8.30 – 12.30
Session 14 Sustainable livestock farming – defining metrics and rationalising trade-offs? – Part 2 Chair: Lee / O'Mara **Session 15** Breeding for climate change – mitigation Chair: Pszczola / Neumann **Session 16** Alternative and new feed resources to reduce feed/food competition and enhance circularity of livestock production Chair: Ranilla García / Stergiadis **Session 17** Climate care dairy farming: follow up – Part 2 Chair: Edouard / Becciolini **Session 18** Carcase grading for quality and efficiency – underpinning future sustainability of beef and lamb Chair: Pethick / Neveu **Session 19** Advances in nutritional modelling and feeding systems (in memory of Daniel Sauvant) Chair: Baumont / Cannas **Session 20** Early career competition 'Innovative approaches to pig and poultry production', supported by Wageningen Academic Publishers Chair: Vigors / Nilsson **Session 21** Insects as feed: sustainability, legislation and safety Chair: Veldkamp / Barbosa **Session 22** Environmental stress: mitigating the adverse effects on animal physiology Chair: Zachut **Session 23** Improving animal welfare in conventional production systems Chair: Minozzi / Trocino **Session 24** New traits and proxies from sensor technologies for breeding purposes Chair: Pocrnic / Lourenco **Session 25** Molecular measures of diversity and their role in monitoring and management of breeds Chair: Danchin / Fernandez	**Session 26** EAAP Plenary Session, Leroy and WAAP Awards Lectures Chair: Casasús

Tuesday 29 August 2023

14.00 – 15.00
Poster session

15.00 – 18.30

Session 27
Breeding programs & strategies
Chair: Egger-Danner / Martin

Session 28
Breeding for improved animal health and welfare
Chair: Kargo / Tiezzi

Session 29
Innovative animal products and farming systems oriented to new consumer demand and trend
Chair: Resconi / Realini

Session 30
Looking back, looking forward – Research on Beef production and quality in China and France (CSBF) and between Australia-France-China (The Triangle Project)
Chair: Liu / Pethick

Session 31
Pasture based solutions to realise a role for ruminants in a sustainable food system
Chair: Boland / Baumont

Session 32
Beyond rumen: role of nutrition in cattle's intestinal health
Chair: Kreuzer / Pinotti

Session 33
Horses in farms and territories: challenges and new solutions
Chair: Vial / Tapprest

Session 34
Sustainable pig and poultry production, including the use of water
Chair: Bee / Brodziak

Session 35
Is there a future for experimental animal research in Europe and, if so, what is it?
Chair: Chalvon-Demersay / Van Milgen

Session 36
Technologies in insect production + controlling quality
Chair: Heckmann / Deruytter

Session 37
Omics and integrative analyses towards understanding inter-organ cross-talk and whole body physiology of livestock
Chair: Louveau

Session 38
Equal opportunities and open communication for early career scientists
Chair: Pszczoła / Sell-Kubiak / Newton / Obsteter / Gislon

Session 39
TechCare and ClearFarm: pilots on PLF tools for monitoring animal welfare
Chair: Morgan-Davies / Llonch

Session 40
The transition from pregnancy to lactation in dairy goats and sheep
Chair: Hernández-Castellano / Salama

Wednesday 30 August 2023
8.30 – 13.00

Session 41
Open the black box: using omics to better understand biological interpretation of complex traits
Chair: Tiezzi / Pocrnic

Session 42
Animal genetics to address food security and sustainability
Chair: Soelkner / Granados

Session 43
Better calves for better farms: how the young stock can be the key to success
Chair: Costa / Keatinge

Session 44
Building on a resilient dairy sector – from animal, farm and regional perspective – Part 1
Chair: Kuipers / Schuenemann

Session 45
Transformation of livestock practices in response to society's expectations on animal welfare, market demand for animal products and environmental issues
Chair: Thenard / Boyle

Session 46
Innovation in ruminant nutrition and feeding
Chair: Stergiadis / Alvarez Rodriguez

Session 47
One-health, one nutrition, one Earth: role of nutrition in livestock production
Chair: Latifa / Weill

Session 48
How can poultry farming systems evolve to meet the major societal and environmental challenges?
Chair: Stadnicka / Meda

Session 49
Monoguthealth – Part 1
Chair: Bee / Trevisi

Session 50
Insect genetics and reproduction
Chair: Sandrock / Maya

Session 51
Leveraging the microbiome for resilience and sustainability in ruminant production – insights from H2020 HoloRuminant
Chair: Kenny / Morgavi

Session 52
Nitrogen management in the dairy supply chain and other tools to improve the efficiency and sustainability of cattle farming
Chair: Righi / Tsiplakou

Session 53
PLF for health and welfare – Part 1
Chair: Maselyne / Calderon Diaz

Session 54
Life Green Sheep and Farm management to adapt goat and sheep farming to climate change: solutions and experiments
Chair: Throude / Satzori

Wednesday 30 August 2023	Thursday 31 August 2023 8.30 – 12.00
14.00 – 15.00 Poster session	**Session 69** Methods & genomic prediction Chair: Sell-Kubiak / El-Ouazizi
15.00 – 18.30	**Session 70** Genetic parameters & GWAS Chair: Egger-Danner / Martin
Session 55 Breeding for climate change – adaptation Chair: Sell-Kubiak / Pszczola	**Session 71** Building on a resilient dairy sector – from animal, farm and regional perspective – Part 2 Chair: Brocard / Klopcic
Session 56 Establishing breeding programs in extensive systems (including developing regions) with climate change in mind Chair: Leroy / Berg	**Session 72** What are the stakeholder and societal expectations of intrinsic and extrinsic quality of animal products? Chair: Martin
Session 57 Grassland management and grassland-based feeding systems for efficient and sustainable milk and meat production Chair: Berard / Probo	**Session 73** Molecules in animal nutrition: over and above building blocks Chair: Van Milgen / Geraert / Mercier
Session 58 The future of animal products: improved quality management, more alternatives or cell-based products? Chair: Ellies-Oury / Chriki	**Session 74** Balancing the feed for the economy, the environment, and society Chair: De Campeneere / Bruggeman
Session 59 Environmental challenges of cattle tropical grazing systems Chair: Dossa / Salgado	**Session 75** Livestock based circularity for health and sustainable land use Chair: Siqueira / De Olde
Session 60 Minerals in animal nutrition: supplementation – Part 1 Chair: Alvarez Rodriguez / Windisch	**Session 76** Humans and horses: from One Health to One Life perspectives Chair: Evans / Montovani
Session 61 Resilient livestock farming systems in the face of climate and other global challenges – adaptation and mitigation Chair: Sturaro / Poulopoulou	**Session 77** Trade-offs between health, production and welfare in pigs and poultry: which tools do we have and which research is needed? Chair: Nilsson / Millet
Session 62 Poultry and pig low-input and organic production systems' welfare Chair: Daş / Holinger	**Session 78** Insect nutrition (substrates) & insect health Chair: Biasato / Lefebvre
Session 63 Monoguthealth – Part 2 Chair: Bee / Trevisi	**Session 79** Improved insight into the reproductive physiology of livestock Chair: Kenny
Session 64 Project session: Farmyng Chair: Kihanguila / Gasco	**Session 80** Animal behaviour Chair: Dantas / Boyle
Session 65 Remote sensing applied to livestock farming: discovering physiological and ethological clues to optimize productive performance and animal welfare Chair: Montanholi	**Session 81** Precision feeding Chair: Druyan / Bloch
Session 66 Innovation in non-ruminant nutrition and feeding Chair: Alvarez Rodriguez / Stergiadis	**Session 82** Product quality and efficiency of small ruminant production systems Chair: Hernández-Castellano / Tsiplakou
Session 67 PLF for health and welfare – Part 2 Chair: Pastell / Stygar	
Session 68 Small Ruminant Technologies (Sm@RT) and Thematic network EuroSheep Chair: Morgan-Davies / Grisot	12.00 – 13.00 **Commission meetings** Cattle / Genetics / Health and Welfare / Horse / Insects / LFS / Nutrition / Physiology / Pig / PLF / Sheep and Goats

Thursday 31 August 2023
14.00 – 18.00

Session 83
Genetic diversity
Chair: Neumann / Schiavo

Session 84
EuroFAANG: genotype-to-phenotype research across Europe and beyond
Chair: Clark / Kuhn

Session 85
Climate care dairy farming: herd management – Part 3
Chair: Cieślak / Kuipers

Session 86
How to address stakeholder and societal expectations of intrinsic and extrinsic quality of animal products?

Session 87
Carbon sources and sinks within global livestock systems
Chair: Salami / Poulopoulou

Session 88
Minerals in animal nutrition: basal diet – Part 2
Chair: Alvarez Rodriguez / Windisch

Session 89
Explainable models and artificial intelligence supporting farming decisions
Chair: Foy / Odintsov

Session 90
Genetics in horses
Chair: Stock / Cox

Session 91
Nitrogen excretion and ammonia emission in pig and poultry production
Chair: Millet / Brodziak

Session 92
Inflammation and energy metabolism in young and adult livestock
Chair: Sauerwein

Session 93
Recent advances on in vitro systems (from cell culture to in vitro digestion methods): relevance as alternatives to animal experimentation for livestock research
Chair: Louveau

Session 94
Free communications in animal health and welfare
Chair: Boyle / Calderon Diaz

Session 95
Happy Delivering innovative and holistic milk based monitoring and decision-making PLF tools – collaboration with HappyMoo
Chair: Gengler / Knijn

Session 96
ERANET-funded 'Grass To Gas' and EU-funded 'SMARTER' projects
Chair: Conington / Vouraki

Scientific programme

Session 01. WAAP Plenary Sessions

Date: Sunday 27 August 2023; 9.00 – 18.00
Chair: Chemineau / Peyraud / Mottet / Rieu / Gauly

Theatre Session 01

vited The real place of livestock in GHG emissions and their control for climate change 149
K. Johnson

vited Global emissions from livestock systems: updated baselines, projections, and mitigation options 149
D. Wisser, G. Tempio, A. Falcucci and G. Cinardi

vited Genetic control of ruminant methane emissions in livestock 150
A.E. Van Breukelen, M.N. Aldridge, R.F. Veerkamp and Y. De Haas

vited Effective strategies to mitigate enteric methane emissions by ruminants 150
C. Arndt

vited Tannins, legumes and methane production by ruminants 151
H.H. Lardner

vited Enteric methane reduction in ruminants by early-life interventions 151
D.P. Morgavi

vited Sustainable intensification of animal production with silvopastoral systems 152
J. Chára

vited Sustainability of low-input livestock systems 152
A. Mottet

vited Agroecological transformation of tropical livestock production through integrating improved forages 153
A.M.O. Notenbaert

vited Livestock-wildlife conflicts: a struggle for land use and conservation in the U.S. Mountain West 153
J.B. Taylor and H.N. Wilmer

vited Community-based breeding programs (CBBP): platforms for genetic improvement and biological discovery 154
J.M. Mwacharo, A. Haile and B. Rischkowsky

Session 02. Sustainable livestock farming – defining metrics and rationalising trade-offs? – Part 1

Date: Monday 28 August 2023; 8.30 – 13.00
Chair: Lee / O'Mara

Theatre Session 02

vited 'Sustainable livestock systems': what does this mean? – welcome and introduction 154
F. O'Mara and M. Lee

vited Friend or Foe? The role of animal-source foods in healthy and environmentally sustainable diets 155
L. Merbold, T. Beal, C. Gardner, M. Herrero, L. Iannotti, S. Nordhagen and A. Mottet

invited The economic sustainability of dairy production systems in the EU and beyond 155
F. Thorne, E. Dillon, T. Donnellan, P. Jeanneaux, R. Jongeneel and L. Latruffe

invited Methods to assess the sustainability of livestock systems: challenges and opportunities 156
E. De Olde

invited SustAinimal – a multi-actor knowledge centre for livestock in future Swedish food systems 156
M. Hetta, S. Agenäs, D.J. De Koning, H. Oscarsson, P. Peetz Nielsen and A. Wallenbeck

invited Breeding livestock for sustainable systems 157
P.W. Knap, K.M. Olson and M.A. Cleveland

invited Panel discussion (morning) 157
A. Granados and A.S. Santos

Use of grazing with dairy goats to design sustainable food systems 158
H. Caillat, E. Bruneteau and B. Ranger

Benefits and limits of farm animals to control herbage mass, pests and weeds in orchards: a review 158
G. Maillet, A. Dufils, F. Angevin, S. Ramonteu, J.L. Peyraud and R. Baumont

A new concept for agro-ecological efficiency at different scales of ruminant production systems 159
H. Nguyen-Ba, P. Veysset and A. Ferlay

Session 03. Up side and down side of 'genomic selection'

Date: Monday 28 August 2023; 8.30 – 13.00
Chair: Miglior / Jornaux

Theatre Session 03

invited Issues and perspectives in genomic selection in limited size breeds: the case of Italian Simmental 159
N.P.P. Macciotta, L. Degano, D. Vicario and A. Cesarani

Impact of genomic selection on genetic diversity in five European local cattle breeds 160
J.J. Winding, T.H.E. Meuwissen, P. Croiseau, G. Restoux and R. Bonifazi

Accounting for sequential genomic selection in broiler breeding 160
J. Hidalgo, D. Lourenco, S. Tsuruta, V. Breen, W. Herring and I. Misztal

Changes in the genome due to genomic selection in two pig populations 161
Y.C.J. Wientjes, M.P.L. Calus, P. Bijma, A.E. Huisman and K. Peeters

invited Practical approaches to managing increased homozygosity caused by genomic selection 161
C. Baes, C. Obari, B. Makanjuola, C. Rochus, C. Maltecca, F. Schenkel and F. Miglior

invited Positive and negative effects of genomic selection 162
I. Misztal

Storing and analysing a million genomes on a desktop computer 162
G. Gorjanc, J. Obsteter, G. Mafra Fortuna, R. Ros-Freixedes, M. Johnsson and I. Pocrnic

Large scale screening for genetic defects in Holstein cattle using transmission disequilibrium test 163
F. Besnard, M. Boussaha, H. Leclerc, J. Jourdain, D. Boichard and A. Capitan

Genomic selection strategies and their potential to maintain rare alleles and de-novo mutations 163
M.F. Schrauf, Y.C.J. Wientjes, H.A. Mulder and J. Vandenplas

Genomic prediction with selected sequence variants in gestation length of New Zealand dairy cattle 164
Y. Wang, K.M. Tiplady, E.G.M. Reynolds, M.A. Nilforooshan, C. Couldrey and B.L. Harris

Novel runs of homozygosity islands in the Finnish Ayrshire population 164
K. Sarviaho, P. Uimari and K. Martikainen

vited Sometimes we win, sometimes we lose: the consequences of genomic selection 165
D. Lourenco, F. Guinan, G. Wiggans, J. Dürr, S. Tsuruta and I. Misztal

Session 04. Adaptation of breeding strategies and genetics to climate change, right animal for right environment

Date: Monday 28 August 2023; 8.30 – 13.00
Chair: Kargo / Sturaro

Theatre Session 04

Old, native cattle breeds; from critically endangered to successfully utilized in niche productions 165
N. Svartedal and O. Vangen

Kenyan pastoralists and climate change: breeding as a possible adaptation strategy? 166
J. Jandl, M. Wurzinger, L. Gichuki, B. Habermann, R. Siamito and T.A. Crane

Adapting climate-smart breeding practices for small ruminants in pastoral communities of Kenya 166
J.M. Ojango, E. Oyieng, J.W. Gitau, N. Ndiwa, J. Gachora and A.W.T. Muigai

Zooarchaeology for conservation biology: introducing ARETI 167
A. Spyrou, G. Hadjipavlou and D. Bradley

A robustness proxy based on variations in energetic allocation to growth in Duroc fattening pigs 167
G. Lenoir, L. Flatres-Grall, R. Muñoz-Tamayo, I. David and N.C. Friggens

Genetic architecture of skin histology with implications for heat tolerance in beef cattle 168
R.G. Mateescu, F.M. Rezende, K.M. Sarlo Davila, A. Hernandez, A.N. Nunez Andrade, G.A. Zayas, E.E. Rodriguez and P.A. Oltenacu

Ewe-lamb allocation trade-offs shape ewe lifetime production depending on dietary energy scarcity 168
M. Hiltpold and F. Douhard

System performance of three dairy-beef genotypes divergent in carcass merit 169
N. Byrne, D. Fahy, M. Kearney and N. McHugh

Characterization and main factors of variation of milk production in suckler cows 169
B. Sepchat, M. Barbet and A. De La Torre

Mediterranean Baladi cattle presents adaptive traits to cope with climate change in the Near East 170
A. Shabtay, E. Lipkin, M. Soller, F. Garcia-Solares, J. Sölkner, E. Shor-Shimoni, A. Bagnato, M.G. Strillacci and M. Cohen-Zinder

Effect of some environmental variables on milk traits in dairy sheep 170
F. Correddu, A. Cesarani, G. Gaspa, I. Peana, G. Fois, C. Dimauro and N.P.P. Macciotta

Genomic tolerance to harsh climate conditions in beef cattle as a result of cross-breeding program 171
R. Tian and H. Asadollahpour Nanaei

Expected selection responses in breeding plans aiming to limit environmental impacts of trout farming 171
S. Pouil, J. Aubin and F. Phocas

Defining resilience traits in sheep from fibre diameter variation of wool 172
E.G. Smith, S.F. Walkom, D.J. Brown and S.A. Clark

Evaluating the impact of uncertainty in estimated breeding values on optimal contribution selection 172
I. Pocrnic, J. Ortiz and G. Gorjanc

Increasing genetic gain without compromising diversity by selection based on mating qualities 173
T.A.M. Niehoff, J. Ten Napel, P. Bijma, T. Pook, Y.C.J. Wientjes, B. Hegedűs and M.P.L. Calus

Session 05. Climate care dairy farming: follow up – Part 1

Date: Monday 28 August 2023; 8.30 – 13.00
Chair: Galama / Rees

Theatre Session 05

invited Holistic perspectives on climate care cattle farming 173
G. Van Duinkerken

Climate care farming aspects in China 174
W. Wang

Producer adoption of climate-smart agriculture practices on U.S. grazing lands 174
A. Blair, K. Cammack, H. Menendez, J. Brennan and K. Ehlert

Assessing impacts of new legislation on Dutch dairy farms using a revised linear programming model 175
A. Van Der Linden, L.M. Alderkamp, C.W. Klootwijk, G. Holshof, N. Van Eekeren, F. Taube and C.E. Van Middelaar

Gaseous emissions (building, storage, pasture) of dairy systems combining or not grazing and housing 175
N. Edouard, X. Vergé, C. Flechard, Y. Fauvel and A. Jacotot

Agri-food and net zero: a science-policy-society perspective 176
R. McGuire, S. Huws, C. Foyer, P. Forster, M. Welham, L. Spadavecchia, D. Curry and N. Scollan

Testing the effects of grassland swards for yield, greenhouse gas emissions and soil health 176
P.R. Hargreaves, D.Z. Kreismane, A. Dorbe, K. Klumpp, J. Bloor, A. Cieślak, M. Szumacher, A. Szejner and R.M. Rees

How to mitigate methane and ammonia emissions at the farm level with innovative approaches 177
P.J. Galama, H.J. Van Dooren, H. De Boer and C. Schep

Effect of mitigation measures on GHG and ammonia emissions of pilot farms in European countries 177
D. Ruska, K. Naglis-Liepa, P. Hargreaves, A. Lenerts, M. De Vries, P. Galma, A. Kuipers, A. Cieslak, J. Bell, R. Rees, V. Juskiene and M. Skorupka

Economic aspects of mitigation practices on pilot dairy farms in Europe 178
K. Naglis-Liepa, D. Ruska, A. Lenerts and D. Kreismane

Specialized or integrated systems: on-farm eco-efficiency of dairy farming in northern Germany 178
R. Loges, T. Reinsch, I. Vogeler, C. Kluss and F. Taube

Modelling methane emission from dairy cows, barn, and storage 179
S. Lashkari, F.R. Dalby and C.F. Børsting

Impact of sorbent application on NH_3 and CO_2 and CH_4 gas emissions from dairy cattle manure 179
V. Juskiene, R. Juska, G. Kadziene, D. Stankeviciene and R. Juodka

Poster Session 05

Greenhouse gas emissions on dairy farms in Slovenia 180
M. Klopčič, T. Perčič, M. Bric, S. Rogina, G. Šen, T. Kramer, J. Klopčič, D. Drašler and A. Dolinar

Session 06. Production of high quality animal products and implementation of new technologies in animal husbandry

Date: Monday 28 August 2023; 8.30 – 13.00
Chair: Georgieva / Hocquette

Theatre Session 06

'Greenanimo' – green future through research 180
S.Y. Georgieva

PLS-DA analysis on hand-held spectrometer for on-line prediction of beef marbling in slaughterhouses 181
M. Kombolo, A. Goi, M. Santinello, N. Rampado, S. Atanassova, J. Liu, P. Faure, L. Thoumy, A. Neveu, D. Andueza, M. De Marchi and J.-F. Hocquette

Application of computed tomography and hyperspectral images for enhanced meat quality evaluation 181
T. Stoyanchev, I. Penchev, S. Georgieva, A. Daskalova and S. Atanassova

Effect of husbandry factors on marbling deposition 182
A. Nicolazo De Barmon, I. Legrand, J. Normand and J.J. Bertron

Effect of birth type on meat quality in Ile-de-France lambs 182
E. Achkakanova, I. Penchev and S. Georgieva

Relationships between lamb feed efficiency, rumen volume and carcass quality measured by CT scanning 183
N.R. Lambe, A. McLaren, K.A. McLean, J. Gordon and J. Conington

Do biomarkers of residual feed intake in beef cattle remain consistent regardless of feeding level? 183
G. Cantalapiedra-Hijar, K. Nedelkov, P. Crosson and M. McGee

Practical application of Observer XT software for behaviour and welfare research in sheep and cattle 184
N.A. Bozakova and M. Halil

Software tools and technologies used to study animal behaviour-benefits and capabilities of Observer 184
M. Halil, N. Bozakova and S. Georgieva

Animal welfare: from research to practice 185
I. Veissier, V. Brunet, D. Ledoux and A. De Boyer Des Roches

Effects of warm climatic periods on dairy cow behaviour and welfare in a maritime European climate 185
M.J. Haskell, I. Sullivan, M. March and L. Shewbridge-Carter

Behavioural and hormonal effects of intensive sheep farming on milking behaviour in a milking parlour 186
I. Nedeva, T. Slavov, V. Radev, S. Georgieva and I. Varlyakov

Combining cattle and sheep in a grassland-based system: which effects for system multiperformance? 186
S. Prache, K. Vazeille, B. Sepchat, P. Note, P. Veysset and M. Benoit

Milking temperament and it relation with test day milk yield in Bulgarian Murrah buffaloes 187
T. Stepancheva, I. Marinov and Z. Gergovska

Poster Session 06

Introduction of lupins in lamb diets: effects on carcass, meat quality and intramuscular fatty acids 187
M. Almeida, S. Garcia-Santos, D. Carloto, A. Arantes, J. Lorenzo, J.A. Silva, V. Santos, J. Azevedo, C. Guedes, L. Ferreira and S. Silva

Lipid composition of dromedary camels (*Camelus dromedarius*) raised in extensive and intensive system 188
M. Lamraoui, J. Pestana, M. Costa, C. Alfaia, Y. Khelef, N. Sahraoui, A.M. Almeida and J.A.M. Prates

Effective application of UAVS in modern livestock agriculture 188
T. Petrova, Z. Petrov and M. Marinov

Effects of dietary beet pulp on rumen fermentation, beef quality, and intramuscular adiposity 189
M. Baik, S.H. Lee and S.Y. Kim

The identification of high-quality *Perilla frutescens* Tan lamb meat by multi-metabolomics analyses 189
Y. Yu, B.Y. Zhang, X.Z. Jiang, Y.M. Cui, H.L. Luo and B. Wang

Effect of feeding garlic-citrus supplement on carcass characteristics and fatty acid profile of beef 190
M.E. Eckhardt, T. Brand, S.A. Salami, B.M. Tas, J.F. Legako, T.E. Lawrence and L.W. Lucherk

Main microclimatic indicators in a milking parlor for dairy cows 190
D. Dimov, T. Penev, I. Marinov, J. Mitev, T. Miteva and M. Stoynov

Estimation of rib composition and intramuscular fat from DXA or smartphone imaging in crossbred bull 191
C. Xavier, B. Meunier, I. Morel, Q. Delahaye, Y. Le Cozler, M. Bonnet and S. Lerch

Perception of beef by the Algerian consumer 191
M. Sadoud and J.F. Hocquette

Effect of intramammary infection on milk protein profile measured at the quarter level in dairy cows 192
V. Bisutti, D. Giannuzzi, A. Vanzin, A. Toscano, M. Gianesella, S. Pegolo and A. Cecchinato

Antioxidant and anti-hypertensive properties of acid whey from Greek yoghurt 192
E. Dalaka, A. Vaggeli, G.C. Stefos, I. Palamidi, A. Savvidou, I. Politis and G. Theodorou

Session 07. Challenges of fish nutrition for sustainable aquaculture

Date: Monday 28 August 2023; 8.30 – 13.00
Chair: Medale / Skiba

Theatre Session 07

invited Recent developments in aquaculture feeds – an industry perspective 193
A. Obach

Nutritive value of soybean meals from different origins fed to Nile tilapia 193
N.V. Nguyen, H. Le, V.K. Tran, D.H. Pham, T.L.T. Tran, J.M. Hooft and D.P. Bureau

EcoXtract® green solvent increases rainbow trout performances and decreases susceptibility to VHSV 194
D. Rigaudeau, N. Abdedaim, T. Rochat, L. Jep, F. Terrier, P. Boudinot, S. Skiba-Cassy, L. Jacques and C. Langevin

Environmental impact of aquafeed for rainbow trout containing alternative protein meals 194
M. Berton, C. Fanizza, V. Stejskal, M. Prokešová, M. Zare, H.Q. Tran, F. Brambilla, E. Sturaro, G. Xiccato, A. Trocino and F. Bordignon

Invited New strategies for fish nutrition 195
S. Skiba-Cassy, J. Roy, L. Marandel, E. Cardona and F. Medale

Rainbow trout hepatic cell lines reveal differences of DL-Methionine and DL-HMTBa function 195
M. Klünemann, K. Pinel, K. Masagounder, I. Seiliez and F. Beaumatin

Millenial salmon – sustainable salmon feed using black soldier fly meal 196
M.L. Lefranc, H.B. Bergoug and K.K. Kousoulaki

Digestibility of ingredients derived from food by-products (insect and single-cell) in Asian seabass 196
R. Le Boucher, W. Chung, J. Ng Kai Lin, L. Tan Shun En, C.S. Lee and C. Wu

Invited Specialty feeds to enhance the health value of farmed fish 197
J. Dias, A. Ramalho-Ribeiro, A. Gonçalves, M.T. Dinis and P. Rema

Poster Session 07

Evaluation of growth performance of pure *Clarias gariepinus* and its hybrid in monosex culture 197
F.O. Olukoya, O.T.F. Abanikannda, K.O. Kareem-Ibrahim, M.S. Hedonukun and S.O. Adebambo

The effect of feeding on dissolved oxygen and turbidity in trout cultivated in an aquaponics 198
I. Sirakov, K. Velichkova, A. Asenov and R. Rusev

Factors affecting feeding of fish cultivated in recirculation aquaculture system 198
K. Velichkova and I. Sirakov

Effect of mycotoxin co-occurrence on zootechnical performance and health indices of *Sparus aurata* 199
C. Papadouli, S. Vardali, T. Karatzinos, M. Maniaki, P. Panagiotaki, D. Kogiannou, C. Nikoloudaki, I. Nengas, M. Henry, G. Rigos and E. Golomazou

Fasting influences on several blood chemistry changes: indicators of malnutrition in Nile tilapia 199
S. Boonanuntanasarn, P. Pasomboon, S. Seedum and W. Auppakhun

Technological developments and 3D imaging for the diagnosis of aquaculture fish diseases 200
M. Merhaz, M. Frétaud, D. Rigaudeau and C. Langevin

Comparative analyses of growth parameters of pure and hybrid African Catfishes fed two feed types 200
O.T.F. Abanikannda, F.O. Olukoya, K.O. Kareem-Ibrahim, S.M. Adebambo and M.S. Hedonukun

Influence of feeding frequency on growth parameters of African Catfish 201
F.O. Olukoya, K.O. Kareem-Ibrahim, A.A. Jimoh, K.O. Bakare, Z.F. Shopeyin, S.A. Shitta and O.T.F. Abanikannda

The impact of a probiotic supplementation in aquaculture – a bioinformatic modelling 201
A. Brzoza, L. Napora-Rutkowski, M. Mielczarek, T. Kaminska-Gibas, T. Suchocki and J. Szyda

Session 08. Various topics in pig production

Date: Monday 28 August 2023; 8.30 – 13.00
Chair: Trevisi / Vigors

Theatre Session 08

Role of daily feeding rhythms in the genetics of feed efficiency in growing pigs 202
H. Gilbert, L. Agboola, A. Foury, M. Teissier, F. Gondret and M.P. Moisan

Using sequence variants to detect SNP and pathways affecting pig gut microbiota and feed efficiency 202
H. Gilbert, M. Teissier, K. Fève, S. Legoueix, Y. Billon, V. Déru, A. Aliakbari, J. Bidanel, O. Zemb, Y. Labrune and J. Riquet

Using digestive efficiency to improve feed efficiency in pig breeding schemes 203
V. Déru, M.J. Mercat, D. Picard, B. Ligonesche, G. Lenoir, L. Flatrès-Grall, F. Ytournel, J. Bidanel and H. Gilbert

Can automatic records of activity predict maternal ability and health in sows? 203
K. Nilsson, A. Bergh and A. Wallenbeck

Cross-breeding strategies to optimize litter size – a way to improve piglet survival? 204
K. Nilsson and A. Wallenbeck

Genetic parameters for genetic variance uniformity in Swiss pigs' birth weight 204
C. Kasper, A. Lepori, J.P. Gutiérrez, N. Formoso-Rafferty, E. Sell-Kubiak and I. Cervantes

A multivariate gametic model for the analysis from reciprocal crosses for two Iberian varieties 205
H. Srihi, D. López-Carbonell, N. Ibáñez-Escriche, J. Casellas, P. Hernández, S. Negro and L. Varona

Impact of vaccination against GnRF on growth performance and meat quality of gilts at market weight 205
N. Quiniou, P. Chevillon and F. Colin

Using different gas mixtures for pig stunning – influence on meat quality and gene expression 206
J. Gelhausen, T. Krebs, T. Friehs, J. Knöll, I. Wilk, D. Mörlein and J. Tetens

Lean tissue grows slower in the belly and loin regions of Australian domestic pork carcases 206
L.T. King, F. Anderson, M. Corlett, A. Williams and G.E. Gardner

Carcass characteristics and boar taint in entire male pigs from commercial French organic farms 207
S. Lombard, A. Roinsard and A. Prunier

Boot swabs to assess efficiency of cleaning and disinfection of slatted floors in pig barns 207
R.H. Mateus-Vargas, N. Kemper, K. Butenholz, N. Volkmann, C. Sürie and J. Schulz

The impact of temperature on transmission of African swine fever in contaminated livestock vehicles 208
Y. Gao, A. Boklund, L. Nielsen, L. Alban and M. De Jong

Porcine ear necrosis is associated with social behavior and blood biomarkers in weaned pigs 208
T. Nicolazo, C. Clouard, C. Teixeira Costa, G. Boulbria, A. Lebret, V. Normand and E. Merlot

Pelleting and botanical source influence starch utilization in growing pigs 209
A. Agouros, M. Le Gall, K. Quéméneur, Y. Lechevestrier, L. Montagne, N. Quiniou and E. Labussière

Effects of a blend of phytogenic feed additives on performance in fattening pigs fed liquid feed 209
A. Samson, W. De Gaiffier, E. Janvier and S. Constantin

Poster Session 08

Genetic parameters of digestive coefficients in three pig breeds 210
C. Carillier-Jacquin, P. Ganier, J. Bidanel, C. Hassenfratz, B. Blanchet, V. Deru, E. Labussiere and H. Gilbert

Breeding for uniformity in piglet birth weight to improve survival 210
C. Kasper, A. Lepori, J.P. Gutiérrez, N. Formoso-Rafferty, E. Sell-Kubiak and I. Cervantes

Genetic lines influence carcass traits and quality of pork 211
X. Li, M. Ha, R. Warner, R. Hewitt, D. D'Souza and F. Dunshea

Boar taint: pedigree-based BLUP and genomic BLUP, and associations with green-ham quality traits 211
S. Faggion, P. Carnier and V. Bonfatti

Diversity of pig breeds reared in the Czech Republic 212
E. Krupa, Z. Krupová, E. Žáková, N. Moravčíková and I. Vrtková

Wild boar and domestic pig identification based on GBS study 212
A. Koseniuk, G. Smołucha, A. Gurgul, T. Szmatoła, M. Oczkowicz and A. Radko

Structural analysis of the HK2 gene in the aspect of pigs performance and meat quality parameters 213
K. Woźniak, A. Terman, D. Polasik, G. Żak, M. Tyra and K. Ropka-Molik

Methods to maximize accuracy of selection in maternal genetic model for genetic evaluation of pigs 213
M. Satoh

Parentage control of pigs based on SNP data 214
E. Žáková, Z. Krupová, E. Krupa, J. Stibal and I. Vrtková

Prediction of piglet survival based only on birth weight or within-litter birth weight variability 214
J.M. Mbuthia, C. Kasper, M. Zenk, G. Bee, C.C. Metges and G. Daş

Effect of fenugreek cotyledons on farrowing sow performances 215
M. Le Bot, J. Laurain, E. Belz and A. Benarbia

Effects of different dietary fibres on constipation in sows during late pregnancy 215
D. Lu, Y. Pi, H. Ye, D. Han, B. Kemp, N. Soede and J. Wang

Effect of transportation condition of pigs during transport for slaughter under different season 216
D.C. Song, S.Y. Chang, J.W. An, S.H. Park, K.H. Jeon and J.H. Cho

Salmonella excretion level by pigs and impact of disinfection on antibiotic resistance of *E. coli* 216
C. Soumet, A. Kerouanton, A. Bridier, B. Anger, V. Dorenlor, V. Rose, F. Eono, C. Houdayer, E. Houard, E. Eveno, F. Souchaud, B. Houry, C. Valentin, P. Houée, I. Attig, N. Haddache, M. Denis and C. Fablet

Climate change adaptation in mixed pigs-beef cattle systems in grasslands areas: key role of organic 217
S. Mugnier, L. Valero, C. Husson, F. Von Kerssenbrock, B. Dounies, H. Rapey and C. Philippeau

Financial implications associated with ante- and *post-mortem* inspection findings in finishing pigs 217
D.L. Teixeira, L.C. Salazar and L.A. Boyle

Training during rearing: effect on body reserves' flexibility & long-term reproduction in rabbit doe 218
J.J. Pascual, E. Martínez-Paredes, L. Ródenas, E. Blas and M. Cambra-López

Session 09. Insect applications

Date: Monday 28 August 2023; 8.30 – 13.00
Chair: Sanchez / Gligorescu

Theatre Session 09

Replacement of poultry by-product meal by black soldier fly larvae meal in diets for dogs 218
B. Agy Loureiro, R.K. Nobrega Cardoso, R. Silva Carvalho, W.A. Zamora Restan, M. Dalim, N. Martin Tome and A. Paul

Hermetia illucens meal in Rainbow trout diet – preliminary data of a commercial scenario 219
S. Bellezza Oddon, I. Biasato, C. Caimi, P. Badino, F. Gai, M. Renna and L. Gasco

Mealworm protein hydrolysate as a novel functional ingredient for aquaculture applications 219
L. Sanchez

Effect of black soldier fly larvae oil on performance and health of lactating sows and piglets 220
C. Omphalius, M. Walraven, S. Juliiand and H. Bergoug

Effects of dietary *Tenebrio molitor* meal and chitosan on health and meat quality of weaned piglets 220
C. Zacharis, E. Bonos, A. Tzora, I. Skoufos, G. Magklaras, I. Giavasis, I. Giannenas, E. Antonopoulou, C. Athanasiou and A. Tsinas

Defatted insect meals: impact on *in vitro* ruminal fermentation and lipid biohydrogenation 221
M. Renna, M. Coppa, C. Lussiana, A. Le Morvan, L. Gasco, L. Rastello and G. Maxin

Can the mixture of *Hermetia illucens* and *Tenebrio molitor* meals improve performance in broilers? 221
I. Biasato, M. Gariglio, E. Fiorilla, V. Bongiorno, E.E. Cappone, S. Bellezza Oddon, L. Gasco and A. Schiavone

Including different products from black soldier fly larvae in the diet of slow-growing broilers 222
A. Rezaei Far, T. Veldkamp, J. Van Harn, S. Naser El Deen, P. Van Wikselaar and I. Fodor

Effect of feeding black soldier fly larvae products to slow-growing broilers on welfare and health 222
A. Doerper, G. Gort, I.C. De Jong, T. Veldkamp and M. Dicke

The immune response of slow-growing broiler chickens fed black soldier fly larvae meal 223
A. Rezaei Far, C.A. Jansen, J. Van Harn, P. Van Wikselaar, S.K. Kar and T. Veldkamp

Does BSF meal have an impact in broilers with a subclinical necrotic enteritis challenge? 223
T. Veldkamp, J.J. Mes, A. Rezaei Far, S. Naser El Deen, P. Van Wikselaar, I. Fodor, R. Van Emous and L. Van Eck

Black soldier fly larval diet eliminate chicken coronavirus at an early infection stage 224
Y. Zhang, C.Y. Yang, C.J. Li, Z.H. Xu, P. Peng, C.Y. Xue, J.K. Tomberlin, W.F. Hu and Y.C. Cao

Black soldier fly larvae meal alter chicken gut microbiota to restricts coronavirus infection 224
C.Y. Yang, Y. Zhao, O. Peng, C. Li, Y. Cao and Y. Zhang

Black soldier fly larvae meal as the main protein source improves feed efficiency in laying hens 225
A. Rezaei Far, I. Fodor, P. Van Wikselaar, S. Naser El Deen and T. Veldkamp

Antibacterial and anticoccidial activities of black soldier fly extract 225
L. Sedano, E. Chambellon, F.I. Bussiere, E. Helloin, M. Vian, C. Guidou, C. Trespuech and A. Silvestre

Poster Session 09

The metabolizable energy value of black soldier fly larvae fat used in broiler chicken diets 226
B. Kierończyk, M. Rawski, K. Stuper-Szablewska, K. Dudek and D. Józefiak

Hermetia illucens larva fat affects broiler chicken breast meat quality 226
B. Kierończyk, M. Rawski, Z. Mikołajczak, P. Szymkowiak, K. Stuper-Szablewska, M. Dudek and D. Józefiak

Evaluation of two black soldier fly products on hens' performance, hatchability, and health traits 227
P. Hristakieva, N. Mincheva, I. Ivanova, K. Velikov and A. Petrova

Insect meals digestibility for rainbow trout aquafeeds: *in vivo* and *in vitro* preliminary results 227
C. Caimi, F. Moyano Lopez, S. Bellezza Oddon, M.J. Aznar, V. Zambotto, F. Gai and L. Gasco

Suitability of full-fat and defatted black soldier fly meal as ingredient for chicken diets 228
D. Murawska, D. Witkowska, M. Gesek, T. Bakuła and J. Błażejak-Grabowska

Protein and amino acids digestibility of two black soldier fly larvae meal forms in broilers 228
A. Cerisuelo, E.A. Gómez, A. Martínez-Talaván, L. Piquer, D. Belloumi, C. Cano, M. Martínez, C. Fernández and S. Rojo

CIPROMED: a PRIMA project on the use of alternative proteins in the Mediterranean value chains 229
C.G. Athanassiou, S. Smetana, A. Tassoni, L. Gasco, F. Gai, A. Shpigelman, D. Pleissner, M. Gastli, L. Conceição, F. Robinson, J.-I. Petrusán and C.I. Rumbos

Replacing fishmeal by *Hermetia illucens* larvae on growth performance and meat quality in broiler 229
K.H. Jeon, D.C. Song, S.Y. Chang, J.W. Ahn, H.A. Cho, S.H. Park, H. Kim and J.H. Cho

Investigation of the suitability of two *Tenebrio molitor* meals in diets of weaned piglets 230
C. Zacharis, G. Magklaras, I. Giavasis, I. Giannenas, E. Antonopoulou, A. Tsinas, C. Athanasiou, A. Tzora, I. Skoufos and E. Bonos

Effect of different black soldier fly frass in plant yield and nutritional value 230
I. Rehan, I. Lopes, D. Murta, O. Moreira and R. Menino

Session 10. Impact of epigenetics and genetics in determining animal physiology

Date: Monday 28 August 2023; 8.30 – 13.00
Chair: Jammes

Theatre Session 10

invited The place of epigenetics in the livestock of tomorrow 231
E.M. Ibeagha-Awemu and M. Wang

A capture-based approach for DNA methylation analysis in cattle 231
V. Costes, A. López-Catalina, O. Gonzalez Recio, M. Ben Sassi, D.E. Rico, G. Foucras, C. Le Danvic, C. Hozé, M. Boussaha, D. Boichard, H. Jammes and H. Kiefer

DNA-methylome analysis of milk somatic cells upon subclinical mastitis in Holstein cattle 232
S. Pegolo, D. Giannuzzi, A. Vanzin, V. Bisutti, A. Toscano, E. Capra, P. Ajmone Marsan and A. Cecchinato

Feed efficiency correlates with the transcriptomic response to feed intake in the pig duodenum 232
G. Devailly, K. Fève, S. Saci, J. Sarry, S. Valière, J. Lluch, O. Bouchez, L. Ravon, Y. Billon, H. Gilbert, M. Beaumont and J. Demars

Epigenetic biomarkers for environmental enrichment and parity in pregnant sows 233
M.M. Lopes, A. Chaulot-Talmon, A. Frambourg, H. Kiefer, C. Gerard, J. Demars, E. Merlot and H. Jammes

A large population study to assess the magnitude of foetal programming in cattle 233
C. Fouéré, C. Hozé, M. Boussaha, H. Kiefer, M.P. Sanchez and D. Boichard

Non-coding exonic microsatellite in bovine Nrf2 gene influences sperm oxidative stress capacity 234
K. Anwar, G. Thaller and M. Saeed-Zidane

Maternal methionine restriction affects liver metabolism and foie gras production in duck offspring 234
C.M.D. Bonnefont, A. Sécula, H. Chapuis, L. Bodin, L.E. Bluy, A. Bonnet, A. Cornuez, M.-D. Bernadet, J. Barrieu, E. Cobo, X. Martin, H. Manse, L. Gress, M. Lessire, F. Mercerand, M.-C. Le Bourhis, A. Collin and M. Morisson

Garcinol promotes hepatic gluconeogenesis by inhibiting PCAF acetyltransferases in late-pregnant Sow 235
W.L. Yao, Q.Y. Fu and J.X. Ning

Early high nutrition affects epigenetics related to lipogenesis in grass-fed Wagyu 235
D. Nishino, S. Khounsaknalath, K. Saito, A. Saito, T. Abe, E. Kobayashi, S. Yasuo and T. Gotoh

Poster Session 10

Construction of a single-cell transcriptome map of sheep horn development 236
X.H. He, Y.Y. Luan, S.J. Wu, Y.H. Ma and L. Jiang

The effect of short and long term cryopreservation on chicken primordial germ cells 236
M. Ibrahim, K. Stadnicka, E. Grochowska, B. Lazar, E. Varkonyi and M. Bednarczyk

The effect of garcinol on muscle *post mortem* glycolysis and meat quality in pigs 237
T.X. Wang, L. Huang and X.M. Liu

The effect of methionine supplementation on DNA methylation during the suckling period of ewe lambs 237
A. Martín, O. Madsen, F.J. Giráldez, J. De Vos and S. Andrés

Effect of supplementary diet at cow-calf phase on DNA methylation in muscle of Angus-Nellore calves 238
R. Curi, L. Rodrigues, R. Cooke, W. Baldassini, L.A. Chardulo and O. Machado Neto

Session 11. Improving pig welfare in conventional production systems

Date: Monday 28 August 2023; 8.30 – 13.00
Chair: Gottardo / Blanco Penedo

Theatre Session 11

invited Improving pig welfare in commercial pig farms 238
K. O'Driscoll

Effects of different nutritional strategies on the prevalence of tail biting in weaned piglets 239
E. Janvier, W. De Gaiffier, J. Piqué, S. Lebas and A. Samson

Validation of an Irish tail biting risk assessment tool 239
R.M. D'Alessio, C.G. Mc Aloon, C. Correia-Gomes, A. Hanlon and K. O'Driscoll

Improving finishing pig welfare by providing rooting and bedding materials 240
M. Ocepek, R.C. Newberry and I.L. Andersen

Impact of enrichment material and space allowance on damaging behaviour in pigs 240
M. Cupido, S.M. Mullan, K.M. O'Driscoll and L.A. Boyle

An investigation of early life indicators in relation to ear necrosis in pigs 241
L. Markland, K. O'Driscoll, F. Leonard and L. Boyle

invited Improving the welfare of sows and piglets: an exploration of teeth resection under the 3S framework 241
L. Boyle

Mitigating the impacts of hot conditions on lactating sows through meal delivery or feed 242
N. Quiniou, J. Dupuis and D. Renaudeau

Effect of comfort and enrichment for pregnant sows on the health and resilience of their offspring 242
M. Lagoda, K. O'Driscoll, M. Galli, J. Cerón, A. Ortín-Bustillo, J. Marchewka and L. Boyle

Use of organic nest-building material in free-farrowing pens with short term fixation 243
J. Wahmhoff, N. Kemper, A. Van Asten and I. Traulsen

Welfare of group-housed sows in the mating area 243
S. Geisler, A. Van Asten, N. Kemper and I. Traulsen

Current knowledge and ways forward to improve welfare of cull sows during transport and in lairage 244
C. Kobek-Kjeldager, L.D. Jensen, L. Foldager, L.M. Gould, K. Thodberg, D.L. Schrøder-Petersen and M.S. Herskin

Exposure times and stunning effectiveness for argon and nitrogen-argon mixture for pigs at slaughter 244
J. Knöll, J. Gelhausen, T. Friehs, T. Krebs, D. Mörlein, J. Tetens and I. Wilk

Poster Session 11

Stability of performance in sows switching between crate and pen from one parity to the next 245
L. Canario, J. Bailly, S. Reignier, M. Poupin, Y. Billon and W. Hébrard

Session 12. Technologies for outdoor systems: which application and opportunities?

Date: Monday 28 August 2023; 8.30 – 13.00
Chair: Maroto-Molina / Gautier

Theatre Session 12

Preliminary results on the use of weather data to predict production output in a Merino flock 245
P.G. Theron, T.S. Brand, S.W.P. Cloete and K. Dzama

GIS-based data integration system as a design tool for rangeland management in productive landscapes 246
M. Odintsov Vaintrub, M. Bianchini, L. Lizzi, F. Paciocco, M. Francioni and P. D'Ottavio

Remotely monitored animal behaviour using sensor ear tags on cattle in Switzerland 246
K. Ueda, U. Heikkilä, N. Gobbo Oliveira Erünlü, J. Bérard, O. Wellnitz and S. Rieder

Differentiating feeding behaviour in grazing cattle based on IMU data and machine learning algorithms 247
N. Tilkens, A.L.H. Andriamandroso and J. Bindelle

Monitoring animal movements by virtual fencing collars – a comparison of different GPS intervals 247
D. Hamidi, M. Komainda, N.A. Grinnell, F. Riesch, J. Horn, M. Hamidi, I. Traulsen and J. Isselstein

The impact of virtual fence use on blood markers linked to stress and acute phase response in cattle 248
J. Ranches, M. Ferreira, G.M.P. Hernandez, A.R. Santos, R.C. O'Connor, D. Johnson, D.W. Bohnert and C. Boyd

Drones to facilitate the monitoring of grazing animal: which application and opportunities? 248
A. Lebreton, A. Demarbaix, F. Demarquet, J. Douhay, P.-G. Grisot, L. Depuille and E. Nicolas

Modular communication gateways, a new opportunity for precision management of agroecological farms 249
M. Odintsov Vaintrub and P. Di Giuseppe

Deploying a LoRa network in mountainous areas to connect animals and shepherds 249
T. Kriszt, O. Benoit, G. Besche, J.D. Guyonneau, E. González-García and J.B. Menassol

Poster Session 12

Effect of successive drone herding on behaviour and spatial distribution of grazing cattle 250
H. Anzai and M. Kumaishi

Evaluation of the RumiWatchSystem to measure activity and grazing behaviour of sheep 250
E. Dunne, N. McHugh and F.M. McGovern

Remote monitoring of cattle using sensor ear tags 251
K. Ueda, T. Kuntzer, U. Heikkilä, N. Gobbo Oliveira Erünlü, J. Bérard, C. Beglinger, O. Wellnitz and S. Rieder

Session 13. Innovations and technological advancements in small ruminant production with particular emphasis in Mediterranean semi-extensive or extensive systems

Date: Monday 28 August 2023; 8.30 – 13.00
Chair: Hadjipavlou / Lauvie

Theatre Session 13

invited Update on sensor technologies for performance recording, management and welfare in small ruminants 251
G. Caja, A. Elhadi, E. González, J.B. Menassol, G. Tessnière, V. Giovanetti, M. Decandia, M. Acciaro, E.M. Sossidou, S.I. Patsios, L.T. Cziszter, L. Grova, G.H.M. Jorgensen, I. Halachmi, A.B. Shamai, T.W.J. Keady, C.M. Dwyer, T. Waterhouse, A. McLaren and C. Morgan-Davis

P@stor-all: a spatialized information system for decision-making in extensive grazing systems 252
U. Kalenga Tshingomba, L. Sautot, M. Teisseire and M. Jouven

Day and night sexual activity of rams as recorded by an automatic heat detector 252
M. Alhamada, E. González-García, N. Debus, A. Lurette, J.-B. Menassol and F. Bocquier

Walking distance and maintenance energy requirements of sheep during mountain pasturing 253
T. Zanon, M. Gruber and M. Gauly

Percutaneous liver biopsy obtention by ultrasonographic guidance in small ruminants 253
S. González-Luna, X. Moll, S. Serhan, B. Chaalia, A.A.K. Salama, X. Such and G. Caja

Modelling milk yields of New Zealand dairy goats undergoing normal and extended lactations 254
N. Lopez-Villalobos and S.A. Turner

Operational use of GPS collars for decision making by agro-silvo-pastoral farmers in Mediterranean areas 254
L. Sautot, I. Candela and M. Jouven

AGRICYGEN: innovative phenotyping and high-throughput genotyping studies of Cyprus sheep and goats 255
G. Hadjipavlou, S. Andreou, T. Christofi, A.C. Dimitriou, A.N. Georgiou, L. Koniali, G. Maimaris, P. Markou and S. Panayidou

Genetic improvement of milk production traits in the Cyprus Damascus goat population 255
S. Andreou, A.N. Georgiou, G. Maimaris, A. Dimitriou, L. Koniali and G. Hadjipavlou

Genetic trends for test day milk yield in White Maritza and Patch-faced Maritza sheep breeds 256
P. Zhelyazkova and D. Dimov

Genetic and environmental effects on the survival and growth rates of Cyprus Chios lambs 256
T. Christofi, S. Panayidou, L. Koniali and G. Hadjipavlou

Genetic and non-genetic factors affecting survival and growth of Cyprus Damascus goat kids 257
L. Koniali, S. Andreou, T. Christofi and G. Hadjipavlou

Feed restriction in dairy ewes decreases milk lipolysis and remodels the milk proteome and lipidome 257
M. Delosière, L. Bernard, C. Hurtaud, H. Larroque, S. Emery, A. Delavaud, M. Taillandier, P. Le Faouder, M. Bonnet and C. Cebo

Poster Session 13

Garlic essential oil supplementation in sheep: effect on fermentation parameters 258
C. Saro, I. Mateos, F.J. Giráldez and M.J. Ranilla

Effect of garlic essential oil supplementation on rumen microbial community of sheep 258
I. Mateos, E. Mata-Nicolás, C. Saro, R. Li, R. Baldwin, F.J. Giráldez and M.J. Ranilla

High-precision scanning system for complete 3D goat udder and teat imaging and analysis 259
G. Coquereau, P.G. Marnet, L. Delattre, J.M. Delouard and P. Martin

Asymmetric udder – a possible indicator of subclinical mastitis in ewes 259
K. Tvarožková, V. Tančin, M. Oravcová, M. Uhrinčať, L. Mačuhová, B. Gancárová and M. Ptáček

Effects of the nature of milk replacer on growth performances and carcass and meat quality of kids 260
J. Normand, C. Boyer, S. Meurisse, A. Pommaret and M. Drouet

Effect of somatic cell count and stage of lactation on milk production 260
L. Mačuhová, V. Tančin, J. Mačuhová, M. Uhrinčať and M. Oravcová

Fit of mathematical models to the lactation curves of dairy sheep 261
L. Guevara, L.S. Gloria, J.C. Angeles-Hernandez, A.M. Fernandes and M.L.C. Abreu

Udder health in dairy ewes in Slovakia: current situation 261
V. Tančin, K. Tvarožková, B. Gancárová, L. Mačuhová, M. Uhrinčať, M. Vršková and M. Oravcová

Evaluation of udder health of dairy goats using electrical conductivity under practical conditions 262
M. Uhrinčať, V. Tančin, K. Tvarožková, B. Gancárová, L. Mačuhová and M. Oravcová

Milk recording protocols and goodness of fit of models applied to dairy sheep lactations 262
L. Guevara, L.S. Gloria, J.C. Angeles-Hernandez, I. Nacarati Da Silva, A.M. Fernandes and A. Peláez Acero

Heritability of milk lipolysis in French Alpine goats 263
H. Larroque, C. Robert-Granié, S. Meurisse, P. Trossat, L. Bernard, C. Hurtaud, C. Huau, T. Fassier, R. Rupp and C. Cebo

Castration method affects the fatty acid profile of subcutaneous and intramuscular fat in sheep 263
H.A. O'Neill, N. Skele, O.B. Einkamerer, A. Hugo and F.W. Neser

Somatic cell count and prevalence of udder pathogens isolated in raw goat milk 264
B. Gancárová, K. Tvarožková, M. Uhrinčať, L. Mačuhová and V. Tančin

Lowering crude protein and supplementing rumen protected methionine and lysine in dairy ewes 264
K. Droumtsekas, A. Plomaritou, I.C.C. Christou, A. Tsigkas, M.A. Belaid, D. Martinez Del Olmo, J. Mateos and A. Foskolos

FASTOChe project: meat sheep grazing on plant species rich in bioactive secondary compounds 265
M. Bernard, H. Hoste, L. Sagot and D. Gautier

Session 14. Sustainable livestock farming – defining metrics and rationalising trade-offs? – Part 2

Date: Monday 28 August 2023; 14.00 – 18.00
Chair: Lee / O'Mara

Theatre Session 14

Multiple mitigation strategies may lead to reduction in emissions for mixed Kenyan dairy systems 265
M.W. Graham, C. Arndt, D. Korir, S. Leitner, L. Merbold, P. Ndung'u, F. Teillard and A. Mottet

Assessing multifunctionality of livestock breeds and species at global level 266
G. Leroy, F. Joly, C. Looft, P. Boettcher and R. Baumung

Predicting sustainable futures: does disease in early life result in trade-offs in ewe productivity? 266
J. Duncan, H. McDougall, G. Mitchell, R. Evans, M. Reeves, L. Andrews, E. Geddes, L. Melville, D. Ewing and F. Kenyon

The welfare of housed and grazing beef cattle assessed by hormones levels and physical check scores 267
M.J. Rivero and A.S. Cooke

Creating a shared framework for standardizing life cycle assessments in livestock production systems 267
P. Goglio, M. Trydeman Knudsen, K. Van Mierlo, N. Röhrig, M. Fossey, A. Maresca, H. Fatemeh, M. Ahmed Waqas, J. Yngvesson, G. Nassy, R. Broekema, S. Moakes, C. Pfeifer, R. Borek, D. Yanez-Ruiz, M. Quevedo Cascante, A. Syp, T. Zylowsky, M. Romero-Huelva and L.G. Smith

Defining operational sustainability and the need for systems approach in assessing livestock farming 268
A.H. Herlin, S. Hellstrand and H.U. Sverdrup

The global roundtable for sustainable beef; multi-stakeholder engagement driving progress worldwide 268
R. Petre and J. Eisele

Developing a decision support system for integral sustainability improvement of Dutch dairy farm 269
M. De Vries, J.W. Reijs, G.J. Doornewaard, A.P. Bos and C.A. Schep

Animal welfare index in sustainability assessments 269
L. Rydhmer and E. Röös

Sustainability analysis of large-scale dairy farming in China from water-energy-labour nexus insight 270
X. Du, Q. Wang and Z. Shi

Characterizing the socio-economic and environmental performance of Irish beef and sheep farms 270
M.C. Ayala, J.C.J. Groot, K. Kilcline, I.J.M. De Boer, C. Grace, J. Kennedy, B. Moran and R. Ripoll-Bosch

Benchmarking the sustainability performance of pastoral dairy-beef production systems 271
M. Kearney, J. Breen, E. O` Riordan and P. Crosson

Measuring, reporting and verifying farm sustainability (Food Futures) 271
R. McGuire, A. Aubry, S. Morrison, J. Birnie and N. Scollan

invited Closing of the 3rd ATF-EAAP LFS symposium 272
F. O'Mara and M. Lee

Poster Session 14

Environmental performance by type of Iberian farm in the Dehesa ecosystem 272
J. García-Gudiño, J. Perea, E. Angón and I. Blanco-Penedo

Ration protein content affects intake, production, efficiency and methane emission of dairy cows 273
F. Schori, C. Kasper and A. Münger

Is protein autonomy synonymous with economic efficiency? 273
E. Bertrand, C. Corre and R. Bore

Development of multicriteria analysis' methodology in an experimental farms network 274
C. Corre and C. Brasseur

Multicriteria analysis on French experimental farm which develops it feed self-sufficiency 274
C. Corre and R. Boré

Partial indicators monitoring in dairy farm: effect of feed efficiency improvement on gross margin 275
G.S.S. Gian Simone Sechi and A.A.S. Alberto Stanislao Atzori

Comparison of methodologies for estimating animal emissions from smallholder cattle systems 275
E. Balcha and C. Arndt

Developing a common framework to measure sustainability in food systems with limited arable land 276
P. Bhérer-Breton, A. Woodhouse, E. Svanes, H.F. Olsen and B. Aspeholen Åby

Partial indicators monitoring in dairy farms: feed efficiency implications on gross margin 276
G.S. Sechi and A.S. Atzori

Session 15. Breeding for climate change – mitigation

Date: Monday 28 August 2023; 14.00 – 18.00
Chair: Pszczola / Neumann

Theatre Session 15

nvited Genomic evaluation to reduce enteric methane emissions in Holstein cattle 277
F. Malchiodi, H.R. Oliveira, S. Narayana, A. Fleming, H. Sweett, S. Shadpour, J. Jamrozik, G. Kistemaker, P. Sullivan, F.S. Schenkel, B. Van Doormaal, C.F. Baes and F. Miglior

Methane phenotyping for genetic selection – can we measure bulls, not cows? 277
L.R. McNaughton, P. Beatson, G. Worth, D.J. Garrick, D. Garrick, R. Handcock and R.J. Spelman

Novel formulas to calculate methane production from concentrations while using sniffers 278
C.I.V. Manzanilla-Pech, M.H. Kjeldsen, T. Villumsen and J. Lassen

Predicting methane production and intensity from milk mid-infrared spectra 278
S. Fresco, A. Vanlierde, R. Lefebvre, M. Gaborit, D. Boichard, R. Bore, S. Fritz, N. Gengler and P. Martin

Genetic parameters for enteric methane emissions of dairy cows using random regression models 279
A.E. Van Breukelen, M.N. Aldridge, R.F. Veerkamp and Y. De Haas

Heritability of methane emission in dairy cows estimated from GreenFeed measures in commercial herds 279
B. Heringstad, K.A. Bakke and G. Difford

Rapid measurements for breeding low methane cattle 280
S.J. Rowe, A. Searle, T. Bilton, S. Sevier, S. Gebbie, A. Jonker, P. Janssen, S. Hickey, N. Amyes, S. Worku and J. McEwan

Collecting and preparing sniffer records for genetic evaluation of enteric methane in dairy cattle 280
T.M. Villumsen, P. Løvendahl, V. Milkevych, R. Krogh, C.I.V. Manzanilla-Pech, M. Bjerring, J. Lassen and G. Sahana

Methods and tools for automated processing of sniffer-based methane emission data from dairy cows 281
V. Milkevych, R.K. Nielsen, T.M. Villumsen, P. Løvendahl and G. Sahana

Indicators of genetic selection using sniffer method to reduce methane emissions from Holstein cows 281
Y. Uemoto, T. Tomaru, M. Masuda, K. Uchisawa, K. Hashiba, Y. Nishikawa, K. Suzuki, T. Kojima, T. Suzuki and F. Terada

Reducing methane emissions: foundations for genetic evaluations for sustainable Irish beef cattle 282
C.V. Ryan, S. Conroy, T. Pabiou, D.C. Purfield and R. Evans

Determining relative values of genetic traits of Chinese dairy cattle in greenhouse gas mitigation 282
Y. Wang, R. Shi, A. Van Der Linden, B.J. Ducro, Y. Wang, Y. Hou, S.J. Oosting and C.E. Van Middelaar

Effects of breeding for feed efficiency on carbon footprint of milk – LCA approach 283
S. Hietala, A. Astaptsev, E. Negussie, M.H. Lidauer and T. Mehtiö

Poster Session 15

Automated CT analysis to identify low emitting ruminants 283
S. Hitchman, L. Grundy, W. Bain, T. Johnson, M. Reis, H. Gunnarsdottir, J. McEwan and S. Rowe

Enteric methane emission of Nellore cattle from two lines selected for yearling weight 284
S.B. Gianvecchio, M.S. Borges, J.A. Silva, J.P.S. Valente, J.O.S. Marcatto, L.G. Albuquerque, J.N.S.G. Cyrillo and M.E.Z. Mercadante

Genetic correlation between enteric methane emission and feed efficiency and growth traits in cattle 284
M.E.Z. Mercadante, J.P.S. Valente, S.B. Gianvecchio, J.O.S. Marcatto, M.S. Borges, J.A. Silva, I.M.C. Garcia, S.F.M. Bonilha and L.G. Albuquerque

Heritability of predicted enteric methane emission in Japanese Holstein cows 285
R. Tatebayashi, A. Nishiura, F. Terada, Y. Uemoto, M. Aihara, T. Suzuki, I. Nonaka and O. Sasaki

Using genomic markers and microbiome data to predict methane emissions from Holstein cows 285
N. Saedi, M.S. Lund and E. Karaman

Session 16. Alternative and new feed resources to reduce feed/food competition and enhance circularity of livestock production

Date: Monday 28 August 2023; 14.00 – 18.00
Chair: Ranilla García / Stergiadis

Theatre Session 16

invited Non-traditional feed ingredients in diets for pigs 286
H. Stein

Effect of genotypes on agronomic and antinutritional traits of *Lupinus albus* L. for livestock 286
M. Oteri, D. Scordia, R. Armone, V. Nava, F. Gresta and B. Chiofalo

Dietary effects of xylanase and flaxseeds on performance, meat quality and bone health in broilers 287
D. Lanzoni, M. Skřivan, M. Englmaierová, M. Marounek, E. Skřivanová, A. Baldi and C. Giromini

The effect of hydrolysed yeast on production performance and gastrointestinal health in broilers 287
N. Irshad, V. Perricone, S. Sandrini, C. Lecchi, P. De Faria Lainetti, A. Agazzi and G. Savoini

Yoghurt acid whey effects on broiler performance, meat oxidative status and caecal barrier integrity 288
V.V. Paraskeuas, G. Papadomichelakis, A. Pappas, P. Simitzis, E.C. Anagnostopoulos, I.P. Brouklogiannis, E. Griela, I. Palamidi, G. Theodorou, I. Politis and K.C. Mountzouris

The effect of *Citrus unshiu* peel on growth performance, digestibility and immunity in weaning pig 288
A.S. Sureshbabu, A.K. Adhimoolam and T.M. Min

The effect of replacing grains with bakery by-products on performance and pigs' diet preference 289
F. Veldkamp, H.M. Vermeer, J. Kater, A. Ten Berge, J.M.J. Rebel and I.C. De Jong

Replacing hexane by 2-methyloxolane for defatting soybean meal does not impair dairy cow performance 289
V. Menoury, A. Ferlay, L. Jacques, V. Rapinel and P. Nozière

In vitro rumen fermentation traits of *Mentha piperita*, and products of oil extraction 290
S. Massaro, J. Andersen, G. Secchi, D. Giannuzzi, S. Schiavon, E. Franciosi and F. Tagliapietra

In vitro rumen fermentation of mushroom-pretreated *Acacia mellifera* simulating ruminant digestion 290
C. Rothmann, E.D. Cason and P.D. Vermeulen

Valorisation of atypical land in cattle production systems in search of self-sufficiency 291
D. Starling, A. Farruggia and T. Puech

Growth of black belly ewe lambs ingesting non-conventional feeds in the form of complete feed blocks 291
D. Barde, E. Traore, R. Arquet, N. Minatchy and H. Archimede

Poster Session 16

Diet effect of *Hermetia illucens* on proximate and fatty acid composition of broiler meat 292
R. Berrocal, E. Molinero, S. Rueda, M.M. Campo and L. Mur

Antioxidant and rumen microbiota shift of Holstein cows by aquatic plant silage in Tibetan plateau 292
B. Li, X. Huang, X. Yang and Y. Wang

Feeding olive cake alters the mRNA of SREBF1 in adipose tissue of cows 293
M.C. Neofytou, D. Miltiadou, E. Sfakianaki, S. Symeou, D. Sparaggis, A.L. Hager-Theodorides and O. Tzamaloukas

Improving gut functions and egg nutrition with stevia Residue in laying hens 293
M.X. Tang, Y.L. W., C. Y., C. P., H.G. L., Y. L., Y.D. C., W. L., X. X. and X.F. K.

In vitro evaluation of olive stones and sorghum used as forage in ruminant diets 294
N. Kellali, N. Merino, E. Mata-Nicolás, I. Mateos, C. Saro, N. Lakhdara and M.J. Ranilla

Preservation of agro-industrial co-products as silage for ruminant feed 294
K. Paulos, C. Costa, J.M.S. Costa, L. Cachucho, P.V. Portugal, J. Santos-Silva, F. Lidon, M.R. Marques, E. Jerónimo and M.T.P. Dentinho

Combination of organic acid and heat treatment decreased ruminal protein degradation in Soybean meal 295
D.K. Yoo, S.W. Jeong and J.K. Seo

Bakery by-products in herbage-based diets for dairy cows: effects on milk yield and reticular pH 295
A.-M. Reiche, M. Tretola, J. Eichinger, A.-L. Hütten, A. Münger, L. Eggerschwiler, L. Pinotti and F. Dohme-Meier

Evaluation of legumes for fermentability and protein fractions using *in vitro* rumen fermentation 296
B.Z. Tunkala, K. Digiacomo, P.S. Alvarez Hess, F.R. Dunshea and B.J. Leury

Dehydration temperature- effect on physicochemical and nutritional characteristics of byproducts 296
M. Alves, K. Paulos, C. Costa, D. Gonçalves, T. Orvalho, J.M.S. Costa, L. Cachucho, J. Santos-Silva, E. Jerónimo and M.T.P. Dentinho

Bacterial ecology and silage quality improvement upon treatment with microbial inoculants 297
I. Nikodinoska, M. Franco, M. Rinne and C.A. Moran

Growth performance, carcass and meat quality of pigs fed *Chlorella vulgaris* supplemented diets 297
J.M. Almeida, M. Parreiras, R. Varino, A. Sequeira, R.J.B. Bessa and O. Moreira

Inclusion of slow-release ammonia-treated barley in concentrates – feed intake and milk production 298
K.S. Eikanger, M. Eknæs, I.J. Karlengen, J.K. Sommerseth, I. Schei and A. Kidane

Reducing aggression in finishing pigs through short application liquid calming herbal blend 298
C. Nicolás-Jorrillo, D. Carrión, D. Escribano, A. Brun and E. Jiménez-Moreno

Evaluating fermentative profile in corn silage using chemical and biological additives 299
C. Mastroeni, S. Van Kuijk, F. Ghilardelli, S. Sigolo, E. Fiorbelli and A. Gallo

Effect of *Chlorella vulgaris* incorporation in pig diets on pork chemical composition and fatty acids 299
J.M. Almeida, A.P. Portugal, R.J.B. Bessa and O. Moreira

Bacillus fermentation products of hog hair waste serve as potential animal feed additives 300
W.J. Chen, K.C. Chang, Z.J. Zhuang, Y.H. Cheng and Y.H. Yu

Effect of coffee grounds on digestibility, ruminal fermentation, and microbial protein synthesis 300
M. Medjadbi, I. Goiri, R. Atxaerandio, S. Charef, C. Michelet, J. Ibarruri, B. Iñarra, D. San Martin and A. García-Rodríguez

Environmental impact of former food estimated trough life cycle assessment 301
M.G.D. Azzena, A. Luciano, L. Pinotti and A.S. Atzori

Effect of coffee grounds on productive performance, milk fatty acid profile and methane production 301
M. Medjadbi, I. Goiri, R. Atxaerandio, S. Charef, C. Michelet, J. Ibarruri, B. Iñarra, D. San Martin and A. García-Rodríguez

Effect of including fresh pears as feedstuff in lactating ewes' diet on the ewes' milk curd strength 302
I. Caro, J. Mateo, S. Kasaiyan, N. Merino, E. Mata-Nicolás, I. Mateos, C. Saro, A. Martín, F.J. Giráldez and M.J. Ranilla

Upcycling of Citrus and Vinification by-products' bioactive compounds for broiler diets 302
A. Mavrommatis and E. Tsiplakou

Session 17. Climate care dairy farming: follow up – Part 2

Date: Monday 28 August 2023; 14.00 – 18.00
Chair: Edouard / Becciolini

Theatre Session 17

Simplified method developed for estimating the on-farm EFGHG and NH_3 – presentation & results 303
X. Vergé, P. Robin, V. Becciolini, A. Cieślak, N. Edouard, L. Fehmer, P. Galama, P. Hargreaves, V. Juškienė, G. Kadžiene, A.-S. Lissy, D. Ruska and M. Szumacher

Simulation of three farming strategies on ammonia and GHG emissions and economics 303
P. Galama and H. Pishgarkomileh

Exploring the effect of dairy cow replacement decisions on feed efficiency and sustainability 304
A. De Vries

Practical monitoring of individual methane production rates in dairy cows: an alternative 304
C.M. Levrault, J.T. Eekelder, P.W.G. Koerkamp, C.F.W. Peeters, J.P.M. Ploegaert and N.W.M. Ogink

Air filtering as alternative approach to combat emissions from cattle facilities 305
A. Kuipers, P. Galama, R. Maasdam, S. Spoelstra and P. Groot Koerkamp

UAV-based approaches for gaseous emissions assessment in cattle farming 305
V. Becciolini, A. Mattia, M. Merlini, G. Rossi, F. Squillace, G. Coletti, U. Rossi and M. Barbari

Reducing N leaching by adapting N application timing and quantities to weather and grass growth 306
L. Bonnard, E. Ruelle, M. O'Donovan, M. Murphy and L. Delaby

Reducing greenhouse gases in dairy cattle farming through innovative technologies 306
A. Svitojus and E. Gedgaudas

Annual quantification of GHG and NH_3 emissions coming from compost-bedded pack housing systems 307
E. Fuertes, G. De La Fuente, J. Balcells and R. Seradj

A study of the building-integrated photovoltaic cattle houses application scenario in China 307
Q. Wang, X. Du and Z. Shi

Effect of manure management and environmental temperature on microbial communities in dairy manure 308
E. Fuertes, R. Seradj, J. Balcells and G. De La Fuente

Poster Session 17

Simultaneous comparison of 5 methods to quantify enteric methane emitted by lactating dairy cows 308
A. Vanlierde, F. Dehareng, O. Christophe and E. Froidmont

Session 18. Carcase grading for quality and efficiency – underpinning future sustainability of beef and lamb

Date: Monday 28 August 2023; 14.00 – 18.00
Chair: Pethick / Neveu

Theatre Session 18

Meeting consumer expectations: a 3G global beef eating quality predictive model for Europe 309
A. Neveu, R. Polkinghorne, R. Watson, H. Cuthbertson and J. Wierzbicki

Development of a carcass grading system for South African beef 309
M.S. Faulhaber, P.E. Strydom and N. Hall

French consumer evaluation of eating quality of Limousin beef 310
M. Kombolo, J. Liu, I. Legrand, F. Noel, P. Faure, L. Thoumy, D. Pethick and J.-F. Hocquette

BeefQ – testing a beef eating quality prediction system for Wales and England 310
P.K. Nicholas-Davies, R. Polkinghorne, A. Neveu, H. Cuthbertson and T. Rowe

Assessment of untrained consumers on eating quality in Zebu and crossbred cattle at aging times 311
A. Barro, M. Coutinho, G. Rovadoscki, A.M. Bridi and J.F. Hocquette

Loin residual glycogen and free glucose do not affect Australian lamb loin eating quality 311
S.M. Moyes, D.W. Pethick, G.E. Gardner and L. Pannier

Combining eating quality and individual cut tracking to maximise value and decrease waste in beef 312
H. Cuthbertson, R. Polkinghorne and A. Neveu

Changing the Australian trading language for beef and lamb to adapt to new measurement technologies 312
G.E. Gardner, S. Stewart, P. McGilchrist, C. Steele, H. Calnan, S. Connaughton, K. Mata and R. Apps

The precision and accuracy of the Q-FOM grading camera predicting rib eye traits in beef carcasses 313
S.M. Stewart, M. Christensen, H. Toft, T. Lauridsen and R. O'reilly

Predicting IMF% and visual marbling scores on beef portion steaks with a Marel vision scanner system 313
L. Pannier, T.M. Van De Weijer, F.T.H.J. Van Der Steen, R. Kranenbarg and G.E. Gardner

Improved analysis of dual energy x-ray absorptiometry images enhances lamb eating quality prediction 314
F. Anderson, L. Pannier, C. Payne, S. Connaughton, A. Williams and G.E. Gardner

Aligning upper and lower commercial beef DEXA images to predict CT composition 314
H.B. Calnan, A. Williams and G.E. Gardner

Calibration of chain-speed DXA units using synthetic phantom blocks in two Australian lamb abattoirs 315
S.L. Connaughton, F. Anderson, A. Williams and G.E. Gardner

Objective carcase measurement from commercial supply chains contributing to genetic improvement 315
D.J. Brown, R. Alexandri, S.F. Walkom, D.W. Pethick, P. McGilchrist, S. Stewart, W.S. Pitchford and G.E. Gardner

Poster Session 18

Potential of deep learning video image analysis for automated SEUROP-classification 316
T. Rombouts, M. Seynaeve and S. De Smet

Hand-held NIR device predicting chemical intramuscular fat% in pork 316
M.T. Corlett, F. Anderson and G.E. Gardner

An optimisation model for operational planning in lamb supply chains 317
G. Wang, F.L. Preston, C. Smith, D. Flanigan, S.M. Miller, W.S. Pitchford and D. Meehan

Predicting P8 fat depth in hot beef carcasses using a handheld microwave system 317
J. Marimuthu, R.K. Abraham and G.E. Gardner

Meat@ppli, a smartphone application to predict fat content of beef 318
J. Normand, B. Meunier, M. Bonnet and B. Albouy-Kissi

Computed tomography vs chemical composition for determination of Australian lamb carcase composition 318
K. Mata, G. Gardner and F. Anderson

Assessing dual-energy X-ray absorptiometry prediction of intramuscular fat content in beef 319
C.L.N. Nunes, R.S.R. Vilela, E.B. Schultz, M.I. Hannas, C.M. Veloso, M.M. Ladeira and M.L. Chizzotti

In-abattoir 3D image parameters of beef carcasses for predicting carcass classification and weight 319
H. Nisbet, N. Lambe, G. Miller, A. Doeschl-Wilson, D. Barclay and C.A. Duthie

Prediction of liveweight from linear conformation traits in beef cattle 320
M. Meister, L. Speiser and A. Burren

Beef marbling assessment by accredited graders and a hand-held camera device 320
M. Kombolo-Ngah, N.S.R. Mendes, A. Neveu, A. Barro, M. Christensen and J.F. Hocquette

Beef on dairy – meat quality and prediction of intramuscular fat on the slaughter line 321
F.F. Drachmann and M. Therkildsen

Pig carcass and cuts lean meat content determined by X-ray computed tomography 321
M. Font-I-Furnols, A. Brun and M. Gispert

Relationships between ultimate pH, weight, conformation and fatness of carcasses 322
N. Silva Rodrigues Mendes, R. Rodrigues Silva, T. Ferreira De Oliveira, P.F. Rivet, M.P. Ellies-Oury, S. Chriki and J.F. Hocquette

Elite Dairy Beef – a pathway for male dairy calves into the premium beef market 322
H. Cuthbertson, R. Polkinghorne, A. Neveu, T. Maguire, O. Catalan and R. Law

Session 19. Advances in nutritional modelling and feeding systems (in memory of Daniel Sauvant)

Date: Monday 28 August 2023; 14.00 – 18.00
Chair: Baumont / Cannas

Theatre Session 19

The contribution of Daniel Sauvant to modelling in animal science 323
J. Van Milgen and P. Faverdin

invited Systemic modelling of nutrient partitioning: from the seed sown by D. Sauvant to future prospects 323
L. Puillet

Mechanistic model of the dynamics of body retention and utilization of phosphorus and calcium in sow 324
J. Heurtault, P. Schlegel and M.P. Létourneau-Montminy

Modelling the digestive utilization of calcium and phosphorus in laying hens 324
F. Hervo, B. Méda, A. Narcy and M.-P. Létourneau-Montminy

A dynamic mechanistic model to forecast the oscillatory feeding behaviour of lactating dairy cows 325
E. Fiorbelli, A.S. Atzori, A.T. Büşra, L. Tedeschi and A. Gallo

A dynamic *in silico* model of rumen microbial fermentation and methane production 325
R. Muñoz-Tamayo, S. Ahvenjärvi, A.R. Bayat and I. Tapio

invited Future directions in ruminant feeding systems: opportunities and challenges for sustainability 326
L.O. Tedeschi

Meta-analysis in animal science: a valuable tool to generate new knowledge from previous experiments 326
M.P. Létourneau Montminy, C. Loncke, P. Schmidely, M. Boval, J.B. Daniel and D. Sauvant

Global sensitivity analysis of INRAtion®V5 for dairy cows: Sobol' indices 327
S. Jeon, S. Lemosquet, T. Senga Kiesse and P. Nozière

Effects of housing or feeding practice in dairy goats estimated by the INRAtion®V5 feeding system 327
S. Giger-Reverdin, V. Berthelot and D. Sauvant

Dynamic models to inform precision supplementation of heifers grazing dormant native grasslands 328
H.M. Menendez Iii, J.R. Brennan, A.K. Dagel and K. Olson

Comparison of measured enteric methane emission with model calculations of commercial dairy diets 328
L. Koning and L.B. Šebek

Poster Session 19

Optim'Al can optimize dairy cow rations on economic cost and protein autonomy with INRA 2018 system 329
P. Chapoutot, G. Tran, V. Heuze and A. Berchoux

Modelling the dynamics of dairy cattle response to heat stress using system dynamics methodology 329
R. Cresci, B. Atamer Balkan, L.O. Tedeschi and A.S. Atzori

Volatolomics for developing non-invasive biomarkers of ruminal function and health in dairy cows 330
E. Jorge-Smeding, C. Martin, L. Volmerange, E. Rispal, M. Bouchon, Y. Rochette, N. Salah, F. Violleau and M. Silberberg

Estimation of total urinary nitrogen excretion in fattening beef cattle from urinary spot samples 330
C. Garcia-Vazquez, P. Nozière, L. Salis, R. Bellagi, J. David and G. Cantalapiedra-Hijar

Advancing mathematical modelling for equines 331
E.M. Leishman, C. Cargo-Froom, P. Darani, S. Cieslar and J.L. Ellis

Predicting equine voluntary forage intake using meta-analysis 331
E.M. Leishman, M. Sahar, P. Darani, S. Cieslar and J.L. Ellis

Score using indicators of reticulo-rumen pH kinetics to characterize subacute ruminal acidosis 332
C. Villot, K. Biard, C. Martin, A. Boudon and M. Silberberg

The INRA mechanistic beef growth model captures feed efficiency ranking in Charolais bulls 332
G. Cantalapiedra-Hijar and S. Lerch

A framework for predicting the environmentally attainable intake of dairy cows 333
J.F. Ramirez-Agudelo, L. Puillet and N.C. Friggens

An integrated tool to apply the INRA2018 feeding system in ruminant rationing: INRAtion®V5 333
P. Noziere, J. Agabriel, R. Baumont, P. Chapoutot, L. Delaby, R. Delagarde, A. De La Torre, P. Faverdin, S. Giger-Reverdin, P. Hassoun, A. Lamadon, G. Maxin, D. Sauvant, B. Sepchat, P. Souvignet and S. Dutot

Milk yield and mammary metabolism in response to increasing amino acids and protein supplies 334
S. Lemosquet, C. Loncke, J.C. Anger, A. Hamon, J. Guinard-Flament and M.N. Haque

Ruminal fatty acid biohydrogenation pathways in lambs and kids fed a high concentrate diet 334
V. Berthelot, Y.D. Yekoye and L.P. Broudiscou

Can high fat concentrates offset the high methane production caused by high forage diets? 335
N. Ayanfe, S. Ahvenjärvi, A. Sairanen and A.R. Bayat

Updates of the expenditures and requirements for growing dairy goats 335
S. Giger-Reverdin and D. Sauvant

Feed efficiency relates to the ability of supporting lactation in dairy cows facing feed restriction 336
L. Barreto-Mendes, J. Pires, A. De La Torre, I. Ortigues-Marty, I. Cassar-Malek and F. Blanc

Effects of rumen protected calcium gluconate on hindgut dynamics and nutrient partitioning 336
M. Hall and J.G.H.E. Bergman

Session 20. Early career competition 'Innovative approaches to pig and poultry production', supported by Wageningen Academic Publishers

Date: Monday 28 August 2023; 14.00 – 18.00
Chair: Vigors / Nilsson

Theatre Session 20

The impact of observation frequency and duration, on resilience traits in pigs 337
W. Gorssen, C. Winters, R. Meyermans, L. Chapard, K. Hooyberghs, S. Janssens, A. Huisman, K. Peeters, H. Mulder and N. Buys

Are tail biting and tail necrosis really only a problem in the domestic pig? 337
I. Czycholl, K. Büttner, W. Baumgärtner, C. Puff and J. Krieter

Innovative rearing strategies for heavy pigs: weight and chemical composition of dry-cured hams 338
A. Toscano, D. Giannuzzi, I.H. Malgwi, V. Halas, P. Carnier, L. Gallo and S. Schiavon

Comparison of different non-linear growth models on post-weaning growth of pigs 338
K.O. Kareem-Ibrahim, O.T.F. Abanikannda and O.A. Folorunsho

Effect of glutamine and/or milk supplementation post-weaning on pig growth and intestinal structure 339
E.A. Arnaud, G.E. Gardiner, S.R. Vasa, J.V. O'Doherty, T. Sweeney and P.G. Lawlor

Effect of lysine level in finisher diets on performance in crossbreds from two terminal sire lines 339
S. Goethals, K. Hooyberghs, N. Buys, S. Janssens and S. Millet

Birth weight effects on performance and gut health status of weaned piglets 340
C. Negrini, F. Correa, D. Luise, S. Virdis, M. Mele, G. Conte, M. Mazzoni and P. Trevisi

Effects of sugary and salty ex-food in pigs' diets on growth performance and pork sensory traits 340
S. Mazzoleni, M. Tretola, C.E.M. Bernardi, P. Lin, P. Silacci, G. Bee and L. Pinotti

Performances, milk quality, and piglet growth in lactating sows fed former food products 341
P. Lin, G. Bee, S. Neuenschwander and M. Girard

Effect of arginine supplementation on the productive performance of gestating sows: a meta-analysis 341
S. Virdis, D. Luise, P. Bosi and P. Trevisi

Are pre-weaning piglet and sow parameters a proxy for uniformity and daily gain? 342
K. Hooyberghs, S. Goethals, W. Gorssen, L. Chapard, R. Meyermans, B. Chakkingal Bhaskaran, S. Millet, S. Janssens and N. Buys

Dietary polyphenol extracts improve the performance of broilers challenged with necrotic enteritis 342
S.A. Salami, L. Chew, B.S. Lumpkins and B. Tas

Effects of *Bacillus subtilis* DSM 32315 or *Bacillus velezensis* CECT 5940 on chicken PBMCs 343
F. Larsberg, M. Sprechert, D. Hesse, G. Loh, G.A. Brockmann and S. Kreuzer-Redmer

Effect of creatine monohydrate *in ovo* feeding on progeny performance of young breeder flocks 343
C.A. Firman, V.K. Inhuber, D.J. Cadogan, W.H.E.J. Van Wettere and R.E.A. Forder

Poster Session 20

Statistical modelling and estimation of pig body weight based on its linear body measurements 344
O.T.F. Abanikannda, K.O. Kareem-Ibrahim and Z.A. Ahmad

Session 21. Insects as feed: sustainability, legislation and safety

Date: Monday 28 August 2023; 14.00 – 18.00
Chair: Veldkamp / Barbosa

Theatre Session 21

invited The potential of insects in sustainable agri-food systems 344
K.B. Barragán-Fonseca

Sustainability potential of insect production systems 345
S. Smetana, A. Bhatia, U. Batta, N. Mouhrim and A. Tonda

Maximizing sustainability in insect production: a multi-objective optimization approach 345
A. Anita, N. Mouhrim, A. Green, D. Peguero, A. Mathys, A. Tonda and S. Smetana

Lowering impacts of chicken meat through *H. illucens* larvae supplementation in the feed? 346
D. Ristic, S. Kechovska, C.L. Coudron, A. Schiavone, J. Claeys, F. Gai and S. Smetana

Use of locusts as feed and food: turning threats into opportunities 346
H.P.S. Makkar, V. Heuzé and G. Tran

Chrysomya chloropyga in broiler feed: valorising abattoir waste for sustainable feed 347
E. Pieterse, E.Y. Gleeson and L.C. Hoffman

Waste side story: a success role for mealworms 347
C. Ricciardi, L. D'Orleans, C. Oundjian, M. Leheup, T. Lefebvre and F. Peyrichou

Ammonia emissions of *Hermetia illucens* larvae grown on different diets 348
C.L. Coudron, S. Berrens, M. Van Peer, J. Claeys and D. Deruytter

Chemical & microbiological safety of insect rearing on yet to be legally authorised residual streams 348
E.F. Hoek- Van Den Hil, A.F.G. Antonis, Y. Hoffmans, K. Van Zadelhoff, R. Deurenberg, O.L.M. Haenen, M.E. Bruins, J.W. Van Groenestijn, A. Borghuis, F. Schaafstra, T. Veldkamp, P. Van Wikselaar and M. Appel

Growth and chemical composition of insects reared on 'yet to be legally authorised' residual streams 349
S. Naser El Deen, P. Van Wikselaar, A. Rezaei Far, I. Fodor, E.F. Hoek - Van Den Hil, Y. Hoffmans, M.E. Bruins, J.W. Van Groenestijn, A. Borghuis, F. Schaafstra and T. Veldkamp

Safety and production performance of insects reared on catering waste 349
E.F. Hoek- Van Den Hil, S. Naser El Deen, K. Van Rozen, P. Van Wikselaar, H.J.H. Elissen, R.Y. Van Der Weide, A. Rezaei Far, I. Fodor, N. Meijer and T. Veldkamp

Mycotoxin accumulation and metabolization by insects: an overview 350
K.C.W. Van Dongen, K. Niermans, E.F. Hoek-Van Den Hill, J.J.A. Van Loon and H.J. Van Der Fels-Klerx

Manure as a substrate for insect rearing: ecological progress or health risk? 350
A. Anselmo, A. Cordonnier, L. Plasman, J. Maljean, C. Aerts, D. Evrard, A. Marien, E. Janssen, V. Ninane, S. Gofflot, P. Veys, M.-C. Lecrenier, V. Baeten and D. Michez

Poster Session 21

Quality characteristics of black soldier fly produced by different substrates 351
A.R. Hosseindoust, S. Ha, J. Mun, H. Tajudeen, S. Park and J. Kim

The value of organic fertilizer from *Hermetia illucens* frass in the sustainable agriculture 351
S. Kaczmarek and K. Dudek

Effects of rearing substrates and processing methods on black soldier fly meal quality 352
S. Ha, A.R. Hosseindoust, Z. Khajehmiri, J. Mun, H. Tajudeen, S. Park and J. Kim

The use of insect meals for broiler feeds with a lower environmental footprint 352
P. Chantzi, I. Giannenas, E. Bonos, E. Antonopoulou, C. Athanasiou, A. Grigoriadou, A. Tzora and I. Skoufos

Diet mediated PUFA accumulation and ώ6-ώ3 ratio improvement in mealworm larvae (*Tenebrio molitor*) 353
S. Yoon, C. Ricciardi, C. Oundjian, M. Leheup, T. Lefebvre and F. Peyrichou

Agro-industry by-products for *Tenebrio molitor* rearing to generate a new aquaculture feed ingredient 353
I. Vieira, D. Murta and M.V. Santos

Transfer of veterinary drug residues from substrate to black soldier fly larvae and frass 354
K.C.W. Van Dongen, E. De Lange, L. Van Asseldonk and H.J. Van Der Fels-Klerx

Session 22. Environmental stress: mitigating the adverse effects on animal physiology

Date: Monday 28 August 2023; 14.00 – 18.00
Chair: Zachut

Theatre Session 22

Drinking heated water improves the growth performance and rumen microbiota of beef cattle in winter 354
T. He, S. Long, X. Wang, T. Liu and Z. Chen

Elevated temperature induces transcriptional and epigenetic changes in the lungs of chickens 355
D. Schokker, J. De Vos, P.B. Stege, O. Madsen, H.J. Wijnen, S.K. Kar and J.M.J. Rebel

Supplementing heat stressed cows with plant extract affects performance welfare and oxidative stress 355
J.R. Daddam, D. Daniel, I. Pelech, G. Kra, H. Kamer, U. Moallem, Y. Lavon and M. Zachut

Immunomodulatory effect of exosomes from the plasma of heat-stressed cows on bovine monocytes 356
L.G. De Matos, S. Dimauro, M. Falco, J.F.S. Filipe, C. Zamboni, F. Ceciliani, V. Martini, A. Agazzi, A. Scarafoni, G.C. Heinzl, G. Sala, A. Boccardo, A. Maggiolino, P. Depalo and C. Lecchi

ıvited Late gestation heat stress: programming effects on three generations 356
J. Laporta

Temporal patterns of environmental heat stress in Holstein heifers: association with age at calving 357
V. Tsiamadis, G.E. Valergakis, A. Soufleri, G. Arsenos, G. Banos and X. Karamanlis

Effect of stocking density on heat stress in fattening pigs 357
L. De Prekel, D. Maes, A. Van Den Broeke and M. Aluwé

Chromium yeast alleviates heat stress in Holstein mid-lactation dairy cows 358
F. Ma, Y. Wo, Q. Shan and P. Sun

May anti-oxidants reduce glyphosate-based herbicides adverse effects on chicken embryo-development? 358
M. Fréville, A. Estienne, C. Ramé, C. Rat, J. Delaveau, P. Froment and J. Dupont

Bromide as a thyroid hormone disruptor 359
H.L. Lucht and N.H. Casey

The impact of THI on milk characteristics, productivity, and metabolism in lactating dairy cows 359
G. Meli, V. Fumo, G. Savoini and G. Invernizzi

Poster Session 22

A selected phytogenic solution improves performance and thermal tolerance of heat stressed pigs 360
A. Morales, P. Sakkas, M. Soto, N. Arce, N. Quilichini and M. Cervantes

Effects of Met, Lys and His supplementation on the metabolic response to heat stress in dairy cows 360
Á. Kenéz, E. Jorge-Smeding, Y.H. Leung, A. Ruíz-González and D.E. Rico

Evaluation of earwax and hair cortisol level in relation to environmental stressors in Hanwoo cattle 361
M. Ataallahi, G.W. Park, E. Nugrahaeningtyas, M. Dehghani, J.S. Lee and K.H. Park

Effects of energy levels and heat stress on growth performance and blood metabolites in Hanwoo calve 361
Y.H. Jo, J.G. Nejad, J.S. Lee, K.K. Park, E.J. Kim and H.G. Lee

The effect of a summer diet on heat stress in lactating sows and their piglets 362
A. Van Den Broeke, L. De Prekel, D. Maes and M. Aluwé

Physiological implication of omega-3 fatty acid supplementation on animal welfare in laying hens 362
S.K. Kang

The effect of different bedding materials on dust and toxic gas concentrations in calf housing 363
D. Witkowska, A. Ponieważ and D. Murawska

Mitigate density stress by using a proprietary orange essential oil 363
A.N. Nalovic, A.A. Auvray and J.G. Gabarrou

The effects of heating drinking water or calcium propionate on fattening beef cattle in winter 364
T. He, Z. Chen and S. Long

Water quality in dairy cattle farms: impact on animal production, reproduction and health 364
V. Resende

Study on the effect of improving the pig gut microbiota of *Rhodobacter sphaeroides* 365
D. Shin, J. Park, S. Son, H. Lee and J. Kim

Algae-based beta-glucan mitigates the adverse effects of heat stress in lactating sows 365
J. Mun, A. Hosseindoust, S. Ha, S. Park, H. Tajudeen, S. Oh and J. Kim

Effect of dietary mineral level and creep feeding on sows performance under high ambient temperature 366
H. Tajudeen, A. Hosseindoust, J.Y. Mun, S. Ha, S. Park, S. Yoon and J. Kim

Session 23. Improving animal welfare in conventional production systems

Date: Monday 28 August 2023; 14.00 – 18.00
Chair: Minozzi / Trocino

Theatre Session 23

invited Pre-weaned dairy calf welfare: from the 'End the Cage Age' movement to the EFSA mandate 366
M. Brscic

Impact of individual pens removal on veal calves' behaviour 367
D. Bastien and V. Lefoul

Quarter level milk yield in dairy cows before and after separation in a cow-calf contact system 367
S. Ferneborg and S. Agenäs

Two different training protocols to encourage dam-reared calves to drink supplemental milk 368
J. Sørby, S. Ferneborg, S.G. Kischel and J.F. Johnsen

Effects of automatic TMR feeding system on the health status of finishing bulls and heifers 368
O. Martinić, L. Magrin, P. Prevedello, G. Fabbri, G. Cozzi and F. Gottardo

Associations between feed push-up frequency, feeding behaviour and milk yield in dairy buffaloes 369
G. Esposito, E. Raffrenato, G. Bosoni, F. Caruso and F. Righi

Comparing dairy cow welfare in loose-housing and in tie-stall systems using animal-based indicators 369
V. Lorenzi, L. Bertocchi, C. Montagnin and F. Fusi

Time-controlled hay racks horses: what effect on behaviour? 370
M. Roig-Pons and S. Briefer

Exploring techniques to mitigate heat stress in feedlot lambs 370
P.G. Theron, T.S. Brand, S.W.P. Cloete and K. Dzama

Management practices and welfare of dairy goats in Greece 371
G. Arsenos, S. Vouraki, V. Papanikolopoulou, L.V. Ekateriniadou and I. Sakaridis

Risk factors for navigation ability of laying hens at housing in an aviary system 371
C. Ciarelli, F. Bordignon, G. Pillan, G. Xiccato and A. Trocino

Busy birds – what do turkey hens use dust baths for? 372
K. Skiba, M. Kramer, P. Niewind, B. Spindler and N. Kemper

Pre-slaughter fasting on stress response and meat quality in rainbow trout (*Oncorhynchus mykiss*) 372
A. De La Llave-Propín, A. Martínez Villalba, R. González Garoz, J. De La Fuente, C. Pérez, E. González De Chavarri, M.T. Díaz, A. Cabezas, M. Villarroel and R. Bermejo-Poza

Poster Session 23

Effects of a new cooled cubicle waterbed on cow behaviour and lameness: a randomized control trial 373
O. Levrad, R. Guatteo, A. Lehébel, L. Bouglé, M. Tourillon, N. Brisseau and A. Relun

Evaluation of milk and residual milk production of dairy Gyr cows 373
R.C. Castanheira, M.I. Paiva, K.T. De-Sousa, M.S.V. Salles, E.A. Silva and L. El Faro

Environmental enrichment and tactile stimulation on the welfare of F1 Holstein × Gir calves 374
C.O. Miranda, A.E. Vercesi Filho, M.L.P. Lima, F.F. Simili, M.S.V. Salles, E.G. Ribeiro, J.A. Negrão and L.E.F. Zadra

Heat stress indicators in dairy cattle breeding provided by automatic milking systems (AMS) 374
M.A. Leandro, J. Stock, J. Bennewitz and M.G.G. Chagunda

Impact of floor type and group size on veal calves' behaviour 375
D. Bastien and V. Lefoul

Session 24. New traits and proxies from sensor technologies for breeding purposes

Date: Monday 28 August 2023; 14.00 – 18.00
Chair: Pocrnic / Lourenco

Theatre Session 24

Heterogeneity of 3D cameras feed intake heritabilities in Holstein, Jersey and Nordic Red cows 375
C.I.V. Manzanilla-Pech, R.B. Stephansen and J. Lassen

Genomic correlations of MIR-predicted and measured feed efficiency traits in Holstein Friesian 376
A. Seidel, C. Heuer, L.M. Dale, A. Werner, H. Spiekers and G. Thaller

Heritability of dry matter intake in Norwegian Red with measures on feed intake in commercial herds 376
K.A. Bakke and B. Heringstad

Assessing the potential of faecal NIR spectra for the prediction of feed efficiency in dairy cows 377
S. Ampuero Kragten, T. Haak, K.-H. Südekum and F. Schori

Genetic parameters for potential auxiliary traits for lameness based on data from PLF technologies 377
K. Schodl, B. Fuerst-Waltl, F. Steininger, M. Suntinger, H. Schwarzenbacher, A. Köck, D4dairy Consortium and C. Egger-Danner

Breeding for resilience based on rumination time data of Chinese Holstein heifers 378
W. Lou, R. Shi, B. Ducro, A. Van Der Linden, H.A. Mulder, S.J. Oosting and Y. Wang

Validation of a high-throughput movable 3D device for the acquisition of the whole cattle body 378
A. Lebreton, C. Allain, L. Delattre, J. Loof, Y. Do and M. Bruyas

MIRS coupled with machine learning for the prediction of cow colostrum immunoglobulins concentration 379
A. Costa, A. Goi and M. De Marchi

Extracting video-based phenotypes in a pig breeding programme 379
C. Coello, Ø. Nordbø, R. Sagevik, F. Cheikh, M. Ullah, K.H. Martinsen and E. Grindflek

Impact of the social effect on feeding behaviour in pigs 380
P. Núñez, S. Gol, J. Reixach and N. Ibañez-Escriche

Assessing lactating sow behaviour using sensor technology and machine learning 380
G. Dumas, L. Maignel, J.-G. Turgeon and P. Gagnon

Relationships between sow activity over the days after farrowing and litter weight gain 381
O. Girardie, D. Laloë, M. Bonneau, Y. Billon, J. Bailly, I. David and L. Canario

Multi-camera tracking of turkeys in large groups using instance segmentation 381
Z. Wang, P. Langenhuizen, B. Visser, H.P. Doekes, P. Bijma and P.H.N. De With

Evaluation of tracking individual broilers using video data 382
J.E. Doornweerd, G. Kootstra, R.F. Veerkamp, B. De Klerk, M. Van Der Sluis, A.C. Bouwman and E.D. Ellen

Poster Session 24

Genetic variation of new sheep traits measured by dual energy Xray absorptiometry 382
C.E. Payne, B. Paganoni, S.F. Walkom, G.E. Gardner and D.J. Brown

Session 25. Molecular measures of diversity and their role in monitoring and management of breeds

Date: Monday 28 August 2023; 14.00 – 18.00
Chair: Danchin / Fernandez

Theatre Session 25

invited Monitoring within-breed genetic variation at global level 383
G. Leroy, R. Baumung, G. Mészáros, I. Curik, J.J. Windig, B. Rosen, Y.T. Utsunomiya, P. Burger, L. Coli, P. Boettcher, A. Stella, C. Looft and J. Soelkner

Real-time monitoring of genetic diversity in animal populations: the RAGEMO project 383
C. Danchin, I. Palhière, B. Servin and G. Restoux

Evolution of the genomic relationship between a local breed and its mainstream sister breed 384
H. Wilmot, M.P.L. Calus and N. Gengler

Fine decomposition of inbreeding into the source of the co-ancestries 384
S. Antonios, S.T. Rodríguez-Ramilo, J.M. Astruc, L. Varona and Z.G. Vitezica

Predicting homozygosity-by-descent to manage inbreeding and diversity 385
N.S. Forneris and T. Druet

GWAS analyses for dairy traits in Cyprus sheep and goat breeds reveal important protein networks 385
G. Maimaris, A.N. Georgiou, A.C. Dimitriou and G. Hadjipavlou

A severely introgressed South China pig breed in urgent need of purification 386
L. Cao, T. Luan, T.H.E. Meuwissen, P. Berg, J. Yang and Z. Wu

Insights into ancient and recent inbreeding patterns in Alpine Grey cattle using ROH analysis 386
G. Gomez Proto, E. Mancin, B. Tuliozi, C. Sartori and R. Mantovani

Study of Bottleneck effect for conservation of Poonchi chicken from international borders of India a 387
M. Azad, D. Chakraborty, K. Kour, D. Kumar and P. Birwal

Up to date results from the ongoing *in situ* genetic management of the local breed Noire de Challans 387
R. Rouger, H. Deloison, M. Teissier, G. Restoux and S. Brard-Fudulea

An empirical approach to assess increase in homozygosity: the Gochu Asturcelta pig 388
K.D. Arias, J.P. Gutiérrez, I. Fernández, I. Álvarez and F. Goyache

Preliminary analysis of Nubian goats and their influence on Old English and Anglo-Nubian goats 388
S.A. Rahmatalla, D. Arends, G.B. Neumann, H. Abdel-Shafy, J. Conington, M. Reissmann, M.K. Nassar and G.A. Brockmann

Towards a flexible definition of core sets based on the haplotype diversity of German sheep 389
J. Geibel, C. Reimer, A. Weigend, Y. Shakya, H. Melbaum, K. Gerdes and S. Weigend

Use of genomic information for the determination of insemination doses per sire in a gene bank 389
C. Reimer, S. Weigend, T. Pook and J. Geibel

Poster Session 25

Is it possible to skip the SNP selection step for breed assignment? 390
H. Wilmot, T. Niehoff, H. Soyeurt, N. Gengler and M.P.L. Calus

Copy number variation regions differing in segregation patterns spanned different sets of genes 390
K.D. Arias, J.P. Gutiérrez, I. Fernández, I. Álvarez and F. Goyache

Genomic characterization as tool in the breed recognition process 391
C. Danchin and D. Duclos

Homozygosity by descent in mice divergently selected for environmental birth weight variability 391
C. Ojeda-Marín, J.P. Gutiérrez, N. Formoso-Rafferty and I. Cervantes

Inbreeding depression is associated with recent HBD segments in Belgian Blue beef cattle 392
M. Naji, J.L. Gualdron Duarte, N.S. Forneris and T. Druet

Assessing the intrapopulation structure of Slovak Spotted cattle by genome-wide data 392
R. Kasarda, N. Moravčíková, J. Candrák, J. Prišťák and I. Pavlík

Genomic selection has aided Nordic Jersey to decrease risks of inbreeding 393
S. Tenhunen, J.R. Thomasen, L.P. Sørensen, P. Berg and M. Kargo

Management of local cattle breed diversity in Croatia: use of molecular and pedigree indicators 393
A. Ivanković, Z. Ivkić, G. Šubara, M. Pećina, J. Ramljak and M. Konjačić

Genomic differences between the domesticated pig breed and its ancestor wild boar 394
N. Moravčíková, R. Kasarda, M. Hustinová, E. Krupa, Z. Krupová and E. Žáková

Genetic polymorphism of kappa-casein in Serbian Holstein-Friesian cattle 394
M. Zorc, M. Šaran, L.J. Štrbac, D. Janković, P. Dovč and S. Trivunović

Estimation of recent and ancestral inbreeding for X-chromosome in Old Kladrub horse 395
L. Vostry, M. Shihabi, J. Farkas, G. Kövér, H. Vostra-Vydrova, B. Hofmanova, I. Nagy and I. Curik

Genetic diversity and population structure of a Peruvian nucleus cattle herd using SNP data 395
F.-A. Corredor, D. Figueroa, R. Estrada, W. Salazar, C. Quilcate, H. Vasquez, J. Gonzales, J. Maicelo, P. Medina and C. Arbizu

Within-breed stratification for across-breed reference population 396
H. Wilmot, T. Druet, I. Hulsegge, N. Gengler and M.P.L. Calus

Session 26. EAAP Plenary Session

Date: Tuesday 29 August 2023; 8.30 – 13.00
Chair: Casasús

Theatre Session 26

invited The livestock sector and research in animal science in France 396
X. Fernandez, J.F. Hocquette and S. Ingrand

invited Solutions to achieve healthy and sustainable diets worldwide 397
H. Guyomard, A. Forslund, A. Tibi, B. Schmitt, P. Debaeke and J.-L. Durand

invited Moving towards bold food systems resilience 397
J. Fanzo

invited A social-ecological agenda for transforming land management toward sustainability 398
T. Plieninger

invited EAAP beyond Europe: establishing and using capacity to feed the world 398
H. Sölkner

Session 27. Breeding programs & strategies

Date: Tuesday 29 August 2023; 15.00 – 18.30
Chair: Egger-Danner / Martin

Theatre Session 27

Effect of genomic selection on allele frequencies of QTL associated to number of teats in pigs 399
C.A. Sevillano, B. Harlizius, M.S. Lopes, M. Van Son and E.F. Knol

Management of genetic defects in breeding programs 399
S.T. Rodríguez-Ramilo, I. Palhière, J. Raoul and J. Fernández

Genetic management of cryptorchidism and horn mutations in Manech tête Rousse dairy sheep breed 400
J. Raoul, F. Fidelle, C. André, M. Ben Braiek, S. Fabre, A. Gouzenes, D. Buisson and I. Palhière

Strategies to improve selection compared to selection based on estimated breeding values 400
T. Pook, T. Niehoff, Y. Wientjes, L. Zetouni, M. Schrauf and M. Calus

The validity of breeding goals 401
R. Wellmann

Beyond scenarios-optimization of breeding program design using evolutionary algorithms (MoBPSopti) 401
A. Hassanpour, J. Geibel and T. Pook

Ecobreed – what is the economically optimal longevity of a cow? 402
S. Schlebusch

Differences in breast milk composition in rabbit lines divergently selected for intramuscular fat 402
P. Hernández, N. Ibáñez, I. Heddi, M. Martínez-Álvaro and A. Zubiri-Gaitán

Combined genomic evaluation of Australian Merino and Dohne Merino sheep populations 403
M. Wicki, D.J. Brown, P.M. Gurman, J. Raoul, A. Legarra and A.A. Swan

Effect of germplasm exchange strategies on genetic gain and diversity in dairy stud populations 403
E.A. Lozada-Soto, F. Tiezzi, J. Jiang, J.B. Cole, P.M. Vanraden, S. Toghiani and C. Maltecca

Local livestock breeds in Flanders are confirmed to be 'at risk'! 404
S. Janssens, L. Chapard, R. Meyermans, W. Gorssen and N. Buys

Probabilistic breeder's equation used for retrospective and prospective analyses 404
D. López-Carbonell, I. Pocrnic, G. Gorjanc and L. Varona

Poster Session 27

Re-organising the Danish dairy cattle sector with alternative breeding goals and crossbreeding 405
J.B. Clasen, R.D. Kring, J.R. Thomasen and S. Østergaard

Session 28. Breeding for improved animal health and welfare

Date: Tuesday 29 August 2023; 15.00 – 18.30
Chair: Kargo / Tiezzi

Theatre Session 28

vited How to balance selection for litter size in pigs with survival, health and welfare 405
P.W. Knap, A. Huisman, C. Sørensen and E.F. Knol

Selection for robustness and welfare in Iberian pig through birth weight uniformity as criterion 406
J.P. Gutiérrez, N. Formoso-Rafferty, F. Sánchez-Esquiliche, M. Muñoz, J.M. García-Casco and I. Cervantes

Resilience parameters in fattening pigs are heritable and associated with tail biting and mortality 406
W. Gorssen, C. Winters, R. Meyermans, L. Chapard, K. Hooyberghs, J. Depuydt, S. Janssens, H. Mulder and N. Buys

The impact of phenotyping, genotyping, and the boar's origin on the genetic gain of organic pigs 407
R.M. Zaalberg, J.B. Clasen, T.M. Villumsen, J. Jensen and T.T. Chu

Genetic correlations between ostrich behavioural traits and slaughter traits 407
P.T. Muvhali, M. Bonato, A. Engelbrecht, I.A. Malecki and S.W.P. Cloete

Genome-wide copy number variants associated with calving ease and retained placenta in Holstein cows408
I.C. Hermisdorff, H.R. Oliveira, G.A. Oliveira Júnior, T.C.S. Chud, S.G. Narayana, C.M. Rochus, A.M. Butty, F. Malchiodi, P. Stothard, F. Miglior, C.F. Baes and F.S. Schenkel

Indirect effects in infection transmission enhance genetic selection and other interventions 408
A.D. Hulst, P. Bijma and M.C.M. De Jong

Optimizing dairy cattle breeding goals to improve production and udder health of crossbred cows 409
A. Bouquet, H.M. Nielsen, V. Milkevych, M. Kargo, J.R. Thomasen and M. Slagboom

Genes and variants involved in resistance to paratuberculosis in Holstein and Normande cattle 409
V. Sorin, A. Boulling, A. Delafosse, M. Boussaha, C. Hozé, R. Guatteo, C. Fourichon, S. Fritz, D. Boichard and M.P. Sanchez

A genome-wide association study for clinical mastitis in Italian Holstein 410
F. Galluzzo, G. Visentin, G. Mészáros, J.B.C.H.M. Van Kaam, R. Finocchiaro, M. Marusi and M. Cassandro

About the genetic connection between milk and health traits in dairy cows within functional regions 410
H. Schneider, J. Heise, A.-M. Krizanac, C. Falker-Gieske, J. Tetens, G. Thaller and J. Bennewitz

Poster Session 28

Frequency of genetic disorders in genomically tested females in the Netherlands 411
C. Schrooten and E. Mullaart

Genome regions and metabolic processes associated with tick resistance in beef cattle 411
P. Martin, T. Hüe, J. Mante, A. Lescane, D. Boichard and M. Naves

Innovative strategy using phenotypes and genetics to optimize partial sampling applied to boar taint 412
A. Markey, J. Wavreille, P. Mayeres, A.C. Huet, D. Duarte and N. Gengler

Genetic susceptibility determined by SNP2 in piglets challenged with F4-Enterotoxigenic *E. coli* 412
A. Middelkoop, H. Kettunen, X. Guan, J. Vuorenmaa, R. Tichelaar, M. Gambino, M.P. Rydal and F. Molist

Reference genes selection for RT-qPCR in chickens after in ovo synbiotic and choline administration 413
E. Grochowska, P. Guz, K. Stadnicka and M. Bednarczyk

Genetic parameters of coronet scores related to hoof disorders in Japanese dairy cattle 413
Y. Saito, A. Nishiura, T. Yamazaki, S. Yamaguchi, O. Sasaki and M. Satoh

Selection for a lower growth rate to avoid reaching bone apposition limits in chickens 414
C. Leterrier

Daily milk yield, variance and skewness of milk deviations in cows' resilience 414
L. Zavadilová, E. Kašná, J. Vařeka and Z. Krupová

Lameness scoring and its use in genetic selection to improve claw health in Austrian Fleckvieh cows 415
A. Köck, C. Fuerst, B. Fuerst-Waltl, M. Suntinger, K. Linke, J. Kofler, K. Schodl, F.J. Auer and C. Egger-Danner

Association between udder health genomic breeding values and dairy and health traits in French cows 415
R. Lefebvre, S. Barbey, F. Launay, M. Gaborit, L. Delaby, P. Martin and D. Boichard

Validating resilience indicators derived from longitudinal performance measurements 416
M. Ghaderi Zefreh, R. Pong-Wong and A. Doeschl-Wilson

Approaching the genetic evaluation models for clinical mastitis in Spanish dairy cows 416
M.A. Pérez-Cabal, I. Cervantes, J.P. Gutiérrez, J. López-Paredes and N. Charfeddine

Milk somatic cell counts in winter and summer are genetically different traits in Norwegian goats 417
H.B. Olsen, J. Jakobsen and T. Blichfeldt

Effects of two different traditional sire breeds on performance and behaviour of rearing pigs 417
A. Lange, M. Wutke, S. Ammer, A.K. Appel, H. Henne, A. Deermann and I. Traulsen

Evaluation of milk parameters and mastitis predisposition in Cyprus Chios sheep 418
S. Panayidou, T. Christofi, A.N. Georgiou and G. Hadjipavlou

Influence of early rearing system on later performance of commercial laying hens 418
Q. Berger, N. Bédère, P. Le-Roy, S. Lagarrigue, T. Burlot and T. Zerjal

Differentially expressed genes in tissues of Pietrain sired pigs challenged with high fiber/fat diet 419
E.U. Nwosu, B. Chakkingal Bhaskaran, E. Kowalski, W. Gorssen, R. Meyermans, S. Janssens, S. Millet, M. Aluwé, S. De Smet and N. Buys

GWAS based on imputed sequence-level genotypes in a huge cow data set: many roles for the GC gene 419
A.M. Krizanac, C. Falker-Gieske, C. Reimer, J. Heise, Z. Liu, J. Pryce, J. Bennewitz, G. Thaller and J. Tetens

GWAS for clinical metritis in Polish HF cows 420
P. Topolski, T. Suchocki and A. Żarnecki

Genetic relationships between mastitis, milk yield traits and somatic cell count in Polish HF cows 420
P. Topolski and W. Jagusiak

Session 29. Innovative animal products and farming systems oriented to new consumer demand and trend

Date: Tuesday 29 August 2023; 15.00 – 18.30
Chair: Resconi / Realini

Theatre Session 29

invited A New Zealand perspective of the role of genetics in underpinning future sheep production systems 421
P. Johnson, D. Scobie, S.-A. Newman, S.J. Rowe, S.M. Clarke and K.M. McRae

Blockchain technology in the beef breeding sector 421
B. Jornet and I. Achour

Using science to optimise production systems to produce a premium product – the LuminaTM Lamb Story 422
P.L. Johnson, S. Saunders and A. McDermott

Factors influencing willingness to pay for red meat with potential to improve consumer wellness 422
R. Zhang, Z. Kallas, M.P.F. Loeffen, M. Lee, L. Day, M.M. Farouk and C.E. Realini

Perception of the use of rangeland and roughage by actors of the organic pig and layer hen sectors 423
L. Montagne, M. Goujon and F. Marie

invited Role of dairy foods in human nutrition 423
C.M. Weaver

Organoleptic properties of fast-growing and slow-growing chicken meat 424
R. Berrocal, L. Mur, J.L. Olleta, V. Resconi, M. Barahona, J. Romero and M.M. Campo

Implementation of sustainable livestock farming principles for generating innovative animal products 424
M. Cohen-Zinder, E. Shor-Shimoni, H. Omri, F. Garcia Solares, S. Kaakoosh and A. Shabtay

Recent changes and future expectations about meat consumption 425
F. Montossi, G. Ares, L. Antúnez, G. Brito, M. Del Campo, C. Saunders, M. Farouk and C.E. Realini

invited Sustainability in the beef sector: aligning consumer and farmer perspectives 425
M. Henchion and V.C. Resconi

Poster Session 29

FT-MIR spectroscopy to discriminate intramuscular fat of beef fed conventional vs sustainable diets 426
S. León-Ecay, I. Goenaga, A. López-Maestresalas, A. Gobeti Barro, M.P. Ellies-Oury, M. Beruete and K. Insausti

Study of the genetic parameters of the destructured meat defect of ham on 3 male lines 426
A. Le Dreau, C. Garcia Baccino, D. Penndu, A. Buchet, P. Doussal and B. Ligonesche

Consumers' perception towards using seaweed as an alternative to antibiotics in rabbits in Spain 427
S. Al-Soufi, J. García, E. Cegarra, A. Muíños, M. Miranda and M. López-Alonso

Perceptions of viticulture-livestock complementarity in Burgundy: first insights 427
R. Ibidhi, O. Aguirre-Saavedra, S. Ambrosino, G. Houndafoche, M. Seignon, Y. Tanguy-Roump and C. Philippeau

Feeding seaweed as an alternative to antibiotics in growing rabbits improves the meat fat profile 428
S. Al-Soufi, J.M. Lorenzo, J. García, E. Cegarra, A. Muiños, M. Miranda and M. López-Alonso

Effects of lamb genotype, primal cuts, and vegetables on baby soups' sensory acceptability 428
M.R. Marques, M. Pimpão, C. Oliveira and J.M. Almeida

Lipid fraction composition of the longissimus thoracis of Lidia beef breed 429
M.M. Campo, M. Barahona, V.C. Resconi, J.V. Romero, S. Romero, S. Zabala, J. Villalon and J.L. Olleta

Innovation in the production and marketing of local breed products, path to long-term sustainability 429
A. Ivanković, G. Šubara, E. Šuran, M. Cerjak, J. Ramljak and M. Konjačić

The ability of the Distell fatmeter to predict intramuscular fat in Yellowtail Kingfish 430
D. Milotic, F. Anderson, G. Partridge, A. Lymbery and G. Gardner

Effect of game species on fatty acids, meat quality and shelf life 430
M.N. Hlohlongoane, U. Marume, O.C. Chikwanha and C. Mapiye

Selection for A2 β-casein genotype in Holstein bulls 431
L. Jiménez-Montenegro, L. Alfonso, J.A. Mendizabal and O. Urrutia

Session 30. Looking back, looking forward – Research on Beef production and quality in China and France (CSBF) and between Australia-France-China (The Triangle Project)

Date: Tuesday 29 August 2023; 15.00 – 18.30
Chair: Liu / Pethick

Theatre Session 30

Consumer perception of the challenges facing livestock production and meat consumption 431
J.J. Liu, S. Chriki, M. Kombolo, M. Santinello, S. Pflanzer, É. Hocquette, M.P. Ellies-Oury and J.F. Hocquette

invited Sustainable development of China's beef industry driven by a successful triangle beef cooperation 432
Q. Meng

nvited The production and quality improvement of mutton in China 432
H.L. Luo

Comparison of sheep meat eating quality thresholds for Chinese, American and Australian consumers 433
R.A. O'Reilly, D.W. Pethick, G.E. Gardner, A.B. Pleasants and L. Pannier

Differences in liver nutrient metabolism contribute to residual feed intake of beef cattle 433
Z.M. Zhou

Marbling in French cattle: average level and factors of variations 434
A. Nicolazo De Barmon, I. Legrand and J. Normand

Effect of high sulphur diet on rumen fermentation and epithelial barrier function in beef cattle 434
H. Wu, Y. Li, Q.X. Meng and Z.M. Zhou

Research progress on quality characteristics and regulation of yak meat in China 435
L.Z. Hao, B.Q. Bai, Y. Xiang, Q.X. Meng, J.Z. Niu, Y.Y. Huang and S.J. Liu

Comparison of beef eating quality terminology databases 435
A. Barro, K. Insausti, M. Kombolo, M.P. Ellies-Oury and J.F. Hocquette

Comparison of slaughter performance and meat quality of Tan sheep under different feeding regimes 436
X.G. Zhao, C. Zhang, M. Liu and H.L. Luo

Effect of *Piper sarmentosum* extract on the growth and nutrient digestion of Hainan Black goat 436
G.D. Ren, Z.Y. Sheng, Y.X. Chen, X.G. Zhao, H. Zhang, H.L. Zhou, W.L. Lv and H.L. Luo

Effects of grazing intensities and supplementary levels on the nutritional composition of lamb meat 437
H. Yang, J. Ji and H.L. Luo

Poster Session 30

Perspectives on consumer attitudes to meat consumption 437
P. Purslow, W. Zhang and J.F. Hocquette

Effects of different grazing intensities and supplementary feeding levels on meat quality of Hulunbu 438
X.H. Ma, Z. Li, X.G. Zhao, H. Yang and H.L. Luo

Progress of yak carcass grading in Qinghai-Tibet Plateau of China 438
S.S. Zhang

Session 31. Pasture based solutions to realise a role for ruminants in a sustainable food system

Date: Tuesday 29 August 2023; 15.00 – 18.30
Chair: Boland / Baumont

Theatre Session 31

vited Future-proofing the sustainability of pasture based beef production systems 439
A.K. Kelly and T.M. Boland

Effect of sward type on growth performance and methane output of lambs in the post-weaning period 439
S. Woodmartin, P. Creighton, T.M. Boland, A. Monaghan and F. McGovern

On farm grass growth prediction in Ireland – 4 year evaluation 440
E. Ruelle, L. Bonnard, D. Hennessy, L. Delaby and M. O'Donovan

Incorporating plantain forage into a grazing dairy system: effect on farm productivity 440
O. Al-Marashdeh and H.M.G.P. Herath

Dry matter production of multispecies swards under dairy grazing in two chemical nitrogen scenarios 441
C. Hearn, M. Egan, M.B. Lynch and M. O'Donovan

Nitrogen herbage yield in grass and grass-white clover swards receiving zero nitrogen 441
Á. Murray, B. McCarthy and D. Hennessy

Grassland-based dairy farms of French Massif Central adapt to climate change with diverse strategies 442
L. Allart, V. Oostvogels, F. Joly, C. Mosnier, N. Gross and B. Dumont

Effect of grazing intensity on nitrogen cycle of alpine summer pasture soil 442
S. Raniolo, A. Squartini, L. Da Ros, F. Camin, L. Bontempo, L. Maretto, D. Gianelle, M. Ramanzin, E. Sturaro and M. Rodeghiero

The visibility of the invisible: analysing heifers reactions while learning the virtual fence system 443
D. Hamidi, N.A. Grinnell, M. Komainda, L. Wilms, F. Riesch, J. Horn, M. Hamidi, I. Traulsen and J. Isselstein

Sheep integration into cropping systems for agroecological transition of farming systems in France 443
F. Stark, J. Ryschawy, R. Mettauer, M. Grillot, I. Shaqura and M. Moraine

Nutrient cycling and efficiency: a comparative flow analysis of meat and dairy sheep farming systems 444
F. Stark, N. Amposta, W. Nasri, M. Lamarque, S. Parisot, P. Salgado and E. González-García

Pasture-based ruminants and biodiversity in contrasting contexts: an account of farmers' narratives 444
V.J. Oostvogels, B. Dumont, H.J. Nijland, L. Allart, I.J.M. De Boer and R. Ripoll-Bosch

Poster Session 31

Meat quality in Podolian young bulls grazing on wood-pasture in a South Italy marginal area 445
F. Giannico, S. Tarricone, A. Caputi Jambrenghi, L. Tedone, D. Campanile and M.A. Colonna

Good management of the pasture makes it possible to reduce the supplementation of autumn calves 445
D. Douhay

Investigating the effects of lactation number, stage, and milking interval on nitrogen use efficiency 446
Y.M. Hu

Plasma urea and diet crude protein concentration relationship in grazing dairy cows 446
R. Delagarde and N. Edouard

An assessment of woody forage resources indigenous to southern Africa 447
M. Trytsman, F.L. Müller, M.I. Samuels, C.F. Cupido and A.E. Van Wyk

The Effect of heterofermentative lactic acid bacteria on quality of King Napier grass 447
P. Lounglawan

Grass chemical composition to predict methane production from grazing sheep at high latitude 448
Q. Lardy, M. Hetta, M. Ramin and V. Lind

Forage height of native grasslands as an indicator of spatial and temporal grazing management 448
P. Aparicio, I. Paparamborda and P. Soca

Improve the efficiency of fattening grazing sheep on rangeland by supplementary feeding 449
Y. Zhang and M. Xu

The use of sheep of native breeds in the control of invasive plants 449
M. Pasternak, M. Puchała, J. Sikora and A. Kawęcka

Session 32. Beyond rumen: role of nutrition in cattle's intestinal health

Date: Tuesday 29 August 2023; 15.00 – 18.30
Chair: Kreuzer / Pinotti

Theatre Session 32

vited Post-rumen health and its implications on health and performance 450
M.V. Sanz-Fernandez

vited Impacts of inflammation and inflammatory diseases on reproduction in dairy cows 450
J.E.P. Santos

vited The role of mycotoxins in the rumen and intestinal health of dairy cows 451
A. Gallo

vited Absorption, metabolism and secretion of tocopherol (vitamin E) stereoisomers in dairy cows 451
S.K. Jensen and S. Lashkari

vited The role of rumen health (acetate) on milk fat synthesis and dietary strategies to increase milk fat 452
K.J. Harvatine and C. Matamoros

Session 33. Horses in farms and territories: challenges and new solutions

Date: Tuesday 29 August 2023; 15.00 – 18.30
Chair: Vial / Tapprest

Theatre Session 33

EUNetHorse-European network to improve the resilience and the performance of equine farms 452
M. Addes

Robustness of equine business structures and global coherence 453
S. Boyer

The equine network, a tool for horse professionals, advisors and teachers 453
G. Mathieu

Does the diversity of the French territories impact farms keeping equines? 454
J. Veslot and G. Bigot

Equestrian practitioners: essential services to support self-organization 454
C. Eslan, C. Vial and S. Costa

Agricultural animal traction in France: characterization of the practice and its users 455
M.M. Miara, M.G. Gafsi and P.B. Boudes

Effects of milking interval and frequency on milk yield in the mare 455
J. Auclair-Ronzaud, M. Bouchet, L. Laschon, E. Lambolez and L. Wimel

Mineral profile of pasture-based mare milk from Basque Mountain horse breed: effect of lactation 456
A. Blanco-Doval, L.J.R. Barron and N. Aldai

Sustainable utilisation of horse manure 456
M. Meyer, C. Eiberger, T. Schilling, D. Winter and L.E. Hoelzle

Use of green waste compost as an alternative bedding to straw in horse husbandry 457
H. Unseld, D. Winter and M. Meyer

Grazing management on practical horse farms 457
C. Siede, M. Komainda, B. Tonn, S.C.M. Wolter, A. Schmitz and J. Isselstein

Use of digital technologies in horse husbandry to increase animal welfare and health 458
M. Pfeiffer, D. Winter, U. Dickhoefer and L.T. Speidel

Poster Session 33

Working time requirements of work processes in the individual keeping of horses 458
L.T. Speidel, D. Winter and U. Dickhoefer

Equine traction in viticulture: what about the workload of the horse-driver pair? 459
B. Pasquiet, N. Delepouve and C. Bénézet

Physiological and behavioural responses of donkeys to effort and discomfort 459
N. Seguin, C. Bonnin, A. Ruet and S. Biau

Offering horse meat in restaurants to develop the demand 460
C. Vial, M. Sebbane and A. Lamy

Relationship between average daily gain and an estimation of consumed grass 460
L. Wimel, C. Lesoudard and J. Auclair-Ronzaud

Comparison of flying and soil invertebrates' biodiversity on meadows with or without horse grazing 461
G. Goudet, F. Reigner and M. Beltramo

Alterations of faecal microbiota in Jeju crossbred weanling horses 461
J.A. Lee, M.C. Shin, J.Y. Choi and S.M. Shin

Colic incidence among horses in different husbandry systems 462
E. Schlotterbeck, L.T. Speidel and D. Winter

Session 34. Sustainable pig and poultry production, including the use of water

Date: Tuesday 29 August 2023; 15.00 – 18.30
Chair: Bee / Brodziak

Theatre Session 34

The effect of feeding a soy-free diet on performance and carbon footprint of the feed intake by pigs 462
S. Millet, M. Lourenço, S. Palmans and C. De Cuyper

Exploring variation in feed efficiency of grower-finisher pigs 463
M. Van Der Heide, J.V. Nørgaard and J.G. Madsen

Former food slightly affect pig faecal microbiota without impairing jejunal integrity and physiology 463
M. Tretola, S. Mazzoleni, P. Silacci, L. Pinotti and G. Bee

Bakery products and legume seeds in the diet of growing-finishing pigs 464
M. Van Helvoort and P. Bikker

Recycling agricultural by-products: using purple carrots as an ingredient in layer quails feeding 464
A.S.-G. Sarmiento-García, A.B.-D. Benito-Diaz, C.V.-A. Vieira-Aller and O.O. Olgun

Effect of camelina cake doses as soybean meal substitutions on growth and gut health of piglets 465
D. Luise, F. Correa, S. Virdis, C. Negrini, G. Cestonaro, L. Nataloni, G. Titton, E. Sattin, E. Costanzo and P. Trevisi

The dietary fibre solubility in the maternal diet does not affect the muscle development of piglets 465
M. Girard, F. Correa, F. Palumbo, P. Silacci and G. Bee

Improving the sustainability of the Australian pork industry 466
G.L. Wyburn

Are tailor-made health plans effective in triggering changes in pig farm management? 466
P. Levallois, M. Leblanc-Maridor, A. Scollo, P. Ferrari, C. Belloc and C. Fourichon

Pork production can contribute positively to human protein supply 467
R.J.E. Hewitt, D.N. D'Souza and R.J. Van Barneveld

Environmental and economic assessment of a French free-range chicken production system 467
B. Méda, E. Barlier, E. Péchernart, J.-Y. Limier and S. Mignon-Grasteau

Effects of values and calculation changes in three IPCC guidelines on greenhouse gas inventories 468
E. Nugrahaeningtyas, J.S. Lee and K.H. Park

Poster Session 34

Bioavailability and repellent activity of phytocompounds against red mites 468
C. Carlu, T. Chabrillat, C. Girard and S. Kerros

Riboflavin yield of common raw ingredients for organic poultry diets 469
V. Decruyenaere, N. Everaert, P. Rondia and J. Wavreille

Sustainable poultry production: biocontrol measures, dietary approaches and packaging strategies 469
G. Maiorano and S. Tavaniello

Genetic correlations between feeding behaviour, meat quality, carcass and production traits in pigs 470
A.T. Kavlak, T. Serenius and P. Uimari

The effect of plant-derived terpenes additive in *Ascaris suum* management in organic fattening pigs 470
H. Bui, E. Belz, J. Ligonniere and M.E.L.A. Benarbia

Effect of fattening and slaughter value of gilts on their lifetime piglet production 471
M. Szyndler-Nędza, M. Tyra and A. Mucha

Are the ethics of pig breeding addressed in literature? 471
M. Van Der Sluis, R.S.C. Rikkers and K.H. De Greef

Benzoic acid reduces environmental impact in fattener pig diets using life cycle assessment 472
E. Perez-Calvo, S. Lagadec, C. Drique, D. Planchenault, S. Potot and C. Valliere

Session 35. Is there a future for experimental animal research in Europe and, if so, what is it?

Date: Tuesday 29 August 2023; 15.00 – 18.30
Chair: Chalvon-Demersay / Van Milgen

Theatre Session 35

Is there a future for experimental animal research in Europe? 472
J. Van Milgen and T. Chalvon-Demersey

Testing minimally invasive blood collection techniques in pigs used for research 473
F.A. Eugenio, F. Gondret, M. Oster and C. Ollagnier

Development of protocols for standard management and recording in pig research facilities 473
A. Wallenbeck, M. Girard, M. Johansen, S. Düpjan, M. Aluwe, C. De Cuyper, E. Labussière, M. Font-I-Furnols, M. Heetkamp and R. Westin

Development of a protocol for administering a capsule to sample small intestine content 474
I. Garcia Viñado, M. Tretola, G. Bee and C. Ollagnier

A modelling approach to investigate metabolic fluxes of amino acids in the small intestine of pigs 474
C.J.J. Garçon, J. Van Milgen, N. Le Floc'h and Y. Mercier

In vitro gas production of pre-treated or untreated feedstuffs using cecum inoculum from horses 475
R.H. Jensen, R.B. Jensen, M.O. Nielsen and A.L.F. Hellwing

Round table discussion 475
J. Van Milgen and T. Chalvon-Demersey

Poster Session 35

UpDown – an R Package to characterize unknown disturbances from longitudinal observations 476
I. David, V. Le and T. Rohmer

Session 36. Technologies in insect production + controlling quality

Date: Tuesday 29 August 2023; 15.00 – 18.30
Chair: Heckmann / Deruytter

Theatre Session 36

Effect of sex ratio and density mating on reproductive performances in *Tenebrio molitor* 476
E. Soucat, K. Paul, Q. Li, A. Masseron, F. Gagnepain-Germain, A. Chauveau, E. Sellem and T. Lefebvre

Behavioural responses of black soldier fly to olfactory stimuli: evaluation of a Y-tube olfactometer 477
A. Spindola and M. Pulkoski

Cold storage: a tool for delayed and stable black soldier fly (*Hermetia illucens*) pupae eclosion 477
D. Deruytter, S. Bellezza Oddon, L. Gasco and C.L. Coudron

Influence of temperature on coloration in honey bees 478
J. Bubnič and J. Prešern

The juvenile hormone analogue, pyriproxifen, alters the body composition of *Tenebrio molitor* larvae 478
V. Hill, T. Parr, J. Brameld and A. Salter

Fate of food pathogens during black soldier fly rearing and processing 479
J. De Smet, D. Van De Weyer, E. Gorrens, N. Van Looveren and M. Van Der Borght

Insect as animal feed: Fourier-transformed spectroscopy to reveal their protein molecular structure 479
J. Ortuño and K. Theodoridou

Proteomics evaluation of the barrier role of insects for the indirect recycling of fast food in feed 480
M.C. Lecrenier, M. Aerts, A. Cordonnier, L. Plasman, O. Fumière and V. Baeten

Scaling up fly mating chambers: lessons learned from operating 4 and 24 m^3 fly mating chambers 480
S.P. Salari and M.L. De Goede

ALL-Yn: An automated solution for phenotyping *Tenebrio molitor* larvae 481
Q. Li, T. Mangin, J. Richard, E. Sellem, A. Masseron, F. Gagnepain-Germain, T. Lefebvre and R. Baude

Does sex matter? The impact of sex on the buffalo pupation phase 481
N. Gianotten, N. Heijmans and N. Steeghs

Assessment of chemical hazards in insect meal production for aquaculture feeds 482
I. Amaral, S.M. Ahmad, M. Ângelo, P. Correia Da Silva, A. Quintas, J.A.A. Brito, D. Murta, M.V. Santos, I. Vieira, T. Ribeiro and L.L. Gonçalves

Poster Session 36

Microbiological risk analysis in insect meal production 482
D. Guerreiro, D. Murta, M.V. Santos, I. Vieira, T. Ribeiro and H. Barroso

Multi directional approach in fly emergence evaluation and forecasting for *Hermetia illucens* 483
B. Grodzki, M.G. Zhelezarova, T. Bruder and M. Tejeda

Application of NIRS on *Tenebrio molitor* protein meal 483
J. Richard, L. Daraï, L. Sanchez and B. Lorrette

Session 37. Omics and integrative analyses towards understanding inter-organ cross-talk and whole body physiology of livestock

Date: Tuesday 29 August 2023; 15.00 – 18.30
Chair: Louveau

Theatre Session 37

State of the art in understanding interorgan crosstalk and physiology in farm animals 484
I. Cassar-Malek, I. Louveau, C. Boby, J. Tournayre, F. Gondret and M. Bonnet

Gene networks controlling functional cell interactions in the pig embryo revealed by omics studies 484
A. Dufour, C. Kurilo, J. Stöckl, Y. Bailly, P. Manceau, F. Martins, S. Ferchaud, B. Pain, T. Fröhlich, S. Foissac, J. Artus and H. Acloque

1H-NMR metabolomic study of Large White and Meishan pigs in late gestation: part 1 – foetal placenta 485
J. Guibert, A. Imbert, N. Marty-Gasset, L. Gress, C. Canlet, L. Canario, Y. Billon, A. Bonnet, L. Liaubet and C.M.D. Bonnefont

Metabolic pathways leading to different intramuscular fat content in rabbit divergent lines 485
P. Hernández, A. Blasco and A. Zubiri-Gaitán

Proteomics profiles of longissimus thoracis muscle of Arouquesa cattle from different systems 486
L. Sacarrão-Birrento, A. Dittmann, S.P. Alves, L. Kunz, D.M. Ribeiro, S. Silva, C.A. Venâncio and A.M. De Almeida

Serum metabolomics of newborn lambs before and after colostrum feeding 486
L. Lachemot, J. Sepulveda, X. Such, J. Piedrafita, G. Caja and A.A.K. Salama

Characterization of milk small extracellular vesicles to study adaptation to lactation in ruminants 487
C. Boby, A. Delavaud, J. Pires, L.E. Monfoulet, S. Bes, S. Emery, L. Bernard, C. Leroux, A. Imbert, M. Tourret, F. Fournier, D. Roux, H. Sauerwein and M. Bonnet

Multi-tissue transcriptome analysis of bovine herpesvirus-1 (BoHV-1) challenged dairy calves 487
S. O'Donoghue, B. Earley, M.S. McCabe, S.L. Cosby, K. Lemon, J.W. Kim, J.F. Taylor, D.W. Morris and S.M. Waters

Transcriptome analysis reveals a different immune response depending on a SOCS2 gene point mutation 488
C. Oget-Ebrad, C. Cabau, S. Walachowski, N. Cebron, J. Sarry, C. Allain, R. Rupp, G. Foucras and G. Tosser-Klopp

Expression quantitative trait loci in whole blood influence putative immune genes in sheep 488
K. Dubarry, M. Coffey and E. Clark

Thigh muscle proteomics revealed key pathways related to feed efficiency in slow-growing chicken 489
P. Kaewsatuan, C. Poompramun, S. Kubota, W. Molee, P. Uimari and A. Molee

Poster Session 37

Multi-breed, multi-tissue systems biology analysis of beef cattle divergent for feed efficiency 489
K. Keogh, D.A. Kenny, P.A. Alexandre, M. McGee and A. Reverter

Contribution of plasma proteins to the phenotypic signature of feed efficiency in Charolais bulls 490
I. Cassar-Malek, A. Imbert, A. Delavaud, H. Sauerwein, R. Bruckmaier, G. Cantalapiedra-Hijar and M. Bonnet

Serum metabolomics of fattening bulls fed dry-total mixed ration (TMR) or corn silage-based TMR 490
C. Stöcker-Gamigliano, M.H. Ghaffari, C. Koch, M. Schönleben, J. Mentschel, N. Göres, P. Fissore, I. Cohrs, S. Schuchardt and H. Sauerwein

Physiological importance of blood vitamin E in production characteristics of Japanese Black steers 491
M. Kim, T. Masaki, K. Ikuta, E. Iwamoto, Y. Uemoto, F. Terada, S. Haga and S. Roh

Age-related changes in ruminal and blood parameters and intramuscular adiposity in Hanwoo steers 491
S.Y. Kim and M. Baik

Post-ruminal urea release impact liver proteomics of beef cows at late gestation 492
M.M. Santos, T.C. Costa, R.D. Araújo, J. Martín-Tereso, I.P. Carvalho, M.P. Gionbelli and M.S. Duarte

Liver transcriptome profiles of dairy cows with different serum metabotypes 492
M. Hosseini Ghaffari, H. Sadri, N. Trakooljul, C. Koch and H. Sauerwein

Comparison of metabolomic profile of the offspring of nulliparous and primiparous Holstein cows 493
M. Terré and M. Tortadès

Transcriptome of the feto-maternal interface in pigs in late gestation: part 1 – foetal placenta 493
A. Bonnet, S. Maman, L. Gress, A. Suin, C. Bravo, G. Cardenas, Y. Billon, L. Canario, N. Vialaneix, C.M.D. Bonnefont and L. Liaubet

Transcriptome of the feto-maternal interface in pigs in late gestation: part 2 – sow endometrium 494
A. Bonnet, S. Maman, L. Gress, A. Suin, S. Legoueix, C. Bravo, Y. Billon, N. Vialaneix, C.M.D. Bonnefont and L. Liaubet

1H-NMR metabolomic study of Large White and Meishan pigs in late gestation: part 2 – sow endometrium 494
A. Imbert, N. Duprat, N. Marty-Gasset, L. Gress, C. Canlet, Y. Billon, N. Vialaneix, C.M.D. Bonnefont, A. Bonnet and L. Liaubet

The effect of dietary *Laminaria digitata* on urinary and kidney proteomes of piglets 495
D.M. Ribeiro, A. Dittmann, D.F.P. Carvalho, L. Kunz, J.P.B. Freire, J.A.M. Prates and A.M. Almeida

Vitamin D3 supplementation has little effect on gut microbiome in pigs 495
M. OcZkowicz, A. Wierzbicka, A. Steg and M. Świątkiewicz

Session 38. Equal opportunities and open communication for early career scientists

Date: Tuesday 29 August 2023; 15.00 – 18.30
Chair: Pszczoła / Sell-Kubiak / Newton / Obsteter / Gislon

Theatre Session 38

Equal opportunities in science 496
I. Adriaens, M. Pszczola and T. Wallgren

The responsibilities of authors, readers and learned-societies in animal science publishing 496
J. Van Milgen, I. Ortigues-Marty, G. Bee, M. Wulster-Radcliffe, J. Sartin, T.A. Davis and P.J. Kononoff

Animal – open space the new member of the animal consortium 497
G. Bee

Artificial intelligence language models in the scientific career 497
M. Pszczola, I. Adriaens and T. Wallgren

Session 39. TechCare and ClearFarm: pilots on PLF tools for monitoring animal welfare

Date: Tuesday 29 August 2023; 15.00 – 18.30
Chair: Morgan-Davies / Llonch

Theatre Session 39

TechCare: exploring the use of precision livestock farming for small ruminant welfare management 498
C. Morgan-Davies, G. Tesniere, C. Dwyer, G. Jorgensen, E. Gonzalez-Garcia and J.M. Gautier

Monitoring water trough attendance in shed: a potential indicator of sheep health or welfare issues? 498
G. Tesnière, U. Jean-Louis, E. Doutart, S. Duroy, C. Douine, M. Rinn, D. Gautier, A. Hardy, A. Aupiais, F. Guimbert, J.-M. Gautier and C. Morgan-Davies

TechCare UK pilots – integrated sheep system studies using technologies for welfare monitoring 499
A. McLaren, A. Waterhouse, F. Kenyon, H. MacDougall, S. Beechener, A. Walker, M. Reeves, N. Lambe, J. Holland, A. Thomson, J. Duncan, A. Barnes, C. Dwyer, F. Gimbert, J.M. Gautier, G. Tesniere and C. Morgan-Davies

An attempt to aggregate pig welfare indicators from sensors: achievements and barriers 499
H.L. Ko, L.J. Pedersen, G. Franchi, M.B. Jensen, M. Larsen, I.J.M.M. Boumans, J.D. Bus, E.A.M. Bokkers, X. Manteca and P. Llonch

Is sensor technology ready for use in animal welfare assessments of gestating sows? 500
I.J.M.M. Boumans and E.A.M. Bokkers

Identification of behavioural pattern associated with mastitis in dairy cows 500
L. Herve, Y. Gómez, K. Chow, A.H. Stygar, G.V. Berteselli, E. Dalla Costa, E. Canali and P. Llonch

Use of PLF sensors to monitoring rumination as an alternative for early clinical mastitis detection 501
G.V. Berteselli, E. Dalla Costa, Y. Gómez Herrera, M.G. Riva, S. Barbieri, R. Zanchetta, P. Llonch, X. Manteca and E. Canali

Monitoring the stress of light goat kids transported over short journeys 501
M. Sort, A. Elhadi, R. Costa, A. Recio, A.A.K. Salama and G. Caja

Monitoring the stress of light suckling lambs transported over short journeys 502
A. Elhadi, J.C. Jesús, R. Costa, A. Recio, A.A.K. Salama and G. Caja

Exploration of indicators measured by PLF sensors to monitor pig welfare during transportation 502
H.-L. Ko, P. Fuentes Pardo, F. Jiménez Caparros, X. Manteca and P. Llonch

Quantifying the value of early warning system (EWS) based on 'smart water trough' in a sheep farm 503
A. Bar-Shamai, I. Shimshoni, A. Godo, J. Lepar and I. Halachmi

Economic feasibility of using sensing technologies for welfare monitoring in the pig value chain 503
A.H. Stygar, M. Pastell, P. Llonch, E.A.M. Bokkers, J.D. Bus, I.J.M.M. Boumans, L.J. Pedersen and J.K. Niemi

Poster Session 39

ORIOLE: a web application for cleaning data from the walk-over-weighing device in livestock systems 504
I. Sanchez, E. González-García, B. Fontez and B. Cloez

Monitoring liveweight in Sarda dairy sheep using a walk-over-weighing system 504
M. Decandia, M. Acciaro, V. Giovanetti, G. Molle, F. Chessa, I. Llach and E. González-García

Stakeholders' attitudes to early warning systems to promote animal welfare in small ruminants 505
E.N. Sossidou, S.I. Patsios, S. Beechener, V. Giovanetti, G. Tesniere, L. Grova, T. Keady, G. Caja, A. Bar-Shamai, L.T. Cziszter and C. Morgan-Davies

Comparison between automatic milking and milking parlour system on dairy cows' welfare 505
G.V. Berteselli, S. Cannas, E. Dalla Costa, R. Zanchetta, G. Pesenti Rossi and E. Canali

Prioritization of welfare issues and precision technologies for welfare monitoring in dairy sheep 506
A. Elhadi, R. González-González and G. Caja

Session 40. The transition from pregnancy to lactation in dairy goats and sheep

Date: Tuesday 29 August 2023; 15.00 – 18.30
Chair: Hernández-Castellano / Salama

Theatre Session 40

invited Update of nutritional requirements of goats for growth and pregnancy in hot environments 506
I.A.M.A. Teixeira, C.J. Härter, J.A.C. Vargas, A.P. Souza and M.H.M.R. Fernandes

Intramammary administration of lipopolysaccharides at parturition affects colostrum quality 507
M. González-Cabrera, A. Torres, M. Salomone-Caballero, N. Castro, A. Argüello and L.E. Hernández-Castellano

Goat kids are not affected by the intramammary administration of lipopolysaccharides at parturition 507
M. González-Cabrera, M. Salomone-Caballero, S. Álvarez, A. Argüello, N. Castro and L.E. Hernández-Castellano

Near-infrared spectroscopy prediction models preliminary results in ewes' colostrum 508
S. González-Luna, E. Albanell, G. Caja and C. L. Manuelian

No kid, no milk? Trials about induction of goat lactation without gestation 508
L. Fito, C. Constancis and M. Bouy

Perinatal rumen microbiota in relation to feed utilization and biochemical parameters in sheep 509
J.X. Chen

Alpine goats divergent for functional longevity differ in metabolic profile during transition period 509
J. Pires, T. Fassier, M. Tourret, C. Huau, N. Friggens and R. Rupp

Effects of sodium butyrate supplementation at late gestation of ewes 510
Y.J. Zhang, X.Y. Zhang, X.H. Duan, R.C. Yang and S.W. Zhang

New artificial teats to improve goat milking device laboratory tests 510
M. Despinasse, C. Bonin, G. Coquereau and J.L. Poulet

invited Olive cake feeding alters the expression of lipogenic genes in mammary and adipose tissue in goats 511
M.C. Neofytou, A.L. Hager-Theodorides, E. Sfakianaki, S. Symeou, D. Sparaggis, O. Tzamaloukas and D. Miltiadou

Poster Session 40

Chemical composition and physicochemical characteristics of mountainous goat milk during lactation 511
E. Kasapidou, I.V. Iliadis, G. Papatzimos, M.A. Karatzia, Z. Basdagianni and P. Mitlianga

Energy balance and body reserve dynamics in early lactation dairy ewes 512
F. Corbiere, C. Machefert, J.M. Astruc, M. El Jabri, B. Fanca, P. Hassoun, C. Marie-Etancelin, A. Meynadier and G. Lagriffoul

Effects of Sulla flexuosa hay as alternative feed resource on goat's milk production and quality 512
S. Boukrouh, A. Noutfia, N. Moula, C. Avril, J.-L. Hornick, M. Chentouf and J.-F. Cabaraux

High DCAD and ascorbic acid modify acid base, oxidative stress and kidney function in dairy goats 513
S. Thammacharoen, S. Semsirmboon, D.K. Do Nguyen, S. Poonyachoti, T.A. Lutz and N. Chaiyabutr

Link between chest circumference at late gestation and milk somatic cell count in Sarda ewe-lambs 513
D. Sioutas, A. Ledda, A. Mazza, A. Marzano, M. Sini and A. Cannas

Session 41. Open the black box: using omics to better understand biological interpretation of complex traits

Date: Wednesday 30 August 2023; 8.30 – 13.00
Chair: Tiezzi / Pocrnic

Theatre Session 41

invited Incorporating high-dimensional omics phenotypes into models for predicting breeding values 514
O.F. Christensen

Biologically informed genomic predictions with the NextGP.jl statistical analysis package 514
E. Karaman, V. Milkevych and L. Janss

Fine mapping QTL associated with fertility in dairy cattle using gene expression 515
I. Van Den Berg, A.J. Chamberlain, I.M. MacLeod, T.V. Nguyen, M.E. Goddard, R. Xiang, B. Mason, S. Meier, C.V.C. Phyn, C.R. Burke and J.E. Pryce

How heritable are metabolomic features? An omic based approach in pigs 515
S. Bovo, G. Schiavo, F. Fanelli, A. Ribani, F. Bertolini, V. Taurisano, M. Gallo, G. Galimberti, S. Dall'olio, P.L. Martelli, R. Casadio, U. Pagotto and L. Fontanesi

Combined transcriptomics and metabolomics in the whole blood to depict feed efficiency in pigs 516
C. Juigné, E. Becker and F. Gondret

Imputation of sequence variants in more than 250,000 German Holsteins 516
A.M. Krizanac, C. Falker-Gieske, C. Reimer, J. Heise, Z. Liu, J. Pryce, J. Bennewitz, G. Thaller and J. Tetens

Longitudinal study on the rabbit's gut microbiota variation through age 517
I. Biada, M.A. Santacreu, A. Blasco, R.N. Pena and N. Ibáñez-Escriche

Studying cattle structural variation and pangenome using whole genome sequencing 517
G.E. Liu

Host genetics affect the composition of the lower gut microbiota in dairy cows 518
L. Brulin, S. Ducrocq, J. Estellé, G. Even, S. Martel, S. Merlin, C. Audebert, P. Croiseau and M.P. Sanchez

A model including host genotype, gut microbiome, and fat deposition measures in swine 518
F. Tiezzi and C. Maltecca

protiPig: microbial analyses of traits related to nitrogen utilization efficiency in pigs 519
M. Schmid, N. Sarpong, M. Rodehutscord, J. Seifert, A. Camarinha-Silva and J. Bennewitz

Rumen microbiome promote host metabolism and microbial bile acids bio-transformation in sheep 519
B.Y. Zhang, Y. Yu, X.Z. Jiang, Y.M. Cui, H.L. Luo and B. Wang

Blending multivariate models to predict feed efficiency and explore multiple omics in meat sheep 520
Q. Le Graverand, F. Tortereau, C. Marie-Etancelin, A. Meynadier, J.L. Weisbecker and K.A. Lê Cao

The effect of dietary *Laminaria digitata* on the muscle proteome and metabolome of weaned piglets 520
D.M. Ribeiro, D.F.P. Carvalho, C.C. Leclercq, S. Charton, K. Sergeant, E. Cocco, J. Renaut, J.A.M. Prates, J.P.B. Freire and A.M. Almeida

Search for new mutations in cattle by systematic whole genome resequencing 521
M. Boussaha, C. Eché, C. Escouflaire, C. Grohs, C. Iampietro, A. Capitan, M. Denis, S. Fritz, C. Donnadieu and D. Boichard

Poster Session 41

Estimation of non-additive genetic effects for semen production traits in beef and dairy bulls 521
R. Nagai, M. Kinukawa, T. Watanabe, A. Ogino, K. Kurogi, K. Adachi, M. Satoh and Y. Uemoto

Evaluating functional properties of feeding black soldier fly larvae in laying hens by FeedOmics 522
A. Rezaei Far, E. Zaccaria, I. Fodor, P. Van Wikselaar, S. Naser El Deen, S. Kar and T. Veldkamp

Characterization of two divergent lines through functional inference of ruminal microbiota 522
J. Guibert, A. Meynadier, V. Darbot, C. Allain and C. Marie-Etancelin

Blood transcriptomic comparison under different THI in Holstein 523
J.-E. Park, H. Kim, J.-H. Cho, H.-G. Lee, W. Park and D. Shin

Metabolome profile of the pectoralis major muscle of red-winged tinamou: a pilot study 523
J.M. Malheiros, C.S.M.M. Vilar, P.F. Silva, L.A. Colnago, J.A.I.I.V. Silva and M.E.Z. Mercadante

Post-mortem muscle proteome of crossbred bulls and steers: carcass and meat quality 524
O.R.M. Neto, W.A. Baldassini, L.A.L. Chardulo, R.A. Curi, G.L. Pereira, B.M. Santiago and R.V. Ribeiro

ITGA6 homozygous splice-site mutation causes junctional epidermolysis bullosa in Charolais cattle 524
C. Grohs, M. Boussaha, A. Boulling, V. Wolgust, L. Bourgeois-Brunel, P. Michot, N. Gaiani, M. Vilotte, J. Riviere and A. Capitan

A sheep pangenome reveals the spectrum of structural variations and their effects on tail phenotypes 525
R. Li, M. Gong, X.M. Zhang, F. Wang, Z.Y. Liu, L. Zhang, S.Q. Gan and Y. Jiang

A Chinese indicine pan-genome reveals novel structural variants introgressed from other Bos species 525
X.L. Dai, P.P. Bian, D.X. Hu, F.N. Luo, Y.Z. Huang, R. Heller and Y. Jiang

Multi-omic analysis reveal epigenetic regulation during muscle growth and development in sheep 526
Q.J. Zhao, Y.H. Ma and Y. Liu

The effect of production system on the muscle proteome of dromedary camel (*Camelus dromedarius*) 526
M. Lamraoui, D.M. Ribeiro, Y. Khelef, N. Sahraoui, H. Osório and A.M. De Almeida

Effect of mitochondrial DNA copy number on productive traits and meat quality in pigs 527
E. Molinero, R.N. Pena, J. Estany and R. Ros-Freixedes

Multi-omics analysis revealed the interaction of microbiota and host in rumen of hanwoo 527
W.C. Park, J.W. Son, M.J. Jang, N.R. An, H.J. Choi, S.A. Jung, J.A. Lim, D.H. Kim, J.H. Cha, S.Y. Choi, Y.J. Lim, S.S. Jang, D.J. Lim and H.H. Chai

Session 42. Animal genetics to address food security and sustainability

Date: Wednesday 30 August 2023; 8.30 – 13.00
Chair: Soelkner / Granados

Theatre Session 42

Challenge session: animal genetics to address food security and sustainability 528
A. Granados and J. Sölkner

Causal structures between gut microbiota and efficiency traits in poultry 528
V. Haas, M. Rodehutscord, A. Camarinha-Silva and J. Bennewitz

Exploring crossbreeding to reduce GHG Emissions and feed-food competition in beef production 529
A. Mertens, L. Kokemohr, E. Braun, L. Legein, C. Mosnier, G. Pirlo, P. Veysset, S. Hennart, M. Mathot and D. Stilmant

Session 43. Better calves for better farms: how the young stock can be the key to success

Date: Wednesday 30 August 2023; 8.30 – 13.00
Chair: Costa / Keatinge

Theatre Session 43

nvited Genetic selection for dairy calf disease resistance traits: opportunities and challenges 529
C. Lynch, F.S. Schenkel, N. Van Staaveren, F. Miglior, D. Kelton and C.F. Baes

Genetic research on colostrum quality traits and passive transfer of immunity in Greek dairy herds 530
A. Soufleri, G. Banos, N. Panousis, G. Arsenos, A. Kougioumtzis, V. Tsiamadis and G.E. Valergakis

Phenotypic and genetic analysis of beef-on-dairy crossbred calves 530
R.H. Ahmed, C. Schmidtmann, J. Mugambe and G. Thaller

Assessing passive immune transfer in newborn calves: salivary and serum IgG association 531
F.G. Silva, E. Lamy, S. Pedro, I. Azevedo, P. Caetano, J. Ramalho, L. Martins, A.M.F. Pereira, J.O.L. Cerqueira, S.R. Silva and C. Conceição

Biochemical predictors of successful transition from milk to solid feed in Holstein calves 531
P. Kazana, N. Siachos, N. Panousis, G. Arsenos and G.E. Valergakis

The impact of dairy farm management on long-term robustness of veal calves 532
F. Marcato, H. Van Den Brand, L. Webb, M. Wolthuis-Fillerup, F. Hoorweg and K. Van Reenen

The assessment of transfer of passive immunity in dairy calves affected by neonatal diarrhoea 532
G. Sala, V. Bronzo, A. Boccardo, A.L. Gazzonis, P. Moretti, V. Ferrulli, A.G. Belloli, L. Filippone Pavesi, G. Pesenti Rossi and D. Pravettoni

Automatic monitoring of calves' behaviour for a precision weaning approach 533
R. Colleluori, D. Cavallini, A. Formigoni and L. Mammi

Can dairy herds be in a positive colostrum stock balance? 533
A. Soufleri, G. Banos, N. Panousis, G. Arsenos, A. Kougioumtzis, V. Tsiamadis and G.E. Valergakis

Dairy-beef calves: current practices and views of calf producers and rearers 534
D.J. Bell, M.J. Haskell, C.S. Mason and C.-A. Duthie

Sodium percarbonate as a preservative in waste milk fed to dairy calves 534
D.J. Wilson, G.M. Goodell, R. Dumm, T. Kelly and M. Bethard

Pre-transport diet affects the physiological status of calves during transport by road and ferry 535
S. Siegmann, L.L. Van Dijk, N.L. Field, G.P. Sayers, K. Sugrue, C.G. Van Reenen, E.A.M. Bokkers and M. Conneely

Clonal dissemination of MDR *Pasteurella multocida* ST79 in a Swiss veal calves 535
J. Becker, J. Fernandez, A. Rossano, M. Meylan and V. Perreten

Effects of feeding milk with antibiotic residues on calf performance during the pre-weaning period 536
A. Flynn, C. Mc Aloon, M. Mc Fadden, J.P. Murphy, S. Mc Pherson, C. Mc Aloon and E. Kennedy

Effects of extending lactation for dairy cows on health, development and production of their calves 536
Y. Wang, R. Goselink, E. Burgers, A. Kok, B. Kemp and A.T.M. Van Knegsel

Poster Session 43

Dam-calf contact rearing in Switzerland: Aspects of management and milking 537
J. Rell, C. Nanchen, P. Savary, C. Buchli and C. Rufener

A survey of colostrum management practices in dairy farms of Piedmont region (Italy): a pilot study 537
G.V. Berteselli, G. Pesenti Rossi, G. Vezzaro, E. Dalla Costa, S. Barbieri and E. Canali

Serum profiles of dairy calves fed a milk replacer or whole milk at two levels of supply 538
T. Chapelain, J.B. Daniel, J.N. Wilms, J. Martín-Tereso and L.N. Leal

Prevalence of foot lesions in French slaughter dairy and beef young bulls housed in indoor feedlot 538
S. Ishak, R. Guatteo, A. Lehébel, N. Brisseau, M. Gall, A. Wache and A. Relun

Effect of direct-fed microbial supplementation on performance and health of pre-weaning dairy calves 539
J. Magalhaes, B.I. Cappellozza, T.C. Dos Santos, F.N. Inoe, M.S. Coelho, V. Soares and J.L.M. Vasconcelos

The effect of astaxanthin on health and calves performance 539
E. Sosin, I. Furgał-Dierzuk, B. Śliwiński and A. Burmańczuk

Strategic grouping of dairy beef calves on arrival at a rearing unit 540
D.J. Bell, C.S. Mason, K.C. Henderson, M.J. Haskell and C.-A. Duthie

Calves sexing and crossbreeding to optimize the destination of the young from French dairy farms 540
S. Dominique

Inclusion of grass silage in finishing total mixed rations for rosé veal calves 541
M. Vestergaard, M. Bjerring, A.L.F. Hellwing, M.B. Jensen, B. Muhlig, L. Mogensen and N.B. Kristensen

Veal calves' housing in France: current situation and investment needs 541
M. Tourtier and C. Martineau

Evolution of the use of antibiotics in the veal calves' sector in France between 2013 and 2020 542
M. Chanteperdrix, A. Chevance, M. Orlianges, D. Urban, M. Tourtier and P. Briand

Session 44. Building on a resilient dairy sector – from animal, farm and regional perspective – Part 1

Date: Wednesday 30 August 2023; 8.30 – 13.00
Chair: Kuipers / Schuenemann

Theatre Session 44

Resilience4Dairy: sharing knowledge to improve sustainability and resilience of the dairy sector 542
V. Brocard, M. Klopcic and J. Boonen

Invited Indicators and influencing factors of livestock resilience 543
I.D.E. Van Dixhoorn, J. Ten Napel, A. Mens and J.M.J. Rebel

Invited Reducing stress of dairy cows and farmers to improve resiliency and welfare 543
M.T.M. King and T.J. De Vries

Invited Building resilience in the dairy sector of China 544
S. Li, W. Wang, W. Du, X. Sun, K. Yao and J. Xia

Invited Building resilience in farming: dairy cattle and workforce management 544
G.M. Schuenemann and J.M. Piñeiro

Working on resilience in the Ukrainian dairy sector 545
L. Stepura

Future scenarios for livestock agriculture in New Zealand 545
C. Vannier, T. Cochrane, L. Bellamy, T. Merritt, H. Quenol and B. Hamon

Invited Breeding approaches to improve robustness and resilience in dairy cows 546
K. May and S. König

Resilience from the perspective of farm economics 546
E. Kołoszycz and A. Wilczyński

Labour: a key factor in the resilience of the European dairy farmer 547
S. Debevere, L. Dejonghe, I. Louwagie, I. Vuylsteke, E. Béguin, S. Fourdin, P. Rondia, L. Boulet, S. Mathieux and G. Elluin

Resilience of dairy farming: the farmers' point of view 547
E. Castellan, C. Bausson and V. Brocard

Poster Session 44

Local breed as an alternative to Holstein-Friesian cows in a farm with low level of milk production 548
M. Sobczuk-Szul, Z. Nogalski, M. Momot and P. Pogorzelska-Przybyłek

Session 45. Transformation of livestock practices in response to society's expectations on animal welfare, market demand for animal products and environmental issues

Date: Wednesday 30 August 2023; 8.30 – 13.00
Chair: Thenard / Boyle

Theatre Session 45

Product quality as a lever to change farming practices to meet society's expectations 548
V. Thenard, S. Couvreur, L. Fortun-Lamothe, B. Méda and T. Petit

Transformation in the dairy sector: a global analysis of sustainability certification standards 549
K. McGarr-O'Brien, J. Herron, L. Shalloo, I.J.M. De Boer and E.M. De Olde

The analysis of a co-design process to develop an eco-citizen dairy cattle farming system experiment 549
J.E. Duval, M. Taverne, M. Bouchon and D. Pomiès

Perceived quality of meat products in short circuits by producers and their customers 550
C. Couzy, G. Haj Chahine, V. Diot, S. Masselin-Sylvin, S. Meurisse, M. Klingler and C. Bièche-Terrier

Value-adding attributes for dairy calves within local beef sector – perceptions among stakeholders 550
L. Schönfeldt, M.G.G. Chagunda and N. Ströbele-Benschop

Farmpédia: how to improve the global acceptability of livestock systems with communication 551
A.-L.J. Thadee and G. Brunschwig

How and why involve citizens in a participatory research project aiming to design livestock farming? 551
P. Coeugnet, J. Labatut, G. Vourc'h and J.E. Duval

Social sustainability: what concepts to approach farmers' satisfaction at work 552
B. Dedieu, J. Duval, P. Girard, N. Hostiou, S. Mercandalli and G. Soullier

Implementing agroecological practices: what are the effects on working conditions of dairy farmers? 552
A.-L. Jacquot, M. Gérard, J.E. Duval and N. Hostiou

Promoting and guiding transformation of French veal calf farms in response to societal expectations 553
C. Martineau, D. Bastien, M. Chanteperdrix, C. Denoyelle, V. Lefoul and M. Orlianges

Designing rabbit breeding systems with access to the outdoors with the innovative design method 553
L. Fortun-Lamothe, M.H. Jeuffroy and L. Le Du

Challenges and opportunities for transitioning to 'low anthelmintics use' in livestock systems 554
M. Sautier and P. Chiron

The rearing of calves, kids and lambs with adults in dairy systems in the AuRA region, France 554
C. Constancis, A. Igier and F. Debrez

Expectations of French suckler farmers in terms of technical support services 555
A. Antoni-Gautier, J. Chambeaud, T. Falcou, C. Galvagnon, N. Lemonnier and J.B. Menassol

Poster Session 45

'Antibiotic-free' strategies in chicken production in France: success factors, assets and limitation 555
N. Rousset and J. Hercule

Veal calves' production: what societal expectations in terms of animal housing for their welfare? 556
D. Bastien, M. Tourtier and A. Warin

Salmon welfare perceived by various stakeholders: where do we stand? 556
C.M. Monestier, L.R.B. Reverchon-Billot, M.S. Stomp and A.W. Warin

A review of sustainable livestock strategies to deal with anti-livestock activists 557
M. Dehghani and K.H. Park

AuthenBeef: use of blockchain technology in beef production to secure authenticity and traceability 557
G. Arsenos, S. Vouraki, V. Papanikolopoulou, A. Argyriadou, V. Fotiadou, S. Minoudi, D. Karaouglanis, N. Karaiskou, P. Fortomaris and A. Triantafyllidis

Structural equation modelling applied to multi-performance objectives in French suckler cattle farms 558
L. Billaudet, I. Veissier, J.J. Minviel and P. Veysset

Systemic enablers and barriers to extending the productive life of Swiss dairy cows 558
A. Bieber, R. Home, M. Rödiger, R. Eppenstein and M. Walkenhorst

Providing outdoor access to pigs: what are the profiles of farmers working in those systems? 559
S.B. Brajon, C.T. Tallet, E.M. Merlot and V.L. Lollivier

The explosion of sustainability indicators in the European livestock sector 559
B. Van Der Veeken, M. Carozzi, C. Barzola Iza and F. Accatino

Pig farmers' and citizens' opinions on outdoor access for livestock 560
S.B. Brajon, C.T. Tallet, E.M. Merlot and V.L. Lollivier

Short food supply chains and food safety issues: HACCP utilization, opportunities and limits 560
R. Rafiq, O. Boutou, M.P. Ellies-Oury and B. Grossiord

Feeding strategy in organic farming as a lever to improve various quality dimensions of pork 561
C. Van Baelen, L. Montagne, S. Ferchaud, A. Prunier and B. Lebret

Body reserve dynamics using metabolites and hormones profiles of Romane ewes in two farming systems 561
A. Nyamiel, D. Hazard, D. Marcon, F. Tortereau, C. Durand, A. Tesnière and E. González-García

Session 46. Innovation in ruminant nutrition and feeding

Date: Wednesday 30 August 2023; 8.30 – 13.00
Chair: Stergiadis / Alvarez Rodriguez

Theatre Session 46

Inclusion of macroalgae in high forage beef cattle diets 562
S.A. Terry, T.W. Coates, R.J. Gruninger, D.W. Abbott and K.A. Beauchemin

Rumen protected potassium gluconate increases average daily gain of beef 562
A. Santos, J.G.H.E. Bergman, J.A. Manzano and M. Hall

Impact of additives and forage levels on performance and enteric methane emissions of Nellore bulls 563
E. Magnani, T.H. Silva, L.B. Tosetti, A. Berndt, E.M. Paula, P.R. Leme and R.H. Branco

Differences in digestive traits of young bulls fed contrasted diets and diverging in feed efficiency 563
M. Coppa, C. Martin, A. Bes, L. Ragionieri, F. Ravanetti, P. Lund, G. Cantalapiedra-Hijar and P. Nozière

Effect of solid feed intake on feeding behaviour and energy metabolism in growing calves 564
E. Labussiere, L. Montagne, Y. Le Cozler, C. Martineau and D. Bastien

Effects of dietary inclusion of willow leaves on feed intake and methane emission in sheep 564
J.J. Thompson, S. Stergiadis, O. Cristobal-Carballo, T. Yan, S. Huws and K. Theodoridou

Effect of supplementing Omega-3 rich oil on ruminal fermentation and dietary digestibility in lambs 565
O. Cristobal-Carballo, F. Godoy-Santos, S. Huws, S. Morrison, A. Aubry, E. Lewis and T. Yan

Essential oils and integral diets to optimize rumen function and decrease methanogenesis in lambs 565
N. Ghallabi, G. Gonzalo, A. Garcia, O. Catalán, P. Romero, M. Hassan, A.I. Martín-García, D.R. Yáñez-Ruiz and A. Belanche

Novel water based delivery of seaweed extracts to improve the sustainability of ruminant production 566
A. Casey, T. Boland, Z. McKay and S. Vigors

Supplementation of Ca gluconate improves fertility and time to peak milk yield in dairy cattle 566
D.J. Seymour, M.V. Sanz-Fernandez, J.B. Daniel, J. Martin-Tereso and J. Doelman

DFM supplementation during the gestation and dry periods on postpartum performance in dairy cows 567
O. Ramirez-Garzon, D.G. Barber, J. Alawneh, L. Huanle and M. Soust

In vivo evaluation of tannins and essential oils mixtures as additives for dairy cows 567
G. Foggi, L. Turini, F. Dohme-Meier, A. Muenger, L. Eggerschwiler, J. Berard, G. Conte, A. Buccioni and M. Mele

Palmitic to oleic ratio in fat supplement influenced the digestibility and production of dairy cows 568
J. Shpirer, L. Lifshitz, H. Kamer, Y. Portnick and U. Moallem

Xylooligosaccharides and enzyme increased milk yield and reduced methane emissions of Jersey cows 568
L.F. Dong and Q.Y. Diao

Defatted black soldier fly larvae meal as substitute of soybean meal in Kenyan dairy cow rations 569
D.J.M. Braamhaar, D.J. List, S.J. Oosting, D. Korir, C.M. Tanga and W.F. Pellikaan

Classifying lipogenic and glucogenic diets in dairy cows based on metabolomics profiles 569
X. Wang, S. Jahagirdar, W. Bakker, C. Lute, B. Kemp, E. Saccenti and A.T.M. Van Knegsel

Poster Session 46

Comparative study of digestibility of safflower varieties 570
M. Besharati, N. Khoshnam and D. Azhir

Effect of different levels of propolis, nitrate, thyme and mint essential oil on digestibility 570
M. Besharati and M. Mousavi

Combination of lactic acid bacteria inoculant in difficult to ensile grass after 15 days of ensiling 571
M. Duvnjak, L. Dunière, B. Andrieu, E. Chevaux and C. Villot

Effect of whey on *in vitro* ruminal methane formation and digestibility in cows 571
H. Luisier-Sutter, L. Isele, M. Terranova, S.L. Amelchanka and M. Schick

Ellagic acid and gallic acid reduced methane and ammonia in an *in vitro* rumen fermentation model 572
M. Manoni, S. Amelchanka, M. Terranova, L. Pinotti, P. Silacci and M. Tretola

Effect of toasted soybean on dairy cows milk performances 572
A. Berchoux, M. Duval, E. Hermant, M. Legris and M. Jouffroy

The addition of dry ice as an attempt to inhibit proteolysis during ensilage of lucerne 573
M. Borsuk-Stanulewicz, C. Purwin and M. Mazur-Kuśnirek

In vitro effects of *Bacillus subtilis* CH201 and *Bacillus licheniformis* CH200 on rumen microbiota 573
R. Gresse, G. Copani, B.I. Cappellozza, A. Torrent, D. Macheboeuf, E. Forano and V. Niderkorn

Effect of nitrate and its interaction with starch levels on methane production in continuous culture 574
Y. Roman-Garcia, S. El-Haddad, S. Van Zijderveld and G. Schroeder

Effect of garlic processing method on *in vitro* methane production and rumen fermentation 574
N.F. Sari, S. Stergiadis, P.P. Ray, C. Rymer, L.A. Crompton and K.E. Kliem

Effects of Red Sorghum and *Rhizoma paridis* on rumen protozoa, fermentation characteristics *in vitro* 575
R. Yi, S. Vigors, L. Ma, J.C. Xu and D.P. Bu

Phytochemicals can modify the rumen fermentation profile as monensin 575
L. Gonzalez, A.C. Dall-Orsoletta, D. Mattiauda, A. Daudet, T. Garcia, P. Chilibroste, A. Meikle, M. Arturo-Schaan, A. Casal and M.A. Bruni

In vitro fermentation of TMR using rumen fluids from cows supplemented with hemp and savory leaves 576
S. Arango, S. Massaro, S. Schiavon, N. Guzzo, M. Montanari, L. Bailoni and F. Tagliapietra

Bacillus licheniformis and *B. subtilis* on *in vitro* ruminal parameters and greenhouse gas emission 576
B.R. Amancio, E. Magnani, T.H. Silva, A.L. Lourenço, B.I. Cappellozza, R.H. Branco and E.M. Paula

Bioproduct from royal palm colonized by *Lentinula edodes* on *in vitro* ruminal fermentation 577
B.M. Rocha, R.L. Savio, G.S. Camargo, K.E. Loregian, A.R. Cagliari, A.C. Casagrande, F. Rigon, E. Magnani, T.G. Timm, L.B.B. Tavares, M.I. Marcondes, T.H. Silva, R.H. Branco, E.M. Paula and P.D.B. Benedeti

Enterococcus faecium and *Saccharomyces cerevisiae* on *in vitro* rumen parameters and greenhouse gases 577
B.R. Amancio, T.H. Silva, G.M. Wachekowski, H. Reolon, T.G. Timm, B.I. Cappellozza, E. Magnani, E.M. Paula and R.H. Branco

Rumen protected calcium gluconate improves milk production of cows 578
J.G.H.E. Bergman, B. Skibba and M. Hall

Rumen protected calcium gluconate improves lactational performance 578
J.G.H.E. Bergman, F. Morisset and M. Hall

Increase milk production by preserving the nutritional value of the dairy ration 579
L.L.C. Jansen, J.G.H.E. Bergman and S.J.A. Van Kuijk

VistaPre-T, a crude fermentation extract to support the sustainable use of forages in dairy rations 579
V. Blanvillain, E. Bungenstab and G. Gomes

Early lactation trial with a blend of fat encapsulated vitamin B 580
A.D.G. Esselink and C. Gordon

The effect of a blend of fat encapsulated vitamin B on milk production of highly productive cows 580
M. Hall and C. Gordon

Evaluation of milk performances and enteric methane emissions on by-products feed base 581
A. Berchoux, M. Jouffroy, A. Laflotte and R. Boré

Supplementing lambs with plant extract supplement enhances growth and improve feed conversion ratio 581
V. Ballard and P.H. Pomport

Effects of *Artemisia annua* residue on rumen microorganisms and antioxidant function of mutton sheep 582
S.S. Wang, C.F. Peng, Y.R. Shao, M.M. Bai, Y.H. Zhang, M. Zhang, X. Xiong and H.N. Liu

Nutrient rich novel feed supplements for grazing lambs 582
A.S. Chaudhry

The order of distribution of two forages affects daily intake and diet composition in dairy goats 583
R. Delagarde, J. Belz and B. Bluet

Feeding frequency has no effect on intake and milk production in dairy goats fed on fresh herbage 583
R. Delagarde, J. Belz and B. Bluet

Effects of dietary energy on late-gestation metabolism in prolific ewes 584
M. Plante-Dubé, C. Sylvestre, R. Bourassa, P. Luimes, S. Buczinski, F. Castonguay and R. Gervais

The effect of feeding with hemp seeds addition on physicochemical and sensory properties of beef 584
P. Pogorzelska-Przybyłek, C. Purwin, M. Modzelewska-Kapituła, M. Borsuk-Stanulewicz and K. Tkacz

Association of feed efficiency with growth and slaughtering performance in Nellore cattle 585
S.F.M. Bonilha, J.A. Muñoz, B.R. Amâncio, J.N.S.G. Cyrillo, R.H. Branco, R.C. Canesin and M.E.Z. Mercadante

Effects of dietary nitrate on performance and enteric methane production in Hanwoo steers 585
R. Bharanidharan, P. Xaysana, R. Ibidhi, J. Lee, B.M. Tomple, J. Oh, M. Baik and K.H. Kim

Impact of additives and forage levels on nutrients digestibility and sorting index of Nellore bulls 586
E. Magnani, T.H. Silva, L.B. Tosetti, E.M. Paula, P.R. Leme and R.H. Branco

Feed efficiency traits calculated at post-weaning and pre-slaughter periods in Nellore cattle 586
S.F.M. Bonilha, J.A. Muñoz, B.R. Amâncio, J.N.S.G. Cyrillo, R.H. Branco, R.C. Canesin and M.E.Z. Mercadante

Environmental protection study on replacing alfalfa with sesbania for feeding ruminants 587
L.Y. Wang and Y.J. Tian

Multi performance analysis of soybean self-consumption on a mixed crop-livestock farm 587
M. Jouffroy, A. Berchoux, M. Duval, M. Weens, E. Hermant and M. Legris

Effects of leaf size and harvesting season on nutritive quality of white clover 588
X. Chen, K. Theodoridou, O. Cristobal-Carballo and T. Yan

Image analysis of feed boluses collected from cows after ingestive chewing 588
B. Delord, M. Berger, R. Baumont, P. Nozière, A. Le Morvan, F. Guillon and M.F. Devaux

Session 47. One-health, one nutrition, one Earth: role of nutrition in livestock production

Date: Wednesday 30 August 2023; 8.30 – 13.00
Chair: Latifa / Weill

Theatre Session 47

invited A documented example of the One Health concept 589
P. Weill, N. Kerhoas and B. Schmitt

Dietary lipid supplements affect milk composition and butter properties in dairy cows 589
M. Landry, Y. Lebeuf, M. Blouin, F. Huot, J. Chamberland, G. Brisson, D.E. Santschi, É. Paquet, D.E. Rico, P.Y. Chouinard and R. Gervais

Decreasing GHG footprint while improving nutrition value in ruminant product: a new challenge 590
S. Mendowski, G. Chesneau, G. Mairesse and N. Kerhoas

Effect of dietary protein source and *Saccharina latissima* on milk fatty acids profiles and bromoform 590
B. Wang, S. Ormston, N. Płatosz, J.K. Parker, N. Qin, D. Humphries, Á. Pétursdóttir, A. Halmemies-Beauchet-Filleau, D. Juniper and S. Stergiadis

Human health markers improvements in clinical trials when animal feed is the only variable 591
N. Kerhoas, P. Weill and B. Schmitt

The impact of combination of Inulin fibre with the 4 major PAHs present in meat on colorectal cancer 591
L. Abdennebi-Najar, M. Zaoui, N. Ferrand, L. Louadj and M. Sabbah

Animal feeding strategy: a move towards giving strategic direction to East African countries 592
H. Makkar, K. Agyemang, D. Balikowa, A. Sebsibe and R. Mondry

invited Macrominerals and trace elements in retail milk: their variation and nutritional implications 592
S. Stergiadis, E.E. Newton, S. Beauclercq, J. Clarke, N. Desnica and Á. Pétursdóttir

Effect of a spice feed additive on behaviour, saliva composition, and ruminal pH in fattening bulls 593
C. Omphalius, J.-F. Gabarrou, G. Desrousseaux and S. Julliand

Role of niacin in regulating intestinal health in piglets 593
H.B. Yi

Effect of maternal nutrition on thymus development in Wagyu (Japanese Black) foetus 594
O. Phomvisith, S. Muroya and T. Gotoh

Poster Session 47

Bacillus sp. strains protect the intestinal barrier from oxidative stress and deoxynivalenol 594
G. Copani, B.I. Cappellozza and E.J. Boll

Production and carcass characteristics of growing-fattening rabbits under three feeding phases 595
V.C. Resconi, M. López, J.L. Olleta, J. Romero and M.M. Campo

Niacin improves intestinal health through up-regulation of AQPs expression induced by GPR109A 595
X. Yang, Y. Qiu, S. Liu and Z. Jiang

Session 48. How can poultry farming systems evolve to meet the major societal and environmental challenges?

Date: Wednesday 30 August 2023; 8.30 – 13.00
Chair: Stadnicka / Meda

Theatre Session 48

invited How can poultry farming systems evolve to meet the major societal and environmental challenges? 596
P. Thobe and P. Van Horne

invited Can we enhance environmental impact without compromising bird welfare in broiler systems? 596
I. Kyriazakis

What is the impact of the farming system on the quality of the chicken breast meat? 597
J. Albechaalany, S. Yilmaz, M.P. Ellies-Oury, M. Bourin, Y. Guyot, J. Saracco, J.F. Hocquette and C. Berri

Improving broiler wellbeing and micro-climate through PLF application 597
S. Druyan, N. Barchilon and I. Halachmi

Effects of heat stress and spirulina on productive performances of two slow growth broiler strains 598
E.A. Fernandes, C.F. Martins, D.F.P. Carvalho, L.L. Martins, A. Raymundo, M. Lordelo and A.M. And Almeida

invited Evaluating services provided by free-range poultry systems 598
G. Chiron

Biosecurity gaps in 7 major poultry producers (breeder and layer farms) in EU: a farmers perspective 599
A. Amalraj, H. Van Meirhaeghe, R. Souillard, A. Zbikowski and J. Dewulf

Reducing environmental impact of broiler production: the role of crude protein and soybean meal 599
T. De Rauglaudre, B. Méda, S. Fontaine, W. Lambert and M.P. Létourneau Montminy

Elevated platforms for broilers on commercial farms: usage and effects on health and performance 600
J. Stracke, F. May, J. Müsse, N. Kemper and B. Spindler

Mutation effects as key driver of maintaining genetic variation for long-term selection in broilers 600
B.S. Sosa-Madrid, N. Ibañez-Escriche, G. Maniatis and A. Kranis

Early interventions during incubation and impact on a muscle of locomotory relevance in broilers 601
T. Kettrukat, A. Dankowiakowska, M. Mangan, E. Grochowska, K. Stadnicka and M. Therkildsen

Summary of the posters in session 48 by the chairs 601
B. Méda and K. Stadnicka

Poster Session 48

Effect of gallic acid supplementation to corn-soybean-gluten meal-based diet in broilers performance 602
J.H. Song, C.B. Lim, S. Biswas, Q.Q. Zhang, J.S. Yoo and I.H. Kim

Veterinary coaching to stimulate biosecurity compliance 602
A. Amalraj, H. Van Meirhaeghe, M. De Gussem and J. Dewulf

Establishing elevated perforated platforms in broiler chicken housing – is hygiene a barrier? 603
B. Sake, J. Müsse, F. May, J. Stracke, N. Kemper, J. Schulz and B. Spindler

What's going on outside? Use of winter gardens by rearing hens of different genetics 603
A. Riedel, N. Kemper and B. Spindler

Metabolomic analysis reveal the molecular mechanism related to leg disease in broilers 604
J. Zheng, G. Zhang, Q. Li and G. Zhao

Housing conditions do not influence the effects of a nutritional challenge for broilers 604
V. Michel, C. Deschamps, N. Regrain, E. Devillard and J. Consuegra

Compliance of biosecurity in poultry farms in France: remaining obstacles and levers for improvment 605
N. Rousset, A. Battaglia, J. Puterflam, J. Marguerie, R. Souillard, S. Le Bouquin-Leneveu and A.-C. Lefort

Untrimmed beaks in turkey hens: Effect on injuries and mortality rate 605
M. Kramer, K. Skiba, P. Niewind, F. Von Rüden, N. Kemper and B. Spindler

Habitat affects egg characteristics of Libyan local pigeon (domestic vs feral pigeon) 606
F. Akraim, M.F. Idrees and M.M. Sghieyer

Effects of heat stress and spirulina on meat traits of two slow growth broiler strains 606
E.A. Fernandes, J.R. Sales, L.L. Martins, M. Lordelo, A. Raymundo and A.M. And Almeida

How does broiler range use impact forage intake, outdoor excretion and gaseous emissions? 607
C. Bonnefous, B. Méda, K. Germain, L. Ravon, T. De Rauglaudre, J. Collet, S. Mignon-Grasteau, M. Reverchon, C. Berri, E. Le Bihan-Duval and A. Collin

Heritability of the number of crossovers as proxy of recombination rate in chicken 607
V. Riggio, E. Tarsani and A. Kranis

The potential of early warning system at health issues in poultry by sound 608
I. Halachmi, T. Lev-Ron, Y. Yitzhaky and S. Druyan

Pre-slaughter fasting changes the *ante mortem* muscle proteolysis levels and *post mortem* meat quality 608
S. Katsumata, M. Kamegawa, A. Katafuchi, A. Ohtsuka and D. Ijiri

Session 49. Monoguthealth – Part 1

Date: Wednesday 30 August 2023; 8.30 – 13.00
Chair: Bee / Trevisi

Theatre Session 49

invited Host-microbiota interactions in swine and poultry: disentangling causes and effects 609
J.F. Pérez

Glutamine and glucose metabolism in suckling low birth weight piglets supplemented with glutamine 609
D. De Leonardis, Q.L. Sciascia, S. Goers, A. Vernunft and C.C. Metges

Creep feeding (dry, liquid) and pen hygiene (low, high) impacts pre-weaning growth in pigs 610
S.R. Vasa, G.E. Gardiner, K. O'Driscoll, G. Bee and P.G. Lawlor

The impact of early incubation temperature on broiler walking ability and final meat quality 610
T. Kettrukat and M. Therkildsen

In vitro and *in vivo* analysis of bioactive substances growth and antioxidant activities 611
M. Mangan, C. Metges and M. Siwek

Prophybiotics, a novel approach for *in ovo* gut microbiome reprograming of broilers 611
R.N. Wishna-Kadawarage, R. Hickey and M. Siwek

Antibacterial plant blends modulate gut microbiota in organic piglets challenged with *E. coli* F18 612
K. Jerez-Bogota, M. Jensen, O. Højberg and N. Canibe

Role of caecal microbiota in flock weight heterogeneity 612
M.Z. Akram, E.A. Sureda, L. Comer and N. Everaert

1H-NMR metabolomics reveals alterations in the metabolism of ascarid-infected laying hens 613
O.J. Oladosu, B.S.B. Correia, B. Grafl, D. Liebhart, C.C. Metges, H.C. Bertram and G. Daş

Validated machine-learning model to detect IUGR piglets 613
R. Ruggeri, G. Bee, P. Trevisi and C. Ollagnier

In vivo validation of a non-invasive tool to collect intestinal content in pigs – CapSa- 614
I. García Viñado, F. Correa, P. Trevisi, G. Bee and C. Ollagnier

Effect of creep feeding (liquid milk, dry and liquid diet) on pig growth and intestinal structure 614
E.A. Arnaud, G.E. Gardiner, M. Chombart, J.V. O'Doherty, T. Sweeney and P.G. Lawlor

In vitro anthelmintic evaluation of Greek oregano against *Ascaridia galli* 615
I. Poulopoulou, E. Sarrou, E. Martinidou, L. Palmieri, D. Masuero, S. Martens and M. Gauly

Session 50. Insect genetics and reproduction

Date: Wednesday 30 August 2023; 8.30 – 13.00
Chair: Sandrock / Maya

Theatre Session 50

invited Genetic diversity and improvement of the black soldier fly 615
C.D. Jiggins and T. Generalovic

Selection for larval weight in the black soldier fly – empirical evidence 616
K. Shrestha, E. Facchini, E. Van Den Boer, P. Junes, G. Sader, K. Peeters and E. Schmitt

Full-sib group records as a practical alternative to individual records in insect breeding 616
L.S. Hansen, A.C. Bouwman, H.M. Nielsen, G. Sahana and E.D. Ellen

Simulating breeding programs based on mass selection in black soldier fly (*Hermetia illucens*) 617
M. Slagboom, H.M. Nielsen, M. Kargo, M. Henryon and L.S. Hansen

Effects of artificial selection in the black soldier fly – a Pool-seq approach 617
E. Facchini, A. Vereijken, D. Bickhart, A. Michenet, K. Shrestha and K. Peeters

Adaptive responses of black soldier fly to simple low-quality diets 618
A. Gligorescu and J.G. Sørensen

Molecular sexing of black soldier flies 618
R.S.C. Rikkers, E. Van Der Valk, A.A.C. De Wit, L. Kruijt, E.D. Ellen, J. Van Den Heuvel and B.A. Pannebakker

Biomarker discovery for the black soldier fly (*Hermetia illucens*) 619
E.M. Espinoza

Improving black soldier fly genetics by CRISPR\Cas9 gene editing 619
I.N.Y. Nevo Yassaf, A.G. Goren, R.A. Adler and I.A. Alyagor

Expected response to selection on larval size and development time in the housefly (*Musca domestica*) 620
H.M. Nielsen, T.N. Kristensen, G. Sahana, S.F. Laursen, S. Bahrndorff, J.G. Sørensen and L.S. Hansen

ŸnFABRE: design of reference populations for genomic selection in *Tenebrio molitor* 620
E. Sellem, A. Donkpegan, Q. Li, K. Paul, A. Masseron, F. Gagnepain-Germain, K. Labadie, B. Vacherie, P. Garrabos, M.A. Madoui and T. Lefebvre

The potential of instrumental insemination for honeybee breeding 621
M. Du, R. Bernstein and A. Hoppe

Genetic analysis of production and behavioural traits of French honeybees 621
T. Kistler, C. Kouchner, C. Dumas, R. Dupain, A. Vignal, F. Mondet, P. Jourdan, B. Basso and F. Phocas

SIMplyBee: R package for simulating honeybee populations and breeding programs 622
J. Obšteter, L.K. Strachan, J. Bubnič, J. Prešern and G. Gorjanc

Demonstrating the principles of genetic inheritance in honeybees using SIMplyBee 622
L. Strachan, J. Bubnič, G. Petersen, G. Gorjanc and J. Obšteter

Poster Session 50

Exploring the potential for artificial selection in the black soldier fly, *Hermetia illucens* 623
T. Generalovic and C. Jiggins

Weight of *Hermetia illucens* eggs from breeding and wild individuals 623
J. Lisiecka, Z. Mikołajczak, M. Dudek, K. Dudek, B. Kierończyk and D. Józefiak

Paternity assignment tool in honey bees (*Apis mellifera*) 624
S. Andonov, G. Aleksovski, B. Dahle, M. Kovačić, A. Marinič, A. Moškrič, J. Prešern, B. Pavlov, Z. Puškadija and A. Uzunov

Single-step genomic BLUP allows for the genetic evaluation of commercial honeybee queens 624
G.E.L. Petersen, F.S. Hely, M. Araujo, P.F. Fennessy and P.K. Dearden

Deformed wing virus quantification: effect of selection and correlation with varroa related traits 625
M.G. De Iorio, S. Ottati, G. Molinatto, D. Bosco and G. Minozzi

Monitoring the distribution of *Apis mellifera* genetic resources in Italy using mtDNA information 625
V. Taurisano, A. Ribani, K.E. Johnson, D. Sami, G. Schiavo, S. Bovo, V.J. Utzeri and L. Fontanesi

Session 51. Leveraging the microbiome for resilience and sustainability in ruminant production – insights from H2020 HoloRuminant

Date: Wednesday 30 August 2023; 8.30 – 13.00
Chair: Kenny / Morgavi

Theatre Session 51

vited How does the cow's microbiome respond to physiological challenges? 626
J. Seifert

Upper respiratory tract microbiota of dairy calves experimentally challenged with BRSV 626
S. O'Donoghue, B. Earley, M.S. McCabe, D. Johnston, K. Ní Dhufaigh, S.L. Cosby, D.W. Morris and S.M. Waters

Differential effects of *Ostertagia ostertagi* vaccination and infection on the rumen microbiome 627
J. Lima, T.N. McNeilly, P. Steele, M. Martínez-Álvaro, M.D. Auffret, R.J. Dewhurst, M. Watson and R. Roehe

vited Novel ruminal microbiome solutions to reducing enteric methane emissions 627
T.A. McAllister, L.L. Guan and R.I. Mackie

Relationship between feed efficiency and rumen microbiota in feedlot bulls fed contrasting diets 628
A. Ortiz-Chura, M. Popova, G. Cantalapiedra-Hijar and D. Morgavi

Establishment and evolution of ruminotypes of lactating Lacaune ewes 628
T. Blanchard, C. Marie-Etancelin, Y. Farizon, C. Allain and A. Meynadier

Effect of colostrum source and calf breed on diarrhoea incidents in pre-weaned dairy calves 629
S. Scully, P.E. Smith, B. Earley, C. McAloon and S.M. Waters

vited Metagenome strain deconvolution and abundance estimation enabled by low-error long-read DNA sequence 629
D. Bickhart, M. Kolmogorov, E. Tseng, P. Pevzner and T. Smith

Stakeholders views regarding new practices to control microbiomes 630
F. Bedoin, A. Ait-Sidhoum, E. Vanbergue, A. Stygar, T. Latvala, Á. MacKen-walsh, S. Waters, P. Smith and J.K. Niemi

Poster Session 51

Temporal establishment of the colon microbiota in angus calves from birth to post-weaning 630
M. Stafford, P. Smith, S. Waters, F. Buckley, E. O'Hara and D. Kenny

Preliminary study of faecal microbiota in a selection experiment for birth weight variability in mice 631
L. El-Ouazizi El-Kahia, N. Formoso-Rafferty, J.P. Gutiérrez, C. Esteban Blanco, J.J. Arranz and I. Cervantes

The effects of crude protein levels on rumen microbiome and CH_4 of the fattening Hanwoo steers 631
H. Kim, H. Cho, S. Jeong, K. Kang, S. Jeon, M. Lee, H. Kang, S. Lee, S. Seo and J. Seo

Rumen metabolites of periparturient cows varying in SARA susceptibility modify fermentation *in vitro* 632
H. Yang, S. Heirbaut, J. Jeyanathan, X.P. Jing, N. De Neve, L. Vandaele and V. Fievez

Prevalence of resistant *E. coli* and their transmission among dairy cattle in Swiss tie stalls 632
B. Köchle, V. Bernier Gosselin and J. Becker

Stability of a *Bacillus*-based DFM following preparation of a milk replacer and premix 633
G. Copani, A. Segura, N. Milora, M. Schjelde and B.I. Cappellozza

Effects of a *Bacillus*-based direct-fed microbial on performance and digestibility of lactating cows 633
M. Terré, N. Prat, D. Sabrià and B. Cappellozza

Stability of a *Bacillus*-based direct-fed microbial post-palletisation under different temperatures 634
B. Cappellozza, C. Galschioet and G. Copani

Microbial signals in peripheral blood mononuclear cells of Australian Angus cattle 634
P. Alexandre, A. Wilson, T. Legrand, R. Farr, S. Denman and A. Reverter

Metabarcoding of milk, faeces and ruminal fluid of Sarda ewes fed with Alfalfa (*Medicago sativa*) 635
A. Vanzin, D. Giannuzzi, G. Zardinoni, A. Cecchinato, N. Macciotta, F. Correddu, A. Atzori, S. Carta, A. Ledda, S. Schiavon, L. Gallo and S. Pegolo

An experimental approach for assessing causal microbes in early life diarrhoea in lambs 635
L. Voland, D. Graviou, K. Vazeille, A. Ortiz-Chura, D.P. Morgavi and M. Popova

Mycotoxin-deactivating feed additive supplementation in dairy cows fed *Fusarium*-contaminated diet 636
A. Catellani, Y. Han, V. Bisutti, F. Ghilardelli, F. Fumagalli, E. Trevisi, A. Ceccinato, H. Swamy, S. Van Kuijk and A. Gallo

Effects of 3-NOP on enteric methane production in growing beef cattle offered a forage based diet 636
S.F. Kirwan, L.F.M. Tamassia, N.D. Walker, A. Karagiannis, M. Kindermann and S.M. Waters

Session 52. Nitrogen management in the dairy supply chain and other tools to improve the efficiency and sustainability of cattle farming

Date: Wednesday 30 August 2023; 8.30 – 13.00
Chair: Righi / Tsiplakou

Theatre Session 52

invited COWFICIENCY project: a field attempt to increase nitrogen use efficiency of dairy cattle 637
A. Foskolos, A. Plomaritou, M.E. Hanlon and D. Kantas

Dairy heifers intake capacity: are estimated values still correct? 637
J. Jurquet, Y. Le Cozler, D. Tremblais, F. Launay and L. Delaby

Current status of feed nitrogen use efficiency in dairy replacement heifers in Greece 638
A. Plomaritou, M.E. Hanlon, K. Gatsas, S. Athanasiadis, K. Droumtsekas, D. Kantas and A. Foskolos

Impact of feed-grade and slow-release ureas on dairy cattle performance and nitrogen efficiency 638
M. Simoni, G. Fernandez-Turren, F. Righi, M. Rodríguez-Prado and S. Calsamiglia

Current nitrogen status and management of dairy farms in Greece 639
M.E. Hanlon, A. Plomaritou, E. Tsiplakou, I. Vakondios, T. Michou, D. Kantas and A. Foskolos

Dairy cattle holistic nutritional management for reduced nitrogen pollution 639
A. Foskolos, A. Plomaritou, M.E. Hanlon and D. Kantas

Manure management practices of dairy farms in Greece 640
L. Makridis, D. Vouzaras, D. Kantas and A. Foskolos

peNDF modulates chewing activity, rumen fermentation, plasma metabolites, performance in dairy cow 640
Y.C. Cao, L.M. Wang and J.H. Yao

Impact on milk production of feeding organic acids during pre and post-partum period 641
L. Jansen and C. Gordon

Peptide profile as fingerprinting of the ripening period of Alpine Asiago cheese 641
S. Segato, S. Khazzar, G. Galaverna, A. Caligiani, G. Riuzzi, L. Serva, F. Gottardo and G. Cozzi

Producing sustainable milk in agroecology with Normande breed and grassland in Normandy 642
B. Rouillé, F. Lepeltier and L. Morin

Hepatic metabolome of grazing dairy cows with or without feed restriction during early lactation 642
M. Carriquiry, M. García-Roche, A.L. Astessiano, D. Custodio, G. Ortega and P. Chilibroste

Nitrogen balance of lactating cows from herds fed hay-based diets in Northern Italy 643
T. Danese, M. Simoni, R.G. Pitino, G. Mantovani, M.C. Sabetti, F. Righi and M.E. Van Amburgh

Assessing sward managements on nitrogen fixation from a white clover high sugar grass mixture 643
M. Verbeeck, C. Segura, A. Louro-Lopez, N. Loick, P. De-Meo-Filho, S. Pulley, J. Hood, B.A. Griffith, L.M. Cardenas and D. Enriquez-Hidalgo

Amino acid intake of hay-based diets and the relationship with milk production and urea content 644
M. Simoni, R. Pitino, T. Danese, G. Mantovani, E. Tsiplakou and F. Righi

Poster Session 52

Does acidification affect urinary creatinine in dairy cattle? 644
T. Danese, M.C. Sabetti, M. Simoni, R.G. Pitino, G. Mantovani and F. Righi

Effect of an alternative to sodium bicarbonate on performance of dairy cows 645
V. Leroux, C. Jaffres, A. Budan and N. Rollet

Session 53. PLF for health and welfare – Part 1

Date: Wednesday 30 August 2023; 8.30 – 13.00
Chair: Maselyne / Calderon Diaz

Theatre Session 53

Evaluation of three measuring methods for the ammonia concentration for practical use in pig houses 645
J. Witt, J. Krieter, K. Schröder and I. Czycholl

The information and communication technologies in livestock production – expectations and concerns 646
S. Opalinski, K. Olejnik, E. Popiela, A. Jankowska-Makosa, D. Konkol, M. Korczynski, D. Knecht, R. Kupczynski, I. Tikasz and T. Banhazi

Implementation of a deep learning based system for monitoring farrowing in sows 646
M. Wutke, C. Lensches, A. Holzhauer, M.A. Lieboldt and I. Traulsen

Who's biting? Detecting pig screams for identifying tail biting events 647
P. Heseker, T. Bergmann, M. Scheumann, S. Ammer, I. Traulsen, N. Kemper and J. Probst

Veterinarians' perceptions of using PLF technologies in pig husbandry in the Netherlands and Germany 647
M.F. Giersberg and F.L.B. Meijboom

Automatic detection and quantification of ear biting in pigs 648
A. Odo, R. Muns, L. Boyle and I. Kyriazakis

Installation modified carbon felt in pig house to control airborne pathogenic microorganisms 648
X.D. Zhao, F. Qi, H. Li and Z.X. Shi

Benefit of caliper use at insemination on different genetic types: impact on farrowing performances 649
C. Teixeira Costa, C. Chevance, T. Nicolazo, G. Boulbria, V. Normand, J. Jeusselin and A. Lebret

Combined effect of genetics, gut microbiota and environment on vaccine responses and welfare in hens 649
A. Lecoeur, F. Blanc, D. Gourichon, N. Meme, T. Burlot, V. Guesdon, V. Ferreira, L. Calandreau, L. Warin, F. Calenge and M.H. Pinard Van-Der-Laan

Using triaxial accelerometers to monitor peripartum behaviour of Purebred Spanish mares 650
M.J. García García, F. Maroto Molina, C.C. Pérez Marín and D.C. Pérez Marín

Relationships between direct and indirect genetic effects of RFI and feeding behaviour traits 650
M. Piles, M. Mora, M. Pascual and J.P. Sánchez

Audio-based event detection and health monitoring in poultry 651
F. Hakansson and D.B. Jensen

Poster Session 53

First approach of using sows' water consumption data to detect the onset of farrowing 651
J. Probst, N. Volkmann, C. Lensches, P. Heseker, G. Thimm, M. Lieboldt, I. Traulsen and N. Kemper

Social network analysis of cattle and horses inferred from sensor ear tag (SET) and GPS based data 652
U. Heikkilä, K. Ueda, N. Gobbo Oliveira Erünlü, M. Baumgartner, M. Cockburn, I. Bachmann, M. Roig-Pons and S. Rieder

Session 54. Life Green Sheep and Farm management to adapt goat and sheep farming to climate change: solutions and experiments

Date: Wednesday 30 August 2023; 8.30 – 13.00
Chair: Throude / Satzori

Theatre Session 54

Modelling adaptation strategies to climate change in Mediterranean small ruminant systems 652
A. Lurette, S. Lobón, F. Douhard, M. Blanco-Alibes, D. Martin-Collado, A. Madrid, M. Curtil-Dit-Galin and F. Stark

Feed365: Early findings for sheep grazing novel year-round forage systems 653
C.E. Payne, D. Real, A. Loi and C. Revell

Quantifying responses to hot conditions in divergent sheep breeds in South Africa 653
S. Cloete, S. Steyn, J. Van Zyl and T. Brand

Impacts of heat peaks in France on the performances of dairy goats housed in an insulated roof shed 654
K. Boissard, A. Fatet, M. Lambert, P. Sales and H. Caillat

Cortisol level in sheep wool underwent to three different pasture management 654
L. Turini, A. Ripamonti, A. Silvi, E. Giua, A. Mantino, G. Conte, F. Bonelli and M. Mele

How sheep sectors do face environmental, social and economic issues in French Pyrenees region? 655
A.-L. Jacquot, P.-G. Marnet and Y. Le Cozler

Aligning carbon footprint estimates from different tools across Europe 655
A.S. Atzori, O. Del Hierro, B. Lyubov, R. Vial, C. Buckley, M.G. Serra, M. Habeanu, R. Ruiz, L. Lanzoni, M. Acciaro, T. Keady and S. Troude

Determining carbon footprint of sheep farms in Europe: first results of the LIFE Green Sheep project 656
S. Throude, M. Acciaro, A. Atzori, R. Ruiz, O. Del Hierro, C. Buckley, L. Bragina, T.W.J. Keady, C. Dragomir, M.A. Gras and J.B. Dollé

Greenhouse gas emission intensity of milk production in three Slovenian goat breeds 656
M. Bizjak, Ž. Pečnik and M. Simčič

No difference in PAC methane emission corrected for milk yield in three Norwegian dairy goat farms 657
J.H. Jakobsen, T. Blichfeldt, K. Dodds and J.C. McEwan

Sustainability of Irish sheep production 657
C. Buckley, L. Bragina and T. Keady

Should animal welfare indicators be integrated into the environmental impact assessment of farms? 658
L. Lanzoni, L. Whatford, K. Waxenberg, R. Ramsey, R.M. Rees, J. Bell, E. Dalla Costa, S. Throude, A.S. Atzori and G. Vignola

Testing mitigation actions to reduce GHG emissions from sheep farming in Europe 658
S. Throude, M. Acciaro, A. Atzori, R. Ruiz, O. Del Hierro, C. Buckley, L. Bragina, T.W.J. Keady, C. Dragomir, M.A. Gras and J.B. Dollé

The effect of forage type on methane production from hill bred lambs grazing alternative forages 659
M. Dolan, T. Boland, N. Claffey, F. McGovern and F. Campion

Poster Session 54

Preliminary results of the LIFE Green Sheep project in Italy 659
M. Acciaro, M. Decandia, M.G. Serra, S. Picconi, V. Giovanetti, A. Atzori, D. Usai and S. Throude

Carbon footprint assessment of a Pecorino cheese produced in central Italy 660
L. Lanzoni, L. Di Paolo, S. Abbate, M. Giammarco, M. Chincarini, I. Fusaro, A. Atzori, D. Di Battista, S. Throude and G. Vignola

Grass management to adapt goat farming to climate change in western France 660
J. Jost, M.-G. Garnier, L. Robin, M. Proust, M. Bourasseau, R. Lesne, A. Villette, V. Tardif, T. Soulard, O. Subileau and O. Prodhomme

Positive effects of grazing mulberry trees in summer on dairy goats' milk and cheese 661
C. Boyer, H. Le Chenadec, F. Noël, A. Stocchetti, J. Jost, A. Pommaret, S. Fressinaud and R. Delagarde

Effects of heat stress and forage quality on feed intake and milk production of Sarda dairy ewes 661
M. Sini, F. Fulghesu, A. Ledda, A.S. Atzori and A. Cannas

Seasonal rainfall patterns modify summer energy balance and nutritional condition of grazing sheep 662
Y. Yoshihara, B. Choijilsuren, T. Kinugasa and M. Shinoda

Effect of the inclusion of rumen-protected amino acids in the diet of high production dairy sheep 662
A. Cabezas, D. Martinez Del Olmo, J. Mateos, J. Matilla, M.T. Díaz, R. Bermejo-Poza, J. De La Fuente and V. Jimeno

Session 55. Breeding for climate change – adaptation

Date: Wednesday 30 August 2023; 15.00 – 18.30
Chair: Sell-Kubiak / Pszczola

Theatre Session 55

Reaction norm model analysis for heat stress tolerance of growth performance in purebred pigs 663
Y. Fukuzawa, S. Ogawa, T. Okamura, N. Nishio, K. Ishii, K. Tashima, K. Akachi, H. Takahashi and M. Satoh

Investigating the slick gene and its effects on heat stress in New Zealand grazing dairy cattle 663
G.M. Worth, E.G. Donkersloot, L.R. McNaughton, S.R. Davis and R.J. Spelman

Use of sensors for the detection and genetic evaluation of heat stress in dairy cattle 664
P. Lemal, M.-N. Tran, M. Schroyen and N. Gengler

Evaluation of heat stress effects on production traits and somatic cell score of Dutch Holstein cows 664
J. Vandenplas, M.L. Van Pelt and H. Mulder

Multi-omics and multi-tissues data to improve knowledge of heat stress acclimation mechanisms 665
G. Huau, D. Renaudeau, J.L. Gourdine, J. Fleury, J. Riquet and L. Liaubet

Should we consider fertility when improving thermotolerance in dairy cattle? 665
M.J. Carabaño, C. Díaz and M. Ramón

Trade-off between fertility and production in French dairy cattle in the context of climate change 666
A. Vinet, S. Mattalia, R. Vallee, A. Barbat, C. Bertrand, B.C.D. Cuyabano and D. Boichard

Genetic analyses of resilience indicator traits in German Holstein, Fleckvieh and Brown Swiss 666
F. Keßler, R. Wellmann, M. Chagunda and J. Bennewitz

Estimating the heritability of nitrogen and carbon isotopes in the tail hair of beef cattle 667
M. Moradi, C. Warburton and L.F.P. Silva

Changes in genetic correlations over generations due to selection and random drift 667
B.C.D. Cuyabano, S. Aguerre and S. Mattalia

Longitudinal study of environmental effects for American Angus beef cattle over 30 years 668
G. Rovere, B.C.D. Cuyabano, B. Makanjuola and C. Gondro

Characterization of environmental impact of 13,000 French dairy farms 668
R. Vial, A. Stocchetti, M. Mevel and C. Brocas

Poster Session 55

Plateau-linear regression analysis of farrowing records on temperature data for pigs reared in Japan 669
S. Ogawa, T. Okamura, Y. Fukuzawa, M. Nishio, K. Ishii, M. Kimata, M. Tomiyama and M. Satoh

Genetic parameter of heat tolerance for reproductive traits in Landrace, Large White and Duroc pigs 669
T. Okamura, Y. Fukuzawa, M. Nishio, S. Ogawa, K. Ishii, H. Takahashi, K. Tashima, K. Akachi and M. Satoh

Breeding soundness evaluation of bulls in extensive systems in interior centre and south of Portugal 670
J. Várzea Rodrigues, L. Pinto De Andrade, S. Dias, J. Carvalho and M. Martins

Transcriptome analysis identifies genes affected by heat stress in hen uterovaginal junction 670
S. Kubota, P. Pasri, S. Okrathok, O. Jantasaeng, S. Rakngam, P. Mermillod and S. Khempaka

Session 56. Establishing breeding programs in extensive systems (including developing regions) with climate change in mind

Date: Wednesday 30 August 2023; 15.00 – 18.30
Chair: Leroy / Berg

Theatre Session 56

ited Establishing and scaling up breeding programs: a challenging, but not impossible task 671
M. Wurzinger

ited Development of a breeding programme for oysters 671
P. Haffray, R. Morvezen, F. Enez, L. Dégremont and P. Boudry

Status of implementation of EU animal breeding legislation for endangered breeds 672
H. Göderz, L. Balzar, J. Wider, C. Danchin, M. Spoelstra, M. Schoon and S. Hiemstra

Spatial modelling improves genetic evaluation of Tanzanian smallholder crossbred dairy cattle 672
I. Houaga, R. Mrode, M. Okeyo, J. Ojango, Z. Nziku, A. Nguluma, A. Djikeng, E. Lavrenčič, G. Gorjanc and I. Pocrnic

Genetic parameters for grazing behaviour traits of Boutsko sheep 673
S. Vouraki, V. Papanikolopoulou, A. Argyriadou, V. Fotiadou, V. Tsartsianidou, A. Triantafyllidis, G. Banos and G. Arsenos

Research and development innovations for climate-smart beef production in subtropical countries 673
M.M. Scholtz, G.M. Pyoos, M.L. Makgahlela, M.C. Chadyiwa, M.D. MacNeil, M.M. Seshoka and F.W.C. Neser

Genetic and environmental factors influencing skin traits of South African farmed ostriches 674
K.R. Nemutandani, A. Engelbrecht, S.W.P. Cloete, K. Dzama and O. Tada

Genomic signatures of adaptive response driven by transhumant pastoralism in native Boutsko sheep 674
V. Tsartsianidou, S. Vouraki, P. Papanikolopoulou, G. Arsenos and A. Triantafyllidis

An Australian sheep genomic reference to meet the evolving breeding objectives of industry 675
S.F. Walkom, D.J. Brown and J.H.J. Van Der Werf

Poster Session 56

Evaluation of the perceptions of the functions of local breeds of domestic ruminants in Mayotte 675
J. Vuattoux, A. Lauvie, A. Giraud, E. Ozarak, J. Janelle, A. Rozier, T.T.S. Siqueira, M. Naves and E. Tillard

Signature of selection in South African Dexter cattle reveal resistance genes and genetic variations 676
E.D. Cason, J.B. Van Wyk, P.D. Vermeulen and F.W.C. Neser

Breed environment interaction and suitability of Dutch cattle breeds for low input systems 676
J.J. Windig, G. Bonekamp, M.A. Schoon, A.H. Hoving and S.J. Hiemstra

Identification of candidate gene variants for the alpaca Suri phenotype by WGS analysis 677
S. Pallotti, D. Pediconi, M. Picciolini, M. Antonini, V. Napolioni and C. Renieri

Session 57. Grassland management and grassland-based feeding systems for efficient and sustainable milk and meat production

Date: Wednesday 30 August 2023; 15.00 – 18.30
Chair: Berard / Probo

Theatre Session 57

invited Extensive permanent grasslands in Europe: multifaceted functions, threats and prospects 677
M. Bassignana

The effect of establishment and grazing management on clover and herb establishment and persistence 678
L. McGrane, N. McHugh, T.M. Boland and P. Creighton

Strategic concentrate supplementation in reducing slaughter age in pasture-based dairy-beef systems 678
J. O' Driscoll, D. Purfield, N. McHugh and N. Byrne

Morphology and body composition of beef-on-dairy heifers along compensatory growth itinerary 679
I. Morel, A. Dieudonné, R. Siegenthaler, C. Xavier and S. Lerch

In France, a new beef × dairy calf to steer production for the out-of-home consumers 679
M.A. Brasseur, C. Fossaert, F. Guy, J.J. Bertron, T. Dechaux and S. Brouard

Behavioural and welfare responses of dairy cows learning a virtual fencing system 680
P. Fuchs, J. Stachowicz, M. Schneider, M. Probo, R. Bruckmaier and C. Umstätter

Age does not affect the learning capacity of virtually fenced cows 680
A. Confessore, C. Aquilani, P. Fuchs, C. Pugliese, C.M. Pauler, M. Schneider, G. Argenti and M. Probo

Integrating multiple data streams and models to inform precision grazing management in the U.S 681
J.R. Brennan, H. Menendez and K. Ehlert

Yearly monitoring of soil ingestion by dairy cows in a grassland system with feed supply 681
C. Collas, A. Laflotte, C. Feidt and S. Jurjanz

Pre-grazing sward height affects enteric methane emission during grazing 682
L. Koning, G. Holshof, A. Klop and C.W. Klootwijk

Forage shortage affects performances, CH_4 emissions and cheese quality in grass- or corn-fed cows 682
M. Bouchon, I. Verdier-Metz, M. Eugene, C. Bord, B. Martin, J. Bloor, M.C. Michalski, B. Graulet and C. Delbès

Variability of economic and GHG performance in dairy-beef systems at different stocking rates 683
M. Kearney, E. O'Riordan, J. Breen, R. Dunne, P. French and P. Crosson

Poster Session 57

Feeding behaviour, methane emission and digestibility of crossbred heifers along compensatory growth 683
B. Hayoz, I. Morel, A. Dieudonné, M. Rothacher, R. Siegenthaler, F. Dohme-Meier and S. Lerch

Involving farmers in the development of a grassland monitoring tool: sunshine's co-design approach 684
D.M. Mathy, C.L. Lucau-Danila, Y.C. Curnel, E.R. Reding, K.D. Dichou and S.L. Lagneaux

Effects of increasing portion of grass-silage in dairy cow diet on carbon footprint of raw milk 684
S. Hietala, A. Vanhatalo, K. Kuoppala, T. Kokkonen, A. Reinikainen, K. Timonen and A.-L. Välimaa

Blood metabolite, hormone and δ13C turnover kinetic during compensatory growth of crossbred heifers 685
S. Lerch, P. Silacci, G. Cantalapiedra-Hijar, R. Siegenthaler, S. Dubois, A. Delavaud, M. Bonnet and I. Morel

Metabolic assessment of parasite dilution and forage niche sharing in sheep/cattle mixed-grazing 685
F. Joly, P. Nozière, P. Jacquiet, S. Prache and B. Dumont

Effects of different additives on the correlation between fermentation characteristics of wilted rye 686
Y.F. Li, L.L. Wang, Y.S. Yu, H.J. Kim and J.G. Kim

Bite item selection by grazing suckler cows in multi-species grasslands 686
C. Siede, W. Pohlmann, A. Juch, D. Hamidi, J. Isselstein and M. Komainda

Nutritional value of intramuscular fat of the muscle of Arouquesa weaners from different systems 687
L. Sacarrão-Birrento, C.A. Venâncio, A.M. De Almeida, L.M. Ferreira, M.J. Gomes, J.C. Almeida, J.A. Silva and S.P. Alves

Prediction of nutritional parameters of naturalized grassland in the dry zone of Chile using NIRS 687
P.M. Toro-Mujica

Effect of cutting length on fermentation dynamics of wilted Italian ryegrass silage 688
J. Kim, Y. Li, L. Waang, Y. Yu and H. Kim

Grazing behaviour of energy-limited dairy cows and development of detection method by deep learning688
Y. Shinoda, S. Asakuma, Y. Ueda, S. Tada and K. Sudo

Evaluation of 3 equations based on grass height measurement for estimating grass stocks in pastures 689
F. Lessire, J.-L. Hornick and I. Dufrasne

Breeding of native breeds as a chance for the development of livestock households in ecological sys 689
P. Radomski and P. Moskala

Conceptual model for the analysis of energy allocation of cow-calf farms native grassland-based 690
V. Figueroa, I. Paparamborda, S. Scarlato and P. Soca

Do botanically diverse pastures effect the meat eating quality of lamb? 690
S. Woodmartin, P. Creighton, T.M. Boland, E. Crofton, A. Monaghan and F. McGovern

Altering milking frequency from 14 to ten milking's per week: effects on milk production of pasture 691
E. Kennedy, K. McCarthy, J.P. Murphy and M. O'Donovan

Robustness of suckling cows at herd level is associated with cows' productive longevity 691
L. Barreto-Mendes, A. De La Torre, S. Ingrand and F. Blanc

Session 58. The future of animal products: improved quality management, more alternatives or cell-based products?

Date: Wednesday 30 August 2023; 15.00 – 18.30
Chair: Ellies-Oury / Chriki

Theatre Session 58

Variability in consumer perception of meat and meat substitutes 692
E. Hocquette, J. Liu, S. Chriki, M.P. Ellies-Oury, M. Kombolo, J.H. Rezende-De-Souza, S.B. Pflanzer and J.F. Hocquette

Bibliometric analysis of scientific articles related to 'cultured meat' 692
J.F. Hocquette, D. Fournier, M.P. Ellies-Oury and S. Chriki

Addressing the challenges of animal-free meat using plant-based tissue engineering 693
M.O.R. Yahav

Comparing the potential of meat alternatives for a more sustainable food system 693
T. Bry-Chevalier

invited IMR3G Foundation, DATAbank software to facilitate collaborative data collection for mutual benefit 694
R. Polkinghorne, H. Cutherbertson, A. Neveu and J. Wierzbicki

Beef processors experience large variation in yield and quality traits on a daily basis 694
W. Pitchford and S. Miller

Introduction to plant-based, cultivated, and fermentation-made meat, eggs, and dairy 695
S. Kell

Environmental impact of dairy alternatives: a case study of Hemp milk and other products 695
B. Queiroz Silva, J. Ferdouse and S. Smetana

The tools of prediction of the sensory quality, the opinion of the French professionals 696
T. Fayet

Limitations and challenges for the successful launch to market of cultured animal protein products 696
J.F. Fuentes-Pila

Implementing advanced characterization methods and building a new reference for alt-meat development 697
M.O.R. Yahav

Poster Session 58

Whey proteins as alternative supplement to FBS in C2C12 muscle cells for cultured meat production 697
T.S. Sundaram, D. Lanzoni, R. Rebucci, F. Cheli, A. Baldi and C. Giromini

German consumers' attitudes towards cultured meat 698
A.-K. Jacobs, M.-P. Ellies-Oury, H.-W. Windhorst, J. Gickel, S. Chriki and J.-F. Hocquette

Development of the beef eating quality management system in Poland 698
G. Pogorzelski, J. Wierzbicki, E. Pogorzelska-Nowicka, A. Jasieniak and A. Wierzbicka

Session 59. Environmental challenges of cattle tropical grazing systems

Date: Wednesday 30 August 2023; 15.00 – 18.30
Chair: Dossa / Salgado

Theatre Session 59

Degradation of Amazonian grasslands by weeds, how to managing this situation? 699
V. Blanfort, C. Favale, S. Bazan, V. Petiot, D. Bastianelli and T. Le Bourgeois

Analysing strategies adopted to cope the climate variability in pastoral zone of Burkina Faso 699
H.P. Yarga, A. Kiema, L. Ouedraogo and S. Ouedraogo

How do changes in crop-livestock integration and specialisation affect farm performance in Vietnam? 700
A. Le Trouher, H. Le Thi Thanh, T. Dinh Khanh, T. Han Anh, C.-H. Moulin and M. Blanchard

Problem of water supply in a context of climate variability in the pastoral zones of Burkina Faso 700
H.P. Yarga H Paul, S. Ouedraogo, A. Kiema and L. Ouedraogo

Local feeding strategies allow to reduce enteric methane emission from cattle in Sahel 701
G.X. Gbenou, M.H. Assouma, C. Martin, D. Bastianelli, L. Bonnal, T. Kiendrebeogo, O. Sib, B. Bois, S. Sanogo and L.H. Dossa

Development of a decision support tool to secure cattle production in chlordecone-contaminated areas 701
A. Fournier, C. Feidt, A. Fourcot, M. Saint-Hilaire, Y. Le Roux and G. Rychen

Can tropical legume grass forage reduce enteric methane yield from suckler cows in the Sahel? 702
M.H. Assouma, A. Baro, G.X. Gbenou, O. Sib, S. Sanogo, H. Marichatou and E. Vall

Agroforestry: an opportunity to improve the sustainability of livestock systems in Vietnam? 702
M. Blanchard, P. Tos, A. Le Trouher, A. Lurette and H. Le Thi Thanh

Selection signatures of the indigenous Sanga cattle of Namibia 703
D.A. Januarie, E.D. Cason and F.W.C. Neser

Diodelle sarmentosa, an invasive plant in the rangelands of the sylvopastoral zone of Senegal 703
E.H. Traore and F. Sow

vited Environmental challenges in dry tropical livestock systems: GHG emissions and carbon storage balance 704
M.H. Assouma, D. Bastianelli and P. Salgado

Poster Session 59

Resilience analysis based on heifer productive and reproductive aspects 704
V.T. Rezende, G.R.D. Rodrigues, A.H. Gameiro, M.E.Z. Mercadante, R.C. Canesin and J.N.G.S. Cyrillo.

Impact of GreenFeed protocols on cattle visitation and methane data collection 705
M.C. Parra, M.H. Dekkers, S.A. Cullen and S.J. Meale

Effect of production system and season on composition of retail cow milk in Greece 705
G. Papatzimos, R.A. Stergioudi, V. Papadopoulos, Z. Basdagianni, M.A. Karatzia, P. Mitlianga and E. Kasapidou

Influence of heifer resilience on the productive performance of calves 706
G.R.D. Rodrigues, V.T. Rezende, C. Raineri, M.E.Z. Mercadante, S.F.M. Bonilha and J.N.S.G. Cyrillo

Feed autonomy and manure's recycling of dairy sheep farming systems in Roquefort (France) 706
W. Nasri, F. Stark, N. Amposta, M. Lamarque, C. Allain, D. Portes, S. Arles, S. Parisot, C. Corniaux, P. Salgado and E. González-García

Metabarcoding study of the microbial dynamics in cheeses as a function of milk tank temperature 707
L. Giagnoni, C. Spanu, A. Tondello, S. Deb, A. Cecchinato, P. Stevanato, M. De Noni and A. Squartini

Adaptive integumentary traits of cattle raised in a silvopastoral system in tropical region 707
A.R. Garcia, A.N. Barreto, M.A.C. Jacintho, W. Barioni Junior, L.N. Costa, F. Luzi, J.R.M. Pezzopane, A.C.C. Bernardi and A.M.F. Pereira

Impact of fodder quality seasonality on enteric methane emission from cattle in Sub-Saharan Africa 708
G.X. Gbenou, M.H. Assouma, C. Martin, D. Bastianelli, L. Bonnal, T. Kiendrebeogo, O. Sib, B. Bois, S. Sanogo and L.H. Dossa

Session 60. Minerals in animal nutrition: supplementation – Part 1

Date: Wednesday 30 August 2023; 15.00 – 18.30
Chair: Alvarez Rodriguez / Windisch

Theatre Session 60

invited Variation in mineral content of feeds for dairy cows and how that can affect ration formulation 708
W.P. Weiss

Evolution of nutritional explorations in herds of dairy and beef cattle between 2013 and 2021 709
L. Reisdorffer

Selenium supply in animal feeds, a powerful nutritional tool against cancer 709
M. López-Alonso, I. Rivas and M. Miranda

Effect of mineral source on 48-h *in vitro* fermentation 710
G.M. Boerboom, C.B. Peterson, L. Jansen, M.M. McCarthy, J.S. Heldt and J. Johnston

Feed phosphates market dynamics: does price influence demand? 710
G. Milochau

Reduction of trace mineral supplementation on performance and mineral status of fattening pigs 711
E. Gourlez, J.Y. Dourmad, F. Beline, A. Monteiro, A. Boudon, A. Narcy, P. Schlegel and F. De Quelen

Optimal level of dietary zinc for pigs between 10 and 30 kg 711
T.S. Nielsen, S.V. Hansen, J.V. Nørgaard and T.A. Woyengo

Administration of potentiated Zn and monovalent Cu in weanling piglets diet 712
L. Marchetti, R. Rebucci, P. Cremonesi, B. Castiglioni, F. Biscarini, A. Romeo and V. Bontempo

Unexpected Cu and Zn speciation patterns in the feed-animal-excreta system 712
S. Legros, M. Tella, A.N.T.R. Monteiro, A. Forouzandeh, F. Penen, S. Durosoy and E. Doelsch

Coarse limestone particles limit the formation of Ca-phytate complexes in laying hens 713
F. Hervo, M.-P. Létourneau-Montminy, B. Méda, M.J. Duclos and A. Narcy

Poster Session 60

The duration of efficacy of a single oral dose of selenium in sheep 713
S.E. Gallimore, E.J. Hall and N.R. Kendall

Dietary manganese impacts on growth, carcass and reproductive traits in angus bulls 714
J.R. Russell, E.L. Lundy-Woolfolk, A.S. Cornelison, W.P. Schweer, T.M. Dohlman and D.D. Loy

Micronutrient supplementation for suckling calves 714
M.S.V. Salles, F.J.F. Figueiroa, A. Saran Netto, C.M. Bittar, F.F. Simili and H.N. Rios

Impacts of trace mineral source and ancillary drench on steer performance during backgrounding 715
K. Harvey, L. Rahmel, J. Cordero, B. Karisch, R. Cooke and J. Russell

Impact of trace minerals and water/feed deprivation on performance and metabolism of grass-fed beef 715
M.J.I. Abreu, I.A. Cidrini, D. Brito De Araujo, F.D. Resende and G.R. Siqueira

The impact of trace mineral sources of copper and zinc on performance and ruminal bacteria diversity 716
I.A. Cidrini, I.M. Ferreira, D. Brito De Araujo, G.R. Siqueira and F.D. Resende

Are inorganic Mn sources soluble and improve rumen fermentation? 716
A. Vigh, C. Gerard and C. Panzuti

Effect of diet type, Cu source and antagonists on rumen *in vitro* fermentation and Cu distribution 717
I. Bannister, J.A. Huntington, L.A. Sinclair, J.H. McCaughern and A.M. McKenzie

Meta-analysis on zinc oxide's mode of action in reducing weaning stress in healthy piglets 717
C. Negrini, D. Luise, F. Correa, P. Bosi, A. Roméo and P. Trevisi

Marine mineral complex reduces nutrient interactions and allows efficient use of Ca and P in broiler 718
M.A. Bouwhuis, R. Casserly, A. Craig, D. Currie and S. O'Connell

Seaweeds in animal nutrition, a valuable source of minerals but in need of fine-tuning 718
S. Al-Soufi, J. García, E. Cegarra, A. Muíños, V. Pereira and M. López-Alonso

Online mislabelling of mineral and complementary feeds available in France 719
F. Touitou, C. Marin, T. Blanchard, A. Meynadier and N. Priymenko

Diversity of practices and advisors in mineral and vitamin supplementation of dairy farms 719
C. Manoli, G. Springer, L. Barbier, C. Chassaing, C. Sibra, G. Maxin, A. Boudon and B. Graulet

Session 61. Resilient livestock farming systems in the face of climate and other global challenges – adaptation and mitigation

Date: Wednesday 30 August 2023; 15.00 – 18.30
Chair: Sturaro / Poulopoulou

Theatre Session 61

vited How to improve resilience, from animal to system level 720
A. Mottet, R. Baumung, G. Velasco Gil, G. Leroy and B. Besbes

On the link between climate change mitigation and adaptation in dairy cow farming in West of France 720
B. Godoc, E. Castellan, A. Madrid and C. Karam

Resilience of ruminant organic systems to climatic hazards: a study model in a French grassland area 721
C. Boivent and P. Veysset

Resilgame: a game to experiment farm adaptation to climate change 721
G. Martel and S. Colombié

Evolution of agroecology and associate indicators – looking for balance in farming systems 722
E. Benedetti Del Rio, A. Michaud and E. Sturaro

Small ruminants farming systems of Spain: challenges and attributes for their resilience 722
J. Lizarralde, B. Soriano, A. Benhamou-Prat, P. Gaspar-García, Y. Mena-Guerrero, J.M. Mancilla-Leyton, A. Horrillo, R. Ruiz, D. Martín-Collado and N. Mandaluniz

Building resilience in drylands' extensive livestock systems under climate uncertainty 723
A. Tenza-Peral, I. Pérez-Ibarra, A. Breceda, J. Martínez-Fernández and A. Giménez

A thirty-year assessment of interactions between weather conditions and sheep milk yield and quality 723
A. Mantino, M. Milanesi, M. Finocchi, G. Conte, G. Vignali, G. Chillemi, L. Turini and M. Mele

Indicators for animal health on agro-ecological dairy farms 724
A. Ceppatelli, M. Crémilleux, A. Michaud and E. Sturaro

Exploring climate change adaptation strategies form the perspective of Mediterranean sheep farmers 724
D. Martin-Collado, S. Lobón, M. Joy, I. Casasús, A. Mohamed-Brahmi, Y. Yagoubi, F. Stark, A. Lurette, A. Abuoul Naga, E. Salah and A. Tenza-Peral

Combining serious games in a process to support sustainable livestock farming systems 725
R. Etienne, S. Dernat, C. Rigolot and S. Ingrand

Assessing how farm features and farmers' profile contribute to farm resilience 725
A. Prat-Benhamou, B. Soriano, D. Ondé, J. Lizarralde, J.M. Mancilla-Leyton, N. Mandaluniz, P. Gaspar-García, Y. Mena-Guerrero and D. Martín-Collado

Poster Session 61

The difference between abnormal climate and extreme climate that cause yield damage to silage corn 726
M. Kim and K. Sung

Causality in climate-soil-yield network for silage corn 726
M. Kim and K. Sung

Can studying the health of livestock systems be a way to improve their resilience? 727
M. Cremilleux, B. Martin and A. Michaud

Effect of heat stress on extensive beef cattle's calving percentage in the Central Bushveld Bioregio 727
S.M. Grobler, M.M. Scholtz, F.W.C. Neser, J.P.C. Greyling and L. Morey

Contrasting rearing and finishing regimens on performance and methane emissions of Angus steers 728
J. Clariget, V. Ciganda, G. Banchero, D. Santander, K. Keogh, D.A. Kenny and A.K. Kelly

Energy and greenhouse gas emissions: tools to discuss sustainability of livestock systems in Amazon 728
D.C.C. Corrêa, R.J.M. Poccard-Chapuis, M. Lenoir, V. Blanfort, J.L. Bochu and P. Lescoat

Microclimate and production of a tropical forage intercropped with pigeon pea 729
J.R.M. Pezzopane, P.P.A. Oliveira, A.F. Pedroso, W. Bonani, V.M. Gomes, C. Bosi, H.B. Brunetti, R. Pasquini Neto and A.J. Furtado

Reducing energy consumption to dry alfalfa using organic acids – a field trial 729
T. Fumagalli and L.L.C. Jansen

Carbon and energy footprint of dehydrated alfalfa production, from planting to factory output 730
D. Coulmier, P. Thiebeau, S. Recous and H. Labanca

Impact of heat waves on the quality of milk and lactic farmhouse goat cheeses in the Aura region 730
S. Raynaud, E. Lemée, H. Le Chenadec, C. Laithier, P. Thorey, C. Boyer, M. Legris, S. Morge, S. Anselmet, V. Béroulle, S. Fressinaud, C. Delbès, M. Brocart, N. Morardet, J. Birkner and Y. Gaüzere

Climate-smart practices can reduce GHG emissions intensity on smallholder dairy farms in Kenya 731
L. McNicol, M. Graham, M. Caulfield, J. Kagai, J. Gibbons, A.P. Williams, D. Chadwick and C. Arndt

Session 62. Poultry and pig low-input and organic production systems' welfare

Date: Wednesday 30 August 2023; 15.00 – 18.30
Chair: Daş / Holinger

Theatre Session 62

invited Welfare barriers and levers for improvement in organic and low-input outdoor pig and poultry farms 731
C. Leterrier, C. Bonnefous, J. Niemi, PPILOW Consortium and A. Collin

Range use relationship with welfare and performance indicators in four organic broilers strains 732
C. Bonnefous, A. Collin, L.A. Guilloteau, K. Germain, S. Mignon-Grasteau, M. Reverchon, S. Mattioli, C. Castellini, V. Guesdon, L. Calandreau, C. Berri and E. Le Bihan-Duval

Case study of a newly-developed genotype for dual-purpose rearing of male chicks 732
H. Pluschke, S. Lombard, B. Desaint, M. Reverchon, A. Roinsard, O. Tavares, A. Collin-Chenot, M. Ferriz, S. Seelig and L. Baldinger

Poultry production: using dual-purpose genotypes to reduce the culling of day-old male chicks? 733
J. Niemi, M. Väre, A. Collin, M. Almadani, M. Quentin, L. Baldinger, S. Steenfeldt, T.B. Rodenburg, F. Tuyttens and P. Thobe

Longitudinal assessment of health indicators in four organically kept laying hen flocks 733
L. Jung, M. Krieger, L. Matoni and D. Hinrichs

vited Alternative pig housing systems with high welfare standards – status quo and perspectives 734
M. Holinger

Characterising outdoor pig systems in Ireland 734
O. Menant, S. Mullan, F. Butler, L. Boyle and K. O'Driscoll

Animal welfare and pork quality of intact male pigs in organic farming according to genotype 735
B. Lebret, S. Ferchaud, A. Poissonnet and A. Prunier

Large White genetics in organic system: breeding for piglet survival 735
L. Canario, S. Ferchaud, S. Moreau, C. Larzul and A. Prunier

Comparing animal welfare assessments by researchers and free-range pig farmers with the PIGLOW app 736
E.A.M. Graat, C. Vanden Hole, S. Nauta, M.F. Giersberg, T.B. Rodenburg and F.A.M. Tuyttens

Session 63. Monoguthealth – Part 2

Date: Wednesday 30 August 2023; 15.00 – 18.30
Chair: Bee / Trevisi

Theatre Session 63

Growth performance and digestive tract parameters in weaned piglets fed *Nannochloropsis limnetica* 736
A.A.M. Chaves, C.F. Martins, D.F. Carvalho, A.R.J. Cabrita, M.R.G. Maia, A.J.M. Fonseca, R.J.B. Bessa, A.M. Almeida and J.P.B. Freire

Synbiotic administration to suckling piglets on health parameters at weaning and postweaning period 737
E.A. Sureda, M. Schroyen, J. Uerlings, F. Fannes, J. Liénart, A. Sabri, P. Thonart, V. Delcenserie, J. Wavreille and N. Everaert

Fine characterisation of a standardized citrus extract and it's effect on weaned piglet performances 737
S. Cisse, J. Laurain and M.E.A. Benarbia

Essential oils and butyric acid effects on growth performance, blood metabolites and health in pigs 738
U. Marume and R.B. Nhara

The impacts of a spectrum of varied lifestyle factors on the porcine gut microbiota 738
L. Comer, E. Arévalo Sureda and N. Everaert

In vitro inhibition of avian pathogenic *Enterococcus cecorum* isolates by probiotic Bacillus strains 739
M. Bernardeau, S. Medina-Fernandez and M. Cretenet

Effects of breed and early feeding on intestinal microbiota, gene expression and welfare indicators 739
F. Marcato, D. Schokker, S.K. Kar, J.M.J. Rebel and I.C. De Jong

Yogurt Acid Whey addition affects broiler caecal microbiota composition and metabolic activity 740
I. Palamidi, V.V. Paraskeuas, I. Politis and K.C. Mountzouris

A richer gut microbiota is related to better feed efficiency and diet adaptability in laying hens 740
M. Bernard, A. Lecoeur, J.L. Coville, N. Bruneau, D. Jardet, S. Lagarrigue, F. Calenge, G. Pascal and T. Zerjal

Gut microbiome variations during the productive lifespan of two high-yielding laying hen strains 741
C. Roth, J. Seifert, M. Rodehutscord and A. Camarinha-Silva

A rabbit nutrition hypothesis to prevent digestive problems: 'Feed the fusus coli' 741
K.H. De Greef and M. Van Der Sluis

Gut microbiota of growing rabbits fed diets with different fibre and lipid contents 742
G. Zardinoni, P. Stevanato, A. Trocino, M. Birolo, F. Bordignon and G. Xiccato

Poster Session 63

Gut microbiota-metabolome response to dietary porcine intestinal mucosa hydrolysate in piglets 742
S. Segarra, A. Middelkoop and F. Molist

Effects of probiotics isolated from faeces of fast-growing pigs on growth performance of weaning pigs 743
Y.H. Choi, Y.J. Min, J.E. Kim, Y.D. Jeong, H.J. Park, C.H. Kim and S.J. Sa

Effect of butyric acid salts on the palatability of feed for piglets 743
W. Kozera, A. Woźniakowska, K. Karpiesiuk, A. Okorski and G. Żak

Synbiotics-glyconutrients enhance growth performance and fatty acid profile in finishing pigs 744
C.B. Lim, Q.Q. Zhang, S. Biswas, J.H. Song, O. Munezero, J.S. Yoo and I.H. Kim

Impact of gut health product on appetite associated hormones using *ex vivo* porcine intestinal cells 744
N. Browne and K. Horgan

Development of antimicrobial peptides against multidrug-resistant enterotoxigenic *Escherichia coli* 745
W.J. Chen, M.Y.W. Kwok, K.C. Wu, K.F. Hua, Y.H. Yu and Y.H. Cheng

Influences of quercetin inclusion to corn-soybean-gluten meal-based diet on broiler performance 745
S. Biswas, Q.Q. Zhang, J.H. Song, C.B. Lim, I.H. Kim and J.S. Yoo

Effects of *Bacillus* species – fermented products on growth performance and gut health in broilers 746
Y.H. Yu, S.H. Hsiao, Y.H. Cheng, W.J. Chen and K.F. Hua

Lactobacillus ingluviei C37 improves gut health in lipopolysaccharide challenged broiler chickens 746
S. Khempaka, M. Sirisopapong and S. Okrathok

Effects of different probiotics on laying performances, egg quality and gut health in laying hens 747
F. Barbe, L. Blanc, A. Sacy and E. Chevaux

Simple methodology to evaluate faeces quality on farm and effect of *S. boulardii* on faecal scoring 747
A. Sacy, F. Barbé, S. Poulain, P. Belloir, C. Paes and E. Chevaux

A comparison of yeast gut health products ability to limit attachment of *Salmonella* to IPEC-j2 cells 748
N. Browne, A. McCormack and K. Horgan

Butyric acid and *Bacillus subtilis* 29784 improve cumulatively the intestinal barrier 748
A. Mellouk, D. Prévéraud, T. Goossens, O. Lemâle, E. Pinloche and J. Consuegra

Session 64. Project session: Farmyng

Date: Wednesday 30 August 2023; 15.00 – 18.30
Chair: Kihanguila / Gasco

Theatre Session 64

Safety of BSF larvae reared on substrates which are spiked or naturally contaminated with aflatoxins 749
K. Niermans, E.F. Hoek- Van Den Hil, N. Meijer, S.P. Salari, M. Gold, H.J. Van Der Fels-Klerx and J.J.A. Van Loon

Transfer of aflatoxin, lead and cadmium from larvae reared on contaminated substrate to laying hens 749
M. Heuel, M. Kreuzer, I.D.M. Gangnat, E. Frossard, C. Zurbrügg, J. Egger, B. Dortmans, M. Gold, A. Mathys, J. Jaster-Keller, S. Weigel, C. Sandrock and M. Terranova

Determination of minimal nutrient requirements of *Tenebrio molitor* larvae 750
B. Tamim, T. Parr, J. Brameld and A. Salter

FlAgship demonstration of industrial scale production of nutrient resources from mealworms 750
W. Kihanguila

Tailored vitamins and minerals premix for *Tenebrio molitor* farming 751
E. Barbier, V. Gerfault, T. Lefebvre, F. Peyrichou, C. Ricciardi and N. Tanrattana

Tenebrio molitor genomics revealed limited molecular diversity among available populations 751
L. Panunzi, E. Eleftheriou, B. Vacherie, K. Labadie, T. Lefebvre and M.A. Madoui

Assessing the quality of insect-derived products: methods and findings from the FARMYNG project 752
S. Gofflot, A. Pissard, F. Debode, A.C. Laplaize, B. Lorrette and J.F. Morin

Authentication of insect-derived products: methods and findings from the FARMYNG project 752
B. Dubois, A. Marien, S. Guillet, J.-F. Morin, B. Lorrette and F. Debode

More than an organic fertilizer: mealworm frass as a substitute to conventional fertilizers 753
E. Bohuon, D. Houben, G. Daoulas, M.-P. Faucon and A.-M. Dulaurent

The environmental life cycle assessment of Ynsect insect based proteins 753
K. Hsu, A. Eiperle and M. Jouy

Insects' nutrients – the Animal Frontiers special issue 754
T. Veldkamp and L. Gasco

Session 65. Remote sensing applied to livestock farming: discovering physiological and ethological clues to optimize productive performance and animal welfare

Date: Wednesday 30 August 2023; 15.00 – 18.30
Chair: Montanholi

Theatre Session 65

Time-course change in lamb composition and reflectance properties: implications for authentication 754
L. Rey-Cadilhac, D. Andueza, A. Prunier and S. Prache

Individual adaptive responses of meat ewes facing an abrupt nutritional challenge after lambing 755
E. González-García, M. Gindri, L. Puillet and N.C. Friggens

Large variation in emission intensities from dual-purpose sheep production system 755
B.A. Åby, S. Samsonstuen and L. Aass

Considering the morphology of cows for comfortable milking 756
J. Fazilleau, S. Guiocheau and J.L. Poulet

Classification of honeybee flight activity patterns reveals impact of recruitment behaviours 756
G.E.L. Petersen, D. Gupta, P.F. Fennessy and P.K. Dearden

Implementation of large-scale climate smart agriculture research on U.S. beef cattle grazing lands 757
K. Cammack, A. Blair, L.O. Tedeschi, H.M. Menendez Iii and J.R. Brennan

Remote monitoring behaviour of bison in captivity: effects of gender, weather and daytime 757
R.R. Vicentini, D. Moya, J. Church, A.C. Sant'anna, W. Balan, W. Squair and Y.R. Montaholi

Artificial intelligence for measuring the respiration rate in dairy cows 758
L. Dißmann, R. Antia, L. Chinthakayala, N. Landwehr, T. Amon and G. Hoffmann

Using millimetre-wave radar for monitoring sow postural activity in individual pen: first results 758
D. Henry, J. Bailly, T. Pasquereau, W. Hebrard, J.F. Bompa, E. Ricard, H. Aubert and L. Canario

The environmental variance on daily feed intake as a measure of resilience in pigs 759
C. Casto-Rebollo, P. Nuñez, S. Gol, J. Reixach and N. Ibáñez-Escriche

Poster Session 65

Factors affecting somatic cell count in ewe milk and its effect on milk yield and composition 759
M. Oravcová, V. Tančin, L. Mačuhová and M. Uhrinčať

Session 66. Innovation in non-ruminant nutrition and feeding

Date: Wednesday 30 August 2023; 15.00 – 18.30
Chair: Alvarez Rodriguez / Stergiadis

Theatre Session 66

Digestibility of soybean meal vs toasted soybeans in newly weaned piglets 760
C. De Cuyper, P. Dubois and S. Millet

Effect of variety and technological treatment on intake of lupin seed in pigs 760
E. Labussiere, H. Furbeyre, M. Guillevic and G. Chesneau

Comparative energy values of 10 forages in finishing pigs 761
D. Renaudeau, J. Pirault, M. Dumesny, S. Lombard and F. Marie

Implication of different amylose/amylopectin ratio on low protein diet on piglet's performance 761
P. Trevisi, D. Luise, F. Correa, S. Virdis, C. Negrini and S. Dalcanale

Effects of two blends of phytoextracts on growth and gut health of weaning pigs to replace zinc oxide 762
D. Luise, F. Correa, C. Negrini, S. Virdis, M. Mazzoni, S. Dalcanale and P. Trevisi

Effect of olive cake in growing pig diets on faecal microbiota fermentation and composition 762
D. Belloumi, P. García-Rebollar, P. Francino, S. Calvet, A.I. Jiménez-Belenguer, L. Piquer, O. Piquer and A. Cerisuelo

Impact of increasing dietary sphingolipids on feed intake and growth performance of piglets 763
S. Chakroun, R. Larsen, J. Levesque, F. Cerpa Aguila, M.-P. Letourneau, J.E. Rico and D.E. Rico

Faeces microbiota and metabolism correlated with reproductive capacity in early parity sow 763
J. Wang, Q. Xie and B. Tan

Dehydrated sainfoin in rabbit feed: effects of high incorporation on the health and performances of 764
C. Gayrard, P. Gombault, A. Bretaudeau, H. Hoste and T. Gidenne

Nutritional value of defatted larvae meal and whole larvae from black soldier fly 764
P. Belloir, F. Hervo, E. Gambier, L. Lardic, C. Guidou, C. Trespeuch, N. Même, E. Recoules and B. Méda

Poster Session 66

Diet taste monotony decreases feed acceptability and preferences in nursery pigs 765
J. Figueroa, E. Huenul, R. Palomo and D. Luna

Dietary taste variety improves performance in nursery pigs 765
J. Figueroa, T. Cabello, E. Huenul, R. Palomo and D. Luna

Impact of some dietary agro-industrial by-products on the gut microbiota of finishing pigs 766
I. Skoufos, A. Nelli, C. Voidarou, I. Lagkouvardos, E. Bonos, K. Fotou, C. Zacharis, I. Giannenas and A. Tzora

Effects of Greek aromatic/medicinal plants on health and meat quality characteristics of piglets 766
G. Magklaras, C. Zacharis, K. Fotou, E. Bonos, I. Giannenas, J. Wang, L.Z. Jin, A. Tzora and I. Skoufos

Effects of mint leaf powder on performance and egg nutrient composition of laying hens 767
Y.H. Zhang, M.M. Bai, H.N. Liu, X.F. Kong and F.C. Wan

Weaned piglets' gut microbiota regulation by dietary olive, winery, and cheese waste by-products 767
A. Tzora, A. Nelli, A. Tsinas, E. Gouva, G. Magklaras, S. Skoufos, B. Venardou, K. Nikolaou, I. Giannenas and I. Skoufos

Effects of medium-chain fatty acids on feed intake, body weight and egg composition of laying hens 768
F. Cerpa Aguila, S. Chakroun, M.P. Letourneau Montminy, J.E. Rico and D.E. Rico

Session 67. PLF for health and welfare – Part 2

Date: Wednesday 30 August 2023; 15.00 – 18.30
Chair: Pastell / Stygar

Theatre Session 67

Open-sourcing behavioural algorithms for ruminant welfare monitoring using raw wearable sensor data 768
D. Foy, T.R. Smith and J.P. Reynolds

Evaluation of an automated cattle lameness detection system 769
N. Siachos, A. Anagnostopoulos, B.E. Griffiths, J.N. Neary, R.F. Smith and G. Oikonomou

Breath analysis in dairy cattle: going beyond methane emission 769
I. Fodor, E. Van Erp-Van Der Kooij and I.D.E. Van Dixhoorn

Dairy cow personality: correlations with age, weight, back fat thickness and activity 770
P. Hasenpusch, T. Wilder, A. Seidel, J. Krieter and G. Thaller

Automatic behaviour assessment of young bulls in pen using machine vision technology 770
A. Cheype, J. Manceau, V. Gauthier, C. Dugué, L.-A. Merle, X. Boivin and C. Mindus

Parity and lactation stage preserve a stable structure in the dynamic social networks in cattle 771
H. Marina, I. Hansson, I. Ren, F. Fikse, P.P. Nielsen and L. Rönnegård

Application of rumen boluses to receive welfare parameters from fattening bulls 771
K. Fromm, J. Heinicke, T. Amon and G. Hoffmann

Stepwise modelling for improved bovine health 772
C.M. Matzhold, K.S. Schödl, C.E. Egger-Danner, F.S. Steininger and P.K. Klimek

Is milking order connected to social interactions? 772
I. Hansson, H. Marina and L. Rönnegård

Combining ultra-wideband (UWB) location and accelerometer data for cattle behaviour monitoring 773
S. Benaissa, F. Tuyttens, D. Plets, L. Vandaele, L. Martens, W. Joseph and B. Sonck

FEMIR report – the new MIR advising tool 773
L.M. Dale, A. Werner, C. Natterer, E.J.P. Strang, Emissioncow Consortium, Remissiondairy Consortium and J. Bieger

Heat stress relief of dairy cows by evaporative cooling under Mediterranean summer conditions 774
S. Pinto, C. Ammon, F. Estellés, A. Villagra, T. Amon and G. Hoffmann

Poster Session 67

Performance of subcutaneous thermochips implanted in dairy cows: preliminary report 774
A.R. Garcia, L.K. Zanetti, T.C. Alves, L.M. Neira, L.F. Pinho, A.N. Barreto, M.J. Moraes, G.G. Ramos, C.E. Grudzinski and G.N. Azevedo

Dairy cow behaviour as proxy for heat stress sensitivity in dairy cows 775
I. Fodor, R.S.C. Rikkers, M. Taghavi and I. Adriaens

Exploring the feeding behaviour of dairy calves: insights from automated milk feeders 775
K.J. Hemmert, M.H. Ghaffari, T. Förster, C. Koch and H. Sauerwein

Characterization of behavioural anomalies of lameness in dairy cows using sensor technology 776
N.L. Mhlongo

Test of Bluetooth low energy localization system for dairy cows in a barn 776
J. Maxa, D. Nicklas, J. Robert, S. Steuer and S. Thurner

Integrating inline milk infrared spectra and genomics to predict metabolic profiling in dairy cattle 777
D. Giannuzzi, L.F. Macedo Mota, H. Toledo Alvarado, S. Pegolo, L. Gallo, S. Schiavon, E. Trevisi and A. Cecchinato

Influence on the water intake of lactating dairy cows 777
J. Heinicke, C. Ammon, T. Amon, G. Hoffmann and S. Pinto

Session 68. Small Ruminant Technologies (Sm@RT) and Thematic network EuroSheep

Date: Wednesday 30 August 2023; 15.00 – 18.30
Chair: Morgan-Davies / Grisot

Theatre Session 68

Sm@RT: Identifying sheep and goats farmers' technological needs and potential solutions 778
C. Morgan-Davies, L. Depuille, J.M. Gautier, A. McLaren, T.W.J. Keady, B. McClearn, L. Grova, P. Piirsalu, V. Giovanetti, I. Halachmi, A. Bar-Shamai, R. Klein, F. Kenyon and I. Llach-Martinez

EuroSheep: increasing flock profitability through improved sheep health and nutrition management 778
P.G. Grisot, B. Fança, A. Carta, S. Salaris, C. Morgan-Davies, I. Beltran De Heredia, R. Ruiz, S. Ocak Yetisign, T.W.J. Keady, B. McClearn, R. Klein, D. Tsiokos and C. Ligda

Sm@RT: Innovative technologies training for small ruminant producers 779
L. Depuille, J.M. Gautier, A. McLaren, T.W.J. Keady, B. McClearn, L. Grøva, P. Piirsalu, V. Giovanetti, I. Halachmi, A. Bar Shamai, R. Klein, F. Kenyon, I. Llach and C. Morgan-Davies

EuroSheep: end-users assessments of flock health and nutrition best practices 779
P.G. Grisot, B. Fança, A. Carta, S. Salaris, C. Morgan-Davies, I. Beltran De Heredia, R. Ruiz, S. Ocak yetisign, T.W.G. Keady, B. McClearn, R. Klein, L. Perucho and C. Ligda

Sm@RT: main lessons from New Zealand on PLF uptake in small ruminants 780
J.M. Gautier, C. Morgan-Davies, L. Depuille, A. McLaren, B. McClearn, L. Grøva, P. Piirsalu, V. Giovanetti, I. Halachmi, A. Bar-Shamai, R. Klein, F. Kenyon, E. Gonzalez-Garcia and T.W.J. Keady

French regional project SO-PERFECTS: project methodology 780
C. Douine, L. Sagot, A.S. Thudor, M. Miquel, M. Bernard, M. Goyenetche and D. Gautier

French regional project SO-PERFECTS: trial results 781
M. Bernard, L. Sagot, A.S. Thudor, C. Douine, M. Miquel, M. Goyenetche and D. Gautier

Use of innovative and precision tools in research stations with small ruminants: the INRAE case 781
I. Llach, H. Caillat, A. Fatet, S. Breton, T. Aguirre-Lavin, D. Dubreuil, A. Eymard, J. Boucherot, T. Fassier, D. Marcon, S. Parisot, C. Durand, G. Bonnafe, D. Portes, C. Morgan-Davies and E. González-García

FEC check: development of an online tool to aid farmer understanding of roundworm faecal egg counts 782
E. Geddes, A. Duncan, K. Lamont, J. Duncan, F. Kenyon and L. Melville

Assessing sheep behaviour in an human-animal interaction test using infrared termography 782
M. Almeida, A. Afonso, C. Guedes and S. Silva

Session 69. Methods & genomic prediction

Date: Thursday 31 August 2023; 8.30 – 12.00
Chair: Sell-Kubiak / El-Ouazizi

Theatre Session 69

Detection of interchromosomal rearrangements in bulls using large genotype and phenotype datasets 783
J. Jourdain, H. Barasc, T. Faraut, C. Grohs, C. Donnadieu, A. Pinton, D. Boichard and A. Capitan

Functional information embedded in the unmapped short reads of whole-genome sequencing 783
G.B. Neumann, P. Korkuć, M. Reißmann, M.J. Wolf, K. May, S. König and G.A. Brockmann

Expanding the capabilities of single-step GWAS with *P*-values for large genotyped populations 784
N. Galoro Leite, M. Bermann, S. Tsuruta, I. Misztal and D. Lourenco

The Life-Functions Ratio: a new indicator trait of trade-offs to go beyond genetic correlations 784
N. Bedere, O. Cado, N.C. Friggens and P. Le Roy

Impact of pedigree errors on the quality of predicted genetic merit from animal models 785
E.C.G. Pimentel, C. Edel, R. Emmerling and K.-U. Götz

Bias in estimated variance components and breeding values due to pre-correction of systematic effect 785
P. Duenk and P. Bijma

Improving computing performance of genomic evaluations by genotype and phenotype truncation 786
F. Bussiman, C. Cheng, J. Holl, A. Legarra, I. Misztal and D. Lourenco

A single-step evaluation of functional longevity of cows including data from correlated traits 786
L.H. Maugan, T. Tribout, R. Rostellato, S. Mattalia and V. Ducrocq

Expected values of genomic prediction validation parameters for non-random validation sets 787
M.P.L. Calus, M. Schrauf, T. Pook, L. Ayres, R. Bonifazi, J. Ten Napel and J. Vandenplas

Unknown parent groups and metafounders in genomic evaluation of Norwegian Red cattle 787
T.K. Belay, A.B. Gjuvsland, J. Jenko, L.S. Eikje and T. Meuwissen

Exploring non linear genetic relationships between correlated traits 788
F. Shokor, P. Croiseau, R. Saintilan, T. Mary-Huard, H. Gangloff and B.C.D. Cuyabano

Single-Step Genomic Prediction in six German Beef Cattle Breeds 788
D. Adekale, H. Alkhoder, Z. Liu, D. Segelke and J. Tetens

Poster Session 69

Incorporating QTL genotypes in the model to predict phenotypes and breeding values 789
J. Yang, T.H.E. Meuwissen, Y.C.J. Wientjes, P. Duenk and M.P.L. Calus

Improving the efficiency of genomic evaluations with random regression models 789
A. Alvarez Munera, D. Lourenco, I. Misztal, I. Aguilar, J. Bauer, J. Šplíchal and M. Bermann

Comparison of two software to estimate breeding value in cattle by single-step approach 790
M. Jakimowicz, D. Słomian, T. Suchocki and J. Szyda

Estimating (co)variance components using Monte Carlo EM-REML in a multi-trait SNPBLUP model 790
H. Gao, M.H. Lidauer, M. Taskinen, E.A. Mäntysaari and I. Strandén

Molecular phenotyping to predict neonatal maturity 791
L. Liaubet, N. Marty-Gasset, L. Gress, A. Bonnet, P. Brenaut and E. Maigné

Comparison of genetic maps from different cattle breeds 791
X. Ding, H. Schwarzenbacher, F.R. Seefried and D. Wittenburg

Genomic relationships across metafounders using partial EM algorithm and average relationships 792
A. Legarra, M. Bermann, Q. Mei and O.F. Christensen

Impact of the correlation between SNP effects in different breeds on the accuracy of predictions 792
P. Croiseau, R. Saintilan, D. Boichard and B. Cuyabano

Early prediction of lactation persistency of multiparous cows managed for extended lactation 793
C. Gaillard, M. Boutinaud, J. Sehested and J. Guinard-Flament

Accuracy of genomic prediction by singular value decomposition of the genotype matrix 793
L. Ayres, M.P.L. Calus, J. Ødegård and T. Meuwissen

Enhancing long-term genetic gain through a Mendelian sampling-based similarity matrix 794
A.A. Musa and N. Reinsch

Mining the convergence behaviour of a single-step SNP-BLUP model for genomic evaluation of stature 794
D.S. Dawid Słomian, J.S. Joanna Szyda and K.Ż. Kacper Żukowski

Increase in prediction accuracy can be achieved by combining multiple populations 795
A. Ajasa, S. Boison, H. Gjøen and M. Lillehammer

Enhancing bovine genome SNP call accuracy with autoencoder analysis of nucleotide impact with AI 795
K. Kotlarz, M. Mielczarek, B. Guldbrandtsen and J. Szyda

Efficient SNP calling: Nextflow vs Bash on the whole genome bovine sequence 796
P. Hajduk, M. Sztuka, K. Liu, K. Kotlarz, M. Mielczarek and J. Szyda

Variant calling and genotyping accuracy of ddRAD-seq: comparison with WGS in layers 796
M. Doublet, F. Lecerf, F. Degalez, S. Lagarrigue, L. Lagoutte and S. Allais

Performing single-step genomic evaluation for superovulatory response traits in Japanese Black cows 797
A. Zoda, R. Kagawa, H. Tsukahara, R. Obinata, M. Urakawa, Y. Oono and S. Ogawa

Production traits in Nellore cattle classified by residual feed intake 797
J.A. Muñoz, R.H. Branco, R.C. Canesin, J.N.S.G. Cyrillo, M.E.Z. Mercadante and S.F.M. Bonilha

Effect of paternal breeding values for residual feed intake on reproductive performance of heifers 798
T. Devincenzi, M. Lema, L. Del Pino, A. Ruggia and E.A. Navajas

Species identification of animal/pig DNA in non-animal food 798
M. Natonek-Wiśniewska and P. Krzyścin

Development of a proxy for feed efficiency prediction in dairy cows based on mid-infrared spectra 799
M. Raemy, T. Haak, F. Schori and S. Ampuero Kragten

Improving taste and flavour in dairy product through analysis of free fatty acid by MIR spectroscopy 799
O. Christophe, R. Reding, J. Leblois, D. Pittois, C. Guignard and F. Dehareng

Modelling growth curves in challenged mice lines divergently selected for birth weight variability 800
V. Mora-Cuadrado, I. Cervantes, J.P. Gutiérrez and N. Formoso-Rafferty

Genomic prediction of commercial layers' bone strength 800
M. Sallam, H. Wall, P. Wilson, B. Andersson, M. Schmutz, C. Benavides, M. Checa, E. Sanchez, A. Rodriguez, I. Dunn, A. Kindmark, D.J. De Koning and M. Johnsson

Variability of daily milk yield during the first 100 days of lactation in Holstein cows 801
E. Kašná, L. Zavadilová and J. Vařeka

Session 70. Genetic parameters & GWAS

Date: Thursday 31 August 2023; 8.30 – 12.00
Chair: Egger-Danner / Martin

Theatre Session 70

Single-step GBLUP for growth and carcass traits in Nordic beef cattle 801
A. Nazari Ghadikolaei, F. Fikse and S. Eriksson

Genetic parameters of pig birth weight variability 802
Y. Salimiyekta, S.B. Bendtsen, K.V. Riddersholm, M. Aaskov and J. Jensen

Disentangling paternal and maternal components of within litter birth weight variability in mice 802
N. Formoso-Rafferty, L. El-Ouazizi El-Kahia, I. Cervantes and J.P. Gutiérrez

(Co)variances between anogenital distance and fertility in Holstein-Friesian dairy cattle 803
M.A. Stephen, C.R. Burke, N. Steele, J.E. Pryce, S. Meier, P.R. Amer, C.V.C. Phyn and D.J. Garrick

Genetic relationships between productive and reproductive traits on Friesian cows 803
S. Abomselem, A. Badr and A. Khattab

Heritabilities of the mid-infrared spectra of sheep milk throughout the lactation 804
C. Machefert, C. Robert-Granié, J.M. Astruc and H. Larroque

Genetic parameters for the composition of milk fatty acid of Holsteins 804
Y. Masuda

External and genetic factors influencing fertility in Latxa dairy sheep breed 805
C. Pineda-Quiroga, I. Granado-Tajada, E. Ugarte and A. Basterra-García

Single-step genome-wide association for milk urea concentration in Walloon Holstein 805
H. Atashi, Y. Chen, C. Bastin, S. Vanderick, X. Hubin and N. Gengler

Genome-wide association study for milk production traits in the Cyprus Chios sheep 806
A.N. Georgiou, G. Maimaris, S. Andreou, A.C. Dimitriou and G. Hadjipavlou

Characterization of additive, dominance, and runs of homozygosity effects inbreeds of dairy cattle 806
H. Ben Zaabza, M. Neupane, M. Jaafar, K. Srikanth, S. McKay, A. Miles, H.J. Huson, I. Strandén, H. Blackburn and C.P. Van Tassell

A genome-wide association study identified a major QTL affecting the red colour in nitrate free hams 807
J. Vegni, M. Zappaterra, R. Davoli, R. Virgili, N. Simoncini, C. Schivazappa, A. Cilloni and P. Zambonelli

Poster Session 70

Genetic relationships between milk fatty acids at early lactation and fertility in Holstein cows 807
T. Yamazaki, A. Nishiura, S. Nakagawa, H. Abe, Y. Nakahori and Y. Masuda

Genetic parameters for early-life racing performance in pigeons 808
P. Duenk and D. Shewmaker

Genetic parameters for milk urea in Swiss dairy cattle breeds 808
A. Burren and S. Probst

Estimation of genetic correlations between predicted energy balance and fertility in Holsteins 809
A. Nishiura, O. Sasaki, S. Yamaguchi, Y. Saito, R. Tatebayashi and T. Yamazaki

Comparison of random regression test-day models for production traits of South African Jersey cattle 809
M.G. Kinghorn, E.D. Cason, V. Ducrocq and F.W.C. Neser

Application of single-step genomic method in routine evaluation of Czech Holstein cattle 810
J. Bauer, J. Šplíchal, D. Fulínová and E. Krupa

A meta-analysis of the genetic parameter estimates for lamb survival 810
S. Fernandes Lazaro, H. Rojas De Oliveira and F. S. Schenkel

Survival analysis in the conservation program of an endangered wild ungulate (*Nanger dama mhorr*) 811
S. Domínguez, I. Cervantes, E. Moreno and J.P. Gutiérrez

Genetic parameters of maturing rate index and asymptotic adult weight in French beef cattle 811
A. Lepers, S. Aguerre, J. Promp, S. Taussat, A. Vinet, P. Martin, A. Philibert, A. Laramee and L. Griffon

Genetic architecture of the persistency of production, quality, and efficiency traits in laying hens 812
Q. Berger, N. Bedere, P. Le-Roy, T. Burlot, S. Lagarrigue and T. Zerjal

On the potential of improving daily milk yield by extending productive lifespan 812
A. Bieber, F. Hediger, F. Leiber, C. Pfeifer and M. Walkenhorst

Genetic trends in a selection process using electronic feeders to improve feed efficiency in rabbit 813
J.P. Sánchez, M. Pascual and M. Piles

Maturity, a heritable trait in French dairy goat 813
M. Arnal, M. Chassier, V. Clément and I. Palihière

Multivariate genome-wide associations for immune traits in two maternal pig lines 814
C. Große-Brinkhaus, K. Roth, M.J. Pröll-Cornelissen, A.K. Appel, H. Henne, K. Schellander and E. Tholen

Estimation of heritability for digital dermatitis in Polish Holstein-Friesian cows 814
M. Graczyk-Bogdanowicz, K. Baczkiewicz and K. Rzewuska

Genetic parameters for β-hydroxybutyrate concentration in milk in Spanish dairy cows 815
I. Cervantes, M.A. Pérez-Cabal, N. Charfeddine and J.P. Gutiérrez

Session 71. Building on a resilient dairy sector – from animal, farm and regional perspective – Part 2

Date: Thursday 31 August 2023; 8.30 – 12.00
Chair: Brocard / Klopcic

Theatre Session 71

Comparison of self-assessment and objective indicators of attributes driving farms resilience 815
D. Martin-Collado, B. Soriano, J. Lizarralde, J.M. Mancilla-Leyton, N. Mandaluniz, P. Gaspar-García, Y. Mena-Guerrero and A. Prat-Benhamou

Inventory and analysis of needs towards resilient dairy farming in 15 EU countries 816
A.M. Menghi and C.S.S. Soffiantini

Assessment of solutions for resilient dairy farming in fifteen European countries 816
A. Kuipers, J. Zijlstra, R. Loges and S. Ostergaard

Knowledge needs and solutions related to resilience in the European dairy sector 817
K. Kuoppala, M. Rinne, N. Browne and V. Brocard

Resilience of contribution to food security of specialized Walloon dairy systems 817
C. Battheu-Noirfalise, E. Froidmont, D. Stilmant and Y. Beckers

Innovative solutions supporting resilience of dairy farms in Netherlands 818
P.J. Galama, J. Zijlstra and A. Kuipers

Needs of the dairy sector: a Hungarian overview 818
L. Czeglédi, B. Béri, I. Komlósi and E. Török

Eco-efficient low-cost pasture based dairy production on a mixed farm in Northern Germany 819
R. Loges and F. Taube

Factors contributing to the financial resilience of spring-calving pasture-based dairy farms 819
G. Ramsbottom, B. Horan, K.M. Pierce, D.P. Berry and J.R. Roche

Planning for resilient dairy farms in the USA 820
A. De Vries

Towards a socially sustainable dairy sector with cow-calf contact systems 820
H.W. Neave, M. Bertelsen, E.H. Jensen and M.B. Jensen

Resilient, healthy or efficient? The ideal animal according to breeders of small ruminants in Europe 821
E. Janodet and M. Sautier

Poster Session 71

Strategies and cases of resilience from dairy farming community in Slovenia 821
M. Klopčič

How farm management influences the longevity of dairy cows: a comparative study of Swiss dairy farms 822
R.C. Eppenstein, A. Bieber, M. Lozano-Jaramillo and M. Walkenhorst

High herd exit rates existence of small herds: a case study from North West Province, South Africa 822
M.D. Motiang and E.C. Webb

Building on a resilient dairy sector- highlights and discussion 823
A. Kuipers, V. Brocard and M. Klopčič

Economic and environmental impacts of cattle longevity extension by altered reproductive management 823
R. Han, A. Kok, M. Mourits and H. Hogeveen

Session 72. What are the stakeholder and societal expectations of intrinsic and extrinsic quality of animal products?

Date: Thursday 31 August 2023; 8.30 – 12.00
Chair: Martin

Theatre Session 72

Economic sustainability of different levels of extensiveness in fattening pig farms 824
P. Ferrari, C. Montanari and L. Giglio

How to characterise the European livestock production systems? 824
E. Bailly-Caumette, S. Moakes, C. Pfeifer and D.R. Yáñez-Ruiz

The IntaQt project's stakeholders' involvement: impact on the research work? 825
F. Bedoin, C. Couzy, C. Laithier, C. Berri and B. Martin

EU policy impacts on the sustainability of the livestock sector, insights from the PATHWAYS project 825
N. Roehrig, A. Sans, K.-E. Trier-Kreutzfeldt, M.A. Arias Escobar, F.W. Oudshoorn, N. Bolduc, E. Regnier, P.-M. Aubert and L.G. Smith

Stakeholders' perception of pig and chicken local breeds – a broad survey by the GEroNIMO project 826
M.J. Mercat, A.J. Amaral, R. Bozzi, M. Čandek-Potokar, P. Fernandes, J. Gutierrez Vallejos, D. Karolyi, D. Laloë, Z. Luković, H. Lenoir, G. Restoux, A. Vicente, V. Ribeiro, T. Rodríguez Silva, R. Rouger, D. Škorput and M. Škrlep

Responding to stakeholder needs and consumer-driven demands in the dairy goat and poultry sectors 826
C. Bonardi and M. Gerevini

The INTAQT project: stakeholders' opinions on future multicriteria scoring tools for animal products 827
I. Legrand, A. Nicolazo De Barmon, F. Albert, M. Berton, M. Bourin, V. Bühl, A. Cartoni Mancinelli, R. Eppenstein, D.A. Kenny, E. Kowalski, S. McLaughlin, G. Plesch, F. Bedoin, C. Couzy, C. Berri, B. Martin and C. Laithier

Co-designing Agroecology farming concepts to increase food sovereignty in the Global South 827
M. Simataa, R. Valkenburg and H. Romijn

Mapping of value chains in the Italian bovine sector 828
M. Finocchi, M. Moretti, A. Mantino, A. Ripamonti, G. Conte and M. Mele

The INTAQT project: stakeholders' expectations on husbandry systems and innovative practices 828
R.C. Eppenstein, V. Bühl, I. Legrand, A. Nicolazo De Barmon, B. Martin, F. Albert, M. Berton, M. Bourin, A. Cartoni Mancinelli, D.A. Kenny, E. Kowalski, S. McLaughlin, G. Plesch, F. Bedoin, C. Couzy, C. Berri and C. Laithier

The INTAQT project: stakeholders' perceptions and points of view on products quality 829
C. Laithier, F. Bédoin, F. Albert, I. Legrand, A. Nicolazo De Barmon, M. Bourin, M. Berton, V. Bühl, R. Eppenstein, A. Cartoni Mancinelli, D.A. Kenny, E. Kowalski, S. McLaughlin, G. Plesch, C. Couzy, C. Berri and B. Martin

Life cycle assessment of different pig production systems around Europe: mEATquality project 829
C. Reyes-Palomo, A. Pignagnoli, S. Sanz-Fernández, P. Meatquality Consortium and V. Rodríguez-Estévez

Poster Session 72

Consumer expectations for beef in the French region Auvergne-Rhône-Alpes 830
S. Chriki, J. Normand, C. Brosse, L. Hallez, L. Vallet, V. Payet and J.F. Hocquette

Analysis of preferences and perception of cheese products by Portuguese consumers 830
V.M. Merlino, M. Renna, M. Tarantola, A. Ricci, A.S. Santos, A. Monteiro and J. Nery

How a risk-based strategy could contribute to a more sustainable agri-food system? 831
B. Grossiord and M.P. Ellies-Oury

Perceptions of meat quality of UK stakeholders: from intrinsic to extrinsic factors 831
S. McLaughlin, F. Bedoin, C. Couzy, I. Legrand, A. Nicolazo De Barmon, C. Laithier and N. Scollan

Transforming livestock practices by federating around common values 832
S. Nade

Session 73. Molecules in animal nutrition: over and above building blocks

Date: Thursday 31 August 2023; 8.30 – 12.00
Chair: Van Milgen / Geraert / Mercier

Theatre Session 73

Genetic response of red yeast supplementation in feed to mycotoxin contamination in laying hens 832
S. Hosseini, B. Brenig, W. Tapingkae and K. Gatphayak

Faecal starch content as an indicator of starch digestibility by fattening Japanese Black cattle 833
M. Matamura, S. Uzawa and M. Kondo

Impact of isoacids on performance and digestibility in gilts fed different dietary fibre sources 833
W. Schweer, B. Kerr, M. Socha, A. Cornelison and L. Rodrigues

The enterotype is associated with the phenotype variation in the pigs treated with dietary fibre 834
H. Li and B.E. Tan

A high rumen degradable starch modulates jejunum microbiota and bile acids in dairy goats 834
L.M. Wang, J.H. Yao and Y.C. Cao

Development of a novel endolysin, RalLys8, for the specific inhibition of *Ruminococcus* albus 835
J. Moon, H. Kim and J. Seo

Dietary microalgae (*Nannochloropsis limnetica*) and probiotic on performance and intestine of piglets 835
M.F. Pedro, D.F.P. Carvalho, A.M. Almeida, R.J.B. Bessa, A.J.M. Fonseca, A.R.J. Cabrita and J.P.B. Freire

Muramidase inclusion reduces gut inflammation in weaned piglets, especially in high protein diets 836
U.M. McCormack, J. Schmeisser, E. Bacou, P. Jenn, F. Amstutz and E. Perez Calvo

Utilization of tryptophane by kynurenine, indoles and serotonin pathways are modified by fructose 836
A. Gual-Crau, M. Jarzaguet, D. Dardevet, D. Rémond, A. Lefèvre, A. Bernalier-Donadille, P. Emond and I. Savary-Auzeloux

Regulation of glucose metabolism in rumen epithelium of cows transitioned from forage to high-grain 837
S. Kreuzer-Redmer, A. Sener, C. Pacífico, F. Dengler, S. Ricci, H. Schwartz-Zimmermann, E. Castillo-Lopez, N. Reisinger and Q. Zebeli

Altering nutrients supply modified dynamics of milk components synthesis in dairy cows 837
J.C. Anger, C. Loncke, R. Bidaux and S. Lemosquet

Phytogenic effects on layer performance, egg quality and cytoprotective response in the ovaries 838
I.P. Brouklogiannis, E.C. Anagnostopoulos, V.V. Paraskeuas, E. Griela, G. Kefalas and K.C. Mountzouris

Poster Session 73

Phytogenic effects on layer performance, and cytoprotective response in the ceca 838
E.C. Anagnostopoulos, I.P. Brouklogiannis, V.V. Paraskeuas, E. Griela, G. Kefalas and K.C. Mountzouris

Effects of guanidinoacetic acid supplementation on growth performance in Nellore cattle 839
O.R.M. Neto, I.M.S.C. Farias, R.N.S. Torres, R.V. Ribeiro, W.A. Baldassini, R.A. Curi, L.A.L. Chardulo and G.L. Pereira

In feed histidine supplementation improves muscle carnosine content and meat quality in pigs 839
M. Paniagua, B. Saremi, B. Matton and S. De Smet

Impact of functional amino acids on performance parameters in post-weaning piglet 840
A. Simongiovanni, K. Fenske, F. Witte, H. Westendarp and T. Chalvon-Demersay

Phytobiotics from *Thymus* species in poultry feed – thymol, carvacrol and rosmarinic acid 840
M. Taghouti

Dietary microalgae (*Chlorella vulgaris*) and probiotic on performance and intestine of piglets 841
M.F. Pedro, D.F.P. Carvalho, A.M. Almeida, R.J.B. Bessa, A.J.M. Fonseca, A.R.J. Cabrita and J.P.B. Freire

Session 74. Balancing the feed for the economy, the environment, and society

Date: Thursday 31 August 2023; 8.30 – 12.00
Chair: De Campeneere / Bruggeman

Theatre Session 74

Effects of linseed oil, a brown seaweed and seaweed extract on methane emissions in beef cattle 841
E. Roskam, D.A. Kenny, V. O'Flaherty, M. Hayes, A.K. Kelly and S.M. Waters

Effect of whey on methane emission and systemic comparison of *in vivo* and *in vitro* methane formation 842
H. Luisier-Sutter, M. Terranova, S.L. Amelchanka, K. Schweingruber, S. Hug, K. Müller, L. Isele, K. Sommer and M. Schick

Effects of feeding fresh white clover on digestibility and methane emission in lambs 842
X. Chen, S. Ormston, S. Stergiadis, K. Theodoridou, O. Cristobal-Carballo and T. Yan

Algae for reducing methane from dairy cows: what expectation with French local resources? 843
B. Rouillé, H. Marfaing, F. Dufreneix, S. Point, M. Gillier, R. Boré, J. Jurquet and N. Edouard

Managing the rumen microbiome to reduce methane emissions through dietary interventions 843
G. Pugh, O. Cristobal Carballo, T. Yan, C. Creevey and S. Huws

Production and enteric methane emissions in grazing dairy cows fed flaxseed-based supplement 844
M.A. Rahman, K.V. Almeida, D.C. Reyes, A.L. Konopoka, M.A. Arshad and A.F. Brito

Effect of feeding brown seaweed and its extract on methane emissions and performance in dairy cattle 844
K.D. Barnes, S. Huws, T. Yan, X. Chen, M. Hayes and K. Theodoridou

Establishing conclusive links between the gastrointestinal microbiota and feed efficiency in cattle 845
M.M. Dycus, U. Lamichhane, C.B. Welch, K.P. Feldmann, T.D. Pringle, T.R. Callaway and J.M. Lourenco

Variations in bulk milk urea content on dairy farms in Flanders, Belgium in 2019-2021 845
J. Vandicke, K. Goossens, Z. Lipkens, T. Vanblaere and L. Vandaele

Impact on environment and performance of the replacement of soybean meal in post-weaning pig diets 846
E. Royer, P. Pluk, J. De Laat, G. Binnendijk, K. Goris and P. Bikker

Simulating pig multiperformance in contrasted breeding systems using an individual-based model 846
E. Janodet, F. Garcia-Launay and H. Gilbert

Environmental impacts of substituting soybean with rapeseed or haemoglobin meal in broiler diets 847
V. Wilke, J. Gickel, A. Abd El-Wahab and C. Visscher

Poster Session 74

Manipulating dietary degradable protein and starch to increase nitrogen efficiency in dairy cows 847
P. Piantoni, Y. Roman-Garcia, C. Canale, M. Messman and G. Schroeder

Substitution of soybean meal with canola meal in dairy cow diets: effect on enteric methane emission 848
C. Benchaar and F. Hassanat

Session 75. Livestock based circularity for health and sustainable land use

Date: Thursday 31 August 2023; 8.30 – 12.00
Chair: Siqueira / De Olde

Theatre Session 75

invited Does better circularity in livestock require a paradigm shift? 848
A. Mottet and M. Benoit

More or better to obtain sustainable food production? 849
H.F. Olsen, H. Møller, S. Samsonstuen, M.T. Knudsen, L. Mogensen and E. Röös

Satisfying meat demand and avoiding excess manure nitrogen at the regional scale in China 849
F. Accatino, Y. Li and Z. Sun

Role of livestock in the nutrient and carbon metabolism of the agri-food system of a tropical island 850
M. Alvanitakis, V. Kleinpeter, M. Vigne, A. Benoist and J. Vayssières

State and regional nitrogen and phosphorus balances to assess manure sheds in New York State 850
O.F. Godber, K. Workman and Q.M. Ketterings

invited Circularity in livestock production: from theory to practice 851
E.M. De Olde, O. Van Hal, A. Groenewoud and I.J.M. De Boer

The perspective of young farmers on circular agriculture: definition, implementation and barriers 851
A.G. Hoogstra, H. Geerse, I.J.M. De Boer, M.K. Van Ittersum, A.G.T. Schut, C.J.A.M. Termeer and E.M. De Olde

Companion modelling approach for collective nitrogen management in a French municipality 852
G. Martel, F. Garcia-Launay and V. Souchère

Historical transformation of crop-livestock integration and its drivers in a French region 852
R. Pedeches, C. Aubron, S. Bainville and O. Philippon

Livestock: option or necessity? Changes of energy flows in an Indian village, 1950-2022 853
C. Hemingway, C. Aubron and M. Vigne

Poster Session 75

Developing livestock-based circularities for healthy and sustainable territories: the CLiMiT project 853
T.T.S. Siqueira, J.-M. Sadaillan, M. Vigne, J. Veyssières, A. Benoist and M. Miralles-Bruneau

Co-design of a scenario of biomass valorisation within a circular approach on Reunion Island 854
R. Youssouf, E. Cavillot, A.-L. Payet, T.T. Da Silva Siqueira and J.-P. Choisis

Barriers for farmers to 'sustainabilize' their farm: the example of nematode control management 854
M. Sautier, A. Somera, R. Rostellato, P. Ly, Y. Labrunne, J.M. Astruc and P. Jacquiet

Benefits and limits of an organic agroforestry system associating rabbits and apple trees 855
D. Savietto, V. Fillon, S. Simon, E. Lhoste, M. Grillot, A. Dufils, L. Lamothe, F. Derbez, M. Fetiveau and S. Drusch

INOSYS livestock farming systems network: benchmarks for advisors, trainers and policy makers 855
P. Sarzeaud, J. Seegers, O. Dupire and T. Charroin

APIVALE scientific consortium: integrated approach for organic effluent recycling and valorisation 856
F. De Quelen, E. Jarde, C. Le Marechal, T. Lendormi, S. Menasseri and F. Beline

Progress made in whole-farm nitrogen and phosphorus mass balances on New York dairy farms 856
O.F. Godber, K. Workman and Q.M. Ketterings

Measuring greenhouse gas emission from pretreatment and liquid composting storage in biogas facility 857
G.W. Park, M. Ataallahi, N. Eska and K.H. Park

Effect of yoghurt acid whey on quality characteristics of corn silage 857
I. Palamidi, V.V. Paraskeuas, I.P. Brouklogiannis, E.C. Anagnostopoulos, I. Politis, I. Hadjigeorgiou and K.C. Mountzouris

Session 76. Humans and horses: from One Health to One Life perspectives

Date: Thursday 31 August 2023; 8.30 – 12.00
Chair: Evans / Montovani

Theatre Session 76

How do horses perceive human emotions? 858
P. Jardat, C. Parias, F. Reigner, L. Calandreau and L. Lansade

Sensorimotor empathy in horse-human interactions: a new way to understand inter-species performance 858
M. Leblanc, B. Huet and J. Saury

Rethinking horse work 859
R. Evans

Humans and equines: shared working conditions 859
V. Deneux - Le Barh and C. Lourd

vited Working horses: humanities and social sciences approach 860
V. Deneux - Le Barh

Web based dissemination of research to support human knowledge and understanding for horse welfare 860
A.-L. Holgersson, K. Lagerlund and G. Gröndahl

Comparison of ethological & physiological indicators in headshakers & control horses in riding tests 861
L.M. Stange, T. Wilder, D. Siebler, J. Krieter and I. Czycholl

Relationship between hay botanical diversity and faecal bacterial diversity in horses 861
C. Omphalius, P. Grimm, V. Milojevic and S. Julliand

vited Thoroughbreds in equine assisted services programmes: selection, living and working conditions 862
C. Neveux, S. Mullan, J. Hockenhull, J. Barker, K. Allen and M. Valenchon

The role of gender in the worldwide Pura Raza Español equestrian sector 862
M. Ripolles, A. Encina, M. Valera and M.J. Sánchez-Guerrero

Anthelmintic activity of chicory (*Cichorium intybus*) in grazing horses 863
J. Malsa, G. Sallé, L. Wimel, J. Auclair-Ronzaud, B. Dumont, L. Boudesocque-Delaye, F. Reigner, F. Guégnard, A. Chereau, D. Serreau and G. Fleurance

Poster Session 76

Does pain condition influence the human-horse relationship? 863
L. Sobrero, M.G. Riva, M. Minero, A. Cafiso, A. Gazzonis and E. Dalla Costa

Saliva steroidome and metabolome in mare during anoestrus, oestrus cycle and gestation 864
S. Beauclercq, C. Douet, A. Piano, L. Haddad, F. Reigner, P. Liere, L. Nadal-Desbarats and G. Goudet

Session 77. Trade-offs between health, production and welfare in pigs and poultry: which tools do we have and which research is needed?

Date: Thursday 31 August 2023; 8.30 – 12.00
Chair: Nilsson / Millet

Theatre Session 77

invited The future of genetic selection in pigs and poultry 864
L. Verschuren and P.W. Knap

invited The future of nutrition research in pigs and poultry 865
S. Vigors and J.V. Milgen

invited Economically sustainable and animal welfare friendly animal production: what is the role of research 865
A. Silvera and H.A.M. Spoolder

Group discussion 866
K. Nilsson and S. Millet

Session 78. Insect nutrition (substrates) & insect health

Date: Thursday 31 August 2023; 8.30 – 12.00
Chair: Biasato / Lefebvre

Theatre Session 78

invited Farmed insects to create a circular bio-economy in the food and feed industry 866
A. Vilcinskas

Amino acid requirements of mealworm and black soldier fly larvae 867
T. Spranghers, A. Moradei and M. Boudrez

Evaluation of the suitability of hemp production side-streams for the rearing of edible insects 867
A. Kolorizos, G. Baliota, C. Adamaki-Sotiraki, I. Malikentzos, C.I. Rumbos and C.G. Athanassiou

Dietary fat sources impact black soldier fly larvae performance 868
R. Zheng, S. Karanjit and A. Hosseini

Hermetia illucens production parameters and the effect of dietary energy source 868
E.Y. Gleeson and E. Pieterse

Starch digestion in *H. illucens* conversion: exploring the role of amylases from larvae and substrate 869
J.B. Guillaume, S. Mezdour, F. Marion-Poll, C. Terrol, C. Brouzes and P. Schmidely

Investigating the nutritional requirements of black soldier fly larvae using artificial substrates 869
L. Broeckx, L. Frooninckx, A. Wuyts, S. Berrens, M. Van Peer and S. Van Miert

Km0 diets for black soldier fly larvae: the link between insect rearing and biogas systems 870
S. Bellezza Oddon, I. Biasato, Z. Loiotine, A. Resconi, C. Caimi and L. Gasco

Sugar processing by-products for black soldier fly farming: does the rearing scale matter? 870
I. Biasato, S. Bellezza Oddon, A. Resconi, Z. Loiotine and L. Gasco

Optimization of a hatchery residues fermentation process to feed black soldier fly larvae 871
M. Dallaire-Lamontagne, G.W. Vandenberg, L. Saucier and M.H. Deschamps

Fish side residues as a substrate for *Hermetia illucens* 871
M.G. Zhelezarova

Poster Session 78

Black soldier fly larvae production is optimized by the presence of HMTBa 872
K. Luyt, G. Crielaard, M. Briens, J.A. Conde-Aguilera and M. Ceccantini

Black soldier fly larvae production is optimized by the presence of a multi-carbohydrase 872
K. Luyt, G. Crielaard, M. Briens, J.A. Conde-Aguilera and M. Ceccantini

Effect of water-soluble complementary feed on performance in nursery of black soldier fly 873
L. Schneider, M. Brake, A. Heseker, W. Westermeier and G. Dusel

Digestibility in *H. illucens* larvae: resolving faeces collection and ingesta quantification issues 873
J.B. Guillaume, S. Mezdour, F. Marion-Poll, C. Terrol and P. Schmidely

Black soldier fly larvae as tools for the bioconversion of sludge from wastewater treatments 874
C. Ligeiro, I. Lopes, T. Ribeiro, I. Rehan, K. Silvério, M. De Fátima and D. Murta

Mediterranean agricultural by-products as insect diet ingredients: the ADVAGROMED perspective 874
C.G. Athanassiou, S. Bellezza Oddon, V. Zambotto, T. Ribeiro, R. Rosa García, A. El Yaacoubi, C. Adamaki-Sotiraki, I. Biasato, A. Resconi, D. Murta, C.I. Rumbos and L. Gasco

Multigenerational and nutritional traits of BSF reared on seaweed or selenium enriched substrates 875
M. Ottoboni, L. Ferrari, A. Moradei, F. Defilippo, P. Bonilauri and L. Pinotti

Bacterial biomass improves performance and antibacterial activity of black soldier fly larvae 875
N. N. Moghadam, K. Dam Nielsen, A. Simongiovanni and T. Chalvon-Demersay

Impact of zinc supply on black soldier fly larvae growth, bioconversion and microbiota 876
L. Frooninckx, L. Broeckx, D. Vandeweyer, C. Keil, M. Maares and S. Van Miert

Session 79. Improved insight into the reproductive physiology of livestock

Date: Thursday 31 August 2023; 8.30 – 12.00
Chair: Kenny

Theatre Session 79

Review: recent outcomes associating time to PSPB increase with pregnancy loss in dairy cows 876
T. Minela, A. Santos and J.R. Pursley

Improving cow fertility by immunizing against inhibin and P4 supplementation 877
Z.D. Shi, F. Chen and R.H. Guo

Characterization of sex chromosomes-linked lncRNAs in Holstein spermatozoa under stress conditions 877
A. Yousif, G. Thaller and M. Saeed-Zidane

Inbreeding affects the freezing ability in sperm samples of Pura Raza Español stallions 878
Z. Peña, M. Valera, A. Molina, N. Laseca and S. Demyda-Peyrás

Sperm transcripts and seminal plasma metabolome reveal the regulatory mechanism of bull semen quality 878
W.L. Li and Y. Yu

Application of an oligo-based FISH method in fertility assessment of boars by chromosomal analysis 879
W. Poisson, J. Prunier, A. Bastien, A. Carrier, I. Gilbert and C. Robert

Artificial insemination success: a new trait for French dairy goats 879
V. Clément, A. Piacère and M. Chassier

The male effect as an alternative to eCG in oestrus induction and synchronization treatment in ewes 880
N. Debus, G. Besche, S. Fréret, A. Hardy, M.T. Pellicer-Rubio, A. Tesniere and J.-B. Menassol

Circulating anti-Müllerian hormone from 5-month old Merino ewe lambs predicts first birthing rates 880
J. Daly, J. Kelly, K. Kind and W. Van Wettere

Poster Session 79

Impact of pinecone oil on reproductive performance, milk composition and serum parameters in sows 881
Q.Q. Zhang, J.H. Song, C.B. Lim, S. Biswas, J.S. Yoo and I.H. Kim

Maternal *Forsythia suspensa* extract alleviated oxidative stress and improved gut health in sows 881
S. Long, Z. Chen and T. He

Effects of PGF2α on luteal tissue morphology, expression of relative genes in Hu sheep 882
Y.Q. Liu, Y. Li, C.H. Duan and Z.P. Song

Relationship between some micronutrients in serum and their effect on the seminal quality of bulls 882
A. Benito-Diaz, A.S. Sarmiento-García, M. Montañes-Foz, R. Bodas-Rodríguez and J.J. García-García

Genetic correlations on kinetic sperm parameters in a closed population of PRE stallions 883
Z. Peña, A. Molina, C. Medina, M. Valera and S. Demyda-Peyrás

Gestation length in Braunvieh cows 883
M. Fanger, A. Burren and H. Jörg

Distribution of adipokines in reproductive tract and embryonic annexes in hen 884
O.B. Bernardi, A.E. Estienne, M.R. Reverchon, A.B. Brossaud, C.R. Rame and J.D. Dupont

The effects of PROK 1, alone or in combination with IFN gamma on endometrial immune response in pig 884
S.E. Song and J. Kim

Session 80. Animal behaviour

Date: Thursday 31 August 2023; 8.30 – 12.00
Chair: Dantas / Boyle

Theatre Session 80

Creating a culture of openness for the responsible use of video observation in animal sciences 885
M.F. Giersberg and F.L.B. Meijboom

Salivary oxytocin and lachrymal caruncle temperature as indicators of anticipation in growing pigs 885
G.A. Franchi, L.R. Moscovice, H. Telkänranta and L.J. Pedersen

Pig oxidative stress model: behaviour as a potential indicator of oxidative stress in pigs 886
R.D. Guevara, J.J. Pastor, S. López-Vergé, X. Manteca, G. Tedo and P. Llonch

Comparison of aversiveness of 8 different inert gas (mixtures) to CO_2 for stunning pigs at slaughter 886
I. Wilk, J. Gelhausen, T. Friehs, T. Krebs, D. Mörlein, J. Tetens and J. Knöll

Detecting onset of farrowing using CUSUM-charts based on sows' activity 887
T. Wilder, B. Baude and J. Krieter

Impact of environmental enrichment on the behaviour and immune cell transcriptome of pregnant sows 887
M.M. Lopes, C. Clouard, J. Chambeaud, M. Brien, N. Villain, C. Gerard, F. Hérault, A. Vincent, I. Louveau, R. Resmond, H. Jammes and E. Merlot

The knowns and unknowns about feather pecking in laying hens 888
A. Harlander and N. Van Staaveren

Lower redness of the facial skin is a marker of a positive human-hen relationship 888
D. Soulet, A. Jahoui, M.-C. Blache, B. Piégu, G. Lefort, L. Lansade, K. Germain, F. Lévy, S. Love, A. Bertin and C. Arnould

Do sheep differentiate emotional cues conveyed in human body odour? 889
I. Larrigaldie, F. Damon, S. Mousqué, B. Patris, L. Lansade, B. Schaal and A. Destrez

Evaluation of inter-observer reliability of dichotomous and four-level animal-based indicators 889
B. Torsiello, M. Giammarino, L. Battaglini, M. Battini, S. Mattiello, P. Quatto and M. Renna

Social network analysis of dairy cows' group structure at the feeding trough 890
T. Wilder, P. Hasenpusch, A. Seidel, G. Thaller and J. Krieter

Roughage and type of dispensers: what consequences on horses' feeding behaviour? 890
M. Roig-Pons and S. Briefer

Poster Session 80

Caudectomy effect on the severity of tail lesions and abscesses' frequency in pig carcasses 891
P. Trevisi, D. Luise, S. Virdis, S. Dalcanale and U. Rolla

Characterising cattle daily activity patterns using accelerometer data 891
S. Hu, A. Reverter, R. Arablouei, G. Bishop-Hurley and A. Ingham

Effects of olfactory exposure to twelve essential oils on behaviour and health of cows with mastitis 892
R. Nehme, C. Michelet, E. Vanbergue, O. Rampin, S. Bouhallab, A. Aupiais and L. Andennebi-Najar

Assessment of stress in dairy cattle industry via analysis of cortisol residue in commercial milk 892
M. Ataallahi, G.W. Park, E. Nugrahaeningtyas, M. Dehghani, J.S. Lee and K.H. Park

The physiological and behavioural effects of COVID-19 restrictions on lactating dairy cows 893
S.J. Morgan and D. Barrett

Effects of the living environment on the behaviour of rabbits 893
M. Fetiveau, M. Besson, V. Fillon, M. Gunia and L. Fortun-Lamothe

Behaviour test and study of sheepdogs' abilities 894
B. Lasserre, B. Ducreux, M. Chassier, L. Joly, T. Le Morzadec, P. Cacheux and C. Gilbert

Innovative evaluation method of behavioural reactivity for 'foie gras' ducks 894
C.M. Monestier, A.W. Warin, S.L. Lombard, S.L.M. Laban-Mele and W.M. Massimino

Use of scaring devices to avoid roe deer fawns getting injured or killed during mowing 895
J. Mačuhová, T. Wiesel and S. Thurner

Session 81. Precision feeding

Date: Thursday 31 August 2023; 8.30 – 12.00
Chair: Druyan / Bloch

Theatre Session 81

Precision feeding, recent advances for gestating sows and dairy cows 895
C. Gaillard, C. Ribas and M. Durand

Individual ingestion time prediction with RGB-D cameras in beef cattle: a machine learning approach 896
P. Guarnido-Lopez, F. Ramirez-Agudelo, S. Tomozyk, A. Kjorvel, A. Carvalho, F. Jezegou-Bernard, P. Gauthier and M. Benaouda

Using image classification to estimate feed intake in weaned piglets under commercial conditions 896
T. Van De Putte, J. Degroote and J. Michiels

invited The value of precision feeding technologies: economic, productivity and environmental aspect 897
J.K. Niemi

Precision feeding in dairy cows: what do professionals in ruminant nutrition think about it? 897
A. Igier, A. Petillon-Pronk, J. Martin, E. Hertault, N. Gaudillière, J. Jurquet, A. Fischer and Y. Le Cozler

Characterizing dairy cow's individual reaction to a decrease in production concentrate distribution 898
A. Fischer, R. Lehuraux and J. Jurquet

Longitudinal patterns for feeding traits in the two parental populations of the mule duck 898
H. Chapuis, C. Ribas, H. Gilbert, M. Lagüe and I. David

Models predicting methane emissions and methane conversion factor of Finnish Nordic Red dairy cows 899
J.S. Adjassin, A. Guinguina, M. Eugène and A.R. Bayat

Use of automated head-chamber systems as exclusive concentrate dispensers during bulls fattening 899
N. Lorant, E. Henrotte, Y. Beckers, F. Forton, A. Vanlierde, M. Mathot and A. Mertens

Poster Session 81

On-farm nutritional supplementation to improve sows and piglets' performances: field report 900
P. Engler, G. Gemo, S. Cissé and J.M. Garcia

Effects of monosaccharides on *in vitro* NDF and starch digestibility, pH and volatile fatty acids 900
E. Raffrenato, G. Esposito and L. Bailoni

Benefit of full matrix application for a novel phytase using a phased-dosing or fixed dosing strateg 901
B.C. Hillen, S. Gilani, R.D. Gimenez-Rico, K.M. Venter, P. Plumstead and Y. Dersjant-Li

Use of fodder beets in French dairy diets: farmers points of views and experimental results 901
V. Brocard, A. Marsault, L. Vivenot, E. Tranvoiz and J. Jurquet

Effect of wilting grass silage on milk production of dairy cows fed two concentrate protein levels 902
M. Grøseth, L. Karlsson, H. Steinshamn, M. Johansen, A. Kidane and E. Prestløkken

Zootechnical performance of dairy cattle fed ensiled Italian ryegrass or winter rye 902
L. Vandaele, J.L. De Boever, T. Van Den Nest and K. Goossens

Performance of NIR spectrometry to predict amino acids content in forages 903
L. Bahloul, V. Larat, P. Riche and B. Sloan

Estimation of feed weight of dairy cows using computer vision 903
M. Taghavi, T. Izquierdo and I. Fodor

Session 82. Product quality and efficiency of small ruminant production systems

Date: Thursday 31 August 2023; 8.30 – 12.00
Chair: Hernández-Castellano / Tsiplakou

Theatre Session 82

The performance of cross fostered lambs from birth to weaning on commercial sheep flocks in Ireland 904
F.P. Campion, J. Molloy and M.G. Diskin

Latent factors analysis of protein profile, composition, and cheese-making traits of goat milk 904
N. Amalfitano, G. Secchi, M. Pazzola, G.M. Vacca, M.L. Dettori, F. Tagliapietra, S. Schiavon and G. Bittante

Pyrenean wools: how improving the organisation of the upstream and downstream sectors in the massif 905
C. Viguié, S. Fichot and G. Brunschwig

The Gentile di Puglia merino sheep breeds preliminary wool quality assessment 905
V. Landi, E. Ciani, G. Molina, R. Topputi, F.M. Sarti, A. D'Onghia, G. Mangini, F. Pilla, S. Grande, A. Maggiolino and P. De Palo

Using indoor and outdoor finishing systems to finish hill lambs to carcass weights between 12-16 kg 906
M. Dolan, T. Boland, N. Claffey and F. Campion

Influence of climatic disturbances on the lactation curve in dairy goats 906
A. Harnois Gremmo, N. Gafsi, F. Bidan, O. Martin and L. Puillet

Genetic parameters for fleece uniformity in Alpacas 907
J.P. Gutiérrez, A. Cruz, R. Morante, A. Burgos, N. Formoso-Rafferty and I. Cervantes

Farmers' competitions: stimulating new management practices by Peruvian alpaca and llama farmers 907
E. Quina, T. Felix, M. Aguilar, G. Gutierrez, J. Candio, J. Gamarra, M. Mamani, A. Mejía and M. Wurzinger

YoGArt project: milk quality and oxidative status of ewes fed microalgae blend 908
A. Mavrommatis, P. Kyriakaki, F. Satolias and E. Tsiplakou

Goat AI programs based on the male effect using less or no hormones progress with suitable results 908
A. Fatet, L. Johnson, L. Jourdain, F. Bidan and P. Martin

Fiber characteristics of cashmere goat in Mongolia 909
S.B. Baldan, M.P. Purevdorj, G.M. Mészáros and J.S. Sölkner

Poster Session 82

Anco fit improves feed efficiency, milk production and milk solids in lactating ewes 909
C. Panzuti, R. Breitsma and C. Gerard

No effect of meslin flattening on dairy ewes 910
B. Fança, A. Hardy, M. Rinn and L. Buisson

Quality of cheeses made from the milk of native breeds of goats 910
A. Kawęcka, M. Pasternak, J. Sikora and M. Puchała

The impact of live-weight and body condition score on reproductive success in ewes 911
P. McCarron, N. McHugh, N. Fetherstone, H. Walsh and F.M. McGovern

Genetic parameters of medullation types in alpaca fibre 911
A. Cruz, Y. Murillo, A. Burgos, A. Yucra, M. Quispe, E. Quispe and J.P. Gutiérrez

Blood and gastrointestinal nematode profile of sheep-fed diets supplemented with fossil shell flour 912
O.O. Ikusika and C.T. Mpendulo

Synchrotron FTIR microspectroscopy investigation on the sperm cryopreservation of Sanan goat 912
S. Ponchunchoovong, F. Suwor, K. Tammanu and S. Siriwong

A need for knowledge on the utilisation of forage by goats according to the distribution methods 913
B. Fança and B. Bluet

Changes in fattening and slaughter traits of old-type Polish merino between 2010 and 2020 913
M. Puchała, A. Kawęcka, J. Sikora and M. Pasternak

Milk fat nutritional indices – a comparative study between retail goat and cow milk 914
G. Papatzimos, P. Mitlianga, M.A. Karatzia, Z. Basdagianni and E. Kasapidou

Mountain caprine milk – the higher the better? 914
M.A. Karatzia, M. Amanatidis, G. Papatzimos, E. Kasapidou, P. Mitliagka and Z. Basdagianni

Results of milk performance of Carpathian goats 915
J. Sikora, A. Kawęcka, M. Puchała and M. Pasternak

Barley in concentrates for dairy goats – effects of alkaline and mechanical treatments 915
A. Martinsen, D. Galméus, H. Volden, K. Hove, M. Silberberg and M. Eknæs

Session 83. Genetic diversity

Date: Thursday 31 August 2023; 14.00 – 18.00
Chair: Neumann / Schiavo

Theatre Session 83

The effect of population history, mutation and recombination on genomic selection 916
D. Adepoju, T. Klingström, A.M. Johansson, E. Rius-Vilarrasa and M. Johnsson

The detection of putative recessive lethal haplotypes in Irish sheep populations 916
R. McAuley, N. McHugh, T. Pabiou and D.C. Purfield

Assessing the impact of inbreeding on mastitis and digital dermatitis in German dairy cattle 917
J. Mugambe

Genomic inbreeding in the Austrian Turopolje pig population 917
G. Mészáros, B. Berger, C. Draxl and J. Sölkner

A genetic diversity study of a local Belgian chicken breed using the new IMAGE SNP array 918
R. Meyermans, W. Gorssen, J. Bouhuijzen Wenger, O. Heylen, J. Martens, S. Janssens and N. Buys

Genetic diversity in Dutch sheep breeds shaped by geography, history, use and genetic management 918
J.N. Hoorneman, W.J. Windig and M.A. Schoon

Genetic diversity and population structure of the Cyprus Chios sheep and Damascus goat breeds 919
A.C. Dimitriou, G. Maimaris, A.N. Georgiou and G. Hadjipavlou

Genome-wide analysis of Greek goats with global breeds revealing population structure and diversity 919
E. Tosiou, V. Tsartsianidou, S. Vouraki, V. Papanikolopoulou, L.V. Ekateriniadou, E. Boukouvala, I.G. Bouzalas, I. Sakaridis, G. Arsenos and A. Triantafyllidis

Genome-wide scan for runs of homozygosity in South Americam camelids 920
S. Pallotti, M. Picciolini, M. Antonini, C. Renieri and V. Napolioni

Exploring global cattle genealogy with tree sequence 920
G. Mafra Fortuna, J. Obsteter, A. Kranis and G. Gorjanc

Which diversity measures are important for small breeds like German Black Pied Cattle (DSN)? 921
G.A. Brockmann, G.B. Neumann, P. Korkuc, M.J. Wolf, K. May and S. König

Characterisation of free-living and exotic animal species at species and individual level 921
M. Zorc, M. Cotman, A. Dovč, J. Zabavnik-Piano and P. Dovc

Comparative analysis of heterozygosity-enriched regions in two autochthonous Italian cattle breeds 922
G. Schiavo, S. Bovo, F. Bertolini, A. Ribani, V. Taurisano, S. Dall'olio, M. Bonacini and L. Fontanesi

Recovering latent population stratification using ADMIXTURE and metafounders in Brown Swiss 922
C. Anglhuber, C. Edel, E.C.G. Pimentel, R. Emmerling, K.-U. Götz and G. Thaller

Poster Session 83

Genomic diversity analysis of the Swedish Landrace goat 923
B. Hegedűs, A.M. Johansson and P. Bijma

Effect of interpopulation distance on average heterosis in crosses 923
A. Legarra, D. Gonzalez-Dieguez, A. Charcosset and Z.G. Vitezica

Genome of the extinct Gotland cattle breed 924
M. Johnsson and A.M. Johansson

Genetic variability in Italian Mediterranean buffalo: implications for conservation and breeding 924
M.M. Gómez, R. Cimmino, D. Rossi, G. Zullo, G. Campanile, G. Neglia and S. Biffani

A comparative genome analysis across pig breeds can help to identify putative deleterious alleles 925
M. Ballan, S. Bovo, G. Schiavo, F. Bertolini, M. Bolner, M. Cappelloni, S. Tinarelli, M. Gallo and L. Fontanesi

The effect of recessive genetic defects on pregnancy loss in Swedish dairy cattle 925
P. Ask-Gullstrand, E. Strandberg, R. Båge, E. Rius-Vilarrasa and B. Berglund

Session 84. EuroFAANG: genotype-to-phenotype research across Europe and beyond

Date: Thursday 31 August 2023; 14.00 – 18.00
Chair: Clark / Kuhn

Theatre Session 84

EuroFAANG – an infrastructure for farmed animal genotype to phenotype research in Europe and beyond 926
E.L. Clark, E. Giuffra, A. Granados-Chapette, M. Groenen, P.W. Harrison, C. Kaya, S. Lien, M. Tixier-Boichard and C. Kuehn

GENE-SWitCH: improving the functional annotation of pig and chicken genomes for precision breeding 926
E. Giuffra, H. Acloque, A.L. Archibald, M.C.A.M. Bink, M.P.L. Calus, P.W. Harrison, C. Kaya, W. Lackal, F. Martin, A. Rosati, M. Watson and J.M. Wells

Genome-wide association studies for body weight in broilers using sequencing and SNP chip data 927
E. Tarsani, V. Riggio and A. Kranis

Using methylation annotation to improve genomic prediction of gene expression in pigs 927
B.C. Perez, J. De Vos, D. Crespo-Piazuello, M.C.A.M. Bink, M. Ballester, M.J. Mercat, O. Madsen and M.P.L. Calus

The path from functional annotation towards improved genomic prediction in cattle 928
C. Kühn

Biology-driven genomic predictions for dry matter intake within and across breeds using WGS data 928
R. Bonifazi, G. Plastow, M. Heidaritabar, A.C. Bouwman, L. Chen, P. Stothard, J. Basarb, C. Li, Bovreg Consortium and B. Gredler-Grandl

Newly annotated genomic features for biology-driven genome selection: the BovReg contribution 929
G.C.M. Moreira, L. Tang, S. Dupont, M. Bhati, H. Pausch, D. Becker, M. Salavati, R. Clark, E.L. Clark, G. Plastow, C. Kühn, C. Charlier and The Bovreg Consortium

AQUA-FAANG: decoding genome function to enhance genotype-to-phenotype prediction in farmed finfish 929
D. MacQueen and S. Lien

Integrating functional annotation data in genomic prediction of VNN resistance in European sea bass 930
S. Faggion, R. Mukiibi, L. Peruzza, M. Babbucci, R. Franch, G. Dalla Rovere, S. Ferraresso, D. Robledo and L. Bargelloni

GEroNIMO (Genome and Epigenome eNabled breedIng in MOnogastrics) 930
S. Lagarrigue, F. Pitel and T. Zerjal

RUMIGEN: new breeding tools in a context of climate change 931
S. Mattalia, A. Vinet, M.P.L. Calus, H.A. Mulder, M.J. Carabaño, C. Diaz, M. Ramon, S. Aguerre, J. Promp, R. Vallée, B.C.D. Cuyabano, D. Boichard, E. Pailhoux and J. Vandenplas

The genetic basis of ruminant microbiomes – contribution of the HoloRuminant project 931
Y. Ramayo-Caldas, I. Mizrahi, P. Pope, C. Creevey, J.P. Sanchez, R. Quintanilla and D. Morgavi

Poster Session 84

The effect of production systems on adipose tissue gene expression in Krškopolje pig 932
K. Poklukar, M. Čandek-Potokar, N. Batorek-Lukač, M. Vrecl, G. Fazarinc and M. Škrlep

A lncRNA gene-enriched atlas for GRCg7b chicken genome using Ensembl, RefSeq and two FAANG databases 932
F. Degalez, M. Charles, S. Foissac, H. Zhou, D. Guan, L. Fang, C. Klopp, F. Lecerf, T. Zerjal, F. Pitel and S. Lagarrigue

Session 85. Climate care dairy farming: herd management – Part 3

Date: Thursday 31 August 2023; 14.00 – 18.00
Chair: Cieślak / Kuipers

Theatre Session 85

Effect of the Rumitech in a high-forage diet on methane production and performance of dairy cows 933
A. Cieslak, H. Huang, B. Nowak, M. Kozłowska, D. Lechniak, P. Pawlak, M. Szumacher-Strabel and J. Dijkstra

vited Sorghum hybrids as viable forage alternative to corn silage when water availability is limited 933
D. Duhatschek, J. Bell, D. Druetto, L.F. Ferraretto, K. Raver, J. Goeser, J. Smith, S. Paudyal, G.M. Schuenemann and J.M. Piñeiro

Feed intake and milk production of dairy cows fed with a ration with ensiled tall fescue 934
M. Cromheeke, L. Vandaele, D. Van Wesemael, J. Baert, M. Cougnon, D. Reheul and N. Peiren

Effect of supplementing live bacteria on methane production in lactating dairy cows 934
O. Ramirez-Garzon, D.G. Barber, J. Alawneh, L. Huanle and M. Soust

Greenhouse gas emissions from dairy cows fed best practice diets 935
M. Managos, C. Lindahl, S. Agenäs, U. Sonesson and M. Lindberg

Effects of cashew nutshell liquid on milk production and methane emission of dairy cows 935
R.J.F. Gaspe, T. Obitsu, T. Sugino, Y. Kurokawa and Y. Kuroki

In vivo testing of a methane-suppressing feed additive that acts by altering rumen redox potential 936
C. O'Donnell, C. Thorn, A.C.V. Montoya, S. Nolan, M. McDonagh, E. Dunne, F. McGovern, R. Friel, S. Waters and V. O'Flaherty

Diurnal variation and repeatability of CO_2 production in GreenFeed studies with lactating cows 936
A. Guinguina and P. Huhtanen

Repeatability and correlations of residual carbon dioxide and feed efficiency in Nordic Red cattle 937
A. Chegini, M.H. Lidauer, T. Stefanski, A.R. Bayat and E. Negussie

Methane emission from purebred Holstein, Nordic Red and F1 crossbred cows of the two breeds 937
A.L.F. Hellwing, M. Vestergaard and M. Kargo

How does a beef × dairy calving affect the dairy cow's following lactation? 938
R.E. Espinola Alfonso, W.F. Fikse, M.P.L. Calus and E. Strandberg

Extended lactation and milk yield – a randomized controlled trial in high yielding older cows 938
A. Hansson, C. Kronqvist, R. Båge and K. Holtenius

Effect of usage BWB sire on Holstein cow performances 939
K. Elzinga, C. Schrooten, G. De Jong and P. Duenk

Combined analysis: genetic and phenotypic trends in German Holstein dairy cattle 939
L. Hüneke, J. Heise, D. Segelke, S. Rensing and G. Thaller

Poster Session 85

Whole-crop maize forage preservation improvement with silage additives 940
I. Nikodinoska, E. Wambacq, G. Haesaert and C.A. Moran

Gallic acid alleviates the negative effect of *Asparagopsis armata* on milk yield in dairy cows 940
R. Huang, P. Romero, C. Martin, E.M. Ungerfeld, A. Demeter, A. Belanche, D.R. Yáñez-Ruiz, M. Popova and D. Morgavi

Milk production in primiparous cows with customized voluntary waiting period 941
A. Edvardsson Rasmussen, E. Strandberg, R. Båge, M. Åkerlind, K. Holtenius and C. Kronqvist

Association of nisin and *Cymbopogon citratus* against *S. aureus* isolated from bovine mastitis 941
L. Castelani, T.M. Mitsunaga, L.C. Roma Jr. and L.E.F. Zadra

Session 86. How to address stakeholder and societal expectations of intrinsic and extrinsic quality of animal products?

Date: Thursday 31 August 2023; 14.00 – 18.00

Theatre Session 86

Quick sustainability scan calculator for intensive and extensive pig farms 942
S. Sanz-Fernández, C. Reyes- Palomo, P. Meatquality Consortium and V. Rodríguez-Estévez

Bulk milk and farms characterization in the Parmigiano Reggiano Consortium area: the INTAQT project 942
M. Berton, M.A. Ramirez Mauricio, N. Amalfitano, L. Gallo, A. Cecchinato and E. Sturaro

Bridging environmental sustainability and intrinsic quality traits of pork 943
M. Gagaoua, F. Gondret, F. Garcia-Launay and B. Lebret

Improving milk intrinsic quality: considering synergies and antagonisms of farming practices 943
L. Rey-Cadilhac, A. Ferlay, M. Gelé, S. Léger and C. Laurent

Imagining futures of husbandry farming: new business models for the sustainable transition? 944
B. Smulders, R. Valkenburg and M. Anastasi

Current prediction and authentication tools used and needed in EU livestock product chains 944
S. McLaughlin, F. Albert, F. Bedoin, C. Couzy, I. Legrand, A. Nicolazo De Barmon, C. Laithier, C. Manuelian, F. Klevenhusen, S. De Smet, E. Sturaro and N. Scollan

Milk quality assessment for intensive and extensive goat farming of the Skopelos breed in Greece 945
Z. Basdagianni, I. Stavropoulos, G. Manessis, C.G. Biliaderis and I. Bossis

Rheological evaluation of rennet-induced curdling of goat milks from different farming systems 945
K. Kotsiou, M. Andreadis, A. Lazaridou, C.G. Biliaderis, Z. Basdagianni, I. Bossis and T. Moschakis

invited Technological tools for authentication assessment of animal products 946
C.L. Manuelian

Potential of milk infrared spectroscopy to discriminate farm characteristics: the INTAQT project 946
M.A. Ramirez Mauricio, D. Giannuzzi, L. Gallo, M. Berton, A. Cecchinato and E. Sturaro

A longitudinal cohort study of health and welfare status in extensively and intensively reared goats 947
V. Korelidou, A.I. Kalogianni and A.I. Gelasakis

New tool for accurate analysis of gas concentrations in barns 947
O. Bonilla-Manrique, A. Moreno-Oyervides, H. Moser, J.P. Waclawek, B. Lendl and P. Martín-Mateos

Poster Session 86

Implementation of husbandry practices improving quality and sustainability: a living lab approach 948
E. Sturaro, C. Berri, D. Berry, R. Eppenstein, C. Laithier, A. Cartoni Mancinelli, B. Martin and F. Leiber

Muscle proteomics towards molecular understanding of colour and water-holding biochemistry in meat 948
S. Yigitturk

Exploring YOLO deep learning model for goat detection in animal welfare and behavioural monitoring 949
A. Temenos, A. Voulodimos, D. Kalogeras and A. Doulamis

The association between body condition score and the udder skin surface temperature in goats 949
V. Korelidou, A.I. Kalogianni and A.I. Gelasakis

Session 87. Carbon sources and sinks within global livestock systems

Date: Thursday 31 August 2023; 14.00 – 18.00
Chair: Salami / Poulopoulou

Theatre Session 87

vited Leveraging the measuring and managing of livestock farms on the journey towards Net Zero 950
J. Gilliland

The role of carbon sequestration in organic dehesas ruminant farms 950
R. Casado, M. Escribano, P. Gaspar and A. Horrillo

Estimating soil carbon sequestration gaps for ruminant systems across the globe 951
Y. Wang, I. Luotto, Y. Yigini, R. Vargas, D. Wisser, T.P. Robinson, U.M. Persson, C. Cederberg, R. Ripoll-Bosch, I.J.M. De Boer and C.E. Van Middelaar

Agronomic and environmental impacts of sheep integration in cover crop management in Wallonia 951
N. Lorant, B. Huyghebaert and D. Stilmant

Carbon footprint assessment of Korean native beef cattle and options to achieve net-zero emissions 952
R. Ibidhi, T. Kim, J. Byun, R. Bharanidharan, Y. Lee, S. Kang and K. Kim

Carbon Footprint of organic beef from dairy male calves 952
L. Mogensen, T. Kristensen, C. Kramer, A. Munk, P. Spleth and M. Vestergaard

Model to calculate the impact of interventions on the carbon footprint of dairy 953
A. Esselink

Tools to optimise the carbon footprint of milk production 953
D. Schwarz

Evaluation of feed additives under varying diets of dairy cows on performances and enteric emissions 954
M.E. Uddin, M.R.A. Redoy, S. Ahmed, M. Bulness, D.H. Kleinschmit, J. Lefler and C. Marotz

Supplementation with a calcium peroxide additive mitigates enteric methane emissions in beef cattle 954
E. Roskam, D.A. Kenny, V. O'Flaherty, A.K. Kelly and S.M. Waters

Methane emission from rosé veal calves feed a corn cob silage-based or grass silage-based ration 955
A.L.F. Hellwing and M. Vestergaard

Feeding a starch and protein binding agent for mitigating enteric methane emissions in sheep 955
P. Prathap, S.S. Chauhan, J.J. Cottrell, B.J. Leury and F.R. Dunshea

Poster Session 87

Neutralization of bovine enteric methane emission by the presence of trees in a silvopastoral system 956
J.R.M. Pezzopane, H.B. Brunetti, P.P.A. Oliveira, A.C.C. Bernardi, A.R. Garcia, A. Berndt, A.F. Pedroso, A.L.J. Lelis and S.R. Medeiros

Dairy goat farms in Extremadura: carbon sources and sinks in several management systems 956
L. Madrid, M. Escribano, P. Gaspar and A. Horrillo

Life cycle assessment of IntelliBond on the carbon footprint of a dairy farm 957
D. Brito De Araujo, K. Perryman and J.G.H.E. Bergman

Effects of supplementation of *Asparagopsis taxiformis* as a means to mitigate methane emission 957
M. Angellotti, M. Lindberg, M. Ramin, S.J. Krizsan and R. Danielsson

Session 88. Minerals in animal nutrition: basal diet – Part 2

Date: Thursday 31 August 2023; 14.00 – 18.00
Chair: Alvarez Rodriguez / Windisch

Theatre Session 88

invited Trace element concentrations in feed ingredients: previously overlooked, now under the spotlight 958
M. López-Alonso

Review on native Se concentrations in feed ingredients 958
D. Cardoso and A. Hachemi

Does USEtox predicts adequately Cu and Zn ecotoxicity in soils amended with animal effluents? 959
E.P. Clement, M.N. Bravin, A. Avadi and E. Doelsch

Challenges in establishing the mineral composition of feed materials 959
G. Tran and V. Heuzé

Swiss feed database: mineral and trace element composition of feedstuffs 960
M. Lautrou, E. Manzocchi and P. Schlegel

Overview of international aquaculture feed formulation database with reference to mineral nutrition 960
D.P. Bureau, L. Manomaitis and F.M. Damasceno

invited Mineral analysis: opportunities and limitations 961
P. Berzaghi, R. Fornaciari, M. Dorigo and G. Cozzi

Determination of forage mineral analysis with portable X-ray Fluorescence device 961
R. Balegi, F. Penen, M. Lemarchand and A. Boudon

A new phosphorus feeding system for the sustainability of swine and poultry production 962
M.P. Létourneau Montminy, M. Lautrou, M. Reis, C. Couture, N. Sakomura, B. Meda, C.R. Angel and A. Narcy

Phosphorus digestibility of porcine processed animal proteins (PAPs) in broiler diets 962
J. Van Harn and P. Bikker

Prediction of magnesium absorption in dairy cows: An update 963
R. Khiaosa-Ard and Q. Zebeli

Se and Co balance in dairy cows: longitudinal study from late lactation to subsequent mid-lactation 963
J.B. Daniel and J. Martín-Tereso

Poster Session 88

Mineral analysis in feed using a new automated analytical method 964
D. Schwarz and D. Aden

Trace mineral trends in British and Irish forages 964
A.H. Clarkson and N.R. Kendall

Does size matter? Comparing liver sample sizes for trace element status 965
A.H. Clarkson, J. Angel and N.R. Kendall

Investigation on the variation in phosphorus of wheat bran: effect in diets of fattening pigs 965
R. Puntigam, P. Riesinger, A. Honig, M. Schaeffler, W. Windisch and H. Spiekers

Swiss Feed Database: a closer look at minerals and trace elements in 10 years of roughage surveys 966
E. Manzocchi, M. Lautrou and P. Schlegel

Mineral forage value in the INRAE feeding system for ruminants 966
A. Boudon and G. Maxin

Minerals in cattle nutrition – meet the needs! 967
A.C. Honig, V. Inhuber, H. Spiekers, W. Windisch, K.-U. Götz, G. Strauß and T. Ettle

Ruminant feed rations rich in clay minerals may induce Zn deficiency 967
M. Schlattl, M. Buffler and W. Windisch

Zn, Cu, Mn and Fe balance in dairy calves fed milk replacer or whole milk at two feeding allowances 968
T. Chapelain, J.B. Daniel, J.N. Wilms, L.N. Leal and J. Martín-Tereso

Impact of drinking water salinity on lactating cows 968
A. Iritz and Y. Ben Meir

Seasonal variation of trace essential and toxic minerals in milk and blood of dairy ewes 969
A. Nudda, M.F. Guiso, G. Sanna, A. Cesarani, M. Deroma, G. Pulina and G. Battacone

Session 89. Explainable models and artificial intelligence supporting farming decisions

Date: Thursday 31 August 2023; 14.00 – 18.00
Chair: Foy / Odintsov

Theatre Session 89

Measuring the behaviour of lambs in an isolated environment with artificial intelligence methods 969
B. Benet and R. Lardy

Estimating sow posture from computer vision: influence of the sampling rate 970
M. Bonneau, J.A. Vayssade and L. Canario

Estimating pig mass in a high-density pig group during transportation 970
V. Bloch, A. Valros, C. Munsterhjelm, M. Heinonen, M. Tuominen-Brinkas, H. Koskikallio and M. Pastell

Assessment of piglet maturity at birth using computer vision 971
L. Maignel, R. Mailhot, A. Carrier and P. Gagnon

Validation of new software (r-Algo) for predicting meat chemical composition from ultrasound images 971
B. Ahmadi, T. Schwarz and P.M. Bartlewski

Coupling a sow herd model with a bioclimatic model of gestation rooms: development and evaluation 972
E. Dubois, F. Garcia-Launay, N. Quiniou, M. Marcon, J.Y. Dourmad, D. Renaudeau and L. Brossard

Supervised machine learning as a tool to improve farrowing monitoring and stillborn rate in sows 972
C. Teixeira Costa, G. Boulbria, C. Dutertre, C. Chevance, T. Nicolazo, V. Normand, J. Jeusselin and A. Lebret

Use of accelerometers to predict the behaviour of growing rabbits 973
M. Piles, J.P. Sánchez, L. Riaboff, I. David and M. Mora

Use of accelerometry data to detect kidding in goats 973
P. Gonçalves, M.R. Marques, A.T. Belo, A. Monteiro and F. Braz

FTIR milk fatty acids quantification for non-invasive monitoring of rumen health in dairy cows 974
F. Huot, S. Claveau, A. Bunel, D. Warner, D.E. Santschi, R. Gervais and E.R. Paquet

Optimizing breeding performance through algorithmic approaches to maximize meat quality in livestock974
J. Albechaalany, M.P. Ellies-Oury, J.F. Hocquette, C. Berri and J. Saracco

Detection of multiple feeding behaviours in calves using noseband and accelerometer sensors 975
S. Addo, K.A. Zipp, M. Safari, F. Freytag and U. Knierim

Numerical detection of productive anomalies induced by heat stress in dairy cows 975
M. Bovo, M. Ceccarelli, C. Giannone, S. Benni, P. Tassinari and D. Torreggiani

Poster Session 89

Using existing slaughterhouse data to assist detection of boar taint 976
B. Callens, M. Aluwé, R. Klont, M. Bouwknegt and J. Maselyne

Evaluation of SARA risk prediction models based on non-invasive measurements in dairy cows 976
V. Leroux, C. Jaffres, A. Budan and N. Rollet

Session 90. Genetics in horses

Date: Thursday 31 August 2023; 14.00 – 18.00
Chair: Stock / Cox

Theatre Session 90

Use of linear data for characterization and selection of sport horses with highest genetic potential 977
K.F. Stock, A. Hahn, I. Workel and W. Schulze-Schleppinghoff

Effect of mitochondrial genetic variability on performance of endurance horses 977
A. Ricard, S. Dhorne-Pollet, C. Morgenthaler, J. Speke Katende, C. Robert and E. Barrey

Early life jumping traits' potential as proxy for jumping performance in Belgian Warmblood horses 978
L. Chapard, R. Meyermans, W. Gorssen, N. Buys and S. Janssens

Genetic covariance components of conformation, movement and athleticism traits in Irish Sport Horses 978
J.L. Doyle, S. Egan and A.G. Fahey

Start status in Swedish Warmblood horses 979
Y. Blom, S. Eriksson and Å. Gelinder Viklund

Genetic analysis of the precocity potential in trotting races of Spanish Trotter Horses 979
M. Ripollés-Lobo, D.I. Perdomo-González, M.D. Gómez, M. Ligero and M. Valera

Is the ability to race barefoot a heritable trait in Standardbred trotters? 980
P. Berglund, S. Andonov, A. Jansson, T. Lundqvist, C. Olsson, E. Strandberg and S. Eriksson

Searching for genomic regions associated with conformation traits in the Pura Raza Español horse 980
N. Laseca, C. Ziadi, D.I. Perdomo-González, M. Valera, P. Azor, S. Demyda-Peyrás and A. Molina

Successful genotype imputation from medium to high density in Belgian Warmblood horses 981
L. Chapard, R. Meyermans, W. Gorssen, B. Van Mol, F. Pille, N. Buys and S. Janssens

Inbreeding in the Belgian equine warmblood population: current degree and evolution 981
B. Van Mol, H. Hubrechts, R. Meyermans, L. Chapard, W. Gorssens, M. Oosterlinck, N. Buys, F. Pille and S. Janssens

Population structure assessment using genome wide molecular information in Martina Franca donkey 982
V. Landi, E. Ciani and P. De Palo

Unravelling genomic regions with transmission ratio distortion in horse 982
N. Laseca, A. Cánovas, M. Valera, S. Id-Lahoucine, D.I. Perdomo-González, P.A.S. Fonseca, S. Demyda-Peyrás and A. Molina

Microsatellite-based detection of transmission ratio distortion in the Pura Raza Española horse 983
D.I. Perdomo-González, S. Id-Lahoucine, A. Molina, A. Cánovas, N. Laseca, P.J. Azor and M. Valera

Poster Session 90

Estimation of the genetic parameters for temperament in Haflinger horses 983
T.H. Zanon, B. Fürst Waltl, S. Gruber and M. Gauly

Estimation of the genetic propensity to suffer hock osteochondrosis in Pura Raza Española horses 984
M. Ripollés, A. Molina, M. Novales, C. Ziadi, E. Hernández and M. Valera

Genetic characterization of white facial marking in Pura Raza Español horses depending on coat colour 984
A.E. Martínez, M.J. Sánchez Guerrero, A. López, M. Ligero and M. Valera Córdoba

Climate effects on foaling event: genotype-by-environment interactions in horse fertility 985
C. Sartori, E. Mancin, G. Gomez-Proto, B. Tuliozi and R. Mantovani

Session 91. Nitrogen excretion and ammonia emission in pig and poultry production

Date: Thursday 31 August 2023; 14.00 – 18.00
Chair: Millet / Brodziak

Theatre Session 91

Impact of raw and digested pig manure on ammonia volatilization after land application 985
F.M.W. Hickmann, R. Rajagopal, N. Bertrand, I. Andretta, M.-P. Létourneau-Montminy and D. Pelster

Influence of housing manure management on ammonia emissions from broiler and laying hen productions 986
E. Caron, P. Le Bras and M. Hassouna

NH_3, N_2O and CH_4 emission from piggeries: comparison of several manure management techniques 986
N. Guingand and P. Le Bras

Feeding pigs with low-protein diets: impact of pig manure nitrogen content on biogas production 987
F.M.W. Hickmann, I. Andretta, L. Cappelaere, B. Goyette, M.-P. Létourneau-Montminy and R. Rajagopal

A tool to reduce worker exposure from ammonia and particles in swine and poultry housing 987
N. Guingand, S. Lagadec, K. Amin, D. Bellanger, A.L. Boulestreau-Boulay, C. Delaqueze, C. Depoudent, L. Gabriel, E. Koulete, V. Le Gall, P. Lecorguille, L. Leroux, G. Manac'h, S. Roffi and M. Ruch

Methodological redesign of the poultry mass-balance excretion and variability of nitrogen excretion 988
V. Blazy, E. Caron, D.D. Djenontin-Agossou and Y. Guyot

Access to bedding and an outdoor run for growing-finishing pigs and their impact on the environment 988
A.K. Ruckli, S. Hörtenhuber, S. Dippel, P. Ferrari, M. Gebska, J. Guy, M. Heinonen, J. Helmerichs, C. Hubbard, H. Spoolder, A. Valros, C. Winckler and C. Leeb

Reduction of protein and potassium to improve welfare and environmental footprint in broilers 989
T. De Rauglaudre, B. Méda, W. Lambert, S. Fournel and M.P. Létourneau-Montminy

Effects of reducing protein content in broiler diets on environmental impacts 989
J. Gickel, V. Wilke, C. Ullrich and C. Visscher

Effect of reduced-crude protein diets on performance, meat yield, and nitrogen production in broiler 990
C. Gayrard, T. Wise, J.D. Davis, J.C. De Paula Dorigam, V.D. Naranjo and W.A. Dozier, Iii

Effect of reducing dietary protein content on performance and environmental impact of pig production 990
E. Gonzalo, A. Simongiovanni, C. De Cuyper, B. Ampe, M. Aluwe, W. Lambert and S. Millet

Impact of genotype×feed-interactions in nitrogen- and phosphorus-reduced ration on of fattening pigs 991
C. Große-Brinkhaus, B. Bonhoff, I. Brinke, E. Jonas, S. Kehraus, K.H. Südekum and E. Tholen

Pig performance, carcass composition and meat quality can be maintained with high CP reduction 991
L. Cappelaere, F. Garcia-Launay, W. Lambert, A. Simongiovanni and M.-P. Létourneau-Montminy

Poster Session 91

Alpha mannan polysaccharide supplement increase growth performance and reduce gas emission in pig 992
J.H. Song, C.B. Lim, M.D.M. Hossain, S. Biswas, J.S. Yoo, I.H. Kim and Q.Q. Zhang

Characteristic of odour substances from broiler farms 992
S.Y. Seo, J.S. Park, S.Y. Park and M.W. Jung

Environmental assessment of layer-type male chicks breeding 993
E. Dubois and M. Quentin

Session 92. Inflammation and energy metabolism in young and adult livestock

Date: Thursday 31 August 2023; 14.00 – 18.00
Chair: Sauerwein

Theatre Session 92

Milk metabolites differ with feed efficiency during early lactation but not during feed restriction 993
J. Pires, T. Larsen, S. Bes, I. Constant, D. Roux, M. Tourret, A. De La Torre, I. Ortigues-Marty, F. Blanc and I. Cassar-Malek

Dairy cow inflammatory status is modulated by physiological stage and feed restriction 994
C. Delavaud, A. De La Torre, D. Durand, S. Bes, D. Roux, A. Thomas, M. Tourret, I. Ortigues-Marty, I. Cassar-Malek, M. Bonnet and J. Pires

Energy balance clusters in relation to metabolic and inflammatory status and disease in dairy cows 994
J. Ma, A. Kok, R. Bruckmaier, E. Burgers, R. Goselink, J. Gross, T. Lam, A. Minuti, E. Saccenti, E. Trevisi and A. Van Knegsel

Effects of parity on metabolism, redox status and cytokines in early lactating dairy cows 995
A. Corset, A. Boudon, A. Remot, S. Philau, P. Poton, O. Dhumez, B. Graulet, P. Germon and M. Boutinaud

Tissue distribution and pharmacological characterization of bovine free fatty acids-sensing GPCRs 995
T.C. Michelotti, M. Bonnet, V. Lamothe, S. Bes and G. Durand

Maternal nutrition carry-over effects on beef cow colostrum but not on milk fatty acid composition 996
N. Escalera-Moreno, B. Serrano-Pérez, E. Molina, L. López De Armentia, A. Sanz and J. Álvarez-Rodríguez

Different types and doses of colostrum to optimize the passive immune transfer and health in lambs 996
A. Belanche, F. Canto and O. Calisici

Oxidative status in female Holstein calves fed with or without transition milk 997
C.S. Ostendorf, M. Hosseini Ghaffari, B. Heitkönig, C. Koch and H. Sauerwein

Effects of concentrate feeding and hay quality on adipogenesis and inflammation in dairy calves 997
R. Khiaosa-Ard, A. Sener-Aydemir, S. Kreuzer-Redmer and Q. Zebeli

Synergistic benefits of marine derived bioactives to maintain the gut barrier in in an *ex vivo* model 998
M.A. Bouwhuis, I. Nooijen, E. Van Der Steeg and S. O'Connell

Effects of milk replacer diets on the intestinal development of sucked piglets 998
G.Y. Duan, C.B. Zheng, J. Zheng, P.W. Zhang, M.L. Wang, J.Y. Yu, B. Cao, M.M. Li, F. Cong, Y.L. Yin and Y.H. Duan

The impact of early thermal manipulation on the hepatic energy metabolism of mule duck 999
C. Andrieux, M. Morisson, V. Coustham, S. Panserat and M. Houssier

Poster Session 92

Feed efficiency and responses of plasma and milk isotopic signatures in Charolais beef cows 999
A. De La Torre, J. Pires, I. Cassar-Malek, I. Ortigues-Marty, F. Blanc, L. Barreto-Mendes, G. Cantalapiedra-Hijar and C. Loncke

How the composition of Holstein cow colostrum differs according to the immunoglobulins G level 1000
A. Goi, A. Costa and M. De Marchi

An educational kit for effective prevention and management of energy deficit in dairy cows 1000
M. Gelé, M. Boutinaud, A. Bouqueau, M. Marguerit, J. Jurquet, Y. Le Cozler and J. Guinard-Flament

Inflammation status weakly modulated by short-term feed restriction in lactating beef cows 1001
C. Delavaud, J. Pires, M. Barbet, I. Ortigues-Marty, I. Cassar-Malek, M. Bonnet and A. De La Torre

Metabolic profiles of grazing dairy cows with or without feed restriction during early lactation 1001
A.L. Astessiano, M. Garcia-Roche, D. Custodio, G. Ortega, P. Chilibroste and M. Carriquirry

Hepatic mitochondrial function in grazing dairy cows with or without feed restriction 1002
M. García-Roche, D. Custodio, A. Astessiano, G. Ortega, P. Chilibroste and M. Carriquiry

On-farm emergency slaughtered dairy cows: causes and haematological biomarkers 1002
F. Fusi, I.L. Archetti, S.M. Chisari, V. Lorenzi, C. Montagnin, R. Salonia, L. Bertocchi and G. Cascone

The effect of early feed restriction of ewe lambs on milk miRNAome of the filial generation 1003
A. Martín, F. Ceciliani, F.J. Giráldez, R. Calogero, C. Lecchi and S. Andrés

Quercetin supplementation in perinatal sows diet influencing suckling piglets and its mechanism 1003
Y. Li, Q.L. Yang, Y.X. Fu, S.S. Zhou, J.Y. Liu and J.H. Liu

Development of muscle injury in weanling piglets under chronic immune stress 1004
Y. Duan

Session 93. Recent advances on *in vitro* systems (from cell culture to in vitro digestion methods): relevance as alternatives to animal experimentation for livestock research

Date: Thursday 31 August 2023; 14.00 – 18.00
Chair: Louveau

Theatre Session 93

invited The great power of being tiny: *in vitro* gut systems in livestock research 1004
S.K. Kar

Comparison of adult intestinal stem cell derived organoids between different ages and sex in pigs 1005
R.S.C. Rikkers, O. Madsen, S.K. Kar, L. Kruijt, A.A.C. De Wit, E. Van Der Valk, L.M.G. Verschuren, S. Verstringe and E.D. Ellen

Butyric glycerides directly and indirectly enhance chicken enterocyte resistance to pathogens 1005
A. Mellouk, N. Vieco-Saiz, V. Michel, O. Lemâle, T. Goossens and J. Consuegra

invited Spheroid and organoid models to study animal reproduction 1006
K. Reynaud, M. Ta, C. Pucéat, M. Billet, C. Mahé, L. Schmaltz, L. Laffont, P. Mermillod and M. Saint-Dizier

3D-cell culture systems; a breakthrough to improve the human and animal disease research 1006
E.S. Smirnova, K.A. Adhimoolam and T.M. Min

Slurry chemical characteristics and ammonia emissions assessed by different methods 1007
L. Sarri, E. Fuertes, E. Pérez-Calvo, A.R. Seradj, R. Carnicero, J. Balcells and G. De La Fuente

In vitro protein fractionation methods for ruminant feeds 1007
B.Z. Tunkala, K. Digiacomo, P.S. Alvarez Hess, F.R. Dunshea and B.J. Leury

Storage conditions of rumen inoculum: impact on gas productions in mini dual flow fermenters 1008
V. Berthelot, M. Charef-Mansouri, A.-M. Davila and L.P. Broudiscou

Dynamics of rumen microbiome and methane emission during *in vitro* rumen fermentation 1008
R. Dhakal, R. Sapkota, P. Khanal, A. Winding and H.H. Hansen

Effects of *Asparagopsis armata* inclusion on methane production in semi-continuous fermenters 1009
P. Romero, S.M. Waters, D.R. Yañez-Ruiz, A. Belanche and S.F. Kirwan

Calcium propionate mitigated adverse effects of incubation temperature shift on *in vitro* fermentation 1009
T. He, S. Long and Z. Chen

Poster Session 93

In vitro generation of ovine monocyte-derived macrophages for SRLV infection studies 1010
E. Grochowska, M. Ibrahim, M. Wu and K. Stadnicka

Resistance of probiotics to antibiotics and antagonism to poultry pathogens 1010
N. Akhavan, K. Stadnicka, D. Thiem and K. Hrynkiewicz

Metabolic footprint of prebiotics and probiotics in Chick8E11 and Caco-2 intestinal cell lines 1011
S. Zuo, W. Studziński, K. Stadnicka and P. Kosobucki

Cellular imagery evaluation of butyric acid biological activities in Caco-2 cell line 1011
D. Gardan-Salmon, B. Saldaña, J.I. Ferrero and M. Arturo-Schaan

Effect of silage sample preparation on rumen fermentation (*in vitro*) 1012
J.A. Huntington, T. Snelling, L.A. Sinclair, H. Warren, D.A. Galway and D.R. Davies

Repeatability of fermentation kinetics using *in vitro* gas fermentation 1012
B. Jantzen and H.H. Hansen

Evaluation of rumen bypass protein in processed soybean meal products 1013
A. Chariopolitou, A. Plomaritou, A. Tzamourani, K. Droumtsekas and A. Foskolos

Effect of rumen inoculum on predicted *in vivo* methane production from barley and oats 1013
P. Fant, M. Ramin and P. Huhtanen

Comparison of *in vitro* and *in vivo* methane production in dairy cows 1014
D.W. Olijhoek, É. Chassé, M. Battelli, M.V. Curtasu, M. Thorsteinsson, M.H. Kjeldsen, W.J. Wang, G. Giagnoni, M. Maigaard, C.F. Børsting, M.R. Weisbjerg, P. Lund and M.O. Nielsen

Session 94. Free communications in animal health and welfare

Date: Thursday 31 August 2023; 14.00 – 18.00
Chair: Boyle / Calderon Diaz

Theatre Session 94

What impairs laying hens' foot health? A retrospective German study 1014
N. Volkmann, A. Riedel, N. Kemper and B. Spindler

Pododermatitis in broilers: prevalence and risk factors present in reused beddings 1015
J. Montalvo, M. Cevallos and M. Cisneros

Digital dermatitis in French young bulls fattening farms 1015
A. Waché, M. Petitprez, E. Dod Ioan, M. Delacroix and C. Guibier

Prevalence and factors associated with teat-end hyperkeratosis in dairy cows 1016
F.G. Silva, C. Antas, A.M.F. Pereira and C. Conceição

Milk calcium content to detect hypocalcaemia in dairy cows at the onset of lactation 1016
T. Aubineau, R. Guatteo and A. Boudon

PRRSV stabilisation programs in French farrow-to-finish farms: a way to reduce antibiotic use 1017
C. Teixeira Costa, G. Boulbria, V. Normand, C. Chevance, J. Jeusselin, T. Nicolazo and A. Lebret

Monitoring of antimicrobial use on French pig farms from 2010 to 2019 using INAPORC panels 1017
A. Poissonnet, I. Correge, C. Chauvin and A. Hemonic

Digital drug registration and its relation to veterinary slaughter findings 1018
H. Görge, I. Dittrich, N. Kemper and J. Krieter

A critical lens to regulations around veterinary antimicrobial use across countries and species 1018
G. Olmos Antillón and I. Blanco-Penedo

An approach of the place of animal health and role of veterinarian in sustainable farming systems 1019
A. Scholly-Schoeller and G. Brunschwig

Developing an animal welfare benchmarking framework for Australian lot-fed cattle 1019
T. Collins, E. Taylor, A. Barnes, E. Dunston-Clarke, D. Miller, D. Brookes, E. Jongman and A. Fisher

H2020 mEATquality: on-farm animal welfare assessment in slaughter pigs 1020
T. Rousing, L.D. Jensen, M.L.V. Larsen and L.J. Pedersen

Variability in pig hair cortisol concentrations at the end of the fattening period 1020
P. Levallois, M. Leblanc-Maridor, S. Gavaud, B. Lieubeau, G. Morgant, C. Fourichon, J. Hervé and C. Belloc

Identification and characterization of IUGR piglets 1021
C. Ollagnier, R. Ruggeri, J. Bellon and G. Bee

Poster Session 94

Investigating the association between sole ulcers and sole temperature in dairy cattle using IRT 1021
A. Russon, A. Anagnostopoulos, M. Barden, B. Griffiths, C. Bedford and G. Oikonomou

Osteopathic manipulative treatment: a complementary approach to promote milk production in cows 1022
C. Omphalius, A. Hardouin, M. Launay Ventelon and S. Julliand

The relationship between stress levels and skin allergy or atopy problem by analysing hair cortisol 1022
G.W. Park, M. Ataallahi and K.H. Park

SECURIVO: self-assessment tools for biosecurity in veal calves' farms 1023
M. Chanteperdrix, M. Drouet, M. Mounaix, M. Tourtier, D. Le Goic, C. Jaureguy, P. Briand, M. Coupin, A. Hemonic and N. Rousset

Temporally synchronized ruminal vs reticular fluid pH and short chain fatty acids in dairy cows 1023
A. Kidane, K.S. Eikanger and M. Eknæs

Antimicrobial use and main causes of treatment in Italian buffalo farms 1024
G. Di Vuolo, F. Scali, C. Romeo, V. Lorenzi, C.D. Ambra, M. Serrapica, G. Cappelli, F. Fusi, E. De Carlo, G.L. Alborali, L. Bertocchi and D. Vecchio

Herd factors associated with levels of parasitism in alternative pig farms 1024
M. Delsart, N. Rose, B. Dufour, J.M. Répérant, R. Blaga, F. Pol and C. Fablet

Biosecurity and Animal Welfare relationship on buffalo farms through ClassyFarm assessments 1025
D. Vecchio, G. Santucci, V. Lorenzi, C. Caruso, C.D. Ambra, M. Serrapica, G. Di Vuolo, G. Cappelli, F. Fusi, E. De Carlo, C. Romeo, F. Scali, G.L. Alborali, A.M. Maisano and L. Bertocchi

Immune function of pre-weaned calves fed a fortified milk replacer under heat stress conditions 1025
A.A.K. Salama, S. Serhan, L. Ducrocq, M. Biesse, A. Joubert and G. Caja

The effect of udder health on milk composition 1026
Z. Nogalski and M. Sobczuk-Szul

Session 95. Happy Delivering innovative and holistic milk based monitoring and decision-making PLF tools – collaboration with HappyMoo

Date: Thursday 31 August 2023; 14.00 – 18.00
Chair: Gengler / Knijn

Theatre Session 95

An attempt to predict dairy cows chronic stress biomarkers using milk MIR spectra 1026
C. Grelet, H. Simon, J. Leblois, M. Jattiot, C. Lecomte, R. Reding, J. Wavreille, E.J.P. Strang, F.J. Auer, Happymoo Consortium and F. Dehareng

Prediction of body condition score for the entire lactation in Walloon Holstein cows 1027
H. Atashi, J. Chelotti and N. Gengler

Conditions to develop successful mid-infrared (MIR) based BCS and BCS change predictions 1027
J. Chelotti, H. Atashi, C. Grelet, M. Calmels, J. Leblois and N. Gengler

Large-scale analysis of chronic stress in dairy cows using hair cortisol and blood fructosamine 1028
H. Simon, C. Grelet, S. Franceschini, H. Soyeurt, J. Leblois, M. Jattiot, C. Lecomte, R. Reding, J. Wavreille, E.J.P. Strang, F.J. Auer and F. Dehareng

Calculation of dry matter intake and energy balance based on MIR spectral data at European level 1028
V. Wolf, L. Dale, M. Gelé, U. Schuler, U. Müller, N. Gengler, J. Leblois and M. Calmels

Validation of HappyMoo MIR energy balance models on external datasets with feeding restriction 1029
M. Calmels, J. Pires, M. Boutinaud, A. Leduc, C. Leroux, J. Chelotti, L. Dale, C. Grelet, A. Tedde, J. Leblois and M. Gelé

Genetic analyses of principal components of milk mid-infrared spectra from Holstein cows 1029
Y. Chen, P. Delhez, H. Atashi, H. Soyeurt and N. Gengler

MastiMIR – MIR prediction for udder health 1030
L.M. Dale, A. Werner, E.P.J. Strang, Happymoo Consortium and J. Bieger

Prediction of lameness and hoof lesions using MIR spectral data 1030
M. Jattiot, L.M. Dale, M. El Jabri, J. Leblois, M.-N. Tran and HappyMoo Consortium

MIR spectral prediction based on heat stress in dairy cattle 1031
L.M. Dale, M. Jattiot, A. Werner, E.J.P. Strang, P. Lemal, Happymoo Consortium, Klimaco Consortium and N. Gengler

Use of MIR spectra-based indicator for genetic evaluation of heat stress in dairy cattle 1031
P. Lemal, L.M. Dale, M. Jattiot, J. Leblois, M. Schroyen and N. Gengler

A unique overview of enteric methane emissions by dairy cows in Bretagne, France 1032
S. Mendowski, O. Garcia, C. Bruand, M. Tournat, T. Viot and L. Meriaux

Genetic parameters analysis of milk citrate for Holstein cows in early lactation 1032
H. Hu, Y. Chen, C. Grelet and N. Gengler

Poster Session 95

Milk fluctuations in daily milk yield associated with diseases in Chinese Holstein cattle 1033
A. Wang, D.K. Liu, C. Mei and Y.C. Wang

Session 96. ERANET-funded 'Grass To Gas' and EU-funded 'SMARTER' projects

Date: Thursday 31 August 2023; 14.00 – 18.00
Chair: Conington / Vouraki

Theatre Session 96

Impact of genomic selection for methane emissions in a high performance sheep flock 1033
S.J. Rowe, T. Bilton, P. Johnson, S. Hickey, A. Jonker, N. Amyes, K. McRae, S. Clarke and J. McEwan

Improving feed efficiency in meat sheep increases CH_4 emissions measured indoor or on pasture 1034
F. Tortereau, J.-L. Weisbecker, C. Coffre-Thomain, Y. Legoff, D. François, Q. Le Graverand and C. Marie-Etancelin

Effects of sire and diet on rumen volume and relationships with feed efficiency 1034
N.R. Lambe, A. McLaren, K.A. McLean, J. Gordon and J. Conington

Evaluating the effect of herbage composition on methane output in sheep 1035
F.M. McGovern, P. Creighton, E. Dunne and S. Woodmartin

Dry matter intake across life stages in sheep 1035
E. O'Connor, N. McHugh, E. Dunne, T.M. Boland and F.M. McGovern

Selecting feed-efficient sheep with concentrates alters their efficiency with forages and behaviour 1036
C. Marie-Etancelin, J.L. Weisbecker, D. Marcon, L. Estivalet, Q. Le Graverand and F. Tortereau

Genome-wise association study of footrot and mastitis in UK Texel sheep 1036
K. Kaseja, S. Mucha, J. Yates, E. Smith, G. Banos and J. Conington

Genetic parameters of nematode resistance in dairy sheep 1037
B. Bapst, K. Schwarz, S. Thüer and S. Werne

Contrasting genetic resistance to GIN on growth performance and feed efficiency of Corriedale lambs 1037
E.A. Navajas, G. Ciappesoni and I. De Barbieri

Poster Session 96

Australian Merino: animal welfare and resilience in extensive systems 1038
M. Del Campo, J.L. De Araújo Pimenta, I. De Barbieri, P. Lorenze, F. Rovira and J.M. Soares De Lima

Prediction of feed efficiency related traits from plasma NMR spectra 1038
A. Marquisseau, F. Tortereau, N. Marty-Gasset, C. Marie-Etancelin and Q. Le Graverand

Metabolism in lambs from two feed-efficiency genetic lines subjected to different early rearing practices 1039
G. Cantalapiedra-Hijar, M.M. Milaon, S. Parisot, C. Durand, M. Vauris, F. Tortereau and C. Ginane

Genetic link between fertility and resilience in sheep and goat divergent selection experiments 1039
R. Rupp, C. Oget-Ebrad, S. Parisot, T. Fassier, G. Tosser Klopp and S. Freret

Phenotypic and genetic variability of health and welfare traits in French dairy goats 1040
I. Palhiere, A. Bailly-Salins, A. Gourdon, M. Chassier, R. De Cremoux, M. Berthelot and R. Rupp

Assessment of phenotypic and genetic variability of rumen temperatures in goats 1040
I. Palhiere, C. Huau, T. Fassier, R. Rupp and L. Bodin

Heritability of novel metabolite-based resilience biomarkers in dairy goat 1041
M. Ithurbide, T. Fassier, M. Tourret, J. Pires, T. Larsen, N.C. Friggens and R. Rupp

Rumen size of sheep: difference between a modern and a native Norwegian sheep breed 1041
B.A. Åby, M.A. Bhatti and G. Steinheim

Diurnal feed intake pattern of two sheep breeds fed different silage qualities 1042
I. Dønnem, B.A. Åby and G. Steinheim

Session 01 Theatre 1

The real place of livestock in GHG emissions and their control for climate change

K. Johnson
Washington State University, Animal Sciences, 126 ASLB, P.O. Box 646351, Pullman, WA 99164-6351, USA; johnsoka@wsu.edu

Climate change presents multiple challenges to the livestock industry including the need to examine animal and environmental management practices to enhance efficiency, reduce greenhouse gas emissions from animals and manure, reduce water and other resource use, and pressure to eliminate meat and livestock products from diets. This pressure is because methane is a relatively short-lived greenhouse gas and reductions in methane emissions now will benefit concentrations in 10 years. This argument does not necessarily take into consideration the role of ruminant livestock across many different cultures across the world and the benefits grazing livestock have to regenerative agriculture. Additionally, the role and benefits livestock provide through up-cycling byproducts of human consumption can be lost in the discussion but will have significant environmental consequences if they are not used. There is a recent body of literature that describes different alternatives for livestock management that address some of the concerns while maintaining the positive attributes ruminants provide. Examination of the ideas in that literature and associated data can benefit livestock managers, scientists who study livestock and society in general as important public policy decisions are debated and enacted.

Session 01 Theatre 2

Global emissions from livestock systems: updated baselines, projections, and mitigation options

D. Wisser, G. Tempio, A. Falcucci and G. Cinardi
Food and Agriculture Organization of the United Nations (FAO), Animal Production and Health Division, Viale di Terme di Caracalla, 00153 Roma, Italy; dominik.wisser@fao.org

Livestock provide valuable nutritional benefits as well as supporting livelihoods and the resilience of families and communities. Though growth has slowed, demand for animal products is projected to grow by 34% by 2050 globally, mostly as a result of demand growth in low and middle income countries. Livestock systems contribute to a considerable fraction of all anthropogenic greenhouse gas emissions but can contribute to climate change mitigation through gains in efficiency along the production chain as well as through specific interventions aiming at reducing greenhouse gases, in particular methane. We present an updated assessment of global emissions associated with livestock using FAO's Global Livestock Environmental Assessment Model (GLEAM). GLEAM adopts a Tier 2 methodology to estimate detailed and spatially explicit emissions for a specified base year and uses a lifecycle assessment (LCA) approach, including both direct emissions from animals as well indirect emissions upstream and downstream. This approach differentiates key stages within livestock agrifood systems, such as land use change, feed production, processing and transport; animal production, animal feeding and manure management; and the processing and transport of products. We will present an overview of the approach and updated estimates of emissions from all species and production systems globally and highlight major sources. These results are made available through a recently released dashboard (https://www.fao.org/gleam/dashboard/en/). Using a business-as-usual scenario, we will then illustrate pathways of future emissions taking into account improvements in productivity as well as specific mitigation interventions. Results suggest that emissions from livestock systems currently account for about 11% of all anthropogenic GHG emission but considerable reductions in emission intensity (emissions per unit of output) and in total emissions can be achieved through efficiency gains along the production chain through interventions in animal health, feed, breeding, and others, while meeting projected future demand for animal products.

Session 01 Theatre 3

Genetic control of ruminant methane emissions in livestock

A.E. Van Breukelen[1], M.N. Aldridge[1], R.F. Veerkamp[1] and Y. De Haas[2]
[1]Wageningen University & Research, Animal Breeding and Genomics, P.O. Box 338, 6700 AH Wageningen, the Netherlands, [2]Wageningen University & Research, Animal Health & Welfare, P.O. Box 338, 6700 AH Wageningen, the Netherlands; anouk.vanbreukelen@wur.nl

Animal agriculture is a large contributor to anthropogenic greenhouse gas emissions. The main source is enteric methane (CH_4) produced by ruminants, with emissions increasing due to growing livestock numbers and increased productivity of individual animals. By mitigating CH_4 emissions, short-term wins can be made towards meeting the 1.5 °C Paris agreement temperature target, due to the short lifetime of CH_4. There is a world-wide interest in applying animal breeding as a mitigation strategy, because the effect is permanent, cumulative, and cost-effective. Initial studies have shown that there is natural variation in CH_4 emissions of cows, that it is heritable, and that theoretical genetic gains will have a significant mitigation effect. Nevertheless, practical implementations that mitigate CH_4 emissions in breeding programs are still limited, which is largely because enteric CH_4 emissions of individual cows are challenging and costly to record. Although phenotyping with sniffers makes large scale recording easier, this technology is still at its infancy and therefore genetic correlations with key production traits are yet still unknown or uncertain, whereas they are essential to ensure that all trait changes are in the desired direction. Furthermore, incentives to apply CH_4 in breeding programs need to be found as the economic value of mitigating CH_4 emissions is uncertain. Nonetheless, many countries are working towards applications in national breeding programs, or have recently started implementing breeding values for CH_4. First implementations confirmed that animal breeding could deliver a substantial mitigation towards Paris agreement targets, ideally when combined with other dietary and management approaches.

Session 01 Theatre 4

Effective strategies to mitigate enteric methane emissions by ruminants

C. Arndt
ILRI, Naivasha Rd, 00100 Nairobi, Kenya; claudia.arndt@cgiar.org

Agricultural methane emissions must be decreased by 11 to 30% of the 2010 level by 2030 and by 24 to 47% by 2050 to meet the 1.5 °C target. A meta-analysis identified strategies to decrease product-based (PB; CH_4 per unit meat or milk) and absolute (ABS) enteric CH_4 emissions while maintaining or increasing animal productivity (AP; weight gain or milk yield). Next, the potential of different adoption rates of one PB or one ABS strategy to contribute to the 1.5 °C target was estimated. The database included findings from 430 peer-reviewed studies, which reported 98 mitigation strategies that can be classified into three categories: animal and feed management, diet formulation, and rumen manipulation. A random-effects meta-analysis weighted by inverse variance was carried out. Three PB strategies – namely, increasing feeding level, decreasing grass maturity, and decreasing dietary forage-to-concentrate ratio – decreased CH_4 per unit meat or milk by on average 12% and increased AP by a median of 17%. Five ABS strategies – namely CH_4 inhibitors, tanniferous forages, electron sinks, oils and fats, and oilseeds – decreased daily methane by on average 21%. Globally, only 100% adoption of the most effective PB and ABS strategies can meet the 1.5 °C target by 2030 but not 2050, because mitigation effects are offset by projected increases in CH_4 due to increasing milk and meat demand. Notably, by 2030 and 2050, low- and middle-income countries may not meet their contribution to the 1.5 °C target for this same reason, whereas high-income countries could meet their contributions due to only a minor projected increase in enteric CH_4 emissions.

Session 01 — Theatre 5

Tannins, legumes and methane production by ruminants

H.H. Lardner
University of Saskatchewan, Department of Animal and Poultry Science, 51 Campus Drive, Saskatoon, Saskatchewan, Canada, S7N 5A8, Canada; bart.lardner@usask.ca

Methane (CH_4) is a greenhouse gas (GHG) with a global warming potential 28 times that of carbon dioxide. Due to methane's contribution to the agriculture emission footprint, investigating mitigation strategies are needed. Condensed tannins (CT) are secondary plant polyphenol compounds, when fed at low levels (5-10 g/kg DM) to ruminants can increase rumen by-pass protein, enhance animal performance and reduce enteric methane production. Several CT legumes in pasture grazing systems include sainfoin (*Onobrychis viciifolia* Scop.), white prairie clover (*Dalea candida* Michx. ex Willd), purple prairie clover (*Dalea purpurea* Vent.), and birdsfoot trefoil (*Lotus corniculatus*). Fermentation of carbohydrates in ruminant diets results in volatile fatty acids, microbial protein synthesis, accompanied by the release of gases such as carbon dioxide and CH_4. Methane is synthesized by anaerobic archaea coupled with bacteria, protozoa and fungi in the rumen ecosystem. Several studies have reported the effects of tannins as alternative feed additives to modify rumen fermentation, improve animal production and mitigate CH_4 production. Opportunities for reducing rumen methanogenesis through dietary inclusion of tannins will be discussed.

Session 01 — Theatre 6

Enteric methane reduction in ruminants by early-life interventions

D.P. Morgavi
INRAE, Centre Auvergne-Rhône-Alpes, Theix, 63122 Saint Genès Champanelle, France; diego.morgavi@inrae.fr

Modulating the microbial colonisation of the developing rumen towards communities that produce less methane is gaining attention as a way to reduce the carbon hoofprint of ruminants. The rationale is that in the first days and weeks after birth, when the gastrointestinal tract is colonised and microbial consortia are installed, inducing changes in the rumen microbiota can have a lasting effect later in life. The rumen microbiota in early life can be modulated by management and targeted interventions that favour phenotypes of interest, such as decreased methane emissions. Modulation options include: allowing contact with dams or mature conspecifics with a desirable microbiota, inoculation with gastrointestinal contents (engraftment) obtained from mature conspecifics, diet and supplementation with direct-fed microbials, prebiotics and feed additives. To modulate methane emissions, the use of nutritional strategies such as feed additives and supplements have been most studied. Two different types of approach can be distinguished in terms of their mechanisms of action: one using feed supplements or additives that affect the microbial ecosystem in a general way, and the other using specific inhibitors of methanogens or the methanogenesis pathway. The presentation will provide an update on the latest information in this field, focusing on its potential and the challenges ahead.

Session 01 Theatre 7

Sustainable intensification of animal production with silvopastoral systems

J. Chára
CIPAV, Centre for Research on Sustainable Agriculture, Carrera 25 # 6-62, 760046. Cali, Colombia; julian@fun.cipav.org.co

Livestock production has the challenge of generating more food and protein to fulfil the global growing demand and, at the same time, reducing its climatic and environmental footprint. Animal production occupies a great portion of the ice-free land both directly in grazing areas, and indirectly to generate feed and other inputs. Silvopastoral systems (SPS) have been proposed as part of nature-based solutions to reverse the negative environmental impact of cattle ranching and increase animal production and economic performance at the same time. SPS are agroforestry arrangements that combine trees, shrubs, forage production and animal grazing on the same land. The trees in SPS provide shade, shelter, and fodder for animals, and generate income from the sale of products, such as fruits, nuts or timber. SPS produce more edible dry matter and nutrients per hectare and increase milk or meat production while reducing the need of chemical fertilizers and concentrate feeds, thus improving profitability. In addition to the improved production, the presence of shrubs and trees in these arrangements have demonstrated effects on biodiversity by creating more complex habitats for wild animals and plants, harbouring a richer soil biota and increasing landscape connectivity. In farmed landscapes, SPS provide food and cover for birds and other organisms, serving as wildlife corridors where unique species assemblages can be found. The tree component has also demonstrated effects on soil physical, chemical and microbiological properties since they provide more layers of vegetation that transform solar energy into biomass that is deposited on the soil in the form of leaves, branches, fruits, legumes and exudates. They contribute also to increase carbon sequestration, reduce enteric methane emissions due to improved nutrition and the presence of fodder legumes, and reduce GHG emission from soil due to the reduction of fertilizer use and enhanced soil microbiota. SPS have gained interest in several parts of the world where they are used for improving animal welfare and comfort of livestock combined with fruit or timber production. The improved efficiency of the system provides an opportunity to release land for restoration and production of other crops thus contributing to sustainability.

Session 01 Theatre 8

Sustainability of low-input livestock systems

A. Mottet
FAO, Viale delle Terme di Caracalla, 00153, Italy; anne.mottet@fao.org

Livestock farming is found in most ecosystems around the world. Thanks to the diversity of animal genetic resources, livestock farmers have been able to live in areas where there is no alternative livelihood. Hundreds of millions of families rely on small-scale and low-input livestock production and pastoralism. They provide food and essential nutrition, manure, animal traction, transport and income. They have a key role to play in particular for women's empowerment and youth employment, both in rural and urban areas. While demand for meat, milk and eggs is increasing, low-input livestock systems have a key role to play for the sustainable development of the sector. Five main areas of environmental priority and inputs can be considered: biodiversity, water, land use, soils and climate change. The type of feeding system is arguably one of the most important drivers for the sustainability of livestock in all five areas. Livestock have the ability to convert grass and swill into protein and the efficiency in this conversion varies between production systems. For example, grazing systems in non-OECD countries need an average 200 g of edible plant protein to produce 1 kg of edible animal protein. The entire livestock sector uses about 2.5 billion hectares, 77% of which are grasslands, and a large part of this can be considered non-cultivable and therefore only valuable if grazed by animals. While low-input livestock systems contribute to GHG emissions, they also have significant potential for climate change mitigation through the adoption of best practices for the reduction of emissions, in feed management but also herd and manure management. However, while most of the sector's growth has so far benefited rather large-scale production systems, access to technologies, services and markets is essential if small holders are to make a bigger contribution. Targeted innovative public policies are required to enhance this access, and social sustainability should be considered as important as economic and environmental sustainability in this development.

Agroecological transformation of tropical livestock production through integrating improved forages

A.M.O. Notenbaert
Alliance of Bioversity and CIAT, Shinyalu Road 88, 00100, Kenya; a.notenbaert@cgiar.org

Livestock production systems play a crucial role in meeting the global demand for animal protein, but they also contribute to global warming, deforestation, biodiversity loss, water use, pollution and land/soil degradation. This paper argues that, although the environmental footprint of livestock production presents a real threat to planetary sustainability, also in the global south, this is highly contextual. Under certain context-specific management regimes livestock can deliver multiple benefits for people and planet. We provide evidence that a move towards sustainable livestock production is possible and could mitigate negative environmental impacts and even provide critical ecosystem services, such as improved soil health, carbon sequestration and enhanced biodiversity on farms. To facilitate the agroecological transformation, it is suggested that cultivated forages – consisting of grasses, legumes, and trees – many of which have been enhanced through selection or breeding, be utilized within integrated crop-tree-livestock systems. However, to realize this, a multi-stakeholder approach is needed, involving farmers, researchers, policymakers, and consumers.

Livestock-wildlife conflicts: a struggle for land use and conservation in the U.S. Mountain West

J.B. Taylor and H.N. Wilmer
Agricultural Research Service, U.S. Department of Agriculture, 19 Office Loop, Dubois, Idaho 83423, USA; bret.taylor@usda.gov

The upper Mountain West of the United States (U.S.) is home to nearly one-third of the national sheep inventory and generates one-third and one-half of the total U.S. lamb crop and wool revenue, respectively. Unfortunately, the Mountain West breeding ewe inventory has declined 37% since 2002 compared with a 26% decline experienced in the remaining states. To date, this sheep depopulation trend has not slowed, indicating obstacles unique to the Mountain West which are not experienced elsewhere. Most sheep production in the Mountain West is transhumant and largely depends on grazing access to U.S. government-owned ('public') rangelands. These rangelands consist of mainly indigenous vegetation, are managed as natural ecosystems, and provide distinct, critical ecosystem services. Specifically, of great interest to the public, is the provisioning of abundant wildlife habitat. However, over the last 100 years, extensive conversion of rangelands to croplands, exurban or energy development, and fire suppression have resulted in a substantial loss of wildlife habitat, thus creating competing ideologies for how remaining rangelands should be managed with regard to wildlife conservation. Some wildlife-focused paradigms embraced by a vocal public minority explicitly exclude livestock grazing on all public lands. Such exclusion negatively impacts the operational capacity of Mountain West sheep ranches and land management capability of government agencies, thus posing serious threats to the U.S. sheep industry and its contributions towards national food security, rural community viability, and rangeland conservation. Unfortunately, in the debate surrounding public land use and wildlife-livestock conflicts, a common foe to all – climate change – is somewhat overlooked. Accordingly, opposing groups must come together to consider the 'real' issues and collectively work towards developing agroecological solutions that perpetuate resilient, multi-use rangelands in U.S. Mountain West.

Session 01 Theatre 11

Community-based breeding programs (CBBP): platforms for genetic improvement and biological discovery

J.M. Mwacharo[1,2], A. Haile[2] and B. Rischkowsky[2]
[1]SRUC and CTLGH, Roslin Institute Building, Midlothian, EH25 9RG, United Kingdom, [2]ICARDA, Small Ruminant Genomics, P.O. Box 5689, Ethiopia; j.mwacharo@cgiar.org

One major constraint facing smallholder farmers in developing countries has been low farmgate productivity attributed to lack of fit-for-purpose breeding programs and technical knowhow. Scientific discovery in smallholder low-input systems has also lagged behind due to dearth of quality data. Conventional breeding programs with centralised nucleus flocks have been implemented as a mitigating strategy but have had limited or no success due to small flock sizes and limited resources, while providing data only from the nucleus flocks. CBBPs are low-cost operations that are implemented by farmers who set their breeding objectives collectively and have proved to be a sustainable alternative to conventional breeding and are gaining popularity among smallholder farmers in developing countries. CBBPs have resulted in increased productivity and profitability in the target breeds without undermining their resilience and genetic integrity, and in household income in the absence of expensive interventions. A key element of CBBPs is that farmers are trained on better and alternative selection approaches, farmers are organised in groups and pool their flocks to broaden the genetic base, a constant interaction between farmers and scientists to evaluate genetic progress and, performance recording that allows for genetic evaluations to select individuals of high genetic merit to sustain genetic progress. CBBPs are making it possible to obtain datasets that have been challenging and almost impossible to collect in smallholder farms. They are also providing the infrastructure that makes it possible to test the application of cutting-edge genomic tools and technologies in smallholder farms in developing countries. CBBPs therefore are offering a blueprint for learning and knowledge exchange between farmers and, between farmers and researchers for the improvement of animal productivity in smallholder farms and in generating information that enrich the existing body of scientific knowledge. In our talk, we shall discuss and demonstrate using real-world data and studies how CBBPs have been a cornerstone of breeding improvement and scientific discovery.

Session 02 Theatre 1

'Sustainable livestock systems': what does this mean? – welcome and introduction

F. O'Mara[1] and M. Lee[2]
[1]Animal Task Force, 149 rue de Bercy, 75012 Paris, France, [2]Harper Adams University, Newport, Shropshire, TF10 8NB, United Kingdom; mail@animaltaskforce.eu

The EU will introduce a Framework Law on Sustainable Food Systems in 2023. In this context, what are sustainable livestock systems, which common ground could they have and how will they differ from region to region? Policymakers, civil society, NGOs, consumers, the agri-food sector and scientists all want to better define and contribute to more sustainable livestock systems. However, there is no clear definition of what a sustainable livestock system is, how to assess the state of existing systems, and the direction in which they must evolve. In the meanwhile, food security has grown in importance because of the recent crisis occurring in Europe. How to combine food security, resilience and sustainability should be at the centre of research and innovation and policy initiatives. This topic will explore the key attributes of sustainable livestock systems, the trade-offs and synergies between different aspects of the systems and their interactions with other parts of the food system. It will also explore how current livestock systems can evolve to be more sustainable and will examine methodologies to assess the sustainability of livestock systems: (1) the benefits and trade-offs of livestock systems from an economic, social and environment viewpoint; (2) Assessing the sustainability of livestock systems and its different levels; (3) barriers for young farmers to enter livestock farming systems; (4) The role of livestock systems in rural economies and job creation; (5) the role of livestock systems in providing nutrient-rich and affordable foods; (6) sustainability and healthy and robust livestock; (7) livestock systems in the context of climate change and the effects of heat and water shortage; (8) intrinsic qualities of livestock products in food and non-food uses; (9) the place of animal welfare concerns in sustainable livestock systems and benefits/trade-offs with the three main pillars of sustainability; (10) contribution to culture; (11) tools to assess and improve sustainability; e.g. development of new metrics, feeding and breeding.

Friend or Foe? The role of animal-source foods in healthy and environmentally sustainable diets

L. Merbold[1], T. Beal[2], C. Gardner[3], M. Herrero[4], L. Iannotti[5], S. Nordhagen[2] and A. Mottet[6]
[1]Agroscope, Reckenholzstrasse 191, 8046 Zürich, Switzerland, [2]GAIN, Rue Varembé 7, 1202 Geneva, Switzerland, [3]University of California, 2201 North Hall, Santa Barbara, USA, [4]Cornell University, 340 Tower Road, Ithaca, USA, [5]Washington University, One Brooking Drive, St. Louis, USA, [6]FAO, Viale delle Terme di Caracalla, 00153 Roma, Italy; lutz.merbold@agroscope.admin.ch

Discussions around the role of animal-source foods (ASFs) in healthy and environmentally sustainable diets are often polarizing. Our study reviews the evidence on the environmental and health benefits and risks of ASFs, focusing on primary trade-offs and tensions. We further aim at summarizing the evidence on alternative proteins and protein-rich foods in such diets. ASFs are rich in bioavailable nutrients commonly lacking globally and can make important contributions to food and nutrition security. Populations in Sub-Saharan Africa and South Asia could benefit from increased consumption of ASFs through improved nutrient intakes and reduced undernutrition. In areas where consumption is high, processed meat should be limited, and red meat and saturated fat should be moderated to lower noncommunicable disease risk – this could also have co-benefits for environmental sustainability. Evidence shows that ASF have a large environmental impact; yet, when produced at the appropriate scale and in accordance with local ecosystems and contexts, ASFs can play an important role in circular and diverse agroecosystems. Under these circumstances, they can be beneficial for restoring biodiversity and degraded land and mitigate greenhouse gas emissions from food production. In summary, the amount and type of ASF that is healthy and environmentally sustainable will depend on the local context and health priorities. This will change over time as populations develop, nutritional concerns evolve, and alternative foods from new technologies become more available and acceptable. Efforts by governments and civil society organizations to increase or decrease ASF consumption should be considered in light of the nutritional and environmental needs and risks. Policies, programs, and incentives are needed to ensure best practices in production, curb excess consumption where high, and sustainably increase consumption where low.

The economic sustainability of dairy production systems in the EU and beyond

F. Thorne[1], E. Dillon[1], T. Donnellan[1], P. Jeanneaux[2], R. Jongeneel[3] and L. Latruffe[4]
[1]Teagasc, Department of Agricultural Economics and Farm Surveys, Ashtown, Dublin. D15 KN3K, Ireland, [2]VetAgroSup, UMR Territoires, 89 avenue de l'Europe, 63370 Lempdes, France, [3]WUR, Department of Social Sciences, Wageningen Economic Research, Hollandseweg 1, 6706 KN Wageningen, the Netherlands, [4]INRAE, Univ. Bordeaux, CNRS, BSE, UMR 6060, 33600 Pessac, France; fiona.thorne@teagasc.ie

The economic sustainability of farming is one of the core objectives of the Common Agricultural Policy (CAP), as set out in the objectives of the Treaty on the Functioning of the EU (TFEU). Whilst a number of TFEU provisions lay down objectives pertaining to the multi-dimensional aspects of sustainability, the purpose of this paper is to examine the current economic sustainability of EU livestock production systems, using the dairy sector as a case study. Whilst economic sustainability is identified as a clear policy objective, the means to measure this term are not clearly identified in policy or literature. Almost all contributions to date, define the concept in a different manner. Hence, the literature relating to defining the theory and practise of measuring farm level economic sustainability across countries or regions will be reviewed in this paper. Based on an agreed set of indicators, harmonised data sources will be consulted to examine the situation with regards to comparative economic sustainability, both within the EU and internationally. A mix of primary and secondary data sources will be reviewed to draw inferences regarding the drivers of economic sustainability. Whilst seminal papers in the literature relating to the drivers of performance will be reviewed, key case study regions where farm level data is available will also be examined using parametric and non-parametric approaches. Finally, the potential role of policy reform in impacting economic sustainability and the evolving role of policy direction will be considered.

Session 02 Theatre 4

Methods to assess the sustainability of livestock systems: challenges and opportunities

E. De Olde[1,2]

[1]*Wageningen University & Research, Wageningen Economic Research, P.O. Box 29703, 2502 LS The Hague, the Netherlands,* [2]*Wageningen University & Research, Animal Production Systems group, P.O. Box 338, 6700 AA Wageningen, the Netherlands; evelien.deolde@wur.nl*

Assessing the sustainability performance of livestock systems is becoming increasingly important, not only in response to pressure from NGOs or consumers, but because of a variety of reasons. First of all, given the substantial environmental impact of the livestock sector, sustainability assessments are seen as a key step to identify possible pathways for improving the environmental as well as economic and social performance of livestock systems. Research into the trade-offs and synergies of (interventions in) livestock systems plays an important role in this. Second, gaining insight into the sustainability practices and performance of farms is increasingly part of corporate sustainability programs and certification schemes which aim to differentiate themselves and establish a market for more sustainable products. In fact, actors in the agri-food sector are increasingly formulating sustainability goals, for instance with regard to the reduction of greenhouse gas emissions. Meanwhile, the use of sustainability indicators in policy (for instance through payments based on sustainability performance) and finance (through discounts on land rent or interest rates) is becoming more prominent. Despite this increased interest, there are certain persistent challenges in the development and use of sustainability methods and indicators. This talk gives insight into these challenges as well as the key characteristics and trade-offs of a variety of methods. Which sustainability issues and indicators are generally included, and which tend to be ignored? What are opportunities for methods to be improved, and options for harmonizing methods? The talk concludes with a reflection on the implication of these challenges and opportunities for future directions in the field of sustainability assessments of livestock systems.

Session 02 Theatre 5

SustAinimal – a multi-actor knowledge centre for livestock in future Swedish food systems

M. Hetta[1]*, S. Agenäs*[2]*, D.J. De Koning*[2]*, H. Oscarsson*[3]*, P. Peetz Nielsen*[4] *and A. Wallenbeck*[2]

[1]*Swedish University of Agricultural Sciences, 901 83 Umeå, Sweden,* [2]*Swedish University of Agricultural Sciences, Box 7023, 750 07 Uppsala, Sweden,* [3]*Vreta Kluster AB, Klustervägen 11, 585 76, Vreta kloster, Sweden,* [4]*Research Institutes of Sweden, Scheelevägen, 17, 223 63 Lund, Sweden; marten.hetta@slu.se*

SustAinimal is a novel knowledge centre created to identify and develop the future role of livestock production for increased sustainable and competitive food production in Sweden. The activities aim to evaluate the potential contributions to ecosystem services and loads as well as enhancing the competitiveness and resilience of the sector. The organization is a resource for decision-makers and stakeholders. The focus areas include grazing and fodder production, digitalization, governance for transition, and regional applications. The partners in the project include academia, research institutes, companies, industry organizations and governing authorities, working together along the value chain from farm to fork. SustAinimal has its own academy with special focus on PhD students and post-docs to ensure that the next generation of livestock scientists are well equipped to contribute to transdisciplinary research processes and knowledge for future challenges. Sweden is a relatively large and long country from North to South and thereby there are different settings for food production, depending on local conditions and differences in proximity to consumers. The work in SustAinimal focuses on three geographical regions for investigations of the importance of local conditions for sustainability and competitiveness in future food systems. An important part of Sweden's agricultural land is currently used for production of animal feed. The intersection between plant cultivation, land use and animal husbandry is therefore highlighted in the regional activities. The centre monitors the supply of food and ecosystem services and evaluate their impact on climate, environment and social sustainability. The project SustAinimal includes different regional living labs for co-creation of innovative future land management practices. The living labs will provide new biodiversity assessment tools and aims to increase the awareness of the role of grazing animals in future food systems.

Session 02 Theatre 6

Breeding livestock for sustainable systems

P.W. Knap[1], K.M. Olson[2] and M.A. Cleveland[2]
[1]GENUS-PIC, Ratsteich 31, 24837 Schleswig, Germany, [2]Genus-ABS, 1525 River Road, DeForest 53532, WI, USA; pieter.knap@genusplc.com

At least five of the UN's 17 Sustainable Development Goals (SDG) can be supported by animal breeding. SDG2: 'achieve food security and improve nutrition': produce more cost-efficient animals, which makes animal source food (ASF) an affordable source of essential amino acids and micronutrients for a wider group of consumers. SDG3: 'Ensure healthy lives for all': produce leaner and more resilient animals that produce food with a lower fat content, and that need fewer antibiotics. SDG12/13: 'ensure sustainable consumption and production patterns; combat climate change and its impacts': improve the efficiency of ASF production by creating animals that are more feed-efficient and produce more high quality protein, and by improving animal health to reduce losses; all of this results in reduced greenhouse gas emission. SDG15: 'halt biodiversity loss': implement responsible management and conservation of genetic resources. From another point of view, sustainability of livestock production follows the same Triple Bottom Line 'PPP' pattern as any other production system: People, Planet & Profit – but there is a fourth element, representing the animals themselves (so, Pigs-Poultry-Puminants-Phish production follows a PPPP pattern). Note that this element does not feature in the UN's SDGs. Animal breeding can contribute significantly to the interests of all P elements, and we discuss two of these: 'Planet' is about: (1) biodiversity; and (2) environmental load and climate change, 'Pigs-etc' is about: (3) animal health; and (4) animal welfare. 'People' is about social justice, with little connection to livestock breeding technology; a possible case would be biopiracy which is more a political and economic issue than a technical one, and more relevant in the plant breeding sector. Influencing 'Profit' by breeding has been covered intensely since selection indexes were designed – there is no need to repeat that here apart from the imminent role of carbon shadow prices in livestock production. It must be borne in mind throughout that 'sustainability will always be a matter of more or less: it can never be an absolute goal'.

Session 02 Theatre 7

Panel discussion (morning)

A. Granados and A.S. Santos
Animal Task Force, 149 rue de Bercy, 75012 Paris, France; mail@animaltaskforce.eu

Panel discussion with speakers and the audience, moderated by Ana Granados & Ana Sofia Santos, ATF Vice-Presidents (45 min.).

Session 02 Theatre 8

Use of grazing with dairy goats to design sustainable food systems

H. Caillat, E. Bruneteau and B. Ranger
INRAE UE FERLus, Les Verrines, 86600 Lusignan, France; hugues.caillat@inrae.fr

France is the largest producer of goat's milk in Europe with about 550 million litres collected. Today, French goat farms have a low feed self-sufficiency (61%), increasing economic risks and feeding costs. To improve sustainability, a larger use of grazing, in particular of legume-based pastures, might be a solution. FERLus-PATUCHEV is an experimental farm of INRAE based in Lusignan (New Aquitaine- 46.43°N, 0.12°E) whose objective is to assess and propose innovative, low input and sustainable goat farming systems by a greater use of grazing. This platform aimed at evaluating 2 independent farmlets about of 60 Alpine goats differing in kidding period: February (F) or September (S). The area for each farmlet was of 10.4 ha and was divided between temporary multi-specific pastures (7.4 ha) and a cereal-protein crops mixture (3 ha). The study period was from 2020 to 2022. Dairy goats number was higher for the F farmlet (62.3 vs 59.8). Average days of grazing was 150 days/year for the F goats, i.e. +19 days/year than the S goats. Average feed self-sufficiency of F farmlet was 83.6%, i.e. +3.9% than S farmlet. Average concentrates quantities were higher for the S (340 vs 286 kg/goat/year), despite a large fodder proportion in the ration for the 2 farmlets (68.4 vs 70.2%). Average milk production per goat and average milk solids per hectare were higher for the F farmlet (818 vs 739 l/year; 365 vs 319 kg). This pasture-based farms are net producers of protein because 82% of the proteins and 81% of the energy consumed by the goats were not edible by humans. Other complementary studies have demonstrated that bulk milk from goats eating fresh grass was richer in nutriments for humans (fatty acids and B6 B12 E vitamins). By limiting the use of resources and producing food for human, these grazing dairy goat farms, in particular with kidding in February, make it possible to realise more sustainable food systems by limiting the feed/food competition. However, the main limit of these systems is the parasitism management because we have observed a continue increase of the annual excretion of strongyles (+24.8% per year since 10 years). The second point is the lack of attractiveness in France for the consumption of kid meat, which would however improve the efficiency indicators of feed-food competition.

Session 02 Theatre 9

Benefits and limits of farm animals to control herbage mass, pests and weeds in orchards: a review

G. Maillet[1], A. Dufils[2], F. Angevin[3], S. Ramonteu[1], J.L. Peyraud[4] and R. Baumont[5]
[1]ACTA, 149 rue de Bercy, 75012 Paris, France, [2]INRAE, UMR ECODEV, 84000 Avignon, France, [3]INRAE, UR Info&Sols, 45000 Orléans, France, [4]INRAE, UMR Pegase, 35590 Saint-Gilles, France, [5]INRAE, UMR Herbivores, 63122 Saint-Genès-Champanelle, France; rene.baumont@inrae.fr

One of the levers of the agroecological transition is the diversification of systems, in particular by reintroducing animals into systems specialised in field crops or arboriculture. In this study, we focus on the introduction of animals into fruit and vineyard plots. One of the assumed benefits of introducing animals into the plots is the management of pests through biological regulation. The objective of our study, which analysed 66 documents (23 scientific and 18 technical articles, 25 other documents), was to make a comparative synthesis of the effects of the introduction of different animals (sheep, goats, cattle, poultry, pigs and rabbits) on the control of herbage mass, weeds and pests in vineyards and orchards. Sheep is a traditional and efficient solution for managing grass cover in orchards and vines; and cows are traditionally used in high-stem orchards like 'pré-vergers' or coconut plantations. Geese are herbivorous poultry that are easy to introduce into vines or orchards and chickens or guinea fowl a possible solution but which can lead to the appearance of bare soil. In contrast, goats and pigs are not very well recommended for managing grass cover because of the damage they cause to trees and soil respectively. Several benefits can be expected from the introduction of animals on pest and disease management. For rodent management, the destruction of rodent burrows by trampling is the most frequently reported mechanism. Animals can also have a direct or indirect effect on the management of insects and other invertebrates. For example, some animals, in particular poultry are heavy consumers of larvae, insects or mollusks. The introduction of animals into orchards can also play a prophylactic role. For example, the consumption or trampling of dead leaves or fallen fruits can reduce the scab inoculum. Introducing animals can be a solution for managing weed, pest and diseases and lead to reduction of chemical treatments. However, not all species can be equally used and some mechanisms of pest and disease control remain unknown.

Session 02 Theatre 10

A new concept for agro-ecological efficiency at different scales of ruminant production systems

H. Nguyen-Ba, P. Veysset and A. Ferlay
Université Clermont Auvergne, INRAE, VetAgro Sup, UMR 1213 Herbivores, Theix, 63122 Saint-Genès-Champanelle, France; hieu.nguyen-ba@inrae.fr

Improving efficiency of ruminant production systems is deemed essential to achieve their sustainability. But does efficiency *per se* really mean sustainability? The efficient high-input systems that intensively breed high-performing animals in confined housing and use nutrient-dense diets are increasingly criticized for accelerating feed-food competition, waterbody contamination, biodiversity loss, animal welfare concerns and rural unemployment, etc. Agroecology has emerged as a promising alternative for current farming. Applying principles of ecology on agriculture such as relying on biodiversity and interactions among components of agroecosystems allows to close nutrient cycle and boost production capacity. It aims to reduce dependencies on synthetic and fossil resources and brings more added values to farmers, society and nature. However, under the simplistic view of current efficiency concept (defined as a mere ratio between outputs and inputs), agroecological ruminant systems are less efficient and have more environmental impacts per kg of products than conventional counterparts. Moreover, numerous studies had showed that it is difficult to maintain at higher scales of ruminant farming systems the efficiency's gains obtained at animal scale. The aim of this study was thus to conceptualize a more holistic criteria for efficiency at different scales of agroecological ruminant systems (i.e. agroecological efficiency). Based on literature review and expert's counsels, we propose that agroecological efficiency at the system scale is a combination of production, environment, economy and farmer's work. More precisely, agroecological efficiency can be improved based on 5 principles: (1) reduce the use of intermediate consumptions (purchased goods and services), (2) reduce pollution and losses by closing the nutrient cycle within the system, (3) reduce feed-food competition, (4) increase the added value and the remuneration of the farm workers in relation to the gross value of the production and (5) increase the 'net' labour productivity. We then adapt this concept to define specific criteria for improving efficiency at herd and animal scales that can potentially avoid the potential losses of efficiency due to changing scales.

Session 03 Theatre 1

Issues and perspectives in genomic selection in limited size breeds: the case of Italian Simmental

N.P.P. Macciotta[1], L. Degano[2], D. Vicario[2] and A. Cesarani[1]
[1]Università di Sassari, Dipartimento di Agraria, Viale Italia 39, 07100 Sassari, Italy, [2]ANAPRI, Via Ippolito Nievo n. 19, 33100 Udine, Italy; macciott@uniss.it

Genomic selection has had a huge impact in the dairy cattle breeding industry, particularly in large breeds as Holsteins, with strong effects on the genetic trend and on the reduction of the generation interval. However, small size breeds like dual purposes or local breeds have already implemented GS programs although they have to cope with specific issues. A case is the Italian Simmental (IS) cattle breed that has started genomic evaluations in 2011. A first main issue was represented by the curse of dimensionality, i.e. the marked unbalance between animals genotyped and markers (usually the medium density 50K panel). In the original two-step approach, this issue has been addressed by reducing the prediction dimensionality through the Principal Component Analysis. This approach allowed to reduce the dimensionality of predictors of about 90% maintaining the same GEBV accuracy of the whole SNP panel. A further strategy to overcome this issue has been the exchange of genotypes with related cattle bred as the German and Austrian Simmental populations. The next step has been the implementation of the single step GBLUP procedure, that was carried out with a project that involved the main cattle breeds farmed in Italy in order to harmonize genomic evaluations. The ssGBLUP is currently implemented in the GS of IS. The GS has impacted the accuracy of GEBV and the genetic trend. Other co-products of the GS breeding program of the IS are biodiversity studies, in particular the comparison of its genomic structure with other breeds of the Simmental group farmed in Europe. Genomic inbreeding has been also investigated. Future challenges will be represented by the consideration of novel phenotypes and the improvement of sustainability. Among them, feed efficiency (i.e. residual feed intake) and environmental adaptation are under investigation. Proxies provided by precision livestock farming, such as milk infrared spectra, could help to predict phenotypes on large scale that can be used in the breeding program. Finally, the role of epigenetics in the mechanisms of adaptation, in particular for thermotolerance, and its possible consideration in the genetic model, is an emerging topic.

Session 03 Theatre 2

Impact of genomic selection on genetic diversity in five European local cattle breeds

J.J. Winding[1], T.H.E. Meuwissen[2], P. Croiseau[3], G. Restoux[3] and R. Bonifazi[1]
[1]Wageningen University & Research, Animal Breeding and Genomics, Droevendaalsesteeg 1, Radix, 6700 AH Wageningen, the Netherlands, [2]Norwegian University of Life Sciences, Department of Animal and Aquacultural Sciences, Oluf Thesens vei 6, 1433 Ås, Norway, [3]Université Paris-Saclay, INRAE, AgroParisTech, Joey-en-Josas, 78350, France; jack.windig@wur.nl

Genomic selection (GS) has revolutionized animal breeding and in general accelerated genetic gains in cattle breeding programs. However, in Holstein and Jersey cattle, the implementation of GS was accompanied by increases in inbreeding rates. Similarly to these popular breeds, GS has been implemented also for smaller populations and local breeds. In this study, we investigated inbreeding trends in five local cattle breeds from three European countries and evaluated the possible impact of implementing GS on breeds' genetic diversity. The five breeds evaluated were: Abondance, Tarantaise, Vosgienne (from France), Norwegian Red (from Norway), and MRIJ (from the Netherlands). We estimated trends in inbreeding and kinship based on pedigree and genomic information (available as medium or high-density SNP genotypes), before and after the introduction of GS. The number of available genotyped animals for each breed was 16,478, 8,589, 4,474, 51,799, and 4,997, for Abondance, Tarantaise, Vosgienne, Norwegian Red, and MRIJ, respectively. Results show that inbreeding trends did not differ between pedigree-based or DNA-based estimates. In all five breeds, periods with higher and lower inbreeding rates occurred across time. However, no clear trend could be observed after the introduction of GS in any of the breeds. Inbreeding rates either slightly increased (MRIJ, Abondance), remained stable (Tarantaise, Vosgienne), or decreased (Norwegian Red). Such small deviations could be due to GS allowing for shorter generation intervals, screening for a larger number of individuals compared to traditional breeding schemes, and preselection of a smaller number of sires. Ultimately, our results suggest that the genetic management of these breeds is more important in determining inbreeding rates than the implementation of GS per se. This project has received funding from the European Union's Horizon 2020 Programme for Research & Innovation under grant agreement n°101000226.

Session 03 Theatre 3

Accounting for sequential genomic selection in broiler breeding

J. Hidalgo[1], D. Lourenco[1], S. Tsuruta[1], V. Breen[2], W. Herring[2] and I. Misztal[1]
[1]The University of Georgia, Animal and Dairy Science Department, 425 River Rd, Athens, GA 30602, USA, [2]Cobb-Vantress Inc., 4703 US-412 E, Siloam Springs, AR 72761, USA; jh37900@uga.edu

Sequential selection occurs in broiler breeding; in the first selection stage, the best individuals for growth are selected for reproduction. These birds are later evaluated for reproductive traits in a separate evaluation for time efficiency; however, excluding data from the broiler phase can yield inaccurate and biased predictions. This study aimed to find the most accurate, unbiased, and time-efficient approach for jointly evaluating broiler and reproductive traits. Broiler data was incorporated sequentially into the evaluation for reproductive traits to assess the impact on accuracy, bias, and dispersion (computed using the Linear Regression method) of predictions for hens from the last two breeding cycles and their sires. Data included a pedigree with 577K birds (146K genotyped), phenotypes for three reproductive (R1, R2, R3; 9K each), and four broiler traits (up to 467K). To find the most time-efficient approach, we evaluated three core definitions for the algorithm of proven and young: a core set with 19K, including parents and young animals, and two random core sets with 7K and 12K animals. The reproductive evaluation (RE) included pedigrees, genotypes, and phenotypes for reproductive traits of selected animals; in RE2, we added their broiler phenotypes; in RE_BR, broiler phenotypes of non-selected animals, and in RE_BR_GE, their genotypes. From RE to RE_BR_GE, accuracy changes were null or negligible for R1 (0.51 in hens, 0.59 in roosters) and R3 traits (0.47 in hens, 0.49 in roosters) that had heritability close to 0.3. For R2 trait (heritability=0.02), accuracy increased in hens (roosters) from 0.4 (0.49) to 0.47 (0.53). The bias of GEBV for hens (roosters), in additive SD units, decreased from 0.69 (0.7) to 0.04 (0.05) for R1, 1.48 (1.44) to 0.11 (0.03) for R2, and 1.06 (0.96) to 0.09 (0.02) for R3. The dispersion was stable in hens (roosters) at ~ 0.93 (~ 1.03) for R1, improved from 0.57 (0.72) to 0.87 (1.0) for R2 and from 0.8 (0.79) to 0.88 (0.87) for R3. Using data from all selection phases is advised in breeding programs with sequential selection. A random core with 7K birds halves computing time, maintaining the quality of predictions.

Session 03 Theatre 4

Changes in the genome due to genomic selection in two pig populations

Y.C.J. Wientjes[1], M.P.L. Calus[1], P. Bijma[1], A.E. Huisman[2] and K. Peeters[2]
[1]Wageningen University & Research, Animal Breeding and Genomics, P.O. Box 338, 6700 AH Wageningen, the Netherlands, [2]Hendrix Genetics B.V., P.O. Box 114, 5830 AC Boxmeer, the Netherlands; yvonne.wientjes@wur.nl

The implementation of genomic selection in breeding programs has resulted in more genetic gain. However, it has also resulted in faster changes in the genome compared to pedigree selection. Our aim here was to investigate the changes in allele frequency and in the results of genome-wide association studies (GWAS) from 2015 to 2021 in two commercial sow lines undergoing genomic selection. For line A, genotypes of 44,054 segregating markers were available for 2,616 to 7,689 animals per birth year, with a total of 40,075 animals. For line B, genotypes of 44,000 segregating markers were available for 921 to 4,995 animals per birth year, with a total of 23,487 animals. Moreover, phenotypes for eight traits under selection were available on a subset of the genotyped animals (Line A: 738-6,423 animals per trait per birth year; Line B: 406-3,965 animals per trait per birth year), including general production and reproduction traits. Over the seven birth years included in the dataset, absolute allele frequency changes up to 0.35 were observed in each line, with a few clear peaks. The regions with the largest changes in allele frequency did not overlap between lines. In each line, GWAS were performed separately for each birth year-trait combination. Results showed that the most significant peaks were present across birth years. This suggests that the changes in the genetic architecture (allele frequencies and allele substitution effects) were relatively limited over the seven years. Some significant GWAS regions overlapped between the two lines. Surprisingly, the largest changes in allele frequency were in regions where no significant markers were found related to any of the traits. Moreover, the correlation between significance levels of markers and changes in allele frequency was close to zero. Additional analyses will be performed to investigate whether the observed allele frequency changes were larger than expected due to drift in those populations. Altogether, those results indicate that the genome changes over the years, however, it is not yet clear to what extent this is due to genomic selection vs drift.

Session 03 Theatre 5

Practical approaches to managing increased homozygosity caused by genomic selection

C. Baes[1,2], C. Obari[2], B. Makanjuola[2], C. Rochus[2], C. Maltecca[3], F. Schenkel[2] and F. Miglior[2,4]
[1]University of Bern, Institute of Genetics, Vetsuisse Faculty, Bremgartenstrasse 109a, 3012 Bern, Switzerland, [2]University of Guelph, Centre for Genetic Improvement of Livestock, Department of Animal Biosciences, University of Guelph, 50 Stone Road East, N1E 2W1 Guelph, Canada, [3]North Carolina State University, Department of Animal Science, Polk Hall, 120 W Broughton Dr, NC 27607 Raleigh, USA, [4]Lactanet Canada, Genetics, 660 Speedvale Ave W, N1K 1E5 Guelph, Canada; cbaes@uoguelph.ca

The mating of related animals is unavoidable in most breeding programs, but the rate of inbreeding in livestock populations has increased sharply since genomic selection, and will continue to increase. Intense directional selection, coupled with novel technologies, have allowed the industry to recognize and identify recessive disorders associated with increased homozygosity. Although the economic gains of intense directional selection strategies currently outweigh negative effects of inbreeding, long-term consequences of this approach are unclear. Several approaches have been proposed for estimating, monitoring and controlling inbreeding, including those using pedigree and genomic information, but these methods do not consider genetic relationships at the herd level. The objective of this study was to investigate the average genetic and genomic relationships of individual sires to the active cow population within each herd (within-herd R-value), and across herds (across-herd R-value). The active sire population had an average relationship ranging from 1 to 8.5% with the active cow population for the across-population pedigree-based R-value. In addition to this, pedigree-based R-values showed the average sires' relationship with cows ranged from 0.43 to 32.38% across herds. By quantifying the R-value on a within-herd basis, more meaningful herd-level information can be used by breeders in selection and mating programs to control inbreeding. Future work will focus on developing a monitoring system to rapidly identify, understand, and manage detrimental haplotypes in dairy. A rapid-response feedback system will be developed in which detrimental haplotypes will be identified before their frequency in the population increases.

Session 03 Theatre 6

Positive and negative effects of genomic selection

I. Misztal
University of Georgia, Animal and Dairy Science, Athens, GA 30605, USA; ignacy@uga.edu

Initial reports on genomic selection indicated strong improvement for major traits and even successful selection for antagonistic traits. Lately, mostly unofficial reports, indicate increased frequency of problems mostly related to fitness traits. The change can be explained by treating the genetic selection as a form of optimization according to a resource allocation theory. Traits in explicit or implicit selection index are improved while other traits change according to genetic correlations among the traits. After long selection, usually production traits improve while fitness traits deteriorate. At the gene level, genes associated with positive effects on major traits tend to fixation, and remaining genes with large effect exhibit pleiotropy. Strong negative effects of genetic selection on fitness traits were noted before the genomic selection, however, adding some of those traits in the index and constantly improving management counterbalanced the deterioration. Under genomic selection, the generation interval declined, and management modifications are not fast enough to alleviate the problems. If problematic traits are recorded in sufficient volume, an obvious solution is a stronger weight for these traits in the selection index. Otherwise, a stronger selection can be implemented for already recorded trait(s) that include problematic traits, e.g. productive life or survival. Setting up an appropriate selection index required to implement the changes requires current genetic parameters. Reports indicate that such parameters are changing: heritabilities for strongly selected traits decline while genetic correlations between production and fitness traits become more antagonistic. An accurate selection index would require new methods to estimate changing genetic parameters with large genomic data as current methods are not applicable. The genomic selection enters a new phase where possible negative effects need to be considered.

Session 03 Theatre 7

Storing and analysing a million genomes on a desktop computer

G. Gorjanc[1], J. Obsteter[1], G. Mafra Fortuna[1,2], R. Ros-Freixedes[3], M. Johnsson[4] and I. Pocrnic[1]
[1]The Roslin Institute, University of Edinburgh, Easter Bush, EH259RG, Edinburgh, United Kingdom, [2]Agricultural Institute of Slovenia, Hacquetova ulica 17, 1000 Ljubljana, Slovenia, [3]Universitat de Lleida, Pl. de Víctor Siurana, 1, 25003 Lleida, Spain, [4]Swedish University of Agricultural Sciences, Ultuna campus, 750 07 Uppsala, Sweden; gregor.gorjanc@roslin.ed.ac.uk

Whole-genome sequence data holds promise for a more informed and accurate study of genetic and phenotypic variation than the established SNP array genotypes. To deliver on this promise, we must efficiently manage and analyse this ultimate source of genomic variation. We describe a plan to manage the whole-genome sequence data from Roslin and Genus/PIC projects – almost a million haploid whole-genome sequences by leveraging pedigrees, SNP array genotypes, strategic sequencing, phasing, and imputation. The data includes 440,610 pigs and 46+ million sequence variants. Storing this data in a rectangular format in double precision requires ~150 TB of memory. While we can lower precision, storing and analysing such data has at least a quadratic complexity. By leveraging pedigree structure and inferred haplotypes, we can succinctly encode the data with tree sequence format in only ~15 GB. The tree sequence contains four key tables: haplotype identifications (nodes), relationships between ancestor-descendant haplotypes (edges), sequence variant loci (sites), and sequence variant alleles (mutations). We estimate the node table will contain ~170M haplotype identifications (~2.5 GB), the edge table will contain ~334M haplotype relationships (~10 GB), and the sites table will contain 46M sequence variant loci (~1 GB). These tables describe whole-genome sequences within the pedigree, while extending the tables with a deep ancestry beyond the pedigree will recapitate the tree sequence up to the most recent common ancestor for every genome region. For this last part, we expect the mutations table to contain ~46M biallelic mutations with ~1.4 GB. While this is a work in progress, we have already successfully inferred tree sequence from the 1000 bull genome data, where we have observed a more than 90% reduction in data storage. This novel way of storing and analysing whole-genome sequence data will pave the way for future biological discovery and more informed breeding.

Session 03 Theatre 8

Large scale screening for genetic defects in Holstein cattle using transmission disequilibrium test

F. Besnard[1,2], M. Boussaha[1], H. Leclerc[3], J. Jourdain[3], D. Boichard[1] and A. Capitan[1]
[1]INRAE, Domaine de Vilvert, 78350 Jouy-en-Josas, France, [2]IDELE, 149 Rue de Bercy, 75012 Paris, France, [3]ELIANCE, 149 Rue de Bercy, 75012 Paris, France; florian.besnard@idele.fr

In this study, we exploit the bovine population structure with large paternal progeny groups to detect loci with transmission disequilibrium, corresponding to multiple determinism defects frequently overlooked by other approaches. Transmission disequilibrium test (TDT) in large genotyped families is an efficient tool to detect these conditions. In a healthy progeny group, both haplotypes are expected to be inherited equally from the parent. In contrast, if the parent carries a deleterious abnormality, a deficit is expected in the surviving progeny carrying the haplotype associated with the mutation. For a recessive defect, this disequilibrium increases with the frequency of the deleterious allele in the population. For a defect with incomplete penetrance or mosaicism, the disequilibrium depends on these parameters. To conduct this study, we selected 401 Holstein bulls with at least 500 genotyped calves, resulting in a total of 532,864 calves. In each progeny, we observed the transmission of 20-SNP haplotypes and selected the haplotypes in significant transmission disequilibrium. The minimum number of calves required for a disequilibrium to be considered significant was determined from a binomial distribution adjusted for multiple testing. Of the 401 bulls tested, 321 bulls had at least one haplotype in disequilibrium. We then grouped all significant identical haplotypes and, for each 10 MB window, selected the most unbalanced haplotypes shared by at least five bulls. This resulted in a subset of 33 significant haplotypes. To detect candidate variants, we used whole genome sequences of 301 bulls. We selected 56 candidate variants (SNPs or structural variants in a 20 MB window around the peak) on the basis of their correlation with the haplotype status of each animal and their Sift score. We propose candidate variants for the previously known QTL in BTA 18 associated to viability and suggest candidate variants in BOLA region that could partly explain early mortality. The functional impacts of these variants are currently under investigation. FB is a recipient of a CIFRE grant with the financial support of ANRT and APIS-GENE.

Session 03 Theatre 9

Genomic selection strategies and their potential to maintain rare alleles and de-novo mutations

M.F. Schrauf, Y.C.J. Wientjes, H.A. Mulder and J. Vandenplas
Wageningen University and Research, Animal Breeding and Genomics, P.O. Box 338, 6700 AH Wageningen, the Netherlands; matias.schrauf@wur.nl

Sustainable breeding programs need to balance short-term genetic improvement with the conservation of genetic diversity. While genomic selection has considerably increased the genetic gain for many breeding programs, consequences on diversity can be less desirable. This is particularly the case for rare alleles and de-novo mutations, as markers used in genomic selection are generally not strongly associated to rare alleles. Moreover, genomic selection allows for the selection of young individuals without records, thereby ignoring the effects of de-novo mutations. To study possible solutions, we simulated populations of 1000 individuals subject to 50 generations of selection. We evaluated four selection strategies to identify the ones which best conserve favourable rare variants and de-novo mutations in the presence of additive and non-additive gene action. The genomic selection strategies represent a variety of approaches to balance between genetic improvement and diversity management and were: truncation selection, which only focuses on short-term gain; optimal contribution selection, which balances that gain with a constraint in the relatedness of the selected individuals; allele-reweighted selection, which upscales the effect of rare alleles in the breeding values; and constrained allele loss selection, a novel strategy which balances short-term gain with a constraint on the reduction in frequency of rare alleles estimated to be favourable. Systematic differences between the strategies were not observed for traits with non-additive gene action. For the trait under additive gene action, allele-reweighted selection obtained a higher genetic gain than truncation selection while preserving a similar level of genetic variance. Meanwhile, constrained-allele-loss selection obtained a similar genetic gain than truncation selection while accumulating a higher number of favourable de-novo mutations. These and similar strategies may contribute to the sustainable long-term use of genomic selection. This project has received funding from the European Union's Horizon 2020 Programme for Research & Innovation under grant agreement n°101000226.

Session 03 Theatre 10

Genomic prediction with selected sequence variants in gestation length of New Zealand dairy cattle

Y. Wang, K.M. Tiplady, E.G.M. Reynolds, M.A. Nilforooshan, C. Couldrey and B.L. Harris
Livestock Improvement Corporation, Research and Development, Private Bag 3016, Hamilton 3240, New Zealand; yu.wang@lic.co.nz

The increasing availability of whole genome sequence (WGS) data has opened up new opportunities for improving genomic prediction in livestock breeding. The aim of the study was to investigate the performance of genomic prediction combining a filtered Illumina50k single nucleotide polymorphism (SNP) panel, with a subset of informative sequence variants selected from a large genome-wide association study (GWAS). The study used gestation length phenotypes for 97,522 New Zealand admixed dairy cattle, where the gestation length for each calf was calculated as the difference between its dam's calving and mating dates in days. The animals were genotyped and imputed up to whole-genome sequence level with 16,122,291 variants. BOLT-LMM software was used for the iterative GWAS analysis, then a subset of significant variants were selected and added to the filtered Illumina50k markers in the prediction models. The prediction analyses were performed using BayesR methods implemented in the GCTB software. Compared to other scenarios, the approach that estimated the effects of selected SNPs in the same training population from which they were chosen resulted in the highest number of significant sequence variants been selected (783), the greatest amount of explained genetic variance (0.518), and the highest prediction accuracy (0.570). However, a relatively high degree of bias (0.887) was also observed. When animals were separated by birth year into GWAS and prediction populations, a relatively high prediction accuracy (0.556) and low bias (0.927) were achieved, compared to the scenarios with either separate animals of equal number for GWAS and prediction, or more animals for GWAS. In conclusion, including informative sequence variants can improve the performance of genomic prediction, offering not only biological insights but also practical applications such as the development of customized SNP arrays or the use of imputation in existing industry SNP panels.

Session 03 Theatre 11

Novel runs of homozygosity islands in the Finnish Ayrshire population

K. Sarviaho, P. Uimari and K. Martikainen
University of Helsinki, Department of Agricultural Sciences, P.O. Box 28, 00014 Helsinki, Finland; katri.sarviaho@helsinki.fi

Artificial selection and inbreeding increase autozygosity and result in long homozygous stretches, runs of homozygosity (ROH), in the genome. In particular, genomic selection may increase ROHs around quantitative trait loci (QTL). High frequency of ROHs at certain genomic regions in a population form 'ROH-islands' – population-specific indicatives of selection pressure. ROH-islands have revealed selection signatures in range of livestock breeds (e.g. cattle, pig, and sheep). The aim of our study was to identify differences in ROH-island patterns, ΔROH-islands, in the Finnish Ayrshire (FAY), that have emerged after genomic selection (GS) was introduced. Our data included genotypes for 45,834 SNPs from 53,469 FAY cows. Cows were divided into two populations; cows born between 1980-2011, before GS was introduced (6,108 cows), and cows born between 2015-2020 (47,361 cows), after GS was introduced. First, ROHs were identified using PLINK version 1.90b6.20. We calculated difference in the occurrence of each SNP in a ROH between the two populations (ΔH-score). Next, the top 1% of SNPs with the largest ΔH-scores were considered significant, and if separated by a maximum of two non-significant SNPs formed a ΔROH-island. We listed genes under positive selection within the ΔROH-islands based on UCSC Table browser (ARS-UCD1.2/bosTau9 assembly, April 2018). Finally, overlap of ΔROH-islands with QTL identified in previous studies were studied using CattleQTLdb version 14 release 47. We detected a total of 1,237 SNPs (including the conjunctive SNPs) within 22 ΔROH-islands in ten *Bos Taurus* autosomes. ΔROH-island length ranged from 45.4 Kbp to 9.23 Mpb and encompassed 61.4 Mbp in total. The ΔROH-islands identified in our study included 424 annotated genes. A total of six ΔROH-islands in BTAs 2, 14, 16, 18, 25, and 26 completely or partially overlapped 777 QTL previously identified in Nordic Red cattle (FAY, Swedish Red, and Danish Red). The QTL were related to milk fat yield, fertility index, somatic cell score, and stature. In conclusion, we identified signatures of selection in genomic regions with previously identified dairy QTL that have appeared after genomic selection was introduced in the Finnish Ayrshire population.

Session 03 Theatre 12

Sometimes we win, sometimes we lose: the consequences of genomic selection

D. Lourenco[1], F. Guinan[1], G. Wiggans[2], J. Dürr[2], S. Tsuruta[1] and I. Misztal[1]
[1]University of Georgia, 425 River Rd, Athens, GA 30602, USA, [2]CDCB, 4201 Northview Drive, Bowie, MD 20716, USA; danilino@uga.edu

Since implementing genomic selection (GS), the dairy industry has seen faster genetic progress because of the rapid turnover of genomic bulls. Generation intervals for Holsteins bulls have decreased from 7 years in 2008 to 2.2 years in 2020, and the genetic gain has more than doubled for some traits in Holstein and Jersey. The yearly changes in average predicted breeding values nine years after the implementation of genomics are greater than in the nine years before for several traits like fat, protein, and daughter pregnancy rate. In fact, the overall increase in genetic gain for Holstein and Jersey resulted from the higher level of implementation and early adoption of genomics compared to breeds like Ayrshire, Brown Swiss, and Guernsey. Besides the increase in genetic gain per generation, GS was supposed to help reduce rates of inbreeding; however, this is not the reality. Right before genomics, the pedigree and genomic inbreeding levels for Holstein bulls were both 6%, but after genomics, they increased to 10.9 and 12.7%, respectively. In Jersey, the increase in inbreeding levels was only 1 to 2%, probably because of different selection decisions. Although pedigree and genomic inbreeding levels are similar, the genetic trends based on pedigree and genomic-based models differ. Once GS started being used, breeding values from BLUP became biased because this method does not account for genomic information that is used for selection; therefore, deregressed breeding values based on BLUP are questionable. Before genomics, predictions based on BLUP were stable and, therefore, trusted by the industry; however, genomic predictions are less stable, and top-ranked animals can easily drop in subsequent evaluations, generating concerns about GS. Another issue related to GS is the ever-increasing computational cost because of the amount of genomic data available. In this talk, we will review the advantages and disadvantages of GS after almost 15 years of its first implementation and discuss how to leverage the positive and overcome the negative aspects.

Session 04 Theatre 1

Old, native cattle breeds; from critically endangered to successfully utilized in niche productions

N. Svartedal[1] and O. Vangen[2]
[1]NIBIO, PB 115, 1431 Aas, Norway, [2]NMBU, PB 5003, 1432 Aas, Norway; nhs@nibio.no

Around 1990 the old cattle breeds native to Norway were critically endangered. There was little information on where these breeds might be found or if breeds had gone extinct. Thus, two actions were taken by the Committee on animal genetic resources. Recording remaining animals of these breeds showed that the population sizes varied between 11 and 113 breeding cows. Additionally, frozen semen from a few bulls from most of the breeds was made available. The semen was collected by Geno, the breeding organisation for the main producing commercial cattle breed, Norwegian Red (NRF). GENO was the only organisation which had facilities to collect and store bull semen. The recorded animals were registered in a pedigree database generated specifically for these breeds. Access to semen, free participation in the pedigree database and public support when establishing breed societies were the only extra support the owners of these breeds had access to up to 2000. Thereafter, subsidies for native endangered cattle breeds were introduced. Slowly the number of breeding animals increased in numbers, and by 2022 the population sizes varied between 307 and 1965 breeding and none of the breeds were any longer characterized as 'critically endangered', only 'endangered'. The different public actions to support the farmers with these breeds have been of crucial importance for the breeds' survival and growth. Still, no cattle breed can be sustainably maintained without being utilized in a production system. Traditionally the breeds were kept in dairy production. However, as the average milk yield of these breeds are appr 4,000 kg/year, only a few farmers manage to make a living as dairy farmers. Some farmers have extended their profit by producing local cheese and other dairy products. However, during the last ten years the growth in population sizes have exclusively been linked to meat production. Direct sale of meat from farmers to consumers, at farmers' market, etc. has been a great success. As seen in many other European countries as well, there is a future for the old, local breeds in alternative low-input production systems, both linked to product quality, local adaptation, different grazing- and plant preferences and local production close to consumers.

Session 04 Theatre 2

Kenyan pastoralists and climate change: breeding as a possible adaptation strategy?

J. Jandl[1], M. Wurzinger[1], L. Gichuki[2], B. Habermann[2], R. Siamito[2] and T.A. Crane[2]
[1]BOKU-University of Natural Resources and Life Sciences, Vienna, Gregor-Mendel-Str. 33, 1180 Vienna, Austria, [2]ILRI-International Livestock Research Institute, P.O. Box 30709, 00100 Nairobi, Kenya; maria.wurzinger@boku.ac.at

Pastoralists in the semi-arid areas of Kenya are grappling with the impacts of climate change, yet livestock is at the centre of the pastoralists' livelihood strategies. Therefore, this study aims at analysing the current management and breeding strategies of Maasai pastoralists to adapt to the changing conditions. The study was carried out in Kajiado county, which has an annual rainfall between 500-700 mm and a bi-modal rainfall pattern. Qualitative interviews and focus group discussions were conducted to gather general information on livelihood strategies and breeding strategies for the different livestock species, cattle, goats and sheep. Data were recorded, transcribed and analysed using NVIVO software. Income diversification by combining livestock keeping with other agricultural activities is a common management strategy. However, livestock remains the primary source of income, provides milk and meat for home consumption and plays an important cultural role. Thus, improving livestock productivity by making strategic breeding decisions is critical to pastoralists' mainstay. In all three species, breeding animals are selected based mainly on their physical appearance. Crossbreeding is the standard approach for cattle. Non-local bulls (Sahiwal, Borana) are mated with local Zebu cattle. Pastoralists aim for large animals, which can fetch a reasonable price on the market. In addition, resistance to drought and diseases, early puberty and good fertility are essential traits. Migration, herd splitting, destocking (applied by some), controlled matings and cultivated pasture on fenced land for breeding bulls are other adaptation mechanisms. Short-term economic aspects are more important in breeding decisions than adaptation to climate change. While this strategy is essentially important, it could be disadvantageous in the long term, as the animals may not have enough adaptive potential for difficult environmental conditions.

Session 04 Theatre 3

Adapting climate-smart breeding practices for small ruminants in pastoral communities of Kenya

J.M. Ojango[1], E. Oyieng[1], J.W. Gitau[1], N. Ndiwa[1], J. Gachora[2] and A.W.T. Muigai[3]
[1]ILRI, Bioscience, 30709, Nairobi, Kenya, [2]Min Agric. Liv. Fish. & Coop, Livestock, 34188 Nairobi, Kenya, [3]National Defence Univ., AA & Research, 370 Nakuru, Kenya; j.ojango@cgiar.org

Changing climatic conditions with high frequencies of droughts have increased the vulnerability of pastoral communities in the Northern rangelands of Kenya and necessitate prompt interventions in sheep and goat breeding practices. Sheep and goats in the arid areas are locally adapted indigenous breeds with low potential for meat and milk production. The animals are reared in large numbers by pastoral families who individually own small flocks (5 to 10 animals) that are herded in large communal groups. Since 2018, the International Livestock Research Institute in collaboration with the Ministry of Agriculture, Livestock and Fisheries and Cooperatives in Kenya commenced a program to build the resilience of pastoralists to the changing climatic conditions using their small ruminant assets rather than depending on humanitarian emergency responses. Interventions necessitated changes in traditional management practices related to sheep and goat production in each community. Core Innovation Groups (CIG), each comprising 30 pastoral households, were established to adopt and model new practices using participatory processes in Isiolo, Marsabit and Turkana Counties of Kenya. The CIG members underwent a three-year phased training program on breeding management practices for more efficient, resilient, and productive animals aligned to a collaboratively predesigned impact pathway. Data collated over three years by extension personnel engaged to monitor the CIG was analysed using logistic regression techniques to assess household-level adoption of livestock breed improvement, feeding and disease control interventions. All CIG adopted more than one of the introduced technologies concurrently. Prevention of diseases was the most readily adopted, followed by crossbreeding using indigenous breeds of sheep and goats from other arid areas. The study shows that pastoral communities are open to adapting technological interventions that will positively impact animal productivity. The adoption of the technologies was enhanced by the experiential capacity development activities adapted to different socio-economic parameters within the communities.

Zooarchaeology for conservation biology: introducing ARETI

A. Spyrou[1], G. Hadjipavlou[2] and D. Bradley[3]
[1]The Cyprus Institute, STARC, 20 Konstantinou Kavafi Street, Nicosia, 2121, Cyprus, [2]The Agricultural Research Institute, Athalassa Area, 1011, Cyprus, [3]Trinity College Dublin, The Smurfit Institute of Genetics, College Green, Dublin 2, Ireland; a.spyrou@cyi.ac.cy

Animals REsilient in TIme (*ARETI*) is an interdisciplinary research project that explores the genetic, economic and cultural history of cattle on the island of Cyprus from the prehistoric times to the present. By weaving together evidence from Zooarchaeology, a subfield of archaeology that studies the remains of animals, cattle iconography and palaeogenomics, ethnography and folklore studies, the project's main aim is to unearth vital information about the last 8,000 years of human-cattle interactions on the island and highlight the historical value of an animal that has accompanied Cypriot rural societies for many centuries. One of the core elements of the project is to demonstrate the physical presence of the thermotolerant *Bos indicus* or zebu cattle (or hybrids) on the island and its potential link to past climatic shifts and human migrations. Our genomic analysis focused so far on 15 prehistoric cattle samples (petrous bone) from key archaeological sites, spanning the early Neolithic-Iron Age, while blood samples from 100 unrelated cattle belonging to the Cyprus local breed have also been selected to be genotyped by using a high density bovine 800K beadchip microarray. Archaeological findings are combined and compared with contemporary DNA mapping of the local Cyprus cattle breed to look for long-lasting signatures of selection, conservation and historical adaptation and contribute to the development, promotion and implementation of sustainable strategies for conserving and further improving the adaptive genetic traits of the Cyprus' indigenous cattle breed. Overall, through the main concepts and methodologies of ARETI as well as our public outreach plan (e.g. documentary, educational activities for children, etc.), we demonstrate what can be achieved when different disciplines such as archaeology, animal genomics and conservation biology join forces.

A robustness proxy based on variations in energetic allocation to growth in Duroc fattening pigs

G. Lenoir[1,2,3], L. Flatres-Grall[2], R. Muñoz-Tamayo[3], I. David[1] and N.C. Friggens[3]
[1]GenPhySE, Université de Toulouse, INRAE, ENVT, Auzeville Tolosane, 31326 Castanet Tolosan, France, [2]Axiom, La Garenne, 37310 Azay-sur-Indre, France, [3]Université Paris-Saclay, INRAE, AgroParisTech, UMR Modélisation Systémique Appliquée aux Ruminants, 22 place de l'Agronomie, 91120 Palaiseau, France; glenoir@axiom-genetics.com

In the context of global warming, livestock, especially pigs, evolve in an increasingly changing and challenging environment. The ability of an animal to cope with these changes while ensuring the expression of its production potential can be associated with its robustness. Our objective was to develop a robustness proxy on the basis of modelling of longitudinal energetic allocation coefficient to growth for fattening pigs, from 75 to 150 days of age. A total of 3,710 pigs from Duroc paternal line were raised at the AXIOM boar testing station (Azay-sur-Indre, France) from 2015 to 2023. This farm was equipped with automatic feeding system, recording individual weight and feed intake at each day t of the fattening period. We used a dynamic linear regression model to characterize the daily evolution of the allocation coefficient (α_t) between cumulative net energy available for growth and cumulative weight gain during fattening period. The cumulative net energy available for growth at day t was estimated as the difference between the net energy intake and the net energy requirement for maintenance. Longitudinal energetic allocation coefficients were analysed using a two-step approach. In a first step, the repeated trait α_t was analysed using a linear random regression animal model (RR). The residuals at day t for each animal of the RR model was used to compute log transformed squared residuals (LSR_t). In a second step, the heritability of LSR was estimated using a multi-traits animal model including LSR and the four traits under selection. The LSR trait, which could be interpreted as an indicator of the response of the animal to perturbations/stress, showed low heritability (0.06±0.01). The trait LSR had high favourable genetic correlations with average daily growth (-0.79±0.07) and unfavourable with feed conversion ratio (-0.54±0.10). In this study, we proposed an approach for characterizing the robustness through the variability in the allocation.

Session 04 Theatre 6

Genetic architecture of skin histology with implications for heat tolerance in beef cattle

R.G. Mateescu[1], F.M. Rezende[1], K.M. Sarlo Davila[2], A. Hernandez[1], A.N. Nunez Andrade[1], G.A. Zayas[1], E.E. Rodriguez[1] and P.A. Oltenacu[1]
[1]University of Florida, Animal Science, 2250 Shealy Dr, Gainesville, FL 32611, USA, [2]ORISE/NADC, 1920 Dayton Ave, Ames, IA 50010, USA; raluca@ufl.edu

Development of effective strategies to improve the ability to cope with heat stress is imperative to enhance productivity of the livestock industry and secure global food supplies. However, selection focused on production and ignoring adaptability results in beef animals with higher metabolic heat production and increased sensitivity to heat stress. Cattle lose heat predominantly through cutaneous evaporation at the skin-hair coat interface when experiencing heat stress. Sweating ability, sweat gland properties, and hair coat properties are a few of the many variables determining the efficacy of evaporative cooling. Sweating is a significant heat dissipation mechanism responsible for 85% of body heat loss when temperatures rise above 86 °F. This is, to our knowledge, the first study reporting an extensive investigation of skin histology properties in a multibreed Angus-Brahman heifer population. Heat loss adaptations at the skin level are anticipated to have a negligible impact on productivity and thus provide an excellent opportunity to select for animals with superior thermal adaption and food production abilities. The breed group had a statistically significant effect on all skin properties in this study except dermis thickness. Brahman cattle had significantly thinner and longer epidermis, thinner dermis, larger sweat gland areas, longer sweat glands closer to the skin surface, smaller sebaceous gland area and more sebaceous glands compared to Angus cattle. Equally important, these differences are also accompanied by significant levels of variation within each breed, which is indicative that selection for these skin traits would improve the heat exchange ability in beef cattle. Single-trait GWAS were used to investigate the relevance of direct additive genetic effects on each trait. Genomic windows explaining more than 1% of direct additive genetic variance were considered to be associated with the analysed trait. Several quantitative trait loci (QTLs) were identified by GWAS for sweat gland area with a large QTL explaining over 1.4% of the genetic variance.

Session 04 Theatre 7

Ewe-lamb allocation trade-offs shape ewe lifetime production depending on dietary energy scarcity

M. Hiltpold and F. Douhard
GenPhySE, Université de Toulouse, INRAE, ENVT, 31326 Castanet-Tolosan, France; maya.hiltpold@inrae.fr

Pasture-based production systems strongly depend on weather and seasonal conditions and are potentially facing limits in climate change adaptation without sufficiently robust animals. Robustness likely involves the balanced allocation of the mother's limited resources between her own survival and future reproduction and her offspring's survival and growth until weaning. Thus, we expect that the optimal resource allocation strategies exist depending on the available resources in the animal's environment. To study the effect of different priorities between mother and offspring traits, we applied a mathematical model of energy acquisition and allocation to suckler sheep. Our dynamic model predicts the whole lifetime production trajectory of an ewe with a given energy allocation strategy between growth, maintenance, reproduction and body reserves in response to dietary energy. Model outputs include the changes of ewe body weight and condition across successive reproduction cycles as well within each cycle, the number of lambs and their growth from birth to weaning. Importantly, the model predictions are based on direct energy transfers from the ewe to her lambs during pregnancy and lactation. The predicted ewe and lamb conditions were assumed to affect ewe and lamb survival, prolificacy and conception rates. We explored the consequences of various allocation strategies on ewe lifetime production for a range of grazing scenarios (schematically described in terms of seasonal energy availability) under fixed management conditions (one lambing per year, winter period). First results suggest that individual variation in energy allocation leads to trade-offs among the components of lifetime production. Comparable lifetime production levels might be achieved with strategies emphasizing either early first lambing and large litter sizes over a short ewe life or moderate litter sizes over a long ewe life. However, prudent allocation strategies were preferred in energy scarce environments. Our work contributes to understanding the optimum ewe performance in contrasting environments by identifying not only the best animal for a given environment, but also the most robust ewe which can deal with a variable environment.

Session 04 Theatre 8

System performance of three dairy-beef genotypes divergent in carcass merit

N. Byrne, D. Fahy, M. Kearney and N. McHugh
Teagasc, Animal & Grassland Research and Innovation Centre, Grange, Dunsany, Co. Meath, C15 PW93, Ireland; nicky.byrne@teagasc.ie

The proportion of beef output originating from the dairy herd in many countries, such as Ireland, is increasing. Compared to suckler origin beef, dairy-beef systems has the potential for greater carbon efficiency; however, in Ireland there is an increasing proportion of dairy-beef animals failing to meet minimum carcass specifications, limiting profitability. The objective of this study was to compare the physical, financial and environmental performance of three dairy-beef genotypes, within an efficient grass-based steer production system. These genotypes consist of male Holstein Friesians (HF) and two Angus (AAX) genotypes, sired by bulls of divert genetic merit for carcass weight and conformation (Low and High AAX). The effect of early-life calf nutrition on lifetime performance was also evaluated, whereby half of each genotype group received either 4 or 8 litres (L) of milk replacer/day from 30 days of age until weaning. Milk feeding level had no effect on lifetime growth and carcass performance (P>0.05), however, calves on the 4 L treatment consumed significantly more concentrate during the rearing phase (+21 kg DM, P<0.01). Genotypes achieved the same lifetime growth performance (P>0.05), except for the first grazing season where HF had significantly higher growth. Terminal performance differed between genotypes for age at slaughter, conformation and carcass value (P<0.001). Although numerically in favour of the High AAX genotype, pairwise comparisons indicated non-significant (P>0.05) differences between AAX genotypes for individual carcass traits. Using coefficients generated from each genotypes physical performance their contribution to overall farm economic and environmental efficiency was modelled using the Grange Dairy Beef Systems Model. The High AAX genotype achieved a €728/ha net margin, €121 and €266 greater than Low AAX and HF genotypes, respectively. Both AAX groups had a reduced carbon footprint, producing 9% less CO_2 eq per carcass kg and were net producers of human edible protein compared to the HF genotype, who consumed 25% more protein than they produced. From this study, dairy-beef genotypes of reduced slaughter age, increased carcass weight and conformation have the best overall farm system performance.

Session 04 Theatre 9

Characterization and main factors of variation of milk production in suckler cows

B. Sepchat[1], M. Barbet[1] and A. De La Torre[2]
[1]INRAE, Herbipôle, Theix, 63122 Saint-Genès-Champanelle, France, [2]UCA, INRAE, VetAgro Sup, UMRH, Theix, 63122 Saint-Genès-Champanelle, France; bernard.sepchat@inrae.fr

The improvement of conformation and muscular development in suckling herd results in a decrease in milk production (MP) leading to an increase of concentrate consumption by calves. In a tense economic, environmental and societal context, it is necessary to maintain or even improve the MP of beef cows to support calves' growth. Assessment of MP is indirect and consists to calculate milk drunk by the calf as the difference of its weight after and before suckling. The objectives of this work were to acquire knowledge on MP and determine the main factors of variation. For that, MP measured in Herbipole unit (more than 12,000 measures over 1000 cows controlled at least 7 times per cow and lactation) over the last 15 years in Charolais (C), Salers (S) and Limousin (L) cattle were analysed. Breed is the main factor of variation of MP (29% difference between S and L): 1,600±313 kg milk / lactation in L, 1,840±355 in C and 2,250±470 in S as well as parity ±10% difference between primiparous and multiparous regardless of breed. In a 'classic' system of weanling's production, characterized by a winter calving and semi-extensive pasture, two peaks are observed across the lactation. The first occurs one-month post-calving and the second largest at turn-out. This peak is ranged from 0.8 to 2.3 kg of milk according to the breed, parity and calving date. In this study, the persistence of lactation and its variability were quantified, (-17 to -27 g drank milk / day). Persistence is higher for low-producing cows (-0.5 against -0.9 kg milk / day less per month) than for high-producing cows. High MP supports high growth rate of calf. Throughout lactation, this gain is 60 g/l of milk drunk in more, either 70 kg of live weight gain for a lactation of 2,300 vs 1,200 kg. The composition of milk from suckler cows is very poorly known, a first approach was performed in a subset of Charolais cows. The average composition (in g/kg of milk) in protein, fat and lactose is respectively 35.4±2.2, 39.1±10.2 and 49.9±1.8. In suckler cows, a better characterization of milk production and its composition is required to improve feed autonomy and contribute to animal selection strategies in a context of climatic deregulation.

Mediterranean Baladi cattle presents adaptive traits to cope with climate change in the Near East

A. Shabtay[1], E. Lipkin[2], M. Soller[2], F. Garcia-Solares[1], J. Sölkner[3], E. Shor-Shimoni[1], A. Bagnato[4], M.G. Strillacci[4] and M. Cohen-Zinder[1]
[1]Agricultural Research Organization, Sustainable Ruminant Production Lab; Model Farm for Sustainable Agriculture, Newe Ya'ar Research Center, P.O. Box 1021, Ramat Yishay 30095, Israel, [2]The Hebrew University of Jerusalem, Dept. of Genetics, Silberman Life Sciences Institute, Givat Ram Campus, Israel, [3]University of Natural Resources and Life Sciences, Gregor-Mendel-Strasse 33/II 1180, Vienna, Austria, Austria, [4]Università degli Studi di Milano, Milan, Italy, Via Dell'Università, 6. Facoltà Medicina Veterinaria 26900 Lodi, Italy; shabtay@volcani.agri.gov.il

The Near East is amongst the driest and water-scarce regions in the world. Moreover, in light of the global climate changes, it is expected to become even hotter and drier, and hence, to experience a reduction in crop yields and quality and turn more vulnerable to geographical redistribution of pests and diseases. It is, therefore, recommended to accelerate the buildup of resilient agri-food systems, which in case of the livestock sector, would lean on breeds under adaptive selection to harsh environment, that can maintain sustainable production. The Baladi is an indigenous cattle, known for its hardiness, disease resistance, and ability to utilize feedstuffs of low nutritious quality. Located at the centre of the three major cattle domestications events, its genome contains relatively similar proportions of the three ancestral components, implying a rich pool of genetic diversity and selection signatures for adaptation to the Near East environment. At physiological level, the capability to maintain lower energy expenditure enable Baladi animals to retain improved energy balance over the productive Simmental breed, under poor diets. Moreover, intrinsic features of Baladi erythrocytes imply an improved resistance to tick borne diseases. Although bearing relatively low dressing percentage comparing to traditional beef breeds, preliminary analysis mark Baladi meat qualitative, in terms of physical and organoleptic characteristics. The current presentation will screen the bio-agricultural advantages described above, and discuss the biological and economic efficiency of Baladi cattle breeding program, as part of future sustainable resilient agri-food system in the Near East.

Effect of some environmental variables on milk traits in dairy sheep

F. Correddu[1], A. Cesarani[1], G. Gaspa[2], I. Peana[3], G. Fois[3], C. Dimauro[1] and N.P.P. Macciotta[1]
[1]Università di Sassari, Dipartimento di Agraria, viale Italia, 39, 07100 Sassari, Italy, [2]Università di Torino, Dipartimento di scienze agrarie, forestali e alimentari – DISAFA, Largo Braccini, 2, 10095 Grugliasco (TO), Italy, [3]Agenzia Regionale per la Protezione dell'Ambiente della Sardegna, Dipartimento MeteoClimatico – Servizio Meteorologico, Agrometeorologico ed Ecosistemi, Viale Porto Torres, 119, 07100 Sassari, Italy; macciott@uniss.it

Due to the climate changes, livestock sector will face the possibility of farming more resilient animals. This is particularly true for the extensive dairy sheep farming systems of the Mediterranean area. Aim of this work was to evaluate the impact of the Temperature Humidity Index (THI) and Wind Chill Index (WCI) on milk traits in Sarda dairy sheep. Data consisted of 2,695 test day records from 555 Sarda breed ewes. Maximum THI and minimum WCI registered the same day (0d), one day (1d), two days (2d) or three days (3d) before each test, respectively, were retrieved from the closest meteorological stations of the ARPAS (Sardinian Regional Agency for the Environment Protection). A total of 7 and 5 classes were created for THI and WCI, respectively. The considered traits were milk yield (MY), fat (FP), protein (PP) and lactose (LC) contents, and somatic cells count (SCC). Each trait was analysed using a linear mixed model with sampling date, days in milk (DIM) class, parity, climate variable (THI or WCI) and the interaction between DMI and climate variable as fixed effects and flock and animal as random effects. Both THI and WCI were negatively correlated with MY and LC, and positively with FP and PP. THI class was always highly significant (P<0.001) for all traits except SCC. According to the F-values, the largest effects of THI were found at 3d, 2d, 1d, and 0d for MY, FP, PP, and LC, respectively. The negative effect of THI on MY was evident at almost each class of DIM, with lower production observed at THI>72. WCI class was significant for MY only at 3d, whereas for LC only at 1d; it was always significant for PP and at 0d, 1d, and 2d for FP. The largest effects of WCI class were observed at 0d for FP and PP and at 1d for LC. Results of the present work evidenced an effect of the climate variables on the temporal evolution of milk production traits in dairy sheep.

Session 04 — Theatre 12

Genomic tolerance to harsh climate conditions in beef cattle as a result of cross-breeding program

R. Tian and H. Asadollahpour Nanaei
Inner Mongolia Academy of Agricultural & Animal Husbandry Sciences, Animal science, 010031, Hohhot, 7521325, China, P.R.; tiannky@163.com

Understanding the evolutionary forces related to climate changes that have been shaped genetic variation within species has long been a fundamental pursuit in biology. In this study, we generated whole-genome sequence (WGS) data from 65 cross-bred and 45 Mongolian cattle. Together with 62 whole-genome sequences from world-wide cattle populations, we estimated the genetic diversity and population genetic structure of cattle populations. In addition, we performed comparative population genomics analyses to explore the genetic basis underlying variation in the adaptation to climate change and immune response in cross-bred cattle located in the cold region of China. To elucidate genomic signatures, we performed three statistical measurements, fixation index (FST), log2 nucleotide diversity ($\theta\pi$ ratio) and cross population composite likelihood ratio (XP-CLR), and further investigated the results to identify genomic regions under selection for cold adaptation and immune response-related traits. Analysis of population structure demonstrated evidence of shared genetic ancestry between studied cross-bred population and both Red-Angus and Mongolian breeds. Among all studied cattle populations, the highest and lowest levels of linkage disequilibrium (LD) per Kb were detected in Holstein and Rashoki populations. Our search for potential genomic regions under selection in cross-bred cattle revealed several candidate genes related with immune response and cold shock protein on multiple chromosomes. We identified some adaptive introgression genes with greater than expected contributions from Mongolian ancestry into Molgolian × Red Angus composites such as TRPM8, NMUR1, PRKAA2, SMTNL2 and OXR1 that are involved in energy metabolism and metabolic homeostasis. In addition, we discovered some candidate genes probably associated with immune response-related traits. The identification of these genes may clarify the molecular basis underlying adaptation to extreme environmental climate and as such they might be used in cattle breeding programs to select more efficient breeds for cold climate regions.

Session 04 — Theatre 13

Expected selection responses in breeding plans aiming to limit environmental impacts of trout farming

S. Pouil[1], J. Aubin[2] and F. Phocas[1]
[1]Université Paris-Saclay, INRAE, AgroParisTech, GABI, Domaine de Vilvert, 78350 Jouy-en-Josas, France, [2]INRAE, Institut-Agro, SAS, 65 rue de Saint-Brieuc, 35042 Rennes, France; simon.pouil@inrae.fr

With the growing societal concerns about the sustainability of food production systems, there is increasing interest in considering not only economic gains but also environmental impacts in the selective breeding of farmed species. In this study, we compared expected selection responses for alternative breeding programs aiming to limit the environmental impacts of the production of rainbow trout, one of the most important farmed fish species in Europe. The consequences of genetic improvement based on optimal election indexes derived to minimize various environmental impacts were investigated in a theoretical rainbow trout farm producing constant annual fry production volumes. A cradle-to-farm-gate life-cycle assessment was performed to evaluate the environmental value (ENV) of each trait that has been used in the breeding goals. The tested breeding goals included three different traits: the body weight (BW), the feed conversion ratio (FCR) measured through the feed conversion ratio and the fry survival rate (SR). Due to a lack of knowledge about the genetic links across these traits, we tested several correlation scenarios between the traits. We explored different impact categories as various environmental breeding goals, such as acidification, climate change, cumulative energy demand, eutrophication, land occupation and water dependence. Annual genetic gains ranged from 0.9 to 1.4% for the different impact categories, while the annual genetic gains ranged from 0.4 to 4.6% for BW, 0.0 to 2.8% for FCR and -11 to 0.9% for SR. We demonstrated interest in using ENV in breeding goals to minimize environmental impacts at the farm level, while maintaining high genetic improvements in growth and feed efficiency-related traits. Nevertheless, another selection strategy should be considered to avoid negative consequences on SR when considering possible negative correlations between survival and production traits. Although our results are promising, their interpretations have to be qualified by the consideration of the economic repercussions of such a selection strategy.

Session 04 Theatre 14

Defining resilience traits in sheep from fibre diameter variation of wool

E.G. Smith[1], S.F. Walkom[2], D.J. Brown[2] and S.A. Clark[1]
[1]University of New England, School of Environmental and Rural Science, University of New England, Elm Avenue, 2351 Armidale NSW, Australia, [2]Animal Breeding and Genetics Unit, University of New England, Elm Avenue, 2351 Armidale NSW, Australia; esmith76@myune.edu.au

The capacity to measure and select livestock that are more resilient to environmental fluctuation is of increasing importance amidst climate change, labour shortages and increasing production demand. Currently, however, there is no consensus on how to quantify resilience, particularly in extensive sheep populations. In this study, we explored the ability to derive resilience indicator traits from fibre diameter variation measured longitudinally (5 mm increments) along the wool staple. Fibre diameter varies in relation to the supply of nutrients to the wool follicles and thereby provides a stable archive of the animal's physiological status across the preceding wool growth period. From this fibre diameter variation, ways to detect and characterise an animal's ability to withstand or be minimally affected by its environment were explored. The heritability estimates of these traits were shown to be low to moderate (0.10 to 0.31), indicating that genetic variation exists for fibre diameter variation measured along the wool staple which may be interpreted as a measure of resilience. The inclusion of such measures in sheep breeding programs has the potential to improve the resilience of sheep to environmental challenges, which may have positive implications for sheep enterprise profitability, health and welfare.

Session 04 Theatre 15

Evaluating the impact of uncertainty in estimated breeding values on optimal contribution selection

I. Pocrnic, J. Ortiz and G. Gorjanc
The University of Edinburgh, The Roslin Institute, Easter Bush Campus, EH25 9RG Edinburgh, United Kingdom; ivan.pocrnic@roslin.ed.ac.uk

Many livestock breeding programmes have a closed nucleus and therefore have to manage the conversion of genetic variation into genetic gain with care. Consequently, these breeding programmes use some methods for managing genetic variation. One such method is the optimal contribution selection (OCS), which aims to maximise genetic gain for a given loss in genetic variation. The accuracy of this optimisation depends on the accuracy of its input parameters; the estimated breeding values (EBV) and the relationship matrix. In OCS, we typically assume that EBV are true values, that is, that they have no associated uncertainty, which is seldom the case. The aim of this study was to evaluate how uncertainty in EBV impacts optimised contributions and success of OCS. To this end, we have stochastically modelled a small breeding programme and obtained 1,000 posterior samples of EBV from either the pedigree (BLUP) or genomic (GBLUP) evaluations. We then ran OCS for each EBV sample (probabilistic OCS), for the posterior mean of EBV (naïve OCS), as well as for the true breeding values (true OCS). We finally evaluated the distribution of optimised contributions for the selected parents and of genetic mean and genetic variation of their progeny. We compared scenarios with OCS that aimed for an effective population size of 50 and 100. Results have shown considerable underestimation of optimised contributions for naïve OCS compared to probabilistic OCS, which were closer to the true OCS. Future work includes the use of robust optimisation that can work with a full distribution of estimated breeding values, thereby enabling probabilistic OCS for routine use.

Session 04 Theatre 16

Increasing genetic gain without compromising diversity by selection based on mating qualities

T.A.M. Niehoff, J. Ten Napel, P. Bijma, T. Pook, Y.C.J. Wientjes, B. Hegedűs and M.P.L. Calus
Wageningen University & Research, Animal Breeding and Genomics, Droevendaalsesteeg 1, 6700 AH Wageningen, the Netherlands; tobias.niehoff@wur.nl

Selection decisions traditionally rely on expected offspring performance, hence breeding values. By also considering the expected Mendelian sampling variance of a mating pair, the probability to produce animals with high performance can be increased, which means a shift in the planning horizon of the breeding objective by one generation. We extended this idea to multiple generations. Thus, by extending the planning horizon, our criterion allows to maximize genetic gain in several generations ahead. We tested our newly developed criterion against previously developed criteria, namely selection based on: (1) breeding values; (2) the probability to select top offspring; (3) the expected breeding value of selected offspring; and (4) a recently published index that describes the linearized probability to produce top offspring. Comparisons were based on a simulated recurrent selection breeding scheme using the software MoBPS. We explored our new criterion in an ideal breeding program with known QTL effects and linkage information to test the theoretical benefit without errors induced by estimating QTL effects and haplotypes. Our criterion achieved higher genetic gain compared to all other criteria while maintaining more genetic variance and achieving lower inbreeding levels relative to selection based on breeding values, both after 5 and 20 generations. This is because our criterion allows to consider the diversity that is present in the current generation to project how it can be turned into genetic gain in a certain generation in the future. In conclusion, our criterion allows faster genetic progress without compromising diversity.

Session 05 Theatre 1

Holistic perspectives on climate care cattle farming

G. Van Duinkerken
Wageningen Livestock Research, De Elst 1, 6708 WD Wageningen, the Netherlands; gert.vanduinkerken@wur.nl

The EU 'Green Deal' includes a commitment to reach a climate neutral economy by 2050. European agreements indicate that greenhouse gas (GHG) emissions from EU agriculture should be reduced with 30% by 2030, compared with 2005 levels. However, with unchanged policies and efforts, and based on projections per member state, it is expected that a reduction of only 2% is expected in the time span 2005 to 2030. Moreover, individual member states are focusing on integrated solutions to mitigate GHG emissions, and simultaneously optimise nitrogen management to reduce ammonia emission and nitrate leaching. Because the cattle chain is a large contributor to EU agricultural emissions, especially methane and ammonia, there is a strong need for applicable measures to reduce emissions along this chain. About two thirds of cattle GHG-emissions have an on-farm origin. For example enteric methane, and emissions related to manure management and crop cultivation. One-third has an off-farm origin, such as the production of fertilizers and concentrates, and processing and transport of farm products. The EU-project Climate Care Cattle Farming ('CCCFarming') focuses on reducing GHG and ammonia emissions while maintaining socio-economic farm performance. Processes within the farm itself are being studied, and emission reducing innovations are being developed and implemented. Potential trade-offs with off-farm emissions are being evaluated. Trade-offs can also occur on-farm. For example an increased maize cultivation, and higher share of maize silage in the animal diet, can reduce enteric methane emissions and increase nitrogen use efficiency at animal level, but the associated reduction of permanent grassland will reduce biodiversity and grazing possibilities and will increase the risk of nitrate leaching from farm land. In this presentation emission mitigation strategies, their interactions, and their relevance for integral sustainable dairy farming will be highlighted.

Climate care farming aspects in China

W. Wang
China Agricultural University, No.2 Yuanmingyuan West Road, 100193, Beijing, China, P.R.; wei.wang@cau.edu.cn

China is the world's largest emitting economy and the world's largest emitter of greenhouse gases from agriculture. President Xi Jinping announced at the 75th United Nations General Assembly in September 2020 that China strives to peak its carbon dioxide emissions by 2030 and become carbon neutral by 2060. The commitment made by President Xi gives national strategic importance to carbon reduction efforts in all sectors. With the growth of China's population and economic level, China's livestock industry is still rapidly developing to meet the growing demand for animal products. Major livestock animals such as pigs, chickens, dairy cattle, and beef cattle are among the top in the world and still maintain a high growth rate. With the continuous improvement of scale, automation, and intelligence, China's livestock industry has undergone radical changes in livestock breeds, breeding methods, and feed types. Under such dramatic changes, some basic parameters of greenhouse gas emissions have also changed a lot. However, compared with developed countries, there is still much room for improvement in China's livestock and poultry production level, livestock and poultry waste treatment technology, and the breeding cycle implementation. There is an urgent need to conduct long-period greenhouse gas emission monitoring of the livestock industry to determine the emissions baseline.

Producer adoption of climate-smart agriculture practices on U.S. grazing lands

A. Blair, K. Cammack, H. Menendez, J. Brennan and K. Ehlert
South Dakota State University, Animal Science, 711 N Creek Dr, Rapid City, SD 57703, USA; amanda.blair@sdstate.edu

Today's livestock producers face increasing public scrutiny because animal agriculture is often cited as a major contributor to greenhouse gas (GHG) emissions. Despite proven ecosystem benefits associated with grazing, range livestock and land managers are typically overlooked in GHG reduction and carbon sequestration incentive programs. South Dakota State University, along with ten external partners, was recently awarded a commodity development grant for grazing beef cattle and bison. A primary goal of this project is to provide economic incentives to beef and bison producers to implement land practices that have carbon and GHG benefits. Furthermore, this project will develop producer-friendly carbon and GHG measuring and monitoring technologies. The long-term goal of this project is to develop sustainable market opportunities for producers that employ climate-smart practices on their operations. Such opportunities can include not only emerging carbon markets, but commodity markets with premiums assigned to responsibly raised beef and bison. Accessible measuring and monitoring technologies are needed by producers to verify the impacts of their practices and successfully enter into these emerging market opportunities. Producer partnership is key to the success of this project and long-term market sustainability. We will work with specific groups of producers to ensure widespread participation, including: (1) established and early practice adopters; (2) late practice adopters; and (3) underserved groups, including indigenous producers. Through this project, we will work with producers that manage more than 970,000 ha of grazing lands. To accomplish this, we have developed a multi-tiered approach between academia, governmental agencies and industry to facilitate sustainable and effective climate-smart practice implementation. Providing economic incentives to producers for adopting climate-smart practices will help overcome transactional entry barriers often associated with new practice implementation and will position producers to merge into new market opportunities aligned with consumer preferences and demand.

Assessing impacts of new legislation on Dutch dairy farms using a revised linear programming model

A. Van Der Linden[1], L.M. Alderkamp[1], C.W. Klootwijk[2], G. Holshof[2], N. Van Eekeren[3], F. Taube[1,4] and C.E. Van Middelaar[1]
[1]Wageningen University, Animal Production Systems group, De Elst 1, 6708 WD Wageningen, the Netherlands, [2]Wageningen Livestock Research, Animal Nutrition, De Elst 1, 6708 WD Wageningen, the Netherlands, [3]Louis Bolk Institute, Kosterijland 3-5, 3981 AJ Bunnik, the Netherlands, [4]Christian Albrechts University, Grass and Forage Science/Organic Agriculture, Hermann-Rodewald Str. 9, 24118 Kiel, Germany; aart.vanderlinden@wur.nl

The Dutch agricultural sector is not eligible for derogation anymore in 2023, which affects dairy farmers because the application room for nitrogen (N) in animal manure is reduced to the European standard of 170 kg/ha. The objective of this study is to quantify the impacts of new legislation, including the loss of derogation, on the economic and environmental performance of Dutch dairy farms. An existing linear programming model for Dutch dairy farms was revised. During revision, the model was written in the R programming language, input parameters were updated, and new legislation was included. As a potential measure to deal with the new legislation, parameters of grass-white clover and grass-white red clover sward were added to the model. Labor income of a typical farm on a sandy soil in the Netherlands (50 ha, Holstein-Friesian cows) was maximized given legislation and available resources on-farm. Due to new legislation, the annual labour income decreased by € 18,350, and nitrogen (N) surplus decreased by 63 kg/ha. The number of cows decreased from 101 to 93, and the percentage maize land increased from 20 to 31% of the total area. Artificial N fertilizer use decreased by 18 kg/ha. New legislation and adoption of grass-white clover (66% land area) decreased income only by € 5,900 and decreased N surplus by 77 kg/ha. Grass with white and red clover (21% land area) decreased income by €8,700 and decreased N surplus by 48 kg/ha. In conclusion, model results indicated that labour income and N surplus will both decrease under the new legislation. Fewer animals are kept and more maize is cultivated. Adopting grass clover-swards could mitigate the decrease in labour income and reduce N surplus depending on the sward type, which shows their potential to improve environmental and economic performance of dairy farms.

Gaseous emissions (building, storage, pasture) of dairy systems combining or not grazing and housing

N. Edouard[1], X. Vergé[2], C. Flechard[3], Y. Fauvel[3] and A. Jacotot[3]
[1]INRAE, Institut Agro, Pegase, 35590 Saint-Gilles, France, [2]Institut de l'élevage, Monvoisin, 35650 Le Rheu, France, [3]INRAE, Institut Agro, SAS, 35000 Rennes, France; nadege.edouard@inrae.fr

Sustainable dairy farms need to make better use of feed resources, reduce the use of inputs and their environmental impacts, particularly in terms of nitrogen (N) losses. Dairy cattle are largely fed on grazed grass in Western Europe but, at certain times of the year, conserved forages and concentrates may be added to the animal diet. Few studies have investigated the consequences of this combination on the animal's N use and manure composition. Moreover, in these situations, animals divide their time between grazing, where urine and solid excreta fall directly onto the soil, and the building, where manure need to be managed and stored, leading to contrasted impacts on the environment. Our project focuses on strategies combining grazed and conserved forages in dairy systems and their consequences on N flows and environmental impacts. Several experiments were conducted to compare animal performance, N use efficiency and gaseous emissions (ammonia and greenhouse gases) of full-housing (FH) vs half-housing-half-grazing (HH-HG) vs full grazing (FG) management systems in spring and autumn 2022. In the FH treatment, cows were housed in mechanically ventilated rooms where they were fed a basic diet of maize silage and concentrates *ad libitum*. Manure was scraped, collected and transferred to controlled pens. Gaseous emissions were measured in the house and during manure storage by spot air samples, with several methods. In the FG treatment, cows grazed a temporary pasture equipped with an eddy covariance flux tower and several trace gas infrared analysers (NH_3, N_2O, CH_4, CO_2, H_2O), and ALPHA passive diffusion samplers for NH_3 coupled with short-range atmospheric dispersion modelling for the determination of field-scale gaseous emissions. Cows on the HH-HG treatment were housed in a mechanically ventilated room at night (receiving 8 kg DM of the basic diet) and grazed on a temporary pasture during the day (8 hours). The results will contribute to the acquisition of new knowledge on these mixed systems especially in terms of gaseous losses over the whole continuum of cattle feeding and manure management.

Agri-food and net zero: a science-policy-society perspective

R. McGuire[1], S. Huws[1], C. Foyer[2], P. Forster[3], M. Welham[4], L. Spadavecchia[5], D. Curry[6] and N. Scollan[1]
[1]Queen's University Belfast, Belfast, BT9 5DL, United Kingdom, [2]University of Birmingham, Birmingham, B15 2TT, United Kingdom, [3]University of Leeds, Leeds, LS2 9JT, United Kingdom, [4]UK Research and Innovation Biotechnology and Biological Sciences Research Council, Swindon, SN2 1FL, United Kingdom, [5]The Department for the Environment, Food and Rural Affairs, London, SW1P 4DF, United Kingdom, [6]House of Lords, London, London, SW1A 0PW, United Kingdom; r.mcguire@qub.ac.uk

Globally, agriculture is responsible for up to 8.5% of all greenhouse gas emissions – whilst in the UK, the agri-food system produces ~23% of UK emissions. Despite this significance, mitigations must be based on scientific knowledge, technological innovation, and the socio-economic capacities of industry and society. Hence, the need for a science-policy-society interface. Using the UK as a case study, through a series of multi-stakeholder workshops, the key policy, political and scientific challenges to a net zero agri-food system were scrutinised, whilst demonstrating farmer's views to a net zero strategy. Overall, in terms of effective mitigation, it is essential that agri-food does not wait for 'perfection'. Indeed, rapid advances to net zero are possible using existing knowledge and technology, supported by targeted policies. The most significant barrier is the lack of plausible road maps that enjoy widespread support and confidence. Such road maps must detail feasible emission mitigation targets for farmers. This sector's ability to translate scientific knowledge into practical application is poor and the knowledge-exchange processes are fragmented. Governments must urgently outline their export strategy, sustain national production standards (particularly animal health and welfare) and outline how they plan to work in partnership with their own farmers. There is a view among some stakeholders that governmental regulation should be the last resort, because it can compromise innovation. The key challenge for a net zero strategy is to create the conditions that initiate industry buy-in, controls for the different capacities of farm systems and delivers the evidence of climate-friendly food production to consumers.

Testing the effects of grassland swards for yield, greenhouse gas emissions and soil health

P.R. Hargreaves[1], D.Z. Kreismane[2], A. Dorbe[2], K. Klumpp[3], J. Bloor[3], A. Cieślak[4], M. Szumacher[4], A. Szejner[4] and R.M. Rees[1]
[1]SRUC, Kings Buildings, West Mains Road, Edinburgh, EH9 3JG, Edinburgh, United Kingdom, [2]Latvia University of Life Sciences and Technologies, Faculty of Agriculture, Institute of Soil and Plant sciences, 2 Liela Street, Jelgava, 3001, Latvia, [3]INRAE, VetAgro Sup, Grassland Ecosystem Research Unit, Université Clermont Auvergne, 5 Chemin de Beaulieu, Clermont Ferrand, France, [4]Poznan University of Life Sciences, Department of Animal Nutrition, 60-637 Poznań, Wołynska 33, Poland; paul.hargreaves@sruc.ac.uk

As part of the Climate Care Cattle Farming mixtures of red clover and other species were tested to assess their contribution to the mitigation of greenhouse gas (GHG) emissions (carbon dioxide (CO_2), nitrous oxide (N_2O) and methane (CH_4)) and the promotion of soil health, whilst retaining yield. Four countries (France, Latvia, Poland, Scotland) established experiments. Treatments were: red clover + perennial ryegrass with fertiliser, red clover + perennial ryegrass, red clover + chicory and red clover + tonic plantain. Plots were cut and the yield calculated as dry matter (t/ha). The experiments ran for two years (2021 to 2022). Soil samples were taken on establishment, at the end of the first year and twice in the second year and analysed for pH, nutrients, N and C. Soil structure was assessed using a visual method and bulk density. Gas samples were taken weekly during the growing seasons, using static chambers and analysed for CO_2, N_2O and CH_4. Results showed variations in dry matter (DM) with the perennial ryegrass + red clover providing similar amounts of DM even with the addition of the fertiliser. The tonic plantain + red clover provided similar yield as the perennial ryegrass mixtures however, the chicory gave the lowest yields for most of the experiments (France, Poland and Scotland). The emissions of N_2O varied across the sites but were lower for the tonic plantain (France and Scotland) and greatest for the perennial ryegrass (fertilised). The tonic plantain gave similar yield to the perennial ryegrass and produced less N_2O emissions. That the two perennial ryegrass mixtures with and without fertiliser produced comparable yield indicated inorganic fertiliser use could be reduced, saving costs on farm and GHG emissions.

Session 05 Theatre 8

How to mitigate methane and ammonia emissions at the farm level with innovative approaches

P.J. Galama, H.J. Van Dooren, H. De Boer and C. Schep
Wageningen University and Research, Livestock research, De Elst 1, 6700 AH Wageningen, the Netherlands; paul.galama@wur.nl

Complying with intensive societal discussions, the Dutch Dairy chain has set goals for 2030 to increase the sustainability on the topics of climate, welfare, grazing, biodiversity, environment, new business model and land based farming. Also, a coalition of several Dairy organizations have set management goals about dilution of manure, grazing and protein in ration together with the Ministry of Agriculture to reduce the nitrogen losses, especially ammonia emission. The potential to reduce ammonia and greenhouse gas emissions with these management measures and investment in housing systems like floor types to separate faeces and urine, daily removal of manure from the barn, different freewalk housing systems, Cow-toilet and air extraction systems will be shown. The challenge is to design a cow barn that improves animal welfare, manure quality and reduces emissions. These indicators were studied in case control studies with groups of 16 cows at research station Dairy Campus. The Cowtoilet is an automatic urinal that cows use voluntarily in a concentrate feeder. It collects 35% of the urine production and reduced the ammonia emission by around 35-45%. A permeable plate on a slatted floor improves the walkability of the cows and collects all the urine underneath the floor. The ammonia emission can be reduced between 35-50% by acidification of the urine, flushing the plates with the urine or by spraying 20 litre water per cow per day on the floor in combination with a urease inhibitor. A freewalk housing system with woodchips bedding material decreased the ammonia emission with 32% but did increase methane emission with 30%. A new development is a freewalk system with sand bedding that separates the urine by drains at the bottom of the bedding. The faeces are picked up by a bedding cleaner behind the tractor. The data of 12,000 dairy farmers using the Annual Nutrient Cycle Assessment tool (ANCA) were analysed and show the importance of fertilizing, feeding and housing systems on the emissions of ammonia and greenhouse gasses. it illustrates that low emissions of ammonia can go hand-in-hand with low emissions of greenhouse gasses.

Session 05 Theatre 9

Effect of mitigation measures on GHG and ammonia emissions of pilot farms in European countries

D. Ruska[1], K. Naglis-Liepa[2], P. Hargreaves[3], A. Lenerts[2], M. De Vries[4], P. Galma[4], A. Kuipers[4], A. Cieslak[5], J. Bell[3], R. Rees[3], V. Juskiene[6] and M. Skorupka[5]
[1]Latvia University of Life Sciences and technologies, Faculty of Agriculture, Institute of Animal Sciences, 2 Liela street, LV-3001, Jelgava, Latvia, [2]Latvia University of Life Sciences and technologies, Faculty of Economics and Social Development, 18 Svetes Street, LV-3001, Jelgava, Latvia, [3]Scotlands Rural College (SRUC), Barony Campus, Parkgate, DG1 3NE Dumfries, United Kingdom, [4]Wageningen Livestock Research, De Elst 1, 6708 WD, the Netherlands, [5]Poznań University of Life Sciences, Department of Animal Nutrition, Wolynska 33, 60-637 Poznan, Poland, [6]Lithuanian University of Health Sciences, R. Zebenkos, 1282317 Baisogala, Lithuania; diana.ruska@lbtu.lv

A large number of solutions to reduce greenhouse gas (GHG) and ammonia emissions have been proposed for the dairy sector, but their suitability and impact may be strongly influenced by local conditions and characteristics of farming systems. The aim of this research, therefore, was to choose mitigation measures with farmers and simulate effects of measures on GHG and ammonia emissions based on actual farm situations. First, a group of experts from eight countries summarised promising mitigation measures for simulation in carbon footprint tools, and chose options from them that were most suitable for the conditions in their country. In total thirteen mitigation measures for dairy farms were chosen. Second, the list of mitigation measures was discussed with project pilot farmers, who chose measures more suitable for their farm. Expectations of the most effective measures for the simulation process from farmers' and experts' points of view were renewable energy sources on farm, methane blockers as a feed additive, and improve of feed efficiency. Third, measures were simulated using two carbon footprint calculation tools. Simulated results were discussed with the farmers involved, and a farm plan prepared from the suggested measures, estimated reductions, and economic evaluation.

Economic aspects of mitigation practices on pilot dairy farms in Europe

K. Naglis-Liepa, D. Ruska, A. Lenerts and D. Kreismane
Latvia University of Life Sciences and Technologies, Svētes iela 18, 3001, Latvia; kaspars.naglis@lbtu.lv

The EU has taken a leading role in mitigating climate change and preserving the environment, where farmers also have an important role to play. EU countries are responsible for implementing measures to reduce GHG emissions, which would ensure the achievement of the EU Green Deal objectives. The practical implementation of GHG reducing measures differs significantly both in terms of practical implementation and their impact on emission reduction potential as well as farm economics. One of the tasks of the Climate Care Cattle Farming Systems study is to assess the socio-economic impact of GHG emission reduction practices. A pilot farm from 8 EU countries participates in the study, where the impact of GHG measures on emission reduction is simulated and the costs of implementing and maintaining the measures are determined. This information is used to construct the Marginal Abatement Cost Curve, which allows the evaluation of the effectiveness of these measures. The results provide interesting insights into the impact of equal GHG and ammonia measures on European farmers.

Session 05 Theatre 11

Specialized or integrated systems: on-farm eco-efficiency of dairy farming in northern Germany

R. Loges, T. Reinsch, I. Vogeler, C. Kluss and F. Taube
Kiel University, Grass and Forage Science/Organic Agriculture, Hermann-Rodewald-Str. 9, 24118 Kiel, Germany; rloges@email.uni-kiel.de

Potential advantages with regards to ecosystem services of alternative systems such as full-grazing (FG) or integrated dairy/cash-crop (IFG) systems compared to intensive all year indoor feeding (IC) systems for dairying are widely discussed. To investigate performance and environmental impacts, we compared four prevailing dairy systems in an on-farm study. The farm types differed in their quantity of resource inputs and access to pasture: (1) an IC with high import of supplements and mineral fertilizers; (2) a semi-confinement (SC) with daytime grazing during summer and moderate import of supplementary feeds (3) FG based on grazed grass-clover with no purchased N-fertilizers and low quantities of supplementary feeds; and (4) an IFG comparable to FG based on grass-clover leys integrated in a cash-crop rotation. Results revealed highest milk productivity (16 t energy-corrected-milk (ECM)/ha) and a farm-N-balance of 230 kg N/ha) in IC; however, the highest product carbon footprint (PCF; 1.2 CO_2eq/kg ECM) and highest N-footprint (13 g N/kg ECM) were found in the SC system. The PCF in FG were comparable to IC (0.9 vs 1.1 kg CO_2eq/kg ECM) but at a lower N-footprint (9 vs 12 g N/kg ECM). However, the farm-N-surplus in the FG system exceeded 90 kg N/ha. A further reduction was possible in the IFG by accounting for a potential N-carry-over from N-rich plant residues to the cash-crop unit, leading to the lowest PCF (0.6 kg CO_2eq/kg ECM) for the IFG. According to this study based on field data, improved integrated grazing systems could provide an important opportunity to increase the ecosystem services from dairy farming.

Session 05 Theatre 12

Modelling methane emission from dairy cows, barn, and storage

S. Lashkari[1], F.R. Dalby[2] and C.F. Børsting[1]
[1]Aarhus University, Department of Animal and Veterinary Sciences, 8830 Tjele, Denmark, [2]Aarhus University, Department of Biological and Chemical Engineering, Environmental Engineering, 8000 Aarhus, Denmark; saman.l@anivet.au.dk

There is a large variation among dairy farms in the productivity, feed intake, feed composition, and slurry management in the barn and in outside storage, which affect emissions. The aim of the present study was to develop a model to quantify enteric methane emission and emission from barn and storage based on diet composition, performance level and farm management. The annual emission was modelled for a dairy farm with 200 cows (11.2 tons milk/cow/year and 8,480 kg DMI/cow/year based on the Danish Manure Normative System). Slurry was removed every 28 d from a barn with a ring-channel slurry storage under a slatted floor. Slurry was stored in an outside slurry storage with a natural crust and without tent cover until field application (avg. ret. time of 3.2 months). Nutrient composition of feedstuffs was obtained from Norfor feed table. Enteric methane was estimated based on DMI, fat (g/kg DM), and NDF (g/kg DM). Nutrient composition of faeces was calculated based on feedstuff composition and digestibilities of organic matter, sugar, starch, protein, and fat. Methane emission from slurry was estimated with the anaerobic biodegradation model, which tracks organic matter conversion to methane in two steps: (1) hydrolysis of organic matter to VFA; and (2) VFA conversion to methane and CO_2 by methanogens. Hydrolysis rate constants were applied individually for each organic matter component in the slurry (protein, fat, fibre, etc.). The hydrolysis rate constants were derived from laboratory experiments with pig slurry incubated at different temperatures and are preliminary. Three diets with a wide range of concentrate to forage ratios (49:51%; diet 1, 70:30%; diet 2, and 91:9%; diet 3) were used to investigate the effect of extremely different diet compositions on emissions. Enteric methane emission was 55, 51, and 47 (tons/year) in diets 1, 2, and 3, respectively, and methane emission from the barn was 6.0, 6.1, and 6.2, and it was 9.8, 9.4, 9.0 (tons/year) from the storage in diets 1, 2, and 3, respectively. These preliminary modelling results indicate that feed composition has an impact on the total emission of methane, mainly due to differences in enteric methane.

Session 05 Theatre 13

Impact of sorbent application on NH_3 and CO_2 and CH_4 gas emissions from dairy cattle manure

V. Juskiene, R. Juska, G. Kadziene, D. Stankeviciene and R. Juodka
Lithuanian University of Health Science, Animal Science Institute, R. Žebenkos 12, Baisogala, 82317, Lithuania; violeta.juskiene@lsmuni.lt

Livestock manure storage is a significant source of gas emissions. Application of various manure additives is one of possible ways to reduce gas emissions. Therefore, this study was conducted to evaluate the effectiveness of sorbents for NH_3, CO_2 and CH_4 emissions from dairy cattle manure. The study was carried out on laboratory scale, using 90-litter tanks. Three treatments, including biochar, peat and dolomite were applied to stored fresh cow liquid manure. The manure was stored for 42 days under a constant ambient temperature of 14 °C and 60-62% relative humidity. The emission reductions were evaluated by the treatments as compared to the control group without sorbents. The results of the study showed that average NH_3 emission without using sorbents during all study period was 9.2±1.6 mg/m²/h, using biochar – 8.7±1.4 mg/m²/h, peat 9.3±21.7 mg/m²/h, and dolomite 9.1±1.5 mg/m²/h. The addition of biochar reduced NH_3 emissions from manure by 5.9%, dolomite – by 0.6%, however, peat increased the emission by 0.8%. The highest average CO_2 emission was fixed from manure with dolomite treatment – 155.1±8.2 mg/m²/h, and the lowest from biochar treatment – 120.5±8, 0 mg/m²/h (P=0.03). CO_2 emissions from manure without sorbents and manure with peat were 133.1±8.2 and 147.5±9.8 mg/m²/h, respectively. It was found that the highest emission values in all groups were observed in the first week, then they gradually decreased and at the end of the study were negligible and the differences between the various treatments were the most insignificant. The highest positive correlation (0.973) between NH_3 and CO_2 emissions was determined for manure treated with dolomite. Sorbents reduced CH_4 emissions from manure on average by 33.6%. In summary, it can be concluded that sorbents most efficiently reduced gas emissions in the first week after adding the sorbents to manure, then this effect decreased. The most effective emission reduction was achieved using biochar. However, at the same time, it can be stated that the ability of sorbents to reduce gas emissions from liquid cow manure is not high, perhaps due to the specific property of cattle manure to form a natural crust during storage.

Greenhouse gas emissions on dairy farms in Slovenia

M. Klopčič, T. Perčič, M. Bric, S. Rogina, G. Šen, T. Kramer, J. Klopčič, D. Drašler and A. Dolinar
University of Ljubljana, Biotechnical Faculty, Department of Animal Science, Groblje 3, 1230, Slovenia; marija.klopcic@bf.uni-lj.si

Greenhouse gas emissions on dairy farms are influenced by various management-related factors such as animal nutrition (feed ratio and feeding regime), housing of dairy cattle, disposal and storage system and application of animal excreta, barn ventilation. In addition, breed of the animal, the level of milk yield, the micro-climate of the barn, and local weather conditions also have an important influence. As part of the EIP-AGRI project 'Innovative environmental-climate farm management systems on dairy cattle farms', greenhouse gas concentrations are measured monthly on 11 farms with dairy cows in Slovenia in connection with different housing systems (tied-in housing system, free barn with cubicles, compost bedded pack barn, barn with artificial floor), with different farming system (conventional / organic) and depending on the season. Measurements of greenhouse gases (CH_4, CO_2, N_2O) and NH_3 concentrations on selected dairy farms has been performed once a month with the use of an FTIR analyser for ambient air measurements, with the possibility of measuring 25 gases at different, predetermined locations inside and outside the barn. Simultaneously with these measurements, we also perform measurements of the micro-climate parameters of the barn (temperature, humidity, air flow) at these same locations. Based on numerous measurements over the past year, we have identified large differences already in the barn itself - different places where the measurements are taken, big differences between farms and large differences between months.

'Greenanimo' – green future through research

S.Y. Georgieva
Trakia University, Faculty of Agriculture, Students campus, 6000 Stara Zagora, Bulgaria; svetlana.georgieva8888@abv.bg

Project 'GREENANIMO' aims to significantly strengthen the competencies of Trakia University, an institution of excellence in Bulgaria, in agro-ecological herbivores productivity and meat quality, by creating a strong collaborative link with two scientific institutions INRAE, France, and SRUC, Scotland that are international leaders in this field. The target areas and fields of tasks of the project are as follows: 'Enhancing meat quality'; 'Increasing feed efficiency'; 'Improving animal welfare'; 'Designing sustainable ruminant farming systems'. The challenge for the Bulgarian researchers is to bridge the gap between sciences, business and education-training, which will enable the livestock sector and rural areas to become more sustainable and competitive. By working with the researchers from the internationally-leading partners INRAE and SRUC, knowledge transfer and successful integration into new networks is being achieved. The project clearly addresses all instruments as active measures in the project implementation: scientific exchanges; expert visits and medium and short-term on-site and virtual training; conference attendance, workshops; training courses for students and farmers, etc. Dissemination activities and promotion of newly-developed modules, knowledge, and experiences are oriented to the young and early career researchers and business through a multidimensional interdisciplinary system for scientific and popular media activity, and publications, as well as social demonstration and programmes. Addressing the problems of the primary production from ruminants and the needs for a thematic University centre interacting with the meat quality and productivity systems is a core challenge for the present project. Integration of the project in the University life and educational future is optimized by the students' activity and post graduate programs and individual courses with a practical training set.

Session 06 Theatre 2

PLS-DA analysis on hand-held spectrometer for on-line prediction of beef marbling in slaughterhouses

M. Kombolo[1], A. Goi[2], M. Santinello[2], N. Rampado[2], S. Atanassova[3], J. Liu[1], P. Faure[4], L. Thoumy[5], A. Neveu[6], D. Andueza[1], M. De Marchi[2] and J.-F. Hocquette[1]
[1]INRAE, UMR1213, Rte de Theix, 63122 Saint-Genès-Champanelle, 63122, France, [2]University of Padova, Viale dell'Università 16, 35020 Legnaro, Italy, Italy, [3]Trakia University, Faculty of Agriculture, Stara Zagora, 6000, Bulgaria, [4]INRAE, Herbipôle, Rte de Theix, 63122 Saint-Genès-Champanelle, 63122, France, [5]IDELE, Boulevard des Arcades, 87000 Limoges, France, France, [6]IMR3GF, Smulikowskiego St. 4/217, 00-389 Warsaw, France; moise.kombolo-ngah@inrae.fr

Few studies have used near infrared (NIR) spectroscopy to assess meat quality traits directly in the chiller. This study aimed to predict marbling scores with a handheld NIR spectrometer operating in the 740-1,070 nm region on intact meat muscles in the chiller in France and Italy. Marbling was assessed according to the 3G (Global Grading Guaranteed) protocol. The scores ranged from 100 to 1,190 with a mean of 330. Five scans were performed at different points of the *Longissimus thoracis* muscle. Two PLS-DA models were used with or without sex and breed included. The first dataset was made of 677 samples (with sex and breed known), the second dataset was made of all of the 829 samples. The models were developed using an external validation set. Both models gave similar outcomes. The models were first evaluated using a confusion matrix which describes the classification performance. The overall accuracy for both confusion matrices was 61%. The model was also assessed with receiver operating characteristic (ROC) curves where AUROC corresponds to the area under a ROC curve and a single value indicates the overall performance of a binary classifier. It ranges from 0.5 to 1 where the lowest value represents a random classifier and the maximum value represents a perfect classifier. AUROC values were higher for the low and high classes (ranging between 0.8 and 0.7). Finally, permutation plots were obtained for each class, using 100 permutations. Values of the permuted R^2 (the explained variance) and Q^2 (the predictive capability of the model) indicated that only the medium class prediction could be built randomly. In conclusion, results did provide a moderate prediction of the marbling scores which can be useful in the European industry context to predict low and high classes of MSA marbling.

Session 06 Theatre 3

Application of computed tomography and hyperspectral images for enhanced meat quality evaluation

T. Stoyanchev[1], I. Penchev[2], S. Georgieva[2], A. Daskalova[1] and S. Atanassova[2]
[1]Trakia University, Faculty of Veterinary Medicine, Students Campus, 6000 Stara Zagora, Bulgaria, [2]Trakia University, Faculty of Agriculture, Students Campus, 6000 Stara Zagora, Bulgaria; todor.stoyanchev@trakia-uni.bg

Meat quality is difficult to determine because it is a combination of microbiological, nutritional, technological, and organoleptic components. Classical methods for measuring meat quality require the destruction of samples. Recently, X-ray computed tomography (CT) and hyperspectral imaging (HSI) have been investigated as tools for non-contact inspection and monitoring of meat and whole carcasses. This study aimed to evaluate the capabilities of CT and HSI to assess the quality of beef cuts. Different steaks (Ribeye steak; Denver steak; Rump steak; Top-sirloin steak) were purchased from the city market in Stara Zagora. The meat samples were analysed by CT tomography Somatom Go (Siemens-Healthcare, Germany) and Hyperspectral camera AVT Goldeye CL-008, (Specim, Spectral Imaging Ltd. Oulu, Finland) in the spectral range 900-1,700 nm. Spectronon software (Resonon Inc. Bozeman, MT, USA} was used for processing hyperspectral images. Fat; protein; dry matter and ash content of meat samples were determined by classical laboratory methods. PLS regression was used for the quantitative determination of the chemical content of meat. The percentage of lean meat and fat was estimated on the base of CT and HSI images. A comparison of the determination accuracy of the two investigated methods was made. Equations for estimation of the chemical composition of meat samples based on spectral information in the near-infrared range provide good accuracy of the determination. In conclusion, results show the capabilities of the CT and HSI for fast and non-destructive estimation of meat quality, which can be useful in the meat industry.

Effect of husbandry factors on marbling deposition

A. Nicolazo De Barmon[1], I. Legrand[1], J. Normand[1] and J.J. Bertron[2]
[1]Institut de l'Elevage, Service qualité des carcasses et des viandes, 149, rue de Bercy, 75012 PARIS, France, [2]Institut de l'Elevage, Service production des viandes, 149, rue de Bercy, 75012 PARIS, France; aubert.nicolazodebarmon@idele.fr

Various studies confirm the positive impact of marbling on the overall palatability of meat. Thus, this criterion has been chosen by the French beef interbranch organization (INTERBEV) as a priority to better answer to consumers expectations. However, husbandry practices that enhance marbling deposition are partially known. The objective of the present study is to identify practices which allow to produce marbled meat from beef breeds females (Limousines and Charolaises). Measurements on carcasses were made in slaughterhouses with the new French marbling grid (from 1: no marbling, to 6: very high marbling). Then, two farm's groups were separated: a group producing carcasses with low marbling (LM – marbling score 2.2±0.8) and another group with high marbling carcasses (HM – marbling score 3.7±0.9). Interviews of breeders were conducted to collect husbandry practices to try to explain marbling levels. First, they had to define the major genetic type of their livestock between: 'beefy type' selected on muscle deposition, 'livestock type' chosen for maternal qualities and body size or 'mixed type', which is intermediate. 'Beefy type' is largely represented in LM group and 'livestock type' in HM group. Results indicate two important nutrition periods that may affect marbling: between 5 and 12 months, a period embracing the 'marbling windows' already identified for Anglo-Saxon breeds, and during finishing. Differences between the two extreme groups are important during these two periods. Between 5 and 12 months, the HM group distribute concentrate during a longer period (5.1±2.4 months) than LM (2.6±2.0 months). Moreover, HM distribute more concentrate than LM (around 30% of HM breeders give it *ad libitum* vs 0% for LM). Fattening periods are longer for HM group (5.7±1.4 months) than for LM (2.9±0.9 months). In addition, energetics levels during fattening are more important in the HM group (+1.3 UFV/d compared to LM group). The efficiency, technical and economical feasibility of the practices identified in this study must be confirmed in experimental farms to make them operational.

Effect of birth type on meat quality in Ile-de-France lambs

E. Achkakanova[1], I. Penchev[2] and S. Georgieva[3]
[1]Institute of Animal Science – Kostinbrod, Spirka Pochivka, 2232, Kostinbrod, Bulgaria, [2]Trakia University, Faculty of Agriculture, Department of Animal husbandry – Ruminant animals and animal products technologies, Students campus, 6000, Stara Zagora, Bulgaria, [3]Trakia University, Faculty of Agriculture, Department of Fundamental sciences in animal husbandry, Students campus, 6000, Stara Zagora, Bulgaria; ivan.penchev@trakia-uni.bg

The aim of the present study was to investigate the influence of the factor 'birth type' in lambs on the performance of the meat obtained from them. The studied lambs were of the Ile de France breed and were divided into two groups: the first group is *Single* and the second group *Multiple*. The lambs were slaughtered at 120 days of age, with 4 lambs from each group slaughtered. Samples were taken at 24 hours *post mortem* from the following muscles: m. longissimus thoracis et lumborum (LTL), m. iliopsoas (IP) and m. semimembranous (SM). The research of the samples for the chemical composition and technological qualities of the meat was carried out 48 h *post mortem*, stored at 4 °C. Regarding the total chemical composition of the meat, no significant difference was observed between the two study groups. There was also no significant difference between the groups in terms of technological qualities (pH value, colour, roasting losses, brittleness). Only a significant difference ($P<0.01$) was found in the indicator of water-holding capacity between the groups in the studied LTL and IP muscles.

Session 06 Theatre 6

Relationships between lamb feed efficiency, rumen volume and carcass quality measured by CT scanning

N.R. Lambe, A. McLaren, K.A. McLean, J. Gordon and J. Conington
SRUC, SRUC Hill and Mountain Research Centre, FK20 8RU, United Kingdom; nicola.lambe@sruc.ac.uk

There is mixed evidence in the literature about the relationships between feed intake or efficiency and body composition of sheep. In cattle there is some evidence that selection for feed efficiency may reduce fatness at a fixed age or weight. These relationships require further investigation before sustainable strategies to breed for improved efficiency and reduce methane can be proposed. Across two years, Texel × Scotch Mule lambs (n=236 in total) from 10 sires were recorded through individual feed intake recording equipment, after weaning, for a total of six weeks (~14-20 weeks old), following a two week adaptation period. Lambs were CT scanned at the end of the feeding trial and a number of carcass quality traits were calculated from the resulting images, as well as reticulo-rumen volume (RRvol; known to be linked to methane emissions). Residual feed intake (RFI) was calculated for each lamb, by adjusting average daily dry matter intake for live weight, average daily liveweight gain and fixed effects (sex, litter size). Residual values for the CT traits were calculated, after adjusting for fixed effects (sex, year, litter size in which the lamb was reared, age of dam) and live weight at CT scanning. Low to moderate negative correlations between residuals imply that reduced RFI is favourably associated with increased carcass muscle weight, eye muscle area and depth. No significant correlations with RFI were observed for fat traits (carcass fat weight or CT-predicted intramuscular fat), spine traits (length or vertebra number), or RRvol. Low to moderate negative correlations were observed between RRvol and most of the carcass traits, suggesting poorer carcass yield and quality (but reduced fatness) in lambs with higher RRvol, which has previously been associated with higher methane emissions. Larger data sets are being amassed to allow genetic relationships among these traits to be further investigated.

Session 06 Theatre 7

Do biomarkers of residual feed intake in beef cattle remain consistent regardless of feeding level?

G. Cantalapiedra-Hijar[1], K. Nedelkov[2], P. Crosson[3] and M. McGee[3]
[1]INRAE, UMR Herbivores, 63122, France, [2]Trakia University, Faculty of Veterinary Medicine, 6000, Stara Zagora, Bulgaria, [3]TEAGASC, Animal & Grassland Research and Innovation Centre, Grange, Dunsany, Ireland; gonzalo.cantalapiedra@inrae.fr

The use of novel blood biomarkers to predict residual feed intake (RFI) has been proposed as a cost-effective technology to identify feed efficient cattle. However, it is unclear whether these biomarkers are linked to RFI because they reflect the metabolic efficiency of the animal or simply co-vary with the inherent differences in feeding level. This study aimed to determine if plasma biomarkers of RFI, identified under *ad libitum* feeding conditions, remain consistent when animals are feed-restricted on the same grass silage-based diet. Sixty Charolais crossbred young bulls divided into two groups of 30 animals were used in a cross-over design study with two 70-day test periods. Group 1 was fed *ad libitum* in period 1 (A1) and then restricted during period 2 (R2), while the opposite occurred for Group 2 (R1 and then A2). Animals in R1 and R2 were restricted at a level of 1.45% of their body weight. Blood samples were collected from the 12 most divergent RFI (6 Low-RFI, efficient; 6 High-RFI, inefficient) animals in both groups at the end of the first test period, and again on the same animals after the second test period (n=48). Plasma samples were analysed by LC-tandem mass spectrometry and colorimetric methods for quantifying a total of 74 targeted metabolites. Repeated measurements analysis was conducted with the fixed effects of RFI, feeding level and their interaction and the random effect of animal; RFI was considered as either a categorical (Group 1; Low vs High) or continuous (Group 2) variable. Fourteen plasma metabolites had a moderate-to-high repeatability ($0.55 \leq r \leq 0.91$) across both feeding levels. In Group 1, the plasma concentration of α-aminoacidic acid was lower in Low-RFI compared to High-RFI cattle for both feeding levels (FDR=0.02). In Group 2, 5-aminovaleric acid concentration was positively correlated (r=0.72) with RFI across both feeding levels (FDR=0.01). These two metabolites belong to the lysine degradation pathway. Results suggest that metabolic regulations associated with RFI are not solely driven by differences in feeding levels.

Session 06 Theatre 8

Practical application of Observer XT software for behaviour and welfare research in sheep and cattle

N.A. Bozakova and M. Halil
Trakia University, Students campus, 6000 Stara Zagora, Bulgaria; nadiab@abv.bg

In modern scientific research with animals, it is especially relevant to synchronize and combine various vital signs to obtain a complete picture of their welfare. Observer XT software provides detailed and refined data on animal behaviour, but also provides the ability to visualize physiological data, as well as export and synchronize ethological data with other physiological indicators. The aim of the present study is to review the practical use of Observer XT software for synchronizing and integrating ethological observations in sheep and cattle with different physiological parameters in relation to their welfare. To achieve the goal, we reviewed over 250 official documents, and scientific publications through electronic networks – PubMed, Research Gate, and Elsevier, related to the use of Observer XT software to integrate data from video recordings of various behavioural reactions of cattle, calves, sheep, lambs, and goats with their physiological, hormonal, biochemical, immunological and other indicators related to their welfare. As a result, we summarized and systematized the scientific data from the practical use of Observer XT software to synchronize and integrate indicators from ethological observations in sheep and cattle with other multimodal data. This makes it possible to establish relationships and regularities between the various vital indicators and, as a result, to obtain a more complete picture of animal welfare, as well as to significantly improve the quality of scientific research. Based on established data, the Observer XT software is an optimal method for integrating and synchronizing ethological data in research with different physiological parameters in sheep and cattle in relation to their welfare.

Session 06 Theatre 9

Software tools and technologies used to study animal behaviour-benefits and capabilities of Observer

M. Halil, N. Bozakova and S. Georgieva
Trakia university, Animal husbandry, Student campus, Stara Zagora, 6000, Bulgaria; mehmed.halil@trakia-uni.bg

Some behaviours can be used to gain insight into the emotional state and welfare of the animals. It is important to be able to recognize abnormal behaviours, equally important to understand which typical behaviours could be indicative of poor welfare when performed in excess, and which behaviours can indicate positive welfare. Using recording methods to quantitatively assess behaviours related to positive and negative welfare can be a powerful tool for professionals working with farm animals. Welfare assessment methods have evolved significantly in recent decades in terms of both behavioural and physiological indicators. Modern ethological studies require a good knowledge and precise measurements and synchronization of video data of behaviour with the physiological parameters of the studied animals in this scientific field. Nowadays, applications are using that support these processes, such as software for studying animal behaviour – Observer XT. The aim of this research is to investigate the benefits and capabilities of Observer XT in analysing and study of farm animal behaviour, as well as its practical application in teaching students. After numerous observations, trainings and specialized literature searches for the application of this unique software, in the present scientific study, we present that the Observer XT software offers a wide range of possibilities for simultaneous integration and synchronization of video – data from ethological observations, with physiological indicators such as heart rate, abdominal movements, respiratory rate in animals. Due to its flexibility, the application can be successfully used in various fields such as Ethology, Zoology, Veterinary Hygiene and Technology, both for scientific purposes and for training of future specialists.

Session 06 Theatre 10

Animal welfare: from research to practice

I. Veissier[1], V. Brunet[1], D. Ledoux[2] and A. De Boyer Des Roches[2]
[1]INRAE, UMR Herbivores INRAE-VetAgro Sup, Centre de Clermont-Ferrand Theix, 63122 Saint Genes Champanelle, France, [2]VetAgro Sup, UMR Herbivores INRAE-VetAgro Sup, 1 avenue Bourgelat, 69280 Marcy l'Etoile, France; alice.deboyerdesroches@vetagro-sup.fr

In the second half of the 20th century, research on animal welfare began. Initially, most research aimed at identifying and reducing suffering; e.g. research helped to define minimum space allowances per animal or to identify the need for social interactions. Then research focused on what would make animals comfortable; e.g. preferences between lying surfaces have been studied. More recently, the concept of positive welfare has been introduced, which goes beyond the mere satisfaction of needs by providing a rich environment and promoting positive emotions (expressed through play behaviour, positive interactions, exploration, etc.). Much is now known about animal welfare, at least in theory. On the basis of this knowledge, legislation has been adapted in many countries, at least in the EU, to guarantee minimum standards of animal welfare, i.e. essentially to avoid poor welfare. Quality standards have also been developed by several production chains, some of which offer a higher level of welfare than the legislation. To bridge the gap between theory and practice, researchers need to work with stakeholders to define best practices that can be applied on farms (or during transport and slaughter) to ensure a high level of animal welfare. These recommendations should not be seen as an addition to farming practices, but as an integral part of those practices. In other words, the latest knowledge on animal behaviour, sensory and cognitive abilities, health, physiology, etc. should be considered when defining husbandry practices that meet the needs of animals and farmers. We believe that good practices should cover animal needs (basic needs, comfort) and best practices should promote positive welfare. We illustrate this approach with the CARE4DAIRY project, which is developing good and best practice guidelines for the dairy cattle sector. The scientific and technical knowledge will be used to define these practices which will be discussed with stakeholders (farmers, farm advisors, policy makers) before final guidelines are produced. The project covers dairy calves, heifers and cows, including cows at the end of their productive lives.

Session 06 Theatre 11

Effects of warm climatic periods on dairy cow behaviour and welfare in a maritime European climate

M.J. Haskell, I. Sullivan, M. March and L. Shewbridge-Carter
SRUC, West Mains Road, EH9 3JG Edinburgh, United Kingdom; marie.haskell@sruc.ac.uk

Global warming is resulting in an overall increase in temperatures and in the frequency of extreme weather events. In dairy cattle, thresholds within the temperature-humidity index (THI) have been used to indicate points at which cattle will likely experience thermal stress (e.g. a THI threshold of 75 predicts thermal stress). However, high-yielding dairy cows that reside in temperate maritime climates may experience some degree of thermal discomfort below this threshold particularly when they are housed. Housing often results in high levels of humidity. The use of technology such as activity monitors and automated intake measures allow us to monitor responses. The aim of this study was to use technological solutions to assess behavioural changes in response to moderate increases in THI levels. Data from dairy cattle on an experimental unit were used. Data on daily lying times, lying bout frequency, step count, feed and water intake were extracted for 10 pairs of warmer (THI<65) and ten matching cooler (THI=43 to 60) periods. Each period was 3 or more days each, and warm and cooler periods were no more than 5 weeks apart to ensure that the data from the same animals were being compared. Results showed that total daily lying time was shorter during warmer periods than cooler periods ($P<0.05$; means and SEMs (h): warm: 11.4±0.04; cool: 12.0±0.04) with a tendency for cows to have more daily lying bouts in warm periods ($P=0.08$; (counts): warm: 12.1±0.1; cool: 11.9±0.1). However, there was no effect of THI level on the no. of steps taken by cows ($P>0.05$ (counts): warm: 858±6; cool: 856±6). Water intake was higher during warm periods ($P<0.05$: (l) = warm: 79.4±0.6; cool: 71.5±0.5). Milk yield was lower during warm periods than cool periods ($P<0.05$; (l): warm: 30.3±0.2; cool: 30.7±0.2). This suggests that behaviour and milk yield are adversely affected even in conditions that are not traditionally regarded as exceeding cows' ability to cope with thermal challenge. Technological solutions aid in detection on cow thermal distress and may be used as a routine monitoring system.

Behavioural and hormonal effects of intensive sheep farming on milking behaviour in a milking parlour

I. Nedeva, T. Slavov, V. Radev, S. Georgieva and I. Varlyakov
Trakia University, Faculty of Agriculture, Department of Fundamental Sciences in Animal Husbandry, Students campus, 6000 Stara Zagora, Bulgaria; ivelina.nedeva@trakia-uni.bg

The aim of the present study was to evaluate how intensive dairy sheep farming influenced their behaviour during milking in a milking parlour. The sheep (n=633) were divided into groups according to milk yield (high- and low-yielding) and stage of lactation (beginning, middle, end). Using video surveillance and data from morning and evening milking of the automated system installed in the milking parlour, an analysis of ethological parameters order of entry in the milking parlour (EMI) and milking parlour side preference (SMI) was made. The calculated EMI and SMI indices served for individual scoring of each animal for evaluation of studied factors. The blood concentrations of thyroid hormones (T_3 and T_4) and cortisol were assayed during three different seasons (summer, autumn, winter). The daily milk yield of studied sheep was 2.451 l, with a peak of 3.967 l in mid-lactation. The sheep built a stable hierarchical order in the group throughout the entire lactation period, manifested with high values of the EMI index: 668.62. The milk yield had no effect on both the order of entry in the parlour and side preference. The established milking parlour side preference (SMI=69.38%) was not accompanied with preference to the milking place, which facilitate the technological process of milking. A statistically significant effect of the season ($P<0.001$ in the autumn) on blood cortisol, triiodothyronine and thyroxine concentrations was demonstrated with no relation with milk yields. Regardless of the established seasonal changes in adrenal and thyroid gland hormones, they had no effect on the health and welfare of sheep reared in intensive systems, but point out to a more difficult adaptation to this farming system.

Combining cattle and sheep in a grassland-based system: which effects for system multiperformance?

S. Prache, K. Vazeille, B. Sepchat, P. Note, P. Veysset and M. Benoit
INRAE, Theix, 63122 Saint-Genès-Champanelle, France; sophie.prache@inrae.fr

The association of beef cattle and sheep shows benefits at the grazing season level, but a comprehensive assessment at system level is lacking. Three grassland-based organic systems were managed for 4 years as separate farmlets, with similar surface area and stocking rate: one mixed system combining beef cattle and sheep (MIX, 60:40 cattle:sheep livestock units (LU)) and two specialised systems, beef cattle (CAT) and sheep (SH). Calving and lambing were adjusted to grass growth to optimise grazing. Calves were pasture-fed from 3 months old until weaning in October, fattened indoors with haylage and slaughtered at 12-15 months. Lambs were pasture-fed from 1 month old until slaughter; if lambs were not ready for slaughter when the ewes mated, they were stall-finished with concentrates. The decision to supplement adult females with concentrate was based on the achievement of a target body condition score (BCS) at key periods. The decision to treat animals with anthelmintics was based on mean faecal egg excretion remaining below a certain threshold. A higher proportion of lambs were pasture-finished in MIX vs SH due to a higher growth rate which led to a lower age at slaughter. Ewe prolificacy and productivity were higher in MIX vs SH. The level of concentrate consumption and number of anthelmintic treatments in sheep were lower in MIX vs SH. Cow productivity, calf performance, carcass characteristics and the level of external inputs used did not differ between MIX and CAT. However, cow BW gain during the grazing season was higher in MIX vs CAT. These outcomes validated our hypothesis that the association of beef cattle and sheep promoted the self-sufficient production of grass-fed meat in sheep. It also promoted better female BCS and BW at key stages of the reproduction cycle and better development of the females used for replacement, which may enhance animal and system resilience. It improved economic and environmental performance and feed-food competition in the sheep enterprise, due to better animal performance and reduced inputs use, but not in the beef cattle enterprise.

Session 06 Theatre 14

Milking temperament and it relation with test day milk yield in Bulgarian Murrah buffaloes

T. Stepancheva, I. Marinov and Z. Gergovska
Trakia University, Faculty of Agriculture, Department of Animal husbandry – Ruminant animals and animal products technologies, Students campus, 6000 Stara Zagora, Bulgaria; ivaylo.marinov@trakia-uni.bg

The aim of the study was to assess temperament during preparation for milking and the milking itself of Bulgarian Murrah buffaloes reared in Bulgaria. The study included 91 buffalo cows that were between 30 and 240 Days in milk (DIM). Cows were housed under the conditions of tie-stall housing system and milked with a milking pipeline. The average milk yield of cows for standard lactation was 2245.37 kg with 7.77% fat and 4.34% protein content in milk. The average score for temperament during attaching the milking cluster was 1.83, and for milking temperament – 1.93. The highest was the percentage of cows that reacted by leg lifting (18.9%), followed by animals that were moving on the stall bed during milking (10%), cows that definitely kick (9.9%), and 13.3% managed to remove the milking cluster during milking. A higher percentage of cows responded by leg lifting and kicking during the milking cluster attaching compared to milking itself, 27.8 and 13.3%, respectively. During milking, 72.2% of the buffaloes stood still or only have stepped from foot to foot (scores 1 and 2), and 14.5% have shown undesirable behaviour (scores 4 and 5). A significant difference between the first and second temperament scores during preparation for milking and during milking was not reported. With the highest LS-means for test day milk yield (TDMY) were cows with the most undesirable behaviour during milking, scores 5 (8.18 kg) followed by those with a score of 4 (7.65 kg). The milk yields of cows with milking temperament scores from 1 to 3 were almost the same and lower than that of aggressive and nervous cows, respectively from 7.21 to 7.37 kg. In cows with scores 1, 2 and 3, the lactation curves were similar in both shape and variation. In all cows, the maximum milk yield was 7.5 to 8.0 kg and was maintained for several months with small variation. The lactation curve in cows with a score 5 had the most fluent shape and the highest maximum milk yield – over 8.5 kg. The lactation curve in cows with a temperament score 4 was with the most undesirable shape – steep reaching the peak and a sharp decrease in milk yield after the peak.

Session 06 Poster 15

Introduction of lupins in lamb diets: effects on carcass, meat quality and intramuscular fatty acids

M. Almeida[1,2,3], S. Garcia-Santos[1,4], D. Carloto[1], A. Arantes[1], J. Lorenzo[5,6], J.A. Silva[1,2,3], V. Santos[1,2,3], J. Azevedo[1,2,3], C. Guedes[1,2,3], L. Ferreira[1,4] and S. Silva[1,2,3]
[1]University of Trás-os-Montes e Alto Douro, Quinta de Prados, 5000-801, Vila Real, Portugal, [2]Associate Laboratory for Animal and Veterinary Sciences (AL4AnimalS), Portugal, Portugal, Portugal, [3]Veterinary and Animal Research Centre (CECAV), Quinta de Prados, 5000-801, Vila Real, Portugal, [4]Centre for the Research and Technology Agro-Environmental and Biological Sciences (CITAB), Quinta de Prados, 5000-801, Vila Real, Portugal, [5]Área de Tecnología de los Alimentos, Facultad de Ciencias de Ourense, Universidad de Vigo, 32004 Ourense, Spain, [6]Centro Tecnológico de la Carne de Galicia, Rúa Galicia N° 4, Parque Tecnológico de Galicia, San Cibrán das Viñas, 32900 Ourense, Spain; mdantas@utad.pt

Over the last decade, the EU has been focused on solving its dependency on imported soybean for livestock feeding, which means Mediterranean legumes species may represent a local solution and possible replacement candidate. The objective of this preliminary study was to evaluate the effects of partial replacement of soybean meal by lupins on portuguese autochtonous Churra da Terra Quente lambs' diets, on the carcass traits, meat characteristics, and meat fatty acid profile. Two trials were conducted: on trial 1, the soybean meal (control; C) was partially replaced by *Lupinus albus* or *Lupinus luteus* (50 g/kg; LA5 and LL5, respectively); on trial 2, lambs were fed four diets with graded levels of *Lupinus luteus* (0, 100, 150 and 200 g/kg; C, LL10, LL15, LL20, respectively). At the end of the feeding trials, animals were slaughtered to evaluate carcass characteristics and meat composition, including fatty acids. Carcass composition in tissues was not affected ($p>0.05$) by diet in both trials. Also, no significant ($P<0.05$) differences were observed in meat quality attributes between diets on trials 1 and 2. Overall, the *Longissimus* muscle's fatty acid content was not affected by diet ($p>0.05$) in both trials. Carcass and meat quality was overall comparable between lambs fed with soybean meal and lupins, indicating the latter as a potential alternative protein source. However, the lack of significant differences could also be attributed to the small sample size. This work was supported by the projects UIDP/CVT/00772/2020 and LA/P/0059/2020 funded by the Portuguese Foundation for Science and Technology (FCT).

Session 06 Poster 16

Lipid composition of dromedary camels (*Camelus dromedarius*) raised in extensive and intensive system

M. Lamraoui[1], J. Pestana[2,3], M. Costa[2,3], C. Alfaia[2,3], Y. Khelef[4], N. Sahraoui[5], A.M. Almeida[6] and J.A.M. Prates[2,3]
[1]LBBBS, Université de Bejaia, 06000, Algeria, [2]Laboratório Associado para Ciência Animal e Veterinária, AL4AnimalS, Universidade de Lisboa, Portugal, [3]CIISA, Faculdade de Medicina Veterinária, Lisboa, Portugal, [4]LBEH, University of El Oued, 39000, Algeria, [5]LBRA, Université Saad Dahlab, 09000, Algeria, [6]LEAF, Instituto Superior de Agronomia, 1349-017, Portugal; messaouda.lamraoui@univ-bejaia.dz

Although camel meat is a valuable source of protein for human consumption, it remains relatively understudied. In this study, we investigated the effect of production systems on the fatty acid profile and fat-soluble compounds of Sahraoui dromedary camel meat. We collected meat samples from 12 healthy male camels of similar age from both extensive (EPS, n=6) and intensive (IPS, n=6) production systems for analysis. Meat from camels raised in IPS had higher lipid content (3.1 g/100 g fresh weight) than meat from those raised in EPS (1.2 g/100 g). The primary fatty acid present in camel meat was oleic acid, which accounted for 28.7 and 23.9% of the fatty acid composition in meat from IPS and EPS, respectively. Additionally, meat from EPS had higher percentages of saturated fatty acids (53.7%) compared to meat from IPS (44.7%). Moreover, polyunsaturated fatty acids content was higher in meat from animals raised in EPS (19.2%) compared to those raised in IPS (6.6%). Interestingly, cholesterol levels were higher in the meat from camels raised in EPS (0.7 mg/g) than those from IPS (0.52 mg/g). However, there were no significant differences in α-tocopherol content between the two production systems. Results suggest that the observed differences in fatty acid profile and fat-soluble compounds in camel meat could be attributed to variations in animal feeding, which can ultimately affect the growth and development of the animals. The insights gained from this study are highly relevant to camel producers and consumers, as it provides an understanding of its suitability for human diets and could help in improving feeding strategies.

Session 06 Poster 17

Effective application of UAVS in modern livestock agriculture

T. Petrova[1], Z. Petrov[2] and M. Marinov[3]
[1]Trakia University, Faculty of Agriculture, Department of Agricultural Engineering, Students campus, 6000 Stara Zagora, Bulgaria, [2]Rakovski National Defence College, Defence Advanced Research Institute, 82 Evlogi i Hristo Georgievi Blvd., 1000 Sofia, Bulgaria, [3]Georgi Benkovski Air Force Academy, 1 Sv. sv. Cyril and Methodius str., 5855 Dolna Mitropolia, Bulgaria; zhpetrov@gmail.com

This article examines the effectiveness of aerial photography using unmanned aerial vehicles (UAVs) in free-ranging livestock. The goal is to analyse the effectiveness of applying algorithms for contrast enhancement and histogram analysis in image processing. The results of the study of three algorithms for contrast enhancement and histogram analysis of digital images are shown. All studied algorithms can also be used on colour images. The results show that it is very important to use appropriate computational image processing methods to achieve the objectives.

Session 06 Poster 18

Effects of dietary beet pulp on rumen fermentation, beef quality, and intramuscular adiposity

M. Baik, S.H. Lee and S.Y. Kim
Seoul National University, Department of Agricultural Biotechnology, College of Agriculture and Life Sciences, Gwanak-ro 1, Gwanak-gu, Seoul, Republic of Korea, 08826, Korea, South; mgbaik@snu.ac.kr

Beet pulp is a byproduct of sugar beet processing and contains abundant neutral detergent fibre and pectin, which are readily fermentable in the rumen. We investigated the effects of partial replacement of corn flake in the diet with beet pulp on growth performance, ruminal volatile fatty acid profiles, lipogenic gene expression, adipocyte cellularity, and carcass traits in Hanwoo (Korean cattle) steers. The eighteen steers (body weight, 636±10.9 kg; age, 25.9±0.25 months) were equally divided into the corn flake (CF) and beet pulp (BP) groups. Approximately 89% of dry matter of the requirement was offered as a concentrate portion, and the remaining 11% was offered as tall fescue hay. The 78 and 72% of concentrate portion was provided by the pelleted basal concentrate to CF and BP groups, and the remaining 22 and 28% were supplemented with corn flake or beet pulp, respectively. Dietary crude protein and energy levels of two groups were similar. The experiment was conducted for 25 weeks, including a 5-week adaptation period. Average daily gain and feed efficiency were not affected ($P \geq 0.79$) by beet pulp feeding. The proportion of ruminal acetate was higher ($P<0.001$) in the BP group than in the CF group, whereas proportion of ruminal propionate was lower ($P<0.001$) in the BP group. The beef yield grade ($P=0.10$), quality grade ($P=0.10$), and beef price per kg ($P<0.001$) were tended to be higher or higher in the BP group than in the CF group. Intramuscular adipocyte size of longissimus thoracis (LT), which was determined by image analysis of histological section of the LT, was larger ($P<0.001$) in the BP group than in the CF group. Fatty acid synthase mRNA levels determined by qPCR were higher ($P=0.03$) in the BP group than in the CF group. The Increased lipogenic gene expression may, in part, contribute to the increased adipocyte size of the LT by beet pulp feeding. In conclusion, beet pulp could be used as a lipogenic energy source for improving beef quality grade without affecting growth performance of cattle.

Session 06 Poster 19

The identification of high-quality *Perilla frutescens* Tan lamb meat by multi-metabolomics analyses

Y. Yu, B.Y. Zhang, X.Z. Jiang, Y.M. Cui, H.L. Luo and B. Wang
China Agricultural University, State Key Laboratory of Animal Nutrition, College of Animal Science and Technology, 2 Yuanmingyuan west road, Haidian District, 100193, Beijing, China, P.R.; wangb@cau.edu.cn

Consumers are paying increasing attention to eating healthily, thus, food enriched with omega-3 polyunsaturated fatty acids (n-3 PUFAs) is admired worldwide. *Perilla frutescens* seeds and *P. frutescens* oil are widely used in the food market and livestock due to their abundance of α-linolenic acid. This study aimed to investigate effects of dietary *P. frutescens* seed supplementation on animal growth, meat quality, muscle fatty acid profiles, and volatile, lipophilic, and hydrophilic metabolome of Tan-lambs. Forty-five Tan-lambs (approximately 6 months old) were randomly divided into three group with HC (a high-grain based control diet), LC (a low-grain based control die) and PFS (a low-grain based control diet with supplemented 3% *P. frutescens* seed). Longissimus dorsi was collected for this study. Both HC and PFS showed a significantly increased growth performance compared to the LC. The PFS modified the 24h meat colour than LC. Compared to the HC and LC, the PFS increased n-3 PUFA content, resulting in the upregulated ratio of PUFA to saturated fatty acid. The ratio of n-6 PUFA to n-3 PUFA was significantly different among the three groups, with the highest value in the HC and lowest value in the PFS. Untargeted high-resolution LC-MS and untargeted GC-MS metabolomics were further used to analyse muscle volatile, lipophilic, and hydrophilic metabolome in the PFS and LC. In total, 68 differential metabolites and 7 volatile compounds were screened between PFS and LC based on LC-MS and GC-MS, respectively. Among them, 18 lipid molecules, 23 hydrophilic molecules and 4 volatiles were significantly increased. Most of the differential metabolites were mainly enriched in the pathways related to antioxidant function and branched-chain amino acid synthesis. Together, this study demonstrated that dietary *P. frutescens* seed supplementation improved the meat colour and fatty acids profiles, and increased the antioxidant capacity of Tan-lambs, indicating the *P. frutescens* seed can be used to enhance lamb meat quality with enriched contents of healthy and flavour compounds for populations.

Session 06 Poster 20

Effect of feeding garlic-citrus supplement on carcass characteristics and fatty acid profile of beef

M.E. Eckhardt[1], T. Brand[2], S.A. Salami[3], B.M. Tas[3], J.F. Legako[4], T.E. Lawrence[1] and L.W. Lucherk[1]
[1]West Texas A&M University, Canyon, TX 79016, USA, [2]Mootral GmbH, Waldseeweg 6, 13467 Berlin, Germany, [3]Mootral Ltd, Roseheyworth Business Park North, Abertillery, United Kingdom, [4]Texas Tech University, Lubbock, TX 79409, USA; ssalami@mootral.com

Several studies have shown that feeding garlic- and citrus-extract-based supplement (GCE; Mootral Ruminant) reduces enteric methane emissions in ruminants. However, there is limited information on the impact of feeding this supplement on beef quality. This study evaluated the carcass characteristics, proximate composition, and fatty acid (FA%) profile of beef from cattle supplemented with GCE. Twenty feedlot cattle were randomly assigned to 4 groups (5 cattle/group). Two groups were fed for 9 months (control vs GCE) and the other two groups were fed for 12 months (control vs GCE). All animals were fed a typical feedlot diet while the GCE-fed cattle were supplemented with 27 g of GCE/head/day and top-dressed on the ration. At the end of each feeding duration, all animals were slaughtered, and carcasses were evaluated according to the USDA beef grading measures. Strip loin steaks and ground beef patties were prepared at 14- and 15-days *post-mortem*, respectively. Data were statistically analysed using SAS and treatment effects were considered significant when $P<0.05$. Marbling score, backfat, hot carcass weight and dressing percentage were similar between the control and GCE groups. Cattle fed for 9 months had higher backfat compared to those fed for 12 months. Relative to the control, the ribeye area was lower in cattle fed GCE for 9 months but did not differ after 12 months of feeding GCE. Proximate analysis of beef steaks and patties showed that fat, moisture, protein, and collagen were similar between the control and GCE diets. In general, feeding GCE and duration of feeding did not exhibit substantial changes in FA% of beef steaks and patties. Dietary GCE decreased C20:2 in steaks and increased C15:0 and C17:1 in patties. Compared to 9-month feeding, the 12-month steaks had higher C18:2 *n*-6, C20:4 *n*-6 and total polyunsaturated FA whereas only higher C18:2 *n*-6 was observed in 12-month patties. Overall, these results indicate that feeding GCE had minimal or no effect on carcass characteristics and FA profile of beef regardless of the feeding duration.

Session 06 Poster 21

Main microclimatic indicators in a milking parlor for dairy cows

D. Dimov[1], T. Penev[1], I. Marinov[2], J. Mitev[1], T. Miteva[1] and M. Stoynov[1]
[1]Trakia University, Faculty of Agriculture, Department of Ecology and Zoohygiene, Students campus, 6000 Stara Zagora, Bulgaria, [2]Trakia University, Faculty of Agriculture, Department of Animal husbandry – Ruminant animals and animal products technologies, Students campus, 6000, Stara Zagora, Bulgaria; toncho.penev@trakia-uni.bg

The study was conducted in the milking parlour of a cattle farm with a capacity of 400 cows of the Holstein-Friesian breed. The milking installation was a double 8 'Herringbone' type without windows, and the roof was constructed of glass. The reporting of temperature, air humidity and temperature-humidity index (THI) was performed three times during each milking (at the start, in the middle and at the end of milking) with measurements repeated during the morning, midday and evening milking. The highest mean and maximum daytime air temperature values were recorded in summer and spring. Although the average values for the spring season were lower than those for summer (by about 4 °C), the maximum values reached were equally high – 31.4 °C. In terms of relative air humidity, the highest mean values were reported for the winter season – 82.39%. For the other seasons, the relative humidity values were on average high and close in value – from 62.51 to 67.46%. At THI, the highest mean daily and maximum values were reported in the summer months – 73.41 and 80, respectively.

Session 06 Poster 22

Estimation of rib composition and intramuscular fat from DXA or smartphone imaging in crossbred bull

C. Xavier[1,2], B. Meunier[3], I. Morel[2], Q. Delahaye[3], Y. Le Cozler[1], M. Bonnet[3] and S. Lerch[2]
[1]INRAE-Institut Agro, PEGASE, 35590 St-Gilles, France, [2]Ruminant Nutrition and Emissions, Agroscope, 1725 Posieux, Switzerland, [3]INRAE, Université Clermont Auvergne, Vetagro Sup, UMRH, 63122 Saint-Genès-Champanelle, France; xaviercaroline@orange.fr

Aim was to compare dual X-ray absorptiometry (DXA) or smartphone-derived picture for estimating the 11th beef rib tissue composition and intramuscular fat content (IMF). Forty-nine beef-on-dairy crossbred bulls (♀ Swiss Brown × ♂ Angus, Limousin or Simmental) were slaughtered at 519.6±8.5 kg body weight. Left 11th rib was DXA scanned (iLunar, GE Med. Syst., 'Right Arm' mode) and DXA total mass, lean and fat proportions were recorded. An open-access computer image analysis method based on smartphone pictures of both faces of the rib was used to estimate the total rib area, from which bone, muscle (longissimus and others) and adipose tissue (subcutaneous, inter- and longissimus intra-muscular) proportions were recorded. As gold standard measures, rib was dissected, and *longissimus* IMF determined by Soxhlet after acid-hydrolysis. Estimative equations of rib composition from DXA or smartphone data were set-up by linear regressions (R 4.2.2). Rib contained 42.6±3.3% longissimus and 63.5±3.3% total muscles, 12.6±1.8% intermuscular and 18.9±3.0% total adipose tissues, and *longissimus* 1.6±0.5% IMF. Rib longissimus and total muscle proportions were precisely estimated by DXA [root mean square error (RMSE) 1.7 and 1.3%, R^2=0.72 and 0.85], as well as intermuscular and total adipose tissues (RMSE=1.0 and 0.9%, R^2=0.69 and 0.90, respectively). Precision was similar for smartphone cranial ribs cross-section estimations (RMSE=1.1, 1.4, 1.0 and 1.1%, R^2=0.88, 0.82, 0.64 and 0.86, for longissimus, total muscles, intermuscular and total adipose tissues). Longissimus IMF was estimated less precisely, but still satisfactorily by either DXA or smartphone (RMSE=0.28%, R^2=0.58 for both). For smartphone, precision was comparable when based on caudal or the average of both faces. At the exception of intermuscular adipose tissue, crossbreed effect was included (P<0.05) in DXA and smartphone models. A single rib DXA scan or cross-section picture seem promising methods to estimate rib composition in a simple, quick, precise, and non-destructive way.

Session 06 Poster 23

Perception of beef by the Algerian consumer

M. Sadoud[1] and J.F. Hocquette[2]
[1]H. Benbouali Chlef University, Faculty of Science, 0200 Chlef, Algeria, [2]INRAE, UMR1213 Herbivores, 63122, Theix, France; mh.sadoud@univ-chlef.dz

Meat consumption is often the symbolic marker of prosperity of a society and/or of specific socio-economic groups. While meat consumption has decreased in the North of the Mediterranean, it has increased in North Africa (Tunisia, Morocco and Algeria) from 23.5 to 39 kg/year/person during the last decade. The development of cattle breeding has always been a priority for Algeria to meet the needs of its population in animal proteins, particularly in the north where meat is mostly consumed. The cattle population of Algeria is around 2 million heads. The beef self-sufficiency rate is 55%, the rest coming from imports. This communication aims to analyse the perception of beef by Algerian consumers according to their socio-demographic profiles. The survey took place in 2018, with 300 consumers from four different age groups. The questionnaire was sent to heads of households, who are mainly responsible for purchasing meat. This survey made it possible to draw up an inventory of consumer preferences in order to know their perceptions of beef. The surveys comprised 24 questions targeting regular consumers of beef. Different factors were studied: household size, family situation, level of education, income, type of the most purchased meat, frequency of purchase, cut, colour, smell, taste and juiciness of the meat, consumption preferences, cooking time, diet, cuts purchased and consumed. Many factors, such as psychological factors and sensory factors, influence consumer behaviour towards beef. First, the majority of consumers go to the butcher for meat purchases. About 60% of them choose beef due to its perceived high nutritional value due to its richness in proteins. Most of them think it is of satisfactory quality. Second, consumers put importance on specific attributes, namely colour, taste and price, and a lower interest in fat content. In addition, this analysis of meat perception by consumers in the Algerian region shows very different behaviours towards beef. In conclusion, beef is well consumed in Algeria despite a great variability mainly due to different socio-demographic and economic conditions of the population in the studied region resulting in specific consumption patterns.

Session 06 Poster 24

Effect of intramammary infection on milk protein profile measured at the quarter level in dairy cows

V. Bisutti[1], D. Giannuzzi[1], A. Vanzin[1], A. Toscano[1], M. Gianesella[2], S. Pegolo[1] and A. Cecchinato[1]
[1]University of Padova, DAFNAE, viale dell'Università 16, 35020, Italy, [2]University of Padova, MAPS, viale dell'Università 16, 35020, Italy; alessio.cecchinato@unipd.it

Subclinical mastitis is an impactful disease affecting, other than animal health and welfare, also milk productivity, composition, and technological traits. This work evaluated the impact of subclinical intramammary infection (sIMI) induced by four pathogens (*Staphylococcus aureus*, *Streptococcus agalactiae*, *Streptococcus uberis* and *Prototheca* spp.) on the detailed milk protein profile of Holstein cows from three Italian dairy farms. After an initial bacteriological screening (T0) performed on all animals (n=450) to detect cows with sIMI, only the positive ones (n=78) were followed up at a quarter level after two (T1) and six weeks (T2) from T0. In total, 529 quarter milk samples were collected, on which a validated RP-HPLC method was used to identify and quantify 4 caseins (CN; κ-, α_{s1}, α_{s2}, and β-CN), and 3 whey proteins (β-LG, α-LA, and lactoferrin). Traits were analysed using a hierarchical linear mixed model including as fixed effects: cows' DIM, parity, herd, somatic cell count (SCC) in classes, bacteriological status (BACT, negative and positive), and the interaction SCC×BACT. As random effect, we used the individual cow/replicate nested within herd, DIM, and parity, that is the error line on which the latter effects were tested. At T1, sIMI decreased β-CN (-6%, $P<0.01$) but increased κ-CN (+5%, $P<0.05$) and α_{s2}-CN (+6%, $P<0.01$) content, while at T2 we observed an increase of lactoferrin (+7%, $P<0.05$), an antimicrobial peptide, in positive animals. Somatic cell count affected most protein fractions, especially at T2. Increases in SCC were associated with a reduction in β-CN ($P<0.001$) and with increases in κ- and α_{s2}-CN ($P<0.001$ and $P<0.05$, respectively). At T2, the interaction SCC×BACT was associated with α_{s1}-CN ($P<0.05$), with the highest proportion of α_{s1}-CN found in positive samples with SCC≥50,000 and <200,000 cells/ml, while the lowest in positive animals with SCC>400,000 cells/ml. This study added new insights on the alteration driven by sIMI of the milk protein profile, at quarter level. Acknowledgments: This study was part of the LATSAN project funded by MIPAAF.

Session 06 Poster 25

Antioxidant and anti-hypertensive properties of acid whey from Greek yoghurt

E. Dalaka, A. Vaggeli, G.C. Stefos, I. Palamidi, A. Savvidou, I. Politis and G. Theodorou
Agricultural University of Athens, Animal Science, Iera Odos 75, 11855, Athens, Greece; gtheod@aua.gr

Greek yogurt has gained immense popularity due to its high nutritional value. However, acid whey (AW) is produced in enormous volumes and its disposal has a negative environmental impact. Fermentation releases antioxidant (AO) and anti-hypertensive (AH) compounds, supporting that AW can be upcycled to develop value-added products. The aim of this study was to evaluate *in vitro* AO and AH activity of AW derived from yogurts before and after a simulated *in vitro* digestion model. AO activity was measured by ORAC, FRAP, ABTS and reducing power assays for AW, digested AW and the <3 kDa fraction. *In vitro* digestion improves the AO activity as assessed by all assays. Within each fraction, animal origin (cow, sheep, goat), season (winter-summer), region and straining method were studied. Regarding animal origin statistical differences were observed in ORAC and FRAP assays. Season influenced the AO properties of samples as assessed by the ORAC method. Regarding region statistical differences were noted in ABTS and FRAP assays. Finally, samples from different straining methods exhibited statistically different AO properties as assessed by ABTS and reducing power assays. AH properties of AW samples and post-digestion fractions of <10 kDa and <3 kDa were evaluated by measuring their ability to inhibit the angiotensin converting-enzyme using the same factors as above. *In vitro* digestion improves the AH activity of samples. No differences were observed between samples of different animal origin and samples of different region. On the other hand, season and straining method affected the AH activity of samples. This research is co-financed by Greece and the European Union through the Operational Programme 'Human Resources Development, Education and Lifelong Learning' in the context of the project 'Strengthening Human Resources Research Potential via Doctorate Research' (MIS-5000432), implemented by the State Scholarships Foundation (IKY). This research is co-financed by the European Regional Development Fund of the European Union and Greek national funds through the Operational Program Competitiveness, Entrepreneurship and Innovation, under the call RESEARCH – CREATE – INNOVATE (project code: T2EDK-00783).

Session 07 Theatre 1

Recent developments in aquaculture feeds – an industry perspective

A. Obach
Skretting, Aquaculture Innovation, Sjøhagen 3, 4016 Stavanger, Norway; alex.obach@skretting.com

Fish meal and fish oil have traditionally been the main ingredients in aquaculture feeds. These are excellent raw materials; however, they are finite. The Food and Agriculture Organisation (FAO) published an update on the state of the World Fisheries and Aquaculture in 2022. According to the FAO, aquaculture is expected to grow by close to 20% in the next 10 years. This means another 18 million tonnes of seafood to be produced in 2030. Today more than 70% of the world production of fish meal and fish oil goes into aquafeeds. In order for aquaculture to grow we need to eliminate our dependence on marine ingredients and to find alternative sources that could provide all the benefits, without the limitations. Fish meal and fish oil replacement have been one of the main focus in fish nutrition research for the last two decades, and several EU-funded projects involving major research institutions and other industry stakeholders have addressed the issue. Due to this extensive research, since 2016, diets without fish meal have become commercially available for Atlantic salmon. A significant proportion of the fish oil in aquafeeds had already been replaced by alternative fat sources, both of vegetable and animal origin. However, complete substitution of fish oil has been difficult due to the lack of alternative sources providing the long-chain n-3 fatty acids, eicosapentaenoic acid (EPA) and docosahexaenoic acid (DHA). Today new ingredients with EPA and/or DHA are available, such as algae meals and oils or genetically modified plant oils, although in relatively small volumes and at a higher cost. In recent years, marine ingredients have been replaced by ingredients of both vegetable and animal origin. Reducing our dependence on fish meal and fish oil has also opened an ocean of opportunities to develop new raw materials that can contribute to make our industry even more sustainable. The so-call novel ingredients such as insect meals and microbial proteins and oils are becoming a commercial reality.

Session 07 Theatre 2

Nutritive value of soybean meals from different origins fed to Nile tilapia

N.V. Nguyen[1], H. Le[1], V.K. Tran[1], D.H. Pham[1], T.L.T. Tran[1], J.M. Hooft[2] and D.P. Bureau[2]
[1]Research Institute for Aquaculture (RIA2), Research Center for Aquafeed Nutrition & Fishery Post-Harvest Technology, 116 Nguyen Dinh Chiet St, District 1, Ho Chi Minh City, Viet Nam, [2]University of Guelph, Dept of Animal Biosciences, 50 Stone RD East, N1G2W1, Guelph, Canada; dbureau@uoguelph.ca

Soybean meal is an important protein source in aquaculture feeds. However, few studies have examined the impact of the origin of the soybeans on the nutritive of SBM to aquaculture species. In this study, batches of soybeans of Argentinian (ARG), Brazilian (BRA) or American (USA) origin were purchased and characterized. USA soybeans contained less damage bean (1.1%) compared to ARG (3.2%) and BRA (4.6%) soybeans. These soybeans were then processed to produce three SBMs and used to produce eight (8) extruded diets formulated to contain 21% (basal diet), 26, 31 or 36% crude protein (CP) using either ARG, BRA or USA SBM. These diets were fed to four replicate groups of Nile tilapia in a 12-week growth trial. A digestibility trial with the same eight (8) diets fed to triplicate groups of tilapia was carried out in parallel. Growth performance of the fish increased linearly ($P<0.0001$) or quadratically ($P<0.05$) with increasing levels of CP in the diet. Moreover, linear ($P<0.0001$) and quadratic ($P<0.001$) decreases in FCR (feed:gain) were associated with increasing CP content of the diet regardless of the origin of the SBM. Fish fed the diets containing USA SBM appeared to have significantly better ($P<0.05$) growth performance than did fish fed the diets containing ARG or BRA SBM. Nutrient retention and nutrient retention efficiencies increased significantly ($P<0.05$) with increasing CP content of the diets. A multiple regression approach was used to estimate apparent digestibility of nutrients of the SBMs of different origins. The apparent digestibility coefficient (ADC) of CP of USA SBM was estimated to be 91% while that of ARG SBM and BRA SBM were observed to be significatively lower at 88 and 85%, respectively. Overall, SBM produced from USA soybeans appeared to support better growth performance of Nile tilapia compared to SBM produced from ARG or BRA soybeans, possibly due to their lower degree of damage and higher nutrient digestibility.

EcoXtract® green solvent increases rainbow trout performances and decreases susceptibility to VHSV

D. Rigaudeau[1], N. Abdedaim[1], T. Rochat[2], L. Jep[3], F. Terrier[3], P. Boudinot[2], S. Skiba-Cassy[3], L. Jacques[4] and C. Langevin[1]
[1]INRAE, IERP, Domaine de Vilvert, 78350 Jouy en Josas, France, [2]INRAE, VIM, Domaine de Vilvert, 78350 Jouy en Josas, France, [3]INRAE, NuMeA, Aquapôle INRAE, 64310 Saint Pée-sur-Nivelle, France, [4]Pennakem Europa, 224 avenue de la Dordogne, Dunkerque 59944, France; christelle.langevin@inrae.fr

The Farm to Fork Strategy of the European Green Deal strategy aims to make food systems fair, healthy and environmentally-friendly. The sustainability of aquaculture production has to face several challenges among which the maintenance of fish health and welfare, the fight against outbreaks of infectious diseases and the need to develop alternative fish feeds protecting the environment and the biodiversity. Soybean meal (SBM) is a common alternative, whose processing is based on hexane extraction. In 2021, 2-methyloxolane (EcoXtract®) was described as an alternative green solvent for production of soybean oil and defatted meal. Our study aims to evaluate the impact of the solvent used and SBM inclusion level on rainbow trout. Fish were monitored for survival, growth performance, digestibility of SBMs and diets, metabolism, gut health and robustness to infectious challenges. The results showed that fingerlings fed EcoXtract-extracted SBM presented statistically higher performance than those fed hexane-extracted SBM, with higher increasing at 15% inclusion rate. No significant effect was detected on the growth in juveniles, although, differences in gene expression were revealed for glycolytic and lipogenesis pathways in fish fed diets containing 40% SBM, likely related to the lower level of starch in these diets. Experimental infections of fingerlings by immersion with *Flavobacterium psychrophilum* did not reveal difference in fish susceptibility for the inclusion rates and the extraction methods. In contrast, fingerlings fed EcoXtract-extracted SBM showed significantly lower susceptibility to the viral haemorrhagic septicaemia virus compared to those fed hexane-extracted SBM, with a stronger phenotype at 15% inclusion rate. Analyses are in progress to evaluate the impact of these diets on the microbiota and the intestinal health of fingerlings by using innovative 3D histology method.

Environmental impact of aquafeed for rainbow trout containing alternative protein meals

M. Berton[1], C. Fanizza[1,2], V. Stejskal[3], M. Prokešová[3], M. Zare[3], H.Q. Tran[3], F. Brambilla[4], E. Sturaro[1], G. Xiccato[1], A. Trocino[1] and F. Bordignon[1]
[1]University of Padova, 35020, Legnaro, Italy, [2]Universitat Politècnica de València, 46022, València, Spain, [3]University of South Bohemia, 37005, České Budějovice, Czech Republic, [4]NaturAlleva (VRM s.r.l.), 37044, Cologna Veneta, Italy; francesco.bordignon@unipd.it

The study analysed the effect on global warming potential (Life Cycle Assessment) of the partial replacement of fishmeal (from by-products) by alternative protein meals in diets for rainbow trout. A total of 1,020 trout (17±7.5 g) were fed four diets (three tanks per diet) with fishmeal being partially substituted by alternative protein meals: a control diet (FM) with 307 g/kg fishmeal and three alternative diets where about 40% of fishmeal was replaced with poultry by-product meal (diet PBM), feather meal (diet FeM), or feather + rapeseed meals (diet FeM+RM). All diets contained also rapeseed oil, soybean protein concentrate and bacterial protein meal, among other ingredients. Final weight (191 g) and feed conversion ratio (1.05) did not differ among diets. The system boundaries included the impact of aquafeed production; 1 kg increase of fish was used as functional unit. Global warming potential, without (GWP) and with (GWP_LUC) emissions due to land-use change were calculated. Impact values were analysed by ANOVA with diet as a fixed effect. No differences among diets were found in term of GWP (1.74±0.07 kg CO_2-eq, on average) and GWP_LUC (2.75±0.10 kg CO_2-eq). In all diets, the major contribution to global warming was due to ingredients different from the four protein meals tested, i.e. rapeseed oil (25 and 31% of the total impact for GWP and GWP_LUC), followed by soybean protein concentrate (10 and 23%) and bacterial protein meal (20 and 12%). Among the protein meals tested, fishmeal provided the highest contribution (14% GWP and 9% GWP_LUC), while the contribution of the three alternative protein meals was limited (0.3 and 0.3% for poultry by-product meal; 2.8 and 1.8% for feather meal; 1.7 and 1.4% for rapeseed meal). In conclusion, the tested alternative protein meals are promising and sustainable ingredients for trout aquafeeds guaranteeing good growth rates and, compared to fishmeal from by-products, a lower contribution to global warming.

New strategies for fish nutrition

S. Skiba-Cassy, J. Roy, L. Marandel, E. Cardona and F. Medale
INRAE, NuMéA, Aquapôle INRAE, 64310 Saint Pée sur Nivelle, France; sandrine.skiba@inrae.fr

Global fish production has been growing rapidly for several decades and is expected to increase by a further 14% over the next ten years, reaching 203 million tons in 2031. This increase will be mainly due to the growth of aquaculture, which is expected to overtake capture fisheries by 2023 and account for 53% of global fish production in 2031. For fish, as for other species of agronomic interest, feed plays a central role, as it determines the performance of the animals and the economic profitability of the farm. The expectations of feed are therefore numerous and go beyond the strict coverage of nutritional requirements. In aquaculture, it is now expected that a feed will be multi-performing, i.e. that it will provide essential nutrients for the growth and the development of the products, but also functional elements that can improve the robustness of the fish in the light of global changes. Aquafeeds are also a potential lever for reducing the environmental footprint of aquaculture. Fish feeds have undergone a strong evolution mainly to reduce the reliance of aquaculture on the use of fishmeal and fish oil. However, total substitution of these marine ingredients by plant protein and oil sources, especially in carnivorous species, has not yet been achieved. One of the limitations rely on the regulation of feed intake. Research efforts have therefore expanded. They are now focusing on the use of new ingredients and micro-ingredients (insects, yeasts, microalgae, by-products of the food industry, pre or probiotics, etc.) and the consideration of the environmental impact of the feed. Selection for adaptation of fish to new aquafeeds is also considered and more exploratory approaches based on the principle of early programming are on ongoing.

Rainbow trout hepatic cell lines reveal differences of DL-Methionine and DL-HMTBa function

M. Klünemann[1], K. Pinel[2], K. Masagounder[1], I. Seiliez[2] and F. Beaumatin[2]
[1]Evonik Operations GmbH, Animal Nutrition Research, Rodenbacher Chaussee 4, 63457 Hanau, Germany, [2]Université de Pau et des Pays de l'Adour, E2S UPPA, INRAE, NUMEA, 173, RD 918 Route de St Jean de Luz, 64310 Saint-Pée-sur-Nivelle, France; martina.kluenemann@evonik.com

The replacement of fishmeal by plant protein sources in aquafeeds requires the use of added methionine (MET) sources to balance the amino acid composition and to meet the metabolic needs of fish of agronomic interest such as rainbow trout (RT; *Oncorhynchus mykiss*). There are different MET products commercially available, DL-MET and DL-2-Hydroxy-4-(methylthio) butanoate (HMTBa), and there is a debate if one product is used more efficiently than another by fish. To address this question, we used hepatic cell lines to control cell growth conditions by fully depleting MET from the media and studied whether DL-Met and HMTBa are capable to restore cell growth and metabolism when supplemented back. We evaluated DL-MET and DL-HMTBa based on their effects on targeted cell proliferation, MET-related intracellular metabolites and specific amino acid sensing pathways. The results of cell proliferation assays, Western blots, qPCR and liquid chromatography analyses from two RT liver-derived cell lines revealed a better absorption and metabolization of DL-MET than DL-HMTBa with the activation of the mechanistic Target Of Rapamycin (mTOR) pathway for DL-MET and the activation of integrated stress response (ISR) pathway for DL-HMTBa. In detail, DL-MET treated cells proliferated significantly more than DL-HMTBa treated cells, independent of cell line ($P<0.05$). We observed an significantly higher expression of the MET related enzymes and a significantly higher concentration of MET related metabolites like glutathione in DL-MET treated cells vs DL-HMTBa treated cells ($P<0.05$). We found no significant differences between DL-HMTBa supplemented cells and cells grown in MET deprived media for any of the measured parameters ($P>0.05$) except intracellular DL-HMTBa concentration, which was higher in DL-HMTBa treated cells. Altogether, our study results in heptic cell lines clearly showed that DL-Met and HMTBa are not biologically equivalent, suggesting similar effects to be found in RT liver and, accordingly metabolic advantages of DL-Met over HMTBa.

Millenial salmon – sustainable salmon feed using black soldier fly meal

M.L. Lefranc[1], H.B. Bergoug[1] and K.K. Kousoulaki[2]
[1]Innovafeed, 85 Rue de Maubeuge, 75010, France, [2]Nofima, Nutrition and feed technology, Kjerreidvika 16, 5141 Fyllingsdalen, Norway; maxime.lefranc@innovafeed.com

Millennial Salmon Project is a strategic partnership along the value chain of salmon production between leading research institutes (Nofima, Sintef Ocean) and commercial actors including leading microalgae (Corbion) and insect producers (Innovafeed), feed producer (Cargill), salmon producer (MOWI), processor (Labeyrie) and retailer (Auchan). This collaboration aims at developing sustainable farmed salmon using novel ingredients. DHA-rich *Schizochytrium limacinum* biomass and black soldier fly insect meal are combined to satisfy large parts of the nutritional requirements of salmon in high quality proteins and long chain omega-3 polyunsaturated fatty acids, allowing for sustainable future growth in salmon industry without the need of further deforestation or compromising wild fish biodiversity. The project focuses on optimally maximising dietary inclusion of heterotrophic microalgae and insect meal substituting large part of fish meal, soy protein concentrate and fish oil without negative effects on salmon health, animal welfare, growth performance and feed physical properties. Life cycle analysis of feed ingredients was also conducted to demonstrate the environmental aspects of the suggested innovation.

Digestibility of ingredients derived from food by-products (insect and single-cell) in Asian seabass

R. Le Boucher, W. Chung, J. Ng Kai Lin, L. Tan Shun En, C.S. Lee and C. Wu
Temasek Life sciences Laboratory, 1 Research Link National University of Singapore, 117604, Singapore; richard@tll.org.sg

Food by-products transformation could become a more sustainable source of ingredients for aquaculture but only little is known yet about their digestibility for marine species. This study gathers the learning of two trials conducted on 2,700 (48.2 g) and 840 (591.3 g) Asian seabass (*Lates calcarifer*) grown in large-scale RAS and fed with experimental diets containing 30% of two types of black soldier fly meal (BSF), and two types of single cell protein meal (SCP). Diets were produced with a twin-screw extruder and 0.1% Yttrium oxide was added to estimate digestibility. In small and large fish trials, thermal-unit growth coefficient (TGC), feed conversion rate (FCR) and nutrient retention efficiency were measured. Fish faeces were collected in each tank, diet and ingredient digestibility were estimated for protein, energy and amino acids. TGC of fish fed with BSFM and SPC was never significantly lower than TGC of fish fed with the control diet ($P<0.05$). FCR of fish fed with BSFM (1.05) was higher ($P<0.05$) than the FCR of fish fed with the control diet. Apparent digestibility coefficients (ADC) of BSFM were high for protein (90.2-91.9), energy (87.7-89.9) when ADC of SCP were moderate. The impact of extrusion parameters and water temperature on digestibility is discussed, together with the role these new ingredients could play in future Asian seabass feed formulation.

Specialty feeds to enhance the health value of farmed fish

J. Dias[1], A. Ramalho-Ribeiro[2], A. Gonçalves[3], M.T. Dinis[4] and P. Rema[5]
[1]SPAROS LDA, Area Empresarial de Marim, Lote C, 8700-221 Olhao, Portugal, [2]Politécnico de Coimbra, ESAC, Bencanta, 3045-601 Coimbra, Portugal, [3]Instituto Português do Mar e Atmosfera, Av. Alfredo Magalhães Ramalho, 6, 1495-165 Algés, Portugal, [4]Centro de Ciencias do Mar do Algarve, Universidade do Algarve, Campus Gambelas, 8005-139 Faro, Portugal, [5]Universidade de Trás-os-Montes e Alto Douro, Qunta dos Prados, 5000-801 Vila Real, Portugal; jorgedias@sparos.pt

Due to the rising demand for nutritious, safe and sustainable seafood products (fish and shellfish), aquaculture is predicted to be a major contributor to the nutritional needs of future generations. A continued research effort led to significant progress on lowering the use of fishmeal and fish oil in aquafeeds. However, this trend is altering the nutritional value of edible fish, conditioning the expected beneficial effects for consumers. Some of the strategies available today to counteract this potential loss of nutritional value in fish comprise the use of novel sustainable feed ingredients and feed fortification with selected healthy nutrients. Several trials undertaken with rainbow trout (*Oncorhynchus mykiss*) and gilthead seabream (*Sparus aurata*) evaluated the use of several emergent raw materials (microalgae, macroalgae, yeasts) as tools to fortify fish fillets with health valuable nutrients. In trout, the dietary incorporation of an iodine-rich macroalgae (*Laminaria digitata*) and a selenised-yeast, at the maximum iodine and selenium permitted levels in feed, resulted in a 6-fold increase for iodine and a 2.9-fold increase for selenium contents in trout fillets, without altering sensorial traits. The fortified trout presented a nutritional contribution of 12.5% DRI for iodine, 78% DRI for selenium and 80% DRI for vitamin D3. One meal portion of seabream fed a diet with 10% *Laminaria digitata* covered 84% DRI of iodine and 60% DRI of vitamin D3. Market-size seabream fed a diet with 2.5% of *Phaeodactylum tricornutum*, a fucoxanthin-rich microalgae, showed a significantly higher lightness in ventral skin and a more vivid yellow pigmentation in the operculum and scored higher than the control, in terms of external appearance and brightness assessed by a consumer panel. Integrating consumers' dietary needs and expectations must be considered when designing aquafeeds.

Evaluation of growth performance of pure *Clarias gariepinus* and its hybrid in monosex culture

F.O. Olukoya[1,2], O.T.F. Abanikannda[2,3], K.O. Kareem-Ibrahim[3], M.S. Hedonukun[2] and S.O. Adebambo[2]
[1]University of Lagos, Department of Marine Sciences, Faculty of Science, Akoka, Lagos, Nigeria, [2]Lagos State University, Department of Zoology and Environmental Biology, Faculty of Science, Badagry Expressway, Ojo, Lagos, Nigeria, [3]Lagos State University, Department of Animal Science, School of Agriculture, Epe Campus, Lagos, Nigeria; otfabanikannda@hotmail.com

This study compared sexual differences in growth parameters of pure and hybrid African catfishes (*Clarias gariepinus* and *Clariabranchus*) raised in a monosex culture. The two species were obtained from an earlier study and grouped by sex separately as male and female in four separate plastic tanks, representing two replicates per sex. Ten each of the pure clarias and its hybrid were reared in each tank. In all, 40 male and 40 female fish comprising 20 each of the pure clarias and its hybrid for each sex were evaluated for the growth parameters. A total of 80 fish were included in the study and the study lasted nine (9) weeks. The mean weight of the fish at the commencement of the experiment was 419.9±21.5 and 429.4±23.9 g respectively for male and female, which was not statistically different ($P>0.05$). The fish were fed commercially compounded feed twice daily throughout the period of the experiment. Length-weight relationship and growth parameters such as mean growth rate (MGR), specific growth rate (SGR), absolute growth rate (AGR), relative growth rate (RGR), condition factor (CF) were computed and evaluated for sex effect. All statistical analyses involving descriptive, general linear model analysis of variance (ANOVA) and post hoc test were done using Minitab® 17 Statistical Software. Sex had significant ($P<0.05$) effects on all growth parameters albeit at different levels, except Absolute Growth Rate (AGR), but specie only had significant ($P<0.05$) effect on Mean Growth Rate (MGR), while the interaction of sex × specie exerted significant ($P<0.05$) effects on all growth parameters except Condition Factor (CF). This study revealed that growth performance is significantly ($P<0.05$) affected by sexual dimorphism and also by interaction of sex and specie.

The effect of feeding on dissolved oxygen and turbidity in trout cultivated in an aquaponics

I. Sirakov[1], K. Velichkova[1], A. Asenov[2] and R. Rusev[3]
[1]Trakia University, Students campus, 6000 Stara Zagora, Bulgaria, [2]University of Ruse Angel Kanchev, 8 Studentska str., 7000 Ruse, Bulgaria, [3]Hach Bulgaria, Kr.Sarafov №45, 1000 Sofia, Bulgaria; ivailo_sir@abv.bg

The control of hydrochemical parameters in recirculation systems in the cultivation of hydrobionts is of the paramount importance. One of the factors that significantly influence them is nutrition. The aim of the present study was to investigate the effect of feeding on two major hydrochemical parameters- dissolved oxygen and water turbidity, during rainbow trout cultivation in an aquaponic recirculating system. Rainbow trout in good health were stocked in an aquaponic system at 90 pc.m-3 stocking density. The fish had an initial mean mass of 43.5±2.2 g. Fish were fed with the trout feed of appropriate size and composition. Hach process probes and Sc 1000 controller were used to monitor the dynamics of the two hydrochemical parameters. During the experimental period over 37,500 data were received. On this basis, the dynamics of the two hydrochemical parameters during the day was determined. The relations of dissolved oxygen and turbidity with the nutrition of the cultured species were analysed.

Factors affecting feeding of fish cultivated in recirculation aquaculture system

K. Velichkova and I. Sirakov
Trakia University, Students campus, 6000 Stara Zagora, Bulgaria; genova@abv.bg

Recirculation systems are characterized by high intensity of production processes and are classified as super-intensive technologies. The normal operation of the recirculation systems is unthinkable without constant monitoring of the main parameters of the water and the technical condition of the various facilities. The main requirement when cultivating fish in a recirculation system is that the water meets certain quality indicators. Of these, the most important are temperature, amount of dissolved oxygen, pH, hardness, biogenic elements, etc. An important element is the quality of the feed to supply the necessary nutrients for the growth of the fish, the size of the granule should be matched to the size of the fish, it is easy to consume the feed and has a low consumption. Of particular importance are the correct amount and the right time to eat. Feeding frequency is affected by the amount of feed consumed in one day and it is very important to manage the feeding process in fish because it affects growth, survival, feed conversion, water quality as well as profit maximization. The role of nutrition is a powerful modulator of the endocrine system governing the feeding behaviour and growth of fish. Food availability and feed composition are two of the most influential external cues modulating hormones, regulating appetite, and modulating growth. The provision and setting of optimal conditions specific to the cultured species are of primary importance for its successful cultivation in aquaculture. It is through the application of scientific developments in this regard that fish farmers will be able to optimize management practices and will provide micro environment for better food intake and its utilization.

Session 07 Poster 13

Effect of mycotoxin co-occurrence on zootechnical performance and health indices of *Sparus aurata*

C. Papadouli[1], S. Vardali[1], T. Karatzinos[1], M. Maniaki[1], P. Panagiotaki[1], D. Kogiannou[2], C. Nikoloudaki[2], I. Nengas[2], M. Henry[2], G. Rigos[2] and E. Golomazou[1]

[1]University of Thessaly, Department of Ichthyology and Aquatic Environment, University of Thessaly, Fytokou str., 38446, Volos, Greece, [2]Institute of Marine Biology, Biotechnology, and Aquaculture, Hellenic Centre for Marine Research, 46.7 km Athens-Sounion, 19013, Attiki, Greece; egolom@uth.gr

This study aimed to investigate the effects of the emerging dietary mixture of mycotoxins deoxynivalenol (DON), fumonisin B1 (FB), and aflatoxin B1(AFB1), at various contamination levels, on the growth and health performance of gilthead seabream (*Sparus aurata*). Fish held in triplicate aquariums received five experimental diets: A (DON:500, FB:1000, AFB1:5 ppb), B (DON:150, FB:650, AFB1:2 ppb), C (DON:3,000, FB:40, AFB1:2 ppb), D (DON:150, FB:40, AFB1:10 ppb), E (DON:150, FB:100, AFB1:2 ppb), and were compared against the control group (CTRL) that was fed a mycotoxin-free diet. The feeding trial lasted 12 weeks. To assess the effect of mycotoxin-contaminated feeds, fish were hand-fed *ad libitum* two meals per day, six days per week. Feed intake was recorded daily, and fish were observed for signs of abnormal behaviour, disease symptoms, and mortalities. Water physiochemical parameters were maintained within the standard levels. Endpoint assessment included the evaluation of haematological parameters, feed intake, growth parameters, and induced genotoxicity. One-way ANOVA was used to determine the variance among treatments, followed by Tukey's post hoc test at a significance level of $P<0.05$. All treatments containing mycotoxins caused significantly lower food consumption compared to the control group. Group C, with the highest concentration in DON, posed the lowest feed intake ($P<0.05$), followed by group D with the highest concentration in AFB1. Similarly, a stunted growth (lower mean weight, length, and total biomass increase) was also recorded in group C. A significantly lower haematocrit was measured in all groups compared to the CTRL group. No differences in induced genotoxicity were recorded between experimental groups, using the single-cell gel electrophoresis method for DNA damage evaluation.

Session 07 Poster 14

Fasting influences on several blood chemistry changes: indicators of malnutrition in Nile tilapia

S. Boonanuntanasarn, P. Pasomboon, S. Seedum and W. Auppakhun

Suranaree University of Technology, School of Animal Technology and Innovation, Institute of Agricultural Technology, 111 University Avenue, Tambon Suranaree, Muang, Nakhon Ratchasima, 30000, Thailand; surinton@sut.ac.th

This study investigated the effect of food deprivation on whole body composition and blood chemistry in of Nile tilapia (*Oreochromis niloticus*). Fish were subjected to starve for 0 (fed fish), 1, 2, 3 or 4 weeks. Body chemical composition and chemical blood chemistry were determined. The results showed food deprivation affected to decrease lipid contents in whole body ($P<0.05$). However, protein and ash contents in the whole body did not change. Food deprivation led to decrease blood triglyceride and phosphorous ($P<0.05$). Food deprivation affected to increase SGOT, SGPT, chloride, iron, and phosphorous ($P<0.05$). Blood glucose decrease during week 2 and 3 of food deprivation, but it increased at week 4. Food deprivation for 3 weeks led to decreased blood cholesterol; however, it induced to increase blood cholesterol at week 4. Food deprivation did not changed blood protein, albumin, blood urea nitrogen and calcium. Taken together, food deprivation influenced whole body contents. Additionally, food deprivation affected several blood chemistry parameters which could be interpret as indicators of malnutrition.

Session 07 Poster 15

Technological developments and 3D imaging for the diagnosis of aquaculture fish diseases

M. Merhaz[1], M. Frétaud[2], D. Rigaudeau[1] and C. Langevin[1]

[1]INRAE, IERP, Domaine de Vilvert, 78350 Jouy en Josas, France, [2]INRAE, VIM, Domaine de Vilvert, 78350 Jouy en Josas, France; christelle.langevin@inrae.fr

The development of cutting-edge technologies such as high-resolution imaging has paved the way for fundamental discoveries in cell biology. The implementation of such techniques has the potential to revolutionise the field of host-microbe's interactions and is likely to generate unprecedented insights into human and animal health. Veterinary medicine does not have the means available in human medicine, despite recent lifting of technological barriers achieved by innovation in imaging instruments. Tissue clearing techniques aim to render fixed biological tissue transparent for 3D visualization of entire organs and organisms at cellular resolution. Various field of biology have taken benefits from these techniques which have contributed to improve the analysis / visualization of complex biological tissue. The co-development of in toto immunohistochemistry and imaging methods applied to large cleared samples (>1 cm) has been an important milestone allowing the broad usage of tissue clearing from mammalian to other animal species. We recently described the adaptation of such clearing techniques to render transparent zebrafish but also trout and carps (4cm long). These developments, when combined to lightsheet deep fluorescence microscopy and data processing with visualization tools provide unprecedent 3D visualization of entire trout, carp and stickleback fish bodies at cellular resolutions. Therefore, in addition to functional imaging (CT-scan /MRI) and classical histology (on thin tissue sections), these 3D imaging technics enabled multiscale analyses of fish organisms assessing the tissue architecture the presence of tissue damages and the activation of immune responses. To go further in the development of novel diagnosis tools, we extend these methods to monitor the intestinal health of various fish species using a so-called '3D histology' method. Such techniques have already been reported for biomedical applications such as 3D histopathology of human tumour microenvironments. Our work demonstrates for the first time their potential for applications in developmental biology, comparative anatomy, toxicology but also diagnosis of aquaculture fish diseases.

Session 07 Poster 16

Comparative analyses of growth parameters of pure and hybrid African Catfishes fed two feed types

O.T.F. Abanikannda[1,2], F.O. Olukoya[1,3], K.O. Kareem-Ibrahim[2], S.M. Adebambo[1] and M.S. Hedonukun[1]

[1]Lagos State University, Department of Zoology and Environmental Biology, Faculty of Science, Badagry Expressway, Ojo, Lagos, Nigeria, [2]Lagos State University, Department of Animal Science, School of Agriculture, Epe Campus, Lagos, Nigeria, [3]University of Lagos, Department of Marine Sciences, Faculty of Science, Akoka, Lagos, Nigeria; otfabanikannda@hotmail.com

Cost of feed has been the biggest challenge confronting aquaculture and efforts are geared towards reducing cost of production occasioned by feed. The standard feed used in aquaculture tanks are the extruded (floating) type which are more expensive than the non-extruded (sinking) feed types common in earthen ponds. This study aimed to evaluate the effect of feed types (extruded and non-extruded) on growth parameters in pure *Clarias gariepinus* and its hybrid. Two strains of 40 fishes each with an initial weight ranging from 134 to 820 g due differences in breed, and an overall mean weight of 414.00±26.4 g were randomly assigned to either of the two feed types, in two replicates each in a randomized complete block design, and reared for eight weeks. Growth parameters such as mean growth rate (MGR), specific growth rate (SGR), absolute growth rate (AGR), relative growth rate (RGR) and condition factor (CF) were computed for all subclasses. Statistical analyses included descriptive and general linear model analysis of variance (ANOVA) using Minitab® 17 Statistical Software. Feed type was significant ($P<0.05$) on all growth parameters except Absolute Growth Rate (AGR), while specie and interaction of feed × specie was not significant ($P>0.05$) on all growth parameters studied. Similarly, feed type was a significant ($P<0.05$) source of variation on weekly gain at weeks 7 and 8 whereas specie was significant ($P<0.05$) on gains at weeks 1, 4, 5 and 8, but the interaction of both was not significant ($P>0.05$) throughout the period of study. This study revealed that growth performance as indicated by the growth parameters and weekly gain is influenced by feed type, with the extruded feed (floating) exerting superior influence over the non-extruded (sinking) feed.

Influence of feeding frequency on growth parameters of African Catfish

F.O. Olukoya[1,2], K.O. Kareem-Ibrahim[3], A.A. Jimoh[2], K.O. Bakare[2], Z.F. Shopeyin[2], S.A. Shitta[2] and O.T.F. Abanikannda[2,3]
[1]University of Lagos, Department of Marine Sciences, Faculty of Science, Akoka, Lagos, Nigeria, [2]Lagos State University, Department of Zoology and Environmental Biology, Faculty of Science, Badagry Expressway, Ojo, Lagos, Nigeria, [3]Lagos State University, Department of Animal Science, School of Agriculture, Epe Campus, Lagos, Nigeria; otfabanikannda@hotmail.com

Fish growth is influenced by feed availability and intake, genetics, age and size, environment and nutrition, and feed intake is perhaps the most prominent factor affecting growth rate of fish. In a bid to surmount this, there is need to strike a balance between rapid fish growth and optimum use of supplied feed by recommending the optimal feeding frequency to reduce wastage and its attendant economic losses, by avoiding the twin traps of overfeeding and underfeeding, that are both detrimental to fish health and may cause marked deterioration in water quality, reduced weight, poor food utilization, and increased susceptibility to infections. A total of 270 African Catfish fingerlings were evaluated for the effects of feeding frequency on growth performance. The fish were hatched and reared to 12 weeks under similar conditions prior to the experiment which lasted for 40 days. Feeding frequency included equally shared feed based on recommended 5% of body weight, equally spaced feed allocation at 6, 8 and 12 hourly intervals. Each of the three treatment groups comprised of six replicates of 15 fish per plastic tank. Body weight, and two linear measurements (Total and Standard Length) were taken twice weekly on each of the fish in all the tanks and the respective length-weight relationship and condition factor (K) were computed. Weekly gain, final weight, average daily gain, absolute, relative, specific and mean growth rates for the three groups were evaluated and compared. Results showed that the 6-hourly treatment group consistently had higher values across all growth parameters and the difference was mostly statistically significant ($P<0.05$) except for total length that was not significant ($P>0.05$). However, there was no statistical ($P>0.05$) difference between the values recorded in the 6-hourly and 8-hourly treatment groups implying that either method can be adopted.

The impact of a probiotic supplementation in aquaculture – a bioinformatic modelling

A. Brzoza[1], L. Napora-Rutkowski[2], M. Mielczarek[1,3], T. Kaminska-Gibas[2], T. Suchocki[1,3] and J. Szyda[1,3]
[1]Wroclaw University of Environmental and Life Sciences, Department of Genetics, The Biostatistics Group, Kozuchowska 7, 51-631 Wroclaw, Poland, [2]Polish Academy of Sciences, Institute of Ichthyobiology and Aquaculture in Golysz, Kalinowa 2, 43-520 Chybie, Poland, [3]National Research Institute of Animal Production, Krakowska 1, 32-083 Balice, Poland; joanna.szyda@upwr.edu.pl

The goal of the research was to study the effect of different commercially available EM (effective microorganism) supplements on the microbial communities of the intestine of Common carp (Cyprinus carpio) and water, as well as on fish growth, by conducting the 94-day experiment. The microbiome composition was identified by sequencing two hypervariable regions (V3 and V4) of the gene encoding the 16S rRNA ribosomal subunit. The experimental setup comprised seven tanks including: a control tank without fish and without probiotic supplementation, two water tanks with fish and no probiotic supplementation, two tanks with fish and the commercially available water supplement W1 and feed supplement F1, two water tanks with fish and with the commercially available water supplement W2 and feed supplement F2. The modelling of abundance diversity was carried out on the family taxonomic level. The overall diversity within each sample quantified by the Shannon index was generally very low and no significant differences in diversities between experimental groups were observed. Neither among water, nor among intestinal samples. Considering changes in abundance of particular families, we observed that 22/13 families significantly increased/decreased their abundance in water during the course of experiment. The highest increase was registered for Acidobacteriaceae (log2FoldChange=-15.5, $P=6\cdot10\text{-}15$) while Devosiaceae was the family with highest decrease in abundance (log2FoldChange=8.3, $P=1\cdot10\text{-}12$). However, no difference in the dynamics of abundance changes could be observed between the experimental and control tanks. All the significant comparisons for the intestinal microbiome after W1F1 and W2F2 supplementation involved cases when sequence reads representing a given family were either not detected in the supplemented individuals, but present in the intestines of some of the control individuals or the opposite.

Session 08 Theatre 1

Role of daily feeding rhythms in the genetics of feed efficiency in growing pigs

H. Gilbert[1], L. Agboola[1], A. Foury[2], M. Teissier[1], F. Gondret[3] and M.P. Moisan[2]
[1]INRAE, GenPhySE, Castanet-Tolosan, 31320, France, [2]INRAE, NutriNeuro, Bordeaux, 33076, France, [3]INRAE, Institut Agro, PEGASE, Saint-Gilles, 35100, France; helene.gilbert@inrae.fr

Feed efficiency is a major driver for sustainable pig breeding, saving feed resources and lowering environmental impacts. In pigs, contrasted feed efficiency are associated to differences in feeding behaviour, with a reduced feeding time in more efficient pigs. Studies in human and rodents have shown that aligning feed intakes with day/night cycles is essential to optimize energy metabolism and body composition. In pigs, the role of daily feeding rhythms on feed efficiency remains to be explored. In our study, temporal feed intake data of about 4,000 pigs from two divergent lines selected for residual feed intake (RFI) were recorded by automatic feeders. Feeding rhythm was mainly distributed during the day, with two peaks, in the morning and in the afternoon. Significant differences were reported between lines for this pattern, with pigs from the more efficient line eating proportionally more during the peaks (+ 73 g/d and +89 g/d, $P<0.001$, respectively) and less in the intervals between the peaks (-119 g/d, $P<0.001$) than pigs from the less efficient line. These traits had significant heritability estimates ($>0.30\pm0.04$), and showed significant genetic correlations with daily feed intake and RFI. The sequence variants detected in the founders of the two lines were then imputed to all breeding pigs of the lines previously genotyped for a 60K medium density SNP chip. The alleles frequencies were computed for the genomic variants segregating in the ten core clock genes regulating circadian rhythm. Changes of allele frequencies with selection were tested separately in each line. Significant responses to selection were pointed out in the ARNTL and CLOCK genes in the less efficient line ($P<0.05$ genome-wide level), suggesting a molecular basis for the differences of feeding rhythm detected between the lines. These results indicate that the genetic variability of feed efficiency in pigs could be related to alterations in circadian rhythms.

Session 08 Theatre 2

Using sequence variants to detect SNP and pathways affecting pig gut microbiota and feed efficiency

H. Gilbert[1], M. Teissier[1], K. Fève[1], S. Legoueix[1], Y. Billon[2], V. Déru[1,3], A. Aliakbari[1], J. Bidanel[4], O. Zemb[1], Y. Labrune[1] and J. Riquet[1]
[1]INRAE, GenPhySE, 31320 Castanet-Tolosan, France, [2]INRAE, UE GenESI, 17700 Surgères, France, [3]FG Porc, La Motte au Vicomte, 35650 Le Rheu, France, [4]IFIP, La Motte au Vicomte, 35650 Le Rheu, France; helene.gilbert@inrae.fr

Only few genomic variants are reported as associated to feed efficiency. To empower our analysis, we used imputation of sequence variants in two Large White pig designs (1,943 pigs and 1,942 pigs, respectively) with records for feed efficiency, and abundances of faecal microbiota genera from 16S rRNA gene sequencing (600 and 1,400 pigs, respectively). Imputation was carried out jointly on the two designs with FImpute, in two steps. First, the 650 K genotypes of 128 founders (Affymetrix Axiom Porcine Array) of the two designs were imputed to animals of the full designs, that had been genotyped with 60K SNP chips. Then, sequence variants, called on 45 founders of the two designs using the nf-core/sarek workflow, were imputed to full designs, leading to a total of 17,033,057 SNP imputed (correlation between true and imputed genotypes per animal = 0.94 ± 0.011). After quality control, association studies were carried out with GEMMA in each design separately, and meta-analyses were used to combine the corresponding outputs with the *metal* package in R. The simpleM approach was used (principal component analysis applied to the matrix of genotypes) to estimate the number of independent tests, leading to a genome-wide threshold of 7.22 for the -log10(*P*-values). In total, 26 QTL regions were significant, and 272 regions were considered as suggestive. An enrichment analysis applied to all annotated genes in these regions pointed out GO terms related to membrane compounds and functions. Focusing on the 10 QTL regions associated with the abundances of 9 microbiota genera, functional candidate genes were suggested, for instance previously reported in colorectal cancer in mice or human. Altogether, the approach combining sizable datasets, imputation of sequence variants and records of traits related to sub-functions of complex traits allowed to pinpoint new pathways and genes related to variation of feed efficiency in pigs.

Using digestive efficiency to improve feed efficiency in pig breeding schemes

V. Déru[1,2], M.J. Mercat[3], D. Picard[3], B. Ligonesche[4], G. Lenoir[5], L. Flatrès-Grall[6], F. Ytournel[7], J. Bidanel[3] and H. Gilbert[1]
[1]INRAE, 24 Chemin de Borde Rouge, 31320 Auzeville-Tolosane, France, [2]Alliance R&D, La Motte au Vicomte, 35651 Le Rheu, France, [3]IFIP – Institut du Porc, La Motte au Vicomte, 35651 Le Rheu, France, [4]Nucléus SAS, 7 rue des Orchidées, 35650 Le Rheu, France, [5]Axiom, La Garenne, 37310 Azay-sur-Indre, France, [6]CAE29 Chrysalide, Kerbastard, 29150 Dinéault, France, [7]Choice Genetics, rue Maryse Bastié, 35170 Bruz, France; vanille.deru@inrae.fr

Selection on feed efficiency (FE) is carried out for 50 years in French pig breeding schemes. Recent work has highlighted the interest of digestive efficiency (DE) measurements to improve FE. Indeed, DE is heritable (h^2>0.26) and genetically favourably correlated (>0.23) with FE, with affordable costs. This study aimed at determining the best strategie(s) to integrate DE into pig breeding schemes to improve FE. For this purpose, simulations of pig breeding schemes for paternal lines were set up with the AlphaSimR package in R. A generic pig breeding scheme was first simulated over 8 generations and 6-months cycles. The initial cohort was composed of 5,000 young boars (40% with FE) and 5,000 gilts (no FE data). The top 1% of boars without FE data, 2% of boars with FE data and 13% of gilts were selected. Then, animals could remain active during 5 cycles, with selection from 42 to 7% of sires, and from 38 to 8% of dams from cycle 1 to 5. The genetic and phenotypic parameters were initialized from data collected on 2,287 Large White pigs, with FE, DE and carcass traits. Then, various scenarios were simulated, including the phenotyping of various proportions of candidates to selection for DE with different strategies: on-farm or on-station candidates, with different family relationships with pigs phenotyped for FE. For all scenarios, indicators including genetic gain and genetic variance for each trait, pig breeding values, inbreeding coefficient and phenotyping cost were computed. They were compared with those of the reference scenario to identify if DE measures can improve genetic gains at a given cost, or can reduce costs while maintaining the genetic gain. Thus, our results contribute to propose strategies to improve genetic progress on FE in French paternal pig populations.

Can automatic records of activity predict maternal ability and health in sows?

K. Nilsson[1], A. Bergh[2] and A. Wallenbeck[3]
[1]Swedish University of Agricultural Sciences, Department of Animal Breeding and Genetics, P.O. Box 7023, 75007, Sweden, [2]Swedish University of Agricultural Sciences, Department of Clinical Sciences, P.O. Box 7054, 75007, Sweden, [3]Swedish University of Agricultural Sciences, Department of Animal Environment and Health, P.O. Box 7068, 75007, Sweden; katja.nilsson@slu.se

Litter size is the major reproduction trait of economic importance in piglet production. However, larger litters generally have a larger proportion of smaller and weaker piglets, and a higher mortality rate. One of the most common causes for pre-weaning mortality are crushing by the sow, and previous research has shown that the risk of crushing is related to the frequency and manner of position changes and movements performed by the sow. Transition from standing or sitting to lying and rolling movements have been shown to be especially risky. Lameness and poor leg quality of the sow is, apart from being painful and causing discomfort to the sow, also associated with an increased risk of crushing of piglets. Our aim was to investigate if records of general activity can predict risk of crushing, and also if activity can serve as an indicator of lameness in sows. Records of activity were collected using accelerometers attached to a collar around the neck of the sow. Activity data have so far been collected from approximately 60 sows at Research centre Lövsta, Uppsala. The accelerometers recorded movements over a period of 2.5-7 days for each sow, around farrowing and during lactation. Sows were loose-housed individually in farrowing pens, with access to straw. Litter size at birth, number of dead piglets and cause of death was recorded by the staff in the stable. Sows varied in total amount of activity, especially around the day of farrowing. However, preliminary analyses showed no correlation between activity and piglet mortality, suggesting that sows that move more do so without putting their piglets at risk. We will continue to analyse the relationship between activity and lameness.

Cross-breeding strategies to optimize litter size – a way to improve piglet survival?

K. Nilsson[1] and A. Wallenbeck[2]
[1]Swedish University of Agricultural Sciences, Department of Animal Breeding and Genetics, P.O. Box 7023, 75007, Sweden, [2]Swedish University of Agricultural Sciences, Department of Animal Environment and Health, P.O. Box 7068, 75007, Sweden; katja.nilsson@slu.se

Larger litters generally have a larger proportion of smaller and weaker piglets. The negative consequences of this are even more pronounced in organic production systems, where piglet mortality is generally higher compared with conventional systems. Organic pig production makes use of the same sow breeds and breed crosses as conventional production systems – breeds that are selected heavily for litter size. Many producers, and organic producers, in particular, would prefer more moderate litter sizes with heavier and more vital piglets. The aim of this project was to investigate the consequences of obtaining a more moderate litter size by crossing so-called sire breeds into breeding sows in piglet-producing herds. Sire breeds are not selected for litter size, but rather for growth rate and meat quality traits. The project was carried out at the research herd at Lövsta Research Center, Uppsala. 16 crossbred sows between Yorkshire and Landrace (YL) were compared with 12 sows crossbred between Yorkshire and Hampshire (YH) and 9 sows crossbred between Yorkshire and Duroc (YD). Sows were loose-housed in individual farrowing pens with access to straw. Litter size, individual piglet growth from birth to weaning, mortality, and cause of death was recorded by the stable staff. YL sows had significantly larger litters at birth, 15 piglets compared with approx. 12 for YH and YD. However, the survival rate was also better for YL sows, resulting in the same advantage of approx. three extra piglets at weaning. The growth rate from birth to weaning was similar for all breed crosses. YL sows lost significantly more weight during lactation, leading us to conclude that they invest more resources in milk production compared with the sire line crosses. Crossing sows with sire lines does result in a more moderate litter size but does not improve survival and pre-weaning growth due to poorer maternal ability in the sows.

Genetic parameters for genetic variance uniformity in Swiss pigs' birth weight

C. Kasper[1], A. Lepori[2], J.P. Gutiérrez[3], N. Formoso-Rafferty[4], E. Sell-Kubiak[5] and I. Cervantes[3]
[1]Agroscope, Animal GenoPhenomics, Tioleyre 4, 1725, Switzerland, [2]Suisag, Allmend 10, 6204 Sempach, Switzerland, [3]UCM, Department of Animal Production, Veterinary Faculty, Avda. Puerta de Hierro s/n, 28040 Madrid, Spain, [4]Universidad Politécnica de Madrid, Department of Agricultural Production, ETSIAAB, Sende del Rey 18, 28040 Madrid, Spain, [5]Poznan University of Life Sciences, Department of Genetics and Animal Breeding, Wojska Polskiego 28, 60-637 Poznań, Poland; claudia.kasper@agroscope.admin.ch

Increasing litter sizes require cross-fostering to regulate litter size, but also to homogenize the weight of piglets growing up together. This practice increases labour costs and poses immunity concerns, but increases preweaning survival and thus farm productivity. Therefore, identifying factors resulting in birth weight (BW) variability is a priority. We estimated the genetic component of residual variance for BW using a data set with 23,313 BW records from 1,748 litters of 813 sows and 26,107 individuals in the pedigree. The heteroscedastic model included the sex (2 levels), farm-month-year (75 levels), litter size (21 levels), breed (4 levels), and age of the sow (300 to 1,956 days) as fixed effects. The litter effect (nested to the sow) was included as random effect in addition to the genetic effect. The data was assigned to the mother, and the same effects were fitted for mean BW and its variability. The model was fitted using a Markov chain Monte Carlo software (GSEVM). The genetic coefficient of variation was 0.189, and the genetic correlation between the mean BW and its variability was 0.398 (SE=0.116). Regarding fixed effects, female piglets were more consistent regarding BW variability despite having lower weight, and litters with more than 9 piglets were less variable than smaller ones. The age of the sow affected mean BW positively, and reduced BW variation. The results are in line with previous genetic parameters estimated for BW uniformity in pigs. Even though the mean birth weight might decrease, selection for uniformity is expected to be beneficial since selection for uniformity has been shown to increase the robustness of the animal. Finally, homogenizing BW might reduce the occurrence of intrauterine growth retardation, thereby improving piglet health and carcass value.

Session 08 Theatre 7

A multivariate gametic model for the analysis from reciprocal crosses for two Iberian varieties

H. Srihi[1], D. López-Carbonell[1], N. Ibáñez-Escriche[2], J. Casellas[3], P. Hernández[2], S. Negro[4] and L. Varona[1]
[1]Universidad de Zaragoza, Instituto Agroalimentario de Aragón (IA2), C. Miguel Servet, 177, 50013, Zaragoza, Spain, [2]Universitat Politècnica de València, Camino de Vera, s/n, 46071, València, Spain, [3]Universitat Autònoma de Barcelona, Plaza Cívica, 08193 Barcelona, Spain, [4]INGA FOOD S.A., Av. de a Rúa, 2, 06200 Almendralejo, Spain; houssemsrihi@unizar.es

Crossbreeding is used in pig breeding schemes to obtain heterosis for reproductive traits in the crossbred maternal lines. In some cases, the performance of reciprocal crosses may be different, and selection in purebred populations may be more effective for one reciprocal cross than for the opposite. This phenomenon can be explained by variability in the correlations between the gametic effects acting as sire or dam in pure and crossbred populations. Therefore, we propose a multivariate gametic model that defines up to four correlated gametic effects for each parental population (sire for purebred, sire for crossbred, dam for purebred, dam for crossbred). The model was applied to a data set of litter size (total number born) from a reciprocal cross between two Iberian pig populations (Entrepelado and Retinto). It consists of 6,933 records from 1,564 purebred Entrepelado (EE) sows, 4,995 from 1,015 Entrepelado × Retinto (ER), 2,977 from 756 Retinto × Entrepelado (RE) and 7,497 from 1,577 purebred Retinto (RR). The gametic effects were calculated by using a pedigree of 6,007 individual-sire-dam entries. The model of analysis also includes the order of parity (6 levels), breed of the sire of service (5 levels), herd-year-season (141 levels) and a random dominance effect for each pure or crossbred population. The analysis was implemented with a Bayesian approach using a Gibbs Sampler with a single long chain of 1,100,000 iterations after discarding the first 100,000. All the posterior estimates of the gametic correlations were positive, and they ranged between 0.05 (Sire for purebred with Dam for crossbred in Retinto) and 0.57 (Sire for purebred with Dam for Purebred in Entrepelado). Moreover, the posterior mean-variance estimates of the maternal gametic effects were larger than the paternal in the four populations.

Session 08 Theatre 8

Impact of vaccination against GnRF on growth performance and meat quality of gilts at market weight

N. Quiniou[1], P. Chevillon[1] and F. Colin[2]
[1]IFIP-Institut du Porc, BP35, 35650 Le Rheu, France, [2]ZOETIS, 10 rue Raymond David, 92240 Malakoff, France; nathalie.quiniou@ifip.asso.fr

Immunization against GnRF (gonadotropin-releasing-factor) is being investigated by the dry-cured ham industry as a solution to increase carcass fatness of entire male pigs in a context of physical castration ban. In gilts, the vaccine temporarily suppresses the ovarian function and the associated secondary effects have been characterized in a trial involving 144 crossbred Piétrain × (Large White × Landrace) group-housed (6/pen) gilts. At 70 d of age (26.3±3.6 kg), full- or half-sisters were randomly allocated to a control non-treated group (C, 12 pens) or to the vaccinated (Improvac®, Zoetis) group (V, 12 pens). Gilts V were vaccinated at 103 (V1) and 132 (V2) d of age. Pigs were harvested either 4 or 5 weeks after V2. A 2-phase *ad libitum* feeding strategy was used, with Phase-1 and -2 diets formulated for 9.75 MJ net energy/kg, and 9.0 and 7.8 g ileal standardized digestible lysine/kg, respectively. Growth before V2 was similar for both groups (P>0.10). Within the 1st week after V2, no difference in average daily feed intake (ADFI) was observed between groups, whereas significant differences were observed afterwards: between V2 and harvest, ADFI was 0.39 kg higher (P<0.001), growth rate was 89 g/d higher (P=0.002) and feed conversion ratio tended to be 0.15 kg/kg higher (P=0.08) in gilts V. They tended to be 2.6 kg heavier at harvest than gilts C (115.6 vs 113.0 kg, P=0.06). The significant (P<0.05) increase in ultimate pH in the Semimembranosus (+0.08 unit) and the Longissimus dorsi (+0.03 unit), and the reduced drip losses (-1.2%) in meat from gilts V suggest an improvement in meat quality for cooking and slicing yields. The increase in backfat thickness (+2 mm, P<0.01) would increase the proportion of carcasses that fit better the expectations of the dry-cured ham industry. In conclusion, vaccination of gilts against GnRF may help producers to improve production performance and meat quality, allowing for more carcasses meeting the industry requirements. The carcass payment grid should be adapted so that it compensates the carcass value for the cost of vaccination and the increase in feed intake when the pig farmer adapts the management of the pigs in response to the quality expectations of the meat industry.

Using different gas mixtures for pig stunning – influence on meat quality and gene expression

J. Gelhausen[1], T. Krebs[1], T. Friehs[1], J. Knöll[2], I. Wilk[2], D. Mörlein[1] and J. Tetens[1]
[1]Georg-August-University, Department of Animal Science, Burckhardtweg 2, 37077 Göttingen, Germany, [2]Friedrich-Loeffler-Institute, Institute of Animal Welfare and Animal Husbandry, Dörnbergstr. 25/27, 29223 Celle, Germany; julia.gelhausen@uni-goettingen.com

Carbon dioxide (CO_2) stunning is the most commonly used method for pig slaughter. It causes a deep unconsciousness and allows group stunning. However, aversions such as hyperventilation or escape attempts can be observed. Inert gases such as Nitrogen (N_2) and Argon (Ar) seem to be promising alternatives to reduce these aversions. In the project for Testing Inert Gases in order to Establish Replacements for high concentration CO_2 stunning for pigs at the time of slaughter (TIGER), we are investigating the effect of ten different gas mixtures on animal welfare and meat quality in a commercial slaughterhouse using a Dip-lift system with a newly developed gas supply technology. From each group samples were taken from 30 [DE × DL] × PI animals. The pigs were either stunned with Ar (100%) or N_2 (70%N_2 /30%Ar) as well as mixtures of those with 10-30% CO_2 or two CO_2 control atmospheres (<1% or <2% O_2). Meat quality traits (pH, temperature, and electrical conductivity) were assessed in loin (*M. longissimus thoracis et lumborum*) and ham (*M. semimembranosus*) 40 min and 24 h *post mortem*, respectively. Corresponding RNA samples were taken from the loin and sequenced. The hams were further examined for petechial haemorrhages and the loin for drip loss, cooking loss, shear force, and colour. Statistically significant differences ($P<0.05$) were found for both, pH_{45min} and pH_{24h} in ham and loin, with no evidence of PSE or DFD meat formation. RNA sequencing showed few differentially expressed genes between gases, which is in line with the results for meat quality. However, notably the conductivity in the ham was increased in the mixtures containing nitrogen, as was the incidence of haemorrhages in the ham with these gases. Concluding from these pilot trails, the alternative gas mixtures investigated here are not inferior to conventional CO_2 stunning concerning loin meat quality. On the other hand, the increased incidence of haemorrhages in the ham requires further investigations.

Lean tissue grows slower in the belly and loin regions of Australian domestic pork carcases

L.T. King, F. Anderson, M. Corlett, A. Williams and G.E. Gardner
Murdoch University, Centre for Animal Production and Health, 60 South Street, Murdoch 6150, Western Australia, Australia; l.king@murdoch.edu.au

Pork supply chains in Australia select for pigs displaying rapid lean growth. Cuts from different carcass regions vary in value; therefore, selection focused on growth in high valued regions and finishing practices that allow their maturation may maximise economic returns. Therefore, we quantified the relative rates of lean growth across carcase regions to benchmark domestic Australian pork. 360 pork carcases from 3 separate processing plants were CT-scanned and simple thresholds were used to differentiate lean/rind from fat, and bone. Huxley's allometric equation ($y = ax^b$) was used to determine the relative rate of growth of carcase lean relative to carcase weight and either the fore, loin, belly, or hind section lean relative to carcase lean weight. Data was converted to natural logarithms (*ln*) to linearise the data and allow subsequent analysis using least squared regression. B values>1 represent slower rates of lean maturation within the carcase, while values <1 represent early rates of lean maturation within the carcase. For *ln* whole carcase lean weight vs *ln* carcase total weight, the b value was 1.02 (R^2=0.87), indicating that carcase lean matured at a similar rate to total carcase weight. For *ln* section weight vs *ln* total lean weight, b values for *ln* loin lean (b=1.15; R^2=0.84) and *ln* belly lean (b=1.64; R^2=0.83) were greater than 1, indicating that the lean in these regions matured slower than total carcase lean. Alternatively, b values for *ln* fore lean (b=0.84; R^2=0.91) and *ln* hind lean (b=0.78; R^2=0.92) were less than 1, indicating more rapid rates of lean maturation than total carcase lean. These results indicate that whole carcase lean grows at a similar rate to carcase weight, and yet varies substantially across different regions of the pork carcase. This implies that production systems focused on optimising the weight of belly and loin cuts will reap benefits from finishing to heavier weights, allowing this lean to grow proportionately more than in other carcase regions.

Session 08 Theatre 11

Carcass characteristics and boar taint in entire male pigs from commercial French organic farms

S. Lombard[1], A. Roinsard[2] and A. Prunier[3]

[1]ITAB, 9 rue André Brouard, 49105 Angers, France, [2]FOREBIO, 117 rue de Charenton, 75012 Paris, France, [3]PEGASE, Institut Agro, INRAe, 35590 Saint-Gilles, France; sarah.lombard@itab.asso.fr

Surgical castration of male pigs is now forbidden without anaesthesia and analgesia in France. Rearing entire male pigs is a way to stop castration and to improve feed conversion and Lean Meat Percentage (LMP) which have important economic values in organic pig production. However, there is a risk of tainted meat which is mainly due to androstenone (A) and skatole (S) stored in fat tissues. This project focuses on the performance and boar taint of entire male pigs in organic farming. Entire male pigs from six organic farms located in the western part of France were followed along one year. Data collection included age at slaughter, carcass weight, LMP (n=849), human nose evaluation (n=622), S and A concentrations in backfat (n=577). For human nose evaluation, carcasses were scored 0 (no boar taint), 1 (suspicious odour), or 2 (boar taint). Most of the boars (84%) were slaughtered before 210 days of age with great variation between farms from 178±1 to 209±2 days (mean ± SE, P<0.001). The average carcass weight varied also between farms (90.1±0.7 to 99.2±1.6 kg, P<0.001) as well as the average LMP (59.2±0.3 to 60.7±0.3 P<0.001). For human nose evaluation, most carcasses (94.5%) were scored 0, 4.0% were scored 1, and 1.4% scored 2. Median S (0.02 to 0.06 µg/g pure fat) and A (0.54 to 1.78 µg/g) concentrations in backfat varied a lot between farms (P<0.001). A positive correlation was depicted between A and S (P<0.0001), and between carcass weight and A (P<0.03). Within farms, analyses revealed a significant variation between trimesters in one farm for A and in a second farm for S and a positive correlation between age at slaughter and A in two farms (P<0.03). Rearing entire male pigs can be a good alternative to castration in organic farming provided the risk of boar taint is under control choosing a genotype with low risk for A, avoiding old and heavy pigs at slaughter, and maintaining a clean and well-ventilated housing.

Session 08 Theatre 12

Boot swabs to assess efficiency of cleaning and disinfection of slatted floors in pig barns

R.H. Mateus-Vargas[1,2], N. Kemper[2], K. Butenholz[2], N. Volkmann[2], C. Sürie[3] and J. Schulz[2]

[1]Department of Animal Sciences, University of Göttingen, Burckhardtweg 2, 37077 Göttingen, Germany, [2]Institute for Animal Hygiene, Animal Welfare and Farm Animal Behavior (ITTN), University of Veterinary Medicine Hannover, Foundation, Bischofsholer Damm 15, 30173 Hannover, Germany, [3]Lehr- und Forschungsgut Ruthe, University of Veterinary Medicine Hannover, Foundation, Schäferberg 1, 31157 Sarstedt, Germany; rafael.mateus-vargas@uni-goettingen.de

Cleaning and disinfection (C&D) of animal houses within and between production cycles are important hygienic measures to control the transmission of pathogens between batches. To adequately assess the level of success of C&D procedures in barns, the use of boot swab sampling may represent an adequate alternative. The aim of this study was to utilize a boot swab sample method to assess the success of C&D in pig barns (farrowing, rearing, and fattening) with slatted floors on two farms located in Northern Germany. Animal houses were cleaned and disinfected by a specialized contractor according to standard protocols. The success of C&D procedures was assessed using a boot swab method before water cleaning (BC), after water cleaning (AC), and after disinfection (AD). Microbiological examinations of swabs included the quantitative determination of indicator bacteria. With one exception, water cleaning alone resulted in statistically significantly reduced hygiene indicators (around 2 log10 cfu per boot swab pair). Additionally, an average decrease of bacterial loads of about 1 log10 cfu per boot swab pair was observed at AD in comparison to AC. Thus, the use of the boot swab sampling method allowed to assess the efficiency of C&D procedures at different time points in pig barns, including the detection of unexpected outcomes, which may indicate incorrectly performed C&D protocols. Such results are valuable for evaluations of C&D practices by animal owner or specialist contractors. However, the boot swab sampling method may only record the microbial contamination on the upper surface, while bacterial loads between and beneath the slots in the floor might not be considered adequately. Combinations with other methods should be further evaluated. This work was supported by the QS-Wissenschaftsfonds.

The impact of temperature on transmission of African swine fever in contaminated livestock vehicles

Y. Gao[1], A. Boklund[2], L. Nielsen[3], L. Alban[2,3] and M. De Jong[1]
[1]Quantitative Veterinary Epidemiology, Wageningen University and Research, Radix, 6700 AH, Wageningen, the Netherlands, [2]Faculty of Health and Medical Sciences, University of Copenhagen, Blegdamsvej 3B, 2200, Copenhagen, Denmark, [3]Danish Agriculture & Food Council, Axelburg, 1609, Copenhagen, Denmark; yuqi.gao@wur.nl

African Swine Fever Virus (ASFV) is the cause of an infectious disease in pigs. The disease is lethal and can bring huge economic losses to an area once introduced. There is neither a vaccine nor a treatment available. Long viability of the virus has been shown for several contaminated materials, especially under low temperature. Therefore, when exposed to a contaminated environment, new infections could occur without the presence of infectious individuals. For example, a contaminated, poorly washed, empty livestock vehicle poses a risk to the next load of pigs. Here we calculated the duration of infectivity of fomites, depending on the temperature in the empty period of a pen/vehicle after contamination. A quantitative stochastic environmental transmission model was applied to both a direct transmission experiment and an environmental transmission experiment to estimate the epidemiological parameters. For a latent period of 5 days, the estimated parameters were as follows: the transmission rate parameter was 1.5 (0.9-2.4) day^{-1}, the decay rate parameter was 1.0 (0.7-1.5) day^{-1} (at 20 °C), and the excretion rate parameter was 2.7 (2.5-3.1) day^{-1}. However, the experiments we referred to were conducted at 20 °C, and data describing such experiments at low temperatures are absent. Thus, we extrapolated from earlier half-life research on urine, faeces, and organ tissues. Based upon this, the hazard level of the environment and the probabilities of infection at 20, 10, 0, and -10 °C were calculated. The results show that the probability of the contaminated environment causing at least one of four recipient pigs (based on the initial study set-up) to become infected was 100% before the start of the empty period and declined to 6.2% after 5 days at 20 °C, compared to 42.9, 94.4, and 100%, at 10, 0, and -10 °C, respectively, all after 5 days. Thus, although in the original experiment the probability of recipients becoming infected was low, the probabilities at commonly occurring lower temperatures are non-negligible.

Porcine ear necrosis is associated with social behavior and blood biomarkers in weaned pigs

T. Nicolazo[1], C. Clouard[2], C. Teixeira Costa[1], G. Boulbria[1], A. Lebret[1], V. Normand[1] and E. Merlot[2]
[1]Rezoolution, ZA De Goheleve, 56920 Noyal-Pontivy, France, [2]INRAE, PEGASE, Le Clos, 35590 Saint-Gilles, France; t.nicolazo@rezoolution.fr

Porcine ear necrosis (PEN) is a worldwide health issue and its aetiology is still unclear. The aim of this study was to describe the prevalence and the severity of PEN in a commercial farm, associated with behavioural changes and health biomarkers measures. On two consecutive batches, PEN prevalence was determined at the pen level. PEN scores, blood haptoglobin concentration and oxidative status biomarkers were measured on two pigs per pen (n=48 pens), 9, 30 and 50 days (D) after arrival to the post-weaning unit. For two to three other pigs per pen, social nosing, oral manipulation, aggression of pen mates and exploration of enrichments were observed twice a week from D9 to D50. At the pen level, the higher the time spent nosing pen mates, the lower the proportion of pigs affected by PEN during both the D9-D30 and the D31-D50 periods ($P<0.002$). On the opposite, the higher the time spent manipulating orally pen mates during the D31-D50 period, the higher the percentage of affected pigs within a pen ($P=0.03$). Mean PEN scores on D50 were higher in pens in which focal piglets spent more time nosing pen mates during the D31-D50 period ($P=0.02$), and to a lesser extent, during the D9-D30 period ($P=0.08$). At the pig level, the higher the increases in hydroperoxides and haptoglobin during the D9-D30 period, the higher the PEN scores on D30 ($P<0.001$). Our study evidenced that an increase in inflammatory and oxidative stress biomarkers might be associated with PEN severity. In addition, this study is the first to suggest that social nosing and oral manipulations have opposite associations with PEN syndrome.

Session 08 Theatre 15

Pelleting and botanical source influence starch utilization in growing pigs

A. Agouros[1,2], M. Le Gall[1], K. Quéméneur[1], Y. Lechevestrier[1], L. Montagne[2], N. Quiniou[3] and E. Labussière[2]
[1]Cargill, Ferchaud, 35320 CREVIN, France, [2]INRAE, Institut AGRO, 16 Le Clos Domaine de, La Prise, 35590 St-Gilles, France, [3]Institut du Porc, La Motte au Vicomte, 35650 Le Rheu, France; etienne.labussiere@inrae.fr

Feed pelleting improves digestibility and metabolic use of energy. As starch provides almost 70% of the dietary energy, the study aimed to test the effects of starch origin and pelleting on metabolism of pig. Six diets were obtained in a 3×2 factorial design differing by starch source (wheat, W; maize, M or barley, B) and presentation (flour, F or pelleted, P). The diets were named as the combination of botanical source and presentation form. Diets were distributed four times a day, providing 2.4 MJ ME/kg $BW^{0.60}$ per day to 72 male pigs (mean body weight (BW): 52.2±4.7 kg) housed in pairs during one week in an open-circuit respiration chamber to measure the dynamics of gas exchanges and energy balance. Pigs were fitted with a catheter in the portal vein to study the dynamics of blood metabolites after a meal. At the end of the experimental period, pigs received two meals with indigestible markers 6 and 1 h before slaughter for quantifying marker recovery in each digestive tract segment. Marker passage in stomach 6-h after intake was higher with WP, MF and MP than with WF, BF and BP (87 vs 73%; $P<0.05$). Passage rate in stomach of marker given 1-h before slaughter were similar with WP, MF, MP and BF but higher than with WF and BP (34 vs 14 and 18%, respectively; $P<0.05$). Dry matter faecal digestibility was greater with WP and BP (+1.9 and +1.0%; $P<0.05$) compared to WF and BF. Preprandial portal insulinemia was higher with MP than MF (54 vs 9 µU/ml; $P<0.05$). From 150 to 180 min after the meal, insulinemia was greater with WP and BP than WF and BF (41, 29, 12.5 and 16.4 µU/ml; $P<0.05$). Mean portal glycaemia was greater with BP than BF from 30 to 150 min after the meal (1.30 vs 1.47 g/l) and tended to be higher with WP compared to WF (1.44 vs 1.66 g/l). Pelleting improved digestive utilization and modified nutrient metabolic dynamics with W- and B- but not M-based diets. Varying responses to digestive and metabolic use between botanical sources have to be taken into account for feed formulation.

Session 08 Theatre 16

Effects of a blend of phytogenic feed additives on performance in fattening pigs fed liquid feed

A. Samson, W. De Gaiffier, E. Janvier and S. Constantin
ADM, Z. A. La Pièce 3, 1180 Rolle, Switzerland; werner.degaiffier@adm.com

Plant extracts may optimize feed efficiency in pigs by increasing the activity of endogenous enzymes involved in digestion and absorption. This study aimed to assess the effect of a blend of phytogenic feed additives on growth performance of fattening pigs fed restricted-feed liquid diets. In total, 270, 68-days-old pigs housed in pens of 6 were fed on a growing diet for the first 35 days, followed by a finishing diet until slaughter. Pigs were divided into 3 experimental groups: a Negative Control (NC) fed diets with low nutritional value (Growing: 9.6 MJ NE/kg, 0.84% SID Lys; Finishing: 9.7 MJ NE/kg, 0.78% SID Lys), a Positive Control (PC) consisting of the same diets as the NC but with a higher NE and SID Lys content (+0.2 MJ NE/kg and +0.02% SID Lys), and a Phytogenic group (PHYT) fed the same diets as the NC but supplemented with 80ppm of phytogenic feed additive containing cinnamaldehyde (3%), capsicum oleoresin (2%) and carvacrol (5%). As pigs were on restricted-feed regimes, the ADFI did not differ significantly between the groups over the study (2.26 kg/d on average, P=NS). As expected, PC pigs significantly improved their performance over the fattening period compared to the NC group (+ 6.3% ADG and -6.6% F/G, $P<0.0001$). The PHYT group also had a significantly improved performance compared to the NC (+ 6.1% ADG and -6.6% F/G, $P<0.0001$) and it was comparable to the PC performance throughout the study. Lean meat percentage was significantly lower for the PC group compared to the NC pigs (-1.5%, $P=0.01$) and intermediate for the PHYT group. This data confirms that this blend of phytogenic feed additives can improve feed efficiency in fattening pigs on restricted-feed liquid diets and can compensate for the reduced nutritional value of the diet administered in the study.

Session 08 Poster 17

Genetic parameters of digestive coefficients in three pig breeds

C. Carillier-Jacquin[1], P. Ganier[2], J. Bidanel[3], C. Hassenfratz[4], B. Blanchet[5], V. Deru[1,6], E. Labussiere[2] and H. Gilbert[1]
[1]GenPhySE, Université de Toulouse, INRAE, ENVT, Chemin de Borde rouge, Auzeville, CS52627, 31320 Castanet-Tolosan, France, France, [2]PEGASE, INRAE, Institut Agro, 35590 Saint-Gilles, 35590 Saint-Gilles, France, [3]France Génétique Porc, 35651 Le Rheu Cedex, France, 35651 Le Rheu Cedex, France, [4]IFIP-Institut du Porc, 35651 Le Rheu Cedex, France, 35651 Le Rheu Cedex, France, [5]UE3P, INRAE, 35590 Saint-Gilles, France, 35590 Saint-Gilles, France, [6]Alliance R&D, 35651 Le Rheu, 35651 Le Rheu, France; celine.carillier-jacquin@inrae.fr

Digestive efficiency predicted via near infrared spectrometry (NIRS) is a new indicator of interest for pig selection schemes, making it possible to target a component of feed efficiency particularly interesting in a context where feed contains more dietary fibres. The aim of this study is to estimate genetic parameters of digestive ability for three French pig breeds: Large White, Landrace and Piétrain. Faecal samples at 21 weeks of age were taken from 629 Large White, 188 Landrace and 213 Piétrain male pigs. All Large White animals were genotyped, and digestibility coefficients (DC) for organic matter, energy and nitrogen were predicted from faecal samples by NIRS. Genetic parameters (genetic variance, residual variance and heritability) of DC and genetic correlations with classical production traits (growth rate, feed efficiency and carcass composition) were estimated using REML algorithms considering genotyped data when it was available. Heritabilities of DC estimated were moderate (between 0.10 for DC of nitrogen and 0.35 for DC of energy in Landrace), which is close to estimations reported in the literature for Large White pigs. Heritability for DC in the three breeds were similar, with slightly lower estimates for DC of organic matter in Landrace and Piétrain breeds compared to Large White. Trends of the genetic correlations with production traits were similar in the three breeds: negative for growth rate, feed intake and null for carcass leanness. These first estimates of genetic parameters for DC in Landrace and Piétrain suggest that the NIRS predictions can be used in most selected pig breeds to further improve feed efficiency.

Session 08 Poster 18

Breeding for uniformity in piglet birth weight to improve survival

C. Kasper[1], A. Lepori[2], J.P. Gutiérrez[3], N. Formoso-Rafferty[4], E. Sell-Kubiak[5] and I. Cervantes[3]
[1]Agroscope, Animal GenoPhenomics, Tioleyre 4, 1725, Switzerland, [2]Suisag, Allmend 10, 6204 Sempach, Switzerland, [3]UCM, Department of Animal Production, Veterinary Faculty, Avda. Puerta de Hierro s/n, 28040 Madrid, Spain, [4]UPM, Department of Agricultural Production, ETSIAAB, Senda del Rey, 18, 28040 Madrid, Spain, [5]Poznan University of Life Sciences, Department of Genetics and Animal Breeding, Wojska Polskiego 28, 60-637 Poznań, Poland; claudia.kasper@agroscope.admin.ch

Selection for uniformity in birth weight (BW) could lead to a more ethical and efficient livestock production because it results in more robust animals, which are easier to manage, more feed-efficient and are more likely to survive to weaning. This study aimed to estimate the genetic component of residual variance for BW and its relationship with piglet survival in a Swiss experimental farm. The data set comprised 43,135 records of BW from 3,163 litters of 986 sows, and pedigree data for 45,737 individuals. A heteroscedastic model was used, including fixed effects such as sex, month-year, litter size, and parity, and the litter effect was added as a random effect in addition to the genetic effect. The data was assigned to the mother, and the same effects were fitted for mean BW and its variability. A threshold homoscedastic model was performed for the probability of stillbirth (SB). A multivariate analysis was performed for BW and SB including the additive genetic effect and the maternal genetic effect using a Markov Chain Monte Carlo software (GSEVM) for the heteroscedastic model and TM software for the homoscedastic one. The genetic coefficient of variation was 0.29, and the genetic correlation between the mean BW and its variability was 0.24 (SE=0.09). The individual and maternal h^2 were 0.04 (0.01) and 0.23 (0.03) and 0.00 (0.00) and 0.05 (0.01), for BW and SB, respectively. The direct genetic correlation between BW and SB was 0.14 (0.31) and the maternal one 0.01 (0.11). In conclusion, our results show that there is potential for selection for the reduction of environmental BW variability, and the correlation between maternal breeding values indicates that it will not negatively affect the survival of piglets.

Session 08 Poster 19

Genetic lines influence carcass traits and quality of pork

X. Li[1], M. Ha[1], R. Warner[1], R. Hewitt[2], D. D'Souza[2] and F. Dunshea[1,3]
[1]School of Agriculture and Food, Faculty of Science, University of Melbourne, 3010, Parkville, VIC, Australia, [2]SunPork Group, SunPork Group, 4009, Eagle Farm, QLD, Australia, [3]Faculty of Biological Sciences, University of Leeds, Leeds, LS2 9JT, United Kingdom; xiyingl@student.unimelb.edu.au

While the Australian swine industry has made efforts in selecting for lean growth, especially in the terminal lines, this may have adverse effects on pork quality. This study aimed to investigate how genetic lines affect carcass traits and pork quality and how chemical composition affects pork texture. Female pigs from 6 lines were used: Pure maternal, Landrace-type (PM-LR, n=18); Pure maternal, Large White-type (PM-LW, n=18); Pure maternal, Duroc-type (PM-D, n=18); Synthetic terminal, large white and Landrace-type (SynT-LWLR, n=12); Pure terminal, Duroc-type (PT-D, n=18); and Pure terminal, Large White-type (PT-LW, n=18). Live weight, dressing percentage, P2 fat depth and ultimate pH were recorded. From each line, 12 carcasses were used to measure the cooking loss, Warner-Bratzler shear force (WBSF), texture profile analysis (hardness, cohesiveness, adhesiveness, chewiness, and springiness), collagen content and solubility and intramuscular fat (IMF) content of Longissimus thoracis et lumborum and Semimembranosus. Results showed that the terminal lines had lower P2 fat depth than the maternal lines (9.26 vs 11.8 mm, $P<0.001$) and their pork showed higher cooking loss (17.2 vs 16.1%, $P=0.002$), hardness (35.7 vs 34.0 N, $P=0.027$) and chewiness (12.4 vs 11.7 N, $P=0.037$) than maternal lines. Considering individual lines, SynT-LWLR showed the highest hardness ($P=0.039$) and cohesiveness ($P=0.023$). The IMF content of SynT-LWLR was lower than the other five lines (1.08 vs 1.69% to 1.73%, $P=0.004$). When all lines were analysed, muscle collagen content was correlated ($P<0.05$) with hardness (r=0.20), chewiness (r=0.27), adhesiveness (r=0.29) and springiness (r=0.30), while collagen solubility was correlated ($P<0.05$) with chewiness (r=-0.18) and springiness (r=-0.22). IMF was correlated ($P<0.05$) with WBSF (r=-0.29), adhesiveness (r=0.25) and springiness (r=0.34). In individual line, no significant correlations were found in PM-D or PT-LW. Genetic selection may adversely influence pork quality and genetic lines affect the contribution of collagen and IMF to pork texture.

Session 08 Poster 20

Boar taint: pedigree-based BLUP and genomic BLUP, and associations with green-ham quality traits

S. Faggion, P. Carnier and V. Bonfatti
University of Padova, Department of Comparative Biomedicine and Food Science, viale dell'Università, 16, 35020, Legnaro, Italy; sara.faggion@unipd.it

Boar taint (BT) occurs in entire male pigs after puberty due to the accumulation in the adipose tissue of three main compounds: androstenone (AND), skatole (SKA), and indole (IND). Slaughtering entire male pigs before sexual maturity is not feasible for pigs intended for protected designation of origin (PDO) dry-cured ham production and, with a perspective future ban on surgical castration in Europe, selecting pigs with reduced ability to accumulate BT compounds in their tissues seems a promising strategy. BT compound concentrations were measured in 1,115 purebred pigs; animals were genotyped using a high-density SNP chip (29,844 SNPs after quality control). Breeding values for BT compounds were computed with pedigree-based BLUP (PBLUP) and genomic BLUP (GBLUP) using the genomic relationship matrix. Model performance was estimated in a 5-fold random cross-validation. The accuracy of PBLUP and GBLUP (defined as the Pearson correlation between predicted EBV and adjusted phenotypes divided by the square root of heritability) was 0.38 and 0.61, respectively, for AND, 0.54 and 0.51 for IND, 0.50 and 0.54 for SKA. Since most slaughter pigs are crossbred, it is important to consider potential effects on commercial traits in crossbreds before selecting against BT in purebreds. Genetic correlations between BT compound concentrations and carcass and ham quality traits from 26,577 crossbred Italian heavy pigs were then estimated. Estimates were obtained in a set of bivariate Bayesian analyses including one BT trait and one production/ham quality trait at a time. Heritability estimates for AND, SKA and IND were 0.41, 0.49 and 0.37, respectively. Genetic correlations between BT compounds were positive (0.40 to 0.85). Negative correlations between SKA and carcass yield (-0.40) and between BT compounds and backfat (between -0.26 and -0.55) were observed. Conversely, positive correlations (from 0.11 to 0.54) between SKA and ham fat thickness traits were detected. Correlations between BT compounds and iodine number ranged from -0.07 (AND) to -0.64 (SKA), whereas those with PUFA ranged from -0.13 (IND) to -0.33 (SKA).

Session 08 Poster 21

Diversity of pig breeds reared in the Czech Republic

E. Krupa[1], Z. Krupová[1], E. Žáková[1], N. Moravčíková[2] and I. Vrtková[3]
[1]Institute of Animal Science, Přátelství 815, 10400 Prague, Czech Republic, [2]Slovak University of Agriculture, Tr. A. Hlinku 2, 94901 Nitra, Slovak Republic, [3]Mendel University, Zemědělská 1, 613 00 Brno, Czech Republic; krupova.zuzana@vuzv.cz

The aim of the study was to compare the diversity parameters of the dam pig breeds: Czech Large White (CLW) and Czech Landrace (CL) based on pedigree and SNP data. Total number of animals in pedigree were 16,438 and 7,266 for CLW and CL, respectively. A slight increase in the inbreeding and the proportion of inbred animals used in the breeding was observed. The average inbreeding rates in 2010, 2015, 2020 and 2022 for the CLW were 0.7, 1.5, 1.7 and 2.0%, respectively. The proportion of inbred animals included in the breeding program in these years was 39.5, 83.1, 97.8 and 97.4%, respectively. The CL breed had slightly higher values in those years. The effective population size ranged from 99 (CLW) to 181 (CL) animals. Selected DNA isolates (603 and 201 animals of CLW and CL, respectively) were applied to a GGP Porcine 50k chip using the Illumina HD Infinium technology protocol. A total of 50,697 SNPs were identified in the selected DNA isolates. The average 'call rate' was 0.99, 0.97 and 0.96 for bristle, insemination batch and ear graft samples used as source, respectively. Thus, data from 590 CLW and 196 CL animals containing 45,065 and 46,257 SNPs, respectively, were used for further evaluation. The genomic inbreeding coefficient was calculated for each ROH class (FROH). The effective population size based on genomic data was calculated using GONE software. The inbreeding coefficient obtained from SNP data was lower than that obtained from pedigree data. It was 0.40% for the CLW breed and 2.23% for the CL breed. The differences were probably due to smaller number of genotyped animals. The effective population size over the last 9 generations exceeded 300 individuals for both breeds, although a decrease in effective population size was observed, especially in the last three years. The results of the SNP analyses are generally consistent with the results from the pedigree analysis in both breeds – a slight increase in inbreeding and a decrease in effective population size. The study was supported by Czech Republic project QK1910217 and MZE-RO0723 – V02.

Session 08 Poster 22

Wild boar and domestic pig identification based on GBS study

A. Koseniuk[1], G. Smołucha[1], A. Gurgul[2], T. Szmatoła[2], M. Oczkowicz[1] and A. Radko[1]
[1]National Institute of Animal Production, Department of Animal Molecular Biology, Krakowska 1, Street, 32-083, Poland, [2]University of Agriculture in Krakow, Center for Experimental and Innovative Medicine, Rędzina 1c, 30-248 Kraków, Poland; anna.koseniuk@iz.edu.pl

According to scientific reports, efforts have also been made to provide an efficient diagnostic tool for distinguishing wild boar (*Sus scrofa scrofa*), domestic pig (*Sus scrofa domestica*), and their hybrids. The issue is getting more complex since both subspecies interbred sporadically over the last few decades and after the last glaciation. Genotyping-by-sequencing (GBS) is a method of identifying genetic variants that do not require knowledge of the genome. Recently, the method has become more attractive due to its relatively short analysis time and low cost. The research aims to identify the DNA regions that underwent strong selection during pig domestication and give an insight into Polish wild boar and domestic pigs' genetic diversity by implementing the genotyping-by-sequencing (GBS) technique. We have DNA extracted from ear fragments of wild boars (n=20) from southern Poland and hair bulbs pigs (n=10). The pig samples belonged to the breeds: PBZ (Polish Landrace, n=2), WBP (Polish Large White, n=4), DUR (Duroc, n=1), PUL (n=3), P (Pietrain, n=2). PstI endonuclease was used for DNA digestion. After the digestion stage, the ligation of DNA fragments with adapters was carried out, and the libraries prepared in this way were subjected to PCR. Fragment sequencing was performed using the HiScanSQ system (Illumina, USA). The genetic distance was calculated in the MEGAX program and the phylogenetic tree was created using the Maximum Likelihood (ML) method. The structure of the population and the degree of hybridization were assessed using the Structure program. Results and conclusions. The conducted analyses showed that both groups of animals are phylogenetically separated from each other. The study of the genetic structure of both populations showed that the most probable number K=2, wild boar samples are genetically uniform, while several subpopulations were identified in the group of domestic pigs. Based on the obtained results, the presence of hybrids in both tested groups of animals cannot be unequivocally stated.

Structural analysis of the HK2 gene in the aspect of pigs performance and meat quality parameters

K. Woźniak[1], A. Terman[1], D. Polasik[1], G. Żak[2], M. Tyra[2] and K. Ropka-Molik[2]
[1]West Pomeranian University of Technology in Szczecin, al. Piastów 17, 70-310 Szczecin, Poland, [2]National Research Institute of Animal Production, ul. Sarego 2, 31-047, Poland; grzegorz.zak@iz.edu.pl

Scientific research in the area of pork quality may in the future enable the selection of animals with a preferred genetic variant, whose meat will be characterized by more favourable meat quality parameters. In this context, the analysis of polymorphic variants of the HK2 gene in terms of pig performance characteristics and pork quality parameters can be considered justified due to the biological role in the regulation of the glycolysis process, as well as due to the molecular nature of the mechanism indicating the variability of the encoded proteins. The aim of this study was to detect polymorphisms in the gene encoding hexokinase (HK2) in domestic pig (*Sus scrofa domestica*) and to determine the potential relationships between the genotypes of the analysed gene fragments and the performance characteristics of pigs (fattening and slaughter) and selected parameters of meat quality. The research covered 722 pigs of 3 breeds: Polish Landrace, Polish Large White and native breed Puławska. The animals were kept at Pig Tests Stations. Feeding and housing conditions were consistent for all animals. Genomic DNA was isolated from longissimus dorsi muscle with the use of A&A Biotechnology (Poland). In the first phase of the research, exons in the gene HK2 had to go through PCR-HRM method. Fragments of PCR amplification selected by this technique, and the obtained products were sequenced using the Sanger method. 15 mutations were found in the analysed HK2 gene fragments. For two identified polymorphisms, SNP type located in the splicing region of exon 7 and in exon 12 an appropriate molecular method was developed (PCR-RFLP and PCR-ACRS, respectively) allowing to determine the frequency of selected mutations on a larger group of animals. The obtained frequencies of the polymorphic variants of the analysed gene show that there is variability within the HK2 gene. The identified polymorphisms show a significant association with selected fattening and slaughter characteristics as well as meat quality in pigs of various breeds.

Methods to maximize accuracy of selection in maternal genetic model for genetic evaluation of pigs

M. Satoh
Tohoku University, Graduate School of Agricultural Science, Aramaki-Aza-Aoba 468-1, Sendai, Miyagi, 980-8572, Japan; masahiro.satoh.d5@tohoku.ac.jp

A mixed model allows us to estimate genetic parameters and easily obtain estimates of direct genetic effects (g_d) and maternal genetic effects (g_m). However, it is difficult to assign appropriate economic weights for g_d and g_m. In this study, we devised a method to obtain an appropriate weighting vector (v) for g_d and g_m. First, for the ith animal, let g_i = [g_{di} g_{mi}]' and its BLUP vector be g_i^. When v is known, we confirmed that the weighting vector w that minimizes the prediction error variance, *var*(v'g_i – w'g_i^), of H_i (= v'g_i) is w = v when H_i^ = v'g_i^. Next, when v is unknown, we determined that the v that maximizes the coefficient of determination ($r_{Hi^Hi}{}^2$) between H_i and H_i^ is approximately the solution of the eigen equation $|G_0^{-1}C^{rr} - kI|v = 0$, where G_0 = *var*(g_i) and C^{rr} is a 2×2 matrix derived from the inverse matrix of the left-hand side of the mixed-model equations. Since the eigenvalue k = v'C^{rr}v/v'G_0v, $(1 - k)^{1/2}$ represents the accuracy of selection. Let $k_1<k_2$ be the two eigenvalues, then k_1 maximizes r_{Hi^Hi} and k_2 minimizes r_{Hi^Hi}. Moreover, the elements of the eigenvector are the weights for g_d and g_m. This method was verified using a Monte Carlo simulation. The genetic parameters used were 0.3 for direct heritability, 0.1 for maternal heritability, and -0.5, 0, and 0.5 for the correlations between them. A closed breeding herd of 20 sires and 100 dams over four generations was simulated. Each litter produced two males and two females. All animals were randomly selected and five females were randomly mated to one male. Let G0 be the base population and animals in G4 are candidates for selection. Data were generated based on an infinitesimal additive genetic model. r_{Hi^Hi} in G4 was 0.614, 0.671, and 0.722 for correlations between direct and maternal heritability conditions of -0.5, 0, and 0.5, respectively. In addition, when the element of eigenvectors v_1 (for g_d) = 1, v_2 (for g_m) was 0.513, 0.546, and 0.549, for the three correlation conditions, respectively.

Session 08 Poster 25

Parentage control of pigs based on SNP data

E. Žáková[1], Z. Krupová[1], E. Krupa[1], J. Stibal[2] and I. Vrtková[3]

[1]Institute of Animal Science, Přátelství 815, 10400 Prague, Czech Republic, [2]Czech Pig Breeders Association, Bavorská 14, 155 41 Prague, Czech Republic, [3]Mendel University, Zemědělská 1, 613 00 Brno, Czech Republic; krupova.zuzana@vuzv.cz

Genotyping of animals included in the national breeding programme CzePig has been applied for genomic evaluation and pedigree assessment. To determine accurate animal evaluation and selection the parentage test is necessary. Therefore, the present study is aimed to exploit the pig genomic data to define an SNP panel that could be used for routine parentage control. Samples were taken from the animals that were parents of the next generation of breeding animals or that had sufficient performance data records in the studbook database. From the 1,489 animal sampled, 68% belongs to maternal and 17% to sire breeds and 15% were of the endangered Prestice black-pied breed. In 869 and 221 animals one and both parents were genotyped, respectively. The ratio of the evaluated samples by sex was 55:45 for sows and boars. Three SNP chips were used for genotyping (Illumina, Inc. Porcine SNP60v2 and PorcineSNP60 BeadChip and GeneSeek® Genomic Profiler™ GGP Porcine 50K). From all of the SNP 32,898 were shared between these chips. The GenCall score (GC ≥70 and ≥80), minor allele frequency (30%≤pM≤50%) and known position of each SNP locus on all autosomes were considered. The evaluation was done separately for each breed. Panels consisting of 435 (GC≥70) and 240 (GC≥80) SNP loci were selected for routine parentage control. The SNPs have to be known for more than 95% of animals. Subsequently, the genotype of an individual was compared with one or with both of parents. Only 3.6% of samples were assessed as incorrect (unreliable sample or pedigree error). These may be refined in the future as data increases. Simultaneously, the panels will be continuously reassessed due to upcoming samples (especially of sire breeds). The study was supported by Czech Republic project QK1910217 and MZE-RO0723 – V02.

Session 08 Poster 26

Prediction of piglet survival based only on birth weight or within-litter birth weight variability

J.M. Mbuthia[1], C. Kasper[2], M. Zenk[1], G. Bee[2], C.C. Metges[1] and G. Daş[1]

[1]Research Institute for Farm Animal Biology (FBN), Institute of Nutritional Physiology, Wilhelm-Stahl-Allee 2, 18196 Dummerstorf, Germany, [2]Agroscope, 'Animal GenoPhenomics' and 'Swine Research Unit', Rte de la Tioleyre 4, Posieux, 1725, Switzerland; gdas@fbn-dummerstorf.de

High piglet mortality is of both economical and ethical relevance. Piglet birth weight (BW) and the birth weight variability within litter (BWvar) highly affect the pre-weaning survival. Further, the identification of low birth weight (LBW) piglets often relies on arbitrary statistical classifications. We established optimal cut-off values for BW and BWvar to predict pre-weaning piglet survival. BW, BWvar and survival data were obtained from FBN and Agroscope experimental pig facilities. The FBN data (2012 to 2021) for the German Landrace breed consisted of records from 28,242 total number of piglets born (TNB) i.e. born dead or alive from 752 sows. The Agroscope data (2004 to 2022) for the Swiss Large White consisted of 43,159 piglet records from 980 sows. The cut-off values for BW and BWvar to predict piglet survival from birth to weaning were estimated by the Receiver Operating Characteristic (ROC) curves analysis using the cutpointr package of R (v 4.0.3). Postnatal piglet mortality was highest within the first 3 days. From the piglets born alive, about 85.2 and 83.7% survived day 3 at FBN and Agroscope, respectively. Respective survival rates at weaning were 81.1 and 78.6%. Overall, the piglets below a BW cut-off value of 1.18 kg (i.e. FBN=1.17 kg and Agroscope=1.19 kg) had a lower survival probability at birth than those piglets>1.18 kg, and can be considered as LBW piglets. With a cut-off value of 0.278 g (AUC=0.55) the BWvar had a lower prediction accuracy than BW (AUC=0.66). Moreover, a BW cut-off value of 1.22 g (AUC=0.74) would increase the probability of piglets to survive until weaning. Although facilities were different in terms of genotypes and environments, highly similar cut-off values were estimated for identification of LBW with smaller survival probability for both facilities. The results also suggest that prediction of piglet survival with only BW is more informative than only with BWvar. The identified LWB piglets may benefit from extra support measures such as e.g. supplemental milk.

Session 08 Poster 27

Effect of fenugreek cotyledons on farrowing sow performances

M. Le Bot, J. Laurain, E. Belz and A. Benarbia
Nor-Feed SAS, 3 rue Amédéo Avogadro, 49070 Beaucouzé, France; amine.benarbia@norfeed.net

Nutritional management of sow from the gestation to the end of lactation is critical for achieving and maintaining optimal sow productivity, longevity and litter growth. Fenugreek (*Trigonella foenum-graecum*) is an annual plant that belongs to the family of the *Fabaceae*. Due to the secondary metabolites, as steroidal saponins providing therapeutic properties and appetite stimulation, fenugreek seeds are commonly used in human and animal diets to stimulate appetite and weight gain. Feeding gestating and lactating sows with fenugreek cotyledons, the saponin-rich part of the seed, could improve optimal sow productivity and consequently litter growth during these periods. In this study, we assessed the effect of a commercial product based on fenugreek cotyledons (Norponin® Cotyl) on sow productivity and the impact on litter. Briefly, 100 sows (experimental farm, France) were randomly divided into two groups from 2 consecutive batches: a control group (CTL) fed with a standard diet (gestation and lactation) and a group (COTYL) supplemented with 1,500 ppm of fenugreek cotyledons. The trial started at the arrival of sows in farrowing unit (9 days before farrowing) and during all lactation (21 days) for a total of 30 days. Results showed that the average daily gain was higher in the COTYL group compared to the CTL group with 6.17 and 5.86 kg/day/sow respectively. This difference of +5.3% ($P<0.05$) could lead to an increase in the milk production of the sows and would explain in part the improvement of the average daily gain of the piglets observed during this trial. Indeed, results corresponded to 3.03 for the CTL group and 3.13 kg/day/litter (+3.3%) for the COTYL group. The results obtained from this trial show that the supplementation with fenugreek cotyledons contributes to significantly increase feed intake and thus to improve piglet's growth during lactation. Further studies are however necessary to confirm these results.

Session 08 Poster 28

Effects of different dietary fibres on constipation in sows during late pregnancy

D. Lu[1,2], Y. Pi[2], H. Ye[1], D. Han[2], B. Kemp[1], N. Soede[1] and J. Wang[2]
[1]Wageningen University&Research, Department of Animal Sciences, Adaptation Physiology Group, Zodiac, De Elst 1, 6700 AH Wageningen, the Netherlands, [2]China Agricultural University, College of Animal Science and Technology, State Key Laboratory of Animal Nutrition, Yuanmingyuan Road, Haidian District, 100193, Beijing, China, P.R.; dongdong.lu@wur.nl

Constipation in sows during late pregnancy increases farrowing duration and thereby increases the number of stillborn piglets. Dietary fibre supplementation can change intestinal microbiota composition, thereby increasing intestinal motility and reducing constipation in many animal models. However, the effects and mechanisms of fibres with different physicochemical properties on constipation have not been fully explored. In this study, 80 sows were randomly allocated to Control (CON, basic corn-soybean meal) and one of three dietary fibre treatments with the same total dietary fibre content (TDF) from day 85 of gestation to delivery: LIG (replace 1.5% of wheat bran with lignocellulose), PRS (replace 2.0% of wheat bran with resistant starch), and KON (replace 2.0% of wheat bran with konjaku flour). Results showed that the defecation frequency and faecal consistency were highest in PRS (2.93/day and 3.03). PRS and KON significantly increased serum levels of gut motility regulatory factors, 5-hydroxytryptamine (5-HT), motilin (MTL), endothelin-1 (ET-1), acetylcholinesterase (AChE) and reduced serum inflammation factors IL-6 and TNF-α. Furthermore, PRS and KON significantly reduced the number of stillborn piglets compared to CON. Microbial sequencing analysis showed that PRS and KON increased short-chain fatty acids (SCFAs) producing genera *Bacteroides*, *Parabacteroides*, and decreased the relative abundance of endotoxin-producing bacteria *Desulfovibrio* and *Oscillibacter*. Besides, the relative abundance of *Turicibacter* was highest in PRS. In conclusion, PRS and KON reduced sow constipation, which was associated with higher levels of gut motility regulatory factors under the genus *Turicibacter* and SCFAs stimulation, thereby increasing gut motility and reducing the number of stillborn piglets.

Session 08 — Poster 29

Effect of transportation condition of pigs during transport for slaughter under different season

D.C. Song, S.Y. Chang, J.W. An, S.H. Park, K.H. Jeon and J.H. Cho
Chungbuk National University, Department of Animal Science, 1, Chungdae-ro, Seowon-gu, Cheongju-si, Chungcheongbuk-do, Republic of Korea, 28644, Korea, South; paul741@daum.net

Animal welfare during transport became an largely issue because of increasing demand for improved animal welfare standards. The welfare of pigs during transportation is impacted by vibration brought on the driver's driving style. In vehicles, animals are exposed to vibration and environmental variations, which can lead to physiological and behavioural disturbances. Also, bedding may affect transport losses as a significant microenvironment component. Thus, the objective of study was to collect and quantify three axis acceleration and determine the effect of rubber type of bedding for transporting pigs from farm to slaughterhouse. A total of 2,553 crossbred pigs of mixed sex with same genetics ([Yorkshire × Landrace] × Duroc) were transported from same commercial farms to same commercial slaughterhouse. A 2×2 completely randomized factorial design was used to investigate the effects of bedding (bedding or non-bedding) and with two levels of driving style (wild or normal). Pigs transported bedding with normal driving style groups had higher ($P<0.05$) than pigs transported bedding with wild driving style groups. Transported with bedding groups showed higher pH, L value and sensory colour than transported with non-bedding groups in winter. Also, transported with bedding groups showed higher ($P<0.05$) standing behaviour but lower ($P<0.05$) lying behaviour than transported with non-bedding groups in winter. Also, transported with bedding groups showed higher ($P<0.05$) standing behaviour but lower ($P<0.05$) lying behaviour than transported with non-bedding groups in winter. Transported with bedding groups showed less ($P<0.05$) aggression behaviour than transported with non-bedding groups in spring, fall and winter. In spring, autumn and winter season, normal driving style groups or bedding groups showed low ($P<0.05$) cortisol level compared to wild driving style or non-bedding groups. In conclusion, driving style and bedding are important part of animal transportation to protect animal welfare and economic losses.

Session 08 — Poster 30

Salmonella* excretion level by pigs and impact of disinfection on antibiotic resistance of *E. coli

C. Soumet[1], A. Kerouanton[2], A. Bridier[1], B. Anger[1], V. Dorenlor[3], V. Rose[2], F. Eono[3], C. Houdayer[2], E. Houard[2], E. Eveno[3], F. Souchaud[2], B. Houry[2], C. Valentin[1], P. Houée[1], I. Attig[4], N. Haddache[4], M. Denis[2] and C. Fablet[3]
[1]Anses, CS 40608, 35306 Fougères, France, [2]Anses, BP 53, 22440 Ploufragan, France, [3]Anses, BP 53, 22440 Ploufragan, France, [4]Anses, 14 rue Pierre et Marie Curie, 94700 Maisons Alfort, France; christelle.fablet@anses.fr

The control of contamination of food by *Salmonella* and the reduction of antibiotic resistance are two major public health issues. The study aimed at acquiring data on the level of *Salmonella* excretion by pigs and assessing the impact of disinfectants on the evolution of antibiotic resistance in *Escherichia coli* strains. Four French pig farms deemed to be *Salmonella* positive were sampled 3 times. At each visit, individual faeces from 10 lactating sows and 20 finishers were collected. The detection and enumeration of *Salmonella* was carried out from faeces by standardized methods. On 3 farms, the pen partitions and floor surfaces of 3 farrowing rooms were swabbed before and after cleaning and disinfection (CD) procedures with quaternary ammonium compounds based disinfectants. Total *E. coli* and *E. coli* resistant to antibiotic were enumerated from swabs on Petrifilm™ Select *E. coli* without and with a concentration of antibiotics respectively. Total bacteria were enumerated on non-selective media. No sow and 21.6% of finishers tested positive for *Salmonella*. A low *Salmonella* excretion level was estimated on most of the positive samples. The excretion level was variable between farms and pigs within a farm. Ten of 36 samples after CD were positive for *E. coli*. CD procedures reduced counts of *E. coli* and total bacteria (by 3 to 4 Log10) and antibiotics resistant *E. coli*. This exploratory study allows collecting for the first time in France quantitative data on the level of *Salmonella* excretion in pigs naturally infected. These data are required to build accurate risk assessment models and ultimately allow better control of the risk associated with *Salmonella* contamination of food. Disinfection of pig premises was effective to reduce total bacteria and *E. coli* counts. Reduction of *E. coli* resistant to tested antibiotics suggested that disinfectant exposure would not have selected antibiotic resistance in *E. coli* strains on these herds at the time of the study.

Session 08 Poster 31

Climate change adaptation in mixed pigs-beef cattle systems in grasslands areas: key role of organic

S. Mugnier[1], L. Valero[1], C. Husson[1], F. Von Kerssenbrock[2], B. Dounies[3], H. Rapey[2] and C. Philippeau[1]
[1]L'Institut Agro Dijon, BP 87999, 21079 Dijon, France, [2]Université Clermont Auvergne, AgroParisTech, INRAE, VetAgro Sup, UMR 1273 Territoires, 9 Av. Blaise Pascal, 63170 Aubière, France, [3]Association Porc Montagne, 9 allée Pierre de Fermat, 63170 Aubière, France; sylvie.mugnier@agrosupdijon.fr

Global climate change may increase the frequency and severity of drought periods which may potentially strongly reduce grass yields due to reduced water and nutrient availability in soils. In this context, the grassland management is essential to ensure the fodder autonomy in cattle farms. Several studies showed that multi-species livestock farming (in particularly, sheep and cattle) increases the resource-use efficiency. However, the potential benefits of systems combining pigs and cattle are less studied. We can suppose that the careful management of the different types of manures may limit variation in plant productivity and, more particularly, grass yield due to climate changes. Our objective was to develop a comprehensive analysis of mixed pigs-cattle farming systems in the French Massif Central area in order to outline and discuss potential benefits and limitations of using different manures to fertilise grasslands and crops. Two studies were performed. Firstly, on-farm 40 surveys were carried out in order to characterize how pigs and cattle effluents were used as organic fertilisers. Most farmers assigned specific fertiliser properties according to each type of effluent. For example, for most farmers, the livestock slurry promotes a rapid growth of grass whereas the application of solid manure enhance organic matter enrichment in soil. Slurry is applied to grass pastures in the spring and solid manure preferentially before the sowing date of cereals. A second survey was carried out on 20 mixed pigs-beef cattle farms which were located in a same pedoclimatic area (North of the French Massif Central) to analyse the variations in fertilisation management of grassland between the last 3 years. This study provided more precise knowledges about the pig manure application as organic fertiliser. Also, farmers seemed aware of the interest of pig slurry in the grassland management in extensive livestock production systems in order to limit variations in grass yields due to climate changes.

Session 08 Poster 32

Financial implications associated with ante- and *post-mortem* inspection findings in finishing pigs

D.L. Teixeira[1], L.C. Salazar[2] and L.A. Boyle[3]
[1]Hartpury University, Department of Animal and Agriculture, Gloucester, GL19 3BE, United Kingdom, [2]Pontificia Universidad Católica de Chile, Departamento de Ciencias Animales, Santiago, Region Metropolitana, Chile, [3]TEAGASC Moorepark, Pig Development Department, Fermoy, Co. Cork, Ireland; laura.boyle@teagasc.ie

This study aimed to investigate the associations between severe ear, tail and skin lesions, hernias, bursitis and rectal prolapses and meat inspection finding in slaughter pigs, including carcass weight and financial implications associated with carcass condemnations at batch level. Data were collected from 13,296 pigs from 116 batches from a single abattoir. Spearman's correlation coefficients were calculated to analyse the degree of association between the prevalence of welfare issues and condemnation findings. The association between batch-level results of carcass weight, batch size and the prevalence of welfare issues was analysed using generalized linear mixed models. The prevalence of tail lesions was significantly associated with both entire ($r=0.224$; $P=0.0432$) and partial ($r=0.276$; $P=0.0120$) carcass condemnation. Batches with pigs affected by more than one welfare issue were 9.9 kg lighter than those without welfare issues ($P<0.05$), which was equivalent to a potential loss of €11.28 per pig. Our findings indicate that *ante-mortem* inspection could be useful to predict *post-mortem* findings at batch level and that welfare issues in pigs represent a financial loss to producers, as they are paid on a per kg basis and have tight margins.

Session 08 Poster 33

Training during rearing: effect on body reserves' flexibility & long-term reproduction in rabbit doe

J.J. Pascual, E. Martínez-Paredes, L. Ródenas, E. Blas and M. Cambra-López
Universitat Politècnica de Valencia, Institute for Animal Science and Technology, Camino de Vera s/n., 46022, Spain; jupascu@dca.upv.es

Animals' flexibility to mobilize and recover body reserves increases with age or parities. As most breeds are selected for early productive criteria, training the body reserves flexibility of young breeds could improve their future reproduction and survival. This work evaluated the effect of a rearing training strategy for young rabbit does, based on 0 to 3 feed restriction schemes, on their body reserves flexibility and long-term reproduction. Each restriction was addressed to simulate the mobilization of reserves around parturition, 6 days of progressive reduction from *ad libitum* to zero and 3 days of progressive recovery until *ad libitum* feeding. At 63 days of age, 120 rabbit females were divided into 4 groups (30 each): A, fed *ad libitum*; 1R, fed *ad libitum* with one restriction scheme from 92 to 101 days of age; 2R, fed *ad libitum* with two restriction schemes from 70 to 79 and 114 to 123 days of age; 3R, fed *ad libitum* with three restriction schemes from 70 to 79, 92 to 101 and 114 to 123 days of age. Females were artificially inseminated (AI) at 137 days of age and at 11 days postpartum thereafter. Live weight (LW), perirenal fat thickness (PFT) and feed intake of females were controlled until the 2nd parturition. Alive and total litter size at birth was controlled until the 9th reproductive cycle. A few days before the first AI, does were challenged with isoproterenol to determine their lipolytic potential. Young rabbit does from the R groups showed clear losses of LW and PFT during the application of restriction schemes but recovered the A group values some weeks after refeeding. At first AI, R females had lower basal blood concentration of non-esterified fatty acids with respect to A females (on av. -13.9±4.5 uEq NEFA/l; P=0.002), but no differences were observed between groups in the increase of NEFA after challenge. There was a linear increase of total born and born alive during 9 reproductive cycles with the number of restrictions applied during rearing (+0.47±0.20 and +0.42±0.19 per restriction, respectively; P<0.05). The restrictions applied during rearing did not affect body reserves' flexibility but improved the prolificacy of the females.

Session 09 Theatre 1

Replacement of poultry by-product meal by black soldier fly larvae meal in diets for dogs

B. Agy Loureiro[1], R.K. Nobrega Cardoso[2], R. Silva Carvalho[2], W.A. Zamora Restan[3], M. Dalim[1], N. Martin Tome[1] and A. Paul[1]
[1]Protix B.V., Industriestraat 3, 5107 NC Dongen, the Netherlands, [2]Universidade Federal da Bahia, Adhemar de Barros, 40170-110, Salvador, Brazil, [3]Universidade Federal da Paraiba, PB 079 km 12, 58.397-000, Areia, Brazil; bruna.loureiro@protix.eu

The study evaluated the use of black soldier fly larvae (BSFL) meal in diets for dogs on digestibility, intestinal fermentation end-products and faecal microbiota. Two kibble iso-nutrient diets were developed using either poultry by-product (PBP) meal or BSFL meal as main protein. Eight beagle dogs were assigned in a cross-over design, with 2 treatments (diets) and 2 periods of 50 days each (with 7 days of wash-out between periods). In the first period, 4 dogs received either the PBP diet or the BSFL diet, while in the second period the diets were inverted. At day 15 of each period, dry matter, organic matter, crude protein and fat digestibility; and metabolizable energy (ME) were determined by total faeces collection method for 5 days. Volatile fatty acids and ammonia were analysed in fresh faecal samples collected on days 21 to 24 of each period. After each period (50 d) fresh faeces were collected for metagenomic analysis using bacterial 16s rRNA marker gene sequence. Nutrients digestibility was similar between the food treatments, except for fat digestibility and diet ME, which was higher when dogs were fed BSFL food (P=0.01). Faecal ammonia was lower (151 vs 94 mmol/g faeces) when dogs were fed BSFL in comparison to PBP (P=0.004). BSFL diet promoted changes in faecal microbiota, with a significant difference in beta diversity, with taxa dissimilarity by Unifrac (P=0.036). BSFL diet promoted a higher relative abundance of *Bacteroides* (P=0.040), responsible to contributes to intestinal permeability; and *Phocaeciola* (P=0.028), considered a biomarker of human health. On the other hand, BSFL reduced the abundance of *Lachnospira* (P=0.003), positively correlated with intestinal butyrate production, despite no diet differences found for volatile fatty acid in faeces. In conclusion, the use of BSFL meal in dog diet didn't affect the use of nutrients, but increased diet fat digestibility and ME; reduced faecal ammonia, and positively modified the faecal microbiome of dogs, favouring some beneficial bacteria genera.

Session 09 — Theatre 2

Hermetia illucens meal in Rainbow trout diet – preliminary data of a commercial scenario

S. Bellezza Oddon[1], I. Biasato[1], C. Caimi[1], P. Badino[1], F. Gai[2], M. Renna[1] and L. Gasco[1]
[1]University of Turin, Largo Paolo Braccini, 2, 10095, Grugliasco, Italy, [2]Institute of Sciences of Food Production, Largo Paolo Braccini, 2, 10095, Grugliasco, Italy; sara.bellezzaoddon@unito.it

Hermetia illucens (HI) meal has already been tested as feed ingredient in rainbow trout under experimental conditions, but few studies have been carried out under commercial farm conditions. The present study aimed to evaluate the inclusion of 0, 2.5, 5 and 10% (named C, HI2.5, HI5 and HI10) of partially defatted HI meal as substitute of fish meal in iso-nutrient diets. 1,560 rainbow trout (117.1±6.4 g) were allotted to 12 tanks (130 fish/tank, 3 replicates/treatment) and fed on a fish tank biomass basis (from 1.4 to 1.1%) with the daily quantity of feed updated every 14 days. Every 38 days (T1, T2 and T3, respectively), blood samples were collected from 6 fish/tank to evaluate the oxidative stress biomarkers in serum (dROM and OXY-Test). At the trial end, all fish were lightly anesthetised and weighed. The following performance parameters were calculated: mortality, individual weight gain, specific growth rate, feed conversion ratio (FCR) and protein efficiency ratio. After slaughter, the following indexes were calculated on 30 fish/treatment: carcass yield, Fulton's condition factor, coefficient of fatness, hepatosomatic and viscerosomatic indexes, and 5 fish/tank were filleted for the evaluation of the fillet physical quality (pH, colour, drip, thawing and cooking losses). Data were analysed by one-way ANOVA (post-hoc: Tukey). The dROM and OXY-Test were influenced by the dietary treatment at T2 and T3. The HI10 group showed lower concentration of dROM than C ($P<0.05$), while OXY-Test values were statistically lower in all insect-based diets than C (HI2.5, $P<0.01$ and HI5-10, $P<0.001$). Among growth performance parameters, only FCR was affected by the diet – with HI10 value being reduced compared to other HI treatments ($P<0.05$). No differences were observed in term of somatic indexes and fillet physical quality ($P>0.05$). These preliminary results confirm that the use of HI meal in rainbow trout diets does not negatively influence production parameters and fillet physical quality, with a positive effect being even observed on oxidative stress.

Session 09 — Theatre 3

Mealworm protein hydrolysate as a novel functional ingredient for aquaculture applications

L. Sanchez
Ÿnsect, R&D Department, 1 rue Pierre Fontaine, 91000 Evry, France; lorena.sanchez@ynsect.com

Animal protein hydrolysates are used in animal nutrition for diverse purposes. They are a source of highly digestible protein and provide bioactive peptides and amino acids, which confer nutritional and physiological functions in animals. In aquaculture, hydrolysates can function as functional feeds that are able to provide superior performance by promoting growth and health when incorporated to conventional feeds. In the alternative protein space, mealworms are rising as a high quality and sustainable source of dietary proteins in animal nutrition. In this work, the role of an enzymatically hydrolysed mealworm protein ingredient in the diet was investigated using two salmonoid species as a model in recirculating aquaculture system (RAS). A 84-day zootechnical test was performed with juvenile rainbow trout to asses the impact of the mealworm hydrolysate on the growth parameters of fish using a low incorporation dosage (1% in dry matter). Compared to a control diet based on fishmeal, the experimental diet supplemented with the protein hydrolysate from *T. molitor* gave higher values for final body weight and final total length ($P<0.05$). In a three-week zootechnical test with juvenile Atlantic Salmon (17 g), the experimental diets supplemented with the *T. molitor* protein hydrolysate (1% in dry matter) improved the proportion of solid faeces from 40% in the fishmeal control diet to 70% in the diets with the functional ingredient. Growth and welfare parameters remained unchanged between the fish under the control and experimental diet. Lower incidence of semi solid stools, makes this ingredient interesting for application in RAS, which need high water quality to avoid negative effects on the biofilters. This novel functional aquafeed, based on a mealworm protein hydrolysate, have the potential to play an important role on the pathway to more sustainable practices with higher fish yields and improved water quality, thereby decreasing environmental footprint.

Session 09 Theatre 4

Effect of black soldier fly larvae oil on performance and health of lactating sows and piglets

C. Omphalius[1], M. Walraven[2], S. Juliiand[1] and H. Bergoug[2]
[1]Lab To Field, 26 bd Dr Petitjean, 21000 Dijon, France, [2]Innovafeed, 85 rue de Maubeuge, 75010 Paris, France; hakim.bergoug@innovafeed.com

Insect oil (IO) is an alternative to imported soybean oil (SO) because of its lower environmental cost. The specific fatty acid (FA) profile of IO could modulate performance and health in pigs. Thus, 36 sows Large White × Youli and their offspring were included in a longitudinal trial to evaluate whether *H. illucens* larvae oil boosts zootechnical efficiency and health of sows and piglets. Two homogeneous groups of sows and then their offspring received feeds adapted to each stage containing either IO or SO as the sole source of oil (1.5% inclusion during lactation, 2% otherwise). The weight and back fat thickness of sows were recorded 7 days before farrowing (birth=d0) and at weaning (d28). At d7, milk was sampled, and the FA profile was determined. Performances (numbers of piglets born alive and stillbirths, deaths until weaning) were recorded. Piglets were weighed at d1, d27, and d63, and scored at d7, d14 and d21 for diarrhoea. The effect of oil source on performances was tested using a MIXED procedure, and a Chi^2 test was performed to compare the prevalence of diarrhoea between groups (SAS). Feed intake and zootechnical performances were similar between groups: no significant difference among treatments was recorded on sows' weight, maternal performances, and daily weight gain and weight of piglets all along the trial. Sows in IO tended to lose less fat (-1.2 mm) during lactation than SO (-2.1 mm), suggesting that they mobilized less stored fat (P=0.083). This could be beneficial to sustain reproductive performances during the next cycle. Milk FA profile highly differed between groups, especially lauric and myristic acids which represented a higher proportion of FA in group IO (P≤0.001). As these FA carry antibacterial properties, this could benefit piglets' health. Due to the low incidence of diarrhoea in this trial, although the prevalence was lower in IO (0.8% at d7; 95% CI: 0-2.0%) during the first week of life, it did not differ significantly (P=0.263) with SO (2.0% at d7; 95% CI: 0.3-3.8%). Further works on the effect of IO over several parities or on gastrointestinal ecosystem will be valuable to characterize the impacts on health and performance in pigs.

Session 09 Theatre 5

Effects of dietary *Tenebrio molitor* meal and chitosan on health and meat quality of weaned piglets

C. Zacharis[1], E. Bonos[1], A. Tzora[1], I. Skoufos[1], G. Magklaras[1], I. Giavasis[2], I. Giannenas[3], E. Antonopoulou[4], C. Athanasiou[5] and A. Tsinas[1]
[1]University of Ioannina, Department of Agriculture, Arta, 47100, Greece, [2]University of Thessaly, Department of Food Science and Nutrition, Karditsa, 43100, Greece, [3]Aristotle University of Thessaloniki, School of Veterinary Medicine, Thessaloniki, 54124, Greece, [4]Aristotle University of Thessaloniki, Department of Biology, Thessaloniki, 54124, Greece, [5]University of Thessaly, Department of Agriculture, Plant Production and Rural Development, Volos, 38446, Greece; ebonos@uoi.gr

The purpose of this trial was to examine the effects of the dietary use of *Tenebrio molitor* meal and chitosan in weaned piglet diets. 48 weaned piglets (34-day-old) were allocated to 4 groups: (1) control; (2) insect meal 100 g/kg; (3) chitosan 0.5 g/kg; (4) insect meal 100 g/kg and chitosan 0.5 g/kg. At the last day of the trial (42nd), fresh stools were sampled from each piglet to determine gut microbial populations. Then, six piglets per group were sacrificed, and meat samples were procured for chemical, microbiological, and oxidative stability analyses. Two-way analysis of variance (ANOVA, insect meal × chitosan supplementation; SPSS) was performed. Insect meal supplementation increased (P≤0.05) body weight and growth rate on the 21st day of the trial, decreased (P≤0.05) pancetta sample ash content, decreased (P≤0.05) *E. coli* and *C. jejuni* populations on shoulder samples and increased (P≤0.05) total phenols on the same samples. Chitosan supplementation decreased (P≤0.05) total aerobic gut populations, increased (P≤0.05) total anaerobic gut populations and decreased (P≤0.05) TBARS counts on shoulder sample. The combined use of insect meal and chitosan increased the final body weight (P≤0.05) and the growth rate (P≤0.05) compared to the single chitosan use, decreased (P≤0.05) total aerobic gut populations compared to the other groups, decreased (P≤0.05) *E. coli* and *Staphylococcus* spp. populations on shoulder samples, and tended to increase (0.05<P≤0.01) total phenols on pancetta meat samples. Acknowledgements: The research has been funded by National Greek Funds. Project code: T2EΔK-02356. Acronym 'InsectFeedAroma'.

Session 09 — Theatre 6

Defatted insect meals: impact on *in vitro* ruminal fermentation and lipid biohydrogenation

M. Renna[1], M. Coppa[2], C. Lussiana[3], A. Le Morvan[4], L. Gasco[3], L. Rastello[1] and G. Maxin[4]
[1]University of Turin, Dept. Veterinary Sciences, L.go Paolo Braccini 2, 10095, Italy, [2]Independent Researcher, INRAE – UMR 1213 Herbivores, Rte de Theix, 63122 Saint-Genès-Champanelle, France, [3]University of Turin, Dept. Agricultural, Forest and Food Sciences, L.go Paolo Braccini 2, 10095, Italy, [4]INRAE, UMR 1213 Herbivores, Rte de Theix, 63122 Saint-Genès-Champanelle, France; manuela.renna@unito.it

Ruminant diets are characterized by low amounts of lipids (<6%), hence defatted insect meals could be an interesting sustainable solution to provide both protein and energy to the rations. The residual ether extract (EE) content of defatted insect meals can vary widely depending on the applied defatting technology. In this study, we evaluated the effects of residual EE of defatted *Hermetia illucens* (HI) and *Tenebrio molitor* (TM) meals on *in vitro* ruminal digestibility and lipid biohydrogenation. Six EE levels for HI (26.9, 19.7, 12.8, 9.2, 7.0 and 4.7 g EE/100 g dry matter – DM) and three EE levels for TM (39.2, 8.1 and 5.7 g EE/100 g DM) were tested. Rumen fluid for the *in vitro* fermentations was obtained from four cannulated sheep. Fermentation parameters and fatty acids (FA) of rumen digesta after 24 h *in vitro* ruminal incubation of the insect meals were measured. A GLM ANOVA was performed to test the effects of the residual EE (regressive factor) and of its interaction with the insect species (fixed factor). We observed a decrease by 0.78 and 0.36% of DM digestibility per 1% increase of EE content for the HI and TM meals, respectively. Irrespective of insect species, a decrease by 12.90% in CH_4 and 15.70% in CO_2 production was also observed. On the contrary, for both HI and TM, the residual EE content had little effect on the FA profile of rumen digesta (e.g. C18:2 *c*9*t*11: +0.01 and +0.02% for HI and TM meals, respectively). One of the major effects in FA was observed for C18:1 *c*9, which decreased by 0.14% for HI and increased by 0.32% for TM. Thus, the use of defatting processes can simplify the inclusion of insect meals in ruminant diets by limiting the negative effects on nutrient digestibility related to a high EE content, with minor effects on lipid biohydrogenation.

Session 09 — Theatre 7

Can the mixture of *Hermetia illucens* and *Tenebrio molitor* meals improve performance in broilers?

I. Biasato, M. Gariglio, E. Fiorilla, V. Bongiorno, E.E. Cappone, S. Bellezza Oddon, L. Gasco and A. Schiavone
University of Torino, Largo Paolo Braccini 2, 10095 Grugliasco (TO), Italy; ilaria.biasato@unito.it

Hermetia illucens (HI) and *Tenebrio molitor* (TM) meals have widely been used in broiler chickens, but their mixture has never been tested. This study investigated the effects of HI and TM meals – alone and as mixture (1:1) – on growth and slaughtering performance of broiler chickens under commercial conditions. A total of 420 1-day-old male broiler chicks were allotted to 7 diets (6 pens/diet, 10 birds/pen, 3 feeding phases): C (control), HI5 (5% HI meal), HI10 (10% HI meal), TM5 (5% TM meal), TM10 (10% TM meal), MIX5 (5% HI-TM mixture), and MIX10 (10% HI-TM mixture). Growth performance were calculated, and, at 38 days of age, 12 birds/diet were slaughtered to record carcass traits. Data were analysed by SPSS software ($P \leq 0.05$). Growth performance were similar in starter phase ($P>0.05$). In grower phase, TM5 birds displayed the highest live weight (LW), average daily gain (ADG) and daily feed intake (DFI), while the lowest ones were observed in HI10, TM10 and MIX10 groups ($P<0.01$). The best and worst feed conversion ratio (FCR) were observed in TM5 and MIX5 birds, and C, HI10 and TM10 groups, respectively ($P<0.05$). In finisher phase, TM5 birds showed the highest LW, whereas the worst performance were highlighted in HI10 (LW, ADG, DFI and FCR) and MIX10 (DFI and FCR) groups ($P<0.001$). Overall, the best performance were recorded in TM5 (ADG, DFI and FCR), and TM10 and MIX5 (FCR) birds, while the worst ones in HI10 group ($P<0.001$). The TM5 and MIX5 birds showed the highest slaughtering weight (SW), ready-to-cook carcass weight, and chilled carcass weight (CCW), while the lowest values were observed in HI10 and MIX10 groups ($P<0.001$). The MIX5 and TM5 birds also displayed the highest CC (%SW) and breast (%CCW) yields, respectively, whereas the lowest ones were highlighted in HI10 and MIX10 groups ($P<0.001$). MIX10 and TM5 birds also showed the highest and the lowest thighs yield (%CCW), respectively ($P<0.05$). In conclusion, the use of TM and MIX meals at lower inclusion levels may improve both the growth and the slaughtering performance in broiler chickens.

Including different products from black soldier fly larvae in the diet of slow-growing broilers

A. Rezaei Far, T. Veldkamp, J. Van Harn, S. Naser El Deen, P. Van Wikselaar and I. Fodor
Wageningen Livestock Research, De Elst 1, 6700 AH Wageningen, the Netherlands; arya.rezaeifar@wur.nl

Despite the crucial role of chicken meat in human food security, there are environmental and social concerns about intensive farming and the use of soybean (SB) meal as a protein source in broiler diets. In response, several innovations have emerged in broiler production, e.g. the use of more sustainable ingredients and the emergence of production systems with slow-growing broilers (SG). In 2021, Dutch supermarkets shifted completely their supply of fresh chicken meat to meat from SG. Of the alternative ingredients, black soldier fly larvae (BSFL) products are valuable sources of protein and energy and their bioactive components, e.g. chitin and lauric acid may provide health benefits for broilers. Although the inclusion of insect products in diets of fast-growing broilers has been studied, little is known about insect products in the diets of SG. This study aims to evaluate the effect of different dietary inclusion levels (1%, 5%) of BSFL meal (BSFLM) or BSFLM with extra chitin (BSFLMc) in exchange for SB meal, and two inclusion levels of BSFL oil (1%, 2.5%) in exchange to SB oil on production performance and carcass characteristics of SG broilers. This study consists of 8 treatments and 8 replicates per treatment. Day-old male Hubbard JA757 were assigned to 64 floor pens (22 bird/pen). From day one, birds were fed either a control diet program or one of the 7 experimental diet programs that were isocaloric and with the same digestible content of essential amino acids in each feeding phase. In general and compared to the control diet program, including insect products did not affect performance parameters, mortality, or relative weight of carcass and carcass parts. Over 0-56 d, including 1% BSFLMc resulted in the highest body weight gain but did not significantly differ from the control diet program and the diet programs with insect oil. Increasing the inclusion level of BSFLMc to 5% reduced body weight (3.8%) and feed intake (4.6%). The results suggest that replacing SB meal and SB oil with insect meal and insect oil is possible in SG broilers without negative effects on performance and carcass characteristics.

Effect of feeding black soldier fly larvae products to slow-growing broilers on welfare and health

A. Doerper[1], G. Gort[2], I.C. De Jong[3], T. Veldkamp[3] and M. Dicke[1]
[1]Wageningen University & Research, Laboratory of Entomology, P.O. Box 16, 6700 AA Wageningen, the Netherlands, [2]Wageningen University & Research, Biometris, P.O. Box 16, 6700 AA Wageningen, the Netherlands, [3]Wageningen University & Research, Wageningen Livestock Research, De Elst 1, 6700 AH Wageningen, the Netherlands; anna.doerper@wur.nl

Insects as feed ingredients for poultry is a topic of increasing interest. Research often focusses on replacing unsustainable feed ingredients. Insects such as black soldier fly larvae (BSFL) might influence also poultry health and welfare. Bioactive compounds have the potential to shape the health of chickens, while the visual attractiveness of live larvae can stimulate natural behaviour. Broilers could benefit from this, since a major issue in broiler production is their increasing inactivity during growth, leading to leg issues. While promoting natural behaviour is likely to be an exclusive function of live BSFL, bioactive compounds might still be present and functioning in processed BSFL such as meal and oil. Bioactive compounds could support broilers, making them more robust against infections. When considering BSFL products in broiler diets there is no consensus on product inclusion levels to achieve health and welfare benefits. In the current experiment, 1,728 one-day-old slow-growing broilers (Hubbard JA757) were housed for seven weeks. In total nine different treatments (T1 to T9) were tested and each treatment had eight replicates (24 broilers/pen of 2.15 m^2). As control (T1), a commercial broiler diet was used. In T2, 5% of the dry matter feed intake (DMFI) was replaced by live BSFL. In T3, 5% DMFI was replaced by 2/3 BSFL meal and 1/3 BSFL oil. The diets of T4 and T5 were based on T3, with diet T4 containing the same amount of BSF meal as only replacement and diet T5 the same amount of BSF oil. The same pattern of diets was applied in T6 to T9 with a replacement of 10% DMFI by live BSFL as a baseline. All diets were isocaloric and balanced for digestible amino acids. During the experiment, each pen was recorded biweekly during the morning, noon and afternoon to evaluate broiler behaviour. Blood was sampled at the end of the trial to evaluate broiler health. The results show that BSFL influence broiler behaviour and that effects on broiler health were limited.

Session 09 Theatre 10

The immune response of slow-growing broiler chickens fed black soldier fly larvae meal

A. Rezaei Far[1], C.A. Jansen[2], J. Van Harn[1], P. Van Wikselaar[1], S.K. Kar[1] and T. Veldkamp[1]
[1]Wageningen Livestock Research, Animal Nutrition, De Elst1, 6708 WD, the Netherlands, [2]Wageningen University, Cell Biology & Immunology, De Elst1, 6708 WD, the Netherlands; soumya.kar@wur.nl

There is no evidence on the effects of feeding black soldier fly larvae meal (BSF) to slow-growing broiler chickens (SGBC) on immunity. In this study, we investigated the effects of feeding a BSF based diet on the systemic immune cells of SGBC. A soybean meal (SBM) based diet served as a reference. Experimental diets were formulated by substituting the corresponding amounts of BSF for SBM in the reference diet, i.e. 1% BSF, 5% BSF, 1% BSF with extra chitin, 5% BSF with extra chitin; or by substituting the appropriate amounts of BSF oil for soybean oil, i.e. 1% BSF oil and 2.5% BSF oil. At 42 days of age, blood samples were collected from 2 birds per pen to measure the effects of feeding BSF and its derived ingredients on the systemic immune response. Using flow cytometry, we measured subsets of T lymphocytes (CD4+, CD8+, CD4+ CD8+), B cells, leukocytes (CD45+), and natural killer cells (NKC) in whole blood samples. In addition, we measured antibody titers to New-Castle Disease Virus (NCDV). Researchers are interested in the low levels of inclusion to examine how this affects SGBC, thus we focused on the results of diets based on 1% BSF. The addition of 1% BSF diet had no significant (P<0.05) effect on zootechnical and performance parameters compared with the reference feed. For systemic immune cells, only the number of B cells was significantly (P<0.05) higher in the 1% BSF treatment compared to 5% BSF with additional chitin. Higher (statistically not significant, P>0.05) antibody titer against NCDV was observed in the 1% BSF group compared to the reference and other treatment groups. These results are interesting as with low inclusion level of BSF in SGBC diet resulted in a higher humoral immunity, especially due to the increased frequency of the B-cell population in the peripheral blood. The results of our study suggest that lower inclusion level of BSF results in better adaptive immunity in SGBC.

Session 09 Theatre 11

Does BSF meal have an impact in broilers with a subclinical necrotic enteritis challenge?

T. Veldkamp[1], J.J. Mes[2], A. Rezaei Far[1], S. Naser El Deen[1], P. Van Wikselaar[1], I. Fodor[1], R. Van Emous[3] and L. Van Eck[3]
[1]Wageningen University & Research, Wageningen Livestock Research, De Elst 1, 6700 AH Wageningen, the Netherlands, [2]Wageningen University & Research, Wageningen Food & Biobased Research, Bornse Weilanden 9, 6708 WG Wageningen, the Netherlands, [3]Cargill Animal Nutrition & Health, Global Innovation Center Velddriel, Veilingweg 23, 5334 LD Velddriel, the Netherlands; teun.veldkamp@wur.nl

Black soldier fly larvae (BSFL; *Hermetia illucens*) and its bioactive components such as chitin, antimicrobial peptides, and lauric acid may result in additional health benefits compared to traditional protein sources in animal feed. Based on the observed *in vitro* inhibitory effects of BSF products on *Clostridium perfringens*, BFSL protein meal and chitin-rich BSFL protein meal were selected for an *in vivo* broiler experiment to study the effect at different dietary inclusion levels (5 and 10%) during mild necrotic enteritis (NE) on growth performance. A standardized model for *C. perfringens*-associated NE was used in combination with an *Eimeria maxima* infection. NE-challenged broilers were inoculated at day 8 with *E. maxima* and at day 14 with *C. perfringens*. The experiment was conducted with 468 NE-challenged and 468 non-challenged Ross 308 male broilers between 0 to 35 d of age. The broiler diets with the two BSFL protein meals at different inclusion levels were formulated isocaloric and isonitrogenous. Soybean meal in the diet was substituted by BSFL protein meals. After *E. maxima* inoculation, between 8 to 14 d of age, challenged broilers in comparison to non-challenged broilers had lower daily feed intake (12.8%), daily weight gain (ADG) (28.7%), and gain:feed (G:F) ratio (18.3%) (all P<0.05). The *C. perfringens* inoculation at 14 d of age exacerbated the impaired performance in challenged treatments. The effect of the NE-challenge on performance alleviated between 28 to 35 days of age and challenged treatments had a higher ADG (5.2%) and G:F ratio (6.0%) (both P<0.05) compared with the unchallenged treatments indicating a compensatory growth. During the presentation, the effects of including BSFL protein sources on performance of NE-challenged and non-challenged broilers will be shown.

Black soldier fly larval diet eliminate chicken coronavirus at an early infection stage

Y. Zhang[1,2], C.Y. Yang[1], C.J. Li[1,3], Z.H. Xu[1], P. Peng[1], C.Y. Xue[1], J.K. Tomberlin[4], W.F. Hu[5] and Y.C. Cao[1]
[1]Sun Yat-sen University, State Key Laboratory of Biocontrol, School of Life Science, Xingang West Road 135, Haizhu district, 510006, China, P.R., [2]Shenzhen Institutes of Advanced Technology, Chinese Academy of Sciences, Brain Cognition and Brian Disease Institute (BCBDI), No. 1068 Xueyuan Road, University Town of Shenzhen, Shenzhen, 518055, China, P.R., [3]Guangzhou Unique Biotechnology Co., Ltd, Kehuijingu K east – 1717, Guangzhou, 510000, China, P.R., [4]Texas A&M University, Department of Entomology, 2475 TAMU, College Station, TX, 77843-2475, United States, 77840, USA, [5]South China Agricultural University, College of Food Science, Wushan Road 483, Guangzhou, 510642, China, P.R.; chujun.li2013@gmail.com

Avian infectious bronchitis virus (IBV), belonging to Gammacoronavirus, is an economically important respiratory virus affecting poultry industry worldwide. The virus can infect chickens at all ages, whereas young chickens (less than 15 day old) are more susceptible to it. The present study was conducted to investigate effects of dietary supplementation of black soldier fly (*Hermetia illucens* L.) larvae (BSFL) on immune responses in IBV infected 10-day-old chickens. BSFL were ground to powder and mixed with commercial fodder (1, 5 and 10% [mass] BSFL powder) to feed 1-day-old yellow broilers for ten days and then challenged with IBV. Our results indicated that commercial fodder supplemented with 10% BSFL [mass] reduced mortalities (20%) and morbidities (80%), as well as IBV viral loads in tracheas (65.8%) and kidneys (20.4%) from 3-day post challenge (dpc), comparing to that of IBV-infected chickens fed with non-additive commercial fodder. Furthermore, at 3-day post challenge (dpc), 10% BSFL [mass] supplemented chickens presented more CD8+ T lymphocytes in peripheral blood and a rise in interferon-g (IFN-γ) at both mRNA and protein levels in spleens, comparing with chickens fed with commercial fodder. Furthermore, the mRNA abundance of *MHC-I*, *Fas*, *LITAF*, and *IL-2* in the spleens of 10% BSFL [mass] supplemented chickens increased at different time points after challenge. The present results suggest that supplemental BSFL could improve $CD8^+$ T lymphocytes proliferation, thus benefit young chickens to defend against IBV infection.

Black soldier fly larvae meal alter chicken gut microbiota to restricts coronavirus infection

C.Y. Yang[1], Y. Zhao[1], O. Peng[1], C. Li[1], Y. Cao[1] and Y. Zhang[2]
[1]Sun Yat-sen University, State Key Laboratory of Biocontrol, School of Life Science, No. 135 Xingang Xi Road, 510275 Guangzhou, China, 510275 Guangzhou, China, P.R., [2]Shenzhen Institutes of Advanced Technology, Chinese Academy of Sciences, Brain Cognition and Brain Disease Institute (BCBDI), No. 1068 Xueyuan Road, University Town of Shenzhen, 518055 Shenzhen, China, 518055 Shenzhen, China, P.R.; chujun.li2013@gmail.com

Insects, containing high quality and quantity of proteins, are novel, alternative feed ingredients for animal nutrition. Among various insect species, black soldier fly larvae (BSFL) meal is the most widely used in poultry feeding. In the current research, we evaluated the effect of diet black soldier fly larvae supplementation on resistance of coronavirus in young SPF chickens. Day-old SPF chickens were randomly assigned to six groups assessing to bean-based diet as control or BSFL diet with 10% [mass] black soldier fly larvae (BSFL) powder. After challenged with chicken coronavirus, chickens fed with BSFL fodder presented less viral loads in different organs including tracheas, kidneys, and ileums. Transcriptional expression of type I interferon (IFN) signalling was also enhanced in the BSFL group. We further found an alternative structure of gut microbiota of the chickens fed with BSFL meal with main differences in Clostridia, Gammaproteobacteria, and Bacilli, comparing with that of the control group. Furthermore, when gut microbiota transplantation was performed in chickens fed with bean-based control fodder using cecum microbes of the BSFL-fed chickens, decreased viral loads and enhanced type I IFN expression were also observed in these transplanted-chickens after coronavirus infection. Our present results indicate that supplemental BSFL could enhance host immunity through affecting gut microbiota, thus benefit young chicken with resistance to coronavirus infection.

Session 09 Theatre 14

Black soldier fly larvae meal as the main protein source improves feed efficiency in laying hens

A. Rezaei Far, I. Fodor, P. Van Wikselaar, S. Naser El Deen and T. Veldkamp
Wageningen Livestock Research, De Elst 1, 6700 AH Wageningen, the Netherlands; arya.rezaeifar@wur.nl

Chicken egg is an inexpensive but highly nutritious animal product, and its global consumption is expected to grow to 102 Mt by 2050. Consequently, the egg production sector requires about 214 Mt of feed by that time. Feed has a major contribution to the environmental impact of egg production, especially as soybean meal (SBM) has been the most common protein source in laying hens diets. Finding more sustainable alternative protein sources for poultry is an important goal. The European Commission recently authorised using insect proteins such as black soldier fly larvae meal (BSFM) in poultry diets. So far, a few studies investigated the effects of including BSFM in laying hen diets, however, there is no consensus about the effects of BSFM inclusion level on production performance yet. This study aims to evaluate the effects of replacing SBM with two inclusion levels of BSFM (5 and 10%) on production performance and egg quality in laying hens in the aviary system. This study consisted of 3 treatments with 9 replicates each. Brown Nick pullets (n=378) at 19 weeks of age and prior to the onset of lay were transported from a commercial farm and were assigned to 27 floor pens (14 birds/pen). For the first 3 weeks, birds were fed the control diet representing the commercial laying hen diet in the Netherlands. For the rest of the experimental period (8 weeks) birds were fed either the control diet (13% SBM w/w) or diets including 5% (SBM 7% w/w) or 10% (0% SBM w/w) BSFM. Diets were isocaloric and with the same digestible content of essential amino acids. Over the experimental period, including BSFM reduced the feed intake ($P<0.05$), while egg weight, laying rate, and body weight of the laying hens were not affected ($P>0.05$). Hence, the feed conversion ratio improved by 0.097 units using 10% BSFL meal compared to the control diet ($P<0.05$). In comparison to the control, including 10% BSFM increased the relative weight and colour of yolk, and shell thickness ($P<0.05$). In conclusion, we demonstrated that replacing SBM with 10% BSFM in the diet of laying hens improves feed efficiency while increasing the thickness of the eggshell.

Session 09 Theatre 15

Antibacterial and anticoccidial activities of black soldier fly extract

L. Sedano[1], E. Chambellon[1], F.I. Bussiere[1], E. Helloin[1], M. Vian[2], C. Guidou[3], C. Trespuech[3] and A. Silvestre[1]
[1]INRAE, Animal Health, Université de Tours, UMR 1282, 37380, France, [2]INRAE, Université Avignon, UMR SQPOV, Avignon, France, [3]MUTATEC, Châteaurenard, France; anne.silvestre@inrae.fr

To improve the protein autonomy and enrichment for animal welfare, the production of insects for animal feed is growing. Black soldier fly (*Hermetia illucens*) larvae (BSFL) are easily raised and can valorise co-products and food wastes. Besides a high nutrient content, BSFL contain compounds (chitin, lauric acid, antimicrobial peptides) of interest for gut microbiota and animal health. A wide variety of bacteria can cause diseases in poultry and foodborne illness in human. Avian coccidiosis is another highly prevalent disease, caused by *Eimeria* protozoan. Disease severity extends from morbidity to mortality. Its economic impact was recently reassessed to 13 billion $/year worldwide. The occurrence of antibacterial and coccidiostat drugs residues in animal-food products may promote: (1) risk of allergic reactions by hypersensitive individuals; (2) bad impact on the dynamics of gastrointestinal flora; and (3) the antibiotic resistance in gut bacteria. The steady increase of occurrence of bacteria resistant to multiple antibiotics has become a global public health threat that is driving the prudent use of antimicrobial in animals and the development of new alternatives. The aim of this study was to evaluate antimicrobial activities of BSFL extracts, comparing protein extracts solubilized in water and lipid extracts solubilized in methanol. We screened a library of bacterial strains that threaten livestock production (poultry, cattle, pigs) and against *Eimeria tenella*, responsible for avian caecal coccidiosis. Although protein extracts had no antibacterial activity, an inhibition of *Eimeria* development was observed at [0.001-0.01 g/l] of dry matter. The lipid extracts were efficient (0.7-5.66 g/l dry matter) against some strains of *Pasteurella multocida, Corynebacterium bovis, Streptococcus suis, Riemerella anatipestifer and Trueperella pyogenes*. They also inhibited *Eimeria* development at 0.6 mg/l dry matter. More research is needed to confirm those results *in vivo*: the antimicrobial effects of BSFL could improve the health and immune response of birds, when facing sanitary or environmental challenges.

Session 09 Poster 16

The metabolizable energy value of black soldier fly larvae fat used in broiler chicken diets

B. Kierończyk[1], M. Rawski[2], K. Stuper-Szablewska[3], K. Dudek[4] and D. Józefiak[1]
[1]Poznań University of Life Sciences, Department of Animal Nutrition, Wołyńska 33, 60-637 Poznań, Poland, [2]Poznań University of Life Sciences, Department of Zoology, Laboratory of Inland Fisheries and Aquaculture, Wojska Polskiego 71C, 60-637 Poznań, Poland, [3]Poznań University of Life Sciences, Department of Chemistry, Wojska Polskiego 38/42, 60-637 Poznań, Poland, [4]HiProMine S.A., Poznańska, 12F, 62-023 Robakowo, Poland; krzysztof.dudek@hipromine.com

The present study aimed to investigate the apparent metabolizable energy (AME) and apparent metabolizable energy corrected to zero nitrogen balance (AMEn) *levels of H. illucens* (BSF) larvae fat for broiler chickens of various ages. A total of 400 1-day-old male Ross 308 birds were randomly assigned to four dietary groups (10 replicate pens per treatment; 10 birds per pen). The following treatments were applied: HI0 – basal diet without dietary fat inclusion, HI03 – basal diet enriched with 30 g/kg BSF larvae fat, HI06 – basal diet enriched with 60 g/kg BSF larvae fat, and HI09 – basal diet enriched with 90 g/kg BSF larvae fat. Broilers had *ad libitum* access to mash form feed and water. Excreta samples were collected on d 14, d 28, and d 35. To establish the AME and AMEn values of BSF larvae fat, the simple linear regression method was used. The results show that the AME and AMEn values of BSF larvae fat for broiler chickens are 9,049 kcal/kg, and 9,019 kcal/kg, respectively. Furthermore, because of the fact that the birds' age significantly affected the AME and AMEn levels, the implementation of BSF larvae fat to broiler diets should be considered in each nutritional period using the recommended regression model AME = 2,559.758 + 62.989 × fat inclusion (%) + 7.405 × day of age and AMEn = 2,543.2663 + 62.8649 × fat inclusion (%) + 7.3777 × day of age. The present data highlighted that the BSF larvae fat metabolizable energy level is similar to that of soybean oil. However, the authors do not recommend using the above-mentioned regression equations to calculate the energy level of BSF larvae fat for young birds, i.e. before 14 d of age. This work was supported by an OPUS-20 grant titled 'The role of *Hermetia illucens* larvae fat in poultry nutrition – from the nutritive value to the health status of broiler chickens' no. 2020/39/B/ NZ9/00237.

Session 09 Poster 17

***Hermetia illucens* larva fat affects broiler chicken breast meat quality**

B. Kierończyk[1], M. Rawski[2], Z. Mikołajczak[1], P. Szymkowiak[1], K. Stuper-Szablewska[3], M. Dudek[4] and D. Józefiak[1]
[1]Poznań University of Life Sciences, Department of Animal Nutrition, Wołyńska 33, 60-637 Poznań, Poland, [2]Poznań University of Life Sciences, Department of Zoology, Laboratory of Inland Fisheries and Aquaculture, Wojska Polskiego 71C, 60-637 Poznań, Poland, [3]Poznań University of Life Sciences, Department of Chemistry, Wojska Polskiego 38/42, 60-637 Poznań, Poland, [4]HiProMine S.A., ul. Poznańska, 12F, 62-023 Robakowo, Poland; monika.dudek@hipromine.com

This study aimed to evaluate the dose-dependent effect of black soldier fly (BFL) larvae fat inclusion in broiler chicken diets on breast meat quality. Four hundred 1-day-old birds were assigned to the following 4 treatments (10 replicates, 10 birds each): HI0, a basal diet without dietary fat inclusion, and HI03, HI06, and HI09, basal diets enriched with 30 g/kg, 60 g/kg, and 90 g/kg of BSF larvae fat, respectively. Principal component analysis showed noticeable differentiation between the selected plant, animal, and insect-origin dietary fats. The BSF fat exhibits a strong relationship with saturated fatty acids (SFAs) resulting in a high concentration of C12:0 and C14:0. The fatty acid (FA) profile in breast muscle obtained from broilers fed diets with increasing insect fat inclusion showed a significant linear effect in terms of C12:0, C15:0, C18:2, C18:3n6, and total FAs. The proportion of dietary insect fat had a quadratic effect on meat colour. The water-holding capacity indices have stayed consistent with the meat colour changes. Throughout the experiment, favourable growth performance results were noticed in HI06. The present study confirmed that BSF larvae fat negatively affects the n3 level in meat. However, the physico-chemical indices related to consumer acceptance were not altered to negatively limit their final decision, even when a relatively high inclusion of insect fat was used. This work was supported by an OPUS-20 grant titled 'The role of *Hermetia illucens* larvae fat in poultry nutrition – from the nutritive value to the health status of broiler chickens' (no. 2020/39/B/ NZ9/00237), which was financed by the National Science Center (Poland).

Session 09 Poster 18

Evaluation of two black soldier fly products on hens' performance, hatchability, and health traits

P. Hristakieva[1], N. Mincheva[1], I. Ivanova[1], K. Velikov[1] and A. Petrova[2]
[1]Agricultural Institute, Agricultural Academy, Stara Zagora, 6000, Bulgaria, [2]NASEKOMO, bul. Dragan Tzankov 8, Sofia, 1164, Bulgaria; adelina.petrova@nasekomo.life

Insect products have a great potential in addressing sustainability in poultry. To understand their impact on hens' homeostasis and performance, differential analyses of the inclusion of two Black soldier fly products in hens' diets were investigated. A total of 180 Rhode Island White hens (Line N; 45 week old) were individually weighed to make uniform groups (60 hens/group;20 hens/pen) and assigned to three dietary treatments: Control (soybean meal), 7% Black soldier fly defatted (BSFd) and 7% Black soldier fly whole larvae (BSFw). All production parameters were measured real time, and at week 6 of the experiment significant differences were noted. The egg production of BSFw-fed hens was significantly higher (85.47%; $P<0.01$) compared to the BSFd (72.26%) and Control (78.79%) groups. Similar trend was observed for the feed conversion ratio (FCR), where the BSFw group had a better performance (2.39; $P<0.05$) compared to the BSFd (3.14) and the Control (2.74) group. Hens from the experimental groups produced smaller eggs, however, statistical significance was recorded for the BSFd (56.32 g; $P<0.05$) compared to the Control (59.12 g). Yolk colour is an important parameter for consumers and its higher intensity is often regarded as healthier eggs. All diets were free of artificial colorants and assessment of this parameter showed higher Roche colour score in egg yolks of BSFw-fed group (4.7; $P<0.001$) compared to the Control (3.83) and BSFd (3.47) groups. The shell thickness increased in both experimental groups, but a significant difference was observed only for BSFd (0.359 mm; $P<0.05$) compared to the Control (0.321 mm) and BSFw (0.339 mm) groups. No differences were observed among egg incubation parameters except the lower weight of hatched chicks in the experimental groups (BSFd-36.70 g and BSFw-36.39 g; $P<0.01$) compared to the Control (38.55 g). Several blood serum parameters were analysed indicating no negative impact on hens' health status. The two BSF products had a different impact on hens' performance with BSFw having superior results. Overall, the study underlines the need for more detailed analyses of insect products.

Session 09 Poster 19

Insect meals digestibility for rainbow trout aquafeeds: *in vivo* and *in vitro* preliminary results

C. Caimi[1], F. Moyano Lopez[2], S. Bellezza Oddon[1], M.J. Aznar[2], V. Zambotto[3], F. Gai[3] and L. Gasco[1]
[1]University of Turin, Largo Paolo Braccini 2, 10095, Italy, [2]University of Almeria, Ctra. Sacramento s/n, 04120 La Cañada de San Urbano, Spain, [3]National Research Center, Largo Paolo Braccini 2, 10095, Italy; laura.gasco@unito.it

The apparent digestibility coefficients (ADCs) of the dry matter (DM), crude protein (CP) and ether extract (EE) of 0, 2.5, 5 and 10% (named C, HI2.5, HI5 and HI10) of partially defatted *Hermetia illucens* (HI) meal, as substitute of fish meal, in iso-nutrient diets for rainbow trout *(Oncorhynchus mykiss)* have been assessed by an *in vivo* digestibility trial. Moreover an *in vitro* digestibility assay, simulating stomach and intestine digestion in rainbow trout, was carried out in order to assess protein hydrolysis of the experimental diets. For the *in vivo* trial, rainbow trout (115±6.4 g) were allotted to 12 tanks (16 fish/tank, 3 replicates/treatment) and fed by hand to visual satiety twice a day, seven days a week. The fish faeces were collected from each tank twice a day during the collection periods for four consecutive weeks, using a continuous automatic device. The faeces were frozen and then freeze-dried prior to the subsequent chemical analyses. *In vitro* digestibility assays were performed using bioreactors where acid and alkaline stages of the hydrolysis were carried out at pH 3.5 and 8.5, respectively. Total Amino Acids (AA) released, expressed as % initial protein, were recorded after 1, 2, 3, 4.5 and 6 hours of incubation and assays were run in triplicate for each diet. Data were analysed by one-way ANOVA (post-hoc: Tukey). *In vivo* trial showed significant results for ADCs of DM and CP. As far as ADC_{CP}, C diet showed the highest value (89.8%) followed by HI2.5 (87.5%) and HI5 and HI10 diets (84.3 and 86.3, respectively). On the other hand, results of total AA released after 6 h of hydrolysis showed that C diet performed better (67.4%) than HI2.5 (65.2%) diet with HI5 and HI10 having similar values (61.9 and 61.7, respectively). In conclusion, experimental data about dietary protein digestibility obtained with both approaches demonstrate a good correlation between the *in vivo* and *in vitro* results.

Session 09 Poster 20

Suitability of full-fat and defatted black soldier fly meal as ingredient for chicken diets

D. Murawska[1], D. Witkowska[2], M. Gesek[3], T. Bakuła[4] and J. Błażejak-Grabowska[5]

[1]University of Warmia and Mazury in Olsztyn, Oczapowski St.2, 10-719, Poland, [2]University of Warmia and Mazury in Olsztyn, Oczapowski St.2, 10-719, Poland, [3]University of Warmia and Mazury in Olsztyn, Oczapowski St.2, 10-719, Poland, [4]University of Warmia and Mazury in Olsztyn, Oczapowski St.2, 10-719, Poland, [5]University of Warmia and Mazury in Olsztyn, Oczapowski St.2, 10-719, Poland; daria.murawska@uwm.edu.pl

Nowadays black soldier fly (*Hermetia illucens* L.) has been widely used as an ingredient of animal diets, but the results of studies on the protein of diet on poultry growth performance, are inconsistent. The aim of this study was to compare selected growth performance effects in broiler chickens fed diets with different full fat (HIF) or defatted (HID) black soldier fly (HI) larvae meal – processed animal proteins (PAP) content. The experiment was run on a total of 420 1-day-old female Ross 308 broilers. Chickens were randomly assigned to 7 dietary treatments (6 replications per treatment, 10 birds per pen). The diets were formulated by including, on as-fed basis, increasing levels of HIF or HID larvae meal (5, 10 and 15%). The birds were raised to 35 d of age and fed *ad libitum*. At 35 d of age, two birds per pen were selected in each feeding group and slaughtered. The results are presented as means and the SEM. The significance of differences in mean values between age groups was determined by Duncan's test. Significance was set at $P \leq 0.05$. The final body weight (BW) of chickens fed diets containing 15% HI-PAP was smaller to BW in the control ($P \leq 0.05$, respectively: 1,780 g HI0, 1,830 g HIF5, 1,806 g HIF10, 1,585 g HIF15, 1,754 g HID5, 1,645 g HID10, 1,411 g HID15). A positive effect of the HIF10 diet on the proportion of pectoral muscles in the body weight of birds was found, and a negative one for the HIF15 and HID15 diets ($P \leq 0.05$). In conclusion, the replacement of soybean protein (5-10%) to full-fat or defatted HI-PAP in the chicken's diet did not deteriorate the final BW and the basic characteristics of the slaughter value. Acknowledgements: this work was supported by the National Science Centre, Grant: 'Development of a strategy for the use of alternative protein sources in animal nutrition enabling the development of its production on the territory of the Republic of Poland', No Gospostrateg1/ 385141/16 /NCBR /2018.

Session 09 Poster 21

Protein and amino acids digestibility of two black soldier fly larvae meal forms in broilers

A. Cerisuelo[1], E.A. Gómez[1], A. Martínez-Talaván[1], L. Piquer[1], D. Belloumi[1], C. Cano[1], M. Martínez[1], C. Fernández[2] and S. Rojo[3]

[1]IVIA, CITA, Pol. Ind. La Esperanza, 100, 12400 Segorbe, Castellón, Spain, [2]Bioflytech, S.L., Ctra. Cementerio, km2,2, 30320 Fuente Álamo, Murcia, Spain, [3]UA, Dep. C. Ambientales y R. Naturales, Ap. 99, 03080 Alicante, Spain; dhekrabelloumi1@gmail.com

A trial was conducted to evaluate the protein (CP) and amino acids (AA) digestibility of two types dehydrated black soldier fly (i.e. *Hermetia illucens*) larvae meal: a defatted meal (DF) and a full fat meal (FF) in broilers. These insects were all fed with agri-food by-products. A total of 140 male broilers of 21 days of age were used. Five experimental feeds were formulated to calculate standardized ileal digestibility (SID) of CP and AA using the regression method. After 7 days of receiving the experimental diets, animals were slaughtered and the terminal ileum content was collected to determine CP and the individual AA digestibility. The FF meal showed a higher amount of crude fat, gross energy and calcium and a lower amount of CP (52.8 vs 43.2% in dry matter basis) compared with DF insect meal. The AA profile was similar between the two insect meals, and the most abundant essential AA were valine, isoleucine, and lysine in both ingredients. The digestibility of CP and essential AA was high (>70%) in both ingredients. The SID of CP was similar in DF and FF (76.6 vs 73.0%, respectively). In general, AA digestibility was lower in FF compared with DF, but the differences were not significant. Among the essential AA, those with a higher SID were methionine, phenylalanine and arginine in both insect meals (95.5 vs 91.8, 88.7 vs 80.5 and 88.9 vs 799.4 in DF and FF, respectively). Regarding the non-essential AA, glutamic acid and tyrosine were the most digestible in both ingredients. In conclusion, in terms of CP and AA, both DF and FF black soldier fly larvae meals can be suitable protein and AA sources for broiler chickens' diets, especially methionine. Thus, insects fed agricultural by-products represent a novel and promising feed ingredient for poultry diets and could potentially be used as substitution ingredients of soybean in broiler diets, especially DF sources. This work was supported by the project Wayst'up, funded by the European union's horizon 2020 research and innovation programme under grant agreement no. 818308.

CIPROMED: a PRIMA project on the use of alternative proteins in the Mediterranean value chains

C.G. Athanassiou[1], S. Smetana[2], A. Tassoni[3], L. Gasco[4], F. Gai[5], A. Shpigelman[6], D. Pleissner[7], M. Gastli[8], L. Conceição[9], F. Robinson[10], J.-I. Petrusán[2] and C.I. Rumbos[1]
[1]University of Thessaly, Phytokou Str., 38446, Volos, Greece, [2]Deutsches Institut für Lebensmitteltechnik e.V., Prof.-von-Klitzing-Str. 7, 49610, Quakenbrück, Germany, [3]Alma Mater Studiorum-University of Bologna, Piazza di Porta San Donato n.1, 40126, Bologna, Italy, [4]University of Turin, largo P. Braccini 2, 10095, Turin, Italy, [5]Italian National Research Council, Via Amendola, 122/O, 70126, Bari, Italy, [6]Technion – Israel Institute of Technology, Technion city, 32000003, Haifa, Israel, [7]Institut für Lebensmittel- und Umweltforschung e.V., Papendorfer Weg 3, 14806, Bad Belzig, Germany, [8]nextProtein, 13 rue des Juges, 2091, Ariana, Tunisia, [9]SPAROS Lda, Área Empresarial de Marim, 8700-221, Olhão, Portugal, [10]AquaBioTech Group, Central Complex, In-Naggar, MST1761, Mosta, Malta; crumbos@uth.gr

To cover the needs for livestock feeds and aquafeeds, as well as for human food, current EU agri-food production systems largely depend on protein imports. However, this dependency renders them unprotected from unforeseen events that disrupt the global supply chains. Therefore, there is an urgent need for efficient, viable and locally produced alternative protein sources. CIPROMED aims to apply, validate and scale up an integrated array of processes, to recover a significant amount of by-products and related sources that can be used as protein for food and feed. These sources are based on insects, microalgae and legumes, as well as agri-industrial side-streams that will subsequently be integrated with animal production and aquaculture, but also directly for human consumption. The integration of insects, microalgae and legumes in food and feed will be based on circular economy attributes, along a wide range of LCA- and CEA-proofed economically and environmentally sustainable extraction, modification and stabilization techniques. The resulting high-value ingredients will be extensively tested on different 'real-world' systems, through feeding trials and clinical studies. This research is supported by the EU-PRIMA program project CIPROMED (Prima 2022 – Section 1).

Replacing fishmeal by *Hermetia illucens* larvae on growth performance and meat quality in broiler

K.H. Jeon, D.C. Song, S.Y. Chang, J.W. Ahn, H.A. Cho, S.H. Park, H. Kim and J.H. Cho
Chungbuk National University, Animal Science, 344, S21-5, 1, Chungdae-ro, Seowon-gu, Cheongju-si, Chungcheongbuk-do, Korea, 28644, Korea, South; jeonkh1222@gmail.com

This study was conducted to investigate the effects of replacing fish meal with black soldier fly larvae (*Hermetia illucens* larvae; HL) as defatted powder form and hydrolysates form in broiler diets on growth performance, blood profiles, nutrient digestibility, faecal bacterial count, meat quality characteristics, and foot-pad dermatitis. The 60 Arbor Acres (AA) broilers with an initial body weight (BW) of 39.52±0.11 g were randomly assigned to a completely randomized three dietary treatments (5 broilers per cage and 4 replicates cage per treatments). For 4 weeks, three experimental diets were provided: (1) CON, basal diet; (2) T1, basal diet without a fish meal and substitute with defatted HL powder; (3) T2, basal diet without a fish meal and substitute with HL hydrolysates. The T2 group had significantly higher ($P<0.05$) BW than the other groups at 4 weeks. Also, the T2 group had significantly higher ($P<0.05$) BW gain and feed intake than the other groups during the overall periods. The T2 groups showed higher ($P<0.05$) Valine, and Leucine digestibility at 2 weeks and Lysine, Methionine, and Tryptophan digestibility at 4 weeks than the other groups. Besides, the T2 groups significantly improved ($P<0.05$) Glycine digestibility at both 2 weeks and 4 weeks. The CP digestibility at 2 weeks showed higher ($P<0.05$) in T2 group than in other groups. In meat quality characteristics, the T1 group showed higher ($P<0.05$) water-holding capacity than the other groups. Regarding pH in meat quality characteristics, substitute HL in the diet showed higher ($P<0.05$) than the basal diet. However, blood profile, faecal bacterial count, and foot-pad dermatitis had no significant effect among the treatment groups. In summary, substitute defatted HL powder induced positive effects on growth performance and nutrient digestibility. Substitute HL hydrolysates exerted beneficial effects on meat quality characteristics. These findings revealed that HL was a potential replacement for fishmeal.

Investigation of the suitability of two *Tenebrio molitor* meals in diets of weaned piglets

C. Zacharis[1], G. Magklaras[1], I. Giavasis[2], I. Giannenas[3], E. Antonopoulou[4], A. Tsinas[1], C. Athanasiou[5], A. Tzora[1], I. Skoufos[1] and E. Bonos[1]
[1]University of Ioannina, Department of Agriculture, Arta, 47100, Greece, [2]University of Thessaly, Department of Food Science and Nutrition, Karditsa, 43100, Greece, [3]Aristotle University of Thessaloniki, School of Veterinary Medicine, Thessaloniki, 54124, Greece, [4]Aristotle University of Thessaloniki, Department of Biology, Thessaloniki, 54124, Greece, [5]University of Thessaly, Department of Agriculture, Plant Production and Rural Development, Volos, 38446, Greece; tzora@uoi.gr

In the present work, two insect meals from *Tenebrio molitor* which were created previously by our team were examined in weaned piglet diets. The insects for the first meal were reared using a conventional substrate, while the insect for the second meal were reared in a substrate enriched with functional ingredients of aromatic and medicinal plants. In total, 36 weaned piglets, 34-days-old were allocated in 3 groups: (A) Control; (B) conventional insect meal; (C) enriched insect meal. The whole experiment lasted 42 days. On the last day of the experiment fresh stool from each pig was sampled to examine gut microbiota. Then six piglets from each group were sacrificed and meat samples from shoulder, ham, pancetta, and boneless steak were procured for chemical analysis, microbiological analysis and oxidative stability analysis. The results of the statistical analysis (ANOVA, using SPSS software) showed that the bodyweight on day 21 increased ($P \leq 0.05$) in Group B compared to the Group A. Concerning gut microbiota, total aerobes were reduced ($P \leq 0.05$) in Group C, while the others examined microbial populations did not differ ($P > 0.10$). On shoulder and pancetta meat samples the microbiological analysis showed a reduction ($P \leq 0.05$) of *E. coli*, *C. jejuni* and sulphite reducing *Clostridium* in Groups B and C. Total phenols were increased ($P \leq 0.05$) on shoulder, pancetta and boneless meat cuts in Groups B and C, while a reduction ($P \leq 0.05$) of TBARs in Group B was observed on shoulder meat samples. Based on the results of our experiment, insect meal can be added in weaned piglet diets without negative effects on growth performance, gut microbiota, and produced meat quality. Acknowledgements: The research has been funded by National Greek Funds. Project code: T2EΔK-02356. Acronym 'InsectFeedAroma'.

Effect of different black soldier fly frass in plant yield and nutritional value

I. Rehan[1,2], I. Lopes[3,4], D. Murta[3,4], O. Moreira[2,5,6] and R. Menino[7]
[1]FCT- UNL, GeoBioTec Research Center, Campus da Caparica, 2829-516, Portugal, [2]INIAV, Polo de Inovação da Fonte Boa, Vale de Santarém, 2005-424, Portugal, [3]CiiEM-Multidisciplinary Research Center of Egas Moniz, Monte de Caparica, Portugal, 2829-511, Portugal, [4]EntoGreen, Ingredient Odyssey SA, Santarém, 2005-079, Portugal, [5]Associate Laboratory for Animal and Veterinary Sciences (AL4AnimalS), Portugal, Lisboa, 1300-477, Portugal, [6]CIISA – Faculty of Veterinary Medicine, Lisboa, 1300-477, Portugal, [7]INIAV, Oeiras, 2780-159, Portugal; iryna.rehan@iniav.pt

Black soldier fly (BSF) larvae have an outstanding capacity of biodigesting organic wastes. The resulting frass can be considered a high-quality organic fertiliser, although it usually has varying composition that is dependent of the nature of the organic waste being converted. This study aimed to evaluate the effect of frass (deriving from food industry coproducts-TFE; and from sludge from wastewater treatment-TFN) on plant yield and quality. A greenhouse study was designed in pots, in which the commercial mixture AVEX® (oats, ryegrass, vetch and annual clovers) was cultivated in three treatments: TFE, TFN and TM (mineral fertilizer), with 5 replicates each. To each of the frass treatments, 10% nitrate N was added. The treatment TFN rendered the highest yield per pot 77±7 g, in comparison to 63±7 g in TM. Similarly, TFN plants had higher Na concentration than TM 3.7±0.4 vs 2.5±0.5 g/kg, respectively. N concentration was higher in TM, followed by TFE and TFN, with respective values of 4.4±1.2, 3.5±0.3 and 3.1±0.2%. Plants from TFN had the lowest content of K 56±3 g/kg while those from TFE showed the highest content of P 8.3±0.5 g/kg. The plants fertilised with frass displayed lower Mn levels comparatively to TM 68±11 mg/kg. There were no significant differences among the three treatments for NDF, ADF, Ca, Mg, Fe and Zn. The frass types evaluated in this study, resulting from the bioremediation by BSF larvae of organic wastes (food coproducts and wastewater sludge) are adequate to be used as fertilizers in crop production.

Session 10 Theatre 1

The place of epigenetics in the livestock of tomorrow

E.M. Ibeagha-Awemu and M. Wang
Agriculture & Agri-Food Canada, Sherbrooke Research & Development Centre, 2000 Rue College, Sherbrooke, QC, J1M 0C8, Canada; eveline.ibeagha-awemu@agr.gc.ca

Epigenome alterations resulting from the intricate interactions of the genome and the exposome have been implicated in the growth, development, health, reproduction and environmental adaptation of livestock through links to improved immune response, feed efficiency, growth performance and livestock products. Thus, epigenetic factors are widely acknowledged as vital players in individual growth, development, health and production ability. Consequently, epigenetics has gained considerable interest due to its potential to provide new insights into the mechanisms underlying complex livestock traits, disease and environmental adaptation. Such new insights have the potential to contribute to the improvement of current breeding programs or the development of new breeding strategies in livestock. As seen with most new technologies, the inclusion of epigenetics information in current livestock breeding programs could lead to further improvements in the rate of genetic gain by increasing the accuracy of genomic selection and selection intensity amongst others. Moreover, knowledge of epigenome alterations is a prerequisite to the application of an emerging technology like epigenome editing which could lead to further improvements in livestock productivity and sustainability. However, significant challenges remain before the wide adoption of epigenetics data in the livestock industry, such as (1) elaborate investigations to identify the causal relationships between epigenetic modifications and complex traits, (2) development of robust tools to support the identification of epigenetic alterations and their associations with livestock traits on a large scale and (3) significant investments in research and development to support realization of the potential of epigenetics in the livestock of tomorrow. In conclusion, epigenetics holds enormous potential to revolutionize the livestock industry by providing additional levels of information for the further enhancement of livestock production efficiency, animal health and welfare, and development of breeding and management practices to support improved and sustainable livestock production to meet the growing demand for high-quality animal products while minimizing the environmental impact of animal agriculture.

Session 10 Theatre 2

A capture-based approach for DNA methylation analysis in cattle

V. Costes[1], A. López-Catalina[2], O. Gonzalez Recio[2], M. Ben Sassi[3], D.E. Rico[4], G. Foucras[5], C. Le Danvic[1], C. Hozé[1], M. Boussaha[6], D. Boichard[6], H. Jammes[6] and H. Kiefer[6]
[1]Eliance, 149 rue de Bercy, 75595 Paris cedex, France, [2]INIA-CSIC, Crta. de la Coruña km 7.5, 28040 Madrid, Spain, [3]Integragen, 5 Rue Henri-Auguste Desbruères, 91030 Evry, France, [4]CRSAD, 1S0, Deschambault, QCG0A, Canada, [5]IHAP-ENVT, Université de Toulouse, 31076 Toulouse, France, [6]INRAE, Domaine de Vilvert, 78350 Jouy en Josas, France; valentin.costes@eliance.fr

DNA methylation is a potential source of phenotypic variation that may help refining genomic prediction in livestock, provided that affordable tools that can be used in routine analyses become available. Several methods have been developed for whole genome DNA methylation analysis, allowing important advances in the field of epigenetics. However, their cost is yet a limitation for their implementation in large populations. A cost-effective alternative is to target a panel of informative CpGs that show methylation variations across individuals. We assessed the potential of a capture-based approach to measure methylation at selected CpGs in the cattle genome. These CpGs were highlighted by reduced representation bisulphite sequencing (RRBS) conducted on 236 bull semen samples and 82 cow blood samples. Their methylation status in semen varied according to fertility, age, breed or early life nutrition; whereas it varied according to age, nutrition, physiological status and inflammatory status in blood samples. Furthermore, Oxford Nanopore Technology (ONT) was applied to 6 semen samples and 12 blood samples; and CpGs with differential methylation status according to fertility, somatic cell scores and heat stress were identified. The RRBS and ONT results were merged into a list of 128,059 unique, non polymorphic CpGs. A custom design was conducted on the ARS-UCD1.2 genome assembly by Twist Bioscience, and 64,237 probes targeting 97,412 CpGs were synthesized. The technology was assessed on 71 semen, 91 blood and 30 mixed samples, for which RRBS and/or ONT data are available. Enzymatic conversion of DNA, capture, library construction and sequencing at a 100× depth on a Novaseq6000 were carried out by Integragen. Results about the repeatability and sensitivity of the method will be showed, as well as the consistency with RRBS and ONT data (H2020 grant 101000226, RUMIGEN).

Session 10 Theatre 3

DNA-methylome analysis of milk somatic cells upon subclinical mastitis in Holstein cattle

S. Pegolo[1], D. Giannuzzi[1], A. Vanzin[1], V. Bisutti[1], A. Toscano[1], E. Capra[2], P. Ajmone Marsan[3] and A. Cecchinato[1]
[1]DAFNAE, University of Padua, Viale dell'università 16, 35020, Legnaro PD, Italy, [2]IBBA, CNR, via Einstein, 26900, Lodi, Italy, [3]DIANA, Catholic University of the Sacred Heart, Via Emilia Parmense 84, 29122, Piacenza, Italy; sara.pegolo@unipd.it

DNA methylation, as a key epigenetic mechanism for the regulation of gene function, is involved in bovine mastitis. We analysed genome-wide DNA methylation profiles of bovine milk somatic cells in Holstein cows naturally infected by subclinical mastitis to identify differences in DNA methylation pattern linked with disease susceptibility. Three experimental groups were defined based on the bacteriological analyses conducted on 188 animals belonging to one farm: healthy animals (H, n=15), animals infected by *Streptococcus agalactiae* (Sa+, n=10) and animals infected by *Prototheca* spp. (P+, n=9). Sequencing was done a HiSeqX platform using a 150 bp PE approach. Bowtie2 was used for aligning sequencing reads to the reference genome (ARS-UCD1.2.108) and MACS2 was used for peaks calling. Peaks annotated by HOMER within genes promoter regions were 3,787 for H, 5,197 for Sa+, and 4,793 for P+. Around 40% of these genes were shared among groups and 20% were shared between Sa+ and P+. The group-specific genes were 472 for H, 812 for Sa+ and 478 for P+. Functional analyses showed that P+ group includes genes mainly involved in immune pathways, especially related to the cytokine production and regulation (e.g. TLR4, IL13) while the Sa+ group includes genes involved in fat metabolism (e.g. FASN) and innate immune pathways such as the complement activation (e.g. C4A). A further step will explore the identification of differentially methylated regions, using the DiffBind R package. The results will enhance the knowledge on the role of DNA methylation in mastitis infection and help providing new target genes and epigenetic markers for mastitis resistance in dairy cattle. Acknowledgements. The research was part of the LATSAN project funded by MIPAAF (Italy). This study was also carried out within the Agritech National Research Center and received funding from the European Union Next-Generation EU (Piano Nazionale di Ripresa e Resilienza (PNRR) – Missione 4 Componente 2, Investimento 1.4 – D.D. 1032 17/06/2022, CN00000022).

Session 10 Theatre 4

Feed efficiency correlates with the transcriptomic response to feed intake in the pig duodenum

G. Devailly[1], K. Fève[1], S. Saci[1], J. Sarry[1], S. Valière[2], J. Lluch[2], O. Bouchez[2], L. Ravon[3], Y. Billon[3], H. Gilbert[1], M. Beaumont[1] and J. Demars[1]
[1]GenPhySE, Université de Toulouse, INRAE, ENVT, 31326, Castanet Tolosan, France, [2]INRAE, US 1426, GeT-PlaGe, Genotoul, 31326, Castanet Tolosan, France, [3]Pig phenotyping and Innovative breeding facility, GenESI, UE1372, INRAE, 17700 Surgères, France; guillaume.devailly@inrae.fr

Feed efficiency is a trait of interest in pigs as it contributes to lowering the ecological and economical costs of pig production. A divergent genetic selection experiment from a Large White pig population was performed for 10 generations, leading to pig lines with relative low- (LRFI, more efficient) and high- (HRFI, less efficient) residual feed intake (RFI). The meals of pigs from the LRFI line are shorter and less frequent as compared to the HRFI line. We hypothesised that these differences in feeding behaviour could be related to differential sensing and absorption of nutrients in the intestine. We investigated the duodenum transcriptomic response and DNA methylation profile to short term feed intake in LRFI and HRFI lines (n=24). We identified 1,106 differentially expressed genes between the two lines, notably affecting pathways of the transmembrane transport activity and related to mitosis or chromosome separation. The LRFI line showed a greater transcriptomic response to feed intake, with 2,222 differentially expressed genes before and after a meal, as compared to 61 differentially expressed genes in the HRFI line. Feed intake affected genes from both anabolic and catabolic pathways in the pig duodenum, such as autophagy and rRNA production. We noted that several nutrient transporter genes were differentially expressed between lines and/or by short term feed intake. Duodenal DNA methylation profiles were not altered by feed intake. However differences in DNA methylation profiles could be identified between LRFI and HRFI. Altogether, our findings highlighted that the genetic selection for feed efficiency in pigs changed the transcriptome profiles of the duodenum, and notably its response to feed intake.

Epigenetic biomarkers for environmental enrichment and parity in pregnant sows

M.M. Lopes[1], A. Chaulot-Talmon[2], A. Frambourg[2], H. Kiefer[2], C. Gerard[3], J. Demars[4], E. Merlot[1] and H. Jammes[2]
[1]INRAE, PEGASE, 16 Le Clos, 35590 Saint-Gilles, France, [2]INRAE, BREED, All. de Vilvert, 78352 Jouy-en-Josas, France, [3]Chambre Régionale d'Agriculture de Bretagne, Maurice le Lannou, 35042 Rennes, France, [4]INRAE, GenPhySE, Chemin de Borde-Rouge, 31326 Castanet Tolosan, France; mariana.mescouto-lopes@inrae.fr

Changes in the blood cells' epigenome, such as variations in DNA methylation, have been proposed as markers of the long-term effects of various factors in humans and in livestock. Therefore, this study aimed to identify pan-genomic DNA methylation variations in association with the well-being of animals submitted to contrasted welfare states. Pregnant sows of mixed parities (low parity (LP) – 2nd and 3rd gestation and high parity (HP) – 4th gestation or higher) were housed in two contrasting conditions throughout gestation (0 to 105 days): in a conventional system on a slatted floor (C: LP, n=9 and HP, n=6) or in an enriched system on accumulated straw with additional space per sow (E: LP, n=6 and HP, n=7). At gestation day 98, pan-genomic DNA methylation from the sows' blood mononuclear cells was analysed by reduced representation bisulphite sequencing (RRBS), following the lab's protocol. Only CpGs sites covered by at least 10 uniquely mapped reads (CpG_{10}) were retained and filtered out using a list of 105,171 known Single Nucleotide Polymorphisms (SNPs). Methylation percentages at each CpG_{10} were calculated and cluster analyses were conducted. Differentially methylated cytosines (DMCs) were identified using methylKit v1.0. ($\Delta_{meth}\geq 25\%$ and adjusted *P*-value <1%) considering the following comparisons: $[LPvsHP]_C$ and $[LPvsHP]_E$ (parity effect); $[CvsE]_{LP}$ and $[CvsE]_{HP}$ (system effect). Cluster analyses revealed a clear separation corresponding to parity groups. [LPvsHP] displayed more DMC in C (5,391) than in E (3,886) with a similar loss of methylation (53 and 55% in C and E, respectively). Regarding the housing effect, the contrast [CvsE] displayed more DMCs in HP (2,769) than in LP sows (2,183), with an equal number of hypo and hyper DMCs. DMC-targeted genes were mostly associated with cell migration and adhesion, and other immune processes. Taken together, these results suggest that parity has a stronger effect on the immune cells' methylome than the housing system.

A large population study to assess the magnitude of foetal programming in cattle

C. Fouéré[1,2], C. Hozé[1,2], M. Boussaha[1], H. Kiefer[3], M.P. Sanchez[1] and D. Boichard[1]
[1]INRAE, AgroParisTech, GABI, G2B, Domaine de Vilvert, 78350 Jouy-en-Josas, France, [2]Eliance, 149 rue de Bercy, 75012 Paris, France, [3]Université Paris-Saclay, INRAE, Ecole Nationale Vétérinaire d'Alfort, BREED, Domaine de Vilvert, 78350 Jouy-en-Josas, France; corentin.fouere@inrae.fr

Prenatal factors may influence the future performance of dairy cows. We therefore investigated the impact of a suboptimal prenatal environment of a cow on the performance of the resulting daughter in the Holstein breed. The factors investigated in this analysis were associated either with the use of assisted reproductive technologies (sexed semen, embryo transfer (ET) or ET combined with *in vitro* fertilization (IVF-ET)) or with different stress factors during dam's gestation (parity, milk fat-to-protein ratio as a proxy for the metabolic status of the dam). The effects on daughter's performance of these factors, occurring before or during different periods of the dam's gestation, were tested in a model including non-genetic effects commonly used in genetic evaluations and the own direct genomic value for the trait considered, as a covariate. Depending on the factor considered, from 10,000 to 200,000 genotyped cows were used. IVF-ET calves were found to have more difficult birth conditions, suggesting a heavier weight, although neither IVF-ET, nor ET had any effect on the stature in first parity and on the milk performance of the resulting cows. Cows born from heifers and derived from sexed semen produced slightly less milk (-0.3%) than their counterparts derived from conventional AI. Low and high fat/protein ratios of the dam's milk, which are indicators of metabolic disorders, were associated with slightly reduced offspring milk production (-1%) and fertility. The parity of the dam was positively associated with the milk performance of the offspring. For all the factors tested in our study the adverse effect was moderate (e.g. less than 1% for milk yield), suggesting that the negative impact of foetal programming appears to be limited. Further work will be carried out to investigate the effects of other situations, on a wider range of traits and breeds. CF is recipient of a CIFRE PhD grant from ANRT and APIS-GENE. This work was part of the POLYPHEME project funded by ANR and APIS-GENE.

Non-coding exonic microsatellite in bovine Nrf2 gene influences sperm oxidative stress capacity

K. Anwar, G. Thaller and M. Saeed-Zidane
Institute of Animal Breeding and Husbandry, Christian-Albrechts-University Kiel, Olshausenstraße 40, 24098 Kiel, Germany; kanwar@tierzucht.uni-kiel.de

Nrf2 signalling plays a crucial role in cellular defence against oxidative stress that could impair bull fertility. In this regard, the current study was conducted to investigate the potential genetic and epigenetic regulations of Nrf2 gene associated with higher oxidative stress capacity and semen quality in bovine. For that, 12 qualified (older bulls, relatively long duration at service, n=6), and non-qualified (young bulls, relatively short duration at service, n=6) Holstein bulls were subjected to semen collection in 4 seasons. The Nrf2 gene was genotyped in all animals for the GCC microsatellite (9 repeats, NCBI Reference Sequence Database) located within the untranslated region of the first exon. Spermatozoa-borne-relative mRNA expression levels of Nrf2 signalling-related genes were analysed for all semen samples. Interestingly, one bull showed 15 GCC repeats in each group, and one bull had 8 GCC repeats. However, 4 bulls in each group had 9 GCC repeats. Furthermore, mRNA expression analysis revealed that the PRDX1 gene was the most abundant sperm-borne-mRNA among all analysed genes, and was significantly higher expressed in qualified bulls. Across the year's four seasons, higher mRNA expression level was found for Nrf2-202 in winter, Keap1, Nrf2-201, NF-κB, GPX1 in spring, Keap1, Nrf2-201, Nrf2-202, Prdx1, NF-κB, GPX1 in summer, and Keap1, Nrf2-202, and NF-κB in autumn semen of qualified bulls. Meanwhile, higher mRNA expression level was found for Keap1, Nrf2-201, NF-κB, and GPX1 in winter, Nrf2-202 in spring, and Nrf2-201, GPX1 in autumn semen of non-qualified bulls. On the other hand, except for spring the mRNA level of the DNA methyltransferase 3A gene (DNMT3A) was higher in non-qualified bulls. Overall results showed that predominantly higher expression of Nrf2 transcripts and the downstream antioxidants (Prdx1 and GPX1) in qualified bulls in stress seasons alleviated the harmful effects of oxidative stress, which is associated with spermatozoa gDNA fragmentation in non-qualified counterparts. Further ongoing work is to investigate the methylation pattern of Nrf2 in bovine sperm cells.

Maternal methionine restriction affects liver metabolism and foie gras production in duck offspring

C.M.D. Bonnefont[1], A. Sécula[1], H. Chapuis[1], L. Bodin[1], L.E. Bluy[1], A. Bonnet[1], A. Cornuez[2], M.-D. Bernadet[2], J. Barrieu[2], E. Cobo[1], X. Martin[2], H. Manse[1], L. Gress[1], M. Lessire[3], F. Mercerand[4], M.-C. Le Bourhis[4], A. Collin[3] and M. Morisson[1]
[1]GenPhySE, Université de Toulouse, INRAE, ENVT, 31326, Castanet Tolosan, France, [2]UEPFG INRAE Bordeaux-Aquitaine, Domaine d'Artiguères, 40280 Benquet, France, [3]INRAE, Université de Tours, BOA, 37380 Nouzilly, France, [4]INRAE, PEAT, 37380 Nouzilly, France; cecile.bonnefont@inrae.fr

Maternal nutrition effects on offspring phenotypes and production performance have been documented in livestock including farmed birds. We thus studied the effects of a reduced dietary methionine level applied to common female ducks on the hepatic metabolism and zootechnical performance of their offspring in relation to 'foie gras' production. Sixty female ducks were divided into 2 groups receiving either 0.25% or 0.40% of methionine during the growing and laying periods. The restriction reduced egg weight (P<0.001), albumen weight (P<0.001) and duckling body weight at hatching (P<0.001). It also affected the ducklings' energy metabolism, as their plasma glucose (P=0.03) and triglyceride (P=0.01) levels were increased, while their plasma free fatty acid level was decreased (P=0.01), as was the alanine aminotransferase activity (P=0.002). The study of the hepatic expression level of 170 target genes identified 38 differentially expressed genes between the 2 groups of ducklings, some being related to energy metabolism and others to one-carbon metabolism and epigenetic mechanisms. This suggested a modulation in the establishment of metabolic pathways in the very early liver development with long-term effects on energy metabolism. Indeed, ducks issued of the restricted dams were leaner, with lower liver lipid (P=0.005) and abdominal fat (P<0.04) at 12 weeks of age. Later, at 14 weeks of age and after force-feeding, 'foie gras' production was reduced by over 10% (P=0.003). Expression analysis of the 170 target genes at 12 et 14 weeks of age is under progress. Thus, maternal methionine restriction resulted in nutritional programming of the hepatic metabolism of their progeny with effects persisting beyond their period of force-feeding and altering their 'foie gras' production.

Garcinol promotes hepatic gluconeogenesis by inhibiting PCAF acetyltransferases in late-pregnant Sow

W.L. Yao, Q.Y. Fu and J.X. Ning
huazhong Agricultural University, Shizishan Street, Hongshan District, Wuhan City, Hubei Province, 430070, China, P.R.; 85755350@qq.com

Disorder of hepatic glucose metabolism is the characteristic of late pregnant sows. The purport of our study was to look into the mechanism of garcinol on the improvement of hepatic gluconeogenic enzyme in late pregnant sows. Thirty second- and third-parity sows (Duroc × Yorkshire × Landrace, n=10/diet) were fed a basal diet (control) or that supplemented with 100 mg/kg (Low Gar) or 500 mg/kg (High Gar) garcinol from day 90 of gestation to the end of farrowing. The livers were processed to measure enzymatic activity. Hepatocytes from pregnant sows were transfected with P300/CBP associating factor (PCAF) siRNAs or treated with garcinol. Dietary garcinol had no effect on average daily feed intake (ADFI), body weight (BW), backfat and BW gain of late pregnant sows. Garcinol promoted plasma glucose levels in pregnant sows and newborn piglets. Garcinol upregulated hepatic gluconeogenic enzymes expression and decreased PCAF activity. Garcinol had no effect on the expression of peroxisome proliferator activated receptor-g co-activator 1 (PGC-1α) and forkhead box O1 (FOXO1), but significantly increased their activity and decreased their acetylation in late pregnant sows. Transfection of PCAF siRNAs to hepatocytes of pregnant sows increased PGC-1α and FOXO1 activities. Furthermore, in hepatocytes of pregnant sows, garcinol treatment also upregulated the activities of PGC-1α and FOXO1 and inhibited the acetylation of PGC-1α and FOXO1. Garcinol improves hepatic gluconeogenic enzyme expression in late pregnant sows, and this may be due to the mechanism of downregulating the acetylation of PGC-1α and FOXO1 induced by PCAF in isolated hepatocytes.

Early high nutrition affects epigenetics related to lipogenesis in grass-fed Wagyu

D. Nishino[1], S. Khounsaknalath[1], K. Saito[2], A. Saito[3], T. Abe[2], E. Kobayashi[2], S. Yasuo[1] and T. Gotoh[4]
[1]Kyushu University, Fukuoka, 819-0395, Japan, [2]National Livestock Breeding Center, Fukushima, 961-8511, Japan, [3]Zenrakuren, Tokyo, 151-0053, Japan, [4]Hokkaido University, Hokkaido, 060-0811, Japan; 0214daichi@gmail.com

The early high plane of nutrition improved meat quality and meat production in Wagyu (Japanese black cattle) fattened on roughage (Khounsaknalath et al, 2021) and altered DNA methylation level significantly. We hypothesized that this was an epigenetic effect of early nutrition on skeletal muscles. In this study, we comprehensively analysed transcriptomics by microarray and DNA methylation level by whole genome bisulphite sequence (WGBS) analysis in the longissimus muscle (LM) of this grass-fed Wagyu model. Wagyu calves were randomly allocated into two groups. EHN (n=12) received intensified nursing (maximum intake of 1.8 kg/day) until 3 months of age (mo) and then fed a high-concentrate diet from 4 to 10 mo. Another ELN (n=11) received normal nursing (maximum intake of 0.6 kg/day) until 3 mo and then fed only roughage *ad libitum* until 10 mo. After 11 mo, both groups were fed roughage *ad libitum* until 31 mo (Khounsaknalath et al, 2021). We analysed the transcriptomics in LM at 3, 10, 14, 20, and 31 mo and DNA methylation levels in LM at 31 mo. Additionally, we integrally analysed gene expression and DNA methylation level of differentially expressed genes from 10 mo, which is immediately after the metabolomic imprinting event, to 31 mo consistently. In EHN, 1 gene was hypo-methylated and highly expressed, and 6 genes were hyper-methylated and consistently under-expressed. We focused on forkhead box O1 (FOXO1) which is involved in fat synthesis by GO analysis because the intramuscular fat content in LM of EHN was significantly high (Khounsaknakath et al, 2021). FOXO1 which inhibits the transcription of peroxisome proliferator-activated receptor γ (PPARG), a master regulator of adipogenesis, was highly methylated and consistently under-expressed after 10 mo in EHN. Additionally, PPARG was highly expressed at 10, 14, and 31 mo in EHN. These results suggested that FOXO1 was affected epigenetically by the early metabolic imprinting event, and it possibly contributes to fat cell differentiation and lipid accumulation by not inhibiting the transcription of PPARG.

Session 10 Poster 11

Construction of a single-cell transcriptome map of sheep horn development

X.H. He, Y.Y. Luan, S.J. Wu, Y.H. Ma and L. Jiang
Institute of Animal Sciences, Chinese Academy of Agricultural Sciences (CAAS), No. 2, Yuanmingyuan West Road, Haidian District, Beijing, China, 100193, Beijing, China, P.R.; hexiaohong@caas.cn

Horns are headgear, which are unique structures of ruminants. As a globally distributed ruminant, the study of horn formation is critical not only for the understanding of natural and sexual selection but also for the breeding of polled sheep breeds to facilitate modern sheep farming. Nevertheless, the underlying genetic mechanisms in sheep horn remain largely unknown. In the present study, we performed scRNA-seq of horn buds and forehead skin during Mongolian sheep the embryonic period. Single cell transcriptome sequencing and analysis was performed on the forehead skin of a 90-day-old sheep foetus and three developmental stages of horn buds (90-day-old, 120-day-old, and 150-day-old). To construct single-cell pseudotime differentiation trajectory, we used Monocle (v 2.10.0) to order single cells along pseudotime according to the official tutorial. Metascape was used to perform gene ontology (GO) analysis to investigate gene functions in each gene cluster. Histological appearances indicate that the horn buds were significant different from the forehead skin, with a thicker epidermis and more layers of epithelial cell. Interestingly, we found that the hair follicles degenerated in the horn region. Single-cell suspensions from the horn buds and forehead skin tissues were prepared for unbiased scRNA-seq. Based on UAMP dimension reduction analysis, we identified 6 distinct cell populations from 41,405 single-cell transcriptomes. Increased proportions were observed for endothelial cells and pericyte cells in horn buds, suggesting that it was related to the development of horn buds. Furthermore, we identified 9 cell populations in the epithelial cell subset and found two cell subsets associated with horn development. GO and KEGG analysis showed that the two cell subsets were related to epidermis development, ECM-receptor interaction, Focal adhesion and PI3K-AKT signalling pathway. By using pseudotime ordering analysis, we constructed the epithelial cell subsets trajectory and revealed the different cell fates. Meanwhile, we found that the proportions for dermal papilla cells were reduced in the dermal cell subset.

Session 10 Poster 12

The effect of short and long term cryopreservation on chicken primordial germ cells

M. Ibrahim[1], K. Stadnicka[2], E. Grochowska[1], B. Lazar[3], E. Varkonyi[3] and M. Bednarczyk[1]
[1]Bydgoszcz University of Science and Technology, Department of Animal Biotechnology and Genetics, Mazowiecka 28, 85-039 Bydgoszcz, Poland, [2]Collegium Medicum, Nicolaus Copernicus University in Torun, Bydgoszcz, Faculty of Health Sciences, Łukasiewicza 1, 85-821 Bydgoszcz, Poland, [3]Institute for Farm Animal Gene Conservation, National Centre for Biodiversity and Gene Conservation, 200 Isaszegi, 2100 Gödöllő, Hungary; miriam.ibrahim@pbs.edu.pl

Chicken primordial germ cells (PGCs) are the precursors of functional gametes, and the only cell type capable of transmitting genetic and epigenetic information from generation to generation. These cells offer valuable starting material for cell-based genetic engineering and genetic preservation. Chicken PGCs can be cultured and cryopreserved without losing their biological features; however, the handling of PGCs is challenging and requires further development. Herein, we sought to compare the effects of different conditions (freezing-thawing and *in vitro* cultivation) on the expression of PGC-specific gene markers. The chicken used in this study was a purebred Green-legged Partridge like (ZS-11). Embryonic blood containing circulating PGCs was isolated after 2.5 days of egg incubation (13-17 HH embryonic development stage). The blood was pooled separately for males and females following sex determination. The conditions applied to the blood containing PGCs were as follows: (1) fresh isolation; (2) cryopreservation for a short duration (2 days); and (3) *in vitro* culture (30 days) with long term cryopreservation of purified PGCs (~2 years). RNA was isolated in order to further characterize PGCs by reverse transcription-polymerase chain reaction (RT-PCR) for the specific germ cell markers (*SSEA1*, *CVH*, *DAZL*), pluripotency markers (*OCT4*, *NANOG*), and the chemokine/receptor axis (*CXCR4*, *SDF1-α*). A validation study using whole genome sequencing was performed on the *in vitro* purified PGCs vs freshly isolated blood containing PGCs to distinguish the specified PGC markers panel from embryonic whole blood. This study may support avian germplasm conservation strategies via culture and cryopreservation of chicken PGCs. Research was funded by National Science Centre, Poland (UMO-2017/27/B/NZ9/01510, and 2020/37/B/NZ9/00497) and The Excellence Initiative-Research University IDUB programme, UMK Poland.

Session 10 Poster 13

The effect of garcinol on muscle *post mortem* glycolysis and meat quality in pigs

T.X. Wang, L. Huang and X.M. Liu
Huazhong Agricultural University, 1 Shizishan Street, Hongshan District, Wuhan City, Hubei Province, 430070, China, P.R.; 535996673@qq.com

The objective of this study was to evaluate the effects of dietary garcinol (0, 200, 400 and 600 mg/kg) on the growth performance, meat quality, *post mortem* glycolysis and antioxidative capacity of finishing pigs. Dietary garcinol increased pigs' average daily gain, pH 24 h, a* and myoglobin content of longissimus dorsi (LM) ($P<0.05$), and decreased feed/gain ratio, the L* 24 h, glycolytic potential, drip loss, shear force, and backfat depth ($P<0.05$). The glutathione peroxidase (GPx), catalase (CAT) and total antioxidative capacity (T-AOC) were significantly increased by garcinol ($P<0.05$), while the activity of lactate dehydrogenase (LDH) and malonaldehyde (MDA) content were decreased ($P<0.05$). Moreover, garcinol decreased the p300/CBP-associated factor (PCAF) activity, the acetylation level and activities of glycolysis enzymes phosphoglycerate kinase 1 (PGK1), glyceraldehyde-3-phosphate dehydrogenase (GAPDH) and 6-phosphofructo-2-kinase/fructose-2, 6-bisphosphatase-3 (PFKFB3) ($P<0.05$). The results of this study showed that garcinol decreased *post mortem* glycolysis, and this may be due to the mechanism of decreasing glycolytic enzyme acetylation induced by PCAF. The present study indicates that garcinol can facilitate the growth performance of pigs and improve pork quality by changing *post mortem* glycolysis and antioxidative capacity.

Session 10 Poster 14

The effect of methionine supplementation on DNA methylation during the suckling period of ewe lambs

A. Martín[1], O. Madsen[2], F.J. Giráldez[1], J. De Vos[2] and S. Andrés[1]
[1]Instituto de Ganadería de Montaña (CSIC-Universidad de León), Finca Marzanas s/n, 24346, Grulleros (León), Spain, [2]Wageningen University and Research, Animal Breeding and Genomics, P.O. Box 338, 6700 AH Wageningen, the Netherlands; alba.martin@csic.es

Nutritional programming events during early life of ewe lambs might program feed efficiency during the whole life time due to stable changes of epigenetic marks caused by methylation of DNA. Nutritional strategies supplying methyl donors, such as methionine, during the early post-natal period might therefore be used to improve animal performance along the whole life. In this study, 26 newborn ewe lambs from the same flock were stratified and distributed into two groups with equal distribution of body weight. The control group (CTRL, n=13) was fed *ad libitum* with a milk replacer, whereas the second group (n=13) received the same milk replacer supplemented with 1 g of methionine/kg of milk replacer on a dry matter basis. After weaning, all animals were raised in exactly the same way, being fed *ad libitum* with a complete pelleted diet during the replacement phase. Blood was sampled from the jugular vein for DNA isolation at 45 days of age, and blood was also sampled at 45 days and 9 months of age using TEMPUSTM Blood RNA tubes for RNA extraction. DNA methylation was assessed by reduced representation bisulphite sequencing (RRBS) in ten animals [5 ewe lambs (MET) vs 5 ewe lambs (CTRL), corresponding to 5 twin births, from which a lamb was assigned to each group]. The first preliminary results indicate that there were no global differences in the level of methylation between the two groups. Differentially methylation analysis (CTRL vs MET lambs) at 45 days of age resulted in 61 genes differentially methylated in either promoter and/or gene bodies. A selection of these genes was validated by RT-qPCR on samples at 45 days and 9 months of age and most notable was the increased expression of KIF4 and RFLN genes, thus corroborating the hypo-methylation observed by RRBS. These genes are involved in mitotic spindle elongation (KIF4) and regulation of chondrocyte development (RFLN). Furthermore, the methionine supplementation of ewe lambs during the suckling period was not sufficient to result in a clear difference in live body weight when animals were 9 months old.

Effect of supplementary diet at cow-calf phase on DNA methylation in muscle of Angus-Nellore calves

R. Curi[1], L. Rodrigues[1], R. Cooke[2], W. Baldassini[1], L.A. Chardulo[1] and O. Machado Neto[1]
[1]São Paulo State University, Campus de Botucatu, SP, 18618-681, Brazil, [2]Texas A&M University, College Station, TX 77845, USA; rogerio.curi@unesp.br

Different strategies are used to increase the intramuscular fat in beef produced in Brazil. The use of Nellore (*Bos indicus*) × *Bos taurus* and creep-feeding, a supplementation system of calves at cow-calf phase with a diet rich in energy and protein, are examples. Nutritional stimuli can modify DNA methylation, gene expression and phenotypes. However, there are no studies evaluating the effect of creep-feeding on the epigenetic state of adipogenic and lipogenic genes in the muscle of beef cattle. The objective was to analyse methylation levels in the genome of the *Longissimus thoracis* (LT) muscle at weaning of crossbred cattle submitted to different nutritional strategies during the lactation phase. Forty-eight F1 Angus-Nellore steers, not castrated, half-sibs, kept from 30 days of age until weaning under two treatments (n=24/treatment): group 1 (G1) – no creep-feeding; group 2 (G2) – creep-feeding, were used. The animals were weaned at 210 days and then finished in a feedlot for 180 days. The groups showed differences ($P<0.05$) for weaning weight and marbling content in the LT, without differing in weights at the beginning of the experiment and slaughter. At weaning, biopsies of LT were collected and differences in methylation patterns of regulatory gene regions were prospected between five individuals per group by RRBS technique. A total of 272 differentially methylated regions were identified (>25%: q-value<0.01), of which 30 were found overlapping the promoter regions of 20 genes. Among these, *HDAC5* hyper-methylated in G2, was highlighted for their potential relationship with the difference in the intramuscular fat content, due to adipogenesis and lipogenesis, observed between animals from G1 and G2. Enrichment analysis showed *HDAC5* related to (FDR<0.05): apelin signalling pathway; regulation of lipid metabolism by PPARα; HDACs deacetylase histones; cell differentiation; and adipogenesis. In the model studied, the effects of observed methylation patterns for *HDAC5* gene can lead to excessive fat accumulation in the muscle tissue and, consequently, an increase in meat quality.

Improving pig welfare in commercial pig farms

K. O'Driscoll
Teagasc, Pig Development Department, Moorepark, Fermoy, Co. Cork, Ireland; keelin.odriscoll@teagasc.ie

The welfare of pigs on commercial farms is of concern to the public, and as such is an area that is increasingly becoming a policy question for governments, non-profit, and commercial enterprises. The most significant indicator that the welfare needs of pigs are not typically met is the widespread occurrence of damaging and inappropriate behaviours that they perform. These can be an indicator of poor welfare, and in the case of tail biting, can also cause poor welfare. In the face of stressors, animals adjust their behaviour as a component of an initial attempt to cope. When the duration or intensity of stressors experienced by pigs surpasses their ability to cope, tail biting often ensues. To prevent this behaviour, docking of pigs' tails is a widespread practice in commercial pig farming. This is the case world-wide, even though in the EU in particular there is legislation in place that prohibits its routine use. In recent years, there has been a spotlight on this legislation, and the lack of enforcement. This has included an increased amount of national and EU level funding being allocated towards research projects aiming to find methods to reduce the stress levels that pigs' experience, and thus reduce the need to dock. In Ireland, pig farms typically consist of systems with fully slatted floors, which makes provision of functional enrichment challenging – a barren environment is considered a key risk for tail biting. Pig farms are large, and the cost of production is high. This means that there is a perceived high level of cost and risk for producers when it comes to changing to management strategies and infrastructure that may better meet the pigs' needs. During the past 10 years we have undertaken a research programme in Teagasc Moorepark which has aimed to elucidate strategies that can be employed to reduce the need to tail dock in a typical Irish pig farm. Our findings have confirmed that it is possible to significantly reduce these risks, even in fully slatted systems.

Session 11 Theatre 2

Effects of different nutritional strategies on the prevalence of tail biting in weaned piglets

E. Janvier[1], W. De Gaiffier[2], J. Piqué[3], S. Lebas[1] and A. Samson[1]
[1]ADM, Animal Nutrition, Talhouët, 56250 Saint-Nolff, France, [2]ADM, Animal Nutrition, A One Business Center, La Pièce 3, 1180 Rolle, Switzerland, [3]SETNA, Rambla d'Egara, 235, 5° AB, 08224 Terrassa (Barcelona), Spain; werner.degaiffier@adm.com

Discussions are ongoing in the EU to assess the feasibility of banning tail docking in pigs, this however, may increase the frequency and severity of tail biting. This study aimed to investigate the potential of nutritional solutions to lower the risk of tail biting. A total of 106 mixed-sex piglets were assigned to three experimental groups during the post-weaning phase (d21-d70). Pigs were housed in pens of 35 or 36 piglets in a single pen per room, each receiving a different dietary treatment. Rooms were otherwise managed similarly. In phase 1 (d21-d42), dietary treatments were as follows: CON: P1 basal diet; ADD: P1 basal diet + feed additives implicated in stress resilience (magnesium, valerian and Passiflora extracts and lavender flavour); ADD+NUT: diet with optimized nutrients implicated in tail biting (energy fraction, protein, tryptophan, crude fibre, sodium) + stress resilience feed additives. In phase 2 (d42-d70), CON and ADD received the P2 basal diet only, and the ADD+NUT group was fed a diet with optimal nutrient levels as in Phase 1, but without the stress resilience feed additives. Piglets were weighed individually, and feed intake recorded at the pen level. Tail lesions were scored daily using the IFIP's score grid (from 0 = no visible lesions to 3 = severe lesions). For the groups ADD and ADD+NUT, a lick block was added as additional enrichment when 10% of the piglets had slight injuries (score 1). The proportion of severe lesions (scores 2+3) was 8.4, 5.4 and 0.8% in the CON, ADD and ADD+NUT treatments, respectively (d21-70, $P<0.001$). There was no effect of sex on the proportion of severe lesions ($P>0.10$). The tail lesions were mainly observed early in Phase 2 when the feed was changed without transition. Piglets' body weights were significantly improved in the ADD and ADD+NUT groups for the overall period ($P<0.01$). This trial suggests the potential of nutritional solutions to reduce tail biting in piglets when all other husbandry practices are well managed.

Session 11 Theatre 3

Validation of an Irish tail biting risk assessment tool

R.M. D'Alessio[1,2], C.G. Mc Aloon[1], C. Correia-Gomes[3], A. Hanlon[1] and K. O'Driscoll[2]
[1]UCD Veterinary Sciences Centre, University College Dublin, University College Dublin, Belfield, Dublin 4, Ireland, D04W6F6, Ireland, [2]Teagasc Moorepark, Pig Development Department, Teagasc Moorepark, Fermoy, Co Cork, Ireland, P61C996, Ireland, [3]Animal Health Ireland, 2-5 The Archways, Carrick on Shannon, Co. Leitrim, Ireland, N41WN27, Ireland; robertamaria.dalessio@teagasc.ie

Tail biting is a complex multifactorial, and farm-specific problem often observed in commercial pig farms. The risk factors contributing to developing this damaging behaviour can be difficult to recognise. Therefore, using a farm-specific husbandry assessment tool, to identify the causes, can help to determine the risk of tail biting on the farm and guide the development of a tail-biting prevention and management plan. This study sought to validate a novel tail-biting risk assessment protocol, by investigating whether the level of risk determined by the tool would be reflected in tail lesion scores recorded at slaughter. A total of 27 farms (six pens/farm) were assessed using the tool by trained assessors, all private veterinary practitioners. The assessment included reporting on animal-based measures such as lesions and behaviour observations. Based on these, the assessors decided whether they considered the pigs in the pen had a risk of tail biting. Following the assessment a batch of pigs per farm was followed to the abattoir, where the severity of tail skin damage (0 [undamaged] - 4 [partial/full loss of tail]) and presence/absence of bruises was scored. Data were analysed using SAS v9.4. Visible injuries, dirty flanks and tucked tails was not associated with the risk of tail biting in pen as reported by the assessors ($P>0.05$). Similarly, the risk of tail biting per pen was not associated with aggressive, damaging and exploratory behaviours ($P>0.05$). At slaughter, 70% of pigs' tails *post-mortem* did not present tail skin damage, and only 4% presented moderate and severe tail skin damage. Additionally, there was no association between the scores reported at slaughter with the risk of tail biting reported by the assessors. These data suggest that this risk assessment tool may not help to correctly evaluate the risk of tail biting on pig farms.

Improving finishing pig welfare by providing rooting and bedding materials

M. Ocepek, R.C. Newberry and I.L. Andersen
Norwegian University of Life Sciences, Faculty of Biosciences, Department of Animal and Aquacultural Sciences, P.O. Box 5003, 1432 Ås, Norway; marko.ocepek@nmbu.no

Norwegian pig producers are required to provide rooting and bedding materials in commercial production. Information about how rooting and bedding material types and management routines around their provision (amount, frequency) influence the pigs has not been well documented. The aim of this field study (n=87 pig farms; n=648 pens; n=5,769 pigs; about 8 pens per farm) was to investigate associations between these factors. We predicted that all of these factors would be associated with bite marks on pigs (≥1 red stripe, including healed wounds). Proportions of pigs with bite marks on the body, tail, and ears were analysed using a generalised linear mixed model with binomial distribution. Rooting material types (1 vs >1 of: chopped or long straw, silage, hay, newspaper), rooting material distribution frequency (≤1 daily vs >1 daily), and amount of bedding (sparse, low, or moderate amount of wood shavings, or high amount of straw bedding), as well as the interactions between them, were included as fixed effects in the model. Provision of newspaper resulted in a lower proportion of pigs with bite marks on the body (P=0.050), and provision of hay reduced the proportion of pigs with bite marks on the tail (P=0.015) and ears (P=0.002), compared to the other rooting material types. With increasing amount of bedding, the proportion of pigs with body (P<0.001), tail (P=0.048) and ear (P=0.031) bite marks was lower. A lower proportion of pigs had body bite marks when given more than one rooting material at a frequency of more than once daily compared with providing only one rooting material type at a frequency of up to once daily (P=0.020). Similarly, the proportion of pigs with body bite marks was lower when providing a combination of more than one rooting material type and a greater amount of bedding (P<0.001). Furthermore, the proportion of pigs with body bite marks was lower when pigs were given more bedding along with more frequent provision of rooting materials (P<0.001). Our results suggest that the pigs receiving more than one type of rooting material, distributed more than once daily, in addition to ample bedding material over the solid-floored resting area, had better welfare.

Impact of enrichment material and space allowance on damaging behaviour in pigs

M. Cupido[1,2], S.M. Mullan[1], K.M. O'Driscoll[2] and L.A. Boyle[2]
[1]University College Dublin, School of Veterinary Medicine, Belfield, Dublin 4, D04 W6F6, Ireland, [2]Teagasc Moorepark, Pig Development Department, Fermoy, Co. Cork, P61 P302, Ireland; lissacupido@gmail.com

The majority of commercial pigs in Ireland are reared in intensive systems with little meaningful enrichment provided, limiting them from expressing instinctive investigative behaviours. Legally permissible stocking densities in countries that do not permit tail docking are all lower than that currently allowed in Ireland, and across the EU. This study compared the effectiveness of four manipulable enrichment materials (Straw, Haylage, Hay and Grass) applied to pigs in similar pens containing either 8 (LOW), 10 (MID) or 12 (HIGH) pigs (weaner: 0.78, 0.62, 0.52 m^2/pig; finisher: 1.2, 0.6, 0.8 m^2/pig, respectively) in reducing damaging behaviour for 10 weeks post weaning. Thirty-two litters containing healthy piglets with undocked tails were assigned at weaning to enrichment and space allowance treatments using a 4×3 experimental design. Every 2 weeks, pens were directly observed continuously for 5 min, 4 times/day and all occurrences of pigs interacting with the enrichment, and performing aggressive (head-knock, bite, fight) and damaging (tail or ear biting, and belly nosing) behaviours were recorded. Pigs were individually scored every week for ear and tail lesions. Data were analysed using SAS v9.4. There was an effect of material on interaction with enrichment (P<0.05), with Grass pigs tending to interact more with it than Hay (P=0.07) and Haylage pigs (P=0.07), but there was no impact on damaging or aggressive behaviour. Pigs in LOW (0.41±0.04 instances/pig/5 min) and MID (0.37±0.05 instances/pig/5 min) performed less aggressive behaviour than HIGH (0.68±0.04 instances/pig/5 min; P<0.001 for both) and LOW also performed less damaging behaviour (0.29±0.03 instances/pig/5 min; P<0.05) than HIGH (0.38±0.03 instances/pig/5 min). More Hay pigs (35.00%) had some tail length reduction (via biting) than Straw (26.25%), Haylage (27.50%), or Grass (12.50%) pigs. Although space allowance had a greater impact on damaging behaviour than enrichment material, Grass pigs has most enrichment interactions, and least damage to their tails. Thus both space allowance and enrichment material could have a part to play in reducing tail biting risk.

Session 11 Theatre 6

An investigation of early life indicators in relation to ear necrosis in pigs

L. Markland[1,2], K. O'Driscoll[2], F. Leonard[1] and L. Boyle[2]
[1]University College of Dublin, School of Veterinary Medicine, Belfield, Dublin, D04 V1W8, Ireland, [2]Teagasc, Animal & Grassland Research & Innovation Centre, Moorepark, Fermoy, P61 C996 Cork, Ireland; lucy.markland@teagasc.ie

Ear necrosis is a poorly understood pig welfare concern. We investigated associations between early life characteristics of piglets and subsequent ear lesion (EL) development. Piglets born to 94 sows in 6 batches from March 2022 to January 2023 were followed from birth until 11 weeks of age. All piglets were handled a min. of 3 times for weighing at birth and at weaning and for teeth clipping and tail docking. Piglets were inspected weekly for EL post-weaning which were scored according to severity from 0 to 5 and collapsed into 1] presence or absence of EL; 2] presence or absence of severe EL (scores 4+5). Batches were involved in 4 experiments investigating pre-weaning (batches 3-6) or post-weaning (batch 1+2) factors. Piglets in three of the batches (3-5) were weighed throughout the suckling period such that they were HANDLED (n=689 piglets) 8 times compared to piglets in batches 1, 2 and 6 that were handled 3 times (NOHAND=589 piglets). Data were analysed using SASv9.4 to investigate the role of parity (young=0 [n=166 gilts]; mid=1-3 [n=687]; old≥4 [n=425]), birthweight, and whether or not piglets received antibiotics or were HANDLED during the suckling period on the likelihood of developing EL. There were no treatment effects on EL within experiment (P>0.05) and birthweight had no effect on the likelihood of developing EL (P>0.05). Offspring of gilts were more likely to develop EL (52% of piglets) compared to mid (27%) and old (28%) sows (P<0.001). More HANDLED piglets developed EL (46 vs 12.7%, P<0.001) and severe EL (12.8 vs 2.2%, P<0.001). Whether or not piglets received antibiotics had no effect on the likelihood of developing EL (P>0.05). However, a higher proportion of HANDLED pigs were treated with antibiotics than NOHAND piglets (P<0.01). This study supports the generalised poorer immunity in piglets from gilts. Further, while ear necrosis is a multifactorial issue, it is clearly exacerbated by frequent handling during the suckling period. The mechanism appears to be partially one of stress induced immunosuppression given the associated finding of higher antibiotic use in frequently handled piglets.

Session 11 Theatre 7

Improving the welfare of sows and piglets: an exploration of teeth resection under the 3S framework

L. Boyle
Teagasc, Pig Development Department, Moorepark, P61 P302, Ireland; laura.boyle@teagasc.ie

In commercial production, piglets' canine teeth are resected (clipped or grinded) to prevent them from injuring one another (piglet facial lesions-FL) when fighting to establish the teat order. Teeth resection is stressful and pain can result from the procedure itself and/or the resultant injuries which can become infected. Hence, the procedure is legally restricted but leaving the teeth intact means piglets may injure the sows (teat lesions-TL) so the practice is widespread and presents a substantial pig welfare issue. In the same way that Russell and Burch's 3R's (reduce, refine and replace) help to improve the treatment and reduce the suffering, of laboratory animals, the 3S's (suppress, substitute and soothe) offers a framework to address painful procedures in animal production. A recent survey indicated several means of suppressing the need for teeth resection including better focus on sow mothering traits and nutrition, improving piglet water intake and nutrition, conducting frequent checks in the farrowing house and using nurse sows. Also respondents using alternative farrowing systems more frequently reported that FL and TL were manageable. Additionally, the recent EFSA report on pig welfare indicated the importance of reduced litter size in suppressing the need for teeth resection. There is some evidence that substituting clipping for grinding can mitigate resection pain and associated injury. However grinding is still painful and because it takes longer to perform, is associated with handling stress. Given emerging findings on the detrimental effects on piglet health and welfare of handling pre-weaning, even in the absence of painful procedures, substituting clipping for grinding is not ideal. The issue of analgesia to soothe pain in piglets following teeth resection is not well studied but there is recent evidence that pain persists for up to 6 weeks. Hence, piglets should be provided with pain relief when subjected to either procedure. General anaesthetic is excessive for teeth resection and local anaesthesia likely impractical so veterinarian intervention is unnecessary but pig producers should administer NSAIDS to piglets undergoing teeth resection to soothe related pain.

Mitigating the impacts of hot conditions on lactating sows through meal delivery or feed

N. Quiniou[1], J. Dupuis[1] and D. Renaudeau[2]
[1]IFIP-Institut du Porc, BP35, 35650 Le Rheu, France, [2]Pegase, INRAE, Institut Agro, 35590 Saint-Gilles, France; nathalie.quiniou@ifip.asso.fr

Six batches of 24 sows were used to study the impacts of the feeding schedule (alternative frequency and time, group M) or the diet (alternative net energy content (NE) and NE to metabolizable energy (ME) ratio, group D) on performance, either under natural summer heat wave situation (trial 1, 2 batches) or chronic heat stress induced with a fan setpoint kept at 25 °C (trial 2, 4 batches). Control sows (group C) and group D were fed 4 times a day between 7:30 am and 8:00 pm and group M 6 times between 6:30 pm and 12:00 am. The diet delivered to groups C and M was formulated at 9.5 MJ NE/kg with a NE/ME ratio of 73.6%. Corresponding values were 10.3 MJ/kg and 76.3% for group D. At farrowing and weaning, Sows and piglets were weighed, and backfat (BT) and muscle (MT) thicknesses were measured by ultrasound. Sows suckled 14.7 and 14.6 piglets on average in trials 1 (P=0.78) and 2 (P=0.77), respectively. Litter growth rate was not influenced by treatments in trial 1 (3.08 kg/d, P=0.42) and 2 (2.99 kg/d, P=0.35). In trial 1, a small number of heat waves of limited intensity occurred either during the first or the second half of the lactation depending on the batch, but the cumulated spontaneous NE intake was significantly higher in groups D and M than in group C (1,616, 1,512 and 1,325 MJ NE/sow, respectively, P=0.04). However, the difference in BW, BT and MT losses was not statistically significant (-39, -38 and -40 kg, P=0.88; -4.5, -5.2 and -4.6 mm, P=0.67; -4.6, -4.9 and -7.1 mm, P=0.56). In trial 2, the NE intake was higher in group D than in groups M and C (1,708, 1,398 and 1,428 MJ NE/sow, P<0.001) and associated to a lower MT loss (-4.1, -9.9 and -8.6 mm, respectively, P=0.007) but no difference in BW loss (-38, -40 and -43 kg, P=0.41) and BT loss (-4.1, -4.8 and -4.7 mm, P=0.34). In conclusion, these results demonstrate that a low thermogenic diet can help to increase NE intake when sows are exposed to heat waves and chronic heat stress, while changing the meal delivery increases feed intake only under heat waves. More data are expected under heat waves to characterize more precisely the interest of alternative feeding strategies on maternal body reserves and piglets.

Effect of comfort and enrichment for pregnant sows on the health and resilience of their offspring

M. Lagoda[1,2], K. O'Driscoll[2], M. Galli[3], J. Cerón[4], A. Ortín-Bustillo[4], J. Marchewka[1] and L. Boyle[2]
[1]Institute of Genetics and Animal Biotechnology, Dept of Animal Behaviour, Jastrzębiec, 05-552, Poland, [2]Teagasc, Pig Development Dept, Fermoy, P61 P302, Ireland, [3]University of Padova, Legnaro, 35020, Italy, [4]Interdisciplinary Laboratory of Clinical Analysis, University of Murcia, Murcia, 30100, Spain; laura.boyle@teagasc.ie

Poor welfare experienced by pregnant sows has negative effects on piglet health mediated by prenatal stress. We compared sow welfare in 2 gestation housing systems (improved, IMP; conventional, CON) and investigated whether improvements to housing would benefit piglet health. Sows were mixed into 12 stable groups (6 groups/treatment, 20 sows/group) 29 d post-service in pens with 20 free-access, full-length stalls. CON pens had fully-slatted concrete floors, with 2 blocks of wood and 2 chains in the group area. IMP pens were the same but with rubber mats and 1 m of rope in each stall, and straw in 3 racks in the group area. Saliva was collected from each sow on d80 of pregnancy and analysed for haptoglobin (Hp). Tear stains (TS) on sows' right and left eyes were scored in mid-lactation and at weaning. Reproductive performance measures included number of piglets born alive, dead or mummified, and total number of piglets born. Piglets were weighed, scored for vitality and intrauterine growth retardation (IUGR) at birth. Presence of scour in farrowing crates was scored every 2nd day during lactation, and scores were summed to give a total score per litter/sow during lactation. Sows in IMP pens had lower Hp (CON 672±54.0, IMP 463±54.1; P=0.007) and TS scores during lactation (P<0.05), and had fewer mummified piglets than CON sows (CON 0.4±0.06, IMP 0.2±0.04; P=0.013). Piglets of IMP sows had lower IUGR scores (P=0.006) and scoured less (CON 5.4±0.2, IMP 3.7±0.2; P<0.001) during the suckling period. Improved housing during pregnancy led to less stressed sows with less inflammation. In turn their offspring were better developed at birth and healthier during the suckling period. Improvements in reproductive performance (fewer mummified piglets) were significant but small, and though arguably not relevant biologically, should be confirmed using larger numbers. Such benefits offer incentives to producers to improve sow welfare during pregnancy.

Session 11 Theatre 10

Use of organic nest-building material in free-farrowing pens with short term fixation

J. Wahmhoff[1], N. Kemper[2], A. Van Asten[3] and I. Traulsen[1]

[1]Georg-August-University, Department of Animal Sciences, Albrecht-Thaer-Weg 3, 37075 Goettingen, Germany, [2]University of Veterinary Medicine Hannover, Institute for Animal Hygiene, Animal Welfare and Farm Animal Behaviour, Bischofsholer Damm 15, 30173 Hannover, Germany, [3]Chamber of Agriculture of North Rhine-Westphalia, Centre for Agriculture Haus Düsse, Ostinghausen, 59505 Bad Sassendorf, Germany; johann.wahmhoff@uni-goettingen.de

The aim of this study was to investigate the use of organic nest-building material on a commercial sow farm in relation to piglet crushing and animal behaviour. The study was carried out on a farm in Northern Germany where farrowing systems with a short-term fixation (2 days a.p. until approx. day 7 p.p.) and a pen area of 7.0 m^2 were installed. In up to now 3 out of 12 planned batches, the sows were provided with hay (n=29) or straw (n=29) as nest-building material within a permanently accessible rack from gestation day 112 until birth. Jute bags were offered in the control pens (n=24). Sows got up to 240 g of material per day (120 g in the morning and 120 g in the afternoon if needed). To determine daily intake, the residues in the racks were reweighed each morning. In addition, crushed piglets were documented daily until day 3 p.p. and at the end of each batch. Focal animals were videotaped. Results showed that on day 3 a.p., the intake of straw (112.7±71.7 g) tended to be higher than the intake of hay (93.4±98.4 g), but no significant difference was detected (P=0.201). On day 1 a.p., the intake of hay (144.1±94.54 g) and straw (143.3±86.5) was very similar (P=0.65). Based on the high standard deviations, a clear animal-specific effect was identified. Independently of the material, the intake of nest-building material tended to increase towards time of birth (P=0.065), based on nest-building behaviour. First results on the number of crushed piglets showed no significant difference between the various materials (jute bags: 5.65%, hay: 7.14%, straw: 7.05%). This work is funded by the German Federal Ministry of Food and Agriculture (BMEL) based on a decision of the Parliament of the Federal Republic of Germany, granted by the Federal Office for Agriculture and Food (BLE; grant number 28N305602).

Session 11 Theatre 11

Welfare of group-housed sows in the mating area

S. Geisler[1], A. Van Asten[2], N. Kemper[3] and I. Traulsen[1]

[1]Georg-August-University, Department of Animal Sciences, Albrecht-Thaer-Weg 3, 37075 Goettingen, Germany, [2]Chamber of Agriculture of North Rhine-Westphalia, Centre for Agriculture Haus Düsse, Ostinghausen, 59505 Bad Sassendorf, Germany, [3]University of Veterinary Medicine Hannover, Foundation, Institute for Animal Hygiene, Animal Welfare and Farm Animal Behaviour, Bischofsholer Damm 15, 30173 Hannover, Germany; swantje.geisler@uni-goettingen.de

Legislative adjustments by the German government are leading to changes in housing conditions for sows in oestrus, requiring a group housing system with short-term fixation and a space allowance of 5 m^2/animal. To assess how sows cope with the modified housing conditions a skin lesion assessment was conducted on two German commercial sow farms that already comply with the new legislative standards. The applied skin lesion assessment, used as an indicator of welfare, was based on the Welfare Quality® assessment for pigs. Skin lesions were recorded three times (day 0, day 2 and day 6/7 after weaning) for each sow (farm A n=205, farm B n=127) in three body areas (front, middle, back) on a scale from a (no/low extent of lesions) to d (very high extent of lesions). A total skin score (TS) was calculated from the individual body area scores on a scale from 0 (low extent) to 2 (high extent). Performance and fertility data were also collected (farm A n=152, farm B n=38). Binary logistic regression was used to investigate the association between skin lesions and insemination success (0-1). On both farms, a significant effect of the recording day on TS was determined (P<0.001). On farm A, a significant increase in the frequency of higher TS was observed between day 0 and day 2 and between day 2 and day 7 (P<0.001), whereas on farm B only a significant increase between day 0 and day 2 was detected (P<0.001). No effect of the skin lesions on the insemination success could be detected. In conclusion, these first results suggest that group housing may affect the magnitude and severity of skin lesions. To improve welfare in group-housing systems, focusing on the management of the social cohabitation could be a control element.

Session 11 Theatre 12

Current knowledge and ways forward to improve welfare of cull sows during transport and in lairage

C. Kobek-Kjeldager[1], L.D. Jensen[1], L. Foldager[1,2], L.M. Gould[1], K. Thodberg[1], D.L. Schrøder-Petersen[3] and M.S. Herskin[1]
[1]Aarhus University, Animal and Veterinary Sciences, Blichers Allé 20, 8830 Tjele, Denmark, [2]Aarhus University, Bioinformatics Research Centre, Universitetsbyen 81, 8000 Aarhus, Denmark, [3]DMRI, Danish Technological Institute, Gregersensvej 9, 2630 Taastrup, Denmark; cecilie.kobek-kjeldager@anivet.au.dk

Across Europe, annual culling rates per sow herd of ~50% means that millions of sows are sent to slaughter each year, while there is a knowledge gap on the welfare of this distinct category of swine according to a recent EFSA Scientific Opinion. In two studies, we investigated effects of journey duration and temperature inside vehicles transporting sows to slaughter on the behaviour during transport and in lairage. In the first experimental study, behaviour was observed during transport of 28 loads of sows in a 3×2 factorial design of journey duration (4 h, 6 h, 8 h), and with or without a pre-planned stop halfway (±Stop). In the second observational study, sows from 23 commercial loads with varying journey duration were observed 1 h in lairage. During transport, sows were upright 89-92% (median) of observations/hour; more in the first hour after departure and less when temperature was higher(P<0.05). With a longer latency until the stop, less sows were upright during the stop (P<0.05). The median events of aggression were 2-3/sow for each duration (0-155). No differences were detected during stop and immediately before or after. Time since departure and temperature in the vehicle had no influence on aggression. In lairage, most sows were active at first, decreasing to 30-46% after ~30 min. Drinking was observed in 36% of the sows. Aggression was initiated by 36% and received by 72%. After short journeys, the level of aggression was higher with increasing temperature, while after long journeys more lying and less aggression and drinking was observed when temperature was higher (P<0.05). Reduced aggression can be interpreted as positive for welfare but also as a sign of fatigue when newly mixed sows do not fight. Overall, the results indicate that the welfare of sows during transport and lairage may be challenged by mixing and resting problems. Physiological and motivational indicators are needed to clarify this. Recommendations on ways forward will be presented.

Session 11 Theatre 13

Exposure times and stunning effectiveness for argon and nitrogen-argon mixture for pigs at slaughter

J. Knöll[1], J. Gelhausen[2], T. Friehs[2], T. Krebs[2], D. Mörlein[2], J. Tetens[2] and I. Wilk[1]
[1]Friedrich-Loeffler-Institute, Institute of Animal Welfare and Animal Husbandry, Dörnbergstr. 25/27, 29223 Celle, Germany, [2]Georg-August-Universität Göttingen, Department of Animal Sciences, Burckhardtweg 2, 37077 Göttingen, Germany; jonas.knoell@fli.de

Despite concerns over its aversiveness, stunning of pigs at the time of slaughter using a high concentration of CO_2 is the most common method in Europe. Inert gases and mixtures of inert gases with low concertation of CO_2 have been proposed as an alternative, but have so far not been considered market-ready due to concerns of gas stability, meat quality, costs and stunning effectiveness. As part of the project for Testing Inert Gases in order to Establish Replacements for high concentration CO_2 stunning for pigs at the time of slaughter (TIGER), experiments were conducted in a commercial Dip-Lift system using a new gassing system, which allowed for residual oxygen concentrations <1%. Pigs were stunned and slaughtered using argon, a nitrogen-argon mixture or high concentrations of CO_2 (control). Stunning effectiveness was closely monitored by checking reflexes, stimulus-response tests and signs of breathing activity (e.g. gasping) for at least three minutes after ejection from the stunning system. Exposure time was dynamically adjusted for each gas mixture using an adaptive staircase procedure in 10 s steps. Initially, exposure time was reduced for every two successfully stunned animals. With each insufficiently stunned animal, exposure time was increased and the number of successfully stunned animals needed for the next reduction was doubled up to a maximum of 200. Using a probit model, exposure times to reach a likelihood of insufficient stuns of <0.5% were estimated. Exposure times in the adaptive method reached more than 4 minutes for both inert gas atmospheres. Stun failures were typically well predicted by the onset of gasping or other obvious signs like a righting reflex. Sufficiently stunned animals typically did not show any breathing activity. We conclude that, while longer exposure times were needed for inert gas atmospheres compared to CO_2 in high concentrations, safe levels for stunning can be reached, with a potentially easy assessment of insufficient stuns based on onset of gasping or righting reflex.

Session 11 Poster 14

Stability of performance in sows switching between crate and pen from one parity to the next

L. Canario[1], J. Bailly[2], S. Reignier[2], M. Poupin[2], Y. Billon[2] and W. Hébrard[2]
[1]INRAE, Animal Genetics, GenPhySE, 31326 Castanet-Tolosan, France, [2]INRAE, UE GenESI, 17700 Surgères, France; laurianne.canario@inrae.fr

Maintaining sows on farm longer contributes to the sustainability of pig production. As guidelines for limiting the use of heavy restraint systems are being developed, another difficulty is to produce piglets at an economically viable level in different housing systems. We set a protocol to characterise a population of Large White sows in a conventional farm equipped with 2 types of farrowing units, with either farrowing crates or individual pens. The aim was to analyse the stability of individual performance in the face of repeated changes of environment. Sows switched from 100% blocked (B) to 100% free (F) – or the opposite – from one parity to the next. Performance of their purebred litters was recorded over the 4 or 5 first parities, with 44 sows alternating as per B1F2B3F4 (number is parity) and 23 sows alternating as per B1B2F3B4F5. The ideal sow is one that maintain a high and stable production level in successive litters regardless of the restraint system. Piglet survival and growth, and maternal ability traits were recorded. Crossfostering was implemented 24 h after farrowing according to the number of functional teats. Models for analysis included the fixed effect of parity, and the random effects of sow identity and farrowing batch. On average, in the B1F2B3F4 sows, number of piglets born alive was similar in B1 and F2 (14.4 and 14.8), increased in B3 (16.6; $P<0.05$) and was 15.6 in F4. The piglet survival rate until weaning was lower in F4 than previous parities (71 vs 80%). The B1B2F3B4F5 sows had similar litter size across parities (14.8) and lower piglet survival rate in F5 than previous parities (62 vs 81%). More piglet crushing occurred in the latest parity. Variation in other traits is under study. Next, groups of sows with highest and lowest variations were identified according to CV values. Finally, the effect of physical enrichment was evaluated in the entire population, considering the mean change of performance from B to F of each sow. Some sows less sensitive than others to change of environment were identified.

Session 12 Theatre 1

Preliminary results on the use of weather data to predict production output in a Merino flock

P.G. Theron[1], T.S. Brand[1,2], S.W.P. Cloete[1] and K. Dzama[1]
[1]University of Stellenbosch, Animal Sciences, Private Bag X1, Matieland, 7600, Stellenbosch, South Africa, [2]Western Cape Department of Agriculture, Directorate: Animal Sciences, 80 Muldersvlei Road, 7600, Stellenbosch, South Africa; pieter.theron@westerncape.gov.za

Extensive small ruminant production systems globally are under threat from climate change and associated changes in weather patterns but little information on the precise relationship between production traits and weather conditions is available. This study aimed to explore the use of weather data as input factors in predictive models for flock production. Weather and (re)production data recorded between 1993 and 2021 on Elsenburg Research Farm were used. Multiple linear regressions were used to relate production traits in two Merino lines divergently selected for number of lambs weaned (NLW) to climate data. Climate data included maximum and minimum temperatures, rainfall, maximum and minimum relative humidity as well as temperature-humidity indices for both day- and night-time. Weather data from the preceding year (y-1) was related to production data from year y to allow for the effect of a full years' weather patterns on production to be studied. Approximately a third of the variation in ewe reproductive performance (conception, lambing and multiple birth rate) in both lines was accounted for by the fitting of multivariate models. Thus producers could experience year-to-year fluctuations in production of more than 30% even when the same management program is applied. Weaning performance was also modelled successfully in the line selected for NLW. Average weaning weight was mainly determined by rainfall. Overall it seemed that weather played a greater role in influencing production in the line selected against NLW, potentially due to the lower genetic reproductive capacity. The study provides empirical evidence of the effect of weather on production output. The models could potentially be used to estimate climate change impacts on production output in extensive sheep farming systems. More work needs to be done to refine the models and elucidate relationships between the various weather parameters and production traits in different environments.

Session 12 Theatre 2

GIS-based data integration system as a design tool for rangeland management in productive landscapes

M. Odintsov Vaintrub, M. Bianchini, L. Lizzi, F. Paciocco, M. Francioni and P. D'Ottavio
Universita Politechnica delle Marche, Agriculture, Piazza Roma, 22, Ancona, 60121, Italy; m.odintsov.vaintrub@gmail.com

'Productive landscape' is a management system used in protected areas (e.g. national parks, and nature reserves) to integrate complex local dynamics. The 'Monte Genzana' Special Areas of Conservation (SAC, IT7110100) is one such area that includes 3 main municipalities (58 km^2 total) each with its own grazing areas for privately own livestock (horses, cattle and sheep). The current work aims to provide a GIS-based expert system to design conservation management strategies for the rangeland productive landscapes of Monte Genzana. Land existing databases (i.e. topographic, geological, soil) were collected and integrated with rangeland biodiversity and livestock management data in a multi-layer geographic information system (QGIS). The dataset was analysed by creating a 10×10 m grid to obtain a map of pastoral value (PV: 0-100) and potential carrying capacity (PCC, expressed in LU/ha/grazing period). Traditional methods of planning (i.e. planimetric area corrected by the average slope) were also implemented. The resulting PCC was compared with the recorded actual stocking rate (LU/ha). Additionally, on-site botanical surveys were performed in 12 sites to assess the impact of the vegetation dynamics on: (1) plant biodiversity conservation; (2) PV and PCC; and (3) land use change. A generalized understocking condition was recorded in most areas, with the PCC of both systems resulted to be similar (1-2% deviations). However, the 10×10 m grid assessment was significantly more reliable in detecting changes in land morphology and local under/over grazing. This was supported by vegetation dynamics observed by Satellite imagery (i.e. encroachment by shrubs and herbaceous species) and on-site botanical surveys which detected large grazing areas with standing dead biomass. In conclusion, the use of an integrated data approach, by using a multi-layer GIS combined with available real-time analysis can significantly reduce the margins for errors in pasture allocation. It can facilitate the monitoring of designated site-specific management strategies in complex environments of rangeland productive landscapes.

Session 12 Theatre 3

Remotely monitored animal behaviour using sensor ear tags on cattle in Switzerland

K. Ueda[1], U. Heikkilä[1], N. Gobbo Oliveira Erünlü[2], J. Bérard[2], O. Wellnitz[2] and S. Rieder[1]
[1]Identitas AG, R&D, Stauffacherstrasse 130A, 3014 Bern, Switzerland, [2]Agroscope, Animal Production Systems and Animal Health, Route de la Tioleyre 4, 1725 Posieux, Switzerland; kosuke.ueda@identitas.ch

Sensor ear tags (SET) serve for the identification, and localization, of individual animals and herds, but are also capable of delivering individual welfare and behavioural metrics. For remote summering pastures, a sound delineation of normal behaviour variations can be key to better identify potential responses to the presence of large predators. The present contribution is based on a comprehensive evaluation of a SET as identification and remote monitoring tool under Swiss conditions, regarding tolerance and technical feasibility, which has been reported elsewhere. Behaviour-related data were collected for two cattle groups under different settings, one a summer long stay on Alpine pastures, and one a mixed barn-pasture husbandry setting that is typical for low-land enterprises. The datasets contain metrics of individual activity and derived alerts for all settings. These were transmitted in an autonomous fashion over a satellite network. In addition, high-resolution raw data were downloaded over Bluetooth in and around the barn in the mixed husbandry setting. For the latter setting, data were both acquired with the SET fixed on the ear and on a collar, separately. Activities exhibit strong diurnal variation that likely reflects the circadian rhythm, for both settings. The pattern matches the distribution of alerts for very high activity. Behavioural metrics computed on the device correspond well to each other when comparing ear and collar fixation, except for the highest activity class, where frequent twitching of the ear in response to insects is suspected to contribute strong accelerations. Geographical location patterns also bear relevance to understanding and establishing herd behaviour baselines. Analyses on herd level behaviour potentially leverage the use of tags on larger herds and are reported separately. Here, a terrain-based time-space distribution survey is reported, in conjunction with weather and SET communication patterns. Analyses of sensor raw data deliver indications on nightly repositioning of individual cattle and establish a footprint of their sleep behaviour.

Differentiating feeding behaviour in grazing cattle based on IMU data and machine learning algorithms

N. Tilkens[1,2], A.L.H. Andriamandroso[2] and J. Bindelle[1,3]
[1]University of Liège, AgricultureIsLife, TERRA, Passage des Déportés 2, 5030 Gembloux, Belgium, [2]University of Lille, Junia, UMRT 1158 BioEcoAgro, 2 rue Norbert Ségard, 59800 Lille, France, [3]University of Liège, Precision Livestock and Nutrition Unit, AgroBioChem, Passage des Déportés 2, 5030 Gembloux, Belgium; nicolas.tilkens@junia.com

A better understanding of the ecological mechanisms is key to empower agroecological approaches increasing the sustainability of grassland-based livestock systems. To do so, the first steps require to monitor how herbivores spend their time on pastures and identify the grazing movement from other types of unitary behaviours. A fine-level description of the time budget use by grazing animals could be enabled using precision livestock farming approaches combining wearable sensors to machine-learning algorithms. We equipped 9 dry red-pied Holstein cattle and 2 Blonde d'Aquitaine × Belgian White and Blue cross-bred with collar-mounted inertial measurement units (IMU) recording 3D accelerometer and gyroscope data at a 100 Hz frequency and set them to graze on ryegrass (*Lolium perenne*) pastures. We also shot 106 videos clips (average 28'07", standard deviation 4'38") to record the cows' behaviour focusing on the identification of grazing moments against other types of behaviours. Quadratic support vector machine (QSVM) and bagged tree (BT) algorithms were explored as Machine Learning classifiers to automatically detect the grazing moment in the time series. Four different time-windows were tested (3 and 5 seconds without overlap, and 10 and 30 seconds with 90% overlap) as well as two different splits of the datasets: split 1 was set to have different breed in the training and testing data, and split 2 was set for an equal representation of 'grazing' and 'other' behaviours in both training and testing data. Results show that for both splits BT algorithms gave the best results, with an accuracy of 91,8% using 5 seconds windows without overlap for split 1 and 94,0% using 30 seconds windows with 90% overlap for split 2. Finally, this paper will explore how the detection of grazing behaviour could be used in future works as a first step to quantify the amount and frequency of bites during each meal and at the scale of a grazing day.

Monitoring animal movements by virtual fencing collars – a comparison of different GPS intervals

D. Hamidi[1], M. Komainda[1], N.A. Grinnell[1], F. Riesch[1,2], J. Horn[1], M. Hamidi[1], I. Traulsen[3] and J. Isselstein[1,2]
[1]University of Goettingen, Crop Sciences, Von-Siebold-Str. 8, 37075 Göttingen, Germany, [2]University of Goettingen, Centre for Biodiversity and Sustainable Land Use, Büsgenweg 1, 37077 Göttingen, Germany, [3]University of Goettingen, Animal Sciences, Burckhardtweg 2, 37077 Göttingen, Germany; dina.hamidi@uni-goettingen.de

The grazing of farm animals, in particular access to pasture for cattle is an essential issue for the design of sustainable future livestock systems. Barriers to the implementation of grazing are the laborious tasks of ground-based-fencing on the one hand and, the difficulties associated with monitoring animals on pasture on the other hand. Using virtual fencing (VF) collars to enclose animals without physical fences can greatly simplify grazing, especially for more complex systems such as rotational grazing. Additionally, VF collars provide GPS data on animal locations, which can be used for calculating walking distances as an indicator to assess livestock vitality. High frequency GPS data from collared animals on pasture can provide continuous data for an improved animal monitoring. However, a lower frequency of GPS relocations reduces the energy demand and prolongs the working life of the precision livestock equipment. An open question regarding the performance of GPS tracking to monitor livestock behaviour is the setup of an appropriate frequency of recorded GPS positions as this likely affects the walking distances. To quantify the information loss using lower frequency data, we calculated the daily walking distances using 5, 10 and 15-min recording intervals, and compared them to the walking distances derived from one-minute data. The GPS recording was performed in a replicated rotational grazing trial with Fleckvieh heifers during 2021 using VF collars (Nofence®, Norway). An average reduction of 55% of the daily walking distances under a 15-min interval compared to a one-minute interval indicated a significant information loss. Moreover, walking distances calculated by using different time intervals differed significantly ($P<0.0001$) (1 min: 2,984±54.8; 5 min: 2,181±54.8; 10 min: 1,658±54.8; 15 min: 1,345±54.8 (mean ± SE)). This is an obvious trade-off between battery life and the accuracy of monitoring cattle behaviour.

The impact of virtual fence use on blood markers linked to stress and acute phase response in cattle

J. Ranches[1], M. Ferreira[1], G.M.P. Hernandez[1], A.R. Santos[1], R.C. O'Connor[2], D. Johnson[1], D.W. Bohnert[1] and C. Boyd[2]
[1]Oregon State University, Eastern Oregon Agricultural Research Center (EOARC), 67826-A Hwy 205, 97720, USA, [2]USDA-ARS, Eastern Oregon Agricultural Research Center (EOARC), 67826-A Hwy 205, 97720, USA; juliana.ranches@oregonstate.edu

This study evaluated the effects of an automated virtual fence (VF) on blood markers associated with stress and inflammatory response in beef cattle. The automated VF employs auditory and electric stimuli to control cattle movement within a predefined virtual boundary, thus possibly raising concerns about cattle welfare. Forty mature Angus × Hereford cows were enrolled in this study. Cows enrolled in the study had never worn VF collars and therefore were considered naïve to the technology. All cows were equipped with VF collars on day 0 (d0). Body weight, body condition score, chute score, chute exit velocity, and blood samples were collected from all cows on d0. Upon collaring and collections, cows were moved to a drylot with predefined VF boundaries that were unknown to the cows. Cows remained in the pasture with VF boundaries for 4 consecutive days. Cows were brought to the working facility on d5 for a second round of collections as performed on d0. Cows were considered the experimental unit for this study, and all data were analysed using day as a repeated measure with the MIXED procedure of SAS (SAS Inst. Inc., Cary, NC, USA). No effects ($P \geq 0.89$) were observed on body weight (595 and 594 kg, respectively for d0 and d5). Chute exit velocity changed ($P=0.001$) from d0 to d5, with cows being slower on d5 (1.20 and 0.94 m/s, respectively). Blood samples were analysed for cortisol (COR), haptoglobin (HP), and ceruloplasmin (CP). No differences ($P \geq 0.11$) in blood COR (2.70 and 2.50 µg/dl, respectively for d0 and d5) or CP (28.3 and 29.3 mg/ml, respectively for d0 and d5) concentrations were observed from d0 to d5. However, HP concentration increased ($P<0.0001$) from d0 to d5 (0.40 and 0.45 mg/ml, respectively). These findings suggest that brief exposure to VF boundaries and the associated stimuli does not have adverse effects on production traits. Moreover, COR concentration appears to remain unaffected by VF use. However, changes in HP concentration may indicate a mild inflammatory response, possibly attributable to local stimuli.

Drones to facilitate the monitoring of grazing animal: which application and opportunities?

A. Lebreton[1], A. Demarbaix[2], F. Demarquet[3], J. Douhay[1], P.-G. Grisot[1], L. Depuille[1] and E. Nicolas[1]
[1]Institut de l'Elevage, 149 Rue de Bercy, 75595 Paris, France, [2]Ferm'Inov, La Prairie, 71250 Jalogny, France, [3]EPLEFPA Carmejane, Route d'Espinouse, 04510 Le Chaffaut-Saint-Jurson, France; adrien.lebreton@idele.fr

Drone technologies are becoming more and more accessible including for farmers. Many authors have seen the potential of drones to automate animal detection and counting. Others have focused on automating drone flights on rangelands. However, there is not much scientific or grey literature on the use and the impact of using 'off-the-shelf' drones on the current legal framework even if many challenges exist: choice of the technology, the understanding of the regulations and the risks, the safety for people and animals. We studied the use of drones as 'eyes in the sky' piloted by farmers themselves on different rangelands (meadows, woody rangelands, summer mountain rangelands). We first analysed the possible applications by crossing an analysis of European drone regulations, technologies performances and farmer's needs on sheep and beef cattle systems. Farmer's needs were recorded through an online survey (19 European farmers), interviews (4 farmers) and a survey during an on-farm demonstration (22 farmers and livestock industry stakeholders). It appears that livestock stakeholders appreciate drone technology, many would recommend it to others, and many could easily implement it on their farms. However, only a few are already equipped due to significant barriers to adoption: (1) most believe it is an unaffordable technology although some would take the risk of investing in it; (2) drones are considered not that easy to use. On-farm demonstrations seem to be an important lever for farmers to have a better understanding of the technologies and can be complemented by other training materials. Defining an appliable framework for the use of drones in grazing-based livestock farming was the first step of the ICAERUS project and will soon be communicated through guidelines and recommendations. Further works will assess risk and interest of using drones on the technical, economic, environmental, and social aspects with an important focus on the impact on the 'work dimensions' described by Hostiou and Fagon.

Session 12 Theatre 8

Modular communication gateways, a new opportunity for precision management of agroecological farms

M. Odintsov Vaintrub and P. Di Giuseppe
Regrowth s.r.l.s, Contrada specola (snc), 64100, Italy; m.odintsov.vaintrub@gmail.com

Agroecology is a farming practice that works in conjunction with natural processes by emulating ecological nutrient dynamics. Agroecological producers usually cater to a niche market of high-value direct sales and clients that appreciate their ecosystem services and values. This makes the farmers particularly interested in technologies useful in showcasing their unique value proposition. However, the farming method presents particular challenges for technology adoption as it usually includes smaller numbers of animals, distributed in different plots frequently located in areas with minimal network coverage. In the current work, we evaluated the feasibility of using a unified data management and communication 'gateway' as a modular solution to address some of the challenges. A custom ESP32 board was used for the initial screening and analysis of data collected with a LoRa receiver from various on-field nodes. The data was then stored both physically on a local removable SD support and on a cloud server via GSM/Satellite communication. This field unit also included its own energy supply (5-12V solar panels and battery modules) and energy efficiency methods were implemented to increase overall resilience. The total costs of the gateway amounted to roughly 100€, with a monthly cost of another 20€/farm. During 2020-2023, three independent trials were conducted in diversified farm settings. In all trials, the gateway was installed where network coverage was unavailable, and managed 3-12 Walk-over-Weight stations at a time. For periods of 30-60 days, the gateway was able to provide a consistent connection with only occasional failure of the external casing. Integration of additional data sources such as VR alarms was also supported by the gateway, with direct streaming activation on demand with a lag time of 5-10 sec. In conclusion, the installation of an affordable and reliable communication system is feasible with the currently available components. It facilitates the installation of IoT systems needed for PLF data collection, significantly reducing the challenges each of them would have faced independently. This modular approach is particularly adapted for the small number, and high diversity typical to agroecological livestock production.

Session 12 Theatre 9

Deploying a LoRa network in mountainous areas to connect animals and shepherds

T. Kriszt[1], O. Benoit[2], G. Besche[2], J.D. Guyonneau[2], E. González-García[1] and J.B. Menassol[1]
[1]SELMET, L'Institut Agro Montpellier, CIRAD, INRAE, Univ Montpellier, 34000, Montpellier, France, [2]SELMET, INRAE, CIRAD, L'Institut Agro Montpellier, Univ Montpellier, 13300, Salon-de-Provence, France; theo.kriszt@supagro.fr

In remote and mountainous areas, the heterogenous and often low quality of the mobile network limits the potential for the introduction of digital technologies requiring connected services for the shepherds or digital tools to monitor livestock movements and activities. Consequently, few digital solutions exist despite numerous possible applications associated for instance with real-time monitoring of the animals (recovery of animals, predator alerts, etc.). As an alternative, we tested the deployment of a LoRa network in the French Alps during the summer period when animals are grazing in altitude. The test area consisted in the summer pasture (from June to September) for one flock (Institut Agro Domaine du Merle) of roughly 1000 sheep guarded by one shepherd and covering 5.77 km^2 (2,109 and 1,505 m of highest and lowest altitudes, respectively). Two LoRa gateways were deployed on pre-determined sites at an altitude of 2,049 and 2,006 meters for GATE-1 and GATE-2 respectively, using the Longley Rice Irregular Terrain Model (ITM) to predict radio coverage given the existing mobile network quality. Each gateway was fitted on a mast of up to 16 metres in height and was powered by a battery and a solar panel to meet the energy requirements for the whole monitoring period. LoRa connectivity was tested based on a virtual grid of 400 m squares. At each vertex (n=57), GNSS modules designed to be embarked on animals were used to transmit their current position through LoRa. Data on network connectivity showed that both GATE-1 and GATE-2 provided a coverage in accordance with the ITM simulation. In our conditions, LoRa network proved to be resilient to highly uneven terrain and showed overall great performances. The coverage simulation allowed to accurately plan the deployment of the gateways and prevented unnecessary on-site efforts. In order to ensure connectivity for animal and human applications at reduced costs, since one gateway can relay on the internet the communications from a large number of low-cost devices, the LoRa network proves to be a good alternative.

Session 12 Poster 10

Effect of successive drone herding on behaviour and spatial distribution of grazing cattle

H. Anzai and M. Kumaishi
University of Miyazaki, Faculty of Agriculture, 1-1 Gakuen Kibanadai Nishi, Miyazaki, 889-2192, Japan; anzai.hiroki@gmail.com

Robotic herding is attracting attention as a new technology for managing grazing animals. We investigated the behavioural responses of cows to drone herding over consecutive days and its performance in manipulating grazing distribution in a pasture. A herd of approximately 30 cows was stocked in a 1.1-ha pasture for 5 consecutive days each month from May to October 2022. The cows were herded by a drone for 10 days during the grazing period in August and September. The pasture was divided into nine plots, with two plots assigned as the herding area in August and three plots in September. When the cows grazed in the herding area, the operator manoeuvred the drone to move them out of the area. The drone was first approached at an altitude of 10 m, and if the cows did not move away, the altitude was gradually lowered to 3 m. The behavioural responses to the drone, the success or failure of the herding (whether cows exited the area or not), and the altitude of the drone were classified based on videos recorded during herding. The behaviour and location of the cows were observed during the time in the pasture. Utilization rates (percentage of grazing time) of the plots were calculated to evaluate the effect of herding. On the first and second days of the herding, the cows responded in 59 and 46% of the cases. On the first day, cows were startled in 23% of responses. From the first to the third day, the herding was successful in 51-75% of cases. However, these percentages declined thereafter. Both when the cows responded to the drone and exited the area, the drone altitude was most frequently 3-4 m, the lowest altitude. On the first day, the utilization rate of the herding area was about half of that on the days without herding. It increased from the second day, and the effect of herding almost disappeared from the third day onward. These findings suggest that cows were not afraid of an approaching drone after one day, but that subsequent habituation also reduced the desired behavioural responses in manipulating grazing distribution. Further studies are warranted to determine the stimuli that persistently elicit the desired responses.

Session 12 Poster 11

Evaluation of the RumiWatchSystem to measure activity and grazing behaviour of sheep

E. Dunne, N. McHugh and F.M. McGovern
Teagasc, Animal & Grassland, Teagasc, H65R718, Ireland; eoin.dunne@teagasc.ie

Animal health and performance are highly influenced by feeding behaviour. The RumiWatchSystem provides an opportunity to quantify grazing practices in livestock farming. Data collected gives potential to optimize grazing practices and make informed management decisions; thus improving animal welfare and increasing the sustainability and efficiency of grazing operations. The objective of this experiment was to evaluate the RumiWatchSystem in comparison to visual observations on animal activity and grazing behaviour. A total of 12 animals were monitored for a total of 6 hours each on two separate occasions. Minute by minute observations were carried out whereby grazing, rumination, eating and other behaviours of the sheep were recorded at pasture by both the RumiWatchSystem sensors and the two observers. The final dataset contained 121 unique 1-minute measurement time points per animal per measurement, providing a dataset of 8,704 unique time point measurements. Analysis were undertaken to compare the RumiWatch System observations to the visual observations. Results showed that the percentage of agreement between the RumiWatchSystem and the visual observations was 87.33%. Strong correlations were also observed between the RumiWatchSystem sensor and visual observations, with a Spearman's rank and concordance correlation coefficient of 0.90 and 0.87, respectively ($P<0.001$). Almost perfect agreement was calculated between the visual observations and the RumiWatchSystem based on the Cohens kappa value of 0.81; with a bias value of 0.08 between both methods. These results indicate a strong relationship between the visual observations and the RumiWatchSystem sensor technology in activity and grazing behaviours of sheep in a pasture based system, enabling further research using this technology to optimise sheep grazing performance, animal health and welfare.

Session 12 Poster 12

Remote monitoring of cattle using sensor ear tags

K. Ueda[1], T. Kuntzer[1], U. Heikkilä[1], N. Gobbo Oliveira Erünlü[2], J. Bérard[2], C. Beglinger[1], O. Wellnitz[2] and S. Rieder[1]
[1]Identitas AG, R&D, Stauffacherstrasse 130A, 3014 Bern, Switzerland, [2]Agroscope, Animal Production Systems and Animal Health, Rte de la Tioleyre 4, 1725 Posieux, Switzerland; kosuke.ueda@identitas.ch

The remote monitoring of animals by means of digital instruments is gaining in importance. This is due, i.a., to evolving production systems, due to structural developments in agriculture, such as ever larger farms and rising costs for personnel, or simply the lack of qualified personnel. On the other hand, digital 'shepherds' collect data on a 24/7 basis. Remote monitoring of individual animals and herds can lead to early insights that would otherwise remain hidden to livestock keepers. In the present study, a sensor ear tag (SET) was used and evaluated for remote monitoring of cattle under Swiss livestock management conditions. The tolerance of the SET regarding animal welfare as well as the suitability of the technology, data acquisition and processing were the focus of the investigation. A SET seems to be particularly attractive for remote monitoring because it is presumably able to combine the mandatory animal identification of cloven-hoofed animals with information obtained directly from the animal. The findings show that the SET in its present form, fixation and placement is not yet suitable for animal identification in Switzerland. Alternatively, the SET was fixed to a neck collar to continue data acquisition via three independent data channels (SATCOM, Bluetooth, RFID). Information was derived, i.a. on animal locations and movements, animal behaviour (daily rhythmicity), and land use. Combining SET data and satellite images, the effect of grazing and draught during summer 2022 was traced. During the winter half-year, or while animals are housed, the energy level of the SET battery, powered by solar panels, showed a rather low level. Hardly any data packages via SATCOM were received when animals were kept indoors. However, the data stream via Bluetooth was maintained throughout. RFID turned out to be a secure, albeit static, data source depending heavily on the placement of antennas. During summering, data flow was exclusively upheld via SATCOM, without additional infrastructure needed. The latter makes the SET particularly attractive for remote monitoring of livestock in pasture-based production systems.

Session 13 Theatre 1

Update on sensor technologies for performance recording, management and welfare in small ruminants

G. Caja[1], A. Elhadi[1], E. González[2], J.B. Menassol[2], G. Tessnière[3], V. Giovanetti[4], M. Decandia[4], M. Acciaro[4], E.M. Sossidou[5], S.I. Patsios[5], L.T. Cziszter[6], L. Grova[7], G.H.M. Jorgensen[7], I. Halachmi[8], A.B. Shamai[8], T.W.J. Keady[9], C.M. Dwyer[10], T. Waterhouse[10], A. McLaren[10] and C. Morgan-Davis[10]
[1]UAB, G2R, 08193 Bellaterra, Spain, [2]INRAE, SELMET, 34060 Montpellier, France, [3]IDELE, CS 52637, 31321 Castanet-Tolosan, France, [4]Agris, Sardegna, 07100 Sassari, Italy, [5]ELGO, Dimitra, 57001 Thessaloniki, Greece, [6]BUAS, BFAR, 300645 Timisoara, Romania, [7]NIBIO, Wildlife and Rangelands, 6630 Tingvoll, Norway, [8]ARO, PLF lab, 7505101 Rishon Lezion, Israel, [9]Teagasc, Athenry, Galway H65 R718, Ireland, [10]SRUC, Kirkton, Crianlarich FK20 8RU, United Kingdom; gerardo.caja@uab.cat

Small ruminants (SR) are numerous livestock species with a low uptake of modern technologies. In the EU, most SR are electronically identified with transponders, which is a key opportunity for the implementation of sensors for PLF. Compared to transponders, which send fixed outputs, sensors send variable signals according to the type and intensity of the input. In practice, sensors are classified as non-wearable and wearable. Among them, their use for: behaviour, animal tracking, virtual fencing, automated weighing, performance recording and health problems detection (lameness, mortality, etc.), with special attention to early warning systems, will be analysed in SR. The study of sensors for SR welfare monitoring is the main aim of the Project TechCare (https://techcare-project.eu/), currently in progress. Advanced results showed that prioritization (by experts and stakeholders) of welfare problems throughout the value chain of SR, varies according to countries, productive purposes and production systems (meat and dairy sheep, dairy goats, suckling and fattening lambs/kids). The tools of interest for the detection of welfare problems were also prioritized and are currently under evaluation to be implemented in large scale trials in commercial farms. Stakeholders showed positive interest on PLF uses in SR and wearable sensors seem to be the ideal solution for animal-based indicators, although non-wearable sensors may be an option of interest in large farms due to their cost-benefit. Further research is needed to support the current opportunities of using sensors in SR. Funded by the EU H2020 program (Contract #862050).

Session 13 Theatre 2

P@stor-all: a spatialized information system for decision-making in extensive grazing systems

U. Kalenga Tshingomba[1,2], L. Sautot[2], M. Teisseire[2] and M. Jouven[1]
[1]Univ Montpellier, Institut Agro, CIRAD, INRAE, SELMET, 2 place Pierre Viala, 34060, France, [2]Univ Montpellier, AgroParisTech, CIRAD, CNRS, INRAE, TETIS, 500 rue Jean-François Breton, 34090, France; lucile.sautot@agroparistech.fr

Agro-silvo-pastoralism is common in Mediterranean regions and provides multiple services. The sustainability of such systems depends on efficient utilization of rangelands. The latter requires good knowledge of the spatial behaviour of the flock, which can be documented with digital tools. The diversity of agro-silvo-pastoral systems complicates both data acquisition and processing, and has slowed down the uptake of precision farming. However, a variety of data is available to document the functioning of the pastoral ecosystem: free satellite data, GPS collars, meteorological data but also technical references in digital format and feedback from farmers. The objective of the P@stor-all project was to group such heterogeneous data in one single tool and combine it in order to provide knowledge and indicators for the improvement of grazing management. Designing such an information system (IS) required to find solutions to integrate structured and unstructured data and produce meaningful indicators for the farmers. A participatory approach was chosen: 7 farmers and 2 research farms representing a variety of systems were partners in the project and were associated in the discussions and decisions about the functionalities of the IS, the conceptual model of the pastoral ecosystem, the spatio-temporal resolution of data, the choice of indicators, the user interface and the terms of use of the IS. GPS collars were deployed in the farms and their data was cross-analysed with free satellite data in order to produce indicators of the spatial utilization of rangelands by the flocks. The proposed indicators were different for the commercial farms (1 commercial GPS collar to localize the flock in real time and record spatial behaviour), and for the research farms (up to 30 GPS collars to test the correlations between spatial behaviour, animal characteristics and environmental conditions). A first prototype of the P@stor-all IS will be tested and refined in spring and summer 2023; by the end of the year, the platform will be freely available to all French farmers.

Session 13 Theatre 3

Day and night sexual activity of rams as recorded by an automatic heat detector

M. Alhamada[1,2], E. González-García[2], N. Debus[2], A. Lurette[2], J.-B. Menassol[3] and F. Bocquier[3]
[1]Istom, 4 Rue Joseph Lakanal, Angers, 49000, France, [2]SELMET, INRAE, CIRAD, L'Institut Agro Montpellier SupAgro, Univ Montpellier, 34000 Montpellier, Fr, 2 Place Viala, Montpellier, 3400, France, [3]SELMET, L'Institut Agro Montpellier SupAgro, CIRAD, INRAE, Univ Montpellier, 34000 Montpellier, Fr, 2 Place Viala, Montpellier, 34000, France; moutaz.alhamada@hotmail.com

The objective was to analyse the day and night-time sexual behaviour of Mérinos d'Arles rams reared under extensive farming conditions. Eight rams were evaluated in their interaction with the ewes (ratio 1♂:40♀) during two consecutive years at the timing of the reproductive period from late April to early May. An automatic heat detection device, already validated in such conditions, was used to monitor the timing and number of mounts for each ram. Data were analysed within two distinct periods i.e. daytime 14 h (from 6 a.m. to 8 p.m.) and nighttime 10 h (from 8 p.m. to 6 a.m.). Means were compared using Tukey's HSD test. Results show that all rams kept a nocturnal sexual activity. The circadian rhythm of ram activity differed from one animal to another ($P<0.05$). One ram displayed 11±6 and 36±25 mounts/h during the night and day, respectively (corresponding to 23 and 77% of its daily activity averaging 25±18 mounts/h). Conversely, another ram displayed 65% of its activity during the night (13±9 mounts/h). No correlation was observed between total daily and diurnal sexual activity ($P>0.05$). Within a same year, the total mounting activity between rams was significantly different ($P<0.001$) and ranged from 1.9 to 89.8 mounts/h. Between years, the total number of mounts for each ram was significantly different ($P<0.01$) but this did not affect the ranking among rams based on this criteria (repeatability 83%; $P<0.05$). Our results suggest that sheep, often classified as diurnal animals in their sexual behaviour, may display a significant nocturnal activity which must be taken into account. We can conclude that studying the animal only during the daytime may lead to a lack of information and/or efficiency in the management of reproduction. In this context, the automatic heat detector developed by our team may be an effective and easy-to-use alternative tool for studying ram reproduction behaviour.

Walking distance and maintenance energy requirements of sheep during mountain pasturing

T. Zanon, M. Gruber and M. Gauly
Free University of Bolzano, Piazza Universitá 5, 39100, Italy; thomas.zanon@unibz.it

Sheep pasturing has become an important means for landscape management and conservation in marginal areas of mountain regions by reducing succession with dwarf shrubs and bushes, thus creating space for valuable forage grasses. Furthermore, mountain sheep pasturing for meat production also contributes to local food security. However, little is known about the energetic expenditures of sheep during mountain pasturing, which is important information for optimizing the productivity and economic efficiency, environmental aspects (e.g. biodiversity) as well as animal welfare. Therefore, the aim of the following study was to estimate the maintenance energy requirements of ewes over the whole mountain pasturing period (transhumance) considering movement patterns assessed by using satellite-based Global Positioning System (GPS) tracking devices. Energy requirements for walking increased rapidly at the beginning of transhumance (May-June) (4.14-4.17 MJ/d), which could be explained by the longer walked distance and by overcoming variable altitude during that phase. Walking speed (2-8 m/min) was slower compared to previous findings due to the difficult terrain of mountain pastures on which sheep moved. Energy demand for walking was correlated with walking distance (0.45, $P<0.001$) and walking speed (0.26, $P<0.001$). Results out of this study contribute in promoting the efficiency and consequently the rentability of alpine sheep pasturing systems. The latter ensures the production of local food and further preserves the ecosystem services linked to this low-input production system.

Percutaneous liver biopsy obtention by ultrasonographic guidance in small ruminants

S. González-Luna[1,2], X. Moll[3], S. Serhan[2], B. Chaalia[2], A.A.K. Salama[2], X. Such[2] and G. Caja[2]
[1]Facultad de Estudios Superiores Cuautitlán, Universidad Nacional Autónoma de México, Departmento de Ciencias Pecuarias, Ctra. Cuautitlán-Teoloyucan km 2.5, 54714 Cuautitlán Izcalli, Mexico, [2]Universitat Autònoma de Barcelona, Group of Research in Ruminants (G2R), Department of Animal and Food Sciences, Campus Universitari de la UAB, 08193 Bellaterra, Spain, [3]Universitat Autònoma de Barcelona, Department of Animals, Medicine and Surgery, Campus Universitari de la UAB, 08193 Bellaterra, Spain; san_dy_sam@hotmail.com

Liver biopsies can provide relevant information about metabolism. However, there is no reference describing a biopsy technique for dairy small ruminants. The objective was to develop a liver biopsy procedure by ultrasonographic guidance for ewes and goats. A total of 20 Murciano-Granadina goats (42.4±1.3 kg BW) in mid lactation (141±4 DIM) and 10 Manchega ewes (68.6±1.2 kg BW) in late lactation (199±8 DIM) were used. Liver biopsies were obtained twice from each animal (6-wk interval) under aseptic conditions. Animals were sedated and placed in left lateral recumbency with previous right flank shaving (7th to 12th intercostal spaces) for local anaesthesia at the biopsy site. Semiautomatic VI Trucut type SuperCore 14G × 9 cm biopsy needles (Argon Medical Devices, Athens, TX) were used. Liver visualization and needle insertion was accomplished by real time B-mode ultrasonography with a convex C60/5-2 MHz transducer (SonoSite Ultrasound System, Vet180 Plus, Bothell, WA). Liver biopsies were snap frozen in liquid N, and stored at -80 °C. Time from antisepsis to stapling averaged 13±2 min in goats and ewes. Milk yield before biopsy averaged 1.78±0.08 in goats and 0.59±0.05 kg/d in ewes. Milk yield declined by 52% (-0.92±0.10 kg/d) the day after biopsy, but recovered after 4 d (1.75±0.09 kg/d) in goats, whereas milk decline was 57% (-0.34±0.05 kg/d) and they recovered after 7 d (0.56±0.09 kg/d) in ewes. The RNA concentration of biopsies averaged 1,313±102 and 246±52 ng/µl in goats and ewes, respectively. In conclusion, percutaneous liver biopsy obtention by ultrasonographic guidance is a recommendable technique that allowed a full recovery of milk production and to be suitable for RNA extraction and RNA-seq analyses in dairy small ruminants.

Session 13 Theatre 6

Modelling milk yields of New Zealand dairy goats undergoing normal and extended lactations

N. Lopez-Villalobos[1] and S.A. Turner[2]
[1]Massey University, School of Agriculture and Environment, 4410 Palmerston North, New Zealand, [2]Dairy Goat Co-operative, 18 Gallagher Drive, 3240 Hamilton, New Zealand; n.lopez-villalobos@massey.ac.nz

Daily milk yields from dairy goats, in a New Zealand herd, undergoing normal and extended lactations, were predicted using a random regression (RR) with 3rd and 5th order Legendre polynomials, respectively. Normal lactations were defined as those lactations with less than or equal to 305 days in milk. Extended lactations were defined as those lactations with more than 305 days, but less than 670 days in milk. Persistency of extended lactation was defined as (B/A)×100 where A was the accumulated yield from day 1 to 305 days and B was the accumulated yield from day 306 to the last day in milk (but only up to 670 days). The relative prediction errors between the actual and the predicted yields using RR were close to 10% and the concordance correlation coefficients were >0.92 indicating that the RR models with Legendre polynomials are an adequate technique to model the normal and extended lactation curves of daily milk yields for dairy goats. Average total milk production in normal and extended lactations were 1,183 kg and 2,473 kg respectively. Average persistency of extended lactation was 117%. Effects of parity were significant ($P<0.01$) on both, 305-day and 670-day totals yields of milk. The average total milk yield of first-parity goats with a normal lactation was 946 kg while the average total milk yield of second-parity goats with a normal lactation was 1,284 kg, a total of 2,230 kg from the two normal lactations. The average total milk yield of first-parity goats with an extended lactation was 2,140 kg. Thus, on average, a goat with two normal lactations following the first- and second-parity produced 90 kg more milk than a first-parity goat did from an extended lactation. However, a second-parity goat produced 43 kg more milk from an extended lactation than the total milk produced by a goat with normal lactations following the second- and third-parity (2,639 kg vs 1,284 kg +1,312 kg). These results indicate that the persistency of extended lactations is high and that goats subjected in a 670-day extended lactation will produce similar amounts of milk compared with goats subjected to two 305-day normal lactations.

Session 13 Theatre 7

Operational use of GPS collars for decision making by agro-silvo-pastoral farmers in Mediterranean areas

L. Sautot[1], I. Candela[2] and M. Jouven[2]
[1]AgroParisTech, UMR TETIS, 500 rue JF Breton, 34090 Montpellier, France, [2]Institut Agro Montpellier, UMR SELMET, 2 Place Pierre Viala, 34060 Montpellier, France; lucile.sautot@agroparistech.fr

In the last decade, GPS collars have increasingly been used in Mediterranean agro-silvo-pastoral farms, especially when animals graze unsupervised in large surface areas of rangeland, with or without fences. Farmers' interest is mainly to reduce the time spent finding and gathering the animals and eventually to identify off-limit behaviours or predation. However, GPS tracks can be analysed further and provide additional useful information. In the P@stor-all project, we explored ways to exploit the tracks recorded with GPS collars to provide operational indicators for decision making in pastoral contexts. Our main objective was to describe the spatial behaviour of the flock in terms of both occupation of the rangeland and grazing effort for the animals. The indicators should interest the farmers and be suited to 'real life' conditions. Seven farmers participated in the P@stor-all project, representing a variety of agro-silvo-pastoral systems and were trusted with a GPS collar, to be deployed in their flock. The frequency of data acquisition was set at 5 min, in order to strike a balance between the accuracy of data acquisition and the feasibility of recharging the batteries. The GPS tracks were recorded in various seasons, grazing conditions and locations. Based on the characteristics of the collected data and the associated free satellite data, but also on repeated interviews with the farmers, we proposed and calibrated a set of 7 indicators: area explored by the flock, occupation density, daily distance covered by the flock, instant speed along the grazing route and daily total ascent. In the P@stor-all platform, each farmer may visualize these various indicators, together with basic information about calculation methods, sensitivity to the quality of GPS data, and suggestions of interpretation. A variety of indicators is needed to account for the diversity of grazing contexts: depending on the farm and season, different indicators might be meaningful to adjust grazing management. The absolute values and relative changes in each indicator also depend on local conditions, thus each farmer will need to build his own local references and make his own decisions.

AGRICYGEN: innovative phenotyping and high-throughput genotyping studies of Cyprus sheep and goats

G. Hadjipavlou, S. Andreou, T. Christofi, A.C. Dimitriou, A.N. Georgiou, L. Koniali, G. Maimaris, P. Markou and S. Panayidou
Agricultural Research Institute, Animal Production, P.O. Box 22016, 1516 Lefkosia, Cyprus; ghadjipavlou@ari.moa.gov.cy

The Project AGRICYGEN (CYprus AGRIcultural Genomics CENtre) has been fully supported by national funds since 2021 and is coordinated and run by the Agricultural Research Institute (ARI) in Cyprus. The project employs a unified approach, through genomics, to improve the productivity and sustainability of the whole system (animals, plants and microbes), and produce novel products, services and recommendations for improving livestock output, crops used for feed and soils cultivated for feed production. In addition, an Aid Scheme through the Cyprus Resilience and Reform Plan was approved for supporting new phenotyping technology implementation and genomic analyses of local Cyprus Chios sheep and Damascus goats in private farms, and has been initiated in 2022 (with funding from the EC-NextGenerationEU), under ARI coordination. The Scheme is connected with the AGRICYGEN research project and aims to upgrade the sheep and goat farming sector in Cyprus by granting targeted subsidies to sheep and/or goat breeders to achieve advanced recording practices on farm as well as ARI-led genetic and genomic evaluation services, improved reproductive and overall livestock and breeding unit management. The long-term goal for the implementation of both initiatives is to implement research outcomes to achieve advanced genetic improvement of an extensive number of sheep and goat animals by 2026 and concurrently a significant increase in farm productivity, mainly in terms of milk production. Research findings from the initial stages of project implementation, are promising and signify the importance of combining various technological approaches to promote the advancement of the Cyprus sheep and goat sector in semi-extensive Mediterranean systems, such as the one predominantly present in Cyprus.

Genetic improvement of milk production traits in the Cyprus Damascus goat population

S. Andreou, A.N. Georgiou, G. Maimaris, A. Dimitriou, L. Koniali and G. Hadjipavlou
Agricultural Research Institute, Animal Production, P.O. Box 22016, 1516, Lefkosia, Cyprus; sandreou@ari.moa.gov.cy

Small ruminant breeding is of great financial and environmental importance around the world. The Damascus goat is of Syrian origin and was imported in Cyprus in early 20th century in order to increase milk and meat yield, and has since then been genetically improved and become a local breed. The Agricultural Research Institute (ARI) is the core research centre in Cyprus for ruminant genetic improvement. Currently, selection of high producing animals in Cyprus is achieved by combining pedigree information with selection indices and with the evaluation of phenotypes on growth rate in young animals and milk production capacity of their ancestors. Research efforts at the ARI Animal Production unit focus on advancing the genetic evaluation of the Cyprus sheep and goat populations under the AGRICYGEN project and Recovery and Resilience Plan (RRP) (NextGenerationEU). Therefore, within the present study, high-throughput genomic mapping techniques were used to pursue for the first time the implementation of genomic evaluations of milk production in Cyprus Damascus goats, combined with a primary investigation of possible genomic regions associated with high levels of milk yield. Records on 2,872 goats born between 2000 and 2022 have been collected and analysed, with the earliest milk quantity and quality information recorded in 2008. In addition, a pool of 891 goat DNA samples were genotyped with the GoatSNP65 BeadChip. Quality control analysis resulted in 55,772 informative SNPs. This is the first case study focusing on Cyprus goat population, exploring genome-wide associations (GWAS) with milk yield volume and quality traits of economic interest such as protein and fat content. In addition, implementation of genomic BLUP models enabled the calculation of individual genomic breeding value and hence more precise genetic improvement for milk quality and quantity in subsequent generations. This study will be further expanded to pursue genomic evaluations and GWAS of genomic and phenotypic data collected from private Cyprus Damascus goat flocks, as part of an Aid Scheme under the Cyprus Resilience and Recovery Plan, funded by the European Commission.

Genetic trends for test day milk yield in White Maritza and Patch-faced Maritza sheep breeds

P. Zhelyazkova and D. Dimov
Agricultural university, Animal sciences, 12, Mendeleev blvrd., 4000 Plovdiv, Bulgaria; angela_pp@abv.bg

Data from milk recording were used to analyse genetic trends of test day milk yields for White Maritza and Patch-faced Maritza sheep breeds in Bulgaria. Both breeds are native to the country and are subject to conservation and improvement through approved breeding programs. A total of 9,556 (White Maritza) and 22,029 (Patch-faced Maritza) test-day records gathered by the Breeding association of native Maritza sheep breeds for the period 1990-2020 were entered in the analyses. The respective pedigree data comprised 4,687 (White Maritza) and 3,961 (Patch-faced) records. Repeatability test day model was used to calculate the heritability and breeding value estimations for test day milk yield (TDMY) and also to estimate the fixed and random effects assumed to affect the TDMY, separately for each breed. The genetic trends of the two breeds were estimated by the weighted regression of the average breeding value estimations of the animals on the year of birth. The average values for the TDMYs in the total database of the White Maritza was 790.27 ml and in the Patch-faced Maritza population was 744.67 ml. The heritability estimates of TDMY were 0.29±0.024 and 0.19±0.045 respectively for the two breeds. In the population of White Maritza breed, the straight line of the genetic trend over the 30-year period was slightly positive, but in the population of Patch-faced Maritza breed was negative. Over the years the graphs show positive and negative fluctuations between generations. Distinctive cyclical patterns which reflected long-time variation in genetic trends of the two breeds were found. Explanations for the different tendencies of genetic trends are given and discussed.

Genetic and environmental effects on the survival and growth rates of Cyprus Chios lambs

T. Christofi, S. Panayidou, L. Koniali and G. Hadjipavlou
Agricultural Research Institute, Animal Production, P.O. Box 22016, 1516, Lefkosia, Cyprus; tchristofi@ari.moa.gov.cy

The Chios sheep has been imported in Cyprus more than 60 years ago and, through genetic improvement, has been adapted to the environmental conditions of the island. It is the main commercial breed in Cyprus, both raised as pure-bred and in crossbreeding, as it is resilient and has high performance for both milk and meat production. Recent research efforts at the Agricultural Research Institute focus on employing genomic breeding methodologies to enhance productivity as well as to include additional traits in the selection objective, such as survival, reproductive success, longevity and others. Within this study, we utilized 16,864 individual lamb records of the Chios breed collected between 1989 and 2022 from the ARI Athalassa breeding nucleus, located in Nicosia district, Cyprus. Pedigree information was available for all animals with records and reproduction data were available for each dam. Lamb survival at various stages of lamb growth was recorded, in addition to pre and post weaning weights (birth, weaning, 90 days). Heritability estimates and coefficient of variation for maternal (litter size born/reared, litter size live, gestation length) and lamb (survival, rearing method, longitudinal weights) traits were estimated. In addition, for ewes with repeated parturitions, estimated breeding values predicted per animal regarding survival and growth were compared to observed breeding values. In addition, for lamb survival from birth to 3 months of age, analyses were conducted using regression/threshold animal models to estimate genetic and environmental effects. Moreover, various non-linear models were used to fit longitudinal weight data for the Chios breed. Besides phenotypic and pedigree data analyses, genomic analysis (GWAS) will also be pursued to dissect genetic associations with growth and survival traits.

Genetic and non-genetic factors affecting survival and growth of Cyprus Damascus goat kids

L. Koniali, S. Andreou, T. Christofi and G. Hadjipavlou
Agricultural Research Institute, Animal Production, P.O. Box 22016, 1516, Lefkosia, Cyprus; lkoniali@ari.moa.gov.cy

Mortality among pre-weaned goat kids represents an important economic loss for farmers. A number of environmental and animal related parameters have been reported to contribute to neonatal mortality in different breeds and locations, with preliminary studies on Cyprus Damascus goats indicating the birth season, the length of gestation and the birth weight to significantly effect kid mortality both at birth and at weaning. Furthermore, maternal age was found to significantly affect neonatal mortality. In an effort to improve productivity of local goats and reduce reproductive losses, the present study aimed to examine both genetic and non-genetic factors affecting the survival and early growth of Cyprus Damascus goats from birth to weaning with a view to developing more effective genetic selection programs. Records on 5,692 Cyprus Damascus goats (2,756 female and 2,936 male) born between 2005 and 2021 were obtained from the Agricultural Research Institute's experimental farm, situated in the Nicosia district. Information of goat kids' gender, birth type (singleton vs twin vs triplet vs quadruplet), kidding season, live weights at birth (BW) and at weaning (WW), the rearing system used (natural suckling vs artificial feeding), lactation number of the dam (1-10) and weight (DW) and goat kids' survival at birth (day 0), at weaning (day 50) and post-weaning period (day 110) were extracted from the ARI database. Pedigree information was also available for all animals with records. The potential effect of early growth-related traits was analysed with the use of linear mixed models, whereas logistic and binomial analysis were performed for survival traits. In parallel, genome-wide association studies using medium-throughput SNP bead chip genotyping data of a second cohort of ARI kids are in progress to further assess genomic factors contributing to neonatal survival and growth potential of kids.

Feed restriction in dairy ewes decreases milk lipolysis and remodels the milk proteome and lipidome

M. Delosière[1], L. Bernard[1], C. Hurtaud[2], H. Larroque[3], S. Emery[1], A. Delavaud[1], M. Taillandier[1], P. Le Faouder[4], M. Bonnet[1] and C. Cebo[5]
[1]Univ Clermont Auvergne, INRAE, VetAgro Sup, UMRHerbivores, St-Genès-Champanelle, 63122, France, [2]INRAE, Institut Agro, PEGASE, Saint Gilles, 35590, France, [3]Univ de Toulouse, INRAE, ENVT, GenPhySE, Castanet Tolosan, 31326, France, [4]MetaboHUB-MetaToul-Lipidomique, MetaboHUB-ANR-11-INBS-0010, Inserm U1297/Univ Paul Sabatier Toulouse III, Toulouse, 31432, France, [5]Univ Paris-Saclay, INRAE, AgroParisTech, GABI, Jouy-en-Josas, 78350, France; mylene.delosiere@inrae.fr

Milk lipolysis is defined as the hydrolysis of triglycerides, the main components of milk fat, resulting in the release of free fatty acids and partial glycerides that alter the taste and functional properties of the milk such as foaming and creaming abilities, respectively. Milk lipolysis is therefore an important criterion of milk quality. Feed restriction was used as a model to study milk spontaneous lipolysis (SL) and its mechanisms in three ruminant species (dairy cows, goats and ewes). The objective of this experiment in dairy ewes was to characterize the effects of dietary restriction on milk SL in relation with the milk proteome and lipidome. For this purpose, a subset of 2 groups of 10 ewes with contrasted milk lipolysis levels were selected from a larger experiment with 2 groups of 24 ewes (102±2.0 DIM) receiving for 5 days either a control diet (100% of DMI *ad libitum*: unrestricted) or the experimental diet (65% of DMI *ad libitum*: restricted) in a 2×2 cross-over design. The feeding restriction caused a large decrease in morning milk SL (-0.43 Meq/100 g of fat) measured by the BDI method, a decrease in milk yield (-0.43 l/d), an increase of milk fat and protein contents (+3.0 and +0.9 g/l) and no variation in milk fat globules diameter. This significant decrease in SL was accompanied by changes in the milk proteome and lipidome that are being studied in depth. In ewe's skim milk with low lipolysis, we displayed a protein inhibitor of the lipoprotein lipase gene expression and a list of proteins that signal an immune process. In conclusion, in contrast to cows, a large decrease of milk spontaneous lipolysis in dairy ewes was observed in response to feed restriction which was associated with substantial variations in milk protein and lipid abundances.

Session 13 Poster 14

Garlic essential oil supplementation in sheep: effect on fermentation parameters

C. Saro[1,2], I. Mateos[1,2], F.J. Giráldez[1] and M.J. Ranilla[1,2]
[1]Instituto de Ganadería de Montaña (CSIC-Universidad de León), Finca Marzanas, s/n, 24346 Grulleros, Spain, [2]Universidad de León, Campus Vegazana, s/n, 24071 León, Spain; cristina.saro@unileon.es

Garlic essential oil is an additive used to modulate ruminal fermentation. Most of the studies using this additive have been conducted *in vitro* but its effects on ruminal fermentation *in vivo* remain unelucidated. The aim of this trial was to evaluate the effects of garlic essential oil on volatile fatty acid (VFA) concentration and profile and ammonia concentration in the rumen of sheep. Eight non-lactating sheep were randomly divided into two groups and received a 50:50 forage:concentrate diet. One of the groups (GO) received daily through the rumen cannula 200 mg of garlic essential oil in two equal doses administered in the morning and in the evening. The other group received no additive (CON). Rumen fluid was collected before starting the administration of the additive and at 1, 2 and 3 days of treatment. At each sampling day rumen was collected before morning feeding and treatment (0 h) and 6 hours after administration of the diet (6 h). Samples were analysed to determine concentration of VFA and molar proportions of acetate, propionate, butyrate, isobutyrate, valerate, isovalerate and caproate by gas chromatography and ammonia concentration by a colorimetric technique. The results were analysed separately for 0 and 6 h. When comparing GO and CON sheep at 0 hours no effects of the additive were observed for total VFA concentration or ammonia concentration in rumen fluid. Concentration of valerate tended to be higher in GO sheep but the rest of the VFA profile was similar for both groups. However, when samples were taken 6 h after the administration of diet, no differences was observed in total VFA concentration, but concentrations of isobutyrate and isovalerate were lower and that of valerate tended to be lower in GO sheep. Ammonia concentration was also lower in GO sheep when rumen parameters were assessed 6 hours after feeding. Under the conditions of the present study, garlic essential oil affected ruminal fermentation, slightly modifying VFA profile and ammonia concentration.

Session 13 Poster 15

Effect of garlic essential oil supplementation on rumen microbial community of sheep

I. Mateos[1,2], E. Mata-Nicolás[1,2], C. Saro[1,2], R. Li[3], R. Baldwin[3], F.J. Giráldez[1] and M.J. Ranilla[1,2]
[1]Instituto de Ganadería de Montaña (CSIC-Universidad de León), Finca Marzanas, s/n, 24346 Grulleros, Spain, [2]Universidad de León, Campus Vegazana, s/n, 24071 León, Spain, [3]USDA ARS, Beltsville, MD, USA; mjrang@unileon.es

Garlic essential oil is an additive used to reduce methane emissions from ruminants presumably due to its broad antimicrobial activity. However, most of the studies have been conducted *in vitro*. The aim of this *in vivo* research was to evaluate the effects of garlic essential oil on microbial communities present in the rumen of sheep. Eight non-lactating sheep were randomly divided into two groups and received a 50:50 forage:concentrate diet. One of the groups (GO) received daily through the rumen cannula 200 mg of garlic essential oil and the other group received no additive (CON). Rumen fluid was collected before starting the treatment and at 1, 2 and 3 weeks of trial. DNA was extracted and bacterial and archaeal community were assessed by high throughput sequencing. The reads generated were processed using FROGS pipeline and the results were analysed using Phyloseq package in R. Shannon diversity index was unaffected by treatment when comparing GO and CON sheep, but it decreased in GO sheep after starting the treatment. When assessing beta-diversity. the principal component analysis showed that samples tended to group according to treatment, but a strongest effect of individual animal was observed. Similarly to alpha-diversity, the abundance of bacterial and archaeal phyla was similar between groups but it varied along the experimental period in treated sheep. Under the conditions of this study, garlic essential oil slightly modified the structure of the rumen bacterial and archaeal community of sheep.

High-precision scanning system for complete 3D goat udder and teat imaging and analysis

G. Coquereau[1], P.G. Marnet[2], L. Delattre[3], J.M. Delouard[3] and P. Martin[4]
[1]Institut de l'Elevage, 2133 Rte de Chauvigny, 86550 Mignaloux-Beauvoir, France, [2]Institut Agro Rennes-Angers, 65 Rue de Saint-Brieuc, 35042 Rennes, France, [3]3D Ouest, 5 rue Louis de Broglie, 22300 Lannion, France, [4]CapGènes, 2135 Rte de Chauvigny, 86550 Mignaloux-Beauvoir, France; gaelle.coquereau@idele.fr

Nowadays, udder scoring for goat udder phenotyping is made by specialized technician. They are trained to score 5 traits for the udder and 4 for teats. Only one trait, the teat length, is being measured. This work is done only once in the life of the goat and some farms have flocks exceeding 400 goats, so it can easily become a hard work. Recent works underlined a clear degradation of udder/teat shape with parity and an increasing proportion of morphological and functional unbalance of half udders. Thus, for a better genetic selection, we must increase the number and quality of quantitative udder traits measured throughout the productive lifespan of dairy goats. To achieve this goal, we aimed to design a 3D Scanner device to scan the entire udder and to produce high-definition numeric images. The device consists of a portable corridor cage in which the goat can enter freely. The size makes it easy to carry in a van. The goat is blocked for a few minutes while the images are taken and then released. We use 3 Intel Realsense depth D 455 cameras mounted on a mobile trolley situated 60 cm under and on the two sides of the cage and sliding from the front to the back of the animal in approximately 10 s. The D455 extends the distance between the depth sensors to 95 mm which improves the depth error to less than 2% at 4 m and an internal software allow an autocalibration that ensure better evaluation of distances. The image resolution is 1,280×800 pixels and real time scene capture could be done at up to 90 frames per second. The reconstruction of the 3D Image is done by a specific software from these 3 sources. The first prototype was tested and produced very accurate images. It is very promising for a close future usage for the Goat selection scheme.

Asymmetric udder – a possible indicator of subclinical mastitis in ewes

K. Tvarožková[1], V. Tančin[1,2], M. Oravcová[2], M. Uhrinčať[2], L. Mačuhová[2], B. Gancárová[1] and M. Ptáček[3]
[1]Slovak University of Agriculture in Nitra, Institute of Animal Husbandry, Tr. A. Hlinku 2, 94976, Nitra, Slovak Republic, [2]NPPC – Research Institute for Animal Production Nitra, Hlohovecka 2, 951 41 Lužianky, Slovak Republic, [3]Faculty of Agrobiology, Food & Natural Resources Czech Univ Life Sci Prague, Kamýcká 129, 165 00 Prague-Suchdol, Czech Republic; xtvarozkova@uniag.sk

Mastitis is one of the serious health and economic problem in dairy ewes farming and may adversely affect milk yield, its quality and increase the somatic cell count (SCC). The reduction of milk yield caused by mastitis could affect also half udder size. The aim of our study was to determine how asymmetric udder is related to the occurrence of subclinical mastitis. The study was performed at four dairy farms. Totally 162 milk samples at half udder level were collected from 81 ewes. Only ewes with asymmetric udder (larger and smaller half) and free of clinical mastitis were included in the study. For bacteriological analysis, 82 milk samples were cultivated. Milk samples were cultured on blood agar (MkB Test a.s., Rosina, SR). MALDI-TOF MS (Bruker Daltonics, Germany) was used to identify pathogens. SCC was determined using the Fossomatic 90 (Foss Electric, Hillerød, Denmark). Somatic cell score (SCS) was used for statistical evaluation: SCS = LOG_2 (SCC/100,000) + 3. The significant lower SCS (3.45±0.27) in larger half udder compared to smaller half udder (5.61±0.27) (P<0.001) were found out. On the other side, 16.46% udders had opposite cells count where higher SCC in larger half was found out but smaller ones still had SCC over 500×10^3 cells/ml with pathogen presence. Additionally, there were detected 66.67% samples with the mastitis pathogens in smaller half udder. The most common pathogens were coagulase-negative staphylococci (CNS) – 87.88% from bacteriological positive samples. *Staphylococcus caprae* was the most frequent CNS (44.83%). *Staphylococcus aureus* was identified in 9.09% of bacteriological positive samples. In conclusion, asymmetric udder could help farmers to identify animals with subclinical mastitis. Supported by the APVV-21-0134, VEGA 1/0597/22 and GA FAPZ 06/2023.

Effects of the nature of milk replacer on growth performances and carcass and meat quality of kids

J. Normand[1], C. Boyer[1], S. Meurisse[1], A. Pommaret[2] and M. Drouet[1]
[1]Institut de l'Elevage, Carcass and Meat Quality, 69007 Lyon, France, [2]EPLEFPA Olivier de Serres, Ferme du Pradel, 07170 Mirabel, France; jerome.normand@idele.fr

In France, goat kid meat production is regarded as a co-product of goat milk. It represents about 3.000 tons of equivalent carcass for nearly 635.000 kids (GEB, 2021). It is poorly valued as it is very seasonal and not well adapted to French consumer demand. The goat sector therefore wanted to revitalize on-farm kids fattening to improve the image, quality and profitability of kid meat production. The 'ValCabri' project was built to investigate different ways aimed at boosting on-farm kids fattening, and especially the optimization of technical fattening itineraries. During this project, a trial was set up in the experimental farm of 'Le Pradel' to evaluate the impact of the nature of milk replacer on zootechnical performances of kids, characteristics of their carcasses and meat qualities. After the colostral period, 60 male Alpine kids, weighing 4.7 kg on average at birth, were divided into 3 groups and fed either goat milk (GM group) or a milk replacer containing 0% or 65% of SMP (Skimmed Milk Powder – groups 0SMP and 65SMP). The kids were slaughtered at 24 d, with 10 kg live weight and 5.6 kg carcass weight, and no significant difference between groups. The average daily gain from birth to slaughter was similar for the 3 groups: 210 g/d. The kids of the 0SMP group consumed more milk replacer than those of the 65SMP one (respectively 7.8 vs 6.2 kg/kid) but the feeding cost remained slightly lower (14.5 vs 17 €/kid, January 2020 prices). The GM kids consumed a total of 37.7 l of milk per kid, or 1.5 l/d on average. The feeding cost of kids fed with maternal milk varies according to the proportion of post-colostral milk used. It can be null if it is only post-colostral milk and can represent from 27 €/kid if it is milk delivered to the cooperative, to 75 €/kid if it is milk transformed on the farm into Picodon PDO. Regarding the quality of the carcasses, no difference in measurements was observed between the 3 lots. However, the carcasses of the GM group were significantly lighter than those of the 2 others. The sensory qualities of the leg (*semimembranosus*) evaluated in the laboratory by a panel of 12 experts were not different between the 3 groups.

Effect of somatic cell count and stage of lactation on milk production

L. Mačuhová[1], V. Tančin[1,2], J. Mačuhová[3], M. Uhrinčat'[1] and M. Oravcová[1]
[1]NPPC, Research Institute for Animal Production Nitra, Hlohovecká 2, 951 41 Lužianky, Slovak Republic, [2]Slovak University of Agriculture in Nitra, Institute of Animal Husbandry, Faculty of Agrobiology and Food Resources, Tr. A. Hlinku 2, 94901 Nitra, Slovak Republic, [3]Institute for Agricultural Engineering and Animal Husbandry, Vöttinger Str. 36, 85354 Freising, Germany; lucia.macuhova@nppc.sk

Somatic cell count is not regularly evaluated in milk of dairy ewes for mastitis diagnostics. Mastitis negatively influences milk yield and its composition. The aim of this study was to evaluate the effect of somatic cell count (SCC) and stage of lactation on milk production in Tsigai (75%) × Lacaune (25%) dairy sheep (n=124). Therefore, in one flock, individual milk production was recorded and individual milk samples were collected once a month from February to July (i.e. from the start of lambing and thus during milking period to the end of lactation). SCC was measured using Fossomatic devices. Ewes were assigned according to current SCC to very low (VLSCC), low (LSCC), middle (MSCC), high (HSCC), and very high (VHSCC) SCC groups with SCC of ≤200,000 cells/ml, between 200,001 and 400,000 cells/ml, between 400,001 and 600,000 cells/ml, between 600,001 and 1,000,000 cells/ml, and >1,000,000 cells/ml, respectively. No significant differences (P=0.7564) were observed between different SCC groups in milk yield. The occurrence frequency of SCC in evaluated SCC groups was 86, 6.5, 1.5, 1.5, and 4.5% in VLSCC, LSCC, MSCC, HSCC, and VHSCC, respectively The month of measurement had a significant influence (P<0.0001) on milk production. Significant differences were observed between all evaluated months except between February and April. The highest milk production was recorded in March (915.35±24.06 ml) and the lowest in July (389.44±22.71 ml). In conclusion, the assessed flock had a very good udder health, and no effect of SCC on milk production is observed within healthy flock. Milk production was affected only by the advancing stage of lactation. This publication was written during carrying out of the projects APVV-21-0134 and SMART 313011W112.

Fit of mathematical models to the lactation curves of dairy sheep

L. Guevara[1], L.S. Gloria[1], J.C. Angeles-Hernandez[2], A.M. Fernandes[1] and M.L.C. Abreu[3]
[1]Universidade Estadual do Norte Fluminense, Pos-graduação em ciência animal, Av. Alberto Lamego, 2000, 28013-602, Campos dos Goytacazes, RJ, Brazil, [2]Universidad Autónoma del Estado de Hidalgo, Instituto de Ciencias Agropecuarias, Rancho Universitario, Av. Universidad km. 1, 43600, Tulancingo, Hgo, Mexico, [3]Universidade Federal de Mato Grosso, Pos-graduação em ciência animal, Av. Fernando Corrêa da Costa, no. 2367, 78060-900, Cuiaba, MT, Brazil; lilian.mvz@gmail.com

Sheep milk production is an incipient activity in Latin America. Due to this, there is a need to establish work plans in the productive activity and to know the biological behaviour of the production to establish strategies within the system that allow the activity to be economical, profitable and sustainable. The use of mathematical models to describe lactation of sheep has been limited, but the best adjustment of these allows predicting and making decisions on production. The objective of this work is to fit different mathematical models that describe the lactation curve of sheep in an experimental farm in Mexico. A total of 477 weekly test day records from 32 dairy crossbred sheep were analysed. The parameters of models were estimated using the 'nlme' package in the R software. A total of 42 models were tested and the goodness of fit was evaluated by means of AICc. Finally, 6 models were selected (Michaelis-Menten, Brody, exponential parabolic form Sikka, Morant and Gnanasakthy and Pollot multiplicative reduced form two parameters and Guo and Swalve modified) and compared by AICc and evidence ratio criteria. The characteristics for lactation peak yield, time of peak yield, total milk yield and persistence were calculated. The Morant & Gnanasakthy model was the best fit to represent dairy sheep lactation in ewes from México. This study represents an advancement in the knowledge of the lactation curve of sheep in Latin America.

Udder health in dairy ewes in Slovakia: current situation

V. Tančin[1,2], K. Tvarožková[2], B. Gancárová[2], L. Mačuhová[1], M. Uhrinčať[1], M. Vršková[1] and M. Oravcová[1]
[1]NPPC – Research Institute for Animal Production Nitra, Hlohovecka 2, 951 41 Lužianky, Slovak Republic, [2]Slovak University of Agriculture in Nitra, Institute of Animal Husbandry, Tr. A. Hlinku 2, 949 76 Nitra, Slovak Republic; xgancarova@uniag.sk

Somatic cell count (SCC) is used for diagnostic of udder health problems – mastitis in dairy cows. However, it is still in high scientific discussion about physiological level of SCC for diagnostic of mastitis in ewes. Therefore, the aim of the study was to review the current situation of udder health in dairy farms and possible importance to use SCC for diagnostic of mastitis through the relationship of SCC with pathogen presence. We have analysed the data obtained during the three years. First data set: in total 9,211 samples in 2019, 11,826 in 2020 and 11,915 in 2021 were available from 26 farms. The samples on the basis of SCC were divided into 5 SCC classes – C1 ($\leq 2.10^5$ cells/ml), C2 (2.10^5-4.10^5 cells/ml), C3 (4.10^5-6.10^5 cells/ml), C4 (6.10^5-10.10^5 cells/ml), C5 ($\geq 10.10^5$ cells/ml). Second data set: in selected farms 1,499 milk samples at half udder level were collected for SCC analysis. Milk samples were cultured on blood agar (MkB Test a.s., Rosina, SR). MALDI-TOF MS (Bruker Daltonics, Germany) was used to identify pathogens. From the first data set there were 52.67, 15.01, 6.55, 7.34 and 18.40% samples in SCC classes, respectively. There were increasing % in C1 (42.16, 51.45 and 62.01%, respectively), and reduction % in C5 from the year to year (22.80, 17.38, 16.00%, respectively). From second data, there were 5.77% samples with mastitis pathogens in both C1 and C2 SCC groups, but in C5 there were 77.23% of samples with pathogens. The most frequent pathogens were coagulase negative staphylococci and among them *S. chromogenes* was with highest frequency, followed by *S. epidermidis* and *S. xylosus*. In conclusion, we could consider SCC below 4.10^5 cells/ml as the proposal level for diagnostic purposes and that on the basis of pathogens detected the improving of housing and milking environment in dairy practice could significantly contribute to better udder health. Supported by project: APVV-21-0134, VEGA 1/0597/22 and SMART 313011W112.

Evaluation of udder health of dairy goats using electrical conductivity under practical conditions

M. Uhrinčat[1], V. Tančin[1,2], K. Tvarožková[2], B. Gancárová[2], L. Mačuhová[1] and M. Oravcová[1]
[1]NPPC-Research Institute for Animal Production Nitra, Hlohovecká 2, 951 41 Lužianky, Slovak Republic, [2]Slovak University of Agriculture, Tr. A. Hlinku 2, 949 76 Nitra, Slovak Republic; michal.uhrincat@nppc.sk

In practice, the portable conductivity meter is used as a rapid method applied to detect bovine mastitis. The aim of the study was to evaluate this method for dairy goats. In the experiment, the milk samples from 484 udder halves of goats at the same farm were evaluated. After forestripping (2 streams of milk), 10 ml were milked into a Milk Checker N-4L and the electrical conductivity (EC; mS/cm) was measured. Afterwards 1 ml of milk was aseptically gathered into a sterile test tube for cytobacteriological analysis and an additional sample of 30 ml was taken for somatic cell count (SSC) analysis. SCC was log transformed to somatic cell score (SCS). Based on the SCC, the samples were divided into classes: (1) SCC<2×10^5 (n=29); (2) $2\times10^5\leq$SCC<4×10^5 (n=53); (3) $4\times10^5\leq$SCC<6×10^5 (n=76); (4) $6\times10^5\leq$SCC<1×10^6 (n=95); and (5) SCC$\geq1\times10^6$ cells/ml (n=211). EC in the classes 1:2:3:4:5 (mean ± SD) 6.34±0.62: 6.38±0.50: 6.46±0.43: 6.63±0.48: 6.82±0.75 had an upward trend, without significant differences. We found only a moderate correlation (r=0.35) between EC and SCS. A difference in EC greater than 0.5 between halves within the same udder indicates a health problem. At difference EC=0 there were 48 goats with SCC difference less than 2×10^5, 23 goats with a difference from 2×10^5 to 4×10^5 and 41 with a difference of more than 4×10^5 cells/ml. At difference EC=0.2-0.4 there were 39 goats with SCC difference less than 2×10^5, 10 goats with a difference from 2×10^5 to 4×10^5 and 45 with a difference of more than 4×10^5. There were only 23 goats with an EC difference of more than 0.5. Eight different pathogens were detected in 59 bacterially positive samples. The most frequent pathogen was *Staphylococcus caprae* (n=28; EC= 6.82±0.75), but *Staphylococcus aureus* caused the highest EC in positive samples (n=4; EC= 7.40±0.51). We concluded that the detection of mastitis by measuring the milk EC of dairy goats proved to be a less reliable method in this case. This study was supported by APVV -21-0134 and SMART 313011W112.

Milk recording protocols and goodness of fit of models applied to dairy sheep lactations

L. Guevara[1], L.S. Gloria[1], J.C. Angeles-Hernandez[2], I. Nacarati Da Silva[1], A.M. Fernandes[1] and A. Peláez Acero[2]
[1]Universidade Estadual do Norte Fluminense, Pos-graduação em ciência animal, Av. Alberto Lamego, 2000, 28013-602, Campos dos Goytacazes, RJ, Brazil, [2]Universidad Autónoma del Estado de Hidalgo, Instituto de Ciencias Agropecuarias, Rancho Universitario, Av. Universidad km. 1, 43600, Tulancingo, Hgo, Mexico; lilian.mvz@gmail.com

The lactation curve can be described by mathematical models, which can never completely represent a biological process, but provide useful information for decision-making at herd level. The ability of mathematical models to represent the lactation curve can be influenced by factors such as the interval between test day records (TDR). The aim of this study was to evaluate the effect of different TDR intervals on the goodness of fit of the empirical (Wood and Wilmink) and mechanistic models (Pollott and Dijkstra) applied to dairy sheep lactation. A total of 4,494 weekly TDRs from 156 lactation of dairy crossbred sheep were analysed. Three new databases were generated from the original weekly TDR data (7D), comprising intervals of 14(14D), 21(21D), and 28(28D) days. Also, the shape of the lactation curve (typical and atypical) was defined. The goodness of fit was evaluated using the mean square of prediction error (MSPE), Root of MSPE (RMSPE), Akaike's Information Criterion (AIC), Bayesian's Information Criterion (BIC), and the coefficient of correlation (r) between the actual and estimated total milk yield (TMY). Most models exhibited the greatest values of r2 (0.56 to 0.99) to 7D interval. We found higher values of r2 for typical curves in comparison with atypical curves (0.91 vs 0.74, respectively). The TDRs interval affected the capability of Wood and Djisktra models to estimate the peak and time to peak lactation, with better estimations for the 28D interval for the Djisktra model. However, the TMY can be adequately estimated by the four models in all TDR intervals evaluated. Therefore, the selection of lactation model and TDR interval must consider the estimation objectives and structure of the database.

Session 13 Poster 24

Heritability of milk lipolysis in French Alpine goats

H. Larroque[1], C. Robert-Granié[1], S. Meurisse[2], P. Trossat[3], L. Bernard[4], C. Hurtaud[5], C. Huau[1], T. Fassier[6], R. Rupp[1] and C. Cebo[7]
[1]GenPhySE, université de Toulouse, INRAE, ENVT, 31326 Castanet-Tolosan, France, [2]Institut de l'Elevage, 149 rue de Bercy, 75595 Paris, France, [3]Actalia-Cécalait, Rue de Versailles, 39800 Poligny, France, [4]INRAE, Université Clermont Auvergne, Vetagro Sup, UMRH, 63112 Saint-Genes-Champanelle, France, [5]PEGASE, INRAE, Institut Agro, 35590 Saint Gilles, France, [6]INRAE, P3R, Domaine de la sapinière, 18390 Osmoy, France, [7]Université Paris-Saclay, INRAE, AgroParisTech, GABI, 78350 Jouy-en-Josas, France; helene.larroque@inrae.fr

Milk lipolysis, i.e. the hydrolysis of milk triglycerides, is not routinely measured and its occurrence and genetic determinism can therefore be hardly studied. In order to overcome this limitation, in France the LIPOMEC project (ANR-19-CE21-0010) has developed an equation for the prediction of goat milk lipolysis from mid-infrared (MIR) milk spectra. This prediction lipolysis equation was applied to non-standardized MIR spectra of milk samples of Alpine goats with a high diversity of αs1 casein genotypes. Finally, 22,745 log-normalized predictions of 1,035 goats were retained. First, environmental factors affecting milk lipolysis were identified. Then, variance components were estimated using a single trait animal model with repeatability and including the same environmental effects. The part of variance explained by the category of αs1 casein genotypes (for their known effect on protein content) was evaluated by including or excluding it in the model. Among the environmental factors identified, longer intervals between milking and milk analysis increased lipolysis in milk, as well as the evening milking compared with the morning milking. Lipolysis decreased with lactation stage, and its level was higher in the second lactation than in subsequent lactations. The lipolysis level was related to the category of αs1 casein genotypes, and reached a maximum for OO genotypes. Heritability of lipolysis was estimated to 0.21±0.03 and decreased to 0.18±0.03 when including αs1 casein genotypes in the model. The αs1 casein genotype would explain 3% of the total phenotypic variance of lipolysis. These first results will need to be verified using data from commercial farms in order to overcome the small size and the particular genetic structure of the population studied.

Session 13 Poster 25

Castration method affects the fatty acid profile of subcutaneous and intramuscular fat in sheep

H.A. O'Neill, N. Skele, O.B. Einkamerer, A. Hugo and F.W. Neser
University of the Free State, Animal Science, 205 Nelson Mandela Drive Park West Bloemfontein, 9301, South Africa; neserfw@ufs.ac.za

This study investigated the effect of lamb castration method on the fatty acid profile of subcutaneous and intramuscular fat of South African Mutton Merino lambs (n=30). This project was approved by the Animal Ethics Committee (UFS-AED2019/0136/2410). Ten (n=10; EC) lambs were castrated at one week of age using rubber rings; ten (n=10; LC) were castrated at 8 weeks of age using a burdizzo and ten (n=10; NC) were kept intact. These methods mimic commercial procedures that are also ethically acceptable. After castration, sheep were kept individually in a metabolic building and fed a finishing diet for 67 days. The animals were slaughtered at a commercial abattoir when they reached an average of ±45 kg live weight on an empty stomach. Muscle and fat samples were collected from the left m. longissimus at the 9th through 11th rib and frozen at -20 °C until analysis by gas chromatography. The method by Folch *et al.* was used to extract the lipid fraction from previously frozen muscle and subcutaneous fat samples. Fatty acid methyl esters were quantified using a Varian 430 flame ionization Gas Chromatography, with a fused silica capillary column, chrompack CPSIL 88 (100 m length, 0.25 mm ID, 0.2 μm film thicknesses). Fatty acids were indicated as the proportion of each fatty acid to the total of all fatty acids present in the sample. Atherogenicity index was calculated as: AI = (C12:0 + 4 × C14:0 + C16:0)/(MUFA + PUFA). For intramuscular fat, NC lambs had higher (P<0.05) PUFA, α-linolenic acid, n-6, linolelaidic acid, linoleic acid, PUFA: SFA ratio, and PUFA: MUFA compared to both EC and LC lambs. Eicosapentaenoic acid (EPA) and total n-3 were higher (P<0.05) for NC lambs than LC lambs. Palmitic acid and CLA were higher (P<0.05) for EC lambs compared to NC and LC lambs. For subcutaneous fat, NC lambs had higher (P<0.05) heptadecenoic acid and linolelaidic acid than LC lambs. The atherogenicity index (AI) of subcutaneous fat was higher (P<0.05) for LC lambs compared to NC lambs. In conclusion, meat and fat from NC lambs are healthier options to consume compared to meat and fat from castrates. It is recommended to keep lambs intact or castrate at a later stage.

Somatic cell count and prevalence of udder pathogens isolated in raw goat milk

B. Gancárová[1], K. Tvarožková[1], M. Uhrinčat[2], L. Mačuhová[2] and V. Tančin[1,2]
[1]Slovak University of Agriculture in Nitra, Institute of Animal Husbandry, Tr. A. Hlinku 2, Nitra, 949 76, Slovak Republic, [2]NPPC-Research Institute for Animal Production Nitra, Hlohovecká 2, Lužianky, 95141, Slovak Republic; vladimir.tancin@uniag.sk

The physiological value of somatic cell count (SCC) representing a healthy udder for goats is still not clearly defined in many countries. Our objective was to assess the health status of the udder at the mid and late lactation stages based on the identification of bacterial pathogens in the raw milk of White Short-haired goats and their relationship to SCC. The goats had kidded from mid February to mid-March. A total of 405 milk samples of half udders were aseptically collected from the dairy goat farm which is located in the northern part of the Orava region (Slovakia). The samples were divided on the basis of SCC: SCC1$<500\times10^3$; SCC2$\geq500<1000\times10^3$; SCC$^3\geq1000<2{,}000\times10^3$; SCC4$\geq2{,}000\times10^3$ cells/ml. The bacteriologically positive samples represented 14.07% of all samples. The main pathogens were coagulase-negative staphylococci (CNS) (80.7%). Interestingly, in the group SCC4 there were only 50.9% samples with identified pathogens. The most prevalent CNS were *Stahylococcus caprae* (40.4%) and *S. epidermidis* (32%). Other CNS like *S. simulans, S. warneri* and *S. equorum* were less represented. A contagious pathogen *S. aureus* was identified only in mid-lactation in two milk samples. Additionally, *Enterobacter kobei* (14.0%), *Enterobacter cloacae* (3.5%) and *Citrobacter braakii* (1.8%) were identified. The prevalence of the pathogens was similar in mid-lactation (29/57) compared with late lactation (28/57). It can be concluded that with increasing SCC, the occurrence of mastitis pathogens also increases, especially in samples with over one million cells. However a high SCC doesn't necessarily indicate the presence of pathogens in goat milk. Thus the threshold level of SCC for mastitis detection is still undefined for dairy goats. Supported by the VEGA 1/0597/22, APVV-21-0134, and by the GA FAPZ 06/2023.

Lowering crude protein and supplementing rumen protected methionine and lysine in dairy ewes

K. Droumtsekas[1], A. Plomaritou[1], I.C.C. Christou[1], A. Tsigkas[1], M.A. Belaid[2], D. Martinez Del Olmo[2], J. Mateos[2] and A. Foskolos[1]
[1]University of Thessaly, Department of Animal Science, Campus Gaiopolis, Larissa, 41222, Greece, [2]KEMIN Animal Nutrition and Health, Toekomstlaan 42, 2200 Herentals, Belgium; afoskolos@uth.gr

In several dairy sheep systems, crude protein (CP) overfeeding has been documented leading to lower nitrogen (N) use efficiency (NUE) and increased N excretion into the environment. This study investigated the effect of lowering the dietary CP level by supplementing rumen protected methionine (RPM; KESSENT®) and rumen protected lysine (RPL; LysiGEM®) at a constant ratio on milk yield and composition. A total of fifty-four (54) Lacaune ewes were involved in a randomized block design with five treatments: a basal diet with CP at 19.3%DM (CTR), a diet with lower CP at 15.1% DM (NEG), and three treatments with decreasing CP compared with CTR, supplemented with RPM and RPL at a fixed ration, namely AA1, AA2 and AA3. The AA1 treatment consisted on the NEG, supplemented with 3 g/d RPM and 5 g/d RPL, AA2 had a CP level at 16.2%DM supplemented with 3 g/d RPM and 4 g/d RPL, and AA3 had a CP level at 16.9%DM supplemented with 3 g/d RPM and 3 g/d RPL. The study lasted for 12 weeks. Each week, for two consecutive days, milk yield, composition and dry matter intake were measured. Dry matter intake was on average 3,26 kg. Lowering CP levels without adding rumen protected amino acids, reduced milk yield by 12.2% (2.36 vs 2.07 kg/d for CTR and NEG respectively; $P<0.001$), but when NEG was supplemented with RPM and RPL milk yield differences were not significant compared with CTR. The AA3 had the highest milk yield but was not different than CTR (2.51 kg/d). However, milk fat composition was significantly higher (6.69 vs 6.25 for AA3 vs CTR, respectively) without affecting milk protein composition. When protein and fat composition were considered calculating fat and protein corrected milk yield (FPCMY), differences between AA3 and CTR became clear leading to improved FPCMY for AA3. In conclusion, lowering CP levels and supplementing rumen protected amino acids is a valid strategy to improve NUE and increase ewes' productivity.

FASTOChe project: meat sheep grazing on plant species rich in bioactive secondary compounds

M. Bernard[1,2], H. Hoste[3,4], L. Sagot[1,2] and D. Gautier[1,2]
[1]CIIRPO, Le Mourier, 87800 Saint Priest Ligoure, France, [2]Institut de l'élevage, 149 rue de Bercy, 75595 Paris, France, [3]ENVT, UMR IHAP, 31076 Toulouse, France, [4]INRAE, 147 rue de l'Université, 75007 Paris, France; mickael.bernard@idele.fr

For several decades, the control of gastrointestinal strongyles has been based exclusively on the use of chemical molecules with broad-spectrum anthelmintic activities. For a long time, these synthetic molecules were applied in an unreasonable way. These practices have led to the increasing development of resistance to these molecules and to a direct impact on lamb performance. In this context, the search for alternatives is a priority to avoid finding ourselves without effective solutions in the coming years. Several *in vitro* and *in vivo* studies have shown that the intake of feeds and forages rich in bioactive secondary metabolites (BSM) by small ruminants disrupts the biology of gastrointestinal strongyles and the dynamics of infestations. The FASTOChe project is studying the anthelmintic and zootechnical interest of three plants rich in BSM, sainfoin, chicory and lanceolate plantain, grazed by meat sheep. Eight trials on experimental farms and agricultural high schools were conducted between 2019 and 2021. Fields were sown in pure plots and grazed either continuously or in 2-3 weeks courses by growing lambs or renewal ewe lambs. Various measurements were carried out on the grasslands (quantity available, quality, etc.) and on the animals, in terms of health (excretion and parasite infestation, evaluation of anaemia, etc.), performance (weighing, body condition score) and animal welfare. In our experimental conditions, chicory and plantain did not reduce the parasite load. Only sainfoin reduced parasite excretion but to a limited level (1000 vs 771 eggs per gram), which did not allow a reduction in the number of treatments. In terms of zootechnical performance, these three plants had moderate positive effects on growth, but here again sainfoin stood out with higher results (30g vs 142g per day) compared to grazing grass. In conclusion, these three plants do not replace anthelmintic treatments but their feed value allowed the animals to grow correctly.

Multiple mitigation strategies may lead to reduction in emissions for mixed Kenyan dairy systems

M.W. Graham[1], C. Arndt[1], D. Korir[1], S. Leitner[1], L. Merbold[2], P. Ndung'u[1], F. Teillard[3] and A. Mottet[3]
[1]International Livestock Research Institute, Mazingira, Box 30709, Nairobi, 00100, Kenya, [2]Agroscope, Agroecology and Environment, Agroscope Reckenholzstrasse 191, Zurich 8046, Switzerland, [3]U.N. Food and Agriculture Organization, Viale delle terme di Caracalla, Rome, 00153, Italy; lutz.merbold@agroscope.admin.ch

Livestock systems are an important source of livelihoods in Africa, but are also a large source of anthropogenic greenhouse gas (GHG) emissions (i.e. CH_4 from enteric fermentation; CH_4 and N_2O from manure) in most African countries. Many African countries, such as Kenya, have prioritized livestock emissions in their Nationally Determined Contributions under the Paris Agreement. However, there are limited data available on GHG emissions from livestock systems in Africa. Scaled livestock emissions in Africa have been estimated using modelling approaches and were not necessarily based on locally appropriate data. To bridge this gap between limited local data and modelling, we used datasets collected from representative smallholder mixed dairy cattle systems in Kenya to up-scale GHG emissions using the Global Livestock Environmental Assessment Model – interactive (GLEAM-i). We evaluated effects of the following previously evaluated mitigation interventions on milk emission intensities (EI) to compare against baseline data: reduced age at first calving; increased fertility rate; sweet potato vine silage (SPVS) supplementation; dairy concentrate feeding; increased feeding level; all interventions combined. EIs for milk were lower than the baseline for all individual intervention scenarios (-3.6 to -11.0%), and the combined scenario reduced EIs additively by 36%. Individual interventions with the highest overall impact on milk EIs were supplementation with SPVS (-11.0%), increased fertility rate (-10.3%), and increasing feeding level (-10.1%). These results indicate the 'many little hammers' approach to interventions can lead to additive reductions in EIs when combined. Further, we demonstrate that in-situ data based mitigation interventions can be captured by the GLEAM-i model. Future work should focus on filling existing data gaps for emissions from livestock in East Africa to allow further upscaling, particularly for pastoralist systems, small ruminants, and manure.

Session 14 Theatre 2

Assessing multifunctionality of livestock breeds and species at global level

G. Leroy[1], F. Joly[2], C. Looft[3], P. Boettcher[1] and R. Baumung[1]
[1]FAO, Viale delle Terme di Caracalla, Roma, Italy, [2]University of Clermont Auvergne, INRAE, VetAgro Sup, UMR Herbivores, Route de Theix, Saint-Genès-Champanelle, France, [3]Neubrandenburg University of Applied Sciences, Brodaer Str. 2, Neubrandenburg, Germany; roswitha.baumung@fao.org

The choice of appropriate livestock species and populations is one of the key levers of the multiple trade-offs that must be considered for the sustainability of food systems worldwide. Although often considered less productive than their exotic counterparts, native and locally adapted breeds are also known to provide a wide range of ecosystem services. Based on information reported by 41 countries on 3,361 national breed populations of 27 species in FAO's Domestic Animal Diversity Information System (DAD-IS), we investigate the factors influencing the recognition of their links to a set of 52 uses and ecosystem services. On average, 4.46 uses and ecosystem services were reported per national breed population. A larger number of cultural services were reported for horses (2.47 vs 0.75 on average), while ruminants were found to be linked to a greater number of provisioning, as well as regulating and maintaining ecosystem services (2.99 and 1.86 vs 2.39 and 1.32 on average, respectively). Compared to European breeds, livestock in Africa were reported to contribute to a larger number of provision services (3.95 vs 1.88). Native and locally adapted breeds were linked to a greater number of services than exotic breeds, either for cultural (0.83 vs 0.69), provision (2.85 vs 1.68) or regulation and maintenance services (1.75 vs 0.54). This highlights the fact that the former tend to be raised in less specialized production systems than the latter, and are recognized to play multiple roles for the livelihoods of rural communities and the environmental sustainability of food systems. This multifunctionality needs to be carefully assessed and taken into account in the development of livestock policies.

Session 14 Theatre 3

Predicting sustainable futures: does disease in early life result in trade-offs in ewe productivity?

J. Duncan[1], H. McDougall[1], G. Mitchell[1], R. Evans[1], M. Reeves[2], L. Andrews[1], E. Geddes[1], L. Melville[1], D. Ewing[3] and F. Kenyon[1]
[1]Moredun Research Institute, Pentlands Science Park, Bush Loan, EH26 0PZ Penicuik, United Kingdom, [2]SRUC, Roslin Institute Building, Easter Bush Campus, EH25 9RG Midlothian, United Kingdom, [3]Biomathematics and Statistics Scotland (BioSS), JCMB, The King's Buildings, Peter Guthrie, Tait Road, EH9 3FD Edinburgh, United Kingdom; jade.duncan@moredun.ac.uk

Livestock encounter numerous challenges in early life; however, we do not yet understand how these affect the long-term productivity of the animal. Here, our aim is to determine the impact of disease in early life on later performance in ewes. We hypothesise that ewe-lambs exposed to high levels of disease in their first 12 months will have reduced productivity when they mature. Female lambs (n=100) born between 28th March-23rd April 2022 at the Moredun Research Institute, Scotland were closely monitored throughout their first grazing season, for a range of natural gastrointestinal parasite (GIN) infections. Records included birthing ease, maternal bond, litter size, weight gain, faecal egg count (FEC) and faecal oocyst count (FOC), presence of other opportunistic diseases, welfare measures, medicines administered, and number of anthelmintic treatments required. A treatment criterion was used to determine if an animal should receive anthelmintic treatment. At weaning ewes were removed and lambs maintained. The ewe-lambs will be monitored throughout their first year until they give birth at 2 years old. We will then monitor their offspring's health and welfare to account for early life experiences. The 100 female twin lambs' average birth weight was 5.2 kg (range 2.5-7.25). Mean strongyle FEC was 186 epg (range 0-2,421), mean *Nematodirus* FEC was 44 epg (range 0-468) and mean coccidia FOC was 7,605 EPG (range 210-411,399). Notable welfare measures were dag score and presence of injury. Initial modelling suggests that there may be relationships between parasite burden, welfare measures and performance. Furthermore, there may be a negative relationship between the weight gain and the strongyle and *Nematodirus* FEC's ($r=-0.24$, $P<0.01$), which are themselves correlated with one another. This study shows that there is variation present in the disease challenge in a group of co-grazing ewe-lambs.

Session 14 Theatre 4

The welfare of housed and grazing beef cattle assessed by hormones levels and physical check scores

M.J. Rivero[1] and A.S. Cooke[2]

[1]Rothamsted Research, Net Zero and Resilient Farming, North Wyke, Okehampton, EX20 2SB, United Kingdom, [2]University of Lincoln, School of Life Sciences, Brayford Pool Campus, Lincoln, LN6 7TS, United Kingdom; jordana.rivero-viera@rothamsted.ac.uk

Housing/grazing time is a key decision with potential impacts on health and welfare. Understanding welfare implications of husbandry strategies is essential for future farming. The objective of this study was to compare the health and welfare of housed and grazing beef cattle using hormone levels and physical health scores. Two herds of beef cattle (30 heads each) were directly compared from housing at weaning in October 2020 to July 2021 across welfare indicators (i.e. physical health, hormones and behaviour). Here we are only presenting the physical and hormonal results. One herd ('HH') was housed this entire time and fed silage and concentrate to finish animals at 14-months, and the other ('HG') was housed until April, fed mainly silage, before being turned out onto pasture and aimed at finishing animals at 18-20 months. Twice during winter and twice during summer, physical inspections of animals were conducted to assess body condition, cleanliness, diarrhoea, hairlessness, nasal discharge, and ocular discharge. At the same time, hair and nasal mucus samples were taken for quantification of cortisol and serotonin. Physical health indicators were broadly comparable between herds except for body condition scores, which were higher in the HH herd (attributed to diet) (P=0.002) and nasal discharge which was more prevalent in the HH herd (P<0.001). Across all indicators there was an impact of time, with scores improving in summer compared to winter. Similarly, hormone data varied mostly across timepoints. However, a difference was found in hair cortisol levels, with the greatest concentrations observed in the HG herd (P=0.011), however such a pattern was not seen for nasal mucus cortisol, or for serotonin levels. In conclusion, animal welfare indicators in this case study varied more with time than with the production system. Animal behaviour will need to be included in the assessment to better understand the impact of housing and grazing on animal welfare. Acknowledgments: Soil to Nutrition (BBS/E/C/000I0320) and the North Wyke Farm Platform (NWFP, BBS/E/C/000J0100), funded by BBSRC.

Session 14 Theatre 5

Creating a shared framework for standardizing life cycle assessments in livestock production systems

P. Goglio[1], M. Trydeman Knudsen[2], K. Van Mierlo[1], N. Röhrig[3], M. Fossey[4], A. Maresca[5], H. Fatemeh[2], M. Ahmed Waqas[2], J. Yngvesson[6], G. Nassy[4], R. Broekema[1], S. Moakes[7], C. Pfeifer[7], R. Borek[8], D. Yanez-Ruiz[9], M. Quevedo Cascante[2], A. Syp[8], T. Zylowsky[8], M. Romero-Huelva[2,9] and L.G. Smith[3,6]

[1]Wageningen Research, Wageningen Economic Research, Bronlan 103, Wageningen 6708 WH, the Netherlands, [2]Aarhus University, Department of Agroecology, Blichers Allé 20, 8830 Tjele, Denmark, [3]University of Reading, School of Agriculture, Policy and Development, Reading, RG66BZ, United Kingdom, [4]Acta – les instituts techniques agricoles, 149 Rue de Bercy, 75012, Paris, France, [5]SEGES Innovation, P/S, Agro Food Park 15, 8200 Aarhus, Denmark, [6]Swedish University of Agricultural Sciences, Department of Biosystems and Technology, Box 190, 2, Sweden, [7]Research Institute of Organic Agriculture (FiBL), Department of Socio-Economics, Frick, Switzerland, Ackerstrasse 113 / Postfach 219, 5070, Switzerland, [8]Institute of Soil Science and Plant Cultivation, Czartoryskich Str. 8, 24-100 Puławy, Poland, [9]Estación Experimental del Zaidín (CSIC), Profesor Albareda 1, 18008 Granada, Spain; l.g.smith@reading.ac.uk

The intensification of animal production has been identified as a major contributor to environmental impacts however livestock products remain crucial for the human diet, and for key ecosystem services. To address the sustainability of livestock systems, Life Cycle Assessment (LCA) has become a vital tool. However, the accuracy and robustness of LCA methods must be improved. To achieve this goal, a participatory harmonization approach was employed, which involved 21 LCA experts in 29 workshops and two surveys. Five research topics were identified to enhance the accuracy of LCA methods for livestock: food, feed, fuel, and biomaterial competition; crop-livestock interaction; the circular economy; biodiversity; animal welfare; nutrition, and GHG emissions. Furthermore, general evaluation criteria were established based on the characteristics of livestock systems: transparency, reproducibility, completeness, fairness, acceptance, robustness, and accuracy. Overall, this participatory method was successful in narrowing down the evaluation criteria for livestock focused LCA methods, providing a holistic framework to assess the impacts of livestock systems on various key topics.

Defining operational sustainability and the need for systems approach in assessing livestock farming

A.H. Herlin[1], S. Hellstrand[2] and H.U. Sverdrup[3]
[1]Swedish University of Agricultural Sciences, Biosystems and technology, Box 190, 234 22 Lomma, Sweden, [2]Nolby Ekostrategi, Tolita 8, 665 92 Kil, Sweden, [3]Inland Norway University of Applied Sciences, Postboks 400 Vestad, 2418 Elverum, Norway; anders.herlin@slu.se

Livestock production has in recent times been accused of harming the environment, and climate and being unsafe to human health. The EAT-Lancet Commission has received a lot of attention for the proposal of a 'planetary diet', which claims to protect the health of people and the planet. The core of the suggestions is that red meat should be reduced by 82-100% globally. The planetary diet is part of a discourse arguing for the need for a food systems transformation that is enforced by a set of soft and firm policy measures. The planetary diet has a substantial influence on policy processes including e.g. land use and climate change by IPCC and the process for sustainable nutrition recommendations developed in the Nordic countries. Organizations at various levels of society have to some extent, adopted policies accordingly. The analysis of the environmental impact of agriculture producing 'the planetary diet' diet is based on LCA (life-cycle assessment) studies. Since LCA according to ISO14 040 and ISO 14 044 cannot capture the complexity of systems, where biology matters, 'the planetary diet' is decoupled from links to known limits of carrying capacity of agri-food systems, given their biophysical and socio-economic context. The internal logic of LCA studies and the lack of links to known characteristics of agricultural systems can lead to starvation. There is an urgent need to generate an operational definition of sustainability in agricultural systems that reflects the knowledge frontier in the disciplines that have the competence of excellence of the systems and issues in focus. The assessment methods to be chosen should be aligned with the sustainability objectives. This means systems analysis where the components are interconnected and capable to measure the sustainability performance of agricultural-food systems. We have developed a suggestion for a definition of operational sustainability in agriculture, and a supporting tool-kit for its evaluation, including a dynamic systems perspective.

The global roundtable for sustainable beef; multi-stakeholder engagement driving progress worldwide

R. Petre and J. Eisele
Global Roundtable for Sustainable Beef, 13570 Meadowgrass Drive, Colorado Springs, 80921, USA; ruaraidh.petre@grsbeef.org

Livestock Sustainability encompasses Environmental, Social and Economic realms and many factors impacting, people, animals and landscapes. Solutions to transform the livestock sector so that it can remain a positive contributor to a thriving food system need to take account of the geography, climate, legislative environment, production system and culture of those involved. Such solutions are therefore rarely simple or directly transferable from one location to another. In order for solutions to deliver sustainability benefits while contributing to a thriving food system, it is essential that all those involved in the livestock sector participate in their design and implementation. The Global Roundtable for Sustainable Beef, and its member National Roundtables are an example of multistakeholder collaboration at scale. These roundtables have brought together cattle producers, processors, retailers, allied industries and input suppliers, civil society and academia along with government and international organisation observers to define beef sustainability firstly at a global level, and then to add national context in 24 countries covered by 12 national / regional groups. Together GRSB members defined Principles and Criteria for Sustainable Beef covering Natural Resources, People & the Community, Animal Health & Welfare, Food, and Efficiency & Innovation. In 2021 GRSB released ambitious 2030 goals for Climate, Nature Positive Production and Animal Health and Welfare, and in the course of 2023 will set an additional Social goal. GRSB develops materials to assist in consistent measurement and reporting of progress. In 2022 GRSB released its Beef Carbon Footprint Guideline that allows sector wide alignment in the calculation of carbon footprint for the beef cattle lifecycle. GRSB is currently working with stakeholders to develop the reporting framework required to present progress against global goals on climate, nature positive production and animal welfare. The global, supply chain wide approach combined with national level, context specific implementation provides GRSB and its members with the most consistent, joined up and effective means available to meet sustainability challenges in the beef industry.

Developing a decision support system for integral sustainability improvement of Dutch dairy farm

M. De Vries[1], J.W. Reijs[2], G.J. Doornewaard[2], A.P. Bos[1] and C.A. Schep[1]
[1]Wageningen Livestock Research, De Elst 1, 6708 WD Wageningen, the Netherlands, [2]Wageningen Economic Research, Prinses Beatrixlaan 582-528, 2595 BM Den Haag, the Netherlands; marion.devries@wur.nl

Dairy production systems are causing severe pressure on diverse sustainability aspects, such as global warming, eutrophication, acidification, biodiversity loss, animal welfare, and food-feed competition. Choosing effective measures to improve integral sustainability performance, however, is difficult for farmers, given the complexity and wide range of potential effects on environmental, economic and social sustainability aspects. Moreover, whether measures can be implemented on farms depends on its specific biophysical and socio-economic context, and farmers' preferences. The aim of this study was to design a decision support system that: (1) assists Dutch dairy farmers in choosing measures suitable for their specific farming system and personal preferences; and (2) gives them insight in potential effects of these measures on environmental, economic and social sustainability. First, a database was developed containing 130 measures potentially contributing to reducing greenhouse gas emissions or improving biodiversity. The measures were assessed by 17 scientific experts and 2 dairy advisors for: (1) suitability for implementation given certain farm characteristics and farmer preferences; and (2) effects on 56 indicators of economic, ecological and social sustainability. Effects on sustainability indicators were scored on a 1-5 Likert scale ranging from very unfavourable to very favourable. An explanation of the effect was added, and, in case of a variable effect, reasons why the effect could vary. Second, a prototype tool was designed allowing farmers and advisors to access information about measures in the database, including functionalities of (automated) user input and weighting of sustainability themes; shortlisting and sorting user-specific measures; and information about sustainability effects. It was concluded that the decision support system designed in this study is a promising method to support farmers and advisors in orienting on measures for integral sustainability improvement of dairy farms. Further usability testing is needed for development and successful implementation of the tool.

Animal welfare index in sustainability assessments

L. Rydhmer[1] and E. Röös[2]
[1]Swedish University of Agricultural Sciences, Department of Animal Breeding and Genetics, Box 7023, 75007 Uppsala, Sweden, [2]Swedish University of Agricultural Sciences, Department of Energy and Technology, Box 7032, 75007 Uppsala, Sweden; lotta.rydhmer@slu.se

Tools for evaluation of sustainability of food consumption are needed in several research projects and development programs. Since animal-sourced food is included in most people's diets, animal welfare aspects should be part of sustainability assessments. Preferably such assessments are based on observations from farm visits in different production systems. If the aim is to evaluate total consumption of all consumers in a country or region, farm visits are often not feasible for practical reasons. Instead, a rough assessment can be performed based on data available in national data bases, reports of various agricultural organisations and authorities and scientific studies. We propose an animal welfare index based on: (1) number of animals needed to produce a given amount of food (e.g. 1 kg meat, milk or egg); (2) animal welfare judgement of the most common production systems; (3) animals' ability to perceive how they are managed. The number of animals takes all involved animals into account, including e.g. male chickens in egg production. One kg food from small animals, such as crickets, involves many more animal lives than one kg meat from e.g. slaughter pigs. The animal welfare judgement of systems is based on key figures of mortality, disease frequency, barren or enriched environment and duration of slaughter process. Pasture is for example regarded as an enriched environment. Wild animals, such as the moose and the shrimp, also live in an 'enriched' environment. We have used a questionnaire answered by researchers in animal science and veterinary medicine to set values for different animal species' relative ability to perceive how they are managed (including the ability to perceive what kind of environment they are kept in). Mammals have higher ability than e.g. blue mussels. Today's typical diets and alternative diets can be compared with the animal welfare index. By including the animal welfare index in broad sustainability assessments, goal conflict between animal welfare and other sustainability aspects can be revealed.

Sustainability analysis of large-scale dairy farming in China from water-energy-labour nexus insight

X. Du, Q. Wang and Z. Shi
China Agricultural University, College of Water Resources and Civil Engineering, No.17, Tsinghua East Road, Haidian District, 100083 Beijing, China, P.R.; xinyidu@cau.edu.cn

With the scale development of dairy farming in China, the cost of water, energy, and labour increased significantly, which are closely related to management and technique. Meanwhile, due to regional characteristics and policy bias, the economic and ecological efficiency varies greatly among provinces. To promote the sustainability of the sector, this study estimates the direct energy and water consumption, and labour use through on-farm production. Based on slack-based measurement data envelopment analysis, the dynamic efficiency of 11 inputs and 6 outputs was calculated from 2012 to 2020. In 2020, the cattle stock bred in large-scale farms (500 heads and above) reached 5.91 million heads, which is 2.41 times the stock in 2011. On the national level, the cost per cow of water, energy, and labour increased by 30, 59 and 106%, respectively. Correspondingly, the energy footprint increased by 22%, the related carbon emission increased by 57%, whereas the water footprint decreased by 12%. As great differences exist among provinces, the development of management practices is unbalanced and inadequate, which is incompatible with the rapid amplification of farm scale. Overall, the comprehensive efficiency of large-scale farms reached its lowest point in 2015 due to poor management, and the circumstance has been significantly improved. In 2020, Jiangsu is the only one that did not reach the production frontier while other provinces are able to allocate resources reasonably, and the number is 8 in 2015. Excessive consumption of coal and diesel is the most serious problem for Jiangsu to deal with, and the utilization of water and electricity is also expected to improve. As environmental requirements become increasingly strict, more efforts should be made to develop a high level of management adapted to regional characteristics, which is expected to present the due advantage of resource utilization in large-scale farms, as well as promote the economic and ecological performance of dairy farming.

Characterizing the socio-economic and environmental performance of Irish beef and sheep farms

M.C. Ayala[1,2], J.C.J. Groot[3], K. Kilcline[4], I.J.M. De Boer[2], C. Grace[1], J. Kennedy[1], B. Moran[4] and R. Ripoll-Bosch[2]
[1]Devenish Nutrition, Dowth hall, A92T2T7, Co. Meath, Ireland, [2]Wageningen University, Animal Production Systems group, DeElst 1, 6708 WD, the Netherlands, [3]Wageningen University, Farming Systems Ecology group, Droevendaalsesteeg 1, 6708 PB, the Netherlands, [4]Teagasc, Rural Economy & Development Center, 3 Cross Street, H65WV00, Co. Galway, Ireland; cecilia.ayala@wur.nl

Although the beef and sheep sectors are important contributors to the Irish economy, at farm level, socioeconomic viability is threatened by low profitability and low rates of generational renewal. In addition, these sectors incur negative environmental externalities (i.e. greenhouse gas – GHG- emissions, air and water pollution, nature loss) that must be addressed for Ireland to reach its environmental targets. However, identifying interventions to manage these socioeconomic and environmental issues is challenging due to the high heterogeneity of farming systems and contexts present in these sectors. We developed a typology of the Irish beef and sheep sectors to unravel their heterogeneity, better understand the socioeconomic and environmental performance of specific farm types and help identify which interventions would be suitable for them. Our preliminary results showed 6 distinct farm types, ranging from extensive cattle-rearing or sheep farms with low profitability and overall low negative environmental impact, to more intensive cattle-finishing farms with better socioeconomic profiles but higher negative environmental impact. For instance, farms in these last groups account for 60% of the total GHG emissions in the sample, which means that any intervention aiming at reducing GHG emissions will have a much larger impact if tailored to these specific farm types. Meanwhile, if the aim is to increase the economic viability of the sector, interventions should be tailored to the first groups of farms which account for 60% of the non-viable farms in the sample. Ultimately, by identifying the varying farm types and their weight in the sector and their impact as a whole, we hope to be able to help further the debate around proper targeted interventions, which could improve the beef and sheep sectors' viability while supporting Ireland in achieving its environmental targets.

Session 14 Theatre 12

Benchmarking the sustainability performance of pastoral dairy-beef production systems

M. Kearney[1,2], J. Breen[1], E. O` Riordan[2] and P. Crosson[2]
[1]University College Dublin, School of Agriculture and Food Science, University College Dublin, Belfield, Dublin 4, Ireland, D04V1W8, Ireland, [2]Teagasc, Animal & Grassland Research and Innovation Centre, Teagasc, Grange, Dunsany, Co. Meath, Ireland, c15pw93, Ireland; mark.kearney@teagasc.ie

Sustainable beef production involves environmental stewardship and farm-level economic viability. It has emerged as an important global policy issue as demand for animal-sourced food increases due to population growth and growing affluence in developing countries. The abolition of the EU milk quota system and subsequent expansion of the dairy herd has increased the proportion of beef derived from dairy dams. Consequently, there is increasing interest in the performance of dairy-beef production systems particularly with respect to farm profitability and greenhouse gas (GHG) emissions. Therefore, the objective of this paper was to identify key economic and GHG performance indicators for dairy-beef systems. The Grange Dairy Beef Systems Model (GDBSM) was used to model three categories of dairy-beef farms representing Irish 'national average' farms (AVE), farms participating in a farm improvement program (IMP), and a research farm system (RES) in order to identify and quantify key performance metrics underpinning differences in farm performance levels. Key sustainability performance indicators identified in this study include live weight and carcass output per ha, age of slaughter, animal live weight performance, inorganic N application rates and proportion of grazed pasture in the animals' feed budget. It is evident from the current study that dairy-beef systems are multifaceted and that there is no one stand-alone key performance indictor that exists to achieve sustainability.

Session 14 Theatre 13

Measuring, reporting and verifying farm sustainability (Food Futures)

R. McGuire[1], A. Aubry[2], S. Morrison[2], J. Birnie[1] and N. Scollan[1]
[1]Queen's University Belfast, Institute for Global Food Security, 19 Chlorine Gardens, BT9 5DL, Belfast, United Kingdom, BT9 5DL, United Kingdom, [2]2Agri-Food and Biosciences Institute, Livestock Production Sciences, Large Park, Hillsborough, United Kingdom, BT366DR, United Kingdom; r.mcguire@qub.ac.uk

Grassland agriculture requires a data-driven tool measuring, reporting and verifying whole-farm sustainability. This 'nature positive' tool must be scientifically informed, adapt to and satisfy emerging policy and enable / stimulate industry 'buy-in'. Co-created with industry, Food Futures (FF) has developed a holistic metric profile (i.e. carbon footprint, profit, well-being, etc.) which is substantiated by sustainability algorithms – functioning as a tool (dashboard) that measures sustainability and supports on-farm decision making. FF integrates digital technology (i.e. geographical positioning systems, light detection and ranging and life cycle analysis) to stream data from multiple sources to improve soil health, water quality, carbon sequestration, precision nutrient application and on-farm habitat management. FF has successfully completed a commercial pilot study with the Livestock Meat Commission Northern Ireland (06/06/22 – 19/08/22). These farms included 21 dairy, 78 beef, 23 sheep and 40 mixed farm enterprise. Additionally, 28 were designated as disadvantaged, 48 Less Favoured Area (LFA), 65 lowland and 20 severely disadvantaged. Initial results indicate that (1) 40% of farmers use low-emission slurry spreading equipment (LESSE), (2) 48% of farms had completed full-farm soil sampling analysis within the last 5 years and (3) only 96% store slurry and manures in a secure and environmentally friendly methods. In addition, the sustainability performance of these farms has been measured, with each farm receiving a targeted two-way knowledge-exchange output (dashboard) including tailored recommendations for improvement. This digital and targeted approach allows farmers to (1) measure their performance and operate within the parameters of sustainability, (2) delivers continuous environmental and climate 'smart' feedback to drive behavioural change, (3) collates evidence of production footprint for national and global markets and (4) contributes to net zero by 2050.

Session 14 — Theatre 14

Closing of the 3rd ATF-EAAP LFS symposium

F. O'Mara[1] and M. Lee[2]
[1]Animal Task Force, 149 rue de Bercy, 75012 Paris, France, [2]Harper Adams University, Newport, Shropshire, TF10 8NB, United Kingdom; mail@animaltaskforce.eu

Closing of the symposium by the co-chairs.

Session 14 — Poster 15

Environmental performance by type of Iberian farm in the Dehesa ecosystem

J. García-Gudiño[1], J. Perea[2], E. Angón[2] and I. Blanco-Penedo[3]
[1]CICYTEX, Animal Production, Finca La Orden, 06187 Guadajira, Spain, [2]University of Córdoba, Animal Production, Campus Universitario de Rabanales, 14071 Córdoba, Spain, [3]University of Lleida, Department of Animal Science, Av. de l'Alcalde Rovira Roure,191, 25198 Lleida, Spain; isabel.blanco.penedo@slu.se

Traditional Iberian pig production is developed in the Dehesa ecosystem, where the Iberian pigs are linked to the sustainable use of natural resources. Currently, Iberian pig production is diversified in the Dehesa ecosystem. As a result, the ecosystem of the Dehesa is threatened by overexploitation, and it is necessary to adopt measures to make traditional Iberian pig production more sustainable, in line with European policies focusing on sustainable food production. The aim of this paper was to identify different typologies of traditional Iberian farms, based on their economic and environmental performance, with the aim of increasing the sustainability of these farms. Sixty-eight Iberian pig farms were evaluated using multivariate statistical tools in order to establish Iberian farm typologies. Factor analysis revealed three factor components related to management, productivity and area yield that characterise Iberian pig farms. Two groups of Iberian farms were identified: Multiple orientation and Montanera orientation. From the analyses carried out, the relationship between the characteristics of the Iberian farm typologies and the economic and environmental performance was determined. The environmental performance of traditional Iberian pig production was analysed for both Iberian farm types as a result of cluster analysis. Overall, montanera farms obtained better environmental values and economic benefits per environmental unit produced than multi-orientation farms. Consequently, the analysis of the different farm types in Iberian traditional pig production can be used to generate best practice guidelines for a more sustainable Iberian pig production from an economic and environmental point of view.

Session 14 Poster 16

Ration protein content affects intake, production, efficiency and methane emission of dairy cows

F. Schori, C. Kasper and A. Münger
Agroscope, Ruminant Nutrition and Emissions, Animal GenoPhenomics, Tioleyre 4, 1725 Posieux, Switzerland; fredy.schori@agroscope.admin.ch

The main N imports in a milk production system are from feedstuffs, fertilizer and atmospheric deposition. Improvements in management and adaptation of the diet composition have a great potential to improve N use efficiency of lactating cows. The goal of the study was to investigate the effects of protein reduced diets on feed intake, milk production, efficiency and methane emission of dairy cows. The trial lasted three weeks, two weeks for adaptation and one measurement week. A total of 30 Holstein cows were allocated based on their lactation number, milk yield before the trial, lactation stage and body weight. Cows in the control group (CONTROL) were offered a total mixed ration (TMR) that was balanced in terms of absorbable protein in the intestine (based on fermentable energy and rumen available nitrogen) and of net energy for lactation (NEL). The TMR of the N reduced group (REDN) was formulated to cover the requirements for crude protein (CP) and NEL. The two TMR, consisting of maize silage, hay, grass silage, two concentrates (energy and protein-rich in different proportions) and minerals, were iso-energetic with 6.1 MJ NEL/kg dry matter (DM). Only the CP content of the TMR was different, CONTROL 152 g/kg DM and REDN 136 g/kg DM. CONTROL cows ate more (23.4 vs 21.8 kg DM, P<0.001) and produced more energy corrected milk (35.1 vs 32.3 kg, P=0.01). Feed conversion rate (0.67 vs 0.68, P=0.67) and residual feed intake (-0.72 vs -0.89 kg, P=0.82) did not differ between the treatments. In contrast, milk urea content (19.7 vs 13.7 mg/dl, P<0.001), residual nitrogen intake (74.9 vs 13.3 g N, P<0.001) and daily nitrogen excretion via urine (170 vs 119 g, P<0.001) were increased in CONTROL cows, indicating more N losses compared to the REDN cows. At the same time, no differences were seen in the daily methane production between the treatments (481 vs 474 g/d, P=0.71), but the CONTROL cows emitted less methane per energy corrected milk (13.7 vs 14.8 g/kg, P=0.02) and per total DM intake (20.5 vs 21.8 g/kg, P=0.03) compared to the REDN cows. The impacts of measures to reduce N losses in milk production systems should be assessed in the broadest sense.

Session 14 Poster 17

Is protein autonomy synonymous with economic efficiency?

E. Bertrand, C. Corre and R. Bore
Institut de l'Elevage, Production laitière, 149, rue de Bercy, 75012 Paris, France; eric.bertrand@idele.fr

The economic performance of Cap Protéines dairy pilot farms was assessed through the calculation of production costs. The COUPROD tool provides a synthetic indicator through the feed cost (corresponding to the cost of purchased feed plus the cost of supplying the land used by the herd) to assess the link between protein autonomy and economy. Analysis of economic data from pilot and reference farms logically shows a relationship between the level of protein autonomy and the cost of feeding. The cost of feed decreases with the increase in protein autonomy and the reduction of feed purchases. In dairy cattle, for example, the feed cost decreases by 48% when the protein autonomy changes from 50 to 90%, going from 158 € to 76 €/1000 l of milk. There is statistically no change in the cost of soil inputs in relation to the level of protein autonomy of the livestock. Regarding the direct expenses, the savings from lower feed purchases exceed the increase in inputs related to GFS or self-consumed crops. Gaining protein autonomy therefore helps to reduce the feed cost of the herd. This relationship is not maintained at the level of the cost of the feeding system of the dairy workshop, whose calculation integrates the indirect loads and structures related to the areas used to feed the herd (mechanization and land). On the other hand, its composition differs with the improvement of protein autonomy: there is a deferral of purchase costs in mechanization costs (harvest and fodder crops, cereals and protein crops own consumption), and working land. In the end, the cost of the feeding system is neither reduced nor increased with the search for autonomy. This is true in all ruminant sectors. The linear correlation between protein autonomy and the economic efficiency indicator (EBE/PB) is low (R^2+0.198) but positive and significant. It is not related to the level of technical productivity or the size of the dairy workshop. Gaining protein autonomy improves the economic efficiency of the production system through mainly load saving.

Development of multicriteria analysis' methodology in an experimental farms network

C. Corre[1] *and C. Brasseur*[2]

[1]*Institut de l'Elevage, 8 route de Monvoisin, 35650 Le Rheu, France,* [2]*Institut de l'Elevage, Boulevard des Arcades, 87000 Limoges, France; clemence.corre@idele.fr*

Today, environmental, economic, and societal context is complex. To pilot resilient livestock systems, technical indicators are no longer sufficient. Other indicators need to be calculated and analysed periodically to take all matters into account. The aim of this study was to develop a multicriteria analysis in an experimental farms' network, in all ruminants' productions (dairy cows, dairy sheep, dairy goats, suckler cows and suckler sheep). Twelve experimental farms were involved. These farms combine production of milk and meat, experimentation for research in real conditions and sometimes instruction for professionals or future ones. About 150 indicators were selected to describe each farm and assess their annual multiperformance, that means their economic, environmental, and social/societal performances. These indicators were calculated by existing tools and related to technical and economical results, feed self-sufficiency, environmental impacts and benefits, competition between feed and food, and animal welfare. This multicriteria analysis focused on the production system of farms. Some economic products and charges linked to experimentation or instruction were not considered. After validation, this analysis was rolled out in the twelve farms. For the first year, analysis was focus on 2020 or 2021. On these years, protein self-sufficiency varied from 68 to 95% in dairy cows' farms and from 76 to 100% in suckler cows' farms. In dairy and suckler sheep farms and dairy goats' farm protein self-sufficiency was respectively 77, 90 and 22%. Carbon footprint, calculated with CAP2ER®, was also various: from 4,050 to 7,410 kg eq. CO_2/ha UAA, depending on farms. This analysis will be repeat every year to analyse indicators' evolution on farms and to pilot choices.

Multicriteria analysis on French experimental farm which develops it feed self-sufficiency

C. Corre[1] *and R. Boré*[2]

[1]*Institut de l'Elevage, 8 route de Monvoisin, 35650 Le Rheu, France,* [2]*Institut de l'Elevage, 42 rue Georges Morel, 49070 Beaucouzé, France; clemence.corre@idele.fr*

Experimental dairy cows farm of Poisy is located in the lowlands of the Alps mountains in eastern France. A multicriteria analysis was made on this farm to assess its technic, economic, environmental, and societal performances, including feed self-sufficiency. More than 150 indicators were calculated with existing tools and some of them were compared to a group of similar farms. In 2021, global feed self-sufficiency was 83% and protein self-sufficiency was 74%, while group's average was respectively 79 and 68%. Many strategies are implemented to improve feed self-sufficiency in a context of climate change: diversification of grasslands composition, growing plant species better adapted to heat and drought like fodder beet and alfalfa, production and use of raw soybean for animal feeding. The carbon footprint calculated with CAP2ER® method was quite high: 0.92 kg eq. CO_2/l of milk produced, because of heifers' breeding for other farmers that increases methane emissions. The farm can feed 19 persons / ha UAA and 93% of proteins fed to cattle are non-digestible by human ! Indeed, 80% of the dry matter fed annually to animals is composed of grass and rapeseed cake is used instead of soybean cake. This analysis enables to have a global vision on the farm system regarding economic, environmental, and societal issues. Strengths and weaknesses were identified and some measures could be implemented to improve some points.

Session 14 — Poster 20

Partial indicators monitoring in dairy farm: effect of feed efficiency improvement on gross margin

G.S.S. Gian Simone Sechi and A.A.S. Alberto Stanislao Atzori
University of Sassari, Department of Agricultural Science, Viale Italia, 39, Sassari, 07100, Italy; giansimone.sechi@iusspavia.it

The use of partial indicators helps to improve technical and economic efficiency of dairy farms. The aim of this work was to propose a practical approach to monitor farm costs and incomes and related indicators of profitability. A methodological framework to calculate farm gross margin was developed and applied to a dairy cattle farm with 227 milking cows from Arborea (Or, Italy). Farm data were gathered at daily scale and elaborated twice per month. Replacement and dry cattle, health and reproduction cost, energy and other cost that included: fuel, electricity and other materials, were monitored separately gathered through the management software Ecostalla®. Monitored variables included: diet formulas, dry matter (DM) and costs of the produced and purchased feeds, milk delivered and price, meat and cattle sold (culled cows, live animals). The farmer was asked daily to input in an excel spreadsheet feed and milk delivered, milking cows, feed orts for lactation groups and then consistency, feed delivered and weekly the unifeed DM. Partial indicators were calculated for feed efficiency as mik/feed (FE), cost in €/kg of DM, IOFC, gross margin (revenues – considered costs). All the calculations were referred to the year 2022, the average milk price and feed cost of lactating cows were 0.465 €/l and 0.39 €/kg of DM, respectively. The average milk production and DMI per cow was 33 l/d and 25.5 kg/d respectively, with a FE of 1.29. Pooling daily data over the whole year a variation of ±0.1 in FE corresponded to ±1.01 €/head daily change in IOFC. At the end of the year the feeding cost were equal to 817,798 € for lactating cows, 229,354 € for heifers, dry cows and calves, health and reproduction cost were 60,852 € and energy and other cost were 212,132 €. The final gross margin was 18,461 €, ranging from a max of 14,282 € in March and a min of – 15,501 € in September. Improving FE from 1.29 to 1.43, thus reaching 26.5 kg/d of DMI and 38 l/d of milk, supposing increased dietary costs of 0.02 €/kg of DM and keeping constant other costs the gross margin would have been 128,707 €. This approach can be useful to stimulate the farm team of technicians and operators to focus on farm data and their economic implications.

Session 14 — Poster 21

Comparison of methodologies for estimating animal emissions from smallholder cattle systems

E. Balcha[1] and C. Arndt[2]
[1]Mekelle University, Mekelle, Ethiopia, [2]ILRI, Naivasha Rd, Nairobi, Kenya; claudia.arndt@cgiar.org

The objective of this study was to compare gross energy intake (GEI; MJ/animal/day) and enteric methane (CH_4) emission factors (EF; CH_4 kg/animal/year) of different cattle categories from smallholder systems in Ethiopia based on Intergovernmental Panel on Climate Change (IPCC) Tier 1, IPCC Tier 2, and International Livestock Research Institute (ILRI) Tier 2 methodology. The ILRI Tier 2 methodology uses calculations that are based on or modified from equations published in 'Nutrient Requirements of Domesticated Ruminants' (CSIRO 2007). In comparison to IPCC, the ILRI methodology is based on metabolizable and not net energy requirements of the animal. Data from Ethiopian smallholder systems were collected four times corresponding to the beginning and end of the three seasons (spring, summer and winter) to account for the effect of seasonality on animal liveweight, diet and performance. There was a high correlation of GEI between IPCC and ILRI Tier 2 methodology (R^2=0.87). However, the IPCC Tier 2 methodology estimated a higher GEI than the ILRI methodology (P<0.05; based on one-sample t-test). The mean difference in GEI was 28 MJ/animal/day (s.d. 18.2), which was 27 and 38% of the average GEI estimated by IPCC and ILRI methodology, respectively. Compared to ILRI methodology, IPCC methodology used a greater methane conversion factor (Y_m; 7.0 vs 6.3% of GE in feed converted to CH_4). Because of the greater GEI and Y_m, the IPCC methodology estimated on average 49% greater EFs than the ILRI methodology (48.4 vs 32.4 CH_4/animal/day, P<0.05). The EFs calculated across all cattle categories (adult females, intact and castrated males, heifers, young makes, and calves) for IPCC Tier 1 (2019) default EFs were found to be 23-38 and 38-59% lower for the IPCC and ILRI Tier 2 methodology, respectively. Similarly, when compared to Tier 1 methodology, IPCC and ILRI Tier 2 methodology estimated 27 and 51% lower total animal emission from smallholder cattle systems, respectively. The observed difference between the methodologies was based on the differences in predicted feed intakes and Ym. Hence, the predictions by Tier 1 and 2 methodologies need to be compared to *in vivo* intake and CH_4 measurements to determine which methodology predicts GEI and Y_m more accurately.

Developing a common framework to measure sustainability in food systems with limited arable land

P. Bhérer-Breton[1], A. Woodhouse[2], E. Svanes[2], H.F. Olsen[1] and B. Aspeholen Åby[1]
[1]Norwegian University of Life Sciences (NMBU), Elizabeth Stephansens v. 15, 1430 Ås, Norway, [2]Norwegian Institute for Sustainable Research (NORSUS), Stadion 4, 1671 Kråkerøy, Norway; paule.bherer-breton@nmbu.no

Food and food systems are important when finding solutions for the future sustainability challenges and to make progress for the sustainable development goals. To enable sustainable food transformations, frameworks that measure food system sustainability are a crucial tool. The Farm to Fork initiative from 2020 describes proposals for front-of-package labelling for both nutrition and sustainability to empower consumers to make knowledge-based decisions. Labelling systems for nutrition is already implemented in several countries, for instance the Nutri-Score. Whereas for sustainability, there is no common labelling framework yet covering several dimensions of sustainability, and themes and sub-themes within these. Creating a common framework for food labelling could represent a threat to local food security and national food sovereignty, if its' conceptualization is not transparent or do not consider the features of all countries within a region (e.g. Europe), and as such prevent sustainable resource utilization. The aim of this work is to identify the gaps between existing frameworks on environmental and social sustainability, and what is needed to foster sustainable food systems. Relevance and feasibility of those frameworks are discussed the case of Norway, a country where the specificities of local ecosystem are unique due to the country's topography, several climatic zones and limited arable land. This study is part of the project NewTools (Research Council of Norway) where the aim is to contribute to food system change through developing frameworks for food labelling focusing public health and environmental and social sustainability.

Partial indicators monitoring in dairy farms: feed efficiency implications on gross margin

G.S. Sechi and A.S. Atzori
University of Sassari, Department of Agricultural Sciences, via E De Nicola 9, 07100 Sassari, Italy; gia.sechi40@gmail.com

The aim of this work was to propose a practical approach to monitor farm costs and incomes and related indicators of profitability. A methodological framework to calculate farm gross margin was developed and applied to a dairy cattle farm with 227 milking cows from Arborea (Or, Italy). Farm data were gathered at daily scale and elaborated twice per month. Replacement and dry cattle, other costs (health and reproduction cost, energy and other cost that included: fuel, electricity and other materials were monitored separately. Considered variables included: diet formulas, dry matter (DM) and costs of the produced and purchased feeds, milk delivered and price, meat and cattle sold (culled cows, live animals). Data were gathered through the management software Ecostalla®. The farmer was asked daily to input in an excel spreadsheet feed and milk delivered, milking cows, feed orts for lactation groups and then consistency, feed delivered and weekly the unifeed DM, for all the groups. Partial indicators were calculated for feed efficiency as mik/feed (FE), cost in €/kg of DM, IOFC, gross margin (revenues – considered costs). All the calculations were referred to the year 2022, the average milk price and feed cost of lactating cows were 0.465 €/l and 0.39 €/kg of DM, respectively. The average milk production and DMI per cow was 33 l/d and 25.5 kg/d respectively, with a FE of 1.29. Pooling daily data over the whole year a variation of ±0.1 in FE corresponded to ±1.01 €/head daily change in IOFC. At the end of the year the feeding cost were equal to 817,798 € for lactating cows, 229,354 € for heifers, dry cows and calves, health and reproduction cost were 60,852 € and energy and other cost were 212,132 €. The final gross margin was 18,461 €, ranging from a max of 14,282 € in March and a min of – 15,501 € in September. Improving FE from 1.29 to 1.43, thus reaching 26.5 kg/d of DMI and 38 l/d of milk, supposing increased dietary costs of 0.02 €/kg of DM and keeping constant other costs the gross margin will increase from 110,246 € to 128,707 € in respect the current situation. This approach can be useful to stimulate the farm team of technicians and operators to focus on farm data and economic implications of feed efficiency.

Session 15 Theatre 1

Genomic evaluation to reduce enteric methane emissions in Holstein cattle

F. Malchiodi[1,2], H.R. Oliveira[3,4], S. Narayana[4], A. Fleming[4], H. Sweett[4], S. Shadpour[2], J. Jamrozik[4], G. Kistemaker[4], P. Sullivan[4], F.S. Schenkel[2], B. Van Doormaal[4], C.F. Baes[2,5] and F. Miglior[2,4]
[1]Semex, 5653 Hwy 6, N1H 6J2 Guelph, Canada, [2]University of Guelph, Centre for Genetic Improvement of Livestock, Department of Animal Biosciences, N1G 2W1 Guelph, Canada, [3]Purdue University, Department of Animal Sciences, 47907 West Lafayette, IN, USA, [4]Lactanet Canada, 660 Speedvale Avenue West, N1K 1E5 Guelph, Canada, [5]University of Bern, Institute of Genetics, Department of Clinical Research and Veterinary Public Health, 3001 Bern, Switzerland; fmalchiodi@semex.com

Together with other major international dairy organizations, Dairy Farmers of Canada has pledged to reach net-zero greenhouse gas emissions from farm-level dairy production by 2050. Consequently, Canada has been building capacity to measure and/or predict enteric methane (CH_4) emissions for both herd monitoring and genetic tools. First, milk mid-infrared (MIR) records have been saved and stored since 2017 for 90% of milk recorded cows in Canada. Then, data collection of CH_4 emissions started in 2016 in two research herds using the GreenFeed System. This data was then used to predict CH_4 production for first parity Holstein cows using MIR spectra data between 120 and 185 DIM. This prediction serves as input for a new genomic evaluation system that has been launched in April 2023 by Lactanet Canada. Research based out of the University of Guelph using a machine learning algorithm has shown great accuracy of predicting individual animal methane emissions for milk-recorded cows using milk MIR spectral data through two research projects, the Efficient Dairy Genome Project and the Resilient Dairy Genome Project, and milk spectral data collected via our milk recording services. Predicted methane had a genetic correlation with collected methane of 0.85 and a heritability of 0.23 (0.01). Lactanet's genomic evaluation for Methane Efficiency was developed using a 4-trait single-step genomic evaluation for predicted methane for the Holstein breed, including milk, fat and protein yields as energy sinks, and defined as genetic Residual Methane Production independent of milk, Fat and Protein via a linear regression approach. Methane Efficiency is an important selection tool, allowing dairy producers to reduce methane emissions without negatively affecting production.

Session 15 Theatre 2

Methane phenotyping for genetic selection – can we measure bulls, not cows?

L.R. McNaughton[1], P. Beatson[2], G. Worth[1], D.J. Garrick[3], D. Garrick[3], R. Handcock[3] and R.J. Spelman[1]
[1]Livestock Improvement Corporation, 605 Ruakura Road, Hamilton, 3286, New Zealand, [2]CRV, 2 Melody Lane, Hamilton 3216, New Zealand, [3]The Helical Company Limited, 17 Appleby Rise, Whakatane 3120, New Zealand; lorna.mcnaughton@lic.co.nz

Most economically important traits relevant to dairy systems can only be measured on lactating females. Methane production and dry matter intake (DMI) are traits that can be measured on dairy bulls before semen collection begins. These traits were recorded on two farms in New Zealand. The purpose of the trial was to access the adequacy of the data for the estimation of genetic parameters. Some 486 Jersey (J), Holstein-Friesian (HF) or Jersey × Holstein-Friesian (Crossbred) bulls, aged 6-15 months, were housed for 35-day periods. Bulls were fed lucerne hay cubes, with up to 10% of the diet as pelleted concentrate bait feed via Greenfeed (GF) devices. Hokofarm Ric2Discover feed intake bins were used for DMI while methane production was measured using GF systems. Bulls were allowed up to 6 visits per day to the GF device. Liveweights were recorded thrice per week. Genetic analysis was performed from a Bayesian bivariate repeatability model fitted by Monte Carlo Markov chain sampling using Julia for Whole-genome Analysis Software (JWAS v1.1.1). A further 300 bulls are being tested in 2023. There were 13,109 daily methane emission phenotypes and 12,687 daily DMI phenotypes from the initial 486 bulls. The methane emission phenotypes consisted of a daily methane yield-equivalent for each visit the bull made to the GF device. To obtain one daily value per bull, visits within a day were pooled to represent the total daily methane and total time in seconds a bull was measured for methane. Heritability estimates and 95% credibility intervals were obtained for methane of 0.1 (0.06, 0.15) and DMI 0.09 (0.05,0.15). Breeding values (BVs) for methane, adjusted for DMI were significantly ($P<0.05$) higher in Crossbred than J or F. Heterosis effects should be investigated. To validate the approach including the use of bulls for phenotyping, 350 daughters are being generated from the 25 highest and 25 lowest BV bulls to determine if the differences in methane observed in bulls are also observed in lactating dairy cattle.

Session 15 Theatre 3

Novel formulas to calculate methane production from concentrations while using sniffers

C.I.V. Manzanilla-Pech[1], M.H. Kjeldsen[2], T. Villumsen[1] and J. Lassen[1,3]
[1]Center for Quantitative Genetics and Genomics, Faculty of Science and Technology, Aarhus University, C. F Møllers allé 3, 8000 Aarhus C, Denmark, [2]Department of Animal Science and Veterinary Science at Aarhus University, Blichers Allé 20, 8830 Tjele, Denmark, Denmark, [3]Viking Genetics, Ebeltoftvej 16, Assenstoft, 8960 Randers, Denmark, Denmark; coralia.manzanilla@qgg.au.dk

Sniffers are widely used to measure methane emissions in dairy cattle. However, sniffers measure concentrations of methane (MeC) and carbon dioxide. To be able to compare animals across countries and other methane recording methods, it is needed to transform MeC to methane production in g/d (MeP) that is the gold standard trait. Madsen *et al.* formula has been widely used to calculate MeP in dairy cattle from MeC and carbon dioxide concentration. Though, this formula was developed based on a limited number of animals, not similar to modern dairy cows. Recently, Kjeldsen *et al.* have developed couple of formulas based on a data set with larger number of animals (n=1,502) in different countries (n=12) and production systems. The first formula uses ECM and BW as the Madsen *et al.* formula, however the coefficients have been re-estimated based on the new data set. The second formula has DMI and BW as main drivers of methane emissions, this formula could be convenient when DMI is available. However, these new formulas have never been tested for genetic purposes, only for nutritional purposes, where their correlations with respiration chamber and GreenFeed data is from 0.68 to 0.76. Thus, the general aim of this paper is to test these formulas in a large database, 24k methane records from 650 cows measured during 7 years in Danish Cattle Research Center, Aarhus University, Denmark. The specific objectives are: (1) to calculate MeP with these updated formulas; and (2) estimate genetic parameters including genetic correlations with the previous formula; (3) to calculate EBV correlation between the formulas. Preliminary results showed similar heritabilities (0.19-0.23) to the previous formula (0.21) and high genetic correlation (0.79-0.82) between the new formulas and the previous formula. The genetic correlation between these two new formulas was close to unity (0.99).

Session 15 Theatre 4

Predicting methane production and intensity from milk mid-infrared spectra

S. Fresco[1,2], A. Vanlierde[3], R. Lefebvre[1], M. Gaborit[4], D. Boichard[1], R. Bore[5], S. Fritz[1,2], N. Gengler[6] and P. Martin[1]
[1]Université Paris-Saclay, INRAE, AgroParisTech, GABI, Domaine de Vilvert, 78350 Jouy-en-Josas, France, [2]Eliance, 149 Rue de Bercy, 75012 Paris, France, [3]Walloon Agricultural Research Centre, Valorization of Agricultural Products, 9 Rue de Liroux, 5030 Gembloux, Belgium, [4]INRAE, UE326 Domaine Expérimental du Pin, l'Ermite, 61310 Exmes, France, [5]Institut de l'Élevage, 149 Rue de Bercy, 75012 Paris, France, [6]ULiège – GxABT, 2 Passage des déportés, 5030 Gembloux, Belgium; solene.fresco@eliance.fr

Selecting dairy cows against methane (CH_4) emissions is one of the solutions to reduce greenhouse gas emissions. However, as large-scale phenotyping is challenging, proxies have been developed to supply the amount of data necessary to genetic analyses. One of them is predicting CH_4 emissions from milk mid-infrared spectra. This study aimed at comparing prediction equations developed from two reference CH_4 for two CH_4 traits: MeP (g/d) and MeI (g/kg of fat- and protein-corrected milk [FCPM]). Methane emissions were recorded using GreenFeed devices, for 278 cows from nine different experiments. These reference CH_4 emissions were averaged over 1 or 2 weeks and were associated to 1,035 and 680, respectively. Equations were derived using Partial Least Square regression. A second dataset including 104 spectra and 1-week or 2-week CH_4 averages from 46 cows was used as an external validation to assess the performance of each equation based on R^2 and % of RMSE/observations mean (E). Performance was similar for both reference CH_4 time periods. Equations calibrated on 2-week CH_4 averages tended to have lower R^2 and E than the one on 1-week CH_4 averages. The average R^2 (E) were 0.28 (17.7) and 0.43 (17.3) for MeP and MeI, respectively. Indirectly predicting MeP by multiplying predicted MeI by FPCM was found to be as accurate as MeI and more accurate than predicting directly MeP. As selection of MeI presents two disadvantages – lower efficiency when selecting a ratio than the two traits separately and a potential increase of milk production at a given MeP, which would be detrimental to high producing cows' fertility and health – using CH_4 predicted from MeI multiplied by FPCM can be of great interest in genetic selection.

Session 15 Theatre 5

Genetic parameters for enteric methane emissions of dairy cows using random regression models

A.E. Van Breukelen[1], M.N. Aldridge[1], R.F. Veerkamp[1] and Y. De Haas[2]
[1]Wageningen University & Research, Animal Breeding and Genomics, P.O. Box 338, 6700 AH Wageningen, the Netherlands, [2]Wageningen University & Research, Animal Health & Welfare, P.O. Box 338, 6700 AH Wageningen, the Netherlands; anouk.vanbreukelen@wur.nl

To reduce the environmental impact of dairy farming, there is considerable interest in applying selective breeding to mitigate enteric methane emissions of cattle. To add methane to breeding programs, models need to be developed that accurately estimate breeding values for individual cows. The objective of this study was to apply a random regression model to long-term recorded methane emissions, which estimated genetic parameters across the lactation curve. Methane concentrations were measured with sniffers in milking robots. The data comprised of 20,768 weekly mean methane concentration measurements from 1,587 cows, with recording periods ranging from 1 to 17 months, on 14 commercial dairy farms in the Netherlands. Genetic parameters were estimated using a single-trait restricted maximum likelihood model in ASReml. The model included fixed effects for farm, year, week of measurement, parity and second order Legendre orthogonal polynomials for days in milk (DIM). In addition, the model included random regressions on the genetic and permanent environmental effect, using second order Legendre orthogonal polynomials, and a heterogeneous residual effect with four classes. The heritability was moderate at the start of the lactation (0.25±0.09, at 5 DIM), peaked in mid lactation (0.41±0.06) and decreased to a low heritability in late lactation (0.06±0.04, at 305 DIM). The repeatability was approaching one at the start of lactation (<10 DIM), and reduced steeply between 15 and 25 DIM, after which it stabilized but remained high at 0.74±0.02. Genetic correlations between the first 25 days of lactation and later lactation stages were moderate to high (ranging from 0.38±0.19 to 1±<0.01). The results confirmed that using a random regression model is more appropriate than using a model that would assume that genetic correlations between different DIM are one. A novel dataset is currently being recorded, which will increase recording to 100 dairy farms with around 15,000 cows. The results will be used to improve the accuracy of the estimations and to confirm the results presented here.

Session 15 Theatre 6

Heritability of methane emission in dairy cows estimated from GreenFeed measures in commercial herds

B. Heringstad[1,2], K.A. Bakke[1] and G. Difford[2]
[1]Geno Breeding and A.I. Association, Storhamargata, Hamar, Norway, [2]Norwegian University of Life Sciences, Department of Animal and Aquacultural Sciences, Faculty of Biosciences, Ås, Norway; bjorg.heringstad@nmbu.no

The aim of this study was to estimate heritability of methane emissions (CH_4) for Norwegian Red dairy cows. Measures of CH_4 from 15 GreenFeed units installed in commercial herds was available. We used data from 2020 and 2021 and the final dataset had a total of 370,642 records of GreenFeed visits from 814 Norwegian Red cows. The trait analysed was CH_4, gram per cow per day, computed as the average of the cow's individual visits each day. The mean (standard deviation) CH_4 was 422 (96) gram per cow per day. A linear animal repeatability model with fixed effects of parity and lactation week, and random effects of herd-testday, animal, and permanent environment was used to estimate variance components. The estimated heritability (standard error) was 0.34 (0.04) and repeatability 0.42. The predicted breeding values for cows with phenotype varied from -123 to 143, with standard errors between 22 and 34. Results so far are promising. The genetic variation for CH_4 in the Norwegian Red breed indicates that breeding for lower CH_4 is feasible.

Session 15 Theatre 7

Rapid measurements for breeding low methane cattle

S.J. Rowe[1], A. Searle[1], T. Bilton[1], S. Sevier[2], S. Gebbie[2], A. Jonker[3], P. Janssen[3], S. Hickey[4], N. Amyes[4], S. Worku[1] and J. McEwan[1]

[1]AgResearch Ltd, Animal Genomics, Invermay Agricultural Centre, 176 Puddle Alley, Mosgiel 9092, New Zealand, [2]AgResearch Ltd, Engineering Department, 1365 Springs Road, Lincoln 7674, New Zealand, [3]AgResearch Ltd, Grasslands Research Centre, Tennent Drive, Manawatu 4410, New Zealand, [4]AgResearch Ltd, Ruakura Agricultural Centre, 10 Bisley Road, Hamilton 3214, New Zealand; suzanne.rowe@agresearch.co.nz

Global warming attributed to methane emissions from livestock presents a significant global challenge to the agricultural sector. Although opportunities in intensive systems for feeding additives exist, the only proven tool for permanent mitigation at a national level, across sectors is breeding. The greatest challenge to incorporating methane emissions into breeding schemes is the difficulty of direct measures on selection candidates. A single measure of methane emissions from a portable accumulation chamber in sheep has been shown rank animals for their lifetime. These measures have successfully been employed to show a 2% reduction in methane emissions per year without affecting productivity. To extend this to cattle, a portable accumulation chamber was designed and built by engineers at AgResearch. To test the design, thirty Holstein-Friesian heifers at approximately 10 months of age and averaging a weight of 250 kg were evaluated. Heifers were fed *ad libitum* bailage for two weeks prior to the start of measures. Heifers were measured over a 5-day period. On the day of measure, a group of 6 heifers were removed from feed. After one-hour heifers were placed one at a time into the chamber for forty-five minutes. Methane (CH_4), Carbon-dioxide (CO_2) and Oxygen (O_2) measures were taken every 5 minutes using an Eagle II analyser. After two weeks the trial was repeated. Results were very similar on a liveweight basis to those seen in sheep per kg liveweight for CH_4 after a 40-minute measure. Importantly, no behavioural issues were observed while the animals were restrained in the chambers. We conclude that the portable accumulation chamber is potentially a suitable measurement technology for the lifetime ranking of cattle for enteric methane emissions for breeding schemes under any management regime. We will present protocols, behaviour observations, variances and phenotypic repeatabilities.

Session 15 Theatre 8

Collecting and preparing sniffer records for genetic evaluation of enteric methane in dairy cattle

T.M. Villumsen[1], P. Løvendahl[1], V. Milkevych[1], R. Krogh[1], C.I.V. Manzanilla-Pech[1], M. Bjerring[2], J. Lassen[1,3] and G. Sahana[1]

[1]Aarhus University, Center for Quantitative Genetics and Genomics, C.F. Møllers Allé 3, 8000 Aarhus C, Denmark, [2]Aarhus University, Animal and Veterinary Sciences, Blichers Allé 20, 8830 Tjele, Denmark, [3]VikingGenetics, Ebeltoftvej 16, 8960 Randers, Denmark; tmv@qgg.au.dk

The EU Green Deal has made the target of no net emissions of GHG by 2050. This requires reduction of GHG from the livestock sector and particularly from dairy. Genetic selection for low methane (CH_4) emitting dairy cows can be one of the important mitigation strategies. The methane sniffer typically installed in an automatic milking system (AMS) makes it possible to record individual CH_4 concentrations from many cows. However, to achieve a high-quality CH_4 phenotype for genetic evaluation, we need a sniffer setup with high operational reliability and an automated data handling pipeline for data quality control. In Denmark, we currently record gas emission from cows with 14 sniffers based on the Guardian NG methane sensor and an additional carbon dioxide (CO_2) sensor. Each sniffer has a multiplex setup with two channels which makes it possible to switch recordings between two AMS units. Our database currently holds CH_4 records of ~8.500 dairy cows from 22 farms. Each day we get ~1.1 mil. new data points from the sniffers currently installed in five farms with ~1,600 dairy cows. The sniffer setup requires daily monitoring of Wi-Fi connections, data transfer, and concentrations as well as technical assistance on farms when irregularities in data flow are observed. The subsequent handling pipeline includes several steps hereunder. (1) A check of time alignment of AMS data and gas data to pair cow and gas concentrations. For this, and quality control of CO_2 concentrations, we use a matched filter approach (doi.org/10.1016/j.compag.2022.107299). (2) Discard data with irregularities in CH_4, e.g. due to instrument breakdowns, duplicated and obvious bad data. (3) Correct for baseline gas concentrations due to diurnal changes and random drift of sniffers. (4) Overall pruning of early and late gas data during milking, to keep part of gas data where concentrations are the highest. These steps are to be integrated and automated for routine genetic evaluation.

Session 15 Theatre 9

Methods and tools for automated processing of sniffer-based methane emission data from dairy cows

V. Milkevych, R.K. Nielsen, T.M. Villumsen, P. Løvendahl and G. Sahana
Aarhus University, Center for Quantitative Genetics and Genomics, C.F. Møllers Allé 3, 8000 Aarhus C, Denmark; vimi@qgg.au.dk

Mitigating methane emission from livestock production is one of the crucial factors for sustainable agricultural food production. To ensure the demanded sustainability level of agricultural food production, a continuous monitoring and efficient implementation of diverse emissions mitigation strategies are required. Hence, a proper emission measurement technique is crucial, which fulfils three basic requirements: cheap and easy-to-operate; scalability and mobility of measurements; and high level of reliability of a measured data and resulting precise emission phenotypes. Overall, the sniffers are cheap, mobile, and relatively easy-to-operate technique. However, due to the intrinsic properties, mostly associated with technique's hardware and the related data acquisition set-up, the quality and reliability of measured gas emission data are of great concern. This leads to the significant difficulties in drawing reliable (statistical) estimates, such as methane phenotypes which, in fact, are highly important for establishing methane emissions mitigation strategies and animal breeding schemes. Hence, there are several major data-related sniffers-technique's issues: alignment (synchronization), reliability detection, and the great level of measurement (embedded) noise. Here we focus on the methods and software aimed to solve these issues. The methods are based on the general approach developed using the linear filtering theory. The algorithmic implementation of the developed approach allows fast and efficient automated data processing. Besides the methodology, we provide its complete software implementation. Our implementations were verified on the massive bulk of the gas emission data from the multiple commercial dairy farms in Denmark. These were analysed and discussed in terms of the methods accuracy and performance. Our findings support the conclusion that the proposed overall solution is robust, applicable to the problem of cattle gas emission data, and convenient for an automated large-scale data processing. The methods and tools are ready for routine phenotyping for use in both management and genetic evaluation to reduce emission from dairy cattle production.

Session 15 Theatre 10

Indicators of genetic selection using sniffer method to reduce methane emissions from Holstein cows

Y. Uemoto[1], T. Tomaru[2], M. Masuda[3], K. Uchisawa[3], K. Hashiba[3], Y. Nishikawa[4], K. Suzuki[4], T. Kojima[4], T. Suzuki[5] and F. Terada[6]
[1]Graduate School of Agricultural Science, Tohoku University, Sendai, Miyagi, 980-8572, Japan, [2]Gunma Prefectural Livestock Experiment Station, Maebashi, Gunma, 371-0103, Japan, [3]Niikappu station, NLBC, Hidaka, Hokkaido, 056-0141, Japan, [4]Head office, NLBC, Nishigo, Fukushima, 961-8061, Japan, [5]Institute of Livestock and Grassland Science, NARO, Nasushiobara, Tochigi, 329-2793, Japan, [6]Institute of Livestock and Grassland Science, NARO, Tsukuba, Ibaragi, 305-0901, Japan; yoshinobu.uemoto.e7@tohoku.ac.jp

Recently, there has been an increase in interest in the reduction of methane (CH_4), and the genetic selection of cows with low CH_4 emissions has drawn attention for sustainable livestock production. Thus, it is necessary to identify selection indicators that can be measured in a large number of samples at low cost in commercial farms. One of the strategies for measuring CH_4 at low cost is the sniffer method, in which the air near the animal's nostrils is sampled through a fixed tube in a feed trough in an automatic milking system (AMS). Here, our objective was to evaluate whether the CH_4 to carbon dioxide ratio (CH_4/CO_2) and methane-related traits obtained by the sniffer method can be used as indicators of genetic selection for lower CH_4 emissions in Holstein cows. First, we investigated the impact of the model with and without body weight (BW) on the lactation stage and parity for predicting methane-related traits using an on-farm dataset (Farm 1; 400 records for 74 Holstein cows). Second, we estimated the genetic parameters for CH_4/CO_2 and methane conversion factor (MCF) using a second on-farm dataset (Farm 2; 520 records for 182 Holstein cows). Farm 1 results revealed that MCF can be reliably evaluated during the lactation stage and parity, even when BW is excluded from the model. Furthermore, the CH_4/CO_2 and MCF could be evaluated consistently during the lactation stage and parity on both farms. Estimates for genetic parameters revealed low heritability for CH_4/CO_2 (0.12) and MCF (0.13) on Farm 2, but moderate repeatability for CH_4/CO_2 and MCF on both farms (ranging from 0.38 to 0.46). This study demonstrated the applicability of the sniffer method for the genetic selection of cows with low CH_4 emissions at the farm level.

Session 15 — Theatre 11

Reducing methane emissions: foundations for genetic evaluations for sustainable Irish beef cattle

C.V. Ryan[1,2], S. Conroy[2], T. Pabiou[2], D.C. Purfield[1] and R. Evans[2]
[1]Munster Technological University, Bishopstown, Co. Cork, T12 P928, Ireland, [2]Irish Cattle Breeding Federation, Ballincollig, Co. Cork, P31 D452, Ireland; cryan@icbf.com

Livestock production globally is a significant contributor to anthropogenic greenhouse gas emissions, particularly methane. Animal breeding provides a potential solution for reducing enteric methane emissions in a cost-effective, permanent, and cumulative manner. Before adding methane to a selection index, correlations with economically important traits already included in breeding goals must be evaluated. This study aimed to estimate genetic parameters, including heritability and genetic correlations between methane and feed intake, using a large quantity of GreenFeed Emission Monitoring system data. The data available in this study consisted of methane measurements of steers, heifers and young bulls in a commercial feedlot environment, with methane measurements recorded via GreenFeed Emission Monitoring units. To date, 1,450 animals, ranging from 368 to 910 days of age have undertaken methane measurement cumulating in 216,921 methane measurements. The test period duration for measurement ranged from 20 to 83 days. Additional measurements available on the population include detailed feed intake records recorded using Insentec feed boxes. Informative animals (n=1,155) were used to derive heritability estimates for methane and feed intake using DMU, ensuring sire representation across contemporary groups. Multiple definitions of the methane trait were analysed due to the varying length of test periods. The resulting six trait definitions included individual observations, 1-day average, 5-day average, 10-day average, 15-day average and a full test average methane trait. Methane heritability estimates ranged from 0.09 to 0.43 depending on trait definition, with larger averaging periods yielding greater heritability estimates. Each of the methane definitions was moderately correlated with dry matter intake ranging from 0.38 to 0.62. The results from this study will aid in the development of breeding values for greenhouse gas traits and the creation of a novel methane trait for future selection of sustainable livestock.

Session 15 — Theatre 12

Determining relative values of genetic traits of Chinese dairy cattle in greenhouse gas mitigation

Y. Wang[1], R. Shi[1,2,3], A. Van Der Linden[1], B.J. Ducro[3], Y. Wang[2], Y. Hou[4], S.J. Oosting[1] and C.E. Van Middelaar[1]
[1]Wageningen University & Research, Animal Production Systems group, P.O. Box 338, 6700 AH Wageningen, the Netherlands, [2]China Agricultural University, College of Animal Science and Technology, Beijing, 100193, China, P.R., [3]Wageningen University & Research, Animal Breeding and Genomics group, P.O. Box 338, 6700 AH Wageningen, the Netherlands, [4]China Agricultural University, College of Resources and Environmental Sciences, Beijing, 100193, China, P.R.; yue3.wang@wur.nl

Breeding is considered a promising greenhouse gas(GHG) mitigation option for dairy sector. Compared with other mitigation options such as nutrition interventions, the effect of breeding is permanent and cumulative. Using a typical Chinese dairy farm as a case, this study builds on a novel method to determine the relative values of genetic traits in dairy cows to reduce GHG emissions. Life cycle assessment was conducted to calculate emissions from the farm(expressed in kg CO_2-eq per ton fat-and-protein-corrected milk(FPCM)). Results were combined with an existing bio-economic model to determine GHG values of six genetic traits, i.e. milk yield(MY), protein yield(PY), fat yield(FY), productive life(PL), calving interval(CI) and clinical mastitis(MAS). Next, an environmental breeding goal(minimizing GHG emissions) and an economic goal(maximizing farm profit) were formed, and relative weights of the traits under different breeding goals were calculated. Then the environmental and economic consequences of two goals were compared. Results showed that, with an improvement of one genetic standard deviation, MY, PY, FY, PL and MAS resulted in a reduction of 27, 6, 2, 6, and 1 kg CO_2-eq per ton FPCM, respectively, while CI resulted in an increase of 10 kg CO_2-eq per ton FPCM. As the goal shifted from economic to environmental focus, relative weight of MY doubled, meanwhile the weights of PY, FY and CI decreased. Under environmental goal, the reduction in emissions is 59% higher than that of economic goal, while the increase in profit is 37% less than that of economic goal. The study provides insights into trade-offs between economic and environmental gains that can be achieved via breeding. Results can be used by breeding organizations to contribute to a more environmentally friendly dairy sector.

Effects of breeding for feed efficiency on carbon footprint of milk – LCA approach

S. Hietala[1], A. Astaptsev[2], E. Negussie[3], M.H. Lidauer[3] and T. Mehtiö[3]
[1]Natural Resources Institute Finland, Bioeconomy and environment, Paavo Havaksen tie 3, 90570 Oulu, Finland, [2]Valio ltd, Meijeritie 6, 00370, Finland, [3]Natural Resources Institute Finland, Production systems, Myllytie 1, 31600 Jokioinen, Finland; sanna.hietala@luke.fi

The climate change impacts of cattle production are well acknowledged. Enteric fermentation and feed crop production are responsible for a large share of the carbon footprint (CF) of milk. Recent studies have shown a medium to high genetic correlation between feed utilization efficiency (FE) and methane (CH_4) emission implying selection for FE lowers CH_4 emission. However, the magnitude of the achievable effects is not clearly known. The aim of this study was to assess the size of correlated response on CH_4 emission due to selection for FE in Nordic Red cattle. The quantified impacts due to breeding for FE were assessed using life cycle assessment (LCA) method. The assessment for the Finnish dairy production ('business-as-usual', BAU) was conducted utilizing data collected from 700 Finnish dairy farms. The data included herd composition, feed composition, own feed crop production details (crops, yields, fertilizer inputs, crop protection), manure management system and the yields of raw milk and meat from cull cows. LCA model was constructed by following product environmental footprint category rules (PEFCR) of dairy products (EC 2018) and IPCC (2006, 2013) guidelines. These frameworks were complemented by national emission prediction models. System boundary was set to farm gate and the functional unit was 1 kg fat and protein corrected milk (FPCM). The CF was assessed for BAU and for the achieved genetic response of 10% improvement in FE but considering BAU for growing replacement cows. The BAU resulted CF of 0.977 kg CO_2eq/kg FPCM. Integration of the 10% improvement in FE due to breeding resulted in CF of 0.899 kg CO_2eq/kg FPCM, which was -8% less than BAU. The reduction of emissions was clearly largest from enteric fermentation (-0.042 kg CO_2eq/kg FPCM). The results indicate that breeding can be considered as powerful tool for climate change mitigation. Breeding for better FE also directly impacted those emission sources which are responsible of largest share of carbon footprint of milk, enteric fermentation and feed crop production.

Automated CT analysis to identify low emitting ruminants

S. Hitchman[1], L. Grundy[2], W. Bain[3], T. Johnson[3], M. Reis[4], H. Gunnarsdottir[3], J. McEwan[3] and S. Rowe[3]
[1]Agresearch ltd, 10 Bisley Road, Hamilton 3214, New Zealand, [2]Agresearch ltd, 1365 Springs Road, Lincoln 7674, New Zealand, [3]Agresearch ltd, 176 Puddle Alley, Mosgiel 9092, New Zealand, [4]Agresearch ltd, Tennent Drive, Fitzherbert, Palmerston North 4410, New Zealand; sam.hitchman@agresearch.co.nz

Genetic selection for reduced methane production has been recognized as an important tool within the agricultural industry's toolbox in achieving international targets for reduced greenhouse gas emissions. Recent research has shown that breeding for lowered methane production is associated with changes in rumen physiology (size, constituents) and feed efficiency. It was found that rumen volume was significantly lower in low methane producing animals in a flock divergently selected for methane yield. Rumen volume and composition can be measured in live animals using computed tomography (CT) of small ruminants (sheep, deer), whereby the entire animal is scanned at 5 mm intervals. CT scanning of animals is already implemented as part of a suite of imaging technologies to generate breeding values associated with desirable traits such as carcass composition. These scans also collect excellent data of the other body constituents such as the rumen, however manual processing of the CT data is a major hurdle for industry uptake. There is a need to automate this process to facilitate further research by reducing analysis costs. Recent developments in Convolutional Neural networks have led to advances in automated semantic (pixel wise) segmentation of images. Medical industry applications include tumour segmentation, classification of disease, and automated segmentation for abdominal composition. Using a dataset of 223 manually processed full body CT scans of sheep and deer, the potential to segment the rumen from the surrounding tissues and analyse its properties has been investigated. Latest models perform with Sørensen-Dice coefficients >0.90 on an independent test set, showing these techniques could soon provide unprecedented data for industry and research. This project is funded by the National Science Challenge: Science for Technological innovation. Data collection was funded through Beef and Lamb Genetics and The Pastoral Greenhouse Gas Research Consortium.

Session 15 Poster 15

Enteric methane emission of Nellore cattle from two lines selected for yearling weight

S.B. Gianvecchio[1], M.S. Borges[1], J.A. Silva[2], J.P.S. Valente[1], J.O.S. Marcatto[3], L.G. Albuquerque[1], J.N.S.G. Cyrillo[2] and M.E.Z. Mercadante[2]
[1]School of Agricultural and Veterinarian Sciences, UNESP, 14884-900, Jaboticabal/SP, Brazil, [2]Institute of Animal Science, 14174-000, Sertãozinho/SP, Brazil, [3]Embrapa Environment, 13918-110, Jaguariúna/SP, Brazil; maria.mercadante@sp.gov.br

Two selection lines, established in 1980, were analysed: Nellore Control (NeC), in which animals with yearling weight (YBW) close to the average of the contemporary group are selected (stabilizing selection); and Nellore Selection (NeS), in which animals with higher YBW are selected. Young bulls (mid-test age of 340±54 days) from NeC (n=132) and NeS (n=238), born from 2017 to 2021, were evaluated (60 roughage:40 concentrate diet). Enteric methane emission (g/day, CH_4) was measured by SF6 tracer gas technique. Least square means ± SE of mid-test body weight (BW), dry matter intake (DMI), average daily gain (ADG), and breeding value of YBW (estimated by ssGBLUP), for NeC and NeS, were 245±7.16 vs 321±6.92 kg ($P<0.01$); 6.80±0.12 vs 8.71±0.11 kg/d ($P<0.01$); 0.729±0.03 vs 1.041±0.03 kg/d ($P<0.01$); and 7.14±1.07 vs 69.2±0.79 kg ($P<0.01$), respectively. Concerning feed efficiency, lsmeans for NeC and NeS were -0.123±0.07 vs 0.053±0.05 kg/d ($P<0.05$) for residual feed intake; 9.61±0.27 vs 8.72±0.25 ($P<0.01$) for feed conversion rate; and 2.82±0.06 vs 2.77±0.06% ($P=0.16$) for DMI as percentage of BW. Comparisons of methane emission traits showed that NeC emitted less CH_4 than NeS (129±10.0 vs 167±9.96 g/d; $P<0.01$). However, NeC and NeS showed similar methane yield, expressed as CH_4/DMI ($P=0.51$), and methane emission intensity, expressed as CH_4/BW ($P=0.96$), although NeC had showed greater methane emission intensity, expressed as CH_4/ADG (183±13.9 for NeC vs 167±13.8 g/kg for NeS; $P<0.01$). When methane emission was expressed as a residual trait (CH_4=DMI+res), it resulted in the lowest value for NeC (-4.51±1.80 vs 2.30±1.34 g/d; $P<0.01$). The results showed that cattle selected for growth are more productive without loss of feed efficiency and without increase of methane per kg of body weight or methane yield. In addition, there is evidence that selection for growth traits decreases methane per kg of average daily gain. Funding: FAPESP (#2017/10630-2 and #2017/50339-5) and CAPES (Finance Code 001).

Session 15 Poster 16

Genetic correlation between enteric methane emission and feed efficiency and growth traits in cattle

M.E.Z. Mercadante[1], J.P.S. Valente[2], S.B. Gianvecchio[2], J.O.S. Marcatto[3], M.S. Borges[2], J.A. Silva[1], I.M.C. Garcia[1], S.F.M. Bonilha[1] and L.G. Albuquerque[2]
[1]Institute of Animal Science, 14174-000, Sertãozinho, SP, Brazil, [2]São Paulo State University, 14884-900, Jaboticabal, SP, Brazil, [3]Embrapa Environment, 13918-110, Jaguariuna, SP, Brazil; mezmercadante@gmail.com

This study aimed to estimate genetic parameters for daily methane emission, feed efficiency, and growth traits in Nellore cattle. Data of daily methane emission (CH_4, n=760 animals, 297±57.3 kg of body weight, 328±40.2 days of age), dry matter intake (DMI, n=2,033), average daily gain (ADG, n=2,033), residual feed intake (RFI, n=2,033), and selection (postweaning) weight (WSel, n=10,393), were analysed. Daily methane emission was also expressed as residual (CH_4res) from linear regression equation of CH_4 on DMI. CH_4 were evaluated using the tracer gas Sulphur Hexafluoride technique, in 20 sampling groups, during feed efficiency tests (GrowSafe Systems). The means of the traits studied were: 151±45.9 g/d, 7.55±1.53 kg/d, 1.022±0.257 kg/d, 0±0.661 kg/d, 303±53.1 kg, and 0±20.5 g/d, respectively for CH_4, DMI, ADG, RFI, WSel, and CH_4res. The (co)variance components were estimated by the average information restricted maximum likelihood method, fitting five two-traits animal models in a single-step GBLUP analysis using the BLUPF90 family of programs. Genotype data of 2,256 animals and a pedigree file with 10,810 animals were included. The heritability estimates (h^2) for CH_4 ranged from 0.21±0.06 (two-trait analysis with WSel) to 0.34±0.07 (with RFI). The h^2 for DMI, ADG, RFI, WSel and CH_4res were: 0.39±0.04, 0.29±0.04, 0.20±0.04, 0.40±0.02, and 0.13±0.07. Positive and high genetic correlations were observed between CH_4 and DMI (0.83±0.12) and between CH_4 and growth traits (0.95±0.12 with ADG, and 0.85±0.17 with WSel). Genetic correlation between CH_4 and its residual (CH_4res) was also positive and high (0.63±0.29). On the other hand, a positive and weak to moderate genetic correlation was observed between CH_4 and RFI (0.31±0.16). Selection for production (higher WSel, ADG and DMI) leads to higher methane emission. However, selection for negative RFI may, in the long term, lead to lower methane emissions. These are the first results in *Bos indicus*. Funding: FAPESP (#2017/10630-2 and #2017/50339-5), and CAPES (Finance Code 001).

Session 15 Poster 17

Heritability of predicted enteric methane emission in Japanese Holstein cows

R. Tatebayashi[1], A. Nishiura[1], F. Terada[1], Y. Uemoto[2], M. Aihara[3], T. Suzuki[4], I. Nonaka[1] and O. Sasaki[1]
[1]Institute of Livestock and Grassland Science, NARO, Tsukuba, Ibaraki, 305-0901, Japan, [2]Graduate School of Agricultural Science, Tohoku University, Sendai, Miyagi, 980-0845, Japan, [3]Livestock Improvement Association of Japan, Koto, Tokyo, 135-0041, Japan, [4]Institute of Livestock and Grassland Science, NARO, Nasushiobara, Tochigi, 329-2793, Japan; tatebayashir014@affrc.go.jp

To achieve sustainable livestock farming, reduction of greenhouse gas emissions in the livestock production is needed. Dairy cattle emit methane, a greenhouse gas, on rumen fermentation. It is required to reduce methane emission genetically in dairy cattle, but it is difficult to collect a large number of data necessary for the accurate genetic evaluation. We developed an equation to predict enteric methane emission in dairy cows based on information such as lactation stage, body weight, and milk composition. This study aimed to predict enteric methane emission and estimate the heritability of methane production using large data sets. Enteric methane production at 6-305 days in milk (DIM) was predicted using 118,402 records of Holstein cows in the 1-5 lactation collected by the Livestock Improvement Association of Japan in 2020-2021. Lactation stage, body weight, energy-corrected milk, milk fat percentage, and milk fatty acid composition were used to predict enteric methane emission. A random regression test-day model was used to estimate heritability of predicted enteric methane emission, including herd-test day-milking frequency, region-calving month and parity-calving age as the fixed effects, and additive genetic and permanent environmental effects as the random effects. The Gibbs sampling method was used to estimate variance components. The mean value of predicted enteric methane emission was like that of measured enteric methane emission in previous reports. The mean values of predicted enteric methane emission were increased around 60 DIM and then gradually decreased from 80 DIM during the lactation period. Heritability was estimated 0.070 to 0.21 from 6 to 305 DIM, which was slightly lower than that previously reported. Heritability estimates were low soon after parturition and in early lactation, and then increased until reaching a maximum at 305 DIM.

Session 15 Poster 18

Using genomic markers and microbiome data to predict methane emissions from Holstein cows

N. Saedi, M.S. Lund and E. Karaman
Aarhus University, Center for Quantitative Genetics and Genomics, C.F. Møllers Allé 3, bld. 1130, 8000 Aarhus, Denmark; naghsa@qgg.au.dk

The emission of Methane, which is produced by microbial activities in livestock, contributes to global warming. Microbiome composition and host genetic markers have an impact on the methane emissions of dairy cows. Our objective is to evaluate genomic prediction accuracy for methane emission, by using microbiome composition together with genetic markers. We used a joint model (GMBLUP), which models methane as a function of microbiome data and genetic markers, and microbial OTUs as a function of genetic markers. The phenotype, genotype, and microbiome data of 692 Holstein cows were used. The microbiome data included relative abundancies of 3,894 bacterial operational taxonomic units (OTUs) and 189 archaeal OTUs, for each individual. We compared prediction accuracies from GMBLUP with pedigree-based (PBLUP) and genomic (GBLUP) best linear unbiased prediction methods. Heritability of methane from GMBLUP, GBLUP, and PBLUP was 0.22, 0.22, and 0.15, respectively. The proportion of phenotypic variance explained by microbiome data (microbiability) using GMBLUP was 0.05. The low microbiability demonstrated that the rumen OTUs did not explain a large proportion of phenotypic variance in methane phenotypes. Both GBLUP and GMBLUP provided a prediction accuracy of 0.17, whereas PBLUP provided a prediction accuracy of 0.13. In conclusion, the joint model (GMBLUP) combining microbial OTUs and genotype data, did not improve genomic prediction of methane, over a standard GBLUP method.

Session 16 — Theatre 1

Non-traditional feed ingredients in diets for pigs

H. Stein
University of Illinois, Department of Animal Sciences, 1207 West Gregory Dr. Urbana, IL 61801, USA; hstein@illinois.edu

Non-traditional feed ingredients that may be used in diets for pigs include, but are not limited to, bakery meal, rice bran, field peas, and yeast products. Bakery meal consist of dried former foods and is produced after collection of non-saleable bread, pastries, cookies, and confectionary products. Generally, companies that produce bakery meal are able to supply meals that are consistent in chemical composition, whereas the variability in digestibility of amino acids and for metabolizable energy is considerable among sources. Bakery meal has low digestibility of lysine, but excellent digestibility of phosphorus, and bakery meal may be included in diets for weanling pigs by up to 25%. Rice bran is the co-product produced when bran is removed from de-hulled paddy rice and the product usually contains 20 to 30% starch, 12 to 15% crude protein, and 20 to 25% total dietary fibre. Rice bran contains 15 to 20% acid hydrolysed ether extract, but the product may be defatted, in which case the fat content is around 2%. The metabolizable energy in defatted rice bran is less than in maize, but full fat rice bran contains more energy than maize. Rice bran may be included in diets for weanling or growing-finishing pigs at up to 30% without negative impacts on growth performance. Field peas contain around 40% starch and 20% crude protein. The protein is high in lysine, but low in sulphur-containing amino acids. However, amino acids and phosphorus in field peas have excellent digestibility and the metabolizable energy is close to that in maize. Field peas may be included in diets for weanling pigs by up to 36% and in diets for growing and finishing pigs, field peas may replace all soybean meal. Yeast is a single cell protein that may be produced on different substrates and there is, therefore, some variability in the nutritional value among sources of yeast. Protein concentration is between 40 and 50%, but digestibility of amino acids vary considerably among sources. High quality sources of yeast may be included in diets for weanling pigs by up to 14%. In conclusion, there are many non-traditional feed ingredients available to the swine feed industry, but it is critical that each ingredient is assessed in terms of nutritional composition, energy and nutrient digestibility and impact on growth performance of pigs.

Session 16 — Theatre 2

Effect of genotypes on agronomic and antinutritional traits of *Lupinus albus* L. for livestock

M. Oteri, D. Scordia, R. Armone, V. Nava, F. Gresta and B. Chiofalo
University of Messina, Department of Veterinary Sciences, Via Palatucci, 98168, Italy; marianna.oteri@unime.it

Lupin species (*Lupinus* spp.) could represent a realistic and sustainable alternative protein source in both monogastric and ruminant feeds, capable of replacing soy without loss of quantity and quality of livestock products. However, a drawback of lupin seed quality are the antinutritional factors (ANFs), namely the alkaloid compounds. Aim of this study was to explore the productive traits and the quinolizidine alkaloids (QAs) in four genotypes of *Lupinus albus* L. grown in the Mediterranean environment. Two recently released varieties (Volos and Luxor) and two ecotypes from South-Italy (Ecotype F and Ecotype G) were compared in a field trial. Seed yield (SY) and thousand seed weight (TSW) were evaluated at seed physiological maturity. QAs analysis was performed by HR-GC/MS (LOQ: 0.01 mg/100 g). Data were subjected to a one-way ANOVA and means were separated by the Tukey HSD test ($p \leq 0.05$). SY was the significantly highest in Luxor and Ecotype G (2.27 and 1.81 Mg/ha), and the lowest in Volos (1.02 Mg/ha). On the contrary, the TSW was the highest in Volos (347.2 g) and the lowest in the two Ecotypes (315.4 g, on average) Luxor (6 mg/100 g) and Volos (5 mg/100 g) showed very low concentrations of total QAs, while Ecotype F (250 mg/100) and G (202 mg/100 g) showed the highest values. Lupanine and angustifoline were significantly higher in Ecotypes F (183 and 39 mg/100 g) and G (146 and 29 mg/100 g) than in Luxor (5 and 0.3 mg/100 g) and Volos (4 and 0.2 mg/100 g). Other QAs (sparteine, multiflorine, 13-alpha-hydroxylupanine, 13-alpha-hydroxymultiflorine and angeloxylupanine) were below the limit of toxicity (0.02%) for animal consumption, nonetheless, the two ecotypes showed the highest content. This study underlines the importance of genotype selection for a safe and sustainable animal nutrition. Among tested genotypes, present findings suggests Luxor well balanced between productive and ANFs traits, while Volos was characterized by a low productivity and the two ecotypes by too high ANFs content. Further effort is needed to breed new white lupin genotypes well adapted to a wide range of climatic conditions and resilient to abiotic adversities.

Dietary effects of xylanase and flaxseeds on performance, meat quality and bone health in broilers

D. Lanzoni[1], M. Skřivan[2], M. Englmaierová[2], M. Marounek[2], E. Skřivanová[2], A. Baldi[1] and C. Giromini[1]
[1]Università degli Studi di Milano, Department of Veterinary and Animal Science (DIVAS), Via dell'Università 6, 29600 Lodi, Italy, [2]Institute of Animal Sciences, Department of Nutritional Physiology and Animal Product Quality, Pratelstvi 815, 104 00 Prague, Czech Republic; davide.lanzoni@unimi.it

The need to increase the nutritional profile of wheat and to study new crops that can ensure both food/feed safety and environmental protection are nowadays primary objectives in the feed industry. Wheat-treatment with xylanase (XL) and the use of flaxseeds (FSs), recognised for their high lipid profile rich in n-3 fatty acids (FAs), are potential candidates. For this, the aim of this experiment was to test the effects of XL and FSs in broiler chickens fed a wheat-based diet, on performance, nutrient retention, quality of breast meat, and bone health. 720 one-day-old broiler Ross 308 were divided into four experimental group: (Control, basal diet), FSs (80 g/kg), XL (0.1 g/kg), and XF received FSs + XL combination. Experimental period lasted 35 days. XL not only reduced feed intake by improving feed conversion and reduced mortality, but also significantly increased (P<0.05) retention of dry matter, crude protein, and ash (56.9; 11.2; 55.1 g/kg, respectively), values not observed following FSs treatment. XL significantly increased (P<0.05) fat in the breast meat (12.1 g/kg) than in the control group (10.6 g/kg), also improving its oxidative stability. At the same time, FSs significantly increased the n-3 FAs content (65.5 mg/100 g) than control (24.3 mg/100 g) and XL treatment (29.6 mg/100 g). Higher values were obtained for XF group (97.2 mg/100 g). HSs resulted in a better n-6/n-3 ratio in the diet (1.96 mg/100 g), also in combination with XL (1.81 mg/100 g), showing statistically significant differences compared to the control (7.66 mg/100 g) and XL treatment (6.96 mg/100 g). FSs and XL treatment significantly improved bone health than the control group (P<0.05). Although further studies are needed to investigate the best level of FSs inclusion in the diet of broilers, these data suggest how their use in combination with XL can achieve positive levels in broilers. These results are enhanced by the sustainable impact of FSs to ensure environmental protection.

The effect of hydrolysed yeast on production performance and gastrointestinal health in broilers

N. Irshad, V. Perricone, S. Sandrini, C. Lecchi, P. De Faria Lainetti, A. Agazzi and G. Savoini
University of Milan, Department of Veterinary Medicine and Animal Sciences, Via dell'università 6, 26900 Lodi, Italy; nida.irshad@unimi.it

The goal of the present study was to ascertain the effects of dietary inclusion of hydrolysed yeast (HY) on the growth performance, meat quality, and gastrointestinal health of broilers. A total of 320 male 1-day-old chicks (ROSS 308) were used in this experiment. The animals were homogeneously separated into two groups and given either the basal diet (CTR) or the basal diet supplemented with yeast (TRT, 500 mg/kg HY). Each experimental group was composed of 160 birds, which were distributed among eight replicates (20 birds per replicate). Growth performance and body lesions (hock burn and foot pad dermatitis) were evaluated at the beginning, at each feeding phase change, and at the end of the trial (i.e. 0, 10, 21, 42). On day 42, all the animals were taken to a slaughterhouse for slaughter and sample collection to further determine meat quality (water holding capacity, pH, or colour) and gastrointestinal health (gene expression). Statistical Analysis System software (SAS version 9.4; SAS Institute Inc., Cary, NC, USA) applying a MIXED procedure, the GLM procedure, and PROC FREQ was used for the analysis of the data. Differences between groups were considered statistically significant at P<0.05, whereas a trend for a treatment effect was noted for 0.05≤P<0.10. The results revealed no difference between the CTR and TRT groups both in terms of production performance and meat quality (P>0.05). Furthermore, the mRNA expression studies on the Adiponectin system genes (AdipoQ, Adipo Receptor 1, and Adipo Receptor 2) and tight junctions (Zonula occludens ZO-1, Occludin, Claudin-3) showed no significant difference (P>0.05) between the CTR and TRT groups. This study was carried out within the Agritech National Research Center and received funding from the European Union Next-Generation EU (Piano Nazionale di Ripresa e Resilienza (Pnrr) – Missione 4 Componente 2, Investimento 1.4 – D.D. 1032 17/06/2022, CN00000022). This manuscript reflects only the authors' views and opinions, neither the European Union nor the European Commission can be considered responsible for them.

Yoghurt acid whey effects on broiler performance, meat oxidative status and caecal barrier integrity

V.V. Paraskeuas, G. Papadomichelakis, A. Pappas, P. Simitzis, E.C. Anagnostopoulos, I.P. Brouklogiannis, E. Griela, I. Palamidi, G. Theodorou, I. Politis and K.C. Mountzouris
Agricultural University of Athens, Animal Science, Iera Odos 75, 11855, Greece; v.paraskeuas@aua.gr

In this research, the investigation of broilers' response to dietary supplementation of yoghurt acid whey (YAW) at 4 different dietary levels (0, 25, 50, and 100 g/kg of diet) was conducted on growth performance, nutrient digestibility, breast meat oxidative stability and quality traits as well as caecal barrier integrity. A total of 300 male 1 d old Ross 308 broilers were randomly allocated in 4 experimental treatments with 5 replicates of 15 broilers each. All experimental treatments received a maize-soybean meal basal diet following a two-phase feeding plan. The 4 experimental treatments were as follows: control was fed with basal diet with no YAW addition (W0), the other three treatments were fed basal diet supplemented with YAW at 25 g/kg of diet (W25), 50 g/kg of diet (W50) and 100 g/kg of diet (W100), respectively. At the starter period (1-10 d) increasing YAW inclusion level resulted in reduction ($P<0.05$) of body weight (BW) and body weight gain (BWG) in treatments W50 and W100, compared with W0. Breast meat oxidative stability was improved ($P<0.05$) during refrigerated storage for 1 and 3 d in all treatments and only at W25 birds at 6 and 9 d. From breast meat quality traits, yellowness appeared to increase ($P<0.05$) in treatments W50 and W100 compared to W0. The relative expression levels of occludin and claudin-1 in caecal mucosa were increased ($P<0.05$) with the addition of YAW at 25 g/kg of diet compared with those of W0 treatment. The results suggest that YAW inclusion at 25 g/kg of diet did not impair the performance, extended meat shelf life by reducing lipid oxidation and enhanced caecal gut barrier. This research has been co-financed by the European Regional Development Fund of the European Union and Greek national funds through the Operational Program Competitiveness, Entrepreneurship, and Innovation, under the call RESEARCH – CREATE – INNOVATE (project code:T2EDK-00783).

The effect of *Citrus unshiu* peel on growth performance, digestibility and immunity in weaning pig

A.S. Sureshbabu[1], A.K. Adhimoolam[2] and T.M. Min[3]
[1]Jeju National University, Department of Animal Biotechnology, 102 Jejudaehak-ro, 63243, Korea, South, [2]Jeju National University, Subtropical Horticulture Research Institute, 102 Jejudaehak-ro, 63243, Korea, South, [3]Jeju National University, Department of Animal Biotechnology, Bio-Resources Computing Research Center, Sustainable Agriculture, 102 Jejudaehak-ro, 63243, Korea, South; anjbio9@stu.jejunu.ac.kr

Mandarin orange (*Citrus unshiu*) is one of the important fruits widely cultivated in Jeju Island, South Korea. *C. unshiu* peels (CUP) account for nearly half of the fruit body and are the main waste used to produce a vast amount of by-products. CUP rich in nutrients/ active compounds, therefore, it is potential feed additives instead of antibiotics for the husbandry of both livestock and poultry. We characterized the biologically active compounds from CUP and studied their beneficial effects (Anti-inflammatory and antioxidant effects). Further, we used it as a feed additive in weaning pigs and examined their growth performance, digestibility, and immunity in weaner pigs. A total of 75 weaning pigs [(Yorkshire × Landrace) × Duroc], with an average body weight of 6.5±0.05 kg, were used in a three weeks feeding trial. Pigs were examined for initial body weight and gender and randomly allotted into experimental treatments (5 replicates pens per treatment, 5 pigs per pen). The dietary treatments were as follows: (1) Control; (2) CUP1; and (3) CUP 2. After three weeks and during the overall period, dietary CUP supplementation showed a positive effect on growth performance, digestibility, and immunity. In another experiment, we also investigated the combined effect of CUP and curcumin nanospheres in weaning pigs and obtained positive results. Taken together, CUP and curcumin nanospheres supplementation can be beneficial in enhancing the performance of weaning pigs.

The effect of replacing grains with bakery by-products on performance and pigs' diet preference

F. Veldkamp[1,2], H.M. Vermeer[2], J. Kater[2], A. Ten Berge[2], J.M.J. Rebel[1] and I.C. De Jong[2]
[1]Wageningen University, Adaptation Physiology Group, De Elst 1, 6708 WD, Wageningen, the Netherlands, [2]Wageningen Livestock Research, Animal Health & Welfare, De Elst 1, 6708 WD, Wageningen, the Netherlands; fleur1.veldkamp@wur.nl

Around one-third of all food produced in the world is lost or wasted. Part of this can be re-used in animal diets to reduce the usage of grains, making it a good diet ingredient for improved livestock sustainability. The aim of this study was to test the effect of replacing feed grains in weaned pig feed with bakery by-products on growth performance, health, welfare and behaviour (data latter three not yet analysed) and pigs' diet preference. The first part of the study was carried out in eight pens with ten weaned piglets per pen (initial starting weight = 7.95±0.9 kg) until a body weight of 25.7±2.5 kg during four batches. Pigs received either standard (PS, based on grains) or circular (PC, based on bakery by-products) feed and were offspring from sows that were fed standard feed (SS, based on grains) during the first two batches or circular (SC, based on bakery by-products) feed during the last two batches. Body weight at pen level was measured at the start and end of each batch. Feed provision was determined daily. Average daily gain (ADG, in grams), average daily feed intake (ADFI, in grams) and feed conversion ratio (FCR) were calculated. The ADG was higher for PC than for PS (451±42 vs 423±57, P=0.03) and the ADFI showed a tendency to be higher for PC than PS (PC=624±63.1 vs PS=583±60.7, P=0.05) whereas FCR (PC=1.50±0.09 vs PS=1.49±0.07, P=0.97) did not differ significantly. No interaction between pig feed and sow feed was found. The second part of the study was carried out in two pens with 40 pigs per pen during four batches. Pigs had the choice between PS or PC feed (two feeding throughs per feed per pen). The results revealed that pigs consumed overall more PC than PS (PC=52.3%±0.03 vs PS=47.7%±0.03, P=0.002). In summary, using bakery by-products as replacements for feed grains in weaned pig feed did not negatively impact growth performance and pigs even preferred feed based on bakery by-products over feed based on grains.

Replacing hexane by 2-methyloxolane for defatting soybean meal does not impair dairy cow performance

V. Menoury[1], A. Ferlay[1], L. Jacques[2], V. Rapinel[2] and P. Nozière[1]
[1]UCA, INRAE, VetAgroSup, UMR Herbivores, Theix, 63122 Saint Genès Champannelle, France, [2]Pennakem Europa (EcoXtract®), 224 av. de la Dordogne, 59944 Dunkerque, France; valentin.menoury@inrae.fr

Soybean meals (SMB) fed to dairy cows are obtained through mechanical and/or chemical extraction of oil from the seed. Chemical extraction consists of solvent extraction, extraction-grade hexane being the most used solvent in the industry. Concerns regarding the health effects of n-hexane residues in food and feed are driving the industry to find alternatives to extraction-grade hexane. 2-methyloxolane (2-meOx) could be one of these. However, process differences between hexane and 2-meOx defatted SBM may lead to differing nutritive values and may affect the performance of dairy cows. A 4×4 Latin-square experiment with 16 primiparous lactating Holstein cows was conducted to study the effects of 2-meOx on milk yield and composition, nitrogen use efficiency for milk production, and solvent residues in milk. Cows were fed a diet containing 14% SBM and 165 g/kg DM crude protein. The 4 treatments differed according to the inclusion level of hexane and 2-meOx defatted SBM (i.e. 100% hexane, 66% hexane – 33% 2-meOx, 33% hexane and 66% 2-meOx, and 100% 2-meOx). Results show that dry matter intake, milk yield, milk protein concentration and excretion, milk protein fractions, and milk fat excretion did not differ among treatments. Milk fat concentration tended to increase with 33% hexane – 66% 2-meOx compared to 100% hexane SBM (P=0.07). Nitrogen intake, milk nitrogen excretion and nitrogen use efficiency for milk production did not differ according to the treatment. N-hexane residues tended to be more often detected in the milk from 100% hexane-fed cows compared to 100% 2-meOX-fed ones (P=0.08). 2-meOx residues were only detected in milk from 2-meOx-fed cows, but the difference with 100% hexane-fed cows was not significant. When detected in the milk, n-hexane and 2-meOx concentrations were below the quantification threshold of 10 ng/g. These preliminary results indicate that 2-meOx-defatted SBM can replace hexane-defatted SBM without affecting milk production performance and nitrogen use efficiency. Further analyses are required to confirm results regarding the detection of solvent residues in milk.

In vitro rumen fermentation traits of _Mentha piperita_, and products of oil extraction

S. Massaro[1], J. Andersen[2], G. Secchi[1,2], D. Giannuzzi[1], S. Schiavon[1], E. Franciosi[2] and F. Tagliapietra[1]
[1]University of Padova, DAFNAE, Viale dell'Università, 16, 35020, Legnaro (PD), Italy, [2]Edmund Mach Foundation, CRI, Via E. Mach, 1, 38098, San Michele all'Adige (TN), Italy; selene.massaro@phd.unipd.it

By-products obtained from essential oil extraction could be valued as ruminant feeds for their fibre and bioactive compounds content. This study aimed to evaluate the *in vitro* rumen fermentable characteristics of *Mentha piperita* (Mp, as dry forage), the residual product achieved from a supercritical CO_2 extraction (residue: rMp), and the residual of extraction plus the extract (erMp, to reconstitute Mp). A sample of 1 gram for each of these 3 materials and a grass hay (control) were tested in 3 repeated incubations, using 3 different rumen fluids, for a total of 36 bottles and 9 blanks. DM degradability (DMd), gas production (GP) kinetic and composition, volatile fatty acid (VFA), and bacterial profile over 24 h of fermentation were investigated. A linear mixed model was run to evaluate the fermentation characteristics of these materials. Orthogonal contrasts were used to compare the control vs Mp, the Mp vs rMp and the Mp vs erMp. The control and Mp in DMd did not differ for total VFA and total GP. However, Mp decreased n-butyric acid and slightly increased acetic acid proportions, increased the GP release between 3 and 9 h, and reduced CH_4 (% v/v) emissions compared to the control. The fermentation parameters of rMp did not differ from those of Mp, except for an increase of GP between 15 and 24 h, and a slightly increase of CH_4 (% v/v) at 24 h. Also, erMp did not affected the fermentative parameters compared to Mp, except for an increase of GP rate in the first 3 h of incubation, a slight increase of branched VFA, and an increase of CH_4 proportions. The Mp caused a modification of the microbial profile in the fermentation fluid with a reduction of the diversity indexes compared to the control. The Mp degradability was similar to that of the grass hay, notwithstanding the signs of some changes in the fermentation pattern and in the rumen microbial population. The extraction of essential oil had only a limited effect on fermentative properties of mint suggesting a potential use of this residual by-product as ruminant's feed. Funded by the project: BIRD213117/21; FoodTraNet 956265.

In vitro rumen fermentation of mushroom-pretreated _Acacia mellifera_ simulating ruminant digestion

C. Rothmann, E.D. Cason and P.D. Vermeulen
University of the Free State, Animal Sciences, Fakulteits Road, Brandwag, Bloemfontein, South Africa; rothmannc@ufs.ac.za

The increasing growth of agro-industrial activity resulting in excessive amounts of agri-waste has led to the accumulation of lignocellulosic residues all over the world, in particular, deforestation initiatives of invasive trees in South Africa. These lignocellulosic residues are rich in energy resources but considered non-digestible by ruminant animals. The use of lignolytic fungi such as mushrooms in solid-state fermentation could sufficiently degrade the indigestible lignocellulosic components and add medicinal and nutritional value to otherwise unusable, high-energy waste material. The current study presents the data of the bioconversion of lignocellulosic substrate using solid-state fermentation with the edible and medicinal mushrooms, *Pleurotus ostreatus*, *Ganoderma lucidum* and a new *Ganoderma* species, to increase digestibility and nutritional value to be applied as ruminant animal feed. Additionally, next-generation sequencing was performed to determine if the use of mushrooms as pre-treatment influenced the microbial diversity of the ruminal fermentation. The solid-state fermentation process was optimised, and the resulting product analysed for the degradation of the lignocellulosic components. Data obtained after 18 weeks of degradation indicated a significant ($P<0.05$) reduction in the acid detergent fibre, acid detergent lignin and neutral detergent fibre fractions of the biomass, with up to 20% reduction in indigestible components. After a 48-hour *in vitro* fermentation using rumen fluid, results indicated a significant ($P<0.001$) improvement in digestibility of all parameters tested namely crude protein, acid detergent fibre, acid detergent lignin, neutral detergent fibre, gross energy, dry matter, and organic matter when compared to an untreated control. It was also found that the species of mushroom used for pre-treatment influenced the microbial community diversity selectively. With the implementation of the AMPTS II testing system, fermentation profiles for all substrates were provided. Results indicated that the use of lignocellulose rich waste substrates, pretreated with mushrooms are suitable for use as ruminant animal feed.

Valorisation of atypical land in cattle production systems in search of self-sufficiency

D. Starling[1], A. Farruggia[1] and T. Puech[2]
[1]INRAE, ACT, UE DSLP, 545 Rue du Bois Mâche, 17450, Saint Laurent de la Prée, France, France, [2]INRAE, ACT, UR ASTER, 662 Av. Louis Buffet, 88500 Mirecourt, France; thomas.puech@inrae.fr

The challenge of feeding nearly 10 billion people by 2050 in a context of climate change and increasing scarcity of natural resources requires rethinking of agricultural systems and the resources they use. Livestock are questioned on the competition for food, land use and their capacity to contribute to the nutrient cycle. This study is based on four cattle breeding systems (two dairy, two suckling) that are not representative of the dominant systems of the Western Europe such as: (1) they use minimal inputs (mineral fertilization, concentrates); (2) they rely on atypical land use providing atypical resources: woody formations, spontaneous vegetation in marshes – reed, duckweed, azolla, grasslands abandoned by conventional livestock. This study aims to understand the different uses of these resources and to quantify, from a metabolic approach of nitrogen flows, their contribution to agroecological and feed performances. Our results show that the self-sufficiency of these systems is mainly based on the use of renewable resources. Although atypical resources represent a small part of the circulating biomass, they contribute significantly to the self-sufficiency of these systems via: (1) animal feeding or nitrogen supplementation during periods of deficit in the production of 'conventional' fodder, particularly during summer droughts; (2) the supply of winter stocks; and (3) the production of biomass for animal bedding, particularly when the land is not well adapted to cereal crops. In addition their composition in antioxidants and micronutrients potentially gives them a role in terms of animal health and tolerance to heat stress. Finally, these resources can offer flexibility of management to systems because they can be used for two purposes: feed or bedding. However, the valorisation of these atypical lands requires harvesting practices that respect the life cycle of the resources generated to ensure their renewal but also the habitats that they provide for many wildlife species.

Growth of black belly ewe lambs ingesting non-conventional feeds in the form of complete feed blocks

D. Barde[1], E. Traore[2], R. Arquet[1], N. Minatchy[1] and H. Archimede[1]
[1]INRAE, Prise d'eau Petit-bourg, 97170, Guadeloupe, [2]ISRA, saint louis, 24005, Senegal; dingamgoto.barde@inrae.fr

The co-products of crops and agro-industry are biomasses that can be used for animal feed. This reduces the need for land for animal feed and the associated carbon footprint of meat production. Various constraints limit the application of this strategy, including seasonal availability, nutritional imbalances and practicality of use. The development of low-tech technologies on the farm, including complete feed blocks, is a solution that should be evaluated considering the socio-economic contexts of the territories and countries. The objective of this study is to compare the feed value of 4 complete feed blocks composed of crop coproducts and agro-industry relative to a conventional feed based on hay grass and concentrate. A fattening trial involving 50 ewes lambs Black belly for 3 months was conducted. Five diets were evaluated: (1) Diet 1, control consisting of a 2-month-old tropical grass hay distributed *ad libitum* and 300 g of concentrate; (2) Diet 2, sorghum straw/molasses/cottonseed; (3) Diet 3, sorghum straw/sorghum grain/cottonseed ration; (4) Diet 4, sorghum straw/sorghum grain/*Leucaena leucocephala*; (5) Diet 5, sorghum straw/molasses/ *Leucaena leucocephala*. The diets were formulated to be isoenergetic and isoproteic. The growth potential of Black belly ewe lambs recorded at INRAE is 150 g /day. Intakes and daily growths (g/d) were 955 and 91, 1,219 and 116, 1,233 and 110, 1,223 and 79, 1,167 and 66 for control, diets 1, 2, 3 and 4 respectively. The differences in intake were not significant. Growth was lower with diets 3 and 4 compared to the other rations ($P<0.05$). The daily growth rates recorded with the control diet are lower than the growth potential of Black belly ewe lambs. This result is attributable to the poor quality of the hay and its low intake. Lowest growth observed with diets 3 and 4 are probably due to the insufficient protein intake with *Leucaena leucocephala* compared to cotton seed.

Session 16 Poster 13

Diet effect of *Hermetia illucens* on proximate and fatty acid composition of broiler meat

R. Berrocal[1,2], E. Molinero[2], S. Rueda[2,3], M.M. Campo[1] and L. Mur[1,2]
[1]Instituto Agroalimentario IA2, Miguel Servet, 177, 50013 Zaragoza, Spain, [2]UVESA, Pol Montes del Cierzo km86, 31500 Tudela, Spain, [3]Venttus, Pierre Gassier, 23C, 50018 Zaragoza, Spain; rberrocal@uvesa.es

Poultry production development in recent decades and increasing costs make it necessary to study new sources of protein. *Hermetia illucens* meal is offered as an alternative to soybean meal due to its high animal protein content (CP=50%) and amino acid profile. But it is important to know the effect on productive parameters and the final product quality. The aim of this study was to assess the influence of including this source of protein in the diet on the chemical composition and the fatty acid profile of the breast in broilers. A total of 1440 one-day-old chicks were housed in 24 boxes (60 birds/pen). Three groups were defined with 8 replicates each one: control group (CG) with soybean meal as the main protein source, a second and third groups with 5% of *Hermetia illucens* meal included in the diet until 21 (H21) and 42 (H42) days, respectively. Animals were slaughtered at 42 days and 12 randomly carcasses were sampled per treatment. Breast proximate composition and fatty acid profile were analysed in frozen samples kept in vacuum packaging. Data were analysed with a General Lineal Model with SPSS (26.0) and differences between treatments were assessed with a Tukey test. No differences were found in the protein or fat content of the breast due to the inclusion of *H. illucens*. The percentage of saturated fatty acids was higher in animals fed with *H. illucens* meal ($P\leq0.05$), especially due to the higher percentage of lauric and myristic acids ($p\leq0.001$). Nevertheless, palmitic acid was lower in H42 because part of the dietary palm oil in the last lifecycle days was replaced by *H. illucens* meal (22.7 vs 23.7% in CG and 24.0% in H21, $P\leq0.001$). Also, the inclusion of *H. illucens* tended to reduce the percentage of monounsaturated acids (39.7 vs 41.3 and 42.3%, $P=0.07$, in H42, H21 and CG, respectively) especially due to the reduction of oleic acid ($P\leq0.05$). However, the evolution of linoleic and α-linolenic acids follow the opposite way, increasing in H42 vs the other two groups ($P\leq0.001$). It is essential to continue researching alternative feed ingredients beyond profitability because they may change the final product composition.

Session 16 Poster 14

Antioxidant and rumen microbiota shift of Holstein cows by aquatic plant silage in Tibetan plateau

B. Li[1], X. Huang[1], X. Yang[1] and Y. Wang[2]
[1]Tibet Academy of Agricultural and Animal Husbandry Sciences, Institute of Animal Husbandry and Veterinary, Lhasa, 850000, China, P.R., [2]China Agricultural University, College of Animal Science and Technology, Beijing, 100000, China, P.R.; xukesuolibin@163.com

The Qinghai-Tibetan Plateau is a region with a unique geographical environment and abundant natural resources. However, due to harsh environmental conditions, forage production in the region is limited. To tackle the issue of feed scarcity, the use of non-conventional feedstuffs as a substitute for conventional feed has emerged as a sustainable solution for animal husbandry. In this study, the aquatic plant silage with optimal performance was selected instead of corn silage and applied in a dairy cow feeding experiment (n=24, 74 days) in the plateau. we evaluated the growth and lactation performance of Holstein cows in the aquatic plant silage treatment groups and the corn silage control group. Milk fatty acid and blood indicators were analysed, along with the rumen microbial response as community, functions, and metabolic profiles through metagenomics. Results showed that the milk fat, lactose, total solids contents and milk fatty acid content varied among different treatment groups. Additionally, the milk somatic cell counts in the formula 1 group was significantly lower than those in the control group, indicating anti-inflammatory activity driven by possible higher bioactive compounds. The blood biochemical indexes of dairy cow among groups also revealed that the addition of aquatic plant silage improved host antioxidant activity and immunity. Metagenomic analysis of the rumen microbiota showed significant enrichment of specific taxa among different groups which is align with rumen metabolism as shown in VFAs and metabolome analysis. These findings suggest that different rumen microorganisms may play important roles in regulating the immune and antioxidant capacity of Holstein cows, as well as affecting the production performance of the animals. Overall, the study provides evidence that substitution of corn silage with aquatic plant silage could improve the production performance, immunity and antioxidant capacity of Holstein cows in the Tibetan plateau, highlighting the potential of using aquatic plants as substitute for ruminant feed.

Session 16 Poster 15

Feeding olive cake alters the mRNA of SREBF1 in adipose tissue of cows

M.C. Neofytou[1], D. Miltiadou[1], E. Sfakianaki[2], S. Symeou[1], D. Sparaggis[3], A.L. Hager-Theodorides[2] and O. Tzamaloukas[1]

[1]Cyprus University of Technology, Department of Agricultural Sciences, Biotechnology and Food Science, P.O Box 50329, Limassol, Cyprus, [2]Agricultural University of Athens, Department of Animal Science, Iera Odos, 11855 Athens, Greece, [3]Agricultural Research Institute, P.O. Box 22016, Nicosia, Cyprus; ouranios.tzamaloukas@cut.ac.cy

This study evaluated the use of ensiled olive cake (EOC) on yield, composition, fatty acid (FA) profile of milk and expression of selected genes involved in mammary and adipose lipid metabolism. Mid-lactating cows were randomly assigned to two feeding treatments (12 animals/ group), named the control (C) and the olive cake diet (OC), where EOC was at 10% DM, for a four-week period. Mammary and perirenal adipose tissues were sampled for gene expression analysis using quantitative RT-PCR at the end of the trial. The expression of 11 genes, involved in FA synthesis (ACACA, FASN, G6PDH), FA uptake and/or translocation (VLDLR, LPL, SLC2A1, CD36, FABP3), FA desaturation (SCD1) and transcriptional regulation (SREBF1, PPARγ), was assessed. The findings showed that EOC feeding increased the milk fat yield (kg/day) and improved the FA profile of bovine milk by significantly decreasing the concentration of de novo synthesized FA and saturated FA, and increasing long – chain, monounsaturated FA concentration as well as the content of specific FA like C18:1 cis-9, the sum of C18:1 trans-10 and trans-11 and CLA cis-9, trans-11 acids. Concerning gene expression results of adipose tissue, only SREBF1 was significantly upregulated in the perirenal fat of OC group in response to the increased level of C18:0 and C18:1 cis-9 in the OC diet, while the mRNA expression of the genes involved in FA synthesis and desaturation was unaffected. Additionally, the decreased proportion of the de novo C<16 FA noticed for the OC group was not accompanied by changes in mammary lipogenic gene expression. The increased secretion of LCFA observed in the milk of OC group did not affect the mammary LPL mRNA abundance, or in mRNA abundance of genes involved in the uptake, transport and trafficking of FA in the cells (CD36, FABP3, VLDLR, SLC2A1). Overall, 10% DM of EOC could be included in cow diets without adversely affecting milk traits or the expression of genes involved in lipid metabolism in cows.

Session 16 Poster 16

Improving gut functions and egg nutrition with stevia Residue in laying hens

M.X. Tang[1,2], Y.L. W.[2], C. Y.[1], C. P.[2], H.G. L.[2], Y. L.[1], Y.D. C.[3], W. L.[3], X. X.[2,3] and X.F. K.[2,3]

[1]Hunan Agricultural University, Hunan Key Laboratory of Traditional Chinese Veterinary Medicine, Changsha, Hunan province, Changsha 410128, China, P.R., [2] Chinese Academy of Sciences, Institute of Subtropical Agriculture, Hunan province, Changsha 410125, China, P.R., [3] Fuyang Normal University, School of Biology and Food Engineering, Anhui Province, Fuyang, 236000, China, P.R.; xx@isa.ac.cn

The objective of this study was to investigate the effect of stevia Residue (STER) on the production performance, egg quality, egg nutritional value, antioxidant ability, immunity, gut health, and microbial populations of laying hens during the peak laying period. To this end, 270 Laying hens (35 weeks of age) of the Yikoujingfen 8 were randomly divided into 5 treatment groups, including the control group and 2, 4, 6 and 8% STER added groups. The results showed that STER had no statistically significant effect on egg quality and production performance of laying hens (P>0.05), but it could significantly improve the yolk colour, amino acid content (alanine, proline, valine, ornithine, asparagine, aspartic acid, and cysteine) of egg whites (P<0.05), and the content of acetate and cis-13, 16-docosadienoic acid in egg whites (P<0.05). Moreover, the addition of STER increased the levels of serum immunoglobulin G (IgG) and interleukins 2, 4, and 10 (P<0.05), improved the antioxidant ability of the liver and oestradiol level in the oviduct (P<0.05), and decreased the level of cortisol (P<0.05). Further research showed that STER could enhance the height of the villi and crypt depth in the jejunum (P<0.05) and increase the abundance of *Actinobacteriota* in the gut (P<0.05), while inhibiting the abundance of harmful bacteria such as *Proteobacteria*, *Desulfobacterota*, and *Synergistota* (P<0.05). Among all the treatment groups, the 6% addition of STER showed the best results. In conclusion, STER has great potential as a plant-based additive in the feed of laying hens.

In vitro evaluation of olive stones and sorghum used as forage in ruminant diets

N. Kellali[1,2], N. Merino[2,3], E. Mata-Nicolás[2,3], I. Mateos[2,3], C. Saro[2,3], N. Lakhdara[1] and M.J. Ranilla[2,3]
[1]Institute of Veterinary Sciences (University of Mentouri Brothers), Road of Guelma, 25100 El Khroub, Algeria, [2]Universidad de León, Campus Vegazana, s/n, 24071 León, Spain, [3]Instituto de Ganadería de Montaña (CSIC-Universidad de León), Finca Marzanas, s/n, 24346 Grulleros, Spain; mjrang@unileon.es

Tannins are secondary compounds with anti-nutritional factors. If they are present in feed ingredients they could affect ruminal fermentation of the diet. A trial was designed to test the effect of tannins present in sorghum and olive stones when theses ingredients are included in diets and incubated *in vitro* for 24 hours. Two diets were formulated including 60% sorghum or olive stones, 22% wheat bran, 10.2% corn, 7.2% soybean meal and 0.4% vitamin-mineral supplement. Diets were incubated either with or without the addition of polyethylene glycol (PEG) to assess fermentation end products. Volume of gas produced was measured. Methane and volatile fatty acids were assessed by gas chromatography and disappearance of dry matter and neutral detergent fibre were determined. When sorghum was included in the diet, ml of gas produced increased by 15.8% ($P<0.05$) when PEG was added to incubation medium. When olive stones were used as ingredient of the diet gas and methane produced increased by 29.7 and 30.4% respectively ($P<0.05$) when PEG was added. No effect of PEG was detected for any of the diets in the volatile fatty acid production or proportions of individual fatty acids with the exception of the proportion of isobutyrate that decreased by 38% ($P<0.05$) when PEG was added and sorghum was included as ingredient of the diet. Disappearance of dry matter or neutral detergent fibre was unaffected by the addition of PEG for both diets. Under the conditions of the present study, the effect of the tannins on ruminal fermentation was strongest in olive stones than in sorghum.

Preservation of agro-industrial co-products as silage for ruminant feed

K. Paulos[1,2], C. Costa[2], J.M.S. Costa[2], L. Cachucho[3,4], P.V. Portugal[2], J. Santos-Silva[2,3], F. Lidon[1], M.R. Marques[2], E. Jerónimo[4,5] and M.T.P. Dentinho[2,3]
[1]Geobiotec, Departamento de Ciências da Terra, FCT Nova, Lisboa, Portugal, [2]INIAV, Fonte Boa, Santarém, Portugal, [3]CIISA, Avenida Universidade Técnica, Lisboa, Portugal, [4]CEBAL, Centro de Biotecnologia Agrícola e Agro-Alimentar do Alentejo, Beja, Portugal, [5]MED, Instituto Mediterrâneo para a Agricultura, Ambiente e Desenvolvimento, Beja, Portugal; katia.paulos@iniav.pt

Every year, considerable amounts of agro-industrial co-products are produced in Portugal that can be used in animal feed, ensuring nutritious and low-cost diets. Sweet potato (SP), brewers' grains (BG), and tomato pomace (TP) are a few examples of co-products that have great nutritional value but must be preserved for usage outside of their production season due to their high moisture content. This research aimed to characterize these co-products chemically and nutritionally and to preserve them as ingredients in silage mixtures. The silages were prepared to contain 40% dry matter (DM) and 14% crude protein (CP). Wheat bran (WB) and grass hay (GH) were added to the mixtures to attain the DM content. Sweet potato has high starch (32.8% DM) and sugar (23.3% DM), and tomato pomace and brewer's grains are good protein sources (19.6 and 29% DM, respectively). So, two silages were prepared: (1) SP/BG silage (30% SP + 35% BG + 20% WB + 15% GH); and (2) SP/TP silage (30% SP + 35% TP + 20% WB + 15% GH). After 4 months of ensiling, well-preserved silages were obtained, with low pH (4.11±0.03 and 3.90±0.02 in SP/BG and SP/TP, respectively), NH_3-N<10% of total N and soluble N<50% of total N. The CP level was 15 and 14% in DM in SP/BG and SP/TP silages, respectively. Organic matter digestibility was higher in SP/TP than in SP/BG silage (55.33±1.23 vs 51.77±0.50%). Ensiling these co-products in mixtures is a good method of preserving them, resulting in stable and nutritionally balanced feed for ruminant nutrition. This work is funded by PRD2020 through the FEADER, project SubProMais (PDR2020-101-030988, PDR2020-101-030993) and by National Funds through FCT – Foundation for Science and Technology projects UIDB/05183/2020 (MED) and UIDP/CVT/00276/2020 (CIISA).

Session 16 Poster 19

Combination of organic acid and heat treatment decreased ruminal protein degradation in Soybean meal

D.K. Yoo, S.W. Jeong and J.K. Seo
Pusan National University, Animal Science, Department of Animal Science, Life and Industry Convergence Research institute, College of Natural R, 50463, Korea, South; fbrjsgud@pusan.ac.kr

This study aims to evaluate the protective effects against rumen degradation when organic acid and heat are combined with soybean meal, which is widely used as a protein feed. The protected soybean meal (PSBM) was developed by adding an organic acid solution with a concentration of 1.5 M to the soybean meal, and then heating the mixture at a temperature of 160 °C for 1 h. To investigate the ruminal degradation of the PSBM developed in this study, both *in situ* and *in vitro* fermentation experiments were conducted. The *in situ* method was conducted to determine the degradability of rumen dry matter (DM) and crude protein (CP), and the results showed that the PSBM had significantly lower effective degradability than the Control group ($P<0.01$). No significant differences were observed in *in vitro* intestinal degradation between the two treatments. Consistent with the *in situ* results, the *in vitro* experimental results showed that the DM and CP degradability were significantly lower in PSBM than in the Control ($P<0.01$). The concentration of ammonia nitrogen was also significantly lower in the PSBM ($P<0.01$). The methane emission was significantly lower in the PSBM than in the Control ($P<0.01$). In conclusion, the PSBM developed in this study could potentially serve as a protein source that reduces ruminal ammonia release while increasing intestinal protein utilization.

Session 16 Poster 20

Bakery by-products in herbage-based diets for dairy cows: effects on milk yield and reticular pH

A.-M. Reiche[1], M. Tretola[1,2], J. Eichinger[1], A.-L. Hütten[1], A. Münger[1], L. Eggerschwiler[1], L. Pinotti[2] and F. Dohme-Meier[1]
[1]Agroscope, Ruminant Nutrition and Emissions, Tioleyre 4, 1725 Posieux, Switzerland, [2]University of Milan, Department of Veterinary Medicine and Animal Sciences, Via Trentacoste 2, 20134 MILANO, Italy; anna-maria.reiche@agroscope.admin.ch

Feeding both recycled former food, such as bakery by-products (BBP) and human-inedible resources such as herbage is a key element of sustainable livestock farming. As the combination of both feed resources was not yet investigated, this work evaluated the effects of including BBP in concentrate (CONC) feed in herbage-based diets on feed intake, milk production and reticular pH of early lactating dairy cows. Twenty-three Holstein cows (days in milk: 33±15 d) were used in two seasons (n=11 in spring and n=12 in autumn). The cows were fed freshly cut grass *ad libitum* and, balanced for parity, DIM and milk yield, they were assigned to two CONC types: a cereal-based control (CON) CONC (n=11) and a CONC containing 55% of BBP (BP, n=12). The CONC were calculated for similar gross energy (19 MJ/kg dry matter (DM) for CON and 20 MJ/kg DM for BP) and crude protein contents (11% of DM). In both seasons, the CONC intake was similar for CON and BP cows. Intakes of herbage DM and ensuing total DM, CP, NDF and NEL were greater for BP than CON cows in spring, whereas they were similar in autumn (season × CONC type interaction: all $P<0.05$). The CON cows ingested more starch, irrespective of the season ($P<0.05$), and less water-soluble carbohydrates (more pronounced in spring than in autumn, interaction: $P<0.05$) than BP cows. Milk yield and composition were not altered by CONC type, except milk lactose percentage which was increased in CON compared to BP cows in both seasons ($P<0.05$). In spring, mean, min and max pH tended to be greater for BP than CON cows, while in autumn the opposite tendency was found (interaction: $P<0.10$). The time below pH 6.04 of CON cows tended to be greater in spring and lower in autumn than that of BP cows. The inclusion of 55% BBP in CONC did not show any negative effects on feed intake and milk production. The season-dependent effects on reticular pH might be related to the varying grass and BBP compositions and need further investigation, as does the relationship between pH range and rumen health.

Session 16 Poster 21

Evaluation of legumes for fermentability and protein fractions using *in vitro* rumen fermentation

B.Z. Tunkala[1], K. Digiacomo[1], P.S. Alvarez Hess[2], F.R. Dunshea[1,3] and B.J. Leury[1]
[1]The University of Melbourne, Agricultural Sciences, Parkville campus, 3010, Parkville VIC, Australia, [2]Agriculture Victoria Research, 1301 Hazeldean Road, 3821, Ellinbank VIC, Australia, [3]Faculty of Biological Sciences, The University of Leeds, Leeds, LS2 9JT, United Kingdom; btunkala@student.unimelb.edu.au

A total of eight forage legumes including *Peltophorum pterocarpum, Neptunia monosperma, Acacia sutherlandii, Gliricidia sepium, Bauhinia hookeri* and three *Desmanthus* species were collected to assess their *in vitro* fermentability, degradable and undegradable protein fractions using *in vitro* fermentation technique. Soybean meal and lucerne hay were used as control. The total gas production ranged from 12.8 ml/g in *P. pterocarpum* to 127.3 ml/g in soybean meal. There was no difference in the total gas and volatile fatty acid (VFA) production between *Desmanthus* cultivars, $P<0.001$. The total VFA concentration from *G. sepium* (117.7 mM/l) and *A. sutherlandii* (111.3 mM/l) was larger than other legumes except for soybean meal (157.1 mM/l) and lucerne hay (130.4 mM/l), $P<0.001$. The methane gas percentage (1.98 and 2.08%) and total VFA (97.0 mM/l and 96.9 mM/l) were not different between *B. hookeri* and *P. pterocarpum*, $P<0.001$. The maximum *in vitro* digestible crude protein (IVDP) was calculated for soybean meal (91%) and lucerne hay (85%), followed by *A. sutherlandii* (73%) and *G. sepium* (58%), $P<0.01$. *B. hookeri* and *P. pterocarpum* showed a negative IVDP after 4 h incubation and reached 25 and 33% after 24 h, respectively. The percentage of fraction 'a' was larger in JCU9 (55.1% crude protein (CP)), and *G. sepium* (47.2%CP), and lower in *B. hookeri* (1.09%CP) and *P. pterocarpum* (1.48% CP), $P<0.001$. Soybean meal (61.3%CP), lucerne hay (44.9%CP) and *A. sutherlandii* (38.6%CP) exceeded others in the proportion of fraction 'b', followed by *G. sepium* (27.2%CP) and JCU5 (20.7%CP), $P<0.001$. The unavailable fraction increased with increasing phenolic content and reached more than 94% for both *B. hookeri* and *P. pterocarpum*. The findings indicate the possibility of using *A. Sutherlandii* and *G. Sepium* as a substitute for lucerne hay for their greater protein potential. However, these legumes need to be tested *in vivo* before promoting for further use to confirm the variability reported here.

Session 16 Poster 22

Dehydration temperature- effect on physicochemical and nutritional characteristics of byproducts

M. Alves[1], K. Paulos[2,3], C. Costa[3], D. Gonçalves[1], T. Orvalho[1], J.M.S. Costa[3], L. Cachucho[4,5], J. Santos-Silva[3,4], E. Jerónimo[5,6] and M.T.P. Dentinho[3,4]
[1]INOV.LINEA, TAGUSVALLEY – Science and Technology Park, 2200-062 Abrantes, Portugal, [2]Departamento de Ciências da Terra, Geobiotec, FCT Nova, 2829-516 Caparica, Portugal, [3]Instituto Nacional de Investigação Agrária e Veterinária, Quinta da Fonte Boa, 2005-424 Vale de Santarém, Portugal, [4]Centro Investigação Interdisciplinar em Sanidade Animal, University of Lisbon, 1300-477 Lisboa, Portugal, [5]Centro de Biotecnologia Agrícola e Agro-Alimentar do Alentejo, IPBeja, 7801-908 Beja, Portugal, [6]Mediterranean Institute for Agriculture, Environment and Development, CEBAL, 7801-908 Beja, Portugal; teresa.dentinho@iniav.pt

The aim of this work was to test the dehydration conditions of sweet potato (SP) and carrot (C) wastes and tomato pomace (TP) with temperature and air speed control for use as animal feed. The effects on colour, water activity (a_W), chemical and nutritional composition and cost were evaluated. Carrot was dehydrated at 60, 70 and 80 °C and SP and TP were dehydrated at 70 and 80 °C. Dehydration was conducted until weight loss stabilized. Freeze-dried samples were used as Control for chemical and nutritional composition. Except SP at 70 °C the moisture in all samples was reduced to values below the critical limit that ensures microbial stability. In SP dehydrated at 70 °C the value was 0.7% above the critical limit. However, all samples had a low a_W ($\leq$0.35), which suggests microbiological and chemical stability. Organic matter digestibility was not affected by temperature but in SP, starch was significantly reduced (24, 19 and 13% in dry matter in Control, 70 and 80 °C, respectively). The energy cost to process 1 kg of fresh by-products was lower at 60 °C for C (€0.39-044), 70 °C for SP (€0.51-0.57) and 80 °C for TP (€0.80), assuming an electricity cost of €0.15/kWh. With these dehydration conditions, it is possible to ensure product stability at a lower cost. In an industrial setting, process, costs can be reduced by increasing the scale of production and with more efficient dehydrators. This work is funded by PRD2020 through the FEADER, project SubProMais (PDR2020-101-030991, 030988, 030993) and by National Funds through FCT-Foundation for Science and Technology projects UIDB/05183/2020 (MED) and UIDP/CVT/00276/2020 (CIISA).

Session 16 Poster 23

Bacterial ecology and silage quality improvement upon treatment with microbial inoculants

I. Nikodinoska[1], M. Franco[2], M. Rinne[2] and C.A. Moran[3]
[1]Alltech European Headquarters, Summerhill Road, A86X006, Dunboyne, Co. Meath, Ireland, [2] Natural Resources Institute Finland (Luke), Tietotie 2, FI-31600 Jokioinen, Finland, [3]ALLTECH SARL, Regulatory Affairs Department, Rue Charles Amand, 14500, Vire, France; ivana.nikodinoska@alltech.com

Microbiota composition in grass silage produced under two fermentation conditions, spontaneous or controlled via microbial inoculant addition, was examined. A second cut timothy (*Phleum pratense*) and meadow fescue (*Festuca pratensis*) grass was harvested (August, 2022) in Jokioinen, Finland, using a farm scale precision chopper. The raw material contained 27.4% dry matter (DM) and 2.95% water-soluble carbohydrates on a fresh matter basis, and thus classified as moderately difficult forage to ensile. The grass was inoculated with silage additive composed of *Lactiplantibacillus plantarum* IMI 507026 and *Pediococcus pentosaceus* IMI 507025 at a dose of 1×10^6 cfu/g of fresh matter, or with water (control). Five replicates per treatment were prepared in 12 l capacity cylindrical silos at a density of 160 kg DM/m^3. After 90 days of ensiling at 20 °C, samples of each replicate were collected. Freeze-dried and ground silages (0.25 g) were subjected to DNA extraction using QIAamp DNA Stool Mini Kit (Qiagen; Sollentuna, Sweden). Amplicon metagenomic sequencing (2×250 bp), targeting 16S rRNA gene V4 region, according to Novogene's pipeline was used for sequencing and bioinformatic analysis. QIIME2's classify-sklearn algorithm was used for species annotation of each amplicon sequence variants (ASV) and Silva database for bacterial ASV taxonomy assignment. T-test was performed to determine species with significant variation between groups (P-value<0.05). Microbiota analysis evidenced *Lactobacillaceae* and *Leuconostocaceae* as dominant families, with *Pediococcus-Lactobacillus* and *Weissella-Pediococcus* as dominant genera in treated and control silage, respectively. T-tests confirmed higher (P<0.05) *Pediococcus-Lactobacillus* and lower (P<0.05) Weissella-*Lactococcus-Leuconostoc-Enterococcus* abundance in treated compared to control silages. These findings suggest that the microbial inoculation used in this study can dominate and modulate silage fermentation, inhibiting the growth of heterofermentative strains such as *Weissella*, which may alter silage losses and quality.

Session 16 Poster 24

Growth performance, carcass and meat quality of pigs fed *Chlorella vulgaris* supplemented diets

J.M. Almeida[1,2,3], M. Parreiras[3], R. Varino[3], A. Sequeira[3], R.J.B. Bessa[1,2] and O. Moreira[1,2,3]
[1]AL4AnimalS, Associate Laboratory for Animal and Veterinary Sciences, FMV-UL, Lisboa, Portugal, [2]CIISA – Centre for Interdisciplinary Research in Animal Health, FMV-UL, Av Universidade Técnica, 1300-477 Lisboa, Portugal, [3]INIAV – National Institute for Agricultural and Veterinary Research, Rua Prof Vaz Portugal, 2005-424 Vale de Santarém, Portugal; joaoalmeida@iniav.pt

Microalgae are being highlighted as new and alternative ingredients for animal feed. *Chlorella vulgaris* has been used to supplement monogastric diets, but its administration could impact growth and meat quality. The current study assessed the impact of adding 1.5 and 3% of dried *C. vulgaris* to an iso-energetic and iso-protein conventional diet on growth performance during the finishing period of 72 pigs F1 [Pietran × (Large White × Landrace)]. Males and females with initial weights of 48.4 and 49.4 kg, respectively, were allocated to 24 pens of 3 animals, and within each pen, allocated to one of the diets. The pens allowed individualised feeding during two 30-minute periods (morning and afternoon), and water access was free all day. Live weight was checked weekly, and animals were slaughtered when they reached 100 kg. Carcass and meat parameters were evaluated immediately after slaughter and at 24 h *post-mortem*, when longissimus lumborum (LL) samples were collected for further analysis. The dietary inclusion of *C. vulgaris* increased the time to reach the target slaughter weight (P<0.05), reduced the voluntary dry matter intake and average daily gain and tended (P<0.10) to decrease hot and cold carcass weight but did not affect (P>0.10) the feed conversion rate. Dietary *Chlorella* supplementation did not affect the weight or proportion of the abdominal adipose deposits nor the backfat layers' thickness. However, females showed significantly lower values for most of these variables. The dietary treatments did not affect the longissimus and semimembranosus muscles' pH, LL weight losses, shear force or meat and backfat colour parameters. Ingestion of up to 3% *C. vulgaris* did not change the carcass or meat quality traits, aside from a slight but considerable impact on intake and slaughter weight. Funding: Project: ALGAVALOR: POCI-01-0247-FEDER-035234; LISBOA-01-0247-FEDER-035234; ALG-01-0247-FEDER-035234.

Session 16 Poster 25

Inclusion of slow-release ammonia-treated barley in concentrates – feed intake and milk production

K.S. Eikanger[1], M. Eknæs[1], I.J. Karlengen[2], J.K. Sommerseth[3], I. Schei[3] and A. Kidane[1]
[1]Norwegian University of Life Sciences, Faculty of Biosciences, Arboretveien 6, 1430 Ås, Norway, [2]Norgesfôr AS, Akershusstranda 27, 0150 Oslo, Norway, [3]TINE SA, BTB – NMBU PB. 5003, 1432 ÅS, Norway; katrinee@nmbu.no

The need to utilize alternative protein sources for sustainable milk production, plus the ability of the rumen microbiome to convert dietary non-protein nitrogen into high-quality microbial protein, has led to an increasing interest in locally produced cereal grains treated with slow-release ammonia as a replacement for conventional protein ingredients. We assessed the effects on feed intake and milk production of two slow-release ammonia (Home'n Dry, FiveF Alka Ltd) treated concentrates with high contents of locally produced barley differing in pre-pelleting particle size (i.e. Alka mjolk-fine=Alka-F, and Alka mjolk-coarse=Alka-C) against a negative control based on the Alka ingredients mixed with urea (Urea-F) and a soy-based positive control (Soya-F). The concentrates were roughly iso-nitrogenous and iso-energetic. Eight multiparous early lactation Norwegian Red dairy cows with initial daily milk yield (MY±SD) of 30.5±5.94 kg were used in a 4×4 duplicated Latin square design experiment (4 diets over 4 periods of 35 d each) accommodated in a tie-stall housing with *ad lib* access to grass silage and water. The concentrates were fed in optimized amounts based on Soya-F (w/w replacement) in 3-split portions per day. Feed intake, MY, milk composition, energy corrected MY (ECM), and ECM efficiency were analysed using a linear mixed model in R (Version 4.2.2) with treatment, period, and week within a period as fixed effects and cow as random effect. Covariance structure was selected based on Bayesian information criterion. Feed intake was not affected ($P \geq 0.05$) by concentrate type. But MY, ECM, and ECM efficiency were higher ($P<0.05$) for Alka-F and Soya-F than Urea-F, while Alka-C was intermediate. Milk fat (%) tended to be lower ($P=0.10$) in Soya-F, and milk protein (%) was significantly lower ($P=0.007$) for Alka-C compared to others. Milk lactose (%) did not differ among treatments ($P=0.27$). In conclusion, slow-release ammonia-treated concentrates, based on local ingredients, can potentially replace soy-based concentrates without adverse effects on milk production.

Session 16 Poster 26

Reducing aggression in finishing pigs through short application liquid calming herbal blend

C. Nicolás-Jorrillo[1], D. Carrión[1], D. Escribano[2], A. Brun[3] and E. Jiménez-Moreno[1]
[1]Cargill Animal Nutrition, Poligono Industrial Riols s/n, 50170 Mequinenza, Spain, [2]University of Murcia, Animal Production, Campus de Espinardo s/n, 30100 Murcia, Spain, [3]IRTA, Food Quality and Technology Program, Finca Camps i Armet, 17121 Monells, Spain; carlos_denicolas@cargill.com

The aim of this study was to evaluate the effects of the inclusion of a calming herbal extract blend in liquid form (BehavePro® L) varying in dosage and in application length on feeding and aggressive behaviours and carcass quality. BehavePro is a nutritional solution containing natural compounds and plant extracts added to drinking water. Four hundred seventy-two [Pietrain × (Landrace × Large White)] pigs, were sexed (entire males and females) and allotted randomly to two treatments (group A; 0.25% solution for 15 days, and group B; 0.8% solution for 5 days period) with twenty-four conventional pens of 17 pigs each and six pens of 12 pigs each with one automatic feeding system per pen. Four days before adding the solution (day 118), pigs weighed 115±2.0 and 116±4.5 kg per pig for the group A and B, respectively. Pigs were slaughtered at day 137 and 146 for the evaluation of carcass quality and skin lesions, and the feeding behaviours after fasting, respectively. After the first load, group A continued receiving 0.25% while group B, was maintained at 0.8% for 2 days following by 0.25% till slaughter. All pigs had free access to water until transportation to the abattoir. Fasting lasted for 21 h and the transportation and lairage time was 4 h. Feed intake of pig was affected for the first 2 days after fasting with higher values in the group B than in the group A ($P<0.05$). Cortisol concentration was reduced with the use of BehavePro and the time of application ($P<0.05$). Final pig BW was not affected by the treatment ($P>0.1$). Hot carcass tended to be 1.5 kg heavier in group B than A ($P=0.08$). The incidence of the skin lesions was lower in the group B than in the group A in the front (68.1 vs 77.3% pigs with less than 2 scratches for the group A and B, respectively) and middle regions of the carcass ($P<0.05$). In conclusion, the addition of 0.8% of BehavePro via water for the last 5 days of fattening reduced the skin damage at slaughter due to its calming effect on aggressive behaviour.

Evaluating fermentative profile in corn silage using chemical and biological additives

C. Mastroeni[1], S. Van Kuijk[2], F. Ghilardelli[1], S. Sigolo[1], E. Fiorbelli[1] and A. Gallo[1]
[1]Università CAttolica del Sacro Cuore, Department of Animal Science, Food and Nutrition (DIANA), Via E. Parmense 84, Piacenza 29121, Italy, [2]Selko, Stationsstraat 77, 3800 AG Amersfoort, the Netherlands, the Netherlands; antonio.gallo@unicatt.it

Silage additives are widely used to reduce dry matter loss, decrease temperature, and enhance quality of ensiled fodder. The objective of this work was to evaluate the effects of chemical and biological silage additives on the fermentative profile of ensiled corn. After harvesting, fresh-chopped corn was ensiled directly (control, CTR) or sprayed with one of two mixes of organic acids before ensiling, Fyvalet Silage (FS, Selko) at a concentration of 3 g/kg of freshly-chopped forage or Selko TMR (ST, Selko), at a concentration of 2 g/kg of freshly-chopped with a combination of *L. hilgardii* CNCM I-4785 and *L. buchneri* NCIMB 40788 (Magniva Platinum 1 or MP1, Lallemand SAS), at recommended concentration of 150,000 cfu/g of freshly-chopped forage. Corn silage was ensiled in 20 l mini-silos, stored at room temperature and opened after 12 or 20 weeks. There were differences in dry matter, pH, lactic acid, acetic acid, propionic acid, and ethanol between treatments ($P<0.05$), but all treatments showed an adequate fermentative profile. The MP1 mini-silos had higher ($P<0.05$) acetic acid content than the other groups, whereas the highest ($P<0.05$) value of ethanol was measured in FS mini-silos. No differences were observed for LAB, mould, and yeast counts, indicating that chemical or biological additives in ensiled corn allow achieving a safe and well-fermented product. Evaluation of fermentative profiles of corn silage with chemical and biological additives included shows that both solutions improve the final quality of the fodder.

Effect of *Chlorella vulgaris* incorporation in pig diets on pork chemical composition and fatty acids

J.M. Almeida[1,2,3], A.P. Portugal[3], R.J.B. Bessa[1,2] and O. Moreira[1,2,3]
[1]AL4AnimalS – Associate Laboratory for Animal and Veterinary Sciences, FMV-UL, Lisboa, Portugal, [2]CIISA – Centre for Interdisciplinary Research in Animal Health, FMV-UL, Av Universidade Técnica, 1330-477 Lisboa, Portugal, [3]INIAV – National Institute for Agricultural and Veterinary Research, Rua Prof Vaz Portugal, 2005-424 Vale de Santarém, Portugal; joaoalmeida@gmail.com

Supplementing diets for finishing pigs with microalgae may affect the nutritional value of their meat. The current study assessed the impact of adding dried *Chlorella vulgaris* to iso-energetic and iso-protein conventional final diet on the nutritional composition and meat fatty acid (FA) profile. The experimental design involved 36 males and 36 females [Pietran × (Large White × Landrace)] to study the three experimental diets with 0%, 1.5 and 3% incorporation of *C. vulgaris*. The animals, separated by sex, were assigned to 24 pens of 3 animals and allocated to one of the diets within each pen. The pens allowed individualised feeding during two 30-minute periods (8:00; 16:00), and water access was free all day. The animals were weighed weekly and slaughtered when they reached 100 kg live weight. Longissimus lumborum muscle samples were collected 24 h *post-mortem*, vacuum packed and stored at -20 °C until analytical procedures. Including *C. vulgaris* in the diets increased linearly ($P<0.001$) the ash content but not the calcium, sodium, phosphorus, potassium or magnesium content. The intramuscular fat (IMF) content was lowest, and the meat protein content was highest with the dietary treatment with 1.5% of *Chlorella* (Quadratic contrast $P<0.05$). Females showed a lower ($P<0.05$) IMF content than males, 1.88 vs 1.38%. The FA composition of the longissimus muscle was little affected by the diets but differed with sex, reflecting the IMF differences. Due to their higher IMF, males had higher saturated FA and lower polyunsaturated FA (PUFA). Neither the omega-3 PUFA nor the omega-6 PUFA were influenced by the inclusion of microalgae in the diet. Similarly, the diets did not affect monounsaturated fatty acids, although 16:1c9 and 18:1c11 showed higher levels in animals fed the 1.5% Chlorella diet. Funding: Project: ALGAVALOR: POCI-01-0247-FEDER-035234; LISBOA-01-0247-FEDER-035234; ALG-01-0247-FEDER-035234.

Session 16 Poster 29

Bacillus fermentation products of hog hair waste serve as potential animal feed additives

W.J. Chen, K.C. Chang, Z.J. Zhuang, Y.H. Cheng and Y.H. Yu
National Ilan University, Department of Biotechnology and Animal Science, No.1, Sec.1, Shennong Rd., 26047, Taiwan; wjchen@niu.edu.tw

We isolated three *Bacillus* strains from livestock waste: *Bacillus licheniformis* (BL), *Bacillus amyloliquefaciens* (BA) and *Bacillus subtilis* 25 (BS25). The crude enzyme produced by BA, BL and BS25 showed protease activity at 40-70 °C and pH 8-12. Heating at 90 °C for 2 hours, the enzyme activity still retains about 90%, and it will not be inhibited by EDTA, iodoacetic acid and PMSF. In zymogram assay, we confirmed that the molecular sizes of these three keratinases were mainly distributed between 17 and 52 kDa. When testing the fermentation conditions, the intact hog hair and hog hair powder were both tested, and BA strain with the best keratinase activity was selected for further study and its optimal fermentation conditions established. We performed acid hydrolysis on the product obtained from intact hog hair fermented for 5 days to analyse the amino acid composition of hog hair before and after fermentation with BA strain. The results showed that in the fermentation products of hog hair, the contents of essential amino acids such as methionine (Met), tryptophan (Trp), phenylalanine (Phe), lysine (Lys) and other non-essential amino acids were all significantly increased. We further established the optimal fermentation conditions for BA to decompose intact hog hairs as follows: fermentation temperature of 37 °C, adding 2% molasses as carbon source, 4% intact hog hair as nitrogen source, and the initial bacterial concentration of 10^{11} cfu/ml. Then, soybean meal, a raw material commonly used in broiler's feed was applied as an adsorbent, and the hog hair fermentation product is sprayed on it using a fluidized bed dryer, serving as broiler feed additive. The results of the animal experiments on broiler showed that 2.5 and 5% substitution of soybean meal with hog hair fermentation product can effectively improve the growth performance of broilers, such as body weight, average daily weight gain and feed conversion rate, in a dose-dependent manner with statistically significant difference. Based on our current findings, BA has the potential for decomposing hog hair waste and the value-added usage as novel feed additives, thus achieving the goal of circularity in agriculture.

Session 16 Poster 30

Effect of coffee grounds on digestibility, ruminal fermentation, and microbial protein synthesis

M. Medjadbi[1], I. Goiri[1], R. Atxaerandio[1], S. Charef[1], C. Michelet[1], J. Ibarruri[2], B. Iñarra[2], D. San Martin[2] and A. García-Rodríguez[1]
[1]NEIKER – Basque Institute for Agricultural Research and Development, Department of Animal Production, Agroalimentary campus of Arkaute, s/n, N 104, km355, 01192 Arkaute, Álava, Spain, [2]AZTI, Food Research, Basque Research and Technology Alliance (BRTA), Technology Park of Bizkaia, Astondo Bidea, Building 609, 48160, Derio, Bizkaia, Spain; mmedjadbi@neiker.eus

Coffee grounds (CG), through their phenolic compounds, can contribute to modulating rumen fermentation towards more efficient fermentation routes and decreasing enteric methane production. This study evaluated the effect of 4 levels of CG inclusion in concentrate on enteric methane production, feed intake, apparent digestibility, ruminal fermentation, and microbial protein synthesis. In a replicated 4×4 Latin square design, 8 non-productive Latxa ewes were randomly assigned to a concentrate that differed in the level of CG: 0, 10, 15, and 20%. In each period, 15 days of adaptation were allowed, followed by 7 days in metabolic cages, and two days in respiration chambers. To avoid a carry-over effect a minimum of 7 days were allotted in which ewes consumed concentrate with 0% CG and forage. Total organic matter intake and CH_4 emissions (g/d) presented a quadratic response ($P<0.050$) to increasing levels of CG in the feed. However, when CH_4 emissions were corrected for a kg of organic matter intake (OMI), a linear decrease was observed with increasing levels of CG in the concentrate ($P=0.009$). This reduction in CH_4 emissions (g/kg OM) could be explained by the linear decrease ($P=0.034$) observed in apparent digestibility with increasing levels of CG in the concentrate. No significant response was found in CH_4 emissions corrected for digestible organic matter input or individual VFA ratios. For microbial N flux, a linear increase in microbial protein supply efficiency ($P=0.008$) was observed with increasing levels of CG in the concentrate. In conclusion, increasing doses of CG in the concentrate linearly reduced methane emissions per kg of OMI, due to a decrease in digestibility, and increased the efficiency of microbial synthesis.

Session 16 Poster 31

Environmental impact of former food estimated trough life cycle assessment

M.G.D. Azzena[1], A. Luciano[2], L. Pinotti[2] and A.S. Atzori[1]
[1]University of Sassari, Department of Agricultural Science, via E De Nicola 9, 07100 Sassari, Italy, [2] University of Milan, Department of Health, Animal Science and Food Safety, VESPA, Via Celoria 10, 20134 Milano, Italy; asatzori@uniss.it

The livestock sector is called to improve resource utilization moreover minimizing the direct food-feed competition in land use. Former food products (FFPs), i.e. food products for human consumption that comply with EU food legislation but cannot be used for human consumption must be found for inclusion as a supplementary food in the diet of farm animals, as an alternative to cereals. Due to their variability in composition, they are classified according to origin into bakery and confectionary products. There is little information on life cycle assessment of FFPs in animal feed. This study aimed to retrieve information on the environmental impact of bakery and confectionery products obtained with the complete life cycle assessment (LCA) from the point of production to incorporation into feed for animal nutrition. Existing environmental product declaration (EPD) of edible foods can be adopted as environmental impact of FFps in animal diets. The EPD system considers: Upstream, which involves calculating the LCA of individual ingredients used in the formulation; Core, which calculates the energy expended, including water, electricity and methane emitted; and Downstream, which considers the LCA analysis from when the product leaves the factory until it arrives on the shelf. Assuming that the final product is used in animal feed, emissions produced as they undergo further processing might increase in energy expenditure and methane produced. Literature evidences gathered from scientific databases showed that variations in emissions depends on the origin of the product, with values ranging on average from 1.323 kg CO_2eq/kg (crackers), 1.577 (rusks), 1.277 (bread), 1.173 (pasta), 1.645 (shortbread), 2.710 (filled cookies), 2.982 (snacks) 2.922 (puff pastry), and 1.478 (breakfast cereals) kg of CO_2eq/kg per 100 g of bakery product. Considering the variations in product emissions, the addition of LCA information would help the feed industry reconsider environmental policies and promote more sustainable feed production by reducing environmental impact.

Session 16 Poster 32

Effect of coffee grounds on productive performance, milk fatty acid profile and methane production

M. Medjadbi[1], I. Goiri[1], R. Atxaerandio[1], S. Charef[1], C. Michelet[1], J. Ibarruri[2], B. Iñarra[2], D. San Martin[2] and A. García-Rodríguez[1]
[1]NEIKER – Basque Institute for Agricultural Research and Development, Department of Animal Production, Agroalimentary campus of Arkaute, s/n, N 104, km355, 01192 Arkaute, Álava, Spain, [2]AZTI, Food Research, Basque Research and Technology Alliance (BRTA), Technology Park of Bizkaia, Astondo Bidea, Building 609, 48160, Derio, Bizkaia, Spain; mmedjadbi@neiker.eus

Coffee grounds (CG) are one of the food wastes that have been studied as an alternative in livestock feeding in the last decade. These studies have proved that the addition of up to 10% of CG in the concentrate does not have a negative effect on the productive performance of dairy ruminants. This study evaluated higher levels of inclusion of CG (0, 10, 15, and 20%) in the concentrate on milk yield and quality, milk fatty acid profile, and methane production. In this trial of 6 weeks of duration, 48 Latxa dairy sheep were used. Ewes were divided into blocks according to parity, milk yield and days in milk. All the concentrates were formulated to be isoenergetic, isoproteic, isofat and to meet the production needs. The concentrate was given in two doses of 450 g of dry matter/milking and fescue hay was offered *ad libitum*. The increase of CG doses in concentrate up to 20% did not affect milk yield or milk quality but affected the milk fatty acid profile. The increase in CG doses resulted in a linear increase in the content of C18:0, C18:1 trans-11, total n-3, total n-6, conjugated linoleic acid, total monounsaturated fatty acids, and total polyunsaturated fatty acids. In contrast, decreased linearly the content of C16:0 and total saturated fatty acids. Methane production (g/d) and methane intensity (g /l of milk) decreased linearly with increasing levels of CG inclusion in the concentrate. In conclusion, increasing the inclusion level of CG up to 20% in concentrate in dairy sheep did not have a negative effect on milk production and its quality, but modified the milk fatty acid profile towards a healthier profile and decreased methane production and intensity.

Effect of including fresh pears as feedstuff in lactating ewes' diet on the ewes' milk curd strength

I. Caro[1], J. Mateo[2], S. Kasaiyan[2], N. Merino[2,3], E. Mata-Nicolás[2,3], I. Mateos[2,3], C. Saro[2,3], A. Martín[3], F.J. Giráldez[3] and M.J. Ranilla[2,3]
[1]Universidad de Valladolid, Av. Ramón y Cajal, 7, 47005, Spain, [2]Universidad de León, Campus Vegazana, s/n, 24071 León, Spain, [3]Instituto de Ganadería de Montaña (CSIC-Universidad de León), Finca Marzanas, s/n, 24346 Grulleros, Spain; nmerb@unileon.es

Spain is the one of major European producers of fruit and vegetables and almost half of the initial production of fruits and vegetable is lost or wasted at different stages of the food chain supply. The inclusion of discarded fruits in animal feed would allow minimizing this impact and reducing the costs of livestock production. The aim of this research is to assess the eventual effect of using pears in the diet of lactating ewes on a relevant milk quality trait. Two groups of ten lactating Assaf ewes on the 2nd third of lactation were fed on two different isoproteic and isoenergetic diets (total mixed rations) for 35 days. One included pear, at a level of 116 g/kg dry matter, as ingredient (PEAR) and the other did not contained pear (CONTROL). The last day of the trial, the milk from each ewe was sampled, skimmed. to avoid the effect of fat, and samples (in duplicate) were used for composition analysis and texture analysis. For the texture, 100 ml of skimmed milk were poured into 150 ml beaker and tempered to 35 °C and 4 ml of diluted (1/10) rennet (1:100,000 force) were added to the milk and mixed. The beaker was incubated at 35 °C for 40 min and then refrigerated at 4 °C for 24 hours. Afterwards, the curd in the beaker was tempered at 22 °C and the gel strength was determined by a penetration text using a texture analyser (TA-XT2, Stable Micro Systems Ltd, Surrey, UK) equipped with metallic cylindrical probe (SMSP/40) which penetrated the curd 15 mm at 0.2 mm/s. The maximum force (firmness) was recorded. Neither the protein content nor the curd firmness was affected by the diet. The protein contents (%) of CONTROL and PEAR ewe milk were (mean ± standard deviation) respectively 5.30±0.43 and 5.50±0.63 (P=0.418), and the curd strengths (N) 2.55±0.80 and 2.88±1.11 (P=0.767). The curd strength was highly correlated with the protein content in the milk from both ewe groups (the regression coefficient, r2, for CON and PEAR was 0.817 and 0.857, respectively).

Upcycling of Citrus and Vinification by-products' bioactive compounds for broiler diets

A. Mavrommatis and E. Tsiplakou
Agricultural University of Athens, Laboratory of Nutritional Physiology and Feeding, Department of Animal Science, Iera Odos 75, 11855 Athens, Greece; mavrommatis@aua.gr

Although carotenoids and polyphenols derived by *Citrus* and vinification by-products, respectively, generally possess vital antimicrobial and antioxidant properties their *in vivo* impact in animals diet have not been thoroughly studied. For this purpose, 300 one-day-old as hatched chicks (Ross 308) were assigned to five treatments, with four replicate pens and 15 birds in each pen. Birds were fed either a basal diet (CON) or a basal diet supplemented with 25 g/kg ground grape pomace (GGP), or 2 g/kg wine lees extract rich in yeast cell walls (WYC), or 1 g/kg starch including 100 mg pure grape stem polyphenolic extract/kg (PE), or 1 g/kg starch including 25 mg carotenoid extract (CCE) for 42 d. Procyanidin B1 and B2, gallic acid, caftaric acid, (+)-catechin, quercetin, and trans-resveratrol were the prevailing compounds in vinification by-products while β-Cryptoxanthin, β-Carotene, Zeaxanthin, and Lutein in *Citrus reticulata* extract. The CCE feed additive exerted inhibitory properties against *Staphylococcus aureus*, *Klebsiella oxytoca*, *Escherichia coli*, and *Salmonella typhimurium in vitro*. Body weight and feed consumption were not significantly affected. The mRNA levels of *GPX1* and *SOD1* tended to increase in the liver of WYC-fed broilers. SOD activity in blood plasma was significantly increased in WYC and PE groups. The total antioxidant capacity showed significantly higher values in the breast muscle of PE-fed broilers, while the malondialdehyde concentration was significantly decreased in WYC-, PE-, and CCE-fed broilers. The mRNA levels of Interleukin 1β and tumour necrosis factor (*TNF*) were downregulated in the livers of the CCE-fed broilers, while *TNF* and interferon γ tended to decrease in the spleens and bursa of Fabricius, respectively. Our study provided new insights regarding the beneficial properties of bioactive compounds present in *Citrus* and vinification by-products in broilers' immune-oxidative status. These promising outcomes could be the basis for further research under field conditions.

Simplified method developed for estimating the on-farm EFGHG and NH_3 – presentation & results

X. Vergé[1], P. Robin[2], V. Becciolini[3], A. Cieślak[4], N. Edouard[2], L. Fehmer[5], P. Galama[6], P. Hargreaves[7], V. Juškienė[8], G. Kadžiene[8], A.-S. Lissy[9], D. Ruska[10] and M. Szumacher[4]
[1]IDELE, Environmental Service, 8 route de Monvoisin, 35652 Le Rheu, France, [2]INRAE, Institut Agro, 35590 Saint Gilles, France, [3]University of Florence, Via san Bonaventura 13, 50145 Firenze, Italy, [4]Poznan University, Wolynska 33, 60-637 Poznan, Poland, [5]Justus Liebig Universität, Ludwigstraße 21 b, 35390 Gießen, Germany, [6]Wageningen Livestock Research, De Elst 1, 6708 WD Wageningen, the Netherlands, [7]SRUC, Barony Campus, Parkgate, DG1 3NE Dumfries, United Kingdom, [8]Lithuanian University of Health Sciences, R. Zebenkos, 1282317 Baisogala, Lithuania, [9]INRAE Transfert – METYS, Bâtiment de Bioclimatologie, Route de la Ferme, 78850 Thiverval-Grignon, France, [10]Latvia University of Life Sciences and technologies, Faculty of Agriculture, Institute of Animal Sciences, 2 Liela street, Jelgava, 3001, Latvia; xavier.verge@idele.fr

Estimating the greenhouse gas (GHG) and ammonia (NH_3) emission factors (EF) of cattle farming is challenging in open barn production systems since ventilation rates are very difficult to do in such situations. To overcome this difficulty, we developed a method based on: (1) indoor/outdoor CO_2, CH_4, N_2O and NH_3 concentration measurements; and (2) a farm questionnaire estimating the carbon mass balance at the building scale. This 'Simplified Method' has been applied in an international study including eight European countries totalizing over 50 dairy farms with seasonal repetitions and the results of the questionnaire were also used to characterize the farm diversity. The dairy population in the barn where the air samplings were done ranged from 8 to 979 cows. The feeding quantity was between 14 and 31 kg DM/animal/day and the milk production between 18 and 50 kg milk/animal/day. Most of them had concrete or slatted floors (40 and 35% respectively) and one fourth were separated urine and faeces. The 'Simplified Method' will be shortly described and examples of the results per countries, farms and seasons will be presented.

Simulation of three farming strategies on ammonia and GHG emissions and economics

P. Galama and H. Pishgarkomileh
Wageningen University and Research, Livestock research, De Elst 1, 6700 AH Wageningen, the Netherlands; paul.galama@wur.nl

Tackling climate change and other environmental concerns is a global challenge. Given the high contribution of livestock sector on the greenhouse gas (GHG) emissions, this industry has come into focus. Efforts have been initiated by introducing many mitigation strategies. However, still there are debates about effectiveness of them. Given the complexity of dairy system, quantifying the impacts is a challenge. In this study some of the available mitigation strategies have been models using a farm model (DairyWise). The assessment had two parts: (1) individual assessment of mitigation strategies; (2) combination of mitigation strategies as different production systems. For individual assessments, five mitigation strategies including increasing milk production level (6,000, 8,500, and 10,000 kg milk/cow), increasing longevity by changing the young stock ratio (5, 6.7, and 8 young stock per 10 dairy cow), increasing area of long-term grasslands (20:20, 33:6, 39:0 grass area:maize area), sowing clover on grasslands, and increasing grazing intensity (900, 1,600, and 3,600 hours/year) were selected and evaluated at three levels. To have a better overview, three production systems including extensive-regular (ER), extensive-organic (EO), and high-tech (HT) dairy systems were compared with the baseline to show the technical, environmental, and economic differences. The production systems differed from the point of view of the stocking rate, milk production level, grazing intensity, cultivation of clover, available land area, and type of stall. For comparisons, environmental (GHG and ammonia emissions), technical (nutrient balance), and economic aspects were considered. Obtained results showed that the total GHG emission of baseline, ER, EO and HT were 1.21, 1.18, 1.14 and 1.20 kg g CO_2-eq per kg milk, respectively. Results of ammonia emission showed the lowest emissions for EO while the highest emissions were reported for HT. Economic results showed that HT is the most attractive system. Results showed that moving towards more extensive and organic production system reduces the environmental impacts and can be attractive for farmers in term of economic aspect.

Session 17 — Theatre 3

Exploring the effect of dairy cow replacement decisions on feed efficiency and sustainability

A. De Vries
University of Florida, 2250 Shealy Drive, Gainesville, FL 32608, USA; devries@ufl.edu

An important metric for the impact of dairy farming on climate sustainability is the amount of milk produced per unit of enteric methane. Enteric methane emissions in dairy cattle are strongly related to dry matter feed intake. Therefore, the amount of milk produced per unit of dry matter intake is relevant. This should include dry matter eaten by all dairy cattle on the farm, including growing replacement heifers, dry cows, and lactating cows. Increasing the productive life of dairy cows implies fewer replacement heifers are needed but may also result in keeping low milk producing cows in the herd longer. The objective of this study was to explore how economic replacement decisions affect feed efficiency and thereby the climate sustainability of dairy farming. We developed and used a stochastic dynamic programming model of a dairy herd consisting of heifers and cows. The model optimizes voluntary replacement and insemination decisions for millions of states of animals but can also simulate scenarios with non-optimal decisions. Animal states are defined by parity, week in milk, week of pregnancy, week of oestrous cycle, and levels of milk production. Each state is associated with relevant performance, such as milk yield, dry matter intake, fertility, and cash flow. Herd inputs were chosen to represent a typical herd in the USA. Initial results showed that a scenario with optimal replacement decisions led to a 36% annual cow replacement rate with profit of $332/cow per year. Average milk yield was 36.5 kg/cow per day. Herd feed efficiency, calculated as all milk yield divided by all dry matter intake in the herd, was 1.42. Lifetime milk yield was 37,278 kg produced in 1,729 days of lifetime, or 21.49 kg per day of life. A scenario where cows were kept longer led to an 28% annual cow replacement rate with a profit of $265/cow per year. Average milk yield was 35.2 kg/cow per day and herd feed efficiency was also 1.42. Lifetime milk yield was 46,548 kg produced in 2023 days of lifetime, or 22.89 kg per day of life. In conclusion, initial results show that trade-offs exist between profitability, longevity, and climate sustainability. Extending the productive life of dairy cows does not necessarily improve climate sustainability.

Session 17 — Theatre 4

Practical monitoring of individual methane production rates in dairy cows: an alternative

C.M. Levrault[1], J.T. Eekelder[1], P.W.G. Koerkamp[1], C.F.W. Peeters[2], J.P.M. Ploegaert[3] and N.W.M. Ogink[1,3]
[1]Wageningen University & Research, Farm Technology Group, P.O. Box 16, 6700 AA Wageningen, the Netherlands, [2]Wageningen University & Research, Mathematical & Statistical Methods Group, P.O. Box 16, 6700 AA Wageningen, the Netherlands, [3]Mathematical & Statistical Methods Group, Livestock & Environment, P.O. Box 338, 6700 AH Wageningen, the Netherlands; cecile.levrault@wur.nl

Over the past decades, the potent greenhouse gas emissions from cattle farming have kept rising, and reducing the impact of the sector on the environment has become crucial. While reduction strategies have already been identified, the evaluation of the performance of these measures remains limited by the available assessment methods. Several devices have already been developed in an attempt to quantify the individual methane production rates (MPR) of dairy cows under barn conditions. Unfortunately, they are often inaccurate, laborious, or expensive. The newly developed Cubicle Hood Sampler (CHS) could offer a way to compensate for this shortcoming. The principle of the CHS is to collect, in a non-intrusive manner, the gas mixtures produced by cows when lying down through the use of extraction hoods that are positioned above the cubicles. Additional RFID sensors allowed each measurement to be assigned to a specific cow. Camera vision was used to monitor cow postures and filter out biased observations. Due to the physiological rhythm of cattle and the location of the system, individual monitoring of up to 12-h per day can be achieved. To investigate the ability of the CHS to estimate MPR and build a ranking index to select cows based on their methane production levels, the methane emissions of 28 lactating Holstein cows were monitored using, alternatively, climate respiration chambers (CRC, 4-d) and CHS (7-d). Using a partially-polled hierarchical model we developed, the estimations made by the CHS were converted from discrete measurements to continuous methane production curves. Comparisons with the MPR estimated by the CRC showed that the CHS can sense variability between cows, and that its estimates are moderately correlated with the CRC ones. These first results position the CHS as a promising system to monitor the MPR of cows in practical conditions.

Session 17 Theatre 5

Air filtering as alternative approach to combat emissions from cattle facilities

A. Kuipers, P. Galama, R. Maasdam, S. Spoelstra and P. Groot Koerkamp
Wageningen University & Research, De Elst 1, 6708 WD, the Netherlands; abele.kuipers@wur.nl

Various strategies can be applied to reduce ammonia and methane emissions. One strategy is to adapt the animal to the environment, like adding methane blockers to the feed or selecting for low methane animals, and the other is to adapt the environment to the animal. We study the simultaneously capturing of methane and ammonia from air, a practice belonging to the 2nd strategy. In some intensive livestock areas, air scrubbers are common in pig houses to capture ammonia from the air. This technique has not been adopted in cattle housings, mostly because these facilities have an open structure. Moreover, capturing of methane from air in cattle houses is complicated due to the very low concentration in barns (in 60 farms spread over Europe ranging from 5 to 80 ppm at 2 m height) and low solubility of methane in water. In a dairy housing about 30% of methane comes from the manure and 70% from the mouth of cows through the process of rumen fermentation. We examined several options to remove ammonia and methane from dairy barns by use of a survey based on scientific literature, contacts with experimenting farmers and firms and experimentation on capturing more concentrated methane (with aid of a hood mounted over the lying place of cows and recirculating of air). We found no feasible technology to capture ammonia and methane simultaneously, because of the low concentrations and chemical and physical inertness of methane. We conclude that two separate technologies need to be applied for capturing ammonia and methane, respectively. When stored manure is a significant source of ammonia and – temporarily- methane, both emissions can be guided to an air scrubber to remove ammonia by acid. For higher levels of methane, a biobed depending on consortia of methanotrophs could be considered. The methane concentration of ventilation air of dairy barns is too low to be treated. No technologies based on activity of methanotrophs or physical-chemical methods involving adsorption and oxidation of methane have been identified until now. Concentration to around 300-500 ppm seems to be a way out. Technologies like catching methane near the head of the animal, e.g. by a hood or designing new ventilation systems, seem a direction to be further explored.

Session 17 Theatre 6

UAV-based approaches for gaseous emissions assessment in cattle farming

V. Becciolini[1], A. Mattia[1], M. Merlini[1], G. Rossi[1], F. Squillace[1], G. Coletti[2], U. Rossi[2] and M. Barbari[1]
[1]University of Florence, Department of Agriculture, Food, Environment and Forestry (DAGRI), via San Bonaventura 13, 50145 Firenze, Italy, [2]Project & Design S.r.l.s., via Livorno 8/28, 50142 Firenze, Italy; valentina.becciolini@unifi.it

The use of unmanned aerial vehicles (UAVs) as in-flight sensing platforms for atmospheric chemical research is rapidly growing, with an increasing number of field studies published in the past 5 years and with the appearance on the market of gas sensors specifically designed for UAVs. Nevertheless, the main applications are connected to air quality and emission monitoring in urban or industrial contexts. Aiming to provide a comprehensive view on the perspectives of application in dairy farming, we investigated and tested UAV-based approaches for gaseous emissions assessment focusing on equipment (drones, sensors), technical solutions and methods for fluxes estimation. While rotorcrafts represent the primary option, due to their manoeuvrability and autonomous GPS-based hovering, drone size choice largely varies depending on the utilized gas sensing equipment. Non-Dispersive Infrared Sensors (NDIR) are considered suitable for measuring CO_2, with small-sized and low-cost options available and providing sufficient accuracy. Measuring CH_4 in farming contexts requires sensors with appropriate limit of detection and accuracy. Finding miniaturized and low-cost options on the market with such performances is currently challenging since most available sensors, both electrochemical and NDIR, have limits of detection of 100 ppm. Methane sensing over livestock facilities, thus, requires high accuracy sensors that can be used in-flight. Ground-based analysers (e.g. CRDS analysers) are also used to measure methane and N_2O concentrations using UAVs as mobile platforms for air sampling. When coming to fluxes estimation using UAVs, current methods rely on mass balance or on the use of plume inversion techniques. Both approaches require the simultaneous collection of accurate gas measurements and atmospheric data as temperature, relative humidity, atmospheric pressure and wind measurements. Together with literature information, the outcomes of an experimental application of a prototype multi-sensor platform for UAV and ground gaseous measurements in cattle dairy farms is presented and discussed.

Reducing N leaching by adapting N application timing and quantities to weather and grass growth

L. Bonnard[1,2], E. Ruelle[2], M. O'Donovan[2], M. Murphy[1] and L. Delaby[3]
[1]MTU, Department of Process, Energy and Transport Engineering, Cork, Ireland, [2]Teagasc, Animal & Grassland Research and Innovation Centre, Moorepark, Fermoy, Cork, Ireland, [3]INRAe, Institut Agro, PEGASE, Saint-Gilles, 35590, France; laeticia.bonnard@teagasc.ie

The EU Green Deal Farm to Fork strategy has set a target to reduce nutrient losses by at least 50% and fertilizer use by at least 20% by 2030. This will require more appropriate management of Nitrogen (N) fertiliser increasing N use efficiency leading to a reduction of N available for loss. A modelling exercise has been conducted, using the MoSt GG model, to highlight the potential for N precision management to reduce N leaching. Four scenarios were simulated over 19 years (2003 to 2021) using weather data recorded by the Met Eireann synoptic weather station, located in Moorepark Co. Cork, Ireland. The base fertiliser application had a total annual N fertilisation of 225 kg N/ha/yr. Then, the following rules were set to adjust the timing and quantity of the fertilisation: The N application had to be delayed if (1) the average temperature recorded over the preceding and/or the coming week dropped below 5 °C, (2) heavy rainfall occurred in the four preceding days or if it was expected in the three coming days, (3) from April onwards, if the predicted grass growth for the following week was below 30 kg DM/ha/d, fertilisation was cancelled. In each case, if the N application was delayed by more than seven days, the N application was reduced (or even cancelled if over 21 days). For each scenario, four paddocks cut every four weeks in rotation with a total of 10 cuts/yr/paddock were simulated. The precision N management led to an average reduction in N leaching of 3%, associated with an average reduction of 10 kg of N applied/ha/yr and a grass dry matter (DM) yield decrease of 1%. While the highest impact on the amount of N leached was the weather (yearly effect), the positive impact of precision fertiliser application has been shown on specific years. In 2014, for example, the N leaching decrease of 7% with 5 kg DM less grass produced. The years with the greatest N leaching reduction due to the reduction of N applied are often associated with a loss of grass DM production, highlighting that trade-offs will have to be found.

Reducing greenhouse gases in dairy cattle farming through innovative technologies

A. Svitojus and E. Gedgaudas
Chamber of Agriculture of the Republic of Lithuania, K. Donelaičio g. 2, 44239, Lithuania; arunas_svitojus@yahoo.com

The presentation will provide an overview of the situation of Lithuanian dairy farms, the capacity of the Lithuanian dairy industry, and the volume of milk produced in Lithuania and its imports from other countries. Information will be provided on milk purchase prices in Lithuania – the different prices paid for raw milk to small and medium-sized dairy producers and large dairy producers. It will also provide information on the average milk procurement prices in the EU in 2022 and the average milk procurement prices in Lithuania. Information will be provided on changes in prices of dairy products sold by Lithuanian dairy companies, structural changes in dairy farms, the main indicators of milk production in Lithuania and the situation of co-operation, as well as measures to reduce the greenhouse effect and increase milk profitability. One of the way to reduce the greenhouse effect is milk production. Lithuanian dairy farms include in selection programs bulls with high milk production indexes, health traits, longevity. In Lithuanian we increase milk production from 5,601 kg in 2017 until 6,425 in 2021 per cow and we still have a lot potential. Farmers have big interest in female sexed semen and beef on dairies management technology.

Session 17 Theatre 9

Annual quantification of GHG and NH_3 emissions coming from compost-bedded pack housing systems

E. Fuertes, G. De La Fuente, J. Balcells and R. Seradj
University of Lleida, Animal Science, Av. Alcalde Rovira Roure, 191, 25198, Lleida, Spain; esperanza.fuertes@udl.cat

Traditionally, dairy cattle in Spain has been housed on freestall cubicles system, where manure is stored on an open air pool. Nevertheless, compost-bedded pack systems (CBP) are gaining popularity due to its benefits on animal health and welfare, and its use is being extended. In this system, the animals rest on a bed made up by their own manure, which is daily composted *in situ*. As emission of pollutant gases such as certain GHG or NH_3 coming from this system have not been quite studied yet, the need appears to evaluate the effect that the composting process has on emissions from CBP dairy systems. Moreover, as gas emission are not constant along the year due to temperature variations, annual emissions were registered to evaluate differences between seasons. Measurements were taken every two months during a whole year on a CBP farm. Two distinct emission phases were differentiated: static emission (SE), corresponding to the period of the day when the compost bed is kept at rest; and dynamic emission (DE), happening during and afterwards the mechanical composting of the bed. To determine emissions, an airtight structure was located on the surface of the bed. An air flux was performed inside the structure, counting with an air entry where clean air come in and an outlet way where the air inside the simulator was taken to an analysis station in order to collect the gases of interest. During DE, surface of the bed was composted by means of a rototiller, simulating in time and depth the composting work performed on the rest of the barn. Emission was calculated as the difference between entry and outlet concentrations. GHG and NH_3 emissions were significantly higher during DE compared with SE for all gases under study (CH_4: 0.55 vs 0.07 g/m2&h; CO_2: 1.5 vs 0.21 g/m2&h; N_2O: 7.6 vs 1.1 mg/m2&h; NH_3: 0.28 vs 0.12 g NH_3/m2&h). NH_3, N_2O and CO_2 showed clearly how their emissions decreased during the cold months of the year and increased constantly as temperatures warmed up, while CH_4 showed the opposite trend. Despite being a positive system when it comes to animal health and welfare, CBP composting phase would present an important contribution to pollutant gas emissions generated from manure.

Session 17 Theatre 10

A study of the building-integrated photovoltaic cattle houses application scenario in China

Q. Wang, X. Du and Z. Shi
China Agricultural University, NO.17 Tsing Hua East Road, Beijing, 100083, China, P.R.; qiwang@cau.edu.cn

For the sake of sustainably meeting the growing demand for livestock products, it is essential to improve resource efficiency and reduce the corresponding emissions, which highlights the significance and necessity of harnessing the renewable energy. The use of building-integrated photovoltaic (BIPV) system can be well optimized on the barn, so that the solar radiation received by the roof of the barn can be fully utilized. This study conducts a feasibility study on the application of BIPV cattle houses in various provinces in China (counting the data for 31 mainland provinces without Hong Kong, Macao and Taiwan). In the study, a virtual standard barn was applied to each province in China, with setting the solar radiation and electricity price differently for each province. The results showed that the application of BIPV cattle houses in all provinces could help to achieve self-sufficiency in the power of the barn. The cattle houses in most provinces could generate a large amount of surplus electricity feeding in the grid, which may reduce most of (over 95%) the annual greenhouse gas (GHG) emissions of the cattle houses in each province. From the perspective of generating electricity power, Tibet had the highest amount of photovoltaic electricity generation, the largest proportion of electricity saved and the biggest number of the gross annual GHG emission reduction, 866,569 kWh, 251% and 546.9 tCO_2, respectively. Seven provinces, including Beijing, Hebei, Shanxi, Inner Mongolia, Jilin, Qinghai and Ningxia, were relatively higher among all provinces, with averages of 665,202 kWh, 193% and 418.8 tCO_2. The financial viability of the above-mentioned results showed that the equity payback of six provinces including Beijing, Hebei, Shanxi, Inner Mongolia, Jilin and Ningxia were shorter, with an average of 3.8 years, while Tibet and Qinghai were longer, with an average of 5.2 years. In general, the photovoltaic electricity generation effect and the income of BIPV cattle houses applied in six northern provinces reached a better result. With the finding, this study certainly provides a preliminary basis for the application of BIPV cattle houses.

Session 17 Theatre 11

Effect of manure management and environmental temperature on microbial communities in dairy manure

E. Fuertes, R. Seradj, J. Balcells and G. De La Fuente
University of Lleida, Av. Alcalde Rovira Roure, 191, 25198 Lleida, Spain; esperanza.fuertes@udl.cat

Lleida region is home to some of the largest dairy farms in Spain; this intense farming generates a great amount of manure to be managed, which is known to act as a source of pollutant gases. The two main dairy housing systems which stand out from others are compost-bedded pack (CBP) and cubicles system (CUB), each one with a different manure management procedure. Since pollutant gas production depends on the activity of certain microbial communities, understand the connection between manure microbiota and housing system might help to characterize the microbial dynamics involved in the emission of pollutant gases in dairy manure as well as the impact of the environmental temperature on such communities. Manure samples (n=81) from both housing systems (CBP, n=3; CUB, n=3) were collected during two seasons: summer and winter. Samples were frozen and the bacterial genomic DNA was extracted, using a commercial kit. Then, high-throughput sequencing of the V3-V4 hypervariable region of 16S rRNA was conducted on Illumina MiSeq system. Taxonomic assignment of phylotypes was performed using a Bayesian Classifier trained with Silva database; both microbial alpha and beta diversity as well as the taxonomic profiles of the samples were analysed using 'vegan' and 'mixomics' packages from R software. CBP showed higher microbial biodiversity than CUB, and both systems presented higher index values during summer season. Community structure was also clustering by system and season although separation between seasons was more remarkable in CPB systems. In CUB systems, higher abundance of methanogenic archaea such as Methanobrevibacter (9%) was found, explained by the favourable anaerobic conditions provided by the storage pool compared to CBP. Some urease producer bacteria, like Corynebacterium and Pseudomonas, were more abundant on CBP systems, especially on summer, results in accordance with literature showing higher NH_3 volatilization from CBP systems compared to CUB. Results obtained show that manure microbiota changes between housing systems and highlights the importance of environmental temperature on microbial activity, and in consequence on the gas emissions produced in these scenarios.

Session 17 Poster 12

Simultaneous comparison of 5 methods to quantify enteric methane emitted by lactating dairy cows

A. Vanlierde, F. Dehareng, O. Christophe and E. Froidmont
Walloon Agricultural Research Center, 5030 Gembloux, Belgium; a.vanlierde@cra.wallonie.be

Different techniques to estimate methane (CH_4) emissions from livestock are available. This study aimed to compare simultaneously 5 methods. Two measurement techniques were considered: SF_6 tracer gas and automated head-chamber system (AHCS – Greenfeed system); as well as three indirect predictive models: based on milk fatty acids obtained by gas chromatography (GC), on milk mid infrared (MIR) spectra and on faecal near infrared (NIR) spectra. The focus was on the ranking of animals in function of their level of CH_4 emissions depending of the technique. Variability is needed between animals (parity, milk yield, level of CH_4 emissions, corresponding spectral information, etc.). Ten animals were chosen based on CH_4 predictions from milk MIR spectra and zootechnical characteristics. The basal diet was a classical RTM. After 3 weeks of adaptation (W1 to W3) to the AHCS device, 4 weeks of measurement with the AHCS were performed (W4 to W8) and an individual representative milk sample (proportional mix of am and pm milking) was collected daily to obtain the MIR spectra. SF_6 tracer technique was used to obtain a double daily measurement of CH_4 emission per cow during 3 days both on W5 and W7. During the same days a spot faecal sample (to obtain NIR spectra) and an additional mix of am/pm milk sample was collected to perform GC analyse. Two cows obtained CH_4 values with SF_6 technique with more than 20% of difference between W5 and W7 which is physiologically questionable as no health trouble was noticed and the diet is stable. Consequently, the cows have been removed of the analyse. The range of CH_4 values obtained by SF_6 tracer technique was about 563 to 872 g/d (mean ± SD=674±82) while it was between 530 and 717 g/d for the AHCS (618±56 g/d). Values obtained from milk MIR spectra were about 25% lower than the AHCS values while the predictions from the very first model based on faecal NIR spectra dedicated to dairy cattle were about 50% lower. However, due to the know errors of the prediction models the variability of CH_4 emissions between animals was too low to conclude about the ranking of animals between techniques.

Session 18 — Theatre 1

Meeting consumer expectations: a 3G global beef eating quality predictive model for Europe

A. Neveu[1], R. Polkinghorne[2], R. Watson[3], H. Cuthbertson[2] and J. Wierzbicki[4]
[1]SARL Birkenwood Europe, Lanon, 03240, France, [2]Birkenwood Pty Ltd, Blandford, 2338 Murrurundi, Australia, [3]Melbourne University, Department of Mathematics and Statistics, University of Melbourne, 3010, Australia, [4]IMR3G Foundation, Smulikowskiego 4, 00-389 Warsaw, Poland; neveu.alix@gmail.com

The meat industry faces several challenges and consistent, predicable eating quality is one of them. In Europe, the actual EUROP commercial grading system for carcass conformation and fat distribution doesn't align with consumer's meal experiences. This was proven over the past 15 years with meat eating quality research trials using untrained consumers across Europe. Since 2017 the collaborative platform hosted by the International Meat Research 3G Foundation (IMR3GF) has compiled the data collected using UNECE Beef Eating Quality protocols. The IMR3GF members comprise leading scientists from several countries who have collaborated extensively to assemble consumer and animal data across many countries and over 25,000 European consumers. The approach was to work with UNECE to create common standards for meat products and sensory evaluation to ensure data were compatible and able to be merged to create sufficient scale from which reliable consumer sensory estimates could be developed. These data have enabled the Foundation to develop a European predictive model based on research trials. The database contains data connecting cattle, carcase treatments, cuts and cooking styles to consumer answers on eating quality. This represents consumer studies conducted in total on 11 muscles over the last decades in European countries (Poland, France, England, Northern Ireland, Wales, Ireland). The database contains a considerable number of variables characterising the carcass using the 3G Chiller Assessment method and the consumers sensory testing according to the UNECE standards. The predictive model is an evolutive tool with further eating quality accuracy and scope developed with greater data. As all industry revenue directly relates to the consumers judgement of value, with the most critical component meal satisfaction, industry profitability can be enhanced by delivering consistent eating quality through strong commercial brands built on a solid scientific foundation.

Session 18 — Theatre 2

Development of a carcass grading system for South African beef

M.S. Faulhaber[1], P.E. Strydom[2] and N. Hall[1]
[1]Cavalier Foods, 83 Performance road, Farm Tweefontein, Cullinan, 1000, South Africa, [2]University of Stellenbosch, Stellenbosch Central, Stellenbosch, 7602, South Africa; michaelaf@cavalier.co.za

The South African red meat industry is diverse in terms of production systems, landscape, climate, breeds, age, sex, feeding regimes, external parasite loads, growth stimulant use (hormonal and beta-agonists) and management practices; each providing its own influence on the quality of meat provided to the consumer. This diversity in conditions often makes it difficult to identify strong correlations between single intrinsic factors and meat quality attributes. Due to this complexity, there is paucity in the understanding the exact factors which influence meat quality. Currently, no method exists which can effectively and consistently predict or ensure the eating quality of the meat which is produced within South Africa. To develop a quality grading model within South Africa, a developing country, the effect of animal inputs unique to South Africa on the eating experience of the consumer, needs to be determined. These unique inputs include local breeds (Sanga types), growth stimulant use (beta-agonists and hormonal growth promotants), sexes (predominantly young bulls) and traditional production methods (grass-fed & feedlot raised) which have not been included in the Meat Standards Australia (MSA), International Meat Research 3G Foundation (IMR3GF) or other international grading model developments, yet. A matrix of animals (n=276), inclusive of the above-mentioned attributes, were slaughtered as per current South African standards and data recorded utilising registered UNECE protocols. A total of 1,980 beef cuts of varying quality were harvested to be subjected to consumer testing to determine eating quality as experienced by the South African population. Subsequent grading prediction scores were compared to the South African consumer's eating experience. The results will update the existing grading models of the International Meat Research 3G Foundation, Meat and Livestock Australia and be utilised to develop the first of its kind grading system for South Africa. Once developed, the model will be implemented within one of the leading red meat value-chain companies and food retailers (the exclusive owner of the South African IP) within South Africa.

French consumer evaluation of eating quality of Limousin beef

M. Kombolo[1], J. Liu[1], I. Legrand[2], F. Noel[3], P. Faure[4], L. Thoumy[2], D. Pethick[5] and J.-F. Hocquette[1]
[1]INRAE, UMR1213, Theix, 63122, France, [2]IDELE, Bd des Arcades, 87000 Limoges, France, [3]IDELE, Rte d'Epinay sur Odon, 14310 Villers-Bocage, France, [4]INRAE, Herbipôle, Theix, 63122, France, [5]Murdoch University, Food Futures Institute, Perth, 6150, Australia; moise.kombolo-ngah@inrae.fr

The aim of this study was to evaluate the eating quality of 2 beef cuts (striploin and rump) from 102 Limousin cows. Carcasses were first graded according to the Meat Standards Australia (MSA) methodology at the 5th and 10th rib 24 h *post-mortem* by 2 chiller assessors. This allowed the prediction of the MSA index for the whole carcass, which is an indicator of beef potential eating quality. Muscle samples were then collected, sliced with a thickness of 2.5 cm, aged for 10 days and finally grilled according to the MSA protocol. A total of 480 consumers scored beef for tenderness, juiciness, flavour and overall liking on a 0 to 100 scale. In addition, consumers were asked to assign a quality rating to each sample: 'unsatisfactory' (2*), 'good everyday' (3*), 'better than everyday quality' (4*) or 'premium quality' (5*) with an average of 3,03 and 2,9 for striploin and rump. Observed MQ4 scores were calculated combining scores of tenderness, juiciness, flavour and overall liking according to the MSA protocol. MQ4 values for the 2*/3*, 3*/4* and 4*/5* boundaries were 54, 62 and 76 respectively. The MSA index ranged from 45.5 to 58.5 and had a median value of 51,33. The marbling scores at the 5th and 10th ribs were positively correlated with the MSA index (0.34 and 0.47 respectively). Finally, beef carcasses were divided into two equal groups (under or above the median MSA index). The MSA index and the marbling score were 13 and 15% higher respectively in the second group of carcasses while ribfat thickness was 88% higher. Scores for tenderness, juiciness, flavour and overall liking as well as MQ4 were on average 9 to 13% higher for the rump and 5-10% higher for striploin in the second group compared to the first one. Overall, the results showed that the Australian MSA model might be relevant for Limousin cows but may require improvements for carcass grading at the 5th rib.

BeefQ – testing a beef eating quality prediction system for Wales and England

P.K. Nicholas-Davies[1], R. Polkinghorne[2], A. Neveu[2], H. Cuthbertson[2] and T. Rowe[3]
[1]Aberystwyth University, Department of Life Sciences, Penglais Campus, SY233DA Aberystwyth, United Kingdom, [2]Birkenwood Pty Ltd, 45 Church Street, Hawthorn, Victoria, 3122, Australia, [3]Celtica Foods, Heol Y Plas Cross Hands Food Park,Cross Hands, SA146SX Llanelli, United Kingdom; neveu.alix@gmail.com

The BeefQ project was implemented to address industry and consumer concerns around beef eating quality consistency. To adapt the Meat Standards Australia (MSA) model for UK conditions, consumer taste testing beef was conducted using the standard protocol developed in Australia. Four cuts (sirloin, tenderloin, salmon and feather blade) from 90 sides of beef were evaluated as grilled steak by 1,200 Welsh and English consumers. Carcase suspension method and 7 and 21 days maturation were also tested. Cuts were collected in two seasons from cattle types typical of the Welsh herd (beef breed steers and heifers, dairy cross steers and heifers, young bulls and cows). Discriminate analysis determined the relationship between the individual consumer scoring scales of tenderness, juiciness, flavour and overall liking and provided a measure of the relative importance of each scale in determining the final quality decision. Overall eating quality scores (an aggregation of scales above (1-100)) were used to determine cut off values for what consumers consider to be good everyday, better than every day and premium quality beef. The relative importance of each variable was 0.3 tenderness + 0.1 juiciness + 0.3 flavour + 0.3 overall liking. These weightings were similar to current MSA 30:10:30:30 ratios indicating that the two consumer populations were similar. The cut off scores for better than every day and premium quality, respectively, were slightly lower (37 vs 41) and slightly higher (79 vs 77) than current MSA values, indicating that BeefQ consumers may discriminate more for both unsatisfactory and premium beef. This is important for brands where any inconsistency may impact value. These analyses provide strong evidence that Welsh and English consumers clearly differentiate eating quality from unsatisfactory to premium and that a universal set of sensory weightings and cut off values can adequately define these categories.

Assessment of untrained consumers on eating quality in Zebu and crossbred cattle at aging times

A. Barro[1,2], M. Coutinho[3], G. Rovadoscki[3], A.M. Bridi[2] and J.F. Hocquette[1]
[1]INRAE Vetagro Sup, UMR1213, Theix France, 63122 Theix, France, [2]Londrina State University, Highway Celso Garcia Cid, 86057970 Londrina, Brazil, [3]Research and Development Department, Brazil Beef Quality Ltd, Piracicaba, Sao Paulo 13415-000, Brazil; amandagbarro@gmail.com

Crossbreeding has been used to improve the sensory quality of Zebu animals, and aging time improves meat tenderness. The objective of this study was to evaluate the effects of aging times on beef eating quality in Zebu and crossbred cattle. Two experiments are reported in this study. The first one was based on carcasses of Nellore and crossbred (Nellore × Angus) animals at 3 aging times (5, 15 and 25 days) for Longissimus thoracis et lumborum (LD). The second experiment involved 6 muscles from Nellore and Brangus at 2 aging times. Sensory analysis with untrained consumers was conducted with a neighbour-balanced Latin square design (6×6) with two block factors, consumer, and assessment order. Consumers rated tenderness, juiciness, and overall liking on a non-structured scale (0-100). The t test was used to evaluate the difference between breeds and aging times in first experiment. A variance analysis was performed for muscle comparison with the post-hoc test including Bonferroni correction for pairwise comparisons. In the first experiment, breed had no effect on sensory traits of the LD muscle. However, beef aged 25 days had on average eating scores 13-20% higher compared to 5-day aged beef. In the second experiment, at 5 days of aging, Brangus cattle presented better scores for all sensory traits (+11-20%). However, increasing ageing time from 5 to 15 days did not improve sensory scores of beef from Brangus. On the opposite, increasing ageing time improved sensory scores of Nellore beef by 11.6-15.4%. At 5 days, muscles showed a difference in tenderness, with the highest grade for rumpsteak. The striploin improved in flavour and the top Sirloin improved in beef eating quality after 15 days of ageing. As a consequence, there was no more difference between cuts in sensory traits at 15 days of ageing. In conclusion, the crossbred animals need shorter aging time compared to Zebu animals to produce good beef. Increasing aging time in Nellore may be a strategy to improve eating quality of beef.

Loin residual glycogen and free glucose do not affect Australian lamb loin eating quality

S.M. Moyes, D.W. Pethick, G.E. Gardner and L. Pannier
Murdoch University, 90 South Street, Murdoch 6150, Australia; s.moyes@murdoch.edu.au

Post-slaughter, muscle pH declines through anaerobic metabolism of muscle glycogen to lactic acid. An ideal pH of at least 5.7 is achieved within 24 hours *post-mortem* when sufficient levels of glycogen are available at slaughter. The inactivation of enzymes that breakdown glycogen occurs as muscle pH lowers, inhibiting further production of lactic acid and thus pH decline, resulting in residual glycogen remaining in muscles. During ageing, concentration of free glucose increases from hydrolysis and diffusion of residual glycogen. Both free glucose and residual glycogen may positively affect lamb eating quality through the Maillard reaction during cooking. However, the link between untrained consumer eating quality and residual glycogen has yet to be demonstrated, nor whether this link varies between ageing times. Loin samples were collected from 139 lamb (Maternal, Merino, Terminal) and 40 yearling (Merino) carcasses 24 hours post-slaughter and aged for either 5, 14 or 21 days, prior to grilled sensory testing for overall liking by untrained consumers. An additional 5 g sample was collected at each ageing time and assessed for residual glycogen and free glucose. Consumer responses (n=1,790) were modelled using ageing time as a fixed effect, and residual glycogen or free glucose tested one at a time as covariates. Unadjusted mean residual glycogen concentration was 14.83, 13.81 and 13.39 µmol/g at 5, 14 and 21 days. Unadjusted mean free glucose was similar across the ageing times ranging from 12.27 at 5 days to 12.22 µmol/g at 21 days ageing. Results showed no association (P>0.05) of loin residual glycogen with loin overall liking scores at any of the ageing times. A similar finding was observed for free glucose on loin overall liking. Despite this lack of effect, further research is warranted to explore the impact of residual glycogen or free glucose on other sensory traits (tenderness, juiciness, flavour liking). Additionally, residual glycogen and free glucose levels vary between muscles due to fibre type and other anatomical differences, therefore their impact on eating quality within other cuts should be explored.

Session 18 Theatre 7

Combining eating quality and individual cut tracking to maximise value and decrease waste in beef

H. Cuthbertson[1], R. Polkinghorne[1] and A. Neveu[2]
[1]Birkenwood International Pty Ltd, Murrurundi, NSW, Australia, [2]SARL Birkenwood Europe, Lanon, 03240, Saint-Sornin, France; holly.cuthbertson1@gmail.com

Typically, processors are limited to how many ways cattle can be sorted as they come into the boning room. In most cases, individual identification is lost as cuts fall onto the belt. This affects the ability to maximise individual carcass value and often results in better quality carcasses or cuts being downgraded because of their pre-assigned batch. Yield based measurement systems are becoming popular in processing facilities. These systems can individually identify and track cuts, whilst collecting multiple yield measurements along the boning line, enabling processors to segregate product of consistent weight and size into specific brands. Eating quality is an important value for calculating consumer satisfaction across multi-cuts, cook types and days ageing on varying cattle populations. Produced from the IMR3GF 3G grading Model, an eating quality matrix provides processors with the ability to differentiate their product into trusted brands based on EQ cut off scores. By facilitating an accurate interpretation of 3G's eating quality matrix and feeding this into a cut tracking system combined with yield results, it enables individual cut sorting to happen in real-time at chain speed. This system would eliminate the need for extensive pre-sorting of carcasses, decrease capture losses and generate increased carcass value through the uplifting of cuts into better valued brands. Using Birkenwood's MEQ Toolkit and Marel's StreamLine system, a case study on generic data reported on the potential benefit of this system over a 3-month period. The processor was previously using basic cut offs for its brands across a small number of cuts. The analysis only looked at the system improving what was currently in place at the processor. Results from the case study revealed an large increase in product capture into the top brand (average 82.5%, range 28-175%) and increased the average $/Hd across the entire dataset from $996.43 to $1,223.14. Potential value and cut capture would only increase if expanded to look at multi-cuts.

Session 18 Theatre 8

Changing the Australian trading language for beef and lamb to adapt to new measurement technologies

G.E. Gardner[1,2], S. Stewart[1,2], P. McGilchrist[1], C. Steele[1], H. Calnan[1,2], S. Connaughton[1,2], K. Mata[1,2] and R. Apps[1]
[1]Advanced Livestock Measurement Technologies, (ALMTech), Australia, [2]Murdoch University, College of Science, Health, Engineering and Education, Murdoch, 6150, Australia; g.gardner@murdoch.edu.au

Since 2016 an Australian project (ALMTech) has accelerated the development and implementation of technologies that predict traits describing eating quality and saleable meat yield %. Initially these technologies were trained to predict traits that are commercially traded upon within industry. However, it soon became apparent that the existing industry trading language was limiting, as evidenced by the following 3 scenarios. The first is an example of where technologies can measure a trait that was not legislated for trading. In this case we introduced intramuscular fat % (IMF%) into the lamb industry trading language – a crucial step given the importance of this trait for predicting eating quality. This was underpinned by a Soxhlet laboratory method as the gold standard. In the second scenario, we introduced an alternative trait for one that poorly describes carcase value. In this case GR tissue depth in lamb was the historical measurement used to reflect saleable meat yield %. We introduced the alternative whole carcase lean % as a better indicator of saleable meat yield %, using computed tomography as the gold standard measurement. In the third scenario we provided an alternative to an existing trait that performed poorly as a calibrating standard for technologies. In this case we introduced IMF% into the beef industry to act in parallel with visual marble scores which have proven too erroneous for accrediting technologies. Conversion equations have also been established enabling these values to be interchanged. These new traits calibrated against robust gold standards have delivered multiple benefits. Firstly, technology provider-companies were instilled with the confidence to commercialise due to the provision of achievable accreditation standards. Secondly, processors had the confidence to invest in these technologies and establish payment grids based upon their measurements, knowing that reliable accuracy standards had been met. And lastly, data flow into genetic databases, industry data systems (MSA), and as feedback to producers was enhanced because of these technologies.

The precision and accuracy of the Q-FOM grading camera predicting rib eye traits in beef carcasses

S.M. Stewart[1], M. Christensen[2], H. Toft[2], T. Lauridsen[2] and R. O'reilly[1]
[1]Murdoch University, 90 South Street, Murdoch 6150, Australia, [2]Frontmatec A/S, Smørum 2765, Smørum 2765, Denmark; r.o'reilly@murdoch.edu.au

The Q-FOM is a non-contact, rechargeable device specifically designed for ribeye grading according to AUS-MEAT and MSA grading standards, either in the chiller or at a grading station. The Q-FOM is able to grade in real-time MSA and AUS-MEAT marble score, AUS-MEAT eye muscle (EMA), AUS-MEAT meat colour, AUS-MEAT fat colour, MSA sub-cutaneous rib fat and chemical intramuscular fat (IMF%).A total of 4,677 commercial carcasses from two Queensland export abattoirs were graded and imaged using 3 cameras. A diverse range of carcass categories (Wagyu, grain fed (long, medium, short), organic, grass-fed, dairy, cull cow) were targeted to achieve a broad phenotypic range in rib eye traits. Visual ribeye grading was performed by two MSA expert graders and one commercial grader according to the AUS-MEAT chiller assessment language guidelines and used as reference. Precision (R2, RMSEC/RMSEP) accuracy (bias) were reported for both calibration and validation analyses. The performance of the camera against the AMILSC accreditation standards were also reported. In addition, a high level of repeatability and reproducibility of Q-FOM Beef cameras was demonstrated across all traits.

Predicting IMF% and visual marbling scores on beef portion steaks with a Marel vision scanner system

L. Pannier[1], T.M. Van De Weijer[2], F.T.H.J. Van Der Steen[2], R. Kranenbarg[2] and G.E. Gardner[1]
[1]Murdoch University, College of Science, Health, Engineering and Education, South Street, 6150 Murdoch, Australia, [2]Marel Meat B.V., Handelstraat 3, 5830 AD Boxmeer, the Netherlands; l.pannier@murdoch.edu.au

Marbling score is an important factor in many beef carcass grading schemes to classify carcass quality grades. Under the current Meat Standards Australia (MSA) and AUS-MEAT grading system, marbling score is visually assessed by accredited graders on the caudal surface of the m. longissmus thoracis et lumborum (loin) at a rib site between the 5th and 13th rib. The marbling score assessed at the grading site, is assumed to adequately represent the entire loin. Contrary to this assumption some studies have indicated that marbling varies within the beef loin. Therefore the grading of individual steaks could enhance classification of marbling grades for individual portions and could underpin an enhanced marbling specification for portion-cut product entering premium markets. This study describes the performance of a Marel conveyer vision scanner system in its ability to predict MSA and AUS-MEAT marbling scores of portion steaks from the cube roll. The capacity to predict the new AUS-MEAT trait of IMF% that has been approved as the gold-standard-trait for accreditation of objective measurement technologies was also assessed. Vision scanner marbling scores were acquired on fresh-cut steaks of beef carcasses (n=102) that represented a wide range in visual marbling score. The vision scanner predictions were tested using a leave-one-out cross validation method, which demonstrated precise and accurate predictions of IMF% (R^2=0.87; RMSEP=1.16; slope=0.09; bias=0.22), MSA (R^2=0.82; RMSEP=70.11; slope=0.09; bias=17.08) and AUS-MEAT marbling (R^2=0.79; RMSEP=0.75; slope=0.16; bias=0.08). The Marel vision scanner prediction of visual grader scores was relatively less precise and accurate than its prediction of IMF%, likely due to random error in subjective grader scores. These results indicate that the Marel Vision Scanner could be used to underpin a market for premium portion-cut steaks where enhanced product description and consistency is required.

Session 18 Theatre 11

Improved analysis of dual energy x-ray absorptiometry images enhances lamb eating quality prediction

F. Anderson[1,2], L. Pannier[1,2], C. Payne[1,2], S. Connaughton[1,2], A. Williams[1,2] and G.E. Gardner[1,2]
[1]Advanced Livestock Measurement Technologies, Murdoch, 6105, Western Australia, Australia, [2]Murdoch University, School of Agricultural Sciences, Murdoch, 6105, Western Australia, Australia; f.anderson@murdoch.edu.au

Dual energy x-ray absorptiometry (DEXA) is used in commercial abattoirs in Australia to predict carcass composition. Images captured using DEXA are at two different energy levels, with the pixels captured at low energy expressed as a ratio (R value) to those captured at a high energy, and relate to the atomic mass of the tissue. The mean R values from DEXA images have previously been used to predict lamb eating quality using isolated bone regions from the images, however this was in a small number of lambs and also demonstrated correlation with intramuscular fat content. Improvements in image analysis have enabled better isolation of the entire lamb skeleton, enabling calculation of the mean bone DEXA R value. This technique has been applied to a larger dataset from 2 abattoirs for lambs that varied in age/maturity (n=198). As the bone DEXA R value increased there was a decrease in the eating quality of some cuts, as evaluated by untrained consumers on a scale from 0 to 100. The relationship between bone DEXA R and eating quality was most consistent in the loin grill where overall liking scores decreased by 10 units ($P<0.05$) at both abattoirs. This relationship was independent of loin intramuscular fat % and carcass lean %, which were tested simultaneously in models alongside bone DEXA R values. Similar associations were also found in other cuts, although the magnitude of effect varied. The improvements to DEXA image analysis enables chain speed DEXA scanning of lamb carcasses to rapidly predict aspects of lamb eating quality. Given the large proportion of the Australian lambs that undergo DEXA scanning at slaughter (≈50%), this represents the ability to provide information on a large proportion of the lambs processed in this country. DEXA eating quality predictions may be incorporated along-side other production traits and carcass measures that could better allow for in-plant carcass sorting and recognition of carcass quality on which to base payment.

Session 18 Theatre 12

Aligning upper and lower commercial beef DEXA images to predict CT composition

H.B. Calnan[1,2], A. Williams[2] and G.E. Gardner[1,2]
[1]Advanced Livestock Measurement Technologies, Murdoch, 6160, Australia, [2]Murdoch University, College of Science, Health, Engineering and Education, Murdoch, 6160, Australia; honor.calnan@murdoch.edu.au

The Australian beef industry is focused on improved measurement of beef carcase composition and valuation along the supply chain. The first commercial dual energy X-ray absorptiometry (DEXA) system has been installed in Rockhampton Australia to measure computed tomography (CT) composition in entire beef sides scanned at line speed. The system is comprised of two DEXA units positioned vertically that scan the entire length of hanging beef sides to produce upper and lower images that overlap centrally. For established DEXA CT prediction equations to be aligned with and outputted by this commercial DEXA system, an automated method of stitching the beef DEXA images is needed that accounts for the duplicated central section and thereby maintains a consistent measure of whole side composition. Two approaches were explored to achieve this: (1) using an anatomical landmark; and (2) based on the number of duplicated pixel rows in carcase DEXA images. The caudal aspect of the 13th rib was identified as an anatomical landmark in the overlapping section of both DEXA images that could guide image stitching. An advantage of this method is that upper and lower DEXA images become a good approximation of the carcase 'forequarter' (lower image) and 'hindquarter' (upper image). However, difficulties were encountered in the ability of an operator to consistently identify the 13th rib landmark in both upper and lower DEXA images, making this method inconsistent and unreliable. Alternatively, the number of pixel rows duplicated in upper and lower DEXA images was determined using scans of plastic calibration blocks. This method of image stitching is advantageous in that it can be consistently applied to beef DEXA images regardless of carcase size or length and can be easily automated. This method has therefore been adopted to produce a single DEXA value for each beef side for use in pre-existing DEXA CT prediction algorithms to output an estimate of carcase composition. However direct calibration of this commercial DEXA system against CT measures of beef composition will further refine this image stitching method.

Calibration of chain-speed DXA units using synthetic phantom blocks in two Australian lamb abattoirs

S.L. Connaughton[1,2], F. Anderson[1,2], A. Williams[1,2] and G.E. Gardner[1,2]
[1]Advanced Livestock Measurement Technologies, Murdoch, 6150, Australia, [2]Murdoch University, College of Science, Health, Engineering and Education, Murdoch, 6150, Australia; s.connaughton@murdoch.edu.au

Dual Energy X-ray Absorptiometry (DXA) technology has been developed for chain-speed predictions of carcass composition in Australian lamb abattoirs, capable of predicting fat %, lean muscle % and bone % with high precision and accuracy. With multiple DXA units in operation across the country, a simple yet reliable method of calibrating them against one another was required. A synthetic phantom constructed of three different plastics (acrylic, high molecular weight polyethylene, and nylon) of varying mixtures and thicknesses is installed at each DXA site and is scanned at the commencement of every production day. The thickness of the plastic sections, as determined by the individual DXA, is compared to the thicknesses at one DXA site in Western Australia operating as the reference. Similarly, the predicted fat % composition of each plastic block is compared to that from the reference site, generating a calibration curve for the R-values of the DXA images. The thickness and R-value differences between the test site and the reference site are applied to all pixels in the DXA image through linear transformations, allowing the test site to calculate the carcass composition using the same algorithm as that applied at the reference site. The predictions of carcass composition can be validated against the computed tomography (CT) determination of fat, muscle, and bone %. The application of this cross-site calibration resulted in an increased precision for the prediction of CT fat % (R^2=0.86, RMSE=1.71%) from the uncalibrated values (R^2=0.82, RMSE=2.77%), while the accuracy was also increased (slope=0.96, bias=-0.12%) from the uncalibrated values (slope=1.24, bias=11.43%). This calibrated precision of CT fat % prediction was well aligned with the precision at the reference site (R^2=0.94, RMSE=1.10%), as was the accuracy (slope=0.97, bias=-1.19%). This method of calibration is suitable for calibration of DXA systems between abattoirs, as it is quickly calculated and applied, and enables DXA to be accredited for trading across Australia.

Objective carcase measurement from commercial supply chains contributing to genetic improvement

D.J. Brown[1], R. Alexandri[1], S.F. Walkom[1], D.W. Pethick[2], P. McGilchrist[3], S. Stewart[2], W.S. Pitchford[4] and G.E. Gardner[2]
[1]University of New England, AGBU, Armidale, 2351, Australia, [2]Murdoch University, School of Veterinary and Life Sciences, Murdoch, 6150, Australia, [3]University of New England, School of Environmental and Rural Science, Armidale, 2351, Australia, [4]University of Adelaide, School of Animal and Veterinary Sciences, Roseworthy, 5371, Australia; dbrown2@une.edu.au

Australia's meat supply-chains are investing in a range of new technologies to improve productivity across the whole value-chain. This has been a collaborative effort aiming to transform industry competitiveness by creating feedback and decision support systems linked to accurate carcase measurements. The Australian sheep industry has also been developing tools for sheep breeders to make simultaneous improvements in the proportion of carcase saleable meat (lean meat yield: LMY) and its eating quality (EQ). Lean meat yield and eating quality attributes are antagonistically related and 'difficult to measure' for seedstock selection purposes. Objective carcase measurements from commercial supply chains offer an important tool for seedstock breeders, and the genetic validation of new technologies for assessment of LMY and EQ can increase opportunities to improve genetic progress. Genetic analysis of both LMY and EQ traits, recorded using several new technologies, demonstrated excellent accuracy. This analysis provided the platform for the new measures to be used in routine genetic evaluation and fast-track genetic progress. Furthermore, our existing genomic selection protocols coupled with objective carcase measurements from commercial lambs offer new opportunities to increase the reference population size and reduce the current reliance on expensive resource populations. New LMY and EQ measuring technologies can also help seedstock breeders by using the collected data and developing platforms to underpin pricing signals for commercial sheep producers. Currently there are limited pricing signals for both LMY and EQ but their development can help seedstock breeders to invest in these traits within their breeding objectives.

Session 18 Poster 15

Potential of deep learning video image analysis for automated SEUROP-classification

T. Rombouts, M. Seynaeve and S. De Smet
Ghent University, Department of Animal Sciences and Aquatic Ecology, Coupure Links 653, 9000, Belgium; toon.rombouts@ugent.be

In cattle slaughterhouses in Europe, video image analysis (VIA) combined with linear regression (LR) modelling is widely used to assist or replace human SEUROP-classifiers. During the authorization procedure of the VBS 2000 (E+V) in Belgium in 2022, a deep learning (DL) algorithm was evaluated as an alternative for LR in terms of correct classification as well as other potential advantages. In Belgium, carcasses are classified into 18 different subclasses for conformation (SEUROP with subclasses +, = and -), 15 subclasses for fat cover (12,345 with subclasses +, = and -) and 7 animal categories (ABCDEZV). For the authorization test, a sample grid of 650 carcasses was made according to the distribution of category, conformation and fat cover in the Belgian carcass population. A jury of 5 assigned classification experts (2 from Belgium and 3 from other EU member states) were positioned next to the VBS 2000 and independently graded 1,123 bovine carcasses, of which 650 were finally used to fill in the grid on a first come, first served basis. The carcass classification was predicted by a previously developed VBS 2000 DL algorithm as well as a LR model, and these scores were compared with the median score of the jury. Correct classification was defined as having the same subclass or one subclass difference compared with the median expert score. For conformation, DL and LR resulted in high and comparable correct classification (90.9 and 92.0% respectively). Differences of 2 and 3 subclasses were observed for 8.9 and 0.2% respectively of the carcasses for DL, and 7.4 and 0.6% for LR. For fat cover, the correct classification was meaningfully larger for DL than for LR (95.7 vs 86.6%). A difference of 2 subclasses was found for 1.3 and 12.0% of the cases respectively, whereas DL had no scores with 3 subclasses difference and LR had 1.4%. Unlike LR, DL does not need the animal category nor the hot carcass weight as input, meaning it can operate separately from other software. In addition, DL was able to grade carcasses as S+, which LR couldn't, and produced less errors due to damage of carcasses. In conclusion, DL offers advantages compared to LR for bovine carcass classification.

Session 18 Poster 16

Hand-held NIR device predicting chemical intramuscular fat% in pork

M.T. Corlett, F. Anderson and G.E. Gardner
Murdoch University, 90 South Street, Murdoch, 6150, Australia; f.anderson@murdoch.edu.au

The use of fast growing, muscular genotypes of pigs in Australia has been shown to result in some genotypes having intramuscular fat (IMF) as low as 1%. The SOMA S-7090 NIR is an AUS-MEAT accredited device for measuring chemical IMF% in sheep meat. This study tested the SOMA using a lamb algorithm to predict pork chemical IMF%. Chemical IMF% and SOMA measures were taken from the same location on the exposed loin eye surface (*M. longissimus lumborum*). Pork carcasses were collected from 3 sites; site 1 in New South Wales (n=120), site 2 in Queensland (n=115) and site 3 in Western Australia (n=104). The SOMA predicted pork IMF% with low precision ($R^{2=}0.10$, RMSEP=0.63) and accuracy (bias=-0.15). This work demonstrated the sheep meat IMF% algorithm currently installed in the device was not transferable to predicting pork IMF%. However, nearly 70% of the chemical IMF data was below 1.5%. This aligns with the national pork IMF% levels which are inherently low due to the muscular, fast growing genotypes utilised. Further work could involve training a pork specific IMF% algorithm which would likely see additional improvements in precision and accuracy.

Session 18 — Poster 17

An optimisation model for operational planning in lamb supply chains

G. Wang[1], F.L. Preston[1], C. Smith[1,2], D. Flanigan[1,3], S.M. Miller[1], W.S. Pitchford[1] and D. Meehan[1,4]
[1]University of Adelaide, Davies Livestock Research Centre, Roseworthy, 5371, SA, Australia, [2]Agbiz Solutions, RMB 539, Natimuk, 3409 Vic, Australia, [3]Meat Data Logic, Horsham, 3400, Vic, Australia, [4]Meehan Consulting, Brisbane, 4000, Qld, Australia; vince.wang@adelaide.edu.au

A Carcase Optimisation Tool was developed for use by the lamb processing industry to sort carcases into boning groups by optimally allocating them to cutting plans using lean meat yield and weight, and to maximise profit. Processors are challenged to design cutting plans and sort carcases into groups to fabricate the most efficient and profitable combination of cuts from each carcase whilst satisfying market specifications. Currently, carcases are grouped using hot weight as the primary sortation criteria. As technologies that precisely and accurately measure carcase yield have become available, such as dual-energy X-ray absorptiometry (DXA), carcase lean meat yield percentage (LMY%) may be used with carcase weight to predict individual cut weights to allocate 'the right carcase' to 'the right cutting plan'. This task is well suited to be solved using mathematical optimisation models. The tool uses an Integer Linear Programming model to allocate carcases of varying weight and LMY% to cut plans to maximise profit. The tool does this by; • specifying multiple cut plans; • using cut weight ranges (min and max) and piece counts to constrain and shape the allocation of carcases to plans to meet market orders; • accounting for the differential cost of labour to process individual carcases based on their weight and LMY%. The potential for the tool to improve profit was tested in a case study in an Australian domestic lamb supply chain. An optimised scenario using carcase weight and LMY% was compared to: (1) carcases randomly allocated to cut plans; and (2) carcases allocated to cut plans by weight; as is the current industry standard practice. Using cut weight alone, profit was 4.1% greater than when carcases were randomly allocated to cutting plans. Optimisation yielded an additional 1% profit when compared to the carcase weight scenario. This work provides a pathway whereby DXA technology, combined with cut weight prediction and optimisation algorithms, and carcase sortation may improve the profitability of lamb supply chains.

Session 18 — Poster 18

Predicting P8 fat depth in hot beef carcasses using a handheld microwave system

J. Marimuthu, R.K. Abraham and G.E. Gardner
Advanced Livestock Measurement Technologies Project, School of Veterinary and Life Sciences, Murdoch University, Murdoch, WA 6150, Australia; jayaseelan.marimuthu@murdoch.edu.au

Non-destructive and non-invasive methods of measuring live animal and carcase fatness on genotypically and phenotypically diverse cattle are crucial as they enable optimised carcase boning, producer feedback, and value-based trading. However, the suitability of the technology that takes these measurements depends upon numerous factors including accuracy, reliability, cost, portability, durability, speed, ease of use, safety, and for *in vivo* measurements the need for fixation or sedation. Working within these constraints a portable microwave system has been developed at Murdoch University to measure fat depth. This study details the accuracy of this system to predict P8 fat depth of hot beef carcases at commercial abattoir chain speeds. Beef carcases (n=1,304) were scanned hot using the Microwave System at the P8-site 30 min *post-mortem* and simultaneously measured for P8 fat depth by the in-house grader. The average carcase weight of these cattle was 476±30 kg and ranged between 156 and 5.48 kg, while the average P8 fat depth was 16±5 mm and ranged between 2 and 35 mm. The microwave system operated at frequencies of 100 MHz to 5.4 GHz with output power of -10 dBm coupled with a prototype broadband Vivaldi patch antenna. The reflected microwave signals were recorded at 10 MHz intervals across 531 frequencies. The magnitude of the frequency domain signals was used to predict the P8 fat depth, with the model trained using an ensemble stacking technique in WEKA and tested using 5-fold cross validation. The average precision was high with average R^2 of 0.81, with root-mean-square-error of the prediction of 2.78. The average bias (difference between the predicted and actual values at the mean of the dataset) was 0.12, while the average slope between predicted and actual values deviated from 1 by 0.02. This study demonstrates that the microwave system can precisely and accurately predict hot P8 fat depth across divergent phenotypes. Given that these measures were captured at commercial chain speed, these results suggest that the hand-held device has the potential for commercial deployment as a simple cost-effect technology for measuring carcase fatness.

Session 18 Poster 19

Meat@ppli, a smartphone application to predict fat content of beef

J. Normand[1], B. Meunier[2], M. Bonnet[2] and B. Albouy-Kissi[3]
[1]Institut de l'Elevage, Carcass and Meat Quality, 69007 Lyon, France, [2]INRAE, Université Clermont Auvergne, VetAgro Sup, UMR Herbivores, 63122 Saint-Genès-Champanelle, France, [3]Institut Pascal, Université Clermont Auvergne, 43009 Le-Puy-en-Velay, France; jerome.normand@idele.fr

Fat has a major economic importance in the beef sector. It affects all the meat food chain steps: from the farmer to the consumer. However, nowadays, the monitoring of fat, and especially marbling, in beef is difficult, due to the lack of a suitable assessment tool, i.e. reliable, simple, fast, non-destructive and inexpensive. The exponential growth of smartphones equipped with high quality imagers and high computing power has provided tremendous opportunities for measuring fat on bovine carcasses. The Meat@ppli project aimed: (1) to predict intramuscular and total fat content of 6th rib from its image captured under non-standardized and uncontrolled conditions, using image analysis methods and deep learning; (2) to embed the algorithms in a smartphone application. For this purpose, cross section images of the 6th rib of 164 carcasses chosen to be representative of the beef marbling variability, were captured with a smartphone Samsung® Galaxy S8 fitted with polarizing filters. The ribs were then removed to determined gold standard measures: total fat content by dissection and weighing, and intramuscular fat content (IMF) by the Soxhlet method. From more than 3,500 images of 6th ribs and gold standard measures, several artificial neural networks were trained to segment the rib, the ribeye, IMF in the ribeye and total fat in the rib. The correlations between the gold standards and the parameters from the image analysis were strong, with correlation coefficients of 0.91 and 0.79 for IMF and total fat content, respectively. The prediction models were then embedded in the Meat@ppli application. The application starts by taking a picture of the cross section of the 6th rib. The captured image is then displayed and submitted for validation. In less than 10 seconds, the application calculates IMF and total fat content. The Meat@ppli application remains a proof-of-concept that, in the future, could be used by the beef industry to route carcasses to the most suitable distribution channels and to perform massive phenotyping for the selection of bovines with appropriate marbling.

Session 18 Poster 20

Computed tomography vs chemical composition for determination of Australian lamb carcase composition

K. Mata, G. Gardner and F. Anderson
Murdoch University, College of Science, Health, engineering and education, 90 South street, 6150, Australia; g.gardner@murdoch.edu.au

Computed tomography (CT) is commonly used as the gold standard measure of carcase composition in Australian lamb. For industry acceptance, experimentation against its predecessor, chemical composition was undertaken to demonstrate its comparability. A group of 30 lambs with a wide phenotypic range were sectioned into fore, saddle and hind and CT scanned at 36 hrs *post-mortem*. The 3 sections including bones were minced post scan through a 5 mm grinding plate until a uniform and homogenous mix was produced. Five samples were obtained in each of the fore(F), saddle(S), and hind(H) sections of the carcase and tested for their chemical composition of protein, lipid, and ash. These values expressed as percentages were then compared to the CT estimates of fat %, lean % and bone %. There was a strong association between chemical protein % and CT lean % with the R^2 and RMSE values for each section 0.86 and 2.82% (F), 0.92 and 2.82% (S) and 0.91 and 2.20% (H). A similar association was found with chemical lipid % and CT fat % across all carcase sections showing R^2 and RMSE values of 0.93 and 2.11% (F), 0.93 and 3% (S), and 0.89 and 2.05% (H). These strong associations demonstrate the relative equivalence between these two values as indicators of carcase composition. There was a weaker association found with chemical ash % and CT bone % showing R^2 values of <0.67 across all three sections and highlighting the well documented difficulty in homogenizing bone in a multi-tissue sample. Chemical analysis for protein, lipid and ash all showed marked sub-sample variation. Despite extensive mixing, the average deviation of each rep from the sample mean was 1.15 protein % units, an amount which represents 3.42% of the average protein value. Comparatively, for CT, the average deviation from the mean for each rep was 0.01 CT lean % units, which represents 0.02% of the average CT lean value. Similar findings were found for lipid and CT fat %, and ash and CT bone %. This variation in chemical sampling makes calibration of objective measurement tools against chemical analysis an unreliable method when compared to CT.

Session 18 Poster 21

Assessing dual-energy X-ray absorptiometry prediction of intramuscular fat content in beef

C.L.N. Nunes[1], R.S.R. Vilela[1], E.B. Schultz[1], M.I. Hannas[1], C.M. Veloso[1], M.M. Ladeira[2] and M.L. Chizzotti[1]
[1]Universidade Federal de Viçosa, Department of Animal Science, Av PH Rolfs, s/n. Campus Universitário, 36570900 Viçosa, Brazil, [2]Universidade Federal de Lavras, Trevo Rotatório Professor Edmir Sá Santos, 37203202 Lavras, Brazil; mariochizzotti@ufv.br

This study assessed the capability of dual-energy X-ray absorptiometry (DEXA) to predict intramuscular fat (IMF) content of beef longissimus steaks against chemical IMF as the gold standard. DEXA performance of fat prediction was assessed using a leave-one-out cross validation method among Angus and Nellore steaks, which generated a chemical fat range of 14.05-36.82% and 2.46-7.84%, respectively. There was a significant positive association between DEXA predicted fat and chemical fat content. However, higher precision was found for pooled data (R^2=0.95, RMSECV=1.95) and Angus (R^2=0.75, RMSECV=2.39) than Nellore (R^2=0.15, RMSECV=1.22) group. Accuracy also had the same response with average slope values close to one for pooled data and Angus and a lower value (0.42) for Nellore group. DEXA precisely predicts IMF content across a wide range of fat content. However, its precision and accuracy of prediction within low-fat content samples are lower than in high-fat content beef steaks. Funding: supported by Fundação de Amparo à Pesquisa de Minas Gerais (FAPEMIG) #APQ-02403–2017; RED-00172-22, Conselho Nacional de Desenvolvimento Científico e Tecnológico (CNPq), # 443718/2018–0; 308241/2022-3, and Instituto Nacional de Ciência e Tecnologia de Ciência Animal (INCT-CA).

Session 18 Poster 22

In-abattoir 3D image parameters of beef carcasses for predicting carcass classification and weight

H. Nisbet[1,2], N. Lambe[2], G. Miller[2], A. Doeschl-Wilson[1], D. Barclay[3] and C.A. Duthie[2]
[1]The University of Edinburgh, The Roslin Institute, Easter Bush, Midlothian, EH259RG, United Kingdom, [2]Scotland's Rural College, The Roslin Institute Building, Easter Bush, Midlothian, EH259RG, United Kingdom, [3]Innovent Technology Ltd., Northern Agri-Tech Innovation Hub, Easter Bush, Midlothian, EH259RG, United Kingdom; holly.nisbet@sruc.ac.uk

Imaging technology can extract measurements from beef carcasses, allowing for objective grading. However, many abattoirs still rely on manual grading due to the infrastructure and cost required, making the technology unsuitable. This study explores 3-Dimensional (3D) imaging technology, requiring limited infrastructure and its ability to extract automated measurements from beef carcasses to predict cold carcass weight (CCW), and EUROP conformation and fat class on the 15-point scale. Time-of-flight cameras captured 3D images of beef carcasses in a commercial abattoir in Scotland over 6-months. Up to 35 frames were captured per carcass, with 74 measurements (lengths, widths, and volumes) extracted from each image using machine vison software. Values were averaged across frames giving one data row per carcass (9,577 steers, 8,323 heifers). The data were randomly split into training and validation datasets (70:30). The training dataset was used to build multiple linear regression and stepwise selection models, using fixed $effects_1$ (sex, breed type, kill date (and CCW for conformation and fat class predictions)) or a combination of fixed effects and 3D $measurements_2$. Including 3D measurements substantially improved the fit of models for conformation (R^2_{Adj}=0.25_1, 0.54_2, RMSE=1.18_1, 0.93_2), fat class (R^2_{Adj}=0.20_1, 0.30_2, RMSE=1.55_1, 1.38_2) and CCW (R^2_{Adj}=0.2_1, 0.74_2, RMSE=31.14_1, 21.02_2). Validation of the best fitting models had low, moderate, and high accuracy, respectively, for fat (R^2=0.3_2, RMSE=1.36_2), conformation (R^2=0.54_2, RMSE=0.93_2), and CCW (R^2=0.74_2, RMSE=21.28_2). Mapping predictions on to the traditional EUROP grid used in the UK showed that 95% of fat classes and 81% of conformation classes were classified within the correct or one neighbouring grade. The 3D measurements were found to add value to the models, improving accuracy, indicating the potential for technology requiring limited infrastructure to predict carcass traits.

Session 18 Poster 23

Prediction of liveweight from linear conformation traits in beef cattle

M. Meister[1], L. Speiser[2] and A. Burren[1]
[1]Bern University of Applied Sciences, School of Agricultural, Forest and Food Sciences HAFL, Länggasse, 85, 3052, Switzerland, [2]Mutterkuh Schweiz, Staperferstrasse 2, 5201 Brugg, Switzerland; marc.meister@students.bfh.ch

In the present study, live weight in beef cattle was modelled using linear conformation traits. For this purpose, Mutterkuh Schweiz provided data from a total of 26,374 beef cattle (♀: 17,920; ♂: 8,454) of 19 different breeds from years 2015-2023. For each animal, the first linear description was used in each case. In the linear description, the live weight of the animal was also determined with a scale in each case. The relationship between live weight and linear conformation traits was investigated using a linear mixed model. The model included the fixed effects sex, breed, age and month at the measurement, height at withers (cm), top line (score:1-9), body length (cm), length of rump (cm), rump angle (score:1-9), chest depth (cm), width at hips (cm), muscularity (score:60-99), BCS (score:1-9), muscularity shoulder side view (score:1-9), thickness of loin (score:1-9), thigh rounding side view (score:1-9), thigh length (score:1-9) and final score (score:60-99), as well as expert as random effects and residual effect. This model resulted in a coefficient of determination of 87.7% and a root mean squared error of 38.26 kg. The largest fixed effects were observed for sex ((♀-♂: 37.97±1.17 kg) and the breeds Grauvieh-Charolais (36.51±2.00 kg), Grauvieh-Aubrac (33.55±1.97 kg), Grauvieh-Dexter (24.51±3.64 kg), Grauvieh-Galloway (16.80±4.00 kg) and Grauvieh-Simmental (16.08±1.85 kg). The main estimated fixed effects for linear conformation traits were 8.51±0.49 kg/score, 7.92±0.49 kg/score, 6.26±0.12 kg/cm, 6.17±0.12 kg/cm, 5.80±0.43 kg/score and 5.55±0.67 kg/score for thigh rounding side view, muscularity shoulder side view, width at hips, chest depth, BCS and rump angle, respectively. In conclusion, linear conformation traits can be used to estimate live weight of beef cattle.

Session 18 Poster 24

Beef marbling assessment by accredited graders and a hand-held camera device

M. Kombolo-Ngah[1], N.S.R. Mendes[1], A. Neveu[2], A. Barro[1], M. Christensen[3] and J.F. Hocquette[1]
[1]INRAE, VetAgroSup, UMRH, 63122 Theix, France, [2]IMR3GF, Smulikowskiego St. 4, 217 00-389 Warsaw, France, [3]Frontmatec A/S, A/S, Smørum 2765, Denmark; moise.kombolo-ngah@inrae.fr

Marbling is one of the most important traits of beef related to eating quality. According to the MSA (Meat Standards Australia) methodology, marbling should be evaluated visually in the chiller by accredited graders although various technologies have been developed. The aim of this study was to evaluate the performance of a hand-held camera (Q-FOM Beef) to predict MSA marbling scores (MSA-MS) between the 5th and the 6th rib. The MSA-MS ranged from 100 to 1,190. Two MSA trained graders were used for this study, one being an expert grader. A total of 285 carcasses were assessed in the chiller by these 2 graders. The R^2 of prediction between scores from the expert grader and the second grader was 0.78 with a RMSE of 47.9 MSA marbling points. Then, 779 images from the same carcasses were acquired with the Q-FOM Beef camera (i.e. between 2 and 3 images per carcass). In a first analysis with the 285 carcasses, a Q-FOM Beef calibration model using expert grader MSA-MS as reference was developed. This calibration model was applied to one image per 285 carcasses resulting in a R^2 of prediction 0.75 and the RMSEP was 44.9 MSA marbling score points. To be accredited in Australia, the requirements are that ≥49% of the samples must be within 50 MSA-MS from the expert grader, ≥79% of the samples must be within 100 MSA-MS and ≥97% must be within 200 MSA-MS. The grader-to-grader comparison showed 75.9% were within 50 MSA-MS, 97.1% were within 100 MSA-MS and 100% were within 200 MSA-MS. This indicates that the graders performed well when comparing their scores. In a second analysis, a subset (n=124) of the 285 carcasses were also assessed on-screen by the expert grader. The R^2 of prediction between in chiller and on-screen MSA-MS was 0.78 with a RMSE of 48.7. The in-chiller and on-screen comparison showed that 79.8% were within 50 MSA-MS, 96.8% were within 100 MSA-MS and 100% were within 200 MSA-MS. This suggests that on-screen assessment would be an acceptable method to develop a marbling calibration model, although more investigations would be required.

Session 18 Poster 25

Beef on dairy – meat quality and prediction of intramuscular fat on the slaughter line

F.F. Drachmann and M. Therkildsen
Aarhus University, Department of Food Science, Agro Food Park 48, 8200 Aarhus n, Denmark; margrethe.therkildsen@food.au.dk

Increased use of beef semen on dairy cows in Denmark is the consequence of focus on sustainable beef production, where beef on dairy has less CO_2 emission than beef from purebred beef breeds. However, this also leads to a larger variation in the carcasses at the slaughterhouse and an interest in optimizing the meat quality from these cross-bred calves. No tools have previously been implemented on the slaughter line to quantify meat quality in Denmark. However, Q-FOM, a handheld camera solution, designed to predict intramuscular fat (IMF) could be a solution to quantify IMF in the carcasses, and a way to generate a large number of meat quality data, which could be used in genomic selection of breeding stock. The aim of the study was to characterize the meat quality of cross-bred calves from Holstein cows sired with Danish Blue (DB), Charolais (CHA) or Angus (ANG) bulls and to evaluate the Q-FOM to predict IMF in the loin on carcasses split between 5th and 6th thoracic vertebra. 335 cross-bred calves were included in the study, representing 126 DB, 100 CHA and 109 ANG including both heifers and bulls. They were slaughtered at the age of approximately 9.5 months. One day *post mortem*, the carcass was split between the 5th and 6th thoracic vertebra and a Q-FOM image was captured of the loin at 5th thoracic vertebra side. At the same time, 7 cm of the loin was removed and aged for additional 2 days before analysis. The Q-FOM prediction model was developed based on the chemical IMF analysis of 261 samples of these cross-bred calves and in addition 111 samples representing loins from cattle with IMF between 6 and 23%. The carcasses from the DB cross-bred calves were the heaviest (221 kg) followed by CHA (216 kg) and ANG (210 kg), and the bulls were on average 15 kg heavier than heifers, although the heifers were 21 days older at slaughter. The amount of IMF measured chemically was highest in ANG (3.5%) vs CHA (3.0%) and DB (2.5%), and with a clear difference between heifers (3.7%) and bulls (2.3%). The Q-FOM model had a RMSEC of 1.5% (R^2=0.89), however if the model was used only on cross-bred calves below 6% IMF, the RMSEC was 1.4%. The study was supported by the GUDP (J. 34009-18-1434), Ministry of Food, Agriculture and Fisheries of Denmark.

Session 18 Poster 26

Pig carcass and cuts lean meat content determined by X-ray computed tomography

M. Font-I-Furnols, A. Brun and M. Gispert
IRTA-Food Quality and Technology, Finca Camps i Armet, 17121 Monells, Spain; maria.font@irta.cat

Currently, the value of the pig carcass is determined by its lean meat content (LMP). At slaughterhouse, LMP is determined objectively by means of manual, automatic or semi-automatic devices. These devices need to be previously calibrated and calibration is carried out by performing a cut-test, using as a reference the manual dissection of the carcass or the virtual dissection by means of a computed tomography (CT) equipment. In Spain an old fixed CTf (GE HiSpeed Zx/I) was calibrated to determine carcass LMP. However, a new mobile CTm unit (Philips Brilliance 16) has recently become available, and it is necessary to validate this new equipment to be used as a reference. Therefore, two different tests have been carried out, (1) adapting the old CTf formula to the new CTm and, (2) calculating a new formula for the CTm. For both approaches, a total of 20 carcasses and its 4 main cuts (ham, belly, shoulder and loin) had been scanned with the new CTm and, after that, they have been fully dissected to determine the LMP. In the first test, when the equation of the old CTf was applied to the images of the new one (CTm) and regressed against the LMP obtained by full dissection, R^2 was 0.97. As a second test, a preliminary and simple new formula for LMP determination with the new CTm has been obtained calculating the volume associated to Hounsfield values between 0 and 120 and divided it by the carcass (or cuts) weight. A correction factor of 1.072 has been found to be applied to adjust the LMP to the dissection with an accuracy of R^2=0.99. This formula has also been applied to the images obtained from the cuts (n=20×4) scanned with the CTm and the relation between LMP obtained by CTm and by dissection is R^2 0.99. Thus, it can be concluded that with these preliminary results, it is possible to have a good determination of pig carcass LMP with the new CTm either by applying the calibration formula of old CTf with a correction factor or by determining the LMP considering the volume associated to lean and the carcass weight, also with a correction factor.

Relationships between ultimate pH, weight, conformation and fatness of carcasses

N. Silva Rodrigues Mendes[1,2], R. Rodrigues Silva[2], T. Ferreira De Oliveira[2], P.F. Rivet[3], M.P. Ellies-Oury[1,4], S. Chriki[1,5] and J.F. Hocquette[1]
[1]INRAE, VetAgro Sup, UMR1213, Theix, 63122, France, [2]Federal University of Goiás-UFG, Campus Samambaia, Rodovia Goiânia, Goiânia, 74690-900, Brazil, [3]Beauvallet, 18 rue de l'abattoir, Limoges, 87000, France, [4]Bordeaux Sciences Agro, CS 40201, Gradignan, 33175, France, [5]ISARA, 23 rue Jean Baldassini, Cedex 07, Lyon, 69364, France; nathaliasrm@gmail.com

We hypothesized that the transportation of cattle prior to slaughter causes stress that results in lower quality carcasses due a higher final ultimate pH. Therefore, the objective of this study was to investigate the effects of stress during transport to slaughter from farms located in different geographical areas in France, specifically the effects on ultimate pH for Limousin cattle. Cold carcass weight, age, and the EUROP conformation and fatness scores were recorded from 3,809 Limousine carcasses from a private slaughterhouse in Limoges, France, which were evaluated 24 h *post-mortem*, from May 2021 to November 2021. We performed statistical analyses of the relationships between ultimate pH, geographical areas (<50 km; between 50 and 150 km; and between 150 and 250 km from the slaughterhouse), cold carcass weight, conformation and fatness of carcasses. European conformation score was correlated with cold carcass weight ($r=0.76$, $P<0.001$) and fat score ($r=0.57$, $P<0.001$). Similarly, fat score was correlated with cold carcass weight ($r=0.57$, $P<0.001$). As expected, the variability in cold carcass weight can be explained in part by the variables studied (mainly age, European conformation, and fat score), explaining approximately 60% of the variability in carcass data. The variables studied (area, age, European conformation, and fat scores) explained only 2.45% of the total variability in pH data. Values of ultimate pH were lower in the geographical area farthest from the slaughterhouse (5.74 vs 5.78, $P<0.05$) because cattle were transported the day before slaughter in contrast to the first two areas. In conclusion, long distances slightly affect beef pH, with consequences on beef eating quality that remain to be studied but are likely to be small.

Elite Dairy Beef – a pathway for male dairy calves into the premium beef market

H. Cuthbertson[1], R. Polkinghorne[1], A. Neveu[2], T. Maguire[1], O. Catalan[3] and R. Law[4]
[1]Elite Dairy Beef Pty Ltd, Hawthorn, VIC, Australia, [2]SARL Birkenwood Europe, Lanon, Saint-Sornin, France, [3]Inzar S.L, Zaragoza, 50018, Spain, [4]Anupro Ltd, Omagh, BT78 1TS, Ireland; hcuthbertson@elitedairybeef.com

The Elite Dairy Beef program is based on consumer eating experience from dairy beef raised under a unique tightly controlled nutritional program. The program is aimed at changing industry perception of 'dairy beef' from the traditional lower quality manufacturing image to a high value premium product that justifies the raising of male dairy calves. A requirement for cattle marketed under this brand requires animals to be antibiotic, hormone and ionophore free. The nutritional program was initially developed in Spain and is widely used in Europe and the UK in the premium beef programs. A locally adapted program utilising key ingredients from Spain in conjunction with local sourcing has since been tested and proven in Australia. The milk replacer and rations are of extremely high quality and specifically targeted at superior early life nutrition to rapidly develop the calf immune system and rumen function, thereby maximising health and avoiding the use of antibiotics. Contrary to conventional rearing systems, this system is based on a low milk replacer intake and immediate concentrate consumption from birth. The critical and interlinked aims are to avoid negative energy balance, optimise gut health, and fast-track rumen development. Calves are purchased at 5 days of age and transported to a rearer. They have *ad lib* access to concentrate Quickstart, plus specialised milk powder InzarMilk, fed at 2 l twice a day for 3-4 weeks. After 2 weeks, calves transition onto a grower ration Papincalf enabling early weaning of the calf. This ration is fed until 14 weeks after which cattle are transitioned onto the final ration Econbeef until slaughter. Cattle are finished at 12-13 months of age with a liveweight of 500-550 kg and a carcass weight of 280-300 kg. Optimal fat cover on these animals is observed for minimal trimming and little waste, with straight Holsteins yielding similar to their beef cross counterparts. Cattle also carry a positive carbon story, with the calf primarily offset by the cow coupled with a highly efficient animal capable of good conversion and fast finishing times.

Session 19 Theatre 1

The contribution of Daniel Sauvant to modelling in animal science

J. Van Milgen and P. Faverdin
INRAE, Institut Agro, Pegase, Le Clos, 35590 Saint-Gilles, France; jaap.vanmilgen@inrae.fr

Mechanistic modelling in animal science started in the 1980s with researchers from the USA, the UK, Australia, and New Zealand leading this. In France, Daniel Sauvant was a 'driving force' in the development of a culture of modelling in research and education. Most animal scientists perceived modelling as somewhat mysterious, more related to complicated mathematics than to biology. An important contribution of Daniel has been the conceptualization of biological systems through an operating system (i.e. the flow of nutrients and metabolites through compartments) and a decisional system (the control of the flows). Furthermore, he integrated concepts proposed by Bauman and Currie of homeorhesis (a long-term trajectory) and homeostasis (short-term controls) in the models. The concept of a long-term trajectory was applied to protein and lipid deposition during the productive life of a growing animal, but also to the management of priorities by the animal during its entire life (e.g. to grow and mature, to reproduce and care for newborns, and eventually to die). In 1994, Daniel stated that 'to progress further in metabolic modelling, it will be necessary to have a better knowledge of the interaction of the metabolic effects of homeorhetic and homeostatic regulations'. He modelled this by comparing the actual situation of an animal with its position on the homeorhetic trajectory, to use the difference (e.g. due to the nutrition or sanitary conditions) as a regulatory mechanism to return towards the trajectory. He applied the concepts an operating and a decisional system not only to animals, but also to livestock production systems in which homeorhesis can be perceived as a long-term goal of a farmer, and homeostasis as his management practices of the herd. Many of us who worked with Daniel may have been frustrated one day because our model did not predict what we expected, to which Daniel could reply "but it goes in the right direction". There is probably not a better way to illustrate the limitations and potential of modelling in animal science. Daniel's approach to structure complicated and complex systems and his skills to makes this accessible have been a great source of inspiration for many of us.

Session 19 Theatre 2

Systemic modelling of nutrient partitioning: from the seed sown by D. Sauvant to future prospects

L. Puillet
Université Paris-Saclay, INRAE, AgroParisTech, UMR Modélisation Systémique Appliquée aux Ruminants, 91120 Palaiseau, France; laurence.puillet@inrae.fr

Predicting how nutrients are channelled to biological functions and eventually converted into animal products has always been a key issue in nutrition models. Many progresses were made to quantify the biochemistry of nutrition and then model metabolic functioning. In this view, the animal is a passive convertor and regulations are short-term homeostasis mechanisms. The role of long-term regulations, that play a key role in coordinating changes in metabolism to support physiological stages, was known but remained a challenge for modellers. A major innovation enabled by the work of Daniel Sauvant was to propose an operational framework for implementing these long-term regulations, with the concept of meta-hormones reflecting priorities among functions. The objective of this presentation is to illustrate how this pioneer systemic proposal was further developed in modelling works. A first type of work is the explicit implementation of animal's priorities. Combining a dynamic theoretical model of priorities among functions (growth, lactation, gestation and reserves) with a model of energy partitioning allows the simulation of multi-traits lifetime trajectories. Interactions among functions are taken into account throughout the succession of reproductive cycles. This consistent biological building-block, proposed as a virtual animal, has contributed to improve herd simulators. A second type of work is the use of systemic view when interpreting time-series data. Pure statistical models are very efficient to capture dynamics but interpreting outputs may be difficult due to their lack of biological meaning. Using a systemic view based on animal's investment into functions is powerful to better interpret trajectories and study individual variability. A third type of work is the implementation of acquisition-allocation framework. This theory, grounded in ecology, has started to be more and more used in animal science, especially with all the issues related to robustness and the need to better account for trade-offs among functions. The examples presented here, not exhaustive, illustrate how powerful the systemic approach is to address challenging issues in animal science, a view that Daniel Sauvant has always defended during his career.

Session 19 — Theatre 3

Mechanistic model of the dynamics of body retention and utilization of phosphorus and calcium in sow

J. Heurtault[1,2], P. Schlegel[1] and M.P. Létourneau-Montminy[2]

[1]Agroscope, Route de la Tioleyre 4, 1725, Posieux, Switzerland, [2]Université Laval, 2425 rue de l'Agriculture, G1V 0A6, Québec, Canada; julien.heurtault.1@ulaval.ca

A better understanding of the fate of dietary phosphorus (P) use by sows will allow its optimization and enhance sustainable practices. Current models for predicting P and calcium (Ca) requirements of sows are extrapolated from growing pig models in which the driving force is the body weight or the protein gain. Furthermore, recent data, of bone mineralization kinetics in sows and its evolution during the production cycle, have shown the ability of the sow to mobilize part of her body P and Ca reserves in lactation to cover her high requirements and also to replenish her body reserves at the beginning of the next gestation period, when her requirement is lower. Thus, a new mechanistic approach could simulate the dynamics of retention and utilization of body P and Ca reserves during lactation. The model includes 3 sub-modules. First, the digestion sub-module considers the sources of dietary minerals including responses to microbial and plant phytase and Ca and P interactions and predicts absorption and faecal excretion. Then, protein and lipid compartments growth sub-module is based on InraPorc model principles. The third one is the body ash sub-module that simulates the partitioning of absorbed Ca and P into the bone, milk, protein, and lipid compartments as well as urinary losses. The P and Ca deposited into soft tissue and excreted into milk have priority over bone deposition and if dietary supply is not sufficient bone are mobilized. This approach aims at precising the understanding and prediction of the fate of dietary P and Ca to then inversed it to calculate precisely the requirements and then optimize the use of P and reduce the losses. The next step will be validation with the generation of a new set of data.

Session 19 — Theatre 4

Modelling the digestive utilization of calcium and phosphorus in laying hens

F. Hervo[1,2], B. Méda[1], A. Narcy[1] and M.-P. Létourneau-Montminy[2]

[1]INRAE, Université de Tours, BOA, 37380, Nouzilly, France, [2]Sciences Animales, Université Laval, G1V 0A6, Québec city, Québec, Canada; fabie.hervo@inrae.fr

Laying hens require a massive daily quantity of calcium (Ca) to form the eggshell. Calcium is provided by the diet and resorption of medullary bone. Calcium mobilization from the skeleton simultaneously leads to the release of phosphorus (P) which is excreted in the urine. Thus, the levels of P and Ca in the feed must be finely balanced to ensure eggshell and bone quality while reducing P excretion in manure. Nevertheless, the fate of P and Ca in the gastrointestinal tract (GIT) and effects of the main modulating factors are not well understood in laying hens. Thus, the aim of this study was to develop a mechanistic model predicting the digestive fate of these minerals in a producing hen (34 weeks of age) considering dietary Ca and non-phytic P (NPP) levels, as well as phytase and coarse limestone particles incorporation. The model consists in 2 compartments with 5 sub-compartments representing the anatomical sections of the GIT: (1) the pre-intestinal compartment with crop and gizzard; and (2) the intestinal compartment with the duodenum, jejunum, and ileum. Because of the expression of a specific appetite for Ca, Ca and P intake were parametrized dependently of the photoperiod. In the pre-intestinal compartment, solubilization of Ca and P and complexation between Ca and phytic P (PP) or NPP are modelled depending on the pH in the compartment. Furthermore, the conversion of PP into NPP depends on the presence of microbial and plant phytases. A longer retention time and a slower solubilization of coarse limestone particles in the gizzard are also represented. In order to consider hormonal regulations occurring at the intestinal level, Ca and P absorption flows are modelled depending of the eggshell formation cycle. In the intestinal phase, the propensity of Ca to form insoluble complexes with PNP or PP as well as the effect of intestinal phytase on PP hydrolysis were parametrized. The sensitivity analysis and external validation of the model are currently in progress. This model provides a promising tool to easily evaluate different feeding strategies to optimise Ca and P utilization in laying hens, and reduce P excretion while maintaining the eggshell and skeleton quality.

Session 19 — Theatre 5

A dynamic mechanistic model to forecast the oscillatory feeding behaviour of lactating dairy cows

E. Fiorbelli[1], A.S. Atzori[2], A.T. Büşra[2], L. Tedeschi[3] and A. Gallo[1]

[1]Università CAttolica del Sacro Cuore, Department of Animal Science, Food and Nutrition (DIANA), Via E. Parmense 84, Piacenza 29121, Italy, [2]Università di Sassari, Dipartimento di Agraria, Viale Italia 39, Sassari 07100, Italy, [3]Texas A&M University, Department of Animal Science, College Station, TX 77843-2471, USA; antonio.gallo@unicatt.it

The aim of this study was to evaluate the ability of the model proposed by Fischer (1996) and reproduced by Tedeschi and Fox (2020) to predict the oscillatory feeding behaviour of lactating dairy cows. The model used the stock and flow diagram (Vensim ver 9.2.4, Ventana Systems) and included compartments for potentially degradable proteins, soluble carbohydrates, slowly digestible fibres, and very slowly digestible fibres fractions. The intake regulation included feedback structures for distention, chemostatics and protein effects. Inputs to the model were limited to protein (CP), 48-hour IVDMD corrected for residual microbial mass, and NDF. In the experimental facility (CERZOO srl, Piacenza, Italy), 20 Holstein dairy cows were randomly selected and the feeding behaviour was monitored for four consecutive days using the Roughage Intake Control System automatic recorder (Hokofarm Group, Marknesse, The Netherlands). The average age was 38±7 months and the average milk yield was 33.3±4.36 kg/cow/day. Oscillation differences in frequency and amplitude (meals per day and amount eaten per meal) and the variability of the observed meals (kg/hour) were discussed in relation to the average meal eaten by each animal during the experimental period. On average, the animal consumed 7±2 meals/cow per day, the intake per meal was 3.60±1.12 kg, the daily dry matter intake (DMI) was 23.81±2.73 kg/cow per day and the time spent eating was 21.0±4.6 minutes per meal. In this experiment, the animal components had a significant influence on the variation in DMI. New feedback loops in the structure of the stock and flow diagram were included to improve the model in simulating the daily oscillatory system of the lactating cow's behaviour and the daily DMI.

Session 19 — Theatre 6

A dynamic *in silico* model of rumen microbial fermentation and methane production

R. Muñoz-Tamayo[1], S. Ahvenjärvi[2], A.R. Bayat[2] and I. Tapio[3]

[1]Université Paris-Saclay, INRAE, AgroParisTech, UMR Modélisation Systémique Appliquée aux Ruminants, 91120, Palaiseau, France, [2]Animal Nutrition, Production Systems, Natural Resources Institute Finland (Luke), 31600, Jokioinen, Finland, [3]Genomics and Breeding, Production Systems, Natural Resources Institute Finland (Luke), 31600, Jokioinen, Finland; rafael.munoz-tamayo@inrae.fr

Mathematical (*in silico*) models have been developed to enhance understanding of rumen function. Model evaluations have shown that there is still room to enhance their accuracy for predicting volatile fatty acids (VFA) concentration and methane production. Previously, we developed an alternative model of rumen fermentation under *in vitro* conditions aimed at better representing the rumen microbiota and hydrogen dynamics. In the present work, we extended the previous model to account for *in vivo* conditions. We performed an experiment with four Nordic-Red dairy cows equipped with rumen fistulas to provide dynamic data to support model construction. The experiment determined animal daily patterns of feed intake, rumen fermentation and enteric methane production. Feed intake and methane emissions were measured in respiration chambers during two days after an adaptation period. Rumen liquid was collected every three hours for determination of VFA concentration. In the model, the rumen microbiota is represented by three functional groups namely sugar-, amino acids- and hydrogen-utilisers. The feed is expressed in three pools as neutral detergent fibre, non-structural carbohydrates and proteins. The model consists of 18 compartments. The model performance was satisfactory as evaluated by the coefficient of variation of the root mean squared error, which was 9% for acetate, 11% for butyrate, 13% for propionate and 21% for methane. Our model has the potential to be used as virtual platform to simulate rumen fermentation dynamics. Acknowledgements. The authors acknowledge funding from the MASTER project, an Innovation Action funded by the European Union's Horizon 2020 research and innovation programme under grant agreement No 818368.

Session 19 Theatre 7

Future directions in ruminant feeding systems: opportunities and challenges for sustainability

L.O. Tedeschi
Texas A&M University, Department of Animal Science, 230 Kleberg Center, 2471 TAMU, College Station, TX 77843, USA; luis.tedeschi@tamu.edu

Ruminants have a specialized digestive system that allows them to extract nutrients from fibrous plant material through microbial fermentation in their rumen. Mathematical models simulate the complex interactions between animal nutrition, digestion, metabolism, and production to optimize feed formulations, predict nutrient requirements, and evaluate the effects of different feeding strategies on animal performance and health. Advancements in decision support tools (DST) must incorporate data from multiple sources (sensors and satellite imagery) to improve predictability and provide real-time recommendations for feeding strategies based on animal performance, feed availability, and environmental conditions. DST will require advanced modelling techniques (machine learning and agent-based models) to enhance the adoption of big data. The real-time decision-making recommendations require multi-objective optimization based on animal performance, environmental impacts, and economic feasibility. An overhaul of the mathematical models is needed to support the DST directives by revising the underlying principles of ruminant nutrition, revamping existing submodels, and incorporating novel concepts. Must-improve topics are nutrient utilization by gut microbes: producing metabolites such as volatile fatty acids, ammonia, and microbial protein; developing environmental impact assessment tools to evaluate the environmental and economic impacts of different feeding strategies and identify opportunities for improvement: genetic/genomics and animals stress/disease/metabolic disorders to enhance prediction of animal performance, feed efficiency, nutrient utilization, and strategic medication. Data quality is the most critical issue in the success of DST and model predictability. The evolution of nutrition models for ruminants will depend on technological advances, data collection, scientific knowledge, and the livestock industry's and society's changing needs and priorities. The increased predictive accuracy of nutrition models for ruminant animals requires a multi-disciplinary approach involving collaboration among scientists, producers, and industry stakeholders and using advanced technologies and modelling techniques.

Session 19 Theatre 8

Meta-analysis in animal science: a valuable tool to generate new knowledge from previous experiments

M.P. Létourneau Montminy[1], C. Loncke[2], P. Schmidely[2], M. Boval[2], J.B. Daniel[3] and D. Sauvant[2]
[1]Laval University, Animal Science, 2425 rue de l'Agriculture, G1V0A6, Canada, [2]AgroParisTech, UMR MOSAR, 22 place de l'Agronomie, CS 20040, France, [3]Trouw nutrition R&D, P.O. Box 299, 3000, Amersfoort, the Netherlands; marie-pierre.letourneau@fsaa.ulaval.ca

Meta-analysis is a useful tool to analyse large sets of heterogeneous data that gain popularity in animal science since the 2000s. Meta-analysis has proven to be efficient to create new knowledge based on already published data through empirical models, allowing progress in both understanding, highlighting research effort and lack of information, and obtain improve prediction. Review on this tool in animal science by Sauvant *et al.* have first focus on good practices namely graphical interpretation, the way of considering the experiment effect (i.e. fixed or random), and then interfering factors. Meta-analysis has then been used for updating feed unit systems (e.g. 'Feeding System for Ruminants'). Another important use of meta-analysis is for the development of mechanistic models to simulate fluxes, parameterization or external validation database. Based on these years of experience, the current review will paint a portrait of the evolution of meta-analysis, followed by a description of the key steps including study objectives well defined to preform exhaustive search of the literature, article filtering to select or discard publications, database construction, coding, and statistical analysis. A focus will also be done regarding the good practices and the potential pitfalls in the conduct of meta-analyses in animal sciences. These include traceability and criteria used to select/discard publications in a way that one could reproduce the final selection of data which is know require by most journals. Then, some key elements for statistical processing such as the coding step of selected data to isolate specific experimental factors, the study of the meta-design, the accurate consideration of study, and the post-analytic study that most often allows the analysis to be re-run to improve the quality of the procedure. Several examples of meta-analysis types will be presented to demonstrate the advantages and the limits of the approach.

Global sensitivity analysis of INRAtion®V5 for dairy cows: Sobol' indices

S. Jeon[1], S. Lemosquet[2], T. Senga Kiesse[3] and P. Nozière[1]
[1]INRAE-VetAgro Sup, UMR Herbivores, Saint-Genès-Champanelle, 63122, France, [2]INRAE-Institut Agro, UMR Pegase, Saint Gilles, 35590, France, [3]INRAE-Institut Agro, UMR SAS, Rennes, 35000, France; seoyoung.jeon@inrae.fr

The effective ruminal degradability (ED6_N) and true intestinal digestibility (dr_N) of nitrogen measured *in sacco* are used to describe the nutritive values of feedstuffs in the INRA 2018 feeding system for ruminant and its rationing software (INRAtion®V5). In the perspective to move from *in sacco* to *in vitro* tests, a global sensitivity analysis (GSA) was conducted to measure the sensitivity of the INRAtionV5 to variations in input variables: gross energy (GE) and crude protein (CP) contents, organic matter digestibility (OMd), ED6_N and dr_N. We used Sobol's indices that calculated input's contribution to each output with interactions among inputs (St_i). The GSA was performed with 6 diets varying according to forages (corn silage, grass fresh/hay/silage) and concentrates (cereals and oil seed meals). These diets were formulated for a multiparous dairy cow at week 14 of lactation, and contained 14 to 18% CP. Then, 10,000 quasi-random samples of the 5 input variables were generated per diet using Sobol' sequences. Variations on predicted dry matter intake were on average mainly due to GE (St_{GE}=53±5.3%), but with grass hay and grass silage diets, OMd was the main contributor (St_{OMd}=52 and 51%, respectively). The contribution of OMd was the highest for milk yield (MY; 63±11.0%), protein yields (MPY; 68±8.7%) and protein contents (MPC; 79±7.2%). With grass silage and fresh forage diets (14% CP), ED6_N was the second contributor to MPY (24 and 16%, respectively) and MY (2 and 18%, respectively). Urinary N flow was most affected by CP (60±8.4%), and faecal N flow (FN) by GE (35±9.6%) and OMd (33±14.4%). The ED6_N was the second contributor to FN in corn silage diet (33%) and grass hay diet (41%). Interaction among the input variables was the highest (>35%) on MPY and MY with grass hay diets with oil seed meal, but was the lowest (<5%) with grass hay diet without oil seed meal. In conclusion, with diets containing 14 to 18% CP, protein-related inputs less contributed than energy related inputs to changes in DMI and milk production. But with 14% CP diets, production responses are sensitive to ED6_N.

Effects of housing or feeding practice in dairy goats estimated by the INRAtion®V5 feeding system

S. Giger-Reverdin, V. Berthelot and D. Sauvant
Université Paris-Saclay, INRAE, AgroParisTech, UMR Modélisation Systémique Appliquée aux Ruminants, 22 Place de l'Agronomie, 91120, Palaiseau, France; sylvie.giger-reverdin@agroparistech.fr

Effects of housing or feeding practice were tested on outputs of the INRAtion®V5 rationing system based on results obtained at MoSAR's unit. Dairy goats were housed either in individual crates (CRA) or in collective pens (PEN). They were fed either separate feed ingredients (SF) or total mixed ration (TMR) with individual measurement of feed intake. Individual weekly mean values of all results were pooled to obtain groups receiving the same basal roughage at similar stages of lactation, housing and feeding practice. Mean values were weighed by the number of observations within a group. The saturation coefficient is the ratio between the dietary fill value and the intake capacity corrected for the level of refusals that is specific to goats, because they exhibit a higher sorting behaviour than other ruminants. Observations with a level of roughage refusal below 7% were discarded. The saturation coefficient of the CRA group (1.04+0.021, ngroups=41) was close to 1 and lower that for the PEN group (1.20±0.024, n=16). The model is accurate for the CRA group and underestimates saturation coefficient for the PEN group. This might be due to data used for the setting up of the model (Caprinut database) obtained with animals in individual crates. In the CRA group, the saturation coefficient was higher for TMR (1.08±0.026, n=23) than for SF (0.96±0.035, n=18). It was quite low (0.80±0.108) for the 8 groups fed corn silage. Standard milk yield did not differ between CRA (3.28±0.179 kg/day) and PEN (3.42±0.203), but predicted milk yield was lower for CRA (2.83±0.174 kg) and higher for PEN (4.35±0.197). The non-producing requirements might be overestimated by the lack of activity for CRA goats. For PEN, an increase in passage rate due to a higher intake might overestimate the feed values. Another explanation might be the evolution of the genetic of the herd, as the trials for CRA were performed between 1978 and 2017 and for PEN after 2018. Animals might exhibit a higher intake capacity. This first test on a large number of rations is quite promising and must be extended to a larger number of diets and in other locations.

Session 19 Theatre 11

Dynamic models to inform precision supplementation of heifers grazing dormant native grasslands

H.M. Menendez Iii, J.R. Brennan, A.K. Dagel and K. Olson
South Dakota State University, Animal Science, 711 N. Creek Drive, 57703, USA; hector.menendez@sdstate.edu

Decreasing supplementation cost while optimizing animal intake is critical to maintaining cow-calf operation viability. Precision livestock technologies (PLT) can be used to target supplementation to meet individual animal nutrient requirements on extensive grasslands. However, real-time animal performance data and animal nutrition models are needed to dynamically adjust precision supplementation levels. The objectives of our study were to: (1) Develop a precision systems model (PSM) that leveraged real-time weight data and nutrient equations to dynamically inform individual heifer supplementation (kg hd/d); and (2) compare differences in predicted and observed rates of gain (kg/d). Angus heifer calves [n=60, initial body weight (BW)=237.6±15.5 kg] were assigned to one of two treatments, control or precision, and grazed dormant native grass. Both treatments were provided 2.27 kg hd/d of pelleted supplement. The control group was supplemented using a bunk feeding method and the precision group was offered supplement via the Super Smartfeed Producer™. Individual daily BW was measured using SmartScales™ and used to calculate daily rate of gain. The PSM was developed in Program R and utilized equations for metabolizable energy intake (MEI), net energy for maintenance (NEm) and gain (NEg) requirements (Mcals/d) at two-week intervals to program precision supplementation to obtain 60% mature BW (381 kg) at time of breeding. Results indicate that of 893 individual animal two week feeding periods, animals were overfed 69% of the time and underfed 31% of the time. Performance gap was calculated between observed and modelled (optimal) rates of gain, resulting in significant differences between the control (0.39±0.05) and precision (0.13±0.05) groups (*P*<0.01). This study suggests precision supplementation can decrease heifer development costs by reducing intake variation and supplement overconsumption without negatively influencing heifer performance. Refining PSM to integrate real-time data with animal nutrition models will be essential to reduce feed costs and meet individual nutrition requirements and performance targets.

Session 19 Theatre 12

Comparison of measured enteric methane emission with model calculations of commercial dairy diets

L. Koning and L.B. Šebek
Wageningen Livestock Research, Animal Nutrition, P.O. Box 338, 6700 AH Wageningen, the Netherlands; lisanne.koning@wur.nl

According to the international climate agreement the Dutch dairy sector is obligated to reduce methane (CH_4) emissions substantially. To determine enteric CH_4 emission on commercial dairy farms model estimations are used. The model used in the Netherlands (Annual Nutrient Cycling Assessment; ANCA) is (indirectly) based on data from respiration chambers and underlying principles of rumen fermentation and it is important to know how well the model fits the wide variation in diets fed on farms. Measurements of enteric CH_4 on commercial farms to validate the model are however limited. This study fills that gap and thereby enabling a comparison between measured CH_4 emission and model estimations by ANCA. On ten commercial dairy farms in the Netherlands enteric CH_4 emission of a subset of lactating dairy cows (between 16 and 53 cows per farm) was measured for twelve weeks between June 2020 to January 2022, divided into three periods of four weeks (two weeks of adaptation and two weeks of measuring period). In each period a different diet was fed (n=29, of one farm only two periods were measured). Enteric CH_4 emission was measured using the GreenFeed Emission Monitoring system (GEM, C-lock Inc. Rapid City, SD, US), an adapted feeding station that measures both continually CH_4 concentration and the quantitative airflow. Additionally, individual milk samples were taken to analyse milk composition and feed intake data on herd level (including feed composition and quality) were collected. Based on feed intake and feed composition CH_4 emission was calculated using ANCA and compared with measured CH_4 in a lack of fit regression analysis. The lack of fit analysis indicated a large fitted bias (F=76.92, *P*<0.001). The measured CH_4 yield (g CH_4/kg DMI) was not significantly related to the calculated CH_4 yield (*P*=0.669). Overall the difference between measured and calculated CH_4 yield was less than 5%. The results suggest that the model is accurate on national level and for a select range of diets, but lacks factors that are necessary to estimate CH_4 more accurately on individual farms in practice.

Session 19 Poster 13

Optim'Al can optimize dairy cow rations on economic cost and protein autonomy with INRA 2018 system

P. Chapoutot[1], G. Tran[1], V. Heuze[1] and A. Berchoux[2]
[1]AFZ, 22 place de l'Agronomie, 91120 Palaiseau Cedex, France, [2]IDELE, 23 rue Jean Baldassini, 69364 Lyon Cedex 7, France; patrick.chapoutot@zootechnie.fr

The low-cost formulation by linear programming (LP) is not widely used in ruminants, unlike in monogastric production. The novelties brought by the INRA 2018 system do not simplify its adaptation to the LP: on the one hand, the calculations implement many iterative processes and, on the other hand, the animals' responses are not linear. Moreover, taking into account protein autonomy of farms is more and more important to increase metropolitan sovereignty for the supply of protein to livestock. In this context, Optim'Al v2 is an optimization tool for dairy cow rations based on LP and integrating nevertheless the principles of the new INRA system, in which feed values and animal requirements depend on the ration characteristics. Different simulations, implementing an iterative calculation process, reveal the interest of a sensitivity analysis of the optimized solutions, useful to reflect on feed valorisation and purchase strategies for dairy farmers. A post-optimization calculation phase takes the non-linear responses of the animals into account and allows a technical-economic diagnosis of dairy cow rations based on feed cost margin and on their environmental values. In addition, a protein-dependency value of feeds according to their supply perimeter, can be used instead of feed cost as an optimization function to obtain 'autonomous' rations. Moreover, both 'economy' and 'protein autonomy' approaches can be combined in an optimization multi-objective function. This tool, designed by AFZ within the framework of the French 'Cap Protéines Program', was based on a prototype proposed by Chapoutot *et al.*, and will be commercialized by Idele.

Session 19 Poster 14

Modelling the dynamics of dairy cattle response to heat stress using system dynamics methodology

R. Cresci[1], B. Atamer Balkan[1], L.O. Tedeschi[2] and A.S. Atzori[1]
[1]University of Sassari, Department of Agricultural Sciences, Viale Italia 39, 07100, Italy, [2]Texas A&M University, Department of Animal Science, 2471 TAMU Kleberg Center, 77843-2471, USA; rcresci@uniss.it

Exposure to prolonged periods of extreme heat, such as heat waves (HW), might negatively affect dairy cows' health, and productive and reproductive indices. Due to climate change, besides a possible change in the pattern of occurrence, HW are expected to rise in intensity, duration, and frequency, making heat dissipation mechanisms less effective and increasing the animal heat load. Several empirical, mathematical, and dynamic models were developed to predict the effects of heat stress (HS) in animals. However, modelling individual animal response under HS conditions is challenging because it exhibits complex system features with dynamic and nonlinear behaviour, including delayed reactions that are often misinterpreted or unaccounted. One formal mathematical modelling approach to understanding the nonlinear behaviour of complex systems over time is the system dynamics (SD) methodology. In this work, we applied SD methodology to model the dynamics of dairy cow response and observed milk yield (MY) under HS. To apply the model, we used MY and temperature-humidity index (THI) data collected from a dairy cattle farm in August 2022. From the case study farm, 20 cows were selected based on the days in milk (DIM; $70 \leq DIM \leq 220$ d), to avoid the effects of the stage of lactation. The model calibration was performed for the relevant parameters related to selected cows' response to HS. The results showed a high accuracy (mean absolute percent error, MAPE<5%; $R^2>0.6$; concordance correlation coefficient, CCC>0.6) of the proposed model structure in capturing the effect of HS in 13 cows, which can be identified as heat-sensitive. The behaviour of the other 7 cows could not be fully captured within the defined parameter space: either they show heat-resistant behaviour, or they may experience different biological delays due to HW events. Our HS model considered nonlinear feedback mechanisms to help farmers and decision-makers quantify the animal response to HS, predict MY under HS conditions, and distinguish the heat-sensitive cows from heat-tolerant cows at the farm level.

Session 19 — Poster 15

Volatolomics for developing non-invasive biomarkers of ruminal function and health in dairy cows

E. Jorge-Smeding[1], C. Martin[1], L. Volmerange[2], E. Rispal[3], M. Bouchon[3], Y. Rochette[1], N. Salah[4], F. Violleau[2] and M. Silberberg[1]
[1]INRAE, VetAgroSup, UMRH, Saint-Genès-Champanelle,France, 63122, France, [2]INP, Toulouse, 31300, France, [3]INRAE, UEH, Saint-Genès-Champanelle,France, 63122, France, [4]Phileo by Lesaffre, Marcq-en-Baroeul, 59700, France; ejorgesmeding@gmail.com

Invasive techniques for studying ruminal function such as fistulae are being banned, and so the development of non-invasive approaches is required. Similar to biological fluids metabolomics, volatolomics propose to phenotype volatile compounds and has been shown to be successful to explore biomarkers either in gaseous samples (e.g. breath) or in the head space generated by liquid and solid samples (e.g. urine) in human. In particular, the use of single-ion flow tube coupled with mass spectrometry (SIFT-MS) has been proved to be a powerful chemometric approach when doing volatolomics due to its ability to measure thousands quickly, without requiring any sample preparation. This study aimed to test, as a proof of concept, the use of volatolomics to follow ruminal functioning. Sixteen Holstein cows (106±11 DIM) were enrolled in 2 balanced groups according to milk yield, composition and BW during a 10 weeks design based on two isoproteic diets (131 g/kg DM on average) either rich in fibre and low in starch (HFibre; 50% NDF, 12% starch) or low in fibre and rich in starch (HStarch; 40% NDF, 30% starch), the latter expected to be at risk of ruminal acidosis as a model a ruminal dysfunction. Between weeks 1 and 5, all cows received the HFibre diet, then half of the cows were progressively switched to HStarch diet (n=8) with the final diet (30% starch) being achieved on weeks 9 and 10. Milk yield and ruminal pH were measured daily, milk composition and DMI were determined twice per week. Breath air, ruminal liquid, sweat, urine, faeces and milk samples were taken on weeks 4, 5, 9 and 10 and stored at -80 °C until SIFT-MS analysis. Samples are currently being analysed and results will be presented in the meeting. In order to explore the ability of the volatolomic fingerprint to discriminate individuals between the diets, data will be submitted to univariate (repeated measure analysis) and multivariate (partial-least square discriminant analysis) analysis.

Session 19 — Poster 16

Estimation of total urinary nitrogen excretion in fattening beef cattle from urinary spot samples

C. Garcia-Vazquez[1], P. Nozière[1], L. Salis[1], R. Bellagi[1], J. David[2] and G. Cantalapiedra-Hijar[1]
[1]INRAR, URMH, Clermont-Ferrand, 63122, France, [2]INRAR, URM, Clermont-Ferrand, 63122, France; cjgvzgz@gmail.com

Livestock farming's impact on the environment is a contentious issue, particularly regarding pollution from N compounds. Matching protein supply to an animal's protein requirement is an effective way to reduce N excretion, evaluated through a N balance trial in metabolic cages. However, animal welfare issues limit this kind of experimental approaches. Therefore, the aim of this study was to develop prediction equations for urinary N excretion in beef cattle based on urinary biomarkers. For this, we used two independent N balance trials conducted with 48 fattening Charolais young bulls. During 10-days, the total amount of urinary N excreted was determined for each animal and spot urine samples were collected on the first and last day of the each 10d-N balance. Spot urinary samples were analysed for N (Dumas method), and creatinine and urea concentrations (colorimetric analysis). The mean urinary concentration of these three metabolites in the spot samples, as well as the weight of the animals and the amount of ingested N, were used as predictors of total urinary N excretion. Linear regression analyses were performed using Minitab® 21.3 software with various combinations of these independent variables. Two prediction equations were selected based on fitting criteria and model errors. The first equation predicted urinary N excretion (g/d) from two variables: the ratio of urinary urea-N to total N ratio in the urinary spot and the total N intake (r^2=0.81 and error=14%). Both predictors were uncorrelated in this first equation (VIF=1.67). The second selected equation predicted urinary N excretion (g/kg $BW^{0.75}$) only from the ratio of urinary urea-N to creatinine in the urinary spot (r^2=0.71 and error=13%). An a posteriori statistical power analysis of the developed prediction equations concluded that a total of around of 44 animals per treatment would be necessary to detect a 10% difference in total urinary N excretion regardless of the equation. The prediction equations that were developed enable practical estimation of urinary N excretion, making it possible to differentiate between treatments aimed at reducing urinary N excretion under field conditions.

Advancing mathematical modelling for equines

E.M. Leishman[1], C. Cargo-Froom[1], P. Darani[2], S. Cieslar[2] and J.L. Ellis[1]
[1]University of Guelph, Animal Biosciences, 50 Stone Road East, N1G 2W1, Guelph, ON, Canada, [2]Mad Barn Inc., 1465 Strasburg Road, N2R 1H2, Kitchener, ON, Canada; eleishma@uoguelph.ca

Mechanistic modelling has a long history of application in animal production systems, where it is used to synthesize knowledge, further understanding of biology, and provide decision-support. Detailed mechanistic models are used globally in dairy, swine, and poultry production, and these models have evolved along with our scientific knowledge. However, this development path has not been paralleled in equine species, and few attempts have been made to develop animal-level nutrient dynamic models in horses. While this may be due to less fundamental research being conducted in equines compared to predominant agriculture species, this gap limits the ability of the equine sector to address complex challenges such as the interactions between equine nutrition, management, health, and welfare. The aim of this large-scale research project is to develop (1) a mechanistic equine metabolism model, (2) a mechanistic equine digestion kinetics model, and (3) integrate these models together to comprehensively describe nutrient intake, digestion, and metabolism in the horse. Ultimately, the model will be expanded to consider breed and sex differences, common metabolic disorders (e.g. insulin resistance), meal and exercise patterns, and gestation and lactation. To support model development, experimental work will be ongoing, such as *in vitro* characterization of feedstuff degradability and fermentability to inform overall model inputs, and *in vivo* experiments on nutrient passage, digestibility, and absorption. This project will accelerate progress in the equine sector by leveraging decades of modelling work from other species, capitalizing on the extant equine literature, and by performing new research to develop a mechanistic model for horses. The development of this model will stimulate and assist further research, inform revisions to the nutrient requirements of horses, increase understanding of digestive and metabolic processes as well as related disorders, reduce waste via precision feeding and support the development of new products and services to improve equine health.

Predicting equine voluntary forage intake using meta-analysis

E.M. Leishman[1], M. Sahar[1], P. Darani[2], S. Cieslar[2] and J.L. Ellis[1]
[1]University of Guelph, Animal Biosciences, 50 Stone Road East, N1G 2W1, Guelph, ON, Canada, [2]Mad Barn Inc., 1465 Strasburg Road, N2R 1H2, Kitchener, ON, Canada; eleishma@uoguelph.ca

To properly formulate diets, the ability to accurately estimate feed intake is critical since the amount of food consumed will influence the amount of nutrients delivered to the animal. Inaccurate intake estimates may lead to under- or over-feeding of nutrients to the animal. Individual differences in equine forage intake are well-known, but predictive equations based on animal and nutritional factors, from information extracted from the body of published literature, are not prevalent. The objective of the present study was to consolidate current knowledge on voluntary forage intake in equines and conduct a meta-analysis to identify driving factors and sources of heterogeneity. Therefore, a systematic literature search was used and identified 99 publications which met the inclusion criteria. From each study, the outcomes of interest (e.g. forage intake), diet composition (e.g. forage information, nutrient composition), and animal factors (e.g. sex, age, breed, bodyweight, exercise level) were extracted. Forage intake was analysed as two different outcome variables: (1) forage intake in kg/d and (2) forage intake in g/kg bodyweight. Linear mixed model analysis using a backward stepping approach was used to identify potential prediction models for both outcome variables where all terms have $P<0.1$. The best fitting models for both outcome variables included similar factors like forage quality (i.e. NDF or crude protein content), forage composition (i.e. grass, legume, or mixed), the animals' size category (i.e. horses vs ponies), and some management factors (i.e. pasture access). As anticipated, forage intake increased when higher quality forages were fed (i.e. lower NDF or higher crude protein), potentially due to improved digestibility. In conclusion, using data gathered from multiple studies, equations to predict equine voluntary forage intake with high accuracy and precision were developed. The results of this meta-analysis confirm that animal traits and forage quality have a significant impact on the voluntary intake of equines and should be accounted for when formulating diets to meet nutritional requirements.

Session 19 Poster 19

Score using indicators of reticulo-rumen pH kinetics to characterize subacute ruminal acidosis

C. Villot[1], K. Biard[2], C. Martin[3], A. Boudon[4] and M. Silberberg[3]
[1]Lallemand SAS, 19 rue des Briquetiers, 31700, France, [2]University Paul Sabatier, Toulouse, Toulouse, France, [3]INRAe, Theix, 63122 Saint-Genès-Champanelle, France, [4]INRAe, Agrocampus Ouest, Saint-Gilles, France; cvillot@lallemand.com

The development of novel farm management technologies are promising strategies to improve welfare, efficiency, and productivity toward sustainable agriculture. The non-invasive and continuous recording of physiological parameters thanks to the use of biosensors, coupled to the development of dedicated data processing, make possible to precisely monitor the kinetics of specific parameters to assess rumen environment and other physiological traits. The objective of this study was to validate a set of novel indicators of reticulo-rumen (RR) pH kinetic that can better define the pattern of RR pH during a SARA challenge. Two different experimental SARA challenge studies (3-period switch-over design including a high starch diet: HSD>30% starch and <30% NDF) were conducted with 10 animals equipped individually with a RR sensor allowing a pH measurement every 10 min. Commonly used pH indicators of SARA (daily average, time spent under RR pH<6.0, area under the curve and daily amplitude) were calculated as well as descriptors modelling the pH curve. Unexpectedly, daily pH mean was not able to predict accurately the change of diet when cows were fed a HSD whereas performance and health indicators were altered confirming the susceptibility of the animals to the SARA challenge. However, indicators derived from pH kinetic significantly responded to when cows were fed a HSD compared to a low starch diet. For instance, the number of pH drops was significantly lower during the SARA challenge compared to a low starch diet (3.05 vs 4.71 drops/day, respectively for study 1 and 2.84 vs 3.27 drops/day respectively for study 2, $P<0.01$). Finally, a significant higher SARA score combining multiple daily pH indicators was able to identify when the animals were fed a HSD (3.8 vs 0.24 for study 1 and 12 vs 5.7 for study 2). Pre and post SARA challenge periods were comparable for these parameters, suggesting a good correlation with animal biology. These new indicators derived from RR pH kinetic carry additional information on pH variation; this makes them good candidates as descriptors of ruminal status.

Session 19 Poster 20

The INRA mechanistic beef growth model captures feed efficiency ranking in Charolais bulls

G. Cantalapiedra-Hijar[1] and S. Lerch[2]
[1]INRAE, UMR Herbivores, 63000, Clermont-Ferrand, France, [2]Ruminant Nutrition and Emissions, Agroscope, 1725, Posieux, Switzerland; sylvain.lerch@agroscope.admin.ch

The INRA dynamic and mechanistic growth model predicts the daily deposition of protein and fat in growing cattle, based on metabolizable energy (ME) intake (main driven force) and animal age. Originally developed for specific animal types (breed and sex), the aim of this study was to calibrate the INRA growth model at the individual level by adjusting coefficients for either rates of protein and lipid synthesis (α and β) or degradation (γ and δ), or modulation of the ME use efficiency (cMEU). The relationship between the adjusted model parameters and residual feed intake (RFI) or feed conversion efficiency (FCE) was further assessed. Individual performance data (daily ME intake and fortnightly recording of body weight) and estimations of body composition (d0, d84, d200) were obtained from an *in vivo* trial on 32 extreme RFI Charolais bulls fed two contrasting silage-based diets (maize vs grass). The three sets of INRA growth model parameters (α and β, γ and δ, and cMEU) were adjusted for each animal using daily ME intake as input, and minimizing model deviation from body lipid and protein masses estimations (Pay-Off procedure, Vensim 7.3.5). The five adjusted parameters were analysed by ANOVA for the effects of RFI, or alternatively by ANCOVA for the covariable FCE. No effect of diet or its interaction with RFI or FCE was found ($P>0.05$). The model accuracy was higher when α and β (average RMSEP of 4.0 kg body lipids or proteins) or γ and δ (3.9 kg) were adjusted, compared to only cMEU (4.4 kg). As the FCE increased, greater synthesis (α and β) and lower degradation (γ and δ) rates for both protein ($P<0.05$) and lipids ($P<0.10$) were found. A trend for lower cMEU (greater efficiency of ME use) as FCE increased ($r=-0.30$; $P<0.10$) was also observed. The RFI ranking only impacted the protein synthesis (greater in efficient; $P<0.05$) and degradation (lower in efficient; $P<0.10$) rates in agreement with observations made during the *in vivo* study. In conclusion, based on ME intake and body composition estimates, metabolic parameters adjusted at the individual level in the INRA growth model were significantly impacted by feed efficiency ranking, especially FCE.

Session 19 Poster 21

A framework for predicting the environmentally attainable intake of dairy cows

J.F. Ramirez-Agudelo, L. Puillet and N.C. Friggens
Université Paris-Saclay | Campus Palaiseau, 22 place de l'Agronomie – CS 80022, 91120, Palaiseau, France; nicolas.friggens@agroparistech.fr

Accurately predicting the intake of dairy cows is essential for evaluating their efficiency. However, measuring individual intake can be challenging and costly, especially in constrained environments. While mathematical models can help predict intake, they are typically designed for normal farm conditions. To overcome this limitation, this article presents a framework for predicting the dry matter intake of dairy cows allowed by constrained environmental conditions, i.e. the environmentally attainable intake (EAI, g DM/d). The framework represents EAI as the product of two main components: maximum sustainable Eating Rate (ER, g DM/min) and Eating Time (ET, min/d). The ER is determined by the physical NDF content of the food and the animal's age, which is a proxy for oral capacity. The daily time available for feeding is divided into four activities: seeking for food, ET, rumination, and additional processing time, influenced by environmental factors such as temperature, photoperiod, farm activities, and food availability. The framework includes assumptions such as a mandatory minimum amount of time for animals to sleep soundly, farm activities that prevent access to food, and the effect of heat stress. To define the most influential parameters, Bayesian inference and a database of 79 treatments from 24 published papers covering grazing and housed cows from several countries, was used for Sensitivity Analysis. The framework is sensitive to six of the 18 inferred parameters. The mean absolute errors for EAI prediction were 2.4 and 4.2 kg DM/d for grazing and housed cows, respectively. The framework separates the animal side of intake from the environmental side, enabling the study of complex phenotypes such as feed efficiency, and genotype by environment interactions. It is simple, suitable for use with resource acquisition-allocation models, and can be readily adapted for use in other livestock species. The framework is extensible to add constraints such as predation pressure, reproductive costs, competition, parasitism, or diseases. We believe this framework could provide a useful tool for designing management strategies to optimize productivity and animal welfare in dairy farms under challenging environmental conditions.

Session 19 Poster 22

An integrated tool to apply the INRA2018 feeding system in ruminant rationing: INRAtion®V5

P. Noziere[1], J. Agabriel[1], R. Baumont[1], P. Chapoutot[2], L. Delaby[3], R. Delagarde[3], A. De La Torre[1], P. Faverdin[3], S. Giger-Reverdin[2], P. Hassoun[4], A. Lamadon[1], G. Maxin[1], D. Sauvant[2], B. Sepchat[5], P. Souvignet[1] and S. Dutot[6]
[1]INRAE, UMR Herbivores, 63122 St-Genès Champanelle, France, [2]INRAE AgroParisTech, UMR Mosar, 91477 Palaiseau, France, [3]INRAE, UMR PEGASE, 35590 St-Gilles, France, [4]INRAE, UMR Selmet, 34000 Montpellier, France, [5]INRAE, UE Herbipole, 63122 St Genès Champanelle, France, [6]SIEL, 42 rue de Châteaudun, 75009 Paris, France; pierre.noziere@inrae.fr

The INRA feeding system for rationing ruminants has been regularly updated for 40 years. Thanks to its support software, INRAtion®, it is widely used in France and in several European, African and South American countries. The recent revision of the system was aimed at developing new approaches to rationing, allowing to calculate rations at an objective different from covering animals' needs at the production potential and to predict the multiple responses of the animals in terms of production, product quality, animal health and environmental emissions. The development of the INRAtionV5 rationing system was achieved through a partnership with the SIEL association, which brings together 40 livestock adviser organizations, representing more than 50% of the dairy herd in France. INRAtionV5 calculates feed values from their laboratory analysis, predicts diet nutritive value considering digestive interactions, feed and nutrients intake including on pasture, and the multiple responses of the animals. An algorithm optimizes the rations according an ‹objective function› that integrates production objectives, management of body reserves, and efficiency of protein use. Other predicted animal responses allow the user to evaluate the calculated rations on several technical, environmental and economic dimensions, as a decision support tool. The current version includes feed value calculations, and ration optimizations for dairy animals (cow, ewe and goat), as well as growing and fattening cattle, and is available under license in French and English (https://www.inration-ruminal.fr/) for industry and academic institutions. The connection to the adviser's information systems will allow a massive feedback of results from the field, which may be used for further development of the INRA feeding system and the rationing tool.

Session 19 Poster 23

Milk yield and mammary metabolism in response to increasing amino acids and protein supplies

S. Lemosquet[1], C. Loncke[2], J.C. Anger[1,3], A. Hamon[1], J. Guinard-Flament[1] and M.N. Haque[1,4]
[1]PEGASE, INRAE, Institut Agro, 35590 Saint-Gilles, France, [2]Université Paris-Saclay, INRAE, AgroParisTech, UMR MoSAR, 91120 Palaiseau, France, [3]Provimi, Cargill Animal Nutrition & Health, 35320 Crevin, France, [4]Department of Animal Nutrition, University of Veterinary and Animal Sciences, Lahore, Pakistan; sophie.lemosquet@inrae.fr

The aim of this study was to understand milk yield variations in response to a balanced profile of 10 essential amino acids (EAA) at 2 levels of metabolizable protein (MP) supplies. Diets and infusions provided 14.5 vs 16.8 g/MJ for MP·NEL^{-1} in INRA 2007 system and 128 vs 145 g/kg DM of CP, respectively in Low vs high MP. Four lactating dairy cows received 2 dietary treatments (low vs high MP) and 2 different duodenal infusions of 10 EAA (AA- and AA+), according to a 2×2 factorial design. The AA- and AA+ treatments provided (in % of MP) 1.7 vs 2.4% of Met, 5.8 vs 7.0% of Lys, 7.9 vs 9.0% of Leu and 1.9 vs 3.0% of His, respectively. Six blood samples were collected from carotid artery and mammary vein to calculate the uptakes of all the energetic nutrients using the Fick principle on Phe + Tyr for mammary plasma flow. Results were analysed using the MIXED procedure of SAS (2009) taking the effect of cows as random. Milk yield increased in AA+ vs AA- treatments (9.1 vs 8.6 kg/12 h·per half udder, P=0.04), with no significant difference in lactose yield between the 2 treatments (442 vs 422 g/12 h·per half udder). However, the water content in milk was higher in the AA+ vs AA- treatments (7.90 vs 7.45 kg/12 h·per half udder, P=0.05) and explained by a parallel increase in minerals yield (69 vs 64 g/12 h·per half udder, P=0.05). Mammary net uptake of glucose (GLC) tended to increase in AA+ vs AA- treatments (1,729 vs 1,562 mmol/h of C, P=0.06) as the difference between GLC minus lactose (504 vs. 391 mmol/h of C, P<0.1). This indicates that there was a shift in GLC partition towards ATP furnishing to support the increased milk protein yield (MPY; from 262 to 290 g/12 h·per half udder; P=0.01). In conclusion, milk volume could vary differently than lactose yield and MPY contrary to the meta-analysis of Daniel and Sauvant and the predictions of INRA (2018) model as it depends on minerals in response to EAA supplies.

Session 19 Poster 24

Ruminal fatty acid biohydrogenation pathways in lambs and kids fed a high concentrate diet

V. Berthelot[1], Y.D. Yekoye[1] and L.P. Broudiscou[1,2]
[1]Université Paris-Saclay, INRAE, AgroParisTech, UMR Modélisation Systémique Appliquée aux Ruminants, 22 place de l'agronomie, 91120 Palaiseau, France, [2]INRAE, UPPA, UMR Nutrition, Métabolisme, Aquaculture, 173 Route de St Jean De Luz, 64310 Saint-Pée-sur-Nivelle, France; valerie.berthelot@agroparistech.fr

Understanding the ruminal biohydrogenation (BH) pathways or lipid metabolism is important to improve the nutritional value of ruminant products. Differences in milk or meat fatty acid (FA) composition between sheep and goat have been shown, possibly due to species-linked differences in lipid BH processes or metabolism. In this study, *in vitro* incubations were used to compare ruminal C18 BH pathways between lambs and kids. After weaning, six lambs and six kids were fed the same concentrate diet based on wheat, rapeseed meal and dehydrated alfalfa. At slaughter, individual rumen digesta were collected to inoculate 6 dual-effluent continuous fermenters for two 8-day periods. The last 3 days, each fermenter provided fermentation medium to inoculate 75 ml culture tubes. Over an 8-hour batch incubation, the ruminal BH process was investigated via the addition of oleic, linoleic or linolenic acids. Kids had lower proportions of anteiso FA (1.62 vs 2.10% total FA, P=0.03) in their rumen but with a tendency to increase the proportion of iso FA (0.41 vs 0.34% total FA, P=0.07). These differences might be linked to a different composition of the ruminal microbiota between these two species. For both species, BH was predominantly via the t10 pathway, with t10 C18:1 being the predominant trans C18:1 isomer (50% trans C18:1). This is consistent with a typical microbiota of a high concentrate diet. Over the 8 hours of incubation, the proportion of C18:0 was significantly higher in lambs compared to kids (14.3 vs 6.5 (% of total C18) respectively, P=0.01) indicating a more intense BH in lambs. Lambs also showed a lower proportion of C18:1 t11 than kids (P=0.05), a higher C18:1 t10/C18:1 t11 ratio (P=0.06) and an increase in C18:2 t10,c12 for linoleic supplementation suggesting a greater orientation of BH pathways towards the t10 pathway and a higher sensitivity of lambs to high starch diets. In conclusion, these findings give evidences that there are interspecies differences in BH pathway between goat and sheep.

Session 19 | Poster 25

Can high fat concentrates offset the high methane production caused by high forage diets?

N. Ayanfe[1], S. Ahvenjärvi[1], A. Sairanen[2] and A.R. Bayat[1]
[1]Natural Resources Institute Finland (LUKE), Tietotie 2C, FI-31600, Jokioinen, Finland, [2]Natural Resources Institute Finland (LUKE), Halolantie 31A, FI-71750 Maaninka, Finland; nisola.ayanfe@luke.fi

Increased dietary forage to concentrate ratio (F:C) would improve the sustainability of dairy production. However, high forage (HF) diets increase enteric methane (CH_4) emissions, but dietary fat supplementation may counterbalance this effect. Therefore, this study examined the effects of supplementing HF or low forage (LF) diets with two concentrates of intrinsically differing fat concentration on enteric CH_4 emission, nutrient digestibility, and milk production in lactating dairy cows. Four multiparous Nordic Red dairy cows were randomly assigned to a 4×4 Latin square design in a 2×2 factorial arrangement of treatments with four 21-d periods each consisting of 5 sampling days in metabolic chambers. The treatments comprised a total mixed ration of grass silage with F:C of 70:30 or 50:50 containing either low or high fat concentrates. Two fat concentrations of 24 and 65 g/kg dry matter (DM) were attained by replacing rapeseed meal and barley with rapeseed cake and oats at 25 and 50% of concentrates, respectively on DM basis. DM intake was higher in LF compared to HF and fat did not affect it. Interactions between F:C and fat concentration were observed for DM, gross energy and fibre digestibility coefficients with high fat concentration increasing fibre digestion ($P<0.01$) when included in LF and decreasing it in HF. Feeding high fat diet increased fat and crude protein digestibility compared to diets with low fat concentration. No interaction between F:C and fat concentration was observed on daily CH_4 production and CH_4 yield showing that the mitigating effect of fat does not depend on the F:C. Fat reduced CH_4 intensity (12.3 vs 10.6 g/kg energy-corrected milk, ECM) in LF but not in HF ($P<0.05$). However, feeding LF and high fat diets decreased CH_4 yield compared to HF (average 23 vs 21 g/kg DMI) and low fat (average 23 vs 21 g/kg DMI) diets, respectively. High fat concentrate increased the proportion of propionate in rumen VFA leading to increased ECM production only in LF. In the current study, high fat concentrate decreased daily CH_4 emissions by 7% regardless of dietary forage to concentrate ratio.

Session 19 | Poster 26

Updates of the expenditures and requirements for growing dairy goats

S. Giger-Reverdin and D. Sauvant
Université Paris-Saclay, INRAE, AgroParisTech, UMR Modélisation Systémique Appliquée aux Ruminants, 22 Place de l'Agronomie, 91120, Palaiseau, France; sylvie.giger-reverdin@agroparistech.fr

A novel approach based on modelling was undertaken to update the expenditures and requirements for growing dairy goats in the INRA2018 system. It replaced former approaches based on feeding trials. Growth, growth rate of the whole body weight (BW), empty BW (eBW), the fat and protein contents of eBW were modelled along growth. The first step was to model the potential growth curves and growth rate from data obtained from the MoSAR's herd in Grignon (Alpine or Saanen breeds). Then the approach used data obtained in Brazil and in France on Alpine or Saanen growing goats using slaughtering techniques. A meta-analysis was conducted on 12 publications, with eight dealing with males, five with castrated males, and only four with females. In the second step, eBW was estimated in function of BW. The third step concerns the modelling of body lipids. The comparison of the 'three sexes' exhibited a significant 'sex' effect. A first approach was done on the males, because they were more numerous and the percentage of lipids/eBW was adjusted with a monomolecular model. Then a coefficient was calculated for the ratio of this percentage between females and males. This enables to model the body lipid component in function of age. In a fourth step, body proteins content was modelled in a similar approach using all the data as the sex effect was non significant. Growth energy requirements for young females were obtained from the body energy content due to lipids or protein deposition. It ranged from 1.55 UFL/kg BW gain around birth to 1.63 UFL (1st gestation) and to 1.80 UFL (2nd lactation). The total energy requirement increases from 0.50 (1st month) to 1.75 (end of 1st gestation) UFL/d. Growth PDI requirements are computed from the protein gain divided by PDI efficiency. It decreases with the age of the animal. Total PDI requirements include growth PDI and those related to nonproductive functions. With 'normal' diets they are around 70 g/d for a one month goat and 130 g/d at the end of 1st gestation. The new values proposed are higher than the previous ones and in agreement with other feeding systems. A more detailed description is in Chapter 21 'Dairy and growing goats'.

Session 19 Poster 27

Feed efficiency relates to the ability of supporting lactation in dairy cows facing feed restriction

L. Barreto-Mendes, J. Pires, A. De La Torre, I. Ortigues-Marty, I. Cassar-Malek and F. Blanc
Université Clermont Auvergne, INRAE, VetAgro Sup, UMR Herbivores, 63122, Saint-Genès-Champanelle, France; luciano.mendes@inrae.fr

We hypothesized that individual differences in prioritization of dairy cows exposed to feed restrictions (FR) are associated with feed efficiency. The objective was to explore clustered associations among adaptive responses to experimental FR and metrics of feed efficiency. 16 Holstein and 17 Montbéliarde cows, starting at 87±6 DIM, underwent four 4d FR periods (50% of individual NEL). Cows were fed *ad libitum* after FR1 (during 10 d), and FR2, FR3, FR4 (during 3 d). Energy-corrected milk yield (ECMY), plasma non-esterified fatty acids (NEFA) and bilirubin concentrations were measured prior, during and after FR. Residual feed intake (RFI) and milk energy conversion efficiency (ECE) were calculated in early and mid-lactation. Individual ECMY pre-FR, and integrated areas under the curve during FR (AUC1) and recovery (AUC2) were calculated and clustered separately for FR1 (or challenge 1, CH1) and the sum of AUC responses during FR2 through FR4 (CH2). Then cluster (CL) differences were tested for RFI and ECE. For both CH1 and CH2, 3 distinct CL, each composed of both breeds, emerged: CL1 with low ECMY pre-FR, low ECMY AUC1 and ECMY AUC2, and low NEFA AUC response; CL2 with high ECMY pre-FR, average ECMY AUC1 and better ECMY recovery in terms of AUC2; and CL3 with high ECMY in pre-FR, high ECMY AUC1 and AUC2, and low bilirubin. RFI_{early} was numerically higher for cows in CL3-CH1. A concordant effect occurred for CL3-CH2 in terms of ECE_{early}, where the low mobilization of body reserves may be explained by better feed efficiency to support lactation. 50% of cows changed CL typology between CH1 and CH2, suggesting that individual adaptation mechanisms are impacted when cows face multiple challenges. In conclusion, feed efficiency seems to have buffered the lower body reserves mobilization of a subgroup of cows during FR, in order to support lactation.

Session 19 Poster 28

Effects of rumen protected calcium gluconate on hindgut dynamics and nutrient partitioning

M. Hall[1] and J.G.H.E. Bergman[2]
[1]Trouw Nutrition GB, Ruminant Department, Blenheim House, Ashbourne DE6 1HA, United Kingdom, [2]Nutreco, Selko, Stationsstraat 77, 3811 MH Amersfoort, the Netherlands; mark.hall@trouwnutrition.com

Rumen protected calcium gluconate (RPCG) fed as a dietary supplement has been consistently shown to increase milk and component yields in dairy cattle under both research and practical conditions. These responses are thought to be mediated by improvements in post-ruminal barrier function upon fermentation in the hindgut. Although its exact mechanism of action is currently unknown, evidence from the available studies point to beneficial effects at different levels. Gluconic acid salts have been shown to have a prebiotic effect in the lower gut of monogastric animals where it serves as a precursor for volatile fatty acids (VFA). In agreement, RPCG supplementation tended to increase both butyrate and total VFA concentrations in cecum digesta of growing lambs. Similarly, an overall increase in VFA concentrations in the hindgut was observed when supplementing RPCG to growing steers. This was accompanied by changes in the microbial population associated with gut homeostasis and health, and the formation of beneficial end products. Interestingly, models of hindgut acidosis show a decrease in milk urea nitrogen (MUN) due to increased N retention in the hindgut for microbial proliferation. This contrasts with the increase in MUN concentrations observed when feeding RPCG, which might reflect a shift in hindgut fermentation. Finally, supplementing RPCG consistently increases circulating NEFA without an increase in blood BHB, suggesting changes in postabsorptive nutrient partitioning. In conclusion, the beneficial effects of RPCG on milk performance might be partially mediated by an improved barrier function, reducing the energetic requirements of intestinal damage and leaky gut, allowing for improved whole-body bioenergetics, and ultimately increased performance.

Session 20 Theatre 1

The impact of observation frequency and duration, on resilience traits in pigs

W. Gorssen[1], C. Winters[2], R. Meyermans[1], L. Chapard[1], K. Hooyberghs[1], S. Janssens[1], A. Huisman[3], K. Peeters[3], H. Mulder[4] and N. Buys[1]
[1]KU Leuven, Center for Animal Breeding and Genetics, Kasteelpark Arenberg 30, bus 2472, 3001, Belgium, [2]KU Leuven, Laboratory for Biological Psychology, Tiensestraat 102, 3000 Leuven, Belgium, [3]Hendrix Genetics, P.O. Box 114, 5830 AC, Boxmeer, the Netherlands, [4]Wageningen University & Research, Animal Breeding and Genomics, P.O. Box 338, 6700 AH Wageningen, the Netherlands; wim.gorssen@kuleuven.be

Recent studies showed pigs' resilience can be measured via deviations in longitudinal data. Recording longitudinal data in pigs has been facilitated by the introduction of automated feeding systems, generating daily records on weight, feed intake and feeding behaviour. However, feeding stations can possibly be used more efficiently. Therefore, we studied the influence of observation frequency and duration on resilience trait. This study used 324,207 daily weight measurements over a 60 day period (95-155 days) on 5,939 Piétrain pigs with known pedigree. The resilience trait *lnvar* was calculated as the natural logarithm of the variance of deviations of predicted vs observed weights after Gompertz modelling. Genetic analyses were done via blupf90 software. The impact of observation frequency was studied by lowering data density from 1 in 4 (~twice a week) up to 1 in 14 (~biweekly) records. The impact of observation duration was investigated by creating three 20 day periods: early (95-115 days), middle (115-135 days) and late (135-155 days). For both scenarios, phenotypic and genetic correlations were estimated between the full dataset and subsets. Heritability of *lnvar* was estimated at h^2=11.0%. We showed that *lnvar* was robust to low observation frequencies with phenotypic correlations r_p>0.76 and genetic correlations r_g>0.96. Moreover, phenotypic and genetic correlations between 20-day periods were moderate for *lnvar* (r_p=0.37-0.50; r_g=0.61-0.78), whereas they were high between 20-day periods and the full dataset (r_p=0.66-0.74; r_g=0.81-0.89). Our findings show that *lnvar* can be used as a resilience trait in finishing pigs with as low as one recording every two weeks. Moreover, daily records on a 20-day period seemed informative to predict *lnvar* over a pigs' finishing period.

Session 20 Theatre 2

Are tail biting and tail necrosis really only a problem in the domestic pig?

I. Czycholl[1], K. Büttner[2], W. Baumgärtner[3], C. Puff[3] and J. Krieter[4]
[1]University of Copenhagen, Department of Animal Welfare and Disease Control, Grønnegårdsvej 8, 1870, Denmark, [2]Justus Liebig University, Unit for Biomathematics and Data Processing, Faculty of Veterinary Medicine, Frankfurter Str. 95, 35392 Giessen, Germany, [3]University of Veterinary Medicine Hannover, Department of Pathology, Bünteweg 17, 30559 Hannover, Germany, [4]Institute of Animal Breeding and Husbandry, Kiel University, Olshausenstr. 40, 24098 Kiel, Germany; ic@sund.ku.dk

Tail biting is a multifactorial problem in pig husbandry that has been linked, amongst others, to deficiencies in the health status. But which pathological findings of domestic pigs can be rated as normal and which might play a subclinical role in the behavioural disorder tail biting? 13 young wild boars, shot during normal hunting practice in Germany, were subjected to pathological examinations. The findings were compared by Chi^2 Tests and t-Tests to pathological findings of 17 rearing pigs from a conventional farm. Moreover, 44 additional tails of wild boars were analysed pathologically. Typical findings in wild boars were linked to parasitic infections, which were not seen in rearing pigs at all (most common findings: nematodes in the lung (11 wild boars affected; $P<0.0001$) and in the gut (8 wild boars affected; $P=0.0002$), milkspots (5 wild boars affected; $P=0.005$). However, significantly more rearing pigs were affected by parameters linked to the gut health (hyperplasia (4 rearing pigs, 0 wild boars; $P=0.05$) and crypt abscesses (7 rearing pigs, 1 wild boar; $P=0.04$) in the colon associated lymphoid tissue and crypt abscesses in the gut (5 rearing pigs, 0 wild boars affected; $P=0.03$). Moreover, significantly more rearing pigs were diagnosed to have a rhinitis (13 rearing pigs, 4 wild boars; $P=0.01$), whereas significantly more wild boars had a purulent inflammation of the mesenteric lymph nodes (3 wild boars, 0 rearing pigs, $P=0.03$) Although some pathological deviations in the tails of wild boars were seen, these can be linked to external parasites (specifically mites) but not to tail necrosis or tail biting occurrence. Overall, it can be concluded that wild boars are not healthier than rearing pigs (or vice versa), but the causes for the diseases differ. There is no scientifically proven evidence for tail necrosis or tail biting in the wild boar population.

Innovative rearing strategies for heavy pigs: weight and chemical composition of dry-cured hams

A. Toscano[1], D. Giannuzzi[1], I.H. Malgwi[1], V. Halas[2], P. Carnier[3], L. Gallo[1] and S. Schiavon[1]
[1]University of Padua, Dept. DAFNAE, viale dell'Università 16, 35020 Legnaro, Italy, [2]Hungarian University of Agriculture and Life Sciences, Dept. Farm Animal Nutrition, Guba Sandor Utca 40, 7400 Kaposvár, Hungary, [3]University of Padua, Dept. BCA, viale dell'Università 16, 35020 Legnaro, Italy; alessandro.toscano@unipd.it

To explore the influence of 4 feeding strategies on the dry-cured ham quality, 336 Goland C21 pigs, barrows and gilts of 90 kg body weight (BW), were divided into 4 groups, assigned to one of four treatments, and housed in 8 pens with automated feeders. In the control group (C), the pigs were fed restrictively medium-protein feeds and slaughtered at 170 kg BW and 257 d of age. With the older age (OA) treatment, the pigs were restrictively-fed low protein feeds and slaughtered at 170 kg BW and 273 d of age. The other two groups were fed *ad libitum* high protein feeds, the younger age (YA) group was slaughtered at 170 kg BW and 230 d of age, the greater weight (GW) at 257 d of age and 193 kg BW. According to the Prosciutto Veneto product specification the hams were dry-cured and seasoned for 607 d. Hams were weighed before and after seasoning, deboned and scored for fat cover depth. Sixty hams were sampled and sliced. The lean and the fat tissues were separated and analysed for proximate composition and fatty acid profile. The model of analysis considered sex, treatment and sex ' treatment as fixed factors. With respect to C: (1) OA lowered the ham weight, the lean protein content, but increased the marbling and decreased the proportion of polyunsaturated fatty acids (PUFA) in lean and fatty tissues; (2) YA hams had thicker fat cover with lower PUFA proportion in lean and fatty tissues; (3) GW increased the deboned ham weight, fat cover depth and marbling, but reduced the PUFA proportion in lean and fatty tissues and the seasoning losses, without alteration of the lean moisture content. Sex had a negligible impact. Innovative rearing strategies had positive influences on the chemical quality of the seasoned hams.
Acknowledgements: Agritech National Research Center, funded by the European Union Next-Generation EU (Piano Nazionale di Ripresa e Resilienza (PNRR) – Missione 4 Componente 2, Investimento 1.4 – D.D. 1032 17/06/2022, CN00000022) and PSR Veneto DGR 2175/2016, code 3682902.

Comparison of different non-linear growth models on post-weaning growth of pigs

K.O. Kareem-Ibrahim, O.T.F. Abanikannda and O.A. Folorunsho
Lagos State University, Department of Animal Science, School of Agriculture, Epe Campus, Lagos, Nigeria; otfabanikannda@hotmail.com

Pig growth curve models are useful tools for identifying optimum management strategies for individual grower-finisher pigs. The use of non-linear growth models in fitting the growth rate of pigs is more effective than linear models. This study aimed at evaluating post weaning growth of pigs as affected by sex and modelling of the weekly weight using various nonlinear curve fitting methods, with a view to assess and compare the efficiency and accuracy of the different models, and recommend the model of best fit for post-weaning weekly growth of pigs. A total of 43 piglets comprising 18 female and 25 male weaners were evaluated. Data was collected for a period of 12 weeks on a weekly basis from the weaners using a digital weighing scale, sensitive to 0.00 g. Preliminary analysis of variance of the effect of sex on the initial weight at weaning was not significant (P>0.05), thus leading to the merger of the weaners across sex. Mean weekly weight of piglets at 0, 7, 14, 21, 28, 35, 42, 49, 56, 63, 70, 77 and 84 days were 5.69±0.17, 6.71±0.17, 7.02±0.19, 8.05±0.22, 8.28±0.24, 9.38±0.28, 10.35±0.34, 11.13±0.37, 12.51±0.40, 14.02±0.47, 15.34±0.53, 15.85±0.55 and 16.51±0.57 kg respectively. Seven non-linear models ranging from two to four parameters (Asymptotic Regression, Logistic, Log- logistic, Gompertz, Richards, Weibull and Mechanistic) were used to fit the growth curves. All modelling and statistical analyses were done using JMP Statistical Software. Test of goodness of fit of the various models included coefficient of determination (R^2), adjusted coefficient of determination (Adj. R^2), root mean square error (RMSE), Akaike's Information Criterion (AIC) and Bayesian Information Criterion (BIC). Out of the seven models used, the Richard growth model had the best fit for the modelling of pig growth and the least root mean square error. It is thus recommended that the Richard be further investigated for growth curve fitting of pigs at the post weaning stage.

Session 20 — Theatre 5

Effect of glutamine and/or milk supplementation post-weaning on pig growth and intestinal structure

E.A. Arnaud[1,2], G.E. Gardiner[1], S.R. Vasa[1,2], J.V. O'Doherty[3], T. Sweeney[3] and P.G. Lawlor[2]
[1]South East Technological University, Dept. of Science, X91K0EK, Waterford, Ireland, [2]Teagasc, Pig Development Dept., Co. Cork, P61R966, Fermoy, Ireland, [3]University College Dublin, School of Agriculture & Food Science, Belfield, D04C7X2, Dublin, Ireland; elisa.arnaud@teagasc.ie

Improving intestinal structure and function in weaned pigs will help to overcome the negative growth effects associated with weaning. This study aimed to determine the effect of providing supplemental milk and/or dietary inclusion of 1% L-glutamine on the growth and intestinal structure of newly weaned pigs. At weaning, 480 pigs were grouped into single sex groups of 10 pigs of even weight (12 pens/treatment) and randomly assigned to one of 4 treatments in a 2×2 factorial arrangement; factors being provision of supplemental milk (yes/no) and dietary supplementation with 1% L-glutamine (yes/no). Supplemental milk (167 g of Swinco Opticare milk powder/l of water) was provided from weaning to day (d)10 post-weaning (pw). L-glutamine was added at 1% to the starter feed and milk powder before reconstitution from weaning to d10pw. After d10pw, all pigs were fed the same standard diet. Group weights and feed disappearance were recorded at intervals up to slaughter. At d7pw, 40 pigs were euthanised and intestinal tissue sampled for histological analysis. Data were analysed using PROC MIXED (v9.4, SAS Institute Inc.). The pen group was the experimental unit. There was a milk × glutamine interaction on body weight (BW) at d10, d20, d28, d47 and d124pw (slaughter); at d10, d20, d28 and d47pw, milk increased pigs BW with and without L-glutamine supplementation ($P<0.05$). At d124pw, milk feeding increased BW when L-glutamine was not fed but did not when L-glutamine was fed ($P<0.05$). There was a milk × glutamine interaction for average daily feed intake (ADFI) from weaning to d10pw, d20-28pw and d28-47pw; milk feeding increased ADFI when L-glutamine was and was not fed ($P<0.05$). At d7pw, milk feeding increased duodenal, jejunal and ileal villous height by 13, 38 and 29%, respectively ($P<0.05$). In conclusion, milk supplementation increased post-weaning feed intake and intestinal villous height, leading to an increase of ~3.0 kg in BW at slaughter.

Session 20 — Theatre 6

Effect of lysine level in finisher diets on performance in crossbreds from two terminal sire lines

S. Goethals[1], K. Hooyberghs[2], N. Buys[2], S. Janssens[2] and S. Millet[1]
[1]ILVO, Flanders Research Institute for Agriculture, Fisheries and Food, Scheldeweg 68, 9090 Melle, Belgium, [2]KU Leuven, Department of Biosystems, Kasteelpark Arenberg 30, bus 2472, 3001 Leuven, Belgium; sophie.goethals@ilvo.vlaanderen.be

Adequate knowledge on lysine requirements and supplementation of free amino acids enables to reduce the dietary protein level, resulting in lower feed costs and reduced nitrogen excretion. However, a deficient supply of amino acids reduces performance. Requirements depend on sex and genotype because of the difference in capability for lean growth and feed intake. The current study evaluated the effect of lysine level in the finisher phase (80-115 kg) on growth performance in crossbred pigs from two terminal sire lines. Gilts and barrows (n=360 in total) with a different paternal genetic background (stress negative sire line selected for growth rate vs stress positive sire line selected for carcass quality) were allocated to a high or low lysine diet (6.3 vs 7.8 g standardized ileal digestible lysine per kg diet, with 12.6 and 14% crude protein, respectively and 9.3 MJ NE/kg), according to a 2×2×2 factorial design. The feed conversion ratio (FCR) in the finisher phase was lower for gilts compared to barrows (2.85 vs 2.93, $P=0.049$). Within sexes, there was only a significant difference in FCR between the low and high lysine diet of the crossbred selected for carcass quality (2.95 vs 2.76, $P=0.023$). Higher serum urea levels in barrows demonstrated the lower amino acid requirements compared to gilts (14.4 vs 10.8 mg/dl, $P<0.001$). In barrows, the high lysine diet resulted in 31% higher serum urea levels compared to the low lysine diet ($P<0.001$), independent of sire line, suggesting an excess of protein in the high lysine diet for all barrows. In gilts, a trend for higher serum urea levels (9% increase, $P=0.06$) was observed for the high compared to the low lysine diet, also independent of sire line. In general, both the low lysine diet and the genetic background selected for carcass quality positively affected nitrogen efficiency in the finisher phase. Only in gilts selected for carcass quality, the low lysine diet did not result in a better nitrogen efficiency compared to the high lysine diet.

Session 20 Theatre 7

Birth weight effects on performance and gut health status of weaned piglets

C. Negrini[1], F. Correa[1], D. Luise[1], S. Virdis[1], M. Mele[2], G. Conte[2], M. Mazzoni[3] and P. Trevisi[1]
[1]University of Bologna, DISTAL, Viale fanin 46, 40127, Italy, [2]University of Pisa, DAFE, Via del Borghetto 80, 56124, Italy, [3]University of Bologna, DIMEVET, Via Tolara di Sopra 43, 40064, Italy; federico.correa2@unibo.it

Over the past few decades, the swine industry has focused on litter size as a key production indicator. However, selecting for greater litter size has resulted in an increase in the number of piglets born with low birth weight (<1 kg). This study aims to assess the impact of birth body weight (BBW) on piglet performance, microbial profile, and gut status in after weaning. At birth, 64 piglets from 13 litters were selected for their BBW and at weaning (d0), piglets were allotted into 2 groups: NBW (>1 kg; 32 piglets) and LBW (<1 kg; 32 piglets). Piglets were weighted weekly until d21 post-weaning. Faecal score and feed intake (FI) were daily recorded. On d9 and d21, a total of 16 piglets/group were slaughtered, and jejunum tissue was collected for morphology and immunohistochemistry analysis, colon content was collected for microbiota and SCFAs analysis, and the pH of jejunum, cecum and colon was measured. Data were analysed using a linear mixed model or a generalized linear mixed model with a Poisson distribution including the class of BW and the litter as fixed and random factors respectively. Piglets from the NBW group had consistently higher BW throughout the study ($P<0.01$). From d0 to d21, the NBW group tended to have a higher average daily gain (ADG) ($P=0.07$), and a higher FI ($P=0.006$). LBW piglets tended to have a lower faecal index from d0 to d21 ($P=0.076$). At d21, the NBW piglets had a higher villus height ($P=0.05$) and tended to have higher VH:CD ratio ($P=0.07$), while LBW piglets had higher number of T-lymphocytes in the epithelium and crypts ($P<0.01$). The LBW piglets had higher concentration of iso-butyrate, butyrate, isovalerate, valerate, and the total amount of SCFAs at d21 ($P<0.05$). Alpha diversity was never different between the two groups, while Beta diversity tended to be influenced by the BW category on d9 ($P=0.068$, $R^2=0.05$) and was significantly different on d21 ($P=0.026$, $R^2=0.05$). In conclusion, the BBW affected the piglet's response to weaning and the gut health status. Selecting piglets for their BBW could represent a valid challenge model to test different feeding strategies.

Session 20 Theatre 8

Effects of sugary and salty ex-food in pigs' diets on growth performance and pork sensory traits

S. Mazzoleni[1], M. Tretola[1,2], C.E.M. Bernardi[1], P. Lin[1], P. Silacci[2], G. Bee[2] and L. Pinotti[1,3]
[1]University of Milan, Department of Veterinary Medicine and Animal Sciences, Via dell'Università 6, 26900 Lodi, Italy, [2]Agroscope, Institute for Livestock Sciences, Rte de la Tioleyre 4, 1725 Posieux, Switzerland, [3]CRC I-WE, Coordinating Research Centre: Innovation for Well-Being and Environment, 20134 Milan, Italy; sharon.mazzoleni@unimi.it

Ex-food also known as Former foodstuffs products (FFPs) are promising alternative ingredients to reduce the waste of natural resources and the environmental impact of food production. These products can be classified into two categories: (1) salty leftovers, mainly obtained by bakery products (pasta, bread and salty snacks); or (2) sugary leftovers, obtained by confectionery products (chocolate, breakfast cereals, cookies). This study investigated the effects of salty and sugary FFPs on growth performance and sensory characteristics of the loins of growing and finishing pigs. Thirty-six Swiss Large White male castrated pigs were assigned to the three grower (G) and finisher (F) experimental diets: (1) standard diets (ST-G; ST-F), 0% FFPs; (2) 30% conventional ingredients replaced by sugary FFPs (SU-G, SU-F); (3) 30% conventional ingredients replaced by salty FFPs (SA-G, SA-F). The grower and finisher diets were iso-energetic and iso-nitrogenous. Body weight was measured weekly. Feed intake (FI) was determined daily. Average daily gain (ADG), average daily feed intake (ADFI), feed conversion ratio (FCR) and gain to feed were calculated for both the growing and finishing phases. The pork loins were assessed for sensory attributes. Feed efficiency in terms of FCR was improved ($P<0.05$) in ST-G than SA-G and SU-G. However, when considering the overall mean of the entire period BW, ADFI, ADG and FCR were similar ($P>0.05$) between the three groups. The loin from SU pigs was perceived as significantly sweeter ($P<0.001$) than the control, while the loin from SA pigs was perceived as significantly sweeter ($P<0.001$) and more tender ($P<0.001$) than the control. This study confirmed that inclusion of FFPs had no detrimental effects on growth performance in growing and finishing pigs and did not impair the sensory characteristics of the loin, further analyses on the chemical composition of the meat and the fatty acid composition of the intramuscular fat are planned.

Session 20 Theatre 9

Performances, milk quality, and piglet growth in lactating sows fed former food products

P. Lin[1,2,3], G. Bee[1], S. Neuenschwander[2] and M. Girard[1]
[1]Agroscope, Route De La Tioleyre 4, 1725 Posieux, Switzerland, [2]ETH Zurich, Institute of Agricultural Sciences, Universitätstrasse 2, 8092 Zurich, Switzerland, [3]Current address: University of Milan, Department of Veterinary Medicine and Animal Sciences, Via dell'Università 6, 26900 Lodi, Italy; peng.lin@unimi.it

Due to their nutrient composition, former food products (FFPs) can replace grains in pig diets. Therefore, we hypothesized that including FFPs in a sow's lactation diet wouldn't affect their lactation performance. However, the FFPs' richness in saturated fatty acid (FA) and scarcity in unsaturated FA will impact the FA profile of sow milk. As unsaturated FAs have immunomodulatory effects, one could expect effects on piglet growth. After farrowing, twenty Swiss Large White sows were allocated to either a diet with 25% FFPs (FFP) or a diet with 15% linseed cake (LIN) during lactation (n=10 per group). The two groups were balanced for body weights (BW), backfat thickness (BF), and parity. Litters were standardized to 13 piglets on average. Individual BW and BF of sows were measured at farrowing and at weaning and feed intake was recorded daily. Milk samples were collected on d17 post-partum to evaluate the chemical composition and the immunoglobulin (Ig) content. Piglet individual BW was measured at several time points until 5 weeks post-weaning. Although FFP sows ingested daily less (P<0.05) feed, BW and BF of FFP and LIN sows at weaning was similar (P>0.05). The contents (g/kg) of dry matter, protein and fat in milk were or tended to be higher (P<0.10) while that of lactose was lower (P<0.001) in the FFP than in the LIN group. The percentage (g/100 g milk FA) of lauric acid was higher (P<0.001) whereas the percentage of linoleic acid, alpha-linolenic acid, and eicosapentaenoic acid was lower (P<0.01) in the FFP group compared to the LIN group. The concentrations of milk IgG and IgA were unaffected (P>0.05) by the diets. The FFP piglets tended (P≤0.10) to grow slower in the pre-weaning and overall period and consumed less feed (P<0.05) from 3 to 5 weeks post-weaning. The inclusion of 25% of FFPs in the lactation diet may be too high as it reduced the daily feed intake, which may have impaired milk yield and ultimately hindered piglet growth.

Session 20 Theatre 10

Effect of arginine supplementation on the productive performance of gestating sows: a meta-analysis

S. Virdis, D. Luise, P. Bosi and P. Trevisi
University of Bologna, Department of Agricultural and Food Sciences, Viale Fanin 46, 40126 Bologna, Italy; sara.virdis3@unibo.it

The supplementation of Arginine (Arg) to gestating sows could play a primary role in litter development. The aim of this study was to investigate the effects of the concentration of dietary standardized digestible Arg (SID Arg) on productive performance of gestating sows using a meta-analytical approach. A total of 19 studies conducted between 2007-2020 were collected form literature. Data regarding backfat thickness, total number of piglets born and born alive, number of stillborn piglets, litter weight at birth and weaning were extrapolated from each experimental group and expressed as a percentage of the value of the control group (CO) within the studies. The SID Arg supplementation was expressed as the percentage of SID Arg in the treated group (TRT) minus that included in the CO group within each study (Additional SID Arg %). Linear and quadratic models were built using a mixed procedure by Minitab® software including Additional SID Arg %, parity and feed intake classes and the period of Arg supplementation as fixed factors, and the study as random factor. The SID Arg in CO groups was higher than the doses suggested by the NRC (2012; 4.20-8.90 g/kg of feed vs 2.02-3.10 g/kg, respectively). A quadratic response of Additional SID Arg % was observed for placental efficiency (PE; P=0.003; a=0.0019), number of total piglets born (P=0.027; a=-0.0003), number of piglets born alive (P=0.005; a=-0.0006) and backfat thickness loss (P=0.057; a=0.0029). A linear increase in plasma concentration of Proline (P=0.004; b=0.120), Ornithine (P=0.002; b=0.284), Arg (P=0.001; b=0.425), and a decrease in the concentration of urea (P=0.037; b=-0.063) was observed with an increasing level of Additional SID Arg %. There was no effect on placenta weight, alive litter birthweight, individual piglet birthweight and parameters at weaning. The estimated requirements of SID Arg for improving PE, reducing backfat thickness loss, increasing the total number of piglets born and born alive were 11.17 g/kg, 13.03 g/kg, 14.83 g/kg and 15.71 g/kg of feed, respectively. This meta-analysis highlighted the importance of redefining the Arg requirements of gestating sows to improve their productive performance.

Session 20 — Theatre 11

Are pre-weaning piglet and sow parameters a proxy for uniformity and daily gain?

K. Hooybergh[1]*, S. Goethals*[2]*, W. Gorssen*[1]*, L. Chapard*[1]*, R. Meyermans*[1]*, B. Chakkingal Bhaskaran*[1]*, S. Millet*[2]*, S. Janssens*[1] *and N. Buys*[1]
[1]*Center for Animal Breeding and Genetics, KU Leuven, Department of Biosystems, Kasteelpark Arenberg 30 bus 2472, 3001 Leuven, Belgium,* [2]*ILVO, Scheldeweg 68, 9090 Melle, Belgium; katrijn.hooyberghs@kuleuven.be*

Body weight uniformity within a batch of pigs is important from different viewpoints: economic purposes but also to lower environmental and social costs. The objective of the UNIPIG project is to identify factors associated with uniformity and to research the evolution of uniformity from birth until slaughter. Data were collected during an observation trial in two Belgian sow farms where 1,177 crossbred pigs born in 70 litters were recorded. Uniformity was assessed as the coefficient of variation (CV; standard deviation/mean) of body weight at several time points between birth and slaughter. CV per litter was calculated at birth (CV= 0.20), weaning (CV= 0.19), 1 (CV= 0.19) and 2 weeks (CV= 0.18) after weaning, at the start (9-11 weeks) (CV= 0.17) and in the middle (17 weeks) (CV= 0.13) of the finishing period. Furthermore, the association between variables such as sow condition (body weight, chest circumference and back fat thickness), order of birth and litter size, and body weight and CV of body weight was studied. Our results show that litters with a higher number of live-born piglets have a higher CV of body weight throughout life, especially at birth (r=0.43, *P*<0.001). Litter size was also moderately correlated to a lower average birth weight per litter (r=-0.40, *P*<0.001) and a higher percentage of gilts at birth (r=0.26, *P*=0.02). Furthermore, the time between the birth of piglets was found to be lowly correlated with birth weight (r=0.11, *P*<0.001). Moreover, other factors that contribute to the variation in body weight in a biological litter and in a batch were evaluated. At this moment, pairwise relationships are reported but in the next step, several factors will be examined together. Our study contributes to more insight into body weight uniformity within a litter and within a batch of pigs and this could help to arrive at a more sustainable and efficient pig production.

Session 20 — Theatre 12

Dietary polyphenol extracts improve the performance of broilers challenged with necrotic enteritis

S.A. Salami[1]*, L. Chew*[2]*, B.S. Lumpkins*[3] *and B. Tas*[1]
[1]*Mootral Ltd, Roseheyworth Business Park North, Abertillery, United Kingdom,* [2]*Mootral SA, Jalan Kiara, Kuala Lumpur, Malaysia,* [3]*Southern Poultry Feed & Research Inc., Roquemore Road, Athens GA, USA; ssalami@mootral.com*

Necrotic enteritis (NE) is a common intestinal disease in poultry and imposes significant economic loss due to the poor productivity of birds. Polyphenols are natural bioactive compounds that can be used in poultry nutrition to improve birds' gut health and productivity. This study investigated the effects of supplementing proprietary polyphenol extracts (IQV10-3) on the performance of broilers challenged with NE. A total of 400-day-old chicks were assigned to five groups (10 replicates/group) that were all challenged with NE: T1, control; T2 received IQV10-3 dose A; T3 received IQV10-3 dose A + coccidiostat; T4 received IQV10-3 dose B; T5 received IQV10-3 dose B + coccidiostat. The trial lasted for 28 d and NE was induced by inoculating all birds with *Eimeria maxima* on d 14 and *Clostridium perfringens* on d 19-21. IQV10-3 was administered via drinking water (dose A: 0.8% from d 0-2 and 0.2% from d 3-28; dose B: 0.8% from d 0-2, 0.4% from d 3-10 and 0.2% from d 11-28) while in-feed coccidiostat (125 ppm Amprolium on d 13-21) was administered to T3 and T5. Feed intake (FI), body weight gain (BWG) and feed conversion ratio (FCR) were evaluated at d 13-21 (challenge phase) and d 13-28 (challenge+recovery phase), and mortality was measured on d 0-28. On d 21, NE lesions were scored, and gut permeability was determined by the levels of serum fluorescein isothiocyanate-dextran (FITC-d). Contrast analysis was used to evaluate the effects of IQV10-3 without (T2+T4) or with (T3+T5) coccidiostat compared to the control (T1). For the challenge phase, T2+T4 and T3+T5 significantly increased BWG and reduced FCR. For the challenge+recovery phase, T2+T4 tended to increase (*P*=0.08) BWG and reduce (*P*=0.06) FCR whereas T3+T5 increased (*P*<0.01) FI and BWG without affecting FCR. No treatment effect on FITC-d was observed. However, T2+T4 and T3+T5 reduced mortality and lesion scores. These results suggest that IQV10-3 without or with coccidiostat had similar performance improvements compared to T1. Overall, IQV10-3 could be an effective nutritional solution for ameliorating the negative impacts of NE on broiler performance.

Session 20 Theatre 13

Effects of *Bacillus subtilis* DSM 32315 or *Bacillus velezensis* CECT 5940 on chicken PBMCs

F. Larsberg[1,2], M. Sprechert[2], D. Hesse[2], G. Loh[3], G.A. Brockmann[2] and S. Kreuzer-Redmer[1]
[1]Veterinärmedizinische Universität Wien, Institute of Animal Nutrition, Nutrigenomics, Veterinärplatz 1, 1210 Vienna, Austria, [2]Humboldt-Universität zu Berlin, Albrecht Daniel Thaer-Institute, Breeding Biology and Molecular Genetics, Unter den Linden 6, 10099 Berlin, Germany, [3]Evonik Operations GmbH, Research, Development & Innovation Nutrition & Care, Kantstr. 2, 33790 Halle, Germany; filip.larsberg.1@hu-berlin.de

Feeding probiotic *Bacillus subtilis* DSM 32315 (BS) and *B. velezensis* CECT 5940 (BV) has been described to promote health in chicken. One of the most plausible mechanisms underlying the beneficial effects is the modulation of the immune system. Here, we explored direct interactions of chicken peripheral blood mononuclear cells (PBMCs) and probiotics *in vitro*. We performed cell culture experiments with PBMCs of Cobb500 broiler chicken in a co-culture with vital BS or BV in a ratio of 1:3 (PBMCs:*Bacillus*) for 24 hours. The effects on PBMCs were analysed using flow cytometry, RNA-sequencing, and chicken cytokine array Q1 (RayBiotech). We found a higher Δ (difference between treatment and control) relative cell count of CD4+ T-helper (Th) cells ($P<0.05$) and CD4+CD25+ activated Th cells ($P<0.01$) after treatment with BS. Accordingly, RNA-sequencing of BS-treated PBMCs revealed higher Δ normalized counts of T cell-related genes CD4, CD28, IL2Ra (CD25), and IL10. Consistently, we found an increased Δ protein concentration of IL10 and decreased levels of pro-inflammatory IL16. However, the RNA-sequencing and cytokine array results were not statistically significant. Furthermore, the Δ relative cell count of CD8+ cytotoxic T-cells (CTLs, $P<0.05$) and CD8+CD25+ activated CTLs ($P<0.05$) was increased in PBMCs exposed to BS. After treatment with BV, we found a higher Δ relative cell count of CD4+ Th cells ($P<0.1$) as well as CD4+CD25+ activated Th cells ($P<0.05$). Correspondingly, the RNA expression of CD4, CD28, and CD25 was elevated. Furthermore, the expression and Δ protein concentration of IL16 decreased after BV treatment. However, those differences were not significant. Our results suggest a modulation of the T cell immune response by BS and BV in chicken PBMCs *in vitro* and, therefore, provide evidence of a direct immunomodulatory effect on the cellular immune response.

Session 20 Theatre 14

Effect of creatine monohydrate *in ovo* feeding on progeny performance of young breeder flocks

C.A. Firman[1], V.K. Inhuber[2], D.J. Cadogan[3], W.H.E.J. Van Wettere[1] and R.E.A. Forder[1]
[1]The University of Adelaide, School of Animal and Veterinary Sciences, Roseworthy Campus, 5371 Roseworthy, South Australia, Australia, [2]Alzchem Trostberg GmbH, Animal Nutrition, Dr.-Albert-Frank-Straße 32, 83308 Trostberg, Germany, [3]Feedworks Pty. Ltd., 13 High st, 3434 Lancefield, Victoria, Australia; vivienne.inhuber@alzchem.com

Broilers hatched from young broiler breeder hens (<30 weeks of age) have higher first week mortalities and lower final bodyweights compared to broiler chickens from older breeder hens. This may be due to the fact that younger breeder hens produce smaller eggs with smaller extraembryonic yolk sacs. Hence, less energy-rich components are available during embryonic development and hatch. The aim of this study was to evaluate if IOF of Creatine Monohydrate, with Creatine (Cr) being a cellular energy storage, positively influences post-hatch growth performance of broilers hatched from young breeder flocks. Four hundred eggs (Ross 308 breeder hens, 27-29 weeks) were collected and at embryonic d14 evenly assigned to three experimental groups: (1) no IOF (CON); (2) 0.75% saline injection (IOF-CON); (3) 8.16 mg CrM in 0.75% saline (IOF-CrM). At hatch, 72 birds (n=24/treatment) and at d42, 96 (n=32/treatment) birds were euthanized and dissected. Birds were reared in a total of 24 pens with 8 pens/IOF treatment until d42. Hatch rate was not statistically different between the three treatment groups. However, a notable difference could be observed between IOF-CrM (93.5%) vs both CON and IOF-CON (88.8 and 88.6%, respectively). Early post-hatch mortality was numerically reduced in the CrM group as compared to both CON and IOF-CON (weeks 4 and 5). *In ovo* feeding of CrM led to increased ($P<0.05$) Cr concentrations on both liver and heart tissue at hatch as compared to both CON and IOF-CON. There were no treatment effects on growth performance over the whole rearing period. Enhanced energy status at hatch may have improved hatching rate in the CrM group, but was only leaving little resources for post-hatch overall performance. This research indicates that improved energy status at hatch enables producers to place more robust chicks and therefore, has immense application to the industry.

Session 20 — Poster 15

Statistical modelling and estimation of pig body weight based on its linear body measurements

O.T.F. Abanikannda, K.O. Kareem-Ibrahim and Z.A. Ahmad
Lagos State University, Department of Animal Science, School of Agriculture, Epe Campus, Lagos, Nigeria; otfabanikannda@hotmail.com

Body weight is an essential index used by producers to determine market-ready pigs, and its accurate estimation before the finishing stage helps reduce losses in pig production. However, since most pig farms at the smallholder level have limited or no access to weighing scales, this study aimed at statistical modelling of linear body measurements to estimate pig body weight. The study involved 83 weaners from a commercial farm at three different ages (4, 5 and 7 weeks) from four breeds comprising Camborough (CB), Large White (LW), and crosses of Camborough with Large White (CB × LW) and Landrace (CB × LR). Body weight was measured using digital weighing scale sensitive to 0.00 g, while the eight linear body measurements were taken using a flexible graduated tape, for the points of reference. Nine parameters including body weight (BW), total body length (BL), thoracic circumference (TC), palette length (PL), shank length (SKL), shank circumference (SC), hip circumference (HC), heart girth (HG) and standard body length (SL) were measured, with mean values of 23.12±1.04 kg, 65.24±3.95 cm, 39.66±3.93 cm, 21.49±0.29 cm, 16.16±1.00 cm, 12.96±0.81 cm, 47.12±2.95 cm, 44.77±2.79 cm and 46.78±2.78 cm, respectively. Breed and age exerted significant (P<0.05) influence on each of the nine variables measured, while multiple correlation of variables was mostly highly significant (P<0.01) except for palette length and other variables which was mostly not significant (P>0.05). Four of the variables, BW, TC, HG and HC had highest loadings in the eigenvalues obtained from the two principal components, which accounted for 99.1% of the total variation. The general model describing body weight in the study was Body Weight = -18.69 – 1.19 TC + 1.46 HG + 0.50 HC which explained 88.26% of the total variation in body weight. The study confirmed the accuracy and reliability of body weight estimation using linear body measurements in pigs, and thus it is recommended that a quick appraisal of the thoracic circumference, heart girth and or hip circumference can give a fair estimate of the pig body weight.

Session 21 — Theatre 1

The potential of insects in sustainable agri-food systems

K.B. Barragán-Fonseca
Universidad Nacional de Colombia, Faculty of Veterinary Medicine and Animal Sciences, Animal Production Department, Carrera 45 # 26-85 Bogotá, 11132, Colombia; kbbarraganf@unal.edu.co

Insects play an important role in sustainability of agri-food systems, as they provide numerous ecosystem services that benefit humans and the environment, including pollination, decomposition, and pest control, as well as serving as a source of food and feed. To evaluate the potential of insects in sustainable agri-food systems, we must consider the following: the objective of insect use; climatic and ecological factors; biological and behavioural traits of the insect species; ways of obtaining the insects; and the scale of their use. Insect farming is rapidly expanding on all continents, and sustainability dimensions vary across small, medium, and large scales of operation. A current societal tendency exists to develop an economic vision that not only focuses growth of the gross domestic product, but also on ecological and social factors. This vision emphasizes the importance of optimizing how insects contribute to various Sustainable Development Goals, including food security, circular economy, and social transformation. While use of insects has the potential to transform societies, not all insect species are suitable for all purposes, and the way in which they are obtained as well as scale of use may vary depending on a variety of factors, including community needs, ecological settings, local regulations, and investment trends. To address these challenges, a comprehensive and interdisciplinary approach is needed to scrutinize the social, economic, and environmental sustainability of insect utilization, assessing various forms of use (gathering, semi-rearing, and production) and their scale. Furthermore, mapping the global use of insects, including current 'hotspots' of their use, can help identify suitable species, forms of production, and scale by region, integrating insects into food systems in an economically, socially, and environmentally sustainable manner.

Sustainability potential of insect production systems

S. Smetana[1], A. Bhatia[1,2], U. Batta[1], N. Mouhrim[3] and A. Tonda[3,4]
[1]DIL German Institute of Food Technologies (DIL e.V.), Professor-von-klitzing-straße 7, 7, 49610, Germany, [2]University of Osnabrück, Neuer Graben 29, 49074, Osnabrück, Germany, [3]INRAE, Université Paris-Saclay, UMR 518 MIA-PS, 91120 Palaiseau, France, [4]Institut des Systèmes Complexes de Paris Île-de-France (ISC-PIF), UAR 3611 CNRS, Paris, France; s.smetana@dil-ev.de

The current food systems face severe sustainability problems, e.g. climate change, deforestation, and biodiversity loss. This study aims to assess the potential of insect production to improve sustainable food systems in Europe. Using a comprehensive approach that models key sustainability aspects as defined in FAO SAFA guidelines, the study determined that environmentally beneficial insect value chains can reduce the impact of livestock production systems by 40-97% in categories of global warming potential, land use, and fossil resource scarcity. This reduction is achievable by substituting meat (beef, pork, and poultry) with efficiently produced insect biomass. However, substituting compound feed production with insect value chains would only be environmentally beneficial in cases where extremely efficient production systems are used (e.g. insects grown on waste or low-cost feeds, relying on side-stream heat and alternative energy sources). In such cases, up to 93% of the impact of compound feed production in categories such as global warming potential, land use, and fossil resource scarcity can be eliminated. Insect biomass did not show benefits in the categories of water footprint and animal welfare. This is partly due to the lack of assessments carried or lack of specialized methods (e.g. animal welfare for insects). It is important to consider various factors, such as social acceptability, economic feasibility, and scalability of production, when evaluating the sustainability of insect production chains to fully evaluate the potential of insect production as a sustainable food system at a European level.
This project received funding from the European Union's Horizon 2020 research and innovation program under grant agreement no. 861976 (Project SUSINCHAIN). This document reflects only the authors' views, and the Commission is not responsible for any use that may be made of the information it contains.

Maximizing sustainability in insect production: a multi-objective optimization approach

A. Anita[1,2], N. Mouhrim[3], A. Green[4], D. Peguero[4], A. Mathys[4], A. Tonda[3] and S. Smetana[2]
[1]University of Osnabrück, Osnabrück, 49074, Germany, [2]German Institute of Food Technologies (DIL e.V.), Quakenbrück, 49610, Germany, [3]UMR 518 MIA-Paris, INRAE, Paris, 75231, France, [4]Laboratory of Sustainable Food Processing, ETH, Zurich, 8092, Switzerland; S.Smetana@dil-ev.de

Determining the sustainability of insect production is a complex and challenging task. The life cycle sustainability assessment (LCSA) of insect production can estimate social, economic, and environmental impacts. To use insects as feed and food, several aspects of sustainability must be considered, including food and feed production regulations, nutrition factors, social acceptance, and greenhouse gas regulation. Multi-objective optimization (MOO) can consider three competing goals: maximization of economic, social benefits and minimize environmental impacts. The MOO-decision support system is based on: (1) modular LCA and scenario-based life cycle inventory (LCI) that analyse environmental, social, and economic performance of insect species (*A. domesticus*, *M. domestica*, *H. illucens*, *T. molitor*); (2) non-dominated sorting genetic algorithm (NSGA-II) to obtain pareto-optimal solutions covering all sustainability indicators to compare insect production scenarios and test the sensitivity of results with different feeds, processing, utilities, end product type, packaging, and scaling options. This MOO-DSS is developed in the scope of SUSINCHAIN project grant agreement no. 861976 to examine the insect production chains in Netherlands, Germany, France, and the United Kingdom. According to findings, *H. illucens* should be the most used insect for food/feed production, followed by *T. molitor*, *M. domestica*, and *A. domesticus*. *H. illucens* is the most versatile species in terms of feed types (poultry feed, fruits, vegetables, and brewer's-spent-grains). *T. molitor* and *A. domesticus* should be fed on fruits, vegetables, and plant residues, whereas *M. domestica* on fruits, vegetables, and brewer's--spent-grains. The study shows that the scaling potential of each species varies, with *H. illucens* having the highest potential for scale-up. The feed types used to rear insects vary depending on region and production scalability. This data can assist farmers and researchers in determining the best feed types and insects for their region and needs.

Lowering impacts of chicken meat through *H. illucens* larvae supplementation in the feed?

D. Ristic[1,2], S. Kechovska[2,3], C.L. Coudron[4], A. Schiavone[5], J. Claeys[4], F. Gai[6] and S. Smetana[2]
[1]University of Nat. Resources and Life Sciences (BOKU), Institute of Food Technology, Muthgasse 18, 1190 Vienna, Austria, [2]German Institute of Food Technologies (DIL e.V.), Prof-von-Klitzing-St 7, 49610 Quakenbrück, Germany, [3]University of Ss. Cyril and Methodius, Faculty of Technology and Metallurgy, Ruger Boskovic 16, 1000 Skopje, Macedonia, [4]Inagro, Ieperseweg 87, 8800 Rumbeke-Beitem, Belgium, [5]University of Turin, Veterinary Sciences, L.go Paolo Braccini 2, 10095 Grugliasco, Italy, [6]National Research Council, Institute of Sciences of Food Production, L.go Paolo Braccini 2, 10095 Grugliasco, Italy; d.ristic@dil-ev.de

Feed production is responsible for most environmental impacts in poultry production chains. Environmentally friendly sources of protein for feed are required to reduce the ecological footprint of poultry production. The slow-growing Label Naked Neck chickens were divided into 2 groups: one, reared on commercial organic feed and 10% BSFL substitution, and the other, reared on commercial organic feed. The modular, attributional life cycle assessment (LCA) was developed to assure a structured and quantitative approach. The results are based on experimental data collected from the project partners, extended by the background data and data from the literature. This study followed the cradle-to-slaughterhouse gate perspective with further extensions to waste treatments, thus including feed production, larvae production, hatchery, poultry production, and slaughterhouse. The endpoint results show major impacts on ecosystem quality and human health, followed by climate change, while resource depletion was the least impacted. Crucially, there was no significant difference in the environmental impact of chicken meat introduced by the inclusion of 10% of BSFL into the chicken diet; the differences can rather be attributed to the difference between the sexes. The midpoint results reveal major impacts in Land occupation (male 1.1725 and 1.1553, female 0.9924 and 0.9187 mPt), Global warming (male 1.1718 and 1.1417, female 1.1095 and 0.9924 mPt) and Respiratory inorganics (male 1.0382 and 1.0171, female 0.938 and 0.8856 mPt). Better results might be expected if insect feed were adjusted to overproduced fruits and vegetables, and if the portion of BSFL in broilers' diets were increased.

Use of locusts as feed and food: turning threats into opportunities

H.P.S. Makkar[1], V. Heuzé[2] and G. Tran[2]
[1]University of Hohenheim, Schloß Hohenheim 1, 70599 Stuttgart, Germany, [2]AFZ, AFZ/AgroParistech, 22 place de l'Agronomie, 91120 Palaiseau, France; valerie.heuze@zootechnie.fr

Locusts cause massive destruction of crops and pastures and adversely affect livelihoods and food security of farmers and pastoralists. Locusts are rich in protein (50-65% in dry matter). Their essential amino acid composition is good. In the diets of poultry, pigs and fish, replacement of up to 25% of the conventional protein-rich feed resources such as soymeal and fishmeal is possible with locusts. Among 10 species of locusts have been traditionally consumed by humans for millennia in 65 countries. Their nutritional composition is comparable or superior to that of conventional meat. Main constraint in their use as food/feed is the presence of insecticides, sprayed for controlling outbreaks. Only insecticide-free locusts must be used as a feed/food. Locusts rearing techniques have not been upscaled; however, harvesting approaches are available: use of nets, light and sound. Upscaling of harvesting techniques could offer opportunities to enhance feed/food availability in many developing countries during the locust plaque. However, use of harvesting approaches has not been fully considered by the locust controlling agencies. A concern has also been raised that people actively collecting locusts during nights puts them at risk from wildlife/other rural population. Communication of advantages of locust harvesting to the community: prevention of animal/human health deterioration because of insecticide spray would help overcoming the resistance. Locust harvesting using light and sound would deter wild animals from coming near. Alternatively, collection can be done in the mornings when the locusts are numb. An integrated approach that strategically uses insecticides and employs mass harvesting techniques must be considered to control these pests. This would decrease use of pesticides and their associated negative effects. The people should be educated on collection of only live locusts for use as food/feed. The dead ones could be used for composting. Advance preparation of the locust control emergency units towards use of the integrated approach would be a triple-win for smallholder farmers, environment, and human/animal health.

Session 21 Theatre 6

Chrysomya chloropyga in broiler feed: valorising abattoir waste for sustainable feed

E. Pieterse, E.Y. Gleeson and L.C. Hoffman
Stellenbosch University, Animal Sciences, Merriman avenue, Stellenbosch, 7600, South Africa; elsjep@sun.ac.za

Insects in animal feed have been extensively researched over the past decade and interest in the field has grown exponentially over the past few years. Larvae from different fly species, most notably *Hermetia illucens*, have gather a great deal of attention as they represent possible avenues for more sustainable feed ingredients by valorising a vast array of waste and by-products. The various species have different dietary preferences and attributes that determines the substrates on which they thrive. Therefore, it is important to determine the best suited species to work with depending on the waste or by-product that would be used as substrate. *Chrysomya chloropyga*, otherwise known as the copper bottom blow fly, is a fly species that is particularly well suited to valorising abattoir waste as it is a carnivorous species commonly associated with carrion. The *C. chloropyga* larvae are rich in protein and represent a promising protein and iron source for broiler nutrition. This paper will present data gathers over several studies where *C. chloropyga* larvae were raised on abattoir waste and processed into a meal. The studies determined digestibility of the meal, evaluated the effect of inclusion level on production parameters, assessed carcass and meat quality characteristics and established the viability of the use of the meal as an iron source for broiler production. It was concluded from these studies that *C. chloropyga* can be successfully used to valorise abattoir waste and that the resultant larvae have a high digestibility, fits the ideal amino acid profile of poultry closely, can serve as an iron source for broilers. It was also established that the use of the larvae in broiler feeds has no negative impact on carcass- or meat quality characteristic. The addition of *C. chloropyga* to the list of insect species used as animal feed ingredients will contribute to creating more dynamic insect production schemes that are aimed at optimizing the valorisation of various wastes and by-products and moving towards circular economies in the animal production sector.

Session 21 Theatre 7

Waste side story: a success role for mealworms

C. Ricciardi, L. D'Orleans, C. Oundjian, M. Leheup, T. Lefebvre and F. Peyrichou
Ynsect, 1, rue Pierre Fontaine – Genopole Campus 3 – Batiment 2, 91000 Evry Courcouronnes, France; carmelo.ricciardi@ynsect.com

Today, one third of all food produced for human consumption is wasted. According to FAO's Food Loss Index (FLI), half of that amount is lost from post-harvest to sale and the other half from sale to consumers. In this context insects can be a powerful resource to reduce and valorise the food wastes. Insects, specifically *Tenebrio molitor*, can be reared in vertical farms, reducing the surface needed, and are able to convert the feed given more efficiently than common livestock. Inclusion of wastes in the diet of *Tenebrio molitor* presents two major benefits: reducing the use of more noble ingredients to provide nutrients of interest, and so decreasing the diet price, and improving the global sustainability of the rearing by reducing the carbon footprint of the feedstock. Ÿnsect conducted several trials in order to assess the effect of including different types of food wastes on the rearing performances. Different diets were formulated according to specific constraints and tested during the exponential growth phase of larval stage, until the harvest. The results obtained were significative: (1) Average reduction of both the developmental time and the FCR when a mixture of humid wastes was provided to the larvae; (2) Stability in the performances for isoproteic and isoenergetic diets formulated incorporating wastes of different origin (Cereals based foods, Vegetables, Processed foods); (3) Reduction of the environmental impact of the formulated diets, estimated as kg of CO_2 equivalent for kg of larvae produced (values obtained from the software Simapro). The trials made proved that the inclusion of different type of food wastes allows to maintain the same, or improved, rearing performances of the standard diet and to decrease the environmental impact and the cost of the diet. The use of food wastes opens new opportunities for feedstock sourcing and outsourcing. These are indeed more environmental friendly and at a cheaper cost than traditional ones. It appears surely feasible and desirable to formulate diets optimized for the industrial rearing of *Tenebrio molitor*.

Session 21 Theatre 8

Ammonia emissions of *Hermetia illucens* larvae grown on different diets

C.L. Coudron[1], S. Berrens[2], M. Van Peer[2], J. Claeys[1] and D. Deruytter[1]
[1]Inagro vzw, Ieperseweg 87, 8800 Rumbeke-Beitem, Belgium, [2]Thomas More University of Applied Sciences, Radius, Kleinhoefstraat 4, 2440 Geel, Belgium; carl.coudron@inagro.be

The impact of ammonia emissions and associated depositions from intensive livestock farming on eutrophication leads in some countries to specific measures to reduce emissions. The production of black soldier fly larvae is a novel activity that could similarly lead to ammonia emissions. Several studies have already examined this topic, but new studies are still needed due to the huge array of potential substrates and nuances in rearing techniques, that could all influence ammonia emissions. Better understanding of this topic can only help clarifying which measures, if any, need to be taken so that local decision makers can make appropriate decisions on the matter. A study was performed using an accumulation chamber, a stainless steel box wherein the concentration of ammonia is monitored. A crate (60×40 cm) with growing larvae (15,000) in their feed is placed inside of the accumulation chambre for a short period (10 minutes up to 1 hour) every 24 h. An exponential function f(t)=a(1-b*exp(-ct)) is fit on the ammonia evolution in the accumulation chamber and derived at time zero to determine the momentaneous ammonia emitted by a crate at the time the larvae were put inside. The combined data throughout larval growth can be used to determine peak and total ammonia emissions. This method was performed on a total of four diets (Gainesville diet, chicken feed, swill and brewers spent grains). In addition, both swill and brewers spent grains were enriched with casein protein or protein was diluted using straw pellets in order to vary the protein content between 12 and 25% crude protein on a dry matter basis. This gives an indication of how protein content influences ammonia emissions. Simultaneous with the ammonia measurement, the nitrogen balance of each crate was determined via chemical analysis of the feed, larvae and frass, the inner crate conditions were monitored by measuring pH and temperature of the substrate daily as well as the insects growth parameters. The combination of these parameters gives better insights on how and when ammonia is produced and how it can be reduced.

Session 21 Theatre 9

Chemical & microbiological safety of insect rearing on yet to be legally authorised residual streams

E.F. Hoek- Van Den Hil[1], A.F.G. Antonis[2], Y. Hoffmans[1], K. Van Zadelhoff[1], R. Deurenberg[2], O.L.M. Haenen[2], M.E. Bruins[3], J.W. Van Groenestijn[3], A. Borghuis[4], F. Schaafstra[4], T. Veldkamp[5], P. Van Wikselaar[5] and M. Appel[1]
[1]Wageningen Food Safety Research (WFSR), Akkermaalsbos 2s, 6708 WB, the Netherlands, [2]Wageningen Bioveterinary Research (WBVR), Houtribweg 39, 8221 RA Lelystad, the Netherlands, [3]Wageningen Food & Biobased Research (WFBR), Bornse Weilanden 9, 6708 WG Wageningen, the Netherlands, [4]HAS Den Bosch University of Applied Science, Onderwijsboulevard 221, 5223 DE 's-Hertogenbosch, the Netherlands, [5]Wageningen Livestock Research (WLR), De Elst 1, 6700 AH Wageningen, the Netherlands; elise.hoek@wur.nl

Insects can convert low quality residual streams into protein-rich ingredients for food and feed. However, chemical residues, pathogens and viruses could be present in these residual streams. Under current European law, it is not permitted to use residual streams that contain animal products or manure for insect rearing. Therefore necessary safety data is needed to provide a basis for enforcement of legalization. Within the project SAFE INSECTS, black soldier fly larvae (BSF, *Hermetia illucens*) and yellow mealworms (YMW, *Tenebrio molitor*) were reared on supermarket mix (food products that passed the expiration date or are unsaleable for other reasons), category 2 meat meal from animal rendering, organic waste from household kitchens, and poultry manure. Chemical and microbiological safety parameters were investigated in the residual streams that were used as substrates, the insects and the frass. Chemical safety parameters were heavy metals, other elements, pesticides, veterinary drugs, dioxins, and PCBs. Residual packaging materials and microplastics were also studied. Microbiological parameters were *Listeria monocytogenes*, *Bacillus cereus*, *Clostridium perfringens*, *Campylobacter*, *Salmonella*, *Staphylococcus aureus*, parasites, the viruses HEV and AIV, and antimicrobial resistant genes. During this presentation, safety results will be shown for both BSF and YMW with a focus on hazards which were present in the residual streams, and on possible accumulation or absence of safety hazards in the insects. With these data, we aim to contribute to possible adaptation of legalization on the permitted use of these residual streams as substrates for insects rearing.

Session 21 Theatre 10

Growth and chemical composition of insects reared on 'yet to be legally authorised' residual streams

S. Naser El Deen[1], P. Van Wikselaar[1], A. Rezaei Far[1], I. Fodor[1], E.F. Hoek - Van Den Hil[2], Y. Hoffmans[2], M.E. Bruins[3], J.W. Van Groenestijn[3], A. Borghuis[4], F. Schaafstra[4] and T. Veldkamp[1]
[1]Wageningen Livestock Research (WLR), Wageningen University & Research, De Elst 1, 6700 AH Wageningen, the Netherlands, [2]Wageningen Food Safety Research (WFSR), Wageningen University & Research, P.O. Box 230, 6700 AE Wageningen, the Netherlands, [3]Wageningen Food & Biobased Research (WFBR), Wageningen University & Research, Bornse Weilanden 9, 6708 WG Wageningen, the Netherlands, [4]HAS Den Bosch University of Applied Science, Onderwijsboulevard 221, 5223 DE 's-Hertogenbosch, the Netherlands; somaya.nasereldeen@wur.nl

Insects can be used to increase the circularity of agriculture. Underutilized residual streams such as food waste, slaughter by-products, and animal manure are a source of nutrients and insects can convert these residual streams into high-quality products for animal and human consumption. It is not permitted to use residual streams that contain animal products or manure for insect rearing according to current EU legislation. In the project SAFE INSECTS the aim is to complete the necessary food safety data to enable the application of legally prohibited residual streams to grow insects. Next to the safety data also the growth performance of the insects and the chemical composition of the insects and frass harvested from these residual streams were studied to see if these residual streams are promising substrates for insect rearing. Black soldier fly larvae (BSF; *Hermetia illucens*) and yellow mealworms (YMW; *Tenebrio molitor*) were reared on supermarket mix (food products that passed the expiration date or are unsaleable for other reasons), category 2 meat meal from animal rendering, organic waste from household kitchens, and poultry manure. The growth performance and the chemical composition of the substrates, the starting and harvested larvae, and the frass were determined in order to calculate the conversion of nutrients such as nitrogen (protein) and fat. These data will indicate if the organic residual streams are promising substrates as such to grow insects or if these should be further optimized to get satisfactory results for insect-rearing companies.

Session 21 Theatre 11

Safety and production performance of insects reared on catering waste

E.F. Hoek- Van Den Hil[1], S. Naser El Deen[2], K. Van Rozen[3], P. Van Wikselaar[2], H.J.H. Elissen[3], R.Y. Van Der Weide[3], A. Rezaei Far[2], I. Fodor[2], N. Meijer[1] and T. Veldkamp[2]
[1]Wageningen Food Safety Research (WFSR), Akkermaalsbos 2, 6708 WB, the Netherlands, [2]Wageningen Livestock Research (WLR), de Elst 1, 6700 AH, the Netherlands, [3]Wageningen Research – ACRRES, Edelhertweg 1, 8219 PH Lelystad, the Netherlands; elise.hoek@wur.nl

Farmed insects intended for food or feed are considered as 'farmed animals', and the EU has established restrictions on the type of feed (substrate) that these insects may be reared on. Consequently, such insects may only be fed with materials of vegetal origin or certain specified animal products such as milk, eggs, etc. – but no meat or fish. Catering waste or former foodstuffs which can contain meat, however, are a very promising substrate compared to other residual streams. These products were food grade and can therefore be considered safe, presuming the risks had been correctly managed in the chain. However, 'intra-species recycling' ('cannibalism') should be avoided: including 'indirect' recycling with insects acting as intermediary. Therefore, the objective of this study was to determine the presence of animal DNA (pig/poultry/bovine) and pathogens in insects reared on substrates containing animal proteins. Black soldier fly larvae (BSFL, *Hermetia illucens*) were reared for one week on catering waste stored until usage at room temperature (18-22 °C) or cool conditions (4-8 °C). The different types of catering waste tested were: (1) fresh French fries, (2) fried French fries, (3) mix location 1 (bread, fries, and meat), (4) mix location 2, and (5) mix location 3. Larval performance in terms of growth rate, waste reduction index, and efficiency of conversion of ingested substrate were determined as well as the chemical composition of the substrate, the larvae, and the frass; to calculate the conversion of nitrogen and fat. Microbiological safety (*Salmonella, B. cereus*, and *S. Aureus*) was assessed. Furthermore, data suggests that gut emptying is an effective measure to reduce the level of animal DNA below detectable limits in the larvae. All results will be presented. With these data we hope to contribute to possible legalization of the use of catering waste for rearing insects.

Mycotoxin accumulation and metabolization by insects: an overview

K.C.W. Van Dongen[1], K. Niermans[1,2], E.F. Hoek-Van Den Hill[1], J.J.A. Van Loon[2] and H.J. Van Der Fels-Klerx[1]
[1]Wageningen Food Safety Research, Akkermaalsbos 2, 6708 WB Wageningen, the Netherlands, [2]Wageningen University, Department of Plant Sciences, Laboratory of Entomology, Droevendaalsesteeg 1, 6708 PB Wageningen, the Netherlands; katja.vandongen@wur.nl

Animal feed production needs novel protein sources, and insects are seen as a promising feed ingredient. In particular when insects can be reared on substrates of low quality, that cannot directly be used as feed ingredient. When insects could upgrade waste or side streams into high quality proteins, insect production would contribute to an economic viable and sustainable food system. When rearing insects on waste and side streams, potential safety issues could arise, and these should be studied in advance, using a safe-by-design approach. In particular, mycotoxin contamination is seen very often, around the world, in all sorts of plant-based commodities. These contaminants can have very severe adverse health effects, so possible accumulation in insects, from the substrate, should be investigated. In this presentation, a state of the art overview will be given on research into the possible accumulation of mycotoxins in insects, in particular black soldier fly and (lesser and yellow) mealworm larvae, as well as impacts of mycotoxins on insect growth and survival. General findings are: (1) mycotoxins have no or very low effects on insect growth and survival; (2) accumulation of mycotoxins was low in black soldier fly and lesser and yellow mealworm larvae; (3) mycotoxins are metabolized within the larvae to some extent. The degree of metabolization depends on the particular insect species and respective mycotoxin. Results of the studies performed so far provide valuable findings to use insects as a means to use commodities contaminated with mycotoxins in a valuable way. However, since full routes of metabolization are not known yet, care must be taken, and more research is needed.

Manure as a substrate for insect rearing: ecological progress or health risk?

A. Anselmo[1,2], A. Cordonnier[2], L. Plasman[2], J. Maljean[2], C. Aerts[2], D. Evrard[1], A. Marien[2], E. Janssen[2], V. Ninane[2], S. Gofflot[2], P. Veys[2], M.-C. Lecrenier[2], V. Baeten[2] and D. Michez[1]
[1]University of Mons, Research Institute for Biosciences, Laboratory of Zoology, Place du Parc 20, 7000 Mons, Belgium, [2]Walloon Agricultural Research Centre, Chaussée de Namur 24, 5030 Gembloux, Belgium; a.anselmo@cra.wallonie.be

Insect rearing represents a food and ecological issue that could help solve future challenges of human societies. Insects are known to be very rich in protein, very easy to produce and require minimal maintenance. These various advantages make them a serious alternative protein for food and feed. However, as the insect sector is not yet sufficiently developed, the price of insect meal is to high. To further develop the insect meal market, the European Union has authorised the use of 8 species in aquaculture since 2017 and, more recently, for pig and poultry feed. Insects are known for recycling organic waste and can transform a low-quality material into a very high protein product. Several studies have tested the use of organic waste substrates to raise insects intended for animal feed. This type of experimentation highlights the crucial issue of the health risk of using waste to produce feed for animals. To address this issue, *T. molitor* larvae were reared on a substrate adulterated with 5, 15 or 25% bovine manure. At the end of the experiment, these larvae were recovered, cleaned and ground to obtain a larval meal. The different insect meals were evaluated by multiple analyses. Firstly, microscopic analyses were carried out according to Annex 6 of Commission Regulation (EC) No 125/2009. Secondly, real-time PCR and proteomic analyses were conducted to check for the presence of ruminant DNA and proteins to assess the potential risk of BSE transmission. Thirdly, microbiological analyses focused on faecal indicator bacteria. Finally, protein and lipid analyses were performed to evaluate a possible modification of the insect nutritional value. The aim of this study was to get a comprehensive overview of the potential health risk associated with the use of organic waste and, if the results are encouraging, to consider circular production in the insect sector. The results of this study will be presented and discussed during the lecture.

Session 21 Poster 14

Quality characteristics of black soldier fly produced by different substrates

A.R. Hosseindoust, S. Ha, J. Mun, H. Tajudeen, S. Park and J. Kim
Kangwon National University, Department of Animal Industry Convergence, Chuncheon, 24341, Korea, South; hosseindoust@kangwon.ac.kr

Insects as a source of livestock feed, such as the black soldier fly (BSF; *Hermetia illucens*), have high accumulations of essential nutrients, particularly protein and fatty acids. BSF is therefore regarded as a new option for sustainable agriculture and a source of animal feed. There were 3 treatments including tofu by-product, food waste, and vegetables. This study was conducted to assess the effectiveness of these botanical waste substrates for BSF growth, conversion efficiency, nutrient accumulation, and fatty acid profiles. Tofu by-product treatment showed the greatest weight at d 14 and the harvest period, compared with food waste and vegetable treatments. Moreover, BSF larval weight was greater in the food waste treatment compared with the vegetable treatment at d 14 and harvest. The larval length was increased in the food waste and tofu by-product treatments. Larvae yield result was greater in the vegetable treatment compared with the food waste. The content of DM and CP were increased in BSF fed with tofu by-product, and the EE content was greater in the food waste and tofu by-product treatments compared with the vegetable treatment. The bioconversion rate was higher in the tofu by-product treatment compared with the vegetable. The waste reduction rate was higher in the tofu by-product treatment compared with the food waste and vegetable treatments. Moreover, the waste reduction rate was higher in the food waste treatment compared with the vegetable. The protein conversion rate was the highest in the food waste treatment. Lipid conversion rate, protein yield, and lipid yield were greatest in the tofu by-product treatment. The content of lauric acid was increased in BSF fed tofu by-product compared with the food waste treatment. The concentration of C16:1 was the highest in the tofu by-product treatment. The content of oleic acid and α-linolenic acid were higher in BSF fed tofu by-product compared with the vegetable treatment. In conclusion, tofu by-product shows benefits of larvae growth and nutrient accumulation, which can improve larval quality for livestock feed ingredients.

Session 21 Poster 15

The value of organic fertilizer from *Hermetia illucens* frass in the sustainable agriculture

S. Kaczmarek and K. Dudek
HiProMine S.A., Poznańska, 12F, 62-023 Robakowo, Poland; krzysztof.dudek@hipromine.com

With the challenges of climate change and the global environmental crisis, sustainable agriculture is becoming increasingly important. In this context, frass from the larvae of the fly *Hermetia illucens*, also known as 'black gold', is gaining importance as an organic fertilizer that can reduce the use of chemical fertilizers, improve soil quality and increase yields. Fertilizer made from *H. illucens*, insect frass, is a valuable source of nutrients for plants. Due to its origin, insect frass contains chitin in its composition, which provides additional application benefits – in addition to improving plant productivity, it can also result in induced plant resistance to abiotic stresses. Chitin can also improve plant resistance to pathogens and diseases. Thus, it can be used to promote crop and soil health on farms, thereby reducing the need for chemical crop protection products. Studies have shown that the use of frass improves soil structure and increases the content of nitrogen, phosphorus, potassium and other elements in the soil. In addition, the use of this fertilizer contributes to reducing greenhouse gas emissions and organic waste. There is also evidence that frass improves the physiological state of plants and their nutritional value, especially for crops under drought stress. Research indicates that the use of frass can contribute to the development of sustainable agriculture, as well as provide a valuable resource for agriculture by using it as a source of plant nutrients and improving soil structure. Insect frass-based organic fertilizer is a product that can be safely applied to all fields, gardens, parks and green spaces, nurseries, hobby gardens, greenhouses, orchards, or ornamental plants. This work was supported by an Szybka Ścieżka grant titled 'Development of a technology for the production of organic fertilizer (in the form of pellets/granules) based on the *H. illucens* frass and testing its impact on selected plants' (no. POIR.01.01.01-00-1503/19), which was financed by the Narodowe Centrum Badań i Rozwoju (Poland).

Session 21 — Poster 16

Effects of rearing substrates and processing methods on black soldier fly meal quality

S. Ha, A.R. Hosseindoust, Z. Khajehmiri, J. Mun, H. Tajudeen, S. Park and J. Kim
Kangwon National University, Department of Animal Industry Convergence, room 201, Animal Resources dep, (409), 24341, Korea, South; hosseindoust@kangwon.ac.kr

Insects as a source of livestock feed, such as the black soldier fly (BSF; *Hermetia illucens*), have high accumulations of essential nutrients, particularly protein and fatty acids. BSF is therefore regarded as a new option for sustainable agriculture and a source of animal feed. The study included six treatments, comprising three different substrates (tofu by-product (TF), food waste (FW), and vegetables (VEG)) and two drying methods (hot air (HA) and microwave (MW)). This study was conducted to assess the effectiveness of these botanical waste substrates for BSF growth, conversion efficiency, nutrient accumulation, and fatty acid profiles. The TF treatment showed a higher crude protein and gross energy content compared with the VEG. There was a tendency for higher ether extract in the FW compared with the VEG. The dry matter was higher in the MW but ether extract was higher in the HA treatment. The *in vitro* pepsin digestibility of BSF meal was the lowest in the VEG. Among processing methods, the *in vitro* pepsin digestibility of BSF was higher in the HA. The thiobarbituric acid reactive species analysis showed higher values in the TF and FW compared with the VEG. The thiobarbituric acid reactive species was decreased in the MW compared with the HA. The content of C12:0 in BSF meal in the TF was higher than in FW. The content of C16:1 was higher in the TF compared with the FW and VEG. The content of C18:1 in BSF meal in the TF was higher than VEG. In conclusion, tofu by-product and food waste show a high potential for larvae growth and nutrient accumulation, which can improve larval quality for livestock feed ingredients.

Session 21 — Poster 17

The use of insect meals for broiler feeds with a lower environmental footprint

P. Chantzi[1,2], I. Giannenas[2], E. Bonos[3], E. Antonopoulou[4], C. Athanasiou[5], A. Grigoriadou[6], A. Tzora[3] and I. Skoufos[3]
[1]Aristotle University of Thessaloniki, Department of Physical & Environmental Geography, Thessaloniki, 54124, Greece, [2]Aristotle University of Thessaloniki, School of Veterinary Medicine, Thessaloniki, 54124, Greece, [3]University of Ioannina, Department of Agriculture, Arta, 47100, Greece, [4]Aristotle University of Thessaloniki, Department of Biology, Thessaloniki, 54124, Greece, [5]University of Thessaly, Department of Agriculture, Plant Production and Rural Development, Volos, 38446, Greece, [6]HAO-Demeter, Institute of Plant Breeding and Genetic Resources, Thessaloniki, 57001, Greece; ebonos@uoi.gr

Recent awareness has led to the transition to alternative sources of protein as feeds and foods. The use of insects in animal nutrition is possible environmentally-friendly approach as insects: (1) do not have a high demand of energy to maintain their body temperature; (2) have a high rate of food conversion into edible meal protein; (3) have great potential of reuse food waste; (4) have lower land use and water requirements than traditional systems. There are different models to study the environmental footprint of insect use such as the Life Cycle Assessment (LCA) of carbon footprint estimation. This initial study evaluated whether the environmental footprint of insect meal production for feed use in broiler systems was improved compared the production of common feeds such as soybean meal. Also, a short literature review on GWP values of broiler, insect, and soybean meal production based on previous LCA studies was conducted to create a dataset for further analysis. LCA results were presented as kg CO_2-eq per kg protein and the FU was nutrient based. The first step was to estimate the GWP of feed ingredients. The mean ratio of soya meal in the basal diet was about 30% and the insect replacement was about 5-10%. The FCR factor was about 1.67. Preliminary data analysis provided a wide range of GWP values without clear evidence yet that GWP of insect protein in broiler systems could result in a reduced environmental footprint. Acknowledgments: Research was co-financed by Greece and EU in context 'Research–Create–Innovate' within the Operational Program Competitiveness, Entrepreneurship and Innovation of the NSRF 2014-2020, Project Code: T2EΔK-02356. Acronym: InsectFeedAroma.

Session 21 Poster 18

Diet mediated PUFA accumulation and ώ6-ώ3 ratio improvement in mealworm larvae (*Tenebrio molitor*)

S. Yoon, C. Ricciardi, C. Oundjian, M. Leheup, T. Lefebvre and F. Peyrichou
Ynsect, 1, rue Pierre Fontaine – Genopole Campus 3 – Batiment 2, 91000 Evry Courcouronnes, France; carmelo.ricciardi@ynsect.com

Edible insects being recognized as a promising alternative ingredient for both animal and human consumption, being able to adjust the final composition of insects through their nutrition may open new products and markets opportunities. As one of the current challenge in human nutrition is the supply of polyunsaturated fatty acids (PUFA), the aim of this study is to optimize the ω6/ω3 ratio of Yellow Mealworm by substituting fixed percentages (5% or 10%) of the standard production diet with oilseeds: chia seeds, flax seeds, rape seeds, soybean, hemp seeds. 9 diets were formulated and tested for 25 days during the exponential growth phase of the larvae until harvest. The results showed that there is a positive correlation between the nutritional composition of the diets and the body composition of larvae, without negatively impacting the survival, growth rate and feed conversion ratio. The PUFA content of the larvae was positively correlated to the PUFA content of the diets. Larvae fed with the standard diet had a ω6/ω3 ratio of 19.07. Chia and flax seeds had the best ω6/ω3 (0.33 and 0.26) among oilseeds. Larvae fed with the diet substituted by 5 and 10% of chia seeds obtained a omega 6 to 3 ratio of 1.4 and 0.9 respectively, approaching the 1:1 ratio, beneficial to human health. By 5 and 10% flaxseed substitution, larvae obtained ratio of 1.6 and 1.0. In addition, rearing performance was positively improved by the addition of those two seeds, compared to the standard diet. It would therefore seem that it is possible to modulate the fatty acid profile of the larvae by accordingly formulate the diets, while improving the production system.

Session 21 Poster 19

Agro-industry by-products for *Tenebrio molitor* rearing to generate a new aquaculture feed ingredient

I. Vieira[1], D. Murta[2] and M.V. Santos[1]
[1]Thunder Foods, Santarém, 2005-332, Portugal, [2]EntoGreen, Santarém, 2005-079, Portugal; marisa.santos@thunderfoods.pt

The progressive increase in world population and animal protein consumption is propelling the growth of animal food production and consequently, the production of compound feeds. For aquaculture, feed formulations, composed of fishmeal, fish oil, soybean and various types of seeds and grains, represent around 60-70% of the production costs and present several sustainability challenges. This situation boosts the search for sustainable alternatives to conventional nutritional sources, essential to support the growth of aquaculture sector. Insects are a possibility, being suitable for large scale production, with positive environmental impact, high feed conversion efficiency and functional properties. Research on this field has recently demonstrated that insect larvae can improve the health of animals, which makes them a promising ingredient for dietary supplementation, with potential to replace the conventional proteins. Within the insect species, *Tenebrio molitor* usually produced using dry grain flours, presents high nutritional quality. The use of other agri-food by-products in insect production could contribute to develop a circular economy approach, encompassing production, consumption and by-products management, which could lead to the development of a secondary market for new raw materials. To test this approach, different agro-industrial by-products, such as olive pomace and beer bagasse, were tested with levels of incorporation varying from 50 to 100%. Positive results were obtained for both by-products used, with surprising production parameters observed for 50 and 60% olive pomace, in the preliminary assays. Considering the water content above 80% in olive pomace and that *T. molitor* is highly impacted by moisture conditions, these results are very promising. The best conditions are being scaled-up to an industrial production environment, to assess the viability of using these agro-industrial by-products for the production of new protein feed ingredients for aquaculture. Acknowledgments: The InFishMix project (PT-INNOVATION-0094) is funded by Iceland, Liechtenstein and Norway through the EEA and Norway grants.

Session 21 Poster 20

Transfer of veterinary drug residues from substrate to black soldier fly larvae and frass

K.C.W. Van Dongen, E. De Lange, L. Van Asseldonk and H.J. Van Der Fels-Klerx
Wageningen Food Safety Research, Akkermaalsbos 2, 6708 WB Wageningen, the Netherlands; katja.vandongen@wur.nl

Insect production for food or feed is increasing in Europe, and is expected to further increase in the near future. To evaluate food and feed safety aspects, quantitative information on transfer of possible contaminants from substrates to larvae is needed, especially when low quality organic substrates or residual streams (e.g. manure) are (to be) applied as substrate to contribute to a more sustainable food system. Veterinary drugs applied to e.g. chicken and pigs, which are regularly detected in manure, were included in this study and evaluated for their transfer to black soldier fly larvae (*Hermetia illucens*). These included three different antibiotics (enrofloxacin, oxytetracycline, sulfamethoxazole), three coccidiostats (narasin, salinomycin, toltrazuril) and one antiparasitic drug (eprinomectin). The veterinary drugs were spiked to the substrate in final concentrations ranging from 0.5 and 5 mg/kg or 5 and 50 mg/kg (depending on the particular drug residue). Black soldier fly larvae were reared for one week on the spiked substrates and the transfer of the veterinary drugs to the respective larvae and frass was quantified using LC-MS/MS. Transfer of the spiked veterinary drugs to the larvae, relative to the spiked amount, was on average 9.5% for narasin, 3.9% for salinomycin, 4.2% for toltrazuril, 12% for enrofloxacin, 19.2% for oxytetracycline, 0.2% for sulfamethoxazole and 8.1% for eprinomectin. Quantified concentrations were compared to the most relevant and available maximum residue limit (MRL). Mass-balance calculations showed that the larvae seem to metabolize the spiked veterinary drugs at least to some extent. As compared to the control, exposure to eprinomectin affected the growth of the larvae significantly, while other veterinary drugs did not affect larvae growth. In conclusion, differences in the transfer of the tested veterinary drugs were observed and their presence in substrates used for insect rearing requires evaluation for feed or food safety purposes as well as for insect growth.

Session 22 Theatre 1

Drinking heated water improves the growth performance and rumen microbiota of beef cattle in winter

T. He, S. Long, X. Wang, T. Liu and Z. Chen
China Agricultural University, College of Animal Science and Technology, No. 2, Yuanmingyuan West Road, Haidian District, Beijing, 100194, Beijing, China, P.R.; hetengfei@cau.edu.cn

Drinking heated water is essential for resisting cold stimulation and improving the growth performance of beef cattle in winter. The effects of drinking different water temperature on growth performance, serum stress hormones, rumen microbiota, and nutrient digestibility of fattening beef cattle was conducted in winter. Fifty Simmental fattening bulls with an average (± SD) body weight of 647±7.25 kg were randomly assigned according to body weight to five drinking water temperature treatments: 4.39±2.54, 10.6±1.29, 18.6±1.52, 26.3±1.74 and 32.5±1.28 °C. The trial lasted for 60 days with an average environmental temperature of 5.65±3.68 °C. Results showed that increasing drinking water temperature linearly and quadratically increased the average daily gain (ADG) and apparent digestibility of neutral detergent fibre (NDF) ($P<0.01$). A quadratic linear-parabolic model revealed that the maximum ADG of 1.23 kg/d ($R^2=0.84$) was obtained at a drinking water temperature of 29.0 °C. The 26.3±1.74 °C group significantly reduced ($P<0.05$) serum cortisol, insulin, and creatinine concentration, enhanced ($P<0.05$) the mRNA expression levels of ZO-1, Occludin, and Claudin-4 in the rumen epithelium, and increased ($P<0.05$) the concentrations of propionate and total volatile fatty acids, as well as the relative abundances of phylum Bacteroidetes and genus *Prevotella*, whereas decreased ($P<0.05$) the relative abundances of genus *Streptococcus* in the rumen fluid compared to the 4.39±2.54 °C group. In addition, the net meat rate trended to increase ($P=0.07$) and the drip loss of longissimus dorsi muscle trended to decrease ($P=0.09$) in the 26.3±1.74 °C group compared to the 4.39±2.54 °C group. Therefore, using heated drinking water at an appropriate temperature (29 °C) in winter is recommended to improve the rumen microbiota composition and function, increase NDF digestibility, and enhance the anti-cold stress ability and growth performance of fattening beef cattle.

Session 22 Theatre 2

Elevated temperature induces transcriptional and epigenetic changes in the lungs of chickens

D. Schokker[1], J. De Vos[2], P.B. Stege[1], O. Madsen[2], H.J. Wijnen[3], S.K. Kar[4] and J.M.J. Rebel[1]
[1]Wageningen Bioveterinary Research, Houtribweg 39, 8221 RA Lelystad, the Netherlands, [2]Wageningen University, Droevendaalsesteeg 4, 6708 PB Wageningen, the Netherlands, [3]HatchTech, Innovatielaan 3, 6745 XW De Klomp, the Netherlands, [4]Wageningen Livestock Research, De Elst 1, 6708 WD Wageningen, the Netherlands; dirkjan.schokker@wur.nl

Health and resilience against respiratory diseases are important features for broiler chicken. In this study, epigenetic and transcriptomic changes in the lungs of broiler chickens of different ages during rearing that were either exposed to elevated egg shell temperature (HIGH) of 38.9 °C during mid-incubation or normal egg shell temperature (control; CON). The objective was to better understand how environmental challenges, such as heat stress during egg incubation, affect the development of the immune system and health of broiler chicken at later age. To this end we generated both epigenetic and transcriptomic data of lung tissue of elevated HIGH and CON chicken, furthermore these chicken were challenged by introducing either an infectious *E. coli* or an IBV vaccination to monitor the respiratory response. Thousands of differential methylated sites were observed at days 15 and 33, when comparing HIGH vs CON. Pathway enrichment analysis of HIGH vs CON showed that differentially expressed genes were mainly involved in cilium, cytoskeleton, and immune processes. These findings provide insight into the underlying biological mechanisms of early life conditions, like elevated EST, and their potential role in health of broilers.

Session 22 Theatre 3

Supplementing heat stressed cows with plant extract affects performance welfare and oxidative stress

J.R. Daddam[1], D. Daniel[2], I. Pelech[3], G. Kra[1,2], H. Kamer[1], U. Moallem[1], Y. Lavon[4] and M. Zachut[1]
[1]Volcani Institute, Ruminant Science, Rishon LeZion, 7528809, Israel, [2]the Robert H. Smith Faculty of Agriculture, Food and Environment, the Hebrew University, Rehovot, 76100001, Israel, [3]Extension Service, Ministry of Agriculture, Rishon Lezion, 7528809, Israel, [4]Israeli Cattle Board Association, Caesarea, 38900, Israel; mayak@volcani.agri.gov.il

We examined the effects of a supplement comprised of plant polyphenols extracts of green tea, capsicum and fenugreek, and electrolytes [(Na+, K+), AXT; Axion ThermoPlus, CCPA, France] during summer heat load on production, welfare, and on oxidative stress proteins in adipose tissue (AT) of dairy cows. Forty-two multiparous mid-lactation cows were divided into 3 groups during summer, and were fed for 2 wks either a standard milking cows' diet (CTL, n=14), or supplemented with 100 g/d of AXT (100AXT, n=14), or 150 g/d of AXT (150AXT, n=14), while being cooled 5 times a day; then, half of the cows from each treatment were cooled (CL) or not cooled (NCL) for 2 wks, after which the CL/NCL were switched for additional 2 wks. Cows were milked 3 times a day and milk composition was analysed at the end of each period. Vaginal temperature (VT) was measured by sensors in each period. Biopsies of subcutaneous AT were taken from 10 NCL (five CTL and five 150AXT) at the end of the period, and examined by proteomics analysis. Data were analysed with PROC MIXED of SAS; the model included the effects of treatment, cooling, and interactions. Proteomics was analysed by t-test. Milk and 4% FCM were higher in 100AXT than in CTL. DMI was higher in 100AXT than in CTL. The effect of cooling was significant for DMI, FCM 4% and Milk/DMI. The percentage of hours that VT was >39 °C was lower in 100AXT and 150AXT than in CTL. Daily rumination time was higher in 150AXT vs CTL, and lying time was increased in 100AXT and 150AXT vs CTL. Proteomics demonstrated increased abundances of peroxidase, microsomal-glutathione-S-transferase-2 and heme-oxygenase-1 in 150AXT vs CTL; the Nrf2-mediated oxidative stress response was enriched in 150AXT. AXT during heat load increased DMI and production, lowered VT, improved welfare indices, and affected the Nrf2-oxidaitve stress response.

Immunomodulatory effect of exosomes from the plasma of heat-stressed cows on bovine monocytes

L.G. De Matos[1], S. Dimauro[1], M. Falco[1], J.F.S. Filipe[1], C. Zamboni[1], F. Ceciliani[1], V. Martini[1], A. Agazzi[1], A. Scarafoni[2], G.C. Heinzl[2], G. Sala[1], A. Boccardo[1], A. Maggiolino[3], P. Depalo[3] and C. Lecchi[1]
[1]Università degli Studi di Milano, Department of Veterinary Medicine and Animal Science, Via dell'Unisità, 6, 26900 Lodi, Italy, [2]Università degli Studi di Milano, Department of Food, Environmental and Nutritional Sciences, via Celoria 2, 20133 Milan, Italy, [3]Università degli Studi di Bari A. Moro, Department of Veterinary Medicine, SP per Casamassima, km 3, 70010, Valenzano, Italy; luiz.dematos@unimi.it

The impact of climate change is expected to negatively affect human and animal health. From the one health point of view, hyperthermia induces impairment of immune cell functions, increasing the risk of infection, morbidity, and mortality, and represents one of the main issues in the livestock industry. Exosomes are nano-sized extracellular vesicles (30-150 nm) involved in cell-to-cell communication, and regulation of immune response and found in many biological fluids, delivering several molecules released into target cells, including immune cells, inducing a change in cellular response. This study aimed to assess the immunomodulatory effects of plasma exosomes purified from cows exposed to thermal comfort conditions (TC, THI<68) and after a natural four-day (heatwave hyperthermia, HS, THImax=84, THImin=69) on bovine monocytes (CD14+). Exosomes were isolated from the blood of three healthy multiparous Holstein cows during TC and HS. Monocytes (CD14+) were isolated from the blood of 6 clinically healthy multiparous Holstein cows during TC, sorted by Magnetic Activated Cell Sorting (MACS), and cocultured with exosomes (200 exosomes/cell) for 20 h at 37 °C and 5% CO_2. To test the influence of exosomes on the immune activity of bovine monocytes, viability, apoptosis, phagocytosis, and chemotaxis assays were performed. Results demonstrated that the phagocytosis ability significantly decreased (P=0.008) in monocytes cultured with HS- compared to TC-exosomes, while exosomes did not affect viability, apoptosis, and chemotaxis of bovine CD14+ monocytes. In conclusion, the exosomes released into the bloodstream during HS significantly impaired monocytes' phagocytosis, exposing cows to a greater risk of developing infections.

Late gestation heat stress: programming effects on three generations

J. Laporta
University of Wisconsin, Department of Animal and Dairy Sciences, 1675 Observatory Dr, 53706, USA; jlaporta@wisc.edu

In dairy cows, the dry period is a 6 to 8-week non-lactating phase between two subsequent lactations intended to regenerate mammary epithelial senescent cells with new ones for the next lactation. The dry period also overlaps with a portion of the last trimester of gestation, a period of exponential foetal growth. Exposure to environmental elevated temperature and humidity leading to heat stress during this period can impact three generations: (1) the dam (F0); (2) the developing daughter (F1), and 3) her germline (future F2). We have previously shown that dry-period heat stress disrupts the highly coordinated cellular processes during mammary gland involution and redevelopment phases (i.e. cellular turnover) and derails the expression of critical genes and pathways (i.e. ductal branching morphogenesis and cell death), resulting in lower milk production during the dam's subsequent lactation. Intrauterine hyperthermia also affects offspring growth, physiology, and performance into adulthood. Experiencing intrauterine hyperthermia for several consecutive weeks affects many tissues, cells, and entire organ systems vital for productivity and maintaining physiological functions. This presentation will highlight the phenotypical, histological, and molecular adaptations of the daughter's mammary gland to in-utero heat stress. We will also discuss the impact of in-utero heat stress on other tissues and organs, including the adrenal glands and the ovaries. Last, carry-over effects on the granddaughters' survival and milk production will be presented. This presentation will provide insights into the molecular underpinnings of disrupted organ structure and function arising from prior exposure to heat stress, leading to lower milk yields for at least three generations.

Session 22 — Theatre 6

Temporal patterns of environmental heat stress in Holstein heifers: association with age at calving

V. Tsiamadis[1], G.E. Valergakis[1], A. Soufleri[1], G. Arsenos[1], G. Banos[1,2] and X. Karamanlis[3]

[1]Faculty of Veterinary Medicine, Aristotle University of Thessaloniki, Laboratory of Animal Husbandry, Box 393, 54124, Thessaloniki, Greece, [2]Scotland's Rural College, Roslin Institute Building, University of Edinburgh, Easter Bush Campus, EH25 9RG Midlothian, United Kingdom, [3]Faculty of Veterinary Medicine, Aristotle University of Thessaloniki, Laboratory of Ecology and Environmental Protection, Box 393, 54124, Thessaloniki, Greece; asoufler@vet.auth.gr

A retrospective study was conducted to evaluate temporal patterns of environmental heat stress (EHS) during in-uterus (IUD) and 3-month post-natal (PN) period of dairy heifers as well as to estimate their association with age at first calving (AFC). Data from 30 dairy herds in Northern Greece including 9,098 heifers were extracted from National Cattle Database. Moreover, 230,100 farm-specific daily temperature and relative humidity records, were obtained from ERA5-Land during 2005-2019. Average monthly THIs were calculated and matched for each heifer to their IUD and PN (IUDM and PN, respectively). They were categorized as: low ≤68 and high>68, and then entered in Hierarchical and Two-Step Cluster Analysis as predictors to allocate heifers to THI clusters. The association of clusters with AFC (in days) was assessed with linear regression (LR) analysis. The model included the fixed effects of herd, age of dam at conception (4 levels, age quartile 1: 292-537, 2: 537-880, 3: 881-1,375 and 4: ≥1,376 days), the THI cluster category of each heifer, and adjusted for the year of heifers' conception (16 levels: 2005-2019). The association of clusters with the risk of AFC>787 d (AFC median) was assessed with binary LR (logit function). Analyses were performed with SPSS 25.0. Eight clusters (HSC-1 to -8) were identified. Compared to HSC-8 (8th-9th IUDM and 1st PNM), heifers of HSC-2 (2nd IUDM and 2nd-3rd PNM), HSC-3 (2nd-3rd IUDM and 3rd PNM), HSC-4 (2nd-5th IUNM), HSC-5 (4th-7th IUDM) and HSC-6 (6th-8th IUDM) calved 14.5, 8.4, 6.4, 13.8, and 17.8 days later, respectively ($P<0.01$-0.001). Furthermore, heifers of HSC-5 and HSC-6 had 1.15 ($P<0.01$) and 1.34 ($P<0.001$) higher risk of an AFC>787 d compared to HSC-8, respectively. Heifers' exposure to THIs>68 during IUD and PN was associated with higher AFC.

Session 22 — Theatre 7

Effect of stocking density on heat stress in fattening pigs

L. De Prekel[1], D. Maes[1], A. Van Den Broeke[2] and M. Aluwé[2]

[1]University Ghent, Unit of Porcine Health Management, Salisburylaan 133, 9820 Merelbeke, Belgium, [2]ILVO, Scheldeweg 68, 9090 Melle, Belgium; lotte.deprekel@ugent.be

Lower stocking densities (SD) may protect fattening pigs from the adverse effects of heat stress (HS) through improved radiant and convective heat losses due to the ability to avoid direct contact with pen mates and greater body-to-floor contact. In the present study, we investigated whether a reduced SD can decrease the effects of HS in fattening pigs during higher heat loads. One hundred and fifty fattening pigs (crossbred Piétrain sire × hybrid sow) were randomly divided into three treatment groups: MIN (minimal SD of 1.3 m^2/pig, n=12 pens), MED (medium SD of 1.0 m^2/pig, n=11 pens) and MAX (maximum SD of 0.8 m^2/pig, n=6 pens). All pens had partially slatted floors with a total pen surface of 4.88 m^2, corresponding with 4, 5 and 6 fattening pigs in the MIN, MED and MAX pens, respectively. An artificial heat wave was induced for seven days when the pigs were 21 weeks of age. In the week before (pre-heat), during (heat) and after the heat wave (post-heat), respiration rate (RR) and rectal temperature (T_{rectal}) were observed on four observation moments per period. Individual animal weight and feed intake per pen were measured weekly, to calculate average daily gain (ADG) and feed intake (ADFI). During the heat weave, the average temperature-humidity index was >75.6, corresponding with a mild heat wave and a warning for HS. There was no effect of the interaction of SD and period or effect of SD on RR or T_{rectal}. However, independent of SD, RR significantly increased with ±20 bpm and T_{rectal} with 0.1 °C during the heat period ($P<0.001$). A significant difference in ADG between the SD was found ($P=0.035$), regardless of period. The ADG of the MIN group was 9% higher than the MED group, possibly due to decreased stress levels as ADFI did not differed between SD. There was no significant SD or SD and period interaction on ADG and ADFI, but ADFI decreased with 15% ($P<0.001$) and ADG with 19% ($P<0.001$) in all groups during the heat period. In conclusion, a lower SD (1.3 m^2/pig) positively affected ADG but did not ameliorate the adverse effects of HS on physiological and performance parameters.

Session 22 Theatre 8

Chromium yeast alleviates heat stress in Holstein mid-lactation dairy cows

F. Ma, Y. Wo, Q. Shan and P. Sun
Institute of Animal Science, Chinese Academy of Agricultural Sciences, No. 2 Yuanmingyuan West Road, Haidian District, 100193, Beijing, China, P.R.; sunpeng02@caas.cn

Two experiments were conducted to identify the optimal dose of CY in the diet and the metabolic mechanisms whereby CY supplementation alleviates the negative effects of heat stress in mid-lactation dairy cows. Cows were experiencing heat stress as the average temperature-humidity index (THI) was greater than 72 in both experiments. In Experiment 1, twenty-four healthy Chinese Holstein mid-lactation dairy cows receiving the same basal diet containing 0.11 mg Cr/kg of DM were divided into 2 blocks according to milk yield (block 1 and block 2 for low- and high-producing cows). Cows of block 1 or block 2 each were randomly allocated to four treatments: a negative control group (without CY supplementation, CON), and groups that received CY at 0.18, 0.36, and 0.54 mg Cr/kg DM, respectively. The experiment lasted 10 wk, including a pre-feeding period of 2 wk. CY supplementation reduced rectal temperature (RT) and respiration rates (RR), and increased dry matter intake and milk lactose content ($P<0.05$). Supplementation with CY improved the antioxidant and immune function ($P<0.05$). The appropriate dose of CY at 0.36 mg Cr/kg DM is recommended. In Experiment 2, twelve Holstein dairy cows were fed the same basal diet containing 0.09 mg of Cr/kg DM. They were allocated randomly to 2 groups: a control group (CON, without CY supplementation) and a CY group (CY, administered 0.36 mg Cr/kg DM). The experiment was performed over 8 weeks. CY supplementation reduced RT, and increased the lactation performance of the heat-stressed dairy cows ($P<0.05$). Supplementation with CY increased the serum glucose and thyroxine concentrations, but reduced the urea nitrogen, insulin, and triiodothyronine concentrations on d 56 ($P<0.05$). Specifically, plasma concentration of nicotinamide was higher after CY supplementation ($P<0.05$). In conclusion, CY supplementation reduces RT, influences metabolism by reducing serum insulin concentration and increasing serum glucose and plasma nicotinamide concentrations, and finally increases lactation performance of heat-stressed dairy cows.

Session 22 Theatre 9

May anti-oxidants reduce glyphosate-based herbicides adverse effects on chicken embryo-development?

M. Fréville, A. Estienne, C. Ramé, C. Rat, J. Delaveau, P. Froment and J. Dupont
INRAe (Institut national de recherche pour l'agriculture, l'alimentation et l'environnement), UMR PRC, 37380 Nouzilly, France; freville.mathias1@gmail.com

Glyphosate is an active molecule massively used in agriculture worldwide for its herbicide properties. Numerous studies have shown that glyphosate-based herbicides (GBH) exposure can adversely affect metabolism and fertility on animal models. In avian species, our previous works have demonstrated that chronic dietary GBH exposure disturbed hens caecal microbiome, and increased the gizzard weight of animals. These effects were associated with an increase in plasma oxidative stress. GBH exposure also dramatically increased the early and late embryo mortality. The aim of the present study was to further investigate the effects of a GBH exposure on chicken embryonic development. We injected or not (non injected) growing doses of GBH (i.e. 0.03 ng/g; 0.3 ng/g; 3 ng/g and 30 ng/g of glyphosate equivalent) or PBS (control) in E13 (i.e. on their 13th day of development) eggs and weighted the organs of the subsequent E17 embryos. Surprisingly, the hearts of the embryos exposed to the weakest dose (0.03 ng of glyphosate equivalent/g egg) were significantly heavier as compared to the other embryos (i.e. all other doses and controls). By using CAM (chorioallantoic membrane) assay, we also observed that the weakest dose of GBH (0.03 ng of glyphosate equivalent /g egg) reduced significantly the neovascularization in E10 eggs. We then performed primary exposed primary cardiac fibroblasts and Gizzard Smooth Muscle Cells cultures from chicken embryos and exposed them to GBH concentrations ranging from 0.03 to 30 ng of glyphosate equivalent/g egg. We observed that the highest dose (30 ng of glyphosate equivalent/g egg) significantly reduced the cell viability and proliferation of gizzard and heart primary cells. We are now investigating the effect of tocopherol or anti-oxidant grape seed extracts on the potential reversibility or attenuation of the negative effect of GBH on the chicken embryo development.

Session 22 Theatre 10

Bromide as a thyroid hormone disruptor

H.L. Lucht and N.H. Casey
University of Pretoria, Animal Science, Private Bag X20, 0028, Hatfield, South Africa; norman.casey@up.ac.za

Literature, which included historical and current research to provide an overview of baseline knowledge and what has since been reported, focused on the potential action of inorganic bromide (Br-) as an endocrine disrupting chemical (EDC). Many substances have been assigned a no observed adverse effect level (NOAEL) but there was none scientifically established for bromide (Br-) until the suggested NOAEL of 0.01 mg/l was validated in 2019. Br- with a concentration even mildly above the (NOAEL) has the potential to be a TDC by interfering most notably with TH and consequent metabolic processes. Evidence suggests that Br- functions as a TDC that can have an effect even at low concentrations with chronic exposure. Once ingested, Br- moves easily through the body water spaces in the body by means of passive and active transport using chloride (Cl-) channels. The hypothalamus-pituitary-thyroid (HPT) axis regulates TH production by means of a negative feedback loop, and energy homeostasis, growth and development depend on TH. Disruption of TH synthesis by Br- interference would potentially disrupt the negative feedback loop causing compensation mechanisms by the thyroid gland, leading to thyroid hypertrophy to mitigate the resultant hypothyroidism. Toxicity is not a sudden effect unless a critical concentration of a highly metabolically disruptive substance appears suddenly in the body. As the concentration in the body increases through further ingestion, the particular susceptible metabolic pathway becomes more exposed and offers more bonding opportunities until a threshold concentration becomes overwhelming and disrupts the metabolic pathway. The hypothesis is that Br- disrupts the deiodination process within the body and prevents the conversion of inactive T4 to the active form, T3. Br- can be considered to be a TDC when outcompeting I- and hampering T4 production in the thyroid gland, decreasing T3 levels in the body and thus disrupting the regulation of TH production in the HPT axis. The synthesis of iodinated thyronines diminishes as Br concentration rises. The implication for livestock production is that the compromised metabolic functioning of the animal will negatively affect milk production, egg production, and protein accretion of animals raised for meat.

Session 22 Theatre 11

The impact of THI on milk characteristics, productivity, and metabolism in lactating dairy cows

G. Meli, V. Fumo, G. Savoini and G. Invernizzi
Università degli studi di Milano, DIVAS, Via dell'Università 6, 26900, Italy; giovanna.meli@unimi.it

The aim of the study was to evaluate the impact of temperature humidity index (THI) on metabolic biomarkers, milk production and milk characteristics in lactating dairy cows. Ten mid lactation (139±40 DIM; average milk production 28.33±5.43 kg/day) primiparous and pluriparous Holstein dairy cows were selected and were housed in a free-stall barn at the Experimental Zootechnical Centre of the Università degli Studi di Milano, in Lodi, Italy. All the experimental animals were fed the same basal diet as total mixed ration. The trial started in May 2019 and lasted for 4 weeks. Temperature and humidity values were continuously recorded and a daily THI was calculated. The threshold for heat stress was considered when THI was above 68: no heat stress (NHS) =THI<68 vs heat stress (HS) =THI>68. Blood samples were collected at day 0 and 28 and milk samples every two weeks. Milk yield was recorded weekly and feed intake was recorded daily. Collected data were analysed by mixed procedure of SAS. Heat stress significantly ($P<0.05$) decreased dry matter intake (DMI; HS= 21.15±1.33 vs NHS= 23.27±1.31 kg/day), milk yield (HS= 25.54±2.15 vs NHS 28.86±2.15 kg/day), protein yield (HS= 0.87±0.07 vs NHS= 1.0±0.07 kg/day) and lactose content (HS= 5.03±0.04 vs NHS 5.1±0.04 g/100 ml). Long chain fatty acids increased significantly (HS= 27.61±0.84 vs NHS= 25.36±0.84 g/100 ml; $P<0.05$) and medium chain fatty acids tended to increase ($0.05<P<0.1$). Plasma glucose decreased significantly (HS= 3.63±0.09 vs NHS= 3.91±0.09; $P<0.05$). DMI reduction could have affected blood glucose concentration and subsequent conversion to lactose in the mammary gland. In conclusion, heat stress negatively impacted DMI, milk yield, milk characteristics and caused several physiological changes in lactating animals. Acknowledgments This study was carried out within the Agritech National Research Center and received funding from the European Union Next-GenerationEU (Piano Nazionale di Ripresa e Resilienza (Pnrr) – Missione 4 Componente 2, INVESTIMENTO 1.4 – D.D. 1032 17/06/2022, CN00000022). This manuscript reflects only the authors' views and opinions, neither the European Union nor the European Commission can be considered responsible for them.

Session 22 Poster 12

A selected phytogenic solution improves performance and thermal tolerance of heat stressed pigs

A. Morales[1], P. Sakkas[2], M. Soto[1], N. Arce[1], N. Quilichini[2] and M. Cervantes[1]
[1]ICA, Universidad Autónoma de Baja California, 21100 Mexicali, Mexico, [2]DELTAVIT, CCPA Group, Z.A. du Bois de Teillay, 35150 Janzé, France; nquilichini@ccpa.com

Pig exposure to high ambient temperature (AT) penalises their performance. Dietary supplementation with phytogenic solutions containing specific plant secondary metabolites, may lower body temperature (BT) and increase the upper critical temperature threshold level, thus improving thermal tolerance. Herein, we investigated the effects of offering a phytogenic solution based on selected and characterized plant extracts including Capsicum spp. (PHY; 0.2%) on performance and thermal tolerance of heat-stressed (HS) pigs. Forty-two individually housed pigs (BW 27±3 kg) were randomly allotted to three treatments: TN-C, thermoneutral (TN, 22±2 °C) pigs fed a control diet based on wheat and soybean meal; HS-C and HS-PHY, pigs housed under HS conditions (29-36 °C) fed control diet without or with PHY, respectively. Following 8 days of adaptation, all pigs were fed *ad libitum* for 8 days (twice daily, 07:00 and 19:00). Thermographs were implanted subcutaneously in five pigs per treatment to record BT every 5 min. Respiration rates were assessed in the morning (07:00) and in the afternoon (17:00). Performance, respiration rates and average BT following morning and evening feedings were analysed using the Statistix software. The TN-C pigs had increased ADG and ADFI in comparison with HS-C (P<0.01) and HS-PHY pigs (P<0.05), and reduced FCR in comparison to HS-C pigs (P<0.01). However, HS-PHY pigs had greater ADFI (P<0.01), ADG (P<0.05) and tended to have improved FCR than HS-C pigs (P=0.07). Respiration rates were higher for HS pigs than TN-C during the afternoon (P<0.01), when AT was higher, while additive supplementation had no effect. Higher BT was observed in HS pigs compared to TN pigs (P<0.01), while the BT increment observed after the evening meal (from 20:00 to 03:00 next day) in HS-PHY pigs was smaller than that of HS-C pigs (P<0.05). In conclusion, dietary phytogenic supplementation partially alleviated the adverse effects of HS on pig performance and reduced BT after the afternoon feeding. Further research should reveal whether observed effects of dietary PHY supplementation are associated with improved intestinal health.

Session 22 Poster 13

Effects of Met, Lys and His supplementation on the metabolic response to heat stress in dairy cows

Á. Kenéz[1], E. Jorge-Smeding[1,2], Y.H. Leung[1], A. Ruíz-González[3,4] and D.E. Rico[3]
[1]City University of Hong Kong, Department of Infectious Diseases and Public Health, Hong Kong, China, P.R., [2]Universidad de la República, DPAP, FAGRO, Montevideo, Uruguay, [3]Université Laval, Quebec, QC, Canada, [4]Centre de Recherche en Sciences Animales de Deschambault, Deschambault, QC, Canada; akos.kenez@cityu.edu.hk

Low-protein diets were shown to decrease heat production, thus providing a potential avenue to alleviate metabolic effects triggered by heat stress (HS). However, Met, Lys and His must be supplemented according to requirements as they are the common limiting AA. Twelve lactating Holstein cows (primiparous, P, n=6; multiparous, M, n=6; 83±28 DIM) were enrolled in a Latin square design with 14-days treatment periods: Heat stress (HS; maximal THI 84; 17% crude protein (CP); 1,715 metabolizable protein (MP), 107 Lys, 34 Met, and 37 His (g/d)); pair feeding in thermo-neutrality (TN; maximal THI 64; CP, MP and AAs supply equal to HS); and HS with a balanced AA diet (HS+AA; maximal THI 84; 17% CP, 1,730 g/d MP, 178 Lys, 64 Met, and 43 His (g/d)). Blood plasma was sampled on the last day of each period and analysed with the AbsoluteIDQ p400 metabolomics assay (Biocrates, Innsbruck, Austria). ANOVA was used to evaluate the treatment (TN, HS, HS+AA), parity (P, M) and their interaction as fixed effects, and the cow as random effect. P-values were adjusted by false discovery rate (FDR) correction. A total of 15 plasma metabolites differed (FDR≤0.048) between treatments but no metabolite was affected by the interaction between treatment and parity. Several of them increased in HS+AA compared with TN and HS (including α-aminoadipic acid, Met, Lys, Val, Phe, Trp and sarcosine). Further, some lipid compounds had lower concentrations in HS than in TN but intermediate concentrations in HS+AA (including phosphatidylcholine (PC) 40:7, PC 40:9, PC-O 34:1, PC-O 40:6, and diglyceride (DG) 39:0). Our results suggest that a relatively greater availability of Met in HS+AA stimulated PC synthesis, recovering it to similar levels to TN. Increased Lys availability in HS+AA likely led to increased oxidation of Lys and to lower catabolism of other essential AA (such as Val, Phe, Trp). The potential signalling function of the affected lipids will be further investigated.

Session 22 Poster 14

Evaluation of earwax and hair cortisol level in relation to environmental stressors in Hanwoo cattle

M. Ataallahi, G.W. Park, E. Nugrahaeningtyas, M. Dehghani, J.S. Lee and K.H. Park
Kangwon National University, College of Animal Life Sciences, Chuncheon, Gangwon, Republic of Korea, 24341, Korea, South; ataallahim@kangwon.ac.kr

Recently, global warming issues in the cattle industry received more attention. The severity of heat stress on cattle can be evaluated through the measurement of cortisol secretion. In this study earwax cortisol concentration (ECC) and hair cortisol concentration (HCC) of native Korean cattle (Hanwoo) were compared to evaluate their usefulness in reflecting the chronic stress from environmental temperature and humidity. Earwax samples (n=70) were collected from the left ear and hair samples (n=70) were collected from the forehead of each cattle with an age ranging from 1 to 72 months in the livestock research farm at the Kangwon National University from April to July 2022. Temperature-humidity index was calculated based on microclimate data from the Korea Meteorological Administration. The ECC ranged from 7.1 to 55.3 pg/mg (average: 29.5±11.1), while the HCC ranged from 1.1 to 25.7 pg/mg (average: 6.7±4.5) via enzyme immunoassay. The ECC was higher in July than April, May, and June ($P<0.05$), whilst it was lower in June in comparison to April and May ($P<0.05$). The HCC was higher in July than May ($P<0.05$). However, the comparison of HCC was not different between April, May, and June, nor between July with April and June ($P>0.05$). Overall, the concentration of cortisol in earwax and hair can be suitable indicators of long-term heat stress in Hanwoo. However, factors such as earwax buildup and earwax collection procedure may affect the production rate of earwax that eventuates inaccuracy of earwax cortisol measurement as a reliable chronic stress indicator.

Session 22 Poster 15

Effects of energy levels and heat stress on growth performance and blood metabolites in Hanwoo calve

Y.H. Jo[1], J.G. Nejad[1], J.S. Lee[1], K.K. Park[1], E.J. Kim[2] and H.G. Lee[1]
[1]Konkuk university, Animal Science & Technology, 120, Neungdong-ro, Gwangjin-gu, Seoul, Republic of Korea, 05029, Korea, South, [2]Kyungpook National University, Department of Animal Science and Biotechnology, 2559, Gyeongsang-daero, Sangju-si, Gyeongsangbuk-do, Republic of Korea, 37224, Korea, South; hglee66@konkuk.ac.kr

This study investigated the effects of dietary energy levels on growth, blood metabolites, and expression of heat shock proteins in Hanwoo calves subjected to heat stress (HS). Twenty-four calves (BW: 221.5±24.9 kg; age: 162±4.8 d) were randomly housed in climate-controlled chambers using a 3×2 factorial design. There were three energy treatments, including low (LE=2.53), medium (ME=2.63), and high energy levels (HE=2.72 Mcal/kg of DM) and two stress levels (threshold: THI=70-73; severe: THI=89-91). The calves were adapted to 22 °C for 7 days, then to the target THI level for 14 days. Energy intake, average daily gain, and gain to feed ratio were determined to decline ($P<0.05$) under severe HS compared with the threshold. Under severe HS, the rectal temperature was increased by 0.67 °C compared with the threshold. Severe HS increased glycine, ammonia, and 3-methylhistidine levels compared with the threshold ($P<0.05$). Gluconeogenic AAs in the blood were increased among the various energy levels regardless of HS. In PBMCs, the expression of the HSP70 gene was increased in the LE group ($P<0.05$), and the HSP90 gene expression was increased in LE and ME groups ($P<0.05$) under severe HS. However, the expression of genes HSP70 and HSP90 in the HE group did not differ under severe HS ($P>0.05$). HE may mitigate ATP depletion in PBMCs. No differences in growth performance were found when increasing energy intake with high protein (CP 17.5%) under HS. However, the increase in energy levels resulted in increased gluconeogenic AAs but decreased urea and 3-methylhistidine in the blood. In conclusion, increased energy levels are thought to improve HS adaptability by inhibiting muscle degradation and glucose production using gluconeogenic AAs, which may improve the health of calves under HS conditions.

The effect of a summer diet on heat stress in lactating sows and their piglets

A. Van Den Broeke[1], L. De Prekel[2], D. Maes[2] and M. Aluwé[1]
[1]ILVO, Scheldeweg 68, 9090 Melle, Belgium, [2]University Ghent, Unit of Porcine Health Management, Salisburylaan 133, 9820 Merelbeke, Belgium; alice.vandenbroeke@ilvo.vlaanderen.be

Heat stress is an emerging problem in pig farming, especially in lactating sows. High temperatures and relative humidity rates can negatively affect the sow's health, productivity, and reproduction. Feeding strategies can help alleviate the effects of heat stress in lactating sows. Apart from changes in the energy, fat and protein levels of the feed, addition of minerals, vitamins and electrolytes can help to maintain the sow's body condition and milk production during times of heat stress. The aim of this trial was to compare the effect of a standard vs a summer lactation diet supplemented with vitamin E (200 ppm), vitamin C (200 ppm), Selenium (organic formula), Betaine (0.1%) and adjusted electrolyte balance on physiological and performance parameters of 50 lactating sows and their piglets during the summer of 2022. To examine the effect of the diet on physiological parameters affected by heat load, observations were performed at control days (temperature-humidity index (THI) below 75) and at days with a high heat load (THI at least 6 hours above 75). Independent of diet, sows had an increase in respiration rate (68 vs 39 breaths per minutes) and rectal temperature (39.6 vs 38.7 °C) on hot vs control days, but there was no significant difference due to diet. Global effects of diet on reproduction parameters like total born piglets, weaned piglets, mortality rate of piglets were also not observed. At the end of the nursery period however, piglets of the standard group had a higher weight (20.7 vs 20.1 kg, P=0.042), higher growth rate (391 vs 368 g/day, P=0.003) and a better feed conversion ratio (1.33 vs 1.37, P=0.030) compared to piglets of the summer diet group, although the latter group had numerically a higher weight at weaning compared to piglets from standard diet sows. To conclude, this study did not observe effects of a summer diet on reproduction or physiological parameters but lactation diet had a significant impact on the performance of the piglets after weaning.

Physiological implication of omega-3 fatty acid supplementation on animal welfare in laying hens

S.K. Kang
Graduate School of International Agricultural Technology, Seoul National University, 1447, Pyeongchang-daero, Daehwa-myeon, Pyeongchang-gun, Gangwon-do, 25354, Korea, South; kangsk01@snu.ac.kr

In general, the lipid mediators which derived from omega-6 fatty acids enhance inflammatory reaction whereas omega-3 fatty acid-derived lipid mediators relieve it. Thus, the balance of omega-6 and omega-3 fatty acid in animal's body is crucial to maintain immune homeostasis. However, most of modern livestock animals considered to be suffered from the chronic inflammation because of the imbalance in omega-6 to omega-3 fatty acid ratio majorly due to the feeding system depending on omega-6 fatty acid-biased resource such as corn. In this study, it is hypothesized that supplementation of dietary flaxseed, an omega-3 fatty acid-rich feed ingredient, to livestock animals could alleviate the chronic inflammation and stress conditions during their feeding by altering the lipid mediators profile. 33-week-old laying hens in their intensive production period were chosen for the investigation, then lipid mediator profiles, indices of inflammation (serum pro-inflammatory cytokines), and stress indices (corticosterone level and heterophils to lymphocytes ratio) in serum have been monitored after feeding 0, 0.9, 1.8, or 3.6% (w/w) dietary flaxseed for 4 weeks. As results, significant increment of several omega-3 fatty acid-derived lipid mediators, reduction of pro-inflammatory cytokine level such as TNF-α, alleviation of stress indices, and improvement of overall laying performance have been observed. So far, in livestock industry, flaxseed supplementation in feed just has been focused on the production of omega-3 fatty acid-fortified animal products. This study newly suggests that the supplementation of omega-3 fatty acids to livestock animals could be helpful to reduce chronic inflammation and stress during their production period associated with animal welfare issues.

Session 22 Poster 18

The effect of different bedding materials on dust and toxic gas concentrations in calf housing

D. Witkowska, A. Ponieważ and D. Murawska
University of Warmia and Mazury, Oczapowski Str. 5, 10-719 Olsztyn, Poland; dorota.witkowska@uwm.edu.pl

The type and quality of the bedding material is an important aspect of the health status of the calves. One of the most popular bedding materials in many countries is straw. Nevertheless, some research indicate that contamination levels can be significantly higher in facilities using straw, in comparison with alternative bedding materials. Given that calves' immunity builds for a relatively long time and that long-term exposure to aerial contaminations predisposes them to respiratory system health disorders, a given type of litter may improve the results of calves breeding. For this reason, the aim of our study was to determine the effect of classic straw and six alternative bedding materials (light pellets, medium sawdust, peat, chopped straw, flax, hemp) on the levels of dust fraction (particulate matter/PM1, PM2.5, PM4, PM10 and total suspended particulate/TSP) and toxic gas (ammonia/NH_3, hydrogen sulphide/H_2S) concentrations in the air of calf housing. The calves were kept in accordance with the EU and Polish standards in identically equipped and separated boxes (8 calves on each type of bedding). Analyses of contaminations were conducted every day at 6 am (before calves handling) for a period of 2 weeks at 5 locations in each box, using a gas detector (Nanosens DP-24 VET) and gravimetric apparatus (DustScan 3020). The data were normally distributed (Kolmogorov-Smirnov test) and analysed by ANOVA (StatSoft). Considering dust concentration, the most contaminated air (by each fraction, apart from PM1) was registered using traditional straw ($P<0.01$; PM2.5=13.02, PM4=50.48, PM10=285.21, TSP=477.78 µg/m3). In case of PM4-10 and TSP the lowest concentrations ($P<0.01$) characterized light pellet and sawdust. Chopped bedding materials: straw, flax, and hemp, and peat had an intermediate effect. Whereas, in the case of PM1-2.5, pellets had the worst impact on the quality of air ($P<0.01$; PM 1=2.91, PM2.5=11.36 µ/m3). The toxic gas concentrations were the lowest using light pellets ($P<0.01$; NH_3=0.25 ppm; H_2S=0.5 ppm) and the highest in case of sawdust ($P<0.01$; NH_3=1.17 ppm; H_2S=1.29 ppm). It could be concluded that alternative bedding materials, especially pellet, significantly improve air quality and hygiene standards in calf housing.

Session 22 Poster 19

Mitigate density stress by using a proprietary orange essential oil

A.N. Nalovic[1], A.A. Auvray[2] and J.G. Gabarrou[2]
[1]Réseau Cristal, 23 Rue Olivier de Serres, 85500 Les Herbiers, France, [2]Laboratoires Phodé, 8 avenue de la Martelle, 81150 Terssac, France; aauvray@phode.fr

Tail biting in pig farm is a major welfare issue that is also impacting farm profitability. Even when on farm policies are taken to reduce risk factors, uncontrolled variables remain like density stress. The aim of this study was to measure the effect of an antistress functional feed additive on gilt stress level and its impact on tail biting. Four successive batches between 120 and 140 gilts each were monitored from post-weaning to fattening, alternating control and experimental batch every 4 weeks: two of them were control batches while the two others received the feed additive, VeO® (Phodé, France) at the dose of 100 ml per 1000 l of drinking water and for the whole duration of the trial. Gilts activity was daily measured through video recording analysed by group scan sampling, using BORIS®. Tail lesions were scored every 15 days, following a scoring system derived from Honeck *et al.* Saliva have been taken on sixteen identified animal 3 times: before and after the transfer (=stressor) and the day after, to measure the cortisol rate. All the data were analysed on SPSS® and using an analysis of variance (General Linear Model) or Khi-square test when appropriate. For the all period, experimental gilts expressed more eating, drinking and digging behaviour ($P<0.001$) and less manipulating objects or other gilts ($P<0.001$). The total number of tail lesions was lower in the experimental group (4.7 vs 13.4%; $P<0.001$) as was the severity of these lesions (14.0 vs 46.9%; $P<0.001$). Regarding the evolution of the tail lesions, we observed a substantial increase 45 days after the entrance in fattening stage 20.6% for the control and 11.2% for the experimental batches supposing a higher density stress. In addition, and just after the stressor, cortisol rate of control group was higher compared to experimental group (8.06 vs 82.67 mg/ml; $P<0.01$). Animals receiving VeO showed a better resilience to stressful situation. Normal behaviour has been maintain longer in experimental group. As a consequence, lower tail biting lesions has been counted. This cerebral approach seems key in an ending of tail docking practise.

Session 22 Poster 20

The effects of heating drinking water or calcium propionate on fattening beef cattle in winter

T. He, Z. Chen and S. Long
China Agricultural University, College of Animal Science and Technology, No. 2, Yuanmingyuan West Road, Haidian District, Beijing, 100194, Beijing, China, P.R.; tenghe@ethz.ch

The effects of heated drinking water (HDW) or calcium propionate (CaP) supplementation on growth performance, nutrient digestibility, and rumen microbiota in finishing beef cattle during winter (Average ambient temperature 6.47±3.52 °C) were evaluated. Ninety-six Simmental finishing bulls with an average body weight of 613±16.5 kg were randomly allocated into four treatments with six replicates per treatment and four bulls per replicate, with a 2×2 factorial arrangement of treatments for 60 d periods. Factors were drinking water temperature (room temperature of 5.12±2.01 °C or heated water temperature of 28.63±3.25 °C) and CaP (0 or 200 g/d per head, dry matter basis). The results showed that HDW significantly increased average daily gain, average daily feed intake, and average daily water intake ($P<0.05$), while the CaP supplementation significantly decreased the feed-to-gain ratio ($P<0.05$). No significant interaction was observed between HDW and CaP supplementation on growth performance and water intake. HDW and CaP supplementation significantly increased the digestibility of neutral detergent fibre and crude protein and significantly interacted on cellulase and xylanase concentration in rumen ($P<0.05$). Moreover, CaP supplementation increased the concentration of total volatile fatty acids, acetate, and propionate in rumen, and had an interactive effect on rumen propionate concentration with HDW ($P<0.05$). HDW significantly reduced serum cortisol and thyroid hormone concentrations ($P<0.05$), while no significant effect was observed on CaP supplementation in serum stress hormone levels. HDW significantly reduced the ratio of Firmicutes to Bacteroidetes and increased the relative abundance of *Prevotellaceae_ucg_003* and decreased the relative abundance of *Lachnospiraceae* ($P<0.05$) in rumen. CaP supplementation did not significantly affect rumen microbial populations but significantly increased the relative abundance of *Succinivibrionaceae* ($P<0.05$). In conclusion, HDW or supplementing CaP in a cold environment could improve growth performance, rumen fermentation function, and nutrient digestibility, enhancing the overwintering ability of Simmental finishing bulls.

Session 22 Poster 21

Water quality in dairy cattle farms: impact on animal production, reproduction and health

V. Resende
Universidade Évora e Instituto Mediterrâneo para a Agricultura, Ambiente e Desenvolvimento, Departamento de Zootecnia, Pólo da Mitra, Apartado 94, 7002-554 Évora, Portugal, Portugal; vjgr33@gmail.com

Climate change is currently one of the great global challenges, affecting in particular the water sector, namely the lack of precipitation and consequent periods of prolonged drought. Drinking water is a scarce resource in many regions of the world. Water is essential for the life of animals, since it intervenes in various metabolic processes. An inadequate water supply could reduce the health and performance of the animals. In dairy farms, the use of quality water is essential to maximize the milk production of animals. The objective of this study is to verify: (1) the importance of water quality on intensive dairy farms; (2) water quality affects animal production, reproduction and health. Presentation of the preliminary results of (2) biochemical analysis and toxicology. Serum samples were tested for Urea, Creatinine, ASL, ALT, Alkaline F. and Manganese. Urine samples were tested for urea, creatinine, and manganese. Milk samples were tested for manganese, calcium and copper. The requirements of cattle are not precisely knows, but Hartmans recommended 25 mg Mn/kg DM for Dutch dairy cattle. Marginal bands for manganese concentrations in diet (10-20 mg/kg DM), blood (µg/l) and serum (µg/l) in cattle. Toxic levels Mn 1000 mg/kg DM and symptoms of toxicity: Slow growth, anaemia, gastrointestinal lesions and occasionally neurological sings. Values were found for the untreated group of ALT=31.5 and 36.2 IU/l (min 6.9 and max 35.3). AST=201.1-269.1 (min 45.3 and max 110.2). GGT= 32.9-135.7 (min 4.9 and max 25.7). GGT – elevated values related to liver disease of toxic origin. This study helps to verify the importance attributed by national dairy cattle producers to the availability and quality of water on their farms.

Study on the effect of improving the pig gut microbiota of *Rhodobacter sphaeroides*

D. Shin[1], J. Park[2], S. Son[1], H. Lee[1] and J. Kim[3]
[1]Jeonbuk National University, Jeonju-si, Jeollabuk-do, Korea, 54932, Korea, South, [2]Jeju National University, Jeju-si, Jeju-do, Korea, 63243, Korea, South, [3]Chung-Ang University, Anseong-si, Gyeonggi-do, Korea, 17546, Korea, South; sdh1214@gmail.com

Pig is attractive to consumers due to its characteristics such as tenderness and supercilious taste. Recently, in the case of pigs, not only the taste of pork but also the importance of animal welfare is increasing. In particular, health is very important in the process of raising pigs. Recently, several previous studies reported that intestinal microbiome was closely related to health and immunity of the host. Herein, we explored a possible relation between pig gut microbiota composition and *Rhodobacter sphaeroides* as feed additive. A feeding experiment was conducted on 20 hybrids (Landrace × Yorkshire × Duroc) with an average weight of 7.73±0.13 kg (10 control group, 10 *Rhodobacter* treatment group), and this experiment was conducted for 2 weeks. After collecting faeces, the pig gut microbiota was characterized by the V3-V4 region of 16S rRNA using an Illumina Miseq. We identified 4,198 OTU-level core bacteria mainly belonging to the kingdom bacteria. Results showed that feeding additives affect the composition of pig gut microbiota. Compared with the control group, *R. sphaeroides* treatment group showed a higher proportion of the putative beneficial bacteria. Also, at the genus level, this feed additive increased Olsenella, whereas Bacteroides were enriched in the control group. Collectively, our study suggested that the *R. sphaeroides* might improve pig health by modulating gut microbiota.

Algae-based beta-glucan mitigates the adverse effects of heat stress in lactating sows

J. Mun, A. Hosseindoust, S. Ha, S. Park, H. Tajudeen, S. Oh and J. Kim
Kangwon National University, Department of Animal Industry Convergence, room 201, Animal Resources dep, (409), 24341, Korea, South; 202016455@kangwon.ac.kr

Heat stress has negative effects on sow performance, but proper nutritional strategies may help alleviate the consequences. This study aimed to investigate the interactive effects of algae-based beta-glucans (BG) on reproductive performance and metabolic response in lactating sows under heat stress. Thirty multiparous sows at day 112 of gestation were randomly assigned to one of three treatments: control (CON), 50 g BG/kg (BG5), and 100 g BG/kg (BG10), for 21 days. Results showed that sows fed BG5 and BG10 had reduced respiratory rates compared to CON. Sows in the BG10 group had higher feed intake ($P<0.01$) than CON. Body weight loss and weaning to oestrus interval tended to be lower in the BG5 and BG10 groups during lactation. Piglet weight ($P<0.05$) was higher in the BG10 group at weaning compared to CON. Sows in the BG10 group had significantly lower hair cortisol levels, and the lowest ($P<0.01$) TNF-α levels in blood. In contrast, a higher ($P<0.01$) concentration of TNF-α was observed in the BG5 group compared to CON. The blood concentration of lipopolysaccharide was decreased ($P<0.01$) in sows fed the BG10 diet compared to CON. The lipopolysaccharide-binding protein concentration in the jejunum tended to be lower ($P<0.01$) in the BG10 group compared to CON. Sows in the BG10 group showed higher ($P<0.05$) superoxide dismutase in the blood. However, there were no significant differences in reactive oxygen species, hydrogen peroxide, total antioxidant capacity, malonaldehyde, and catalase concentrations. Overall, the supplementation of BG at a dose of 10 g/kg was more effective than 5 g/kg, suggesting that higher doses of BG should be considered in future studies.

Session 22 — Poster 24

Effect of dietary mineral level and creep feeding on sows performance under high ambient temperature

H. Tajudeen, A. Hosseindoust, J.Y. Mun, S. Ha, S. Park, S. Yoon and J. Kim
Kangwon National University, Room 201-2, Department of Animal Industry Convergence dep, (409), 24341, Korea, South; 202016455@kangwon.ac.kr

Heat stress is an important issue, and the dietary mineral level and supplementation of creep feed for suckling piglets may be a solution to reduce the adverse effects of heat stress. This study evaluated the interactive effect of different mineral levels and creep feeding on reproductive performance, stress level, and immune status of lactating sows during heat stress. There were a total of 4 treatments including 2 mineral levels (CON, NRC-based; HM, 200% of NRC recommendation) with or without creep feeding (CF) in a 2×2 factorial arrangement during high ambient temperature. Sows in the CF treatment showed a lower respiratory rate at days 8, 10, 11, and 14 of lactation. The rectal temperature of lactating sows was mainly over 38.5 °C indicating relatively high heat stress. Sows in the CF treatments showed a lower body weight loss during lactation. Backfat thickness, farrowing duration, weaning to oestrus interval, born alive, weaned number, weaned weight, and litter weight was unaffected. The CF treatment showed a higher litter uniformity compared with suckling pigs without creep feeding. The survivability of piglets was increased in the CF treatment. Sows in the CF treatment showed a lower hair cortisol concentration. However, hair cortisol was not affected in the HM group. Sows in the CF treatments showed a lower tumour necrosis factor-α during lactation. The concentration of lipopolysaccharide and lipopolysaccharide-binding protein were not different in the CF and HM treatments. In conclusion, creep feeding is recommended for suckling pigs during heat stress due to reducing weight loss, stress level, and inflammatory response in sows.

Session 23 — Theatre 1

Pre-weaned dairy calf welfare: from the 'End the Cage Age' movement to the EFSA mandate

M. Brscic
University of Padova, Department of Animal Medicine Production and Health (MAPS), Viale dell'Università 16, 35020 Legnaro, Italy; marta.brscic@unipd.it

Aim of this contribution is to analyse the outcomes of the European Food Safety Authority (EFSA) mandate in regards to pre-weaned dairy calf and veal calf welfare and the risks associated with individual housing and insufficient space, in particular. Before an in-depth analysis of the contents of the EFSA scientific opinion on the welfare of calves adopted in February 2023, an overview of the state of the art at European level will be given starting with the 'End the Cage Age' movement, the large European Citizens' Initiative that in June 2021, that has obtained the response from the Commission to phase out cages for a number of farm animal species and categories and has risen the lack of science-based knowledge to fully respond to it. The campaign and the EU Farm to Fork Strategy, core of the Green Deal, are promoting rapid changes in the EU food production chains with implications as raising standards of animal welfare to improve animal health and food quality, reduce the need for medication and promote biodiversity preservation. The contribution will progress with a SWOT analysis on the importance of implementing a whole sector approach from farm to fork in order to guarantee higher levels of animal welfare still maintaining food safe and accessible, and the need to support farmers to upgrade their farms to more animal-friendly facilities meeting needs of several: animals, consumers and citizens, farmers and stakeholders. In this context, some major concerns related to dairy calves farming as the abrupt early calf-cow separation at birth (within minutes to 1-2 days), the actual housing systems used for pre-weaned dairy calves and veal calves and their space allowances will be analysed and the respective recommendations from the EFSA scientific opinion will be described. The contribution will end with a discussion on the potential impact that recommendations might have on dairy and veal calves sectors.

Session 23 Theatre 2

Impact of individual pens removal on veal calves' behaviour

D. Bastien and V. Lefoul
Institut de l'Elevage, Qualité des carcasses et des viandes, Monvoisin, 35650 Le Rheu, France; virginie.lefoul@idele.fr

The European Commission plans to phase out the use of cages for farm animals from 2025. In this context, a trial was conducted to study the impact of the removal of individual pen in veal calf production. 60 male Prim Holstein calves aged 20 days and weighing 49.0 kg were divided into 3 batches and housed for the first 28 days in individual pens (control, IP28), by pair (pair housing, PH28), or in groups of 5 calves (GROUP). The calves were then fattened in collective pen of 5 calves. The total fattening period last 24 weeks. Drinking from buckets in collective pens requires the presence of headlocks that allow the calves to be blocked during milk drinking. Continuous sampling observations were carried out from 6 am to 8 pm on 4 days (D13, D27, D112 and D156). The use of headlocks for the PH28 and GROUP caused stress in the calves (35% of calves stressed in the first two days and 20.7% at 2 weeks). At D13, cross-sucking of the foreskin was more frequent in PH28 and GROUP (8.2 and 11.1% of their daytime vs 0.4% for IP28, $P<0.01$). 65% of this activity occurred between 1.5 hours before and 1.5 hours after milk drinking. At D27, this cross-sucking behaviour was still very present for PH28 (3.2% of their daytime, with 65% calves affected compared to 0.0% for IP28, $P<0.01$, and 1.6% for GROUP being intermediate with 56% of calves affected, *NS*). In the GROUP batch, 2 calves with high cross-sucking behaviour had to be isolated (at D54 and D133) because they were no longer drinking their milk and were losing weight. Positive interactions such as grooming, or muzzle and ear sucking were equivalent between the 3 batches (representing 1.3% of the day's activities). Nevertheless, IP28 spent more time expressing PICA behaviour on D13 (7.8 vs 1.5% for the PH28 and GROUP batches, $P<0.01$). These differences are balanced out during the rest of the fattening period. Furthermore, no significant differences were observed on the number of health treatments per batch (1.7 treatment/calf, *NS*) or on live weight at slaughter, despite 10 kg difference between IP28 vs PH28 and GROUP (IP28=262.7 kg, PH28=252.5 kg, GROUP=251.9 kg). These results show that collective management of male calves at their arrival at the fattening farms can lead to fatal decline of their welfare.

Session 23 Theatre 3

Quarter level milk yield in dairy cows before and after separation in a cow-calf contact system

S. Ferneborg[1] and S. Agenäs[2]
[1]Norwegian University of Life Sciences, Department of Animal and Aquacultural Sciences, P.O. Box 5003 NMBU, 1432 Aas, Norway, [2]Swedish University of Agricultural Sciences, Department of Animal Nutrition and Management, P.O. Box 7070, 75007 Uppsala, Sweden; sabine.ferneborg@nmbu.no

There is an increased interest among consumers, farmers and other stakeholders, to facilitate cow-calf contact (CCC) systems in dairy production, where cow and calf are kept together for an extended period of time. However, research on the effects of the system on milk yield before and after separation is scarce. Farmers report milk ejection difficulties in CCC cows during the time together with their calves and also low milk yield after separation. The aim of this study was to investigate quarter level milk yield prior to and after separation among cows kept in a CCC system with automatic milking. In total 35 cows were enrolled in the trial and assigned to CCC (n=17) until the calves were 127±6.6 d old, or control (n=18) that were separated from their calves within 12 h after calving. All cows were milked in an AMS and individual quarter level records were obtained from each milking during one lactation. Analysis of variance was performed using the mixed model function in SAS Studio, with a model including the fixed effects of treatment, separation, quarter and the random effect of cow Milk yield delivered to the robot was higher in control cows per quarter and milking than in CCC cows before separation ($P<0.001$) but not after ($P=0.92$). In CCC cows, average milk yield per quarter and milking increased from 2.1±1.12 before separation to 2.7±0.93 kg/milking (mean ± SD) after separation, while control cows decreased from 3.9±1.10 to 3.0±1.00 kg/milking over the same period. Within-cow variation was higher in CCC cows before (CV=58%) than after separation (CV=38%). Collectively, these results indicate that milk yield per quarter and milking is similar between control and CCC cows after separation and that in turn suggests that total yield is similar between the groups also before separation. The lower milk yield recorded by the robotic milking is therefore more likely an effect of calf milk intake than of impaired milk ejection.

Session 23 Theatre 4

Two different training protocols to encourage dam-reared calves to drink supplemental milk

J. Sørby[1], S. Ferneborg[1], S.G. Kischel[2] and J.F. Johnsen[3]

[1]Norwegian University of Life Sciences, Oluf Thesens vei 6, 1433, Norway, [2]TINE SA, BTB-NMBU, Pb. 5003, 1432, Norway, [3]Norwegian Veterinary Institute, Arboretveien 57, 1433, Norway; johanne.sorby@nmbu.no

At debonding, cow-calf contact (CCC) calves' performance may be improved, and stress alleviated through encouraging the calf to drink supplemental milk (SM, e.g. automatic feeder). However, calves often struggle learning how to drink milk from a new source. This study describes the success (intake of ≥1.5 l SM more than once) of two different training protocols to encourage CCC calves to drink SM from an automatic feeder during debonding. In training protocol 1 calves (n=16) were pushed into the feeder where they involuntarily received whole SM from a teat bottle the first two d, thereafter assisted in drinking from the feeder the following three d regardless of success with the teat bottle. In training protocol 2, we familiarized the calves (n=14) with artificial teats and the feeder by placing two dummy teats in the creep and moving the salt lick within the feeder (i.e. race-gate). In addition, we enticed the calves with a focus on positive reinforcement during the 5 d training period using something sweet; calves could voluntarily choose to approach a teat bottle with approx. 2 dl of electrolytes and teat greased with molasses, before enticing them towards the feeder. Upon success with this first step, electrolytes were substituted with whole SM from a bottle within the feeder. Finally, calves received individual assistance in the feeder *if* successful with the bottle. Protocol 2 appeared gentler for the calves, and descriptive analysis showed a higher success rate compared to protocol 1 (71 and 50%, respectively). In addition, successful calves trained by protocol 2 had on average 10.4±4.58 success days out of 38 possible (amounting 50.8±42.7 l/calf for the period) compared to successful calves trained by protocol 1 having on average 3.4±2.34 success days (18.4±21.08 l/calf). Experiences from this study confirm that CCC calves struggle to learn how to drink SM during debonding, but familiarity with an artificial teat and enticement during training can be promising tools.

Session 23 Theatre 5

Effects of automatic TMR feeding system on the health status of finishing bulls and heifers

O. Martinić, L. Magrin, P. Prevedello, G. Fabbri, G. Cozzi and F. Gottardo

University of Padova, Department of Animal Medicine, Production and Health, Viale dell'Università 16, 35020 Legnaro, Italy; oliver.martinic@phd.unipd.it

The aim of this study was to evaluate the effects on the health status of Limousine finishing bulls and heifers of automatic feeding systems (AFS) with dynamic total mixed ration (TMR) distribution in comparison to the conventional feeding system (CFS). Two feeding systems (FS) prepare and distribute the same composition TMR: the CFS consisted of a TMR prepared and distributed by a mixer wagon in a single daily distribution; AFS consisted of an automatic system that prepared and delivered TMR dynamically several times daily according to animals' intake observed by feeding delivery robot sensor which measured feed leftovers in the bunk. A total of 1,440 bulls and 1,129 heifers regularly imported in 75 batches from France were monitored during the experimental period from January 2020 to March 2022 in a farm located in the province of Treviso, Italy. After an adaptation period, animals were separated into groups of 10 animals based on sex, batch, and initial weight, and were transferred to two identical fattening barns with different FS. Animals were slaughtered when they reached a suitable finishing status evaluated by a slaughterhouse expert. The individual health status of the animals was daily checked by the farm veterinarian throughout the fattening period. The trained veterinarian inspected the rumen mucosa of a random sample of animals of each batch at the slaughterhouse. The percentage of treated animals was greater for CFS heifers ($P<0.01$) and bulls ($P<0.05$) than AFS. The number of treatments for respiratory disorders in heifers was greater for CFS than for AFS ($P<0.05$). Furthermore, the number of treatments received per bull for locomotor disorders was greater for CFS than AFS ($P<0.05$). The FS affected the prevalence of hyperkeratosis, plaques, and signs of ruminitis both in bulls and heifers, being lower for AFS than CFS. Rumens' mucosa belonging to AFS heifers showed a lower prevalence of star scars than those belonging to CFS ($P<0.005$). Collected data showed that an increasing frequency of feed deliveries over the day could improve the rumen health status of bulls and heifers.

Session 23 Theatre 6

Associations between feed push-up frequency, feeding behaviour and milk yield in dairy buffaloes

G. Esposito[1], E. Raffrenato[2], G. Bosoni[1], F. Caruso[1] and F. Righi[1]

[1]University of Parma, Veterinary Medicine, via del Taglio,10, 43126 Parma, Italy, [2]University of Padua, Comparative Biomedicine and Animal Nutrition, viale dell'Università, 35020 Legnaro, Italy; giulia.esposito@unipr.it

The objective of this study was to assess the effects of feed push-up frequency on the feeding behaviour and milk yield of Italian Mediterranean dairy buffaloes. Lactating buffaloes (n=95±6.2, parity = 4±1; DIM: 107±8.2) were housed in a freestall, milked twice daily, and offered *ad libitum* access to water and a total mixed ration containing, on as fed basis: 25% corn flour, 17%distillers, 15% wheat silage, 15% soy, 14% haylage, 4% alfalfa and 10%mineral supplement), provided daily. Buffaloes within the pen were exposed to each of 2 treatments in a crossover design with 15-d treatment periods and a 10-d adjustment periods between treatments; treatment 1 had 3 feed push-up a day (at 7:30, 11.30 and 18.30), whereas treatment 2 had 2 feed push-up a day (7.30 and 12.30). Dry matter intake and milk yield were recorded daily whereas every five days feed samples of fresh feed and orts were collected for particle size analysis. Two separators were used: the Penn State Particle size separator, fitted with 3 screens (19, 8, and 1.18 mm) and a bottom pan, resulting in 4 fractions (long, medium, short, fine) and a Z-box fitted with a 3.18 mm screen (for determination of the physical effectiveness factor (pef) of the TMR). Sorting was calculated as the actual intake of each particle size fraction expressed as a percentage of the predicted intake of that fraction. Treatment had no effect on milk yield (13.26±0.04 vs 13.021±0.05 kg/d for 3 vs 2 push-up) or feed sorting. Buffaloes sorted against long particles (76.4±31.96%) and for short (99.38±0.72%) and fine (100.06±1.09%) particles. The pef of the TMR was not affected by treatment. Dry matter intake increased with increased push-up frequency (41.71±0.03 kg vs 37.11±0.03 kg for 3 vs 2 push-up). These preliminary results suggest that altering feed push-up frequency increases intake but doesn't affect feed sorting or milk yield in dairy buffaloes. Further studies are needed to evaluate long term effects of feeding frequency on milk yield and composition.

Session 23 Theatre 7

Comparing dairy cow welfare in loose-housing and in tie-stall systems using animal-based indicators

V. Lorenzi, L. Bertocchi, C. Montagnin and F. Fusi

Istituto Zooprofilattico Sperimentale della Lombardia e dell'Emilia Romagna 'Bruno Ubertini', Italian Reference Centre for Animal Welfare, Via Bianchi 9, 25124 Brescia, Italy; francesca.fusi@izsler.it

Due to the geographical features of the territory, both tie-stall and loose-housing systems for dairy cows coexist in Italy. Tie-stall barns are more common in the mountainous areas, while loose-housing systems (free-stall or deep litter barns) are widespread in lowland areas. During the year 2022, using the ClassyFarm system of the Italian Ministry of Health (www.classyfarm.it), 10 animal-based indicators (ABIs) were collected in 1,385 tie-stall farms (range 10-656 heads, mean 51 heads) and in 4,123 loose-housing farms (range 10-700 heads, mean 223 heads). In each farm, 4 out of the 10 ABIs were collected separately for lactating cows (LC), dry cows (DC) and heifers (HF). These 4 ABIs were: avoidance distance test, body-condition score, coat cleanliness (flank, leg and udder) and integument alterations. The other 6 collected ABIs were: lameness in adult cows, milk somatic cell count, antimicrobial usage for mastitis, annual mortality rate of adult cows, annual mortality rate of calves and mutilations. For each ABI a score was assigned to the farm, based on whether or not the set warning threshold was exceeded. Results obtained in loose-housing farms and tie-stall farms were compared. Loose housing farms achieved better results for the ABIs: avoidance distance test in HF; body condition score in LC, DC and HF; coat cleanliness in LC and DC; integument alterations in LC, DC and HF; milk somatic cell count and antimicrobial usage for mastitis. Tie-stall farms showed a better score for the ABIs: lameness, annual mortality rate of lactating cows, annual mortality rate of calves and mutilations. No differences were found for the ABI 'avoidance distance test' in LC and DC and for the ABI 'coat cleanliness' in HF. In terms of animal welfare, the tie-stall housing system of dairy cows is strongly debated. On one hand, the tied system restricts the voluntary movement possibilities and the social behaviour of dairy cows; on the other it limits competition for the resources. As stated by some authors, the adverse effects of tie-stall could be reduced, when periods of exercise are possible.

Time-controlled hay racks horses: what effect on behaviour?

M. Roig-Pons and S. Briefer
Agroscope, HNS, Les Longs Prés 1, 1580 Avenches, Switzerland; marie.roig-pons@agroscope.admin.ch

Under domestic conditions, horses are typically fed roughage 2-3 times a day in limited quantities, which differs from their natural feeding behaviour. To better regulate feeding intervals without increasing the workload, caretakers can rely on time-controlled hayracks (rack with roll-up doors opening and closing at programmed times). We studied (1) the association between lying behaviour (essential for REM-sleep) and the openings of the rack at night and (2) the evolution of aggressiveness and risk of injury between close (HC) and open (HO) hayrack. The study was conducted on 3 groups of 5 mares from the SNSF in loose housing with bedded resting area. Straw was *ad libitum* and hay was accessible for 9 hours, split into 5 meals of 1.30-2 hours using time-controlled rack. Continuous observations of social interactions were carried out for 10 hours per group and social status, aggression towards conspecifics (Specific Index of Aggression – SIA) and social tension endured (Social Tension Index – STI) were calculated for each mare. Lying behaviour was recorded with MSR accelerometers. (1) Results highlighted a significant increase of agonistic behaviours in HO compared to HC ($NbAgoBehav_{HO}$/h=175.5, $NbAgoBehav_{HC}$/h=122.2, t-test, P<0.05). In HO, mares displayed in average more threats (64.9% of agonistic behaviours in HO vs 53.8% in HC) and less passive displacements (27.5% HO vs 39.4% HC) (χ^2, P<0.05). The average level of aggressiveness therefore increases in HO (SIA_{HF}=3.5±2.4, SIA_{HO}=5.3±4.8 paired Wilcoxon test, P<0.05) but injury risk behaviours (kicks, bite and chase) remained stable. STI_{HO} and STI_{HC} were similar, meaning that the mares retain their 'initiatior' or 'recipient' status. Lying behaviour obtained was below the 10% of daily time reported in the literature (%DailyTimeSpentLying = 4±2.7). A clear association between HO and lying behaviour was found ($TotNbLyingBouts_{HC}$ = 706, $TotNbLyingBouts_{HO}$ = 8), even though these time slots are indicated in the literature as preferable for lying phases. Five mares laid down after 8am, suggesting that some dominated mares prefer to shift their sleeping bouts to take advantage of feeding slots. These results emphasize the need of further research to find feeding management that optimizes the welfare of horses housed in groups.

Session 23 Theatre 9

Exploring techniques to mitigate heat stress in feedlot lambs

P.G. Theron[1], T.S. Brand[1,2], S.W.P. Cloete[1] and K. Dzama[1]
[1]University of Stellenbosch, Animal Sciences, Private Bag X1, Matieland, 7600, Stellenbosch, South Africa, [2]Western Cape Department of Agriculture, Directorate: Animal Sciences, 80 Muldersvlei Road, 7600, Stellenbosch, South Africa; pieter.theron@westerncape.gov.za

The increased demand for food from a growing population, linked to the projected increase in drought frequency and mean annual temperatures over Southern Africa is likely to force the predominantly extensive sheep farming industry to intensify production. Feedlotting is the commonest intensification technique but information on how to keep the method sustainable under extreme temperatures must be gathered proactively. This study explored the use of shearing and varying dietary energy levels as tools to mitigate heat stress in lambs during feedlot finishing. Fifty-six Dohne Merino ram lambs were finished on a higher (10.6 MJ ME/kg) or lower (9.1 MJ ME/kg) energy grain-based feedlot diet during summer and half of each dietary group were randomly selected to be shorn. Lambs were fed individually and production traits (growth, backfat depth, feed intake and feed conversion) and heat stress indicators (respiration rate, surface and rectal temperature) were monitored throughout the finishing period. Dietary energy level did not affect production traits or heat stress indicators, potentially due to the difference between diets not being large enough. Shorn lambs exhibited lower rectal temperatures (39.77 vs 39.91 °C) and respiration rates (140 vs 158 breaths per minute), indicating that they were more effective at coping with the high (>30 °C) environmental temperatures present during the study. Surface temperature did not differ between groups. Backfat depth was the only production trait that differed between shearing groups with unshorn lambs having greater fat depth (+0.4 mm). Growth and feed intake did not differ significantly between shearing groups, suggesting that shearing will not affect feedlot production or profitability. Shearing is thus a viable technique to improve welfare in feedlot animals as it will reduce heat stress without negatively affecting production output.

Management practices and welfare of dairy goats in Greece

G. Arsenos[1], S. Vouraki[1], V. Papanikolopoulou[1], L.V. Ekateriniadou[2] and I. Sakaridis[2]
[1]School of Veterinary Medicine, Aristotle University, University Campus, 54124 Thessaloniki, Greece, [2]Hellenic Agricultural Organization DIMITRA, Campus of Thermi, 57001 Thessaloniki, Greece; vipapani@vet.auth.gr

The objective was to assess management practices related to animal welfare in dairy goat farms in Greece. A random sample of 18 dairy goat farms located in Northern and Central Greece was used. The average farm comprised 390±448 goats with an average milk yield of 230±143.9 kg/milking period. The prevailing farming system was semi-extensive. The average grazing time and walking distance of goats was 5.9±2.35 hours and 5.5±2.88 km, respectively. There was also one intensive and one semi-intensive farm. Reared goats belonged to six indigenous Greek breeds (Eghoria, Skopelos, Chalkidikis, Paggaiou, Aridaias, Serres; n=11 farms) and three foreign breeds (Damascus, Anglo-Nubian, Murciano-Granadina; n=5 farms) whereas two farms had composite populations of crossbred goats. Housing and management practices were recorded by a group of veterinarians using a designated questionnaire. Records included information on flock size, housing conditions and shed dimensions, milking procedures, vaccinations, and antiparasitic treatments; available space per goat was calculated. Descriptive statistical analysis was performed. Results showed that in 39 and 16% of farms, available space per goat was less than 2 and 1.5 m^2, respectively. In most farms (68%), machine-milking was performed; use of gloves by milkers and post-dipping were reported in 28 and 5% of cases, respectively. Moreover, 21, 42 and 84% of farmers did not vaccinate against *Clostridium perfringens*, *Mycoplasma agalactiae* and *Chlamydophila abortus*, respectively. Routine treatments for endoparasites and ectoparasites were performed in 79 and 84% of farms, respectively. Overall, results indicate that milking procedures, lack of vaccination against common diseases and limited space availability are management practices associated with the welfare status of dairy goats in Greece. Further improvement of those practices is needed to increase animal productivity and welfare. This research has been co-financed by the European Regional Development Fund of the European Union and Greek National Funds through the Operational Program Central Macedonia 2014-2020 (KMP6-0083632; GRAEGA CHEESE).

Risk factors for navigation ability of laying hens at housing in an aviary system

C. Ciarelli, F. Bordignon, G. Pillan, G. Xiccato and A. Trocino
University of Padova, Department of Agronomy, Food, Natural Resources, Animals and Environment, viale dell'Università 16, 35020, Italy; angela.trocino@unipd.it

To evaluate the risk factors for navigation ability of two genotypes of laying hens at housing in an aviary system, 1,800 pullets, half Lohmann White-LSL and half Hy-line Brown, were randomly allocated at 17 weeks of age in 8 pens of an experimental aviary (3 tiers), according to a bi-factorial arrangement, 2 genotypes (brown vs white hens) × 2 types of pens (enriched or not enriched with additional perches). Data collected by video-recording at 17 and 20 weeks of age were used to assess the number of successful and unsuccessful landings from any part/equipment of the aviary to the floor. Risk factors related to failed landings were evaluated by multivariate logistic regression analysis through a forward stepwise selection using the PROC LOGISTIC of SAS. The regression coefficients were expressed as odds ratio (OR) with 95% confidence interval (CI). Compared to brown hens, white hens performed a significantly higher number of landings per hour in the observation interval (80.7 vs 35.4; $P<0.001$) with a higher success rate (94.8 vs 88.6%; $P<001$). Brown hens had higher odds of failed landings compared with white hens (OR: 6.65; 95% CI: 4.36-10.1). The logistic regression analysis showed significantly higher odds (OR: 1.90; $P<0.001$) of experiencing failed landings at 17 weeks compared to 20 weeks. No significant difference in the number or in the rate of success of landings was recorded between enriched and not enriched pens, where the logistic regression analysis only measured a trend (OR: 1.29; $P=0.09$). The odds of failed landings were greatly higher when comparing long with medium flight distance to floor (i.e. hens starting from the third vs the second tiers of the aviary) (OR: 31.1; $P<0.001$) and lower comparing short (i.e. hens starting from the first tiers of the aviary) with medium flight distance (OR: 0.17; $P<0.001$). In conclusion, under the condition of the present study, white hens exhibited higher navigation activity and ability compared to brown hens since the first week after housing; the navigation ability improved four after housing weeks; the enrichment with additional perches played a minor role. Long-term effects on space use should be evaluated over the laying period.

Busy birds – what do turkey hens use dust baths for?

K. Skiba[1], M. Kramer[1], P. Niewind[2], B. Spindler[1] and N. Kemper[1]
[1]University of Veterinary Medicine Hannover, Foundation, Institute for Animal Hygiene, Animal Welfare and Farm Animal Behaviour (ITTN), Bischofsholer Damm 15, 30173 Hannover, Germany, [2]Agricultural Chamber of North Rhine-Westphalia, Haus Duesse 2, 59505 Bad Sassendorf, Germany; karolin.skiba@tiho-hannover.de

In this study, six groups of 71 turkey hens were observed for one hour after renewing the dust baths using video cameras. Each group was kept in a 17.5 m^2 compartment. Next to straw pellets, which were used as bedding material, each compartment included one dust bath (DB, 1 m^2). The DB was available at all times during the fattening phase (5th-16th week of life (WL)) and renewed once weekly, using either rock flour, sawdust or sand for two groups each. The occupancy of the DB was analysed via scan sampling every five minutes within one hour after renewing the baths, five times during the fattening phase (6th, 8th, 10th, 12th and 14th WL). Next to counting the animals inside the DB, the behaviour of each animal was recorded. In order to categorise each behaviour correctly, the previous 20 seconds to each scan were watched. Results showed an average number of 2.2 turkey hens per hour and DB throughout the fattening phase. In 8.9% of all samples, the process of dust bathing was identified. The remaining behaviours shown in the DB were exploration (37.6%), rest/inactivity (32.3%), agonistic behaviour (15.0%), and comfort behaviour (6.2%). The peak of bird numbers in the DB was found in the 10th WL (n=2.5), which matched with the peak for dust bathing activity (n=0.4; 15.5%). Over the course of the fattening phase, next to all other behaviours, dust bathing occurred most frequently in rock flour (n=0.3; 12.9%), followed by sand (n=0.2; 9.0%) and then sawdust (n=0.1; 4.8%). In this study, turkey hens used dust baths primarily for exploration and rest. Dust bathing frequencies varied between substrate type and age of the animals and should be investigated further. This work (Model- and Demonstration Project for animal welfare #Pute@Praxis) is financially supported by the Federal Ministry of Food and Agriculture (BMEL) based on a decision of the Parliament of the Federal Republic of Germany, granted by the Federal Office for Agriculture and Food (BLE); grant number 'FKZ 2817MDT611'.

Session 23 Theatre 13

Pre-slaughter fasting on stress response and meat quality in rainbow trout (*Oncorhynchus mykiss*)

A. De La Llave-Propín[1], A. Martínez Villalba[2], R. González Garoz[2], J. De La Fuente[2], C. Pérez[3], E. González De Chavarri[2], M.T. Díaz[2], A. Cabezas[2], M. Villarroel[1] and R. Bermejo-Poza[2]
[1]ETSIAAB, Universidad Politécnica de Madrid, Departamento de Producción Agraria, Avenida Complutense 3, 28040, Madrid, Spain, [2]Facultad de Veterinaria, Universidad Complutense de Madrid, Departamento de Producción Animal, Universidad Complutense, Avenida Puerta de Hierro s/n, 28040, Madrid, Spain, [3]Facultad de Veterinaria, Universidad Complutense de Madrid, Departamento de Fisiología Animal, Universidad Complutense, Avenida Puerta de Hierro s/n, 28040, Madrid, Spain; alvaro.delallave.propin@alumnos.upm.es

The stress response in fish is conditioned by water temperature. The objective of this study was to evaluate the effects of the application of different pre-slaughter fasting on the stress response and meat quality in rainbow trout (*Oncorhynchus mykiss*) in summer. 270 fish (initial biomass = 17±0.10 kg/tank) were subjected to different pre-slaughter fasting in terms of degree days (no fasting -0 °C d-, 3 days fasting -50 °C d- and 6 days fasting -100 °C d-) in summer conditions (water mean temperature = 22 °). After slaughter by ikejime, biometric (slaughter weight, standard length, body condition and hepato-somatic index) and flesh quality parameters (muscle and liver colour, muscle pH, liver glycogen, liver weight and *rigor mortis*) were measured. Fasting at 50 °C d showed intermediate results over the groups of 0 and 100 °C d, with the exception of liver colour, exhibiting lower values than the rest of the groups. Fasting at 100 °C d had a greater effect, decreasing biometric parameters and presenting lower meat quality. Muscle pH, rigor mortis and liver glycogen were lower in the fasting groups compared to the non-fasted. In conclusion, it was observed that fasting had a significant effect on the parameters analysed and that the fish subjected to fasting, especially at 100 °C d, showed a worse productive state and meat quality, so a pre-slaughter fasting of less than 50 °C d is recommended in rainbow trout.

Session 23 Poster 14

Effects of a new cooled cubicle waterbed on cow behaviour and lameness: a randomized control trial

O. Levrad, R. Guatteo, A. Lehébel, L. Bouglé, M. Tourillon, N. Brisseau and A. Relun
Oniris, INRAE, BIOEPAR, Chantrerie, 44300 Nantes, France; olivier.levrard@oniris-nantes.fr

In Europe, farmers commonly use fans or sprinklers to mitigate heat stress in dairy cows. Innovative strategies are needed to reduce water and energy use. A randomized controlled trial was implemented in summer 2021 to assess the effects of a new cooled waterbed on the behaviour and lameness of dairy cows, assuming that cows on the cooled waterbed would spend less time standing and more time feeding, and therefore develop fewer claw lesions and lameness. At calving, 219 cows were randomly allocated to one of two adjacent barns, with cubicles equipped with either a conventional or a water-cooled cubicle mat (Louisiane vs Aquaclim; Bioret-Agri, France). The temperature-humidity index (THI) was recorded every 15 minutes. Posture and location of cows in the barns were continuously monitored using computer vision (Alherd; France). Cows' feet were trimmed by a hoof trimmer at the beginning and end of the trial. Gait was scored every fortnight. Behavioural response variables and incidence of lameness were analysed by multiple regression (linear and Cox models respectively). At baseline, cows were primarily affected by digital dermatitis, with no difference in lameness score between mats (P=0.25). Between 14 June and 30 September, 17 days were slightly hot (68≤THI<72) and 8 days moderately hot (72≤THI<79). 83 cows calved during this period. Cows in the waterbed group spent significantly more time feeding (+ 46 min/d), lying down (+1 h 47 min/d) and less time drinking (-6 min/d), standing in the alley (-63 min/d) and in the stall (-67 min/d), regardless of average daily THI. Consequently, cow comfort and stall use indexes were significantly improved in the waterbed group (74 and 54% respectively vs 63 and 44% in the control group). Of the 58 cows that were sound at their first post-calving assessment 41 developed lameness before 150 DIM (75 and 65% in the control and waterbed groups, P=0.42), at a mean of 68 DIM (P=0.83). Incidence of lameness did not differ between mats. The water-cooled cubicle mat is an interesting tool to improve cow comfort. Further studies are needed to draw a firm conclusion on its impact on lameness and to find the optimal settings to mitigate heat stress.

Session 23 Poster 15

Evaluation of milk and residual milk production of dairy Gyr cows

R.C. Castanheira[1], M.I. Paiva[1], K.T. De-Sousa[1], M.S.V. Salles[1], E.A. Silva[2] and L. El Faro[1]
[1]Animal Science Institute, Beef Cattle Center, Rod Carlos Tonani, km 94, 14174-000 Sertaozinho-SP, Brazil, [2]Epamig Oeste – Campo Getulio Vargas, Dairy Cattle, R. Afonso Rato, 1301, 38060-040, Uberaba, MG, Brazil; lenira.zadra@sp.gov.br

We aimed to evaluate the milk production (MY) and residual milk (RM) of 40 dairy Gyr cows. The cows calved in Uberaba, Minas Gerais, Brazil, in two periods: PC1- October and November/2021, and PC2- December/2021 and January/2022. Milking was performed twice a day, and the calves had access to one teat from their dam during each milking, until 90 days of age. The calves were distributed into two treatments according to the feeding management: conventional (n=20) and optimized (n=20). For optimized management, the calves receive 4 or 2 l of milk (depending on age) in bottles after milking. MY and RM were recorded at 30, 60, 90 e 120 days-in-milk, in both milkings. To measure the RM, 1 ml of oxytocin was applied right after milking. The variables were analysed using a linear model, with measures repeated over time using the SAS MIXED procedure. The fixed effects were treatment, period of calving (PC=1, 2), parity order (PO, primiparous and multiparous), days-in-milk (DIM), and interactions (Treatment×PC; Treatment×PC×DIM; PO×DIM). Means estimated by least squares were compared by Tukey's test. For MY, there was an effect (P<0.05) of PO (primiparous: 7.9±0.88 kg; multiparous: 10.2±0.83 kg) and DIM (d30: 8.1±0.61 kg; d60: 8.2±0.61 kg; d90: 7.5±0.61; d120: 9.3±0.61 kg). There was an effect (P<0.05) of the PC and DIM on RM. The least square means for RM in P1 was 1.9±0.17 kg, and in P2 was 1.3±0.21 kg. The estimated RM varied over days-in-milk (d30: 1.3±0.34 kg; d60: 1.2±0.34 kg; d90: 2.0±0.34; d120: 2.4±0.34 kg). The lower value of RM in P2 can be explained by the fact that calves in P2 consumed more additional milk and food than the calves in P1 and, possibly sucked smaller amounts of milk from their mothers. Our results showed that residual milk may be related to the amount of feed the calf receives. Based on this, future studies can evaluate strategies for feeding calves, in order to improve milk production and reduce residual milk.

Session 23 Poster 16

Environmental enrichment and tactile stimulation on the welfare of F1 Holstein × Gir calves

C.O. Miranda[1], A.E. Vercesi Filho[1], M.L.P. Lima[1], F.F. Simili[1], M.S.V. Salles[1], E.G. Ribeiro[1], J.A. Negrão[2] and L.E.F. Zadra[1]

[1]Animal Science Institute, Beef Cattle Center, Carlos Tonani, 94, 14.174-000 Sertaozinho Sao Paulo, Brazil, [2]Universidade de São Paulo, FZEA, Duque de Caxias, 225, Campus Fernando Costa, 13635-900, Pirassununga, Sao Paulo, Brazil; lenira.zadra@sp.gov.br

This study aimed to measure the physiological traits related to the welfare of F1 Gyr × Holstein calves. The experiment was carried out at Instituto de Zootecnia, Ribeirao Preto, SP, Brazil, with 34 calves with 30 days of age, housed for 90 days in collective barns. A 2×2 factorial design, with two environments (with and without environmental enrichment with playful objects, (EE and WE) and two managements (with and without tactile stimulation, TS and WS). During 31 days of experiment (up to 60 days of age), heifers received TS (brushing). Weaning occurred at 60 days of experiment (90 days of age). Serum levels of cortisol, lactate, glucose and oxytocin, as well as the flight distance (FD) were analysed using mixed linear models, with repeated measures over time. The model of analysis for the variables (cortisol, lactate, glucose and oxytocin) included the fixed effects of management group class (enriched environment and/or tactile stimulation and the interaction between them), day of record (1, 30, 55, 60, 65 and 90 days). For FD, the same fixed effects were considered, but the collection days at 10, 20 and 22 days after weaning. There was no interaction between EE and ST for all traits (P>0.05). The effect of animal within treatment accounted the repeated records. Both tactile stimulation and EE were important to reduce the animals' stress, but as the TS was only offered for 30 days, the EE effect was more visible throughout the experimental period. Animals that received EE had shorter flight distance (2.19+0.22 vs 3.09+0.22 m), lower serum lactate levels (15.49+0.79 vs 20.06+0.81 mg/dl) and higher oxytocin levels (39.7+0.28 vs 18.8+0.29 uUI/ml) than WE animals. For tactile stimulation management only oxytocin level was significantly different, with higher estimated means for ST (36,84±0,31 uUI/ml) than that WS (20,88±0,27 uUI/ml). The use of EE together with TS promoted important in the improvement of the welfare of the animals.

Session 23 Poster 17

Heat stress indicators in dairy cattle breeding provided by automatic milking systems (AMS)

M.A. Leandro, J. Stock, J. Bennewitz and M.G.G. Chagunda

University of Hohenheim, Garbenstraße 17, 70599, Stuttgart, Germany; miguel.ribeiroleandro@uni-hohenheim.de

Heat stress in dairy cattle represents a major economic loss and animal welfare problem. Apart from direct effects on animals, heat stress has an indirect effect on the increased proliferation of diseases such as mastitis and claw disease. Automatic milking systems (AMS) are becoming common on dairy farms and allow for the non-invasive measuring of many animal parameters without disturbing the cow's normal behaviour. While common for predicting the onset of disease, research on prediction and management of heat stress in dairy cows using AMS is sparse. This review examines the associations of milk- and animal behaviour-based parameters with heat stress. It also examines the availability and accessibility of such parameters in the AMS and infers their potential to be used as indicators of heat stress in dairy cattle breeding. This review also evaluates the potential of these traits to be used in breeding programmes through examination of the heritability values. In terms of animal behaviour-related traits, the literature showed that heat stress manifests through reduced feed intake (up to 10%), higher respiratory rate (panting) and reduced activity. It was also shown that animals reduced number of visits to the AMS, as they spent more time laying down and at water sources. For milk-based indicators, the literature has shown reduced milk yield (up to 21%), correlated with an increase in somatic cells in the milk as well as reduced protein (up to 2.7%) and fat content (up to 9.5%). Milk temperature has also been found to be correlated to body temperature. Furthermore, the AMS also provides information on its interactions with the cows, such as the number of successful milkings per individual, and milking efficiency. From an animal breeding perspective, most of these traits have been reported to have moderate heritability with milk yield=0.27, milk fat content=0.44, milk protein content=0.55, respiratory rate 0.04 and body temperature=0.15-0.3. This review develops a framework examining the potential of generating non-invasively measured indicators of heat stress phenotypes from automatically milked cows.

Session 23 Poster 18

Impact of floor type and group size on veal calves' behaviour

D. Bastien and V. Lefoul

Institut de l'Elevage, Qualité des Carcasses et des Viandes, 8 route de Monvoisin, 35650 Le Rheu, France; didier.bastien@idele.fr

Calf housing is at the heart of societal and regulatory issues. In this context, a trial was conducted to study the impact of floor type and group size in veal calf production. 80 male Prim Holstein calves aged 20 days and weighing 52.7 kg were divided into 4 batches: wooden floor with 5 calves per pen (control, W5), wooden floor with 10 calves per pen (W10), rubber floor with 10 calves per pen (R10), and wooden and rubber mixed floor with 10 calves per pen (M10). The calves were individually housed the first 28 days and were fattened for 24 weeks using the same feeding plan and health protocol. Each batch had the same density (1.8 m^2/calf). Scan sampling and continuous sampling observations were carried out from 6 am to 8 pm on 3 days (D22, D75 and D145), supplemented by activity measurements from pedometers placed on 5 calves per batch. W10 calves walked more than W5 calves (+20% at mid fattening, NS and +40% at the end of fattening, $P<0.05$). However, the group size had no effect on lying time (68.2% of the day spent lying for both batches at the beginning of fattening period, 66.9% at mid-fattening and 65.5% at the end of fattening), nor on ruminative or stereotypic activities. It also had no impact on the number of health treatments per calf (W5=3.2 treatments vs W10=3.7 treatments, NS). Total growth rates were also identical (W5=1,244 g/d vs W10=1,236 g/d, NS). The calves preferred the rubber floor for lying: 90% of the lying in M10 was done on the rubber floor. However, the lying time was identical between W10, M10 and R10. We measured more steps for R10 than W10 (334 steps/day vs 288 at the beginning of fattening, 524 vs 427 at mid-fattening and 459 vs 414 at the end of fattening), but these differences are NS (M10 being intermediate). 33% of R10 calves were scored dirty during fattening vs 18% for M10 and 1% for W10. This state of cleanliness had an impact on calf grooming (W10=4.8% of daily activity in mid fattening vs M10=7.1% and R10=5.7%). In terms of zootechnical performances, floor type had no effect on calf growth or carcass weight. Nevertheless, it had an impact on carcass colour (W10=5% of coloured carcasses vs M10= 35% and R10=30%).

Session 24 Theatre 1

Heterogeneity of 3D cameras feed intake heritabilities in Holstein, Jersey and Nordic Red cows

C.I.V. Manzanilla-Pech[1], R.B. Stephansen[1] and J. Lassen[1,2]

[1]Aarhus University, C. F Møllers allé 3, 8000 Aarhus C, Denmark, [2]Viking Genetics, Ebeltoftvej 16, Assenstoft, 8960 Randers, Denmark, Denmark; coralia.manzanilla@qgg.au.dk

Recording sparse phenotypes as feed intake on a large scale is becoming possible with the incorporation of recently developed new technologies, as the use of 3D cameras. With the availability of dry matter intake (DMI) records from such systems, it becomes possible to estimate heritabilities across lactation with complex models as random regression. The objective of this study was to estimate heritabilities for DMI measured by 3D cameras across lactation week in first parity Holstein (HOL), Jersey (JER), and Nordic Red (NR) cows. Data was collected on 18 commercial farms in Denmark (HOL=7, JER=5, NR=6) in the period from 2020 to 2022. A total of 69,971 weekly records in 3,572 first parity cows were analysed (HOL=28,966 records/1,441 cows, JER=23,234 records /1,086 cows, NR= 17,771 records /1,045 cows). In average, 17 to 21 weekly DMI records were available per cow per breed. Fixed effects included herd-year-season, week of lactation (1-44, modelled with 3rd order Legendre polynomial) and age at first calving as covariate. Random effects included genetic and permanent environmental effects, which were modelled using a 2nd order Legendre polynomial. Heterogeneous residual variance was modelled per month of lactation (1-11). Preliminary results indicate estimated heritabilities for DMI ranging from 0.18 to 0.35 for Holstein, from 0.11 to 0.21 in Jersey and from 0.38 to 0.53 in Nordic Red. These heritabilities values are within the range of previously reported in literature for DMI across the lactation measured by traditional methods as feed bins.

Genomic correlations of MIR-predicted and measured feed efficiency traits in Holstein Friesian

A. Seidel[1], C. Heuer[1], L.M. Dale[2], A. Werner[2], H. Spiekers[3] and G. Thaller[1]
[1]Christian-Albrechts-University, Institute of Animal Breeding and Husbandry, Olshausenstr. 40, 24098 Kiel, Germany, [2]Regional association for performance testing in livestock breeding of Baden-Wuerttemberg, Heinrich-Baumann Str. 1-3, 70190 Stuttgart, Germany, [3]Bavarian State Research Centre, Institute for Animal Nutrition and Feed Management, Prof.-Dürrwaechter-Platz 3, 85586 Poing, Germany; aseidel@tierzucht.uni-kiel.de

Feed is a major component of variable and environmental costs associated with dairy systems and is therefore a crucial parameter for future breeding programs. However, feed related traits are complex and costly to measure at the population level. Thus, individual estimations based on routinely measured mid-infrared spectroscopy (MIR) data from milking cows could serve as a potential indicator for phenotypic and genetic purposes. Therefore, the aim of this study was to compare predictions for the traits energy balance (EB NEL) energy efficiency (EE NEL) according to GfE (2001) as well as feed efficiency (FE ECM) from the MIR spectra to on-farm measurements. All MIR models were built in R software with the help of 'pls' and 'glmnet' packages with a RPD higher than 2. Pairwise bivariate (one trait, on-farm vs MIR) random regression models were established to estimate SNP based genomic correlations. For the analyses, 13,797 measured as well as MIR predicted observations of 671 Holstein Friesian cows from four German research farms were used. For all animals 50k SNP genotypes were available. In the models, herd-year-season and lactation number (1, 2, 3, 4, ≥5) were considered as fixed effects. Parameters were estimated from lactation days 10 to 300. The additive genetic and the permanent environmental effect were both modelled using the third degree polynomial distribution of lactation day, allowing for heterogeneous variances. SNP based correlations were calculated as the average correlation (20,000 replicates) of posterior breeding values for each lactation day. Average genomic correlations within trait and across days in milk varied between 0.34-0.95. Specifically, for the traits EB NEL, EE NEL and FE ECM, 0.82 to 0.95, 0.34 to 0.51 and 0.52 to 0.55 respectively. The results demonstrate high potential integrating MIR-estimated efficiencies as indicator traits for innovative breeding purposes.

Heritability of dry matter intake in Norwegian Red with measures on feed intake in commercial herds

K.A. Bakke[1] and B. Heringstad[1,2]
[1]Geno A.I and Breeding organisation, Storhamargata 44, 2317 Hamar, Norway, [2]Norwegian University of Life Sciences, Oluf Thesens vei 6, 1430 Ås, Norway; karoline.bakke@geno.no

An ongoing project for monitoring direct feed-intake in commercial dairy herds enables longitudinal measurements on cows direct feed intake. The aim of the current study was to perform the first genetic analyses of daily dry matter intake in Norwegian Red cows. Equipment for recording of roughage intake has been installed in collaborating dairy herds. Data from 7 of these herds from 2022 was included in the analyses. After editing the data and combining data from concentrate intake, roughage intake, and the monthly feed analyses of roughage a total of 58,370 records of daily total dry matter intake from 484 Norwegian Red cows entered this study. The mean (standard deviation) daily dry matter intake in the period from January to October 2022 were 20.5 (4.3) kg. Variance components for daily dry matter intake were estimated with a linear animal repeatability model. The fixed effects of herd, days in milk and parity were included. For first parity cows, age at calving in months were included. As random effects we modelled herd-test-day, additive genetic effect, and permanent environment effect of animal. The heritability (se) for daily dry matter intake in Norwegian Red was 0.18 (0.05) with a repeatability (se) 0.33 (0.02). Predicted breeding values (BLUP) varied from -3.49 to 3.81 (1.22) for the 484 animals with data. Results are promising for Genos further work on feed efficiency and the amount of data are steadily increasing. The future work will include genetic analyses of energy-corrected feed-intake and its correlation to milk yield and to estimate genomic breeding values for all cows in the 14 herds with equipment for recording feed intake.

Session 24 Theatre 4

Assessing the potential of faecal NIR spectra for the prediction of feed efficiency in dairy cows

S. Ampuero Kragten[1], T. Haak[2,3], K.-H. Südekum[2] and F. Schori[3]

[1]Agroscope, Method Development and Analytics, 1725 Posieux, Switzerland, [2]University Bonn, Institute of Animal Science, Bonn, Germany, [3]Agroscope, Ruminant Nutrition and Emissions, 1725 Posieux, Switzerland; silvia.ampuero@agroscope.admin.ch

Optimizing the feed efficiency (FE) of ruminants might reduce greenhouse gas emissions while limiting production costs. Thereby, reducing the environmental footprint of dairy production. This paper assesses the potential of faecal NIR spectra prediction models to estimate: feed conversion ratio [FCR = dry matter intake (kg DMI/d)/energy corrected milk yield (kg/d)], nitrogen use efficiency [NUE = milk-N yield (g/d)/feed-N intake (g/d)], residual feed, energy and nitrogen intakes [actual – predicted values (RFI: kg DMI/d), (REI: MJ NEL/d), (RNI: g N/d) resp.]. Within the frame of larger projects, FE traits were evaluated for 86 lactating cows (60 Holstein, 16 Swiss Fleckvieh), (grazing and stall-fed), primiparous and multiparous, in mid and late lactation. The n-alkane marker method for grazing cows and weighing consumed rations for stall-fed cows were used for DMI determination together with Swiss feed recommendations for predicted intake, NEL and N values. Freeze-dried faecal samples were scanned in diffuse reflectance mode with a NIRFlex N-500 (Büchi, CH-Flawil), with 3 replicate spectra per sample, 21 scans per replicate, in the range of 4,000 to 10,000 cm^{-1}. 132 samples were a 7-days mass ratio pool of a one-week measurement period; 32 of those were also scanned as 3-days pool (1, 4 and 7 d) and as single-day sample (1 d). NIRS models were developed with PLS algorithms including: snv (normalization standard normal variate), mf (normalization MSC Full) and db1g2 (1st derivative BCAP Gap 2). NIRS models (independent calibration (cal) and validation (val) sets, ~70 and 30% of samples resp.) showed moderate to good predictive potential for FE traits [R^2 (cal/val): 0.75/0.76, 0.84/0.84, 0.58/0.58, 0.67/0.65 and 0.88/0.82; residual prediction deviation (SD calibration-set/standard error of prediction): 2.2, 2.5, 1.6, 1.9 and 2.8 for FCE, NUE, RFI, REI and RNI resp.], which is interesting, especially for selection of dairy cows. These models should be incremented with samples from diverse origins (breeds, diets, production and environmental conditions) to improve their robustness.

Session 24 Theatre 5

Genetic parameters for potential auxiliary traits for lameness based on data from PLF technologies

K. Schodl[1,2], B. Fuerst-Waltl[2], F. Steininger[1], M. Suntinger[1], H. Schwarzenbacher[1], A. Köck[1], D4dairy Consortium[1] and C. Egger-Danner[1]

[1]ZuchtData EDV-Dienstleistungen GmbH, Dresdner Str. 89, 1200 Vienna, Austria, [2]Univ.Nat.Res.Life Sci. Vienna, Gregor-Mendel-Str. 33, 1180 Vienna, Austria; schodl@zuchtdata.at

Lameness in dairy cattle presents a widespread animal welfare and economic issue. Lameness may be associated with changes in milk performance or behaviour. Integrating lameness into routine genetic evaluation is important for improving dairy health and welfare. However, like many other direct health traits large scale phenotyping is challenging. Precision livestock farming (PLF) technologies such as sensor and automatic milking systems (AMS) provide a valuable opportunity for large scale phenotyping. Within the D4Dairy project we investigated the potential of PLF data for deriving auxiliary traits for lameness, which may be used in genetic evaluation. Data from 32 farms equipped with Lely AMS and sensors measuring activity, rumination and feeding time were collected between January 2020 and March 2021. Lameness was assessed by employees of the milk recording organisations during their routine visits and data from 78 farms were included. The lameness assessment scale ranged from 1 (not lame) to 5 (severely lame). Daily mean, median, and standard deviation (SD) were calculated for the sensor parameters and daily milk yields were derived from the AMS. Variables for genetic parameter estimation comprised lameness scores, daily milk yields and sensor parameters on days 0, -5, and -10 to the lameness assessment. Bivariate linear animal models were fitted to 33,977 observations of 4,888 Simmental cows (13,742 observations and 1,993 animals including sensor and AMS data) to estimate heritabilities (h^2) and genetic correlations (r_g). The h^2 was highest for feeding behaviour (0.11-0.44) followed by rumination (0.04-0.30) and activity (0.7-0.18) and lameness (0.12) with low standard errors for all traits. Low r_g were found for mean feeding time (0.07), for rumination SD during days -10 (0.21) and for median activity on day 0 (-0.13). Heritabilities show that sensor parameters have potential for use in breeding. Low correlations to lameness were found for some traits, but the use as proxies for lameness require further work on trait definition.

Session 24 Theatre 6

Breeding for resilience based on rumination time data of Chinese Holstein heifers

W. Lou[1,2], R. Shi[1,2], B. Ducro[2], A. Van Der Linden[2], H.A. Mulder[2], S.J. Oosting[2] and Y. Wang[1]
[1]China Agricultural University, No.2 Yuanmingyuan West Road, Haidian District, 100193 Beijing, China, P.R., [2]Wageningen University & Research, P.O. Box 338, 6700 AH Wageningen, the Netherlands; wenqi.lou@wur.nl

Breeding for resilient cows could result in lower impact of stressors and reduce extra costs for labour and antibiotics. Resilience indicators can be derived when it is possible to record perturbations in the cows' rumination. These indicators can be used in farm management and breeding provided these are inherited traits. Therefore, the aims of this study were to define resilience indicators from daily rumination time (DRT) in heifers, and to investigate if they can be used to genetically improve resilience. After quality control and outlier detection, 113,813 records on 2,246 Chinese Holstein heifers were used in this study. The resilience indicators were defined based on deviations from a quantile polynomial regression model on DRT. Genetic parameters for the resilience indicators were estimated using an animal model. Results showed that DRT had a moderate heritability (0.17-0.44) and repeatability (0.21-0.66), while resilience indicators had low a to moderate heritability (0.03-0.16). The top 10% heifers with highest estimated breeding value (EBV) for resilience had more often a successful first insemination (57 vs 53%), an easy first calving more often (76 vs 66%), and a lower disease incidence (26 vs 30%) than the 10% heifers with the lowest EBV for resilience. The top 10% sires with the highest EBV for resilience had a higher genomic EBV for feed efficiency (71.92 vs 61.13) and feed saved (79.09 vs -48.83), productive life (1.54 vs 0.78), and body confirmation score (0.15 vs -0.29) than the 10% sires with lowest EBV for resilience. These findings lay a foundation for future work to select more resilient dairy cows based on resilience indicators from rumination time.

Session 24 Theatre 7

Validation of a high-throughput movable 3D device for the acquisition of the whole cattle body

A. Lebreton[1], C. Allain[1], L. Delattre[2], J. Loof[1], Y. Do[1] and M. Bruyas[3]
[1]Institut de l'Elevage, 149 Rue de Bercy, 75595 Paris, France, [2]3D Ouest, 5 Rue Louis de Broglie, 22300 Lannion, France, [3]Eliance, 149 Rue de Bercy, 75595 Paris, France; adrien.lebreton@idele.fr

In the past ten years, proofs of concept have flourished around the use of three-dimensional imagery to assess cattle morphology, but devices were not adapted for on-farm phenotyping. Indeed, most of them were sensitive to animal movements, sunlight or dust. Our objective was to upscale a 3D image scanner adapted to on-farm phenotyping and movable from one commercial farm to another. The 3D scanner is a dismountable gantry (3×2.5×0.7 m) embedding 10 depth sensors which combine their data acquisition to provide a 3D image of the whole body. Animals are scanned in motion by walking under the device allowing high throughput phenotyping. Embedded algorithms clean the image automatically allowing a direct processing of the image. The scanner was tested in a variety of situations: inside and outside, positioned at the end of a scale or between 2 pens in a fattening barn. To validate the accuracy of the 3D image acquisition, we used the method proposed by Le Cozler *et al.* comparing the values measured on the live animals (REF) to those extracted from the 3D images manually (MAN) and automatically by image processing (PROC). The traits used in the validation process were hip width (HW), heart girth (HG), chest depth (CD), wither height (WH) and sacrum height (SH). REF measures were collected in 2 sessions on 33 animals (crossbreed calves 6-12 mo). Animals were scanned in the different conditions described above. Correlations with REF for both MAN and PROC methods were higher than 0.9 on WH and SH measures, higher than 0.8 for CD and HW measures and higher than 0.7 on HG. RMSEs and coefficients of variation of repeatability were also assessed. The accuracy of this new device allows on farm phenotyping. Automatic processing of the image is already promising and will be further developed into the PHENO3D project that aims to predict body weight and morphological trait scores used in the beef cattle genetic indexes.

Session 24 Theatre 8

MIRS coupled with machine learning for the prediction of cow colostrum immunoglobulins concentration

A. Costa[1], A. Goi[2] and M. De Marchi[2]
[1]University of Bologna, via Tolara di Sopra, Ozzano dell'Emilia 40064, Italy, [2]University of Padova, Viale dell'Università, Legnaro 35020, Italy; angela.costa2@unibo.it

The concentration of immunoglobulins G (IgG, g/l) is the key parameter for the quality of colostrum and IgG below 50 g/l indicates insufficient quality due to the concrete risk of failure of passive transfer in calves. Mid-infrared spectroscopy (MIR), routinely used for milk official analyses worldwide, can be exploited for quality screening of colostrum for different purposes on both portable or benchtop devices. In the present study we evaluated how MIR spectra coupled with machine learning perform for the prediction of colostrum quality based on IgG. 531 colostrum samples were harvested within 6 h from calving from 531 Italian Holstein cows in different parities (1 to 8) and farms (n=9) between 2019 and 2020. Cows were not vaccinated before calving. The RID assay was adopted for IgG concentration determination, whereas MIR spectra were collected through a Milkoscan FT7 (FOSS A/S, Hillerød, Denmark). Samples were flagged as 'low quality' if IgG was either below 50 or 70 g/l. The 1,060 spectral wavelengths, in absorbance, were used as predictors of: (1) IgG in regression; (2) and quality level in classification. The 70% of data were adopted to train the model using bootstrapping and testing was carried out on the remaining 30% of data. Regardless of the approach used, MIR coupled with machine learning provided promising prediction accuracy. As regards the classification ability, when the threshold at 70 g/l was considered, the amount of true positive (96.7%) and true negative samples (81.8%) in the testing set were greater compared to the other threshold, i.e. 50 g/l. In training, the out-of-bag errors were 0.05 (50 g/l) and 0.07 (70 g/l). Satisfactory accuracies were observed in regression: the coefficient of determination in training (0.97) and testing (0.74) resulted promising. The ten most important wavenumbers were similar between regression and classification with a slight re-ranking and were located in milk protein absorption region. Based on our findings, it would be possible in future to accurately assess colostrum quality exploiting MIR-based devices.

Session 24 Theatre 9

Extracting video-based phenotypes in a pig breeding programme

C. Coello[1], Ø. Nordbø[1], R. Sagevik[1], F. Cheikh[2], M. Ullah[2], K.H. Martinsen[1] and E. Grindflek[1]
[1]Norsvin AS, Storhamargata 44, 2317, Norway, [2]NTNU Gjøvik, Department of Computer Science, Teknologivegen 22, 2815 Gjøvik, Norway; christopher.coello@norsvin.no

In breeding and management of pigs there are needs to improve welfare of pigs raised in groups, for example to minimize biting and damage to pen mates, minimize stress and fair of the animal itself, or avoid aggression towards farmers. Behaviour phenotypes are, however, difficult to obtain in an objective, repetitive and cost-effective way for breeding companies. Video cameras, together with analysis pipelines extracting the phenotypes of interest, have been proposed in order to achieve this. One critical aspect of this analysis pipeline is the necessity to identify the animal from which the behaviour is detected. The robustness of this identification is necessary in order to define the genetic component for the behaviour. Pigs' similarity makes this step complex and prone to error. In addition, other factors like animal density in the pen, pig's willingness to lie very close to each other or frequent occlusions occurring due to the top-down acquisition geometry of the video camera. Therefore, additional information needs to be combined to the video feed in order to achieve this identification robustly. The objective of this work is to set up a baseline pipeline to measure activity budget phenotypes like walking distance, time spent lying, etc. To achieve this, we have implemented a state-of-the-art activity budget detection pipeline based on a detection module (Yolov8), a tracking module (StrongSORT) and an identification module (colocalization with feeding station). We have looked at how the identification of the tracks can be successful using additional information like colour eartag information. Manually labelled video tracks were used to validate this automatic phenotype extraction method. Use-case specific validation metrics, like the fraction of the video feed that an animal is correctly identified, were measured in parallel to the more classic detection metrics (e.g. F_1-score=0.965 and mAP@.5=0.976) and tracking metrics (e.g. identity switches). From our initial experiments, robust identification of animals (correct identity>50% of the time) cannot be achieved with the current tracking algorithm by only using colocalization with feeding station as identification procedure.

Session 24 Theatre 10

Impact of the social effect on feeding behaviour in pigs

P. Núñez[1], S. Gol[2], J. Reixach[2] and N. Ibañez-Escriche[1]
[1]Universitat Politècnica de València, Institute for Animal Science and Technology, Cami de Vera, 46022, Spain, [2]Selección Batallé S.A., Riudarenes, 17421, Spain; pnuez@posgrado.upv.es

Automatic feeders allow the measurement of individual feeding behaviour in porcine. Social interactions within a group of growth pigs could influence feeding behaviour. This study aims to estimate the percentage of phenotypic variance explained by the social effect on feed consumption and duration per visit. Data analysed had 843,605 visits from 1,608 purebred Pietrain pigs (Batalle S.A. selection nucleus) housed in groups of 12.7±1.8 pigs with a 1.5 m^2 per pig density. The feed intake and duration were measured with automatic feeders (PPT, Nedap, Netherlands). Bayesian analyses were carried out using a classical animal model y= Xb+Za+e, and a 'social model' y= Xb+Za+Z's+e, which included the social effect (Z's) of the pig that visited the feeder (follower) immediately after the visit under analysis. The analyses were performed with the complete database, as well as for two subgroups classified as competitive visits (time<60 seconds between visits, n=336,914) and no competitive visits (>600 seconds, n=129,180). Fixed effects included the contemporary group, the hour of the day, and the age in days. The assumed distributions were a ~ N (0,A σ^2) and s ~ N (0,I σ^2) for the additive genetic and social effect, respectively. Models' effects and variance components were obtained from the posterior marginal distribution generated with Gibbs1f90. The estimated heritabilities with both models showed similar values for the three databases (from 0.21-0.25 for the duration and 0.15-0.16 for consumption). The social effect explained less than 1% of both traits in the complete database and the 'no competitive' database. However, this percentage increased to 2.3% in the competitive visits. These results are similar to those found in Angarita *et al.*, where the same models were evaluated for duration but with fewer animals. As we expected, the social effect was more important in the competitive visits compared with no competitive visits. However, this effect explained a small portion of the phenotypic variance (2.3%). These results indicate that in these housing conditions and genetic line, the social effect did not have a relevant impact on the duration and feed intake per visit.

Session 24 Theatre 11

Assessing lactating sow behaviour using sensor technology and machine learning

G. Dumas[1], L. Maignel[2], J.-G. Turgeon[1] and P. Gagnon[1]
[1]Centre de développement du porc du Québec, Place de la Cité – Tour Belle Cour 450 – 2590, boulevard Laurier, Quebec City, Quebec, G1V 4M6, Canada, [2]Canadian Centre for Swine Improvement, 960 Carling avenue Building 75, Ottawa, Ontario, K1A 0C6, Canada; laurence@ccsi.ca

Sow behaviour during lactation is a key factor affecting pre-weaning survival. Maternal behaviour is a combination of several complex traits, including sow temperament and nervousness after farrowing. The frequency of postural changes in lactating sows is often used as a predictor of their stress level. Calm sows tend to wean more piglets, however, they have to be alert to raise up quickly to avoid crushing piglets, and stand up to eat and drink enough to cover their high energy requirements during lactation. Frequent standing/laying transitions lead to increased risk of piglet crushing. Vision systems allowing for 24/7 tracking of lactating sows allow for real-time behaviour tracking as potential management tools and provide opportunities to record activity levels as potential selection criteria. Since it might not be possible to install cameras in all farms, a simplified system using a set of infrared sensors was designed by the *Centre de développement du porc du Québec* (CDPQ) and tested to collect information to identify and quantify basic sow postures over time. Ten farrowing pens were equipped with cameras and sensors at the CDPQ Sow Research and Training Barn (Armagh, Québec) and a total of 59 commercial sows were tracked from entry to weaning. Specific 1-hour video sequences were annotated using standard methods to be used as a training set. The data collected by the sensors was used to predict basic sow postures including standing, sitting and laying down using machine learning approaches such as random decision forests. Preliminary results show that the sensors can classify sow postures and detect sow postural changes, however the prediction accuracy varies across sensors, and the algorithm should be improved before we can use it to compute indicators of interest such as the amount of time spent in each posture and the frequency of postural transitions for each day in lactation. The newly developed sensor system is a promising easy to install, inexpensive, contactless and accurate solution to track sow behaviour during lactation.

Session 24 Theatre 12

Relationships between sow activity over the days after farrowing and litter weight gain

O. Girardie[1], D. Laloë[2], M. Bonneau[3], Y. Billon[4], J. Bailly[4], I. David[1] and L. Canario[1]
[1]INRAE, GenPhySE, 31320 Auzeville-Tolosane, France, [2]INRAE, GABI, 78350 Jouy-en-Josas, France, [3]INRAE, ASSET, 97170 Petit-Bourg, Guadeloupe, [4]INRAE, GenESI, 17700 Surgères, France; oceane.girardie@inrae.fr

Litter growth during the first week of lactation depends on sow maternal ability. The objective of the study was to identify which behaviours of the sow are related to the average daily gain of her litter (ADGL) considering 3 periods after farrowing: D0 to D1, D1 to D3 and D3 to D7. Behaviour of 21 Large White and 22 Meishan primiparous sows, kept in individual pens, was extracted from the analysis of video records. Convolutional Neural Networks were used to obtain sow daily time-budgets for posture activity and standing activity. Sow postures included standing, sitting, lying ventrally, lying laterally with/without udder exposed and the number of posture changes per day (total and those hiding the udder). Sow standing activity included eating, drinking and other (e.g. exploring, rooting). Daily feed intake of each sow was also recorded and grouped with standing activity behaviours. Piglets were individually weighted at D0, D1, D3 and D7. To dissect the temporal evolution of data structure, a special type of PCA, the Partial Triadic Analysis (PTA) was applied to litter ADGL values, and either posture activity or standing activity, considering the 3 periods. Such analysis permits to study the temporal evolution of the association between behaviour and litter growth in comparison to a compromise. Since breeds differed greatly in piglet growth, data was adjusted for the breed effect. On this compromise, ADGL showed a slightly positive correlation with sow feed intake and time spent eating, according to axis 1 which explained 51% of variability. This association was much stronger on the first period (D0-D1) than subsequent periods. The ADGL was positively correlated to number of posture changes and negatively correlated to time spent lying ventrally according to axis 2 (27% of variability explained), markedly at D0-D1 but not at subsequent periods. In conclusion, sows that were eating more and were more restless on D0 had a higher litter weight gain during D0-D1 than other sows. Further explanatory analysis of ADGL according to behavioural data is in progress.

Session 24 Theatre 13

Multi-camera tracking of turkeys in large groups using instance segmentation

Z. Wang[1], P. Langenhuizen[2], B. Visser[3], H.P. Doekes[1], P. Bijma[1] and P.H.N. De With[2]
[1]Wageningen University and Research, Animal Breeding and Genomics, P.O. Box 338, 6700 AH Wageningen, the Netherlands, [2]Eindhoven University of Technology, Centre for Care & Cure Technology, P.O. Box 513, 5600 MB Eindhoven, the Netherlands, [3]Hendrix Genetics Research, Technology & Services B.V., Spoorstraat 69, 5830 AC Boxmeer, the Netherlands; zhuoshi.wang@wur.nl

Aggressive feather pecking in poultry can result in poor plumage, skin damage and mortality. Feather pecking is a multidimensional problem, relating to e.g. housing conditions, feed, management and genetics. Although the trait is heritable, breeding against feather pecking is still limited due to a lack of automated phenotyping methods for large groups of birds. To develop an automated phenotyping method, there are two main challenges: (1) detecting pecking events, and (2) identifying the birds involved. Here, we focus on the latter. We set up a system based on computer vision and radio-frequency identification (RFID) to track over 300 turkeys in a pen of 120 m^2. We installed 35 cameras, such that each part of the pen was covered by two cameras. We equipped each bird with a RFID tag and placed RFID antennas in the feeders for (re-)identification. Between a feather pecking event and re-identification at a RFID antenna, birds need to be tracked through computer vision. For this purpose, we here present a multi-camera tracking algorithm. First, the 35 fields of view were calibrated and transformed to a global coordinate system. Second, a Yolov7 instance segmentation model was trained to segment the turkeys in each field of view. Third, to track birds across cameras, the overlap of instance segments from neighbouring cameras was calculated. If two segments overlapped more than 60%, they were considered to be the same bird. Fourth, to track birds over time, the overlap of segments between subsequent frames was calculated. Again, if overlap was more than 60%, the segments were considered to be the same bird. Here we present results of the tracking accuracy. The current tracking time is not enough yet to bridge the time gap between a pecking event and re-identification at the feeder. However, we believe that improvement of this model will enable tracking of animals for longer time periods in the near future.

Evaluation of tracking individual broilers using video data

J.E. Doornweerd[1], G. Kootstra[2], R.F. Veerkamp[1], B. De Klerk[3], M. Van Der Sluis[1], A.C. Bouwman[1] and E.D. Ellen[1]
[1]Wageningen University and Research, Animal Breeding and Genomics, P.O. Box 338, 6700 AH, Wageningen, the Netherlands, [2]Wageningen University and Research, Farm Technology, P.O. Box 16, 6700 AA, Wageningen, the Netherlands, [3]Cobb Europe BV, Koorstraat 2, 5831 GH, Boxmeer, the Netherlands; janerik.doornweerd@wur.nl

Tracking individual broilers on video could provide valuable information on the health, welfare, and performance on many animals with one sensor. These individual broiler tracks would enable breeders to phenotype for novel or indicator traits that aid genetic improvement. However, tracking group-housed individual broilers is a challenge due to the large number of similar looking animals in the barn. The aim of this study was to measure how long broilers could be tracked in terms of time and distance in our experimental pen. The experimental pen (1.80×2.61 m) housed 39 male broilers, of which 35 were non-colour marked. Only data at 18 days of age were used. A YOLOv7 model was trained (n=140), validated (n=30), and tested (n=30) on two-hundred annotated frames to detect broilers within frames. On the test set, YOLOv7 had a precision and recall of 0.99 and an average precision (@0.75 – Intersection over Union threshold) of 0.98, which showed that individual animals were reliably detected within frames. The tracker (SORT) – to link individual detections between frames – was evaluated on human annotated ground-truth tracks of eleven white broilers based on the first frame of every minute (136 frames) over an approximately 2-hour period. Broilers had an average of 10 ID-switches (IDSW) per ground-truth track, with a maximum of 20 IDSW and a minimum of 5 IDSW. Broiler tracklets were on average 12 minutes, with the longest tracklet lasting 51 minutes. The average tracking distance was 1.83 meters per tracklet, with a maximum of 17.07 meters. Shorter tracklet durations and distances seem to coincide with more IDSW, and thus a shorter time to phenotype the animal. Almost all broilers were detected (YOLOv7), but the tracker (SORT) could be improved to reduce the number of IDSW. Overall, tracking group-housed individual broilers on video shows promise in enabling breeders to phenotype for novel or indicator traits to aid genetic improvement.

Session 24 Poster 15

Genetic variation of new sheep traits measured by dual energy Xray absorptiometry

C.E. Payne[1], B. Paganoni[1], S.F. Walkom[2], G.E. Gardner[3] and D.J. Brown[2,3]
[1]Department of Primary Industries and Regional Development WA, Nash Street, Perth WA, 6000, Australia, [2]Animal Genetics & Breeding Unit, University of New England, Armidale NSW, 2351, Australia, [3]Advanced Livestock Measurement Technologies, Murdoch University, Murdoch WA, 6150, Australia; claire.payne@dpird.wa.gov.au

Dual energy Xray absorptiometry (DEXA) is regularly used to measure body composition and bone characteristics in human medicine. Medical grade DEXA can accurately measure the proportion of fat, lean and bone in production animals, as well as bone minerals in lambs. Recently, DEXA modified for lamb carcase scanning at abattoir-chain-speed precisely predicted CT fat and lean % over a range of lamb genotypes. There is opportunity to measure new traits and identify the genetic variation of these traits on live sheep using a medical grade DEXA. Lambs from the Genetic Resource Flock (n=400) are live DEXA scanned at the Katanning Research Station and subsequently DEXA scanned as carcases at the processing facility. DEXA technology scans an object at two different energy levels. The ratio (R value) between the high and low energy level for each pixel in the image relates to the atomic mass of the tissue. Analysis will include whole body and sub-regional lean and fat %, and bone %. Individual pelvic bones are also analysed on the live DEXA scans to measure pelvic area and bone content. This data will provide the opportunity to identify correlations between live DEXA and carcase DEXA measures of body composition within and across sire types. Furthermore, there is potential to link individual bone analysis to important genetic reproduction traits such as lamb birthweight and lamb survival.

Session 25 Theatre 1

Monitoring within-breed genetic variation at global level

G. Leroy[1], R. Baumung[1], G. Mészáros[2], I. Curik[3], J.J. Windig[4], B. Rosen[5], Y.T. Utsunomiya[6], P. Burger[7], L. Colli[8], P. Boettcher[1], A. Stella[9], C. Looft[10] and J. Soelkner[2]
[1]FAO, Viale delle Terme di Caracalla, Roma, Italy, [2]BOKU, Gregor-Mendel-Straße 33, Vienna, Austria, [3]University of Zagreb, Svetošimunska cesta 25, Zagreb, Croatia, [4]WUR, P.O. Box 338, Wageningen, the Netherlands, [5]USDA, 10300 Baltimore Avenue bldg, Beltsville, USA, [6]UNESP, Av. Prof. Dr. Orlando Marques de Paiva, 87, São Paulo, Brazil, [7]University of Veterinary Medicine, Vetmeduni Savoyenstrasse 1, Vienna, Austria, [8]UCSC, Via Emilia Parmense 84, Piacenza, Italy, [9]CNR, Via Edoardo Bassini, 15, Milano, Italy, [10]University of Applied Science, Brodaer Str. 2, Neubrandenburg, Germany; gregoire.leroy@fao.org

Despite the recognised importance of maintaining genetic variation, the inclusion of related indicators in global monitoring systems has been neglected in recent decades. Especially when considering molecular indicators, there are many challenges related to technological and organisational capacities: Sampling requirements, the choice of the marker sets and methods' properties limit the feasibility of many indicators. A review of demographic, pedigree and genomic parameters that could complement the existing indicators in Domestic Animal Diversity Information System (DAD-IS) shows that the main constraint is the lack of availability of pedigree and molecular data, especially in developing countries. In this context effective population size (Ne) appears to be an appropriate and practical candidate, as it is a direct measure of how well genetic diversity in a population is maintained, and therefore of the genetic risk status of a population. It can be compared across populations and species and is straightforward in terms of interpretation. Moreover, it can be estimated from demographic, pedigree and genomic information sources. Molecular tools provide the opportunity to obtain estimates of Ne that are considerably more accurate compared to demographic indicators and pedigree based measures, especially if pedigree information is of low quality. Given the approximations and potential biases for specific methodologies and data sources, details should be carefully documented in respective information systems, especially for long-term monitoring standardization and comparison.

Session 25 Theatre 2

Real-time monitoring of genetic diversity in animal populations: the RAGEMO project

C. Danchin[1], I. Palhière[2], B. Servin[2] and G. Restoux[3]
[1]IDELE, 149 rue de Bercy, 75012 Paris, France, [2]INRAE – GenPhyse, 24 chemin de Borde-Rouge, 31326 Castanet Tolosan, France, [3]INRAE – GABI, Domaine de Vilvert, 78350 Jouy-en-Josas, France; gwendal.restoux@inrae.fr

Genetic diversity is essential to maintain genetic gain and adaptation abilities for new breeding goals or conditions. However, in selected populations, it is often threatened due to either or both selection on the one hand and genetic drift and inbreeding on the second hand. Thus, a careful and frequent monitoring of the level of variability is necessary. While pedigrees are widely available in France for most breeds, many local breeds have still limited genealogies. Thus, SNP arrays used for genomic selection appear as an affordable solution to estimate the levels of genetic diversity in these cases. Moreover, they allow to estimate additional information about the structure and history of the population like the level of admixture between breeds, the age of the co-ancestry and the presence of deleterious mutations. The RAGEMO project funded by Apis-Gene proposes to develop an automated analysis framework from the deposit of the genotype to the design of a tailor-made report. We worked together with the stakeholders including breed associations (CAPGENES), to select the appropriate indexes and to design the reports depending on the target audience and their uses. The chosen indexes include: (1) within population indexes: (kinship, inbreeding, effective population size, etc.); (2) population structure (PCA, admixture, etc.); and finally (3) the identification of deleterious mutations. The output of these analysis should allow a precise management of breed by making it possible to check the quality of pedigrees, retrieve genealogical relationships, measure the level of diversity and define the limit of the breed (e.g. inclusions of external individuals). As a proof of concept, we first focused on French goats, since they have interesting features including numerous breeds of different population sizes with various status (conservation, selection, with more or less complete pedigrees, etc.). The first results and the analysis pipeline will be presented. The ultimate goal is to extend this monitoring to all French ruminant species by automatically including on the fly all available genotypes.

Session 25 Theatre 3

Evolution of the genomic relationship between a local breed and its mainstream sister breed

H. Wilmot[1,2], M.P.L. Calus[3] and N. Gengler[1,2]
[1]ULiège-GxABT, TERRA Teaching and Research Centre, Passage des Déportés, 2, 5030 Gembloux, Belgium, [2]National Fund for Scientific Research (F.R.S.-FNRS), Rue d'Egmont, 5, 1000 Brussels, Belgium, [3]Wageningen University & Research, Animal Breeding and Genomics, Droevendaalsesteeg, 1, 6700 AH Wageningen, the Netherlands; Helene.Wilmot@uliege.be

In Benelux countries, there is a genomic continuum of red-pied cattle breeds. Some of them are still rather common, like the Meuse-Rhine-Yssel (MRY), from the Netherlands, while others have the endangered European status like the Red-Pied of Ösling (RPO), from Luxembourg. Survival of a small population like RPO is based on finding a good balance between keeping genetic specificities and collaborating with their sister breeds. We therefore tried to estimate the historical evolution of the genomic relationship between RPO and MRY. For this purpose, we divided the MRY samples into four cohorts depending on the birthdate of animals: before 1990 (cohort 1), between 1990 and 1999 (cohort 2), between 2000 and 2009 (cohort 3) and after 2009 (cohort 4). Results from principal component analysis, F_{st} values and the genomic relationship matrix showed that current RPO animals were on average slightly more related to MRY animals from cohorts 1 and 4 (born before 1990 or after 2009) than to MRY animals from cohorts 2 and 3 (born between 1990 and 2009). These results confirmed the known history of RPO. Breeders in Luxembourg imported MRY animals or semen on a large scale before 1990. After 1990, breeding in Luxembourg was more focused on collaboration with Germany and German dual-purpose red-pied (RDN) cattle were not collaborating a lot with MRY at that time. This changed during the last 20 years and many recent RDN sires have MRY ancestors which could explain increasing relationships with RPO too. Even if this study did not contain RDN animals, it shows the evolution of the complex relationships between some of these sister breeds. Genome assisted mating advice should be used to help breeders in their choice of an appropriate MRY (or other related breed) animal for mating. These animals should be genomically close to the current RPO population which will strengthen the collaboration of these breeds and, finally, support the further across-breed genomic evaluations.

Session 25 Theatre 4

Fine decomposition of inbreeding into the source of the co-ancestries

S. Antonios[1], S.T. Rodríguez-Ramilo[1], J.M. Astruc[2], L. Varona[3] and Z.G. Vitezica[1]
[1]INRAE, GenPhySE, 24 chemin de Borde Rouge, 31326 Castanet Tolosan, France, [2]Institut de l'Elevage, 24 chemin de Borde Rouge, 31321 Castanet Tolosan, France, [3]Universidad de Zaragoza, Instituto Agrolimentario de Aragón (IA2), Facultad de Veterinaria, 50013 Zaragoza, Spain; simona.antonios@inrae.fr

Offspring from related individuals yield to a higher degree of homozygosity across the whole genome, also called inbreeding. Several decompositions of inbreeding have been proposed. For instance, inbreeding can be partitioned into effects attributed to founders by following the pathways back to the origin of the alleles that might be identical by descent in inbred animals. In this study, inbreeding is decomposed into the sources of co-ancestry between the parents of an individual. This decomposition is called the Mendelian decomposition of inbreeding and involves founders and the Mendelian sampling of non-founders. The objective of this work is to decompose inbreeding into the source of the co-ancestries. Real data of three French dairy sheep breeds (Basco-Béarnaise, BB; Manech Tête Noire, MTN; and Manech Tête Rousse, MTR) were used. The pedigree data included 190,276, 166,028 and 633,655 animals in BB, MTN, and MTR, respectively, born between 1985 and 2021. In BB, 41% of the individuals were inbred with an average inbreeding coefficient of 0.013. In MTN, inbreeding affected 24% of the individuals, and the average inbreeding coefficient was 0.007. In MTR, 56% of the individuals were inbred and the average inbreeding coefficient was 0.014. The partial inbreeding coefficients from the Mendelian decomposition of inbreeding were 7,044,811, 1,501,362, and 54,758,519 coefficients generated from 3,268, 2,563, and 11,078 ancestors in the BB, MTN, and MTR breeds, respectively. The average partial inbreeding coefficients (±standard deviation) were 5.82×10^{-5} ($\pm1.04\times10^{-3}$) in BB, 1.912×10^{-4}($\pm2.06\times10^{-3}$) in MTN, 2.33×10^{-5} ($\pm8.01\times10^{-4}$) in MTR. Most of the partial inbreeding coefficients were very low; 99.9 and 96.4%, 99.6 and 92.9%, and 99.9 and 98.7% were lower than 0.01 and 0.001 in the BB, MTN, and MTR breeds, respectively. Our findings contribute to understand the state of inbreeding in those breeds and will be helpful for future studies to predict the inbreeding depression load of any individual in the pedigree.

Session 25 Theatre 5

Predicting homozygosity-by-descent to manage inbreeding and diversity

N.S. Forneris and T. Druet
Unit of Animal Genomics, GIGA-R & Faculty of Veterinary Medicine, University of Liège, Avenue de l'Hôpital 1, 4000, Belgium; nforneris@uliege.be

Prediction of homozygous-by-descent (HBD) segments in the future offspring of a given mating is a valuable tool for the management of inbreeding and diversity as it provides information on expected inbreeding levels, at both genome-wide and local scales, and has proven to be efficient. Such prediction amounts to identify identical-by-descent (IBD) segments in the parents. Here we took advantage of 98 whole-genome sequenced trios (sire-dam-offspring) of a Holstein pedigree (264 animals, 84 couples) to evaluate different HBD prediction methods that we classified as follows: (1) hidden Markov models (HMM) that model IBD along the four parental chromosomes, with 15, 9 or 3 IBD classes; (2) ruled-based haplotype-matching methods that scan for identity-by-state segments; (3) methods designed to estimate inbreeding within individuals (ZooRoH or the rule-based ROH) applied on pairs of parental haplotypes; (4) SNP-by-SNP measures; and e) pedigree-based. Medium-density (50K) genotypes from the parents were used to predict inbreeding in the offspring. Scenarios with lower marker density data, genotyping-by-sequencing (GBS) data or smaller sample sizes were also investigated. The methods were compared in terms of correlations between predicted and reference genome-wide inbreeding levels or between locus-specific values (HBD probabilities). Reference inbreeding levels were defined with respect to different based populations (recent vs ancient inbreeding). Two HMM approaches (relying respectively on a 15 IBD class model and on ZooRoH) proved accurate, flexible and robust to predict recent inbreeding across the different tested scenarios, both at the genome-wide (r=0.74-0.75) and locus-specific levels (r=0.48-0.49). An efficient haplotype-matching method proved fast and accurate for genome-wide prediction. Use of GBS (both globally and locally) and lower marker density (at the locus-specific level) data resulted in lower performances than those obtained with the medium density array. Lower sample size did not affect the relative behaviour of the predictors, when results were averaged across subsets. We are currently running simulations, with two demographic scenarios, to further characterize the different methods.

Session 25 Theatre 6

GWAS analyses for dairy traits in Cyprus sheep and goat breeds reveal important protein networks

G. Maimaris, A.N. Georgiou, A.C. Dimitriou and G. Hadjipavlou
Agriculture Research Institute, Animal Production, P.O. Box 22016, 1516, Lefkosia, Cyprus; gmaimaris@ari.moa.gov.cy

The importance of food securing and hence of investing in the Primary Production Sector was repeatedly highlighted in the past few years because of the climate change impacts, the globally long-lasting economic crisis and the fast-changing geopolitics, especially in Eastern Europe. The island of Cyprus is located in the southeastern part of Europe and due to unfavourable environmental conditions, crop production is limited while at the same time sheep and goat milk demand is constantly increasing. This need is not only attributed to the local growing population but also to the recent registration of halloumi cheese as a product with Protected Designation of Origin. Halloumi cheese consumption is one hundred times more abroad than within the island's borders. Therefore, given the increasing milk demand, the arid climate and the limited size of the country, investing in the genetic improvement of the small ruminant population, is of utmost importance for Cyprus. Towards this direction, the Agriculture Research Institute (ARI) has initiated a four-year project including 15,000 animals from the local Cyprus Chios sheep and Cyprus Damascus goat breeds. Given their constant presence on the island for more than 70 years, these breeds are highly adapted to the local environmental conditions. Therefore, they are ideal candidates for genetically improving commercial-value genetic-traits including milk-, protein- and fat- yield or protein- and fat content, along with weight gain, adaptation and disease resistance, in the context of optimum cost-effective conditions. Using commercially available SNP microarrays (OvineSNP50v3, ICSG_SNP65k_v2), 1000 individuals from each breed have already been genotyped and identified polymorphisms have been incorporated in GWAS analyses. The resulting gene sets and their associated proteins were then incorporated in network and pathway analysis, revealing enrichment in important Biological Processes and Molecular Functions related to the genetic traits of interest.

A severely introgressed South China pig breed in urgent need of purification

L. Cao[1], T. Luan[2], T.H.E. Meuwissen[2], P. Berg[2], J. Yang[1] and Z. Wu[1]
[1]South China Agricultural University, College of Animal Science, Wushan Street 483, Tianhe District, 510642, Guangzhou, China, P.R., [2]Norwegain University of Life Sciences, Faculty of Biosciences, P.O. Box 5003, 1432 Ås, Norway; cao.lu@scau.edu.cn

Yuedong Blck pig (YDH) only distribute in Guangdong province in China. It is one of the four national indigenous pig breeds that are originated in South China region and currently under conservation status. The breed features of black hair and excellent meat quality make YDH popularly embraced by the local market with over 100 million consumers, leading to a considerably higher price than other pig breeds. In addition, its rough feeding resistance, tameness, good fecundity, long service life of sows, and good performance in crossbreeding draw increasing concern from breeders and producers on its potential for development and utilisation. Therefore, we studied the genetic diversity, phylogeny, and population structure of 360 representative YDH pigs using their SNP chip data, in combination of 782 publicly downloaded genome-wide SNP data from other 42 Eurasian or American breeds and wild boars. The final clean dataset contained 1,142 individuals with 34,579 variants. Our results showed the present inbreeding level, expected and observed heterozygosity, ROH, and LD ($r^2_{0.3}$) of YDH population was respectively <0.254, 0.265~0.370 and 0.266~0.331, 0.09~1.15 Mb, and 1.06~7.44 kb, depending on the subgroups of 360 YDH. These results indicated relatively rich genetic diversity in the existing YDH population possibly due to recent introgression into YDH. Moreover, huge genetic variation within YDH population was observed from scattered clustering in principal component analysis. Neighbour-joining tree constructed by identical-by-state-based genomic distance matrix captured 56 YDH pigs far from the major cluster of YDH population, suggesting some severely admixed YDH pigs. Admixture analysis after excluding 56 severely admixed YDH revealed 19.08~43.96% introgression rate in current YDH population. For further study, advanced optimum contribution selection strategy considering genomic information will be mainly explored to search for the possibility of purifying YDH, with aim to provide insight to YDH conservation program.

Insights into ancient and recent inbreeding patterns in Alpine Grey cattle using ROH analysis

G. Gomez Proto, E. Mancin, B. Tuliozi, C. Sartori and R. Mantovani
UNIPD, DAFNAE, Viale dell'Università 16, 35020, Italy; guido.gomezproto@studenti.unipd.it

The maintenance of genetic diversity is a major concern in livestock populations therefore the study of inbreeding is a determining factor to guarantee this genetic background, especially in local breeds such as the Alpine Grey cattle. The use of runs of homozygosity (ROH) as a method for exploring inbreeding has gained popularity due to its ability to precisely identify genome regions that have undergone recent or ancient inbreeding. This approach allows even knowledge of the inbreeding pattern, as well as to make conservation and breeding decisions. The aim of this study was to evaluate genomic inbreeding using ROH analysis and identify regions of genome with ancient and recent inbreeding patterns, using 1,194 Alpine Grey individuals genotyped with medium-low (33K), and high-density (150K) panels. The ROHs were detected using the sliding-window methodology within the 'detectRUNS' R package. Genes potentially related to disease resistance and fertility were found in BTA6 and BTA7. Additionally, two sets of ROHs were established based on their length: short or ancient ROH (ROHA), and long or recent ROH (ROHR), with a cut-off of 8 Mbp of length. The inbreeding coefficient based on ROH (FROH) was calculated as the ratio between the total length of all ROH and the autosomal genome length. The Alpine Grey showed a higher prevalence of short ROHs than long ROHs, with an average of 35.9 ROHA and 4.3 ROHR per individual, meaning that inbreeding in this breed is mainly ancient. Additionally, the regions containing short ROH had a higher FROH than regions containing long ROH with values of 0.04 and 0.02. The correlation between FROHA and FROHR was weakly positive, with a value of 0.05. The maintenance of genetic diversity in Alpine Grey cattle, as well as its productivity and adaptability, should be the focus on future conservation and selection efforts. Achieving this goal will require the development of strategies that mitigate the negative effects of inbreeding on these selected traits. The study was done within Agritech NRC and received funding from the EU Next-GenerationEU (PNRR – Missione 4 Componente 2, Investimento 1.4–D.D.1032 17/06/2022, CN00000022) and by the Dualbreeindg project (CUP J51J18000000005).

Study of Bottleneck effect for conservation of Poonchi chicken from international borders of India a

M. Azad[1], D. Chakraborty[1], K. Kour[1], D. Kumar[1] and P. Birwal[2]
[1]Sher-E-Kashmir University Of Agricultural Sciences &Technology Of Jammu, Animal Genetics and Breeding, Main Campus, Chatha Jammu, Jammu and Kashmir, 180009, India, [2]Punjab Agriculture University, Food Engineering, Ludhiana, 141004, India; mandeepsinghazad@gmail.com

In conservation studies population bottlenecks attributes to the loss of genetic variation in the population in due course of time due to drastic climate changes, predation, limited resources that lead to random fluctuation in the population size. Bottleneck effect within Poonchi chicken of Jammu and Kashmir was evaluated using ten number of microsatellite markers. The microsatellite markers namely ADL0268, MCW026, MCW0081, ADL0278, MCE0069, MCW0111, MCW0222, MCW0016, LEI0094 were analysed in Poonchi chicken. A total of 72 alleles were observed with maximum alleles (9) contributed by locus MCW069, MCW0016 & and the lowest alleles (6) by ADL0268, ADL0278, MCW022. The average observed heterozygosity was 0.8520 with standard deviation of 0.1331, whereas the average expected heterozygosity was 0.8110 with standard deviation of 0.0471. The expected numbers of loci with heterozygosity excess in Poonchi chicken were 5.97 ($P<0.05$), 5.93 ($P<0.05$), 5.90 ($P<0.05$) for Infinite Allele Model (IAM), Two Phase Model of Mutation (TPM) model and Stepwise Mutation Model (SMM), respectively in Sign test. The IAM, TPM and SMM values for one tail for heterozygosity excess in Wilcoxon rank test revealed significant values of ($P<0.05$) deviation indicated all the loci deviates from mutation-drift equilibrium. The qualitative test for allele frequency showed a slight shift from normal L-shaped curve suggesting the recent bottleneck effect in the population. The presence of genetic bottleneck might have affected the number of alleles and resulted in loss of several effective alleles. Furthermore, loss of several effective alleles suggests urgent need of designing effective breeding policy to conserve this unique native germ plasm of Jammu and Kashmir. This study on native chicken of Poonch region is the first report indicating genetic bottleneck effect.

Up to date results from the ongoing *in situ* genetic management of the local breed Noire de Challans

R. Rouger[1], H. Deloison[2], M. Teissier[3], G. Restoux[4] and S. Brard-Fudulea[1]
[1]SYSAAF, Centre INRAE Val de Loire, UMR BOA, 37380 Nouzilly, France, [2]Association pour la sauvegarde et la valorisation de la poule Noire de Challans, 52 Le Broussais, 44460 Saint Nicolas de Redon, France, [3]INRAE GenPhySE, 24 chemin de Borde-Rouge – Auzeville Tolosane, 31326 Castanet Tolosan, France, [4]INRAE GABI, Domaine de Vilvert, 78530 Jouy-en-Josas, France; romuald.rouger@inrae.fr

Poultry local breeds represent a genetic reservoir so large that it is difficult to measure exhaustively. Contrarily to other livestock species, individual identification of local breed animals is uncommon and pedigrees are rarely recorded. As a result, these breeds are at risk of genetic dilution and high level of inbreeding. In this work, we present up to date results about the use of molecular tools to investigate the level of genetic diversity in the French local breed 'Noire de Challans' and reconstruct the pedigree of the animals. One hundred and ninety two SNP markers were tested on a reference population of 54 individuals (generation G0: 15 males and 39 females). After having discarded markers of poor quality, the average minor allele frequency (MAF) of G0 was 0,36. The F_{IS} was significantly positive due to a genetic substructure in G0. A final panel of 96 markers was designated by keeping those with the highest MAF among the markers for which the F_{IS} is comprised between -0.25 and 0.25. The probabilities of identity (1.3×10^{-37}) and exclusion (>0.9999) of this final panel were compatible with its use for parentage assignment purposes. A simulated annealing algorithm was used to design breeding pens aiming to maximise the genetic distance between sires and dams. Ninety-five individuals from the next generation (G1) were genotyped using the panel of 96 markers. Parentage assignment only permitted to designate unambiguously the parents of 47 G1 individuals because of the use of non-genotyped mother to compensate zootechnical issues in G0. Next generations should permit to validate this proof of concept of an *in situ* management of genetic diversity in a local poultry breed. This project has received funding from the European Union's Horizon 2020 research and innovation programme under grant agreement N°101000236 (GEroNIMO project).This project is part of EuroFAANG.

An empirical approach to assess increase in homozygosity: the Gochu Asturcelta pig

K.D. Arias[1], J.P. Gutiérrez[2], I. Fernández[1], I. Álvarez[1] and F. Goyache[1]
[1]SERIDA-Deva, Camino de Riosedo 1225, 33394 Gijón, Spain, [2]UCM, Dpto. de Producción Animal, 28040 Madrid, Spain; kathyah18@gmail.com

Despite the availability of SNP array data, differentiation between the observed homozygosity and that caused by matings between relatives (autozygosity) has major difficulties. This research investigates the usefulness of two recent estimators of increase in homozygosity to monitor the genetic diversity in small populations. Up to 526 Gochu Asturcelta pig parent-offspring trios (534 individuals; 76 families) were genotyped using the Axiom-PorcineHDv1 (Affymetrix). Pedigree depth varied from 0 (founders) to 4 equivalents to discrete generations (t). Four homozygosity parameters (Runs of Homozygosity, H_{ROH}, Heterozygosity-Rich Regions, H_{HRR}, Li and Horvitz's, H_{LH}, and Yang *et al.*'s, H_{YAN}) were computed for each individual, adjusted for the variability in the Base Population (BP; six individuals) and further jackknifed over autosomes. Individual increase in homozygosity (depending on t) and increase in pairwise homozygosity (i.e. increase on the parents' mean) were computed for each individual; effective population size (N_e) was computed for five different cohorts. Genealogical parameters were used for comparisons. Mean inbreeding was 0.120±0.074; mean BP-adjusted homozygosity varied from 0.099±0.081 (H_{LH}) to 0.152±0.075 (H_{YAN}). After jackknifing, mean values were slightly lower. Increase in pairwise homozygosity tended to be two-fold higher than the corresponding individual increase in homozygosity values. When compared with genealogical estimates, N_e estimates based on increase in pairwise homozygosity using both H_{ROH} and H_{HRR} took the lower root-mean-squared errors. Our results suggest that parameters characterizing homozygosity may have difficulties in depicting losses of variability in real small populations in which breeding policy avoids matings between close relatives. Assuming that increase in homozygosity depends only on pedigree depth lead to underestimations in populations with shallow pedigrees. Increase in pairwise homozygosity computed from either H_{ROH} and H_{HRR} can be a promising approach to characterize autozygosity. Project PID2019-103951RB/AEI/10.13039/501100011033. Grant PRE2020-092905.

Preliminary analysis of Nubian goats and their influence on Old English and Anglo-Nubian goats

S.A. Rahmatalla[1,2], D. Arends[3], G.B. Neumann[1], H. Abdel-Shafy[4], J. Conington[5], M. Reissmann[1], M.K. Nassar[1,4] and G.A. Brockmann[1]
[1]Humboldt-Universität zu Berlin, Albrecht Daniel Thaer-Institute, Invalidenstr. 42, Berlin, Germany, [2]University of Khartoum, Faculty of Animal Production, Shambat str., Khartoum, Sudan, [3]Northumbria University, Dep. of Applied Sci., Newcastle upon Tyne, NE7 7XA, United Kingdom, [4]Cairo University, Faculty of Agriculture, Uni. str., Giza, Egypt, [5]SRUC, W Mains, Rd, Edinburgh, United Kingdom; siham.rahmatalla@hu-berlin.de

Nubian goats found in the countries of Northeastern Africa, such as Sudan and Egypt. They play a crucial role in providing self-sufficiency and food security, due to their ability to provide a rich source of milk and meat. In the latter half of the 19th century, Nubian goats were exported from Africa to Britain and crossed with native Old English goats to create the modern Anglo-Nubian breed. We characterized the genetic diversity of Nubian goats from Sudan (n=124) and Egypt (n=11) and their relationship to Old English (n=40) and Anglo-Nubian (n=13) British goats. Animals were genotyped with the Axiom Caprine 60K SNP chip. Nucleotide diversity, observed and expected heterozygosity, and excess of homozygosity were used to assess the genetic diversity. Fixation index (F_{ST}), hierarchical clustering, and principal component analysis (PCA) were used to compare the relationship between populations. The nucleotide diversity in Anglo-Nubian (2.16×10^{-05}) was the lowest compared to the other breeds. Consistently, the highest observed heterozygosity was found in Nubian from Sudan (35.97) and the lowest in Anglo-Nubian goats (24.62). Excess of homozygosity showed the reverse pattern. Four distinct clusters were detected with the hierarchical clustering (Nubian from Sudan, Old English, Anglo-Nubian, and Nubian from Sudan and Egypt), which were confirmed by PCA and F_{ST} analyses. The first two PCs explain 14.3% (PC1) and 6.7% (PC2) of the genetic variation between all individuals. The SNPs that contributed strongly to PC1, and therefore, mostly differentiate Nubian from Old English goats, are located across 16 chromosomes. Candidate genes for adaptation, immune response, reproduction, pigmentation, metabolic process, and goat evolution were found in these regions. Further analysis of these regions will enhance our understanding of breed adaptation.

Session 25 Theatre 13

Towards a flexible definition of core sets based on the haplotype diversity of German sheep

J. Geibel[1,2], C. Reimer[1,2], A. Weigend[2], Y. Shakya[3], H. Melbaum[4], K. Gerdes[5] and S. Weigend[1,2]
[1]University of Goettingen, Center for Integrated Breeding Research, Carl-Sprengel-Weg 1, 37075 Göttingen, Germany, [2]Friedrich-Loeffler-Institut, Institute of Farm Animal Genetics, Höltystraße 10, 31535, Germany, [3]University of Goettingen, European Master of Animal Breeding and Genetics, Albrecht-Thaer-Weg 3, 37075 Göttingen, Germany, [4]Landes-Schafzuchtverband Weser-Ems e.V., Mars-la-tour-Str. 6, 26121 Oldenburg, Germany, [5]Chamber of Agriculture in Lower Saxony, Mars-la-tour-Str. 6, 26121 Oldenburg, Germany; johannes.geibel@fli.de

In recent years, attitudes towards gene banks changed from being a pure backup of a breed for the case of emergency towards being also an active tool in breed conservation. This requires setting scientifically derived demarcations between a 'core' collection (to serve as long term backup) and a 'working' collection of the gene bank. Usual recommendations thereby suggest the optimization of the core collection to preserve a certain amount of the diversity of a breed of interest. However, common definitions of the diversity of populations do not represent something like a total amount of diversity a population has and of which a certain proportion is available in the gene bank, a number which would be possible to communicate to stakeholders. We use simulations based on deep pedigrees to provide a control of the historic breeding background as well as genotypes of East Friesian Milk Sheep and Bentheimer Landschaf being already part of the German Gene Bank together with selected individuals from the field to evaluate the ability of haplotypes (defined as haploblocks) segregating in the population to serve as proxy of the founder genome content of the animals. This allows for the definition of a 'proportion of total diversity' based on the share of all population-wide haploblocks that is present in the collection and by this the derivation of reasonable required sizes of the core collection. This number then allows to be visually communicated to stakeholders. It further enables derivation of conservation decisions independent from the knowledge of potential future relevance of certain genomic variants, as a haploblock should preserve all potential additive and local epistatic effects of single variants.

Session 25 Theatre 14

Use of genomic information for the determination of insemination doses per sire in a gene bank

C. Reimer[1,2], S. Weigend[1,2], T. Pook[1,3] and J. Geibel[1,2]
[1]University of Goettingen, Center for Integrated Breeding Research, Carl-Sprengel-Weg 1, 37075 Göttingen, Germany, [2]Friedrich-Loeffler-Institut, Institute for Farm Animal Genetics, Höltystraße 10, 31535 Neustadt, Germany, [3]Wageningen University and Research, Animal Breeding and Genomics, Droevendaalsesteeg 1, 6700 AH Wageningen, the Netherlands; christian.reimer@fli.de

The overarching aim of a gene bank is to cryo-conserve reproducible material of breeding animals for a future moment when the in-situ population became inviable. Strategies to restore a breed from semen through a back-crossing scheme exist for long and support the prominent assumption of required 25 sires with 100 insemination doses each. Recently, the role of gene banks has been interpreted in a broader sense. Rather than just ensuring the restoration of an extinct breed, they can be active elements in the breeding program. Individual animals could be re-introduced into the population in time in order to restore genetic diversity and advantageous traits. If the relevant traits cannot be genomically characterized, what we expect to be the normal case, and specifically be introgressed, this consequently means ensuring that the majority of the donor genome is transferred to subsequent generations. We have derived a simple equation to determine how many inseminations are necessary to pass on a certain proportion of the donor genome, taking into account the number of offspring, degree of inbreeding, fertilisation success and the number of chromosomes, and support this by simulating appropriate scenarios on the genomic level. Upon the simplistic estimation, it showed that in cattle, due to the high efficiency of artificial insemination and in pigs, due to the high number of offspring, around ten doses, are needed to transfer 95% of the diploid donor genome. In small ruminants, small litter sizes and low efficiency of artificial insemination coincide, and similar transfer rates as in the aforementioned species require already 35 insemination portions. Autozygosity of an animal can reduce the number of required doses, but only with moderate effect unless it is up to a high degree. In view of the fact that this number is sometimes close to the technical maximum to be collected, there is an urgent need for improvements in efficient reintroduction strategies.

Session 25 Poster 15

Is it possible to skip the SNP selection step for breed assignment?

H. Wilmot[1,2], T. Niehoff[3], H. Soyeurt[1], N. Gengler[1,2] and M.P.L. Calus[3]
[1]ULiège-GxABT, TERRA Teaching and Research Centre, Passage des Déportés, 2, 5030 Gembloux, Belgium, [2]National Fund for Scientific Research (F.R.S.-FNRS), Rue d'Egmont, 5, 1000 Brussels, Belgium, [3]Wageningen University & Research, Animal Breeding and Genomics, Droevendaalsesteeg, 1, 6700 AH Wageningen, the Netherlands; helene.wilmot@uliege.be

Breed assignment of animals generally relies on three main steps: (1) selection of a SNP panel that allows to distinguish between the breeds of interest; (2) training of a classification algorithm; and (3) validation of this classification on new animals. In recent years, to select the most informative SNPs, more complex methodologies have been developed, e.g. by the combination of several methods. There is no consensus about the protocol to follow to select the best SNPs and many questions stay unanswered. How to select the most informative SNPs? What is the optimal number of SNPs to select? To solve these issues, we proposed to skip the SNP selection step by using a genomic relationship matrix based on all available SNPs to assign animals to their breed of origin. Three new methodologies were developed for three cattle breeds: (1) breed assignment based on the highest mean relatedness of an animal to the reference population of each of the three breeds; (2) breed assignment based on the highest standard deviation (SD) of the relatedness of an animal to the reference population of each of the three breeds; and (3) a methodology that combines the values of mean and SD of the relatedness of each animal in a linear support vector machine model (SVM). These new methodologies were compared with a control methodology: a previously developed model based on a reduced SNP panel. The linear SVM achieved a similar percentage of correct assignment as the control methodology, and was substantially faster to compute. The main advantage of using the new methodology, based on the linear SVM, is to bypass the SNP selection step.

Session 25 Poster 16

Copy number variation regions differing in segregation patterns spanned different sets of genes

K.D. Arias[1], J.P. Gutiérrez[2], I. Fernández[1], I. Álvarez[1] and F. Goyache[1]
[1]SERIDA-Deva, Camino de Rioseco 1225, 33394 Gijón, Spain, [2]UCM, Dpto. de Producción Animal, 28040 Madrid, Spain; kathyah18@gmail.com

Copy Number Variations Regions (CNVR) are genomic alterations influencing performance frequently used as informative markers in association studies for economically important traits in livestock. However, CNVR segregation patterns usually depart from Mendelian rules and there is no consensus on how CNVR mirror useful genomic variation. This research aims to identify the genes spanned in CNVR differing in segregation patterns using a pedigree of the Gochu Asturcelta pig. Up to 478 parent offspring-trios belonging to 61 different families were genotyped using the Axiom_PigHDv1 Array. CNVR were identified following Arias *et al.* and classified as singleton de novo CNVR (sdnCNVR; n=20), recurrent de novo CNVR (rdnCNVR; n=255) and segregating CNVR (sCNVR; n=69). Gene-annotation enrichment and functional annotation analyses were performed using BioMart Software (Ensembl Genes 99 database) and DAVID Bioinformatics Resources 6.8. Previous studies show that sdnCNVR, occurring in one individual only, are probably Allele-Drop-In loci caused by technological issues whereas rdnCNVR are present in various individuals and families and may represent real genomic variation. Enrichment analyses informed that both the rdnCNVR (identified in 5 individuals or more) and sCNVR spanned 1,292 and 970 candidate genes, respectively. As expected, both CNVR classes gathered genes mainly involved in immunity and control of cellular function. Interestingly, sCNVR in the Gochu Asturcelta pig breed spanned genes putatively involved in prolificacy of sows and piglets' survival via the oxygen transport function, whereas rdnCNVR spanned growth-promoting genes. Genomic variation gathered by sCNVR may be of importance at the whole population level whereas that of rdnCNVR may explain differences in individual performance. Project PID2019-103951RB/AEI/10.13039/501100011033. Grant PRE2020-092905.

Session 25 Poster 17

Genomic characterization as tool in the breed recognition process

C. Danchin and D. Duclos
IDELE, 149 rue de Bercy, 75012 PARIS, France; coralie.danchin@idele.fr

The most prominent French goat breed is the Alpine, which originated in the northern part of the French Alps. For 50 years the breed has been intensively selected for dairy traits. The main selection nuclei are now located in the west and centre of France. Meanwhile, some farmers kept the traditional population in the Alps and claimed to have limited exchange of animals or germ plasm with the rest of the breed. A breeders' association was created 20 years ago to preserve this population which they called 'Savoie'. The breeders' association set up pedigree recording with the help of the French Livestock Institute (IDELE) however the pedigree information remained quite scarce. Finally, from a phenotyping point of view, the Alpine and the Savoie are sharing some colour patterns since the Alpine is not completely standardized with the 'bezoar' colour pattern. The association struggled to achieve an official recognition by the Ministry of Agriculture since it was difficult to prove that the population was significantly different from the Alpine breed. In the mid-2010s, a medium SNP goat chip (by Illumina Inc.) became available in France for a reasonable price (about 40 € /genotype). The breeders' association, in collaboration with IDELE, took this opportunity to sample and genotype about 40 goats to compare it with available Alpine genotypes. The goats were chosen based on pedigree and phenotyping information to maximize the sample diversity. First IDELE performed a principal component analysis based on a kinship genomic matrix that showed that the Savoie breed is indeed a close relative to the Alpine breed, but that the population differs now from it. A second analysis was performed by using the ADMIXTURE software to detect animals that were clearly crossbred recently with the Alpine, leading to their withdrawal from the conservation programme. The results of the genomic analysis were provided to an expert commission and were one of the factors that led to the official recognition of the Savoie breed by the French Ministry of Agriculture in 2020.

Session 25 Poster 18

Homozygosity by descent in mice divergently selected for environmental birth weight variability

C. Ojeda-Marín[1], J.P. Gutiérrez[1], N. Formoso-Rafferty[2] and I. Cervantes[1]
[1]UCM, Producción Animal, Avda Puerta del Hierro, s/n, 28040, Spain, [2]UPM, E.T.S.I.A.A.B, Producción Agraria, Avda Puerta del Hierro, s/n, 28040, Spain; candelao@ucm.es

Inbreeding could lead negative effects as inbreeding depression. Purging process produced that deleterious alleles are mostly young and maintain at low frequencies in populations. As a result of divergent selection experiment of birth weight environmental variability in mice two lines, high variability line (H-Line) and low variability line (L-Line), were created. L-Line presented advantages in robustness, animal welfare and production. The aim of this study was to analyse genomic inbreeding in both selected lines and to compare it with pedigree inbreeding (F_{PED}) to ascertain if there were differences between lines. A total of 844 individuals of H-Line and 855 individuals of L-Line belonging to 26 generations of selection were genotyped using a high density genotyping array. Homozygosity by descent segments (HBD) were analysed using a multiple classes Hidden Markov Model with 9 HBD classes (2,4,8,16,32,64,128,256 and 512) and one non HBD class (512). Hence, global HBD inbreeding (F_{HBD}) and specific class HBD inbreeding (F_{HBD2}, F_{HBD4}, F_{HBD8}, F_{HBD16}, F_{HBD32}, F_{HBD64}, F_{HBD128}, F_{HBD256} and F_{HBD512}) were computed. Genomic inbreeding computed using runs of homozygosity (F_{ROH}) and F_{PED} were also computed. Moreover, F_{PED} was decomposed in recent inbreeding (F_N) and old inbreeding (F_O) using as generation thresholds 4, 5, 6, 7 and 8 generations. Correlations between different inbreeding coefficients were computed. The evolution of F_{PED}, F_{HBD} and F_{ROH} were positive across the generations. F_{HBD} and F_{PED} presented greater differences between lines than F_{PED}. High correlations were observed between F_{HBD}, F_{PED} and F_{ROH}. The greatest correlations were found between F_{HBD}-F_{ROH} in both lines. H-Line presented greater contribution of younger HBD classes than L-Line and F_{HBD16} presented higher correlations with F_N and F_O in H-Line. Differences between lines were observed in the contribution of HBD classes to total F_{HBD} that were not observed with other inbreeding coefficients. Further studies should analyse local HBD to clear up if there were genomic differences in homozygosis between lines.

Inbreeding depression is associated with recent HBD segments in Belgian Blue beef cattle

M. Naji[1], J.L. Gualdron Duarte[2], N.S. Forneris[1] and T. Druet[1]
[1]Unit of Animal Genomics, GIGA-R & Faculty of Veterinary Medicine, University of Liège, Avenue de l'Hôpital, 1, 4000 Liège, Belgium, [2]Walloon Breeders Association (awe group), Ciney, 5590, Belgium; mnaji@uliege.be

Intensive use of high-merit bulls is common in cattle populations and results in high inbreeding levels that can lead to outbursts of genetic defects or to inbreeding depression (ID). We herein used data from 14,205 Belgian blue beef cattle cows, genotyped at medium marker density (30K SNPs) and phenotyped for eleven linear classification traits, to study ID. We first computed several estimators of the inbreeding coefficient F, relying either on the pedigree information (F_{PED}), on the correlation between the uniting gametes (F_{UNI}), on the diagonal elements from the genomic relationship matrix (F_{GRM}), or on homozygous-by-descent (HBD) segments identified with the RZooRoH package (F_{HBD}). We observed that F_{UNI} and F_{GRM} were highly sensitive to allele frequencies used in their estimation (e.g. founder vs sample allele frequencies), whereas F_{HBD} was more stable. Using a linear mixed model, we detected significant ID for four traits, related to height and length; F_{UNI} and F_{HBD} had the strongest association, confirming that both are efficient estimators of F. The regression coefficient associated with ID was for instance equal to -21 cm for height when using F_{HBD}. RZooRoH classifies HBD segments in different age-based groups (the length of HBD segment being inversely related to the number of generations to the common ancestors). We took advantage of this feature to test whether HBD segments associated with more recent common ancestors were more deleterious than shorter HBD segments associated with more remote ancestors. Recent HBD classes (ancestors present approximately up to 15 generations in the past) presented stronger ID than more ancient HBD classes. Simulations indicated that these observations were not due to lack of variation in more ancient HBD classes. Similar results were obtained with higher-marker density, suggesting that lack of ability of the model to capture shorter HBD segments was not an issue neither. Overall, our results suggest that mutational load decreases with haplotype age, and that mating plans should consider mainly the levels of recent inbreeding.

Assessing the intrapopulation structure of Slovak Spotted cattle by genome-wide data

R. Kasarda[1], N. Moravčíková[1], J. Candrák[1], J. Prišťák[1] and I. Pavlík[2]
[1]Slovak University of Agriculture in Nitra, Institute of Nutrition and Genomics, Tr. A. Hlinku 2, 94976 Nitra, Slovak Republic, [2]Research Institute of Animal Production – NPPC Slovakia, Hlohovecká 2, 95141 Lužianky, Slovak Republic; radovan.kasarda@uniag.sk

Artificial selection, especially the preferential use of particular sire lines and maternal families, simultaneously affects the diversity status and genetic structure of livestock populations across generations. Therefore, this study aimed to assess the intrapopulation genetic structure and diversity of the Slovak Spotted cattle by analysis of genome-wide data. Slovak Spotted cattle gene pool represented 37 sires born in 1972-2011, 50 dams and 89 cows from active populations. The two most commonly accepted diversity parameters, genomic inbreeding (F_{ROH}) and effective population size (N_{eLD}), were used to describe the current genetic status of the breed. The intrapopulation structure was tested by principal component analysis. A total of 9,719 runs of homozygosity (ROHs) with an average length of 8.73 Mbp and 3,5487 ROHs with an average length of 8.96 Mbp were identified in the genome of sires and cows, respectively. The observed level of inbreeding ($F_{ROH8\text{-}16Mbp}$=1.65±1.30% and $F_{ROH>16Mbp}$=1.43±1.27%) pointed to a sufficient level of variability in the current population, but there were significant differences among the individuals, as confirmed by the analysis of the individual inbreeding coefficient depending on the sex. The current effective population size of 75.49 individuals indicated that the preferential use of sire lines and a general decline in breeders' interest led to a relatively high intergeneration decline (7.76 individuals). PCA showed that sires and dams created a common genetic cluster, while cows were distributed depending on the sire lines preferred by a particular farmer. The distribution of genetic variability outside the breed nucleus creates the potential for the selection of suitable genotypes of future sires, whose inclusion in breeding can positively impact the development of effective population size. This study was supported by the Slovak Agency for Research and Development (grants no. APVV-17-0060 and APVV-20-0161).

Genomic selection has aided Nordic Jersey to decrease risks of inbreeding

S. Tenhunen[1,2], J.R. Thomasen[1], L.P. Sørensen[1], P. Berg[3] and M. Kargo[2]
[1]VikingGenetics, Ebeltoftvej 16, 8960 Randers SØ, Denmark, [2]Aarhus University, QGG, C. F. Møllers Allé 3, bld. 1130, 8000 Aarhus, Denmark, [3]Norwegian University of Life Sciences, NMBU, Universitetstunet 3, 1433 Ås, Norway; saten@vikinggenetics.com

Recently, many dairy cattle populations across the world have demonstrated a rapid increase in inbreeding which may lead to inbreeding depression. The main reasons for this have been reported to be the changes in breeding practices due to genomic selection (GS). This research aimed to identify and study past events which have influenced the rate of inbreeding (ΔF) and coancestry (ΔC) in the Nordic Jersey (NJ) population. Retrospective analysis was performed of 99,083 NJ animals with pedigree information and 41,896 SNP imputed genotypes. Animals were divided into two groups: before (2009-2014) and after (2015-2021) GS implementation in the NJ population. Pedigree (PED) was used to calculate yearly ΔF and ΔC and runs of homozygosity (ROH) was used to calculate genomic ΔF where shared segments (SEG) between pairs were used to calculate genomic ΔC. Effective population size (Ne) was calculated from ΔF and ΔC was used to calculate future effective population size (FNe) for each animal group. The yearly average generation interval (L) decreased from 4.7 to 3.2 after GS. PED ΔF was at the same level before and after GS, but there was a distinct increase in ROH ΔF from 0.0035 to 0.0052. PED ΔC had decreased from 0.0031 to 0.0017 between the periods and no changes in SEG ΔC were seen. Ne PED increased from 32 to 55 and Ne ROH was 30 in both periods. FNe PED increased significantly after GS implementation from 34 to 91, and FNe SEG saw a more moderate increase from 41 to 57. Increases in Ne ROH and FNe SEG were mainly caused by a decrease in L between the periods. From the yearly estimates we could identify a period before GS when ΔC increased rapidly in the NJ. In this period, heavy use of related sires was observed. This may explain the increased ΔC in the population which may have caused the higher ΔF after GS. Ne is currently on the increase in the NJ population and FNe indicates that this trend will continue in the coming years. In conclusion, GS seems to have been instrumental in decreasing the risk for inbreeding depression now and in the future.

Management of local cattle breed diversity in Croatia: use of molecular and pedigree indicators

A. Ivanković[1], Z. Ivkić[2], G. Šubara[3], M. Pećina[1], J. Ramljak[1] and M. Konjačić[1]
[1]University of Zagreb Faculty of Agriculture, Svetošimunska cesta 25, 10000, Croatia, [2]Croatian agency for agriculture and food, Vinkovačka cesta 63 c, 31000 Osijek, Croatia, [3]Agency for Rural Development of Istria, Tugomila Ujčića 1, 52000 Pazin, Croatia; aivankovic@agr.hr

In addition to conventional cattle breeds, three local breeds are also preserved in Croatia. They are classified as endangered breeds and are included in a program for conservation and economic revitalization. Breeding associations carry out breeding programs with the help of public and scientific institutions and optimize breeding and economic measures. There is constant monitoring of breeding activities, optimization of genetic evaluation methods and harmonization of population indicators (*molecular, pedigree, exterior, productivity*). Three local cattle breeds in Croatia were studied under different population situations (*size, distribution and competitiveness*). Molecular characterization using microsatellites is performed continuously, supported by regular pedigree testing. Genotyping of the breed using SNP-s is performed only sporadically and on a smaller number of individuals, mainly for scientific purposes. The indicators of pedigree analysis are used daily in direct selection (*maintenance of lines, preparation of mating plans, etc.*). Experience in implementing the management program shows that pedigree indicators, supported by molecular information, are the basis for efficient management of population diversity. Pedigree and molecular markers synergistically show that the greatest preserved genetic diversity is found in the Buša population, while the Slavonian-Syrmian Podolian population has the highest level of inbreeding or bottleneck. The indicators for the Istrian cattle population are more favourable, and the sustainability of the breed is more likely, especially considering the increasing demand for 'Istrian beef' The example of local cattle breeds in Croatia illustrates the need for synergistic and continuous use of molecular and pedigree indicators to maintain the diversity of local breeds, as this increases their informative value and applicability. It should be noted that pedigree indicators are easier and more understandable for breeders and can be used in daily herd and population management.

Genomic differences between the domesticated pig breed and its ancestor wild boar

N. Moravčíková[1], R. Kasarda[1], M. Hustinová[1,2], E. Krupa[3], Z. Krupová[3] and E. Žáková[3]
[1]Slovak University of Agriculture in Nitra, Tr. A. Hlinku 2, 94976 Nitra, Slovak Republic, [2]Slovak Hunting Chamber, Štefánikova 10, Bratislava 81105, Slovak Republic, [3]Institute of Animal Science, Přátelství 815, 104 00 Praha-Uhříněves, Czech Republic; nina.moravcikova@uniag.sk

This study aimed to identify the genomic differences between the autochthonous Czech pig breed Přeštice Black-Pied and wild boar by screening selection signals distribution in the autosomal genome. Analysis of selection signals was based on the assumption that SNPs extremely related to the population structure can be responsible for biological adaptation. We used genomic data from 265 Přeštice Black-Pied pigs and 75 wild boars collected in Czechia and Slovakia, respectively. Animals were genotyped by two platforms, Illumina Porcine SNP60 and GGP Porcine 50k. After SNP pruning, the database included 30,704 informative autosomal SNP markers. Selection signals were defined by outlier SNPs resulting from the principal component analysis while accounting for population structure. Above the cut-off value set based on the Bonferroni correction were found 234 outliers. Selection signals were distributed in different areas on autosomes 1, 2, 4, 8, 9, and 14. The strongest signal was found on autosome 1 close to the *CTIF* gene included in pathways resulting in the formation of proteins by the translation of mRNA. In addition, *TMEM161B*, *ZNF697*, *HSD3B1*, *EDNRA* and *CTNNA3* genes involved in pathways controlling the frequency or rate of heart contraction, cardiac muscle cell action potential, reproduction, vasoconstriction, regulation of blood pressure, fertility was identified close to the top of signals on chromosomes 2, 4, 8, and 14. The results of this study can contribute to research related to the explanation of the effect of domestication and livestock breed development on their genome structure compared to wild ancestors. This study was supported by the Slovak Agency for Research and Development (grants no. APVV-17-0060 and APVV-20-0161) and Ministry of Agriculture of the Czech Republic (project no. MZE-RO0723 V02 and QK1910217).

Genetic polymorphism of kappa-casein in Serbian Holstein-Friesian cattle

M. Zorc[1], M. Šaran[2], L.J. Štrbac[2], D. Janković[2], P. Dovč[1] and S. Trivunović[2]
[1]University of Ljubljana, Biotechnical Faculty, Department of Animal Science, Jamnikarjeva 101, 1000 Ljubljana, Slovenia, [2]University of Novi Sad, Faculty of Agriculture, Department of Animal Science, Trg Dositeja Obradovića 8, 21000 Novi Sad, Serbia; minja.zorc@bf.uni-lj.si

Milk proteins have been thoroughly studied in cattle. Considerable genetic variation has been identified and characterized in the genes coding milk proteins, mainly due to the effects of milk protein variants on milk composition and cheese-making characteristics. Previous research has determined that kappa-casein polymorphism affects the total milk protein content and cheese yield. In order to examine the polymorphism of kappa-casein, a total of 1,600 Serbian Holstein-Friesian cows were genotyped with a GGP Bovine 100K SNP array. Nine markers from the SNP array targeting the kappa-casein gene (CSN3) were examined. Out of the total number of cows, the genotype was successfully determined for 1,591 cows. It was found that the most frequent allele was allele A (63%), followed by allele B (28%) and allele E (9%). Kappa-casein genotype frequencies were: 35.3% AA, 38.0% AB, 17.8% AE, 8.5% BB, and 0.4% EE. Using the chi-square test, it was determined that the population is not in Hardy-Weinberg equilibrium for the studied gene ($P<0.05$), which may indicate the existence of evolutionary processes or selection in the studied population. The observed heterozygosity (Ho) and expected heterozygosity (He) were 0.56 and 0.52, respectively. Constant monitoring of milk protein variation in different breeds of cattle is an essential practice to avoid an increase in the frequency of mutations with unfavourable effects on cheese production. Acknowledgments: This research is funded by Ministry of Science, Technological Development and Innovation of the Republic of Serbia (Contract No. 451-03-47/2023-01/ 200117). The data used in this research were collected during the realization of PROMIS project: 'A Bioinformatics Approach to Dairy Cattle Breeding Using Genomic Selection', No. 6066512, funded by the Science Fund of the Republic of Serbia.

Estimation of recent and ancestral inbreeding for X-chromosome in Old Kladrub horse

L. Vostry[1], M. Shihabi[2], J. Farkas[3], G. Kövér[3], H. Vostra-Vydrova[1], B. Hofmanova[1], I. Nagy[3] and I. Curik[2]
[1]CZU Prague, Kamycka 129, 165 00 Prague, Czech Republic, [2]University of Zagreb, Faculty of Agriculture, Svetošimunska 25, 10000 Zagreb, Croatia, [3]Hungarian University of Agricultural and Life Sciences, Guba S 40, 7400 Kaposvar, Hungary; vostry@af.czu.cz

A number of pedigree inbreeding coefficients (F_{ped}) have been developed to analyse inbreeding depression or its purging, such as the new Kalinowski inbreeding coefficient (F_{new}), the Kalinowski ancestral inbreeding coefficient (Fa_Kal), the Ballou ancestral inbreeding coefficient (F_{a_Bal}), and the Ancestral History coefficient (AHc). Estimation of inbreeding coefficients between sex and autosomal genes is different and depends on the order of male and female individuals in their pedigree. Unfortunately, there is no programme to calculate sex-specific inbreeding coefficients (e.g. F_{new}, F_{a_Kal}, and/or F_{a_Bal}) needed for purging analysis. In this study, we estimated and analysed a large number of sex-specific inbreeding coefficients calculated by the algorithm implemented in the GRainX software using a long and complete pedigree of Old Kladrub horses. The ratio of sex-specific to autosomal inbreeding estimates (sex/autosomal) was higher for F_{ped} (1.3), F_{ped} five-generation (1.46), F_{a_kal} (1.4), and F_{new} (1.3). In contrast, a lower ratio (sex/autosomal) was observed for F_{a_Bal} (0.93) and AHc (0.71). At the same time, correlations between autosomal and sex inbreeding coefficients ranged from 0.37 (AHc and AHcX) to 0.65 (F_{ped} and F_{pedX}). Our results showed that sex-specific inbreeding coefficients obtained with GRainX software can be used to estimate inbreeding depression and its purging for sex chromosomal genes. This study was supported by projects QK1910156 and ANAGRAMS-IP -2018-01-8708.

Genetic diversity and population structure of a Peruvian nucleus cattle herd using SNP data

F.-A. Corredor[1], D. Figueroa[1], R. Estrada[1], W. Salazar[1], C. Quilcate[1], H. Vasquez[2], J. Gonzales[3], J. Maicelo[2], P. Medina[1] and C. Arbizu[1]
[1]Instituto Nacional de Innovación Agraria, Lima, 15024, Peru, [2]Universidad Nacional Toribio Rodríguez de Mendoza de Amazonas, Chachapoyas, 01001, Peru, [3]Universidad Nacional de Frontera, Piura, 20103, Peru; corredor@alumni.iastate.edu

New-generation sequencing technologies, among them SNP chips for massive genotyping, are useful for the effective management of genetic resources. To date, molecular studies in Peruvian cattle are still scarce. For the first time, the genetic diversity and population structure of a reproductive nucleus cattle herd from a Peruvian institution were determined. Samples from Brahman, Braunvieh, Gyr, Simmental, and creole cattle (AFB) were incorporated. Female individuals were genotyped with the GGPBovine100K and males with the BovineHD. Quality control, and the proportion of polymorphic SNPs (Pn), minor allele frequency (MAF), expected heterozygosity (He), observed heterozygosity (Ho), and inbreeding coefficient (Fis) were estimated for the five breeds. Admixture, principal component analysis (PCA), and discriminant analysis of principal components were performed. Also, a dendrogram was constructed using the Neighbour-Joining clustering algorithm. The genetic diversity indices in all breeds showed a high Pn, varying from 51.42% in Gyr to 97.58% in AFB. Also, Braunvieh possessed the highest Ho (0.43±0.01), while Brahman the lowest (0.37±0.02), indicating that Brahman was less diverse. The pairwise genetic differentiation estimates between breeds showed values that ranged from 0.08 (Braunvieh vs AFB) to 0.37 (Brahman vs Braunvieh). Similarly, pairwise Reynold's distance ranged from 0.09 (Braunvieh vs AFB) to 0.46 (Brahman vs Braunvieh). The dendrogram, similar to the PCA, identified two groups, showing a clear separation between *Bos indicus* (Brahman and Gyr) and *B. taurus* breeds (Braunvieh, Simmental and AFB). Simmental and Braunvieh grouped closely with the AFB cattle. Similar results were obtained for the population structure analysis with K=2. The results from this study would contribute to the appropriate management, avoiding loss of genetic variability in these breeds and for future improvements in this nucleus. Additional work is needed to speed up the breeding process in the Peruvian cattle system.

Within-breed stratification for across-breed reference population

H. Wilmot[1,2], T. Druet[2,3], I. Hulsegge[4,5], N. Gengler[1,2] and M.P.L. Calus[5]
[1]ULiège-GxABT, TERRA Teaching and Research Centre, Passage des Déportés, 2, 5030 Gembloux, Belgium, [2]National Fund for Scientific Research (F.R.S.-FNRS), Rue d'Egmont, 5, 1000 Brussels, Belgium, [3]ULiège-GIGA-R & Faculty of Veterinary Medicine, Unit of Animal Genomics, Avenue de l'Hôpital, 1, 4000 Liège, Belgium, [4]Wageningen University & Research, Centre for Genetic Resources, Droevendaalsesteeg, 1, 6700 AH Wageningen, the Netherlands, [5]Wageningen University & Research, Animal Breeding and Genomics, Droevendaalsesteeg, 1, 6700 AH Wageningen, the Netherlands; nicolas.gengler@uliege.be

In endangered breeds, it is difficult to have large reference populations needed for accurate genomic predictions. To solve this issue, one solution can be to build an across-breed reference population. In that case, candidate animals should however be genetically close enough to the reference population. Therefore, special attention must be paid to the animals from other breeds that are selected to be part of the across-breed reference population. In our study, we explored different measures of genetic diversity, namely principal component analysis (PCA) and ADMIXTURE clustering based on genotypes, to detect stratification within Dutch red-pied cattle breeds with the aim to build an across-breed reference population for two genomically related endangered red-pied cattle breeds from Belgium and Luxembourg. The PCA analysis showed that the Dutch, Belgian and Luxembourgish breeds were part of a genomic continuum. We detected stratification within the Dutch breeds and consequently also genomically related animals to the Belgian and Luxembourgish breeds. Although empirical validation of across-breed genomic prediction is needed to confirm which reference population yields the best predictions, our results do suggest that stratification within breeds should be considered when construction across-breed reference populations.

The livestock sector and research in animal science in France

X. Fernandez[1], J.F. Hocquette[2] and S. Ingrand[1]
[1]INRAE, Animal Physiology and Livestock Systems division (Phase), 37380 Nouzilly, France, [2]INRAE, VetAgroSup, INRA Herbivores, UMR1213, 63122 Theix, France; xavier.fernandez@inrae.fr

In 2022, the numbers of farm animals in France were, in million heads, 17.8 for cattle, 13.7 for pigs, 7.2 for sheep, 1.4 for goats and 157.6 for poultry meat. In 2020, 145,000 French farms were devoted to livestock farming, i.e. 37% of all farms (meat, milk, mixed cattle; sheep; pigs; poultry). The consumption of animal products (per capita / year) in France is the following: for pig meat, 31 kg; for poultry, 28 kg; for cheese, 26 kg; for beef, 22 kg; for eggs, 13 kg; for butter, 8 kg. France produces 24 billion litres of cow's milk. This is the 2nd European production with 51 dairy protected designations of origin (PDO). There are also 226 products with the 'label rouge' quality sign in the poultry/egg sector (from https://agriculture.gouv.fr/infographie-lelevage-francais). The different animal sectors in France are represented by interprofessional organizations such as INTERBEV for the red meat sector, INAPORC for the pig sector, ANVOL for the poultry sector, CNIEL for the dairy sector and FGE for the genetic sector. According to scientific papers in the Web of Science, France is the 7th country in terms of research in dairy and animal science (after the USA, Canada, Brazil, India, Germany and Australia) with 50.4% of the papers published in open access. INRAE has co-authored 85% of French scientific articles of the Web of Science in dairy and animal science. The main research areas in animal science (genetics, health, physiology, livestock systems) will be described orally. After INRAE, the other main actors of French research in dairy and animal science are universities, 'grandes ecoles' for engineers and veterinarians, other research institutes (such as CIRAD) and also technical institutes for herbivores (IDELE), pigs (IFIP) and poultry (ITAVI) production in charge of research transfer to the industry. Coordination between all these actors is achieved through the scientific group of interest 'Avenir Elevage' for research and the French Association for Animal Production (AFZ) for dissemination.

Session 26 Theatre 2

Solutions to achieve healthy and sustainable diets worldwide

H. Guyomard, A. Forslund, A. Tibi, B. Schmitt, P. Debaeke and J.-L. Durand
INRAE, SDAR, La Motte au Vicomte, 35650 Le Rheu, France; herve.guyomard@inrae.fr

Word food systems are neither healthy nor sustainable. In that context, numerous studies aim at assessing the ability of world agriculture to feed an increasing world population in a healthy and sustainable way. On the final consumption side, the presentation adopts a normative assumption for diets that would evolve in all world regions towards healthy diets by 2050. We first describe what these healthy regional diets are by 2050 and the required changes relative to current regional diets. In a second part, we use the GlobAgri biomass model to analyse cropland and pastureland induced needs based on complementary assumptions for crop and forage yields, cropping intensities, and livestock feed efficiencies. Yields and maximum cultivated areas are impacted by climate change. In addition, we adopt a second normative assumption by excluding any deforestation since the latter is a key factor of greenhouse gas emissions and biodiversity loss. Simulation results allow us to determine cropland and pastureland needs in the world divided into 21 regions as well as impacts on agri-food trade. Several regions in the world would be constrained by their maximum available cultivated areas. In addition, pastureland needs will be very high in several regions, notably in Sub-Saharan Africa. As a consequence, the last part of the presentation analyses to what extent it would be possible to reduce agricultural land needs by activating supply drivers (sustainable intensification of yields, sustainable improvements in feed efficiencies, reduction of pre-harvest losses, etc.) as well as demand drivers (reduction in post-harvest losses, replacement of red meat by white meat in diets, replacement of animal proteins by plant proteins in diets, etc.). We conclude by implications for public policies.

Session 26 Theatre 3

Moving towards bold food systems resilience

J. Fanzo
Berman Institute of Bioethics, Nitze School of Advanced International Studies (SAIS), and Bloomberg School of Public Health, Johns Hopkins University, 1717 Massachusetts Ave NW 730, Washington DC 20036, USA; jfanzo1@jhu.edu

With climate change, the COVID-19 pandemic, and ongoing conflicts, food systems are facing increasing fragility. In a turbulent, hot world, threatened resiliency and sustainability of food systems could make it all the more complicated to nourish a population of 9.7 billion by 2050. Climate change is having adverse impacts across food systems with more frequent and intense extreme events that will challenge food production, storage, and transport, potentially imperilling the global population's ability to access and afford healthy diets. At the same time, the way food is grown, processed, packaged, and transported is having adverse impacts on the environment and finite natural resources further accelerating climate change, tropical deforestation, and biodiversity loss. Food system policies and actions can contribute to climate adaptation and mitigation responses and, at the same time, improve human and planetary health, however multi-lateral cooperation, action and investments must be bold. While there is significant urgency in acting, it is also critical to move beyond the political inertia and bridge the separatism of food systems and climate change agendas that currently exists among governments and private sector actors.

Session 26 Theatre 4

A social-ecological agenda for transforming land management toward sustainability

T. Plieninger[1,2]

[1]Faculty of Organic Agricultural Sciences, University of Kassel, Steinstr. 19, 37213 Witzenhausen, Germany, [2]Department of Agricultural Economics, University of Göttingen, Platz der Göttinger Sieben 5, 37073 Göttingen, Germany; plieninger@uni-goettingen.de

Recent high-level reports (for instance, the Global Assessment of the Intergovernmental Panel on Biodiversity and Ecosystem Services, IPBES, and the Global Sustainable Development Report, GSDR) have highlighted that dominant research modes in the agricultural sciences and related research fields are insufficient to guide the transformative change of society that is needed to achieve the Sustainable Development Goals. Accordingly, new approaches toward a transformative land sustainability science are currently on the rise. This keynote presentation takes up a framework of transformative sustainability science to: (1) analyse the systemic interactions in land-management systems that lead to synergies or trade-offs between sustainability outcomes; (2) identify competing values and interests of actors in land management that help or hinder sustainability; and (3) help understand transformations toward sustainable land management in concrete contexts. A social-ecological research agenda will be illustrated with empirical studies from Europe, Asia, and North Africa that focus on multiple innovative agricultural and other land-use practices and systems that are in need of and with potential for sustainable development. The keynote will include cases of agroforestry systems, extensive livestock grazing systems, conservation agriculture, and collaborative agri-environmental governance.

Session 26 Theatre 5

EAAP beyond Europe: establishing and using capacity to feed the world

H. Sölkner

University of Natural Resources and Life Sciences, Vienna, Gregor Mendel Str. 33, 1180 Vienna, Austria

The European Federation of Animal Science (EAAP) provides very strong glue for animal scientists in Europe, helping them thinking outside their narrow disciplinary boxes and doing interdisciplinary research, so valuable for society at large. Over the last two decades, EAAP has evolved a science communication and dissemination platform far beyond its original scope. It is communication partner of virtually all EU funded research projects in the field of animal science. Very recently, EAAP established a partnership program, connecting doctoral students in Europe with doctoral students from around the world. What a great effort! It is my strong belief that animal scientists in Europe can play a crucial role in improving livestock populations and their productivity across the globe, enhancing the livelihoods of many people in developing countries and reducing malnutrition, particularly of infants. The animal breeding research group at BOKU has embarked on this mission around 25 years ago and I have good news to share. Together with local partners as well as the international institutions ICARDA and ILRI, we developed community based breeding programs (CBBP) for small ruminants, cattle and New World camelids in Africa and Latin America. Open nucleus breeding programs, establishing and multiplying breeding stock on station and distributing male breeding animals to farmers used to be the way to go in many developing countries. This approach almost invariably failed, due to many reasons, including the animals not fitting the smallholder system or the type of animal not being accepted by farmers. Establishment of CBBP starts with finding the breeding objectives of farmers with participatory tools and agreeing on simple on farm recording of traits of farmer interest. During the first generations of CBBP implementation, young male animals ready for breeding are ranked based on breeding objectives and corresponding records and farmer committees choose among the candidates. Ways of using breeding animals are agreed on and associations are formed, providing by-laws for that. The most impressive use case of CBBP is that of small ruminants in Ethiopia, where this system was decreed the breeding method of choice by the national government. The key to this success has been the capacity and commitment of a few persons leading the programs. This is very similar to the farmer-led cattle and pig breeding programs in Austria, where technically sophisticated and well trained staff of breeding associations play that role. International collaboration, including its North-South and South-South variants, has very many facets, joint supervision of doctoral students is one of them. The EAAP international student partnerships are a new device of that collaboration. May that work well and may the student partners take their respective supervisors on their journey. Good luck to all of you!

Session 27 Theatre 1

Effect of genomic selection on allele frequencies of QTL associated to number of teats in pigs

C.A. Sevillano[1], B. Harlizius[1], M.S. Lopes[1,2], M. Van Son[3] and E.F. Knol[1]
[1]Topigs Norsvin Research Center, P.O. Box 43, 6640 AA Beuningen, the Netherlands, [2]Topigs Norsvin, Visconde do Rio Branco 1310, 80.420-210 Curitiba, Brazil, [3]Norsvin, Storhamargata 44, 2317 Hamar, Norway; claudia.sevillanodelaguila@topigsnorsvin.com

Number of teats (NTE) is an important trait for pig welfare because it influences directly the milk production of the sow and indirectly the survival and weight gain of piglets. NTE is a heritable trait and shows considerable variation between and within breeds. NTE have been included in the breeding goal of Topigs Norsvin maternal lines in the last decades, while genomic selection (GS) was adopted in 2012. It is known that GS has increased the rate of genetic gain in the short term but its impact on allelic diversity is not known. In this study we aimed to better understand the impact of GS on NTE by following the changes in allele frequency across generations on different QTL regions affecting NTE in two maternal lines. NTE was recorded at birth on 187,867 L (Landrace) and 173,288 LW (Large-White) animals from both sexes born between 2011 and 2022. All animals were genotyped on Illumina GeneSeek custom SNP chips (25K, 50K or 80K) or Axiom porcine 660K array from Affymetrix and imputed within population to 660K SNPs. The mean value for NTE is around 1 teat higher in the L compared to LW (16.89 vs 15.96). L line has gained 1.8 extra teat in the last 11 years, while the LW line has only gained 0.7. Heritability was 0.39 for L and 0.34 for LW. Results from a single-SNP GWAS performed within population clearly show the *VRTN* promoter SNP (VRTN_RS709317845) located on chromosome 7 as the most significant SNP in both lines. In addition, 2 QTL regions overlap between the two populations on chromosomes 10 and 12. The frequency of the allele C of the *VRTN* promoter SNP increasing NTE is almost five times higher in the L line compared to LW (0.79 vs 0.17). Allele frequency trend across the last 11 years shows that the frequency has been increased in L line, whereas this favourable allele is decreasing in LW line. These results show that although the breeding objective of increasing NTE has been achieved, the selection pressure on the *VRTN* promotor is different in both lines. The reasons for this difference, such as unfavourable correlations with other traits, needs to be further evaluated.

Session 27 Theatre 2

Management of genetic defects in breeding programs

S.T. Rodríguez-Ramilo[1], I. Palhière[1], J. Raoul[1,2] and J. Fernández[3]
[1]GenPhySE, Université de Toulouse, INRAE, ENVT, Castanet-Tolosan, 31326, France, [2]IDELE, Castanet-Tolosan, 31326, France, [3]Departamento de Mejora Genética Animal, INIA – CSIC, Madrid, 28040, Spain; jmj@inia.csic.es

Allele frequencies of genetic defects can increase in selected populations mainly due to a couple of reasons. Firstly, hitchhiking effect linked to genes controlling traits included in the selection objective. Secondly, genetic drift caused by the reduced number of candidates selected as parents of the next generation. Breeding programs for populations with individuals carrying genetic defects need considering the reduction of their allele frequencies, besides the typical aims of maximisation of genetic gain, and minimisation of genetic diversity loss. The objective of this study was to evaluate the evolution of allele frequencies of genetic defects, the genetic gain and diversity when managing genetic defects through a selection index combining the GEBVs with the individual weighted gene content. Stochastic simulations were used to generate a population with reasonable levels of linkage disequilibrium and expected heterozygosity using a mutation-drift equilibrium approach. Around 50,000 biallelic loci were assumed to be SNPs for management purposes and another 50,000 non-genotyped loci to determine the effect on the rest of the genome. An additive quantitative trait with a heritability of 0.4 was defined. One, 3 or 10 loci (with initial allele frequencies between 0.1-0.2) controlled the expression of a monogenic genetic defect each. One hundred individuals were randomly sampled from the population at equilibrium constituting the founders of the breeding program. Truncation selection on the index values was performed for ten non-overlapping generations. Averaged results for one hundred replicates shows that both the genetic gain and genetic diversity were affected by the number of genetic defects and the weight given for each of them. Removal of the genetic defects from the population was fast with high weights on the genetic defects. However, mean breeding values were lower when weights were higher and a high number of genetic defects were considered. Genetic diversity remained relatively stable across simulated scenarios. Extension of this study will involve the management of these genetic defects in an optimum contribution selection strategy.

Session 27 — Theatre 3

Genetic management of cryptorchidism and horn mutations in Manech tête Rousse dairy sheep breed

J. Raoul[1,2], F. Fidelle[3], C. André[3], M. Ben Braiek[2], S. Fabre[2], A. Gouzenes[2], D. Buisson[1] and I. Palhière[2]
[1]Idele, CS 52637, 31321 Castanet-Tolosan, France, [2]GenPhySE, Université de Toulouse, INRAE, ENVT, 31326 Castanet-Tolosan, France, [3]CDEO, 140 Route Ahetzia, 64130 Ordiarp, France; jerome.raoul@inrae.fr

Since the 90s, the 'horned' phenotype has been counter-selected in the Manech tête Rousse (MTR) dairy sheep breed. Among the elite dams, around 30% are horned and thus excluded from matings that produce candidate males for artificial insemination. In the meantime, breeding society managers have observed an increase in the number of cryptorchid young males among the candidates to selection. Recent work has highlighted the genetic determinism of these two traits and their interrelationship in the MTR breed. This determinism can be formalized by the segregation of three haplotypes: the allele for horns (H), the allele for being polled and non-cryptorchid (Pn) and the allele for being polled and cryptorchid (Pc). Only [HH] females are horned while all [HH] males and part of [HPn] males are horned (incomplete penetrance). [HPc] and [PcPc] males are polled, and only [PcPc] males are cryptorchids. Using stochastic simulations that mimics the current breeding program, we evaluated strategies combining: (1) selection of young AI males (that are genotyped for genomic selection) based on genotypes for the 'horn-cryptorchid' haplotype; and (2) the selection based on 'horned/polled' phenotype for other categories: young male candidates to be genotyped, elite dams, and ewe lambs. Strategies were compared over a 15-year period for both the genetic gain for milk yield and the changes in allelic and genotypic frequencies of the 'horn-cryptorchid' haplotype of ewe lambs. Selection based on genotype of AI males was the most efficient for the increase of the Pn allele frequency (f(H)=0. 12; f(Pn)=0.85; f(Pc)=0.03) and the control of genotypes associated with horned or cryptorchid phenotypes (f(HH)=0.01; f(HPn)=0.22; f(HPc)=0.01; f(PnPn)=0.71; f(PnPc)=0.05; f(PcPc)=0.00). This strategy is associated with a loss of genetic gain, -14%, compared to the reference strategy where horn phenotype is not considered. Combined strategies, as the selection based on genotype of young males and the exclusion of horned elite dams, limited the loss of gain (12%) whereas some other strategies induced higher losses (-22%).

Session 27 — Theatre 4

Strategies to improve selection compared to selection based on estimated breeding values

T. Pook, T. Niehoff, Y. Wientjes, L. Zetouni, M. Schrauf and M. Calus
Wageningen University & Research, Animal Sciences, Droevendaalsesteeg 1, 6700 AH Wageningen, the Netherlands; torsten.pook@wur.nl

Selecting animals based on estimated breeding values based on additive effects is shown to maximize short-term response to selection when using random mating. As rare beneficial alleles are oftentimes difficult to identify, have small estimated effects, and are in linkage with unfavourable alleles, this can lead to the loss of those alleles by drift and hitchhiking. To offset this, there are various strategies to reduce inbreeding, increase genetic diversity and thus allow for higher long-term genetic gains at the cost of slightly lower short-term genetic gain. We compared several of these strategies in a simulation study, using the software MoBPS. Applying a weighting factor to loci based on the frequency of the beneficial allele when calculating the estimated breeding value reduced the short-term genetic gain by -0.14 genetic standard deviations (gSD) while resulting in an increase of the long-term genetic gain of 2.02 gSD after 50 generations and reducing inbreeding rates by 22% compared to selection according to traditional estimated breeding values. Secondly, we considered including an additional trait in the selection index that represents the uniqueness of a given individual for the breeding program. As uniqueness criteria, we considered the number of estimated beneficial rare alleles, the total number of rare alleles, the average kinship to other animals, inbreeding rates, and the position in a principal component analysis. Putting 5% of the index weight on the average kinship to top animals led to virtually no short-term losses (-0.01 gSD) while reducing inbreeding rates by 6% and yielding 0.61 gSD more long-term genetic gain. The presented results probably represent a conservative estimate as the simulations did not include any deleterious variants and / or inbreeding depression. All considered strategies do not result in additional costs or extensive computations and thus implementation into practical breeding programs should be very straightforward and come with virtually no downsides. Considered strategies could also be combined with each other, with methods for optimum contribution selection, or with approaches to account for mendelian sampling variance.

The validity of breeding goals

R. Wellmann
University of Hohenheim, Department of Animal Genetics and Breeding, Garbenstraße 17, 70599, Germany; r.wellmann@uni-hohenheim.de

Breeding goals of livestock breeds are commonly derived from profit calculations, while breeding goals of companion breeds are defined by expert assessments. In both cases, the question arises whether the obtained breeding goals are valid. This study characterizes the validity of breeding goals and proposes a method for optimising them. A breeding goal can be called valid if there exists a group of owners who would prefer the improved breed over other breeds, and if this group of owners remains sufficiently large for the long-term survival of the breed. The environments that are provided by this group of owners can be called the breed's niche. A breeding goal can thus be called valid if there would exist a niche for the improved breed that is sufficiently large for the long-term survival of the breed. As a small population of the breed that would be placed into that niche would increase in size, this definition of a niche extends the definition of an ecological niche towards domestic animal breeds. The definition of the validity of a breeding goal allows to evaluate the breeding goals of arbitrary domestic breeds. Furthermore, it provides a sound base for the optimisation of breeding goals. The proposed optimisation method consists of defining an envisaged niche for the breed and of determining the conformation, performance and behaviour that maximises the adaptation of the breed to its envisaged niche. The general framework for the optimisation of breeding goals can be applied to both, livestock breeds and companion breeds. For example, the breed standards of many dog breeds changed only very little over long periods of time, even though their population sizes are decreasing, and their historical uses have changed. This observation suggests that the validity of their breeding goals is questionable. The same could apply to some endangered livestock breeds. Their breeding goals should be evaluated and adjusted not only with regard to animal welfare aspects, but also with regard to their validity.

Beyond scenarios-optimization of breeding program design using evolutionary algorithms (MoBPSopti)

A. Hassanpour[1], J. Geibel[1,2] and T. Pook[1,3]
[1]University of Goettingen, Center for Integrated Breeding Research, Animal Breeding and Genetics Group, Albrecht-Thaer-Weg 3, 37075, Goettingen, Germany, [2]2Friedrich-Loeffler-Institut, Institute of Farm Animal Genetics, Höltystraße 10, 31535 Neustadt, Germany, [3]Wageningen University & Research, Animal Breeding and Genomics, P.O. Box 388, 6700 AH Wageningen, the Netherlands; azadeh.hassanpour@uni-goettingen.de

Optimization of breeding program design requires accounting for multiple objectives along with many interdependent factors. Improving resource allocation facilitates breeders' decisions on what combinations and to what extent different breeding objectives are attainable. In a previous study, we used stochastic simulations and a kernel regression model to optimize the allocation of resources in a dairy cattle scheme. Our goal was to balance gain and inbreeding within a specified budget. Although kernel regression has aided in testing a broad range of potential breeding programs, the large number of needed simulations ultimately weakens its usability. This study aims to decrease the number of required simulations till convergence for the same problem via an evolutionary algorithm model. To achieve this, initial simulations were carried out with randomly generated parameter settings, and their fulfilment of the target criteria was evaluated. The most promising parameters will be selected as parents for the next iteration. Offspring parameterizations are generated through linear combination (recombination) and with minor changes to individual parameters (mutations) to explore different combinations of parameters in the search space. Our iterative process has an initial search phase with higher mutation rates to identify promising parameter combinations, followed by a refinement phase with a lower mutation rate to reduce parameter variations and thus narrow done the actual optima even further. We utilized Snakemake to automate all steps for efficient execution. The method proved robust by converging faster and stabilizing more quickly than kernel smoothing.

Session 27 Theatre 7

Ecobreed – what is the economically optimal longevity of a cow?

S. Schlebusch
Agroscope, Dürntnerstrasse 9, 8340, Switzerland; simon.schlebusch@agroscope.admin.ch

The question when to cull a dairy cow is frequently asked by farmers as well as by the scientific community. In Switzerland, there is the general opinion that cows should be kept longer to reach the economical optimum, though in reality this is almost never realized. Hence, are longer living cows really more profitable or are the farmers culling decisions sub optimal due to lack of complete economic information. The goal of the project Ecobreed is to provide to the farmers complete economic information about their culling decisions. This is done with a model that calculates the expected monetary value of a cow in the future. The first part of the model consists of a Markov chain simulation based on herd data from 2010 to 2018. In this simulation the life of a cow is spilt up in 4 different states, lactation, month in milk, month pregnant and the culled state. The model then calculates with which probability a cow moves from one state to another until the cow reaches the culled state. The time between birth and the cow reaching the culled state corresponds to the expected lifetime of the cow. This life expectancy is then combined with a monthly profit equation of the cow's revenue and cost, such as milk yield and price, feed cost, veterinarian cost, etc. The profit equation combined with the life expectancy results then in the cow own worth. In the simulation based on the previously described model the optimal length of life of a cow is calculated. This is done for different farm types based on typical farms in Switzerland. In Switzerland there is a vivid debate about implementing direct payments to farmers for cows with a certain minimum number of lactations. Therefore, the goal of the project is also to find the effects of such a payment on the economically optimal longevity of a dairy cow. The results of this project are meant to be implemented in an online tool that can be given to farmers. This tool should be able to help the farmers to make informed culling decisions.

Session 27 Theatre 8

Differences in breast milk composition in rabbit lines divergently selected for intramuscular fat

P. Hernández, N. Ibáñez, I. Heddi, M. Martínez-Álvaro and A. Zubiri-Gaitán
Institute for Animal Science and Technology. Universitat Politècnica de València, Camino de Vera s/n, 46022 Valencia, Spain; phernan@dca.upv.es

An experiment of divergent selection for intramuscular fat (IMF) in Longissimus thoracic and lumborum (LTL) muscle was performed in rabbits over ten generations. The aim of the study is to estimate the correlated response to selection for IMF of LTL muscle on does' milk composition since previous studies have pointed out the importance of maternal effects for IMF. Milk samples were collected on day 15 of lactation from 38 does from the high-IMF (H) line and 34 from the low-IMF (L) line, and its composition and fatty acid profile were analysed. A linear model was fitted to estimate the differences in milk composition between the H and L lines, solved using Bayesian inference. The model included the line, sampling date (month), and parity order as fixed effects, and the number of lactating kits (LK) as a covariate. The marginal posterior distribution of the difference between lines was obtained using the Rabbit program (https://github.com/VLabUPV/runRabbit). The median (DH-L) of the marginal posterior distribution of the difference between lines and its probability of differing from 0 (P0) were calculated. The milk of the H-line showed greater fat content (P0=0.98) than the L- line with DH-L of 0.9%. In contrast, L-line milk had a higher percentage of lactose than H-line, with DH-L of 0.2% (P0=1). No difference was found in urea (DH-L=11.6, P0=0.62) and protein (DH-L=-0.05, P0=0.63) content. In general, H-line milk showed greater saturated fatty acid (SFA) and lower polyunsaturated fatty acid (PUFA) percentage than L-line, with no differences in monounsaturated fatty acids (MUFA). Among the SFA, the H line had a higher percentage of C10:0, C12:0, and C16:0 and a lower percentage of C8:0, C15:0, and C17:0 than the L line. The C18:3n6 and CLA9t11t were present in higher percentages in the H line. As for PUFA percentage, C18:2n6, C18:3n3, and C20:3n3 percentage was lower in the H line. The two lines divergently selected for IMF content presented substantial differences in breast milk composition confirming a correlated response to selection for IMF, suggesting that the milk composition is one of the factors explaining the maternal effects for IMF.

Session 27 Theatre 9

Combined genomic evaluation of Australian Merino and Dohne Merino sheep populations

M. Wicki[1,2], D.J. Brown[3], P.M. Gurman[3], J. Raoul[1,2], A. Legarra[1,4] and A.A. Swan[3]
[1]INRAE, INP, UMR 1388 GenPhySE, Animal Genetics, 24 Chemin de Borde Rouge, 31320 Castanet-Tolosan, France, [2]Institut de l'Elevage, Animal Genetics, 24 Chemin de Borde Rouge, 31320 Castanet-Tolosan, France, [3]AGBU, a joint venture of NSW Department of Primary Industries and University of New England, AGBU Building, 2350 Armidale, Australia, [4]Current address: Council on Dairy Cattle Breeding, 4201 Northview Dr, Bowie, 20716 MD, USA; marine.wicki@inrae.fr

Historically, the Merino has been the dominant sheep breed in Australia having first been imported to the country over 200 years ago. In addition, starting in the late 1990s, sheep producers began importing Dohne Merino embryos from South Africa to improve on attributes such as reproduction and carcase composition. Since then, this breed has continued to expand in Australia but the number of genotyped and phenotyped purebred individuals remains low, calling into question the feasibility of genomic selection in this breed. The Australian Merino on the other hand has a very large reference population in a separate genomic evaluation. Many studies have shown that combined genomic evaluation of several populations can be advantageous in terms of accuracy when it involves genetically close populations. In particular, combined evaluations can be very advantageous for small breeds that can benefit from the large reference population of another breed. This study was based on 27,632 Australian Merino (M), Dohne Merino (D) and crossbred (C) genotyped individuals and a pedigree of more than 4 million animals and more than 5.2 million phenotypes for two wool traits (fibre diameter and greasy fleece weight) and one weight trait (yearling liveweight). The first objective of this study was to characterize the genomic structure and relationships between these populations. A Principal Component Analysis of the genomic relationship matrix as well as computations of Hudson's fixation index (Fst), that were below 0.08, revealed a low genetic differentiation between M, D and C populations. This suggests that crossbred or combined predictions may be feasible. In the following step of this work, we will investigate the accuracy of genomic evaluation in a Dohne validation population based on purebred D, purebred M, crossbred C or the combination of these three reference populations.

Session 27 Theatre 10

Effect of germplasm exchange strategies on genetic gain and diversity in dairy stud populations

E.A. Lozada-Soto[1], F. Tiezzi[2], J. Jiang[1], J.B. Cole[3], P.M. Vanraden[4], S. Toghiani[4] and C. Maltecca[1]
[1]North Carolina State University, Department of Animal Science, Raleigh, NC, 27607, USA, [2]University of Florence, Department of Agriculture, Food, Environment and Forestry (DAGRI), Florence, 50144, Italy, [3]URUS Group LP, Madison, WI, 53718, USA, [4]USDA, Animal Genomics and Improvement Laboratory, Beltsville, MD, 20705, USA; cmaltec@ncsu.edu

This study aimed to determine the effect of different germplasm exchange strategies on long-term genetic gain, homozygosity, and inbreeding. Genotypes from U.S. Holstein cattle born between 2010 and 2020 and belonging to 3 distinct stud populations were used as a basis for a simulated breeding program. Ten generations of selection were simulated (using the R AlphaSimR package) within each population by mating 200 bulls with 5,000 cows. Three traits were simulated (one for each population) with the same underlying genetic architecture (either 10 or 100 QTL/per chromosome) and an additive genetic correlation between traits that varied according to a parameter ε (ε=0.10, 0.50, or 0.90). Animals within a given population were selected either randomly, using true breeding value (TBV), estimated breeding value (EBV), or an estimated breeding value penalized for kinship (pEBV) with females. To evaluate the effect of germplasm exchange between populations, we simulated scenarios where sires (25 or 100) from the other two populations were used for mating instead of a portion of within-population sires. These sires were chosen either randomly, using pEBV, or using genomic future inbreeding (GFI). The study monitored genetic progress, homozygosity, and inbreeding. The simulation was replicated ten times. A mixed-linear model was fitted using the fixed effects of within- and across-population selection scenarios, population, number of QTL, ε parameter, all the two-order interactions, and the random effect of replicate. Least-squares means were obtained for each level of the main effects. Using sires from other populations (specifically the pEBV and GFI scenarios) provided higher long-term genetic gain (pEBV_100 = 2.30 genetic SD vs no exchange = 1.77 genetic SD), a decrease in homozygosity (GFI_100 = -1.16% vs no exchange = 0.28%), and a lower average inbreeding increase (pEBV_100 = 0.26% vs no exchange = 0.38%).

Local livestock breeds in Flanders are confirmed to be 'at risk'!

S. Janssens, L. Chapard, R. Meyermans, W. Gorssen and N. Buys
Center for Animal Breeding and Genetics, KU Leuven, Department of Biosystems, Kasteelpark Arenberg 30, Box 2472, 3001 Leuven, Belgium; steven.janssens@kuleuven.be

Within the EU, member states can support their local breeds if they are classified with a status 'at risk'. Granting such a status should be supported by advice, which is issued by an experienced body. Here, we report the assessment of the status of 17 breeds of cattle (4), sheep (8), goat (3) and pigs (2) in Flanders which was requested by the Flemish government, Department Agriculture and Fisheries. We investigated demographic and genealogic information, as well as genomic information for 8 breeds. The primary source of information was pedigree data, containing id's (animal, sire and dam), birth year, the owner of the animal The cattle pedigree was preselected and contained the active animals for the years 2017-2022 and all ancestors of these animals. Sheep, goat and pig pedigrees were obtained 'unselected' from the responsible organizations and from these, the cohorts of animals born and registered in 2017-2022 and all their known ancestors were selected. Active breeding females and males (at least one offspring in a particular year) were counted per year. The number of active females and its trend over time were retained as separate criteria and converted in scores (0=no problem; 5=high risk) based on FAO guidelines. All pedigree files were cleaned and were used to infer effective sizes (N_e) following the decision tree implemented in POPREPORT. Genotypic data was available for 4 breeds of cattle and 4 sheep breeds and was used to compute genomic inbreeding and Ne. Ne estimates were converted in scores from 0 to 5, 5 corresponding to N_e<45. N_e was retained as the third criterion. A score of 4 or 5 for any of the 3 criteria was considered sufficient to classify a breed as endangered. Following this rule, all 17 breeds in the study were categorized as being at risk. Most breeds fall short on more than one criterion and their risk status is often aggravated by low and decreasing numbers of active breeders (2017-2022). Our across-species approach, applying formal criteria helps stakeholders to better monitor the local breeds in Flanders and calls for urgent action in the field.

Probabilistic breeder's equation used for retrospective and prospective analyses

D. López-Carbonell[1], I. Pocrnic[2], G. Gorjanc[2] and L. Varona[1]
[1]Instituto Agroalimentario de Aragón (IA2), Universidad de Zaragoza, Departamento de Anatomía, Embriología y Genética Animal, Calle Miguel Servet, 177, 50013, Zaragoza, Spain, [2]The Roslin Institute and Royal (Dick) School of Veterinary Studies, The University of Edinburgh, Easter Bush Campus, EH25 9RG, Edinburgh, United Kingdom; davidlc@unizar.es

The breeder's equation (BE) is the most important equation in breeding and also key in quantitative genetics. It is traditionally used to predict the response to selection in a population. This prediction is based on the intensity of selection, the accuracy of selection criteria, and the additive genetic standard deviation. However, in non-experimental populations, the genetic contributions, accuracies and genetic variances are unknown variables inferred from the data. This entails uncertainty in predictions with the BE, which is commonly ignored. While the BE is mainly used for a prospective analysis/prediction, it also has application in retrospective analysis. To this end, García-Cortés *et al.* developed a method to analyse past genetic trends by partitioning breeding values of individuals into parent contribution terms and Mendelian sampling and allocating these terms to different paths of selection. Here we combine the principles of the BE and the partitioning method for probabilistic retrospective analyses. By doing this, the posterior distribution of the elements of the BE for user-defined paths of selection is obtained. These results provide an understanding of the past genetic changes due to selection within and between different paths of selection and how each of these paths contributed to the overall change. To test the method, we first simulated 5 generations of selection on true breeding values with variable male and female contributions. The results showed that genetic changes are proportional to the genetic contribution of male and female paths and their respective selection intensities following the BE. Second, we simulated selection on breeding values estimated from the data with uncertainty. We used the developed method to obtain the posterior distribution of selection intensity, accuracy, and genetic variance in each generation. This work will give further insight into past genetic changes in breeding programmes and provide probabilistic predictions.

Session 27 Poster 13

Re-organising the Danish dairy cattle sector with alternative breeding goals and crossbreeding

J.B. Clasen[1,2], R.D. Kring[1], J.R. Thomasen[3] and S. Østergaard[1,2]
[1]SimHerd A/S, Niels Pedersens Allé 2, 8830 Tjele, Denmark, [2]Aarhus University, Animal and Veterinary Science, Blichers Allé 20, 8830 Tjele, Denmark, [3]VikingGenetics, Ebeltoftvej 16, 8960 Randers SØ, Denmark; julie.clasen@anivet.au.dk

In this simulation study, we analysed the economic impact of re-organising the Danish dairy cattle sector inspired by the pig sector. The system involves herds with purebred animals acting as multipliers for production herds with crossbred animals. Hereby, utilising the strengths of the major dairy breeds in Denmark in a terminal crossbreeding system, potentially increasing the overall profitability of the sector. The selection of female breeding candidates will be intensified by only using embryo technologies. The proportions of cows in multiplier and production herds were calibrated to meet the minimum number of breeding animals for the highest genetic gain with a reasonable inbreeding rate. It resulted in 7% of the cows being purebreds in multiplier herds, and 93% of the cows being crossbreds in production herds. A scenario illustrating the current dairy sector was simulated for economic comparison. In the economic analyses we assumed same market prices and costs associated with breeding in the current and re-organised sector. We also assumed a zero-profit of production of crossbred heifers in the multiplier herds. The preliminary results show, that multiplier herds in will gain a 3% higher contribution margin than an average herd in the current sector, while production herds will gain 19%. The additional contribution margins are expected to cover the cost of producing crossbreds by using embryo technologies in the breeding programme of the re-organised sector. Furthermore, the preliminary results show that the re-organised sector is able to deliver the same amount of milk as the current sector, but with a 5% larger cow population. On the other hand, the number of young stock for replacement (both purebred and crossbred) will be reduced by 35%. We expect that including genetic gains will increase the economic potential of crossbreeding the re-organised dairy sector. Furthermore, the economic potential of having alternative breeding goals aimed to specialise the breeds used in the crossbreeding system will be analysed.

Session 28 Theatre 1

How to balance selection for litter size in pigs with survival, health and welfare

P.W. Knap[1], A. Huisman[2], C. Sørensen[3] and E.F. Knol[4]
[1]Genus-PIC, Ratsteich 31, 24837 Schleswig, Germany, [2]Hendrix Genetics, P.O. Box 30, 5830 AA Boxmeer, the Netherlands, [3]Danish Agriculture & Food Council, Agro Food Park 15, 8200 Aarhus N, Denmark, [4]Topigs Norsvin Research Center, Schoenaker 6, 6641 SZ Beuningen, the Netherlands; pieter.knap@genusplc.com

Animal populations adapt to their environments in order to survive. Adaptation to farming conditions can take place by natural selection (when the best-adapted individuals produce more offspring), but targeted genetic selection results in larger changes. To steer genetic change in the desired direction, livestock genetic programs need to record many characteristics on individual animals, ideally this should be done under the farming conditions of the future. In the EU, societal discussions will soon lead to clearer definition of those conditions, e.g. free farrowing for sows. Another EU- and worldwide development is the increasing shortage of labour, especially on (pig) farms. Both these issues require an increased self-sufficiency of sows and piglets. Breeding goals that emphasize pig meat production efficiency lead to a genetic increase in litter size. Litter size has negative (unfavourable) genetic correlations to piglet survival, mediated by changes in weight and physiological maturity at birth, gestation length, and farrowing duration. Such unfavourable correlations can be neutralized by balanced index selection based on a proper understanding of the underlying biology; this leads to favourable trends in litter size and piglet survival at the same time. This involves appropriate recording of piglet survival itself (at farrowing and during lactation), birth weight and its within-litter variation, teat number, weaning weight, gestation length, farrowing duration (relevant for colostrum intake), and/or the various types of maternal behaviour and mothering ability. We present: (1) realized phenotypic and genetic trends of various reproduction traits and the way they relate to each other; (2) past and future breeding goals; and (3) developments in data recording, of four leading pig breeding organizations. We show that it is possible to balance selection for litter size in pigs with survival, health, and welfare. And by doing so improve both animal welfare and efficiency within the pork value chain.

Selection for robustness and welfare in Iberian pig through birth weight uniformity as criterion

J.P. Gutiérrez[1], N. Formoso-Rafferty[2], F. Sánchez-Esquiliche[3], M. Muñoz[4], J.M. García-Casco[4,5] and I. Cervantes[1]
[1]Dpto. Producción Animal. Facultad de Veterinaria, UCM, 28040 Madrid, Spain, [2]Dpto. Producción Agraria. E.T.S.I.A.A.B, UPM, 28040 Madrid, Spain, [3]Dpto. Agropecuaria, Sánchez Romero Carvajal-Jabugo, 21290 Jabugo, Huelva, Spain, [4]Dpto. Mejora Genética, INIA-CSIC, 28040 Madrid, Spain, [5]Centro de I+D del cerdo Ibérico-Zafra, INIA-CSIC, 06300 Zafra, Badajoz, Spain; icervantes@vet.ucm.es

After the genetic selection for increased litter size (LS), the pig industry has noticed an increase in birth weight (BW) variability within litter. Within-litter variation of BW has a high importance in breeding practice for animal survival and from an economic perspective. The objective of this study was to analyse the data from a selection experiment for BW variability in Iberian pigs (Sánchez Romero Carvajal population). Two replicates of the generation 0 have been performed during 2022. A total of 243 litters belonging to nulliparous sows (345-458 days of age) and 28 boars were used, totalling 1,544 BW piglets. A heteroscedastic model was used to select animals, including mother age, piglet sex, LS at birth and replicate as systematic effects. Genetic effects (mean and variability) were assigned to the mother. The pedigree contained 1,026 animals. Weight at 21 days (W21), average daily gain to 21 days (ADG) and survival (at birth and at 21 days) were also studied to evaluate the impact of selection. The genetic variance for the BW variability was 0.39 (genetic variation coefficient of 0.62) with a genetic correlation of -0.20 between BW and its variability, although not significant. The mean BW was 1.17 kg and 1.18 kg for the selected and the total sows. LS was 7.3 averaging the number of individuals per litter in which each piglet was born, being 7.8 in the selected sows for uniformity. The intermediate LS values were less variable than the extremes. The correlation between the mean trait and its variability was close to 0 for W21 and ADG. Correlations between BW variability and other traits were favourable for BW, W21 variability, ADG variability, survival at birth and at 21 days. However, they were slightly unfavourable for W21 and ADG. A reduction of 16% in the present BW variability could be achieved without negatively modifying the BW neither LS.

Resilience parameters in fattening pigs are heritable and associated with tail biting and mortality

W. Gorssen[1], C. Winters[2], R. Meyermans[1], L. Chapard[1], K. Hooyberghs[1], J. Depuydt[3], S. Janssens[1], H. Mulder[4] and N. Buys[1]
[1]KU Leuven, Center for Animal Breeding and Genetics, Kasteelpark Arenberg 30, bus 2472, 3001, Belgium, [2]KU Leuven, Laboratory for Biological Psychology, Tiensestraat 102, 3000 Leuven, Belgium, [3]Vlaamse Piétrain Fokkerij vzw, Aardenburgkalseide 254, 9990 Maldegem, Belgium, [4]Wageningen University & Research, Animal Breeding and Genomics, P.O. Box 338, 6700 AH, Wageningen, the Netherlands; wim.gorssen@kuleuven.be

Previous research in pigs showed that body weight deviations in longitudinal data are heritable and can be used as a proxy for pigs' general resilience. However, only a few studies investigated their relationship with specific resilience-related traits, such as lameness, bite wounds and mortality. Moreover, most studies focused on purebred pigs, which are often less exposed to challenges than commercial pigs. In this study, 1,865 fattening pigs with known pedigree (135 Piétrain sires and 266 crossbred dams) were weighed every 2 weeks (~8 recordings per pig). During weighing, pigs were also scored for physical abnormalities, such as ear swellings, umbilical hernia, lameness and ear and tail biting wounds. Recordings took place in the same stable between July 2020 and July 2021. The resilience trait *lnvar* was calculated as the natural logarithm of the variance of deviations of predicted vs observed weights after Gompertz modelling. Heritabibilites, and phenotypic and genetic correlations between *lnvar*, physical abnormalities and mortality were estimated using remlf90 (BLUPF90 software). The resilience trait *lnvar* was moderately heritable (h^2=31.4%). Moreover, *lnvar* was phenotypically and genetically correlated with tail biting wounds (r_p=0.22; r_g=0.39±0.25), lameness (r_p=0.21; r_g=0.44±0.05) and mortality (r_p=0.27; r_g=0.43±0.07). Low correlations were found between *lnvar* and umbilical hernia (r_p=-0.01; r_g=-0.02±0.81), ear biting wounds (r_p=0.02; r_g=-0.02±0.26) and hematomas (r_p=-0.02; r_g=-0.22±0.21). Our findings show that deviations in longitudinal weight data are heritable and positively associated with tail biting wounds, lameness and mortality. These findings are valuable for pig breeders, as they offer evidence that these deviations are an indication of animals' general health and resilience.

Session 28 Theatre 4

The impact of phenotyping, genotyping, and the boar's origin on the genetic gain of organic pigs

R.M. Zaalberg[1], J.B. Clasen[2,3], T.M. Villumsen[1], J. Jensen[1] and T.T. Chu[1]
[1]Aarhus University, Center for Quantitative Genetics and Genomics, C. F. Møllers Allé 3, 8000 Aarhus, Denmark, [2]Aarhus University, Department of Animal and Veterinary Sciences, Blichers Alle 20, 8830 Tjele, Denmark, [3]SimHerd Inc., Niels Pedersens Alle 2, 8830 Tjele, Denmark; roos.zaalberg@qgg.au.dk

Pig welfare and organically produced pork are popular among consumers. Yet, for organic pigs, there are no breeding programs that aim specifically at improving organic pigs according to the organic principles. Instead, organic pig breeding programs depend on genetic material from conventional boars. Little is known about how organic pig breeding programs could be optimized. This study investigated how genetic gain in an organic pig population is impacted by phenotyping and genotyping of different types of pigs in the pig production chain. Furthermore, we considered the biology of traits and the origin of the boars. We simulated breeding schemes that included three types of pigs, namely nucleus pigs (L), crossbred pigs (YL), and finisher pigs (DYL). The scenarios used four factors: phenotyping of DYL, genotyping of the three types of pigs, the origin of the L-sire, and the genetic correlation between the purebred- and crossbred performance. The results show that using boars from an external breeding program stagnated the genetic gain of both L- and DYL-pigs. On the other hand, using organic boars resulted in a significantly higher and positive genetic gain in both L- and DYL-pigs. Genotyping of L-pigs increased genetic gain in both the L- and DYL-pigs, whereas genotyping of YL- and DYL-pigs made no difference. Phenotyping of DYL-pigs had a positive impact on the genetic gain in DYL, but only when the correlation between the purebred and crossbred performance was strong. In conclusion, to genetically improve the organic pig population, different factors should be considered. Organic boars should be used instead of external boars, and genotyping should only be done for the L-pigs. Phenotyping DYL may also be considered, under the condition that the correlation between the purebred and crossbred performance is strong.

Session 28 Theatre 5

Genetic correlations between ostrich behavioural traits and slaughter traits

P.T. Muvhali[1,2], M. Bonato[2], A. Engelbrecht[1], I.A. Malecki[2,3] and S.W.P. Cloete[1,2]
[1]Western Cape Government, Directorate Animal Sciences, Private Bag X1, 7607, Elsenburg, South Africa, [2]University of Stellenbosch, Department of Animal Sciences, Private Bag X1, 7602, Matieland, South Africa, [3]The University of Western Australia, School of Agriculture and Environment, 35 Stirling Highway, WA 6009, Crawley, Australia; pfunzo.muvhali@westerncape.gov.za

Behavioural responses of animals towards humans in livestock industries are vital to improve animal welfare, productivity and occupational health and safety. In the ostrich industry, it is unknown if selection of less temperamental animals that are willing to associate with humans, influence productivity and other traits of economic importance on a genetic level. The aim of this study was to estimate genetic correlations, first among behavioural response traits of juvenile ostriches towards humans and then of juvenile ostrich behavioural traits with bird slaughter weight and skin traits (i.e. skin size, nodule size, nodule shape, presence of hair follicles). Behavioural traits, namely willingness of birds to approach a human, keeping a distance away from the human, allowing touch interactions with a human, excessive pecking, beak clapping and wing flapping were recorded for 1,092 juvenile ostriches. The results of this study revealed moderate to high heritability estimates ranging from 0.12 for excessive pecking to 0.48 for willingness to approach and keeping a distance from the human. Willingness to approach, keeping a distance and allowing touch interactions were all highly correlated genetically with each other at respectively -0.99, 0.87 and -0.87. No unfavourable genetic correlations between behavioural traits and slaughter traits were recorded for slaughter weight, skin size, nodule size score and nodule shape score. There were unfavourable genetic correlations approaching an absolute value of 0.50 of approach, keeping a distance and allowing touch interactions with the prevalence of hair follicles, though the underlying biology of these unfavourable correlations is unknown at present. In conclusion, our results suggested that temperament of ostriches could be improved by selecting birds that demonstrate willingness to associate with humans. Such selection is unlikely to compromise slaughter weight and most skin traits.

Session 28 Theatre 6

Genome-wide copy number variants associated with calving ease and retained placenta in Holstein cows

I.C. Hermisdorff[1], H.R. Oliveira[1,2], G.A. Oliveira Júnior[1], T.C.S. Chud[1], S.G. Narayana[1], C.M. Rochus[1], A.M. Butty[3], F. Malchiodi[1,4], P. Stothard[5], F. Miglior[1,6], C.F. Baes[1,7] and F.S. Schenkel[1]
[1]Department of Animal Biosciences, University of Guelph, Guelph, ON, N1G2W1, Canada, [2]Department of Animal Sciences, Purdue University, West Lafayette, IN, 47907, USA, [3]Qualitas AG, Zug, 6300, Switzerland, [4]Semex, Guelph, ON, N1H6J2, Canada, [5]University of Alberta, Edmonton, AB, T6G2R3, Canada, [6]Lactanet, Guelph, ON, N1K1E5, Canada, [7]Institute of Genetics, Vetsuisse Faculty, University of Bern, Bern, 3012, Switzerland; idacosta@uoguelph.ca

Calving difficulty and retained placenta (RP) can lead to significant economic losses for dairy producers due to increased veterinary costs, reduced milk production, and increased risk of culling. Copy number variants (CNVs) have the potential to explain part of the phenotypic variation of quantitative traits not explained by other genetic markers. However, despite their potential importance, there is a lack of studies in the literature investigating the impact of CNVs on RP and calving ease (CE). Here, genome-wide association studies using CNVs and de-regressed breeding values for RP and CE were performed using data from 2,465 Canadian Holsteins. A total of four genomic regions containing CNVs (CNVR) were found to be significantly associated (FDR<0.05) with RP (located on BTA4, BTA5, BTA6 and BTA23), while one CNVR was found to be significantly associated with CE (BTA7). The CNVR detected on BTA7, BTA23 and BTA4 had deletions in 36.3, 16.3 and 1.2% of the evaluated animals, respectively. On the other hand, the CNVR detected on BTA5 and BTA6 had duplications in 2.2 and 1.2% of the animals, respectively. A total of 50 (RP) and 93 (CE) positional candidate genes were retrieved from the significant CNVR identified. Among them, two important candidate genes found for CE (*CCDC105* and *SLC1A6*) have been previously associated with fertility traits in dairy cattle. Additionally, the gene *MCM3*, found for RP in a CNVR located on BTA23, plays an important role in cell cycle pathway and is a functional candidate gene for the regulation of follicular maturation. Our findings provide new insights into the genetic architecture of RP and CE in Holstein cattle and may help reveal new strategies for future genetic improvement for these traits.

Session 28 Theatre 7

Indirect effects in infection transmission enhance genetic selection and other interventions

A.D. Hulst[1,2], P. Bijma[1] and M.C.M. De Jong[2]
[1]Wageningen University & Research, Animal Breeding and Genomics, P.O. Box 338, 6700 AH Wageningen, the Netherlands, [2]Wageningen University & Research, Quantitative Veterinary Epidemiology, P.O. Box 338, 6700 AH Wageningen, the Netherlands; dries.hulst@wur.nl

Genetic selection of livestock to lower the prevalence of infectious diseases has long attracted interest as an addition or alternative to other interventions, but its potential has been seen as limited because of the low heritability of binary disease status. The usual quantitative genetic methods used to estimate this heritability, however, treat disease status simply as an individual trait and thereby ignore the transmission dynamics of infectious diseases. Here we show that the effect of transmission dynamics on the results of genetic selection are considerable. We use simulations of an established epidemiological model (SIS) with genetic variation in individual susceptibility. Feedback effects in the transmission process rapidly lead to a certain degree of herd immunity, which might eventually result in eradication of the infection. Feedback in transmission occurs because individuals that have a low susceptibility are not only less likely to get infected themselves, but thereby also contribute to a lower number of infectious individuals in the population. This leads to indirect (genetic) effects, the effects of the genes of an individual on the infection status of its herd mates. Calculations show that these indirect effects result in a total effect of selection that is a factor of the inverse of the prevalence larger than the direct genetic effect (the breeding value for individual disease status). Thus, the total effect increases strongly with decreasing prevalence, implying that genetic selection will become more and more effective as prevalence goes down. This effect is not limited to genetic selection. If other interventions are applied that reduce prevalence, genetic selection will become more effective as well. Consequently, it can be very beneficial to combine genetic selection with other interventions, and not do only one or the other. One should, for example, not reduce hygienic measures once genetically resistant animals enter the herd, nor should breeding for increased resistance be rejected because hygienic measures are already applied.

Optimizing dairy cattle breeding goals to improve production and udder health of crossbred cows

A. Bouquet[1], H.M. Nielsen[1], V. Milkevych[1], M. Kargo[1], J.R. Thomasen[2] and M. Slagboom[1]
[1]Aarhus University, Center for Quantitative Genetics and Genomics, C.F. Møllers Allé 3, 8000 Aarhus C, Denmark, [2]VikingGenetics, Ebeltoftvej 16, 8960 Randers SØ, Denmark; alban.bouquet@qgg.au.dk

Crossbreeding is known to increase economic profitability by benefiting from heterosis and breed complementarity. Breeding goals (BG) used to select dairy cattle breeds have so far been aimed at improving purebred performance which may be sub-optimal when the goal is crossbred performance. The aim of this study is to compare various purebred BGs in a two-way terminal crossbreeding system for dairy cattle over several generations of crossbreeding. We postulated that differentiating BGs in purebred parental lines could be more efficient to increase crossbred performance. Breeding schemes of Danish Jersey (DJ) and Nordic Holstein (NH) cattle were simulated using real haplotypes of 200 founders sampled from each breed. We simulated a founder population where NH had a higher breeding value for milk production and a lower breeding value for udder health compared to DJ. First, genomic reference populations were constituted for 20 years to predict genomic breeding values with appropriate reliabilities. In year 21-30, a genomic selection scheme for NH and DJ aimed at crossbred performance was simulated. Each year, 2,000 young bulls and 2,000 heifers were genotyped. Out of the young bulls, 50 were selected as sires. Three scenarios were set up to test the hypothesis: (1) a reference scenario resembling the economic values in the present Nordic total merit index for DJ and NH; (2) the same BG in NH as in 1, but a higher value on health in the BG of DJ; and (3) BGs only weighing production in NH and only health in DJ. In year 25-30, a population of crossbred cows was simulated alongside the two pure breeding schemes by mating Jersey cows with Holstein bulls. Purebreds in both nuclei were evaluated both for purebred and crossbred performance. For evaluation of crossbred performance, SNP-BLUP breeding values were estimated by using allele frequencies of the purebred population that an animal was bred to, whereas GBLUP was used for purebred evaluation. Outcomes of this study will be useful to assess the importance of updating BGs in purebred breeding schemes to optimize crossbred performance.

Genes and variants involved in resistance to paratuberculosis in Holstein and Normande cattle

V. Sorin[1], A. Boulling[1], A. Delafosse[2], M. Boussaha[1], C. Hozé[3], R. Guatteo[4], C. Fourichon[4], S. Fritz[3], D. Boichard[1] and M.P. Sanchez[1]
[1]Université Paris Saclay, INRAE, AgroParisTech, GABI, 78350 JouyenJosas, France, [2]GDS Orne, 61000 Alençon, France, [3]Eliance, 75012 Paris, France, [4]Oniris, INRAE, BIOEPAR, 44300 Nantes, France; marie-pierre.sanchez@inrae.fr

Bovine paratuberculosis, or Johne's disease, is a contagious and incurable disease, caused by *Mycobacterium avium* subsp. *paratuberculosis* (MAP), with adverse effects on animal welfare and serious economic consequences. The recent implementation of genomic evaluation for paratuberculosis resistance in France, effective in Holstein and ongoing in Normande, provides access to a large database of cows with disease status estimated from blood serological data, i.e. likely to be infected (with or without clinical signs), or not to be infected. Using a subset of 4,677 Holstein and 4,845 Normande cows with chip genotyping data, we performed sequence-based genome-wide association studies (GWAS) to identify genes and variants influencing resistance to paratuberculosis. Genotypes from the whole genome sequence were imputed using a multibreed reference population from the 9th run of the 1kBG consortium and in-house data comprising 1,414 Holstein and 160 Normande animals. The GWAS was carried out with a mixed linear model, testing the individual effect of ~14M variants after filtering for low imputation accuracy ($R^2<0.2$) or frequency ($MAF<0.005$), and accounting for the population structure through polygenic effects estimated from a 50k-based genomic relationship matrix. A total of 3,026 and 5,063 variants had significant effects ($7.3\leq -\log(P)\leq 22$) on resistance to MAP infection in Holstein and Normande, respectively. They were located on chromosomes 12, 13, 18, and 23 in Holstein and on chromosomes 3, 6, 12, and 23 in Normande. The genomic region with the most significant effect was located on chromosome 13 in Holstein (~ 63 Mb) and on chromosome 23 in Normande (~ 23 Mb). Based on their position, we identified *PEAR1*, *ELOVL5*, *HS6ST3*, *SNTA1*, and *BOLA-DRA* as the most plausible candidate genes. This study confirms the genomic regions initially identified in Holstein for resistance to MAP infection and reveals novel regions, candidate genes and variants in both Holstein and Normande breeds.

Session 28 Theatre 10

A genome-wide association study for clinical mastitis in Italian Holstein

F. Galluzzo[1,2], G. Visentin[1], G. Mészáros[3], J.B.C.H.M. Van Kaam[2], R. Finocchiaro[2], M. Marusi[2] and M. Cassandro[2,4]
[1]University of Bologna, DIMEVET, via Tolara di Sopra 50, 40064 Ozzano dell'Emilia, Italy, [2]Associazione Nazionale Allevatori della Razza Frisona, Bruna e Jersey Italiana, Via Bergamo 292, 26100 Cremona, Italy, [3]University of natural resources and life sciences, NUWI, Gregor-Mendel-Straße 33/II, 1180 Wien, Austria, [4]University of Padova, DAFNAE, Viale dell'università, 16, 35020, Legnaro, Italy; ferdinandogalluzzo@anafi.it

Clinical mastitis (CM) has a huge detrimental impact on animal welfare, farms' net profit and antimicrobial resistance. Nowadays, on-farm CM diagnoses are available in Italy from the Livestock Environment Opendata (LEO) project of the Italian Breeders Association (AIA). The aim of this study was to identify the SNPs associated to CM resistance and investigate its genetic background performing a genome-wide association study (GWAS). CM was considered a binary trait. Data came from 4 North-Italian regions (Lombardia, Emilia-Romagna, Veneto and Piemonte), recorded from 2014 to 2020: only first records per lactation, from 1 to 305 days in milk, were included. The maximum parity considered was 5. Lower and upper bounds of incidence per herd per year were set to 0.15 and 0.75 respectively. Herd-year-season of calving groups with less than 10 animals were discarded. The edited dataset had 113,289 records; the prevalence was 0.32. The 4 generations pedigree was composed of 140,385 animals. Heritability was previously calculated with a single-trait repeatability threshold animal model: (co)variances were converted to the observed scale in order to fit the trait as linear. Heritability on the observed scale was 0.04. GWAS analysis was performed and, in the intersection of the highest thousand SNPs for explained variance and the thousand most significant SNPs, we identified 32 SNPs. There were 47 genes in their vicinity considering windows of 200 kb, mostly located on chromosomes 5, 7, 25 and 27. Among others, PLXNC1, associated with mastitis resistance in sheep, SEMA7A, associated with immune response in humans and CEP83, associated with paratuberculosis pathological outcomes in Holstein cattle. Results confirmed the complexity of the trait and its highly polygenic nature and provide a better understanding of CM resistance in Italian Holstein.

Session 28 Theatre 11

About the genetic connection between milk and health traits in dairy cows within functional regions

H. Schneider[1], J. Heise[2], A.-M. Krizanac[3], C. Falker-Gieske[3], J. Tetens[3], G. Thaller[4] and J. Bennewitz[1]
[1]Institute of Animal Science, University of Hohenheim, Garbenstr. 17, 70599 Stuttgart, Germany, [2]Vereinigte Informationssysteme Tierhaltung w.V., Heinrich-Schröder-Weg 2, 27298 Verden (Aller), Germany, [3] Georg-August-University Göttingen, Division of Functional Breeding, Department of Animal Sciences, Burckhardtweg 2, 37077 Göttingen, Germany, [4]Institute of Animal Breeding and Husbandry, Christian-Albrechts University of Kiel, Olshausenstr. 40, 24098 Kiel, Germany; helen.schneider@uni-hohenheim.de

Many genomic mutations, all having a small impact, shape most complex traits in mammals. Intensive attempts were made to detect the underlying causal mutations, not only to scrutinize the biological mechanisms underlying these traits but also to enhance genomic selection. Unfortunately, difficulties arise because of linkage between variants and their usually small effect sizes. However, implementing functional and evolutionary information in genomic analyses enhances the detection of potential causal mutations. Thus, the intention of this study was to investigate the contribution of 27 functional and evolutionary genome partitionings to the connection between milk production and health status in dairy cows. De-regressed proofs for ~30,000 German Holstein cows with ~17 million imputed variants were available. Next to milk yield, the observed traits were mastitis, interdigital hyperplasia, digital dermatitis, claw ulcers, retained placenta, metritis and cystic ovaries. We performed a bivariate analysis to obtain genetic correlations between milk yield and the eight health traits on the other side. Each of the eight trait combinations was analysed within each of the 27 partitionings. Thereby, two relationship matrices were included in the model, one consisting of variants on the 50K array and the other one consisting of the respective partitioning. Our results indicate a significant negative connection between milk yield and health traits, which can be assigned to the noncoding part of the genome. This agrees with recent studies observing pleiotropy of causal variants. Our results confirm that variants altering gene expression play an important role for the variation of complex traits and reveal that this holds true also for the trait's genetic correlation.

Session 28 Poster 12

Frequency of genetic disorders in genomically tested females in the Netherlands

C. Schrooten[1] and E. Mullaart[2]
[1]CRV Cooperative, Animal Evaluation Unit, P.O. Box 454, 6800 AL Arnhem, the Netherlands, [2]CRV B.V., Global Genetics, P.O. Box 454, 6800 AL Arnhem, the Netherlands; chris.schrooten@crv4all.com

In 2007, CRV started to use genomic chips to obtain genotypes of a large number of SNPs. The chips that have been used since 2014 contained SNP for the most relevant genetic disorders, with a few disorders added on later versions of the chip. Some of the disorders are lethal when animals carry both recessive alleles (for example BLAD, Bovine Leucocyte Adhesion Deficiency), whereas others are less detrimental. In this research, the development of the frequency of the disorders among genomically tested females in the Netherlands was analysed, for female animals born from 2008 onwards. Frequencies were based on real genotypes, imputed genotypes were not taken into account. Number of genotypes used in the analyses ranged from 79,000 (CVM, patent expired end of 2021) to around 500,000 for the disorders that were present on the chip since 2014. For the lethal haplotypes, HH1 and HH3-HH7, highest frequency of carriers was found for HH3 in 2012 (8.2%), HH4 in 2009 (8.3%), and HH5 in 2016 (4.8%). Frequency of carriers decreased for all HH-haplotypes, highest frequency for animals born in 2023 was observed for HH3 (2.6%). The frequency of BLAD carriers gradually decreased from 1.8% in 2006 down to 0.10% in 2022. For cholesterol deficiency haplotype (CDH), which was discovered more recently, carrier frequency decreased from 6.6% in 2015 to 1.9% in 2022. Highest carrier frequency for CVM was for animals born in 2013 (4.2%), carrier frequency for CVM in recent years was less than 1%. For other disorders, like mulefoot, citrullinemia and dumps, carrier frequencies were always lower than 0.4%. The development in frequency of carriers was favourable for all disorders studied, resulting in better health and viability, due to actively selecting against these disorders by AI organizations worldwide.

Session 28 Poster 13

Genome regions and metabolic processes associated with tick resistance in beef cattle

P. Martin[1], T. Hüe[2], J. Mante[3], A. Lescane[4], D. Boichard[1] and M. Naves[5]
[1]Université Paris-Saclay, INRAE, AgroParisTech, GABI, Domaine de Vilvert, 78350 Jouy-en-Josas, France, [2]Institut Agronomique néo-Calédonien, ARBOREAL, Port Laguerre, 98890 Païta, New Caledonia, [3]France Limousin Sélection, Pôle de Lanaud, 87220 Boisseuil, France, [4]Unité Néo-Calédonienne de sélection et de promotion des races bovines, Port Laguerre, 98890 Païta, New Caledonia, [5]INRAE, ASSET, Domaine Duclos, 97170 Petit-Bourg, French West Indies, France; pauline.martin@inrae.fr

Ticks cause significant production losses in cattle and consequences of infestation can go up to the animal death. The geographical areas of the parasite keep expending due to climate change. For these reasons, a better understanding of the genetic control and the metabolic processes involved in host resistance appears of primary importance. To address this question, regular phenotyping have been implemented in 11 commercial farms of French New Caledonia from 2014 to 2021. Six different traits were considered: score of adult female ticks, score of juvenile ticks, total score, and the logarithm of the three previous traits. A total of 556 Limousin animals and 302 Charolais animals were phenotyped, with 1 to 8 visits per animal for the Limousin breed (mean=2.3, s.d.=1.6, total=1,273), and 1 to 12 visits for the Charolais breed (mean=4.2, s.d.=2.8, total=1,281). All the animals were genotyped with the 50K EuroGMD SNPchip. First, a generalized linear model was used to correct performances by the effects of sex, age, herd, technician and period. Then, genome-wide association studies were performed with the GCTA software using an average corrected performance per animal. Five genomic regions in Charolais and 13 regions in Limousin were found to be associated to at least one trait, with two regions in Limousin being highly significant (*P*<10-11) and shared between traits. Finally, all the significant regions have been included in a Gene Ontology analysis, using the Cytoscape software. Results shows that 37 terms from 9 functional groups are overrepresented, with the three most overrepresented group being the linoleic acid metabolic process, the negative regulation of insulin secretion and the excitatory extracellular ligand-gated ion channel activity. These processes appears as a promising candidate that requires deeper investigation.

Session 28 Poster 14

Innovative strategy using phenotypes and genetics to optimize partial sampling applied to boar taint

A. Markey[1], J. Wavreille[2], P. Mayeres[3], A.C. Huet[4], D. Duarte[1,3] and N. Gengler[1]
[1]ULiège – GxABT, Passage des Déportés 2, 5030 Gemboux, Belgium, [2]Walloon Agricultural Research Centre, Rue du Liroux 8, 5030 Gembloux, Belgium, [3]Elevéo, Rue des Champs Elysées 4, 5590 Ciney, Belgium, [4]CER Groupe, Rue de la Science 8, 6900 Aye, Belgium; alice.markey@uliege.be

Avoiding castration of male piglets improves their welfare but creates the risk of boar tainted meat. Detecting and reducing boar taint is an important issue in the context of entire male fattening. The incidence of boar taint corresponds to 3-10% of declared carcasses by the human nose method at abattoir as carrying this defect. This detection is however controversial in relation to its reliability. An more reliable detection alternative, but costly and slow, therefore unfeasible in the slaughterhouse, is chemical analyses of compounds known as skatole and androstenone. Reliable genetic and genomic evaluation of boar taint risk needs the most complete phenotyping. However, extending chemical analyses to the whole population is not economical sustainable. An innovative strategy to optimize partial sampling was developed to select samples for chemical analyses after culling. The goal was to obtain a sampling that would be considered 50% tainted and 50% untainted originating from all the tested sire boar lines. A genetic boar taint risk ranking of Pietrain × Landrace crossbred boars based on available phenotypes (792 human noses (one from the abattoir and one from the lab) of which 213 had chemical analyses) in 16 contemporary groups and adjusted for their metabolic weight was computed through a multi-trait model considering a 4-generation pedigree. For the 6 posteriors contemporary groups (i.e. not yet analysed) 15% of samples from animals that were at the top of the list (i.e. most also detected by human nose) or at its bottom were chosen. Diversity of half-sire sib families were guaranteed by avoiding repeating boars from the same family. After chemical analyses allowing to compare observed vs expected status, 15% of errors were found, leading to 57% of tainted and 43% of untainted selected samples. Even if this strategy needs to be improved, it could be used to identify relevant samples to complete human nose phenotyping.

Session 28 Poster 15

Genetic susceptibility determined by SNP2 in piglets challenged with F4-Enterotoxigenic *E. coli*

A. Middelkoop[1], H. Kettunen[2], X. Guan[1], J. Vuorenmaa[2], R. Tichelaar[1], M. Gambino[3], M.P. Rydal[3] and F. Molist[1]
[1]Schothorst Feed Research, Meerkoetenweg 26, 8218 NA Lelystad, the Netherlands, [2]Hankkija Oy, Peltokuumolantie 4, PL 390, 05800 Hyvinkää, Finland, [3]University of Copenhagen, Grønnegårdsvej 1, 1870 Frederiksberg C, Denmark; amiddelkoop@schothorst.nl

In pig production, post-weaning diarrhoea (PWD) is a frequent gastro-intestinal problem for weaned piglets, commonly associated with Enterotoxigenic *E. coli* (ETEC) as pathogen. Reducing PWD caused by ETEC and associated antimicrobial usage is crucial to improve piglet welfare and to restrict antimicrobial resistance spread among ETEC isolates. Genome-wide association studies have suggested an important role of SNP2 (ASGA0091537), positioned between the genes HEG homolog 1 and mucin 13, in ETEC adherence to intestinal cells. However, the effect of SNP2 on F4-ETEC susceptibility has not been tested yet. The aim of this study was therefore to investigate genetic susceptibility determined by SNP2 in piglets challenged with F4-ETEC. A DNA marker test targeting SNP2 identified 75 suckling piglets (Tempo × TN-70) as susceptible (heterozygous) or resistant (homozygous) to developing F4-ETEC. A total of 50 piglets, half SNP2+ and half SNP2-, were selected and weaned at 30 days of age (7.8 kg, mixed gender). The trial consisted of 10 pens with 5 piglets/pen. At day 10 post-weaning, all piglets were orally infected with F4-ETEC. Fresh faecal samples were collected from individual pigs to determine faecal consistency and the faecal concentration of F4-ETEC using quantitative PCR targeting the inoculated F4-ETEC strain. In addition, the body weight of the piglets was monitored until the end of the study, i.e. day 22 post-weaning. The concentration level of faecal F4-ETEC shedding, the percentage of pigs that developed F4-ETEC diarrhoea (72 vs 32%, $P<0.01$) as well as diarrhoea duration (2.6 vs 0.8 days, $P<0.001$) following infection were higher in SNP2+ compared to SNP2- piglets. SNP2+ piglets also had a reduced growth performance compared to SNP2- piglets (302 vs 399 g/d, $P=0.03$). To conclude, the reported association between SNP2 and ETEC F4ac susceptibility suggests that SNP2 can be a suitable marker for screening piglets to determine F4ac susceptibility for experimental and potential breeding purposes.

Session 28 Poster 16

Reference genes selection for RT-qPCR in chickens after in ovo synbiotic and choline administration

E. Grochowska[1], *P. Guz*[1], *K. Stadnicka*[2] *and M. Bednarczyk*[1]
[1]*Bydgoszcz University of Science and Technology, Department of Animal Biotechnology and Genetics, Mazowiecka 28, 85-084, Poland,* [2]*Collegium Medicum Nicolaus Copernicus University, Faculty of Health Sciences, Łukasiewicza 1, 85-821, Poland; grochowska@pbs.edu.pl*

Synbiotics are bioactive substances, which can have positive effect on animal health and performance. Synbiotics and choline are considered potential epigenetic factors. They can modulate gene expression in different organs and tissues. Gene expression can be measured using RT-qPCR. One of the most important steps in RT-qPCR experiments is the selection of optimal reference genes. This study aimed to select the two most stable reference genes, which can be used for gene expression determination in five spcific chicken tissues after *in ovo* synbiotic and choline administration. The experiment was conducted on Green-legged Partridgelike chickens. Synbiotic PoultryStar® (Biomin) and choline were injected *in ovo* on the 12th day of egg incubation. Three groups were established: (1) control (C); (2) synbiotic (S); and (3) combined synbiotic and choline (SCH). Jejunum, spleen, cecum, cecal tonsils and liver were sampled from 21-week-old birds. The expression of 11 commonly used reference genes (*ACTB, GAPDH, YWHAZ, SDHA, B2M, RPL13, RPL30, G6PDH, HPRT1, PPIA, TBP*) was determined by RT-qPCR. The stability of the genes was evaluated by RefFinder, which integrates geNorm, Normfinder, BestKeeper, and the comparative Δ-Ct method. The comprehensive analysis in RefFinder showed two most stable genes for each specific tissue: *GAPDH* and *YWHAZ* for jejunum, *B2M* and *RLP13* for cecal tonsils, *ACTB* and *HPRT1* for spleen, *TBP* and *YWHAZ* for liver, *ACTB* and *YWHAZ* for cecum. The *ACTB* and *YWHAZ* genes were ranked by the comprehensive analysis method in RefFinder as the two most suitable reference genes for all five tissues. In conclusion, different reference genes are recommended for normalization of gene expression results in specific chicken tissues after *in ovo* administration of synbiotic and choline. *ACTB* and *YWHAZ* are the most suitable reference genes for the gene expression normalization across all 5 tissues in the experiment. This research was supported by National Science Centre, Poland (grant no. 2020/37/B/NZ9/00497).

Session 28 Poster 17

Genetic parameters of coronet scores related to hoof disorders in Japanese dairy cattle

Y. Saito[1,2], *A. Nishiura*[2], *T. Yamazaki*[3], *S. Yamaguchi*[4], *O. Sasaki*[2] *and M. Satoh*[1]
[1]*Tohoku University, Graduate School of Agricultural Science, Sendai, Miyagi, 980-8572, Japan,* [2]*NARO, Institute of Livestock and Grassland Science, Tsukuba, Ibaraki, 305-0901, Japan,* [3]*NARO, Hokkaido Agricultural Research Center, Sapporo, Hokkaido, 062-8555, Japan,* [4]*Livestock Improvement Association of Japan, Koto, Tokyo, 135-0041, Japan; saitoy865@affrc.go.jp*

Hoof disorders are one of the factors that lead to the culling of dairy cattle. They cause serious economic losses and are also an animal welfare problem as they are difficult to detect until they have become severe. The coronet score is used as an indicator of the hoof health of cattle in the Dairy Herd Improvement Program of Japan. It is determined on a scale of 1 (normal) to 5 (severe) based on the degree of redness and swelling around the coronary band, dew claw, and heel bulb. The objective is to investigate the genetic characteristics of the scores to genetically improve the hoof health of cattle. The data used for the analysis were the monthly test-day records of Holstein cows collected by the Livestock Improvement Association of Japan in 2014-2021. The numbers of records used for the analysis of the coronet scores were 606,417, 491,283, and 336,461 on first, second, and third lactation, respectively. Genetic parameters were estimated from a repeatability model of first to third lactation and a multi-trait model by assuming different traits at each lactation time. Herd year, lactation age group, test month, and lactation stage were considered as fixed effects. Additive genetic and permanent environmental effects were included as random effects. Estimated heritabilities were 0.02 for all parities, 0.007 for first and second lactation, and 0.01 for third lactation. Compared to previously reported estimates for the other traits associated with hoof disorders, e.g. locomotion scores, the heritability of the coronet score was estimated to be similar or slightly lower. Genetic correlations between first and second lactation and first and third lactation were 0.78 and 0.70, respectively. In contrast, those in the second and third lactation were 0.92. This suggests that the expression of genotype effects for coronet scores differs between the first and later lactations.

Selection for a lower growth rate to avoid reaching bone apposition limits in chickens

C. Leterrier
CNRS, IFCE, INRAE, Université de Tours, PRC, 37380 Nouzilly, France; christine.leterrier@inrae.fr

Broiler chickens have been selected for high growth rate. This selection was accompanied by an increase in non-infectious leg troubles. These disorders are related with low bone stiffness due to high porosity of the diaphysis cortex. Comparisons of chickens from genetic strains or crossbreeds with different growth rates were used to understand the pathogenesis of this bone porosity. The results of the histological and biomechanical studies show that the growth of the external diameter of tibiotarsi is linked to the allometric rules. Thus, at the same age, the chickens from the fast-growing animal strains had a tibial diameter greater than the slow-growing chickens. This increase in tibial diameter induces an increase in the external area of the bone, where the osteons are located, and an increase in the diameter of these osteons. The osteoblasts completely covered the internal surface of these osteons and the mineral apposition rate was very high, indicating that the osteon infilling is maximum. These observations have shown that bone apposition in the centre of the osteons reaches the physiological limits, and that the porosity of the bone tissue observed is related to the increase in the size of the osteons imposed by the external growth of the diaphysis, itself imposed by the increase in body weight of fast growing chickens. This work on bone tissue indicates that the selection of high growth rate has led to a growth unbalance: diaphysis cannot get less porous since its strengthening cannot be achieved at the same rate as its external growth. Such limitations are also observed in terms of muscle fibre growth, which indicates that the use of crossbreeds with a lower growth rates is the only way to keep an healthy locomotor apparatus in meat-type chickens of broilers healthy and to preserve their welfare. The importance of genetic factors in determining the diaphyseal growth rate was reinforced by the identification of many QTL related to different parameters describing the diaphysis structure in the chicken.

Daily milk yield, variance and skewness of milk deviations in cows' resilience

L. Zavadilová, E. Kašná, J. Vařeka and Z. Krupová
Institute of Animal Science, Genetics and Breeding of Farm Animals, Přátelství 815, 10400, Czech Republic; zavadilova.ludmila@vuzv.cz

Daily milk yield observations (DMY) were analysed from 484 lactations of 300 Holstein and 84 lactations of 61 Czech Fleckvieh cows monitored via Affimilk software from 1994 to 2017. The database included 167,112 DMY observations between 1 and the maximum ranging from 242 to 305 days in milk. The highest completed lactation was the eighth. Using a linear model with sixth-order Legendre polynomials, the lactation curve predicted a cow's expected milk yield on each lactation day. The deviations of DMY from the predicted milk yield, their variance and skewness (Skew) were determined. The variance of deviations was log-transformed (LnVar) to achieve their normal distribution. There were estimated least square means for DMY, deviations, LnVar and Skew, and the significance of fixed effects and correlations between traits analysed by statistical programme SAS 9.4® using proc GLM and proc CORR. DMY ranged between 5.1 and 54.9 kg, with a mean of 31.3 kg of milk per cow and day. DMY were higher in the Holstein breed than in the Fleckvieh breed and increased with parity. In both breeds, the deviations were significantly influenced by year and calving season and increased with parity. While differences between breeds were found for both LNVar and Skew, LnVar was mainly influenced by parity and Skew by calving year and calving season. The total number of recorded cases of clinical mastitis was 1,253. Lactation incidence for the whole lactation was 95%, in early and middle lactation 50% and in late lactation 34%. The lactation occurrence of clinical mastitis was high. Nevertheless, we could not find the effect of mastitis on the increase of deviations in any breed. The significant effect of mastitis occurred on Skew for Holstein. For LnVar, the effect of mastitis was found in the early lactation of Fleckvieh cows. The correlation between Skew and LnVar was 0.15 ($P<0.001$) for Holstein cows. Supported by the Ministry of Agriculture of the Czech Republic, institutional support MZE-RO0723 and by the project QK22020280.

Session 28 Poster 20

Lameness scoring and its use in genetic selection to improve claw health in Austrian Fleckvieh cows

A. Köck[1], C. Fuerst[1], B. Fuerst-Waltl[2], M. Suntinger[1], K. Linke[1], J. Kofler[3], K. Schodl[1], F.J. Auer[4] and C. Egger-Danner[1]
[1]ZuchtData EDV-Dienstleistungen GmbH, Dresdner Str. 89, 1200 Vienna, Austria, [2]University of Natural Resources and Life Sciences, Vienna, Institute of Livestock Sciences, Gregor-Mendel-Straße 33/II, 1180 Vienna, Austria, [3]University of Veterinary Medicine, Vienna, Veterinärplatz 1, 1210 Vienna, Austria, [4]LKV-Austria, Dresdner Str. 89, 1200 Vienna, Austria; koeck@zuchtdata.at

This study was part of the D4Dairy project. This project aimed, amongst others, to investigate the potential of sensor data and other farm and cow-specific data for disease prediction and genetic improvement of metabolic, udder and claw health. The specific objective of this study was to estimate genetic parameters for lameness and to investigate relationships with claw diseases and culling reasons. A total of 53,490 data records from 6,383 Fleckvieh cows from 91 farms were available from September 2019 to June 2021. Data collection was performed by the regional milk recording organizations. Lameness was scored for each animal at each milk recording using the scoring system of Sprecher *et al.* with 1 = normal, 2 = slightly lame, 3 = moderately lame, 4 = lame, and 5 = severely lame. For genetic analyses, a linear animal model was fitted with fixed effects of herd, lactation, and calving year-month and random effects of animal and permanent environment. The frequency of lame cows was 15.1% (lameness score≥3). Lameness frequency was lowest in first lactation cows (8%) and increased with lactation number (25.4% in fifth and higher lactations). The estimated heritabilities for lameness ranged from 0.06 to 0.09 depending on the trait definition. Relationships between lameness and claw diseases and culling reasons were investigated. The results confirmed the usefulness of lameness scoring to improve claw health.

Session 28 Poster 21

Association between udder health genomic breeding values and dairy and health traits in French cows

R. Lefebvre[1], S. Barbey[2], F. Launay[2], M. Gaborit[2], L. Delaby[3], P. Martin[1] and D. Boichard[1]
[1]Université Paris-Saclay, INRAE, AgroParisTech, GABI, 78350, Jouy-en-Josas, France, [2]INRAE UE326, Domaine Expérimental du Pin, 61310 Gouffern en Auge, France, [3]PEGASE, INRAE, Institut Agro, 35590 Saint-Gilles, France; rachel.lefebvre@inrae.fr

Mastitis is a major issue in dairy cows. Although environmental effects are preponderant, the genetic variability of mastitis resistance is important. A divergent genetic selection experiment on mastitis resistance was carried out at INRAE Le Pin experimental unit in Holstein and Normande breeds, yielding females of resistant and control lines, based on their sire breeding values. The aim of this study was to estimate the impact of udder health genomic breeding values on dairy and health traits throughout the lactation. This study involved 584 Holstein and 343 Normande cows from the experimental facility. Cows were genotyped and characterized by their direct genomic value (DGV) for all traits routinely evaluated in France, including somatic cell score (SCS) and clinical mastitis (CM) summarized in a udder health genomic breeding values (UHg). Own performances were not included in the genomic evaluation. Milk yield (MY) and composition were recorded daily and twice a week respectively, all along their lactation, and all health events were recorded. CM and metritis data were pooled by 2 weeks-period and limited to the first 6 months of lactation for CM and 2 months of lactation for metritis. The model included the date as a fixed effect, the DGV for the analysed trait as a covariate, a within-parity Wilmink model of lactation curve and a regression, function of UHg and the standardized stage of lactation. Trajectories of all traits were found to be dependent on UHg in both breeds. In the beginning of lactation, animals with high UHg (i.e. genetically more resistant) showed lower MY and SCS (-1.5 kg and -0.7 point per UHg genetic standard deviation, respectively), higher fat and protein contents (+1.2 and +0.4 g/kg, respectively) and less frequent CM and metritis. The effect of UHg was constant all along lactation for SCS, whereas it varies with days in milk for other traits. Some of these effects were unexpected and need further investigations.

Session 28 Poster 22

Validating resilience indicators derived from longitudinal performance measurements

M. Ghaderi Zefreh, R. Pong-Wong and A. Doeschl-Wilson
University of Edinburgh, The Roslin Institute and R(D)SVS, Midlothian, EH259RG, United Kingdom; mghaderi@ed.ac.uk

Resilience is defined as the ability of an animal to be minimally affected or quickly recover from a disturbance. Thanks to an increasing abundance in automated on-farm monitoring systems, longitudinal performance measures of individual animals become more routinely available. Several studies have proposed that statistical measurements of the deviation of an animal from its target trajectory (i.e. performance in ideal condition) provide useful resilience indicators (RI) and may be suitable for genetic selection. The aim of this study was to assess the ability of these RI to discriminate between different response types, and their dependence on the quality of available data. The RI considered in this study were skewness, autocorrelation, integral, mean of squares and log variance of performance deviations. Performance trajectories of three broad response types with respect to a short-term challenge were simulated which were, *Fully Resilient* (not affected by the challenge), *Partially Resilient* (affected but recovered after a period), and *Non-Resilient* (permanently affected by the challenge). The simulations included individual variation within response types. The ability of the RI to discriminate correctly between the response types was assessed assuming that target trajectories were unknown and using different methods to estimate these. Across all simulated scenarios, it was found that all RI could correctly distinguish *Fully Resilient* from *Partially or Non-Resilient animals*. However, only log variance, integral and mean of squares correctly identified the *Partially Resilient* response type as more resilient than the *Non-Resilient* type, and this required data both within and outside the perturbation period. The results of this study highlight the potential risk of misclassifying animals based on the diverse RI, and method and data requirements to overcome these. Furthermore, response to selection for the diverse RI on actual disease resilience of animals is currently evaluated using an existing mechanistic host-pathogen interaction model for gastro-intestinal nematode infections in lambs. This study has received funding from the European Union's Horizon 2020 research and innovation program under grant agreement No 772787 (SMARTER).

Session 28 Poster 23

Approaching the genetic evaluation models for clinical mastitis in Spanish dairy cows

M.A. Pérez-Cabal[1], I. Cervantes[1], J.P. Gutiérrez[1], J. López-Paredes[2] and N. Charfeddine[2]
[1]Faculty of Veterinary Science, Complutense University of Madrid, Avda. Puerta de Hierro, 28040 Madrid, Spain, [2]Spanish Holstein Association (CONAFE), Valdemoro, 28340 Valdemoro, Spain; mapcabal@vet.ucm.es

Clinical mastitis (CM) is one of the most important health problems on dairy farms due to the costs it causes and the decrease in animal welfare. In the last years, breeding programs are focused on selecting animals less susceptible to udder health problems, among others. The objective of this work was to determine the best model for estimating genetic parameters of CM traits in Spanish Holstein population following the recommendations of the Eurogenomics Golden Standard. We studied six CM traits after differentiating among lactations and periods. Clinical mastitis data collected by CONAFE in first lactation (L1), second lactation (L2) and third or more lactations (L3) have been considered. Each lactation was split into two periods: up to day 68 postpartum (P1) and from day 69 up to 305 days in milk (P2). The final dataset comprised 89,225 records from 57,857 cows calved from 2013 to 2022. Traits were transformed to a normal distribution using the Snell transformation by herd-year strata. The models always included lactation-age (as fixed effect with up to 44 levels), calving month (as fixed effect, 12 levels), and animal and permanent environmental effect (only for L3) as random effects. The tested effects were ratio of days at risk in each period or standardized ratio by herd (as a covariate), herd-year (as fixed effect with up to 481 levels, or as random). The pedigree contained a total of 156,961 individuals. Genetic parameters were estimated using ASReml v.4. and model election was based on Bayesian Information Criterium (BIC). The prevalence of CM in P1 ranged from 8.9 to 16.0%, and in P2 from 17.8 to 34.9%. Regardless lactation and period, the best model should include the standardized ratio. However, in P1 the model should consider herd-year as random while in P2 should be included as fixed. The univariate analyses leaded to estimated heritabilities in P1 of 0.01, 0.02 and 0.01, for L1, L2 and L3 (in P2 were 0.07, 0.04 and 0.03, respectively). The estimated repeatability for L3 was 0.22 and 0.47 (for P1 and P2, respectively). Further studies should approach bivariate analyses among CM traits.

Milk somatic cell counts in winter and summer are genetically different traits in Norwegian goats

H.B. Olsen, J. Jakobsen and T. Blichfeldt
The Norwegian Association of Sheep and Goat Farmers, Breeding and genetics, Box 104, 1431 Ås, Norway; hbo@nsg.no

High milk somatic cell count (SCC) is associated with mastitis infections and poor milk quality, but compared to dairy cows, the dairy goats are more prone to elevated SCC due to environmental factors. A recent study using data from Norwegian Dairy Goats (NDG) found an effect of calendar month on SCC with the highest values in summer on pasture. The aim of the present study was to determine if SCC recorded during winter and summer is genetically the same trait. We extracted individual SCC records obtained in 2018-2022 from 5,357 one year old goats from the goat recording system. The goats started lactating between mid-December and through March and belonged to 32 herds. Test day was used to determine if the SCC records should belong to winter (SCC1) or summer (SCC2). Winter was defined as January through March, and summer as June and July. For goats having more than one record within season, the first record was kept for winter and the record closest to July 1st was kept for summer. Individual daily milk yield (DMY) observations obtained the same test date as the SCC1 (DMY1) and SCC2 (DMY2) were included in the analysis with SCC traits. Pedigree was traced as far back as possible on animals with data. We used a multivariate linear animal model to estimate genetic correlations between SCC1, SCC2, DMY1 and DMY2 with year and stage of lactation as fixed effects, and herd-testday, additive genetic effect of animal and residual as random effects. The genetic correlation between SCC1 and SCC2 was 0.49 (0.12) and the genetic correlation between DMY1 and DMY2 was 0.83 (0.05), indicating that SCC during winter and summer are different traits genetically, while milk yield during winter and summer to a large extent are determined by the same genes. Due to seasonal kidding, the winter trait is recorded earlier in the lactation than the summer trait, which makes it difficult to separate lactation stage and lactation season. Nevertheless, somatic cell count recorded in winter and summer is not the same trait in Norwegian goats.

Session 28 Poster 25

Effects of two different traditional sire breeds on performance and behaviour of rearing pigs

A. Lange[1], M. Wutke[1,2], S. Ammer[3], A.K. Appel[4], H. Henne[4], A. Deermann[5] and I. Traulsen[1]
[1]Georg-August-University, Livestock Systems, Albrecht-Thaer-Weg 3, 37075 Göttingen, Germany, [2]Georg-August-University, Breeding Informatics Group, Margarethe von Wrangell-Weg 7, 37075 Göttingen, Germany, [3]Bavarian State Research Center for Agriculture, Prof.-Dürrwaechter-Platz 2, 85586 Poing, Germany, [4]BHZP GmbH, An der Wassermühle 8, 21368 Dahlenburg-Ellringen, Germany, [5]ASB GmbH, Boschstraße 9, 49770 Herzlake, Germany; martin.wutke@uni-goettingen.de

As the breeding of fast-growing, lean-meat-producing hybrid pigs and at the same time the frequency of tail-biting has been increasing since the 1950s, this study investigated the effects of different local, traditional sire breeds on the behaviour and performance of rearing pigs. In total, 1,561 piglets were weaned from hybrid sows (Landrace × Large White) that were paired with either Swabian-Hall (SH), Bentheim Black Pied (BB) or Piétrain (Pi) boars. Tails of the piglets were left intact (43.5%) or docked (56.5%), and male piglets were castrated. Piglets were conventionally reared on fully slatted plastic flooring in mixed-sex groups. Starting one day after weaning, skin lesions were scored once per pig, and tail lesions and losses were scored weekly until the end of rearing. The average daily gain was documented for the suckling and rearing period. The activity behaviour of eight focal pens was analysed using video recordings. Differences between modern and traditional breeds were found in this study for skin and tail lesions and tail losses. Significantly fewer BB pigs had severe skin lesions on the front body than SH or Pi pigs ($P<0.05$). In the first half of the rearing period, significantly more BB pigs were assessed without tail lesions and tail losses than SH and Pi pigs ($P<0.01$). However, these differences disappeared in the second half of rearing. Either docked or undocked Pi pigs had significantly higher average daily gains than SH and BB pigs ($P<0.05$). The activity of the focal pens was not influenced by the sire breed or tail lesion class. To conclude, the use of the traditional sire breed BB showed the genetic potential to reduce injurious behaviour in the offspring. Selective breeding for suitable behaviour traits of performance-orientated lines might be promising to further reduce the incidences of tail lesions and losses.

Evaluation of milk parameters and mastitis predisposition in Cyprus Chios sheep

S. Panayidou, T. Christofi, A.N. Georgiou and G. Hadjipavlou
Agricultural Research Institute, Animal Production, P.O. Box 22016, 1516, Lefkosia, Cyprus; spanayidou@ari.moa.gov.cy

Sheep is one of the most important livestock species around the world and different breeds have a significant variation in diverse phenotypic traits. The Cyprus Chios breed is a productive and prolific dual-purpose sheep breed and is the main commercial breed of Cyprus, imported into the country and genetically improved under local conditions for more than 60 years. As part of extensive research activity at the Agricultural Research Institute, data on individual daily milk yield, monthly milk quality traits, and other milk parameters such as individual milk sample electrical conductivity, pH, and somatic cell count have been collected on average for 500 ewes per year, for more than ten years. We also have complete pedigree information for all animals and genomic data on about 1,200 ewes from 2020 onward, based on a 50K single-nucleotide polymorphism (SNP) genotyping array. The aim of this study is to identify genetic factors that affect milk yield, milk quality traits (fat, protein, lactose, total solids), and susceptibility to mastitis for the Cyprus Chios breed, as well as the genetic correlation between traits. Our hypothesis is that milk yield, electrical conductivity, pH, and somatic cells of milk can be predictive of milk composition and indicators for the health status of the mammary gland. Genetic effects and correlations between genetic and phenotypic data for these factors are estimated across all animals, and the genetic difference is examined separately for high- and low-yielding groups of animals for different lactation periods. Estimation of genomic breeding values for mastitis susceptibility indicators shall be incorporated within the existing genetic improvement program of the Chios breed, with the use of more phenotypic and genomic information from about 40 private Chios sheep farms, taking part in a recently initiated research project within an Aid Scheme of the Cyprus Resilience and Recovery Plan, funded by the European Commission.

Influence of early rearing system on later performance of commercial laying hens

Q. Berger[1], N. Bédère[2], P. Le-Roy[2], S. Lagarrigue[2], T. Burlot[3] and T. Zerjal[1]
[1]Université Paris-Saclay, INRAE, AgroParisTech, GABI, 78350, Jouy-en-Josas, France, [2]PEGASE, INRAE, Institut Agro, 35590, Saint-Gilles, France, [3]NOVOGEN, 5 rue des Compagnons, 22960 Plédran, France; tatiana.zerjal@inrae.fr

Poultry production is undergoing rapid changes from cage systems to cage-free environments such as floor housing systems. In this study, we examined the effects of early life rearing environment on late performance (from 70 to 93 weeks of age). We used data from 1,024 purebred hens from the nucleus of Novogen, reared in either a floor housing system (120 hens) or in collective cages (904 hens) during the first laying period (FLP) from 18 to 55 wks of age, and then transferred to individual cages for the late laying period (LLP) from 56 to 93 wks fo age. More specifically, we recorded daily egg production throughout the measurement period and measured individual feed intake twice a week for three weeks, starting at 70, 80, and 90 wks of age, as well as body weight at the beginning and end of each feed intake recording period. Egg quality was assessed at 70 and 90 wks and body organ weight at slaughter. Random regressions models were used to study trait variation on a trajectory of time and genotype by time interactions. To study the impact of the rearing system of the first laying period on the performance of the late laying period, a fixed effect was included in the model, with two levels: floor or collective cage. The results showed that the rearing environment of the FLP did not affect body weight, feed conversion ratio, weight gain, yolk percentage, or abdominal fat content recorded in the LLP. Instead it affected daily feed intake, egg weight, laying rate, egg mass, Haugh unit, and residual feed intake. Specifically, floor hens had a 9.4% higher daily feed intake, 4% higher egg weight, 3.8% higher laying rate, 7.3% higher egg mass, 6.8% higher Haugh unit, and 133% higher residual feed intake compared to those reared on collective cages in the FLP. They also presented increased liver (+ 15%) and breast muscle (+9%) weight proportion. Although more data are needed to confirm this observation, our results suggest that floor rearing in the first laying period may have an impact on performance at later laying stages, perhaps reflecting a change in the animals' energy allocation.

Session 28 Poster 28

Differentially expressed genes in tissues of Pietrain sired pigs challenged with high fiber/fat diet

E.U. Nwosu[1], B. Chakkingal Bhaskaran[1], E. Kowalski[2,3], W. Gorssen[1], R. Meyermans[1], S. Janssens[1], S. Millet[2], M. Aluwé[2], S. De Smet[3] and N. Buys[1]
[1]Centre for Animal Breeding and Genetics, KU Leuven, Department of Biosystems, Kasteelpark Arenberg 30, Box 2472, 3001 Leuven, Belgium, [2]ILVO, Scheldeweg 68, 9090 Melle, Belgium, [3]Laboratory for Animal Nutrition and Animal Product Quality, Ghent University, Coupure Links 653, Block F, 9000 Ghent, Belgium; emmanuelauchenna.nwosu@kuleuven.be

The use of more by-products in animal feeding is one possibility to close nutrient cycles. They are often higher in fibre and/or fat content and lower in starch and monosaccharides than regular feed ingredients, and may be more variable in nutrient digestibility and composition. In this study, Quantseq was used to identify transcriptomes in liver, duodenum and pancreas, in order to determine the differential expression of genes between a control vs a high fibre feed group in Pietrain crossbred progeny. Thirty two animals were included in a 2×2×2 experiment with diet (based on soybean-cereals vs based on byproducts), sex (castrated male vs female), and genetic background (high vs low estimated breeding value for feed intake) as factors. Compared to the standard diet, the challenge diet was lower in starch and higher in fibre and fat. Pigs were reared for 15 weeks and until 14 weeks all the pigs received the first feeding phase which was similar to the control diet and at 14 week half of the pens received the challenge diet. One pig per pen was euthanised at a liveweight of approx. 50 kg. Tissues were snap frozen in liquid nitrogen, thereafter stored at -80 °C until total RNA extraction and subsequent Quantseq analysis. Total RNA was extracted from liver, duodenum and pancreas using the RNAeasy Mini kit. The RNA quality was quantified using Nanodrop and Agilent Bioanalyser. Quantseq was performed on 96 RNA samples and differential expression of genes were tested between the 3 main factors: comparison of high and low EBV for feed intake, comparison of control and challenge feed, and comparison of barrow and gilt using R package DESeq2. The result of the comparisons showed that several genes were expressed differentially. These data will allow a better understanding of the gene functions regulating metabolism of high fibre and fat feeds.

Session 28 Poster 29

GWAS based on imputed sequence-level genotypes in a huge cow data set: many roles for the GC gene

A.M. Krizanac[1], C. Falker-Gieske[1], C. Reimer[2], J. Heise[3], Z. Liu[3], J. Pryce[4], J. Bennewitz[5], G. Thaller[6] and J. Tetens[1]
[1]Göttingen University, Burckhardtweg 2, 37077 Göttingen, Germany, [2]Friedrich-Loeffler-Institut, Höltystr. 10, 31535 Neustadt, Germany, [3]Vereinigte Informationssysteme Tierhaltung w.V. (VIT), Heideweg 1, 27283 Verden, Germany, [4]Agriculture Victoria Research, Bundoora, Victoria 3083, Australia, [5]Hohenheim University, Garbenstr. 17, 70599 Stuttgart, Germany, [6]Kiel University, Hermann-Rodewald-Str. 6, 24118 Kiel, Germany; jens.tetens@uni-goettingen.de

GWAS for 57 milk, health, fertility, and conformation traits was performed in a dataset of 252,285 chip genotyped cows imputed to WGS level (presented in another contribution). Such huge data set allows for the identification of highly significant lead variants even for lowly heritable traits. In order to identify genes and variants affecting metabolic stability, we analysed clinical ketosis (KET) and body condition score (BCS) as a potentially correlated trait using deregressed proofs as phenotypes. Initially, we analysed SNP-based heritability and genomic correlation of the two traits based on medium density chip data. The chip heritability for KET and BCS were estimated as 0.07±0.004 and 0.29±0.007, respectively. The genomic correlation was 0.24±0.03. In the GWAS for BCS, 32,309 variants showed genome-wide significance after Bonferroni correction; for KET we obtained 3,940 significant variants. For both traits, the most significant variants were located in a major QTL on BTA 6 containing 14,008 and 3,504 of the significant SNPs, respectively. Next, we partitioned the genomic correlation into a component stemming from BTA 6 and the rest of the genome in a bivariate GREML-run and found that a major part of the correlation stems from this chromosome (rG_{BTA6}= 0.73±0.25, rG_{Rest}= 0.14±0.09). The top variants on BTA6 for both traits as well as for milk performance traits and from a multi-trait GWAS were around the *GC* gene, which encodes the Vitamin D binding protein. This gene has previously been implicated in milk fever, mastitis and milk traits. Here, we demonstrate its major association with traits related to metabolic stability and thus animal health in early lactation.

GWAS for clinical metritis in Polish HF cows

P. Topolski[1], *T. Suchocki*[1,2] *and A. Żarnecki*[1]
[1]*National Research Institute of Animal Production, Department of Cattle Genetics and Breeding, Krakowska 1, 32-083 Balice, Poland,* [2]*Wroclaw University of Environmental and Life Sciences, Department of Genetics, Kozuchowska 7, 51-631 Wroclaw, Poland; piotr.topolski@iz.edu.pl*

Metritis is one of the most common causes of declining fertility of cows in dairy farming, associated with a significant economic implications and animal welfare. The low heritability of the trait causes that selection for resistance to metritis is possible however, genetic response to traditional selection is small. Consequently, new genomic methods are being used to help a better understand the genetic variability underlying metritis and present options for direct selection of such traits. In our study, we performed a genome-wide association study (GWAS) to identify genetic loci associated with clinical metritis in Polish HF cows. Data for the study were drawn from the data base PLOWET used for veterinary-recording health traits in 4 experimental dairy farms of the National Research Institute of Animal Production and including a total of 1,812 cow genotypes; the incidence of clinical metritis has been recorded in 517 cows. Each cow was genotyped or imputed to Illumina BovineSNP50K beadchip in version 2. As a final dataset we used 53,557 SNPs with MAF≥1% and minimal quality of genotyping ≥99%. The GWAS data were analysed using single SNP mixed model. We used FDR multiple testing correction. Finally, we found 3 statistical significant SNPs that may be associated with clinical metritis located on 3 chromosomes: 18, 8 and 22. These SNPs were surveyed to their corresponding genes within a distance of 100 kbp. The pathway analysis showed that part of these SNPs were located in or very near genes associated with immune system.

Genetic relationships between mastitis, milk yield traits and somatic cell count in Polish HF cows

P. Topolski[1] *and W. Jagusiak*[1,2]
[1]*National Research Institute of Animal Production, Department of Cattle Genetics and Breeding, ul. Krakowska 1, 32-083 Balice near Krakow, Poland,* [2]*University of Agriculture in Krakow, Department of Genetics, Animal Breeding and Ethology, Al. Mickiewicza 24/28, 30-059 Krakow, Poland; piotr.topolski@iz.edu.pl*

Genetic correlations between Mastitis (MA), Milk Yield Traits and Somatic Cell Count (SCC) at the first lactation in the population of Polish Holstein-Friesian (PHF) cows were estimated. The dataset for health traits (e.g.MA) was created on the basis of PLOWET database developed by the National Research Institute of Animal Production and included phenotypic data collected in 4 experimental dairy farms. The dataset for Milk Yield Traits and SCC trait was created on the basis of SYMLEK database belonging to the Polish Federation of Cattle Breeders and Dairy Farmers. Phenotypic evaluation of health traits of the cows was carried out between 2016 and 2022 and was based on veterinary registration. The dataset used in our study comprised 6,049 records of cows. Gibbs sampling algorithm implemented in the BLUPF90 package was used for the computations. The linear observation models included additive random effects of animal, fixed effects of herd-year-season of calving subclass (HYS) and lactation stage, fixed regressions on cow age at calving and fixed regressions on cow age at the time the disease was diagnosed. Cow and bull pedigrees contained two generations of ancestors i.e. parents and grandparents. Genetic correlation coefficients were ranged from -0.11±0.019 [MA vs Milk yield (kg)] and -0.12±0.086 [MA vs Protein(%)] to -0.39±0.012 [MA vs Fat(kg)] and 0.47±0.016 [MA vs SCC]. The magnitude of the obtained coefficients of genetic correlation are in the range of coefficients published in the papers of other authors and indicate the possibility of including these results in the national cattle improvement program e.g. to optimize selection indexes.

Session 29 Theatre 1

A New Zealand perspective of the role of genetics in underpinning future sheep production systems

P. Johnson[1], D. Scobie[2], S.-A. Newman[1], S.J. Rowe[1], S.M. Clarke[1] and K.M. McRae[1]
[1]AgResearch Invermay, Animal Genomics, Puddle Alley, Mosgiel, 9053, New Zealand, [2]Lincoln University, 85084 Ellesmere Junction Road, Lincoln 7647, New Zealand; tricia.johnson@agresearch.co.nz

New Zealand (NZ) has a world-leading sheep genetics improvement programme, underpinned by research. Traditionally, genetic improvement in the NZ sheep industry focussed on production traits, maximising the amount of product produced, but sheep farming operations in NZ (and worldwide) are facing an increasing number of economic, environmental, cultural and social challenges. These challenges include the need for adaptation to overcome the impact of climate change on production systems, a need to contribute towards mitigation of greenhouse gases which contribute towards climate change, together with consumer expectations about their food quality and the systems in which it is produced e.g. animal welfare. As a result, future production systems need to evolve to tackle these challenges. Genetic improvement is a proven method to optimise production systems provided genetic variation in the required traits exists. This has led to numerous research programmes identifying traits that will allow the NZ sheep industry to overcome these challenges and work is underway investigating their genetic control and relationships between these novel traits and other traditional traits. Examples, which will be discussed, include those associated with climate mitigation (methane), climate adaptation (disease traits, heat tolerance and feed efficiency), meat quality (intramuscular fat) and animal welfare (disease traits, tail length and wool cover). For traits where genetic control is being identified, genomic selection is being implemented to increase rates of genetic gain. In addition to traditional host DNA, novel sources of variation that may influence the identified traits, including the rumen microbiome and host genome methylation profiles, are also being explored as additional targets for selection. Cumulatively these research programmes are seeking to provide solutions to ensure that products produced from NZ sheep production systems address the challenges confronting the NZ sheep industry and as such match the expectations consumers and society are increasingly placing on such production systems.

Session 29 Theatre 2

Blockchain technology in the beef breeding sector

B. Jornet and I. Achour
Label Food Chain, 23 Rue Jacques Cottin, 93500 Pantin, France; benjamin.jornet@gmail.com

The initial observation is that there is a loss of information, mainly due to insufficient collaboration between professionals of the beef industry, who only partially communicate traceability information amongst themselves. The result is a loss of information at each stage of the chain, and therefore a loss of added value for the product vis-à-vis the final consumer. This insufficient transparency does not create the conditions necessary to establish a relationship of trust with the final consumer who is obliged to 'believe' the partial information communicated by the actor at the end of the chain (i.e. the distributor in most cases); This article describes how blockchain technology can respond to this problematic. The principle is to establish a decentralized architecture associated with blockchain technology. Indeed, loss or falsification of information is technically impossible in a blockchain system with a satisfying level of decentralization. In practice, multiple entry points and an electronic signature system allow each actor of the chain to provide and certify information that he/she disposes of. The system allows actors to better guarantee the integrity of the data and their origin and thus to self-regulate. For example, a processor could directly access information coming from the farm, without having to go through an intermediate party such as the slaughterhouse. In addition, the blockchain makes it possible to move from a 'declarative' system of information to a system of information 'certified' by each link in the chain. At the end of the chain, the consumer can, thanks to a QR code, access different sources of complete and certified information provided directly by each of the actors of the chain, from the breeder to the final distributor via the slaughterhouse and the processor. Link: https://www.viandesetproduitscarnes.fr/index.php/fr/1135-la-blockchain-dans-le-secteur-elevage-viande.

Using science to optimise production systems to produce a premium product – the LuminaTM Lamb Story

P.L. Johnson[1], S. Saunders[2] and A. McDermott[3]
[1]AgResearch Invermay, Puddle Alley, Mosgiel, 9092, New Zealand, [2]Stag Valley Station, RD2, Lumsden, 9792, New Zealand, [3]AgriFood Strategy, P.O. Box 59, Collingwood, 7054, New Zealand; tricia.johnson@agresearch.co.nz

Science has underpinned the development of a premium meat product in New Zealand, Lumina™ lamb. The journey, that resulted in the Lumina product, started with the establishment of a novel breeding programme to breed sheep of composite genetics suited to the hard-hill country of New Zealands South Island (Headwaters) with an emphasis on standard productivity traits. With the additional aim to identify, through research, a point of difference under genetic control that could be selected for, to produce a marketable differentiated premium lamb product. The research identified intramuscular fat and omega-3 fats as traits that could be improved through genetic improvement which could lead to a differentiated premium lamb product. It was, however, recognised that genetics was only one part of the equation towards a premium product, and commercial production systems that could complement the breeding programme were also required. This part of the research investigated aspects of the commercial production system including the sex of the lamb, different types of finishing forages and the length of time the lambs required on the finishing forage identified to produce a consistent, differentiated premium product. The resulting integrated value chain for Lumina lamb involves stud breeders producing genetically superior rams which are used on commercial farms where the lambs produced are finished on a chicory/clover diet for a minimum of 35 days, which further increases levels of omega-3 in the final product. All commercial male lambs are castrated to reduce potential pH issues, that can occur at certain times of the year in ram lambs. The resulting combination has been demonstrated in independent research to generate a product identified as most preferred by New Zealand and Chinese consumers when compared with lamb from other standard New Zealand production systems. The programme has also jointly invested in the development of in-line hyperspectral imaging technologies to estimate the levels of intramuscular fat in lamb meat for use within the breeding programme and ultimately for commercial assessment of the product.

Session 29 Theatre 4

Factors influencing willingness to pay for red meat with potential to improve consumer wellness

R. Zhang[1], Z. Kallas[2,3], M.P.F. Loeffen[4], M. Lee[5], L. Day[1], M.M. Farouk[1] and C.E. Realini[1]
[1]AgResearch Ltd, Te Ohu Rangahau Kai, Cnr Riddet Rd & University Ave, Massey University Manawatu, 4474, New Zealand, [2]DEAB, Department of Agrifood Engineering and Biotechnology, Baix Llobregat Campus, Building D4 (Campus Baix Llobregat). C. Esteve Terradas, 8, 08860, Spain, [3]CREDA-UPC-IRTA, Centre for Agro-food Economy & Development, Parc Mediterrani de la Tecnologia Edifici ESAB Esteve Terradas 8, 08860, Spain, [4]Delytics Ltd, Waikato Innovation Park, 1 Melody Lane, Hamilton, 3216, New Zealand, [5]Meat & Livestock Australia Donor Company, Coca-Cola Place, 1/40 Mount St, North Sydney, 2060, Australia; carolina.realini@agresearch.co.nz

Consumers are shifting towards a more holistic view of health and well-being and are not only interested in food consumption for physical fitness but also mental wellness. This study investigated consumers' attitudes towards physical and mental wellness improvements through red meat consumption and their willingness to pay (WTP) premium price for the meat. Online surveys were conducted in the US (n=1,000) and Australia (n=523) using commercial platforms. Results showed that over 90% of respondents indicated their interest in purchasing red meat that would improve their wellness status. Respondents from the two countries indicated different interests in the aspects of wellness that they would like to improve through food consumption. Additionally, about 85% indicated their WTP for red meat with potential to provide wellness benefits, with stronger interest by American than Australian respondents. The strong interest consumers showed about eating red meat for improved physical and mental wellness was the key factor contributing to consumers' WTP a price premium. Other favourable factors for increasing WTP were also observed, such as higher frequency of meat consumption, more exercise, good sleep quality and better economic position. Positive consumer responses obtained in this study, could provide the meat industry with a unique opportunity to position meat outlining qualities that support improved health and mental well-being. Providing science-based evidence that meat consumption can result in immediate and/or longer-term wellness benefits, would represent a major opportunity to create and capture more values for the red meat industry.

Session 29 Theatre 5

Perception of the use of rangeland and roughage by actors of the organic pig and layer hen sectors

L. Montagne[1], M. Goujon[2] and F. Marie[3]
[1]PEGASE, INRAE, Institut Agro, Saint-Gilles, 35590, France, [2]CAPL, Angers, 49105, France, [3]IBB, Rennes, 35700, France; lucile.montagne@institut-agro.fr

Organic specification imply to add roughage to the ration of monogastric animals. This can be done through access to rangeland, compulsory for laying hens, or the distribution of roughage. This study aimed to cross the points of view of farmers and others stakeholders on the use of rangeland and roughage in organic laying hens and pigs. Farmers in Bretagne or Pays de la Loire (136 for hens, 87 for pigs) and 18 French stakeholders (from research to marketing) participated to a semi-structured interview. This allowed identifying 326 motivations, 266 barriers and 108 levers to the use of rangeland or roughage. The topics included aspects related to technic, animal health and welfare, economy, environment, nutrition, regulation, work, and social dimensions. Barriers and levers were cited in equivalent numbers for both sectors. More motivations (215/326) were cited by actors of the egg sector. The farmers described more motivations (218/326) and levers (66/108) than other stakeholders that reported more brakes (169/266). The main motivation was the improvement of animal welfare: expression of natural and reduction of deleterious behaviours (116/326). All the actors of both sectors shared this point of view that answer to consumer expectations. The barriers mostly evocated were technical aspects (84): maintenance of vegetation cover in rangelands, difficulty for fodders distribution), animal health (44): increased risk of parasitism and infectious diseases, and degradation of working conditions (28/266). Stakeholders also evocated a lack of knowledge (26) in particular in animal nutrition (23). The topic economy included brakes (25/266, cost of fence and equipment) but also motivations (22/326) especially for the pork sector as access to rangeland would be a marketing argument. Levers concerned mainly technic (59/108, rangeland management) and the feed value of forages (23/108). This study highlighted shared elements but also differences of perception of the use of rangeland and roughage between actors and sectors. It has been a prerequisite to research projects aiming to furnish references on the nutritive values of roughage and on rangeland management.

Session 29 Theatre 6

Role of dairy foods in human nutrition

C.M. Weaver
San Diego State University, 5500 Campanile Drive, 92182-7251, USA; cmweaver@sdsu.edu

Consumer demand for milk and milk products is driven by nutrition as well as taste, affordability, and sustainability. Adequate dairy intake is promoted by most countries throughout the lifespan, primarily for bone health. Most countries recommend 2 to 3 servings daily for most age groups. This recommendation is because of the huge contribution of dairy to intake of essential nutrients. On a global scale, milk consumption is ranked very high compared to other food sources for bone building nutrients i.e. first for calcium, second for phosphorus and potassium, third for protein, and fourth for magnesium. Milk also ranked among the top three foods contributing to the global intake of essential amino acids such as lysine, threonine, methionine, and tryptophan. The dairy matrix includes the complex organizational structure of these essential nutrients and other constituents. The dairy matrix may promote health and lower the risk of many chronic diseases beyond the contribution of essential nutrients. Modelling shows that replacing dairy foods with non-dairy foods to meet nutrient intakes is not feasible. Dairy foods are sustainable as well as nutrient dense and are more affordable than plant-based beverage alternatives to milk.

Organoleptic properties of fast-growing and slow-growing chicken meat

R. Berrocal[1,2], L. Mur[1,2], J.L. Olleta[2], V. Resconi[2], M. Barahona[2], J. Romero[2] and M.M. Campo[2]
[1]UVESA, Nutrición animal, Polígono Montes del Cierzo A68, Km86, 31500 Tudela, Spain, [2]Instituto Agroalimentario IA2, Universidad de Zaragoza-CITA, Producción animal, Calle de Miguel Servet, 177, 50013 Zaragoza, Spain; rberrocal@uvesa.es

The use of intermediate-growing strains appears in the market as an alternative production to reduce fast-growing welfare consequences. It also responds to the new trends in European countries towards longer lifecycles and improved farm conditions. But it is necessary to know the characteristics of the final product and the consumers' opinion. Breast organoleptic characteristics were compared between one of these alternative genetics and a conventional fast-growing broiler through a 7-member trained taste panel. Genetics (Ross 308 vs Ranger Classic) and sex (males vs females) were compared at two ages associated to two commercial fast-growing (44 d) and slow-growing (56 d) nutritional programs at the same experimental farm. Twelve carcasses from each treatment were dissected 24 hours after each slaughter age and the breasts were vacuum packaged and kept frozen until analyses were performed. Meat was grilled until the internal temperature reached 70 °C and attributes were assessed on a 0-10 cm lineal scale from no intense to extreme intense. Genetics, sex, age and their interactions were analysed with a General Lineal Model (SPSS 26.0). Ross breast was more fibrous, firmer and juicier than Ranger breast (6.61 vs 6.17, $P \leq 0.001$; 6.85 vs 6.27, $P \leq 0.001$; 6.99 vs 6.61, $P \leq 0.05$; respectively) and had more intense chicken odour (7.98 vs 7.82; $P \leq 0.05$) and flavour (7.69 vs 7.21; $P \leq 0.001$). Nevertheless, Ranger breast was more tender (7.93 vs 7.47; $P \leq 0.001$) and had a greater metallic taste intensity (4.65 vs 4.35; $P \leq 0.05$). Sex did not have a significant effect in the studied assessors, although meat from males tended to be firmer than from females (6.69 vs 6.64, $P<0.1$). Meat from younger animals showed more tenderness (7.94 vs 7.46, $P \leq 0.05$) and chewiness (6.76 vs 6.63, $P \leq 0.05$) than older ones which had more fibrousness (6.51 vs 6.28, $P \leq 0.05$) and firmness (6.77 vs 6.35, $P \leq 0.05$). Although all scores were high, it is important to continue with an improving genetics to equal current meat characteristics ensuring consumer preferences.

Implementation of sustainable livestock farming principles for generating innovative animal products

M. Cohen-Zinder, E. Shor-Shimoni, H. Omri, F. Garcia Solares, S. Kaakoosh and A. Shabtay
Sustainable Ruminant Production Lab; Model Farm for Sustainable Agriculture, Newe Ya'ar Research Center, Agricultural Research Organization, P.O.B 1021, Ramat Yishay 30095, Israel; mirico@volcani.agri.gov.il

It is well evident that intensive livestock farming imposes undesired impacts on the environment, animal welfare and public health. These trends have initiated a growing consumer demand for sustainable, environmentally friendly, economically viable for farmers, socially acceptable livestock systems, that are considerate of animal welfare and offer high quality animal food products, as an outcome of transparent and traceable practices along the food chain. To this end, we study and implement new practices of sustainable ruminant production, in order to produce innovative animal-based food products, some of which will be highlighted herein. In a Farm to Table approach, we: (1) demonstrate superiority of the local Holstein cattle by means of meat quality (e.g. tenderness, IMF% and fatty acid profile) over imported beef animals; (2) select towards a new cattle breed (Simmental × Baladi; *SimBal*), adapted among others to climate change and meagre feeds, with unique organoleptic signature; (3) grow a novel crop (*Moringa oleifera*) to be used as animal feed, to improve their production performance and meat quality characteristics. Specifically, our findings highlight the potential of Moringa provided as food additive to increase the accumulation of health promoting compounds in animal food products, and to prolong its shelf life. These innovative sustainable livestock farming practices may promote future production of animal products with longer shelf life, as well as improved edible properties and health advantages for livestock and humans.

Session 29 Theatre 9

Recent changes and future expectations about meat consumption

F. Montossi[1], G. Ares[2], L. Antúnez[2], G. Brito[1], M. Del Campo[1], C. Saunders[3], M. Farouk[4] and C.E. Realini[4]
[1]INIA URUGUAY, Brigadier Gral. Fructuoso Rivera Km 386, 45000, Uruguay, [2]UDELAR, By Pass de Rutas 8 y 101, Pando, Uruguay, [3]Lincoln University, 85084 Ellesmere Junction Road, Lincoln 7647, Canterbury, New Zealand, [4]AgResearch Limited, Te Ohu Rangahau Kai, Massey University Campus, Grasslands, Tennent Drive, 4474 Palmerston North, New Zealand; fmontossi@inia.org.uy

A telephone survey was carried out in Uruguay in 2021 with 601 participants over 18 years of age. Participants answered a series of questions about the frequency of recent changes and future expectations in meat consumption. Responses were analysed according to gender, age, place of residence, and educational and socioeconomic level. The percentage of participants who increased (6%), maintained (50%), decreased (35%), and did not consume (9%) meat consumption was not affected by the sociodemographic variables studied ($P>0.05$). In the future, 65% of participants do not intend to alter their meat consumption, while 24% intended to reduce it, mostly women and people with a higher educational level ($P<0.01$). Thirty-three percent of the interviewees were willing to substitute meat for proteins of vegetable origin, mainly women and people with a higher educational level ($P<0.01$). Only 17% of respondents would be willing to try meat analogues produced in laboratories or factories, mainly young people (under 30 years old) and those with higher education level. The proportion of people willing to pay more for meats that have labels/brands and are certified for animal welfare, grass-fed, organic, sustainable, traceable, without the use of antibiotics and growth promoters, and grain-fed were 65, 61, 60, 55, 50, 45, 43, and 18%, respectively. The attributes of organic production, care for animal welfare, and the non-use of antibiotics and growth promoters during production were particularly relevant for participants under 30 years old and those with a higher educational and socioeconomic level ($P<0.05$). Sustainable production is more relevant for older participants ($P<0.05$). Recent and future changes in meat consumption are differentially influenced by sociodemographic variables, and their better understanding can be used in designing public information strategies and promotion campaigns to promote meat consumption.

Session 29 Theatre 10

Sustainability in the beef sector: aligning consumer and farmer perspectives

M. Henchion[1] and V.C. Resconi[2]
[1]Teagasc, Agrifood Business and Spatial Analysis, Ashtown, D15 KN3K, Ireland, [2]University of Zaragoza, Department of Animal Production and Food Sciences, C. Miguel Servet 177, 50013 Zaragoza, Spain; maeve.henchion@teagasc.ie

Consumer demands from meat products, and associated production systems, can be complex and conflicting, particularly when seeking to address different elements of sustainability. Following a very brief overview of currents trends in meat production in Europe, this presentation will draw on the results of a pan-European consumer survey to identify some key motivations and factors influencing consumer demands with regards to beef. It will complement this perspective with knowledge gathered within an EU network that was established to address sustainability issues within the beef sector. Specifically, it will identify farmers' concerns relating to sustainability and provide an overview of some of the solutions farmers across Europe are putting into practice, as well as other solutions from research that have yet to be put into practice. It will conclude with some observations about the challenges and opportunities in aligning within and across these different perspectives in order to achieve more sustainable production systems.

FT-MIR spectroscopy to discriminate intramuscular fat of beef fed conventional vs sustainable diets

S. León-Ecay[1], I. Goenaga[1,2], A. López-Maestresalas[1], A. Gobeti Barro[3,4], M.P. Ellies-Oury[4,5], M. Beruete[1] and K. Insausti[1]
[1]Universidad Pública de Navarra, Arrosadia, 31006 Pamplona, Spain, [2]TRASA, Milagro, 31320 Milagro, Spain, [3]Londrina State University, Highway Celso Garcia Cid, 86057970 Londrina, Brazil, [4]INRAE, Vetagro Sup, UMR1213, Theix, France, 63122 Theix, France, [5]Bordeaux Sciences Agro, CS 40201, 33175 Gradignan, France; kizkitza.insausti@unavarra.es

Large amounts of vegetable by-products are generated by the agri-food industry every day. Giving them a second life can be positive for food and beef companies, because while improving their sustainability indicators they can add value to their products. Besides, the meat industry requires non-destructive, sustainable, and rapid methods that can provide objective and accurate quality assessment with little human intervention. In this context, the objective of this research was to study the usefulness of FT-MIR technology to discriminate the intramuscular fat (IMF) of beef fed a sustainable diet vs a conventional diet. To do so, a total of 23 samples of IMF (Soxhlet extraction) were analysed. 11 came from beef fed vegetable by-products, and 12 from beef fed concentrate and straw. To obtain their infrared spectra, a Fourier-transform infrared (FTIR) Vertex 80v spectrometer with an A225/Q Platinum Attenuated Total Reflectance (ATR) accessory was used. 10 replicates per sample were taken from 32 scans in the 4,000-400 cm^{-1} spectral range, with a resolution of 4 cm^{-1}, carrying out a calibration before each experiment. To understand the behaviour and find patterns between diets, a decomposition by Principal Component Analysis (PCA) with the software PLSToolbox under MATLAB ® R2020a was done. The best resolution was obtained using the pretreatment Standard Normal Variate with 9 Principal Components, recording a RMSEC of 0.01 and a RMSECV of 0.03. The errors were very low and manageable, and for computational issues it shows fast processing. Thus, the potential of FTIR-MIR as a non-destructive and easy to handle methodology to discriminate among the IMF of beef fed on different diets was demonstrated. This research was supported by Government of Navarra, project 0011-1365-2020-000288 'BEEF+: Carne saludable a través de la economía circular'.

Study of the genetic parameters of the destructured meat defect of ham on 3 male lines

A. Le Dreau[1], C. Garcia Baccino[1], D. Penndu[2], A. Buchet[2], P. Doussal[2] and B. Ligonesche[1]
[1]NUCLEUS, 7 rue des orchidées, 35650 Le Rheu, France, [2] Cooperl Innovation, 1 rue de la gare, 22640 Plestan, France; a.ledreau@nucleus-sa.com

The destructured meat defect (also called PSE-like defect) occurs during the transformation procedure into cooked ham, which leads to loss of yield and therefore to an economic loss. This defect is quantified using a visual scoring grid that requires deboning the ham. The genetic origin of this defect has not been extensively studied, and the few studies on the subject have always focused on crossbred animals. A study was therefore set up by Cooperl and Nucléus to study the genetic determinism of the PSE-like defect on its purebred populations Pietrain, Duroc and the synthetic line Kador. More than 5,000 hams from animals coming from 3 selection farms have been noted. The scoring consists of assigning a score during the deboning of the ham by a unique trained operator. At the slaughterhouse, meat quality measurements were carried out, namely the measurement of the ultimate pH and the rate of exudate in the loin. Piece weight predictions were also obtained using information from an ultrasonic scanner, the Autofom III ™. The traits of interest were measured on these 5,000 animals and their contemporaries. These data involved information about the growth (age standardized at 100 kg), the conformation (back fat thickness and muscle thickness, both standardized at 100 kg) and for some animals, their individual food consumption (to calculate the feed conversion ratio). Due to the fact that the 'destructured' notation is categorical, a part of the study consisted in fitting an adequate model to these data. The pedigree information consisted on 10 generations for each breed and the analysis was carried out using AIREMLF90 software. The first results on Pietrain show a significant heritability of 0.19, with favourable genetic correlations with other meat quality traits. Consequently, selection against the PSE-like defect is possible, and is already made indirectly by selection for the meat quality traits considered in the selection objective. The same analysis on Duroc and Kador will permit to confirm and compare these results.

Session 29 Poster 13

Consumers' perception towards using seaweed as an alternative to antibiotics in rabbits in Spain

S. Al-Soufi[1], J. García[2], E. Cegarra[3], A. Muiños[4], M. Miranda[5] and M. López-Alonso[1]
[1]University of Santiago de Compostela, Animal Pathology, Veterinary Faculty, 27002, Spain, [2]Polytechnic University of Madrid, Dept Producción Agraria, 28040 Madrid, Spain, [3]DeHeus España, Plaza de Mina 1, 15004 A Coruña, Spain, [4]Porto Muiños SL, Cerceda, 15185 A Coruña, Spain, [5]University of Santiago de Compostela, Anatomy, Animal Production and Veterinary Clinical Sciences, Veterinary Faculty, 27002, Spain; sabelaalsoufi.novo@usc.es

The rabbit meat industry has faced critical challenges in the last few years, during which the ban on the prophylactic use of antibiotics in animal feed has added to the weakness of the production system and a decrease in consumption of rabbit meat, mainly within the younger population, due to the lack of rabbit processed products that fit modern types of consumption and to the perception of rabbit as a pet. Seaweed have demonstrated to be a good alternative to antibiotics in animal nutrition by improving the gut health; moreover, some seaweed species enhance the fat quality of the meat. Within the project TIRAC, that aims to reduce the use of antibiotics in growing rabbit farming in Spain by using seaweed, this study seeks to obtain data about the consumers' perception of and attitudes towards consuming rabbit meat if seaweed were included in the feed to reduce the use of antibiotics. A semi-structured questionnaire was carried out using a computer-assisted web interviewing (CAWI) in the city of Lugo (NW Spain). Responders (n=383) were selected with a non-probablistic sampling by quota based on their gender and age. Most responders declared to be interested in a meat produced using seaweed in the diet of the animals to reduce the use of antibiotics, the positive attitude being related not only to their own health, but also to animal welfare and the environment. Our results indicate that the inclusion of marine algae in rabbit food could potentially become a commercial marketing strategy to attract new consumers concerned about environmental sustainability and who are looking for different, high-quality foods.
The project TIRAC was co-financed by the European Agricultural Fund for Rural Development (EAFRD) of the European Union (80%) and the Ministry of Agriculture, Fisheries and Food (20%), within the framework of the National Rural Development Program 2014-2020.

Session 29 Poster 14

Perceptions of viticulture-livestock complementarity in Burgundy: first insights

R. Ibidhi[1], O. Aguirre-Saavedra[2], S. Ambrosino[2], G. Houndafoche[2], M. Seignon[2], Y. Tanguy-Roump[2] and C. Philippeau[2]
[1]Agroécologie, INRAE, Institut Agro, Univ. Bourgogne, Univ. Bourgogne Franche-Comté, 17 rue Sully, Dijon, 21000, France, [2]L'Institut Agro Dijon, 26, bd Docteur Petitjean, BP87999, 21079 Dijon cedex, France; ridha.ibidhi@agrosupdijon.fr

Unlike over specialized farming systems, the integration of crop and livestock systems has been recognized for its potential to improve ecosystem services. The integration of livestock in viticulture production is described as a promising example of an agroecological integrated crop-livestock system. While some studies of the integration of livestock especially sheep into vineyards are already available for other French areas, there is still scarce research on its implementation in Burgundy viticulture systems, a wine region that produces prestigious wines. This study aimed to explore winegrowers and agricultural stakeholders' perceptions on viticulture and livestock synergy and identify potential initiatives implemented in Burgundy. Semi-structured interviews were conducted within 5 winegrowers, 2 winegrowers who use animals in their vineyards and 2 viticulture stakeholders. We find that seasonal integration of sheep in the period of vine dormancy is the most common form of synergy. Indeed, sheep grazing is important for winegrowers to manage cover crop growth. Overall, winegrowers perceive significantly more benefits than challenges with the integration of sheep into vineyards, particularly reduce mowing, improve soil health and organic matter from sheep faeces and reduce chemical costs for weed control. In addition, grazing in vineyards offers an additional supply of forage for sheep. Some practitioners reported an image enhancement for the winegrowers with new marketing options. However, some of the interviewees mentioned that this form of synergy is a labour cost practice and they can also cause compaction in waterlogged soils. Finally, this study suggests a wide-scale adoption of seasonal integration of sheep into vineyards in Burgundy. Where, the integration of livestock into vineyard may contribute to ecological benefits. However, this practice is not yet been sufficiently scientifically validated.

Session 29 Poster 15

Feeding seaweed as an alternative to antibiotics in growing rabbits improves the meat fat profile

S. Al-Soufi[1], J.M. Lorenzo[2], J. García[3], E. Cegarra[4], A. Muiños[5], M. Miranda[6] and M. López-Alonso[1]
[1]University of Santiago de Compostela, Animal Pathology, Veterinary Faculty, 27002 Lugo, Spain, [2]CETECA, San Cibrao das Viñas, 32900 Ourense, Spain, [3]Polytechnic University of Madrid, Dept Producción Agraria, 28040 Madrid, Spain, [4]DeHeus España, Plaza de Mina, 1, 15004 A Coruña, Spain, [5]Porto Muiños SL, Cerceda, 15185 A Coruña, Spain, [6]University of Santiago de Compostela, Anatomy, Animal Production and Veterinary Clinical Sciences, Veterinary Faculty, 27002 Lugo, Spain; marta.lopez.alonso@usc.es

The rabbit meat industry has faced critical challenges in the last few years, during which the ban on the prophylactic use of antibiotics in animal feed has added to the weakness of the production system and a decrease in consumption of rabbit meat. Seaweed are a good alternative to antibiotics in animal nutrition by improving the gut health and enhancing the meat stability, the antioxidant capacity as well as the fatty acid profile by increasing the polyunsaturated and reducing the saturated fatty acids concentrations in the main livestock species; however information in rabbits is very sparce. Within the project TIRAC, that aims to reduce the use of antibiotics in growing rabbit farming in Spain by using seaweed to improve the gut health, this study evaluates the effect of different seaweed supplements (whole *Saccharina latissima*; aqueous extract of *S. latissima*, *Ulva lactuca*, *Himanthalia elongata*, given at 1%) on the proximate composition, antioxidant capacity and the fatty acid profile on the meat. The supplementation with *U. lactuca* showed a statistically significant effect on the fatty acid profile compared to the control group by increasing the concentration of mono-unsaturated, polyunsaturated, n-3 fatty acids and n-6 fatty acids. The potential of seaweed to enhance the fatty acid profile of the meat of rabbits could become a commercial marketing strategy to attract new consumers concerned about environmental sustainability and looking for high quality foods. The project TIRAC was co-financed by the European Agricultural Fund for Rural Development (EAFRD) of the European Union (80%) and the Ministry of Agriculture, Fisheries and Food (20%), within the framework of the National Rural Development Program 2014-2020.

Session 29 Poster 16

Effects of lamb genotype, primal cuts, and vegetables on baby soups' sensory acceptability

M.R. Marques[1], M. Pimpão[1], C. Oliveira[1] and J.M. Almeida[1,2,3]
[1]INIAV – National Institute for Agricultural and Veterinary Research, Rua Prof Vaz Portugal, 2005-424 Vale de Santarém, Portugal, [2]AL4AnimalS – Associate Laboratory for Animal and Veterinary Sciences, FMV-UL, Lisboa, Portugal, [3]CIISA – Centre for Interdisciplinary Research in Animal Health, FMV-UL, Av Universidade Técnica, 1300-477 Lisboa, Portugal; joaoalmeida@iniav.pt

Due to the lambs' diet based on sheep's milk, this red meat has a high nutritional quality. Thus, it's one of the first types of animal protein recommended being incorporated into a baby's diet. However, the consumption of this meat, due to its characteristic flavour, is not spontaneous among the new generations of consumers. This study aimed to evaluate the sensory properties of carrot and pea soups with 10% of lamb meat from two Portuguese pure breeds, Saloia (S), Merino Branco (MB) and Ile France × Merino Branco (IF×M) crossbreed. Lambs received the same farm management and were slaughtered at four months of age, which provided carcasses with an average weight of Saloia 10 kg; MB 12.5 kg; IF×M 13.5 kg. It was only used meat from less valued retail cuts, and each soup formula had a lean meat content of 10% (P/P), which ranged from 40 to 50% and 60% of shoulders (S) and breast+neck (BN) in equal parts (40S/60BN; 50S/50BN; 60S/40BN). A Thermomix Vorwerk and standard formula vegetables were used to make 48 soups, which were then evaluated by a panel of experts. Lambs' genetics influenced all sensory parameters ($P<0.001$), with more intense meat odour and flavour and globally less acceptable from the Saloia meat breed. Conversely, lamb meat from the M and IL×M breeds has given rise to soups with less aromatic intensity, less meaty flavour, and less lamb flavour, which were overall more acceptable. The intensity of lamb flavour was the only characteristic influenced ($P<0.001$) by the meat proportion from the various primal cuts (S or BN). The base vegetable used significantly influenced global acceptance, with panel preference for peas that originated soups with less intense lamb and meat flavour and higher acceptability. Less valued retail lamb cuts can give highly appreciated baby soups with reduced lamb and meat taste and flavour. Funding: Project: ALGAVALOR: POCI-01-0247-FEDER-035234; LISBOA-01-0247-FEDER-035234; ALG-01-0247-FEDER-035234.

Lipid fraction composition of the longissimus thoracis of Lidia beef breed

M.M. Campo[1], M. Barahona[1], V.C. Resconi[1], J.V. Romero[1], S. Romero[2], S. Zabala[2], J. Villalon[3] and J.L. Olleta[1]
[1]Instituto Agroalimentario IA2, Universidad de Zaragoza-CITA, 50013 Zaragoza, Spain, [2]IMIDRA, 28070 Madrid, Spain, [3]RUCTL, 28010 Madrid, Spain; marimar@unizar.es

Lidia breed is reared to achieve certain characteristics and behaviour in the bullring. Meat production is not the main aim for farmers, but it has commercial interest. The husbandry system is clearly extensive, reared under natural environments such as dehesa meadows, which results in particular characteristics of the meat in comparison with fattening units. This work aimed to differentiate meat from different commercial products from Lidia breed. At the processing plant, the loin of 20 bulls coming from the bullring (5.4 years old), 10 heifers (2.6 yo) and 6 cows (15.2 yo) were obtained 24-48h after slaughtering. At 7 days of ageing under vacuum conditions at 4 °C, a slice of *longissimus thoracis* was obtained, vacuum packaged and kept al -18 °C until lipid content (ISO 1443:1973) and fatty acid composition (extraction with chloroform:methanol and methylation with KOH) were assessed. Data were analysed with SPSS 26.0 with a GLM procedure, and Tukey test was used to differentiate mean values. Fat content of the cows (4.99%) almost doubled that of the heifers (2.59%) and three-folded that of the bulls (1.57%; $P<0.001$). The older the animal, the fatter the muscle becomes, but in this breed males are very lean, even at 5 years of age. The saturated fatty acid (SFA) profile from bulls was less saturated than that from females ($P<0.001$). No significant differences were found in the monounsaturated profile, but the percentage of linoleic and arachidonic *n*-6 polyunsaturated fatty acids (PUFA) was significantly higher in bulls than in females, especially in cows; whereas heifers showed the highest α-linolenic acid percentage and bulls the smallest. These data make the ratio PUFA/SFA very favourable (0.44; $P<0.001$) but the *n*-6/*n*-3 PUFA very unfavourable (18.9; $P<0.001$) for the bulls. This reflects the nutrition pattern during the last rearing period. Even while suffering a drought that reduces the availability of pastures, heifers are not supplemented. However, a more valuable animal such as the bull receives concentrates with cereals that modifies the fatty acid profile. These differences support the differentiation of products at retail.

Innovation in the production and marketing of local breed products, path to long-term sustainability

A. Ivanković[1], G. Šubara[2], E. Šuran[2], M. Cerjak[1], J. Ramljak[1] and M. Konjačić[1]
[1]University of Zagreb Faculty of Agriculture, Svetošimunska 25, 10000 Zagreb, Croatia, [2]Agency for Rural Development of Istria, Tugomila Ujčića 1, 52000 Pazin, Croatia; aivankovic@agr.hr

The development of society (community) is also reflected in the changes in consumption habits. Depending on the development of the society and the region, there is a stratification of consumers in terms of willingness to pay for animal products with nutritional and gastronomic added value. Local breeds are suitable for the design of innovative animal products, as they can highlight the uniqueness of the breed, the territory, the production technology and the innovative approach to the preparation of animal products. The effect (potential) of introducing innovations into the gastronomic offer, i.e. the production of donkey milk, was studied using the example of products made from Istrian beef and milk from Istrian donkeys. An insight into the connection of stakeholders in the 'short supply chains' of Istrian beef (farmer – butcher – chef – consumer) showed a significant influence of gastronomic innovation, which unites the '*terroir*' of beef in the original territory and maintains the interest of consumers. Gastronomic innovation makes the breeding of Istrian cattle more profitable and sustainable. It has been estimated that all stakeholders in the path of animal products (from farmer to consumer) achieve a higher income of 15-25% and the region has recognizable local products. The introduction of innovations in the production and processing of Istrian donkey milk has significantly increased consumer interest, diversified the local food offer and created '*short supply chains*' that ensure the sustainability of Istrian donkey households. The introduction of innovations in the production and supply of animal products from Istrian cattle and Istrian donkeys highlights the '*terroir*' of animal products, arouses consumer interest and significantly increases the economic viability of local endangered breeds.

Session 29 Poster 19

The ability of the Distell fatmeter to predict intramuscular fat in Yellowtail Kingfish

D. Milotic[1], F. Anderson[1], G. Partridge[2], A. Lymbery[1] and G. Gardner[1]
[1]Murdoch University, Agriculture, 90 South St, Murdoch WA, 6150, Australia, [2]Harvest Road, Harvest Road Oceans, P.O. Box 3155, Broadway Nedlands, 6009, Australia; dino.milotic@murdoch.edu.au

The aquaculture of Yellowtail Kingfish (YTK) is rapidly developing in Australia. This fast-growing species reaches market size in 12 to 18 months and is not demanding to maintain. Additionally, the YTK can target premium sashimi markets internationally if it contains an optimal amount of intramuscular fat (IMF). Early detection of IMF would inform breeding programs, optimise the grow-out period and thus reduce production costs and enable better market allocation of fish based on quality. A non-invasive measurement of IMF can be acquired using the Distell fatmeter which is a microwave-based technology that estimates crude lipid content within tissues based on the inverse relationship between moisture and lipids. This device was used to scan 240 fish ranging in weights from 300 to 1,500 g. Fish were scanned whole and alive, then after evisceration with head removed (bulleted), and again after skin-off fillets were produced. Scans were captured at four locations (2 scans on each side, above the lateral line), and larger fish were scanned six times (3 per side). Scan locations of right skin-off fillets were matched to previous scans. Chemical analysis consisted of freeze drying the entire right skin-off fillet, homogenizing it using a blender, and extracting crude fat using a chloroform Soxhlet method. Fatmeter values were then associated with chemically extracted lipid %. Distell fatmeter values of whole fish had a moderate association with chemically extracted fillet fat %, with R^2=0.39, RMSE=41.1. For skin-off fillet this association was stronger, with R^2=0.49, and RMSE=36.6. These results demonstrate that the Distell fatmeter has some capacity to predict fillet IMF. The skin-on results suggest that it could be used *in vivo* on YTK to identify breeding stock or finishing condition prior to slaughter.

Session 29 Poster 20

Effect of game species on fatty acids, meat quality and shelf life

M.N. Hlohlongoane[1], U. Marume[2], O.C. Chikwanha[3] and C. Mapiye[4]
[1] North West University, Mafikeng Campus, Animal Science, Corner of Albert Luthuli and University Drive, Mmabatho, 2745, South Africa, [2] North West University, Mafikeng Campus, Animal Science, Corner of Albert Luthuli and University Drive, Mmabatho, 2745, South Africa, [3]Stellenbosch University, Animal Science, Mike de Vries Building, Merriman Avenue, 7600, South Africa, [4]Stellenbosch University, Animal Science, Mike de Vries Building, Merriman Avenue, 7600, South Africa; matsheponthabeleng@gmail.com

The study determined the effect of species on fatty acids, meat quality, and shelf life of impala (*Aepyceros melampus*), mountain reedbuck *(Reduncula fulvorufula*), and springbok (*Antidorcas marsupialis*). Eighteen, 12-14 months old impala, mountain reedbuck, and springboks (n=6) were culled from a game estate. They were then transported to the slaughterhouse on the farm. After 24 hours of chilling (4 °C), the Longissimus thoracis et lumborum was removed for meat fatty acid, meat quality, and shelf life analysis. All data were analysed using generalised linear model procedure of SAS v 9.4. Total saturated fatty acids was affected by species with mountain reedbuck and springbok having higher ($P\leq0.05$) proportions than impala. Impala and springbok had higher ($P\leq0.05$) proportions of total monounsaturated fatty aids than mountain reedbuck. Overall, polyunsaturated fatty acids was higher ($P\leq0.05$) for impala and mountain reedbuck than springbok. Impala and springbok carcasses had higher ($P\leq0.05$) ultimate pH values than the mountain reedbuck. Meat colour was affected by species ($P>0.05$). Cooking loss and tenderness were influenced by species ($P\leq0.05$). Impala and springbok meats had higher ($P\leq0.05$) cooking loss than the mountain reedbuck. Springbok species had more tender ($P\leq0.05$) meat than the impala and mountain reedbuck. Both lipid and protein oxidation had species × day interactive effects ($P\leq0.05$). Overall, it was observed that species affected the total proportions of fatty acids with regard to game meat. As an outcome of this study, meat from impala, mountain reedbuck, and springbok shows desired physical meat quality traits and fatty acids. The findings may be valuable in increasing the marketing of these three game species as a source of meat.

Session 29 Poster 21

Selection for A2 β-casein genotype in Holstein bulls

L. Jiménez-Montenegro, L. Alfonso, J.A. Mendizabal and O. Urrutia
Public University of Navarre, Department of Agronomy, Biotechnology and Food, Campus de Arrosadia 31006 Pamplona, Navarre, 31006, Spain; lucia.jimenez@unavarra.es

In the last few years, an increasing interest in the use of A2A2 semen bulls has been experienced. This specific genotype belongs to the protein β-casein (β-CN) which is encoded by a single gene located on chromosome 6 called *CSN2*. From the several alleles that are codified from this gene, the A1 and the A2 are the most common. The A1 β-CN has been related to risk factor in aetiology of some human disorders such as digestive discomfort, but not A2 β-CN. This reason, together with the increase in both the consumption of plant-based substitutes and the milk production costs, has made that some dairy producers moved towards new products development with higher added value such as the A1 β-CN free milk, which only contains A2 β-CN protein. However, it is needed to know if the genetic selection of cows and bulls for A2A2 genotype to produce this type of milk could lead to possible future genetic implications in dairy herds. Thus, the aim of this work was to study the possible association between the genomic selection for A2A2 and estimated breeding values for milk performance traits. Six bovine semen companies were selected for the analysis and statistical analysis using ANOVA was carried out. The results showed a high frequency of the A2 allele, and that the general strategy of the analysed companies seems to be the increase in the number of bulls with the A2A2 genotype. The associations between β-casein genotypes and breeding values for milk production and composition traits varied among companies and there was not a clear common pattern that suggests there is some physiological mechanism that can explain them. Finally, breeding values for milk performance traits for bulls with A2A2 and A1A2 genotypes compared to A1A1 were generally either better or similar, but not worse. This suggest that genetic selection for A2A2 genotype would not have future negative implications in dairy herds in the analysed traits.

Session 30 Theatre 1

Consumer perception of the challenges facing livestock production and meat consumption

J.J. Liu[1], S. Chriki[2,3], M. Kombolo[3], M. Santinello[4], S. Pflanzer[5], É. Hocquette[2], M.P. Ellies-Oury[3,6] and J.F. Hocquette[3]
[1]Teagasc Ashtown Food Research Centre, Department of Food Quality and Sensory Science, Scribbletown, Dublin 15, D15 KN3K, Ireland, [2]Isara, 23 rue Jean Baldassini cedex 07, 69364 Lyon, France, [3]INRAE, UMR1213, Saint-Genès-Champanelle, 63122, France, [4]University of Padova, Department of Agronomy, Food, Natural Resources, Animals and Environment, VEN, 35020 Legnaro, Italy, [5]University of Campinas, Department of Food Engineering and Technology, Rua Monteiro Lobato, Campinas 13083-862, SP, Brazil, [6]Bordeaux Sciences Agro, 1 cours du Général de Gaulle, CS 40201, 33175 Gradignan, France; liujingjing1003@126.com

With the global meat market growing and intensive livestock farming systems increasing, the impacts of livestock are a growing concern among consumers, further influencing their meat consumption. Therefore, it is essential to understand consumer perception of livestock production. This study was conducted with 16,803 respondents in China, France, Brazil, Cameroon and South Africa to investigate the different perceptions of the ethical and environmental impacts of livestock among consumer segments depending on their sociodemographic characteristics. On average, the current respondents in China and Brazil and/or who are females, work outside the meat sector, low meat eaters and/or more educated, were more likely to believe that livestock meat production causes serious ethical and environmental problems; while those who from China, France and Cameroon and/or who are women, younger, outside the meat sector, low meat eaters and/or more educated, agree more that reducing meat consumption could be a good solution to these problems. Additionally, an affordable price and sensorial quality are main drivers of food purchase for the current respondents. In conclusion, sociodemographic factors have significant effects on consumer perception of livestock meat production and meat consumption habits. Perceptions of challenges facing livestock meat production differ between countries from different geographical regions depending on social, economic, cultural contexts and dietary habits.

Session 30 Theatre 2

Sustainable development of China's beef industry driven by a successful triangle beef cooperation

Q. Meng
China Agricultural University, The Center for Sino-French Beef Research & Development, College of Animal Science and Technology, Yuanmingyuan Ws. Rd., Haidian District, Beijing 100193, China, 100193, China, P.R.; qxmeng@cau.edu.cn

China's beef cattle industry has developed rapidly in recent years, which is inseparable from a successful international beef cooperation between China, France, Australia and other countries. By the end of 2021, China had 98.17 million inventory cattle population, of which 82.54 million were beef cattle. The total beef output was 6.98 million tons, the beef imports were 2.33 million tons, and the per capita beef consumption was 6.58kg. China's beef cattle inventory and beef output have jumped to the third place in the world, and its annual beef import has ranked the first place in the world, indicating that China has developed into a major beef producer and consumption country in the world. The international cooperation in China's beef industry has experienced three stages: the early stage from 1949 to 1979, the steady development stage from 1980 to 2000, and the comprehensive development stage from 2001 up to now. Among them, the establishment of the Center for Sino-French Beef Research and Development and the successful Triangle beef cooperation between China, France and Australia are the most remarkable, which have become an important supporting force to promote the right development of China's beef industry. In the future, China's beef industry will develop towards the large-scale, health and sustainability.

Session 30 Theatre 3

The production and quality improvement of mutton in China

H.L. Luo
China Agricultural University, College of Animal Science and Technology, 2 Yuan Mingyuan Xilu, Haidian District, Beijing, 100193, China, P.R.; luohailing@cau.edu.cn

China has been the world's largest producer in mutton from 4.08 million ton in 2013 to 5.14 million ton in 2021, meanwhile the quality has been improved. In China, there are three kinds of feeding methods for mutton production including grazing, restricted grazing time plus supplementary feeding and house-feeding. The different feeding methods led to the meat quality change, such as intramuscular fat deposits increasing in lamb meat, specifically over-rich in ω-6 polyunsaturated fatty acids but low in ω-3 polyunsaturated fatty acids. Generally, the ratio of ω-6 to ω-3 fatty acids in meat increases from 1 to 20 when lambs are switched from a grazing pasture to house-feeding with concentrate. It is impressed that the ω-6/ω-3 ratio with restricted grazing time of 0 hr, 2 hr, 4 hr, 8 hr and 12 hr was decreasing gradually with 20.23, 9.74, 6.18, 4.99 and 4.28 for the weaned male Tan lambs, and with 5.75, 3.8, 3.67, 3.94, 3.45 and 1.26 in the weaned male Ujumuqin lambs, since the different kinds of grass were intake in different pasture between them. However, the ω-6/ω-3 ratio in the muscle of Hu sheep was significantly lower than that of Tan sheep and Duper sheep by feeding the same diet. The fatty acid composition of lamb can be improved by feed additives. The addition of vitamin E to the diet reduce the drip loss of meat and the content of stearic acid and branched-chain fatty acids related to the 'mutton odour', but improve the content of CLA and unsaturated fatty acids in muscle. Feeding different types of oil in diet had a significant effect on the fatty acids composition in lamb meat. The soybean oil and canola oil increased the ω-6/ω-3 ratio from 9.59 to 14.23, but fish oil decreased the ratio to 1.70 significantly. The addition of alfalfa saponins, a compound that regulates lipid metabolism, reduced the ω-6/ω-3 ratio from 6.45 to 5.60. In summary, there has been an adverse effect on the composition change of fatty acids in lamb meat by changing grazing system to housing feeding. However, this adverse effect can be improved by either grazing regime plus concentrate feeding or house feeding by different additives, or special sheep breed with the high preferred fatty acid composition.

Comparison of sheep meat eating quality thresholds for Chinese, American and Australian consumers

R.A. O'Reilly[1,2], D.W. Pethick[1,2], G.E. Gardner[1,2], A.B. Pleasants[3] and L. Pannier[1,2]
[1]College of Science, Health, Engineering and Education, Murdoch University, WA, 6150, Australia, [2]Australian Cooperative Centre for Sheep Industry Innovation, NSW, 2351, Australia, [3]Massey University, Al-Rae Centre for Animal Breeding and Genetics, Hamilton, 3214, New Zealand; r.oreilly@murdoch.edu.au

A new cuts-based Meat Standards Australia (MSA) model has been developed within Australia to predict eating quality grades of sheep products. It is based on untrained Australian consumers therefore it is important to examine whether the expectations of consumer groups in other countries align with Australian perceptions. The objective of this study was to examine American, Australian, and Chinese consumer allocation of Australian lamb and yearling products into MSA quality grades and determine whether differences exist in the eating quality grade thresholds between these consumer groups. Untrained consumers were recruited across Australia, China and the USA (720 per country). Each consumer tasted and scored 6 grilled sheep meat samples (3 loin and 3 topside muscles) collected from 164 lambs and 168 yearlings. Samples were scored on tenderness, juiciness, liking of flavour and overall liking on a scale line of 0-100, in addition to assignment of a quality grade: 2 star (fail), 3 star, 4 star, or 5 star. Linear discriminate analyses were used to weigh the importance of the four sensory traits and determine quality thresholds for each consumer group. Chinese consumers had lower quality grade thresholds than American and Australian consumers, ranging from 6 to 11 points lower. Within each consumer group, there was minimal difference between the thresholds for lamb or yearling products. The combined lamb and yearling thresholds for 2-3 star, 3-4 star and 4-5 star were: 37, 56, 74 for China, 43, 65, 81 for Australia, and 46, 67, 82 for the USA. Thresholds reflected the pattern observed in allocation of samples to quality grades, with Chinese consumers assigning a larger proportion of samples to higher quality grades, while American and Australian consumer allocations were more critical. These findings suggest that quality perceptions can vary for consumer groups, thus assignment of sheep meat products to quality grades may need adjustment to ensure quality expectations are met within export markets.

Differences in liver nutrient metabolism contribute to residual feed intake of beef cattle

Z.M. Zhou
China Agricultural University, No. 2 Yuanmingyuan West Road, Haidian District, Beijing, 100193, China, P.R.; zhouzm@cau.edu.cn

Residual feed intake (RFI) is a good measure of feed efficiency, which is defined as the difference between the actual dry matter intake (DMI) and the predicted DMI based on body size and growth. RFI is a trait independent of growth performance. Compared with the high RFI (HRFI), the low RFI (LRFI) animals can reduce feed consumption without affecting growth performance. As an important metabolic and immune organ, liver plays important roles in physiological processes. The changes of metabolism and gene expression may lead to the variation of feed efficiency. Ninety Angus heifers (410±25 kg, 15 months) were fed with the same diet for 144 days, and all conditions were consistent. All heifers had *ad libitum* access to water and feed. Daily feed intake of individual animals was obtained from an automatic feed intake recording system, and body weight was obtained at the beginning and end of the experiment and at 14-day intervals. Individual RFI value was calculated through dry matter intake, average daily gain, and middle metabolic weight. The liver samples of heifers were collected with the highest (n=6) and the lowest (n=6) RFI values, and stored at -80 °C until subsequent analysis. The results showed that a total of 47 differential metabolites were identified in the liver with different RFI groups ($P<0.05$), which enriched in the following KEGG pathways: protein digestion and absorption, D-Glutamine and D-glutamate metabolism, and aminoacyl-tRNA biosynthesis ($q<0.05$). And a total of 495 differentially expressed genes were enriched in the following KEGG pathways: glutathione metabolism, PPAR signalling pathway, protein processing in the endoplasmic reticulum, B cell receptor signalling pathway ($q<0.05$). For proteomic analysis, a total of 411 differentially expressed proteins ($P<0.05$) only significantly enriched in the glycerophospholipid metabolism pathway ($q<0.05$). After further analysis, we found that PNPLA6, PTDSS1, DGK, CDS2, PNPLA8, LYPLA2 and GPAT4 were up-regulated in the LRFI group. Compared with the HRFI group, the LRFI group had stronger glycerolphospholipid synthesis, cell proliferation, cell membrane transport, cell signal transduction, and immune function, but fat transport was accelerated.

Session 30 Theatre 6

Marbling in French cattle: average level and factors of variations

A. Nicolazo De Barmon, I. Legrand and J. Normand
Institut de l'Elevage, Service qualité des carcasses et des viandes, 149, rue de Bercy, 75012 PARIS, France; aubert.nicolazodebarmon@idele.fr

The importance of fat infiltration or 'marbling' in meat for organoleptic quality has been proved. Thus, this is a research axis of the French bovine meat interbranch organization (Interbev) to improve beef quality for consumers. In this respect, the French Livestock Institute (IDELE) developed a grid with 6 levels of marbling (from 1 (no marbling), to 6 (very high marbling)) for the French meat industry to be used in slaughterhouses. The objective of this study is to make a clear picture of marbling levels from French breeds and to see if professional opinions regarding possible factors affecting meat marbling would merit to be further studied. Marbling of 3,218 carcasses from various breeds, categories and slaughterhouses have been evaluated by different graders. Most carcasses were from cows, they can be divided in three classes: 1/3 poorly marbled (1 or 2), 1/3 with a medium level (3) and the last third marbled or extremely marbled (4, 5 or 6). Different carcass traits seem to affect marbling levels. Once again, breed impact has been confirmed: early maturing breeds are more marbled (around 50% of dairy cows graded equal or more than 4) than late maturing ones (only 25% of beef cows). The category also affects marbling level: as previously seen in different studies, young beef bulls are poorly marbled (70 to 80% graded 1 or 2) whereas females have higher marbling score (only 25 to 30% graded 1 or 2). Moreover, carcasses with high marbling level are the more fatty, heavier and with the best conformation. This observation is linked to fattening which affects these characteristics and marbling deposition. However, a large variability exists at the same level of weight, conformation and fatness. Therefore, a given conformation, weight or fatness score doesn't guarantee a marbling level. Age effect on marbling is not really clear and has to be more studied. Thus, these results give references to the French meat industry to better answer to consumers' demands, especially from an organoleptic aspect.

Session 30 Theatre 7

Effect of high sulphur diet on rumen fermentation and epithelial barrier function in beef cattle

H. Wu, Y. Li, Q.X. Meng and Z.M. Zhou
China Agricultural University, College of Animal Science and Technology, 2 Yuanmingyuan West Road, Haidian District, Beijing, 100193, China, P.R.; wu2213@cau.edu.cn

The aim of the study was to investigate the mechanism of high-sulphur diet inducing rumen epithelial injury and inflammation. Eight 24-month-old Angus steers (350±43 kg) fitted with permanent rumen fistulas were used in a repeated 4×4 Latin square design: Cattles in the control group were fed a basal diet (sulphur content was 0.4%, CON), and those in the experimental groups were fed a diet with sulphur content of 0.6% (LSD), 0.8% (MSD) and 1.0% (HSD) with sodium sulphate, respectively. Total gas production and methane gas concentrations were linearly decreased with dietary sulphur content increasing ($P<0.01$), while hydrogen sulphide gas concentration was significantly increased ($P<0.05$). Compared with CON, the concentration of TVFA and the proportion of butyric acid of HSD were significantly increased ($P<0.01$), but the proportion of propionic acid was significantly decreased ($P<0.01$). The concentration of NH_3-N was not significantly different between LSD and MSD, but was significantly lower than that of CON and HSD ($P<0.01$). Total thickness of rumen epithelium in HSD was significantly higher than that in CON and LSD ($P=0.01$), but not significantly different with MSD. Specifically, the thickness of spinous layer and basal layer increased ($P=0.02$), while there was no significant difference between corneum and granular layer. The degree of peeling and keratinization on the rumen papilla surface was more obvious in the HSD. At the same time, the high-sulphur diet led to the formation of large cracks and sprouting of rumen epithelium. The bacteria attached to the rumen epithelium surface were mainly coccus and bacillus in CON, while the high-sulphur diet showed more fusarium and actinomycetes. qPCR results showed that the mRNA expression levels of TJP1 and CLDN-1 genes in rumen epithelial cells in HSD were significantly lower than those in CON ($P<0.05$). According to the above results, it can be inferred that dietary excess sulphur can affect rumen fermentation, change the type of rumen fermentation, and damage the integrity of rumen epithelium morphological structure, thereby increasing the permeability of rumen epithelium and destroying barrier function of beef cattle.

Session 30 Theatre 8

Research progress on quality characteristics and regulation of yak meat in China

L.Z. Hao[1], B.Q. Bai[1], Y. Xiang[1], Q.X. Meng[2], J.Z. Niu[1], Y.Y. Huang[3] and S.J. Liu[1]
[1]Qinghai University, Qinghai Academy of Animal Science and Veterinary Medicine, Key Laboratory of Plateau Grazing Animal Nutrition and Feed Science of Qinghai Province, Xi'ning, 810016, China, P.R., [2]China Agricultural University, Beef Cattle Research Center, Beijing, 100193, China, P.R., [3]INRAE, UMR PEGASE, 16 le Clos, 35590 Saint Gilles, France, France; qxmeng@cau.edu.cn

The yak (*Poephagus grunniens*), an indigenous herbivore raised at 3,000 to 5,000 meters above sea level across the Asian highlands, is one of the main livestock grazing on the Qinghai-Tibetan Plateau (QTP). At present, there are approximately 17.6 million yaks in the world, of which more than 95% are raised in China. Yaks play a vital role in ecosystem stability, livelihood security, socio-economic development, and ethnic cultural traditions on the QTP. Yak meat is the main product of yak industry, and its quality characteristics have always been unclear. This research systematically reviewed the eating quality, processing quality, nutrition quality, health safety quality and cultural quality of yak meat by collecting yak meat samples from different species, ages, genders, feeding methods and regions in the QTP. The quality characteristics, advantages and disadvantages of yak beef were demonstrated. On this basis, the research progress of yak meat quality through nutrition regulation was systematically introduced, which provided the basis for the construction of the world's largest green organic yak meat export land and the development of yak industry.

Session 30 Theatre 9

Comparison of beef eating quality terminology databases

A. Barro[1,2], K. Insausti[3,4], M. Kombolo[2], M.P. Ellies-Oury[2,4] and J.F. Hocquette[2]
[1]Londrina State University, Highway Celso Garcia Cid, 86057-970 Londrina, Brazil, [2]INRAE, Vetagro Sup, UMR1213, 63122 Theix, France, [3]Public University of Navarre, ETSIAB-ISFOOD, Arrosadía Campus, 31006 Pamplona, Spain, [4]Bordeaux Sciences Agro, CS 40201, 33175 Gradignan, France; amandagbarro@gmail.com

The standardisation of the evaluation of beef eating quality is a prerequisite to set up a global meat research database and to drive continuous improvement in meat quality prediction. This is the case of the Meat Standards Australia (MSA) methodology, which is based on common definitions of carcass characteristics for carcass graders to maintain skills and accreditation to ensure data consistency. In this line, the objective of this project was to extract and compare relevant terms related to beef eating quality present in pre-existing ontologies and terminology databases. The technical terms used in the MSA methodology were considered as the reference. Their equivalence in different languages (French, Portuguese and Spanish) including minor languages (e.g. Basque) was recorded. A semi-automatic search was carried out in existing databases. This was followed by a manual part to search for the available definitions and establish the equivalence of the terms in the different databases. A total of 19 databases freely accessible online were consulted, including specific ontologies for animal production, (ATOL, NAL USDA, and GACS), the Meat thesaurus available in the AGROPORTAL ontology, online dictionaries, and materials from world-renowned institutions (MLA, ICAR, USDA, IBEEF, AMSA). In total, we identified 56 terms used in the meat industry, in animal production, carcass quality and sensory characteristics. No database could find all the terms used in this research. Besides, some terms (marbling, subcutaneous fat thickness, carcass weight, ribeye area, etc.) are easy to find but are described differently in each database. For example, marbling is measured differently in some countries, which is important to point out in the database. In conclusion, among the consulted sources and ontologies for animal production already established, some of the more specific terms used in the MSA methodology are still missing. Conversely, carcass classification databases lack information on pre-slaughter factors that influence beef quality, mainly for the MSA methodology.

Comparison of slaughter performance and meat quality of Tan sheep under different feeding regimes

X.G. Zhao, C. Zhang, M. Liu and H.L. Luo
China agriculture University, college of animal science and technology, animal nutrition and feed science, 2 West Yuanmingyuan Road, 100193, China, P.R.; 1404010216@cau.edu.cn

Grazing with time limited is a burgeoning production technology which improves production efficiency and maintains the balance between forage and animal. The change of feeding regime, however, had additional effects on the quality of the meat. The aim of this study is to accurately compare the slaughter performance and meat quality of the Tan Sheep under different feeding regimes. In this study, three groups of three-month-old Tan Sheep were raised in three regimes, grazing(G), grazing with time limited + supplementary feeding (GT) and indoor feeding (F). The sheep of three groups were slaughtered after three months. The results showed the final body weight of the GT was significantly higher than the G and F group. The proportion of subcutaneous fat to carcass of the F group was significantly higher than GT and G groups. The proportion of tail fat to carcass of the F group was significantly higher than GT group, which was significantly higher than G groups. The Longissimus thoracis et lumborum (LTL) muscle pH45min of the G group was significantly higher than the F group, but pH24h was lower than F group. The initial and 24 hours after slaughter lightness of LTL muscle of the F group were significantly higher than the GT group and G group. The lightness of 48 hours after slaughter of LTL muscle of the F group was significantly higher than the GT group. The yellowness of 24 hours after slaughter of LTL muscle of the F group was significantly higher than the G group. The drip loss of the G group was significantly lower than the F group and the GT group. The cooking loss of the F group was significantly lower than the GT groups, which was lower than the G group. In summary, Tan Sheep under grazing with time limited + supplement feeding regime had higher meat production than grazing regime, and the meat quality was more similar to grazing sheep.

Effect of *Piper sarmentosum* extract on the growth and nutrient digestion of Hainan Black goat

G.D. Ren[1,2], Z.Y. Sheng[1,2], Y.X. Chen[1,2], X.G. Zhao[2], H. Zhang[2], H.L. Zhou[3], W.L. Lv[3] and H.L. Luo[1,2]
[1]China Agricultural University, Yazhouwan Science and Technology City, 57200, China, P.R., [2]China Agricultural University, 2 Yuan Mingyuan Xilu, Haidian District, Beijing, 100193, China, P.R., [3]Chinese Academy of Tropical Agricultural Sciences, 4 Xueyuan Road, Longhua District, Haikou, 571101, China, P.R.; guodongren630@cau.edu.cn

Piper sarmentosum as a naturally occurring medicinal plant in the tropics, whose extracts have a wide range of bioactive substances due to their anti-inflammatory, antioxidant and insecticidal activities. This study was conducted to investigate the effects of *P. sarmentosum* extract (PSE) addition on the growth performance and meat quality of Hainan black goat. Thirty-six goats (body weight=9.48±0.25 kg, age=90±10 day; mean ± SD) were fed a 50:50 concentrate: roughage basal diet and randomly divided into four groups: 0 (control), 200 (200 PSE), 400 (400 PSE), or 600 mg/kg DM (600PSE) PSE, respectively. The experimental period was 105 days, with 15 days for adaptation and 90 days for data collection. The results showed that feed efficiency responded linearly ($P<0.05$) with the highest values for the 400PSE group. Average daily gain and dry matter gain were not affected by dietary supplementation with PSE ($P>0.05$). Dietary supplementation with PSE linearly increased the pH24h of longissimus dorsi (LD) muscle ($P<0.05$) but decreased the value of L*24 h ($P<0.05$). As supplementation with the PSE increased, the value of b* in LD muscle responded linearly ($P<0.05$), and 600PSE group had lower values of b*45 min and b*24h than other treatment groups ($P<0.05$). The carcass weight, net meat weight and internal organ development indexes were not influenced by dietary supplementation with PSE ($P>0.05$). Goats receiving 400 mg/kg DM PSE addition had higher digestibility of DM, NDF and ADF ($P<0.05$). In addition, the serum content of glucose, IL-2, IL-4 and IL-6 responded linearly ($P<0.05$) with addition of PSE. The blood activity of glutathione peroxidase and level of total antioxidant capacity were linearly ($P<0.05$) increased in the PSE addition groups whereas decreased in malondialdehyde ($P<0.05$). Therefore, the results indicated that dietary addition of 400 mg/kg DM PSE improved feed efficiency, nutrient digestion and antioxidant capacity in Hainan black goats.

Effects of grazing intensities and supplementary levels on the nutritional composition of lamb meat

H. Yang, J. Ji and H.L. Luo
China Agricultural University, State Key Laboratory of Animal Nutrition, College of Animal Science and Technology, 100193, Beijing, China, P.R.; yanghuan2021@cau.edu.cn

Our research objective is to determine the effects of different grazing intensities and supplementary levels on the nutritional composition of lamb meat. Six treatments were compared, with 2 grazing intensities and 3 supplementary levels investigated at both grazing intensities. The 2 grazing intensities were moderate grazing (MG, the utilization rate of grassland is 80%, the plot area is 0.2 ha) and heavy grazing (HG, the utilization rate of grassland is 40%, the plot area is 0.4 ha); The 3 supplementary levels were as follows: 0% supplement (NS), 1% supplement (LS) and 2% supplement (HS) of lamb weight. 72 healthy three-month-old male Hulunbeier lambs were used in a randomized complete block design and divided in 6 groups for the entire experiment, which had 12 lambs (3 plots for each treatment, and 4 lambs in each plot) in each treatment group. After 90 days of grazing, the slaughter experiment was conducted. The results showed that the GR value of lambs decreased with the increase of grazing intensity as well as the decrease of supplementary level ($P<0.05$). The net meat percentage increased with the increase of supplementary level ($P<0.05$). The water content of lumborum muscle in of NS group was higher than that of LS and HS groups ($P<0.05$) and crude protein content was increased in LS group compared to NS group ($P<0.05$). Decreasing the intensity of grazing and increasing the level of supplementary led to a linear increase the proportion of intramuscular fat ($P<0.05$). The n-3 PUFA, including C18:3n3 and C20:5n3 increased in muscle of NS group, and the ratio of n-6/n-3 decreased from 4.76 to 3.03 with decreased supplement level, while the ratio of MUFA to SFA increased in HS group ($P<0.05$). The HG group and NS group increased the contents of EAA (e.g. threonine, valine, isoleucine, leucine, phenylalanine and lysine) and sweet amino acids (e.g. asparagine, glutamicacid, alanine and arginine) ($P<0.05$), compared to MG group and supplementary groups. In conclusion, moderate grazing and supplementary concentrate improved the yield and quality of lamb meat, while the meat content of fatty acids and amino acids under NS or HG was more beneficial to human health.

Perspectives on consumer attitudes to meat consumption

P. Purslow[1], W. Zhang[2] and J.F. Hocquette[3]
[1]The University of Melbourne, 3010, Victoria, Australia, [2]Nanjing Agricultural University, College of Food Science and Technology, Nanjing, China, P.R., [3]INRAE, UMR Herbivores, 63122, France; jean-francois.hocquette@inrae.fr

The production and consumption of meat are regularly discussed in the public sphere. As the first widely recognized journal in its field, the scientific journal Meat Science must provide objective data concerning evolution in perception of meat production and consumption. This is the purpose of a special issue focusing on the range, consensus and diversity of consumer attitudes to meat in the world. Several of the 24 papers on this issue highlight different consumer attitudes between countries. For instance, whereas debates around health, environment and welfare issues are quite strong in the USA, there is a sustained demand for meat in this country, despite success of meat alternatives for mostly young, highly-educated and rich consumers. In China, increasing income is the main factor explaining meat consumption, while the biggest concern is safety. Other countries also express their own specific drivers and concerns. An affordable price and safety but also health, animal welfare and environmental issues are among the factors which explain the decrease in meat consumption. Conversely, the pleasure of eating meat, culinary culture, cultural aspects and national traditions are among the main drivers of meat consumption in relationship to the reported customs and social attitudes. Based on this complexity, a comprise has sometimes to be found, for instance between sensory traits and ethical issues related to pig castration and other practices rejected by consumers in many countries. Whereas plant-based products are already commercialized, there are significant technical, ethical and regulatory issues to fix before getting 'cultured meat' available in the market. Indeed, information on production processes and product composition are not publicly available, making it impossible to check product characteristics and sustainability. However, consumer reactions are dominated by affective, rather than by cognitive factors. To sum up, the current market is disrupted. Meat producers and also the promoters of meat substitutes must adapt their strategies according to these general trends, nuanced by local specificities, while being more transparent.

Session 30 Poster 14

Effects of different grazing intensities and supplementary feeding levels on meat quality of Hulunbu

X.H. Ma, Z. Li, X.G. Zhao, H. Yang and H.L. Luo
China Agricultural University, No.2 Yuanmingyuan West Road, Haidian District, Beijing, 100193, China, P.R.; 2095602677@qq.com

Many studies have shown that meat quality is better in sheep under grazing conditions. However, the changes in meat quality under different grazing levels and supplementary feeding levels are not clear. In this study, 72 Hulunbeier sheep were selected for testing under different grazing and supplementary feeding levels. The results showed that moderate grazing and high supplementary feeding significantly increased the pre-slaughter live weight and carcass weight of the sheep ($P<0.05$), but had no effect on their slaughter rate. Low levels of supplementation significantly ($P<0.05$) increased the greater retinal lipid index in sheep. There was no significant effect on other fat indices. The pH of Hulunbeier lamb ranged from 6.53 to 6.91 at 45 min after slaughter and from 5.70 to 6.23 at 24 h. The pH at 24 h decreased overall compared to the pH at 45 min. This is in line with the pattern of decreasing pH after slaughter. The pH decreased the slowest at the low re-feeding level compared to 45 min. The pH range of 5.8 to 6.2 measured after 24 h of slaughter at low temperatures. L values at 0 h were significantly higher in the no and low supplementation meat colours than in the high supplementation level, while L values at 24 h were higher in the high supplementation level than in the other two groups. The fact that the yellowness values at 0 h were higher in the no- and low-feeding levels than in the high-feeding level, but not in the 24 h level, also indicates that the no- and low-feeding levels were favourable to the development of flesh colour. Moderate grazing significantly reduced the yellowness values at 24 h. The a-values at 24 h were significantly higher at no and low levels of supplementation than at high levels of supplementation. Drip loss was significantly reduced by no and low levels of supplemental feeding. Overall, no and low levels of supplemental feeding were more favourable to the development of good meat quality under moderate grazing conditions.

Session 30 Poster 15

Progress of yak carcass grading in Qinghai-Tibet Plateau of China

S.S. Zhang
Institute of Animal Science Chinese Academy of Agricultural Sciences, 2# Yuanmingyuan West Road, Haidian District, Beijing, 100193 Beijing, China, P.R.; 474002029@qq.com

Yak, as a precious species resource in the Qinghai-Tibet Plateau of China, has great commercial development potential due to its high protein content, low fat content, high Ca, Fe, Zn, Mg and other minerals, and yak meat is the main resource of animal protein for Tibetans. Traditional yaks are fat in autumn and lean in winter because of their grazing conditions which makes them grows slowly. The yaks are mainly slaughtered at the age 5-7. The growth characteristics and industry periodicity of yak are different from those of ordinary beef cattle, so that its quality evaluation system should be different from that of ordinary beef cattle. The current beef grading standard is not applicable to yak meat grading, and the absence of national or industrial yak carcass grading standard makes the evaluation of yak carcass a confusion. So, carrying out yak carcass grading is conducive to the high-quality development of yak industry in the Qinghai-Tibet Plateau.

Session 31 Theatre 1

Future-proofing the sustainability of pasture based beef production systems

A.K. Kelly and T.M. Boland
University College Dublin, Agriculture and Food Science Centre, Belfield, Dublin 4, Ireland; alan.kelly@ucd.ie

Beef is a high-quality source of protein and demand is increasing globally. However, population growth and environmental constraints will continue to pressure beef producers and wider industry to improve productivity, efficiency and sustainability. Most notably, beef producers and beef industries face significant environmental and sustainability challenges around greenhouse gas emission, water quality, arresting and reversing environmental degradation, land use (food vs feed) and biodiversity loss. Future proofing the sustainability of beef farm production systems will require a combined approach supporting high on-farm efficiency and productivity, paramount to farm profitability and succession, coupled with financial incentives that rewards producers in achieving target-market specifications for higher-value premium quality beef, using suitable genetics and appropriate management principles, that contribute to reduced environmental and animal welfare impacts necessary for provenance and social licence to farm. This review provides an overview of the principles and practices that support efficient, high output, productive pastoral-based beef production systems, including rule of genetics, nutritional strategies and other management practices. Reducing the environmental footprint of pasture-based beef cattle production systems will undoubtedly be a challenge and a secondary objective of this review GHG mitigation solutions, both currently available and under development, for temperate pasture-based beef cattle production systems.

Session 31 Theatre 2

Effect of sward type on growth performance and methane output of lambs in the post-weaning period

S. Woodmartin[1], P. Creighton[1], T.M. Boland[2], A. Monaghan[1] and F. McGovern[1]
[1]Teagasc, Animal and Grassland Research and Innovation Centre, Mellows Campus, Athenry, Galway, Ireland, H65R718, Ireland, [2]School of Agriculture and Food Science, University College Dublin, Belfield, Dublin 4, Ireland, D04FX62, Ireland; sarah.woodmartin@teagasc.ie

In Irish sheep systems, a large proportion of lambs grazing perennial ryegrass swards (*Lolium perenne L.*;PRG) in the post-weaning period have live weight gains below target. Subsequently, forage inclusion is gaining popularity alongside PRG to enhance animal performance. The aim of this study was to assess the effect of sward type on lamb performance and methane (CH_4) output in the post-weaning period. Five sward types were investigated, PRG only, PRG plus white clover (*Trifoluim repens L.*;PRG+WC), PRG plus red clover (*Trifoluim pratense L.*;PRG+RC), PRG plus chicory (*Chicorium intybus L.*;PRG+Chic) and PRG plus plantain (*Plantago lanceolate L.*;PRG+Plan). A complete randomised block design experiment was used creating five farmlets each representing a dietary treatment. Lambs were weaned at 14 weeks and a leader-follower grazing system was implemented. Methane measurements were obtained using portable accumulation chambers (PAC) on 120 lambs over two production seasons (n=24 per treatment). Lambs were weighed fortnightly and drafted accordingly, targeting a 20 kg carcass. Days to slaughter (DTS) was calculated as the number of days from birth to slaughter. Results show that lambs grazing swards containing a forage had reduced DTS of between 16 days (PRG+Plan) and 47 days (PRG+RC) (P<0.0001) relative to those grazing PRG only swards. The lambs grazing PRG+RC (10.3±0.225) ranked lowest for CH_4 production (g/day) in comparison to all other dietary treatments. Lambs grazing PRG+WC (11.6±0.225) and PRG+Plan (11.8±0.225) ranked intermediately and lambs grazing PRG+Chic and PRG only (12.9±0.225) ranked the highest (P<0.0001). The addition of forages to the grazed sward offered to lambs post-weaning gives the potential to reduce DTS and lower the ranking of CH_4 (g/day) output when compared to lambs grazing PRG monocultures. Lambs finished on the PRG+RC sward had superior growth performance coupled with the lowest CH_4 output. Forage inclusion can benefit the environmental and economic performance of lamb finishing systems.

On farm grass growth prediction in Ireland – 4 year evaluation

E. Ruelle[1], L. Bonnard[1,2], D. Hennessy[1], L. Delaby[3] and M. O'Donovan[1]
[1]Teagasc, Moorepark, Fermoy, Co. Cork, Ireland, [2]MTU, Cork, Co. Cork, Ireland, [3]INRAe, Institut Agro, PEGASE, 35590 Saint-Gilles, France; elodie.ruelle@teagasc.ie

In pasture-based systems farmers need to make daily management decisions to ensure that their livestock have good quality feed in adequate quantities. 2018 was a challenging year in Ireland due to a late start to grass growth in the spring and a drought in the summer. This situation highlighted the need to be able to predict grass growth weekly to help farmers manage their grass and better anticipate grass shortages in a changing climate. The grass growth modelling program at Moorepark, based on the MoSt Grass Growth Model, increased from 39 farms in 2019 to 78 in 2022 allowing a good representation of grass growth across Ireland. To predict grass growth, data from the farms participating in the project are extracted from PastureBase Ireland (PBI – a grassland decision support tool). Data includes number of paddocks and their area, grazing and cutting dates, number of grazing animals and feed supplementation, and nitrogen fertiliser application date and rate. Other data required by the model include the soil type for each paddock, and historical and forecast weather data (provided by Met Éireann, the Irish Meteorological Service). This paper will present the evaluation of four years of grass growth prediction by the latest version of the MoSt model compared to PBI farmer data using historical weather data. The number of available data points possible for comparison was 191,021 at the paddock level and 9,695 when averaged at farm level. At farm level, the average RMSE was of 13.5 (39 farms), 17.5 (56 farms), 15.7 (78 farms) and 14.3 kg DM/ha (78 farms) for 2019, 2020, 2021, and 2022, respectively. At paddock level (n=121,021), the error was higher and was 24.2 (1,295 paddock), 27.3 (1,880 paddock), 26.0 (2,471 paddock) and 24.8 kg DM/ha (2,478 paddock) for 2019, 2020, 2021, and 2022, respectively. The lower accuracy at paddock level is expected due to the difficulty in accurately replicating inter-paddock variability and the low reliability of the data entered into PBI at paddock level. While the predictions are not perfect, the accuracy at farm level is sufficient to make the MoSt model a useful tool for grassland management on Irish farms.

Incorporating plantain forage into a grazing dairy system: effect on farm productivity

O. Al-Marashdeh and H.M.G.P. Herath
Lincoln University, Faculty of Agriculture and Life Sciences, Lincoln, Canterbury 7647, New Zealand; omar.al-marashdeh@lincoln.ac.nz

Plantain (*Plantago lanceolata* L.) has been exploited as a natural mitigation strategy to reduce nitrate leaching from pastoral dairy systems. However, the productivity of dairy farm system that incorporated plantain into a ryegrass/white clover mixed swards has not been reported yet. Therefore, in a replicated farm system study and over full production season, the objective of this study was to investigate the effect of the increasing proportion of plantain in a ryegrass-white clover (RGWC) mixed sward on farm productivity. The study was conducted at Lincoln University Research Dairy Farm, Lincoln, New Zealand using a total of 108 Jersey × Friesian dairy cows during 2021/2022 production season (start calving date 27th July 2021 and dry-off date 11th May 2022). Cows were blocked into replicated herds of 12 cows (total of nine herds), and randomly allocated into one of three pasture treatments (n=3) sown with an increasing plantain seed rate: (1) RGWC with nil plantain (PL0); (2) RGWC + 3 kg/ha plantain seed rate (PL3); or (3) RGWC + 6 kg/ha plantain seed rate (PL6). A total of 32 ha of 0.3-ha paddocks were allocated to farmlets in this study, with 12 paddocks (total area 3.6 ha) per farmlet. During the period of study, farmlets were maintained on their treatments and managed individually. Individual cow's milk yield was recorded at each milking event and fortnightly herd test was conducted for determining milk composition. Individual paddock herbage mass was estimated weekly by a calibrated rising plate meter, and daily pasture growth rate was calculated. Annual pasture production was similar (13.0±0.61 t DM/ha) across treatments. Total milk yield (average 4,052±111.9 kg) and milk solid production (fat + protein; 406.9±11.93 kg/cow) were not different between treatments. Total milk fat (232±9.2 kg/cow) and protein (175±3.4 kg/cow) yields of cows did not differ between treatments. Integrating plantain in a mixed sward of ryegrass-white clover does not seem to affect productivity of dairy system, and if environmental benefits are confirmed at farm scale, plantain could be used as a mitigation strategy to reduce environmental impact of grazing dairy systems with no impact on farm profitability.

Session 31 Theatre 5

Dry matter production of multispecies swards under dairy grazing in two chemical nitrogen scenarios

C. Hearn[1,2], M. Egan[2], M.B. Lynch[1,3] and M. O'Donovan[2]

[1]University College Dublin, School of Agriculture and Food Science, University College Dublin, Belfield, Dublin 4, D04 V1W8, Ireland, [2]Teagasc, Animal and Grassland Research and Innovation Centre, Moorepark, Fermoy, Co. Cork, P61 P302, Ireland, [3]Teagasc, Environmental Research Centre, Johnstown Castle, Co. Wexford, Y35 Y521, Ireland; michael.odonovan@teagasc.ie

Irish dairy farmers have become increasingly interested in the use of multispecies (MS) swards for intensive grazing systems. A grazed plot experiment was established to investigate the annual dry matter (DM) yield and botanical composition of MS swards containing three plant functional groups; grass, legumes and herbs under two nitrogen (N) fertiliser scenarios; zero N (N0) and a more conventional 200 kg N/ha/year (N200). Five sward types were established which included a perennial ryegrass (PRG, *Lolium perenne* L.) monoculture and sward mixtures of the following species: PRG, white clover (WC, *Trifolium repens* L.), chicory (CH, *Chicorium intybus* L.) and ribwort plantain (PL, *Plantago lanceolate* L.); with monoculture PRG N200 as the control. Plots were grazed by lactating dairy cows on eight, nine and eight occasions in years one, two and three, respectively. Swards in the N200 treatment produced a higher level of DM with an average of 2,497.4 kg DM/ha/year more than N0 swards ($P<0.001$). The sward mixture of PRG, WC, CH & PL produced the highest level of DM; 2,184.3 kg DM/ha more than the PRG monoculture sward on average across both N scenarios. Swards including WC produced an average of 1,599.8 kg DM/ha/year more than those sown without WC ($P<0.001$); the level of WC was influenced by N fertiliser where MS swards in the N200 protocol had a WC content of 15.4% less than that of those in the N0 protocol ($P<0.001$). These results indicate that the inclusion of WC in grazed MS swards is crucial to increased DM production and the level of WC in grazed MS swards is influenced by the application of chemical N fertiliser.

Session 31 Theatre 6

Nitrogen herbage yield in grass and grass-white clover swards receiving zero nitrogen

Á. Murray, B. McCarthy and D. Hennessy

Teagasc, Animal and Grassland Research and Innovation Centre, Paddy O Keeffe Building, Moorepark Teagasc, P61C996, Ireland; aine.murray@teagasc.ie

Grazed grass is considered the cheapest feed available for dairy cows in temperate regions but is highly reliant on nitrogen (N) inputs. Nitrogen fertilisers are a major contributor to ammonia (NH_3) and greenhouse gas (GHG) emissions through their inputs in grazed herbage systems. Nitrogen use efficiency (NUE) can be increased through reduced N inputs to grazed herbage systems but production needs to be maintained. The objective of this study was to evaluate the nitrogen uptake yield from perennial ryegrass (*Lolium perenne* L.; PRG) and PRG-white clover (*Trifolium repens* L.; WC) herbage receiving zero nitrogen inputs. The study consisted of zero N plots that were set up in 2020 and 2021 within paddocks grazed by dairy cows of either PRG-only or PRG-WC sward types at Clonakilty Agricultural College, Cork, Ireland. The plots received no chemical or organic N and were not grazed by animals for the duration of the grazing season. Plots were relocated within paddock between grazing seasons. Herbage yield was measured within the plots at the same time as herbage yield for the paddock was measured prior to each grazing. The PRG-only sward plots yielded 7,950 kg DM/ha per year and 203 kg N/ha within this herbage. The PRG-WC swards yielded significantly higher herbage (11,499 kg DM/ha per year) and N (306 kg N/ha) than the PRG-only swards ($P>0.001$). The results provide an indication of background N mineralization and biological N fixation within PRG-only and PRG-WC swards receiving zero nitrogen. This study warrants further measurements to be carried out to gain a better understanding of the N dynamics and responses within grazed swards to allow for better utilization of this valuable resource and in turn reduce N use and increase NUE.

Grassland-based dairy farms of French Massif Central adapt to climate change with diverse strategies

L. Allart[1], V. Oostvogels[2], F. Joly[1], C. Mosnier[1], N. Gross[3] and B. Dumont[1]
[1]Université Clermont Auvergne, INRAE, VetAgro Sup, UMR Herbivores, 63122 Saint-Genès-Champanelle, France, [2]Wageningen University & Research, Animal Production Systems group, De Elst 1, 6708 WD Wageningen, the Netherlands, [3]Université Clermont Auvergne, INRAE, Unité de Recherche sur les Ecosystèmes Prairiaux, 5 Chemin de Beaulieu, 63100 Clermont-Ferrand, France; lucie.allart@inrae.fr

Grassland-based dairy systems from upland areas, which are based on the grazing or mowing of semi-natural grasslands, are threatened by climate change, e.g. more frequent and severe summer droughts. The choices of adaptation of farmers can include the preservation of semi-natural grasslands or their conversion to other land-use as fodder crops. We surveyed 15 dairy farmers from French Massif central and investigated their perceptions of climate change, biodiversity, and grassland function in their adaptive capacity by semi-structured interviews. Using framework analysis and Multiple Component Analysis (MCA), we built four groups based on axes of the MCA explaining 71% of the variance. The first group (7 farms) is made of farmers who have a clear adaptive strategy that largely relies on permanent grassland management (e.g. building stocks by decreasing or optimising fodder consumption, changing harvesting methods to stabilise fodder yields and quality to cope with climate variability) among other levers. A second group of farmers (2 farms) also has clear strategies to adapt, based on building stocks from fodder crops and temporary grasslands, and preservation of hedgerows and trees. Farmers from the third group (3 farms) are still developing their adaptive strategy. So far, they give limited value to permanent grasslands in their adaptation strategy. The fourth group (3 farms) is characterised by a low intention to adapt. Grassland appreciation of the first group was not related to specific systems, as farm area and annual milk yield did not distinguish the first group from the others. Only the share of permanent grasslands in farm area was significantly higher in the first group than in all three others (89 vs 57%, $P<0.05$). These preliminary results reveal how dairy farmers from the same area have contrasted perceptions on how grassland diversity allows adapting to climate change.

Effect of grazing intensity on nitrogen cycle of alpine summer pasture soil

S. Raniolo[1], A. Squartini[1], L. Da Ros[2], F. Camin[3,4], L. Bontempo[3], L. Maretto[1], D. Gianelle[3], M. Ramanzin[1], E. Sturaro[1] and M. Rodeghiero[3]
[1]University of Padova, DAFNAE, Viale dell'Università 16, 35020, Italy, [2]Free University of Bolzano, Faculty of science and technology, piazza Università 5, 39100, Italy, [3]Fondazione Edmund Mach di San Michele all'Adige, Via Edmund Mach 1, 38098, Italy, [4]Center Agriculture Food Environment, Agriculture Food Environment Centre, Via Edmund Mach 1, 38098, Italy; salvatore.raniolo@phd.unipd.it

Alpine pastures are relevant agro-ecosystems classified as High Nature Value Farmland thanks to their high biodiversity. The aim of this study is to characterize the agroecological relationships among grazing animals, soil conditions, and microorganisms in an alpine summer pasture located in the Dolomites, Eastern Italian Alps, at an elevation around 2,000 m asl, grazed by dairy cows from mid-June to mid-September, with an area of 180 ha and a stocking rate of 0.84 livestock unit/ha. The grazing intensity was characterized at a fine spatial scale by combining GPS tracking of grazing animals and farmer's interviews. To assess the impact of grazing on soil conditions and microbial functions, topsoil cores were sampled at two different depths in three pasture areas at different grazing intensity during September, the end of the grazing period. Topsoil samples were analysed for bulk density, content of ammonium, nitrate, and the N isotopic ratio ($\delta 15N$), as an index of relative rates of N inputs and losses. In addition, the functional potentials of microbial communities were assessed by quantifying with real-time PCR the abundances of target genes *nirK* and *nosZ* for denitrification, *amoA* Archea and Bacteria for nitrification, and *nifH* for nitrogen fixation. Nitrogen content and bulk density were significantly affected by both grazing intensity and soil depth and their interaction, while microbial functions was significantly affected only by grazing intensity, showing a likely and diversified impact of grazing on nitrogen cycle into topsoil. These results could be used to develop minimally invasive biophysical indicators of regulating ecosystem services in grasslands soils. These tools can inform sustainable management of grazing livestock systems in accordance with agroecology principles.

The visibility of the invisible: analysing heifers reactions while learning the virtual fence system

D. Hamidi[1], N.A. Grinnell[1], M. Komainda[1], L. Wilms[1], F. Riesch[1,2], J. Horn[1], M. Hamidi[1], I. Traulsen[3] and J. Isselstein[1,2]

[1]University of Goettingen, Crop Sciences, Von-Siebold-Str. 8, 37075 Göttingen, Germany, [2]University of Goettingen, Centre for Biodiversity and Sustainable Land Use, Büsgenweg 1, 37077 Göttingen, Germany, [3]University of Goettingen, Animal Sciences, Burckhardtweg 2, 37077 Göttingen, Germany; dina.hamidi@uni-goettingen.de

An interesting approach to optimising the role of ruminants in sustainable food systems is the use of virtual fencing (VF) to graze land that cannot be managed with ground-based fencing. VF is an innovative technology with the potential to substitute common fencing systems. However, animal welfare concerns are a general problem for the wider implementation. When the animals approach the VF boundaries an acoustic signal is emitted by the collar followed by a short time electric pulse if the animal continues moving forward. Both signals are coupled, as the electric pulse is emitted by the last note of a melody rising in pitch. The heifers' reaction that led to the stopping of the coupled signal was used to analyse the learning success. We defined a reaction score (RS) to classify the heifers' reactions while turning around causing the signal to stop. RS 1: continue grazing/walking slowly <three steps, RS 2: walking >three steps/jumping with front legs, RS 3: running (trot/canter), RS 4: escaping. Data collection was conducted in a 12-day training trial in northern Germany (2021) with 16 Fleckvieh heifers fitted with VF collars (Nofence®, Norway), equally divided into two groups. Four 'focus' heifers per group were observed by one observer. The heifers were naive with VF prior to the study. Each RS was observed for each of the heifers' except RS 4 (observed twice). On the first day of the trial, 10.13 reactions per hour were observed (RS 1: 18%; RS 2: 60%; RS 3: 22%). On the second day, proportions for the respective RS changed to 61% (RS 1), 29% (RS 2) and 10% (RS 3). On the last day, 100% RS 1 was observed. By observing the heifers, it was possible to analyse the behavioural change in the animals' reactions to the invisible VF signals as a visible sign of learning success.

Sheep integration into cropping systems for agroecological transition of farming systems in France

F. Stark[1], J. Ryschawy[2], R. Mettauer[3], M. Grillot[2], I. Shaqura[1] and M. Moraine[4]

[1]SELMET, Univ Montpellier, INRAE, CIRAD, L'Institut Agro, Montpellier, France, [2]AGIR, Univ Toulouse, INPT, INRAE, Toulouse, France, [3]SAS, INRAE, L'institut Agro, Rennes, France, [4]INNOVATION, Univ Montpellier, INRAE, CIRAD, L'Institut Agro, Montpellier, France; fabien.stark@inrae.fr

Limits of the specialization of agriculture are widely documented in the scientific literature in terms of environmental, social and economic impacts. At the same time, ecological issues and agroecological transition of agri-food systems must be considered both at farm and territorial levels. Crop-livestock integration at both levels seems a promising avenue for building sustainable systems. In this study, we elaborate scenarios of agroecological transitions based on innovative crop-livestock synergies. We studied sheep integration into organic wines and cereals cropping systems in South-western France. We firstly assessed the technical potential for crop-livestock integration at a territorial level with a seasonal supply-demand model. Secondly, we designed acceptable future scenarios from real situation through a participative process and simulated them. These scenarios are related to climate change impact on resources, use of new fodder resources, and adaptation of animal requirements to fodder resources availability. Our results show that in heterogeneous landscape mosaic as it is the case here, sheep herds are fed with a wide range of resources, available in a complementary way throughout the seasons. Under climate change, current situation won't be sustainable due to insufficient fodder resources at certain critical times (in summer specifically). However, additional resources at territorial scale (legume fodder, rangeland, vineyard inter-row) could be used at different seasons to maintain or even increase the number of sheep bred. Moreover, transhumance during summer could decrease animal needs when fodder resources are missing and transhumance during winter could increase the number of sheep bred when the demand for weed control in the vineyard is high. Finally, we discuss brakes and levers resulting from the analysis to assess to what extent and in what terms this innovative initiative draws a model of agroecological transition of farming systems at territorial level.

Session 31 — Theatre 11

Nutrient cycling and efficiency: a comparative flow analysis of meat and dairy sheep farming systems

F. Stark[1], N. Amposta[2], W. Nasri[1,3], M. Lamarque[2], S. Parisot[2], P. Salgado[1,3] and E. González-García[1]
[1]SELMET, Univ Montpellier, INRAE, CIRAD, L'Institut Agro, Montpellier, France, [2]La Fage, INRAE, Roquefort-sur-Soulzon, France, [3]CIRAD, UMR SELMET, Montpellier, France; fabien.stark@inrae.fr

Nutrient cycling and efficient use of resources are issues of concern for the future of sustainable livestock systems. A deep analysis of nutrient flows involved in the functioning of meat and dairy sheep farming systems (FS) was undertaken using the Ecological Network Analysis (ENA) method. Farm autonomy, efficiency and productivity performances were assessed. As part of the TrustFarm project (ERANET), the La Fage INRAE experimental farm (Aveyron, France) was the case studied. It is composed of two sub-systems comprising: (1) a dairy sheep flock (600 Lacaune ewes), under semi-intensive conditions and mostly fed with on-farm fodder; and (2) a meat flock (280 Romane ewes) extensively reared on rangelands and complemented in winter with on-farm fodder. A historical dataset (2015-2019) was used to represent nutrient flows (i.e. nitrogen, N) between FS components (i.e. flocks, feeds, effluents, crops, grasslands and rangeland). Outcomes on annual farm' balance revealed that internal N flows between components (excluding inputs, outputs and losses) are responsible for half of the nutrient cycling, which is mainly concerned by forage distribution, forage autonomy representing around 80%. In a counter-intuitive way, on-farm fodder distributed to meat flock covers almost half of requirements despite grazing on rangelands all year round. In the same line, dairy flock' N outputs are mostly determined by lamb exports, rather than milk production which is supposed to be the objective of dairy production (46 vs 14%). The farm' N use efficiency was 52%, with high variability between system components, meat flock presenting a better efficiency than dairy flock (139 vs 90%), whereas rangeland and effluent components showed the lowest values (23 and 43%). These preliminary results on nutrient balance provide a different, complementary approach to analyse performances at different scales of the FS. It contributes to deeper analyses of nutrient cycling aiming to adjust management practices for designing more sustainable and agroecological systems.

Session 31 — Theatre 12

Pasture-based ruminants and biodiversity in contrasting contexts: an account of farmers' narratives

V.J. Oostvogels[1], B. Dumont[2], H.J. Nijland[3], L. Allart[2], I.J.M. De Boer[1] and R. Ripoll-Bosch[1]
[1]Wageningen University & Research, Animal Production Systems group, De Elst 1, 6708 WD Wageningen, the Netherlands, [2]Université Clermont Auvergne, INRAE, VetAgro Sup, UMR Herbivores, Theix, 63122 Saint-Genès-Champanelle, France, [3]Hanneke J. Nijland Research & Consultancy – Antennae, Beeklaan 48, 6869 VH Heveadorp, the Netherlands; vincent.oostvogels@wur.nl

In debates about roles for pasture-based ruminants in sustainable food systems, biodiversity is a central theme. Different narratives occur as to how pasture-based ruminants and biodiversity relate and which courses of action should be taken. To arrive at effective and just solutions, it is essential to include in these debates the narratives of local stakeholders such as farmers. However, there is still little insight in their narratives, and how these may vary within and between individuals and areas, considering differences in socioecological context. Therefore, we conducted a study to unravel farmers' narratives in two contrasting areas with pasture-based dairy farming: an area in Massif Central in France and a fen peat area in the Netherlands. These areas differ, among others, in land use intensity, environmental conditions, and value chains. We followed a qualitative approach, based on interpretative analysis of 30 one-hour in-depth interviews (15 in each area). We encountered diverse and sometimes conflicting narratives, with different problem definitions and proposed ways forward. We found that these differences were grounded in underlying differences in biodiversity conceptualizations and values. The two areas showed similarities in the narratives' basic lines of reasoning. For instance, in both areas we encountered 'productivist' as well as 'post-productivist' narratives. Nevertheless, there were differences in the specific topics that were linked to the narratives. For example, in the Dutch area more links were made with religion, and in the French area with food culture. Our results highlight the pluralism and – on certain points – strong context-specificity in the encountered narratives. We argue that better understanding and inclusion of stakeholder's narratives, with more attention for this pluralism and context-specificity, can benefit the debate around biodiversity in pasture-based production.

Session 31 — Poster 13

Meat quality in Podolian young bulls grazing on wood-pasture in a South Italy marginal area

F. Giannico[1], S. Tarricone[2], A. Caputi Jambrenghi[2], L. Tedone[2], D. Campanile[3] and M.A. Colonna[2]
[1] University of Bari Aldo Moro, Department of Veterinary Medicine, Valenzano (BA), 70010, Italy, [2]University of Bari Aldo Moro, Department of Soil, Plant and Food Sciences, Bari, 70125, Italy, [3]Puglia Region, Department of Agriculture, Rural and Environmental Development, Bari, 70121, Italy; francesco.giannico@uniba.it

In dry marginal areas, pasture plays an important role in land preservation and wildfire prevention. The aim of the study was to evaluate meat production and quality in autochthonous Podolian young bulls grazing on spontaneous grass and wood-pasture vegetation in South Italy marginal areas. Sixteen Podolian male calves grazed with their dams until they were 10 months old (±15 days). Afterwards, two groups of calves homogeneous for weight (310±10 kg) were made. The control group (C) was kept in a loose barn with a straw-bedded resting area and an open exercise area with a total space allowance of 15 m^2/subject. They received a commercial feed (B/82 – Bovi mix, Galtieri, Italy; 6.0 kg/day/subject) and wheat straw (2 kg/day/subject). The second group grazed during the day on a spontaneous 10 ha wooded pasture (WP); at housing, in the evening, the calves received 2 kg/subject of the same commercial feed and straw *ad libitum*. Pasture availability and composition were monitored during the experiment and the nutritive value was assessed by *in vitro* gas production (IVGP). All the calves were slaughtered at 18 months of age. In WP bulls, the average daily gain (1.25 vs 0.95) and final weight (520 vs 485) were lower ($P<0.05$) as compared with control. Grazing worsened meat colour in terms of lightness (32.8 vs 33.9) but it increased ($P<0.05$) red (21.98 vs 20.87) and yellow (2.04 vs 1.96) indices. Meat from the WP group showed a lower cooking loss as compared to C bulls, while tenderness (WBS) was markedly lower ($P<0.05$). The concentration of conjugated linoleic acid (CLA) in the Longissimus dorsi muscle was significantly higher in WP bulls (0.28 vs 0.18; $P<0.05$). Results of the consumer test showed that although panellists detected differences in meat tenderness and chewiness, they recognized and tended to prefer ($P<0.05$) meat from the WP bulls for its taste and flavour.

Session 31 — Poster 14

Good management of the pasture makes it possible to reduce the supplementation of autumn calves

D. Douhay
Institut de l'Elevage, Saône-et-Loire, La Prairie, 71250 Jalogny, France; jeremy.douhay@idele.fr

The production of weanlings born in the fall, a practice not very widespread in the Charolais suckling area, was set up at FERM'INOV, the experimental farm of Jalogny in Saône-et-Loire as part of a system study aimed at producing heavy lean males of 400-420 kg live at destined for the Italian market during low production period (June-July). To meet this commercial objective, experimental work aims to determine the optimal level of supplementation under the mother and to study the impact of good spring grazing management on the growth of calves until weaning. The growth target is 1,400 g/d between birth and sale. The latest round of work conducted in 2020/2021 shows promising results. On average, over the 'birth-sale' period, the average daily gain levels of the two groups of calves of around 1,550 g/d are not different and are above the growth target. The group not supplemented on pasture was weaned 15 days later to ensure an identical selling weight between the two groups. The absence of supplementation on pasture allows a saving of 100 kg of concentrates per calf, i.e. around €25/calf for a feed price of between €250 and €270/T (2021 economic situation). Weaning and sale are deferred for 15 days with no impact on the selling price per kg of weanling (€2.7/kg live for the two groups of calves). In 2021 grazing conditions, a difference in the feed cost margin of €41/grazing in favour of the group not supplemented with grazing is observed. A winter supplement of 1 kg of concentrates per 100 kg of live weight is sufficient to produce weanlings born in the fall of 400 kg of live weight at weaning at the end of June, when the feeding of suckler cows allows the expression of their full dairy potential. In the spring, optimized rotational grazing management, with control of grass heights at each entry and exit from the plot, has a positive impact on the growth performance of calves that do not need to be supplemented. It allows a gain of autonomy and a saving of work related to the distribution of the concentrates.

Session 31 Poster 15

Investigating the effects of lactation number, stage, and milking interval on nitrogen use efficiency

Y.M. Hu
Gembloux Agro-Bio Tech (GxABT)/ University of Liège, Precision Livestock and Nutrition, Passage des Déportés, 2, 5030 Gembloux, Belgium; yumei.hu@student.uliege.be

Nitrogen use efficiency (NUE) is considered one of the most important indicators of ruminant feed utilization. The objective of this study was to describe the relationship of NUE with lactation number, stage, and milking interval in Holstein dairy cows using an automatic milking system. Data was collected between May 2021 and June 2022 from a dairy farm equipped with an automated milking system (AMS), with a total of 466 dairy cows (260 primiparous cows and 206 multiparous cows). The AMS management system recorded the date and time of milking during the 24 h, and the milk yield of each cow as measured by the milk meters installed on each AMS unit. AMS concentration intake was recorded at each milking and the partial mixed ration (PMR) was recorded every day. Compared with primiparous cows, the multiparous cows had higher NUE, AMS concentration N intake, PMR N intake, and Milk N ($P<0.001$). In contrast, Manure N was higher in primiparous cows ($P<0.001$). There were significant differences were found in the NUE ($P<0.001$) between different lactation stages, and cows in DIM 0 to 100 days has the highest NUE (31.74%). The milking frequency had a significant influence on NUE ($P<0.001$). Meanwhile the NUE of cows increases with milking frequency and is most frequent between the milking intervals of 7 to 12 hours. We recorded the highest NUE with the multiple parties, in the early lactation stage (DIM 0 to 100 days) at a 3× milking frequency/d. Overall, detailed knowledge of these factors associated with increasing NUE using AMS, such as lactation number, stage, and milking interval, will help guide future recommendations to producers for maximizing NUE in the dairy industry.

Session 31 Poster 16

Plasma urea and diet crude protein concentration relationship in grazing dairy cows

R. Delagarde and N. Edouard
PEGASE, INRAE, Institut Agro, 16 Le Clos, 35590 Saint-Gilles, France; remy.delagarde@inrae.fr

It is well known that plasma urea (PU, mg/dl) is a good indicator of protein status in dairy cows fed conserved forages, and also that this relationship may differ between conserved forages and fresh herbage. The objective of this study was to relate plasma urea of grazing dairy cows to herbage, supplements, diet, grazing and cows characteristics. A dataset of 8 experiments carried out between 1993 and 2014, in which herbage intake was measured individually, was used. Herbage intake was measured from the faecal output (Cr2O3 or Yb2O3)/faecal index (N and ADF) method, or by the n-alkanes method. Diet crude protein (CP) concentration (dietCP, g/kg DM) was determined from measured intake and CP concentration of each feed, including grazed herbage. Supplements (conserved forages and/or concentrates) represented 12% of the diet (0 to 34%). Multiple regression analyses including cows, feeds, diet and grazing variables, either with individual cow × period data (n=709), or with data averaged per herd and period (n=114), showed that dietCP explained almost all the PU variability. Addition of cows (weight, milk production, daily DM intake), supplements (amount, nature, proportion of diet), or pasture and grazing (fibre concentration, digestibility, sward height, herbage allowance, daily access time to pasture) characteristics did not improve the model. Plasma urea and dietCP averaged 21 mg/dl (6 to 43) and 171 g/kg DM (106 to 246), respectively. At herd level, PU may thus be predicted from dietCP only: PU = 0.236 dietCP – 20, R^2=0.82, s.d.=3.8. This equation means that any increase of 10 mg/dl of PU originates from an increase of 42 g CP/kg DM of the diet. Plasma urea may also be used as an indicator of dietCP, with dietCP = 3.47 PU + 99, R^2=0.82, s.d.=14.6. This equation suggests a total urea recycling when dietCP is close to 100 g/kg DM, with no more urea in plasma. When measured simultaneously in the morning before feeding, PU and milk urea were strongly correlated, with a slope of one and an origin of zero. It is concluded that plasma urea or milk urea may be good indicators of diet crude protein concentration in grazing dairy cows, and vice versa.

Session 31 Poster 17

An assessment of woody forage resources indigenous to southern Africa

M. Trytsman[1], F.L. Müller[1], M.I. Samuels[1], C.F. Cupido[1] and A.E. Van Wyk[2]
[1]Agricultural Research Council, Animal Production: Range and Forage Sciences, Private Bag X2, 0062 Irene, South Africa, [2]University of Pretoria, Department of Plant and Soil Sciences, Private Bag X20, 0028 Hatfield, South Africa; mtrytsman@arc.agric.za

The South African National Forage Genebank (SA-NFG) mandated to identify, collect, conserve, and evaluate important indigenous forage genetic resources, initiated a project to identify valuable indigenous South African woody (trees and shrubs) forages. To date, a total of 722 species and infraspecific taxa from 76 families have been recorded, which were statistically grouped into seven distinct phytochoria with clear geographical patterns. The aim of this work was to prioritize woody species within each phytochorion, based on relevant forage attributes documented in, among others, scientific publications and textbooks. These include identifying key families and species, functional traits (degree of deciduousness, seed-bearing structure, and growth form), utilization by game and livestock (plant part and browser type), nutritive values (crude protein, ADF and NDF), and the presence of secondary compounds. Preliminary findings show that Asteraceae and Leguminosae are important families, the former mostly in the Succulent Karoo Biome and the latter in the Savanna Biome. Furthermore, Leguminosae contains the most species with crude protein values higher than 7%, however, contains many species with relatively high condensed tannin levels. This browse forage database, together with the Leguminosae and Poaceae databases, will continuously be updated, and accessed by the SA-NFG for collection, conservation, screening, and characterization strategies of these indigenous forage genetic resources.

Session 31 Poster 18

The Effect of heterofermentative lactic acid bacteria on quality of King Napier grass

P. Lounglawan
Suranaree University of Technology, School of Animal Technology and Innovation, Suranaree University of Technology, Nakhon Ratchasima, Thailand, 30000, 30000, Thailand; pipat@sut.ac.th

This research aimed to study the chemical change of Napier grass (*Pennisetum purpureum* × *Pennisetum americanum*) silage after using lactic acid producing bacteria in heterofermentative lactic acid bacteria group. The experiment aims to select the type of Heterofermentative lactic acid bacteria suitable for Napier grass silage. The results showed that *L. buchneri* at the level of 1×10^6 cfu/g of fresh forage resulted in the higher acetic acid content of Napier grass silage than other bacteria. However, the use of this bacterium was found to negatively affect the quality of Napier grass silage. In addition, the use of *P. pentosaceus* at 1×10^6 cfu/g of fresh forage gave the fermentation plants a higher lactic acid content than those of other bacteria.

Grass chemical composition to predict methane production from grazing sheep at high latitude

Q. Lardy[1,2], M. Hetta[1], M. Ramin[1] and V. Lind[2]
[1]Swedish University of Agricultural Sciences, Department of Animal Nutrition and Management, SLU, Umea, Sweden, [2]Norwegian Institute of Bioeconomy Research, Department of Grassland and Livestock, NIBIO, Steinkjer, Norway; marten.hetta@slu.se

Our study aimed to investigate the effects of seasonal changes in the chemical composition of temperate pasture and their impact on the potential enteric methane (CH_4) production in grazing sheep. The cultivated pasture was located 100 km south of the artic circle. The experimental periods represent contrasting vegetative conditions under a Norwegian conditions. The month of July is used to produce silage while the animals are in the mountain grazing and august is usually dedicated to grazing to finish the lambs. Twelve grass samples were collected with an automatic clipping machine, then dried, grinded and pooled together every two days during 10 days at each period. The five samples per period were analysed by NIRS for chemical composition and incubated for 48 hours using the automated *in vitro* gas production technique for estimating *in vivo* CH_4 production from ruminants. The average pasture height was 13.5±1.4 cm in July and 28.5±3.23 cm in August. The chemical composition (g/kg DM) of the pasture for July and August, respectively, was CP 104±4.1 and 150±5.7; NDF 572±20.5 and 501±5,5; sugar 209±20.1 and 183±10.9; OM 930±1.3 and 918±1.1. The estimated potential *in vivo* CH_4 production of the grass were significantly higher in July (34.6±2.19 ml CH_4/g DM) than in August (31.0±1.02 ml CH_4/g DM). Organic matter (3.19; 0.77), NDF (0.60; 0.59) and sugar content of the grass (0.48; 0.61) have a positive impact on the potential CH_4 production (coefficient; r^2). On the other hand, CP (-0.77; 0.59) had a negative corelation with the potential CH_4 production. We observed a difference in chemical composition and nutritive values of the grasses between the two periods. Our findings suggest that the chemical composition of the herbage in combination with recordings with the automated *in vitro* gas production technique may be useful tools to predict enteric CH_4 production from grazing sheep. However, estimates of herbage intake is needed to quantify the CH_4 emission from the grazing animals.

Session 31 Poster 20

Forage height of native grasslands as an indicator of spatial and temporal grazing management

P. Aparicio, I. Paparamborda and P. Soca
Universidad de la República, Faculty of Agronomy, Animal Production and Pastures, Av. E Garzón 780, 12900, Sayago, Montevideo, Uruguay; patriciaam@live.com

In the Pampa Biome, cow-calf systems on native grasslands are pivotal in giving competitiveness to the meat chain, providing ecosystem services and biodiversity conservation. Campos grasslands are composed of C4 species; therefore, there is a period of forage production and another utilization. We postulate that forage height in autumn, winter and spring, together with its intrafarm variability, is an indicator of spatial and temporal grazing management. Our study aimed to quantify the forage height of the native grasslands, forage allowance and the Body Condition Score (BCS) of cows in production systems of the Sierras del Este region in Uruguay. In 8 cow-calf systems during the period March 2021- March 2022, forage height (cm), forage mass (kg DM/ha), and forage allowance (kg DM/kg BW) was measured every 30 days, and BCS was quantified four times during the year. The study cases were selected in a paired manner considering grazing intensity, available resources, soil types and location. In this way, 2 groups were formed (G1 and G2), one of moderate and the other of low grazing intensity. With a matrix of dimensions 92×6 made up of monthly records of height (cm), stocking rate (kg PV/ha) and forage supply (kg DM/kg PV) for each of the 7 systems, an analysis of variance was carried out considering the categorical variable. Explanatory to the grazing intensity group to which it corresponded and to the other variables as a response. For body condition, a database with 570 records was used at 3 important moments and the same procedure was applied as for the other variables. In this analysis, G1 presents an average forage height of 5 +1.4 cm of deviation, a stocking rate of 327+45.0 kg PV/ha and 5+0.8 points of BCS. Instead, G2 had an average height of 6+1.4 cm, a load of 354+45.5 kg PV/ha and 5.2+0.7 BCS points. The forage height has a significant effect on the formation of groups with a *P*-value of 0.01, the same as the BCS. The stocking rate affects significantly with a *P*-value of 0.01. These differences between G1 and G2 would be indicating that the spatio-temporal management of the grazing intensity would improve the forage stock and the body condition of the herd.

Improve the efficiency of fattening grazing sheep on rangeland by supplementary feeding

Y. Zhang and M. Xu
China Agricultural University, 2 Yuan Ming Yuan Xilu, Beijing, 100193, China, P.R.; zhangyj@cau.edu.cn

We performed an experiment to investigate the efficiency improvement of fattening sheep grazing on open rangeland. This experiment was set as a factorial design with two factors (grazing intensity and supplement feeding level). Grazing intensity has two levels: 80% (heavy) and 40% (moderate) of grassland utilization. Supplementary feeding included no supplementation low supplementation (1% of body weight), and high supplementation (2% of body weight). This experiment was conducted at the Tenihe Ranch in Inner Mongolia, China (49°26′47″N, 120°9′30″E). Our results showed that low levels of supplementary feeding increased intake of *Leymus chinensis*. and *Artemisia tanacetifolia*, but decreased selectivity by high levels of supplementary feeding. Grazing intensity decreased the sheep intake of organic matter, crude protein, NDF and ADF. Dry matter intake was increased by supplementary feeding. For example, compared with no supplementation low and high supplementary feeding increased dry matter intake by 214.53 and 325.82 g/d, respectively. There was a significant positive correlation between the levels of concentrate supplementation and dry matter intake of sheep, with regression equations of $DMI = 0.743 \times Sup + 895.38$, $R^2=0.547$. Grazing intensity decreased dry matter intake by 135.23 g/d. Grassland livestock productivity increased with increasing grazing intensity and supplementary feeding levels. For instance, livestock productivity was increased by 46.89 and 107.57 kg/ha, respectively with low and high levels of supplementary feeding. Grazing intensity increased livestock productivity by 110.28 kg/ha. Feeding conversion rate reached 8.45±1.49, 10.73±2.18 and 13.97±3.28%, respectively, for the no, low and high feeding supplementary levels. Organic matter digestibility and crude protein digestibility was significantly increased with increasing levels of supplementary feeding, while NDF and ADF digestibility was decreased. Low and high supplementary feeding decreased methane emissions per unit of sheep weight gain by 26 and 48%, and methane emissions per unit of digestible organic matter intake by 24 and 34%, respectively. In conclusion, supplementary feeding can improve sheep fattening efficiency and decrease greenhouse gases emission.

The use of sheep of native breeds in the control of invasive plants

M. Pasternak, M. Puchała, J. Sikora and A. Kawęcka
National Research Institute of Animal Production, Department of Sheep and Goat Breeding, 2 Sarego Street, 31-047 Kraków, Poland; michal.puchala@iz.edu.pl

Sheep grazing can be used to care for valuable natural landscapes and help maintain the specific nature of ecosystems and biodiversity of wild species. In addition to the protective function, it can also be used to combat and limit the expansion of invasive plants. An invasive species is an alien species whose introduction causes a threat to local biodiversity and human management. There are many species of invasive plants in Poland, including Heracleum sosnowskyi and Solidago canadensis L. These plants have a negative impact on the environment by transforming natural habitats and displacing native species as a result of competition or limiting the food base. They also cause economic damage and may be dangerous for human and animal health. None of the chemical and mechanical methods of control is fully effective, therefore an attempt was made to use native sheep grazing for these purposes. The Department of Sheep and Goat Breeding of the National Research Institute of Animal Production in Balice participates in the implementation of the Norvegian project entitled 'The use of sheep grazing to reduce the occurrence of invasive plants'. The project is financed from the European Economic Area Financial Mechanism under the 'Environment, Energy and Climate Change' Program. The activities include sheep grazing in two locations: Ojców National Park (30 ewes of Olkuska sheep, reducing Canadian goldenrod) and Cieszyńskie Źródła Tufowe (30 ewes of Polish Pogórza sheep, reducing Heracleum sosnowskyi), both are the area of Natura 2000. Two rounds of grazing, approximately one month long, were carried out in May and July, separated by the period needed for the regrowth of the pasture sward. Sheep were on the pasture 24 hours a day. After the first year of the experiment, better results were found in the case of fighting goldenrod, which was very eagerly eaten by sheep. In the case of hogweed, the intended effect was not achieved, which was probably due to the late start of grazing. The stems were too hard and reluctantly eaten by the sheep. In the next year of the experiment, it is planned to modify the date of the first grazing and repeat the botanical inventory, which will allow to assess the effects obtained.

Session 32 — Theatre 1

Post-rumen health and its implications on health and performance

M.V. Sanz-Fernandez
Trouw Nutrition R&D, P.O. Box 299, 3800 AG Amersfoort, the Netherlands; victoria.sanz-fernandez@trouwnutrition.com

The gastrointestinal (GI) tract is a highly sensitive organ that can be negatively impacted by multiple factors during the productive cycle of cows. Diet management and characteristics, feed deprivation, local and systemic inflammation, and psychological and environmental stress, can all disrupt the barrier function. Further, several of these factors occur simultaneously at specific times, like around dry-off and calving. In healthy conditions, the GI barrier function maintains a regulated separation between the luminal content and the host. However, when disrupted, material from the lumen infiltrates through the enterocyte lining resulting in hyperpermeability or leaky gut. In this context, luminal content of the GI tract becomes a source of inflammatory stimuli and contributes to systemic immune activation. Inflammation constitutes an animal welfare issue, as well as an economic burden for producers, as it diverts nutrient and energy resources away from production toward the immune cells. Further, immune activation might play a role in metabolic maladaptation during the transition period, as an enhanced inflammatory status has been associated with an increased incidence of metabolic diseases. In cattle, rumen health has received the most attention in the context of subacute ruminal acidosis, however, it is increasingly evident that other sections of the GI tract can be negatively affected by similar factors. For instance, the lactation ration is associated with a large fraction of bypass starch that can induce hindgut acidosis. Despite acidosis developing similarly in both the rumen and the hindgut, the latter might be more vulnerable based on its structure, buffer capacity and immunity. It has been demonstrated that inducing hindgut acidosis independently of the rumen results in leaky gut with systemic consequences. Further, nutritional strategies intended to improve hindgut health improve performance, demonstrating that the energetic cost of poor hindgut health is diverting resources away from productivity.

Session 32 — Theatre 2

Impacts of inflammation and inflammatory diseases on reproduction in dairy cows

J.E.P. Santos
University of Florida, Animal Sciences, 2250 Shealy Drive, Gainesville, FL 32611, USA; jepsantos@ufl.edu

Calving and the onset of lactation are associated with increased risk of diseases, many of which are linked with inflammatory responses. Approximately 30 to 35% of the dairy cows develop at least one clinical disease in the first month of lactation, and postpartum diseases often result in local as well as systemic signs of inflammation. A conserved mechanism during disease is suppression of appetite, presumably to attenuate the inflammatory response; however, acute inflammatory diseases also alter the partition of nutrients with increased used by immune cells and the hepatic tissue for immune activation and the acute phase response. Collectively, inflammatory processes further exacerbate the negative nutrient balance dairy cows experience in early lactation, and the degree of negative nutrient balance is linked with the ability of the cow to resume oestrous cyclicity and become pregnant. Nevertheless, inflammation and inflammatory mediators have direct effects on oocyte competence, granulosa cell function, follicle development, the uterine environment, and on embryo and conceptus development that ultimately compromise the establishment and maintenance of pregnancy in dairy cattle. Pathogens that colonize the reproductive tract damage the superficial and glandular epithelia and increase the expression of inflammatory genes, and endometrial inflammation alters the composition of the uterine histotroph and disrupt conceptus development. Cows that develop uterine diseases have reduced fertilization, compromised early embryo development, impaired conceptus elongation, and increased risk of pregnancy loss. Inducing endometritis in cows promoted an acute phase response with increased concentrations of haptoglobin, suppressed appetite, and transiently reduced milk yield. Conceptuses from cows subjected to induced endometritis synthesized less interferon tau and had molecular signatures compatible with inflammation and reduced anabolic pathways needed for cell proliferation. Induced endometritis resulted in endometrial changes with reduced presence of immune cells required for tolerance to conceptus alloantigens. Diseases and the resulting inflammatory responses compromise nutrient balance and result in systemic and localized responses that compromise reproduction in dairy cows.

Session 32 Theatre 3

The role of mycotoxins in the rumen and intestinal health of dairy cows

A. Gallo
Università Cattolica del Sacro Cuore, Department of Animal Science, Food and Nutrition – DIANA, Via E. Parmense 84, 29122, Italy; antonio.gallo@unicatt.it

With an increased knowledge of the mechanism of action of mycotoxins, the concept that these substances are deleterious only for monogastric species is obsolete. Indeed, most mycotoxins can be converted into less toxic compounds by the rumen microflora of healthy animals. However, mycotoxin absorption and its conversion to more toxic metabolites, as well as their impact on rumen and intestinal functionality, immune response and subsequently animal welfare, reproductive function, and milk quality, during chronic exposure should not be neglected. Among the Fusarium-produced mycotoxins, the most studied are deoxynivalenol (DON), zearalenone (ZEN), and fumonisins from the B class (FB). Further, a lot of emerging mycotoxin can be detected in feed used in ruminant diets, including stored forage, and they are starting to be studied by scientific community. Anyway, it is remarkable that there is a paucity of *in vivo* researches, with a low number of studies on nutrient digestibility and impact of these substances on rumen and intestinal function. Moreover, most of the *in vitro* studies are related to the reproductive function or are restricted to rumen incubation. When evaluating the production performance, milk yield is used as an evaluated parameter, but its quality for cheese production is often overlooked. Here, the most recent findings regarding the adverse effects of these mycotoxins on dairy cattle are presented, with special attention to their performance and the effects of mycotoxin on rumen and intestinal functionality.

Session 32 Theatre 4

Absorption, metabolism and secretion of tocopherol (vitamin E) stereoisomers in dairy cows

S.K. Jensen and S. Lashkari
Aarhus University, Department of Animal and Veterinary Sciences, Blichers Alle 20, 8830 Tjele, Denmark; skj@anivet.au.dk

Tocopherols (Toc) in cattle feed are often a combination of synthetic Toc (all-rac-α-Toc) from vitamin supplements and natural tocopherols (RRR-α-Toc) and tocotrienols (T-3) from the feed. Among different stereoisomers of all-rac-α-Toc, RRR-α-Toc has the highest biological activity. Pre-absorption metabolism of Toc and T-3 in ruminants differs from monogastrics due to the extensive microbial fermentation in the anaerobic rumen. Two experiments were conducted. The first study examined ruminal metabolism and intestinal digestibility of Toc and T-3 in four ruminal and intestinal cannulated Danish Holstein cows. Cows were fed a TMR consisting of oat, fava beans, grass silage and a mineral/vitamin mixture with all-rac-α-Toc acetate. This gave a daily intake of 1,862 mg total α-Toc (819 mg natural α-Toc and 1,043 mg all-rac-α-Toc), 63 mg α-T-3 and 224 mg γ-Toc. Total α-Toc (102, mg/day) and synthetic α-Toc (279 mg/day) was degraded in the rumen, in contrast RRR-α-Toc was formed in the rumen (221 mg/day). The small intestinal and feed-ileum digestibility ranked in the following order: RRR > synthetic 2R > 2S-α-Toc. Results showed the first evidence for RRR-α-Toc formation under anaerobic condition in rumen. In addition, synthetic α-Toc stereoisomers, γ-Toc and α-T-3 were degraded in the rumen, and small intestinal absorption seems to favour of RRR-α-Toc absorption. The second experiment studied the pharmacokinetics of stereoisomers of all-rac-α-Toc and investigated the discrimination and distribution of α-Toc stereoisomers in plasma and milk as well as quantitative secretion into milk with lactating Holstein cows after a single intramuscular injection of 2.50 g of all-rac-α-Toc acetate. The highest half-life (2.92/h) and lowest elimination rate (0.36/h) were found for RRR-α-Toc. Highest maximal daily increase (8.36 mg/day) and accumulated secretion (50.8 mg) were observed for milk RRR-α-Toc. The majority of the 2S stereoisomers was found in the liver 36 h post injection, while the 2R stereoisomers were more equally distributed between liver and plasma. The present findings showed a clear discrimination between RRR-α-Toc, synthetic 2R stereoisomers and 2S stereoisomers in milk and plasma of dairy cows.

Session 32 Theatre 5

The role of rumen health (acetate) on milk fat synthesis and dietary strategies to increase milk fat

K.J. Harvatine and C. Matamoros
Penn State University, Department of Animal Science, University Park, PA 16802, USA; kjh182@psu.edu

Major advances have been made over the past 25 years in our understanding of milk fat synthesis. Nutritionally, milk fat depression (MFD) can decrease milk fat by up to 50% and garnered most of the research interest for years. Bioactive intermediates of ruminal fatty acid biohydrogenation were identified as the causative factor and provided insight into the regulation of mammary lipogenesis and led to targeted approaches to manage milk fat. Large changes in milk fat are undoubtedly due to these bioactives, but recent work has shown that other nutritional factors can have smallerimpacts on milk fat and are economically important. Milk fat synthesis requires acetate as a source of carbon and reducing equivalents. We initially ruminally infused acetate to investigate the effect of acetate spared during MFD on adipose tissue metabolism and observed an unexpected increase in milk fat yield. We then conducted a dose titration experiment and observed a maximal response of 220 g of milk fat with a ruminal infusion of 600 g of acetate. The increase in milk fat has also been observed when feeding sodium acetate in a total mixed ration. The response has been consistent, and further experiments have found no interaction with dietary starch, fibre digestibility, or unsaturated fatty acids despite the potential for these factors to change rumen fermentation. Acetate also did not interact with *trans*-10, *cis*-12 CLA, indicating that the mechanism is likely independent of bioactive FA that cause MFD. It is still unclear if acetate increases milk fat simply by increasing substrate to the mammary gland or if it increases metabolic capacity by changing cellular physiology. Although sodium acetate supplementation is not practical, this work highlights the importance of fibre digestibility and maintaining optimal rumen function. Genetic selection is increasing the potential for milk fat synthesis and acetate supply will likely be increasingly important to meet genetic potential. Thus, maximizing milk fat yield requires not just minimizing MFD, but also optimizing acetate supply and other dietary and management factors.

Session 33 Theatre 1

EUNetHorse-European network to improve the resilience and the performance of equine farms

M. Addes
French Horse and Riding Institute, La Jumenterie du Pin, 61310 Exmes, France; marlene.addes@ifce.fr

The overall objective of EUNetHorse is to establish an active multi-stakeholder network in France, Portugal, Spain, Belgium, Germany, Romania, Poland, and Finland, in order to increase the resilience and performance of their equine farms to face environmental, social, health, economic or political crises by widely disseminating practices, tools and solutions that improve (1) their resilience and socio-economic performance, (2) animal welfare and health on farms, and (3) the environmental sustainability of the sector. To achieve this, EUNetHorse will: (1) increase the flow of practical information between farmers in these 8 countries in a geographically balanced way and taking into account the differences between territories by structuring National- Horse AKISs in each country; (2) collect and assess grassroots solutions and practices, disseminate best practices and solutions on the three thematic areas mentioned above replying to specific needs; (3) achieve a greater user acceptance of the collected solutions and best practices, through cross-fertilisation between all actors of the sector (horse breeders, advisors, trainers, technical experts, scientists, policy makers, sector representatives, etc.) and all levels (local, regional, national and European), during exchange activities allowing peer-to-peer learning, such as workshops, demonstration days, training and cross visits; (4) maintain the practical knowledge in the long term – beyond the project period – by sharing the full set of project results on the EUFarmbook platform accessible to all, by training advisors and trainers during the project on these topics, by setting up a sustainable network of trainers and advisors who can continue to train equine farmers and disseminate these solutions using the training kit made available through their activities with equine farmers.

Session 33 Theatre 2

Robustness of equine business structures and global coherence

S. Boyer
IDELE, Route de troche, 19230 Arnac Pompadour, France; sophie.boyer-lafaurie@idele.fr

Whatever the hazards encountered by farm operators, whether external to the farm (health crisis) or internal to the operation (health problems), some always manage to bounce back and maintain good economic results. We advance the idea that beyond the techno-economic results and the skills of the farm operator, the robustness of the operation lies in the human dimension. It is the interpersonal skills, the ability to listen and to challenge oneself, to have the courage to make those changes which will enable a farm operator to develop his/her operation in such a way as it remains consistent with his/her values and resources. Our study combined an analysis of the technical and economic data of 87 equine business structures, which had been monitored within the Equine Network, and ten surveys, referred to as 'trajectory analysis', exploring the way farm operators had implemented the sociological dimension in the running of their operations. The ten qualitative surveys were conducted in France: four in riding schools, five in livery stables and one in a horse farm. The aim of these semi-structured interviews was to identify the levers used to overcome any difficulties which may have been encountered. Six main types of levers were identified: knowledge, know-how, interpersonal skills, strategic elements, available resources, and finally professional and personal objectives and values. We will present these results in the form of an allegory. The levers are represented as a living and ever-growing plant: the flowers symbolizing knowledge; the leaves, the means; the roots, the resources; the stem, the strategy and the objectives of the operator, all of which is nourished by the soil which represents the values. We will present a synthetic analysis of the trajectories analysed, highlighting the importance of the ability to react to hazards, to grow one's business in such a way as it is always in line with one's aspirations. These initial results provide us with a wealth of information and would appear to be particularly useful, for future equine project leaders and for established farm operators undergoing a transitional period and who need to give new-found meaning to their project. Before presenting these results, a short overview will be given of the French equine systems covered in this study.

Session 33 Theatre 3

The equine network, a tool for horse professionals, advisors and teachers

G. Mathieu
Institut de l'élevage, 149 rue de Bercy, 75012 PARIS, France; guillaume.mathieu@idele.fr

The Equine Network is a French partnership between the livestock Institute, the Chambers of Agriculture, the French Institute for Horse and Riding and the Federation of Horse Councils. It is based on the INOSYS livestock network method. The Equine Network provides technical and economics guides and tools for advice and planning. All of these tools are intended for experts, advisors, horse professionals, institutional decision-makers, teachers and students. In total, 130 organizations are monitored, representing the different equine sectors: riding schools, equestrian tourism centers, boarding stables, trotter equine stables, draught horse breeders, horse drafting providers. To that, additional organizations on specific and least well-known activities such as mare's milk, shared stables or innovative stables are monitored. In the 130 organizations, technical, economic, environmental and social data are collected. The collected data is analysed and used to produce various supports with dashboards, technical and economic summaries, typical cases, and books. The network aim is to understand the operating logic of each of the farming systems that are monitored. These results can be used in the field to help installing a new horse professional, to have references to compare economic results, to improve horse professional practices and to obtain economic data that decision makers can rely on.

Session 33 Theatre 4

Does the diversity of the French territories impact farms keeping equines?

J. Veslot and G. Bigot
INRAE, UMR Territoires, 9 avenue Blaise Pascal, 63170 Aubière, France; jacques.veslot@inrae.fr

In spite of professional and institutional databases on French horse activities, it remains difficult to have a national panorama of this industry, and its specificities according to the diversity of territories. To answer this question, we analysed the French Farm Structural Survey (FSS), where more than 54,000 farms with equines in 2010 kept almost half of the national herd. This study resulted in a typology of equine herds in farms, using herd size, type of breeds (warmblood, coldblood and donkeys) and the presence of mares, and resulting in 18 groups (8 'specialized' structures, and 10 groups with very small herds). A clustering of the 90 French departments (areas equivalent to European NUTS3) was carried out using numbers of farms in these 18 groups, leading to 9 types of areas with different local characteristics of horse keeping. The density of farms and predominant equine types raised, differed largely between regions according to altitude, the importance of grassland in agricultural area and the presence of large cities (more than 150,000 inhabitants). For example, 'Manche' (in Normandy) and 'Pyrénées Atlantiques' presented both high densities of farms with equines, but differed in their predominant type of breed: warmblood in Manche and coldblood in Pyrénées Atlantiques. The analysis also highlighted noticeable patterns: most areas in uplands (Pyrénées, Massif Central, Vosges) with a medium density of farms and mainly coldblood horses; and clusters of areas in lowlands, mainly located in the centre and in the South-East of France, with low densities of farms with equines. This classification of territories based on the diversity of farms keeping equines could be interesting for future studies on the territorial impacts of changes in the horse industry.

Session 33 Theatre 5

Equestrian practitioners: essential services to support self-organization

C. Eslan[1,2,3], C. Vial[2,3] and S. Costa[3]
[1]FFE, Parc fédéral, 41600 Lamotte-Beuvron, France, [2]IFCE, Pôle Développement Innovation et Recherche, La jumenterie, 61310 Exmes, France, [3]INRAE, MoISA, Place Viala, 34000, France; celine.vial@ifce.fr

Sporting practices are increasingly breaking away from the federal model, leading to a shift from clubs to more independent organizations. Consequently, since the 1990s, self-organized people have been steadily turning towards private service providers for support. A wide range of services is on offer: initiation, teaching, rental, training or even advice. The French equestrian sector is mainly organized around professional structures such as riding schools or livery stables. Nevertheless, it is facing a decrease in the number of federal memberships and an increase in independent practices. Consequently, the French equine industry is interested in people who organize themselves to take care of their equines outside professional structures. They are often located on the outskirts of cities or in rural areas even though the service offer is a fortiori closer to densely populated areas. Based on these observations, our research analyses the service needs of self-organized equestrian users for the practice of their sport and the daily management of their horses. A quantitative online survey conducted in 2021 among self-organized equestrian users shows that needs for services differ according to the type of organization. Nevertheless, more than a quarter of the respondents have difficulties finding service providers adapted to their needs (19%), and advice (9%). Advice or training on agricultural issues, access to large machinery such as a rotary slasher or to products for the maintenance of land seem essential for 17% of the self-organized people. Still, they do not have these services available. 41% of them find arrangements with neighbouring farmers. For the practice of their sport, self-organized equestrian users do not find coaches who meet their expectations. Respondents who value more than everything else the luxury of having their horses close to home cannot entertain the idea of renewing membership in a club-like structure. These results show the importance of adapting the offer of services to the needs of self-organized equestrian users. Moreover, the development of this market can be instrumental in the sustainable development of rural and suburban areas.

Agricultural animal traction in France: characterization of the practice and its users

M.M. Miara[1], M.G. Gafsi[2] and P.B. Boudes[3]
[1]INET, 17, cours Xavier Arnozan, 33000 Bordeaux, France, [2]LISST, 5 All. Antonio Machado, 31100 Toulouse, France, [3]ESO, 65 Rue de Saint-Brieuc, 35000 Rennes, France; maurice.miara@gmail.com

Faced with the challenges that emerge in our modern societies, many injunctions to achieve the agro-ecological transition and the energy transition are being raised. Animal traction seems to meet these two objectives and is benefiting from an emerging dynamic in France with an increase in the use of working equids, whether in communities, in the forest, or in agriculture. However, despite its growth, this practice remains little studied and its characterization remains unclear. From this perspective, our study aims to understand what is animal traction in agriculture in France, who are its users, what are its networks and how it is mobilized in practice. To answer our questions, we used an interview methodology to question the practice in the field. Thus, 32 semi-directive surveys were conducted with users of animal traction in France during the year 2022. 16 market gardeners, 10 wine growers, 4 nurserymen and 2 cereal growers were interviewed. Our paper will detail the profile of these users and their farms through a sociological analysis. Thus, our analysis enabled us to understand the life paths of animal traction users, as well as the motivations that led them to this practice. Our study enabled us to characterize the practice and the farms that use it by identifying farmer typologies and associated farm management strategies. Thus, the farms are oriented towards autonomy and short circuits, on relatively small surfaces and a high added value per hectare. Motivations for using animal traction vary according to the respondents. We find passion for the horse, political choice, or technical interest as the main modality. There is a high proportion of neo-rural users, but a significant proportion have a family link to agriculture and horses. Although heterogeneous, animal traction is part of a redesign of agriculture, a societal paradigm shift, where the animal has a central place in the peasant project carried out by these farmers.

Effects of milking interval and frequency on milk yield in the mare

J. Auclair-Ronzaud, M. Bouchet, L. Laschon, E. Lambolez and L. Wimel
IFCE – Plateau Technique de Chamberet, 1 Impasse des Haras, 19370 Chamberet, France; juliette.auclair-ronzaud@ifce.fr

Mare milk production is expending in France but practices vary widely from one farm to another, especially due to the lack of technical information provided to the producers. This study aims to explore the effect of milking interval and frequency on milk yield. Twenty-three mares were studied with a mean age of 9.6±4.2 years old. Mares and foals were kept in group in pasture and were weighed every two weeks. Two groups were studied, a group of 13 mares milked once (1M) with 5 primiparous and 8 multiparous, and a group of 10 mares milked twice (2M) with 3 primiparous and 7 multiparous. Milking protocol was performed at week 5, 9 and 13 of lactation. These weeks, mares were milked during 5 consecutive days according to the following procedure. At 8:30, foals were equipped with muzzle to prevent them from suckling. After one hour, 2M mares were milked for the first time. Then, two hours from the muzzling of the foals, 2M mares were milked a second time, and 1M mares were milked. Milkings were performed using a manual milker, starting by the right udder. For each udder, collected milked was weighed. The effect of mare parity, age and weight, foal gender and weight, day of milking (from 1 to 5), week of milking and group on milk yield was studied using a linear mixed model with individual as random effect. No effect of mare weight nor parity, foal gender and day of milking was shown. Lack of significance regarding parity, however, has to be taken with caution as few primiparous mares were studied. Milk yield was significantly influenced by the week of milking, mares producing more milk at the 5th week (326.70±140.63 g) compared to the 9th (309.76±141.36 g) and 13th week (290.59±156.77 g) and more during the 9th week than during the 13th. In addition, foal weight is positively correlated with milk yield, mares caring for larger foals producing more milk. Regarding group, 2M mares produced significantly more milk than 1M mares (377.59±148.25 g and 255.52±122.11 g, respectively). Thus, one hour interval between two milking is sufficient to increase daily milk production. This result highlight the importance of milking interval on milk yield and more research on this theme would benefit the equid milk industry.

Session 33 Theatre 8

Mineral profile of pasture-based mare milk from Basque Mountain horse breed: effect of lactation

A. Blanco-Doval, L.J.R. Barron and N. Aldai
University of the Basque Country (UPV/EHU), Pharmacy and Food Science, Paseo de la Universidad 7, 01006 Vitoria-Gasteiz, Spain; ana.blancod@ehu.eus

Minerals are essential elements that play a key role in human physiology and need to be acquired through diet. Milk and dairy products are a good source of some minerals, e.g. calcium and phosphorous, although the mineral composition of milk differs among mammal species. In this respect, mare milk is very similar to human milk, both with lower overall mineral contents than ruminant milk. However, studies on mare milk minerals are particularly scarce. In this study, the content of 7 macrominerals (Ca, K, P, S, Na, Mg and Cl) and 4 trace elements (Zn, Fe, Cu and Mn) in mare milk from Basque Mountain Horse breed was analysed. This is an autochthonous equine breed from northern Spain and its milk has never been studied before. Furthermore, the effect of lactation and farm management on the milk mineral composition was studied. Milk samples (n=310) were collected all along the lactation period (six months) from 18 mares belonging to three different commercial farms with different management system. Inductively coupled plasma mass spectrometry (ICP-MS) was used for mineral analysis, and the ANOVA Linear Mixed Model was used for the statistical analysis. Major macrominerals and trace elements in mare milk were Ca, K and P, and Zn and Fe, respectively. Mn was only quantifiable in 11% of the samples. In comparison to milk from other horse breeds, milk from Basque Mountain Horse breed presented similar Ca, K, P, Na, Mg, Zn and Cu contents, but lower Cl and S and higher Fe contents. Lactation stage significantly affected the content of all minerals: Ca, K, P, S and Mg decreased along lactation, whereas Na, Zn and Cu contents increased and Cl and Fe contents fluctuated. Significant differences were also found in Ca, P, S and Na contents between milk from mares with long and short grazing period, probably due to grazing activity in agreement with previous findings in cow milk. These results substantiate the quality of mare milk from Basque Mountain Horse breed from a human consumption perspective, contributing to the valorisation of the food produced (milk), the breed and extensive equine grazing systems.

Session 33 Theatre 9

Sustainable utilisation of horse manure

M. Meyer[1], C. Eiberger[2], T. Schilling[1], D. Winter[3] and L.E. Hoelzle[1]
[1]University of Hohenheim, Institute of Animal Science, Department Livestock Infectiology and Environmental Hygiene, Garbenstraße 30, 70593 Stuttgart, Germany, [2]Marbach State Stud, Gestütshof 1, 72532 Gomadingen, Germany, [3]University of Applied Science Nuertingen-Geislingen, Neckarsteige 6-10, 72622 Nürtingen, Germany; madeline.meyer@uni-hohenheim.de

Horse manure is a challenge for everyone who has responsibility for at least one horse and is therefore in charge of proper storage and utilisation of produced manure. Increasingly, horse farms, especially in urban areas and farms with no or only small agricultural land, have serious problems with the utilisation of horse manure. Experts assume that the possibilities for manure utilisation decrease and the associated costs will even increase in the future. In summary, approaches associated with storage and especially sustainable utilisation of horse manure are currently of enormous importance and highly demanded. In this research project, light is shed on different utilisation options of horse manure with special regard to storage capacity, nutrient availability, labour management as well as hygiene parameters. Advantages and disadvantages of the respective possibilities are determined to deliver practical solutions for a sustainable utilisation of horse manure. The project has four different research goals. The first focus involves the analysis of the composting process of horse manure. Within the scope, four differently treated experimental muck heaps remain on the manure plate for six weeks, two heaps are turned over after three weeks. The muck heaps are spiked with temperature loggers as well as parasite and *Salmonella* samples. Furthermore, nutrients samples are taken at the beginning and at the end of the research trail. In the second subproject, the impact of the usage of a combi-mulcher and rotting accelerator is tested in a field and an exact trail. Furthermore, the third focus deals with the influence of the biogas process on the hygienic status of the digestate. Various hygiene samples are examined in the laboratory for this purpose. The last research field analyses the use of a rotting accelerator in loose housing and its effect on the composting process of horse manure in the stable.

Session 33 Theatre 10

Use of green waste compost as an alternative bedding to straw in horse husbandry

H. Unseld[1], D. Winter[1] and M. Meyer[2]
[1]University of Applied Sciences Nürtingen-Geislingen, Applied agricultural research, Neckarsteige 6-10, 72622 Nürtingen, Germany, [2]Principal and State Stud Marbach, Gestütshof 1, 72532 Gomadingen, Germany; harald.unseld@hfwu.de

Horse care is associated with high demands for owners and boarding stables. A demand-oriented and high-quality supply of feed and bedding materials also represents a high cost item for horse-keeping operations. Animal welfare means that high-quality feed must not be skimped on. Cost savings and working time represent important challenges. With regard to animal welfare, the bedding used must fulfil multiple functions. For example, the materials must be hygienic, dry and low in dust, and have good binding of urine and noxious gases. In addition, they should be easy to muck out to save labour time, produce a small volume of manure to limit manure storage space, and be inexpensive to purchase. In addition, they should provide good lying comfort for the horses. This paper addresses the question if composted green manure meets these requirements. Green compost and wheat straw were investigated for their water absorption capacities in laboratory tests. In addition, practical tests were conducted to determine the working times for manure removal and the necessary material requirements, for re-spreading various bedding materials. The cost rates were shown and compared in each case. The stable air factors ammonia and particulate matter, were measured for the different bedding variants. In addition, the lying behaviour of horses was investigated. The results showed, that about 50% of the bedding costs could be saved by using green waste compost, compared to straw bedding. In particular, the material costs of green waste compost are, on an annual average, 80% lower than those of wheat straw. The manure quantity of green compost litter is reduced by 50% compared to straw. The ammonia load in the trial is lower than with straw, but the dust load is higher than with straw. In summary, it can be said that green waste compost has an enormous savings potential in terms of material and labour costs compared to straw. Green waste compost at a bedding height of 20 cm provides a good indoor climate with non-perceivable ammonia values that are far below the official limits.

Session 33 Theatre 11

Grazing management on practical horse farms

C. Siede[1], M. Komainda[1], B. Tonn[2], S.C.M. Wolter[1], A. Schmitz[1] and J. Isselstein[1,3]
[1]University of Goettingen, Deparment of Crop Science, Von-Siebold-Str. 8, 37075 Göttingen, Germany, [2]FiBL Switzerland, Department of Livestock Science, Ackerstr. 113, 5070 Frick, Switzerland, [3]University of Goettingen, Centre for Biodiversity and Sustainable Land Use, Büsgenweg 1, 37077 Göttingen, Germany; caroline.siede@uni-goettingen.de

Grassland offers horses the opportunity to act on their natural foraging and grazing behaviour and allows them movement and social contacts which all enhances their wellbeing. However, little is known about pasture management and pasture-based nutrition of horses on farms. Therefore, we investigated the actual grazing management and to what extent aboveground herbage (AGH, kg/ha) changes throughout the grazing season on six practical horse farms around Göttingen, Germany, from May till October 2019. Compressed sward height (CSH) was measured monthly on all pastures that were used for horse grazing. On four focus pastures per farm in two short-grazed (CSH<7.2) and two tall-rejected (CSH≥7.2) grass sward areas AGH was cut close to ground level of known CSH to develop a linear model for prediction of standing AGH from CSH values. The proportions of each area were calculated per pasture and month. The farmers had to fill out a grazing diary on a daily basis to obtain a measure of stocking intensity, as expressed in terms of livestock unit grazing days (LUGD) per month and pasture. There was a significant interaction between farm×month×area on the AGH (P=0.002). As the proportion of tall grass sward areas declined progressively over the season, short grass sward areas increased correspondingly. This coincided with an increase of AGH in short grass sward areas per ha of horse pasture. In addition, we found no correlation between AGH per pasture and the LUGD. LUGD were relatively constant on almost all farms throughout the grazing season, whereas only one farm showed a pattern of LUGD adapted to the grass growth. On most practical horse farms, grassland is rather used for movement and social contacts than nutrition. With adjusting the stocking, grassland could provide sufficient herbage and thus reduce inputs of supplemental feeding. There is a need to inform horse famers about improved grassland management.

Session 33 Theatre 12

Use of digital technologies in horse husbandry to increase animal welfare and health

M. Pfeiffer[1], D. Winter[1], U. Dickhoefer[2] and L.T. Speidel[1]
[1]Nuertingen-Geislingen University, Institute of Applied Agricultural Research (IAAR), Neckarsteige 6-10, 72622 Nuertingen, Germany, [2]Kiel University, Institute of Animal Nutrition and Physiology, Hermann-Rodewald-Strasse 9, 24118 Kiel, Germany; linda.speidel@hfwu.de

Even if horses have been domesticated, their needs concerning their living conditions have hardly changed. However, under today's husbandry conditions, the daily routine of horses partly deviates from the horses' natural time budget, which can have negative psychological and physical effects on the horses. The recording of long-term activities serves to control the horses' time budget. Furthermore, it can also serve as a basis for the evaluation of management, feeding and husbandry decisions. The collection of animal-related indicators by humans is very time-consuming, subjective and reflects a snapshot. Digital technologies can continuously and objectively collect indicators and ensure their documentation. The aim of the present work was to show different ways in which horse farms can use digital technologies to ensure animal welfare and health and to record activities in the long term. The studied indicators include feeding, resting and exercise behaviour. These were recorded using a video-based system supported by artificial intelligence. In addition, horses' reactions to the use of digital technologies were recorded when the feeding process was changed from manual to automated for a total of six horses in two farms. During manual concentrate feeding, statistically significant higher heart rates of horses were found compared to automated feeding. For roughage feeding, horse heart rates were slightly higher with automated feeding than with manual roughage feeding. Digital technologies can support farm management in the continuous and objective recording and documentation of indicators. In addition, animal welfare can be improved, especially in the area of feeding, through the possibility of automated feed presentation. Possible challenges related to the digitalisation in horse husbandry must be assessed together with the farm managers in the future.

Session 33 Poster 13

Working time requirements of work processes in the individual keeping of horses

L.T. Speidel[1], D. Winter[1] and U. Dickhoefer[2]
[1]Nuertingen-Geislingen University, Institute of Applied Agricultural Research (IAAR), Neckarsteige 6-10, 72622 Nuertingen, Germany, [2]Kiel University, Institute of Animal Nutrition and Physiology, Hermann-Rodewald-Strasse 9, 24118 Kiel, Germany; linda.speidel@hfwu.de

Working time requirements in horse husbandry depend on the type of husbandry and stabling, the associated mechanisation as well as the arrangement of the buildings and the services offered. Current data from the literature can be used for personnel planning and the calculation of labour costs on horse farms. However, these data are more than ten years old. Therefore, current data on essential areas of activity in horse husbandry were collected in the present work. In general, the working time per horse and day varies between 1-10 minutes per horse for the individual activities. Most of the time is spent on manure and pasture care, the least time is needed for the manual feeding of concentrated feed. The use of technology saves time, especially for manure removal, and should be considered in operational planning. The required working times vary considerably for the different activities and depend on many factors, such as the experience and motivation of the employees, the distances travelled, and the technology used. This should always be considered when calculating labour costs and the associated pension prices. A possible addition to this study is the comparison with group husbandry of horses and automated work processes.

Session 33 Poster 14

Equine traction in viticulture: what about the workload of the horse-driver pair?

B. Pasquiet[1], N. Delepouve[1] and C. Bénézet[2]
[1]Institut français du cheval et de l'équitation, Plateau technique de Saumur, Avenue du Cadre Noir, BP207, 49411 Saumur, France, [2]Institut français du cheval et de l'équitation, Plateau technique d'Uzès, Mas des tailles, 30700 Uzès, France; benoit.pasquiet@ifce.fr

Animal welfare is a growing social concern. In this context, the mobilization of equids for the realization of various urban, agricultural or even natural sensitive works can raise questions not only concerning their daily management (feeding, housing, care) but also concerning their working conditions. A return to the use of equine traction in vineyards is currently observed, mainly for soil maintenance. In this context, we are interested in the workload of the horse-driver pair, with the Caract-Equivigne project. Our goal is to provide first references on the effort represented by this work according to the operation carried out, the type of worked plot and the tool used. For this study, 14 equine-driver pairs, working for vineyards spread over different regions, have been selected. Each pair will be followed for two years. For each vineyard, we observe two contrasting plots in terms of effort (variation in slope, soil type, etc.). Two cultivation operations are measured each year (ridging and plowing the ridges). We analyse the locomotion of the horse with an accelerometer. We monitored both horse and driver heart rate. The traction effort is determined by a dynamometer placed between the hitch lines and the collar. In addition, the horse is filmed throughout the measurement session: the videos are then used to analyse its behaviour. For the traction and cardiac data, different effort thresholds have been considered and allow to estimate the time spent above and below each threshold. These data will allow to characterize the effort levels provided by the equids in different soil maintenance situations and will be compared to references in other work situations. The references built at the end of the project could be used to create educational resources for driver training.

Session 33 Poster 15

Physiological and behavioural responses of donkeys to effort and discomfort

N. Seguin[1], C. Bonnin[1], A. Ruet[2] and S. Biau[2]
[1]INET, 17 Cours Arnozan, 33000 Bordeaux, France, [2]IFCE, Avenue de l'ENE BP207, 49411 Saumur, France; noseguin@hotmail.fr

In a preliminary study, the physiological solicitation and behaviours expressed by donkeys were described in different work situations. The objective of this new study was to analyse the effects of two loads being pulled, with or without physical discomfort, on cardiac physiology and behaviour of donkeys in a standardised context. Twelve donkeys (6±4 years) were tested with no load (F0), with a load being pulled of 10±2% (F10) and 20±2% (F20) of weight, with and without discomfort (C). This last consisted of a piece of rope placed unilaterally between the collar and the neck at the point of traction. Modalities order was randomised. Mean heart rates (HR) and occurrences of 10 behaviours were recorded for each modality. For behavioural analyses, four Z-scores were calculated: Z-score 'locomotion' (spontaneous stops, gait changes, trampling, lowering of hips), Z-score 'mobility' (head and ear movements, chewing, snorting), Z-score 'tail movement', Z-score 'defecation'. Non-parametric Friedman's tests and post-hoc tests with Bonferroni correction were performed, with a significance level ≤0.05. HRs were significantly different (X^2=13.4, P<0.01) between F0 (101,6±10,8 bpm) and F20+C (131,7±29,0 bpm; P<0,01). The Z-score 'locomotion' was significantly different (X^2=21.4, P<0.001) between F0 (-0,36±0,33) and F20 (0,51±0,40; P<0,01), and F10 (-0,25±0,33) and F20 (P<0,01). The Z-score 'mobility' was significantly different (X^2=13.9, P<0.01) between F0 (0,60±0,74) and F20 (-0,22±0,41; P<0,05), and F0 and F20+C (-0,40±0,40; P<0,05). The other two Z-scores did not show significant differences between modalities. The presence of discomfort did not have a significant effect on HR and Z-scores. These results show an increase in HR, indicating increased physiological effort, with the presence of loads being pulled. The donkeys also appeared to express more difficulty in moving and showed an overall decrease in forehand movements with increasing loads. Physical discomfort was not expressed in this study, either because the rope did not induced discomfort or the effort masked the discomfort. This standardised test with a sample of donkeys allowed to identify physiological and behavioural indicators enabling the working effort to be monitored.

Session 33 Poster 16

Offering horse meat in restaurants to develop the demand

C. Vial[1,2], M. Sebbane[1,3] and A. Lamy[1,3]
[1]MoISA, Univ Montpellier, CIHEAM-IAMM, CIRAD, INRAE, Institut Agro, IRD, Montpellier, France, [2]IFCE, pôle développement innovation et recherche, 61310 Exmes, France, [3]Centre de recherche, Institut Paul Bocuse, 69131 Ecully, France; celine.vial@inrae.fr

Although historically the consumption of horse meat in France has never been very high, it is currently declining significantly. Out of all French people who do not eat horse meat, 15% would be ready to do so if they had the opportunity, mainly with friends and relatives, or in restaurants. In this context, our research focuses on the obstacles and levers regarding the development of the horse meat offer in commercial catering. Qualitative interviews were conducted with 12 chefs, questioning their relationship to meat in general and their representations and knowledge about horse meat. According to the chefs, horse meat is rarely on the menu in restaurants because it would not appeal to many consumers, its consumption poses a problem of acceptability, its price is high, there is a lack of recipes and no teaching is given on this subject to chefs-in-training. Chefs also lack knowledge about the horse meat sector. Nevertheless, they present characteristics suggesting that they could be interested in this product: meat occupies an important place in general for them, they are open to innovation and new produce, they are interested in healthy menus, animal welfare, the environment, and local produce. Concerning the produce itself, they mention its advantages: colour, taste, adaptation to certain types of restaurants, and they are curious about the French breeding sector of draft horses for butchery. Finally, three profiles stand out among the chefs interviewed: the connoisseurs, those who are pragmatic, and those who are reluctant. Thus, first thoughts are emerging to develop the horse meat market in France. On the one hand, they rely on enhancing the nutritional, organoleptic and environmental qualities of this produce, which can meet the current challenges of food sustainability. On the other hand, it is necessary to increase the visibility and availability of the offer, for consumption at home, but also in commercial catering that appears to be relevant to familiarize new potential consumers with this produce. This involves overcoming the lack of knowledge of chefs regarding horse meat and the lack of a recipe index.

Session 33 Poster 17

Relationship between average daily gain and an estimation of consumed grass

L. Wimel, C. Lesoudard and J. Auclair-Ronzaud
IFCE – Plateau Technique de Chamberet, 1 Impasse des Haras, 19370 Chamberet, France; laurence.wimel@ifce.fr

Grassland, as base of equine alimentation, is not used consistently by owners as the resource management is difficult. Thus, to broaden the use of grassland it is important to give owners information regarding horse performances when they are fed exclusively with grass. Relationship between average daily gain (ADG) and disappeared grass quantity (DGQ; expressed as kilograms of dry matter of grass consumed by horse per day) was studied on 3 grazing seasons (2018; 2019 and 2020). Horses were kept in herds and grassland was managed with rotational grazing. Grazing seasons started between the begging of April and mid April and ended between begging of October and begging of November. In total 188 horses were studied (80 in 2018; 111 in 2019 and 104 in 2020) some being studied for several years. DGQ was estimated using grass height measured before and after grazing (from 75 to 150 points depending on the plot size). Dry matter was estimated by sampling grass on 0.5m^2 surface. Grass height was measured before and after cutting with a mini-mower. Cut grass was then weight after and before drying at 60 °C for 72 hours. Relationship between ADG and DGQ was studied using a generalized mixed model with ADG as the variable to explain, DGQ, age of the horses, year, month of grazing and interactions (DGQ:age, DGQ:year and DGQ:month) as fixed variables, and individual as random variable. Surprisingly, a significant negative correlation appeared between ADG and DGQ, which might reflect the impossibility for horses already reaching a high weight to gain more. Moreover, needs increasing with weight, heavier horses will have to ingest more grass to cover the needs and not inducing weight gain. A significant effect of month of grazing on ADG was also demonstrated, being a reflect of grass growth which depends on environmental factors. Moreover, ADG was lowered by age of the horses which might be explained by ingestion capacity. The latter is influenced by body weight and growing horses are lighter than adults. As a relationship was established between ADG and DGQ, the later might be a good indicator of the former, even though more studies are need as several factors seem to influence the correlation.

Comparison of flying and soil invertebrates' biodiversity on meadows with or without horse grazing

G. Goudet[1], F. Reigner[2] and M. Beltramo[1]
[1]INRAE, PRC, 37380 Nouzilly, France, [2]INRAE, PAO, 37380 Nouzilly, France; ghylene.goudet@inrae.fr

By grazing preferentially some spots, horses create patches of short grass and tall vegetation in which they concentrate their dungs. This leads to habitat heterogeneity that could increase grassland animal biodiversity. This pilot study aimed to test a method to compare, using non-lethal protocols, invertebrates' biodiversity on meadows undergoing or not horse grazing. This 2-years study was performed in our experimental farm facilities using one-hectare paddocks with meadows 5-years-old and similar bordering environment. Two paddocks were grazed from April to November by 6 pony mares each and 2 paddocks were not grazed. For bees monitoring, 2 trap nests of 32 tubes each were placed at paddocks' edge. Nest occupancy was monitored by counting sealed tubes from March to October. Butterflies were monitored once per month from May to September by recording all butterflies in a 5×5×5 m cube around the observer during a 10-min long transect. To monitor soil invertebrates, 3 wooden boards of 30×50 cm were laid on the ground, 2 at the edge and 1 at the centre of the paddock. Once per month from March to November all invertebrates present under the board were identified and assigned to 3 categories: predators (carabid beetle and spiders), herbivores (mollusks), and detritivores (earthworms, millipedes, woodlouse, ants). The nests were occupied by solitary bees in the genus Osmia and Megachile, and the number of sealed tubes was different between grazed and non-grazed meadows. Butterflies from Hesperiidae, Pieridae, Lycaenidae and Nymphalidae families were observed. Their abundance and diversity tend to be higher in grazed meadows. The abundance of predators was similar between grazed and non-grazed meadows, but the abundance of mollusks and detritivores tend to be different. These tendencies should be considered with caution because of the short period of observation and the limited number of meadows included in the study. Collection of additional data is in progress to consolidate the data and confirm the impact of horse grazing. In spite of this limitation, this pilot study shows that wild biodiversity can be studied using non-lethal protocols on horse-grazed meadows and that horse grazing may influence some invertebrates groups.

Alterations of faecal microbiota in Jeju crossbred weanling horses

J.A. Lee, M.C. Shin, J.Y. Choi and S.M. Shin
National Institute of Animal Science, Subtropical Livestock Research Institute, 593-50, Sanrokbukro, 63242, Korea, South; amasss@korea.kr

This study was conducted to discover changes in taxonomic composition and diversity of faecal microbiota in Jeju crossbred weanling horses, by 16S rRNA gene amplicon sequencing. Faecal balls were collected from 27 Jeju crossbred (Jeju × Thoroughbred) horses, from 4 months and 8 months of age. Genomic DNA was extracted from each faecal ball using QIAamp PowerFecal DNA kit according to manufacturer's instructions. The faecal microbiota was characterized on the basis of the V3-V4 hypervariable region of the 16S rRNA gene. At 4 months, *Firmicutes* (45%), *Bacteroidetes* (27%), *Spirochaetes* (7%) and *Fibrobacteres* (6%) were identified as the major phyla. On the other hand, the major phyla were changed at 8 months in the order of *Firmicutes* (40%), *Fibrobacteres* (18%), *Bacteroidetes* (18%) and *Spirochaetes* (14%). Interestingly, *Fibrobacter* was observed as the most dominant genus at 4 and 8 months, relative abundances of *Fibrobacter* were higher at 8 months. The results of α-diversity analysis showed that the richness (Chao1, ACE, Jacknife) in microbial communities were statistically significant ($P<0.001$), and the evenness (Shannon) was similar ($P=0.21$) between 4 and 8 months. The PCoA (principal coordinates analysis) results indicated that 4 and 8 months were divided into two groups with PC1(48%) and PC2(24%) at the species level. The PERMANOVA (permutational multivariate analysis of variance) showed that the microbiota composition differences between 4 and 8 months were statistically significant ($q<0.001$) with Unifrac distance matric. LEfSe (Linear discriminant analysis effect) was performed to select taxonomic makers between 4 and 8 months. At the genus level, EF436358_g ($q<0.001$) and AB494828_g ($q<0.001$) were identified as markers in 4 months, whereas Treponema ($q<0.01$), Streptococcus ($q<0.01$) were selected in 4 months. We also performed PICRUST (phylogenetic investigation of communities by reconstruction of unobserved states) analysis to predict metagenome functional content between 4 and 8 months. Purine metabolism, starch and sucrose metabolism were predicted in 4 months, Steroid hormone biosynthesis was predicted at 8 months, respectively.

Session 33 — Poster 20

Colic incidence among horses in different husbandry systems

E. Schlotterbeck, L.T. Speidel and D. Winter
Nuertingen-Gesilingen University, Institute of Applied Agricultural Research (IAAR), Neckarsteige 6-10, 72622 Nürtingen, Germany; linda.speidel@hfwu.de

Colic is considered one of the most common diseases of domestic horses and is influenced and caused by many different factors. Due to the small stomach volume compared to the body size, the digestive system of horses is particularly prone to disorders. The present study was initiated to find out whether there is a correlation between the frequency of colic and the type of husbandry. To answer the research question, a quantitative survey was conducted. For this purpose, an online survey was created and published via Google Forms. The questionnaire consisted of 35 questions and was distributed over a period of eight weeks via social media and horse magazines. 2,039 horse owners and farm managers participated in the survey. The survey asked about the demographics of the participants, the keeping of the horses, the frequency of feeding, the quality of feed and bedding and the colic of the horses. The 'Numbers' and 'R' programmes were used for the analysis. The study shows that there is a significant correlation between the type of housing and the frequency of colic. Horses kept individually (56%) suffer more frequently from colic than those kept in groups (44%), whereby the frequency of colic decreases with increasing access to pasture, regardless of the type of housing. In the present study, no significant correlation between feeding management and colic frequency could be found. As an explanation for this result, it is assumed that the feeding management of the participating horses can be assessed as optimal. The most common form of colic in this study was constipation. Further studies on colic symptoms could examine the individual types of husbandries in more detail and include the type of use and training of the horses.

Session 34 — Theatre 1

The effect of feeding a soy-free diet on performance and carbon footprint of the feed intake by pigs

S. Millet[1], M. Lourenço[1], S. Palmans[2] and C. De Cuyper[1]
[1]ILVO, Scheldeweg 68, 9090 Melle, Belgium, [2]PVL, Kaulillerweg 3, 3950 Bocholt, Belgium; sam.millet@ilvo.vlaanderen.be

Because of its balanced digestible amino acids content, soybean meal is a common ingredient in pig feed. However, its production is associated with high greenhouse gas emissions, especially related to deforestation. Therefore, interest in alternative protein sources for pig diets is growing. We describe a trial where pigs were fed either a control diet with soy or a soy-free test diet with alternative protein sources. Three-phase feeding was used and diets were formulated in consultation with Flemish stakeholders. Diets were isocaloric (9.6, 9.4, and 9.25 MJ NE/kg diet, for phase 1, 2 and 3, respectively) and were low in crude protein (155, 140 and 125 g/kg). Soy-free diets were more expensive (15.8, 8.4 and 0.9 euro/ton for phase 1, 2 and 3, respectively) than the control diets. Performance, nitrogen efficiency and carbon footprint of the feed intake were calculated. The daily gain (773±30 vs 773±30 g/day), daily feed intake (1,826±99 vs 1,818±64 g/day), feed conversion ratio (2.36±0.1 vs 2.35±1 g/g), carcass gain per kg feed (351±15 vs 354±12 g/kg) or nitrogen efficiency (56.5±2.7 vs 56.0±2.1) did not differ significantly between control and test diet (P>0.05). Carcass yield was significantly higher in the control than the treatment group (79.4+0.2 vs 78.6+0.2, P<0.01), while meat percentage did not differ significantly (64.3+0.2 vs 63.9+0.2, P>0.05). The carbon footprint of the feed intake, expressed as CO_2eq per kg of carcass gain was significantly higher in the control group (2,448±100 CO_2eq) compared to the soy-free group (1,752±56 CO_2eq). The experiment showed that it is possible to obtain similar technical results when feeding a growing-finishing diet without soy. This may lead to a lower climate impact of pig production but it comes with an increased cost. This research was funded by the Flemish Government and the European Agricultural Fund for Rural Development.

Session 34 Theatre 2

Exploring variation in feed efficiency of grower-finisher pigs

M. Van Der Heide[1], J.V. Nørgaard[1] and J.G. Madsen[2]
[1]Aarhus University, Animal and Veterinary Sciences, Blichers Allé 20, 8830 Tjele, Denmark, [2]University of Copenhagen, Veterinary and Animal Sciences, Grønnegårdsvej 2, 1870 Frederiksberg C, Denmark; johannes.g.madsen@sund.ku.dk

Improving feed efficiency (FE) of grower-finisher pigs is of great importance considering its impact on production economy, nutrient excretions to the environment as well as carbon-footprint. Feed efficiency estimated on pen level, does take individual feed intake per day (FID), time spent feeding (TSF) and meal frequency (MF) into account, and thus variation between pen mates is impossible to establish. Thus, by using advanced feeding stations enabling individual measurements of these parameters, this study aimed to explore the variation in FE between pen mates fed one of three levels of crude protein (CP). The study included 60 gilts divided in two series, housed as 10 pigs/pen. From 30-115 kg BW, pigs were fed either a low (L; 113 CP g/kg), standard (S; 121 CP g/kg) or high (H; 124 CP g/kg) CP diet, and traits BW, AWG, FCR (weekly based), and FID, TSF and MF (daily based) were estimated on individual basis. The data were analysed in R as a linear mixed model with dietary treatment (DT; L, S or H), week (W; 7-14 from 30 kg BW) or day (D; 50-100 from 30 kg BW) and DT × W or DT × D as fixed, and series, station within series and animal (repeated measurement) as random effects. Crude protein level had no effect on any traits, whereas W had a significant effect on all traits measured. There was a DT × W interaction where BW was greater ($P<0.05$) in H-compared with S- and L-CP during W12-14 (W14; 82.5, 76.9 and 76.0 kg BW), slope for AWG varied significantly ($P<0.05$) as from W12 on slope decreased for H-CP and increased for L-CP, whereas slope FCR between W12-W14 increased steeper ($P<0.05$) than S-CP and L-CP (decreasing slope). With respect to FID, TSF and MF there were also several DT × W interaction effects, notably, H-CP had on D70, D90 and D100 steeper increasing ($P<0.05$) slope FID than L-CP, H-CP had on D70 and D100 lower ($P<0.05$) slope TSF than L-CP, and L-CP had higher ($P<0.05$) absolute MF than H-CP on D60-70. Together the results suggest that there is a change in efficiency of nutrient utilization over time, indicating that the response to dietary CP level is of dynamic nature during the grower-finisher phase.

Session 34 Theatre 3

Former food slightly affect pig faecal microbiota without impairing jejunal integrity and physiology

M. Tretola[1,2], S. Mazzoleni[1], P. Silacci[2], L. Pinotti[1] and G. Bee[2]
[1]University of Milan, Department of Veterinary Medicine and Animal Science, Via dell'Università 6, 26900 Lodi, Italy, [2]Agroscope, Institute for Livestock Sciences, La Tioleyre 4, 1725 Posieux, Switzerland; marco.tretola@agroscope.admin.ch

Food industry leftovers, also called former foodstuffs products (FFPs) are promising alternative ingredients to improve the circular economy in food production. This study investigated the effects of salty and sugary FFPs on gut microbiota and intestinal integrity of grower-finisher pigs. Thirty-six Swiss Large White barrows were assigned to three grower and finisher diets: (1) standard diet (ST); (2) 30% conventional ingredients replaced by sugary FFPs (SU); and (3) 30% conventional ingredients replaced by salty FFPs (SA). At the beginning of the trial (T1), at the end of the growing period (T2) and 1 day before the slaughter (T3), faeces were collected from the rectal ampulla, snap-frozen, and used for next-generation sequencing to analyse the composition and the alpha and beta diversity indexes of the microbial population. At slaughter, a minimum of 7 pigs per group were randomly selected for electrophysiology measurements. For each pig, two jejunal samples were collected and mounted in Ussing chambers to evaluate the trans-epithelial resistance (TEER) and the active uptake of D-glucose, L-Arginine, L-Methionine and L-Glutamate. For each time point, no difference in the faecal microbiota alpha and beta diversity indexes were found between the dietary treatments. At T3, the core microbiota of the SU pigs was composed by 12 ASVs, while in the ST and SU groups it was characterized by 9 and 8 ASVs, respectively. The Linear Discriminant analysis of effect size (LefSe) showed that at T1, no differential taxa between groups were found. However, differences ($P<0.05$) at the genus level between ST, SU and SA groups were found at T2 and T3. The dietary treatments did not affect ($P>0.05$) the TEER nor the D-Glucose uptake. Among the selected AAs, only the L-Glutamate uptake differed ($P<0.05$), with a higher uptake in SU compared to SA (+366%). When used to partially replace traditional ingredients in growing-finishing pig's diet, both SU and SA FFPs can slightly modulate specific bacterial taxa in faecal microbiota but without impairing the intestinal integrity and physiology.

Bakery products and legume seeds in the diet of growing-finishing pigs

M. Van Helvoort[1] and P. Bikker[2]
[1]De Heus Animal Nutrition, Rubensstraat 175, 6171 VE, the Netherlands, [2]Wageningen University & Research, Wageningen Livestock Research, P.O. Box 338, 6700 AH Wageningen, the Netherlands; mhelvoort@deheus.com

It is fundamental for future pig production to raise the animals as part of a circular food system. The aim of this study was to determine the effect of total or partial replacement of soybean meal, palm kernel expeller and cereal grains by bakery products and legume seeds in the diet of growing-finishing pigs to improve the circularity of food production. In total 736 pigs (26.0±4.5 kg) were randomly allocated to 8 dietary treatments by mixing 3 diets on pen level with the computerised feeding system. Three grower diets for week 1-3 and three finisher diets for week 4 onwards were produced: (1) a control diet: high in cereal grains, with soybean meal and palm kernel meal; (2) a diet high in bakery products with approximately 50% bread meal; and (3) a diet high in legume seeds with 22.5% peas and 25% faba beans. The dietary treatments were fed to slaughter at 125 kg of live weight. Efficiency of nitrogen (N) and phosphorus (P) utilisation was calculated using reference values for N and P content in pigs. Carbon footprint of the feeds was based on list version 5 of Nevedi (2022). Data were analysed with analysis of variance using Genstat with pen as the experimental unit. Dietary treatments had limited effects on growth performance and feed efficiency of the pigs. The efficiency of N and P utilisation was reduced by approximately 4-5 and 8-9% units, respectively ($P<0.001$) due to the higher N and P content of the diets with bakery products and legume seeds. The carbon footprint of the feeds expressed in g CO_2 eq. per kg of carcass gain was reduced by inclusion of bakery products with 64% including and 38% excluding land use change (LUC) ($P<0.001$). In contrast, inclusion of legume seeds reduced the carbon footprint including LUC with 23% but increased the carbon footprint excluding LUC with 23% ($P<0.001$). Overall, these results indicate that cereal grains and soybean meal can be replaced by bakery products and legume seeds with minor effects on growth performance of the pigs, but efficiency of N and P utilisation is reduced. Replacement by bakery products and legume seeds reduced the carbon footprint including LUC, however legume seeds increased the carbon footprint excluding LUC.

Recycling agricultural by-products: using purple carrots as an ingredient in layer quails feeding

A.S.-G. Sarmiento-García[1,2], A.B.-D. Benito-Diaz[1], C.V.-A. Vieira-Aller[1] and O.O. Olgun[3]
[1]Instituto Tecnológico Agrario de Castilla y León, Estación Tecnológica de la Carne, Avenida de Filiberto Villalobos, 5, 37770 Guijuelo, Salamanca, Spain, [2]Universidad de Salamanca, Área de Producción Animal, Departamento de Construcción y Agronomía, C/Filiberto Fillalobos1 119, 37007, Salamanca, Spain, [3]University of Selçuk, Department of Animal Science, Faculty of Agriculture, Ardıçlı, Selçuk Ünv. Alaaddin Keykubat Kampüsü No:371, 42250, Selcuk, Konya, Turkey; asarmg00@usal.es

The carrot (*Daucus carota L.*) is an interesting ingredient due to its chemical composition. In the last years, there has been an increase in the intake of coloured carrots due to apparent health benefits. Nevertheless, it has resulted in increasing amounts of agricultural waste by-products. The goal of the current experiment was to establish the effect of dietary concentrations of purple carrot powder (PCP) waste by-products on performance, egg production, eggshell quality, and yolk colour of layer quails (*Coturnix coturnix Japonica*). A total of one hundred and fifty 22-week-old Japanese laying quails were allotted to 5 dietary treatments each with 6 replicates of 5 quails. Purple carrots were provided by local markets (Konya, Turkey), which had been discarded for human sale. The carrots were dried in an oven (100 °C until reaching 20% humidity) and subsequently ground. The PCP was incorporated into the basal diet (0%) by replacing the proportional part of corn as follows: 0.1, 0.2, 0.3, and 0.4% PCP which were supplied *ad libitum* for 10 weeks. No differences were detected between dietary treatments for any of the performance parameters or egg production. Eggshell weight ($P<0.01$) and eggshell thickness ($P<0.01$) were linearly affected by dietary PCP supplementation, reaching maximum levels at 0.4% of PCP supplementation, while the percentage of damaged egg and egg-breaking strength remained similar for all experimental groups ($P>0.05$). Above the aforementioned results, using PCP, an agricultural by-product that is both safe and easily available, as am ingredient for laying quail diets was found to be effective without negatively affecting quail production or egg quality.

Effect of camelina cake doses as soybean meal substitutions on growth and gut health of piglets

D. Luise[1], F. Correa[1], S. Virdis[1], C. Negrini[1], G. Cestonaro[2], L. Nataloni[2], G. Titton[2], E. Sattin[3], E. Costanzo[2] and P. Trevisi[1]
[1]University of Bologna, Viale G Fanin, Bologna, 4127, Italy, [2]Ceral Docks S.p.A, Camisano Vincentino, 36043, Italy, [3]BMR genomics, Via della republica, Padova, 35131, Italy; diana.luise2@unibo.it

The camelina cake (CAM) is a co-product proposed as an alternative protein source; however, data on piglets are still limited. This study aimed to evaluate the effect of different doses of CAM in substitution of soybean meal on growth and gut health of weaned pigs. At 14 days post-weaning (d0), 64 piglets balanced for body weight and litter, were assigned to one of the four following isoenergetic and isoproteic diets: standard diet (CO) or a diet with the inclusion of 4% (C4), of 8% (C8) or 12% (C12) of CAM. Pigs were weighed weekly. Faeces and blood were collected at d7 and d28 for microbiota (v3-v4 of 16s rRNA gene) and ROMs analysis. At d28, pigs were slaughtered; pH was recorded on gut contents, organs were weighed and jejunum was used for morphological and gene expression analysis. A mixed-model including diet as a fixed factor and pen and litter as random factors was applied. From d0-d7, the CAM linearly reduced the average daily gain (ADG) ($P\leq 0.01$). No effect was observed in the subsequent weeks, but, from d0-d28, the CAM linearly reduced the ADG ($P=0.01$). From d0-d7, the feed intake linearly decreased ($P=0.04$), and the feed to gain linearly increased ($P=0.004$) with the increase of CAM. The liver weight linearly increased with the rise in CAM ($P<0.0001$). The diet did not affect ROMs, intestinal pH and morphology, and jejunal gene expression except for the expression of zonulin-1 ($P=0.07$, quadratic effect). The CAM, at all doses, increased the alpha diversity indices at d28 ($P<0.05$). The C4 diet promoted the abundance of Butyricicoccaceae_UCG-008 and Erysipelatoclostridiaceae_UCG-004, which are usually correlated with resilient piglets' gut microbiome. In conclusion, despite reducing the ADG of piglets, the CAM did not affect gut health and enhanced resilient gut microbiome. CAM could be evaluated as a potential alternative protein source in weaned pigs. Acknowledgements: FA&AF project (ID 10288429), POR FESR 2014-2020, Veneto Region, Italy and Agritech National Research Center (PNRR project D.D. 1032 17/06/2022, CN00000022).

The dietary fibre solubility in the maternal diet does not affect the muscle development of piglets

M. Girard[1], F. Correa[2], F. Palumbo[2], P. Silacci[1] and G. Bee[1]
[1]Agroscope, Tioleyre 4, 1725 Posieux, Switzerland, [2]University of Bologna, Department of Agricultural and Food Sciences, viale G Fanin 44, 40127 Bologna, Italy; marion.girard@agroscope.admin.ch

Within litter, piglets born with the same birthweight may have very contrasting growth in the suckling period. Investigating the underlying mechanism behind this is important to harmonize weaning weights. Piglet weaning weight is affected by the insoluble (IDF)-to-soluble (SDF) dietary fibres ratio in sow diet. The present study investigated whether the IDF/SDF ratio in the maternal diet can affect the myofibre development and the lifetime growth of slow (S)- and fast (F)-growing piglets. From 85 days of gestation to weaning, 32 sows were fed a gestation and a lactation diets containing either 4% of IDF or 4% of SDF. The birth order as well as the birth weight and the weight at one day were recorded to calculate colostrum intake. On day 17, one or two pairs of S and F piglets per litter with similar birth weights were selected. On day 25, the semitendinosus muscle was collected on 32 piglets. The 48 remaining piglets were fattened until 110 kg. There was a maternal diet × growth interaction ($P<0.05$) for colostrum intake with IDF-F piglets having a greater ($P<0.05$) colostrum intake than IDF-S and SDF-S with intermediate values for SDF-F. The maternal diet had no effect on the birth weight, ponderal index or myofibre development but IDF piglets had a greater ($P<0.05$) growth from birth to 17 days. However, the growth until slaughter and the age at slaughter were similar between IDF and SDF pigs. Regardless of the maternal diet, F and S piglets had similar birth characteristics. Nevertheless, besides a similar number and proportion of type I and II myofibres, S piglets had smaller ($P<0.01$) myofibres than F piglets. In addition, S pigs were on average 13 days older ($P<0.001$) at slaughter due to a slower ($P<0.001$) growth compared to F pigs. This study shows that S piglets have a reduced myofibre hypertrophy at 25 days and do not have compensatory growth until slaughter. In addition, the positive effect of the maternal diet on piglet growth fades during the fattening period and is not related to muscle development on day 25.

Session 34 — Theatre 8

Improving the sustainability of the Australian pork industry

G.L. Wyburn
Australian Pork Limited, 2 Brisbane Avenue, Barton 2600 Australian Capital Territory, Australia; gemma.wyburn@australianpork.com.au

In recent years, the Australian pork industry has embraced the sustainability challenge and implemented a program of research, development and extension to help industry understand, measure and improve their performance. Australian Pork Limited (APL), the national industry body has developed a Sustainability Framework that has goals across the categories of people, pigs, planet and prosperity. The framework goals with respect to 'Planet' cover carbon and nutrient accounting, natural resource stewardship and reducing waste. Two industry roadmaps have been developed – one on activities to help reduce a farms' carbon footprint and the other to reduce waste on farm. In addition, a full lifecycle assessment (LCA) of the Australian pork industry's key environmental impacts based on the 2020 financial year was conducted. The study looked at the footprints of 32 farrow to finish operations representing 44% of the total Australian pork industry based on sow numbers. This study included all major Australian pork processors. Outcomes of the LCA showed that manure, feed production, land use and direct land use change and farm services were the biggest carbon emissions contributors. The average emission was calculated at 3.7 kg of CO_2-e per kg of liveweight with a range from 1.4 to 6.4 kg CO_2-e for scope 1,2 and 3 emissions at farm gate. This represents a reduction of 13% since the last measurement taken in 2010. When post- farm processing is included, total emissions had an average of 5.9 kg CO_2-e per kg of wholesale pork. Water and energy usage have reduced more than expected and the report also quantifies for the first time the eutrophication impact for phosphorus (P) and nitrogen (N) across the industry for both direct and indirect impacts. Apart from outdoor producers, P results were low and N results varied depending on which grain growing region they sourced feed from. High industry participation and the nation-wide scope of the life-cycle assessment project, together with the industry roadmaps has provided the Australian pork industry with a clear position on the progress made to date and the sustainability challenges and opportunities that lie ahead. The author wishes to acknowledge the work of Integrity Ag and Environment in undertaking these studies.

Session 34 — Theatre 9

Are tailor-made health plans effective in triggering changes in pig farm management?

P. Levallois[1], M. Leblanc-Maridor[1], A. Scollo[2], P. Ferrari[3], C. Belloc[1] and C. Fourichon[1]
[1]Oniris, INRAE, BIOEPAR, 44300 Nantes, France, [2]University Torino, Grugliasco, 10095 Torino, Italy, Italy, [3]CRPA, 42121 Reggio Emilia, Italy; christine.fourichon@oniris-nantes.fr

A tailor-made health plan is a set of recommendations for a farmer to achieve and maintain a high health and welfare status. Tailored to each farm, it is intended to be an effective way of triggering change and meeting new expectations. This study aimed to evaluate the effectiveness of tailor-made health plans in pig farms, designed in various situations after a systematic biosecurity and herd health audit. An intervention study was carried out on 20 farrow-to-finish pig farms. An initial standardised audit and discussion between the farm veterinarian and the farmer resulted in a specific plan. Compliance with recommendations was monitored 8 months. Changes in health, performance and antimicrobial use were monitored. We defined two categories of plans: (1) 14 plans targeting a given health disorder present in a farm; (2) 17 plans to improve prevention, not targeting a specific disorder (one farm could have both types of plans). A small number of priority recommendations were made per farm. In 18 farms, farmers implemented 1 to 4 recommendations (none in 2 farms). Of the 17 non-disorder-specific plans, 11 were considered effective (>50% recommendations implemented), 3 intermediate (at least one but less than half of the recommendations implemented) and 3 ineffective (no implementation). Of the 14 disorder-specific plans, 9 were followed with full or good compliance (>50% recommendations implemented), 2 with intermediate compliance (1 recommendation implemented out of 2) and 3 with no compliance (no recommendation implemented). When at least one recommendation was implemented, change in clinical, performance and antimicrobial use indicators was assessed if a biological association with the disorder was deemed plausible and if their initial value showed room for improvement. Improvement was evidenced 4/9, 1/6 and 1/6 times for these indicators, respectively. Independently, veterinarians concluded in effectiveness for 8/14 plans. Overall, tailor-made health plans were effective in triggering changes in farm management. Several approaches were combined to investigate their effects, in the perspective of multicriteria assessment.

Pork production can contribute positively to human protein supply

R.J.E. Hewitt, D.N. D'Souza and R.J. Van Barneveld
SunPork Group, 6 Eagleview Place, Eagle Farm, 4009, Australia; robert.hewitt@sunporkfarms.com.au

The respective sustainability of our foods systems is increasingly under focus as we are challenged to produce more food from diminishing resources to feed a growing global population. Historically, livestock produced nutrient-rich food from lower value inputs that were not consumed by humans. Ruminants consumed inedible fodder from generally non-arable land, whilst monogastrics were fed undesired human food by-products such as excess grain, dairy co-products and waste streams from milling or other food processing. Livestock production today has become reliant on feed crops grown on arable land to improve efficiency, reduce the cost of production or meet premium market requirements. Livestock feed consumption now competes directly with human food, with one-third of global grain production being fed to livestock. The simple solution would be to divert human-edible feedstuffs directly to humans, but all proteins are not the same, they vary in nutritional profile, digestibility, environmental implications, and consumer acceptance. Livestock can bank poorer quality proteins and make them available for humans. Net Protein Contribution (NPC) describes the contribution of a production system to meeting human nutritional requirements, with values greater than one indicating a system is contributing more human edible protein than it consumes. In determining a systems NPC, we must understand the attributes of the protein and its quality, the efficiency of its conversion and the contribution it makes to meeting the needs of the consumer. The Australian pork production system is characterised by the utilisation of co- and by-products of other agricultural systems, including the utilisation of non-human edible processed animal proteins that are notable omissions from most pig diets in Europe and to some extent North America. An intensive Australian pork supply chain is estimated to contribute over three times the human edible protein that it consumes (NPC=3.26), which compares favourable to current European NPC estimates of pork, and other livestock, production systems. Focussing livestock systems on the utilisation of waste streams, co-products and human-inedible feedstuffs ensures they can make a sustainable contribution to human food supply.

Environmental and economic assessment of a French free-range chicken production system

B. Méda[1], E. Barlier[1], E. Péchernart[2], J.-Y. Limier[3] and S. Mignon-Grasteau[1]
[1]INRAE, Université de Tours, BOA, 37380 Nouzilly, France, [2]ITAVI, 7 Rue du Faubourg Poissonnière, 75009 Paris, France, [3]CAFO, 41 B Av. Chanzy, 41240 Beauce-la-Romaine, France; bertrand.meda@inrae.fr

French 'Label Rouge' (LR) chicken production is particularly appreciated by consumers for its animal-friendly production conditions (outdoor access, lower animal density and growth rate) and the organoleptic qualities of its meat. Yet, it is also facing various challenges, such as a poorer feed efficiency and a larger exposure to climate change than in conventional production. Analysing LR value chains is a prerequisite to identify progress margins for poultry operators. The aim of this study was thus to assess the environmental/economic performances of one representative chicken LR value chain in the centre of France, the '*Poulet de l'Orléanais*'. Environmental and economic performances were assessed using Life Cycle Analysis and by calculating production costs (PC), respectively. For 1 kg of live weight at farm gate, PC was estimated to be 2.07 € and environmental impacts were: 2.58 kg CO_2-eq for climate change (CC), 26.7 MJ for cumulated energy demand (CED), 0.048 kg SO_2-eq for acidification (AC), 0.019 kg PO_4^{3-}-eq for eutrophication (EU), and 4.38 m^2.y for land occupation (LO). Using an economic allocation approach, the same indicators were calculated at slaughter house gate for 1 kg of ready-to-cook chicken and 1 kg of breast meat. For these two products, results were respectively: 3.40 and 12.46 € for PC, 3.79 and 8.93 kg CO_2-eq for CC, 42.8 and 100.8 MJ for CED, 0.068 and 0.161 kg SO_2-eq for AC, 0.027 and 0.064 kg PO_4^{3-}-eq for EU, and 6.33 and 14.92 m^2.y for LO. Regardless of the functional unit, environmental impacts differed between the two genetics strains used in this production (S757N and G657N), with higher values for S757N (CC: +8.8%; CED: +4.5%; AC: +5.3%; EU: +0.9%), except for LO (-5.0%). Despite a better feed efficiency, the S757N strain had greater feed impacts, because of higher corn and soybean dietary contents. These results suggest that improvements should be made regarding genetic selection and nutrition in order to decrease the environmental impacts of LR chicken products. In particular, much attention should be paid to cut parts such as breast meat, since consumer demand is rapidly increasing.

Session 34 Theatre 12

Effects of values and calculation changes in three IPCC guidelines on greenhouse gas inventories

E. Nugrahaeningtyas, J.S. Lee and K.H. Park
Kangwon National University, Department of Animal Industry Convergence, 1 Gangwondaehakgil, Chuncheon-si, 24341, Korea, South; eskanugrahaeningtyas@kangwon.ac.kr

The increase in anthropogenic activities since the industrial era has increased GHG emissions (CH_4, N_2O, etc), resulting in climate change and global warming. Based on the Life Cycle Assessment (LCA), livestock represents 14.5% of total anthropogenic emissions. Despite emphasizing how livestock production affects the environment, it ignores how livestock contributes to food security (i.e. nutrient-dense food and the transformation of non-edible food into edible food) as well as other industries (i.e. pharmaceuticals). LCA identifies the impacts on the environment from the whole process but does not display Transparency, Accuracy, Completeness, Comparability, Consistency (TACCC), as does GHG inventory (GHGI). GHGI as the main tool for GHG reporting targets the main sources of livestock emissions, enteric fermentation (CH_4) and manure management (CH_4, N_2O). It accounts for 4.96% of total carbon dioxide equivalent (CO_2-eq) emissions and 51.04% of total agricultural emissions. Calculation of GHGI is based on Intergovernmental Panel on Climate Change Guidelines (IPCC GL), which are available in 1996 GL, 2006 GL, and 2019 Refinement. There is a difference in the emissions among guidelines due to different default values in the guidelines. Consequently, guidelines should be used carefully to prevent overestimating or underestimating the calculation which may be used to gain an advantage. GHG emissions from Korea's livestock were calculated to explain guidelines differences. The result showed that the estimated emission was different with each guideline. The emissions using 1996 GL and 2019 Refinement were nearly similar and the emission using 2006 GL was higher than that in others. Although the emissions were similar in 1996 GL and 2019 Refinement, the contribution of emission's sources differed. This difference may result in uncertainty when changing a GL and can be used to gain some benefits. Hence, changing to the most recent guidelines is the best practice regardless of the results of estimated emissions. Since an improved method is used to calculate emissions, the 2019 Refinement may better represent livestock conditions in the country.

Session 34 Poster 13

Bioavailability and repellent activity of phytocompounds against red mites

C. Carlu, T. Chabrillat, C. Girard and S. Kerros
Phytosynthese, 57 avenue Jean Jaurès, 63200 Mozac, France; claire.carlu@phytosynthese.fr

Dermanyssus gallinae, known as the 'red mite,' is a blood feeding ectoparasite, commonly found in laying hens and is one of the most important epidemiological and economic problem. An uncontrolled red mite population affects animal performance and welfare. Adding plant bioactives to diets may be an efficient and natural alternative to chemical repellents. *D.gallinae* repellent activity appears to come from neurotoxic effects, blocking γ-aminobutyric acid and reducing the ability of neurons in the nervous system. This *ex vivo* experiment measures the repellent activity of PhytoDerm against red mites. PhytoDerm is a natural solution composed of monoterpenes from essential oils and plants known for their activity against arthropods. Broilers from Phytosynthese experimental farm were separated in 2 groups of 2,200 birds JA657; one group received PhytoDerm from 64 to 70 days-old through drinking water at 1 ml/l. At 70 days of age, blood samples were collected from 6 broilers from the control group and 6 broilers from the supplemented group. In a partner lab, 3 groups of blood were realized: (1) C made with the blood from the 6 control birds; (2) A made with the blood from the control birds with an artificial supplementation of PhytoDerm at 1 ml/l at the lab; (3) B made with the blood from the 6 birds supplemented at the farm. Each blood sample was put into contact with 200 unfed mites via an artificial feeding model using skin of broilers and maintained at 38 °C. After one hour in the dark, red mites and dark mites were visually counted; mites were considered fed according the red colour. Trial was performed with 6 repetitions. The number of red-fed-mites were significatively reduced in A and B groups, respectively -34 and -28% (P<0.001). No statistical difference was observed between blood samples from groups A and B (P=0.3). This *ex-vivo* trial was able to evaluate, the effect of PhytoDerm based on plant bioactives to reduce *D. gallinae* feeding behaviour and the positive effect of oral supplementation of PhytoDerm implying a good bioavailalibity of its bioactives in poultry blood. Further trials in farming conditions are carried out to evaluate how this preliminary repellent activity is influencing the mites' population dynamic.

Session 34 Poster 14

Riboflavin yield of common raw ingredients for organic poultry diets

V. Decruyenaere[1], N. Everaert[2], P. Rondia[1] and J. Wavreille[1]
[1]Walloon agricultural Research Centre, Animal production Unit, 8 rue de Liroux, 5030 Gembloux, Belgium, [2]KU Leuven, Department of Biosystems, 30 Kasteelpark Arenberg, 3001 Leuven, Belgium; v.decruyenaere@cra.wallonie.be

The supply in vitamin B2 or riboflavin is strictly necessary for poultry. Prolonged deficiency leads to serious health problems and reduced performances. The riboflavin previously used in organic poultry feed was derived from a process using GMOs, although this is strictly prohibited. The research for alternative sources is therefore becoming a necessity for the organic poultry industry. The riboflavin recommended level is 3.6 mg/kg feed from 0-6 weeks and 3 mg/kg of feed from 6-8 weeks for broilers. For laying hens, at the pullet stage, 3.4 to 3.6 mg/kg feed is needed. From 8 to 12 weeks of age, requirements are covered by 1.7 to 1.8 mg/kg of feed and from 18 weeks to the first egg, 1.7 to 2.2 mg/kg of feed. During laying, the recommended riboflavin content is 3.1 mg/kg of feed. To determine the riboflavin content of feedstuffs included in poultry diets, a bibliography research was conducted. It appeared that milk powder (20 mg/kg), algae (20-40 mg/kg), insect flours (>15 mg/kg), beer yeasts (12 mg/kg) and forages (12-17 mg/kg) had riboflavin contents higher than cereal (1-2 mg/kg) and oilseed cakes (2-4 mg/kg), or protein grains (2 mg/kg). Feed samples were collected on Walloon farms and from manufacturers and their riboflavin content was determined ($n \approx 170$ feedstuffs, Eurofins method EN 14152 2006 mod. [CN Food]). Within product category, the variability in vitamin B2 levels was high. For cereals, the dosed values were lower than those announced in feed tables. Among cereals, maize had the lowest values (0.67 mg/kg) followed by wheat (0.82 mg/kg). Milk powder, oilseed cakes and protein grains had values comparable to those of feed tables. Many feedstuffs included in poultry diets did not meet the riboflavin recommended level. Only few ingredients rich in riboflavin can be used in poultry formulations (yeast, milk powder, dehydrated forage feeds, insect flours). Their rate of incorporation in formulas is limited and their cost sometimes high (milk powder for example). The search for alternative forms of vitamin B2 for organic poultry diets must continue.

Session 34 Poster 15

Sustainable poultry production: biocontrol measures, dietary approaches and packaging strategies

G. Maiorano and S. Tavaniello
University of Molise, Department of Agricultural, Environmental and Food Sciences, Via F. De Sanctis snc, 86100 Campobasso, Italy; siria.tavaniello@unimol.it

The aim of the project is to develop, implement and transfer concrete innovations into practice based on a farm-to-fork approach to reduce the use of antibiotics in poultry production. It is an integrated approach along the entire food chain, acting: (1) at embryonic stage by *in ovo* synbiotic injection; (2) during the rearing period by nano-encapsulated polyphenol extracts (PE) (from olive leaves and olive mill wastewater) administered in feed; (3) at packaging using antioxidant-releasing film. The *in ovo* delivery of synbiotics, as the earliest method to stimulate intestinal environment, combined with dietary supplementation of nano-encapsulated PE, could be a promising strategy to eliminate all non-prudent use of antibiotics, while optimizing productivity. Moreover, the incorporation of PE into film packaging could be an effective method to extend shelf-life of meat, reducing food waste. The project includes pilot and field trials for the characterisation of the modes of action and for evaluating the robustness and effectiveness of the results in practice. The first part of the project includes: the set-up of the synbiotic formulation and the extraction of different polyphenolic compounds from olive by-products. Thereafter, *in vitro* studies will be carried out to evaluate the antimicrobial activity of PE against the major poultry pathogens. Based on the combined information coming from *in vitro* results and computer modelling by computational chemistry, the best PE and the relative dose will be chosen for the encapsulation process. A field trial will be carried out to characterize the overall response of chickens to synbiotic and PE administration on: gut microbiota; gene expression pattern; growth performance; carcass and meat quality traits. The last step will be the development, by experimental and theoretical methods, of packaging with the incorporation of PE, as a method to extend meat shelf-life. The integrated approach proposed in this project will give the scientific support to develop an innovative and sustainable poultry production chain. Acknowledgments PRIN 2020 project (Prot. 2020ENLMHA), Ministry of University and Research (MUR), Italy.

Session 34 Poster 16

Genetic correlations between feeding behaviour, meat quality, carcass and production traits in pigs

A.T. Kavlak[1], T. Serenius[2] and P. Uimari[1]
[1]University of Helsinki, Koetilantie 5, 00790 Helsinki, Finland, [2]Figen Oy, Kuusisaarentie 1, 68600 Pietarsaari, Finland; alper.kavlak@helsinki.fi

The Finnish pig breeding programmes have been very successful in the genetic improvement of economically important traits such as average daily gain (ADG), feed conversion rate (FCR) as well as carcass traits during the last decades. Even though the main goal for selection would be to improve feed efficiency, the genetic association within feeding behaviour and other traits among pigs could also be used in a selection program for modifying the improvement in efficiency for specific environments which could have a direct effect on pork production. Therefore, the objective of this study is to estimate the heritability of these traits and the genetic correlations between feeding behaviour, production, meat quality, and carcass traits. In this study, the data consists of purebred Finnish Yorkshire and Finnish Landrace pigs and their F1-crosses raised in a controlled test station environment located in Längelmäki (Figen Oy). The data were collected from 7,405 pigs that had entered the test station between 2011 and 2016. Pigs arrived at the test station at an average age of 89±10 days (mean ± standard deviation) and an average weight of 34.7±6.4 kg. The slaughter age is 186±10 days, and the slaughter weight is 121.2±12.9 kg. In this study, the most important feeding behaviour traits (FBT) are considered as number of visits per day (NVD, counts), time spent in feeding per day (TPD, min), daily feed intake (DFI, g), time spent in feeding per visit (TPV, min), feed intake per visit (FPV, g) and feed intake rate (FR, g/min). The production traits (PT) are considered in pig breeding programmes are average daily gain (ADG, g) and feed conversion rate (FCR, g/g) while the meat quality traits (MQT) are considered as pH of loin and ham ($Loin_{Ph}$ and Ham_{Ph}) and the lightness (L), redness (a) and yellowness (b) of loin and ham ($Loin_L$, $Loin_a$, $Loin_b$). Lastly, the carcass quality traits (CQT) include meat percentage (M%, kg) as well as backfat thickness (BF, mm) and muscle depth (MD, mm). The restricted maximum likelihood method (REML) and the DMU software are going to be used for the variance component estimations. The results will be presented and discussed during the conference.

Session 34 Poster 17

The effect of plant-derived terpenes additive in *Ascaris suum* management in organic fattening pigs

H. Bui, E. Belz, J. Ligonniere and M.E.L.A. Benarbia
Nor-Feed, R&D, 3 rue Amedeo Avogadro, 49070, France; hoa.bui@norfeed.net

Ascaris suum is the most prevalent intestinal parasite in pig production. The infection causes significant economic losses due to health and production issues. Today, within the high risk of resistance development and the residual concern, sustainable alternatives to conventional anthelmintics are under demand. The objective of this study was to assess the effect of the plant-based feed additive (PB-A) on *A. suum* and compare it to flubendazole in fattening pigs. PB-A is standardized in terpenes, composed mainly of natural essential oils (EOs) including *Cymbopogon nardus* and *Citrus scinensis* 160 pigs, at 10- weeks age, were randomly divided into 2 groups. PB-A group: 80 pigs received a standard diet supplemented PB-A at 10g/ 100 kg of body weight (bw), during 5 consecutive days, at 16 and 21 weeks of age. Control group (CT): 80 pigs had standard diet supplemented flubendazole at 10 g/100 kg bw during 5 consecutive days, at 16 week of age. *A. suu*m infection was evaluate before and after trials using ELISA method to quantify the specific serum antibody derived from *A. suum* infective larvae 3 (L3). 8 random blood samples were taken per group for every evaluation. Statistical analysis was performed by using ANOVA. Obtained results showed that both tested groups were negative to L3 at the beginning of trial. There was no significant difference in the antibody titer between the two group before supplementation. However, at the end, pigs in the control group were tested positive to L3 while pigs from PB-A group remained negative. The average antibody titer in CT pigs (0.3) were significantly higher (P=0.0004) compared to the one of STPA pigs (0.18). In conclusion, PB-A supplementation in feed during fattening period resulted in significantly restricting the proliferation of *A. suum* pressure compared to flubendazole in such experimental condition. It was also observed that the single flubendazole treatment during fattening in accordance with French organic farming regulation may be not effective enough. This protocol was adapted to be close to farm practical condition so that the negative control group was not possible to install. Further study with controlled conditions should be carried out to reinforce the results.

Session 34 Poster 18

Effect of fattening and slaughter value of gilts on their lifetime piglet production

M. Szyndler-Nędza, M. Tyra and A. Mucha
National Research Institute of Animal Production, Department of Pig Breading, ul. Krakowska 1, 32-083 Balice, Poland; magdalena.szyndler@iz.edu.pl

The purpose of this study was to examine how the performance traits of young female pigs (gilts) influence the number of piglets born and raised during their reproductive life. The data for the study was gathered from the entire population of Puławska sows in breeding herds that were part of a genetic resources conservation program. The study included 2,722 Puławska gilts with complete information on their fattening and slaughter evaluation and reproductive performance. The data collected from the fattening and slaughter evaluation included body weight, daily gain, backfat thickness, average backfat thickness, loin eye height, and carcass meat percentage. This data was paired with information on how many purebred piglets the sows had given birth to and reared in successive litters. The collection contained information on 12,665 purebred litters from 71 breeding herds and parities 1 to 15. In order to determine the impact of certain factors on sow production performance, four groups of factors were created based on parity, body weight, daily gain, and meatiness. The study found that young gilts evaluated at a body weight of up to 55.9 kg and with daily gains of up to 519 g/day had the poorest production parameters in both the first and subsequent parities. Females with excessive carcass meat content (≥60%, class S) had fewer piglets born and reared mainly in the first parities, while a high meatiness value in the second and third parities contributed to a reduction in the number of piglets reared. The work financed from the National Research Institute of Animal Production (Task no. 01-11-05-11).

Session 34 Poster 19

Are the ethics of pig breeding addressed in literature?

M. Van Der Sluis, R.S.C. Rikkers and K.H. De Greef
Wageningen Livestock Research, P.O. Box 338, 6700 AH Wageningen, the Netherlands; malou.vandersluis@wur.nl

Animal breeding activities may raise ethical concerns, especially because of the technologies used and the (potential) effects on the animals. This study examined to what extent ethics of genetic selection in pigs are discussed in the scientific literature, and which themes are focused on. A literature search was performed in Scopus, based on titles, key words and abstracts, using a combination of three search term groups: (1) an animal part; (2) a breeding part; and (3) an ethics part. The search was limited to publications from 1990 to 2022, resulting in 130 publications retrieved on December 15th 2022. Subsequent manual selection resulted in 31 publications that met the criterium of mentioning ethics of pig genetic selection. Of these, 14 only used the term ethics without elaborating on it, whereas 17 addressed ethics explicitly. The majority of these publications were not pig-specific, but addressed multiple species or farm animals in general. Ethical analysis of explicit pig-related themes that raise concerns (such as for example litter size) was scarce. Overall, three main ways of addressing ethics were identified: (1) the term 'ethics' being used as a synonym for societal acceptance, for example regarding genome editing, where ethics sometimes appear to be considered as a hurdle to overcome before implementation; (2) identification of ethical aspects of (consequences of) breeding, largely focusing on animal welfare; or (3) an ethical analysis of choices or dilemmas in breeding was posed, for example the balance between production economics (e.g. efficiency traits) and non-market value traits (e.g. animal welfare) in breeding programs. Overall, ethical analysis of specific pig breeding issues appears to be scarce in literature. An inventory of issues surrounding animal breeding across species (and in general) will likely be more informative for insight in ethical considerations. However, this will probably lack attention for specific pig issues.

Session 34 Poster 20

Benzoic acid reduces environmental impact in fattener pig diets using life cycle assessment

E. Perez-Calvo[1], S. Lagadec[2], C. Drique[2], D. Planchenault[3], S. Potot[1] and C. Valliere[1]
[1]DSM Nutritional Products, 576 Wurmisweg, 4303, Kaiseraugst, Switzerland, [2]Chambre d'Agriculture de Bretagne, 56 Rue de la Fontaine, 56300 Pontivy, France, [3]DSM Nutritional Products, 71 Boulevard National, 92250 La Garebbe-Colombes, France; estefania.perez-calvo@dsm.con

The objective of this study was to calculate the environmental impact of raising pigs from 30 to 115 kg fed a commercial diet supplemented with benzoic acid (BA) vs unsupplemented (C) using life cycle assessments (LCA). One hundred and forty-four pigs were randomly placed into 24 pens and assigned to treatments. Basal diets included corn, barley, wheat and sunflower meal as main ingredients and feed enzymes (phytase and xylanase). BA was included at 0.5%. A Two-phase feeding was used including 14% CP (grower) and 12.2% CP (finisher) and animals' weight and feed intake were monitored. Data were analysed by one way ANOVA in JMP. Overall performance showed that BA pigs increased ADG compared to C pigs (905 vs 939 g/d, $P=0.04$) and FCR was also improved in BA pigs compared to C pigs (2.56 vs 2.45, $P=0.04$). LCA were performed including the impact of feed production, housing and slurry management using an attributional approach. The results were given according to the 19 impact categories of the EU PEF methodology using Sustell™ LCA software (ISO: 14040/44). Over the 19 environmental impact categories and after normalization, 4 categories, including Global Warming Potential (GWP100 – fossil, biogenic, land use and land transformation), respiratory inorganics (RI), marine eutrophication (MA), and freshwater eutrophication (FE), were found to be the most relevant impact categories for this study. Feed production was identified as the main contributor for climate change (GWP 100), (up to 44%), as well as MA (up to 67%) and FE (up to 73%) in both groups. Inclusion of BA in the diet helped to reduce by 6, 7, 9 and 11% in GWP100, EF, EM and RI, respectively, mainly through improved performance. Additionally, BA intervention reduced emissions on ammonia by 13% and methane from the manure management process by 3%. In conclusion, 0.5% BA supplementation to GF pig diets resulted in improved performance and consequently in a reduced environmental impact especially in climate change and ammonia emissions.

Session 35 Theatre 1

Is there a future for experimental animal research in Europe?

J. Van Milgen[1] and T. Chalvon-Demersey[2]
[1]INRAE, Institut Agro, Pegase, Le Clos, 35590 Saint-Gilles, France, [2]METEX Animal Nutrition, 32, rue Guersant, 75017 Paris, France; jaap.vanmilgen@inrae.fr

Research involving experiments with animals is increasingly questioned by society. In 2021, the European Parliament asked the European Commission to 'Develop plans and actions to accelerate the transition to innovation without the use of animals in research, regulatory testing and education'. The resolution was adopted with 664 votes in favour, 4 against, and 16 abstentions. In February 2022, The European Commission responded that 'The legal obligation to replace the use of animals when new non-animal methods become available is firmly embedded in EU legislation, providing a step-wise approach as science advances'. The Commission also indicated that it is impossible to predict when these new methods become available and that non-animal alternatives are most efficiently developed in clearly defined contexts of use. Later that year, a European Citizens' Initiative was launched asking the European Commission to 'Modernize science in the EU' and to 'Commit to a legislative proposal plotting a roadmap to phase-out all animal testing in the EU before the end of the current legislative term'. The initiative collected more than one million signatories (from at least seven EU countries with a threshold for each country), which means that the European Commission has to consider it and a formal reply from the Commission is expected by July 2023. The 'legal obligation to replace the use of animals when new non-animal methods become available' and the 'clearly defined contexts of use' means that we, as an animal science community are not only concerned by, but also have a responsibility on how research on livestock production will be done in the future. This includes the development and training of future generations of animal scientists on using alternative methods (such as *in vitro* methods and computer modelling), and in engaging in an open and constructive debate with legislators and society. It also means that the data that we have generated in the past and that we generate now from animal experiments will have a great scientific value in the future. Open Data will allow us to bring new life in old data.

Session 35 Theatre 2

Testing minimally invasive blood collection techniques in pigs used for research

F.A. Eugenio[1], F. Gondret[2], M. Oster[3] and C. Ollagnier[1]
[1]Swine Research Unit, Agroscope, Posieux, 1725 Fribourg, Switzerland, [2]PEGASE, INRAE, Le Clos, 35590 Saint-Gilles, France, [3]Research Institute for Farm Animal Biology (FBN), German, Wilhelm-Stahl-Allee 2, 18196 Dummerstorf, Germany; florence.gondret@inrae.fr

The collection of blood to monitor circulating concentrations of metabolites is an important tool for research to better describe the nutritional and health status of animals. In pigs, venepuncture requires the animal to be restrained, which can be stressful and painful, and large veins are also not very visible which complicates the procedure. An alternative is catheterization, but this involves a surgical procedure which induces an inflammatory response that can mask certain metabolic responses, and needs the housing of pigs in isolated pens. This study aimed to test alternative blood collection procedures and associated scaled down methods to assess circulating metabolites in growing pigs. In Exp 1, four blood collection methods were tested in pigs (n=6 females, 61 kg body weight): (1) catheterization; (2) venepuncture with the aid of a handheld infra-red imaging tool; (3) ear pricking with a lancet to collect drops of blood in filter papers; and (4) blood sucking *Dipetalogaster maxima* parasites. Blood collection was done at four time points using each method in a randomised order after feeding the pigs a liquid glucose meal. Methods were compared for the relative ease of collection, animal stress indicators, and obtained concentrations of cortisol and glucose. In Exp 2, glucose sensors with an intradermal needle were secured on the neck of pigs (n=8 females, 50 kg body weight); pigs were also equipped with a jugular catheter to serve as a control. Circulating glucose concentrations analysed by the sensor were obtained during five days with two diets (high starch then a high fat). Both basal concentrations and postprandial glucose curves during four hours after two meal test procedures were also compared with those obtained by blood sampling through the catheter. Ongoing laboratory and data analysis will reveal the best suitable approach in minimising the negative impact of blood collection in pigs used in experimental research. The PIGWEB project has received funding from European Union's Horizon 2020 under Grant Agreement No 101004770.

Session 35 Theatre 3

Development of protocols for standard management and recording in pig research facilities

A. Wallenbeck[1], M. Girard[2], M. Johansen[3], S. Düpjan[4], M. Aluwe[5], C. De Cuyper[5], E. Labussière[6], M. Font-I-Furnols[7], M. Heetkamp[8] and R. Westin[1]
[1]SLU, Box 234, 523 32 Skara, Sweden, [2]Agroscope, Rte de la Tioleyre 4, 1725 Posieux, Switzerland, [3]AU, S20, 3310, 8830 Tjele, Denmark, [4]FBN, Wilhelm-Stahl-Allee 2, 18196 Dummerstorf, Germany, [5]ILVO, Scheldeweg, 68, 9090 Melle, Belgium, [6]INRAE, Saint, Gilles, France, [7]IRTA, Finca Camps i Armet, 17121 Monells, Spain, [8]WUR, P.O. Box 338, 6700 AH Wageningen, the Netherlands; anna.wallenbeck@slu.se

European pig research infrastructures have different focus areas, but share common challenges. All perform basic pig management and often the same standard traits are recorded, providing possibilities for (improvement of) standard operating procedures (SOPs) within and across facilities. Researchers are not always aware of SOPs and their impact on the quality of their studies and data, while facility staff might not be fully aware of the potential impact of deviation from standard procedures. To safeguard good research and data quality, this awareness needs to be improved in all involved parties In the PigWeb project, one of the aims is to improve and harmonise protocols for standard management and recording in pig experimental facilities of eight partners in the network. The process to reach this aim included three main steps: (1) Identification of key areas of standard management and recording; (2) Development of improved protocols based on compilation of current practises, and (3) SWOT analyses on implementation of improved protocols. We conclude that procedures varied between facilities but key areas with potential for improvements and harmonisation could be identified. Many facilities had no written SOPs in place, even though adequate routines were applied. The primary suggestion for improvement is that if SOPs are not in place, the first important step is to develop SOPs on current procedures, leading to harmonisation within the facility. The suggested improved protocols developed in the process described above are not to be used strictly, but as templates to facilitate and promote development of SOPs that suit the specific facility and initiate communication between facility staff and researchers. Important activities in further harmonisation over facilities are knowledge exchange on SOP development.

Development of a protocol for administering a capsule to sample small intestine content

I. Garcia Viñado, M. Tretola, G. Bee and C. Ollagnier
Agroscope, Rte de la Tioleyre 4, 1725, Switzerland; catherine.ollagnier@agroscope.admin.ch

Pig's microbiome has recently been the focus of research, as modulating microbiota has profound effects on health. So far, stool samples have been the only non-invasive method to study gut microbiota, but the microbiome composition and diversity are different along the entire gastrointestinal tract (GIT) and faeces. The Capsule for Sampling (CapSa) is a swallowable device that transits through the natural digestive pathway, collects small intestine content, and is excreted with the faeces. First administration of capsules to pigs revealed that the capsules were chewed, got stuck in the stomach, or were lost to the environment after excretion in the faeces. A total of eight pilot studies have been conducted to test and optimize the procedure to administer the capsule and to improve capsules' recovery rate in the faeces. To validate the procedure, 93 Swiss Large White pigs (ranging from 6.45 to 67.1 kg) were administrated 2 capsules each and monitored the following 3 days to search for the CapSas in the faeces. Subsequently, the pigs were euthanized to locate the missing capsules. In all trails, the pH of the capsules' content was assessed and a pH level of 6 or higher was considered to rule out a stomach sampling. The capsules could be administered to 97% of the pigs, and 44% were found in the faeces within 72 hours of administration, 31% were retrieved in the stomach after euthanasia, while 25% could not be located. There was a correlation between the pigs' BW and the ability of the capsule to pass the stomach ($P<0.05$) and to be retrieved in the faeces ($P=0.03$). Capsules were not able to pass the stomach in pigs with a body weight lower than 12 kg. Overall, 72% of capsules retrieved from faeces had a pH>5.5, indicating they had sampled gut content. It is necessary to conduct additional studies to validate the administration protocol and to study the microbiome in order to verify that Capsa samples the content of the small intestine.

A modelling approach to investigate metabolic fluxes of amino acids in the small intestine of pigs

C.J.J. Garçon[1,2], J. Van Milgen[2], N. Le Floc'h[2] and Y. Mercier[1]
[1]Adisseo SAS, 2 rue marcel Lingot, 03082 Commentry, France, [2]INRAE, UMR PEGASE, 16 Le clos, 35590 St-Gilles, France; clement.garcon@inrae.fr

Studies on animal metabolism often require invasive interventions and may be limited in the coming years because of ethical concerns. This is the case to study *in vivo* metabolic processes linked to digestion, which is done with catheterized animals. Several *in silico* models of digestion or post-absorptive nutrient utilization have been developed, however little attention has been given to the intestinal tract itself. Therefore, our aim was to aggregate current knowledge to better understand the dynamics of metabolic fluxes of amino acids (AA) in the small intestine of pigs. A mechanistic model was built as a series of differential equations representing the metabolism of an unspecific AA in the intestine of pigs. The model was built with a series of functional intestinal segments, with an identical structure. Each segment included four state variables representing dietary proteins, luminal free AA, free AA in intestinal cells, and protein-bound AA. These state variables were linked by fluxes representing the main metabolic pathways of AA metabolism, namely hydrolysis of dietary protein, absorption of resulting AA, synthesis and degradation of cellular protein, endogenous secretions, and exchanges with blood. To parametrize the model, data were obtained from the literature and, if not available, values assumed as reasonable were used. A simulation was done with 1000 intestinal segments during a period of 24 h in which the animal received five meals followed by an overnight fast. During periods of feed intake, the model simulated the export of AA to the blood, while during fasting it simulated the importation of AA from the blood to maintain cellular homeostasis. About 30% of the absorbed AA did not appear in the blood due to their usage for synthesis of intestinal proteins. Part of these proteins were redirected to the intestinal lumen as endogenous secretions driven by the presence on intestinal content. In this version of the model, recycling of endogenous proteins was not yet considered, which led to an increase in intestinal sequestration of AA. This first model constitutes a basis for the further development as well as a tool to test hypotheses about metabolic fluxes of AA in the intestine.

Session 35 — Theatre 6

In vitro gas production of pre-treated or untreated feedstuffs using cecum inoculum from horses

R.H. Jensen[1], R.B. Jensen[2], M.O. Nielsen[1] and A.L.F. Hellwing[1]

[1]Aarhus University, Department of Animal and Veterinary Sciences, Blichers Allé 20, 8830 Tjele, Denmark, [2]Norwegian University of Life Sciences, Department of Animal and Aquacultural Sciences, Universitetstunet 3, 1433 Ås, Norway; annelouise.hellwing@anivet.dk

Horses are hindgut fermenters, and the composition and digestibility of feedstuffs reaching the hindgut might have an impact on the gas/methane production. Sugar, starch, protein and fat are degraded and absorbed in the small intestine so the nutrient content in digesta entering the hindgut differs from the original feedstuff. The aim of the study was to compare *in vitro* gas production from feedstuffs pre-treated to remove sugar, starch, protein and fat or untreated, using cecum fluid from horses as inoculum. Twelve feedstuffs (early or late cut grass hay, alfalfa haylage, maize silage, seed grass straw, barley straw, dried carrots, oat, maize, dried sugar beet pulp, dried apple pulp and peas) were investigated. Pre-treated feedstuffs were exposed to α-amylase and pepsin at 38 °C and washed in acetone. Both untreated and pre-treated feedstuffs were incubated for 48 hours (h) in buffered cecum fluid in the ANKOM RF gas production system. Data were fitted to a monophasic model and maximum gas production (MGP) ml/g dry matter (DM), time point of half of the MGP (T½) in h and the rate constant (C) were calculated. The data were analysed in PROC GLM with the effect of pretreatment and feedstuff. Pre-treatment removed from 17% (seed grass straw) to 78% (carrots) of the DM. Pre-treatment affected the MGB and T½, and the average MGP was 167 and 180 ml/g DM for pre-treated and untreated feedstuffs, respectively. T½ was 13.7 and 16.0 h for the pre-treated and untreated feedstuffs. There was a significant effect of feedstuffs on MGB, T½ and C. The highest MGB was found on peas (311 ml/g DM) and the lowest barley straw (69 ml/g DM). T½ was highest on grass seed straw (28 h) and lowest for dried sugar beet pulp (7 h). The rate constant was highest for maize silage and lowest for late cut grass hay. The pre-treatment decreased the gas production due to removal of dietary components with high degradability. This highlights the importance of pre-treating feedstuffs when evaluating the gas/methane production in *in vitro* systems to reflect the horse.

Session 35 — Theatre 7

Round table discussion

J. Van Milgen[1] and T. Chalvon-Demersey[2]

[1]INRAE, Institut Agro, Pegase, Le Clos, 35590 Saint-Gilles, France, [2]METEX Animal Nutrition, 32, rue Guersant, 75017 Paris, France; jaap.vanmilgen@inrae.fr

A round table discussion with early-career scientists will be organized.

Session 35 Poster 8

UpDown – an R Package to characterize unknown disturbances from longitudinal observations

I. David[1], V. Le[1,2] and T. Rohmer[1]
[1]INRAE, GenPhySE, Castanet Tolosan, 31320, France, [2]Alliance R&D, Le Rheu, 35650, France; ingrid.david@inrae.fr

The response of an animal to a disturbance depends on its resilience and the characteristics of the disturbance experienced (intensity and duration). As the latter are generally not registered on the farm, the R package UpDown was developed in order to detect and characterize unknown disturbances from the analysis of longitudinal individual observations that are now available in many species thanks to the development of electronic devices. Depending on their nature, disturbances are likely to affect one or more animals organised in hierarchical groups. For instance, in pig farming, the animals are grouped in pens which are themselves grouped in batches leading to a 3-levels hierarchical organization (individual, pen, batch). The UpDown method consists in analysing the summarised longitudinal observations at these different group levels in order to gain power to detect disturbances and facilitate their characterisation. The Up step identifies elements undergoing a disturbance at the different hierarchical levels that is validated and characterised in the down step. The UpDown package can consider as many hierarchical levels as desired and different validation options in the down step, which allow it to be adapted to different farming systems and environmental conditions. Applied to simulated data mimicking 100 days of observations in a pig farming system, the UpDown package showed a sensibility to detect elements undergoing a disturbance that increased with the hierarchical level (from 43% at the individual level to 93% at the batch level) and was associated with a good specificity for all levels (>95%). The quality of the characterization of the disturbances increased with their duration. The correlations between the estimated and the true intensities were large (>0.72 for the group scales). The median gap between the estimated and true starting (ending) date was lower than 3 days. Different information on individual trajectories can be obtained from the output of the UpDown package and used to analyse the resilience of animals after correction by the disturbances identified and characterized by the UpDown package.

Session 36 Theatre 1

Effect of sex ratio and density mating on reproductive performances in *Tenebrio molitor*

E. Soucat, K. Paul, Q. Li, A. Masseron, F. Gagnepain-Germain, A. Chauveau, E. Sellem and T. Lefebvre
Ynsect, R&D Biotech, 1 rue Pierre Fontaine, 91000 Evry, France; katy.paul@ynsect.com

Ynsect is leading an ambitious R&D program called YnFABRE, aiming to develop a genetic and genomic selection for *Tenebrio molitor*. In this way, reproduction skills belong to a trait of interest, which may be improved through genetics and environmental ways. This work aims to study the effect of sex ratio deviation and breeders' density on each female's egg quality and production. We set up an experimental design, with equivalent-aged individuals from the Ynsect population, containing 15 conditions with 2 replicates for 5. We examined the influence of sex ratio on reproductive traits with 7 different sex ratios tested (polyandry: 0.25, 0.33, 0.5; ex-aequo: 1; polygamy: 2, 3, 4); and also the effect of 8 different densities (2 to 10 animals per cup of 16 cm^2). We followed the egg production per cup twice weekly during the two first weeks. Five weeks after the start of the experimentation, egg quality was assessed by measuring the hatching rate. The sex ratio does not affect significantly the females' performance (number of eggs). Indeed, during the 2 first weeks, each female produced on average 223, 219 and, 194 eggs in the various sex ratio groups (i.e. polyandry, ex-aequo, and polygamy, respectively). It is, therefore, possible to increase the number of females to males (polygamy) in a tray to improve egg production per tray with the same breeders' number. However, on average hatching rate was significantly lower under polygamy (68%;$P<0.005$) than under polyandry or with equal breeders' sex ratio (79 and 77%). Furthermore, limiting the spawner density (~less than 0.056 g/cm^2) should be necessary, as this parameter seems negatively correlated to individual female egg production. These preliminary results provide guidelines for modifying density and sex ratio in mealworm breeding to improve egg production. Further analysis is needed to confirm these first results on a larger number of density and sex ratio with more replicates. In addition, to better understand the variation in performance due to sexual behaviour under various sex ratios and density environments, breeder tracking will be also developed.

Session 36 — Theatre 2

Behavioural responses of black soldier fly to olfactory stimuli: evaluation of a Y-tube olfactometer

A. Spindola and M. Pulkoski
Enviroflight, Animal Biology, 2100 Production Drive, 27539, USA; aline.malawey@enviroflight.net

Secondary colonizers are sensitive to microbial volatile organic compounds (MVOCs) that relay information about resources, potential mates, oviposition site, and habitat suitability. These colonizers' responses are partially governed by the sex and physiological state (e.g. gravid and non-gravid) of the fly and concentration of the MVOCs. Two days-old black soldier flies (BSF) virgin and mated males and females and gravid females were examined, by group, in a Y-tube olfactometer. The results showed a relationship between sex and physiological state related to dose-dependent attraction to MVOCs, suggesting that specific MVOCs provide distinct types of information to the flies with differing foraging interests. Specific MVOCs at a given concentration could prevent oviposition outside of egg collectors and the nuisance presence of conspecific males and non-gravid females within the oviposition site.

Session 36 — Theatre 3

Cold storage: a tool for delayed and stable black soldier fly (*Hermetia illucens*) pupae eclosion

D. Deruytter[1], S. Bellezza Oddon[2], L. Gasco[2] and C.L. Coudron[1]
[1]Inagro, Insect Research Centre, Ieperseweg 87, 8800, Belgium, [2]University of Turin, Department of agricultural, Forest and Food Sciences, Via Verdi 8, 10124 Turin, Italy; david.deruytter@inagro.be

The black soldier fly (BSF) is known and revered for its rapid development. Although this comes with its drawbacks. The fast and single egg clutch deposit by BSF females may be quite a challenge especially for small scale productions and or research colonies. One of the ways to delay and stabilize the production is to ensure control of when the eclosion of the pupae occurs. In this study, BSF (pre)pupae of different ages (from prepupae to 5 day old pupae) were stored at different cold temperatures (15 and 18 °C) for a range of exposure times (2-35 days). After cold storage, the pupae were transferred back to control conditions (28-30 °C). Finally, the 18 °C exposure experiment was done for 2 populations (Belgium and Italy) to assess any differences. During the experiments the eclosion time was assessed on a daily basis and the mortality was recorded. Based on this, the 50% eclosion time was then calculated. The results indicate that lower temperatures slowed the development of the pupae and, surprisingly, this decrease was irrespective of the pupae age or the exposure time. At 15 °C the development slowed down to 15% of the control rate and at 18 °C this increased to 31-33%. This means that it is easy to predict the eclosion of the flies when only 3 parameters are known: the age of the pupae, the storage-time in the cooler and the normal development time. The latter may differ between populations as the normal development rate at Inagro (Belgium) was 8 days, this was only 5.4 days at Unito (Italy). Finally, although storing them in a cold environment is good for delaying pupation it may significantly increase the mortality. At Inagro this was observed when storing at 15 °C for more than 14 days and near complete mortality after 35 days, but no mortality occurred at 18 °C. At Unito up to 60% mortality was observed at 18 °C. The apparent combined faster development with a lower cold tolerance does mean that cold storage is less effective for Unito compared to Inagro. Further research is needed to elucidate what causes the different observations.

Influence of temperature on coloration in honey bees

J. Bubnič and J. Prešern
Agricultural Institute Of Slovenia, Animal production department, Hacquetova 17, 1000, Slovenia; jernej.bubnic@kis.si

Abdominal coloration of honeybees was one of the first traits used to describe subspecies. The main disadvantage in using abdominal coloration as an indicator of the honeybee subspecies is the lack of knowledge of the genetic background of the trait and the subjectivity of coloration grading. Coloration seems to be strongly subjected to the environmental impact: previous studies showed that a change in abdominal coloration in workers of *A. cerana* can be induced by manipulating the environmental temperature to which workers are exposed as pupae. Similar results were obtained in *A. mellifera* queens. We incubated honey bee (*Apis mellifera carnica*) brood obtained from two colonies, one exhibiting yellow marks on the abdomen of workers and the other not (workers had only gray abdomens; labels gray or yell) at two different temperatures (30 and 34 °C; labels 30 or 34). We collected hatched workers and photographed abdomens under the stereo microscope. Images were analysed using custom written R script to obtain single value that represents the coloration of abdomen and vectors of average pixel values. We tested influence of brood origin and temperature on coloration using ANOVA. We used vectors for UMAP analysis and to train and test the support vector machine model. The mean value for coloration in 34gray group was 446.7±19.3and 458.1±19.2 in 34yell group. In 30gray group the mean value was 441.9±23.9 and 454.0±21.1 in 30yell group. ANOVA did not reveal differences between 30gray and 34gray (P=0.477) nor between 30yell and 34yell (P=0.742). Differences were confirmed between groups 34yell and 34gray (P=0.031) and between 30yell and 30gray (P=0.009). UMAP analysis did not position individual abdomens according to experimental group. The performance of trained model was: area under ROC was 0.776 classification accuracy was 0.543 F1 is 0.0541 precision was 0.547, and recall was 0.543. We observed differences in abdominal colour between different temperature regimen, however those differences were not significant. We developed a useful tool for quantifying abdominal coloration. The outputs of the tool could be used in various statistical analysis for scientific purposes as well as on field when assessing subspecies purity.

The juvenile hormone analogue, pyriproxifen, alters the body composition of *Tenebrio molitor* larvae

V. Hill, T. Parr, J. Brameld and A. Salter
University of Nottingham, School of Biosciences, Sutton Bonington, LE12 5RD, United Kingdom; victoria.hill1@nottingham.ac.uk

Tenebrio molitor larvae (yellow mealworms, MW) are a potential alternative protein source, but they contain relatively high concentrations of fat (25-36%). The aim of the study was to determine if the juvenile hormone analogue, pyriproxifen, could alter MW body composition. MW were fed wheat bran (WB) containing 3 ml acetone vehicle control (Vcont) or 2 mg pyriproxifen/kg WB (JH-L) or 15 mg pyriproxifen/kg WB (JH-H) with 4 replicate groups per treatment, each with 300 MW per group. MW were housed in an incubator in the dark at 25 °C and 60% humidity, and fed *ad libitum* feed and water. Dead and pupated MW were removed throughout. After 28 days, MW were culled and proximate nutrient and amino acid analysis conducted, along with protein analysis using SDS-PAGE. Data were analysed by one or two-way ANOVA using Genstat (20th Edition). There was a significant time × treatment interaction for average individual MW body weight (P<0.05), which was not different until day 28 when JH-L treated MW were significantly heavier than JH-H (P<0.05), but not heavier than the Vcont treatment group. After 28 days of treatment, both JH-L and JH-H treated MW had significantly (P<0.001) decreased fat contents, with JH-H decreasing by 68% compared to Vcont. Crude protein (CP) was significantly increased (P<0.001) in both JH-L and JH-H, the latter being increased by 46% compared to Vcont. SDS-PAGE indicated JH-H induced changes in the types of protein present, with the apparent induction of a distinct 150 kDa protein in JH-H, which proteomics identified as the egg-yolk protein, vitellogenin. This was accompanied by a significant (P<0.001) 17.5% increase in lysine content in JH-H compared to Vcont. The shift in protein to fat ratio demonstrates the potential to manipulate MW body composition, with the CP (70.26% of dry matter (DM)) and fat (8.88% of DM) contents of JH-H treated MW being similar to fishmeal (CP, 70.7% of DM; fat, 10% of DM). The change in CP is accompanied by changes in protein composition, which may be responsible for the increase in lysine content. However, it remains to be established to what extent the overall change in body composition is due to increased protein or reduced lipid deposition.

Fate of food pathogens during black soldier fly rearing and processing

J. De Smet, D. Van De Weyer, E. Gorrens, N. Van Looveren and M. Van Der Borght
KU Leuven Campus Geel, Department of Microbial and Molecular Systems, Kleinhoefstraat 4, 2440 Geel, Belgium; jeroen.desmet@kuleuven.be

The black soldier fly (*Hermetia illucens* L.) is a commonly used insect for the bioconversion of organic waste on an industrial scale. One of the reasons explaining its success is the ability of the larvae of *H. illucens* to thrive on a wide range of organic materials, ranging from rotting fruit to manure. While these niches can be a rich source of nutrients, they are also known to contain a high microbial load. It is thus important to evaluate the prevalence and risks associated with the presence of food pathogens during both the rearing cycle and downstream processing of the larvae. Our group has worked on this topic in the frame of three projects: Entobiota, Upwaste and SUSINCHAIN. Here, we provide an overarching view of the main findings in these projects, some of which have been published. By focusing on the fate of *Salmonella* spp. and *Staphylococcus aureus* during rearing, we observed a more nuanced pathogen reduction by the larvae, than sometimes reported in literature. Based on these findings, we advise to be careful with claims on the ability of the black soldier fly larvae to reduce pathogens. Therefore downstream processing steps are crucial to ensure sufficient reduction of food pathogens that may persist in the larvae. We have investigated the effectiveness of three innovative strategies for processing: extrusion, microwave, and radio-frequency drying. Another observation is that some food pathogens are only rarely associated with insects (e.g. *Clostridium perfringens*), making it difficult to assess the effectivity of the current processing steps against these food pathogens. In the future, we will therefore also use pathogen supplementation during processing to fully understand the final microbiological risk for the end-consumer of insects.

Session 36 Theatre 7

Insect as animal feed: Fourier-transformed spectroscopy to reveal their protein molecular structure

J. Ortuño and K. Theodoridou
Institute of Global Food Security, Queen's University Belfast, 19 Chlorine Gardens, Belfast, BT9 5DL, United Kingdom; k.theodoridou@qub.ac.uk

The aim was to characterize for first time the protein molecular structure of four insect species (black soldier fly larvae-BSF, Mealworms-MW, Field Crickets-FC and Banded Crickets-BC) in correlation with their nutritive value, for monogastric animals. Three batches of those species were purchased in three consecutive weeks from a commercial supplier. Insects were sieved to remove frass and litter, fasted for 12 h at room temperature, washed, freeze-dried until constant weight, milled and sieved (<1 mm). In total 12 samples were analysed using 3 different techniques: FTIR, wet chemistry and *in vitro* nutrient digestibility (DMd). No significant difference for CP or EE content between the two types of crickets, while BSF showed the lowest ($P<0.001$) CP compared with the rest of the species. Both crickets had higher ($P<0.05$) chitin content than BSF and MW. NDF was similar between BSD and MW but lower ($P<0.001$) than FC and BC. MW showed the highest DMd (81.2%) ($P<0.05$) than the other species. This is the first study exploring *in vitro* the kinetics of the different DMd phases in insects. The pancreatin DMd in both cricket species was higher (>30%) ($P<0.05$) than for MW (<20%) and BSF (<10%). The protein molecular structure was identified and the AmideII height was greater ($P<0.05$) in BSF, followed by the two crickets and MW. AmI band used to predict protein secondary structure, was lower ($P<0.001$) for BSF (0.311) compared to MW, FC and BC (0.644, 0.810, 0.712, respectively). BSF contained lower ($P<0.001$) percentage of a-helix (0.307) and β-sheet (0.303), compared to MW, FC, and BC (0.609, 0.778, 0.679 and 0.599, 0.756, 0.653, respectively). Positive correlation (r>0.95; $P<0.001$) between the AmI height, α-helix and β-sheet with the CP was found. Chitin showed only a modest correlation (r=0.58; $P<0.05$) with the a-helix/β-sheet ratio. Digestibility during the pancreatin phase was correlated ($P<0.01$) with CP (r=0.91) and NDF (r=0.80). Insect powders showed differences in their nutritive value which was reflected in their protein molecular structure. In conclusion, FTIR has the potential to become a rapid technique to predict nutritive value of insect powders included in animal diet and prevent fraud and adulteration.

Session 36 Theatre 8

Proteomics evaluation of the barrier role of insects for the indirect recycling of fast food in feed

M.C. Lecrenier[1], M. Aerts[2], A. Cordonnier[1], L. Plasman[1], O. Fumière[1] and V. Baeten[1,2]
[1]Walloon Agricultural Research Centre (CRA-W), Chaussée de Namur, 24, 5030 Gembloux, Belgium, [2]University of Louvain-la-Neuve (UCLouvain), Croix du Sud, 2, 1348 Louvain-la-Neuve, Belgium; m.lecrenier@cra.wallonie.be

Since 2021, insect meal is authorised in feed intended for pig and poultry, in addition to that for fish. Insect meal is considered as one of the most promising alternative ingredient totally in line with the EU goal aiming to reduce its dependency on critical feed materials like soya. Even these regulatory changes seem to increase the insect market, its use in feed remains limited and still more expensive that other protein sources. One of the reasons for this high price is linked to the limitation of authorised substrates for breeding. Indeed, from a legislation point of view, edible insects are considered as farmed animals and therefore they must follow EU animal by-products regulation. Following the same sustainability strategy, EU has also adapted legislation to promote the use of another feed source, the former foodstuffs (FFS). FFS retain a significant nutritional value and their use is completely in line with the current trend of circular economy. However, FFS containing meat or fish remain prohibited. Their use as a substrate for insect breeding could in the future be a contributing way to tackle food waste and an indirect use of these products for non-ruminants by having a role as health barrier. But is it really a sufficiently strong barrier? At the end of the insect rearing, insects are separated from feed media and incorrect procedure may lead to the presence of residual feed materials. The objective of this study was to develop a UHPLC mass spectrometry-based proteomics method to detect the presence of animal proteins in insect meal. Non-ruminant meat, raw or cooked or under its fast food form, were used to identify specific peptides markers. In order to evaluate the sensitivity of the protocol, insect substrate adulterated or not with meat as well as insect reared on this substrate were analysed. Analyses were performed by liquid chromatography (Acquity UHPLC system, Waters) coupled with a triple quadrupole mass spectrometer (Xevo TQ-XS, Waters). Results will be presented and discussed during the meeting.

Session 36 Theatre 9

Scaling up fly mating chambers: lessons learned from operating 4 and 24 m^3 fly mating chambers

S.P. Salari and M.L. De Goede
InsectoCycle, Bronland 10, 6708 WH, the Netherlands; seppe.salari@insectocycle.nl

InsectoCycle has been dedicated to developing large-scale black soldier fly reproduction units. These units feature a fly mating chamber (light cages) that is either 4 or 24 m^3 and have been in operation for the past two years. Compared to typical laboratory scale experiments, these units are notably larger, resulting in new interactions and diverse microclimates within the cage. In-depth research has been conducted to comprehend the key factors that impact fly mating and egg deposition, as well as how to measure and influence these factors. Some of the factors that have been examined include air temperature, surface temperature differences between the floor and walls, humidity at the floor vs the ceiling, airflow through the cage, and the necessary refreshment rate to regulate CO_2 and ammonia levels for optimal performance. However, controlling these elements simultaneously can be challenging and determines the success and stability of the reproduction process.

ALL-Yn: An automated solution for phenotyping *Tenebrio molitor* larvae

Q. Li[1], T. Mangin[2], J. Richard[1], E. Sellem[1], A. Masseron[1], F. Gagnepain-Germain[1], T. Lefebvre[1] and R. Baude[2]
[1]Ynsect, R&D Biotech and Innovations, 1 rue Pierre Fontaine, 91000 Evry, France, [2]Aprex-Solutions, 2 allée André Guinier, 54000 Nancy, France; qi.li@ynsect.com

Tenebrio molitor is a promising alternative to produce high quality protein products with low environmental impact. The industrial scale production handles an important number of insects, putting forward the emphasis on genetic diversity and performances. The high number of individuals leads to the development of automated solutions for insect phenotyping. To that end, All-Yn has been developed to automatically count the number of larvae and assess the biomass of both larvae and feed leftover by associating artificial intelligence state of the art and parametric computer vision algorithm. 50 samples, containing in total 969 larvae, have been simultaneously counted and weighted manually and with All-Yn. According to the sample, different larval weights were assessed (min=23.8 mg, max=247.5 mg). The feed leftover were also quantified. Anova analysis has been used to validate these comparisons. Out of 969 larvae, 939 are detected and segmented for the first batch of samples, with no false positives counted. A subset of 224 larvae was assessed with the laboratory set up to phenotype the weight. The first results show no significant (P>0.05) differences between the predicted biomass and the reference (respectively 1,958.5±999.2 mg and 1,951.2±975.6 mg on average). The feed leftover biomass evaluation is still ongoing. These first outcomes are promising and will empower the AI learning. The dataset has to be completed with more samples aiming to reach higher detection rate. Through the cooperation of neural network and segmentation, we will be able to propose a highly robust and flexible set up for daily laboratory work by reducing the bench lab duration (minutes per sample) and by increasing the data accuracy and objectivity. These technological approaches, applied to other phenotypes will improve and facilitate the traits assessments in large population for the development of a genetic or genomic selection scheme.

Does sex matter? The impact of sex on the buffalo pupation phase

N. Gianotten, N. Heijmans and N. Steeghs
Ynsect Netherlands, R&D Biotech, Harderwijkerweg 141B, 3852 AB, the Netherlands; natasja.gianotten@ynsect.com

In insect rearing, the number of individuals per tray or container can be very large. Depending on the type of rearing system, tens of thousands or even hundreds of thousands of individual insects per tray are not uncommon. In lesser mealworms (*Alphitobius diaperinus* or buffalo) the sex of an individual is not easy to distinguish. Currently, in the large-scale lesser mealworm rearing at Ynsect Netherlands, the sex is not considered, both in the larval and the adult stage. Previous publications have shown an impact of sex on the individual weights of pupae and beetles, with females being bigger than males. These experiments were executed more than 25 years ago, which is very long in this field of work. It was therefore decided to repeat this experiment with our current line of *A. diaperinus*, which is already artificially reared for feed and food purposes for more than 40 years. Additional to individual weights of female and male pupae and beetles, also the impact of sex on moment of pupation, and on length of pupal period is investigated. The possible impact of biological parameters like individual weight, length of pupation period and male/female ratio in ovipositing beetles on the large-scale rearing of the *A. diaperinus* will be discussed.

Assessment of chemical hazards in insect meal production for aquaculture feeds

I. Amaral[1], S.M. Ahmad[1], M. Ângelo[1], P. Correia Da Silva[1], A. Quintas[1], J.A.A. Brito[1], D. Murta[1,2], M.V. Santos[3], I. Vieira[3], T. Ribeiro[2] and L.L. Gonçalves[1]
[1]CiiEM (Centro de Investigação Interdisciplinar Egas Moniz, Egas Moniz School of Health and Science, Monte de Caparica, 2829-511 Caparica, Portugal, [2]Entogreen – Ingredient Odyssey, S.A., R. Eng. Albertino Filipe Pisca Eugénio no.140, 2005-079 Santarém, Portugal, [3]Thunder Foods, Rua Santo António Lt 9, Zona Industrial Santarém, 2005-002 Santarém, Portugal; lgoncalves@egasmoniz.edu.pt

Insect meals are sustainable alternatives to the ingredients currently used in formulation of aquaculture feeds. However, the use of insects as feed raises safety concerns, stemming from a combination of factors: the nature of the substrate used to feed the insects, the specific production methods, the harvest stage, the insect species as well as the methods used for further processing. In fact, the presence of a wide variety of chemical contaminants, including heavy metals (Hg, Cd, Pb), metalloids (As), and POPs (persistent organic pollutants) such as Dioxins, PCBs, pesticides, can be found in these products. However, published data on hazardous chemicals in captive bred insects and their possible transfer from different substrates to insects and from insects to other animal species are scarce. Thus, the present study aims to assess the abovementioned chemical risks in insect meals and in fish feed, throughout the food chain. For that purpose, 6 agricultural by-products selected for the growth of *Tenebrio molitor* (TM) and *Hermetia illucens* (BSF) and fish feed formulations obtained from the mentioned insects were screened for elements of toxicological concern by wavelength dispersive X-ray fluorescence spectrometry (WDXRF), using a 4 kW commercial system (Bruker S4 Pioneer), according to published methods. Pesticide residues, such as organochlorides, were monitored in the same samples, by GC-MS (Agilent 6890N with a 5973 Network GC/MS system at SIM mode), using a QuEChERS method from the literature. Preliminary results revealed that the agri-food by-products, as well as the insect meals tested, did not present chemical risks that could compromise their future use in feed formulations for aquaculture.
Acknowledgments: The InFishMix project (PT-INNOVATION-0094) is funded by Iceland, Liechtenstein and Norway through the EEA and Norway grants.

Microbiological risk analysis in insect meal production

D. Guerreiro[1], D. Murta[1,2], M.V. Santos[3], I. Vieira[3], T. Ribeiro[2] and H. Barroso[1]
[1]CiiEM, IUEM, Egas Moniz School of Health and Science, Campus Universitario, Qta da Granja, Monte de Caparica, 2829-511 Caparica, Portugal, [2]Entogreen, Zona Industrial de Santarém, 2005-079 Santarém, Portugal, [3]Thunder Foods, Zona Industrial de Santarém, 2005-332 Santarém, Portugal; mhbarroso@egasmoniz.edu.pt

Currently, the ingredients used in the manufacture of aquaculture feeds are expensive and their use pose a risk to sustainability, which leads aquaculture food producers to look for other, friendlier alternatives. The circular economy is a key to tackle climate change and redefine the economy. With the world population increase, it becomes necessary to look for alternative sources of protein. Insects are a viable alternative with their production having a positive environmental impact. This study is part of the InFishMix project where the main goal is the development of a new ingredient for aquaculture feed, mixing different species of insects produced from agroindustry by-products. Throughout the production process, the quality and safety of by-product substrates, insects, feed and fish will be monitored, with the objective of guaranteeing safe fish for human consumption. So far, microbiological analyses were performed for: (1) the substrates used for the insects feed, namely, vegetable by-products mixture, olive pomace, soy pomace and wheat bran; (2) insects (black soldier fly and *Tenebrio molitor* larvae) fed with these substrates, and (3) processed black soldier fly and *Tenebrio molitor* larvae. The presence of different potentially pathogenic microorganisms of importance to human health were assessed (by conventional methods), as well as mycotoxins (by ELISA assays). The results showed that the processed insect larvae obtained are microbiologically safe. Of the different substrates studied, it was found that olive pomace was the one with the lowest microbial contamination. The use of olive pomace as feed for the larvae may contribute to the resolution of a serious environmental problem. We conclude that the use of the studied agroindustry by-products to feed insects is safe, resulting in a safe ingredient for aquaculture feed in terms of microbiological evaluation. Acknowledgments: The InFishMix project (PT-INNOVATION-0094) is funded by Iceland, Liechtenstein and Norway through the EEA and Norway grants.

Multi directional approach in fly emergence evaluation and forecasting for *Hermetia illucens*

B. Grodzki, M.G. Zhelezarova, T. Bruder and M. Tejeda
Nasekomo AD, Saedinenie 299, 1151, Bulgaria; bartosz.grodzki@nasekomo.life

Various studies have been conducted on intrapuparial BSF development aiming at establishing the concrete development phases from a pupa into fly. Having a clear knowledge about the stages sequence and length can enable prediction of insects emergence time. Furthermore, as synchronicity in development is crucial in black soldier fly farming, this basic knowledge of the fly biology can bring benefit in large scale production operations, by reducing loss due to delayed emergence and facilitate the animals maintenance. The late larva-prepupa-pupa-adult transition were studied simultaneously with three different approaches. A thermal model of development, physiological stage tracking by pupa dissections and an object detection algorithm approach. Production batches were grown into two temperature controlled rearing rooms and monitored from late pupa till emergence, dissections were performed at different stages of pupa development, photographed under the stereomicroscope, and analysed with a computer algorithm. The different approaches combined information allows Nasekomo future production to forecast the batches time of emergence from environmental or dissection data. Furthermore, the algorithm was able to detect specific structures like legs and compound eye of the pharate insect and thus could also be used to determine stage of development and probably, quality of production. Described models and tools can be used for precise management of fly emergence in reproduction sector of BSF colony. Proper evaluation and forecasting of adults emergence allow to deliver constant (in time and numbers) outputs of flies to the system and plan production according to the population's dynamic.

Application of NIRS on *Tenebrio molitor* protein meal

J. Richard[1,2], L. Daraï[1], L. Sanchez[1] and B. Lorrette[1]
[1]Ynsect, R&D Animal Nutrition & Health, 1 rue Pierre Fontaine, 91000 Evry, France, [2]Ynsect, R&D Biotech, 1 rue Pierre Fontaine, 91000 Evry, France; lorena.sanchez@ynsect.com

The potential of near infrared spectroscopy (NIRS) was studied to evaluate the ŸnMeal composition of total fat and protein, moisture and minerals. ŸnMeal, a highly digestible protein concentrate from *Tenebrio molitor* larvae, is perfectly suitable for the nutrition of farmed fish and pets. This study presents the development of calibration models of ŸnMeal content with NIRS technology and it's daily use potential during routine applications for quality control. Analyses were done with the NIRS XDS (400 to 2,500 nm; Metrohm) in a moving cell by reflectance with a spot size of 17,25 mm and an average of 32 scans. All the spectra and calibrations have been analysed with Vision®, the dedicated software developed by Metrohm. Each sample involved in calibration, have been analysed by NIRS and sent for external analysis (sub-contractor using COFRAC methods) to obtain reference values. Calibrations are made with PLS method and different mathematic treatments have been used to determine the best parameters for each calibration. Calibrations have been created for moisture, protein, lipid, and ash content from 290 samples, and varying from 0.1 to 12.0%, 62.1 to 77.0%, 8.6 to 15.2% and from 3.1 to 5.5% respectively. Low uncertainty of measurement have been observed (1.0, 1.6, 1.0 and 0.3% for moisture, protein, lipid and ash calibration respectively). Every uncertainty obtained are significantly lower than the standard deviation of the data set. These NIRS calibrations are used in routine quality control since 2018. They are monthly controlled with 5 samples analysed in parallel with an external control and they are fed with new data every year making them more accurate. The accumulated results obtained in our lab conditions, highlight the suitability of NIRS for ŸnMeal content control and pave the road for the development of other product quality controls.

Session 37 Theatre 1

State of the art in understanding interorgan crosstalk and physiology in farm animals

I. Cassar-Malek[1], I. Louveau[2], C. Boby[1], J. Tournayre[1], F. Gondret[2] and M. Bonnet[1]
[1]Clermont Auvergne University, INRAE, VetAgro Sup, UMR Herbivores, 63122 Saint-Genès-Champanelle, France, [2]PEGASE, INRAE, Institut Agro, 35590 Saint-Gilles, France; isabelle.cassar-malek@inrae.fr

One of the challenges for livestock production is to manage phenotypic traits associated with health and production performance across diverse production systems in a wide range of climatic conditions and agro-ecological practices. Such a management requires a better understanding of the physiology of key tissues and organs involved in these traits. Therefore, it is important to determine the mechanisms behind the traits (feed efficiency, body composition, meat quality) and their regulation along physiological states (lactation, reproduction, growth) and in changing environments (robustness), and to get an integrated view at the level of the animal. Inter-organ communication, so called organ crosstalk, is required to regulate the different physiological functions, and to orchestrate multiple signalling molecules to maintain body homeostasis. In this review, we will show that tremendous opportunities are now offered for ruminants and pigs with the recent advances in: (1) omics (transcriptomics, proteomics, lipidomics) technologies for the profiling of organs and biofluids, and the knowledge of mediators in extracellular communication (e.g. microvesicles/ exosomes); (2) *in silico* analysis with bio-informatics prediction tools (secretome/surfaceome, interaction networks); (3) integration of heterogeneous data (including multivariate analyses, clustering, network inference, graphs); (4) biosensing technologies; and (5) organ- on- chip systems. This will be helpful to get a holistic view for production and adaptation traits (incl. symbiotic interactions with resident microbiota), and to discover biomarkers to develop minimally invasive phenotyping tools to qualify these traits and assess the impact of rearing practices.

Session 37 Theatre 2

Gene networks controlling functional cell interactions in the pig embryo revealed by omics studies

A. Dufour[1], C. Kurilo[2], J. Stöckl[3], Y. Bailly[4], P. Manceau[4], F. Martins[2], S. Ferchaud[4], B. Pain[5], T. Fröhlich[3], S. Foissac[2], J. Artus[6] and H. Acloque[1]
[1]Université Paris Saclay, INRAE, AgroParisTech, GABI, 78350 Jouy en Josas, France, [2]Université de Toulouse, INRAE, ENVT, GenPhySE, 31326 Castanet-Tolosan, France, [3]LMU, Genzentrum, Feodor-Lynen-Str. 25, 81377 München, Germany, [4]INRAE, GenESI, 86480 Rouillé, France, [5]Université de Lyon, Inserm, INRAE, SBRI, 69500 Bron, France, [6]Université Paris Saclay, Inserm, UMRS1310, 7 rue Guy Moquet, 94800 Villejuif, France; herve.acloque@inrae.fr

Pig embryonic development differs from that of humans and mice from the blastocyst stage and is characterised by a much later implantation. This particular period is associated with a lengthening and significant growth of the extra-embryonic tissues and is still poorly understood. To better understand the biology of the pig blastocyst, we generated a large dataset of single-cell transcriptomics (scRNAseq) and multi-omics (paired RNAseq and scATAC-seq) at different embryonic stages (early, late, ovoid and elongated blastocysts) and the associated proteomic dataset from the corresponding uterine fluids. These data were cleaned, filtered and represent a total of 35,000 cells. Using these data, we first characterised embryonic and extra-embryonic cell populations and their evolution, and identified specific markers of these populations. We then used a pig-adapted version of the SCENIC package to infer gene regulatory networks (regulons) and selected those that were specifically active in each embryonic population. Meta-analysis of other scRNAseq publications on the preimplantation embryo in pigs and humans increased confidence in the regulons we identified. We then linked these regulons to signalling pathways and biological processes. To do this, we used the CellCom software to build signalling networks from ligands (expressed by cells or present in the uterine fluids), receptors, intermediary players, to transcription factors. Thanks to this integrative study, we have identified key new players controlling the biology of the three main cell lineages and identified novel stage-specific subpopulations. Taken together, our work provides new insights into the functional interactions of the first cell lineages within the conceptus prior to implantation.

1H-NMR metabolomic study of Large White and Meishan pigs in late gestation: part 1 – foetal placenta

J. Guibert[1], A. Imbert[1], N. Marty-Gasset[1], L. Gress[1], C. Canlet[2], L. Canario[1], Y. Billon[3], A. Bonnet[1], L. Liaubet[1] and C.M.D. Bonnefont[1]
[1]GenPhySE, Université de Toulouse, INRAE, INPT, ENVT, 31326 Castanet Tolosan, France, [2]Axiom Platform, MetaToul-MetaboHUB, National Infrastructure for Metabolomics and Fluxomics, 31027 Toulouse, France, [3]GenESI, INRAE, Le Magneraud, 17700 Surgères, France; cecile.bonnefont@inrae.fr

The most critical period for piglet survival is the first three days after birth. The risk of piglet mortality is increased if the piglets have a lower degree of maturity at birth, resulting from intra-uterine growth retardation. Foetal development is highly dependent on the functioning of the placenta, and is accelerated in the later stages of gestation. Therefore we decided to study the placental metabolism at the end of gestation. To better understand the relationship between foetal development and piglet maturity at birth, we chose to compare 14 Large White (LW) sows, representing the European commercial breed that is highly affected by piglet mortality before weaning, with 14 Meishan (MS) sows, representing the Asian breed with lower piglet mortality. All sows were inseminated with mixed semen from both breeds and develop purebred and crossbred foetuses. Placental samples were collected from 224 foetuses at 90 or 110 days of gestation (dg). Hydrophilic metabolites were extracted and ^{1}H-NMR spectra were acquired. Raw spectra were processed using the ASICS R package to identify and quantify 48 metabolites. PLS-DAs were performed at 90 dg and 110 dg. Both separated the LW and MS purebred foetuses on the 1st latent variable (LV), explaining 22 and 16% of the variability, respectively. The crossbred foetus were intermediate. There were 19 and 21 metabolites important (VIP>1) for the 1st LV at 90 and 110 dg, respectively. The metabolites were mainly involved in amino acid metabolism and in energy metabolism. Metabolites were then correlated with morphological parameters. In particular, placental myo-inositol concentration was negatively correlated with foetal body mass index (BMI, r=-0.4) and positively correlated with foetal brain weight/body weight ratio (r=0.4), indicating stunted growth. In conclusion, differences in metabolism between the foetal genotypes may be involved in their differences in survival at birth.

Metabolic pathways leading to different intramuscular fat content in rabbit divergent lines

P. Hernández, A. Blasco and A. Zubiri-Gaitán
Institute for Animal Science and Technology. Universitat Politècnica de València, Camino de Vera S/N, 46022 Valencia, Spain; phernan@dca.upv.es

A divergent selection experiment for intramuscular fat content (IMF) was developed in rabbits at the Universitat Politècnica de València. These divergent lines were used to perform an untargeted metabolomic analysis with the aim of identifying the metabolic pathways involved in their differentiation. Plasma samples were obtained from 24 animals of the high IMF (H) and 24 of the low IMF line (L) at 9 weeks of age, after fasting for 4 h. The metabolomic profile was obtained by UPLC-MS/MS, detecting 997 metabolites. Metabolites with more than 10% of zeros in at least one line were removed, and the remaining zeros were imputed using random forest. ALR transformation was applied to account for the compositional nature; then the data was standardized and analysed using two multivariate approaches: PLS-DA to find the differences between the lines, and PLS to find the metabolites that adjusted linearly to IMF. Double cross-validation was performed to select the relevant metabolites without overfitting. Metabolites with VIP≥0.8 and Jack Knife confidence interval that did not include the 0 were selected. A total of 322 metabolites were relevant in the discrimination between the lines and adjusted linearly to the IMF, being mostly involved in the metabolism of lipids and amino acids. The L line had greater abundance of lipids including, among others, several free fatty acids and acylcarnitines, and greater abundance of metabolites from the arginine metabolism. The H line showed greater abundance of free carnitine, branched-chain amino acids (BCAA) and BCAA-associated species like butyrylcarnitine and propionylcarnitine, and metabolites from the aromatic amino acids and the histidine metabolisms. The changes in lipids metabolism evidenced a reduced uptake in the muscle and adipose tissues and an impaired β oxidation of fatty acids of the L line, indicating a limited capacity to obtain energy from lipids. The H line showed greater BCAA catabolism, which could partially explain its greater IMF content. Additionally, the changes in the metabolisms of BCAA, aromatic amino acids, arginine, and histidine suggest the relevant role of microbiome activity in the development of the trait.

Session 37 Theatre 5

Proteomics profiles of longissimus thoracis muscle of Arouquesa cattle from different systems

L. Sacarrão-Birrento[1], A. Dittmann[2], S.P. Alves[3], L. Kunz[2], D.M. Ribeiro[1], S. Silva[4,5], C.A. Venâncio[5,6] and A.M. De Almeida[1]

[1]LEAF, ISA, Tapada da Ajuda, Lisboa, Portugal, [2]Functional Genomics Center Zürich, University of Zurich, Zurich, Switzerland, [3]CIISA and Al4Animals, FMV, Lisboa, Portugal, [4]CECAV, UTAD, Vila Real, Portugal, [5]ECAV, UTAD, Vila Real, Portugal, [6]CITAB, UTAD, Vila Real, Portugal; laurasvbirrento@isa.ulisboa.pt

Beef is one of the most important protein sources in human nutrition, so it is essential to guarantee its quality. Local breeds, such as the Arouquesa from Northern Portugal, produced under extensive systems lead to high quality and healthier products. The aim of this work was to compare the protein abundance profiles of beef from the Arouquesa traditional production system with improved systems, using label-free proteomics. Sixty animals were allocated to five systems: TF (n=11), the traditional where the animals were weaned and slaughtered at 9 months; TFS1 (n=13), the addition of an initiation supplement (S1); S1S2 (n=15), animals eating S1 until the weaning at 5 months and then a growing supplement (S2) until 9 months; TFS3 (n=10) and S3 (n=11), produced like TFS1 and S1S2, with the addition of a rearing period until 12 months and a finishing supplement (S3). We used label-free quantification to determine protein profile in the Longissimus thoracis muscle (10 samples per group). The meat quality had only differences on the exudative and cooking losses being higher in TF and TFS1 groups. We identified 3,774 proteins in all comparisons. The proteins were considered differentially abundant with $P<0.05$: 18 in the TFxTFS1 comparison, with TF beef showing higher accumulation of proteins involved in actin (fascin and GC) and calcium binding (hyaluronoglucosaminidase) pathways; 28 in the TFxS1S2 comparison, with actin, ATP and calcium binding metabolic pathways being affected; 65 in the TFxTFS3 comparison that were involved in proteolysis (Tripeptidyl-peptidase 2), muscle contraction (Myosin heavy chain 4) and proteasomal protein processes; and finally, 36 in the TFxS3 comparison involved in glucose metabolic process (Phosphoglucomutase-1 and fibrillin-1) and in protein transport (exportin-2). Some potential biomarkers of meat quality when comparing the traditional group with the new systems could be established.

Session 37 Theatre 6

Serum metabolomics of newborn lambs before and after colostrum feeding

L. Lachemot[1], J. Sepulveda[1,2], X. Such[1], J. Piedrafita[1], G. Caja[1] and A.A.K. Salama[1]

[1]Universitat Autonoma de Barcelona, Ruminant Research Group (G2R), Campus UAB, 08193 Bellaterra, Spain, [2]Universidad Nacional de Colombia, Facultad de Medicina Veterinaria y Zootecnia, Edificio Uriel Gutiérrez, Bogotá D.C., Colombia; ahmed.salama@uab.cat

Colostrum feeding to the neonatal lamb is essential for health and survival. Nevertheless, little is known about the changes in blood metabolomics in neonatal lambs during the first hours of life. Ten lambs were separated from their mothers just after birth, and blood samples were collected. Thereafter, lambs received 2 colostrum meals (200 ml each) at 1 and 5 h after birth. A second blood sampling was done at 6 h after birth. Serum was obtained and frozen at -80 °C until the metabolomic analyses. Blood serum was ultra-filtrated (3 KDa cutoff), mixed with DSS (0.5 mM), and analysed using ^1H nuclear magnetic resonance spectroscopy at 600 MHz. Metabolites were quantified using Bayesil (bayesil.ca). Enriched pathways were identified by KEGG using MetaboAnalyst (v. 5.0). Feeding colostrum resulted in an increment ($P<0.05$) in the blood concentrations of D-glucose (+71%), acetone (+28%), 3- hydroxybutyrate (+61%), isobutyric acid (+111%), formate (+15%), and 8 amino acids (Glu, Ile, Leu, Lys, Met, Phe, Tyr, and Val) that increased by 47 to 189%. On the other hand, colostrum consumption caused decreased ($P<0.05$) levels of lactate (-42%), acetate (-27%), glycerol (-54%), succinate (-67%), betaine (-31%), carnitine (-34%), choline (-38%), creatinine (-25%), and 5 amino acids (Ala, Arg, Asp, Gly, Thr) that were reduced by 51 to 74%. Additionally, there were 8 altered pathways (FDR<0.05) after colostrum feeding. These pathways included aminoacyl-tRNA biosynthesis, Phe-Tyr-Trp biosynthesis, Phe metabolism, Gly-Ser-Thr metabolism, Ala-Asp-Glu metabolism, and glyoxylate and dicarboxylate metabolism. In conclusion, the metabolomic blood profiling of neonate lambs revealed several changes in the intermediates of TCA and lipid metabolism to satisfy their energy needs during the first hours of life. Amino acid metabolism pathways were the most affected pathways by colostrum ingestion. Acknowledgments: Funded by the Spanish Ministry of Science and Innovation (Project RTA #PID2020-113913RR).

Session 37 Theatre 7

Characterization of milk small extracellular vesicles to study adaptation to lactation in ruminants

C. Boby[1], A. Delavaud[1], J. Pires[1], L.E. Monfoulet[2], S. Bes[1], S. Emery[1], L. Bernard[1], C. Leroux[1], A. Imbert[1], M. Tourret[1], F. Fournier[3], D. Roux[3], H. Sauerwein[4] and M. Bonnet[1]

[1]Université Clermont Auvergne, INRAE, VetAgro Sup, UMR Herbivores, Centre INRAE CARA, 63122 Saint-Genès-Champanelle, France, [2]Université Clermont Auvergne, INRAE, Human Nutrition Unit, Centre INRAE CARA, 63122 Saint-Genès-Champanelle, France, [3]INRAE, Herbipôle, Centre INRAE CARA, 63122 Saint-Genès-Champanelle, France, [4]University of Bonn, Institute of Animal Science, Physiology Unit, Katzenburgweg 7, 53115 Bonn, Germany; celine.boby@inrae.fr

Small extracellular vesicles (EV) are secreted into the extracellular space by all cells. Due to the diversity of their cellular origin and the molecules they contain, small EVs have the ability to ensure extracellular communication and carry molecular signatures of their tissue of origin and its physiological state. In ruminants, early lactation is characterized by profound changes in energy balance and metabolic status. To explore physiological adaptations during early lactation, this study investigated the use of non-invasive milk EVs to identify specific indicators of inter-organ signalling. Milk samples were collected from 8 cows on weeks 2 and 7 postpartum, corresponding to negative and neutral energy balance, respectively. Small EVs were isolated by ultracentrifugation coupled with size exclusion chromatography and characterized by morphological, biophysical and biochemical criteria. Labelled-free shotgun quantitative proteomics was performed by nanoLC-MS/MS. Electron microscopy revealed cup-shaped vesicles with a diameter of about 100 nm, characteristic of small EVs. The diameter was confirmed by Tunable Resistive Pulse Sensing and the specificity of small EVs isolation by the presence of cytoplasmic (TSG101) and membrane (CD63) markers. A total of 508 proteins were identified in milk EVs at weeks 2 and 7 of lactation. Multilevel PCA analysis showed a clear separation between the 2 time points, indicating a strong effect of lactation stage on the protein composition of milk EVs. This proteomic dataset will be analysed to identify small EV molecular signatures of tissue interactions that coordinate nutrient partitioning and adaptation during early lactation.

Session 37 Theatre 8

Multi-tissue transcriptome analysis of bovine herpesvirus-1 (BoHV-1) challenged dairy calves

S. O'Donoghue[1,2], B. Earley[2], M.S. McCabe[2], S.L. Cosby[3], K. Lemon[3], J.W. Kim[4], J.F. Taylor[4], D.W. Morris[1] and S.M. Waters[2]

[1]University of Galway, Discipline of Biochemistry, Galway city, Galway, Ireland, [2]Teagasc, Animal and Bioscience Research Dept, Grange, Co. Meath, Ireland, [3]Agri-food and Biosciences Institute, Veterinary Sciences Divison, Stormont, Belfast, United Kingdom, [4]University of Missouri, Division of Animal Sciences, Columbia, MO, USA; stephanie.odonoghue@teagasc.ie

BoHV-1, a key virus associated with the onset of bovine respiratory disease (BRD), is a leading cause of morbidity and mortality in cattle. Previously, changes in gene expression were identified in the whole blood of BoHV-1 infected dairy calves, but no study has examined responses of key respiratory or immune associated tissues. Our study objective was to examine the gene expression profiles of multiple respiratory and immune associated tissues in BoHV-1 infected dairy calves. Holstein-Friesian bull calves (mean age(SD) 149.2 days(23.8); mean weight(SD) 174.6 kg(21.3)) were challenged with either BoHV-1 inoculate (6.3×10^7/ml × 1.35 ml)(n=12) or sterile phosphate buffered saline (n=6). Animals were euthanised on day 6 post-challenge and tissue samples (lesioned (LL) and healthy right cranial lobe (HL), mediastinal (MLN) and bronchial lymph nodes (BLN) and pharyngeal tonsil (PGT)) were collected. Total RNA was extracted, RNA quality determined, libraries prepared using the TruSeq stranded mRNA kit and then sequenced on an Illumina NovaSeq 6000 (paired-end, 100 bp (MLN, PGT) and 150 bp (HL, LL and BLN). Sequence reads were quality assessed, trimmed, and aligned to the ARS-UCD1.2 bovine reference genome. Differential expression analysis was conducted using EdgeR. Differentially expressed genes (DEGs) were input to DAVID for pathway and gene ontology (GO) analysis and DEGs identified between control and infected animals in all tissues (P<0.05, FDR<0.1, FC>2). The PGT had the greatest number of DEGs with 1,833 while 67 were identified in the LL. Across all tissues, KEGG pathways for Influenza A and Hepatitis C and the GO terms defence response to virus, innate immune response and negative regulation of viral genome replication were enriched (P<0.05, FDR<0.05). The herpes simplex infection pathway was enriched in the MLN, LL and LH. These findings provide further insight into the host immune response to BoHV-1 infection.

Session 37 Theatre 9

Transcriptome analysis reveals a different immune response depending on a SOCS2 gene point mutation

C. Oget-Ebrad[1], C. Cabau[1], S. Walachowski[2], N. Cebron[2], J. Sarry[1], C. Allain[3], R. Rupp[1], G. Foucras[2] and G. Tosser-Klopp[1]

[1]GenPhySE, Université de Toulouse, INRAE, 24 chemin de Borde Rouge, 31326 Castanet-Tolosan, France, [2]IHAP, Université de Toulouse, ENVT, INRAE, 23 Chemin des Capelles, 31076 Toulouse, France, [3]UE Domaine de La Fage, INRAE, La Fage, 12250 Saint-Jean et Saint-Paul, France; gwenola.tosser@inrae.fr

The suppressor of cytokine signalling 2 (*SOCS2*) gene encodes a protein involved in the immune response. In sheep, a point mutation in SOCS2 has been associated with higher susceptibility to mastitis in commercial conditions. The objective of this study was to investigate the underlying role of the SOCS-2 protein in response to bacterial infection in a dairy sheep model under controlled experimental settings. We performed an intramammary inoculation of 14 ewes belonging to contrasted genotypes at the *SOCS2* R96C point mutation (7 T/T and 7 C/C) with the *Staphylococcus aureus* SA9A strain. We monitored these ewes during 152 hours post inoculation (hpi) with extensive sample collection and measures including clinical observation, milk and blood metabolites and blood transcriptome. The clinical mammary response was more pronounced in susceptible T/T ewes when compared to wild type C/C ewes, as indicated by higher ruminal temperature and detrimental mammary clinical scores. Blood transcriptional analysis revealed large numbers of differentially expressed genes (DEG) from 16 hpi until the end of the challenge with *S. aureus*. DEG between genotypes were detected almost only at 56 hpi (n=177). The most represented pathways according to the Ingenuity Analysis were the interferon signalling and STAT-3 pathways. All genes involved in the latter pathways were overactivated in T/T genotype. At 56 hpi, the modular analysis allows a near perfect clustering according to the genotype groups, showing Inflammation-related modules are more activated in the T/T group. Altogether these data support that the higher susceptibility to mastitis in *SOCS2* mutated T/T ewes is not a higher predisposition to infection, but rather a lower capacity of these ewes to master the mammary inflammation leading to the development of a chronic inflammatory state, involving the interferon JAK/STAT3 pathway. This study was funded by the Agence Nationale de la Recherche (REIDSOCS project: ANR-16-CE20-0010).

Session 37 Theatre 10

Expression quantitative trait loci in whole blood influence putative immune genes in sheep

K. Dubarry[1], M. Coffey[2] and E. Clark[1]

[1]The Roslin Institute and Royal (Dick) School of Veterinary Studies, Easter Bush, Edinburgh, United Kingdom, [2]SRUC, Easter Bush, Edinburgh, United Kingdom; katie.dubarry@ed.ac.uk

Sheep are livestock animals of global importance for the production of meat, milk, and fibre. Sheep breeding, health, and welfare can be improved through a better understanding of immune function and resilience to disease. Molecular techniques, such as RNA-Sequencing, allow us to interrogate the regulation of the immune system at the gene level, and it has been proposed that integrating molecular phenotypes into breeding algorithms could improve genetic gain and performance. Gene expression (GE) is one such molecular phenotype, and we can investigate the genomic regions driving variation in GE through Expression Quantitative Loci (eQTL) discovery. eQTL are genomic markers that explain a proportion of variance in the transcriptome. Analysing eQTL can identify candidate genes of interest for immunity or other production relevant traits. However, to date only a few eQTL studies have been published in sheep. We performed an eQTL analysis using RNA sequence data generated from whole blood and matched single nucleotide polymorphism (SNP) genotypes produced on the Ovine 50k chip, for a population of 57 Texel × Scottish Blackface lambs. This allowed us to identify 611 genomic variants associated with gene expression patterns (*FDR<0.05*). Significant eQTL were found across the genome with most related to 'housekeeping' pathways. Two of the top eQTL were associated with STING1 and ENO1, which are known to play a role in immune function in humans. According to the sheep gene expression atlas, STING1 is ubiquitously expressed across tissues but ENO1 is highly expressed in macrophages, indicating it is involved in the innate immune response. In total, STING1 had 15 variants with a significant association, while ENO1 had 2. Our approach demonstrates that eQTL in whole blood have the potential to reveal novel genomic regions of interest underpinning complex immune molecular phenotypes. In particular, these findings indicate that the genomic variants associated with ENO1 and STING1 are possible targets for further investigation, validation, and potentially inclusion in genomic selection algorithms for sheep.

Session 37 Theatre 11

Thigh muscle proteomics revealed key pathways related to feed efficiency in slow-growing chicken

P. Kaewsatuan[1], C. Poompramun[1], S. Kubota[1], W. Molee[1], P. Uimari[2] and A. Molee[1]
[1]Suranaree University of Technology, Institute of Agricultural Technology, Nakhon Ratchasima, 30000, Thailand, [2]University of Helsinki, Department of Agricultural Sciences, Helsinki, 00790, Finland; d5930029@g.sut.ac.th

Korat chicken (KR) is a slow-growing chicken developed in Thailand. It has recently become a popular alternative for meat-type chicken providing tasty meat with a high nutritional value. However, its feed efficiency (FE) is low causing high production costs and low competitiveness compared to commercial breeds. Understanding the metabolic pathways related to FE will allow the identification of potential biomarkers that can be used in the selection for improving feed utilization efficiency chicken. Therefore, this study aimed to investigate the proteome differences of the thigh muscle between male KRs with either high FE (HFE) or low FE (LFE) using a label-free liquid chromatography-mass spectrometry (LC-MS) proteomic approach. At ten weeks of age, thigh muscle samples from six animals (three HFE and three LFE chickens) were collected for differential abundant proteins (DAPs) analysis. A total of 75 proteins were different between the two groups ($P<0.05$), of which 26 proteins had high abundance and 49 proteins had low abundance in the LFE group compared to the HFE group. Furthermore, a functional enrichment and pathway analysis of DAPs revealed that the glycolysis/gluconeogenesis, biosynthesis of amino acid, pyruvate metabolism, pentose phosphate pathway, and fructose and mannose metabolism were significantly enriched within DAPs. These findings provide new insights into understanding the molecular mechanism in the thigh muscle that contributed to FE in slow-growing chickens.

Session 37 Poster 12

Multi-breed, multi-tissue systems biology analysis of beef cattle divergent for feed efficiency

K. Keogh[1,2], D.A. Kenny[2], P.A. Alexandre[1], M. McGee[3] and A. Reverter[1]
[1]CSIRO, Agriculture & Food, Queensland Bioscience Precinct, 306 Carmody Rd., St. Lucia, Brisbane, QLD 4067, Australia, [2]Teagasc, Animal and Bioscience Research Department, Grange, Dunsany, Co. Meath, C15PW93, Ireland, [3]Teagasc, Livestock Systems Research Department, Grange, Dunsany, Co. Meath, C15PW93, Ireland; david.kenny@teagasc.ie

Provision of feed in beef production systems is a major determinant of profitability. Thus, the identification of genes implicated in regulating feed efficiency may allow for the selection and subsequent breeding of more feed efficient cattle, with obvious benefits for sustainability. It is also crucial that gene markers identified as contributing to feed efficiency are robust across various factors including breed type as well as environmental influence such as diet. In this study, gene co-expression network analysis was undertaken on RNAseq data generated from *Longissimus dorsi* and liver tissue samples collected from steers of two contrasting breed types (Charolais and Holstein-Friesian) divergent, within breed, for residual feed intake (RFI), across contrasting dietary phases: ((1) high-concentrate; (2) zero-grazed grass; (3) high-concentrate). Differentially expressed genes (DEG) based on the contrasts of breed, diet and RFI phenotype were utilised as nodes of the gene co-expression networks. Significant network connections were identified using an algorithm that exploits the dual concepts of partial correlation and information theory (PCIT). PCIT network analysis resulted in the formation of three RFI specific clusters of co-expressed genes which were separated by tissue type. Pathway analysis revealed enrichment ($P<0.05$) of biological processes related to fatty acid biosynthesis in both liver and muscle clusters as well as immune-related pathways in a separate muscle specific cluster. Genes contained within the immune-related cluster were also breed specific DEGs highlighting a potential role for these genes as robust biomarkers for RFI across varying breed type. Acknowledgement: This research was funded by the Irish Department of Agriculture, Food and the Marine (RSF13/S/519). Kate Keogh received funding from the Research Leaders 2025 programme (co-funded by Teagasc and the European Union's Horizon 2020, Marie Skłodowska-Curie grant agreement number 754380).

Contribution of plasma proteins to the phenotypic signature of feed efficiency in Charolais bulls

I. Cassar-Malek[1], A. Imbert[1], A. Delavaud[1], H. Sauerwein[2], R. Bruckmaier[3], G. Cantalapiedra-Hijar[1] and M. Bonnet[1]
[1]Clermont Auvergne University, INRAE, VetAgro Sup, UMR Herbivores, 63122 Saint-Genès-Champanelle, France, [2]University of Bonn, Institute of Animal Science, Physiology Unit, 53115 Bonn, Germany, [3]University of Bern, Vetsuisse Faculty, Veterinary Physiology, 3012 Bern, Switzerland; isabelle.cassar-malek@inrae.fr

Improving the efficiency of feed utilization by reducing the amount of feed intake while maintaining the same level of production has significant economic and environmental impacts in beef cattle. One major challenge is to easily phenotype this multifactorial animal trait for the purpose of genetic selection and precision feeding. Biomarkers and proxies of feed efficiency (FE) may contribute to this challenge. In order to identify phenotypic signatures that discriminate FE groups, we characterized 17 Charolais young bulls fed grass silage diets on their FE parameters (Residual Feed Intake: RFI, and feed conversion efficiency: FCE), carcass composition, some targeted blood parameters (hormones: GH, IGF, insulin, adiponectin, leptin, FT4; metabolites: glucose, NEFA, alpha amino nitrogen, BOH butyrate, urea), and plasma proteome. Labelled-free shotgun proteomics enabled identifying 229 plasma proteins (with 2 peptides, 1% FDR). Univariate, multivariate, and a multiblock sPLSDA analyses were applied to obtain a signature combining omics and carcass traits that discriminates the groups of FE animals. For that, we integrated our multiple datasets while explaining their relationship with FE variables (DIABLO method, mixOmics package). For RFI, the first latent variable of plasma protein abundances was strongly correlated with the one of carcass traits ($|r|= 0.63$) and with the one of blood parameters ($|r|=0.68$) and enabled a clear discrimination of the groups. The key variables that drove such discrimination were identified as 25 proteins together with 3 carcass parameters and 8 hormones and metabolites. Inefficient bulls (with positive RFI) were characterized by: (1) bigger bladder and heart; (2) higher GH levels, as well as lower insulin and IGF1 levels, higher alpha amino nitrogen and NEFA levels; (3) higher abundance in 20 plasma proteins (incl. ITIH3, IPSP, ASPN and PSA6) and lower abundance in 5 plasma proteins (incl. MBL2, ANGPTL3, CO3).

Serum metabolomics of fattening bulls fed dry-total mixed ration (TMR) or corn silage-based TMR

C. Stöcker-Gamigliano[1,2], M.H. Ghaffari[2], C. Koch[1], M. Schönleben[3], J. Mentschel[3], N. Göres[3], P. Fissore[3], I. Cohrs[4], S. Schuchardt[5] and H. Sauerwein[2]
[1]Hofgut Neumühle, Alsenz, 67728, Germany, [2]University of Bonn, Bonn, 53111, Germany, [3]Sano, Loiching, 84180, Germany, [4]University of Gießen, Gießen, 35392, Germany, [5]Fraunhofer ITEM, Hannover, 30625, Germany; morteza1@uni-bonn.de

In Europe and developed countries, fattening bulls are fed corn silage and starch-rich feeds and climate change could affect yields and drive the use of straw and other by-products. The objective of this study was to compare serum metabolites of Simmental bulls fed two diets: a conventional total mixed ration based on corn silage (CONVL) and a dry TMR based on straw and concentrates (DRY) (n=12/group). The main difference between the two rations was the absence or presence of corn silage. Blood samples collected at an average age of 420±11 days underwent targeted metabolomics analysis with the MxP® Quant 500 kit. MetaboAnalyst was used for statistical analysis. After performing multivariate analyses on 282 serum metabolites, it was found that the data differed significantly by treatment. Volcano plot analysis identified 29 significantly different serum metabolites (FDR=0.05 and fold change ≥1.5) from the 282 detected in all samples. The DRY group had lower serum levels of several metabolites compared to the CONVL group, including 3-indolepropionic acid, and hippuric acid. Gut microbiota-derived metabolites with health benefits include 3-indolepropionic acid, a potent neuroprotective antioxidant derived from tryptophan, and hippuric acid, a phenylalanine metabolite that increases with gut development and rumen function and decreases with digestive disorders like displaced abomasum. The DRY diet impacted lipid metabolism and cellular functions, as evidenced by lower levels of three cholesterol esters and five phosphatidylcholine species and higher levels of 17 triacylglycerols. Higher levels of trimethylamine N-oxide (TMAO) were observed in the DRY group compared to the CONVL group. Increased levels of TMAO have been associated with adverse health conditions in human studies. The results of this study suggest that feeding Simmental bulls CONVL or DRY TMR during the fattening period results in changes in serum metabolites related to gut microbiota-derived metabolites, lipid metabolism, and cellular functions.

Session 37 Poster 15

Physiological importance of blood vitamin E in production characteristics of Japanese Black steers

M. Kim[1], T. Masaki[2], K. Ikuta[2], E. Iwamoto[2], Y. Uemoto[1], F. Terada[1], S. Haga[1] and S. Roh[1]
[1]Tohoku University, Lab of Animal Physiology, Graduate School of Agricultural Science, Aramakiaza-Aoba 4681, Aobaku, Sendai, 980-8572, Japan, [2]Hyogo Prefectural Technology Center of Agriculture, Forestry and Fisheries, Hyogo, 679-0198, Japan; sanggun.roh@tohoku.ac.jp

Vitamin E plays a critical role in liver function, which is essential for maintaining the health and productivity of beef cattle. The objective of this study was to investigate the physiological features between blood vitamin E and the production characteristics in Japanese Black steers bred and raised in Japan. Twenty-one Japanese Black steers aged 12 months were reared until 30 months of age. The experimental period was divided into early fattening (12-14 mo of age; T1), middle fattening (15-22 mo of age; T2), and late fattening phases (23-30 mo of age; T3). Vitamin E concentrations increased from T1 to T2, decreased to T3, and notably increased from 16 months, reaching their peak at 20 months of age. To demonstrate the physiological importance of vitamin E, cattle were divided into two groups (High vs Low) according to the difference from 16 months to 20 months of age, and the relationship with carcass traits was analysed. Five Japanese Black cattle with high vs low vitamin E were identified. The levels of blood metabolites and hormones were investigated in High and Low cattle groups. Feed intake in T2 was significantly higher ($P<0.01$) in the High group. This High group in T2 showed high growth performance and carcass traits. The concentrations of blood total cholesterol increased ($P<0.1$) in the High group, whereas those of blood total protein decreased ($P<0.01$). Blood AST (asparagine amino transferase) and γ-GTP (gamma-glutamyl transpeptidase) levels were significantly higher ($P<0.01$) in the Low group. Additionally, this study determined differentially expressed genes (DEGs) in the liver tissues of both High and Low groups in T2 phase. A total of five DEGs were identified and were mainly involved in physiological functions of liver. These results suggest that elevated vitamin E levels in the middle fattening phases depending on high-energy intake may contribute to the mitigation of oxidative stress and inflammation, thereby promoting the enhanced liver health and function of Japanese Black steers.

Session 37 Poster 16

Age-related changes in ruminal and blood parameters and intramuscular adiposity in Hanwoo steers

S.Y. Kim and M. Baik
Seoul National University, Department of Agricultural Biotechnology and Research Institute of Agriculture and Life Sciences, 1, Gwanak-ro, Gwanak-gu, Seoul, 08826, Korea, South; ksy3773@snu.ac.kr

Understanding age-associated changes in growth, metabolism, and intramuscular adiposity are important for determining beef yield and quality characteristics. We investigated age-related changes in growth rate, ruminal and blood metabolic parameters, and adiposity of longissimus thoracis (LT) of Hanwoo (Korean cattle) steers. Nineteen Hanwoo steers (initial body weight, 352±5.37 kg; age, 12±0.25 months) were used. The blood and rumen fluid were collected at 12, 15, 18, 21, 24, 27, and 30 months of age, and the LT samples were biopsied between the 11th and 12th rib at 12, 18, and 24 months of age. Steers were slaughtered at 31 months of age (772±5.37 kg), and the LT samples were collected. The LT samples were used for histological observation of adipocyte size. Average daily gain and feed efficiency decreased ($P<0.001$) with age. The ruminal ammonia concentrations increased ($P<0.001$) with age from 12 to 21 months, thereafter those were unchanged until 30 months of age. Ruminal volatile fatty acid concentrations were unchanged from 12 to 24 months of age, and increased ($P<0.001$) at 27 months of age. Ruminal acetate proportion decreased with age, whereas propionate proportion increased with age, resulting in decreased acetate: propionate ratio with age (all $P<0.001$). Serum glucose concentrations decreased with age, whereas serum triglyceride concentrations increased with age (all $P<0.001$). Serum free fatty acid concentrations increased ($P<0.001$) with age from 12 to 24 months, thereafter those were unchanged until 30 months of age. Insulin concentrations increased ($P<0.001$) with age. The adipocyte size, which was measured by image analysis of histological section of the LT, increased ($P<0.001$) with age. In conclusion, the increased lipogenic parameters including ruminal propionate proportion and serum free fatty acid concentrations and insulin levels may in part contribute to the increased beef adiposity with age.

Session 37 Poster 17

Post-ruminal urea release impact liver proteomics of beef cows at late gestation

M.M. Santos[1], T.C. Costa[1], R.D. Araújo[1], J. Martín-Tereso[2], I.P. Carvalho[2], M.P. Gionbelli[3] and M.S. Duarte[4]
[1]Federal University of Viçosa, Av. PH Rolfs, S/N, 36570-900, Brazil, [2]Trouw Nutrition Research & Development, Stationsstraat 77, 3800, the Netherlands, [3]Universidade Federal de Lavras, Av. Sul UFLA – Aquenta Sol, 37200-000, Brazil, [4]University of Guelph, Animal Biosciences, 50 Stone Road East, N1G-2W1, Canada; mduarte@uoguelph.ca

We evaluated the effects of post-rumen released urea (PRU) on liver metabolism of pregnant beef cows at late gestation. Twenty-four pregnant cows weighing 545 kg±23 kg were assigned into one of two dietary treatments from 174±23 of gestation up to 270 d of gestation: Control (CON, n=12), consisting of a basal diet supplemented with conventional urea; and PRU (PRU, n=12), consisting of a basal diet supplemented with a urea coated to extensively prevent ruminal degradation. Liver samples were collected and analysed through a NanoAquity high-performance liquid chromatographer (HPLC) coupled with a maXis 3G high-resolution Q-TOF mass spectrometer. The data was then screened for quality control where samples with less than 1% of the proteins identified and proteins represented in less than 10% of samples were removed from the dataset, which resulted in a final dataset with a total of 382 proteins. The data were normalized to library size using the Trimmed Mean of M-values method, via the TMM package using R software. Enriched analyses revealed proteins involved in pathway related to carbohydrate digestion and absorption, glycolysis, pyruvate metabolism, oxidative phosphorylation, pentose phosphate pathway, glutathione synthesis and biosynthesis of amino acids of the exclusively expressed proteins in PRU cows. Enriched pathways were differentiated related to carbohydrate digestion and absorption, glycolysis, pyruvate metabolism, oxidative phosphorylation, pentose phosphate pathway, and biosynthesis of amino acids of the exclusively expressed proteins in PRU cows. The proteomic data and the protein-protein interaction analysis revealed an increase in protein and energy metabolism in the liver of PRU cows. The data also suggest an enhancement in glutathione synthesis due to PRU supplementation, which likely to counteract the oxidative stress from the higher metabolic rates of the liver, since the treatment increased oxidative phosphorylation and glycolysis pathways.

Session 37 Poster 18

Liver transcriptome profiles of dairy cows with different serum metabotypes

M. Hosseini Ghaffari[1], H. Sadri[2], N. Trakooljul[3], C. Koch[4] and H. Sauerwein[1]
[1]University of Bonn, Bonn, 53115, Germany, [2]University of Tabriz, Tabriz, 516616471, Iran, [3]Research Institute for Farm Animal Biology (FBN), Dummerstorf, 18196, Germany, [4]Hofgut Neumühle, Münchweiler an der Alsenz, 67728, Germany; morteza1@uni-bonn.de

The objective of this study was to examine the changes in the hepatic transcriptome of cows with different serum metabotype in early lactation. We performed hepatic transcriptome analysis by RNA sequencing (RNA-seq) for cows in 3 different metabolic clusters: high body condition score (BCS) predicted high BCS (HBCS-PH, n=8), HBCS predicted normal BCS (HBCS-PN, n=6), and normal BCS predicted normal BCS (NBCS-PN, n=8) on day 21 postpartum. A total of 13,118 genes were detected aligned with the bovine genome. Using PCA and heat map analyses, we found that the liver transcriptomes of cows with different metabotypes were largely overlapped. A total of 48 differentially expressed genes (DEG; ~0.4%; FDR≤0.1 and fold-change >1.5) were found between NBCS-PN and HBCS-PH cows, whereas a total of 24 DEG (~0.18%) were found between HBCS-PN and HBCS-PH cows using DESeq2. Reverse transcription-quantitative real-time PCR (RT-qPCR) was performed for seven DEG and showed consistent results with the sequencing results in terms of relative expression across the comparisons. The study identified 31 down-regulated and 17 up-regulated genes between NBCS-PN and HBCS-PH cows. The downregulated DEG in NBCS-PN cows were involved in biosynthesis, cellular organization, localization, catabolism, and response to external stimuli compared to HBCS-PH cows. The upregulated DEG in NBCS-PN compared to HBCS-PH cows were involved in signal transduction, biological quality, cell motility, and molecular function, indicating altered metabolic adaptations to lactation. Furthermore, in the liver tissue of HBCS-PN vs HBCS-PH cows, DESeq2 analysis detected 14 down-regulated and 10 up-regulated genes involved in various processes, including cellular organization, cell adhesion, and motility. Overall, the slight alterations observed in the liver transcriptome may suggest a reduced role in the development of distinct metabotypes but post-transcriptional processes may have hidden their contribution.

Session 37 Poster 19

Comparison of metabolomic profile of the offspring of nulliparous and primiparous Holstein cows

M. Terré and M. Tortadès
IRTA, Ruminant Department, Torre Marimon, 08140, Spain; marta.terre@irta.cat

Foetal imprinting can influence further metabolism of the progeny. Several studies have examined the foetal development among nulliparous, primiparous and multiparous cows and its relationship with reproductive or performance traits of their offspring. The objective of this study was to evaluate metabolic imprinting at the beginning of the first lactation of the offspring from nulliparous or primiparous cows. Sixteen coetaneous primiparous Holstein cows, eight of which were born from nulliparous cows (heifers), and eight from primiparous cows, were raised under the same conditions in a commercial dairy farm. Plasma samples were collected at the beginning of lactation (DIM=32±8.9 days) for a non-targeted metabolomics analysis using a quadrupole time-of-flight mass spectrometry. Data were analysed by principal component analysis, hierarchical clustering heatmaps, random Forest analysis and comparisons of the observed metabolite abundances between treatments by t-test. Neither principal component nor random Forest were able to classify samples by experimental groups. However, t-test found differences ($P<0.01$) in fifteen plasma metabolites. From those 2 were identified: 4-alpha-carboxy-4-beta-methyl-5-alpha-cholesta-8-en-3betaol was down-regulated, and spermine was up-regulated in the offspring from primiparous compared to nulliparous cows, suggesting modulatory effects of cow parity on offspring's cholesterol biosynthesis (4-alpha-carboxy-4-beta-methyl-5-alpha-cholesta-8-en-3betaol) and cellular oxidative stress (spermine).

Session 37 Poster 20

Transcriptome of the feto-maternal interface in pigs in late gestation: part 1 – foetal placenta

A. Bonnet[1], S. Maman[1], L. Gress[1], A. Suin[2], C. Bravo[1], G. Cardenas[1], Y. Billon[3], L. Canario[1], N. Vialaneix[4], C.M.D. Bonnefont[1] and L. Liaubet[1]
[1]GenPhySE, Université de Toulouse, INRAE, ENVT, Auzeville, 31326 Castanet, France, [2]GeT-PlaGe, INRAE, Genotoul, Auzeville, 31326 Castanet, France, [3]GenESI, INRAE, Le Magneraud, 17700 Surgères, France, [4]Université de Toulouse, INRAE, UR MIAT, Auzeville, 31326 Castanet, France; agnes.bonnet@inrae.fr

In pigs, selective breeding has led to a rise in neonatal mortality, in relation to a reduced maturity of newborn piglets. Piglet maturity and growth accelerate at the end of gestation and this requires considerable nutritional support. The process of maturation varies with foetal genome and uterine environment and depends upon placenta, the tissue that regulates feto-maternal allocation of resources. To better understand the foetal maturation process, purebred and crossbred foetal genotypes were produced from 14 Large White (LW) and 14 Meishan (MS) sows differing for piglet maturity at birth and piglet survival. All the sows were inseminated with mixed semen from the two breeds. We compared the transcriptome in 224 placentas collected according to this four foetal genotypes (FG) and two days of gestation (D90-D110, term at 114 days). The transcriptome was obtained by RNAseq technology and processed by nf-core/rnaseq pipeline. Statistical analysis was performed using mixed linear models with sow as random effect followed by a correction for multiple testing (FDR<0.001). We documented the expression profile of 8,407 differential transcripts (DEGs) for gestational age, FG and their interaction. These DEGs reflect an important transcriptomic change that occurs between D90 and D110 in the endometrium (2,901 DEGs) and underline genes involved in pathways essential for normal placental development such as the insulin-signalling pathway. In addition, 1,543 DEGs were identified according to FG. Gene Ontology enrichment analysis of the DEGs between FG highlighted the placental fatty acid b-oxidation pathway, which is proposed to be related to maternal nutrition and body composition. Other transcripts (561 DEGs) showed interactions between foetal genotype and gestational age. In conclusion, our study identified pathways that may be associated with differences in maturity at birth. This research is part of the ANR-20-CE20-0020-01 project CO-LOcATION.

Session 37 Poster 21

Transcriptome of the feto-maternal interface in pigs in late gestation: part 2 – sow endometrium

A. Bonnet[1], S. Maman[1], L. Gress[1], A. Suin[2], S. Legoueix[1], C. Bravo[1], Y. Billon[3], N. Vialaneix[4], C.M.D. Bonnefont[1] and L. Liaubet[1]

[1]GenPhySE, Université de Toulouse, INRAE, ENVT, Auzeville, 31326 Castanet, France, [2]GeT-PlaGe, INRAE, Genotoul, Auzeville, 31326 Castanet, France, [3]GenESI, INRAE, Le Magneraud, 17700 Surgères, France, [4]Université de Toulouse, INRAE, UR MIAT, Auzeville, 31326 Castanet, France; agnes.bonnet@inrae.fr

In pigs, genetic progress has led to a rise in perinatal mortality, mostly due to a reduced piglet maturity. Piglet maturity acquisition and greater foetal growth occur at the end of gestation and require considerable nutritional support. These processes are closely related to the physiology and nutrition of the dam, as nutrients are transmitted from the endometrium to the foetus via the placenta. The endometrium is critical for implantation, placentation, embryonic and foetal development; many critical morphological and secretory changes occur throughout gestation. In order to understand the role of endometrium during the foetal maturation process, we compared two breeds at two days of gestation (D90-D110, term at 114 days): Large White (LW) and Meishan (MS) with high and low neonatal mortality, respectively. The transcriptome of 224 endometrial samples from the two breeds at two gestational stages was obtained by RNAseq technology and processed by Nextflow nf-core/rnaseq pipeline. Statistical analysis was performed using mixed linear models with sow as random effect (limma package) followed by a correction for multiple testing (FDR<0.001). We documented the expression profile of 10,080 differential transcripts (DEGs) for gestational age, mother breed and their interactions. These DEGs reflect an important transcriptomic difference between LW and MS breeds (2,721 transcripts) and underlined genes involved in catalytic activity and metabolic process. A smaller transcriptomic change was observed between D90/D110 in the endometrium (607 transcripts), mainly reflecting a change in cell communication. Moreover, 3,210 DEGs showed interactions between gestational age and breed which are grouped into eight clusters of expression profiles. Results from this study suggest endometrial remodelling at D90/D110 and breed differences in catabolism status and strategy to mobilize reserves for foetal growth. This research is part of the ANR-20-CE20-0020-01 COLOcATION project.

Session 37 Poster 22

1H-NMR metabolomic study of Large White and Meishan pigs in late gestation: part 2 – sow endometrium

A. Imbert[1], N. Duprat[1], N. Marty-Gasset[1], L. Gress[1], C. Canlet[2], Y. Billon[3], N. Vialaneix[4], C.M.D. Bonnefont[1], A. Bonnet[1] and L. Liaubet[1]

[1]GenPhySE, Université de Toulouse, INRAE, INPT, ENVT, Auzeville, 31326 Castanet, France, [2]Axiom Platform, MetaToul-MetaboHUB, Toxalim, INRAE/INP/UPS, route de Tournefeuille, 31027 Toulouse, France, [3]GenESI, INRAE, Le Magneraud, 17700 Surgères, France, [4]Université de Toulouse, INRAE, UR MIAT, Auzeville, 31326 Castanet, France; laurence.liaubet@inrae.fr

The risk of piglet mortality is highest in the first few days after birth. Therefore, late gestation development is of high importance for piglet survival. The endometrium is the maternal tissue that is in direct contact with the placenta of each foetus and thus is the matrix for feto-maternal interactions. To better understand the relationship between foetal development and piglet maturity, endometrial tissues from Large White (LW) sows, with high piglet mortality, were compared to that from Meishan (MS) sows, a breed with less neonatal mortality. Endometrial samples (n=224) were collected from 28 sows (14 LW and 14 MS) at 90 or 110 days of gestation (dg, birth at 114 days), each sample being in direct contact with each of the 224 placentas. Hydrophilic metabolites were extracted and 1H-NMR spectra were acquired. Raw spectra were processed using the ASICS R package to identify and quantify 46 metabolites. A multivariate analysis revealed a main effect of gestational age (13.1% on 1st axis). Mixed models were used to identify 21 metabolites with differential concentrations between the two following conditions, day of gestation and genotype, their interaction, and with the sow as a random effect (one model fitted to each metabolite independently followed by a correction for multiple testing, FDR<0.05). As expected, fructose was more abundant at 90 dg than at 110 dg. Conversely, L-glutathione-reduced was more concentrated at 110 dg than at 90 dg and more concentrated in MS than in LW sows. Citrate and L-glycine were more concentrated in MS sows than in LW sows at both stages of gestation. These results suggest important metabolic differences between the endometrium of the two breeds. This research is part of the ANR-20-CE20-0020-01 project COLOcATION.

Session 37 Poster 23

The effect of dietary *Laminaria digitata* on urinary and kidney proteomes of piglets

D.M. Ribeiro[1], A. Dittmann[2], D.F.P. Carvalho[1], L. Kunz[2], J.P.B. Freire[1], J.A.M. Prates[3,4] and A.M. Almeida[1]
[1]Instituto Superior de Agronomia – Universidade de Lisboa, LEAF – Linking Landscape, Environment, Agriculture and Food, Tapada da Ajuda, 1349-017 Lisboa, Portugal, [2]ETH Zürich/University of Zurich, Functional Genomics Centre Zürich, Winterthurerstrasse 190, 8057 Zürich, Switzerland, [3]Laboratório Associado para Ciência Animal e Veterinária (AL4AnimalS), Avenida da Universidade Técnica, 1300-477 Lisboa, Portugal, [4]Faculdade de Medicina Veterinária – Universidade de Lisboa, CIISA – Centro de Investigação Interdisciplinar em Sanidade Animal, Avenida da Universidade Técnica, 1300-477 Lisboa, Portugal; davidribeiro@isa.ulisboa.pt

Laminaria digitata is a brown seaweed with prebiotic properties, despite having high levels of ash and a recalcitrant cell wall that prevents an efficient digestion. We hypothesized that enzymatic supplementation in monogastric diets can increase nutrient availability of this feedstuff. The objective of this study was to evaluate if the seaweed affects the kidney and urinary proteomes due to disruption of mineral metabolism homeostasis. Weaned piglets (male, aged 35 days) were randomly assigned to one of three diets (n=10): control (wheat-maize-soybean meal based), LA (10% *L. digitata* replacing control) and LAL (LA+0.01% alginate lyase). Each piglet was housed individually in a metabolic cage with free access to water. The trial lasted for two weeks, after which they were slaughtered, and kidney and urine samples were harvested. Samples were then analysed using a label-free LC-MS/MS approach. The dietary treatments had less effects in the urine proteome, where 52 proteins had differences ($P<0.05$, $\log_2(FC)>1$) compared to 253 in the kidney. LAL piglets increased the urinary excretion of complement factor proteins, which have been suggested as a marker for kidney inflammation. LAL vs control yielded the highest number of differences in the kidney, with LAL increasing the abundance of HMOX1 (involved in inflammatory response) and lowering that of ASH1L (negative regulation of inflammatory response). These findings suggest that dietary *L. digitata* may disrupt kidney homeostasis, which is worsened by enzymatic supplementation. This could stem from dietary availability of the mineral fraction, requiring an adaptative response by the piglet.

Session 37 Poster 24

Vitamin D3 supplementation has little effect on gut microbiome in pigs

M. OcZkowicz[1], A. Wierzbicka[1], A. Steg[1] and M. Świątkiewicz[2]
[1]National Research Institute of Animal Production, Department of Animal Molecular Biology, ul Krakowska 1, 32-083, Poland, [2]National Research Institute of Animal Production, Department of Animal Nutrition and Feed Science, ul Krakowska 1, 32-083, Poland; maria.oczkowicz@iz.edu.pl

Dietary supplementation with vitamin D3 (VD3) is widely used in humans, but there are still disputes about the recommended doses. In farm animals, the effects of supplementation are still at the stage of research and the exact cost-benefit balance has not yet been determined. According to the recommendations of the European Union, the dose of VD3 in the feed should not exceed 2,000 U/kg of feed for domestic pigs. However, higher doses seem to be needed to achieve a health-promoting effect. Our study aimed to evaluate the effect of supplementation with high doses of VD3 (5,000 and 10,000 U/kg) in the pig diet on the gut microbiome. The research covered 30 castrated boars, which were divided into three equally numerous groups: Group I: no VD3, Group II: 5,000 U/kg VD3, and Group III: 10,000 U/kg VD3. VD3 supplementation turned out to be effective, the groups differed significantly in the content of 25OH vitamin D in the blood. Caecal digesta samples were collected at slaughter and bacterial DNA was isolated. DNA was sent to weSeq.it company, where V3 and V4 16sRNA libraries were created and next-generation sequencing and bioinformatics processing were performed. After sequencing, an average of 80,013 raw reads were gained per sample, and consequently an average of 55,499 operational taxonomic units 'OTU' at the 'genus' level. Alpha Diversity analysis showed no statistically significant differences between the groups in any of the parameters studied (Chao1, Ace, Shannon, Simpson, Coverage, PD_whole_tree). Similarly, beta diversity analysis revealed no differences among the analysed groups. Moreover, differential analysis between groups, using ANOVA did not show any differences at all taxonomic levels. The only difference ($P<0.05$) was observed after Metastats analysis at some rare microorganisms belonging to *Gemmatimonadota*, *Methylomirabilota*, and *Myxococcota* phylum. In conclusion, our results suggest that VD3 supplementation has little effect on the gut microbiome in pigs, however, we can not exclude its influence on certain rare microorganisms.

Session 38 — Theatre 1

Equal opportunities in science

I. Adriaens[1], M. Pszczola[2] and T. Wallgren[3]

[1]Livestock Technology, Department of Biosystems, Kleinhoefstraat 4, 2440 Geel Belgium, Belgium, [2]Poznan University of Life Sciences, Genetics and Animal Breeding, Wolynska 33, 60-637 Poznan, Poland, [3]Swedish University of Agricultural Sciences (SLU), Department of Animal Environment and Health, Box 7068, 750 07 Uppsala, Sweden; marcin.pszczola@puls.edu.pl

Early career scientists often doubt what the best (academic) path is to partake, and which decisions along the way can help them reach their goals. One hugely uncertain factor in this context is whether a decision can negatively impact this road, for example, because one chooses to start a family or focus on a path more directed toward teaching rather than research. Additionally, many people experience having to work more or harder than colleagues because they do not come from a similar background, religion, or ethnicity or did not have a similarly high-level yet expensive education. Both in business and in academics, this has been discussed in the past as complying with the 'seven ticks': if you have wealthy or educated parents, are born in a western country, are male, heterosexual, white, and are highly educated, in high school and in university, you have all of them. These people are considered the most favoured people in society that are seldom discriminated against and often reach the top functions rather easily with not that many setbacks. In this Young EAAP session, we invited several people with different backgrounds to talk about 'equal opportunities' and their experience with things like starting a family (and being female), coming from different backgrounds, are not white, etc. The session will include a round of introductions, a panel discussion, and room for questions from the audience.

Session 38 — Theatre 2

The responsibilities of authors, readers and learned-societies in animal science publishing

J. Van Milgen[1], I. Ortigues-Marty[2], G. Bee[3], M. Wulster-Radcliffe[4], J. Sartin[4], T.A. Davis[5] and P.J. Kononoff[6]

[1]INRAE, Institut Agro, Pegase, Le Clos, 35590 Saint-Gilles, France, [2]INRAE, Université Clermont Auvergne, VetAgro Sup, UMRH, 63122, Saint-Genès-Champanelle, France, [3]Agroscope, Rte de la Tioleyre 4, 1725 Posieux, Switzerland, [4]American Society of Animal Science, P.O. Box 7410, Champaign, IL 61826-741, USA, [5]USDA/ARS Children's Nutrition Research Center, Baylor College of Medicine, 1100 Bates Street, Houston, TX 77030-2600, USA, [6]University of Nebraska-Lincoln, Department of Animal Science, Lincoln, NE 68583-0908, USA; jaap.vanmilgen@inrae.fr

Scientific publishing was relatively straightforward in the past. Scientific journals, often owned and managed by learned-societies and research organizations, used the expertise in their community to peer-review articles, which was perceived as the gold standard of quality. Peer-review served to weed out low-quality research, but sometimes stringent peer-review standards resulted in work not being published, simply because authors could not find a journal that would accept their sound, but perhaps little-less-novel research. This has changed considerably in the last decade with Open Access publishing. The number of publications in peer-reviewed animal science journals has doubled in just five years (Journal Citation Report, Clarivate). Open Access, combined with the increase in the number of publications, has also resulted in a shift in responsibility from the journals to the reader. Virtually any author with funding can now find a journal willing to publish their research results. But how serious do journals take the peer-review process if they can promise a very rapid time-to-first decision? Of course, we all like to see that our papers are published rapidly and without hassle, but is this really what we want from the journals that we own and manage? The future of animal science publishing has many uncertainties but, as an animal science community, we can certainly shape it. Should peer-review remain the standard to assess 'quality and novelty' or should we, as journals of learned societies, explore new publishing models? How do you see your future as an author and reader in the rapidly changing publishing landscape and what should the role of the journals of learned-societies be in this?

Session 38 Theatre 3

Animal – open space the new member of the animal consortium

G. Bee
Agroscope, Swine Research Group, Route de la Tioleyre 4, 1725 Posieux, Switzerland; giuseppe.bee@agroscope.admin.ch

animal – open space is a new publishing initiative of the animal Consortium, a collaboration between the British Society of Animal Science, INRAe and the EAAP. The journal animal – open space is part of a family of journals including the flagship journal animal and animal – science proceedings. The journal has a wider scope than the flagship animal. The journal fully embraces Open Science and its philosophy is that all carefully conducted reproducible research, the data linked to that research and the associated points of views of the authors contribute to knowledge gain. This knowledge deserves to be rapidly published and open for further discussion from readers and the authors once published. The journal publishes articles that relate to farmed or other managed animals, leisure and companion animals and the use of insects for animal feed and human food. Articles will be accepted from all species if they are in, or contribute knowledge to the aforementioned categories (e.g. cattle, sheep, pigs, poultry, horses, rabbits, fish, cats, dogs). Lack of novelty, negative results or lack of significant treatment differences are not a barrier for publication. The journal publishes three types of articles: data papers, method articles and research articles. The target audiences of animal – open space are animal scientists, stakeholders and policy makers interested in agricultural, biomedical, veterinary and environmental sciences with expected impacts on animal performance and productivity, animal welfare, animal health, food security, environment, climate change, product quality, human health and nutrition, sustainability of animal agriculture, livestock systems and methodology. The impacts can be either of local or international relevance. In conclusion, the final objective of animal – open space is to reinforce the idea of Open Science and by that becoming a source of knowledge of reproducible data obtained from research with farmed or other managed animals, leisure and companion animals and on the use of insects for animal feed and human food.

Session 38 Theatre 4

Artificial intelligence language models in the scientific career

M. Pszczola[1], I. Adriaens[2] and T. Wallgren[3]
[1]Poznan University of Life Sciences, Genetics and Animal Breeding, Wolynska 33, 60-637 Poznan, Poland, [2]Livestock Technology, Department of Biosystems, Kleinhoefstraat 4, 2440 Geel Belgium, Belgium, [3]Swedish University of Agricultural Sciences (SLU), Department of Animal Environment and Health, Box 7068, 750 07 Uppsala, Sweden; marcin.pszczola@puls.edu.pl

Recent developments in artificial intelligence (AI) language models, such as chatGTP, are revolutionizing many different fields of life. These tools can change how some tasks are performed by enabling quick, effortless, and prompt generation of human-like written text; they can assist in coding tasks or data analysis or produce many other types of content. The generated outcome is difficult to distinguish from the work of human-generated work. Next, to the outstanding capabilities of the developed tools, using these technologies raises many concerns in the scientific community, education, and everyday life. Moreover, blind use of AI language models can produce false information based on imperfections of the currently available tools and low-quality information on the internet. Therefore, AI should be perceived as assistance for humans using them rather than as a replacement for expertise and critical thinking. Another important ethical and legal aspect is the authorship of the outcomes generated by AI language models. Currently, scientific journals do not accept co-authorship of AI language models, and this is an issue that needs more attention shortly. Therefore, using AI language models has to ensure alignment with ethical principles and standards thoughtfully.

Session 39 Theatre 1

TechCare: exploring the use of precision livestock farming for small ruminant welfare management

C. Morgan-Davies[1], G. Tesniere[2], C. Dwyer[1], G. Jorgensen[3], E. Gonzalez-Garcia[4] and J.M. Gautier[2]
[1]SRUC, West Mains Road, EH9 3JG, United Kingdom, [2]IDELE, Campus INRAe, 31321 Castanet Tolosan, France, [3]NIBIO, Grassland and livestock Division of Food and society, P.O. Box 115, 1431 Ås, Norway, [4]INRAe, SELMET, Montpellier, France; claire.morgan-davies@sruc.ac.uk

Small ruminant production systems are found in diverse contexts where conditions can be harsh and day to day supervision of animals challenging. Implementing Precision Livestock Farming (PLF) and other new or innovative technologies could help to manage or monitor animal welfare (AWE). The H2020 TechCare project explores such opportunities. Following a series of prioritisation of AWE issues in the 9 partners' countries (France, UK, Ireland, Norway, Israel, Greece, Romania, Italy and Spain), potential PLF and innovative technologies have been identified as promising to help manage, monitor and/or improve AWE. To explore those potentials, a series of pilot studies have been set up in 5 partner countries (France, UK, Norway, Italy and Israel), where the application of near-market technologies is being assessed in different conditions and environments, varying from northern grasslands (UK and Norway) to Mediterranean climates (France, Italy and Israel). Different production purposes are also involved, from dairy sheep to dairy goats and meat sheep, both indoors and outdoors. Alongside those pilots, prototyping and adapting PLF and innovative tools is also performed, where trials under more controlled conditions are being carried out, in order to evaluate the potential of specific tools while adopting adequate approaches for monitoring and/or improving different small ruminant AWE. Those trials are being held in France, UK, Norway, Spain and Italy. Both approaches will inform on the best innovative and PLF technologies, and the most adapted ways to use them, that can most answer AWE priorities in a diversity of small ruminant farming systems. Some of the tools (e.g. Ultra High Frequency tags, proximity loggers) studied are being presented later in this session.

Session 39 Theatre 2

Monitoring water trough attendance in shed: a potential indicator of sheep health or welfare issues?

G. Tesnière[1], U. Jean-Louis[1], E. Doutart[1], S. Duroy[1], C. Douine[2], M. Rinn[3], D. Gautier[1,2], A. Hardy[3], A. Aupiais[1], F. Guimbert[4], J.-M. Gautier[1] and C. Morgan-Davies[5]
[1]Institut de l'élevage, CS 52637, 31321 Castanet-Tolosan, France, [2]CIIRPO, Ferme expérimentale du Mourier, Le Mourier, 87800 Saint-Priest-Ligoure, France, [3]EPLEFPA de la Cazotte, Ferme du lycée agricole, Route de Bournac, 12400 Saint-Affrique, France, [4]Page Up, 13 Rue Marguerite Yourcenar, 21000 Dijon, France, [5]SRUC, West Mains Road, Edinburgh, EH93JG, United Kingdom; germain.tesniere@idele.fr

As part of the TechCare project (2020-2024), various new technologies are tested in seven pilot farms (5 countries) to assess their use as potential early warning of animal welfare issues or of risk factors for small ruminant sectors. Among different welfare indicators, we are interested in water intake as one of the pillars of good health and well-being in sheep. Our objective is to study the watering behaviour of sheep in sheds to potentially detect early welfare or health problems. Our device is composed of three main components. (1) Each animal wears an ultra high frequency (UHF) Radio frequency identification (RFID) tag, and a reading antenna is installed above each water trough. The data is collected using a reader connected to the internet. (2) Each water trough is equipped with a connected water meter with automatic data transmission on a web platform. (3) The device is completed by a motion detection camera to validate our data and observations. In France, this device has been set up in: (1) a meat sheep shed to monitor fattening lambs (91) for 4 weeks in 2022; and in (2) a dairy sheep shed to monitor a batch of lactating ewes (40) for 6 weeks in 2022 and 12 weeks in 2023. A welfare assessment protocol has been designed alongside to collect individual observations and measurements on a series of welfare indicators. These attendance data are studied at the individual scale, at the flock scale, and in relation to other data characteristic of the animals' environment (temperature and humidity of the shed, for example), and results on how they tally with the welfare measurements will be presented in this paper. Monitoring water trough attendance could be a relevant indicator to design an early warning system for sheep health or welfare.

Session 39 Theatre 3

TechCare UK pilots – integrated sheep system studies using technologies for welfare monitoring

A. McLaren[1], A. Waterhouse[1], F. Kenyon[2], H. MacDougall[2], S. Beechener[1], A. Walker[1], M. Reeves[1], N. Lambe[1], J. Holland[1], A. Thomson[1], J. Duncan[2], A. Barnes[1], C. Dwyer[1], F. Gimbert[3], J.M. Gautier[4], G. Tesniere[4] and C. Morgan-Davies[1]

[1]SRUC, West Mains Road, Edinburgh, EH9 3JG, United Kingdom, [2]MRI, Bush Loan, Penicuik, EH26 0PZ, United Kingdom, [3]PageUp, 13 rue Marguerite Yourcenar, 21000 Dijon, France, [4]IDELE, Campus INRAe, 31321 Castanet Tolosan, France; ann.mclaren@sruc.ac.uk

Small ruminant production systems are found in diverse contexts, including extensive grazing regions, where environmental conditions can be harsh and day to day supervision of animals challenging. In addition, farmers within these systems may lack access to, or not be aware of, technologies that may help to improve performance, efficiency and welfare. As part of the H2020 TechCare project, a series of pilot studies have been set up in 5 partner countries to explore the application of technologies to address different welfare priorities, as identified by industry stakeholder groups in each country. In the UK, these priorities included nutritional issues (under/malnutrition), poor maternal relationships and health problems (particularly lameness, mastitis and endoparasites). Two pilot studies have been set-up one at SRUC's Hill & Mountain Research Farm in the West Highlands and the other at MRI's Firth Mains Farm in Midlothian. The technologies investigated on these units include the use of low frequency electronic identification (EID) enabled equipment (e.g. weigh crates and stick readers), proximity and ID loggers (Bluetooth beacons and ultra-high frequency (UHF) tags and reader systems), accelerometers, GNSS trackers, and electronic weather stations. In addition to live weight and body condition score data collection, and in-field observation work including measures of gut parasitic infections (for MRI pilot), regular welfare assessments of the animals are being carried out using the Animal Welfare Indicator (AWIN) protocols. Through a series of different studies covering in combination most of the sheep production cycle (pregnancy; lambing; summer rearing period; weaning and post weaning of lambs) and which will be detailed in this paper, the suitability of these technologies to improve our understanding of the welfare issues highlighted, and potentially to act as an early warning indicator, are being assessed.

Session 39 Theatre 4

An attempt to aggregate pig welfare indicators from sensors: achievements and barriers

H.L. Ko[1], L.J. Pedersen[2], G. Franchi[2], M.B. Jensen[2], M. Larsen[2], I.J.M.M. Boumans[3], J.D. Bus[3], E.A.M. Bokkers[3], X. Manteca[1] and P. Llonch[1]

[1]Universitat Autònoma de Barcelona, Department of Animal and Food Science, Campus UAB, 08190, Cerdanyola del Vallès, Spain, [2]Aarhus University, Department of Animal and Veterinary Sciences, Tjele, 8830, Denmark, [3]Wageningen University and Research, Animal Production Systems, Wageningen, 6700 AH, the Netherlands; pol.llonch@uab.cat

Precision livestock farming (PLF) is developing as a powerful on-farm tool to monitor animal welfare. ClearFarm aims to contribute to this by developing an algorithm that can assess animal welfare continuously, based on the five Welfare Domains. This study summarises the main achievements and barriers experienced in the development of the algorithm. PLF-derived indicators that are relevant to the related Welfare Domains were validated in pig farms, including activity, feeding and play behaviour, body weight, climatic measurements, and respiratory health. Each indicator was classified as: (1) having a solid threshold based on literature; (2) having a range as a threshold based on literature; (3) being unsuitable for a fixed threshold; or (4) having no existing literature to support a threshold. Next, the ClearFarm platform algorithm to aggregate the indicators was meant to be fed with the different indicators extracted from sensor data and the predefined thresholds, providing a score (0-100) for each Welfare Domain. However, several barriers were encountered. First, few indicators from PLF technology have been validated to assess pig welfare. Second, no sensors are currently available to monitor relevant indicators for the mental state domain. Also, other domains (e.g. health) are underrepresented because only one indicator is available (i.e. respiratory health). Third, the pig production chain has different phases (e.g. gestating sows, weaners, fattening pigs) with different durations and welfare threats. Also, the common practice to mix pigs between units and farms makes tracking pigs across these phases difficult. Yet, it is not feasible to propose an algorithm that meaningfully aggregates welfare indicators across domains and production stages in pigs. Rather, ClearFarm proposes how welfare indicators can be achieved from sensors and how they can contribute to assessing pig welfare in different domains.

Session 39 Theatre 5

Is sensor technology ready for use in animal welfare assessments of gestating sows?

I.J.M.M. Boumans and E.A.M. Bokkers
Wageningen University & Research, Animal Production Systems group, P.O. Box 338, 6700 AH Wageningen, the Netherlands; eddie.bokkers@wur.nl

Sensor data is expected to be valuable for continuous and real-time animal welfare assessment. The availability of robust and reliable sensor technologies that are already applied in farms, however, is limited and their value for animal welfare assessment has not been shown in commercial farms. Therefore we aimed to study to what extent commercially available sensors can be used for welfare assessment, with gestating sows as a case study. Two sensor technologies were studied: climate sensors (temperature, relative humidity, CO_2 and NH_3; dol-sensors, Denmark) and electronic sow feeders (ESF, Nedap, The Netherlands). Sensor data was collected on a conventional and an organic farm, for a period of a year. Potential welfare indicators were selected based on scientific literature and calculated and compared between the two different farming systems. Examples of selected welfare indicators are days above a critical temperature indicative of risk on heat stress, days above a critical NH_3 level indicative of risk on respiratory problems and discomfort, average feeder occupation indicative for feed competition and aggression, mean days per sow gestation of low feed intake indicative of reduced appetite and impaired health. Sensor technology can have added value for welfare assessment and lead to new indicators to assess welfare, however, several challenges were identified, including the availability and quality of data, functioning of technology and influence of farm management practices. The availability and quality of results relates strongly on technical functioning of sensors, presence of erroneous data and data cleaning. In particular data cleaning poses challenges as it requires different standards per farm. Outliers in data can corrupt results, but they cannot simply be removed as they might also reflect reality, and as such can contain important information related to welfare. For correct interpretation of results also background information is required about the farming system, e.g. farm management practices, such as use of bedding and (in case of ESF) changing ESF settings. Sensor data can provide valuable additions to welfare assessments, however, the interpretation should be done carefully and further validation is required.

Session 39 Theatre 6

Identification of behavioural pattern associated with mastitis in dairy cows

L. Herve[1], Y. Gómez[2], K. Chow[3], A.H. Stygar[4], G.V. Berteselli[5], E. Dalla Costa[5], E. Canali[5] and P. Llonch[2]
[1]PEGASE, INRAE, Institut Agro, 35590 Saint-Gilles, France, [2]Universitat Autònoma de Barcelona, Department of Animal and Food Science, 08193 Cerdanyola del Vallès, Spain, [3]Universitat Autònoma de Barcelona, Department of Information and Communications Engineering, 08193 Cerdanyola del Vallès, Spain, [4]Natural Resources Institute Finland (Luke), Latokartanonkaari 9, 00790 Helsinki, Finland, [5]Università degli Studi di Milano, Dipartimento di Medicina Veterinaria e Scienze Animali, 26900 Lodi, Italy; lucile.herve@inrae.fr

Mastitis is major health issue compromising the welfare of dairy cows. Individual behaviour of dairy cows few days before mastitis is a potential tool for higher detection accuracy. However, little is known about the existence of behavioural patterns usually expressed by cows in healthy conditions that are associated with the subsequent onset of mastitis. This study aimed to determine whether cows that will subsequently be affected by mastitis express distinctive behavioural patterns compared to the ones that stay healthy throughout lactation. Daily hours spent standing, lying down, walking, ruminating and eating were recorded from calving to 15 d prior mastitis using accelerometer collars in two commercial farms in Italy. The mastitis group (n=16) was constituted with the cows suffering from one mastitis during lactation. A control group of healthy cows (n=16) balanced for parity and lactation stage has been constituted afterwards. The group effect (mastitis vs no mastitis) on the intercept and the slope of the regression curve of each individual cows for all behaviours was analysed using ANOVA. Cows from the mastitis group spent less time lying (9.25 vs 9.66 h/d) and more time walking (2.69 vs 2.40 h/d) and ruminating (8.56 vs 8.14 h/d) than control cows 100 days before the onset of mastitis ($P<0.001$). Behavioural changes over time before mastitis also differed as shown by the lower slope coefficient for lying, walking and ruminating in mastitis than in control cows ($P<0.001$). This study may suggest that cows suffering from mastitis exhibited different lying, walking and ruminating durations, months before the onset of the disease. Further investigations are needed to confirm if differences in behavioural patterns can be considered as promoters of disease.

Use of PLF sensors to monitoring rumination as an alternative for early clinical mastitis detection

G.V. Berteselli[1], E. Dalla Costa[1], Y. Gómez Herrera[2], M.G. Riva[1], S. Barbieri[1], R. Zanchetta[1], P. Llonch[2], X. Manteca[2] and E. Canali[1]

[1]University of Milan, Department of Veterinary Medicine and Animal Science, via dell'Università 6, 26900 Lodi, Italy, [2]Universitat Autònoma de Barcelona, Department of Animal and Food Science, Campus UAB, 08193 Cerdanyola del Vallès, Barcelona, Spain; greta.berteselli@unimi.it

Mastitis is a costly disease and remains a major welfare concern in the dairy industry. Rumination is a vital behaviour for dairy cattle, decreased rumination time has been linked with acute stress, anxiety, or disease. The aim of this study was to clarify whether a decreased rumination time, monitored with accelerometery collars, can predict the onset of mastitis in lactating dairy cows. Two groups of dairy cows balanced for stage of lactation and parity order, were recruited: mastitis group (n=14) and control group (n=14). Mastitis was diagnosed through somatic cell count in milk and clinical examination. For each cow, daily hours spent ruminating was monitored for a total of 21 days using accelerometery collars: 10 days prior, the day of diagnosis and 10 days after the diagnosis. The same time-frame was considered for the control group. A repeated-measures ANOVA determined that mean Rumination time and the interaction between Days and Group differed significantly across the days (F= 4.200 and P=0.030, F=5,518 and P=0.014 respectively). A post hoc pairwise comparison using the Bonferroni correction showed differences between the mastitis and control group at D-9, D-3, D-2, D1 and D10. The results show that a decrease in rumination time can be used as an early indicator of clinical mastitis few days before the onset. PLF technology allows an early detection of decrease in rumination time, helping the farmer to early identify clinical cases, thus improving the welfare of dairy cows and reducing the negative effects of mastitis (e.g. costs, antibiotic treatment).

Monitoring the stress of light goat kids transported over short journeys

M. Sort, A. Elhadi, R. Costa, A. Recio, A.A.K. Salama and G. Caja

Universitat Autonoma de Barcelona, Group of Ruminant Research (G2R), Animal and Food Sciences, Av. dels Turons, 08193 Bellaterra, Spain; gerardo.caja@uab.es

Despite the large number of suckling goat kids harvested in the EU, the EC 1/2005 regulation for animal transport does not mention kid requirements. With this aim, 25 suckling goat kids (9.5±0.4 kg LW) of Murciano-Granadina breed (male, n=10; female, n=15) were used to evaluate the effects of 2 space densities for road transport over a short journey (2 h), using a trailer (1.8×1.0×1.3 m; 2 floors and 4 compartments of 0.9 m^2), equipped with sensors (temperature, humidity, sound and acceleration). A second accelerometer was placed in the driver's cabin. Kid densities were: (1) low (0.018 m^2/kg; n=10); and (2) high (0.013 m^2/kg; n=15). Trip was done on national roads, at moderate speed and under mild weather conditions. Kids were weighed, temperature measured, and blood sampled at h 0 (uploading), 2 (downloading) and 24 (resting) from departure. Acceleration peaks in the z-axis during transport were much higher (P<0.001) in the trailer (11.8×g) than in the cabin (1.5×g). Similarly, high noise peaks were detected in the trailer (volume >800 dB and frequency >1,400 Hz) which should be considered as stressful. All kids decreased weight during transport (-3.5% LW; P>0.05), but recovered after 24 h. Moreover, no differences were detected by sex or density. Despite the mild ambient conditions during the journey (27 °C and 44% RH; THI=74), goat kids lost rectal temperature (-0.2 °C; P<0.001), which recovered at h 24, without effect of density. All metabolic indicators increased after transport, except urea, which indicated a situation of metabolic stress aggravated by the cold (12% of the kids showed shivering at arrival). High density during transport only increased serum creatine kinase (71%; P=0.031) and tended to increase lactate dehydrogenase (15%; P=0.13), which recommended using the low density. In conclusion, a density of 0.018 m^2/kg LW may be adequate for suckling kids, although deficiencies in temperature, noise and acceleration were detected. The use of closed trailers with conditioned ambient and efficient shock absorbers, is recommended. Acknowledgements: Funded by the EU H2020 program (Contract #862050, TechCare Project).

Monitoring the stress of light suckling lambs transported over short journeys

A. Elhadi, J.C. Jesús, R. Costa, A. Recio, A.A.K. Salama and G. Caja
Universitat Autonoma de Barcelona, Group of Ruminant Research (G2R), Animal and Food Sciences, Av. dels Turons, 08193 Bellaterra, Spain; abdelaali.elhadi@uab.cat

Space allowance for lambs <20 kg LW in the EC 1/2005 regulation for animal transport must be under 0.2 m^2/lamb (density, 0.010 m^2/kg), but it is unknown if this is adequate for suckling lambs. To assess this, 20 lambs (13.9±0.4 kg LW) of 2 breeds (Manchega, MN; n=10; Lacaune, LC; n=10) were submitted to a transport experiment over a short journey (2 h) with 2 densities: 0.016 m^2/kg (n=8) and 0.011 m^2/kg (n=12). A trailer (1.8×1.0×1.3 m; 2 floors and 4 compartments of 0.9 m^2), equipped with sensors (temperature, humidity, sound and acceleration) and a second accelerometer placed in the driver's cabin, were used. The trip was on national roads, at moderate speed and under mild weather conditions. Lambs were weighed, temperature measured, and blood sampled at h 0 (uploading), 2 (downloading) and 24 (resting). Only MN lambs were slaughtered after arrival and their carcasses evaluated. Acceleration peaks (z-axis, 11.9×*g*) and noise recorded (volume, 491±98 dB; frequency, 395±98 Hz) in the trailer were much greater (P<0.001) than in the cabin and considered as stressful. Transport produced a non-significant weight loss (-3.2%; P>0.05), which was recovered in the LC lambs after resting, without differences by breed, sex or transport density. Despite the mild ambient conditions (15.6 °C and 69% RH; THI=74), rectal temperature decreased during transport, which was greater in LC than MN lambs (-0.68 vs -0.26 °C; P=0.031) agreeing their fleece traits, but without effect of density. All serum indicators increased during transport (P<0.001), indicating metabolic stress, but only lactate dehydrogenase increased in the high density lambs (+14%; P=0.032). Carcasses of MN differed in colour, which was darker (P<0.05) in low density lambs. In conclusion, >0.010 m^2/kg LW seems to be adequate for suckling lambs transport over short journeys, although stressing effects of cold, noise and acceleration were detected. The use of closed trailers, with conditioned ambient and efficient shock absorbers avoiding noise and jumps, is recommended in light lambs. Acknowledgements: Funded by the EU H2020 program (Contract #862050, TechCare Project).

Exploration of indicators measured by PLF sensors to monitor pig welfare during transportation

H.-L. Ko[1], P. Fuentes Pardo[2], F. Jiménez Caparros[3], X. Manteca[1] and P. Llonch[1]
[1]Universitat Autònoma de Barcelona, Department of Animal and Food Science, Campus UAB, 08193 Cerdanyola del Vallès, Spain, [2]CEFU, S.A., Department of I+D+i, Paraje de la Costera, s/n, 30840 Alhama de Murcia, Spain, [3]ElPozo Alimentación, S.A., Department of I+D+i, Avenida Antonio Fuertes, no. 1, 30840 Alhama de Murcia, Spain; henglun.ko@uab.cat

Most pigs experience transportation at least once in their lives. However, the welfare of the pigs during transportation is often overlooked. The aim of the current study was to investigate potential welfare indicators measured by PLF technologies during truck transportation to the slaughterhouse. Two trucks were used in the study. One GPS tracker and two climate sensors, DOL 114 and DOL 19, were installed on each truck. The GPS tracker recorded the journey distance (km), journey duration (minute), and number of stops during the journey. DOL 114 measured temperature (°C) and humidity (%), and DOL 19 measured CO_2 concentration (ppm), both every 15 minutes. A modified Welfare Quality® assessment was conducted during unloading, with the following parameters recorded: number of dead pigs on arrival, fasting duration, and number of sick/lamed/falling/reluctant-to-move/turning-back pigs. A total of 25 journeys were followed with the averages of 173 pigs on the truck, 51 km and 53.6 minutes per journey, and a fasting duration of 16.2 hours. Data were analysed with correlation tests and general liner models. Preliminary results showed that journey distance and duration were correlated with number of reluctant-to-move pigs during unloading (r=0.62 and r=0.63 respectively, P<0.05). Journey distance was also correlated with humidity (r=0.62, P<0.05) and temperature (r=0.71, P=0.07). Additionally, number of falling pigs during unloading was associated with fasting duration (P=0.06) and CO_2 concentration (P<0.05). Our preliminary results suggest that longer journey increases temperature, humidity, and CO_2 concentration. Together with long fasting duration, the aversive micro-climate could affect animals' status on arrival and therefore lead to difficult handling during unloading.

Session 39 Theatre 11

Quantifying the value of early warning system (EWS) based on 'smart water trough' in a sheep farm

A. Bar-Shamai[1,2], I. Shimshoni[1], A. Godo[2], J. Lepar[2] and I. Halachmi[2,3]
[1]Haifa University, Dept. of Information systems, 65 Hanamal St., Haifa, Israel, [2]Agricultural Research Organization (A.R.O.) – Volcani Institute, Precision Livestock Farming (PLF) Lab., 68 Hamaccabim Road, P.O. Box 15159, Rishon Lezion 7505101, Israel, [3]Ben Gurion University of the Negev, Dept. of Industrial Engineering and Management, P.O. Box 653, Be'er Sheva 8410501, Israel; halachmi@volcani.agri.gov.il

This work describes the calculated economic value of an early warning system (EWS) for meat sheep production. An EWS was developed including a multi-sensor monitored drinking station. Weight, drinking behaviour and water consumption data was collected from 53 Assaf-breed sheep, two to five months old, over a period of three months. The financial benefit model comprised of calculating alternative costs for cases of ineffective growth, and optimal selection for marketing. The EWS uncovered significant potential profit increase of up to 50% by employing a decision support system on farms. In addition, a reduction in labour and feed costs and identification of welfare changes is evident.
Acknowledgements: Funded by the EU H2020 program (Contract #862050, TechCare Project).

Session 39 Theatre 12

Economic feasibility of using sensing technologies for welfare monitoring in the pig value chain

A.H. Stygar[1], M. Pastell[1], P. Llonch[2], E.A.M. Bokkers[3], J.D. Bus[3], I.J.M.M. Boumans[3], L.J. Pedersen[4] and J.K. Niemi[1]
[1]1Natural Resources Institute Finland (Luke), Latokartanonkaari 9, 00790 Helsinki, Finland, [2]Universitat Autònoma de Barcelona, Campus UAB, Barcelona, 08193 Cerdanyola del Vallès, Spain, [3]Wageningen University and Research, P.O. Box 338, 6700 AH Wageningen, the Netherlands, [4]Aarhus University, Blichers Allé 20, 8830 Tjele, Denmark; anna.stygar@luke.fi

The use of sensors provides an opportunity for continuous monitoring of animal welfare status that can potentially benefit the entire pig value chain. However, the economic rationale for using sensor technology to provide welfare information is rarely explored. The aim of this study was to assess the economic benefits of different sensor technologies used to monitor pig welfare and discuss economic implications of comprehensive welfare monitoring throughout the pig value chain. Bio-economic models, based on dynamic programming, were used to determine the effect of sensor application to monitored welfare (good feeding, good housing, good health and appropriate behaviour) at various production stages (sows, piglets, growing pigs). Studied examples concerned application of sensor technologies to monitor tail biting, pen fouling, respiratory diseases in growing pigs, body weight in sows and growing pigs, thermal comfort in sows and playing behaviour in piglets. Model parameters were set to represent average costs and benefits in the Netherlands, Spain and Denmark. The simulated results indicated that the economic benefits of using sensors may not outweigh the costs, particularly on farms where good husbandry practices are already in place. However, our analyses indicated that the use of technologies could bring additional benefits in terms of reducing the environmental impact of pig production (e.g. reduced sow replacement rate). Therefore, it is necessary to integrate all sustainability indicators (social, economic and environmental) in order to provide an overview of the benefits that can be obtained from the application of sensors. Business models based on the sharing of sustainability data should be developed.
This study was conducted within the ClearFarm project which received funding from the European Union's Horizon 2020 research and innovation programme under grant agreement No. 862919.

ORIOLE: a web application for cleaning data from the walk-over-weighing device in livestock systems

I. Sanchez, E. González-García, B. Fontez and B. Cloez
INRAE, 2 place pierre viala, 34060, France; eliel.gonzalez@inrae.fr

The use of the walk-over-weighing (WoW), which automatically records animal live weight (LW) in an automated, non-invasive manner, involves filtering the primary datasets produced by this technology. Removing outliers allows the correct data to be retained for a more consistent interpretation of individual daily physical activity progression. However, the standard methods used so far to perform this cleaning were impractical, time-consuming and required minimal mastery of the methods used. This limits the adoption of WoW by farmers and other end users. Our team previously developed a Kalman filter with impulse noised outliers algorithm for the automatic detection of outliers generated by the WoW (kfino; https://arxiv.org/abs/2208.00961). Once the kfino algorithm was tuned, the ORIOLE web-application was developed and deployed by our team (for OutlieRs detectIOn waLk wEighing; https://oriole.sk8.inrae.fr/). The Shiny library of the R software which enables to easily create user-friendly interactive web apps straight from R was used (https://shiny.rstudio.com/). Our web application allows users to import raw data measured from the WoW and through simple settings to perform outlier detection and weight prediction during the experiment. Descriptive statistics are then available such as number of daily weighing, evolution of weight per animal, evolution of the flock weight, 24 h kinetics of individuals. The web app is a dashboard composed of a menu of several subsets offering a user-friendly experience: (1) a 'Welcome' section; (2) the 'Genesis' of the technology and the web-app project; (3) the heart of the app with a section for the import and analysis of 'WoW data' and producing useful reports; (4) a 'How to' section documenting how to use the app. Users can analyse their data using full advantage of descriptive and statistics plots and download reports for communication and decision making.

Monitoring liveweight in Sarda dairy sheep using a walk-over-weighing system

M. Decandia[1], M. Acciaro[1], V. Giovanetti[1], G. Molle[1], F. Chessa[1], I. Llach[2] and E. González-García[2]
[1]AGRIS Sardegna, Loc. Bonassai S.S. 291 Sassari-Fertilia, 07100 Sassari, Italy, [2]SELMET, INRAE, Montpellier SupAgro, CIRAD, Université Montpellier, 34060 Montpellier, France; mdecandia@agrisricerca.it

Live weight (LW) monitoring is important in livestock management to check nutritional, reproduction and welfare status of animals. In dairy sheep system LW measurement is less frequent because it is time and labour consuming. The use of automated weighing system could facilitate this practice. Within the H2020 Techcare project, aiming to use innovative technologies to improve welfare management for small ruminant systems, a short-test was run to evaluate the precision and accuracy of LW measures carried out using an automated walk-over-weighing (WoW) scale in comparison with a static scale. Thirty-six dry Sarda dairy ewes (LW, mean ± SE 55.49±0.92) were used in a 3-day session. Ewes were daily weighed 3 times following the same circuit, which included a first static LW measurement; once a static position was achieved and LW recorded, the animals continued and traversed the WoW scale for the WoW LW recording. The ewes were previously accustomed to the circuit during a pre-experimental week. Raw static and automatic LW data were first filtered for removing misbehaviours and outlier with Kalman filter algorithm. The concordance correlation coefficient (CCC) has been then calculated (Model Evaluation System) to simultaneously account for precision and accuracy. The two component of CCC (=0.98), correlation coefficient estimate (r, that measure precision) and the bias correction factor (Cb, that indicates the accuracy) were respectively 0.98 and 0.99 indicating high agreement between measures. The WoW system evaluated here is an alternative to the static scales conventionally used on dairy sheep farms. If sound filtration of raw data is applied, WoW could contribute to the close (daily) monitoring of individual LW without operator intervention (i.e. voluntary weighing), taking animal welfare into account (i.e. no stress related to the weighing session on static scales), and potentially detecting nutrition issues highlighted by LW changes.

Session 39 Poster 15

Stakeholders' attitudes to early warning systems to promote animal welfare in small ruminants

E.N. Sossidou[1], S.I. Patsios[1], S. Beechener[2], V. Giovanetti[3], G. Tesniere[4], L. Grova[5], T. Keady[6], G. Caja[7], A. Bar-Shamai[8], L.T. Cziszter[9] and C. Morgan-Davies[2]

[1]Veterinary Research Institute, Ellinikos Georgikos Organismos-DIMITRA, ELCO Campus, 57001 Thermi, Thessaloniki, Greece, [2]Scotland's Rural College (SRUC), Hill & Mountain Research Centre|Kirkton, Crianlarich, FK20 8RU, Scotland, United Kingdom, [3]AGRIS SARDEGNA – Agenzia per la ricerca in agricoltura, Via Giovanni Antonio Carbonazzi, 10, 07100 Sassari SS, Italy, [4]Institut de l'Elevage Service Capteurs, Equipements, Bâtiments, Campus INRAe, CS 52637 – 31321 Castanet Tolosan, France, [5]NIBIO, Norsk institutt for bioøkonomi, Norsk institutt for bioøkonomi, Gunnars vei 6, 6630 Tingvoll, Norway, [6]TEAGASC, TEAGASC, Oak Park, Carlow, R93 XE12, Ireland, [7]UAB, Plaça Cívica, 08193 Bellaterra, Barcelona, Spain, [8]ARO – Agricultural Research Organization of Israel, The Volcani Centre, 68 HaMacabim Road, 7505101 Rishon LeZion, Israel, [9]FBIRA Universitatea de Științele Vieții, Aleea Mihail Sadoveanu nr. 3, 700490 Timisoara, Romania; sossidou.arig@nagref.gr

TechCare project aims to demonstrate innovative approaches to monitor and improve welfare management in small ruminants' farming systems. An integral part of this approach is the early warning systems (EMS) for the timely alert on welfare issues on an individualized base. An integrated multi-actor approach is applied to constantly collect feedbacks from stakeholders and co-design the EWS through a series of national workshops. The 3rd series of TechCare national workshops engaged all relevant stakeholders and collected their feedback on the critical aspects of developing the EWS. Despite the different background, the stakeholders from all countries recognized specific features of the EWS as highly favourable. These features include: the ability to track and trend history and past trendlines of critical data; the integration of novel EWS with existing tools and data management applications; the option to manually add data recordings and other information (i.e. mortality data, diets' information); the ability to group animals; that the EWS should be customizable to the specific needs of each farming system; that the EWS should be easily expandable to include new precision livestock farming (PLF) tools or welfare indicators; and that the interface should be simple and concise.

Session 39 Poster 16

Comparison between automatic milking and milking parlour system on dairy cows' welfare

G.V. Berteselli, S. Cannas, E. Dalla Costa, R. Zanchetta, G. Pesenti Rossi and E. Canali

University of Milan, Department of Veterinary Medicine and Animal Science, via dell'Università 6, 26900 Lodi, Italy; greta.berteselli@unimi.it

The use of robot milking in dairy farming has led to numerous advantages (e.g. increasing hygiene, health; reducing labour) but the impact on animal behaviour and welfare is not fully explored. The aim was to compare behaviours' time-budget, animal-based measures related to welfare, and productivity in two different milking systems: automatic milking and milking parlour. Two groups of healthy lactating cows of two intensive farms (located in a region of Northern Italy) were recruited. Automatic milking group (AM) milking parlour group (MP): 44 and 50 cows respectively. A welfare assessment was performed including the evaluation of animal cleanliness (udder, legs, flank); presence of lesions/swelling and presence of hairless patches; nasal, ocular, vulvar discharge; diarrhoea; locomotion score, BCS, human-animal relationship (avoidance distance test). Furthermore, for each cow, daily hours spent lying, ruminating, walking and eating were recorded using accelerometery collars. A time window of 15 day was considered (7 days prior the assessment, the day of assessment, 7 days after assessment) as well as the milk production. A significant difference was found in the time budgets between AM and MP groups as well as the milk production (Mann Whitney U-test; $P<0.05$): AM group spent more time lying (10.88±1.58 h), walking (2.97±0.84 h), whereas MP group spent more time ruminating (8±0.84 h), and eating (5.01±1.31 h); the milk production was higher in MP-group. A significant difference between groups was also found for udder cleanliness and the presence of lesion/swelling (Chi-square test; $P<0.05$). In particular, MP-group showed a higher % of animals with dirty udder, whereas most of the injured animals belonged to AM group. The AM group showed a better human-animal relationship, with only a 23% of animals showing avoidance behaviours towards unfamiliar person (Chi-square test; $P=0.066$). The results show that the milking system can have an impact on different aspects of dairy cows' life in particular on time-budget and welfare. The human-animal relationship could be positively affected by the automatic milking, but this requires further studies.

Session 39 Poster 17

Prioritization of welfare issues and precision technologies for welfare monitoring in dairy sheep

A. Elhadi[1], R. González-González[2] and G. Caja[1]
[1]Universitat Autonoma de Barcelona, Group of Ruminant Research (G2R), Animal and Food Sciences, Av. dels Turons, 08193 Bellaterra, Spain, [2]Gestión Empresarial de Ovino (GEO), Av. Americas 7, 49600 Benavente, Spain; abdelaali.elhadi@uab.cat

According to the planned activities of WP1 in the TechCare project, 2 National Workshops (NW) were held in 9 countries to discuss and to prioritize (top 3) the welfare issues and sensor technologies for welfare monitoring in sheep (meat and dairy) and goats (dairy), according to species and systems. In Spain, NW were: NW1 (47 participants from 10 Autonomous Communities: 22 scientists, 10 veterinarians, 10 technicians and 5 farmers), which prioritized the welfare issues from an exhaustive list (n=437) based on bibliographic references (WP2), and NW2 (46 part. from 12 A.C.: 16 sci., 8 vet., 15 tech. and 7 farm.), which prioritized a list (n=116) of PLF sensors based on bibliographic references and available market products (WP3). Suckling and fattening lambs and goat kids were also included. Because the COVID-19 pandemic, both Consortium and Spanish meetings were telematic (by Zoom). Additionally, to validate the NW1 and NW2 prioritizations, a presential and *in situ* workshop (Medina del Campo, Valladolid), with focus on dairy farmers, was held late in 2022: NW3 (40 part.: 4 vet., 10 tech. and 26 farm.) before the tools implementation in large-scale trials on commercial farms. The NW3 prioritizations, according to percentage of votes, were: Welfare issues (1st, mastitis and milking management 70%; 2nd, abortion and perinatal mortality 55%; 3rd, housing conditions 43%), and Sensors (1st, weather stations 60%; 2nd, automatic milk meters 53%; 3rd, behaviour accelerometers 43%). The results of NW3 remarkably agreed with those obtained in the NW1 of the TechCare consortium-9 (256 part.) and in the specific case of NW1 for dairy sheep in Spain. The coincidence was greater in the sensor prioritization, which was similar to the NW2 of the TechCare consortium-9 (150 part.) and in the Spanish NW2. In conclusion, Spanish farmers showed interest in welfare monitoring for dairy sheep and prioritized the use of weather stations (with external and internal sensors) to record the environmental conditions of intensive dairy farms. Acknowledgements: Funded by the EU H2020 program (Contract #862050, TechCare Project).

Session 40 Theatre 1

Update of nutritional requirements of goats for growth and pregnancy in hot environments

I.A.M.A. Teixeira[1], C.J. Härter[2], J.A.C. Vargas[3], A.P. Souza[4] and M.H.M.R. Fernandes[5]
[1]University of Idaho, 315 Falls Avenue Evergreen Buiding, 83303-1827 Twin Falls, ID, USA, [2]Universidade Federal de Pelotas, Rua Gomes carneiro, 01, 96160-000 Pelotas, RS, Brazil, [3]Universidade Federal Rural da Amazônia, PA-275 s/n, 68515-000 Paraupebas, PA, Brazil, [4]Universidade Federal do Sul e Sudeste do Para, R. Alberto Santos Dumont, 68557-335 Xinguara, PA, Brazil, [5]UNESP-Universidade Estadual Paulista, Via de Acesso Paulo Donato Castellane, 14884-900 Jaboticabal, SP, Brazil; izabelle@uidaho.edu

Goats are an important source of income in the agricultural business, such as milk and meat. They are also the preferred livestock for rural households due to their inherent resiliency and adaptability to many environments and suitability in sustainable production systems. In the last decades, the nutrition of goats, particularly their nutritional requirements, has received special attention. Different research groups around the world have dedicated efforts to updating feeding systems for goats. In this invited talk, our objective is to present the recent findings on the energy and nutrient requirements of growing and pregnant goats in hot environments. The literature and studies considered herein comprise growing goats of different sexes (females, intact and castrated males) and genotypes (dairy, meat, and indigenous). In short, the energy and protein requirements (maintenance and growth) of growing goats are influenced by sex and genotype only when mature weight is not considered in the models. The maintenance and growth requirements of major minerals are not affected by sex and genotype. Sex and genotype affect the efficiency of energy use for growth but did not affect the efficiency of protein use. The literature findings also suggest that losses in urine and methane in goats are lower than the ones reported for bovine and sheep. Regarding requirements for pregnancy, there was no effect of days of pregnancy on the energy or protein requirements. The efficiency of metabolizable energy utilization for pregnancy increased with the progress of pregnancy. Mineral accretion for pregnancy differs between single and twin pregnancies and irrespective of pregnancy the mineral requirements increased as pregnancy progressed. New approaches and limitations to goat requirements are also presented.

Session 40 Theatre 2

Intramammary administration of lipopolysaccharides at parturition affects colostrum quality

M. González-Cabrera[1], A. Torres[2], M. Salomone-Caballero[1], N. Castro[1], A. Argüello[1] and L.E. Hernández-Castellano[1]
[1]Universidad de Las Palmas de Gran Canaria, Institute of Animal Health and Food Safety, Trasmontaña s/n, 35413 Arucas, Spain, [2]Canary Islands Institute for Agricultural Research, Animal Production, Pasture, and Forage in Arid and Subtropical Areas, Finca El Pico, s/n, 38260 La Laguna, Spain; marta.gonzalezcabrera@ulpgc.es

In this study, 20 Majorera dairy goats were used. The TRT group (n=10) received an intramammary administration (IA) of saline (2 ml) containing 50 µg of lipopolysaccharides (LPS) from *Escherichia coli* (O55:B5) in each quarter at parturition. The CON group (n=10) received an IA of saline (2 ml) without LPS. Rectal temperature (RT) was recorded, and a blood sample collected at parturition (before the IA). In addition, RT was recorded, and blood and colostrum/milk samples were collected at 3 and 12 hours, and on days 1, 2, 4, 7, 15 and 30 relatives to IA. Plasma immunoglobulin (Ig) G and M and serum β-hydroxybutyrate, glucose, calcium, free fatty acids, lactate dehydrogenase and total proteins concentrations were determined. Colostrum/milk yield as well as chemical composition, somatic cell count (SCC) and IgG and IgM concentrations were measured. The MIXED procedure (SAS 9.4) was used, and the model included the IA, time, and the interaction between both fixed effects. Statistical significance was set as $P \leq 0.05$. Goats from the TRT group increased RT after the IA, while the CON decreased RT ($P_{IA} \times T=0.007$). Serum biochemical parameters, and plasma IgG and IgM concentrations were not affected by IA. Colostrum and milk yield as well as chemical composition were not affected by IA, except for milk lactose that was lower in the TRT group compared to the CON group ($P_{IA}=0.026$). Colostrum SCC was higher in the TRT group than in the CON group (3.5±0.09 and 3.1±0.09 cells×10^6/ml, respectively; $P_{IA}=0.009$). Similar results were observed for milk SCC ($P_{IA}=0.004$). The TRT group showed higher IgG ($P_{IA}=0.044$) and IgM ($P_{IA}=0.037$) concentrations on colostrum than the CON group (31.9±4.8 and 19.0±4.5 mg/ml,0.7±0.08 and 0.4±0.08 mg/ml, respectively). No differences on milk IgG and IgM concentrations between groups were observed. In conclusion, the IA of LPS at parturition increases RT, SCC and IgG and IgM concentrations in colostrum without affecting either yield or chemical composition.

Session 40 Theatre 3

Goat kids are not affected by the intramammary administration of lipopolysaccharides at parturition

M. González-Cabrera[1], M. Salomone-Caballero[1], S. Álvarez[2], A. Argüello[1], N. Castro[1] and L.E. Hernández-Castellano[1]
[1]Universidad de Las Palmas de Gran Canaria, Institute of Animal Health and Food Safety, Trasmontaña s/n, 35413 Arucas, Spain, [2]Canary Agronomic Research Institute, Animal Production, Pasture, and Forage in Arid and Subtropical Areas, Finca El Pico, s/n, 38260 La Laguna, Spain; marta.gonzalezcabrera@ulpgc.es

This study evaluated the effect of an intramammary administration (IA) of lipopolysaccharides (LPS) from *Escherichia coli* (O55:B5) to dairy goats at parturition, on performance, biochemical parameters (calcium, LDH, glucose and total proteins) and immune status (IgG and IgM) of goat kids. Twenty dairy goats were used. The treatment (TRT) group (n=10) received an IA of saline solution (2 ml) containing 50 µg LPS per gland at parturition. Similarly, the control (CON) group (n=10) received an IA of saline solution (2 ml) without LPS. At birth, goat kids (n=45) were weighted (day 0) and immediately allocated into either the TRT group (n=19) or the CON group (n=26) based on the experimental group of the dam. They were bottle-fed dam colostrum equivalent to 10% of the birth body weight (BW) divided in two meals (3 and 12 h after birth), and then fed twice daily with milk replacer. Individual milk intake (MI) and BW were recorded on days 7, 15, 21 and 30 of life. Blood samples were taken on days 0, 1, 2, 4, 7, 15, 21 and 30 after birth. Data was analysed using the MIXED procedure of SAS (9.4). The model included IA, time (T) and the interaction (IA × T). Both groups showed similar MI, except for day 7 as the TRT group showed higher MI than the CON group (911±73.9 ml/day and 652±64.6 ml/day, respectively; ($P_{IAxT}=0.001$)), although no differences were observed on BW during the experiment ($P_{IA}=0.367$). The TRT group showed higher calcium ($P_{IA}=0.001$) and total protein ($P_{IA}<0.001$) concentrations than the CON group (12.8 [12.5-13.1] and 12.2 [12.0-12.5] mg/dl, respectively). The CON group showed higher LDH activity than the TRT group (626±20.51 U/l and 561±20.51, respectively; $P_{AI}=0.016$). Glucose and plasma IgG and IgM concentrations were not affected by IA ($P_{IA}=0.217$, $P_{IA}=0.151$ and $P_{IA}=0.157$, respectively). In conclusion, the IA of LPS to dairy goats at parturition does not affect performance or immune status of goat kids but affects calcium, total protein and LDH activity on serum.

Session 40 Theatre 4

Near-infrared spectroscopy prediction models preliminary results in ewes' colostrum

S. González-Luna[1,2], E. Albanell[2], G. Caja[2] and C. L. Manuelian[2]
[1]Facultad de Estudios Superiores Cuautitlán, Universidad Nacional Autónoma de México (UNAM), Departamento de Ciencias Pecuarias, Ctra. Cuautitlán-Teoloyucan km 2.5, 54714 Cuautitlán Izcalli, Mexico, [2]Universitat Autònoma de Barcelona (UAB), Group of Ruminant Research (G2R), Department of Animal and Food Sciences, Campus Universitari de la UAB, 08193 Bellaterra, Spain; carmen.manuelian@uab.cat

Colostrum is key for nutrients provision and passive immunity transfer in newborns. Thus, rapid methods to determine its quality at farm level is crucial. The aim was to investigate near-infrared spectroscopy feasibility to prediction colostrum gross composition, IgG, and insulin. A total of 45 colostrum samples collected <6 h post-partum from Lacaune (n=21) and Manchega (n=24) ewes were scanned in duplicate by using NIR System 5000 equipment (FOSS Electric A/S, Hillerød, Denmark) from 1,100 to 2,500 nm wavelength every 2 nm. Absorbance was recorded as log(1/Transflectance). Reference data were matched with the sample spectra, and modified partial least squares regressions prediction models were developed applying a 5-fold cross-validation. Outlier were discarded before building the final model. Average mean ± SD for fat, protein, true protein, casein, total solids, IgG, and insulin of the retained samples were 7.35±3.45%, 15.84±4.89%, 15.22±4.73%, 5.81±1.84%, 26.17±7.16%, 25.10±10.96 mg/ml, and 13.32±6.12 µg/l, respectively. Calibration models were considered excellent for total solids, total protein, and true protein with a coefficient of determination in cross-validation (R2cv) of 0.99, and a RPD (SD/SEcv) that ranged from 9.40 (true protein) to 15.84 (total solids). Very good calibration model applicable for quality control was achieved for total fat (R2cv, 0.96; RPD, 3.58), and poor calibration applicable for a rough screening was obtained for casein (R2cv, 0.80; RPD, 2.29). However, very poor calibration models resulted for IgG (R2c, 0.88; R2cv, 0.39; RPD, 1.27) and insulin (R2c, 0.68; R2cv, 0.26; RPD, 1.14). In conclusion, these preliminary results suggested the potential use of near-infrared spectroscopy to estimate sheep colostrum quality, and confirm the need for further investigation for IgG.

Session 40 Theatre 5

No kid, no milk? Trials about induction of goat lactation without gestation

L. Fito, C. Constancis and M. Bouy
FiBL France, Research Institute of Organic Agriculture, 150 avenue de Judée, 26400 Eurre, France; caroline.constancis@fibl.org

In dairy production, female small ruminants usually give birth every year to produce milk and be profitable. However, repeated gestation leads to exhaustion of the animals and early culling. Furthermore, consumer demand for goat kid meat is very low, impeding farmers to retrieve satisfying economic and social value from these repeated births. In response to these issues, the practice of 'long or extended lactation' appears like an interesting solution, but raises concerns about animal and farmer welfare. Alternative methods consist of inducing lactation in non-gestating females by subjecting them to intense hormonal treatments which simulate the prepartum period, but these also come with animal welfare and environmental drawbacks. Some farmers do this on one or two non-gestating goats each year but very little research has been done on this practice. The goal of the study was to assess if natural induction of lactation in non-gestating goats was possible, and to evaluate the impact on milk production and on endocrinology. trials on 6 farms has been conducted to investigate the induction of lactation in non-gestating goats by manual stimulation of the teats and by keeping them in presence of gestating ones. twenty-eight non-gestating goats were manually stimulated by farmer in the milking parlour, for one month. Prolactin, progesterone and oestradiol levels, involved in the lactation process, were measured before and after parturition. Kruskal-Wallis tests were used using R software version 3.5.3. Lactation was induced in 61% of the stimulated non-gestating goats. Milk production level of these non-gestating goats gradually increased during the first month, but they reached the same production level and the same milk composition as the others at the end of the lactation. Overall production was 30% lower in in non-gestating goat. We evidenced that the propensity for non-gestating goats to induce lactation was correlated to a higher prolactin level prior to the onset of stimulation (*P*-value<0.001). In conclusion, induction of lactation therefore seems to be promising, and would allow valuable improvements of animal welfare and farm sustainability, but the underlying hormonal mechanisms still need better understanding.

Session 40 — Theatre 6

Perinatal rumen microbiota in relation to feed utilization and biochemical parameters in sheep

J.X. Chen
Hebei Agricultural University, college of animal science and technology, No.2596, Lekainan Street, Lianchi District, 071000 Baoding, China, P.R.; chenjiaxin1226@163.com

This study aimed to investigate the relationship between rumen microflora and nutrient digestion, rumen fermentation and biochemical parameters in sheep. Blood, rumen fluid and faeces samples (n=10) were collected on days -21, -14, -7, 3, 7 and 14 (Q21, Q14, Q7, H3, H7 and H14) relative to expected parturition. Dynamic changes of average daily feed intake (ADFI), nutrient digestibility, rumen fluid and serum parameters were detected. The 16S rRNA sequencing was performed for rumen microflora and the Random Forest method was used to predict Perinatal time-related changes. All data were evaluated by one-way ANOVA. The results showed that the digestibility of dry matter (DMD), crude protein (CPD) and neutral detergent fibre (NDFD) in Q7 d was higher than that in H3 d (P<0.05); Compared with Q21 d, ADFI and total volatile fatty acids (T-VFA) increased after lambing, while acetate to propionate ratio (A/P) decreased (P<0.05); Serum concentrations of glucose, β-hydroxybutyric acid (BHBA), 25-hydroxy vitamin D3 (25HVD3) and neuropeptide Y (NPY) were increased in H14 d compared to Q21 d; Total cholesterol (TC), high density lipoprotein cholesterol (HDL-C), glucagon and leptin were the highest at Q21 d and decreased after lambing (P<0.05). The 16S rRNA gene sequencing revealed that rumen microflora composition also differed in rumens before and after lambing (P<0.05). Nineteen perinatal time-related bacterial genera were predicted, in which the relative abundance of *Anaeroplasma* and *Lachnospiraceae_ND3007_group* was the highest at Q21 d, and was positively correlated with rumen pH, acetate and TC, negatively correlated with propionate (P<0.05), *Endomicrobium, Suttonella, M2PT2-76_termite_group* decreased after parturition and were positively correlated with HDL-C and TC, negatively correlated with T-VFA, 25HVD3 and NPY (P<0.05); *Ruminococcus* and *UCG-005* is increasing in H3 d, and *Ruminococcus* is positively correlated with T-VFA, 25HVD3 and NPY (P<0.05), while *UCG-005* is negatively correlated with NDFD, CaD, PD, CPD and DMD (P<0.05). These results suggest that rumen bacterial community in the perinatal period of ewes affect feed utilization and rumen fermentation patterns, and play an important role in shaping biochemical parameters.

Session 40 — Theatre 7

Alpine goats divergent for functional longevity differ in metabolic profile during transition period

J. Pires[1], T. Fassier[2], M. Tourret[1], C. Huau[3], N. Friggens[4] and R. Rupp[3]
[1]INRAE, UCA, VetAgro Sup, UMRH, Theix, 63122, France, [2]Domaine de Bourges, INRAE, Osmoy, 31326, France, [3]GenPhySE, Université de Toulouse, INRAE, Castanet Tolosan, 31320, France, [4]UMR 0791 MoSar, INRAE, AgroParisTech, Université Paris-Saclay, Paris, 75005, France; jose.pires@inrae.fr

The objective was to study associations among functional longevity and plasma indicators of metabolic adaption in Alpine goats during periparturient period. Two Alpine goat strains divergent for longevity (LGV+ and LGV-) were produced by AI selecting for extreme functional longevity but nondifferent milk yield. A total of 174 primiparous goats were studied in 2018, 2019, 2020 and 2021. Jugular plasma collected on wk -4, -3, -2, -1 relative to expected parturition, and wk 1, 2, 4, 13, 24, 33 of lactation was analysed for NEFA, BHB, glucose, urea and bilirubin. Data were analysed using SAS mixed models with repeated measures, including strain, litter size (LS), wk, and interactions as fixed effects, and year and goat (year) as random effects. Significant wk effects were observed for all metabolites. LGV- goats had greater plasma NEFA on wk-3 (181 vs 123 μM; strain × wk prepartum: P=0.05), and greater BHB prepartum (0.45 vs 0.41 mM; strain effect: P=0.04) than LGV+, which denotes greater fat mobilization and partial oxidation in late gestation LGV- goats. 35% of goats carried multiple foetus (LS2+) and LS not differ with LGV. Prepartum plasma NEFA, BHB and bilirubin were greater for LS2+ compared to single (LS1; P<0.001; 260 vs 174 μM; 0.51 vs 0.39 mM; 0.069 vs 0.056 mg/dl, respectively), whereas glucose was lower for LS2+ (P<0.001; 49.8 vs 54.7 mg/dl). Conversely, plasma NEFA was greater for LS1 during wk 1 and wk 2 postpartum (P<0.05; 558 vs 442 on wk 1, and 421 vs 330 μM on wk 2, respectively), reflecting greater availability of body reserves to support lactation in LS1. Prepartum incidence of BHB>0.80 mM was significantly greater for LS2+ than LS1 (30 vs 3.7%), and for LGV- carrying LS2+ than LGV+ carrying LS2+ (42 vs 28%). Marked LS effects were observed in plasma metabolite profiles in primiparous goats. LGV strains differ in their metabolic adaptations peripartum. This study has received funding from the EU H2020 research and innovation program under grant agreement No 772787 (SMARTER) and from APISGENE (ACTIVEGOAT).

Session 40 Theatre 8

Effects of sodium butyrate supplementation at late gestation of ewes

Y.J. Zhang, X.Y. Zhang, X.H. Duan, R.C. Yang and S.W. Zhang
Hebei Agricultural University, College of Animal Science and Technology, Lekai Street, Baoding, 071000, China, P.R.; zhangyingjie66@126.com

The purpose of this experiment was to study the effects of different levels of sodium butyrate on growth performance, nutrient apparent digestibility and body immunity during late gestation of ewes. Sixty female Hu sheep with similar age, body weight and parity were randomly divided into 4 groups (n=15), and the basal diet was supplemented with 0 g/d (control group), 2.5 g/d (group A), 5 g/d (group B) and 7.5 g/d (group C) sodium butyrate for each sheep, respectively. The trial lasted from day 90 to day 15 before lamb birth. Feed intake and body weight (BW) were recorded and blood samples from jugular vein were collected at day 90 and day 15 before lamb birth. Three ewes in each group were randomly selected for digestion and metabolism test. The results showed that the dry matter intake (DMI) in group B was significantly higher than that of control group and group A ($P<0.05$), but there was no significant difference between group B and group C ($P>0.05$). The BW of ewes in group B was significantly higher than that of control group and group A ($P<0.05$), but there was no significant difference between group B and group C ($P>0.05$). The apparent digestibility of dry matter, crude protein, gross energy and crude fibre in groups B was significantly higher than that in control group ($P<0.05$). At 60 days before lamb birth, the serum level of immunoglobulin G (IgG), interleukin-2(IL-2), interleukin-4(IL-4) and interleukin-6(IL-6) in experimental groups had no differences compared to the control group. At day 15 before lamb birth, IgG in group B was significantly higher than that in control group and group C ($P<0.05$), but there was no significant difference between group A and group B ($P>0.05$). The serum level of IL-2 and IL-6 in groups B and C were significantly lower than those in control group and group A ($P<0.05$). In conclusion, dietary sodium butyrate supplementation at late gestation of ewes can increase feed intake, nutrient apparent digestibility and improve the body immunity. The optimal supplemental level of sodium butyrate was 5 g/d per sheep.

Session 40 Theatre 9

New artificial teats to improve goat milking device laboratory tests

M. Despinasse, C. Bonin, G. Coquereau and J.L. Poulet
Institut de l'Elevage, 149 rue de Bercy, 75595 Paris Cedex, France; jean-louis.poulet@idele.fr

In general, milking devices are evaluated first in laboratories by applying standardized 'wet tests' (ISO 3918:2007 and ISO 6690:2007), with a single design of goat artificial teats (AT). However, studies previously conducted (CASDAR RT Mamovicap, 2017) have stigmatized the shapes and dynamic characteristics (deformation, milk flow) diversities of these teats. CapriMam3D (CASDAR RT, 2020-2023) aims to investigate teat-liner interactions, using new technologies to develop and evaluate new artificial teats. For this project, goat teats of our 2 partner experimental farms (Experimental Farm of Le Pradel and INRAe Experimental Station of Bourges). 3 shapes of teats have been selected: 'M' (Medium, L=60 mm and Ø=38 mm), with dimensions very close to those of standard AT (ISO 6690:2007), 'S' (Small and short, L=40 mm and Ø=30 mm), close to ewe's teat, and 'C' (large and Conical, L=90 mm and Ø=55 mm), finally quite present on farms. For these 3 shapes, different constitutions of AT were evaluated: RAT (Reference Artificial Teat), made with hard plastic and standardized size, SAT (Soft Artificial Teat), obtained by casting silicone in moulds made by 3D printing, and SAAT (Soft Anatomic Artificial Teat), produced in an identical way to the previous ones but with a cavity simulating the teat cistern, obtained using a counter-mould. Milking Liners (ML), among the most widely used in France (analysis of an extraction from Logimat 3, Milking Machine Control input and monitoring software of the French Interprofessional Committee for Milk Production Techniques) were tested with this new AT panel. It emerges from an initial analysis of the results that the characterization of the operation of the AT/ML interface (vacuum fluctuations) can differ greatly depending on the AT used (and possibly the underlying methods). The use of RAT sometimes leads to laboratory test results that are quite different from farm findings (experimental and commercial), while the use of SAAT leads to characterizations of the AT/ML interface closer to these, despite sometimes more complex implementations. The work is currently continuing, by deepening the comparisons between characterization in laboratory and in experimental farms of 2 ML, common but opposed in terms of design.

Session 40 Theatre 10

Olive cake feeding alters the expression of lipogenic genes in mammary and adipose tissue in goats

M.C. Neofytou[1], A.L. Hager-Theodorides[2], E. Sfakianaki[2], S. Symeou[1], D. Sparaggis[3], O. Tzamaloukas[1] and D. Miltiadou[1]

[1]Cyprus University of Technology, Department of Agricultural Sciences, Biotechnology and Food Science, P.O Box 50329, Limassol, Cyprus, [2]Agricultural University of Athens, Department of Animal Science, Iera Odos, 11855 Athens, Greece, [3]Agricultural Research Institute, P.O. Box 22016, Nicosia, Cyprus; despoina.miltiadou@cut.ac.cy

This study assessed the effect of dietary inclusion of ensiled olive cake (OC) on milk yield, composition, fatty acid (FA) profile and the expression of selected genes involved in lipid metabolism in the udder and adipose tissue of goats. Three iso-nitrogenous and iso-energetic diets containing 0% (OC0), 10% (OC10), and 20% (OC20) of EOC (DM) were formulated and offered to mid-lactating Damascus goats (3 pens of 8 animals each) for 42 days. At the end of the trial, mammary and perirenal adipose tissue samples were collected from 6 animals per treatment from OC0 and OC20 groups for gene expression analysis by quantitative RT-PCR. The expression of 10 genes, involved in FA synthesis (ACACA, FASN, G6PDH), FA uptake and/or translocation (VLDLR, LPL, SLC2A1, CD36, FABP3), FA desaturation (SCD1) and transcriptional regulation (PPARγ), was evaluated. The results showed that, among milk traits, milk fat percentage was risen with increasing OC inclusion rates in the diets, while milk protein percentages were elevated in both OC groups. Although, the content of medium-chain FA was reduced in the milk of OC groups, the mammary expression levels of ACACA, FASN, and SCD1 was unaffected. In contrast, the significant increase of the long-chain FA content observed in the milk of OC20 group, coincided with elevated expression of SLC2A1 ($P<0.05$), VLDLR ($P<0.01$), FABP3 ($P<0.01$) in the mammary gland of OC goats. In the adipose tissue, significant increments of C18:1 trans isomers found in the milk of OC group may have affected the mRNA abundance of FASN ($P<0.01$) which was increased in the perirenal fat of OC cows. Similarly, the OC20 diet upregulated the SLC2A1 in the adipose tissue of goats that could likely increase LCFA and glycose uptake in caprine adipose tissue. Overall, OC can be used up to 20% (DM) in goats' diets, without adversely affecting milk traits or the expression of genes involved in lipid metabolism.

Session 40 Poster 11

Chemical composition and physicochemical characteristics of mountainous goat milk during lactation

E. Kasapidou[1], I.V. Iliadis[2], G. Papatzimos[1], M.A. Karatzia[3], Z. Basdagianni[4] and P. Mitlianga[2]

[1]University of Western Macedonia, Department of Agriculture, Florina, 53100, Greece, [2]University of Western Macedonia, Department of Chemical Engineering, Koila, 50100 Kozani, Greece, [3]Research Institute of Animal Science, HAO-Demeter, Paralimni, 58100 Giannitsa, Greece, [4]Aristotle University of Thessaloniki, School of Agriculture, Department of Animal Production, Thessaloniki, 54124, Greece; ekasapidou@uowm.gr

Milk composition affects its capability to produce superior quality dairy products. Additionally, the favourable composition of milk from upland origin has been reported in several studies. The study aimed to record changes in mountainous goat milk's composition and physicochemical characteristics during lactation. Bulk tank milk samples (n=134) were collected every fortnight between March and October 2022 from 10 dairy goat farms following the semi-intensive production system located in an area with an average altitude of 772 m. Daily milk yield and the population of lactating goats were recorded. Milk composition and physicochemical characteristics were determined. One factor variance analysis examined differences at monthly level. The average number of lactating goats per farm was 151 whereas mean daily milk yield (171 l) differed significantly ($P<0.001$) during lactation. The average fat, protein, lactose and solids non-fat content were 4.39, 3.63, 9.14 and 4.49 g/100 ml respectively. Highly significant differences ($P<0.001$) were observed in gross composition during lactation. Higher protein and fat contents were found at the end of the lactation period whereas lactose content was variable throughout lactation. Mean values for physicochemical characteristics were 1.3481, 5.30 mS/cm and 6.75 for refractive index, electrical conductivity and pH respectively. Electrical conductivity and pH were significantly affected ($P<0.001$) during lactation whereas there was no effect ($P>0.05$) on refractive index. Lower pH and higher electrical conductivity values were recorded in late spring and summer months. Results confirm that lactation stage affects milk composition. Information about milk composition during lactation could be beneficial for producers of mountain dairy products enabling standardization of product composition and quality. Support: AGROTOUR Project (MIS 5047196), NSRF 2014-2020.

Session 40 Poster 12

Energy balance and body reserve dynamics in early lactation dairy ewes

F. Corbiere[1], C. Machefert[2], J.M. Astruc[3], M. El Jabri[3], B. Fanca[3], P. Hassoun[4], C. Marie-Etancelin[2], A. Meynadier[2] and G. Lagriffoul[3]

[1]INRAE, UMR INRAE-ENVT IHAP, 23 Chemin des Capelles, 31076 Toulouse, France, [2]INRAE, UMR GenPhyse, Campus INRAE Toulouse, 31326 Castanet-Tolosan, France, [3]IDELE, Campus INRAE Toulouse, 31326 Castanet-Tolosan, France, [4]INRAE, SELMET, Campus international de Baillarguet, 34060 Montpellier, France; fabien.corbiere@envt.fr

The objective of this study conducted on two experimental dairy sheep flocks and eight commercial farms was to evaluate the use of two blood metabolites, beta-hydroxy-butyrate (BHB) and non-esterified fatty acids (NEFA), to assess energy balance and body reserve mobilization in dairy sheep. On 9 farms, body condition scoring (BCS) was performed on 3 occasions (one month before lambing, 3 weeks after lambing and after weaning at the time of the first official milk recording) for a total of 2,125 ewes. In parallel, blood BHB and NEFA concentrations were measured on a subset of 466 2nd lactation ewes in these flocks. On two farms, the diurnal kinetics of blood BHB and NEFA during the day was also studied. Comparisons between BHB and NEFA values measured at different time points or between different groups were performed by Wilcoxon rank tests (for paired data if necessary) and by multivariable linear mixed regression models, with a Holm correction applied for multiple comparisons. Results confirm that the composition of the ration and the postprandial delay are major factors of diurnal variation in BHB and NEFA values. While plasma NEFA values measured 3 weeks after lambing were found to be positively correlated with the extent of BCS loss since late gestation, no significant relationship could be demonstrated for blood BHB. When measured at the first official milk recording (52±9.9 days after lambing), neither BHB and NEFA were correlated to body reserve mobilization. Further, the overall milk yield throughout lactation was strongly correlated with BCS dynamics between late gestation and early lactation. These results support the importance of exploring the early lactation period in dairy ewes, during which lambs are kept under their dams.

Session 40 Poster 13

Effects of Sulla flexuosa hay as alternative feed resource on goat's milk production and quality

S. Boukrouh[1,2], A. Noutfia[1], N. Moula[2], C. Avril[3], J.-L. Hornick[2], M. Chentouf[1] and J.-F. Cabaraux[2]

[1]National Institute of Agricultural Research, Regional Center of Agricultural Research of Tangier, 10090 Rabat, Morocco, [2]ULiège, Department of veterinary management of animal resources, FARAH, Faculty of Veterinary Medicine, Quartier Vallée 2, Avenue de Cureghem, 6, Bât. B43, 4000 Liège, Belgium, [3]HEPH Condorcet, Agronomy category, Rue de la Sucrerie, 10, 7800 Ath, Belgium; jfcabaraux@uliege.be

Sulla flexuosa (*Hedysarum flexuosum* L.) is an endemic legume growing in some Mediterranean areas in rainfed and cold mountainous conditions. It could be used in goat diets as an alternative protein source instead of alfalfa to supplement forest rangeland. This study aimed to test the effects of incorporating Sulla flexuosa (SF) hay in the diet of Beni Arouss goats on their milk production and quality. The hay was introduced at two levels, i.e. 35 or 70% (SF70), on a DM basis; it partially or totally replaced the alfalfa hay of the control diet. The Data were analysed using the PROC MIXED function, including the random effect of goats, the fixed effects of the diet, the sampling week and their interaction. Post hoc analyses were performed using the Tukey test when the results for a parameter were significantly different according to the diets. SF incorporation did not affect milk production or physicochemical composition. However, milk FA content varied in proportion to the percentage of SF incorporation. The SF70 diet was associated with increased milk levels in C18:1n-9, C18:2n-6, C18:3n-3, and C22:6n-3 and total monounsaturated, polyunsaturated, and n-3 fatty acids. As a consequence, atherogenic and thrombogenic indices were improved. Tannins present in SF hay probably had a protective effect on fatty acid biohydrogenation in the rumen and an impact on fatty acid desaturating enzymes in the mammary gland. Additionally, better antioxidant capacity in milk was observed in SF70. The data of this study suggest that SF hay should be incorporated with no less than half of dry matter intake to show noticeable effects. Therefore, SF should be suggested as an alternative forage and protein resource in the lactating goat diet.

High DCAD and ascorbic acid modify acid base, oxidative stress and kidney function in dairy goats

S. Thammacharoen[1], S. Semsirmboon[1], D.K. Do Nguyen[1], S. Poonyachoti[1], T.A. Lutz[2] and N. Chaiyabutr[1]
[1]Faculty of Veterinary Science, Chulalongkorn University, Department of Physiology, Pathumwan, Bangkok, 10330, Thailand, [2]Vetsuisse Faculty, University of Zurich, Veterinary Physiology, Zurich, 8057, Switzerland; sprueksagorn@hotmail.com

Acid-base imbalance and systemic oxidative stress are the consequences of the high ambient temperature (HTa) in dairy animals. This is the important condition that down regulate the mammary gland function. We have shown in dairy goats that the negative effect of HTa can be alleviated by enhancing heat dissipation with high dietary cation and anion difference (DCAD) regimen. In addition, high-dose ascorbic acid (AA) could improve oxidative stratus. The present study aimed to evaluate the effect of both supplements on acid-base balance, oxidative stress and the kidney function in dairy goats. Data containing 2 factors were analysed with general linear model using repeated two-way analysis of variance. When compared with control diet (n=6), high DCAD (n=6) had no effect on urine flow rate (V_U) and endogenous creatinine clearance (C_{cr}) throughout the experiment. High DCAD significantly decreased net acid excretion (NAE) and increased fractional excretion of potassium (FE_K). The results suggested that high DCAD regimen influenced tubular functions without changing urine synthesis. During the period of high DCAD regimen, supplementation with high dose AA significantly increased V_U, however, could not change endogenous C_{cr}. This effect of AA on V_U apparently came from the AA effect that could increase in NAE. high dose of AA decreased in FE_K. Moreover, both high DCAD and AA influenced urine pH. High-dose AA decreased plasma glutathione peroxidase activity and malonaldehyde. The effect was enhanced by a high DCAD. Taken together, it can be concluded that high DCAD regimen and high dose of AA influence kidney function mainly by changing the tubular function. The tubular handling of bicarbonate and acid molecules is activated and partially explains the effect of both treatments on systemic acid-base balance. Finally, it seems likely that body water is maintained at least from the insignificant effect of the present high DCAD regimen on urine synthesis rate.

Link between chest circumference at late gestation and milk somatic cell count in Sarda ewe-lambs

D. Sioutas, A. Ledda, A. Mazza, A. Marzano, M. Sini and A. Cannas
University of Sassari, Agricultural Sciences, viale Italia 39/a, Sassari, 07100, Italy; dsioutas@uniss.it

The objective of this study was to evaluate in Sarda ewe-lambs the association between chest circumference (CC) at 120 d of gestation and udder health, as indicated by somatic cell count (SCC) during the successive lactation. In previous studies conducted on ewes of the Sarda breed, CC has been proven an interesting and accurate alternative method for the assessment of body condition score (BCS) of the animals at the farm level, eliminating the subjectivity factor related to the evaluation of body reserves by trained operators. A total of 26 pregnant Sarda ewe-lambs were selected. The animals presented an average BCS of 2.97±0.12, and CC of 91.15±4.66 cm at 120 d in gestation, while the average age at first partum was 413±26 d. For the determination of the SCC, milk samples were recorded individually during the morning and afternoon milking, once a week, from day 1 to day 105 of lactation. Taking into consideration that milk SCC does not follow the normal distribution, the milk SSC of each animal was log transformed. Average values for SCC during lactation were 1.125×10^6 cells/ml, which corresponds to log SCC equal to 2.37. The effect of CC at late gestation on average log SCC during lactation was analysed using the General Linear Model (GLM) procedure of R software. The results of our study showed that CC at late gestation had a significant effect on average lactation values of log SCC ($P<0.01$). Moreover, CC presented a high and significant correlation with log SCC (r=0.59; $P<0.01$), as well as a moderate but significant correlation with BCS (r=0.43; $P<0.05$). Interestingly, average values for log SCC during lactation increased proportionally to the values of CC at late gestation, i.e. a positive linear relationship was found between CC and log SCC ($R^2=0.35$). These findings imply that animals with relatively higher levels of body reserves close to parturition have a higher probability of presenting an impaired overall udder health status during successive lactation, probably because extensive mobilization of body reserves leads to ketone-body synthesis, which induce a state of sub-ketosis and thus of immunosuppression. However, further studies are needed to confirm this hypothesis.

Session 41 Theatre 1

Incorporating high-dimensional omics phenotypes into models for predicting breeding values

O.F. Christensen
Aarhus University, Center for Quantitative Genetics and Genomics, C.F. Møllers Alle 3, 8000 Aarhus C, Denmark; olef.christensen@qgg.au.dk

Incorporating high-dimensional omics phenotypes into models for predicting breeding values Ole F. Christensen, Center for Quantitative Genetics and Genomics, Aarhus University High dimensional phenotypes like metabolomics and transcriptomics are becoming available in larger quantities and at decreasing costs. Since they are both heritable and associated to phenotypes of interest, they may be relevant to incorporate into a genetic evaluation system. In this presentation I will provide an overview on models and methods for incorporating such high-dimensional omics into genetic evaluation. The primary modelling framework will consist of two parts, where the first part is model is a model for how the phenotype of interest depends on omics phenotypes and genetic effects that are not mediated by the omics phenotypes, and the second part is a model for how omics phenotypes depend on genetic effects. Definition of true breeding value, prediction of breeding values, and to how evaluate predictive performance of predicted breeding values will be discussed. An analysis of metabolomic data and mating quality traits in barley will presented.

Session 41 Theatre 2

Biologically informed genomic predictions with the NextGP.jl statistical analysis package

E. Karaman, V. Milkevych and L. Janss
Aarhus University, Center for Quantitative Genetics and Genomics, C. F Møllers allé 3, 8000 Aarhus C, Denmark; emre@qgg.au.dk

Genomic prediction and genome-wide association studies have now become the backbone of quantitative genetics. Bayesian multiple-regression methods are predominant in genomic research with complete genomic data, i.e. when all phenotyped individuals are also genotyped. As many other genomic data analysis software, the NextGP.jl Julia package implements a collection of widely used 'Bayesian alphabet' methods, such as BayesB and BayesR. One of its key contributions to the field is that it also implements sophisticated Bayesian multiple-regression methods to incorporate prior biological information in genomic predictions. Such prior information can be in the form of functional annotations, which can be either categorical or continuous covariates. Most functional information is coded as 0/1 to indicate membership of a marker in one or more annotation categories. To incorporate such information, it implements recently developed methods, i.e. BayesRC+ and BayesRCπ. Marker-association results such as inclusion probabilities, *P*-values, or *t*-statistics can be used when partially overlapping and nested functional categories exists. NextGP.jl implements a log-linear functional genomic prediction model, BayesLV, to handle such continuous covariates. Although the package was primarily developed for multiple-regression methods, traditional mixed linear models based on relationship matrices (genomic or non-genomic) can also be fitted. This work describes the capabilities of NextGP.jl in genomic prediction of complex traits, with a focus on biologically informed genomic predictions. It uses Bayesian approach in parameter estimations, in most cases, a Gibbs sampling procedure is run. The NextGP.jl package is written purely in an open source language, Julia, and the source code is available at https://github.com/datasciencetoolkit/NextGP.jl.

Fine mapping QTL associated with fertility in dairy cattle using gene expression

I. Van Den Berg[1], A.J. Chamberlain[1,2], I.M. MacLeod[1], T.V. Nguyen[1], M.E. Goddard[1,3], R. Xiang[1,3], B. Mason[1], S. Meier[4], C.V.C. Phyn[4], C.R. Burke[4] and J.E. Pryce[1,2]
[1]Agriculture Victoria, 5 Ring Road, 3082 Bundoora, Australia, [2]School of Applied Systems Biology, La Trobe University, 3082 Bundoora, Australia, [3]Faculty of Veterinary and Agricultural Science, University of Melbourne, 3010 Parkville, Australia, [4]DairyNZ Limited, 605 Ruakura Rd, 3240 Hamilton, New Zealand; irene.vandenberg@agriculture.vic.gov.au

While the inclusion of fertility in dairy cattle breeding objectives has helped to improve female reproductive performance, further improvements are desirable. Detecting genomic regions associated with fertility may help to increase our understanding of the biology underpinning fertility traits, and may lead to the identification of markers to increase the accuracy of genomic prediction of fertility. However, the relatively low heritability and highly polygenic nature of conventional female fertility traits in dairy cattle reduces the power to detect such regions. Recently, gene expression data has been used to facilitate fine mapping of quantitative trait loci (QTL) associated with various quantitative traits. An alternative approach is to select cows for high and low genetic merit for fertility and subsequently investigate phenotypic and genomic differences between these divergent groups, as has been done in a selection experiment in New Zealand. Our aim was to combine gene expression data, fertility phenotypes and allele frequencies in high and low fertility cows to fine map QTL associated with fertility in New Zealand and Australia. We performed a genome wide association study on calving interval using an Australian dataset containing imputed whole genome sequences from 60,395 dairy cattle. Using a New Zealand dataset that had 197 high and 168 low fertility cows, we detected gene expression QTL (eQTL) and QTL for fertility phenotypes. Variants that were significantly associated with calving interval were highly enriched for eQTL. Many of the overlapping QTL were in regions with a high number of copy number variants. We detected 671 genes that were significantly differentially expressed between high and low fertility cows. In conclusion, our results demonstrate how multiple data sources, including expression data, can be used to detect candidate genes for fertility.

How heritable are metabolomic features? An omic based approach in pigs

S. Bovo[1], G. Schiavo[1], F. Fanelli[2], A. Ribani[1], F. Bertolini[1], V. Taurisano[1], M. Gallo[3], G. Galimberti[4], S. Dall'olio[1], P.L. Martelli[5], R. Casadio[5], U. Pagotto[2] and L. Fontanesi[1]
[1]University of Bologna, Department of Agricultural and Food Sciences, Viale Giuseppe Fanin 46, 40127 Bologna, Italy, [2]University of Bologna, Department of Surgical and Medical Sciences, Via Massarenti 9, 40138 Bologna, Italy, [3]Associazione Nazionale Allevatori Suini, Via Nizza 53, 00198 Roma, Italy, [4]University of Bologna, Department of Statistical Sciences 'Paolo Fortunati', Via Belle Arti 41, 40126 Bologna, Italy, [5]University of Bologna, Department of Pharmacy and Biotechnology, Via San Giacomo 9/2, 40126 Bologna, Italy; giuseppina.schiavo2@unibo.it

Dissecting the complexity of production traits in their biological basic components may contribute to define novel phenotypes that can be useful to design new selection strategies in livestock. Metabolites are simple phenotypes that are the products of many different biological processes that underline more complex phenotypes. Metabolites cover a broad range of transient and stable levels over different environmental conditions and physiological stages. Therefore, before metabolites can be used as phenotypes in animal breeding, it is compulsory to estimate their heritability. In this study, we evaluated the heritability of about 200 plasma metabolites analysed in ~900 Italian Large White and ~400 Italian Duroc pigs using a targeted metabolomic platform. All pigs were also genotyped with a high-density single nucleotide polymorphism panel. Pedigree-based and genomic heritabilities were estimated in both pig breeds. The two estimates were highly correlated across different metabolite families. Correlation between breeds was however lower. Heritability was also evaluated in relation to the number of carbons and the degree of unsaturation of different metabolite classes, showing some relation derived by the biochemical features of different biomolecules. According to the obtained results, many metabolites can be considered as proxy phenotypes to substitute more complex phenotypic parameters in estimating breeding values in pigs, opening new avenues to improve complex traits.

Session 41 Theatre 5

Combined transcriptomics and metabolomics in the whole blood to depict feed efficiency in pigs

C. Juigné[1,2], E. Becker[1] and F. Gondret[2]
[1]Univ Rennes, Inria, CNRS, IRISA – UMR 6074, Campus de Beaulieu, 35000 Rennes, France, [2]PEGASE, INRAE, Institut Agro, 16, le clos, 35590 Saint Gilles, France; camille.juigne@irisa.fr

Feed efficiency is a research priority to support a sustainable meat production, but it is also recognized as a complex trait that integrates multiple biological pathways orchestrated in and by various tissues. This study aims to determine the relationships between biological entities underlying the inter-individual variation of feed efficiency in growing pigs. The feed conversion ratio (FCR) was calculated from a total of 47 Large White pigs from a divergent selection for residual feed intake and fed high starch or high-fat high fibre diets during 58 days. We considered transcriptomics (60 k porcine microarray) and metabolomics (1H-NMR analysis and target gas chromatography) datasets obtained in the whole blood of pigs at the end of the trial. We studied whether connecting these two levels of biological organization will help for a better understanding of the key biological processes driving the phenotypic difference in feed efficiency. We identified 33 weighted gene co-expression networks (WGCNA) in the whole blood (13 to 2,491 unique genes). The eigengenes of six modules were correlated to FCR ($P<0.05$). Most of these modules were enriched in genes participating in immune and defence-related processes. Profiles in circulating metabolites and fatty acids were summarized by weighed linear combinations from principal component analyses, and correlated to the eigengenes of the co-expressed genes modules. A significant association was found between a module of co-expressed genes participating to T cell receptor signalling and cell development process and related to FCR, and the circulating concentrations of omega-3 fatty acids. Finally, we provided a new method to integrate those experimental data and the public knowledge from metabolic pathway networks databases, to show where and how these molecules at different levels of biological organization participate and interact with each other. Databases of biological pathways in BioPAX format, represented by graphs, were queried to identify participants of interest in interactions. Altogether, these different approaches will help to depict biological systems and understand complex traits such as feed efficiency.

Session 41 Theatre 6

Imputation of sequence variants in more than 250,000 German Holsteins

A.M. Krizanac[1], C. Falker-Gieske[1], C. Reimer[2], J. Heise[3], Z. Liu[3], J. Pryce[4,5], J. Bennewitz[6], G. Thaller[7] and J. Tetens[1]
[1]University of Göttingen, Department of Animal Sciences, Burckhardtweg 2, 37077 Göttingen, Germany, [2]Friedrich-Loeffler-Institut, Institute of Farm Animal Genetics, Höltystraße 10, 31535 Neustadt, Germany, [3]Vereinigte Informationssysteme Tierhaltung w.V. (VIT), Heinrich-Schröder-Weg 1, Verden 27283, Germany, [4]School of Applied Systems Biology, La Trobe University, 5 Ring Road, Bundoora, Victoria 3083, Australia, [5]Agriculture Victoria Research, AgriBio, Centre for AgriBioscience, 5 Ring Road, Bundoora, Victoria 3083, Australia, [6]University of Hohenheim, Institute of Animal Science, Garbenstraße 17, 70599 Stuttgart, Germany, [7]Christian-Albrechts-University, Institute of Animal Breeding and Husbandry, Hermann-Rodewald-Straße 6, 24118 Kiel, Germany; ana-marija.krizanac@uni-goettingen.de

Imputation allows us to obtain sequence-dense genomic marker information in a cheap and efficient way. The usage of whole-genome sequence (WGS) data helps to improve the power of genome-wide association studies (GWAS) and consequently facilitates the detection of causal variants. A large number of individuals is usually needed to obtain reliable GWAS results. The combination of WGS data and large samples enables the identification of highly significant lead variants that may point to pleiotropic loci in a large, multi-trait GWAS and are probably the underlying quantitative trait nucleotides. In this study, we aimed to identify significant trait-specific and trait-shared variants between 57 milk, health, and conformation traits in German Holstein. For that purpose, a large dataset of 252,285 SNP-chip genotyped cows was imputed to the WGS level using a two-step imputation approach. After quality control and applying different GWAS approaches to infer the best method for correction of genomic inflation, both single-trait and multi-trait GWAS were carried out using the deregressed proofs as phenotypes. Single-trait and multi-trait GWAS led to the discovery of numerous known but also novel quantitative trait loci. Our results will aid in the development of novel breeding strategies for the improvement of animal welfare.

Session 41 — Theatre 7

Longitudinal study on the rabbit's gut microbiota variation through age

I. Biada[1], M.A. Santacreu[1], A. Blasco[1], R.N. Pena[2] and N. Ibáñez-Escriche[1]
[1]Universitat Politècnica de València, Instituto de Ciencia y Tecnología Animal, Camí de Vera, s/n, 46022 València, Spain,
[2]Escuela Técnica Superior de Ingeniería Agraria, Av Rovira Roure, 191, 25198 Lleida, Spain; ibiada@posgrado.upv.es

Unlike the animals' genome, which is static, the microbiome's profile is inherently dynamic, and the host's age is one of the main factors influencing it. Furthermore, it is well known that the microbiome can influence the host's health and possibly longevity, by modulating different physiological functions, like the host's immune system. The aim of this study is to conduct a longitudinal study to investigate the evolution of gut microbiota of two rabbit lines with different longevity (LP, a line founded with longevity criteria, and A, a standard commercial line). To do so, the gut microbiome's profile was analysed using the 16S rRNA gene. A total of 368 soft faeces samples collected from 195 does (78 of line A and 117 of line LP) were sequenced using Illumina MiSeq. After quality control, the sequences were analysed with DADA2 in R. Taxonomic annotation, alpha, and beta diversity were computed within QIIME2. Alpha diversity indices (Shannon and Pielou evenness) were used to compute the ranking Spearman correlation with age, and also as response variables in a repeatability linear model with the parity order (OP) grouped into four categories (OP1, OP 2, OP 3-8, OP>9) and the rabbit line with two levels (LP and A) as fixed effects, age as a covariate, and the animal effect as random. Principal Coordinate Analysis (PCoA) based on beta diversity distance matrices (Bray-Curtis, Jaccard, and UniFrac) were computed. The results evidenced changes in microbial diversity while age increases. Shannon and Pielou's evenness alpha showed a negative Spearman correlation (-0.7 and -0.4, respectively) with age. Moreover, the group OP>9 showed smaller diversity than the rest of the OP groups, with a probability of at least 97%. Finally, PCoA graphs showed clear gradual separations between samples with increasing age. Preliminary alpha and beta diversity results showed clear changes in the rabbits' microbiota with increasing age. Further studies will carry out to investigate how microbial abundance changes over time between the two lines A and LP and its possible implication in longevity.

Session 41 — Theatre 8

Studying cattle structural variation and pangenome using whole genome sequencing

G.E. Liu
USDA ARS, AGIL, Building 306 Room 111, Beltsville, MD, 20705, USA; george.liu@usda.gov

A cattle pangenome representation was created based on the genome sequences of 898 cattle representing 57 breeds. The pangenome identified 83 Mb of sequence not found in the cattle reference genome, representing 3.1% novel sequence compared with the 2.71-Gb reference. A catalogue of structural variants developed from this cattle population identified 3.3 million deletions, 0.12 million inversions, and 0.18 million duplications. Estimates of breed ancestry and hybridization between cattle breeds using insertion/deletions as markers were similar to those produced by single nucleotide polymorphism-based analysis. Hundreds of deletions were observed to have stratification based on subspecies and breed. For example, an insertion of a Bov-tA1 repeat element was identified in the first intron of the APPL2 gene and correlated with cattle breed geographic distribution. This insertion falls within a segment overlapping predicted enhancer and promoter regions of the gene, and could affect important traits such as immune response, olfactory functions, cell proliferation, and glucose metabolism in muscle. The results indicate that pangenomes are a valuable resource for studying diversity and evolutionary history, and help to delineate how domestication, trait-based breeding, and adaptive introgression have shaped the cattle genome.

Session 41 Theatre 9

Host genetics affect the composition of the lower gut microbiota in dairy cows

L. Brulin[1,2], S. Ducrocq[2,3], J. Estellé[1], G. Even[2,3], S. Martel[2,3], S. Merlin[2,3], C. Audebert[2,3], P. Croiseau[1] and M.P. Sanchez[1]

[1]Université de Paris-Saclay, INRAE, AgroParisTech, Animal Genetics, Domaine de Vilvert, 78350 Jouy-en-Josas, France, [2]GD Biotech – Gènes Diffusion, 1 Rue du Professeur Calmette, 59019 Lille, France, [3] PEGASE-Biosciences, Institut Pasteur de Lille, 1 Rue du Professeur Calmette, 59019 Lille, France; louise.brulin@inrae.fr

Symbiotic microorganisms are organized into specific ecosystems along the gastrointestinal tract of animals. In ruminants, studies showed a complex crosstalk between the ruminal microbiota and the host: beyond its role in the digestive process, the microbiota modulates the host phenotype but is also under the influence of the host genome. Concerning the intestinal microbiota, its role on numerous traits, in particular health maintenance, is increasingly described and could make it a key component of a sustainable breeding. However, its genetic control remains elusive. In this context, the present work will aim to perform a comprehensive genetic study of the lower gut microbiota. Faecal samples were collected between 2020 and 2022 from a population of 1930 Holstein cows, reared on 140 French commercial farms. Microbiota 16S rRNA analyses were performed on the samples and the data were processed to obtain amplicon sequence variant (ASV) tables. Genetic parameters were estimated using animal models for (1) the samples' diversity expressed as the Shannon index, and (2) the abundance of the most frequently observed ASVs and genera (with a prevalence threshold of 60%). The 152 ASVs and 87 genera analysed showed low to moderate heritability estimates. Indeed, 32% of ASVs had a heritability between 0.05 and 0.23 (s.e. from 0.04 to 0.10), with an ASV of the *Negativibacillus* genus being the most heritable taxon. Almost half of the genera had heritabilities ranging from 0.05 to 0.21 while the Shannon diversity index appeared to be poorly heritable (h^2=0.04, s.e.=0.05). Overall, our results demonstrate that the host genetics shapes the composition of the cow faecal microbiota. These first encouraging results will be complemented to assess the genetic correlations between the microbiota composition and production traits, and to identify genomic regions involved in the genetic determinism of the bovine gut microbiota through GWAS analyses.

Session 41 Theatre 10

A model including host genotype, gut microbiome, and fat deposition measures in swine

F. Tiezzi[1] and C. Maltecca[2]

[1]University of Florence, DAGRI, Piazzale delle Cascine 18, 50144 Firenze, Italy, [2]North Carolina State University, Animal Science, 120 W Broughton dr, 27695 Raleigh, NC, USA; francesco.tiezzi2@unifi.it

We propose a systematic approach to elicit the host genome control over microbial composition and tissue deposition while accounting for these two components' effects on each other. We developed a mediation test in a Structural Equation Model (SEM) and applied it to measured and latent dependent variables describing fat deposition in swine (*Sus scrofa*). Host genotype (G) contribution was determined using a 60k SNP beadchip. Gut microbiome composition was assessed using 16S sequencing of faecal swabs at 18 and 26 weeks of age (M1 or M2, respectively). Back fat accumulation was measured with ultrasound scan at 18 weeks of age (P1). In addition, a latent variable was constructed (P2), compounding ultrasound and mechanical backfat measures at slaughter, together with the weight of the belly cut. We focused our analysis on this measure that we considered the relevant target endogenous variable. We implemented a SEM and a standard model used in Genome-Wide Association Studies (sGWAS). Both included common effects to account for environmental variation and family stratification. The SEM aimed at estimating the sign and magnitude of the following mediated paths: Path1: G->P1->P2; Path2: G->M2->P2; Path3: G->M1->P2; Path4: G->P1->M2->P2; Path5: G->M1->M2->P2; together with the direct path Path0: G->P2. The sGWAS model, without using microbial information, estimated the total effect of G on P2, such as Path00: G->P2. The two models were run for each 42,546 SNP included in the analysis. The significance of the estimates was assessed using bootstrapping, which allowed to obtain an empirical distribution of the estimates by sampling with replacement. Compared to the sGWAS model, the SEM identified additional SNPs with significant mediated effects. Notably, these did not overlap with those identified as having a significant effect with the standard GWAS model. While the loci showing stronger total effects were primarily located on SSC1 and SSC5, the loci with the strongest indirect effects were found on SSC9, SSC6, SSC7, and SSC13. Results suggest that gut microbial information can be successfully included in (more complex) models.

Session 41 Theatre 11

protiPig: microbial analyses of traits related to nitrogen utilization efficiency in pigs

M. Schmid, N. Sarpong, M. Rodehutscord, J. Seifert, A. Camarinha-Silva and J. Bennewitz
Institute of Animal Science, University of Hohenheim, Garbenstr. 17, 70599 Stuttgart, Germany; markus_schmid@uni-hohenheim.de

In pig production, improving protein efficiency is required to reduce emissions and water pollutions caused by the excretion of nitrogenous compounds as well as to ease the competition between protein sources fed to the animals but also edible for humans. Besides an adequate diet, breeding for an improved protein utilization efficiency and its equivalent nitrogen utilization efficiency (NUE) is a promising approach. As such traits are hard to measure, proxy traits are required for routine animal evaluation. Blood urea nitrogen (BUN) has shown to be a heritable indicator for NUE. Studies revealed that the animal genome as well as the gut microbiome, which is also mediated by genetics, impact efficiency traits in pigs. Results directly using the trait NUE are rare and most of the experiments focus one single sampling timepoint. Within the frame of the protiPig project, around 450 Piétrain × Large White crosses were phenotyped for traits related to NUE and faecal microbiota composition was determined in the grower and finisher phase, respectively. Genomic analyses in this dataset showed that the traits were moderately heritable but the contribution of the genetics to the phenotypic variance differed between grower and finisher phase particularly for NUE. We investigated the role of the faecal microbiome with regard to nitrogen intake and retention, NUE and BUN. The microbial variance, microbiabilities and the microbial correlation between traits were quantified and microbiome-wide association studies (MWAS) were conducted. The influence of the microbial composition on the traits varied strongly within each phase and was substantially larger in the finisher phase. The microbiability.was significant for all traits except for NUE in the grower phase. For NUE in the grower phase, the microbiability could not be estimated (model not converged) and was m^2=0.162 in the finisher phase. Estimated microbiabilities were m^2=0.103 (m^2=0.157) for BUN in the grower (finisher) phase. MWAS revealed polymicrobial trait architectures with the majority of microbial effects being small. We demonstrated that the microbiome mediates NUE and related traits, which might be utilized in a breeding effort.

Session 41 Theatre 12

Rumen microbiome promote host metabolism and microbial bile acids bio-transformation in sheep

B.Y. Zhang, Y. Yu, X.Z. Jiang, Y.M. Cui, H.L. Luo and B. Wang
China Agricultural University, State Key Laboratory of Animal Nutrition, College of Animal Science and Technology, 2 yuanmingyuan west road, Haidian District, 100193, Beijing, China, P.R.; wangb@cau.edu.cn

Bile acids are not only essential for gut fat metabolism but act as signalling molecules and mediums through the interaction between gut microbiota and host. This study aimed to determine the potential efficient roles of rumen microorganisms in bile acid biotransformation, the exogenous porcine bile acids was added to the rumen in both *in vivo* and *in vitro* sheep model. In the *in vivo* experiment, ten male Tan-lambs (approximately 6-month-old) with similar body weights were selected and randomly divided into two groups: control (CON) and supplementation of 0.04% bile acids in the diet (dry matter basis, HCA). In the *in vitro* experiment, control group (no addition, C) and bile acids group (16 mg per 75 ml incubation, CB) were conducted. Rumen fluid was collected from the *in vivo* experiment after 70 d consecutive feeding and from the *in vitro* experiment after 24 h incubation. Exogenous bile acids significantly affected the ruminal microbial fermentation activity and microbiome. We annotated 907 bile acid metabolism-related genesets in 20 rumen fluid samples and found 77 secondary bile acid metabolism pathways among these genesets. Potential primary and secondary bile acid metabolism genes and some specific microbial bile acid metabolic genes were discovered through metagenome-wide association studies *in vivo* and *in vitro* based on KEGG and eggNOG databases. Mediation analysis established 12,010 mediation linkages: 3,540 for the microbial taxonomy, 1,452 for the genes annotated by KEGG and 7,018 for the genes annotate by eggNOG impact on the serum metabolites through rumen bile acids. Finally, significant correlations were established between the serum hyocholic acid, hyodeoxycholic acid, glycohyocholic acid and host phenotypes including body fat (assessed as GR value in sheep), heart ratio, kidney ratio, tail fat ratio, kidney fat ratio. In conclusion, our study provides insights into the potential new roles of rumen microorganisms in bile acid biotransformation and their impact on host metabolism.

Blending multivariate models to predict feed efficiency and explore multiple omics in meat sheep

Q. Le Graverand[1], F. Tortereau[1], C. Marie-Etancelin[1], A. Meynadier[1], J.L. Weisbecker[1] and K.A. Lê Cao[2]
[1]GenPhySE, Université de Toulouse, INRAE, ENVT, Castanet Tolosan, 31326, France, [2]School of Mathematics and Statistics, University of Melbourne, VIC 3010, Australia; quentin.le-graverand@inrae.fr

Selecting sheep for feed efficiency would improve the sustainability of sheep farming by decreasing feeding needs. However, due to the costs of recording feed intake, feed efficiency is rarely selected in sheep. Identifying feed efficiency biomarkers could help resolve this issue. A total of 258 Romane male lambs were phenotyped in the growing period for Residual Feed Intake (RFI) – in three different batches. Rumen fluid and blood were sampled as potential sources of biomarkers for feed efficiency. Multivariate analyses were performed with six distinct 'blocks' of predictors: fixed effects and covariates (FC), genotypes (SNPs), plasma NMR spectra (NMR), ruminal volatile fatty acids (VFAs), long-chain fatty acids (LFAs), bacteria and archaea abundances (16S amplicon sequencing). We modified a Partial Least Square regression approach (PLS) to account for the three batches while selecting biomarkers of feed efficiency. Cross-validation was repeated to fit one model per block on our training data (60% of the samples). Then, predictions for the validation set (30% of the samples) were obtained by using a weighted aggregation – based on the performance on each validation set. Testing data (10%) were independently used to assess the overall prediction accuracy based on Pearson correlations. When RFI was predicted from separate blocks, the average accuracy was low to moderate: 0.08 (standard deviation: 0.17) from VFAs to 0.44 (0.13) from SNPs. When RFI was predicted with our approach combining different omics, accuracy increased and reached an average of 0.55 (0.11). Based on weights attributed to blocks of predictors, we were able to rank the most predictive blocks to explain RFI: SNPs, FC, NMR, 16S, LFA and VFA. Furthermore, within each block we identified variables that were highly associated with feed efficiency RFI, including β-hydroxyisovaleric acid and a SNP located on the chromosome 3. To conclude, blending models is useful to integrate heterogeneous omics data: from predicting efficiency, to identifying associations between multi-omics predictors.

The effect of dietary *Laminaria digitata* on the muscle proteome and metabolome of weaned piglets

D.M. Ribeiro[1], D.F.P. Carvalho[1], C.C. Leclercq[2], S. Charton[2], K. Sergeant[2], E. Cocco[2], J. Renaut[2], J.A.M. Prates[3,4], J.P.B. Freire[1] and A.M. Almeida[1]
[1]Instituto Superior de Agronomia, Universidade de Lisboa, LEAF – Linking Landscape, Environment, Agriculture and Food, Tapada da Ajuda, 1349-017 Lisboa, Portugal, [2]LIST – Luxembourg Institute of Science and Technology, Biotechnologies and Environmental Analytics Platform, Environmental Research and Innovation Departme, 41 rue du Brill, L-4422 Belvaux, Luxembourg, Luxembourg, [3]Laboratório Associado para Ciência Animal e Veterinária (AL4AnimalS), Avenida da Universidade Técnica, 1300-477 Lisboa, Portugal, [4]Faculdade de Medicina Veterinária, Universidade de Lisboa, CIISA – Centro de Investigação Interdisciplinar em Sanidade Animal, Avenida da Universidade Técnica, 1300-477 Lisboa, Portugal; davidribeiro@isa.ulisboa.pt

Laminaria digitata is a brown seaweed with bioactive micronutrients, such as n-3 PUFA and minerals. Its recalcitrant cell wall polysaccharides can cause antinutritional effects for monogastrics. Carbohydrase supplementation is a putative solution. The objective of this study is to evaluate the impact of dietary *L. digitata* and alginate lyase supplementation on the muscle proteome and metabolome of weaned piglets. Weaned (♂, Large White × Duroc) piglets were randomly assigned to three diets (n=10): control (standard diet), LA (10% *L. digitata* replacing control) and LAL (LA + 0.01% alginate lyase). After two weeks of trial, they were slaughtered, and samples of *longissimus lumborum* taken. The muscle proteome and metabolome were analysed using label free LC-MS and GC-MS platforms, respectively. The diets had no effect on growth performance. A total of 126 proteins were differentially abundant ($P<0.05$, $0.67<FC>1.5$) in the muscle proteome of these piglets. Proteins involved in energy metabolism (e.g. PHKG1, ATP5F1A, GPI) were of lower abundance in LA piglets, possibly reflecting a lower dietary energy availability. Accordingly, the abundance of proteins involved in gluconeogenesis (TRAPPC10, CYB5A, IVD) increased in LA piglets. LA and LAL piglets downregulated oxidative metabolism (NEBL), possibly in relation to reduced lipid oxidation compared to control. These diets caused incipient differences on the metabolome. Our study demonstrates that digestion of *L. digitata* influences the metabolism of piglet muscle.

Search for new mutations in cattle by systematic whole genome resequencing

M. Boussaha[1], C. Eché[2], C. Escouflaire[3], C. Grohs[1], C. Iampietro[2], A. Capitan[1], M. Denis[2,4], S. Fritz[3], C. Donnadieu[2] and D. Boichard[1]

[1]Université Paris-Saclay, INRAE, AgroParisTech, GABI, Jouy-en-Josas, 78350, France, [2]INRAE, US1426, GeT-PlaGe, Genotoul, France Genomique, Université Fédérale de Toulouse, Castanet-Tolosan, 31326, France, [3]Eliance, Paris, 75012, France, [4]GenPhySE, Université de Toulouse, INRAE, INPT, ENVT, Castanet-Tolosan, 31326, France; mekki.boussaha@inrae.fr

Systematic whole genome sequencing provides a rapid and powerful method to identify recent novel mutations on a cattle population scale. It helps farmers to detect early carriers of new genetic anomalies and more generally to boost genomic selection. Regarding genetic defects, the aim is to identify new candidate mutations that may have deleterious potential before they are widely spread in the population to inform the breeders. A major drawback is the high rate of false positives, ie of variants with a strong annotation but without any effect. Therefore the annotation is the most critical step. We focussed only on still unknown variants (ie de novo or likely very recent mutations) in highly conserved sequences in other species and/or with effects predicted to be similar to those described in OMIM or MGI databases. We applied this strategy on whole genome sequences on 571 artificial insemination bulls from to 14 dairy and beef breeds. In large breeds, recently marketed bulls were selected. In smaller breeds, influential ancestors not yet sequenced were chosen. In total, we identified 1,548 novel genomic variants with a potential link to certain quantitative traits of interest or possible genetic abnormalities. These variants were investigated by searching for carrier or homozygous descendants, and by adding these variants on the custom part of the EuroGenomics SNP chip for an easy screening of the population. This information is also returned back to bull's owners, with a specific alert only for the few most critical variants. Although the annotation still lacks accuracy, we believe that whole genome sequencing of all artificial insemination bulls is an early, rapid and always cheaper method for identifying genetic defects before they disseminate in the population. The SeqOccIn project was funded by the Occitanie region, FEDER, and Apis-Gene.

Estimation of non-additive genetic effects for semen production traits in beef and dairy bulls

R. Nagai[1], M. Kinukawa[2], T. Watanabe[2], A. Ogino[2], K. Kurogi[3], K. Adachi[3], M. Satoh[1] and Y. Uemoto[1]

[1]Graduate School of Agricultural Science, Tohoku University, Laboratory of Animal Breeding and Genetics, Sendai, Miyagi, 980-8572, Japan, [2]Livestock Improvement Association of Japan, Inc., Maebashi Institute of Animal Science, Maebashi, Gunma, 371-0121, Japan, [3]Livestock Improvement Association of Japan, Inc., Cattle Breeding Department, Koto-ku, Tokyo, 135-0041, Japan; rintaro.nagai.q7@dc.tohoku.ac.jp

Non-additive genetic effects are considered as a key factor to understand reproductive traits, such as semen production traits, in cattle. In order to evaluate the relevance of non-additive genetic effects for semen production traits in beef and dairy bulls using the Illumina BovineSNP50 BeadChip, we performed genome-wide association studies (GWAS) to detect non-additive quantitative trait loci (QTLs). We also evaluated non-additive polygenic effects by estimating non-additive genetic (i.e. dominance and epistatic) variance components. In total, 65,463 records for 615 genotyped Japanese Blacks (JB) as beef bulls and 50,734 records for 873 genotyped Holstein (HOL) as dairy bulls were used to detect QTLs and to estimate non-additive genetic variance components for five semen production traits: semen volume (VOL), sperm number (NUM), sperm motility (MOT), MOT after freeze-thawing (aMOT) and sperm concentration (CON). Additive QTL was detected on *Bos taurus* autosome (BTA) 24 for MOT in JB and on BTA17 for MOT in HOL. Non-additive QTLs were detected on BTA2 for MOT and on BTA26 for CON in JB and BTA6 for MOT, BTA11 for aMOT and BTA17 for NUM, MOT, aMOT and CON in HOL. In HOL, a non-additive QTL on BTA17 has pleiotropic effects on these traits. Assuming non-additive genetic variance components, the broad-sense heritability (0.17 to 0.43) was more than twice as high as the narrow-sense heritability (0.04 to 0.11) for all traits and breeds. In addition, the difference between the broad-sense heritability and repeatability were very small for VOL, NUM and CON, because a large proportion of permanent environmental variance was explained by epistatic variance. In conclusion, our present study suggest that non-additive QTLs and polygenic effects play important roles in semen production traits in beef and dairy bulls.

Evaluating functional properties of feeding black soldier fly larvae in laying hens by FeedOmics

A. Rezaei Far, E. Zaccaria, I. Fodor, P. Van Wikselaar, S. Naser El Deen, S. Kar and T. Veldkamp
Wageningen Livestock Research / Wageningen University & Research, Animal Nutrition, De Elst 1, 6708 WD, Wageningen, the Netherlands; edoardo.zaccaria@wur.nl

Insect proteins, such as black soldier fly larvae meal (BSFM), are a more sustainable alternative protein source than soybean meal (SBM). We engage multi-omics techniques to characterise the potential functional properties of protein-containing feed ingredients that we often refer to as the FeedOmics approach. In this study, we evaluated the effects of replacing SBM with two inclusion levels of BSFM, i.e. 5 and 10%, on production performance and egg quality in laying hens in an aviary system. Measurements of production performance in laying hens have shown that replacing SBM with 10% BSFM significantly ($P<0.05$) improves feed efficiency while increasing eggshell thickness. To further explain the response of laying hens to a partial and complete replacement of SBM with BSFM, we employed a comprehensive multi-omics approach on a wide range of data obtained from various biological samples. We have collected ileum and jejunum digesta to perform bacterial compositional and diversity analysis, jejunum, ileum, liver and uterus for genome-wide transcriptomics, tissues, and blood plasma for global metabolites profiling. The data are being generated from the multi-omics approaches, and results will be presented showing the functional effects of feeding BSFM in laying hens. Such multiomics dataset is the first of its kind to our knowledge that can provide a deep insight into the biological systems of laying hens, revealing new insights into the effects of different inclusion levels of BSFM on laying hen health, production performance and egg quality.

Session 41 Poster 18

Characterization of two divergent lines through functional inference of ruminal microbiota

J. Guibert[1], A. Meynadier[1], V. Darbot[1], C. Allain[2] and C. Marie-Etancelin[1]
[1]INRAE- UMR GENPHYSE, GA, chemin de Borde Rouge, 31326 Castanet Tolosan Cedex, France, [2]INRAE – UE La Fage, Domaine Expérimentale de La Fage, 12250 Roquefort sur soulzon, France; christel.marie-etancelin@inrae.fr

The ruminal microbiota plays an important role in the nutrition of its host, having a direct impact on milk production and animal health. However, as the links between bacteria abundances and host's traits are often difficult to interpret, we proposed to infer the abundances of bacteria functions to characterize divergent lines on somatic cell score (SCS) or on milk persistency (PERS). From 2015 to 2019, we sampled the rumen juice of 700 adults dairy Lacaune ewes belonging to either SCS line (94 SCS+/204 SCS-) or PERS line (200 PERS+/202 PERS-). DNA was extracted from ruminal juice and sequenced for the 16s rRNA gene. We analysed microbiota sequences with FROGS pipeline to obtain relative abundances of bacteria and then to infer functions. The inference consisted of using PICRUSt2 bioinformatic approach to predict the functional composition of a metagenome using marker gene sequences. For the 700 ewes, abundances of 2,059 ASVs were computed. After filtering to ensure a good quality of inference (Nearest Sequenced Taxon Index<0.4 and identity/coverage percentage>90%), abundances of 1,146 ASVs were kept enabling us to quantify abundances of 330 bacterial functions. A mixed model was computed on the function's abundances (after GBM imputation and CLR transformation to consider their compositional nature) within SCS and PERS lines respectively, with fixed effects such as the line, run of sequencing, number of sequences, date of sampling, lactation stage and, for SCS line only, litter size. This analysis revealed functions significantly different for either the SCS (n=14) or PERS (n=60) lines (P-value<0.05), responsible for branched amino acids (SCS) or glycan/lipopolysaccharide (PERS) biosynthesis. Finally, a sparse PLS discriminant analysis (with MixOmics R package) was performed on residual of functional abundances obtained from the previous model (without line effect), considering the divergent line as discriminant factor. While balanced error rates (BER) were poor for both SCS (0,52±0.08) and PERS (0.49±0.04) lines, interesting discriminant functions, needed for vitamin (SCS) or glycan (PERS) biosynthesis, were found.

Session 41 Poster 19

Blood transcriptomic comparison under different THI in Holstein

J.-E. Park[1], H. Kim[2], J.-H. Cho[3], H.-G. Lee[3], W. Park[4] and D. Shin[2]
[1]Jeju National University, 102, Jejudaehak-ro, Jeju-si, Jeju-do, 63243, Korea, South, [2]Jeonbuk National University, 567, Baekje-daero, Deokjin-gu, Jeonju-si, Jeollabuk-do, 54896, Korea, South, [3]Konkuk University, 120, Neungdong-ro, Gwangjin-gu, Seoul, 05029, Korea, South, [4]National Institute of Animal Science, 1500, Kongjwipatjwi-ro, Iseo-myeon, Wanju_gun, Jeollabuk-do, 55365, Korea, South; jepark@jejunu.ac.kr

Recently, the average temperature of the atmosphere has been rising due to rapid climate change, which causes environmental stress such as high temperatures in livestock. Environmental changes due to temperature and humidity in cows are accompanied by decreased milk production, decreased immunity, and changes in body metabolism, which lead to economic losses for farmers due to decreased productivity and death. This study attempted to understand the biological mechanisms of the gene level by comparing the gene expression patterns of cows according to the temperature and humidity index (THI). The specification experiment was carried out for total 14 days which are 3 days of initial adaptation, 4 days of normal level (THI=70-71, temperature 22 °C, humidity 50-60%), 7 days of critical level (THI=72-73, 25-26 °C, 35-50). Differential expression and functional analysis were performed on RNA-seq data with a total of six blood samples collected on the 7th and 14th days of the experiment. 795 differentially expressed (FDR<0.05, FC>1.0) genes (DEGs) were discovered between the two experimental groups. Among them, 441 were expressed upward and 354 were expressed downward. These DEGs were found to have 40 biological processes, 17 Molecular Functions, and 13 Cellular Component functions of Gene Ontology (GO) term. A total 15 of KEGG pathway were identified, including Rap1 and PPAR signalling pathways activated in response to thermal stress. The results suggested that these DEGs and functions are related with climate stress and give the clue for their reaction mechanisms for climate change in dairy cows.

Session 41 Poster 20

Metabolome profile of the pectoralis major muscle of red-winged tinamou: a pilot study

J.M. Malheiros[1], C.S.M.M. Vilar[1], P.F. Silva[2], L.A. Colnago[3], J.A.I.I.V. Silva[4] and M.E.Z. Mercadante[1]
[1]Animal Science Institute, Sertãozinho, SP, Brazil, [2]University of São Paulo, São Carlos, SP, Brazil, [3]Embrapa Instrumentation, São Carlos, SP, Brazil, [4]São Paulo State University, Botucatu, SP, Brazil; jehmalheiros@gmail.com

This study was conducted to reveal the differences in the metabolome profile of *pectoralis major* muscle of red-winged tinamou (*Rhynchotus rufescens*) selected for growth trait. Phenotypic selection index composed of growth traits (live weight and chest and thigh circumference) and male reproductive traits (semen volume and sperm concentration) were used to classify two experimental groups: selection group with a higher index (TinamouS) and commercial group with a lower index (TinamouC). Couples were housed separately for reproduction. Subsequently, eggs were collected, artificially incubated and 20 males of second generation (n=10/group) slaughtered with 350 days of life. Next, pectoralis major muscle sample were collected for metabolomics assays by Proton Nuclear Magnetic Resonance Spectroscopy (^{1}H NMR). Sixty-five polar metabolites were identified and quantified; however, 29 metabolites showed significant difference ($P<0.05$) between the experimental groups. TinamouS group exhibited higher concentrations ($P<0.05$) of anserine, arginine, aspartate, betaine, carnosine, creatine, creatine phosphate, creatinine, glutamate, leucine, proline, threonine, 3-methylhistidine, ADP, AMP, ATP, adenosine, IMP, inosine, NAD+, acetate, 1,3-dihydroxyacetone, glucose, pyruvate and taurine. However, the concentrations of lactose, cysteine, beta-alanine, and choline were lower ($P<0.05$) in TinamouS group. Correlations between metabolites showed 197 significant positive (≥0.5; $P<0.05$) and 79 significant negative (≥-0.5; $P<0.05$). Eight significant pathways were detected ($P<0.05$; impact value>0.35): phenylalanine, tyrosine and tryptophan biosynthesis; alanine, aspartate and glutamate metabolism; D-glutamine and D-glutamate metabolism; β-alanine metabolism; glycine, serine and threonine metabolism; taurine and hypotaurine metabolism; histidine metabolism; phenylalanine metabolism. In conclusion, these findings provide convincing evidence that selection index for muscle growth in the second generation led to differences metabolome in red-winged tinamou.

Post-mortem muscle proteome of crossbred bulls and steers: carcass and meat quality

O.R.M. Neto, W.A. Baldassini, L.A.L. Chardulo, R.A. Curi, G.L. Pereira, B.M. Santiago and R.V. Ribeiro
São Paulo State University, School of Veterinary Medicine and Animal Science, 3780 Universitária Ave, 18610160, Botucatu, São Paulo, Brazil; otavio.machado@unesp.br

Beef cattle producers routinely use castration to simplify the rearing of stock, which helps reducing unwanted breeding and to modify the quality of the carcasses. However, the biochemical and molecular mechanisms that control and regulate these traits are not fully understood. Few studies used young animals and feedlot finishing of 180 days, which support the need for more studies using such a biological model. This study investigated the muscle proteome of crossbred bulls and steers with the aim of explaining the differences in carcass and meat quality traits. Therefore, 640 post-weaning Angus-Nellore calves were fed a high-energy diet for 180 days. In the feedlot trial, comparisons of steers (n=320) and bulls (n=320) showed lower ($P<0.01$) performance (1.38 vs 1.60±0.05 kg/d), slaughter weight (547.4 vs 585.1±9.3 kg), which resulted in lower hot carcass weight (298.4 vs 333.7±7.7 kg) and ribeye area (68.6 vs 81.0±2.56 cm^2). Steers had higher ($P<0.01$) carcass fatness, meat colour parameters and lower meat pH. Lower ($P<0.01$) Warner-Bratzler shear force (WBSF) were observed in steers compared to bulls (WBSF=3.68 vs 4.97±0.08 kg; and 3.19 vs 4.08±0.08 kg). The proteomic approach using two-dimensional electrophoresis, mass spectrometry and bioinformatics procedures revealed several differentially expressed proteins between steers and bulls ($P<0.05$). Interconnected pathways and substantial changes were revealed in biological processes, molecular functions, and cellular components between the *post-mortem* muscle proteomes of animals. Steers had increased ($P<0.05$) abundance of proteins related to energy metabolism (CKM, ALDOA, and GAPDH), and bulls had greater abundance of proteins associated with catabolic processes (PGM1); oxidative stress (HSP60, HSPA8 and GSTP1); and muscle structure and contraction (TNNI2 and TNNT3). The better carcass and meat quality traits of steers were associated with higher abundance of key proteins of energy metabolism and lower abundance of enzymes related to catabolic processes, oxidative stress, and proteins of muscle contraction. São Paulo Research Foundation (FAPESP) – Grant number – 2019/11028-0.

ITGA6 homozygous splice-site mutation causes junctional epidermolysis bullosa in Charolais cattle

C. Grohs[1], M. Boussaha[1], A. Boulling[1], V. Wolgust[2], L. Bourgeois-Brunel[1], P. Michot[1,3,4], N. Gaiani[1], M. Vilotte[1], J. Riviere[1] and A. Capitan[1,4]
[1]Université Paris-Saclay, INRAE, AgroParisTech, GABI, 4 rue Jean Jaurès, 78350 Jouy-en-Josas, France, [2]Unité de Pathologie du Bétail, ENVA, 7 av. du Général-de-Gaulle, 94700 Maisons-Alfort, France, [3]Herd Book Charolais, Agropôle du Marault, 58470 Magny-Cours, France, [4]Eliance, 149 rue de Bercy, 75012 Paris, France; cecile.grohs@inrae.fr

Junctional epidermolysis bullosa (JEB), characterized by mechanically induced blisters of the skin and mucous membranes, is one of the most painful and life threatening recessive genetic disorders described in humans and other animal species. Congenital skin fragility resembling JEB was recently reported in three Charolais calves born in two distinct herds from unaffected parents. Phenotypic and genetic analyses were carried out to describe this condition and its molecular etiology. Genealogical, clinical and histological examinations confirmed the diagnosis of recessive JEB, even if affected calves showed milder symptoms than those of another form of JEB, previously reported in the same breed, and due to a homozygous deletion of ITGB4. Homozygosity mapping followed by analysis of the whole genome sequences of 2 cases and 5,031 control individuals enabled us to prioritize a splice donor site of ITGA6 (c.2160+1G>T; Chr2 g.24112740C>A) as the most compelling candidate variant. This substitution showed a perfect genotype-phenotype correlation in the two affected pedigrees and was found to segregate only in Charolais, and at a very low frequency ($f=1.6\times10^{-4}$) after genotyping 186,154 animals from 15 breeds. Finally, RT-PCR analyses revealed increased retention of ITGA6 introns 14 and 15 in a heterozygous mutant cow as compared with a matched control. The mutant mRNA is predicted to cause a frameshift (ITGA6 p.I657Mfs1) affecting the assembly of the Integrin $\alpha6\beta4$ dimer and its correct anchoring to the cell membrane. This dimer is a key component of the hemidesmosome anchoring complex, which ensures the attachment of basal epithelial cells to the basal membrane. In conclusion, we report a rare example of partial phenocopies observed in the same breed and due to mutations affecting two members of the same protein dimer, as well as the first evidence of an ITGA6 mutation causing JEB in livestock species.

Session 41 Poster 23

A sheep pangenome reveals the spectrum of structural variations and their effects on tail phenotypes

R. Li[1], M. Gong[1], X.M. Zhang[1], F. Wang[1], Z.Y. Liu[1], L. Zhang[1], S.Q. Gan[2] and Y. Jiang[1]
[1]Northwest A&F University, Xinong Road 22, Yangling, Shaanxi, 712100, China, P.R., [2]Xinjiang Academy of Agricultural and Reclamation Sciences, Wuyi Road 221, Shihezi, Xinjiang, 832000, China, P.R.; gongmian2767@126.com

Structural variations (SVs) are a major contributor to genetic diversity and phenotypic variations, but their prevalence and functions in domestic animals are largely unexplored. Here we generated high-quality genome assemblies for fifteen individuals from genetically diverse sheep breeds using PacBio HiFi sequencing, discovering 130.3 Mb non-reference sequences, from which 588 genes were annotated. A total of 149,158 biallelic insertions/deletions, 6,531 divergent alleles and 14,707 multiallelic variations with precise breakpoints were discovered. The SV spectrum is characterized by an excess of derived insertions compared to deletions (94,422 vs 33,571), suggesting recent active LINE expansions in sheep. Nearly half of the SVs displayed low to moderate linkage disequilibrium with surrounding SNPs and most SVs cannot be tagged by SNP probes from the widely used ovine 50K SNP chip. We identified 865 population stratified SVs including 122 SVs possibly derived in the domestication process among 690 individuals from sheep breeds worldwide. A novel 168-bp insertion in the 5′ UTR of HOXB13 was found at high frequency in long-tailed sheep. Further genome-wide association study (GWAS) and gene expression analyses suggest that this mutation is causative for the long-tail trait. In summary, we developed a panel of high-quality de novo assemblies and presented a catalogue of structural variations in sheep. Our data captures abundant candidate functional variations that were previously unexplored and provides a fundamental resource for understanding trait biology in sheep.

Session 41 Poster 24

A Chinese indicine pan-genome reveals novel structural variants introgressed from other Bos species

X.L. Dai[1], P.P. Bian[1], D.X. Hu[1], F.N. Luo[1], Y.Z. Huang[1], R. Heller[2] and Y. Jiang[1]
[1]Northwest A&F University, No.22 Xinong Road, 712100, China, P.R., [2]University of Copenhagen, Ole Maaløes Vej 5, 2200, Denmark; daixuelei2014@163.com

Chinese indicine cattle harbour a much higher genetic diversity compared to other domestic cattle, but their genome architecture remains uninvestigated. Using PacBio HiFi sequencing data from 10 Chinese indicine cattle across southern China, we assembled 20 reference-quality partially phased genomes, and integrated them into a multi-assembly graph containing 148.5 Mb (5.6%) of novel sequence. We identified 156,009 high-confidence non-redundant structural variants (SVs) and 206 SV hotspots spanning ~195 megabases of gene-rich sequence. We detected 34,249 archaic introgressed fragments in Chinese indicine cattle covering 1.93 Gb (73.3%) of the genome. We inferred an average of 3.8%, 3.2%, 1.4%, 0.5%, of introgressed sequence originating respectively from banteng-like, kouprey-like, gayal-like, gaur-like *Bos* species, and 0.6% of unknown origin. We therefore show that the high genetic diversity of Chinese indicine cattle is due to substantial introgression from multiple donors. Altogether, this study highlights the contribution of interspecies introgression to the genomic architecture of an important livestock population, and demonstrates how exotic genomic elements can contribute to the genetic variation available for selection.

Session 41 Poster 25

Multi-omic analysis reveal epigenetic regulation during muscle growth and development in sheep

Q.J. Zhao, Y.H. Ma and Y. Liu
Institute of Animal Science, Chinese Academy of Agricultural Sciences, NO.2 Yuanmingyuan, Haidian district, Beijing, 100193, China, P.R.; zhaoqianjun@caas.cn

Skeletal muscle is the muscle tissue attached to the bone, accounting for about 40% of the total body weight in mammals. It has important functions such as maintaining the state of the body, generating heat and protecting organs. Skeletal muscle has been developed in the embryonic stage and is accompanied by a series of complex regulatory processes. Although there have been a lot of studies on muscle growth and development, as an important agricultural animal, the three-dimensional genomic structure and epigenetic regulation of sheep skeletal muscle has never been revealed. we generated datasets of profiling expression dynamics (RNA-Seq), chromatin accessibility (ATAC-Seq), 3D genome (Hi-C), and histone modifications (ChIP-Seq) in ovine skeletal muscle at different stages. We found that the number of accessible regions identified at each stage of muscle development varied, mainly reflecting a gradual increase in the number of peaks in the early stages of embryonic development and a gradual decrease in the number of peaks in the later stages. In addition, there were difference in compartment switch, the number and size of TAD and Loop during ovine skeletal muscle growth. The three-dimensional chromatin interaction map showed that there were abundant gene remote regulation relationships in sheep skeletal muscle. The dynamic change of TAD boundary leads to the change of gene expression, which in turn regulate muscle development. A total of 110 genes were found in the altered TAD structure, which were enriched in the pathways related to muscle development, such as Ras signalling pathway, Citrate cycle (TCA cycle), TNF signalling pathway, Glycolysis / Gluconeogenesis. Several long-range enhancers were identified, which modulate muscle development through the loops. Overall, our study provides a global genome-wide resource of chromatin dynamics that define unrecognized regulatory networks and the epigenetic regulation of ovine skeletal muscle growth and muscle. The present study also provide novel insights into the mechanisms underlying myogenesis and muscle development.

Session 41 Poster 26

The effect of production system on the muscle proteome of dromedary camel (*Camelus dromedarius*)

M. Lamraoui[1], D.M. Ribeiro[2], Y. Khelef[3], N. Sahraoui[4], H. Osório[5] and A.M. De Almeida[2]
[1]Laboratoire de Biophysique, Bio Mathématiques, Biochimie et Scientométrie, Université de Bejaia, 06000, Béjaia, Algeria, [2]LEAF-Linking Landscape, Environment, Agriculture and Food Research Center, Associated Laboratory, Instituto Superior de Agronomia, Tapada da Ajuda, 1349-017, Portugal, [3]Laboratory of Biology, Environment and Health, University of El Oued, 39000, El Oued, Algeria, [4]Laboratoire de Biotechnologies Liées à la Reproduction, Veterinary Institute, Université Saad Dahlab, 09000, Blida, Algeria, [5]Instituto de Investigação e Inovação em Saúde, University of Porto, 4200-135, Porto, Algeria; messaouda.lamraoui@univ-bejaia.dz

Proteomics has been a common tool used to study domestic animal science over the last decade, being employed in a wide range of topics, namely on meat science. Unlike conventional species, studies concerning dromedary meat proteomics are lacking. This work characterizes the longissimus lumborum muscle proteome of male Sahraoui dromedary camel as affected by the production system (extensive vs intensive) by using a label-free proteomics approach. Samples were taken from six camels of similar body weight and nutritional condition from the two production systems in southern Algeria. In total, 1,405 proteins were identified in the muscle proteome; with 40 showing differential accumulation. Proteome characterization reveals similarities with domestic ruminants, most biological process annotations were comprised between catalytic activity (32.8%) and binding (16.7%). Production systems affected several metabolic pathways. The extensive system group had an increased accumulation of energy metabolism and contractile apparatus proteins such as MYBPC1 and MYL2. This could be explained by the need for dromedaries to move in search of food in the extensive rangeland. On the contrary, the intensive system group showed an increase in proteins related to amino acid and lipid metabolisms (ACAT1 and ACAA2). This is likely a consequence of the higher growth rates and nutritional levels of the animals in the intensive system. These results contribute to differentiate the traditional extensive production system from the intensive production system.

Session 41 Poster 27

Effect of mitochondrial DNA copy number on productive traits and meat quality in pigs

E. Molinero, R.N. Pena, J. Estany and R. Ros-Freixedes
University of Lleida – Agrotecnio-CERCA Center, Department of Animal Science, Av Alcalde Rovira Roure, 191, 25198, Spain; eduard.molinero@udl.cat

Mitochondria are essential for the regulation of cellular energy metabolism. Mitochondrial DNA copy number (mtDNA_CN) is an indicator of mitochondrial number and size, as well as the cellular capacity to generate energy. Studies in livestock species showed that mtDNA_CN is a heritable trait (0.33-0.65) associated with productive traits such as birth and weaning weight, in cattle, and carcass and breast meat yield and abdominal fat, in chickens. Our aim was to evaluate the effect of mtDNA_CN on productive and meat quality traits in pigs, and to identify which genes underlie this association. We sequenced 280 Duroc pigs from different batches of a commercial line at an average coverage of 7.9× (SD 2.4×). The reads were processed using a standard bioinformatic pipeline and aligned to the Sscrofa11.1 reference genome. We calculated mtDNA_CN as the natural logarithm of the ratio between mitochondrial and nuclear DNA coverages. We studied the correlation between mtDNA_CN and different productive traits (growth: weight at 180 and 210 d, and carcass weight; carcass quality: backfat and loin thickness; and meat quality: fat content and composition, pH and colour), accounting for the batch effect. We performed a genome-wide association study for the mtDNA_CN using GEMMA to fit a linear mixed model that accounted for the batch effect and the kinship matrix. Variants with a P-value<10-5 were taken as significant and we explored the genes located ±50 kb from those. In contrast with what has been observed in other livestock species, no correlation between mtDNA_CN and growth traits was observed in our population, nor did we observe associations with fat content and composition traits. Yet, mtDNA_CN was correlated with meat quality traits such as pH (+0.15; P=0.02) and colour parameter L* (-0.29; P=0.04). The genome-wide association study detected 29 associated variants in 12 regions of chromosomes 3, 4, 5, 7, 10, 12, 13, 14 and 17. These regions contained candidate genes MTHFD2, which contributes to the folate mitochondrial pathway, and the KRTs gene family, which are associated with mitochondrial homeostasis and activity. Such candidate genes could influence mitochondrial function and energy metabolism.

Session 41 Poster 28

Multi-omics analysis revealed the interaction of microbiota and host in rumen of hanwoo

W.C. Park[1], J.W. Son[1], M.J. Jang[1], N.R. An[1], H.J. Choi[1], S.A. Jung[1], J.A. Lim[1], D.H. Kim[1], J.H. Cha[1], S.Y. Choi[1], Y.J. Lim[1], S.S. Jang[2], D.J. Lim[1] and H.H. Chai[1]
[1]National Institute of Animal Science, Animal Genomics and Bioinformatics, Wanju, Jeollabuk-do, 55365, Korea, South, [2]National Institute of Animal Science, Hanwoo Research, Daegwallyeong-myeon, Pyeongchang-gun, Gangwon-do, 25340, Korea, South; wcpark1982@korea.kr

Hanwoo is Korean native cattle breed. These cattle have been subjected to intensive artificial selection for last 70 years to improve its meat production traits. However, studies on the mechanism of interaction between the host and the microorganisms in the cattle rumen are insufficient. In this study, we performed multi-omics analysis using rumen tissue (RNA-seq) and microorganism (16S rRNA) at 10,12, 14, 18, 22, 26 and 30 months to confirm changes between host gene and microbial expression patterns according to growth stages. We used the DIABLO method from the mixOmics package. This is because it uses two datasets (RNA-seq and 16S rRNA) and one variable(Month). As a result, our multi-omics dataset consists of two components, 1 and 121, 1 and 161 were identified in genes and microorganisms in component 1 and component 2, respectively. In addition, we identified that the GP5(glycoprotein V platelet) and ENSBTAG00000047411(novel) gene was significantly correlated with 3 and 13 microorganisms respectively according to the growth stage of Hanwoo. Except for the negative correlations of ENSBTAG00000047411 and Oscillospira (genus level), all correlations between genes and microorganisms were positive. Moreover, GP5 and ENSBTAG00000047411 gene have a high contribution at 18 months and 26 months, respectively. These results suggest that interaction between host and microorganisms in rumen according to the growth stage of Hanwoo were identified. Therefore, there is a need to understand the genetic approach of microorganisms and host interaction of rumen according to the growth stage of Hanwoo, our findings will help in developing a muti-omics approach to improve growth performance.

Session 42 Theatre 1

Challenge session: animal genetics to address food security and sustainability

A. Granados and J. Sölkner
EFFAB-FABRETP, Rue de Trèves 61, 1040, Belgium; ana.granados@effab.info

Food security exists when populations have physical and financial access to sufficient, safe, and nutritious food that meets their dietary needs and food preferences for a healthy life. We all need to produce and consume food sustainably while reducing food waste, GHG emissions and the use of resources and improving animal health, welfare, and food quality. Current crises due to the aggression of Russia on Ukraine have shown how access to affordable, healthy, and sustainable food could be compromised. The importance of farmed animals is vital not only for diets, manure, or diversification in agricultural systems but also for the livelihoods of millions of people and the vitality of rural areas across the globe. In its many forms, animal production plays an integral role in the food system by turning inedible crops for humans into highly nutritious, protein-rich food too. Available technologies and innovation in animal genetics open the door to more sustainable production systems. Societal and ethical concerns related to animal welfare standards and the use of technology are increasing due to the association of genomic technics with enhancing animal productivity. Transparency of animal breeding strategies and the responsible use of technologies are needed. Modern animal breeding is the responsible combination of different breeding goals and traits balanced between them, including novel breeding goals and traits. The self-regulative initiative, Code EFABAR(CE), recognises the role of resilience and sustainability in safeguarding food security by showing a commitment to practising responsible use of technologies and balanced breeding. In this FABRE TP challenge session, we will discuss these topics and the application of animal genetics to alleviate the pressure. We want to invite experts worldwide to have constructive debates on the importance of animal genetics, diversity and techniques that could be considered in future breeding programmes. The regulatory framework, ethics, and societal context of food production with all the necessary and relevant stakeholders in this debate are essential. CE, more knowledge gained on animal genomes and traits of vital importance could go towards a progressive regulatory approach to look at the potential of NGTs and other technological advances in the animal breeding and reproductive sector.

Session 42 Theatre 2

Causal structures between gut microbiota and efficiency traits in poultry

V. Haas, M. Rodehutscord, A. Camarinha-Silva and J. Bennewitz
Institute of Animal Science, University of Hohenheim, Garbenstraße 17, 70599 Stuttgart, Germany; valentin.haas@uni-hohenheim.de

Livestock farming struggles with social acceptance, environmental impact, and animal welfare. From these points of view, consideration of feed and phosphorus efficiency is becoming increasingly important in poultry production. Efficiency of nutrients can be divided into digestive and metabolic efficiency. The gastrointestinal tract harbours a large community of microbes. These microbial settlers can influence the digestive efficiency and thus affect the expression of efficiency traits of the animals. It is well known that efficiency traits and gut microbiota composition are partly under the control of the host genome. Thus, the gut microbiota composition can be seen as a mediator trait between the host genome and the efficiency traits. However, causal relationships are not well understood and were to be separated in the present study. A data set of 750 F2 cross Japanese quail was used as a model species for important poultry species. All birds were genotyped for 4k SNPs and trait recorded for phosphorus utilization (PU), phosphorus retention (PR), body weight gain (BWG) and feed per gain ratio. In addition to the genotypic and phenotypic information, the ileum microbiota composition was characterized using targeted amplicon sequencing and the alpha diversity was calculated as the Pielou`s evenness index (J'). A Hill-climbing learning algorithm was applied to construct a stable network with the efficiency traits and the alpha diversity index. The network identified direct and indirect links via the traits used and placed the J' as the most upstream trait and BWG as the most downstream trait. Structural equation models quantified the direct and indirect effects between J', PU, and PR. Three genome-wide significant quantitative trait loci (QTL) with 49 trait-associated SNPs within the QTL regions were identified with QTL linkage mapping. The extension of structural equation model (SEM) to the SEM association analysis separated the total SNP effect for a trait into a direct effect and indirect effects mediated by upstream traits. The method applied allows for the detection of shared genetic architecture of quantitative traits and microbiota diversity.

Exploring crossbreeding to reduce GHG Emissions and feed-food competition in beef production

A. Mertens[1], L. Kokemohr[2], E. Braun[3], L. Legein[1], C. Mosnier[3], G. Pirlo[4], P. Veysset[3], S. Hennart[1], M. Mathot[1] and D. Stilmant[1]

[1]Walloon Agricultural Research Centre, Agricultural Systems, rue du serpont, 100, 6800 Libramont, Belgium, [2] University of Bonn, Institute for Food and Resource Economics, Nußallee 21, 53115, Bonn, Germany, [3]Université Clermont Auvergne, INRAE, VetAgro Sup, UMR Herbivores, 63122 Saint-Genès-Champanelle, France, [4]Council for Agriculturale Research and Economics, Via Antonio Lombardo 11, 26900 Lodi, Italy; a.mertens@cra.wallonie.be

In the context of a growing population, beef production is expected to reduce its consumption of human edible food, and its contribution to global warming. We hypothesize that implementing the innovations of fast rotational grazing alone and with redesigning existing production systems using crossbreeding and sexing may reduce these impacts. In this research, recently accepted in Animals, the bio-economic model FarmDyn is used to assess the impact of such innovations, taking into account farmer's decisions, on farm profit, workload, global warming potential and feed-food competition. The innovations are tested in (1) a Belgian system composed of a Belgian Blue breeder and a fattener farm, (2) a French-Italian system where calves raised in a French suckler cow farm and fattened in a farm in Italy, (3) a German dairy farm that fattens its male calves. The practice of fast rotational grazing with a herd of dairy-to-beef crossbred males from the (1) system is found to have the best potential of greenhouse gases reduction because of the share of the environmental load with milk production while implementing fast rotational grazing alone induced only marginal GHG emission reduction. The reduction of the use of human edible food is also observed when by-products are available and at low stocking rate. The crossbreeding with early maturing beef breeds shows a good potential to produce grass-based beef with little feed-food competition if the stocking rate is adapted to the grassland yield. The results motivate field trials in order to validate the findings.

Genetic selection for dairy calf disease resistance traits: opportunities and challenges

C. Lynch[1], F.S. Schenkel[1], N. Van Staaveren[1], F. Miglior[1,2], D. Kelton[3] and C.F. Baes[1,4]

[1]Centre for Genetic Improvement of Livestock, University of Guelph, 553 Gordon St, N1G 1Y2, Guelph, Ontario, Canada, [2]Lactanet Canada, 660 Speedvale Ave W, N1K 1E5, Guelph, Ontario, Canada, [3]Department of Population Medicine, University of Guelph, 50 Stone Road East, N1G 2W1, Guelph, Ontario, Canada, [4]Institute of Genetics, University of Bern, Bremgartenstr.109a, Postfach, 3001, Bern, Switzerland; clynch@uoguelph.ca

The dairy industry needs to ensure animal health standards are continuously being reviewed and improved. One tool to achieve this is genetic selection, however, limited research has been conducted on the genetics of calf-hood diseases. Therefore, this study aimed to understand the current impact of calf diseases on Canadian farms by investigating incidence rates, estimating genetic parameters, and providing industry recommendations. Producer recorded calf disease data comprised of 69,695 Holstein calf disease records for respiratory problems (RESP) and diarrhoea (DIAR) from 62,361 calves collected on 1,617 Canadian dairy herds from 2006 to 2021. Two combined morbidity traits were also analysed, a binary trait and a categorical trait. Single and multiple trait analyses using a linear animal model for all traits were evaluated. Furthermore, each trait was analysed using two scenarios with respect to minimum herd-year disease incidence threshold criterion (1 and 5%), to highlight the impact of different filtering thresholds on selection potential. Heritability estimates for RESP, DIAR and the two morbidity traits ranged from 0.02 to 0.08 across analyses, while estimated genetic correlations between RESP and DIAR ranged from 0.50 to 0.62. Genetic correlations between both disease traits and production traits were low ranging from 0.03 to 0.08. Furthermore, sires were compared based on their estimated breeding value and their daughter diseased incidence rates. On average, calves born to the bottom 10% of sires were 2.2 times more likely to develop DIAR, 1.8 times to develop RESP, and 1.6 times more likely to exhibit morbidity compared to daughters born to the top 10% of sires. Results from the current study are promising, however, industry outreach on the value of recording and the standardization of collection practices is required for effective genetic evaluation.

Genetic research on colostrum quality traits and passive transfer of immunity in Greek dairy herds

A. Soufleri[1], G. Banos[1,2], N. Panousis[1], G. Arsenos[1], A. Kougioumtzis[1], V. Tsiamadis[1] and G.E. Valergakis[1]
[1]Faculty of Veterinary Medicine, School of Health Sciences, Aristotle University of Thessaloniki, Box 393, 54124 Thessaloniki, Greece, [2]Scotland's Rural College, Roslin Institute Building, EH25 9RG Midlothian, United Kingdom; asoufler@vet.auth.gr

The genetic component of colostrum quality traits and passive transfer of immunity was investigated in 10 Greek dairy herds. Colostrum fat (F), protein (P), lactose (L) and total solids (TS) content records on 1,074 Holstein cows were available. Trait heritability estimates were 0.21, 0.19, 0.15 and 0.27, respectively ($P<0.05$). The genetic correlation of TS measured on-farm with a Brix refractometer with colostrum P and F was unity and zero, respectively. Data on serum total protein content, a proxy for passive transfer of immunity, on 1,013 calves were analysed and a heritability estimate of 0.21 ($P<0.05$) was derived. A strong positive genetic correlation of 0.99 ($P<0.05$) was detected between calf serum total protein content and colostrum P. Estimated breeding values (EBVs) for colostrum traits of 67 Holstein sires were derived using data from 699 cows. Number of daughters (purebred Holsteins with full pedigree) per sire ranged from 5 to 49. The EBVs difference for colostrum TS between the 10th and 90th percentile was 3.20% (-1.50 to +1.70, respectively). The EBVs difference for colostrum F between the 10th and 90th percentile was 2.23% (-1.56 to +0.67, respectively). The EBVs difference for colostrum P between the 10th and 90th percentile was 2.32% (-0.92 to +1.40, respectively). The EBVs difference for colostrum L between the 10th and 90th percentile was 0.33% (-0.21 to +0.12, respectively). Three bulls had positive EBVs, and five negative EBVs for all four colostrum traits studied. When considering TS, F and P only, 13 bulls had positive EBVs and 19 had negative EBVs for all three traits. The EBVs difference between the 13 'positive' bulls and 19 'negative' bulls for TS, F and P were 2.15, 1.10 and 1.28%, respectively. Colostrum traits are amenable to improvement with genetic selection. Correlations of colostrum and calf traits with other productive and functional traits should be investigated.

Phenotypic and genetic analysis of beef-on-dairy crossbred calves

R.H. Ahmed, C. Schmidtmann, J. Mugambe and G. Thaller
University of Kiel, Institute of Animal Breeding and Husbandry, Hermann-Rodewald-Str. 6, 24118 Kiel, Germany; rahmed@tierzucht.uni-kiel.de

Beef semen has been increasingly used in dairy herds since a few years. This trend is driven by volatile milk markets, higher revenues for crossbred calves compared to purebred dairy calves but also due to innovative reproductive technology and management in dairy herds, which allows to inseminate surplus cows with beef semen. However, the production strategy Beef-on-Dairy might increase the risk for calving difficulty and stillbirth compromising animal welfare and resulting in economic losses. In this study, a dataset consisting of 3,500 records for traits such as birthweight, chest circumference, gestation length and calving ease from crossbred calvings was analysed to identify differences among beef breeds used for insemination in dairy herds. Data was analysed using mixed linear regression models for weight of calves and binary logit regression models for calving difficulty. The results showed that calves sired by Angus breed had highest rate of calving difficulty, which might be due to the fact that these crossbred calves had significantly higher chest circumference compared to calves sired by other breeds in the dataset (Belgian Blue & Limousin). Moreover, the results revealed differences in gestation length for the dams inseminated with beef semen (280 d for Angus, 282 d for Belgian Blue and 287 d for Limousin). Additionally, genotypic information for all crossbred and most of the sires were available. This data was utilized for principal component analysis to investigate the genetic structure among crossbred calves within the same breed of sire. The results revealed closest genetic relatedness among calves sired by Belgian Blue. On the other hand, Angus crossbred calves showed highest genetic variability.

Session 43 Theatre 4

Assessing passive immune transfer in newborn calves: salivary and serum IgG association

F.G. Silva[1,2], E. Lamy[1], S. Pedro[1], I. Azevedo[1], P. Caetano[3], J. Ramalho[3], L. Martins[3], A.M.F. Pereira[1], J.O.L. Cerqueira[2,4], S.R. Silva[2] and C. Conceição[1]
[1]Mediterranean Institute for Agriculture Environment and Development & Change, University of Évora, Department of Zootechnic, Largo dos Colegiais 2, 7004-516 Évora, Portugal, [2]Veterinary and Animal Research Centre & Al4AnimalS, University of Trás-os-Montes e Alto Douro, Department of Animal Science, Quinta de Prados, 5000-801 Vila Real, Portugal, [3]Mediterranean Institute for Agriculture Environment and Development & Change, University of Évora, Department of Veterinary Medicine, Polo da Mitra, Apartado 94, 7004-516 Évora, Portugal, [4]Polytechnic Institute of Viana do Castelo (IPVC) – Agrarian School of Ponte de Lima, Rua D. Mendo Afonso, 147 Refóios do Lima, 4990-706 Ponte de Lima, Portugal; fsilva@uevora.pt

The calf is born with very low levels of antibodies, which makes colostrum, abundant in antibodies, crucial for passive immunization. The efficiency of passive immune transfer is assessed by the IgG concentration in the calf's blood 24 h after birth. However, this method undergoes blood collection, which needs specialized training and a good animal constraint and can be a source of physical injury and stress. Thus, measuring IgG in the saliva is a more accessible, simpler, and non-invasive alternative method. This study evaluated the association between total protein and IgG in the serum and saliva of calves before and after passive immunization. Total protein and IgG concentration was measured in eighty saliva and blood samples from twenty calves collected at 4-time points: at birth (approximately 30 min before colostrum consumption), 24 h, 48 h and at day 7 after birth. Total protein was measured by the Bradford method, and IgG was quantified using the ELISA technique. The correlation between IgG between blood and saliva was assessed and will be presented during the EAAP 2023 Annual Meeting.

Session 43 Theatre 5

Biochemical predictors of successful transition from milk to solid feed in Holstein calves

P. Kazana, N. Siachos, N. Panousis, G. Arsenos and G.E. Valergakis
Faculty of Veterinary Medicine, Aristotle University of Thessaloniki, Laboratory of Animal Production, Aristotle University Campus Box 393, 54124 Thessaloniki, Greece; geval@vet.auth.gr

The objective of the study (9 commercial dairy farms, 249 healthy Holstein calves) was to establish tests predicting: (1) which calves are ready to be weaned; and (2) which calves successfully adapted to the solid feed diet. At three time-points relative to the day of weaning (-7, 0 and 7 d), for each calf we performed the following: (1) a clinical examination; (2) blood sampling for determination of serum concentration of β-hydroxybutyrate acid (BHB) and non-esterified fatty acids (NEFA); (3) sampling of rumen fluid to determine the concentration of total volatile fatty acids (VFA, gas chromatography); and (4) measurement of ruminal fluid pH. At each time-point, critical thresholds of total VFA concentrations were identified, below or over which there is increased probability of low or high pH values (multiple ROC curves), using total VFA concentrations as quantitative variables and pH as test variable. Thresholds of total VFA<80-85 mmol/l and >116-134 mmol/l for 'low' and 'high' pH values, respectively, yielded the best predictive values, based on the area under the curve (AUC). Using these thresholds, we divided calves into 3 categories, indicative of 'low', 'adequate' and 'high' solid feed consumption, respectively. Subsequently, multinomial logistic regression (MLR) models were fit within each time-point with BHB and NEFA concentrations as predictors of calves' classification, accounting for the effects of farm and age at weaning. When statistically significant effects of either BHB and/or NEFA were detected, ROC curves were used to determine thresholds dichotomizing calves into 'adequate/high' vs 'low' categories. As BHB and NEFA concentrations on -7 d increased, the odds of a calf being classified as having 'adequate/high' solid feed consumption pre-weaning increased. A BHB concentration of >260 µmol/l and a NEFA concentration of >175 µmol/l predicted the categorization of calves into the 'adequate/high' solid feed consumption with sensitivities of 94.1 and 83.5%, respectively. On 7 d, a NEFA concentration of <245 µmol/l predicted the categorization of calves into the 'low' solid feed consumption with a specificity of 79.2%.

The impact of dairy farm management on long-term robustness of veal calves

F. Marcato[1,2], H. Van Den Brand[2], L. Webb[3], M. Wolthuis-Fillerup[1], F. Hoorweg[1] and K. Van Reenen[1]
[1]Wageningen Livestock Research, P.O. Box 338, 6700 AH Wageningen, the Netherlands, [2]Wageningen University & Research, Adaptation Physiology Group, P.O. Box 338, 6700 AH Wageningen, the Netherlands, [3]Wageningen University & Research, Animal Production Systems Group, P.O. Box 338, 6700 AH Wageningen, the Netherlands; francesca.marcato@wur.nl

Dam parity and dam-rearing are known to have an impact on calf health and welfare, but little is known about the effects on long-term measures of robustness in veal calves. The aim of this study was to investigate the impact of dam parity and dam-rearing on veal calf robustness. Calves (n=683) were transported from 13 dairy farms to 8 veal farms at 2 or 4 weeks of age. Within the large study on 13 dairy farms, the study on dam-rearing was conducted on one of these farms where half of the calves were directly separated from their dam at birth (n=39 of 76), and half of the calves (n= 37 of 76) were reared with their dam during their entire stay at the dairy farm and abruptly separated on the day of transport (2 or 4 weeks). At all farms, dam characteristics were recorded, body weight (BW) of calves was recorded at birth, weekly at the dairy farm, upon arrival at the veal farm and at slaughter. A blood sample was collected from the dam a week before calving and in calves a week after birth, a day before transport and in week 2 and 10 post-transport. These samples, together with colostrum samples were analysed for IgG. Health of calves was scored weekly at the dairy farm and in weeks 2, 6, 18 and 24 at the veal farm. Calves born from first parity cows had the lowest BW until slaughter (day before transport: Δ= -4.7 kg, P=0.01; at slaughter: Δ= -6.1 kg, P=0.07). First parity cows produced lower quality colostrum and their calves' serum-IgG was lower up to week 2 post-transport (P<0.01). Dam-reared calves had a higher BW and more signs of disease (58 vs 27%, P=0.05) at the dairy farm, and higher BW upon arrival at the veal farm (70 vs 63 kg, P=0.01) than calves raised without their dam, but no differences in serum IgG or carcass weights were present. This study suggests that calves from first parity dams may require extra attention, while more research is necessary to understand whether or not dam-rearing affects veal calf robustness at a transport age of 2 or 4 weeks.

The assessment of transfer of passive immunity in dairy calves affected by neonatal diarrhoea

G. Sala[1], V. Bronzo[2], A. Boccardo[2], A.L. Gazzonis[2], P. Moretti[2], V. Ferrulli[2], A.G. Belloli[2], L. Filippone Pavesi[2], G. Pesenti Rossi[2] and D. Pravettoni[2]
[1]University of Pisa, Department of Veterinary Sciences, Via Livornese s.n.c, 56122, San Piero a Grado, Italy, [2]University of Milan, Department of Veterinary Medicine and Animal Science, Via dell'Università 6, 26900, Lodi, Italy; gaia.pesenti@unimi.it

The assessment of immune status has practical implications in calves' management and health. However, the presence of hemodynamic alterations like dehydration caused by neonatal calf diarrhoea (NCD), makes the diagnosis of failure of transfer of passive immunity (FTPI) challenging. This study aimed to evaluate the diagnostic performances of serum total protein refractometry (STP) and gamma-glutamyl-transferase activity (GGT) for assessing FTPI in dairy calves affected by NCD. 91 Friesian calves aged 1 to 10 days were enrolled for this study. Among them, 72 were admitted to the Veterinary Teaching Hospital of the University of Milan as affected by NCD, while 19 were healthy. Each calf underwent a complete clinical examination and dehydration assessment. Blood sampling was performed and serum was tested for STP concentration, GGT activity and IgG concentration. The effect of dehydration status and age on the correlation between the two methods under study and the gold standard (IgG) was investigated with Spearman's correlation index R for ranks. Receiver operating characteristic (ROC) curve analysis was performed to identify the optimal cut-off point to distinguish between diarrheic calves with or without FTPI (IgG<10 g/l). Significant differences in IgG concentration (P= 0.009), STP concentration (P= 0.040), and age (P= 0.042) were found between healthy and NCD calves, while no statistical difference was found in GGT activity (P=0.081). GGT was affected by age (P= 0.003), while STP was influenced by dehydration (P= 0.013). The cut-offs for identifying FTPI were 52 g/l of STP in normohydrated calves, 58 g/l of STP in dehydrated calves, and 126 UI/l of GGT in calves with more than 3 days of age. In normohydrated diarrheic calves STP refractometry showed better diagnostic accuracy, whereas in dehydrated calves its accuracy dropped, and it resulted advisable to use GGT activity. This study confirms the hypothesized better performance of GGT activity in diagnosing FTPI in dehydrated calves.

Session 43 — Theatre 8

Automatic monitoring of calves' behaviour for a precision weaning approach

R. Colleluori, D. Cavallini, A. Formigoni and L. Mammi
University of Bologna, Department of Veterinary Medical Sciences, Via Tolara di Sopra, 50, 40064, Ozzano dell'Emilia (BO), Italy; riccardo.colleluori2@unibo.it

Aim of this study was to evaluate the potential of an accelerometer ear tag monitoring system (AMS) (SenseHub Dairy, MSD Animal Health) for a precision management of pre-weaned dairy calves. This AMS records times of suckling, activity, rumination and intake of calves and, based on these patterns, develop a health index (HI; 0-100 scale) that generates a health alarm (HA) when lower than 86. Twenty-eight Holstein female calves were monitored for the first 90 days of life. After colostrum feeding, animals were fed bulk milk twice daily, pelleted calf starter and long hay with *ad libitum* access to water. For the first 2 months, calves were housed in individual pens, and individual milk, feed, and hay intake was recorded. At d 66.5±13.3 calves were moved to multiple pens. In both housing system, calves were weighed fortnightly and clinically assessed 3 times a week. Moreover, health checks (HC) were carried out at day 0, 1, 2 and 3 from the HA sent by the AMS. Body weight (BW) of calves was 36.68±5.44 kg at birth and 89.65±12.14 kg at the weaning age (74.56±8.08 d). The average daily gains (ADG) recorded at 60 and 90 days were respectively 0.65±0.13 and 0.76±0.14 kg/d. Total number of HA received for each calf varied from 0 to 7, and for the 63.2% of HA mild (25.0%) or severe symptoms (38.2%), were detected at the HC. Animals were retrospectively divided into 3 clusters (C1, C2, C3) based on number of the HA received in the first 90 days (0-1 HA for C1, 2-3 HA for C2, >3 HA for C3), and intakes and growth rate were compared between clusters. Milk and hay intakes showed no differences between the 3 clusters, while starter intake resulted lower (P<0.001) for C3 (198.1±39.8 g/d) and C2 (293.5±35.4 g/d) compared to C1 (388.3±33.1 g/d). BW recorded at 60 and 90 days was lower (P<0.001) for C3 compared to C1 (-16.78 and -22.53 kg, respectively), as well as ADG (0.68±0.04 and 0.52±0.05 kg/d, respectively at 60 d, and 0.82±0.05 vs 0.64±0.05 kg/d at 90 d). According to this data, the automatic monitoring of calves' behaviour reveals the potential to promptly detect weaker calves even though the complete potential of these systems is far from being explored.

Session 43 — Theatre 9

Can dairy herds be in a positive colostrum stock balance?

A. Soufleri[1], G. Banos[1,2], N. Panousis[1], G. Arsenos[1], A. Kougioumtzis[1], V. Tsiamadis[1] and G.E. Valergakis[1]
[1]Faculty of Veterinary Medicine, School of Health Sciences, Aristotle University of Thessaloniki, Box 393, 54124 Thessaloniki, Greece, [2]Scotland's Rural College, Roslin Institute Building, EH25 9RG Midlothian, United Kingdom; asoufler@vet.auth.gr

The importance of colostrum for passive transfer of immunity to newborn calves is indisputable. Keeping a stock of frozen high-quality colostrum, is a well-established practice to cater the needs when cows have low quality/quantity. In this simulation study we investigated whether it is practically feasible for all farms to have adequate stock of high-quality colostrum and the factors associated with it. Data from 9 dairy herds in Greece including 1,067 Holstein cows were used. In each farm, sampling covered on average a period of 8-9 months. Only first milking colostrum was considered. Colostrum yield (CY) was recorded, and quality was determined using a digital Brix refractometer. Cow parity and calving calendar season (CS) were also recorded. The simulation included the following steps: (1) 4 l of good quality colostrum (Brix≥22%) were 'fed' to each calf born (twins included) and any excess quantity was stored; (2) if colostrum quality was poor (Brix<22%), it was 'rejected' and 4 l from the colostrum stock were used; and (3) if quality was good but quantity was <4 l per calf, it was supplemented from stock. These steps created a 'colostrum stock balance' after every calving that was either positive or negative (continuous variable). Effect of CS, parity, CY and quality and the interactions between them on 'colostrum stock balance' was analysed with a univariate general linear model. Four farms were continuously in positive stock balance, four others were in negative stock balance for one or two periods during the study and one farm was continuously in negative stock balance. CS (spring and summer were favourable), CY and the interaction of colostrum CY × quality and of CS × quality, had a statistically significant (P<0.05) effect on colostrum stock balance. The simulation was repeated with a hypothetical supply of 6 l per calf and a quality threshold set at Brix ≥26%, but in both cases, none of the farms were ever in a positive stock balance. Colostrum replacers should be used in case of negative 'colostrum stock balance' for parts or the whole year.

Session 43 Theatre 10

Dairy-beef calves: current practices and views of calf producers and rearers

D.J. Bell, M.J. Haskell, C.S. Mason and C.-A. Duthie
SRUC, Peter Wilson Building, West Mains Road, Edinburgh. EH9 3JG, United Kingdom; david.bell@sruc.ac.uk

In the United Kingdom, there has been a year-on-year increase in the number of dairy-beef cross calves being born as a result of the increasing use of sexed dairy semen in dairy herds. These dairy-beef cross calves tend to be sold pre-weaned for further rearing on another unit and enter the beef production cycle. As part of the study, a small selection of semi-structured qualitative interviews was carried out with dairy farmers (calf producers; n=5) and calf rearers (n=7). The aim of these interviews was to explore the current practices carried out by each group in terms of calf sales, calf purchasing decisions, transfer of information and the information they would be willing to share/receive. The analysis of the interviews highlighted that the calf rearers tended to select calves for purchase by visual assessment based on the appearance of the calves for their age. Apart from the statutory requirements, no additional information, such as immunity status, prior health events and treatments, was formally transferred between parties. Any information that was transferred between calf producers and calf rearers was usually gathered via informal dialogue. All the calf rearers conveyed a desire to obtain as much information as possible about the calves they were purchasing. There was an expression of interest from the calf producers in receiving feedback about the subsequent performance of their calves whilst on the rearing unit but none of them actively pursued this type of information. Calf producers also highlighted the need to establish a good personal reputation as a producer of calves that met the market requirements. Some calf rearers said that they were now selecting calves with genetics that suited their production system. Calf producers all believed that such calves were a valuable additional source of financial income to their business and questioned such calves being referred to as 'surplus'. The calf rearers interviewed stated that they preferred to group calves by source farm when they received them at their rearing unit with health being quoted as the main reason for this practice. Overall, the interviews illustrated the themes of appropriate terminology, selection of the correct genetics, good communication, developing relationships and reputation.

Session 43 Theatre 11

Sodium percarbonate as a preservative in waste milk fed to dairy calves

D.J. Wilson[1], G.M. Goodell[2], R. Dumm[3], T. Kelly[2] and M. Bethard[2]
[1]Utah State University, Veterinary Clinical and Life Sciences, 955 E 700 N, Logan, UT 84322, USA, [2]The Dairy Authority, 8215 W. 20th St., Unit A, Greeley, CO 80634, USA, [3]Dairy Tech, Inc., 10027 County Road 70, Windsor, CO 80550, USA; david.wilson@usu.edu

The objective was evaluation of sodium percarbonate (SP), a preservative used in milk and drinking water, as an inhibitor of bacterial growth in pasteurized waste milk to be fed to dairy calves. Bacteria standard plate counts (SPC) in cfu/ml were performed using standard methods for the examination of dairy products procedures. After pasteurization at 63 °C for 30 min, dairy farm waste milk was incubated at 32 °C (100 °F) in a 0.4 m^3 (14.3 ft^3) incubator. SPC were calculated for 296 aliquots from 7.6 l (2 gal) batches of pasteurized milk at times 0 (as soon as milk was cooled to 49 °C for safe handling) and at 1, 2, 3, 4, 5, 6, 7, 8 and 24 hr after pasteurization. Concentrations of SP added to milk were 0 (untreated control), 200 and 400 mg/l. Statistical significance between SPC for the SP concentrations within each time point was tested using ANOVA. SPC means at the times following pasteurization for the 3 concentrations of SP were as follows: Untreated: 4,499; 4,634*; 5,485*; 3,344*; 85,282*; 58,424*; 88,663*; 245,050; 476,679*; 193,743,750**. 200 mg/l: 3,702; 2,814; 428; 857; 21,706; 5,036; 2,465; 4,188; 4,127; 444,000**. 400 mg/l: 4,062; 1,720; 845; 1,714; 4,008; 5,085; 7,533; 3,722; 2,738; 15,324**. * = untreated milk SPC higher than for either SP treatment, $P<0.05$, ANOVA. ** = SPC were different among all 3 treatments, $P<0.0001$, ANOVA. SP added to pasteurized milk at 200 or 400 mg/l was associated with SPC remaining significantly lower according to ANOVA, usually <8,000 cfu/ml, with all counts <22,000 cfu/ml, compared to those in untreated milk in milk at 32 °C. In contrast, SPC increased markedly beginning 4 h after pasteurization in untreated milk, reaching nearly 500,000 cfu/ml after 8 hr. SP has potential as a preservative for milk fed to calves; further studies are being conducted.

Session 43 — Theatre 12

Pre-transport diet affects the physiological status of calves during transport by road and ferry

S. Siegmann[1,2], L.L. Van Dijk[2,3], N.L. Field[2], G.P. Sayers[3], K. Sugrue[2], C.G. Van Reenen[1], E.A.M. Bokkers[1] and M. Conneely[2]
[1]Wageningen University & Research, Animal Production Systems, Wageningen, the Netherlands, [2]Teagasc, Animal & Grassland Research Centre, Fermoy, Ireland, [3]Munster Technological University, Dept. of Biological and Pharmaceutical Sciences, Tralee, Ireland; susanne.siegmann@wur.ie

Extended feeding intervals during long-distance transport require strategies to be developed that reduce the potential for dehydration, negative energy balance and stress. We studied the effects of providing different amounts of feed pre-transport on the physiology of unweaned dairy calves undergoing long-distance transport by road and ferry from Ireland to the Netherlands. At the Irish assembly centre, calves (n=116) were either fed 2 l of milk replacer the morning of transport (STD) or 3 l of milk replacer the evening before and the morning of transport (ALT). Calves were divided equally over two commercial trips and blood samples were taken before transport at the assembly centre (t=0 h), after roll-on-roll-off ferry transport at a French lairage, after road transport upon arrival at a Dutch veal farm (t=56 h) and one week later. Blood was analysed for parameters indicating energy balance (glucose, NEFA, BHB), dehydration (urea, haematocrit) and stress (cortisol, creatine kinase). Preliminary results show that ALT calves had higher levels of plasma glucose after ferry transport than STD calves (3.7 vs 3.2 mmol/l, P=0.02), but not on arrival at the veal farm. ALT calves had lower BHB (0.06 vs 0.13 mmol/l, P=0.001) and NEFA (0.09 vs 0.24 mmol/l, P=0.0007) levels at the assembly centre than STD calves. BHB was lower for ALT calves than for STD calves after ferry transport (0.28 vs 0.37 mmol/l, $P<0.0001$), whereas NEFA was higher for ALT calves on arrival at the veal farm (0.83 vs 0.66 mmol/l, P=0.008). ALT calves had lower urea concentrations after ferry transport than STD calves (2.7 vs 3.7 mmol/l, P=0.0001). Haematocrit and stress parameters showed no differences between treatments. There were no effects of treatment or transport one week after arrival. Overall, feeding larger volumes of milk replacer before long-distance transport has positive effects on calf energy balance and hydration status that are not sustained for the entire duration of the journey.

Session 43 — Theatre 13

Clonal dissemination of MDR *Pasteurella multocida* ST79 in a Swiss veal calves

J. Becker, J. Fernandez, A. Rossano, M. Meylan and V. Perreten
Vetsuisse Faculty, Bremgartenstrasse 109a, 3012 Bern, Switzerland; jens.becker@unibe.ch

In the framework of a large field study where prevalence of and antimicrobial resistance in *P. multocida* in calves was investigated in 38 Swiss farms, we observed that 20 isolates exhibited an MDR profile. Minimum inhibitory concentrations were determined using Thermo Scientific™ Sensititre™ EUST2 plates for tetracycline, streptomycin, sulfamethoxazole and trimethoprim, and BOPO6F plates for other tested antimicrobial agents, following CLSI recommendations. The complete genome of the *P. multocida* isolates was obtained from Nextera DNA Flex libraries sequenced on Illumina MiSeq (2×150 bp paired-end). Several parenteral treatments with tulathromycin, tylosin and oxytetracycline, and oral administration of amoxicillin were recorded during the study year and the previous year. Clonal dissemination of *P. multocida* sequence type (ST) 79 was observed throughout a period that was distinctly longer than the lifespan of a calf, indicating that *P. multocida* ST79 was maintained within the herd by circulating among calves that were present on the farm at the same time. The frequent use of macrolides, tetracyclines and β-lactams and the presence of *P. multocida* ST79 exhibiting resistance to these antimicrobial agents in different calves over several months indicate that antimicrobial selective pressure was responsible to select and maintain the bacterium over time. Antimicrobial resistance in animal pathogens and the use of antimicrobial agents, particularly of the most critically important ones, need to be strictly monitored to identify farms acting as reservoirs to limit the further spread of such MDR bacteria in livestock.

Session 43 Theatre 14

Effects of feeding milk with antibiotic residues on calf performance during the pre-weaning period

A. Flynn[1,2], C. Mc Aloon[1], M. Mc Fadden[2], J.P. Murphy[2], S. Mc Pherson[2], C. Mc Aloon[1] and E. Kennedy[2]
[1]University College Dublin, School of Veterinary Medicine, Belfield, Dublin, Ireland, [2]Teagasc, Grassland Department, Grassland Dept., Teagasc, Animal & Grassland Research and Innovation Centre, Moorepark, Fermoy, Cork, Ireland; anna.flynn@teagasc.ie

The practice of feeding waste milk, containing antibiotic residues, to pre-weaned calves is not recommended as it is associated with increased risk of diseases such as diarrhoea, and can result in poorer growth rates. However, there is evidence that many farmers feed this non-saleable milk to calves. As such, more knowledge on the effects of duration of exposure to milk with antibiotic residues is needed. This study aimed to investigate the effects of feeding milk replacer (MP) containing antibiotic residues on calf health, and growth across the pre-weaning period (PWP). The study was a randomised block design, dairy heifer calves (n=87) were balanced by birth weight, breed, and birth date. Treatments included: (1) long-term; calves that were fed antibiotic spiked MP for the entire duration of the PWP until weaning (LTA); (2) short-term; calves that were fed the antibiotic spiked MP for a duration of two weeks from the age three to five weeks of age (STA); (3) the negative controls fed MP free from antibiotics (CONT). Calves were fed MP containing neomycin (2.28 mg/l) and amoxicillin (1.68 mg/l), based on previous literature and reflected milk collected from cows during the withdrawal period, following intramammary treatment for mastitis. Measurements included health scores, treatment incidence, and weekly weights (kgs). All calves were gradually weaned by 12 weeks old. Weights data were analysed in linear mixed models, fixed effects included treatment and breed. Health data were analysed by logistic regression, reported as odds ratios (OD). Weights were similar in all treatment groups (P=0.301) across the PWP. Results showed faecal scores of LTA (OD 1.356) and STA (OD 1.561) calves were more likely to be higher than CONT during the PWP, STA calves were also more likely to have nasal discharge than CONT (OD 1.349). A similar number of calves required intervention with antibiotics in all treatments. The findings of this study will help to better inform farmers as to the risks associated with feeding waste milk.

Session 43 Theatre 15

Effects of extending lactation for dairy cows on health, development and production of their calves

Y. Wang[1], R. Goselink[2], E. Burgers[1,2], A. Kok[1], B. Kemp[1] and A.T.M. Van Knegsel[1]
[1]Wageningen University & Research, Animal Science group, Adaptation Physiology group, De Elst 1, 6708 WD Wageningen, the Netherlands, [2]Wageningen University & research, Wageningen Livestock Research, De Elst 1, 6708 WD Wageningen, the Netherlands; yapin.wang@wur.nl

Extending the voluntary waiting period for insemination (VWP) in dairy cows is of interest to reduce the frequency of calving events and inseminate at a moment with less fertility problems. Little is known about the impact of extension of the dams' VWP on the calf. The aim of this study was to evaluate the effect of extending VWP of dams on growth, metabolites, and production performance of their offspring. Holstein Friesian dairy cows (n=154) were blocked according to parity, milk yield, somatic cell count and randomly assigned to a VWP of 50, 125, or 200 days. For the current study, Holstein-Friesian heifer calves (n=61) from cows with different VWP were monitored from birth until 100 days in milk (DIM) after their first calving. Birth weight did not differ among heifer calves of the 3 VWP groups. During the rearing phase, body weight (BW) of heifers was not different among VWP groups. During the first 100 DIM, heifers in VWP50 had a greater BW (557 kg, P<0.01) and a greater fat and protein corrected milk yield (FPCM, 29.01 kg/d, P<0.01), compared with heifers in VWP125 (BW: 533 kg; FPCM: 27.04 kg/d). And both VWP50 and VWP 125 did not differ from VWP200 in terms of BW (549 kg) and FPCM (28.78 kg/d). When heifers were regrouped according to their mothers' real calving interval (CInt; CInt_1: <409 days; CInt_2: 409-468 days; CInt_3: >468 days), heifers born to mothers with CInt_3 had a greater BW after calving compared with heifers in CInt_2 (559 vs 537 kg, P=0.01). Heifers in CInt_1 had higher FPCM than heifers in CInt_2 (29.13 vs 27.23 kg/d, P<0.01). No difference in CInt_1 for BW (544 kg) and CInt_3 for FPCM (28.21 kg/d) compared with other two groups was found. In conclusion, extending VWP of dams did not affect growth of heifer calves during the rearing phase, but affected body condition and milk performance of heifers during the start of their first lactation.

Session 43 Poster 16

Dam-calf contact rearing in Switzerland: Aspects of management and milking

J. Rell[1], C. Nanchen[2], P. Savary[2], C. Buchli[1] and C. Rufener[2]
[1]Centre for Dam-Calf Contact Rearing, Postfach 363, 8903 Birmensdorf ZH, Switzerland, [2]Agroscope, Animal Production Systems and Animal Health, Tänikon, 8356 Ettenhausen, Switzerland; julia@mu-ka.ch

The interest in dam-calf contact rearing systems, where dairy calves remain on their farm of birth and are nursed by their dam, is increasing among consumers and farmers. By means of telephone interviews (n=16), we investigated the current status quo of a large proportion of the existing dam-calf contact farms in Switzerland descriptively. We identified views on practical chances and challenges of the system, individual farming approaches including husbandry, separation and weaning management, milking and health. Furthermore, a controlled on-farm study was conducted to evaluate milkability of nursing cows. Interview data revealed that seven (44%) farmers provide whole-day contact (W), while two (12%) operate with half-day contact (H) and seven (44%) provide short-time contact before (SB) or after (SA) milking. Weaning occurred between two and 12 months of age. Mostly, male calves were weaned abruptly at slaughter while female calves were weaned gradually. None of the farms reported regular calf diarrhoea but most farms had calves with diarrhoea recovering without intervention occasionally. Cow health was reported to be unproblematic on 12 (75%) of the farms. Perception of milkability problems differed widely among farms, ranging between <10 and 80% of cows affected. Eleven (69%) of the farms had taken measures to address poor milkability. Avoiding early separation and better calf health were the main decision criteria to change to a dam-calf contact system. Twenty of the 701 examined milkings (on 10 dam-calf contact farms and 5 control farms with artificial rearing) met the criteria for clear ejection disorder, with 85% of them occurring on W farms and 0% on control farms. Stripping milk fat contents were lower in nursing cows (P<0.01). Milk yield and average milk flow during the main milking phase were higher on SA farms than on W and SB farms (P<0.01). In summary, dam-calf contact systems are divers and may offer an alternative to artificial calf rearing with a high welfare standard and good calf health. Nursing after milking had the least negative influence on milkability and milk production of all contact types.

Session 43 Poster 17

A survey of colostrum management practices in dairy farms of Piedmont region (Italy): a pilot study

G.V. Berteselli, G. Pesenti Rossi, G. Vezzaro, E. Dalla Costa, S. Barbieri and E. Canali
University of Milan, Department of Veterinary Medicine and Animal Science, Via dell'Università 6, 26900 Lodi, Italy; gaia.pesenti@unimi.it

Inadequate transfer of immunoglobulins from dam to calf via colostrum remains a challenge in dairy farming. Failure of transfer of passive immunity has been related to increased morbidity and mortality in calves, lower productivity, and increased risk of culling. The aim of this pilot study was to investigate colostrum administration and calves' management in dairy farms of Piedmont region (Northern Italy). The questionnaire included 28 multiple choice questions and it was administered by a single interviewer to 15 farmers. The 28 items regarded calving management and care of the new-born, calf-dam separation, colostrum management, calf feeding, weaning, and calf housing. Descriptive analysis was performed. In 40% of farms surveyed, the disinfection of the calf's navel was not done. The separation from dam occurred immediately after birth in most of farms (67%). All farms used single pens, where calves were housed up to 60 days of age in most of the farms (80%). Colostrum was administered within 3 hours of life in all farms included. In 80% of farms 2 meals of colostrum were provided. Colostrum bank was present in 80% of farms. The quality of colostrum was assessed in almost half of the farms (47%) but only one farmer checked serum IgG. Weaning occurred at 60 days of age in most of the farms (67%). These findings provide preliminary insight into calves' management and colostrum administration in dairy farms of an Italian region, highlighting critical issues for calf welfare. Despite a prompt administration of adequate quantity of colostrum is carried out by the farmers interviewed, the evaluation of the quality is still partially performed. Moreover, the evaluation of passive immunity in calves remains lacking. Correct management and feeding of high-quality colostrum can reduce calf mortality, strengthen immunity, and increase animal life span.

Session 43 Poster 18

Serum profiles of dairy calves fed a milk replacer or whole milk at two levels of supply

T. Chapelain, J.B. Daniel, J.N. Wilms, J. Martín-Tereso and L.N. Leal
Trouw Nutrition R&D, P.O. Box 299, 3800 AG, Amersfoort, the Netherlands; tchapela@uoguelph.ca

The objective of this study was to compare the effect of feeding fresh whole milk and a milk replacer at two different levels of supply on calf metabolism. Forty-eight newborn Holstein calves (45.0±4.37 kg body weight; 2±1.0 d of age) were enrolled and blocked by age and arrival date and randomly assigned to one of four treatments at: 9.0 l/d of milk replacer (MR-H) or whole milk (WM-H), and 4.5 l/d of MR (MR-L) or WM (WM-L). Dry matter (DM), protein, fat and lactose content (% of DM) in MR and WM were: 16.2 vs 13.9, 23.7 vs 25.8, 17.5 vs 32.7, and 48.8 vs 32.9, respectively. Starter feed and straw were introduced at week 6. Calves were gradually weaned from week 6 to 10 and studied up to week 13. Blood samples were taken weekly, and data was analysed using PROC MIXED (SAS 9.4). Calves fed MR had higher blood glucose concentrations preweaning (P=0.01), but lower during weaning (P=0.02), compared to WM-fed calves. Preweaning blood concentrations of β-hydroxybutyrate (BHB), non-esterified fatty-acids (NEFA), triglycerides (TG) and cholesterol ($P \leq 0.06$) were greater for WM-fed calves. However, during weaning and postweaning, only TG remained greater ($P \leq 0.04$). Concentrations of total protein, urea, albumin, and globulin were higher for WM-fed calves during preweaning ($P<0.01$). Total protein and albumin remained higher during weaning ($P \leq 0.01$). Calves fed high milk allowance had greater glucose (P=0.02) and lowered cholesterol ($P<0.01$) concentrations in the blood during preweaning. During weaning, blood concentrations of glucose, NEFA, total protein, albumin ($P<0.01$) and TG (P=0.04) were higher for calves on high milk supply, whereas BHB ($P<0.01$) and urea (P=0.06) were lower. Postweaning, calves fed high milk allowances had greater NEFA (P=0.04), total protein and albumin ($P<0.01$) concentrations and lower TG concentrations ($P<0.01$). The differences observed in blood metabolites during preweaning and weaning confirm the significant influence of feeding WM or MR and their supply on calf metabolism. However, most metabolites did not differ with the liquid diet source provided during postweaning.

Session 43 Poster 19

Prevalence of foot lesions in French slaughter dairy and beef young bulls housed in indoor feedlot

S. Ishak[1,2], R. Guatteo[2], A. Lehébel[2], N. Brisseau[2], M. Gall[1], A. Wache[1] and A. Relun[2]
[1]French Livestock Institute, Beaucouzé, 49071, France, [2]Oniris, INRAE, BIOEPAR, Nantes, 44300, France; sarah.ishak@idele.fr

Almost half of beef calves and 14% of dairy calves are fattened indoor as feedlot cattle in France. Lameness is increasingly reported in French feedlot cattle, particularly at the end of the fattening period. Previous studies suggest that most cases are due to feet lesions. However, the lack of basic knowledge limits the establishment of cost-effective control measures. The aim of this study is to estimate the prevalence of foot lesions at the end of the finishing period in French dairy and beef young bulls. A cross-sectional study will be conducted in the 3 regions with the highest proportion of feedlot cattle slaughtered in spring 2023. The 4 feet of nearly 1,500 animals will be trimmed and examined *post-mortem* by two trained persons, and the type, severity and location of foot lesions will be recorded according to the ICAR Claw Health Atlas and national scoring methods. Descriptive analysis will be conducted to estimate foot, animal and within-batch level prevalence, to describe the severity and location of foot lesions and to estimate the distribution (e.g. breed, geographical area, etc.) of foot lesions. This study will provide an accurate estimate on the prevalence of foot lesions in a large sample of finishing young bulls. It will provide original data for indoor feedlot cattle, including the frequency of lesions on the front feet, association between foot lesions and associations between foot lesions and several dairy and beef breeds.

Session 43 — Poster 20

Effect of direct-fed microbial supplementation on performance and health of pre-weaning dairy calves

J. Magalhaes[1], B.I. Cappellozza[2], T.C. Dos Santos[1], F.N. Inoe[1], M.S. Coelho[3], V. Soares[3] and J.L.M. Vasconcelos[1]
[1]Sao Paulo State University, School of Veterinary Medicine and Animal Science, Department of Animal Production, Fazenda Lageado, 18618-150 Botucatu, SP, Brazil, [2]Chr. Hansen A/S, Bøgle Allé, 10-12, 2970, Denmark, [3]Fazenda Santa Luzia, Fazenda Santa Luzia, 37902-377 Passos, MG, Brazil; brbrie@chr-hansen.com

The health of pre-weaning dairy calves dictates long-term performance and profitability of the operation and technologies that support pre-weaning health and performance are warranted. We hypothesized that direct-fed microbial (DFM) supplementation would improve health and performance of pre-weaning dairy calves. Our objective was to evaluate the effects of two DFM combinations on performance and health of pre-weaning dairy calves. At birth, 90 female crossbred Gyr × Holstein calves were assigned to: (1) Control: no DFM (CON; n=30); (2) 1 g/head per day of a *Bacillus*-based DFM (BOVACILLUSTM; Chr. Hansen A/S; BAC; n=30); or (3) BAC plus 1 g/head per day of a lactic acid bacteria-based DFM (LACTIFERM®; Chr. Hansen A/S; MIX; n=30). The BAC and MIX treatments were mixed daily in whole milk throughout the 77-day experimental period. All animals were daily observed for signs of adverse health events. On day 77 of the study, weaning body weight (BW) was recorded for average daily gain (ADG) and feed efficiency (FE) calculation. Contrast analysis were performed: (1) DFM effect: CON vs DFM (BAC + MIX); and (2) DFM type: BAC vs MIX. At weaning, DFM-supplemented calves, regardless of type (P=0.68), were heavier than CON calves (P=0.04; 81.7, 88.0, and 86.7 kg for CON, BAC, and MIX, respectively). Moreover, ADG and FE tended to be greater (P=0.06) for DFM vs CON, but no differences were observed between BAC or MIX (P>0.88). Pneumonia occurrence was greater (P=0.05; 47.0, 23.0, and 29.0% for CON, BAC, and MIX, respectively), whereas diarrhoea tended to be greater for CON (P=0.08; 50.0, 29.0, and 32.0% for CON, BAC, and MIX, respectively). Moreover, no differences were observed on number of days sick and costs of pharmacological intervention per calf (P>0.21), but total cost of the interventions was numerically greater for CON calves. In summary, DFM supplementation, regardless of the type, supported health and overall performance of pre-weaning Holstein calves.

Session 43 — Poster 21

The effect of astaxanthin on health and calves performance

E. Sosin[1], I. Furgał-Dierzuk[1], B. Śliwiński[1] and A. Burmańczuk[2]
[1]National Research Institute of Animal Production, Department of Animal Nutrition and Feed Science, Krakowska St. 1, 32-083 Balice, Poland, [2]University of Life Sciences in Lublin, Department of Pharmacology, Faculty of Veterinary Medicine, Akademicka 12, 20-033 Lublin, Poland; ewa.sosin@iz.edu.pl

The aim of the study was to determine the effect of astaxanthin on health indicators and calves performance. The experiment was carried out with 16 Polish Holstein-Friesian bull calves aged between 4±1 and 90 days were assigned based on the analogue principle to 2 equal groups. Calves were given colostrum and whole milk before the experiment, and milk replacer from the beginning of the experiment to 56 days of age. Grains in the mixtures were given in rolled form. The control group (C) was fed with the mixture contained corn grain (56%), oats (18%), soybean pressed cake (22%) and minerals (4%) and in the experimental group (AXT) there were the addition of astaxanthin (0.020 g/kg feed). During the experiment the body weight, feed intake, daily gain, haematological and biochemical and immunological parameters were determined. The daily gain of calves from the AXT group was higher than from the control group before the weaning (766 vs 633 g) and after (1,428.3 vs 1,341.5 g) ($P \geq 0.05$). There were no significant differences between the group in haematological and biochemical parameters except the total cholesterol and LDL fraction which were lower for AXT group. The iron content in the blood were also higher for AXT group. There were no statistical differences between IgA and IgM and IgG at the end of the experiment but IgG (13.39 vs 11.46 mg/g) and IgA (1.29 vs 0.97 mg/g) was higher for AXT group before the weaning. There was found positive influence of astaxanthin on calves health and immunity up to 56 d of age.

Session 43 Poster 22

Strategic grouping of dairy beef calves on arrival at a rearing unit

D.J. Bell, C.S. Mason, K.C. Henderson, M.J. Haskell and C.-A. Duthie
SRUC, Peter Wilson Building, West Mains Road, Edinburgh. EH9 3JG, United Kingdom; david.bell@sruc.ac.uk

There is an increasing trend in the number of dairy beef cross calves being born on dairy farms in the United Kingdom. The majority of these calves are sold off farms pre-weaned for further rearing on dedicated calf rearing units where they enter the dairy-beef production cycle. On these rearing units, such calves are exposed to new environmental and disease challenges as well as often being mixed with other calves from multiple farms. The aim of this preliminary study was to strategically group calves when they arrived on a rearing unit and assess their health and performance after a period of time on the rearing unit. One hundred and forty calves were sourced from seventeen dairy farms within a seventy-mile radius of the rearing unit. The calves ranged from 10 to 67 days of age (mean 26 days). On arrival, the calves were lung ultrasound scanned (LUS) and scored on a scale from 0 (normal aeriated lungs) to 5 (consolidation observed in three or more lung lobes). The calves were also Wisconsin health scored (WS) and weighed. Using this information, calves were then assigned to one of five groups: (1) 'High health' (HH) (LUS 0/1 & WS≤2); (2) 'Low health' (LH) (LUS≥3 & WS≥3); (3) 'Intermediate health' (INT) (LUS 2/3 & WS≤2); (4) 'Mixed health' (MIX) (a proportion of calves (50%) LUS≤2 & WS≤2 along with a proportion of calves (50%) with LUS 4/5 & WS≥2); and (5) 'Normal farm practice' (NFP). The NFP for grouping calves was to keep calves from the same source farm together in the same pen. All calves were weighed, LUS and WS again after 22 days of being on the rearing unit to coincide with the farms routine for handling calves, thus allowing a daily liveweight gain (DLWG; kg/d) to be calculated. A relatively large proportion of the calves (62%) were observed with signs of lung consolidation (LUS≥3) on arrival at the rearing unit. Younger calves had lower LUS on arrival at the rearing unit. Calves grouped as LH had a higher DLWG compared to calves grouped as HH but lower than NFP. The results of this preliminary study highlight some areas of interest that require further investigation.

Session 43 Poster 23

Calves sexing and crossbreeding to optimize the destination of the young from French dairy farms

S. Dominique
Institut de l'Elevage, 23 rue Jean Baldassini, 69364 Lyon cedex 07, France; sandra.dominique@idele.fr

Destination of calves from dairy farms is subject of societal polemics. Indeed, for some dairy breeds, economic value of young male calves is minimal while the export in live of these calves to European countries is contested. In 2022, 370.000 French dairy male calves were exported in Europe to be fatten as young bulls. This number keeps increasing every year: number of French dairy calves exported increased by 58% between 2017 and 2022. In parallel, the number of slaughtered calves in veal calves sector decreased by 6,7% from 2021 to 2022 due to the reduced activity of French integrators and the increase of production costs. Using sexed semen can be a lever for farmers who wants to structure differently calves' destination from his dairy farm. Used in addition with meet crossbreed inseminations and genotyping, it becomes an asset to manage the renewal and calves' valorisation: farmers can choose the sex of calves born. We assessed the evolution of the use of sexed semen in France since 2010 by using the data from the French information genetic system. Our results show a quick expansion from 2010 till 2015, followed by a slow decreased till 2019, and finally a new impulse in its deployment. Data from French suckler farms show a less use of animal inseminations. The proportion of sexed semen used in suckling farms represents only 3% of all sexed semen inseminations in France. Initially used as a last attempt for a non-pregnant dairy cow after several inseminations, the number of first crossbreed animal inseminations with a suckler breed sire as doubled from 2014. Contrary to sexed semen, meat crossbreeding inseminations did not know a decrease in activity. Recent data confirm that volume is still growing. Detailed analysis showed a connection between size of the farms and use of sexed semen. This reproductive tool, associated with meat crossbreed inseminations is useful to focus selection effort on targeted cows and valorise other calves by meat production. Prospects for the evolution of these practices are subjects to various factors. It is not easy to pronounce on the situation tomorrow, but this is a useful tool to optimize valorisation of necessary calves born in dairy sector and a lever to limit exported volumes calves.

Session 43 Poster 24

Inclusion of grass silage in finishing total mixed rations for rosé veal calves

M. Vestergaard[1], M. Bjerring[1], A.L.F. Hellwing[1], M.B. Jensen[1], B. Muhlig[1], L. Mogensen[1] and N.B. Kristensen[2]
[1]Aarhus University, Dept Anim & Vet Sci, Foulum, 8830 Tjele, Denmark, [2]SEGES P/S, Agro Food Park, 8200 Aarhus N, Denmark; mogens.vestergaard@anivet.au.dk

Finishing rations in rosé veal calf production are often pelleted feed combined with straw or total mixed rations with maize-silage as sole forage source. Cereals are the dominating starch source and canola meal the protein source. These rations can compromise ruminal health, due to sub-acute rumen acidosis, rumen wall damages and establishment of liver abscesses. Regarding climate issues, these rations keep the methane production low, but the Carbon Footprint of the feed production is often higher than for grass-based rations. The present study aimed at comparing a grass silage-based TMR (Gr) with a corn-cob-based TMR (Ye) on performance, eating behaviour, methane production and carcass quality. Ye consisted of corn-cob silage, rolled barley and rapeseed meal whereas Gr consisted of rolled barley, 1 cut grass silage (25-35% of DM) and untoasted milled fava beans. DM%, crude protein, NDF, starch, and NE were similar but differed in fill value and physical structure. Holstein bull calves (155 kg) were housed in pens of 8 with two automated feeders per pen. Within each pen, 4 calves had access to Gr and 4 to Ye. A total of 8 pens was used. Calves performed equally well on both rations ($P>0.10$), with Gr being numerically best. Feed intake was similar, but Ye calves ate more meals than GR and spend more time eating from the straw-bedding. Accelerometers showed higher rumination activity on Gr vs Ye but similar lying time and health index. Methane production was 13% higher for Gr than Ye ($P<0.10$). When slaughtered 1 year old, at approximately 500 kg, LW was numerically 3% higher for Gr than Ye ($P>0.10$) and carcass characteristics were similar including lean-fat colour. In Gr calves, two liver abscesses were detected and 5 in Ye. In conclusion, both grass silage and fava beans can replace corn-cob silage and canola meal in TMR for rosé veal calves with a tendency to alleviate the high prevalence of liver abscesses.

Session 43 Poster 25

Veal calves' housing in France: current situation and investment needs

M. Tourtier and C. Martineau
Institut de l'Elevage, Qualité des carcasses et des viandes, 8 route de Monvoisin, 35650 Le Rheu, France; christophe.martineau@idele.fr

A survey on veal calves' housing was conducted by the French Livestock Institute, commissioned by INTERBEV Veaux. The main objective was to establish an inventory of the age, state of disrepair and consistency of veal calves' housing in France. A second objective was to estimate the cost needed to modernize the veal calves' housing according to different scenarios that meet the needs expressed by the veal sector. In 2019, the survey was deployed to all French veal calves' breeder via a questionnaire accessible on internet. The answers of 405 farms (i.e. 20% of the French farms) were analysed, representing 583 farm buildings for 122,000 fattening places. The average of size of the farms surveyed was 351 places, with notable differences depending on the region. The farms had an average of 1.6 buildings with a capacity of 213 places. The buildings' average age was 20.7 years. The answers obtained in this survey and those collected in the INOSYS network farms make it possible to provide quantified cost for the modernization of the 'France' farm according to 5 scenarios: biosecurity, health/welfare, work/automation, environment and outdoor access for animals. According to the assumptions, the cost of modernizing buildings extrapolated to all 2,391 French farms in 2019 ranges from 35.8 million euros for 'biosecurity' scenario to 237.2 million euros for 'environment' scenario. 'Outdoor access for animal' scenario has not been qualified, as none of the farms surveyed could evolve in this direction without calling into question the very organisation of the existing buildings or without incurring disproportionate investments. This scenario should therefore only be considered in the case of construction of new buildings. In conclusion, it is clear that veal calves' housing must evolve by taking into account societal expectations in terms of environmental protection, animal health and welfare and landscape quality. It seems equally important to consider the way in which the 'French' farm must adapt to preserve or even improve the organisation of work and ensure a fair remuneration for breeders, and thus guarantee the sustainability of the veal sector.

Session 43 Poster 26

Evolution of the use of antibiotics in the veal calves' sector in France between 2013 and 2020

M. Chanteperdrix[1], A. Chevance[2], M. Orlianges[3], D. Urban[2], M. Tourtier[1] and P. Briand[4]
[1]Institut de l'Elevage, Qualité des Carcasses et des Viandes, 8 route de Monvoisin, 35650 Le Rheu, France, [2]Agence Nationale du Médicament Vétérinaire (Anses-ANMV), 14 rue Claude Bourgelat, 35133 Javené, France, [3]INTERBEV, 207 rue de Bercy – TSA 21307, 75564 Paris Cedex 12, France, [4]Chambre Régionale d'Agriculture de Bretagne, Rue Maurice Le Lannou – CS 74 223, 35042 Rennes Cedex, France; manuel.tourtier@idele.fr

The veal sector participates in Ecoantibio2017 plan by implementing an ambitious program to know the practices and identify levers for effective action to reduce the use of antibiotics on farms. ANSES-ANMV and the French Livestock Institute were commissioned by INTERBEV Veaux to set up a permanent observatory to estimate the quantities of antibiotics in veal calf farms. This national system was set up in 2016 and is based on active collaboration of breeders, integrators, producer groups and veterinarians and on the support of engineers of the Chamber of Agriculture in Brittany. The observatory is made up of 30 volunteer breeders located in 6 departments of western France. They represent nearly 15,000 calf places, raised in production systems representative of national feeding and housing practices. Farms work for 9 integrators or producers' groups. In 2020, data from 47 calf batches (16,553 animals) were analysed. ALEA is the main standardized indicator used in France for the annual sales reporting of antibiotics and reached 3.14 in 2020. In the survey carried out by ANSES-Lyon in 2013 on 186 batches of calves, the average ALEA was 5.86. Based on this indicator, the exposure of calves to antibiotics decreased by 45.3% between 2013 and 2020.

Session 44 Theatre 1

Resilience4Dairy: sharing knowledge to improve sustainability and resilience of the dairy sector

V. Brocard[1], M. Klopcic[2] and J. Boonen[3]
[1]Institut de l'Elevage, 8 route de Monvoisin, 35650 Le Rheu, France, [2]University of Ljubljana, Kongresni trg 12, 1000 Ljubljana, Slovenia, [3]LTA, Kréiwénkel, 9374 Gilsdorf, Luxembourg; valerie.brocard@idele.fr

R4D – which stands for Resilience for Dairy – is an EU-funded project which aims at improving the European dairy sector's sustainability and resilience. To achieve this goal, the 18 partners of R4D have created a network enabling the exchange of practical and scientific knowledge among European dairy farmers, researchers and other relevant stakeholders. They will focus on 3 Knowledge Areas: economic & social resilience, technical efficiency and environment, animal welfare and society friendly production systems. Those issues are often addressed separately. However, they are interconnected and depend on the livestock farming system, rearing management, people involved in the production process, feeding and material resources, and level of use of innovation. R4D aims to build bridges between them in an innovative cross-fertilization, crowd-innovation and transdisciplinary approach, focusing on Best Practices allowing optimal benefits to be achieved in all three. The overall objective of R4D is to develop and to strengthen a self-sustainable EU Thematic Network on 'resilient and robust dairy farms' designed to stimulate knowledge exchanges and cross-fertilization among a wide range of actors and stakeholders of the dairy industry. The six specific objectives of the project are the following: (1) structuration of the networks by selecting and connecting innovative resilient farms and relevant Operational Groups into National Dairy AKIS; (2) prioritization of farmers' needs to improve resilience and robustness of dairy farms; (3) matching farmers' needs with the inventory of best practices & setting-up of fine-tuned Knowledge Work Plans; (4) implementing collective evaluation and assessment of solutions (including cost-benefit analysis); (5) adaptation and translation of the Best Practices for practitioners; and (6) communication, dissemination and demonstration of the results and best-practices. The aim of R4D is to widely disseminate relevant ready-to-use best practices based on innovations, facilitating knowledge exchange from farmers to farmers.

Session 44 — Theatre 2

Indicators and influencing factors of livestock resilience

I.D.E. Van Dixhoorn[1], J. Ten Napel[1], A. Mens[1] and J.M.J. Rebel[2]
[1]Wageningen Livestock Research, P.O. Box 2176, 8203 AD, the Netherlands, [2]Wageningen Bioveterinary Research, Houtribweg 39, 8221 RA Lelystad, the Netherlands; ingrid.vandixhoorn@wur.nl

Current livestock production systems in Europe have been developed on the premise of a need for supplying more food with the greatest efficiency and least risks. Therefore, these systems have been designed to maximize productivity under well-controlled conditions. This has contributed to highly productive and efficient food production systems. However, we now realize that this strategy has led to vulnerabilities, especially concerning animals in livestock production systems. One of the big questions is how we shift focus from maximized production efficiency towards a system that maximizes animal resilience instead of making the animals more dependent on well-controlled conditions. Then impact of diseases can be reduced simultaneously with the use of antibiotics and the need for animal mutilations, aiming for a more sustainable production. Animal farming utilizing animal resilience starts with farming conditions that are acceptable to society and a level of management that the majority of qualified stockmen can provide. In such farming systems, animals can be bred and prepared for common day-to-day disturbances. Animal resilience can only be explored in an environment that is representative of commercial farming systems in terms of health, climate, housing conditions and diet quality. Differences in resilience between animals can be quantified by measuring variation in longitudinally recorded traits, such as daily milk production, body weight gain, activity patterns, heart rate or body temperature. Resilience at animal level is defined as the capacity of an individual to be minimally affected by disturbances or to quickly recover. A shift in strategy towards more resilient production animals therefore requires indicators of animal resilience as well as knowledge of what factors can be used to influence it. We present concepts of resilience that can be applied in research and give examples of indicators of animal resilience as well as influencing factors in dairy cows and other animals.

Session 44 — Theatre 3

Reducing stress of dairy cows and farmers to improve resiliency and welfare

M.T.M. King[1] and T.J. De Vries[2]
[1]University of Manitoba, Department of Animal Science, 12 Dafoe Road, R3T 2N2 Winnipeg, Canada, [2]University of Guelph, Department of Animal Biosciences, 50 Stone Road East, N1G 2W1 Guelph, Canada; tdevries@uoguelph.ca

The dairy industry has, in many ways, always had to be resilient and adaptable to be able to survive and thrive in changing circumstances. The fact that the dairy industry has been able to continue to produce and succeed through rapidly emerging challenges is a testament to the resiliency of the cows, farmers and other professionals who make up the sector. In recent years, however, many of those changes and challenges have become more pronounced and may threaten the sustainability of the industry. One of those challenges that continues to face the dairy industry is stress, both on dairy cows as well as those who take care of them. Stressors can have both physiological and psychological effects on dairy cows, which can lead to changes in cow behaviour and health status, negatively impacting cow production, reproduction, and welfare. Interestingly, there is growing evidence that the welfare of dairy cows, including their health, is related to the well-being of dairy farmers. There is evidence that farmers may experience higher levels of stress, anxiety, and depression than the average citizen. In such a high stress occupation, physical and mental stress may wear down an individual's ability to cope with complex problems; this may affect how dairy farmers deal with the animals in their care. Therefore, when considering ways to reduce stress for dairy cattle and improve their welfare, focus should also be on improving the well-being of the farmer. Combined efforts to reduce stress, and improve welfare, both of dairy cows and their caretakers, will contribute to maintaining resiliency in the dairy industry.

Building resilience in the dairy sector of China

S. Li, W. Wang, W. Du, X. Sun, K. Yao and J. Xia
China Agricultural University, Beijing, 100193, China, P.R.; lisheng0677@163.com

The Chinese dairy sector has undergone a fundamental transition in recent decades, with traditional smallholders being largely replaced by industrial and specialized farming systems. By 2022, the proportion of dairy cattle kept on farms with more than 100 head reached 73%. The intensification of the dairy sector has been accompanied by rapid improvements and the application of modern industrial technology, particularly in feeding strategies and farm management. For instance, there has been a significant improvement in the nutritional conditions of roughage, as TMR (Total Mixed Ration) feeding technology and high-quality whole-plant corn silage have become widely used. Additionally, almost all large-scale dairy farms in China have adopted mechanized milking. According to data published by the National Dairy Industry and Technology System, the average annual milk production per cow on large-scale dairy farms in China reached 10.1 tons in 2022. Furthermore, manure management has greatly improved with the implementation of more advanced treatment facilities, such as those that produce biogas. These advancements have contributed to the promotion of overall industry quality and resilience. Despite these achievements, the Chinese dairy sector faces numerous challenges in areas such as industry chain integration, self-sufficiency in milk and feed raw materials, improvement of cow breeding, reducing production costs, carbon mitigation management, animal welfare, and environmental pollution. China's per capita dairy consumption is only 60% of that in developed Asian countries. Given the prediction that there will be mild growth in national milk consumption in China over the coming decade, it is crucial to address these challenges and work towards a more resilient and sustainable Chinese dairy sector.

Building resilience in farming: dairy cattle and workforce management

G.M. Schuenemann[1] and J.M. Piñeiro[2]
[1]The Ohio State University, Dept. of Veterinary Preventive Medicine, Columbus, OH, USA, [2]Texas A&M University, Dept. of Animal Science, Amarillo, TX, USA; schuenemann.5@osu.edu

To ensure long-term sustainability of dairy farming, it is essential to build a resilient workforce. Dairy farms in US are consolidating at a faster rate today than any other agricultural commodity. In January 2022, the dairy cattle inventory was 9.3 million cows distributed among 30,000 farms, with an average annual productivity per cow of 11,000 kg of milk. About 50% of licensed dairy farms, mostly with less than 500 cows, ceased operation in the past 20 years at a rate of about 2,300 farms per year without changing the national milk cow inventory. Today, about 5% of US dairy farms are milking 60% of cows with an average herd size of about 3,300 cows. The farm workforce shortages have been an issue for the past few decades and 75% are immigrant workers. This is influenced by demanding work schedules and physically demanding jobs, limited access to educational opportunities or simply individual preferences to live near larger urban communities with greater access to developed infrastructure (e.g. internet, health services, entertainment, schools). Some dairy farms have been able to reverse this labour shortage by providing their employees with housing, covering expenses associated with transportation, offering educational opportunities and work-related clothing. The dairy community faces challenges (e.g. extended drought, increased regulations, changing consumer preferences, workforce shortages) as well as opportunities (e.g. growing demand and precision technology). The adoption of advanced technologies such as automation and data analytics can help reduce labour costs and improve the overall productivity and efficiency of farm operations. However, these solutions include investing in education and training programs, enhancing workplace safety and health, and promoting diversity and inclusion in the workforce. Today, leading dairy operations are completely integrated with an established network of suppliers and professionals, actively engaged with their local communities with a logistic that goes from the farm to consumers. Farms have been integrating best animal welfare and sustainability practices with strong emphasis on prevention and continuous improvement.

Session 44 Theatre 6

Working on resilience in the Ukrainian dairy sector

L. Stepura
Scientific and Methodological Center for Higher and Pre-Higher Vocational Education, Smiliamska str. 11, 03151 Kyiv, Ukraine; ludmila.stepura@gmail.com

The resilience of the dairy sector in Ukraine depends firstly on the cooperation with producers of raw milk. This is especially the case under the conditions of the war. The resilience of raw materials production and adaptation to the real conditions are impacted by many factors: transport logistic, (un)interrupted supply of electricity, use of generators (that leads to production cost increase), mobilization of professionals to the Ukrainian army, etc. The large-scale invasion of the Russian occupiers on the Ukraine territory affected both, producers of dairy raw materials and their processors and consumers. The supply chains between farms, processing plants and trade networks have been disrupted. In order to improve the economy of feeds, farmers started to feed cows only twice per day, which reduced milk yield. In the occupied territories, farms were completely or partially destroyed. The number of cows in all categories of farms decreased by 13.1% in 2022; as a result, the industry did not receive 1 million tons of milk (volume was 8.7 million tons in 2021). Dairy farms in the de-occupied and frontline areas are / will not be able to produce fodder for livestock without demining of their fields and will be forced to reduce or completely close dairy production. Nevertheless, some time after the start of the large-scale war, the country's dairy sector managed to adapt to the new conditions. However, in autumn of 2022, the industry also experienced problems caused by the deliberate damage to the energy infrastructure, but Ukrainian producers also later adapted to these challenges. Nowadays, the dairy sector meets the needs of the domestic market despite considerable industry losses. Due to significant migration of citizens to abroad, the domestic market of dairy products consumption reduced. According to the Association of Milk Producers, the production of milk in 2022 compared to 2021 decreased by 12.1 to 7.66 million tons. The figures for the fall in milk production could have been much higher, but thanks to the efficient functioning of dairy farms in safer regions and the relocation of livestock from war-affected areas, losses were minimized. In particular, the central and western regions of Ukraine increased milk production due to business relocation and increased demand for raw materials.

Session 44 Theatre 7

Future scenarios for livestock agriculture in New Zealand

C. Vannier[1], T. Cochrane[2], L. Bellamy[2], T. Merritt[2], H. Quenol[3] and B. Hamon[2]
[1]Landcare research, 54 Gerald Street, 7608 Lincoln, New Zealand, [2]University of Canterbury, 20 Kirkwood Avenue, 8041 Christchurch, New Zealand, [3]CNRS, place du recteur Henri le Moal, 35043 Rennes, France; vannierc@landcareresearch.co.nz

Agriculture in New Zealand (NZ) faces disruptions from climate change, increasingly stringent environmental regulations, and emerging technologies. Given the importance of agriculture to the NZ economy, government and industry need to develop policies and strategies to respond to the risks and opportunities associated with these disruptors. Livestock production in NZ represents 64% of the agricultural export revenue and 85% of agricultural GHG emissions, mainly from methane and nitrous oxide. There is a research gap in understanding how future disruptions from climate and technology could impact on the environment and productivity. To address this gap, we have developed and applied an assessment tool of NZ livestock farming systems to explore pathways and interventions for increasing agricultural resilience, sustainability, and profitability over the next 5-30 years. A systems dynamic model developed using Stella Architect was designed as a Decision Support Tool (DST) to bring together production, market values, land use, water use, energy, fertiliser consumption, and emissions for each of the main agricultural sectors (dairy, beef, sheep, cereals, horticulture, and forests). The parameters are customisable by the user for scenario building. With stakeholder consultation, scenarios were designed to assess pathways and interventions to underpin strategy initiatives in the arable, dairy, and beef & lamb sectors related to food security and climate change adaptation and alternative protein production. Simulation results suggest potential synergistic opportunities between sectors to enhance productivity and reduce emissions. Achieving food security and cereal self-sufficiency in NZ through a multi-sectoral approach and the development of an alternative protein market (100% NZ branding) could lead to carbon emission reductions and improvement of carbon sequestration. Furthermore, the interactive DST and simulations improved stakeholder engagement, which can facilitate future land planning and policy formulation.

Breeding approaches to improve robustness and resilience in dairy cows

K. May and S. König
Institute of Animal Breeding and Genetics, Justus-Liebig-University Gießen, Ludwigstraße 21B, 35390 Gießen, Germany; katharina.may@agrar.uni-giessen.de

Dairy cows are exposed to environmental challenges including rising and fluctuating temperatures, or increasing pathogen infection pressure due to resistances against available drugs. Against this background, it is imperative to develop breeding strategies for improved robustness and resilience. Robustness reflects the cow's adaptation ability to environmental challenges, implying improved disease resistance and tolerance to environmental stressors by maintaining high production (e.g. milk). A resilient cow is robust and recovers quickly from disease. Genetically, breeding for robustness suggests the selection of genotypes with stable genetic values in different environments, i.e. animals being robust against the impact of genotype-by-environment interactions. Classical breeding approaches focus on the analysis of simple production traits in dependence of environmental descriptors including herd management or climate characteristics by applying reaction norm and random regression models. Such approaches can be enhanced by considering novel functional traits reflecting health, metabolic stability or resource efficiency, and by a more detailed description of the farm environment including feeding or emission aspects. Hence, the general approach of genotype-by-environment interactions can be extended. Insights into the causal genetic mechanisms of cow health are possible when considering genomic data, i.e. dense marker genotypes, genome sequences or gene expression data. Approaches in this regard are genome-wide association studies with the ongoing annotation of potential candidate genes, up to the study of gene expressions in different environments. In ruminants, a further genetic contribution is due to the genetics of the microbiome, addressing novel studies on genotype-by-genotype interactions. Similar methods are suggested when studying the genetic mechanisms of disease resistance, i.e. simultaneously considering the genotype of the host (cow) and the genotype of a pathogen (e.g. endoparasite). Setting-up genetic relationships or similarity matrices for both species (host and pathogen) allow deeper insights into the functional diversity, being a main driver for selection response in a long-term perspective.

Session 44 Theatre 9

Resilience from the perspective of farm economics

E. Kołoszycz and A. Wilczyński
West Pomeranian University of Technology in Szczecin, Department of Management and Marketing, al. Piastów 17, 70-310 Szczecin, Poland; ewa.koloszycz@zut.edu.pl

In an era of increasingly emerging economic, social, environmental and institutional shocks, the importance of farm capacity for survival and further development is increasing. The capacity of farms for robustness, adaptability and transformability serves to build resilience to various types of shocks in the perspective of short- and long-term changes in the environment. The three abilities mentioned are correlated and mutually reinforcing. This means, for example, that robustness in the short term is essential for transformability in the long term. The economic aspect is one of the three most important dimensions of resilience beyond the social and environmental dimensions. Economic resilience can be defined as the ability of a farm to maintain economic viability in the face of emerging turbulence (e.g. market and production risks and policy changes), including the ability to transition to a new equilibrium state. The main objective of this research is to assess the economic resilience of selected European dairy farms and diagnose the life cycle phase they are in. Two types of economic resilience were used for comparative analysis: short-term resilience and long-term resilience. The subjects of the research are groups of farms specializing in dairy farming located in the observation field of the European FADN (type 45 in the FADN Public Database). The research area consists of producers operating in the fifteen countries of the European Union represented in the Resilience for Dairy (R4D) project. The research was conducted in groups of farms classified by economic size, adopting the ES6 typology. The research covers the years 2011-2020, when farms operated under different legal conditions, i.e. during milk quotas and after the abolition of milk production limits. Preliminary results showed that farms with a larger scale of production have higher short-term economic resilience, but also lower long-term resilience compared to farms with a smaller economic size. Of all the farm groups analysed, most of them were in the survival phase. This means that the farms surveyed had problems covering the costs of unpaid labour.

Session 44 Theatre 10

Labour: a key factor in the resilience of the European dairy farmer

S. Debevere[1], L. Dejonghe[1], I. Louwagie[1], I. Vuylsteke[1], E. Béguin[2], S. Fourdin[2], P. Rondia[3], L. Boulet[3], S. Mathieux[3] and G. Elluin[4]

[1]Inagro vzw, Ieperseweg 87, 8800 Rumbeke-Beitem, Belgium, [2]IDELE, 54-56 avenue Roger Salengro, BP 80039, 62051 St Laurent Blangy, France, [3]Centre wallon de Recherches Agronomiques, Rue de Liroux 8, 5030 Gembloux, Belgium, [4]Chambre d'agriculture du Nord Pas de Calais, 54-56 avenue Roger Salengro, BP 80039, 62051 Saint Laurent Blangy, France; sandra.debevere@inagro.be

In the Horizon 2020 project 'Resilience for Dairy', 82% of questioned people in the European Dairy sector indicated that work-life balance is an important factor for farmers to be resilient. Indeed, today, dairy farmers are under increasing pressure due to rapid herd growth and increased competitiveness. Rising labour demands are affecting the quality of life of livestock farmers. The Franco-Belgian Interreg project 'CowForme' aimed to create jobs and increase labour efficiency on dairy farms. The aim was to reduce the workload and increase quality of life. By means of a survey, CowForme investigated what were the biggest challenges for farmers to work with staff. Also the motivation and obstacles of jobseekers to work on a dairy farm were explored. The results of the survey will be presented and discussed. In focus groups, several advisors and dairy farmers shared their experiences and possible solutions how to optimize the workload on the farm. In the presentation, some practical examples will be given that were proposed in the focus groups. During the project, many ways were used to spread information to farmers: study tours, videos, webinars, fact sheets, etc. All this information can be consulted at www.cowforme.eu. By small, but efficient solutions, farmers can save a lot of time and improve their work-live balance and resilience.

Session 44 Theatre 11

Resilience of dairy farming: the farmers' point of view

E. Castellan[1], C. Bausson[2] and V. Brocard[1]

[1]Institut de l'Elevage, 149 rue de Bercy, 75012 Paris, France, [2]Chambre d'agriculture de Normandie, 6 rue des Roquemonts – CS 45346, 14053 Caen Cedex 4, France; elisabeth.castellan@idele.fr

The concept of 'resilience' in not a familiar and concrete one for breeders. Within the frame of EU H2020 Eurodairy program, the French dairy farmers involved in the project proposed a definition from their point of view; then, they imagined a methodology to evaluate it on their farms. Hence a tool named 2MAINS (tomorrow, or 2 hands) was created to simply evaluate the resilience of the farms considering 5 topics (strategy, technical efficiency, economics, social issues, environment). This tool is mainly a good support so start discussions and thought processes on a farm. This process was first implemented with a group of dairy farmers from Hauts de France, then nationally widened with farmers from Normandy, Rhône-Alps and Brittany, to make sure that the steps proposed could fit with various production backgrounds. The second stage consisted in identifying practices considered as resilient for each of the 5 topics, and to share them thanks to video testimonies. These practices were gathered on French farms but also abroad, thanks to the exchange trips organized during the project. These exchanges of innovative best practices are now going on within Resilience4Dairy EU project. The dissemination process about the notion of resilience has also been realized in agricultural schools towards the dairy farmers of the future.

Session 44 — Poster 12

Local breed as an alternative to Holstein-Friesian cows in a farm with low level of milk production

M. Sobczuk-Szul, Z. Nogalski, M. Momot and P. Pogorzelska-Przybyłek
University of Warmia and Mazury in Olsztyn, Department of Animal Nutrition, Feed Science and Cattle Breeding, Oczapowskiego 5, 10-719 Olsztyn, Poland; monika.sobczuk@uwm.edu.pl

Local breeds becomes an interesting option for livestock breeding especially for small producers and in organic systems. Developing local breeds usually requires added value, which could be for example health – promoting ingredients in milk. Although, it is important to generate new approaches to milk production from these breeds and determine the quality of milk from farms with reduced supplementation with concentrates. This study aimed to evaluate the milk quality of black-and white and Holstein Friesian cows under system with reduced use of concentrate, simulating organic production conditions. The experimental material were milk samples collected from 32 cows – 16 of each breed. Cows were in the same barn and similar age and calving date. Milk samples were collected twice in winter season. Average milk production level from this farm was 6,500 kg. Local breed cows had lower milk yield compared to the Holstein-Friesian cows. There were no differences between the content of basic ingredients in milk and between breeds, which may indicate a good use of fodder by black-and-white cows. However the milk produced by local breed proved to be a more valuable source of compounds such as desirable fatty acids or proteins than that of the Polish Holstein-Friesian breed. The results of this study are pave the way for future research, within the pasture feeding period. Project financially supported by the Minister of Education and Science under the program entitled 'Regional Initiative of Excellence' for the years 2019-2023, Project No. 010/RID/2018/19, amount of funding 12.000.000 PLN.

Session 45 — Theatre 1

Product quality as a lever to change farming practices to meet society's expectations

V. Thenard[1], S. Couvreur[2], L. Fortun-Lamothe[3], B. Méda[4] and T. Petit[2]
[1]INRAE, UMR AGIR, centre INRAE Occitanie Tulouse, 31327 Castanet-Tolosan, France, [2]ESA, URC URSE, ESA Angers, 49007 Angers, France, [3]INRAE, UMR GenPhySe, centre INRAE Occitanie Toulouse, 31327 Castanet-Tolosan, France, [4]INRAE, UMR BOA, L'Ofrasière, 37380 Nouzilly, France; vincent.thenard@inrae.fr

Product quality has long been the focus of taste improvement in livestock products. Today, new approaches induced by stakeholders aim to introduce new practices into specifications that meet society's expectations. These changes may emerge at different levels of scale and initiative (private operators, groups of players or entire sector). Based on examples, we show how the evolution of specifications or new approaches meet expectations of society and how practices are being modified. The PDO-PGI labels increase the exigence about environmental issues and local feed use by animals. In milk sector, the cheese production now involves improving the sustainability of farming practices. A commercial approach with the BBC brand aims at modifying the animals' diet to improve the nutritional aspect, mainly by increasing grass for ruminants, or using linseed for other animals. For poultry, the 'European Chicken Commitment' aims to improve welfare of conventional chicken production. This initiative focuses on several farming practices to improve animal welfare, and many operators have agreed to evolve them by 2026. In the rabbit sector, cooperatives and the largest French rabbit slaughterhouse has developed a new system for raising rabbits in large pens on the ground ('Lapin et Bien' brand). The aim was to take better account of animal welfare, and opposition with cage farming. In the beef sector, a new demand for red, tender and marbled meat is born especially in restaurants and butchers. To date, this type of meat is mainly imported. A wholesale butcher company has developed an integrated local chain with cattle farmers to adapt breeds, and practices to improve sustainability. These approaches are evaluated with criteria to asses sustainable livestock farming for tomorrow (feed-food competition, climate change mitigation, nutritional intake and animal welfare). In addition to the co-construction of the technical system between the stakeholders, the approach is also based on shared governance, and a better price for the farmers.

Session 45 Theatre 2

Transformation in the dairy sector: a global analysis of sustainability certification standards

K. McGarr-O'Brien[1,2], J. Herron[1], L. Shalloo[1], I.J.M. De Boer[2] and E.M. De Olde[2]
[1]Teagasc, Animal and Grassland Research and Innovation Centre, Moorepark West, Fermoy, County Cork, P61 P302, Ireland, [2]Wageningen University & Research, Animal Production Systems, P.O. Box 338, 6700 AH Wageningen, the Netherlands; keeley.mcgarr@wur.nl

In the transformation of the livestock sector, private certification standards are increasingly used to meet demand for sustainable dairy production. Research into these standards is, however, lacking. In this study we characterized sustainability certification standards currently used in dairy production globally. A literature search for dairy sustainability initiatives yielded 116 results. Based on selection criteria, 19 of these initiatives qualified as 'sustainability certification standards'. We analysed the 19 standards from three perspectives: (1) the general characteristics; (2) the sustainability themes addressed within each standard; and (3) and the credibility, accessibility, and continuous improvement (also referred to as the 'devil's triangle reflecting the trade-offs in standards). Examining general characteristics of the 19 standards revealed variation in governance, verification processes as well as the demands the standards place on farmers. The variation was recognized also in the sustainability focus between standards, and comprehensiveness of sustainability focus. The most frequently and comprehensively addressed sustainability pillar is the environmental; the least frequently and comprehensively addressed is the economic pillar. The accessibility and credibility of standards are most clearly described and declared within most standards' documents. Whereas continuous improvement is an infrequent area of focus for standards. The variability revealed by this study demonstrates the challenge of using certification standards as proof of sustainability. This could negatively impact upon consumer trust in sustainability certification, as well as impairing the potential use of certification standards as proof of sustainable development in the dairy sector.

Session 45 Theatre 3

The analysis of a co-design process to develop an eco-citizen dairy cattle farming system experiment

J.E. Duval[1], M. Taverne[1], M. Bouchon[2] and D. Pomiès[3]
[1]INRAE, Université Clermont Auvergne, AgroParisTech, INRAE, Vetagro Sup, UMR 1273 Territoires, site de Theix, 63122 Saint-Genes-Champanelle, France, [2]INRAE, UE 1414 Herbipole, site de Theix, 63122 Saint-Genes-Champanelle, France, [3]Université Clermont Auvergne, INRAE, VetAgro Sup, UMR Herbivores, site de Theix, 63122 Saint-Genes-Champanelle, France; julie.duval@inrae.fr

The implementation of step-by-step design approaches of farming systems experiments could be a promising way to produce knowledge useful for the transition to new systems. Moreover, it is hypothesized that innovation towards sustainable farming can benefit from combining scientific and expert knowledge. However, co-innovation research processes remain often abstract concepts without a shared understanding of how to carry it out. The objectives of this study are to characterize the step-by-step co-design process of a farming system experiment. The interactions during the first 2 years of the co-design process aiming at designing an 'eco-citizen' dairy cattle farming system were analysed. The process involved livestock farmers, actors of the local dairy value chain and scientists of the public sector. By analysing the co-design process, we show that the format of the co-design activities strongly affected participation rates. Different formats allowed different frequencies of interactions. The experiment leaders rapidly focused the co-design process on the design object 'farming system' and on operational practices (e.g. ideas on practices to manage animal health). However, participants did not always limit themselves the design object or the conception levels by the leaders targeted. Participants also targeted, for example 'the way the group works together during the co-design process' or focused the attention again on the clarification of the objectives. In addition, participants carried out more collaborative design activities than asked from them, by participating in cognitive synchronisation activities for example. These results provide certain elements to understand how co-innovation processes can be enacted, and characterize challenges and possible lessons learned for future co-design experiences.

Session 45 Theatre 4

Perceived quality of meat products in short circuits by producers and their customers

C. Couzy[1], G. Haj Chahine[1], V. Diot[2], S. Masselin-Sylvin[1], S. Meurisse[1], M. Klingler[1] and C. Bièche-Terrier[1]
[1]Idele – Institut de l'Elevage, 149 rue de Bercy, 75012 Paris, France, [2]IFIP, La motte au Vicomte, 35651 Le Rheu, France; ghida.haj-chahine@idele.fr

Within the framework of the VICTOR project, 40 cattle and pig farmers working in short circuits, 77 consumer clients and 9 managers of commercial or collective catering services were questioned about the quality of meat and meat products during in-depth interviews in four different regions in France. For the farmers, the organoleptic quality of their products is essential. Moreover, they strongly emphasize the interest of their farming practices and the services they offer to their customers. It is essential for them to make the link between their farm and the product, to distinguish themselves from the industry by showing the artisanal aspect of their products. The 77 consumers indicated that they were happy with the varied offer. The relationship of trust they have with the farmers, with whom they talk directly and whose animals they can see, is a major factor in their positive perception of product quality. Many elements of the farmers' discourse are also found among their costumers' answers. The geographical proximity of the farmer is a determining factor for the consumer and the more functional role of short circuits was partly observed in two of the study territories with low population density and more distant commercial basins. Regarding the chefs in the catering industry, they have difficulty finding farmers who can meet their requirements, particularly in terms of securing supply volumes. They and their customers are satisfied with the quality of their products and globally there is less waste. Around one hundred online surveys, conducted throughout France, broaden the panel of farmers' answers. They show that farmers working in short circuits are most concerned by controlling production and manufacturing hygiene, and by guaranteeing their products' origin. They do not feel completely at ease with the legal constraints of the activity such as regulations, labelling rules, setting products' shelf life, etc. For farmers, the short circuit activity requires a lot of manpower. They often call upon service providers, employees, or volunteers for different tasks such as cutting, processing, packaging, delivering, and selling.

Session 45 Theatre 5

Value-adding attributes for dairy calves within local beef sector – perceptions among stakeholders

L. Schönfeldt[1,2], M.G.G. Chagunda[1] and N. Ströbele-Benschop[2]
[1]University of Hohenheim, Animal Breeding and Husbandry in the Tropics and Subtropics, Garbenstr. 17, 70599 Stuttgart, Germany, [2]University of Hohenheim, Applied Nutritional Psychology, Fruwirthstr. 12, 70593 Stuttgart, Germany; lea.schoenfeldt@uni-hohenheim.de

Value-added inclusion into the local beef sector is a strategy to improve ethical and economic appreciation for surplus dairy calves. This study aimed to investigate the role of animal welfare and sustainability, on improving dairy calf's marketability within a potential dairy beef value chain. For successful inclusion, awareness of the dairy calf issue, relevance, and willingness to pay for value-adding attributes need to be better understood at both ends of the value chain. A survey on perceptions of calf welfare and sustainability within dairy production was carried out in the state of Baden-Wuerttemberg, Germany. Data from 127 dairy farmers and 288 consumers on relevance of selected attributes for calf welfare and sustainability were captured using rating scales (mean ± standard deviation). Participants rated physical calf health (1.18±0.55) as most important attribute, followed by a fair compensation for farmer's work (1.26±0.63), calf's space requirement (1.29±0.63), regional food production (1.52±0.91) and avoidance of long-distance transport (1.61±0.97). Contrary to consumers, farmers rated attributes of calf welfare (2.43±0.84) as less important compared to sustainability attributes (1.63±0.66). Dietary habits and awareness of the dairy calf issue influenced relevance of certain calf welfare attributes positively. Practicing organic dairy farming was positively correlated with relevance of sustainability and welfare attributes within the farmers' group. Consumers declared highest willingness to pay a premium for milk (0.43€/litre) and beef (2.63€/kg) derived from cattle, raised under increased welfare standards. This study indicated consensus among dairy farmers and consumers regarding existing importance of the calf's wellbeing and aspects of sustainability within dairy production. Dietary habits and individual understanding of welfare and sustainability might shape farmers and consumers perception. Stakeholder analyses, investigating the feasibility of the inclusion into the local beef sector are proposed throughout the whole dairy beef value chain.

Session 45 Theatre 6

Farmpédia: how to improve the global acceptability of livestock systems with communication

A.-L.J. Thadee and G. Brunschwig
VetAgro Sup, UMR Herbivores, 89 avenue de l'Europe, 63370 Lempdes, France; anne-laure.thadee@vetagro-sup.fr

Farm and animal products and technics have in common to be both used and unknow from the vast majority of citizen. Life in the city, far from the countryside, cuts children off from the rudiments of nature. High school and college teachers often neglect those topics. This situation is all the more marked as the information available to young citizens comes first from television, second from their entourage and third from school. We have therefore chosen to produce resources for teachers (Farmpédia) and short videos (2-5 mn) for young people. Farmpédia is a multimedia encyclopaedia, designed by researcher to tackle this issue. It offers a series of thematic chapters, enriched with data, from science article to press article, videos and real-life testimonials. We start with a French version in 2020 and develop an English version in 2023. To interest a younger audience, and embody the principles of ranching, we have decided to offer a series of experts to answer societal questions, in addition to offering clear definitions of certain points of jargon. At this point, the results are a better understanding of some of the key point of ranching for the people in reach of our media. The work began by defining the questions to be asked of the specialists, which correspond to questions of interest to non-specialist citizens. On the next step we have to record the researchers' answers and illustrate them with images to propose a fun and instructive video. We would like to expand the social diffusion of our work, without falling into cheap advertising like buzz or clickbaits. To enlarge our audience and get theses video to international levels, we add subtitles in English and also produce video in English with French subtitles. The motivation for this action is based on providing information adapted to the citizen, which seems essential to strengthen the link between society and the actors of animal production. We consider that this project requires the participation of researchers and we hope that Farmpedia and our videos will contribute adequately.

Session 45 Theatre 7

How and why involve citizens in a participatory research project aiming to design livestock farming?

P. Coeugnet[1], J. Labatut[2], G. Vourc'h[1] and J.E. Duval[1]
[1]INRAE, Site de Theix, 63122, Saint Genès Champanelle, France, [2]INRAE, 5 Boulevard Descartes, 77454, Marne-la-Vallée, France; philippine.coeugnet@inrae.fr

Various studies have highlighted the interest of collaboration between researchers and citizens in the same project in the light of responsible research and innovation. This has led to an increase of participatory research in different sectors. In agriculture, if many cases of collaboration between agricultural professionals and researchers exist, there are few cases of participation of citizens, within participatory research projects. Citizens are defined as actors who are neither researchers nor agricultural professionals. However, the agricultural sector, and in particular the livestock sector, is the object of various criticisms from society. The participation of citizens in a participatory research project on the design of future livestock systems could be a way of recreating a connection between livestock and society, provided that appropriate participatory methods are mobilized. We set up an innovative co-design process aiming to include citizens at the same level as researchers and livestock professionals within a participatory research project aimed at producing knowledge and solutions for future livestock systems in the French mountain area 'Massif Central'. We adapted the method DKCP (Diagnosis, Knowledge, Concept, Proposal) that was initially developed to support a co-design process between experts within the industrial sector to our study context. The 22 participants were invited to work on the topic of 'the future of dairy calves' and were asked to design dairy farming systems that respect the health and welfare of calves, farmers and society. We analysed the video recordings of the exchanges between the participants and conducted semi-structured research interviews with participants. Our results show that the participation of citizens within this co-design process has made it possible to integrate different types of knowledge, to explore innovations, to democratize science and to generate different types of learning. For this, rigorous participatory methods are necessary in order to ensure the effective participation of citizens within a group of actors with heterogeneous expectations and knowledge.

Social sustainability: what concepts to approach farmers' satisfaction at work

B. Dedieu[1], J. Duval[1], P. Girard[2], N. Hostiou[1], S. Mercandalli[2] and G. Soullier[2]
[1]INRAE, ACT, Theix, 63122 Saint-Genes-Champanelle, France, [2]Cirad, E&S ART-Dev, 73 rue Jean-Francois Breton, 34000 Montpellier Cedex, France; benoit.dedieu@inrae.fr

The social pillar of sustainability is, compared to the economic and environmental ones, considered less often in the existing models used to study livestock production systems. With a generation of farmers soon to retire (at least in the EU), the farming sector – and notably the livestock one – is facing the challenge to ensure that a new generation of farmers emerges despite working conditions considered as difficult and unattractive. In Sub-Saharan Africa, agricultural activities are often labour-intensive but generate little value. The attractiveness of the agricultural sector is also a challenge to absorb a growing rural population. We deepen here one domain of social sustainability, referring to farmers' satisfaction at work. Several concepts exist to capture what 'satisfaction' means for the farmers: decent work, job satisfaction, working conditions, working rationalities, farmers' well-being, quality of life at work, etc. Building on a comparative review of these concepts and their definitions, this communication aims to explicit them in order to highlight their common and divergent topics of interest, their underlying frameworks, methodologies and their set of indicators either based on individual factors or on factors related to the nature of the job and its environment. We then analyse: (1) the way they are connected to farming practices and farming styles (agroecology, sustainable intensification, etc.); (2) to what definition of 'work' they refer (a set of tasks; an occupation with its professional norms; a job); (3) how they incorporate gender and age-related issues as well as local, cultural dimensions; (4) what is the balance between objective and subjective indicators. We present some of our operational applications of 'satisfaction at work' approaches in both Northern and Southern contexts. We open the debate to the consideration of other types of workers (family members, wage-earners, volunteers). We conclude by a reflexion toward a shared vision of the future of livestock farming combining high economic and environmental performances, fulfilness and attractivity for the persons.

Implementing agroecological practices: what are the effects on working conditions of dairy farmers?

A.-L. Jacquot[1], M. Gérard[1,2], J.E. Duval[3] and N. Hostiou[3]
[1]PEGASE, INRAE, Institut Agro, 16 rue le Clos, 35590 Saint Gilles, France, [2]Smart-Lereco, INRAE, Institut Agro, 65 rue de Saint-Brieuc, 35000 Rennes, France, [3]Université Clermont Auvergne, INRAE, AgroParisTech, VetAgro Sup, UMR Territoires, campus des cézeaux, 63178 Aubière, France; anne-lise.jacquot@agrocampus-ouest.fr

To limit their impacts on the environment, farmers are encouraged to adopt agroecological practices, which can affect their working conditions. This study aims to explore effects of the adoption of such practices on farmers' working conditions. During fall and winter 2019-2020, 17 dairy farmers located in Western France were individually surveyed. The interview guide was designed to explore the relationships between the adopted agroecological practices and the effects on farmers working conditions. Dairy farmers were invited to make a list of their agroecological practices adopted and to describe how they experienced their working conditions. The latter were analysed with a broad framework allowing to depict farmers' workload and work organisation, and to understand farmer's perceptions of their job (physical and mental workload, skills, social relations, sense of coherence, etc.). More than thirty agroecological practices were identified concerning the crop, livestock system or landscape infrastructures. All farmers declared an effect of those practices on their working conditions. They claimed various impacts on workload, work organization and the need for special equipment, depending on the nature of the production system and the applied agroecological practices, mainly through: (1) a crop diversification or an increase in the share of grasslands; (2) change of technical operations on the cropping system. Farmers report a greater seasonality of tasks to perform. More generally, they all expressed a positive effect due to an improvement of the physical workload, their skills and relationship with the society, and their job satisfaction. This survey highlighted different effects of the adoption of agroecological practices on farmer's working conditions that may help to design an appropriate support for farmers engaging in the agroecological transition. This study is part of the LIFT project funded from the European Union's Horizon 2020 research and innovation programme (No 770747).

Session 45 Theatre 10

Promoting and guiding transformation of French veal calf farms in response to societal expectations

C. Martineau[1], D. Bastien[1], M. Chanteperdrix[1], C. Denoyelle[1], V. Lefoul[1] and M. Orlianges[2]
[1]Institut de l'Elevage, Qualité des carcasses et des viandes, 8 route de Monvoisin, station expérimentale veau de boucherie, 35650 Le Rheu, France, [2]INTERBEV, 207, rue de Bercy, 75587 PARIS Cedex 12, France; virginie.lefoul@idele.fr

Production and consumption of veal calves is a French specificity. France is the largest consumer in the world of veal calves, with 3.2 kgce per capita. In 2022, approximately 1.1 million calves were slaughtered in France, which is the second largest producer in the world behind the Netherlands. More than 60% of male calves from the French dairy farms are used as veal calves, which plays a major part in regulating the milk and bovine-meat markets. The French veal sector is coveted by countries that have implemented policies of eliminating dairy calves at birth due to the lack of solutions that allow a better valorisation. In this context, the French veal sector must adapt to confront new society's challenges such as environmental protection, animal welfare or animal health. How can the veal production systems converge to meet these societal expectations while preserving the working conditions and a fair remuneration of farmers? INTERBEV Veaux and IDELE proposed to answer these questions through a collective project called 'Le Veau Durable' (= Sustainable Calves) and composed of 3 parts: (1) a research programme to evaluate innovative calves' production methods that meet these societal expectations; (2) the construction of a new collective calves' innovation and research centre (CIRVEAU), to produce references on these innovative breeding methods; (3) a demonstration and communication programme to spread and to promote the results to professionals in the sector, students, and the scientific community. This collective project aims to guide the French veal industry in its reorganisation for the next 20 years by accompanying the transformation of its breeding practices. It involves IDELE as a research and development organisation and INTERBEV Veaux as an interprofessional organisation that federates all the links in the French veal sector, from upstream to downstream.

Session 45 Theatre 11

Designing rabbit breeding systems with access to the outdoors with the innovative design method

L. Fortun-Lamothe[1], M.H. Jeuffroy[2] and L. Le Du[2]
[1]GenPhySE, INRAE, 24 Chemin de Borde Rouge, 31326 Castanet-Tolosan, France, [2]AgroParisTech-Innovation, INRAE, IDEAS, CS 20040, 91123 Palaiseau, France; laurence.lamothe@inrae.fr

Today, most rabbit farming (>90%) is carried out in confinement in small unenriched cages, which is inconsistent with animal welfare. Allowing animals access to the outdoors meets a pressing societal demand solving this problem, but requires breakthrough innovations. To this end, we implemented an innovative design process, using the KCP® method. Two design workshops brought together 14 people: veterinarians, advisors, livestock professionals including farmers, and researchers. Each of the workshops was organised in 3 phases: a knowledge sharing, an exploration phase and a debriefing. By exploring the concept of the 'rabbit's health paddock', the first workshop showed that: (1) it is necessary to design several systems to take into account the diversity of breeders and consumers; (2) it is possible to design a win-win system for animals and breeders; (3) there is a strong need to design systems that are attractive to young breeders; (4) it is necessary to have adaptive management rules (weather, feed resources, state of the rabbits); (5) professional skills will have to be renewed to deal with new uncertainties; (6) the management of these innovative systems requires the evolution of performance indicators that are currently dominant in the sector. These are new properties for rabbit farms. By exploring the two concepts of 'adaptive, efficient and welfare-friendly rabbit farming' and 'attractive and renewed rabbit farming', the second workshop aimed to generate coherent combinations of practices. This workshop allowed us to outline 5 coherent contrasting rabbit farming systems, based on different drivers: flexible, integrated into the territory, multi-skilled and enhancing natural resources, adapted to a diversity of markets, extensive and natural. They share common features (e.g. more robust genetics, flexible breeding management) but also have specificities (e.g. living environment, feeding). These results confirm that innovative design method fosters collective creativity and open innovation processes. The results will be used to design a serious game that aims to support the agroecological transition of rabbit farms.

Session 45 Theatre 12

Challenges and opportunities for transitioning to 'low anthelmintics use' in livestock systems

M. Sautier[1] and P. Chiron[2]
[1]INRAE, GenPhySE, Chemin de Borderouge, 31320, France, [2]Univ. Grenoble Alpes, CNRS, School of Political Studies, Pacte, Grenoble, 38000 Grenoble, France; marion.sautier@inrae.fr

Today's level of anthelmintic use in pasture-based livestock production is a major threat to the environment and the livestock industry. In this context, the research community is looking for alternative ways to equip farmers with preventive and treatment strategies that could help decrease livestock-industry dependence on anthelmintics and have beneficial impacts on ecosystems. Production practices for a sustainable control of parasites have been advocated for almost forty years, but farmers' uptake of these practices has been too slow to address the issues at stake. In this presentation, we examine the rationales behind the under-adoption of alternative worm control practices in grassland-based livestock systems. This research builds on 34 semi-structured interviews with dairy sheep farmers, veterinarians, and agricultural extension agents in southwestern France. The interview material was analysed via qualitative discourse analysis. We highlight farmers' representations and rationales underpinning adoption or non-adoption of the 'low anthelmintics use' innovation. We identify six profiles for nematode control according to the way each farmer included treatment and coprology in their on-farm practice. We analyse the divergent set of ideals and representations around animal health management found among the farmer population. We identify socio-technical barriers and opportunities for transforming livestock systems towards lower anthelmintic use. We then articulate our results around the conclusion that the 'low anthelmintics use' innovation has low potential for adoption, and discuss ways to facilitate transformations, such as increased communication, training and farm visits involving farmers, extension agents and veterinarians.

Session 45 Theatre 13

The rearing of calves, kids and lambs with adults in dairy systems in the AuRA region, France

C. Constancis[1], A. Igier[1] and F. Debrez[2]
[1]FiBL France, 150 avenue de Judée, 26400 Eurre, France, [2]INRAE, GenPhySE, 24 Chemin de Borde Rouge, 31326 Auzeville Tolosane, France; caroline.constancis@fibl.org

Animal welfare is at the heart of concerns about the future of farming. In particular, mother and offspring separation from birth in dairy systems is increasingly criticised by society for many reasons. However, some innovative alternative practices to this separation are emerging in France but are still not well known. They appear to break with dominant practices. The study aims to describe the rearing calves, kids, lambs under adults (dam or foster adult) practice in the Auvergne-Rhône-Alpes region (France). To do so, 45 farmers who have implemented such rearing systems have been interviewed (15 famers by species). Using a semi-structured interview guide in order to analyse (1) its implementation (discovery, motivations and challenges encountered), (2) its day-to-day management and (3) the benefits and constraints of the practice for the animals, the farmer and finally the farming system. The interviews, lasting 2 hours on average, has been recorded and transcribed. The collected data has been analysed by Excel to establish typical cases of farmers and practices. Principal component analyses has been carried out using the R software. Our first results show that practices are very diverse and adapted to the constraints of each dairy sector and farm. The rearing with a foster adult is more common in cows than in small ruminants. In the Auvergne-Rhône-Alpes region, foster cow rearing seemed to be more similar to the dam rearing in the milking herd (foster cows are both suckled and milked) than the foster cow rearing in the north-western France where foster cows are no milked and separated to the milking herd. In general, farmers were satisfied and consider that these rearing systems improve animal welfare and meet societal expectations. They are more or less profitable and takes more or less time depending on the managements and species. Finally this practice, which is still confidential, deserves to be known and studied according to the farmers because of the benefits it brings, both for the animal health and welfare.

Session 45 Theatre 14

Expectations of French suckler farmers in terms of technical support services

A. Antoni-Gautier[1], J. Chambeaud[2], T. Falcou[3], C. Galvagnon[4], N. Lemonnier[5] and J.B. Menassol[2]
[1]L'Institut Agro Dijon, 26 bd Docteur Petitjean, BP 87999, 21079 Dijon Cedex, France, [2]L'Institut Agro Montpellier, 2 place Pierre Viala, 34060 Montpellier, France, [3]VetAgro Sup, 89 Avenue de l'Europe, 63370 Lempdes, France, [4]Bordeaux Sciences Agro, 1 cours du Général De Gaulle, 33175 Gradignan Cedex, France, [5]L'Institut Agro Rennes-Angers, 65 rue de Saint-Brieuc, CS 84215, 35042 Rennes, France; jean-baptiste.menassol@supagro.fr

In France, support and advice for beef farms have long been used to improve both the technical and economic performances of livestock, while at the same time meeting the needs of the livestock stakeholders and consumers. But these services are increasingly questioned, with a percentage of breeder's members of support structures in constant decline for several years. Consequently, two support structures, Races de France and Eliance, commissioned students from 5 agricultural engineering schools to better understand this evolution. A survey was conducted from October 2022 to February 2023 for which 230 suckler farmers were contacted (whatever the breed) and whether or not they are members of these structures. The final sample consisted of 137 farmers (i.e. a positive response rate close to 60%) that were surveyed using semi-structured interviews, resulting in 134 usable interviews. The analyses revealed that the main reasons for not joining the program were lack of interest (59/106 answers), the cost of the services (57/106) or the time required to prepare the animals for the test (51/106). When questioned about their expectations for such services, breeders mostly asked for advice on specific subjects (legislation changes, evolution of practices, future prospects, valorisation of livestock products, etc.) (40/121 answers) and specific interventions on aspects related to herd management (33/121). A reorganization of the service is also expected by the farmers: more visits, better communication, etc. It is concluded that to help beef farmers and the suckling sector to face new challenges such as climate change, societal evolution and a fluctuating economic context, the consulting offer must evolve and go beyond technical and/or scientific frameworks, by enriching itself with new skills and approaches.

Session 45 Poster 15

'Antibiotic-free' strategies in chicken production in France: success factors, assets and limitation

N. Rousset and J. Hercule
TAVI, 7 rue du faubourg Poissonnière, 75009 PARIS, France; rousset@itavi.asso.fr

In the early 2010s, in France, different claims 'raised without antibiotic treatment' have appeared for chicken meat, in a context of changing regulatory framework, and the steering of a national program to reduce use by the Ministry of Agriculture. These claims are based on different poultry production sectors that have implemented strategies to reduce the use of antibiotics. Interviews with different actors of the poultry sector were conducted, to understand how these private standards were built, in the particular case of standard/certified broiler production, how they work, and to analyse their success factors, their assets and limits. Although limited to a part of the production, these strategies seem to have benefited the whole poultry production. In fact, training, technical and moral support was carried out on the long term, for all farmers, in order to improve their technical skills. The information collected also emphasizes the importance of formalizing objectives and the commitment of all actors of the sector for the success of these strategies. Nevertheless, 'zero antibiotics' can be questioned considering its limitations in terms of animal welfare and its relevance to the fight against antibioresistance. The economic equilibrium of these sectors, which are positioned on products that are affordable for consumers, is also questioned, as the revaluation of additional production costs remains limited. An evolution towards a more systemic approach seems to be an avenue to be explored, going beyond the unique indicators of care, to take into account the overall health of the flock.

Session 45 Poster 16

Veal calves' production: what societal expectations in terms of animal housing for their welfare?

D. Bastien[1], M. Tourtier[1] and A. Warin[2]
[1]Institut de l'Elevage, Qualité des carcasses et des viandes, 8, route de Monvoisin, 35650 Le Rheu, France, [2]Bureau BANKIVA, 4 impasse de la fin de chêne, 21410 Gergueil, France; didier.bastien@idele.fr

Calf housing is at the heart of societal and regulatory issues. In this context, the 'RenouVeau' project, funded by the Ministry of Agriculture, was launched to identify main societal expectations for this production. Three types of audience were questioned: (1) consumers through an online survey; (2) welfarist NGOs through their public statements; and (3) students in agricultural schools through an idea competition. 26 student groups (100 students in 11 schools) responded to the idea competition and described their vision for calf breeding in the future. The top issue was the possibility of opening the buildings, with 22 projects, 14 of which provided outdoor access for calves. Projects also highlighted breeding on litter (21 projects), increasing the space allowance per calf (12 projects), providing natural light (7 projects), environmental enrichment (7 projects) and hay in feed (8 projects). The main claims of French welfarist NGOs (apart from specific problems linked to transport and slaughter) mainly concern the fact that the animals are kept in individual housing for less than 8 weeks, the absence of litter on the floor and unsuitable feed (insufficiently rich in iron to meet the market demand for white meat). Almost 2,600 consumers responded to the online survey. 58% of them admitted that they did not know how veal calves are breed. Those who said they know this production give an average score for its image of 5.5 out of 10 (from the most negative to the most positive). Among the negative's points, the notions of claustration, animal welfare and intensive breeding were mentioned. For housing, 83% of consumers want calves to graze in summer and 65% want them to have a farm building or shelter in winter, but with outdoor access. 49% of consumers want a production on litter floor. In conclusion, the main expectations for veal calves are therefore more outdoor access. In addition, societal expectations also relate to group housing of calves from their arrival to fattening farms, with more space, as well as the provision of litter for bedding and a more iron-rich diet.

Session 45 Poster 17

Salmon welfare perceived by various stakeholders: where do we stand?

C.M. Monestier, L.R.B. Reverchon-Billot, M.S. Stomp and A.W. Warin
Bureau Bankiva, Department of Research & Development, 4 impasse de la fin de chêne, 21410 Gergueil, France; aurelia.warin@bankiva.fr

Salmon production has been the subject of heated debate, in which current controversies focus on the need to take animal welfare into account, notably supported by the demonstration of sensory capacities (such as pain or positive emotions) and cognitive abilities (such as learning or personalities). In this light, we drew up an inventory of current scientific findings and requirements on the basis of a literature review covering 317 scientific papers, analysing the expectations of 7 animal welfare NGOs, and screening the standards of 10 labels encompassing salmon welfare, implemented in a wide array of countries worldwide. The resulting critical synthesis aims at assessing the headway made, highlighting the shortcomings on some issues, and charting avenues worth exploring. Most label specifications and animal welfare NGOs thus define clear requirements regarding the handling of fish out of water and the most controversial stunning and killing methods, for which the scientific literature details the negative impacts on salmon. However, even though NGOs and some labels (such as 'Label Rouge', the European Organic Farming label, Demeter, RSPCA and Friend of the sea) limit the duration of fasting before slaughter to a few days depending on water temperature, little research has been carried out on the issue, only drawing partial conclusions. Conversely, although various environmental enrichments have shown positive effects on several species of marine fish, NGOs and labels provide little or no guidelines in this area. Such cross-analyses prove to be crucial to provide optimum and joint guidance to both producers and policy-makers.

Session 45 Poster 18

A review of sustainable livestock strategies to deal with anti-livestock activists

M. Dehghani and K.H. Park
Kangwon National University, College of Animal Life Sciences, Kangwon National University, Chuncheon, Gangwon, Republic of Korea, 24341, Korea, South; kpark74@kangwon.ac.kr

The livestock industry plays an important role in culture, societal well-being, livelihoods, and food security. According to the Food and Agriculture Organization (FAO), this industry supports about 1.3 billion livelihoods and food security by accounting for 40% of the global value of agricultural output in developed countries and 20% in developing ones. During the last decades, rising population, income, and urbanization have all been associated with increasing demand for food products derived from livestock, which has led to increased livestock output and population. Since livestock activities have significant impacts on all aspects of the environment, including air and climate change, land and soil, water and biodiversity, this growth in animal products demand has raised concerns of anti-livestock activists who are against activities of livestock industry without considering the vital role of it in humans life. So far, various solutions have been provided to mitigate and control the destructive effects of this industry, which can promote it towards sustainable development goals such as reduction of enteric fermentation in ruminants through dietary manipulation and improving manure management through acidification. Concerning to sustainability strategies, one of the most effective solutions is circular bio-economy, which has been discussed in this study, and from the results obtained, it can be concluded that activity of the livestock industry is not only an inseparable part of human life, but also for the continuation of other industries it is absolutely vital.

Session 45 Poster 19

AuthenBeef: use of blockchain technology in beef production to secure authenticity and traceability

G. Arsenos[1], S. Vouraki[1], V. Papanikolopoulou[1], A. Argyriadou[1], V. Fotiadou[1], S. Minoudi[2,3], D. Karaouglanis[2,3], N. Karaiskou[2,3], P. Fortomaris[1] and A. Triantafyllidis[2,3]
[1]School of Veterinary Medicine, Aristotle University, University Campus, 54124 Thessaloniki, Greece, [2]School of Biology, Aristotle University, University Campus, 54124 Thessaloniki, Greece, [3]Epigenomics Translational Research, Center for Interdisciplinary Research and Innovation, Balkan Center, 57001 Thessaloniki, Greece; svouraki@vet.auth.gr

The traceability and authenticity system of livestock products is becoming increasingly important from a political, consumer and environmental perspective, promoting the concept of 'farm to fork'. Research has shown that the main attributes consumers value in addition to quality when buying meat are provenance, traceability and quality. The objective of AuthenBeef is to implement a fully traceable beef supply chain using blockchain technology combined with genetic markers both up the chain to market and back down the chain to farm. The key data entry points include the entire production process from feedlots to the final marketing of meat and its products. A designated protocol for the genetic analysis of 17 microsatellite markers, recommended by the International Society for Animal Genetics (ISAG), has been developed. In the farm, blood samples will be collected from recently slaughtered animals and the full set of markers will be genotyped, building a genetic database of the farm. To test the genetic tracing, system meat samples will be collected, at the selling point stage. Individual identification will depend on the obtained genetic profile for each meat sample. To cater the needs of the above protocols, a recording system, applicable to the beef sector, has been developed providing accurate and traceable data of the whole processes. Moreover, to harness blockchain technologies and to allow verification of data a decoding system using QR-coding has been developed to provide real time information to consumers. AuthenBeef offers a marketing strategy making the most of the provenance and traceability of beef and hence ensuring greater financial returns in the value chain. This research has been co-financed by the European Regional Development Fund of the European Union and Greek National Funds through the Operational Program Central Macedonia 2014-2020 (KMP6-0219662; AuthenBeef).

Session 45 Poster 20

Structural equation modelling applied to multi-performance objectives in French suckler cattle farms

L. Billaudet, I. Veissier, J.J. Minviel and P. Veysset
INRAE, UMR Herbivores, Route de Theix, 63122 Saint-Genes-Champanelle, France; larissa.billaudet@inrae.fr

Faced with increasing input prices coupled with consumers' concern about purchasing power, farmers have to produce goods at affordable prices while ensuring the survival of their farm. Society has in addition many other expectations of livestock farming beside mere production: improve animal welfare, preserve biodiversity, reduce greenhouse gases, etc. In the literature, clear references are lacking to guide farmers on how to manage these multi-performance goals simultaneously. The objective of this work is to give insight on the relationship between the three dimensions – animal welfare, economic performance, and environmental performance. To address this question, we sought to analyse which decision strategies – defined as a combination of practices and rearing conditions – lead to synergies or, on the contrary, to antagonisms between relevant indicators for the three dimensions. Structural equation modelling (SEM) was used as an analytical framework. The resulting model relates latent variables – which are concepts that are not directly observable – to measurable variables – which are directly observable. In our case, the latent variables are the three dimensions under investigation. The measurable variables are either indicators of practices used and rearing conditions offered to animals (input variables) or performance indicators (output variables), derived directly or calculated from a multi-year database, which includes more than 300 suckler cattle farms in mainland France. The specification of the theoretical model, i.e. the description of the relationships between the variables (latent and measurable), was derived from knowledge established in the literature between practices/conditions and each dimension separately. The first results suggest that under certain conditions it is possible to obtain good economic, environmental and animal welfare performances at the same time. The results need to be confirmed by refining the assessment of animal welfare e.g. by using measurements of the Welfare Quality protocol.

Session 45 Poster 21

Systemic enablers and barriers to extending the productive life of Swiss dairy cows

A. Bieber, R. Home, M. Rödiger, R. Eppenstein and M. Walkenhorst
FiBL, Ackerstr. 117, 5070 Frick, Switzerland; anna.bieber@fibl.org

The economically and environmentally optimum length of productive life of European dairy cows, which is commonly described in terms of the number of completed lactations or productive life days (milking days) before a cow is replaced, is between six and seven lactations. However, the current average length in Switzerland, and many other countries, is approximately half that figure, which is both inefficient and ethically questionable. The aim of this study is to investigate systemic barriers to, and enablers for, the extension of the length of productive life of dairy cows in Switzerland. Data were collected by means of 29 qualitative expert interviews. Interviewed experts were selected by purposive sampling to cover a broad range of different stakeholder groups within the Swiss dairy farming system. The analysis of the interviews was done by classifying statements to inductive codes based on their content. The results show that breeder associations encourage extended service life by advertising that longer life increases milk yield per day of life. Information on the benefits of extended service life is readily available, but is not a focus of farmer education so farmers are not motivated to use the available calculation tools. Milk prices are low, so farmers try to maximise production while minimising costs so make breeding decisions based on production rather than robustness. Swiss vets are highly educated and familiar with herd management so could potentially motivate a change in the mindset of dairy farmers. However, vets are expensive so few farmers seek advice from veterinarians on herd management. In conclusion, the existing system is so deeply entrenched that no actors feel they can bring about comprehensive change on their own. Any actor who could initiate change in their part of the system fears that they would then no longer be in harmony with the rest of the system. Therefore, they are committed to behaviours that they know are not optimal and overcoming these lock-ins can only take place slowly. Collaborative reflection at industry level, led by breeder's associations, vets, and advisors, and the further development of the existing decision support tools, may lead to optimisation of the system, and to a longer productive life of dairy cows.

Session 45 Poster 22

Providing outdoor access to pigs: what are the profiles of farmers working in those systems?

S.B. Brajon, C.T. Tallet, E.M. Merlot and V.L. Lollivier
INRAE, Institut Agro Rennes-Angers, PEGASE, Le clos, 35590 Saint-Gilles, France; sophie.brajon@agrocampus-ouest.fr

Societal expectation for the provision of outdoor access to farm animals is growing, mainly for questions of natural behaviour expression. It is thus important to understand the farmers motivational factors to engage in this type of farming in order to increase the job attractiveness and develop strategies to encourage and support farmers in moving towards such systems. This study aimed at identifying the motivational profiles of farmers choosing a trajectory of pig farming with outdoor access. A total of 24 pig farmers providing outdoor access (from indoor with outdoor access to free-range) participated in a semi-structured interview. Questions concerned historical context, farm and practices description and perception of the impact of outdoor access on the farmer, pigs, technico-economic performances, environmental and societal aspects. Qualitative data were analysed using thematic analysis which identified four motivational profiles. Farmers providing 'Outdoor access for the animals' were particularly concerned by the animals' welfare and mainly based their decision to provide an outdoor access on the importance of allowing them to express their natural behaviour and have control over their environment. Farmers providing 'Outdoor access for environmental principles' were motivated by the search of low-input, circularity, and autonomy. Those two profiles included farmers working in full outdoor and indoor farms with outdoor access. Farmers developing 'Outdoor access for economic reasons' were able to seize the opportunity to go toward a differentiated market along with welfare quality scheme to meet societal expectations. This profile was exclusively constituted by farmers managing indoor farms with outdoor access. Finally, the last profile concerned farmers giving 'Outdoor access by tradition' for whom the outdoor access was the result of maintaining traditional familial and regional breeding practices, and this concerned farmers rearing local breeds. The present study highlighted the motivational factors that guided pig farmers in their choice to provide an outdoor access, and the relation to the type of access given.

Session 45 Poster 23

The explosion of sustainability indicators in the European livestock sector

B. Van Der Veeken[1], M. Carozzi[2], C. Barzola Iza[3] and F. Accatino[2]
[1]Science, Management & Innovation, Radboud University, Redboud, 1081 AR, the Netherlands, [2]UMR SADAPT INRAE, AgroParisTech, Université Paris-Saclay, Palaiseau, 91120, France, [3]Wageningen University and Research, Wageningen, 6705, the Netherlands; francesco.accatino@inrae.fr

Sustainability of the European livestock sector needs indicators to be measured. Over the last decades, several studies defined sustainability indicators covering the three different pillars of sustainability, namely the environmental (ENV), economic (ECO) and social (SOC) pillar. However, due to an increasing number and diversity of indicators, it is possible to speak about an 'indicator explosion' and the development of indicators was not balanced over the three pillars. By means of a systematic literature review, we aimed at studying the quantity and diversity of indicators used in each pillar and for different livestock systems defined in the last decades. We started from a search result and, after a systematic screening, we selected 44 studies that defined and measured sustainability indicators in Europe at the farm scale. Indicators used in these studies were harmonized (in case they named differently but describing the same thing) and classified into pillars and themes (sub-pillars). They were counted as total (T) or unique (U) (according to weather the same indicator, used in different studies, was counted multiple times or once), quantitative or qualitative. Out of a total of 736 indicators, most were used in ENV (n=282), followed by SOC (n=271) and ECO (n=183). ENV showed a low diversity of indicators used (U/T=0.52), with the most representing themes being emissions, nutrient balance, and non-renewable resources; ECO showed greater diversity of indicators used (U/T=0.64) but the great majority (65.5%) of them regarded profitability and were mostly quantitative (94%). SOC showed the higher diversity of indicators (U/T=0.72), most of them in the animal welfare theme, with the highest percentage among the pillars of qualitative indicators (50%). The review showed that most indicators (38.9%) were used for dairy cattle systems. This study enhanced that more efforts should be used to expand the study of indicators of positive impacts of the livestock system and to make the social indicators more uniformly assessed.

Session 45 Poster 24

Pig farmers' and citizens' opinions on outdoor access for livestock

S.B. Brajon, C.T. Tallet, E.M. Merlot and V.L. Lollivier
INRAE, Institut Agro Rennes-Angers, PEGASE, Le clos, 35590 Rennes, France; sophie.brajon@agrocampus-ouest.fr

The proportion of French citizens in favour of outdoor access for farm animals is increasing, but farmers are sometimes hesitant or reluctant to do so. Two surveys based on semi-structured interviews were performed among 32 French citizens and 36 pig farmers (who worked in all types of farming systems, from conventional to free-range) (1) to better understand citizens' expectations and specify acceptable practices of outdoor access for livestock and (2) to identify the barriers and motivations of farmers to provide outdoor access to pigs. The main theme addressed with the citizens concerned their perceptions of livestock farming and outdoor access. The main theme addressed with farmers concerned perceptions about the impact of outdoor access for pigs on farmers, animals, technico-economic performance and society. Qualitative data were analysed using thematic analysis. Our study reaffirms that citizens interviewed are generally in favour of outdoor access which is synonymous of freedom in vast fields, although they believe that the ability for animals to choose whether to shelter or go outside is important. For farmers, the coherence of the breeding system with their values is crucial. Farmers providing outdoor access have pleasure to work outside and consider the life conditions better for animals. Most indoor farmers do not imagine breeding pigs outdoor mainly due to their job conception and they often mention the disconnection between citizens perception and farming reality. Despite their lack of knowledge on the topic, citizens were aware about main challenges of outdoor access such as the risk of disease transmission by wildlife, land availability and farmers working conditions. In addition, a main limitation for farmers with outdoor access concern the lack of support and technical references to manage the system and face climatic hazards, but experienced farmers mention that knowledge exist and should be shared. Finally, citizens are not able at characterising practical modalities of outdoor access due to a lack of knowledge of livestock but they trust farmers. Communicating about their work and receiving a positive feedback from citizens is a source of pride for farmers and this could be a lever to promote farms with outdoor access acceptable for society.

Session 45 Poster 25

Short food supply chains and food safety issues: HACCP utilization, opportunities and limits

R. Rafiq[1], O. Boutou[2], M.P. Ellies-Oury[1,3] and B. Grossiord[1]
[1]Bordeaux Sciences Agro, Feed &Food, 1 cours du Général de Gaulle, 33170 GRADIGNAN, France, [2]Groupe AFNOR, Délégation Nouvelle-Aquitaine, 3 Avenue Rudolph Diesel, 33700 Mérignac, France, [3]INRAE, UMR1213 Recherches sur les herbivores, VetAgroSup Clermont-Ferrand, 63122 Saint Genès Champelle, France; benoit.grossiord@agro-bordeaux.fr

In order to maintain and develop their activities, but also to bring an appropriate answer to the expectations of the consumers and the society, farmers develop new ways to valorise their productions. One of those consists in shortening the distance between producers and consumers by developing new strategies allowing a direct producer-consumer relation, like the so called short food supply chains (SFSCs). When developing SFSCs, producers have to reinvent their own job as new skills are necessary to manage and control all the steps involved in the processing, storing, marketing or transport of the end products they put on the market. If considering animal husbandry, the slaughter is a key step in the process. Throw the whole process, from living animals to end products, the producer who becomes an entrepreneur has to face many stakes related to those new activities. One of the main stake is the management and control of the food safety all along the chain. Like all the other food business operators, producers involved in SFSCs must respect the EU regulation on food safety (Food Law). Here we focus on Article 6 of the Food Law and its application in SFSCs. Article 6 deals with risk analysis, which consists in three successive actions: risk assessment, risk management and risk communication. The HACCP method (Hazard Analysis and Critical Control Point) is universal and we report here on its utilization in SFSCs. The HACCP allows a full analysis of all the steps involved in the elaboration of the end products, from the production of the raw materials to the commercialization. It is therefore a very powerful tool in order to make the operators be conscious of the many hazards (physical, chemical, biological) they have to manage all along the food chain and to help them identify and apply the proper good hygiene practices (GHPs). A better understanding and utilization of the HACCP could then contribute to the improvement of the food safety culture (FSC), as it is expected all along the food chain.

Session 45 Poster 26

Feeding strategy in organic farming as a lever to improve various quality dimensions of pork

C. Van Baelen[1], L. Montagne[1], S. Ferchaud[2], A. Prunier[1] and B. Lebret[1]
[1]PEGASE, INRAE, Institut Agro, 35590 Saint-Gilles, France, [2]INRAE, GenESI, 86480 Rouillé, France; chloe.van-baelen@inrae.fr

Organic farming is the official quality label in which consumers have the most confidence with. Its specifications evolved in 2022 to reinforce animal welfare and enhance the link to the soil for feeding resources. We considered the feeding strategy as a lever to improve various quality dimensions of organic pork. Pork quality includes intrinsic dimensions (commercial, nutritional, organoleptic, sanitary) and extrinsic dimensions related to animal farming (image). The experiment was conducted with 77 organic non-castrated male pigs (Piétrain NN (non-carrier of the n allele) × Large White) distributed in two batches. Within batch, littermates were divided into two groups at 33 kg body weight (BW). One group received a control feed (C) corresponding to the organic specifications. The other group received a test feed (Bio+) mainly based on French raw materials and contained more fibre (faba bean) and omega-3 fatty acids (linseed, camelina) and had access to fodder (rack). Each group was reared in a pen from the same building on deep straw bedding (1.3 m^2/pig) with free outdoor access (1.0 m^2/pig). Pigs were fed *ad libitum* until slaughter at about 125 kg BW. Average daily gain, feed intake and feed efficiency, and carcass weight and lean meat content were similar between C and Bio+ pigs. Compared to C, loin meat (longissimus muscle) from Bio+ pigs had higher ultimate pH associated to lower glycolytic potential ($P<0.05$), and tended to have lower drip loss ($P<0.10$), meaning an improvement in technological quality. Meat from Bio+ vs C pigs was less light ($P<0.05$) and had a more intense red colour ($P<0.01$). It was judged with a lesser aromatic persistency ($P<0.05$) by a trained sensory panel, whereas odour intensity, tenderness and juiciness did not significantly differ. With Bio+ diet, loin had lower n-6:n-3 fatty acid ratio ($P<0.001$). Backfat skatole and androstenone concentrations were low in most C and Bio+ pigs. However, skatole was slightly lower in Bio+ than C pigs ($P<0.05$) whereas androstenone did not differ. Altogether, the Bio+ diet had positive impacts on nutritional, technological and some sensory qualities of organic pig meat from entire male pigs, while contributing to the relocation of feed resources.

Session 45 Poster 27

Body reserve dynamics using metabolites and hormones profiles of Romane ewes in two farming systems

A. Nyamiel[1], D. Hazard[1], D. Marcon[2], F. Tortereau[1], C. Durand[3], A. Tesnière[4] and E. González-García[4]
[1]INRAE, UMR1388 GENPHYSE Université de Toulouse, ENVT, 31326 Castanet-Tolosan, France, [2]INRAE, UEP3R Bourges, 18390 Osmoy, France, [3]INRAE, UE321 La Fage, 12250 Roquefort-sur-Soulzon, France, [4]INRAE, CIRAD, SELMET Institut Agro Montpellier, Univ Montpellier, 34060 Montpellier, France; agnes.nyamiel@inrae.fr

The objective was to monitor the main effects affecting body reserve (BR) mobilization and accretion, using metabolites and hormones profiles, in Romane ewes reared under two contrasting farming systems (FS; (indoor, IND; extensive, OUT)). The ewes (n=173 IND; n=234 OUT) belonged to two genetic lines selected for low or high residual feed intake. They were monitored during their two first productive cycles at five key physiological stages (Mating, M; mid-Pregnancy, P; 2 weeks Pre-Lambing, bL; 3 weeks Post-Lambing, aL; Weaning, W). Parameters included body condition score (BCS) and metabolic profiles for plasma concentrations on non-esterified fatty acids (NEFA), β-hydroxybutyrate (BHB), Triiodothyronine (T3) and insulin (INS). The relevant fixed effects and their interactions were investigated through analyses of variance using R. Physiological stage, parity, cohort, genetic line, and litter class were fixed effects while ewe and residuals were random effects. Depending on the parameter evaluated, the fixed effects and their interactions, with the exception of the genetic line, were statistically significant ($P<0.05$). Regardless of the FS, results showed that BCS increased until P and declined thereafter. Highest NEFA concentrations were found from bL until W and at W in ewes reared IND or OUT, respectively. Significantly higher BHB levels than those found in M, P, and W were found in bL and aL ewes in both FS. T3 displayed a similar high trend throughout the stages with a peak at aL, and INS increased from P to aL before declining thereafter whatever the FS. These findings indicate that BR mobilization was displayed between P and W, as evidenced by BCS and blood parameters, whereas BR accretion occurred between W to P. Overall, there seems to be a consistent trend in Romane ewes' capacity to mobilize and recover their BR irrespective of the FS. Plasma concentrations of metabolites and hormones at different physiological status can be an indicator of the ewe's metabolic plasticity.

Inclusion of macroalgae in high forage beef cattle diets

S.A. Terry, T.W. Coates, R.J. Gruninger, D.W. Abbott and K.A. Beauchemin
Agriculture and Agri-Food Canada, 5403 1 Ave S, T1J4B1, Lethbridge, Alberta, Canada; stephanie.terry@agr.gc.ca

Supplementing ruminant diets with macroalgae has gained global interest because bromoform-containing macroalgae (e.g. *Asparagopsis* sp.) have been shown to be highly effective enteric methane (CH_4) inhibitors. Due to concerns with production costs, animal and human health, as well as regulatory restrictions, alternative macroalgae were examined for their impact on ruminant CH_4 production. Study 1 examined 33 Atlantic macroalgae on *in vitro* fermentation, and gas and CH_4 production. Study 2 examined the effect of two macroalgae extracts on rumen fermentation, nutrient digestibility, and CH_4 emissions from beef cattle fed high forage diets. All macroalgae were included at 2% of diet dry matter (DM) for each study. Most of the novel macroalgae examined in Study 1 did not have an inhibitory effect ($P>0.05$) on *in vitro* CH_4 production when included in a high forage diet; however, *Battersia* sp. and a shore blend (*Fucus vesiculosus, Furcellaria lumbricalis, Laminaria longicruris, Chondrus crispus*) linearly decreased ($P\leq0.05$) *in vitro* CH_4 production expressed on a dry matter degraded basis. There was no effect ($P\geq0.35$) of macroalgae extract on dry matter intake, feeding rate, or enteric CH_4 production (Study 2); however, apparent total tract digestibility of neutral and acid detergent fibre was increased ($P<0.01$) by both extracts compared to the control. In conclusion, these studies demonstrated the ability of the rumen environment to effectively degrade various macroalgae, with only limited species showing potential to decrease CH_4 production. Further screening efforts and testing using *in vivo* experiments are needed to continue the discovery of novel macroalgae with the potential to inhibit enteric CH_4 production.

Rumen protected potassium gluconate increases average daily gain of beef

A. Santos, J.G.H.E. Bergman, J.A. Manzano and M. Hall
Nutreco, Selko, Stationsstraat 77, 3811 MH Amersfoort, the Netherlands; alberto.santos@trouwnutrition.com

Calcium gluconate is a prebiotic that improves hindgut health. Trials in dairy cattle have shown that feeding rumen protected calcium gluconate (RPCG) improved hindgut health and increases milk production. A trial in steers sowed that feeding RPCG increases the amount of volatile fatty acids in the caecum and colon. In this trial, the impact of RPCG (LactiBute, Selko) on faecal starch levels and pH and on ADG gain in beef was tested. Beef calves on a commercial beef farm in Spain, housed in 6 pens of 20 calves each were included in the trial. Half of the animals (1 pen with females, 2 with males) were fed RPCG, the other half (1 pen with females, 2 with males) served as untreated controls. The fattening period lasted 111 days, RPCG was fed for 43 days. Bodyweight and starch levels and pH of faeces were measured at the start of the trial and at day 43. Faecal pH was reduced from 6.2 in the controls to 5.8 in the RPCG group. Average faecal starch declined from 19.7% in the controls to 15.6% for the RPCG group. ADG in the females increased by 8% or 94 g/day, ADG in males increased by 7% or 95 g/day.

Session 46 — Theatre 3

Impact of additives and forage levels on performance and enteric methane emissions of Nellore bulls

E. Magnani[1], T.H. Silva[1], L.B. Tosetti[1], A. Berndt[2], E.M. Paula[1], P.R. Leme[3] and R.H. Branco[1]

[1]Institute of Animal Science, Beef Cattle Research Center, Rod. Carlos Tonani km94, 14.174-000, Brazil, [2]Empraba Pecuaria Sudeste, Beef cattle, Rod. Washington Luiz, Km 234, 13560-970, São Carlos, SP, Brazil, [3]University of Sao Paulo, Department of Animal Science, R. Duque de Caxias, 225, 13635-900, Pirassununga, SP, Brazil; rhbarnandes@gmail.com

This study was carried out to evaluate the effect of additives and increasing levels of neutral detergent fiber from forage (NDFf) on performance and enteric methane (CH_4) emissions in finishing Nellore bulls. Sixty Nellore bulls [408.4±14.4 kg of body weight (BW) and approximately 24 months of age] were allocated into 6 diets in a factorial 2×3 arrangement during 105 days in an individualized feedlot system. The treatments were the combination of monensin (MON; 30 mg/kg DM), essential oils (EO; 500 mg/kg DM; essential oil was a mixture of castor oil and cashew nut) and 3 levels of dietary FDNf (6, 9, and 15% total diet, DM basis). The diet was composed by *Brachiaria brizantha* cv. Marandu hay, finely ground corn, soybean meal, ground citrus pulp, urea, potassium chloride, sodium chloride and trace minerals + vitamins. For CH_4 emission measurements, 5 random bulls per treatment were evaluated. Methane emissions were measured using the sulphur hexafluoride (SF6) tracer technique. Additives and dietary NDFf levels were analysed in a complete randomized design according to a factorial arrangement. When NDFf levels were statistically significant, linear and quadratic responses were assessed (software SAS 9.4). Greater dry matter intake (DMI) was detected for MON vs EO treatments. Also, DMI increased linearly as NDFf levels increased. Monensin treated bulls had greater final BW and ADG compared to EO. Essential oils reduced CH_4 emission (g/d) compared to MON. In addition, increased CH_4 emission was detected with increasing dietary NDFf levels. These effects reflected on CH_4 emission by carcass unit, where EO reduced CH_4 emissions and increasing NDFf levels increased. Monensin was effective in improving the performance of finishing Nellore bulls compared to EO, in which EO reduced CH_4 emission at performance expense. Further, dietary NDFf may increase enteric CH_4 emission in a level-dependent manner.

Session 46 — Theatre 4

Differences in digestive traits of young bulls fed contrasted diets and diverging in feed efficiency

M. Coppa[1], C. Martin[2], A. Bes[2], L. Ragionieri[3], F. Ravanetti[3], P. Lund[4], G. Cantalapiedra-Hijar[2] and P. Nozière[2]

[1]Independent Researcher, Turin, 10100, Italy, [2]INRAE, UMR-Herbivores, Saint-Genès-Champanelle, 63122, France, [3]Universit° di Parma, Departement of Vererinary Science, Parma, 43126, Italy, [4]Aarhus Universitet, Nordre Ringgade 1, Aarhus, 8000, Denmark; mauro.coppa@inrae.fr

This research aimed to: (1) identify digestive traits that differed on bulls that phenotypically diverged in residual feed intake (RFI) and their relationships with RFI; and (2) explore the hierarchy among digestive traits in discriminating divergent RFI bulls fed contrasted diets. Extreme and phenotypically divergent RFI animals were selected within 100 Charolais growing bulls fed either grass silage (GS, n=50), or maize silage (MS, n=50) diets (8 efficient RFI- and 8 inefficient RFI+ bulls selected per diet). The 32 bulls selected on a first feed efficiency test (84 days) were measured for digestibility, digestive transit rate, rumen pH, behaviour and gas emissions (CH_4, CO_2, H_2). They were also measured for visceral organ weight and reticulo-omasal orifice (ROO) size, rumen particle size, and rumen and ileum histology after slaughter. A feed efficiency test was performed on the whole trial (180 days); RFI were recalculated and regressed by GLM to digestive traits. Irrespective of the diet, efficient bulls (RFI-) tended to have higher CH_4 and CO_2 emissions (g/kg DMI), spent longer time of rumen pH below 5.8 and had lower rumen size, proportion of small particle size in the rumen, less ileum nuclei, than non-efficient ones. The RFI was negatively related to time spent in activity other than ingestion, rumination, and resting, and positively related to abomasum size, and number of cells in the ileal crypts. Diet-dependent relationships were noted: with GS, RFI- bulls showed a slower transit rate, whereas with MS, RFI- bulls tended to have shorter resting events and smaller ROO than RFI+. Transit rate was positively related to RFI only in GS diet. Among digestive traits monitored, rumen size appeared as the major discriminating variable between RFI divergent bulls. These results should be validated on a larger population.

Session 46 Theatre 5

Effect of solid feed intake on feeding behaviour and energy metabolism in growing calves

E. Labussiere[1,2], L. Montagne[2], Y. Le Cozler[2], C. Martineau[3] and D. Bastien[3]
[1]UE3P, INRAE, Le Clos, 35590 Saint-Gilles, France, [2]PEGASE, INRAE, Institut Agro, Le Clos, 35590 Saint-Gilles, France, [3]Idele, Monvoisin, 35650 Le Rheu, France; etienne.labussiere@inrae.fr

In growing calves reared for the production of veal meat, milk replacers (MR) and solid feeds (SF) are fed simultaneously to support high growth rate and address welfare issues. The combination of these diets is associated with modified feeding behaviour, digestion and metabolic processes that may change the utilization of dietary energy for growth compared with the use of MR alone. Present experiment aimed to determine the effects of increasing levels of SF as substitutes to MR on feeding behaviour, physical activity and energy metabolism in growing calves. Twenty Holstein male calves were affected to four dietary treatments in which SF were substituted to MR from 10 to 100%. SF were composed of 87% concentrates, 5% chopped wheat straw and 8% chopped hay. The calves were first adapted to their respective treatments during 100 days until their body weight reached 159±10 kg. MR was given twice daily whereas SF were provided once in the morning. Water and SF were available 18 h 45 min per day. Thereafter, calves were housed individually in an open-circuit respiration chamber during one week to measure their feeding behaviour, physical and energy balance from total collection of faeces and urine, and measurements of gas exchanges (CH_4, O_2 and CO_2) to calculate heat production (HP). The latter was partitioned between components due to basal metabolism, physical activity and thermic effect of feeding (TEF). Water evaporation was measured to determine latent heat losses. Intake of SF increased from 404 to 3,630 g dry matter (DM)/d while MR intake decreased from 2,029 to 0 g DM/d between dietary treatments. When increasing SF intake, number of meals, feeding rate and time spent standing while eating increased whereas total time spent standing decreased. Total HP increased with increasing SF intake because of increased TEF. The proportion of HP lost via latent route decreased from 45 to 32% when SF increased. Energy balance was not affected suggesting that the lower metabolic efficiency of SF was compensated for by behavioural adaptations. These results suggested that calf behaviour and nutrition can adapt to contrasted sources of dietary energy.

Session 46 Theatre 6

Effects of dietary inclusion of willow leaves on feed intake and methane emission in sheep

J.J. Thompson[1], S. Stergiadis[2], O. Cristobal-Carballo[3], T. Yan[3], S. Huws[1] and K. Theodoridou[1]
[1]Institute of Global Food Security, Queen's University Belfast, 19 Chlorine Gardens, Belfast, BT95DL, United Kingdom, [2]University of Reading, School of Agriculture, Policy and Development, Earley Gate, P.O. Box 237 Reading, RG6 6AR, United Kingdom, [3]Agri-Food and Biosciences Institute, Large Park, Hillsborough, BT266DR, United Kingdom; k.theodoridou@qub.ac.uk

The present study aimed to investigate the effects of feeding two varieties of fresh-cut willow leaves on feed intake and enteric methane (CH_4) emissions in sheep. Twelve ewe hoggets (Texel, 12 months old) were allocated in a 3 (diets) × 3 (periods) Latin Square design study with 4 weeks/period. The diets were (DM basis): 100% grass silage (Control), 85:15 grass silage:Terranova leaves (TL) and 85:15 grass silage:Beagle leaves (BL). Each sheep also received concentrate at 180 g/day offered through a GreenFeed unit (5 visits/d). Sheep were housed as a single group with 6 auto-feeders (2 feeders/treatment) with CH_4, H_2 and CO_2 emissions and O_2 consumption measured using a GreenFeed unit during day 14 to 21 of each period. Sheep were then placed in individual metabolism crates for 7 days, with total faeces and urine outputs recorded in the final 6 days. Willow leaves were harvested daily and chopped to a length of 3 cm, totally mixed with grass silage for TL and BL diets and offered to lambs twice a day with a residual rate of 5%. Willow treatments had no effect on CH_4 (g/d) (P=0.933), CO_2 (P=0.898) and O_2 production (P=0.925). H_2 production increased, during period 1 for BL compared to the other two treatments. Dietary willow did not affect (P= 0.061) urinary nitrogen (N) excretion (g/d), but faecal N excretion (g/d) was higher (P<0.01) for both willow treatments compared to control. In conclusion, feeding fresh willow leaves of Terranova or Beagle at 15% (DM basis) had no effect on feed intake and methane emissions of lambs. Further animal trials are needed to quantify the effect of willow fodder of different varieties and at different inclusion rates.

Session 46 Theatre 7

Effect of supplementing Omega-3 rich oil on ruminal fermentation and dietary digestibility in lambs

O. Cristobal-Carballo[1], F. Godoy-Santos[2], S. Huws[2], S. Morrison[1], A. Aubry[1], E. Lewis[3] and T. Yan[1]
[1]AFBI, Hillsborough, BT26 6DR, United Kingdom, [2]Queen's University of Belfast, Belfast, BT9 5DL, United Kingdom, [3]Devenish Nutrition, Belfast, BT3 9AR, United Kingdom; omarcristobal.carballo@afbini.gov.uk

This study assessed the effect of feeding microalgae oil, rich in Omega-3, at different levels on ruminal fermentation, methane emissions, and feed digestibility in finishing lambs. A group of 24 male lambs, Texel/Scottish black face, was selected from a performance study of 64. All lambs were balanced by body weight (BW), and randomly assigned into one of four treatments. Treatments were fed TMR diets of 50:50 grass silage and concentrate. Concentrate meals included 0.00% (Control), 0.54% (Low), 1.08% (Medium) and 1.62% (High) of microalgae oil (DM basis); oil inclusion was balanced with soyabean oil. Daily DMI and weekly BW were recorded. At 70 d of treatment, lambs were gradually introduced at 4 d intervals in 4 groups of 6 into digestibility crates for 7 d and respiration chambers for 4 d. Urine and faeces were collected daily during the final 6 d of crates. Methane emissions were measured in the final 48 h of chambers. Rumen fluid was collected at slaughter (95 d of treatment). Response variables were analysed using linear mixed models via the REML method, with treatment as fixed effect and Run and Chamber as random effects. Pairwise differences between treatments were examined using Fisher's LSD test at P=0.050. GenStat (version19) was used to carry out all analyses. DMI tended to decrease (P=0.068) as the microalgae oil dose increased in the diet. Digestibility of DM, OM, NDF, ADF and Energy did not differ (P>0.255) between treatments. Total VFA concentration was greater (P=0.018) in High and Low when compared to Medium lambs. Acetate and propionate proportions were higher (P<0.001) and lower (P=0.003), respectively, in Control when compared to microalgae oil treatments. Methane production (P=0.009) and yield (CH_4/DMI; P=0.041) were reduced in 21.5 and 20.3%, respectively, in High when compared to Control lambs. It is concluded that microalgae oil can be included in the diet of finishing lambs at doses of 1.62% without affecting the nutrient and energy digestibility, while increasing the total VFAs and propionate proportions, and reducing the acetate proportions and CH_4 emission.

Session 46 Theatre 8

Essential oils and integral diets to optimize rumen function and decrease methanogenesis in lambs

N. Ghallabi[1], G. Gonzalo[1], A. Garcia[2], O. Catalán[3], P. Romero[4], M. Hassan[4], A.I. Martín-García[4], D.R. Yáñez-Ruiz[4] and A. Belanche[1]
[1]University of Zaragoza, Animal Production and Food Science, Miguel Servet 177, Zaragoza, 50013, Spain, [2]Los Chengos SL, Maxima Ladron de Guevara 5, Mula, 30170, Spain, [3]INZAR, Julio García Condoy 41, Zaragoza, 50018, Spain, [4]Estación Experimental del Zaidín (CSIC), Profesor Albareda 1, Granada, 18008, Spain; belanche@unizar.es

This study aims to explore if feeding lambs with fibrous concentrate feeds without forage and their supplementation with feed additives could help to minimize the labour costs and environmental footprint. After receiving artificial milk feeding, a total of 56 male lambs were divided into 28 pens (in pairs) and allocated to 4 experimental treatments (n=7) following a 2×2 factorial design. Lambs were fed *ad libitum* with a conventional concentrate feed and barely straw (CTL) or with a fibrous concentrate feed without forage (FIB), either non-supplemented (-) or supplemented (+) with a commercial blend of essential oils (200 mg/kg DM). Concentrate feed intake was similar across treatments but CTL lambs tended to have higher total DMI than FIB lambs, the difference being the straw consumption. No differences were noted in the feeding pattern. The FIB diet promoted lower glucose / β-hydroxybutyrate ratio during the post-weaning period suggesting an accelerated rumen physiological development. The FIB diet also promoted higher rumen fermentation activity characterized by higher AGV concentration and methane yield and lower rumen pH and acetate/propionate ratio. On the contrary the CTL lambs had higher blood urea concentration suggesting a lower N utilization by the host. Essential oils supplementation tended to increase rumen propionate proportion and decreased the blood urea, blood glucose and methane yield indicating a lower environmental footprint. No differences were noted across treatments in productive parameters or carcass performance. In conclusion, the use of integral diets without forage can represent a successful strategy to increase the rumen fermentation and simplify the animal feeding labour without having negative productive implications. The supplementation with essential oils led to a propionic fermentation and could be considered as a strategy to minimize the environmental footprint in fattening lambs.

Session 46 Theatre 9

Novel water based delivery of seaweed extracts to improve the sustainability of ruminant production

A. Casey, T. Boland, Z. McKay and S. Vigors
University College Dublin, Animal and Crop Sciences, School of Agriculture and Food Science, University College Dublin, Belfield, Dublin 4, Ireland, D4, Ireland; staffordvigors1@ucd.ie

The use of seaweed derived additives has potential to improve the sustainability of ruminant production but dosage rates and delivery need further evaluation. One of the major issues constraining the incorporation of rumen modifiers in grass based production systems is issues in relation to delivery. The objective of this study was to evaluate the inclusion of an extract derived from *Ascophyllum nodusum* on nitrogen use efficiency (NUE) and methane emissions delivered through water. The extract was included through the water at 0.75 and 1.5% of total feed intake and fed to 10 Spring calving mid-late lactation Holstein Friesian dairy cows (182 days in milk, 27.5 kg/day, 3.86% fat, 3.64% protein) in a 3×3 Latin Square design. The experiment compromised three 21 d experimental periods (16 days dietary acclimatization and 5 days data and sampling collection). The inclusion of the seaweed had no effect on dry matter intake or water intake (P>0.05). The inclusion of the seaweed extract at 1.5% increased milk solids compared to the control (P>0.05), while milk yield was unaffected by seaweed inclusion. Of the constituents milk protein was increased in the cows fed seaweed extract at 1.5% (P<0.05) while milk fat and milk lactose were unaffected (P>0.05). The seaweed extracts had no effect on urinary or faecal nitrogen (N) but milk N was increased in the cows supplemented with seaweed included at 1.5% compared to the control group (P<0.05). This increase in milk N led to an improvement of NUE in the group fed the high dose rate compared to the control. The seaweed extracts had no impact on rumen fermentation measures such as ammonia and volatile fatty acids and had no impact on methane production. This research provides interesting data on the mode of action of brown seaweeds with further research ongoing on their potential benefit and use in pasture based production systems.

Session 46 Theatre 10

Supplementation of Ca gluconate improves fertility and time to peak milk yield in dairy cattle

D.J. Seymour, M.V. Sanz-Fernandez, J.B. Daniel, J. Martin-Tereso and J. Doelman
Trouw Nutrition R&D, P.O. Box 299, 3800 AG Amersfoort, the Netherlands; dave.seymour@trouwnutrition.com

Transitioning from pregnancy through parturition to lactation imposes a significant amount of physiological stress and inflammation on dairy cattle, which can, when excessive, negatively impact health and performance. The gastrointestinal tract is a key site of inflammation during the transition period and represents a potential target for prophylactic dietary interventions. Increased milk productivity has been reported in dairy cattle when a fat-embedded prebiotic compound, Ca gluconate (HFCG), was supplemented to mid-lactating dairy cows. The objectives of this study were to evaluate responses in health, fertility, and lactation performance in multiparous dairy cattle supplemented with HFCG during the transition period and early lactation. Holstein cattle in a commercial herd (n=164) were randomly assigned to a 2×2 factorial, receiving either HFCG or control incorporated into a pelleted feed for approximately 21 d prepartum and/or 100 d postpartum. Incidence of culling was monitored as a proxy for animal health, while days to first observed heat, first service, and confirmed pregnancy were used to evaluate fertility. Individual lactation curves were modelled, and the fitted parameters were compared statistically to evaluate production response. No differences in incidence of culling were observed. Animals supplemented with HFCG, both prepartum and postpartum, tended to return to oestrus earlier and reach peak lactation earlier relative to animals that switched to negative control postpartum. No other effects on production performance were observed. Animals supplemented with HFCG postpartum, regardless of prepartum treatment, tended to be serviced earlier and were confirmed pregnant earlier than those receiving control postpartum. While the precise mode of action of HFCG remains unclear, we hypothesize that these findings are driven in part by a combination of reduced inflammation and improved energy status during the transition period and early lactation.

DFM supplementation during the gestation and dry periods on postpartum performance in dairy cows

O. Ramirez-Garzon[1], D.G. Barber[2], J. Alawneh[3], L. Huanle[1] and M. Soust[1]

[1]Terragen Biotech Pty Ltd, Unit 6/ 39-41 Access Cres, Coolum Beach QLD, 4573, Australia, [2]DairyNEXT, Marburg, 4346, Australia, [3]GCP Veterinary Consulting Ltd Pty., Kenmore, Brisbane, 4069, Australia; orlandor@terragen.com.au

This study aimed to test the hypothesis that daily supplementation of dairy cows with a direct-fed microbial (DFM) during gestation and the dry period would have a carry-over effect on milk production and reproductive performance in the first 100 days in milk (DIM) of the subsequent lactation. The study was conducted at a commercial dairy farm in Harrisville (Queensland, Australia) from October 2021 to January 2023 with 150 multiparous Holstein cows randomly selected based on parity (2.0±0.8) and days in milk (126.9±55.4 DIM). Group A (control n=75) received the basic diet, while Group B (DFM, n=75) received a basic diet supplemented with a DFM (top dressed) consisting of three strains of *Lactobacillus* sourced from a commercial product (Mylo®, Terragen Biotech, Australia). During the calving season (March to July 2022) a subset of 82 cows (40 control, 42 DFM) was monitored to assess postpartum performance. Serum metabolic profile and uterine health were evaluated before calving (-4 w to -2 w prepartum) and after calving, (week 1, week 3, and week 6 postpartum). Results showed that cows supplemented with DFM had higher average milk yield and peak milk than the control group (24.0 vs 22.0 l/cow; $P<0.007$ and 34.0 vs 31.2 l; $P<0.014$, respectively). Moreover, the days to first insemination were lower in the DFM group (64.7 d vs 74.3 d; $P=0.03$). Serum levels of albumin, glucose, NEFA, and Mg were higher in DFM cows ($P<0.05$) while total protein, urea, and BOH were higher in the control group ($P<0.05$). No significant differences were observed in uterine health. In conclusion, supplementing cows with DFM during gestation and dry periods can improve their subsequent productive and reproductive performance after calving, providing evidence for the use of DFM as a potential strategy to enhance the performance of dairy cows.

***In vivo* evaluation of tannins and essential oils mixtures as additives for dairy cows**

G. Foggi[1], L. Turini[1], F. Dohme-Meier[2], A. Muenger[2], L. Eggerschwiler[2], J. Berard[2], G. Conte[1], A. Buccioni[3] and M. Mele[1]

[1]University of Pisa, Department of Agriculture, Food and Environment, via del Borghetto, 80, 56124, Pisa, Italy, [2]Agroscope, Ruminant Nutrition and Emissions, Route de la Tioleyre 4, 1725 Posieux, Switzerland, [3]University of Florence, Dipartimento di Scienze e Tecnologie Agrarie, Alimentari, Ambientali e Forestali, Piazzale delle Cascine 18, 50144 Florence, Italy; giulia.foggi@phd.unipi.it

Several studies have focused on the use of feed additives to modulate ruminal fermentation and metabolism. Most of these studies adopted *in vitro* or *in vivo* approach considering pure substances, whereas only a minor part reported the effect of additive mixtures. This study aimed to evaluate the *in vivo* low dosage effects of 3 additives formulated with tannins and/or different blends of essential oils compounds (BEOC) previously tested in RUSITEC. The trial was conducted in a free-stall barn at Agroscope, Posieux (Switzerland), for 12 weeks. A total of 32 lactating Holstein cows were assigned to 4 groups balanced by milk yield (35±5) and days in milk (62±27): negative control, K; positive control, PC; group A; and B. The Basal diet was common to all groups and contained corn and grass silages, grass hay and a concentrate mix (forage:concentrate=62:38). The PC was supplemented with Agolin Ruminant®; A, with a chestnut extract and a BEOC_A; B, with a quebracho extract and a BEOC_B. Gross composition of milk and feed were analysed weekly, feed intake and milk production daily, and ruminal pH and ammonia on weeks 2, 5, 8, 10 and 12. All data were analysed with a mixed model with repeated measures. Cows consumed on average 22.4±0.8 kg/day of DM regardless of group. The fat-corrected milk yield did not differ between groups and averaged 36.5±1.8 kg/day. The ruminal pH (7.04±0.04) and ammonia (9.97±0.60 mM) were not affected by the additive addition, even though ruminal ammonia decreased by 25% ($P<0.001$) in weeks 10 and 12 compared to week 8. Regarding milk composition, there were significant interactions of group and period for milk urea, protein and fat (all $P<0.001$). In conclusion, the use of the selected additives did not affect animal performance and most of the parameters considered, but further evaluations, particularly on ruminal fermentation profile are needed.

Palmitic to oleic ratio in fat supplement influenced the digestibility and production of dairy cows

J. Shpirer[1,2], L. Lifshitz[2], H. Kamer[2], Y. Portnick[2] and U. Moallem[2]
[1]The Hebrew University of Jerusalem, 2Department of Animal Science, Rehovot, Rehovot, Israel, [2]ARO, Volcani Institute, Department of Ruminant Science, HaMaccabim Road, 7505101, Israel; uzim@volcani.agri.gov.il

The form of fat supplements, degree of saturation, and the fatty acids (FA) profile influence the cows' productive response. The objective was to examine the effects of supplemental fats in the form of calcium salts of fatty acids (CSFA) in different ratios between palmitic (PA) and oleic (OA) acids on nutrient digestibility and cows' performance. Forty-two dairy cows were assigned into 3 groups and fed for 13 wks rations contained 2.2% of CSFA (on DM basis) consisting of: (1) CS45:35-45% PA and 35% OA; (2) CS60:30-60% PA and 30% OA; and (3) CS70:20-70% PA and 20% OA. Rumen and faecal samples were taken for VFA and digestibility measurements, respectively. Production data were analysed with PROC MIXED, and rumen and digestibility data with GLM models of SAS. Milk yields were the highest in the CS45:35 (52.0 kg/d), intermediate in the CS60:30 (51.1 kg/d), and lowest in the CS70:20 cows (47.3 kg/d; $P=0.002$). Milk fat content was ~0.35 percentage units lower in the CS45:35 cows than in the other two groups (3.55, 3.94, and 3.87% in the CS45:35, CS60:30, and CS70:20 groups, respectively; $P=0.001$), and fat yields were higher in the CS60:30 than in other groups ($P=0001$). The FCM 4% and ECM yields were higher in the CS60:30 than in other groups. Feed intake was highest in the CS60:30 group (33.5 kg/d) and lowest in the CS70:20 group (31.3 kg/d; $P=0.001$). The milk-to-DMI ratio was the highest in the CS45:35 ($P=0.001$), with no differences in the efficiency for 4% FCM or ECM production. In conclusion, increasing the PA proportion in the fat supplements greatly increased the milk-fat content, and a high OA ratio increased the milk yields. The digestibility of most nutrients was lower in the CS70:20 than in other groups; however, the total fat digestibility was similar between all groups, indicating that the form more than the FA profile influences the fat digestibility. The different impacts of the PA-to- OA ratio in the fat supplements indicates that in the future the FA profile of the provided supplement will be determined according to the defined goal: milk or milk-fat.

Xylooligosaccharides and enzyme increased milk yield and reduced methane emissions of Jersey cows

L.F. Dong and Q.Y. Diao
Institute of Feed Research, Chinese Academy of Agricultural Sciences, No. 12 Zhongguancun South Street, 100081, China, P.R.; donglifeng@caas.cn

Sustainable strategies for enteric methane (CH_4) mitigation of dairy cows have been extensively explored to improve production performance and alleviate environmental pressure. The present study aimed to investigate the effects of dietary xylooligosaccharides (XOS) and exogenous enzyme (EXE) supplementation on milk production, nutrient digestibility, enteric CH_4 emissions, energy utilization efficiency of lactating Jersey dairy cows. Forty-eight lactating cows were randomly assigned to one of 4 treatments: (1) control diet (CON), (2) CON with 25 g/d XOS (XOS), (3) CON with 15 g/d EXE (EXE), and (4) CON with 25 g/d XOS and 15 g/d EXE (XOS + EXE). The 60 d experimental period consisted of a 14-d adaptation period and a 46-d sampling period. The enteric CO_2 and CH_4 emissions and O_2 consumption were measured using the GreenFeed system, which were further used to determine the energy utilization efficiency of cows. Compared with CON, XOS and EXE synergistically ($P<0.05$) increased milk yield, fat concentration, and energy-corrected milk yield (ECM)/DM intake, which could be reflected by the significant improvement ($P<0.05$) of dietary NDF and NDF digestibility. The gaseous results showed that XOS and EXE synergistically ($P<0.05$) reduced CH_4 emission and CH_4 emissions intensities (e.g. CH_4/DM intake, CH_4/milk yield, and CH_4/ECM yield), whereas CO_2 emission and O_2 consumption remained similar among the treatments ($P>0.05$). Furthermore, the synergistic effects of XOS and EXE was observed ($P<0.05$) for metabolizable energy intake and CH_4 energy output as a proportion of gross energy intake, whereas lowest values ($P<0.05$) of CH_4 energy output and CH_4 energy output as a proportion of gross energy intake was observed for cows fed XOS compared with the remaining treatments. Dietary supplementary of XOS and EXE contributed to the improvement of lactation performance, nutrient digestibility, and energy utilization efficiency, as well as reduction of enteric CH_4 emissions of lactating Jersey cows. More research is also needed to investigate the long-term effect and mode of action of these additives for Jersey cows.

Defatted black soldier fly larvae meal as substitute of soybean meal in Kenyan dairy cow rations

D.J.M. Braamhaar[1], D.J. List[2], S.J. Oosting[1], D. Korir[3], C.M. Tanga[4] and W.F. Pellikaan[2]
[1]Wageningen University & Research, Animal Production Systems Group, P.O. Box 338, 6700 AH Wageningen, the Netherlands, [2]Wageningen University & Research, Animal Nutrition Group, P.O. Box 338, 6700 AH Wageningen, the Netherlands, [3]International Livestock Research Institute, Livestock, System and Environment, P.O. Box 30709, 00100 Nairobi, Kenya, [4]International Centre of Insect Physiology and Ecology, P.O. Box 30772, 00100 Nairobi, Kenya; dagmar.braamhaar@wur.nl

Soybean meal (SBM) is worldwide the most used protein source in livestock production. However, SBM is expensive and the availability is limited in Kenya. Additionally, this feed has a limited contribution to circular food systems, since it is associated with feed-food competition. Therefore, there is a demand for alternative feed ingredients. Black soldier fly larvae (BSFL) can be considered an alternative feed because of its contribution to the circular food system by converting bio-waste into high quality feed ingredients. To the best of our knowledge, to date no *in vivo* study has been done with lactating dairy cows. The aim of the experiment was to investigate the effect of replacing SBM with defatted black soldier fly larvae meal (DBSFLM) in diets of lactating Holstein-Friesian (HF) cows on milk production and composition, feed and nitrogen use efficiency, and apparent total-tract digestibility. The experiment was conducted at a commercial dairy farm in Molo, Kenya. Twelve lactating HF cows were blocked by milk production, parity and days in milk into 4 squares in a replicated 3×3 Latin square trial. The basal diets contained 52% corn silage, 4.6% barley straw, 17% maize germ, 8.6% wheat bran, 2.4% sunflower meal, 0.4% urea, 3.6% molasses, 1.6% mineral and premix, and 9.4% protein treatment. The protein treatments were: (1) 92% SBM and 8% soybean oil; (2) 46% SBM, 4% soybean oil and 50% DBSFLM; and (3) 100% DBSFLM. The amount of concentrate was adjusted to milk yield per Latin square and basal diet mixtures were offered *ad libitum*. Data were analysed using the Mixed procedures of SAS. No treatment effect was found for milk production and milk quality (i.e. protein, fat and lactose). Lab results to calculate feed and nitrogen use efficiency, and apparent total-tract digestibility are expected soon.

Classifying lipogenic and glucogenic diets in dairy cows based on metabolomics profiles

X. Wang[1,2], S. Jahagirdar[1], W. Bakker[3], C. Lute[2], B. Kemp[2], E. Saccenti[1] and A.T.M. Van Knegsel[2]
[1]Wageningen University & Research, Department of Agrotechnology and Food Sciences, Systems and Synthetic Biology, Helix, 6700 EJ Wageningen, the Netherlands, [2]Wageningen University & Research, Department of Animal Sciences, Adaptation Physiology Group, Zodiac, 6700 AH Wageningen, the Netherlands, [3]Wageningen University & Research, Department of Agrotechnology and Food Sciences, Toxicology, Helix, 6700 EA Wageningen, the Netherlands; xiaodan.wang@wur.nl

The objective of this study was to investigate the ability of plasma metabolomic profiles to differentiate between early lactation cows based on their diet. Holstein-Friesian cows were randomly assigned to a glucogenic (n=15) or lipogenic (n=15) diet in early lactation. Plasma was collected in week -2, 2, and 4 relative to calving. Metabolomics profiles in plasma were detected using liquid chromatography-mass spectrometry (LC-MS). A total of 37 metabolites were identified. For each cow per timepoint two metabolomics profiles were available: (1) extinction values for metabolites (extinction dataset); and (2) ratios between extinction values for metabolites (ratio dataset). Classification performance of dairy cows to the two diets was done using the XGBoost algorithm. The ratio dataset resulted in better classification performance compared with the extinction dataset for cows fed a lipogenic diet and cows fed a glucogenic diet. Model performance as measured by the area under the ROC curve (AUC) of lipogenic diet and glucogenic diet improved from 0.606 to 0.753 and from 0.696 to 0.842 in week 2 and 4, respectively. The top features to classify the lipogenic and glucogenic treatment were the ratio of arginine to tyrosine and the ratio of aspartic acid to valine in week 2 and 4, respectively. For cows fed the lipogenic diet, choline and the ratio of creatinine to tryptophan were top features to classify cows in week 2 and 4. For cows fed the glucogenic diet, methionine and the ratio of 4-hydroxyproline to choline were top features to classify cows in week 2 and 4. Carnitine and the ratio of asparagine to carnitine were top features to classify week -2 and 4 both for cows fed a lipogenic diet or a glucogenic diet. This study shows that the ratios of metabolites better classify diets in dairy cows in early lactation compared with absolute extinction values.

Session 46 — Poster 17

Comparative study of digestibility of safflower varieties

M. Besharati, N. Khoshnam and D. Azhir
University of Tabriz, Animal Science, 29th Bahman Blv., 51666, Iran; m_besharati@hotmail.com

The purpose of this project was to compare the laboratory digestibility of safflower varieties and the effect of radiation on digestibility and gas production parameters. Experimental treatments include: 1:seed of Gol Mehr variety, 2:seed of Parnian variety, 3:seed of Goldasht variety, 4:seed of Sefa variety, 5:Gol Mehr variety Meal, 6:Parnian Meal, 7:Goldasht Meal, 8:Sefe Meal, 9:Gulmehr fodder, 10:Parnian fodder, 11:Goldasht fodder, 12:Sefa variety fodder, 13:irradiated seed of Gol Mehr variety, 14:irradiated seed of Parnian variety, 15:irradiated seed of Goldasht variety, 16:irradiated seed of Sefa variety, 17:irradiated flour Gol Mehr variety, 18:irradiated meal of Parnian variety, 19:irradiated meal of Goldasht variety, 20:irradiated meal of Sefa variety. Gas production of different safflower cultivars at different incubation times showed that there is a significant difference between the treatments in terms of gas production at different incubation times. The amount of gas production of different varieties of safflower flour in different hours of incubation showed that Mehr flower meal has the highest amount of gas production among different varieties at 2,6,8,12,6 and 24 hours of incubation. Also, the irradiated Gol Mehr flour cultivar had the lowest amount of gas production in the rumen liquid during the mentioned hours of incubation, and it has a significant difference with other cultivars. In the subsequent times of incubation, the irradiated Pernian meal variety had the highest gas production rate compared to other varieties of safflower meal. The results showed that irradiation has improved the percentage of laboratory digestibility of dry matter and organic matter of different varieties of safflower seeds and flour compared to varieties without irradiation, although this performance improvement is different in different varieties and times. The results show that the irradiated Goldasht seed has the highest NDF digestibility in all incubation hours compared to other treatments. According to the obtained results, it can be concluded that radiation can increase the nutritional value of food, which has been proven in several studies, and among them, factors such as different types of food and method Radiations can be considered important factors in achieving effective results.

Session 46 — Poster 18

Effect of different levels of propolis, nitrate, thyme and mint essential oil on digestibility

M. Besharati and M. Mousavi
University of Tabriz, Animal Science, 29th Bahman Blv., 51666, Iran; m_besharati@hotmail.com

The aim of this study was to investigate the effect of using different levels of bee propolis, calcium nitrate, thyme and mint essential oil on *in vitro* gas production and digestibility by using 3 experiments (gas production, *in vitro* degradability, three-step digestibility) was performed in a completely randomized design with 9 treatments. The experimental treatments included: 1. diet (control), 2.diet containing 500 mg propolis/kg DM 3. diet containing 1000 mg propolis/kg DM 4. diet containing 1.5 g calcium nitrate/100 g DM 5. diet containing 3 g calcium nitrate/100 g DM 6. diet containing 100 mg thyme essential oil/kg DM 7. diet containing 200 mg thyme essential oil/kg DM 8. diet containing 100 mg mint essential oil/kg DM 9. diet contained 200 mg mint essential oil/kg DM. The results showed that the treatment containing thyme essential oil (100 mg) caused a decrease and the propolis (1000 mg) and calcium nitrate (1.5 g) treatments significantly increased the amount of gas produced ($P<0.05$). The treatment containing mint essential oil (200 mg) had a significant effect on digestibility in that it reduced the digestibility of dry matter and insoluble fibres in acidic detergent, but in the kinetics of degradability in the rumen, it increased the degradability of dry matter, insoluble fibres in neutral and acidic detergent ($P<0.05$). Treatments containing propolis increased the degradability of dry matter, insoluble fibres in neutral and acidic detergents, and the treatment containing propolis (1000 mg) increased the degradability of dry matter and insoluble fibres in neutral detergents in the total gastrointestinal tract. Treatments containing thyme essential oil (200 mg) and calcium nitrate (1.5 g) significantly reduced the degradability of insoluble fibres in neutral and acidic detergents in the rumen and the total gastrointestinal tract ($P<0.05$).

Session 46 Poster 19

Combination of lactic acid bacteria inoculant in difficult to ensile grass after 15 days of ensiling

M. Duvnjak[1], L. Dunière[2], B. Andrieu[2], E. Chevaux[2] and C. Villot[2]
[1]University of Zagreb, Department of General Agronomy, Faculty of Agriculture, Zagreb, Croatia, [2]Lallemand SAS, Blagnac, 31700, France; cvillot@lallemand.com

The main objective was to evaluate the impact of an inoculant combining homofermentative and heterofermentaive bacteria on the microbiota and fermentation profile of a sugar rich grass ensiled for only 15 days. Pure Italian ryegrass (1st cut, DM=46.8%, WSC=33.4%) was chopped and ensiled in macro-silos of 40 l. Five replicates were ensiled without inoculant (CTRL), whereas another 5 replicates were inoculated with *P. pentosaceus, L. buchneri*, and *L. hilgardii* (INOC). A faster acidification in INOC compared to CTRL samples was observed (4.18 vs 4.58) thanks to the accumulation of lactic acid (103.2 vs 56.0 g/kg DM). A higher level of acetic acid was reported in CTRL compared to INOC (15.0 vs 9.0 g/kg DM) confirming the heterofermentative activity at early stage in non-inoculated silage. A large diversity of LAB (*Lentilactobacillus* and *Latilactobacillus*) was reported in the CTRL silage illustrating the natural fermentation process whereas INOC silage was dominated by *Pediococcus* and *Lentilactobacillus* supporting the fastest pH drop observed. Amplicon sequence variants (ASV) specific of inoculated strains were identified, with inoculated *L. hilgardii* and *L. buchneri* being higher in the INOC silages compared to CTRL. Non-inoculated silage was characterized by two wild ASVs of *L.buchneri/parabuchneri* partly responsible for the highest acetic acid level although no reduction in the total mould development was noted (5.2 vs 4.3 log10 cfu/g). In addition, undesirable bacteria *E. coli* was observed in significantly greater abundance in the CTRL silage than in the INOC samples (0.08 vs 0.03%). The dominant homolactic pattern of the inoculated silages supports lower fermentation losses (3.1 vs 9.9% of DM). The use of homolactic bacteria in difficult to ensile grass helps minimizing fermentation losses early on and may balance the development of dominant heterofermentative epiphytic flora that slows the acidification process without clear effects on antifungal metabolite production. Not all *L. buchneri* strains are metabolically similar and inoculation with selected strains can limit less efficient wild species to drive fermentation profile and silage quality.

Session 46 Poster 20

Effect of whey on *in vitro* ruminal methane formation and digestibility in cows

H. Luisier-Sutter[1,2], L. Isele[1], M. Terranova[3], S.L. Amelchanka[3] and M. Schick[1,2]
[1]University Hohenheim, Institute for Agricultural Engineering, Garbenstrasse 9, 70593 Stuttgart, Germany, [2]Strickhof, Division Animal Husbandry& Dairy Production, Eschikon 21, 8315 Lindau, Switzerland, [3]ETH Zurich, AgroVet-Strickhof, Eschikon 27, 8315 Lindau, Switzerland; helene.luisier@strickhof.ch

The agricultural sector is responsible for a large part of methane emissions worldwide, mainly because of cattle detention. Methane is produced by fermentative processes in the digestive tract of ruminants. Hence, feeding strategies might lower methanogenesis. Whey is a byproduct of caseation. It is available in large quantities, cheap and has a high nutritive value. Whey is rich in water-soluble carbohydrates since it mainly consists of lactose. Water-soluble carbohydrates do modify the fermentative processes in the rumen and might affect methanogenesis. A previous study from Agroscope (unpublished) on whey feeding in heifers observed a methane reduction of 37%. However, further research is needed to investigate the potential of whey as an inhibitory agent on methanogenesis. To this aim, an *in vitro* study was conducted using the Hohenheim Gas Test (HGT) method. A basal diet, consisting of grass-silage and hay (66% : 33%) was supplemented either with Emmentaler cheese whey, whey powder or pure lactose. The latter was added to test if the water-soluble carbohydrates in whey could be responsible for methane inhibition. Ruminal fluid was taken from 3 rumen-canulated Original Brown Swiss cows and incubated in 4 consecutive runs. For every run the roughage-based diet served as basal diet. Whey, whey powder and lactose replaced the basal diet in 3 different dosages: 5, 15 and 30% of diet DM for whey and whey powder and 1, 10 and 20% of diet DM for lactose to test for an effect of dosage as well. In total, there were 20 replicates per treatment and dosage. Against our expectations, preliminary results suggest that the addition of whey increased *in vitro* ruminal methanogenesis. The same effect, on a smaller scale, was found for whey powder and lactose. *In vitro* organic matter digestibility was also elevated with whey, whey powder and lactose. This finding is in line with previous studies which suggested that whey is an interesting option for replacing concentrate due to its high digestibility and feeding value.

Session 46 Poster 21

Ellagic acid and gallic acid reduced methane and ammonia in an *in vitro* rumen fermentation model

M. Manoni[1], S. Amelchanka[2], M. Terranova[2], L. Pinotti[1], P. Silacci[3] and M. Tretola[1,3]
[1]University of Milan, Department of Veterinary Medicine and Animal Science, Via dell'Università 6, 26900, Lodi, Italy, [2]ETH Zurich, AgroVet-Strickhof, Eschikon 27, 8315, Lindau, Switzerland, [3]Agroscope, Posieux, Rte de la Tioleyre 4, 1725, Posieux, Switzerland; michele.manoni@unimi.it

Livestock production must meet the growing demand for animal-source food and reduce the impact on the environment. Different strategies are studied to lower greenhouse gas (GHG) emissions in ruminants, and one of them is tannin supplementation. The dietary supplementation with tannins can affect the enteric fermentation, reducing methane emissions. Following a previous screening, we investigated the effect of two hydrolysable tannins, ellagic acid (EA) and gallic acid (GA), in a long-term *in vitro* rumen fermentation. EA and GA were supplemented to a control diet (CTR: ryegrass hay and barley concentrate, 10 g DM/day) in an 8-fermenter rumen simulation technique (Rusitec), for 10 days. Three experimental conditions were investigated: (1) EA 75 mg/g DM; (2) GA 75 mg/g DM; (3) EA 75 mg/g DM + GA 75 mg/g DM. The data were collected in the last 5 days of the incubation time. Total gas production was not altered over the last 5 days, whereas daily methane (CH_4) production was significantly decreased by EA (-45%) and EA+GA (-60%), compared to control. CH_4 production per unit of dietary organic matter (OM) and short-chain fatty acids (SCFA) was also reduced by EA (-48% and -32%) and EA+GA (-65% and -58%), and less by GA (-19% and -22%). Ammonia formation was significantly reduced by EA (-46%), GA (-19%) and EA+GA (-86%). Total SCFA production was decreased by EA and EA+GA (-26%, -16%), but not by GA. Similarly, EA and EA+GA, but not GA, reduced rumen degradability of OM, crude fibre (CF) and crude protein (CP). All the treatments increased the bacterial count and decreased the protozoal count (except for GA). EA and EA+GA modulated the relative abundance of selected rumen bacterial taxa. These data proved that both EA and GA decreased GHG emissions and ammonia formation, with EA being most effective than GA. Nevertheless, GA showed a lower interfering effect on nutrient rumen degradability. Further details on rumen microbiota dynamics and hydrolysable tannins metabolism will complete the outcome of the study.

Session 46 Poster 22

Effect of toasted soybean on dairy cows milk performances

A. Berchoux[1], M. Duval[1], E. Hermant[2], M. Legris[1] and M. Jouffroy[1]
[1]Institut de l'Elevage, 149 rue de bercy, 75012, France, [2]EPLEFPA Rethel, route de novion, 08300, France; alice.berchoux@idele.fr

Protein crops can be a way to reduce the use of imported soya meal. However, most of the protein content in crude pea/bean crops deteriorate very quickly in rumen. Heating protection of proteins can be a solution to increase the proportion of feed protein in the intestine (PDIA). This process was tested and evaluated on a mixed-crop dairy farm in French Ardennes. The trail was conducted in 2022 on two groups (control and testing) of twenty-seven dairy cows in complete and balance blocks. The control group received crude soybean and the testing group received toasted soybean. The trail started with a pre-testing period of two weeks and followed by a testing period of ten weeks. An analysis of enzymatic degradability at 1 hour (DE1) was done with crude and toasted soybean. For each cow, milk production, composition (protein and fat contents) and urea rate were analysed each week. Ingestion of dry matter per group was measured each day. The statistical analysis was done on the covariance on the mean data of the testing period, with covariate from the pre-testing period. Toasted soybean had a DE1 of 32% against 83% for crude soybean. PDI increased by 49% and PDIA by 123%. Toasting process did not have significantly impact on milk production and fat contents. Protein contents of the testing group significantly increased of 19 g/day ($P<0.05$) but protein concentration (g/kg of milk) did not increased. Urea rate did not vary between each group. DM intake was similar between each group (18.3 kg of DM/d). In regard of feeding system INRA07, the toasting process allowed to enhance PDIE intake of 109 g/d for the testing group. With a marginal efficiency of 20%, the response of protein contents (g/d) was in coherence with the result of our trial, but the increase was not enough to impact protein concentration (g/kg of milk) which is in coherence with previous results. The feeding cost of the testing group was higher of 4,1€/1000 l, with a stable milk performance the feeding margin was lower.

Session 46 Poster 23

The addition of dry ice as an attempt to inhibit proteolysis during ensilage of lucerne

M. Borsuk-Stanulewicz, C. Purwin and M. Mazur-Kuśnirek
University of Warmia and Mazury in Olsztyn, Department of Animal Nutrition, Feed Science and Cattle Breeding, Oczapowskiego 5, 10-719 Olsztyn, Poland; marta.borsuk@uwm.edu.pl

Lucerne (*Medicago sativa* L.) is an important crop plant of economic importance in animal and human nutrition, with a species susceptibility to proteolysis during ensilage. The hypothesis assumes that the addition of carbon dioxide as dry ice would modify the gaseous environment of the ensiled lucerne and improve the fermentation parameters, limiting proteolysis. The aim of the study was to investigate the effect of different levels of dry ice addition on the fermentation profile and protein fractions of ensiled lucerne. Lucerne was harvested in the first cut at the beginning of bud formation stage, at a height of 5 cm. The herbage was cut to a particle with an average length of 25 mm and the one part was allowed to wilted for 12 hours. Lucerne was ensiled in micro silos (1 dm^3) with the addition of: 0 g – control, 0.5, 1 and 2 g dry ice. After 50% of the micro silos volume was filled, dry ice was added as granules of a diameter of 16 mm and densiting was continued. The micro silos equipped with fermentation tubes were closed after the densiting was ended. The proximate chemical composition, fermentation parameters and nitrogen fractions were performer according to standard methods described in the literature. According to the literature, the parameters of fermentation were correct. Dry ice (2 g) significantly reduced ammonia nitrogen in both groups of silage (fresh and wilted) and reduced (1 and 2 g) the degradation of true protein in silage regardless of wilting. In unwilted lucerne silage (2g dry ice) the fraction of protein insoluble in neutral detergent increased. According to the CNCPS, the addition of dry ice significantly reduced the PA1 fraction and the proportion of fibre-bound protein (PB2) in wilted lucerne silage. The addition of 1 and/or 2 g dry ice significantly changed the fermentation parameters, limiting ammonia nitrogen and degradation of true protein. This research was funded in whole or in part by the National Science Centre, Poland grant no. 2021/41/N/NZ9/01881.

Session 46 Poster 24

***In vitro* effects of *Bacillus subtilis* CH201 and *Bacillus licheniformis* CH200 on rumen microbiota**

R. Gresse[1], G. Copani[2], B.I. Cappellozza[2], A. Torrent[1], D. Macheboeuf[1], E. Forano[3] and V. Niderkorn[1]
[1]UCA, INRAE, VetAgro Sup, UMRH, Route de Saint Genes, 63122 Saint Genes Champanelle, France, [2]Chr. Hansen, Animal and Plant Health & Nutrition, Boege Alle 10/12, 2970 Hørsholm, Denmark, [3]UCA, INRAE, UMR 454 MEDIS, Route de Saint Genes, 63122 Saint Genes Champanelle, France; raphaele.gresse@inrae.fr

Direct-fed microbials has improved livestock's performances and overall health status. Several research studies have been undertaken to find new potential probiotics and to gain knowledge about their mechanisms of action in the gut. *Bacillus* species are multifunctional spore-forming bacteria surviving to harsh conditions and thus having numerous potential applications in feed industry and livestock production. The present study aimed to investigate the mode of action of *B. lichenliformis* CH200 and *B. subtilis* CH201 (BOVACILLUS™, Chr. Hansen, Denmark) in the rumen using diverse *in vitro* techniques. *In vitro* inoculation of spores showed that both strains were able to germinate and grow in rumen and intestinal contents obtained from dairy cows. Gas composition analysis of *in vitro* cultures in a medium containing 40% rumen juice revealed that germination of CH200 and CH201 strains could reduce O_2 level, thus favouring an anaerobic environment beneficial for rumen microbes. Additionally, we observed that a cocktail containing CH200 and CH201 spores in a rumen fluid medium could survive and grow with a commercial dose of monensin sodium. The effects of CH200 and CH201 on rumen fermentative activity and microbiota were studied using an *in vitro* batch fermentation assay. In fermenters that received a combination of CH200 and CH201, less CO_2 was produced while dry matter degradation and CH_4 production was similar compared to the control condition, indicating a better efficiency of dry matter utilization by the microbiota. Investigation of the microbiota composition in fermenters showed that although no significant effect was observed on alpha and beta diversity, the differential analysis highlighted changes in several taxa in the supplementation condition compared to the control condition. Altogether, these data suggest that the administration of a cocktail of CH200 and CH201 could have a beneficial impact on rumen function and consequently on health and performance of ruminants.

Session 46 Poster 25

Effect of nitrate and its interaction with starch levels on methane production in continuous culture

Y. Roman-Garcia, S. El-Haddad, S. Van Zijderveld and G. Schroeder
Cargill Animal Nutrition and Health, Innovation Campus, 10383, 165th Ave NW, Elk River, MN, 55330, USA; yairy_romangarcia@cargill.com

The objective of this study was to evaluate the effect of calcium nitrate (SilvAir®, Cargill Inc.) supplementation and its interaction with levels of starch on methane production and rumen fermentation in *in vitro* continuous culture. Twelve fermenters were used in a randomized block design experiment of 2 periods (10-d each) consisting of 2 diets (21% or 28% starch; LST or HST, respectively) with or without nitrate supplementation at 1% DM in a factorial arrangement. Fermenters (2 l capacity) were fed (~ 100 g) a lactating cow diet once a day. Treatments were made isonitrogenous by adding urea or nitrate and starch levels were modified by replacing fine ground corn with soyhulls. Methane in the headspace was monitored continuously in all fermenters using a Micro-Oxymax Respirometer (Columbus Instrument Inc., Columbus OH). Data was analysed with lmer in R with fixed effects of diet, treatment, and their interaction, and the random effects of period and fermenter. Overall, nitrate supplementation reduced ($P<0.01$) methane production rate by 21% (3.67 vs 2.89 g/d). There was a tendency for an interaction ($P=0.06$) starch level by nitrate supplementation, suggesting that reduction of methane by nitrate was higher in HST than LST (-30 vs -6%, respectively). Treatments with nitrate had lower methane production rate on all 24 hours post-feeding. There was no effect of treatments on starch degradation (98%). Replacing urea by nitrate did not affect NDF degradability, but HST reduced NDF degradation (49 vs 41%; $P=0.03$). Supplementing nitrate did not affect total VFA concentration, but that was higher in LST than HST (111 vs 121 mM, $P<0.01$). Both nitrate and LST increased ($P<0.01$) acetate production and decreased ($P<0.01$) propionate production, resulting in higher acetate:propionate ratio (2.30 vs 2.61, $P<0.01$). Mean daily pH was not affected by nitrate, but the LST diet had a tendency ($P=0.08$) to decrease mean daily pH (6.03 vs 5.99) mostly explained by lower pH between 12 to 24 h after feeding. This study validates that nitrate is an effective strategy to reduce methane production and that the magnitude of the reduction can vary with the type of diet.

Session 46 Poster 26

Effect of garlic processing method on *in vitro* methane production and rumen fermentation

N.F. Sari[1,2], S. Stergiadis[2], P.P. Ray[2,3], C. Rymer[2], L.A. Crompton[2] and K.E. Kliem[2]
[1]National Research and Innovation Agency, Cibinong, 16911, Indonesia, [2]University of Reading, Reading, RG6 6EU, United Kingdom, [3]The Nature Conservancy, Arlington, VA 22203, USA; nurulfitri.sari@pgr.reading.ac.uk

The effect of garlic processing method (freeze-dried, FD; oil) and origin (China; Spain) at different dietary inclusion rates (3, 6, 12% on dry matter (DM) basis) on gas, methane (CH_4) and volatile fatty acid (VFA) production was assessed in a batch culture *in vitro* system. Garlic was chopped, and stored at -20 °C until processing. One batch from each origin was freeze-dried and milled to <2 mm, and another batch underwent diethyl ether extraction of the essential oil. A total mixed ration (forage:concentrate, 50:50 DM basis) without (Control, CON) or with garlic at each of the three inclusion rates was incubated for 72 h with rumen fluid: incubation medium at 1:9 v/v. Gas pressure was recorded and gas samples were analysed for CH_4 concentration by gas chromatography at intervals throughout the incubation. DM degradability (DMD) and VFA concentrations in the medium were assessed after 72 h. The gas and CH_4 production profiles were fitted to the model of France *et al*. Data were analysed by linear mixed models, using processing method, origin, and their interaction as fixed, and run as random, factors. DMD (g/kg DM) was higher ($P<0.01$) in FD, than in CON, across all inclusion rates (807, 797, and 791 vs 783 at 12, 6 and 3% DM, respectively); and the same was observed for OIL at 12% DM (811) and 6% DM (802). Total CH_4 production (ml/g DM) was higher ($P<0.05$) in OIL, than in CON (10.9 ml), at 6% (12.4 ml) and at 12% (11.9 ml) but not at 3% DM ($P>0.05$). Acetate:propionate ratio (A:P) was lower ($P<0.001$) in both garlic treatments across all inclusion rates (2.4-2.9) compared with CON (3.0). Garlic origin did not affect DMD and A:P ($P>0.05$). Garlic from China resulted in higher total CH_4 production (12.7 ml/g DM) than garlic from Spain (10.0 ml/g DM) at 3% DM ($P<0.05$) but origin had no effect at 6 and 12% DM ($P>0.05$). Garlic from China resulted in higher terminal pH than garlic from Spain across all inclusion rates (6.8 vs 6.5). Adding garlic did not reduce *in vitro* CH_4 production but improved overall diet DMD.

Session 46 Poster 27

Effects of Red Sorghum and *Rhizoma paridis* on rumen protozoa, fermentation characteristics *in vitro*

R. Yi[1,2], S. Vigors[1], L. Ma[2], J.C. Xu[2] and D.P. Bu[2]
[1]University College Dublin, School of Agriculture and Food Science, Belfield, Dublin 4, Ireland, [2]Institute of Animal Sciences, Chinese Academy of Agricultural Sciences, State Key Laboratory of Animal Nutrition, Yuanmingyuan West Road NO.2, Haidian District, Beijing, 100193, China, P.R.; ran.yi@ucdconnect.ie

The inclusion of plant-derived supplements in ruminant diets is one potential strategy to modify fermentation and reduce emissions. Rumen protozoa play an important role in lowering nitrogen utilization efficiency (NUE) and are therefore, a target for alterations both in numbers and functionality. This research aimed to examine the potential of *Sorghum bicolor* L. Moench (RS) (0, 10.4, 20.8, 41.6, 83.2 mg/g) and *Rhizoma paridis* (RP) (0, 2.6, 5.2, 10.4, 20.8 mg/g) to inhibit rumen protozoa *in vitro* and impact NUE. Fresh rumen fluid was collected from four cannulated dairy cows and filtered. The effect of RS and RP powder (n=6) was tested on rumen protozoal counts and fermentation characteristics following 18 hours of *in vitro* culturing. Rumen protozoal cells were morphologically identified and counted using microscopy. Ammonia nitrogen (NH_3-N) and microbial protein (MCP) concentrations in the *in vitro* cultures were tested using colorimetric determination, and concentrations of volatile fatty acids (VFAs) were determined using gas chromatography. Data were analysed in a complete randomized design using RStudio (version 1.4.1717). The results demonstrated that, compared with the control group, RS and RP stem powder at 83.2 and 5.2 mg/g decreased *Entodinium* cells by 35.33 and 26.04%, respectively (P<0.05). NH_3-N was decreased by 32.15 and 15.27% at 83.2 and 5.2 mg/g inclusion of RS and RP stem respectively. In addition, methane production was predicted to be reduced by 15% with inclusion of 83.2 mg/g of RS, while total VFA production was unchanged. Moreover, after correlation analysis, the number of total protozoa and *Entodinium* are positively correlated with NH_3-N, isobutyrate, A:P ratio and methane, while, negatively correlated with Propionate and MCP. In conclusion, RS and RP may be potential feed additives to enhance NUE in ruminants without impairing fermentation characteristics, but future research is warranted to identify the target phytochemicals and evaluate in animal trials.

Session 46 Poster 28

Phytochemicals can modify the rumen fermentation profile as monensin

L. Gonzalez[1], A.C. Dall-Orsoletta[2], D. Mattiauda[1], A. Daudet[1], T. Garcia[1], P. Chilibroste[1], A. Meikle[3], M. Arturo-Schaan[2], A. Casal[1] and M.A. Bruni[1]
[1]Facultad de Agronomía, Ruta 3 km 363, 60000, Paysandú, Uruguay, [2]Deltavit, CCPA Group, ZA du Bois de Teillay, 35150, JANZÉ, France, [3]Facultad de Veterinaria, Ruta 8 Km 18, 13000, Montevideo, Uruguay; acdallorsoletta@ccpa.com

To investigate the influence of phytochemicals (PE; containing Trans cinnamaldehyde, Flavonoids, Curcuminoids, and Piperine) and a monensin (MO) on the rumen fermentation parameters rumen liquid samples were collected from 9 fistulated cows 7 d prepartum, 30 and 60 d postpartum from cows that had either received a PE (n=3) or MO (n=3) from 3 weeks antepartum on feed, or a control diet (n=3). The samples were analysed for pH, volatile fatty acid (VFA), ammonia concentrations, and protozoal counts. Before calving cows ate a total mixed ration (70% forage+30% concentrate on DM basis) after calving cows had a daily grazing shift (Medicago sativa) and were supplemented with a partial mixed ration (45% forage+55% concentrate on DM basis). In the prepartum ruminal sample, there were no differences in ruminal pH, ammonia, and total VFA as well as the proportion of the main VFA (acetate, propionate, butyrate). However, PE increase the branched-chain fatty acids compared to the control, and MO and PE treatments decreased the number of protozoa by 39% (P<0.01). During the postpartum period, there were no differences in ruminal pH, acetate proportion, and acetate: propionate. Ammonia concentration was lower on PE (P<0.01). Total VFA is similar between PE and MO, but PE decreases is comparable to the control (P=0.09). PE and MO modified the VFA proportion. Propionate proportion was not affected by MO, but was increased by PE treatment (P=0.08). PE reduced butyrate and isobutyrate compared to MO and control (P<0.01). MO and PE treatments decreased the number of protozoa on average by 38 and 64%, respectively (P<0.01). The rumen fermentation profile was affected by physiological conditions, diets, and PE or MO supplementation, additionally, the results suggest that PE can be useful as a rumen fermentation modifier and, considered as a potential alternative to monensin.

Session 46 Poster 29

In vitro fermentation of TMR using rumen fluids from cows supplemented with hemp and savory leaves

S. Arango[1], S. Massaro[1], S. Schiavon[1], N. Guzzo[1], M. Montanari[2], L. Bailoni[1] and F. Tagliapietra[1]
[1]University of Padova, BCA/DAFNAE, V. dell'Universitá 16, Legnaro, 35020, Italy, [2]Council for Agricultural Research and Economics, V. di Corticella 133, 40128 Bologna, Italy; sheylajohannashumyko.arangoquispe@phd.unipd.it

Industrial hemp (*Cannabis sativa* L.) leaves could be a good source of protein and fibre for ruminants. Summer savory (*Satureja hortensis* L.) is an aromatic plant with antimicrobial and antioxidant activities. The aim was to assess the fermentation parameters during *in vitro* incubations of rumen fluids collected from cows fed a total mixed ration (TMR) supplemented with hemp and savory leaves. Hemp leaves were harvested at the Center for Cereal and Industrial Crops (CREA-CI, Rovigo, Italy) and savory was provided by Agripharma srl (Padova, Italy). Six Italian Simmental lactating cows were divided into 3 groups on a 3×3 Latin square design. Each experimental period lasted 2 weeks. A TMR, based on grass and sorghum silages, was used for the experiment. During the first week, the control group (CTR) was fed with TMR, the hemp leaves group (HL) and savory group (SS) were fed TMR plus increasing amounts of leaves until 1.5 and 1.0 kg/d were reached, resp. Rumen fluid samples were collected after the first week of each period by an oesophageal probe and analysed for volatile fatty acids (VFA). Three rumen fluids were used as inocula on *in vitro* incubations with an automated gas production (GP) equipment. Each run (48 h) used TMR as substrate to test the microbial activity. The fluids were analysed after fermentation for pH, dry matter degradability (DMd), ammonia nitrogen, VFA, GP kinetics and gas composition (CH_4, CO_2, H_2). Results showed that the acetic acid proportion in the rumen fluid was lower in the HL than the CTR group (57.4 vs 59.3% of VFA; $P<0.10$). The HL rumen fluid did not affect DMd of the TMR or GP kinetics but, it increased the acetic acid proportion (50.0 vs 48.1% of VFA; $P<0.10$). Despite these, methane production was similar in both groups. Savory supplementation did not show any effect on rumen fluid composition or *in vitro* parameters. To conclude, fresh hemp and savory leaves did not have an inhibitory effect on rumen TMR degradability. It appears that HL have a slight effect on rumen microbial activity by modulating VFA production. Funded: BIRD213117/21.

Session 46 Poster 30

Bacillus licheniformis and *B. subtilis* on *in vitro* ruminal parameters and greenhouse gas emission

B.R. Amancio[1], E. Magnani[1], T.H. Silva[1], A.L. Lourenço[1], B.I. Cappellozza[2], R.H. Branco[1] and E.M. Paula[1]
[1]Institute of Animal Science, Beef Cattle Research Center, Rodovia Carlos Tonanni, km 94, 14.160-970, Sertãozinho, SP, Brazil, [2]CHR Hansen A/S, Research and Development, Boege Alle 10-12, 2970 Horsholm, Denmark; renata@sp.gov.br

Three *in vitro* experiments were conducted to evaluate the effects of increasing levels of *Bacillus licheniformis* and *Bacillus subtilis* (BB) on *in vitro* ruminal parameters and enteric greenhouse gas (GHG) emissions in three different scenarios. For Exp. 1, the basal diet consisted of 25:75 roughage: concentrate ratio (R:C) and was composed by 3 treatments: control (no additive) and 2 levels of BB (3.8 mg and 19 mg). The Exp. 2 consisted of a 40:60 R:C diet and was composed by 3 treatments: control (no additive) and 2 levels of BB (5.6 and 28 mg). The Exp. 3, consisted of a 100:0 R:C diet [*Brachiaria* (syn. *Urochloa brizantha*)] and the same treatments as in Exp. 1. The BB product contained *Bacillus licheniformis* + *Bacillus subtilis* (3.2×10^9 cfu/g). An *in vitro* gas production (GP) system with 12 bottles (AnkomRF) was used in four consecutive fermentation batches each experiment to evaluate total GP (TGP), volatile fatty acids (VFA), enteric methane (CH_4) and carbon dioxide (CO_2). Supplementation levels of each experiment were analysed for linear and quadratic responses using the software SAS 9.4. For Exp. 1, there were no effects of BB on ruminal fermentation and GHG emissions. For Exp. 2, after 48 h of fermentation, TGP and organic matter digestibility (OMD) increased quadratically with increasing BB supplementation. In addition, propionate increased quadratically with increasing doses of BB, which reflected on reduced acetate to propionate ratio. Ammonia-N and CH_4 and CO_2 emissions related to OMD also reduced quadratically according to BB addition. Finally, for Exp. 3, after 48 h of fermentation, TGP increased linearly as BB increased. In addition, butyrate proportion and ammonia-N reduced linearly and quadratically, respectively, as BB increased. Overall, BB supplementation may improve ruminal fermentation and reduce GHG emissions in a dose-dependent manner, but the outcomes are dependent upon diet type.

Bioproduct from royal palm colonized by *Lentinula edodes* on *in vitro* ruminal fermentation

B.M. Rocha[1], R.L. Savio[1], G.S. Camargo[1], K.E. Loregian[2], A.R. Cagliari[1], A.C. Casagrande[1], F. Rigon[1], E. Magnani[2], T.G. Timm[3], L.B.B. Tavares[3], M.I. Marcondes[4], T.H. Silva[2], R.H. Branco[2], E.M. Paula[2] and P.D.B. Benedeti[1]
[1]Universidade do Estado de Santa Catarina, Beloni Trombeta Zanin street, 680E, 89815-630, Chapecó, Santa Catarina, Brazil, [2]Institute of Animal Science, Carlos Tonani Road, km 94, 14160-970, Sertãozinho, São Paulo, Brazil, [3]Fundação Universidade Regional de Blumenau, Antônio da Veiga street, 140, 89030-903, Blumenau, Santa Catarina, Brazil, [4]Washington State University, ASLB, 120, 99164-6310, USA; dasilvath2@gmail.com

The agroindustrial by-products use in ruminant nutrition have a great potential to promote both profitability and sustainability. The king palm (*Archontophoenix spp.*) residue is a high fibre content. Thus, its treatment with fungal (*Lentinula edodes*) colonization produces a more digestible bioproduct that can be an interesting option for ruminant nutrition. Thus, the aim of this study was to evaluate the effect of king palm bioproduct on ruminal fermentation parameters in comparison with corn silage using an *in vitro* gas production system. The treatments were the two feeds (corn silage and king palm bioproduct), which were individually evaluated in three consecutive 48-hour incubations. Each treatment had 8 replicates, plus 3 blanks, totalling 19 observations per incubation. Sixteen 250 ml flasks were used to analyse fermentation parameters, total gas production, and rumen kinetics. Simultaneously, sixteen 100 ml flasks were used for enteric CH_4 and CO_2 evaluation. Data were analysed using the PROC NLMIXED (for gas production models estimation) PROC MIXED (remaining variables) in SAS, with α=0.05. King palm bioproduct had lower total gas production, metabolizable energy, organic matter digestibility, and greenhouse gases profile. Regarding the kinetic parameters, king palm bioproduct presented lowest value for the first pool, compared with corn silage. However, treatments did not differ for the other kinetic parameters. These results suggest that king palm bioproduct has as sustainable potential to be used as roughage in feedlot diets.

***Enterococcus faecium* and *Saccharomyces cerevisiae* on *in vitro* rumen parameters and greenhouse gases**

B.R. Amancio[1], T.H. Silva[1], G.M. Wachekowski[1], H. Reolon[1], T.G. Timm[1], B.I. Cappellozza[2], E. Magnani[1], E.M. Paula[1] and R.H. Branco[1]
[1]Institute of Animal Science, Carlos Tonani Road, km 94, 14160-970, Sertãozinho, São Paulo, Brazil, [2]Chr. Hansen, Boege Alle 10-12, 2970, Denmark; dasilvath2@gmail.com

Three *in vitro* experiments were conducted to evaluate the effects of increasing levels of *Enterococcus faecium* and *Saccharomyces cerevisiae* (ES) on *in vitro* ruminal parameters and enteric greenhouse gases emissions in three different diets. For Exp. 1, basal diet consisted of 25:75 roughage: concentrate ratio (r:c) and was composed by 3 treatments: control (no additive) and 2 levels of ES (1.9 mg and 9.0 mg). The Exp 2 consisted of a 40:60 r:c diet and was composed by 3 treatments: control (no additive) and 2 levels of ES (3.8 and 19.0 mg). Finally, Exp 3. consisted of a 100:0 r:c diet [*Brachiaria* (syn. *Urochloa brizantha*)] and the same treatments as Exp. 1. The ES contained 2.5 and 1.5×10^9 cfu/g of *Enterococccus faecium* and *Saccharomyces cerevisiae*, respectively. An *in vitro* gas production (GP) system with 12 bottles (AnkomRF) was used in four consecutive fermentation batches each experiment to evaluate total GP (TGP), volatile fatty acids (VFA), enteric methane (CH_4) and carbon dioxide (CO_2). Supplementation levels of each experiment were analysed for linear and quadratic responses using the software SAS 9.4. For Exp. 1, after 24 and 48-h of fermentation, TGP and organic matter digestibility (OMD) increased quadratically with increasing ES. Further, ammonia-N, as well as CH_4 and CO_2 related to OMD decreased quadratically. For Exp. 2, after 24 and 48-h of fermentation, TGP and OMD increased quadratically with increasing ES inclusion. Also, butyrate proportion, ammonia-N, and CH_4, and CO_2 emissions related to OMD reduced quadratically according to ES addition. Finally, for Exp. 3, similar quadratic increase of TGP and OMD were detected after 48 and 72-h of fermentation. Further, quadratic increase of pH and branched-chain volatile fatty acids were detected as ES increased. Ammonia-N was quadratically reduced with increasing doses of ES. Overall, ES supplementation may improve rumen fermentation as well as reduce GHG emissions of different diets in a dose-dependent manner with no detrimental effects on VFA profile.

Session 46 Poster 33

Rumen protected calcium gluconate improves milk production of cows

J.G.H.E. Bergman, B. Skibba and M. Hall
Nutreco, Selko, Stationsstraat 77, 3800 AG Amersfoort, the Netherlands; jac.bergman@selko.com

Calcium gluconate (CG) is a prebiotic that increases butyrate in the hindgut, reducing the risk of leaky gut. Trials have shown that abomasal infusion of CG increases milk production and feeding rumen protected calcium gluconate (RPCG) increases fat corrected milk (P=0.056) and energy corrected milk (P=0.086). Purpose of this trial was to test the effect of RPCG (LactiBute, Selko) on 2 commercial farms. The first farm had 550 cows, 65 were randomly divided in 2 groups, immediately after calving until 120 days into lactation. Both groups were fed TMR with 270 g starch and 180 g crude protein/kg DM. RPCG was added to the TMR of the test group. Milk production and solids were recorded on day 0, day 30, day 60, day 90 and day 120. Average milk production over the entire period was 2.35 kg/cow/day higher in cows fed RPCG (P<0.05). Average production of energy corrected milk by 2.77 kg/cow/day (P<0.05). The second farm had 240 cows, 71 cows were randomly allocated to one of 2 groups and included at 30 days after calving until they were 120 days into lactation. Both groups were fed TMR with 260 g starch and 170 g crude protein/DM. RPCG was added to the TMR of the test group. Milk production was recorded on day 30, day 60, day 90 and day 120. Milk solids were not measured on this farm. Average milk production over the entire period was 1.85 kg/cow/day higher in the group of cows fed RPCG (P<0.05). Conclusion: milk production increased in cows kept under commercial conditions as a result of feeding rumen protected calcium gluconate. These results are in line with the results of earlier trials.

Session 46 Poster 34

Rumen protected calcium gluconate improves lactational performance

J.G.H.E. Bergman, F. Morisset and M. Hall
Nutreco, Selko, Stationsstraat 77, 3800 AG Amersfoort, the Netherlands; jac.bergman@selko.com

Rumen protected calcium gluconate (RPCG) has a prebiotic effect in the hindgut, increasing volatile fatty acid production, which reduces the risk of leaky gut and improves milk production. Earlier trials have shown that RPCG increases milk production of dairy cows. Purpose of the current trial was to test RPCG (LactiBute, Selko) in 3 French farms. RPCG was fed for 4 months. Milk recordings were compared to predicted production based on genetic potential of each of the 3 dairy herds included in the trial, based on data from all cows using the same supplier of bull semen that also recorded production data on all their farms. Farm 1 had 120 cows, 155 days in milk at the start of the trial. Persistence of milk production was improved as a result of feeding RPCG with a difference in milk production of up to 1.7 kg of milk. Farm 2 had 90 cows, 138 days in milk when the trial started. RPCG resulted in a better persistence of milk production with a difference of up to 1.8 kg of milk per day. Farm 3 had 200 dairy, 170 days in milk when the trial started. Persistence of milk production in cows fed RPCG increased. When the trial was 2 months underway, respiratory disease occurred, leading to a drop in production. When the herd recovered, milk production in the cows on RPCG recovered faster, with a difference of up to 2.7 kg of extra milk per day. Conclusion: before the start of the trial, production on all 3 farms was equal to the predicted production based on genetic potential. Persistence of milk production improved during the period cows were fed RPCG. Once RPCG was stopped, milk production dropped to a level at or even below the predicted genetic potential.

Session 46 Poster 35

Increase milk production by preserving the nutritional value of the dairy ration

L.L.C. Jansen, J.G.H.E. Bergman and S.J.A. Van Kuijk
Selko, Feed Additives, Stationsstraat 77, 3811 MH Amersfoort, the Netherlands; lonneke.jansen@trouwnutrition.com

Growth of micro-organisms in a Total Mixed Ration (TMR) for dairy cows reduces both the palatability and feeding value of TMR. Applying a synergistic blend of organic acids to reduce the growth of microbes in TMR is a critical step in proper silage management. The objective of this study was to test the impact of TMR preservation with an organic acid based blend on milk production of dairy cows. The study was carried out with 120 lactating, of which 8 fistulated, dairy cows on the Trouw Nutrition dairy research centre (Kempenshof, the Netherlands). Cows were blocked based on parity and days in milk (DIM). After an adaptation period of 1 week, both groups of 60 dairy cows each were fed the same diet based on a mixture of grass and corn silage, mixed into a TMR. The treatment group received TMR including 2kg/mt of the organic acid blend (Selko® TMR, Selko®, Tilburg, The Netherlands). Both groups were fed once a day for a total study period of 7 weeks. Compound feed was fed according to a standard curve based on DIM. Environmental temperature and humidity were monitored. Individual dry matter intake (DMI) of the TMR was measured daily using automatic feeding bins. Milk production including milk solids was recorded weekly for each animal and the volatile fatty acid (VFA) profile of rumen fluid was measured weekly in the 8 fistulated cows. Average daily temperatures were between 20 and 30 °C during the study. The Temperature Heat Index was above 72 regularly, which suggests that the cows were suffering from heat stress. DMI and VFA profile of rumen fluid did not differ significantly during the study. At the start of the study, there was an insignificant difference of 0.3 kg/day (P=0.52) of milk production between the treatment groups. This difference numerically increased towards the end of the study with the cows fed the organic acid blend producing 1.7 kg/day more milk compared to the control (P=0.07). Milk fat and protein levels (g/kg) did differ significantly. Although feed intake was not different in this study, dairy cows fed the treated TMR tended to produce up to 1.7 kg/day more milk compared to the controls. This increase of milk production is most likely related to a higher nutritional value of the TMR treated with the organic acid blend.

Session 46 Poster 36

VistaPre-T, a crude fermentation extract to support the sustainable use of forages in dairy rations

V. Blanvillain, E. Bungenstab and G. Gomes
AB Vista, Woodstock Court, Marlborough, United Kingdom; virginie.rivera@abvista.com

Enteric methane corresponds to approximately 40% of the carbon emissions coming from livestock. While feeding greater dietary starch or fat levels contribute to reducing enteric methane emissions per animal, greater NDF levels have the opposite effect. Under this perspective, forages do not appear to be an alternative to sustainable livestock farming, and the high variability of their quality and nutritional content makes it even more challenging. However, forages have a role to play in the biogenic cycle of methane, and in circular economy models. Along with an accurate forage characterization, nutritional strategies which maximize nutrient utilization from forages are a means to support the use of these inputs while remaining sustainable. VistaPre-T (VPT, AB Vista, UK), a crude fermentation extract, significantly increased NDF and ADF digestibility, by increasing the utilization of hemicellulose and cellulose of forages in an *in vitro* rumen fermentation model (P<0.05). It was assumed that the greater digestibility could result in a beneficial impact on the global warming potential (GWP) of milk, thanks to a greater milk yield or a decrease in feed GWP. Two studies were performed in commercial farms to measure and compare the effect of VPT on GWP to a hundred years (GWP100), per kg of fat and protein corrected milk (FPCM) at farm gate. In the first study, VPT was added on top of a ration containing 70% of grass silage. Dry matter intake (DMI) decreased by 9%, with no significant impact on milk yield or composition. In the second study, the ration was reformulated to balance metabolizable energy content of the total mixed ration, by assigning an energy value of 800 kJ/kg of DMI from forage. A 5% reduction in DMI was observed, along with a 9% increase in milk yield. Overall, GWP100 was reduced by 4 and 15% per kg of FPCM, in the first and second study, respectively. Because no experimental data has yet been published, these estimates did not account for the potential beneficial effect of VPT on enteric methane emissions, as fiber digestibility is increased. Measuring GWP at farm gate helped understand the effect of feeding strategies on productivity, ingredient usage, and overall farm sustainability.

Session 46 Poster 37

Early lactation trial with a blend of fat encapsulated vitamin B

A.D.G. Esselink[1] and C. Gordon[2]

[1]Trouw Nutrition, Global Innovation, Stattionsstraat 77, 3811 MH Amersfoort, the Netherlands, [2]Trouw Nutrition Canada, Dairy Technology Application, 7504 McLean Rd E, Puslinch, ON N0B 2J0, Canada; allard.esselink@trouwnutrition.com

Several researchers have shown synthesis of B vitamins can be insufficient in highly productive cows. Supplementation with B-vitamins can improve milk yield and composition in highly productive dairy cows. The effect of a blend of fat encapsulated B vitamins (FEBV, Vivalto, Selko) was studied in an early lactation trial with 54 cows, divided into 2 groups. One group was fed RBBV between day 4 and day 200 of lactation, the other group served as controls. Supplementing dairy cows with FEBV increased milk yield between 75 and 200 DIM by 3 kg/d ($P<0.05$) and protein yield by 0.08 kg/d when fed between 4 and 200 DIM ($P<0.05$). Feed efficiency increased (+8%; $P<0.03$) from 1.63 kg milk/kg DMI to 1.77 kg milk/kg DMI throughout the 200 day experiment.

Session 46 Poster 38

The effect of a blend of fat encapsulated vitamin B on milk production of highly productive cows

M. Hall[1] and C. Gordon[2]

[1]Nutreco, Selko, Stattionsstraat 77, 3811 MH Amersfoort, the Netherlands, [2]Trouw Nutrition Canada, Dairy Technology Application, 7504 McLean Rd E, Puslinch, ON N0B 2J0, Canada; mark.hall@trouwnutrition.com

Changes in rumen fermentation can have an effect on apparent synthesis of B vitamins. As a result, diets of highly productive cows, particularly if they have a high cereal content do sometimes not meet the demand for B vitamins. Several researchers have shown that B-vitamin supplements can improve milk yield, composition, and metabolic efficiency in highly productive dairy cows. The effect of a blend of fat encapsulated B vitamins (FEBV, Vivalto, Selko) was studied in a 2×2 factorial design experiment including 260 multiparous cows and 140 heifers. Three groups of cows were fed RBBV from 21 days pre-partum to calving, from calving to 150 DIM or from 21 days pre-calving until 150 DIM respectively. A control group was not fed RBBV. Group dry matter intake and daily milk yield and milk composition were recorded. Supplementing multiparous dairy cows with FEBV increased milk yield by 1.37 kg/d and milk protein yield by 47 g/d($P<0.05$) with no effect on milk protein concentration. Milk fat concentration tended to decrease ($P=0.067$) due to dilution effects when FEBV was fed post-partum, but there was no change in milk fat yield. There was no significant increase in milk production in heifers fed FEBV.

Session 46 Poster 39

Evaluation of milk performances and enteric methane emissions on by-products feed base

A. Berchoux[1], M. Jouffroy[1], A. Laflotte[1] and R. Boré[2]
[1]Institut de l'Elevage, 149 rue de bercy, 75012 Paris, France, [2]Université de Lorraine, ferme de la Bouzule, 54280 Laneuvelotte, France; alice.berchoux@idele.fr

By-products are produced by agrifood industries during process of transformation of raw materials. They have high energetic and protein values and represent a local resource for farmers to feed cattle. Moreover, they are not in competition with human food. By-products are very interesting for the feeding of ruminants and 75% of them are used in cattle feed in France. In the North-East of France, in Grand-Est region, most of by-products generated are wet (<30% of humidity rate) and need to be mixed with other by-products and sometimes with fodders to reduce matter loses and sanitary issues. A trial has been carried out in the experimental farm of the University of Lorraine from the 4th of January to the 29th of March 2023 to evaluate the effects of two types of by-products storage on dairy cows milk performances (quantity and quality) and enteric methane emissions. The trial is conducted on two groups (control and testing) of twenty-three dairy cows in complete and balanced blocks. The control group received a Total Mixed Ration (TMR) mixed every day with fodder and by-products stored separately. The testing group received the same TMR but totally mixed before the beginning of trial and stored in the same silo. The trial has started with a pre-experimental period of two weeks and has been followed by a ten weeks experimental period. For each cow, milk production, milk composition (protein and fat contents) and urea rate are analysed each week and enteric methane emissions are collected continuously by two GreenFeed systems. An analysis of silo conservation was done during the pre-experimental period and at the end of the experimental period. Dry matter intake (DMI) per group is registered each day. At this time, milk performances and methane emissions of two groups seems similar. Statistical analysis will be done in May with R software.

Session 46 Poster 40

Supplementing lambs with plant extract supplement enhances growth and improve feed conversion ratio

V. Ballard[1] and P.H. Pomport[2]
[1]CCPA, ZA du Bois de Teillay, 35150 JANZE, France, [2]Ferme de Grignon, Route de la ferme, 78850 Thiverval Grignon, France; vballard@ccpa.com

The plant extract has been employed in ruminants to optimize the diet formulation and improve the energy and protein utilization in the rumen. The aim of this study was to optimize the diet of lamb by reducing the crude protein (CP) by combing 3 different plant extracts supplements (CCPA, France) on performance, meat quality, and health. For this, 172 male lambs 81 days old Romane breed were randomly distributed among the 4 treatments. The treatments consisted of: without plant extracts supplementation (CTL), supplementation with tannins and spices (Grp A); supplementation with tannin, essential oils and spices (Grp B), and supplementation with essential oils and spices (Grp C). Lambs were fed concentrate and straw *ad libitum*. The plant extracts were added to the concentrate and the treatments differed in crude protein content (CP), 17.5 and 16.8% of CP for the CTL and plant extract supplementation treatments, respectively. The CP content in the concentrate was achieved by reducing the soybean meal inclusion to 16,8% in the Grp A, B, and C instead of 19.8% in the CTL. The concentrate intake, average daily gain (ADG), feed conversion ratio (FCR) and protein conversion ratio was followed. On average lambs were slaughtered 114 days old. The morbidity and mortality did not differ between groups. The concentrate intake was 1.43, 1.47, 1.47, and 1.54 kg/day for CTL, Grp A, Grp B, and Grp C, respectively. The plant extract supplementation increased the ADG during the first 14 days of fattening, on average 437, 460, 443 g/day (Grp A, Grp B, and Grp C, respectively) compared to CTL (340, $P<0.001$). The total fattening period ADG tended to be greater for Grp A, B, C (426, 432, 419 g) compared to CTL (390 g, $P<0.15$). No difference was observed between groups for carcass weight at slaughter and carcass classification. The FCR was higher in the Grp A and B. The protein conversion ratio was improved in Grp A and Grp B (62.4, 62.7 g of CP/kg of growth, respectively) when compared to Grp C (64.7 g of CP/kg) and CTL (67.3 g of CP/kg). This suggests that the diet optimization with specifics plants extracts permits reducing CP content and soya utilization improving ADG, FCR, and protein utilization.

Session 46 Poster 41

Effects of *Artemisia annua* residue on rumen microorganisms and antioxidant function of mutton sheep

S.S. Wang[1], C.F. Peng[2], Y.R. Shao[1], M.M. Bai[1], Y.H. Zhang[2], M. Zhang[3], X. Xiong[1] and H.N. Liu[1]
[1]Institute of Subtropical Agriculture, Chinese Academy of Sciences, Institute of Subtropical Agriculture, Chinese Academy of Sciences, 410125, 644 Yuanda 'er Road, Furong District, Changsha City, Hunan Province, China, China, P.R., [2]Hunan Agricultural University, Hunan, Changsha, 410128, 1 Nongda Road, Furong District, Changsha City, Hunan Province, China, China, P.R., [3]Vinsce Bio-Pharm Co., Ltd, Jiangsu, suzhou, 215600, 2 Nanjing Middle Road, Yangzijiang Chemical Industrial Park, Jiangsu Province, China, P.R.; liuhn@isa.ac.cn

Artemisia annua L. (AA) has been widely recognized for its extract artemisinin in the treatment of malaria. At present, the application of AA and its by-products in other fields has gradually attracted people's attention. The purpose of this study is to explore the effects of *A. annua* residue (AAR) as roughage on in ruminant farming. Eighteen Hu sheep with similar body weight (30.50±0.31 kg) and the same body condition were divided into two groups at random, with three replicates per group, and three animals per replicate. The control group was fed a conventional total mixed ration diet and the treatment group was fed 10% AAR instead of partial roughage. The pre-feeding period was 10 days and the experimental period was 60 days. This study showed that the addition of 10% AAR in the diet significantly increased the average daily feed intake (ADFI), muscle storage loss and decreased the ACE index (P<0.05). The levels of aspartate amino transferase (AST) and low density lipoprotein-C3 (LDL-C3), and the activities of glutathione peroxidase (GSH-Px) and superoxide dismutase (SOD) in the serum significantly increased in the treatment group (P≤0.05). AAR treatment significantly increased the rumen empty weight and the relative abundance of *Firmicutes* and *Rikenellaceae_RC9_gut_group* in the rumen contents (P≤0.05). The relative abundance of *Verrucomicrobiota*, *Fibrobacterota*, and *Fibrobacter* significantly decreased in the treatment group (P<0.05). In conclusion, our study provides data in support of feeding ruminants with AAR instead of roughage and demonstrates that AAR can reduce feeding costs and improve the economic benefits of sheep farming without affecting the growth performance.

Session 46 Poster 42

Nutrient rich novel feed supplements for grazing lambs

A.S. Chaudhry
Newcastle University, School of Natural and Environmental Sciences, Agriculture Building, NE1 7RU, United Kingdom; abdul.chaudhry@ncl.ac.uk

Globally, grazed grass is accepted as an affordable and efficient feed resource for ruminant animals. Nevertheless, it is inconsistent in supply and nutritional value depending upon its variety, region, and season. Thus, grass consuming sheep must receive supplements that match their nutrient requirements to achieve variable production targets and vitality. This completely randomised study compared the impact (P<0.05) of *ad libitum* access of grazing lambs to feed blocks containing either Sffmannan (SafM) or no SafM (NSafM) or no blocks (CNT) on their intake, liveweight (LW), LW gain (LWG) and condition score (CS). Two hundred forty weaned Suffolk and Texel cross lambs of both genders with initial LW of 30.6 kg (+0.34) and CS (1.8+0.038) were split into six uniform groups of 40 lambs each. These lambs were dewormed before group housed on a ryegrass paddock (11.2 MJ ME, 215 g CP /kg DM) that was divided into six uniform fields of 2 hectares each by fencing. The lamb groups were randomly distributed into these replicated fields with plentiful grass and fresh drinking water for 84 days. The lambs had access to either grazed grass alone (CNT) or with either SafM or NSafM blocks enriched with molasses, soybean meal and vitamin-mineral premix. The grass quality and quantity remained uniform across the fields and treatments. However, its quantity and nutrient value during the sampling periods of grass, declined for all treatment groups towards the end of the trial. All lambs remained fit while adapted easily to consume feed blocks with more variable intake of SafM than NSafM. The treatments differed significantly (P<0.05) for their effects on LW, LWG but not CS (P>0.05). While both feed blocks showed better lamb growth than the CNT, the NSafM gave greater increase in lamb growth (P<0.01). The Breed and Gender effect or their interaction was not significant (P>0.05) for any of the parameters. It appeared that the nutrient rich feed blocks as supplements were beneficial for grazing lambs. However, their effect did vary depending upon their constituents and the quality and quantity of grazed grass. Further studies should be conducted over a longer duration while examining their effect on fertility, carcass and meat quality of lambs.

Session 46 Poster 43

The order of distribution of two forages affects daily intake and diet composition in dairy goats

R. Delagarde[1], J. Belz[1] and B. Bluet[2]
[1]PEGASE, INRAE, Institut Agro, 16 Le Clos, 35590 Saint-Gilles, France, [2]IDELE, Route de Chauvigny, 86550 Mignaloux-Beauvoir, France; remy.delagarde@inrae.fr

In practice, dairy goats are often fed daily with two separate forages in order to organise the daywork, diversify the diet and stimulate intake. According to their preferences, the order of distribution of these forages could affect goat performance. The objective of this experiment was to investigate the effect of the order of distribution of two forages: fresh herbage cut once daily, and a partial mixed ration (PMR) based on maize silage and soyabean meal (ratio 85:15 on a DM basis). Twenty four multiparous Alpine goats in mid lactation were used in a reversal switchback design ABA-BAB with three 14-d periods in spring 2022. The forage fed in the morning was available during 6h30 (09:45 to 16:15), and the forage fed in the evening was available during 15h45 (17:15 to 09:00). The two forages were given *ad libitum*, with, in both treatments, 12% of refusals for PMR and 14% of refusals for fresh herbage. Each goat also received 400 g/d of a pelleted concentrate based on maize grain (200 g/milking). Feeding behaviour (cameras) and daily intake were measured daily at herd level, and milk production was measured daily at goat level. Fresh herbage and PMR intakes were 1.36 and 0.81 kg DM/d (38% of PMR in forages), respectively, when fresh herbage was fed in the morning. The corresponding figure was 1.70 and 0.39 kg DM/d (19% of PMR in forages) when fresh herbage was fed in the evening. Total DM intake (+0.08 kg DM/d) and milk production (+0.08 kg/d) were greater when fresh herbage was fed in the morning, which is partly a response to the increase in soyabean meal intake (+0.07 kg DM/d). Time spent eating fresh herbage was 235 and 358 min/d when fresh herbage was fed in the morning and in the evening, respectively. The corresponding figure for PMR was 50 and 126 min/d, respectively. In conclusion, the order of distribution affected marginally daily DM intake but more significantly feeding behaviour pattern and diet composition, affecting milk production. Goats showed a clear preference for fresh herbage in both treatments, despite a lower crude protein concentration (102 vs 128 g/kg DM) and a lower intake rate (292 vs 423 g DM/h) for fresh herbage than for PMR.

Session 46 Poster 44

Feeding frequency has no effect on intake and milk production in dairy goats fed on fresh herbage

R. Delagarde[1], J. Belz[1] and B. Bluet[2]
[1]PEGASE, INRAE, Institut Agro, 16 Le Clos, 35590 Saint-Gilles, France, [2]IDELE, Route de Chauvigny, 86550 Mignaloux-Beauvoir, France; bertrand.bluet@idele.fr

According to many goat farmers and their advisors, increasing feeding frequency could be one solution to motivate ruminants to eat more forage and increase milk production. Is this management strategy useful when goats are fed on fresh herbage? The objective of this study was to investigate the effect of the number of fresh herbage distributions (1 vs 2 per day) on dry matter intake, milk production and feeding behaviour. Twenty four multiparous Alpine goats in mid lactation were used in a reversal switchback design ABA-BAB with three 14-d periods in spring 2022. Herbage was cut once daily and fed *ad libitum* either only in the morning (10:00 h) or in the morning (10:00 h) and in the evening (17h30 h). Goats fed once daily were in the milking parlour during the evening feeding period to avoid stimulating them at this moment. Fresh herbage was accessible for the goats all the time without human intervention, except during 1 h in the morning (refusals weighting at 09:00 h). Each goat received as a supplement 500 g/d of a pelleted concentrate based on maize grain, with 250 g at each milking. Feeding behaviour (cameras) and daily intake were measured daily at herd level, and milk production was measured daily at goat level. Herbage was characterised by an unexpected low crude protein concentration (102 g/kg DM). Herbage intake (1.76 kg DM/d), total DM intake (2.16 kg DM/d), milk production (2.58 kg/d), milk fat (34.1 g/kg) and true protein (27.4 g/kg) concentrations, and live weight (53.1 kg) were unaffected by feeding frequency. Goats fed once daily tended to spend more time eating (443 vs 403 min/d, $P<0.07$), and particularly during the afternoon (13:00 to 17:00 h): + 61 min ($P<0.01$). At the reserve, goats fed twice daily ate 30 min longer ($P<0.05$) in the evening (17:00 to 23:00 h). Provided that fresh herbage is accessible throughout the day, increasing the feeding frequency affected daily feeding pattern of goats but not their total dry matter intake, milk production or composition, suggesting possible simplifications of feeding management for dairy goats fed on fresh herbage.

Session 46 Poster 45

Effects of dietary energy on late-gestation metabolism in prolific ewes

M. Plante-Dubé[1], C. Sylvestre[1], R. Bourassa[2], P. Luimes[3], S. Buczinski[4], F. Castonguay[1] and R. Gervais[1]
[1]Laval University, Department of animal sciences, 2425 rue de l'Agriculture, Québec, G1V0A6, Canada, [2]Veterinary Hospital of Sherbrooke, 1771 rue King Est, Sherbrooke, J1G5G7, Canada, [3]University of Guelph, 120 Main St E, Ridgetown, N0P2C0, Canada, [4]University of Montreal, Faculty of Veterinary Medicine, 3200 rue Sicotte, St-Hyacinthe, J2S2M2, Canada; marguerite.plante-dube.1@ulaval.ca

In late gestation, to prevent metabolic disorders in prolific ewes, dietary supplies must be in line with the rapid increase in energy requirements to support the growth of multiple foetuses. The current study was conducted to evaluate the effects of the dietary energy supply and source on the dynamic response of energy metabolism during the last 6 weeks prepartum in F1 prolific breed ewes. Six weeks prior to lambing (wk -6), all 48 mature, crossbred (Dorset × Romanov) ewes were offered grass silage (2.17 Mcal/kg dry matter (DM)) *ad libitum*. Then, ewes were randomly assigned to 1 of 3 dietary treatments (2.62 Mcal/kg DM): a total mixed ration (TMR) based on grass silage and ground corn from wk 4 (GC4) or wk 2 (GC2) or a TMR based on corn silage from wk 4 (CS4). Dry matter intake (DMI) and body condition score (BCS; 1 to 5) were respectively determined daily and twice weekly. Blood concentrations of beta-hydroxybutyrate (BHB) and glucose were monitored twice weekly for wk -6 to -4, then thrice weekly. Treatment affected DMI by wk -4, as DMI for GC2 was lower (1.91^{b}) than GC4 (2.42^{a}) with intermediate values for CS4 (2.00^{ab}kg/d). Starting on wk -3, DMI increased for GC2 and CS4 resulting in similar DMI among treatments by wk -2 (2.39 kg/d). No difference over time or among treatments was observed for BCS (3.65). Glucose concentrations were greater for GC4 and CS4 ewes (3.49^{a}) than GC2 (2.72^{b}mmol/l) at wk -4 and -3. These differences were no longer significant by wk -2 (3.68 mmol/l). By wk -4 BHB concentrations were greater for CS4 (1.04^{a}) than GC2 (0.54^{b}), whereas GC4 presented intermediate values (0.77^{ab}mmol/l). During the last 3 weeks, BHB concentrations for GC2 evolved (0.69^{b}) to be similar to GC4 by wk -2 (0.70^{b}) but lower than CS4 (1.21^{a}). These results suggest that dynamic responses of energy metabolism in late gestation of prolific ewes are influenced by the dietary energy supply and source.

Session 46 Poster 46

The effect of feeding with hemp seeds addition on physicochemical and sensory properties of beef

P. Pogorzelska-Przybyłek[1], C. Purwin[1], M. Modzelewska-Kapituła[2], M. Borsuk-Stanulewicz[1] and K. Tkacz[2]
[1]University of Warmia and Mazury in Olsztyn, Department of Animal Nutrition, Feed Science and Cattle Breeding, ul. M. Oczapowskiego 5/248, 10-719, Poland, [2]University of Warmia and Mazury in Olsztyn, Department of Meat Technology and Chemistry, Plac Cieszyński 1, 10-719, Poland; paulina.pogorzelska@uwm.edu.pl

Hemp by-products have the potential to be used as feed ingredients for ruminant. Some studies have already included hemp by-products as protein sources in finishing diets for goats, sheep, cattle. However, there is a gap in understanding the impact of hemp by-products on the quality and sensory properties of meat. The aim of this study was to determine the effect of whole hemp seeds (*Cannabis sativa* L.) inclusion to the diet of Polish Holstein-Friesian bulls (PHF) on the physicochemical and sensory properties of longissimus lumborum muscle (LL). The experiment was performed on 3 groups of PHF bulls (of 8 animals each), characterized by a similar age and body weight. Over a 91-day finishing fattening, bulls were fed *ad libitum* with maize silage and a concentrate (triticale grain, rapeseed meal, mineral-vitamin premix) with a different proportion of *Cannabis sativa L.* seeds, depending on the feeding group: C – without hemp seeds (HS), HS10 – 10% addition of HS, HS17 – 17% addition of HS. After slaughter samples of LL were removal from the left half-carcass of each animal and stored at -20 °C until analyses. According to the standard methods described in the literature, the colour, chemical composition, physicochemical and sensory properties were determined. The addition of hemp seeds in the rations of PHF bulls had a neutral effect on the chemical composition of meat but affected its colour and sensory properties, which is important for consumers. Meat from the HS10 group was the product with the most desirable aroma, juiciness, tenderness and the highest overall acceptability. This study was conducted within the research project No. 2021/05/X/NZ9/00799 financed by the National Science Center, Poland. Project financially supported by the Minister of Education and Science under the program entitled 'Regional Initiative of Excellence' for the years 2019-2023, Project No. 010/RID/2018/19.

Session 46 Poster 47

Association of feed efficiency with growth and slaughtering performance in Nellore cattle

S.F.M. Bonilha, J.A. Muñoz, B.R. Amâncio, J.N.S.G. Cyrillo, R.H. Branco, R.C. Canesin and M.E.Z. Mercadante
Institute of Animal Science, Beef Cattle Research Center, Rodovia Carlos Tonani, km 94, 14.174-000 Sertãozinho/SP, Brazil; sarah.bonilha@sp.gov.br

Improving feed efficiency (FE) is of major interest for livestock production. This is the case for studies using residual feed intake (RFI). RFI is designed to quantify between-animal variation in FE, with the limitation of being calculated based on short time interval, evaluated in the beginning of the animals' productive life. This study aimed to evaluate growth and slaughtering performance of Nellore males, based on RFI calculated at post-weaning (PW) and pre-slaughter (PS) periods. Sixty-seven animals were evaluated twice: PW with diet formulated for growth; and PS with diet formulated for finishing. RFI was calculated as the residual of dry matter intake (DMI) regression equation as a function of weight gain (ADG), metabolic body weight (MBW) and subcutaneous fat thickness (FT), being animals classified as negative (RFI<0; efficient) or positive RFI (RFI>0; non-efficient) for PW and PS periods. Feed conversion ratio (FCR) and FE were calculated as the ratio between DMI and ADG, and ADG and DMI, respectively. Data was analysed using SAS GLM procedure, considering RFI class as fixed effect and probability of 5%. Pearson's correlation coefficient was estimated for RFI considering both periods. PW RFI values were compared to PS RFI values to determine change in classification. No significant effect was detected for RFI class on ADG, MBW and FT in both periods. However, it was observed that negative RFI animals, when compared to the positive RFI ones, had lower DMI (6.10 vs 7.06 kg/d; P=0.0001) and FCR (7.05 vs 7.77 kg/kg; P=0.0001), and higher FE (0.14 vs 0.13 kg/kg; P=0.0001) at PW period; and lower DMI (7.72 vs 8.30 kg/d; P=0.0069) and similar FCR and FE at PS period. The correlation coefficient between PW and PS RFI was 0.22 (P=0.0740), indicating no linear association in RFI between both periods. From the 67 evaluated animals, 37% changed RFI classification (13 animals changed from negative to positive, and 12 from positive to negative RFI). RFI can interact with FCR and FE mainly at PW period. DMI is influenced by RFI until slaughter, even with animals being reclassified. Acknowledgments: FAPESP Process 2022/12347-4.

Session 46 Poster 48

Effects of dietary nitrate on performance and enteric methane production in Hanwoo steers

R. Bharanidharan[1], P. Xaysana[2], R. Ibidhi[3], J. Lee[4], B.M. Tomple[1], J. Oh[5], M. Baik[4] and K.H. Kim[1,2]
[1]Institutes of Green Bio Science and Technology, Seoul National University, 25354, Pyeongchang, Korea, South, [2]Graduate School of International Agricultural Technology, Seoul National University, 25354, Pyeongchang, Korea, South, [3]Agroécologie, INRAE, Institut Agro, Univ. Bourgogne, Univ. Bourgogne Franche-Comté, 21000, Dijon, France, [4]College of Agriculture and Life Sciences, Seoul National University, 08826, Seoul, Korea, South, [5]Cargill Animal Nutrition, 13630, Seongnam, Korea, South; bharanitharshan76@gmail.com

Nitrate (NO_3^-) has been recognized as an effective dietary additive to reduce enteric methane (CH_4) production in ruminants. The current experiment was aimed to evaluate the effects of NO_3^- on performance, enteric CH_4 production and rumen fermentation characteristics in Hanwoo steers. Twenty Hanwoo steers (348±18 kg) were divided into two groups of 10 steers balanced for body weight and CH_4 yield, and were assigned to one of two diets: (1) Control, 500 g/kg timothy hay and 500 g/kg commercial concentrates; (2) NO_3^-, control diet supplemented with 2.5% calcium ammonium nitrate (1.875% NO_3^- in the dietary dry matter) for 28 days in a completely randomized design. Rumen fluid samples were collected on the last day of the feeding trial. CH_4 measurement was performed for 3 days starting from day 24 using whole body respiratory chambers (RC). Ruminal pH, ammonia concentration and total volatile fatty acid concentration did not differ between treatments (p > 0.05), but NO_3^- containing diets increased and decreased (P<0.0001) acetate and butyrate proportions, respectively. No difference (P>0.05) in dry matter intake (DMI) was noted. However, an increase in feed conversion ratio (10.9 vs 13.6; P<0.05) caused by the decrease in average daily gain (0.73 vs 0.58 kg; P<0.05) was found in NO_3^- fed animals located in barn. In RC, animals fed NO_3^- diet exhibited decreased DMI (6.85 vs 5.31 kg/d; P<0.005), CH_4 production (160.7 vs 91.5 g/d; P<0.0001) and yield (22.4 vs 17.7 g/kg DMI; P<0.05) compared to the control. Future studies are aimed at performing haematological analyses and dosage optimization of NO_3^- in order to minimize the negative effects of NO_3^- on animal performance, and also perform untargeted urinary volatiles profiling to identify factors influencing DMI in RC.

Session 46 — Poster 49

Impact of additives and forage levels on nutrients digestibility and sorting index of Nellore bulls

E. Magnani[1], T.H. Silva[1], L.B. Tosetti[1], E.M. Paula[1], P.R. Leme[2] and R.H. Branco[1]
[1]Institute of Animal Science, Beef Cattle Research Center, Rod. Carlos Tonani km94, 14.174-000, Brazil, [2]University of Sao Paulo, Department of Animal Science, R. Duque de Caxias, 225, 13635-900, Pirassununga, SP, Brazil; dasilvath2@gmail.com

This study was carried out to evaluate the effect of additives and increasing levels of neutral detergent fibre from forage (NDFf) on digestibility of nutrients and sorting index in finishing Nellore bulls. Thirty Nellore bulls, 5 bulls per treatment, [409.6±6.36 kg of body weight (BW) and approximately 24 months of age] were allocated into 6 diets in a factorial 2×3 arrangement during 105 days in an individualized feedlot system. The treatments were the combination of monensin (MON; 30 mg/kg DM), essential oils (EO; 500 mg/kg DM, essential oil was a mixture of castor oil and cashew nut)), and 3 levels of dietary FDNf (6, 9, and 15% of total diet, DM basis). The diet was composed by *Brachiaria brizantha* cv. Marandu hay, finely ground corn, soybean meal, ground citrus pulp, urea, potassium chloride, sodium chloride and trace minerals + vitamins. The particle size distribution of diet and orts were determined weekly using the Penn State Particle separator containing 3 sieves. Additives and dietary NDFf levels were analysed in a complete randomized design according to a factorial arrangement. When NDFf levels were statistically significant, linear and quadratic responses were assessed (software SAS 9.4). The digestibility of dry matter, organic matter, crude protein, and starch reduced linearly as NDFf level increased. For sorting index outcomes, there was sorting against 1.18 mm particles when 15% NDFf level was combined with EO supplementation compared to other levels within EO treated bulls. Sorting against <1.18 mm particles was also detected for EO vs MON treated bulls. Overall, increasing dietary NDFf levels of finishing Nellore bulls reduce the digestibility of nutrients. Further, EO may induce to sorting against particle ≤1.18 in a level-dependent manner.

Session 46 — Poster 50

Feed efficiency traits calculated at post-weaning and pre-slaughter periods in Nellore cattle

S.F.M. Bonilha, J.A. Muñoz, B.R. Amâncio, J.N.S.G. Cyrillo, R.H. Branco, R.C. Canesin and M.E.Z. Mercadante
Institute of Animal Science, Beef Cattle Research Center, Rodovia Carlos Tonani, km 94, 14174-000, Sertãozinho/SP, Brazil; sarah.bonilha@sp.gov.br

Feed efficiency traits have been commonly evaluated in trials performed after weaning with diets formulated for growth, however, its repeatability across stages of production cycle or using different diets is not well known. This study aimed to evaluate the correlations between performance traits of Nellore bulls during two periods (post-weaning-PW and pre-slaughter-PS), and to verify animals' reclassification based on residual feed intake (RFI). Sixty-seven animals were evaluated twice: PW with diet formulated for growth; and PS with diet formulated for finishing in three different years. RFI was calculated as the residual of dry matter intake (DMI) regression equation as a function of weight gain (ADG), metabolic body weight (MBW) and subcutaneous fat thickness (FT), being animals classified as negative (RFI<0; efficient) or positive RFI (RFI>0; non-efficient) for PW and PS periods in each year. Pearson correlation coefficients were estimated between DMI, ADG, MBW, FT and RFI values in both, PE and PS periods. PW RFI values were plotted against PS RFI values to determine change in classification. The correlation coefficients found between PW and PS in year 1, 2, and 3, respectively, were 0.67 (P=0.005), 0.36 (P=0.067), and 0.48 (P=0.016) for DMI; 0.56 (P=0.020), 0.12 (P=0.561), and 0.21 (P=0.311) for ADG; 0.91 (P=0.001), 0.89 (P=0.001) and 0.75 (P=0.001) for MBW; 0.02 (P=0.930), 0.40 (P=0.040) and -0,04 (P=0.851) for FT; and 0.67 (P=0.005), 0.35 (P=0.079) and 0.55 (P=0.005) for RFI. High correlation coefficients within years 1 and 3 indicated that 31 and 36% of animals have changed RFI classification after PW period. For year 2, it was found a trend of 39% of animals to change their RFI classification. There is reclassification of animals based on RFI when changing age of evaluation and/or type of diet. Nelore bulls classified by RFI at PW period can change the efficiency class when tested again at PS period. However, knowing that most of animals keep their PW RFI classification can be decisive for determinations on the animal's productive life. Acknowledgments: FAPESP Process 2022/12347-4.

Environmental protection study on replacing alfalfa with sesbania for feeding ruminants

L.Y. Wang and Y.J. Tian
Tianjin Agricultural University, Tianjin Key Laboratory of Agricultural Animal Breeding and Healthy Breeding, College of Animal Science, Tianjin Agricultural College (East Campus), 22 Jinjing Road, Xiqing District, Tianjin, China, 300384, China, P.R.; wly6888@163.com

Alfalfa has a high protein content and moderate cellulose content, and is generally widely used for feeding ruminants. However, alfalfa has a long growth cycle and requires a large amount of water and nutrients during breeding, which is easy to lead to soil erosion, and a large amount of nitrogen fertilizer is required during growth, which is easy to cause soil degradation. Therefore, Farmers need to find better alternatives to alfalfa. Sesbania has been shown to be a good substitute for ruminants; Sesbania has the characteristics of cold tolerance, drought tolerance, vigorous growth, and high protein content, low fibre content, and the leaves have the effect of clearing heat and detoxifying. Sesbania can grow in saline-alkali land and survive rough breeding, which greatly reduces the cost of artificial breeding, so it becomes an ideal alternative. Some studies have shown that the experimental group of ruminants added sesbania in the total mixed diet, the control group did not add, and the milk yield was recorded every day. After statistical analysis, the average daily milk production of cows in the experimental group was about 1 litre higher than that of cows in the control group. In addition, researchers also found that the use of sesbania to feed cows had no negative effect on the health status of cows. As a feed source, sesbania has a significantly lower carbon footprint than alfalfa. In conclusion, according to the existing scientific research results, alfalfa sesbania has great potential to replace alfalfa as a feed source for ruminants. In addition, growing sesbania can also improve land and protect the environment, reduce greenhouse gas emissions and save economic benefits. More research is therefore needed to more fully understand the substitution potential of alfalfa and the environmental and economic impacts.

Multi performance analysis of soybean self-consumption on a mixed crop-livestock farm

M. Jouffroy[1], A. Berchoux[1], M. Duval[1], M. Weens[2], E. Hermant[3] and M. Legris[1]
[1]Institut de l'élevage, 149 rue de Bercy, 75012 Paris, France, [2]Chambre Régionale d'Agriculture Grand-Est, Route de Suippes, 51000 Chalons en Champagne, France, [3]Lycée Agricole de Rethel, Route de Novion, 08300 Rethel, France; mathilde.jouffroy@idele.fr

Breeders are currently working on the rise of their protein autonomy to reduce their dependence on imported feed with unstable market prices. The objective of this study was to evaluate the consequences of soybean self-consumption on the multi-performance of a mixed crop-livestock dairy farm. A trail was carried out on two groups of 26 dairy cows complemented with rapeseed meal (control) or toasted soybeans (testing). Animal performances (ingestion, milk production and composition) were measured, and economic and environmental performances were evaluated. It was set up on the herd of the agricultural college of Rethel for 9 weeks of testing period. No significantly differences were observed on ingestion, milk production and fate rate between these two groups. However, cows complemented with toasted soybeans had a milk protein rate of 2 g/kg lower than the control group. The decrease of protein rate and the additional cost of feed decreased the feed profit per cow for the testing group (-15€/1000 l). A new system of production with a self-consuming soybean strategy was modelized on this farm. A total of 10 ha of soybeans was needed to feed the herd during the winter period (150 days), replacing 10 ha of wheat. Effect of soybeans as preceding crop enabled a lower nitrogenous fertilization on subsequent crops (-20 units). Introduction of soybeans in cows feeding generated lower purchases of rapeseed meal (-5,772€) but self-consumption of soybean generates extra-costs for toasting and crushing (+2,000€). The simulation of self-production of soybeans with a stable milk production and fate rate and with a lower protein rate of 2 g/kg leaded to a loss of Gross Operating Surplus of -16,445€ (-13%). The decrease of charges (-3,020€) did not balanced the decrease of products (-19,465€). Economy of fertilization and decline of rapeseed meal purchases enabled to reduce farm-scaled emissions of greenhouse gas (-5%), nitrogen balance (-16%), and the carbon footprint (-3%) at milk-scaled. In this case, self-consumption of soybeans leads to economic lost but improves environmental performances.

Session 46 Poster 53

Effects of leaf size and harvesting season on nutritive quality of white clover

X. Chen[1], K. Theodoridou[2], O. Cristobal-Carballo[1] and T. Yan[1]
[1]Agri-Food and Biosciences Institute, Hillsborough, BT26 6DR, United Kingdom, [2]Institute of Global Food Security, Queen's University Belfast, Belfast, BT9 5DL, United Kingdom; xianjiang.chen@afbini.gov.uk

The present study aimed to evaluate variations in nutritive values of white clover varieties with different leaf sizes and harvesting seasons. Thirteen varieties of white clover were selected from recommendation lists in the Republic of Ireland and Northern Ireland, including 4 large leaf sizes (Alice, Aran, Barblanca and Gabby), 5 medium leaf sizes (Aberherald, Buddy, Chieftain, Crusader and Iona) and 4 small leaf sizes (Aberace, Coolfin, Galway and Rivendel). Each was sown in a single paddock in 2018 and harvested on 8 June, 23 July and 5 September 2020, respectively. All 39 samples obtained were analysed using 3 techniques: near infra-red spectroscopy (NIRS), wet chemistry and *in vitro* gas production. Results were analysed as a 3 (leaf size) × 3 (harvesting season) factorial design with the variety fitted as a random factor. There was no significant interaction on any variable between leaf size and harvested season except for wet chemistry NDF content with which the interaction was significant (P=0.034). Leaf size had no significant effect on any nutrient content determined by NIRS (dry matter (DM), crude protein (CP), acid detergent fibre (ADF), water soluble carbohydrates (WSC) or metabolizable energy (ME)), wet chemistry techniques (DM, Ash, CP, GE, ADF, neutral detergent fibre, lipid or WSC), or *in vitro* total gas production. Organic matter digestibility or ME content was similar between the 3 leaf size categories. However, the majority of variables evaluated were influenced by harvesting date. For example, in comparison with early and middle harvesting dates, late harvesting produced a significantly higher CP, lipid and energy contents while lower ADF and WSC contents when measured by wet chemistry techniques. The NIRS DM, CP and WSC contents each were positively related to their counterparts measured by wet chemistry (P<0.001, R^2=0.57, 0.55 and 0.61, respectively). In conclusion, the nutritive value of white clover is similar between leaf size categories although influenced by grazing season. The NIRS technique has potential to be used for predicting DM, CP and WSC contents of white clover.

Session 46 Poster 54

Image analysis of feed boluses collected from cows after ingestive chewing

B. Delord[1], M. Berger[1], R. Baumont[2], P. Nozière[2], A. Le Morvan[2], F. Guillon[3] and M.F. Devaux[3]
[1]Limagrain, Europe, 63360 Saint-Beauzire, France, [2]INRAE, UMR Herbivores, 63122 Saint-Genès-Champanelle, France, [3]INRAE, UR1268 BIA, 44300 Nantes, France; rene.baumont@inrae.fr

Forage feed value depends on its digestibility and ingestibility, i.e. the quantity of dry matter voluntary ingested by the animal. Digestibility first depends on the quantity of indigestible cell-walls in the forage. However, for equivalent digestibility, differences are observed in forage ingestibility and consequently in milk production. Forage ingestibility depends primary on its fill effect in the rumen that is linked to the forage residence time. The latter is determined by the time needed to reduce forage particle size through chewing and microbial degradation to the threshold size allowing feed residues to leave the rumen. Here, we propose to quantify the reduction in particle size due to the ingestive chewing by image analysis. An INRAE homemade macrovision device, the LightBox which allows a large field of view and thus, the imaging of objects several cm long was used, together with image texture analysis to estimate fragment size from the greyscale images. Four Holstein dairy cows hosted at INRAE were fed experimental meals according a Latin square design with four lines of maize selected according to their chemical composition and friability measured by a grinding test. The animals received the maize plants without ear harvested at the silage stage. The feed boluses were collected manually during the meal at the cardia through the rumen canula. Both the offered meals and the boluses were imaged in bulk using the LightBox. The field of view was 113×151 mm^2, with a pixel size of 94 µm. Fragment size was quantified using grey level granulometry that provides with granulometric curves, characteristic of the size of objects even if they are observed in bulk. As expected, the mean fragment size of the offered meals differed between maize lines and a decrease in particle size after chewing was observed. After chewing, the size of the fragments depended first on the cow and then on the maize line. Finally, between maize lines, the size of the fragments after chewing were ranked accordingly to their friability measured by a grinding test, that is promising for further phenotyping of maize lines for friability and ingestibility.

Session 47 Theatre 1

A documented example of the One Health concept

P. Weill, N. Kerhoas and B. Schmitt
Université de Rennes, rue du Tabor, 35000 Rennes, France; pierre.weill@bleu-blanc-coeur.com

One Health concept is based on the principle that human health is linked to the health of its ecosystems. Initially based on the link between animal and human, supported by pandemics and antibiotic resistance, this concept covers very broad fields, sometimes without rigorous demonstration. The synthesis and roles of polyunsaturated fatty acids (PUFA) from soil & plant to man are a good example of One Health that provides measurable markers. Indeed, the precursors of PUFA are exclusively plant-based and participate to plant immunity. Then, only animals elongate these PUFA and incorporate them in their tissues. In animals and in humans, the balance between these PUFA drives inflammation and health prevention. Numerous trials in which the ratio between n-6 and n-3 PUFAs was reduced (from 15 to 5 in average) in animal feed (all other things being equal), provide measurements of environment, animal health and human health impact. n-6 PUFA is mainly provided by corn, n-3 PUFA by grass and linseed At different levels: environmental. The main effect is the reduction of enteric methane emissions from ruminants. These effects are measurable in milk and meat. Animal health and fertility. A decrease in the frequency of metabolic diseases and an improvement in fertility has been measured in cows and sows. Composition of animal products They (eggs, meat, milk) contain less saturated FA and n-6 PUFA and more monounsaturated FA and n-3 PUFA. Human health We note an improvement in the composition of serum and red blood cells in healthy volunteers. A significant improvement of insulin resistance in diabetic patients, an improvement of cardiovascular markers and a lesser weight regain after a hypocaloric diet in obese volunteers are also measured. The decrease of the n-6/n-3 ratio in animal diets is an excellent measurable example of what should be a one health strategy built on the strength of evidence and measurement. Other examples could be built on this measurement scheme from soil to man.

Session 47 Theatre 2

Dietary lipid supplements affect milk composition and butter properties in dairy cows

M. Landry[1,2], Y. Lebeuf[1,2], M. Blouin[1,2], F. Huot[1,2], J. Chamberland[1,2], G. Brisson[1,2], D.E. Santschi[3], É. Paquet[2], D.E. Rico[4], P.Y. Chouinard[1,2] and R. Gervais[1,2]
[1]Institute of Nutrition and Functional Foods, 2440 Bd Hochelaga, G1V0A6, Québec, Canada, [2]Université Laval, 2425 Rue de l'Agriculture, G1V0A6, Québec, Canada, [3]Lactanet, 555 Bd des Anciens-Combattants, H9X3R4, Ste-Anne-de-Bellevue, Canada, [4]CRSAD, 120-A Chemin du Roy, G0A1S0, Deschambault, Canada; myriam.landry.8@ulaval.ca

This trial aimed to evaluate the effects of 4 lipid supplements on milk composition and butter manufacture. In a replicated 5×5 Latin square design, 10 multiparous Holstein cows (64±21 days in milk) received a basal diet without supplementation (CTL) or with 2% (dry matter basis) fatty acids (FA) provided as soybean oil (SO), calcium salts of palm FA (CS), hydrogenated/hydrolysed tallow (HT) or palmitic acid-enriched supplement (PA). Treatment periods lasted 21 days and the last 5 days were used for data and sample collection. At each period, milk from consecutive milkings of days 18 and 19 was collected from each cow, pooled by treatment in refrigerated bulk tanks, and transferred to the pilot plant for butter manufacture with a Stephan mixer. Milk yield was increased with SO compared with CTL (+8%; $P<0.01$). Milk fat content was decreased with SO compared with any other treatment (-11 to -17%; $P<0.01$) and increased with PA compared with CS (+7%; $P=0.04$). Milk crude protein content was greater for PA and CTL compared with CS or SO (+4%; $P\leq0.04$). Casein micelle size in pooled milk was similar among treatments (163±4 nm; $P=0.99$). Milk free FA content was not affected by treatment (0.46±0.07 mEq/100 g fat; $P=0.67$). Fat globule diameter ($D_{4,3}$) in cream before maturation tended to be greater for CS than SO (4.29 vs 3.80 µm; $P=0.10$). The width of the fat globules size distribution after overnight cream maturation was greater for CS than CTL (1.26 vs 1.11; $P=0.03$). Churning time tended to be greater for PA compared with HT and CS (23 vs 11 min; $P=0.08$). At 20 °C, butter made from PA milk was harder and its spreadability lower than any other treatment ($P\leq0.01$), whereas at 4 °C, PA butter was only harder and less spreadable than CS and SO ($P\leq0.03$). Compared with CTL, SO butter was softer at 20 °C ($P<0.01$) and its spreadability was higher at both temperatures ($P<0.01$).

Decreasing GHG footprint while improving nutrition value in ruminant product: a new challenge

S. Mendowski[1], G. Chesneau[1], G. Mairesse[1,2] and N. Kerhoas[2]
[1]Valorex, La Messayais, 35210 Combourtillé, France, [2]VALOREX, La Messayais, 35210 Combourtille, France; s.mendowski@valorex.com

Dairy products are the main source of saturated fatty acids (SFA) in human nutrition. Consumed in excessive amounts, SFA may increase the risk of cardiovascular diseases. Moreover, enteric methane (CH_4) emissions from dairy cows is the most important greenhouse gas in agriculture. Despite other nutritional benefits of dairy products, these two defaults are a brake for dairy industry. As milk fatty acid (FA) profile is linked to dairy cow diets, and CH_4 can also be mitigate by changing diets, is animal feeding a solution to improve both nutritional and environmental quality of dairy products? Two technical webtools, Visiolait and Lait Durable, designed to monitor dairy herds performances and health for the first, and performances and environmental impact for the second, were used to answer this question. The data extracted from these tools gathers 44,217 milk FA profiles analysed between 2021 and 2022 and their associated predicted CH_4 emissions (thanks to the equation used by Bleu-Blanc-Coeur). Data were divided in 6 groups according to the content of linolenic acid (ALA) in milk, which reflects the content of ALA in dairy cow diets. Results show that both SFA and CH_4 decrease in milk when ALA increase. Indeed, for milks containing less than 0.5% of total FA of ALA, SFA and CH_4 average 73.0% of total FA and 16.1 g/l of milk, respectively; while for milks containing more than 0.9% of total FA of ALA, SFA and CH_4 reach averages of 63.6% of total FA and 13.5 g/l of milk, respectively. In conclusion, changing dairy cow diets by increasing the amount of ALA in their feed (spring grass, alfalfa, linseed, etc.) lead to an improvement of both nutritional and environmental milk quality, by increasing the level of unsaturated FA while decreasing both SFA amount and the greenhouse gas footprint.

Effect of dietary protein source and *Saccharina latissima* on milk fatty acids profiles and bromoform

B. Wang[1,2], S. Ormston[2], N. Płatosz[3], J.K. Parker[4], N. Qin[2], D. Humphries[2], Á. Pétursdóttir[5], A. Halmemies-Beauchet-Filleau[6], D. Juniper[2] and S. Stergiadis[2]
[1]China Agricultural University, College of Animal Science and Technology, 100193, Beijing, China, P.R., [2]University of Reading, School of Agriculture, Policy and Development, Reading, RG6 6EU, United Kingdom, [3]Polish Academy of Sciences, Institute of Animal Reproduction and Food Research, Olsztyn 10-748, Poland, [4]University of Reading, Department of Food and Nutritional Sciences, Reading RG6 6DZ, United Kingdom, [5]Matis, Vínlandsleið 12, Reykjavík, 113, Ireland, [6]University of Helsinki, Department of Agricultural Sciences, Helsinki, 00014, Finland; b.wang6@reading.ac.uk

16 Holstein cows were allocated in four 4×4 Latin squares with four 4-week periods (7 d washout, 14 d adaptation, 7 d measurements). Diets (forage:concentrate 75:25 in dry matter (DM)) were: (1) wheat distillers' grains (WDG)-based without seaweed (C-WDG); (2) WDG-based with *Saccharina latissima* (S-WDG); (3) rapeseed meal (RSM)-based without *S. latissima* (C-RSM); (4) RSM-based with *S. latissima* (S-RSM). *S. latissima* was offered at 38.6 g/cow/d. Fresh-cut grass or grass silage (as buffer) was fed *ad libitum* and concentrates were iso-nitrogenous (16% CP in DM). Linear mixed models used protein source, seaweed supplementation, and their interaction as fixed effects; and animal ID and period as the random effects. Compared to WDG, RSM milk had higher ($P<0.05$) concentrations of total SFA (+1.7%), C12:0 (+4.7%), C14:0 (+3.3%), C16:0 (+3.8%) in milk fat. Compared to RSM, WDG milk had higher ($P<0.05$) concentrations of C18:1 t11 (+16.9%), C18:1 c9 (+5.2%), C18:2 c9c12 (+30.7%), C18:2 c9t11 (+12.9%), C18:3 c9,c12,c15 (+6.5%), total MUFA (+3.6%), *trans* MUFA (+7.4%), PUFA (+13.3%), *cis* n-3 PUFA (+4.9%), and *cis* n-6 PUFA (+25.3%) in milk fat. Atherogenicity and thrombogenicity indices were lower ($P<0.05$) in WDG than in RSM. Seaweed did not affect the nutritionally-relevant milk FAs ($P\geq0.05$); while bromoform concentrations were negligible and unaffected by the dietary treatments ($P\geq0.05$). Feeding WDG, instead of RSM, enhanced concentrations of nutritionally-beneficial FA and reduced the nutritionally-undesirable SFA, while feeding *S. latissima* at 38.6 g/cow/d does not pose any risk around bromoform contamination of milk.

Session 47 Theatre 5

Human health markers improvements in clinical trials when animal feed is the only variable

N. Kerhoas, P. Weill and B. Schmitt
Bleu-Blanc-coeur, 8 rue J Maillard de la Gournerie, 35000 Rennes, France; pierre.weill@bleu-blanc-coeur.com

A lot of trials measure the impact of differences in human diets on human health biomarkers (lipidic, glycaemic or anthropometric markers) for prevention of chronic diseases such as cardiovascular diseases, obesity or diabetes. We present results of another type of trial when human diet remains unchanged, but the diet of animal that produce eggs, milk and meat were changed. In all trials, animal diet composition was the only experimental variable, and we introduced 5% of extruded linseed in experimental diets in substitution of corn and soy on an iso caloric, iso proteic basis In the first trial, human volunteers are healthy volunteers and we measure cardiovascular risks markers. In the second trial, human volunteers were diabetic. In the third trial, we recruited obese volunteers. The first trial measured an improvement in cholesterol markers, in correlation with a decrease of stroke risk by 9%. The second trial measured a significant decrease in insulin resistance. In the third one, with obese volunteers, a 90 days hypocaloric regimen caused weight loss of 3 kg in both groups, but 5 months later, we measured a weight re-gain divided by 5 in the experimental group. The nature of animal diets is associated with the quality of animal products and in particular their lipid composition. It is also associated with human health prevention by improving, (all other things being equal), the value of recognized markers of chronic diseases prevention, when the n-6 / n-3 ratio of animal feed is reduced.

Session 47 Theatre 6

The impact of combination of Inulin fibre with the 4 major PAHs present in meat on colorectal cancer

L. Abdennebi-Najar[1,2], M. Zaoui[3], N. Ferrand[3], L. Louadj[3] and M. Sabbah[3]
[1]CRSA, Sorbonne University, INSERM, UMR_S_938, 184 rue du faubourg Saint-Antoine, 75571 Paris cedex 12, France, [2] IDELE Institute, Quality and Health Department, 149 Rue de Bercy, 75012, Paris cedex 12, France, [3]CRSA, Sorbonne University, INSERM, UMR_S_938, 184 rue du faubourg Saint-Antoine, 75571 Paris cedex 12, France; latifa.najar@idele.fr

Evidence suggests that processing and heat treatment of meat may increase cancer risk through exposure to potentially carcinogenic compounds, the polycyclic aromatic hydrocarbons (HAPs) and heterocyclic aromatic amines (AAHs). Inulin-rich foods is known to be potent preventive agents against colorectal cancer. The current study aims to examine the effect of the 8 major (4-HAPs, 4 AAHs) meat contaminants in colorectal cancer and to test whether supplementation with inulin limits cancerogenic process in mice model. Healthy colon cell line (HCEC-1T) and colon cancer cell lines showing different aggressiveness (HT-29, HCT116 and LS174T) were incubated 24 and 48 hours in the presence of the 4-PAHs, 4-AAHs and the mixture. We measured cell proliferation and the expression of genes involved in the metabolism of xenobiotics (CYP1A1, CYP1A2, CYP1B1). We Evaluated in BALB/c DSS mice receiving 3 times/week 50, 100 and 150 mg of the 4-HAPs/kg of body weight, the effect of inulin (1mg/ml in drinking water) on the initiation and development of colon cancer. The 4-PAHs have no effect on healthy colon cells but were able to decrease the viability of tumour colorectal cells and activate the CYP1A1 and CYP1B1 genes involved in the metabolization of xenobiotics. *In vivo*, the 4-PAHs group mice supplemented with inulin did not reveal any changes in carcinogeneis as compared to the one without inulin. However, we did observe a significant ($P<0.05$) decrease of the loss of body weight in mice receiving PAHs+ inulin as compared to the HAPs group. As conclusion, PAHs b are capable of significantly modifying the activity of several colic cancer cell lines and the expression of key genes related to colorectal cancer. In mice, inulin supplementation did not protect against HAPs induced colorectal carcinogenesis. All these results underscore the importance of considering biological association between HAPs exposure at low doses and diet-related tumours in the colon.

Animal feeding strategy: a move towards giving strategic direction to East African countries

H. Makkar[1], K. Agyemang[2], D. Balikowa[3], A. Sebsibe[4] and R. Mondry[5]
[1]Sustainable Bioeconomy, Lammaschg, Vienna, Austria, [2]Agriculture, NRM & Policy, 85233, Gilbert, USA, [3]EAC Secretariat, EAC Road, Arusha, Tanzania, [4]ICPALD, Kapenguria Road, Nairobi, Kenya, [5]Subregional Office for Eastern Africa FAO, CMC Road, Addis Ababa, Ethiopia; ricarda.mondry@fao.org

Inability to feed animals adequately to meet their nutrient requirements round-the-year is a major constraint in East Africa. Despite animal feed and feeding being the foundation of livestock systems, and food and nutrition security being heavily interlinked with feed security, it has received limited attention so far. Most challenges in feed and feeding area are common to countries of the Intergovernmental Authority on Development (IGAD) and East African Community (EAC). Concerted efforts are needed to address them. This prompted the Subregional Office for Eastern Africa of the Food and Agricultural Organization (FAO) in collaboration with IGAD and EAC, to take lead in developing an Eastern Africa Livestock Feed and Feeding Strategy (2023-2037). It has been developed through multi-stakeholders participation from the region, and covers technological, institutional and policy dimensions under 4 pillars in the form of 4 strategic objectives: (1) take stock of the feed and water availability and accessibility, and formulate and put in practice technical solutions to enhance their availability and accessibility; (2) develop and implement appropriate feed processing, feeding strategies, and water provision approaches; (3) develop and strengthen agri-feed businesses; and (4) develop and strengthen institutional, policy-formulation and-research and human capacities on feed production and feeding. The strategy also presents 3 to 6 strategic outcomes under each 4 strategic objectives, with activities under several key focal areas of action. It embraces all the prevalent livestock production systems in East Africa: pastoral, agro-pastoral, mixed-crop livestock and intensive systems; and is inclusive of smallholder, pastoralists and industrial farmers. Recently countries in the region validated this Strategy. It will guide FAO and IGAD to update the East Africa Animal Feed Action Plan, and the countries to develop country-specific feed-focused action plans and embed them in their livestock development agenda and action plans.

Macrominerals and trace elements in retail milk: their variation and nutritional implications

S. Stergiadis[1], E.E. Newton[1], S. Beauclercq[1], J. Clarke[1], N. Desnica[2] and Á. Pétursdóttir[2]
[1]University of Reading, School of Agriculture, Policy and Development, Reading, RG6 6EU, United Kingdom, [2]Matis, Vínlandsleið 12, Reykjavík, 113, Iceland; s.stergiadis@reading.ac.uk

This study investigated the effect of dairy production system (conventional, CON; organic, ORG; channel island, CHA) and seasonal variation (January-December) on the concentrations of macrominerals and trace elements in retail milk and assessed the potential nutritional implications on mineral intakes for consumers. 42 milk brands (26 CON, 12 ORG, 4 CHA) were sampled monthly from retail outlets between January-December 2019 (n=473) and analysed for mineral concentrations by ICP-MS. Data were analysed by linear mixed models using production system, month and their interaction as fixed factors and brand ID (nested within the production system) as a random factor. Compared with CON and ORG, CHA milk contained (respectively) more Ca (+111 and +116 mg/kg milk), Mg (+9 and +12 mg/kg milk), P (+48 and +60 mg/kg milk), Cu (+7.3 and +8.6 µg/kg milk), Mn (+6.7 and +7.0 µg/kg milk), and Zn (+0.67 and +0.32 µg/kg milk); and less K (-127 and -125 mg/kg milk) and I (-109 and ORG -100 µg/kg milk). Although macrominerals did not show a clear seasonal pattern (with both lowest and highest values been within the cows' indoor period), trace elements were at lower concentrations in summer, when cows are typically grazing. For example, the months with the higher milk concentrations for Ca, P, I and Zn had 1.7, 1.4, 1.6, and 4.8 times more of these minerals, respectively, than the months with the lower concentrations. In terms of nutritional implications, milk was an excellent source of Ca, P, I, and Mo across all UK demographics, and a very good source of K, Mg, and Zn for children. Seasonal variation was greater than variation due to milk production system. The seasonal variation, and to a lesser extent milk production system, may have implications for Ca and P supply in children; and I and Zn supply across all consumer demographics. Targeted mineral supplementation in production systems and seasons with reduced milk concentrations in certain minerals (i.e. I during the grazing season), could minimise variation and ensure an optimum supply of minerals to the population throughout the year.

Effect of a spice feed additive on behaviour, saliva composition, and ruminal pH in fattening bulls

C. Omphalius[1], J.-F. Gabarrou[2], G. Desrousseaux[2] and S. Julliand[1]
[1]Lab To Field, 26 bd Dr Petitjean, 21000 Dijon, France, [2]Phodé, ZI Albipôle, 81150 Terssac, France; cleo.omphalius@lab-to-field.com

The acid buffering capacity (aBC) of saliva contributes to maintaining ruminal homeostasis in fattening bulls fed acidogenic diets. Both production and aBC of saliva can be modulated by feeds. Eight fattening bulls (427±40 kg BW) were distributed in a 4×4 Latin square design, with 21-d experimental periods separated by 7-d wash-out periods to evaluate the impact of a spice feed additive (SFA) on behaviour, saliva composition, and ruminal pH. Bulls were fed a controlled diet (in DM: 62% corn silage, 22% barley, 16% soybean meal, 1% minerals, at 1.88% BW DMI) in 2 equal meals (8:30am and 4:15pm) without (CTRL) or with the SFA included at 0.5 g/d (SFA0.5), 1.0 g/d (SFA1.0), or 2.0 g/d (SFA2.0). Ruminal pH was continuously recorded with boluses. During experimental periods, feeding duration and ingestion speed were determined from video records at d16 and d17. At d18, rumination behaviour was recorded from 7am to 7pm (2-min scan sample intervals). Saliva was sampled at d21 (3-h after morning meal) to measure aBC and minerals and HCO_3^- concentrations. Spice effect on saliva composition was assessed using multidimensional descriptive analysis (PCA, R) and through analysis of variance and regressions (SAS). Correlations between variables were calculated (SAS). Saliva composition seemed to differ gradually with the dose on the PCA graph, and [HCO_3^-] and [Na^+] were numerically higher with SFA2.0. These concentrations were positively correlated to saliva aBC ($r \geq 0.87$; $P<0.001$). The higher aBC was associated with numerically lower amplitude of ruminal pH over 24-h (0.54 for SFA2.0 vs 0.60 for other diets). Ruminal pH drop after morning meal was longer when bulls were fed the CTRL diet compared to the SFA-containing diets (6.0h vs 3.7-4.2h; $P=0.080$). The shorter the drop, the faster the ingestion ($r=-0.32$; $P=0.034$) and the longer the rumination ($r=-0.37$; $P=0.013$). Yet, ingestion speed rather followed a quadratic than a linear regression with SFA dose ($P=0.051$). It was faster for CTRL and SFA2.0 than SFA1.0 (9.7, 9.6, and 10.9 min/kg DM, respectively). The bell curve suggests that the effect of this SFA could be optimal with 1.0 g/d in fattening bulls although saliva was more buffered with 2.0 g/d.

Role of niacin in regulating intestinal health in piglets

H.B. Yi
Institute of Animal Science of Guangdong Academy of Agricultural Sciences, 1 Dafeng 1st street, Guangzhou, 510640, China, P.R.; yihongbo@gdaas.cn

How to effectively enhance the expression of antimicrobial peptides(AMPs) without triggering intestinal inflammation is currently an challenge in nutritional regulation in weaned piglets. We found that niacin (NA) could induce AMPs expression in the intestinal epithelium of weaned piglets without increasing the expression of pro-inflammatory cytokines, thereby enhancing its own resistance to disease and alleviating diarrhoea, intestinal damage and inflammation caused by *E. coli* infection. Furthermore, the expression of AMPs was associated with activation of SIRT1 and inhibition of HDAC7, while NA enhanced the levels of pH3S10, H3K9ac and H3K27ac modifications in intestine. We further explored the potential mechanisms by which histone modifications induce the expression of APMs. Firstly, we found that NA combined with sodium butyrate increased piglet daily weight gain, reduced feed-to-weight ratio and diarrhoea significantly improved growth performance of piglets. On the one hand the combination of the two increased the expression of ZO-1, Occludin, MUC2 and AMPs, improved the barrier function in weaned piglets. On the other hand, it regulated the changes of colonic metabolites and increased the concentration of butyric acid and propionic acid in the colon. Acetylation modification results suggested that differential proteins affected by the combination of NA and NaB were significantly enriched in the TCA pathway. We have verified it. The results showed that the combination of NA and sodium butyrate significantly increased the enzymatic activities of HK, PK, and SDH as well as the concentration of NADH, and increased the modification level of H3K27ac in piglet intestine. In addition, Chip-qPCR results demonstrated that the combination of NA and sodium butyrate significantly enhanced the modification level of H3K27ac in the promoter region of porcine intestinal epithelial cells, thereby inducing the expression of antimicrobial peptides in epithelial cells. This study further revealed the mechanism of niacin-induced intestinal AMPs expression in piglets. In conclusion, the addition of NA to the diet can be a safe and efficient way to promote intestinal healthy of weaned piglets.

Effect of maternal nutrition on thymus development in Wagyu (Japanese Black) foetus

O. Phomvisith[1], S. Muroya[2] and T. Gotoh[1,3]
[1]Hokkaido University, Faculty of Environmental Science, Sapporo, Hokkaido, 060-0811, Japan, [2]NARO, Animal Products Research, Tsukuba, Ibaraki, 305-0901, Japan, [3]Kagoshima University, Faculty of Agriculture, Kagoshima, 890-0065, Japan; ouanh80@gmail.com

The rapid growth of the global population involves accumulated pressure on animal-source food. Enlargement of meat and milk production inevitably engage to increase the risk of animal epidemics. The strength of immunity at birth greatly affects the subsequent health of calves. Understanding the biological mechanism of cattle immune development is imperative to support ideal tools to combat existing and potential pathogens for enhancing global food security. Maternal nutrition is crucial for the growth trajectory and development of the structure and physiology of the visceral organs in the foetus. The thymus is a primary lymphoid organ, it has important functions in immune homeostasis during foetal development and after birth. The study aimed to determine the effect of different maternal nutrition on the alteration of immune-related properties in the thymus of foetuses in Wagyu which is a fatty breed. After the conception was proved, 12 pregnant cows were assigned to 2 groups (n=6 for each group), the low-nutrition (LN) group received 60%, and the high-nutrition (HN) group was fed 120% of their nutrient requirement (NARO, 2008). At 260 days of gestation, foetuses were removed; thymus samples were collected and frozen with liquid nitrogen and stored at -80 °C until used for analyses. Relatively, the LN group revealed lighter foetal thymus weight than the HN group ($P<0.01$). The haematoxylin and eosin staining found a larger section of the interlobular segment, and immature Hassall's corpuscles in the LN group, whereas the HN group exhibited a clear zone of cortex and medulla, and apparent Hassall's corpuscles panels. The immunohistochemical analysis displayed a higher CD8 detection in the LN foetal thymus. Furthermore, concentrations of His, Ser, and Thr were higher in the LN foetal thymus compared to the HN group ($P<0.05$) in the metabolome analysis. In summary, the levels of maternal nutrition influenced foetal development, which affected the immunohistochemical properties and tissue metabolism in the foetal thymus of wagyu cattle.

***Bacillus* sp. strains protect the intestinal barrier from oxidative stress and deoxynivalenol**

G. Copani, B.I. Cappellozza and E.J. Boll
Chr. Hansen A/S, Animal and Plant Health & Nutrition, Boege Alle 10/12, 2970 Hoersholm, Denmark; dkgico@chr-hansen.com

Oxidative stress can have a major impact on performance and health of livestock animals. In ruminants, there is a potential negative impact of oxidative stress during transportation and weaning. Mycotoxins present in the feed can cause decreased nutrient utilization/performance and impairment of the intestinal barrier. These stressors can cause intestinal barrier damage (leaky gut), which in turn may facilitate the passage of potential toxins and other luminal content into the bloodstream. The objective of this study was to evaluate *in vitro* beneficial effects of probiotic *Bacillus licheniformis* (CH200) and *B. subtilis* (CH201) (part of BOVACILLUS™, Chr. Hansen) on gut barrier integrity using Caco-2 cell monolayers challenged with: (1) hydrogen peroxide (5 mM of H_2O_2) to simulate oxidative stress (Exp. 1); or (2) the mycotoxin deoxynivalenol (75 µM of DON) (Exp. 2). The FITC-dextran-20 kDa (FD) was added to the apical side of the Caco-2 cells at the same time as the probiotics. The amount of FD translocated to the basolateral side was quantified upon the termination of the TEER measurements (after 20 h for Exp. 1 and after 14 h for Exp. 2) by measuring the fluorescent signal. In Exp 1, H_2O_2 caused an 82% TEER decrease compared to unstimulated Caco-2 cells over 20 h ($P<0.005$). CH200 and CH201 both significantly alleviate the H_2O_2-induced TEER decrease ($P<0.0002$) and FD translocation ($P<0.05$). In Exp 2, DON caused an 85% TEER decrease compared to unstimulated Caco-2 cells over the 14 h ($P<0.0001$) while CH200 and CH201 significantly reduced the DON-induced TEER decrease to 50% ($P<0.0001$) and 37% ($P<0.0001$), respectively. The addition of both CH200 and CH201 significantly reduced the amount of DON-induced FD translocation (79 and 60% respectively; $P<0.0001$). In conclusion, the tested probiotic strains confer protection against oxidative stress caused by H_2O_2 and alleviate the *in vitro* leaky gut condition caused by DON by counteracting its damaging effect on intestinal barrier integrity.

Session 47 Poster 13

Production and carcass characteristics of growing-fattening rabbits under three feeding phases

V.C. Resconi, M. López, J.L. Olleta, J. Romero and M.M. Campo
Universidad de Zaragoza, Animal Production and Food Science, Miguel Servet 177, 50013 Zaragoza, Spain; resconi@unizar.es

Traditionally, rabbit farms use one type of feed after weaning and include a withdrawal feed if medication has been applied previously. On some farms more than one type of feed is used, but their composition, period and mode of administration and names of the phases are not well established. In this study, a three-feeding phase plan was tested on a Spanish farm: from 21 to 35 d old (transition phase), from weaning at 35 to 49 d (post-weaning phase), and from 49 d to slaughter (63-65 d, finishing phase). Percentages of fibre decreased and energy components increased as the feeding phase progressed, while protein was highest in the last feed type and lowest in the second. Daily growth and feed efficiency were calculated weekly in 18 litters with 8 rabbits each allocated in a cage until 63 d. Carcass characteristics were assessed at 21, 35, 49 and 65 d old in 10 animals per age. Growth was higher and feed efficiency lower in the post-weaning phase than in the finishing phase (50.9 vs 45.9 g/d and 2.9 vs 5.0; $P<0.001$; respectively). Carcass yield was maximum at 21 d, when previous feeding was based on lactation; minimum at 49 d for the high proportion of the digestive tract; and increased to 55.4% at 65 d with 2.1 kg liveweight. Considering the muscle in the carcass at 65 d as a reference, 22.1% of the weight was achieved from 21 to 35 d, 20.1% from 35 to 49 d, and 46.6% in the finishing period. In terms of fat (scapular plus kidney fats), 24.5% of the final weight was achieved in the transition period, only 7.0% during post-weaning and 29.6% in the third phase. Most the liver (40.0%) and kidney (28.4%) weight was also gained in the finishing phase, but the highest percentage of the head was reached at 21 d. These changes reflect the modification of feed composition in the different phases and the development of tissues and organs along the growth of the rabbits. The feeding phases studied are feasible to apply under the conditions of this trial (i.e. animal genetics, slaughter weight and sanitary conditions). All three phases are important, but specially the third because, although daily gain and feed efficiency are disadvantageous, most of the edible weight of the carcass is attained.

Session 47 Poster 14

Niacin improves intestinal health through up-regulation of AQPs expression induced by GPR109A

X. Yang, Y. Qiu, S. Liu and Z. Jiang
Institute of Animal Science, Guangdong Academy of Agricultural Sciences, No. 1, Dafeng 1st Street, Wushan, Tianhe District, Guangzhou, 510640, Guangzhou, China, P.R.; yangxuefen@gdas.cn

Changes in the expression of aquaporins (AQPs) in the intestine are proved to be associated with the attenuation of diarrhoea. Diarrhoea is a severe problem for postweaning piglets. Therefore, this study aimed to investigate whether niacin could alleviate diarrhoea in weaned piglets by regulating AQPs expression and the underlying mechanisms. 72 weaned piglets (Duroc × (Landrace × Yorkshire), 21 d old, 6.60±0.05 kg) were randomly allotted into 3 groups for a 14-day feeding trial. Each treatment group included 6 replicate pens and each pen included 4 barrows (n=24/treatment). Piglets were fed a basal diet (CON), a basal diet supplemented with 20.4 mg niacin/kg diet (NA) or the basal diet administered an antagonist for the GPR109A receptor (MPN). Additionally, an established porcine intestinal epithelial cell line (IPEC-J2) was used to investigate the protective effects and underlying mechanism of niacin on AQPs expression after *Escherichia coli* K88 (ETEC K88) treatment. Piglets fed niacin-supplemented diet had significantly decreased diarrhoea rate, and increased mRNA and protein level of ZO-1, AQP 1 and AQP 3 in the colon compared with those administered a fed diet supplemented with an antagonist ($P<0.05$). In addition, ETEC K88 treatment significantly reduced the cell viability, cell migration, and mRNA and protein expression of AQP1, AQP3, AQP7, AQP9, AQP11, and GPR109A in IPEC-J2 cells ($P<0.05$). However, supplementation with niacin significantly prevented the ETEC K88-induced decline in the cell viability, cell migration, and the expression level of AQPs mRNA and protein in IPEC-J2 cells ($P<0.05$). Furthermore, siRNA GPR109A knockdown significantly abrogated the protective effect of niacin on ETEC K88-induced cell damage ($P<0.05$). Niacin supplementation increased AQPs and ZO-1 expression to reduce diarrhoea and intestinal damage through GPR109A pathway in weaned piglets.

Session 48 Theatre 1

How can poultry farming systems evolve to meet the major societal and environmental challenges?

P. Thobe[1] and P. Van Horne[2]
[1]Thuenen Institute, Thuenen-Institute of Farm Economics, Bundesallee 63, 38116 Brausnchweig, Germany, [2]Wageningen University & Research, the Netherlands, P.O. Box 9101, 6700 HB Wageningen, the Netherlands; p.thobe@thuenen.de

Global poultry meat and egg production showed an impressive increase over the last 10 years. FAO data show that poultry meat production was 131 m ton in 2020 compared to 97 m ton in 2010. In this period, North America and Europe lost market share and Asia and South America increased their share. The largest producers of poultry meat are USA (17%), China (13%), Brazil (12%) and the EU (9%). According to the FAO, egg production in 2020 was 87 m ton compared to 64 m ton in 2002. In egg production, China is by far the largest producer (34%), followed by USA (8%) and the EU (7%). In 2020, total production of pig meat was 110 m ton and of cattle meat was 68 m ton. In 2017, poultry meat surpassed the production volume of pig meat. OECD projection for 2030 show a further increase in eggs and poultry meat. The outlook for 2030 shows a higher annual growth for poultry meat compared to pig meat and cattle meat. As a result of population growth and increasing income of a growing middle class in many developing countries, meat consumption will further grow with poultry meat often being the favoured one. Tradition, no religious barriers, a low price and convenient preparation are factors in favour of poultry meat. The most efficient production system with specific breeds and indoor housing is dominating world-wide. Housing of layers in cage and the use of fast-growing broiler breeds is criticized in the EU for animal welfare reasons. As poultry production is highly cost and price driven, costs incurred from adopting welfare practices are therefore of great importance. In terms of sustainability, besides animal welfare, social as well as environmental and economic aspects need to be included, while minimizing potential trade-offs. For bringing poultry farming to a significantly higher level of animal welfare in a socially and economically sustainable manner, a centrally guided concept becomes indispensable. Policy and business actors are working together to ameliorate animal production systems in the EU to comply with higher animal welfare standards. Several new public and private animal welfare labelling initiatives emerged in the past decade.

Session 48 Theatre 2

Can we enhance environmental impact without compromising bird welfare in broiler systems?

I. Kyriazakis
Queen's University, Institute for Global Food Security, 19 Chrlorine Gardens, BT9 5DL Belfast, United Kingdom; i.kyriazakis@qub.ac.uk

As there is a widely recognized need to enhance all aspects of sustainability of livestock systems, the debate about potential trade-offs between reduction of environmental impacts and enhancement of animal welfare has inevitably come into focus. This conundrum is best reflected in the desire to move towards slower growing birds, to overcome some of the health and welfare concerns associated with fast growth and efficiency. Because feed-associated activities, such as feed ingredient production, and manure contribute almost 90% to the various metrics of environmental impact, systems that use slower growing birds which are less efficient in utilizing their feeding resources, will by definition be more environmentally impactful. The reductions in environmental impact that arise from welfare enhancements, such as reductions in mortality, culling, and use of medication will never offset such feed utilization-related impacts. For this reason, instead of entering into comparisons between broiler systems that utilize either fast or slow growing birds, it may be more fruitful to concentrate on practices that have the potential to reduce environmental impacts within a system. In this presentation, examples of such practices and their potential to affect environmental impacts will be given: (1) As the environmental impact of feed ingredient production is associated with land use and land use change, such as in the case of soya production, using home-growing protein sources may reduce the environmental impact of broiler systems. Furthermore, because of the lower amino acid requirements of slower grown strains, in theory it may be possible to replace more of the soya inclusion in the diet with alternative protein sources than in the diets of fast-growing birds and reduce their environmental impacts. (2) A reduction in the stocking density is accompanied with an increase of several environmental impact metrics (e.g. carbon footprint and energy use). However, the use of a heat exchanger applied for ventilation air is able to reduce any increased heating requirements due to low stocking density and offset environmental impact increases.

Session 48 Theatre 3

What is the impact of the farming system on the quality of the chicken breast meat?

J. Albechaalany[1,2], S. Yilmaz[2], M.P. Ellies-Oury[1,2], M. Bourin[3], Y. Guyot[3], J. Saracco[4], J.F. Hocquette[1] and C. Berri[5]
[1]INRAE, UMRH, 63122 Saint-Gènes-Champanelle, France, [2]Bordeaux Sciences Agro, F&F, 33170 Gradignan, France, [3]ITAVI, INRAE CVL, 37380 Nouzilly, France, [4]INRIA, IMB, 33400 Talence, France, [5]INRAE, BOA, 37380 Nouzilly, France; john.albechaalany@doctorant.uca.fr

The study aimed to assess the quality characteristics of the four main types of chicken fillets produced in France. The research was conducted on a set of 7,843 fillets collected from commercial slaughterhouses as part of a survey conducted by ITAVI and INRAE. The four main French production systems – Label Rouge, Certified, Standard and Heavy Broiler – were included in the sample, which varied according to animal type, age and/or weight at slaughter, and rearing practices. These systems have coexisted in France for many years to meet different demands in terms of taste quality and societal expectations, including animal welfare. The study assessed technological quality by ultimate pH, CIELAB colour parameters and curing-cooking yield, sensory quality by juiciness and tenderness after cooking, and nutritional quality by measuring dry matter, protein and fat content. Principal component analysis was performed to identify significant variables that explain chicken fillet quality, and analysis of variance (ANOVA) to investigate relationships between categorical variables. The data also included the time between slaughter and boning of the carcass, which has a significant impact on breast meat tenderness. Label Rouge fillets had the lowest pH and lipid content, but the highest redness and yellowness. The Certified broiler produced the lightest meat. Compared to standard and heavy broilers, Label Rouge and Certified broilers had a higher shear strength, indicating a tougher meat. Heavy broiler breast meat had the highest fat content, but the lowest cooking and processing losses. In addition, the study identified an abattoir effect for each type of broiler, highlighting the significant impact of the slaughter process on breast meat quality. In conclusion, the study highlights the significant impact of the rearing system and genetics on the quality of chicken fillets and the importance of taking into account the characteristics of the animals to optimize the process and ensure optimal meat quality.

Session 48 Theatre 4

Improving broiler wellbeing and micro-climate through PLF application

S. Druyan[1,2], N. Barchilon[1,2] and I. Halachmi[1]
[1]PLF Lab, Agricultural Research Organization, Volcani Institute, 68 HaMakkabbim Road, Rishon Le Ziyyon, 7528809, Israel, [2]Institute of Animal Science, Agricultural Research Organization, Volcani Institute, 68 HaMakkabbim Road, Rishon Le Ziyyon, 7528809, Israel; shelly.druyan@mail.huji.ac.il

The welfare and productivity of broilers are affected by micro-climate conditions. Unsuitable environmental conditions lead to reduced performance, increased morbidity, and even death, resulting in animal welfare concerns and financial losses for farmers. Modern broilers struggle to balance energy expenditure and body water balance, limiting their ability to maintain steady-state mechanisms under sub-optimal environmental conditions. Climate control systems in modern broiler houses are equipped with sensors measuring environmental parameters around the broilers, rather than their actual needs. This study aimed to test and validate a new system that measures individual broiler body temperature using a low-cost infrared camera calibrated by a thermistor, an algorithm for IR image processing, and a Lasso regression model predicting actual body temperature. The system successfully detected a departure from the thermoneutral zone, with both systems detecting broilers crossing the upper and lower critical temperature thresholds on different days. The study identified 16 occurrences where the broilers' body temperature fell below the lower critical threshold, and five cases were detected where the broilers' body temperature exceeded the upper critical threshold. All were identified despite the ambient temperature sensors of the climate control system indicating that the temperature was within the desired range effecting broilers body weight (2,100 grams compared to an average weight of 2,500 as reported by the slaughterhouses for 35-day-old broilers). The results of this study suggest that monitoring individual broiler body temperatures is crucial for ensuring optimal micro-climatic conditions. By using this system as a temperature sensor in the climate control loop in broiler houses, farmers can improve broiler productivity and welfare, leading to better economic outcomes. Overall, this study highlights a practical, low-cost method for achieving optimal micro-climatic conditions for broilers.

Effects of heat stress and spirulina on productive performances of two slow growth broiler strains

E.A. Fernandes[1], C.F. Martins[1,2], D.F.P. Carvalho[1], L.L. Martins[1], A. Raymundo[1], M. Lordelo[1] and A.M. And Almeida[1]
[1]Instituto Superior de Agronomia, University of Lisbon, LEAF – Linking Landscape, Environment, Agriculture and Food Research Centre, Associated Lab. TERRA, Tapada da Ajuda, 1349-017 Lisbon, Portugal, [2]Faculdade de Medicina Veterinária, University of Lisbon, CIISA – Centre for Interdisciplinary Research in Animal Health. AL4AnimalS, Avenida da Universidade Técnica, 1300-477 Lisbon, Portugal; eafernandes@isa.ulisboa.pt

Poultry production faces two problems: heat stress (HS) and lack of sustainable feedstuffs. HS affects productive performance and health. Mainly due the high protein content Spirulina are becoming increasingly interesting in poultry nutrition. The aim of this study was to evaluate the effect of 15% Spirulina (SP) dietary inclusion and the impact of HS on growth performance and organ measurements in two slow- growth broiler strains: FF (normally feathered) and Na (naked neck). 40-one-day-old male chicks (20/strain) were individually housed at 30 °C with *ad libitum* access to water and feed. Each strain received either a Control diet (C) or SP diet. Feed intake (FI) and body weight (BW) were monitored weekly and feed conversion ratio (FCR) and average daily gain (ADG) were calculated. After 84 days of trial, all broilers were slaughtered and organs collected, weighed, and measured. The final BW of the animals was 2,741, 2,536, 2,527 and 2,142 g for CNa, SPNa, CFF and SPFF groups respectively. At the end of the trial, FI and ADG were significantly affected ($P<0.05$). SPNa animals had the highest FI and ADG (145 g/animal/d – 53.34 g/d) and the CFF had the lowest (111 g/animal/d – 39.53 g/d). However, FCR were not significantly affected. Relative crop and gizzard weight ranged between 1.97-2.67 and 10.0-13.0 respectively, highest in SP animals ($P<0.05$). The relative length (cm/kg) of the ileum was increased by the incorporation of SP, 31.5, 30.1, 27.5 and 25.7 for SPFF, SPNa, CFF and CNa groups respectively. The HS negatively influenced growth performance of broilers especially the normally feathered animals. The Na groups could tolerate HS even when fed with SP. Future information provided by digestibility coefficients and meat quality traits will complement these results, allowing a better knowledge of the nutritional values of Spirulina for broilers under HS.

Evaluating services provided by free-range poultry systems

G. Chiron
ITAVI, 7 Rue du Faubourg, 75009 Paris, France; chiron@itavi.asso.fr

In free-range poultry production, outdoor run is a key element for the multiperformance of the farming system. However, managing the outdoor run can be seen more as a constraint when its potential to provide more global agroecological solutions and meet society's expectations, beyond animal welfare, is not perceived. Thus, it is necessary to provide tools for farmers, adapted to different knowledge levels and context, allowing them to think about an optimized management of the outdoor run or to better take into account the diversity of services it provides. The objective of this study was to develop and test a generic method to assess the services provided by different free-range poultry systems (i.e. species, location, or outdoor run characteristics). First, a conceptual framework was developed through a participatory approach by experts before being challenged by futures users (farmers, advisors, outdoor run management structures, etc.) in three poultry production regions (West, South-East and South West of France). The conceptual framework finally considers of 13 services, being allocated in five different service categories (production of resources and value, farmer's life quality, farmer-consumer-citizen interconnection, territorial integration and environmental quality). For each service, the experts also selected relevant indicators to assess each service in an easy, quick, and cheap way in commercial farms. These indicators are based on either on-farm surveys (e.g. labour time or economic performance) or *in situ* measurements (e.g. assessing animal welfare). In order to validate the feasibility and genericity of the method in practical conditions, 21 commercial houses were studied. These farms differed in their location (West, South-East and South West of France), the production (chicken meat, eggs, duck meat) and the shape, size, and design (trees, bushes, woods, etc.) of the outdoor run. The first results indicates that the method is enough sensible to highlight different bundles of services according to the outdoor run characteristics. Therefore, this method could be used by farmers and their advisors, to evaluate the actual provision of services by a free-range poultry system, and further discuss about the most adequate management of the outdoor run for a maximal provision of services.

Session 48 Theatre 7

Biosecurity gaps in 7 major poultry producers (breeder and layer farms) in EU: a farmers perspective

A. Amalraj[1], H. Van Meirhaeghe[2], R. Souillard[3], A. Zbikowski[4] and J. Dewulf[1]
[1]Faculty of Veterinary Medicine, Ghent University, 9820, Belgium, [2]Vetworks BV, Knokstraat, Aalter, Belgium, [3]ANSES, French Agency for Food, Environmental and Occupational Health & Safety, Epidemiology, Health, France, [4]Institute of Veterinary Medicine, Warsaw University of Life Sciences, Nowoursynowska, Poland; arthi.amalraj@ugent.be

Good practices prevent diseases and economic losses, and ensure high production standards in poultry farming. A qualitative assessment was done in poultry breeder (n=46), enclosed layer (n=37) and free-range layer (n=21) farms of the NetPoulSafe project, with a semi-closed questionnaire with focus on 46, 45 and 54 biosecurity measures (BM) respectively in each production. The farmers responded to the frequency of implementation and the reasons for non-compliance. Most measures (75% of all the answers collected) were frequently implemented due to regulatory control. However, a compromise was noticed on implementation of some measures. Wheel disinfection (BM1), cleaning and disinfection of the rendering tank after each collection (BM2) and egg storage room (BM3) after each collection, showering of personnel (BM4), visitors (BM5), and farm-specific clothing and shoes for egg transport drivers (BM6) were less implemented compared to other measures. A descriptive analysis of the collected data showed that from the total respondents, only 78% breeder, 59% enclosed layer and 62% free-range layer farms practiced BM1; only 46% breeder, 67% enclosed layer and 42% free-range layer farms practiced BM2; only 47% breeder, 63% enclosed layer and 69% free-range layer farms practiced BM3 and only 75% breeder, 67% enclosed layer and 56% free-range layer farms followed BM6. In only 65% breeder, 11% enclosed layer and 19% free-range layer farms, farm personnel showered before entering into the poultry house and in only 72% breeder, 24% enclosed layer and 24% free-range layer visitors showered before entering into the poultry house, often reported 'excessive measures' in some productions. The main reasons for non-compliance were 'lack of time', 'not knowing risks/advantages', 'expenses incurred' or 'not considered useful'. This study suggests that intervention is needed in order to further improve biosecurity compliance in the participating farms.

Session 48 Theatre 8

Reducing environmental impact of broiler production: the role of crude protein and soybean meal

T. De Rauglaudre[1,2], B. Méda[1], S. Fontaine[3], W. Lambert[3] and M.P. Létourneau Montminy[2]
[1]INRAE, BOA, Nouzilly, 37380, France, [2]Université Laval, Département des sciences animales, Québec, G1V 0A6, Canada, [3]METEX Animal Nutrition, Paris, 75017, France; theophane.de-rauglaudre.1@ulaval.ca

Lowering crude protein (CP) content in broiler feeds has the potential to reduce environmental burdens related to nitrogen (N) volatilization. As the main source of CP is generally soybean meal (SBM), this strategy also allows to reduce the environmental impacts (e.g. climate change) associated to feed production, when SBM production is associated with deforestation. Yet, in most low CP trials, the effects of CP and SBM are confounded. Thus, the aim of this study was to investigate the effect of decreasing CP content at various SBM levels on growth performance, and N volatilization. A total of 3,384 one-day-old Ross 308 males were randomly distributed in 63 pens. After receiving the same starter feed, 7 experimental diets were used fed during growing (G; 11-22 d) and finishing (F; 23-34 d) periods: A control diet (C; 20.4 and 19.5% CP in G and F periods, respectively) and 6 experimental diets formulated with 3 levels of SBM (100, 50 and 0%) and two CP reduction levels (-15 and -30 g/kg) were randomly assigned. The body composition used to calculate the N body retention was measured by scanning under anaesthesia one bird per pen (9 per treatment) with dual-X ray absorptiometer at days 23 and 35. During the G period, compared to C, average daily gain and feed intake were linearly decreased with CP independently of SBM content (P<0.001), but feed efficiency was not impacted. In the F period, no effect on performance was observed. On the total trial period (11-34 d), only the growth and feed intake in the soy-free diet at -30 g CP/kg were lower than C (-5%; P<0.01). Reducing CP linearly decreased N intake and excretion independently of SBM content while N retention efficiency was linearly increased (P<0.001). This decrease in N excreted led to a decrease of N lost through volatilization (N volatized = -0.44 + 0.58 * N excreted; RMSE=0.20; P<0.001). The current results show that it is possible to simultaneously decrease CP content by 30 g/kg and SBM content by 50% without impairing animal performance. CP and SBM reduction strategies can therefore be used together to improve the sustainability of broiler production.

Elevated platforms for broilers on commercial farms: usage and effects on health and performance

J. Stracke[1], F. May[2], J. Müsse[3], N. Kemper[2] and B. Spindler[2]
[1]University of Bonn, Farm Animal Ethology, Institute for Animal Science, Endenicher Allee 15, 3115 Bonn, Germany, [2]University of Veterinary Medicine Hannover, Foundation, Institute for Animal Hygiene, Animal Welfare and Farm Animal Behaviour, Bischofsholer Damm 15, 30173 Hannover, Germany, [3]Chamber of Agriculture Lower Saxony, Mars-la-Tour-Straße 1-13, 26121 Oldenburg, Germany; jenny.stracke@itw.uni-bonn.de

Barren barns as standard housing systems for broiler chicken can lead to serious welfare problems. Environmental enrichment does not only allow the animals to perform species-specific behaviour, but might also improve other welfare aspects such as health. In the present study, one of two barns in each of three broiler farms was equipped with an elevated platform (90×1.20 m). Video cameras were installed above the elevated platforms, recording the number of animals in three locations (front, middle, back) during fattening (about 42 days). For one day per week (n=6) of the fattening period (n=3), screen shots were taken every 30 minutes for a time period of 24 hours. The number of animals on the elevated platforms was evaluated. Furthermore, each farm was visited three times (beginning, middle end) per fattening period (n=5). Data of animals' performance (weight) and health (food pad dermatitis (FPD), hock burns (HB), injuries and plumage pollution (PP)) was collected per barn for 50 animals each. Generalized linear mixed models were calculated. The animals used the elevated platform frequently with a maximum of 14 broilers per m^2. Significant effects of farm, week and interaction between both (week and farm) was found. Effects of treatment were found for HB and PP, and a significant effect of the interaction between farm and treatment was revealed (HB, FPD, PP). The results of this study indicate the benefits of elevated platforms for broilers' well-being. However, positive effects on health depend on the respective management and need further research. This work (Model- and Demonstration Project for animal welfare MaVeTi) is financially supported by the Federal Ministry of Food and Agriculture (BMEL) based on a decision of the Parliament of the Federal Republic of Germany, granted by the Federal Office for Agriculture and Food (BLE); grant number 'FKZ 2817MDT301'.

Mutation effects as key driver of maintaining genetic variation for long-term selection in broilers

B.S. Sosa-Madrid[1,2], N. Ibañez-Escriche[1], G. Maniatis[3] and A. Kranis[2,3]
[1]Institute for Animal Science and Technology, Universitat Politècnica de València, P.O. Box 2201, Valencia 46071, Spain, [2]The Roslin Institute, University of Edinburgh, EH25 9RG, Midlothian, United Kingdom, [3]Aviagen Ltd., Newbridge, Midlothian, EH28 8SZ, United Kingdom; v1bsosa@exseed.ed.ac.uk

Maintaining genetic variation in a population is important for long-term genetic gain. Theoretically, selection process leads to reducing genetic variation over time in populations undergoing intense directional selection pressure. Mutations are variation sources arising in the genome randomly, and those contributing to the genetic variance of the traits under selection would partly compensate for the loss of genetic variation. In broiler breeding, the breeding goals have been expanding through time to include traits related to biological performance and welfare. Body weight is a trait that has been part of broiler breeding goals historically. The aim of this study was to quantify the contribution of mutational variance for body weight trait in a broiler line, over 23 years of selection and 2 million animal records. Bayesian inference was carried out to estimate the variance components using two models: one with- and other without-mutational effects. The parameterization of Wray's relationship matrix by Casellas and Medrano was used to model the mutational matrix. Mutation variance is regarded as the variation originating from a mutation effect arising from an animal until the last generation. Namely, the genetic effect build-up segregating from a mutation event to the youngest relative is deemed a mutation effect and contribute separately to additive genetic variance. The trajectory of variances was inferred according to Sorensen's method clustering animals by egg hatching period (3 weeks). Results showed that mutational variance (0.71) and mutational heritability (0.0027) estimates are lower in the base population, compared to the additive genetic variance (100.96) and heritability (0.39). However, mutational variance bears an important role over time, as the mutational effects accumulate and thus, their contribution tends to increase. Overall, albeit the contribution of mutation effects is small in the first generation of selection, its role has been essential to ensure genetic variation in a long-term broiler selection programme.

Session 48 Theatre 11

Early interventions during incubation and impact on a muscle of locomotory relevance in broilers

T. Kettrukat[1], A. Dankowiakowska[2], M. Mangan[3], E. Grochowska[3], K. Stadnicka[4] and M. Therkildsen[1]
[1]Aarhus University, Dept of Food Science, Agro Food Park 48, 8200 Aarhus N, Denmark, [2]University of Science and Technology Bydgoszcz, Dept of Animal Physiology & Zoophysiotherapy, ul. Mazowiecka 28, 85-084 Bydgoszcz, Poland, [3]University of Science and Technology Bydgoszcz, Dept of Animal Biotechnology & Genetics, ul. Mazowiecka 28, 85-084 Bydgoszcz, Poland, [4]Nicolaus Copernicus University, Faculty of Health Sciences, ul. Łukasiewicza 1, 85-821 Bydgoszcz, Poland; tobias.kettrukat@food.au.dk

Early intervention through temperature manipulation and *in ovo* microbiome programming can potentially alleviate challenges in broiler production like heat stress and gut health. In contrast to the breast muscle, the impact on leg muscles has rarely been examined. A positive effect on such muscles could improve locomotory ability in broiler chickens, a persistent concern brought on by singular selection on breast muscle growth that compromises welfare. We investigated the effects of temperature manipulation and microbiome programming *in ovo* on *M. gastrocnemius* in Ross 308 broilers. In experiment one, temperatures ranging from 36.5 to 39.0 °C were applied during embryonic day 4 to 7, and *M. gastrocnemius* was sampled for histology at hatch and 35 days of age. In experiment two, a combination of an IBD vaccine with a prebiotic (Astragalus polysaccharides), a probiotic (*B. lactis*) or a synbiotic (both) was applied *in ovo* on embryonic day 18. *M. gastrocnemius* of 42-day-old females was sampled for histological analysis. Fibre dimensions, fibre type and intramuscular fat were determined by H+E, myosin ATPase, and Oil Red O – staining. Absolute and relative muscle weights were influenced by the interventions, but the fibre dimensions seem to be largely unaffected, suggesting no negative effect on the development of leg muscle fibres through the interventions. The findings could be used to develop these early interventions into tools to alleviate negative effects of the intensive selection for body weight. The experiments were funded by the European Union's Horizon 2020 research and innovation program (grant N° 955374), the National Science Centre, Poland (grant N° 2019/35/B/NZ9/03186, OVOBIOM) and the Polish National Agency for Academic Exchange (grant N° PPI/APM/2019/1/00003).

Session 48 Theatre 12

Summary of the posters in session 48 by the chairs

B. Méda[1] and K. Stadnicka[2]
[1]INRAE, Université de Tours, BOA, 37380 Nouzilly, France, [2]Nicolaus Copernicus University, Collegium Medicum, Lukasiewicza 1, 85-821 Bydgoszcz, Poland; katarzyna.stadnicka@cm.umk.pl

A summary of the posters in session 48 will be presented by the chairs of the session. The aim is to increase the visibility of the relevant research topics that address challenges and solutions related to poultry farming systems in the context of major societal and environmental challenges.

Session 48 Poster 13

Effect of gallic acid supplementation to corn-soybean-gluten meal-based diet in broilers performance

J.H. Song[1], C.B. Lim[1], S. Biswas[1], Q.Q. Zhang[1], J.S. Yoo[2] and I.H. Kim[1]
[1]Dankook University, Department of Animal Resource and Science, Cheonan, 31116, Korea, South, [2]Daehan feed company, 13 Bukseongpo-gil, Jung-gu, Incheon, 22300, Korea, South; thdwnsh@naver.com

Gallic acid (GA) is an endogenous plant polyphenol found in fruits, nuts, and plants that has antioxidant, antimicrobial, and health-promoting effects. Broilers were fed graded doses of GA to assess growth efficiency, nutrient retention, footpad lesion score, tibia ash, and meat quality. In a 32-days feeding trial, 576 1-day-old Ross 308 broiler chicks (mixed sex) with an average body weight of 41±0.5 g were arbitrarily assigned to 1 of 4 dietary treatments with 8 repetitions of 18 birds per cage and a basal diet administered with 0, 0.02, 0.04, and 0.06% of GA. Linear and quadratic effects were used to examine the polynomial contrast of increasing level of GA inclusion. Feeding broilers with a graded dose of GA increased body weight gain (BWG) (P=0.003) and feed intake (P=0.012) linearly on days 9-21. BWG also tended to linearly increase on days 21-32 (P=0.092) and the overall period (P=0.051), respectively. In addition, the nutrient digestibility of dry matter (P=0.003) and gross energy (P=0.014) were linearly increased. However, the faecal score, footpad lesion score, and tibia ash reveal no significant effect (P>0.05). By feeding broilers a graded level of GA, relative organ weights including breast muscle and spleen tended to increase linearly (P=0.081 and 0.067, respectively), meat colour of yellowness increased linearly (P=0.005), and pH value tended to decrease (P=0.068). Briefly, adding GA at increasing doses to broiler diets increased growth efficiency, nutritional absorption, and meat quality, so we infer that adding 0.06% of GA could be beneficial for broiler performance.

Session 48 Poster 14

Veterinary coaching to stimulate biosecurity compliance

A. Amalraj[1], H. Van Meirhaeghe[2], M. De Gussem[2] and J. Dewulf[1]
[1]Faculty of Veterinary Medicine, Ghent University, 9820, Belgium, [2]Vetworks BV, Knokstraat, Aalter, Belgium; arthi.amalraj@ugent.be

Farmers' readiness to accept changes is pivotal to biosecurity application and therefore improving biosecurity requires changes in the attitudes and behaviour of farmers. Coaching is a method to help farmers find their short term and long term goals through non-directional questioning with more interaction and is different from previously used advisory techniques where the process is more unidirectional. Veterinary 'coaching' was tested to assess their efficacy in stimulating compliance. A longitudinal study is ongoing on 18 pilot farms (part of NetPoulSafe project) in Belgium: broiler (n=5), enclosed layer (n=2), free-range layer (n=2), turkey (n=2), breeder (n=4), and hatcheries (n=3). The hygiene evaluation tool – Biocheck.Ugent (www.biocheck.ugent.be) and ADKAR® change management model were used as supporting tools for coaching. For ADKAR, whenever the farmers scored low (<3) for the element 'Awareness', the risk factors arising from poor hygiene was discussed. For the element 'Desire' (<3), to provoke an interest, the benefits of the change were explained. For the element 'Knowledge', depending on the specific problem in the farm, an educative approach was used in the form of visual aids and PowerPoint presentations. For the element 'Ability' lower scores were dealt with by discussing topics such as making structural changes and investments towards better biosecurity. An action plan is drawn for the following 6 months and is reviewed by the coach and the veterinarian through a phone call. The biosecurity status was re-evaluated after 6 months. The total biosecurity score obtained by the pilot farms were: broiler (65%), enclosed layer (65%), free-range layer (68%), turkey (70%), and breeder (71%). Breeder farms in Belgium scored highest for external biosecurity (70%) and turkey and free-range layer farms had the highest internal biosecurity scores (80%). The mean ADKA scores for the participating farms were 3.8, 4, 3.8 and 3.9 for the elements Awareness, Desire, Knowledge, and Ability, respectively. Farm biosecurity scores and profiling helped the coach and veterinarian to design an intervention plan specific to each pilot farm.

Session 48 Poster 15

Establishing elevated perforated platforms in broiler chicken housing – is hygiene a barrier?

B. Sake[1], J. Müsse[2], F. May[1], J. Stracke[3], N. Kemper[1], J. Schulz[1] and B. Spindler[1]

[1]University of Veterinary Medicine Hannover, Foundation, Institute for Animal Hygiene, Animal Welfare and Farm Animal Behaviour (ITTN), Bischofsholer Damm 15, 30173 Hannover, Germany, [2]Chamber of Agriculture Lower Saxony, Mars-la-Tour-Straße 1-13, 26121 Oldenburg, Germany, [3]University of Bonn, Farm Animal Ethology, Endenicher Allee 15, 53115 Bonn, Germany; bjoern.sake@tiho-hannover.de

In Europe, broiler chickens are usually raised in littered barns without structuring elements. Elevated perforated platforms (EPP) might have positive effects on animal welfare by providing additional space, and elevated structures to rest. This study aimed to analyse the efficiency of the cleaning and disinfection process (C&D) during the service period. Therefore, microbiological samples were taken during two to three service periods from three farms with barns equipped with an EPP. For sample collection, sterile sponges were taken, and wiped over a defined area at three different locations: the top of the perforated plastic floors of the platform, the metal rack and the barn floor for comparison. At each location five samples were taken and pooled. Three different time-points (TP) at the service period were considered (TP1: before cleaning, TP2: after cleaning and drying, TP3: after disinfection). Samples were analysed microbiologically for the occurrence of *Enterococc*i (ET), *Enterobacteriaceae* (EB), and the total bacterial counts (TBC). The results showed that samples from the EPP at both locations contained less TBC compared to the barn floor. At TP3, neither ET nor EB were detected at both locations of the EPP and all tested locations showed a TBC below 1,000 per cm^2, indicating a reduction of 4 to 6 log levels. Although C&D of the EPP took considerably more time and required additional cleaning tools, the procedure was performed efficiently. It can be concluded that the applied hygiene measures are suitable to reduce the risk of pathogen transmission to the next production cycle.
This work (Model- and Demonstration Project for animal welfare MaVeTi) is financially supported by the Federal Ministry of Food and Agriculture (BMEL) based on a decision of the Parliament of the Federal Republic of Germany, granted by the Federal Office for Agriculture and Food (BLE); grant number 'FKZ 2817MDT301'.

Session 48 Poster 16

What's going on outside? Use of winter gardens by rearing hens of different genetics

A. Riedel[1,2], N. Kemper[2] and B. Spindler[2]

[1]University of Veterinary Medicine Hannover, Foundation, WING (Science and Innovation for Sustainable Poultry Production), Heinestraße 1, 49377 Vechta, Germany, [2]University of Veterinary Medicine Hannover, Foundation, Institute for Animal Hygiene, Animal Welfare and Farm Animal Behaviour, Bischofsholer Damm 15, 30173 Hannover, Germany; anna.katharina.riedel@tiho-hannover.de

In German laying hen husbandry, it is required that pullets are accustomed during rearing to the husbandry system in the laying hen farm. For organic farms, the offer of a winter garden (WG) as a transit area between henhouse and free-range is required, to accustom the pullets to the outdoor climate. In this on-farm study, it was examined how frequently this area was visited by pullets in three flocks of different genetics (H1 NB: NOVOgen Brown, H2 LB: Lohmann Brown-Lite and H3 DW: Dekalb White). They were reared over the same period of 18 weeks of life (LW). The flocks H1 NB and H2 LB were housed in groups of about 4,100 animals each, with 2 WG of about 69 m^2 each. The flock H3 DW was reared in another farm, including 2,400 animals, with one WG of approximately 68 m^2. A wildlife animal camera was installed in one WG of each group, automatically taking photographs in a 15-minute interval over the course of the day. These photographs were evaluated for three days in LW 12 and LW 16/18, and the number of animals present in an area of 35 m^2 (H3 DW), 26.8 m^2 (H1 NB) and 19.5 m^2 (H2 LB) was counted. The results showed a good acceptance of the WG in all three flocks with on average of 2.26 (±1.86) hens/m^2 in the examined area. All flocks showed a significant increase in hens/m^2 from the first (LW 12) to the second (LW 16/18) observational period ($P<0.001$). During the first observational period, no significant differences between the flocks were found, while at the end of the rearing period, H1 NB showed a significantly higher use of the WG (3.8±1.8 hens/m^2) than H3 DW (2.9±0.5 hens/m^2; $P<0.01$). The results for H2 LB were intermediate (3.1±2.4 hens/m^2). This work is financially supported by the Federal Ministry of Food and Agriculture based on a decision of the Parliament of the Federal Republic of Germany, granted by the Federal Office for Agriculture and Food (grant number FKZ 2817MDT301/311).

Metabolomic analysis reveal the molecular mechanism related to leg disease in broilers

J. Zheng, G. Zhang, Q. Li and G. Zhao
Institute of Animal Sciences, Chinese Academy of Agricultural Sciences, Yuanmingyuan West Road No.2, Haidian, Beijing, 100193, Beijing, China, P.R.; liqinghe@caas.cn

White feather broilers are characterized by fast growth and high production efficiency. Rapid weight gain causes the body metabolism to be in an overload state, triggering leg disease, which has economically affected the welfare and development of broilers. However, the relationship between changes in metabolites during the dynamic metabolism and leg disease remains unclear. The results of serological biochemical indices in normal and leg bone deformed broilers showed that the serum calcium to phosphorus ratio was highly reduced in diseased broilers. In addition, serum lipid indices were abnormal, indicating a disturbance in lipid metabolism. Based on extensive targeted metabolomics analysis, 1,005 and 997 metabolites were detected in serum and cartilage tissue samples, respectively. Orthogonal partial least squares discriminant analysis and weighted gene co-expression network analysis were used to screen for important metabolites associated with leg disease, serum calcium and phosphorus levels. As a result, a total of nine key metabolites were screened, which included carnitine C16:0, carnitine C18:1, 3-oxymethyl-L-tyrosine, cis-4-hydroxy-D-proline, cis-L-3-hydroxyproline, and trans-4-hydroxy-L-proline. The top 20 metabolic pathways enriched in the KEGG pathway for differential metabolites in the two tissue samples included fatty acid degradation and arachidonic acid metabolic pathways. Metabolite analysis on these two pathways showed that prostaglandin D_2, prostaglandin J_2, prostaglandin A_2, 15- keto- prostaglandin $F_{2\alpha}$ were significantly different between normal and disease groups. Collectively, the results suggest that the occurrence of leg disease in broiler chickens is associated with metabolic disorders in the organism. Serum calcium to phosphorus ratio can be used as an *in vivo* predictor of leg disease in broiler chickens.

Housing conditions do not influence the effects of a nutritional challenge for broilers

V. Michel, C. Deschamps, N. Regrain, E. Devillard and J. Consuegra
Centre of Expertise and Research in Nutrition, Adisseo, 03600, France; virginie.michel@adisseo.com

In vivo experimentation is essential for animal science research. However, the large number of animals and repetitions required to obtain biological and statistical differences limits the flexibility of designs and heavily impacts animal welfare. For a sustainable animal experimentation, it is pivotal to develop new designs and experimental structures allowing both the reduction of animal use and biological events. Here, we tested the possibility of reducing the number of animals in nutritional challenge thanks to an adaptable system composed of 24 modular broilers cages. This system allows a flexible experimental design in terms of birds per cage and repetitions. A total of 108 one-day-old male Ross 308 chicks were randomly allocated to 4 experimental conditions with a 2×2 factorial arrangement. Two diets were tested: Control (CTL, standard diet with wheat/soybean meal) and Dysbiosis (DYS, an unbalanced feed enriched in rye and saturated fat). Two types of cages were used: small (S) with 3 birds/cage and large (L) with 9 birds/cage. The experimental conditions in the S cages were repeated 9 times and those in L cages were repeated 3 times to obtain 27 birds per treatment. Body weight (BW), feed intake (FI) and feed conversion ratio (FCR) were measured on days 14 and 28. No interaction between diets and cage sizes was observed. At 28 d, DYS significantly decreased broilers body weight by 5.6% in S cages and by 7.4% in L cages compared to CTL ($P<0.05$). Moreover, FCR was significantly degraded with DYS diet by 8% on 0-14 d and 6% on 0-28 d periods ($P<0.01$). However, housing conditions did not affect these parameters. The FI during 0-14 d and 0-28 d wasn't impacted by any experimental factor. Altogether, our results show that both tested housing conditions allow to reproduce the DYS negative impacts on BW, FI and FCR. Moreover, the deterioration of performances is at the same level as the one previously reported on the same challenge using around 2,000 animals on floor (at least -6% of BW at 28 d). Thus, our S cages model allows to increase the number of experimental replicates and to reduce the numbers of animals which offers a larger panel of experimental designs and improves the sustainability of our experimental facilities.

Session 48 — Poster 19

Compliance of biosecurity in poultry farms in France: remaining obstacles and levers for improvment

N. Rousset[1], A. Battaglia[1], J. Puterflam[1], J. Marguerie[2], R. Souillard[3], S. Le Bouquin-Leneveu[3] and A.-C. Lefort[1]
[1]ITAVI, 7 rue du faubourg Poissonnière, 75009 PARIS, France, [2]SNGTV, 5 rue Moufle, 75011 Paris, France, [3]Anses, BP 53, 41 rue de Beaucemaine, 22440 Ploufragan, France; rousset@itavi.asso.fr

The successive outbreaks of avian influenza in France underline more than ever the importance of complying with biosecurity measures in poultry farms. Although the concepts of biosecurity are increasingly well known, it is nevertheless observed that they are not always well applied. The European NetPoulSafe network, launched in 2020, aims to sustainably improve compliance of biosecurity in the poultry sector. In order to make a state of the art of current compliance of these measures, the technical and psychosocial obstacles that remain, and to identify ways of strengthening support, a study was conducted in France with 25 broiler, turkey, laying hen and duck farmers, with different production systems (indoor, outdoor, different sales ways), as well as with 24 technical and veterinary advisors. The study reveals that only a minority of measures are actually applied by all farmers. As for the rest of the measures, a very high degree of heterogeneity was observed depending on the species produced and the production system. The obstacles mentioned by the farmers are often the inconvenience in the work, or even in the organisation of the work that some measures can generate. This is the case in particular for complex production systems in farms with direct sales, with the presence of several species, and multiple flocks. Structural obstacles were also mentioned, e.g. the presence of specific equipment (aviaries, cages) can generate difficulties in applying complete cleaning and disinfection protocols. Finally, many farmers still have doubts about the effectiveness or relevance of some measures, which indicates a need to continue raising awareness and training on the risks. Personal advice, discussion groups within farmers and training based on concrete cases were widely favoured by the interviewees. Although an improvement in biosecurity compliance was generally observed, this study confirms the need to strengthen personalised support, by proposing very operational solutions to overcome the specific obstacles of some production systems.

Session 48 — Poster 20

Untrimmed beaks in turkey hens: Effect on injuries and mortality rate

M. Kramer[1], K. Skiba[1], P. Niewind[2], F. Von Rüden[2], N. Kemper[1] and B. Spindler[1]
[1]University of Veterinary Medicine Hannover, Foundation, Institute for Animal Hygiene, Animal Welfare and Farm Animal Behaviour (ITTN), Bischofsholer Damm 15, 30173 Hannover, Germany, [2]Agricultural Chamber of North Rhine-Westphalia, Haus Duesse, Haus Düsse 2, 59505 Bad Sassendorf, Germany; Marie.Kramer@tiho-hannover.de

Beak trimming is a common non-curative intervention to minimize severe pecking injuries in poultry. This study aimed to show the effect of untrimmed beaks on injuries and mortality rate of turkey hens in an optimized housing environment. Therefore, turkey hens with untrimmed (UT: 5,400 hens) and trimmed beaks (T: 5,400 hens) were housed in one barn each, on one farm and were observed during two fattening batches. To minimize injurious pecking, the UT herd received enrichments such as elevated platforms, hay baskets, metal mobiles and feed dispensers. In the 1st, 5th, 9th and 13th week of life, in both batches, 60 birds per group were scored to detect injuries. Losses due to mortality and separation were recorded daily. Overall, there were 28% hens with injuries in the UT sample and 6% in the T group. The highest prevalence of injured birds in both batches (UT: 72% | 52%; T: 13% | 10%) was recorded in the 13th week of life. The injuries occurred mainly to the snood and head area (80%). Animals with fresh injuries were removed from the herd to a separation compartment. Thus, in the first run, 1,546 animals (29%) were separated in the UT group, while 331 animals (6%) were separated in the T group. In the second run, 718 hens (13%) were separated in the UT group, while 203 hens (4%) were separated in the T group. The cumulative mortality rate in UT herds was 8.65% in the first, and 4.39% in the second run, while in T herds it was 3.98% in the first, and 2.26% in the second run. Concluding, keeping UT hens showed mortality rates nearly twice as high and a higher prevalence of injuries despite the enriched housing environment and the extensive separation management. This work (Model- and Demonstration Project for animal welfare #Pute@Praxis) is financially supported by the Federal Ministry of Food and Agriculture (BMEL) based on a decision of the Parliament of the Federal Republic of Germany, granted by the Federal Office for Agriculture and Food (BLE; grant number *FKZ 2817MDT611*).

Session 48 Poster 21

Habitat affects egg characteristics of Libyan local pigeon (domestic vs feral pigeon)

F. Akraim, M.F. Idrees and M.M. Sghieyer
Omar Al-Mukhtar University, Animal Production Department, University campus, 1118 Al Bayda, Libyan Arab Jamahiriya; fowad.akraim@omu.edu.ly

Domestic pigeon is descended from rock pigeons (*Columba livia*), that inhabit North Africa. The majority of the pigeon population in Libya are domestic or feral pigeons, in addition to a smaller number of wild pigeons. This study aimed to investigate the effect of habitat on some external and internal quality characteristics of eggs collected from feral and domestic Libyan pigeons. Measurements were done on 62 feral pigeon eggs collected from different urban locations and 37 eggs collected from a domestic pigeon farm. Across habitats, the mean egg weight was 15.72 g. Yolk, albumin, and shell represent 67.85, 23.72 and 8.22% of egg weight, respectively. Egg shape index was 74.88%, and shell thickness was 0.26 mm. Feral pigeon eggs were heavier (16.73 vs 15.16 g), bigger (15.82 vs 14.76 cm^3), wider (28.57 vs 27.46 mm), and had a more favourable shape index (76.34 vs 72.30%) than domestic pigeon eggs ($P<0.05$). The yolk percentage was lower (21.51 vs 26.09%) and the albumin percentage was higher (71.07 vs 64.36%) in feral compared with domestic pigeon eggs ($P<0.05$). Those differences could be attributed to nutritional factors linked to the diverse feeding regimes of feral pigeons. A study of multiple domestic pigeon farms is needed to confirm this conclusion.

Session 48 Poster 22

Effects of heat stress and spirulina on meat traits of two slow growth broiler strains

E.A. Fernandes, J.R. Sales, L.L. Martins, M. Lordelo, A. Raymundo and A.M. And Almeida
Instituto Superior de Agronomia – University of Lisbon, LEAF – Linking Landscape, Environment, Agriculture and Food Research Centre, Associated Lab. TERRA, Tapada da Ajuda, 1349-017 Lisboa, Portugal; eafernandes@isa.ulisboa.pt

Heat stress (HS) and lack of sustainable feedstuffs are two of the most severe problems faced by poultry production systems. Heat stress negatively affects productive performance and health. Due to its high protein content, the incorporation of spirulina (SP) is becoming increasingly interesting in poultry nutrition and can play an important role in minimizing the problems associated with HS. The objective of this study was to evaluate the impact of 15% SP dietary inclusion and the effect of HS on different meat quality parameters in normally feathered (FF) and naked neck (Na) slow growth broiler strains. Forty-day-old chicks were distributed in 4 different groups of ten animals: Control diet + Naked neck (CNa), Diet with 15% SP + Naked neck (SPNa), Control diet + Normally feathered (CFF) and Diet with 15% SP + Normally feathered (SPFF). Each animal had *ad libitum* access to water and feed. During the 84 days of the experiment, the temperature was kept at 30 °C. At the end of the trial, all broilers were slaughtered, and breast muscle was collected, weighed, and used for subsequent analysis. Carcass yield was 69.0, 67.3, 66.6 and 65.8% for CFF, CNa, SPNa and SPFF groups, respectively. Carcass yield was decreased ($P<0.05$) with the incorporation of SP regardless the strain. Breast muscle yields were not significantly different between groups ($P>0.05$). SPFF animals had the highest percentage of drip loss (40.7%), while CNa had the lowest (34.8%). Thawing and cooking losses were not significantly affected ($P>0.05$). Regarding colour parameters, lightness (L*) and redness (a*) were not significantly affected ($P>0.05$) but yellowness (b*) was significantly highest ($P<0.001$) in SP groups, when compared to control diet groups. This alteration is associated with a more intense yellow in the groups fed with SP, resulting from the presence of high levels of carotenoids. The incorporation of Spirulina in the diet didn't affect the majority of the meat traits, independently of the strain used.

Session 48 Poster 23

How does broiler range use impact forage intake, outdoor excretion and gaseous emissions?

C. Bonnefous[1], B. Méda[1], K. Germain[2], L. Ravon[2], T. De Rauglaudre[1], J. Collet[1], S. Mignon-Grasteau[1], M. Reverchon[3], C. Berri[1], E. Le Bihan-Duval[1] and A. Collin[1]

[1]INRAE, Université de Tours, BOA, 37380 Nouzilly, France, [2]INRAE, UE EASM, Le Magneraud, CS 40052, 17700, Surgères, France, [3]SYSAAF, 37380, Nouzilly, France; anne.collin@inrae.fr

Free range enables broiler chickens to express natural behaviours, eat grass, worms or insects outdoors, but also induces excretion outside the poultry house. To understand how the outdoor run is impacted by range use, we studied three strains of intermediate to slow-growing chickens: JA757 (734 animals, average daily gain (ADG): 36 g/day, outdoor density (OD): 0.26-0.29 chicken/m^2, rearing duration (RD): 86 days), S757N (735 animals, ADG: 26 g/day, OD: 0.26-0.29 chicken/m^2 and RD: 100 days) and a dual-purpose (DP) crossbreed (771 animals, ADG: 16 g/day, OD: 0.27-0.31 chicken/m^2 and RD: 121 days). Birds had access to a grassy outdoor run with mature trees from 36 days of age. We collected data on the time spent outdoor by broilers using Radio Frequency IDentification (RFID) chips, on the composition and amount of manure and on grass growth to evaluate the impact of range use on grass consumption and outdoor nitrogen (N) / phosphorus (P) excretions. RFID showed that S757N chickens spent about twice as much time outdoor per day than the JA757 and DP ones. Similarly, the S757N chickens also consumed about twice as much grass per day than JA757 and the DP ones. Consistently, S757N chicken outdoor excretions of N and P per day of outdoor access were about twice as much as that of the DP chickens while only about 30% greater than that of JA757. Therefore, we confirmed the relationship between range use and foraging, but we showed strain-dependant outdoor excretion in relation to growth rate and feed intake. A finer understanding of how genetics impact foraging behaviour can help maximize the use of the outdoor run without compromising resource intake and excretion on outdoor areas. The project PPILOW has received funding from the European Union's Horizon 2020 research and innovation programme under grant agreement N°816172.

Session 48 Poster 24

Heritability of the number of crossovers as proxy of recombination rate in chicken

V. Riggio[1], E. Tarsani[1] and A. Kranis[1,2]

[1]The Roslin Institute and R(D)SVS, University of Edinburgh, Easter Bush, EH25 9RG Midlothian, United Kingdom, [2]Aviagen Ltd., Newbridge, EH28 8SZ Midlothian, United Kingdom; valentina.riggio@roslin.ed.ac.uk

Recombination has a direct impact on evolution by shuffling genetic diversity. Avian species appear to have higher rates of recombination than mammals. The variability in recombination observed between chicken breeds is consistent with extensive findings in other species that show it is a heritable trait. However, in mammals, the genetic control of recombination is better studied and it has been shown that some genes have a major effect, whereas in avian species, there is less information about the genetic architecture of the trait. In this study, we aimed at estimating the heritability of the number of crossovers, as proxy of recombination rate. The data consisted of 39,108 genotyped chicken, of which 299 were sires and 1,780 were dams. The pedigree file included 41,189 individuals. Chickens were genotyped with a customised 50k genotyping array, for a total of 48,367 SNP across 28 chromosomes. Sex chromosomes were not included in the analysis. The number of crossovers per offspring for both sires and dams were calculated using the FindHap software v.3 (https://aipl.arsusda.gov/software/findhap/), and then analysed as repeated measurements of the parents. Variance components, heritability and repeatability were estimated using ASReml v.4, fitting a repeatability animal model with sex as fixed effect, age (calculated as difference in weeks between parents and their offspring) as covariate, and the animal (i.e. either the sire or the dam) and the permanent environmental effect as random, using the pedigree. Both heritability and repeatability estimates were moderate and significant, being equal to 0.25±0.02 and 0.55±0.01, respectively, confirming that the recombination rate, expressed as number of crossovers, is heritable and it can potentially be used to make predictions of their incidence and impact on genetic variation. Further studies are been carried out using random regression models to investigate how the recombination rate is affected by the individuals ageing (i.e. whether the more an individual ages, crossovers increase, hence being a proxy of recombination rate).

The potential of early warning system at health issues in poultry by sound

I. Halachmi[1], T. Lev-Ron[1,2,3], Y. Yitzhaky[3] and S. Druyan[1,2]
[1]PLF Lab, Agricultural Research Organization (A.R.O.), Volcani Institute, 68 HaMakkabbim Road, Rishon Le Ziyyon, 7528809, Israel, [2]Animal Sci Inst., Agricultural Research Organization (A.R.O.), Volcani Institute, Poultry and Aquaculture Department, 68 HaMakkabbim Road, Rishon Le Ziyyon, 7528809, Israel, [3]School of Electrical And Computer Engineering, Ben-Gurion University of the Negev, Be'er Sheva, 8410, Israel; halachmi@volcani.agri.gov.il

Sound analysis has a potential to detect health issues such as a respiratory disease. To test the feasibility of the above hypothesis, poultry sound analysis was performed. Two consecutive experiments were conducted, each with 80 chickens exposed to different stressors (sub-optimal acute environmental conditions – cold, heat, and windy conditions), and a control group kept under recommended environmental regime. Data were collected using audio recordings, and ANN models were trained and tested on two different datasets. Among three ANN model sizes, ranging from 6 million to 88 million parameters. The 'base' model (which is the largest) resulted in the highest mAP scores of 0.97, 0.80, and 0.96 for the first dataset, the second dataset, and the combined datasets. The 'small' model yielded mAP scores of 0.96, 0.78 and 0.95 for the first, second and combined experiments accordingly. 'Tiny' model resulted in mAP scores of 0.95, 0.75, and 0.93 for the first, second, and combined experiments, respectively. Further research is recommended applying induced-illness, not only cold, heat and wind stressors.

Pre-slaughter fasting changes the *ante mortem* muscle proteolysis levels and *post mortem* meat quality

S. Katsumata[1], M. Kamegawa[2], A. Katafuchi[2], A. Ohtsuka[2] and D. Ijiri[2]
[1]Okayama University, Graduate School of Environmental and Life Science, 1-1-1, Tsushima-naka, Okayama, 7008530, Japan, [2]Kagoshima University, Faculty of Agriculture, 1-21-24, Korimoto, Kagoshima, 8900065, Japan; s.katsumata@okayama-u.ac.jp

Pre-slaughter fasting is a routine practice in commercial broiler production and it elevates stress levels with muscle protein degradation (proteolysis). The aim of this study was to investigate the effects of pre-slaughter fasting times on the relationship between *ante mortem* muscle proteolysis level and *post mortem* breast meat quality of broilers. Twenty-four broiler chicks (Ross 308) at 0 days of age were provided water and a semi-purified diet with no animal protein *ad libitum* until 27 days of age. At 27 days of age, the chicks were assigned to four treatments: 0-hour of fasting (0H), 8-hours of fasting (8H), 16-hours of fasting (16H), or 24-hours of fasting (24H). Then, they were slaughtered at 28 days of age. Blood samples were collected from the wing vein before the fasting and just prior to slaughter. A portion of the left breast muscle was immediately stored in liquid nitrogen, and the right breast muscle was stored at 4 °C for 48 hours. Plasma N^{τ}-methylhistidine concentration, which is an index of muscle proteolysis level, and muscle free amino acids contents were analysed by the UHPLC system. To evaluate the taste characteristics, chicken soup was made from each breast muscle, and its taste traits were determined by the taste sensing system. *Ante mortem* change in individual plasma N^{τ}-methylhistidine concentration was significantly increased in 8, 16, and 24H compared to that in 0H ($P<0.05$). After 48 hours of *post mortem* storage, glutamic acid (Glu) content in breast muscles was increased with the length of fasting time ($P<0.05$), and umami taste (the fifth basic taste) of chicken soup in the fasting group (8, 16, 24H) was higher than 0H ($P<0.05$). The *ante mortem* change plasma N^{τ}-methylhistidine concentrations were correlated with the Glu contents in breast muscles ($r=0.57$, $P<0.05$), and umami taste ($r=0.66$, $P<0.05$). These results suggest that *post mortem* changes in the Glu contents of chicken meat during storage reflect *ante mortem* muscle proteolysis levels exerted by pre-slaughter fasting.

Session 49 Theatre 1

Host-microbiota interactions in swine and poultry: disentangling causes and effects

J.F. Pérez
Universitat Autònoma de Barcelona, Animal and Food Science, Travessera dels Turons. Edifici V, 08193, Spain; josefrancisco.perez@uab.es

Main concerns of the swine and poultry production reside on the high morbidity and mortality of animals during early periods, and the high within lot BW variability. Mortality and slow growing animals are particularly relevant for the industry, as these animals bring inefficiencies in term of feed and resources, increasing the final variability of the market group. The progressive movement towards a more sustainable livestock industry, with a required lower use of medicines, also means a strong interest for understanding host microbiota interactions aimed to improve overall wellbeing and health. Many solutions are being studied, including the use of additives with prebiotic or probiotic functions, but there is no 'silver bullet' to guarantee animal health in this new scenario. The failure of these strategies may likely derive from our difficulties on understanding the complex relation and interactions established among causes and consequences involved on the health, and productivity of the animals. Characterizing the intestinal microbiota of highly feed-efficient (FE) pigs could help to define an 'optimal' microbial profile for improved FE. In pigs, low residual feed intake was associated to: higher fermentation of carbohydrates, higher release of butyrate, lower mucin secretion. In poultry, it has been also described a genetic control by the host of its microbiota. However, several studies suggest that early life experiences and environmental conditions have a critical role in the growth and development of mammals. On the one side, the early life microbial colonization of the gut, shaped by factors such as host genetics, environment, diet, immunological pressure, and antibiotics, may influence the programming of the mucosal immune response and the development of the gut barrier function, conditioning the microbiota profile and their propensity to develop certain health disorders. Evidence suggests that a disruption of microbial colonization, and events occurring during the early periods, might also cause lifelong deficits in growth and development. Better knowledge of these interactions will allow the design of a holistic approach aimed to improve overall wellbeing, health and feed efficiency in swine and poultry.

Session 49 Theatre 2

Glutamine and glucose metabolism in suckling low birth weight piglets supplemented with glutamine

D. De Leonardis, Q.L. Sciascia, S. Goers, A. Vernunft and C.C. Metges
Research Institute for Farm Animal Biology (FBN), Wilhelm-Stahl-Allee 2, 18196 Dummerstorf, Germany; de-leonardis@fbn-dummerstorf.de

Glutamine (Gln) supplementation has been shown to be beneficial in growing piglets. To study the effect of glutamine supplementation on metabolic pathways in suckling piglets, male German Landrace piglets with low (L; 0.8-1.2 kg) and normal birthweight (N; 1.5-1.9 kg) were selected. At 24 h after birth, 10 L and N piglets/group were allocated to daily Gln (1 g/kg BW/d; L-Gln, N-Gln) or water (W, 6 ml; L-W, N-W) supplementation. At age 14 d, piglets received orally Gln (0.33 g/kg BW) plus $^{13}C_5$ Gln (10 mg/kg BW), and at 16 d, glucose (Glc; 0.4 g/kg BW) plus $^{13}C_6$ Glc (10 mg/kg BW). Blood was collected before (-15 min, basal) and half-hourly until 300 min after tracer administration via a jugular catheter. Mass spectrometry was used to measure red blood cell $^{13}CO_2$ enrichment (E) derived from oxidation of $^{13}C_5$ Gln and $^{13}C_6$ Glc, plasma $^{13}C_5$ Gln E, and plasma $^{13}C_3$ Glc E, newly synthetized from $^{13}C_5$ Gln tracer carbon. Area under the ^{13}C enrichment-time-curve (AUC), maximum enrichment (Emax) and time to maximum enrichment (Tmax) were computed by curve fitting. Statistical evaluation was performed by Student t-tests. Preliminary results show that $^{13}CO_2$ E Tmax from $^{13}C_5$ Gln and $^{13}C_6$ Glc oxidation was greater in N-Gln than in L-Gln and N-W ($P<0.05$). Plasma $^{13}C_5$ Gln E AUC tended to be lower (141.4 vs 174.4 mole % excess (MPE)*min; $P=0.1$) and Emax was lower (0.95 vs 1.24 MPE; $P<0.05$) in N-Gln than in N-W. Plasma $^{13}C_3$ Glc E AUC derived from $^{13}C_5$ Gln metabolism tended to be greater in L-Gln than in L-W and N-Gln piglets (14.9 vs 8.5 vs 9.2 MPE*min; $P=0.1$). The $^{13}C_3$ Glc Tmax was greater in N-Gln than in L-Gln and N-W groups (91.1 vs 57.9 vs 55.9 min; n=3/group; $P<0.05$). Our data suggest that L piglets supplemented with Gln oxidized Gln and Glc faster than N piglets. This agrees with greater utilization of glutamine carbon for Glc de novo synthesis in L-Gln piglets. The lower plasma $^{13}C_5$ Gln E in N-Gln piglets might indicate a greater dilution by endogenous Gln production. These results must be confirmed by further investigations of Gln and Glc metabolism. This project has received funding from the European Union's Horizon 2020 research and innovation program under grant agreement no. 955374.

Session 49 Theatre 3

Creep feeding (dry, liquid) and pen hygiene (low, high) impacts pre-weaning growth in pigs

S.R. Vasa[1,2], G.E. Gardiner[1], K. O'Driscoll[2], G. Bee[3] and P.G. Lawlor[2]
[1]South East Technological University, Dept. of Science, Waterford X91Y074, Ireland, [2]Teagasc, Pig Development Dept., Fermoy P61R966, Ireland, [3]Agroscope, Swine Research Unit, Posieux 1725, Switzerland; shivramveer.vasa@teagasc.ie

Increasing pre-weaning creep feed intake can increase pig weaning weight (WW) and better prepare pigs for weaning. The objective was to evaluate the effect of providing creep feed in dry or liquid form to suckling pigs housed in a low or high hygiene environment, on their growth and intestinal structure. Eighty seven sows, blocked by parity, number of pigs weaned and live-weight, were randomly allocated to one of the four treatments in a 2×2 factorial arrangement. The factors were creep feeding (dry or liquid) and pen hygiene (low or high). Pigs were provided with dry pelleted starter diet from day (d) 10-28, or a mixture of liquid milk and starter diet from d3-28. Either a sub-standard cleaning protocol (water wash, no detergent or disinfectant and no drying) or an optimal cleaning protocol (detergent application, water wash, disinfectant application and thorough drying) was used to obtain a low or high hygiene environment, respectively in the farrowing rooms. Pigs were weighed and feed disappearance recorded on d4 and 28 (weaning) of age. On d4 post-weaning (PW), 10 pigs/treatment were euthanized to collect tissue samples for histological analysis. Data were analysed using PROC MIXED (v9.4, SAS Institute Inc.). There was a creep feeding × hygiene interaction on WW. Liquid feeding increased WW in both high and low hygiene environments ($P<0.05$). Liquid-fed pigs had 0.5 kg heavier WW than dry-fed pigs ($P<0.05$). There was a creep feeding × hygiene interaction on ADFI from d3-28. Liquid feeding increased ADFI in the high hygiene but not in the low hygiene environment ($P<0.05$). High hygiene pigs had 43% higher ADFI than low hygiene pigs ($P<0.05$). From d4-28, ADG was increased by liquid feeding ($P<0.05$) but was not affected by pen hygiene ($P>0.05$). On d4 PW, jejunal villus height and crypt depth were increased by high hygiene in farrowing rooms ($P<0.05$) but were not affected by creep feeding ($P>0.05$). In conclusion, high pen hygiene increased pre-weaning feed intake and improved intestinal structure, whereas liquid creep feeding increased growth and WW. Funded by EU Horizon 2020 under grant agreement No 955374.

Session 49 Theatre 4

The impact of early incubation temperature on broiler walking ability and final meat quality

T. Kettrukat and M. Therkildsen
Aarhus University, Department of Food Science, Agro Food Park 48, 8200 Aarhus N, Denmark; tobias.kettrukat@food.au.dk

With a rearing time of only 5-6 weeks from hatch until slaughter for fast-growing broiler strains, the portion of the life spent in the egg becomes increasingly important for interventions. The incubation temperature has been shown to affect muscle development, bone development, performance, meat quality and possibly walking ability. The latter is a persistent animal welfare issue in commercial broiler flocks around the world leading to concern among consumers and economic losses. Studies on the relationship between these aspects and the possibility to develop temperature manipulation into a tool to improve walking ability and subsequent welfare without compromising meat quality are limited. In the studies presented here, we changed the temperature of ROSS 308 fertile eggs during embryonic day 4 to 7, testing 36.5, 38.5 and 39.0 °C against 37.5 °C. Then, we followed the development of the locomotory system by gene expression analysis and histology in different muscles, as well as bone strength analysis, while documenting performance parameters and meat quality. The walking ability was assessed by gait scoring and its association with different parameters of the locomotory system was examined. Temperature manipulation affected the walking ability of 5 weeks-old broilers: a temperature of 36.5 °C led to higher gait scores than all other treatments. Furthermore, a higher temperature seems to favour the growth of *M. pectoralis* over *M. gastrocnemius* compared to a lower temperature, as indicated by the ratio between the muscles. Analysis of gene expression in the muscles by qPCR suggests more changes in the breast muscle than in the leg muscle. The sex influenced almost all parameters measured, most notably performance, muscle weights and bone dimensions, with males reaching higher weights. An effect on final meat quality is suggested by differences in muscle glycogen content and drip loss. The results give a comprehensive overview about the changes in the locomotory system of broilers induced by early temperature manipulation and their subsequent effects on performance and welfare. This project received funding from the European Union's Horizon 2020 research and innovation program under grant agreement no. 955374.

Session 49 Theatre 5

In vitro and in vivo analysis of bioactive substances growth and antioxidant activities

M. Mangan[1], C. Metges[2] and M. Siwek[1]
[1]Bydgoszcz University of Science and Technology, 85-084 Bydgoszcz, Poland, Department of Animal Biotechnology and Genetics, Mazowiecka 28, Bydgoszcz, Poland, 85-084, Poland, [2]Research Institute for Farm Animal Biology, Wilhelm-Stahl-Allee 2, Dummerstorf, 18196, Germany; modou.mangan@pbs.edu.pl

Heat stress is a major problem in the poultry industry, causing severe economic loss due to its adverse effects on chickens' health and performance. Thus, the main aim of this study was to screen *in vitro* bioactive substances to find those most suitable for *in ovo* modulation of chicken microbiota and heat stress mitigation in chickens. To achieve this, we first determined the kinetic growth curve of the selected probiotics (*Lacticaseibacillus casei*, *Lactiplantibacillus plantarum*, *Limosilactobacillus reuteri*, *Lacticaseibacillus rhamnosus*). Subsequently, these probiotics were combined with prebiotics (raffinose, galactooligosaccharide (GOS), long-chain inulin) and plant extracts (green tea, turmeric, garlic extract). The growth curve of the probiotics in the presence of prebiotics or plant extracts was determined by optical density (OD) at 600 nm. Finally, the antioxidant activities of these bioactive substances were assessed using the 2,2-diphenyl-1-picrylhydrazyl DPPH) assay. From the results obtained, *L. plantarum* and *L. rhamnosus* had the highest growth curve (OD 2). From the DPPH results, *L. plantarum* exhibited antioxidant activity of 69% and thus was selected for *in ovo* injection on day 12 of embryonic development. The second bioactive used for *in ovo* injection and microbiota modulation was prebiotic GOS which was selected due to its positive effects in mitigating heat stress in poultry. These bioactive substances were delivered *in ovo* to test their effects on hatchability and zootechnical parameters on day-old chicks. Preliminary results show that the hatchability was higher in the negative control (not injected) and GOS treatment. The day-old chick body weight was higher ($P<0.05$) when treated with *L. plantarum* compared to the other groups. The day-old chick length was lowest in the negative control. There was no difference between the treatments for Pasgar score. This research is funded by the European Union's Horizon 2020 research and innovation program, grant N° 955374.

Session 49 Theatre 6

Prophybiotics, a novel approach for in ovo gut microbiome reprograming of broilers

R.N. Wishna-Kadawarage[1], R. Hickey[2] and M. Siwek[1]
[1]Bydgoszcz University of Science and Technology, Department of Animal Biotechnology and Genetics, Faculty of Animal Breeding and Biology, Mazowiecka 28, 85-084 Bydgoszcz, Poland, [2]Teagasc Food Research Centre, Department of Food Biosciences, Moorepark, Fermoy, Co. Cork P61 C996, Ireland; ramesha.wishna-kadawarage@pbs.edu.pl

Bioactives administered *in* ovo may reprogram the gut microbiome of chickens before they are exposed to environmental pathogens. Although the antimicrobial and gut microbiome modulation potential of probiotics and phytobiotics alone have been studied intensively, their combined use in *in ovo* model has yet to be investigated. We coined the term Prophybiotics (probiotic + phytobiotics) to describe such a combination. The current study aims to screen and validate the effects of prophybiotics in an *in ovo* model to determine if their application can mitigate pathogenic stress in broilers. Six lactic acid bacteria, *Lactiplantibacillus plantarum*, *Lacticaseibacillus casei*, *Limosilactobacillus reuteri*, *Lacticaseibacillus rhamnosus*, *Leuconostoc mesenteroides* and *Pediococcus pentosaceus* and three plant extracts, turmeric, green tea and garlic were included for *in vitro* screening. Growth curves with plant extract supplementation and antimicrobial assays against *Salmonella* and *Campylobacter* were performed to select the most effective anti-pathogenic and synergistic combination for *in ovo* validation. *L. mesenteroides* (LM) with garlic (G) presented as the most promising prophybiotic *in vitro*. Therefore, this prophybiotic and its probiotic component alone were injected to ROSS308 broiler hatching eggs on day 12 of incubation to validate the beneficial effects on the gut microbiome *in vivo*. The hatchability of LM+G and LM alone treated groups were higher and lower, respectively than that of positive control (injected with physiological saline). The weight of chicks on day one was highest in LM group followed by LM+G group ($P<0.05$). The chick length and Pasgar score were not significantly different between the groups although LM+G treatment resulted in the highest chick quality (Pasgar score 9). Moreover, the impact of LM+G and LM on the gut microbiome and physiology of the chickens will be presented. This project is funded by European Union's Horizon 2020 research and innovation program (Grant agreement N° 955374).

Antibacterial plant blends modulate gut microbiota in organic piglets challenged with *E. coli* F18

K. Jerez-Bogota[1,2], M. Jensen[2], O. Højberg[1] and N. Canibe[1]
[1]Aarhus University, Department of Animal and Veterinary Sciences, Tjele, 8830, Denmark, [2]Aarhus University, Department of Food Science, Aarhus N, 8200, Denmark; jerezbogota@food.au.dk

The study investigated the use of combinations of garlic and apple pomace, or blackcurrant as potential in-feed alternatives to antibiotics and zinc oxide in combating postweaning diarrhoea caused by enterotoxigenic *E. coli* (ETEC) in organically raised piglets. These blends had previously demonstrated *in vitro* synergistic antibacterial activity against ETEC. Here we present the effects on ETEC shedding, diarrhoea incidence, gut microbiota composition (16S rRNA), and oxidative stress markers. For 21 days, 32 piglets (7-weeks old) were randomly assigned to one of four groups: non-challenge (NC); ETEC-challenged (PC); ETEC-challenged receiving garlic and apple pomace (3%+3%; GA); ETEC-challenged receiving garlic and blackcurrant (3%+3%; GB). A strain of ETEC F18 was administered (8 ml; 10^9 cfu/ml) on days 1 and 2 after weaning. Faecal samples were collected daily the first week, and every other day thereafter for diarrhoea assessment; microbiota composition was analysed on days 1, 3, 7, 14 and 21. At the end of the experiment, digesta and mucosa samples from the gastrointestinal tract (GIT) were collected. NC pigs had no ETEC F18 shedding nor diarrheal symptoms. PC group had diarrhoea and ETEC shedding, which were reduced in the GA and GB groups. On day 7, the GA, GB, and NC pigs had a greater (P<0.05) Shannon and I-Simpson α-diversity index than the PC pigs. The PC group had a greater (P<0.05) microbiota volatility (compositional change) than all other groups on days 7 and 14. The *Escherichia*, *Campylobacter*, and *Erysipelothrix* genera were less abundant in the NC, GA, and GB than in the PC group (log2FC>2; P<0.05), whereas *Catenibacterium*, *Dialister*, and *Mitsoukella* were more abundant (log2FC>2; P<0.05). The GB group had the highest abundance of *Prevotella* and *Lactobacillus* (log2FC>2, P<0.05). Overall, adding GA or GB to the weaning feed of organic piglets maintained a healthier gut microbiota profile and overall gut health, that reduced the incidence of diarrhoea. Results on microbiota along GIT and gut mucosa oxidative stress markers will be further presented. The study received funding from the EU H-2020 program (grant # 955374).

Role of caecal microbiota in flock weight heterogeneity

M.Z. Akram, E.A. Sureda, L. Comer and N. Everaert
KU Leuven, Department of Biosystems, Kaastelpark Arenberg 20, 3001, Leuven, Belgium; muhammad.akram@kuleuven.be

Despite improved genetics and farm practices, body weight (BW) heterogeneity yet exists within broiler flocks, causing operational inefficiencies and challenges in processing and marketing. Chicken performance is also linked to the gut microbiota, however specific microbiota affecting BW within a flock remains elusive. This study examined caecal microbiota's role in BW differences among birds from two hatching systems (HS: HH, hatch in hatchery and HOF, hatch on-farm). The 454 male broiler chicks for each HS were raised together until day (d) 7 before being split into low (L, n=147) and high (H, n=140) BW groups, made two factorial design (HS × BW) and four groups (HH-L, HH-H, HOF-L, & HOF-H). On d 7, 14 and 38, ten birds from each group were sampled for 16s rRNA gene sequencing of the caecal microbiota. α-diversity was analysed by 2-way ANOVA and β-diversity with 2-way PERMANOVA. The differential abundance of taxa was analysed using LEfSe. α-diversity was largely influenced by BW with low BW chickens having higher α-diversity on day 38. β-diversity was solely influenced by BW, not HS, and the PCoA plot had distinct BW group clusters on day 7. The interaction of HS × BW had no significant effect on α- and β-diversities. LEfSe analysis revealed that high BW chickens had caecal microbiota enriched with SCFA-producing and health-promoting genera e.g. *Unclassified Lachnospiraceae*, *Faecalibacterium* and *Clostridia family*. Interestingly, *Ruminococcaceae* members and *Lactobacillus* were abundant in low BW chickens along with *Escherichia-Shigella*, *Enterococcus*, *Streptococcus* and *Akkermansia*. Little HS-related differences in caecal microbial composition were observed, including an increased level of *Escherichia-Shigella* in HH system chickens. The study revealed that caecal microbiota diversity and composition were less influenced by HS and differences in taxa abundances were more associated with chicken BW differences. This project has received funding from the European Union's Horizon 2020 research and innovation programme under grant agreement N° 955374.

Session 49 Theatre 9

1H-NMR metabolomics reveals alterations in the metabolism of ascarid-infected laying hens

O.J. Oladosu[1], B.S.B. Correia[2], B. Grafl[3], D. Liebhart[3], C.C. Metges[1], H.C. Bertram[2] and G. Daş[1]

[1]Research Institute for Farm Animal Biology (FBN), Institute of Nutritional Physiology, Wilhelm-Stahl-Allee 2, 18196 Dummerstorf, Germany, [2]Aarhus University, Department of Food Science, Agro Food Park 48, 8200, Aarhus, Denmark, [3]University of Veterinary Medicine Vienna, Clinic for Poultry and Fish Medicine, Veterinärplatz 1, 1210, Vienna, Austria; oladosu@fbn-dummerstorf.de

Nematode infections with *Ascaridia galli* and *Heterakis gallinarum* are associated with impaired performance in laying hens. Furthermore, *H. gallinarum* is a vector of *Histomonas meleagridis*, and is often co-involved in the infections. Here, we provide the first insight into alterations in plasma and liver metabolome of laying hens due to ascarid infections with concurrent histomonosis. ^{1}H nuclear magnetic resonance (^{1}H-NMR) metabolomics was applied to explore the variation in the metabolite profiles of the liver (n=105) and plasma (n=108) from laying hens experimentally infected with *A. galli* and *H. gallinarum* (+ *H. meleagridis*), and compared with uninfected hens at 2, 4, 6, 10, 14 and 18 weeks post infection (wpi). The hens were killed after a 3-hour feed withdrawal for sample collection. In total 31 and 54 metabolites were quantified in plasma and liver extracts, respectively. Data analysis showed no significant difference ($P>0.05$) in any of the 54 liver metabolites between the two groups. In contrast, 20 plasma metabolites showed significantly elevated concentrations in the infected hens ($P<0.05$). Alterations of plasma metabolites occurred in wpi 2, 6 and 10, covering the pre-patent period of worm infections. Plasma metabolites with the highest variation at these time points included glutamate, succinate, trimethylamine-N-oxide, myo-inositol, acetate, and glucose. Pathway analysis suggested that infection induced changes in: (1) phenylalanine, tyrosine, and tryptophan metabolism biosynthesis; (2) alanine, aspartate and glutamate metabolism; and (3) arginine and proline metabolism. In conclusion, ^{1}H-NMR metabolomics revealed significant alterations in the plasma metabolome of infected hens. The alterations suggested upregulation of key metabolic pathways mainly during the patency of infections. This project has received funding from EU-Horizon 2020 under the Marie Sklodowska-Curie grant agreement No 955374.

Session 49 Theatre 10

Validated machine-learning model to detect IUGR piglets

R. Ruggeri[1,2], G. Bee[1], P. Trevisi[2] and C. Ollagnier[1]

[1]Agroscope, Pig Research Unit, Animal Production Systems and Animal Health, Tioleyre 4, 1725 Posieux, Switzerland, [2]University of Bologna, Agricultural and Food Sciences, G Fanin 44, 40127 Bologna, Italy; roberta.ruggeri@agroscope.admin.ch

Intrauterine growth restriction (IUGR) is defined as the impaired development of the foetus during gestation. Piglets affected by IUGR have prioritized brain development as part of an adaptive reaction to placental insufficiency. This mechanism results in a higher brain-to-liver weight ratio (BrW/LW). The aim of this study was to develop a machine-learning model to predict the BrW/LW from a piglet's image and accurately diagnose IUGR. Two days (±1) after birth, brain and liver weight of each piglet were assessed with computed tomography scan (n=299) or by weighting the organs after euthanasia (n=65). A threshold value of 0.94±1 (mean + SD) was chosen to divide the piglets into NORM (BrW/LW<0.94) and IUGR (BrW/LW≥0.94). Videos of the piglets were taken using a RealSense camera. Selected frames of piglets were then used to predict the IUGR status through a convolutional neural network (CNN) developed in Python. The available data was split in two datasets. One dataset was used for training (80% of the data) and the other to validate the model and assess its performance (remaining 20% of the data). The CNN was trained five times and the results were expressed as average recall, precision and F1 score. Recall represents the percentage of IUGR piglets the CNN correctly predicted, over all the IUGR cases. Precision is a measure of how many of the IUGR predictions made were correct. F1 score is the harmonic mean of precision and recall. The CNN performed in the training dataset with a recall, precision and F1 score equal to 97, 53 and 65%, respectively. In the validation phase, recall, precision and F1 score were reduced to 88, 50 and 64%, respectively. The present results showed that the CNN was able to identify most of the IUGR piglets in both the training and the validation phase. However, 32 and 27% of the NORM piglets were classified as IUGR in the training and in the validation dataset, respectively. In conclusion, the model is highly sensitive in detecting the IUGR cases but precision could still be improved. This project has received funding from the European Union's Horizon 2020 research and innovation program under grand agreement N°955374.

In vivo validation of a non-invasive tool to collect intestinal content in pigs – CapSa-

I. García Viñado[1,2], F. Correa[1], P. Trevisi[1], G. Bee[2] and C. Ollagnier[2]
[1]University of Bologna, Department of Agricultural and Food Sciences (DISTAL), Viale Giuseppe Fanin 40-50, 40127 Bologna, Italy, [2]Agroscope, Pig Research Unit, Rue de Tioleyre 4, 1725 Posieux, Switzerland; ines.garciavinado@agroscope.admin.ch

Pig microbiome has become a focus of intense research in recent years, because its modulation can greatly impact pigs' health. So far, stool samples have been the only non-invasive mean to study the gut microbiome. However, microbiome in the gut differs from the one in the stool. Capsule for Sampling (CapSa) is a swallowable device that transits through the natural digestive pathway, collects small intestine content, and is excreted in faeces. In this validation study, 93 pigs (ranging from 6 to 67 kg bodyweight (BW)) were administrated 2 CapSas each and monitored for 3 days until recovery of the CapSas in the faeces. Upon retrieval from faeces, CapSa contents were extracted, their pH were measured, and they were stored at -80 °C until microbiota analysis. Three days after administration, pigs were euthanized and content from small and large intestine, and faeces were collected for microbiome analysis. In total, 38 CapSas were retrieved from 28 pigs (n=9; 12±3.4 kg BW fed a starter diet; n=19; 50.5±12.3 kg BW fed a standard grower diet). Bacterial composition of CapSas were compared with the microbiota extracted from 3 segments of the small and large intestine, and the faeces of the same pig. Pairwise Adonis contrast demonstrated that CapSa sample composition was different ($P<0.001$) from the large intestine and faeces, while it was more similar to the composition of the segment 1 and 2 of the small intestine for lighter pigs fed with a starter diet (P adj=0.32 and $R^2=0.10$; P adj=0.06 and $R^2=0.12$) but not (P adj=0.001 and $R^2=0.12$; P adj=0.001 and $R^2=0.14$) for heavier pigs, fed with a grower diet. The CapSa samples were consistently characterized by the abundance of certain bacteria like *Terrisporobacter* (LDA=4.42, $P=0.002$ and LDA=4.61 and $P<0.01$; in lighter and heavier pigs respectively). CapSa has been validated for repeated collections of the gut microbiota in postweaning pigs, but needs to be confirmed in grower pigs with more precise *post-mortem* samplings. This project has received funding from the European Union's Horizon 2020 research and innovation program under grand agreement N° 955374.

Effect of creep feeding (liquid milk, dry and liquid diet) on pig growth and intestinal structure

E.A. Arnaud[1,2], G.E. Gardiner[1], M. Chombart[2], J.V. O'Doherty[3], T. Sweeney[3] and P.G. Lawlor[2]
[1]South East Technological University, Dept. of Science, X91Y074, Waterford, Ireland, [2]Teagasc, Pig Development Dept., Animal & Grassland Research & Innovation Centre, Moorepark, P61R966, Fermoy, Co. Cork, Ireland, [3]University College Dublin, Belfield, D04C7X2, Dublin, Ireland; elisa.arnaud@teagasc.ie

Increasing early post-weaning (PW) feed intake in pigs is necessary to improve intestinal structure and function and maximise growth. This study aimed to determine the effect of creep feeding (dry pelleted starter diet, liquid milk replacer, liquid starter diet) suckling pigs on pre- and post-weaning growth and PW intestinal structure. After farrowing, 104 sows were blocked on parity, number of pigs weaned in previous parity and body weight (BW) and randomly assigned to one of 4 treatments: (1) CTL; no creep; (2) DRY; pelleted starter diet provided to piglets from day (d)10-28; (3) MILK; liquid milk [150 g of Swinco Opticare milk powder/l of water] provided from d3-28; and (4) MIX; mixture of liquid milk and starter diet which increased as a proportion of the mixture as lactation progressed (152 g DM/l of water) from d3-28. At weaning, 566 pigs were grouped by sow treatment and formed into single sex groups of 10-12 pigs of even weight (12 pens/treatment). Group weights and feed disappearance were recorded at intervals up to slaughter. At d7 PW, 40 pigs were euthanized and intestinal tissue sampled for histological analysis. Data were analysed using the mixed models procedure in SAS (Version 9.4; SAS Institute Inc). The litter was the experimental unit prior to weaning and the pen group PW. Weaning weight was 7.87^c, 8.10^b, 8.27^a and $7.99^{b,c}$ (SEM 0.064 kg; $P<0.05$) for CTL, DRY, MILK and MIX, respectively. From d0-6 PW, average daily gain was 208^a, 198^a, 159^b and 191^a (SEM 12.2 g/pig/day; $P=0.03$), gain to feed was 1.00^a, 0.94^a, 0.83^b and 0.90^a (SEM 0.035 g/g; $P<0.001$) and slaughter weight was 120.8^b, 120.1^b, 119.5^b and 122.7^a (SEM 0.70 kg; $P<0.01$) for CTL, DRY, MILK and MIX, respectively. Villous height in the ileum was 242^B, 284^A, 248^B and 283^A (SEM 14.0 µm; $P=0.07$) for CTL, DRY, MILK and MIX, respectively. In conclusion, providing a liquid mixture of milk and starter diet to suckling pigs increased PW villous height, likely explaining the heavier BW found at slaughter.

Session 49 Theatre 13

In vitro anthelmintic evaluation of Greek oregano against _Ascaridia galli_

I. Poulopoulou[1], E. Sarrou[2], E. Martinidou[3], L. Palmieri[3], D. Masuero[3], S. Martens[3] and M. Gauly[1]
[1]Free University of Bolzano, Faculty of Agricultural, Environmental and Food Sciences, Piazza Università 5, 39100 Bolzano, Italy, [2]Institute of Plant Breeding and Genetic Resources, Hellenic Agricultural Organization DEMETER, Thermi, 57001 Thessaloniki, Greece, [3]Edmund Mach Foundation, Food Quality and Nutrition Department, Via E. Mach 1, 38010 San Michele all'Adige Trento, Italy; ioanna.poulopoulou@unibz.it

This study aims to assess the *in vitro* anthelmintic activity of five different *Origanum vulgare* (OV) sbsp. hirtum accessions with different composition of carvacrol, and thymol from two different years (2019 and 2020). *Ascaridia galli* eggs isolated from the worm uterus were exposed *in vitro* to OV essential oils (EO), in dimethyl sulfoxide (1%) from the accessions originated from Evia 1 (OvH4) Taygetos-Peloponnese (OvH14), Gokceada (Turkey) (OvH15), Evia 2 (OvH17), Thessaloniki (OvH19). Untreated control, negative control petri dishes with 0.5% formalin and positive controls, flubendazole, thymol and carvacrol were also tested at 0.500 mg/ml in duplicate. Eggs' embryonic development (ED) was evaluated every third day (160 eggs/replicate) from the day of egg isolation until day 28 resulting in the examination of 12,160 eggs. The examined eggs were classified as either undeveloped or developed and the percentage of the eggs corresponding to each development class was calculated. Analysis performed using generalized linear mixed model, stating a negative binomial distribution, having plant accession, and year as fixed effect. OvH14 showed significant lower ($P<0.05$) ED compared to all other samples estimated at approximately 38%. The samples OvH14, OvH15 and OvH19 from the year 2020 had lower ($P<0.05$) ED, compared to the year 2019. The phytochemical analysis showed that the sample OvH14 had a high concentration of thymol for both years that estimated approximately at 790 mg/ml EO while carvacrol concentration was low (64 mg/ml EO) while taxifolin a flavonoid with antiparasitic activity was also determined. EO extracts obtained from the reported OV species have promising results in inhibiting ED, contributing to the identification of alternative anthelmintic treatments against *A. galli*.

Session 50 Theatre 1

Genetic diversity and improvement of the black soldier fly

C.D. Jiggins and T. Generalovic
University of Cambridge, Downing Street, CB2 3EJ, United Kingdom; c.jiggins@zoo.cam.ac.uk

To improve food security, sustainable agricultural practices are needed due to threats posed by climate change and population growth. Insect livestock, specifically the black soldier fly, is a promising organism for circular food production by feeding on organic waste. However, knowledge of the fly's evolution, genetics, and potential for genetic improvement is lacking. We have been studying the domestication and genomics of *H. illucens*, revealing evidence of considerable genetic diversity around the globe, and for selective sweeps in the genome associated with domestication. Domestication has likely occurred independently multiple times but is associated withe genetic changes in the same regions of the genome. We have used experimental evolution to select for increased pupal size, with correlated improvements in many industrially important traits. We have also developed CRISPR/Cas9 approaches for genetic modification that hold considerable promise for the future. This work lays the foundation for *H. illucens* as an important global agricultural organism.

Session 50 Theatre 2

Selection for larval weight in the black soldier fly – empirical evidence

K. Shrestha[1], E. Facchini[2], E. Van Den Boer[1], P. Junes[1], G. Sader[1], K. Peeters[2] and E. Schmitt[1]
[1]Protix. B.V., Research and Development_Genetics, Industriestraat 3, 5107 NC, Dongen, the Netherlands, [2]Hendrix Genetics Research, Technology & Services B.V, Spoorstraat 69, 5831 CK, Boxmeer, the Netherlands; kriti.shrestha@protix.eu

Black soldier fly (BSF; *Hermetia illucens*) industrial farming has increased in the last ten years. Interest in the species arises from its ability to convert low-grade organic materials into high-value ingredients. In addition to its use in aquaculture and pet food diets, the European Commission has recently authorized the use of insects derived processed animal proteins in monogastric livestock diets. The demand for insect protein is expected to grow 50 times more by 2030 globally. Therefore, there is a need to improve the efficiency of production further. One means of improvement is selective breeding, which has been widely applied in plants and other animal species. Results on other instances reflecting a focused, multi-year selective breeding effort in BSF is yet to be seen. In 2019, a genetic improvement program for increased larval weight of BSF larvae was started. We present the outcomes of this breeding program after 16 generations of selection. The performance of the selected body weight line was compared to the base population line over six experimental rounds under different environmental conditions. Under automated production settings, an average improvement of +39% in larval weight, +34% in wet crate yield, +26% in dry matter crate yield, +32% in crude protein per crate and +21% crude fat per crate was achieved in the selected line compared to the base population line. The selection in the context of the genetic improvement program translated into improvements in industrial production settings. This research demonstrates the potential contribution of selective breeding to improve production when farming BSF.

Session 50 Theatre 3

Full-sib group records as a practical alternative to individual records in insect breeding

L.S. Hansen[1], A.C. Bouwman[2], H.M. Nielsen[1], G. Sahana[1] and E.D. Ellen[2]
[1]Aarhus University, Center for Quantitative Genetics and Genomics, C.F. Møllers Allé 3, 8000 Aarhus, Denmark, [2]Wageningen University & Research, Animal Breeding and Genomics, P.O. Box 338, 6700 AH Wageningen, the Netherlands; lsh@qgg.au.dk

Selective breeding in insects has predominantly been utilising information on phenotypic variation which does not require records on relatedness. Selection based on estimated breeding values would require: (1) tracking a pedigree over multiple generations; and (2) tracking individuals over their development to link phenotypic records for multiple traits. Most farmed insect species experience several life stages of occasionally distinct morphology, and the entire exoskeleton including any exterior identifier is shed regularly. These aspects of the insect metabolism make tracking of individuals essentially impossible unless they are reared individually. However, rearing insect species such as black soldier fly or housefly in isolation is impractical for large-scale production, and suboptimal for larval development. When tracking and isolation are impossible, we cannot correlate individual phenotypic records collected throughout development, or relate early life phenotypic records to mature selection candidates. To overcome these limitations, an approach would be to use group records. For flies, the genetic relationship of individuals grouped together cannot be tracked unless they are all full-siblings. This has consequences for the accuracy of breeding values estimated using family-level information, since phenotypic covariance will be caused by both genetic similarity and common environmental effects, and this environment could vary dramatically between families. In this study, we use stochastic simulation to investigate the potential for genetic improvement in insects using family records for breeding value estimation in a two-trait breeding scheme. We simulate scenarios with varying ratio of genetic to common environmental variance, and consider the effect of using full-sib subgroups to estimate and correct for the non-genetic sources of variance. The group-based breeding scheme is not limited to applications in multi-trait systems, but also applicable when breeding for traits that cannot be measured at the individual level.

Simulating breeding programs based on mass selection in black soldier fly (*Hermetia illucens*)

M. Slagboom[1], H.M. Nielsen[1], M. Kargo[1], M. Henryon[2] and L.S. Hansen[1]
[1]Aarhus University, Centre for Quantitative Genetics and Genomics, C. F. Møllers Allé 3, 8000 Aarhus, Denmark, [2]Danish Agriculture and Food Council, Axeltorv 3, 1609 København, Denmark; margotslagboom@qgg.au.dk

Interest in black soldier fly production for feed has increased greatly in recent years. Selective breeding can improve chosen traits in the desired direction. Mass selection is the selection on phenotypic performance, i.e. something that is measurable on the individual itself. In this study, we simulated mass selection breeding programs aimed at improving larval body weight (LBW). We tested if more complicated breeding program designs that increase manual labour increase genetic gain while maintaining inbreeding. Specifically, we looked at housing families separately after a female lays eggs, housing larvae individually after weighing, and controlling mating, which resulted in 8 different breeding programs. Each time step, 35,000 larvae were produced, of which 3,000 were phenotyped for LBW. These larvae were either randomly selected from the whole population, or a fixed number per family was selected when families were housed separately. The 400 heaviest larvae were kept alive to emerge as adult flies (when they could be sexed), and they were either individually or group housed. When group housed after phenotyping, it was impossible to connect the phenotype for LBW to individual flies. Therefore, the selection of 200 females and 100 males for breeding was random in this case (selected from the 400 heaviest larvae that were kept alive). Alternatively, larvae were housed individually until development into flies to connect the phenotype for LBW to individual flies. In this case, breeding animals were selected based on LBW. Mating was either controlled in a vial with 2 males and 1 female, or in a group with all breeding animals. Preliminary results show that housing families separately after egg-laying did not have a large effect on either inbreeding or genetic gain, housing larvae individually after weighing to connect the phenotype for LBW to individual flies increased genetic gain by up to 8%, and controlled mating decreased inbreeding at the same level of genetic gain.

Effects of artificial selection in the black soldier fly – a Pool-seq approach

E. Facchini[1], A. Vereijken[1], D. Bickhart[1], A. Michenet[1], K. Shrestha[2] and K. Peeters[1]
[1]Hendrix Genetics Research, Technology & Services B.V., Villa 'de Körver', Spoorstraat 69, 5831 CK Boxmeer, the Netherlands, [2]Protix Biosystems B.V., Industriestraat 3, 5107 NC Dongen, the Netherlands; elena.facchini@hendrix-genetics.com

Black soldier fly (*Hermetia illucens*) is used to convert low-grade organic waste materials into high-quality products. The European Commission has recently authorized the use of insects derived processed animal proteins in monogastric livestock diets, expanding the destination of use that already included aquaculture and pet food diets. In 2019, we started a genetic improvement program to increase larval body weight. Results show how genetic improvement can help the sector improve production. Under commercial conditions, an average gain of +39% in larval weight was achieved after two years of artificial selection. This study aimed to evaluate the effects of artificial selection on genetic variability. The flies enrolled for this study were sampled at the beginning of the program and after eight generations to form three study populations. Population A represents the starting base population at generation zero, population B represents the selection line for body weight at generation eight, and population C is the control line at generation eight. The control line (population C) was derived from the same starting population and was used to benchmark the genetic progress in the selection line. DNA was individually extracted from 30 random flies and aliquoted in equal amounts to form one pooled DNA sample per population. The three DNA pools were sequenced using paired-end next-generation sequencing with a target coverage of 30 Gbp (i.e. 30× coverage). Allele frequencies at single nucleotide polymorphisms and Wright's Fixation index were estimated on a genome-wide scale. Based on differences in allele frequencies, the impact of the breeding program on genetic variability was evaluated. Differences in allele frequencies were detected between the three studied populations, and fixation indexes showed moderately low genetic differentiation. Pool sequencing is an affordable and viable approach for investigating allele frequency changes due to natural or artificial processes.

Session 50 Theatre 6

Adaptive responses of black soldier fly to simple low-quality diets

A. Gligorescu and J.G. Sørensen
Aarhus University, Department of Biology, Section for Genetics, Ecology and Evolution, Ny Munkegade 116, Building 1540, 8000, Aarhus, Denmark; angl@bio.au.dk

The production of black soldier fly larvae (BSFL) is rapidly expanding because of the ability of BSFL to bioconvert cheap and low-quality substrates into high quality insect biomass. Despite this, the consequences of growing BSFL on low-quality substrates during multiple generations are largely unknown. We used three lines of BSF kept on a low-quality diet (wheat bran, WB) or on a high-quality diet (chicken feed, CF), respectively, for more than 30 generations. BSF lines were maintained on either the original or the interchanged diets for three generations, during which we investigated the response in a series of performance parameters (i.e. larval weight at harvest, larval growth rate, survival rate, pupa weight, sex ratio and egg production). As expected, the BSF was found to perform best when kept on high quality diets, irrespective of original maintenance diet. The BSF lines originally maintained on a low-quality diet had a higher growth rate than BSF lines reared on high quality, but shifted to a low quality diet during the first generation. Subsequently, this difference disappeared during the following two generations. Our study has implications for understanding the evolutionary and phenotypic consequences of maintaining BSF lines on single sourced, low-quality substrates. Differences between original diets seems to rely mainly on phenotypic plasticity rather than evolutionary adaptation, and low-quality diets have limited implications for BSF performances.

Session 50 Theatre 7

Molecular sexing of black soldier flies

R.S.C. Rikkers[1], E. Van Der Valk[1], A.A.C. De Wit[1], L. Kruijt[1], E.D. Ellen[1], J. Van Den Heuvel[2] and B.A. Pannebakker[2]
[1]Wageningen Livestock Research, Droevendaalsesteeg 1, 6700 AH Wageningen, the Netherlands, [2]Laboratory of Genetics, Droevendaalsesteeg 1, 6700 AH Wageningen, the Netherlands; elianne.vandervalk@wur.nl

The use of black soldier flies (BSF; *Hermetia illucens*) has exponentially increased the past decade because of their ability to efficiently convert waste products into high value proteins for humans as well as feed for livestock. There is a need to further improve the efficiency of BSF production to meet the rising demands for high-value proteins in human food and livestock feed industries. To set up well functioning breeding programs it is crucial to sex the BSF at an early stage, however in BSF it is only possible to determine the sex phenotypically once they reached the adult phase. Additionally, female larvae are heavier than male larvae, thus selecting larvae on body weight would result in a biased sex ratio. Therefore, the objective of this study was to develop a tool to sex BSF, using molecular sexing. BSF have six autosomes, and additionally females have two sex chromosomes and males have one sex chromosome. DNA from six male and six female adult BSF was isolated and with the use of quantitative Polymerase Chain Reaction (qPCR) we tested nine sex genes and three autosomal genes (reference genes). For males, the qPCR product of the sex genes should be half of the amount relative to the females. Furthermore, the qPCR product of the reference genes should be equal for both sexes. Raw qPCR results were corrected for DNA quantity differences with the Pfaffl method. Results showed that (within this small sample size) we could identify the male and female BSF with a 100% accuracy, based on the differences in qPCR product between the sex genes. Of the nine sex genes, five genes resulted in 100% accurate identifications. Three genes showed higher variation, but also resulted in a 100% accuracy, whereas one gene gave inconclusive identifications. To conclude, we identified genes and developed a tool that could be used to molecular sex BSF which could be used for future applications on e.g. exuviae of larvae. However, validation in a higher sample size and on exuviae is necessary.

Session 50 Theatre 8

Biomarker discovery for the black soldier fly (*Hermetia illucens*)

E.M. Espinoza
EnviroFlight, Genetics R&D, 2100 Production Dr, 27539, USA; eespinoza@enviroflight.net

One of the overarching goals at EnviroFlight Genetics R&D is to create elite black soldier fly (BSF) (*Hermetia illucens*) populations that meet target values for the commercial production of our insect ingredients. Identifying desirable traits throughout the life history of the BSF will help us track phenotypes and aid in the selection of individuals for breeding. We strive to identify potential biomarkers for studying various biological processes important in the farming of BSF. Defining these measurable indicators, we hope to better understand breadth of limiting factors critical to the improved production of elite BSF populations.

Session 50 Theatre 9

Improving black soldier fly genetics by CRISPR\Cas9 gene editing

I.N.Y. Nevo Yassaf, A.G. Goren, R.A. Adler and I.A. Alyagor
FreezeM, Genetics, Nachshonim 1, Nachshonim, 7319000, Israel; inbar@freeze-em.com

The shortage of protein resources is becoming more and more severe as the global population is increasing. Finding new environment-friendly and sustainable protein sources is a necessity to replace the traditional ones. The black soldier fly (BSF), *Hermetia illucens* (Diptera: Stratiomyidae), is one of the most efficient insects in converting organic waste into biomass. To generate improved genetic strains with beneficial features, we utilized the CRISPRIL platform to develop an efficient BSF genome editing platform. Applying a unique gRNA design system developed during the CRISPRIL consortium, we generated 400 genome editing events with an average efficacy of 74%. Additionally, we generated 10 deletion events by applying CRISPR on two adjacent editing sites simultaneously. Later, this technique was applied to generate modified BSF strains with enhanced industrial performance. One of these strains resulted in significantly larger larvae than the wild-type origin. This strain has a better feed conversion ratio (FCR) while maintaining its nutritional values. Our study provides valuable technical and genomic resources for improving BSF lines for industrialization.

Session 50 Theatre 10

Expected response to selection on larval size and development time in the housefly (*Musca domestica*)

H.M. Nielsen[1], T.N. Kristensen[2], G. Sahana[1], S.F. Laursen[2], S. Bahrndorff[2], J.G. Sørensen[3] and L.S. Hansen[1]
[1]Aarhus University, Center for Quantitative Genetics and Genomics, C.F. Møllers Allé 3, 8000 Aarhus, Denmark, [2]Aalborg University, Section for Bioscience and Engineering, Department of Chemistry and Bioscience, Fredrik Bajers Vej 7H, 9220 Aalborg, Denmark, [3]Aarhus University, Department of Bioscience, Ny Munkegade 116, 8000 Aarhus C, Denmark; hannem.nielsen@qgg.au.dk

The demand for protein will increase dramatically in the future. At the same time, we face major challenges related to e.g. greenhouse gas emissions causing climate changes, where the agricultural sector is a large contributor. Proteins from insects can be produced more sustainable compared to other animal protein sources because they generally have higher nutritional value, higher feed conversion efficiency, and produce less ammonia and greenhouse gasses than conventional livestock. However, commercial production of insects for food and feed is still in its early stage, and little has been done to improve quantity and quality of products from insects through selective breeding. The expectation is that implementation of selective breeding programs for traits of economic importance in insect production can lead to tremendous genetic progress and further optimize the production. In this study, we present expected selection response in the houseflies Musca domestica. The traits were; larval size and development time for which we have estimated heritabilities of 0.1 and 0.3, respectively. In our simulations we select 80 dams and 40 males and each dam produced 10 offspring. Candidates were evaluated purely on sib information. Because the genetic correlation between larval size and development time is presently unknown but expected to be unfavourable, the correlation was varied from 0 to 0.6. With equal economic values on the two traits, it was possible to improve both traits simultaneously. With a correlation of e.g. 0.3, response in larval size was 0.72 mm^2, whereas response in development time was -10.4 hours. This study shows that there is a large potential for selecting for important production traits in insects such as the housefly. The knowhow from this study will help to design breeding plans for insect species used for food and feed production.

Session 50 Theatre 11

ŸnFABRE: design of reference populations for genomic selection in *Tenebrio molitor*

E. Sellem[1], A. Donkpegan[2], Q. Li[1], K. Paul[1], A. Masseron[1], F. Gagnepain-Germain[1], K. Labadie[3], B. Vacherie[3], P. Garrabos[4], M.A. Madoui[5] and T. Lefebvre[1]
[1]Ynsect, R&D Biotech, 1 rue Pierre Fontaine, 91000 Evry, France, [2]SYSAAF, UMR BOA, Centre INRAE Val de Loire, 37380 Nouzilly, France, [3]CEA, Genoscope, Institut de biologie François Jacob, Université Paris-Saclay, 91057 Evry, France, [4]Thermo Fisher Scientific, 16 Av. du Québec, 91941 Villebon-sur-Yvette, France, [5]CEA, Service d'Etude des Prions et des Infections Atypiques (SEPIA), Institut François Jacob, Fontenay-aux-Roses, France; qi.li@ynsect.com

Tenebrio molitor is a promising alternative to produce high-quality protein products with low environmental impact. Industrial scale production concentrates an important number of individuals, leading to consider population management to account for genetic diversity conservation and performance improvement. ŸnFABRE project was designed to address the required innovations in genetics/genomics and automated phenotyping for insect industries. This project aims to develop all the required tools to implement a genomic selection scheme adapted to industrial farming conditions. Four main traits will be selected: reproduction, growth, food efficiency, and disease resistance. A reference population of 4,000 phenotyped individuals was generated over 3 generations of unrelated single-pair matings (700 pairs were mated per generation). Several phenotypes have been assessed, a part of them characterized the couple (daily or total egg laying count, hatching rate, larval growth, or feed efficiency), and the other part described the individual (pupae-imago-adult weight, sex, or developmental time egg to pupae). A subset of 136 adults has been sequenced, identifying all the SNP candidates to produce the genotyping HD array (~700K), and then the rest of the adults will be genotyped. The genetic study quantified the part of the phenotype variability explained by the additive genetic effect. Heritabilities ranged between 0.21 to 0.45 for reproduction phenotypes, 0.12 to 0.57 for growth parameters and 0.20 to 0.46 for developmental time according to the parameter studied. These results show the potential for the selective breeding of these various traits, which will allow us to establish the first selected yellow mealworm lines of interest.

Session 50 Theatre 12

The potential of instrumental insemination for honeybee breeding

M. Du, R. Bernstein and A. Hoppe
Institute for Bee Research Hohen Neuendorf, Friedrich-Engels-Str. 32, 16540 Hohen Neuendorf, Germany; manuel.du@hu-berlin.de

Mating control constitutes a crucial part of honeybee breeding. In Germany, it is currently mostly ensured by the use of isolated mating stations, where geographic remoteness allows only drones from drone producing queens with a selected dam to participate in the mating process. However, for many burgeoning breeding programs in Europe, this method of mating control is not feasible, because suitable locations are scarce. Instrumental insemination appears as a viable alternative that is independent of geography and comes with several theoretical advantages. Due to increased control over the mating process, breeding values can be estimated more accurately. Furthermore, a greater variety of sires can be selected, thus enhancing the genetic variance, and, finally, by the direct selection of drone producing queens, the paternal generation interval between selected individuals can be reduced from five to four years. We used computer simulations to quantify the benefits of instrumental insemination for different population sizes and traits with different heritabilities. In instrumental inseminations, queens were paired with 12 drones that all came from the same drone producing queen. When we did not reduce the generation interval, we found that instrumental insemination could indeed generate higher genetic gain at lower inbreeding rates. Only five insemination stations, each with 8 drone producing queens, were needed to conserve as much genetic variance as could be maintained with 20 isolated mating stations of the same size. In contrast, when reducing the paternal generation interval, we observed substantially higher inbreeding rates. By selecting dams and sires from the same generation, the probability of pairing closely related queens and drones was greatly increased. The ideal strategy thus appeared to allow sires to be selected from two consecutive birth cohorts, thereby reducing the paternal generation interval only for a part of the population. In conclusion, we think that instrumental insemination can serve as a great alternative and that its theoretical aspects should be further investigated.

Session 50 Theatre 13

Genetic analysis of production and behavioural traits of French honeybees

T. Kistler[1,2], C. Kouchner[1,3], C. Dumas[1,4], R. Dupain[1,4], A. Vignal[1,5], F. Mondet[1,4], P. Jourdan[1,3], B. Basso[1,4] and F. Phocas[1,2]
[1]UMT PrADE, Avignon, 84000, France, [2]Université Paris-Saclay, INRAE, AgroParisTech, GABI, Jouy-en-Josas, 78350, France, [3]ADAPI, Aix-en-Provence, 13100, France, [4]Abeilles & Environnement, INRAE, Avignon, 84000, France, [5]GenPhySe, INRAE, ENVT, Université de Toulouse, 31320, Castanet-Tolosan, France; tristan.kistler@inrae.fr

Genetic parameters were estimated for traits of interest in a honeybee population reared by professional beekeepers in south-eastern France. Around eight hundred open mated queens from 32 maternal families of instrumentally inseminated dams were tested on 14 apiaries. In 2021 and 2022, respectively 429 and 379 honey production records were collected. Hygienic behaviour was measured by the pin-killed brood method in spring (May). During this visit and a second one in early summer (June), the total sealed brood surface, gentleness, calmness, and the phoretic varroa load, were also measured. BLUP animal models were used considering the workers' genetic effects and the fixed effects of apiaries×year, the open mating area, and observation technicians×year to describe performances. The software BLUPf90 was used with a bee specific inverse relationship matrix. Heritabilities were high to very high for honey production (0.50±0.21), hygienic behaviour (0.68±0.40), phoretic varroa load measured in summer (0.66±0.24), and both sealed brood surface measures (spring: 0.74±1.06, summer: 0.53±0.24). Moderate heritabilities were obtained for gentleness, calmness and phoretic varroa load in spring (from 0.20±0.17 to 0.33±0.24), except for gentleness in summer with a null heritability. No significant genetic correlations could be estimated (due to a small dataset), but strong positive correlations were found between repeated measurements during the season. These results were obtained thanks to the Carnot France Futur Elevage project 'BeeMuSe' and promise that the ADAPI starting breeding plan can bring significant genetic progress in the future.

Session 50 Theatre 14

SIMplyBee: R package for simulating honeybee populations and breeding programs

J. Obšteter[1], L.K. Strachan[2], J. Bubnič[1], J. Prešern[1] and G. Gorjanc[2]
[1]Agricultural Institute Of Slovenia, Hacquetova Ulica 17, 1000, Slovenia, [2]The Roslin Institute and Royal (Dick) School of Veterinary Studies, University of Edinburgh, Easter Bush, Midlothian, EH25 9RG, Edinburgh, United Kingdom; jana.obsteter@kis.si

The Western honeybee is a globally important species, contributing to pollination and food production, but has been experiencing colony losses that lead to economical damage and decreased genetic variability. This situation has increased interest in honeybee breeding and conservation programs. Stochastic simulators are essential tools for rapid and low-cost testing of breeding programs and methods, yet no existing simulator allows for a detailed simulation of honeybee populations. To close this gap, we have developed a holistic simulator of honeybee populations and breeding programs, SIMplyBee. SIMplyBee is an R package, freely available for installation from CRAN at http://cran.r-project.org/package=SIMplyBee and SIMplyBee.info. SIMplyBee builds upon the stochastic simulator AlphaSimR that simulates individuals with their corresponding genomes and quantitative genetic values. To enable a honeybee-specific simulation, SIMplyBee extends AlphaSimR by developing: (1) a class for modelling honeybee colonies, Colony, that holds multiple castes of honeybee individuals with their individual- and colony-level quantitative genetic values; and (2) a class for modelling multiple colonies, MultiColony. SIMplyBee also includes functions to address major specificities of the honeybees: honeybee genome, haplo-diploid inheritance, social organisation, complementary sex determination, polyandry, colony events, and quantitative genetics of honeybees. We have additionally addressed simulating the spatially dependent mating of honeybees. SIMplyBee is a holistic simulator of honeybee populations and breeding programs, including individual honeybees with their genomes, colonies with colony events, and selection on individual- or colony-level quantitative values. SIMplyBee provides a research platform for testing breeding and conservation strategies and their effect on future genetic gain and variability.

Session 50 Theatre 15

Demonstrating the principles of genetic inheritance in honeybees using SIMplyBee

L. Strachan[1], J. Bubnič[2], G. Petersen[3], G. Gorjanc[1] and J. Obšteter[2]
[1]Edinburgh Univeristy, Roslin Institute, Edinburgh, United Kingdom, [2]Agricultural Institute of Slovenia, Hacquetova ulica 17, Ljubljana, Slovenia, [3]AbacusBio Ltd, Moray Place, Dunedin, New Zealand; s2122596@ed.ac.uk

Recently, there has been an increased interest in honeybee genetic variation and in establishing breeding programmes. A key metric in studying genetic variation is the level of genetic relationships between individuals of a population. Here we used SIMplyBee to demonstrate the principles of honeybee genetic inheritance and show the genetic relatedness amongst individual honeybees. SIMplyBee is a stochastic simulator dedicated for the design and optimisation of honeybee breeding programmes. It is available as an open-source R package from CRAN at http://cran.r-project.org/package=SIMplyBee and at http://SIMplyBee.info. A base population of 800 *Apis mellifera carnica* (Car) and 1,600 *Apis mellifera mellifera* (Mel) founder queens underwent a 10-year cycle, simulating a closed Car and a closed Mel population as well as a semi-open MelCross hybrid population by continuous mating of a proportion of Mel queens with drones from Car colonies. Various genetic relationships across the whole genome and at the complementary sex determiner (*csd*) locus were computed between individuals within a single colony, queens of the same population, and queens of different populations. We computed three types of relationships: expected identity-by-descent (IBD, pedigree) and realised IBD (genomic) relationships relative to the founder population, and identity-by-state (IBS) relationships. For the latter, we used different allele frequencies. Summarising the expected IBD relationships we recovered the well-known pedigree expectations, while realised IBD showed variation due to recombination and segregation. Summarising the IBS relationships showed even more variation due to older ancestries. IBS relationships varied substantially in values depending on which allele frequencies we used, suggesting that we must be careful when comparing genomic relatedness between studies. Our comparison of closed and hybridised populations can guide such comparisons. The csd locus showed smaller relationship values than the whole genome due to balancing selection. This work will help interpret the regional population genetic diversity of honeybees in many populations.

Session 50 — Poster 16

Exploring the potential for artificial selection in the black soldier fly, *Hermetia illucens*

T. Generalovic and C. Jiggins
University of Cambridge, Zoology, Downing Street, Cambridge, CB2 3EJ, United Kingdom; tng23@cam.ac.uk

Artificial selection has dramatically shaped the genomes of many species resulting in phenotypic diversification across the tree of life. In an agricultural context, the rising insect livestock industry shows great potential for artificial selection due to large population sizes and short generation times. The black soldier fly, *Hermetia illucens*, has been the focus of this industry over the past decade due to its efficient bioremediation and nutritional properties. Recent studies have identified significant genetic diversity in *H. illucens* populations around the globe. However, capturing and shaping this diversity for optimal industrial application requires further understanding on the phenotypic variation within strains. Here, we genome sequenced domesticated populations of *H. illucens* and explored phenotypic variation within and between strains across several diets. Our experimental assays of domesticated *H. illucens* revealed substantial variation between strains across several life-history traits, suggesting that there may be benefits to using specific strains for waste applications. Exploring genetic and phenotypic variances within strains and families enabled estimates of heritability suggesting several avenues for genetic improvement through artificial selection. We also performed experimental evolution for increased pupal size. Response to selection in *H. illucens* was strong and repeatable over time. Body size showed complex interactions with life-history traits vastly improving biomass yields and larval growth rates over nine months. However, increasing pupal body size had the unintended consequence of significantly reducing the development time of female pupae, revealing a previously undetected trade-off. This work highlights the importance for genetic and phenotypic characterisation of *H. illucens* strains to be incorporated into insect farming practices. We also demonstrate the potential impact of artificial selection in this growing sector. However, further assessment for trade-offs should be explored further prior to mass uptake of selection regimes in this novel industry.

Session 50 — Poster 17

Weight of *Hermetia illucens* eggs from breeding and wild individuals

J. Lisiecka[1], Z. Mikołajczak[1], M. Dudek[1], K. Dudek[1], B. Kierończyk[2] and D. Józefiak[1,2]
[1]HiProMine S.A., Poznańska, 12F, 62-023 Robakowo, Poland, [2]Poznań University of Life Sciences, Department of Animal Nutrition, Wołyńska 33, 60-637 Poznań, Poland; monika.dudek@hipromine.com

The aim of the research was to define the differences between *Hermetia illucens* eggs weight from breeding and wild individuals. Additionally, differences between the time of eggs harvest were compared. In current research, individuals from two lines were compared: breeding, which was HiproMine (HPM) standard genetic material, and wild, from Malaysia (M line). In the beginning of experiment, the generation of lines in production condition was: F26 for HPM line and F1 for M line. During the cycle, eggs were collected on different days. Eggs were placed on microscope slides tared on scale, and weighed (PR223/E, OHAUS Corporation, USA). After that, a microscope preparation using physiological saline was prepared. Prepared slides were observed using a microscope (Stemi 508, Carl Zeiss, Jena, Germany) and ZEISS ZEN Microscopy Software, and the pictures of eggs were taken using Axiocam 208 colour (Carl Zeiss, Jena, Germany). Eggs counting was performed using the ImageJ Software v.1.53. Egg weight (g) for HPM line was estimated to 0.00002528, while for M line to 0.00002598. Statistical analysis showed no significant differences between the groups. However, both lines showed a significant decrease in egg weight with harvesting in the later days of the reproductive cycle (HPM line P=0.007; M line P<0.001). This work was supported by an grant titled 'InnHatch: Innovative Technology for Industrial Insect Reproduction' no POIR.01.01.01-00-0898/19. The project is co-financed from European Funds by the National Center for Research and Development, under Measure 1.1 'R&D projects of enterprises', Sub-measure 1.1.1 'Industrial research and development works carried out by enterprises' of the Innovative Economy Operational Program.

Paternity assignment tool in honey bees (*Apis mellifera*)

S. Andonov[1], G. Aleksovski[2], B. Dahle[3], M. Kovačić[4], A. Marinič[5], A. Moškrič[5], J. Prešern[5], B. Pavlov[2], Z. Puškadija[4] and A. Uzunov[6]
[1]Swedish University of Agriculture, Dep. of Animal Breeding and Genetics, Ulls Vag 26, 750 07, Sweden, [2]CARPEA, 16-ta Makedonska brigada No 3, 1000, Skopje, Macedonia, [3]NBA, Norway, Dyrskuevegen 20, 2040,Kløfta, Norway, [4]Fakultet Agrobiotehničkih Znanosti, Vladimira Preloga 1, 31000 Osjek, Croatia, [5]Kmetijski Inštitut Slovenije, Hacquetova ulica 17, 1000, Ljubljana, Slovenia, [6]Faculty of Agricultural Sciences and Food, 16-ta Makedonska brigada No 3, 1000, Skopje, Macedonia; andonov.sreten@slu.se

Mating control, a key element in honey bee (*Apis mellifera*) breeding programs, is often not, or only partially implemented. Consequently, the genetic gain is not as desired, prompting beekeepers to consider purchasing queens from non-local populations or even commercial hybrids. BeeConSel project aims to establish tailor-made effective mating control in Croatia, Macedonia, and Slovenia. Several approaches to obtain controlled natural matings have been tested for suitability, e.g. geographical isolation, isolation by drone saturation (biological isolation), and temporal isolation. The analysis of patrilines and paternity assignment explores genetic tools in estimating patriline composition. In every approach, we have collected samples of drone-producing queens and their drone brood, and young mated queens, and their brood. In these samples, the DNA was isolated, and 5 micro-satellites (A0007, A0013, Ap043, Ap055, and B0124) were genotyped, resulting in a total of 81 alleles. The readouts were analysed with a custom-developed Linux script. The analysing process had three steps: data verification by confirming that all brood samples originated from the queen of the colony; reference genome composition by calculating all possible combinations of the five loci in combinations of the queen and present known drones at the spot; paternity assignment by matching brood genotypes with the combinations in the reference genotypes. It also includes logical controls and gives summary statistics. Later the script was upgraded to calculate different alleles frequencies for each subpopulation. The tool was tested on *A. m. carnica* and *A. m. macedonica* in mating seasons 2021 and 2022.

Session 50 Poster 19

Single-step genomic BLUP allows for the genetic evaluation of commercial honeybee queens

G.E.L. Petersen[1,2], F.S. Hely[2], M. Araujo[2], P.F. Fennessy[2] and P.K. Dearden[1,3]
[1]FutureBees NZ Ltd, 442 Moray Place, 9016 Dunedin, New Zealand, [2]AbacusBio Ltd, 442 Moray Place, 9016 Dunedin, New Zealand, [3]University of Otago, Biochemistry, 710 Cumberland Street, 9016 Dunedin, New Zealand; gertje_petersen@gmx.de

Although honeybees play a large role in modern food production systems through their role as an abundant and readily manageable generalised pollinator, beekeeping systems in most western countries have remained small scale, with a strong influence of traditional management techniques and low uptake of new technologies. This has led to a generalised absence of historically grown structures for structured genetic improvement (such as the breed societies often present in other livestock species), so the majority of honeybee queens is currently still selected based on ad hoc quality assessments. However, use of a centralised genetic evaluation for European honeybees (Beebreed) has shown that selection based on BLUP-based estimated can significantly increase genetic gain. Simultaneously, where large-scale beekeeping businesses are present (e.g. in North America and Oceania), their selection decisions impact on the population structure and genetic diversity of the honeybee population. These commercial beekeeping businesses are often operating under a cash- and time-poor business models and can only support the most basic of performance and pedigree recording systems. Through the use of genomics, commercial queen selection schemes can be developed that support large-scale beekeeping and contribute significantly to the sustainable improvement of honeybee populations. By blending a historic pedigree-based relationship matrix with a genomic relationship matrix based on non-invasive drone pool genotypes, we were able to calculate preliminary genomic estimated breeding values (gEBVs) for 1,200 honeybee queens in commercial beekeeping businesses in New Zealand. Although prediction accuracy for these is still quite low (as expected for such an early-stage genetic evaluation), these gEBVs will enable more informed selection decisions and empower beekeepers to improve their own populations without having to rely on queen suppliers who often operate in very different environments and under less demanding conditions than commercial honey producers.

Session 50 Poster 20

Deformed wing virus quantification: effect of selection and correlation with varroa related traits

M.G. De Iorio[1], S. Ottati[2], G. Molinatto[2], D. Bosco[2] and G. Minozzi[1,3]

[1]University of Milan, Via dell'Università 6, 26900, Italy, [2]Università degli Studi di Torino, Grugliasco, DISAFA, Largo Paolo Braccini, 2, 10095, Grugliasco, Italy, [3]CNR-IBBA, Via Alfonso Corti 12, 20133, Italy; giulietta.minozzi@unimi.it

Together with the parasitic mite *Varroa destructor*, viral infections are threatening the health of honeybee hives. One of the main viruses studied today is the deformed wing virus (DWV). DWV is an RNA virus that affects honeybees and other hymenopterans often in latent form. Disease outbreaks may occur when virus replicates at high levels and/ or in presence of other stressors, such as heavy varroa infestations. Furthermore, varroa mite is the main vector of DWV. Therefore, selecting for varroa resistance traits should indirectly reduce the DWV presence in hives. In this study we quantified the viral load by rtPCR in 69 hives belonging to a population of bees under selection with the aim to test correlation of DWV load with different varroa related traits and with the genetic lineage. The selected population belonged to a well melted stock including Ligustica, Carnica and Buckfast selected over several years by the same beekeeping farm located in North Italy. Selection was based on three traits: honey yield, gentleness and hygienic behaviour. Since 2015, phenotypic data from 90 hives have been annually registered. In 2021, after six years of selection, 69 out of 90 hives have been tested for DWV load. In the same year other varroa related traits were measured. DWV was present in all the hives showing a good variability in load, which ranged from 2.32E+11 to 5.44E+14 viral genome copies/bee. We found high correlations between DWV load and varroa growth rate (0.80) and between DWV load and the total quantity of phoretic mites (0.84). Furthermore, out of the 8 genetic lines used in the selected population, we identified a higher viral load in line 1 compared to line 8. Further research will be conducted in the next years on the genetics underlining DWV infection in this population.

Session 50 Poster 21

Monitoring the distribution of *Apis mellifera* genetic resources in Italy using mtDNA information

V. Taurisano, A. Ribani, K.E. Johnson, D. Sami, G. Schiavo, S. Bovo, V.J. Utzeri and L. Fontanesi

University of Bologna, Department of Agricultural and Food Sciences (DISTAL), Viale Fanin 48, 40127, Italy; luca.fontanesi@unibo.it

The genetic integrity of *Apis mellifera* subspecies and populations is a matter of concern all over Europe. Honey contains environmental DNA (eDNA) traces from all organisms that directly or indirectly were involved in its production, including the DNA of the honey bees that produced it. Specific mitochondrial DNA (mtDNA) lineages (i.e. mitotypes) characterize several *A. mellifera* subspecies. Different mitotypes can be detected using honey as source of honey bee DNA, providing approximate population genetic information useful to estimate the diffusion and frequency of honey bee mitotypes. In this study, we combined two approaches based on the analysis of: (1) honey eDNA; and (2) DNA extracted from honey bee workers of many different colonies to monitor the distribution of *A. mellifera* mtDNA lineages all over Italy. DNA was extracted from a total of about 2,300 honey samples produced over five years (2018-2022) in several regions of the Italian peninsula and Sicily. In addition, worker bees from 1,100 colonies were analysed to compare the distribution map of the mtDNA lineages in a region of the North of Italy (Emilia-Romagna). PCR products were analysed using a fragment size-based assay followed by Sanger sequencing. Results from the two approaches confirmed that the C lineage was the most frequent mitotype all over Italy except in Sicily where the A lineage was highly represented across all years. Other mtDNA haplotypes (A and M lineages) were present in almost all Italian regions. The obtained results are useful for designing conservation strategies for *A. mellifera* genetic resources in Italy and evaluating the efficacy of previously adopted measures. Moreover, we demonstrated that eDNA from honey can be exploited to design cost-effective non-invasive and simple methods to obtain information on the genetic distribution of honey bee mitotypes at large geographic scale.

How does the cow's microbiome respond to physiological challenges?

J. Seifert[1,2]
[1]*University of Hohenheim, HoLMiR – Hohenheim Center for Livestock Microbiome Research, Leonore-Blosser-Reisen Weg, 70593 Stuttgart, Germany,* [2]*University of Hohenheim, Functional Microbiology of Livestock, Emil-Wolff-Str. 10, 70593 Stuttgart, Germany; jseifert@uni-hohenheim.de*

Dairy cows face several outstanding challenges during their production life span. This includes the calving period with painful parturition and a significant impact on energy metabolism and nutrient requirements at the start of the lactation period. In addition, infections can induce inflammatory reactions such as fever, reduced rumination activity and feed intake. Although cows are of similar genetic background and receiving same feed at equal housing conditions, the response to the described challenges varies individually. The changes of the rumen and intestinal microbiome are studied in respect to feeding changes, etc. but can we link the microbiome composition and its functional response to the individual challenges described above? And does this correlate with individual animal performance? These questions were addressed in a previous study, where the individual variability and long-term changes of the microbiome and metabolome of dairy cows were studied. We found a significant impact in the rumen microbiome after parturition characterized by a bloom of *Bifidobacteria*. Fibrolytic bacteria seem to support the animal with energetic metabolites during the acute phase of an induced inflammatory event. Overall animal microbiomes were clustered into three microbial clusters (enterotypes) based on faecal microbiome sequence information. They are characterized by different diversity indices and dispersion pattern over the complete experimental period. Correlation analyses showed clear links between microbiome clusters and performance, residual energy intake, and health of the animals. This indicates that a certain microbial composition should be considered for future selection and breeding strategies.

Upper respiratory tract microbiota of dairy calves experimentally challenged with BRSV

S. O'Donoghue[1,2]*, B. Earley*[2]*, M.S. McCabe*[2]*, D. Johnston*[2]*, K. Ní Dhufaigh*[2]*, S.L. Cosby*[3]*, D.W. Morris*[1] *and S.M. Waters*[2]
[1]*University of Galway, Discipline of Biochemistry, Galway city, Co. Galway, H91 W2TY, Ireland,* [2]*Teagasc, Animal and Bioscience Research Department, Animal & Grassland Research Centre, Grange, Co. Meath, C15PW93, Ireland,* [3]*Agri-Food and Biosciences Institute, Veterinary Sciences Division, Stormont, Belfast, Northern Ireland, BT4 3SD, United Kingdom; stephanie.odonoghue@teagasc.ie*

Bovine respiratory disease (BRD) is a leading cause of morbidity and mortality in cattle of all ages. The bovine respiratory microbiome plays a critical role in respiratory health. Despite current research, a lack of data exists surrounding the impact of viral infection on the bovine nasal microbiota. A key BRD virus, Bovine Respiratory Syncytial Virus (BRSV) is an enveloped, single stranded RNA virus of the order Mononegavirales, and a significant viral agent in BRD cases in Ireland. The study objective was to characterise the nasal microbiota of dairy calves following experimental challenge with BRSV. Holstein-Friesian bull calves (mean age(SD)120.7 days(14.15); mean weight(SD)154.7(13.6 kg)) were administered either BRSV inoculate ($10^{3.5}$TCID$_{50}$/ml × 15 ml)(n=12) or sterile phosphate buffered saline (n=6). Nasal swab samples were collected prior to euthanasia on day (d) 7 post-challenge. Microbial DNA was extracted from swabs using the Powerlyzer PowerSoil extraction kit. Sequencing libraries were prepared using the Nanopore 16S barcoding kit (SQK-16S024) and amplicons purified using AmPureXP beads. Libraries were pooled and sequenced on a MinION flow cell (9.4 chemistry) using the Mk1C Device. Raw FASTQ files were uploaded to the EPI2ME platform and analysed using the Fastq 16S workflow(v2022.01.07). The top genera identified in BRSV challenged animals were *Pasteurella*, *Moraxella* and *Mannheimia* spp. with *Pasteurella multocida*, *Moraxella nonliquefaciens* and *Mannheimia varigena* identified as the most common species. *Moraxella*, *Pasteurella* and *Mannheimia* spp. were the most common genera identified in controls on d7, while *Moraxella nonliquefaciens*, *Pasteurella multocida* and *Mannheimia varigena* were the most common species identified. In conclusion, differences were identified in the top genera between BRSV infected and non-infected calves, suggesting that BRSV infection causes a dysbiosis of the nasal microbiome.

Session 51 Theatre 3

Differential effects of *Ostertagia ostertagi* vaccination and infection on the rumen microbiome

J. Lima[1], T.N. McNeilly[2], P. Steele[2], M. Martínez-Álvaro[3], M.D. Auffret[4], R.J. Dewhurst[1], M. Watson[5] and R. Roehe[1]
[1]SRUC, West Mains Road, EH9 3JG, United Kingdom, [2]Moredun Research Institute, Pentlands Science Park, EH26 0PZ, United Kingdom, [3]Universitat Politècnica de València, Camí de Vera s/n, 46022, Spain, [4]Agrifirm Belgium, Baarleveldestraat 8, 9031 Drongen, Belgium, [5]Royal (Dick) School of Veterinary Studies, Easter Bush Campus, EH25 9RG, United Kingdom; joana.lima@sruc.ac.uk

The abomasal parasitic nematode *Ostertagia ostertagi* leads to lower performance and welfare in cattle, but its impact on the rumen microbiome is unclear. Two groups of 10 animals received a vaccine against the parasite (V) or an adjuvant-only control (AD) and were then orally administered 1000 *O. ostertagi* L3 larvae 5 times a week for 5 weeks. We sampled the rumen digesta of AD animals identified based on high (H, n=4) or low (L, n=4) cumulative parasite faecal egg counts (FC), 4 VC animals with low FC, and 4 unvaccinated uninfected animals (UN) at *post-mortem* 3 weeks after the last infection, and shotgun sequenced to obtain abundances of microbial genera (MT) and genes (MG). Microbial profiles were centred log-ratio transformed and pairwise compared (H, L, V vs UN) using permutations within discriminant partial least squares models, and the most important MT/MG were identified by the variable importance in projection. A total of 36, 38 and 21 MT and 91, 31, and 57 MG in H, L, and V, respectively, differed from UN. Of these, 26, 29 and 13 MT and 74, 19 and 41 MG exclusively differed between UN and H, L, or V, respectively, suggesting that the parasite influence on the microbiome depends on the degree of infection. MT enriched in H, L, and V included e.g. opportunistic pathogen *Listeria* (H), *Tsukamurella* (L) and *Pseudopropionibacterium* (L, V), which include pathogens *T. paurometabolum* and *P. propionicum*. Methanogens e.g. *Methylocystis* (H, L) and *Methyloversatilis* (L, V) were enriched in H, L, or V. The microbiome functional potential in H, L and V also differed from UN. MG enriched in H, L and V were involved in, e.g. bacterial secretion system (*gspD*), biosynthesis of nucleotide sugars (*evaA*) and fructose and mannose metabolism (*rhaB*), respectively. These results suggest that the abomasal parasite changed the composition of MT and MT in the rumen microbiome.

Session 51 Theatre 4

Novel ruminal microbiome solutions to reducing enteric methane emissions

T.A. McAllister[1], L.L. Guan[2] and R.I. Mackie[3]
[1]Agriculture and Agri-Food Canada, Lethbridge Research and Development Centre, 5403 1st Ave South, Lethbridge, AB T1K 4R5, Canada, [2]University of Alberta, Department of Agricultural Food and Nutritional Science, 410 Agriculture Forestry Center, Edmonton, AB T6G 2P5, Canada, [3]University of Illinois, Animal Science, 1207 W. Gregory Drive, 61801 Urbana, USA; tim.mcallister@agr.gc.ca

Methane (CH_4) is the second most important greenhouse gas (GHG) contributing to climate change, and the most significant emission source from ruminants. Presently, 150 countries have signed the Global Methane Pledge, which targets reducing global CH_4 emissions by 30% from 2020 by 2030. Furthermore, the comparatively short-lived atmospheric duration of CH_4 makes it a logical target if global warming is to be limited to 1.5 °C. The rumen microbiome can have both beneficial and detrimental impacts on cattle productivity and GHG emissions. The rumen microbiome plays a vital role in carbohydrate, protein and fat metabolism and the production of volatile fatty acids, microbial protein and vitamins for use by the cattle host. Changes in the presence/absence/abundance and activity of select microbes within the rumen affects fermentation, alters nutrient supply and enteric CH_4 emissions and thereby the efficiency of meat and milk production. Despite considerable effort, few CH_4 mitigation strategies are available for immediate implementation on farm. The paucity of mitigation solutions suggests that we still lack a fundamental understanding of the complexity and nature of hydrogen (H2) transfer within the rumen ecosystem. Recent studies suggest that even if all known enteric CH_4 mitigation strategies were implemented, they would be insufficient to limit global warming to current targets. Consequently, new strategies that reduce enteric CH_4 emissions are needed, but they must be anchored by a understanding of the fundamentals of H2 flow in the rumen. The current presentation will describe some of the gaps in our current understanding of fermentation and how they may be potentially filled. Tools such as genomics, metagenomics, metatranscriptomics, proteomics and metabolomics of both the host and the rumen microbiome, with integration through machine learning, will play a key role in identifying new approaches to reducing enteric CH_4 emissions in ruminants.

Session 51 Theatre 5

Relationship between feed efficiency and rumen microbiota in feedlot bulls fed contrasting diets

A. Ortiz-Chura, M. Popova, G. Cantalapiedra-Hijar and D. Morgavi
Université Clermont Auvergne, INRAE, VetAgro Sup, UMR Herbivores Theix, 63122 Saint-Genes-Champanelle, France; abimael.ortiz-chura@inrae.fr

Improving feed efficiency in ruminants contributes to the environmental and economic challenges facing the livestock sector. Although the rumen ecosystem plays an essential role in feed digestion, there is no defined pattern between rumen microbiota and feed efficiency. The goal was to find if there is a relationship between divergent residual feed intake (RFI) phenotypes (RFI–= efficient and RFI+= non-efficient) and the rumen microbiota in feedlot Charolais bulls fed contrasting diets. In this experiment, we used the 32 most extreme bulls based on their RFI from 100 animals fed corn-silage (CS; n=50) or grass-silage (GS; n=50) based diets. Rumen fluid samples were obtained at slaughter for fermentation analysis, bacteria and methanogens quantification by qPCR and meta-taxonomic analysis. In this study, total VFA concentration and profiles showed no differences between diets and RFI groups ($P>0.05$). Total bacteria and methanogen populations did not differ between RFI groups ($P>0.05$), although methanogens expressed per total weight of rumen content tended to decrease in efficient bulls compared to non-efficient ones ($P=0.10$). Total bacteria did not differ between diets, but methanogens increased in the GS compared to the CS ($P<0.05$) diet. Rumen microbial community structure differed between diets (Adonis test, $R^2=0.18$, $P<0.001$), but also between RFI groups within each diet. Although we found no RFI × Diet interaction on microbial abundance, different microbial families were associated with the RFI phenotype in a diet-dependent manner. Efficient bulls receiving the CS diet had a higher abundance of the *Lachnospiraceae* and *Veillonellaceae* families (q<0.05), while in the GS diet, the *Acidobacteriaceae* family was more abundant compared to non-efficient ones (q<0.05). In conclusion, diet-related factors were the main drivers of rumen microbiota variation. Furthermore, the RFI phenotype in each diet revealed the establishment of different microbial consortia. However, whether these microbiota rearrangements originate from the RFI phenotype or are associated with the interaction of the multiple variables involved in rumen fermentation remains unclear.

Session 51 Theatre 6

Establishment and evolution of ruminotypes of lactating Lacaune ewes

T. Blanchard[1], C. Marie-Etancelin[1], Y. Farizon[1], C. Allain[2] and A. Meynadier[1]
[1]INRAE, ENVT, GenPhySE, 24 Chem. de Borde Rouge, 31326, Castanet Tolosan, France, [2]INRAE, Experimental Unit of La Fage, 12250, Saint-Jean et Saint-Paul, France; tiphaine.blanchard@inrae.fr

Ruminal microbiota of sheep is being studied to evaluate its impact on animals' energy balances in early lactation stage. The establishment of ruminotypes could be a useful tool to summarize the large dataset of ruminal microbiota abundances. The feasibility of directional selection of gut microbiota enterotypes has been demonstrated in pigs, proving the existence of genetic determinism of the host on enterotypes. Sixty-five Lacaune dairy sheep were studied and their ruminal contents and blood were sampled at 2 weeks (D1), 3 weeks (D2), 4 weeks (D3) and 6 weeks (D4) after lambing. Ewes were weighed on each date and were fed the same diet with hay, grass silage and concentrates. During the first month, lambs remained with the ewes: milk recording (milk quantity, fat and protein contents, somatic cell count) were performed at D4. Ruminal microbiota was analysed by sequencing the 16s rRNA gene for bacteria and sequences were computed with the FROGS pipeline to obtain relative abundances of ASVs which were then labelled according to genus level. Clustering of microbiota genera abundances using Jensen-Shannon divergence and partition around medoids clustering algorithm has led to 2 groups. Spls-da of these two clusters revealed a first ruminotype caracterised by *Mogibacterium*, *Lachnospiraceae NK3A20 group* and *Acetitomaculum* genera (RuM) and a second one by *Prevotella*, *Prevotella_9* and *Lachnospiraceae NK4A136 group* genera (RuP). Over time, a significant shift towards RuP is noted (D1: 22%, D2: 34%, D3: 58%) which becomes a minority again after weaning (D4: 14%). At D1, RuM is associated with higher non esterified fatty acids ($P=0.04$), beta-hydroxybutyrate ($P<0.01$), isobutyrate ($P=0.01$) and lower total volatile fatty acids ($P<0.01$) compared with RuP. At D2, RuM is associated with lower weekly weight loss ($P=0.03$) compared to RuP. At D3, ruminotypes were not associated with zootechnical traits. At D4, RuM was associated with lower milk protein and fat contents ($P<0.02$). Characterization of the specific metabolic pathways of these 2 clusters is underway via a functional inference approach, in order to make links with the zootechnical traits.

Session 51 Theatre 7

Effect of colostrum source and calf breed on diarrhoea incidents in pre-weaned dairy calves

S. Scully[1,2], P.E. Smith[2], B. Earley[2], C. McAloon[1] and S.M. Waters[2]

[1]University College Dublin, School of Veterinary Medicine, Belfield, Dublin, Ireland, [2]Teagasc, Animal and Bioscience Research Department, Teagasc Grange, Dunsany, Co. Meath, Ireland; sabine.scully@teagasc.ie

Neonatal calf diarrhoea is a multifactorial condition and a major cause of economic loss to producers. Several pathogens are reported to cause or contribute to calf diarrhoea. Other factors include environmental and management practices that influence disease outcomes. The study objective was to evaluate effect of colostrum source (CS) and calf breed (CB) on diarrhoea incidents in 51 spring born Holstein (HO; n=29, birth weight (BW) 34.7 (SE 0.69) kg) and Jersey (JE; n=22, BW 25.9 (SE 0.81) kg) heifer calves from birth (day (d) 0) to weaning (d83 (SE 1.04)). Calves were fed 8.5% BW in colostrum, from either, the dam (n=28) or a pooled source of colostrum (≤2 cows, ≤1d; n=23) within 2 hours of birth. A modified Wisconsin-Madison calf health scoring system was used and rectal temperature (RT) measured for clinical assessment at d0, d7, d21, and day-of-diarrhoea incident, and day-of-weaning. Live weights were recorded at d0, d21, and weaning. Diarrhoea incident was assessed using faecal scores (0=normal, 1=semi-formed, 2=moderate, 3=severe diarrhoea), and health status was defined as calves having diarrhoea (n=27), or healthy (n=24). Faecal scores were analysed using the Wilcoxon Mann-Whiney U-test and RT data were analysed using the PROC MIXED procedure in SAS (9.4). The model included CS, CB, and health status and their interactions as fixed effects, and shed as random effect. The mean day post-birth for diarrhoea was d23 (SE 1.04) and d22 (SE 0.88) for HO and JE calves, respectively; 53% of calves had a diarrhoea incident. There was no CS × CB interaction ($P>0.05$) for diarrhoea incident from d0 to weaning. On the day of diarrhoea detection, faecal scores were greater ($P<0.0001$) for diarrhoeic calves (median score 3; range 2-3) than healthy calves (0; 0-1) while RT of diarrhoeic calves was elevated (+0.37 °C (SE 0.01); $P=0.004$). Health status had no effect ($P>0.05$) on ADG from d0 to weaning. Faecal samples from healthy and diarrhoeic calves at d7, d21 and the day-of-diarrhoea incident, and day-of-weaning are currently undergoing microbial metataxonomic analysis using 16s rRNA amplicon sequencing.

Session 51 Theatre 8

Metagenome strain deconvolution and abundance estimation enabled by low-error long-read DNA sequence

D. Bickhart[1], M. Kolmogorov[2], E. Tseng[3], P. Pevzner[4] and T. Smith[5]

[1]Hendrix Genetics, Spoorstraat 69, 5831 CK Boxmeer, 5830 AC Boxmeer, the Netherlands, [2]National Institute for Health, Center for Cancer Research, Building 41, Room A100C, 20892 Bethesda, Maryland, USA, [3]Pacific Biosciences, 1305 O'Brien Drive, 94025 Menlo Park, California, USA, [4]University of California San Diego, Jacobs School of Engineering, 9500 Gilman Drive, 92093 La Jolla, California, USA, [5]United States Department of Agriculture, Meat Animal Research Center, State Spur 18D, 68933 Clay Center, Nebraska, USA; derek.bickhart@hendrix-genetics.com

The sequencing of microbial isolates is considered the gold standard means to generate high quality reference genomes for individual species in lieu to metagenome sequence assembly. We present new methods and tools to generate isolate-quality reference genomes for metagenome assembled genomes (MAG) using the latest in DNA sequencing technologies that achieve lower error-rates (<1%) with longer read sizes (>9 kbp). Using 255 Gbp of PacBio HiFi sequence, we assembled 428 MAGs that had completeness estimates higher than 90%. We then developed a novel strain deconvolution tool, MAGPhase, that identifies SNP haplotypes within assembled MAGs that are representative of lower abundance strains by using read-phasing. Within our sheep MAG dataset, MAGPhase predicted that 220 of our MAGs (51%) were devoid of any phased SNP haplotypes and represented lineage- or strain-resolved reference genomes. We also identified strain haplotypes in several MAGs that had >99% predicted nucleotide identity with the reference MAG. This suggests that a closely related strain was present in the same sample as the MAG and that the proportion of reads that contain the SNP haplotypes could be used to derive relative abundance of the strain to the reference assembly haplotype. Such insights were difficult to derive from MAGs created using shorter or error-prone DNA sequencing reads, as such assemblies were more likely to collapse strain-level variation into chimeric assemblies. We demonstrate that metagenome assemblies generated with lower-error, long-read sequencing methods are most likely to generate lineage-resolved MAGs. Coupled with the use of our MAGPhase algorithm, these reads can be used to identify cohabiting strains prevalent in the sample or in other orthologous samples of the same type.

Stakeholders views regarding new practices to control microbiomes

F. Bedoin[1], A. Ait-Sidhoum[2], E. Vanbergue[1], A. Stygar[2], T. Latvala[2], Á. MacKen-walsh[3], S. Waters[3], P. Smith[3] and J.K. Niemi[2]
[1]Institut de L'elevage, 149 Rue de Bercy, 75012 Paris, France, [2]Natural Resources Institute Finland (Luke), Latokartanonkaari 9, 00790 Helsinki, Finland, [3]Teagasc, Oak Park, Carlow R93 XE12, Ireland; jarkko.niemi@luke.fi

Microbes play an important role when farmers are aiming at sustainable and viable livestock production, and farmers and other stakeholders play a key role in adopting practices that utilize microbiomes. The aim of this study was to gain understanding on the perceptions and expectations farmers and other supply chain stakeholders regarding microbial ecosystems and adoption of innovations that benefit from microbiome. A review of decision-making to explain the adoption of innovations and farming practices and five focus group discussions involving farmers, advisors and other stakeholders in five European countries were carried out. The themes considered were: (1) identification of microbiomes on farms; (2) Stakeholders' knowledge; and (3) opinion on the role of microbiomes in animal production, health and greenhouse gas emissions; and (4) Opinions on innovations relating to early life, dietary transition and environmental issues. Although financial aspects are strong drivers for the adoption of practices, endogenous factors such as the perceived impact of diseases, the lack of knowledge, and technical skills were found barriers for the adoption of new practices. The importance of early establishment of a 'good' microbiome for young animals was understood well. However, many participants were reluctant towards feed additives for the reduction of methane emissions especially because of doubts about their long term effects.

Session 51 Poster 10

Temporal establishment of the colon microbiota in angus calves from birth to post-weaning

M. Stafford[1,2], P. Smith[2], S. Waters[2], F. Buckley[1], E. O'Hara[3] and D. Kenny[2]
[1]University College Cork, School of Biological, Earth, and Environmental Sciences, Distillery Fields, North Mall, University College Cork, T23 N73K, Ireland, [2]Teagasc Grange, Animal and Bioscience Research Department, Teagasc Grange, Meath, C15 PW93, Ireland, [3]University of Alberta, Department of Agricultural, Food, & Nutritional Sciences, Department of Agricultural, Food, & Nutritional Sciences, University of Alberta, Edmonton, AB T6G 2P5, Canada; michelle.stafford@teagasc.ie

During the peri- and early post-partum period, the gastrointestinal tract (GIT) of the calf is colonised by a combination of commensal and pathogenic microbes. Current understanding of the temporal microbial establishment of neonatal colon is, however, limited. This study focused on the ontogeny of colon microbiota establishment in Angus Beef calves from birth through weaning. Colon digesta contents were collected from Aberdeen Angus calves reared on two farms. Calves were euthanized at birth (n=7), day (D) 7 (n=7), D14 (n=6), D21 (n=7), D28 (n=6), and D96 (n=7) of life. Microbial DNA was extracted from colon samples and subjected to 16S rRNA amplicon sequencing. The effect of time on alpha (GLM procedure of SAS; 9.4) and beta (PERMANOVA) diversity of the colon microbiota was assessed with age and farm included as fixed effects. Differential abundance (*Maaslin2*) analyses were conducted with the same fixed effects. The alpha-diversity of the bacterial community progressed from birth onwards (D0 H' mean=0.59 and D96 H' mean=5.33) with apparent stabilisation between D14 v D21 (P=0.21). Weaning had a further statistically significant effect on the diversity of the colon microbiota (D28 v D96; P<0.001). Similarly, PERMANOVA analysis indicated stabilised colonisation of the colon microbiota from D14 to D21 (P=0.22), albeit weaning altered the colon microbiota (D28 v D96; P<0.001). This study is amongst the first to describe the ontogeny of colonisation of the colon microbiota, from birth to weaning in beef calves and provides fundamental information to support the establishment of beneficial microbes in the lower GIT of ruminants.

Session 51 Poster 11

Preliminary study of faecal microbiota in a selection experiment for birth weight variability in mice

L. El-Ouazizi El-Kahia[1], N. Formoso-Rafferty[2], J.P. Gutiérrez[1], C. Esteban Blanco[3], J.J. Arranz[3] and I. Cervantes[1]
[1]Facultad de Veterinaria, UCM, Producción Animal, 28040 Madrid, Spain, [2]E.T.S.I.A.A.B, UPM, Producción Agraria, 28040 Madrid, Spain, [3]Facultad de Veterinaria, ULe, Producción Animal, 24007 León, Spain; lailaelo@ucm.es

A divergent selection experiment for environmental variability of birth weight in mice was carried out successfully for 31 generations. Two divergent lines were created having different variability at birth weight, the high variability line (H-line) and the low variability line (L-line). Further studies showed that the L-line is more advantageous in fertility with higher reproductive longevity. Also, it showed benefits in traits considered welfare and robustness indicators such as feed efficiency and longevity. It has been reported that the microbiota plays a key role in the host phenotypes through the interactions that occur with the host by influencing several physiological functions. Thus the microbiota can have beneficial or detrimental effects leading to differences in animal productivity and welfare. The aim of this study was to evaluate the differences in the microbial composition from the two lines obtained by a divergent selection experiment for birth weight environmental variability. In order to analyse the faecal microbiota, samples from 77 pregnant females from four generations of selection (22-25) were collected,43 of which correspond to the L-line and 34 to the H-line. The V4 region of the 16S rRNA was sequenced, the amplicon sequence variants (ASVs) were determined using the DADA2 pipeline, and the taxonomic assignment was performed using the SILVA database. After the trimming and quality control, 11.5 million sequences were used for further analysis. These sequences were grouped in 45,783 ASVs corresponding to 12 phyla, being *Firmicutes* (48%) the most representative, and 143 genera. There were significant differences in relative abundance at the genus level when comparing the two divergent lines. We observed a marked increase in *Lactobacillus* and *Helicobacter* and a decrease in *Turicibacter* in the L-line compared to the H-line. The selection experiment led to differences in the microbiota composition between lines; studying these three genera could be the first step to analysing the microbiota linked to higher welfare and robustness.

Session 51 Poster 12

The effects of crude protein levels on rumen microbiome and CH_4 of the fattening Hanwoo steers

H. Kim[1], H. Cho[2], S. Jeong[2], K. Kang[2], S. Jeon[2], M. Lee[2], H. Kang[2], S. Lee[3], S. Seo[2] and J. Seo[1]
[1]Pusan National University, Department of Animal Science, 3339, 1268-50 Samrangjin-ro, Miryang, 50463, Korea, South, [2]Chungnam National University, Division of Animal and Dairy Sciences, 99, Daehak-ro, Yuseong-gu, Daejeon, 34134, Korea, South, [3]National Institute of Animal Science, Animal Nutrition and Physiology Division, 1500, Kongjwipatjwi-ro, Iseo-myeon, Wanju_gun, Jeollabuk-do, 55365, Korea, South; khb3850@pusan.ac.kr

Enteric methane (CH_4) emissions are a significant concern in livestock science, due to their substantial impact on global warming potential. The CH_4 is generally produced by diverse microbiota in the rumen. In this study, we aimed to evaluate the effects of crude protein (CP) levels in concentrate on rumen microbiome and CH_4 emissions of Hanwoo steers during the early fattening period. Twenty-four Hanwoo steers (average 27-month-old, 504±33.0 kg) were randomly allocated to one of four treatment groups and fed their respective diets for 16 weeks: (1) low crude protein (12.00%; LCP); (2) lower middle CP (15.50%; LMCP); (3) higher middle CP (17.25%; HMCP); and (4) High CP (19.00%; HCP) levels in concentrate. In the final week of the experimental period, rumen fluid was collected and 16S rRNA gene amplicon sequencing was performed using PacBio Sequel II. There was no significant difference in CH_4 concentration (ppm-m; $P>0.10$). Based on the weighted UniFrac distance, HCP group showed a significant difference compared to LCP group ($P<0.05$). At the family level, Succinivibrionaceae was significantly decreased with increasing levels of CP ($P<0.05$), whereas unclassified family Bacteriodales tended to increase ($P<0.10$). At the species level, *Succinivibrio dextrinosolvens* tended to decrease with increasing levels of CP ($P<0.10$). Functional modules related to amino acids biosynthesis (Cysteine biosynthesis [M00021], tyrosin biosynthesis [M00025], methionine salvage pathway [M00034]) were linearly decreased ($P<0.05$) with increasing levels of CP. Our findings provide a fundamental understanding of the relationship between CP levels and rumen microbiota in beef cattle during the finishing period.

Rumen metabolites of periparturient cows varying in SARA susceptibility modify fermentation *in vitro*

H. Yang[1,2], S. Heirbaut[2], J. Jeyanathan[2], X.P. Jing[2,3], N. De Neve[2], L. Vandaele[4] and V. Fievez[2]
[1]School of Animal Science and Technology, Jiangsu Agri-animal husbandry vocational college, Taizhou, China, P.R., [2]Laboratory for Animal Nutrition and Animal Product Quality, Department of Animal Sciences and Aquatic Ecology, Ghent University, Coupure Links 653, 9000, Gent, Belgium, [3]State Key Laboratory of Grassland and Agro-Ecosystems, International Centre for Tibetan Plateau Ecosystem Management, School of Life Sciences, Lanzhou University, Lanzhou, China, P.R., [4]Animal Sciences Unit, Flanders Research Institute for Agriculture, Fisheries and Food, Scheldeweg 68, 9090 Melle, Belgium; jeyamalar.jeyanathan@ugent.be

The rumen metabolic environment from either SARA susceptible (S) or unsusceptible (U) cows, was hypothesized to alter the fermentative capacity and bacterial community composition. This was evaluated *in vitro* at pH 5.8 and 6.8 through exposure for a period of 24 h of the microbial inoculum of three donor cows to sterile ruminal supernatant obtained from 15 early lactating dairy cows which were either SARA S (n=8) or U (n=7) (2×2 design). Compared to exposure to sterile supernatant from S cows, microbiota exposed to sterile supernatant from U cows produced more total volatile fatty acids, irrespective of the *in vitro* pH (6.8 or 5.8). Specifically, branched-chain volatile fatty acids (BCVFA), such as iso-butyrate and total BCVFA were elevated/tended to be elevated in incubations with the sterile supernatant from U cows. Net lactate accumulation in incubations with U cows' supernatant at pH 6.8 was higher than with S cows' supernatant, whereas no difference was observed between groups when microbiota were exposed to a pH 5.8 buffer. Nevertheless, the gene copy numbers of bacteria were higher at pH 6.8 than at pH 5.8, irrespective of the origin of the sterile supernatant. Also, the bacterial richness and diversity did not differ between incubations exposed to S or U sterile supernatants. Nevertheless, a lower relative abundance of Lachnospiraceae_UCG-008, and *Mogibacterium*, as well as a higher relative abundance of *Prevotella* were observed after incubation with sterile supernatant from S cows as compared with U cows.

Prevalence of resistant *E. coli* and their transmission among dairy cattle in Swiss tie stalls

B. Köchle, V. Bernier Gosselin and J. Becker
Vetsuisse Faculty, Bremgartenstrasse 109a, 3012 Bern, Switzerland; jens.becker@unibe.ch

Dairy cattle are the production branch of Swiss animal husbandry where most antimicrobials are administered. This study was conducted to: (1) describe the prevalence of rectal *E. coli* resistant to antimicrobials; and (2) to investigate whether there are indications for transmission of resistant *E. coli* among tie-stalled dairy cattle that are tied next to each other. During farm visits, triplets of cows were swab sampled to isolate *E. coli* from the rectal mucosa and perform susceptibility testing using microdilution. Triplets included treated and untreated cows: cow A: treated parenterally with antimicrobials within 3-7 days prior to sampling; cow B: untreated neighbouring cow of cow A; cow C: untreated cow not neighbouring cow A or B. A total of 281 farms were visited. Prevalence of resistant *E. coli* was moderate to low in untreated and moderate in treated cows. Overall, resistance was exhibited most frequently to sulfamethoxazole (31.8%), tetracycline (8.5%) and ampicillin (6.3%). Being cow B, i.e. being exposed to treated cow A was not associated with higher prevalence of resistant *E. coli* in comparison to cow C. In our study population, there is no indication that parenteral antimicrobial treatment in tied dairy cows increases the risk of carrying resistant *E. coli* for cows tied at the neighbouring stand. For practitioners, separation of cows that undergo treatment may not be necessary to avoid spread of resistant bacteria and therefore to comply with the standards of prudent use of antimicrobials.

Session 51 Poster 15

Stability of a *Bacillus*-based DFM following preparation of a milk replacer and premix

G. Copani, A. Segura, N. Milora, M. Schjelde and B.I. Cappellozza
Chr. Hansen A/S, Animal and Plant Health & Nutrition, Boege Alle 10/12, 2970 Hørsholm, Denmark; dkgico@chr-hansen.com

The ability of spore-forming bacteria, such as *Bacillus* spp., to tolerate challenging environments allows its inclusion in different types of supplements often fed to calves and mature animals. We hypothesized that the inclusion of a *Bacillus*-based direct-fed microbial (DFM) in milk replacer and mineral-vitamin premix would provide adequate recovery following supplement preparation. Hence, our objective was to evaluate the recovery of a *Bacillus*-based DFM following its inclusion in milk replacer after 1 hour (Exp. 1) and premix over 12 months (Exp. 2). In Exp. 1, a commercial milk replacer was dissolved in water either at 37 or 50 °C and a *Bacillus*-based DFM (*Bacillus licheniformis* and *B. subtilis*; BOVACILLUS™; Chr. Hansen A/S) was included in both mixtures at 1.0×10^6 cfu/ml of milk replacer (n=3 samples/temperature). Following DFM inclusion, milk replacers were incubated for 1 hour at room temperature (25 °C), and samples were enumerated after 30 and 60 min. In Exp. 2, two independent batches of the *Bacillus*-based DFM were included at a commercial premix at a rate of 3.2×10^7 cfu/gram of premix. Samples were stored at 25 °C and analysed at 1, 3, 6, and 12 months following DFM inclusion into the premix. In Exp. 1, no differences were observed in the recovery of the *Bacillus*-based DFM (log cfu/gram of premix, $P=0.68$) vs time 0 ($P>0.49$) following its mixture in milk replacers at 37 or 50 °C ($P=0.62$). Moreover, no differences were observed between the temperatures evaluated herein. In Exp. 2, the recovery of the *Bacillus* spores in the premix averaged (log cfu/g) 99.5, 100.8, 100, and 98.6% at 1, 3, 6, and 12 months, respectively, and did not differ from time 0 ($P>0.17$). In summary, our results support the hypothesis that this Bacillus-based DFM thrives and survives under different environments for feed manufacture, including the ones observed for milk replacer preparation (different temperatures) and mineral-vitamin premix. These results provide additional evidence of the range of opportunities for *Bacillus*-based DFM that could be used in ruminant production systems.

Session 51 Poster 16

Effects of a *Bacillus*-based direct-fed microbial on performance and digestibility of lactating cows

M. Terré[1], N. Prat[1], D. Sabrià[1] and B. Cappellozza[2]
[1]Institute of Agrifood Research and Technology (IRTA), Finca Camps i Armet, Edifici D, 17121 Monells, Girona, Spain, [2]Chr. Hansen A/S, Commercial Development, Animal and Plant Health and Nutrition, Bøge Allé, 10-12, 2970, Denmark; marta.terre@irta.cat

Direct-fed microbials (DFM) have been used to support health and performance of dairy calves, and to improve health, nutrient utilization, milk production, and milk production efficiency in lactating dairy cows. Therefore, we hypothesized that a *Bacillus*-based DFM would improve performance and nutrient digestibility in lactating dairy cows. Hence, our goal was to evaluate the effects of supplementing a *Bacillus*-based DFM on milk production traits and nutrient digestibility of lactating dairy cows. Seventy-six early-lactating Holstein cows (days in milk 43±6) were blocked by calving date and parity in 1 of 2 treatment groups: (1) partial mixed ration (PMR) without DFM supplementation (n=38; CON) or (2) PMR with the addition of 3 g/head per day of a *Bacillus*-based DFM (BOVACILLUS™, Chr. Hansen A/S, Hørsholm, Denmark; n=38; DFM) for a 105-day experiment. The PMR contained ryegrass hay, wheat silage, corn silage, ryegrass silage, alfalfa hay, and concentrate feed. A pelleted soybean-based supplement was used as a carrier for DFM. Milk yield, composition (%), dry matter intake (DMI), and milk production efficiency were evaluated daily. Moreover, faecal samples were collected on week 15 of the experimental period for nutrient digestibility determination. No treatment effects were observed on DMI and milk yield ($P>0.16$), but DFM-supplemented cows had a greater milk production efficiency when compared with CON cohorts (1.54 vs 1.59 kg milk/kg DMI; SEM=0.02; $P=0.03$). No differences were observed on milk composition (% or yield; $P>0.12$), as well as on energy- or fat-corrected milk production and efficiency ($P>0.18$). Lastly, DFM supplementation tended ($P=0.10$) to increase DM digestibility (71.4 vs 72.9%, respectively; SEM=1.13), without further treatment effects being detected ($P>0.24$). In summary, feeding BOVACILLUS improved milk production efficiency, and tended to improve dry matter digestibility in lactating dairy cows. These data highlight the opportunity of using BOVACILLUS to promote milk production efficiency in dairy production settings.

Stability of a *Bacillus*-based direct-fed microbial post-palletisation under different temperatures

B. Cappellozza[1], C. Galschioet[2] and G. Copani[2]
[1]Chr. Hansen A/S, Commercial Development, Animal and Plant Health and Nutrition, Bøgle Allé, 10-12, 2970, Denmark, [2]Chr. Hansen A/S, Innovation, Animal and Plant Health and Nutrition, Bøgle Alle 10-12, 2970, Denmark; brbrie@chr-hansen.com

Spore-forming bacteria, such as *Bacillus* spp., are known to support challenging environments, such as high-temperature. We hypothesized that the inclusion of a *Bacillus*-based direct-fed microbial (DFM) in pellet supplements would provide the expected recovery under three temperatures. Hence, our goal was to evaluate the recovery of a *Bacillus*-based DFM in pellets prepared under 75, 85, and 95 °C. Three independent batches of a *Bacillus*-based DFM (BOVACILLUS™; Chr. Hansen A/S) were evaluated. The supplement was composed of (as-fed basis) 29.0% wheat, 20.0% barley, 15.0% corn, 10% beet pulp, 10% rapeseed cake, 9.5% high-protein soybean, 3.8% soybean oil, 1.0% limestone, and 1.7% vitamin-mineral premix. The *Bacillus*-based DFM was included at a recommended commercial dose (9.6×10^9 cfu/head per day). Its recovery was evaluated following the pellet preparation under 75, 85, and 95 °C. Mash samples (pre-pellet) were taken from the horizontal mixer in 10 different locations. Moreover, meal with the product was mixed for 10 minutes before sampling and pelleting. For each temperature, pellet samples were taken approximately between 9 and 11 minutes post-pelleting to ensure enough time to get a stable environment in the conditioner and to give a homogeneous sample. The expected cfu/g was 3.80×10^5. Following pellet manufacturing, cfu/g (3.89, 3,91, and 4.46×10^5 cfu/g for 75, 85, and 95 °C, respectively) and deviation from expected (102, 103, and 117% for 75, 85, and 95 °C, respectively) were not affected by the temperatures evaluated herein (P>0.21). When the results were transformed to log cfu/g, no differences were observed by the temperature of the pellet preparation (P=0.25), given that the mean counts represented 100, 100, and 101% of the expected count for 75, 85, and 95 °C, respectively. In summary, these data support our hypothesis that BOVACILLUS™ is thermoresistant, given that its recovery in a pellet supplement prepared under different temperatures (75, 85, and 95 °C) was achieved when calculated in cfu/g, log cfu/g, or deviation from expected (%).

Microbial signals in peripheral blood mononuclear cells of Australian Angus cattle

P. Alexandre[1], A. Wilson[2], T. Legrand[1], R. Farr[2], S. Denman[1] and A. Reverter[1]
[1]CSIRO, Agriculture & Food, St Lucia, Queensland, 4067, Australia, [2]CSIRO, Health & Biosecurity, Geelong, Victoria, 3220, Australia; pamela.alexandre@csiro.au

A key aspect of developing sustainable livestock production systems is selecting for improved productivity while considering welfare outcomes. Immune competence is critical to animal health and welfare and can be described as an animal's ability to mount both a cell-mediated (cell-IR) and an antibody-mediated (Ab-IR) immune response. While the beneficial impacts of microbiomes on livestock health and performance are becoming increasingly apparent, major knowledge gaps remain around how their function can be exploited to deliver beneficial outcomes. With that in mind, 51 Angus heifers and steers were evaluated for cell-IR and Ab-IR at weaning and had blood samples collected by jugular venipuncture. Peripheral blood mononuclear cells (PBMC) were isolated using density gradient centrifugation and RNA was extracted using a Qiagen miRNeasy Mini kit. Ribosome depleted RNA libraries were sequenced in an Illumina NovaSeq 2×150 bp. Filtered reads were aligned to the bovine reference genome ARS-UCD1.2 using STAR. Unmapped reads corresponding to ribosomal RNA (rRNA) were identified using SortMeRNA with the reference databases SMRv4.3sensitive and Greengenes2. Considering all samples, an average of 113M±22M reads were sequenced and around 16% of those were unmapped. Unmapped reads can be attributed to technical sequencing artifacts, unknown transcripts, RNA editing, trans-splicing, gene fusion, circular RNAs, and of particular interest, the presence of non-host RNA sequences (e.g. bacterial, fungal, and viral organisms). Most of the identified rRNA across the samples (~2.9% of unmapped reads) aligned to mammals and other eukaryotes. However, the most present bacteria were Proteobacteria. In humans, Proteobacteria represents the most frequent phylum present in blood and an increasing number of studies identifies them as a possible microbial signature of disease. We also identified a small portion of archaea present in our samples. Our ongoing research is now focused on understanding the relationship between microbial rRNA in PBMCs and immune competence measures to identify possible biomarkers that can inform both animal selection and management decisions.

Session 51 Poster 19

Metabarcoding of milk, faeces and ruminal fluid of Sarda ewes fed with Alfalfa (*Medicago sativa*)

A. Vanzin[1], D. Giannuzzi[1], G. Zardinoni[1], A. Cecchinato[1], N. Macciotta[2], F. Correddu[2], A. Atzori[2], S. Carta[2], A. Ledda[2], S. Schiavon[1], L. Gallo[1] and S. Pegolo[1]
[1]University of Padua, DAFNAE, Viale dell'Università 16, 35020, Italy, [2]University of Sassari, Dep. of Agricultural Sciences, Viale Italia 39/a, 07100, Italy; diana.giannuzzi@unipd.it

Precision livestock feeding can decrease livestock environmental impacts by optimizing the use of dietary nutrients and animal nutrient utilization efficiency. Diet is also a key factor influencing rumen and gut microbiota compositions, which have a strong relationship with animal health and welfare. The present study aims to investigate the effects of different diets on rumen, faeces and milk microbiota of Sarda ewes. A group of 24 Sarda ewes with low milk cheese-making ability was divided into two groups fed for 22 days with two diets similar for net energy (1.48 Mcal/kg DM), crude protein (175 g/kg DM) and NDF (343 g/kg DM), but differing for alfalfa hay which was introduced (534 g/kg DM) in partial replacement of soybean meal, meadow hay and corn grain. Individual milk, faeces and ruminal fluid samples were collected at day one (T0) and at day 22 (T1) from the beginning of the trial. The microbiota of these matrices was characterized by sequencing the V2-4-8 and V3-6, V7-9 hypervariable regions of the 16S rRNA gene with the Ion GeneStudio S5 technology. To process amplicon sequencing data from raw reads to taxon abundance tables, DADA2 was used. Then, α- and β-diversity indices were calculated using the Phyloseq package. The results obtained from the ruminal fluid did not show significant differences in relative abundances between the control group (low – alfalfa, L-Alf) and the treated one (high – alfalfa, H-Alf). Alpha-diversity analysis showed a tendency for a higher taxonomic diversity in rumen fluid of animals belonging to the H-Alf diet. The predominant phyla were Bacteroidetes, Proteobacteria and Firmicutes. The most abundant families were Prevotellaceae and Succinivibriaceae, which are associated with feed conversion efficiency. The upcoming integration with milk and faeces data will allow a more complete picture, which will shed new lights on the relationships diet – animal health and welfare – quality of animal products. Acknowledgments: the research was part of the PRECISLAT project funded by MIPAAF (Italy).

Session 51 Poster 20

An experimental approach for assessing causal microbes in early life diarrhoea in lambs

L. Voland[1], D. Graviou[1], K. Vazeille[2], A. Ortiz-Chura[1], D.P. Morgavi[1] and M. Popova[1]
[1]Université Clermont Auvergne, INRAE, VetAgro Sup, UMR Herbivores, Rte de Theix, 63122 Saint-Genès-Champanelle, France, [2]Unité Expérimentale Herbipôle 1414 Les Razats, Laqueuille, 63820, France; laurianne.voland@inrae.fr

In sheep production, separating lambs from dams at birth is common. However, this causes a late and random colonization of the gastrointestinal microbiota in newborn lambs, leading to a higher incidence of digestive disorders like diarrhoea, and neonatal mortality. Understanding the etiology of digestive dysbiosis will allow proposing better prevention and management plans, achieving a balance between welfare and production. We hypothesized that the incidence of diarrhoea would be linked to unbalanced colonization of the gastrointestinal tract in early life. To infer causality of early life scouring, we used triplet lambs allotted into three groups, M: lambs reared with their ewes, AA: artificially reared, no contact with adults, AAC: AA with some contact with adults. All lambs were sampled weekly for rumen contents, with additional sampling if diarrhoea was detected. The sampling times intended to address the first criteria of temporality for establishing causality. The three rearing conditions were designed to obtain different sequences of microbial colonization and thus address the non-spurious criterion of causality (comparing the microbial communities between groups, before and after diarrhoea). In this study, the average daily gain in the M group was 3 times higher than AA and AAC groups in the first week of life; but decreased to a 1.5 fold in week 8,. Throughout the study, scouring episodes were recorded in 83, 66 and 35% of the lambs in the AA, AAC and M groups, respectively. The incidence of severe diarrhoea was more frequent at 4 weeks of age and at weaning in the three groups; however, symptoms were recorded in only 5% of M lambs compared to 21% in AA and 28% in AAC. The experimental setup allowed us to obtain different disease phenotypes per group (AA had more diarrhoea than AAC and M). Currently, the genomic DNA of rumen contents, fermentation parameters and immunity status (blood IgG) are under analysis. Integration of these data should allow determining factors driving diarrhoea occurrences in early life.

Session 51 Poster 21

Mycotoxin-deactivating feed additive supplementation in dairy cows fed *Fusarium*-contaminated diet

A. Catellani[1], Y. Han[2], V. Bisutti[3], F. Ghilardelli[1], F. Fumagalli[1], E. Trevisi[1], A. Ceccinato[3], H. Swamy[2], S. Van Kuijk[2] and A. Gallo[1]
[1]Università Cattolica del Sacro Cuore, Department of Animal Science, Food and Nutrition – DIANA, Via E. Parmense 84, 29122 Piacenza (PC), Italy, [2]Nutreco R&D, Stationsstraat 77, P.O. Box 299, 3800 AG Amersfoort, the Netherlands, [3]University of Padova, Department of Agronomy, Food, Natural Resources, Animals and the Environment – DAFNAE, Agripolis, Viale dell'Università 16, 35020 Legnaro (Pd), Italy; antonio.gallo@unicatt.it

Limited scientific evidence on Fusarium mycotoxins impact is available in dairy cow performance and health, especially after longer exposition time. Also information on the effects of these mycotoxins on milk cheese-making parameters is very poor. The objective of this study was to evaluate a commercially available mycotoxin deactivating product (MDP, TOXO® XXL, Selko, Tilburg, the Netherlands) in lactating dairy cows fed a Fusarium-mycotoxin contaminated diet, and the repercussions on cows' health status and feeding behaviour, milk yield and quality, in term of cheese-making traits. The MDP contains smectite clays, yeast cell walls and antioxidants. In the study, 36 lactating Holstein cows were grouped based on days in milk, milk yield, body condition score and randomly assigned to specific treatments. The study ran over 2 periods (March/May – May/July 2022). In each period, six animals/treatment were considered. Experimental periods consisted of 9 days of adaptation and 54 days of intoxication. Physical activity, rumination time, daily milk production and milk quality was measured. Cows were fed once daily with the same TMR mixed composition. Experimental groups consisted of CTR diet, TMR with low contaminated high moisture corn (HMC), C-B mix and beet pulp; MTX diet, TMR with high contaminated HMC, C-B mix and beet pulp; TOXO XXL diet, MTX diet supplemented with 100 g/cow/day of TOXO XXL. MDP TOXO XXL reduced mycotoxin negative effects on milk yield and quality (protein, casein, lactose, and clotting parameters). The MTX diet had a lower milk yield and feed efficiency than the CTR and TOXO XXL diets. These results provide a better understanding of mycotoxin risk on dairy cows' performances and milk quality. Further analyses should be carried out to evaluate MDP outcome on immune-metabolic responses and diet digestibility.

Session 51 Poster 22

Effects of 3-NOP on enteric methane production in growing beef cattle offered a forage based diet

S.F. Kirwan[1], L.F.M. Tamassia[2], N.D. Walker[2], A. Karagiannis[2], M. Kindermann[2] and S.M. Waters[1]
[1]Teagasc, Animal Bioscience Research Centre, Grange, Dunsany, Co. Meath, C15PW93, Ireland, [2]DSM Nutritional Products, Animal Nutrition and Health, Wurmisweg 576, 4303 Kaiseraugst, Switzerland; stuart.kirwan@teagasc.ie

There is an urgent requirement internationally to reduce enteric methane (CH_4) emissions from ruminants. Enteric CH_4 from ruminants accounts for 6% of global anthropogenic greenhouse gas emissions. Numerous nutritional interventions including feed additives have been suggested as potential CH_4 mitigating strategies, however, the efficacy of some of these strategies varies and are difficult to implement at farm level. The CH_4 inhibitor 3-nitrooxypropanol (3-NOP) is seen as a potential strategy at reducing methane emissions in ruminants, but to date the effects have mainly been observed in adult beef cattle and dairy cows mainly fed high energy concentrate diets. Therefore, the objective of this study was to investigate the effects of 3-NOP inclusion in growing beef cattle offered a forage based diet. A total of 68 Dairy × Beef male calves (≤6 months; BW: 147±38 kg) were assigned to 1 of 2 treatments over a 12-week period, in a completely randomized block design. Dietary treatments were: one of two groups: (1) control, placebo (no 3-NOP), and (2) 3-NOP applied at 150 mg/kg dry matter (DM). Methane, H_2 and CO_2 were measured using the GreenFeed system. Total weight gained, dry matter intake, average daily gain, and animal health variables were not affected by the inclusion of 3-NOP ($P>0.05$). The inclusion of 3-NOP decreased ($P<0.001$) CH_4 emissions: (g/d; g/kg DMI) by 29.87 and 26.41% respectively, and was consistent throughout the course of the study. This reduction in methanogenesis resulted in a 223.2% increase in H_2 emissions (g/d) ($P<0.001$). Concentrations of ruminal total VFA and individual VFA were not affected by the inclusion of 3-NOP ($P>0.05$). However, ruminal acetate:propionate tended to be lower in animals that received 3-NOP ($P<0.10$). Incorporating 3-NOP into growing beef cattle diets is a potential solution to decrease CH_4 emissions in growing cattle offered a forage based diet.

Session 52 Theatre 1

COWFICIENCY project: a field attempt to increase nitrogen use efficiency of dairy cattle

A. Foskolos, A. Plomaritou, M.E. Hanlon and D. Kantas
Department of Animal Sciences, University of Thessaly, 41500 Larisa, Greece; andreasfosk@hotmail.com

Nitrogen (N) extended use in the food chain led to the N cascade phenomenon with considerable environmental impacts. The European Union has been taken several initiatives at either research or legislative level to reduce N pollution, including applied projects to further implement these strategies at the dairy sector. The CowficieNcy project aims to further develop and implement the mathematical models Cornell Net Carbohydrate and Protein System (CNCPS; a cow-based model) and Catttle Nitrogen Efficiency (CNE; a herd-based model) to improve N use efficiency (NUE) in dairy cattle farming. The first step involved training staff on the CNCPS feed fractionation scheme, chemical analysis of feedstuffs, and models use. The second step described the current situation at the farm level in specific areas in each country, namely Thessaly in Greece (GR), Emilia Romagna in Italy (IT), Catalonia in Spain (SP), and Wales in the United Kingdom (UK). An intensive outreach program was performed to describe the N status in 64 dairy farms (20 in GR, 20 in IT, 14 in UK and 10 in SP). Following farm sampling and management evaluation, data collected for each farm were inputted into the CNCPS to further evaluate dietary performance. Rations at the feeding group level were evaluated and studied to highlight their limitations in terms of NUE, that were discussed with the farmers and/or the nutritionists to inform on solutions to improve their diets. A third step included the implementation of the suggested changes that was at the discretion of the farmer. Farmer's acceptability was on average 20.5%. A significant improvement of milk nitrogen use efficiency (MNE = N in milk / N intake, %) from 30.0 to 34.0% was recorded. In conclusion, the CowficieNcy project developed a set of pilot farms with improved NUE and consequently decreased environmental impact. Furthermore, it revealed the importance of developing links between Universities and the dairy and feed industries to promote trust and increase farmers' acceptability to new strategies. This project received funding from the European Union's Horizon 2020 research and innovation program, under grant agreement No 777974.

Session 52 Theatre 2

Dairy heifers intake capacity: are estimated values still correct?

J. Jurquet[1], Y. Le Cozler[2], D. Tremblais[1], F. Launay[3] and L. Delaby[2]
[1]Institut de l'Elevage, 149 rue de Bercy, 75595 Paris, France, [2]INRAE, L'Institut Agro, UMR PEGASE, 16 le Clos, 35590 Saint Gilles, France, [3]INRAE, Unité expérimentale du Pin, Borculo, Exmes, 61310 Gouffern en Auge, France; julien.jurquet@idele.fr

On-farm heifers feeding practices often challenge on-going recommendations, some advisors hypothesizing that equations developed in early 80's to predict heifers intake capacity (IC) are no longer suitable for current animals. Three complementary approaches were performed to test this hypothesis. First, a survey on 21 advisors in 2020 indicated that 16 of them considered that algorithms they were using adequately predicted heifers IC. These algorithms were based on information regarding body weight and ration composition, but advisors indicated that most farmers do not weigh heifers or analyse feed. In a second approach, feed intake from 58 groups of heifers, from 4 experimental farms, was analysed. The groups size averaged 15 heifers (±7), weighing 400 (±84) kg and aging 15 (±3.5) months. The main forages used were corn silage, hay and wrapped haylage. Differences between predicted dry matter intake (DMI) and measured DMI was in average -0.3 kg, difference being less than 1 kg DMI/day in 2/3 of the cases. Finally, in a third trial, individual DMI were measured during autumns 2019 and 2020, on respectively 47 and 42 heifers of Holstein (n=54) and Normande (n=35) breeds, aged 17 months on average. They were fed *ad libitum with* grass silage in 2019 and hay in 2020. IC was evaluated thanks to the determination of fill value of forage, through the DMI of the heifers or from an alternative method using DMI measured on adult sheep fed *ad libitum*. IC was also determined thanks to the equation $IC = aBW^b$. The 'a' coefficient was adjusted in both approaches by setting the coefficient 'b' to 0.9, as previously recommended. In the first approach (heifers intake), 'a' value was similar to previous published results (0.03915, vs 0.03729 and 0.03749 for silage and hay, respectively). Similar results were observed in the second approach with the sheep intake and the forages fill value estimations. It was concluded that the intake prediction equation used since the 80's remains valid. The lack of heifers weighing and/or of knowledge of forages values probably resulted in wrong initial hypothesis.

Session 52 Theatre 3

Current status of feed nitrogen use efficiency in dairy replacement heifers in Greece

A. Plomaritou, M.E. Hanlon, K. Gatsas, S. Athanasiadis, K. Droumtsekas, D. Kantas and A. Foskolos
University of Thessaly, Department of Animal Science, Campus Gaiopolis, Larissa, 41222, Greece; anplomaritou@uth.gr

The aim of the study was to assess management conditions and define the current nitrogen use efficiency (NUE) status of dairy replacement heifers in Greece. Data were collected from 14 dairy farms in the region of Thessaly, Macedonia and Thrace. Information about the overall farm and heifer rearing management were obtained during a 2-day visit. Three data sets were developed on: (1) reproduction performance (RP), (2) growth performance (GP) and (3) ration evaluation (RE). For the RP data set, reproduction records of all cows and heifers of the herd were retrieved to determine time of first service (T1stSer), services/heifer and time of first calving (T1stCal). For the GP dataset, 15-20% of all heifers and cows were tape measured to estimate body weight (BW) and average daily gain (ADG) in different stages and mature BW (MBW) of cows in each herd. Growth targets based on farm's MBW were set as follows: (1) double birth weight at weaning, (2) 55% MBW at first service, (3) 85% MBW at calving (Target85) and (4) 95% MBW (Target95) at calving and the respective ADG targets for each growth stage were set. For the RE data set, a description of the total mixed ration was recorded and feed samples were collected and chemically analysed for nutrient composition. The rations were evaluated with the Cornell Net Carbohydrate and Protein System (v6.5). It was found that T1stSer and T1stCal was 15.4 (±2.4) and 25,3 months (±2.65), respectively and pregnancy was achieved with 1,54 services/heifer. Heifer BW at T1stSer and T1stCal was 437.1 kg and 692.8 kg, respectively that was within the targets of 55 and 85% of MBW (404 and 625 for T1stSer and T1stCal, respectively) but achieved with a delay of 1.3 months. In all farms, Metabolizable Energy (% of requirements) was the first limiting factor, while Metabolizable Protein was supplied above heifers' requirements and was progressively increasing along with heifers age ($P<0.0001$). Regarding NUE, was higher in the period from 3 to 12 mo (on average 25.1%); and was decreasing as the growing period, from 15 to 24 mo) was progressing (on average12.5%).

Session 52 Theatre 4

Impact of feed-grade and slow-release ureas on dairy cattle performance and nitrogen efficiency

M. Simoni[1], G. Fernandez-Turren[2], F. Righi[1], M. Rodríguez-Prado[3] and S. Calsamiglia[3]
[1]University of Parma, Department of Veterinary Science, Via del Taglio, 10, 43126 Parma, Italy, [2]Universidad de la República, Departamento de Producción Animal y Salud de los Sistemas Productivos, Instituto de Producción Animal, Facultad de Veterinaria, Ruta 1 km 42, CP 80100 San José, Uruguay, [3]Universitat Autònoma de Barcelona, Animal Nutrition and Welfare Service (SNiBA), Departament de Ciència Animal i dels Aliments, 08193 Bellaterra, Spain; federico.righi@unipr.it

The aim of the present study was to assess the effects of feeding feed-grade urea (FGU) or slow release urea (SRU) as a replacement for true protein sources (control; CTR) in high-producing dairy cattle diets. The selected research papers (n=44) were published between 1971 and 2021 and included dairy breed, detailed description of the isonitrogenous diets fed, provision of FGU or SRU (or both), high-producing cows (>25 kg/cow/d), and milk yield as minimum output. Most of the studies compared only 2 treatments, thus a generalized linear mixed model network meta-analysis was employed to compare the effects among CTR, FGU, and SRU. Milk production was 32.9±5.7 l/d with 1.18±0.03 and 1.05±0.02 kg/d of fat and protein while dry matter intake was 22.1±3.45 kg. Average diet composition was 16.4±1.45% CP, 30.8±5.91% NDF, and 23.0±4.62% starch. Average supply of FGU and SRU were 209 and 204 g/cow/d respectively. No differences were detected on performance. The SRU and FGU tended to reduce the MUN ($P=0.09$) and tended to reduce the 4% fat corrected milk ($P=0.10$), protein ($P=0.07$) and lactose ($P=0.06$) yield, probably for a lower energy intake in the 2 treatments. A further analysis was conducted on a subset of articles (n=7) reporting both CP intake and milk CP yield. No difference in milk nitrogen efficiency was found between treatments ($P=0.122$). Among the treatments the use of FGU is less expensive and can be justified. This project received funding from the European Union's Horizon 2020 research and innovation program under the Marie Skłodowska-Curie grant agreement no. 777974.

Session 52 Theatre 5

Current nitrogen status and management of dairy farms in Greece

M.E. Hanlon[1,2], A. Plomaritou[2], E. Tsiplakou[1], I. Vakondios[2], T. Michou[2], D. Kantas[2] and A. Foskolos[2]
[1]Agricultural University of Athens, Department of Animal Science, Iera Odos 75, 11855 Athens, Greece, [2]University of Thessaly, Department of Animal Science, Campus Gaiopolis, Larissa, 41222, Greece; mikenziehanlon@gmail.com

Dairy farming plays a major role in nitrogen (N) emissions causing environmental impacts. Nitrogen emissions and losses can vary largely between farms due to differences in N management practices. Nutritional strategies can be developed to mitigate N losses; however, management and efficiency performance of farms must be determined prior. The objective of this study was to determine the current nitrogen status of dairy farms in Greece. Sixteen dairy farms (202±105 lactating cows, 690±25 kg of BW, 33±6 kg of milk yield, 23±3 kg of DMI) were sampled using a 2 d sampling period. Sampling consisted of a detailed description and collection of the total mixed ration (TMR) offer and orts, feed ingredients, milk yield and composition of each group and a farm questionnaire. To determine the nitrogen efficiency on an animal level, milk nitrogen use efficiency (MNE; milk nitrogen/nitrogen intake) was calculated. Each diet was further evaluated using the Cornell Net Protein and Carbohydrate system (CNCPS) to determine model predicted nitrogen performance. Offered TMRs on dairy farms had a Crude Protein (CP) ranging from 14 to 23% of DM (M=18; SD=3). Nitrogen intake ranged from 441 to 845 g/d (M=610; SD=111) and MNE from 25 to 34% (M=31; SD=3). Model evaluation suggested that the first limiting nutrient for 62% of farms was metabolizable energy. Moreover, rumen ammonia levels, as defined by the CNCPS, ranged from 104 to 319% requirements (M=153; SD=46), suggesting that in some cases CP was oversupplied loading the rumen with soluble N. For all the TMRs limited in metabolizable protein, rumen ammonia levels averaged 145±17%, oversupplying ruminal N. Thus, in Greek dairy farms, N is being oversupplied or accurate sources are not fed. Therefore, indicating potential to improve nitrogen efficiency.

Session 52 Theatre 6

Dairy cattle holistic nutritional management for reduced nitrogen pollution

A. Foskolos, A. Plomaritou, M.E. Hanlon and D. Kantas
University of Thessaly, Department of Animal Science, Campus Gaiopolis, Larissa, 41222, Greece; afoskolos@uth.gr

Nitrogen (N) pollution from agricultural activities has been a central social and scientific issue for at least the last three decades. Within the N cascade theory, there is a sequential transfer of reactive N through environmental systems, and particularly to the transformations of N that make it either to move from one system to another or to be stored within each system. For the European Union (EU) we calculated that: (1) the overall involvement of lactating dairy cattle in N consumption is 26.3% of total N directed to animal feed, and (2) approximately 10% of the virgin N entering into the EU ecosystem is deposited in cattle manure. Several strategies have been proposed to mitigate N pollution referring to the system, the herd or/and the animal. Within these strategies, nutrition has a key role at the animal and herd levels and it is mainly focused on improving milk N use efficiency (MNE = N in milk / N intake). Even though the theoretical maximum MNE is estimated at 40-43%, farm surveys around the world suggest that it is currently between 25 and 30%. To overcome this limitation at the farm level, we developed a Holistic Nutritional Management (HNM) scheme that includes: (1) dairy cattle group formation strategies based on farm's equipment and facilities, (2) diet formulation with sensitive to low crude protein diets systems, such as the Cornell Net Carbohydrate and Protein System, (3) on farm data collection protocols relative to model inputs, and (4) a feed preparation and delivery monitoring scheme to assure that the formulated diet is actually fed. First step evaluation of 20 dairy cattle farms in Greece suggested that current MNE is on average 30.4%, while crude protein overfeeding is in most cases overcame. However, this was done by limiting cows' productivity due to inadequacy of metabolizable protein intake and by supplementing sources of protein that the cow couldn't use efficiently. In a second set of five pilot farms, the HNM was implemented and resulted in MNE improvement (from 30.2 to 34.0%; *P*=0.02). In conclusion, an improved nutritional management as suggested by the HNM scheme may be applied in dairy farms to improve MNE and consequentially to reduce N excretion into the environment.

Session 52 Theatre 7

Manure management practices of dairy farms in Greece

L. Makridis, D. Vouzaras, D. Kantas and A. Foskolos

University of Thessaly, Department of Animal Science, Campus Gaiopolis, Larissa, 41222, Greece; leandros.makridis@gmail.com

Manure management practices play a significant role in the environmental impact of dairy farming. Prior to developing key mitigation strategies, a detailed description of dairy farms and current practices must be identified. The objective of this study was to determine current manure management practices of dairy farms in Greece with the purpose of collecting the baseline information required for the development of further tools and evaluation methods. This study consisted of an on-farm questionnaire of 48 dairy farms describing herd size, bedding used, manure collection and storage type, land owned, manure treatment with anaerobic digestion (AD) and digestate usage for fertilization. Data collected was first categorized based on farm size (small = 0 to 100 lactating cows; medium = 101 to 200; large ≥201). Herd size of lactating cows averaged 67±21 for small farms (n=13), 148±30 for medium farms (n=24), and 335±109 for large farms (n=11). The majority of all farms (81%, n=48) did not provide any bedding for the lactating cows and (82% n=39) of them used bedding only for the dry cows and heifers, creating the need for both liquid and solid manure handling and storage techniques. The farms with access to mechanical manure separator was 8% for small, 38% for medium and 64% for large. The data was then categorized based on availability of storage type to slurry, solid and a combination of both. The most common storage system was the combination of both with 62, 75, 91% followed by solid 23, 17, 0%, and slurry 15, 8, 9%, for small, medium, large farms, respectively. For land possession, 70, 86, and 73% of small, medium, and large farms, owned land for cultivation, while the hectares to number of lactating cows averaged, 0.40±0.36, 0.37±0.45, 0.17±0.05 respectively. AD was used by 64% of all farms as a treatment method for manure, and only 33% of the farms utilized the digestate for land fertilization on their land.

Session 52 Theatre 8

peNDF modulates chewing activity, rumen fermentation, plasma metabolites, performance in dairy cow

Y.C. Cao, L.M. Wang and J.H. Yao

Northwest A&F University, Yangling, Shaanxi, 712100, China, P.R.; caoyangchun@126.com

Subacute ruminal acidosis (SARA) continues to be a common and costly metabolic disorder in high-producing cows worldwide. In order to evaluate if increasing physically effective neutral detergent fiber (peNDF) in diet can prevent SARA in cows fed high concentrate diets. Thirty second-parity Holstein cows were randomly allocated to three treatment groups: H-peNDF8.0, M-peNDF8.0, and L-peNDF8.0, which were prepared by mixing the same total mixed ration for 10, 18, or 60 min, respectively. The peNDF8.0 intake was positively correlated with the peNDF8.0 contents in the diets. Total chewing and ruminating time was lower for the L-peNDF8.0 diet than for the H-peNDF8.0 and M-peNDF8.0 diets (P<0.05). Rumen pH was higher in the H-peNDF8.0-fed cows than in the other two groups (P<0.05). The H-peNDF8.0 and M-peNDF8.0 diets corresponded with higher acetate concentration, acetate:propionate ratio than the L-peNDF8.0 diet (P<0.05), while H-peNDF8.0 and M-peNDF8.0 resulted in lower propionate and valerate concentrations than L-peNDF8.0 (P<0.05). Lowering the peNDF8.0 content decreased the activities of ruminal carboxymethyl cellulase, avicelase, and β-glucanase (P<0.05). H-peNDF8.0 resulted in lower total plasma antioxidant capacity, γ-glutamyl transpeptidase, albumin, and creatinine compared to M-peNDF8.0 and L-peNDF8.0 (P<0.05). Somatic cell counts in milk were positively correlated with the dietary peNDF8.0 content. The feed and milk energy efficiencies were unaffected by the treatments. In conclusion, increasing the content of peNDF8.0 in diet could help alleviate SARA and improve animal health among early lactation cows fed a high concentrate diet by increasing peNDF8.0 intake, chewing activity, and rumen pH.

Session 52 — Theatre 9

Impact on milk production of feeding organic acids during pre and post-partum period

L. Jansen[1] and C. Gordon[2]

[1]Nutreco, Selko, Stationsstraat 77, 3811 MH Amersfoort, the Netherlands, [2]Trouw Nutrition Canada, Dairy Technology and Application, 7504 McLean Rd E, Puslinch, ON N0B 2J0, Canada; lonneke.jansen@trouwnutrition.com

Organic acids have been used in dairy diets to stabilize ruminal pH, reduce methanogenesis, and stimulate ruminal bacterial growth and activity, which results in enhanced ruminant performance. The impact of a mixture of organic acids (MOA, Renergy, Selko) on milk production was tested in a trial with 4 groups of 18 cows. MOA was fed to a pre-partum group from day -21 until calving, to a post-partum group from calving until 28 days in milk (DIM) and continuously from -21 to 28 DIM, whereas 1 group served as controls. There was a significant (P<0.05) increase of 5.3 kg of energy corrected milk as a result of feeding MOA continuously from -21 to 28 DIM. Feeding MOA in pre- or post-partum periods did not show a significant benefit in energy corrected milk yield. Treatment did not affect dry matter intake in the first 28 days in milk. Liveweight and body condition score were used to assess whether the additional milk was coming from cows losing more bodyweight or body-condition. Similarly, the milk fat and protein concentrations, together with somatic cell count were assessed, but no statistically significant differences were observed.

Session 52 — Theatre 10

Peptide profile as fingerprinting of the ripening period of Alpine Asiago cheese

S. Segato[1], S. Khazzar[1], G. Galaverna[2], A. Caligiani[2], G. Riuzzi[1], L. Serva[1], F. Gottardo[1] and G. Cozzi[1]

[1]Padova University, Dept. of Animal Medicine, Production and Health, Viale dell'Università, 16, 35020 Legnaro (PD), Italy, [2]Parma University, Dept of Food and Drug, Parco Area delle Scienze, 27/A, 43124 Parma, Italy; severino.segato@unipd.it

The biochemical cheese ripening process is characterized by an intensive microbial proteolytic activity on milk curd caseins (CN) originating a broad range of oligopeptides. The study was undertaken to evaluate the proteolysis process with the main goal of identifying a potential pool of oligopeptides to be used as molecular markers of cheese ripening time. Samples (n=32) of protected designation of origin (PDO) Asiago were manufactured with mountain raw bulk milk according to the rules defined by the European Community Regulation and then aged for 6 or 12 months (12±2 °C, 82±4% of relative humidity). The detection of peptides in the water-soluble extracts of cheese samples was carried out by a liquid chromatography/electrospray ionization mass spectrometry (LC-ESI-MS) analysis. The peptides were identified according to their molecular weight and significant fragment ions. All identified peptides were semi-quantified by comparison with a Phe-Phe internal standard as ratio between areas in the extract ion chromatogram. A total of 76 peptides were identified and most of them derived from the N-terminal or C-terminal part of $_{\alpha s1}$-CN and $_{\beta}$-CN, while only a small number from $_{\alpha s2}$-CN and none from $_{k}$-CN. A principal component analysis showed that the most representative peptides were $_{\beta}$-CN f(10-14), $_{\alpha s1}$-CN f(17-21), $_{\beta}$-CN f(8-14), $_{\alpha s1}$-CN f(10-16) and $_{\alpha s1}$-CN f(3-13) at 6-mo of ageing, and $_{\beta}$-CN f(193-209), $_{\alpha S2}$-CN f(189-207), $_{\alpha S1}$-CN f(6-28) and $_{\alpha S2}$-CN f(183-207) at 12-mo. The outcomes of this research confirmed that the peptides profile could be a useful molecular benchmark to discriminate cheeses according to their ageing time.

Session 52 Theatre 11

Producing sustainable milk in agroecology with Normande breed and grassland in Normandy

B. Rouillé[1], F. Lepeltier[2] and L. Morin[2]

[1]Institut de l'Elevage, 149 rue de Bercy, 75595 Paris, France, [2]Association de la ferme expérimentale de La Blanche Maison, La Blanche Maison, 50880 Pont-Hébert, France; benoit.rouille@idele.fr

Normandy is an important cattle breeding region in western France. It ranks second among French regions with 17% of dairy cows. Its territory is largely covered with meadows and is therefore valued primarily by ruminants. La Blanche Maison experimental farm is located in the heart of this territory with an annual rainfall of approximately 1,100 mm. A dairy system, based on the principles of agroecology, was studied for 5 years (2017 to 2021). The system is built with 90 dairy cows of the Normande breed and an area of 97 hectares dedicated solely to fodder production for the herd. A specificity of the system is the conduct of reproduction in 4 periods of artificial insemination of 6 weeks each (one every three months) to rationalize the labour demand. The ambition of the research program was to improve the production system on three major axes: (1) zootechnics by increasing the delivery of milk from the farm with a constant number of cows; (2) the environment by reducing the footprint carbon of the milk produced; and (3) labour and the economy by improving income and work organization for farmers. Over the 5 years of the study, milk production increased by 1,500 kg of milk/cow/year, in particular by improving feed strategy with better grazing management, better quality for harvested fodder and an additional supply of concentrates. This also led to improve milk composition in fat and protein. On the environment, the efficiency of the use of nitrogen in the system has increased from 39 to 46%. At the same time, the carbon footprint calculated with CAP2ER® has decreased by 12% in 5 years. Finally, the cost of milk production went down from €653/1,000 l to €524/1,000 l, thus improving the economic performance of the farm. The system has therefore evolved favourably on zootechnics, the environment and the economy, thanks to simple levers accessible to breeders.

Session 52 Theatre 12

Hepatic metabolome of grazing dairy cows with or without feed restriction during early lactation

M. Carriquiry, M. García-Roche, A.L. Astessiano, D. Custodio, G. Ortega and P. Chilibroste

Facultad de Agrononomía, UdelaR, Department of Animal Sciences and Pastures, Ave. Garzón 780, 12400, Uruguay; mcarriquiry@fagro.edu.uy

Six multiparous early lactation (29±10 days in milk (DIM)) Holstein cows (477±73 kg body weight and 2.75±0.16 body condition score (BCS) units) were selected to assess the effect of a five-day feed restriction on the hepatic metabolome. From 29 to 38 DIM cows went through an adaptation period where they were *ad libitum* fed (36.5 kg/DM/cow/d pasture allowance and 5.3 kg/DM/cow/d of an energy concentrate), from 38 to 43 DIM cows were fed restricted (16.5 kg/DM/cow/d pasture allowance and 2.6 kg/DM/cow/d of an energy concentrate) and from 43 to 54 DIM cows went through a flushing period when they were again *ad libitum* fed. Milk yield was recorded daily and milk composition was analysed twice during every period while liver biopsies were collected at the end of every period. Milk yield and components were analysed using a model with period as a fixed effect and metabolomic data analysis including Partial Least Squares Discriminant Analysis (PLS-DA) and Metabolite Set Enrichment Analysis (MSEA) was done using Metaboanalyst 5.0. Energy corrected milk yield was lower during the feed restriction than the adaptation or flushing periods (27.1 vs 33.4 and 33.7±1.9 kg/d, $P<0.01$). The PLS-DA model showed that the periods that differed the most were the feed restriction and flushing periods as components 1 and 2 accounted for 47% of the variation. The PLS-DA model plots showed there were 28 metabolites with Variable of Importance Projection scores >1.2. Differentially abundant metabolites were determined using a volcano plot with fold change >1.5 and $P<0.01$, relative concentrations of beta-hydroxybutyrate, D-glucose-1-phosphate and ribose were higher while relative concentration of malate was lower during the restriction period. Results of MSEA showed that the tricarboxylic acid (TCA) cycle was differently expressed (false discovery rate (FDR) adjusted $P<0.05$) and one-way ANOVA identified that only malate had FDR<0.05. Our results suggest that feed restriction changed the availability of TCA cycle intermediates in the hepatocyte and non-intake energy precursors such as beta-hydroxybutyrate and D-glucose-1-phosphate were used.

Nitrogen balance of lactating cows from herds fed hay-based diets in Northern Italy

T. Danese[1], M. Simoni[1], R.G. Pitino[1], G. Mantovani[1], M.C. Sabetti[1], F. Righi[1] and M.E. Van Amburgh[2]
[1]University of Parma, Veterinary Science, Strada del Taglio 10, 43126, Italy, [2]Cornell University, College of Agricutural and Life Science, Morrison Hall, 14850, NY, USA; tommaso.danese@unipr.it

The aim of this study was to measure nitrogen balance (NB) of Holstein dairy herds fed hay-based diets in Northern Italy. For this purpose, 20 dairy herds (60-367 animals, 34±3 kg MY, 26±1.7 kg DMI, 678±33 kg BW), located in Emilia Romagna's Parmigiano cheese region and fed a total mixed ration (TMR), were involved in this study. Group feed intakes were measured twice a day, and representative TMRs and bulk milk samples were collected, while individual samples of faeces and urine were collected from 10% of the lactating cows. TMRs and faeces were analysed to determine the following parameters: DM, N, aNDFom, ADFom, NDIN and ADIN. Undigested NDF was determined at 240 h *in vitro* fermentation and used as a marker to estimate total faecal output. Urine samples were collected by perineal massage from the same cows as faecal sampling and acidified to obtain a pH 2, pooled per TMR group and then analysed for creatinine content, employed as marker to calculate the total urine output. Faecal, urine and bulk milk samples were analysed for N content. Milk nitrogen, urine nitrogen, and faecal nitrogen outputs ranged between 147 and 211 g/d (mean 180±14), 191 and 423 g/d (mean 272±57), 128 and 422 g/d (mean 235±64) respectively, while nitrogen intake resulted in the interval between 448 and 768 g/d (mean 627±72). Therefore, NB was on average -60±93 g/d. Overall, NB appears to be consistent with ranges reported in literature but shows a wider variability, probably related to the type and quality of the forage employed. The negative values of NB are likely due to partial collection of faeces and urine and the reliance on markers to estimate total excretion as the negative value is within the standard deviation of the measurement and spans zero however the high end of the excretion for urine and faeces is high compared to literature values. Thus, the cattle are likely excreting excess nitrogen, and this approach was not sensitive to those differences. This project has received funding from the European Union's Horizon 2020 research and innovation program, under grant agreement No 777974.

Assessing sward managements on nitrogen fixation from a white clover high sugar grass mixture

M. Verbeeck[1,2], C. Segura[2], A. Louro-Lopez[2], N. Loick[2], P. De-Meo-Filho[2], S. Pulley[2], J. Hood[2], B.A. Griffith[2], L.M. Cardenas[2] and D. Enriquez-Hidalgo[2,3]
[1]Soil Service of Belgium, Willem De Croylaan 48, 3001 Heverlee, Belgium, [2]Rothamsted Research, Net zero and resilient farming, EX20 2SB North Wyke, United Kingdom, [3]University of Bristol, Bristol Veterinary School, BS405DU Langford, United Kingdom; daniel.enriquez@bristol.ac.uk

Biological nitrogen fixation (BNF) in mixed swards can be reduced by poor management or unfavourable soil conditions. In this study, the BNF of a grass-white clover pasture was examined under set-stocking conditions with different grazing management. A split-plot block experiment was run at the North Wyke Farm Platform. Plot treatments were: (1) grazing in unfenced plots; (2) pasture for silage; and (3) simulated grazing, the latter two in fenced plots. The split-plot treatments were either or not application of 14 ton/ha of farmyard manure (FYM). Three replicate blocks were with ewes and lambs grazing from April until October, the other three blocks with pasture reserved for silage production with a cut in June, afterwards cattle grazed from July until October. Samples were taken monthly to assess herbage production, species composition and BNF (assessed with the ^{15}N natural abundance method). The herbage was analysed for macro- and micronutrients in July. The herbage production ranged between 4-10 ton dry matter (DM)/ha, higher in the silage compared to the grazed and simulated grazing treatments (9.1 vs 6.5-6.1 ton DM/ ha). The clover content in the herbage ranged from a few percentages to 40% (DM basis), was lower in the grazing treatments than in the other treatments (10 vs 17-18% DM). The estimated %N derived from the atmosphere (%Ndfa) ranged between 84-93%, but was not determined by management or FYM addition. The BNF ranged between 7-76 kg N/ha, was higher in the silage treatments compared to the other treatments (52 vs 32-23 kg N/ha). The BNF was strongly determined by herbage production (ρ=0.72) and clover content (ρ=0.68) but not by the %Ndfa. The FYM addition had a positive effect on the mineral N concentration in the soil and on the K and Mo concentration in the herbage, however the FYM at the tested rate did not affect the BNF nor the herbage mass of the white clover-grass pasture under set-stocking conditions.

Session 52 Theatre 15

Amino acid intake of hay-based diets and the relationship with milk production and urea content

M. Simoni[1], R. Pitino[1], T. Danese[1], G. Mantovani[1], E. Tsiplakou[2] and F. Righi[1]
[1]University of Parma, Department of Veterinary Science, Strada del taglio, 10, Parma, 43126, Italy, [2]Agricultural University of Athens, Laboratory of Nutritional Physiology and Feeding, Department of Animal Science, School of Animal Bi, Iera Odos, 75, Athens, 11855, Greece; giorgia.mantovani@unipr.it

This study aimed at evaluating the effects of amino acid intake (dAA) of Holstein dairy herds' hay-based diets on milk yield (MY) and composition. A total of 20 dairy herds located in the Parmigiano-Reggiano area were involved in the trial (185±89 cows; 678±33 kg BW). The diets were collected in 2 consecutive d, their composition and AA profile was determined. At the same time, dry matter intake (DMI) and MY were measured and 2 bulk milk samples were collected every 12 h during the 2nd monitoring day. The diets were on average 15.1±0.7% CP, 38.1±3.3% NDF and 21.8±2.2% starch (DM basis). The dAA were determined after acid hydrolysis, except for sulphur AA which were pre-oxidized with performic acid. Barium hydroxide was used for Trp. The procedure was optimized for sample homogenization and substrate to solvent ratio. Samples were analysed by RP-UPLC/ESI-MS in the SIR scan mode. The AA content was estimated using the internal standard method (Trp), and an external calibration curve (all other AA). Casein, fat and urea yield (CY, FY, UY) were determined by MilkoscanTM, milk total protein yield (CPY) was determined by Dumatherm. Correlations were tested between dAA and milk composition parameters using SPSS v.28. The CPY was related to Ala, Val, Thr, Ile, Leu, Asp, Glu, Arg (g/d; r=0.60, r=0.47, r=0.44, r=0.59, r=0.47, r=0.50, r=0.51, r=0.62, P<0.05); ECM to Pro, Glu, Arg and Met (g/d, r=0.45, r=0.49, r=0.46, r=0.52, P<0.05). The milk nitrogen efficiency was correlated with Val intake (g/d, r=-0.47, P=0.04). The FY was correlated with Pro, Ile, Leu, Glu, Arg, Met, Trp (g/d, r=0.47, r=0.50, r=0.48, r=0.54, r=0.46, r=0.59, r=0.50, P<0.05). The UY was found to be affected only by Arg (g/d, r=0.47, P=0.03). This project received funding from the European Union's Horizon 2020 research and innovation program, under grant agreement No 777974.

Session 52 Poster 16

Does acidification affect urinary creatinine in dairy cattle?

T. Danese, M.C. Sabetti, M. Simoni, R.G. Pitino, G. Mantovani and F. Righi
Università di Parma, Dipartimento di Scienze Medico-Veterinarie, Strada del taglio 10, 43126, Italy; tommaso.danese@unipr.it

Creatinine is a marker commonly employed to quantify the amount of urine produced by dairy cattle. Combined with urine nitrogen content, the latter allows to measure the urinary nitrogen excretion which is needed when nitrogen balance must be calculated. However, when urine is sampled, a quota of ammonia nitrogen tends to volatilize potentially leading to an underestimation of the content of this element. This issue is normally prevented by sample acidification with sulphuric acid which, along with diluting the sample, could chemically alter the creatinine content, leading to bias in N excretion quantification. The objective of this trial was to evaluate the impact of urinary samples acidification on creatinine concentration and analytical detection in dairy cattle urine. A total of 60 urine samples were collected from lactating dairy cows and randomly divided in three groups (n=20), and immediately treated through acidification with H_2SO_4 to obtain pH<2 (group 1), addition of the same volume of distilled water to analyse the dilution effect (group 2) or stored without acid nor water (group 3). Creatinine analyses were conducted with the Jaffe colorimetric method. Kruskal-Wallis test was used to compare urine groups for treatments (P<0.05). Bland-Altman's test was used to analyse inter-assay agreement between measurements (group 1 vs group 3). Results show that urinary creatinine it is not statistically different between groups 1 (median 48.5 mg/dl; range 36.9-83.0), group 2 (median 47.5 mg/dl; range 36.5-80.7) and group 3 (median 48.9 mg/dl, range 37.2-84) (overall P=0.7). Bland-Altman's analysis showed agreement between the standard method (group 3) and the experimental method (group 1). The measurement of urinary creatinine by Jaffe method is not influenced by sample acidification; therefore, the use of creatinine as a marker for total urine output is still to consider viable under acidification of the samples. This project has received funding from the European Union's Horizon 2020 research and innovation programme, under the grant agreement No 777974.

Effect of an alternative to sodium bicarbonate on performance of dairy cows

V. Leroux, C. Jaffres, A. Budan and N. Rollet
Neolait, Cargill Animal Nutrition, Voie de la ville Louze, 22120 Yffiniac, France; noemie_rollet@cargill.com

Adding sodium bicarbonate (SB) to the diet of dairy cows is a common practice to prevent digestive disorders induced by diets dense in fermentable energy. Acidomix (AC) is a blend of mineral ingredients that have shown to increase rumen pH *in vitro* from 6 to 24 h compared to sodium bicarbonate in a 1:1.5 ratio. A trial was conducted to evaluate the effect of AC compared to SB in 1:1 and 1:1.5 ratios on the performance of dairy cows fed with a diet based on corn silage containing 47.7% fermentable energy. Compared to SB in a 1:1 ratio, AC decreased milk protein (-0.02%, $P<0.01$), improved body condition score (+0.40 during 45 days, $P<0.01$) and margin over feed cost (+0.06€/c/d, $P=0.05$). In a 1:1.5 ratio, milk fat was increased (+0.23%, $P=0.02$) with AC. Dung pH was higher for AC compared to SB with the 1:1 ratio (+0.08, $P<0.01$) and with the 1:1.5 ratio (+0.09, $P<0.01$). These results suggest that compared to BC, AC is a better economical solution with a 1:1 ratio and increase milk fat with a 1:1.5 ratio.

Evaluation of three measuring methods for the ammonia concentration for practical use in pig houses

J. Witt[1], J. Krieter[1], K. Schröder[1] and I. Czycholl[2]
[1]Institute of Animal Breeding and Husbandry, Olshausenstr. 40, 24098 Kiel, Germany, [2]Department of Veterinary and Animal Sciences, University of Copenhagen, Grønnegårdsvej 15, 1870 Frederiksberg C, Denmark; jwitt@tierzucht.uni-kiel.de

Ammonia negatively affects pig health and productivity. A value of 20 ppm should therefore not be exceeded. Thus, a simple and reliable control of the ammonia concentration (AC) is necessary. This study examines the use of two hand-held measuring devices, the ammonia meter 'basic ammonia meter' (AM) and the 'Dräger accuro®' gas detection pump (DA) as an alternative to the continuously measuring sensor 'DOL53 Ammonia Meter' (DOL) and the factors that influence these measuring methods. The AC on three different farms (farm 1: organic; farm 2,3: conventional) over three to four batches from farrowing to the end of fattening were measured weekly with the AM (n=284) and the DA (n=142). In addition, the DOL (n=77,170) measured the AC every 15 minutes and showed the daily course of the AC. The influence of the farm, which was investigated with the use of a linear mixed model, on the AC was significant ($P\leq0.05$) for all three measurement methods. The freely ventilated farm 1 showed the lowest AC (mean DOL 0.69 ppm), which could be caused by a lower animal density and lower stable temperatures compared to the mechanically ventilated farms 2 (mean DOL 12.4 ppm) and 3 (mean DOL 10.6 ppm). The season was a further significantly ($P\leq0.05$) influencing factor for the mechanically ventilated farms 2 and 3, which is reflected in higher AC in the winter compared to the summer months. This could be due to a lower ventilation rate, due to the lower outdoor temperatures in winter. The DOL was also influenced by the barn, higher AC were measured as the production stage of the pigs progressed, with highest values in the fattening barn (farm 2 up to 60.8 ppm, farm 3 up to 58.5 ppm). Moreover, the DOL showed an increase of AC by feeding and activity phases of the pigs. The comparison of the AM and DA with the DOL did not achieve acceptable or good agreements (limits of agreement >5% or 10% difference, respectively). It can therefore be concluded that the hand-held meters cannot replace the sensor under practical conditions.

The information and communication technologies in livestock production – expectations and concerns

S. Opalinski[1], K. Olejnik[1], E. Popiela[1], A. Jankowska-Makosa[2], D. Konkol[3], M. Korczynski[3], D. Knecht[2], R. Kupczynski[1], I. Tikasz[4] and T. Banhazi[5]

[1]Wroclaw University of Environmental and Life Sciences, Department of Environmental Hygiene and Animal Welfare, Chelmonskiego 38C, 51-630, Poland, [2]Wroclaw University of Environmental and Life Sciences, Institute of Animal Husbandry and Breeding, Chelmonskiego 38C, 51-630 Wroclaw, Poland, [3]Wroclaw University of Environmental and Life Sciences, Department of Animal Nutrition and Feed Management, Chelmonskiego 38C, 51-630 Wroclaw, Poland, [4]Institute of Agricultural Economics Nonprofit Kft., Zsil u. 3-5, 1093, Budapest, Hungary, [5]AgHiTech LTD, Budapest, Hungary; sebastian.opalinski@upwr.edu.pl

The use of digital innovations in the animal production sector enables the prediction of animal diseases in advance (animal welfare improvement), optimises production processes, develops the environmental sustainability of farms, and reduces the need for manual labour. Despite all these positive aspects resulting from implementing information and communication technologies (ICT) in livestock production, farmers are not open to new technologies. The main goal of the LivestockSense project was to identify the barriers to introducing modern technologies and to support farmers through education and mentoring. The focus group discussion (FGD) with the representatives related to the pig and poultry sector (scientists, policymakers, technology providers, ICT developers and farmers) has been used to identify critical characteristics of farmers' attitudes towards ICT tools. Participants of the FGD stated that ICT tools are a time and money-saving solution, enabling production optimisation and filling staff shortages. These are the main motivations and purposes for using smart technologies. However, the main barriers to ICT adoption were the high financial risk, especially for small farms. Among solutions that might motivate the spread of ICT tools, the most frequently mentioned were subsidies on the cost of installation and maintenance, governmental support programs, pressure from the side of food contractors (fast food chains or supermarkets) and demonstration farms. Research was funded by the National Centre for Research and Development (NCBR), no. ICTAGRIFOOD/I/LIVESTOCKSENSE/01/21, within the EU's Horizon 2020, No 862665 ERA-NET ICT-AGRI-FOOD.

Implementation of a deep learning based system for monitoring farrowing in sows

M. Wutke[1], C. Lensches[1], A. Holzhauer[1], M.A. Lieboldt[2] and I. Traulsen[1]

[1]Georg-August University Göttingen, Animal Sciences, Albrecht-Thaer-Weg 3, 37075 Göttingen, Germany, [2]Chamber of Agriculture Lower Saxony, Division Agriculture, Mars-la-Tour-Straße 6, 26121 Oldenburg, Germany; martin.wutke@uni-goettingen.de

Monitoring the farrowing process plays an important role in pig production systems, as complications during parturition pose a potential risk to the sow's health status as well as to the newborn piglets. In this study, we propose a system using computer vision and deep learning techniques to monitor the farrowing process in temporarily crated and free farrowing sows. By first determining the sow's orientation and a potential maternity area, the times of birth can then be determined by detecting newborn piglets in the target area. In addition, by analysing the locomotion behaviour of the detected piglets, an early assessment of their respected health status can be computed. We evaluated the performance of our framework using a two-stage approach and two distinct test sets for the detection of target objects and the subsequent identification of critical events during farrowing. As a result, we achieved a Precision, Recall and MAP score of 0.982, 0.989 and 0.993, respectively, for our object detection model and an Accuracy, Recall and Precision score of 0.9, 0.8 and 1, respectively, for the corresponding determination of important events. The findings of this study demonstrate the potential of artificial intelligence systems and computer vision techniques as a useful addition to the farm management process and could contribute to animal welfare through the early identification of critical situations during farrowing.

Session 53 Theatre 4

Who's biting? Detecting pig screams for identifying tail biting events

P. Heseker[1,2], T. Bergmann[3], M. Scheumann[3], S. Ammer[1], I. Traulsen[1], N. Kemper[2] and J. Probst[2]
[1]Georg-August-University Göttingen, Department of Animal Sciences, Livestock Systems, Albrecht-Thaer-Weg 3, 37075 Göttingen, Germany, [2]University of Veterinary Medicine Hannover, Foundation, Institute for Animal Hygiene, Animal Welfare and Farm Animal Behaviour (ITTN), Bischofsholer Damm 15, 30173 Hannover, Germany, [3]University of Veterinary Medicine Hannover, Foundation, Institute for Zoology, Bünteweg 17, 30559 Hannover, Germany; philipp.heseker@uni-goettingen.de

Early identification of tail biting and intervention are necessary to reduce tail lesions and its impact on animal health and welfare. Removal of biters has become an effective intervention strategy, but finding them can be difficult and time-consuming. The aim of this pilot study was to investigate if tail biting could be identified by detecting pig screams. The study included 288 undocked weaner pigs that were housed in six pens in two batches. Pigs had coloured ear-tags and each pen was equipped with a video camera and a microphone. Individual tail examinations were carried out biweekly. Once a tail biter (n=7) was identified and removed by the farm staff, the previous days of video recordings were analysed for pig screams (sudden increase in loudness with frequencies from 1 to 4 kHz) and tail biting events until no biting was observed. In total 2,846 screams were detected in four tail biting pens, from which 53.6% were caused by tail biting, 24.9% originated from other pens, 8.7% were not assignable and 12.8% occurred due to other pig manipulation. Biters were identified between 1 and 9 days prior to their removal from the pen. Tail lesions decreased up to 52.3% one week after removal. Detecting pig screams has the potential for identifying tail biting events. Earlier removal of biters is possible and can reduce tail lesions and thus increase animal health and welfare. Possibilities of an automatic evaluation will be analysed in the next steps. The study is part of the project DigiSchwein, which is supported by funds of the Federal Ministry of Food and Agriculture (BMEL) based on a decision of the Parliament of the Federal Republic of Germany. The Federal Office for Agriculture and Food (BLE) provides coordinating support for digitalisation in agriculture as funding organisation, grant number 28DE109E18 and 28DE109F18.

Session 53 Theatre 5

Veterinarians' perceptions of using PLF technologies in pig husbandry in the Netherlands and Germany

M.F. Giersberg and F.L.B. Meijboom
Utrecht University, Veterinary Medicine, Yalelaan 2, 3584 CM Utrecht, the Netherlands; m.f.giersberg@uu.nl

As important players in livestock production, veterinarians are confronted with emerging technologies in the field, such as precision livestock farming (PLF). PLF allows for automated monitoring of animals and related environmental processes on commercial farms. Despite their potential for improving animal health and welfare, these technologies may attract public criticism, for instance through substituting human care and transforming the human-animal relationship. The aim of this study is to make sense of how veterinarians perceive the use of PLF in the context of current pig production. Therefore, semi-structured interviews were carried out with pig veterinarians located in the Netherlands and Germany. The interview data were analysed by using an inductive and semantic approach to reflexive thematic analysis. Based on the developed themes, we draw the following main conclusions: (1) Veterinarians agree on broad concepts, such as defining their own role as advisory and framing PLF technologies as supporting tools. What these concepts entail in detail, however, varies within the group of pig veterinarians. (2) The veterinarians' elaborations on relationships among different stakeholders, including themselves, show that they share positions with different groups of society. They are aware of and reflect on competing interests of these groups. (3) Some veterinarians see the development and provision of PLF as part of their professional activities. This demonstrates that they play an active role in the emerging field of sensor technologies in pig production. (4) The veterinarians also have a clear idea of the chances and challenges PLF may pose. The use of PLF technologies in pig husbandry must not ignore the present distance between agriculture and society. According to the veterinarians, PLF should not reinforce the misconceptions resulting from this distance. Applying PLF to improve pig welfare, a topic in which part of society is interested, is seen as a starting point to reconcile the sector and the public. However, external obstacles, such as financial dependencies, may prevent veterinarians to take an active stance in this process in practice.

Automatic detection and quantification of ear biting in pigs

A. Odo[1], R. Muns[2], L. Boyle[3] and I. Kyriazakis[1]
[1]Institute for Global Food Security, School of Biological Sciences, Queen's University Belfast, 19 Chlorine Gardens, BT9 5DL, United Kingdom, [2]Agri-Food and Biosciences Institute, 18a Newforge Ln, Belfast, BT9 5PX, United Kingdom, [3]Teagasc, Teagasc, Animal & Grassland Research Centre, Moorepark, Fermoy, Co. Cork, P61 C996, Ireland; a.odo@qub.ac.uk

Ear biting is considered a main risk factor for ear necrosis, a growing and highly prevalent problem in weaned pigs associated with health and welfare challenge, and poor performance in pig production systems. We aimed to develop an automated method for the quantification of ear biting. We collected data samples comprising videos from two farms (experimental and commercial) with diverse management and monitoring settings. The data samples were from different pens and at various times of the day to ensure good representation of pen conditions and scenarios. Video frames with ear biting events were manually annotated. Behaviours targeting the ear of pen mates (gentle manipulation, quick bites, ear pulling, and chewing) were labelled as ear biting. We defined the region of interest using a bounding box that overlay the heads of the interacting pigs. The size of the bounding boxes varied as the orientation of the biter and bitten pig change. We used a deep learning-based object detection network (YOLOv4) to localize these regions and track them over multiple frames. Multiple objects trackers namely the Centroid and DeepSORT associate detected instances to event episodes. The association of events enabled the computation of frequency and duration of ear biting outbreaks. YOLOv4 detected ear biting at average precision (AP@0.5) of 0.98 and 0.86 on the respective datasets, and 0.92 on the combined dataset. Results showed that automated detection and tracking of ear biting was possible. There are 2,113 events on the test set. The Centroid tracker identified 2,040 events but also resulted in higher false-alarm rate of 34% compared to the DeepSORT-based system (14%). Our method quantified ear biting at group-level which means that frequency and duration of ear biting was not considered for individual performer. Identifying individual bitter will facilitate fine-grained intervention such as isolation of the offender. We consider this as the first step in the development of an on-farm early warning system for ear biting detection.

Installation modified carbon felt in pig house to control airborne pathogenic microorganisms

X.D. Zhao, F. Qi, H. Li and Z.X. Shi
China Agricultural University, College of Water Resources & Civil Engineering, No.17, Tsinghua East Road, Haidian District, 100083, Beijing, China, P.R.; zhaoxd723@cau.edu.cn

The pathogenic microorganisms in the air have a significant impact on pig health and even biosecurity of pig industry. Carbon material shows great potential in microbial adsorption to reduce the misuse of chemical disinfectants in pig houses. Herein, modified carbon felt was selected for installation in pig houses to control the levels of airborne microorganisms. The distribution law and initial level of air microorganisms in the pig house were determined first, and carbon felts were placed in the areas with high microbial levels in different ways. High-throughput pyrosequencing of the 16S rRNA was used to detect bacterial composition in the pig house before and after the work of carbon felts. The results show that the modified carbon felt was competent to adsorb several pathogenic bacteria commonly found in pig farms, such as *Escherichia coli*, *Staphylococcus aureus*, *Salmonella cholerae*, and *Clostridium perfringens*. It can also achieve a 90.2% removal of mixed flora. The control efficiency to air microorganisms was higher at 2×2 m coverage area per carbon felt in the pens. Specifically, the relative abundance of *Enterobacteriaceae*, *Streptococcaceae*, and other pathogenic bacteria showed a progressively decreasing trend over 18 d. This study provides a promising strategy to control air microorganisms based on modified carbon felt, thus enhancing biosafety and animal welfare in pig farms.

Session 53 Theatre 8

Benefit of caliper use at insemination on different genetic types: impact on farrowing performances

C. Teixeira Costa, C. Chevance, T. Nicolazo, G. Boulbria, V. Normand, J. Jeusselin and A. Lebret
Rezoolution, ZA De Goheleve, 56920 Noyal-Pontivy, France; t.nicolazo@rezoolution.fr

Backfat thickness (BFT) is widely measured and used to adapt the feed ration of sows during gestation to increase the sustainability of pig production and optimize performances. Recently, a tool called Sow Caliper (SC) was developed to assess the physical structure of sows. The aim of this study was to evaluate the interest of the SC to classify the sows according to their body condition to optimize performances and reduce worktime. The SC was tested on six farms of different genetic types (mean of 284 sows per farm). Five categories (C1 to C5) were used to classify the 1,701 measurements realized at insemination. In parallel, BFT of the same sows was measured with an ultrasonic tool (Renco Lean-Meater®). To determine an optimal category of SC measurement, all farrowing performances were collected. Category means were compared for each genetic type. Correlation coefficients between the measurements with the SC and the BFT were calculated by Spearman test correlation and the effects of the SC category on the farrowing performances by litter rank were tested by analysing the variance otherwise using a non-parametric test (R Studio). The SC measurements were positively correlated with BFT, especially for sows of parity 3 or more ($r=0.65$, $P<0.001$). The correlation was weaker for parity 1 and 2 ($r=0.41$, $P<0.001$). Based on performances at next farrowing, category C3 was considered as ideal for sows of parity 3 or more (on average, approximatively +0.5 total born and -0.13 stillborn than other categories). On the opposite, category C4 was optimal for gilts and to a lesser extent for sows of parity 2, depending on the genetic type (for example, for gilts +0.5 total born). For gilts, SC measurement was a better predictor of farrowing performances than BFT alone. According to these results, under certain conditions, SC measurements at insemination can replace BFT particularly for older sows. However, especially for gilts, combining SC and BFT measurements seems more appropriate.

Session 53 Theatre 9

Combined effect of genetics, gut microbiota and environment on vaccine responses and welfare in hens

A. Lecoeur[1], F. Blanc[1], D. Gourichon[2], N. Meme[2], T. Burlot[3], V. Guesdon[4], V. Ferreira[5], L. Calandreau[5], L. Warin[6], F. Calenge[1] and M.H. Pinard Van-Der-Laan[1]
[1]Université Paris-Saclay, INRAE, AgroParisTech, GABI, GeMS, 78350 Jouy-en-Josas, France, [2]INRAE, PEAT, 37380 Nouzilly, France, [3]Novogen, 5 rue des compagnons, 22960 Plédran, France, [4]Comportement Animal et Systèmes d'Elevage, JUNIA, 59000 Lille, France, [5]CNRS, IFCE, INRAE, Université de Tours, PRC, 37380 Nouzilly, France, [6]ITAVI, ITAVI, 37380 Nouzilly, France; alexandre.lecoeur@inrae.fr

Vaccination is one of the most effective strategies for preventing infectious diseases. However, most vaccines vary in their efficacy. Our objective was to investigate the effect of genetic background, caecal microbiota and housing on vaccine responses and on welfare. We studied a cohort of 400 animals from 2 commercial laying hen lines from the Novogen breeding company: Rhode Island Red (RIR) and White Leghorn (LEG). To study the impact of the microbiota, we modified its composition by giving hens a cocktail of three antibiotics. To study the housing, half of the animals had an access to an outdoor yard from 12 weeks to the end of the experiment while the others not. The vaccine responses were monitored by an indirect ELISA test for 3 vaccines and by hemagglutination inhibition assays for one of them. The behaviour was studied by group evaluation (image analysis and EBENE® method), individual behavioural evaluations and RFID tagging to record animal exits. The caecal microbiota (16S rRNA gene sequences) was performed at the end of the experiment and on subsets of animal batches at 4 time points to study microbiota dynamics through time. For all the vaccines tested, RIR harboured higher antibody levels compared to LEG with a longer persistence. Microbiota perturbation and rearing conditions also moderately altered vaccine responses. Behaviours differed according to the line: LEG individuals were more exploratory and expressed more resting behaviours, whereas RIR individuals expressed more behaviours related to fear and vigilance. The microbiota perturbation also affected the exploratory behaviour, and differently according to the line. These results confirm the triple impact of genetics, microbiota and rearing environment on the vaccine responses and the hen behaviour.

Using triaxial accelerometers to monitor peripartum behaviour of Purebred Spanish mares

M.J. García García, F. Maroto Molina, C.C. Pérez Marín and D.C. Pérez Marín
Universidad de Córdoba, Department of Animal Production, Campus de Rabanales. Ctra. Madrid-Cádiz km. 396, 14071 Córdoba, Spain; g42gagam@uco.es

Behavioural monitoring can provide relevant information on horse health and welfare, as well as on the occurrence of events of interest, such as foaling or oestrus. Direct observation of animal behaviour requires lots of time and effort and often involves subjective decisions. The use of accelerometers to monitor behaviour limits the influence of human presence and makes it possible to record unusual behaviours that would otherwise be difficult to observe. But reliable classification models are needed to convert raw acceleration data into meaningful behaviour classifications. The aim of this study was to develop algorithms able to automatically monitor mare behaviour from acceleration data. Tri-axial accelerometer collars gathering data at 10 Hz were used on 9 Purebred Spanish mares around foaling. Video recordings and BORIS software were used to tag mare behaviour as: 'standing-head down', 'standing-head up', 'lying on belly' and 'lying on side'. Data were processed and analysed using the statistical software R and the package rabc. A total of 26 features were calculated from acceleration data, which were divided into 20-second fragments. A combination of filtering and wrapping methods was used to select the best subset of features. Finally, an XGBoost supervised machine learning model was trained with the selected features. The best subset of features was: mean acceleration on the X axis, mean of absolute values on the Y axis, mean angle on the Y axis, maximum acceleration on the Y axis, variance on the Z axis, variance on the Y axis and standard deviation on the X axis, with XYZ being the longitudinal, transverse, and perpendicular axes to the neck of the mare. The classification model obtained presented an overall accuracy of 82.7%. In terms of class statistics, the model obtained a F1 score of 0.86 for standing-head down, 0.82 for standing-head up, 0.76 for lying on belly and 0.94 for lying on side. The model obtained was able to classify mare behaviour appropriately and may be used to detect behavioural changes around reproductive events, such as foaling.

Relationships between direct and indirect genetic effects of RFI and feeding behaviour traits

M. Piles, M. Mora, M. Pascual and J.P. Sánchez
IRTA, Torre Marimon, Caldes de Montbui, Barcelona, 08140, Spain; miriam.piles@irta.cat

Records from electronic feeders (EF) allow for the definition of new phenotypes that can be used to improve the response to selection for feed efficiency, as well as the behaviour of the animals in the group towards less competitive/aggressive animals. By fitting a set of two-traits social animal models to residual feed intake (RFI) and one feeding behaviour trait (FB) we aimed at identifying: (1) FB traits whose direct genetic effects (d) are genetically associated with d of RFI and can be used to improve the prediction of genetic evaluations for this trait; (2) FB traits whose d are related to the indirect genetic effects (s) of RFI and therefore can be considered as part of the mechanisms involved in the effects that an animal exerts on the RFI of its group mates. Data came from 1,086 rabbits of a line selected for growth rate under feed restriction at fattening (35 to 59 d). They comprised records of body weight at 35 and 59 d, feed intake and 15 FB variables: number of visits to the EF, occupation time, consumption, feeding rate and ratios indicating how those FB traits aggregated per hour are distributed between cage mates. The models included the effects of sex, batch, parity order, litter size at birth, litter and cage, as well as d and s effects. Correlations between d of RFI and ratio traits were very high (range=0.89 [0.04] to 0.93 [0.03]) as it was the correlation with the mean number of meals (0.70 [0.16]), number of visits per meal (0.56 [0.23]) and time interval between meals (-0.66 [0.18]). These correlations indicate that, genetically, the most efficient rabbits are those that eat fewer meals, wait more time between meals, eat the least amount of food consumed in an hour in the cage and occupy the EF the least. The variables whose d were most correlated with s of RFI were the variables related to the distribution of time, feed and feeder visits occurring in each hour between cage mates (-0.85 [0.08] to -0.90 [0.05]), indicating that animals that occupy the EF longer, eat the most feed, and visit the EF more times than their cage mates are the ones that cause their cage mates to be more efficient. Thus, there is a substantial genetic variation on FB traits which is correlated with d and s effects of feed efficiency.

Session 53 Theatre 12

Audio-based event detection and health monitoring in poultry

F. Hakansson and D.B. Jensen
University of Copenhagen, Department of Large Animal Sciences, Groennegaardsvej 2, 1870, Denmark; fh@sund.ku.dk

With the continuous growth of intensive livestock production systems, health and welfare impairments are increasing challenges in poultry production. Timely identification of developing challenges can enable farmers to take immediate management actions, thus limiting potential negative implications. Previous research has shown that audio data can be utilized as biomarker for health conditions in poultry, but studies and applications of bioacoustics monitoring in intensive livestock systems are rare. This study aims at developing a tool for automated and continuous monitoring of audio data collected within poultry barns, to detect audible signs of stress and disease, such as sneezing, alarm calls or others sounds. This study will utilize video and audio data collected at a single poultry production site, and signal processing and machine-learning models will be implemented to detect and count instances of undesired sounds/ noises. In an initial step, it will be explored if transfer learning from existing models to detect coughing in pigs can be utilized to detect signs of respiratory diseases in poultry. Furthermore, combinations of signal processing models such as wavelet filtering, or machine-learning models accounting for temporal features in the data (such as long-short term memory- LSTM models) will be scrutinized to create novel models to detect early signs of diseases or stress. This work was funded by Code Re-farm Horizon 2020 project: 'Consumer-driven demands to reframe farming systems' (Grant No: 101000216; https://coderefarm.eu/wp-coderefarm/).

Session 53 Poster 13

First approach of using sows' water consumption data to detect the onset of farrowing

J. Probst[1], N. Volkmann[1], C. Lensches[2], P. Heseker[2], G. Thimm[3], M. Lieboldt[4], I. Traulsen[2] and N. Kemper[1]
[1]University of Veterinary Medicine Hannover, Foundation, Institute for Animal Hygiene, Animal Welfare and Farm Animal Behaviour (ITTN), Bischofsholer Damm 15, 30173 Hannover, Germany, [2]Georg-August-University Göttingen, Department of Animal Sciences, Livestock Systems, Albrecht-Thaer-Weg 3, 37075 Göttingen, Germany, [3]Johann Heinrich von Thünen Institute, Institute of Agricultural Technology, Bundesallee 47, 38116 Braunschweig, Germany, [4]Chamber of Agriculture Lower Saxony, Mars-la-Tour-Str. 6, 26121 Oldenburg, Germany; jeanette.probst@tiho-hannover.de

To offer sows individual and timely support at parturition, the present study aimed to predict the onset of farrowing by measuring the sows' water consumption (WC). In a first approach, the hourly WC of five sows (parity 2-5, mean=3.6) was recorded by a magnetic-inductive waterflow meter. WC was analysed 96 hours before and 72 hours after start of farrowing (determined by video analyses) using the sows' individual cumulative sum (CUSUM) control chart. Daily WC ranged between 18.7 and 91.4 litres. All sows retrieved the highest amount of water 24 hours before and the lowest 24 hours after the hour the first piglet was born. Analysing each CUSUM control chart with allowance value (k=0.1) and individual smoothing value (h=4.1 to 6.5), the onset of farrowing was detected five to nine hours before it started by analysing sows' hourly WC. Further investigations will be carried out to analyse a larger data set and consider potential influences on animals' WC (e.g. fixation, health problems). This first study on using WC data to detect the onset of farrowing in sows shows the potential of this approach. The study is part of the project DigiSchwein, which is supported by funds of the Federal Ministry of Food and Agriculture (BMEL) based on a decision of the Parliament of the Federal Republic of Germany. The Federal Office for Agriculture and Food (BLE) provides coordinating support for digitalisation in agriculture as funding organisation, grant number 28DE109E18.

Session 53 Poster 14

Social network analysis of cattle and horses inferred from sensor ear tag (SET) and GPS based data

U. Heikkilä[1], K. Ueda[1], N. Gobbo Oliveira Erünlü[2], M. Baumgartner[3], M. Cockburn[3], I. Bachmann[3], M. Roig-Pons[3] and S. Rieder[1]

[1]Identitas AG, Staffacherstrasse 130A, 3014 Bern, Switzerland, [2]Agroscope, Posieux, 1725 Posieux, Switzerland, [3]Agroscope, Avanches, 1580 Avenches, Switzerland; ulla.heikkilae@identitas.ch

Solar powered sensor ear tags (SET) have been applied on cattle and horses within an accompanying study. Animal social networks give insights of an individual's social integrity within a herd and thus give indications on its welfare state. As energy supply is a limiting factor in animal sensors, it is of interest to see whether social networks can be detected with the low temporal reporting frequency data of the SET. This study thus attempts to analyse the social structures of the herd by building a contact network of the animals by using the location data delivered by the SET (up to 4 data points per day). We define a 'contact' where two animals have dwelled within 5 meters of each other within 1 hour. These constrictions are fairly loose, being restricted by the data transfer frequency of the ear tags. These contacts are then assessed over the period of 3 months and a network is built. Most contacts occur on feeding at a shared haystack and could be coincidental. We therefore conduct a verification study on horses that are equipped with both the SET and a GPS tracker with a recording frequency of 10 s during a period of one month. The feeding spots are ignored in the verification step. Furthermore, we observe the animals in both studies in the field to compare the theoretical networks with the actual behaviour. The main features of the contact network can be confirmed and show that the social contacts in the herd follow a fairly fixed structure.

Session 54 Theatre 1

Modelling adaptation strategies to climate change in Mediterranean small ruminant systems

A. Lurette[1], S. Lobón[2], F. Douhard[3], M. Blanco-Alibes[2], D. Martin-Collado[2], A. Madrid[4], M. Curtil-Dit-Galin[1] and F. Stark[1]

[1]UMR SELMET, INRAE, Campus de La Gaillarde, 2 place Pierre Viala, 34060 Montpellier, France, [2]CITA, Avda. Montañana 930, 50059 Zaragoza, Spain, [3]UMR GENPHYSE INRAE, 24, chemin de Borde-Rouge - Auzeville Tolosane, 31321 Castanet Tolosan, France, [4]IDELE, CS 52637, 31321 Castanet Tolosan, France; amandine.lurette@inrae.fr

Mediterranean livestock farming systems have specific characteristics, such as heterogeneity of animals, diversity in land use and flock mobility, that make them particularly sensitive but also adaptive to climatic hazards. Combining resilient herds and an efficient use of various feed resources is central to develop adapted livestock farming systems to climate change. In the PRIMA AdaptHerd project, this study aims at evaluating the multi-level implications of adaptation levers that can be mobilized by Mediterranean small ruminant farmers. An approach which combined a participatory design of adaptation strategies and a simulation of these strategies on four Mediterranean French and Spanish farm types was used. The farm types differed in their level of herd intensification and in their feeding practices. For the four contrasting situations, groups of experts were consulted to design: (1) the projected impact of climate change on vegetation; and (2) several adaptation strategies. The resulted adaptation levers to climate change were different between farm types and flock management (sedentary or transhumant). We therefore tested the effects of three levers: (1) increasing the part of pastoral area; (2) shifting the grazing periods; and (3) decreasing the number of lambings or the age at first mating to better match with resources availability. The assessment of the effect of adaptation levers on livestock systems was based on their forage autonomy, grazing rate and the number of days when needs are not covered. The simulation of the designed adaption levers showed that they allowed the impacts of climate change to be mitigated, but they did not allow a return to an initial situation with needs covered all year round and a high autonomy and grazing rate.

Session 54 — Theatre 2

Feed365: Early findings for sheep grazing novel year-round forage systems

C.E. Payne, D. Real, A. Loi and C. Revell
Department of Primary Industries and Regional Development WA, Nash Street, East Perth WA, 6000, Australia; claire.payne@dpird.wa.gov.au

Grazing systems in southern Western Australia are challenged by climate change with increasingly hotter, drier, and more variable seasons. New forage systems need to be identified and re-designed to sustain livestock grazing all-year-round with limited supplementary feeding. Grazing plots (n=48) of 0.5 ha each are spread across 2 paddocks of varying soil types at the Katanning Research Station. Sheep (merino hoggets) grazing time and density is dependent on total forage mass, estimated pasture growth rate and type of forage available. In winter and spring sheep are grazing plots that contain mostly annual legumes and grasses, whilst in summer and autumn sheep are grazing dry residues, winter crop stubbles, summer crops, and perennial legumes and grasses. A total of 20 forage/ forage mixes were trialled consisting of 30 different forage species. Sheep are weighed and condition scored every three weeks whilst grazing plots. Sheep are not given any supplementary feeding whilst grazing plots. During the 2022 winter spring period (June-November) sheep on average gained weight and gained or maintained condition on all plots. The exception being one plot of dry slashed oat stubble where sheep lost between a quarter and half of a condition over the period of early October to late November. During the 2021/2022 summer autumn period sheep also gained or maintained weight on all grazing treatments. Condition was maintained on most plots during the summer months (January-February) however lost condition in the autumn months (March-April), which was expected during the first establishment year. The ability to graze and monitor sheep performance on multiple forage types and mixes will identify forage systems that can successfully maintain livestock productivity all-year-round in a changing climate whilst minimising supplementary feeding.

Session 54 — Theatre 3

Quantifying responses to hot conditions in divergent sheep breeds in South Africa

S. Cloete[1,2], S. Steyn[2], J. Van Zyl[2] and T. Brand[1,2]
[1]Western Cape Department of Agriculture, Directorate Animal Sciences, Private Bag X1, 7607, South Africa, [2]Stellenbosch University, Private Bag X1, 7602, South Africa; schalkc2@sun.ac.za

Maximum temperatures are expected to increase in future. Although the ovine species is generally resilient to high temperatures, it is important to know the extent of breed differences in heat tolerance. Therefore, 10 ewes each of 2 terminal sire breeds (Dormer and Ile de France), 2 wool breeds (Merino and Dohne), a dual-purpose breed (SA Mutton Merino or SAMM), 2 composite breeds with indigenous content (Dorper and Meatmaster) and an unselected indigenous fat-tailed breed (Namaqua Afrikaner) were studied for their short-term responses to heat. The former 5 breeds were all from temperate regions. Rectal temperature and respiration rate of the ewes, as indicators of heat stress, were recorded over three days, involving 3 morning and 3 afternoon sessions. The mornings were appreciably cooler than the afternoons with a mean thermo-humidity index of 19.8, as compared to 29.1 during the afternoons. Rectal temperature ranged in a narrow band of 38.5°C (SAMM) to 38.8°C (Dohne and Meatmaster) in the mornings. These temperatures increased to between 39.0°C (Dorper) and 39.3°C (Ile de France) in the afternoons. Morning respiration rates of the breeds with indigenous content ranged from 51 (Meatmaster) to 60 (Dorpers). The corresponding range for the temperate breeds were from 77 (SAMM) to 91 (Merino). Afternoon respiration rates were appreciably higher: 132 for the Namaqua Afrikaner, 135 for the Meatmaster, 157 for the Dorper and from 167 (Dohne) to 192 (Dormer and Ile de France) for the temperate breeds. Rectal temperatures were low-moderately repeatable across sessions at 0.16. All between-ewe variance partitioned to a term accounting for the reranking of ewes from the mornings to the afternoons for respiration rate. All breeds maintained their rectal temperature in a relatively narrow band, but the breeds from temperate regions needed a higher respiration rate to do this. If heat stress conditions persist, this additional energy need could compete with energy required for other production functions. It was therefore contended that breeds with indigenous content would adapt better to the anticipated hotter conditions expected in future.

Session 54 Theatre 4

Impacts of heat peaks in France on the performances of dairy goats housed in an insulated roof shed

K. Boissard[1], A. Fatet[1], M. Lambert[2], P. Sales[3] and H. Caillat[1]
[1]INRAE UE FERLus, Les Verrines, 86600 Lusignan, France, [2]Institut de l'élevage, 23 rue Baldassini, 69007 Lyon, France, [3]Chambre d'Agriculture de l'Aveyron, Carrefour de l'Agriculture, 12000 Rodez, France; hugues.caillat@inrae.fr

Thermal comfort of animals and breeders is decisive in ensuring the sustainability of goat and sheep farming systems. Climate projections show that summer periods will be increasingly hot and will lead to changes in feeding and breeding management. The objective of this study is to measure impacts of heat peaks on dairy goats housed in an insulated roof shed. Data were collected from June 1st to September 15th, 2022 in the INRAE FERLus Experimental Unit based in Lusignan (New Aquitaine- 46.43°N, 0.12°E). Monitoring was conducted on 3 batches of animals: 1 group is grazing and breeds during sexual season, 1 group is grazing and breeds during the off-season and 1 group is housed inside year-round and breeds in the off-season. Recorded data included precise description of housing and zootechnical data: feed (distributed and refusal), milk production (collective and individual quantity and quality), watering, sanitary, behaviour, grazing time, panting score (16 goats per batch). Sensors were used to record hourly temperatures and humidity in 2 places in the building, then calculate the temperature hygrometry index (THI). Meteorological data were collected by a 300 m-distant station. In conditions defined like hot (>27 °C), the temperature inside the shed was lower by 1.6 °C than outside, and when the temperature was >36 °C, the temperature inside was lower by 3.2 °C. In neutral condition (<27 °C), the temperature inside was higher by 3.7 °C which means that this building buffers the impact of high temperatures but also when the temperature drops. In 2022, 3 periods of 1 or 2 days with a THI above 77 were observed. Milk production was 6.6% lower during 3 days following the THI>77 period than the preceding 3 days (3.76 vs 3.51 kg/d). This drop was still significant over a period of 5 days. The mean THI was not different (72±2.4) between the 3 or 5 day-periods before or after THI>77, and no differences were observed on the dry matter intake (2.74±0.24 kg/d) or the water consumption (7.3±2.4 l/d). This study will be repeated in 2023 after the addition of a ventilation system.

Session 54 Theatre 5

Cortisol level in sheep wool underwent to three different pasture management

L. Turini[1], A. Ripamonti[1], A. Silvi[1], E. Giua[1], A. Mantino[1], G. Conte[1], F. Bonelli[2] and M. Mele[1]
[1]University of Pisa, Dipartimento Scienze Agrarie, Alimentari, Agro-ambientali, Via del Borghetto 80, 56124 Pisa, Italy, [2]University of Pisa, Dipartimento Scienze Veterinarie, Via Livornese snc, 56122 Pisa, Italy; luca.turini@unipi.it

Cortisol has been used as a biomarker to evaluate different sources of stress in livestock (i.e. heat stress, lameness, water restriction, and tail docking). In sheep, wool cortisol concentration is a more reliable predictor of stress compared to serum cortisol and is not influenced by acute stress as sampling. The aim of this study was to evaluate cortisol concentration in the wool of 20 healthy sheep bred into 3 dairy sheep farms with different characteristics of pasture management (Farm A, B, C). Pasture area of Farm A had no shade allowance, neither drinking water sources, whereas farm B had water sources, but no shade and farm C had both water sources and shade allowance. Twenty animals in early lactation were randomly chosen from each flock and underwent to a clinical examination for health evaluation. Wood samples were monthly collected from March to June (T1, 2, 3, 4, respectively). Wool cortisol was analysed by an ELISA kit. The average temperature humidity index (THI) was recorded for each sampling time and classified as 'absence of heat stress', 'moderate heat stress', 'severe heat stress', and 'extreme severe heat stress'. Data about wool cortisol concentrations showed a non-Gaussian distribution so a Log10 transformation was applied to normalize it. A linear mixed model was used to evaluate the difference in wool cortisol throughout T1 to T4, between Farm A, B, C and the interaction between them. All the farms showed a 'no stress' THI at T1 and T2, 'severe heat stress' at T3 and 'extreme severe heat stress' at T4. The highest wool cortisol levels were recorded at T4, while the lowest at T1 (P<0.001). The effect of season in increasing wool cortisol levels could be related to the rise of THI values from March to June, suggesting that sheep may suffer of heat stress during the warm season. Farm B showed lower wool cortisol concentrations (P<0.001), suggesting that management factors at pasture different to water and shade availability, such as predator attacks and the nutritional quality of the pasture concur to affect wool cortisol levels.

Session 54 Theatre 6

How sheep sectors do face environmental, social and economic issues in French Pyrenees region?

A.-L. Jacquot[1], P.-G. Marnet[2] and Y. Le Cozler[1]

[1]PEGASE, INRAE, Institut Agro, 35590 Saint Gilles, France, [2]Institut, Agro, 35000 Rennes, France; anne-lise.jacquot@agrocampus-ouest.fr

French Pyrenees Mountains are traditional territory for sheep production with two different areas: Pyrenées-Atlantiques dedicated to milk and Hautes-Pyrénées to meat production. As other livestock sectors, the French sheep sector is facing environmental, economic, and social issues (climate changes, limitation of environmental impacts, renewal of farmers, lack of attractiveness, etc.). In present study, we explored the responses and adaptations of this sector to these multiple issues, thanks to a survey performed during fall 2022 by 23 'master 2' students. In practice, they individually surveyed 33 stakeholders located in Pyrenees, including 13 farmers. The interview guide consisted of exploring major themes (environmental impacts, climate changes, remuneration, market influences, farm transfer, sector relations, societal expectations, working conditions, etc.) and aimed at designing to better understand actual situation and precise stakeholders' perceptions concerning the those major issues. Results indicated that sheep systems are considered as environmentally friendly, due low green gas emissions and grasslands-use. Sheep sector has an important role in the patrimonial identity of both territories, mainly through the preservation of pastoral lands and landscapes, highly supported by subsidies, and the presence of quality label products. Transhumance, i.e. moving flocks towards other available feed resources, such mountain pastures, is considered to be of importance to adapt to climate changes and is in line of societal expectations, but labour and predation issues might negatively affect this practice. Transmission to new generation is a major concern for all stakeholders, especially for meat farms. Quality labels such as protected designation of origin (PDO), have a key-role in the structuration of milk sector, enabling to maintain high-prices, sufficient incomes and support stakeholder activities. Both sectors are highly depending on subsidies and export market (feed and lamb to Spain), and this is considered as a huge threat for the sustainability. For all stakeholders, regardless of their production and activity, communication is crucial for the future of these productions.

Session 54 Theatre 7

Aligning carbon footprint estimates from different tools across Europe

A.S. Atzori[1], O. Del Hierro[2], B. Lyubov[3], R. Vial[4], C. Buckley[3], M.G. Serra[5], M. Habeanu[6], R. Ruiz[2], L. Lanzoni[7], M. Acciaro[5], T. Keady[3] and S. Troude[4]

[1]University of Sassari, Sassari, Italy, via DeNicola 9, 07100, Italy, [2]Neiker, Victoria, Spain, [3]5Teagasc, Athenry, Co Galway, Ireland, [4]6Institut de l'Elevage, France, France, [5]AGRIS Sardegna, Loc. Bonassai, Sassari, Italy, [6]Institutul National de Cercetare-Dezvoltare, Romania, [7]University of Teramo, Italy; asatzori@uniss.it

LIFE Green Sheep (LIFE19 CCM/FR/001245) has been targeting a common Carbon Footprint (CF) methodology at European level. The aim of the study was to aling estimates of 4 tools, already available in Europe CAP'2ER (C2E; Institute de l'Elevage, France); ArdiCarbon (AC; Neiker, Spain); CarbonSheep (CS; Univ. of Sassari, Italy) and Sheep LCA (S-LCA; Teagash; Ireland), for carbon footprint of dairy and meat sheep farms. The comparison was performed gathering data from 12 dairy and 12 meat sheep farms (from France, Spain, Ireland, Romania, Italy). Data included a broad range of characteristics in semi-extensive farming systems. The tools performed a simplified life cycle assessment with different approaches to estimate enteric methane, N excretion while different emission coefficients were used for feed purchased, crop and fertilizers, energy and fuel. Enteric emissions were aligned using the same methane emission formula (dry matter intake, DMI * Ym/55.65; DMI as from Pulina *et al.*), the emission coefficients for each hotspot were standardized. Evaluations consisted of 24×4 model runs before and after tool alignment. After a first comparison the mean farm CF estimated in Dairy farms was 4.69, 3.20 and 3.43, and 3.34 kg CO_2eq/kg of milk and 31.1, 16.4, 21.3 and 17.2 kg CO_2eq/kg of meat for C2E, AC, CS, S-LCA, respectively. Large differences (more the 30% among tools) were observed among tools often due to enteric methane and manure emissions. After the building app phase the mean CF was equal to 4.1±1.3 kg CO_2eq/kg of milk, being 4.21, 3.98, 4.14, and 4.1 kg CO_2eq/kg of milk for C2E, AC, CS, S-LCA, respectively. For meat farms it was equal to 22,9±7.1 kg CO_2eq/kg of carcass weight (CW) with small variations among farms being 22.9, 26.2, 22.1, 21.9 and 21.3 kg CO_2eq/kg of CW.

Determining carbon footprint of sheep farms in Europe: first results of the LIFE Green Sheep project

S. Throude[1], M. Acciaro[2], A. Atzori[3], R. Ruiz[4], O. Del Hierro[4], C. Buckley[5], L. Bragina[5], T.W.J. Keady[5], C. Dragomir[6], M.A. Gras[6] and J.B. Dollé[1]
[1]Institut de l'Elevage, 149 rue de Bercy, 75012 Paris, France, [2]AGRIS, SS 291 Sassari-Fertilia, Loc. Bonassai, Italy, [3]University of Sassari, Viale Italia, 39, 07100 Sassari, Italy, [4]Neiker, Parque tecnológico de Bizkaia, 48160 Derio, Spain, [5]Teagasc, Mellows Campus, Athenry Co. Galway, Ireland, [6]National R&D Institute for Animal Biology and Nutrition, Calea Bucuresti N1, Balotesti 077015, Romania; sindy.throude@idele.fr

Small ruminants farming generates greenhouse gas (GHG) emissions accounting for 6.5% of livestock emissions. To reduce the environmental impact of this sector, it is important to know and understand its level of emissions in order to identify areas of improvement. This is one of the objectives of the LIFE GREEN SHEEP project which aims at reducing the carbon footprint of meat and milk sheep by 12% while ensuring farms' sustainability. One of the project's actions is to create an observatory of the carbon footprint of sheep production as well as its sustainability performances. Data from 1,355 sheep farms (935 meat and 420 dairy) representing various rearing systems in 5 European countries (France, Ireland, Italy, Romania, Spain), the project provides a good overview of the average EU sheep milk and sheep meat carbon footprint. Preliminary results indicate an overall GHG emissions from milk and meat sheep of 3.98 kg CO_2e/kg FPCM (fat and protein corrected milk) and 21.5 kg CO_2e/kg carcass, respectively. These results differ between countries and rearing systems. Nevertheless, the variability is mainly observed within each type of system. For instance, within the French grassland-based dairy sheep system, the average GHG emissions of the 10% farms with the lowest emissions is 71% lower (2.13 kg CO_2e/kg FPCM) than the average emissions of the group (2.58 kg CO_2e/kg FPCM). This reflects the farms' differences in productive efficiency. The variations on the carbon footprint are strongly related to herd management: the farms with the lowest emissions have the highest productivity (prolificacy and fertility) rates, the lowest consumption of concentrates and spend more days grazing. These first results enable a better understanding of the correlation between carbon footprint and farms' practices and to identify areas of improvement.

Greenhouse gas emission intensity of milk production in three Slovenian goat breeds

M. Bizjak[1], Ž. Pečnik[2] and M. Simčič[1]
[1]University of Ljubljana, Biotechnical faculty, Department of animal science, Groblje 3, 1230 Domžale, Slovenia, [2]The Agricultural Institute of Slovenia, Hacquetova 17, 1000 Ljubljana, Slovenia; marko.bizjak@bf.uni-lj.si

The aim of the study was to determine the intensity of greenhouse gas (GHG) emissions from milk production of three goat breeds in Slovenia, to identify the trends, and to determine the main impacts on greenhouse gas emissions. Based on information on milk yield, protein and fat content, average body mass of each breed, litter size, and kidding interval, we estimated methane and nitrous oxide emissions for the period 2008-2021. Emissions were estimated for 9,890 lactations. GHG emissions were expressed in carbon dioxide equivalents. Emission intensity was expressed as emissions per kg of milk produced. Differences in GHG emission intensity among goats were fourfold, ranging from about 0.4 to more than 1.5 kg of CO_2 equivalent per kg of milk. On average, the emission intensity expressed in kg CO_2 equivalent per kg milk was 0.756 for the Slovenian Saanen goat, 0.740 for the Slovenian Alpine goat, and 0.873 for the autochthonous Drežnica goat. The highest emission intensity was found for goats at the first parity (0.809 kg CO_2 equivalent per kg milk), while goats at the fifth parity had the lowest emission intensity (0.733 kg CO_2 equivalent per kg milk). Emission intensity decreased with increasing litter size. The average emission intensity expressed in kg CO_2 equivalent per kg milk was 0.811 for goats delivering single kids, 0.726 for goats delivering twins, and 0.680 for goats delivering triplets. Emission intensity increased with increasing kidding interval. For goats with extended kidding interval, total milk yield was not high enough to compensate for the higher maintenance requirements of these goats. The intensity of GHG emissions from milk production in herds with controlled goats varied over the years (ranging from 0.803 to 0.688 kg of CO_2 equivalent per kg of milk), but decreased overall by about 14% during the study period. We concluded that some fertility traits are correlated with milk production and consequently with the intensity of GHG emissions as well. In particular, relatively short kidding interval could reduce the intensity of GHG emissions through higher daily milk production.

No difference in PAC methane emission corrected for milk yield in three Norwegian dairy goat farms

J.H. Jakobsen[1], T. Blichfeldt[1], K. Dodds[2] and J.C. McEwan[2]
[1]The Norwegian Association of Sheep and Goat Breeders, Box 104, 1431 Ås, Norway, [2]Research Limited, Invermay Agr. Centre, PB 50034, Mosgiel 9053, New Zealand; jj@nsg.no

Use of portable accumulation chambers (PAC) for capturing enteric methane (CH_4) emissions in sheep is spreading globally, while the usage for goats has been limited. Technically the equipment is equally suitable for both species. The Norwegian agricultural sector has made an agreement with the government to reduce the greenhouse gas emissions from agriculture by a cumulative five-million-ton CO_2-eq by 2030, relative to 2020. The agreement is expressed as per kg product. The Norwegian goat milk industry is small and absolute contribution to the national emission is relatively minor. However, dairy goats may act as a model for dairy cattle in dose-response mitigation trials. The aim of the current study was to compare weight corrected CH_4 and milk corrected CH_4 for three farms. Thirty goats participated in each trial, repeated on three different farms. The thirty goats were divided into three lots of ten, with four goats from first, three goats from second and three goats from third parity. All goats were in their early lactation, on average 70 days from kidding and fed indoor on silage and concentrate. Each lot of 10 goats were measured simultaneously three times during 24-hours and at the same time of the day during the three consecutive days, yielding a total of nine PAC visits per goat. Sixty-minute CH_4 concentrations were captured for each visit. Goats were weighed prior to measurement and ID's merged with latest official milk recording. Average CH_4 emission was 1.94, 1.37 and 1.34 g/hr for each of the three farms respectively with corresponding average body weights of 64.1, 49.9, and 49.5 kg, indicating higher emission at higher body weight, and significant differences between farms. Average daily milk yield for the goats with methane emission on the three farms were 3.75, 2.73 and 2.88 kg/day. CH_4 emissions g/hr / kg-milk/day were 0.52, 0.51, and 0.50, respectively, with no significant difference between farms. Based on this study and experimental design, milk yield explains a larger part of the between farm variation in CH_4 emission than live weight in dairy goats and this should be considered as part of on farm calculations.

Sustainability of Irish sheep production

C. Buckley, L. Bragina and T. Keady
Teagasc, Mellows Campus, Athenry, Ireland; lyubov.bragina@teagasc.ie

The paper assesses the economic, environmental and social sustainability of sheep farming across the Republic of Ireland using data from the EU Farm Accountancy Data Network. Results are based on a sample of 111 specialist sheep farms, which are population weighted to represent 13,979 on a national basis. *Economic Sustainability Indicators* The average gross output per hectare for sheep farms was €1,476 in 2021, and the average gross margin was €623 per hectare. Across all sheep farms, 39% were defined as economically viable. The average income per labour unit on sheep farms was €18,725. Approximately 66% of output was generated from the market, with the remaining 34% derived from direct payments. The average family farm income per hectare on sheep farms was €445 in 2021. *Environmental Sustainability Indicators* In 2021, the average sheep farm produced 166 tonnes CO_2 equivalent of Agricultural GHG emissions. In all, 54.1% of these emissions were generated by the sheep enterprise, with the remaining emissions (45.5%) generated by a cattle enterprise present on specialist sheep farms, with the remainder coming from other sources (minor arable enterprise). On average, sheep farms emitted 4.1 tonnes of CO_2 equivalent per hectare. The average sheep farms emitted 0.17 tonnes of energy based CO_2 equivalent per hectare. *Social Sustainability Indicators* On average, 21% of all specialist sheep farms were classified as being at risk of isolation. The proportion of all specialist sheep farms with a high age profile was 37%. Sheep farmers worked an average of 1,541 hours per year on farm in 2021 (or 29.6 hours a week) and 2,085 hours in between on and off-farm work (approximately 40.1 hours per week). On average, 29% of sheep farmers reported experiencing business related stress in 2021 and 45% indicated stress levels had increased over the previous 5 years. Only 52% of sheep farmers indicated having daily human contact outside the farm household. *Trade-offs* Higher GHG emissions per farm and per hectare were associated with the more profitable sheep farms. Whereas, better economic performance was linked with lower GHG and ammonia emissions per kg of live-weight output.

Should animal welfare indicators be integrated into the environmental impact assessment of farms?

L. Lanzoni[1], L. Whatford[2], K. Waxenberg[3], R. Ramsey[3], R.M. Rees[3], J. Bell[3], E. Dalla Costa[4], S. Throude[5], A.S. Atzori[6] and G. Vignola[1]

[1]University of Teramo, Dep. of Veterinary Medicine, Località Piano d'Accio, 64100, Teramo, Italy, [2]Royal Veterinary College, North Mymms, AL97TA, Hertfordshire, United Kingdom, [3]Scotland's Rural College, King's Buildings, EH93JG, Edinburgh, United Kingdom, [4]University of Milan, Dep. of Veterinary Medicine and Animal Sciences, Via dell'Università 6, 26900, Lodi, Italy, [5]Institut de l'Elevage, Rue Jean Baldassini 23, 69007, Lyon, France, [6]University of Sassari, Dep. of Agricultural Sciences, Viale Italia 39, 07100, Sassari, Italy; llanzoni@unite.it

Animal welfare (AW) is fundamental priority for a sustainable transition of the livestock sector. Understanding to what extent AW affects the environmental performances of farms can promote the integration of animal welfare (AW) indicators into environmental assessments (EA). The aims of the present project were to (1) quantify the climatic cost of impaired AW in sheep farms and to (2) summarise existing methods that integrate AW indicators with farm EA. A scenario study was developed using Agrecalc© to test the change in emission intensity (EI, kgCO_2/kg fat and protein-corrected milk) from a baseline dairy sheep farm to six scenarios of impaired welfare, modelled from literature data. Moreover, a scoping review was conducted to summarise the studies that combined AW with Life Cycle Assessment (LCA). Impaired welfare caused an increase in EI for all the scenarios compared to the baseline: overstocking (+10.0%), mastitis (+6.8%), gastrointestinal nematodes (+6.5%), lameness (+2.1%,), water deprivation (+1.6%), and thermal discomfort (+1.0%). The scoping review, which included 24 articles, showed that while LCA was performed with a fairly homogeneous approach, AW was assessed rather heterogeneously, mainly due to differences in the number and type of indicators used. Sixteen papers aggregated the AW indicators into a final score. Within these, four proposed to associate the welfare scores to the LCA functional unit, while the others used either an AW scale (n=10) or a traffic light system (n=2). These results show that although the inclusion of AW indicators in the EA of farms can help identify the best mitigation strategies, guidelines are needed to harmonise their systematic integration.

Testing mitigation actions to reduce GHG emissions from sheep farming in Europe

S. Throude[1], M. Acciaro[2], A. Atzori[3], R. Ruiz[4], O. Del Hierro[4], C. Buckley[5], L. Bragina[5], T.W.J. Keady[5], C. Dragomir[6], M.A. Gras[6] and J.B. Dollé[1]

[1]Institut de l'Elevage, 149 rue de Bercy, 75012 Paris, France, [2]AGRIS, SS 291 Sassari-Fertilia, Loc. Bonassai, Italy, [3]University of Sassari, Viale Italia, 39, 07100 Sassari, Italy, [4]Neiker, Parque tecnológico de Bizkaia, 48160 Derio, Spain, [5]Teagasc, Mellows Campus, Athenry Co. Galway, Ireland, [6]National R&D Institute for Animal Biology and Nutrition, Calea Bucuresti N1, Balotesti, 077015, Romania; sindy.throude@idele.fr

Climate change is an existential threat to Europe and the world. To overcome this challenge, the Europe commission aims at reducing net greenhouse gas (GHG) emissions by at least 55% by 2030, compared to 1990 levels, and becoming the first climate-neutral continent by 2050. The livestock sector plays an important role in climate change and has to reduce its GHG emissions. In 2013, FAO shows that there was a potential to significantly reduce global GHG emissions by 33%. In line with this, the LIFE GREEN SHEEP project aims to identify, demonstrate and disseminate innovative good practices on sheep farms to reduce GHG emissions by 12% and to increase carbon storage. In Green Sheep, 282 innovative farms are monitored with an environmental and sustainability assessment, and some mitigation actions have been identified and tested to target the objective of the project. These meat and dairy sheep farms are located in 5 EU countries (France, Ireland, Italy, Romania, Spain) and represent different sheep systems, and have implemented mitigation actions dealing with flock, feed, manure and crop management. Impacts of these actions varies according to their contribution to GHG emissions. Indeed, flock management actions enable the reduction of enteric methane emissions which represent on average 55% of farm GHG emissions. Thus, mitigation actions tested enable the reduction of GHG emission by -1% to -10%. Most of the mitigation actions facilitate an improvement in flock technical and economic performance. Furthermore, one of the main mitigation actions is carbon storage, which can compensate for around 50% of emissions of sheep farming. It is a real asset for this sector to reduce its GHG emissions. As sheep farming enhances a large proportion of grasslands and pastoral areas, it has a strong capacity in preserving and increasing absorption of carbon in soil.

Session 54 — Theatre 14

The effect of forage type on methane production from hill bred lambs grazing alternative forages

M. Dolan[1], T. Boland[2], N. Claffey[1], F. McGovern[1] and F. Campion[1]
[1]Teagasc, Animal & Bioscience Department, Animal & Grassland Research & Innovation Centre, Mellows Campus, Athenry, Galway., H65R718, Ireland, [2]University College Dublin, School of Agriculture and Food Science, University College Dublin, Belfield, Dublin 4, Ireland, D04 V1W8, Ireland; mark.dolan@teagasc.ie

Ruminant diets are an important factor in enteric methane (CH_4) production as changes in CH_4 emissions from ruminants are closely correlated to forage quality and digestibility. The objective of this experiment was to investigate the effect of forage type offered post-weaning on CH_4 production from Scottish Blackface (SB) ram and castrate lambs and Texel cross Scottish Blackface (TXSB) ram lambs offered one of six dietary treatments from October to January; forage rape (*Brassica napus* L.;FR) hybrid brassica (*Brassica napus* L.; HB) kale (*Brassica oleracea* L.; K) perennial ryegrass reseed (*Lolium perenne*; RS) permanent pasture (predominantly *Lolium perenne*; PP) grazed in-situ and *ad libitum* concentrates indoors (ALC). The study was a 6×2×2 factorial design with lambs blocked by weight and balanced for breed and sex. The study was carried out over two separate years with 16 entire males and 8 castrates per treatment per year. Lambs were removed from feed and weighed a minimum of one hour and maximum of three hours prior to entering portable accumulation chambers (PAC). Methane measurements were obtained using individual chambers over a 50 minute time period at three specific time points (0, 25 and 50 minutes) on two separate occasions 14 days apart. Lambs offered ALC had the lowest CH_4 emissions compared to all other dietary treatments (5.13±0.719 g/day; $P<0.01$). Lambs grazing kale had higher CH_4 emissions than lambs offered FR and HB (10.87 g/day, 8.82 g/day and 9.19±0.534 g/day; $P<0.01$) but these were lower than the CH_4 emissions from lambs offered RS and PP (14.34 g/day and 15.80±0.579 g/day; $P<0.01$). Breed had no effect on CH_4 emissions ($P>0.05$). However, lamb weight and time off feed had a significant effect on CH_4 produced ($P<0.001$). Results from this study show that while lambs offered ALC diets had the lowest CH_4 emissions, the inclusion of forage brassica crops has the potential to reduce CH_4 emissions from fattening lambs in the autumn winter period compared to perennial ryegrass swards.

Session 54 — Poster 15

Preliminary results of the LIFE Green Sheep project in Italy

M. Acciaro[1], M. Decandia[1], M.G. Serra[1], S. Picconi[1], V. Giovanetti[1], A. Atzori[2], D. Usai[3] and S. Throude[4]
[1]AGRIS Sardegna, Loc Bonassai, 07100, Italy, [2]University of Sassari, Department of Agricultural Sciences, Sassari, 07100, Italy, [3]LAORE Sardegna, Cagliari, 09123, Italy, [4]Institut de l'Elevage, Paris, 75012, France; macciaro@agrisricerca.it

LIFE Green Sheep project (LIFE19 CCM/FR/001245) aims at reducing the carbon footprint of milk and meat sheep by 12%, while ensuring farms' sustainability, in five European countries (France, Ireland, Italy, Spain, and Romania). One of the project's action is to carry out a large-scale assessment of GHG emissions and carbon storage in 1,355 'demonstrative' sheep farms (885 in France, 185 in Ireland, 100 in Italy, 100 in Romania and 90 in Spain) using a simplified life cycle inventory tool, in order to create a European observatory of the environmental and sustainability performances of this sector. This paper summarizes the first results for 64 dairy sheep farms in Italy (only Sardinia) using CAP'2ER® tool level 1 (C2E; 'Institute de l'Elevage', France) that allows a simplified estimate of farms' environmental impact. The preliminary assessment reported an average GHG emissions from Italian milk sheep of 3.70±1.37 kg CO_2-eq/kg FPCM (fat-protein corrected milk), with 29±36% of GHG emission offset from carbon sequestration. Therefore, the net carbon footprint turns out to be 2.67±1.68 kg CO_2-eq/kg FPCM. The largest contribution to GHG emissions comes from the enteric fermentation (41±9% of GHG emissions). CAP'2ER also enabled to estimate: (1) the contribution of sheep farms to maintain biodiversity, expressed as hectares equivalent (ha-eq) of biodiversity that is on average 0.52±0.47 ha-eq/ha TSA (total sheep area, total surface used by sheep flock); (2) the carbon sequestration, estimated to be 173±201 kg C/ha TSA. The high variability observed is related to the different extensification/intensification level of dairy sheep farms. This is also confirmed by the variability in milk productivity (247±95 kg FPCM/animal) and number of heads per farm (429±296 animals/farm). Finally, the data analysis confirmed the relationship between animal productivity (expressed as kg FPCM/sheep) and total GHG emissions, evidenced by the good Pearson's correlation coefficient (r=0.49, P-value<0.001).

Carbon footprint assessment of a Pecorino cheese produced in central Italy

L. Lanzoni[1], L. Di Paolo[2], S. Abbate[2], M. Giammarco[1], M. Chincarini[1], I. Fusaro[1], A. Atzori[3], D. Di Battista[2], S. Throude[4] and G. Vignola[1]
[1]University of Teramo, Dep. of Veterinary Medicine, Località Piano d'Accio, 64100 Teramo, Italy, [2] University of L'Aquila, Dep. of Industrial and Information Engineering and Economics, Piazzale Ernesto Pontieri 1, 67100 L'Aquila, Italy, [3]University of Sassari, Dep. of Agricultural Sciences, Viale Italia 39, 07100 Sassari, Italy, [4]Institut de l'Elevage, Rue Jean Baldassini 23, 69007 Lyon, France; llanzoni@unite.it

To mitigate the environmental impact of small ruminant farming it is important to define tailored mitigation solutions, which require a thorough understanding of the different systems. The following study aimed to assess the environmental impact of three different sheep farms located in central Italy, oriented towards the production of local pecorino cheese. The three farms were classified according to the different management strategies adopted as 'extensive' (1 mating group, lambs with the mothers), 'intermediate' (2 mating groups, lambs weaned at 1 month) and 'intensive' (3 mating groups, lambs artificially reared). Inventory data were collected on the farm to perform the Life Cycle Assessment (LCA) and 'CarbonSheep', developed by the University of Sassari, was used to evaluate the climate change impact to produce 1 kg of pecorino cheese with an economic allocation. The carbon footprint of the three farms was 20.04 ('extensive'), 17.4 ('intermediate') and 16.64 ('intensive') kgCO_2eq/kg pecorino, reflecting the farms' differences in productive efficiency. However, farms had a different relative contribution to the farm sectors, associated with their management differences. The extensive farm, which was the less efficient, had a higher relative impact (RI) derived from enteric fermentation and manure management (83 vs 76%-intermediate and 74%-intensive). The 'intensive' farm had a higher RI from the purchased feed (15 vs 9%-extensive and 6%-intermediate) while the 'intermediate' farm from the use of fuel (18 vs 8%-extensive and 11%-intensive) due to a higher farmed land extension. The present results indicate that, even though the same production process and geographical area are involved, mitigation solutions need to be studied on a case-by-case basis as the types of management can change considerably.

Grass management to adapt goat farming to climate change in western France

J. Jost[1,2], M.-G. Garnier[3], L. Robin[3], M. Proust[4], M. Bourasseau[5], R. Lesne[6], A. Villette[7], V. Tardif[8], T. Soulard[8], O. Subileau[9] and O. Prodhomme[10]
[1]BRILAC, Réseau REDCap, CS 45002, 86550 Mignaloux-Beauvoir, France, [2]Institut de l'Elevage, CS 45002, 86550 Mignaloux-Beauvoir, France, [3]Saperfel, 228 rue Androlet, 79410 Echire, France, [4]Innoval, CS 80038, 35538 Noyal Sur Vilaine Cedex, France, [5]Civam du Haut-Bocage, 2 Place du Renard, 79700 Mauléon, France, [6]Ardepal, Boulevard Des Arcades, 87100 Limoges, France, [7]Chambre d'agriculture de la Dordogne, 295 Boulevard des Saveurs, 24660 Coulounieix-Chamiers, France, [8]Seenovia, 141 boulevard des Loges, 53940 Saint-Berthevin, France, [9]GAB 72, 16 avenue Georges Auric, 72000 Le Mans, France, [10]CRA Pays de la Loire, 9 rue André-Brouard, 49105 Angers, France; jeremie.jost@idele.fr

Ten groups of goat farmers (4 to 8 farmers per group and their advisor) in western France (Nouvelle-Aquitaine and Pays de la Loire) have been working since 2020 to identify solutions to adapt their forage systems to climate change. Grassland management (choice of species and varieties, sowing techniques, fertilisation strategy) and stocking grass (hay, silage) were identified as the primary levers. It will also be necessary to plan for a greater carryover of fodder stock, which will also secure the system. Diversification of fodder resources and the method of valorisation are also favourable, as is the establishment of grasslands under a cover of sunflower, meslin, oats, barley, etc., alfalfa in the spring or grass/multi-species meadows in the autumn. For non-grazing systems, the first cut will be earlier (wrapping, grass silage, barn drying) to ensure a quality harvest and regrowth before the summer droughts. Grazing and green feeding can also help to make the most of this early spring grass. The spring yield of grassland will be greater in the future, with a reduced period of grass growth. It will be necessary to ensure that there is enough work to harvest the grassland at the best stage, sometimes with a diversity of harvesting methods. For grazing farms, the main issue will be linked to summer droughts (and the capacity of the grassland to grow back in the autumn). The choice of species and varieties will be important. Other forage crops are interesting for summer/autumn, such as sorghum or rape.

Session 54 Poster 18

Positive effects of grazing mulberry trees in summer on dairy goats' milk and cheese

C. Boyer[1], H. Le Chenadec[1], F. Noël[1], A. Stocchetti[1], J. Jost[1], A. Pommaret[2], S. Fressinaud[2] and R. Delagarde[3]
[1]Institut de l'elevage, 149 avenue de Bercy, 75012 Paris, France, [2]EPLEFPA Olivier de Serres, Ferme du Pradel, 950 chemin du pradel, 07170 Mirabel, France, [3]PEGASE, INRAE, Institut Agro, 16, Le Clos, 35590 Saint-Gilles, France; jeremie.jost@idele.fr

With more frequent and intense summer droughts linked to climate change, fodder trees appear to be a solution allowing farmers to offer green fodder to their animals at this period. The objective of this project was to study the effect of grazing white mulberry (Morus alba L.) young trees on dairy goats' production and cheese making capacity as Picodon PDO. Grazing of white mulberry trees was tested during 16 days in July 2021 and 12 days in July 2022, at the goat farm of Pradel (Ardèche, France). Each year, 2 homogeneous groups of 24 goats were used. The Experimental group grazed the mulberry trees between 8:00 am to 3:00 pm (7 h /d), while the Control group was fed alfalfa hay *ad libitum* in the barn at that time. After evening milking, both groups were fed alfalfa hay *ad libitum* indoor. Each goat was individually fed 250 g/d of a commercial concentrate and 500 g/d of whole maize grain. After 7 days of grazing, individual milk production and milk fat and protein contents were measured on 4 days. Faeces were sampled from 4 representative subgroups of goats every 4 days to estimate the proportion of mulberry leaves in the diet from n-alkanes profile in faeces and feeds. With only 30% of the daytime spent outdoor, mulberry leaves (140 g of crude protein/kg DM) represented 61±4 and 67±7% of the goats' forage ration in 2021 and 2022, respectively, on a DM basis. Milk production (3.5 kg/d) was similar between groups. Milk fat content was much greater when goats grazed mulberry trees (39 vs 30 g/kg of milk) as was milk protein content (32 vs 30 g/kg of milk), with lower milk urea (300 vs 450 mg/l of milk). Due to the greater milk fat and protein contents, cheese yield increased by 1.5 kg of cheese per 100 kg of milk, without problems in draining. The Picodon cheeses were tasted after 12 or 14 days of ripening and no difference was found between the two groups.

Session 54 Poster 19

Effects of heat stress and forage quality on feed intake and milk production of Sarda dairy ewes

M. Sini, F. Fulghesu, A. Ledda, A.S. Atzori and A. Cannas
Università degli studi di Sassari, Agricultural Science, Viale Italia 39, 07100, Italy; msini1@uniss.it

The persistency of lactation of dairy ewes is positively affected by the utilization of highly degradable fibre sources. At the same time, mid-late lactation often occurs during spring-summer, periods with high temperatures. Thermal stress can markedly affect the performances of dairy ewes, especially when fed fibre rich feeds, which highly thermogenic. Thus, this study evaluated the impact in lactating ewes of three different forage sources, fed *ad libitum*, and of thermal stress (assessed measuring the temperature-humidity index, THI). Sarda dairy ewes (n=21; 150, days in milk) were divided into three groups balanced for dry matter intake (DMI), milk yield (MY), body weight (BW) and BCS (mean ± SD; DMI 2.22±0.05 kg/d MY 2.00±0.03 kg/d; BW 53.9±2.04 kg; BCS 2.98±0.08). Two groups received dehydrated chopped alfalfa hay of medium quality (AMQ:19.7% CP, 43.5% NDF, 6.48% ADL, DM basis) or high quality (AHQ: 23.5% CP, 39.0% NDF, 5.8% ADL, DM basis) fed *ad libitum* and a supplementation of 528 g of DM/d of whole corn grains. The third group received dehydrated chopped oat hay (OAT: 7.3% CP, 63.9% NDF, 4.78% ADL, DM basis) *ad libitum* and a supplementation, DM basis, of 176 g/d of whole corn grains and 360 g/d of soybean meal. The experimental period lasted 21 d. A factorial design with feeding treatment, THI classes, time and their interactions as fixed factors, animals as random effect was applied. Voluntary DMI was affected by the diet (OAT=1.289 vs AMQ= 2.085 vs AHQ= 2.733 kg/d, P<0.001) and its interaction with THI, showing a decrease in DMI with THI>76 (-17% of ingestion compared to THI=70; P<0.05) in OAT group, while the AHQ and AMQ were not affected by heat stress. Milk yield was affected by the diets (OAT=1.491 vs AMQ=1.755 vs AHQ=2.028 kg/d, P<0.001) and by its interaction with the THI when comparing OAT vs AHM group (P<0.05), while no effects was observed when OAT was compared to the AMQ group and also between AHQ and AMQ. This study highlighted the importance of feeding high quality forage during the mid-lactation phase in order to maintain the persistency of lactation, sustain the voluntary dry matter intake and increase the tolerance response to heat stress by the animal.

Session 54 Poster 20

Seasonal rainfall patterns modify summer energy balance and nutritional condition of grazing sheep

Y. Yoshihara[1], B. Choijilsuren[2], T. Kinugasa[3] and M. Shinoda[4]
[1]Mie university, kuirmachoyacho1577, 5148507, Tsu city, Japan, [2]Institute of Veterinary Medicine, Zaisan, PO 53/2, Ulanbaatar, Mongolia, [3]Tottori university, Koyamacho 4-101, Koyamacho 4-101, Tottori city, Japan, [4]Nagoya University, Furocho, Nagoya city, Japan; marmota.sibirica@gmail.com

Recent spring and summer rainfall patterns in Mongolian grasslands have shown high annual fluctuations and are thought to affect sheep energy balance and nutritious condition via plant community changes. To test this, we obtained climatic and vegetation data in Mongolian semi-arid grasslands from spring to summer for 3 years. Sheep energy intake and expenditure were calculated, and the energy balance and nutrition indicators were compared among years with different seasonal rainfall patterns. In 2019 and 2022, rainfall patterns were characterized by the presence or absence of an early summer drought, respectively. Compared to 2019, plants were tall and abundant in 2022, and thus mean bite size, energy intake, and body weights were greater. Estimated energy intake and expenditure were 4.56 and 2.10 Mcal in 2019, and 6.75 and 1.63 in 2022, respectively, demonstrating that rainfall timing and amount affected vegetation assemblage, plant height, herbage nutrition, grazing behaviour, animal energy balance, and nutrition conditions. Insufficient rainfall events before early summer in semi-arid grazing lands resulted in low energy intake and balance, and the delayed body weight gain could not be recovered. The findings can inform management of grazing land to ensure optimal livestock condition.

Session 54 Poster 21

Effect of the inclusion of rumen-protected amino acids in the diet of high production dairy sheep

A. Cabezas[1], D. Martinez Del Olmo[2], J. Mateos[2], J. Matilla[3], M.T. Díaz[1], R. Bermejo-Poza[1], J. De La Fuente[1] and V. Jimeno[4]
[1]Facultad de Veterinaria. UCM, Av. Puerta de Hierro s/n, 28040 Madrid, Spain, [2]Kemin Animal Nutrition and Health, Herentals, 2200 Herentals, Belgium, [3]Oceva Sdad.Coop, Mirador del Duero 5, 49017 Zamora, Spain, [4]E.T.S.I.A.A.B. UPM, Av. Puerta de Hierro 2, 28040 Madrid, Spain; diego.martinez@kemin.com

The aim of the study was to examine the effect of including rumen-protected amino acids in the diet of high-production dairy sheep, on performance and milk composition. A total of 188 Lacaune dairy sheep were divided into two homogeneous groups in lactation number (3) and located in separated sheep sheds. Treatments were a control diet (CON) and a diet with rumen-protected amino acids LysiGEM™ and KESSENT® M*e* (Kemin Animal Nutrition and Health, Belgium) (APR). Diets were supplied after milking (morning and afternoon) for 60 days. Both diets were isoproteic (17% CP and 11.4% PDI) but with different energy levels (1.02 and 1.00 Milk Forage Units, MFU/kg DM, in CON and APR diets, respectively) LysDI and MetDI (7.4 and 2.0% PDIE in CON vs 7.9 and 2.8% PDIE in APR diets). Twenty-five sheep of each group were daily controlled in milk production and a sample of milk was weekly taken to determine milk composition. Blood samples were collected from venipuncture of the jugular vein at 1, 4, 7, and 9 weeks from the start of the trial and after the morning milking in the selected sheep. Data were analysed using the PROC GLM procedure of SAS with diet as a fixed factor. Milk yield was similar in CON and APR diets (3.95 vs 3.79 l/d respectively; P=0.47), however, milk fat content was higher (P=0.0003) in CON (5.53%) than in APR (5.23%), while milk protein content was higher (P=0.0149) in APR (4.77%) than in CON (4.69%). CON and APR showed significant differences (P=0.0013) in plasma D-β-hydroxybutyrate (mmol/l), being lower in APR (0.61) than CON (0.70) while plasma glucose (mg/dl) was higher (P=0.0001) in APR (50.78) than CON (46.78), which means that energy metabolism was more efficient in APR than in CON. Using rumen-protected amino acids, it is possible to formulate diets with lower levels of energy density (MFU/kg DM) and achieve greater efficiency in the use of nitrogen from the diet to synthesize protein in milk without affecting daily milk production.

Session 55 Theatre 1

Reaction norm model analysis for heat stress tolerance of growth performance in purebred pigs

Y. Fukuzawa[1,2], S. Ogawa[2], T. Okamura[2], N. Nishio[2], K. Ishii[2], K. Tashima[3], K. Akachi[3], H. Takahashi[3] and M. Satoh[1]
[1]Tohoku University, Graduate School of Agricultural Science, Sendai, Miyagi, 980-8572, Japan, [2]National Agiculture and Food Research Organization, Institute of Livestock and Grassland Science, Tsukuba, Ibaraki, 305-0901, Japan, [3]Global Pig Farms, Inc., Shibukawa, Gumma, 377-0052, Japan; fukuzaway282@affrc.go.jp

Heat stress due to global warming has increased the loss of productivity among essential livestock such as pigs. However, genetic improvement may provide a solution. Therefore, reaction norm model analysis was performed, to develop a methodology to evaluate the genetic ability for heat tolerance of purebred Landrace (L), Large White (W), and Duroc (D) pigs in Japan. The number of records of lifetime average daily gain (LADG) from birth to the end of testing collected from five farms was approximately 35,000 per breed. First, accumulated thermal load (TL) as an indicator of heat stress on LADG was developed via evaluation over a suitable period (from 1 to 16 weeks before the day at the end of testing), and a threshold temperature (10-30 °C) that showed best-fit performance was ascertained. Next, reaction norm model analysis was performed using the developed TL as an environmental variable (linear covariate). The threshold temperature and period of study were 15 °C and 6 weeks, 18 °C and 5 weeks, and 20 °C and 9 weeks for L, W, and D, respectively. The maximum TL values were 19, 16, and 13 for L, W, and D respectively and varied according to the threshold. Values of heritability estimated from reaction norm model analysis ranged between 0.50-0.55, 0.46-0.56, and 0.35-0.47 in L, W, and D, respectively. Minimum values of estimated genetic correlation were 0.75, 0.56, and 0.46 for L, W, and D, respectively. These results suggest a genotype-TL interaction effect; therefore, growth performance under different TL conditions could be genetically improved. All breeds showed negative estimated genetic correlations between the intercept and slope terms, specifically, -0.45 for L, -0.56 for W, and -0.64 for D, implying that some pigs could be genetically more sensitive to heat stress. Further studies investigating the approach to ranking selection of candidates are necessary for efficiently improving growth performance while considering the genetic ability of heat tolerance.

Session 55 Theatre 2

Investigating the slick gene and its effects on heat stress in New Zealand grazing dairy cattle

G.M. Worth, E.G. Donkersloot, L.R. McNaughton, S.R. Davis and R.J. Spelman
Livestock Improvement Corporation, 605 Ruakura Road, Hamilton 3286, New Zealand; gemma.worth@lic.co.nz

In New Zealand pastoral grazing systems, dairy cattle are exposed to increased heat load via environmental temperature and solar radiation combined with the effects of walking to and from the milking parlour. We investigated the use of the slick gene, a naturally occurring deletion in the prolactin receptor gene found in Senepol cattle introgressed into a New Zealand dairy genetic background. This results in a slick, short hair coat and an improved ability to regulate body temperature under heat load. Our objective was to compare thermotolerance of slick and control cattle when observed under heat stress conditions. Data was collected across 3 studies on a commercial dairy herd in (1) non-lactating (slick n=9 and control n=9) heifers (exercise induced heat challenge), (2) first lactation cows (slick n=9 and control n=9) with rumen bolus temperature and vaginal temperature loggers at pasture and (3) first lactation cows (slick n=7 and control n=7) with indoor heat challenge plus sweating rate measurement. The vaginal temperature differential between slicks and control cattle were similar to the temperature differential seen with the rumen bolus. While at pasture and a temperature-humidity index (THI)<70 temperatures were similar between groups. When exposed to heat load (walking) at a THI of >74, differences in temperatures between the two groups were significant (t-test, $P<0.001$) in both non-lactating heifers and lactating cows rising to a 0.8-1.0 °C difference. Additionally, slick animals returned to baseline temperatures more rapidly than control animals. Hair length differed (t-test, $P<0.05$) between slicks and controls when sampled at the neck (4.3±0.46 vs 13.8±1.16 mm, respectively), pin bone (6.9±0.61 vs 15.2±1.42 mm, respectively) and shoulder (6.5±0.69 vs 14.9±1.5 mm, respectively). Previous research has identified a difference in in sweating rates, no difference between groups was identified in this study. These findings demonstrate that cattle with the slick genotype are better able to maintain temperature homeostasis under heat load in grazing conditions. A breeding programme is continuing to improve the genetic merit of slick dairy cattle.

Session 55 Theatre 3

Use of sensors for the detection and genetic evaluation of heat stress in dairy cattle

P. Lemal[1], M.-N. Tran[2], M. Schroyen[1] and N. Gengler[1]
[1]ULiège – GxABT, Passage des Déportés, 2, 5030 Gembloux, Belgium, [2]Elevéo, Rue des Champs Elysées, 4, 5590 Ciney, Belgium; pauline.lemal@uliege.be

Heat stress (HS) detection and its genetic evaluation is still a challenge. Indeed, the low frequency of milk recording drastically limits the number of records available during hot days. A solution could be to add the information of data from sensors. The objectives of this study were thus to evaluate the usability of sensor data to detect HS and to assess the possible gain of adding sensors data for HS genetic evaluation. SenseHub™ collars daily records were obtained from October 2019 to July 2022 in six herds. A total of 453 Walloon Holstein cows were followed during this period for activity and rumination time. Meteorological data and milk yield records were also obtained from 2015 to 2022 for 1,740 cows from the same herds. The thresholds at which the different traits start to be affected by heat stress were estimated at a temperature-humidity index (THI) of 63, 64 and 66 respectively for milk yield, activity, and rumination time with multi-traits models. Sensor data showed clear thresholds with a good variation along the THI scale. A three-trait random regression reaction norm model was fitted with these different thresholds. Heritability values were 0.19, 0.14 and 0.21 respectively for the three traits at the thresholds and these values were similar at high THI. Regarding reaction norm effects, activity and rumination time showed a positive phenotypic correlation with milk yield (0.37 and 0.50 respectively) and activity time also showed a positive genetic correlation with milk yield (0.51). If confirmed, results seem to validate that daily available sensors data could be a great help to detect HS animals because it allows to follow the cows in real time. Moreover, the drop is important (good variation along the THI scale), sudden (clear thresholds) and the same cows will probably also suffer from milk yield drop (positive phenotypic correlations with milk yield regression on THI). Activity time could also be a valuable tool for genetic evaluation of HS because the daily recording gives information for all events of HS, it is heritable including at high THI and the genetic correlations between the effect on milk yield and activity time is positive.

Session 55 Theatre 4

Evaluation of heat stress effects on production traits and somatic cell score of Dutch Holstein cows

J. Vandenplas[1], M.L. Van Pelt[2] and H. Mulder[1]
[1]Wageningen University & Research, Animal Breeding and Genomics, P.O. Box 338, 6700 AH, the Netherlands, [2]CRV, P.O. Box 454, 6800 AL, the Netherlands; jeremie.vandenplas@wur.nl

Global climate changes are expected to have an impact on livestock production throughout the world. For example, heat waves should increase in frequency and intensity, resulting in temperate regions facing regularly hot periods. During these hot periods, dairy cows will suffer from heat stress, reducing their milk production, reproductive performance and welfare. Therefore, selecting heat-tolerant dairy animals is of interest to cope with future heat waves. Among several indicators for heat stress, one of them is the temperature-humidity index (THI) that combines the effects of temperature and humidity on animals. The first aim of this study is to evaluate heat stress effects on milk production traits and somatic cell score (SCS) of Dutch Holstein cows using a THI indicator. The second aim is to estimate genotype by THI interactions for the same traits to evaluate the magnitude of the individual variability in heat tolerance. Data included more than 1.5 million test-day records for milk, fat, and protein yields, and somatic cell score (SCS). Data were collected from around 500 thousand first and second-parity cows in 1,581 Dutch herds between 2010 and 2022. All test-day records were associated with 3-day average THI values computed from publicly available meteorological records. For both parities, decreases in milk yield are observed from a value of THI around 65 (i.e. between 18 and 23 °C), and decreases in fat and protein yields are observed from a value of THI around 50. For SCS, increasing THI results in increasing SCS for the first parity (especially from a value of THI around 40), but does not impact SCS in the second parity. Individual responses to THI and to days-in-milk were estimated simultaneously using random regression models. Most genetic correlations along the THI gradient were higher than 0.90 at the 150th day-in-milk. Based on these results, genotype by THI interactions are weak for production traits and SCS of Dutch Holstein cows.
This project has received funding from the European Union's Horizon 2020 Programme for Research & Innovation under grant agreement n°101000226. This project is part of EuroFAANG.

Session 55 Theatre 5

Multi-omics and multi-tissues data to improve knowledge of heat stress acclimation mechanisms

G. Huau[1,2], D. Renaudeau[2], J.L. Gourdine[3], J. Fleury[4], J. Riquet[1] and L. Liaubet[1]
[1]INRAE, GenPhySE, 24, Chemin de Borde-Rouge, 31326 Castanet, France, [2]INRAE, UMR PEGASE, 16 Le clos, 35590 Saint Gilles, France, [3]INRAE, URZ, Domaine Duclos, 97170 Petit-Bourg, France, [4]INRAE, PTEA, Domaine Duclos, 97170 Petit-Bourg, France; guilhem.huau@inrae.fr

Elevated temperature is one of the main stressors with a huge impact on performance in the pig industry, and warming climate will enhance these concerns in the future. Long-term stress lead to acclimation of the pigs to cope with the effect of heat stress (HS) but underlying physiological mechanisms remains poorly described. This study aims to use multi-omics and multi-tissues data to understand the physiological pathways involved in heat tolerance in pigs. Transcriptomic data from seven different tissues (muscle, adipose tissue, liver, blood, pituitary, thyroid, adrenal gland), metabolomics data from four tissues (muscle, liver, plasma, urine), and blood parameters where obtained in an experiment involving 36 pigs from 3 genotypes slaughtered before (n=18) or after a 5-d exposure to 32 °C (n=18). Respiratory rate, rectal and skin temperatures significantly increased within the first 24 to 48 h of exposure to 32 °C when compared to the thermoneutral (TN) conditions (P<0.01). The average daily voluntary feed intake was significantly lower in HS conditions than in TN conditions (1,560 vs 1,707 g/d; P<0.01). With the aim to identify differentially expressed genes (DEG) and differentially produced metabolites, data were analysed with a mixed model with the effects of HS conditions, breed and sire origin. We identified 4,248 DEG in all tissues, with differences in the expression pattern between regulatory tissues (pituitary, thyroid, adrenal gland) and the other tissues. With a pathway enrichment analysis querying KEGG database, we have identified tissue-specific and recurrent pathways. Integrating metabolome with transcriptome data from muscle and sub-cutaneous adipose tissue, we have produced correlation networks about oxidative stress and thermogenesis, two important mechanisms related to HS. Integration of transcriptome data between regulatory tissues and an effector tissue allowed identifying hub genes that could have an importance regarding HS acclimation. This research was part of the MP ACCAF project P10157 (PigChange).

Session 55 Theatre 6

Should we consider fertility when improving thermotolerance in dairy cattle?

M.J. Carabaño[1], C. Díaz[1] and M. Ramón[2]
[1]INIA-CSIC, Mejora Genética Animal, Ctra de La Coruña km 7,5, 28040 Madrid, Spain, [2]IRIAF- CERSYRA, Departamento de Investigación-Area de Reproducción y Mejora Genética Animal, Av. del Vino s/n, 13300 Valdepeñas, Spain; mjc@inia.csic.es

Breeding for adaptation to heat stress (HS) has been normally approached by the use of indicators derived from the decline in production traits (quantity and quality) under high temperatures in dairy species. However, indicators of impairment of other functions associated with HS, such as reproductive performance, which have been less studied, may be advantageous because the reproductive function is most affected by stress and because the antagonistic relationship between productive level and productive decline under HS. In this study, different indicators of heat tolerance through the use of artificial insemination (AI) outcome and meteorological data have been analysed and the relationship with traditional productivity indicators quantified. Data for conception rate (CR, measured as 0/1) in first lactation, first insemination from 913,493 Holstein cows in 7,858 herds in Spain were used to select the optimal phenotypic indicator for HT derived from fertility data. Ridge regression considering average temperatures for the 5,10,15 and 30 days pre and post AI were used to compare the alternative indicators and decide in which period the AI outcome is more sensitive to the effects of heat. Substantial loss in CR was observed for all indicators beyond 20 °C of average temperature. Post AI indicators showed better goodness of fit than pre AI values. Genetic evaluations under a threshold random reaction norm model on CR using average temperature for 7 days post AI showed relevant variability in the slope of CR decay and an antagonistic relationship fertility level and slope of CR decay under HS. Correlations with heat tolerance derived from productive traits where close to null, indicating that selection for heat tolerance through maintaining productivity will not result in improving reproductive performance under HS and vice versa. This result points at the consideration of including both productive and reproductive declines when aiming at improving adaptation to high temperatures in dairy cattle. This project has received funding from the European Union's Horizon 2020 Programme for Research & Innovation under grant agreement n°101000226.

Trade-off between fertility and production in French dairy cattle in the context of climate change

A. Vinet[1], S. Mattalia[2], R. Vallee[2], A. Barbat[1], C. Bertrand[3], B.C.D. Cuyabano[1] and D. Boichard[1]
[1]Université Paris Saclay, INRAE, AgroParisTech, GABI, Jouy-en-Josas, 78350, France, [2]Idele, UMT eBIS, Paris, 75012, France, [3]INRAE, CTIG, Jouy-en-Josas, 78350, France; aurelie.vinet@inrae.fr

Climate change will induce harsher environments for European cattle production. One likely consequence is a growing antagonism between production and functional traits. To address this question, we investigated the evolution of trade-offs between fertility and production in French dairy cattle across a range from 15 to 75 of the temperature-humidity index (THI). Conception rate (CR) at first insemination (AI) and test-day protein yield (PY) from Holstein (HOL) and Montbeliarde (MON) cows were analysed. Only first lactation performances, recorded between 2010 and 2020, were considered. CR and PY were modelled according to the average THI of the 8 days after the AI and of the 3 days before the test-day, respectively. In total, we analysed 3,351,068 and 649,814 CR and 10,245,692 and 1,966,985 PY records from 3,368,605 and 656,164 HOL and MON cows, respectively. The evolution of trade-offs between CR and PY according to THI values was estimated with bivariate random regression models which included the fixed effects of herd × year, conventional vs sexed semen, weekday, age at AI, days in milk at AI for CR, and herd-test-day, age at calving, and days in milk for PY, and two random effects, the additive genetic effects of the sire for both traits and the permanent environment effect of the cow for PY. A sire model was chosen to allow very large analyses and thus ensure accurate estimates. The random effects were modelled with THI-dependent third-order Legendre polynomials. These models estimate genetic variances all along the THI trajectory as well as genetic correlations between traits across increasing THI conditions. The genetic correlations between PY and CR were moderately negative and evolved only slightly in the range of the observed THI. They increased with THI in HOL (around -0.1 for high THI), whereas they decreased in MON (around -0.15 for high THI). This study received funding from the European Union's Horizon 2020 research and innovation program under grant number 101000226 (Rumigen) and from APIS-GENE (CAICalor). The authors thank Meteo-France for the Safran database.

Genetic analyses of resilience indicator traits in German Holstein, Fleckvieh and Brown Swiss

F. Keßler, R. Wellmann, M. Chagunda and J. Bennewitz
University of Hohenheim, Animal Genetics and Breeding, Garbenstraße 17, 70599 Stuttgart, Germany; franziska.kessler@uni-hohenheim.de

Modern livestock farming faces various challenges such as climate change and new evolving pathogens, associated with new animal stressors. To meet these challenges, high performance and resilient animals are needed that are able to cope with disturbances. Animal breeding has been dealing with the possibilities of breeding for resilience since several years. In this context, the indicator traits variance of performance (Var) and variance (LnVar), autocorrelation (r_{Auto}) and skewness (Skew) of deviation of the observed from the expected performance were introduced in different animal species. With the long-term objective of establishing resilience as a breeding goal, we used in our study daily milk yields to apply these resilience indicators to 9,373 lactations from 4,426 dairy cows of German Holstein, Fleckvieh, and Brown Swiss breeds. A breed comparison and the effect of different degrees of smoothing of the expected lactation curves were investigated. We estimated heritabilities, phenotypic- and genetic correlations between resilience indicators and resilience indicators of different smoothing degrees. We also examined the correlations between resilience indicators and performance traits as well as functional traits. Resilience indicators showed low to moderate heritabilities in all breeds e.g. approximately 0.09 in Var, 0.14 in LnVar, 0.06 in r_{Auto} and 0.02 in Skew. Significant phenotypic and genetic correlations between resilience indicators were found e.g. approximately 0.43 Var/LnVar, -0.21 LnVar/r_{Auto} and 0.63 LnVar/Skew. The highest correlations between smoothing levels were found for LnVar and the lowest correlations were identified for Skew. The strongest correlations between performance traits and resilience indicators were found for LnVar e.g. 0.23 with milk yield. Estimation of genetic correlations with further functional and health traits are in progress, as well as a detailed investigation of external triggers (e.g. heat waves) that causes the resilience traits to be relevant.

Session 55 Theatre 9

Estimating the heritability of nitrogen and carbon isotopes in the tail hair of beef cattle

M. Moradi, C. Warburton and L.F.P. Silva
Queensland Alliance for Agriculture and Food Innovation, The University of Queensland, Gatton, QLD, 4343, Australia; m.moradi@uq.edu.au

The natural abundance of nitrogen isotopes ($\delta^{15}N$) in animal tissues is related to the ability of cattle to lose less nitrogen in the urine and has been used to estimate nitrogen use efficiency and feed efficiency. Carbon isotope enrichment ($\delta^{13}C$) has also been used to estimate feed conversion ratio and grazing behaviour in beef cattle. There are significant variations between the individual beef cattle in the same environment for $\delta^{15}N$ and $\delta^{13}C$ of the tail hair. Thus, the objective was to determine the heritability of $\delta^{15}N$ and $\delta^{13}C$ in the tail hair of tropically adapted beef cattle to further validate their use in genetic breeding programs. A total of 497 *Bos indicus* steers (269 Brahman and 228 Droughtmaster) were used, and the steers had reliable pedigree information for three generations as they were part of the beef cattle genetic evaluation system in Australia. Two single, mixed breed, contemporary groups of steers are used including the steers weaned in 2019 (n=254) and steers weaned in 2020 (n=243). Samples of tail switch hair representing hair segments growing during the dry season were collected and analysed for $\delta^{15}N$ and $\delta^{13}C$. Fixed effects fitted in the linear regression model were age, breed, year (as contemporary groups), and average daily gain during the different seasons. The heritabilities of $\delta^{15}N$ and $\delta^{13}C$ in the tail hair were estimated at 0.43±0.14, and 0.41±0.15, respectively, which indicates both are moderately heritable traits. The genetic correlation between $\delta^{15}N$ and $\delta^{13}C$ was negative at -0.78±0.16, and the phenotypic one was -0.40±0.04. The negative correlation can be explained by the contrasting isotope fractionation in the urine of cattle, as urine has less $\delta^{15}N$ and more $\delta^{13}C$ than the diet. The findings of this study have shown for the first time that both $\delta^{15}N$ and $\delta^{13}C$ in the tail hair of cattle are heritable traits. These results support the feasibility of utilizing tail hair isotopes as a promising approach for identifying cattle with greater nitrogen use efficiency, which is a key trait in low-protein diets. Moreover, genetic selection for nitrogen-efficient cattle can reduce the environmental impact of beef production.

Session 55 Theatre 10

Changes in genetic correlations over generations due to selection and random drift

B.C.D. Cuyabano[1], S. Aguerre[2] and S. Mattalia[2]
[1]INRAE, GABI, Domaine de Vilvert, 78350 Jouy-en-Josas, France, [2]IDELE, Domaine de Vilvert, 78350 Jouy-en-Josas, France; beatriz.castro-dias-cuyabano@inrae.fr

The evolution of genetic variance over generations is a topic that has been widely studied in quantitative genetics. In many breeding populations, both selection (resulting in the Bulmer effect) and limited population sizes (which builds up co-ancestry, resulting in random drift and inbreeding) reduce a trait's genetic variance. Until recently most research related to this topic remained theoretical due to limitations of data availability with sufficient generations for many species. Moreover, studies regarding the consequences on genetic correlations between traits under selection (and on their evolution over generations) are limited in the theoretical perspective, and to our knowledge inexistent empirically. It is however, reasonable to expect that both the Bulmer effect and random drift impact genetic correlations between traits under selection. Furthermore, if changes are significant when compared to a base population (typically used for inferences of genetic parameters), they should be accounted for when performing genetic evaluations comprising the most recent generations of a breeding population. A preliminary study on simulated data that mimics a dairy cattle population with respect to production and fertility (h2=0.3 and h2=0.01, respectively, and genetic correlation -0.18) indicated that when the population evolved through ten generations under random mating, although genetic variances decrease due to random drift, the genetic correlation remains unchanged. However, through ten generations under selection for either one of the traits individually, the genetic correlation was significantly attenuated. This attenuation was systematically progressive when selection was performed for production; when selection was performed for fertility, an immediate and relatively large attenuation is attained, then the genetic correlation is partially recovered and its trend follows the same as that of when selection was performed for production. Such attenuations are expected to be observed in real populations, a work that is currently undergoing. This study received funding from the European Union's Horizon 2020 research and innovation program under grant number 101000226 (Rumigen).

Session 55 Theatre 11

Longitudinal study of environmental effects for American Angus beef cattle over 30 years

G. Rovere[1], B.C.D. Cuyabano[2], B. Makanjuola[1] and C. Gondro[1]
[1]Michigan State University, Animal Science, East Lansing, 48824-1225, MI, USA, [2]INRAE, Domain de Vilvert, 78350 Jouy-en-Josas, France; roveregabriel@gmail.com

Beef cattle production in the USA is mostly pasture-based, thus, animals spend most of their life outdoors. In such a vast country, herds are exposed to different degrees of climate variability across location, each potentially influencing animals' performance differently. We analysed the trends of the American Angus population on birth weight (BW; ~6.7Mill records), weaning weight (WW; ~6.6Mill records), and post-weaning gain (PWG; ~3.4Mill records) over 30 years, and used historical records of climate variables to associate those trends with the climate variability in the locations of the herds. A single-step-multi-trait genetic evaluation was performed, with fixed contemporary group (CG) effect, maternal effect for BW and WW, and a permanent effect for WW. Using historical climate records from 1980-2020, we defined 5 classes of locations (LOC) in the USA, according to their frequency of higher than usual temperature-humidity-index (THI) during the breeding season. This classification of LOC was then assigned to the herds in our data through their GPS coordinates. From 1990, phenotypic and genetic trends were both decreasing for BW, and increasing for WW and PWG, with a smaller phenotypic slope observed after 2005 on WW and PWG. As for the effects obtained for CG (comprising climate+management), the trend was of decreasing effects over the years for all traits, while the variance of the CG increased during the same period. Interestingly, when studied per LOC, genetic trends were equivalent irrespective of LOC; phenotypic trends, although in the same direction for all LOC, presented lower values in LOC with more unusual THI events (LOC 4 and 5), and slopes of their trends were significantly different. CG effects were also consistently lower and with higher coefficient of variation in LOC 4 and 5 for all traits. From a population perspective, these results might suggest that, although genetic progress was equally achieved for herds in all LOC, in regions where unusually high THI events are becoming more frequent, management practices might not be sufficient to countervail the depressor effect of the climate on the animals' performance.

Session 55 Theatre 12

Characterization of environmental impact of 13,000 French dairy farms

R. Vial[1], A. Stocchetti[2], M. Mevel[3] and C. Brocas[3]
[1]Institut de l'Elevage, 9 rue de la Vologne, 54520 Laxou, France, [2]Institut de l'Elevage, 149 rue de Bercy, 75595 Paris Cedex 12, France, [3]Institut de l'Elevage, Monvoisin, 35652 Le Rheu, France; remi.vial@idele.fr

The French dairy sector is committed to the fight against global warming thanks to the low carbon dairy farm initiative. Indeed, several projects have been carried out to build the milk carbon footprint reduction road map. As part of these projects, a national life cycle assessment tool named CAP'2ER® was developed to measure the milk carbon footprint considering GHG emissions and carbon sequestration. Since the creation of the tool, 13,000 audits in dairy farm have been made. These assessments are a significant source of information on environmental impact of French dairy farms. A multicriteria analysis have been carried out to access the contribution of production parameters to the carbon footprint. The results are expressed in kg CO_2 equivalent per litre fat and protein corrected milk and per ha. There is a big variability in the results among farms. The average carbon footprint is 1 kg CO_2eq per litre and the farms which are among the 10% lowest emitters have carbon footprint 20% lower than the average. The production system has a small impact on the milk carbon footprint and practices explain a big part of the variability. Practices with the largest impact are milk yield, age at first calving, quantity of concentrate, N-fertilizer used, and direct energy consumed.

Session 55 Poster 13

Plateau-linear regression analysis of farrowing records on temperature data for pigs reared in Japan

S. Ogawa[1], T. Okamura[1], Y. Fukuzawa[1,2], M. Nishio[1], K. Ishii[1], M. Kimata[3], M. Tomiyama[3] and M. Satoh[2]
[1]Institute of Livestock and Grassland Science, NARO, Division of Meat Animal and Poultry Research, 2 Ikenodai, Tsukuba, Ibaraki, 305-0901, Japan, [2]Graduate School of Agricultural Science, Tohoku University, Sendai, Miyagi, 980-8572, Japan, [3]CIMCO Corporation, Koto-ku, Tokyo, 136-0071, Japan; ogawas897@affrc.go.jp

In Japan, national-scale genetic evaluation for female reproductive traits has been conducting with data collected throughout the year at multiple farms distributed around Japan. Farrowing performance of sows is affected by the time of year such as season, and heat stress induced by high temperature would be a contributing factor in shaping the effect of time of year. Here, for robust modelling of heat load, a plateau-linear regression analysis was performed using records of number born alive (NBA) from purebred Landrace, Large White, and Duroc female pigs reared in six farms in Japan obtained during 2007 and 2022 and public weather data. Daily maximum temperature data measured by Automated Meteorological Data Acquisition System (AMeDAS) at Japan Meteorological Agency (JMA) meteorological stations nearest from the farms were downloaded. The number of NBA records analysed was 52,668 for Landrace, 42,668 for Large White, and 5,459 for Duroc pigs. Coefficient of determination of the plateau-linear regression model was obtained with changing the breakpoint value from 10.0 to 25.0 °C at 0.1 °C intervals, and the value showing the greatest model fitting was reported as the threshold temperature. Daily maximum temperature data at mating day was used. For all breeds, values of coefficient of determinations showed trajectories like upward convex parabolas and estimated values of plateau-linear regression coefficient were negative, indicating the adverse effects of heat load by high temperature on sows reared in Japan. Our results show the possibility of utilizing publicly available weather data for considering the effects of heat load on NBA of pigs reared in Japan, which might contribute more flexible statistical modelling for national genetic evaluation of female reproductive performance to cope with ongoing global warming.

Session 55 Poster 14

Genetic parameter of heat tolerance for reproductive traits in Landrace, Large White and Duroc pigs

T. Okamura[1], Y. Fukuzawa[1,2], M. Nishio[1], S. Ogawa[1], K. Ishii[1], H. Takahashi[3], K. Tashima[3], K. Akachi[3] and M. Satoh[2]
[1]Institute of Livestock and Grassland Science, National Agriculture and Food Research Organization, 2 Ikenodai, Tsukuba, Ibaraki, 3050901, Japan, [2]Graduate School of Agricultural Science, Tohoku University, 468-1, Aramaki Aza Aoba, Aoba-ku, Sendai, Miyagi, 9808572, Japan, [3]Global Pig Farms, Inc., 800, Kamihakoda, Hokkitsu-machi, Shibukawa-shi, Gunma, 3770052, Japan; okamut@affrc.go.jp

This study aimed to estimate the genetic variance components reacting to heat stress in five reproductive traits, i.e. number of born alive; NBA, number of weaning piglets; NW, number of death piglets during suckling; NDW, total weaning weight; WW and average weaning weight; AWW. The data consisted of 26,770, 25,973, and 12,544 litters of Landrace, Large White, and Duroc sows, respectively, from five farms in Japan. The heat loads for each trait and breed as heat stress indicator were defined based on the daily maximum ambient temperature with optimal thresholds and periods for each trait and breed. The reaction norm model was fitted with the heat load for each breed and traits. Although the ratio of variances of heat tolerance to that of performance under comfortable condition were from 18 to 32%, these results indicated heat tolerance of these traits were heritable traits. The genetic correlations between heat tolerance and performance under comfortable condition were almost zero or positive (-0.01 to 0.79) for NW, NDW, WW and AWW, although these correlations were negative (-0.98 to -0.13). This suggested improving the heat tolerance by genetic selection could not interfere with the performance under comfortable condition for except of NBA. These results demonstrated that the heat tolerance was heritable and could be improved by genetic selection, and that can achieve to adapt pork production under global warming.

Breeding soundness evaluation of bulls in extensive systems in interior centre and south of Portugal

J. Várzea Rodrigues[1], L. Pinto De Andrade[1,2], S. Dias[1], J. Carvalho[1] and M. Martins[1,3]
[1]Instituto Politécnico de Castelo Branco, ESA, Av Pedro Álvares Cabral, no. 12, 6000-084, Portugal, [2]CERNAS, Av Pedro Álvares Cabral, no. 12, 6000-084, Portugal, [3]QRURAL, Av Pedro Álvares Cabral, no. 12, 6000-084, Portugal; luispa@ipcb.pt

Reproductive efficiency in beef cattle is a fundamental objective for the economic viability of farms and it is known that the bull contributes significantly in defining the profile that the fertility of a herd presents. Between January 2012 and December 2022, 1147 animals aged between 13 and 151 months were evaluated through breeding soundness evaluation (BSE) on 85 beef cattle farms in the interior centre (1) and in the south (2) of Portugal. BSE was carried out throughout the year with a higher prevalence in the months of September to January (71.58%), the time of greatest demand for this service by farmers, to prepare for the subsequent mating season. Considering the hottest months (June to October), the probability of bull failure is almost twice as high as in the rest of the year (OR=1.9867: IC 95%, 1.2841-3.0737). In the coldest months (November-March), the probability of failure is between 1.39-3.36 lower than in the rest of the year (OR=0.4622: IC 95%, 0.2974-0.7183). Of the evaluated animals, 80.21% were satisfactory, 6.28% unsatisfactory, and 13.51% deferred. Among the deferred animals (n=155), only 58 were reassessed and of these, 74.71% were reclassified as satisfactory. Assessing Body Condition Score (BCS) scale 1-9, we found that in the satisfactory animals, 11.62% had BCS≤4, 62.73% had BCS between 5 and 6 and 25.65% had BCS≥7. Considering the regions 1 and 2, 86.99 and 94.71% males were considered satisfactory and within these, the percentage of animals aged ≤24 months satisfactory was 24.28 and 78.57% respectively. The andrological examination should be considered an essential tool for the elimination of males characterized by subfertility or infertility, optimizing reproductive results, however it should be complemented with libido tests. The re-evaluation of animals classified as deferred is pertinent and necessary, recovering many males and avoiding the elimination of bulls that may be necessary for mating and may have a good genetic value.

Session 55 Poster 16

Transcriptome analysis identifies genes affected by heat stress in hen uterovaginal junction

S. Kubota[1], P. Pasri[1], S. Okrathok[1], O. Jantasaeng[1], S. Rakngam[1], P. Mermillod[2] and S. Khempaka[1]
[1]Suranaree University of Technology, School of Animal Technology and Innovation, 111 University Avenue, Suranaree, Muang, 30000 Nakhon Ratchasima, Thailand, [2]National Research Institute for Agronomy, Food and Environment (INRAe), UMR de Physiologie de la Reproduction et des Comportements, INRAE, 37380, Nouzilly, France; skubota@sut.ac.th

Female birds have the potential to modulate the motility of resident sperms. Heat stress decreases the reproductive ability of broiler breeder hens. However, its effects on the sperm storage tubules (SSTs) in the uterovaginal junction (UVJ), the primary site of sperm residence, remain unclear. In this study, we used RNA-sequencing to identify the differentially expressed genes (DEGs) in UVJ tissues containing SSTs of breeder hens affected by heat stress (36 °C for 6 h). A total of 561 genes, including 181 upregulated and 380 downregulated DEGs, were found to be differentially expressed compared to those in chickens raised under thermoneutral conditions (23 °C). Gene Ontology analysis revealed 209 significantly enriched terms involving heat shock proteins (HSPs). Kyoto Encyclopaedia of Genes and Genomes analysis identified nine significant pathways, including the protein processing in endoplasmic reticulum, neuroactive ligand-receptor interaction, biosynthesis of amino acids, ferroptosis, and nitrogen metabolism pathways. Protein-protein interaction network analysis of DEGs revealed two large networks. Our findings indicate that heat stress inhibits innate immunity in UVJ tissues of broiler chickens and that heat-stressed chickens protect their cells by increasing the expression levels of HSPs. Our results suggest that DEGs, including *HSP25*, *HSPA5*, *HSPA8*, *GKN2*, *IL4I1*, *PDK4*, *TAT*, *CA*, *LHCGR*, *GPX*, and interferon-stimulating genes may be used to further investigate SSTs in UVJ tissues of heat-stressed hens.

Session 56 — Theatre 1

Establishing and scaling up breeding programs: a challenging, but not impossible task

M. Wurzinger
BOKU-University of Natural Resources and Life Sciences, Vienna, Department of Sustainable Agricultural Systems, Gregor-Mendel-Str. 33, 1180 Vienna, Austria; maria.wurzinger@boku.ac.at

Establishing breeding programs is challenging, especially in extensive systems and developing regions. Breeding programs are complex and require good planning, but they also require constant monitoring during implementation. Many logistical problems, complex communication, and the coordination of different actors (farmers, agricultural extension service, research organizations, ministries) must be overcome. In recent years, a new approach to implementing breeding programs has been developed. So-called community-based breeding programs (CBBP), which focus on the needs and objectives of livestock farmers, have been successfully established. These programs usually start in pilot regions with fewer farmers with a common interest in improving their animals. Once the program is established, the next critical step is up-scaling. Expansion with new members also requires rethinking and reorganization, as existing decision-making structures often no longer meet the requirements. All stakeholders must be aware of and play their role in successful implementation. Scientists must work in an inter- and transdisciplinary approach and have a good understanding of the livestock production system, a profound technical understanding of animal breeding, and organizational management and development knowledge. Programs must be institutionally based and not dependent on individuals. Local and national governments must fulfil their mandate, create an enabling environment and ensure long-term support for farmers. Different examples of breeding programs are presented, and the specific challenges and possible solutions are discussed. Breeding programs are complex, very dynamic and must constantly adapt to new conditions (e.g. market requirements, and policies). But at the same time, this dynamic makes them so appealing and makes the work so exciting for animal breeders.

Session 56 — Theatre 2

Development of a breeding programme for oysters

P. Haffray, R. Morvezen, F. Enez, L. Dégremont and P. Boudry
INRAE/SYSAAF, Campus de Baulieu, 35042 Rennes, France; romain.morvezen@inrae.fr

In this presentation, we report how the French hatcheries invested since 1990 in polyploidization and selective breeding of the Pacific oyster *Crassostrea gigas*. After its massive introduction in 1971 to replace previous oyster species hit by a virus and two parasites, the production was hit again mid 80's by two new pathogens: OsHV-1 herpes virus and *Vibrio aesturianus*. A collective initiative to increase genetic resistance to OsHV-1 virus by sib-selection was not supported as this innovation could compete with the production of wild spats. Commercial hatcheries invested then in breeding program with SYSAAF expertise initially based on mass selection of surviving candidates to OsHV-1. Research works, developed in partnership with Ifremer and INRAE research organizations, improved practices by the development and application of reproductive and genomic tools to ease family productions (reproductive extenders), to manage inbreeding (DNA-parentage assignment) and since the last ten years by sib and genomic selection of the diploid lines (high-through put phenotyping, genomic indexation). The hatchery production of nearly 3 billion of mostly triploid spats (>50% of the national production) is making the application of breeding technologies and genomic selection as a case study for low trophic and extensive productions.

Session 56 Theatre 3

Status of implementation of EU animal breeding legislation for endangered breeds

H. Göderz[1], L. Balzar[1], J. Wider[1], C. Danchin[2], M. Spoelstra[3], M. Schoon[3] and S. Hiemstra[3]
[1]BLE, Deichmanns Aue 29, 53179 Bonn, Germany, [2]IDELE, 149 rue de Bercy, 75595 Paris Cedex 12, France, [3]WUR, P.O. Box 338, 6700 AH Wageningen, the Netherlands; holger.goederz@ble.de

With the Implementing Regulation (EU) 2022/2077 the European Commission designated the European Union Reference Centre for Endangered Animal Breeds (EURC-EAB). The EURC-EAB is led by Wageningen University & Research (WUR). Partners in the consortium are the Institut de l'Elevage (IDELE) and the Federal Office for Agriculture and Food (BLE). The EURC-EAB shall work with breed societies and third parties designated by them, competent authorities and other authorities of EU Member States to facilitate the preservation of endangered breeds and genetic diversity existing within those breeds. Within its Work Programme, EURC-EAB initiated a European wide survey and mapping process to show the status of the implementation of Regulation (EU) 2016/1012 concerning breeding programmes for endangered breeds. The mapping process will identify challenges and obstacles in the development and implementation of breeding and conservation programmes for endangered breeds. The regulation gives the opportunity to implement special derogations for endangered breeds in the design and implementation of their breeding programmes. The survey will show to what extent these derogations are used by breed societies. Breed societies will also be asked about the status of implementation of relevant elements of their breed specific breeding programmes, including performance recording, genetic evaluation, monitoring of genetic diversity and their collaboration with national gene banks. The survey will give insight in differences between countries, in particular their respective policies and procedures at national level, such as definitions and endangerment classification systems. Through this survey both breed societies and competent authorities will also be able to indicate specific challenges and obstacles in relation to the implementation of the breeding legislation and the development of sustainable breeding programs. The results of the study will be presented to the Standing Committee on Zootechnics for discussion among the Member States.

Session 56 Theatre 4

Spatial modelling improves genetic evaluation of Tanzanian smallholder crossbred dairy cattle

I. Houaga[1,2], R. Mrode[3,4], M. Okeyo[3], J. Ojango[3], Z. Nziku[5], A. Nguluma[6], A. Djikeng[2], E. Lavrenčič[7], G. Gorjanc[1] and I. Pocrnic[1]
[1]University of Edinburgh, The Roslin Institute and Royal (Dick) School of Veterinary Studies, EH25 9RG, Edinburgh, United Kingdom, [2]The Roslin Institute and Royal (Dick) School of Veterinary Studies, Centre for Tropical Livestock Genetics and Health, EH25 9RG, Edinburgh, United Kingdom, [3]International Livestock Research Institute, P.O. Box 30709, Nairobi 00100, Kenya, [4]Scotland's Rural College, The King's Buildings, EH9 3JG Edinburgh, United Kingdom, [5]Tanzania Livestock Research Institute, P.O. BOX 5016, TANGA, Tanzania, [6]Sokoine University of Agriculture, P.O. Box 3000, Chuo Kikuu, Morogoro, Tanzania, [7]University of Ljubljana, Biotechnical Faculty, Kongresni trg 12, 1000 Ljubljana, Slovenia; ihouaga@ed.ac.uk

African smallholder dairy production systems are characterized by small and scattered herds with low genetic connectedness between them. Coupled with the lack of an appropriate recording system, genetic evaluations are challenged to deliver meaningful accuracy, and hence typical genetic progress is low. One key underlying mechanism of genetic selection is an accurate separation of environmental and genetic effects. Thus, this study aimed to assess the impact of accounting for spatial distance/relationships on variance components estimation and the accuracy of genetic evaluation for daily milk yield in Tanzanian smallholder crossbred dairy cattle. To this end, we applied GBLUP-based model on 19,494 test-day milk yield records of 1,906 genotyped crossbred dairy cows from 1,394 herds. Genomic information consisted of 664,822 SNP markers after quality control. We modelled herds either as a random independent effect, a random spatially-correlated effect between the wards, or a random spatially-correlated (distance-based) effect between the herd locations. The results show large amount of spatial variation as well as of genomic estimated breeding values. We conclude that spatial modelling of herd effect more accurately separates genetic and environmental effects than independent herd effect and thus increases the accuracy of genomic evaluations. Further studies that integrate the genotype-by-environment interactions are needed to further modelling of the staggering variation in African smallholder crossbred dairy production systems.

Session 56 Theatre 5

Genetic parameters for grazing behaviour traits of Boutsko sheep

S. Vouraki[1], V. Papanikolopoulou[1], A. Argyriadou[1], V. Fotiadou[1], V. Tsartsianidou[1,2], A. Triantafyllidis[1,2], G. Banos[3] and G. Arsenos[1]

[1]Aristotle University, University Campus, 54124 Thessaloniki, Greece, [2]Epigenomics Translational Research, Center for Interdisciplinary Research and Innovation, Balkan Center, 57001 Thessaloniki, Greece, [3]Scotland's Rural College, Easter Bush, EH25 9RG Midlothian, United Kingdom; svouraki@vet.auth.gr

The objective was to estimate genetic parameters for grazing behaviour traits of Boutsko sheep reared semi-extensively in mountainous regions. A total of 296 ewes were randomly selected from three flocks (n=198, n=49 and n=49, respectively) in Epirus, Greece. Rotational monitoring of animal grazing behaviour was performed from June to September 2021, using global positioning system devices (GPS, n=50) attached on designated collars. GPS tracking of each animal was performed for 4-10 days at 2-60 minutes intervals. Traits measured included duration of daily grazing, distance, speed, altitude difference and elevation gain. Blood samples of the ewes were obtained from the jugular vein and DNA was extracted; samples were genotyped with the OvineSNP50 Genotyping Bead Chip v2. (Co) variance components and genetic parameters of grazing behaviour traits were estimated with mixed model analyses. Statistically significant ($P<0.05$) heritability estimates were reported for duration of grazing and speed (0.75 and 0.78, respectively). The latter traits and elevation gain were significantly repeatable (0.85, 0.86, and 0.14, respectively). Significant ($P<0.05$) negative genetic and phenotypic correlations were found between grazing duration and speed (-0.99 and -0.96, respectively) and positive ones between duration and distance (0.28 and 0.24, respectively). Such correlations indicate that practices for higher grazing duration are expected to decrease speed and favourably increase distance. Overall, results suggest that genomic selection practices could be implemented to improve grazing behaviour in the context of a multi-trait breeding programme. Moreover, there is significant between-animal variation for most grazing behaviour traits studied to support management practices aiming to improve adaptation to extensive rearing conditions. This work was funded by the SMARTER Horizon 2020 project (Grant No: 772787), https://www.smarterproject.eu/

Session 56 Theatre 6

Research and development innovations for climate-smart beef production in subtropical countries

M.M. Scholtz[1,2], G.M. Pyoos[1,2], M.L. Makgahlela[1,2], M.C. Chadyiwa[2], M.D. MacNeil[1,2], M.M. Seshoka[1,3] and F.W.C. Neser[1]

[1]University of the Free State, Department of Animal Science, P.O. Box 339, 9301 Bloemfontein, South Africa, [2]Agricultural Research Council, Animal Production, Private Bag X2, 0062, Irene, South Africa, [3]Northern Cape Department of Agriculture, Environment, Land Reform and Rural Development, Vaalharts Research Station, Private Bag X9, 8550 Jan Kempdorp, South Africa; neserfw@ufs.ac.za

The impact of climate change and the release of greenhouse gases influence the ruminant livestock industry in two ways. Firstly, the continuous increase in temperature will have both direct and indirect effects on the animal. The direct effects are mostly associated with heat and the indirect effects with feed sources, ecosystem changes and diseases. Secondly, ruminant production has the responsibility to limit the release of greenhouse gases to ensure future sustainability. The focus should be on adaptation, mitigation and resilience. Some research and innovations to achieve this in beef cattle in subtropical developing countries, will be discussed. The utilization of adapted and indigenous genotypes and the development of early warning systems should get attention. This could result in maintained levels of production in spite of adverse weather conditions. Any mitigation strategy should be aimed at improving efficiency of production, which will have positive effects on sustainability, while reducing the carbon footprint. The focus should thus be to improve cow-calf efficiency, selection for alternative measures of efficiency as well as the effective use of crossbreeding. It is recommended that, when selecting for cow-calf efficiency the use of the calf/cow weight ratio should be replaced by an index that includes both weights and reproductive performance. Residual traits such as residual feed intake, residual daily gain or a combination of both should be used when selecting for efficiency. Effective crossbreeding can have a small to medium effect on the reduction of the carbon footprint, while increasing the efficiency of production. Lastly resilience can be achieved through the breeding of less plastic or more climate resilient genotypes (identify genes for function / adaptation that make them more resilient).

Session 56 Theatre 7

Genetic and environmental factors influencing skin traits of South African farmed ostriches

K.R. Nemutandani[1], A. Engelbrecht[2], S.W.P. Cloete[3], K. Dzama[3] and O. Tada[1]

[1]University of Limpopo, Agricultural economics and animal production, Private Bag X1106, Sovenga, 0727 Polokwane, South Africa, [2]Western Cape Government, Directorate Animal Sciences, Department of Agriculture, P.O. Box 351, 6620 Oudtshoorn, South Africa, [3]Stellenbosch University, Department of Animal Sciences, Private Bag X1, 7602 Matieland, South Africa; khetho.nemutandani@ul.ac.za

Genetic and environmental parameters for ostrich skin traits were estimated in this study. Slaughter records of 2,660 South African Black ostriches were used, with an age range of 210 to 596 days. Fixed effects included were contemporary group, sex, dam age and their interactions. ASReml program was used for analysis. Age influenced all traits, except for neckline traits. Sex was significant for crown length, hair follicle score, skin weight, skin thickness, nodule size score and nodule shape score, with males producing thicker skins than females. Males had significantly higher scores for nodule size (4.6 vs 4.3), nodule shape (4.4 vs 4.1) and hair follicles (3.9 vs 3.5) than females. Direct single-trait heritability for skin size, slaughter weight, skin weight, skin thickness, hair follicle score and pitting score were 0.41±0.06, 0.37±0.06, 0.27±0.06, 0.20±0.05, 0.42±0.06 and 0.08±0.04, respectively. The significant genetic correlations of slaughter weight with crown width, crown length and crown shape were moderate to high at 0.58, 0.81 and 0.79, respectively. The genetic correlation of slaughter weight with skin size was high (0.92), moderate with skin weight (0.48), and low with skin thickness (0.15). Slaughter weight was also positively and significantly correlated to nodule size on the genetic level. High genetic correlation (0.71) was observed between nodule shape and nodule size scores. It was concluded that appreciable levels of genetic variation existed in most quantitative skin characteristics and genetic progress appears to be feasible through indirect selection as all traits are genetically favourably correlated to slaughter weight.

Session 56 Theatre 8

Genomic signatures of adaptive response driven by transhumant pastoralism in native Boutsko sheep

V. Tsartsianidou[1,2], S. Vouraki[3], P. Papanikolopoulou[3], G. Arsenos[3] and A. Triantafyllidis[1,2]

[1]Center for Interdisciplinary Research and Innovation, Genomics and Epigenomics Translational Research, Balkan Center, 57001 Thessaloniki, Greece, [2]Aristotle University of Thessaloniki, School of Biology, University Campus, 54124 Thessaloniki, Greece, [3]Aristotle University of Thessaloniki, School of Veterinary Medicine, University Campus, 54124 Thessaloniki, Greece; tsarvale@bio.auth.gr

Current and future climate fluctuations combined with selected livestock populations for production purposes raise major concerns about the preservation of animal genetic resources and food security. This study aimed to identify signatures of adaptive selection related to the impact of environmental heterogeneity on the genome of Boutsko, a Greek native transhumant sheep. Global positioning system (GPS) devices were used to monitor a total of 296 purebred Boutsko sheep reared semi-extensively during summer in three flocks located in Northwestern Greece. These populations were georeferenced based on the farm coordinates for lowland winter locations and genotyped with the Ovine SNP50K Bead Chip v2. A genotype-environment association (GEA) analysis was conducted with high and low-altitude grazing locations implementing logistic regressions. Fifteen SNPs were significantly associated with the bioclimatic variable of precipitation seasonality (bio15) representing the variation of monthly precipitation totals over the course of the year. Five candidate SNPs located on chromosomes 6, 15 and 22 were positioned within annotated genomic regions and their spatial genotype distribution coincided with geographic variation of precipitation seasonality. Functional enrichment analyses detected gene sets related to climate response, such as hypoxia response, energy metabolism, cell proliferation and differentiation and body conformation traits. The association between specific genomic loci and precipitation variation in the contrasting eco-climatic regions could be attributed to water and relevant pasture availability and composition within grazing areas seasonally and altitudinally. These results can enhance the incorporation of adaptive-related SNPs in selective breeding programs to face projected climate change. This work was funded by the SMARTER H2020 project (Grant No: 772787).

Session 56 Theatre 9

An Australian sheep genomic reference to meet the evolving breeding objectives of industry

S.F. Walkom[1], D.J. Brown[1] and J.H.J. Van Der Werf[2]
[1]Animal Genetics and Breeding Unit, University of New England, Armidale, 2351, Australia, [2]School of Environmental and Rural Science, University of New England, Armidale, 2351, Australia; swalkom@une.edu.au

The Australian sheep industry has used genomically enhanced breeding values since 2012. The ability to incorporate genomic information was only made possible by industry investment in a national genomic reference. Initiated in 2007, the Sheep CRC Information Nucleus Flock and its later incarnation the MLA Resource Flock has provided growth, carcase, and wool phenotypes on ~40k genotyped individuals across Merino, Terminal and Maternal breed types. Australian Wool Innovation's Merino Lifetime Productivity project and the Australian Merino Sire Evaluation sites also provided valuable data for the Merino breed. By maintaining a successful genomic reference that directly contributes to the national genetic evaluation producers were shown the direct value of genotyping. Currently, across the core national Sheep Genetic analyses, there are over 500k genotyped animals from industry and research contributing to the evaluation. As a result, the genomic reference is expanding beyond the research population and the role of industry levies to fund the reference population is declining. Future funding will depend on co-investment by breeders beyond levy contributions, with investment dependent on breeders perceiving value in a genomic reference. New and hard to measure traits continue to require recording in research flocks. Consequently, the next iteration of the resource flock will focus on underrepresented and future traits of importance. Providing the core population for the recording of methane and feed intake as well as resilience and maternal behaviour traits. The role of the reference population can be further enhanced by creating stronger linkages between different maternal and Merino ram sources, needed for prediction of breeding values across breed types and of crossbred animals. This would achieve more reliable genomic prediction reliability for key traits such as reproduction and meat eating quality across the commercial breeding flocks.

Session 56 Poster 10

Evaluation of the perceptions of the functions of local breeds of domestic ruminants in Mayotte

J. Vuattoux[1], A. Lauvie[2], A. Giraud[3], E. Ozarak[3], J. Janelle[4], A. Rozier[3], T.T.S. Siqueira[1], M. Naves[5] and E. Tillard[6]
[1]CIRAD, 7 chemin de l'IRAT, ligne paradis, 97410 Saint Pierre, France, [2]INRAE, 2 place Viala, 34060 Montpellier Cedex 01, France, [3]CIRAD, 69 Rue Moussa Oili, 97660 Tsararano, Dembéni, Mayotte, [4]CIRAD, Chem. de Baillarguet, 34980 Montferrier-sur-Lez, France, [5]INRAE, Centre de Recherche Antilles-Guyane, Domaine Duclos – Prise d'eau, 97170 Petit-Bourg, Guadeloupe, [6]CIRAD, Campus agronomique de KOUROU, Avenue de France, BP 701, 97387 Kourou cedex, French Guiana; anne.lauvie@inrae.fr

In Mayotte, local populations of ruminants are well adapted to the difficult breeding conditions of the island. The importation of animals and artificial insemination with European breeds, has led to a policy of crossbreeding and a decrease of the size of the local breeds herds. The work carried out by CIRAD, INRAE and the Chamber of Agriculture enabled the characterization of local bovine, ovine and caprine breeds. These three populations are a reservoir of adaptive genes. The conservation of domestic breeds is often considered from a genetic and zootechnical perspective. However, the choice of breeds follows along socio-economic dimensions. Two studies questioned the perception that breeders have of local breeds. The first study aimed at understanding the factors that influence the preservation or abandon of the Mahoran zebu breed. Sixty farmers were surveyed through semi-structured interviews. The data collected made it possible to describe the diversity of cattle farms in five types: (T1) small traditional farms; (T2) small precarious farms; (T3) farms specialized in dairy production; (T4) intermediate farms between tradition and intensification; (T5) farms in transition. The zebu is used for its ease of breeding and its resistance, in addition to its heritage and cultural value (T1, T2). It is of economic interest for crossbreeding (T3, T4) and ¾ of the breeders note the superior quality of its meat. The second study which is based on comprehensive interviews carried out on 19 farms on the island, explore the link between the functions assigned by the farmer to his small ruminant herd and the choice of its genetic composition. Both of these studies help to define the design of the conservation programs of the ruminant breeds in Mayotte, based on farmers perspectives.

Session 56 Poster 11

Signature of selection in South African Dexter cattle reveal resistance genes and genetic variations

E.D. Cason[1], J.B. Van Wyk[1], P.D. Vermeulen[2] and F.W.C. Neser[1]
[1]University of the Free State, Department of Animal Sciences, 205 Nelson Mandela Drive, Bloemfontein, 9301, South Africa, [2]University of the Free State, Office of the Dean, 205 Nelson Mandela Drive, Bloemfontein, 9301, South Africa; CasonED@ufs.ac.za

The environment shapes the genome of animals and drives them to carry sufficient genetic variations to adapt to changes in temperature. The Dexter breed is a small, dual purpose, breed of cattle that originated in the Southwestern region of Ireland and was imported to South Africa in 1917. These animals are farmed intensively and were under artificial and natural selection for adaptation to the semi-arid South African climate. Due to the small initial population, these animals were also crossbred with other local breeds. Selection should thus have left detectable signatures on the genome of these Africanised cattle. However, the selection response for production and adaptation traits in the Dexter genome is not explored. In this study, we provide the first overview of the selection signatures in the Dexter genome. The cattle were genotyped using the BovineSNP50 BeadChip and the SNP genotype data of its European contemporaries and constituent breeds were obtained from affiliates. Signatures of selection revealed genomic regions subject to natural and artificial selection that may provide background knowledge to understand the mechanisms that are involved in economic traits and adaptation to the South African environment. Principal component analysis (PCA) and genetic clustering emphasised the genetic distinctiveness of Dexter cattle relative to other South African cattle breeds. HapFLK results identified selection sweep regions distributed across 3 chromosomes, 1, 14 and 27. The identified regions overlapped with regions previously associated with quantitative trait loci (QTL) which included tick resistance, *Mycobacterium paratuberculosis* susceptibility, bovine tuberculosis susceptibility and bovine respiratory disease susceptibility. Selection for production traits were also present, such as carcass weight, milk yield, malving ease, fertility index as well as marbling and tenderness score.

Session 56 Poster 12

Breed environment interaction and suitability of Dutch cattle breeds for low input systems

J.J. Windig[1,2], G. Bonekamp[2], M.A. Schoon[1,2], A.H. Hoving[2] and S.J. Hiemstra[1]
[1]Wageningen UR, Centre for Genetic Resources the Netherlands, P.O. Box 338, 6700 AH, the Netherlands, [2]Wageningen UR, Wageningen Livestock Research, P.O. Box 338, 6700 AH Wageningen, the Netherlands; jack.windig@wur.nl

Dutch dual-purpose cattle breeds have been replaced for 98% by international transboundary high input/high output breeds, mostly the Holstein Friesian cattle breed. However, there is renewed interest in native dual purpose cattle breeds and their potential suitability for more 'nature-inclusive' production systems that require less external inputs. We performed a national inventory of dairy herds with Dutch dual purpose cattle breeds using data on production, health and fertility, combined with farming system characteristics and input and output parameters across production environments and regions in the Netherlands. Results were contrasted with performance of Holstein Friesian herds neighbouring the Dutch dual purpose cattle herds. Milk production varied considerably within breeds but was on average of a higher level in Holstein compared to the Dutch dual purpose breeds. Across the range of farming systems and environments milk, fat and protein levels were lower in grass based systems, on smaller farms and in more 'nature-inclusive' low input systems within the Holstein breed. For the dual purpose breeds changes across environments and systems were less clear, mainly because these breeds were more confined to specific regions and farm systems. Fat and protein and reproduction traits of Dutch dual purpose breeds were comparable and in some cases higher than levels in Holstein. For example, the Groningen White Headed had higher protein levels in grass based systems and the MRY in systems where more maize and concentrates were fed compared to the Holstein. These results help to identify future opportunities for the Dutch native dual purpose breeds in the context of transitions in Dutch livestock farming systems.

Session 56 Poster 13

Identification of candidate gene variants for the alpaca Suri phenotype by WGS analysis

S. Pallotti[1], D. Pediconi[2], M. Picciolini[3], M. Antonini[4], V. Napolioni[1] and C. Renieri[2]
[1]School of Biosciences and Veterinary Medicine, University of Camerino, Via Gentile III Da Varano, 62032 Camerino (MC), Italy, [2]School of Pharmacy, University of Camerino, Via Gentile III Da Varano, 62032 Camerino (MC), Italy, [3]SYNBIOTEC Laboratori s.r.l., Località Torre del Parco, 62032 Camerino, Italy, [4]Italian National Agency for New Technologies, Energy and Sustainable Development (ENEA), Casaccia, Via Anguillarese, 301, 00123 Galeria (RM), Italy; stefano.pallotti@unicam.it

Alpaca is a South American camelid bred for meat and fibre production. Two different phenotypes are described for this species: the short and crimped hair phenotype, known as Huacaya, and the long, straight and luster hair phenotype, known as Suri. To date, the genetic background behind the two phenotype is still unknown, however, segregation analysis suggest the Suri phenotype as dominant trait. In this research, whole-genome sequencing (WGS) analysis was used to uncover the genetics variant behind the Suri phenotype. Sample consisted of 19 WGS from Huacaya alpacas retrieved from public available repository (NCBI-SRA), 3 new WGS of Huacaya alpacas and 4 new WGS of Suri alpacas. Single reads were aligned to the most updated alpaca reference genome (VicPac3.1) and the genomic joint variant calling was performed. 37,421,914 variant were called, classified as single nucleotide polymorphism (30,749,986), insertions (3,547,803) and deletions (3,124,125). The called variants were then annotated in order to predict the phenotypic effects. 81,859,906 phenotypic effects were predicted, of which 61,094 were of 'high impact', 312,519 were of 'moderate impact', 545,541 were of 'low impact' and lastly, 80,940,752 were classified as 'modifier impact' influencing genomic untranslated regions. Finally, a case-control filtering were performed assuming the Suri phenotype as dominant mutation. 258 out of 35.371.788 variants remains. Five variants located on four loci were missense mutations while the remaining 253 were annotated to untranslated genomic regions. The five missense variants identified are promising candidate for the alpaca Suri phenotype.

Session 57 Theatre 1

Extensive permanent grasslands in Europe: multifaceted functions, threats and prospects

M. Bassignana
Institut Agricole Régional, Reg. La Rochère, 1/A, 11100 Aosta, Italy; m.bassignana@iaraosta.it

Permanent grasslands represented in 2020 more than 30% of the 157 million hectares of land used for agricultural production in the EU, but they were incurring a declining trend, with a reduction of 2.0 million hectares in a decade. The EU Biodiversity Strategy for 2030 explicitly highlights the preservation of low-intensive permanent grasslands as one of the sustainable practices to be implemented, in accordance with the Farm to Fork Strategy, in the framework of the new Common Agricultural Policy. The first and evident function of extensive permanent grasslands, with particular reference to mountain ones, is to support sustainable animal productions, given that the breeding of domestic ruminants is the only way to obtain food for humans in vast areas where, at the present time, no other type of food crops would be possible. At the same time, their aptitude to deliver multiple ecosystem services, in addition to provisioning ones, is unique and peculiar, ranging from biodiversity and supporting services, to regulating and cultural ones. Mountain grasslands, in fact, are essential elements of centuries-old breeding systems which have been maintaining over time local breeds – often on the verge of extinction – as well as the continuity of mountain civilization, whose distinctive characteristics are absolutely original and peculiar. The durability of these systems is increasingly jeopardized by the combination of various factors, such as low economic benefits, workload and scarcity of manpower, climate change and drought, leading to the abandonment of marginal areas. In order to promote the proper preservation of these resources and of the farming systems associated with them, multidisciplinary research is essential to highlight their significance – for animal production but more generally for the environment and for people, be aware of the risks they are facing and, finally, discern what prospects can be expected for their future.

Session 57 Theatre 2

The effect of establishment and grazing management on clover and herb establishment and persistence

L. McGrane[1,2], N. McHugh[2], T.M. Boland[1] and P. Creighton[2]
[1]University College Dublin, School of Agriculture and Food Science, Belfield, Dublin 4, D04V1W8, Ireland, [2]Teagasc, Animal and Grassland Research and Innovation Centre, Mellows Campus, Athenry, Co. Galway, H65R718, Ireland; Lisa.McGrane@teagasc.ie

The addition of clovers and herbs to grazed swards has shown many benefits including increased performance of grazing ruminants, increased herbage production and reduced dependence on chemical nitrogen inputs. Poor plant persistency under grazing is one of the main challenges associated with these diverse sward types. Although there has been an increase in the use of diverse sward types in recent years, there is a lack of grassland management advice available for their use. Two adjacent sheep grazed plot studies were established using a complete randomised block design in 2019 and measured in the years 2019 to 2022 inclusive, investigating the effect of: (1) establishment method (n=60 plots); and (2) post-grazing sward height (n=36 plots), on binary sward mixtures of perennial ryegrass plus one companion forage (white clover, red clover, plantain or chicory). Post removal of the pre-existing sward the four establishment methods implemented were: (1) conventional (plough, till, sow); (2) Disc (disc harrow cultivation, power harrow cultivation, sow); (3) Power Harrow (Power harrow cultivation, sow); and (4) Direct Drill (direct drill with no cultivation of the soil). The three post-grazing sward heights implemented were: (1) 4.0 cm; (2) 4.75 cm; (3) 5.5 cm. Results show that establishment method had no effect on white clover, red clover or chicory content in the sward, however there was a higher plantain content in swards sown using the Direct Drill method compared to the Conventional method (P<0.05). In the third full production year, white clover content was similar across all post-grazing sward heights, however there was a higher red clover content in swards grazed to 4.75 cm than both the 4.0 and 5.5 cm treatments (P<0.05). There were higher chicory and plantain contents in swards grazed to 4.0 cm than those grazed to 5.5 cm (P<0.05). In conclusion, establishment method had an impact on plantain content in the sward and appropriate grazing management is vital to maximise forage persistency in diverse sward types.

Session 57 Theatre 3

Strategic concentrate supplementation in reducing slaughter age in pasture-based dairy-beef systems

J. O'Driscoll[1,2], D. Purfield[1], N. McHugh[2] and N. Byrne[2]
[1]Munster Technological University, School of Biological Sciences, Bishopstown, Cork, T12 P594, Ireland, [2]Teagasc, Animal & Grassland Research and Innovation Centre, Grange, Dunsany, Co. Meath, C15 PW93, Ireland; jamie.odriscoll@teagasc.ie

Increased efficiency within livestock production systems is critical to reducing greenhouse gas emissions, with reduced age at slaughter cited as a key mitigation policy. Beef systems of reduced slaughter age must maximise the use of grazed grass over the animal's lifetime due to its lower cost of production and lessened environmental impact compared to other feeds. The objective of this study was to investigate the strategic use of concentrate within a pasture-based dairy-beef steer production system, to establish its effectiveness in reducing slaughter age, using Holstein Friesian (HF; n= 80 animals) and Angus × Holstein Friesian (AAX; n= 160 animals) genotypes. Contrasting supplementation strategies were analysed: (1) grass only (GO), pasture only diet in the first and second grazing season; (2) intermediate (INTER), concentrate supplementation throughout the first grazing season, pasture only thereafter; and (3) high concentrate supplementation (HIGH), concentrate supplementation throughout the first grazing season, and from July until slaughter/housing of the second grazing season. Drafting for slaughter across each strategy occurred at a body condition score ≥3.75, on a 5-point scale. The age at slaughter of AAX steers was 80 days earlier than HF (631 vs 712±2 days: P<0.001) across concentrate treatments. HIGH treatment AAX steers were slaughtered 43 (±3.8) days earlier (P<0.001), and produced a lighter carcass (294 vs 307±4.8 kg, P<0.05) than both GO and INTER AAX steers. However, HIGH AAX steers consumed more concentrates (+122±21.75 kg, P<0.001) over the finishing period than the GO and INTER steers (278±21.75 kg), which did not differ. Irrespective of concentrate treatment, slaughter age did not differ for HF steers (P>0.05). HIGH HF steers produced a heavier carcass (326.2 vs 308.2±6.3 kg, P<0.05), and consumed more concentrate (+334±32.2 kg, P<0.001) than GO and INTER HF steers, which did not differ. Age at slaughter of AAX steers, can be reduced by strategic concentrate use during the second grazing season, allowing earlier finish at pasture.

Morphology and body composition of beef-on-dairy heifers along compensatory growth itinerary

I. Morel[1], A. Dieudonné[1], R. Siegenthaler[1], C. Xavier[1,2] and S. Lerch[1]
[1]Ruminant Nutrition and Emissions, Research Contracts Animals, Agroscope, 1725 Posieux, Switzerland, [2]INRAE-Institut Agro, PEGASE, 35590 St-Gilles, France; isabelle.morel@agroscope.admin.ch

The aim was to evaluate if finishing heifers fed a grass-based diet during a compensatory growth period (CG) would recover the effects of a previous restrictive feeding (RF). Sixty-six crossbred heifers (♀ Swiss Brown × ♂ Angus, Limousin or Simmental) were used from 271 to 527 kg BW. Thirty-three grew discontinuously (DI), 111 d on mountain pasture, followed by 80 d CG at barn (65:20:8:7 grass silage/hay/maize silage and concentrates, DM basis). The remaining 33 were fed continuously (CO) the barn-diet during 191 d. Body morphological measurements were performed on 12 heifers per treatment by full body 3D imaging, dorsal and rump ultrasounds, and trained operator grading (CHTAX), at the beginning and the end of RF, at the end of CG and before slaughter. At the same dates (except end of CG), direct *post mortem* measurements of empty body (EB) chemical composition were performed on 6 additional heifers per treatment and date. The RF induced lower average daily gain (ADG, 0.45 vs 1.06 kg/d), leading to a weight difference of 84.4 kg at the start of CG ($P<0.001$), lower fat cover score (1.34 vs 3.27) and skin and adipose tissue thickness ($P<0.05$) for DI vs CO heifers. Accordingly, lipid content in EB (9.0 vs 17.3%) was lower and protein content higher at the end of RF ($P<0.001$). Morphologically, the DI heifers were significantly lower in body length, area, and volume, but not in height. Over 80 days of CG, the DI heifers ingested more (8.7 vs 7.8 kg DM/d), had a higher ADG (1.46 vs 1.02 kg/d) and valorised the ration more efficiently (0.17 vs 0.13 kg ADG/kg DMI; $P<0.01$) than CO heifers, achieving a BW compensation index of 43%. Increase in body height was however strongly slowed down during CG, thus prioritizing width gain. At slaughter after 160 d of refeeding, full compensation of the gap due to RF between DI and CO heifers was observed, in carcass grading, morphological measurements, as well as for EB lipid and protein contents ($P>0.10$). This trial highlights wide adaptive processes in body morphological and composition traits when growing cattle had to cope with nutritive challenges along CG itinerary.

In France, a new beef × dairy calf to steer production for the out-of-home consumers

M.A. Brasseur[1,2], C. Fossaert[1,2], F. Guy[1,2], J.J. Bertron[2], T. Dechaux[2] and S. Brouard[2]
[1]CIRBEEF, 4 La Touche Es Bouvier, 56430 Mauron, France, [2]Institut de l'élevage, 149 rue de Bercy, 75595 Paris, France; marc-antoine.brasseur@idele.fr

Following the decline in veal and young bull productions, French dairy calves are increasingly oriented towards export, resulting in long distance transport for young calves and rising questions from consumers. However, these calves could produce carcasses adapted to the out-of-home market, which is mainly supplied by imported meats from dairy herd. In this context, trials have been realized on the CIRBEEF farm in Mauron (56), with the aim of developing beef × dairy steer production (16-17 months) with light carcass weight (280-300 kg carcass), well fattened and maximizing grazed or conserved grass. Four groups of 56 beef × dairy calves have been studied with two modalities depending on finishing mode (indoor/pasture) linked to the birth period (fall/winter). Weaned 56 days after their arrival, then fed a maize silage-based ration, the four groups showed an average daily gain (ADG) of 736 g/day on the arrival-weaning period and 1,011 g/day on the weaning-120 days of presence period. The two very contrasting climatic years had a strong impact on the growth of grass and therefore on the feeding practices (housing in summer, supplementation, etc.). However, the global growths are 1,070 g/day for the group finished indoor and 973 g/day for the group finished on pasture. With 5.4 and 7.4 months of pasture, respectively for the groups finished indoor and on pasture, the grass account for 54 and 61% of the lifetime feed consumption. This main grass part in the diet explains the protein self-sufficiency level of 81% which is better than 51% for our full indoor production based on maize silage. Of the 222 animals slaughtered, the average age at slaughter and the average carcass weight were 17.2 months and 304 kg. Slaughterhouses are looking for conformations O+ or better which is the case of 71% of carcasses. Moreover, the carcass yields were relatively high for dairy origin animals (on average 54%) and reflect well the diversity of the potentialities of the meat breeds used for the crossbreeding. The meat produced with these systems have gross emissions of 6.4 kg eq. CO_2 /kg LW and a net carbon footprint of 6.2 kg eq. CO_2 /kg LW (Calculated using the CAP'2ER® method).

Behavioural and welfare responses of dairy cows learning a virtual fencing system

P. Fuchs[1,2], J. Stachowicz[2], M. Schneider[2], M. Probo[2], R. Bruckmaier[3] and C. Umstätter[4]
[1]GCB, University of Bern, Mittelstr. 43, 3012 Bern, Switzerland, [2]Agroscope, Rte de la Tioleyre 4, 1725 Posieux, Switzerland, [3]University of Bern, Veterinary Physiology, Bremgartenstr. 109a, 3001 Bern, Switzerland, [4]Thünen-Institute of Agricultural Technology, Bundesallee 47, 38116 Braunschweig, Germany; patricia.fuchs@agroscope.admin.ch

Virtual fencing (VF) offers the possibility of replacing physical fences with a virtual system. A proper use of VF aims to condition animals to an audio tone (AT, 82 dB for 5-20 s) to avoid a light electric pulse (EP, 0.2 J for 1 s). The present study investigated the learning behaviour of 10 lactating cows under VF and its effects on cow behaviour and welfare compared to 10 cows managed with electric fences (EF). Both treatments were split into groups of 5 (2× VF, 2× EF) balanced by age and lactation stage. Each group grazed half-day in a separate, EF paddock during a 3 d lead-in period (P0) with inactive VF, followed by 4 periods (P1-4) with active VF. All cows were equipped with a VF collar (Nofence AS, Batnfjordsøra, Norway) and an IceQube pedometer (Peacock Technology Ltd., Stirling, UK). We continuously monitored cow activity and daily milk yield, body weight, and feed intake. Further, we recorded milk cortisol and the frequency of agonistic interaction, vocalization and excretion at the start, middle, and end of P0-P4. During 59 d, the cows received a mean of 108±54 AT and 7±3 EP. At Day 1, the number of EP was highest (18) but decreased to 5 at Day 3 (-72%) and remained below that threshold for the rest of the trial. With each paddock change, cows were successively conditioned to the AT, which was reflected in a decreasing ratio of EP/AT from Week 1 (22%) to Week 4 (2%), 6 (1%), and 8 (0%). Milk yield, milk cortisol, feed intake, body weight, and activity were similar in VF and EF groups. Throughout the trial, we observed a mean of 11.2 vocalizations and 5.9 displacements more per cow in the VF groups compared to the EF groups ($P<0.05$). However, their frequency did not differ between P0-P4. Our results indicate that all cows learned to cope with the VF system without lasting behavioural changes or negative effects on animal welfare. The conditioning of the cows to the AT of the VF system succeeded at herd level after 3 repetitions on a new virtual fence.

Age does not affect the learning capacity of virtually fenced cows

A. Confessore[1], C. Aquilani[1], P. Fuchs[2], C. Pugliese[1], C.M. Pauler[2], M. Schneider[2], G. Argenti[1] and M. Probo[2]
[1]University of Florence, Department of Agriculture, Food, Environment and Forestry (DAGRI), Piazzale delle Cascine, 18, 50144 Firenze (FI), Italy, [2]Agroscope, Rte de la Tioleyre, 4, 1725 Posieux, Switzerland; andrea.confessore@unifi.it

Virtual Fencing (VF) can be a helpful technology in managing herds in pasture-based systems. In VF systems, paddock boundaries are set in GIS only. Animals wear a GPS collar, which emits a sound signal of increasing pitch as soon as an animal crosses the virtual fence, followed by a weak electrical impulse if the animal does not return. The stimuli sequence is repeated up to 3 times if the animal continues to walk forward. It is well known that animals can learn a VF system, but it is unknown if learning capacity decreases with animal age. The study aimed to investigate whether old learn the system worse than young ones and whether VF impacts grazing activities and milk production. The study was conducted in the Swiss lowlands on 4 strip-grazing paddocks, comparable in forage biomass and botanical composition. 20 lactating Holstein-Friesian cows were divided into 4 groups of 5 animals, with each group grazing a separate paddock. Groups differed in terms of age: 2 old age groups (average of 5.2 and 4.8 lactations) and 2 young age groups (first lactation). After a 7-d training period, each paddock was gradually increased in size by VF during 5 consecutive grazing periods, based on forage biomass availability. Each cow was equipped with VF collars (Nofence AS, Batnfjordsøra, Norway) and leg pedometers (Peacock Technology Ltd., Stirling, UK) to record the daily number of sounds and electrical pulses as well as step count respectively. Data were analysed by generalized mixed-effect models demonstrating that age had no significant impact on animals' overall response to the VF system. However, there was a clear effect of time: independent of age, both electrical pulses and sounds significantly decreased within days of each period. Moreover, there were no significant differences in daily step count between ages. Finally, no changes in milk production were detected, comparing before, during, and post-VF treatment. In conclusion, results suggest that age does not affect animals' learning capacity, grazing activities, and milk production as well.

Integrating multiple data streams and models to inform precision grazing management in the U.S

J.R. Brennan, H. Menendez and K. Ehlert
South Dakota State University, 711 N Creek Dr., Rapid City, SD 57703, USA; jameson.brennan@sdstate.edu

Precision livestock management (PLM) technology has the potential to increase efficiency and sustainability of animal agriculture. Although any specific piece of technology requires unique steps for processing and turning data into insights, key to the advancement of PLM will likely be the integration of multiple technologies and data streams into animal nutrition models to better inform management decisions. Stocking rate (the number of animal units on an area of land per unit of time) is an essential component for setting forage utilization targets, minimizing overgrazing, and meeting land management objectives. However, setting appropriate stocking rates requires information on current forage production (kg/ha), and animal body weight (kg), variables which are often unknown to livestock managers at relevant time scales. The objectives of our study were to: (1) develop machine learning models to predict forage quantity and quality using open source climate and satellite imagery datasets; (2) develop data pipelines to process real-time information on individual animal and herd weight; and (3) develop a precision grazing model that integrates forage predictions and animal weights to dynamically adjust grazing rotations using virtual fence technology. Random forest regression trees were used to predict forage quality and quantity across the two sites in South Dakota, U.S. based on climate and imagery metrics derived from Google earth engine. R2 values for predicted vs measured data in 2020-2021 were 0.91 for forage production, 0.78 for CP, 0.55 for NDF, and 0.77 for ADF. SmartScalesTM were deployed in grazing areas, and an automatic program interface was developed to estimate daily average herd weight. Weight and forage predictions were incorporated into a decision grazing model to estimate when to rotate animals based on forage utilization targets under three stocking rate scenarios (heavy, moderate, and light). The decision grazing model was then used in conjunction with virtual fencing technology to inform grazing area rotations and is a critical next step to maximizing precision technologies and mathematical animal nutrition modelling.

Yearly monitoring of soil ingestion by dairy cows in a grassland system with feed supply

C. Collas[1], A. Laflotte[2], C. Feidt[1] and S. Jurjanz[1]
[1]Université de Lorraine, INRAE, URAFPA, 54000 Nancy, France, [2]ENSAIA, Université de Lorraine, Centre R&D La Bouzule, 54280 Laneuvelotte, France; claire.collas@univ-lorraine.fr

Soil may be a vector of contaminants, such as trace metals or organic pollutants, from the environment to the animal. This can result in exceeding regulatory thresholds in food of animal origin and questioning the livestock farming sustainability in contaminated areas. Quantifying soil ingestion and determining its variation factors allow to characterise and adapt farming practices to reduce the free-range animal exposure to contaminants. Soil ingestion has been few studied in dairy cows, once in different barns in the USA and once with strip grazing in France. The present study monitored the soil ingestion in a dairy herd in East France from July 2019 to June 2020. During the winter period from December to March, the cows were housed without outside access and fed with a total mixed ration based on corn silage. Otherwise, cows had access to permanent grassland with the same ration in lower proportions depending on grass availability. Soil ingestion was estimated collectively every fortnight when the cows had access to pasture and every month when they were housed. A composite faecal sample was taken at each time point from a pool of 12 cows. Individual estimates of soil ingestion were made on 30 cows on four dates. Grass, feed, soil and faeces samples were analysed using titanium as a soil marker. Data were analysed by mixed models. Daily soil ingestion averaged 0.5% of total intake (125 g dry soil). The highest level (1.5% of total intake, 300 g dry soil) was obtained in late autumn when the soil was very wet which could amplify grass soiling. Low or not-detectable soil ingestions were observed in winter (housing conditions) and in summer when the soil was dry. Animal characteristics (age, lactation stage, milk yield) affected poorly soil ingestion. Compared to the literature, this farm management resulted in relatively low soil ingestions compared to cattle on exclusive grazing, but consistent for grazing with supplementation. Feed supply during difficult pasture conditions has shown to be an efficient practice to maintain low levels of soil ingestion and ensure animal health and food safety.

Session 57 Theatre 10

Pre-grazing sward height affects enteric methane emission during grazing

L. Koning, G. Holshof, A. Klop and C.W. Klootwijk
Wageningen University and Research, Wageningen Livestock Research, De Elst 1, 6708 WD Wageningen, the Netherlands; cindy.klootwijk@wur.nl

In order to reduce national greenhouse gas emissions, the Dutch dairy sector aims to reduce the enteric methane (CH_4) emission of dairy cattle. Grazing strategies could facilitate in achieving this goal as recent research showed that enteric CH_4 might be lower during grazing compared to grass silage diets. One of the strategies expected to influence CH_4 emission during grazing is the height of the sward prior to grazing. The pre-grazing sward height is related to the number of growing days thereby impacting the quality of the grass. The objective of this study was to compare enteric CH_4 emission based on a grazing system with a low and a high pre-grazing sward height. An experiment was conducted in 2020 and 2021 in Leeuwarden, The Netherlands. Two groups of fifteen dairy cows were randomly assigned to two treatment groups: a low pre-grazing sward height (8 cm; LS) and a high pre-grazing sward height (15 cm; HS). The experiment was divided into two weeks of adaptation and two weeks of measurement period and was repeated three times during the season: Apr-May (period 1; P1), Jun-Jul (period 2; P2), and Aug-Sept (period 3; P3). Cows had outdoor access to a pasture during the day and received grass silage indoors during the night in a 50:50 ration. In addition, they were fed 5.5 kg concentrates per cow per day. Enteric CH_4 emission was measured using GreenFeed® stations. Per period no significant differences were found in CH_4 emission, with exception of the CH_4 intensity (per kg FPCM) in P1 of 2020 and P3 of 2021, which was lower for LS compared to HS. The meta-analysis of the CH_4 emission across periods and years showed a significantly ($P<0.001$) lower CH_4 yield, intensity and production for LS (resp. 20.0 g CH_4/kg DM, 15.2 g CH_4/kg FPCM and 378 g CH_4/cow/day) compared to HS (resp. 20.9 g CH_4/kg DM, 16.2 g CH_4/kg FPCM and 397 g CH_4/cow/day). Based on these results we can conclude that the pre-grazing sward height, which is related to the number of growing days, can influence enteric CH_4 emission. This study highlights that grazing strategies can contribute to mitigate enteric CH_4 emission in dairy farming systems.

Session 57 Theatre 11

Forage shortage affects performances, CH_4 emissions and cheese quality in grass- or corn-fed cows

M. Bouchon[1], I. Verdier-Metz[2], M. Eugene[3], C. Bord[2], B. Martin[3], J. Bloor[4], M.C. Michalski[5], B. Graulet[3] and C. Delbès[2]
[1]INRAE, UE Herbipole, 63122 Saint-Genes-Champanelle, France, [2]UCA, VAS, INRAE, UMR Fromages, 15000 Aurillac, France, [3]UCA, INRAE, VAS, UMR Herbivores, 63122 Saint-Genes-Champanelle, France, [4]UCA, INRAE, VAS, UMR UREP, 63000 Clermont-Ferrand, France, [5]INRAE, Inserm, Univ-Lyon, UMR CarMeN, 69310 Pierre-Bénite, France; matthieu.bouchon@inrae.fr

Grass-based dairy systems are considered to contribute to the agroecological transition and resilience of farming compared with more intensive corn silage-based systems, but the impacts of drought-induced forage shortages on services provided by these different systems are unclear. This study aims at evaluating the impact of a reduction of grass availability of -25% of the total intake during two months, as might be expected in case of a summer drought, for groups of cows fed either a pasture-based diet (AE) or a corn silage-based diet (IN). In each group, only half of the cows were subjected to the reduction in grass allowance (75 to 50% in AE, replaced by a hay-based diet; 25 to 0% in IN, replaced by corn-silage based TMR). Milk yield (MY) was higher for IN cows but the reduction in grass allowance did not affect MY nor milk fat content. MY decreased more over time for AE cows, in line with seasonal decreases in grass nutritional value. Milk protein content was the highest for IN cows with no more access to pasture. Dry matter intake was higher for IN cows and increased by the reduction in grass allowance. Raw CH_4 emissions were estimated using INRAE 2018 equations and were higher for IN cows, particularly for those with no grass in the diet. However, CH_4 emissions calculated on a MY basis were similar for all cows but expressed on a DMI basis, they were lower for IN groups and decreased in response to forage shortage for both AE and IN cows. Cheeses made with raw milk from AE groups were more yellow and softer while cheeses from IN groups had a lesser intense global odour. Within each group, cheeses were less yellow and firmer for cows subjected to a forage shortage. This study suggests that reducing access to pasture in a grazing system has less effect than removing pasture in a corn-silage based system on the services provided by the farming system.

Session 57 — Theatre 12

Variability of economic and GHG performance in dairy-beef systems at different stocking rates

M. Kearney[1,2], E. O'Riordan[1], J. Breen[2], R. Dunne[3], P. French[4] and P. Crosson[1]
[1]Teagasc, Animal & Grassland Research and Innovation Centre, Teagasc, Grange, Dunsany, Co. Meath, Ireland, C15PW93, Ireland, [2]University College Dublin, School of Agriculture and Food Science, University College Dublin, Belfield, Dublin 4, Ireland, D04V1W8, Ireland, [3]Teagasc, Johnstown Castle, Co Wexford, Y35Y521, Ireland, [4]Teagasc, Livestock Systems Department, Animal & Grassland Research and Innovation Centre, Moorepark, Fermoy Co. Cork, Ireland., P61C997, Ireland; mark.kearney@teagasc.ie

The capacity of dairy-beef systems to run at high stocking rates (SR) has been identified as a key driver in increasing farm profitability and reducing GHG emissions. However, excessive SR may restrict animal performance with potential implications for GHG emissions and farm profitability. Therefore the objectives of this study was to parameterize a farm-level model using data from an experimental field study to model the effects of contrasting stocking rates on the economic and GHG emission performance of dairy-beef production systems. A total of 216 spring-born early and late-maturing dairy-beef steer and heifer calves were purchased and blocked by breed, birth date and weight on arrival and assigned to one of three treatments. The three treatments were based on SR and classified as low SR (LSR; 2.6 Livestock Units (LU)/ha), medium SR (MSR; 2.9 LU/ha) and high SR (HSR; 3.2 LU/ha). Each treatment consisted of 36 steers and 36 heifers. Slaughter weight and carcass weight were heavier and better conformed for LSR than MSR and HSR. Steers and heifers finished in HSR had greater live weight and carcass output per hectare and subsequent net profit per ha than LSR and MSR. GHG emissions per kg of product were lower for HSR. Increases in stocking rate maximise profit per hectare and reduce GHG emissions per kg of product provided it is undertaken via improvements in pasture productivity and utilisation.

Session 57 — Poster 13

Feeding behaviour, methane emission and digestibility of crossbred heifers along compensatory growth

B. Hayoz, I. Morel, A. Dieudonné, M. Rothacher, R. Siegenthaler, F. Dohme-Meier and S. Lerch
Research Contracts Animals; Ruminant Nutrition and Emissions, Agroscope, 1725 Posieux, Switzerland; bastien.hayoz@agroscope.admin.ch

Compensatory growth (CG) occurs along refeeding after a period of feed restriction, and is characterized by high feed intake. Aim was to study feeding behaviour, enteric methane (CH_4) emission and digestibility during CG in beef-on-dairy crossbreeds. Heifers [♀ Brown Swiss × ♂ Angus (AN), Limousin (LI) or Simmental (SI), n=66] were used from 271 to 527 kg body weight (BW). Thirty-three grew discontinuously (DI), 111 d on mountain pasture, followed by refeeding 80 d at barn (65:20:8:7 grass silage/hay/maize silage and concentrates, DM basis). The remaining 33 were fed continuously the barn-diet (CO). Individual intake and feeding behaviour were obtained using automatic weighing troughs. Enteric CH_4 emission was measured with greenfeed system and digestibility over 5 d using insoluble ashes as indigestible marker. Individual means were computed over refeeding for ANOVA analysis (package GLM, RStudio 4.1.3) with fixed effects of dietary treatment, crossbreed and their interaction, and pen as random effect. Correlations between variables were explored (package RCORR). During CG, DI had higher DM intake (DMI) than CO heifers (9.2 vs 8.2 kg/d, P<0.01) with concomitant increases in meal duration and DMI per meal (P≤0.05). Nevertheless, DI had a lower organic matter digestibility (dMO) than CO (79.8 vs 81.9%, P<0.01), and tended to have lower CH_4 yield (28.4 vs 29.8 g/kg DMI, P=0.09). Compared to LI and SI, AN heifers had higher DMI but lower dMO (P≤0.01), whereas none difference was observed on CH_4 yield (g/kg DMI) between crossbreeds (P=0.34). Methane (g/d) was positively correlated with DMI and DMI per meal (r=+0.60, +0.39, respectively; P≤0.05) and negatively with dMO (r=-0.42, P<0.01). The dMO was also negatively correlated with DMI (r=-0.50, P<0.001). Only correlation between DMI and CH_4 remained significant when DI and CO groups are considered separately (r≥+0.48, P≤0.01). This trial highlights broad adaptive processes in feeding behaviour, concomitant with increase in feed intake along CG, with further consequences on CH_4 emission and digestibility.

Session 57 Poster 14

Involving farmers in the development of a grassland monitoring tool: sunshine's co-design approach

D.M. Mathy[1], C.L. Lucau-Danila[1], Y.C. Curnel[1], E.R. Reding[2], K.D. Dichou[3] and S.L. Lagneaux[1]
[1]Walloon Agricultural Research Center, Rue du Serpont, 100, 6800 Libramont, Belgium, [2]Walloon Breeders Association (Elevéo), Rue des Champs Elysées, 4, 5590, Ciney, Belgium, [3]Gembloux Agro-Bio Tech (ULG), Passage des Déportés, 2, 5030, Gembloux, Belgium; d.mathy@cra.wallonie.be

To help farmers managing their grasslands more efficiently, researchers around the world have been working on the development of new technologies such as biomass assessment out of satellite imaging and grass growth modelling. The Sunshine project aims at adapting these technologies for Wallonia (Belgium) and making them available in a dedicated decision support tool (DST). However, latest research revealed a low utilization rate of other existing grassland-related DSTs by Walloon farmers. To understand the reasons of this low adoption rate and to overcome it, a participatory approach including farmers in a co-design process has been set-up. In the first stage, we led semi-directive interviews with dairy and beef farmers spread across Wallonia, which allowed us to collect their design ideas, unveil their practical constraints and build a list of the ideal features for the software. In the second stage, we submitted this list to evaluation by a group of technicians and a group of farmers in order to identify the most relevant features. In the third stage, we developed a prototype with a core selection of basic features. Then, we engaged the farmers actively in the design process by organizing focus groups where they could try the prototype and exchange in an advanced co-design session. Using this approach allowed us to refine a typology of grassland management practices. Subsequently, this made possible to identify a core of features that would match to a greatest range of farmers and at the same time to avoid overspecific features that would inevitably lead to poor utilization rate of the software. This approach also generated novel ideas coming from the farmers, such as the way information should be addressed (e.g. eating-days left on a given paddock) or by reversing the purpose of a given feature. Lastly, our case study allowed us to discuss the critical points that raised our attention when it comes to involving farmers in the development of new DSTs.

Session 57 Poster 15

Effects of increasing portion of grass-silage in dairy cow diet on carbon footprint of raw milk

S. Hietala[1], A. Vanhatalo[2], K. Kuoppala[3], T. Kokkonen[2], A. Reinikainen[1], K. Timonen[1] and A.-L. Välimaa[3]
[1]Natural Resources Institute Finland, Bioeconomy and environment, Paavo Havaksen tie 3, 90570 Oulu, Finland, [2]University of Helsinki, Department of Agricultural sciences, Koetilantie 5, 00014 Helsinki, Finland, [3]Natural Resources Institute Finland, Production systems, Paavo Havaksen tie 3, 90570 Oulu, Finland; sanna.hietala@luke.fi

Finnish dairy cows' diet consists of 55-60% of the grass silage (GS). Aim of this study was to find the impact of increasing GS content of dairy cows' diet to the carbon footprint (CF) of milk. We conducted a simulation study using experimental data from lactating dairy cows. Different diets were based on GS or equal mixture of red clover / GS and supplemented with barley and oats. Current results are for different proportions of forages on GS based diets, GS 55 or 65%. The CF of milk was assessed with life cycle assessment (LCA). IPCC approach was used as described in Hietala *et al.* and it was adjusted for the diet composition, as well as milk yield and composition according to the simulation results. Milk yield was converted to fat and protein corrected milk (FPCM). Average number of calvings per cow was assumed to be 3.3 and slaughter age was 1,825 d. Equal slaughter weight was assumed for both diets. The '55% GS' diet included 4,000 kg dry matter (DM) GS/year, cereals 1,800 kg DM/year (1:1 barley and oat) and rapeseed meal 650 kg DM/year. The milk yield was 10,735 kg FPCM/year. The '65% GS' diet included 4,413 kg DM GS/year, cereals 1,298 kg DM/year (1:1 barley and oat) and rapeseed meal 469 kg DM/year. The predicted milk yield was 10,092 kg FPCM/year. Increasing the GS proportion in the diet led to lowered milk yield (-1,769 kg FPCM). The demand for arable land was reduced (-2,353 m^2), when GS portion was increased. The total emissions of the dairy cow were lowered by 1,819 $kgCO_2eq$ with the increased GS content of the diet. Yet, when the emissions were allocated to the products, the higher FPCM yield of the 55% GS diet levelled out the difference at product level and carbon footprint of FPCM was 2% higher with 65% GS diet. Acknowledgments This study was conducted in Leg4Life project (2019-2025) funded by the Strategic Research Council at the Academy of Finland (grant numbers 327700 and 327698).

Blood metabolite, hormone and δ13C turnover kinetic during compensatory growth of crossbred heifers

S. Lerch[1], P. Silacci[1], G. Cantalapiedra-Hijar[2], R. Siegenthaler[1], S. Dubois[1], A. Delavaud[2], M. Bonnet[2] and I. Morel[1]
[1]Ruminant Nutrition and Emissions, Animal Biology, Research Contracts Animals, Feed Chemistry, Agroscope, 1725 Posieux, Switzerland, [2]INRAE, Université Clermont Auvergne, Vetagro Sup, UMRH, 63122 Saint-Genès-Champanelle, France; sylvain.lerch@agroscope.admin.ch

Compensatory growth occurs when cattle are refed after feed restriction and is characterized by changes in metabolic and hormonal profiles. Aim was to explore relationships between feed efficiency or compensation intensity, and blood metabolites, hormones and ^{13}C natural abundance ($\delta^{13}C$). Sixty-six beef-on-dairy heifers (♀ Swiss Brown × ♂ Angus, Limousin or Simmental) were used from 270 to 527 kg body weight (BW). Thirty-three grew discontinuously (DI), 111 d on mountain pasture, followed by 80 d refeeding at barn (65:20:8:7 grass silage/hay/maize silage and concentrates, DM basis). The remaining 33 were fed continuously the barn-diet (CO). Blood serum non-esterified fatty acids (NEFA), glucose, creatinine, τ-methyl-hystidine (τMH, CAS:332-80-9), and plasma insulin, IGF-1 and $\delta^{13}C$ were determined at d 0, 2, 4, 8, 16, 35 and 70 of refeeding. Relationships (Proc CORR, SAS 9.4.) were explored with feed conversion efficiency [FCE=average daily gain (ADG)/DM intake] and compensatory index (CI; reduction of BW differences between DI and CO treatments over 80 d refeeding). When compared to CO, ADG of DI heifers was lower at pasture (0.45 vs 1.06 kg/d), but higher during compensation (1.46 vs 1.02 kg/d), as for FCE (0.17 vs 0.13; P<0.001). CI was 43% whatever the crossbreed (P=0.52). Plasma $\delta^{13}C$ at d 0 was negatively correlated with ADG at pasture (r=-0.77), but positively over compensation (r=+0.76), as for FCE (r=+0.63); the reverse being observed for d 0 IGF1 (r=+0.46, -0.76, -0.51) and d 70 τMH (r=+0.34, -0.65, -0.59; $P\leq0.10$). The FCE was positively correlated with increases (by difference) in glucose from d 0 to 70 and creatinine from d 0 to 4 (r=+0.59, 0.56, respectively; $P\leq0.06$), but negatively with d 2 to 16 τMH and d 16 to 70 insulin increases (r=-0.62, -0.58; $P\leq0.05$). The CI was positively correlated with d 0 to 4 creatinine increase (r=+0.60) and $\delta^{13}C$ turnover (r=+0.57; $P\leq0.06$). Feed efficiency and CI were associated with blood metabolites, hormones and $\delta^{13}C$ kinetic following a refeeding in beef-on-dairy heifers.

Metabolic assessment of parasite dilution and forage niche sharing in sheep/cattle mixed-grazing

F. Joly[1], P. Nozière[1], P. Jacquiet[2], S. Prache[1] and B. Dumont[1]
[1]Université Clermont Auvergne, INRAE, VetAgro Sup, UMR Herbivores, 63122 St Genes Champanelle, France, [2]IHAP, INRAE, ENVT, Université de Toulouse, 31076 Toulouse, France; frederic.joly@inrae.fr

Mixed-grazing by sheep and cattle is the simultaneous or sequential grazing of a pasture by both species. It can improve lamb liveweight gain through parasite dilution (PD) and/or forage niche sharing (FNS). Here, we assessed the relative strengths of the two mechanisms through a novel metabolic approach. We used recently published equations to model the infection cost of gastrointestinal nematodes in metabolizable energy (ME) and crude protein (CP). By comparing infection levels in mixed and monospecific grazing, we quantified the gains of PD in ME and CP. We also used feed value tables to assess the gains in ME and CP, resulting from sheep diet improvement through FNS. We applied this approach to the dataset of an experiment, comparing sheep monospecific grazing to simultaneous mixed sheep-cattle grazing. We also applied it to a generic situation where we studied the relative gains in ME and CP, along gradients of increasing strength of PD and FNS. The approach applied to ewe lamb in our experimental data revealed that: (1) infection by gastrointestinal nematodes can represent 100% of ME and 75% of CP requirements in monospecific grazing; (2) mixed-grazing can reduce these costs to 25 and 15% of requirements, respectively; and (3) PD was more important than FNS in terms of ME gains, whereas it was the opposite for CP. However, meeting CP requirements was less constraining than meeting ME requirements in our experimental conditions, which puts into perspective the importance of CP gains. With the generic approach, most of the situations modelled also identified PD as the main mechanisms of ME gain (79%), whereas it was FNS for CP (70%), with the same observation that CP requirements were less difficult to meet. Both our experiment and generic approach thus suggest that PD matters more than FNS in mixed-grazing, owing to the greater difficulty in meeting ME requirements. We proposed a novel approach to assess the roles of two contrasting mechanisms through common metrics. It can help improve the comprehension of the biological processes involved in agroecological practices, such as mixed-grazing.

Session 57 — Poster 18

Effects of different additives on the correlation between fermentation characteristics of wilted rye

Y.F. Li[1], L.L. Wang[1], Y.S. Yu[1], H.J. Kim[2] and J.G. Kim[1,2]

[1]Seoul National University, Graduate School of International Agricultural Technology, 1447 Pyeongvhang-Ro, Daehwa, Pyeongchang, Gangwon, Korea, 25354, Korea, South, [2]Seoul National University, Institute of GreenBio Science and Technology, Seoul National University, 1447 Pyeongvhang-Ro, Daehwa, Pyeongchang, Gangwon, Korea, 25354, Korea, South; 2019-20640@snu.ac.kr

The primary objective was to explore whether chemical or biological additives have effective effects on the correlation between the fermentation characteristics of wilted rye. The whole rye was harvested at the early heading stage and wilted in the field for 24 h. Wilted rye was divided into 6 treatments: no additive (C), 3 g/kg sodium diacetate (SDA3), 6 g/kg sodium diacetate (SDA6), *Lactobacillus plantarum* (Lp), *L. buchneri* (Lb), or their equal mixture (Lbp) at 1×10^6 cfu/g fresh matter. After 60 days ensiling, it showed that lactic acid (LA) was negatively correlated with acetic acid (AA) except for the C, where the negative correlation of Lp was the highest (-0.994, $P<0.05$). There was a negative correlation between LA and NH_3-N in all groups with Lp (-0.990) having the highest negative correlation and S3 (-0.198) the lowest ($P<0.05$). LA in C and S6 showed a positive but extremely low correlation with its pH while the other groups showed a negative correlation. LA in C, Lp and Lbp was negatively correlated with WSC while the other three groups were positively correlated and Lbp was the highest (0.997, $P<0.05$). Interestingly, the correlation between AA and WSC was just the opposite. NH_3-N increased with AA in all silages except the C. AA in all silage showed a positive correlation with pH but a negative correlation with CP. WSC concentration of C, Lp and Lbp was negatively correlated with NH_3-N and pH. There was a positive correlation between NH_3-N and LAB counts only in Lp and Lbp and Lp (0.927)>Lbp (0.144). The NH_3-N concentration was low, the pH was high except for the C. AA and pH were higher, CP was lower. CP content was positively correlated with LA but negatively correlated with NH_3 apart from C. Additives significantly altered the correlation between fermentation characteristics and LP was optimal.

Session 57 — Poster 19

Bite item selection by grazing suckler cows in multi-species grasslands

C. Siede[1], W. Pohlmann[1], A. Juch[1], D. Hamidi[1], J. Isselstein[1,2] and M. Komainda[1]

[1]University of Goettingen, Deparment of Crop Science, Von-Siebold-Str. 8, 37075 Göttingen, Germany, [2]University of Goettingen, Centre for Biodiversity and Sustainable Land Use, Büsgenweg 1, 37077 Göttingen, Germany; caroline.siede@uni-goettingen.de

Bite item (BI) choice of grazing cattle depend on various interacting factors including BI diversity and availability in the grass sward. Multi-species grasslands provide a spatio-temporally varying availability of potential BI. It remains an open question if cattle utilize the available BI diversity by choice or not and if that depends on the grazing intensity (GI). The BI availability and choice of suckler cows was studied in a long-term replicated cattle grazing experiment under two GI (moderate, M; lenient, L) (in total six 1 ha paddocks). The GI paddocks were stocked with three or two non-lactating Fleckvieh cows. During two grazing periods in 2022 (spring, autumn) the actual available BI were assessed pre grazing along two transects (200 points per paddock) using a modified sward stick (10×10 cm steel frame) to simulate the bite size area of adult cattle. At each point, the height as well as the botanical composition (functional groups, colour and the phenological stage) were measured. The diversity of unique BI per pasture and period was calculated subsequently. The realised BI choice was assessed by video recording and observation of each cow in the morning and afternoon (4×2 minute intervals; in total 16 min/cow and period) using a mobile phone app for recording and subsequent video analysis. The diversity of chosen BI was calculated. In total 51 unique BI were found. Available BI diversity in the grass sward was significantly higher in spring (M: 22.8±2.15 vs 13.5±2.15; L: 22.3±2.15 vs 17.0±2.15). Chosen BI diversity increased with greater available BI diversity in the grass sward and was affected by the interaction of GI×period. Generally, it was greater under lenient grazing (spring: 12.1±0.08 vs 14.9±0.09; autumn: 1.4±0.03 vs 4.5±0.05). More BI are selected in pastures with greater available BI diversity.

Session 57 Poster 20

Nutritional value of intramuscular fat of the muscle of Arouquesa weaners from different systems

L. Sacarrão-Birrento[1], C.A. Venâncio[2,3], A.M. De Almeida[1], L.M. Ferreira[3,4], M.J. Gomes[2,3], J.C. Almeida[2,3], J.A. Silva[2] and S.P. Alves[5]

[1]LEAF, ISA, Lisboa, Portugal, [2]CECAV, UTAD, Vila Real, Portugal, [3]ECAV, UTAD, Vila Real, Portugal, [4]CITAB, UTAD, Vila Real, Portugal, [5]CIISA and Associate Laboratory for Animal and Veterinary Sciences, FMV, Lisboa, Portugal; laurasvbirrento@isa.ulisboa.pt

The awareness about the fatty acids (FA) content in beef is increasing due to the relation of some FA with several diseases. Autochthonous breeds like the Arouquesa from Northern Portugal, are associated with healthier beef due to a more extensive way of production. This work aimed to compare the lipid nutritional value of Longissimus thoracis from Arouquesa weaners produced under different systems: TF, traditional system with weaning and slaughtering at 9 months; TFS1, the addition of an initiation supplement (S1); S1S2, animals fed with S1 until weaning at 5 months and then a growing supplement (S2); TFS3, addition of a rearing period until 12 months with a finishing supplement (S3); and S3, animals produced like the S1S2 but with the rearing period. Total lipids were extracted from freeze-dried muscle with dichloromethane and methanol (2:1), and intramuscular fat (IMF) was quantified gravimetrically by weighting the final lipid residue. FA methyl esters were prepared and quantified through gas chromatography with flame ionization detection. The IMF differed between the 5 groups being highest in the TFS1 group and lowest in TFS3 group. The omega-6/omega-3 ratio in meat was lowest in the TF and TFS1 groups (3.53 and 4.52), as expected since the omega-3 FA were higher, and highest in the TFS3 and S1S2 groups (11.37, and 10.0). About the ratio PUFA/SFA, it was lowest in TF, TFS1 and S3 groups (0.35, 0.30 and 0.36) and highest in TFS3 (0.56). The h/H index (hypocholesterolemic/hypercholesterolaemic fatty acids) was higher in groups S1S2 and TFS3 (2.24 and 2.35). In conclusion, Arouquesa meat produced under the traditional system seems to be healthier for human consumption when considering the omega-3 FA. Regarding the h/H ratio and PUFA/SFA, that are associated with lowest proportions of hypercholesterolaemic fatty acids, the group TFS3 produced the meat nutritionally more favourably, however, this effect could be associated with the lowest IMF content.

Session 57 Poster 21

Prediction of nutritional parameters of naturalized grassland in the dry zone of Chile using NIRS

P.M. Toro-Mujica

Instituto de Ciencias Agroalimentarias, Animales y Ambientales. Universidad de O'Higgins, Ruta I-90 km 1, 3070000, Chile; paula.toro@uoh.cl

The naturalized grassland in the dry central zone of Chile presents a wide variety of species whose presence is conditioned by climatic, edaphic, and management factors. The nutritional quality of the pasture, in addition to the species present, depends on its phenological state. In this way, the nutritional quality varies widely between farms and throughout the year. The objective of the work was to develop calibration curves based on near infrared (NIR) spectra of parameters related to the nutritional quality of the naturalized grassland in the dry central zone of Chile. One hundred thirty eight grass samples were obtained from 32 exclusion plots on four farms from August to December 2021 and 2022. The spectra of the samples were obtained through a FOSS NIRS™ DS2500 to later determine dry matter (DM), crude protein (CP), acid detergent fibre (ADF), and neutral detergent fibre (NDF) by chemical analysis. Modified partial least squares calibration equations were developed in different spectral regions, applying several mathematical treatments. The values (mean ± standard deviation) obtained for the parameters were 60.9±28.6, 8.4±4.7, 40.2±5.7, and 57.4±10.7% for DM, CP, ADF, and NDF, respectively. Models showed good predictive values for the prediction. Coefficients of determination of cross validation (1-VR) of the selected equations presented values between 0.8 and 0.9 and standard error of cross validation (SECV) between 1.47 and 4.39. These results support the viability of NIRS technology to predict the composition parameters of naturalized grassland in the dry central zone of Chile.

Session 57 Poster 22

Effect of cutting length on fermentation dynamics of wilted Italian ryegrass silage

J. Kim[1,2], Y. Li[2], L. Waang[2], Y. Yu[2] and H. Kim[1]
[1]Seoul National University, Institute of GreenBio Science and Technology, #1447, Pyeongchang-ro, Pueongchang, Kangwon, 25354, Korea, South, [2]Seoul National University, Graduate School of International Agricultural Technology, #1447, Pyeongchang-ro, Pueongchang, Kangwon, 25354, Korea, South; forage@snu.ac.kr

Italian ryegrass silage is currently quite popular in Korea and the cutting length plays an important role in the success of the fermentation in making silage. The harvested Italian ryegrass was wilted in the field for 2 days and divided into 3 parts, which were chopped into 10 (CL10), 20 (CL20) and 30 (CL30) mm respectively. The ensiled IRG was sampled at 1, 2, 3, 5, 10, 20, 30, and 45 days indicated that different cutting length had significant effect on CP content for 2-days wilted IRG silage. CL30 had the highest CP from 5 to 45 days (P<0.05), and showing an upward and downward trend in the first 10 days, and then an upward trend. CP changes in CL20 was more stable than CL10 and CL30 and CL10 was the greatest (P<0.05). NDF and ADF content of CL10, CL20 and CL30 had similar trends during ensiling but different cutting length had no significant effect on the NDF content changes (P>0.05). For ADF, CL20 was significantly lower than CL10 and CL30 (P<0.05). IVDMD was negatively correlated with NDF in all treatments. WSC as a whole was declining during ensiling. After 45-days ensiling, CL10 was the lowest (11.44 g/kg DM) but at day 10 CL20 and CL30 showed the opposite change from CL10 (P<0.05). Cut length significantly affected lactic acid production and CL10 was the highest, followed by CL20, and CL30 was the lowest (P<0.05). pH value and lactic acid have the opposite trend of change, and CL10 was lower than CL20 and CL30 (P>0.05). pH value had an overall slow decline trend.

Session 57 Poster 23

Grazing behaviour of energy-limited dairy cows and development of detection method by deep learning

Y. Shinoda, S. Asakuma, Y. Ueda, S. Tada and K. Sudo
Hokkaido Agricultural Research Center, NARO, Sapporo, Hokkaido, 062-8555, Japan; shinoday080@affrc.go.jp

Understanding the exact energy status of grazing dairy cows is important for health management. However, it is difficult to accurately estimate the amount of dry matter intake in pasture. Recent advances in sensors have enabled long-term and high-density monitoring of animal behaviour. The purpose of this research is to clarify the behavioural characteristics caused by energy deficient by analysing behavioural data from sensors and to develop a method to automatically detect the energy-deficient of cows. From 2019 to 2021, a total of 24 lactating Holstein-Friesian cows, eight each year, were grazed in Hokkaido region for one month during the summer. Two treatment groups were established according to the amount of supplementary feed supplied at the barn: a control group (n=12) with a TDN sufficiency of 110% and a limited group (n=12) with a TDN sufficiency of 80%. The behaviour of individual was monitored using a global navigation satellite system (GNSS) logger and an accelerometer attached to the cow's collar. To assess the impact of energy deficient on behaviour, a generalized linear mixed model was used with collected behavioural data as the explanatory variable, treatment group and grazing duration as fixed effects and individual and day and year as mixed effects. Total eating time, eating bout duration, distance travelled during eating bout and total moving distance were significantly longer in the limited group than in the control group. We developed an energy deficient detection model using a convolutional neural network (CNN) using all 441 movement trajectory images. Of all movement trajectory images, 70% were divided into data for model creation and 30% were divided into data for testing. Moreover, of the former, we randomly split 70% of the data as training data and 30% as validation data to develop a detection model. As a result, the highest accuracy of energy deficient detection model using movement trajectory images was about 65.1%. Although there is a need for improvement, it was suggested that the movement trajectory could be used to detect the energy deficient of grazing dairy cows.

Session 57 Poster 24

Evaluation of 3 equations based on grass height measurement for estimating grass stocks in pastures

F. Lessire, J.-L. Hornick and I. Dufrasne
Faculty of Veterinary Medicine, FARAH, Avenue de Cureghem, 6 B43, 4000 Liège, Belgium; flessire@uliege.be

Despite its benefits, grazing is decreasing, in southern Belgium because of its difficult management. In fact, farmers lack confidence in the amount of fodder available on the grasslands. The herbage mass (HM – kg DM/ha) is an essential data to estimate grass stocks. To obtain it, the reference method is to clip grass from delimited quadrats in the paddocks. The weight of the sample (kg DM/quadrat) allows an estimate of the stocks (kg DM/ha) on the paddock. However, this method needs labour to get a valuable overview of the grassland stocks. Another method is based on measuring the sward height (H-cm) using rising plate meters (RPM) and then to convert it into HM. Using the RPM is easy but needs a calibration equation to convert the compressed height (CSH, cm) to HM. Yet, the rising plate meters available on the market are calibrated by manufacturers under conditions somewhat different from those where they will be used. This study aims thus to evaluate the accuracy of different calibration equations from commercial RPM. Measurements of CSH and HM were carried out in 5 Walloon farms on permanent grasslands from 2013 to 2015. In total, 299 data were collected. We compared three calibration equations. Equation 1 developed in New-Zealand is proposed at use of Jenquip EC20®. Equation 2 developed in Ireland is proposed at use of GrassHopper® and the Equation 3 is developed in France. Calculated HM was faced to the collected field data. The coefficient of determination (R^2), the mean square error (MSE) and the relative prediction error (RPE) were determined. The HM and CSH of field data were 1,599±674 kg DM/ha and 9.73±2.86 cm (mean ±SD). The values of HM obtained from Equ.1 to 3 were: 1,862±400 kg DM/ha (Equ. 1), 1,489±339 kg DM/ha (Equ. 2) and 2,315±680 kg DM/ha (Equ. 3). The R^2 values ranged from 0.41 (Equ.2) to 0.42 for Equ.1-3. The Equ.1 and 3 led to a recurrent over-evaluation of stocks (74% in Equ.1 and 90% in Equ. 3). This over-estimation reached 503 kg DM/ha. Conversely, the RPE of Equ. 2 was the lowest (33.6%) and the errors were more equally balanced (55% under vs 45% over-estimation). It seems thus that the Irish equation could be fairly used.

Session 57 Poster 25

Breeding of native breeds as a chance for the development of livestock households in ecological sys

P. Radomski and P. Moskala
National Research Institute of Animal Production, Sarego 2, 31-047 Krakow, Poland; pawel.radomski@iz.edu.pl

Modern methods of breeding and production of meat and other products have, on the one hand, reduced prices for consumers, but on the other hand, they have made the food market less diversified. The increasingly widespread ecological awareness of consumers reinforces the ever-growing tendency to eat products derived from animals traditionally kept, i.e. in natural conditions, with free access to open space and grasslands. The growing demand for traditional (without enhancers) food products should contribute to an increase in the consumption of meat, cheese and other products derived from farm animals fed and reared on traditional (self-sufficient) farms. It is therefore necessary to draw the farmer's/producer's attention to the development of agricultural production using environmentally friendly methods and ensuring food safety. This is a great opportunity for the development of breeding and the use of the potential of native breeds kept in low-input production systems of family farms. Of great importance is the fact that global sales of organic food and drinks are growing every year and in 2019 reached over EUR 106 billion. The leading country is the United States with €44.7 billion, followed by Germany (€12 billion) and France (€11.3 billion). With regard to spending on organic products per capita, Denmark (EUR 344) and Switzerland (EUR 318) are in the lead, while in Poland the average is EUR 8. In terms of the area of ecological agricultural land, Australia is invariably the world leader (35.7 million ha). Argentina comes second (3.6 million ha) and Spain comes third (2.4 million ha). In Europe, organic production covered almost 16.5 million hectares (as at the end of 2019), which is almost 1 million hectares more compared to 2018. Spain boasts the largest area of organic farming in Europe (2.4 million hectares), France (2.2 million ha) and Italy (2 million ha). Poland ranks 9th in this ranking – 0.5 million ha. Currently, Poland ranks in the second ten in organic animal production compared to other European countries in each of the basic groups of animals: cattle breeding – 20th place, pig breeding – 14th place, sheep breeding – 20th place, and the highest place is only in poultry farming – 11th.

Session 57 Poster 26

Conceptual model for the analysis of energy allocation of cow-calf farms native grassland-based

V. Figueroa, I. Paparamborda, S. Scarlato and P. Soca
Universidad de la República, Departament of Animal Production and Pastures, EEBR., Ruta 26, Km. 408, Bañados de Medina, Cerro Largo, 37000, Uruguay; vfigueroa@fagro.edu.uy

Reproductive and productive outcome of cow-calf farms on native grassland is controlled by the energy nutritional management in breeding herds. Farmers' techniques are necessary to manage energy nutrition of livestock. The index of bovine breeding techniques (ITB) condenses in a hierarchical manner the techniques recommended by research and their mode of application in beef cows systems. We hypothesize that an increase in the value of ITB is positively related to the beef cows pregnancy rate and calf weight at weaning. The 12 cases were analysed, participants of the 'Livestock and Climate' Project (L&C). Distributed in the East and North of Uruguay, from March 2020 to March 2023. L&C is a project with approach the Co-innovación, developed cycles of analysis and design in 60 farms to achieve a systemic rethinking of the farm through ecological intensification. The sum of the value of each technique is the final result of the ITB, therefore farms with a higher value of the ITB correspond to a greater and better way of applying strategic and tactical management. The variables production of bovine meat per animal unit (PBMAU), calf weight at weaning (CWW), weight of weaning calf per mating cow (kgWC/MC) and bovine pregnancy rate (%BP) were analysed using a linear model. PBMAU was calculated as total kilos of bovine meat referred to the animal stock in the year (kg) and %BP is the ratio between pregnant cows and matching cows. Statistical model included (ITB) and project phases (beginning and end) as fixed effects. The results showed that ITB increased from 57,0 to 73,3 between project phases ($P<0.0001$). The PBMAU improved from 127.2 to 135.0 kg between project phases ($P<0.0001$), finding a positive relationship between ITB and %BP, CWW, so contribute to explain the improve of PBMAU and kgWC/MC. The results suggesting, ITB constitute robust conceptual model for the analysis and comprenhens of energy allocation and use of cow-calf farms. The co-innovation was favourable to encourage the ITB, a necessary condition to the ecological intensification of cow-calf farms on native grassland.

Session 57 Poster 27

Do botanically diverse pastures effect the meat eating quality of lamb?

S. Woodmartin[1], P. Creighton[1], T.M. Boland[2], E. Crofton[3], A. Monaghan[1] and F. McGovern[1]
[1]Teagasc, Animal and Grassland Research and Innovation Centre, Mellows Campus, Athenry, Galway, H65R718, Ireland, [2]School of Agriculture and Food Science, University College Dublin, Belfield, Dublin 4, D04FX62, Ireland, [3]Teagasc, Ashtown Food Research Centre, Food Quality and Sensory Science Department, Ashtown, Dublin, D15 DY05, Ireland; sarah.woodmartin@teagasc.ie

Efficient utilization of pasture is Ireland's competitive advantage in sheep meat production. The aim of this study was to assess the effect of binary swards on slaughter data and meat quality. Five treatments were investigated, perennial ryegrass (*Lolium perenne* L.;PRG), PRG plus white clover (*Trifoluim repens* L.;PRG+WC), PRG plus red clover (*Trifoluim pratense* L.;PRG+RC), PRG plus chicory (*Chicorium intybus* L.;PRG+Chic) and PRG plus plantain (*Plantago lanceolate* L.;PRG+Plan). Sixty lambs (n=12 per treatment) were selected from the main cohort of lambs born. Males were castrated within 24 hours of birth. Groups were balanced for sex and weight. Post-weaning, weight was recorded from all lambs fortnightly. Lambs were drafted for slaughter once they reached the desired weight of 45 kg. Meat samples were obtained from the longissimus dorsi muscle. Sensory analysis was carried out by a descriptive panel. Quality attributes were scored on a scale of one (low) to ten (high) under red filtered light in isolated booths. There was no difference in pre-slaughter live-weight, carcass weight or kill out percentage with mean values of 45.4 kg, 21.1 kg and 46.3%, respectively. Average values of pH5.6 and 12.0°C were recorded for ultimate pH and temperature of the *longissimus dorsi* with no difference between treatments. Sensory results show meat from lambs finished on PRG+Plan scored lower for tenderness and higher for chewiness when compared to meat from lambs finished on PRG+WC or PRG+Chic ($P<0.05$). Diets with herb inclusion, PRG+Chic or PRG+Plan, produced lambs with lower meat juiciness scores than the legume treatments, PRG+WC and PRG+RC ($P<0.05$). Diverse pastures did effect meat eating quality. Meat from lambs grazing PRG+WC swards was deemed more tender and juicy than meat from lambs grazing PRG+Plan. This gives scope for future work to further explore the relationship between diverse pastures and red meat eating quality.

Session 57 Poster 28

Altering milking frequency from 14 to ten milking's per week: effects on milk production of pasture

E. Kennedy, K. McCarthy, J.P. Murphy and M. O'Donovan
Teagasc, Animal & Grassland Research and Innovation Centre, Moorepark, Fermoy, Co. Cork, Ireland; emer.kennedy@teagasc.ie

One of the biggest factors in attracting and retaining people to work on dairy farms is that they must be desirable places to work. Milking is the most labour demanding task on Irish dairy farms and one, which necessitates farmers being present on the farm twice daily, if milking twice-a-day (TAD). The objective of this study was to alter weekly milking frequency by milking ten times per week and compare their milk production to cows milked 14 times per week, the industry standard. Pre-calving, cows were balanced based on calving date, breed, PD for milk, fat and protein and assigned to one of three treatments: (1) milked ten times per week for the full lactation (F107); (2) milked TAD for the first half of lactation and ten times per week for the second half of lactation (P107); and (3) milked TAD for the full lactation (TAD; control). When cows were milked ten times per week the milking schedule was TAD milking on a Monday, Wednesday and Friday and once-a-day milking on Tuesday, Thursday, Saturday and Sunday. The TAD cows were milked at 7am and 4pm daily; when the F107 and P107 were milked once-a-day milking time was approximately 10am, it was 7am and 4pm on TAD days. Milk yield was recorded at each milking and milk composition and quality were determined across four days each week. Data were analysed using mixed models in SAS. Completing ten milkings per week for the full lactation reduced milk yield ($P<0.01$) by 10% in comparison to TAD cows (5,210 kg) while milk solids yield (kg fat + kg protein) was reduced by 11% when milking ten times per week in comparison to TAD (463 kg; $P<0.001$). Interestingly, switching to ten milkings per week half way through lactation (approx. 20 weeks) resulted in no difference in total lactation milk or milk solids yield compared to cows milked TAD for the entire lactation (5,204 and 458 kg, respectively). There was no difference in somatic cell score between any of the three treatments across the entire lactation (1.99). The results from this study suggest there are options to reduce the labour associated with milking, particularly for the second half of lactation, without reducing production.

Session 57 Poster 29

Robustness of suckling cows at herd level is associated with cows' productive longevity

L. Barreto-Mendes, A. De La Torre, S. Ingrand and F. Blanc
Université Clermont Auvergne, INRAE, VetAgro Sup, UMR Herbivores, 63122 Saint-Gen`es-Champanelle, France; luciano.mendes@inrae.fr

Reproductive failure is one of the main reasons for involuntary culling in beef cattle herds. Consequently, individual cows' productive longevity (PL, 1st calving to culling interval), a proxy of their robustness, is highly determined by their ability to conceive and recalve within the time window defined by the farmer. One indicator to measure this ability is the calving-to-calving interval (CCI). At the herd level, robustness can be defined as the ability to maintain performance stable across time. The objective was to verify whether the herds' robustness was associated to cows' robustness. We used time-series from 89 herds (min. size of 20 cows) of Charolais suckling cows with individual dates of birth, calving and culling for at least 10 years (max. 36 years). Primiparous and multiparous cows that did not manage to complete pregnancy were attributed a theoretical CCI calculated as the max. herd CCI at a given year plus 1.5 times inter-quartile range. For each herd, robustness was assessed by the standard error of mean PL (SE_PL) and mean CCI (SE_CCI) across time. Links between herd and cows' robustnesses were studied via the correlations between SE_PL and the average PL of cows present in a farm across the studied period and between SE_CCI and the average CCI. Between herds time averaged PL and CCI were negatively correlated (-0.28, $P=0.008$). Correlations between log(SE_CCI) and log(SE_PL) with PL and CCI were negative and positive (-0.18, $P=0.09$; 0.35, $P=0.0009$), respectively. PL stability (log(SE_PL)) was positively associated with CCI stability (log(SE_CCI)) as well (0.40, $P=0.0001$). When herds were grouped based on these four variables, two clusters (CL1 and CL2) emerged that presented distinct properties as follows: CL1 had significantly higher PL than CL2 (PL_{CL1}=2,213±35 d; PL_{CL2}=2,004±47 d; $P=0.0005$), associated with significantly lower SE_CCI and higher SE_PL ($P<0.0001$), respectively. On the other hand, CL2 presented significantly higher CCI than CL1 (CCI_{CL1}=389±1 d; CCI_{CL2}=401±1 d; $P<0.0001$) and presented significantly higher and shorter SE_CCI ($P<0.0001$), respectively. In conclusion, robustness traits observed at the individual scale seem to have been transmitted to the herd scale.

Session 58 Theatre 1

Variability in consumer perception of meat and meat substitutes

E. Hocquette[1], J. Liu[2], S. Chriki[1,2], M.P. Ellies-Oury[2,3], M. Kombolo[2], J.H. Rezende-De-Souza[4], S.B. Pflanzer[4] and J.F. Hocquette[2]
[1]ISARA, 23 rue Jean Baldassini, 69364 Lyon Cedex 07, France, [2]INRAE, VetAgro Sup, UMR1213, 63122 Theix, France, [3]Bordeaux Sciences Agro, 1 cours du Général de Gaulle, 33175 Gradignan, France, [4]University of Campinas, Rua Monteiro Lobato, 80, Campinas 13083-862, SP, Brazil; ehocquette@etu.isara.fr

This study, conducted with more than 16,000 respondents in 5 countries (Brazil, Cameroon, China, France, South Africa) was aimed at analysing the consumption of meat and meat substitutes according to sociodemographic factors. For this, we asked for the criteria to choose food products at purchase time and for the proportion of people consuming meat substitutes and willing to consume 'cultured meat'. The most important criteria when purchasing food products are the following: sensory quality (67%), price (56%), food safety (47%), origin/traceability (45%), ethics (42%), nutritional value (35%), environmental impact (33%), and then appearance (24%) and presence of a label (22%). Men place less importance on food safety (44 vs 50% for women, $P<0.01$). There is also an age effect ($P<0.01$), people over 51 years of age putting less importance on price (40 vs 52-69% than younger respondents). Respondents who rarely consume meat place price first, vegans/vegetarians place ethical and environmental concerns first, unlike meat consumers who consider sensory quality to be the most important ($P<0.01$). These results also depend on countries ($P<0.01$): sensory quality, food safety, origin/traceability and price are more important in Brazil, China, France and then two African countries respectively. On average, 45% of respondents eat meat substitutes. This result depends on gender (50% for women vs 39% for men), country (70% in China vs 29% in Brazil) and dietary habits, with flexitarians and vegetarians being 59-60% to consume meat substitutes. Thirty nine percent of the respondents would be regularly willing to eat cultured meat (43% of women and 36% of men; 46% among 18-30 year-old respondents vs 33-36% for the oldest). This proportion is higher for flexitarians and vegetarians (47-49%). The French are the least ready to consume 'cultured meat' (17%) vs 54% in Brazil. To conclude, perception of meat and meat substitutes depends on sociodemographic factors, mainly countries and dietary habits.

Session 58 Theatre 2

Bibliometric analysis of scientific articles related to 'cultured meat'

J.F. Hocquette[1], D. Fournier[2], M.P. Ellies-Oury[1,3] and S. Chriki[1,4]
[1]INRAE, VetAgro Sup, UMRH 1213, 63122 Theix, France, [2]INRAE, SDAR, 34000 Montpellier, France, [3]Bordeaux Sciences Agro, 1 cours du Général de Gaulle, CS 40201, 33175 Gradignan, France, [4]ISARA, 23 rue Jean Baldassini, Cedex 07, 69364 Lyon, France; jean-francois.hocquette@inrae.fr

'Cultured meat' aims to produce large quantities of 'meat' from muscle cell culture to feed humanity while slaughtering fewer animals. It is a hot topic, but which is much less present in academic research. Indeed, a first study found a total of 327 scientific publications only on this topic though the first cultured meat was approved in 2020 for commercialisation in Singapore. The purpose of this work was therefore to analyse the recent evolution of the scientific literature as of February 13, 2023. Thus, 826 scientific publications are present on the Web of Science (108 in 2020, 180 in 2021 and 242 in 2022) including 159 reviews. Although the number of scientific papers on this topic has increased over the last three years, the total number of scientific articles remains modest and mainly on technological aspects. While a bibliometric search was carried out with more than 20 keywords, it appears that 'cultured meat' is present in the title in about 30% of the articles from 2020. More than a third of scientific articles concern the 'Food Science and Technology' section. The top three journals publishing articles on this subject are Foods (39 articles), Frontiers in Sustainable Food Systems (two recent publishers) and Fleischwirtschaft (a technical international magazine for the meat industry) (24 articles each). Authors originate mainly from the USA (197 articles), UK (93 articles), China (73 articles), Germany (59 articles) and The Netherlands (55 articles). The two authors who published the most are Prof. Mark Post from The Netherlands (16 articles), who trusts the technology, and JF Hocquette (15 articles) from France who has a more critical view. More generally, the network of authors is very fragmented with more than 15 groups of authors who do not publish together, which may reflect various approaches on this topic. In conclusion, the scientific literature on cultured meat is limited but originates mainly from countries with an Anglo-Saxon or Germanic culture, and from China, which tend to support this innovation.

Session 58 Theatre 3

Addressing the challenges of animal-free meat using plant-based tissue engineering

M.O.R. Yahav
Redefine Meat, Technology and innovation, Openhaimer 10 rehovot, 7670110, Israel; morya@redefinemeat.com

Addressing the Challenges of Animal-Free Meat using Plant-Based Tissue Engineering By Mor Yahav, Redefine Meat Ltd. Today the necessity of replacing animal meat with animal-free alternatives is commonly accepted in our society. This concept has matured along with the growing awareness among the public of the extent of the ecological damage that animal farming and industrial meat production are causing to our habitat and to the biodiversity of our planet. While it is clear that today's society is not ready to give up the meat-eating experience, the food industry and food science are facing the challenge and the opportunity of making meat without animals. The degree to which this alternative meat should imitate its animal twin varies according to the market segments. While vegans and vegetarians can be content with even a rough approximation of meatiness, the flexitarians, who replace only part of their diet with meat alternatives, set a much higher bar. Indeed, the introduction of new generation of meat replacements in the last decade seemed to drive penetration to new market segments and expand the total market size. However, over the last year, this market has shown stagnation, possibly caused by the lack of progress in culinary and organoleptic performance of alternative meat. There is much work done to further disrupt the alternative meat industry and new ways are proposed for bringing the performance of new meat a significant leap forward. In this work, we disclose some novel ways for reconstructing meat using plant-based ingredients, while acknowledging for its anatomical origins and its composite tissue structure. We show that most of the organoleptically-relevant meat features can be mimicked by first modelling the composite structure of meat and its biochemical building blocks and reconstructing them separately via plant-based formulations. Then, an automatized process is employed for assembling these blocks and recreating the meat tissue elements and its dynamical and thermal behaviour.

Session 58 Theatre 4

Comparing the potential of meat alternatives for a more sustainable food system

T. Bry-Chevalier
Université de Lorraine, BETA, 23 rue Baron Louis, 54000 Nancy, France; tom.bry-chevalier@chaireeconomieduclimat.org

A growing body of scientific works document how high meat consumption is incompatible with a sustainable food system through a disproportionate use of resources. The negative externalities of livestock farming are not limited to its impact on the environment but extend to health, antibiotic use and the risk of epidemics. In this paper, I investigate how alternative proteins perform regarding the positive and negative externalities of the current food system. Most studies on alternative proteins focus solely on one environmental dimension when they discuss their relative benefits compared to livestock farming. In this paper, I stress the importance of having a multi-dimensional approach to fully assess the relevance of meat alternatives in mitigating the externalities of the food system. Moreover, alternative proteins are rarely compared with each other. In this paper, I compare the relative merits of different meat alternatives not only on environmental dimensions but also on their overall public health, scalability and acceptability potentials. Although some alternative proteins may be complementary it is not impossible that they also compete with each other for funding or purchases in shops. For example Slade finds that preferences for plant-based burgers and cultivated meat are broadly, though not perfectly, correlated. Overall, I find that the most promising category of alternative proteins for achieving a sustainable food system is plant-based meats and proteins produced by fermentation. Cultivated meat may be an interesting addition if it appeals to a different category of consumers (people with a strong attachment to meat), but it cannot be considered a solution for the immediate climate issues given the remaining challenges it has to face to achieve mass production at an affordable price. Insects probably have the lowest potential because of the difficulties in maintaining their environmental benefits on a large scale as well as their very low acceptability compared to other alternative proteins.

Session 58 Theatre 5

IMR3G Foundation, DATAbank software to facilitate collaborative data collection for mutual benefit

R. Polkinghorne[1], H. Cutherbertson[1], A. Neveu[2] and J. Wierzbicki[3]
[1]Birkenwood Pty Ltd, Blandford, 2338, Australia, [2]Birkenwood Europe, Lanon, 03240, France, [3]IMR3GF, Warsaw, 00-389, Poland; rod.polkinghorne@gmail.com

Development of consumer prediction models requires extensive high-quality research data. The International Meat Research 3G Foundation has developed a collaborative software system to facilitate integrated project design and delivery across multiple international partners who maintain ownership of their individual data with underlying protocols to ensure data compatibility. The DATAbank software supports experimental design through sequential processes that assist statistical balance. After specifying the number and type of livestock to be acquired for an experiment these are allocated to primary treatment-based groups and with further treatments progressively assigned through to sensory samples. While developed for specific cattle and sheep use the base design is adaptable to other species of any size with the principal being that the live animal is progressively converted to component portions to final sensory samples. For bovine and ovine use, the carcase can be assigned sides with allocation of side-based treatments to achieve parsimonious allocation to the smallest number of animals needed to achieve treatment balance. UNECE Bovine language codes define the carcase portions collected from each side and which individual muscles are available from each portion and, from carcase weight, the expected muscle mass, designated within-muscle positions and feasible sample numbers for evaluation by 10 consumers. Parsimonious allocation of multiple treatments including cooking methods, ageing periods, further treatments, packaging and sample destinations can be overlaid on the base sample plan. The final design is then supported by automated production of labelling and control files to facilitate data collection, with the completed samples and their related information stored in the DATAbank. Further routines assist in assignment of samples to consumer sensory test sessions and associated production of cooking, serving and data collection protocols. It is intended that the software be made widely available at minimal cost and hoped that it will prove valuable in facilitating highly compatible data across research projects thereby increasing the value of data through extensive linkage.

Session 58 Theatre 6

Beef processors experience large variation in yield and quality traits on a daily basis

W. Pitchford and S. Miller
University of Adelaide, Davies Livestock Research Centre, Roseworthy Campus, 5371, Australia; wayne.pitchford@adelaide.edu.au

Increasingly beef processors are able to extract more value from higher quality carcasses which will increase demand for quality and be reflected in price. Pitchford *et al.* examined various pricing strategies based on yield and quality and concluded that the majority of variation was associated with yield even when high premiums were placed on quality. However, concerns were raised by processors that the data set used had less variation in quality than they commonly experienced. This work was conducted using a subset of the Meat Standards Australia (MSA) database, covering a period of 4 years from start of 2010 to end of 2013, totalling 1,159 days. This subset covers a range of different lots from across Australia, processed at nine different plants that slaughter a total of approximately 1.7 million carcases with data from 35 variables. The carcase weight and traits associated with yield and eating quality variables utilised for this section of work were Hot Standard Carcase Weight (HSCW, kg), Eye Muscle Area (EMA, cm^2), Ossification Score (OSS, score out of 590), MSA Marbling Score (MARB, score out of 1,190), MSA Index Score (MSA, index), P8 fat depth (P8, mm), and Rib fat depth (RIB, mm). Variance was partitioned by differences and presented graphically. The variance for each trait was HSCW 2,506 kg2, EMA 101 cm4, OSS 3,416 scores2, MARB 8,945 scores2, MSA 11.1 index2, P8 18.3 mm^2 and RIB 12.7 mm^2. The largest proportion (49-73%) of the variation was between and within lots so processors experience the bulk of the variation in carcass quality on a daily basis. Processors experience much greater variation in ossification and marbling than that observed within genetics trials where animal age and growth path are more consistent. The result of this is that genetics type trials will underestimate the importance of quality relative to yield when modelling price effects.

Session 58 Theatre 7

Introduction to plant-based, cultivated, and fermentation-made meat, eggs, and dairy

S. Kell
GFI Europe, ASBL Drève du Pressoir 38, 1190 Forest Belgium, Belgium; serenk@gfi.org

Animal agriculture causes 20% of global greenhouse gas emissions – equivalent to all the planes, trucks, cars, trains and ships on Earth. And research cited by the Intergovernmental Panel on Climate Change shows it will be impossible to meet the Paris Agreement targets without a reduction in conventional meat production. Additionally, intensive animal agriculture is a leading driver of antimicrobial resistance, environmental and habitat destruction and – to feed a population of 10 billion by 2050 – we need a system less sensitive to climate shocks and global supply chain vulnerabilities. Yet global demand for meat will have grown by 52% by 2050. People from all walks of life want our food system to be sustainable, secure and just. But most people's day-to-day food choices are based on taste, price and convenience, and alternative sources of protein can not yet compete on these terms: people are unlikely to move away from animal products unless they're presented with sustainable food that looks and tastes as good as the conventional products they love. This talk will cover the growing need for transforming our global food system in order to sustainably feed the world by 2050, introducing the role that plant-based, fermentation-made, and cultivated meat can play in this transition. It will also overview the present commercial and investment landscape across these sectors, as well as the scientific and industrial challenges presently preventing these solutions from achieving large-scale market uptake. Finally, it will outline the opportunities for researchers and scientists from multiple disciplines (including agricultural and animal sciences) to help to tackle these challenges, and how they can contribute towards this flourishing global research community.

Session 58 Theatre 8

Environmental impact of dairy alternatives: a case study of Hemp milk and other products

B. Queiroz Silva, J. Ferdouse and S. Smetana
DIL Deutsches Institut für Lebensmitteltechnik e.V., Food Data Group, Professor-von-Klitzing-Straße 7, 49610 Quakenbrück, Germany; b.silva@dil-ev.de

Milk consumption in humans lasts longer than in other mammal species. Today consumers' awareness of the environmental burden that some products carry keeps growing. Thus, they look for alternatives that are more environmentally friendly and nutritionally similar. One example of growing demand can be found in plant-based beverages, as these have a lower environmental impact, though not nutritionally comparable to bovine milk. Though these beverages bring benefits (e.g. overall lower environmental impact), there are also disadvantages (e.g. high water consumption in almond beverages). One promising plant is industrial hemp, as all parts of the plant can be used in different ways (e.g. fibres for paper production, seeds for animal feed or milk production). This type of plant does not require many inputs and is capable of mitigating soil desertification. Though some studies show the potential of hemp milk as a more sustainable alternate plant-based beverage, there are no LCA studies. This research aims to explore the lack of data using fat and protein-corrected beverages (1 kg of FPC beverage). The boundaries are from farm to industry gate, leaving out transportation, retail and consumer-related impacts, and data was collected from licensed agricultural farmers, databases and literature. The methods used were ReCiPe 2016 Midpoint (H)(V1.06) and Cumulative Energy Demand (V1.01). It was found that hemp milk has a high impact on human carcinogenic (7.61E02 kg 1,4-DCB) and non-carcinogenic (3.04E05 kg 1,4-DCB) toxicity, marine (7.90E02 kg 1,4-DCB) and freshwater (6.67E02 kg 1,4-DCB) ecotoxicity when compared to bovine milk (5.44E-1,4.39E-2, 4.02E-2 and 3.23E-2 kg 1,4-DCB respectively). On the other hand, bovine milk had higher values for global warming potential (1.54E00 kg CO_2 eq.), land use (7.69E00m^{2}a) and water consumption (3.86E-2 m^3). In conclusion, overall, the environmental impacts associated with this plant-based beverage are lower than animal milk, as reported for other similar beverages in literature.

Session 58 Theatre 9

The tools of prediction of the sensory quality, the opinion of the French professionals

T. Fayet
Ecole supérieure d'agricultures, 55 Rue Rabelais, 49000 Angers, France; thomas-fayet@outlook.fr

The French beef industry is structured by two types of consumption: everyday purchases oriented towards economical products in tender portions, often processed (such as chopped steak) and pleasure purchases oriented towards a search for gustatory pleasure and the satisfaction of societal and environmental criteria. However, it is difficult for the industry to guarantee regular and homogeneous products to satisfy consumers. These inadequacies stem from the current carcass grading systems. Thus, the professionals we met appear to be in favour of a change in the grading system based on a sensory quality prediction system that could be inspired by foreign systems such as 'Meat Standards Australia' for butchered cuts. Such a system, through its segmentation, could meet the expectations of both types of consumption, daily and pleasure, allowing to generate an added value for the whole sector as it is the case in Australia. However, the diversity of organizations with sometimes divergent interests makes it very unlikely, in the short term, to set up a prediction system on a sector-wide scale. Thus, the implementation of a carcass prediction system would more likely be the result of an individual initiative. The links where an individual initiative is most likely are, on the one hand, mass distribution for which the triggering lever lies in the dissemination of knowledge and, on the other hand, meat companies independent of livestock farming that wish to ensure a regular and qualitative supply. In addition, economic, operational, political and knowledge barriers make it unlikely that a sensory quality prediction system for beef will be developed collectively or by the upstream sector. However, a low probability exists, depending on the perception of a possible socio-economic opportunity by an innovative organization or on the evolution of European regulations.

Session 58 Theatre 10

Limitations and challenges for the successful launch to market of cultured animal protein products

J.F. Fuentes-Pila
Technical University of Madrid (UPM), Agricultural Economics, Statistics, and Business Management, Av. Puerta de Hierro, 2-4, 28040 Madrid, Spain; joaquin.fuentespila@upm.es

Sustainable development goals 2 (zero hunger), 3 (good health and well-being), and 13 (climate action) require healthier, safer, and more sustainable diets. High-quality protein diets are essential for reaching goals 2 and 3, but meat products alone will not be able to assure food security, safety, and sustainability at a global scale in the medium term. In this scenario, cultured animal protein products are becoming a key alternative protein to meat. I will provide, in this presentation, an analysis of the major limitations and challenges for the successful launch to market of cultured animal protein products in a way that allows to reach the SDGs 2, 3, and 13 globally. The main limitations and challenges identified are: (1) immortalized cell lines vs non-immortalized cells lines; (2) tridimensional growth with scaffolds vs growth in suspension without scaffolds; (3) serum-free cell culture media and growth factors; (4) development of cultured animal protein and fat products; (5) bioreactors design and size; (6) cultured meat vs hybrids of plant-based protein and cultured animal protein and fat; (7) financial limitations and business model; (8) sustainability and transparency in the value chain; (9) consumers' acceptance and willingness to pay. Possible solutions to these limitations and challenges will be discussed.

Session 58 — Theatre 11

Implementing advanced characterization methods and building a new reference for alt-meat development

M.O.R. Yahav
Redefine meat, technology and innovation, Openhier 10 Rehovot, 7670110, Israel; morya@redefinemeat.com

Implementing Advanced Characterization Methods and Building a New Reference For Alt-Meat Development By Mor Yahav, Redefine Meat Ltd. Recently, with the increasing demand for meat analogues, we have become acutely aware of the gap between the performance of plant-based meats vs livestock meat products. The alt-meat industry is making a substantial effort to close this gap by introducing new ingredients, processes, and manufacturing technologies. However, to make a tangible impact, the various academic and industry players must develop an in-depth understanding of the product that they strive to mimic, optimally in a quantitative form. Like to other players, Redefine Meat is facing this challenge, while implementing additive manufacturing technology for recreating beef whole- muscle cuts. and is addressing this task with a major investment in meat science, combining the collection and organization of existing literature data with its own studies aimed to populate the white spaces in this data gamut. In this talk we will review the Redefine Meat's custom-made Meat Knowledge Center and two unique studies that were recently conducted for understanding specific aspects of beef whole-muscle cuts. In one study, synchrotron radiation was employed for micro-CT scanning of bovine muscle and understanding the effect of cooking process on its morphology and structure. In the second study, a robotic tool was built for automatic measurement of multiple areas of a steak and creating 2D maps of characteristics that illustrate the sophisticated, heterogeneous structure of livestock meat.

Session 58 — Poster 12

Whey proteins as alternative supplement to FBS in C2C12 muscle cells for cultured meat production

T.S. Sundaram, D. Lanzoni, R. Rebucci, F. Cheli, A. Baldi and C. Giromini
Università degli studi di Milano, Veterinary Medicine and Animal Science, via dell'università 6, 29600 Lodi, Italy; carlotta.giromini@unimi.it

Lab-cultured meat has gained worldwide attention as a potential sustainable alternative for conventionally farmed meat. Unlike traditional meat, it doesn't involve animal cruelty, emits less greenhouse gas, and importantly reduces human diseases associated with antibiotic resistance. The complex structure of livestock muscle is recreated in lab-cultured meat by cultivating cells in artificial medium consisting foetal bovine serum (FBS), and other essential nutrients. However, since cultured meat's goal is to decrease animal slaughter, the primary challenge lies in the acquisition of FBS from calves' blood. Whey is a by-product of the dairy industry, which has become an interest of research due to its bioactive and nutritional properties. Presently, we accessed the suitability of high-hydrolysed whey (HW), beta-lactalbumin, and lactoferrin as an alternative to FBS in the mouse C2C12 muscle cells during its proliferative stage. To this aim, the colorimetric assays such as MTT cell viability and lactate dehydrogenase (LDH) cytotoxicity were performed after 24, 48, and 72 h of treatment with 0.03-1% HW and beta-lactalbumin while, 3.125-200 µg lactoferrin in DMEM medium. Statistical analysis was performed in GraphPad Prism 9.3.1 for repeated measures one-way ANOVA with Tukey's post-hoc test. We observed that HW did not significantly affect the cell viability and LDH activity until 72 h compared to the control (0%) ($P<0.05$). Conversely, only 1% of beta-lactalbumin significantly ($P<0.05$) enhanced the cell viability until 72 h, while it did not affect the LDH activity. Further, 6.25-200 µg of lactoferrin after 48 h and 200 µg after 72 h significantly improved the cell viability compared to the control (0 µg) ($P<0.05$). This study shows beta-lactalbumin and lactoferrin could be a promising alternative to FBS as a growth supplement for utilization in cell culture systems. The data need to be confirmed in further studies, considering not only the proliferation stage but also the full differentiation process. (Project funded under the National Recovery and Resilience Plan (NRRP),'ON Foods – Research and innovation network on food and nutrition Sustainability, Safety and Security–Working ON Foods').

German consumers' attitudes towards cultured meat

A.-K. Jacobs[1], M.-P. Ellies-Oury[2,3], H.-W. Windhorst[1], J. Gickel[1], S. Chriki[2,4] and J.-F. Hocquette[2]
[1]University of Veterinary Medicine Hannover, WING, Heinestraße 1, 49377 Vechta, Germany, [2]INRAE, Université d'Auvergne, Vetagro Sup, UMR Herbivores, Theix, 63122 Saint Genès Champanelle, France, [3]Bordeaux Sciences Agro, 1 cours du Gal De Gaulle, 33175 Gradignan, France, [4]Isara, 23 rue Jean Baldassini, 69364 Lyon, France; jean-francois.hocquette@inrae.fr

Meat plays an important role in German nutrition, but recent surveys reveal a growing interest in plant based meat alternatives. The aim of this study was to document similarities and differences regarding the attitudes of potential German consumers towards other meat alternatives such as cultured meat. For this purpose, the responses of 3,558 German participants of an online survey were evaluated. More than 94% of the respondents were familiar with cultured meat technology. Nearly 63% of them thought that this novel food is promising/acceptable and 22% indicated that it is absurd/disgusting. Most respondents believed that cultured meat is both a more ethical (67%) and environmentally friendly (58%) solution than conventional meat. In terms of future, almost 75% of respondents believed that cultured meat production and consumption will be commercialised in more than 5 years. The vast majority (70%) would be willing to try this new product, while around 57% only would be willing to eat it regularly. Among them, respondents could imagine a regular consumption especially at home (47%), and in equal shares in restaurants and ready-to-eat meals (37%). Around 40% would prefer to pay the same price as for conventional meat. Only 27% would be willing to pay more or much more whereas 33% want to pay less or much less. There were significant impacts of demographic factors on the willingness to try, regularly eat, or pay for cultured meat. For example, a high willingness to try and to eat this new product was found among male respondents who were young (18-30 years), rarely meat's consumers or with a low income (<1,500€). This also applies to the female respondents, who, however, belonged to higher income classes. Males with the highest income were only willing to pay much less/less for cultured meat. But females with a low income would like to pay the same/more. These results are important for the discussion of a paradigm change in global meat production.

Development of the beef eating quality management system in Poland

G. Pogorzelski[1], J. Wierzbicki[2], E. Pogorzelska-Nowicka[3], A. Jasieniak[4] and A. Wierzbicka[1]
[1]Institute of Genetics and Animal Biotechnology of the Polish Academy of Sciences, Department of Biotechnology and Nutrigenomics, Postępu 36A, Jastrzębiec, 05-552 Magdalenka, Poland, [2]Polish Beef Association, Smulikowskiego 4, 00-389 Warsaw, Poland, [3]Warsaw University of Life Sciences, Department of Technique and Food Development, Faculty of Human Nutrition and Consumer Sciences, Nowoursynowska 159C, 02-776 Warsaw, Poland, [4]Food and product law centre, Wiejska 17 street, suite 6, 00-480 Warsaw, Poland; grzegorz.t.pogorzelski@gmail.com

Since 2010, work has been started in Poland on the assessment of beef quality other than only commercial quality assessment according to the EUROP system. During the ProOptiBeef project, various beef cuts were subjected to four cooking methods and tested by almost 10,000 consumers. This made it possible to build a prototype of eating quality predicting model for Polish beef. The construction of the model was based on the Australian approach to eating quality management. For this purpose, a significant amount of beef carcasses was chiller assessed and then samples were collected from selected carcasses for consumer evaluation. Meat Standards Australia (MSA), an eating quality system based on consumer evaluation, uses an interactive prediction model to provide objective descriptions of beef meal outcomes for 33 carcass muscles by multiple cooking methods. The Australian beef eating quality prediction system is proving to be a useful tool for quality management. It not only supports quality management but is also responsible for the creation of new beef brands and maintains their stability. Therefore, it was decided to continue work on the development of the Polish model. The next step in the construction of the Polish quality management system will be the identification of gaps in the existing data to obtain the accuracy of the tool sufficient for use in the production plant and the development of a research plan on this basis. The following step will be to conduct a chiller assessment, select the appropriate beef carcasses, collect samples and conduct consumer tests. It is planned to use approximately 6,000 consumers to evaluate the eating quality of various beef cuts.

Session 59 Theatre 1

Degradation of Amazonian grasslands by weeds, how to managing this situation?

V. Blanfort[1], C. Favale[1], S. Bazan[1], V. Petiot[2], D. Bastianelli[1] and T. Le Bourgeois[1]
[1]CIRAD, Mediterranean and Tropical Livestock Systems Unit, University of Montpellier, France, Campus international de Baillarguet – TA C-112/ A, 34398 Montpellier Cedex 5, France, [2]Chamber of Agriculture of French Guiana, 1 avenue des Jardins de Sainte-Agathe, 97355 Macouria-Tonate, French Guiana; vincent.blanfort@cirad.fr

Tropical grasslands face two major challenges: maintaining food security in the face of growing needs and coping with global change. In French Guyana, pastures are not very profitable for farmers in a constrained climatic context and are criticized for their potential contribution to Amazon deforestation. This work aims to study the sustainability of pasture cover in French Guiana and to characterize the weed flora. On the other hand, this work seeks to determine the relationship between agronomic practices and pasture degradation, mainly with the evolution of pasture flora 4 years after the implementation of dynamic rotational grazing. Flora surveys of 83 stations in 21 exploitations, distributed from Regina to Saint-Laurent-du-Maroni, have been done. 132 species were referenced as weeds. 27% of the stations are characterized by a degradation index greater than or equal to 25%, and 23% by a degradation index greater than or equal to 50%. The weed flora mainly is mainly composed of *Cyperaceae*, *Mimosa pudica* and *Spermacoce verticillata* are the two most frequent and abundant weeds. *Cyperus aromaticus*, a recently recorded alien species, is of concern. No environmental factors seem to be linked to the level of degradation, but agronomic practices are the most likely to explaining it. This work contributed to the collaborative portal on weeds in tropical agrosystems, WIKTROP, on Guianese pastures.

Session 59 Theatre 2

Analysing strategies adopted to cope the climate variability in pastoral zone of Burkina Faso

H.P. Yarga, A. Kiema, L. Ouedraogo and S. Ouedraogo
INERA, Gestion des ressources naturelles, 01 BP 910 Bobo Dioulasso, 910, Burkina Faso; yargapaul@yahoo.fr

The objective of this study is to identify the strategies implemented by agro-pastoralists to cope with the effects of climate variability in two representative pastoral zones in Burkina Faso (Gadeghin in the Sudanian zone and Sideradougou in the northern Sudanian zone). The study methodology consisted of a one-pass survey using a questionnaire with 290 agropastoralists in the two pastoral zones. The questionnaire focused on livestock and pastoral resource management practices in relation to climate variability. The data collected were analysed with SPSS 22 software. The results show that producers have adaptive practices to climate change ranging from pastoral resource management to herd management. A total of 10 pastoral resource management methods were identified. In order of importance, these are: digging deep wells (31%), using organic manure (20%), planting trees (18%) and applying water and soil conservation methods (14%). In terms of herd management, 9 techniques were identified. The most important are, in order of importance: reducing the number of animals by destocking (27%), dividing the herd into sub-groups for management (20%), purchasing animal species that are resistant to climatic hazards (17%) and purchasing animal feed (13%). Climate variability is a global phenomenon. Therefore, for these strategies to be more efficient, the government and development actors must support these endogenous initiatives developed by producers through capacity building and by providing adequate resources.

Session 59 Theatre 3

How do changes in crop-livestock integration and specialisation affect farm performance in Vietnam?

A. Le Trouher[1], H. Le Thi Thanh[2], T. Dinh Khanh[2], T. Han Anh[2], C.-H. Moulin[3] and M. Blanchard[1]
[1]CIRAD – UMR SELMET, Campus de Baillarguet, 34398 Montpellier Cedex 5, France, [2]NIAS, 9 P. Tân Phong, Thuỵ Phương, Từ Liêm, Hà Nội, Viet Nam, [3]L'institut Agro Montpellier, INRAE – UMR SELMET, 2 Place Pierre Viala, 34060 Montpellier, France; alice.le_trouher@cirad.fr

In Northwest Vietnam, mixed farms integrating crop and livestock are undergoing major changes driven by a multitude of factors: lack of labour force, reforestation law, limited land resources, and fluctuating markets. The changes in farming systems and practices raise questions about the future performance of farms and their role in the sustainability of the territory. This study aimed to assess the effects on farm performance of changing crop-livestock integration practices and farming systems at the farm level. A participatory workshop with local stakeholders was held to define prospective scenarios based on available knowledge of farming systems and their past evolution. The participants identified the drivers of change in practices and their impacts. A modelling tool has been used to simulate the effects of the scenarios on a diversity of mixed farms. The model used considers all biomass flows on the farm and between the farm and the other farms, the market and the environment (livestock and crop products, crop residues, forage and manure). Based on a set of indicators, the model assesses the effect of each scenario on the productivity, economic, environmental and social performances of farms. The five scenarios simulated were: the continuity of current practices (scenario A), the development of local feed self-sufficiency (scenario B), the implementation of environmental protection measures (scenario C), the development of intensive livestock farming (scenario D) and the development of quality livestock and crop products (scenario E). As expected, the simulation results show contrasting differences between farms. The results show which scenarios have the most positive effects on farm performances and highlight which farms are most affected by resource access limitations. These results could be used to support dialogue with local stakeholders on the expected development of crop and livestock systems and the sustainability of the agricultural sector.

Session 59 Theatre 4

Problem of water supply in a context of climate variability in the pastoral zones of Burkina Faso

H.P. Yarga H Paul, S. Ouedraogo, A. Kiema and L. Ouedraogo
INERA, Gestion des Ressources Naturelles, 01 BP 910 Bobo Dioulasso, 910, Burkina Faso; yargapaul@yahoo.fr

This study aims to analyse the modalities of water supply under the effect of land pressure and climate variability in two representative pastoral areas in Burkina Faso, Gadeghin in the northern Sudanian zone and Sidéradougou in the southern Sudanian zone. The methodology consisted of processing geospatial data (Landsat 8 OLI-TIRS images) and a cross-sectional survey conducted with a questionnaire in 2022 among 290 herders in 10 villages in the two pastoral zones studied. Cartographic processing was done using Qgis 3.22.6 software to produce thematic maps. Data analysis was done using SPSS 22 software to produce descriptive statistics. The results showed an unequal distribution of hydraulic infrastructures in the pastoral zones. Thus, around 70% of the livestock move more than 2.5 km to have access to a water point. Furthermore, 80% of the tracks leading to a water source are occupied by crops during the rainy season. In the dry season, water sources are limited to water drillings, 38% of which are frequently out of service during this period. In response to these difficulties, producers rationalize the water requirements of livestock (37%) and use deep wells (25%). However, these strategies are not very efficient because they partly affect the animal's regular need for water. The solution for better access to water in these areas relies in the implementation of a development plan that will enable a good distribution of water points, the protection of pastoral paths by planting trees and a good management of hydraulic facilities by well qualified committees in this matter.

Session 59 Theatre 5

Local feeding strategies allow to reduce enteric methane emission from cattle in Sahel

G.X. Gbenou[1,2,3], M.H. Assouma[2,4], C. Martin[5], D. Bastianelli[6], L. Bonnal[6], T. Kiendrebeogo[3], O. Sib[2,4], B. Bois[6], S. Sanogo[4] and L.H. Dossa[1]
[1]FSA-UAC, Cotonou, BP 526, Benin, [2]SELMET, Univ Montpellier, CIRAD, INRAE, Institut Agro, Montpellier, 34398, France, [3]CNRST, INERA, Bobo-Dioulasso, 01 BP 910, Burkina Faso, [4]CIRDES, Bobo-Dioulasso, 01 BP 454, Burkina Faso, [5]INRAE, UMRH, Clermont-Ferrand, 63122, France, [6]CIRAD, UMR SELMET, 34398, Montpellier, France; gerard_xavier_djidjo.gbenou@cirad.fr

In Africa, a large diversity of diets (forage + staple food crop residues or other agricultural byproducts) is used by cattle herders in different production systems. It is reported that feeding strategies could help to reduce enteric methane (eCH_4) in ruminants by up to 55%. In order to test different feeding strategies, a trial was carried out (CIRDES, Burkina Faso) on 10 Sudanese peulh zebu steers (32 months of age) with live weight (LW) 147±5.8 kg (i.e. 0.59 Tropical Livestock Unit). Animals were kept in individual box and were fed at 3.5% LW with natural pasture and crop byproducts (75:25, DM basis). Individual eCH_4 (GreenFeed system) and total tract digestibility were measured. The 6 main crop byproducts used in cattle diet were identified with a survey from farmers: legume haulm-based diets (cowpea and peanut haulms), and cereal straw-based diets (maize, sorghum, millet and rice straws). For each crop byproduct, the experiment consisted in 3 weeks, including 2 weeks of feed adaptation and 1 week of data collection. Basal diets consisted in natural forage hay (CTL hay) to test cereal crops, and Panicum maximum hay (CTL Pm) to test legume crops. Dry matter intake (DMI, g/kg LW) were on average 18.0±1.92 in cereal straw diets vs 16.4±2.06 in CTL hay; and 25.5±1.69 in legume haulm diets vs 21.7±2.29 in CTL Pm. The DM digestibility (%) were 48.1±3.01 in cereal straw diets vs 46.0±3.05 in CTL hay; and in legume haulm diets 49.8±1.4 vs 48.7±6.1 in CTL Pm. Intake content was lower in fibre for cereal straw and legume haulm diets compared to their control (NDF, g/kg DMI, 664±26.7 and 611±8.5 vs 734±13.6 and 712±14.0, respectively). Amounts of eCH_4 (g/kg DMI) emitted were 26.5±6.85 in cereal straw diets vs 30.6±7.17 in CTL hay; and 24.0±4.26 in legume haulm diets vs 27.9±4.66 CTL Pm. Complementation with cereal and legume crops reduced by 13 and 21% eCH_4 yield.

Session 59 Theatre 6

Development of a decision support tool to secure cattle production in chlordecone-contaminated areas

A. Fournier[1], C. Feidt[1], A. Fourcot[2], M. Saint-Hilaire[3], Y. Le Roux[1] and G. Rychen[1]
[1]Université de Lorraine, INRAE, UR AFPA, 2 avenue de la Forêt de Haye, 54000 Nancy, France, [2]GIP-FCIP, 4 rue du Père Delawarde Desrochers, 97200 Fort-de-France, France, [3]Institut Pasteur de Guadeloupe, Morne Jolivière, 97139 Les Abymes, France; agnes.fournier@univ-lorraine.fr

Soil historical contamination by chlordecone (CLD) is weakening the Caribbean livestock production in contaminated areas. The first impact is concerning farmers who are threatened by providing non-compliant carcasses. Indeed, some breeders have decided to stop their activity without any new facilities taking over, reducing, thus, the volume of local production. Furthermore, the existence of contaminated carcasses affects consumer confidence in local animal products. Such elements impair societal objectives of maintaining ecosystem services provided by livestock farming and producing healthy local food. Animal contamination depends on environment contamination and on farming practices. Its determinism remains therefore complex. However, it is essential for farmers to know the contamination status of their animals before deciding whether or not to send them to the slaughterhouse. These elements of context made it necessary to develop a decision support tool (DST) to guarantee the conformity of animal products. Thus a DST, based on a PBPK model, was created and developed on the fate of CLD in adult cattle. This DST is based on two main elements: (1) the ability of the adult bovine to eliminate CLD; (2) the predictability of edible tissue concentrations from serum levels. Thus, from an initial blood sample, the decontamination period required to respect the Maximum Residue Limit in perirenal fat (regulated target tissue) can now be estimated. This DST has recently been validated *in situ*. This validation process was possible due to a close collaboration between stakeholders: volunteer farmers recruited, local livestock organizations (GDSM, Sanigwa), administrative authorities taking charge of CLD analyses (DAFFs), and analytical laboratories (IPG, LDA). A steering committee including the DGAL and the two Prefectures was also set up. In the next step, local actors will be trained to implement the DST on a larger scale.

Session 59 Theatre 7

Can tropical legume grass forage reduce enteric methane yield from suckler cows in the Sahel?

M.H. Assouma[1,2,3], A. Baro[3], G.X. Gbenou[2,3], O. Sib[1,2,3], S. Sanogo[1], H. Marichatou[4] and E. Vall[2,3]
[1]CIRDES, Bobo-Dioulasso, 01 BP 454, Burkina Faso, [2]SELMET, Univ Montpellier, CIRAD, INRAE, Institut Agro, Montpellier, 34398, France, [3]CIRAD, Bobo-Dioulasso, 01 BP 454, Burkina Faso, [4]AGRHYMET, Niamey, BP 11011, Niger; habibou.assouma@cirad.fr

Enteric methane emissions (eCH_4) from ruminants are the main source of greenhouse gas from the livestock sector in sub-Saharan Africa. In this region, very few references with *in vivo* measurements are available on local breeds, particularly on suckler cows. With the development of the dairy sector in Africa, it is important to integrate the evaluation of enteric methane emissions into the various feeding strategies. The aim of this study was to evaluate the impact of protein supplementation with a legume grass fodder (*Stylosanthes hamata*) on enteric methane emissions in Sudanese Zebu suckler cows. The study was conducted with Greenfeed in stall at CIRDES (Burkina Faso). The experiment was carried out on 10 animals: 5 steers (38 months old with liveweight (LW) of 179±20.3 kg, i.e. 0.72 Tropical Livestock Unit (TLU), and 5 suckler cows (75 months old with LW of 204±13.3 kg, i.e. 0.82 TLU). Animals were kept in individual boxes. The measurements of eCH_4 and milk production were made on the cows, and *in vivo* digestibility test was made simultaneously on the steers. Two trials of 4 week including 2 rounds of one-week data collection were successively conducted. The animals were fed respectively with *Bracharia ruziziensis* (100) (R1) and *Bracharia ruziziensis* + *S. hamata* (75:25) (R2). The results showed that the daily dry matter intake (DMI) is 15.67±2.70 g/kg LW for R1 and 20.33±4.28 g/kg LW for R2. *In vivo* DM digestibility (%) was 52.72±1.4 for R1 and 58.25±1.43 for R2 when milk production was 3.0±1.2 kg/d for R1 and 3.46±0.98 kg/d for R2. Amounts of eCH_4 (g/kg DMI) emitted were 25.05±8.27 for R1 and 20.99±7.30 for R2. Our results show that protein supplementation with *S. hamata* L. increases the DM intake (g/kg LW) for 30%, the digestibility (%) for 11%, and reduces eCH_4 yield (g/kg DMI) for 16%.

Session 59 Theatre 8

Agroforestry: an opportunity to improve the sustainability of livestock systems in Vietnam?

M. Blanchard[1,2], P. Tos[2], A. Le Trouher[1,2], A. Lurette[3] and H. Le Thi Thanh[1]
[1]NIAS, Hanoi, 01, Viet Nam, [2]CIRAD, UMR SELMET, Univ Montpellier, CIRAD, INRAE, Institut Agro, Montpellier, France, 34000, France, [3]INRAE, UMR SELMET, SELMET, Univ Montpellier, CIRAD, INRAE, Institut Agro, Montpellier, France, 34000, France; melanie.blanchard@cirad.fr

In the mountainous areas of Vietnam, agroforestry is being developed in response to new market opportunities, the need to improve forest cover, and to reduce soil erosion. These resulted in changes in land use and affect the animal mobility and feeding patterns of ruminants, which represents an economic opportunity and contributes to the trade balance on meat. Using the example of the Dien Bien district in northwestern Vietnam, and based on a modelling of crop livestock integration and biomass flows in the area, this study assesses the effect of these changes on livestock systems, helping to define strategies for the sustainable development of livestock systems. The development of agroforestry contributes to the intensification of feeding systems, or leads to a cessation of livestock activities. Agroforestry systems intercropping forage within tree plantations increase the areas dedicated to forage production and the quantities and quality produced. Depending on their objectives, farmers maintain livestock activities to maintain soil fertility from avoiding over grazing and have access to manure as an organic fertilizer source for crops, trees and forage. They adapt their livestock systems, by stalling the animals, recycling grasses produced elsewhere and ensure feed supply in the whole year, thanks to silage. Animal housing, harvesting, transport and forage conservation technics can also represent a significant workload for farmers who then abandon livestock farming. The increase in forage production with agroforestry offers opportunities to develop local forage and silage markets and to participate in local livestock development, while providing opportunities to improve the livelihoods of farms without livestock. Agroforestry with forage production, combined with forage conservation, offers opportunities for sustainable intensification of livestock systems in the Mountainous area of Vietnam.

Session 59 — Theatre 9

Selection signatures of the indigenous Sanga cattle of Namibia

D.A. Januarie, E.D. Cason and F.W.C. Neser
University of Free State, Animal Sciences, P. O. Box 339, 9300, Bloemfontein, South Africa; deidre.januarie@gmail.com

The majority of the indigenous Sanga cattle in Namibia are found in the Northern Communal Areas (NCA) of Namibia. Tradition, culture and environment play an important role in the selection of the animals. The indigenous Sanga cattle are well adapted to the harsh environment of Namibia resulting in identifying possible candidate genes that could explain performance and adaptability of these animals. Hair samples from the Kavango- (n=124), Caprivi- (n=135), Ovambo- (n=269), Kunene- (n=118), Herero (n=17) Sanga ecotypes as well as Nguni (n=148) cattle were collected and analysed using a 150K Single Nucleotide Polymorphism (SNP) chip. HapFLK results identified selection sweep regions over chromosomes 6, 4, 14 and 5. Selection sweep regions used Quantitative Trait Loci (QTL) to determine the possible candidate genes under selection. QTL for hair pigmentation, reproduction traits (gestation length, calving ease, age at first calving, heifer pregnancy and stayability), tick resistance, milk traits (yield, fat and protein percentage, somatic cell score), structural soundness traits (foot angle, rump width & length, teat placement), growth traits (average daily gain & residual feed intake) body weight and carcass traits (tenderness, eye muscle area, dressing percentage, marbling, weight) were identified. These results underline traditional selection habits for traits such as milk production, walking ability, coat colour, tick resistance, ease and age of calving were being applied.

Session 59 — Theatre 10

***Diodelle sarmentosa*, an invasive plant in the rangelands of the sylvopastoral zone of Senegal**

E.H. Traore and F. Sow
Institut Sénégalais de Recherches Agricoles (ISRA), Laboratoire National de l'Elevage et de Recherches Vétérinaires, ISRA, PRH, Hann, 3120, Senegal; elhadji.traore@isra.sn

Livestock occupies an important place in Africa and particularly in Senegal where, the ruminant size is about 3,541millions cattle, 6,678 millions sheep, 5,704 millions goats and 5 thousand camels. This livestock essentially conducted on extensive mode, has as a feed resources the natural pasture which is becoming rara due to a strong agrarian and urban pressure, but also, a negative influence of global changes, which affect the rainfall, on which depends the development of fodder. The effects of climate change can sometimes lead to the invasion of pastures by new plant species, especially herbaceous species that are more or less consumed. Thus, since some years, the pastoral space of the sylvopastoral zone of Senegal, has been colonized by an invasive species: *Diodelle sarmentosa*, (diodia) which has caused indigestion problems especially in straw state in ruminants, especially cattle and sheep. Then, the species although fleeting, begins to be palatable. The present work was devoted to study the bromatology of diodia, completed by *in vivo* ingestion and digestibility tests. The trials took place at the Centre de recherches zootechniques (ISRA-CRZ) of Dahra Djollof and involved 16 sheep male (08 *Peul-peul* and 08 *Touabir* breeds), aged about 22 months, divided into two batches (4 Touabir and 4 Peul-peul per batch) and kept in digestibility cages. Daily rations consisted of 1,500 g of *D. sarmentosa* straw alone for batch 1 and 1,500 g of a mixture of 50% *D. samentosa* and 50% other natural forages for batch 2. Refusal and faeces were assessed each morning. The bromatological analysis shows that the ration 2 was richer in MAT than the one made of diodia straw alone. Also, the negative values of the MAT digestibility coefficients of both rations, show their low nitrogen content. However, a significant difference in the digestibility of MAT, NDF and ADF in favour of ration 2 (diodia and other forages) is noted. It appears that the feed value of *D. sarmentosa* is low. However, this is a preliminary work that should be continued to better determine the best methods to valorise this invasive plant if not to eliminate it.

Environmental challenges in dry tropical livestock systems: GHG emissions and carbon storage balance

M.H. Assouma, D. Bastianelli and P. Salgado
CIRAD, TA C-112/A, Campus International de Baillarguet, 34398 Montpellier, France; habibou.assouma@cirad.fr

The effects of livestock on global warming and more generally on the environment are widely reported. It is estimated that the livestock sector accounts for 14.5% of global anthropogenic greenhouse gas (GHG) emissions of which 39% is due enteric methane (CH_4). Sub-Sahara African (SSA) livestock (pastoral and agropastoral) systems suffer from a criticism because their environmental impact appears high when GHG emissions are expressed per kg of product (milk, meat), due to their limited productive efficiency. Enteric emission rates are also high because of the low overall quality of forage resources and their seasonal variability. But these assessments do not account for the animal's role in organic matter recycling and carbon, nitrogen, and phosphorus cycling in grazed ecosystems. Moreover, most calculations are based on default emission factors which are not adapted to local conditions, and to approximative regional livestock populations. One of the major challenges of SSA pastoral and agropastoral livestock systems is to modify this negative perception of their impact, by working to recognize these systems as a mode of production and lifestyle and acknowledging their environmental contribution. Through research work and production of reference data, CaSSECS project is promoting an original approach to assess the environmental impact of livestock systems in SSA. This original approach, called an ecosystem approach of the carbon balance, aims to evaluate over a complete annual cycle all the GHG emission (animals, soil and water surfaces) and carbon storage (trees and soil) by using local references. To calculate the carbon balance at the national level, FAO's GLEAM tool will be updated with novel GHG emission references and an additional module for carbon storage of land management related to the livestock sector assessment in soil and trees.

Resilience analysis based on heifer productive and reproductive aspects

V.T. Rezende[1], G.R.D. Rodrigues[2], A.H. Gameiro[1], M.E.Z. Mercadante[3], R.C. Canesin[3] and J.N.G.S. Cyrillo.[3]
[1]University of São Paulo, Faculty of Veterinary and Animal Science, 13635-900, Pirassununga, SP, Brazil, [2]Uberlândia Federal University, BR-050, 38410-337, Uberlandia, MG, Brazil, [3]Institute of Animal Science, Beef Cattle Research Center, 14174-000, Sertãozinho, SP, Brazil; jgcyrillo@sp.gov.br

Extensive beef cattle production on tropical areas has the dry seasons as a challenge in all phases of breeding because of low forage availability. This study aimed to assess the influence of heifer resilience at dry season on pregnancy rates. Data from Nellore heifers born between 2008 and 2017 belonging to a selection experiment started in 1980 were analysed. Heifers from selection line (NeS, n=283) and Control line (NeC, n=129) selected for higher and mean yearling weight (YW), respectively, were submitted to the breeding season at two years of age. Resilient animals were those with higher values of the difference (DFW770) between observed weight at 770 days old (W770, 347±54 kg), measured at the beginning of the breeding season, and the standardized weight at the same age (W770E, 296±43 kg), obtained using the ADG of the rainy season (391 to 550 days of age) as a growth pattern for the studied period. Animals were classified as susceptible (S, 300.07±4.40 kg, n=82), average (A, 291.34±3.11 kg, n= 214), or resilient (R, 301.66±3.73 kg, n= 107) based on the first, second, and third quartiles of DFW770, respectively. A logistic model included the fixed effects: RES, breeding season year, selection lines, and body condition score, using R 4.2.2 software. The significance of the odds ratio was given by the confidence interval of 95% (CI). All factors showed significant effects on pregnancy rates, except the selection lines (P<0.05). Heifers with the highest body score (7) had a 74.16% (CI: 70.45-77.87%) chance of becoming pregnant than heifers with a low body score (4) (P=0.0048). R animals had a 32.41% (CI: 30.78-34.02%) higher chance of becoming pregnant than S animals (P=0.0201). There were no significant differences between animals S and A (P=0.4708) regarding the chance of becoming pregnant. Resilient animals have a greater chance of pregnancy compared to non-resilient ones, regardless of selection for yearling weight and the body weight on breeding season.

Impact of GreenFeed protocols on cattle visitation and methane data collection

M.C. Parra[1], M.H. Dekkers[2], S.A. Cullen[2] and S.J. Meale[1]
[1]The univeristy of Queensland, School of Agriculture and Food Science., 5391 Warrego Highway, 4343, Australia, [2]The univeristy of Queensland, Queensland Animal Science Precint., 5391 Warrego Highway, 4343, Australia; m.parramunoz@uq.edu.au

Reliable methane emission data from a GreenFeed emission measuring (GEM) unit depends on cattle visitation. Six drought-master steers (411.83±48.29 kg BW) housed in a 3-ha paddock with *ad libitum* pasture were offered 2 kg/d of alfalfa pellets for 20 d, followed by grain-based pellets for 25 d to test attractiveness and subsequent number of visits to the unit. Two GEM protocols were used with Alfalfa pellets: (1) six daily visits (240 g/ea) with 8 drops/visit; (2) six daily visits (180 g/ea) with 6 drops/visit. While four protocols were used with energy-based pellets, (3) six daily visits (180 g/ea) with 6 drops/ visit; (4) five daily visits (300 g/ea) with 10 drops/visit; (5) Five daily visits (390 g/ea) with 13 drops/ visit and; (6) four daily visits (510 g/ea) with 17 drops/ visit. All drops released 30 g of pellet at intervals of 30 sec, except protocol 2 which dropped pellets at 45 sec intervals. Daily average visits per head and pellet intake were measured. Visits >3 mins in length were considered to provide reliable methane (CH_4) data. Protocol 1 resulted in more visits to the GEM and a greater pellet consumption (3.2× daily and 0.7 kg/d/h), compared to protocol 2 (2.4× daily and 0.4 kg/d/h). Protocol 2 aimed to increase the time spent at the feeder, but steers left the GEM unit prior to consuming all available drops and increased the number of visits unable to capture reliable CH_4 data to 32.06%, compared to 10.43% for protocol 1. Grain-based pellets increased steers visits, where protocol 3 and 5 showed the highest daily visits per head (4.8 and 4.8 times, respectively), compared to protocol 4 and 6 (4.2 and 3.8 times, respectively). Pellet intake was greater with protocol 5 and 6 (2.2 and 1.8 kg/d/h, respectively), compared to protocol 3 or 4 (0.8 and 1.3 kg/d/h, respectively). Protocol 6 had more visits providing reliable CH_4 data (98.1%), followed by protocol 5 or 6 (~96.1%) and protocol 4 (94.9%). Therefore, to maximise emissions data collection we recommend the use of grain-based pellets with protocol 5 or 6 for future grazing trials using a GEM unit.

Effect of production system and season on composition of retail cow milk in Greece

G. Papatzimos[1], R.A. Stergioudi[2], V. Papadopoulos[1], Z. Basdagianni[3], M.A. Karatzia[4], P. Mitlianga[2] and E. Kasapidou[1]
[1]University of Western Macedonia, Department of Agriculture, Florina, 53100, Greece, [2]University of Western Macedonia, Department of Chemical Engineering, Kozani, 50100, Greece, [3]Aristotle University of Thessaloniki,School of Agriculture, Department of Animal Production, Thessaloniki, 54124, Greece, [4]Research Institute of Animal Science, HAO-Demeter, Paralimni, 58100, Greece; papatzimos@hotmail.com

Information on cow milk nutritional quality at retail level is limited even though the effect of breed, diet, production system and season on milk composition has been well documented. Regarding Greece, there are no studies on the nutritional value of retail fluid milk. This study investigated the effect of the production system and season on the composition of cow milk produced in Greece. Milk samples (n=108) were collected monthly from four major supermarket retailers between November 2019 and October 2020. All samples were full-fat homogenized milk, either conventionally (n=7 brands) or organically produced (n=2 brands). Milk composition (fat, protein, lactose and solids non-fat content) was determined with a milk analyser, and fatty acid profiles were analysed by GC flame ionisation detection. General linear models were used to investigate differences in milk composition due to production system and season. Production system did not affect ($P>0.05$) milk chemical composition, whereas a highly significant ($P<0.001$) season effect was observed in protein content. Lactose and solids non-fat contents were also affected ($P<0.05$) by season. Higher protein levels of protein, lactose and solids non-fat were observed in autumn and winter. The lower protein content observed in summer months is related to the impact of high temperatures on lactating cows. Season did not affect ($P<0.05$) milk lipid classes, i.e. saturated, monounsaturated and polyunsaturated fatty acids, while a significantly higher content ($P<0.01$) of polyunsaturated fatty acids was found in organically produced milk. Differences in fatty acid composition in relation to the production system are attributed to diet-related parameters such as differences in the forage to concentrates ratio, type of flora in the grazing site and type of grass (fresh or silage).

Session 59 Poster 15

Influence of heifer resilience on the productive performance of calves

G.R.D. Rodrigues[1], V.T. Rezende[2], C. Raineri[1], M.E.Z. Mercadante[3], S.F.M. Bonilha[3] and J.N.S.G. Cyrillo[3]
[1]Uberlândia Federal University, BR-050, 38410-337, Brazil, [2]University of São Paulo, Faculty of Veterinary and Animal Science, 13635-900, Brazil, [3]Institute of Animal Science, Beef Cattle Research Center, 14174-000, Brazil; jgcyrillo@sp.gov.br

This study aimed to assess the relationship between heifer's resilience to the dry season challenge on calves' weights at birth and weaning. Data from Nellore heifers born between 2008 and 2017, belonging to a selection experiment started in 1980 were used. Heifers from selection line (NeS, n=283) and control line (NeC, n=129), selected for higher and mean yearling weight, respectively, were submitted to the breeding season at two years of age. Resilient animals (DFW770) were those with higher observed weights at 770 days old (W770, 347±54 kg), measured at the beginning of the breeding season, when compared to the standardized weight at the same age, obtained using the ADG of the rainy season as a growth pattern (W770E, 296±43 kg). Heifers were classified as susceptible (S) (n=82), average (A) (n=214), or resilient (R) (n=107) based on the first, second, and third quartiles of DFW770. Calves data corresponded to birth weight (BW) and adjusted weight at 210 days of age (WW) from NeS (n=183) and NeC (n=89). Analyses were carried out in R 4.2.2. The model of analysis included the effects of heifer resilience (RES), month of birth (M), calf gender, selection line (SL), and interaction RES×SL. Differences between factors were evaluated by SNK test at 5% significance. NeS calves were born and weaned heavier (32.08±0.32 kg, 198.17±2.87 kg) than NeC calves (26.00±0.38 kg, 161.58±2.77 kg) (P<0.001). Calves from R heifers had higher BW (31.31±0.49 kg) than calves from A and S animals (30.01±0.44 kg × 28.69±0.65 kg) (P<0.001). Calves from R heifers had higher WW (191.26±2.83 kg) than calves from A and S animals (180.26±2.33 kg × 161.35±3.44 kg) (P<0.001). Calves from R NeS (198.17±2.87 kg) and R NeC (161.58±2.77 kg) heifers had higher WW than calves from S NeS (166.94±4.97 kg) and S NeC (156.06±4.65 kg) heifers (P=0.013). Results suggest that the dry season challenge on heifers impacts calves' birth and weaning weight; R heifers wean heavier calves than S heifers in both selection lines.

Session 59 Poster 16

Feed autonomy and manure's recycling of dairy sheep farming systems in Roquefort (France)

W. Nasri[1,2], F. Stark[1], N. Amposta[3], M. Lamarque[3], C. Allain[3], D. Portes[3], S. Arles[3], S. Parisot[3], C. Corniaux[1,2], P. Salgado[1,2] and E. González-García[1]
[1]UMR SELMET, Univ Montpellier, CIRAD, INRAE, L'Institut Agro, 34060 Montpellier, France, [2]CIRAD, UMR SELMET, 34398 Montpellier, France, [3]INRAE, UE0321 La Fage, 12250 Roquefort-sur-Soulzon, France; waad.nasri@cirad.fr

The aim of this work was to identify the main gaps and opportunities for increasing the efficiency and autonomy in the use of resources (i.e. natural, farm- or locally-produced, external or imported), as well as the resilience to climate change in dairy sheep farming systems (DSFS). The DSFS from the INRAE Experimental Farm La Fage (43°54′54.52″N; 3°05′38.11″E) was chosen as a typical case study, representative from the Roquefort region, the main production basin for sheep milk in France. Firstly, a deep characterization was carried out with the aim of diagnosing the global functioning of the DSFS; then its nutrient flows were analysed using Ecological Network Analysis. A second step was looking to identify the main opportunities for building future DSFS through a multi-stakeholder platform. A historical dataset for the period 2015-2019 was collected, organized, processed, analysed and interpreted. Detailed features of each DSFS component and interactions between them were detailed in the deep characterization phase. The main components of the DSFS are: the flock (n=608 females; 70% ewes and 30% hoggets), feeding resources, infrastructures (buildings, sheds, etc.) and manure stockpile. The feeding system is based on improved (96%) and native (4%) grasslands both used for direct grazing and producing roughage. The grazing rate is 12% while the indoors feeding covers 88% of the yearly flock consumption. Main gaps are related to the feeding system and manure use. In fact, the quantity of purchased feed keeps increasing (from 7 to 30% during 2015-2019), contrarily to the quantity of on-farm produced feedstuffs (from 92 to 70%). There are considerable nitrogen losses mainly by leaching and by volatilization during manure's storage and spreading. There are opportunities for increasing feed and forage autonomy of the farm as well as for optimising processes for organic matter recycling, conservation and use. Acknowledgements: This work is part of the TRUSTFARM project carried out under the ERA-Net Cofund FOSC (Grant N° 862555).

Session 59 Poster 17

Metabarcoding study of the microbial dynamics in cheeses as a function of milk tank temperature

L. Giagnoni[1,2], C. Spanu[2], A. Tondello[1], S. Deb[1], A. Cecchinato[1], P. Stevanato[1], M. De Noni[3] and A. Squartini[1]
[1]University of Padova, DAFNAE, Viale dell'Università 16, 35020, Italy, [2]University of Sassari, Department of Veterinary Medicine, Via Vienna 2, 07100, Italy, [3]Latteria Montello S.p.A., Via Fante d'Italia 26, 31040, Italy; l.giagnoni@studenti.uniss.it

Milk tank temperature variations and fresh cheese seasoning timing can affect the shelf-life and the sensory properties of cheese as final product. Milk refrigeration temperature before pasteurization plays a crucial role as it can affect especially the proportions of psychrotrophic taxa abundance over the total milk bacterial population. Generally, 4 °C is the standard choice, due to its general growth control effect. However, some cold-tolerant genera featuring a proteolytic attitude, upon proliferation, could negatively affect curd clotting and regular cheese maturation at 4 °C, becoming an inconvenience. The present study was commissioned by a major fresh cheese producing company with an in-factory trial. The aim of this study was to evaluate the microbial communities of cheese preparations derived from raw milk stored at different temperatures, before pasteurization and cheese making, by DNA extraction and 16S-matabarcoding sequencing. Working at actual dairy facility level, raw milk was stored at 4, 7 and 9 °C, performing a gradual shift of the same tank across the above values. Milk was then pasteurized and used to prepare a caciotta-style fresh cheese. The temporal dynamics of the microbial communities at different times of maturation, from time zero up to two months, were evaluated. The taxa alpha and beta diversity across the different communities and their relative shifts in abundance, with particular emphasis on the complementary dynamics of psycrotrophic groups vs temperature-favoured enteric ones, will be discussed. Relevant clues for a better anticipation of the effects of thermal abuse and overall process parameters variation, have been gathered, to the benefit of an improved handling of the technical conditions by the cheese manufacturing industry.

Session 59 Poster 18

Adaptive integumentary traits of cattle raised in a silvopastoral system in tropical region

A.R. Garcia[1,2], A.N. Barreto[1], M.A.C. Jacintho[2], W. Barioni Junior[2], L.N. Costa[3], F. Luzi[4], J.R.M. Pezzopane[2], A.C.C. Bernardi[2] and A.M.F. Pereira[5]
[1]UFPA, Rod BR 316, 68740, Castanhal, Brazil, [2]Embrapa Pecuária Sudeste, P.O. Box 339, 13560, São Carlos, Brazil, [3]Università di Bologna, Viale Fanin, 46, 40127, Bologna, Italy, [4]Università degli Studi di Milano, Via Celoria, 10, 20133, Milan, Italy, [5]Universidade de Évora, R Cardeal Rei, s/n, 7000, Évora, Portugal; alexandre.garcia@embrapa.br

This work aimed to investigate whether the environment of afforested pastures or in a silvopastoral system (SPS) determines adaptive differences in the morphophysiological characteristics of the integument of beef cattle. The experiment was conducted for 12 months in a tropical climate (São Carlos, Brazil, 21°57'42"S 47°50'28"W). Steers previously raised together in conventional pastures (n=64 Nelore and Canchim; 31.4±1.6 mo; 482.6±47.7 kg) were placed in rotational grazing systems with *Urochloa brizantha* in full sun (FS, n=32) or in shaded area with *Eucalyptus urograndis* (SPS, n=32). Animals were evaluated monthly, during periods of thermal challenge (11:00am-2:00pm). Skin microbiopsies were performed in winter and summer for histomorphometric evaluation (Bioethical Prot. CPPSE 07/17). No significant interaction (treat×seas) was found. Overall means were compared using Tukey's test ($P \leq 0.05$). Rectal temperature (FS=39.45±0.05 vs SPS=39.29±0.05 °C; $P<0.05$) and the coating thickness differed (FS=1.60±0.04 vs SPS=1.42±0.04 mm; $P<0.05$), suggesting that steers raised in SPS had an external adaptive feature that favoured heat loss. Histological analysis revealed greater epidermal epithelial thickness in animals raised in FS (FS=76.4±42.2 vs SPS=64.4±42.2 µm; $P<0.05$), which may be attributed to greater outer cell proliferation in the integumentary layer of the animals most exposed to ultraviolet radiation. The distance between the epidermis and sweat glands was smaller in SPS animals (FS=805.9±25.1 vs SPS=727.1±25.1 µm; $P<0.05$), indicating their greater glandular activity and lumen filling, which increases the sweating efficiency. Integrated production systems have been pointed out as a strategy to mitigate and adapt to the effects of climate change. Indeed, animals raised in SPS showed greater plasticity and adaptability of skin tissue, favouring cutaneous convective and evaporative thermolysis.

Session 59 Poster 19

Impact of fodder quality seasonality on enteric methane emission from cattle in Sub-Saharan Africa

G.X. Gbenou[1,2,3], M.H. Assouma[2,4], C. Martin[5], D. Bastianelli[6], L. Bonnal[6], T. Kiendrebeogo[3], O. Sib[2,4], B. Bois[6], S. Sanogo[4] and L.H. Dossa[1]

[1]FSA-UAC, Cotonou, BP 526, Benin, [2]SELMET, Univ Montpellier, CIRAD, INRAE, Institut Agro, Montpellier, 34398, France, [3]CNRST, INERA, Bobo-Dioulasso, 01 BP 910, Burkina Faso, [4]CIRDES, Bobo-Dioulasso, 01 BP 454, Burkina Faso, [5]INRAE, UMRH, Clermont-Ferrand, 63122, France, [6]CIRAD, UMR SELMET, Montpellier, 34398, France; gerard_xavier_djidjo.gbenou@cirad.fr

In Sahel, pastures feed resources drastically decline both in quantity and quality from rainy to dry season. Several models to estimate enteric methane (eCH_4) emission were developed but have low accuracy in Sahelian production systems. In order to improve these models, direct reference measurements are necessary. We aimed to measure *in vivo* eCH_4 emission in Sudanese peulh zebu during each season (rainy: RN, cold dry: CD, and hot dry: HD) of the year. The experiment was carried out (CIRDES, Burkina Faso) on 10 steers of 32 months old with liveweight (LW) of 147±5.8 kg (i.e. 0.59 Tropical Livestock Unit-TLU). Animals were kept in individual boxes. Individual eCH_4 (GreenFeed system) and total tract digestibility were measured. The animals were fed with natural forage harvested from rangelands in each season. Different levels of forage were offered by mimicking the gradient of forage availability on pastures during a year: 7% LW during the rainy season; 3.5% LW during the cold dry season; and 3.5, 2.5 and 1.5% LW during the hot dry season. Each trial lasted 3 weeks including 2 weeks of feed adaptation and 1 week of data collection. In RN, CD and HD, the average content of intake in crude protein (CP, g/kg DMI) was 65±1.6, 48±1.6, and 28±1.1, respectively. Average content of intake in fibre (NDF, g/kg DMI) was respectively 619±29.0, 660±17.4, and 712±17.8. Dry matter intake (DMI, g/kg LW) was respectively 22±1.4, 23±1.5, and 16±1.8 in RN, CD, and HD. The average DM digestibility (%) was respectively 56.9±3.04, 49.9±2.12, and 46.2±3.08. Amounts of eCH_4 (g/kg DMI) emitted was different with HD compared to others seasons: (33±7.9 vs 21±4.4 in RN and 25±4.7 in CD). A global calculation over a year revealed that in extensive systems of Sub-Saharan Africa, local zebus cattle fed on natural pastures emitted 44.9±3.34 kg of eCH_4/TLU/year.

Session 60 Theatre 1

Variation in mineral content of feeds for dairy cows and how that can affect ration formulation

W.P. Weiss

Ohio State University, 2145 Riverside Drive, Cincinnati, OH 45202, USA; weiss.6@osu.edu

Sources of variation in the mineral concentrations of feedstuffs include those produced by the observer (sampling and analytical variation) and the true variation caused by soil, climate, manufacturing method, etc. Partitioning variation correctly requires large amounts of data with multiple points of replication. From a previous experiment, observer variation for macrominerals (Ca, P, Mg, and K) and trace minerals (Cu, Fe, Mn, Zn) was about 10% and 60-80% of total variation, respectively. Data on observed variation in mineral composition can be obtained from various databases, but this observed variation will include both sampling and analytical variation. For this abstract, the database from NASEM (2021) Dairy Requirements was used. That database was derived from $>10^6$ records obtained from commercial testing labs that used approved methods. The data were screened using a robust statistical method to remove mislabelled feeds and partition feeds into subpopulations when appropriate. Several exceptions exist but some general conclusions can be reached. Concentrations of minerals were more variable (based on CV) than the concentration of total ash. For most feeds, concentrations of ash, P, Mg, and S approximated a normal distribution but for all TM and Ca, distributions were skewed with long tails at higher concentrations. Within feeds, concentrations of P were least variable (CV averaged 15%) and Fe was most variable (CV >50%). As a comparison crude protein had CV generally <10. Within a feed, variation in Cu and Zn concentrations were similar. Variation in mineral concentrations was generally greater for forages than concentrates, but differences between those feed classes were less for macrominerals than for TM. However, sampling variation may have caused much of these differences. Although variation for many minerals is quite high in feeds, well-made mixed diets will vary less because the composition of feeds vary independently. Using a simple 5 ingredient diet with no supplemental minerals, means and SD from NASEM, and Monte Carlo simulations, diet CV for crude protein was 2.9. For P, Mg, S, Cu, and Zn the CV were 4.7, 9.5, 9.1, 15.9, and 10.9. Variation in basal concentrations of minerals is greater than for macronutrients, but sampling error is likely a major contributor to that variation.

Session 60 Theatre 2

Evolution of nutritional explorations in herds of dairy and beef cattle between 2013 and 2021

L. Reisdorffer
OBIONE, 71, 239 Rue Fernand Leger, 71000 Macon, France; lr@obione.fr

Nutritional explorations are helpful in understanding the problems encountered in farming. For several years the company Obione in partnership with the Labeo laboratory have been offering Nutritional Explorations (NE) to French breeders and veterinarians. Based on their data between 2013 and 2021, the results of the NE are analysed by elements, by year and by month. The analysis of these data shows variations which are explained both by the analysis technique but also by season and therefore at the same time the fodder availability, by the type of housing which conditions the intake of minerals in the ration of animals and by drought which can limit the availability of vitamins for animals. Prevalences are still quite high since about 50% of the samples that are processed show by deficiencies. The hierarchy within the deficient elements shows a fairly marked evolution with for the Trace elements, Copper is the leading deficient element and for Vitamins, it is Vitamin A. These data should lead veterinarians and breeders to not forget these two elements significant and often loss-making.

Session 60 Theatre 3

Selenium supply in animal feeds, a powerful nutritional tool against cancer

M. López-Alonso[1], I. Rivas[1] and M. Miranda[2]
[1]University of Santiago de Compostela, Animal Pathology, Veterinary Faculty, 27002 Lugo, Spain, [2]University of Santiago de Compostela, Anatomy, Animal Production and Veterinary Clinical Sciences, Veterinary Faculty, 27002 Lugo, Spain; marta.lopez.alonso@usc.es

Selenium (Se) is an essential trace element for the proper functioning of all organisms. It is a cofactor of many essential enzymes, as glutathione peroxidase, selenoprotein P, with a well-known role on the protection against oxidative stress as a free radical scavenger. Selenium deficient areas are very frequent around the world, including Europe, and given the low concentrations of Se in the feed ingredients, animal diets are regularly supplemented with Se to maintain health and production. In this areas, human food ingredients also contain inadequate levels of Se, and Se deficiency is associated with a compromised immune system and increased susceptibility to many diseases, including arthritis, cardiovascular disease, cataracts, cholestasis, cystic fibrosis, diabetes, immunodeficiency, lymphoblastic anaemia, macular degeneration, muscular dystrophy, stroke and some others, even though the most compelling evidence exists in relation to the cancer-protective effects of Se. In epidemiological observations and prospective studies, an inverse correlation between Se levels in food and blood and the risk of cancer and cancer mortality was observed. Particular interest in Se has been generated as a result of clinical studies showing that dietary supplementation with organic Se decreased cancer mortality two-fold. As a result, finding solutions to this problem is now on the agenda of many government health bodies. Results derived from various research studies conducted over the last few years have indicated that the enrichment of animal-derived foods with Se via supplementation of animal feeds can be an effective way of increasing human Se status in countries where Se consumption falls below the recommended daily allowances. This paper reviews how Se concentrations of animal products (meat, eggs and milk) can be manipulated to give increased levels, especially when organic Se is included in diet, which offers the opportunity to produce tailored Se-enriched animals products well adapted to the needs of the population.

Session 60 — Theatre 4

Effect of mineral source on 48-h *in vitro* fermentation

G.M. Boerboom[1], C.B. Peterson[2], L. Jansen[1], M.M. McCarthy[2], J.S. Heldt[2] and J. Johnston[3]
[1]Selko, Stationsstraat 77, 3811 MH Amersfoort, the Netherlands, [2]Selko USA, 3905 Vincennes Rd, Indianapolis, 46268, USA, [3]Fermentrics Technologies Inc, 961 Campbell Drive, Arnprior K7S0E1, Canada; carlyn.peterson@selko.com

A total of 4 trials were completed to evaluate the effect of trace minerals (TM) sources on *in vitro* fermentation. In the 1st trial the inclusion of copper (Cu), zinc (Zn) and manganese (Mn) was evaluated when comparing sulphate (SU) to IntelliBond (IB). In the 2nd trial, the inclusion of Mn from Mn oxide (MO), Mn sulphate (SM), organic Mn (OM) and IntelliBond M (IBM) was compared. Targeted supplemental TM were 222 mg/d and 555 mg/d for Mn and 1,035 mg/d for Zn, for a dairy cow consuming 25 kg DM with 120 l rumen volume. In the 3rd trial, 7 different sources of Mn were tested at 500 mg/d: Control (CM, no TM), MO, SM, IBM, Vistore Mn (VM), Availa Mn (AM), and Maintrex Mn (MM). In the 4th trial, 7 different sources of Zn were tested at 750 mg/d: Control (CZ, no TM), Zn oxide (OZ), Zn sulphate (SZ), IntelliBond Z (IBZ), Vistore Zn (VZ), Availa Zn (AZ), and Mintrex Zn (MZ). Fermentation vessels contained a mix of KSU buffer (80%) and rumen fluid (20%) with 6 mm ground dairy TMR, as fermentation substrate, in 5×10 cm bags. Vessels were incubated for 48-h in a 39.5 °C water bath. Data were analysed using PROC MIXED (SAS institute Inc, Cary, NC). A P-value $\leq$0.05 was considered significant. The 1st trial indicated reduced apparent organic matter disappearance (aOMD) with SU compared with the control ($P<0.01$), whereas IB did not differ from control. The 2nd trial indicated that MO and SM reduced aOMD relative to the control ($P<0.001$), whereas IBM and OM were similar to the control. IBM and OM had higher apparent microbial biomass production (aMBP) than the control, SM and MO treatments ($P<0.01$). In the 3rd trial, IBM increased aMBP compared to MM, SM and AM ($P<0.05$) but was no different from VM or the control. Compared to MO a trend for an improved aMBP was observed for IBM ($P<0.10$). In the 4th trial, IBZ resulted in greater aMBP compared to the control, MZ, VZ and SZ ($P<0.05$). A trend was observed for an improved aMBP compared to AZ ($P<0.10$). IBZ also increased aOMD compared to all other treatments ($P<0.05$). Overall, results indicated that trace minerals can affect *in vitro* fermentation, with IntelliBond reducing the negative effects.

Session 60 — Theatre 5

Feed phosphates market dynamics: does price influence demand?

G. Milochau
Praxed, 1 rue Napoléon Ancelin, 78400 Chatou, France; guillaume.milochau@praxed.fr

For people new to the inorganic feed phosphate (IFP) market, it's hard to find any logic! This market is highly volatile, and it is hard to find a good proxy or benchmark to determine when it's the right time to close a deal. For example, when it comes to Polish imports, the price range of monocalcium phosphate (MCP) is wide: from below €200/tonne up to €1,400/tonne! Of course, if we compare feed phosphate prices with prices for phosphate fertilizers such as diammonium phosphate (DAP), we can see they follow the same basic trend, but there is no clear correlation between fertilizer and feed phosphates. These markets have different dynamics, although they share the same fundamentals. Access to phosphorous is limited to a few regions in the world, with most of the reserves in North Africa. Many key agricultural regions depend on the import of phosphorous in one form or another (phosphate rock, phosphoric acid, DAP, TSP, etc). Production of phosphates is increasingly concentrated within vertically-integrated producers that are not only expanding their production capacities but also investing in their own distribution systems. This is what we mean when we talk about mine-to-market strategies. We hear from the field that consumption has been heavily reduced, and soon feed phosphates may become niche products in pig feed, for instance. Are feed phosphates going to become niche products with the introduction of last phytase generation? Does the recent price of IFP will have destroyed demand for good?

Session 60 Theatre 6

Reduction of trace mineral supplementation on performance and mineral status of fattening pigs

E. Gourlez[1,2,3], J.Y. Dourmad[3], F. Beline[2], A. Monteiro[1], A. Boudon[3], A. Narcy[4], P. Schlegel[5] and F. De Quelen[3]
[1]Animine, rue Léon Rey Grange, 74960 Annecy, France, [2]INRAE, OPAALE, Av. de Cucillé, 35000 Rennes, France, [3]INRAE, PEGASE, Le Clos, 35590 Saint-Gilles, France, [4]INRAE, BOA, Centre INRAE Val de Loire, 37380 Nouzilly, France, [5]Agroscope, Rte de la Tioleyre 4, 1725 Posieux, Switzerland; emma.gourlez@inrae.fr

Excreted Cu and Zn can accumulate in the upper soil layer and harm its fertility. Thus, managing dietary contents in pig diets should reduce such risk. However, a limited supply of these essential trace elements to pig requires a better knowledge. This study aimed to evaluate the effect of three levels and two sources of Cu and Zn on performance and mineral status of fattening pigs. Four dietary treatments were compared including a negative control (NC) corresponding to the basal growing and finishing diets, without Zn and Cu supplementation (5 and 29 mg/kg Cu and Zn respectively); an intermediate level (O1) supplemented with Cu and Zn oxides (Cu_2O and ZnO; CoRouge® and Hizox®, Animine, France), which provided on average 7.4 mg/kg Cu and 47.5 mg/kg Zn; two diets were supplemented with oxides (O_2) or sulphates (PC, positive control) at maximal EU regulation levels (25 and 120 mg/kg for total Cu and Zn, respectively). Ninety-six pigs (24.3±3.3 kg BW) were allocated to one of the four experimental treatments and raised in individual pens during 14 weeks (up to 110.3±8.9 kg). Animal performance were measured and samples of plasma (at d1, d41 and d90), bones and liver (at slaughter) were collected. Every third week, samples of faeces were obtained to determine the dynamic of Cu and Zn excretion. Over the whole experimental period feed intake, body weight and feed conversion ratio were not affected by the level nor the source of Cu and Zn. Plasma Cu was not affected neither by treatment nor period, whereas plasma Zn increased with period but did not differed between treatments. Hepatic Cu increased ($P<0.05$) with dietary content, and a tendency for increase of bone Zn with dietary content ($P<0.10$) was observed. Faecal Cu and Zn decreased significantly ($P<0.01$) with dietary content. Although it seems possible to reduce dietary Cu and Zn without affecting performance, these results require to be validated in commercial farms with more challenging health constraints.

Session 60 Theatre 7

Optimal level of dietary zinc for pigs between 10 and 30 kg

T.S. Nielsen, S.V. Hansen, J.V. Nørgaard and T.A. Woyengo
Aarhus University, Animal and Veterinary Sciences, Blichers Allé 20, Box 50, 8830 Tjele, Denmark; tinas.nielsen@anivet.au.dk

Zinc (Zn) is an essential nutrient but excessive dietary Zn is excreted via faeces by pigs and spread to the environment. To establish the optimal level of dietary Zn from 10-30 kg, we conducted a Zn dose-response experiment (80, 92, 117, 189 and 318 ppm total Zn) in 150 pigs, 2 to 6 weeks post-weaning (PW). The basal diet (including 1000 phytase units/kg feed) without added Zn contained 80 ppm Zn, and added Zn was high purity zinc oxide. From weaning at 28 days of age to 2 weeks PW, pigs were supplied 1,500 ppm total Zn, which we previously showed was the optimal level of dietary Zn in this period. From 2 to 6 weeks PW, pigs were housed individually and feed intake, BW and faecal scores were recorded. Zinc status was evaluated in blood (weekly), and in liver, small intestinal mucosa and bone by the end of the trial. Average daily feed intake (ADFI) and average daily gain (ADG) was unaffected by dietary Zn content ($P=0.99$ and $P=0.60$, respectively) but there was a tendency for feed conversion ratio (FCR) to be better ($P=0.07$) for pigs fed 80 ppm Zn (1.34) compared to 318 ppm Zn (1.39), with the other dietary Zn levels being in between. The probability of a diarrheic faecal score was lowest (0.35-0.36) at 80, 189 and 318 ppm Zn and highest (0.45) at 92 ppm Zn ($P<0.001$). Serum Zn status was on average 873 µg/l at the beginning of the experimental period, but decreased ($P<0.001$) by 13-24% after one week, depending on dietary Zn level. After 3 and 4 weeks, 80 and 92 ppm Zn still resulted in serum Zn levels lower ($P<0.001$) that at the beginning of the experimental period. Zinc status in liver, bone and intestinal mucosa was affected ($P<0.001$) by dietary Zn level but for liver, there was no difference between dietary Zn levels. Intestinal mucosa Zn increased from the lowest to the highest dietary Zn level ($P<0.05$), whereas there was a quadratic effect of dietary Zn level on bone Zn content. Based on ADFI, ADG, FCR and probability of diarrhoea, we conclude that 10-30 kg pigs can do without added Zn in the diet when a high level of phytase is included and pigs are initially supplemented with 1,500 ppm Zn. However, serum Zn status decline gradually when Zn is not added to the diet and other Zn status parameters may also be lower at 30 kg.

Session 60 Theatre 8

Administration of potentiated Zn and monovalent Cu in weanling piglets diet

L. Marchetti[1], R. Rebucci[1], P. Cremonesi[2], B. Castiglioni[2], F. Biscarini[2], A. Romeo[3] and V. Bontempo[1]
[1]Università degli Studi di Milano, Department of Veterinary Medicine and Animal Science, Via dell'Università, 6, 26900, Lodi, Italy, [2]Institute of Agricultural Biology and Biotechnology, Via Einstein, 26900, Lodi, Italy, [3]Animine, 10 rue Léon Rey Grange, Annecy 74960, France; luca.marchetti1@unimi.it

The aim of this study was to determine the suitability of potentiated Zn combined with a monovalent Cu as a replacement for pharmacological ZnO supplementation in piglets after weaning. A total of 120 piglets (7.143±0.924 kg) were weaned at 27 d and divided into 4 experimental treatments. A positive control (PC, 2,500 ppm of Zinc through standard ZnO) was compared to 3 treatments in which Cu and Zn were supplemented through potentiated Zn (HiZox®) and monovalent Cu (CoRouge®) at European and non-European levels of inclusion: EU (120 ppm of Zn; 140 ppm of Cu), non-EU^+ (300 ppm of Zn; 200 ppm of Cu) and non-EU^- (300 ppm of Zn; 140 ppm of Cu). Performance data were analysed using the MIXED procedure of SAS. Serum diamine oxidase (DAO) and D-lactate were analysed through a MIXED procedure of SAS at 0 d and 14 d. SIgA was evaluated with a GLM procedure of SAS (28 d). Faecal samples (14 d and 28 d) and intestinal content (28 d) were collected and the V3-V4 hypervariable regions of the bacterial 16S gene were sequenced in one MiSeq (Illumina) run. Performances were not affected by different Zn and Cu dosages. Plasma DAO was negatively ($P<0.05$) affected in non-EU^- treatment. The level of sIgA was increased in PC group (157.73±19.58 ng/ml) compared to EU (143.81±12.96 ng/ml; $P<0.05$) and non-EU^- group (138.01±16.88 ng/ml; $P<0.01$). non-EU^- contributed to a significant decrease in biodiversity in faecal microbiota as shown by the Shannon diversity and Simpson evenness indexes ($P<0.05$). A total of 18 differentially abundant genera ($P<0.05$) were identified in the caecal content with the presence of genera linked to the disruption of the gut barrier (*Campylobacter*) with non-EU^-. These results suggest that a more balanced supplementation of the two trace elements, through more bioavailable sources, could represent a valid tool to enhance gut health of weaning piglets and reduce environmental impact.

Session 60 Theatre 9

Unexpected Cu and Zn speciation patterns in the feed-animal-excreta system

S. Legros[1], M. Tella[1], A.N.T.R. Monteiro[2], A. Forouzandeh[3], F. Penen[2], S. Durosoy[2] and E. Doelsch[4]
[1]CIRAD, Avenue Agropolis, 34398 Montpellier, France, [2]Animine, 10 rue Léon Rey Grange, 74960 Annecy, France, [3]Universitat Autònoma de Barcelona, Department of Animal and Food Science, Edifici V, 08193 Bellaterra, Spain, [4]CIRAD, Recycling and risk research unit, CEREGE Europole de l'Arbois, 13545 Aix-en-Provence Cedex 4, France; doelsch@cirad.fr

Trace minerals such as copper (Cu) and zinc (Zn) are animal nutrition supplements necessary for livestock health and breeding performance, yet they also have environmental impacts via animal excretion. Here we investigated changes in Cu and Zn speciation from the feed additive to the animal excreta stages. The aim of this study was to assess whether different Cu and Zn feed additives induce different Cu and Zn speciation patterns, and to determine the extent to which this speciation is preserved throughout the feed-animal-excreta system. Two types of animals were considered, broilers and pigs. Synchrotron-based X-ray absorption spectroscopy (XAS) was used for this investigation. The principal findings were: (1) In feed, Cu and Zn speciation changed rapidly from the feed additive signature (Cu and Zn oxides or Cu and Zn sulphates) to Cu and Zn organic complexes (Cu phytate and Zn phytate). (2) In the digestive tract, the results depend on the animal studied. In broilers, we showed that Cu and Zn phytate were major Cu and Zn species while Cu sulphide and Zn amorphous phosphate species were detected but remained minor species. In pigs, we showed that the Cu and Zn dominant species were Zn phytate and Cu sulphide. (3) In fresh excreta, the results depend on the element studied. For Cu, Cu sulphide is always the major species regardless of the animal studied. For Zn, Zn amorphous phosphate is the major species for broilers while Zn sulphide is the major species for pigs. These results should help to: (1) enhance the design of future research studies comparing different feed additive performances; (2) assess Cu and Zn bioavailability in the digestive tract; (3) gain further insight into the fate of Cu and Zn in cultivated soils when poultry manure is used as fertilizer.

Session 60 Theatre 10

Coarse limestone particles limit the formation of Ca-phytate complexes in laying hens

F. Hervo[1,2], M.-P. Létourneau-Montminy[2], B. Méda[1], M.J. Duclos[1] and A. Narcy[1]
[1]INRAE, Université de Tours, BOA, 37380, Nouzilly, France, [2]Sciences Animales, Université Laval, G1V 0A6, Québec city, Québec, Canada; fabie.hervo@inrae.fr

Reducing phosphorus (P) excretion in laying hen production is a challenge. In the digestive tract, calcium (Ca) can form de novo unavailable complexes with phytate, impairing phytase efficiency to release non-phytic P (NPP) from phytate. In broilers, it has been shown that coarse limestone particles (CL) limit the formation of Ca–phytate complexes. Thus, the effect of CL and microbial phytase was investigated in laying hens between 31 and 35 wk of age to study the potential beneficial effect of CL incorporation on P utilization. Seventy-two Lohmann Tradition laying hens were randomly assigned to one of the four experimental diets. A 2×2 factorial arrangement was used with two levels of phytase and basal available P (aP); 0 FTU/kg with 0.30% aP or 300 FTU/kg with 0.15% aP and two limestone particle sizes (LmPS); fine particles (FL, <0.5 mm) or a mix (MIX) of 75% CL (2-4 mm) and 25% FL. Diets contained equivalent levels of Ca (3.5%), phytic P (PP, 0.18%) and, considering the phytase P equivalency, equivalent levels of aP (0.30%). Egg production and average daily feed intake (ADFI) were measured weekly throughout the experimentation. Tibia ash (%), apparent pre-caecal digestibility (APCD) of P and Ca (%) and PP disappearance at the ileal level (%) were measured. No differences were observed between treatments in ADFI, FCR and tibia ash. Phytase and CL together increased the APCD of Ca by 7.3 percentage points (Phytase × LmPS, $P<0.001$). Hens fed with FL and phytase exhibited a lower APCD of P compared with hens fed FL without phytase, while no differences were observed in hens fed with CL (Phytase × LmPS, $P<0.001$). Additionally, hens fed with FL and phytase showed a lower PP disappearance at the ileal level than hens fed with CL and phytase (Phytase × LmPS, $P=0.005$). These results may be explained by a lower formation of Ca-phytate complexes when limestone was provided as CL, and thus a higher availability of minerals for the animal. This strategy may therefore help improving the utilization of P and Ca and reducing P excretion in laying hens.

Session 60 Poster 11

The duration of efficacy of a single oral dose of selenium in sheep

S.E. Gallimore[1], E.J. Hall[1] and N.R. Kendall[2]
[1]Rumenco Ltd, T/A Nettex, Stretton House, Derby Road, Burton-on-Trent, DE13 0DW, United Kingdom, [2]University of Nottingham, School of Veterinary Medicine and Science, Sutton Bonington campus, Loughborough, LE12 5RD, United Kingdom; sgallimore@rumenco.co.uk

Selenium is commonly deficient in the diets of grazing sheep and can limit immune responses, growth, fertility and flock profitability. Oral drenches are widely used to supplement Se and research on duration of efficacy of these products is required to inform best-practise mineral supplementation. This study aimed to determine the response and duration of efficacy of a single dose of oral Se in grazing lambs. Two groups of weaned lambs (ewe n=339, wether n=347) were randomly allocated to 3 treatments; 1.55 or 3.5 mg of Se administered on 0 d via a sodium selenite drench, or a control group which received no drench. Lambs were weighed on 0 and 54 d; between these days they were run as a single group and then separated into 2 groups (ewes/wethers) rotating around the same block of pasture. From each treatment group, 6 ewes and 6 wethers were sampled by jugular venepuncture on 0 d prior to drench administration and subsequently sampled on days 12, 26, 33, 40, 47, 54, 68, 89 and 108. Samples were analysed for plasma Se and Co by ICP-MS, for erythrocyte glutathione peroxidase (GSHPx) by colorimetric assay and for plasma vitamin B_{12} by immunoassay. Lambs across all treatments began the trial with adequate Se status. Results showed both drench treatments elevated plasma Se for 6-7 weeks. The 3.5 mg dose had a slightly longer duration with significance ($P<0.05$) compared to controls on 40 d present for this dose only. The 3.5 mg dose also displayed a higher peak in plasma Se compared to 1.55 mg on 12 d ($P<0.05$). No other plasma Se differences between doses were found. GSHPx activity declined over the duration of the study in both treatment and control groups, which could be attributed to diminishing effects of maternal Se transfer. However, both Se drenches partially mitigated this with elevated activity ($P<0.05$) present from 33 to 108 d compared to unsupplemented controls. There were no treatment effects for weight, plasma Co or vitamin B_{12}. In lambs with adequate Se status, a single dose of oral Se elevated plasma Se for 6-7 weeks and subsequently enhanced GSHPx activity for 108 days post treatment.

Session 60 Poster 12

Dietary manganese impacts on growth, carcass and reproductive traits in angus bulls

J.R. Russell[1], E.L. Lundy-Woolfolk[2], A.S. Cornelison[1], W.P. Schweer[1], T.M. Dohlman[2] and D.D. Loy[2]
[1]Zinpro Corporation, Eden Prairie, MN, USA, [2]Iowa State University, Ames, IA, USA; jrussell@zinpro.com

Although minerals like zinc and copper have been substantially researched in beef cattle diets, manganese (Mn) remains poorly understood. A pooled analysis revealed improved growth and carcass traits when zinc amino acid complex was provided in diets at twice the NASEM (2016) requirement. Therefore, the current objective was to evaluate dietary Mn sources at twice the NASEM requirement. Sixty weaned angus bulls received a common starting ration containing Mn sulphate while acclimating to feed bunks capable of measuring individual feed intake. Bulls then transitioned to a finishing diet in which a pellet was included to provide one of two dietary treatments (TRT) for 182 d. The TRT provided 40 ppm supplemental Mn on a total dietary basis, consisting of either inorganic Mn hydroxychloride (INORG; Intellibond M, Micronutrients LLC, US) or Mn amino acid complex (AVMN; Zinpro Availa Mn, Zinpro Corporation, US). Starting on d 60, bulls were electro-ejaculated (EJAC) monthly to collect semen for puberty determination. Following harvest (d 183), carcass traits were measured, and one testicle was collected to measure testicular degradation (TD) and other traits. Utilizing PROC Mixed of SAS for all continuous variables, bull was experimental unit (n=30/TRT) and TRT was a fixed effect. Initial BW (267 kg) served as a covariate for final bodyweight and carcass weight. PROC Glimmix was utilized to evaluate the pubertal data (binomial) and TD data (multinomial). Growth and carcass traits did not differ (P>0.39) due to TRT. Though TRT did not affect (P>0.3) puberty across timepoints, more AVMN bulls met or exceeded puberty prerequisites than INORG (20 vs 16, respectively) on the final EJAC day (d 171). Pubertal headcount for INORG regressed and it is unclear why, though TD score was greater (P=0.01) in post-harvest testicles from INORG vs AVMN bulls. No other differences (P>0.15) were detected for testicular measurements post-harvest. Although Mn source had no measurable impact on growth and carcass traits in developing bulls, Mn source effects on reproductive outcomes should be further explored.

Session 60 Poster 13

Micronutrient supplementation for suckling calves

M.S.V. Salles[1], F.J.F. Figueiroa[2], A. Saran Netto[2], C.M. Bittar[3], F.F. Simili[1] and H.N. Rios[1]
[1]Animal Science Institute, A. Bandeirantes, 2419, 14030-670, Brazil, [2]FZEA/USP, R. Duque de Caxias, 225, 13635-900, Brazil, [3]ESALQ/USP, Av. Pádua Dias, 11, 13418-900, Brazil; marcia.saladini@gmail.com

Micronutrient supplementation for suckling calves is important to support metabolism and the immune system, thus reducing the incidence of diseases and improving performance in this early stage of life. The objective of this research was to evaluate the supplementation of selenium, iron, and vitamin E on metabolism, the incidence of diarrhoea, and the performance of calves in an immunological challenge. Holstein male newborn calves (n=42) up to 60 days old, in a randomized block design, were allocated in three treatments: C (control milk substitute); SeVitE (milk substitute supplemented with 0.6 mg organic selenium/kg + 100 IU vitamin E); SeVitEFe (milk substitute supplemented with 0.6 mg organic selenium/kg + 100 IU vitamin E + 200 mg Fe chelate/kg). The calves received 6 litres of milk substitute daily until 30 days of age and then 4 litres until weaning at 60 days, and received concentrate *ad libitum*. The animals were inoculated with *Anaplasma marginale* at 40 days of age. Food intake was monitored daily. Blood was collected for nutrient analysis and the weight and growth of the animals were monitored. The daily faecal score was used to monitor the incidence of diarrhoea. The results were analysed using the SAS PROC MIXED. Intakes of the nutrient selenium, vitamin E, and iron were higher in the supplemented animals compared to the control treatment. There was an increase in selenium in the serum of supplemented calves, with no change in vitamin E and iron. Animals with SeVitEFe tended to have lower diarrhoea scores and a lower frequency of diarrhoea. The performance of the animals did not differ between treatments. Supplementation with the micronutrient selenium, vitamin E, and iron in the diet helped to control the incidence of diarrhoea without altering the performance of the calves during the suckling phase.
Financial support: FAPESP 2017/04165-5 CNPQ: 301990/2022-0.

Session 60 Poster 14

Impacts of trace mineral source and ancillary drench on steer performance during backgrounding

K. Harvey[1], L. Rahmel[1], J. Cordero[1], B. Karisch[1], R. Cooke[2] and J. Russell[3]
[1]Mississippi State University, Starkville, MS, USA, [2]Texas A&M University, College Station, TX, USA, [3]Zinpro Corporation, Eden Prairie, MN, USA; ks3114@msstate.edu

Crossbred steers (120 hd) were purchased at auction and transported to the experimental facility. Steer BW was recorded at arrival (d -1; initial shrunk BW=227.7±1.3 kg). On d 0, steers were ranked by BW and allocated to 1 of 8 groups (8 steers/group) and housed in dry lot pens (6×12 m) equipped with GrowSafe automated feeding systems (Model 8000; 2 bunks/pen; d 0 to 60). Groups were randomly assigned to receive a total mixed ration (TMR; 47% corn) containing: (1) sulphate sources of Cu, Co, Mn, and Zn (INR; n=40); (2) organic complexed sources of Cu, Mn, Co, and Zn (AAC; Availa 4; Zinpro Corp., Eden Prairie, MN; n=40); or (3) AAC and an organic trace mineral drench (30 ml/hd; ProFusion, Zinpro Corp.) on d 0 and ancillary to any morbidity treatment (APF; n=40). Diets were formulated to provide the same daily amount of energy, protein, macro minerals and trace minerals based on 7 g/steer daily of Availa4. Steers were assessed for bovine respiratory disease (BRD) signs daily and vaccinated for BRD pathogens on d 0 and 21. Individual feed intake was evaluated daily (d 0 to 60) using GrowSafe 8000 software. Liver biopsy was performed on d 0 (used as covariate), 28, and 60. Final BW was recorded on d 60 and 61. Blood samples were collected on d 0, 2, 6, 10, 13, 21, 28, and 45. No treatment differences were detected ($P\geq0.59$) for TMR intake, final BW, ADG, feed efficiency, or BRD incidence. Mean liver Co concentrations were greater ($P=0.02$) in AAC and APF compared to INR steers. Mean liver Cu was greater ($P=0.02$) in APF compared to AAC steers. Liver Zn tended to be greater ($P=0.10$) on d 28 but less ($P=0.05$) on d 60 for INR compared to AAC and APF steers. Plasma haptoglobin was lowest ($P=0.05$) for steers supplemented with AAC on d 6, whereas AAC steers tended to have greater ($P=0.09$) plasma haptoglobin on d 13 compared with APF. Plasma cortisol tended to be greater ($P\leq0.10$) for INR steers on d 28 and 45 compared to AAC and APF. While supplementing cattle with AAC or INR resulted in similar animal performance and clinical disease, AAC and APF improved stress and acute phase protein responses.

Session 60 Poster 15

Impact of trace minerals and water/feed deprivation on performance and metabolism of grass-fed beef

M.J.I. Abreu[1], I.A. Cidrini[1], D. Brito De Araujo[2], F.D. Resende[3] and G.R. Siqueira[3]
[1]Sao Paulo State University, School of Veterinary & Agricultural Sciences, Via de acesso Professor Paulo Donato Castellane s/n, 14884-900, Jaboticabal, Brazil, [2]Nutreco Nederland, Selko Feed Additives, Stationstraat 77, 3811 MH, Amersfoort, the Netherlands, [3]Sao Paulo State Agency of Technology & Agribusiness, Department of Research & Innovation, Avenida Rui Barbosa s/n, 14470-000, Brazil; davi.araujo@selko.com

Two studies investigating how sources of Cu/Zn and 48-hours water/feed deprivation affect performance and metabolism of grass-fed beef cattle. Study 1: 20 castrated-canulated Nellore steers (350±132 kg) were blocked and randomly distributed in individual pens, in 2×2 factorial: supplemental Cu/Zn sources from inorganic (ITM; sulphate) vs hydroxy (HTM, Selko® IntelliBond®); and 48-hours deprivation (WFD) vs unrestricted (WFU) access to water/feed. The study was divided in adaptation (-21 to -1 d) and evaluation (0 to 36 d) periods. Deprivation did not affect ($P>0.10$) DMI but deprivation × period interaction was detected ($P<0.05$) for digestibility of DM, OM, NDF and ADF. HTM increased DMI ($P=0.075$) and NDFD in 1.6%. VFA concentration decreased from 62.3 to 20.1 mMol/dl, while rumen pH increased from 7.3 to 8.2 ($P<0.05$). After the deprivation, butyrate increased ($P=0.065$) and VFA ($P=0.004$) decreased in WFU. Study 2: 84 intact Nellore males (260±35 kg) were blocked and randomly assigned to *Urochloa brizantha* cv. Marandu paddocks for 131 d. Animals were randomly distributed among same treatments as describe in study 1. WFD animals increased urea and NEFA ($P<0.05$) and increased AST and total protein on 2 d ($P=0.080$). HTM lowered urea peak on 2 d and remained lower on 12 and 105 d ($P<0.05$). Liver Cu was higher in WFU/HTM animals ($P<0.05$). Interaction between deprivation × period ($P<0.05$) was detected for BW and ADG. WFD lost 34.5 kg, resulting in lower BW on 2 d, but increased ADG after 12 d, recovering BW lost. In conclusion, water/feed deprivation were able to impact nutrient digestibility and ruminal fermentation parameters. After deprivation, animals were able to compensate, recovering BW faster, resulting in no performance differences between WFD and WFU. Additionally, HTM increased liver Cu and increased DMI and NDFD, although no changes in performance observed when compared to ITM.

The impact of trace mineral sources of copper and zinc on performance and ruminal bacteria diversity

I.A. Cidrini[1], I.M. Ferreira[1], D. Brito De Araujo[2], G.R. Siqueira[3] and F.D. Resende[3]
[1]Sao Paulo State University, School of Veterinary & Agricultural Sciences, Via de Acesso Professor Paulo Donato Castellane s/n, 14884-900, Jaboticabal, Brazil, [2]Nutreco Nederland, Selko Feed Additives, Stationsstraat 77, 3811 MH, Amersfoort, the Netherlands, [3]Sao Paulo State Agency of Technology & Agribusiness, Department of Research & Innovation, Avenida Rui Barbosa s/n, 14470-000, Colina, Brazil; davi.araujo@selko.com

Research indicates the solubility of trace mineral (TM) can vary between sources and impact rumen function, bioavailabity and performance. Two studies were carried out to evaluate the effect of TM source on performance and rumen microbial diversity of grazing beef cattle. In the first study, 120 intact Nellore males (BW=349.72±24 kg; 24 m) were blocked by BW and randomly assigned to 12 paddocks (06 paddocks/treatment; 10 animals/paddock). Animals were grazing *Urochloa brizantha* cv. Marandu and offered 01 of 02 supplements (0.5% BW; 25% CP; 65% TDN). Supplements were formulated for 40 mg Cu and 148 mg Zn/kg DM from either inorganic (ITM; Cu-sulphate and Zn-oxide) or hydroxy (HTM; Selko® IntelliBond ®) during 90 d. BW were collected every 30 d. Data were analysed by ANOVA using PROC MIXED. Each 30 d period was used as REPEATED measurement. HTM-fed animals increased ADG (0.469 vs 0.506 kg/day, P=0.01) and increased final BW (391.97 vs 397.11 kg P=0.03) compared to ITM. In the second study, 08 cannulated Nellore steers (541±18 kg) were blocked by BW and randomly distributed in individual paddocks and supplemented during 101 days with the same diet supplements from study 1. Ruminal content samples were collected on d97 for total DNA extraction using commercial kit. The V3/V4 regions of 16SrRNA gene was sequencing using Illumina MiSeq and QIIME (v.1.9.1) to filter reads, determining OTUs. A total of 293 OTUs were identified at genus level. HTM supplementation resulted in a higher ruminal abundance of several groups such as *Prevotella* 1 (P=0.01), *Ruminococcaceae* (P=0.01) and lower abundance of *Fibrobacter* (P=0.04) and others. These results suggests that HTM source of Cu and Zn improves animal performance and ruminal abundance of bacteria, i.e. *Firmicutes* phylum with important structural/non-structural carbohydrates degradation functions.

Are inorganic Mn sources soluble and improve rumen fermentation?

A. Vigh, C. Gerard and C. Panzuti
ADM Animal Nutrition, Talhouet, 56250, France; clemence.panzuti@adm.com

The aim of this study was to compare the ruminal solubility and bioavailability of two inorganic sources of manganese (Mn) and their effect on ruminal activity. *In vitro* fermentations, using a low-Mn substrate, a Mn-free buffer and a rumen juice-based inoculum were conducted for 70 h. The substrate was tested solely (CON), or with an addition of 0.06% DM of Mn oxide (MnO) or Mn sulphate (MnS), introduced at the start (Mn_0h), after 24 h (Mn_24h) or 48 h (Mn_48h) of incubation. Fermentation activity was measured through total gas production and substrate dry-matter degradability (dDM%). Mn ruminal bioavailability and solubility were assessed by measuring the Mn concentration in fractions obtained after successive centrifugations of the final fermentation juice, allowing to separate the remaining big particles, containing feed particles, insolubilized minerals, protozoa, a bacteria concentrated fraction, containing bioavailable minerals and a final supernatant, containing the solubilized minerals. Concerning the fermentation activity, the Mn_0h (regardless of source) tended (P<0.10) to decrease the total gas production compared to the CON (224±3.9 and 234±5.4 ml/gDM, respectively). dDM% was significantly lower (P<0.01) with MnO_24h compared to CON (86.5±2.3 and 88.6±2.4 dMD% respectively), while all the other treatments tended (P<0.10) to decrease dDM%. Analysis of the Mn content of the centrifugation fractions showed that 84-93% of the additional Mn was found solubilized in the supernatant, 4-11% in the big particles, and only 2-4% in the bacteria concentrated fraction. Mn concentration of the supernatant was significantly increased (P<0.05) in all treatments compared CON, with no differences between MnO and MnS source. Mn concentration of the bacteria concentrated fraction was significantly higher (P<0.001) in all treatments compared to CON (+238%, +358%, +354%, +307%, +353% and +338% for MnO_0h, MnO_24h, MnO_48h, MnS_0h, MnS_24h and MnS_48h, respectively). These results indicate that both inorganic sources of Mn are soluble in the rumen and can be assimilated by the rumen bacteria. However, as the addition of these soluble Mn forms had no positive effects on the gas production or dry matter fermentability, it seems that the rumen microbiota does not require Mn for its fermentative activity.

Session 60 Poster 18

Effect of diet type, Cu source and antagonists on rumen *in vitro* fermentation and Cu distribution

I. Bannister[1], J.A. Huntington[2], L.A. Sinclair[2], J.H. McCaughern[2] and A.M. McKenzie[2]
[1]SRUC Aberdeen, Department of Veterinary and Animal Sciences, Craibstone Estate, AB21 9YA, United Kingdom, [2]Harper Adams University, Harper Adams University, Edgmond, TF10 8NB, United Kingdom; jhuntington@harper-adams.ac.uk

Effects of molybdenum and sulphur on copper status in dairy cattle have recently been shown to be influenced by diet type. Diets high in maize silage or high in starch resulted in an increased copper status compared with a high grass silage or a diet low in starch. A 2×3 factorial rumen *in vitro* study was designed to further investigate the interaction between copper, molybdenum and sulphur and diet type. The diets had either a Barley : Hay ratio of 80:20 or 20:80 on a DM basis. All cultures were supplemented with 7.5 mg/kg DM Mo and 4.0 g/kg S. Copper was supplemented at 10.0 mg/kg DM either as $CuSO_4$ or Cu_2O (CoRouge, Anime) or received no supplemental copper. Diet (2.0 g DM) was incubated in 200 ml buffered (Mould's buffer) rumen fluid (20%) at 39 °C for 48 h. Cumulative increase in headspace gas pressure was recorded and gas profile was measured for 48 h. After 48 h, culture pH was recorded and the cultures were split into solid, liquid and microbial pellet samples. These were than analysed by ICP-MS for mineral levels. There was a significant increase in cumulative gas pressure in the high barley diet ($P<0.001$) but there was no effect of copper source. Final rumen pH was 5.84 for the high barley diets compared with 6.26 for the hay diets (sed=0.016; $P<0.001$). H_2S gas levels were significantly greater in the high barley diet compared with the hay diet (3,400 vs 2,182 ppm; sed=171.6; $P<0.001$). Diets supplemented with copper had significantly higher levels on copper in the solid fraction but there were no effects of Cu source on the liquid or microbial fractions. It was concluded that the effect of diet type on rumen pH and H_2S gas levels were a major influence on the interaction between copper, molybdenum and sulphur in the rumen.

Session 60 Poster 19

Meta-analysis on zinc oxide's mode of action in reducing weaning stress in healthy piglets

C. Negrini[1], D. Luise[1], F. Correa[1], P. Bosi[1], A. Roméo[2] and P. Trevisi[1]
[1]University of Bologna, DISTAL, Viale fanin 46, Bologna, 40127, Italy, [2]Animine, 10 rue Léon Rey Grange, Annecy, 74960, France; federico.correa2@unibo.it

Pharmacological concentrations of zinc oxide (ZnO) are known to improve growth performance and reduce diarrhoea incidence in piglets. However, to address environmental concerns, the EU banned the use of ZnO at pharmacological dosage (2,500-3,000 ppm). This meta-analysis aims to examine the effect of different doses of ZnO on post-weaning pig productivity and gut health. Forty-one peer-reviewed articles from 2001 to 2021 were selected and assigned to four ZnO levels: LZn (0-200 ppm), M1Zn (201-630 ppm), M2Zn (631-1,600 ppm), and HZn (1,601-3,000 ppm). Statistical analyses were based on ANOVA mixed models in Minitab. Results showed that the M2Zn and HZn groups had significantly higher average daily gain (ADG) values than the LZn group ($P<0.05$). HZn also had the highest gain-to-feed (G:F) values compared to M1Zn and LZn ($P<0.05$). Higher liver Zn concentrations were observed in the HZn group ($P=0.01$), consistent with increased dietary Zn levels. However, blood Zn concentration was higher in M2Zn ($P=0.003$). Villus height and width increased with increasing Zn level ($P<0.05$). HZn increased the expression of ZO-1 ($P<0.05$) and decreased the expression of IFN-γ ($P<0.05$), while also tending to increase occludin and TNF-α expression ($P=0.06$ and $P=0.09$, respectively). The presence of *Escherichia coli* was higher in the HZn group ($P<0.05$), while the highest presence of *Lactobacillus* content was found in the LZn group ($P=0.05$). In summary, the positive effects of HZn were confirmed, but medium dosages (M1Zn and M2Zn) yielded comparable results. Our findings suggest a mode of action for HZn and can aid in the development of alternative feeding strategies.

Marine mineral complex reduces nutrient interactions and allows efficient use of Ca and P in broiler

M.A. Bouwhuis[1], R. Casserly[2], A. Craig[3], D. Currie[3] and S. O'Connell[1,4]
[1]Celtic Sea Minerals, Strand Farm, Carrigaline, Co. Cork, Ireland, [2]Iernevation Ltd, Clonfower, Lanesboro, Co. Longford, Ireland, [3]Roslin Nutrition, Gosford Estate, Longniddry, East Lothian, United Kingdom, [4]Munster Technological University, Plant Biostimulant Group, Shannon ABC, Clash Road, Tralee, co. Kerry, Ireland; m.bouwhuis@celticseaminerals.com

Enhancing performance through reducing calcium (Ca) levels in broilers has been well investigated. Lowering Ca will reduce interactions with phosphorus (P) and phytate. However, bone issues have been reported which impact implementation. Novel strategies are required to allow this. A Ca rich marine mineral complex (MMC) derived from red seaweed *Lithothamnion sp.* is assembled on a scaffold of polysaccharide hydrogel, reducing the interaction between Ca, available P (avP) and phytate and making them more bioavailable. In this study, the potential to reduce Ca and avP levels in broiler diets was evaluated with MMC inclusion. The study was set up as a central composite design with 3 avP levels and 3 Ca levels, resulting in 9 treatments. The avP levels were 0.17-0.22, 0.34 and 0.45% and Ca levels were 0.5, 0.7, 0.90-0.98% in the starter period (day 0-20) and were reduced in the grower phase (day 21-35). All diets contained 0.2% MMC. The Ca and avP levels were obtained by differing levels of monocalcium phosphate, limestone and phytase. At day 0, 1,260 Broilers were divided over 9 treatments, resulting in 7 replicates containing 20 birds per treatment. Bird performance was measured by weight gain (ADG), feed intake (ADFI) and feed conversion (FCR). The data was analysed as a complete randomised design by JMP® PRO 14.2.0. An effect of dietary treatments was observed for all parameters ($P<0.05$). The observed trends were similar for ADG, ADFI and FCR. The highest performance was observed at high avP levels, with the lowest performance observed at the low avP levels ($P<0.001$). A high Ca:avP ratio reduced performance ($P<0.001$). Hence, absolute avP level had the highest influence on performance, while Ca:avP ratio had a smaller influence. This study concludes that inclusion of MMC enhances bird performance by allowing a reduced Ca level, while maintaining a high avP level, thereby reducing the optimal Ca:avP ratio to 1.6:1.

Seaweeds in animal nutrition, a valuable source of minerals but in need of fine-tuning

S. Al-Soufi[1], J. García[2], E. Cegarra[3], A. Muíños[4], V. Pereira[1] and M. López-Alonso[1]
[1]University of Santiago de Compostela, Animal Pathology, Veterinary Faculty, 27002 Lugo, Spain, [2]Polytechnic University of Madrid, Dept Producción Agraria, 28040 Madrid, Spain, [3]DeHeus España, Plaza de Mina 1, 15004 A Coruña, Spain, [4]Porto Muiños SL, Cerceda, 15185 A Coruña, Spain; sabelaalsoufi.novo@usc.es

Seaweeds are a rich source of biologically active compounds such as polysaccharides, proteins and essential amino acids, minerals, polyunsaturated fatty acids, antioxidants, vitamins and pigments, among others, with beneficial antioxidative, anti-inflammatory, antibacterial or antiviral properties. These properties give them a great potential as innovative and eco-friendly feed ingredients/additives to provide animals with high nutritive value, promote their health and productivity, and obtain high-quality products. Although very variable depending on the species, seaweeds contained higher amounts of both macrominerals and trace elements than those reported for edible land plant, making them, particularly brown and red spp., an interesting source of minerals that are oftentimes deficient in European populations. In turn, high concentrations of some elements, such as iodine, need to be carefully addressed when evaluating seaweed consumption, since excessive intake of this element was proven to have negative impacts on health, and could limit the amount of seaweed used in animal feeds. Another weakness of seaweed consumption is their capacity to accumulate several toxic elements, particularly arsenic, which can pose some health risks. This study evaluates the macro and micro element profile of 17 seaweed species and seaweed extracts with the potential to be used in rabbit farming as an alternative to antibiotics. Overall, iodine concentration can be as high as 6,850 mg/kg DM in Saccarina Lattisima, which limit the amount of algae used in the complete feed. Moreover, although the arsenic concentration is higher than in most other feedstuffs (up to 35 mg/kg DM) their low level of inclusion (generally 1-2%) make not a limitation when used in animal feed. The project TIRAC was co-financed by the European Agricultural Fund for Rural Development (EAFRD) of the European Union (80%) and the Ministry of Agriculture, Fisheries and Food (20%), within the framework of the National Rural Development Program 2014-2020.

Session 60 Poster 22

Online mislabelling of mineral and complementary feeds available in France

F. Touitou[1], C. Marin[1], T. Blanchard[1], A. Meynadier[1] and N. Priymenko[2]
[1]INRAE, ENVT, GenPhySE, 23 Chemin de Borde Rouge, 31326 Castanet-Tolosan, France, [2]INRAE, ENVT, Toxalim, 180 Chemin de Tournefeuille, 31027 Toulouse, France; florian.touitou@envt.fr

Many complementary feedstuffs for horses are currently available on the market in order to balance diets and meet horses' mineral requirements. Producers' websites are the main source of information available. A database of feed producers selling in France was obtained using veterinary product catalogues as well as listing members of the French national association of equine feed producers (CNEF) and producers appearing on the main online shopping sites. Only complementary feed that are supplemented to the diet daily and in a quantifiable form were retained resulting in 83 brands and 1,164 complementary feeds. Among these, mineral feeds (i.e. complementary feeds containing at least 40% crude ash, Regulation (EC) 767/2009), complementary feeds that were presented as 'mineral feeds' and complementary feeds containing at least 10% crude as and 5% calcium were retained since they are all likely to be used to balance horses base diets. It resulted in a database containing 235 complementary feeds from 71 brands: 132 were mineral feeds, 81 were complementary feeds containing less than 40% of crude ash and 22 were presented as complementary feeds but did not mention their crude ash content as mandatory. Systematic analysis of the data available for each of these products on the internet highlighted mistakes in the composition (e.g. a percentage of calcium higher than the advertised content of crude ash) and translation mistakes (e.g. feeds identified as feed material in French that were actually compound feeds). Furthermore, many claims, that should legally be objective, verifiable by the competent authorities and understandable by the user, were found to be quite abusive and sometimes dealt with particular nutritional purposes when feeds were not labelled as feed intended for particular nutritional purposes. In conclusion, a large choice of mineral and other complementary feed enriched in minerals is available on the market but quasi-systematic mislabelling and inaccuracies on producers' websites impair the capacity for the veterinarian or the owner to choose an adequate mineral supplementation.

Session 60 Poster 23

Diversity of practices and advisors in mineral and vitamin supplementation of dairy farms

C. Manoli[1], G. Springer[1], L. Barbier[1], C. Chassaing[2], C. Sibra[2], G. Maxin[2], A. Boudon[3] and B. Graulet[2]
[1]ESA, USC URSE, 55 rue Rabelais, 49007 ANGERS, France, [2]Université Clermont Auvergne, INRAE, VetAgro Sup, UMR Herbivores, Rte de Theix, 63122 St-Genès-Champanelle, France, [3]INRAE, UMR PEGASE, 16 Le Clos Domaine de, La Prise, 35590, Saint Gilles, France; c.manoli@groupe-esa.com

In a context of agroecological transition, reducing inputs and maintaining animal health are important issues for livestock activities. In French dairy farms, addition of minerals and vitamins to rations is a common practice to prevent health disorders or production losses, but knowing and covering cows' requirements is a complex process and the farmers' practices are poorly described. Moreover, scientific recommendations adapted to feeding systems with a large part of fodder are scarce and it is difficult to know how do farmers take decisions on mineral and vitamin supplementation. Our objective was to decipher the way practices and recommendations are performed for mineral and vitamin supplementation in the context of herb-based livestock, such as largely encountered in Auvergne-Rhône-Alpes region (France). A study based on semi-structured interviews was conducted with 24 dairy farmers and 13 of their advisors. A qualitative and thematic analysis of discourses was conducted and led to identify 5 main variables characterizing the diversity of practices. A multiple correspondence analysis and ascending hierarchical classification allowed to classify the farmers into 3 types according to their mineral and vitamin supplementation practices: (1) farmers expressing indifference toward mineral and vitaminic supplementation; (2) farmers expressing a high need to be advised; (3) farmers looking for autonomy. The first type gathered more simplified practices than the other 2 types. Interviews with advisors revealed 3 types of advisors: feed sales representatives; nutritionists of performance monitoring associations, and veterinarians. Each advisor is characterized by a particular message. Our results are consistent with sociological studies on the relationship between farmers and their advisors and confirm the extreme complexity and diversity of minerals and vitamins supplementation practices, in a context of lack of scientific knowledge on this subject.

Session 61 Theatre 1

How to improve resilience, from animal to system level

A. Mottet, R. Baumung, G. Velasco Gil, G. Leroy and B. Besbes
FAO, Viale delle Terme di Caracalla, 00153, Italy; anne.mottet@fao.org

Climate change impacts, such as droughts and floods, market crisis and disease outbreaks are the main shocks affecting the global livestock sector and regularly threatening the livelihoods of the most vulnerable populations in the world. Building resilience means not only reducing vulnerability but also increasing adaptive capacity, at all levels, animal, farm, system and community. We provide examples from a diversity of contexts and countries. The presence of livestock can be in itself a climate change adaptation strategy, as a way to buffer climate or market variability through the management of animal mobility, herd structure and feed resources. The management of animal genetic resources, but also of feed and fodder species, are among the most efficient ways to support the necessary adaptation of livestock to climate change, for example through breeding strategies for better resilience or the choice of adapted and robust livestock populations. However, at farm and household level, resilience can and should be built differently depending on the type of systems considered. Examples are available from recent work on the economics of pastoralism in Argentina, Chad and Mongolia, as well as from a cross-country analysis of the results from the FAO Tool for Agroecology Performance Evaluation. Changes to production systems to increase resilience can be both incremental (e.g. growing drought resistant fodder, improving water management, breeding towards resilience animals) or transformational (e.g. areas where extensive grazing replaces partially mixed crop-livestock farming, diversification in terms of species and breeds). Social protection mechanisms, such as cash transfers and school meals, or insurances are also key to reduce vulnerability. Policies for increasing resilience also include disaster risk reduction and management (DRRM), which can take the form of disease prevention and outbreak containment plans or early warning systems (for example drought monitoring).

Session 61 Theatre 2

On the link between climate change mitigation and adaptation in dairy cow farming in West of France

B. Godoc, E. Castellan, A. Madrid and C. Karam
Institut de l'Elevage, French Livestock Institute, Monvoisin, 35650 Le Rheu, France; brendan.godoc@idele.fr

The dairy cow sector faces two intertwined challenges at the same time: while there is a growing necessity to lower the greenhouse gas emissions from dairy farming, farmers need to adapt their systems and practices to harsher climate conditions. Our work studied the relationship between the concepts of adaptation and mitigation through the design of future dairy systems using a participatory approach. First, four groups of dairy farmers in the French Loire region co-constructed resilient dairy systems using a serious board-game called 'forage rummy'. Farmers started with four representative dairy systems with contrasting share of maize silage in their total forage area. Choices upon forages, grassland and herd management were made to ensure the balance between year-round forage production and animal feeding requirements with the support of the computerized model of the serious game. The four resulting systems manage to maintain production performances in a projected 2050 climate implying new crop yields and grass yield repartition. Climate data of the regionalised model were used to predict grass growth through the STICS crop model under the RCP8.5 business-as-usual scenario. Secondly, the life cycle assessment model CAP'2ER Level-1 allowed us to estimate the greenhouse gas emissions of the four co-constructed systems. The four simulated dairy farms all reduced their net carbon footprint when comparing past and future greenhouse gas emissions and soil organic carbon sequestration. Results show that strategies to adapt dairy farming systems to climate change can lead to mitigate the contributions of these systems to climate change.

Resilience of ruminant organic systems to climatic hazards: a study model in a French grassland area

C. Boivent and P. Veysset
INRAE, UMR Herbivores, 63122 Saint-Genès-Champanelle, France; patrick.veysset@inrae.fr

Climate change and the increase of frequency of climatic perturbations could have a severe impact on organic farms in grassland areas. Years of severe drought such as 2003 or more recently 2022 prove the need to study the adaptations of ruminant farms to climate perturbations. Resilience provides a framework for analysing livestock systems in this context. Based on structural, technical, and economic data from thirty-six specialized ruminants (beef and dairy cattle, dairy sheep and sheep for meat) organic farms in the Massif Central (a mountain area) monitored between 2014 and 2020, and meteorological data enabling to calculate agro-climatic indicators for each farm and each year, we studied the determinants of the resilience of the systems. Resilience was measured by the stability or increase in the gross value added (gross output without subsidies minus intermediate consumption) of the farm. We used a Partial Least Square (PLS) path modelling and then a hierarchical ascending classification (HAC). In the farms studied, the management of the forage system is primarily based on maximizing grazing at the expense of the constitution of fodder stocks. Maintaining herd production, even at the expense of fodder autonomy, is essential for the resilience of the farms. Good management of the fodder system with fodder stocks also seem to be decisive for the sustainability of farms. Finally, the climatic variation of the seven years of observations did not impact the structures of the farms. Farmers managed the climatic perturbations by maximizing grazing and by purchasing fodder to compensate forage deficits. However, more concentrates were not used to compensate climatic perturbations. More intense hazards could, however, call into question the capacity of the systems to manage climatic variability, mainly if organic fodder become scarce on the market.

Resilgame: a game to experiment farm adaptation to climate change

G. Martel[1] and S. Colombié[2]
[1]INRAE, UMR BAGAP, 55 rue Rabelais, 49100 Angers, France, [2]Chambre d'Agriculture des Pays de la Loire, 9 rue André-Brouard, 49000 Angers, France; gilles.martel@inrae.fr

The issue of climate change is particularly important for the agricultural sector, whose production is directly linked to the climate and where the effects are already observed. Although farmers are aware of this, they only make marginal adjustments to mitigate these effects. Adaptation levers to improve the resilience of farming systems have already been identified within the different compartments of the farm. But all this knowledge is fragmented and difficult to integrate at the same time by both farmers and their advisors. This is why we have designed a serious game to confront farmers and student, with future climates and put them in a decision-making situation. To build the game, we brought together people involved in projects related to adaptation to climate change, i.e. teachers, trainers, advisors, and researchers in various disciplines: agronomy, meteorology, animal science, economics, and sociology. The game is made up of two parts: a common board where players choose adaptation levers (for crops, pastures and animals), react to hazards and observe environmental performances; and an individual board where they decide on the cropping plan and animal numbers. The goal of the game is to reach objectives (income, reduction of phytosanitary treatments or work time) according to the character played. Crops can be sold or used to feed the animals. Crops, animals, but also technical choices and responses to hazards affect greenhouse gases, crop protection treatments, net margin and labour time. The game includes climatic and economic hazards that affect the characteristics of the crops and animals. Their effects can sometimes be countered by an adaptation lever. Some of these levers can be activated in reaction to the hazard, while others must be anticipated for the effect to be active. The first tests of the game have confirmed its ability to simulate the consequences of climate changes that force players to change their practices. It must be combined with knowledge inputs to help players understand the mechanisms and the effects of global changes and the levers of adaptation. The next step is to evaluate the players' ability to implement the adapted practices on their farms.

Session 61 Theatre 5

Evolution of agroecology and associate indicators – looking for balance in farming systems

E. Benedetti Del Rio[1], A. Michaud[2] and E. Sturaro[1]

[1]University of Padova, DAFNAE, Viale dell'Università 16, 35020 Legnaro, Italy, [2]VetAgro Sup, UMR-1213-Herbivores, 89 Avenue de l'Europe, 63370 Lempdes, France; elena.benedettidelrio@phd.unipd.it

Agroecology can be defined as an integrated approach that applies ecology and sociology to agricultural production systems. The agroecological transition has to be accounted through multi-disciplinary approaches, in fact in the last 20 years it has been studied coupling agricultural or animal sciences with social and economic approaches. Since 2014, the growth of case studies was exponential and produced several indicators to measure its applicability. This study aims to identify a set of indicators for monitoring agroecological transition in grassland-based farming systems with a systematic literature review approach, through Scopus search engine. Selection criteria are the following: (1) published since 2000; (2) concerning agroecology as a discipline. The database consists of 74 papers, 35 of them related to our objective. Through article's aim we sorted indicators, based on the principles from Dumont *et al.* and Wezel and Peeters. The results showed that for Dumont at least 20 articles cover 3 principles out of 5: 'decreasing input', 'decreasing pollution' and 'preserving biodiversity in agroecosystems'. Only 5 articles relate to 'increasing animal health' and 9 to 'enhancing diversity in animal production'. Based on Wezel and Peeters at least 21 articles reach out 4 principles out of 6: 'resources' and 'system management', 'biodiversity conservation' and 'knowledge, culture and socio-economics of farmers'. The principles of the authors overlap on biodiversity and resources management. Both indicators and principles suggest that most of the articles still relate to the agricultural sector more than livestock. Moreover, it appears that the two sectors are still taken into consideration in a separate way. The results of this research evidenced that the indicators can be used to highlight threats and opportunities of different contexts, to enhance the resilience of grassland-based farming systems, both from an economical and practical point of view, and favour the agroecological transition. The outputs of this study will be used to define a multicriteria approach for grassland-based case studies in Italy and France.

Session 61 Theatre 6

Small ruminants farming systems of Spain: challenges and attributes for their resilience

J. Lizarralde[1], B. Soriano[2], A. Benhamou-Prat[3], P. Gaspar-García[4], Y. Mena-Guerrero[5], J.M. Mancilla-Leyton[5], A. Horrillo[4], R. Ruiz[1], D. Martín-Collado[3] and N. Mandaluniz[1]

[1]Basque Institute for Agricultural Research and Development, Campus Agroalimentario de Arkaute, 01192 Arcaute, Álava, Spain, [2]Research Centre for the Management of Agricultural and Environmental Risks of the Polytechnic Univer, P.° de la Senda del Rey, 13, 28040 Madrid, Spain, [3]Agrifood Research and Technology Centre of Aragon, Av. de Montañana, 930, 50059 Zaragoza, Spain, [4]University of Extremadura, Av. de la Universidad, 10003 Cáceres, Spain, [5]University of Sevilla, C. San Fernando, 4, 41004 Sevilla, Spain; jlizarralde@neiker.eus

Small ruminants farming systems (SRFS) is facing an important crisis (linked to high input prices, low meat and milk prices, social issues, etc) which ends up in a significant lack of generational turnover. The aim of this paper is to analyse the current challenges with different origin (economic, environmental, institutional or social) and temporality (short-term, ST; long-term, LT) faced by SRFS in Spain and to identify the attributes that promote or limit their resilience. To collect data, 24 depth interviews with farmers have been conducted in 4 case studies of SRFS in Spain: 8 in dairy goats in Andalucía, 6 in meat sheep in Aragón, 4 in dairy sheep in Extremadura and 6 in the Basque Country and Navarra. The interviews were analysed using deductive content analysis coding them according to the bibliographic compilation of challenges and attributes. Results have shown that the SRFS have common challenges referred to institutional ST bureaucracy and administrative workload. However, particularities between the systems have been observed: for example, Andalusian system highlights economic difficulties; Aragon and Extremadura systems highlight LT and ST institutional aspects; and Basque Country and Navarra system highlights ST social challenges. In terms of the attributes that strengthen their resilience, farmers differ according to the specificities of each system, but they agree that resilience is strengthened by aspects such as the autonomy, functional diversity and human capital, and it is weakened by the lack of economic capital. The knowledge about the attributes that promote the resilience will help to define the political measures to support the SRFS to deal with current and future challenges.

Session 61 Theatre 7

Building resilience in drylands' extensive livestock systems under climate uncertainty

A. Tenza-Peral[1], I. Pérez-Ibarra[1], A. Breceda[2], J. Martínez-Fernández[3] and A. Giménez[4]
[1]University of Zaragoza, Department of Agricultural Sciences and the Natural Environment, Calle de Miguel Servet 177, 50013 Zaragoza, Spain, [2]Center for Biological Research of the Northwest, Instituto Politécnico Nacional s/n, 23096 La Paz, Mexico, [3]New Water Culture Foundation, Pedro Cerbuna 12, 50009 Zaragoza, Spain, [4]Miguel Hernández University, Avenida de la Universidad s/n, 03202, Spain; atenza@unizar.es

Extensive livestock farming systems in drylands ensure food security, support social well-being and provide vital ecosystem services. They have evolved to cope with the spatiotemporal variability of natural resources and frequent disturbances. However, climate change poses significant challenges to the adaptive capacity of these social-ecological systems. Here, using the dynamic simulation model SESSMO about a case study (the oasis of Comondú in the Sonoran Desert, Mexico), we assessed the contribution of management strategies proposed by local stakeholders to the resilience and long-term sustainability of these social-ecological systems under the uncertainty of climate change. Management strategies were classified according to their contribution to: (1) strengthening robustness (e.g. improving veterinary technical assistance); (2) improving adaptive capacity (e.g. changes in marketing channels); and (3) fostering transformation (e.g. local tourism development). We explored outcomes under various climate scenarios (i.e. favourable and unfavourable). Our simulation results showed the great sensitivity of these social-ecological systems to climate variability. Contrary to what was expected, fostering transformation showed limited effects. Tourism development is not a panacea. Strengthening robustness showed a very similar impact on systems dynamics, and the effect was reinforced when the robustness and the adaptive capacity were enhanced together, even under unfavourable climate conditions. Here we identified a clear synergy between these strategies that do not imply structural changes. Nevertheless, to cope with climate uncertainty, the best strategy was to act simultaneously on robustness, adaptive capacity, and transformation.

Session 61 Theatre 8

A thirty-year assessment of interactions between weather conditions and sheep milk yield and quality

A. Mantino[1], M. Milanesi[2], M. Finocchi[1], G. Conte[1], G. Vignali[2], G. Chillemi[2], L. Turini[1] and M. Mele[1]
[1]University of Pisa, Department of Agriculture, Food, and Environment, via del Borghetto 80, 56124, Pisa, Italy, [2]University of Tuscia, Department for Innovation in Biological, Agro-Food and Forest Systems, via San Camillo de Lellis snc, 01100, Viterbo, Italy; alberto.mantino@unipi.it

The Mediterranean area is a climate change hot spot. Locally, up to 40% of winter precipitation could be lost and, generally, summer will be warmer and drier. Semi-extensive mixed farming systems are threatened by the reduction and the variation of rainfall distribution rather than intensive systems because the increase of frequency of dry spells negatively affects pasture mass and quality. Moreover, elevated temperatures combined with intense solar radiation lead to the decrease of intake and animal welfare due to heat stress. A study focusing on the interaction of climate and animal production was performed in the Maremma region (central Italy) on milk sheep farms. Firstly, a thirty-year (1993-2022) assessment of the spatial and temporal patterns of dry spells and thermal comfort indexes was conducted on a daily dataset. Secondly, the collection of milk yield and quality daily data was carried out on more than 200 farms on the same time span. Lastly, the analysis of interactions between weather conditions and sheep milk production was performed. In the Mediterranean, climate changes seem to negatively affects milk production in semi-extensive farms and, moreover, dairy sheep productivity is being threatened by several factors such as the predation by wild carnivores and the volatility of diet supplement costs. The outcomes of the analysis will be presented. This study was carried out within the Agritech National Research Center and received funding from the European Union Next-GenerationEU (Piano Nazionale di Ripresa e Resilienza (Pnrr) – Missione 4 Componente 2, Investimento 1.4 – D.D. 1032 17/06/2022, CN00000022). This manuscript reflects only the authors' views and opinions, neither the European Union nor the European Commission can be considered responsible for them.

Session 61 Theatre 9

Indicators for animal health on agro-ecological dairy farms

A. Ceppatelli[1], M. Crémilleux[2], A. Michaud[2] and E. Sturaro[1]
[1]UNIPD, DAFNAE, Viale dell'Università 16, 35020 Legnaro, Italy, [2]Université Clermont Auvergne, INRAE, VetAgro Sup, UMR Herbivores, Saint-Genès-Champanelle, 63122, France; andrea.ceppatelli@phd.unipd.it

The agroecological transition, supported by European policies and strategies, can strengthen the sustainability and resilience of farming systems in the face of climate and socio-economic change challenges. This approach considers the farm as a key part of the ecosystem in which it is embedded. Thus, the farm contributes to regulate the health of the system, which is influenced by all the actors involved: environment, farm, productions. Making the most of these interactions and improving their level of health contributes to achieving the highest possible level of sustainability. The present research aims to identify a set of indicators to assess animal health on agroecological dairy farms, adopting a Global Health approach. 18 farms located in the Auvergne-Rhone-Alpes Region (France) rearing dairy cows, goats, and sheep, were monitored over two years. Four rounds of on-site visits were conducted to carry out measurements on the animals (behaviour, nutritional and health status, milk parameters, parasitism). Two lists of animal health indicators were selected by combining statistical analysis on the collected data (Principal Component Analysis) with expert knowledge. For dairy cows, a list with 16 indicators related to animal health, housing, behaviour, feeding, production and reproduction was performed. For sheep and goats, another list with 16 indicators referring to animal health, housing, feeding and production was performed. These indicators were then evaluated on the 18 farms, to assess their level of health. On average, veterinary interventions were low (0.18/animal/year for cows and 0.07 for small ruminants), as was the incidence of lesions (8%), lameness (5%) and dirty animals (7%). However, small ruminants had higher levels of parasitism (Eimeria and Strongles) than dairy cows (Parampistome). Based on the results obtained, the selected indicators can contribute to the definition of agroecological practices within the farms and to the global assessment of their sustainability and resilience. Other farms located in similar territorial contexts (Italian Eastern Alps) will be involved in the network to validate the results.

Session 61 Theatre 10

Exploring climate change adaptation strategies form the perspective of Mediterranean sheep farmers

D. Martin-Collado[1], S. Lobón[1], M. Joy[1], I. Casasús[1], A. Mohamed-Brahmi[2], Y. Yagoubi[2], F. Stark[3], A. Lurette[3], A. Abuoul Naga[4], E. Salah[4] and A. Tenza-Peral[5]
[1]CITA-IA2, Avda. Montañana 930, Avda. de Montañana 930. CIF: Q5000823D, 50059, Spain, [2]ESAK-INRAT, Rue Hédi Karray, 1004 Menzah 1 Tunisie, Tunisia, [3]INRAE-UMR SELMET, Campus International de Baillarguet, 34398 Montpellier, France, [4]APRI, Nadi El Said, Dokki, Giza, Egypt, [5]UNIZAR, Calle Miguel Servet, 177, 50013 Zaragoza, Spain; dmartin@cita-aragon.es

Sheep production systems in the Mediterranean region are particularly vulnerable to the impact of climate change (CC) due to their relatively high use of local food resources. Most studies analysing farm adaptation strategies tend to follow top-down approaches without considering the feasibility of their implementation on farms. In this work, which is part of the PRIMA project Adapt-Herd, we investigated farmer perceived CC impact on-farm and their view about the best adaptation strategies to future CC scenarios in representative farming systems in Egypt, France, Spain, and Tunisia. Scenarios of CC were developed for each region based on IPCC projections. Two hundred-five farmers were surveyed face-to-face and asked to point out the actions they would take to adapt to the CC scenarios. Possible actions covered feed, grazing, reproduction and flock management, breeding, and machinery and facilities. Strategies were identified using K-modes cluster analysis. Discriminant analysis was used to determine their preferred strategies across countries, farming systems, farm features, and farmer profiles and perceptions. Five general strategies were identified: (1) farm machinery and facilities (high preference in Egypt, France. and Spain); (2) feed intensification (Egypt and Spain); (3) flock management and feed extensification (France); (4) feed optimization (Tunisia); and (5) general farm adaptation. Besides differences among countries, farmers with irrigation systems perceive the impact of feed shortage, and heat stress on farms to be lower than other farmers and tend to prefer strategies focus on machinery and facilities and feed intensification. Our results show that both farm profile and CC impact perception factors influence farmer's views on adaptation strategies.

Session 61 Theatre 11

Combining serious games in a process to support sustainable livestock farming systems

R. Etienne[1], S. Dernat[1], C. Rigolot[1] and S. Ingrand[2]

[1]INRAE, ACT, UMR Territoires, campus des Cezeaux, 63000 Aubiere, France, [2]INRAE, PHASE, Theix, 63000 Saint Genès Champanelle, France; rebecca.etienne@inrae.fr

Farmers have to articulate individual and collective objectives, to face global challenges but need support to address these issues. The objective of this study is to propose a methodological approach to support this process, based on a combination of serious games. This method was proposed to a group of farmers involved in a French cheese PDO area ('Fourme de Montbrison'), in order to improve their fodder autonomy in the context of climate change. Three serious games were combined at different scales in order to (1) choose levers of adaptation (Lauracle); (2) simulate and design their effects on the farm systems (Forage Rummy); (3) enhance collective decision at territorial scale (Dynamix). These serious games were chosen because they address forage system adaptation and help to collaboratively apprehend the trade-offs between individual and collective scales. Other forms of interventions are also set up (farmers lead on-farm experiments, trainings, farm visits). The support method is evaluated along the way and *a posteriori* with an evaluation model adapted for the case study. It allows to follow four levels of evaluation (reactions, learning, behaviours and results) which are detailed in the presentation. Different tools are combined before, during and after game sessions such as participant observation; in-game observations and debriefings of game sessions; interviews and technical diagnosis. The results highlight contributions of the serious games on future changes of practices, at individual and collective level, either technical or organizational innovations. We could observe that farmers decide the levers of adaptation, then simulate some of them and finally, set up on-farm experiments (long-lasting multi-species grasslands adapted to drought) but also impulse the creation of a machine to brush grassland seeds. These learnings and changes on behaviours consider the articulation of individual and collective objectives. Those results will ensure the development of an operational method for agricultural extension services to articulate both farm and territorial scales towards sustainable livestock farming systems.

Session 61 Theatre 12

Assessing how farm features and farmers' profile contribute to farm resilience

A. Prat-Benhamou[1], B. Soriano[2], D. Ondé[3], J. Lizarralde[4], J.M. Mancilla-Leyton[5], N. Mandaluniz[4], P. Gaspar-García[6], Y. Mena-Guerrero[5] and D. Martín-Collado[1]

[1]Agrifood Research and Technology Centre, Animal Science, Avda. Montañana 930, 50059 Zaragoza, Spain, [2]Polytechnic University of Madrid, C/ Senda del Rey 13, 28040 Madrid, Spain, [3]Complutense University of Madrid, Campus de Somosaguas, s/n, 28223 Madrid, Spain, [4]Basque Institute for Agricultural Research and Development, Berreaga Kalea 1, 48160 Bizkaia, Spain, [5]University of Sevilla, Ctra. de Utrera 1, 41013 Sevilla, Spain, [6]University of Extremadura, Av. de Adolfo Suárez, 06007 Badajoz, Spain; abenhamou@cita-aragon.es

Under current global change situation, strengthening farming systems' resilience is an aim of agricultural policy institutions. Resilience is defined as 'the ability to ensure the provision of the system functions in the face of increasingly complex and accumulating shocks and stresses, through capacities of robustness, adaptability and transformability'. In this sense, we test the hypothesis that farms' features (farm attributes) and farmers' profile (personal attributes) influence the resilience capacities in different manner. Based on this hypothesis, we used farmer's resilience self-assessments to evaluate resilience drivers. We conducted face-to-face surveys to farmers of 4 different small ruminant farming systems in Spain, that resulted in 160 questionnaires. Data analyses were based on multiple regression models. Results showed that farmer's optimism and the availability of different management options enhance the three resilience capacities. However, most attributes influenced differently to each capacity. Farmers' who are proud of their achievements and bounded to local traditions, lead more robust farms, farms with a higher autonomy and better farm infrastructures are more adaptable, and proactiveness and cooperation with other economic sectors are positive for transformability. Finally, some attributes showed trade-offs for the resilience capacities. We found the use of natural resources influenced positively farm robustness but negatively adaptability due to the complexity of natural processes. In conclusion, we argue that both farm features and farmer personal attributes are key for the understanding of farm resilience and should be considered in resilience assessments.

Session 61 Poster 13

The difference between abnormal climate and extreme climate that cause yield damage to silage corn

M. Kim[1] and K. Sung[2]

[1]Institute of Animal Life Scinece, Gwangwondaehak gil 1, 24341, Korea, South, [2]Kangwon National University, Gwangwondaehak gil 1, 24341, Korea, South; lunardevil@kangwon.ac.kr

This study aimed to compare the characteristics by estimating the impacts of abnormal climate and extreme climate on silage corn. The silage corn metadata (n=3,232) were collected from cultivation experiments that conducted by the Rural Development Administration of Korea. To define abnormal climate and extreme climate, climate data including daily weather information were collected from the weather information system of the Korean Meteorological Administration. The variables were dry matter yield (DMY, kg/ha), mean temperature (MT, °C), lowest temperature (LT, °C), highest temperature (HT, °C), maximum precipitation (MP, mm/hr), accumulated precipitation (AP, mm), maximum wind speed (MW, m/s), mean wind speed (WS, m/s), sunshine duration (SD, hr). Firstly, to define the extreme climate, the median-interquartile method was used instead of the mean-standard deviation method. Secondly, the abnormal climate was defined by comparing the characteristics of favourable and poor conditions for the growth and development of silage corn. For doing this, principal component analysis and discriminant analysis were used. Finally, the t-test and ANOVA with the 5% significance level were carried out to confirm the difference in DMY. For the extreme climate in mid-June, the low-extreme LT, high-extreme AP, and high-extreme MW were detected in the trends. However, the year records were different for each point. Thus, the damage in DMY was estimated based on high-extreme AP, as a representative scenario. As a result, the damage in DMY caused by the extreme climate in mid-June was estimated to be 927.2 kg/ha. Meanwhile, for the abnormal climate in mid-June, the characteristics of early monsoons without typhoons were confirmed as poor. In particular, MP in abnormal climate (20.54 mm/hr) was three times greater than that in normal climate (6.61 mm/hr). The damage in DMY caused by the abnormal climate in mid-June was estimated to be 1,155.9 kg/ha. Hence, the magnitude of yield damage to silage corn caused by extreme climate and abnormal climate, as well as characteristics of high precipitation, were similar.

Session 61 Poster 14

Causality in climate-soil-yield network for silage corn

M. Kim[1] and K. Sung[2]

[1]Institute of Animal Life Scinece, Gwangwondaehak gil 1, 24341, Korea, South, [2]Kangwon National University, Gwangwondaehak gil 1, 24341, Korea, South; lunardevil@kangwon.ac.kr

This study aimed to confirm causality of climatic and soil physical factors on yield of silage corn based on climate-soil-yield network in Korea. The climatic variables were GDD, high temperature, low temperature, surface temperature, rainfall, relative humidity, wind speed, sunshine duration in the before and after silking stages. The soil physical variables were effective soil depth, slope and drainage classes. The yield variables were total digestible nutrients, dry and fresh matter yields. The network was constructed by structure equation modelling and neural network modelling. In the result of networking, three causalities were remarkable. First, all longitudinal climatic causality between before and after silking stages was significant. It implies that the effect of climates in vegetative stage reaches to yield through themselves in reproductive stage. Second, there was the causality between climatic and soil physical factors based on indirect effects. Thus, it is likely to lead to offset between the direct and indirect effects of soil physical factors. Finally, the effects of drought and heavy rainfall were clear in before and after silking stages, respectively. It indicates the stress can damage yield of silage corn. Here, the damage caused by the drought could be recovered due to various indirect effects, while the damage caused by heavy rainfall was fatal because there was a lack of an indirect path to recover from. This study contributed to identifying how various climatic and soil physical factors can affect production in the network. Furthermore, the climate-soil-yield network for silage corn of this study will be helpful to extend the structure with various factors in the future study. This study was supported by the National Research Foundation of Korea funded by the Ministry of Science and ICT (NRF-2023R1C1C1004618).

Session 61 Poster 15

Can studying the health of livestock systems be a way to improve their resilience?

M. Cremilleux, B. Martin and A. Michaud
Université Clermont Auvergne, INRAE, VetAgro Sup, UMR Herbivores, Theix, 63122 Saint Genès Champanelle, France; maeva.cremilleux@vetagro-sup.fr

Livestock systems are facing multiple challenges, particularly economical, zoonosis and climatic hazards. These crises, caused by human activities and population growth, remind humanity that human, animal and ecosystem health are interdependent (developing notions of One Health, Eco Health, planetary health, etc.). The farm scale seems particularly suited to a comprehensive approach to health in agriculture. A farm constitutes a system as such and it is on this scale that farmers make decisions, experiment and establish interactions between humans and non-humans (animals, plants, environment, etc.), which have an impact on health. The aim of this study is to characterize farm health (definition and indicators) and to show to what extent this approach can be used to reason about the resilience and sustainability of livestock systems. We first built a framework for analysing the health of a farm, which was then tested in the field. For this purpose, 18 sheep, cow and goat dairy farms – claiming to be agroecological- were monitored in the Massif Central (France), for 2 years (2021-2022). We assessed the soil health (physical, chemical, and biological indicators), grassland health (biodiversity, yield, resilience, products quality, etc.), animal health (production, reproduction, feeding, behaviour, buildings, and health) and farmer health (survey on psychological health) on each farm. We proposed an assessment of the health status of the monitored farms and highlighted their good health. We also discussed the value of health approaches in promoting the sustainability and resilience of farming systems. Moreover, the analysis of the management practices implemented by the farmers and the functioning of the system indicates that it is the coherence of the system that promotes health. This work also highlighted the interest of the notion of health for studying farms. Health, thanks to its holistic vision, allows the study of the decision-making system, the integration of the farmers' relationship with nature and to show the coherence of the systems.

Session 61 Poster 16

Effect of heat stress on extensive beef cattle's calving percentage in the Central Bushveld Bioregio

S.M. Grobler[1], M.M. Scholtz[1,2], F.W.C. Neser[2], J.P.C. Greyling[2] and L. Morey[3]
[1]Agricultural Research Council, Animal Production, Irene, Pretoria, 0062, South Africa, [2]University of the Free State, Department of Animal, Wildlife and Grassland Sciences, Bloemfontein, 9300, South Africa, [3]Agricultural Research Council, Biometry department, Hatfield, Pretoria, 0028, South Africa; mgrobler@arc.agric.za

Little research has been done locally, on the effect of heat stress in the sub-tropical summer rainfall area of the Central Bushveld Bioregion, even though the mating season coincides with the warmest months of the year. A study was conducted over a six-year period to investigate the effect of heat stress just before and during the mating season on the subsequent calving percentage of extensively managed beef cattle. Average monthly minimum- and maximum-temperature (°C) and average monthly minimum- and maximum relative humidity (%) were used to calculate a monthly discomfort index. An index value above 90 was considered very uncomfortable, relating to heat stress. The highest calving percentage of 89.6% was obtained when the discomfort index never rose above 89 before, during, or after the mating season. The lowest calving percentage of 60.0% was obtained with an average monthly discomfort index above 90 within the month just before mating and during the first two months of the mating season. Average monthly minimum- and maximum- temperature (°C), average monthly minimum- and maximum relative humidity (%), total monthly precipitation (mm), and discomfort index were taken into account performing forward stepwise regression procedures for the dependent variable calving percentage. Maximum relative humidity one month prior to the mating season had a high negative Pearson's correlation coefficient of -0.95. Results suggest that heat stress before and during the mating season has a negative impact on the calving percentage of extensively managed beef cattle in the sub-tropical summer rainfall area of the Central Bushveld Bioregion.

Session 61 Poster 17

Contrasting rearing and finishing regimens on performance and methane emissions of Angus steers

J. Clariget[1,2,3], V. Ciganda[3], G. Banchero[3], D. Santander[3], K. Keogh[2], D.A. Kenny[2] and A.K. Kelly[1]
[1]UCD, School of Agriculture and Food Science, Belfield, Dublin 4, Ireland, [2]Teagasc, Animal & Bioscience Research Department, Dunsany, Co. Meath, Ireland, [3]INIA, Colonia, 70000, Uruguay; jclariget@inia.org.uy

Beef cattle production contributes to global warming mainly through the emission of methane (CH_4) generated during the normal process of feed digestion. Compensatory growth could provide a means to reduce CH_4 emissions, since feed efficiency improves during refeeding and an increase in feed digestibility could explain this improvement. Steers are commonly fattened on pasture or feedlot, so the aim of the experiment was to evaluate the effect of dietary restriction in Angus steers on DM digestibility (DMD) and CH_4 emission under two contrasting refeeding diets. Eighty steers with an average live weight (LW) of 444±39 kg and age of 18±1 months, were blocked, and randomly assigned to 1 of 4 treatments, in a 2×2 factorial arrangement: severity of dietary restriction (moderate vs mild) and fattening system (pasture vs feedlot). During the 97 days period of dietary restriction target growth rates were 0.3 and 0.6 kg/d, for moderate and mild groups, respectively. The dietary restriction period was followed by a subsequent fattening period of 84 days. Methane emissions and DMD were measured at the end of fattening period in 36 and 80 steers, using SF_6 and acid insoluble ash technique, respectively. During fattening, previous moderate restricted steers had higher LW gain compared to mild steers (1.2 vs 1.0 kg/d; $P<0.01$). Feedlot steers had higher LW gain (1.3 vs 0.9 kg/d; $P<0.01$) than pasture steers. Irrespective of previous dietary restriction treatment, methane emission and methane yield were lower for feedlot steers ($P<0.05$; 230 vs 313 g CH_4/d, and 22.2 vs 25.8 g CH_4/kg DMI, respectively). No difference in DMD was evident between previously restricted steers, whereas feedlot steers achieved higher DMD than pasture steers (82.1 vs 69.2%; $P<0.01$). Moderate steers had lower methane intensity than mild steers (246 vs 321 g CH_4/LW gain; $P<0.01$). In conclusion, for both finishing systems, the lower methane intensity of steers initially subjected to a moderate dietary restriction is due to their relatively higher performance during fattening, rather than lower methane emissions or yield *per se*.

Session 61 Poster 18

Energy and greenhouse gas emissions: tools to discuss sustainability of livestock systems in Amazon

D.C.C. Corrêa[1,2], R.J.M. Poccard-Chapuis[2], M. Lenoir[2], V. Blanfort[2], J.L. Bochu[3] and P. Lescoat[1]
[1]AgroParisTech, INRAE, Université Paris-Saclay, UMR Sadapt, 91120, Palaiseau, France, [2]CIRAD, UMR Selmet, 34980, Montferrier-sur-Lez, France, [3]SOLAGRO, CS27608, 75 voie du Toec, 31076 Toulouse, France; caroline.da_cruz_correa@cirad.fr

Greenhouse gas (GHG) emissions and energy efficiency are key parameters to consider more sustainable farming systems. Management practices on farms can either improve or deteriorate energy efficiency and GHG emissions. It is therefore essential to identify relevant agricultural practices. This study determined the energy and GHG 'eco'-efficiency of cattle breeding, using energy balance and GHG emissions of 31 farms in two breeding area in the Brazilian Amazon, using a diagnostic tool (AgriClimateChange Tool) adapted to the region. The analysed dairy and meat farms are representative of the diversity of livestock systems in the region: suckling herd, fattener and suckling-fattener and of their degree of intensification. The energy efficiency is on average 16.27 GJ/t produced live weight/year (min= 2.35 GJ/t, max= 45.45 GJ/t) which is 47% lower than studies conducted in mainland France for similar systems. This increased efficiency could be linked to pasture feeding which makes better use of natural resources (highly productive grasses and edaphoclimatic factors) compared to other systems in the world which use more inputs and infrastructure. Results showed that fertilisers/amendments and the purchase of young animals accounted for high percentages of total energy consumption and that fuel constituted the majority of the direct energy used on farms. GHG emissions are on average 24,47 teqCO_2/t produced live weight/year (min= 3.78 teqCO_2/t, max= 65.59 teqCO_2/t) more important than in mainland France (14.20 teqCO_2/t produced LW/year) and similar to French Guiana (30.00 teqCO_2/t produced LW/year). These farms can be also considered as a carbon sink when the forests on their surface are included in the balance, with a production cost in terms of emissions close to neutral. This study shows the interest of this diagnostic tool, which has been calibrated for the tropics, and opens up prospects for the adoption of this tool as an indicator of 'eco'-efficiency of livestock farming systems.

Session 61 Poster 19

Microclimate and production of a tropical forage intercropped with pigeon pea

J.R.M. Pezzopane[1], P.P.A. Oliveira[1], A.F. Pedroso[1], W. Bonani[1], V.M. Gomes[2], C. Bosi[3], H.B. Brunetti[1], R. Pasquini Neto[1] and A.J. Furtado[1]

[1]Embrapa Pecuaria Sudeste, Rod. Washington Luis, Km 334, 1356-3970, São Carlos, Brazil, [2]UNICEP, Rua Miguel Petroni, 5111, 13563-470, São Carlos, Brazil, [3]UFPR, Rua Getúlio Vargas, 1642, 87704-010,Paranavaí, Brazil; jose.pezzopane@embrapa.br

Pasture degradation is the main problem in Brazilian livestock systems. Several strategies can be used for pasture recovery, including the use of shrub legumes, which can change the environment (microclimate and soil water content) and pasture productive potential. This research aimed to evaluate the microclimate, soil water dynamics, and production of a tropical pasture intercropped with pigeon pea. The study was carried out in pasture areas of *Urochloa decumbens (syn. Brachiaria decumbens)* Stapf cv. Basilisk in São Carlos, SP, Brazil, from 2020 to 2022. Different pasture production systems were evaluated: Degraded pasture (DP); Pasture recovered with application of nitrogen fertilizer (200 kg of N/ha/year) (RP); and pasture intercropped with pigeon pea (*Cajanus cajan*). cv. Mandarin (INT). The microclimate (photosynthetically active radiation and wind speed) and soil moisture characterization were assessed. The effect of the presence of pigeon pea on the vegetative and productive characteristics of the pastures was also evaluated. The results showed that the pigeon pea plants intercropped with the pasture did not promote significant changes in the soil water content in the layer 0-60 cm deep but reduced the transmission of photosynthetically active radiation to the pasture (from 0 to 65%), in addition to decreasing the incidence of winds (from 0 to 60% of reduction). During two experimental years, pigeon pea showed a high potential for biomass production (average of 12,615 Mg/ha/year), serving as forage supply to the animals during the dry season of the year. The recovery of the pasture through the input of nutrients conventionally in the recovered pasture (RP) or by intercropping with pigeon pea (INT) had an impact on forage accumulation compared to degraded pasture (PD). Additionally, the pasture intercropped with pigeon pea showed higher crude protein content compared to the other systems. Acknowledgment: Fapesp 2017/20084-5 and CNPq 421788/2018-6.

Session 61 Poster 20

Reducing energy consumption to dry alfalfa using organic acids – a field trial

T. Fumagalli and L.L.C. Jansen

Selko, Feed Additives, Stationsstraat 77, 3811 MH, the Netherlands; tancredi.fumagalli@trouwnutrition.com

Italian hay's exporter harvest alfalfa on the field and then dry it in an industrial dryer. The dryer temperature is manually adjusted aiming to achieve the desired humidity. Targeting for a higher moisture level to reduce energy consumption for drying while maintaining shelf life can be a challenge but can lower the costs. The aim of this study was to reduce dryer's energy consumption (source: methane gas) while preserving the product's shelf life by proper moisture management in a challenging condition like during a foggy day. The harvested alfalfa had an average moisture content of 40% (NIR analysis). During a processing period of about 2 hours, 14 bales (700 kg each) of untreated alfalfa (avg moisture 10.0%) and 14 bales of treated alfalfa (avg moisture 12.1%) were produced. The dryer temperature was decreased by 10 °C of the treatment group to reach a higher final moisture content. The treatment consisted of 2 kg of a liquid organic acid-based blend (Fylax® Forte-HC, Selko®, Tilburg, The Netherlands) per ton of alfalfa. At the end of the test, 4 samples of treated and untreated alfalfa each were collected at day 0 and after 36 days. A qualitative mould and yeast analysis was performed with a rapid test (Hygicult® Y&F) 4 days after sample collection. After 36 days of storage samples were recollected to perform a quantitative yeast and mould analysis in order to obtain an accurate count of fungal present. The average difference in final moisture was +2.2% at day 0 (P=0.0019) of the treated alfalfa. On day 4, untreated alfalfa showed on average high mould levels while treated alfalfa showed moderate to low mould levels. Yeasts counts were low for both treatments (<1000 cfu/ml). These findings were partially confirmed on day 36 by the quantitative analysis, where on average equal low levels of moulds (<10,000 cfu/g) and yeasts (<50,000 cfu/g) were detected in both treatments. The study showed that with the use of the organic acid-based blend it is possible to preserve the hay's shelf life even with higher moisture. This benefits in term of potential increase in volumes produced, and of consumption of energy and relative cost, due to a lower need for methane gas to support a drying temperature of -10 °C, which can correspond to up to 23 kg/h less of methane gas consumption.

Session 61 Poster 21

Carbon and energy footprint of dehydrated alfalfa production, from planting to factory output

D. Coulmier[1], P. Thiebeau[2], S. Recous[2] and H. Labanca[3]
[1]DESIALIS, Mont Bernard, 51000 Chalons en Champagne, France, [2]INRAE, UMR FARE, Esplanade R. Garros, 51100 Reims, France, [3]La Coopération Agricole, 43 rue Sedaine, 75011 Paris, France; didier.coulmier@desialis.fr

Agriculture accounts for a significant part of the factors of global warming through the consumption of non-renewable energy (NRE). Dehydration as a means of preservation of the harvested fodder is a great energy-consuming process. Reducing the use of NRE in this area is an important lever for mitigating the impacts of agricultural activity on the climate and can be complemented by the carbon storage achieved by crops. This study updates the energy and carbon balance of the French dehydrated alfalfa (DA) production sector with respect to the efforts made in recent years. The inventory is built with elements carried in the liabilities for both the carbon and energy part: fuel consumption required to the cultivation operations until the entrance of the factory, energy consumption (primary energy renewable or not, electricity) related to the dehydration in factory. The carbon asset consists of the atmospheric carbon fixed by the crop (harvested aerial biomass and root biomass). The energy asset consists of the energy valorisation of dehydrated alfalfa (LD) by the dairy cow. Over the period under consideration (2016-2019 – P2), the energy balance becomes positive with +1.2 GJ/t DM compared to the previous study period (2006-2009 – P1). The carbon balance also improves, reaching +390 kg C/t DM of DA. When entered into the Agribalyse/Ecoalim v8 reference database, these data show a 57% reduction in the impact of DA on the 'climate change (CC)' criterion compared to the pre-existing data (1.150 kg eq. CO_2 in P1 vs 0.494 kg eq. CO_2/kg DM of DA in P2). The environmental footprint of the sector therefore improved between the two periods studied. This improvement results from the generalisation of the flat wilting of alfalfa in the field to all the production sites, reducing the quantities of water to be evaporated, and from the implementation of less energy-consuming furnaces (250 vs 750 °C) able of incorporating biomass as an energy source.

Session 61 Poster 22

Impact of heat waves on the quality of milk and lactic farmhouse goat cheeses in the Aura region

S. Raynaud[1], E. Lemée[1], H. Le Chenadec[1], C. Laithier[1], P. Thorey[1], C. Boyer[1], M. Legris[1], S. Morge[2], S. Anselmet[3], V. Béroulle[4], S. Fressinaud[5], C. Delbès[6], M. Brocart[7], N. Morardet[8], J. Birkner[9] and Y. Gaüzere[9]
[1]French Livestock Institute, 149 rue de Bercy, 75595 Paris Cedex 12, France, [2]Chambre d'Agriculture de l'Ardèche, Le Pradel, 07170 Mirabel, France, [3]Chambre d'Agriculture de l'Isère, ZA Centr'Alp 34 Rue du Rocher de Lorzier, 38430 Moirans, France, [4]Syndicat caprin de la Drôme, 70 route de Choméane Est, 26400 Divajeu, France, [5]Ferme caprine expérimentale du Pradel Eplefpa, Le Pradel, 07170 Mirabel, France, [6]INRAE UMRF, 20, côte de Reyne, 15000 Aurillac, France, [7]Association Nationale Interprofessionnelle Caprine (ANICAP), 42 rue de Châteaudun, 75314 Paris cedex 09, France, [8]Auvergne-Rhône-Alpes Elevage, Agrapole, 23 rue Jean Baldassini, 69364 Lyon Cedex 07, France, [9]ENILBIO, Rue de Versailles, 39800 Poligny, France; cecile.laithier@idele.fr

In France, heat waves are becoming increasingly frequent, intense and long. Effects of these heat waves on the forage system and animal feeding are explored in different project but less is known on the effects on milk and milk products quality. The aim of this study was to characterise the consequences of these heat waves on milk and farmhouse lactic cheese quality and technology. Nine farms in the Auvergne-Rhône-Alpes region were monitored over 3 periods: before, during and after a high heat episode. The period has an effect on milk yield and composition (fat and protein rate, fatty acid percentage) and on cheese yields and acidification. These effects can be linked to the decrease of animal intake during periods of heat stress, especially concerning forage, and also to the temperature differences. The importance of these performance losses is nevertheless variable on the different farms in the study. This initial exploratory study on commercial farms allowed to identify issues and methodologies for future experimental or larger-scale studies.

Climate-smart practices can reduce GHG emissions intensity on smallholder dairy farms in Kenya

L. McNicol[1], M. Graham[2], M. Caulfield[2], J. Kagai[2], J. Gibbons[1], A.P. Williams[1], D. Chadwick[1] and C. Arndt[2]
[1]School of Natural Sciences, Bangor University, Bangor, LL57 2UW, United Kingdom, [2]International Livestock Research Institute (ILRI), Nairobi, 00100, Kenya; lsm20fqj@bangor.ac.uk

Milk consumption in Kenya is projected to rise. To meet projected demand in a sustainable manner, Kenya must focus on upscaling climate-smart agricultural (CSA) practices to increase productivity. The Kenya Climate-Smart Agriculture Project has been supporting smallholder farmers to adopt integrated climate-smart Technology, Innovation and Management Practices (TIMPs). The aims of this study were to investigate the effects of these TIMPs on milk production and GHG emission intensity (EI; farm gate GHG emissions per unit of product). Survey data were collected from four counties in Kenya: Baringo, Bomet, Kericho and Laikipia. Emissions estimates were calculated for 566 farms using Agrecalc and data were analysed using multiple linear regressions accounting for variability in geographical context and production system (no graze, semi-intensive, and extensive). Mean daily milk yields ranged from 0.5 to 15.7 l/cow/day. EIs ranged from 0.6 to 12.0 kg CO_2e/kg fat and protein corrected milk (FPCM), highlighting the opportunity for efficiency gains. The lowest EIs were found in Laikipia (2.4 kg CO_2e/kg FPCM) and the highest in Bomet (3.1 kg CO_2e/kg FPCM). This could be due to the difference in milk yields between counties and the prevalence of more extensive systems in Bomet. Dairy production system was the most important explanatory variable for the variability in milk production and EI (P<0.001). County also had a significant effect on milk production (P<0.001) and EI (P=0.017). Increasing adoption of TIMPs led to increased milk production (P=0.068) and reduced EI (P=0.117). The largest gains in milk yield and reductions in EI were seen in extensive systems. While this was not significant, there was a strong visual trend in extensive systems, but not in intensive or semi-intensive systems. Our results show that adoption of TIMPs generally increased milk yields and decreased GHG EIs. Therefore, adoption of CSA practices could allow Kenya to increase milk production to meet projected demand, whilst keeping associated GHG emission increases below business as-usual predictions.

Welfare barriers and levers for improvement in organic and low-input outdoor pig and poultry farms

C. Leterrier[1], C. Bonnefous[2], J. Niemi[3], PPILOW Consortium[4] and A. Collin[2]
[1]CNRS, IFCE, INRAE, Université de Tours, PRC, 37380 Nouzilly, France, [2]INRAE, Université de Tours, BOA, 37380 Nouzilly, France, [3]Natural Resources Institute Finland, Luke, Kampusranta 9, 60320 Seinäjoki, Finland, [4]PPILOW, www.ppilow.eu, https://www.ppilow.eu/wp-content/uploads/2023/03/List-authors.pdf, France; christine.leterrier@inrae.fr

The PPILOW project aims to co-construct innovations to improve Poultry and Pig Welfare in Low-input outdoor and Organic farming systems through a multi-actor approach. Its first step was to sum up animal welfare challenges observed in these systems and levers of improvement, from a review of literature data and research projects. Data were completed with information from key informants of the supply chains of poultry meat, eggs and pork in Italy, France, the United Kingdom and Finland. The interviews indicated that the main issues in poultry were: feeding, biosecurity, lack of range use and range management, feather pecking, weather, regulation, flock size or density, predation, bone fractures, lack of robustness, parasitism, pododermatitis, arthrosis, nervousness, water quality, catching and time spent by farmers. The main issues in pig were: feeding, tail biting, mortality, weather, predation, lack of robustness, lack of range use, castration, animal aggressiveness and competition, water quality, range management, human welfare, biosecurity issues, flock size or density, parasitism, insolation burns, joint abnormalities, parturition in freedom and pollution. This information has implemented a participatory approach for proposing welfare-improvement levers. Some issues and potential solutions were included in PPILOW experiments (phytotherapy against parasitism, involvement of animal personality in range use, rearing of entire pig males, genetic selection for reduced piglet mortality, improved farrowing huts for sows and piglets reared on range, avoiding feather pecking in laying hens with intact beaks, avoiding the killing of day-old male chicks, etc.), and solution costs evaluated. The results will provide a combination of practical solutions for welfare improvement in Europe. The PPILOW project has received funding from the European Union's Horizon 2020 Research and Innovation Programme under grant agreement N°816172.

Range use relationship with welfare and performance indicators in four organic broilers strains

C. Bonnefous[1], A. Collin[1], L.A. Guilloteau[1], K. Germain[2], S. Mignon-Grasteau[1], M. Reverchon[3], S. Mattioli[4], C. Castellini[4], V. Guesdon[5], L. Calandreau[6], C. Berri[1] and E. Le Bihan-Duval[1]

[1]INRAE, Université de Tours, BOA, 37380 Nouzilly, France, [2]INRAE, UE EASM, Le Magneraud CS 40052, 17700 Surgères, France, [3]SYSAAF, 37380, Nouzilly, France, [4]University of Perugia, Department of Agricultural, Environmental and Food Science, Borgo XX Giugno 74, 06124 Perugia, Italy, [5]France Junia, Comportement Animal et Systèmes d'Elevage, 59000, Lille, France, [6]INRAE, CNRS, IFCE, Université de Tours, PRC, 37380, Nouzilly, France; anne.collin@inrae.fr

Free range allows chickens to express more behaviours such as foraging and locomotion in a different environment from the poultry house. To understand how range use is related to welfare and performances of chickens, we studied four intermediate- to slow-growing strains with outdoor access until slaughter at 71 to 106 days of age depending on the strain. Males from JA757, S757N, White Bresse and a dual-purpose crossbreed were classified according to their range use and divided in two extreme groups of 25 high- and low-rangers. We did not observe any significant relationship between range use and welfare indicators collected at the slaughterhouse (hock burn, pododermatitis scores and struggling activity on the slaughter line) in the four studied strains. Leg health of JA757 chickens improved significantly with range use, with tibia being 4% shorter and 2% stronger in high-rangers compared to low-rangers. In both JA757 and S757N strains, range use also reduced the immune and inflammatory responses of birds. However, greater range use was negatively related to performances in all strains, including a 12, 8 and 7% reduction in carcass weight in JA757, S757N and the White Bresse strains, respectively, and a significant 2% reduction in thigh yield in the dual-purpose strain. Overall, greater range use in slow-growing birds did not affect welfare, health and meat quality indicators, but we confirmed a negative relationship with performance. The project PPILOW has received funding from the European Union's Horizon 2020 research and innovation programme under grant agreement N°816172.

Case study of a newly-developed genotype for dual-purpose rearing of male chicks

H. Pluschke[1], S. Lombard[2], B. Desaint[2], M. Reverchon[3], A. Roinsard[2], O. Tavares[2], A. Collin-Chenot[4], M. Ferriz[5], S. Seelig[6] and L. Baldinger[1]

[1]Thünen-Institute of Organic Farming, Trenthorst 32, 23847 Westerau, Germany, [2]ITAB, 9 rue André Brouard, 49100 Angers, France, [3]SYSAAF, Centre INRAE, Val de Loire, 37380 Nouzilly, France, [4]INRAE, Université de Tours, 37380 Nouzilly, France, [5]La Bassecour Bio, 41 Chem. de Chaponnay, 69970 Chaponnay, France, [6]Wendland Geflügel, Diahren 3, 29496 Waddeweitz, Germany; h.pluschke@thuenen.de

The culling of male layer chicks has been subject to widespread disapproval and led to its ban in Germany and France. One approach to divert from this practice is the use of dual-purpose genotypes (DPG) with a balanced performance in egg and meat production. Practitioners and stakeholders selected males of a DGP with a focus on laying (C) for on-farm evaluation under organic conditions in France and Germany. In Germany, the medium-growing JA757 (D) while in France the naked neck strain S757N (F) were reared as control groups. Data collection included mortality, feed consumption, live weight, welfare indicators, behaviour observations and carcass characteristics. In Germany, C cockerels were slaughtered at 16 while D at 13 weeks (wks) of age. In France there were two slaughter dates for C and F: 13 and 15 wks of age. In Germany, the carcass weight of C was 1.8 kg and that of D 2.4 kg. Carcass weights (incl. necks) at 13 and 15 wks of age were 2.9 and 3.3 kg for F while C weighed 2.0 and 2.5 kg in France. The FCR of DPG C until wk 13 was 3.7 for both countries, and 2.6 for F and 2.7 for D. In Germany, D showed dirtier breasts and more footpad lesions than C; on behavioural aspects, D spent more time resting than C while C spent more time foraging. DPGs could be an alternative to end the practice of chick culling, and thus fulfil the societal demand for a shift towards welfare-oriented production. A longer fattening period with higher FCR of DPG cockerels may be economically feasible if their meat is sold at higher price than that of usual genotypes. A perspective is to decipher whether they could valorise side products of the food industry to decrease feeding cost. Furthermore, the productivity of the females should be considered for a complete economical analysis of DPG. This project received funding from EU's Horizon 2020 research & innovation program under grant agreement N°816172.

Session 62 Theatre 4

Poultry production: using dual-purpose genotypes to reduce the culling of day-old male chicks?

J. Niemi[1], M. Väre[1], A. Collin[2], M. Almadani[3], M. Quentin[4], L. Baldinger[3], S. Steenfeldt[5], T.B. Rodenburg[6], F. Tuyttens[7] and P. Thobe[3]

[1]LUKE, Kampusranta, Seinäjoki, Finland, [2]INRAE, Université de Tours, Nouzilly, France, [3]Thünen-Institute, Farm Economics and Organic Farming, Bundesallee, BS, Germany, [4]ITAVI, Cunicole, Nouzilly, France, [5]Aarhus University, Tjele, 88333, Denmark, [6]Utrecht University, Yalelaan, the Netherlands, [7]EV ILVO, Animal and Veterinary Sciences, Scheldeweg, Melle, Belgium; jarkko.niemi@luke.fi

Existing studies show that consumers consider methods that avoid the killing of male day-old chicks (DOC) from layer genotypes as desirable practices. However, only 24% of producers argue that these measures are currently being applied. The use of dual-purpose breeds or *in ovo* sexing are two methods that can be used to address the ethical issues associated with the killing of male DOC. In this context, our study aims to develop a business case that would provide incentives to farmers to adopt these methods. More specifically, using the CANVAS method and insights from expert workshops, a business model outline was developed through reviewing the features, advantages and disadvantages of *in ovo* sexing and dual-purpose breeds. The results show that the promise of more ethical production is the key value proposition of these two methods. The key target groups were found to be consumers who are aware of current practices or are ethically conscious and/or better-informed. The results reveal that the use of dual-purpose breeds is considered a good option to avoid killing of male DOC. However, this option is considered to adversely affect broiler production, as male birds are slow-growing, have low feed efficiency and meat yields, and thus higher production costs, compared with conventional broilers. Hence, the business case of dual-purpose birds was based on addressing both animal welfare and ethical concerns, which in turn, could widen consumer group and promote demand toward compensating production costs. Raising consumers' awareness of culling one-day-old male chicks gives growing potential for keeping dual-purpose birds. The PPILOW project has received funding from the European Union's Horizon 2020 Research and Innovation Programme under grant agreement N°816172. www.ppilow.eu.

Session 62 Theatre 5

Longitudinal assessment of health indicators in four organically kept laying hen flocks

L. Jung[1], M. Krieger[2], L. Matoni[2] and D. Hinrichs[1]

[1]University of Kassel, Animal Breeding, Nordbahnhofstr. 1a, 37213 Witzenhausen, Germany, [2]University of Kassel, Farm Animal Behavior and Husbandry, Nordbahnhofstr. 1a, 37213 Witzenhausen, Germany; lisa.jung@uni-kassel.de

Good health is an important precondition for good welfare. Health disorders like feather damage, skin lesions, keel bone damage or footpad lesions are known to be present in most laying hen flocks regardless of the housing system. The aim of this study was to find out whether there is: (1) a difference between genetics in terms of disease prevalence; (2) a specific lifetime for disease occurrence; (3) a temporal relationship between disease occurrences; and (4) a cumulation and subsequence of diseases in individual animals. Health indicators were assessed in four successive organically kept laying hen flocks of Lohmann Brown (flock 1 and 2) and the dual-purpose breed Coffee&Cream (flock 3 and 4) in the years 2018 to 2022. In the first flock 60 hens and in the three following flocks 150 hens were individually marked and assessed at 4-week intervals from housing (17-18 weeks) to depopulation (61-78 weeks). First results show that the prevalence of all indicators was comparable to those found in conventional laying hen flocks. There was no marked difference in terms of general health status between Lohmann Brown and Coffee&Cream chickens. Keel bone damage was the most prominent welfare problem in all flocks. Descriptive statistics showed that all flocks were affected by comb lesions from housing until depopulation and by footpad ulcerations from peak of lay until depopulation. Differences between flocks were found in relation to feather and skin condition. Contrary to the hypothesis, that hens with feather damage are disabled in navigation, this leading to more collisions and thus more keel bone damages, the latter generally occurred prior to plumage damage. Likewise, no general order with respect to the occurrence of plumage damage and skin lesions could be determined from individual animal data.

Session 62 Theatre 6

Alternative pig housing systems with high welfare standards – status quo and perspectives

M. Holinger

Research Institute of Organic Agriculture (FiBL), Livestock Sciences, Ackerstrasse 113, 5070 Frick, Switzerland; mirjam.holinger@fibl.org

The term 'alternative housing system' is an umbrella term for systems that encompass welfare and/or sustainability while also ensuring productivity. 'Conventional housing' usually describes a barren indoor system that strongly limits animals' behavioural freedoms. Alternative systems such as organic, free-range, pasture-based, agroforestry or regenerative generally meet the social, exploratory and locomotory needs of the pigs. Organic production is an absolute niche, with a market share in pigs of less than 1% in Europe. Legislation and private standards for organic livestock production differ between countries. In some countries organic pigs are kept predominantly free-range, in other countries they have access to a concrete outdoor run but not to soil. These differences influence pig health and welfare. Free-range pig husbandry (organic or non-organic) is associated with less (e.g.) respiratory and gastrointestinal problems. As bedding material is provided, organic pigs in general have a lower prevalence of auxiliary bursae. The risk for parasite infestation, however, is increased because of restrictions on preventive deworming. Newer, innovative approaches aim at integrating various essential resources within the housing system. Provision of a rooting area filled with earth-like materials and a pool enable display of rooting and wallowing behaviour, both of which are impeded in most housing environments. While both resources are intensively used by pigs, the soiling of these areas is a management challenge. Targeted temporary restriction on the use of these resources or distribution of feed are measures that have been tested in on-farm settings to reduce soiling. Future developments will aim at further increasing the variability of environments and providing pigs with more choice. More variability could for example be implemented in feeding (location, type, and accessibility) or installations for cognitive enrichment. The ideas for such new approaches originate either from farmers or from researchers. The implementation needs to be accompanied by research to analyse the desired effects on health and welfare, but also to optimize practicability.

Session 62 Theatre 7

Characterising outdoor pig systems in Ireland

O. Menant[1], S. Mullan[2], F. Butler[3], L. Boyle[1] and K. O'Driscoll[1]

[1]Teagasc, Pig Development Department, Moorepark, P61P302, Ireland, [2]School of Veterinary Medicine, University College Dublin, Belfield, Dublin 4, Ireland, [3]School of Biological, Earth and Environmental Sciences, University College Cork, T12 K8AF, Ireland; ophelie.menant@teagasc.ie

OneWelPig is a project to develop a roadmap for expansion of outdoor pig systems in Ireland not only to meet societal and policy demands, but also through the agro-ecological role that pigs can play in farming systems. As little is known about outdoor pig systems in Ireland, the objective of this study was to characterize current husbandry and management practices. Between Dec 2022 and Feb 2023, 57 owners of outdoor pigs (in 21 out of the 32 Irish counties) completed an anonymous online survey. They had 8±6 years of experience (range: 1-30) and mainly kept pigs for personal meat consumption (62.3%), followed by 'meat for sale' (56.6%), breeding pigs for sale (45.3%), conservation of traditional Irish and/or rare breeds (32.1%), hobby (30.2%) and land management (28.3%). Pigs were kept in a variety of land types: pasture (74.5%), scrubland (33.3%) and woodland (37.3%), with additional straw (23.6%); and up to 18.5 animals/ha. Meat was mostly sold directly at the farm (66.6%), farmers markets (45%), in online shops (41%) and in supermarkets (23.3%). Pigs were mainly bred on farm (60.4%) or came from other outdoor farms (41.5%) with fewest from conventional farms (13.2%). Breeds were mainly Duroc (35.8%), Oxford Sandy and Black (32.1%), Tamworth (22.6%), Gloucester old spot (20.8%) and few conventional Large White (7.5%). Participants were all operating at a small-scale (average 7 sows/gilts, 1 boar, 18 grower pigs, 17 piglets at any one time in the year and managed by 2 people) and had a communication network separate from the conventional pig industry. At least 51% of the participants were members of the Irish Pig Society (and 7.5% from others) which promotes 'non-intensive' pig farming. When they are facing pigs health issues, 61.7% of the owners consult veterinarians (34.5% with strong experience in outdoor animals, 44.8% generalist large animals, and 10.3% conventional pig specialist), but also farmers that raise pigs outdoors (31.9%). The preliminary survey results demonstrate that a diverse range of outdoor pig keeping systems emerged in Ireland in recent years.

Session 62 Theatre 8

Animal welfare and pork quality of intact male pigs in organic farming according to genotype

B. Lebret[1], S. Ferchaud[2], A. Poissonnet[3] and A. Prunier[1]
[1]PEGASE, INRAE, Institut Agro, 35590 Saint-Gilles, France, [2]INRAE, GenESI, 86480 Rouillé, France, [3]IFIP, Institut du Porc, 35650 Le Rheu, France; benedicte.lebret@inrae.fr

In organic pig farming, genotype was tested as a lever to improve welfare and pork quality traits (tenderness, processing ability) of intact males while controlling the risk for boar taint. A total of 81 organic intact males from two genotypes: Large White × Duroc (D, n=47) or Large White × Piétrain NN (P, non-carrier of the n allele, n=34) were involved in two batches, each including one group of pigs per genotype. Each group was reared in a pen from the same building on deep straw bedding (1.3 m^2/pig) with free outdoor access (1.0 m^2/pig) from 27 to 125 kg live weight. All pigs received *ad libitum* the same organic growing and finishing diets and had free access to hay (rack). Overall, health and welfare indicators in live pigs (mortality, lameness, skin scratches and tail lesions) showed few problems, but mortality, severe lameness or number of skin scratches were lower in D than P pigs. On carcasses, the number of skin scratches was lower in D than P pigs (P<0.05). Average growth rate, feed intake and feed efficiency (per pen) were similar in both genotypes. Carcass weight was similar but lean meat content was lower (P<0.001) in D than P pigs. Compared to P, loin (Longissimus) meat from D pigs had lower drip loss and lightness (P<0.05), higher lipid content (P<0.001) and tended to have lower shear force (P<0.10). Backfat androstenone was higher in D than P pigs (P<0.01) but skatole did not differ. Even if not significant, the risk of rejection by consumers of tainted carcasses was higher for D (17.4%) than P (8.8%) pigs. Androstenone content increased with slaughter weight (P<0.02) but not skatole. Altogether, raising D instead of P intact males in organic farming seemed favourable for welfare and some pork quality traits (technological, colour, texture) but impaired carcass value. The higher boar taint risk could be reduced by decreasing slaughter weight. Solutions can thus be proposed to better satisfy the needs of farmers and stakeholders according to their priorities for organic pork. The project PPILOW has received funding from the European Union's Horizon 2020 research and innovation programme under grant agreement N°816172.

Session 62 Theatre 9

Large White genetics in organic system: breeding for piglet survival

L. Canario[1], S. Ferchaud[2], S. Moreau[2], C. Larzul[1] and A. Prunier[3]
[1]INRAE, Animal Genetics, GenPhySE, 31326 Castanet-Tolosan, France, [2]INRAE, UE GenESI, 86480 Rouillé, France, [3]INRAE, PEGASE, Agrocampus Ouest, 35590 Saint Gilles, France; laurianne.canario@inrae.fr

A multi-generation selection to improve the survival of piglets in organic condition was initiated with the POWER project and is ongoing with the PPILOW project (Horizon 2020 EU's Research and Innovation programmes No. 727495 and No. 816172). The experiment is carried out using Large White animals. In 2019, daughters of sows (G0) with good performance from the national scheme (kept in farrowing crate during lactation) were recruited as G1 for the organic farm. All sows are inseminated with semen from Large White boars with higher genetic merit for litter survival rate and a not too high genetic merit for litter size. Sow performance and behaviour are used to select future breeders. In G1 sows, we studied the effect of temporary crating around farrowing on performance, with comparison of two groups genetically similar. In one batch (n=4 per parity), pairs of sisters were inseminated with the same boar. One sister was 100% free to move in an individual pen (L), while the other was restricted to a crate around farrowing (B). The 24 pairs of sisters were evaluated over their first 3 parities. Causes of piglet death, piglet growth and sow behaviour were analysed. Prolificacy tended to increase between the 1st and 3rd litter in B sows. Piglet survival rate was 75-88% in the 48 h after farrowing and 64-75% until weaning; it varied greatly between sows and sisters. The number of piglets weaned in parity 1 was higher in L sows than in B sows. In parity 2, prolificacy was higher in B sows than in L sows (16.8 vs 14.4 live born piglets, P=0.01), but survival rate tended to differ (60.2 vs 69.7%) which led to equivalent litter sizes at weaning (9.9 vs 9.6). Mortality seemed to be slightly delayed in crated compared to loose-housed sows. Sows that were more maternal (at return of their piglets on day 1) had higher litter survival, especially those temporarily crated. First results indicate that G2 sows produced slightly fewer piglets in their two first litters as compared to G1 sows. Survival rate until 48h after farrowing was slightly higher in G2 sows than G1 sows (88.1 vs 83.7%). Response to selection will be further investigated.

Session 62 Theatre 10

Comparing animal welfare assessments by researchers and free-range pig farmers with the PIGLOW app

E.A.M. Graat[1,2], C. Vanden Hole[2], S. Nauta[3], M.F. Giersberg[3], T.B. Rodenburg[3] and F.A.M. Tuyttens[1,2]
[1]Ghent University, Salisburylaan 133, Merelbeke, Belgium, [2]ILVO, Scheldeweg 68, Melle, Belgium, [3]Utrecht University, Yalelaan 2, Utrecht, the Netherlands; evelien.graat@ilvo.vlaanderen.be

The PIGLOW app was developed for free-range and organic farmers to conduct welfare assessments of their own pigs. The goal of the app is to stimulate farmers to take a closer look at their animals and perhaps gain a different view of certain welfare aspects. A study is being conducted in which 12 free-range pig farmers use the PIGLOW app regularly for two years. While the main goal is to see if this leads to improved animal welfare on the farms, another aim is to compare welfare assessments with the app by researchers and by farmers. For this purpose, one of two researchers visits each farm at the start and end of the study to conduct the first and last welfare assessment simultaneously with the farmer. For all first welfare assessments, the group means for farmers ($\bar{x}_F$) and researchers ($\bar{x}_R$) and the average absolute differences in the percentages of animals/groups that they scored as 'positive' for a welfare indicator were calculated for 18 animal-based indicators. For group indicators, the largest differences were for 'coughing/sneezing' ($|\bar{x}_F-\bar{x}_R|=14.6$, $\bar{x}_F=12.5$, $\bar{x}_R=27.1$) and 'difficulty accessing water' ($|\bar{x}_F-\bar{x}_R|=37.5$, $\bar{x}_F=8.3$, $\bar{x}_R=45.8$). For individual indicators, the most striking differences were for 'scratches' ($|\bar{x}_F-\bar{x}_R|=6.5$, $\bar{x}_F=0.5$, $\bar{x}_R=6.8$) and 'too small' ($|\bar{x}_F-\bar{x}_R|=2.4$, $\bar{x}_F=3.7$, $\bar{x}_R=2.1$). Both 'coughing/sneezing' and 'scratches' require close and focused observation, which perhaps the researchers were more skilled at. The difference for 'difficulty accessing water' suggests a difference in judgement of the water facilities, where researchers consistently judged more severely. 'Too small' was one of few indicators scored more severely by farmers, possibly because body size is an important performance parameter. These results show that farmers score the welfare of their pigs differently and often more positively than researchers. These results will be complemented by those of the final welfare assessments, which will be completed by early August 2023. This project has received funding from the European Union's Horizon 2020 research and innovation programme under grant agreement N°816172.

Session 63 Theatre 1

Growth performance and digestive tract parameters in weaned piglets fed *Nannochloropsis limnetica*

A.A.M. Chaves[1], C.F. Martins[1,2], D.F. Carvalho[1], A.R.J. Cabrita[3], M.R.G. Maia[3], A.J.M. Fonseca[3], R.J.B. Bessa[2], A.M. Almeida[1] and J.P.B. Freire[1]
[1]Instituto Superior de Agronomia, Universidade de Lisboa, LEAF Linking Landscape, Environment, Agriculture and Food, Associated Laboratory TERRA, Tapada da Ajuda, 1349-017 Lisboa, Portugal, [2]Faculdade de Medicina Veterinária, Universidade de Lisboa, CIISA – Centre for Interdisciplinary Research in Animal Health, AL4AnimalS, Av. da Universidade Técnica, 1300-477 Lisboa, Portugal, [3]Instituto de Ciências Biomédicas de Abel Salazar, Universidade do Porto, REQUIMTE, LAQV, R. Jorge Viterbo Ferreira 228, 4050-313 Porto, Portugal; andreia_ac_22@hotmail.com

Presently, it is necessary to find novel and more sustainable alternatives to complement conventional feedstuffs, such as corn or soybean meal. Due to their rich nutritional profile, microalgae such as *Nannochloropsis limnetica* (NCH) can be an interesting solution for monogastric feeding. The aim of this study was to evaluate the effect of dietary NCH on growth performances and digestive tract development of weaned piglets. Animals were individually housed in metabolic cages and randomly allocated to 4 dietary treatments (n=6): control, 5, 10 and 15% dietary incorporation of NCH as a replacement of the basal diet. Water was always available. The experiment lasted 14 days (with a previous adaptation period of 4 days) and feed intake was monitored daily and equalized between experimental diets. Piglets were weighed weekly. At the end of the experiment, animals were slaughtered and stomach, pancreas, vesicle, small and large intestine weighed and measured. Polynomial contrasts were performed to test the effect of NCH incorporation levels in the diet. No differences were observed in average daily intake, average daily gain and feed conversion ratio among dietary treatments ($P>0.05$). The weight of full ($P<0.05$) and empty ($P<0.001$) small intestine and liver ($P<0.05$) linearly increased with dietary NCH inclusion. The group fed with 10% incorporation of NCH had the heaviest pancreas ($P<0.05$, quadratic). Diet did not affect weight of stomach, vesicle and large intestine and length of large intestine. Further information, to be provided by digestibility coefficients analysis, will complement the results herein described, allowing further knowledge of the nutritional value of NCH for the weaned piglet.

Session 63 Theatre 2

Synbiotic administration to suckling piglets on health parameters at weaning and postweaning period

E.A. Sureda[1], M. Schroyen[2], J. Uerlings[2], F. Fannes[3], J. Liénart[3], A. Sabri[4], P. Thonart[4], V. Delcenserie[2], J. Wavreille[5] and N. Everaert[1]

[1]KU Leuven, Kasteelpark Arenberg 30, 3001 Leuven, Belgium, [2]Liège University, Av. Cureghem 7, 4000 Liège, Belgium, [3]Vésale Pharma SA, Rue L. Allaert 9, 5310 Eghezée, Belgium, [4]Artechno SA, Rue H. Meganck 21, 5032 Isnes, Belgium, [5]Walloon Agricultural Research Centre, Rue Liroux 9, 5030 Gembloux, Belgium; ester.arevalosureda@kuleuven.be

In pig production, probiotics, prebiotics and their combination are being widely explored to improve intestinal health. This study investigated if synbiotic (Syn) supplementation to suckling piglets could have health benefits up to the post-weaning (PW) period. From birth to 25 days of age (d25), piglets received Syn1 (*Clostridium butyricum*, oligosaccharides), Syn2 (*Bifidobacterium animalis lactis Intellicaps*®, 2'fucosyllactose) or water [n=6/7]. Growth performance parameters were evaluated weekly. Diarrhoea incidence and behaviour were evaluated for the suckling and the PW period. Faeces (d25) were analysed for detection of the probiotics, beneficial microbial groups and gene expression of butyrylCoA:acetylCoA transferase(BCO). At day 25, 1- and 5-weeks after weaning, a piglet per litter/pen was selected for sampling. Colon content and faeces were analysed for lactate and short-chain fatty acids. Colon tissue was analysed for changes in gene expression. Syn-supplemented piglets had lower weight at the end of the PW period compared with the control group. Yet, syn-supplementation significantly reduced diarrhoea prevalence during suckling (Syn1) and PW periods (Syn2), which was accompanied by an increase of piglets' active behaviour. Microbial communities and metabolites showed no major differences, and BCO expression was reduced in Syn2 piglets' faeces. However, syn-supplemented piglets showed lower expression of MUC2, which could lead to reduced mucus production at d25, and lower expression of inflammation related genes (CXCL10, IFNα, IL1β, IL6, IL8, ILRN1) at all times. Syn-supplementation to piglets had health and well-being benefits even if not resulting in improved piglet's performance. Observed differences in weight of piglets could be caused by higher energy expenditures on syn-supplemented piglets due to increased active behaviour. Synbiotics had anti-inflammatory effects on gene expression in colonic tissue.

Session 63 Theatre 3

Fine characterisation of a standardized citrus extract and it's effect on weaned piglet performances

S. Cisse[1,2], J. Laurain[1] and M.E.A. Benarbia[1,2]

[1]Nor-Feed SAS, R&D, 3 rue Amédéo Avogadro, 49070, France, [2]FeedInTech, R&D, 42 rue Georges Morel, 49070, France; sekhou.cisse@norfeed.net

Plant extracts such as Citrus extract have more and more interest in animal nutrition, as growth performances enhancer. However, their effects on animals' growth performances vary, depending on the used product. A possible explanation of these variation may be the composition. The lack of composition data of plant extracts makes the understanding of their mechanisms of action unclear. This study aimed to evaluate the composition of a standardized natural citrus extract (SNCE). The effect of SNCE dietary supplementation has also been and assessed on weaned piglet growth performances. Apolar and polar compounds of SNCE were characterized by GC-MS and LC-MS (dereplication) respectively. In parallel, 864 piglets weaned at day 28 (Duroc × Large-White × Landrace) were divided into two groups. Each group consisted of 12 replicates of 36 piglets: A control (CTL) group fed with a standard diet and a SNCE group fed with a standard diet supplemented with 250 ppm of SNCE. All diets had been formulated without antibiotics, organic acids or enzymes, or probiotics. The trial duration was in two steps, post-weaning phase during 42 days and growing-finishing phase from day 70 until slaughtering at day 168. Growth performances were recorded and statistical analysis (T-test) was performed using GraphPad prism V7.SNCE characterization allowed to identify pectic oligosaccharides as major compounds as well as 30 secondary metabolites, including hesperidin, naringinin, and eriocitrin. Moreover, no variation has been observed in the composition and the concentration of SNCE active compounds from 5 different SNCE batches manufactured over 2 years. Regarding piglets' performances, supplementation with SNCE increased ADG between day 28 and day 70 compared to CTL group (333.9 vs 322.4, P=0.0426). A positive effect of SNCE was also observed on the FCR (2.179 vs 2.051, P=0.0016). Supplementation with SNCE also improved FCR (2.83 Vs 2,70, P=0.0003) during the growing and fattening period. Data evidenced that Citrus extract could be effective solution as growth performances enhancer for weaned piglets. SNCE characterisation revealed its standardization and the trial on weaned piglets confirm its interest for pig industry.

Session 63 Theatre 4

Essential oils and butyric acid effects on growth performance, blood metabolites and health in pigs

U. Marume and R.B. Nhara
North-West University, Animal Science, P. Bag X2046, Mmabatho, 2735, South Africa; upenyu.marume@nwu.ac.za

The study was conducted to determine the effect of Kalahari melon essential oil (EO), butyric acid (OA) and their blend on performance, blood metabolites and faecal health in growing pigs. Forty growing pigs were weighed, stratified, and randomly allocated to five dietary treatments, NC-Negative control (no antibiotic), PC-Positive Control (zinc bacitracin), EO (0.4%), OA- (0.6%) and EOOA- EO (0.4%) + OA (0.6%). Each dietary treatment had eight replicate pigs regarded as the experimental units. Treatment significantly affected growth performance. Pigs fed the OA and EOOA diet had higher ADFI (1.07 kg and 1.10±0.01 kg) and ADG (0.60 and 0.62±0.01 kg/d) compared with the control. Those fed the NC diet had the lowest values for all growth parameters. The pigs fed the EOOA diet also had lower FCR value compared with the control diet. With regards to the protein efficient ratio which measures the food values of different proteins, the EO and butyric acid inclusion in diets appeared to improve the protein utilisation in the diets with pigs fed the EO, OA and EOOA having the highest values for protein consumed (PC) protein efficient ratio (PER), specific growth rate (SGR) and growth efficiency (GE) compared with the control. Although there were no significant effects on haematology, EO and OA inclusion in diets appeared to slightly improve the haematological parameter. With regards to non-cellular metabolites, the inclusion of OA and EO in diets significantly reduced liver enzyme activity as reflected by lower values for alanine aminotransferase, aspartate transferase and alkaline phosphatase. Nevertheless, the inclusion of OA and EO appeared to deplete the levels of total serum protein probably due to its effect use by the animals. Throughout the study, the pigs fed the EOOA diet consistently had the lowest faecal score with no incidences of diarrhoea, A decline in faecal scores was observed in all other treatment as the trial progressed apart from the NC fed pigs. The findings indicate that essential oils, butyric acid and their blend can positively influence growth performance and blood metabolites and reduce diarrhoea incidence in growing pigs. They can therefore be used as alternatives to the conventional antibiotics.

Session 63 Theatre 5

The impacts of a spectrum of varied lifestyle factors on the porcine gut microbiota

L. Comer, E. Arévalo Sureda and N. Everaert
KU Leuven, Biosystems Department, Kasteelpark Arenberg 30, 3001 Leuven, Belgium; luke.comer@kuleuven.be

Given piglets' vulnerability to gastrointestinal disease during the weaning period, it is vital that we better understand how varied factors influence the gut microbiota, so as to be able to promote the establishment of a healthy microbiota. Studies hitherto have highlighted the importance of a plethora of different factors in shaping the microbiota. Yet, these studies tend not to deviate from small-scale experimental setups, and neglect the interplay of varied lifestyle factors (diet, environment, breed) seen in the real world. We hypothesised that a study analysing the microbiota across a range of varied lifestyles would show more variation in microbial composition, and highlight the dominance of a few factors such as diet and environment-type. To address this, we have established a microbiota repository for which we are obtaining faecal samples from >300 pigs and other members of the Suidae family. Subjects have very varied lifestyle factors including diet, environment-type and age. As an ongoing project, we are continuing to gather more samples. After sample collection, V1-V9 16S rRNA gene sequencing is performed and data is processed using QIIME2 with DADA2. Early analysis indicates a particularly strong influence of several factors (PERMANOVA $P<0.05$) including inside/outside environment and age which drive marked clustering on a PCoA. A number of genera were also significantly different by location, with *Treponema* and *Cellulosilyticum* among the many taxa enriched in pigs with outdoors environments or access to straw. *Lactobacillus* on the other hand was associated with animals fed a more commercial diet. These initial results provide trends to explore and target through further sample collection in the immediate future, as well as providing an insight into the microbiota responses to many lifestyle factors.

Session 63 Theatre 6

In vitro inhibition of avian pathogenic _Enterococcus cecorum_ isolates by probiotic Bacillus strains

M. Bernardeau[1,2], S. Medina-Fernandez[1] and M. Cretenet[1]
[1]Normandy University, UNICAEN, ABTE, Campus 1, Bât M, Esplanade de la Paix, 14032 Caen, France, [2]IFF, Danisco Animal Nutrition and Health, Innovation, Willem Einthovenstraat 4, 2342 BH Oegstgeest, P.O. Box 2300 AE, Leiden, the Netherlands; marion.bernardeau@iff.com

Enterococcus cecorum is a commensal bacteria and opportunistic pathogen that can cause outbreaks of Enterococcal spondylitis in poultry, with a growing concern worldwide. Numerous *Bacillus*-based probiotic strains are commercially available with proven effects in supporting gut health and growth performance, but efficacy against pathogenic *E. cecorum* is unknown. This study compared the *in vitro* inhibitory potential of Cell-Free Supernatants (CFSs) of 18 *Bacilius* strains (14 being commercial probiotic strains) on the growth of 9 clinical *E. cecorum* isolates. Standardized biomass cultures of live *Bacillus* were harvested and filtered to obtain CFSs. Inhibitory potential against *E. cecorum* isolates was assessed via a microdilution assay in which the final pathogen concentration was $\approx 10^4$ cfu/ml. Absorbance (OD) was measured every 15 min for 15 h and used to calculate percentage growth inhibition at an OD equivalent to 0.4 in the positive control (PC) (pathogen but no CFS), and growth delay vs PC. Growth kinetic responses of pathogen isolate-*Bacillus* strain combinations ranged from total pathogen inhibition to partial inhibition, lag in growth, no effect, or increased growth vs PC. Percentage inhibition of individual isolates varied markedly among *Bacillus* strains, from 100% (*B. amyloliquefaciens* CFSs BS8, 15AP4, 2084, #10B/4 and CFS #1/1) to -100% (growth promotion) (*B. amyloliquefaciens* DSM7T). Five *B. amyloliquefaciens* CFSs (3AP4, 2084, ABP278, #1/1 and #10B/1) produced higher average inhibition rates (>75%) than 2 out of 3 *B. licheniformis* CFSs (#12/1 and #10/4, -2.5% and -8.39% vs PC, respectively) and 1 out of 2 *B. subtilis* CFSs (#11/1, 7.3% vs PC) ($P<0.05$). Commercial strain 3AP4 exhibited the highest average percentage inhibition vs PC (85.0%, 7.862SD) and the most consistent inhibitory effect across pathogen isolates. The findings indicate that some commercially available poultry probiotic *Bacillus* strains are effective at inhibiting pathogenic *E. cecorum in vitro*, but effects are highly strain- and pathogen-isolate- dependent.

Session 63 Theatre 7

Effects of breed and early feeding on intestinal microbiota, gene expression and welfare indicators

F. Marcato[1], D. Schokker[2], S.K. Kar[1], J.M.J. Rebel[2] and I.C. De Jong[1]
[1]Wageningen Livestock Research, P.O. Box 338, 6700 AH Wageningen, the Netherlands, [2]Wageningen Bioveterinary Research, P.O. Box 65, 8200 AB Lelystad, the Netherlands; francesca.marcato@wur.nl

Recently the Netherlands has shifted towards more welfare-friendly broiler production systems using slower-growing broiler strains. Early post-hatch feeding is a dietary strategy that is currently used in commercial broiler production to modulate the gut microbiota, increase performance and improve broiler welfare. However, there is a knowledge gap on the effect of both breed and early feeding strategies and their interplay on microbiota composition and diversity, which in turn can affect the release of endotoxins in the excreta of broilers. Endotoxins are a major concern as these can be harmful for both animal and human health. The main aim of this study was to investigate the effects of breed and early feeding on gut microbiome with the ultimate goal to reduce endotoxin concentration in faeces of broilers. The study also investigated the impact on these factors on relative gene expression and welfare indicators of broilers. A total of 624 Ross 308 and 624 Hubbard JA757 day-old male broiler chickens were included until day 37 and day 51 of age, respectively. Within each breed, one half of the chickens received early post-hatch feeding and the other half not. A total of 2 chickens per pen were euthanized at two time points, i.e. target body weight (TBW) of 200 g and 2.5 kg, and gut samples were collected. Jejunum content samples (n=96) were analysed for the microbiome, whereas the jejunum tissue (n=96) was used for gene expression (data not yet analysed). Welfare assessment was conducted on 10 chickens/pen at TBW of 2.5 kg. Results showed that breed affected the microbiome at a TBW of 2.5 kg ($P=0.04$), with Hubbard chickens having a higher mean alpha-diversity compared to Ross chickens, respectively 209 vs 157. Early feeding did not affect the microbiome at both TBW's. Breed and early feeding impaired the cleanliness score and the gait score of Hubbard JA757 at TBW of 2.5 kg ($P<0.01$). These results indicate that breed modulates the intestinal microbiota and affects welfare of chickens, whereas early feeding only affected the welfare at a late TBW.

Yogurt Acid Whey addition affects broiler caecal microbiota composition and metabolic activity

I. Palamidi, V.V. Paraskeuas, I. Politis and K.C. Mountzouris
Agricultural University of Athens, Department of Animal Science, Iera odos 75, 118 55 Athens, Greece; inpalamidi@aua.gr

In Greece strained yoghurt is a high nutritional and consumer-desirable product. The production process yields Yoghurt Acid Whey (70% v/v of the milk used) as a byproduct that can pose an environmental problem if left untreated. In this respect and given the composition of the Yoghurt Acid Whey its utilization in animal nutrition merits investigation. The aim of this study was to investigate the addition of yogurt acid whey powder (YAWP) in four dietary levels on broiler performance, caecal microflora composition and metabolic activity. A total of 300 male one-day-old male Ross broiler chickens were randomly assigned into 4 treatments of 5 replicates each. Broilers were fed maize-soybean meal basal diets following a 2- phase feeding program. Depending on YAWP inclusion level, treatments were no YAWP addition (W0), YAWP at 25 g/kg (W25), YAWP at 50 g/kg (W50) and YAWP at 100 g/kg (W100). Overall body weight gain, feed intake and feed conversion ratio were not affected by YAWP addition. In the ceca, a linear decrease in total bacteria counts and a linear increase in *Lactobacillus* spp. was noted with increasing YAWP dietary level. A linear reduction of *Clostridium* cluster I and *E. coli* species were noted with increasing YAWP dietary level. A linear increase in total volatile fatty acids concentration, acetic acid and butyric acid molar ratios were shown with increasing YAWP level. In conclusion, the addition of YAWP did not affect overall performance, and positively modulated caecal microbiota and metabolic activity of 35-day-old broiler chickens. This research has been co-financed by the European Regional Development Fund of the European Union and Greek national funds through the Operational Program Competitiveness, Entrepreneurship and Innovation, under the call RESEARCH – CREATE – INNOVATE (project code:T2EDK-00783).

A richer gut microbiota is related to better feed efficiency and diet adaptability in laying hens

M. Bernard[1], A. Lecoeur[1], J.L. Coville[1], N. Bruneau[1], D. Jardet[1], S. Lagarrigue[2], F. Calenge[1], G. Pascal[3] and T. Zerjal[1]
[1]Université Paris-Saclay, INRAE, AgroParisTech, GABI, Domaine de Vilvert, 78350 Jouy-en-Josas, France, [2]INRAE, Institut Agro, UMR PEGASE, 16, le clos, 35590 Saint-Gilles, France, [3]GenPhySE, Université de Toulouse, INRAE, ENVT, 24, chemin de Borde-Rouge, Auzeville Tolosane, 31326 Castanet Tolosan, France; maria.bernard@inrae.fr

In the egg industry, feed cost represents the majority of total production costs and breeding efforts are ongoing to improve feed efficiency of laying hens. The gut microbiota is known to play an important role in energy harvest and is likely to affect feed efficiency. In this study, we analysed the composition of caecal microbiota of 31 week old hens by 16S metabarcoding sequencing to characterise its composition, interactions with the host and influence on phenotypes of interest. As an animal model, we used hens of the R+ and R- lines divergently selected for high (low feed efficiency) and low (high feed efficiency) residual feed intake values respectively, that were fed either a commercial wheat-soybean diet (CTR) or a low-energy corn-sunflower diet (LE). Our results show a significant line effect on the microbiota richness and composition with the CTR diet, whereas with the LE diet, the microbiota was primarily affected by the diet change. A line × diet interaction was observed: the high efficient R- line presented a greater microbial richness and a reduced impact of the diet change compared to the low efficient R+ line. Interestingly, common taxonomies and/or predicted functions were highlighted between R+ and CTR diet, and between R- and LE diet, which could suggest that common microbiota mechanisms between feed efficiency and adaptation to nontraditional feedstuffs exist. OTUs of *Actinobacteriota* were more abundant in R+ birds and in the CTR diet, whereas OTUs of *Bacteroidota* were preferentially abundant in R- birds and/or LE diet. At functional level, carbohydrates and fatty acid metabolisms on the one hand, and short-chain fatty acids and amino acids metabolisms, on the other hand, were enriched in birds with low or high feed efficiency, respectively. These results provide insight into the role of the microbiota in laying hen feed efficiency and the impact of diet composition on microbiota.

Gut microbiome variations during the productive lifespan of two high-yielding laying hen strains

C. Roth, J. Seifert, M. Rodehutscord and A. Camarinha-Silva
Hohenheim Center for Livestock Microbiome Research, University of Hohenheim, Emil-Wolff-Str. 10, 70599 Stuttgart, Germany; amelia.silva@uni-hohenheim.de

Gut microbiota affects nutrient digestion, pathogen inhibition, endocrine activity, and interaction with the gut-associated immune system. In laying hens, previous studies focused on excreta and specific gastrointestinal sections, and analysed single time points in the hen's life and their response to specific conditions. This study aimed to characterize the active gut microbiota of a large cohort of two laying hen strains: Lohmann Brown-classic and Lohmann LSL-classic, during their productive life span. All 100 birds were fed the same diet and offered identical raising conditions within each age period. Digesta from the crop, gizzard, duodenum, ileum, and caeca were collected after 10, 16, 24, 30, and 60 weeks of life to represent the complete productive period. RNA from 500 samples was extracted and analysed by target amplicon sequencing. Phylogenetic analysis of the bacterial sequences was assessed using Mothur, followed by multivariate statistical analysis. A statistical significance was obtained for the breed, gastrointestinal section, age, and the combination of all factors ($P<0.05$). Depending on the strain, the detected genera differed in abundance within gut sections or productive periods. A significant shift in the active microbiota of both strains was observed with the laying onset transition between weeks 16 and 24. Metagenomic shotgun sequencing was further used to reveal the functional shifts between these weeks. Functional profiling showed differences between the strains besides up- and downregulated functions with the onset of the laying phase. The inositol phosphate metabolism: 5-keto-L-gluconate epimerase (iolO) (K22233) was significantly downregulated in the ileum of LSL hens at week 16 compared to 24. Functions K22231-22233 and the inositol-phosphate transport system substrate-binding protein (inoE) (K17237) were downregulated in the caeca of LB vs LSL hens of week 24. The strain and age impacted gut microbiome dynamics. Even though birds have been offered the same diet and were housed under standardized conditions, it remains unclear if the organism's needs caused microbiome variations during the lifespan or if the microbiome adapted to the host alterations.

A rabbit nutrition hypothesis to prevent digestive problems: 'Feed the fusus coli'

K.H. De Greef and M. Van Der Sluis
Wageningen Livestock Research, P.O. Box 338, 6700 AH Wageningen, the Netherlands; karel.degreef@wur.nl

Rabbits are vegetarian feed-to-food converters, equipped to digest fibrous diets at a high passage rate. As hind gut digesters, their prime nutrient utilisation is dependent on fibre digestion in the caecum. Modern breeding, feeding and production methods have practically treated the rabbit as a semi-monogastric animal: high levels of easily digestible nutrients are fed to enhance growth. Existing work (especially in France) has illustrated the need for adequate levels of fibre (especially ADF, acid detergent fibre) alongside this to prevent diarrhoea. However, nowadays, this appears inadequate to prevent digestive problems. High productive animals may suffer from diarrhoea, subsequent obstipation and related problems. This aberration is seen as a production disease, and is the main reason for antibiotics use for growing rabbits in the meat production sector. We have reviewed, analysed and combined existing views on the mechanism of derailment of the digestive system of the rabbit, focusing on the onset of the problems. This has led to a more integrated view on the cause and onset of the problem. In short, a rabbit-specific selection mechanism (the *fusus coli*) in the colon decides to send digesta from the colon back to the caecum, or to send the digesta further down the gastrointestinal tract for excretion. This selection mechanism is, among other factors such as stress, influenced by particle size (generally non-digested fibre). The hypothesis is that especially the lack of large particles hinders the signal to pass on digesta. This results in a too high rate of sending back digesta to the caecum, derailing the digestive processes as illustrated by overloaded caeca seen in affected rabbits (diseased or dead). Studies are currently being performed to validate this hypothesis: low digestible particles (of any kind) should prevent the 'traffic jam' in the gastrointestinal tract. If this hypothesis is indeed confirmed in practice, it opens up opportunities to better compose feeds that combine high productivity (addressing the rabbit as a monogastric animal) with good digestive health (respecting the rabbit as a fibre converter).

Session 63 Theatre 12

Gut microbiota of growing rabbits fed diets with different fibre and lipid contents

G. Zardinoni[1], P. Stevanato[1], A. Trocino[1,2], M. Birolo[1], F. Bordignon[1] and G. Xiccato[1]
[1]University of Padova, Department of Agronomy, Food, Natural resources, Animals and Environment (DAFNAE), Viale dell'Universita' 16, 35020 Legnaro, Italy, [2]University of Padova, Department of Comparative Biomedicine and Food Science (BCA), Viale dell'Universita' 16, 35020 Legnaro, Italy; giulia.zardinoni@phd.unipd.it

Fibre with its different fractions, both insoluble and soluble, is the main dietary component guaranteeing the normal functioning of the rabbit digestive physiology and gut health. Thus, the present study aimed to evaluate the effect of an increase of ADF from 18.1 to 18.8% associated with a decrease in dietary starch from 14.3 to 13.8% and an increase in dietary fat from 2.9 to 3.8% on the microbiota composition of caecal content and hard faeces. To this purpose, 576 crossbred rabbits (Hypharm, Groupe Grimaud, Roussay, France) were weaned at 31 d, assigned to the two dietary treatments, and fed the experimental diets until slaughtering (73 d of age), when hard faeces and caecal contents were sampled from 20 rabbits (10 per diet) in the afternoon (h.15:00-16:00) and analysed using a 16S rDNA multi-amplicon sequencing approach. Firstly, sequencing results showed that the microbial diversity and the bacterial community structure of the hard faeces barely differed from that of the caecal content ($P>0.05$). The overall microbial composition was dominated by the phylum of Firmicutes, the Clostridia and Bacilli classes, followed by Ruminococcaceae and Lachnospiraceae as dominant families. Then, as for the diet effect, no differences in alpha and beta diversity of microbiota were detected in rabbits fed the two diets. However, twelve genera, mostly belonging to the family of Lachnospiraceae, increased (Wald test, $P<0.05$) in rabbits fed the diet with the highest fibre and fat contents. Overall, these findings enhance our understanding about gut microbiota in growing rabbits and indicate that even small changes in fibre and fat of the diet may affect the composition of gut microbiota.

Session 63 Poster 13

Gut microbiota-metabolome response to dietary porcine intestinal mucosa hydrolysate in piglets

S. Segarra[1], A. Middelkoop[2] and F. Molist[2]
[1]R&D Bioiberica S.A.U., Esplugues de Llobregat, 08950, Spain, [2]R&D Schothorst Feed Research, NA Lelystad, 8218, the Netherlands; ssegarra@bioiberica.com

In piglets, a healthy gut microbiota contributes to regulating the immune system and enhancing resistance to pathogens. The use of porcine intestinal mucosa hydrolysate (PIMH) as an alternative to other protein sources has been reported to improve performance and profitability in piglets. The present study aimed to assess the effects on gut microbiome and metabolome of incorporating PIMH in the diet in piglets. The trial consisted of two treatment groups (n=16 pens/treatment; 6 piglets/pen). Pigs in the negative control (NC) group received a weaner I diet containing 3.5% skimmed milk powder and a weaner II diet containing 2.5% soy protein concentrate. In the PIMH group, these protein sources were replaced by 5% PIMH (Palbio 50RD®, Bioiberica S.A.U.) in weaner I and 2.5% PIMH in weaner II. Diets were iso-energetic and iso-protein and with a similar lactose content. At day 34 post-weaning, ileum content from the last 1.5 m before the cecum was collected from ten pigs (1:1 gender ratio) from each group. Digesta was homogenized and snap-frozen and used for microbiota analysis with 16S ribosomal RNA (rRNA) sequencing as well as metabolomics using ultra-high-performance liquid chromatography coupled with a time-of-flight (UHPLC-TOF). A permutational multivariate analysis of variance (PERMANOVA) was used for microbiota data, and an ANOVA comparison was done for metabolomics. Microbiota alpha- and beta-diversity were not significantly different ($P>0.10$) between study groups, but a significantly decreased (FDR<0.05) relative abundance of amplicon sequence variants (ASVs) of opportunistic pathogens from the Streptococcus genera was seen with PIMH. Metabolomics showed significantly ($P<0.05$) higher concentrations of spermine (protein-derived biogenic polyamine) and lower amino acid concentrations of N-acetyl-L-methionine and L-glutamine in the PIMH group, which could be related to an antioxidant action, and a positive modulation of metabolism, intestinal barrier and maturation, and protein digestibility. In conclusion, the use of PIMH may improve gut health in piglets through modulation of their gut microbiota and metabolome, which could explain the improvements in performance observed previously.

Session 63 Poster 14

Effects of probiotics isolated from faeces of fast-growing pigs on growth performance of weaning pigs

Y.H. Choi, Y.J. Min, J.E. Kim, Y.D. Jeong, H.J. Park, C.H. Kim and S.J. Sa
Rural Development Administration, Sinbang 1gil, Seonghwan, 31000 Cheonan, Korea, South; cyh6150@korea.kr

The study aimed to evaluate the effect of *Lactobacillus pentosus*, *Lactobacillus petauri*, *Mitsuokella multacida* isolated from faeces of fast-growing pigs given to weanling piglets. The forced feeding procedure was used for transferring the supplements to weaned pigs stomach. Eighty 25-d-old, female crossbred (L×Y×D) piglets were assigned to two experiments (40 pigs per experiment) with four dietary treatments and 10 replicates. Experiment 1 was included a control (CON; no probiotic), *L. pentosus* (LP; 10^{10} per d), *L. petauri* (LPC; 10^{10} per d), and *M. multacida* (MM; 0.1% of diet). Experiment 2 was based on the dose response to *L. pentosus* including 0, 5×10^9, 10^{10}, and 5×10^{10} cfu/d for each pig. In experiment 1, the LP treatment led to higher ($P<0.05$) body weight gain and ADG compared to the CON, while the diarrhoea score was reduced ($P<0.05$) in the LP, LPC, and MM treatments at day 14 and overall. Villus height in the duodenum was improved ($P<0.05$) in the MM compared to the CON, and villus height in the ileum was increased ($P<0.05$) in the MM and LP treatments. The concentrations of acetic acid and butyric acid in the cecum digesta were increased ($P<0.05$) in the LP, LPC, and MM treatments. The concentrations of acetic acid and total short-chain fatty acids in the faeces were increased ($P<0.05$) in the LP, LPC, and MM treatments, while a higher ($P<0.05$) concentration of butyric acid, isobutyric acid, and isovaleric acid in the faeces was observed in pigs fed MM compared to the CON. In experiment 2, supplementation with *L. pentosus* resulted in a linear positive response ($P<0.05$) in average daily gain (ADG) and gain to feed (G:F) at day 14 and 28. The overall growth performance also showed a positive linear response ($P<0.05$) to an increase in final body weight, ADG, and G:F, with a quadratic response ($P<0.05$) of overall G:F to *L. pentosus* supplementation. The diarrhoea score was linearly reduced ($P<0.05$) at day 14, 28, and overall. In conclusion, dietary supplementation with LP improved intestinal integrity and digesta fermentation, leading to improved growth performance during the critical weaning period for piglets.

Session 63 Poster 15

Effect of butyric acid salts on the palatability of feed for piglets

W. Kozera[1], A. Woźniakowska[1], K. Karpiesiuk[1], A. Okorski[1] and G. Żak[2]
[1]University of Warmia and Mazury in Olsztyn, ul. Oczapowskiego 5, 10-719 Olsztyn, Poland, [2]National Research Institute of Animal Production, ul. Sarego 2, 31-047 Kraków, Poland; grzegorz.zak@iz.edu.pl

The post-weaning period causes severe stress in piglets, which often manifests as low feed intake and gastrointestinal dysfunction. The negative effects of the weaning period can be counteracted by using organic acids in piglet mixes. Butyric acid salts influence the development of the piglets' intestinal epithelium, as well as a decrease in the number of Clostridium bacteria in the colon, which reduces the incidence of diarrhoea. They also increase piglet immunity. The inclusion of sodium butyrate in the diet of weaned piglets can affect feed palatability, which consequently improves feed intake and increases daily gain. The aim of this study was to investigate the effect of butyric acid salts on feed palatability and growth rate in weaned piglets. The experiment started when the piglets reached the age of 35 days. In the 8-day experiment, 32 piglets (homogenous in terms of weight and sex) were divided into 4 treatment groups (n=8), with 4 replicates per group and 2 piglets per replicate, to determine the effect of sodium butyrate on feed palatability. The animals received one of the four experimental diets according to the 4×4 Latin Square design. The diets were fed to piglets 4 times a day, at 8.00 a.m., 11.00 a.m., 2.00 p.m. and 5.00 p.m., by the single stimulus method, with free access to one of the diets for 20 minutes during each feeding session (*ad libitum* access to water). Feed intake was monitored daily. Piglets were weighed on days 1 and 8. The analysed additives had no effect on feed palatability. Sodium butyrate in the amount of 4,000 mg/kg of feed significantly influenced the final body weight of piglets ($P\leq0.05$). The highest total body weight gains were observed in all piglets receiving diets supplemented with sodium butyrate and a blend of coated calcium butyrate and sodium butyrate, with no effect on average daily feed intake. The addition of 4,000 mg/kg of sodium butyrate had a more beneficial influence ($P\leq0.05$) on total body weight gains than the addition of 2,000 mg/kg of coated sodium butyrate.

Session 63 Poster 16

Synbiotics-glyconutrients enhance growth performance and fatty acid profile in finishing pigs

C.B. Lim[1], Q.Q. Zhang[1], S. Biswas[1], J.H. Song[1], O. Munezero[1], J.S. Yoo[2] and I.H. Kim[1]
[1]Dankook University, Department of Animal Resource and Science, Cheonan, 31116, Korea, South, [2]Daehan feed company, 13 Bukseongpo-gil, Jung-gu, Incheon, 22300, Korea, South; coqlsdlek@naver.com

Glyconutrients are thought to be beneficial to the body by helping in cell communication. In addition, synbiotics appear to be a promising option for improving immune function. Therefore, we hypothesized that combining synbiotics and glyconutrients could enhance pig nutrient utilization. The experiment utilized 150 pigs (Landrace × Yorkshire × Duroc), initially weighing 58.85±3.30 kg of live body weight (BW) to see how finishing pigs supplemented with synbiotics-glyconutrients (SGN) affected growth performance, nutrient digestibility, gas emission, meat quality, and fatty acid profile. The pigs were matched by BW and sex and randomly assigned to one of three diet treatments: control = Basal diet; TRT1 = Basal diet + SGN 0.15%; TRT2 = Basal diet + SGN 0.30%. The trials were conducted in two phases (weeks 1-5 and weeks 6-10). ADG increased linearly from weeks 6 to 10 in pigs fed a basal diet with SGN (P=0.016). Moreover, pigs fed with SGN 0.30% tended to improve the BW at week 10 (P=0.058), ADFI, and FCR from weeks 6 to 10 (P=0.058, P=0.068, respectively). However, the ATTD of DM, N, and GE did not differ between treatments (P>0.05). Dietary treatments had no effect on NH_3, H_2S, methyl mercaptans, acetic acids, and CO_2 emissions. SGN 0.30% inclusion linearly improved cooking loss and drip loss on day 7 (P=0.025 and P=0.019, respectively). The SGN supplemented group had higher levels of margaric acid (C17:0), omega 3 and 6 fatty acids, and their ratio in both fat and lean tissues. Furthermore, the inclusion of SGN improved palmitoleic acid (C16:1) and linoleic acid (C18:2n6c) levels in fat and lean tissues, respectively. Additionally, SGN 0.3% supplementation improved unsaturated fatty acid (USFA) and monounsaturated fatty acid (MUFA). Thus, supplementing SGN improved growth performance, meat quality, and fatty acid profiles of finishing pigs.

Session 63 Poster 17

Impact of gut health product on appetite associated hormones using *ex vivo* porcine intestinal cells

N. Browne and K. Horgan
Alltech, Research, Summerhill Rd, Dunboyne, Co. Meath, A86X006, Ireland; nbrowne@alltech.com

In piglets it is observed that early weaning can lead to poor weight gain due to an underdeveloped gastointestinal (GI) tract which is unsuitable for efficient absorption of nutrients. Short chain fatty acid (SCFA) butyrate has demonstrated an ability to improve intestinal development by increasing cell proliferation which is vital during this transition period when the small and large intestinal tracts are rapidly growing. The palatability of butyrate makes it difficult for animal consumption which has led to encapsulation with more desirable feed components. Gut health product (GHP), a proprietary mix of yeast, SCFA and Zn proteinate was assessed here to determine its impact on cellular growth, metabolism and appetite associated hormones in *ex vivo* small intestinal pig cells. Pig intestinal cell densities were enumerated after trypsin, centrifugation, and resuspension in equal volumes. Total protein concentrations determined by Qubit and Hormone secretion of Peptide YY (PYY), and Ghrelin determined by ELISA on cell free media from cultured cells in the presence of (150 ppm) GHP, butyrate or the control. Statistics performed by One-way ANOVA (n=3). Intestinal cells had greater cell densities with GHP addition compared to the control ($P \leq 0.05$) and butyrate ($P \leq 0.01$) at 24 h, and a significantly higher cell density at 48 ($P \leq 0.05$) and 72 h ($P \leq 0.01$) over the butyrate treatment. Protein uptake in cells receiving GHP was significantly ($P \leq 0.05$) higher than control cells at 48 and 72 h hours. GHP inclusion resulted in a greater ($P \leq 0.05$) Ghrelin (appetite inducing hormone) release over the control and a numerically greater level than butyrate treated intestinal cells at 48 h. PYY (satiety hormone) was significantly lower ($P \leq 0.05$) than the control following GHP treatment at 72 h. Ghrelin was significantly elevated ($P \leq 0.05$) over the butyrate treated cells at 48 h while PYY was significantly lower ($P \leq 0.05$) for GHP treated cells over the control at 72 h and GHP had the lowest amount of PYY secretion from cells. Higher levels of Ghrelin and lower PYY secretion may drive the uptake of protein in cells receiving GHP. Contributing to the higher cell growth density observed with the inclusion of GHP which may enhance pig performance as observed in previous trials.

Session 63 Poster 18

Development of antimicrobial peptides against multidrug-resistant enterotoxigenic *Escherichia coli*

W.J. Chen, M.Y.W. Kwok, K.C. Wu, K.F. Hua, Y.H. Yu and Y.H. Cheng
National Ilan University, Department of Biotechnology and Animal Science, No.1, Sec.1, Shennong Rd., 26047, Taiwan; wjchen@niu.edu.tw

Post-weaning diarrhoea due to enterotoxigenic *Escherichia coli* (ETEC) is a common disease of piglets and causes great economic loss for the swine industry. Over the past few decades, decreasing effectiveness of conventional antibiotics has caused serious problems because of the growing emergence of multidrug-resistant (MDR) pathogens. Various studies have indicated that antimicrobial peptides (AMPs) have potential to serve as an alternative to antibiotics owing to rapid killing action and highly selective toxicity. Our previous studies have shown that AMP GW-Q4 and its derivatives possess effective antibacterial activities against the Gram-negative bacteria. Hence, in the current study, we evaluated the antibacterial efficacy of GW-Q4 and its derivatives against MDR ETEC and their minimal inhibition concentration (MIC) values were determined to be around 2~32 μg/ml. Among them, AMP Q4-15a-1 with the second lowest MIC (4 μg/ml) and the highest minimal haemolysis concentration (MHC, 256 μg/ml), thus showing the greatest selectivity (MHC/MIC=64) was selected for further investigations. Moreover, Q4-15a-1 showed dose-dependent bactericidal activity against MDR ETEC in time-kill curve assays. According to the cellular localization and membrane integrity analyses using confocal microscopy, Q4-15a-1 can rapidly interact with the bacterial surface, disrupt the membrane and enter cytosol in less than 30 min. Minimum biofilm eradication concentration (MBEC) of Q4-15a-1 is 4× MIC (16 μg/ml), indicating that Q4-15a-1 is effective against MDR ETEC biofilm. Besides, we established an MDR ETEC infection model with intestinal porcine epithelial cell-1 (IPEC-1). In this infection model, 32 μg/ml Q4-15a-1 can completely inhibit ETEC adhesion onto IPEC-1. Quantitative real-time PCR confirmed the anti-inflammatory ability of Q4-15a-1. Overall, these results suggested that Q4-15a-1 may be a promising antibacterial candidate for treatment of weaned piglets infected by MDR ETEC.

Session 63 Poster 19

Influences of quercetin inclusion to corn-soybean-gluten meal-based diet on broiler performance

S. Biswas[1], Q.Q. Zhang[1], J.H. Song[1], C.B. Lim[1], I.H. Kim[1] and J.S. Yoo[2]
[1]Dankook University, Department of Animal Resource and Science, Cheonan, 31116, Korea, South, [2]Daehan feed company, 13 Bukseongpo-gil, Jung-gu, Incheon, 22300, Korea, South; sarbani.dream@gmail.com

Quercetin (prevalent flavonoid) is considered to have antimicrobial and antioxidant properties. This trial was conducted to evaluate the impact of graded doses of quercetin (QS) on growth efficiency, nutrient retention, faecal score, footpad lesion score, tibia ash, and meat quality. In a 32-day feeding test, a total of 576 1-day-old Ross 308 broilers (male) were allocated arbitrarily with an average body weight of 41±0.5 g. The trial had four dietary treatments with eight repetitions of 18 birds per cage and a basal diet incorporating with 0, 0.02, 0.04 and 0.06% of QS. As the QS dosage increased, body weight gain increased linearly (P=0.069) on days 9-21 and tended to increase linearly (P=0.079) during the overall period. Similarly, feed intake increased (P=0.009) linearly with the increasing doses of QS on days 9-21. Likewise, there was a linear improvement in dry matter (P=0.002) and energy (P=0.016) digestibility after QS administration. Moreover, the inclusion of QS supplement linearly increased (P=0.011) tibia ash in broilers. However, the faecal score and footpad lesion score showed no significant outcome (P>0.05). By giving broilers a graded amount of QS, the relative organ weight of breast muscle (P=0.009) and spleen (P=0.001) improved linearly, meat colour lightness increased (P=0.015) and redness tended to improve (P=0.065) linearly, and drip loss decreased (P=0.015) linearly. The inclusion of QS in the graded-level diet led to improvements in growth efficiency, nutrient absorption, meat quality, and tibia ash, which recommended it as a beneficial feed additive for the broiler.

Session 63 Poster 20

Effects of *Bacillus* species – fermented products on growth performance and gut health in broilers

Y.H. Yu, S.H. Hsiao, Y.H. Cheng, W.J. Chen and K.F. Hua
National Ilan University, Department of Biotechnology and Animal Science, No.1, Sec. 1, Shennong Rd., Yilan City, Yilan County 26047, Taiwan; yuyh@niu.edu.tw

This study was aimed to assess the comparative effects of *Bacillus subtilis*-fermented products (BSFPs) and *Bacillus licheniformis*-fermented products (BLFPs) on the growth performance, intestinal morphology and gene expression, and caecal microbiota and microbial virulence factor gene composition in broilers. A total of 160 one-day-old unsexed Arbor Acres broiler chicks were randomly allocated to 4 treatments, with 8 replicates each and 5 chicks per replicate: C = basal diet, E = basal diet plus 10 mg/kg enramycin, BSFP = basal diet plus BSFPs (1×10^8 colony-forming unit (cfu) *B. subtilis* spore/g of feed), and BLFP = basal diet plus BLFPs (1×10^8 cfu *B. licheniformis* spore/g of feed). One-way analysis of variance with Tukey's honestly significant difference test was used for analysing the statistical differences. The significance was defined as $P\leq0.05$. BLFP treatment resulted in a higher body weight at 35 days of age ($P\leq0.01$) and average daily gain at 15 to 35 days ($P\leq0.05$) and 1 to 35 days ($P\leq0.05$) of age than did the control and BSFP treatments. The average villus heights in the jejunum of the broilers in the BLFP group were higher than those in the BSFP group ($P\leq0.05$). *MUC2* mRNA expression levels in the jejunum and cecum of the broilers in the BLFP group were higher than those in the BSFP group ($P\leq0.05$). Principal coordinate analysis of caecal microbiota and microbial virulence factor gene composition showed distinct clustering among the groups. The abundance of *Lactobacillus crispatus* in the caecal digesta was higher in the BSFP group than those in the C and BLFP groups ($P\leq0.01$). Microbial virulence factor gene analysis showed that the abundance of *clpC* gene in *Bacteroides* species was lower in the BLFP group than in those the C group ($P\leq0.05$), whereas the abundance of *clpC* gene in *Faecalibacterium* species was lower in the BSFP group than those in the C and E groups ($P\leq0.01$). These results demonstrated that BSFP and BLFP treatment exhibit differential effects in improving growth performance, modulating intestinal morphology and gene expression, and regulating caecal microbiota and microbial virulence factor gene composition in broilers.

Session 63 Poster 21

***Lactobacillus ingluviei* C37 improves gut health in lipopolysaccharide challenged broiler chickens**

S. Khempaka, M. Sirisopapong and S. Okrathok
Suranaree University of Technology, School of Animal Technology and Innovation, Nakhon Ratchasima, 30000, Thailand; khampaka@sut.ac.th

Lactobacillus ingluviei C37 (LIC37) is Lactobacillus strains isolated from the caecal content of chickens are safe and possess probiotic properties including tolerance to acid and bile salt, antibacterial activity, adhesion activity, antibiotic resistance and cholesterol removal. However, few studies have examined the probiotic properties of *L. ingluviei* on gut health in chickens. Therefore, this study aimed to investigate the effects of LIC37 on gut health through intestinal barrier gene expression and microbial populations in broilers induced with lipopolysaccharide (LPS). Four treatments consisted of: (1) control (orally administered PBS); (2) probiotic LIC37 10^8 cfu/bird/day administered orally; (3) probiotic LIC37 10^9 cfu/bird/day administered orally; and (4) negative control (orally administered PBS). A total of forty, one-day-old male chicks (Ross 308) were distributed to 4 treatments. At 14 days, treatment 2, 3 and 4 chickens were injected intraperitoneally with *Escherichia coli* O55:B5 LPS (1 mg/kg body weight), and the control was injected with sterile PBS. After 24 hours of LPS injections, the chickens were randomly selected in each group, euthanized by exsanguination and their caecal content was immediately collected for analysis of the caecal microbial population. Intestinal mucosa was collected for intestinal tight junction proteins measurement. It was found that LIC37 can increase the population of LAB and *Bifidobacterium* spp. while reducing *Enterobacteria* spp. and *E. coli* in the caecal content of chickens ($P<0.05$). LPS-challenged LIC37 group showed a significant increase of intestinal TJ proteins and MUC2 mRNA ($P<0.05$). The results indicate that LIC37 can improve gut health during LPS-mediated immunological challenge states.

Session 63 Poster 22

Effects of different probiotics on laying performances, egg quality and gut health in laying hens

F. Barbe[1], L. Blanc[2], A. Sacy[1] and E. Chevaux[1]
[1]Lallemand SAS, 19 rue des briquetiers, 31702, France, [2]Aveyron Labo, 195 rue des artisans, 12031 Rodez cedex, France; echevaux@lallemand.com

Probiotic supplementation in layers is already well documented to improve laying performance, while maintaining optimal egg quality and robust general health. The current field interrogation is the existence of differences between probiotics technologies in a context of long cycle production where the challenge of maintaining good performances and optimal shell quality is sought. The aim of this trial was to benchmark 2 commercial live bacteria probiotics on laying performances, faeces quality, egg quality and general health of old laying hens (71 wks old) against a negative control (NC): *Pediococcus acidilactici I-4622* (PA) at 10^{12} cfu/ton of feed or *Bacillus subtilis C-3102* (BS) at 3×10^{11} cfu/ton of feed (as per the producers' recommended dose). 104 individually housed Lohmann Tradition laying hens were involved in an 8-week trial where the first 2 weeks were set as adaptation and reference period without any supplementation. Overall, the 2 probiotics improved several parameters related to laying performances, egg quality and hen's general health, with higher beneficial effects observed for some parameters with PA: the comparison intra-group (before/after supplementation) showed a persistence of the laying rate for PA (+4.2%; NS) compared to a significant reduction for the BS hens (-4.1%; P=0.013). Eggs weight was significantly improved for both probiotic groups (PA: +2.1%, P<0.001; BS: +1.2%, P<0.01). FCR was not altered with PA whereas significantly degraded with BS (NC: 3.3; PA: 2.6; BS: 3.1. Proportion of defective eggs was significantly lower with probiotics (NC: 7.9%; PA: 1.8%; BS: 3.8%; P<0.001). Concerning general health, faeces scorings (from 1 (firm faeces) to 4 (watery faeces)) were improved with both probiotics (NC: 2.0; PA: 1.7; BS: 1.91; P=0.058), suggesting a positive gut microbiota modulation. This trial also showed that every probiotic does not display the same effect looking at the 'lactobacillus/coliforms' ratio in faeces with a significantly higher ratio for PA (1.459; P=0.032) compared to BS (1.255). In conclusion, probiotics solutions perfectly fit the need of performance maintenance in extended production cycle, however the expected effects may vary according to the probiotic strains.

Session 63 Poster 23

Simple methodology to evaluate faeces quality on farm and effect of *S. boulardii* on faecal scoring

A. Sacy[1], F. Barbé[1], S. Poulain[2], P. Belloir[3], C. Paes[3] and E. Chevaux[1]
[1]Lallemand SAS, 31702, Blagnac, France, [2]Aveyron Labo, 12000, Rodez, France, [3]Ecole d'Ingénieurs de Purpan, 31076, Toulouse, France; echevaux@lallemand.com

Intestinal health is currently widely investigated in poultry: preservation of intestinal barrier is recognized as a major pillar to prevent pathogenic translocation and mature and healthy villi are important components of feed efficiency. However, investigation of intestinal health often required invasive or expensive analyses. Faecal condition can provide an easy insight into how a diet is being digested by the bird and the state of gastrointestinal health. This current study aimed at describing a simple scale to evaluate faeces consistency on farm. A 4-score scale was created: 1: well-formed faeces, firm; 2: softer, more rounded; 3: soft, mushy and losing shape; 4: watery. For the different scores, faeces were analysed for dry matter. There was a significant correlation between the 2 factors (humidity in faeces = 2.97 × faecal scoring + 73.14; n=48; R^2=0.57). Probiotics are an accepted solution to improve intestinal health and faeces quality is often described as a good visible sign of their effects. The live yeast *Saccharomyces boulardii* CNCM I-1079 (SB) was tested in 2 studies for faeces consistency against a negative control (NC). The first study was done with 96 Bovans GoldLine layers. Faeces scoring and dry matter were determined after 14 weeks of supplementation and showed a significant improvement of both criteria: NC: 2.54 vs SB: 2.04 (P<0.001) and NC: 19.01 vs SB: 21.73% (P<0.001), respectively, with firmer faeces for SB group. The absence of time by treatment interaction in the first trial indicates a strong and consistent effect of the probiotic on digestive comfort of the hens, from the first sampling starting 4 weeks after the supplementation. The second trial involved 204 Ross PM3 broilers for 35 days. Faecal dry matter was increased with SB (26.89% vs NC: 25.84%; P<0.05). Based on dry matter, faeces were significantly drier with SB compared to control group at the end of both trials. This effect has potential direct consequences on litter quality and the reduction of pododermatitis risk. Other investigations on birds' welfare and microbiota analysis on litter would be interesting to measure to confirm these benefits.

Session 63 Poster 24

A comparison of yeast gut health products ability to limit attachment of *Salmonella* to IPEC-j2 cells

N. Browne, A. McCormack and K. Horgan
Alltech, Research, Summerhill Rd, Dunboyne, Co. Meath, A86X006, Ireland; nbrowne@alltech.com

Salmonella is a leading cause of foodborne contamination for human consumers. This has raised concerns amongst governing bodies for a means to limit the establishment of *Salmonella* strains such as *S. Heidelberg*, *S. Enteritidis* and *S. Typhimurium* in pigs. *Salmonella* contamination of meat can lead to food spoilage and the risk of cross contamination of these harmful strains into the food chain. The supplementation of a yeast cell wall product composed of a mannan rich fraction (MRF), and other yeast cell wall products were assessed for their ability to interfere with the attachment of *Salmonella* strains to IPEC-j2 cells. Adhesion of *S. Heidelberg*, *S. Enteritidis* and *S. Typhimurium* to the surface of IPEC intestinal cells was carried out at a 500:1 ratio in the presence of Salmonella strains alone or with MRF, Product A, B or C for 1 hour at 37 °C. Unattached bacteria were washed away, IPEC cells were lysed and diluted prior to plating on MacConkey's agar. Bacterial colonies were enumerated after incubating over night at 37 °C. TNFα (PRFI00158), IL-1β (SBR5185) and IL-8 (SBR51590) secretion were measured on cell free media by ELISA. Three independent biological replicates were performed, with One-way ANOVA used to test for statistical significance, *$P \leq 0.05$, unless stated otherwise. Attachment of *S. Heidelberg*, *S. Enteritidis* and *S. Typhimurium* to IPEC-j2 cells was consistently impaired with addition of MRF while product A resulted in a greater attachment of *Salmonella* strains compared to the control. Product B and C demonstrated a reduction in attachment by the three strains of *Salmonella* compared to the positive control though higher than with MRF addition. MRF treatment resulted in lower levels of inflammatory markers IL-1β, IL-8 and TNFα in both *S. Enteritidis* and *S. Heidelberg* challenged intestinal cells compared to the control, and lower IL-8 and TNFα following *S. Typhimurium* challenge. MRF and product B and C impaired the attachment of three *Salmonella* strains to IPEC-j2 cells that led to marked reductions in inflammatory markers. MRF and to a lesser extent products B and C may offer a means to limit the abundance of foodborne strains of *Salmonella* found on pig intestinal cells.

Session 63 Poster 25

Butyric acid and *Bacillus subtilis* 29784 improve cumulatively the intestinal barrier

A. Mellouk, D. Prévéraud, T. Goossens, O. Lemâle, E. Pinloche and J. Consuegra
Adisseo, Research in Nutrition, 8 route noire, 03600 Malicorne, France; amine.mellouk@adisseo.com

We have previously reported the beneficial effects of probiotic *Bacillus subtilis* strain 29784 (BS) on gut integrity and acute inflammation. Butyric acid is a short-chain fatty acid known as primary source of energy for colonocytes and enterocytes that improves gut integrity and maturity. Here we aim to study whether butyric acid and BS have synergistic effects on gut epithelial tight junctions and inflammation under inflammatory stress. Tight junction strength was measured *in vitro* through the transepithelial electrical resistance (TEER) in human derived enterocytes Caco-2 cells. The initial TEER of Caco-2 (753 Ohm/cm^2) significantly increased after a preincubation with sodium butyrate or BS reaching 866 and 1,162 Ohm/cm^2, respectively ($P<0.01$). When exposed to IL-1, IFN-α and IFN-γ proinflammatory cytokines, Caco-2 TEER decreases from 709 to 594 Ohm/cm^2. The preincubation with sodium butyrate (2 mM) or BS (2.5×10^5 cfu/well) were able to restore or enhance their TEER (693 and 1,113 Ohm/cm^2, respectively, $P<0.01$). Interestingly, the simultaneous incubation with sodium butyrate and BS significantly amplified Caco-2 TEER (1,193 Ohm/cm^2 in stressful 1,259 Ohm/cm^2 in normal conditions) showing an additional effect on epithelial cells tight junctions. The Caco-2 treatment with the inflammatory cytokines cocktail induces IL-6 and IL-8 secretion (3.2 and 90 ng/ml respectively). Treatments with butyrate or BS significantly reduced the secretions of IL-6 and IL-8 with a maximal effect in presence of BS (0.3 and 27 ng/ml respectively, $P<0.01$). The anti-inflammatory effects of butyrate and BS was not cumulative. Finally, multiplex ELISA allowed to confirm our result on inflammatory cytokines and to identify other significantly modulated biomarkers by butyric acid and/or BS like IL-18, HSP60 and Fabp3 known to be involved in inflammatory process, stress response and metabolism. In summary, our data show that the probiotic BS29784 and butyric acid cumulatively improve gut integrity by strengthening the epithelial enterocytes tight junctions in normal and stressful conditions. Treatment with either Bs29784 or butyric acid can decrease acute inflammatory factors in the gut and improve the resilience of animals under stressful conditions.

Session 64 Theatre 1

Safety of BSF larvae reared on substrates which are spiked or naturally contaminated with aflatoxins

K. Niermans[1,2], E.F. Hoek- Van Den Hil[1], N. Meijer[1], S.P. Salari[3], M. Gold[4], H.J. Van Der Fels-Klerx[1] and J.J.A. Van Loon[2]

[1]Wageningen Food Safety Research, Akkermaalsbos 2, 6708 WB Wageningen, the Netherlands, [2]Wageningen University, Laboratory of Entomology, Droevendaalsesteeg 1, 6708 PB Wageningen, the Netherlands, [3]InsectoCycle, Bronland 10, 6708 WH Wageningen, the Netherlands, [4]ETH Zurich, Sustainable Food Processing, Schmelzbergstrasse 9, 8092 Zurich, Switzerland; kelly.niermans@wur.nl

Aflatoxins are contaminants of agricultural crops in many regions around the world and innovative aflatoxin management strategies are needed. Previous research showed that larvae of the black soldier fly (BSFL) could provide such an opportunity as exposure to aflatoxin B1 (AFB1) did not affect growth or survival, and accumulation did not occur. In the current study, BSFL were exposed to substrates spiked or naturally contaminated with aflatoxins. Spiking was performed with either regular or isotopically labelled aflatoxin B1 (AFB1) at the maximum limit (ML) concentration set for feed materials by the European Commission (20 µg/kg). Additionally, BSFL were exposed to two types of substrates naturally contaminated with aflatoxins, peanut press cake and maize. Concentrations of the aflatoxins or known metabolites were determined by an LC-MS/MS-based method and were examined in the feed offered, the larvae and residual substrate material. Additionally, for the spiked diet, tracking of the isotopic label via HRMS analyses was performed with the aim to detect novel metabolites. Data obtained for growth, survival, and accumulation were in accordance with previously conducted studies, and, when spiked, no isotopic label was found back in unknown compounds. The molar mass balance of AFB1 revealed a missing fraction of 38% when spiked whereas a full mass balance was obtained for natural contamination. Formation of the AFB1 metabolite AFP1 was observed in all situations. Overall, results obtained from feeding studies in which the substrate was spiked may not be fully representative for when naturally contaminated side-streams are used. However, transfer of AFB1 from the feed substrate into the BSFL is limited/absent, and provides a promising outlook for the safety of rearing BSFL with the purpose of further using them as feed.

Session 64 Theatre 2

Transfer of aflatoxin, lead and cadmium from larvae reared on contaminated substrate to laying hens

M. Heuel[1,2], M. Kreuzer[2], I.D.M. Gangnat[2,3], E. Frossard[2], C. Zurbrügg[4], J. Egger[4], B. Dortmans[4], M. Gold[4,5], A. Mathys[5], J. Jaster-Keller[6], S. Weigel[6], C. Sandrock[7] and M. Terranova[8]

[1]Agridea, Eschikon 28, 8315 Lindau, Switzerland, [2]ETH Zurich, Institute of Agricultural Sciences, Eschikon 27, 8315 Lindau, Switzerland, [3]HAFL, Länggasse 85, 3052 Zollikofen, Switzerland, [4]Eawag, Department Sanitation, Water and Solid Waste for Development, Überlandstr. 133, 8600 Dübendorf, Switzerland, [5]ETH Zurich, Laboratory of Sustainable Food Processing, Schmelzbergstr. 9, 8092 Zürich, Switzerland, [6]German Federal Institute for Risk Assessment BfR, Max-Dohrn-Str. 8-10, 10589 Berlin, Germany, [7]FiBL, Livestock Sciences, Ackerstr. 113, 5070 Frick, Switzerland, [8]ETH Zurich, AgroVet-Strickhof, Eschikon 27, 8315 Lindau, Switzerland; melissa-terranova@ethz.ch

The use of low-grade substrates can improve the sustainability of insect-based feed production but also poses food safety risks. These include mycotoxins and heavy metals that may be present in substrates for insects. Thereby they might pass the entire production chain and lead to contaminated foods. We studied the transfer of three contaminants to black soldier fly larvae (BSFL) as well as eggs and poultry meat. Four poultry diets were formulated including four partially defatted BSFL meals (200 g/kg diet) produced at two different facilities. In Indonesia, BSFL were reared on not EU-approved meat-containing food waste, either non-spiked or spiked with environmentally relevant concentrations of Cd (1.9 mg/kg) and Pb (19 mg/kg) or aflatoxin B1 (1.5 mg/kg). As an additional control, in Switzerland, BSFL were reared on EU-approved substrates. Nine late-laying hens per treatment were fed the experimental diets for 4 weeks. Only the diet including BSFL reared on Cd contaminated substrate exceeded the EU-threshold for Cd for complete feed (1.7 vs 0.5 mg/kg). No diet affected laying performance or egg quality. Feeding the heavy-metal contaminated diet doubled Cd concentrations in breast meat and elevated Cd concentrations in kidneys and liver compared to the control. However, all eggs, meat and tissues (except kidneys) ranged below permitted limits for food. Our results show that, under certain conditions, even contaminated material can provide a suitable substrate to produce BSFL for use as feeds for poultry.

Session 64 Theatre 3

Determination of minimal nutrient requirements of *Tenebrio molitor* larvae

B. Tamim[1,2], T. Parr[1], J. Brameld[1] and A. Salter[1]
[1]University of Nottingham, Sutton Bonington, School of Biosciences, LE12 5RD, United Kingdom, [2]King Abdulaziz University, Faculty of Science, Biochemistry Department, 21589, Saudi Arabia; alybt1@nottingham.ac.uk

Tenebrio molitor larvae (yellow mealworms, MW) are often fed on the milling by-product, wheat bran (WB). However, some studies have suggested that they may be able to grow on waste materials, such as polystyrene. The aim of the study was to determine the minimal nutrient requirements of MW. There were six diet groups for the 24-days-trial: 100% wheat bran (WB); 85% cellulose, 15% casein (C+P); 85% cellulose, 15% casein, minerals/vitamins premix (C+P+MV); 85% cellulose, 15% casein, 5% fatty acid (C+P+FA); 85% cellulose, 15% casein, minerals/vitamins premix, 5% fatty acid (C+P+MV+FA); 70% cellulose, 15% casein, 15% glucose, minerals/vitamins premix, 5% fatty acid (C+P+G+MV+FA). MW were divided into 4 replicate containers per diet group, with 250 MW per container. MW were housed at 27 °C and 60% humidity and food replenished every 10 days. Dead or pupated MW were removed and weights were determined every 3 to 4 days. Data were analysed by 2- or 1-way ANOVA (Genstat 21st edition). There was a significant time × diet interaction for average individual MW body weight ($P<0.001$). Overall, MW fed WB achieved the greatest final body weight, while those fed diets not supplemented with MV failed to grow. Of the remaining groups, those with all of the added macronutrients (C+P+G+MV+FA) showed the highest final body weight, though this was still 17% less than those fed WB ($P<0.05$). However, mealworms fed WB showed the highest level of pupation ($P<0.001$), with 59% pupating by the end of the trial (compared to 30% in the C+P+G+MV+FA group). As a result, when considered on the basis of the weight of live MW remaining/ plate, the highest values were actually seen in the C+P+G+MV+FA group, with the total weight being 61% higher than in WB fed larvae. The data shows that MW have an absolute requirement of, at least some, vitamins and minerals. It is also shown that growth can be enhanced by the addition of lipids and carbohydrates. However, larvae fed on the 'complete' synthetic diet still do not pupate at the same rate as those fed on WB. The reasons for this remain to be established. These results cast doubt on whether MW can be reared on minimal substrates, such as polystyrene.

Session 64 Theatre 4

FIAgship demonstration of industrial scale production of nutrient resources from mealworms

W. Kihanguila
Ynsect, 207 Rue de Bercy, 75012 Paris, France; whitley.kihanguila@ynsect.com

The FARMYNG project, co-funded within the Horizon 2020 Public-Private Partnership Bio-Based Industries Joint Undertaking framework(topic BBI.2018.F2 – 'Large-scale production of proteins for food and feed applications from alternative, sustainable sources') aims to establish the largest global fully-automated flagship industrial plant to produce premium proteins from insects (*Tenebrio molitor*) for animal nutrition. The world faces a major challenge with the sharp increase in demand for meat and fish by 2050, to which current modes of production, whether through agriculture, aquaculture or fisheries, cannot be sustained. An innovative solution to this predicted shortage of resources, especially protein, lies in insect production and processing which can provide a serious response. However, up to date, no industrial scale production of insect-derived products has been realized. Coordinated by Ynsect, the FARMŸNG project has the ambition to develop, on an industrial and automated scale, the breeding and transformation of insects for the production of ingredients, with the strong participation of 19 key actors all along the value chain, from the feedstock supply to the final insect transformation. The FARMYNG project will demonstrate a large-scale, first-of-its-kind bio-based value chain producing sustainable, safe and premium feed products from an innovative origin: the *T. molitor* insect (mealworm). The plant developed for FARMYNG will exploit the physiological capabilities of mealworms physiology to efficiently convert vegetal by-products in mealworm biomass and will transform those mealworms into sustainable proteins and lipids for fish feed and pet food end markets. In parallel, manure will be recovered for soil fertilization applications. The project, responding to the increasing worldwide proteins demand, will transfer the technology from a demo plant to the Industrial Flagship Plant able to produce almost 1,500 tons of proteins and 400 tons of oil per month, reaching production rate that has never been achieved by any other insect-protein production plant in the world.

Session 64 Theatre 5

Tailored vitamins and minerals premix for *Tenebrio molitor* farming

E. Barbier[1], V. Gerfault[1], T. Lefebvre[2], F. Peyrichou[2], C. Ricciardi[2] and N. Tanrattana[1]
[1]MG2MIX, R&D, ZA La Basse Haye, 35220 Chateaubourg, France, [2]YNSECT, R&D Biotech, 1 Rue Pierre Fontaine, 91000 Évry-Courcouronnes, France; n.tanrattana@mg2mix.fr

Premixes are complex mixtures of vitamins, minerals, and trace elements incorporated at low levels in the compound feed. These mandatory micro-nutrients permit to meet the nutritional needs of animals, and to improve the efficiency of a diet. A deficit in even one vitamin can lead to decrease the rearing performances in the main animal productions. The literature about the vitamins and minerals requirements in *Tenebrio molitor* (TM) is limited but existing which is notable for a farmed insect. It dates from 50's-70's, mainly with the pioneer works of Fraenkel and Leclercq, then has not been investigated further since. The objective of this study is therefore to consolidate the understanding of the TM's micronutritional requirements, to then formulate a TM-specific premixes as a supplement of its diet. In that way, an experimental design was performed based on 10 premixes of different levels of vitamins, trace-elements and minerals, and regarding the optimization of both physiological and economical performances of TM production. This study focused on the larval stage in the exponential phase of the growth curve. In order to do this, the micronutrients were grouped. The three investigated groups were the following: Core group (consisting of graded levels of B-group vitamins, considered mandatory), Accessory Vitamins (A, D3, E, K3), and Trace-elements (S, Cu, Mn, Se, I). The results showed that the Core group composition has a significative effect on the feed conversion ratio, whereas none among the Accessory Vitamins and the Trace elements groups showed a significant effect on the rearing performances. Besides, high contents of Mg seem to have a significative detrimental effect on the rearing performances. This is in contradiction with studies on other species where the inclusion of Mg increased animal performances by improving the anti-oxidant status linked to the catalase activity. Thanks to these key information, Ynsect and MG2MIX have thus succeeded in designing a new tailored premix for mealworm farming, increasing the performances of conversion by 13%, and finally achieving a world premiere in the industry of insects for feed and food.

Session 64 Theatre 6

***Tenebrio molitor* genomics revealed limited molecular diversity among available populations**

L. Panunzi[1], E. Eleftheriou[2], B. Vacherie[3], K. Labadie[3], T. Lefebvre[4] and M.A. Madoui[1]
[1]), Institut François Jacob, Commissariat à l'Energie Atomique et aux Energies Alternatives (CEA), 18 Rte du Panorama, 92260 Fontenay aux Roses, France, [2]Génomique Métabolique, Genoscope, Institut François Jacob, Commissariat à l'Energie Atomique (CEA), 2 rue Gaston Crémieux, 91000 Evry, France, [3]2Genoscope, Institut de biologie François Jacob, CEA, Université Paris-Saclay, 2 rue Gaston Crémieux, 91000 Evry, France, [4]Ynsect, 1 Rue Pierre Fontaine, 91000 Evry, France; mohammedmain.madoui@cea.fr

Tenebrio molitor is now a promising species for insect farming to produce high protein products with low environmental impact. However, the molecular diversity and the genetic structure in available *T. molitor* populations are understudied. To identify possible sources of molecular diversity that may help future breeding programs, we analysed the genomic structure of 18 populations from 12 countries using Illumina pool-seq data. More than 4 millions SNPs were selected to compute allele frequencies, then multivariate and pairwise F_{ST} analyses were used to estimate the genetic distances between the 18 *T. molitor* populations. Our results showed low to very low genetic distance between 16 *T. molitor* populations. Two very distant populations, one wild population sampled in Serbia and one from Germany presented very distinct genomic structure. These results suggest that more populations should be sampled to increase the molecular diversity that may help future breeding programs. A specific attention should be made on the sampling of wild populations that have more chance to present higher molecular diversity than laboratory or producer's populations. Also, further investigations will use the data generated by the current study to focus on the detection of loci under natural selection.

Session 64 Theatre 7

Assessing the quality of insect-derived products: methods and findings from the FARMYNG project

S. Gofflot[1], A. Pissard[1], F. Debode[1], A.C. Laplaize[2], B. Lorrette[3] and J.F. Morin[2]
[1]Walloon Agricultural Research Centre (CRA-W), chaussée de Namur, 24, 5030 Gembloux, Belgium, [2]Eurofins Food & Feed France (EAF), Rue Pierre Adolphe Bobierre, 44300 Nantes, France, [3]Ynsect, 1 Rue Pierre Fontaine, 91000 Évry-Courcouronnes, France; jeanfrancois.morin@ftfr.eurofins.com

Insects can serve as an alternative source of protein in the current economic and environmental context of European livestock. Several insect rearing industries have emerged in the EU to produce insect meals. The FARMYNG project (BBI-H2020, 2019-2024) aims to develop methods for quality testing of insect-derived products. Quality testing is critical for assessing the insect meals' nutritional value, and it's important to have tools for rapid characterization of insects or insect meals. The primary benefit of insect meals is their high protein content, but caution is necessary when estimating this parameter because the exoskeleton of insects is made of chitin, which contains nitrogen that could be erroneously included in the protein content estimation. Within FARMYNG, two alternative methods to the one used by the insect company for chitin determination were tested, the ADF-ADL and cellulose methods. It was concluded that the Ynsect method, which is more a lab-scale chitin extraction, is likely the most accurate method for chitin determination. However, it is time-consuming (4-5 days) and must be used for precise chitin determination. Alternative methods such as ADF-ADL (2-3 days) or the Cellulose method (1.5 day) could be used for a rapid estimation of chitin content. The analysis time can be significantly reduced with the development of Near Infra-Red (NIR) approaches and appropriate models. NIR also has the potential to predict other parameters related to the quality of insect-derived products. Two approaches were followed: the 'specific' approach, which was only based on insect products collected during the FarmYng project, and the 'global' approach, which included other animal proteins in the database to increase variability in the values. Both approaches showed similar performance. It can be concluded that NIR coupled methods are a valuable tool for predicting humidity, protein, fat, cellulose, and chitin content with relatively low prediction errors.

Session 64 Theatre 8

Authentication of insect-derived products: methods and findings from the FARMYNG project

B. Dubois[1], A. Marien[1], S. Guillet[2], J.-F. Morin[2], B. Lorrette[3] and F. Debode[1]
[1]CRA-W, chaussée de Charleroi, 234, 5030 Gembloux, Belgium, [2]EAF, rue P.A. Bobierre-BP42301, 44323 Nantes, France, [3]Ynsect, rue Pierre Fontaine, 1, 91000 Evry, France; f.debode@cra.wallonie.be

Eight insect species are currently authorized by regulation in the European Union as feed for aquaculture, pigs, and poultry. Authenticating insect species is necessary to protect producers and users. When aiming at insect species identification, a morphological examination is not adapted as insects are processed into meals. Genomic methods were then considered. Real-time PCR has proven to be effective in detecting insect species, and methods have been developed and published for *Tenebrio molitor*, *Hermetia illucens*, and *Alphitobius diaperinus*. Methods for *Acheta domesticus*, *Gryllus assimilis*, and *Bombyx mori* are also ready. The method validation is based on performance criteria recommended in international guidelines as specificity, sensitivity, applicability and robustness. It was observed that some commercially available insect species were mislabelled, as *A. laevigatus* instead of *A. diaperinus*. However, real-time PCR is limited to detecting only the targeted species and does not allow for an untargeted approach to identify a mixture of different insect species with potentially contaminating insect species. To address this, we used metabarcoding approaches based on high-throughput sequencing. To select the most appropriate barcodes for this purpose, we designed, compared and tested 34 barcodes based on 12S, 16S, COI, COII, CytB, 18S, and 28S on mock communities, which are artificial samples composed of DNA from multiple species mixed in known proportions, here consisting of 10 insect species. Qualitatively, several primer pairs were found to be effective in detecting all insect species. However, in terms of quantitative analysis, significant differences were observed among the different targets, and none of them could accurately detect proportions of 10% for every species. To analyse the insect composition of various samples, a QIIME2 bioinformatics pipeline was also developed. Multiple datasets were processed using this pipeline to validate its efficiency. The bioinformatics treatment was adapted to handle information from both classical Illumina and Oxford Nanopore technologies.

Session 64 Theatre 9

More than an organic fertilizer: mealworm frass as a substitute to conventional fertilizers

E. Bohuon[1], D. Houben[2], G. Daoulas[1], M.-P. Faucon[2] and A.-M. Dulaurent[2]
[1]Ÿnsect, 1 rue Pierre Fontaine, 91000, France, [2]UniLaSalle, AGHYLE, 19 rue Pierre Waguet, 60026, France; emilien.bohuon@ynsect.com

With the rapid growth of the insect production sector, frass (insect excreta) has the ability and capacity to reduce our reliance on conventional fertilizers and improve our sustainability. The main objectives of this study were to investigate mealworm (*Tenebrio molitor* L.) frass (MF) potential to sustain crops needs, improve soil life/quality and reduce our environmental footprint. Its well-balanced nutrient composition (NPK ratio 4-3-2) and its quick mineralization makes MF suitable to be applied on any kind of crop and can be as effective as a conventional fertilizer. Analysis of the MF surface also revealed an even distribution of P, K and Ca within the organic matter which implies an homogeneous soil nutrient distribution after mealworm frass application. These findings suggest a very good MF potential to be used as substitute to conventional fertilizers. Unlike conventional fertilizers, a high organic matter content in MF (around 80%) and composition has the ability to increase soil microbial activity and maintain soil functional diversity. Studies also revealed a synergetic effect between MF and earthworms[3] which improves plants capacity to uptake MF nutrients when earthworms are present. Moreover, after MF application, water soluble P is five times lower compared to conventional fertilizers which prevent P loss and sorption improving the P use efficiency. Thus, MF seems to be a better asset to maintain or even increase soil quality and micro/macro organisms biodiversity. Given its capacity to capture carbon into the soil and to reduce agriculture environmental footprint, MF is promising fertilizer. Life cycle assessment revealed that the use of MF as a substitute to conventional fertilizer can significantly reduce crop production environmental footprint (Internal Ÿnsect source). More than an organic fertilizer, MF has a great potential to stimulate soil life, reduce the use of conventional fertilizer and improve environmental footprint. Considering the exponential or rapid growth of the insect industry, frass will become a commonly used organic fertilizer which has a great potential to ensure a more sustainable agriculture.

Session 64 Theatre 10

The environmental life cycle assessment of Ynsect insect based proteins

K. Hsu[1], A. Eiperle[1] and M. Jouy[2]
[1]Quantis, Badenerstrasse 141, 8004 Zürich, Switzerland, [2]Ynsect, 207, rue de Bercy, 75012 Paris, France; karl.hsu@quantis-intl.com

In order to investigate the environmental impacts caused by Ynsect's insect-based protein products, an environmental Life Cycle Assessment (eLCA) will be performed. The objectives of this eLCA is to show the environmental impact across Ynsect's products (main products and by-products) using a number of indicators, such as climate change, land use, water use, and fossil resource use, over the entire value chain for the reference year 2027 due to the production plant currently being constructed. Data was collected for their products (YnMeal, YnFrass, and three by-products) in order to calculate their eLCA. The environmental impact (for all indicators) is expected to be concentrated on the main products YnMeal and YnFrass compared to the by-products, as an economic allocation approach will be used. Further work post eLCA will be carried out in order to compare the impact of using Ynsect products within fertilisers, pet feed, and aqua feed as a replacement to conventional ingredients.

Session 64 Theatre 11

Insects' nutrients – the Animal Frontiers special issue

T. Veldkamp[1] and L. Gasco[2]
[1]Wageningen University & Research, Wageningen Livestock Research, De Elst 1, 6700 AH Wageningen, the Netherlands, [2]University of Turin, Department of Agricultural, Forest and Food Sciences, Largo P. Braccini 2, 10095 Grugliasco, Italy; teun.veldkamp@wur.nl

With a growing world population and rising prosperity, the demand for animal-derived product is increasing. To meet this increasing demand, the global feed production has to increase, and new protein sources are being sought worldwide. These proteins must have a good nutritional value and must be produced sustainably. Insects can be part of the solution and the last decade has seen a growing interest in using insects as a sustainable and nutritious source of raw material for animal feed. Insects are high in protein and other essential nutrients and are part of the natural diet of many animal species. In perfect agreement with the circular economy principles, insects can be grown on organic waste, reducing the environmental impact of food waste and converting residual products into high-quality nutrients. The main aim of the special issue of Animal Frontiers is to provide an overview of the development of insects in the global food chain. This special issue tackle different aspects. How these insects should be reared and fed and which is the status of the legislation in the different countries. How to processed insects into protein and lipids and what is the impact of the use of these insect-derived products in animal feed? How to guarantee the safety in the food chain and how sustainable it is to include insects as an important link in the food chain.

Session 65 Theatre 1

Time-course change in lamb composition and reflectance properties: implications for authentication

L. Rey-Cadilhac[1], D. Andueza[1], A. Prunier[2] and S. Prache[1]
[1]UCA, INRAE, VetAgro Sup, UMR Herbivores, 63122 St-Genès-Champanelle, France, [2]INRAE, Institut Agro, UMR PEGASE, 35590 St-Gilles, France; lucille.rey-cadilhac@inrae.fr

Differences in the chemical composition of meat from lambs fed contrasted diets, for example pasture vs a concentrate-based diet indoors, are well known. They are the basis of authentication methods to discriminate between pasture-fed and concentrate-fed lamb meat. However, the question of the timing of appearance of markers of pasture-feeding in lamb meat relative to the start of pasture-feeding is not known. The objective of the present study was therefore to study the kinetics of change in lamb meat composition and to determine whether it was possible to discriminate between different grazing durations in order to authenticate lamb meat production conditions. Four groups of 55 Romane lambs were used: L0, concentrate-fed lambs, no grazing; L21, L42 and L63, with lambs grazing alfalfa for 21, 42 and 63 days before slaughter, respectively. At slaughter, measurements of the commercial and sensory qualities of the carcass and meat were performed, as well as spectral analysis of two fat tissues (perirenal and dorsal fat, PF and DF) and of the longissimus dorsi (LD) muscle using visible and infrared spectroscopy. Regarding authentication purposes: (1) PF skatole concentration (skatole is a compound both responsible for off-flavour and proposed as a marker of grass feeding in some publications) did not allow for a sufficiently reliable classification of pasture-fed lambs vs concentrate-fed lambs; (2) the combination of PF skatole concentration and an indicator of PF carotenoid concentration, using a decision tree, enabled to correctly discriminate L0 lambs from the pasture-fed (L21, L42, L63) lambs, but not to authenticate the duration of pasture-finishing; (3) discriminant analysis on the visible spectrum of PF or DF also allowed to discriminate L0 lambs from the pasture-fed (L21, L42, L63) lambs, but not to authenticate the duration of pasture-finishing. The LD spectral data did not allow to discriminate L0 lambs from L21 lambs and L42 lambs from L63 lambs, but they enabled to reliably discriminate L0 and L21 lambs from L42 and L63 lambs. The perspectives are to expand the spectral range explored using infrared spectroscopy.

Session 65 Theatre 2

Individual adaptive responses of meat ewes facing an abrupt nutritional challenge after lambing

E. González-García[1], M. Gindri[2], L. Puillet[2] and N.C. Friggens[2]
[1]INRAE, PHASE, SELMET, Univ Montpellier, CIRAD, INRAE, L'Institut Agro Montpellier SupAgro, 34060 Montpellier, France, [2]INRAE, PHASE, Université Paris-Saclay, INRAE, AgroParisTech, UMR MoSAR, 75005, Paris, France; eliel.gonzalez-garcia@inrae.fr

Simulating a climate change event, responses of Mediterranean meat ewes when facing an abrupt nutritional challenge (NC; i.e. fed with cereal straw of very low nutritional value only) were studied at a very sensitive physiological stage (i.e. just after lambing). Forty Romane ewes were chosen at early-mid pregnancy (around 2 mo) according to parity (20 primiparous, PRIM; 20 multiparous, MULT); feed efficiency genetic line [residual feed intake (RFI); inefficient, RFI-, n= 10 per parity; efficient, RFI+, n= 10 per parity); litter size (i.e. bearing twins, diagnosed by ultrasonography); and BW and body condition score (BCS) [initial BW and BCS (mean ±SD): 51.6±7.41 kg; 2.5±0.20, respectively; representing average BW and BCS of their parity in the flock]. Effects on intake, ewes' BW and BCS, subcutaneous back-fat thickness (BFT), energy metabolism [plasma NEFA, β-OHB, glucose, urea, tri-iodothyronine (T3)], and lambs' growth were examined before, during and after NC. Individuals' profiles of the response-recovery of each ewe to NC were described using a piecewise mixed-effects model and clustered using principal components analysis and Euclidean distance. MULT presented sharper β-OHB recovery from NC than PRIM ($P \leq 0.05$). Parity or genetic line did not affect the other evaluated traits. Clusters of individuals' response-recovery to NC suggested three different adaptive strategies to NC (i.e. adaptation on acquisition, allocation or trade-off between acquisition and allocation of energy. Interestingly, ewes' response-recovery to NC demonstrated also to be related to lamb average daily gain (ADG, g/d), especially plasma β-OHB and NEFA ($r \geq 0.50$). Results provide new insights in how such short and abrupt NC affect some key physiological parameters, and to what extent the impacts of NC and the ewes' potential response-recovery are influenced by the individual nature of the animals (i.e. observed inter-individual differences in the responses).
This work was financed by the PRIMA ADAPTHERD project (https://www.adapt-herd.eu/).

Session 65 Theatre 3

Large variation in emission intensities from dual-purpose sheep production system

B.A. Åby, S. Samsonstuen and L. Aass
NMBU, Department of Animal and Aquacultural Sciences, Box 5003, 1432, Ås, Norway; bente.aby@nmbu.no

Sheep currently account for approx. 4% of global GHG emissions from livestock and the demand for sheep products is likely to increase due to human population growth. Thus, to limit GHG emissions from sheep production, it is essential to reduce emission intensities, i.e. GHG emissions from producing one unit of product, e.g. sheep and lamb carcass and wool. Due to large variability in resource base, management practices, and production between individual farms, emission intensities must be calculated at farm level and farm-specific mitigation options need to be implemented. Whole-farm models take into account interactions and trade-offs between different GHG sources; therefore, these models are powerful tools to estimate total farm GHG emissions, assess mitigation options at farm level, and tailor mitigation strategies to specific farms or production systems. HolosNorSheep is such a whole-farm model, based on IPCC methodology, modified to Norwegian conditions. The model estimate GHG emissions from dual-purpose sheep meat and wool production and considers direct CH_4 from enteric fermentation and manure management, direct and indirect N_2O from manure management and soils, and CO_2 emissions from production and use of farm inputs. Emissions are allocated to meat and wool using a biophysical allocation on the basis of net energy requirements for growth and wool production. In this study, emission intensities for sheep and lamb carcass and wool were estimated for 38 farms distributed across Norway with varying climate, natural resource base, feeding and management practices. The data were from year 2019-2021 and several farms had data for more than one year, giving a total of 65 calculations. Animal performances varied considerably across farms and years, e.g. average number of weaned lambs per adult ewe varied from 1.4 to 3.0 while the carcass weight of lambs ranged from 14.5 to 27 kg. Average emission intensity per kg sheep and lamb carcass, and wool was 20.4 (±5.0) and 22.2 (±7.2) kg CO_2-eq, respectively. The most important emission source for all farms was enteric methane. The large variation in calculated emission intensities demonstrates the potential to reduce emission intensities from Norwegian dual-purpose sheep production.

Session 65 Theatre 4

Considering the morphology of cows for comfortable milking

J. Fazilleau[1], S. Guiocheau[2] and J.L. Poulet[1]
[1]Institut de l'Elevage, 149 rue de Bercy, 75595 Paris Cedex, France, [2]Chambre Régionale d'Agriculture de Bretagne, Aéroport, 29600 Morlaix, France; jean-louis.poulet@idele.fr

ErgoTraite (CASDAR IP project 2021-2023) aims to ensure the sustainability of conventional bovine milking. This requires an overall improvement of milking, by proposing innovative organization models and milking equipment designs. Within this project, measurements were carried out in 7 milk farms in the West of France, with diversified profiles, to verify the adequacy of the recommendations for the design of milking parlours for current herds. 726 udders were characterized, with 2 main dimensions: Udder Floor Height (UFH: vertical distance between highest teat attach on the udder and the platform) and Horizontal Reach Distance (HRD: horizontal distance between the farthest teat and the platform border). Platforms Heights (PH: vertical distance between milker floor and cow plateform) had been also measured. Vertical Reach Distance (VRD = PH + UFH) has been calculated by combining 2 previous indicators. Main anatomical characteristics of milkers were also recorded: Milker Height (MH), Elbows (EH) and Shoulders Heights (SH) and length of arms (AL). Indeed, these dimensions make it possible to qualify the areas of intervention for the milkers (adapted from INRS, 1999). 'Comfort zones' are in terms of vertical movements, at EH<VRD<SH, and horizontally, at HRD<2/3 AL. From VRD>EH+10 cm or HRD>AL, it is considered that intervention is done in a 'painful zone'. In average UHF being at 63.8 cm, VRD were thus quite significant, with extreme teats over 1.85 m, especially in parallel parlours, even for tall milkers (1.80 m) and HRD were at 41.8 cm. By comparing these data with the recommendations, teats can be considered in uncomfortable areas for 88% of VRD and 86% of HRD. HRD is difficult to modify, unless if the stall is adaptable to each animal. It seems essential to consider UFH when dimensioning milking parlours (in addition to milker height, often already considered), as also reported by Cockburn & Al. (2015), current platform height recommendations being no longer adapted to cow morphologies. By doing so, rear stall elements (located at 70 cm from the platform to prevent cow from falling) must be modified to be sure they won't become obstacles to udders visibility and accessibility.

Session 65 Theatre 5

Classification of honeybee flight activity patterns reveals impact of recruitment behaviours

G.E.L. Petersen[1,2], D. Gupta[2], P.F. Fennessy[2] and P.K. Dearden[1,3]
[1]FutureBees NZ Ltd, 442 Moray Place, 9016 Dunedin, New Zealand, [2]AbacusBio Ltd, 442 Moray Place, 9016 Dunedin, New Zealand, [3]University of Otago, Biochemistry, 710 Cumberland Street, 9016 Dunedin, New Zealand; gpetersen@abacusbio.co.nz

Remote sensing technologies are likely to greatly improve our understanding of the livestock species we rely on for the production of food and fibre for human consumption. From walk-over weighing to wearable motion sensors, new technologies are constantly being developed. For the European Honeybee, *Apis mellifera*, numerous hive telemetry systems are available today and can support the decryption of bee behaviours and their impact on hive productivity. Stationary hive telemetry can lend great insight into the flight activity, production and health of honeybee colonies, especially where observations are continuous. Here, we monitored IN activity (No of bees entering the hive over a 10-minute interval), OUT activity (No of bees leaving the hive) and total hive weight over the 3-month nectar flow period on 140 beehives in the Canterbury region of New Zealand. Daily activity patterns were analysed by checking for the presence of local maxima indicating that a spike in flight activity occurred, and then classified into predominantly early activity, predominantly late activity, and two-phase activity, indicating that foraging efforts had been reduced over a mid-day period. Classification results showed that while daily changes in environmental conditions had the biggest impact on flight activity patterns, individual honeybee colonies that were found to express an 'early' or 'late' pattern were more likely to show this pattern again at a later time. Two-phased activity patterns were more common during times that resulted in higher hive weight gain, indicating that the mid-day break present in the activity data is the result of recruitment behaviours between food handler bees inside the hive and returning foragers. The high influx of nectar and pollen results in higher demand for worker bees to move food deep into the hive and make room for a second wave of returning foragers in the afternoon. Colonies with a higher proportion of days with a two-phased activity patterns were found to be more successful in terms of weight gain over the nectar flow period.

Session 65 — Theatre 6

Implementation of large-scale climate smart agriculture research on U.S. beef cattle grazing lands

K. Cammack[1], A. Blair[1], L.O. Tedeschi[2], H.M. Menendez Iii[1] and J.R. Brennan[1]
[1]South Dakota State University, Animal Science, 711 N. Creek Drive, 57703, USA, [2]Texas A&M University, Animal Science, 474 Olsen Blvd, College Station, Texas, 77843, USA; kristi.cammack@sdstate.edu

Grazing lands are globally crucial for their greenhouse gas (GHG) mitigation and sequestration potential. However, livestock producers are often neglected in enteric GHG reduction and carbon sequestration incentives because cost-effective methods to determine and measure carbon and GHG sinks and sources are lacking. Emerging technologies have bridged that gap, paving the way to develop, implement, and monitor climate-smart agricultural (CSA) practices. Successful implementation and monitoring of CSA practices have the potential to translate to climate-smart livestock commodities that sustainably assign economic incentives to resilient livestock grazing land production systems. Nearly 380 million ha in the U.S. are rangelands dedicated to supporting forage-based livestock production. This land area is more than double that dedicated to row crops and has enormous potential to contribute to U.S. climate change commitments. Recently, South Dakota State University, with 13 other partners, was awarded a CSA commodity development grant to develop markets for Climate Smart Beef in the U.S. Northern Great Plains (NGP), one-fifth of U.S. beef production, through the USDA-NRCS. To achieve this, the SDSU team utilizes an innovative measuring, monitoring, reporting and verification (MMRV) method to evaluate the effectiveness of climate-smart agriculture practices such as prescribed grazing and planting native grasses. This project will integrate precision technologies and data analytics for a large-scale sampling (971,245 ha) of grazing lands. Key metrics include enteric methane emissions and soil carbon sequestration for cow-calf and stocker production phases. The data collected in our project will be utilized to further enhance the COMET-Farm model, a whole ranch carbon and greenhouse gas accounting tool, by improving or validating coefficients regarding practice emission or mitigation potentials across grazing management styles, soil types, and climates; a critical next step in enhancing global sustainability of livestock farming systems.

Session 65 — Theatre 7

Remote monitoring behaviour of bison in captivity: effects of gender, weather and daytime

R.R. Vicentini[1,2], D. Moya[3], J. Church[2], A.C. Sant'anna[4], W. Balan[1], W. Squair[1] and Y.R. Montaholi[1]
[1]Lakeland College, School of Agricultural Sciences, 5707 College Drive, Vermilion, AB, T9X 1K5, Canada, [2]Thompson Rivers University, 805 Tru Way, Kamloops, BC, V2C 0C8, Canada, [3]The University of Saskatchewan, 52 Campus Dr, Saskatoon, SK, S7N 5B4, Canada, [4]Federal University of Juiz de Fora, Rua José Lourenço Kelmer, Juiz de Fora, MG, 36036-900, Brazil; yuri.montanholi@lakelandcollege.ca

Studying bison in zoo can provide valuable insights about the behaviours of this species with application in the monitoring and management of commercial herds. Besides this biological characterization, we hypothesize that gender, weather parameters, and daytime periods may affect the behaviours of the bison. The objective of this research was to employ remote video surveillance to identify and quantify the behaviours of bison. Four bisons (*Bison bison*; female: 2; male: 2), provided by Zoo Aquarium de Madrid (Madrid, Spain), were video monitored during 26 consecutive days. Three solar-powered cameras with motion sensitive tracking (Reolink Argus PT, Reolink Digital Technology Co. Ltd., Hong Kong, China) were installed surrounding the enclosure of the bisons. The footage collected was computed considering five behaviours: laying, walking, standing, drinking and eating. The behaviours were assessed and quantified by a judge as percentage of observation time behaviour and used to perform the means analysis. Preliminary results indicate that the most frequently observed behaviour of the herd was laying (65.8%), followed by standing (16.2%), moving (7.7%), eating (9.4%) and drinking (0.7%). The analyses of the behaviours by gender indicated that males spent more time laying (21.72%, 15.87%) and less time standing (3.57%, 5.73%) than females. The time spent walking (2.04%, 2.40%), eating (2.79%, 2.60%) and drinking (0.31%, 0.09%) were similar in both sexes. Potential effects of weather parameters (i.e. temperature, humidity, sunlight duration, etc.) and daytime periods remain to be evaluated. Further analysis may also evidence the existence of circadian cycle related to behaviours evaluated. The parametrization in development will be of great relevance to monitor bison behaviour in confined spaces, such as observed in commercial feedlots, which will assist optimize the husbandry practices.

Session 65 Theatre 8

Artificial intelligence for measuring the respiration rate in dairy cows

L. Dißmann[1], R. Antia[2], L. Chinthakayala[3], N. Landwehr[3], T. Amon[1] and G. Hoffmann[1]
[1]Leibniz Institute for Agricultural Engineering and Bioeconomy (ATB), Department Sensors and Modeling, Max-Eyth-Allee 100, 14469 Potsdam, Germany, Germany, [2]dida Datenschmiede GmbH, Hauptstraße 8, 10827 Berlin, Germany, [3]University of Hildesheim Foundation, Data Science, Universitätsplatz 1, 31141 Hildesheim, Germany; ldissmann@atb-potsdam.de

The respiration rate (RR) of dairy cows is a very sensitive parameter that can indicate a stress situation (e.g. heat stress, anxiety) as well as pathological processes at an early stage. The common method for determining the RR of cattle is to visually count the flank movements in breaths per minute. However, this method is very time consuming and the close contact with the animal affects the cow's behaviour. Therefore, our approach to record the RR is based on a non-contact method. The novelty here is the use of two imaging techniques. On the one hand, we used infrared thermography (FLIR AX5 series, Oregon, USA) to record the RR in the area of the nostrils based on the temperature difference between inhaled and exhaled air. First, the artificial Intelligence (AI) detected the area of interest (AOI) and extracted afterwards from the average temperature in the AOI a time series from which we derived the RR. On the other hand, we used a depth camera (IntelRealsense D455, California, USA) to record the respiration based on the flank movement of the cow. Here, the AI extracted a RR curve from the maximum expansion difference from the flank area. The gold standard for both methods was the visual counting of the RR due to the flank movements. We conducted multiple tests to assess the accuracy of these image-processing techniques for an automatic detection of RR from various camera positions and angles as well as the feasibility under agricultural conditions. The results showed that it is generally possible to record the RR with both camera systems automatically, but at this stage, the recording is limited to individual animals, as the recordings can only capture 2-3 lying cubicles in the barn with one camera. The project KAMI is supported by funds of the Federal Ministry of Food and Agriculture (BMEL) based on a decision of the Parliament of the Federal Republic of Germany via the Federal Office for Agriculture and Food (BLE) under the innovation support program.

Session 65 Theatre 9

Using millimetre-wave radar for monitoring sow postural activity in individual pen: first results

D. Henry[1], J. Bailly[2], T. Pasquereau[1], W. Hebrard[2], J.F. Bompa[1], E. Ricard[1], H. Aubert[3] and L. Canario[1]
[1]INRAE, GenPhySE, Castanet-Tolosan, 31326, France, [2]INRAE, GenESI, Surgères, 17700, France, [3]CNRS, LAAS, Toulouse, 31400, France; dhenry@laas.fr

A millimetre-wave radar is tested in an INRAE experimental unit to monitor sow postural activity in presence of her piglets. A total of 16 sows of the Large White breed with piglets aging from 7 to 15 days are monitored inside farrowing pens with measurement sequences lasting 2 to 3 hours at different dates. The radar is attached to the farrowing pen entrance at a distance to ground of 1.8m, and the sow's position is remotely estimated from the backscattering of electromagnetic waves. The radar-based detection technique does not require equipping animals with radiofrequency tags. The automatized system records 3D images are built from the simultaneous azimuth (digital) and elevation (mechanical) radar beam scannings with a time resolution of 3 seconds. By applying an algorithm based on Constant False Alarm Rate, undesirable radar echoes from the pen are mitigated and only radar detections of the sow are recorded over time. A clustering algorithm is applied to the detections to obtain the 3D position of the sow. A classification of sow postures is performed from a Quadratic Discriminant Analysis of the 3D positions. Ground-truth postures of sow are annotated manually from video recordings. The two following classes of postures are finally analysed: the 'standing and transition' and the 'lying' classes ('transition' refers here to both 'sitting' and 'kneeling' postures). The training data is composed of radar detections of 4 sows for the total monitoring duration of around 8 hours, or equivalently for 7,097 detections. The tested data is composed of 34,356 radar detections of 12 other sows for a total monitoring duration of 38 hours. The precision and sensitivity are 88.3 and 90% for the 'standing and transition' class, and 97.9 and 98% for the 'lying' class. Precision and sensitivity of the classification may vary from one sow to another due to the difficulty to classify correctly 'transition' postures for some of them. Based on these first encouraging results, future work will be devoted to further develop the radar detection method, and detect the motion of the sow according to changes in her location in the pen.

The environmental variance on daily feed intake as a measure of resilience in pigs

C. Casto-Rebollo[1], P. Nuñez[1], S. Gol[2], J. Reixach[2] and N. Ibáñez-Escriche[1]
[1]Institute for Animal Science and Technology, Universitat Politècnica de València, 46022 València, Spain, [2]Batallé SA, Riudanares, 17421 Girona, Spain; pedronunez136@gmail.com

To achieve a more sustainable livestock system, knowing how animals cope with environmental factors is crucial. Identifying the best indicators of resilience is challenging, and identifying which animals are resilient remains an active area of research. The environmental variance (VE) of traits, specifically their residual variance, is under genetic control and has been proposed as a promising indicator of overall resilience. However, its biological impact on the health status and welfare of the animals is still unclear. This study aimed to disclose the biological mechanisms of resilience in Pietrain pigs. For this purpose, we used the VE of the daily feed intake (DFI) as a novel indicator of resilience. 780,112 feed intake (FI) records were collected from automated feeders of 1,628 sires aged between 80 and 180 days. VE was calculated using the pre-corrected DFI for batch, feeder, and body weight effects. Animals were classified into three resilience categories using the 25% (Q1) and 75% (Q3) percentiles of VE: resilient (VE less than Q1), moderately resilient (VE between Q1 and Q3), and non-resilient (VE greater than Q3). Survival data from 230,504 offspring were used to calculate the progeny mortality rate per sire (at least 40 offspring). Differences in progeny mortality were found between resilient (mean 36.18%) and non-resilient (mean 43.22%) animals, with a *P*-value of 0.01. Bayesian statistical analysis indicated that the differences in this mortality between the non-resilient and resilient animals averaged 0.44 standard deviations, with a probability of greater than 0 of 99%. These results suggest that the VE of DFI may act as a measure of progeny survival and thus may account for differences in resilience between animals. Therefore, to unravel the biological mechanism underlying animal resilience and to understand its impact on feed efficiency and critical animal health traits, genomic studies using VE of DFI are ongoing.

Factors affecting somatic cell count in ewe milk and its effect on milk yield and composition

M. Oravcová, V. Tančin, L. Mačuhová and M. Uhrinčať
National Agricultural and Food Centre, Hlohovecká 2, 95141 Lužianky, Slovak Republic; marta.oravcova@nppc.sk

The objective of the presented study was to analyse milk traits (daily yield and composition) in dependence on somatic cell count (five SCC classes defined) in purebred and crossbred Tsigai and Lacaune ewes. Variations due to parity, month in milk/month of measurement, genotype, flock, control year and ewe were also taken into account. Ewes (906 heads, with 7,608 records available) were bred in two flocks; they produced between 2019 and 2022. Estimates of daily milk yield in individual levels of SCC class decreased as follows: 0.835±0.012 l (<200 ths/ml), 0.817±0.014 l (200 ths/ml.<400 ths/ml), 0.816±0.018 l (400 ths/ml<600 ths/ml), 0.811±0.018 l (600 ths/ml<1 mil/ml) and 0.774±0.015 l (≥1 mil/ml). Estimates of lactose content in dependence of SSC class decreased from 4.577±0.008% through 4.540±0.011%, 4.508±0.015%, 4.991±0.015%, to 4.406±0.011%. Estimates of fat content increased from 7.481±0.028 to 7.726±0.042%; estimates of protein content increased from 5.820±0.017 to 6.043±0.023%. Vice versa, the analysis of logarithmic SCC (same factors with exception of SCC class considered) showed their influence on this parameter (increasing trend along with increasing parity, decreasing trend along with month in milk/month of measurement and unclear pattern in SCS across control years revealed). The findings indicate that tissue changes due to udder inflammation were in progress throughout lactation and mastitis (clinical or subclinical) was partly responsible for reduced milk yield. Regular SCC monitoring, which could contribute to improve flock management and milk production efficiency, and further research aimed at prevention of mastitis, are needed. The support of the Ministry of Education, Science, Research and Sports of the Slovak Republic/The Slovak Research and Development Agency (APVV-21-0134 and APVV-15-0072) is gratefully acknowledged.

Session 66 Theatre 1

Digestibility of soybean meal vs toasted soybeans in newly weaned piglets

C. De Cuyper[1]*, P. Dubois*[2] *and S. Millet*[1]
[1]*ILVO, Scheldeweg 68, 9090 Melle, Belgium,* [2]*Danis N.V., Knijffelingstraat 15, 8851 Koolskamp, Belgium; sam.millet@ilvo.vlaanderen.be*

In feed evaluation, digestibility coefficients are usually determined in growing pigs between 40 and 100 kg. The same values are used when formulating piglet diets. However, digestive capacity is lower in piglets than in growing pigs and digestibility may therefore be overestimated. In piglet diets, toasted soybeans are often preferred over soybean meal as feed ingredient. Still, feeding tables indicate a higher crude protein (CP) digestibility for soybean meal than for toasted soybeans. We hypothesized that in piglets, CP digestibility is higher for well processed and monitored toasted soybeans compared to soybean meal. Therefore, an experiment was designed to analyse and compare the digestibility of CP and crude fat (CFAT) of both feed ingredients in young piglets. At four weeks of age, a total of 180 piglets were selected and divided over pens of 6 animals (3 barrows and 3 gilts). Pens were randomly assigned to one of the four dietary treatments: a low protein (soy-free) control diet, 2 diets with control + toasted soybeans and 1 diet with control + soybean meal + soy oil. The soybean meal originated from Brazil and contained 46.5% CP. Soybeans were sampled from commercially available batches. An 18-day adaptation period allowed the animals to get used to the test feed. This was followed by a four-day collection period during which faecal samples were collected. After this collection period, pigs were euthanized to sample ileal content. Apparent ileal digestibility (AID) of CP was significantly ($P<0.05$) higher for toasted soybeans (77±8%) compared to soybean meal and soy oil (67±8%), confirming our hypothesis. Moreover, apparent total tract digestibility (ATTD) of CFAT tended ($P=0.076$) to be higher for toasted soybeans (84±4%) compared to soybean meal and soy oil (81±3%). This study also confirmed that digestibility values may be lower in weaned piglets compared to table values, especially in soybean meal. The difference in digestibility according to bodyweight may be ingredient specific. The research was supported by Danis NV through financial funding and providing a batch of toasted soybeans (Danex®).

Session 66 Theatre 2

Effect of variety and technological treatment on intake of lupin seed in pigs

E. Labussiere[1]*, H. Furbeyre*[1]*, M. Guillevic*[2] *and G. Chesneau*[2]
[1]*PEGASE, INRAE, Institut Agro, Le Clos, 35590 Saint-Gilles, France,* [2]*Valorex, La Messayais, 35210 Combourtillé, France; etienne.labussiere@inrae.fr*

The utilization of lupin seed in pig diets is limited because of anti-nutritional factors that decreased dietary intake when inclusion is higher than 15%. Nevertheless, lupin seed is a feedstuff with a high content in protein which is of interest to diversify protein sources in pig diets. The experiment aimed to test whether differences in variety (n=3, *Lupinus albus, Lupinus angustifolius, Lupinus luteus*) or technological treatment (n=3, none, extrusion, dehulling and extrusion) increase feed intake in pigs. The experimental feedstuffs were included in nine diets at an inclusion rate of 25%. Soybean meal, wheat bran, soybean hulls, sugar beet pulp and sunflower oil were used to balance the diets for their content in crude protein (from 21.9 to 23.3% DM), crude fibre (from 6.1 to 6.5% DM) and fat (from 5.3 to 5.5% DM). These diets were offered by pairwise comparisons to two groups of 25 male pigs (mean body weight = 48.5 kg). Each group had access to four automatic feeders filled with two different experimental diets. Each pig can eat to each feeder and feeding behaviour was recorded, allowing to calculate the proportion of each diet in the daily intake as an indicator of pig preference. Each week, the two diets tested for each group changed and the experiment lasted 7 consecutive weeks. It was not possible to test *L. luteus* with no technological treatment because of lacking of feedstuff. Feed intake decreased by 50% the first week the pigs received the diets containing lupin seed but recovered thereafter. Whatever the technological treatment, *L. albus* was the less preferred lupin seed, whereas *L. luteus* was the most preferred (account for 92 to 98% of intake). *L. luteus* and *L. angustifolius* were preferred after dehulling and extrusion than after extrusion only but dehulled and extruded *L. angustifolius* was less preferred than crude seed. Extruding *L. albus* increased intake over crude seed and dehulled and extruded seed. The increase in the intake of lupin-based diets would be related to the reduction of antinutritional factors by the technological treatments. These treatments therefore contribute to increase the interest of these seeds.

Session 66 Theatre 3

Comparative energy values of 10 forages in finishing pigs

D. Renaudeau[1], J. Pirault[2], M. Dumesny[3], S. Lombard[4] and F. Marie[5]
[1]INRAE, UMR PEGASE Institut Agro, 35590 St Gilles, France, [2]INRAE, Domaine de la Motte, 35650 Le Rheu, France, [3]DESHYOUEST, 11 rue Louis Raison, 35113 Domagné, France, [4]ITAB, 9, rue André Brouard, 49105 Angers, France, [5]IBB, 2, Square René Cassin, 35700 Rennes, France; david.renaudeau@inrae.fr

In specific production systems, forage can make a valuable contribution to pig nutrition and reduce feed costs. Little is known about the energy value of forage in swine. The objective of the present study was to conduct 3 trials to measure the apparent digestibility coefficient (DC) of nutrient and energy and the energy value of 10 forages from different botanical origins and/or with different conservation methods: 3 fresh fodders (chicory, red clover and perennial ryegrass) (T1), 2 dehydrated forages legumes (lucerne and red clover) obtained different drying conditions (130 vs 50 °C at the dehydrator outlet) (T2), 2 hays (lucerne and a mix of 30% red clover and 70% grasses) and a mixed forages legumes (lucerne + red clover) preserved by wrapping (T3). Forages were mixed with a control wheat-soybean mash diet (C diet) and water. The DC of DM, organic matter (OM) and gross energy (GE) of each experimental diet were evaluated with 4 to 5 80 kg BW males pigs. Pigs were adapted to the diet and digestibility cage for 14 d before total collection of faeces and urine over a period of 7 d. The DC of OM and GE was significantly reduced in forages than in C diet ($P<0.001$) due to the higher ash and fibre contents as well as the lignification degree of fibre. On a DM basis, the digestible energy (DE) values were 28 to 61% lower in forages than in C diet (15.3 MJ/kg DM). In T1, the DE values of fresh chicory, red clover and perennial ryegrass averaged 11.05, 10.39 and 10.66 MJ/kg of DM, respectively. In T2, the reduction of dehydration temperature did not affect the DE value of dehydrated lucerne (9.74 MJ/kg DM on average) whereas it negatively affected the DE value of red clover (10.63 vs 8.69 MJ/kg DM; $P<0.05$). In T3, hays had lower DE values than forage conserved in wrapped big bale (6.67 vs 10.41 MJ/kg DM). In conclusion, the study provide original information regarding the energy value of forages according to their conservation method. These data have to be complemented by similar results on the nitrogenous fraction.

Session 66 Theatre 4

Implication of different amylose/amylopectin ratio on low protein diet on piglet's performance

P. Trevisi, D. Luise, F. Correa, S. Virdis, C. Negrini and S. Dalcanale
University of Bologna, Department of Agricultural and Food Sciences, Viale G. Fanin 46, 40127, Italy; paolo.trevisi@unibo.it

In pigs, reducing the dietary crude protein (CP) increase the environmental sustainability and reduce the risk for gut health impairment. Anyway, the high rate of free amino acids (AAs) inclusion in low CP diet could increase the asynchronism between the blood picks of amino acids and glucose that results in a low protein synthesis and increased NH_3 excretion. This study aims in maximizing the growth of pigs fed a low CP diet enriched with amylopectin, which would increase the synchronism between the AAs and glucose picks. At weaning (d0), 90 pigs were divided into 3 groups balanced for body weight (BW) and litter (10 box/diet): (1) control diet (CO): standard diet (Phase1: 18%; Phase2: 16.6%; Phase3: 16.7%); (2) low CP diet (LP) (Phase1: 16%; Phase2: 14.7%; Phase3: 14.5%); (3) as LP but with Waxy corn characterized by 99% of amylopectin (LPW). The feeding phases were d0-d14; d15-d28; d29-d49. BW and feed intake (FI) were weekly recorded. Faecal samples were collected at d14 and d49, faecal score (FS) and lesion score index (LSI) for tail and ears were calculated. Data on BW and ADG, faecal NH_3 and calprotectin were analysed using an ANOVA model considering group, box and litter of origin as factors using piglets as the experimental unit. While the ANOVA model for FI, FCR, LSI and faecal score considers group as factor and box as experimental unit. The BW at d7 and d14 did not differ between groups. From d21 to d49, the LP had lower BW than CO ($P<0.01$). BW of LPW did not differ from the CO until d28, then it was reduced ($P<0.05$). The ADG d0-14 tended to be reduced in the LP ($P=0.06$) and was not reduced in the LPW. From d15-d28, d29-d49 and d0-d49, the CO group had a higher ADG compared with LP and LPW groups ($P\leq0.01$). The FI never differs between the groups. The FCR did not differ from d0-d14 but it was lower in the CO group on d14-d28 ($P\leq0.01$). The faecal NH_3 and calprotectin, FS and LSI did not differ between the groups. Despite the reduction of the ADG, the reduction of the CP did not affect pigs' FI and health and the LPW diet do not reduce the performance of pigs in the Phase1 Acknowledgment: RDP-Emilia Romagna Region, Piano di Innovazione 'MELioR DIET', 16.2.01, ID:5404635.

Session 66 Theatre 5

Effects of two blends of phytoextracts on growth and gut health of weaning pigs to replace zinc oxide

D. Luise[1], F. Correa[1], C. Negrini[1], S. Virdis[1], M. Mazzoni[2], S. Dalcanale[1] and P. Trevisi[1]
[1]Univerisity of Bologna, Distal, Viale G. Fanin, 40127, Italy, [2]Univerisity of Bologna, Dimevet, Ozzano dell'emilia, 40064, Italy; diana.luise2@unibo.it

Phytoextracts (PY) and essential oils (ES) and their mixture with organic acids (OS) are potential alternatives to pharmacological doses of zinc oxide (ZnO) for their antimicrobial, anti-inflammatory and antioxidant abilities. This study aims to evaluate, compared to ZnO, the effect of two blends of PY containing ES and OS on performance and gut health of weaned pigs. At weaning (d0), 96 piglets (7,058±895 g), balanced for body weight and litter, were assigned to one of 4 groups: CO (control group), ZnO (2,400 ppm ZnO from d0 to d14); Blend1 (cinnamaldehyde, ajowan and clove essential oils, 150 g/100 kg feed); Blend2 (cinnamaldehyde, eugenol and short and medium chain fatty acids, 200 g/100 kg feed). Pigs were weighed weekly. Faeces were collected at d13 and d35 for microbiota (v3-v4 of16s rRNA gene) and *Escherichia coli* count analysis. At d35, pigs were slaughtered; pH was recorded on gut contents, jejunum was collected for morphological and gene expression analysis. From d0-d7, Blend2 had a lower average daily gain (ADG) (P<0.05). ZnO and Blend1 never differ in ADG and feed intake. At d14, Blend1 and Blend2 had lower pH in caecum and colon than ZnO (P<0.05). CO had higher number of haemolytic *E. coli* than Blend1 (P=0.01). At d13, ZnO group had lower alpha diversity (P<0.01) and a different microbial beta diversity (P<0.001). At d13, pigs from the ZnO group were characterized by a higher abundance of Prevotellaceae_NK3B31_group (LDA score=4.5, P=0.011), Parabacteroides (LDA score=4.5, Padj.=0.005), Blend1 by Megasphaera (LDA score=4.1, Padj.=0.045) and Ruminococcus (LDA score=3.9, Padj.=0.015) and Blend2 by Christensenellaceae_R-7_group (LDA score=4.6, Padj.<0.001) and Treponema (LDA score=4.5, Padj.<0.001). Jejunal expression of NFKB2 (P=0.05) and IALP (P=0.08) was higher in the Blend2 group than in the CO group. In conclusion, Blend1 allowed obtaining the same performance as ZnO through modulation of the gut pH and microbiome and the reduction in haemolytic *E. coli* and inflammatory status of the mucosa. Acknowledgements: Agritech National Research Center (PNRR project D.D. 1032 17/06/2022, N00000022).

Session 66 Theatre 6

Effect of olive cake in growing pig diets on faecal microbiota fermentation and composition

D. Belloumi[1], P. García-Rebollar[2], P. Francino[3], S. Calvet[4], A.I. Jiménez-Belenguer[4], L. Piquer[1], O. Piquer[5] and A. Cerisuelo[1]
[1]CITA-IVIA, Pol. Esperanza, 100, 12400 Segorbe, Spain, [2]UPM, Dept. Produccion Agrarias, ETSIAAB, Av. Puerta de Hierro, 2-4, 28040 Madrid, Spain, [3]FISABIO, Salud pública, Av. de Catalunya, 21, 46020 Valencia, Spain, [4]UPV, ICTA, Camí de Vera, s/n, 46022 Valencia, Spain, [5]CEU-Cardenal Herrera, C. Tirant lo Blanc, 7, 46115 Valencia, Spain; belloumi_dhea@gva.es

The intestinal microbiota plays a critical role in the metabolism and health of the host. The present study investigated the impact of two types of olive cake on short-chain fatty acid (SCFA) concentrations and the faecal microbiota composition of pigs. A total of 30 pigs (Landrace × Large white) with an initial body weight (BW) of 47.9±4.21 kg were divided into three groups according to the feed they received: control feed (C), feed with 200 g/kg of partially defatted olive cake (PDOC) or feed with 200 g/kg of cyclone olive cake (COC). Faecal samples were collected from each animal after 3 weeks of the feeding trial. Microbiota composition was analysed by sequencing the V3-V4 region of the 16S rRNA gene. The results showed that the COC group had more total SCFAs, acetic acid, butyric acid, and caproic acid (P<0.05) than the other two groups. No significant differences were found among groups in alpha diversity (P>0.05). At the phylum level, *Firmicutes* and *Bacteroidota* were the predominant phyla across the three groups, with more than 93% of the total community. However, *Spirochaetota* was significantly (P<0.05) more present in the C group than in the PDOC group, suggesting a severe possibility of gut inflammation in this group. *Plantomycetota* was significantly more abundant (P<0.05) in the PDOC group than in the C group. The relative abundances of *Eggerthellaceae* and *Allisonella* were significantly (P<0.05) enriched in the COC group compared to the C group. However, *dgA-11_gut_group* was more abundant (P<0.05) in the C group than in the COC group. These results suggested that supplementing pigs' diets with olive cakes may beneficially affect pigs' gut health without altering the diversity of microbial communities. This work was supported by the project IVIA-GVA 52201L from IVIA (co-financed by the EU through the ERDF Program 2021-2027 Comunitat Valenciana). BD received PhD scholarship (GRISOLIAP/2020/023).

Impact of increasing dietary sphingolipids on feed intake and growth performance of piglets

S. Chakroun[1], R. Larsen[1], J. Levesque[2], F. Cerpa Aguila[1], M.-P. Letourneau[1], J.E. Rico[3] and D.E. Rico[1,2]
[1]Université Laval, 2425 Rue de l'agriculture, Quebec, QC, G1V0A6, Canada, [2]CRSAD, 120-A Chemin du Roy, G0A1S0, Canada, [3]University of Maryland, 8127 Reagents drive, College Park, MD, 20740, USA; daniel.rico@crsad.qc.ca

The weaning period in piglets is associated with impaired growth and increased oxidative stress and inflammation. Milk fat globular membranes are a rich source of anti-inflammatory polar lipids which may improve performance of animals when included in the diet. Weaned male piglets (n=240; 21 days of age; 6.3±0.5 kg of BW) were blocked by initial weight and distributed into 48 pens of 5 animals in a complete randomized block design with a factorial arrangement, where animals received either a soybean lipid-based (n=24 pens; SD) or a polar lipid-based diet (n=24 pens; PD) from weaning to day 42 of the nursery phase. Within each diet group animals received 1 of 3 milk replacers (MR; 0.5 l/d) for the first 7 d: (1) commercial MR (CO; Control); (2) soybean lipids-based MR (S); (3) polar lipid-based (PO; Polar). Animals were switched to a common diet from d 21 to 42. Feed intake was recorded weekly from d 14 and body weight (BW) on d 0, 7, 14, 21, 28 and 42. Data were analysed in a mixed model with the random effects of pen and block, and the fixed effects of diet, MR, time, and their interactions. A diet by time interaction was observed for feed intake (P=0.004) as it was 17% lower in PD relative to SD on d21, but not different at any other timepoint. A 3-way interaction was observed for BW (P=0.003), which increased over time and was 14% lower in PO compared with CO in animals fed the PD diet, but not different between other treatments. Similarly, average daily gain was 9% lower in PO compared with CO in animals fed the PD diet only (Diet × MR P=0.01). The impact of polar lipids on growth performance of piglets is dependent on the method of feeding.

Faeces microbiota and metabolism correlated with reproductive capacity in early parity sow

J. Wang, Q. Xie and B. Tan
Hunan Agricultural University, 1st Nongda street, Changsha, 410128, Hunan, China, P.R.; jingwang023@hunau.edu.cn

The metabolic status and sow fertility are tightly connected and reciprocally regulated. The aim of this study was to analyse the potential correlation among reproductive performance, colostrum composition and faeces microbiota and metabolites. A number of 32 first and second-parity sows were selected to measure litter size and litter weight at farrowing, and to collect faeces and blood on d 110 of gestation (d110G), as well as colostrum (24 h after farrowing). The live litter size of primiparous sows was lower than those of second parity sows (P<0.05). The IgA, IgG, IgM concentrations and protein, lactose, urea nitrogen contents in colostrum of primiparous sows were lower than those in second parity sows (P<0.05). Consistently, the concentrations of secretory IgA (sIgA) in serum and faeces of primiparous sows on d110G were lower than those in second parity sows (P<0.05). The colostrum sIgA and IgM concentrations were positively correlated with the live litter size (sIgA: Pearson r=0.259; IgM: Pearson r=0.361) and initial litter weight (sIgA: Pearson r=0.265; IgM: Pearson r=0.376) (P<0.05). Significant differences were observed in the alpha and beta diversities of faeces microbial community on d110G. The Shannon index in second parity sow faeces was higher than in primiparous sow (P<0.05), and second parity sow showed difference in faeces microbiota composition (P<0.05). The faeces abundance of *Lactobacillus* genus was highly positive correlation with IgM levels in faeces, serum and colostrum and faeces sIgA level (P<0.05). A total of 1,923 metabolites were detected in faeces on d110G, of these, 252 metabolites were up-regulated and 260 metabolites were down regulated in second parity sow. KEGG showed that the significantly up-regulated metabolites were mainly involved in protein digestion and absorption pathway. Citrulline, proline, palmitoleic acid, and valerophenone were positively correlated with sIgA, IgA, IgM, IgG levels in faeces, serum and colostrum, as well as the abundance of *Lactobacillus* genus in faeces (P<0.05). In conclusion, faeces microbial composition and metabolic status might affect the immunoglobulin levels of colostrum and used as a predictor of reproductive capacity in sows.

Dehydrated sainfoin in rabbit feed: effects of high incorporation on the health and performances of

C. Gayrard[1], P. Gombault[2], A. Bretaudeau[3], H. Hoste[4] and T. Gidenne[1]
[1]INRAE – Occitanie Toulouse, Animal physiology and farming systems, GenPHySE, 31326 Castanet Tolosan, France, [2]Multifolia, Viapres-le-petit, 10380 Viapres-le-petit, France, [3]Arrivé-Bellanné, Nueil-les-Aubiers, 79250 Nueil-les-Aubiers, France, [4]INRAE Occitanie Toulouse, Interaction Hôte-Agents Pathogènes, ENVT, 31300 Toulouse, France; thierry.gidenne@inrae.fr

In rabbit farming, coccidiosis and fibre deficient diets (especially lignins) can increase the risk of digestive disorders. Sainfoin is a legume rich in lignins, and contains polyphenols and tannins with anti-parasitic properties. Sainfoin also contained a high level of digestible protein (79g/kg) and energy (8.84 MJ/kg). Here, we studied the effects of a high level (26%) incorporation of dehydrated sainfoin (Perly cultivar, first cut, 'DSp') in rabbit feeding on the performances and health of reproductive doe and growing rabbits were analysed in a sub-optimal (with coccidiosis for the two previous batches) professional breeding environment and over two non-consecutive reproductive cycles (2 replicates). Performances and health of does and growing rabbits were compared for 2 groups of 194 does and associated litters, when fed iso-nutritive feeds containing either 0 or 26% DSp. Dietary DSp incorporation had no effect on doe live weight (4,675 g), fertility rate (88.4%), mortality (6.9%), culling rate (0.8%) and doe coccidia excretions levels (P>0.05). In replicate 1, kits growth before weaning was similar (24.8 g/d) among the two groups, but was 12% lower for the DSp26 group in the 2nd replicate (significant interaction). The post-weaning growth rate was improved by 4% (37.9 vs 39.3 g/d, P=0.02) for DSp26 group. Before weaning, a higher mortality was observed for DSp26 (3.3 vs 1.8%) in replicate 1, while in replicate 2 it was lower (2.1 vs 4.4%). After weaning, the mortality rate decreased (4.5 vs 7.1%, P<0.001) for DSp26 group (mean of two replicates). Coccidia excretions of growing rabbits were not affected neither by dietary sainfoin nor by replicates. Thus, a high incorporation of dehydrated sainfoin in the feed can be recommended to improve the health and performance of growing rabbits, without affecting does performances. Sainfoin is thus a good alternative to the alfalfa meal in the rabbit feeding.

Nutritional value of defatted larvae meal and whole larvae from black soldier fly

P. Belloir[1], F. Hervo[2], E. Gambier[2], L. Lardic[2], C. Guidou[3], C. Trespeuch[3], N. Même[4], E. Recoules[2] and B. Méda[2]
[1]Ecole d'Ingénieurs de PURPAN, 75 Voie du Toec, 31076 Toulouse, France, [2]INRAE, Université de Tours, BOA, 37380 Nouzilly, France, [3]MUTATEC, 1998 Chem. du Mitan, 84300 Cavaillon, France, [4]INRAE, PEAT, 37380 Nouzilly, France; bertrand.meda@inrae.fr

Black soldier fly (BSF) larvae is a novel protein source to feed poultry. However, only nutritional values for BSF meals are available in the literature, whereas the use of whole larvae could be a promising strategy, for instance when used as an environmental enrichment material. Thus, this study aimed to investigate the nutritional value of three different BSF products: defatted larvae meal (LM) and whole larvae, either dried (DL) or fresh (FL). On a dry matter (DM) basis, crude protein (CP) and crude fat (CF) contents were respectively of 57 and 10% for LM vs 39 and 34% for whole larvae. A total of 48 Ross 308 male boilers were randomly assigned to individual cages for a digestibility trial. Four treatments were used: a control diet (C), a LM diet containing 75% C + 25% LM and two larvae diets containing, on a DM basis, 75% C + 25% of FL or DL larvae. Whole larvae were distributed on top of pellets in the same feeder. For the three products (LM, DL, and FL), the metabolizable energy (AMEn) and the apparent total tract digestibility (ATTD) of DM, CF, CP, and gross energy (GE) were measured, as well as the standardised ileal digestibility (SID) of amino acids (AA). The ATTD of DM, GE, and CP were significantly higher (P<0.001) for DL and FL compared to LM (DM: 83 vs 60%; GE: 85 vs 63%; CP: 68 vs 53%). TTAD of CF was significantly higher for DL (98.2%) compared to LM (94.6%; P<0.01) and FL (95.7%; P<0.05). The AMEn was significantly lower for LM compared to DL/FL (2,730 vs 4,950 kcal/kg DM; P<0.001). The SID of all AA were significantly lower for LM (52-86%) compared to DL and FL (77-98%) with respective values of 84 vs 93% for lysine and 85 vs 95% for methionine+cysteine. These differences could be explained by the higher chitin content in LM, due to defatting process. At high concentration, chitin is indeed known to impair nutrients digestibility. These results provide the first data on the nutritional value of whole BSF larvae and confirm that BSF larvae are a highly digestible source of nutrients for poultry.

Session 66 Poster 11

Diet taste monotony decreases feed acceptability and preferences in nursery pigs

J. Figueroa[1], E. Huenul[2], R. Palomo[3] and D. Luna[3]
[1]Universidad de O'Higgins, Ruta 90 km 3, San Fernando, 3070000, Chile, [2]Universitat Autònoma de Barcelona, Facultat de Veterinària, Bellaterra 08193, Spain., 08193, Spain, [3]Universidad de Chile, Facultad de Ciencias Veterinarias y Pecuarias, Av. Santa Rosa 11735, La Pintana, Santiago, 8820000, Chile; jaime.figueroa@uoh.cl

When satiated animals have access to a diet with different sensory properties, they respond with an increase in consumption. However, most commercial diets in swine industry lack sensory variety. Thirty-two nursery pigs (43 days old, 10.9±1.63 kg) allocated in pairs into 16 pens were used to explore the effect of taste variety on their feeding behaviour. Animals were daily exposed for eight consecutive days, to acceptability (n4) and preference tests (n4), to assess sensory-specific satiety for sweet and umami tastes. Pigs were offered for 10 min a commercial feed with an inclusion of 8% sucrose (Suc-feed) or 300 mM of monosodium glutamate (MSG-feed) after being exposed to the same or different feed for the same time (Suc-Suc, MSG-MSG, Suc-MSG, MSG-Suc). Palatability was estimated by consumption patterns (consumption time/n° approaches to the feeder). Preference between Suc-feed and MSG-feed was estimated for 10 minutes after a previous exposure, for the same time, to Suc-feed or MSG-feed. Data was analysed by using the statistical software SAS®. No intake differences were observed between Suc-feed and MSG-feed intake during the acceptability test (P=0.345). Feed intake and palatability were affected by the interaction between the taste consumed and the taste previously exposed, where pigs consumed more and perceived the feed more pleasant when a different taste was offered (P<0.05). During the preference test, pigs preferred to consume MSG-feed than Suc-feed (P<0.001). An interaction between the taste consumed during the preference test and the taste previously delivered was observed (P<0.001) where pigs preferred MSG-feed over Suc-Feed, but the magnitude of this differences was lower if pigs that previously consumed MSG-feed (54 vs 23 g) than Suc-feed (72 vs 13 g). Results support that taste monotony may decrease feed acceptability, palatability and preferences in nursery pigs. Therefore, taste variety is relevant to implement in diets to approach the natural feeding behaviour of pigs.

Session 66 Poster 12

Dietary taste variety improves performance in nursery pigs

J. Figueroa[1], T. Cabello[1], E. Huenul[2], R. Palomo[3] and D. Luna[3]
[1]Universidad de O'Higgins, Ruta 90 km 3, San Fernando, 3070000, Chile, [2]Universitat Autònoma de Barcelona, Facultat de Veterinària, Bellaterra 08193, Spain, [3]Universidad de Chile, Facultad de Ciencias Veterinarias y Pecuarias, Av. Santa Rosa 11735, La Pintana, Santiago, 8820000, Chile; jaime.figueroa@uoh.cl

The lack of dietary sensory variety decreases hedonism during a consumption episode, being able to reduce food intake and the welfare of domestic animals. Thirty-two nursery pigs allocated in pairs into 16 pens were used to explore the feed taste variety's effect on their performance. From the second week after weaning (28 days old, 6.73±0.45 kg), half of the pens (n8) were exposed for 6 weeks to a commercial feed with the inclusion of 300 mM of monosodium glutamate (MSG-feed; Monotony Group). The other half of the pens (n8) were exposed to a daily rotation between 3 different feeds; MSG-feed, a commercial feed with an inclusion of 8% of sucrose and a standard commercial feed (Variety Group). Pig's body weight, average daily gain (ADG), average daily feed intake (ADFI) and feed conversion ratio (FCR) were weekly estimated. Data were analysed with the statistical software SAS®. Pigs in the variety group tended to have a higher body weight at the end of the experiment (16.3 vs 15.1 kg; P=0.088) and presented an overall higher ADG (0.232 vs 0.194 kg; P=0.008), observing significant differences in weeks 3, 4 and 6 (P<0.05). The ADFI was higher in Monotony group at the beginning of the experiment, observing a significant difference at week 3 (P=0.038). However, the Variety Group presented higher ADGs at the middle-end of the experiment, reflected in a significant difference at week 6 (P=0.018). Finally, FCR was better in the Variety Group from week 3 until the end of the experiment, with significant differences at weeks 3 and 4 (P=0.01), also observing an overall difference (2.45 vs 2.82; P=0.003). These results support that a varied diet in taste between days improves performance in nursery pigs, especially from week 4 after weaning compared with animals fed with a sensory monotony diet supplemented with a palatable taste such as MSG. This could improve the productive performance of the swine industry if strategies are developed to promote dietary taste variety.

Session 66 Poster 13

Impact of some dietary agro-industrial by-products on the gut microbiota of finishing pigs

I. Skoufos[1], A. Nelli[1], C. Voidarou[1], I. Lagkouvardos[1], E. Bonos[1], K. Fotou[1], C. Zacharis[1], I. Giannenas[2] and A. Tzora[1]
[1]University of Ioannina, Department of Agriculture, Arta, 47100, Greece, [2]Aristotle University of Thessaloniki, School of Veterinary Medicine, Thessaloniki, 54124, Greece; jskoufos@uoi.gr

The optimum composition of the gut microbiota (GM) not only protects the health of finishing pigs but also improves their production performance. This dietary intervention study assessed the effect of a novel silage composed by a mixture of olive mill wastewater, grape pomace, and deproteinized feta cheese whey on the GM of finishing pigs. Eighteen finishing crossbred pigs, were penned individually and randomly assigned to three different dietary treatments: control, 5% silage and 10% silage. The duration of the trial was 40 days, after which all animals were humanely killed and ileal and colonic digesta were collected for bacterial DNA extraction. The 16s RNA gene sequencing was performed using Illumina MiSeq. The bioinformatic analysis was carried out using IMNGS pipeline and Rhea platform. A sharp separation of the group B microbial cluster from the control group was observed in both ileum (P=0.063) and colon (P<0.05). In the ileum, the relative abundance of *Clostridium* genus (*C. celatum*) was increased in group B compared to the control (P<0.05). In the colon, the relative abundance of *Clostridium* genus (*C. celatum*) was increased, and the relative abundance of *Streptococcus* genus (*S. alactolyticus*) was decreased in groups B and C compared to the control (p≤0.05). Interestingly, the relative abundance of *Bifidobacterium* genus (*B. pseudolongum*) increased in group B compared to the other two treatments (P<0.05). These results showed that 5% was the optimum inclusion level of the innovative silage in the diet of finishing pigs associated with beneficial changes in the GM. Therefore, low-cost agro-industrial by-products are promising dietary supplements for pigs lead to value-added final products. This work has been co-financed by Greece and European Regional Development Fund, Greece-China 2014-2020, project code: T7ΔKI-00313(MIS-5050735), acronym Green Pro.

Session 66 Poster 14

Effects of Greek aromatic/medicinal plants on health and meat quality characteristics of piglets

G. Magklaras[1], C. Zacharis[1], K. Fotou[1], E. Bonos[1], I. Giannenas[2], J. Wang[3], L.Z. Jin[4], A. Tzora[1] and I. Skoufos[1]
[1]University of Ioannina, Department of Agriculture, Arta, 47100, Greece, [2]Aristotle University of Thessaloniki, School of Agriculture, Thessaloniki, 54124, Greece, [3]Nanjing Agricultural University, Jiangsu Key Laboratory of Gastrointestinal Nutrition and Animal Health, Nanjing, CN-210095, China, P.R., [4]Guangzhou Meritech Bioengineering Co. Ltd, Guangzhou, 510300, China, P.R.; jskoufos@uoi.gr

The aim of the study was to investigate the effects of aromatic/medicinal plants, on pig health and meat oxidation. Two herbal mixtures consisting of oregano (*Origanum vulgare* subsp. hirtum) essential oil, crithmum (*Crithmum maritimum*) essential oil, garlic flour (Allium sativum) and camellia flour (*Camelina sativa*) were used in piglet diets at two different proportions. Three groups of weaned piglets were fed either the control diet (A) or one of the enriched diets (B or C, 2 g/kg). After 43 days, meat tissue cuts (biceps femoris, external abdominal, triceps branchii) were collected for chemical analysis and oxidative stability testing. Microbiological analysis of intestinal digesta from the ileum and colon was conducted. Statistical analysis revealed no differences (P>0.05) in the body weights and growth rates among the groups, as well as in meat chemical analysis. Haematological and biochemical parameters did not differ between the groups. An increase (P<0.05) of total aerobic bacteria was detected in the ileum of group B, while *Escherichia coli* counts were reduced (P<0.05) in group C. In the colon, reduction of *E. coli* and Lactobacilli counts were observed in groups B and C. Concentrations of malondialdehyde (MDA) as indicator of lipid peroxidation, were significantly reduced (P<0.05) in triceps branchii and biceps femoris for both groups B and C (day 0). A reduction (P<0.05) of MDA was noticed in triceps branchii and external abdominal meat samples (day 7) for groups B and C. The two investigated mixtures could be used in weaned piglets' diets with positive results on intestinal *E. coli* count reduction and oxidative stability of the meat. Funded by E.U. and National Greek Funds, 'Greece – China'. Project code: T7ΔKI-00313. Acronym: 'Green Pro'.

Session 66 Poster 15

Effects of mint leaf powder on performance and egg nutrient composition of laying hens

Y.H. Zhang[1], M.M. Bai[2], H.N. Liu[2], X.F. Kong[2] and F.C. Wan[1]
[1]Hunan Agricultural University, 410128, 1 Nongda Road, Furong District, Changsha City, Hunan Province, China, P.R., [2]Institute of Subtropical Agriculture, Chinese Academy of Sciences, 410125, 644 Yuanda 'er Road, Furong District, Changsha City, Hunan Province, China, P.R.; liuhn@isa.ac.cn

This experiment was conducted to investigate the effects of mint leaf powder on performance and egg nutrient composition of laying hens in late laying period. A total of 216 laying hens with similar laying rate in late laying period were randomly divided into 4 groups. The control group was fed with basic diet, and the other three groups were added with 0.1, 0.2 and 0.4% mint leaf powder respectively in the basic diet. Each group has 9 replicates, and each replicates has 6 chickens. The whole experiment lasted for 35 days, including 7 days of pre-test and 28 days of trial. Compared with the control group, the addition of mint leaf powder in the diet had no significant difference in the feed to egg ratio, egg production rate and egg weight of laying hens at the later stage of laying (P>0.05). At the same time, the addition of 1 and 4% mint leaf powder can reduce the average daily feed intake (ADFI) of laying hens at the later stage of laying (P<0.05). The three treatment groups significantly increased the IgG content in serum of laying hens (P<0.05). The 0.1% mint leaf powder group significantly increased the number of large white follicles in ovary (P<0.001). There was no significant difference in egg weight, egg shape index and Huff unit among the three mint leaf powder supplementation groups (P>0.05), and there was a trend to increase the total content of L-phenylalanine, L-tryptophan, L-isoleucine and essential amino acid in egg white (0.05<P<0.10). The 0.1 and 0.4% mint leaf powder supplementation groups had no significant effects on fatty acid content in egg yolk (P>0.05).In conclusion, dietary supplementation of 0.1 and 0.4% mint leaf powder can reduce the feed intake of laying hens in late laying period, increase the IgG content in the serum of laying hens, increase the weight of oviduct and the number of follicles in laying hens, and have a trend to increase the content of amino acids in egg white, but have no significant effect on the content of fatty acids in egg yolk.

Session 66 Poster 16

Weaned piglets' gut microbiota regulation by dietary olive, winery, and cheese waste by-products

A. Tzora[1], A. Nelli[1], A. Tsinas[1], E. Gouva[1], G. Magklaras[1], S. Skoufos[1], B. Venardou[1], K. Nikolaou[1], I. Giannenas[2] and I. Skoufos[1]
[1]University of Ioannina, Department of Agriculture, Arta, 47100, Greece, [2]Aristotle University of Thessaloniki, School of Agriculture, Thessaloniki, 54124, Greece; tzora@uoi.gr

Gut microbiota (GM) plays a vital role in the nutrition and metabolism of weaned piglets and therefore feed additives have been developed to adjust the composition of GM in order to improve their health and performance. The purpose of this study was to evaluate the effect of dietary use of a novel silage in weaners, created by combining three agro-industrial waste products namely olive mill wastewater, grape pomace, and deproteinized feta cheese whey, on the GM composition. Forty-five crossbred weaned piglets (34 day-old) were randomly allocated to 3 pens, one per dietary treatment. Group A was the control treatment (no added silage), while a diet enriched with 5 and 10% of the novel silage was fed to groups B and C respectively. After a 40-day dietary supplementation, 6 pigs per treatment were humanely killed and ileal and colonic digesta were collected for bacterial DNA extraction. The 16s RNA gene sequencing was performed using Illumina MiSeq. The bioinformatic analysis was carried out using IMNGS pipeline and Rhea platform. In the ileum, no significant differences were observed between treatments. In the colon, the relative abundance of Prevotellaceae family was increased in group B compared to the control (P<0.05). At the genus level, the relative abundances of Coprococcus and Alloprevotella were increased in group B compared to the control (P<0.05 and 0.635 respectively) and group C (P<0.05 for both). The innovative silage at 5% inclusion level in the diet of weaned pigs altered beneficially the composition of the GM indicating its ability to promote their gastrointestinal health as well as production parameters. This work has been co-financed by Greece and European Regional Development Fund, Greece-China 2014-2020, project code: T7ΔKI-00313(MIS-5050735), acronym Green Pro.

Session 66 Poster 17

Effects of medium-chain fatty acids on feed intake, body weight and egg composition of laying hens

F. Cerpa Aguila[1], S. Chakroun[1], M.P. Letourneau Montminy[1], J.E. Rico[2] and D.E. Rico[1,3]
[1]Université Laval, 2425 Rue de l'agriculture, Quebec, QC, G1V0A6, Canada, [2]University of Maryland, 8127 Reagents drive, College park, 20740, USA, [3]CRSAD, 120-A Chemin du Roy, Deschambault, QC, G0A1S0, Canada; daniel.rico@crsad.qc.ca

Low-protein, high-energy (LPHE) diets may reduce environmental impact of egg production, but may affect animal performance. Medium-chain fatty acids are rapidly oxidized in the liver and could help prevent fatty liver in laying hens fed LPHE diets. In addition, these fatty acids (FA) have been shown to alter egg yolk FA profile. Lohman-white layers (n=100; 25 wk of age; 1.5±0.2 kg of body weight; BW) were blocked by weight and randomly allocated to 1 of 4 diets for 77 days in a randomized complete block design. Treatments were: (1) High-protein, low-energy diet (Control; 2,600 Kcal ME/kg of DM; 17% CP; 3.5% fat); (2) LPHE diet (3,000 Kcal ME/kg of DM; 13% CP; 7% fat) based on medium-chain triglycerides (MCT; 8:0 10:0); (3) LPHE diet based on coconut oil (Coco; 12:0, 14:0); or (4) LPHE diet based on long-chain FA from soybean oil (LCFA; 56% 18:2 n-6). Feed intake (FI) and BW were recorded on d 0, 4, and weekly thereafter. Data were analysed as repeated measures in a mixed model with random effects of block and hen, and the fixed effects of treatment, time, and their interaction. There was a treatment by time interaction for FI, which decreased progressively in all LPHE diets, reaching a nadir on d 42, and being -19% lower than Control ($P<0.001$). Protein intake was reduced progressively in LPHE diets, being 32% lower than in Control (trt × time $P<0.001$). Energy intake of animals was higher 12% higher on d 4 in LPHE relative to Control, but decreased progressively and was 7% lower on d 56, 63, and 70 (trt × time $P<0.001$). There was a treatment by time interaction for body weight ($P<0.001$) as it was 8.9% lower in LPHE diets from day 35 to 77. Egg weight remained constant in the LPHE groups, whereas it increased over time in Control (trt × time $P<0.001$). Treatment had no impact on egg composition, but the proportion of egg yolk was increased whereas that of albumen and shell decreased over time ($P<0.001$). Dietary FA profile had no impact on the overall hypophagic effects of high-fat diets in laying hens.

Session 67 Theatre 1

Open-sourcing behavioural algorithms for ruminant welfare monitoring using raw wearable sensor data

D. Foy[1], T.R. Smith[1] and J.P. Reynolds[2]
[1]AgriGates, Philadelphia, 19146, USA, [2]Western University College of Veterinary Medicine, 309 E. Second St. Pomona, 91766-1854, CA, USA; d.foy@agrigates.io

Wearable sensors have immense potential in monitoring ruminant welfare metrics, but their Black-Box Algorithms (BBA) and siloed data limit their usage beyond specific insights like oestrus, lameness detection, and rumination. We propose open-source behavioural algorithms and collaborations based on quality raw data (RD) output from sensors that capture individual biometrics and behavioural markers for assessing welfare measurements. Data quality is critical for accuracy, value, and understanding the RD from the available sensor adds to the validity and veracity of data used towards ruminant ML/AI and welfare metrics. We explore the need for a gold standard method to evaluate RD output from 3, 6, and 9-axis sensors and the data quality for monitoring the behaviours of food animal livestock. We discuss how the academic community, in concert with commercial enterprises, can collaborate to define acceptable RD, sensor bias, error rates, edit rates, and key biometric and behavioural markers for welfare by applying open-source behavioural algorithms to quality RD. We suggest that the sector collaborates through entities like the Linux Foundation to overcome the challenges of BBA at the farm level and commercial IP issues that arise, providing the wider sector with a standard so that an agile and scalable platform of quality data for behavioural algorithms to the individual animal of greater value at the farm level and supply chain. This approach can be extended to more ruminant species under varying production systems. Unlocking raw wearable sensor data in livestock farming for welfare monitoring that allows the application to each farm. No two farms or animals are the same, ML/AI needs to be able to adapt and learn within the context of the specific farming system and animal. We conclude that validating food animal livestock wearable RD outputs on and off-animal platforms, independent of the supplier, is crucial for improving decision support, improving ML/AI, and meeting target metrics for welfare monitoring of animals in food animal agriculture, while coordinating collectively as a sector, in an initiative to reduce, understand and map BBA critical for improving food animal welfare.

Session 67 Theatre 2

Evaluation of an automated cattle lameness detection system

N. Siachos, A. Anagnostopoulos, B.E. Griffiths, J.N. Neary, R.F. Smith and G. Oikonomou
University of Liverpool, Department of Livestock and One Health, Leahurst Campus, CH64 7TE, United Kingdom; nektarios.siachos@liverpool.ac.uk

Our aim was to evaluate the performance of an automated lameness detection system (CattleEye Ltd) which is using a 2D surveillance camera mounted over an exit race. A total of 29 whole-herd mobility scoring sessions in 8 medium to large-size herds were performed by 4 experienced veterinarians (VETs) using the 4-grade (0-3) AHDB mobility scoring method. The weekly average score for each cow provided by the system (CE) was also stored and analysed after the end of the study. A total of 27,082 mobility scores were collected and matched to the weekly average CE scores. Agreement between CE and each VET was assessed for the binary transformed scores (0: 0,1; 1: 2,3) using percentage agreement (PA), kappa (κ) and Gwet's coefficients (AC). Moreover, the same VET was present in 17 foot trimming sessions in 3 farms and recorded the presence and severity of sole haemorrhage (SH), sole ulcer (SU), white line disease (WL), toe ulcer (TU), digital dermatitis (DD) and interdigital phlegmon (IP) cases. Lesion records were then matched with the weekly average CE scores, resulting in a dataset of 991 cows. The same VET also mobility scored a subset of 340 cows in 2 farms 1-3 days before foot trimming. Accuracy (ACC), sensitivity (SE) and specificity (SP) were calculated for both CE and the VET; presence (binary) of at least one case of SU grade ≥2, WL grade 3, TU, stage M2 of DD and IP grade 2, was used as the gold standard. Overall PA, κ and AC ranged from 81.5 to 86.3%, from 0.23 to 0.41, and from 0.76 to 0.83, respectively (when agreement between CE and VETs in binary mobility scores was assessed). ACC, SE and SP of CE and VET varied notably among farms, yielding an overall combination of 0.83, 0.40 and 0.88, and 0.80, 0.53 and 0.83, respectively. Based on our results, the agreement between CE and VETs is within the moderate and substantial range, in concordance with that reported between experienced assessors in an initial validation study. Further investigation of farm and cow-level factors that potentially influence the predictive ability of CE in identifying cows bearing foot lesions is needed.

Session 67 Theatre 3

Breath analysis in dairy cattle: going beyond methane emission

I. Fodor[1], E. Van Erp-Van Der Kooij[2] and I.D.E. Van Dixhoorn[1]
[1]Wageningen University and Research, De Elst 1, 6700 AH Wageningen, the Netherlands, [2]HAS green academy, Department of Animal Husbandry, Onderwijsboulevard 221, 5223 DE 's-Hertogenbosch, the Netherlands; istvan.fodor@wur.nl

Real-time, non-invasive monitoring tools are continuously being investigated in research and deployed in practice in the dairy cattle industry. So far, breath composition of dairy cattle has been typically monitored for methane emissions only. Although, a growing body of evidence suggests that the potential applications of breath analysis go beyond monitoring methane emissions. The results from our systematic review on using breath analysis to detect diseased cattle showed that research focused on cows before 2000, however, calves have been studied more frequently since then. Ketosis has been the most studied cattle disease using breath analysis so far. Our experimental results showed potential of breath analysis to follow ketosis status in postpartum cows, because a rise in serum β-hydroxybutyrate was significantly related to a rise in breath acetone concentration. We also showed that longitudinal records would be necessary to detect increasing breath acetone levels within individual cows, instead of single spot measurements. In another experiment, we found that a lower respiratory exchange ratio (CO_2/O_2, V/V%) was linked to a larger decrease in body condition score and higher levels of serum β-hydroxybutyrate in the first six weeks postpartum in dairy cows. We also found a link between breath composition and feed efficiency, using data from automated emissions monitoring stations. As an outlook for the future, breath analysis has potential to be used as a practical, non-invasive, real-time monitoring tool on dairy farms, with a broader scope than monitoring methane emissions. Although, more research is needed to further explore the capabilities of this tool, and engineering solutions are also needed to implement this technique on dairy farms.

Session 67 Theatre 4

Dairy cow personality: correlations with age, weight, back fat thickness and activity

P. Hasenpusch, T. Wilder, A. Seidel, J. Krieter and G. Thaller
Christian-Albrechts-University, Institute of Animal Breeding and Husbandry, Olshausenstraße 40, 24098 Kiel, Germany; phasenpusch@tierzucht.uni-kiel.de

Personality traits determine dairy cow reactions to potentially stressful situations such as management procedures and social interactions in group housed dairy cattle. Therefore, identification of personality traits may help to improve animal welfare. A common approach to assess those traits are behavioural tests. Accordingly, the reaction of 200 lactating Holstein-Friesian cows stressed by a novel object and human approach in a known environment was observed. Latencies and quantities of vocalisation, urination, defecation, approaches and contacts as well as the duration of every object contact were measured in the novel object test (NOT). In the human approach test (HAT), the minimal distance and the cow's reaction to three subsequential approaches were documented. Due to the assumption that personality traits could be described by certain combinations of the observed variables, Bartlett's Test of sphericity and Kaiser-Meyer-Olkin Criteria were applied to validate if and which variables could be used for principal component analysis. Correlations between extracted components and age at testing, daily activity (measured via pedometer) and body weight, were estimated ($P<0.05$). Kaiser Criteria suggested the extraction of four distinct components for the NOT. According to correlations with measured variables those components were named: Explorative (factor loading of 0.6 quantity of contacts; 0.99 contact duration), bold (-0.97 latency of approach; 0.96 latency to contact), nervous (-0.97 latency to urinate/defecate; 0.97 quantity of urinations/defecations) and sociable (-0.89 latency to vocalise; 0.89 quantity of vocalisations). For the HAT, one component named trusting (-0.85 first approach; -0.87 second approach; -0.87 third approach) was extracted. Age (0.14), weight (0.21) and back fat thickness (0.2) correlated with trusting. Variation coefficient of daily activity correlated with being explorative (0.16) and trusting (-0.17). Further behavioural studies are currently underway considering individual and group interactions by precise indoor positioning via ultra-wideband.

Session 67 Theatre 5

Automatic behaviour assessment of young bulls in pen using machine vision technology

A. Cheype[1], J. Manceau[2], V. Gauthier[2], C. Dugué[3], L.-A. Merle[4], X. Boivin[5] and C. Mindus[1]
[1]Institut de l'élevage, 149 rue de Bercy, 75012 Paris, France, [2]NeoTec-Vision, 7 allée de la Planche Fragline, 35740 Pacé, France, [3]France Limousin Selection, Pole de Lanaud, 87220 Boisseuil, France, [4]Ferme des Etablières, route du Moulin-Papon, 85000 la Roche-sur-Yon, France, [5]Université Clermont Auvergne, INRAE, VetAgroSup, UMR1213 Herbivores, 63122 Saint-Genès Champanelle, France; agathe.cheype@idele.fr

Changes in animal's behaviour may be good indicators of health and welfare variations. However, human observation is time-consuming and labour-intensive. Development of video technology and image processing is a non-invasive method which may offer the opportunity for a better prevention by detecting behavioural and welfare issues continuously and automatically and therefore at an early stage. We are developing deep learning algorithms to analyse routinely the behaviour of young bulls. In the present work, performances of algorithms developed to automatically detect the different activities of bulls on images are evaluated. Bulls originating from 2 different breeds (Limousine, 6 bulls/pen; Charolais, 13±1 bulls/pen) were housed accordingly to the standard management conditions of their respective stations (Pôle de Lanaud, Ferme des Etablières). Two cameras 2 D colour were installed above each pen with different angular views. Nine postures (standing, lying) and behaviours (eating, drinking, moving, autogrooming, fighting, standing up and lying down) were labelled on 1,108 images extracted from the videos. Annotations are evenly distributed with an average of 123 sequences per type of posture or activity and a standard deviation of 37.0. This annotated set of images was used to train the algorithm, an object detection model that uses convolutional neural networks to detect and classify objects in an image. Preliminary training of the algorithm with 419 standing bulls and 373 lying bulls' pictures is promising with 88% sensitivity and 79% precision. Complementary results of algorithm's performances will be presented by valuing the full dataset. This project BeBoP will contribute to the current need for on-farm, operational behavioural welfare indicators that can be easily used to assess not only the individual welfare but also the welfare of the whole group.

Session 67 Theatre 6

Parity and lactation stage preserve a stable structure in the dynamic social networks in cattle

H. Marina[1], I. Hansson[1], I. Ren[1], F. Fikse[2], P.P. Nielsen[3] and L. Rönnegård[1,4]
[1]Swedish University of Agricultural Sciences, Box 7023, 750 07 Uppsala, Sweden, [2]Växa, Ulls väg 26, 756 51 Uppsala, Sweden, [3]RISE Research Institute of Sweden, RISE Ideon, 223 70 Lund, Sweden, [4]Dalarna University, Högskolegatan 2, 791 88 Falun, Sweden; hector.marina@slu.se

Social interactions between cows are a reflection of animal welfare. Positive interactions between dairy cows promote animal welfare and hence increase milk production. A large number of scientific observational studies focused on social interactions in cattle have been published. These types of studies are limited to short periods of time and to a reduced number of animals. Precision Livestock Farming allows the automated monitoring of animals over long periods of time. In this regard, Real-time location systems provide the opportunity to monitor dyadic social interactions of dairy cattle. Spatial proximity between dairy cows has been associated with affiliative social interactions. In this study, we applied the ultra-wideband location technology to explore dyadic social interactions between dairy cows in a commercial free-stall barn with approximately 210 lactating Holstein Friesian cows. Positioning data were collected every second for 14 days. We assumed that a social interaction occurred when a pair of cows spent at least 10 minutes per day within 2.5 m of each other. The spatial interactions observed in the two different functional areas (the feeding area and the resting area) were used to build daily social networks. The main objective of the study was to determine the role of parity and lactation stage in the maintenance of a stable base structure in the dynamic social networks in cattle. Additionally, we analysed the repeatability of the role of the animals in the social networks to improve our understanding of the dynamics of social networks in dairy cattle. Our study enhanced the role of the parity and lactation stage in the preservation of a stable structure in the dynamic social networks in cattle and reveals insights into how the role of animals in social networks evolves over time. Understanding the evolutionary dynamics of social networks in dairy cattle could contribute to the stability of the social structure, promoting animal welfare and production, and to the comprehension of disease transmission in dairy cattle.

Session 67 Theatre 7

Application of rumen boluses to receive welfare parameters from fattening bulls

K. Fromm[1], J. Heinicke[1], T. Amon[1,2] and G. Hoffmann[1]
[1]Leibniz Institute for Agricultural Engineering and Bioeconomy e.V. (ATB), Department Sensors and Modelling, Max-Eyth-Allee 100, 14469 Potsdam, Germany, [2]Institute of Animal Hygiene and Environmental Health, Department of Veterinary Medicine, Freie Universität Berlin, Robert-von-Ostertag-Str. 7-13, 14163 Berlin, Germany; kfromm@atb-potsdam.de

Despite the immense importance of beef production to our economy, there is only a minor amount of publications about physiological and behavioural responses of feedlot cattle. In comparison to female bovines, the gap of data collection from bulls is even larger. Especially in regard to the welfare situation of bulls, the tools to collect these parameters are restricted. Due to safety grounds, it is not always possible to get in close contact to the animals. Therefore, researchers and farmers need techniques which can be operated from a distance e.g. scores, behavioural analysis, surrounding conditions. The last decades several new sensors were brought on the market that allow us to get physiological information from bulls without the necessity of exposing people to greater risks. For several reasons, they have not found place yet in ours fields. We will discuss one of these systems and investigate the advantages and disadvantages compared to other animal monitoring devices. For our current study, we equipped 60 fattening bulls (initial BW=491±53 kg) with rumen boluses (smaXtec SX 2. EU, smaXtec animal care GmbH, Graz, Austria) that are commonly used for fertility prediction in dairy cows, recording rumen temperature, rumination, activity and daily amount of water intake. Fifty subjects obtained classic boluses, which collected the mentioned parameters and ten bulls were equipped with pH plus boluses, which additionally recorded rumen pH. The animals were observed for 180 days. The investigation showed that the use of the boluses is possible and useful in fattening bulls. They enable an early prediction about health status and subacute ruminal acidosis. Acknowledgements The project InnoRind is supported by funds of the Federal Ministry of Food and Agriculture (BMEL) based on a decision of the Parliament of the Federal Republic of Germany via the Federal Office for Agriculture and Food (BLE) under the innovation support program.

Session 67 — Theatre 8

Stepwise modelling for improved bovine health

C.M. Matzhold[1], K.S. Schodl[1], C.E. Egger-Danner[1], F.S. Steininger[1] and P.K. Klimek[2]
[1]ZuchtData EDV-Dienstleistungen GmbH, Dresdner Str. 89, 1200 Wien, Austria, [2]Complexity Science Hub Vienna, Josefstädter Str. 39, 1080 Vienna, Austria; matzhold@zuchtdata.at

The digitisation of livestock farming is generating multiple streams of data that have the potential to significantly improve bovine health when linked and analysed. However, developing models to transform complex integrated datasets into smart decision-making tools often requires a trial-and-error approach. Therefore, we propose a stepwise modelling approach in which specific dimensions of a farm are first analysed separately to reduce data complexity. We analysed over 50 variables from 8,282 farms and an additional 200 for 457 farms. To ensure quality of our analysis, only farms with regular health data reporting are included. The data will be enriched with dynamic information derived from DHI assessments, the National Weather Service, and AMS systems for 1,025 farms, allowing the development of tools to identify at-risk cows based on a score. Our stepwise process allows for more detailed analysis of the effects of individual trait clusters, as demonstrated by our preliminary results on farm traits. These findings indicate that farms with access to alpine pasture exhibit a 50% lower risk of lameness. Leveraging this insight, we developed predictive models for animal health by enriching the dataset with dynamic variables, resulting in a model predicting lameness with high sensitivity and specificity (F1=0.74). We believe that our approach will improve the accuracy of animal health models by enabling individual and collective analysis of datasets, allowing a deeper understanding of their interactions and the development of more accurate models.

Session 67 — Theatre 9

Is milking order connected to social interactions?

I. Hansson, H. Marina and L. Rönnegård
Swedish University of Agricultural Sciences, Animal Breeding and Genetics, Ulls väg 26, Box 7023, 750 07, Uppsala, Sweden; ida.hansson@slu.se

Information on the social interplay between cows in a dairy herd can be essential to improve herd management and enhance animal welfare and health. Cows' movement around the barn and their interactions with other herd mates have shown to be non-random. Cows differ in their tendency to stay close to other individuals, and some cows seem to create preferential bonds with individuals with similar attributes. This study aims to investigate if milking order to a milking parlour is connected to the cow's social role in the herd, with a further aim to examine the relationship to milk yield. The social behaviour of dairy cattle and dyadic interactions can be measured using advanced sensor technology and social network analysis. Positioning data of dairy cattle were collected from an ultra-wideband indoor positioning system in a herd with approximately 200 cows. Cow positions were recorded every second, with an accuracy of 50 cm, and were used to see which cows were spending time in proximity to one another. If cows were within a distance of 2.5 m for at least 10 min per day, they were assumed to interact socially. The interactions were predicted in two functional areas within the barn: the feeding and resting areas. Topological network parameters such as betweenness centrality, closeness centrality, degree and eigenvector centrality scores were used to describe the individual cow's role within the social network. Cows were divided into four groups depending if they entered the milking parlour in the first, second, third, or last group. Ordinal logistic regression models were used with the milking order group as the response variable and the topological network parameters as different fixed effects in each model. The results showed positive associations between all network parameters in the resting area and entering the parlour early. These results indicate that cows that contribute more to the network's connectivity, are more closely connected to the rest of the individuals, and have more interactions in the resting area are more likely to enter the milking parlour at an early stage. Further analyses will be conducted on whether milking order and cows' social interactions affect milk yield.

Session 67 Theatre 10

Combining ultra-wideband (UWB) location and accelerometer data for cattle behaviour monitoring

S. Benaissa[1,2], F. Tuyttens[1,3], D. Plets[2], L. Vandaele[1], L. Martens[2], W. Joseph[2] and B. Sonck[1,4]
[1]Flanders Research Institute for Agriculture, Fisheries and Food (ILVO), Animal sciences, Scheldeweg 68, 9090 Melle, Belgium, [2] Ghent University/imec, Department of Information Technology, iGent-Technologiepark 126, 9052 Ghent, Belgium, [3] Ghent University, Faculty of Veterinary Medicine, Department of Veterinary and Biosciences, Heidestraat 19, 9820, Merelbeke, Belgium, [4]Ghent University, Faculty of Bioscience Engineering, Department of Animal Sciences and Aquatic Ecology, Coupure links 653, 9000 Ghent, Belgium; said.benaissa@ugent.be

Precision livestock farming (PLF) technologies build on the collection and processing of reliable real-time data at the individual level to efficiently classify animal behaviours for improved monitoring and management systems. The current accelerometer-based systems can monitor only a limited number of behaviours, such as feeding and ruminating (neck collar sensors) or lying and standing (leg sensors). Moreover, the accuracy of detection of behaviours that are less frequently expressed in animals, such as walking or drinking, remains a challenge. The aim of this study was to present a novel efficient method to incorporate indoor location and accelerometer data for improved cattle behaviour monitoring systems. In total, 30 dairy cows were fitted with Pozyx tracking tags (Pozyx, Ghent, Belgium) on the upper (dorsal) side of the cow's neck. A total of 123 hours of video recordings were used for validation. Bland-Altman plots for the correlation and difference between the estimated behaviours (sensors) and the ground-truth behaviours (video) were computed for the performance analysis. In overall, the location-based classification into the correct functional areas was very high (R^2=0.99, P<0.001). The combination of location and accelerometer data improved the RMSE per bout of the feeding time and ruminating time compared to the accelerometer data alone (2.6 to 1.4 min). Moreover, the combination of location and accelerometer enabled accurate classification of additional behaviours that are difficult to detect using the accelerometer alone, such as eating concentrates and drinking (R^2=0.85, and 0.90, respectively). This study demonstrates the potential of combining accelerometer and UWB location data for designing a robust monitoring system for dairy cattle.

Session 67 Theatre 11

FEMIR report – the new MIR advising tool

L.M. Dale[1], A. Werner[1], C. Natterer[1], E.J.P. Strang[1], Emissioncow Consortium[2], Remissiondairy Consortium[3] and J. Bieger[1]
[1]Regional association for performance testing in livestock breeding of Baden-Wuerttemberg (LKVBW), Heinrich Baumann Str. 1-3, 70190, Germany, [2]https://www.emission-cow.de/, Adenauerallee 174, 53113 Bonn, Germany, [3]https://remission-dairy.de/, Erlenweg 23, 49324 Melle, Germany; ldale@lkvbw.de

As part of the ReMission Dairy and eMissionCow project, a report called FeMIR has been developed that provides a good description of the energetic situation of the herd over all lactation stages. Moreover, this report is a new tool that allows a better and easier monitoring of the animals' metabolism. In the last years FeMIR report has been tested in the field by four field workers and three consultants, a field test was carried out in which the new parameters i.e. energy efficiency, feed efficiency, nitrogen efficiency and the fatty acids: DeNovo and Preform were compared with the livestock on the trial farms. As part of the field test farm visits in different sides of Baden Württemberg took place in order to set the guidelines for the new parameters. Here, differences became clear, which made it possible to determine that the report is very helpful as a management tool for feeding and monitoring the metabolism of the animals. The report of the respective farm was performed in the MIR spectral data from the monthly milk recording and as well feed samples was taken and conspicuous animals according to the FeMIR were examined. In addition to the handling and use of the report, the limits of each parameter were determined and established to be able to define an optimal framework in which a farm should be. At this step, the respective physical constitution of the animals, as found on site, confirmed the experts' expectations, which they had derived from the efficiency and energy parameters in the report. On all three farms this assessment could be found, which is why the FeMIR report was also rated by all participants as a valuable and suitable management tool for feeding and monitoring the animals' metabolism.

Heat stress relief of dairy cows by evaporative cooling under Mediterranean summer conditions

S. Pinto[1], C. Ammon[1], F. Estellés[2], A. Villagra[3], T. Amon[1,4] and G. Hoffmann[1]
[1]Leibniz Institute for Agricultural Engineering and Bioeconomy e.V. (ATB), Department Sensors and Modelling, Max-Eyth-Allee 100, 14469 Potsdam, Germany, [2]Universitat Politècnica de València, Camino de Vera, Valencia, Spain, [3]Centro de Tecnología Animal (CITA-IVIA), |, Segorbe, Spain, [4]Freie Universität Berlin, Veterinary Medicine, Institute of Animal Hygiene and Environmental Health, Robert-von-Ostertag-Str. 7-13, Berlin, Germany; severinop@atb-potsdam.de

The warm and humid climate during summer in Mediterranean countries presents a particularly high risk for dairy cows in terms of heat stress. The present study aimed to investigate the effectiveness of evaporative cooling on respiration rate (RR) of lactating dairy cows, considering individual animal factors (standing vs laying). A total of 18 Holstein dairy cows (1st to 5th lactation; mean ± SD; 41.64±4 kg milk per day and 150±4.7 days in milk) were randomly selected in July 2016 from the high-yielding group of a farm. The cows were housed in a loose naturally ventilated barn, located in Bétera, Spain. Cooling sessions were implemented three times a day (05:15, 13:15, and 21:15 h) for 45 minutes each, with side and ceiling fans and sprinklers. RR (bpm; breath per minute) of the cows was hourly measured by visually counting of flank movements, and the body posture (standing vs lying) was documented during the data collection. Data were analysed for differences between factor levels via a linear mixed model with repeated measurements at an overall significance level of 0.05. During the experimental period the temperature-humidity index (THI) inside the barn was recorded (76.6±4.1). Standing cows decreased ($P<0.001$) the RR in the cooling (54±11.8 bpm) compared to the cows in the barn (60±18.9 bpm). In contrast, lying cows (71±14.8 bpm) inside the barn showed higher RR ($P<0.001$) than standing cows. Regardless of cow body posture, implementing measures of heat relief as evaporative cooling reduced RR in lactating dairy cows under pressure from hot climate conditions. The project was funded by the FACCE-ERANET Plus Initiative 'Climate Smart Agriculture' in Brussels; by the Federal Office for Agriculture and Food (BLE), and a scholarship from the Coordination for Improvement of Higher Education Personnel (CAPES), Brazil.

Performance of subcutaneous thermochips implanted in dairy cows: preliminary report

A.R. Garcia[1], L.K. Zanetti[2], T.C. Alves[1], L.M. Neira[2], L.F. Pinho[3], A.N. Barreto[3], M.J. Moraes[3], G.G. Ramos[4], C.E. Grudzinski[4] and G.N. Azevedo[2]
[1]Brazilian Agricultural Research Corporation, Rod Washington Luiz, km 234, 13560, São Carlos, Brazil, [2]UNICEP, R Miguel Petroni, 5111, 13563, São Carlos, Brazil, [3]Federal University of Pará, Rod BR 316, 68740-970, Castanhal, Brazil, [4]University of São Paulo, Av Duque de Caxias, 225, 13635, Pirassununga, Brazil; alexandre.garcia@embrapa.br

The standard technique for measuring the internal temperature of cattle is clinical thermometry (CT), but this does not meet the requirement for automation of data collection in the herds. In this context, the use of microchips for monitoring the temperature of animals has been studied but has its limitations. Therefore, our objective was to evaluate the performance of thermochips for predicting internal temperature. Three ½Holstein × ½Jersey cows (85±41 m, 592±132 kg) were used (Bioethical Prot PRT 06/2022). Thermochips (2.1×12 mm, ISO 11784/85, 134.2 kHz) were implanted subcutaneously in the rear udder attachment (RUA), ear base (EB), tail head (TL), and armpit (AP) of each animal. Rectal temperature (RT) was determined using CT as the gold standard. At the same time, the thermochips were read with a universal scanner and the local microclimate data were used to calculate the temperature and humidity index (THI). Evaluations were conducted during the summer at random times of day on eleven alternate days. The average RT was 38.7±0.57 °C. The average temperatures recorded by the microchips were: TL (38.1±0.69 °C), RUA (37.8±0.73 °C), EB (37.6±0.98 °C), AP (37.1±0.99 °C), indicating that the absolute values of TL and RUA were closer to RT. Spearman correlation ($P<0.05$) between thermochip temperatures and RT were: RUA (0.93), TL (0.88), AP (0.84), and EB (0.78). Significant correlations ($P<0.05$) with THI were: EB (0.46), TL (0.39), and RUA (0.38). The lower the correlation with THI, the less influence the thermal environment has on the anatomic region. Conversely, the higher the correlation with RT, the more appropriate the indicator would be, since RT is the most important parameter of homeothermy in cattle. In view of this, implantation of a thermochip in the RUA or TL seems to be a promising strategy for digital monitoring of temperature in dairy cows. Since our results are preliminary, further research on this topic is recommended.

Session 67 Poster 14

Dairy cow behaviour as proxy for heat stress sensitivity in dairy cows

I. Fodor, R.S.C. Rikkers, M. Taghavi and I. Adriaens
Wageningen University and Research, Animal Breeding and Genomics, Droevendaalsesteeg 4, 6708 PB Wageningen, the Netherlands; istvan.fodor@wur.nl

The frequency and severity of extreme weather events are increasing as a result of climate change. To improve and maintain sustainable animal production, it is crucial to quantify and understand the effect of these weather events to our animals. In our research, we aimed to quantify heat stress sensitivity in dairy cows based on its effect on (1) milk production, (2) fertility and (3) health, and predict this sensitivity based on the behavioural changes of the cows. To this end, we quantified individual cow heat stress sensitivity by modelling both the group and the individual production time series of behaviour and production in function of the temperature-humidity index and covariates such as parity and lactation stage with a linear mixed model. By looking at the group trends and the individual trends in relation to the group, we could separate animals that are more or less affected by heat stress. Additionally, re-arranging the residuals of our mixed model in function of the time, also allowed to quantify time-lagged (the so-called recovery or overcompensation period) effects. We found for example that older cows and animals in mid lactation suffer more from heat stress, and that this translates in increased activity, together with more time spent at the drinking troughs and less at the feeding rack. Also within peer-groups, we found that variability across cows exists in how severe individual cows are affected by the events. As with PLF technologies continuous and automated monitoring and comparison with herd-peers is possible, these results offer opportunities for selecting animals that are best adapted to their individual management environment, and for tailoring the management to the individual cows' needs, e.g. by separating animals that are sensitive in better insulated barns or barns sections with fans.

Session 67 Poster 15

Exploring the feeding behaviour of dairy calves: insights from automated milk feeders

K.J. Hemmert[1,2], M.H. Ghaffari[2], T. Förster[1], C. Koch[3] and H. Sauerwein[2]
[1]Förster-Technik GmbH, Gerwigstraße 25, 78234 Engen, Germany, [2]University of Bonn, Institute of Animal Science, Katzenburgweg 7-9, 53115 Bonn, Germany, [3]Hofgut Neumühle, Neumühle 1, 67728 Münchweiler an der Alsenz, Germany; hemmert@uni-bonn.de

Monitoring the feeding behaviour of dairy calves using automated milk feeders (AMF) became common whereby milk intake and drinking rate are used as indicators of satiety, development, and health. Further detailed information about feeding behaviour can be recorded by modern AMF, such as nudging at the teat, corresponding to natural nudging at the dam's udder. This nudging behaviour has hardly been investigated. In the current study, we aimed at comparing this behaviour between the preweaning phase with high feeding level and the weaning period. Female Holstein calves (n=56) group-housed from d 16 of life has access to 12 l/d of milk replacer (MR, 140 g/l) until d 56 when the daily MR allowance was linearly decreased funtil d 98 of life per d. Records of feeding behaviour comprised milk consumption [ml/d], drinking speed [ml/min], number of rewarded and unrewarded visits, and power of nudging bouts [per min./visit]) via the AMF (VARIO smart and add-on HygieneBox, Förster-Technik GmbH, Engen, Germany). Statistical descriptive analysis and paired sample tests (Wilcoxon) of sensor data were performed using Python 3 and JASP 0.16.4. We evaluated 12,659 rewarded visits from the preweaning period and 11,562 from weaning. The average frequency of AMF visits was 8.5×/d preweaning, 13×/d weaning, entire period: 11×/d and resulted in average intakes of 8.2 l/d. The mean drinking speed was 1.24-fold increased during the weaning time as compared to the preweaning period. Nudging activities during and after a meal were increased ($P<0.001$) during the weaning period as compared to the preweaning time, both in average over all calves, and when using the individuals' averages. The results indicate that calves, beside drinking faster, also increased their natural nudging behaviour during weaning in an effort to increase the MR flow. Despite a relatively long milk feeding period of 14 weeks in total, a high daily allowance for 8 weeks, and early *ad libitum* access to concentrate and forage, the calves seemed to have an urge to satisfy their suckling behaviour and hunger with MR.

Session 67 Poster 16

Characterization of behavioural anomalies of lameness in dairy cows using sensor technology

N.L. Mhlongo
Wageningen University & Research, Environmental Sciences, Room L-C-107 Droevendaalsesteeg 3, 6708 PB Wageningen, the Netherlands; nokuthulalorraine.mhlongo@wur.nl

Immediate or gradual behavioural alterations can be important indicators of lameness in dairy cows. Sensor technology offers potential tools to detect these alterations in behaviour. However, to use sensed behavioural patterns as indicators of lameness, differences between lame and healthy cattle need to be known and quantified. Therefore, this study aimed to use sensors to provide distinctive behavioural patterns that are indicative of lameness in dairy cows. Six Dutch dairy farms were visited in 2021 and 2022 from spring to early autumn. Approximately 70 cows that were either healthy or lame were equipped with an accelerometer and GPS for a week to obtain data on mobility characteristics. Visual behavioural observations (15 minutes per observation) were carried out for behavioural classification. During subsequent statistical analyses, descriptive features were computed and later linked to specific behaviours and health status. The results indicate distinct behavioural differences between lame and healthy cows. As expected, lame cows tended to lie down more and had longer individual lying bouts compared to healthy cows. Moreover, the duration lame cows grazed and the frequency at which it occurred decreased in comparison to the healthy cows. The frequency at which lame cows lay down was also higher than those of healthy cows. Overall, lame cows tended to expectedly move at a slower speed compared to healthy cows. Sensor technology is promising and can potentially be used to distinguish the behavioural characteristics of lame from healthy cows. Considering specific behavioural traits expressed by lame cows could further produce clearer lameness indicators in dairy cows.

Session 67 Poster 17

Test of Bluetooth low energy localization system for dairy cows in a barn

J. Maxa[1], D. Nicklas[2], J. Robert[3], S. Steuer[2] and S. Thurner[1]
[1]Bavarian State Research Centre for Agriculture, Institute of Agricultural Engineering and Animal Husbandry, Vöttingerstr. 36, 85354 Freising, Germany, [2]University of Bamberg, Chair of Mobile Systems, An der Weberei 5, 96047 Bamberg, Germany, [3]University of Erlangen-Nürnberg, Institute of Information Technology, Am Wolfsmantel 33, 91058 Erlangen, Germany; jan.maxa@lfl.bayern.de

Sensor-based cattle monitoring in a barn is widely used nowadays, especially on large-scale farms. However, there is a lack of animal monitoring systems that reliably provide data on the location of the cows with high frequency, both in the barn and on the pasture. Therefore, we develop a cost-effective and energy-efficient localization system for cattle in combined barn and pasture conditions within the project 'WeideInsight'. The main aim of this study was to test and evaluate the localization accuracy of the new localization system in the dairy barn of the Staatsgut Almesbach (BaySG) in Bavaria. It consisted of Bluetooth tags (Safectory) attached in a housing (Cattle Data) mounted on collars worn by the cows and Bluetooth Low Energy (BLE) beacons installed in the barn and recording cow's position every 5s. During the trial conducted in 2022, ten Simmental cows from the automatic milking system herd were included in the study. Positions of cows equipped with Bluetooth tags, together with their behaviour, were recorded based on video observations and used as a localization reference. Due to data noise and irregular occurrence of less than three beacons close to the cow, trilateration was impossible to determine the cow's exact position in the barn (e.g. cubicle). Therefore, an alternative approach of proximity-based localization was applied where only the closest beacon was returned. Signal strength indicator (RSSI) filters of -50 and -80 dBm were then applied to positioning data. Applied RSSI of -50 dBm resulted in the highest position accuracy compared to raw data and other RSSI thresholds. Positioning data without the application of RSSI threshold resulted in higher inaccuracy due to reflections based on barn facilities and obstacles, cow bodies and other factors. Even though BLE localization of cows in a barn provides less accurate information, its low cost and energy efficiency can be advantageous for certain cattle housing systems.

Session 67 Poster 18

Integrating inline milk infrared spectra and genomics to predict metabolic profiling in dairy cattle

D. Giannuzzi[1], L.F. Macedo Mota[1], H. Toledo Alvarado[2], S. Pegolo[1], L. Gallo[1], S. Schiavon[1], E. Trevisi[3] and A. Cecchinato[1]
[1]University of Padua, DAFNAE, Legnaro, PD, Italy, [2]National Autonomous University of Mexico, Genetics and Biostatistics, Mexico City, CDMX, Mexico, [3]Catholic University of Sacred Heart, DIANA, Piacenza, PC, Italy; diana.giannuzzi@unipd.it

Metabolic disorders in dairy cattle have negative effects on cows' health and welfare, impairing farm efficiency and sustainability. Diverse serum metabolites are known to be valuable indicators of cows' health status and could be used to identify stress-resilient animals. In the era of high-throughput genotyping and phenotyping, the possibility of blood profiling using prediction from milk samples obtained during daily milking routine would be an attractive strategy. We developed calibration equations for 28 blood metabolites related to energy, liver function, oxidative stress, inflammation, and minerals using milk near-infrared (NIR) spectra collected by the AfiLab instrument and applying diverse machine learning (ML) methods. The dataset comprised 385 Holstein cows reared in 1 herd in Northern Italy, fully equipped with PLF tools. Cows were genotyped using the Geneseek Genomic Profiler Bovine 100K SNP Chip assay. Two models (M) were fitted: M1 considered the NIR spectra together with individual animal data (DIM and parity), M2 considered predictors in M1 plus the genomic information. Model performance was evaluated using a 10-fold random cross validation. Considering the best performing ML method (stacking ensemble), the correlation (r) between the observed and predicted phenotypes ranged from 0.48 for paraoxonase to 0.76 for Na. Respect to M1, the integration of genomic information in M2 increased on average the r of 13% for metabolites related to energy, 11% for liver function, 15% for oxidative stress, 9% for inflammation, and 15% for minerals. In conclusion, our study indicates that integrating NIR milk spectra with individual and genomic data ameliorates predictability and could be considered for the prediction of metabolic profile during daily milking procedure. This study was carried out within the Agritech National Research Center and received funding from the European Union Next-GenerationEU (Piano Nazionale di Ripresa e Resilienza (Pnrr)–Missione 4 Componente 2, Investimento 1.4–D.D. 1032 17/06/2022, CN00000022).

Session 67 Poster 19

Influence on the water intake of lactating dairy cows

J. Heinicke[1], C. Ammon[1], T. Amon[1,2], G. Hoffmann[1] and S. Pinto[1]
[1]Leibniz Institute for Agricultural Engineering and Bioeconomy e.V., Sensors and Modelling, Max-Eyth-Allee 100, 14469 Potsdam, Germany, [2]Institute of Animal Hygiene and Environmental Health, Freie Universität Berlin, Department of Veterinary Medicine, Robert-von-Ostertag-Str. 7-13, 14163 Berlin, Germany; spinto@atb-potsdam.de

The statement 'drink enough' applies to humans as well as animals. Water is essential, it is involved in all life processes. In dairy cows, water deficiency can lead to reduced feed intake, decreased performance and heat tolerance, and increased susceptibility to diseases. For the production of one litre milk, a cow needs 3-4 l water intake. The recording of individual water intake is often difficult because the entire herd drinks from large collection tanks. A technology (rumen bolus) of smaXtec GmbH (Graz, Austria) enables to record the absorbed amount of water per cow and day. The sensor continuously measures the body temperature in the cow's rumen and registers sudden strong drop as a drinking event. Artificial intelligence is used to determine how much water was consumed. The data show that a lactating dairy cow drinks about 80 to 170 l/d. The presented study investigated the influence of milk yield, lactation stage and climate regarding the water intake. The study was conducted in a naturally ventilated dairy barn in Germany. The herd consisted of 160 lactating Holstein Friesian cows from the 2nd to 7th lactation and had an average daily milk yield of 41.52±5.22 kg. Data from January to December 2022 were analysed. Ambient temperature and relative humidity were recorded within the barn. The temperature-humidity index was calculated to assess the heat stress condition. Cows with a milk yield of approx. 30 l/d showed an increase in water intake from 80 to 130 l per cow and day with a temperature change from 5 to 30 °C. While cows with a higher milk yield (approx. 50 l/d) even had an increase in water intake from 120 to 165 l/d. In conclusion, the exact amount of water intake is important to know for farmers and scientists to identify early warnings and research gaps around animal health, performance and heat tolerance. The project is supported by funds of the Federal Ministry of Food and Agriculture based on a decision of the Parliament of the Federal Republic of Germany via the Federal Office for Agriculture and Food.

Sm@RT: Identifying sheep and goats farmers' technological needs and potential solutions

C. Morgan-Davies[1], L. Depuille[2], J.M. Gautier[2], A. McLaren[1], T.W.J. Keady[3], B. McClearn[3], L. Grova[4], P. Piirsalu[5], V. Giovanetti[6], I. Halachmi[7], A. Bar-Shamai[7], R. Klein[8], F. Kenyon[9] and I. Llach-Martinez[10]
[1]SRUC, Hill & Mountain Research Centre, Kirkton farm, Crianlarich, FK20 8RU, United Kingdom, [2]IDELE, Campus INRAe, 31321 Castanet Tolosan, France, [3]Teagasc, Athenry, Co Galway, Ireland, [4]NIBIO, Gunners veg 6, 6630, Tingvoll, Norway, [5]EULS, Fr.R. Kreutzwaldi 1, Tartu 51006, Estonia, [6]AGRIS, Viale Adua, 07100 Sassari, Italy, [7]ARO, The Volcani Centre, 7505101, Rishon LeTsiyon, Israel, [8]UNIDEB, Egyetem Ter 1, Debrecen 4032, Hungary, [9]MRI, Bush Loan, Penicuik, EH26 0PZ, United Kingdom, [10]INRAe, SELMET, Montpellier, France; claire.morgan-davies@sruc.ac.uk

Small ruminant farming systems are important to the sustainability of many European rural communities. Despite recent advances in digital technologies to improve farm practices, uptake by small ruminant stakeholders has been low. Sm@RT (Sm@ll Ruminant Technology) is a Horizon 2020 funded project, involving 8 countries. Sm@RT established focus groups in each country to assess the technology/innovative tool needs of sheep and goats producers regarding 6 topics and proposed solutions. The main needs identified by topic were: (1) grazing/feeding: issues of forage/pasture quality, fencing; (2) for health and welfare: early detection of health issues and diseases and early diagnosis of mastitis; (3) for reproduction: how to optimise AI, animal selection and early pregnancy diagnosis; (4) for herd/flock management: issues of batch management; (5) for fattening: lamb weighing; and (6) for milking, milking machine maintenance. There were differences in the identified needs between countries and system of production (dairy sheep, dairy goats and meat sheep). The 8 Sm@RT countries identified 50 solutions that were subsequently voted by stakeholders during a transnational workshop. Some of the preferred solutions included EID weigh-crate and auto-sorter, milk feeders for kids/lambs, data recording system, automated grass measuring, and milk meters and milking management software. Sm@RT has identified farmers' needs, and identified many existing tools that could help them if adopted and will encourage uptake.

Session 68 Theatre 2

EuroSheep: increasing flock profitability through improved sheep health and nutrition management

P.G. Grisot[1], B. Fança[1], A. Carta[2], S. Salaris[2], C. Morgan-Davies[3], I. Beltran De Heredia[4], R. Ruiz[4], S. Ocak Yetisign[5], T.W.J. Keady[6], B. McClearn[6], R. Klein[7], D. Tsiokos[8] and C. Ligda[8]
[1]Institut de l'Elevage, 570 avenue de la libération, 04100 Manosque, France, [2]AGRIS, Localita Bonassai Ss, 07100 Sassari, Italy, [3]SRUC, West Mains, EH93JG Edinburgh, United Kingdom, [4]NEIKER, Campus Agroalimentario de Arkaute s/n 01192, 01192 Arkaute, Spain, [5]Ondokuz Mayıs University, Department of Animal Science, Faculty of Agriculture, 55139 Samsun, Turkey, [6]Teagasc, Athenry, Co Galway, Ireland, [7]Unideb, Egyetem Ter 1, 4032 Debrecen, Hungary, [8]HAO, Thessaloniki, 57100 Thessaloniki, Greece; pierre-guillaume.grisot@idele.fr

The objective of the thematic network EuroSheep is to increase profitability of sheep production through improved health and nutrition management. EuroSheep involves 8 countries and used multi actor and transdisciplinary approaches, to exchange experience and knowledge among sheep farmers, veterinarians, technicians, advisors and researchers. The project is built on 6 steps starting with the identification of the end users needs and finishing with the definition of a dissemination strategy and a research exploitation. To identify the needs, a survey was developed and launched asking stakeholders what were their main needs regarding the nutrition and health of 3 categories of sheep (adults, lambs and replacements). Nearly 1,300 surveys have been completed across Europe and Turkey. To address the identified needs, 96 solutions (47 on nutrition and 49 on health) were collected as proposed by stakeholders in the different countries. During national workshops, the stakeholders within each country selected the 10 solutions they preferred. In the nutrition category, the 2 solutions preferred were 'nutrition plan of ewe lambs from weaning to mating' and 'guidelines for interpretation of milk urea concentration in sheep milk'. In the health category, the 3 solutions which were preferred were 'design and strategy of the hoof bath', 'booklet on how to recognize lameness' and 'mixed grazing of cattle and sheep to limit parasite infestation'. Following this choice, 44 solutions have been further assessed by the end users in different countries. Persistent gaps to the main needs (needs without a proposed solution) have been identified for further research exploitation.

Sm@RT: Innovative technologies training for small ruminant producers

L. Depuille[1], J.M. Gautier[1], A. McLaren[2], T.W.J. Keady[3], B. McClearn[3], L. Grøva[4], P. Piirsalu[5], V. Giovanetti[6], I. Halachmi[7], A. Bar Shamai[7], R. Klein[8], F. Kenyon[9], I. Llach[10] and C. Morgan-Davies[2]

[1]IDELE, Campus INRAe, 31321 Castanet Tolosan, France, [2]SRUC, West Mains Road, Edinburgh, EH9 3JG, United Kingdom, [3]Teagasc, Athenry, Co Galway, Ireland, [4]NIBIO, Gunnars veg 6, 6630 Tingvoll, Norway, [5]EULS, Fr.R. Kreutzwaldi 1, Tartu 51006, Estonia, [6]AGRIS, Viale Adua, 07100 Sassari, Italy, [7]ARO, The Volcani Centre, 7505101, Rishon LeTsiyon, Israel, [8]UNIDEB, Egyetem Ter 1, Debrecen 4032, Hungary, [9]MRI, Bush Loan, Penicuik, EH26 0PZ, United Kingdom, [10]INRAe UMR SELMET, 2 place Viala, 34060 Montpellier, France; laurence.depuille@idele.fr

Across Europe, there is a low uptake of digital and precision livestock farming technologies and innovative solutions by small ruminant producers. The thematic network Sm@RT (Sm@ll Ruminant Technology), which involves 8 countries, aims to improve this level of uptake by farmers by identifying existing tools and technologies that can help farmers. To date, the network has selected 50 relevant tools. To encourage uptake, Sm@RT is organising training/ demonstration days on digifarms (research/demonstration farms) and innovative commercial farms. During training days, stakeholders can work, and evaluate different tools and technologies in real situations. Farmers attending the events complete questionnaires before and after evaluating a technology, to gauge if their opinions change after the training. The tools vary from simple electronic identification (EID) stick readers to more complex milking machines and virtual fence collars. Demonstration days are held in a different context, relying on peer-to-peer exchanges between farmers. The innovative farmers show other farmers how they use tools or technologies, in practice, on their farm. The demonstration days allow for more discussion between farmers on the benefits and problems of using the tools. A series of online videos are also available showing how tools and technologies are used and can be accessed by those who did not attend the on-farm training days. These approaches will yield invaluable information regarding barriers and drivers to uptake of technologies on small ruminant farms.

EuroSheep: end-users assessments of flock health and nutrition best practices

P.G. Grisot[1], B. Fança[1], A. Carta[2], S. Salaris[2], C. Morgan-Davies[3], I. Beltran De Heredia[4], R. Ruiz[4], S. Ocak yetisign[5], T.W.G. Keady[6], B. McClearn[6], R. Klein[7], L. Perucho[8] and C. Ligda[8]

[1]Institut de l'Elevage, 570 avenue de la libération, Institut de l'Elevage, 04100, Manosque, France, [2]AGRIS, Localita Bonassai Ss, 07100 Sassari, Italy, [3]SRUC, West Mains, EH93JG Edinburgh, United Kingdom, [4]NEIKER, Campus Agroalimentario de Arkaute s/n 01192, 01192 Arkaute, Spain, [5]Ondokuz Mayıs University, Department of Animal Science, Faculty of Agriculture, 55139 Samsun, Turkey, [6]Teagasc, Athenry, Co Galway, Ireland, [7]Unideb, Egyetem Ter 1, 4032 Debrecen, Hungary, [8]HAO, Thessaloniki, 57100 Thessaloniki, Greece; pierre-guillaume.grisot@idele.fr

A series of 96 Best Practices (BP) addressing sheep health and nutrition issues suggested by the 8 partner countries in the EuroSheep thematic network were identified. During the 3rd national workshops series undertaken in each EuroSheep country, stakeholders selected their 10 most preferred best practices from other countries. These 51 preferred BPs have then been the subjects of cost benefit analyses and the sustainability impacts of their implementation have been evaluated. A total of 147 assessments on 44 different best practices (24 health and 20 nutrition) were subsequently completed by farmers who implemented the practice or by farmers, veterinarians, advisors or other stakeholders who experienced the outcomes of the implementation of the practice. Commercial farmers, research farm staff and veterinarians completed 98, 15 and 34 assessments, respectively. Completed assessments indicated whether the respondent implemented the practice, and the equipment, cost, specific labour and prerequisites required to implement it, as well as the overall acceptance, observed benefits, implementation facility and potential limits to implementation. Overall, most of the BPs implemented received a good satisfaction rating (68%) and only 2 received a poor satisfaction rating. For some of them (27%), assessments were contradictory depending on the farming systems.

Session 68 Theatre 5

Sm@RT: main lessons from New Zealand on PLF uptake in small ruminants

J.M. Gautier[1], C. Morgan-Davies[2], L. Depuille[1], A. McLaren[2], B. McClearn[3], L. Grøva[4], P. Piirsalu[5], V. Giovanetti[6], I. Halachmi[7], A. Bar-Shamai[7], R. Klein[8], F. Kenyon[9], E. Gonzalez-Garcia[10] and T.W.J. Keady[3]

[1]Institut de l'Elevage, BP42118, Castanet Tolosan, 31321, France, [2]SRUC, West Mains Road, Edinburgh, EH9 3JG, United Kingdom, [3]Teagasc, Athenry, Co Galway, H65, Ireland, [4]4NIBIO, Gunnars veg 6, Tingvoll, 6630, Norway, [5]EULS, Fr.R. Kreutzwaldi 1, Tartu, 51006, Estonia, [6]AGRIS, Viale Adua, Sassari, 07100, Italy, [7]ARO, The Volcani Centre, Rishon LeTsiyon, 7505101, Israel, [8]UNIDEB, Egyetem Ter 1, Debrecen, 4032, Hungary, [9]MRI, Bush Loan, Penicuik, EH26 0PZ, United Kingdom, [10]INRAe, UMR SELMET, Montpellier, 34000, France; jean-marc.gautier@idele.fr

Sm@RT (Sm@ll Ruminant Technology) is a thematic network, involving 8 countries, with the objective of improving the uptake of digital and precision livestock farming (PLF) technologies by sheep and goat producers, for labour efficiency and farm profitability. In 2023, representatives of the nine partners undertook a fact-finding tour to New Zealand (NZ) to study (1) the use and uptake of innovative PLF approaches; (2) barriers to PLF uptake; and (3) means to facilitate PLF uptake, for the small ruminant sectors. Similar barriers exist in NZ as in EU, namely cost, perceived lack of return on investment, lack of producer interest, additional management input, ease to use, lack of follow-up support, data interoperability and network coverage. A further issue identified in NZ is the absence of compulsory electronic identification (EID) in sheep. Simple tools instead are used at flock level (not at animal level), e.g. weight crate and manual drafting systems to sort lambs according to weight. As in EU, PLF uptake in NZ is higher in the dairy sector (sheep and goat) both for genetic and flock management. For the emerging dairy sheep sector, some companies propose a PLF package that includes digital tools, advice and training for uptake by their suppliers. Findings from the study tour suggest the following steps for PLF uptake: (1) identify issues at farm level; (2) determine if they can be solved without PLF; (3) if not, identify potential relevant PLF tools; (4) purchase and transfer the technology. The use of digital technologies by NZ researchers is important and can inspire EU researchers for digital, landscape management and agroecological studies.

Session 68 Theatre 6

French regional project SO-PERFECTS: project methodology

C. Douine[1], L. Sagot[1,2], A.S. Thudor[1], M. Miquel[2], M. Bernard[1,2], M. Goyenetche[1] and D. Gautier[1,2]

[1]CIIRPO, Le Mourier, 87800 Saint Priest Ligoure, France, [2]Institut de l'élevage, 149 rue de Bercy, 75595 Paris, France; mickael.bernard@idele.fr

Future sheep farms have to combine efficient production with resilience production and adapted production to societal expectations. The attractiveness of the sheep farming profession and adaptation to climate change are also at the heart of this project. The SO-PERFECTS project contributes to providing solutions to these issues by implementing a working method based on operational groups in Nouvelle Aquitaine region. These groups are made up of actors such as breeders, technicians, teachers, researchers, sharing the same issues. In total, 17 partners participated to promote agro-ecological sheep farming systems that combine economic, environmental and social performance. The objective is to guide farmers in answering the set challenges by relying on the experience of innovative farmers. Indeed, many of them are changing their technical and financial choices in order to adapt to climatic and economic hazards. The objective is to share them. In order to identify innovative themes, the partners met and listed all the themes related to agroecology and innovation. 3 groups of themes were selected: forage resource, herd health, adaptation of breeding systems. Ideas related to ongoing projects, such as tannin plants, were excluded. When the themes were chosen, bibliographic research was carried out in order to identify technical and scientific knowledge. Trials were then set up in livestock farms, agricultural high schools and experimental sites. In total, 15 themes were worked on with 122 farmers who participated in the construction of innovative solutions and the testing of solutions before their transfer. 93 studies were carried out on farms or in agriculture high schools. The results were synthesized in the form of technical sheets, slide shows, panels, press articles and a motion design.

Session 68 Theatre 7

French regional project SO-PERFECTS: trial results

M. Bernard[1,2], L. Sagot[1,2], A.S. Thudor[1], C. Douine[1], M. Miquel[2], M. Goyenetche[1] and D. Gautier[1,2]
[1]CIIRPO, Le Mourier, 87800 Saint Priest Ligoure, France, [2]Institut de l'élevage, 149 rue de Bercy, 75595 Paris, France; mickael.bernard@idele.fr

Various topics were studied in this project. Two of them are presented here: identification of success factor for spring mating and the interest of shearing grass-fed lambs returned to the sheepfold. In spring, some sheep breeds have the ability to reproduce in natural mating, but success rates are subject to wide variations. In 2020 and 2021, 3,459 ewes of different breeds from 9 farms in the Nouvelle-Aquitaine region were monitored. Various measurements, related to the management of the herd, were carried out on animals, including a body condition score. The average fertility rate was 71% with two farms below 50% and five above 80%. Three factors seem to be of prime importance for successful mating: the interval between lambing and mating must be sufficient, with 80% success rate when it is greater than 160 days. The ewes must be in a dynamic weight gain or in good body condition and the duration of the mating must be at least 3 cycles (51 days). Some farmers shear their grass-fed lambs when they return to the sheepfold, the main reasons being to save on concentrate feed and animal welfare. Few recent references exist on this topic. Thus, 7 trials were carried out between 2020 and 2022 in an experimental station and in French farmers in Haute-Vienne (87) and Creuse (23). In total, 448 lambs were divided into two groups, sheared or not at the entrance to the sheepfold. Various measurements were made on animals, regarding performance, feeding and animal welfare. On average, shearing improves performances little with a 3-day reduction in fattening time for sheared lambs, but there were large differences between farms. During high heat (30-35 °C), shearing significantly improves lamb comfort with a 50% reduction in panting time and the fleeces are much cleaner.

Session 68 Theatre 8

Use of innovative and precision tools in research stations with small ruminants: the INRAE case

I. Llach[1], H. Caillat[2], A. Fatet[2], S. Breton[3], T. Aguirre-Lavin[4], D. Dubreuil[4], A. Eymard[5], J. Boucherot[6], T. Fassier[6], D. Marcon[6], S. Parisot[7], C. Durand[7], G. Bonnafe[7], D. Portes[7], C. Morgan-Davies[8] and E. González-García[1]
[1]INRAE, UMR SELMET, 34060 Montpellier, France, [2]INRAE UE1373 FERLus, Les Verrines, 86600 Lusignan, France, [3]INRAE, UE1277 PFIE, 37380 Nouzilly, France, [4]INRAE UE1297 PAO, CR Tours, 37380 Nouzilly, France, [5]INRAE UMR0791 MoSAR Chèvrerie expérimentale, Route de la ferme, 78850 Thiverval-Grignon, France, [6]INRAE UE0332 P3R, La Sapinière, 18390 Osmoy, France, [7]INRAE, UE0321 La Fage, 12250 Roquefort-sur-Soulzon, France, [8]SRUC, West Mains Road, EH9 3JG Edinburgh, United Kingdom; eliel.gonzalez-garcia@inrae.fr

An extensive survey was carried out in experimental units (EU) of INRAE using small ruminants (SR), to get insights in current and historical uses of innovative technologies in their facilities, and staff viewpoints. Ten EU use SR in INRAE (in France and overseas); 6 were visited in 2022 (addresses in the abstract; 3 with sheep -2 meat, 2 both meat and dairy-; and 3 with dairy goats). A detailed questionnaire was prepared. A total of 78 technologies were inventoried. From that, ~10% are invented or co-produced by INRAE, from which 7 are appreciated i.e. automate weighing device (Baléa); sorting gates (3 exits); DH20 (water consumption monitoring and weighing indoor); DAC (automatic distributor of concentrate); DAF (automatic distributor of forage); Gély test tube (individual milk yield monitoring); and Walk-over-Weighing (WoW). Five tools are used by 100% of EU i.e. EID for individual identification; Baléa for weighing; PDA (Personal Digital Assistant); temperature and humidity sensors (mandatory); and sorting gates. Interviewed staff are favourable to techs' use, but mostly for research purposes and they unanimously agreed in positive effects to alleviate workload and routine. Internet connectivity was revealed however as a serious constraint in certain areas. Four techs are recommended for farmers, recognising price may limit adoption: conveyor belt for feeding supply indoor; mixer (with tractor) for preparing total mixed rations; milk tank weighing; Combi clamp (to ease handling). The P3R EU is the best example of phenotyping EU with promising and effective techs for both research and management purposes. Information will be completed, with further upcoming visits to 100% of EUs.

Session 68 Theatre 9

FEC check: development of an online tool to aid farmer understanding of roundworm faecal egg counts

E. Geddes[1], A. Duncan[2,3], K. Lamont[3], J. Duncan[1], F. Kenyon[1] and L. Melville[1]
[1]Moredun Research Institute, Edinburgh, EH260PZ, United Kingdom, [2]University of the Highlands and Islands, Inverness, IV2 5NA, United Kingdom, [3]Scotland's Rural College (SRUC), Centre for Epidemiology and Planetary Health, Inverness, IV2 5NA, United Kingdom; eilidh.geddes@moredun.ac.uk

Faecal egg counts (FECs) are a simple, inexpensive, and accessible tool for sheep producers to monitor the gastrointestinal nematode (GIN) challenge facing their livestock. GINs are the cause of significant production losses to the industry, and now with increasing anthelmintic resistance and pressure to reach environmental goals, sustainable and effective control is key. FECs are being increasingly utilised by producers to guide treatment timings, estimate pasture contamination and test anthelmintic efficacy. However, interpretation can be challenging. To support the effective interpretation of FEC results, a free web-based application ('FEC Check') was co-designed with stakeholders. A prototype tool was developed using R Shiny which provides farmers with a visual representation of the clinical importance of their FEC results. This is accompanied by guidance to support effective decision-making. Subsequently, to ensure the app is tailored to stakeholder needs, the prototype was trialled by farmers, veterinary clinicians, and animal health advisors in four focus groups across two geographically distinct locations in Scotland. The app was trialled by 33 stakeholders, composed of 17 farmers, and 16 vets/advisors. Most farmers (87%) currently used FECs on their farm, however agreed that the level of interpretation and clinical guidance provided with the results varied substantially between test providers. Upon introduction to the app, both the farmers and advisors liked the simplicity of the design. They also highlighted the benefits of being able to download the results to build a picture of the parasite challenge and anthelmintic efficacy over time, information which could be used for future health planning and breeding stock selection. With further development to optimise the app for smartphones and the ability to handle longitudinal data, all participants agreed that this tool could improve utilisation and understanding of FEC results within the industry.

Session 68 Theatre 10

Assessing sheep behaviour in an human-animal interaction test using infrared termography

M. Almeida[1,2,3], A. Afonso[1], C. Guedes[1,2,3] and S. Silva[1,2,3]
[1]University of Trás-os-Montes e Alto Douro, Quinta de Prados, 5000-801, Vila Real, Portugal, [2]Associate Laboratory for Animal and Veterinary Sciences (AL4AnimalS), Portugal, Portugal, Portugal, [3]Veterinary and Animal Research Centre (CECAV), University of Trás-os-Montes e Alto Douro, Quinta de Prados, 5000-801, Vila Real, Portugal; mdantas@utad.pt

The objective of this work was to evaluate behaviour and stress in sheep of the Ile de-France (IF), and Churra da Terra Quente (CTQ) breeds subjected to a human-animal interaction test (arena test). Thermographic imaging analysis was performed as a non-invasive stress assessment tool. The avoidance distance in the pen (ADP) and ocular temperature (IRT) were assessed before and after the presence of an operator. Several behaviours were also registered using an ethogram. It was found that, in general, IF sheep are more reactive to the presence of the operator, which translates into a higher ADP (3.41 vs 2.66, $P<0.05$, respectively). The IRT also shows that IF ewes show lower eye temperature than CTQs (34.47 vs 34.60 ° C, $P<0.05$, respectively). These indicators were also associated with higher time to resume activity, showing that IF ewes might be more reactive to human presence. It was possible to conclude that in general there are differences between the two breeds of sheep, for behavioural indicators, which supports the theory that IF sheep are more reactive than CTQ. Finally, it is possible to conclude that thermography may be a useful non-invasive tool to assess sheep behaviour. This work was supported by the projects UIDP/CVT/00772/2020 and LA/P/0059/2020 funded by the Portuguese Foundation for Science and Technology (FCT).

Session 69 Theatre 1

Detection of interchromosomal rearrangements in bulls using large genotype and phenotype datasets

J. Jourdain[1,2], H. Barasc[3], T. Faraut[3], C. Grohs[1], C. Donnadieu[4], A. Pinton[3], D. Boichard[1] and A. Capitan[1,2]
[1]Université Paris-Saclay, INRAE, AgroParisTech, GABI, G2B, Domaine de Vilvert, 78350 Jouy en Josas, France, [2]Eliance, 149, Rue de Bercy, 75012 PARIS, France, [3]GenPhySE, Université de Toulouse, INRAE, ENVT, 23 Chemin des Capelles, 31320 Castanet-Tolosan, France, [4]INRAE, US 1426, Université Fédérale de Toulouse, GeT-PlaGe, Genotoul, France Génomique, 24 chemin de borde rouge, Auzeville, 31326 Castanet-Tolosan, France; jeanlin.jourdain@inrae.fr

Interchromosomal rearrangements (IR), which result from the transfer of genetic material from one chromosome to another, can have severe phenotypic consequences due to gene dosage defects. Artificial insemination (AI) bulls are not currently screened for IR before their semen is used. Most IR have been detected by surveillance for the t(1;29) Robertsonian fusion or by targeted controls of low fertility bulls. Here, we developed a method to detect IR using linkage disequilibrium (LD) across chromosomes and applied it to 5,571 paternal halfsib families genotyped for genomic evaluation. Thirteen progeny groups (0.23%) showed significant LD, and 12 were confirmed by cytogenetic analyses: one Robertsonian fusion, 10 reciprocal translocations, and the first case of insertional translocation reported in cattle. Using national databases, reciprocal translocation carriers were all found in the worst percentile of their breed for male fertility and their carrier daughters were subfertile. The insertional translocation carrier was the worst bull for mortality with 44% of daughters dying before 365 days of age. We used long-read sequences for 7 bulls to characterize breakpoints and detect genes putatively affected in their expression, possibly leading to deviating phenotypes in balanced progeny, such as the observed delayed growth and high death rate in some balanced daughters. In addition to systematic karyotyping of bulls to avoid IR diffusion, our study highlights the importance of computing LD in halfsib groups with our method to detect remaining smaller IR and to manage carriers in the population. This study is the most comprehensive scan of the cattle population scan and paves the way for follow-up studies on the origins and consequences of IR. JJ is a recipient of a CIFRE PhD grant with the financial support of ANRT and APIS-GENE.

Session 69 Theatre 2

Functional information embedded in the unmapped short reads of whole-genome sequencing

G.B. Neumann[1], P. Korkuć[1], M. Reißmann[1], M.J. Wolf[2], K. May[2], S. König[2] and G.A. Brockmann[1]
[1]Humboldt-Universität zu Berlin, Albrecht Daniel Thaer-Institute for Agricultural and Horticultural Sciences, Invalidenstrasse 42, Ostbau, 10115 Berlin, Germany, [2]Justus-Liebig-Universität, Institute of Animal Breeding and Genetics, Ludwigstr. 21, 35390 Gießen, Germany; guilherme.neumann@hu-berlin.de

Livestock genomics involves resequencing genetic information from additional individuals of the same species with an established reference genome. Short-read sequencing generates DNA sequences of 150-300 bp, which are mapped to the reference genome. Unmapped reads are typically discarded, but they may contain information about pathogenic DNA and structural variants (SVs) not present in the reference genome. In this study, we analysed unmapped reads from whole-genome sequencing of 302 German Black Pied cattle (DSN) to explore these hypotheses. The unmapped reads retrieved from the 302 DSN animals were assembled into scaffolds and blasted against the NCBI's database for reference and representative genomes of all available species. Scaffolds mapping against those genomes covering at least 10% of their respective reference genomes were kept for further analysis. SVs were detected for both mapped and unmapped short reads using Delly and SvABA, respectively. Among the unmapped reads of 302 sequenced DSN animals, 116 contained assembled DNA sequences covering >10% of the genome of eleven different species of bacteria and six viruses. Of those species, Mycoplasma and bovine parvovirus 3 are known pathogens infecting cattle. DNA sequences covering more than 10% of foreign eukaryotes were not found in the assembled unmapped reads. All 302 DSN animals had assembled unmapped reads aligned to Bos species with an average length of 341.3 kb per animal. Using unmapped reads, 26,593 SVs were detected in addition to 19,410 SVs that had been detected with mapped reads. While metagenomics is typically used to detect pathogens, unmapped reads of whole-genome sequencing can also provide valuable information on the presence of pathogens which is crucial for epidemiologists. Furthermore, the examination of unmapped reads revealed additional SVs that could not be identified through the alignment to the *Bos taurus* reference genome. This emphasizes the need for high-quality long sequence reads to accurately detect SVs.

Session 69 Theatre 3

Expanding the capabilities of single-step GWAS with *P*-values for large genotyped populations

N. Galoro Leite, M. Bermann, S. Tsuruta, I. Misztal and D. Lourenco
University of Georgia, 425 River Rd, 30602, Athens GA, USA; mbermann@uga.edu

With the availability of genomic information, there is an increasing interest in genome-wide association studies (GWAS). Most of the methods used for GWAS can account for population structure but do not consider phenotypes for non-genotyped individuals. Single-step methods can combine information on genotyped and non-genotyped individuals because of the use of a joint pedigree and genomic relationship matrix. Single-step GWAS can be utilized in large genotyped populations; however, only SNP effects and variance explained were possible in such a case. A significance test based on *P*-values was available only for small genotyped populations, i.e. up to 50k, depending on the model. This is because *P*-values rely on the prediction error variance (PEV) for each SNP effect, which is backsolved from the prediction error covariance for animals, requiring the inverse of the left-hand side (LHS) of the mixed model equations. When using more than 50k genotyped animals, single-step genomic BLUP (ssGBLUP) methods rely on a sparse representation of the inverse of the genomic relationship matrix (G^{-1}) computed with the APY algorithm. In APY, genotyped animals are split into core and noncore, and recursions on the core animals are used to compute G^{-1} at a low cost. With APY, the PEV of SNP effects relies only on the information of core animals, varying from 4k to 6k in pigs and chickens and 10k to 15k in cattle. Instead of using the inverse of the LHS, we approximated the PEV for core animals, then computed PEV and obtained *P*-values for SNP effects. We obtained similar GWAS resolution between the inverse and approximation using 50k genotyped animals, 1.5M animals in the pedigree, and 850k phenotypes. We then ran ssGWAS with *P*-values using 450k genotyped animals. As expected, a better resolution was observed when using 450k genotyped animals compared to 50k. Having single-step GWAS for large genotyped populations is feasible and allows including all available data in association studies.

Session 69 Theatre 4

The Life-Functions Ratio: a new indicator trait of trade-offs to go beyond genetic correlations

N. Bedere[1], O. Cado[1], N.C. Friggens[2] and P. Le Roy[1]
[1]PEGASE, INRAE, Institut Agro, 35590, Saint-Gilles, France, [2]Université Paris-Saclay, INRAE, AgroParisTech, UMR Modélisation Systémique Appliquée aux Ruminants, 91120, Palaiseau, France; nicolas.bedere@inrae.fr

We aim to study the adaptation abilities of a genotype in terms of both the ability to be consistent across different environments (i.e. robustness) and the ability to cope with perturbations of the environment (i.e. resilience). This means making trade-offs between life functions, especially when resources are limited. Usually, to study those trade-offs between functions, geneticists calculate genetic correlations, which indicate the common genetic share of two traits, and thus common requirements for biological processes and resources. The main drawback of genetic correlations is that they are population parameters. However, we know that individuals in a population will make different trade-offs. To study the inter-individual variability of trade-offs, we have developed an individualized indicator based on resource acquisition and allocation concepts: the Life-Functions Ratio (LFR). We focused on the trade-off between two traits first, to establish the link between LFR and the genetic correlation between both traits. LFR is defined as a product of the transformed traits. It is mathematically related to the conversion efficiencies of both traits and the resource allocation coefficient (alpha). To describe this new indicator trait, LFR between backfat and egg number was calculated on a pedigreed purebred layers' population from the breeding company Novogen, late in the laying period (80 wk of age), when a trade-off occurs between body reserves and egg production. Traits displayed moderate heritabilities: 0.23 for egg number, 0.52 for backfat, 0.45 for LFR, and 0.31 for alpha. LFR was favourably genetically correlated with backfat (+0.79) and with egg number (+0.76). LFR was genetically very different from alpha with a genetic correlation of 0.02. We are currently exploring the genetic architecture of LFR, expecting to find cryptic quantitative trait loci for backfat and egg number. We are also exploring the effect of selecting for LFR on the initial genetic correlation across generations through simulations. Future works will focus on integrating more traits into the equation.

Session 69 Theatre 5

Impact of pedigree errors on the quality of predicted genetic merit from animal models

E.C.G. Pimentel, C. Edel, R. Emmerling and K.-U. Götz
Bavarian State Research Center for Agriculture, Institute of Animal Breeding, Prof.-Dürrwaechter-Platz 1, 85586 Poing-Grub, Germany; eduardo.pimentel@lfl.bayern.de

Pedigrees used in genetic evaluations contain errors. Because of such errors, assumptions regarding the relatedness among individuals in genetic evaluation models are wrong, which may have an impact on prediction quality. In the Single-Step model, which combines both pedigree and genomic information, pedigree errors also cause inconsistencies between the pedigree-based relationship matrix A and the genomic relationship matrix G. The objective of this work was to investigate the effects of pedigree errors on the quality of predicted genetic merit. We used a real pedigree (n=361,980) and real genotypes (n=25,950) of Fleckvieh cattle, sampled in a way to provide a good consistency between A and G. Given the real pedigree and genotypes, true breeding values were simulated to have a covariance structure equal to the matrix H assumed in a Single-Step model. Based on true breeding values, phenotypes were simulated with a heritability of 0.25. Genetic evaluations were conducted with a conventional animal model (i.e. without genomic information) and a Single-Step animal model under scenarios using either the correct pedigree or a pedigree containing 5, 10 or 20% of wrong records. Prediction quality was assessed in terms of correlation and regression of true on estimated breeding values, as well as standard deviation of estimated breeding values. These metrics were calculated for all animals and for different groups of animals, such as progeny tested bulls and young selection candidates. The increasing rates of pedigree errors led to decreasing correlations between true and estimated breeding values, larger deviations of the regression coefficients from the expected values and lower standard deviations of predictions. We further investigated the impact of data truncation on the measures of prediction quality under the different simulated scenarios.

Session 69 Theatre 6

Bias in estimated variance components and breeding values due to pre-correction of systematic effect

P. Duenk and P. Bijma
Wageningen University and Research, Animal Breeding and Genomics, Droevendaalsesteeg 1, 6708 PB Wageningen, the Netherlands; pascal.duenk@wur.nl

In animal breeding, the size of data sets for estimation of heritabilities and breeding values is growing all the time, also because of the increasing availability of genomic data. To be able to analyse such large data sets efficiently, two-step procedures are common. In the first step, systematic effects are estimated as fixed or random effects, for e.g. the sex, herd or litter of the individual. The data is then pre-corrected by subtracting the estimated effects from the raw phenotypes. In the second step, heritabilities or (genomic) breeding values are estimated from the pre-corrected phenotypes. This two-step procedure simplifies the analysis of big data. Even though pre-correction is widely used in animal breeding, the consequences for estimated variance components and breeding values have received little attention. Our results show that pre-correction creates bias in the estimated variance components. For balanced data, we provide simple mathematical expressions for the bias, showing that variance components are underestimated unless the reliability of the pre-correction is one. Bias occurs both with fixed and random effect pre-correction, albeit the use of fixed effects resulted in stronger bias. These theoretical results were consistent with the results from simple simulations. For realistic reliabilities of systematic effects, bias of estimated variance components was about 10% when systematic effects were fitted as random, and about 16% when they were fitted as fixed. We will present further results on the consequences of pre-correction for estimated variance components and (genomic) breeding values. Those results include simulation of more realistic data sets with family structure and genotypes, and the use of an actual data set from a Dutch breeding company. Our research will help to better understand the effects of data pre-correction, and will ideally provide a solution for the resulting bias.

Improving computing performance of genomic evaluations by genotype and phenotype truncation

F. Bussiman[1], C. Cheng[2], J. Holl[2], A. Legarra[3,4], I. Misztal[1] and D. Lourenco[1]
[1]University of Georgia, Animal and Dairy Science, 425 River Rd, 30602, USA, [2]Pig Improvement Company, 100 Bluegrass Commons Blvd, Ste 2200, 37075, USA, [3]Council on Dairy Cattle Breeding, 4201 Northview Dr, 20716, USA, [4]INRA, GenPhySE, 23 Avenue des Capeles, 31076, France; fob@uga.edu

The use of historical data collection is a common practice for genomic evaluation; however, the more data, the more computing power is needed, and fitting the same model to both current and historical data may be inappropriate. This study investigated the use of data truncation to reduce computing costs of genomic prediction with large datasets. Data truncation is usually done for phenotypes and pedigree; however, the increasing number of genotyped animals raises a question on whether all genotyped animals should be included in the evaluations. Data from terminal and maternal pig lines were analysed through different truncation scenarios: TARGET, which removed genotyped animals without own and progeny phenotypes; AGE, where old genotyped animals were removed; TARGET + AGE, which combined the two scenarios above; and ALLGEN, which kept all genotyped animals. Phenotypes were removed (in all scenarios) based on birth year, and pedigree depths were 2, 3, or all generations. We analysed growth and mortality on a terminal line and preweaning and reproductive traits on a maternal line using single-step GBLUP. Validation was based on the linear regression method, and the focal animals were the selection candidates born in 2019. Reliability did not change among scenarios except when truncating the phenotypes in less than two generations of data. The phenotype/genotype truncation or the pedigree depths did not affect the dispersion, except when using the entire pedigree with truncated phenotypes and genotypes. Tracing up to three generations of pedigree is enough for reliable predictions. Data truncation caused a slight drop in reliability when genotyped animals had no phenotypes. Keeping all genotyped animals for truncated datasets did not increase the reliability and resulted in poor convergence. Removing unneeded genotypes, phenotypes, and pedigree increases computing efficiency by up to 90% without compromising predictions for selection candidates.

A single-step evaluation of functional longevity of cows including data from correlated traits

L.H. Maugan[1], T. Tribout[1], R. Rostellato[2], S. Mattalia[3] and V. Ducrocq[1]
[1]GABI, Université Paris Saclay, INRAE, AgroParisTech, 78350 Jouy-en-Josas, France, [2]Geneval, 3 rue du Petit Robinson, 78350 Jouy-en-Josas, France, [3]Institut de l'Elevage, UMT eBIS, 78350 Jouy-en-Josas, France; vincent.ducrocq@inrae.fr

A routine genetic evaluation of dairy bulls based on the length of productive live (LPL) of their daughters corrected for milk production was developed in France in 1997. Initially, only the functional longevity (FL) breeding values of males were available. The FL genetic evaluation was then progressively improved by accounting for time-dependent changes in environmental factors and by computing approximate FL evaluations for cows. It was also found during the following decade that there were routinely collected traits related to fertility, conformation or udder health which could serve as predictors of FL, using an approximate multiple trait approach. Combining these direct and indirect sources of information led to more robust FL genetic evaluations of bulls and cows. With strong selection on production, the economic importance of functional traits such as FL and its predictors gradually increased. At the same time, genomic evaluations and selection became central. However, genomic evaluations of low heritability traits such as FL were usually characterized by poor reliabilities. We propose a 'combined' Single-Step (CSS) approach to obtain genomic evaluations (GEBV) of FL mimicking multiple-trait evaluations. This is illustrated in the context of the Montbéliarde breed, using or not information from predictor traits. Survival curves of a complete cohort of cows born in 2014-2015 were calculated, showing a large superiority of the CSS evaluation: among genotyped cows, 50% of the best 10% cows with the univariate SS reach a productive life of about 1,400 days, i.e. 402 days more than the worst 10% cows. This difference reaches 501 days with the CSS evaluation. The corresponding figures for ungenotyped animals are substantially smaller (187 vs 258 days). In other terms, CSS evaluations allow farmers to detect at birth the genotyped heifers that are more likely to have a long productive life in their herd. In contrast, CSS evaluations of males are much less pertinent to select young bulls, because selection intensity of males is much higher in selection programs.

Session 69 Theatre 9

Expected values of genomic prediction validation parameters for non-random validation sets

M.P.L. Calus, M. Schrauf, T. Pook, L. Ayres, R. Bonifazi, J. Ten Napel and J. Vandenplas
Wageningen University & Research, Animal Breeding and Genomics, P.O. Box 338, 6700 AH Wageningen, the Netherlands; mario.calus@wur.nl

Dispersion bias of genomic estimated breeding values (GEBV) measures whether GEBV contain too much or too little variation. Dispersion bias of GEBV is evaluated by regressing a variable that reflects true breeding values (TBV) on the GEBV being validated. For GEBV without dispersion bias, the expected value for the regression coefficient (b1) estimator is usually considered to be 1, while it may deviate from 1 if the validation set is not a random sample of animals. We aimed to compute expected b1 values for final GEBV with accuracies ranging from 0.4 to 0.9, for a set of validation animals that was progressively more intensely selected (100 to 10%) based on initial GEBV with an accuracy of 0.3. TBV, initial GEBV and final GEBV were drawn from a trivariate normal distribution, where information used to estimate initial GEBV was assumed to also be used to estimate final GEBV (dependent final GEBV), or not (independent final GEBV). For independent final GEBV of unselected validation groups, b1 values were 1 as expected. With 500 animals in the validation group, standard deviations (SD) of b1 were 0.11 and 0.02 for final GEBV with accuracies of 0.4 and 0.9. With 5,000 validation animals the SDs reduced to 0.03 and 0.01. For a selected validation set of 10% of the full data, b1 values were 0.93 (0.98) for final GEBV with an accuracy of 0.4 (0.9), while SDs of b1 increased up to three-fold compared to validating on the full data. For dependent final GEBV, b1 values were all ~1, while SDs were similar as for independent final GEBV. Accuracies, computed as correlations between final GEBV and TBV, were more reduced with more intense selection based on initial GEBV; this reduction was relatively small with independent final GEBV, and considerably larger for dependent final GEBV. Considered scenarios will be extended to other practical situations where validation is based on comparing GEBV based on partial and whole data. It is concluded that this approach is useful to efficiently compute expected values for parameters used in validation of genomic prediction.

Session 69 Theatre 10

Unknown parent groups and metafounders in genomic evaluation of Norwegian Red cattle

T.K. Belay[1], A.B. Gjuvsland[2], J. Jenko[2], L.S. Eikje[2] and T. Meuwissen[1]
[1]Norwegian University of Life Sciences, Animal and Aquacultural Sciences, Oluf Thesens vei 6, 1433 ÅS, Norway, [2]GENO SA, Storhamargata 44, 2317 Hamar, Norway; tesfaye.kebede.belay@nmbu.no

Appropriate ways of accounting for missing parents and incompatibility between the base populations of the pedigree-based (A) and the genomic relationship (G) matrix are crucial for unbiased single-step genomic prediction. Here, we tested the effects of alternative ways of fitting unknown parent groups (UPG) on biases (level-bias and inflation) and stability (ratio of accuracies) of genomic predictions and genetic trends in Norwegian Red cattle. We fitted the UPG as random effects accounting for relationships among all the 52 UPGs, calculated using a recursive method (UPGγ) and among 14 metafounders (MF). We also fitted UPG as in the recently published method, 'Q-Q+', which fits fixed group effects corrected for the part that can be explained by genotypes. The models were evaluated using cross-validation by masking phenotypes of 5,000 young-genotyped cows. Inclusion of relationships among the UPGs or MF introduced biases to the genomic predictions compared to other models where these relationships were not considered. The biases were further increased when G both weighted by 10% A and scaled by alpha (mean difference between A and G) was used in the UPGγ models. The Q-Q+ models performed best in terms of inflation and stability but had higher level bias than the routine model where relationships among UPGs were not considered. Stability of genomic predictions was comparable across the other genomic models with the exception for MF which showed less stability. The Q-Q+ models predicted the highest genetic trends while the MF model predicted the lowest genetic trends, and the genetic trends for the UPGγ models were in between the Q-Q+ and MF models. The genetic trend for routine model was slightly below the Q-Q+ models. In conclusion, the routine method performed best in terms of level bias while the Q-Q+ model with 0.5 allele frequency and unweighted G produced the lowest inflation and most stable genomic predictions. Inclusion of relationships among the UPGs introduced biases especially when scaled G was used and reduced predicted genetic trends. The latter is probably due to the assumed correlations between early and recent UPGs.

Session 69 Theatre 11

Exploring non linear genetic relationships between correlated traits

F. Shokor[1,2], P. Croiseau[1], R. Saintilan[1,2], T. Mary-Huard[1], H. Gangloff[1] and B.C.D. Cuyabano[1]
[1]Université Paris Saclay, INRAE, AgroParisTech, GABI, Domaine de Vilvert, 78350 Jouy-en-Josas, France, [2]Eliance, 149 Rue de Bercy, 75012 Paris, France; fatima.shokor@inrae.fr

Genetic evaluation has emerged as a crucial tool in livestock breeding, enabling decisions about which individuals to keep in a breeding program, based on a selection index that oftentimes considers multiple traits of commercial interest, and the accuracy of selection indexes depends on the accuracy of breeding values (BV) predicted by the model. Different traits may be affected by a same QTL region, resulting in a genetic correlation between such traits, and the current models used to perform genetic evaluations impose a linear genetic correlation between traits. However, if the genetic relationship between two traits is non-linear, the current models fail to comprise this assumption, resulting in less accurate predicted BV. A preliminary study with simulated data indicated that when using gaussian models (i.e. GBLUP) to evaluate two traits with the same heritability (0.3) and genetic correlation of 0.5, the estimated genetic correlation captured by the model was of 0.53 (SD=0.03) when the true genetic correlation was linear. However, when this correlation was non-linear (and in fact quadratic in this preliminary study), the captured correlation decreased to 0.46 (SD=0.02), and GBLUP failed to identify the non-linear genetic relationship. Methods able to account for non-linear genetic relationships between traits may increase the accuracy of predicted BV, and improve the individuals' final selection index, enabling more efficient breeding strategies. The non-parametric nature of machine learning (ML) methods provide the necessary flexibility to account for non-linear genetic correlations, and may thus provide new insights into the underlying genetic architecture of correlated traits. Our preliminary study indicated that, although prediction accuracy obtained with a two-trait ML model was comparable to that of GBLUP for both linearly and non-linearly correlated traits (0.52; SD=0.3 and 0.48; SD=0.03 respectively), our ML approach was able to reveal the non-linear relationship between traits, an information of relevance to improve the selection index accounting for multiple traits.

Session 69 Theatre 12

Single-Step Genomic Prediction in six German Beef Cattle Breeds

D. Adekale[1,2], H. Alkhoder[2], Z. Liu[2], D. Segelke[2] and J. Tetens[1]
[1]Georg-August-Universität Göttingen, Department für Nutztierwissenschaften (Abteilung Functional Breeding, Burckhardtweg 2, 37077 Göttingen, Germany, [2]vereinigte informationssysteme tierhaltung w.v, Biometrie, Heinrich-Schröder-Weg 1, 27283 Verden (Aller), Germany; damilola.adekale@vit.de

The estimation of breeding values for production traits in German Beef cattle breeds is carried out routinely by a multi-trait pedigree BLUP (PBLUP) method. This study aimed to investigate the potential of implementing single-step SNPBLUP (ssSNPBLUP) genomic evaluation in six German Beef cattle breeds. Following linear regression methods, the (G)EBV of the validation animals from the single-step full evaluation was regressed on the (G)EBV obtained from the truncated evaluation. The correlations of (G)EBVs obtained between the full and truncated evaluations ranged between 0.72 and 0.83 for the PBLUP evaluation. The maximum improvement was a 3% increase in correlation with the ssSNPBLUP evaluation. In general, ssSNPBLUP only showed a 1-2% increase in correlation across all traits and breeds. This indicates slightly more stable breeding values with the inclusion of genotypes. The b1 values only show a slight deviation from the expectation of 1 for the PBLUP and ssSNPBLUP evaluations. This suggests that an evaluation with PBLUP or a ssSNPBLUP is neither inflated nor deflated. The SNP effect estimates from the truncated evaluation were highly correlated with the full evaluation, ranging from 0.79 to 0.94. The correlation of the SNP effects is influenced by the number of genotyped animals shared between the full and truncated evaluations. The regression coefficients of the SNP effect of the full evaluation on the truncated evaluation were all close to the expected value of 1, indicating unbiased estimates of the SNP markers. Based on the (G)EBV of the validation population, the single-step model only showed a slight improvement relative to a PBLUP. This is not an unexpected, given the low number of animals with genotype and phenotype records. However, given that the results were slightly better with the inclusion of genotype data, this study can recommend adopting a single-step SNPBLUP evaluation. Furthermore, with increasing number of genotype and phenotype records, the ssSNPBLUP will result in higher accuracy, less bias, and more stability for future genomic evaluations.

Session 69 Poster 13

Incorporating QTL genotypes in the model to predict phenotypes and breeding values

J. Yang[1], T.H.E. Meuwissen[2], Y.C.J. Wientjes[1], P. Duenk[1] and M.P.L. Calus[1]
[1]Wageningen University & Research, Animal Breeding and Genomics, Droevendaalsesteeg 1, 6708 PB Wageningen, the Netherlands, [2]Norwegian University of Life Sciences, Box 5003, 1432 Ås, Norway; jifan.yang@wur.nl

With the accumulation of genotyped individuals, increasingly more QTL are being detected. Including genotypes of those known QTL in genomic prediction models can help to improve the accuracy of predicted breeding values and phenotypes. However, it remains unknown how much genetic variation a QTL should explain in order to see a benefit when it is included in the model. The aim of this study is to investigate at which proportion of genetic variance explained by QTL included in the model the highest prediction accuracy is achieved. A population under selection was simulated, with per generation 20 breeding males being mated with 2,500 breeding females. We considered that QTL effects followed either a gamma or a normal distribution. A standard GBLUP model based on ~58,000 SNPs was used as a benchmark. We then extended the standard GBLUP by including QTL as a separate variance component (hereafter we call it 2GBLUP). The proportion of genetic variance explained by the QTL included in the model was varied from 5 to 99%. This was achieved by adding increasingly more QTL in the model, after sorting them by decreasing genetic variance explained. In addition, weighted 2GBLUP was considered which assigns to each QTL its known genetic variance as a weight, instead of making the prior assumption that each QTL explains the same amount of genetic variance. The results showed that: (1) the highest prediction accuracy was obtained when 80% of the total genetic variance was explained by the QTL included in the model, and prediction accuracies increased approximately linearly as long as the proportion <80%; (2) using weighted 2GBLUP could further increase the prediction accuracy when more than 80% of the genetic variance was explained by the QTL; and (3) with Gamma distributed QTL effects both higher accuracies and stronger bias were obtained than with normally distributed QTL effects. As a next step, we will investigate the potential to include QTL in machine learning models, while still including all SNPs as was done in the 2GBLUP model. The aim of this analysis is to investigate if machine learning models can compete with the 2GBLUP models.

Session 69 Poster 14

Improving the efficiency of genomic evaluations with random regression models

A. Alvarez Munera[1], D. Lourenco[2], I. Misztal[2], I. Aguilar[3], J. Bauer[4], J. Šplíchal[4] and M. Bermann[2]
[1]Universidad Nacional de Colombia, Cra. 65 #59a-110, 4309000, Medellin, Colombia, [2]University of Georgia, Department of Animal and Dairy Science, 420 River Road, 30602, Athens, GA., USA, [3]Instituto Nacional de Investigación Agropecuaria (INIA), Ruta 48 km 10, Rincon del Colorado, 90100, Montevideo, Uruguay, [4]Czech Moravian Breeders' Corporation, Benešovská 123, 252 09 Hradištko, Czech Republic; mbermann@uga.edu

Random-regression models (RRM) are used worldwide to model longitudinal traits in dairy cattle breeding. Due to a more complex structure, RRM present more challenges than other models for longitudinal traits like multiple-trait repeatability models. These challenges include the convergence of the solvers, and approximating accuracies of genomic estimated breeding values (GEBV). These problems might worsen when genomic information is included by single-step genomic best linear unbiased predictor (ssGBLUP). The objectives of this study were to test efficient methods for implementing RRM with ssGBLUP for national dairy evaluations with the BLUPF90 software suite and to develop a method to approximate accuracies in RRM with genomic information. We used Czech dairy data to test the proposed methodologies. Data comprised 30 million test-day records for milk yield across three lactations. The pedigree had 2.5 million animals, from which 55,000 were genotyped. A block-diagonal preconditioner including correlated random traits was used to improve the solver's convergence. To approximate accuracies for 305-days GEBV, 305-days accuracies without genomic information were back-solved to effective record contributions (ERC), which were used as weights in a GBLUP model. Final reliabilities were calculated with ERC obtained from the last model while removing double counting of information. The proposed preconditioner reduced the number of iterations by half. Approximated accuracies were compared with those calculated with the inverse of the coefficient matrix. The correlation between them was 0.98, while the slope and the intercept of the regression were 1.01 and 0.01, respectively. Further research will focus on including the fixed effects block to the preconditioner, adding genetic groups or metafounders to the model, and testing different methods for including external information in the evaluation.

Comparison of two software to estimate breeding value in cattle by single-step approach

M. Jakimowicz[1], D. Słomian[2], T. Suchocki[1,2] and J. Szyda[1,2]
[1]Wrocław University of Environmental and Life Sciences, Department of Genetics, Kożuchowska St. 7, 51-631 Wrocław, Poland, [2]National Research Institute of Animal Production, Krakowska 1, 32-083, Balice, Poland; tomasz.suchocki@upwr.edu.pl

The main goal of the presented study was to compare predictions of breeding values for stature, using single step SNP-BLUP and G-BLUP models respectively implemented in the MiXBLUP and BLUPF90 software. In both programmes the same variance components were used, where genetic was equal to 5,50058 and residual to 4,63406. The data compromised 134,960 genotyped animals, and 1,098,611 phenotyped cows, and 141,686 bulls with pseudo-phenotypes expressed by Interbull-MACE (Multiple across country evaluation). The complete pedigree information used in the analysis included 8,451,809 animals and 36 phantom parent groups that have been defined based on birth year, sex and country of origin. The Data came from the national routine polish genetic evaluation for stature (December 2021) and was provided by the National Research Institute of Animal Production. To compare the results obtained by both of the models, we create rankings of the 100 of the best individuals and 1000 of the best individuals. The rankings were created separately for each sex, based on, the type of data available for animals (genotypes and phenotypes, and non-genotyped animals), and year of birth. We used Venn diagrams and Pearson correlation coefficients to check if there was any overlap between models from both programmes.

Estimating (co)variance components using Monte Carlo EM-REML in a multi-trait SNPBLUP model

H. Gao, M.H. Lidauer, M. Taskinen, E.A. Mäntysaari and I. Strandén
Natural Resources Institute Finland (Luke), Myllytie 1, 31600 Jokioinen, Finland; hongding.gao@luke.fi

The analytical REML-based methods typically used for (co)variance component (VC) estimation require elements from the inverse coefficient matrix of the mixed model equations (MME), i.e. prediction error variances (PEV). Making and inverting a dense and large MME is computationally challenging when genomic information has been used extensively in the breeding programme. To overcome this issue, we implemented the EM-REML method combined with Monte Carlo (MC) algorithm for multi-trait SNPBLUP. The PEV was approximated without explicitly making or inverting the MME coefficient matrix using solutions from MME having observations generated from the same distributions as the original data and current VC estimates. The MME solving step used the preconditioned conjugate gradient iteration and iteration on data, allowing fast computations and high memory scalability. The aims of this study were: (1) to estimate VCs using the MC EM-REML approach in a multi-trait SNPBLUP model, and (2) to assess its scalability in terms of computational time with different numbers of genotyped individuals. Data were simulated based on 10,000 markers and three traits with heritabilities of 0.44, 0.32, and 0.34 and genetic correlations of 0.80, 0.71, and 0.98. Simulated phenotypes had general means, genetic effects based on simulated marker effects, and random residual effects. We investigated six scenarios based on the number of genotyped individuals: 5,000, 10,000, 15,000, 20,000, 25,000, and 30,000. Computations in a three-trait SNPBLUP model used MC EM-REML with three MC samples generated within each REML round. The final VC estimates were in line with the simulated values. The computing time increased linearly with respect to the number of genotyped individuals, given a constant number of markers. Thus, MC EM-REML is well-suited to dense systems and can serve as a viable alternative for estimating VCs for large-scale genomic datasets.

Molecular phenotyping to predict neonatal maturity

L. Liaubet[1], N. Marty-Gasset[1], L. Gress[1], A. Bonnet[1], P. Brenaut[2] and E. Maigné[3]
[1]GenPhySE, Université de Toulouse, INRAE, INPT, ENVT, 24 chemin de Borde Rouge, 31326 Castanet-Tolosan, France, [2]IFIP-Institut du Porc, Le Rheu Cedex, France, La Motte au Vicomte, 35650 Le Rheu, 35650 Le Rheu, France, [3]Université de Toulouse, INRAE, UR MIAT, 24 chemin de Borde Rouge, 31326 Castanet-Tolosan, France; laurence.liaubet@inrae.fr

Improved piglet survival during the suckling period is a strong expectation for breeders. This notably involves taking into account the maturity of the piglet at birth. An immature piglet, which has not reached its full development, will have a greater risk of early death. These piglets have a characteristic morphology: A steep, dolphin-like forehead, bulging eyes and head/body asymmetry. Based on image analysis, the Pic'Let project (CASDAR-RT 2019) aims to offer breeders an innovative tool for phenotyping piglet maturity. On this study, 298 newborns (99 Landrace, 98 Large White, 98 LR×LW) were classified for their maturity level. Furthermore, a metabolomic analysis was also performed by 1H-NMR on blood sample (serum) collected on piglet. Raw spectra were analysed with the R package ASICS and 55 metabolites with non-zero variance have been used in following statistics. A first analysis with PCA suggested a common metabolic profile may be identified whatever the genotype. Next, two predictive models were developed to explain immaturity (23%, severe or light) vs maturity (77%). The first model uses the 55 available metabolites and is based on Random Forests and Lasso methods for the prediction. The second model uses a subset of 14 metabolites selected with the Lasso method and is based on random Forests and GLM methods. The imbalance characteristic of the dataset was adjusted by down sampling and model aggregation. The two models predict 100% of the severe immaturity status in the training and the test samples. Some piglets morphologically determined as mature are expected to be immature with a strong confidence [80-100%] according to metabolic data. Altogether, we identified a molecular signature based on metabolic data able to predict the neonatal maturity status. To validate this method, the next step will be to apply it to another dataset of newborns, where the same metabolomic analysis has been performed, and to study the relationship between the prediction and proxy variables such as piglet birth weight.

Comparison of genetic maps from different cattle breeds

X. Ding[1], H. Schwarzenbacher[2], F.R. Seefried[3] and D. Wittenburg[1]
[1]Research Institute for Farm Animal Biology (FBN), Wilhelm-Stahl-Allee 2, 18196 Dummerstorf, Germany, [2]ZuchtData GmbH, Dresdner Straße 89/B1/18, 1200 Vienna, Austria, [3]Qualitas AG, Chamerstrasse 56, 6300 Zug, Switzerland; wittenburg@fbn-dummerstorf.de

The proximity of loci on a genome can be measured in physical (base pairs) or in genetic distance units (Morgan), where the latter is of special importance for animal breeders. It is expected that one crossover event occurs on average during meiosis at one Morgan distance. Hence, the genetic distance between loci allows drawing inferences, for instance, on genetic variation of not yet born progeny and provides valuable information when searching for top breeding animals. Genetic diversity among cattle breeds and different breeding objectives for meat, dairy or dual purpose require breed-specific genetic maps. We analysed genotype data from seven commercial cattle breeds with sample size ranging from 4,181 to 298,850. Since various assays were used for genotyping the animals, we standardised the data preparation and analysed the data with our pipeline 'hsrecombi'. We investigated the frequency of paternal recombination events and derived genetic-map coordinates of about 50K SNP markers. Additionally, estimates of recombination rate between intra-chromosomal pairs of markers enabled the localisation of further putatively misplaced markers or regions in the bovine genome assembly ARS-UCD1.2. Estimates of map length varied from 23.99 M to 27.36 M between breeds. Two to 49 misplaced candidates were detected in each breed, mostly overlapping among breeds. Furthermore, a genomewide association study on the number of recombination events among progeny revealed 13 significant SNPs in total. A subset of these hits was located in or near two genes with known impact on recombination activity providing options to counteract genetic erosion in breeding populations in future. To explore recombination activity interactively and to evaluate differences between breeds, we implemented all results in an R Shiny app 'CLARITY'.

Session 69 Poster 19

Genomic relationships across metafounders using partial EM algorithm and average relationships

A. Legarra[1], M. Bermann[2], Q. Mei[3] and O.F. Christensen[4]

[1]CDCB, 4201 Northview Drive, 20716 Bowie MD, USA, [2]University of Georgia, Animal and Dairy Science, 425 River Rd, 30602 Athens GA, USA, [3]Huazhong Agricultural University, No.1,Shizishan Street, Hongshan District, 430070 Wuhan, China, P.R., [4]Aarhus University, Center for Quantitative Genetics and Genomics, C. F. Møllers Allé 3, bld. 1130, 8000 Aarhus C, Denmark; andres.legarra@uscdcb.com

Genomic relationships describe relationships among animals previously assumed as unrelated through pedigree, either within or across-breeds. Missing relationships can be modelled using the theory of metafounders, where relationships within and across base populations (metafounders) are encapsulated in a matrix Gamma. Values in Gamma are often hard to estimate, because, first, founder individuals are too far from genotyped individuals, second, the use of several metafounders within-breed to model missing pedigree, and third, many individuals are mixtures of several metafounders. Here we propose a hybrid method to estimate gamma within and across breeds. We use a partial EM maximum likelihood algorithm to estimate Gamma across breeds. We decompose the 'complete' likelihood of markers given Gamma and pedigree into a part that is a direct function of Gamma and a part that is a function of Mendelian sampling variance. We then approximate the first derivative by ignoring the Mendelian sampling variance. The following approximated EM algorithm consists in (1) postulate an initial value of Gamma (2) set up the H-inverse matrix as a function of G-inverse and A-inverse, with A-inverse including rows and columns for metafounders (3) invert the block of H-inverse corresponding to metafounders to obtain H(1:MF,1:MF) (4) set Gamma to the block H(1:MF,1:MF); iterate again. At convergence, we obtain an estimate of Gamma. The algorithm is completed by a check that the (total) log-likelihood is maximized at each iteration. Tests using simulated data show that the algorithm is accurate if the metafounders are not too distant from genotyped animals. For metafounders within breed along time, Gamma can be inferred using a structure that models the increase of relationships and relies on (1) the initial gamma parameter at foundation of the breed and (2) the increase of average relationships within breed along time, measured through pedigree analyses.

Session 69 Poster 20

Impact of the correlation between SNP effects in different breeds on the accuracy of predictions

P. Croiseau[1], R. Saintilan[1,2], D. Boichard[1] and B. Cuyabano[1]

[1]Université Paris-Saclay, INRAE, AgroParisTech, GABI, Domaine de Vilvert, 78350 Jouy-en-Josas, France, [2]Eliance, 149 rue de Bercy, 75012, Paris, France; pascal.croiseau@inrae.fr

Rotational crossbreeding schemes in dairy cattle are an efficient way for breeders to obtain more adaptable and robust animals, as well as more sustainable breeding systems. It takes advantage of breed complementarities and of heterosis. Its practice is currently expanding among cattle breeding systems. Aiming to achieve a correct representation of crossbred animals in genomic evaluation, methods to account for the breed of origin of alleles (BOA) have been developed (BOA-GBLUP). Their results regarding the prediction accuracy of genomic estimated breeding values (GEBV), however, have not yet overcome those obtained with a standard GBLUP, which ignores the BOA, especially when true QTL-effects are similar across breeds. Using simulations, we studied the impact of the correlation between QTL-effects in different breeds, on the accuracy of the predicted GEBV using the standard GBLUP and the BOA-GBLUP. 50k SNPs in linkage disequilibrium were simulated for three different breeds; among these simulated SNPs a random subset of 200 were assigned as QTL for all breeds, and ten scenarios varying the levels of correlation, from high (0.9) to low (0), between the QTL effects on the different breeds were generated. Four traits were investigated with heritabilities ranging from 0.01 to 0.65, and each breed of purebred animals was selected for a different trait, while crossbred animals were selected for all four with a selection index weighting all traits equally. After evolving 10 generations under this scheme, A training population of 20,000 animals (6,000 purebred animals from 3 breeds and 2,000 crossbred animals) was used to predict the GEBV of 2,000 crossbreds in the final generation. Finally, the accuracy of the predicted GEBV and of the estimated SNP effects within breed allowed us to assess under which conditions of QTL-effects correlation, the use of BOA-GBLUP is beneficial for genetic evaluations in rotational crossbreeding schemes. This project has received funding from APIS GENE and from the European Union's Horizon 2020 research and innovation program – GenTORE – under grant agreement No. 727213.

Session 69 Poster 21

Early prediction of lactation persistency of multiparous cows managed for extended lactation

C. Gaillard[1], M. Boutinaud[1], J. Sehested[2] and J. Guinard-Flament[1]
[1]Institut Agro, PEGASE, INRAE, 16 Le Clos, 35590 Saint Gilles, France, [2]ICROFS, Aarhus University, Blichers Allé 20, 8830 Tjele, Denmark; charlotte.gaillard@inrae.fr

Early prediction of lactation persistency will potentially be an important management tool influencing decisions on reproduction and feeding management. Primiparous cows have a relatively stable and high persistency while it is quite variable for multiparous cows. Therefore, the objective was to predict multiparous cows' lactation persistency, within the first 6 weeks of lactation, based on two or three variables. The dataset used contained 19 production data and blood biomarkers of 36 multiparous Holstein cows managed for extended lactation (insemination at 8 months) with access to a milking unit. The persistency was calculated as the ratio of total milk produced during a defined period of 100 days (P1: from 100 to 200 days, P2: from 200 to 300 days, or P3: from 300 to 400 days) over the first 100 days of production (Method 1); as the quantity of milk produced at day 100, 200, 300, or 400 over the quantity produced at day 60 of lactation (Method 2); and as the slope of the smoothed milk yield curve at day 100, 200, 300, and 400 (Method 3). For each method, the cows were divided into 3 groups (12 cows per group) with a low, medium or high persistency. A clustering method (k=3) was then applied on a pair or trio among the data measured at week 1, 3, or 5. Using a trio of variables to predict persistency only improved the maximum accuracy of 3.7% (average over the methods, weeks and periods) compared to a pair of variables, and the highest accuracy obtained with a pair of variables varied from 43 to 70%. With a pair of variables, the highest accuracy was obtained with Method 1, considering milk yield and lactose in milk measured at week 1 to predict the persistency over P3. The second best pair belonged to Method 3 and involved dry matter intake and milking frequency measured at week 5 to predict persistency over P1. To conclude, Method 1 gave the highest accuracies and with this specific data set, a pair of variables in early lactation may be used to predict the persistency of the lactation at different stages of the lactation with a moderate accuracy.

Session 69 Poster 22

Accuracy of genomic prediction by singular value decomposition of the genotype matrix

L. Ayres[1], M.P.L. Calus[1], J. Ødegård[2,3] and T. Meuwissen[3]
[1]Wageningen University & Research, Animal Breeding and Genomics, Droevendaalsesteeg 1, 6700 AH Wageningen, the Netherlands, [2]AquaGen AS, Postboks 1240, 7462 Trondheim, Norway, [3]Norwegian University of Life Sciences, Department of Animal and Aquacultural Sciences, Oluf Thesens vei 6, 1433 Ås, Norway; lucas.ayres@wur.nl

Reference populations for genomic prediction are continuously growing, and there is a tendency to use more single nucleotide polymorphisms (SNPs) from animal genomes. Both developments lead to increasing the dimensions of the genotype matrix, which poses a computational challenge for genomic prediction. To reduce the dimensionality, we applied singular value decomposition (SVD) to the genotype matrix. The objective of this study was to evaluate the effect of the number of SVD components on the accuracy of genomic prediction. To predict the breeding values, we simulated the phenotypes by generating 1000 quantitative trait loci (QTL) randomly drawn from the SNPs of chromosome 1 of Atlantic Salmon. We employed SVD-based principal component ridge regression (PCRR) to estimate breeding values. Accuracies increased steeply for the first few principal components until stabilizing around 100-300 components, and no meaningful gain was obtained with additional components. Maximum accuracies were obtained at around 50-250 components, marginally higher than when the original genotype matrix was used. Our results indicate that, within an appropriate range of components, SVD can be used in genomic prediction to reduce computational burden, while not compromising prediction accuracy.

Session 69 Poster 23

Enhancing long-term genetic gain through a Mendelian sampling-based similarity matrix

A.A. Musa and N. Reinsch
Research Institute for Farm Animal Biology (FBN), Institute of Genetics and Biometry, Wilhelm-Stahl-Allee 2, 18196 Dummerstorf, Germany; musa@fbn-dummerstorf.de

Indices combining the expected breeding value and Mendelian sampling variance (MSV; variability in full-sibs breeding values) of a parent have been recently proposed as alternative selection criteria to expected breeding value to sustain long-term genetic gain. However, such an index tends to select similar parents with a high MSV potential, resulting in the loss of favourable haplotypes and jeopardizing long-term genetic gain. Through simulation, we show that long-term genetic gain can be enhanced using a Mendelian sampling-based similarity matrix which restricts the selection of parents with MSV caused by the same chromosomal segments. Here, we compare several recurrent selection schemes derived by combining breeding value or index with truncation selection (TS) or optimum mate allocation using the similarity matrix (OMA). Our simulation consisted of 500 males and 500 females per generation, and we performed 50 generations of selection, allocating one male to 50 females. This ratio was sometimes lower in schemes involving OMA. Finally, we assumed random mating and known additive marker effects for a trait with a heritability of 0.25. We found that selection schemes involving OMA outperformed TS. Compared to TS on the index, for example, combining the index and similarity matrix realized up to 7% more genetic gain, preserved up to 2,177% more genetic variability, and reduced inbreeding by up to 30% in the terminal generation. Furthermore, up to 25% more favourable quantitative trait loci alleles and up to 24% more single nucleotide polymorphisms were retained in the terminal generation. While further studies are needed, we believe that the inherent benefits of including the similarity matrix in the genomic selection apply to many practical breeding programs and represent a significant step towards efficient long-term genomic selection.

Session 69 Poster 24

Mining the convergence behaviour of a single-step SNP-BLUP model for genomic evaluation of stature

D.S. Dawid Słomian[1], J.S. Joanna Szyda[1,2] and K.Ż. Kacper Żukowski[1]
[1]National Research Institute of Animal Production, Sarego 2, 31-047 Cracow, Poland, [2]Wroclaw University of Environmental and Life Science, Biostatistics Group, Department of Genetics, Kożuchowska 7, 51-631 Wroclaw, Poland; dawid.slomian@iz.edu.pl

The single-step model is becoming increasingly popular for national genetic evaluations of dairy cattle, offering several benefits such as joint breeding value estimation for genotyped and ungenotyped animals. However, the model's high parameterization and correlations among millions of effects can lead to significant computational challenges, especially in terms of the accuracy and efficiency of the preconditioned conjugate gradient method used for the estimation. This study aimed to investigate the effect of pedigree depth on the overall convergence rate of the single-step SNP-BLUP model and the convergence of its different components. The results showed that the data set with a truncated pedigree converged twice as fast as the full data set. Both data sets had very high Pearson correlations between predicted breeding values. The study also compared the top 50 ranking bulls between the two data sets and found a high correlation in predictions of their genetic merits. The convergence patterns of different animal groups and SNP effects were analysed, revealing heterogeneity in convergence behaviour. Pedigree depth influenced the convergence rate of the single-step SNP-BLUP model, with SNP effects converging the fastest and phantom parent groups converging the slowest, reflecting the difference in information content available in the data set for those effects. Among different animal groups, genotyped animals with phenotype data converged the fastest, while non-genotyped animals without their own records required the most iterations. In conclusion, data structure markedly impacts the convergence rate of the optimization, and the truncated data set is more efficient than the full data set.

Session 69 Poster 25

Increase in prediction accuracy can be achieved by combining multiple populations

A. Ajasa[1,2], S. Boison[3], H. Gjøen[1] and M. Lillehammer[2]
[1]Norwegian University of Life Sciences, Department of Animal and Aquacultural Sciences, Arboretveien 6, 1430 Ås, Norway, [2]Norwegian institute of Food, Fisheries and Aquaculture research, Breeding and Genetics, Osloveien 1, 1432 Ås, Norway, [3]Mowi Genetics AS, Sandviksboder 77AB, 5035 Bergen, Norway; afees.ajasa@nofima.no

The accuracy of genomic prediction is in part determined by the size of the reference population. In aquaculture breeding programs, the generation interval is usually three or four years, and thus three or four parallel populations (year-classes) usually exist at any point in time. A reasonable strategy to increase the reference population size can then be to combine multiple year-classes. However, results from terrestrial species have in general indicated limited or no impact on prediction accuracy when combining multiple populations, partly because of the inconsistency of linkage disequilibrium phase between markers and quantitative trait loci (QTLs) across populations. In this study, we evaluated the impact on prediction accuracy when selecting SNPs that have the same phase with the QTLs across populations. This was achieved by including a genome wide association analysis step before the genomic prediction was done. This was to identify markers with similar signs of allele substitution effect across populations (without recourse to whether these markers were significant or not). The data set utilized in this study was gill score records of Atlantic Salmon infected with amoebic gill disease from three year-classes genotyped with a 55K SNP, and the genomic evaluation model was a Genomic best linear unbiased prediction. Utilizing the whole genotype data, combining multiple year-classes, resulted in no or limited increase in prediction accuracy when compared to within year-class prediction accuracy, whereas utilizing selected markers resulted in about 10-20% increase in prediction accuracy, with bias ranging from 0.93-0.98. Our findings suggest that including multiple populations in genomic prediction may enhance accuracy, provided that appropriate marker selection methods are employed to account for the inconsistency of linkage disequilibrium phase between populations.

Session 69 Poster 26

Enhancing bovine genome SNP call accuracy with autoencoder analysis of nucleotide impact with AI

K. Kotlarz[1], M. Mielczarek[1,2], B. Guldbrandtsen[3] and J. Szyda[1,2]
[1]Wroclaw University of Environmental and Life Sciences, Department of Genetics, Kozuchowska 7, 51-631 Wroclaw, Poland, [2]National Research Institute of Animal Production, Krakowska 1, 32-083 Balice, Poland, [3]University of Copenhagen, Department of Veterinary and Animal Sciences, Grønnegårdsvej 8, 1870 Frederiksberg C, Denmark; krzysztof.kotlarz@upwr.edu.pl

A critical step in the analysis of NGS data is variant calling, which involves comparing the sequences to a reference genome to identify variants like, e.g. single-nucleotide polymorphisms (SNPs). However, variant calling is a challenging operation and errors happen for several reasons. Our research focused on the identification of the effects of nucleotide context around an identified variant on incorrect SNP calls. For that purpose, an anomaly detection procedure was implemented via an autoencoder (AE) model. Whole-genome sequences (WGS) of four Danish Red Dairy Cattle bulls were sequenced on the Illumina HiSeq2000 platform and genotyped with the Illumina BovineHD Bead Array. Incorrect SNPs were defined as mismatches between these two technologies. For the classification, the variables from a VCF file with 3 downstream and 3 upstream reference nucleotides were used. The training data set composed data from three bulls and consisted of 2,227,995 correct (97.93%) and 47,093 incorrect SNPs, with a stratified 30% validation subset, while the test data set consisted of data from one bull with 749,507 correct (97.85%) and 16,468 incorrect SNPs. The AE algorithm was implemented via the Keras library with mean square error (MSE) loss function. The training process resulted in a final MSE of 0.147 and 0.239 in the validation dataset. The reconstruction loss of incorrect SNPs was significantly higher ($P=1.19\times10^{-8}$) than the reconstruction loss of correct SNPs. To further evaluate the effectiveness of the AE, the trained model was evaluated on a test dataset that archived MSE of 0.229 and a significant ($P=1.00\times10^{-3}$) difference between correct and incorrect SNP subsets. Our results demonstrated: (1) there is still a considerable number of incorrectly identified SNP genotypes originating from WGS data; (2) by mining the underlying pattern of their explanatory variables in a post-calling analysis one can identify a large proportion of incorrect SNPs.

Efficient SNP calling: Nextflow vs Bash on the whole genome bovine sequence

P. Hajduk[1], M. Sztuka[1], K. Liu[1], K. Kotlarz[1], M. Mielczarek[1,2] and J. Szyda[1,2]
[1]Wroclaw University of Environmental and Life Sciences, Department of Genetics, Kozuchowska 7, 51-631 Wroclaw, Poland, [2]National Research Institute of Animal Production, Krakowska 1, 32-083 Balice, Poland; 121687@student.upwr.edu.pl

Variant calling is a process that involves comparing sequences to a reference genome, allowing for the detection of single-nucleotide polymorphisms (SNPs). It is especially useful when working with NGS data, which has become more popular in recent years. However, due to the sheer size of the data used in this process, time and memory usage become a concern when creating variant calling pipelines. This study compares those aspects for different approaches to SNP-calling pipeline parallelization with the use of Nextflow. Data used in this study included the genomic DNA of five Polish Holstein-Friesian cows, which were sequenced with the Illumina HiSeq2000 platform in the paired-end mode. The bioinformatic pipeline consisted of (1) quality control, (2) alignment to the reference genome, (3) post-alignment processing step, and (4) SNP calling. Three different approaches were compared for various levels of parallelisation expressed by the numbers of cores. The first was a plain bash script in which calculations for each cow were executed in parallel. The second was a bash script wrapped in a single NextFlow process. Third was a NextFlow script with each step of the pipeline treated as a separate process. Results demonstrated that on average the multi-process NextFlow script performed the fastest, no matter the number of CPUs used, with the biggest increase in time when each process was on 10 cores, where the average time difference was 47%. When comparing RAM usage, the only significant differences occurred between scripts given 1 core per process and slight differences when given 5 cores per process. For all other constellations, no differences were found. Nextflow is an efficient computing environment not only for creating pipelines but also for managing them. It offers a lot of tools for visualization during processes as well as after completion, which helps with monitoring each process during its run-time. Even though there were no significant differences regarding RAM usage, there was a noticeable decrease in execution time between NextFlow and plain Bash pipelines.

Variant calling and genotyping accuracy of ddRAD-seq: comparison with WGS in layers

M. Doublet, F. Lecerf, F. Degalez, S. Lagarrigue, L. Lagoutte and S. Allais
UMR 1348 PEGASE INRAe, Institut Agro, 16 Le Clos, 35590 Saint-Gilles, France; mathilde.doublet@inrae.fr

Since the advent of Next Generation Sequencing, sequencing has been used for many applications, including obtaining genotypes for animal genomic selection. However, as full depth whole genome sequencing (WGS) is either unaffordable or inappropriate for certain applications, alternative approaches have been developed. Amongst them, double digested Restriction-site associated DNA sequencing (ddRAD-seq) offers a high sequencing depth on a targeted part of the genome. This type of approach reduces costs compared to full depth WGS and limits the inter-individual variability of a low depth WGS approach. A method such as ddRAD-seq could benefit the laying hen industry by providing low-cost genotyping, assuming it is of sufficient quality. The objective of this study was to evaluate the suitability of ddRAD-seq sequencing to detect variants and obtain bi-allelic SNP genotypes, using different sets of quality control filters. The study was conducted on a population of 50 males of a Rhode Island laying line for which sequences were obtained by ddRAD-seq with the Taq1/Pst1 enzyme couple and by 20X sequencing. First, the variant calling results were compared between 20X and ddRAD-seq. Then, the genotype concordance rate between ddRAD-seq and WGS 20X was calculated on the common SNPs. Finally, the variation of this concordance rate was calculated based on different filter combinations on the average sequencing depth (DP) per SNP and the SNP call rate (CR). 9,270,491 SNPs were genotyped in 20X and 350,490 in ddRAD-seq. The average SNP CR (55%) and the mean SNP DP value (11X) in ddRAD-seq were lower than in 20X (99% and 16X respectively). 327,364 SNPs were detected in both methods and were distributed on all the chromosomes. For these common SNPs, the mean of the concordance rate per SNP (CcR) between the genotypes obtained in ddRAD-seq and those obtained in 20X was on average 82%. The relationship between the CcR, the CR, the DP and the percentage of retained SNPs should allow future users of ddRAD-seq to choose the best filtering thresholds for their analyses.

Session 69 Poster 29

Performing single-step genomic evaluation for superovulatory response traits in Japanese Black cows

A. Zoda[1], R. Kagawa[1], H. Tsukahara[1], R. Obinata[1], M. Urakawa[1], Y. Oono[1] and S. Ogawa[2]
[1]Research and Development Group, Zen-noh Embryo Transfer Center, Kamishihoro, Hokkaido, 080-1407, Japan, [2]Division of Meat Animal and Poultry Research, Institute of Livestock and Grassland Science, NARO, Tsukuba, 305-0901, Japan; zouda-atsushi@zennoh.or.jp

Japanese Black cattle are famous for their excellent meat quality. The value of Japanese Black calves is much higher than that of calves of other beef breeds. Therefore, Japanese Black embryos, which have been produced by superovulation treatments, are widely used for embryo transfer. Previously, using pedigree data, we reported that genetic improvement of superovulatory response traits seems possible. Here, we performed genomic prediction for superovulatory response traits in Japanese Black cows via a single-step approach. Records of the total number of embryos and oocytes (TNE) and the number of good embryos (NGE) per flush were obtained from 1,874 Japanese Black donor cows during 2008-2022. The number of records was 25,332 per trait. Overall, 575 out of the 1,874 cows had genotype information on 36,426 single-nucleotide polymorphisms (SNPs) on autosomes, according to ARS-UCD1.2. A two-trait repeatability animal model was exploited. Variance components were estimated using the BLUPF90+ software with the method VCE option. Pedigree-based and genome-based relationship matrices (A and H matrices) were fitted. Estimated heritabilities of TNE and NGE were slightly lower when using the H matrix than when using the A matrix, although the estimated repeatabilities were similar to each other. When the variance components were fixed in breeding value prediction, the mean reliability was higher when using the H matrix than when using the A matrix. This advantage seems more prominent for cows with low reliability when using the A matrix. The results imply that introducing a genomic prediction scheme could boost the rate of genetic improvement.

Session 69 Poster 30

Production traits in Nellore cattle classified by residual feed intake

J.A. Muñoz, R.H. Branco, R.C. Canesin, J.N.S.G. Cyrillo, M.E.Z. Mercadante and S.F.M. Bonilha
Instituto de Zootecnia, Rodovia Carlos Tonanni, km 94, Sertãozinho, SP, 63-14160-900, Brazil; julianmunoz@alumni.usp.br

In beef cattle, there is considerable variation in feed intake that is independent of size and growth rate and can be calculated as residual feed intake (RFI), which is a measure of the efficiency of feed utilization by the animals. We aimed to evaluate the effects of RFI classification on productive traits of Nellore cattle with different growth potential (GP) finished in the feedlot. A total of 45 non-castrated Nellore males with 528±30 days of initial age and 377±8.87 kg of initial body weight were evaluated in a 2×2 factorial scheme: RFI (low or high) and GP (HG: high growth potential; LG: low growth potential) to determine performance and carcass traits. RFI classification of animals was based efficiency test performed after weaning. Dry matter intake (DMI), weight gain (ADG), body weight (BW), hot carcass weight (HCW), carcass yield (CY), rib eye area (REA), backfat thickness (BF), and rump fat thickness (RF) were analysed by SAS MIXED procedure considering as fixed effects RFI and GP, as covariate age at slaughter, and as random effect diet. Differences between means were verified using the t-Student test with α=0.05. RFI class did not affect the carcass traits of the animals (*P*>0.05). However, there was a trend towards lower DMI in feedlot (10.0 vs 10.6 kg/d; *P*=0.099) in low RFI animals, compared to high RFI animals, regardless of their GP. There was a lower RF (3.50 vs 6.30 mm; *P*=0.022), as well as trends towards lower BW (424 vs 458 kg; *P*=0.074) and lower HCW (240 vs 260 kg; *P*=0.095) in the comparison between HG and LG. It is noteworthy that ADG had an interaction between the factors (RFI and GP), having LG and low RFI animals lower ADG than LG and high RFI animals (1.10 vs 1.56 kg; *P*=0.033) and than HG and low RFI animals (1.10 vs 1.48 kg; *P*=0.049), respectively. In this context, Low RFI Nelore animals, possibly with lower feed intake during the feedlot, had performance equivalent to commercial standards without any difference in carcass traits. Animals with high growth potential (selected for growth) had better performance in the feedlot. However, genetic associations with RFI did not affect traits of economic importance.
Acknowledgments: FAPESP Process 2022/12347-4.

Session 69 Poster 31

Effect of paternal breeding values for residual feed intake on reproductive performance of heifers

T. Devincenzi, M. Lema, L. Del Pino, A. Ruggia and E.A. Navajas
Instituto Nacional de Investigación Agropecuaria, Av. Italia 6201 Edificio Los Guayabos, 11500, Uruguay; tdevincenzi@inia.org.uy

We evaluated the effect of sires' efficiency for RFI on productive and early reproductive performance of Hereford heifers grazing native pastures in Uruguay. We studied an offspring of 98 females born from 7th September to 25th November 2020 in an experimental herd which belong to a Hereford genetic information nucleus. Animals were classified into two groups according to their paternal estimated breeding values (EBV) for RFI (percentiles ≤20% for high; H; n=41, ≥70% for low; L; n=57). Animals were managed together from weaning to 12 m and then were split into two contemporaneous groups. Body weights were recorded at birth, weaning, and then monthly until 25 m of age before heifers first insemination. Average daily gain (ADG) was calculated by seasons (fall/winter-2021, spring/summer-2021, fall/winter-2022 and spring/summer-2022). Ovarian activity (OA) was assessed by ultrasonography at 12,15,18 and 25 m of age and heifers presenting a corpus luteum were registered. Growth performance data were subjected to ANOVA and the frequency distribution of OA was compared by Qui2 test. The statistic model included EBV for RFI group as a class fixed effect. Dam age and the split group were tested but later excluded from the model because they were not significant. The sire EBV group affected birth weight (P=0.00306). Animals from H group were lighter than L group (38.6±5.06 and 40.8±5.01 kg, for H and L groups respectively). EBV group did not affected heifers' performance (P>0.05). Weaning weight was 184.4±19.23 kg at ADG from birth to weaning was 0.773±0,085 g/day. Animals weighted 236.3±25.37, 270.2±23.55, 313.8±26.92 and 384.9±31.65 kg at 12, 15, 18 and 25 m respectively. ADG was 0.218±0.133, 0.520±0.122, -0.014±0.132 and 1.166±0.291 g/day during fall/winter-2021, spring/summer-2021, fall/winter-2022 and spring/summer-2022 respectively. There was no effect of the EBV group on OA (P>0.05). The frequencies of OA were 2.06, 21.4, 51.5 and 73.9% at 12, 15, 18 and 25 months, respectively. In our study paternal EBV for RFI did not affect the productive and early reproductive of female performance, however efforts must be done to include a larger number of animals from different generations in future studies.

Session 69 Poster 32

Species identification of animal/pig DNA in non-animal food

M. Natonek-Wiśniewska and P. Krzyścin
The National Research Institute of Animal Production, Department of Animal Molecular Biology, Krakowska 1 Street, 32-083 Balice, Poland; malgorzata.natonek@iz.edu.pl

Resignation from the consumption of pork or animal products in general is most often related to health care, religious beliefs or cultural affiliation. Regardless of the reasons for the phenomenon, representatives of the above-mentioned social groups are interested in consuming products devoid of animal or, in particular, pork ingredients. Although in Europe there is an obligation to declare the composition of food products (EU 1169/2011), there are cases of food adulteration – intentional or accidental due to poor production or storage conditions. For these reasons, there is a need to control food for the potential addition of undesirable ingredients in the food that is assumed to contain only plant components. The work aimed to develop a method for identifying animal and pork DNA in fruit juice and chocolate. DNA obtained from juice and chocolate was fortified (1, 0.1 and 0.01%) with pork DNA obtained from meat and analysed by PCR with animal and pork primers. DNA from juice and chocolate without the addition of foreign DNA was used as blank controls. No PCR products were obtained for blank controls, while for fortified samples a product indicating the presence of animal/pig DNA was obtained. The obtained results indicate the biological specificity of the method in the tested matrix and the lack of inhibitors. The LOD of the method is 0.01%.

Session 69 Poster 33

Development of a proxy for feed efficiency prediction in dairy cows based on mid-infrared spectra

M. Raemy[1], T. Haak[2,3], F. Schori[3] and S. Ampuero Kragten[1]
[1]Agroscope, Method Development and Analytics, 1725 Posieux, Switzerland, [2]University Bonn, Institute of Animal Science, Bonn, Germany, [3]Agroscope, Ruminant Nutrition and Emissions, 1725 Posieux, Switzerland; marlyse.chatton@agroscope.admin.ch

The present study is an evaluation of the potential of mid-infrared (MIR) spectra of milk to predict feed efficiency (FE) of lactating cows. Feed efficiency can be expressed as: feed conversion ratio (FCR), residual feed intake (RFI), nitrogen use efficiency (NUE), residual nitrogen intake (RNI) and residual energy intake (REI). The advantage of milk MIR spectra is that they can be readily used to predict FE traits since they are routinely collected as part of milk recording of cattle. Fleckvieh and Holstein lactating cows (76) were used, from three feeding trials. In the first trial, the cows grazed almost exclusively, in the second they ate fresh herbage in the barn, and in the third they were offered total mixed rations with two different protein contents. Milk samples were scanned with a MilkoScan RM® from FOSS®, from 929 cm^{-1} to 3,000 cm^{-1}. Different data pretreatments were applied to the raw spectra: multiplicative scatter correction, standard normal variate, Savitzky-Golay smoothing and derivative. Partial least squares regressions with leave one out cross validation (CV) were used to develop the prediction models with all available data except 10 samples kept for independent validation (val) on all FE traits. The prediction models showed moderate to good results with R^2_{CV} of 0.90, 0.81, 0.80, 0.75 and 0.59 for RNI, FCR, NUE, REI and RFI respectively. The R^2val showed lower values 0.83, 0.69 and 0.77 for RNI, FCR and NUE respectively and even lower values for REI (0.27) and RFI (0.31). The residual prediction deviation (standard deviation of calibration-set / root mean square error of prediction of validation-set) showed 1.98 for RNI, 1.63 for NUE, 1.57 for FCR, 1.21 for REI and 1.07 for RFI. These results indicate that milk MIR spectra are a promising tool to predict FE traits, probably with the exception of REI and RFI. It is therefore important to expand the variability of the dataset to further improve the robustness of the prediction models in order to be able to use them for the selection of dairy cows. This would be easier through international collaborations.

Session 69 Poster 34

Improving taste and flavour in dairy product through analysis of free fatty acid by MIR spectroscopy

O. Christophe[1], R. Reding[2], J. Leblois[3], D. Pittois[4], C. Guignard[4] and F. Dehareng[1]
[1]Walloon Agricultural Research Center (CRA-W), 24, Chaussée de Namur, 5030 Gembloux, Belgium, [2]Convis, 4, Zone artisanale et commerciale, 9084 Ettelbruk, Luxembourg, [3]Elevéo asbl, 4, rue des Champs Elysées, 5590 Ciney, Belgium, [4]Luxembourg Institute of Science and Technology, Department Environmental Research & Innovation (ERIN), 41, Rue du Brill, 4422 Belvaux, Luxembourg; o.christophe@cra.wallonie.be

The dairy sector deals with a recurring issue: a taste alteration due to degradation of fat, commonly called lipolysis. Lipolysis happens after the milking, through the physical shocks induced by freezing, pumping, transfer and storage of the milk. Physical break of fat globules makes triglycerides accessible to enzymes and degraded into free fatty acids (FFA). Among them, the volatile short chain FFA lead to organoleptic issues through undesired tastes. An easy quantification of these individual short chain FFA, responsible of taste alteration, is very difficult. Historically, the lipolysis was quantified with the BDI methods by the measurement of the fat acidity. On the other hand, the analysis of a wide range of FFA is now possible by Gas Chromatography coupled with tandem mass spectrometer (GC-MS/MS). This analysis is time consuming, expensive and difficult to apply for routine analysis on a large set of samples. In order to bring a new way of preventive and corrective action for dairies and farmers, this project aims to develop predictive models based on milk mid Infrared spectroscopy (FT-MIR) to quantify FFA. For this purpose, milk samples from four different countries were collected and analysed by MIR spectroscopy as well as GC-MS/MS. The different models provided moderate R^2 for long-chain FFA and relatively low R^2 for short-chain FFA. Indeed, most of short chain FFA were under the limit of quantification. The lack of short-chain FFA concentration was solved by testing different mechanical induced lipolysis without interfering with the MIR spectrum. Among them, time milk homogenization has demonstrated a clear increase of short chain FFA value leading to better predictive models. More than 600 analyses were collected (classical and mechanical induced lipolysis included) to setup this model. R^2 is ranged from 0.3 to 0.8 for the different FFA (C4 to C18).

Session 69 Poster 35

Modelling growth curves in challenged mice lines divergently selected for birth weight variability

V. Mora-Cuadrado[1], I. Cervantes[1], J.P. Gutiérrez[1] and N. Formoso-Rafferty[2]
[1]Facultad de Veterinaria, UCM, Animal Production, Avda. Puerta de Hierro s/n, 28040 Madrid, Spain, [2]ETSIAAB, UPM, Agronomic Production, C/Senda del Rey 18, 28040 Madrid, Spain; nora.formosorafferty@upm.es

Selection for homogeneity can affect feed efficiency, and may benefit productivity and animal welfare, even when feed restrictions are applied. This fact has been proved in a selection experiment in mice for birth weight (BW) variability. After more than 30 generations of selection the Low line (L-line) showed benefits over the High line (H-line) regarding growth, feed efficiency, litter size, survival, and longevity. The objective of this study was to model the growth curves during a feeding restriction challenge in both lines. A total of 80 females (40 from each line) from generation 14 of the experiment were analysed. During the growth stage from weaning (21 days) to 63 days, the mice were separated in two groups per line (20 females per line and group), and received two types of diet: *ad libitum* and a restricted one consisting of 85% of the food consumption *ad libitum*. Animals were weekly weighed and the data were used to estimate the growth curves of each animal using the Brody, Gompertz, Logistic, and Von Bertalanffy growth functions. These functions provided coefficients for the asymptotic weight of the animals and their growth rate. AIC, BIC and r^2 were used as criteria for the selection of the models. A GLM was performed for BW, for the asymptotic BW and for growth rate of each function. The litter size, diet, line, diet×line and days (linear and quadratic only for BW) were included in the models. Differences between the four functions were small. However, the Logistic function was the one with the worst fit in all cases. All effects were relevant showing that the influence of the diet effect was very different between lines, showing a difference of 1.98 g in H-Line and 0.27 g in L-line. L-line animals were more robust and resilient, having a better support of the stress caused by feed restriction, while H-line animals proved to be less homogeneous and susceptible to stress, despite reaching higher weights.

Session 69 Poster 36

Genomic prediction of commercial layers' bone strength

M. Sallam[1], H. Wall[1], P. Wilson[2], B. Andersson[3], M. Schmutz[3], C. Benavides[4], M. Checa[4], E. Sanchez[4], A. Rodriguez[4], I. Dunn[2], A. Kindmark[5], D.J. De Koning[1] and M. Johnsson[1]
[1]Swedish Uni. of Agricultural Sciences, Ulls väg 26, Uppsala, Sweden, [2]Roslin Institute, Edinburgh EH25 9RG, Scotland, United Kingdom, [3]Lohmann Breeders, Am Seedeich 9-11, Cuxhaven, Germany, [4]Uni. de Granada, Mineralogia y Petrologia, Granada, Spain, [5]Uppsala Uni., Akad. sjukhuset, Uppsala, Sweden; mohammed.abdallah.sallam@slu.se

The high prevalence of bone damage is frequently reported on commercial layers (hybrids). The direct selection of hybrids for stronger bone is not possible because they never participate in breeding practices. One possible strategy is to estimate SNP effects of hybrids' bone strength, then use them as parameters for selecting grand-parents of hybrids. We hypothesize that accurate SNP effects of hybrids' bone strength can be obtained with a relatively small reference population. Genotypes on 50k SNPs and tibia strength phenotypes were available from two brands of hybrids (White Bovans and Leghorn Selected Lohmann Classic) housed in two housing systems (cages and pens). This resulted in 4 classes (hybrid-housing combinations) of bone strength: Bovans kept in cages (n=220), Lohmann kept in cages (n=218), Bovans kept in pens (n=217), Lohmann kept in pens (n=218). In addition, we included the tibia strength of a Rhode Island Red breeding line (n=924). Each hybrid-housing combination was fitted separately into single-trait GBLUP. Then multi-trait GBLUP approaches were used to fit hybrid-housing combinations simultaneously: of the same hybrid but housed in different systems, of different hybrids but housed in the same system, of different hybrids housed in different systems, or the latter in combination with RIR data. Single-trait GBLUP has an average accuracy of 0.29. Combining data simultaneously across hybrids resulted in an average accuracy of 0.34. Tibia strength showed genomic-based heritability estimates of 0.1-0.64 and approximately 261-1,891 independent chromosomal segments. Genomic predictions of commercial layers' bone strength can be obtained with a relatively small reference population. Further study is required to test whether the proposed marker effects can be used as parameters for selecting breeding lines with an expected genetic gain in the forward batches of the hybrids.

Session 69 — Poster 37

Variability of daily milk yield during the first 100 days of lactation in Holstein cows

E. Kašná, L. Zavadilová and J. Vařeka
Institute of Animal Science, Genetics and Breeding of Farm Animals, Přátelství, 10400 Prague, Czech Republic; zavadilova.ludmila@vuzv.cz

Fluctuations in daily milk yield (DMY) can be used as an indicator of the health and resilience of dairy cows. We evaluated the variability of DMY during the first 100 days of lactation of cows from three cooperating farms. The dataset included 75,000 DMY records of 548 Holstein cows. The data were collected between October 2022 and January 2023. Cows were required to have records for at least 20 successive days and the first day in milk (DIM) must have been lower than 50. DMY ranged between 1 and 78 kg with a mean of 36 kg of milk per cow and day. The variances of DMY for each cow were log-transformed (LnVar) to achieve their normal distribution. The estimation of heritability (h^2=0.15) and breeding values (BV) for LnVar was based on a single trait linear animal model and performed with the BLUPF90 family of programs. The model included fixed effects of the farm, the month of calving, the lactation order and the first observed DIM, the random effect of animals included in the three-generation pedigree and the random residual. We calculated the correlations of sires' BV for LnVar with their BV for production, fertility, exterior and health traits to indicate their possible genetic correlations. The higher BV(LnVar) were associated with the higher BV for milk, fat and protein yield (kg), and the lower BV for milk protein and fat content (%). The higher BV(LnVar) were related to the deep and wide rear udder and negatively correlated with locomotion, a final class for legs and feet and body condition score. The lower LnVar was associated with better longevity and fertility traits, but the correlations were non-significant. The lower BV(LnVar) were also favourably correlated with health traits BV (somatic cells count, retained placenta, metritis, ovarian cysts, clinical mastitis, ketosis, feet disorders) but the correlations were weak and non-significant. The study was supported by the Ministry of Agriculture of the Czech Republic, project QK22020280 and institutional support MZE-RO0723.

Session 70 — Theatre 1

Single-step GBLUP for growth and carcass traits in Nordic beef cattle

A. Nazari Ghadikolaei[1], F. Fikse[2] and S. Eriksson[1]
[1]Swedish University of Agricultural Sciences (SLU), P.O. Box 7023, 75007 Uppsala, Sweden, [2]Växa, P.O. Box 288, 75105 Uppsala, Sweden; anahit.nazari@slu.se

Single-step genomic prediction (ssGBLUP) has been applied for more accurately estimating genomic breeding values (GEBVs) using pedigree, phenotype and genotype information. The aim of this study was to investigate the use of ssGBLUP in the Nordic beef cattle evaluation of Charolais and Hereford. Genotypes for about 44,000 SNP for in total 4,321 Charolais and 4,532 Hereford were analysed together with phenotype information for in total 358,198 Charolais and 406,857 Hereford animals provided by the Nordic Cattle Genetic Evaluation. Because the number of genotyped Danish animals was low, we here present results for Swedish and Finnish animals. The BLUPF90 family of programs was used for estimation of breeding values for 10 growth and carcass traits. We compared two different weights, alpha values of 0.95 ($a_{0.95}$) and 0.70 ($a_{0.70}$), for genomic information while building the H matrix. The Legarra-Reverter (LR) method was used to estimate relative accuracy improvement (RAI) and dispersion (b1) values. Reduced data was created by removing observations made after 2018. The ssGBLUP and pedigree BLUP (PBLUP) was considered as the whole and partial model respectively. RAI was calculated as the accuracy ratio obtained from ssGBLUP and PBLUP minus 1, and the b1 value was defined as the regression coefficient GEBVs from ssGBLUP on EBVs from PBLUP in the validation set (animals born after 2018). The estimated RAI ranged from 5% for Europe conformation class (SCONF) in Finnish Charolais animals to 71% for post weaning gain (PWG) in Swedish Hereford animals when using $a_{0.95}$. Using $a_{0.70}$ resulted in lower RAI (on average for all traits 15.3 and 30.3% in Charolais and Hereford, respectively) compared to using $a_{0.95}$ (average RAI 23.6% in Charolais and 43.0% in Hereford). The b1 values ranged from 0.80 to 1.05 in both Charolais and Hereford using $a_{0.95}$ and were only slightly, and not consistently, influenced by the alpha value used. We conclude that introducing ssGBLUP in Nordic beef cattle breeding would be feasible and beneficial to improve the accuracy of breeding values especially for young, genotyped animals.

Genetic parameters of pig birth weight variability

Y. Salimiyekta[1], S.B. Bendtsen[1], K.V. Riddersholm[1], M. Aaskov[1] and J. Jensen[2]
[1]Danish Genetics, Lysholt Alle 10, 7100 Vejle, Denmark, [2]Aarhus University, quantitative genetics and genomics, C.F Møllers Alle 3, 8000 Aarhus, Denmark; yas@danishgenetics.dk

The objective of this study was to estimate genetic variance and heritability of birth weight variability in the Landrace and Yorkshire pig populations in Denmark. The data set consisted of 1,352 litters from 1,155 Landrace sows and 1,758 litters from 1,432 Yorkshire sows farrowing in the period from 2021 and 2023. Variance components and breeding values were estimated using single-step genomic BLUP (ssGBLUP) method and the average information restricted maximum likelihood approach (AI-REML). The heritability of birth weight variability was estimated to be 0.100 and 0.114 for Landrace and Yorkshire, respectively. These results show that birth weight variability is heritable and minimizing the total within litter variance through selection is possible. Lower variability in the size of piglets can lead to less preweaning mortality and better growth during sucking in the pig population of Denmark.

Disentangling paternal and maternal components of within litter birth weight variability in mice

N. Formoso-Rafferty[1], L. El-Ouazizi El-Kahia[2], I. Cervantes[2] and J.P. Gutiérrez[2]
[1]ETSIAAB, UPM, Producción Agraria, Senda del Rey 18, 28040, Madrid, Spain, [2]Facultad de Veterinaria, UCM, Producción Animal, Avda. Puerta de Hierro s/n, 28040, Madrid, Spain; nora.formosorafferty@upm.es

Animal production is searching for homogeneity, as it has been related to robustness, that improves functional traits whilst maintaining high production potential. Selection for homogeneity has resulted in more robust animals resilient to environmental challenges improving the animal welfare. A divergent selection experiment for birth weight (BW) environmental variability in mice has been successfully performed during 32 generations. Selection for low variability (L-line) was beneficial for traits usually considered robustness indicators such as birth weight homogeneity, growth, feed efficiency, litter size, survival and longevity. In this mice experiment, it was not possible to differentiate the within litter variability attributable to the father or to the mother because of the one to one mating design. The objective of this study was to ascertain the maternal or paternal nature of the genetic component of birth weight variability crossing mice selected lines between them and with the control line. H-line, L-line and C-line were mated simultaneously all with all, giving 9 different groups. Females were allowed to have up to two parturitions. Observed values for variance (V), standard deviation (SD) and coefficient of variation (CV) of BW, litter size (LS), mean birth weight (MBW), and individual pup birth (BW) and weaning (WW) weights were analysed. The model included the maternal and paternal lines, the interaction between them as an heterosis indicator, parturition number, the sex (for BW and WW) and the litter size (except when it was used as trait) as fixed effects. The model was solved by using TM software. Maternal significant differences between L-line and H-line were 19.2, 8.5, 7.0, 4.6, 1.3, 1.3 and 2.6 times the paternal differences between lines for respectively V, SD, CV, LS, MBW, BW and WW, being the paternal differences significant only for MBW, BW and WW. C-line performed intermediate between lines for all the traits except for LS in which there were no differences with L-line. Heterosis was not relevant. These results showed that variability within litter is a maternal trait.

Session 70 — Theatre 4

(Co)variances between anogenital distance and fertility in Holstein-Friesian dairy cattle

M.A. Stephen[1,2], C.R. Burke[2], N. Steele[2], J.E. Pryce[3,4], S. Meier[2], P.R. Amer[5], C.V.C. Phyn[2] and D.J. Garrick[1]
[1]Massey University, AL Rae Centre for Genetics and Breeding, Ruakura, Hamilton, 3214, WKO, New Zealand, [2]DairyNZ, NZAEL, 605 Ruakura Road, 3240, WKO, New Zealand, [3]La Trobe University, School of Applied Systems Biology, Bundoora, 3083, Victoria, Australia, [4]AgriBio, Agriculture Victoria Research, Bundoora, 3083, Victoria, Australia, [5]AbacusBio, Central Dunedin, 9016, Dunedin, New Zealand; melissa.stephen@dairynz.co.nz

Reproductive performance is an economically important trait for dairy farmers, particularly in pasture-based, seasonal farm systems that use an annual calving pattern to align pasture growth with the feed demands of the herd. Traditional fertility phenotypes are lowly heritable and expressed late in an animal's life, which contributes to relatively slow progress from genetic selection for those traits. Anogenital distance (AGD) is a moderately heritable candidate trait for predicting fertility estimate breeding values (EBVs) in dairy cattle. The objectives of this study were twofold. First, to estimate the genetic and phenotypic (co)variances between AGD measured at two ages and subsequent fertility traits expressed during lactation. Second, to estimate the genetic (co)variances between AGD and body conformation traits. We measured AGD, shoulder height, body length and body weight at 11±0.5 months of age in a population of 5,010 Holstein-Friesian and Holstein-Friesian × Jersey cows, born in 2018 and farmed across 54 seasonal, pasture-based herds. We also measured AGD at 29±0.7 months of age in a subset of 17 herds (n=1,956 cows). Calving, breeding, and conception dates were recorded for all available animals in first and second lactations, which commenced when animals were approximately 24 and 36 months of age, respectively. We report moderate heritabilities of 0.23 and 0.29 for 11-month and 29-month AGD, respectively. Both AGD measures exhibited moderate genetic correlations between 0.19 and 0.63 with calving, breeding, and conception traits. Genetic correlations between both AGD measures and body stature traits were weak (≤0.16). We conclude that AGD is a promising candidate predictor for fertility EBVs in dairy cattle, and genetic selection for shorter AGD in female cattle should result in improved fertility outcomes during lactation.

Session 70 — Theatre 5

Genetic relationships between productive and reproductive traits on Friesian cows

S. Abomselem, A. Badr and A. Khattab
Faculty of Agriculture, Tanta Univ., Animal Production Department, Campus of Siberibia, 35712, Egypt; adelkhattab@yahoo.com

A total of 2166 normal lactation records of Frisian cows, kept at Sakha Farm, belonging to Animal Production, Research Institute, Ministry of Agriculture were used to study the relationship between productive and reproductive traits. Variables studied were total milk yield (TMT), 305 day milk yield (305 d MY), lactation length (LL), number of service per conception (NSC), first service period (FSP), days open (DO) and calving interval (CI). Means of TMY, 305 d MY, LL, NSC, FSP, DO and CI were 2,875 kg, 2,680 kg, 302 d, 3.1, 76.2 d, 159.0 d and 433.2 d, respectively. Direct heritability (h2) for TMY, 305 d MY, LL, NSC, FSP, DO and CI were 0.20, 0.19, 0.09, 0.03, 0.04, 0.05 and 0.06, respectively. Genetic correlations between TMY, 305 d MY and LL were positive and ranged from 0.936 to 0.944. While, genetic correlation among TMY, 305 d MY, LL and NSC, FSP, DO and CI were negative and ranged from -0.10 to -0.30. Phenotype correlations among different traits are similar to genetic correlation. The present results indicated that the genetic improvement in milk production can be achieved through selection breeding program while, low h2 estimates for LL, NSC, FSP, DO and CI indicated that improving the managerial techniques should lead to a considerable decrease in length of lactation length, number of service per conception, first service period, days open and calving interval.

Heritabilities of the mid-infrared spectra of sheep milk throughout the lactation

C. Machefert[1], C. Robert-Granié[1], J.M. Astruc[2] and H. Larroque[1]
[1]GenPhySE, Université de Toulouse, INRAE, ENVT, 31326 Castanet-Tolosan, France, [2]Institut de l'Elevage, CNBL, 75595 Paris, France; coralie.machefert@inrae.fr

Fourier transform infrared (FTIR) milk spectral data were routinely used to predict milk component concentrations and could be considered as predictors reflecting the physiological status and performance of animal. The genetic variability of milk FTIR spectra is well documented in cattle and goat. This study aimed to explore genetic variability of FTIR spectra in sheep milk by estimating heritability at each individual spectral point in the mid-infrared region from French Lacaune dairy sheep's milk obtained in SMARTER European project. The FTIR spectrum of milk provided by spectrometers contains 1,060 points, called wavenumbers (5,012 to 926 cm^{-1}), based on transmittance. Milk was sampled during two years (2020-2021) in 8 commercial farms. After data editing, the number of records available was 41,143 FTIR spectra from 5,281 ewes with a mean of 8 records per ewe for the whole testing period. 1,794 ewes were genotyped with low-density chip then imputed to SNP50 Bead-Chip. Heritability at each wavenumber was estimated, with all records and at each lactation stage, by a single-trait animal model using AI-REML including pedigree and genomic relationships. Parity and a vector of flock-year-stage of lactation were included as fixed effects and an additive genetic effect on the animal, permanent environmental effect and residual as random effects. The pattern and variability of the FTIR spectrum was similar to those estimated in dairy cattle and goat. Heritability estimates ranged from 0 in water absorption regions to 0.42 in regions linked to milk composition. A moderate repeatability of FTIR spectra (0 to 0.53) was observed. Two different heritability patterns were observed between the first three and last lactation stages considered. These results will be analysed regarding the evolution of the farming conditions during lactation. Additional analyses will allow to characterize groups of ewes with similar patterns and genetic correlations between wavenumbers and milk production traits will be estimated. This information will contribute to propose less invasive and cheaper predictors for the characterization of adaptability of dairy ewes. Study financially supported by INRAE and Occitanie region.

Genetic parameters for the composition of milk fatty acid of Holsteins

Y. Masuda
Rakuno Gakuen University, Ebetsu, Hokkaido, 0698501, Japan; yutaka@rakuno.ac.jp

A negative energy balance of milking cows may lead to infertility and metabolic disorders. Fatty-acid (FA) content in milk represents the energy balance, and the measurement can be used to monitor the health condition of a cow. The FA content is now measured by Fourier-transform mid-infrared spectroscopy (FT-MIR) in the dairy milk recording program in Japan. This study aimed to estimate genetic parameters for the FA traits in the first three lactations of Japanese Holsteins. Data included 1,448,373 test day records from 211,462 Holstein cows in the 1st lactation, calving in Hokkaido, Japan, between April 2021 and October 2022. Two groups of FA, de novo FA (between C4 and C14) and preformed FA (equal to or greater than C18), were measured with FT-MIR for a milk sample. The de novo FA percentage (DNF%) and preformed FA percentage (PFF%) were calculated as the ratio of the predicted FA content to the total FA content. The dataset was divided by lactation stage, defined as every 30 days in milk (DIM) in each lactation, i.e. DIM 6 to 35 as stage 1, DIM 36 to 65 as stage 2, and so on. Variance components were estimated with an animal model. For each trait, (co)variances were estimated with a two-trait model for phenotypes between two different stages. In the first lactation, heritability at stage 1 for DNF% was 0.24, for PFF%, 0.30, and the value decreased up to stage 7 (DIM 215) in the first lactation. For both traits, stages 1 and 2 showed lower genetic correlations with later stages, e.g. for DNF%, 0.21 between stages 1 and 5. The genetic correlations of stage 3 or later were high (>0.90) up to stage 7. Similar trends in heritability and genetic correlations were observed in the second and third lactations. Further research will focus on genetic relationships between health and reproductive traits and the FA content, particularly at stages 1 and 2, where the negative energy balance may cause an issue in a milking cow.

Session 70 Theatre 8

External and genetic factors influencing fertility in Latxa dairy sheep breed

C. Pineda-Quiroga, I. Granado-Tajada, E. Ugarte and A. Basterra-García
NEIKER, Animal Production, Campus Agroalimentario de Arkaute, 01080, Spain; igranado@neiker.eus

Pregnancy rate at artificial insemination (AI) is largely variable and dependent on genetic and non-genetic factors. In Latxa breed, herds within the breeding program make use of high genetic value rams to inseminate their ewes once per year. Despite the relevance of the success of AI on the breed's genetic progress and on farms' productivity, this trait has not been up to now explored in this sheep population. To this aim, 135,351 edited AI records from 63,480 Latxa Cara Negra from Euskadi ewes, collected between 2000 and 2021, were used. The outcome of an AI event was treated as a binary response of either success or failure in becoming pregnant. To identify the environmental variables influencing the AI outcome, a multiple logistic regression was first calculated. After that, using the relevant factors identified in the previous step, a threshold model was used to estimate the genetic components of the trait in females and males. Results show that the AI success is higher in ewes having their first parturition at 1 year of age than at 2 or 3 year of age (Odd Ratio [OR] 1.04 and 1.08, respectively), in those with high prolificacy in the lambing previous to the AI event (OR 1.04 and 1.13 by having 2 or 3 lambs, respectively, vs 1 lamb), in those with a larger lambing-AI interval (OR 1.07 and 1.35 for intervals between six months or more, respectively, vs three months interval), as well as in those who had the previous parturition from an AI event compared to natural mating (OR 1.39). In counterpart, the higher the milk produced in the nearest milking AI date, the poorer the AI results (OR 0.92 when producing more than 450 ml vs producing up to 300 ml). Moreover, ewes older than five years and six lambings have less AI success with respect to younger ewes (OR 0.78). Regarding genetic parameters, the heritability and repeatability were 0.090±0.007 and 0.200±0.007 in females and 0.013±0.006 and 0.025±0.002, respectively, in males. These results indicate that there are extensive and diverse external factors influencing the AI success, while the additive component is low. Although the trait shows genetic variability susceptible of selection under the breeding scheme, the genetic progress would be slow.

Session 70 Theatre 9

Single-step genome-wide association for milk urea concentration in Walloon Holstein

H. Atashi[1,2], Y. Chen[2], C. Bastin[3], S. Vanderick[2], X. Hubin[3] and N. Gengler[2]
[1]Shiraz University, Department of Animal Science, Shiraz, 7144113131 Shiraz, Iran, [2]University of Liège, GxABT, Pass. des Déportés 2, 5030 Gembloux, Belgium, [3]Elevéo asbl Awé Group, Ciney, 5590 Ciney, Belgium; hadi.atashi@uliege.be

Milk urea (MU), a normal non-protein nitrogen component in milk, is a by-product of the protein metabolism. It has been shown that MU is correlated with milk production, milk composition, cheese-making properties, nitrogen utilization efficiency and reproductive performance in dairy cattle. Therefore, selection for reducing MU can result in reducing N pollution, improving feed efficiency and animal health, and increasing milk quality and reproductive performance. The aims of this study were to estimate genetic parameters for MU and to conduct a single-step genome-wide association (ssGWAS) to identify candidate genes associated with MU in Walloon Holstein. The used MU data have been collected from 2014 to 2020 on 78,073 first-parity (485,218 test-day records), and 48,766 second-parity (284,942 test-day records) Holstein cows distributed in 671 herds in the Walloon Region of Belgium. Data of 730,539 single nucleotide polymorphisms (SNP), located on 29 *Bos taurus* autosomes (BTA) of 6,617 animals (1,712 males) were used. The proportion of the total additive genetic variance explained by windows of 50 consecutive SNPs (with an average size of ~ 216 Kb) was calculated, and the top-three genomic regions explaining the largest rate of the total additive genetic variance were considered promising regions and used to identify potential candidate genes. Mean (standard deviation) MU was 25.38 (8.02) mg/dl and 25.03 (8.06) mg/dl in the first and second lactation, respectively. Mean heritability estimates for daily MU were 0.21 and 0.23 for the first and second lactation, respectively. The top-3 regions combined explained 1.22 and 1.04% of the total additive genetic variance of MU in the first and second lactations, respectively. The identified regions were located from 80.61 to 80.74 Mb on BTA6, 103.26 to 103.41 Mb on BTA11, and 1.59 to 2.15 on BTA14. Genes including *PAEP*, *SOHLH1*, *GLT6D1*, *LCN9*, *DGAT1*, *CYHR1*, *CPSF1*, *SCX*, *SCRT1*, and *SPATC1* were identified as positional candidate genes for the MU. The findings of this study provide a better understanding of the genomic architecture underlying MU in Holstein cattle.

Genome-wide association study for milk production traits in the Cyprus Chios sheep

A.N. Georgiou, G. Maimaris, S. Andreou, A.C. Dimitriou and G. Hadjipavlou
Agricultural Research Institute, Animal Production, P.O. Box 22016, 1516, Nicosia, Cyprus; ageorgiou@ari.moa.gov.cy

The identification of genetic markers affecting traits of economic interest such as milk performance provide practical benefits for the dairy sheep industry, enabling genomic prediction of the breeding value, based on single nucleotide polymorphisms (SNPs). The present study focuses on the identification of genetic variants associated with total milk yield in Cyprus Chios sheep, which is the most commercially important dairy breed in Cyprus. In this pilot study, total milk yield measurements for the first lactations were recorded over a period of 11 years (2011-2022) for 393 Chios ewes. All animals were genotyped using Illumina Ovine 50k Genotyping Beadchip. A principal component analysis (PCA) was performed to assess genetic diversity among individuals while a genome-wide association analysis (GWAS) was conducted aiming to identify candidate SNPs related to available phenotypes. After quality control, a total of 39,115 markers (over 26 autosomal chromosomes) and 386 ewes were retained for further analysis. The PCA analysis revealed a substructure within the dataset, identifying two principal clusters in the population. The GWAS analysis identified two SNPs, that were associated with total milk yield at a genome-wide significance ($P<1\times10^{-4}$). The two SNPs were located in the non-coding region of *KIAA1217* (chromosome 13) and *CAMD1* (chromosome 15) genes respectively. More than 4,000 ewes across various flocks in Cyprus will be genotyped in the near future to further validate the primary findings of the GWAS analysis, and to enable the identification of further promising biomarkers for milk production (quantity and quality) in the Cyprus Chios sheep. The findings of the current study are anticipated to improve sheep productivity in Cyprus through the estimation of individual genetic merit and subsequently genomic evaluation and improvement of milk performance through the AGRICYGEN (CYprus AGRIcultural GENomics Centre) project.

Characterization of additive, dominance, and runs of homozygosity effects inbreeds of dairy cattle

H. Ben Zaabza[1], M. Neupane[2], M. Jaafar[3], K. Srikanth[3], S. McKay[1], A. Miles[2], H.J. Huson[3], I. Strandén[4], H. Blackburn[5] and C.P. Van Tassell[2]
[1]USDA, Animal Genomics and Improvement Laboratory, ARS, Beltsville, MD, USA, [2]University of Vermont, Department of Animal and Veterinary Sciences, Burlington, VT, USA, [3]Cornell University, Department of Animal Science, Ithaca, NY, USA, [4]Natural Resources Institute Finland, Jokioinen, FI, Finland, [5]National Center for Genetic Resources Preservation, Fort Collins, CO, USA; hafedh.ben-zaabza@uvm.edu

The global application of genomic selection in dairy cattle has raised interest in characterizing dominance effects for a better understanding of inbreeding depression (IB). We believe that a richer understanding of additive (ADD), dominance (DOM), and runs of homozygosity (ROH) effects in purebred and crossbred dairy cattle will help to understand the impact of these factors on IB and heterosis. To identify and localize genomic regions associated with ADD, DOM, and ROH effects we performed a single-SNP GWAS analysis, where SNPs were fit as fixed effects for the ADD, DOM, and ROH one locus at a time. The current analysis has been performed using 125,000 US Holstein cows genotyped on 79,294 SNP markers. We have analysed 3 production traits, 3 fertility traits, and somatic cell score (SCS). For the production traits, SNPs on BTA14 had the largest ADD effects. For fertility traits, SNPs on BTA1 had the most significant ADD effects. SCS had no significant DOM effects. Few DOM effects were detected for production traits and these effects had lower significance than ADD effects. Estimates of fertility trait DOM were detected with similar statistical significance as the ADD effects. The ROH revealed less prominent $-\log_{10}(P)$ peaks than the ADD effects for all traits. Genome-wide inbreeding depression will be examined for all traits. Correlations between ROH and ADD effects were close to 0. Correlations between ROH and DOM effects calculated by ADD-DOM model were 0.006. However, correlations between ROH and DOM effects calculated by ADD-DOM-ROH model were 0.31. These findings suggest that confounding exists between ADD and ROH effects – and that DOM and ROH components capture the same variance. This study will be further extended to a genomic dataset comprising over 4 million (1 million) genotyped purebred Holstein (Jersey) cows and all available crossbreds.

Session 70 Theatre 12

A genome-wide association study identified a major QTL affecting the red colour in nitrate free hams

J. Vegni[1], M. Zappaterra[1], R. Davoli[1], R. Virgili[2], N. Simoncini[2], C. Schivazappa[2], A. Cilloni[3] and P. Zambonelli[1]
[1]Alma Mater Studiorum University of Bologna, Scienze e Tecnologie Agroalimentari (DISTAL), section: Animal Science, Viale Fanin 46, Bologna, 40127, Italy, [2]Stazione Sperimentale per l'Industria delle Conserve Alimentari (SSICA), Viale Faustino Tanara 31/A, Parma, 43121, Italy, [3]Fratelli Galloni S.p.A., Via Roma 84, Langhirano (Parma), 43013, Italy; jacopo.vegni2@unibo.it

Meat colour is an important quality parameter in dry-cured protected designation of origin (PDO) hams where the use of any additives, including nitrates, are not allowed. The typical colour of dry-cured hams, like Parma ham, is due to the formation of the red pigment Zinc protoporphyrin (ZnPP) in which Zn(II) is inserted in protoporphyrin IX. The synthesis of ZnPP has been suggested to be totally or partially enzymatic, and a role was assigned to the endogenous ferrochelatase enzyme. In this work, we carried out a genome-wide association study (GWAS) to identify candidate gene regions affecting meat colour in nitrate-free dry cured hams. A total of of 238 commercial hybrid pigs were genotyped using the GeneSeek® Genomic Profiler genome-wide porcine genotyping array. The activities of ferrochelatase (FECHA) enzyme (responsible for meat red colour) in *Semimembranosus* muscle were determined through fluorimetric assays. The GWAS identified 12 single nucleotide polymorphisms (SNPs) that were significantly associated with FECHA activity. These SNPs were located in the chromosome region that harbours the FECHA gene, suggesting a direct role of variants in this gene with the investigated parameters related to meat colour. Additional studies will be needed to validate the obtained results and identify the causative mutation of this relevant QTL for meat quality parameters.

Session 70 Poster 13

Genetic relationships between milk fatty acids at early lactation and fertility in Holstein cows

T. Yamazaki[1], A. Nishiura[2], S. Nakagawa[3], H. Abe[3], Y. Nakahori[3] and Y. Masuda[4]
[1]Hokkaido Agricultural Research Centre, NARO, Sapporo, 062-8555, Japan, [2]Institute of Livestock and Grassland Science, NARO, Tsukuba, 305-0901, Japan, [3]Hokkaido Dairy Milk Recording and Testing Association, Sapporo, 060-0004, Japan, [4]Rakuno Gakuen University, Ebetsu, 069-8501, Japan; yamazakt@affrc.go.jp

To address the negative energy balance in early lactation is important for improving fertility of dairy cows, and milk fatty acid (FA) compositions are thought to be related to energy status. Our objective here was to investigate the genetic relationships between FA compositions during the early stage of lactation and fertility traits in the first and second lactations of Japanese Holstein cows. We used the insemination and monthly test-day records of Holstein cows that calved between 2019 and 2021 in the Hokkaido region of Japan (33K cows for first lactation and 21K cows for second). We analysed three types of test-day milk FA compositions (de novo FA (C4 to C14) composition on a milk basis (DNM), de novo FA composition on a total FA basis (DNF) and preformed FA (≥C18) composition on a total FA basis (PRF)) during 95 days in milk (DIM) as FA composition traits and conception rate at first insemination (CR) and days open (DO) as fertility traits. We estimated the genetic correlations between FA composition and fertility traits within each lactation by using two-trait (each FA composition traits and CR or DO) linear model, with month and age of calving for FA composition traits, and, month, age, and DIM at first insemination for fertility traits as fixed effects. We also included herd-year, service sire at first insemination (for fertility traits), additive animal, and residual effect as random effects. The genetic correlations of DNM, DNF, and PRF within 35 DIM with fertility traits in the first lactation were 0.31, 0.44, and -0.25 for CR, and -0.26, -0.27, and 0.25 for DO, respectively. These results suggest that genetic improvements in milk FA composition within 35 DIM could help to increase fertility in the first lactation.

Session 70 Poster 14

Genetic parameters for early-life racing performance in pigeons

P. Duenk[1] and D. Shewmaker[2]
[1]Wageningen University and Research, Animal Breeding and Genomics, Droevendaalsesteeg 1, 6708 PB Wageningen, the Netherlands, [2]Shewmaker Genetics, 5575 Pyracantha Dr, 95682 Shingle Springs, USA; pascal.duenk@wur.nl

Pigeon racing is a popular sport in many countries. Races range from local club competitions to international One Loft Races where prize money can be over one million US dollars. Although breeders have been successful in improving racing performance over the years by means of selection, selection decisions are usually based on observations that may be weakly correlated with racing performance. In addition, racing performance traits are not clearly defined, and accurate estimates of genetic parameters for such traits are scarce. Finally, inbreeding is often not monitored, while mating of close relatives is quite common. Breeders may therefore benefit from data driven, structured breeding programs that improve racing performance while monitoring genetic diversity. Before such programs can be designed, however, some questions need to be answered. The objectives of this study are therefore to: (1) define the traits of interest in a breeding program for racing pigeons; (2) estimate genetic parameters of these traits; and (3) calculate historical rate of inbreeding. Data will be collected on young birds (born in 2023) from a single loft, between March and July 2023. During this period, the birds will be trained to return back to the loft, where the distance between release location and the loft will be incrementally increased up to about 450 km. The data will consist of approximately seven contemporary groups, each consisting of 100 birds. Across and within contemporary groups, there may be half- and full-sibs. We will record distance and direction of release location, individual total flight time, and weather conditions. In addition, we will collect pedigree data of up to seven ancestral generations. Genetic parameters will be estimated using a linear mixed model that corrects for environmental and contemporary group effects. Inbreeding rate will be computed from the pedigree. Overall, this study will provide knowledge on the importance of genetics for racing performance in pigeons, and may contribute to the development of more sustainable and efficient racing pigeon breeding and evaluation programs.

Session 70 Poster 15

Genetic parameters for milk urea in Swiss dairy cattle breeds

A. Burren and S. Probst
Bern University of Applied Sciences BFH, School of Agricultural, Forest and Food Sciences HAFL, Länggasse 85, 3052 Zollikofen, Switzerland; alexander.burren@bfh.ch

The objective of this study was to estimate repeatability scores and variance components (VC) for milk urea in the Swiss cattle breeds Braunvieh (BV), Original Braunvieh (OB), Holstein (HO), Montbéliard (MO), Red Holstein (RH), Swiss Fleckvieh (SF) and Simmental (SI). For this purpose, milk urea results from 2,315,437 standard lactations yields (BV=955,231; OB=80,016; HO=265,775; MO=44,590; RH=474,289; SF=367,720; SI=127,816) with totally 2,335,267 pedigree records (BS=676,368; OB=73,927; HO=416,634; MO=70,635; RH=526,760; SF=431,080; SI=139,863) were used. The data were collected in the years 2010-2019. Genetic variances and covariances were estimated by REML using the software ASReml 4.2. The model to estimate VC and repeatabilities included the fixed effects number of lactation and milk yield, the random effects farm and sire, as well as a random additive genetic component and a permanent environmental effect. Estimated heritabilities and standard errors were 0.37±0.004, 0.52±0.012, 0.32±0.007, 0.20±0.013, 0.27±0.005, 0.34±0.006, 0.38±0.009 in BV, OB, HO, MO, RH, SF and SI breed, respectively. The repeatability coefficients varied between 0.30±0.008, 0.34±0.003, 0.40±0.003, 0.40±0.003, 0.44±0.006, 0.45±0.002 and 0.56±0.006 in MO, RH, SF, HO, SI, BV and OB, respectively (±standard error). The results are consistent with other studies and show that genetic modification of milk urea by selection is possible.

Session 70 Poster 16

Estimation of genetic correlations between predicted energy balance and fertility in Holsteins

A. Nishiura[1], O. Sasaki[1], S. Yamaguchi[2], Y. Saito[1], R. Tatebayashi[1] and T. Yamazaki[3]
[1]Institute of Livestock and Grassland Science, NARO, Tsukuba Ibaraki, 3050901, Japan, [2]Livestock Improvement Association of Japan, Koto Tokyo, 1350041, Japan, [3]Hokkaido Agricultural Research Center, NARO, Sapporo Hokkaido, 0628555, Japan; akinishi@affrc.go.jp

It is important to improve the energy balance (EB) of dairy cows in lactation to prevent the deterioration of fertility and health. Our objective was to investigate relationships between predicted EB (PEB) from milk traits and fertility. The datasets we used to predict EB consisted of test-day milk records for the first three lactations of Holstein cows calved in 2015-19. There were 1,185,020, 966,742 and 676,416 records of first, second and third lactation cows each. We calculated average PEB of cows in each class of first conception rate (FCR), number of insemination (NI) and days open (DO) in early lactation. We estimated genetic correlations between PEB and FCR, NI and DO using random regression model. We included herd-test day, region-calving month and parity-calving age as the fixed effects in the model for PEB. We included herd-year, month and parity-age at first insemination as the fixed effects in the model for FCR, NI and DO. Additive genetic and permanent environmental effects were also included in both models as the random effects. The PEB of cows not conceived at first insemination were lower than those of cows conceived especially in multiparous cows. The PEB of cows with higher NI were lower than those of cows with lower NI especially in multiparous cows. The PEB of cows with longer DO were lower than those of cows with shorter DO especially in multiparous cows. Genetic correlations between PEB and FCR were estimated 0.16 to 0.19 in early lactation and became smaller in mid to late lactation. Genetic correlations between PEB and NI were estimated -0.14 to -0.10 in early lactation and became smaller in mid to late lactation. Genetic correlations between PEB and DO were estimated -0.26 to -0.21 in early lactation and became positive in late lactation, 0.10 at 305 days in milk. There were desirable genetic correlations between PEB and fertility traits in early lactation. It would be possible to improve fertility by improving EB in early lactation and PEB could be used as the index of EB.

Session 70 Poster 17

Comparison of random regression test-day models for production traits of South African Jersey cattle

M.G. Kinghorn, E.D. Cason, V. Ducrocq and F.W.C. Neser
University of the Free State, Animal, Wildlife and Grassland Sciences, P.O. Box 339, Bloemfontein 9300, South Africa; neserfw@ufs.ac.za

Fixed regression test-day (TD) models are currently implemented for genetic evaluation of South African (SA) Jersey Cattle. The lactation curve is determined by fixed effects only and assumes that random effects are constant throughout the lactation, implying a genetic correlation of one between TDs. Random regression TD models (RRTDM) don't make the same assumption and split the lactation curve into two parts: a fixed part that accounts for similarities of lactation curves within specific groups and a random animal part that provides information on individual deviations from the fixed lactation curve. Furthermore, inferences can be made about persistency (PER). Our objective was to identify an appropriate RRTDM for SA Jersey cattle using Legendre polynomials (LP) to model the lactation curve by estimating genetic parameters with single- (S) and multiple- (M) lactation RRDTMs. TD records from 3 lactations were analysed using AIREMLF90. Goodness of Fit was assessed using Bayesian information Criterion (BIC). Varying orders of fit for LP were tested with S-RRTDMs, the resulting (co)variance components were used as starting values in the M-RRDTM for each trait (milk (MY), fat (FY) and protein (PY)). S-RRTDMs with a 3rd order LP fit best. For M-RRTDM a 2nd order LP was the best fit. Genetic correlations for total production (TP) between lactations were high for all traits (0.90-0.98), but lower for PER (0.54-0.97). Heritability estimates obtained for TP ranged between 0.20-0.25 (MY), 0.13-0.15 (FY) and 0.19-0.21 (PY) and were similar to those obtained by Interbull for the current SA Jersey model. The current model does not have a comparative heritability estimate for PER. PER estimates obtained in this study were low; 0.08-0.14 (MY), 0.07-0.10 (FY) and 0.08-0.12 (PY), but highest in the 2nd lactation. Stronger relationships for TP and PER in later lactations suggest that more effective selection could be made using a M-RRTDM that accounts for performance in later lactations. Similar heritability estimates (Interbull) for TP, plus the addition of a heritable measurement for PER suggest that the M-RRTDM proposed in this study is an appropriate alternative model to the current SA model.

Session 70 Poster 18

Application of single-step genomic method in routine evaluation of Czech Holstein cattle

J. Bauer, J. Šplíchal, D. Fulínová and E. Krupa
Czech Moravian Breeders` Corporation, Genetic Evaluation, Benešovská 123, 25209 Hradištko, Czech Republic; bauer@plemdat.cz

Single-step genomic evaluation of Holstein Cattle in Czech Republic was in development since 2011 with support of prof. Ignacy Misztal of University of Georgia, USA. At first, the original procedures and programs for prediction of genomic breeding values (GEBVs) were exclusively written inhouse in Cobol and Rexx languages. However, with higher number of genotypes the computational time was considerably increasing and limitations of optimizations of these programs became more apparent. For this reason, the Blupf90 family software was adopted for prediction of GEBVs by single-step method. In 2015 all traits participating in national index for Holstein Cattle (SIH) had prediction of GEBVs which allowed to publish SIH together for all genotyped and non-genotyped animals. In the year 2015 Interbull validated and used Czech single-step GEBVs for milk production as input into genomic Multiple Across Country Evaluation (genomic MACE). It was first time when GEBVs predicted by single-step GBLUP were included at all. Because of the intensive exchange of genetic material between the Czech Republic and other countries, the eventual high genetic quality of imported bulls was often low rated in GEBVs due to lack of information. The issue was solved by development of integration of breeding values from MACE performed by Interbull. The data from international evaluation of bulls is since 2019 integral part of official GEBVs published in Czech Republic for milk production traits, udder health, linear traits, and fertility. Since then, the linear traits were also included in genomic MACE. The current routine genomic evaluation uses advanced techniques such as Algorithm for Proved and Young (APY) and further enhancements like utilization of weighted single-step BLUP are tested. Additionally, more traits will be soon implemented with cooperation with Institute of Animal Science Prague. The highest priority was given to health traits (e.g. mastitis resistance, infectious and non-infectious claw disorders). However, GEBVs of other traits are also in implementing process or are already published such as GEBVs for gestation length making total of 37 traits with predicted routine GEBVs.

Session 70 Poster 19

A meta-analysis of the genetic parameter estimates for lamb survival

S. Fernandes Lazaro[1], H. Rojas De Oliveira[2] and F. S. Schenkel[1]
[1]University of Guelph, Department of Animal Biosciences, 50 Stone Rd E, Guelph, ON, N1G 2W1, Canada, [2]Purdue University, Department of Animal Sciences, 610 Purdue Mall, West Lafayette, IN, 47907, USA; sirlene@uoguelph.ca

Genetic factors are known to play a role in Lamb survival (LS) and knowing its heritability and genetic correlations with other economically important traits is required to optimize the inclusion of LS into sheep breeding programs. The goal of this study was to perform a meta-analysis on the genetic parameters for LS. The dataset comprised 24 heritabilities and 33 genetic correlations between LS and other important traits, which included birth weight, BWT; lambs born per ewe bred; NLB; lambs weaned per ewe bred, NLW; weaning weight, WWT; and lambs born per lambing ewe, LL. All genetic parameters included in this study were estimated using animal linear models. The meta-analysis was performed using a random-effects model, in which the parameter estimates were assumed independent, and normality distributed. The I2 index was used to quantify the degree of heterogeneity among studies, and the effects of breed group (maternal, terminal, wool, and dual purpose), year of publication (from 1994 to 2010 and from 2010 to 2022), and continent (America, Africa, Asia, Europe, and Oceania) were tested. High variability among estimates from different studies were observed (I2>99.58%), which was mostly associated with the continent. The pooled heritability estimated for LS was 0.05 (95% confidence interval (CI): 0.02;0.07), which confirms that genetic improvement through selection is possible. Low to moderate genetic correlation estimates were observed between LS and BWT (0.16, CI: -0.07;0.39), LS and NLB (-0.16, CI: -0.30;-0.01), LS and WWT (0.39, CI: 0.15;0.63), LS and NLW (0.67, CI: 0.56;0.77), and LS and LL (-0.17, CI: -0.34;-0.01), suggesting the importance to directly include LS in the breeding programs. Overall, LS seems lowly heritable and shows low to moderate genetic correlation with the other traits analysed. Genetic parameters reported here support genetic evaluation for LS when reliable population specific parameter estimates are available.

Session 70 Poster 20

Survival analysis in the conservation program of an endangered wild ungulate (*Nanger dama mhorr*)

S. Domínguez[1], I. Cervantes[2], E. Moreno[1] and J.P. Gutiérrez[2]
[1]Estación Experimental de Zonas Áridas-CSIC, Ctra. De Sacramento s/n, 04120 La Cañada de San Urbano, Spain, [2]Faculty of Veterinary, UCM, Department of Animal Production, Avda. Puerta de Hierro s/n, 28040 Madrid, Spain; sdominguez@eeza.csic.es

Juvenile survival is a relevant aspect in the *ex situ* breeding programmes of threatened species, since it directly affects the viability of these populations in the short and long term. It is known that dams and sires can have a great influence on the survival of their offspring, both for their genetic and environmental effect. The objective of this study was to evaluate the genetic component of juvenile survival in the captive population of the critically endangered mhorr gazelle (*Nanger dama mhorr*). The analysed dataset included records of 2,222 calves from 502 different females and a total pedigree of 2,739 individuals. Juvenile survival was studied separately for different ages: 0-15 days, 16-30 days and 31-180 days of life. Sex, type of birth (primiparous or multiparous), year and season of birth were included as systematic effects in all models. As random effects, combination of sire, dam and calf genetic effects were included, beside the residual effect. Models were solved with a continuous and a threshold approach, and then compared using logCPOP values. All analysis were performed in a Bayesian frame using the TM program. The logCPO values indicated that the models with the highest predictive power were the sire-calf threshold model for survival in the first 15 days of life, the dam-calf threshold model for survival in the second fortnight of life and the dam-calf threshold model for survival from one month of life to weaning. The heritability values resulting from these models ranged between 0.02 and 0.50. In the period of 0-15 days, the heritability was 0.07 (SD 0.04) for the sire and 0.50 (SD 0.08) for the calf; with 16-30 days, it was 0.02 (SD 0.01) for the dam and 0.03 (SD 0.02) for the calf; and with 31-180 days, it was 0.05 (SD 0.03) for the dam and 0.08 (SD 0.04) for the calf. These preliminary results suggest the existence of genetic component in the juvenile survival in different ages in the mhorr gazelle. It could be a starting of a selection program as part of the conservation strategy of the species.

Session 70 Poster 21

Genetic parameters of maturing rate index and asymptotic adult weight in French beef cattle

A. Lepers[1,2], S. Aguerre[1,2], J. Promp[1,2], S. Taussat[1,3], A. Vinet[1], P. Martin[1], A. Philibert[2], A. Laramee[2] and L. Griffon[2]
[1]Université Paris-Saclay, INRAE, AgroParisTech, GABI, rue de la manufacture, 78350 Jouy-en-Josas, France, [2]Institut de l'Elevage, 149 rue de bercy, 75012 Paris, France, [3]ELIANCE, 149 rue de bercy, 75012 Paris, France; armance.lepers@idele.fr

In beef cattle, animals that grow and reproduce the fastest are desirable to maximize productivity while limiting the environmental footprint. In order to investigate the genetic basis of this phenomenon, genetic parameters were estimated for both adult weight and maturing rate index. Weights from birth to post-weaning and slaughter weights were collected for commercial suckling females from five different breeds: Charolaise (n=27,341), Limousine (n=56,142), Blonde d'Aquitaine (n=21,529), Parthenaise (n=16,275) and Aubrac (n=1,797). Based on these weights, asymptotic adult weight and maturing rate index phenotypes were estimated using Brody equation. The five breeds studied show slightly different earliness growth curves with the Aubrac breed being the most precocious with a Brody coefficient for maturing rate index of 0.0017 (corresponding to almost 56% of adult weight reached at 15 months). The Parthenaise breed was the less precocious, with only 44% of its adult weight reached at the same age (equivalent to 0.0012 for brody's coefficient of maturing rate index). Genetic parameters of these two traits were estimated for the four breeds with enough data using a bivariate animal model. Results were highly similar across breeds, with heritability estimates close to 0.41 for asymptotic adult weight and 0.24 for maturing rate index. The two traits were highly correlated with an average genetic correlation of -0.64 and a phenotypic correlation of -0.67. The coefficients of genetic variance associated to adult weight and maturing rate index had a mean value of 6.2 and 9.1% respectively. In conclusion, this study shows a significant variability in growth precocity between breeds and confirms that animals with a high adult weight have a lower maturing rate index. The studied phenotypes are heritable and show high genetic standard deviations, which enables their possible consideration for genomic selection.

Session 70 Poster 22

Genetic architecture of the persistency of production, quality, and efficiency traits in laying hens

Q. Berger[1], N. Bedere[2], P. Le-Roy[2], T. Burlot[3], S. Lagarrigue[2] and T. Zerjal[1]
[1]Université Paris-Saclay, INRAE, AgroParisTech, GABI, 78350, Jouy-en-Josas, France, [2]INRAE, Institut Agro, PEGASE, 35590, Saint-Gilles, France, [3]NOVOGEN, 5 rue des Compagnons, 22960, Plédran, France; quentin.berger@inrae.fr

The laying hen industry aims to extend the laying production career to 90 weeks or more to increase profitability and promote ethical and environmental benefits. However, this can be challenging due to declining egg production and quality as well as reduced efficiency in aging hens. To investigate persistency in egg production, quality, and feed efficiency, we studied 1,024 purebred hens from the nucleus of Novogen between 70 and 90 wk of age. We recorded daily egg production throughout the period and measured individual feed intake twice a week for three weeks, starting at 70, 80, and 90 wks of age, as well as body weight at the start and at the end of each feed intake recording period. Egg quality was assessed at 70 and 90 wk Random regressions models were used to study trait variation on a trajectory of time and genotype by time interactions. Among the measured traits, egg weight, feed conversion ratio, weight gain, and Haugh unit showed persistency over the measured period. Daily feed intake, egg mass, residual feed intake, and laying rate decreased over time, while body weight and yolk percentage increased during the late period of production. To assess the viability and ways of selecting for trait persistency (i.e. stability over time), we estimated the genetic variance of the slope and its correlation with the intercept. We found that for laying rate and egg weight, the genetic variance of the slope was negligible, indicating that selection for persistency on these traits requires other means. On the contrary, for other traits such as residual feed intake, there is significant additive genetic variance in the slope. In addition, the genetic correlation between the intercept and the slope was -0.19, meaning that already decreasing residual feed intake level would also reduce the negative slope and help improve persistency. This can be further improved by integrating the slope in the selection criteria. Projects funded by the European Union's Horizon 2020 research and innovation programme under grant agreement N°101000236 and the National research agency under the decision code ANR-20-CE20-0029.

Session 70 Poster 23

On the potential of improving daily milk yield by extending productive lifespan

A. Bieber, F. Hediger, F. Leiber, C. Pfeifer and M. Walkenhorst
Research Institute of Organic Agriculture, Ackerstrasse 113, 5070 Frick, Switzerland; anna.bieber@fibl.org

Productive lifespan (PL) is a key factor for the sustainability of dairy farming. A long productive lifespan amortizes the rearing costs over a longer period of time and dilutes resource consumption and emissions. In addition, beef and milk production can be combined to a greater extent, which is advantageous in terms of emissions compared to specialized (separate) production of meat and milk. Nevertheless, the productive lifespan of dairy cows has been decreasing across main milk producing countries for many years. We aimed at describing the development of production level, productive lifespan and culling reasons from 1999 to 2019 by analysing herdbook data of 2.60 Mio. cows of the breeds Braunvieh (BV, n= 1.011.192), Swiss Fleckvieh (SF, n= 652.299), Holstein from swissherdbook (HO_SHB, n= 497.467), Holstein from Holstein Switzerland (HO_HOS, n=262.359), Simmental (SI, n=128.920), and Original Braunvieh (OB, n=50.063). We compared the increase in daily lifetime milk yield achievable over an extended PL with the increase in daily lifetime milk yield achieved over time from 1999 to 2019. Average milk yield per day of life continuously increased in all studied breeds, ranging from 7.5±4.1 (SI) to 12.1±5.5 (HO_HOS) kg/d in 2019. Average length of productive lifespan in 2019 ranged from 3.0±2.1 (HO_HOS) to 3.8±2.8 (SF) years. In contrast to international reports it increased (and stabilized) in all studied breeds, apart from OB where it decreased. Culling rates during first and second lactations were high, ranging from 40 (SF) to 51% (HO_SHB). Main culling reasons were fertility, udder health and leg/ claw problems in pronounced dairy breeds, while insufficient milk production was relevant in dual-purpose breeds. The comparison revealed that increase of daily lifetime milk yield during 20 years had the same size as could be achieved through extension of the PL by 1.3 years. Therefore, efforts to increase the number of completed lactations per cow may be expected to pay off with greater dynamics than the combined developments in selection for milk yield, improved feeding and housing over time.

Session 70 Poster 24

Genetic trends in a selection process using electronic feeders to improve feed efficiency in rabbit

J.P. Sánchez, M. Pascual and M. Piles
IRTA, Animal Breeding and Genetics Program, Torre Marimon, 08140 Caldes de Montbui, Spain; juanpablo.sanchez@irta.es

An improvement of the efficiency in the use of the feed will have positive consequence on both the economic balance of the rabbitries and on their environmental fingerprint. With the aim to generate animal material with enhanced feed efficiency we have conducted a selection experiment for 7 generations -34 batches- to improve feed efficiency in 3 rabbit lines. ADGR, RFI and GRP lines have been selected for growth under feed restriction, individual residual feed intake and cage-average (4 animals) residual feed intake, respectively. In spite of animals in lines ADGR and RFI were raised in groups (6 animals), as we used electronic feeders an individual recording of the feed intake was possible. The studied traits were average daily gain (ADG), average daily feed intake (ADFI) and feed conversion ratio (FCR=ADFI/ADG). In the line GRP they were cage-average records (n=873) while in RFI (n=2,482) and ADGR (n=2,343) lines they were individual records. The reported genetic trends are the regression coefficients of the within year-of-birth average predicted breeding values on the year-of-birth. Breeding value predictions were obtained with REML. The estimated genetic trends in RFI and ADGR lines were 0.03 ((g/d)/year) (P>0.05) and 1.20((g/d)/year) (P<0.01) for ADG; -1.79 ((g/d)/year) (P<0.01) and 0.34 ((g/d)/year)) (P<0.05) for ADFI; and -0.09 ((g/d)/g/d))/year) (P<0.01) and -0.01 (((g/d)/(g/d))/year) (P<0.05) for FCR. In the line GRP the genetic trends were 0.04 ((g/d)/year) (P>0.05), -1.67 ((g/d)/year) (P<0.01) and -0.07 (((g/d)/(g/d))/year) (P<0.01), for ADG, ADFI and FCR, respectively. The genetic trends on ADG were never companied with environmental trends; however, environmental trends of opposite sign were observed for ADFI and FCR which is an evidence of up-bias in the genetic trends. The environmental trends can be explained by changes over the selection process in the configuration parameters of the electronic feeders. We can conclude that the selection to improve feed efficiency in rabbits raised in collective cages seems to be effective. These trend-based response results are being further validated with experiments using an unselected control population.

Session 70 Poster 25

Maturity, a heritable trait in French dairy goat

M. Arnal[1,2], M. Chassier[2], V. Clément[2] and I. Palihière[1]
[1]GenPhySE, Université de Toulouse, INRAE, ENVT, 24 chemin de Borderouge, 31320 Castanet Tolosan, France, [2]Institut de l'élevage, 24 chemin de Borderouge, 31320 Castanet Tolosan, France; mathieu.arnal@idele.fr

Maturity can be defined as the ratio of first parity milk yield on third parity milk yield and can be, according to the literature, associated with poor longevity. As longevity has been declining for several years in French dairy goats, maturity seems to be interesting to investigate. The objective of this work was to estimate the phenotypic and genetic coefficient of variation of maturity, its heritability, and genetic correlations with milk yield at first, second and third parity in dairy goat, in order to evaluate the feasibility of a future selection on this trait. The data set consisted of 32,807 first parities, 19,215 second parities, 9,386 third parities, and 9,386 maturity phenotypes of Saanen dairy goats. 66,300 animals were in the pedigree. The model developed was a multi-trait model with four traits: maturity, milk yields at first, second and third parities. The heritability of maturity was moderate and equal to 0.12±0.02, suggesting that this trait could be selected. The genetic correlations with production were equal to 0.30±0.07, -0.17±0.08 and -0.40±0.07 for first, second and third parities respectively. The genetic correlations of milk yield between parities were close to those estimated in a previous study: 0.87±0.02 between first and second parity, 0.75±0.03 between first and third parity, and 0.95±0.02 for the correlation between second and third parity. These correlations show that the production capacity of the first and third parity is not the same and that there is variability captured by the maturity trait. The phenotypic coefficient of variation was higher for maturity than for milk yield: 0.26 for maturity, 0.19 for the first parity, 0.21 for the second parity and 0.23 for the third parity. The genetic coefficients of variation were close whatever the trait: 0.09 for maturity, 0.11 for first parity, 0.11 for second parity and 0.12 for third parity. The correlation between maturity EBVs and longevity EBVs of AI bucks with at least 15 offspring was equal to 0.4. This correlation makes this trait interesting for use as a predictor of longevity. A future study will implement different predictors of longevity to improve the accuracy of longevity.

Session 70 Poster 26

Multivariate genome-wide associations for immune traits in two maternal pig lines

C. Große-Brinkhaus[1], K. Roth[1], M.J. Pröll-Cornelissen[1], A.K. Appel[2], H. Henne[2], K. Schellander[1] and E. Tholen[1]
[1]Institut of Animal Science, University of Bonn, Endenicher Allee 15, 53115 Bonn, Germany, [2]BHZP GmbH, An der Wassermühle 8, 21368 Dahlenburg-Ellringen, Germany; cgro@itw.uni-bonn.de

The genetic foundation of immune traits, which determines the immune competence of piglets has not been fully uncovered, so far. Because of the well-known high genetic correlations among immune traits, pleiotropic or linked genetic markers can be expected. Usually, GWAS explores phenotypes in a univariate (uv), trait-by-trait manner, whereas multivariate (mv) methods, which should have a higher QTL detection power, have not been commonly used. Here, two uv GWAS methods and four mv GWAS approaches were applied on sets of 22 immune traits for Landrace (LR) and Large White (LW) pig lines. In total 433 (LR: 351, LW: 82) associations were identified with the uv approach implemented in PLINK and a Bayesian linear regression uv approach (BIMBAM) software. The identification of causal relationships among immune traits before performing mv GWAS helps to reduce extensive computation effort impaired by the realization of all possible mv combinations for all available immune phenotypes. Immune trait combinations of interest were created by performing Bayesian network analyses and principal component analyses. Mv GWAS approaches detected 647 associations for different mv immune trait combinations comprising 133 QTL. SNPs for different mv trait combinations (n=66) were observed with more than one mv method. Most of these SNPs are associated with red blood cell-related immune trait combinations. Functional annotation of these QTL revealed 453 immune-relevant protein-coding genes. With uv methods shared markers were not observed between the breeds, whereas mv approaches were able to detect two common SNPs for LR and LW. In conclusion, the mv methodology outperformed compared with uv methods. Our results indicate that one single test is not able to detect all the different types of genetic effects in the most powerful manner and therefore, the methods should be applied complementary. The study was supported by funds from the German Government's Special Purpose Fund held at Landwirtschaftliche Rentenbank (FKZ 28-RZ-3-72.038).

Session 70 Poster 27

Estimation of heritability for digital dermatitis in Polish Holstein-Friesian cows

M. Graczyk-Bogdanowicz, K. Baczkiewicz and K. Rzewuska
Polish Federation of Cattle Breeders and Dairy Farmers, Centre for Genetics, 79A Dabrowskiego, 60-529 Poznan, Poland; m.graczyk@cgen.pl

Claw disorders are listed as one of the top three reasons for culling in dairy cows. Claw health traits are being included in an increasing number of selection indices to improve genetic resistance to hoof lesions. The high incidence of dermatitis digitalis (DD) has been the main reason for the focus on improving genetic resistance to this disease. The aim of this study was to estimate genetic parameters of DD for the Polish Holstein-Friesian cows. In 2017 a Polish web-based application was launched to record claw health data. Hoof trimmers involved in the 'CGen trimming' project have recorded information on cow health status and hoof disorders during routine farm visits. Identification of lesions was based on 'ICAR Claw Health Atlas'. Data edits and the model were prepared according to the EuroGenomics Golden Standard guidelines for claw health. Data collected by hoof trimmers from 2017 to 2022 were used for this analysis. Data included 166,911 records from 63,131 cows. Multi lactation animal model was used for estimation of genetic parameters. The fixed effects of: age at calving×year, herd-trimming×year, trimmer×year, stage of lactation×year and month of calving×year were included. The variance components were estimated by Bayesian methods with Gibbs sampling, using GIBBS2F90. Results showed that DD was the most frequent hoof disease in Poland, equal to almost 42% of new cases. The estimated heritability of resistance to DD is not high but stays on the typical level for this trait. Heritability for the DD was 4.9, 4.7 and 4.5% for lactations 1, 2, 3+, respectively. The genetic correlation between lactations 1 and 2 was 84%, between 2 and 3+ equal to 80% and the least correlated were lactations 1 and 3+ with 70%. Genetic variation exists which suggests that genetic selection can be performed in this population. Results of this study serve as a basis for routine genetic evaluation for claw health in Polish Holstein-Friesian cows which is currently being developed. Further studies are planned to assess the genetic correlations with other collected lesions and to establish a Polish hoof health index.

Session 70 Poster 28

Genetic parameters for β-hydroxybutyrate concentration in milk in Spanish dairy cows

I. Cervantes[1], M.A. Pérez-Cabal[1], N. Charfeddine[2] and J.P. Gutiérrez[1]
[1]Universidad Complutense de Madrid, Dpto. Producción Animal. Facultad de Veterinaria, 28040 Madrid, Spain, [2]Spanish Holstein Association, (CONAFE), 28340 Valdemoro, Madrid, Spain; icervantes@vet.ucm.es

Nowadays, breeding for disease resistance is a priority in most breeding programs across species. Subclinical ketosis is diagnosed by an increase of β-hydroxybutyrate (BHBA) concentration in blood, due to a negative energy balance at the early lactation. Spanish Holstein Association (CONAFE) is collecting BHBA concentration in milk since 2013. The objective of this study was to evaluate the genetic component of β-hydroxybutyrate concentration in milk as a proxy for a potential selection criterion against subclinical ketosis. Dataset contained a total of 872,302 records of 317,522 different cows from 5 to 68 days in milk (DIM). The analyses were performed separately for first lactation (L1), second lactation (L2) and third or more lactations (L3+). The BHBA was log transformed. The tested models included DIM (as covariate or as discontinuous variable), year (10 levels), region (8 levels), herd (between 1,886 and 1,740 levels), herd-year (between 7,678 and 6,700 levels), region-year (between 35 and 37 levels), age at first calving in days and parturition number (14 levels) as fixed effects in different combinations. Herd-year and environmental permanent effects were included as random variables besides residual and additive genetic effects. The ASReml v.4.2 and VCE6.0 software were used to estimate genetic parameters. The pedigree contained a total of 665,876 individuals. Model election was based on Akaike Information Criterium (AIC). Best model included age at first calving, DIM (as covariate for L2 and as discontinuous effect for L1 y L3+), region-year and parturition number (L3+) as fixed effects. The heritabilities ranged between 0.044 (0.003) for L1 to 0.027 (0.002 for L3+). The ratio of herd-year was higher that the genetic component, being 0.06 for L1 and 0.05 for the rest of traits. Environmental permanent effect ratio ranged between 0.08 for L2 and 0.03 for L3+. These preliminary results indicated that the genetic component of BHBA concentration is low, but enough to include the Spanish Holstein breeding programs, and a genetic evaluation could be performed in following steps for this trait.

Session 71 Theatre 1

Comparison of self-assessment and objective indicators of attributes driving farms resilience

D. Martin-Collado[1], B. Soriano[2], J. Lizarralde[3], J.M. Mancilla-Leyton[4], N. Mandaluniz[3], P. Gaspar-García[5], Y. Mena-Guerrero[4] and A. Prat-Benhamou[1]
[1]Agrifood Research and Technology Centre of Aragon, Av. Montañana, Zaragoza, Spain, [2]Polytechnic University of Madrid, P.° de la Senda del Rey, Madrid, Spain, [3]Basque Institute for Agricultural Research and Development, Campus Agroalimentario, Arkaute, Spain, [4]University of Sevilla, Ctra. de Utrera, Sevilla, Spain, [5]University of Extremadura, Av. de Elvas, Badajoz, Spain; dmartin@cita-aragon.es

Strengthening farming systems' resilience is on the top of the EU and national political agendas. Given the relative novelty of resilience research field, several approaches have been proposed to develop this concept, but there is a lack of methodological consensus to assess resilience at a farm level. Commonly, resilience assessments have been based on objective or subjective measures, assuming that each approach have different strengths and weaknesses. However, very little is known about how subjective self-assessment and objective indicators-based resilience measures compare. Our study aims to fill this research gap in the farming system's field by providing a comparison of both measurement approaches. We understand resilience as result of farm and personal attributes contributing differently to farm robustness, adaptability, and transformability, through the resilience principles set by the Resilience Alliance: system reserves, diversity, tightness of feedbacks, openness, and modularity. Based on this, we identified attributes as specific factors measuring each principle and we did an evaluation of them using self-assessed statement and objective indicators separately. Finally, statistical analyses were based on 149 face-to-face farmer surveys covering four representative small ruminant farming systems in Spain. Results show a moderate correlation between indicators and self-assessed measures, which vary across attributes (-0.27/0.45), suggesting that the two types of measures are not interchangeable. However, in most cases, a positive correlation exists. The clearest alignment between the two measurements is found in attributes of social capital, redundance, and knowledge networks, while the highest discrepancy is found in farmer life quality. We argue that our results have new highlights for the understanding of farm resilience assessments.

Session 71 Theatre 2

Inventory and analysis of needs towards resilient dairy farming in 15 EU countries

A.M. Menghi and C.S.S. Soffiantini
Centro Ricerche Produzioni Animali (CRPA), Economics, Viale Timavo 43/2, Reggio Emilia, 42122, Italy; a.menghi@crpa.it

Within the R4D – Resilience for Dairy H2020 UE founded thematic network, an inventory of needs for dairy farmers to be resilient have been created. The inventory is a result of an online questionnaire spread by R4D partners across 15 EU Countries, 535 answers have been collected. In the survey, a list of 43 needs have been selected and proposed to stakeholders, asking them to assign a score from 0 to 5 to each of them according to the potential to improve farm resilience. The key areas explored across the survey were: (1) technical efficiency; (2) environment, animal welfare and society friendly production systems; and (3) economic efficiency and social resilience. Responders have also been asked to add other relevant needs, with an open-ended question. Moreover, information about their background and, when applicable, their farm has been asked. The results have been elaborated according to geographic distribution, sex, age, level of education, profession and dimension of the farm. Regardless regional specificities, it is clear that the improvement of work-life balance and the necessity of a transparent and effective communication with civil society are in the top 10 issues that farmers have to face, to be resilient in the future, just on the same level of other more technical challenges, like animal health/welfare and energy self-sufficiency.

Session 71 Theatre 3

Assessment of solutions for resilient dairy farming in fifteen European countries

A. Kuipers[1], J. Zijlstra[1], R. Loges[2] and S. Ostergaard[3]
[1]Wageningen University and Research, De Elst 1, 6708 WD, the Netherlands, [2]Kiel University, Grass and Forage Science/ Organic Agriculture, Hermann-Rodewald-Strasse 9, D 24118 Kiel, Germany, [3]Aarhus University, Department of Animal and Veterinary Sciences, Blichers Alle 20, Dk-8830 Tjele, Denmark; abele.kuipers@wur.nl

An assessment scheme was developed based on 5-scale questions related to the sub-categories social resilience (less to more); economic resilience, technical efficiency, environment, animal welfare, societal perception items, readiness and acceptability (low to high). This assessment scheme was applied to 185 practices, techniques and tools (named solutions), which were collected in 15 European countries as part of the EU-Resilience for Dairy project. 62 experts from universities and research institutes from 15 EU-countries scored these solutions with in total 3,300 assessments. The same sub-categories, with focus on readiness and acceptability, were also scored by farmers and stakeholder in local workshops in all 15 countries. The scoring took place with in mind farm types or systems where the solution is applicable and attractive. When answering the question about the impact of the solution, the average dairy farm in the region was taken as a reference. Practices and techniques were analysed within the four composed categories: socio-economic, technical, animal welfare and environment, and readiness. It appeared that top ranking solutions by the experts different in several cases from those chosen by the farmers and stakeholders. Moreover, the choice of solutions was locally coloured.

Session 71 Theatre 4

Knowledge needs and solutions related to resilience in the European dairy sector

K. Kuoppala[1], M. Rinne[1], N. Browne[2] and V. Brocard[3]
[1]Natural Resources Institute Finland, Production systems, Tietotie 2, 31600 Jokioinen, Finland, [2]Teagasc, Animal & Grassland Research and Innovation Centre, Moorepark, Fermoy, Co. Cork P61 P302, Ireland, [3]Idele, Dairy Unit, 8 route de Monvoisin, 35650 Le Rheu, France; valerie.brocard@idele.fr

The EU dairy sector is currently facing many challenges, which leads to multiple needs both within the farm gate and across other stakeholders. In the project Resilience for Dairy (R4D, funded by European Commission), the knowledge needs of 15 European countries were derived from an on-line survey and meetings of the National Dairy AKIS (Agricultural Knowledge and Innovation Systems) of the R4D project partners. Several solutions were also identified that could improve the resilience of the dairy sector. The most urgent needs were categorised within: (1) ecological and environmental footprint/mitigation of climate change/inputs efficiency; (2) financial needs; and (3) social issues: building society friendly dairy systems. Solutions were most often sought within: (1) ecological and environmental footprint/mitigation of climate change/inputs efficiency; (2) labour conditions; (3) dairy cattle management; and (4) animal nutrition and grassland management. Category 'Ecological and environmental issues' scored highest in terms of both needs and solutions. The reasons for that are probably two-fold: the pressure from the society, and the dependency of individual farm success on local weather and biotic resources. The second highest topic was related to financial and labour issues, which are the core of running resilient dairy businesses. Practical management issues related to dairy cow care, nutrition and feed production were emphasized as solutions, which is logical as they can be controlled at farm level. There were clear differences in the top scoring areas between regions. Direct comparisons of the needs and solutions across regions is hampered by the non-standardized format, as they were presented as open questions to NDA representatives, but on-line survey was standardized. Although animal welfare scored high in the on-line survey, it was not emphasized in the NDA outputs, which could mean that although an important topic regarding the image of the dairy chain, it is not experienced as a matter limiting the resilience of dairy farms.

Session 71 Theatre 5

Resilience of contribution to food security of specialized Walloon dairy systems

C. Battheu-Noirfalise[1,2], E. Froidmont[2], D. Stilmant[2] and Y. Beckers[1]
[1]ULiège GxABT, Passage des Déportés 2, 5030 Gembloux, Belgium, [2]Walloon Agricultural Reseach Centre, Rue du Serpont 100, 6800 Libramont, Belgium; c.battheu@cra.wallonie.be

The contribution to food security of dairy systems was mainly addressed through its dimension of food availability, including ('net indicators') or not ('gross indicators') the penalizing use of human-edible feeds. Here, we approach its dimension of stability of supply for both cases. Using accounting data, we calculated the annual gross (GP) and net protein productivity (NP) of 80 dairy farms of the Walloon region (Belgium) over a ten-year interval (2011-2020). We clustered the farms based on nine management parameters using a kmeans algorithm. The difference between the clusters was tested with a mixed model. We calculated the production resilience indicator ($mean^2/sd^2$) for each cluster and estimated the significance of their differences with a Monte-Carlo procedure. Last, we calculated those indicators on two periods separated by the end of milk quotas (2015) in order to analyse its impact on GP and NP resilience. Three clusters were identified. The first farm type was intensive and grass-based (IG). It reached a high GP (302 kgCP/ha) and NP (269 kgCP/ha) associated with the highest resilience for both indicators (81 and 53 for GP and NP, respectively). The second type was also intensive but rather maize-based (IM). It used the same amount of concentrates as the first type but with a higher CP-content. This type reached similar GP (302 kgCP/ha) as the first type but its NP (232 kgCP/ha) was lower. Moreover, its resilience was lower for both GP (55) and NP (21). The last type was extensive and grass-based (EG). It used less concentrates than the two previous types with similar CP-content compared to IG. Its GP (185 kgCP/ha) and NP (202 kgCP/ha) as well as its resilience of GP (49) were the lowest. However, it reached a higher resilience in terms of NP (41) than IM but still lower than IG. Between 2011-2015 and 2016-2020, all types showed a decrease in the resilience of NP and GP linked with type-specific management changes. IG showed an increase in milk production and a decrease in concentrate use and CP-content of concentrates. IM showed an increase in milk production and fodder yield. EG showed a decrease in the use of maize silage.

Session 71 Theatre 6

Innovative solutions supporting resilience of dairy farms in Netherlands

P.J. Galama, J. Zijlstra and A. Kuipers
Wageningen University and Research, Wageningen Livestock Research, De Elst 1, 6700 AH, the Netherlands; paul.galama@wur.nl

Resilience for Dairy (R4D) will tackle urgent sustainability challenges faced by dairy producers by bringing together dairy farmers, farming organisations, advisors, researchers and all relevant actors across 16 member states to close the divide between research and innovation in Europe. R4D is built around the multi-actor approach to implement more intense cooperation between researchers, advisors, farmers and relevant actors to facilitate greater exchange and acceptance of co-created solutions. R4D draws on the EU, national and regional connections of the 17 consortium members to appropriate networks, on the three related key themes of economic and social resilience, technical efficiency and environmental and society friendly production systems. In each of the 16 countries meetings with farmers and stakeholders have been organized to identify needs and solutions to make the dairy sector more resilient. These relate to the farm, the farmer and the dairy chain. The solutions are focusing on the resilience items of robustness, adaptation or transformation. A list of most important shocks and challenges for the dairy sector are made and solutions are selected. The results of the meetings in Netherlands will be shown. Topics are development of new revenue models like credits for CO_2 reduction, CO_2 storage and energy production, or care farming or valorisation of dairy products like A2 milk; increasing soil fertility; grassland management in relation to biodiversity; new housing systems in relation to cow comfort, less emissions and manure quality; personal development in relation to communication with society and cooperation between farmers. It is noticeable that the Dutch farmers and stakeholders present in the meetings had a relatively strong interest in a 'nature kind of approach' to farming. This implies a lot of focus on solutions in the area of quality of soil and crop and grassland farming. There was also quite some interest expressed in socio-economic topics, like communication with society and cooperative forms of manure management along the chain, instead of more detailed techniques and practices at farm level. The group of farmers was forward looking, but deeply concerned about the present discussions and policies in the public domain in the country.

Session 71 Theatre 7

Needs of the dairy sector: a Hungarian overview

L. Czeglédi, B. Béri, I. Komlósi and E. Török
University of Debrecen, Department of Animal Science, 138. Böszörményi Street, 4032, Hungary; czegledi@agr.unideb.hu

The current distribution of the Hungarian cow population is 60% dairy, 35% beef, and 5% dual-purpose. The Holstein-Friesian is the dominant dairy breed (97%) having an average of 10,804 kg milk production in a 305-day lactation. The Hungarian dairy sector can be characterized as intensive milk production and farms of 440 cows in average. 46% of cows are kept on farms with more than 500 cows and 31% on farms with 300-500 cows. Deep bedding, laying boxes, two or three milking a day, and a high proportion of concentrate in the daily ratio are common in most of the farms. Grazing of dairy cows is not common due to the low yields of pastures. An online survey was conducted where respondents, Hungarian dairy farmers and dairy stakeholders, were asked to select their level of interest in a range of needs within three different fields: technical efficiency; environment, animal welfare, and society friendly production systems; economic efficiency and social resilience. Concerning all areas, the innovative systems (milking strategies; feeding system; analysis for early detection of diseases), the improvement of welfare conditions and the effective communication to the general public of agricultural practices were the main issues identified by the respondents. Innovative devices for measuring grass growth and techniques for grazing management, environmental footprint assessment techniques as well as feed additives to mitigate methane emissions were not identified as crucial needs. In the field of technical efficiency, the main needs were relationship with the application of innovative milking, feeding as well as animal health monitoring systems. The improvement of welfare conditions of calves and cows, and the effective communication to the general public of agricultural practices, and the role of agriculture in society proved to be the main issues within the field of environment, animal welfare, and society friendly production systems. Within the economic efficiency and social resilience field, the main requirements were the using of innovative and reliable information channels as well as the salary.

Session 71 Theatre 8

Eco-efficient low-cost pasture based dairy production on a mixed farm in Northern Germany

R. Loges and F. Taube
Kiel University, Grass and Forage Science/Organic Agriculture, Hermann-Rodewald-Str. 9, 24118 Kiel, Germany; rloges@email.uni-kiel.de

Recent intensification in European agricultural production is accompanied by serious environmental trade-offs questioning the sustainability of current specialized production systems for both all arable cash crops and animal products. Under the temperate conditions of North-West Europe, ruminant-based integrated crop-livestock systems are considered as a strategy towards ecological intensification. This is the background for the interdisciplinary project: 'Eco-efficient pasture-based milk production' established 2016 at the organic research farm Lindhof in Northern Germany. The project aims at fulfilling relevant ecosystem services linked to dairy systems: high quantity and quality of agricultural commodities; low nutrient surpluses, a low carbon footprint and contributions to agrobiodiversity. Data are presented based on a 98 spring-calving Jerseys/crossbred dairy herd on an organic former arable farm as an alternative to traditional specialized systems. Measurements include: productivity, production costs, nitrate-losses and product-carbon footprint compared to typical regional dairy farms. The results illustrate the capability of a rotational ley grazing system to provide both a high milk performance per ha combined with low environmental footprints and additionally offer significant yield benefits for the arable crops in the crop rotation. Lindhof acts as pilot-farm in the EU-Horizon2020-project R4D – Resilience for Dairy.

Session 71 Theatre 9

Factors contributing to the financial resilience of spring-calving pasture-based dairy farms

G. Ramsbottom[1], B. Horan[2], K.M. Pierce[3], D.P. Berry[4] and J.R. Roche[5]
[1]Teagasc, Animal and Grassland Research and Innovation Programme, Teagasc, Oak Park, Carlow, R93 XE12, Ireland, [2]Teagasc, Animal and Grassland Research and Innovation Programme, Teagasc, Moorepark, Co. Cork, P61 C996, Ireland, [3]University College Dublin, School of Agriculture and Food Science, Belfield, Dublin 4, D04 R7R0, Ireland, [4]Teagasc, Animal and Grassland Research and Innovation Programme, Teagasc, Moorepark, Co. Cork, P61 C996, Ireland, [5]University of Auckland, School of Biological Sciences, Private Bag 92019, Auckland, 1142, New Zealand; george.ramsbottom@teagasc.ie

The objective of this study was to identify factors contributing to the financial resilience of spring-calving pasture-based dairy farms when ranked by average operating profitability (i.e. net profit/ha). A dataset of 315 Irish pasture-based dairy farms with complete records for 8 consecutive years was used in this analysis. The farms were characterized by expansion and intensification during the 8-year study period, as evidenced by the annual increase in milk fat and protein yield per cow (+15%; $P<0.001$); mean annual pasture DM consumed/ha also increased linearly (+19%; $P<0.05$); production costs increased linearly ($P<0.01$) while net profit was highly variable between years. The 8-year average net farm profit/ha was €1,611/ha, €1,189/ha, €937/ha and €630/ha for the highest, second highest, second lowest and lowest profit quartiles respectively ($P<0.001$). The highest profit quartile contained, on average, smaller farms (59 ha) with greater technical efficiency (stocking rate 2.42 LU/ha; 5,511 litres milk/cow; and, 9.9T pasture DM/ha utilized) ($P<0.001$) than the other profit quartiles. When affected simultaneously by a combination of milk price reduction and adverse weather, they experienced the greatest nominal reduction but highest nadir farm profitability ($P<0.001$). The highest 8-year average net profit quartile experienced a reduction of €850/ha and nadir profit of €763/ha in the adverse year but profit recovery the following year of €743/ha. The second highest, second lowest and lowest profit quartiles had reductions of €730/ha, €625/ha and €562/ha; nadir profits of €478/ha, €311/ha and €46/ha; and, net profit recoveries the following year of €618/ha, €533/ha and €478/ha respectively.

Session 71 Theatre 10

Planning for resilient dairy farms in the USA

A. De Vries
University of Florida, 2250 Shealy Drive, Gainesville, FL 32608, USA; devries@ufl.edu

Resilience implies to many dairy farmers in the USA the capacity to withstand unplanned changes to their production environment with as little impact as possible. The production environment may include physical aspects of dairy production, such as weather, crops, animal disease, energy, and labour availability. It can also include prices for milk, feed, and access to buyers and suppliers. In the presentation I will give examples of common practices that are used on dairy farms in the USA. The need for resilient dairy farms and a more resilient dairy sector was highlighted during the COVID pandemic when milk plants were shut down and feed could not be delivered. This led to dumping of milk, unwanted changes in rations, and forced reductions in herd sizes. An example of resilience related to changing weather is the increase in the number of barns equipped with heat abatement and raising dairy heifers indoors year-round. Dairy farmers also maintain their own power generators in case the power goes off due to storms. Related to cropping, dairy farmers may grow different varieties of corn, the main forage source, depending on maturity, yield, and disease resistance. Many farmers also maintain a one-year supply of corn silage, so that disappointing crop yields do not lead to the need to purchase forages. Forward contracting of feed stuffs is also common, as is hedging of milk, such that future prices are known and variation in prices is reduced. Dairy farmers may also maintain several suppliers of the same goods to not become reliable on a single supplier. Most dairy farmers keep ownership of their own heifers, even when they may be raised by a heifer grower in another state. This practice reduces their risk to high heifer purchase prices and biosecurity. To account for unplanned culling, dairy farmers often raise a buffer of 5 to 10% more heifers than they need. On average, this practice reduces cow longevity. Cross training of farm labour is common so that different people can do the same jobs. Some dairy farmers even pay potential workers to be on stand-by in case their help is needed. In summary, resilience is on the mind of many US dairy farmers, and they use a large variety of practices to keep a resilient dairy farm.

Session 71 Theatre 11

Towards a socially sustainable dairy sector with cow-calf contact systems

H.W. Neave, M. Bertelsen, E.H. Jensen and M.B. Jensen
Aarhus University, Department of Animal and Veterinary Science, Blichers Alle 20, 8830 Tjele, Denmark; heather.neave@anivet.au.dk

The intensification of the dairy sector has led to social sustainability challenges. To address concerns around animal welfare and societal values, alternative systems that provide cow-calf contact have been proposed. Such systems have been shown to increase work satisfaction for farmers, which also relates to social sustainability. However, cow-calf contact systems also present challenges with reduced saleable milk and high animal stress during later separation. Some proposed solutions include reducing daily contact duration and novel weaning methods. We conducted two studies (48 and 56 cow-calf pairs each) providing either full-time (23 h/d) or part-time (10 h/d during daytime) cow-calf contact for 8 wk. We used three methods to wean off milk and separate calves from their dams (using a fence-line): two-step (milk and dam removal separated by a week) or gradual (time with the dam reduced to 50% then 25%, over 2 wk), compared to simultaneous (milk and dam removal occurred together), Our findings suggest that cow and calf behaviour in part-time contact systems also foster strong bonds between cows and calves. However, calves become hungry during daily separation periods, and the vocal response to separation is not reduced, compared to full-time contact systems. A two-step (vs simultaneous) weaning process reduced behavioural and vocal responses of calves to separation, but both cows and calves were similarly vocal during gradual weaning (vs simultaneous). Thus, dividing the weaning and separation process into two steps may be one strategy to reduce the negative behavioural responses of calves at weaning. Further scientific exploration is needed to address welfare, economic and staff labour concerns regarding cow-calf contact systems. In doing so, there can be transformative, proactive development of alternative management systems that support social sustainability and thereby a more resilient dairy sector.

Session 71 Theatre 12

Resilient, healthy or efficient? The ideal animal according to breeders of small ruminants in Europe

E. Janodet and M. Sautier
INRAE, GenPhySE, Chemin de Borderouge, 31320, France; estelle.janodet@inrae.fr

Health, resilience and efficiency of herds is a major issue for the sustainability and profitability of farms. Genetic selection is one of the means for increasing the performance of animals for those three traits. Our study aimed at better understanding the diversity of breeders' preferences with respect to the relative importance of traits related to animal resilience, health, and efficiency (SMARTER H2020 project). Data were obtained through preference surveys based on choice experiment that were administered to small ruminants breeders in 5 countries and for 14 breeds (n>600) with the decision-making software 1000minds. Results highlight differences in preferences between the proposed traits with a lower degree of importance for feed efficiency ($P<0.05$) and prolificity ($P<0.05$) compared to the other 6 traits (dairy production, dry matter in milk, mastitis resistance, parasitism resistance, mortality at weaning, functional longevity). We present divergent preference profiles identified through cluster analysis and characterize them. We investigate the correlations between preferences profiles and: (1) livestock systems; and (2) breeder's profile. Our results may contribute to inform small ruminants industries' experts, especially on the revision of breeding objectives for more resilient and efficient animals.

Session 71 Poster 13

Strategies and cases of resilience from dairy farming community in Slovenia

M. Klopčič
University of Ljubljana, Biotechnical Faculty, Dept. of Animal Science, Groblje 3, 1230 Domžale, Slovenia; marija.klopcic@bf.uni-lj.si

Milk production is the most important production sector of Slovenian agriculture. Dairy cattle farming has been concentrated and specialized since the mid-1990s, which is reflected in the reduction of the number of dairy farmers, the increase in the average size of the herd per farm, the increase in the milk yield of cows and the higher quality of milk. The milk production sector in Slovenia is characterized by family farms with an average of 19 dairy cows per herd, with a large proportion of tied-in housing systems (70%) and with a very unfavourable land structure (a large number of small plots (35), which complicates the development and efficiency of Slovenian dairy farms. Within the R4D – Resilience for Dairy H2020 EU founded thematic network, Slovenian dairy farmers together with researchers, advisers and stakeholders (NDA) discussed about future challenges and solutions for dairy sector in Slovenia. Given that Slovenia is dominated by permanent grassland, they think that 'Increase grazing vs indoor feeding to meet customer desires and added milk value' can be one of the most suitable solutions. In order to improve the well-being of the animals and to make their work easier, there is a distinct desire for freewalk housing system to improve animal welfare and animal health. They also think that personal development on a wide spectrum of topics to increase resilience skills of dairy farmers is very important in this uncertain time. Given that the ownership structure of agricultural land is very unfavourable, there is a clear need for land consolidation to be more efficient. In Slovenia, diversification and the search for additional sources of income through added value are very important, because we have many hilly and mountain dairy farms with small herds (e.g. PDO/PGI/AOP, organic, hay milk/meat, A2A2 milk, agro-tourism, forestry, etc.) to increase income for farmers. Also the topic about reducing emissions of ammonia and GHG is fully in discussion.

Session 71 Poster 14

How farm management influences the longevity of dairy cows: a comparative study of Swiss dairy farms

R.C. Eppenstein, A. Bieber, M. Lozano-Jaramillo and M. Walkenhorst
Research Institute of Organic Agriculture FiBL, Animal Science, Ackerstrasse 113, 5070, Switzerland; rennie.eppenstein@fibl.org

Increasing the productive lifespan of dairy cows is an important means to lowering the environmental impact of dairy production. Farm characteristics, such as location, production type and breed are fix characteristics for most farms. However, farm management strategies can influence the longevity of their dairy herds in the medium- and short-term. Within the framework of the research project 'Longevity of Swiss Dairy Cows' (Nutzungsdauer Schweizer Milchkühe), we aimed at identifying management choices that affect the productive lifespan of dairy cows. Based on data from the Swiss census and the major breeding organizations, we built a database of 142 farms. We defined 15 farm types that best represent the diversity of Swiss dairy production with regard to geographic regions, production zone, breed and production type (organic vs conventional). We allocated 10 dairy farms per farm type. Five of the 10 farms were chosen for having a low average productive lifespan (APL) of their dairy herd. The other five farms were chosen for having a high APL. APL was defined as the average lactation number of all cows culled 5 years in retrospective. From the initial 142 farms, 68 farms participated in a survey to assess the differences in management practices. From these participating farms, 30 were further clustered into matched pairs and were visited on-farm. Farms with low APL did not differ from those with high APL regarding their milk production and average dairy herd size. However, they significantly differed with regard to their APL, thus confirming a successful selection strategy of matched pairs. On average, dairy cows from farms with low APL were culled 2 lactations earlier than cows from farms with comparable characteristics, but with a high APL. Compared to farms with low ALP, farms with high APL were characterized by a higher percentage of loose housing systems, a higher percentage of energy rich feed rations, better fertility and more animals being inseminated with meat breeds. No statistical differences were found in relation to antibiotic treatment incidences and other health parameters.

Session 71 Poster 15

High herd exit rates existence of small herds: a case study from North West Province, South Africa

M.D. Motiang[1] and E.C. Webb[2]
[1]Agricultural Research Council, Animal Production, Old Olifantsfontein Road, Irene, 0062, South Africa, [2]University of Pretoria, Department of Animal and Wildlife Sciences, University of Pretoria, Pretoria, 0002, South Africa; dan@arc.agric.za

The aim of this paper was to compare herd balance between different herd sizes. Data were collected from 308 randomly selected cattle farmers from RSM District North West Province in 2012. Data were divided into four herd size categories of 1-10, 11-30, 30-70 and >70. Data were analysed using IBM SPSS statistics 22 (2013). The GLM multivariate analysis was performed to test effect of herd size category on calving, herd mortality and off-take rates. Means were separated using least significant differences (LSD) tests. Results show that the average calving rate of 55% and cumulative exit of 27.3% resulting from 10% mortality, 15% offtake and 2.3% slaughter rates. Although calving rates did not differ significantly across herd size categories, small herds of 1-10 had significantly higher values for herd exit rates than larger ones. Herd mortality rates ranged from 6% in large herds of >70 to 18% in small herds of 1-10 while offtake rates were 16 and 23%, respectively. Small herds also herd significantly ($P<0.05$) higher slaughter rates (3%) than large herds (1%). Overall herd exit for small herds was 43% compared to 20-21% for other herd categories. On average small herds of 1-10 had 5.9 head of cattle consisting of 3.84 cows that produce 2.06 calves but lose 2.57 head of cattle through sale, slaughter and mortality, which results in a negative herd balance of -0.5 as compared to the average of 0.85 for all herds. This negative herd balance implies that small herds will shrink and ultimately disappear in 12 years if the status quo continues. Herd size categories of 11-30, 30-70- and >70 recorded herd balance of 1.08, 0.95 and 12.02, respectively. Although small herds recorded the highest herd exit values, their average offtake rate of 23% was within the national average of between 25 and 30%. However, 18% mortality rate for these herds was higher than previously reported of approximately 10%. It is recommended that farmer development programmes focus on herd mortality in order to stimulate growth of small herds in the study area.

Session 71 Poster 16

Building on a resilient dairy sector- highlights and discussion

A. Kuipers[1], V. Brocard[2] and M. Klopčič[3]
[1]Wageningen University & Research, De Elst 1, 6708 WD Wageningen, the Netherlands, [2]Institut de l'Elevage, 8 route de Monvoisin, 35650 Le Rheu, France, [3]University of Ljubljana, Biotechnical Faculty, Dept. of Animal Science, Groblje 3, 1230 Domžale, Slovenia; abele.kuipers@wur.nl

This seminar intends to contribute to sharing insights on how best to support dairy farmers in coping with change to achieve a more resilient operation and peace of mind. The concept of resilience and indicators from a technical, environmental and socio-economic perspective, and analysis of needs and solutions, gathered in 15 EU countries in the context of resilience, will be presented, as well as interesting cases of resilience in the field. The focus will be on herd, farm management and communication with society, with aim to answer two key questions: How does the dairy farmer/farm family cope with shocks, barriers and uncertainty in market and environmental policies in the management of the farm? This time slot will be utilized to summarize the findings and farm cases, and discuss the various presentations of this seminar.

Session 71 Poster 17

Economic and environmental impacts of cattle longevity extension by altered reproductive management

R. Han, A. Kok, M. Mourits and H. Hogeveen
Wageningen University and Research, Wageningen, 6706 KN, the Netherlands; ruozhu.han@wur.nl

Introduction: Prolonging dairy cattle longevity is regarded as one of the options to contribute to a more sustainable milk production. Since reproduction failure is the primary reason of culling, this study investigates the effect of extending cattle longevity on farm's gross margin and greenhouse gas emissions (GHG) by altered reproductive management. Materials and Methods: An adapted model of Kok *et al.* is used to stochastically simulate the dynamics of a Dutch dairy herd of 100 cows, by modelling individual cow lactations and calving intervals, while accounting for culling for fertility reasons, mastitis, lameness and other reasons (i.e. general culling). Moreover, the model computes the GHG emissions using a life cycle approach. To extend cattle longevity, two altered strategies for reproduction management were evaluated: (1) the insemination extension strategy, in which the maximum number of inseminations (AI) per cow before she is culled for infertility was raised from 4, to 5 or 6 times; and (2) the reduction in subfertility culling standard strategy, in which the milk production threshold for culling non-pregnant cows was reduced from 20 kg to 15 or 10 kg/day. The model was run for 500 herds of 100 cow places for each reproductive management strategy alternative. Results and Discussion: Age of culled cows increased with the increased maximum number of AI from 2,040 to 2,195 days. The change was larger from 4 to 5 times AI (108 days) than from 5 to 6 times AI (47 days). Annual gross margin increased from €165,847 to €167,570, while GHG decreased from 0.926 to 0.915 CO_2-equivalents per kg FPCM. With the decrease in the subfertility culling standard from 20 to 10 kg/day, the age of culled cows increased from 1,968 to 2,132 days. Annual gross margin decreased with € 168,188 to minimal €161,210, while GHG increased with 0.002 CO_2-equivalents per kg FPCM. Implications: The increased maximum number of AI and subfertility culling standard can benefit a dairy farm economic and environmental sustainable development.

Session 72 Theatre 1

Economic sustainability of different levels of extensiveness in fattening pig farms

P. Ferrari, C. Montanari and L. Giglio
CRPA Research Centre for Animal Production, Economics and Agriculture Engineering, Viale Timavo 43/2, 42021, Italy; p.ferrari@crpa.it

Consumers generally believe that extensive livestock farming systems is synonymous to better meat products quality. This holds particularly for livestock species such as pigs that are kept at relatively high stocking density. The mEATquality project aims to develop novel solutions that address societal demands, environmental concerns and economic needs on farm and in the chain. The 'extensiveness' of production is the key issue, and will be developed in a stepwise approach during the project. The first step surveys extensive husbandry factors in relation to intrinsic meat quality, through data collection on conventional, free-range and organic farms. Standardized sustainability assessment protocols are developed for on-farm data collection across 80 pig farms in Denmark, Italy, Poland and Spain. The protocols include: (1) general farm description; (2) environmental issues; (3) animal welfare assessment; and (4) farms productivity and economics information that will be collected. The starting point for each section was the protocol developed and used by the ERA-net SusAn project 'SusPigSys' (https://suspigsys.fli.de/en/home/) which assessed the sustainability of pig farms in 8 EU Member States. Each protocol was adapted to focus on the extensiveness of production. Farm's economic resilience will be estimated through indicators related to profitability, labour productivity, animal performance, entrepreneurship, risk management and resilience of resources. Economic pig farm protocol includes information about volumes of inputs and outputs both at pig barn level and at crop level used for own produced pig. The focus will be on the degree of 'extensiveness' of the farms surveyed, in particular in relation to their economic results. Data collected will be used to gain insight in the economic performance and its variability according the different degree of pig farms extensiveness. To this end, economic indicators will be calculated such as revenues minus feed costs per animal place, production costs, profit/loss accounts, farm income per labour unit and returns to capital. In addition, technical efficiency of the production process is of great importance and will be included.

Session 72 Theatre 2

How to characterise the European livestock production systems?

E. Bailly-Caumette[1], S. Moakes[2], C. Pfeifer[3] and D.R. Yáñez-Ruiz[4]
[1]ENS Lyon, Biology, Lyon, 69342, France, [2]Aberystwyth University, Aberystwyth, SY23 3FL, United Kingdom, [3]FiBL, Frick, 5070, Switzerland, [4]CSIC, Granada, 18008, Spain; elea.bailly@ens-lyon.fr

In the context of the European Green Deal, the EU PATHWAYS project aims to identify practices and innovations that will ensure a sustainable evolution of European livestock systems, that also address societal expectations. This process requires a characterisation and assessment of existing livestock systems to support development of improved ones. Whilst describing a livestock system is straightforward, difficulties lie in differentiating between livestock systems within the high diversity within Europe, and a single framework or database describing livestock systems in Europe does not exist. We have therefore developed a methodology to utilise and blend data from different existing sources, including Eurostat and FADN databases at NUTS2 level, combined with an expert survey conducted through a questionnaire. Variables linked to farm practices and economy were identified from the databases and analysed through hierarchical clustering to cluster NUTS2 regions with similar livestock system attributes. In parallel, a survey of experts and stakeholders was conducted to gain quantitative (e.g. feed, pasture) and qualitative data (e.g. importance of system, location), resulting in 171 described systems from 12 countries. The combination of these data sources identified specific systems for 8 livestock categories: dairy cows, suckler cows, finishing beef, meat and dairy sheep, goats, breeding and finishing pigs, laying hens and broilers. Maps and a dataset were generated to highlight the systems' attributes and distribution at NUTS2 level, which were further discussed with experts and stakeholders to refine and validate the outcomes. This holistic approach moves beyond existing databases and explores context specific husbandry practices, livestock management and economic performance, highlighting regional tendencies at European level. Although limitations remain, this could serve as an improved knowledge base for livestock's role and impacts, highlighting diversity and complexity of livestock systems and allowing more nuanced sustainability assessments that can inform current political decisions and debates regarding livestock production.

Session 72 Theatre 3

The IntaQt project's stakeholders' involvement: impact on the research work?

F. Bedoin[1], C. Couzy[1], C. Laithier[1], C. Berri[2] and B. Martin[3]
[1]Institut de l'Elevage – French Livestock Institute, 149, rue de Bercy, 75012 Paris, France, [2]INRAE, Université de Tours, BOA, Nouzilly, France, [3]INRAE, Université Clermont Auvergne, VetAgro Sup, UMR Herbivores, Saint-Genès-Champanelle, France; florence.bedoin@idele.fr

Multi-actor research projects are still a recent trend, putting at work together researchers with non-academic partners, like representants of food chain actors. The INTAQT project (INnovative Tools for Assessment and Authentication of chicken meat, beef and dairy products' QualiTies) has chosen to play the game fully: the project was set up with stakeholders' consultations at the heart of the project and with their recommendations implemented in the research. Stakeholders from the beef, dairy or poultry sectors in seven European countries are involved from producers to retailers, including processors and other relevant actors in each context. They were consulted at three levels: first through individual qualitative interviews then invited to national group discussions and some of them to European group meetings. Their role was to give their opinions and suggestions for the choice of farming systems to be studied in the project and on the choice of quality criteria to take into account and on analysis to be conducted. They will be involved all along the project life with their contribution next year to the co-construction of the multi-criteria scoring tool. This presentation is based on analysis of interviews with the project's Work Package leaders and Task Leaders in order to confirm or refute the following hypotheses: (1) there was an a priori reticence of some of the researchers about the importance and relevance of including stakeholders' voices in a research project; (2) after almost of 2 years in the process, there is an interest in the world of research, but also difficulties in effectively integrating positions that are difficult to reconcile (temporality, decision-making methods, governance, etc.). Globally we reflect on how multi-actor approach like the one in INTAQT questions a research project.

Session 72 Theatre 4

EU policy impacts on the sustainability of the livestock sector, insights from the PATHWAYS project

N. Roehrig[1], A. Sans[2], K.-E. Trier-Kreutzfeldt[3], M.A. Arias Escobar[3], F.W. Oudshoorn[3], N. Bolduc[2], E. Regnier[2], P.-M. Aubert[2] and L.G. Smith[1,4]
[1]University of Reading, School of Agriculture, Policy and Development, Earley, Reading RG6 6BZ, United Kingdom, [2]Institut du Développement Durable et des Relations Internationales, 41 Rue du Four, 75006 Paris, France, [3]Innovationscenter for Økologisk Landbrug, Agro Food Park 26, 8200 Aarhus, Denmark, [4]Swedish University of Agricultural Sciences, Department of Biosystems and Technology, Box 190, 234 22 Lomma, Sweden; n.roehrig@reading.ac.uk

How can EU policies be modified to enhance the sustainability of livestock production systems? To find answers to this question, which is central to achieving environmental goals while maintaining adequate production levels, the PATHWAYS project is working towards formulating policy trajectories for sustainable livestock production in Europe by 2050. As part of this work, we have conducted a review of the scientific evidence on the impacts that EU policy has had on the sustainability of livestock systems and identified reasons why they were not successful in promoting sustainability in various areas. This is combined with an analysis of the current and prospective policy framework to determine what already has been done and how EU policies can be expected to shape the sustainability of livestock systems in the future.

Session 72 Theatre 5

Stakeholders' perception of pig and chicken local breeds – a broad survey by the GEroNIMO project

M.J. Mercat[1], A.J. Amaral[2], R. Bozzi[3], M. Čandek-Potokar[4], P. Fernandes[5], J. Gutierrez Vallejos[1], D. Karolyi[6], D. Laloë[7], Z. Luković[6], H. Lenoir[1], G. Restoux[7], A. Vicente[2], V. Ribeiro[8], T. Rodríguez Silva[9], R. Rouger[10], D. Škorput[6] and M. Škrlep[4]

[1]Ifip-Institut du Porc, BP35104, Le Rheu, France, [2]Univ. of Évora, Largo dos Colegiais 2, Évora, Portugal, [3]DAGRI-UNIFI, Via delle Cascine, Firenze, Italy, [4]KIS, Hacquetova 17, Ljubljana, Slovenia, [5]ANCSUB, Largo do Toural, Vinhais, Portugal, [6]UniZG, Fac. of Agriculture, Svetosimunska 25, Zagreb, Croatia, [7]INRAE, GABI, Jouy-en-Josas, France, [8]AMIBA, Rua Domingos Marques, Vila Verde, Portugal, [9]FEUGA, Santiago de Compostela, A Coruña, Spain, [10]SYSAAF, INRAE Val-de-Loire, Nouzilly, France; marie-jose.mercat@ifip.asso.fr

As part of the European Union project GEroNIMO, a broad survey with actors (breeders, managers and processors) involved in the preservation of chicken and pig local breeds (LB) was conducted. The aim was to characterize the LB conservation or breeding programmes and to collect the stakeholders' views. The survey was designed in seven languages including English and conducted mostly online. More than 550 stakeholders from 12 countries, representing 31 pig and 94 chicken LB participated in the survey, with unbalanced representation of breeds and countries. The responses collected showed similarities between species such as free-range and pure-bred breeding on small farms. Although breeding animals are selected mainly on external phenotypes like the breeds' standard, participants showed interest in selection, especially for productive or reproductive traits. However, many barriers to the implementation of breeding programmes exist in both species. Moreover, preserving genetic diversity is the main motivation of the surveyed actors. Leisure activity is the second most important motive for the chicken breeders who have marginal economic activities related to LB. In contrast, with a median income percentage of 42% linked to LB, economic activities constitute the second most important motive of the pig LB actors. Finally, participants expressed concern about the sustainability of LB, especially for economic reasons, and expect more support from public authorities. The project is funded by the European Union's Horizon 2020 research and innovation programme under grant agreement N°101000236.

Session 72 Theatre 6

Responding to stakeholder needs and consumer-driven demands in the dairy goat and poultry sectors

C. Bonardi and M. Gerevini

Tecnoalimenti S.C.p.A., Research, Via Gustavo Fara 39, 20124 Milan, Italy; c.bonardi@tecnoalimenti.com

Code: Re-farm project aims to understand the relationship between intrinsic product quality and husbandry systems in goat and poultry production systems, through consumer driven demands. As a first step, stakeholders' future changes and challenges were identified together with consumer demand drivers. The industries expect tougher regulations on animal welfare and health, supply chain transparency, and food safety along with challenges related to environmental issues and innovation in emissions reduction, feed footprint, and genetics. Local production and animal welfare standards currently affect consumers' behaviours, with a strong impact on alternative product pricing. There are growing concerns about the impact of food products on the environment. Locally made products, high product quality and food chain traceability, are some of the current drivers of the consumer-demands, expected to intensify and reshape the ways consumers choose products. Based on the stakeholders' and consumer inputs, Code: Re-farm tests several novel tools for product quality monitoring methodologies, specialized instrumentation and benchmarking tools capable of reliably indicating differences between intensive and extensive farming systems without bias. As part of the project's visualised data and responding to the consumers' requests for better transparency in the dairy milk and poultry values chains, a new App is developed that will enable product quality monitoring and producer data, animal welfare and health conditions. This work was funded by EU (HORIZON H2020) under the 'Code Re-farm: Consumer-driven demands to reframe farming systems' (Grant No: 101000216).

Session 72 Theatre 7

The INTAQT project: stakeholders' opinions on future multicriteria scoring tools for animal products

I. Legrand[1], A. Nicolazo De Barmon[1], F. Albert[1], M. Berton[2], M. Bourin[3], V. Bühl[4], A. Cartoni Mancinelli[5], R. Eppenstein[4], D.A. Kenny[6], E. Kowalski[7], S. McLaughlin[8], G. Plesch[4], F. Bedoin[1], C. Couzy[1], C. Berri[9], B. Martin[9] and C. Laithier[1]

[1]IDELE, 149, rue de Bercy, 75595 Paris cedex 12, France, [2]University of Padova, DAFNAE, Viale dell'Università 16, 35020 Legnaro, Italy, [3]ITAVI, L'Orfrasière, 37380 Nouzilly, France, [4]FiBL, Ackerstrasse 113, 5070 Frick, Switzerland, [5]University of Perugia, Borgo XX Giugno 74, 06121 Perugia, Italy, [6]TEAGASC, Moorepark, Fermoy, Co. Cork, P61 C996, Ireland, [7]Ghent University, Coupure Links 653, 9000 Gent, Belgium, [8]The Queen's University of Belfast, 19 Chlorine Gardens, Stranmillis, United Kingdom, [9]INRAE, 147 rue de l'Université, 75338 Paris, France; isabelle.legrand@idele.fr

Agri-food chain actors (AFAs) lack reliable information to meet consumer expectations in relation to multiple facets of intrinsic quality of chicken meat, beef, and dairy products from the various European livestock systems. One of the challenges of the INTAQT project is to build, with AFAs, multi-criteria scoring tools related to products global quality. This tool should combine safety, sensory, and nutritional results obtained during the project-based on collection of poultry/beef/dairy samples and possibly other quality criteria. Multi-actor participatory approach was applied to present and discuss the concept of a multicriteria scoring tool, collecting opinions, fears, and expectations on this tool. On a consumer side, it was felt that such a multi-criteria score placed on products could be a clear and simple representation of a complex reality. However, some disadvantages were expressed about its reliability or implementation. For their part, AFAs had varying opinions on the tool's target, mainly about its possible use as an internal tool or for consumer information, with different pros and cons expressed on both aims. Fears dealt with the building of the tool, and its relevance, representativeness, practical use, and the potential dangers, especially if safety aspects were included. However, both consumers and AFAs agree on the need to include in this tool extrinsic criteria such as farming system sustainability and animal welfare.

Session 72 Theatre 8

Co-designing Agroecology farming concepts to increase food sovereignty in the Global South

M. Simataa, R. Valkenburg and H. Romijn

Eindhoven University of Technology, Den Dolech 1, 5612 AE Eindhoven, the Netherlands; m.simataa@tue.nl

Food sovereignty encompasses a community's right to produce, access, and control food. It faces severe challenges in the Global South. Even though intensified peri-urban farming improves food system resilience, poverty reduction, and nutrition security in low- and middle-income countries, concerns rise about the potential adverse effects on health and the environment. A crucial question arises: 'is there a future for small-holder farmers in the Global South?' Applying the principles of *Agroecology* through a *One Health* approach can help address these challenges. Agroecology mimics natural systems, reduces reliance on external inputs and promotes local resilience and food sovereignty. The unique historical, social, and economic circumstances and the limitations of traditional development models should be recognized, and an inclusive approach that considers the agency and perspectives of local actors should be applied to address agroecology transition in Africa. Stakeholder engagement and co-creation are critical to designing and implement sustainable business models. This study aims to understand the current state of the African agro-food sector by starting from the needs, priorities, and challenges of all different stakeholders and recognizing the everyday struggles of African food sovereignty. We will conduct case studies in six countries: Morocco, Senegal, Nigeria, Burkina-Faso, Benin, and Ghana. Data will be gathered through interviews with agro-food actors in these countries and relevant reports and articles. The study has two objectives: (1) to identify stakeholders and their roles; and (2) to diagnose opportunities for value creation towards sustainable transition. Using the value framework Den Ouden developed, we will assess the potential for sustainable business models in agroecological farming. We will showcase our perspective on possible solutions to make agriculture more sustainable in Africa. The European Union's Horizon 2020 program supports the research under grant agreement No. 101059232 – Project Urbane.

Session 72 Theatre 9

Mapping of value chains in the Italian bovine sector

M. Finocchi, M. Moretti, A. Mantino, A. Ripamonti, G. Conte and M. Mele
University of Pisa, Dipartimento di Scienze Agrarie, Alimentari e Agro-ambientali, Via del Borghetto, 80, 56124 Pisa, Italy; matteo.finocchi@phd.unipi.it

The transformation of Europe into a zero-pollution economy by 2050 can be achieved through innovation in agro-industrial systems. Globally, animals provide more than 30% of the protein and substantial employment of rural populations. Livestock farming is the most important agri-food sector for greenhouse gases emissions, accounting for 60-70% of emissions, generated mainly by enteric fermentation, deposition of manure on pasture and its storage. In this scenario, the whole European livestock sector is challenged by the need to continue to produce while limiting environmental impacts and respecting the environment and animal welfare. The mapping of value chains in the dairy and beef cattle sector is fundamental to understand the general picture of national production, from farm to processing industry, distribution, and final consumption. The main aim of the present work is the identification of leverage points along the value chain to foster sustainable transformation in the Italian dairy and beef cattle sector. The analysis of the governance structure of dairy and beef cattle value chains in Italy was investigated through surveys and semi-structured interview with various actors of the chains aimed at describing the power dynamics, the transmission of prices between the different nodes and the policies of both sectors. The main results revealed different physical-economic aspects of the value chain and can be summarized in: (1) main characteristics of the production structure; (2) national and international trade relationship; (3) main actors and governance structure of the value chains. The analysis of the production structure included the volume of production and the localization of the farms, as well as the historical trend in the Italian territory and the regional differences with GPS representations. Results allowed to investigate different scenarios for the ecological transition of the Italian dairy and beef cattle sectors. For example, it provides the info needed to estimate the emissions of both sector through the adoption of predictive models to each single part of the whole value chain since the beginning of the last decade.

Session 72 Theatre 10

The INTAQT project: stakeholders' expectations on husbandry systems and innovative practices

R.C. Eppenstein[1], V. Bühl[1], I. Legrand[2], A. Nicolazo De Barmon[2], B. Martin[3], F. Albert[2], M. Berton[4], M. Bourin[5], A. Cartoni Mancinelli[6], D.A. Kenny[7], E. Kowalski[8], S. McLaughlin[9], G. Plesch[1], F. Bedoin[2], C. Couzy[2], C. Berri[10] and C. Laithier[2]
[1]FiBL, Animal Science, Ackerstrasse 113, 5070, Switzerland, [2]IDELE, 149, rue de Bercy, 75595 Paris cedex 12, France, [3]INRAE, Université Clermont Auvergne, VetAgro Sup, UMR Herbivores, 63122 Saint-Genès-Champanelle, France, [4]University of Padova, DAFNAE, Viale dell'Università 16, 35020 Legnaro, Italy, [5]ITAVI, L'Orfrasière, 37380 Nouzilly, France, [6]University of Perugia, Borgo XX Giugno, 74, 06121 Perugia, Italy, [7]Teagasc, Oak Park Rd, Pollerton Little, Carlow, R93 XE12, Ireland, [8]Ghent University, Coupure Links 653, 9000 Gent, Belgium, [9]The Queen's University of Belfast, 19 Chlorine Gardens, Belfast BT9 5DL, United Kingdom, [10]INRAE, Université de Tours, BOA, 37380 Nouzilly, France; rennie.eppenstein@fibl.org

The INTAQT project aims to establish the relationship between intrinsic and extrinsic quality criteria of animal products and European husbandry systems. Another aim is to evaluate the impact of innovative husbandry practices on the quality of the products. In order to determine, which husbandry systems and practices the project should study, a European-wide stakeholder consultation of the chicken, beef and dairy value chains was conducted. Stakeholders were presented with a preselection of husbandry systems and were requested to modify or add systems according to their interest, and to name innovative practices to be tested. In total, 161 face-to-face interviews took place between October 2021 and March 2022, followed by 12 national and 3 European group meetings. Overall, stakeholders of all value chains were satisfied with the initially selected systems. Suggestions covered both mainstream systems as well as new, innovative ones, which are emerging due to the growing societal rejection of intensive farming practices. These included systems that improve the farms' self-autonomy, use local resources or a circular economy, or improve animal welfare and environmental impact. Similarly, actors in all three value chains suggested innovative practices that centre on increased animal welfare and environmental sustainability, thus reflecting a willingness to respond to societal demands.

Session 72 Theatre 11

The INTAQT project: stakeholders' perceptions and points of view on products quality

C. Laithier[1], F. Bédoin[1], F. Albert[1], I. Legrand[1], A. Nicolazo De Barmon[1], M. Bourin[2], M. Berton[3], V. Bühl[4], R. Eppenstein[4], A. Cartoni Mancinelli[5], D.A. Kenny[6], E. Kowalski[7], S. McLaughlin[8], G. Plesch[4], C. Couzy[1], C. Berri[9] and B. Martin[9]
[1]Institut de l'Elevage, 149, rue de Bercy, 75595 Paris cedex 12, France, [2]ITAVI, L'Orfrasière, 37380 Nouzilly, France, [3]University of Padova, DAFNAE, Viale dell'Università 16, 35020 Legnaro, Italy, [4]FIBL, Ackerstrasse 113, 5070 Frick, Switzerland, [5]University of Perugia, Borgo XX Giugno, 7406121 Perugi, Italy, [6]Teagasc, Oak Park Rd, Pollertoon Little, Carlow, R93 XE12, Ireland, [7]Ghent University, Coupure Links 653, 9000 Gent, Belgium, [8]The Queen's University of Belfast, 19 Chlorine Gardens, Stranmillis, United Kingdom, [9]INRAE, 147, rue de l'Université, 75338 Paris, France; cecile.laithier@idele.fr

The INTAQT project aims to characterize the links between husbandry systems and the quality of poultry meat, beef and dairy products after consulting at national and European levels the actors of each sector on their expectations. The first step was to identify their perceptions and points of view in terms of product quality and this communication focuses on results obtained with producers, processors, retailers and some representatives of citizens' associations. In addition to the intrinsic quality criteria already foreseen in the project (health, nutrition, organoleptic), stakeholders spontaneously expressed the importance of considering extrinsic criteria related to sustainability (animal welfare, environment, socio-economic aspects) as well as technological quality. The other criteria were mentioned in a variable way depending on the type of stakeholder, the country and the sector concerned. These results are consistent with consumers' views and they have been taken into account to include other quality criteria in the project.

Session 72 Theatre 12

Life cycle assessment of different pig production systems around Europe: mEATquality project

C. Reyes-Palomo[1], A. Pignagnoli[2], S. Sanz-Fernández[1], P. Meatquality Consortium[3] and V. Rodríguez-Estévez[1]
[1]Universidad de Córdoba, Ctra. Madrid-Cádiz km 396, 14071, Spain, [2]Centro Ricerche Produzioni Animali, V.le Timavo, 43/2, 42121, Reggio Emilia, Italy, [3]Project mEATquality, Consortium, https://meatquality.eu/, Denmark; v22repac@uco.es

One of the main objectives of the H2020 mEATquality project is to generate information about producing more sustainably through extensive husbandry practices, meeting animal welfare and environmental concerns. The life cycle assessment (LCA) will be the main tool used for the environmental sustainability of the farms evaluated. This LCA study will collect data from 80 fattening pig farms from 4 countries (Denmark, Italy, Poland, Spain) and covering a wide range of models and husbandry practices, ranging from large intensive farms to small, organic, and extensive ones. The enteric fermentation and manure handling emissions will be calculated using Tier 2 equations on IPPC guidelines. The not differentiable emissions such as C sequestration in pasture-based systems will be assessed by economic approach, and environmental impacts will be calculated with Agribalyse database. The limit of the system will be 'cradle to gate'. It is expected that the presence of different pig breeds, with different productive ranges, diverse farming systems and level of extensification will lead to differentiate the environmental impact of the systems and link it to the extensiveness practices. This work will cover a large representation of the different pig husbandry systems around Europe. It is supposed that the inclusion of C sequestration makes the difference in the extensive farms, even though the intensive systems have better performances. This project has received funding from the European Union's Horizon 2020 research and innovation program under Grant Agreement No 101000344.

Session 72 — Poster 13

Consumer expectations for beef in the French region Auvergne-Rhône-Alpes

S. Chriki[1,2], J. Normand[3], C. Brosse[2], L. Hallez[2], L. Vallet[2], V. Payet[2] and J.F. Hocquette[1]
[1]INRAE, VetAgro Sup, UMR1213 Herbivores, Theix, 63122, France, [2]Isara, 23 rue Jean Baldassini, 69364, France, [3]Institut de l'Elevage, Agrapole, 23 rue Jean Baldassini, 69364, France; schriki@isara.fr

In France, beef consumption has been steadily decreasing for many years. The reasons are multifactorial, linked to the: (1) controversial image of beef (its environmental impact, the competition between feed and food, the respect of animal welfare); (2) the modification of consumption patterns (decrease of daily time dedicated to cooking and eating); (3) its high price; and finally (4) its variable sensory quality. Thus, consumers are not always satisfied with the organoleptic quality of beef. In this context, the scientific project OABov-AURA (2022-2024) was conducted. An online survey was carried out to study consumer expectations regarding the intrinsic and extrinsic qualities of beef, in the region Auvergne-Rhône-Alpes (AURA) (the 2nd largest beef cattle breeding region in France). This study involved 509 respondents, mostly female (62%), young (52%<35 vs 32%>45 years old), students (28%) and executives (38%), with a good knowledge of farming (57%), and preferring to buy beef in butcher shops (31%). Among those surveyed, 52% say that eating beef for pleasure is their primary motivation. The majority of consumers are not disappointed with raw (86%) or cooked (58%) beef. Origin (89%), proximity (83%) and SIQO (Identification signs of quality and origin) (82%) are the most important extrinsic criteria. However, taste, tenderness and freshness are the most intrinsic important criteria of beef for the respondents. Consumers also prefer meat that is bright red (50.7%) and rather fatty (57%). The two main reasons for the decrease in beef consumption are economic (28% of responses) and environmental (27%). In conclusion, it is important that consumers in the AURA region have access to a varied supply of meat in butcher shops, of French origin and under quality signs.

Session 72 — Poster 14

Analysis of preferences and perception of cheese products by Portuguese consumers

V.M. Merlino[1], M. Renna[2], M. Tarantola[2], A. Ricci[2], A.S. Santos[3], A. Monteiro[3] and J. Nery[2]
[1]University of Turin, Dept. Agricultural, Forest and Food Sciences, L.go Braccini 2, 10095, Italy, [2]University of Turin, Dept. Veterinary Sciences, L.go Braccini 2, 10095, Italy, [3]FeedInov CoLab, Qta da Fonte Boa, 2005-048, Santarém, Portugal; manuela.renna@unito.it

In a critical dairy chain context, small producers must understand the consumers' needs and the strategies for a winning marketing communication. The aim of this study was to assess consumer preferences and profiles to better define marketing stimuli of cheese consumers in Portugal. A total of 218 Portuguese consumers filled in a structured questionnaire developed to collect information about: (1) socio-demographic traits; (2) cheese purchasing habits and preferences; and (3) attitudes and familiarity towards new cheese products and marketing stimuli. Data were analysed using Principal Component (PC) analysis to assess different preference patterns of consumers towards different cheese attributes (i.e. origin, quality, and certification). Cluster analysis of the PCs was used to classify consumers according to their choice orientation patterns towards cheese quality. Clusters composition was examined using the consumer characteristics as independent variables. Correspondence analysis was performed to identify the association between clusters and the suggested marketing stimuli to improve new product visibility and acceptability. Two PCs were defined as 'Tradition is sustainable' and 'Local is better'. Respondents were divided into 3 groups: 'Terraphilia' (57%), 'Sensitive to local origin' (38%) and 'Undecided' (5%). The 'Terraphilia' group is encouraged to try new cheese products through appealing packaging and detailed information on traceability and sustainability of the supply chain. The 'Sensitive to Local Origin' group is primarily attracted to products with lower price point, clear indications of origin, and a longer shelf-life. The 'Undecided' cluster instead showed the most negative attitude toward new products, being attracted by visual messages conveyed by brands. Portuguese cheese consumers are sensitive to sustainable local production chains. Enhancing the marketing of cheese in this context could be achieved through packaging improvement and information on product origin, sustainability and traceability. Funded by EIT FOOD, 2021(project number: 21324).

Session 72 Poster 15

How a risk-based strategy could contribute to a more sustainable agri-food system?

B. Grossiord[1] and M.P. Ellies-Oury[1,2]

[1]Bordeaux Sciences Agro, Feed &Food, 1 cours du Général de Gaulle, 33170 GRADIGNAN, France, [2]INRAE, UMR1213 Recherches sur les herbivores, VetAgroSup Clermont-Ferrand, 63122 Saint Genès Champelle, France; benoit.grossiord@agro-bordeaux.fr

Nowadays, members of the agri-food systems are the targets of much criticism but also the bearers of many hopes with regard to the challenges facing our society when considering sustainable development. Their contribution to the achievement of the Sustainable Development Goals is crucial and a key element of the coming transition. Their responsibility is engaged more than ever and they must lead the change by knowing the expectations of their stakeholders and managing the risks related to their activities. In order to reach that goals, they can use tools or methods which are available and could contribute to the improvement of the quality of their products, practices and organization. Such tools are successfully used by operators of the food chain, especially those with an industrial size. Small producers, due to harder work conditions and lack of time and means, are not familiar with such approaches although very useful for the improvement of their overall performance and long term viability. In relation to the growing expectations of the consumers and the society, they must consider diversifying their activity at multiple levels and propose new products, follow different quality schemes or change their modes of production. Short food supply chains are part of the many answers they can develop in order to improve the value of their products both at the intrinsic and extrinsic levels. Here we show that many issues around governance, social benefits, environmental impacts, animal welfare, health and nutrition dimensions can be integrated in a global approach of the short food supply chains. For that purpose, the concomitant utilization of tools, methods, guidelines or standards already available is recommended. Using management systems requirements with a consideration to intrinsic and extrinsic quality attributes allows the construction of an integrated approach with a continual improvement spirit. Such an organization will mobilize risk management tools applied to several stakes, like quality of the products, occupational health and security, environmental impacts, and so on, therefore allowing a better global performance.

Session 72 Poster 16

Perceptions of meat quality of UK stakeholders: from intrinsic to extrinsic factors

S. McLaughlin[1], F. Bedoin[2], C. Couzy[2], I. Legrand[2], A. Nicolazo De Barmon[2], C. Laithier[2] and N. Scollan[1]

[1]Queen's University Belfast, School of Biological Sciences, Institute for Global Food Security, 19 Chlorine Gardens, Stranmillis, BT9 5DL Belfast, United Kingdom, [2]IDELE, 149, rue de Bercy, 75595 Paris cedex 12, France; s.mclaughlin@qub.ac.uk

Aim: To understand attitudes and perceptions of livestock product quality held by supply chain actors within the agri-food sector across the UK. Methods: 15 interviews with actors across the UK beef supply chain, from farmers, processors, retailers, animal welfare, research, professional and quality assurance bodies. Qualitative analysis using inductive thematic analysis procedures. Interviews conducted December 2021-Februrary 2022. Results: For UK stakeholders extrinsic attributes of product quality were most important. These include animal welfare and the price of the product. Age at slaughter and carbon footprint of production were also emerging characteristics of quality. Grass fed production and use of native breeds, such as Aberdeen Angus and Hereford breeds were associated with better quality. In terms of intrinsic quality, taste, fat and intramuscular fat, tenderness, appearance, consistency, and flavour were the most cited attributes. Eating quality was often used synonymously with product quality. Conclusions: Both intrinsic and extrinsic of livestock product quality are important to actors within the beef supply chain in the UK.

Session 72 Poster 17

Transforming livestock practices by federating around common values

S. Nade
Obione, 239 rue Fernand Léger, 71000 Mâcon, France; sarahnd57@gmail.com

Today, environmental and animal welfare issues are real concerns for the marketing of animal products. In the image of One Welfare, the environment, animal welfare and, above all, the farmer's welfare are the three essential and inseparable elements for obtaining 'healthy' livestock and for meeting societal expectations. Obione therefore imagined a project in which farmers, veterinarians, processors and consumers would be united around a common objective and shared values. First of all, activities and meetings will be held between farmers, processors and veterinarians to improve communication between them. Indeed, Obione will offer training courses on value creation for and by farmers, on animal and farmer welfare and on the importance of each actor in the chain and the link between them. To improve and facilitate the dissemination of information and communication between the four parties, a website will be created. This site will include each producer registered in the process as well as the people involved in the structure (with their agreement). The aim is to enable consumers to know and understand the design of the products they consume. It will then be possible to directly involve consumers in the tasks carried out on the farms thanks to meetings organised via a connected diary. The Happy system, created by Obione to promote 'happy' farms, will then be offered to all breeders and associated veterinarians, thus enabling them to be valued, to take pride in their work and to report on the well-being of their farms. Indeed, Happy is an approach that evaluates the well-being of the animals and the breeder as well as respect for the environment. This project aims to strengthen the links between the players in the sector and the consumer by informing, sharing and enhancing the value of each of them, while showing the importance of livestock farming for human nutrition, biodiversity, the environment and even within a territory, for its development.

Session 73 Theatre 1

Genetic response of red yeast supplementation in feed to mycotoxin contamination in laying hens

S. Hosseini[1], B. Brenig[1], W. Tapingkae[2] and K. Gatphayak[2]
[1]University of Goettingen, Department of Animal Sciences, Burckhardtweg 2, 37077 Goettingen, Germany, [2]Chiang Mai University, Department of Animal and Aquatic Sciences, 239 Huay Kaew Road, 50200 Chiang Mai, Thailand; shahrbanou.hosseini@uni-goettingen.de

Feed contaminated with toxic fungi in farm animals can lead to impaired immune function, renal function, hepatotoxicity, and genetic mutations. Consumption of contaminated animal products is a global concern for human health. Mycotoxins are toxigenic chemical products of fungi that play a crucial role in liver disease, resulting in reduced production performance and high mortality in animal husbandry system. To minimize the harmful effects of mycotoxins, red yeast, an organic compound, was used as a mycotoxin binder in the poultry industry. Hence, the main goal of this study was to investigate the genetic response underlying feeding with red yeast supplementation in interaction with the mycotoxin in the liver of laying hens. For this purpose, the animals were fed for 63 days with four different diets: control diet (CON), CON with red yeast supplementation 1.0 g/kg (RY1.0), CON with contaminated corn containing 100 µg/kg mycotoxin (MT100), and CON with a combination of RY1.0 and MT100 (RY1.0+MT100). The liver tissue of four animals from each experimental group were collected after the feeding period and the total RNA was isolated for RNA sequencing using Illumina platform. The standard pipeline for RNA read alignment and counting, differential gene expression and functional enrichment analysis was performed for all samples. The results showed a high number of significant differentially expressed genes (DEGs, $p_{adj}<0.05$) in MT100 vs CON group (1,553). This number of DEGs was considerably reduced to 858 genes, when RY was added to the RY1.0+MT100 diet compared to the CON group. As expected, the number of significant DEGs in the diet with RY1.0 was negligible compared to the CON group (8 genes). We identified a set of genes in MT100 and in RY1.0+MT100 diet, which play key roles in phase I (e.g. *CYP2C23a, CYP2C23b*) and phase II (e.g. *UGT1A1, GSTO1*) detoxification process of xenobiotics. Additionally, other genes involved in antioxidant mechanisms (e.g. *CHAC1, SOD1*) and immune response (e.g. *IL2RA, SERPINB10B*) were also detected in the liver of hens in this study.

Session 73 Theatre 2

Faecal starch content as an indicator of starch digestibility by fattening Japanese Black cattle

M. Matamura, S. Uzawa and M. Kondo
Mie University, Graduate School of Bioresources, 1577 Kurimamachiya,Tsu-city, Mie-Pref, 514-8507, Japan; masaya_matamura0@yahoo.co.jp

A total of 2.6 million fattening cattle in Japan are fed approx. 10,000 t of starch every day. However, we do not know how much starch the cattle digest efficiently. To estimate starch digestibility in fattening cattle, a single linear regression equation based on the faecal starch (FS) content has been reported in the USA. As the digestion of nutrients is affected by several factors, regression equations developed under different feeding systems may not be applicable to Japanese Black cattle. Thus, the aim of this study was to develop a method for estimating starch digestibility from the faeces of fattening Japanese Black cattle. Moreover, we compared the developed equation with those obtained from studies in other countries. Total-tract starch digestibility (TTSD) was measured in a total of 116 fattening Japanese Black cattle. A linear regression model was developed to estimate TTSD from the FS content. In addition, research papers on fattening cattle published from 1993 to 2022 conducted in other countries were collected. Based on these data, a linear model between FS and TTSD was also developed. As a result, TTSD ranged from 83.9 to 99.8% and FS ranged from 0.2 to 15.3% in Japanese cattle. A negative linear relationship was observed between these variables, and the following equation was obtained to estimate TTSD: TTSD (%) = 99.7 (%) – 0.85 FS (%DM), n=116, R^2=0.89, RMSE=1.00. The slope of the regression equation obtained from fattening Japanese Black cattle (-0.85) was significantly greater (P<0.001) than that from fattening cattle in other countries (-0.47). Mathematically, the slope of the regression equation represents the ratio of DM indigestibility to dietary starch content. The mean values of DM indigestibility and dietary starch content obtained in this study were 29.5 and 32.5%, respectively, which are significantly different (P<0.001) from those reported by studies conducted in other countries (23.3 and 48.7%, respectively). Therefore, the greater slope of the equation obtained from Japanese Black cattle compared to cattle from other countries could be attributed to the higher DM indigestibility and lower dietary starch content.

Session 73 Theatre 3

Impact of isoacids on performance and digestibility in gilts fed different dietary fibre sources

W. Schweer[1], B. Kerr[2], M. Socha[1], A. Cornelison[1] and L. Rodrigues[1]
[1]Zinpro Corporation, Eden Praire, MN, USA, [2]USDA-ARS, Ames, IA, USA; wschweer@zinpro.com

Isoacids are short, branched chain volatile fatty acids produced by microbial degradation of branched chain amino acids. In ruminants, isoacids are growth factors for cellulolytic bacteria which leads to improved digestibility of fibrous ingredients. The objective of this study was to determine if feeding isoacids to pigs could improve performance and digestibility of high fibre diets. Two groups of 45 gilts (BW=139 kg) were individually penned and fed 1 of 9 dietary treatments for 28 days (n=10 per treatment). Diets were arranged in a factorial manner consisting of diet type: corn-soybean meal (CSBM), a CSBM diet with 40% inclusion of distillers dried grains with solubles (DDGS), or a CSBM diet with 40% inclusion of sugar beet pulp (SBP); in combination with no isoacid (NO) or the addition of 0.50% isobutyrate (IB) or 0.88% inclusion of isoacid mixture (MX). IB and MX diets were formulated to have equal levels of isobutyrate. Gilts were limit fed 2 kg per day. Performance metrics (ADG, ADFI, G:F) were determined for the 28 day period. Faecal samples were collected on day 26 to determine apparent total tract digestibility (ATTD) of gross energy (GE) and nitrogen (N), and a faecal VFA profile. There was no diet type by isoacid interaction and no main effect of isoacids on performance (P≥0.22). Diets with DDGS or SBP reduced ADFI and increased G:F compared to pigs fed CSBM diets (P=0.01). Interactions existed for ATTD (P=0.01), where IB increased ATTD of GE and N in diets containing DDGS compared to DDGS diets with no isoacid. In contrast, MX decreased ATTD of GE and N compared to NO in DDGS diets. In diets containing SBP, IB or MX both decreased ATTD of GE and N compared to no isoacid addition. In CSBM diets, no differences in ATTD of GE or N were noticed between IB and NO (P>0.05); however, MX addition decreased ATTD of GE and N compared to NO. Total faecal VFA tended to be increased by MX in CSBM and SBP diets, but not DDGS diets, leading to an interaction of diet by isoacid (P=0.06). Faecal isobutyrate was increased by IB, but not MX, compared to NO (P=0.01). These results suggest that IB may have an impact on digestibility in diets higher in insoluble fibre compared to soluble fibre.

The enterotype is associated with the phenotype variation in the pigs treated with dietary fibre

H. Li and B.E. Tan
Hunan Agricultural University, No.1 Nongda Road, Changsha, 410128, China, P.R.; bietan@hunau.edu.cn

The gut microbiota degraded DF in hindgut and fermented to producing short chain fatty acids (SCFAs) that provide energy or produce health benefits for pigs. Enterotype was a method of dividing individuals based on the composition of gut microbial community and could be a key factor to determining the phenotype of DF-treated pigs. Therefore, the current study was conducted to analysis the linkages between the dietary fibre (DF) utilization and enterotype in pigs. Based on 8,872 16S rRNA gene sequencing data from fresh faeces or rectal contents from around the world, pigs were stratified into three main intestinal types, named ETL, ETP, ETC according to the feature genera of the gut microbiome. The ETL and ETP were dominated by *Lactobacillus* and *Prevotellacae* members, respectively. Whereas ETC had the microbial communities formed with *Christensenellaceae_R_7_group* as the core genus. The α diversity index of ETC and ETP was generally higher than that of ETL. It was verified that changing the DF source or breed of pigs did not affect the grouping effect of enterotype on pig individuals. Diets with higher soluble dietary fibre (SDF) increased the proportion of ETC and decreased the proportion of ETL in pigs compared to the diet with lower SDF. Additionally, enterotype significantly affected the apparent total tract digestibility (ATTD) of DF and the coefficient values of the DF ATTD in pigs. Further, it was identified that the ratio of SCFAs, mainly produced by fermentation of dietary fibre, also varied due to enterotype changing. ETC was more inclined to acetate production, while ETC may have more butyrate production. In summary, it was expected to provide more accurate nutritional strategies for the application of dietary fibre when considering the proportion of different enterotypes in pig herds.

A high rumen degradable starch modulates jejunum microbiota and bile acids in dairy goats

L.M. Wang, J.H. Yao and Y.C. Cao
Northwest A&F University, Yangling, Shaanxi, 712100, China, P.R.; wanglamei1216@126.com

Fat is the major energy component in milk and accounts for many of the physical properties, manufacturing characteristics, and organoleptic qualities of milk and milk production. Milk fat can be affected by factors that influence the processes from dietary lipogenic precursors to milk fat, in the rumen, intestine, liver, and mammary tissues and so on. This study uncovered the effect of digestive processes and liver metabolism of lipogenic precursors on milk fat synthesis in dairy goats. Eighteen lactating goats were allocated equally into low RDS (LRDS=20.52%), medium RDS (MRDS=22.15%), and high RDS (HRDS=24.88%) diet groups. After 5 weeks of feeding, the HRDS diet increased the relative abundance of Firmicutes and Ruminococcus_2 in jejunum. Firmicutes, which are gram-positive bacteria, are the predominant bacteria able to deconjugate and dehydroxylate primary bile acids into secondary bile acids. *Ruminococcus* perform epimerization during the conversion from primary to secondary bile acids. The expression of bile acid receptor FXR in jejunum and ileum and TGR5 in jejunum were decreased by HRDS treatment, indicating the negative feedback regulation of bile acid synthesis was inhibited in liver. The increase expression of CYP7A1, the rate-limiting enzyme of the bile acid biosynthetic from cholesterol, indicate the increased bile acids synthesis in liver. We also found the HRDS group increased the bile acids TCDCA and TDCA, and disordered the phosphatidylcholines in liver, as well as upregulated TNFα expression. We further measure the genes expression related to lipid metabolism, and found that HRDS group increased the expression of the transcription factor peroxisome proliferator-activated receptor α (PPARα) and its downstream target gene CPT1. This study demonstrated that HRDS diet feeding modulates jejunum microbiota and alters enterohepatic circulation of bile acids, and promote lipid oxidation in dairy goats.

Development of a novel endolysin, RalLys8, for the specific inhibition of *Ruminococcus* albus

J. Moon, H. Kim and J. Seo
Pusan National University, Department of Animal Science, 1268-50 Samrangjin-ro, Miryang, Gyeongsangnam-do, 50463, Korea, South; mantis0044@pusan.ac.kr

Ruminococcus albus (*R. albus*) is the most important cellulolytic bacterium in the rumen. It ferments one mole of glucose, producing ethanol, acetate, carbon dioxide, and a high concentration of hydrogen. Ruminal methanogens use hydrogen to reduce carbon dioxide to methane, which is a potent greenhouse gas, and also a loss of energy for ruminants. Endolysins, bacteriophage-encoded peptidoglycan degrading enzymes, have emerged in recent years as a novel agent to kill target bacteria specifically. Thus, the study hypothesized that if a novel endolysin could specifically kill *R. albus*, it could potentially reduce hydrogen production and mitigate methane emissions. Whole genome sequence of *R. albus* strains and related bacteriophages were collected from the National Center for Biotechnology Information database, and the candidate gene for RalLys8 was isolated based on amino acid sequences and conserved domain database (CDD) analysis. The lytic activity of RalLys8 was evaluated under various conditions (dosage, pH, temperature, NaCl, and metal ions) to determine the optimal lytic conditions. CDD analysis revealed that RalLys8 possesses only the catalytic domain of PGRP superfamily which includes a zinc dependent N-acetylmuramoyl-L-alanine amidase located at the N-terminal without cell wall binding domain. The RalLys8 reduced the optical density of the *R. albus* dose-dependently above a concentration of 3.125 μg/ml, and the activity of RalLys8 was the highest at a concentration of 25 μg/ml. The highest lytic activity of RalLys8 against *R. albus* was observed at pH 9.0 and 31.25 mM NaCl, and this activity was maintained over a temperature range of 16 to 60 °C. The lytic activity of RalLys8 was recovered by treatments with Mg^{2+} and Co^{2+}. However, treatments with Ca^{2+} and Zn^{2+} reduced the activity of RalLys8 by approximately 40 and 20%, respectively. In conclusion, we have successfully developed a novel endolysin, RalLys8, which exhibits potent lytic activity against *R. albus*. Therefore, further studies are needed to evaluate the effect of RalLys8 against *R. albus* and its impact on rumen microbiota under *in vitro* anaerobic condition.

Dietary microalgae (*Nannochloropsis limnetica*) and probiotic on performance and intestine of piglets

M.F. Pedro[1], D.F.P. Carvalho[1], A.M. Almeida[1], R.J.B. Bessa[2], A.J.M. Fonseca[3], A.R.J. Cabrita[3] and J.P.B. Freire[1]
[1]LEAF, TERRA, ISA, Universidade de Lisboa, 1349-017 Lisboa, Portugal, [2]CIISA, AL4AnimalS, FMV, Universidade de Lisboa, 1300-477 Lisboa, Portugal, [3]REQUIMTE, LAQV, ICBAS, Universidade do Porto, 4050-313 Porto, Portugal; mfpedro@isa.ulisboa.pt

Study of alternative feeds can provide solutions that consider both current issues and future requirements, with resource usage and their ecological implications. Feedstuffs often directly compete with human-edibles and are produced with arable land. Microalgae have interesting nutritional and industrial properties but require further study. This work evaluated the effects of *Nannochloropsis limnetica* dietary inclusion and a commercial probiotic (LBC ME10) on growth performance and intestinal physiology of weaned piglets. 120 piglets (Large White × Landrace × Piétrain), with 7.7±0.9 kg live weight (28 d) were blocked by weight in trios and randomly assigned to one of the experimental diets: C – Control diet (wheat/maize/soybean meal); N2.5 – 2.5% *Nannochloropsis*; N5 – 5% *Nannochloropsis*; N2.5P – 2.5% *Nannochloropsis* and Probiotic; and N5P – 5% *Nannochloropsis* and Probiotic. Piglets were housed in three per pen, in a total of 40 pens (8 pens per treatment) with *ad libitum* feeding and water. After 3 days of adaptation, piglets were fed for 5 weeks, with liveweight and feed intake determined weekly and faecal scores daily. At week 5, one piglet from each pen was slaughtered and gastrointestinal contents collected. Performance indicators, viscosity and pH of intestinal contents were determined and data analysed by ANOVA. Diets did not affect average daily gain (498±88 g/d), average daily feed intake (648±103 g/d), nor feed conversion ratio (1.32±0.23). Faecal consistency, viscosity of intestinal contents in upper small intestine (4.2±0.6 cP) and ileum (4.6±0.4 cP), as well as pH of stomach (4.2±0.9), upper small intestine (5.7±0.3), ileum (6.6±0.4), cecum (5.6±0.2) and colon (6.1±0.2) contents were similar within each sampling site for all treatments. Overall, results indicate that *N. limnetica* inclusion in weaned piglet diets up to 5% does not affect growth performance nor faecal consistency, suggesting it may be effectively used in piglet diets with good feed acceptance. Further work is needed to understand effects on digestive physiology.

Session 73 Theatre 8

Muramidase inclusion reduces gut inflammation in weaned piglets, especially in high protein diets

U.M. McCormack[1], J. Schmeisser[2], E. Bacou[2], P. Jenn[2], F. Amstutz[2] and E. Perez Calvo[1]
[1]DSM Nutritonal Products, Wurmisweg 576, 4303 Kaiseraugst, Switzerland, [2]DSM Nutritional Products, 1 Bd d'Alsace, 68128 Village-Neuf, France; estefania.perez-calvo@dsm.com

Weaning compromises gastrointestinal functionality (GIF), including changes in digestion, absorption, microbiology and immunology, resulting in clinical signs like post-weaning diarrhoea (PWD) and reduced growth performance. High protein diets elevate proteolytic fermentation, altered microbial communities and intestinal physiology increasing PWD incidence. Muramidases (MUR) are enzymes that hydrolyse peptidoglycans (PGN) from bacterial cell debris present in the gut, which play an important role in maintaining intestinal homeostasis by down-regulating inflammatory responses. This study assessed dietary MUR inclusion on GIF in early weaned piglets fed either low (LP) or high (HP) protein diets. Sixty-four male 21 d old pigs (initial BW 6.1±0.6 kg) were used in a 14 d study, randomized into 4 treatments: LP:167 g of CP/kg, LP+: LP supplemented with MUR, HP: 210 g of CP/kg and HP+: HP supplemented with MUR. Bodyweight was recorded on d0 and 14. On d14, plasma urea nitrogen (PUN) and concentrations of Vit A and E were measured. Faecal dry matter (DM), and lipocalin as a biomarker of inflammation were evaluated. Additionally, inflammatory gene expression was analysed by PCR array targeting 42 genes in mesenteric lymph nodes, ileal peyer patches and lamina propria. Treatment differences were determined by two-way ANOVA (JMP® software) considering protein level and MUR inclusion as main effects. HP pigs had higher ADG and plasma Vit A (P<0.05) and higher faecal DM (P=0.06) compared to LP pigs. Adding MUR increased faecal DM and plasma Vit A (P>0.05) and reduced PUN (P<0.05) which could explain the trend to increase ADG (P=0.10). No interaction was detected, except for lipocalin (P<0.05) which was reduced in HP, only when MUR was included. Moreover, a downregulation of inflammatory genes (CCL2, CXCL12, CXCL9, IL10RA, IL18, IL27; P<0.05) was found in the different tissues from MUR pigs. A downregulation in IL17B indicating reduced inflammation was also found in HP+. In conclusion, MUR supplementation for 14 d post-weaning reduces gut inflammation, improves nitrogen utilization and Vit A absorption, resulting in lower PWD, especially in HP diets.

Session 73 Theatre 9

Utilization of tryptophane by kynurenine, indoles and serotonin pathways are modified by fructose

A. Gual-Crau[1], M. Jarzaguet[1], D. Dardevet[1], D. Rémond[1], A. Lefèvre[2], A. Bernalier-Donadille[3], P. Emond[2,4] and I. Savary-Auzeloux[1]
[1]UCA, INRAE, UNH, 63000 Clermont-Ferrand, France, [2]iBrain, Université de Tours, Inserm, 37000 Tours, France, [3]UCA, INRAE, MEDIS, 63000 Clermont-Ferrand, France, [4]CHRU de Tours, MEDIS, 63000 Clermont-Ferrand, France; didier.remond@inrae.fr

Tryptophane (Trp) is an essential amino acid used for protein synthesis and in 3 metabolic pathways: Kynurenine (KYN), indoles and serotonin. If Trp deficiencies are known to reduce growth and appetite in monogastrics, less is known on Trp's fate in overfeeding, when animals develop insulin resistance (IR). In humans, increased levels of Trp and metabolites occur in type 2 diabetes. Our aim is to determine if Trp can mark the onset of IR-related metabolic perturbations and what can be the consequences on Trp requirements in a monogastric model: mice. C57BL/6 mice were fed one month with a control diet (CON) (n=8) or CON diet with 10% starch substituted by fructose (n=8) (FRU). Trp and metabolites (from the 3 pathways) were assayed in arterial serum, jejunum, colon, faecal content and liver (LC-HRMS) in the fasted state. Insulin and glucose were measured in serum (ELISA and enzymatic method). Comparisons between groups: ANOVA, post hoc: Tukey. Although weight gain, intake and serum insulin /glucose were not differentially altered in the two groups, Trp increased (P<0.05, +21.8%) in FRU vs CON as well as KYN, 3-OH-KYN, picolinic acid and KYN/Trp ratio. This suggests a stimulation of KYN pathway by fructose. On the contrary, the indole pathway, controlled by gut microbiota activity, is inhibited in FRU (indole-3-sulfate (-25.6% – P<0.05). Lastly, 5-OH Try, a precursor of serotonin, is increased in FRU (P<0.05) but other are unaltered. The activation of KYN pathway in FRU also occurs in jejunum but not colon for 3-OH KYN and 3-OH Anthranilic acid (P>0.05), suggesting an intense activity of KYN pathway in this tissue. Additionally, quinolinic acid (QA), endproduct of KYN pathway, is increased in faeces and liver (P<0.05), preventing an accumulation of QA in blood as QA is neurotoxic. Other pathways are under investigation. We show that Trp's fate in serum and tissues can be strongly altered during the onset of IR. In fast fattening animals, this altered Trp metabolic fate may occur and should require further investigation.

Regulation of glucose metabolism in rumen epithelium of cows transitioned from forage to high-grain

S. Kreuzer-Redmer[1], A. Sener[1], C. Pacifico[1], F. Dengler[2], S. Ricci[1], H. Schwartz-Zimmermann[3], E. Castillo-Lopez[1], N. Reisinger[4] and Q. Zebeli[1]
[1]Institute of Animal Nutrition, Vetmeduni, Veterinärplatz 1, 1210 Vienna, Austria, [2]Institute of Physiology, Vetmeduni, Veterinärplatz 1, 1210 Vienna, Austria, [3]IFA-Tulln, iBAM, BOKU, K.L.-Str. 20, 3430 Tulln, Austria, [4]DSM, BIOMIN Research Center, Technopark 1, 3430 Tulln, Austria; susanne.kreuzer-redmer@vetmeduni.ac.at

The bovine rumen epithelium has a crucial role in the uptake and metabolism of fermentation products. Yet, little is known about the uptake of hexoses and their metabolism. Thus, we have examined the transcriptome of rumen papillae from cows when transitioned from a forage-based to a high-grain diet. Rumen biopsies were collected from ruminally cannulated non-lactating Holstein-Friesian cows fed a forage diet (FD; 0% concentrate, n=9) as baseline or a high-grain diet for 4 weeks (HG; 65% concentrate, n=9). Transcriptome analysis identified 9,481 differentially expressed genes (DEGs) between FD and HG. Within the top 5 most significant DEGs, we detected SLC7A8, which is associated with the transport of glucose and other sugars. A more in-depth analysis of known glucose transporters revealed that *GLUT2, GLUT3, GLUT4, GLUT9, GLUT10, SGLT2* and *SGLT3* were all up-regulated when cows were transitioned to the HG diet, suggesting an increased uptake of glucose, which might be used for the increased energy demands of cells to proliferate and enlarge papillae, or simply for storage. Targeted anion exchange chromatography-HRMS-based metabolomics detected carboxylic acids, sugars and sugar-phosphates in rumen fluid and Spearman correlation analysis revealed positive correlations of hexoses with the differentially expressed glucose transporter genes. Particularly strong and significant correlations were found for *GLUT3, GLUT9, SGLT3* and *SLC7A8* (R^2>0.6, P<0.05). We were able to confirm our NGS results via qPCR on transcript level. In addition, we were able to prove the ruminal abundance of GLUT2, GLUT3, SGLT1, SGLT3 and SLC7A8 on protein level by Western blotting in ruminal papillae. Our data confirm a direct involvement of the rumen epithelium in glucose metabolism. Further experiments are essential to evaluate and deepen our understanding of glucose metabolism within the rumen epithelium.

Altering nutrients supply modified dynamics of milk components synthesis in dairy cows

J.C. Anger[1,2], C. Loncke[3], R. Bidaux[1] and S. Lemosquet[2]
[1]Provimi, Cargill ANH, 35320 Crevin, France, [2]PEGASE, INRAE, Institut Agro, 35590 Saint-Gilles, France, [3]Université Paris-Saclay, INRAE, AgroParisTech, UMR MoSAR, 91120 Palaiseau, France; jean-charles.anger@inrae.fr

This study aimed to analyse the dynamics of milk component synthesis in response to increasing starch and balancing amino acids (AA) during mid-lactation in dairy cows. We hypothesized that altering the nutrient supply could modify the persistency of lactose, protein, and fat differently. Thirty-two multiparous dairy cows were involved in a randomized complete block design. They received 2 levels of starch (LS: low starch; HS: high starch) and 2 levels of AA supplies (AA- vs AA+) through rumen-protected AA according to a factorial design from 56 to 180 days in milk (DIM). The LS and HS diets provided (in g/kg DM) 155 vs 233 of starch, 384 vs 380 of NDF, 138 vs 142 of CP, 89.5 vs 94.5 of MP, and 6.54 vs 6.20 MJ/kg, respectively. The AA- vs AA+ diets were formulated to provide [in % of MP]: 6.4 vs 7.1% Lys, 2.0 vs 2.4% Met, and 2.1 vs 2.4% His, respectively. Milk protein and fat contents were measured twice a week, and lactose content once a week. Milk lactose, protein, and fat yields were modelled by linear regression with DIM, starch, AA, and covariates as predictors, and cow as a random effect. The slope of each milk component model was assumed to be an estimator of persistency. Lactose and protein yields were higher on HS than on LS diets (1,708 vs 1,895 g; P<0.001; 1,085 vs 1,201 g; P<0.001, respectively). The slope of lactose, protein, and fat yields models were higher on HS than on LS diets (-3.41 vs -1.41 g/d, P<0.001; -1.31 vs 0.29 g/d; P<0.001; -2.43 vs -0.43 g/d, P<0.001, respectively). This effect could be due to the stimulation of mammary synthesis by starch or by slightly higher supply of MP in the HS diet. Balancing AA increased the slope of lactose yield (-2.78 vs -2.04 g/d, P=0.01), however the effect tended to be not additive (at LS level: -4.05 vs -2.77 g/d; at HS level: -1.50 vs -1.31 g/d; AA × starch on slope: P=0.06). With the LS diet, balancing AA tended to increase the slope of fat yield (at LS level: -2.90 vs -1.96 g/d, AA × starch on slope: P<0.05). These results suggest that increasing starch or better balancing AA increased the persistency of milk components.

Session 73 | Theatre 12

Phytogenic effects on layer performance, egg quality and cytoprotective response in the ovaries

I.P. Brouklogiannis[1], E.C. Anagnostopoulos[1], V.V. Paraskeuas[1], E. Griela[1], G. Kefalas[2] and K.C. Mountzouris[1]
[1]Agricultural University of Athens, Animal Science, Iera Odos 75, 11855, Athens, Greece, [2]Nuevo SA, Schimatari Viotias, 320 09, Viotia, Greece; gbrouk@outlook.com

The aim of this work was to investigate the inclusion level effects of a natural phytogenic blend (PB) on layer production performance, egg quality and underlying detoxification (aryl hydrocarbon receptor; AhR), antioxidant (Nuclear factor erythroid 2-related factor 2; Nrf2) and inflammatory (nuclear factor-kappa B; NF-κB) responses in the ovaries. Laying hens (n=385; 21-wk-old; Hy-Line Brown) were allotted to 5 dietary treatments with 7 replicates of 11 hens each, for a 12-week feeding trial. Treatments received a corn-soybean meal basal diets with no PB (CON) or supplementation with PB at 250 (PB250), 750 (PB750), 1000 (PB1000) and 1,500 mg/kg diet (PB1500), respectively. The PB consisted of selected Mediterranean plants having olive oil polyphenols, carvacrol and thymol among its main bioactive components (NuPhoria®; Nuevo SA, Greece). Performance and egg quality parameters were determined weekly for a 12-week period (i.e. 33rd wk of layers age) and reported as overall. Ovarian samples from 33-wk-old layers were collected and stored deep frozen, until gene expression analysis with qPCR. Data were analysed by ANOVA and statistical significance was determined at $P<0.05$. Linear and quadratic patterns of biological responses to PB inclusion levels were studied via polynomial contrasts analysis. Increasing PB inclusion, enhanced quadratically egg laying rate and egg mass, with PB750 birds being higher ($P<0.05$) compared to CON. Albumen height and Haugh unit increased quadratically with increasing PB inclusion and peaked at PB750 ($P<0.05$). In the ovaries, increasing PB inclusion level down-regulated ($P<0.05$) the expression of most of AhR pathway genes (80%), while increased the expression of most of the Nrf2 pathway genes (87.5%) assessed. Additionally, most of the genes related to NF-κB pathway (75%) were down-regulated ($P<0.05$) with increasing PB inclusion level. Conclusively, phytogenic inclusion beneficially enhanced the birds' adaptive cyto-protection and anti-inflammatory capacity and documented further production and egg quality improvements, with PB750 displaying the optimal benefits.

Session 73 | Poster 13

Phytogenic effects on layer performance, and cytoprotective response in the ceca

E.C. Anagnostopoulos[1], I.P. Brouklogiannis[1], V.V. Paraskeuas[1], E. Griela[1], G. Kefalas[2] and K.C. Mountzouris[1]
[1]Agricultural University of Athens, Animal Science, Iera Odos 75, 118 55 Athens, Greece, [2]Nuevo SA, Schimatari Viotia, 320 09 Viotia, Greece; vagosanagn@gmail.com

The aim of this work was to evaluate the effect of a natural phytogenic blend (PB) inclusion level on production performance and underlying detoxification (aryl hydrocarbon receptor; AhR) and antioxidant (nuclear factor erythroid 2-related factor 2; Nrf2) molecular responses in layer ceca. The PB consisted of selected Mediterranean plants having olive oil polyphenols, carvacrol and thymol among its main bioactive components (NuPhoria®; Nuevo SA, Greece). A total of 385 21-wk-old Hy-Line Brown layers were placed into 5 dietary treatments with 7 replicates of 11 hens each, for a 12-week feeding trial. Treatments received a maize-soybean meal basal diet without PB (CON) or with PB supplementation at 250 (PB250), 750 (PB750), 1000 (PB1000) and 1,500 mg/kg diet (PB1500), respectively. Laying rate, egg mass, feed intake and feed conversion ratio were determined weekly and presented for periods: 1-6 wks and 7-12 wks. Caecal intestinal samples were collected in the 6th and 12th week of the experiment. The data were analysed by ANOVA procedure and statistical significance was determined at $P<0.05$. Results revealed that PB inclusion, improved ($P<0.001$) egg laying rate, egg mass and feed conversion ratio in the period 7-12 wks, mainly at PB750 treatment, compared to control. Furthermore, dietary PB inclusion up-regulated ($P\leq0.05$) the expression of the majority of the Nrf2 pathway antioxidant genes in the ceca, both in 6th and 12th week of the experiment, compared to CON. Additionally, the AhR pathway related genes were down-regulated ($P<0.05$) with PB inclusion level at both sampling points. Conclusively, including PB in layers' diets beneficially modulated the caecal antioxidant and detoxification responses along with a performance enhancement, with PB750 diet being the most prominent inclusion level.

Effects of guanidinoacetic acid supplementation on growth performance in Nellore cattle

O.R.M. Neto[1,2], I.M.S.C. Farias[1,2], R.N.S. Torres[2], R.V. Ribeiro[1,2], W.A. Baldassini[1,2], R.A. Curi[1,2], L.A.L. Chardulo[1,2] and G.L. Pereira[1,2]

[1]São Paulo State University, School of Agricultural and Veterinarian Sciences, Prof.Paulo Donato Castellane, 14884900, Jaboticabal, São Paulo, Brazil, [2]São Paulo State University, School of Veterinary Medicine and Animal Science, 3780 Universitária Ave, 18610160, Botucatu, São Paulo, Brazil; otavio.machado@unesp.br

Guanidinoacetic acid (GAA) has been used as a feed additive to improve growth performance, feed efficiency and meat quality in non-ruminant animals. However, there is little information about its use in feedlot beef cattle. In this context, the aim was to evaluate the effect of the inclusion of GAA in the diet of non-castrated Nellore cattle finished in feedlot, on the dry matter intake (DMI), performance (ADG), feed efficiency (FE) and carcass traits (CT). We used 120 Nellore bulls, with initial body weight of 345±18.31 kg, blocked by weight and distributed in 24 collective pens (n=12/treatment), receiving two treatments: Control = without GAA and treatment with the inclusion of 1.0 g/kg of DMI of GAA. The experimental period lasted 101 days. In the first 20 days, the animals were submitted to the step-up adaptation protocol, with an increase in the concentrate:roughage ratio in the diet from 64:36 (step 1), 72:28 (step 2), 80:20 (step 3), until the final ratio of 87:13. The final diet contained 13.60% CP; 20.90% NDF and 9.87% Forage-NDF. The inclusion of GAA in the diet in the adaptation phase (1-20 days) did not affect dry matter intake (8.73 vs 8.90 kg/d; $P>0.05$), however it improved FE (0.17 vs 0.20; $P=0.018$). In addition, a trend towards an increase in ADG (1.50 vs 1.76 kg/d; $P=0.067$) was observed with the inclusion of GAA. The inclusion of GAA in the finishing diet (21-101 days) did not affect ($P>0.05$) DMI, ADG, FE and CT. The improvement in FE and the trend towards greater daily weight gain only during the adaptation phase can be attributed to the higher NDF content in the diet during this period, since, according to the literature, GAA has stimulatory effects on cellulolytic microbial growth. The inclusion of GAA in the diet improves FE during the adaptation period, demonstrating the potential of this additive in diets of feedlot cattle.

In feed histidine supplementation improves muscle carnosine content and meat quality in pigs

M. Paniagua[1], B. Saremi[2], B. Matton[2] and S. De Smet[3]

[1]Quimidroga SA, R&D Department Food and Feed, tuset 26, 08006, Spain, [2]CJ Europe GmbH, Research Center, Main Airport Center, Unterschweinstiege 2-14, 60549, Germany, [3]Ghent University, Laboratory for Animal Nutrition and Animal Product Quality, Coupure Links 653, block F, 9000, Belgium; b.matton@cj.net

Nowadays low crude protein (CP) diets in swine are mandatory to improve sustainability by reducing nitrogen emission. Low CP diets, although following the ideal protein concept, reduce the availability of some amino acids such as histidine (His). Thus, carnosine synthesis and meat quality could be compromised. Herein, we test His impact in low CP diets in fattening pigs on meat quality and carnosine content. On a commercial farm, Pietrain genetic pigs (n=2,440) were assigned to control (CON) or HIS (3 kg of BestAmino™ L-His HCl – CJ BIO for the last 12 weeks before slaughter) group. Animals received the same feed formulas based on barley, corn and soybean meal during the growing (CP 14.6%) and finishing (CP 14.1%) phase, where SID His:Lys ratio was 0.37 in CON diet and 0.60 in HIS group. Animals were slaughtered at approximately 130 kg of final BW in 4 consecutive weeks (305 animals per treatment per week). Carcass weight was measured at the slaughterhouse, and samples from the longissimus dorsi (LD) and the semitendinosus (ST) muscles were obtained to analyse the carnosine content and meat quality parameters, such as pH and drip loss. Carcass weight was not affected by treatment (105.5±2.0 and 105.4±1.3 kg for CON and HIS, respectively). Carnosine content was greater ($P<0.05$) in the LD of HIS (307±4 mg / 100 g muscle) compared with CON animals (269±7 mg / 100 g muscle), whereas only a numerical increase was observed in ST of HIS pigs. Additionally, pH at 24 h *post-mortem* was higher ($P<0.01$) in HIS group (5.68±0.02) than in CON pigs (5.58±0.03), and drip loss was lower ($P<0.05$) (2.49±0.19 and 1.96±0.12% for CON and HIS, respectively). The improvement in both parameters might be explained by the higher carnosine content in LD and ST. In conclusion, supplementing L-His in fattening pigs increased muscle carnosine content and improved meat quality, whereas it did not affect carcass weight. Thus, more research is needed to elucidate the effects of low CP diets on meat quality.

Session 73 Poster 16

Impact of functional amino acids on performance parameters in post-weaning piglet

A. Simongiovanni[1], K. Fenske[2], F. Witte[2], H. Westendarp[2] and T. Chalvon-Demersay[1]
[1]METEX Animal Nutrition, 32 Rue Guersant, 75017 Paris, France, [2]Hochschule Osnabrück, Fakultät Agrarwissenschaften und Landschaftsarchitektur, Am Krümpel 31, 49090 Osnabrück, Germany; aude.simongiovanni@metex-noovistago.com

Weaning is a critical period during which piglets are faced with many changes causing stress, with negative effect on feed intake. Commonly, weaned piglets are offered high quality proteins and supplemental amino acids (AA) to balance the diet. Functional AA and polyphenols may provide additional benefits to secure a good start after weaning. Therefore, 192 piglets weaned at 27 days of age were allocated to 16 pens (12 piglets/pen; 50:50 castrated males:females). The diets were offered during 3 post-weaning (PW) phases: from weaning (d0) to d14 PW, from d15 to d28 PW and from d29 to d49 PW. Two dietary treatments were compared: a control commercial diet based on wheat, barley, corn (broken), and soybean meal in varying concentration, or the same control diet supplemented 'on-top' with 0.1% of a mixture of glutamine, arginine, cystine and grape extract polyphenols. Piglets were individually weighted at weaning, d14, d28 and d49 and the average daily gain (ADG) was calculated for the periods between weighting dates. Average daily feed intake (ADFI) and feed conversion ratio (FCR) were calculated per pen for the same periods. No significant differences were found during the first period, though ADG (+12%), ADFI (+8%), and FCR (-4%) were numerically improved for the supplemented group. During phase 2, a significant better growth (+35 g/d ADG; P=0.005) translated in a significant higher body weight (BW) at d14 (+664 g; P=0.022) in the supplemented group. A trend for a better ADFI for the supplemented group can be as well underlined for this period (+29 g/d; P=0.077). If we consider the total period of the trial (d0-49), piglet ADG was significantly improved in the supplemented group (+24 g/d; P=0.026) which may be explained by the trend towards an increased feed intake (+30 g/d ADFI; P=0.068). The final BW at d49 was significantly higher for the supplemented group with +1.2 kg BW more compared to the control group (P=0.026). Collectively, the supplementation with a mixture of functional AA and polyphenols improved piglet performance during the PW phase.

Session 73 Poster 17

Phytobiotics from *Thymus* species in poultry feed – thymol, carvacrol and rosmarinic acid

M. Taghouti
FeedInov CoLab, Integrated production systems, Estação Zootécnica Nacional, Qta da Fonte Boa, Rua Prof. Dr. Vaz Portugal, 2005-424, Portugal; myriam.taghouti@feedinov.com

Phytogenic feed additives (PFAs), also known as phytobiotics are herbal extracts, essential oils (EO) and other plant-derived products that have health promoting potential such as antioxidant, antimicrobial and anti-inflammatory activities. With the increasing interest in using safe alternatives to conventional antibiotics in livestock farming, more focus was given to the in-feed inclusion of PFAs, especially in poultry and swine. Polyphenols are among the molecules that were highly studied and used as feed natural additives such as rosmarinic acid (RA), thymol (TH) and carvacrol (CA). (RA) is a water soluble phenolic acid and is the major component of aqueous and hydro-ethanolic extracts from several *Thymus* species such as *T. vulgaris*, while (TH) and (CA) are monoterpene phenols and are the major compounds in chemotypes of different *Thymus* spp. essential oils. Several studies investigated the *in vivo* effect of inclusion of *Thymus* spp. plants and/or their extracts and EO in the broilers and laying hens diets. General observations confirm the beneficial outcome in animals' welfare, growth performance, meat and egg quality. Described mechanisms behind these improvements are mainly the stimulation of gastrointestinal (GI) beneficial flora, the contribution to a more balanced GI ecosystem by the inhibition of pathogenic microorganisms and modulation of immune response. Effects are dose-dependent which requires an adequate inclusion in the diet taking into consideration the bioavailability of the molecule in the target tissues. Synergy between the different compounds, such as TH and CA relevant synergistic antimicrobial effect, is also an important factor that should be investigated for market-scale formulations. Novel techniques for administration of bioactive molecules, such as encapsulation, enhanced the effectiveness of supplementation resulting in better growth performance, lower feed conversion rates and improved health status. Also, *in ovo* injections are earning more interest especially for the protection from post-hatching infections and early health disorders.

Session 73 Poster 18

Dietary microalgae (*Chlorella vulgaris*) and probiotic on performance and intestine of piglets

M.F. Pedro[1], D.F.P. Carvalho[1], A.M. Almeida[1], R.J.B. Bessa[2], A.J.M. Fonseca[3], A.R.J. Cabrita[3] and J.P.B. Freire[1]
[1]LEAF, TERRA, ISA, Universidade de Lisboa, 1349-017 Lisboa, Portugal, [2]CIISA, AL4AnimalS, FMV, Universidade de Lisboa, 1300-477 Lisboa, Portugal, [3]REQUIMTE, LAQV, ICBAS, Universidade do Porto, 4050-313 Porto, Portugal; mfpedro@isa.ulisboa.pt

Study of alternative feeds can provide solutions that consider both current issues and future requirements, with resource usage and their ecological implications. Feedstuffs often directly compete with human-edibles and are produced with arable land. Microalgae have interesting nutritional and industrial properties but require further study. This work evaluated the effects of *Chlorella vulgaris* dietary inclusion and a commercial probiotic (LBC ME10) on growth performance and intestinal physiology of weaned piglets. 120 piglets (Large White × Landrace × Piétrain), with 7.4±1.0 kg live weight (28 d) were blocked by weight in trios and randomly assigned to one of the experimental diets: C – Control diet (wheat/maize/soybean meal); CV2.5 – 2.5% *Chlorella*; CV5 – 5% *Chlorella*; CV2.5P – 2.5% *Chlorella* and Probiotic; and CV5P – 5% *Chlorella* and Probiotic. Piglets were housed in three per pen, in a total of 40 pens (8 pens per treatment) with *ad libitum* feeding and water. After 3 days of adaptation, piglets were fed for 5 weeks, with liveweight and feed intake determined weekly and faecal scores daily. At week 5, one piglet from each pen was slaughtered and gastrointestinal contents collected. Performance indicators, viscosity and pH of intestinal contents were determined and data analysed by ANOVA. Diets did not affect average daily gain (562±56 g/d), average daily feed intake (706±74 g/d), nor feed conversion ratio (1.27±0.12). Faecal consistency, viscosity of intestinal contents in upper small intestine (3.8±0.8 cP) and ileum (4.9±1.4 cP), as well as pH of stomach (4.3±0.7), upper small intestine (5.7±0.2), ileum (6.6±0.2), cecum (5.7±0.3) and colon (6.1±0.2) contents were similar within each sampling site for all treatments. Overall, results indicate that *C. vulgaris* inclusion in weaned piglet diets up to 5% does not affect growth performance nor faecal consistency, suggesting it may be effectively used in piglet diets with good feed acceptance. Further work is needed to understand effects on digestive physiology.

Session 74 Theatre 1

Effects of linseed oil, a brown seaweed and seaweed extract on methane emissions in beef cattle

E. Roskam[1,2], D.A. Kenny[2,3], V. O'Flaherty[1], M. Hayes[4], A.K. Kelly[3] and S.M. Waters[1,2]
[1]University of Galway, School of Natural Sciences and Ryan Institute, Galway, H91TK33, Ireland, [2]Teagasc Grange, Animal and Bioscience Research Department, Dunsany, Co. Meath, Ireland, [3]University College Dublin, School of Agriculture and Food Science, Dublin, D04 C1P1, Ireland, [4]Teagasc Ashtown, Food Research Centre, Ashtown, Dublin, Ireland; emily.roskam@teagasc.ie

Enteric methane (CH_4) is a by-product of fermentation of feed in ruminants, accounting for two-thirds of Irish agricultural GHGs. Nationally, Ireland are legally bound to reducing agri-GHGs by 25% by 2030, including a 10% reduction in CH_4, thus research into strategies to reduce enteric CH_4, such as anti-methanogenic feed additives particularly those which are naturally derived, has increased. Linseed oil (LSO) is high in PUFAs and can be grown in the European climate. Ascophyllum nodosum (ASC), an indigenous brown seaweed that is found in abundance on Irish coastlines, and a treatment of ASC (TR1) generated to improve palatability and shown to include phlorotannins, were assessed for their ability to reduce CH_4 in beef cattle. Seventy two beef bulls were used in this study. The basal diet was 60:40 (w:w) forage:concentrate. Fresh silage was allocated each morning (09:00), and concentrate supplementation containing the feed additive, twice daily (08:00 and 15:00). Following a 7 d covariate period, animals were offered 1 of 4 dietary treatment groups for a 10 week feeding period; Control (CON), LSO, ASC and TR1 (n=18 per group), included at 0, 4, 2 and 2% of DMI respectively. Gaseous emissions were measured using four GreenFeed systems. A mixed model, including fixed and covariate effects of the treatments and a random effect of block, was applied using SAS 9.4. No effects on DMI or FCR were observed (P>0.05), however TR1 reduced ADG relative to CON (P=0.05). LSO reduced CH_4 g/d and CH_4 g/kg DMI by 18 and 17% respectively (P<0.05). TR1 reduced CH_4 g/d by 7% (P<0.05), while ASC had no effect on CH_4 emissions (P>0.05). Twice daily supplementation of LSO and TR1 are effective strategies to reduce CH_4 production *in vivo*, and have potential for use in pasture-based scenarios with concentrate supplementation at grass.

Session 74 Theatre 2

Effect of whey on methane emission and systemic comparison of *in vivo* and *in vitro* methane formation

H. Luisier-Sutter[1,2], M. Terranova[3], S.L. Amelchanka[3], K. Schweingruber[2], S. Hug[2], K. Müller[2], L. Isele[1], K. Sommer[2] and M. Schick[1,2]

[1]University Hohenheim, Institute for Agricultural Engineering, Garbenstrasse 9, 70539 Stuttgart, Germany, [2]Strickhof, Division Animal Husbandry& Dairy Production, Eschikon 21, 8315 Lindau, Switzerland, [3]ETH Zurich, AgroVet-Strickhof, Eschikon 27, 8315 Lindau, Switzerland; helene.luisier@strickhof.ch

Methane is an important greenhouse gas whose global warming potential is 25-times greater than that of carbon dioxide. Ruminants, through their digestive processes, are one of the main sources of anthropogenic methane emissions worldwide. Thus, adaptations in feeding strategies can inhibit methanogenesis. Whey is a by-product of caseation that is abundantly available and cheap. It is rich in water-soluble carbohydrates and proteins and thus an energy-dense feedstuff interesting for animal production. In a former study by Agroscope (unpublished), supplementing grass-fed heifers with whey reduced methane emissions by 37%. However, the mitigating potential of whey in dairy cows has not yet been examined. Therefore, in this study, the inhibitory potential on methanogenesis of whey from Emmentaler cheese production and whey powder was investigated more closely. Four Original Swiss brown dairy cows were used in a 3×3 cross-over design. All 4 cows received a roughage-based ration consisting of grass-silage, grass pellets, hay and maize-silage and were consecutively supplemented with Emmentaler whey or whey powder. The animals were adapted to each diet for a minimum of 2 weeks and then placed in respiration chambers for 2 days methane emission measurements. After exiting the chambers, rumen fluid was extracted from each cow and incubated *in vitro* using the Hohenheim Gas Test (HGT) method. The diets used for incubation were exact replicates of those fed to the cows previously *in vivo*. Preliminary results did not show an effect on *in vivo* methanogenesis of either Emmentaler whey or whey powder. The *in vitro* methane production behaved similarly. Interestingly, when *in vitro* methane production was correlated to the digestive organic matter, the addition of Emmentaler whey produced an inhibitory effect, but only in two of the four animals. Therefore, our *in vitro* results suggest that the effect of whey is dependent on animal-individual characteristics.

Session 74 Theatre 3

Effects of feeding fresh white clover on digestibility and methane emission in lambs

X. Chen[1], S. Ormston[2], S. Stergiadis[2], K. Theodoridou[3], O. Cristobal-Carballo[1] and T. Yan[1]

[1]Agri-Food and Biosciences Institute, Hillsborough, BT26 6DR, United Kingdom, [2]School of Agriculture, Policy and Development, University of Reading, Reading, RG6 6AR, United Kingdom, [3]Institute of Global Food Security, Queen's University Belfast, Belfast, BT9 5DL, United Kingdom; xianjiang.chen@afbini.gov.uk

This study aimed to investigate the effects of dietary inclusion of white clover on feed intake, dry matter (DM) digestibility and enteric methane (CH_4) emissions in sheep fed zero-grazed herbage. Twelve ewe hoggets (Texel, 12 months old) were allocated in a 3 (diets) × 3 (periods) Latin Square design study with 3 weeks/period. The 3 diets were (DM basis): 100% perennial ryegrass (PRG, Control), 70% PRG and 30% medium leaf size white clover (Chieftain, MLS), and 70% PRG + 30% large leaf size white clover (Barblanca, LLS). In addition, each sheep received concentrate pellets at 158 g/day. All lambs were housed as a single group with 6 auto-feeders with CH_4 and hydrogen emissions measured using a GreenFeed unit during day 1 to 14 of each period. Afterwards, sheep were placed in individual metabolism crates for 7 days, with faeces and urine outputs collected in the final 6 days. Pure grass and pure white clover herbages were harvested daily and chopped to a length of 5 cm, totally mixed (for MLS and LLS) and offered to lambs for *ad libitum* twice a day (5% refusal). In each period, concentrate pellets were offered daily to lambs using a GreenFeed unit (5 visits/d) during day 1 to 14 and as top dressing in 2 equal amounts in the am and pm feeding during day 15 to 21. Results showed dietary treatments (Control, MLS and LLS) had no significant effect on DM intake (1.33, 1.45 and 1.46 kg/day), final body weight (68.5, 68.6 and 68.8 kg) or DM digestibility (790, 788, 783 g/kg). Average daily CH_4 emissions for Control, MLS and LLS treatments were 35.6, 37.6 and 37.0 g/day (SED=0.90, P=0.140), respectively. When expressed per kg of DM intake or body weight, CH_4 emissions were not affected significantly by dietary treatments. However, hydrogen emissions were lower (P<0.001) for Control (56.0 mg/day) than for MLS (68.3 mg/day) and LLS (76.1 mg/day). In conclusion, feeding white clover had no significant effect on feed intake, DM digestibility or CH_4 emissions in sheep.

Algae for reducing methane from dairy cows: what expectation with French local resources?

B. Rouillé[1], H. Marfaing[2], F. Dufreneix[2], S. Point[3], M. Gillier[4], R. Boré[1], J. Jurquet[1] and N. Edouard[5]
[1]Institut de l'Elevage, 149 rue de Bercy, 75595 Paris, France, [2]Centre d'Etude et de Valorisation des Algues, 83 Rue de Pen Lan, 22610 Pleubian, France, [3]Agro Innovation International TIMAC Agro, 18 av. Franklin Roosevelt, 35400 Saint-Malo, France, [4]SAS Les Trinottières, La Fûtais, 49140 Montreuil sur Loir, France, [5]INRAE, Institut Agro PEGASE, 16 Le Clos, 35590 Saint-Gilles, France; benoit.rouille@idele.fr

METH'ALGUES project aims at evaluating the anti-methanogenic capacity of algae available on the French coasts with *in vitro* and *in vivo* trials. From a review of the literature, the species Ascophyllum nodosum, *Chondrus crispus*, *Fucus vesiculosus*, Ulva sp. and *Arthrospira* sp. were identified as candidate to reduce CH_4 emissions thanks to their bromoform and polyphenol concentrations. *In vitro* fermentation tests on rumen juice were carried out on the identified species to select those with the highest methanogenic power and the lowest impact on ruminal fermentations. *F. vesicolosus* and *C. chrispus* were selected to proceed the *in vivo* trials. Indeed, *F. vesiculosus* decreased by almost 100% the production of CH_4 but had an impact on fermentations. *C. crispus* reduced CH_4 production and contain bromoforme. Based on these results, an *in vivo* experiment was carried out with *C. crispus* and *F. vesiculosus* as well as a mix of additives (Roullier Group), compared to a control diet, on the performance of dairy cows and enteric methane emissions. The experiment was conducted in a Latin square design with 20 mid-lactation Holstein cows during four 3-week periods. The basal diet consisted of maize silage, dehydrated alfalfa and concentrates. Enteric methane emissions were recorded with a Greenfeed system. Individual DM intake (25.6±0.3 kg DM/cow/day) and milk production (34.6±0.8 kg/cow/day) did not differ between treatments, suggesting that there were no palatability problems with the addition of algae. There was also no difference in enteric CH_4 emission between treatments (533±4.5 gCH_4/cow/day). However, the inclusion rate was lower than expected due to higher-than-expected DM intake levels. As a complement to the first *in vivo* experiment, a second experiment is ongoing testing higher algae inclusion rates for longer periods with two groups of 28 Holstein cows and will end in April 2023.

Session 74 Theatre 5

Managing the rumen microbiome to reduce methane emissions through dietary interventions

G. Pugh[1], O. Cristobal Carballo[2], T. Yan[2], C. Creevey[1] and S. Huws[1]
[1]Queen's University Belfast, Institute for Global Food Security/ School of Biological Sciences, 19 Chlorine Gardens, Belfast BT9 5DL, United Kingdom, [2]Agri-Food and Biosciences Institute, Large Park, Hillsborough BT26 6DR, United Kingdom; gpugh01@qub.ac.uk

Methane is a natural product resulting from enteric fermentation in the rumen forestomach of ruminants. The majority of CH_4 produced is the rumen derives from the conversion of carbon dioxide and hydrogen. The latter is a consequence of microbial digestion of plant material resulting in formation of volatile fatty acids (VFAs), which are a source of energy for the animal. Increasing concerns over greenhouse gases produced by livestock and their subsequent heating potential has led to an increase in research into mitigating enteric methane from ruminants. Feed additives, such as 3-NOP, have been shown to reduce methane emissions, but apart from 3-NOP, other potential additives have not been studied to a great extent. This study investigated the methane reducing potential of a peroxide-based oxidising inhibitor (OI). *In vitro* studies using a rumen-simulating batch culture system were used to incubate varying inclusion levels of the additives with artificial saliva, rumen fluid and grass silage. Bottles were incubated until the measuring time points (4, 24, 48h). At each time point, pH was measured, gas samples were taken for methane analysis using gas chromatography and liquid samples were analysed for VFAs and ammonia analysis. The OI significantly reduced methane emissions compared with the 0% control when added at 0.375, 0.75, 1.5, 2.25 and 3% (P=0.014). The additive is now under investigation at 2.25 and 3% levels in an animal experiment using 60 ewes. Greenfeed data will be available within the next month and individual animal methane emissions can be analysed. The second animal trial will involve feeding the OI to 70 ewes with twin lambs, to investigate the additive effect on early-life programming of the rumen microbiome. For both animal trials, shotgun sequencing of rumen fluid samples will elucidate the changes occurring at microbiome level and mechanisms of action of this compound. Continuation of laboratory analysis, paired with both animal experiments will offer a novel intervention for methane reduction in ruminants.

Session 74 Theatre 6

Production and enteric methane emissions in grazing dairy cows fed flaxseed-based supplement

M.A. Rahman, K.V. Almeida, D.C. Reyes, A.L. Konopoka, M.A. Arshad and A.F. Brito
University of New Hampshire, Department of Agriculture, Nutrition, and Food Systems, 129 Main St, Durham, NH 03824, USA; mdatikur.rahman@unh.edu

The effects of an extruded flaxseed-based supplement (LinPRO-R) on animal production and enteric methane emissions during the grazing season were studied using a randomized complete block design with 18 multiparous and 2 primiparous Jersey cows (128±52 DIM). Cows grazed a mixed grass-legume pasture (herbage allowance = 15 kg DM/cow daily) overnight and were fed partial total-mixed ration (pTMR) during the day. The pTMR were formulated to comprise (DM basis) 37.5% legume baleage and 62.5% of soybean meal/ground corn-based concentrate. Cows were randomly enrolled to 1 of 2 diets: (1) pasture plus pTMR (control = CTRL) or (2) pasture, pTMR, and 6% LinPRO-R (LIN). Ground corn and soybean meal were replaced with LinPro-R in LIN diet. The experiment lasted 12 weeks with 2-week covariate period and 3 sampling periods at weeks 4, 7, and 10. Two GreenFeed units were used to measure gaseous emissions throughout the study. Individual herbage intake was estimated using Cr_2O_3 and *in vitro* DM digestibility of feeds. Faecal grab samples were collected 8 times over 5 d and ruminal fluid was taken once via stomach tubing in each sampling period. Data were analysed using MIXED procedure of SAS with repeated measures over time. Cows on LIN diet were observed to have a lower herbage intake (7.52 vs 6.92 kg/d; $P<0.01$) compared with CTRL, but pTMR intake tended to be greater with feeding LIN (14.9 vs 14.5 kg/d; $P=0.07$). Intake and apparent total tract digestibilities of nutrients were not affected by treatments. Similarly, treatments had no effects on milk yield (mean=27 kg/d), and milk components. However, MUN concentration was lower ($P<0.001$) in LIN (8.38 mg/dl) than CTRL (11.0 mg/dl). No treatment effects were observed for total VFA concentration (mean=90.8 m*M*), and the molar proportions of acetate, propionate, butyrate, and the acetate:propionate ratio (mean=4.68). Production of CO_2 (mean=10,885 kg/d) and enteric CH_4 (mean=350 g/d), CH_4 yield (mean=15 g/kg of DMI) and CH_4 intensity (mean=10.5 g/kg of ECM) did not differ with feeding CTRL vs LIN. In summary, LinPRO-R fed at 6% diet DM did not affect production and enteric methane emissions in grazing dairy cows.

Session 74 Theatre 7

Effect of feeding brown seaweed and its extract on methane emissions and performance in dairy cattle

K.D. Barnes[1], S. Huws[1], T. Yan[2], X. Chen[2], M. Hayes[3] and K. Theodoridou[1]
[1]Queens University Belfast, School of Biological Sciences, BT9 5DL, United Kingdom, [2]Agri-Food and Biosciences Institute, Large Park, BT26 6DR, United Kingdom, [3]Teagasc Food Research Centre, Ashtown, D15 DY05, Ireland; kbarnes06@qub.ac.uk

Dependent on the species, bioactive components and the dietary inclusion level, seaweed has been shown to reduce enteric methane (CH_4) emissions in ruminants. The study aim was to assess the effect of a brown seaweed species and its extract on dairy cow performance, CH_4 emissions and end-product quality. Hypothesizing that the seaweed extract would elicit a greater response due to the concentrated bioactive components. Fifteen late lactation multiparous barren Holstein-Friesian cows were evaluated in a 3 (diet) × 3 (period) Latin square design experiment with 21 days/period. The 3 groups of cows (5 cows/group) were balanced by parity, bodyweight, and average milk yield. Three total mixed ration diets each contained (dry matter; DM basis) 40% concentrates and, respectively, (1) 60% grass silage (GS) (Control), (2) 56% GS and 4% whole seaweed and (3) 56% GS and 4% seaweed extract. Gas emissions (CH_4 and hydrogen (H_2)) were recorded from day 9-14 of each period via GreenFeed. After day 15, 9 cows (3 cows/treatment) were placed in the metabolism unit individually for 6 days for digestibility measurements. Feed intake, milk production and faecal and urine outputs were recorded, and samples were collected for chemical composition analysis. No significant differences were observed for DM intake ($P=0.067$) and milk yield ($P=0.116$) between and within treatments. Dietary inclusion of seaweed and seaweed extract had no significant effect on CH_4 output in g/day ($P=0.897$), g/kg DMI ($P=0.568$) and g/kg milk ($P=0.240$). The seaweed extract tended to produce lower CH_4 results expressed as g/d and g/kg milk level. No response on H_2 output (g/day) was observed ($P=0.706$). Treatment seaweed samples need to be analysed for chemical composition, polyphenol and phlorotannin content as variations in seaweed quality could have contributed to the lack of response. Additional research needs to be carried out on the bioactive compounds involved and quantity required for consistent and substantial CH_4 reduction in ruminants.
SEASOLUTIONS project. European Union's Horizon2020, under grant agreement No 696356.

Establishing conclusive links between the gastrointestinal microbiota and feed efficiency in cattle

M.M. Dycus, U. Lamichhane, C.B. Welch, K.P. Feldmann, T.D. Pringle, T.R. Callaway and J.M. Lourenco
University of Georgia, Animal and Dairy Science, 425 River Road, 30602, USA; jefao@uga.edu

The gastrointestinal tract of ruminant animals hosts a myriad of microorganisms that can ferment feedstuffs to produce energy for the host animal. The efficiency of this gastrointestinal microbial fermentation can differ between animals and impact animal productivity. Previous findings have indicated that there are associations between the host microbiota and feed efficiency; however, the scope of conclusions drawn are limited due to sample size constraints. Thus, this study was performed to strengthen the understanding regarding the link between the gastrointestinal microbiome and feed efficiency in ruminants, through utilizing a large sample size. Ruminal and faecal samples were collected from 1,480 Angus bulls located at 8 different farms in the United States. Every bull had records of performance and feed intake, which allowed calculation of their individual feed efficiency performance. The collected samples were subjected to extraction/purification and sequencing of microbial DNA in order to determine their microbial composition, evenness, richness, and diversity. As expected, the ruminal and faecal environments had distinct microbial richness, diversity, and presence of specific taxa (e.g. *Fibrobacter* and *Prevotella*). However, despite these substantial differences between the ruminal and faecal environments, both contained taxa that were significantly associated with feed efficiency, indicating that the microbiota of those two environments play important biological roles in the conversion of feedstuffs into usable products for the animal. Furthermore, it suggests that the manipulation of the gastrointestinal microbiota can be a viable approach for improving feed efficiency, and potentially decrease greenhouse gas emissions by cattle.

Variations in bulk milk urea content on dairy farms in Flanders, Belgium in 2019-2021

J. Vandicke[1], K. Goossens[1], Z. Lipkens[2], T. Vanblaere[2] and L. Vandaele[1]
[1]Flanders Research Institute for Agriculture, Fisheries and Food (ILVO), Scheldeweg 68, 9090 Melle, Belgium, [2]Melkcontrolecentrum Vlaanderen vzw (MCC), Hagenbroeksesteenweg 167, 2500 Lier, Belgium; jonas.vandicke@ilvo.vlaanderen.be

Animal production is viewed as one of the main drivers of the nitrogen excess and eutrophication of nature reserves in Flanders, Belgium. Ammonia from manure plays a crucial role and is formed when urea in urine and urease in faeces react. Hence, lowering the urea excretion in urine can reduce ammonia emissions from manure. This can be achieved by reducing the amount of crude protein in the ration of dairy cattle. Feeding low-protein diets could therefore be a valid method for dairy farmers to prove that they are reducing ammonia emissions, if diet composition can be regularly monitored and controlled. However, today, that is not the case. Alternative ways of evaluating the protein content of the ration are therefore needed. Milk urea could be used to this end, since milk urea content is known to be largely driven by the protein content of the ration. Furthermore, milk urea content of every bulk milk collection can be analysed in Flanders by the Milk Control Centre Flanders (MCC), so data can readily be made available for the farmer. In this research, we determined the inter- and intra farm variability in bulk milk urea content on Flemish dairy farms and evaluated its potential as a predictive tool for the protein content of dairy rations. We analysed data from MCC from 1,652,470 milk collections from 3,881 Flemish dairy farms throughout 2019, 2020 and 2021. Mean bulk milk urea content over these three years was 229.6±51.8 mg/l, with higher average values in summer (244.6±53.1 mg/l in August) compared to winter (208.0±51.9 mg/l in December), and brief peaks in milk urea content as a response to heatwaves. However, most of the variation of the milk urea content on individual farm level could not be explained by known variables such as date or temperature. We conclude that milk urea content cannot be used as a standalone predictive tool for the protein content of dairy rations. Further research will determine which additional parameters or analyses are needed to easily monitor the protein content of the ration in a cost-effective, practical and reliable manner.

Session 74 Theatre 10

Impact on environment and performance of the replacement of soybean meal in post-weaning pig diets

E. Royer[1], P. Pluk[2], J. De Laat[2], G. Binnendijk[1], K. Goris[2] and P. Bikker[1]
[1]Wageningen University & Research, Wageningen Livestock Research, P.O. Box 338, 6700 AH Wageningen, the Netherlands, [2]Cargill Animal Nutrition, Global Innovation Center, Veilingweg 23, 5334 LD Velddriel, the Netherlands; eric.royer@wur.nl

Imported soybean meal (SBM) is more widely used in post-weaning than fattening pig feeds. A study tested the hypothesis that a more circular formulation could improve the environmental print of feed but would affect piglet performance and gut health, nutrient digestibility and nitrogen (N) and phosphorus (P) efficiency. 240 weaned pigs (6.5±0.9 kg) were allocated to 6 dietary treatments. A control diet (C) with 14% SBM was compared to diets in which SBM was totally replaced by oilseed meals, i.e. 12% sunflower meal, 12% rapeseed meal and 3% linseed meal (Oil), or legume seeds, i.e. 14% peas, 13% lupines and 13% faba beans (LS), or a mixture of equal amounts of two of these three diets (C-Oil, C-LS, Oil-LS). Diets contained 9.4 to 9.5 MJ NE/kg and 1.12 g dig lys/MJ NE and were given *ad libitum* from 7 d to 40 d after weaning. Performance and faecal consistency were registered. For treatments C, Oil and LS, faeces were collected on d 35 and 36 to determine apparent total tract digestibility (ATTD), and two piglets per pen were slaughtered at d 41 or 42 to collect contents and segments of the digestive tract. The carbon footprint (CFP) was calculated using the Nevedi-GFLI database, including land use. CFP was lower with circular diets (1,011, 778 and 867 g CO_2eq /kg for C, Oil and LS diets, respectively). Dietary treatments had no effect on growth performance and faecal consistency (P>0.05). The LS diet enhanced the weight of empty stomach, full caecum, and full or empty colon. Oil, LS and Oil-LS diets had lower ATTD than C diet for OM and crude protein (P<0.001). But LS diet had lower apparent ileal digestibility than C diet for crude protein (P=0.03). Non-starch polysaccharide ATTD was the highest for LS diet and the lowest for C diet (P<0.001). LS and Oil diets had the highest and the lowest ATTD for P and ash, respectively (P<0.001). N excretion per kg of gain was similar among treatments (P>0.05). In conclusion, the replacement of SBM by oilseed meals or legume seeds can improve the sustainability of post-weaning diets without reduction in piglet performance.

Session 74 Theatre 11

Simulating pig multiperformance in contrasted breeding systems using an individual-based model

E. Janodet[1,2], F. Garcia-Launay[1] and H. Gilbert[2]
[1]INRAE, PEGASE, 35590 Saint-Gilles, France, [2]INRAE, GenPhySE, 31320 Auzeville-Tolosane, France; estelle.janodet@inrae.fr

Feed accounts for more than half of production costs and environmental impacts of the pig industry. The use of a greater diversity of feedstuffs, often more local but less digestible is one lever to mitigate these impacts. However, current pigs value better energy-rich and easy to digest diets. Thus, the multiperformance of different pig profiles in production systems differing for their feeding constraints should be characterized. The objective of this study is to propose a modelling approach to evaluate how different production systems affect the multiperformance of different individual pig profiles. Four contrasted systems were considered: (1) conventional system (CS) representative of standard practices in French pig farms, (2) CS excluding raw materials with high environmental impacts and including low-cost opportunity feedstuffs, (3) CS with local feedstuffs, and (4) an organic system with fodder. A dataset of 1,586 InraPorc® pig profiles was established from a previous trial. Feeds were formulated to cover individual nutritional requirements given the constraints of each system. Two formulation objectives were applied: a least-cost formulation, as commonly used, and a multi-objective formulation that minimizes feed environmental impacts under an economic constraint on feed price. We simulated, in each system and for each formulation objective, responses from these 1,586 pig profiles, using a previously published individual-based model of the pig fattening unit. The model provides animal performance and environmental impacts evaluated through a cradle-to-farm-gate life cycle assessment (LCA): climate change, resource use (fossils, minerals and metals), acidification, marine, freshwater and terrestrial eutrophication and land occupation. Based on these indicators, the 1,586 pigs can be ranked on their technical and environmental performances in each system, to identify which individual profiles fit best in each breeding system, and if G×E occurred. To conclude, combining system and individual level modelling provides a strategy to predict the multiperformance of pigs in a variety of systems, that could later be used to identify individual profiles adapted to each system.

Session 74 Theatre 12

Environmental impacts of substituting soybean with rapeseed or haemoglobin meal in broiler diets

V. Wilke[1,2], J. Gickel[1], A. Abd El-Wahab[2] and C. Visscher[2]
[1]University of Veterinary Medicine Hannover, Foundation, Science and Innovation for Sustainable Poultry Production (WING), Heinestraße 1, 49377 Vechta, Germany, [2]University of Veterinary Medicine Hannover, Foundation, Institute of Animal Nutrition, Bischofsholer Damm 15, 30173 Hannover, Germany; julia.gickel@tiho-hannover.de

As soybean meal (SBM) shows negative impacts on the environment, a partial substitution has a benefit on the environment. In this study, 120 broilers where fed with three different diets (similar energy content) containing only SBM (32.5%) or SBM and an alternative protein source (APS) (14.5% rapeseed meal [RSM] or 4.5% haemoglobin meal [HBM]) from day 8 to 44. The initial and final body weight and feed intake was measured for the trial period. The assessment of the environmental impact of 1 t feed and 1 t carcass weight (according to the ISO 14040) was performed using the online software application Opteinics® (BASF Lampertsheim GmbH, Lampertsheim, Germany), which is based on the Global Feed LCA Institute (GFLI) database. The impact on climate change was 1.9 t CO_2 eq per t feed and was about 20% lower with a partial reduction of SBM by RSM or HSM (1.5 t CO_2 eq per t for both). A lower impact of the diets with APS was also found on eutrophication (marine/freshwater), land use and resource use. The impact of 1 t feed on acidification, eutrophication (terrestrial), particulate matter and water use was higher for the diet with RSM and lower for the diet with HBM, for ozone depletion the impact was higher for both diets with alternative protein sources (compared to SBM). The body weight of the broilers of all diets was similar at day 8 but different at day 44 (SBM: 2.8 kg; RSM: 2.9 kg; HBM: 2.5 kg). The FCR was 1.68 (SBM/HBM) and 1.71 (RSM). The impact on climate change related to carcass weight was 4.9 t per t (excluding day 1-7) and was reduced by 16% with a partial substitution of SBM by RSM or HBM. A lower impact of the meat produced with APS was also found on marine eutrophication, land use, particulate matter and resource use. The impact per t meat on acidification, eutrophication (freshwater/terrestrial) and ozone depletion was higher with an inclusion of RSM or HBM. The impact on water use per t meat was higher using the diet with HBM and lower using the diet with RSM (each compared to SBM).

Session 74 Poster 13

Manipulating dietary degradable protein and starch to increase nitrogen efficiency in dairy cows

P. Piantoni, Y. Roman-Garcia, C. Canale, M. Messman and G. Schroeder
Cargill Animal Nutrition and Health, Innovation Campus, 10383, 165th Avenue Northwest, Elk River, MN 55330, USA; paola_piantoni@cargill.com

The objective of this experiment was to evaluate the effect of decreasing degradable protein (RDP) and increasing starch content as strategies to increase N efficiency (NE). Eighteen multiparous Holstein cows (51±8 kg/d milk yield; 138±78 DIM; mean ± SD) were used in a split-plot design experiment with a randomized complete block on the whole plot and a 3×3 Latin square on the sub-plot. Main plot was dietary starch level (20 and 28% of DM), and sub-plot was RDP level (8.5, 9.75, and 11% of DM). Treatment periods were 28-d long, with the last 7 d used for data collection. The starch content difference was obtained by replacing soyhulls with dry finely ground corn. Dietary CP increased with RDP and ranged from 14.2 to 16.2% of DM. RDP differences were obtained by replacing blood meal and rumen-protected amino acids with urea and soybean meal. All diets covered estimated requirements of essential amino acids. No interaction was observed between main effects for NE. Regardless of starch content, increasing RDP did not affect milk yield but 9.75% RDP decreased intake and increased feed efficiency (both $P_{quad.}<0.01$). Increasing RDP decreased NE (32.4, 31.6, and 28.0%; $P_{quad}<0.01$) but did not affect milk protein yield. Increasing RDP linearly decreased milk fat content (3.65, 3.51, and 3.48%) and yield (1.67, 1.59, and 1.58 kg/d), and increased MUN (9.99, 11.9, and 13.9 mg/dl; all $P<0.01$). Consequently, increasing RDP linearly decreased energy-corrected milk yield (47.1, 46.1, and 45.9 kg/d; $P=0.03$). Regardless of RDP content, increasing dietary starch did not affect intake or energy-corrected milk yield but tended to increase milk yield ($P=0.08$), and increased NE (32.7 vs 28.6%; both $P=0.01$). Increasing starch decreased milk fat content (3.25 vs 3.84%) and MUN (11.3 vs 12.6 mg/dl; both $P=0.02$) but did not affect milk fat yield or protein content and yield. The 23% decrease in RDP increased NE by 16%, and the 40% increase in starch increased NE by 14%. Therefore, decreasing RDP, while balancing for amino acids, was a more effective strategy to increase NE compared with increasing dietary starch content. Results also suggest that combining both strategies increased NE the most.

Session 74 Poster 14

Substitution of soybean meal with canola meal in dairy cow diets: effect on enteric methane emission

C. Benchaar[1] and F. Hassanat[2]

[1]Agriculture and Agri-Food Canada, Sherbrooke R&D, 2000 rue college, Sherbrooke, QC, J1M 1Z3, Canada, [2]Agriculture and Agri-Food Canada, Québec R&D, 2560 Bd Hochelaga, Québec, QC, G1V 2J3, Canada; chaouki.benchaar@agr.gc.ca

A number of studies have shown that replacing solvent-extracted soybean meal (SBM) with solvent-extracted canola meal (CM) in dairy cow diets increases dry matter intake (DMI) and milk production. However, information of such dietary changes on enteric methane (CH_4) emissions are scarce. Accordingly, this study was undertaken to determine the effects of iso-nitrogenous replacement of SBM with CM on CH_4 emissions and intensity and milk production. Sixteen multiparous lactating Holstein cows (DIM=116±23; milk yield=47.5±4.9 kg) were used in a replicated 4×4 Latin square (35-d periods; 14-d adaptation) and fed (*ad libitum*) either a control diet (0%CM) or diet supplemented with 8, 16, or 24% CM, on a dry matter (DM) basis. The forage:concentrate ratio was 52:48 (DM basis) and was similar among the experimental diets. Canola meal was included in the diet at the expense of SBM and soybean hulls, whereas the proportions of the other diet ingredients were kept the same. Methane production (respiration chambers) and milk production were determined over 5 and 7 consecutive days, respectively. Linear and quadratic contrasts (Proc MIXED; SAS 9.4) were used to determine effects of CM levels on variable responses. Significance was declared at $P \leq 0.05$. Dry matter intake increased linearly with increasing CM proportion and the increase was more pronounced for cows fed 24% CM (+1.9 kg/d) compared with cows fed 8% CM (+1.1 kg/d) or 16% CM (+1.5 kg/d) diets. Production of energy-corrected milk (ECM) increased linearly as CM proportion increased in the diet (+1.0, +1.6, and +2.2 kg/d for 8, 16 and 24% CM, respectively). Daily CH_4 emission decreased linearly with increasing CM proportion in the diet (489, 475, 463, and 461 g/d for 0, 8, 16 and 24% CM, respectively). Also, CH_4 yield (% gross energy intake), and CH_4 emission intensity (g CH_4/kg ECM) decreased linearly by up to 14 and 10%, respectively, at the highest CM dietary inclusion level. This study demonstrates that partial or full substitution of SMB with CM can successfully mitigate enteric CH_4 emission and intensity while enhancing animal productivity.

Session 75 Theatre 1

Does better circularity in livestock require a paradigm shift?

A. Mottet[1] and M. Benoit[2]

[1]FAO, Viale delle Terme di Caracalla, 00153 Rome, Italy, [2]INRAE, UMR Herbivores, Herbipôle, 63122 Saint-Genès-Champanelle, France; anne.mottet@fao.org

Livestock contribute to food security by supplying essential macro- and micro-nutrients, providing manure and draught power for agriculture, and generating income for households. But they also consume food edible by humans and graze on pastures that could be used for crop production. Livestock, especially ruminants, are often seen as poor converters of feed into food products. However, about 86% of the dry matter consumed in the global livestock sector are made of materials that are currently not eaten by humans. Livestock therefore play a key role in the bio-economy by converting forages, crop residues and agricultural by-products into high-value products and services. The production of global feed requires 2.5 billion ha of land, which is about half of the world agricultural area. Most of this area, 2 billion ha, is grassland, of which about 1.3 billion ha cannot be converted to cropland (rangeland). This means that 57% of the land used for feed production is not suitable for food production. However, the current global energy scarcity is leading to a sharp increase in its price and indirectly in the price of feed. Therefore, animal production that relies on cereals, pulses and cultivated forage is experiencing a sharp loss of competitiveness, which can lead to increased prices for milk, meat and eggs, a drop in consumption and a loss of incomes for farmers, as the increased price of outputs won't compensate the increase in input prices. To avoid this scenario, we proposed that two consequences seem unavoidable for livestock farming systems: (1) the reduction of arable land dedicated to the production of animal feed; and (2) a switch to feeding strategies based on low opportunity land and raw materials from which livestock production is most likely to benefit, i.e. low-quality resources that are difficult to harvest. This would result in a reduction in animal numbers and a redistribution of livestock in agricultural landscapes, a change in the types and traits of farm animals, an adaptation of supply chains and a rebalancing of diets. Such an evolution of livestock farming should also respond to other major challenges, such as climate change and feeding humanity.

Session 75 Theatre 2

More or better to obtain sustainable food production?

H.F. Olsen[1], H. Møller[2], S. Samsonstuen[1], M.T. Knudsen[3], L. Mogensen[3] and E. Röös[4]
[1]NMBU, Box 5003, 1430 Ås, Norway, [2]NORSUS, Stadion 4, 1671 Kråkerøy, Norway, [3]AU, N. Ringgade 1, 8000 Aarhus, Denmark, [4]SLU, Box 7070, 750 07 Uppsala, Sweden; hanne.fjerdingby@nmbu.no

To ensure future food security, each country should utilize its resources to provide food to society. In current food systems, the focus is on improving the technological and biological production efficiency to produce as much food as possible to cover the increasing demand for food. Due to the risk of depleting nutrients and creating imbalance in the natural ecosystems, it is necessary to discuss the long-term effects of food production strategies. In this study, we explored three different scenarios for livestock production in Norway, which is a country with limited arable land: business as usual (BAU), maximizing the livestock production on domestic feed resources (MaxProd), and best possible utilization of domestic feed resources, including waste and by-products (BestProd). The aim was to discover how the scenarios performed in terms of use of arable land, protein production, environmental impacts including biodiversity, and resource efficiency. The scenarios were discussed with year 2040 as a time frame. MaxProd and BestProd included novel feed ingredients, such as yeast and insects, but did not allow for imported feedstuff as in BAU. An important difference between BestProd and the two others was that BestProd required using all suitable, arable land to produce plants for human consumption to reduce the feed-food competition and that BestProd also prioritized circular food production. The BAU scenario produced the largest amount of animal protein, whereas BestProd produced the lowest. In addition, the BestProd scenario had the highest climate impact due to the high share of ruminants. On the other hand, the BestProd scenario had the lowest land occupation per year but utilized most of the available biomass. Also, the land use ratio indicated that BestProd had the highest level of resource efficiency and the lowest potential loss of biodiversity. In total, we conclude that livestock production using domestically available biomass in best possible way, using livestock for recycling waste and by-products, will produce somewhat less animal protein and give a trade-off for climate change but will overall contribute to a future sustainable food production system.

Session 75 Theatre 3

Satisfying meat demand and avoiding excess manure nitrogen at the regional scale in China

F. Accatino[1], Y. Li[2] and Z. Sun[2]
[1]INRAE, AgroParisTech, Université Paris-Saclay, UMR SADAPT, 22 place de l'agronomie, 91120, France, [2]Institute of Geographic Sciences and Natural Research Research, Chinese Academy of Sciences, Beijing, 100101, China, P.R.; francesco.accatino@inrae.fr

Circularity at the regional scale requires nitrogen (N) excess avoidance through crop-livestock. This could contrast with satisfying local meat demand. We aimed at: (1) classifying 261 eastern China regions according to their ability/inability to satisfy local meat demand and their deficiency/excess of manure N; (2) analysing scenarios of decreased livestock numbers. For 1) we compared meat demand (based on healthy diet recommendations) with meat production (based on livestock abundance and productivity data) and manure production (based of livestock abundance and productivity data) with manure demand (based on crop and nitrogen crop demand data). For (2) we set livestock quantities in each region to some key levels: the level meeting local meat self-sufficiency (scenario A), the level meeting local manure demand (scenario B), the level meeting local meat self-sufficiency in case manure N is not in excess, otherwise manure N excess avoidance is prioritized (scenario C). Concerning (1), our scenario showed that 15% of the analysed regions could meet meat self-sufficiency without manure N excess, however in 76% of the regions meat self-sufficiency was met but while causing manure N excess. Concerning (2) scenario C was the best performing, with 50% of the regions that could meet meat self-sufficiency (according to healthy diet recommendations) while avoiding manure N balance. This scenario achieved, at the level of the whole eastern China a global meat self-sufficiency and avoidance of N excess. Our results suggested that, while a trade-off exists between achieving meet self-sufficiency and avoiding excess of manure N, livestock is in excess in many Chinese regions and, in order to improve the allocation of livestock quantities in the regions, it is a good strategy to prioritise the avoidance of manure N excess. Improving circularity requires solutions for increasing the use of manure nitrogen as substitute of synthetic fertilizer in order to soften the trade-off. Further perspectives should also take into account the self-sufficiency of feed.

Session 75 Theatre 4

Role of livestock in the nutrient and carbon metabolism of the agri-food system of a tropical island

M. Alvanitakis[1], V. Kleinpeter[1], M. Vigne[2], A. Benoist[3] and J. Vayssières[1]
[1]CIRAD, Selmet, 7 chemin de l'IRAT, 97410 Saint-Pierre, Reunion, [2]CIRAD, Selmet, BP 319, 110 Antsirabe, Madagascar, [3]CIRAD, BioWooEB, 40 Chem. Grand Canal, 97490 Saint-Denis, Reunion; manon.alvanitakis@cirad.fr

Increasing nutrient and carbon circularity and use efficiency is a leading topic in the search for more sustainable agri-food-waste systems (AFWS). In this study, we analyse the nitrogen (N), phosphorus (P), and carbon (C) metabolism of the AFWS in Reunion tropical Island to assess the livestock contribution to AFWS circularity, self-sufficiency, and use-efficiency. The livestock sector on Reunion Island relies heavily on imported feeds: 72, 83 and 56% of the N, P, C consumed by animals come from external sources. As a result, livestock is responsible for 36, 44 and 39% of N, P, C imports at the AFWS level. It is also responsible for 17 and 13% of respectively N & C atmospheric losses from the AFWS. On the other hand, livestock is involved in circularity as it plays a recycling supplier role. Among flows of secondary products between all sub-systems, 27, 35 and 17% of N, P and C flows come from livestock. However, this recycling is inefficient. Manure spreading is responsible for 40% of N losses from croplands to the environment. Manure application substitutes only partially the import of mineral fertilizers, which leads to nutrient surpluses at crop level (+90 kg N/ha and +40 kg P/ha) and low nutrients use efficiency of the crop sector (37 and 30% respectively for N and P). Moreover, only 36, 22 and 17% of the N, P and C contained in slaughterhouse and meat cutting wastes are recovered by the waste sector. The rest is burned or sent to landfill. Several livestock-based levers of improvement were identified, including: (1) recycling material burned or sent to landfill as feeds, bedding materials or substrates for manure co-composting; (2) limiting N losses during manure management and spreading; and (3) adjusting feed and fertiliser inputs to livestock and crop needs. Those reorganization of nutrient and carbon flows can potentially reduce the AFWS impacts on the environment. The nutrient and carbon metabolism approach will be helpful to feed a life-cycle assessment of the AFWS with and without improvement levers.

Session 75 Theatre 5

State and regional nitrogen and phosphorus balances to assess manure sheds in New York State

O.F. Godber[1], K. Workman[2] and Q.M. Ketterings[1]
[1]Cornell University, Nutrient Management Spear Program, Department of Animal Science, Ithaca, NY, 14853, USA, [2]Cornell University, PRO-DAIRY, Ithaca, NY, 14853, USA; ofg6@cornell.edu

New York (NY) has a large dairy industry that produces considerable amounts of manure that is applied to cropland. This provides multiple benefits, opportunities, and challenges for attaining a climate resilient, environmentally sustainable, and economically viable agricultural sector. State and regional nitrogen (N) and phosphorus (P) balances illustrate the circularity of the NY agricultural sector, help to identify the feasibility of future improvements to reduce environmental impact, and increase the sustainability of agriculture. Here, we present N and P balances for agricultural cropland in NY, derived from crop and livestock data from the most recent (2017) United States (US) Census of Agricultural survey, and fertilizer sales data from the US Geological Survey. We present an evaluation, at the state, regional and county level, of (1) the most recent (2017) N and P balances, (2) the impact of improved nutrition on N and P excretion rates by dairy cows on N and P balances, and (3) the impact of manure application to legumes on N and P balances. Results show NY balances of 11.7-40.1 kg N/ha, and 2.6-10.8 kg P/ha, depending on the amount of legume acres that receive manure. State, regional and county-based balances reflect that manure is used as a nutrient source, offsetting N and P fertilizer use, and, with a few exceptions, there is sufficient land base to recycle manure.

Session 75 Theatre 6

Circularity in livestock production: from theory to practice

E.M. De Olde[1,2], O. Van Hal[2], A. Groenewoud[2] and I.J.M. De Boer[2]
[1]Wageningen Economic Research, Wageningen University & Research, P.O. Box 29703, 2502 LS The Hague, the Netherlands, [2]Animal Production Systems group, Wageningen University & Research, P.O. Box 338, 6700 AA Wageningen, the Netherlands; evelien.deolde@wur.nl

Our global food system has a substantial impact on the environment through the use of natural resources, emissions to soil, air and water and the contribution to biodiversity loss. To safeguard the health of our (agro)ecosystems and avoid losses and waste, a transition towards a more circular food system has been proposed in science and policy. A recent publication presented five ecological principles for a circular bioeconomy (i.e. safeguard, avoid, prioritize, use and entropy). As such, circularity does not only imply the effective use of resources such as land, but starts with safeguarding the health of (agro)ecosystems to protect its regenerative capacity. This talk will discuss the implications of the concept of circularity and its principles for the livestock sector. For instance, how can the livestock sector contribute to healthy (agro)ecosystems and what is the impact of prioritizing biomass for human needs? The circularity concept and its principles have so far been used mostly theoretical, while (policy) pressure to measure the circularity performance of individual farms is increasing. Translating circularity to the farm level, however, raises many questions such as: What indicators can be used to gain insight into circularity at farm level? What data is available, or needed, to measure circularity? How does circularity at farm level relate to circularity at different spatial and organisational levels? Can indicators for circularity be integrated in environmental impact assessments? We will reflect on these questions using our first findings from a study on evaluating circularity in dairy farming. We will conclude this talk with some reflections on barriers and opportunities in moving towards a more circular and sustainable food system.

Session 75 Theatre 7

The perspective of young farmers on circular agriculture: definition, implementation and barriers

A.G. Hoogstra, H. Geerse, I.J.M. De Boer, M.K. Van Ittersum, A.G.T. Schut, C.J.A.M. Termeer and E.M. De Olde
Wageningen University & Research, Animal production systems, Droevendaalsesteeg 4, 6708 PB Wageningen, the Netherlands; anne.hoogstra@wur.nl

Both in policy and science, the concept of circular agriculture has gained popularity as a way to improve the sustainability of agriculture. In the transition to circular agriculture, farmers and especially young farmers play a key role, however, insight into their perception of circular agriculture is lacking. Therefore, the aim of this study is to examine the perspective of young farmers in the North of the Netherlands with regard to the transition to circular agriculture. We conducted an online survey and gathered insights into their view and definition, interest in circular practices and perceived barriers with regard to circular agriculture. In this study, the focus is on dairy, mixed and arable farmers, who (co)own a farm in the provinces of Friesland, Groningen and Drenthe with a maximum age of 35 years old. In total, there were 53 respondents of which 32 arable, 16 dairy and 5 mixed farmers. The results of the survey show that young farmers vary in their perception of circular agriculture as a solution for agricultural issues in the Netherlands. This variation was also observed in their definition of circular agriculture, although there was a consensus about the themes of the recycling of nutrients in animal manure and stimulating soil vitality. The results concerning the application and interest of farm practices that are associated with circular agriculture, showed that many farmers already apply or are interested in circular farming practices. The most applied practices were precision agriculture and the production of protein crops and the most interest was for low input management of nutrients and chemicals and application of field borders. Multiple barriers were highlighted in the transition towards circular agriculture, mainly (uncertainty) regarding regulations. The insights of this paper provide valuable information for the transition to circular agriculture as it is important to offer a clear perspective for young farmers and stimulate them to adopt circular practices.

Companion modelling approach for collective nitrogen management in a French municipality

G. Martel[1], F. Garcia-Launay[2] and V. Souchère[3]
[1]INRAE, UMR BAGAP, 55 rue Rabelais, 49100 Angers, France, [2]INRAE, UMR PEGASE, 16, le clos, 35590 Saint-Gilles, France, [3]INRAE, UMR SADAPT, Campus Agro Paris Saclay, 91120 Palaiseau, France; gilles.martel@inrae.fr

The specialization of agriculture has led to a disruption of nutrient cycles. Many studies ask for interactions between farms at the territorial level. But the implementation of collective solutions at this level requires the involvement of a diversity of stakeholders who are not all in direct interaction. This is why we have carried out a companion modelling approach to support a municipality to think about the collective management of nitrogen, particularly from effluents. The support approach involved elected officials, advisory and support organizations, associations, cooperatives and agri-food industries and a variety of farmers. From individual interviews and a collective restitution we produced a representation of the actors, resources, dynamics and interactions necessary to talk about nitrogen management in the territory. On this topic, the stakeholders individually identify 38 types of people involved and 53 resources mobilized. Two dynamics stand out: the presence of imported nitrogenous elements in the territory and the decrease in the number of livestock in favour of cultivated surfaces on increasingly large farms. These dynamics make the stakeholders of the territory put forward the solution of the biogas production facility which can convert these inputs into digestate usable on the cultures, even if several fears around the composition of these digestates exist. Nevertheless, with the decrease in the number of animals, farms that do not use all their effluents are rare, limiting the possibilities of direct interactions between farmers and cereal growers on the territory. This work therefore highlights the importance of off-farm actors for nitrogen management at the territorial level, but it also questions the role of farmers in collective management and the resilience of these tools in the event of a decrease in livestock production. These aspects will be the objectives of the role-playing game currently being prepared, which will allow farmers and elected officials to reflect on possible options for better nitrogen management on the territory.

Historical transformation of crop-livestock integration and its drivers in a French region

R. Pedeches[1], C. Aubron[2], S. Bainville[2] and O. Philippon[2]
[1]Université de Montpellier, 163 rue Auguste Broussonet, 34060 Montpellier, France, [2]Institut Agro Montpellier, 2 place Viala, 34060 Montpellier, France; remi.pedeches@laposte.net

The combination of crops and livestock can help reduce the negative impact of agriculture on the environment. Forage crops can be included in the rotation of non-fodder crops and play an important agronomic role, thus reducing the use of chemicals; if animals consume locally produced feed, nutrient flows can be more closed. The decline in the number of mixed crop-livestock farms and the factors behind this decline have been documented in many regions. However, some authors advocate going beyond the criterion of mixed crop-livestock farming and taking into account the functional interactions themselves. In this study, we aim at studying the recent transformations of crop-livestock interactions at the scale of a small agricultural region and at understanding their origin. To trace these interactions and their ecological roles, we had to reconstruct the overall functioning of past and present farms in a region of southwestern France. The required data were collected through 80 interviews to retired and active farmers. In 1955, most of the farms were mixed-crop-livestock farms, self-sufficient in feed and bedding, with forages included in the rotations. From 1960 onwards, the specialization of farms into livestock farms and grain farms isolated crops from livestock, which on the one hand separated forage and non-forage crops in distinct farms and on the other hand created open flows of concentrates between farms and even to beyond the region. However, crop-fodder rotations vanished on many mixed farms: indeed, some of them no longer have ruminants, but only monogastric animals. Besides, in farms with cattle, fodder and cereal crops are often produced on separate lands. We then explain these changes by the evolution of prices, policies, the development of irrigation infrastructures, and the soil and climate characteristics of the area.

Session 75 Theatre 10

Livestock: option or necessity? Changes of energy flows in an Indian village, 1950-2022

C. Hemingway[1], C. Aubron[1] and M. Vigne[2]

[1]L'Institut Agro Montpellier, SELMET Research Unit, 2 Place Pierre Vialla, 34060 Montpellier, France, [2]CIRAD, SELMET Research Unit, Campus international de Baillarguet ou Avenue Agropolis, 34398 Montpellier Cedex 5, France; charlotte.hemingway@supagro.fr

Since the 1950's, agriculture in India has undergone many 'revolutions'. In Anantapur, a dryland area in Southern India, the Yellow, Green and White revolutions focused on boosting the production of groundnut, rice and milk respectively. Those revolutions have not impacted farms in the same way but the role of livestock has changed substantially, modifying the circularity of energy at different levels. We combine the comparative agriculture and the territorial metabolism approaches to study the energy transition of the agriculture of a village in Southern India from the 1950's to today. For both time steps we look at changes of energy flows at farm and territorial levels, assessing the size and nature of the flows of energy (human and animal labour, gross energy of biomass and fossil energies) entering, circulating and exiting the farm and the territory. We rely on a cycling index to estimate the proportion of total internal flows among the total flows involved, and on a fossil energy consumption index to assess the dependency of farms on non-renewable external resources. Circularity of energy within farms has declined but has been reshaped at the scale of the territory. While energy flows are at both periods driven to a large extent by livestock, the nature of circularity has been modified by the use of fossil fuels. In the 1950's circularity and livestock was a sine qua non for agricultural production: livestock served for ploughing, lifting water for irrigation, producing food and manure. Today flows of energy are an outlet of agricultural production: livestock is no longer a necessity but rather an option for farms aiming to specialise in either milk or meat production. In the current configuration, livestock is fed partly on co-products thus fuelling the circularity of the territory. But as the cultivation of crops requires fuel and electricity, the current circularity depends on 'pulses' of external fossil energy to be maintained. We therefore show that livestock-based circularity is not always an indicator of sustainable land-use.

Session 75 Poster 11

Developing livestock-based circularities for healthy and sustainable territories: the CLiMiT project

T.T.S. Siqueira[1], J.-M. Sadaillan[1], M. Vigne[2], J. Veyssières[1], A. Benoist[3] and M. Miralles-Bruneau[1]

[1]Cirad – Mediterranean and Tropical Livestock Systems Unit (SELMET), Environments and Societies, 7 chemin de l'IRAT, 97410, Reunion, [2]Cirad – Mediterranean and Tropical Livestock Systems Unit (SELMET), Environments and Societies, Antsirabe 110, FIFAMANOR, 319, Madagascar, [3]Cirad- Biomass, Wood, Energy, Bioproducts (UPR BioWooEB), Environments and Societies, 40, chemin de Grand Canal, 97443 Saint Denis Cedex 9, Reunion; siqueira@cirad.fr

The implementation of livestock-based circularity strategies contributes to creating healthy and sustainable territories thanks to the: (1) reduction of environmental impacts through synergies in nutrient and energy flows, soil carbon storage in grasslands, organic matter recycling, the transformation of non-edible biomass in food, etc.; and (2) socio-economic development and food security by the creation of wealth and jobs. Despite the recognition of the place of livestock in sustainable development, actors seem to be sparsely or poorly equipped with tools to identify potential and to co-design circular economy projects adapted to their territories. The main scientific objective of the CLiMiT project is to develop and test a toolbox to support actors in the construction of livestock-based circularity projects considering territorial specificities and actors' preferences. Our project adopts a research-intervention approach with a transformative goal. We place the circularity paradigm as the primary vector of territorial development compatible with a reduction in greenhouse gas emissions. We propose an interdisciplinary and comprehensive 3-step approach: (1) characterizing the socio-ecological functioning of territories to better understand them; (2) modelling the different biomass and nutrient flows to assess the environmental and socio-economic impacts of the territories; (3) co-design with local stakeholders technical and organizational solutions for the implementation of circular economy projects including livestock farming.

Co-design of a scenario of biomass valorisation within a circular approach on Reunion Island

R. Youssouf[1,2], E. Cavillot[2], A.-L. Payet[2], T.T. Da Silva Siqueira[1] and J.-P. Choisis[3]
[1]CIRAD, 7, chemin IRAT, 97410, Saint-Pierre, France, [2]ILEVA, 9, chemin Joli Fond, 97410, Saint-Pierre, France, [3]INRAE, 7, chemin IRAT, 97410, Saint-Pierre, France; siqueira@cirad.fr

In outermost islands the current increase of imported agricultural input costs questions the viability of farms and of the whole agricultural sector. The need, for these territories, to be less dependent and to mobilize local resources in a circular way is rising. Nevertheless, technical and organizational lock-in makes changes difficult to achieve. From 2017 to 2020, a Research and Development project has been caried out on Reunion Island to implement a territorial approach of biomass in agriculture based on circular economy. Five case-studies were undertaken with development, agricultural training and research actors and policy makers for the co-conception of organizational biomass scenarios. This project has created a strong momentum and the co-construction of solutions. However, as in many prospective works with actors, the implementation – which is then delegated to stakeholders – is rarely applied. To overcome this pitfall, we built a new project with some of the previous actors, on the territory of a municipality, starting from the insights of two case-studies (the levers to lower manure spreading constraints and the evaluation of the interest of composting manure with green waste). After a first diagnostic phase of the territory, we carried out two kinds of actions. We first organized focus-groups with local actors to conceive an agroecological transition for the territory. Second, we set up on-farm experiments with 5 farmers to develop composting of manure mixed with green waste and to test the compost on meadows and market gardening. The transition scenario confirms the need to improve biomass management to reduce inputs, jointly with the transformation of systems toward more agroecological ones. The development of farm composts along with their use on crops and meadows were successful. But, if the project allows the production of references and learnings, we observed within the project period various dynamics of actors, more or less competing, for treating and valorising manure resources on the territory.

Barriers for farmers to 'sustainabilize' their farm: the example of nematode control management

M. Sautier[1], A. Somera[2], R. Rostellato[1,3], P. Ly[1], Y. Labrunne[1], J.M. Astruc[3] and P. Jacquiet[4]
[1]Université de Toulouse, INRAE, ENVT, GenPhySE, Chemin de Borderouge, 31320 Castanet Tolosan, France, [2]Centre Départemental de l'Elevage Ovin, Quartier Ahetzia, 64130, Ordiarp, France, [3]IDELE, Chemin de Borderouge, 31320 Castanet Tolosan, France, [4]Ecole Nationale Vétérinaire de Toulouse, UMT Santé des Petits Ruminants, UMR INRAE/ENVT 1225 IHAP, 23 Chem. des Capelles, 31300 Toulouse, France; marion.sautier@inrae.fr

Current health management practices in livestock farming are not sustainable, mostly because they select pathogens resistant to treatments. Integrated health management is put forward as a way to manage trade-offs between production, animal health and the environment. But, it embeds in so few livestock farming practices that farmers do not recognize and advocate it per se. In this context, research and development is needed: (1) to identify and design sustainable livestock systems and management tools in line with integrated health management principles; and (2) to better understand adoption and innovation dynamics towards integrated health management. This presentation contributes to the latter. The aim of our study was to explore how knowledge and information circulate among farmers, and between farmers and non-farmer stakeholders around the theme of parasitism control. For this purpose, we carried out a questionnaire-based survey among 536 dairy-sheep farmers in southwestern France. We analysed knowledge networks for parasitism control by listing whom farmers talk to when dealing with parasitism control. We identified the kind of individuals likely to be contacted by farmers depending on the farming system and the farmers' representations. Results are discussed in terms of implications for developing integrated health management programs that take into account the diversity of health management actors and farmers identities.

Session 75 Poster 14

Benefits and limits of an organic agroforestry system associating rabbits and apple trees

D. Savietto, V. Fillon, S. Simon, E. Lhoste, M. Grillot, A. Dufils, L. Lamothe, F. Derbez, M. Fetiveau and S. Drusch
INRAE, GenPhySE, UERI, LISIS, AGIR, Ecodéveloppement, 24 Chemin de Borde Rouge, 31326, France; davi.savietto@inrae.fr

Intensive farming and animal production are facing a crisis of legitimacy due to the use of pesticides, antibiotics and the lack of respect for animal needs. In contrast, organic agriculture and agroforestry systems are models contributing to a more sustainable agriculture. Organic farming bans the use of synthetic inputs, and agroforestry combining trees and animals provides various ecosystem services, among which nutrient cycling. However, those systems are not well developed nor studied. To fill this gap, we designed an organic agroforestry system to raise a small herbivore – the rabbit – in an apple orchard (AO) and evaluated the benefits and limits of this association for both plants (weeding, fertilisation and sanitation) and animals (food, microclimate, protection against predation and welfare). We started by designing prototypes relevant to this association that were improved following the proposals of farmers made during a workshop. We obtained a functional system: mobile-pens (18 m^2 for 6 rabbits) placed within the apple tree rows after leaf fall. We then tested this association with 144 rabbits divided in two groups (72 in the AO and 72 in a grassland (GL) with no trees), resulting in three combinations: rabbits with apple trees, rabbits without apple trees, and apple trees without rabbits. Compared to GL, rabbits in the AO had a more propitious microclimate (thermal amplitude: 11.6 vs 13.1 °C), ingested a similar amount of grass (39 vs 36 g of dry matter/rabbit/day) and pelleted feed (147 vs 154 g/rabbit/day) and grew alike (33 vs 32 g/day). Moreover, rabbits in the AO ingested 58 g of apple/rabbit/day. There was no predation, but one rabbit died of myxomatosis at the AO. Besides the weeding service (bare ground in one week), rabbits gnawed 35 out of 215 trees, despite the presence of physical protection. This association worked well in the autumn: little orchard management, plenty of food resources and mild weather for rabbits. Data on soil fertility and rabbit hair cortisol are under analysis. Globally AO and GL improved the welfare of rabbits, as we observed behaviours (grazing, hopping, digging, etc.) not expressed in caged systems.

Session 75 Poster 15

INOSYS livestock farming systems network: benchmarks for advisors, trainers and policy makers

P. Sarzeaud[1], J. Seegers[1], O. Dupire[2] and T. Charroin[1]
[1]French Livestock Institute, 149 Rue de Bercy, 75595 Paris Cedex 12, France, [2]Chambre d'Agriculture de France, 9 Av. George V, 75008 Paris, France; patrick.sarzeaud@idele.fr

The INOSYS – French Livestock Farming Systems Network has been established during the 80s and illustrates an innovative and long term AKIS cooperation to construct technical and economic benchmarks on farming systems. This multi actor organisation is leaded by the IDELE and the Chambers of Agriculture. Its originality relies on the collaboration of farmers and advisors to design farming systems from the farm monitoring. It makes it possible to produce technical and economic benchmarks adapted to the systems diversity and regional contexts. In 2023, this Network hold 1100 farms monitored by more than 200 chamber- of-agriculture agents and 25 project leaders from the IDELE. It is financed and supported by public authorities and the professional agriculture bodies. Its action can be resumed in three missions: (1) observe the livestock farming systems in place in the regions; (2) identify and support innovative systems; and (3) transfer and disseminate the productions in the form of tools, methods, training and publications. Based on an operational classification, this panel is representative of the systems variety and informs about the main issues they are facing, such as sustainability, competitiveness, changes in working conditions or environmental impacts. In the context of global and competitive economy, the French Livestock Institute proposed a national method and its informatic application COUPROD, to calculate costs of production. INOSYS network is also involved in providing references to several decision tools such as CAP2er (environmental evaluation) or APIBOV (beef production monitoring). The regional and national enhancement of the Network are dedicated to different audiences: farmers, advisers, teachers (for advice or training), individuals or collectives, and local and national decision makers to measure the impact of new farming policies and lead their implementation. To enlarge the dissemination of markers and references, INOSYS network built a new web site. This platform enables visitors to consult and extract updated references covering a wide range of domain (technical, environmental, economic, social, etc.) for more than 200 farming systems models.

Session 75 Poster 16

APIVALE scientific consortium: integrated approach for organic effluent recycling and valorisation

F. De Quelen[1], E. Jarde[2], C. Le Marechal[3], T. Lendormi[4], S. Menasseri[5] and F. Beline[6]
[1]INRAE Institut Agro UMR PEGASE, Saint Gilles, 35590, France, [2]CNRS Université de Rennes UMR Geosciences, Rennes, 35000, France, [3]ANSES, Laboratoire de Ploufragan-Plouzané-Niort, Ploufragan. 22440, France, [4]IRDL Université Bretagne Sud, Lorient, 56100, France, [5]INRAE InstitutAgro UMR SAS, Rennes, 35000, France, [6]INRAE UR OPAALE, Rennes, 35000, France; francine.dequelen@inrae.fr

About 400 million tons of organic waste is produced each year in France with a major contribution of animal production (300 million tons) and agroindustry (45 million tons). Agriculture is thus at the heart of organic waste recycling and valorisation (organic matter, energy, nutrients, etc.). This challenge requires the production of scientific knowledge, the development of technical or organizational innovation and a more holistic approach to better consider the possible synergies on the territories. Organic effluents are subjected to numerous biological, chemical and physical processes that modify their composition, generate emissions to the environment and finally affect the availability of nutrient to plants and soil fertility. An improved knowledge of these different processes is thus required to quantify more precisely the emissions (for environmental evaluation) as well as for their reduction (for mitigation strategies). Different research institutes located in Western France including INRAE, ANSES, CNRS, UBS, Institut Agro, Université de Rennes and ENSCR, have decided to share their skills, experimental facilities and equipment in a scientific consortium named 'APIVALE' in order to develop an integrated approach of organic effluent recycling and valorisation. The scientific consortium provides skills and facilities to perform integrated studies over the whole chain of production and valorisation of organic effluent, possibly in combination with other sources of organic waste (urban or agro-industrial).

Session 75 Poster 17

Progress made in whole-farm nitrogen and phosphorus mass balances on New York dairy farms

O.F. Godber[1], K. Workman[2] and Q.M. Ketterings[1]
[1]Cornell University, Nutrient Management Spear Program, Department of Animal Science, Ithaca, NY 14850, USA, [2]Cornell University, PRO-DAIRY, Ithaca, NY 14853, USA; ofg6@cornell.edu

The whole farm nutrient mass balance (NMB) is an assessment tool that farms can use to calculate their nitrogen (N), phosphorus (P) and potassium (K) use efficiency at the farm level. By calculating the difference in the amount of nutrients imported into and exported out of the farm in a given calendar year, the amount of nutrients remaining on the farm or lost to the environment can be estimated per unit of farmland (hectares), and per unit of output produced (a product of the farm, for example kg of milk). The balance per hectare indicates how well a farm is putting nutrients to use on that farm's land base, and the risk of losing nutrients to the environment. The balance per unit of output indicates how efficiently nutrients are being used to produce that output, and the environmental footprint of that product in terms of nutrient use. A substantial positive P and K balance per hectare indicates soil build-up and potential for nutrient loss over time. A large portion of positive N balances will be lost to the environment as N is more difficult to retain from one year to the next. Some of these losses will be nitrous oxide (N_2O), a potent greenhouse gas (GHG). Here, we show the progress made in reducing N and P balances on dairy farms in New York (NY) State between 2005 and 2020. The data indicate great improvement in P balances per unit of milk produced as well as an improvement in balance per hectare over this time period, despite an increase in animal density. The data also indicate improvements in N balances per unit of milk produced but an increase in balance per hectare due to higher animal densities in 2020 compared to 2005. Findings from a subset of 10 farms with three years of data suggest that strategies that reduce N balances also have the potential to reduce GHGs at the whole farm and product level. Work is ongoing to identify opportunities for improvements in N management.

Measuring greenhouse gas emission from pretreatment and liquid composting storage in biogas facility

G.W. Park, M. Ataallahi, N. Eska and K.H. Park
Kangwon national university, Animal life sciences, Kangwondo, Chuncheonsi, Kangwondaehakgil 1, Dept of animal life sciences 1st building Room#420, Kang, 24343, Korea, South; eskanugrahaeningtyas@gmail.com

Growing number of biogas plants in Korea has brought new assignments for making the new country-specific emissions factor to be included in national greenhouse gas inventory to quantify, monitor, and reduce methane (CH_4) and nitrous oxide (N_2O) gas emissions. The objective of this research focused on comparing the emissions between site demonstration data and the 2006 IPCC guideline. Therefore, gas measurement was the most important for quantifying the greenhouse gas emissions from biogas facility. This facility has treated 70 ton of swine manure from 20,000 heads and 30 ton of food waste per day and generate 3,000 m^3 /day of biogas. In the biogas plant, swine manure and food waste were stored in pretreatment with agitation, and moving into anaerobic digester. From anaerobic digester the biogases such CH_4, CO_2, etc. were moving into biogas storage and providing gas to generator, and treated manure and food waste were moving into liquid composting storage. This research measured CH_4 and N_2O from pretreatment storage which is included slurry, and liquid composting storage. All gases in this biogas facility were vented by negative pressure from the odour treatment tower, except the generator and its blowers. Quantification of the total greenhouse gas emissions from the biogas facility was conducted on 24 hours every 2 weeks, from July to October, by using CH_4/N_2O analyser (Los Gatos Research, San Jose, CA, USA). The air inlet sample was collected from air injector to aerate for the liquid compost and external air. The air outlet sample was collected from odour treatment tower before treated by acid and base. Field demonstrated data for CH_4 and N_2O emission in biogas facility were 9,112.08±3,179.57 kg CH_4/year and 91.07±63.50 kg N_2O/year. But if methane and nitrous oxide emission were calculated by 2006 IPCC guideline Tier 2, it was 48,200 kg CH_4/year and 2,016.70 kg N_2O/year, which means respectively 5.29, 22.14 times higher than field emission data. The result concluded that the country-specific emission factor of biogas plant is necessary to produce more accurate national greenhouse gas inventory as a tool for policy-making for CH_4 and N_2O reduction.

Effect of yoghurt acid whey on quality characteristics of corn silage

I. Palamidi, V.V. Paraskeuas, I.P. Brouklogiannis, E.C. Anagnostopoulos, I. Politis, I. Hadjigeorgiou and K.C. Mountzouris
Agricultural University of Athens, Department of Animal Science, Iera odos 75, 11855 Athens, Greece; inpalamidi@aua.gr

Yogurt acid whey, a by-product of stained yogurt production, presents a serious environmental problem if left untreated, as for every 100 litres of milk used, 70 litres of yogurt acid whey are produced. Given that yogurt acid whey is rich in lactose and lactic acid, its use as a silage additive merits investigation. The aim of this study was to determine the effect of yogurt acid whey powder (YAWP) addition on nutritive, microbial and fermentation characteristics of corn silage. For this purpose, YAWP was added at 0, 2.5, 5 and 10% w/w to corn grass, in a total final weight of 7kg, which was then sieved into appropriate vacuum bags. All bags were then stored at ambient temperature (18±3 °C) for 10 weeks. Representative samples were taken at the beginning and end of ensilation and stored at -30 °C until analysed. At the end of the ensiling, dry matter, crude protein and ash were linearly increased, whereas crude fibre, neutral detergent fibre and acid detergent fibre were linearly decreased with YAWP increasing level. The pH was quadratically decreased with increasing YAWP level, with the 5% YAWP being the lowest. Concentrations of total aerobes were linearly decreased, whereas *Clostridial* counts did not differ. Acetic acid content was linearly decreased with increasing YAWP incorporation level. With increasing YAWP level, propionic acid, lactic acid and the ratio of lactic to acetic acid were increased linearly. Butyric acid content was only detected in 0 and 2.5% YAWP. Ammonia content decreased linearly by YAWP incorporation level. In conclusion, the incorporation of 5 and 10% YAWP to silage preparation improved the nutritive and fermentative characteristics of corn silage. This research has been co-financed by the European Regional Development Fund of the European Union and Greek national funds through the Operational Program Competitiveness, Entrepreneurship and Innovation, under the call RESEARCH – CREATE – INNOVATE (project code:T2EDK-00783).

How do horses perceive human emotions?

P. Jardat[1], C. Parias[1], F. Reigner[2], L. Calandreau[1] and L. Lansade[1]
[1]CNRS, IFCE, INRAE, Université de Tours, PRC, l'Orfrasière, 37380 Nouzilly, France, [2]UEPAO, INRAE, l'Orfrasière, 37380 Nouzilly, France; plotine.jardat@inrae.fr

Recently, unexpected sociocognitive abilities toward humans have been discovered in domestic mammals. We will focus on horses and present recent findings on their perception of human emotions. Studies have shown that horses distinguish human facial expressions as well as vocalizations of joy and anger, and that they can recognize these emotions cross-modally; i.e. they can associate a human vocalization of joy or anger to the right facial expression. In a recent experiment, to broaden the spectrum of emotions studied, we focused on human expressions of sadness and joy. We conducted a cross-modal experiment in which horses were presented with two simultaneous soundless videos of a sad and a joyful face, while sad or joyful vocalizations were played. The results suggest that horses differentiated between the facial and vocal expressions of sadness and joy, and that they may also recognize these emotions cross-modally. Moreover, horses looked less at the sad faces than the joyful faces and seemed less aroused by sad voices than joyful faces. We conclude that emotional contagion may take place between humans and horses when the former express sadness or joy, and that it could be beneficial for horsemen and horsewomen to express joy when interacting with horses. In addition to the visual and auditory channels, we can wonder whether horses can perceive human emotions *via* the olfactory channel. We implemented a habituation-discrimination protocol to examine horses' ability to discriminate between human odours produced in fear vs joy contexts. Sweat odours collected on human participants watching a horror movie or comedy were presented to horses. First, odour A (from the fear or joy context) was presented twice in successive trials (habituation); then, the same odour A and a novel odour B (from the same human in the other context) were presented simultaneously (discrimination). Results show that horses discriminated between the two odours and may have separate emotional processing for each of them. Overall, these findings highlight horses' sensitivity to various human emotions via different sensory channels, which can affect the interactions between horses and their owners, riders or caretakers.

Session 76 Theatre 2

Sensorimotor empathy in horse-human interactions: a new way to understand inter-species performance

M. Leblanc, B. Huet and J. Saury
Nantes Université, MIP, UR 4334, 25bis boulevard Guy Mollet, 44322 Nantes, France; marine.leblanc@etu.univ-nantes.fr

Inter-species performances are complex. They require a mutual understanding which highly implies the actors' body. To perform, the *écuyers*, French elite riders from Cadre noir of Saumur, and the horses both develop a disposition: sensorimotor empathy. This allows them to connect and form complex synergies. The study shows how humans and horses manage to synchronise to achieve high-level collective performances. We conducted this study with an enactivist epistemological approach, relying on the theoretical framework of the Course of action research program, which, from a phenomenological perspective, gives full place to experience. To analyse the activity of two *écuyers* who are experts in work in hand, we filmed 36 training sessions with the horses, and we confronted the *écuyers* with the videos by conducting self-confrontation interviews in order to reconstruct their course of experience. We also kept a field diary to collect ethnographic notes. To analyse the activity of the horses, we used behavioural indicators from scientific ethology in order to infer their experience. The results show that emergence of moments of mutual connection between the *écuyer* and the horse is required, to synchronise and achieve high-performance synergy. The moments of connection are characterised by pleasant sensations for the *écuyer*, which result of an intersubjective agreement between him and the horse. This latter is built in the context of a long-term relationship during which they construct a common cultural practice. It allows them to understand each other and to realise efficient synergies. Through the construction of their common cultural practice, *écuyer* and horse develop a sensorimotor empathy which is a disposition to feel, to perceive and understand the other through the body and adjust finely. The study of sensorimotor empathy helped to understand the development of a common cultural practice between the actors. This allowed them to bring together the appropriate conditions and exploit them to form complex synergies. More globally, the study of sensorimotor empathy in human-non-human interactions seems to be fruitful for understanding how two species that do not share the same 'world' manage to understand each other finely.

Session 76 Theatre 3

Rethinking horse work

R. Evans
Hogskulen for gronn utvikling, Arne Garborgsveg 22, 44340 Bryne, Norway; rhys@hgut.no

This presentation will challenge us to rethink what we mean by 'Horse Work'. It will touch historical milestones which contributed to the traditional view that Working Horses pull heavy things. It will then examine the changing uses of horses in peoples lives and suggest some situations which equally qualify as equine work. Based upon this, it will examine some consequences of these changes for how we see, use, and heal working horses in the 21st Century.

Session 76 Theatre 4

Humans and equines: shared working conditions

V. Deneux - Le Barh[1,2] and C. Lourd[3]
[1]INRAE, UMR Innovation, 2 place Pierre Viala, 34000 Montpellier, France, [2]Institut Français du cheval et de l'équitation, Pôle Développement Innovation Recherches, Avenue de l'école nationale d'équitation, 49111 Saumur, France, [3]Institut Français du cheval et de l'équitation, Equiressources, Le Haras du Pin, 61310 Le Pin au Haras, France; vanina.deneux@inrae.fr

In a context of generalisation of the watchword 'well-being': well-being and personal development, well-being at work, animal welfare, etc. Contemporary society, weighs the weight of an injunction to well-being and welfare and the equine sector is not excluded from it. In the context of work relations, the anthropo-equine professions are all the more conducive to these reflections as they associate two different species: the human and the equine. We hypothesise that there is a link between human and equine working conditions and thus seek to show that the organisation and working conditions can be understood in a multi-factorial manner, when they are positive they are a source of fulfilment whereas they can be deleterious when they are bad. To this end, we propose to explore this questioning based on the results of two doctoral research projects, carried out between 2017 and 2021. The first one, in sociology of work, studied the life and work relationships between professionals and their animals in the different worlds of the horse. For this, the discourse of 108 semi-directive interviews was analysed. The second, in education sciences, focused on understanding the construction of professional careers in the equestrian and horse-riding sectors. This was based on the analysis of three types of sources: 1,117 curricula vitae, 226 surveys and 12 interviews. Although they come from different disciplinary fields, we obtain common results. The communication we propose is the presentation of our three main results. Firstly, we will show that the understanding of the role of organisation and working conditions on the mental and physical health of human and equine individuals depends on the individuals themselves and the geographical and economic environments. Secondly, we will show how poor working conditions have deleterious effects on both humans and equines. We will conclude with a discussion of ways of work organisation that are conducive to the fulfilment of both humans and equines.

Session 76 Theatre 5

Working horses: humanities and social sciences approach

V. Deneux - Le Barh[1,2]

[1]*INRAE, UMR Innovation, 2 place Pierre Viala, 34000 Montpellier, France,* [2]*Institut Français du cheval et de l'équitation, Pôle Développement Innovation Recherches, Route de l'école nationale d'équitation, 49111 Saumur, France; vanina.deneux@inrae.fr*

Wherever we live, whatever our culture, we are linked to animals. There are no societies without animals, we are hybrid societies (Midgley, Lestel). In Western societies, we have been producing goods and services with domestic animals for several millennia, i.e. we are working with animals. However, the relationship of animals to work has been poorly studied. In the 19th century, the disciplinary construction of the life sciences and the humanities and social sciences established a distinction whereby animals would fall within the scope of the first, while work, which is by definition human, would be dealt with by the second. However, over the last fifteen years, we have seen the humanities and social sciences reintegrate animals into their studies and there is a desire to do sociology with animals. The Animal's Lab research group led by J. Porcher in Montpellier analyses the relationships between animals and work. Numerous fields have been conducted on dogs, elephants, horses, performing animals, and farm animals. Based on the theoretical frameworks from the labour sciences and field investigations, Animal's Lab researchers have theorised animal work, i.e. the subjective engagement of animals in work with humans. This engagement is not natural; it requires that the individuals of the species involved mobilise their senses, their intelligence and their experiences. This is learned and this requires the construction of a common language. My proposal of communication is part of this new disciplinary field. Using the case of the horse, I will present the definition and the eight key points of the theory of animal work. I will show the particularities of professions employing humans and horses compared to professions employing only humans. I will conclude the presentation by discussing the deontological and ethical issues involved in working with horses in terms of the living and working conditions of the equines.

Session 76 Theatre 6

Web based dissemination of research to support human knowledge and understanding for horse welfare

A.-L. Holgersson, K. Lagerlund and G. Gröndahl

Swedish University of Agricultural Sciences, Equine Sciences, P.O. Box 7046, 75007 Uppsala, Sweden; anna-lena.holgersson@slu.se

Internet today is overflowing with information and it is not easy for horse owners to find evidence-based, reliable content, based on research and proven experience. The website HästSverige ('HorseSweden'; www.hastsverige.se) is a free-to-use knowledge site in Sweden that communicates independent and objective basic knowledge as well as research and development in the field of horses. All texts are reviewed by researchers or experts from universities or well-known institutes. Through this, HästSverige offers both beginners and qualified professionals access to quality-assured, up-to-date and fact-based knowledge about horses and horse activities. The goal is to raise the level of knowledge among the hundreds of thousands of people who are in daily contact with horses in Sweden. In this way, HästSverige can contribute to better horse care and increased animal welfare. The Swedish University of Agricultural Sciences is responsible for the site and it is financed and operated in collaboration between several organizations representing researchers, university teachers, the Swedish horse sector, authorities, research financiers, sports, breeding and animal welfare. The main collaborators are the National Veterinary Institute and the Swedish Horse Industry Foundation. There are no commercial ads, branded content or integrated advertising on the site. Since its launch in 2011, over 800 pages have been published on the site. The main structure is: The horse, Diseases and injuries, The horse's environment and Horses and people. There is a special section for short basic courses and quizzes. The free program for calculating feed rations is the most visited page on the site. It uses individual variables for the horse and available feed, or default values. Business-related content, such as legislation, management and advice, is available to actors in the equestrian industry. The site has been well received with about 400,000 visitors and 600,000 sessions in 2022 and about 15,500 followers of the attached Facebook page. HästSverige has become an established and valued source of horse-related factual information, which is used by the general public as well as schools, media and veterinarians.

Session 76 Theatre 7

Comparison of ethological & physiological indicators in headshakers & control horses in riding tests

L.M. Stange[1], T. Wilder[1], D. Siebler[1], J. Krieter[1] and I. Czycholl[2]
[1]Institute for Animal Breeding and Husbandry, Ohlshausenstraße 40, 24118 Kiel, Germany, [2]Department of Veterinary and Animal Sciences, Grønnegårdsvej 15, 106 91 Frederiksberg, Denmark; lstange@tierzucht.uni-kiel.de

Equine headshaking syndrome (EHS) is associated with severe headshaking in absence of external stimulus, which affects horses' welfare, handling and rideability. Since EHS mainly occurs under rider, the aim of this study was to investigate different effects in standardised riding tests. It was hypothesised that riding would influence occurrence of EHS and that headshaker would express more signs of discomfort. Therefore, riding tests (n=227) were carried out on 16 headshakers and 12 corresponding control horses in order to compare physiological parameters, rein tension and ethological indicators between the two groups. Statistical analysis was performed using separate generalised linear mixed models for heart rate, rein tension as well as behaviour. In particular, the study tested for the presence of headshaking (PoHS), gait and other fixed effects depending on the model. For the study, 19 geldings and 9 mares were available, of which 19 horses were Warmbloods and nine were ponies. The average age of the headshakers was 12.1 years (control group: 14.3 years). The heart rate dataset revealed that PoHS has no significant impact on the heart rate; whereas, it is significantly influenced by gait ($P \leq 0.05$). The rein tension dataset indicated that this is not influenced by PoHS, but a meaningful influence was caused by the equipment used (saddle: $P \leq 0.05$) and environmental influences (wind: $P \leq 0.05$). The results of the behavioural parameters indicate that there is no difference between headshakers and control horses in terms of defensive reactions. Wearing a nose cover significantly reduced headshaking symptoms (head flicking: $P \leq 0.05$). In summary, no connection between the heart rate as a physiological parameter as well as rein tension and PoHS could be proven. Thus, it can be concluded that riding has a negligible effect on the EHS. Using a nose cover leads to a reduction in certain headshaking movements. This study is a first step to investigate the influence of riding on horses affected by EHS and, in general, contributes to a better understanding of idiopathic EHS.

Session 76 Theatre 8

Relationship between hay botanical diversity and faecal bacterial diversity in horses

C. Omphalius[1], P. Grimm[1], V. Milojevic[2] and S. Julliand[1]
[1]Lab To Field, 26 bd Dr Petitjean, 21000 Dijon, France, [2]Sandgrueb-Stiftung, Sandgrueb, 8132 Egg b. Zürich, Switzerland; samy.julliand@lab-to-field.com

Greater biodiversity in ecosystems leads to greater stability. Horses grazing natural grassland with higher plant variety show a higher hindgut bacterial diversity, as compared to horses in captivity with access to poorer grazing resources. Yet if such relationship between forage and hindgut bacterial diversity is also valid for horses that only have access to hay, remains unknown. Thus, the objective of this study was to evaluate the relationship between hay botanical diversity and equine faecal microbial diversity and activity. Sixteen adult horses were included in a 2×2 Latin Square design to test 2 iso-DM, -energy, -NDF, and -protein diets during two 21 d experimental periods. The high diversity diet (HD) was composed of 100% hay from a natural meadow containing 29 different identified species, and the low diversity diet (LD) of 86% timothy and 14% alfalfa hay. Faeces was collected on d21 of each period. Bacterial composition was determined using 16S rRNA sequencing. Richness (observed OTUs), evenness (Pielou), and diversity (Shannon and inverse Simpson) indices were calculated from faecal samples. Numbers of OTUs were also categorized into 5 abundance classes (>1% of relative abundance, 0.1-1%, 0.01-0.1%, 0.005-0.01%, or <0.005%). Microbial activity was assessed through measurement of volatile fatty acids (VFA) concentrations and pH. A MIXED procedure was used to test the effect of the diet, and a Chi^2 test was performed to evaluate whether the proportion of each abundance class differed between diets (SAS). Microbial activity was not altered by hay botanical diversity, as faecal VFA concentrations and pH were similar with the 2 diets. Bacterial richness did not differ between diets ($P=0.164$), but the evenness and diversity indices were lower when horses were fed the HD diet ($P \leq 0.027$). The lower evenness observed with HD diet was due to the increased number of rare OTUs (between 0.005-0.01%, and <0.005% of relative abundance; $P<0.001$) which might reflect the provision of more diverse substrates to the ecosystem. It is unknown whether this confers the horse a health benefit, and it would be interesting to study if this provides the hindgut microbiota a better resilience to stress.

Thoroughbreds in equine assisted services programmes: selection, living and working conditions

C. Neveux[1], S. Mullan[2], J. Hockenhull[1], J. Barker[3], K. Allen[1] and M. Valenchon[4]
[1]University of Bristol, Bristol Veterinary School, Langford, BS40 5DU, United Kingdom, [2]University College Dublin, School of Veterinary Science, Belfield, Dublin 4, Ireland, [3]Racing to Relate, 82 St John Street, London, EC1M AJN, United Kingdom, [4]Centre INRAE Val de Loire, UMR Physiologie de la Reproduction et des Comportements (PRC-INRAE, CNRS, IFCE, Université de Tours), INRAE, 37380 Nouzilly, France; claire.neveux@bristol.ac.uk

Post-racing Thoroughbreds (TB) are often retrained for sport or leisure purposes and are increasingly being considered for equine assisted services (EAS) but have not yet been widely evaluated. To evaluate the selection criteria of EAS horses as well as their current living and working conditions, we distributed a detailed online survey aimed at EAS practitioners. In total, 129 people responded from 13 different countries. Participants provided detailed information about the 5 horses they use the most and we obtained such data on 427 EAS equids: 13.5% of TB, 77.3% of various other breeds (OB), 8.5% with unknown origins and 0.7% of donkeys. Our sample was mainly composed of geldings (TB=78.6%, OB=59.6%). The TB were significantly younger than OB (median age: TB=14yo, OB=17yo; $P<0.01$) even if the age when acquired did not differ significantly (median age: TB=OB=9yo, $P>0.05$). The three most commonly selected criteria as being important for EAS horses were: showing a good temperament, absence of kicking/biting propensity, enjoying the work. Participants declared that most of the horses participated in less than 7 EAS sessions a week (TB=97.6%, OB=80.7%). Over two third of the horses were housed in collective settings (TB=71.9%, OB=84.4%), had unrestricted access to forage (TB=77.2%, OB=67.3%) and had social contacts the majority of the time (TB=60.8%, OB=75.8%). These results move towards an understanding of the selection criteria as well as the living and working conditions of EAS horses. Some differences were highlighted between TB and OB and would need to be further investigated. The next phase of the project will compare EAS horses and TB characteristics and will involve qualitative interviews, horse personality and reactivity to human tests as well as a 12-month monitoring period.

The role of gender in the worldwide Pura Raza Español equestrian sector

M. Ripolles, A. Encina, M. Valera and M.J. Sánchez-Guerrero
Universidad de Sevilla, Animal Production, Carretera de Utrera KM, 41013, Spain; anaencmar@gmail.com

One of the least studied factors in the equestrian sector, and never carried out in a global breed, is the relevance of the role of gender in owners, riders and morphological appraisers in the Pura Raza Español (PRE) equestrian sector worldwide. The objective of this study was to evaluate the role of gender in the Pura Raza Español equestrian sector, considering: (1) the owner genders of the studs, (2) the different morphological scores and the use of the hole scale according to the appraiser gender and (3) the differences in Dressage scores according to gender of the rider. This study was carried out in 8,180 PRE studs (7,002 were run by men and 1,178 by women); 36,673 PRE appraised for its morphology (23,095 appraised by a man and 13,578 by a woman); 3,499 Dressage riders with different number of participations (1,633 were men and 1,866 were women). A descriptive analysis and comparison of means was carried out to describe the role of gender in this sector. Women run the 15.09, 12.03 and 14.69% in big, medium and small studs respectively. In the European area (leaving Spain out due to this large number of studs) women own the 68.68% of the PRE studs, while Spain, Central and South America and North America had a large predominance of men owners. In percentages, men in general use extremes more, although not in a significant way. Trot and hock lateral view were not influenced by the appraiser gender. The scale is fully used by both genders with the exception of the Ewe Neck. Men had an average of 28 participations and an active period of 3.66±4.62 years and women have an average of 19.96 participations and 2.67±3.48 years. The maximum active period has been 17 years in both genders. The Dressage scores obtained by men and women had not significant differences with the only exception of Submission. It had a significantly higher value for men. So, it could be concluded that women haven underrepresentation as a horse breeder in the Latin countries and acting as morphological appraisers. This is not the case when women role as Dressage rider. So, it is important to create women's networks to address their specific needs to advance in their leadership and increase their public visibility in achieving social and economic change especially in Latin countries.

Session 76 Theatre 11

Anthelmintic activity of chicory (*Cichorium intybus*) in grazing horses

J. Malsa[1], G. Sallé[1], L. Wimel[2], J. Auclair-Ronzaud[2], B. Dumont[3], L. Boudesocque-Delaye[4], F. Reigner[5], F. Guégnard[1], A. Chereau[1], D. Serreau[1] and G. Fleurance[3,6]
[1]INRAE, Université de Tours, UMR 1282 Infectiologie et Santé Publique, Nouzilly, 37380, France, [2]IFCE, Plateau technique de Chamberet, Chamberet, 39170, France, [3]INRAE, Université Clermont Auvergne, VetAgro Sup, UMR 1213 Herbivores, Saint-Genès-Champanelle, 63122, France, [4]Université de Tours, Eq. 7502 Synthèse et Isolement de Molécules Bioactives, Tours, 37020, France, [5]INRAE, EUPAO, Nouzilly, 37380, France, [6]IFCE, Pôle développement, innovation et recherche, Saint-Genès-Champanelle, 63122, France; joshua.malsa@inrae.fr

Cyathostomins are the most prevalent parasitic nematodes of grazing horses. They are responsible for colic or diarrhoea in their hosts. After several decades of exposure to synthetic anthelmintics, they have evolved to become resistant to most compounds. In addition, the drug associated environmental side-effects question their use in the field. To face these challenges, alternative control strategies, like bioactive forages, are needed. Among these, chicory (*Cichorium intybus,* Puna II variety) is known to harbour anthelmintic compounds. We aimed to test its *in vivo* efficacy against cyathostomins. Two groups of 2-year-old infected saddle horses (n=10) grazed rotationally either: (1) a pasture sown with chicory; or (2) a mesophile grassland at the same stocking rate (2.4 LU/ha). The study took place in summer during 45 days to prevent horse reinfection. Chicory group horses mostly grazed chicory (89% of the bites) while horses of the control group grazed mainly grasses (73%) and 4% *Plantago lanceolata*, a plant with known anthelmintic properties in small ruminants. Cyathostomins egg excretion decreased in both groups throughout the experiment (P=0.002; average EPG falling from 2,169±802 EPG to 508±450 EPG). Accounting for this trajectory, the chicory regime was however associated with a significant 93% drop in parasite egg excretion compared to the control group, i.e. similar to the efficacy of a chemical anthelmintic drug. In addition, larval development was significantly reduced in horses grazing chicory (-39.5% between D16 and D45, P=0.02). *In vitro* tests are in progress to analyse the role of terpenes. The grazing of chicory (var. Puna II) by horses could thus constitute a promising strategy to reduce the use of synthetic anthelmintics.

Session 76 Poster 12

Does pain condition influence the human-horse relationship?

L. Sobrero, M.G. Riva, M. Minero, A. Cafiso, A. Gazzonis and E. Dalla Costa
Università degli Studi di Milano, Dipartimento di Medicina Veterinaria e Scienze Animali (DIVAS), Via dell'Università, 6, 26900 Lodi, Italy; emanuela.dallacosta@unimi.it

Today horses are mainly considered as companions for sports and leisure activities. It is well known that being used for recreational purposes or sport competitions can predispose horses to health disorders causing pain. Pain condition can play a role in the development of behavioural problems, such as escape attempts and aggressions, that can be dangerous for both horses and riders. The aim of this study was to investigate the relationship between the presence of signs of pain and the human-horse relationship measured with 3 tests. A total of 137 adult sport and leisure horses (14±6 years), housed in 7 stables, were assessed by two trained assessors. No horses were reported to have major health issues at the moment of the evaluation. Pain was evaluated using the Horse Grimace Scale, while the quality of the H-H relationship was assessed using three tests: Avoidance Distance (AD), Voluntary Human Approach (VAA) and Forced Human approach (FHA) as described in the AWIN welfare protocol. Data were analysed with descriptive analysis and a Chi-Square test (X^2). On the assessed sample, the proportion of horses that showed signs of pain (HGS≥3) was 30.7%. During the VAA test, the proportion of horses without pain (HGS<3) approaching an unknown human was significantly higher than those of horses with pain (HGS≥3), 60.7 and 23.1%, respectively (X^2=6.234; P=0.044), meaning that horses with no pain seek and enjoyed interacting with humans. No differences were found for AD and FHA tests. The results of this study confirm that pain is often under recognised in sport and leisure horses, thus leading to a possible deterioration of the human-horse relationship. Specifically, our results show that, when in pain, horses do tent to reduce positive voluntary interactions to unknown humans. Further research is suggested to investigate whether an effective pain recognition and treatment pain can also re-establish a good human-horse relationship.

Session 76 Poster 13

Saliva steroidome and metabolome in mare during anoestrus, oestrus cycle and gestation

S. Beauclercq[1], C. Douet[2], A. Piano[3], L. Haddad[3], F. Reigner[4], P. Liere[3], L. Nadal-Desbarats[5] and G. Goudet[2]
[1]INRAE, BOA, 37380 Nouzilly, France, [2]INRAE, PRC, 37380 Nouzilly, France, [3]INSERM Université Paris Saclay, U1195, 94276 Kremlin Bicêtre, France, [4]INRAE, PAO, 37380 Nouzilly, France, [5]INSERM Université de Tours, UMR-1253 iBrain, 37000 Tours, France; ghylene.goudet@inrae.fr

Precision livestock farming using omics approach such as metabolomic and steroidomic to acquire precise and real-time data can help farmers in individual animal management and decision making. Moreover, saliva collection is a non-invasive, painless and easy sampling method. Thus, this prospective study proposes a metabolomic and steroidomic analysis in mare saliva during reproductive stages, in order to identify salivary biomarkers to detect their reproductive stage in a welfare friendly production system. Saliva samples from 6 mares were collected in anoestrus, in the follicular phase 3, 2 and 1 day before ovulation and the day when ovulation was detected, in the luteal phase 6 days after ovulation and in gestation 18 days after ovulation and insemination. Metabolome and steroidome analysis were performed by 1H-nuclear magnetic resonance spectroscopy and gas chromatography coupled to tandem mass spectrometry, respectively. We identified 59 metabolites and 25 steroids in saliva. The salivary concentrations of metabolites were significantly different between the anoestrus stage and another stage (n=11 metabolites or group of metabolites), between follicular phase and gestation (n=1), between the day of ovulation and the luteal phase (n=1), during the 4 days until ovulation (n=4). The salivary concentration of pregnenolone during gestation was significantly higher than during anoestrus or follicular phase and tended to be higher than during luteal phase. Most of the 5alpha-reduced metabolites of progesterone showed higher salivary concentrations during the luteal phase and gestation compared to anoestrus and follicular phase. These metabolites and steroids could be potential salivary biomarkers of the reproductive stage of the mare. They could allow to easily detect their reproductive stage for real-time decision making at the individual animal level. Further studies with a greater number of animals are in progress to confirm the reliability of these candidate biomarkers.

Session 77 Theatre 1

The future of genetic selection in pigs and poultry

L. Verschuren[1] and P.W. Knap[2]
[1]Topigs Norsvin, P.O. Box 86, 5268 ZH Helvoirt, the Netherlands, [2]Genus-PIC, Ratsteich 31, 24837 Schleswig, Germany; pieter.knap@genusplc.com

Over the past decades, the efficiency of pig and poultry production strongly increased due to improved technology for animal management, nutrition and breeding. In general, when a system becomes more efficient, it also becomes more sensitive to external and internal disturbance factors. This leads to trade-offs between animal production efficiency (with the carbon footprint as an important component) and animal welfare (with animal health as an important component). Such trade-offs must be neutralized, and this has implications for animal science: the breeding sector needs better technology to: (1) record hard-to-measure traits in the fields of greenhouse gas emission (mainly via nitrogen excretion, in monogastrics), disease resilience, physical robustness, and several types of behaviour (all of these at low cost on very large numbers of animals); (2) process that data into information that can be converted to meaningful breeding value estimates; (3) exploit metabolomics to generate biomarkers for too-hard-to-routinely-measure traits (e.g. disease tolerance); and (4) exploit genomics to generate fully annotated genome sequences with massive discovery of causal variants, to make the EBVs of point (2) more robust and more accurate. We describe a few trade-off systems that became apparent in pig production in the 1970s (carcass lean content vs meat quality), in the 1980s (growth rate vs leg soundness), and in the 2000s (litter size vs piglet survival), and the strategies that were employed to neutralize them. Coming back to the R&D needs of above, we also describe: (1) technology that is currently being developed to record physical robustness (leg soundness) and behaviour (tail biting, and maternal behaviour) in pigs; (2) a case where proper data processing can exploit routinely recorded data to quantify a resilience trait (the volatility of longitudinal individual feed intake); and (3) cases where metabolomics may play a significant role (tail biting, and boar taint).

Session 77 Theatre 2

The future of nutrition research in pigs and poultry

S. Vigors[1] and J.V. Milgen[2]
[1]School of Agriculture and Food Science, University College Dublin, Belfield, Dublin 4, Ireland, [2]INRA, Institut Agro, Pegase, Le Clos, 35590 Saint-Gilles, France; staffordvigors1@ucd.ie

The pig and poultry industries have substantially improved productivity through improvements in nutrition, health and genetics. However, due to environmental, economic and social pressures there is further need for evaluation and improvement which will coincide with an increased demand to feed a global population of 9 billion in 2050. The goal therefore is to produce sustainable food from sustainable feed (and feeding) and ensure minimal trade-offs with the environmental and health aspects of production. Trade-offs means that there is not a 'win-win', but a 'win-lose' dilemma, and the decision where to win and where to lose is made by different stakeholders (e.g. legislators, producers, consumers). Nutrition is a key component of the economic feasibility with over 70% of total costs of production related to feed. An example of a trade-off between animal health and production is the ban on the use of antibiotics and ZnO. This is a 'win-win' for public health and the environment, but a 'lose-lose' for animal (gut) health and production. The environmental impact predominantly relates to nitrogen and phosphorus excretion with strategies such as reduced crude protein, increased use of synthetic amino acids and optimization of phytase rates requiring further industry integration. The future industry will require a focus on precision animal nutrition, validation of alternative feed ingredients (insects, algae, microalgae, seaweed etc), optimization of current enzymatic use and an increased usage of food waste and co-products. This precision nutrition will focus on the interaction of nutrition with animal physiology, microbiology, immunology (immunometabolism), host genetics (nutrigenomics). The tools that we need should quantify the 'win' and the 'lose' of the dilemma, so that each stakeholder can make a well-informed decision. We are now in an era that large-scale non-invasive animal monitoring is feasible and research should focus on how these tools can be used to assess health, production, and welfare. Also, for complex issues such as gut health, targeted experimental research will be required to identify the mechanisms behind promising (nutritional) levers that ensure gut health.

Session 77 Theatre 3

Economically sustainable and animal welfare friendly animal production: what is the role of research

A. Silvera[1] and H.A.M. Spoolder[2]
[1]Svensk Fågel, Sustainability dept, Sweden, [2]Wageningen Livestock Research, P.O. Box 338, 6700 AH Wageningen, the Netherlands; anna.silvera@svenskfagel.se

The European Commission is about to publish legislative proposals to further improve the welfare of livestock. The proposals will be based on extensive scientific opinions from EFSA on the welfare requirements of various livestock species, and a number of impact assessments carried out to determine the likely consequences of any proposed new legislation. There are several areas in pig and poultry husbandry which may be addressed. For pigs it is very probable that there will be proposals on housing lactating sows and their piglets. A ban on farrowing crates will greatly improve the behavioural freedom of the sow, but it may also have implications for e.g. the cost of production, the welfare of the piglets, labour requirements and perhaps even emissions to the environment. Similarly, a ban on surgical castration of piglets also has its trade-offs: entire males may cause welfare problems when they are older, but will immune-castration be accepted by consumers? For broilers a reduction in growth rate together with lower stocking densities, are highly debated in the European Union. Focus from EFSA has been purely on animal welfare, but the interaction between genotype and environmental factors such as management and nutrition as well as the effects of a balanced breeding programme are lacking in the discussion. To make the issue even more complex and relevant in practice, environmental and economical sustainability must be added to the debate. In this 'challenge session' we will invite you to look at the trade-offs that are associated with these proposed changes, and ask you what the role of science can be in providing solutions to the deal with them.

Session 77 Theatre 4

Group discussion

K. Nilsson[1] and S. Millet[2]
[1]Swedish University of Agricultural Sciences, Department of Animal Breeding and Genetics, Box 7023, 75007, Sweden, [2]ILVO, Scheldeweg 68, 9090 Melle, Belgium; sam.millet@ilvo.vlaanderen.be

---

Session 78 Theatre 1

Farmed insects to create a circular bio-economy in the food and feed industry

A. Vilcinskas
Justus Liebig University of Giessen, Institute for Insect Biotechnology, Heinrich-Buff-Ring 26-32, 35392 Gießen, Germany; andreas.vilcinskas@agrar.uni-giessen.de

Farmed insects, such as the black soldier fly *Hermetia illucens*, are considered as a missing link for the circular bio-economy in the food and feed sector because they can mediate the industrial bioconversion of agricultural or industrial side streams into feed for aquaculture and livestock. In turn, the leftovers of farmed insects, the so-called frass (excrements, chitinous exuvia of the moulting larvae and feed leftovers), represents a valuable biofertilizer, which can replace chemical fertilizer in agriculture and vertical farming. The ability of *H. illucens* larvae both to utilize almost all organic substrates, even liquid manure, as a diet and to produce a potent biofertilizer has been attributed to their beneficial microbes in the gut, which can also mediate the detoxification of plant-derived secondary metabolites. The presentation highlights both the characterization and dynamics of the *H. illucens* gut microbiota and it prominent role for the circular bio-economy. The avenues for revenues of industrial insect farming can be expanded beyond the production of food, feed and bio-fertilizer to encompass the development of higher-added value products from the generated protein, lipid and chitin fractions. Taken together, farmed insects can help to create a sustainable and circular bio-economy in the food and feed industry.

Session 78 Theatre 2

Amino acid requirements of mealworm and black soldier fly larvae

T. Spranghers[1], A. Moradei[2] and M. Boudrez[1]
[1]VIVES University of Applied Sciences, Centre of expertise for agro- and biotechnology, Wilgenstraat 32, 8800 Roeselare, Belgium, [2]University of Milan, Department of Veterinary Medicine and Animal Sciences (DIVAS), Via dell'Università 6, 26900 Lodi, Italy; thomas.spranghers@vives.be

In order to achieve optimal growth and health of farm animals, a different amount/ratio of essential amino acids in the feed is necessary in each phase of life. Knowledge about which amino acids are essential and in which quantities/ proportions they are best administered has already greatly advanced the pig and poultry sector and is still being generated today. For the still young, rapidly developing insect sector, this knowledge could mean a major step forward towards rearing optimization. Therefore, in this project the needs of the 2 most commercially reared insect species, namely mealworms (*Tenebrio molitor*) and black soldier fly larvae (*Hermetia illucens*, BSF), are investigated. The focus is on needs for lysine, methionine, threonine, phenylalanine and tryptophan based on indications from the literature and the needs of pigs. Semi-artificial diets based on at least 25% wheat bran (mealworm) and chicken feed/ water (30/70) (BSF) (basic feed important for structure and certain micronutrients) supplemented with sugar and synthetic amino acids (a mixture of 14: the 10 essential for mammals + glutamic acid, glycine, alanine and aspartic acid) were tested. The diets were isoenergetic and isonitrogenous. Of the 5 amino acids studied, one was administered in different doses per experiment. The non-essential glutamic acid was used as a substitute at lower doses. Growth, as measured by weight gain, and survival were observed. For lysine, from a content of 0.35 g/100 g on, no extra mealworm growth was noticeable. For methionine this was from 0.13 g/100 g on. Each time more threonine was given, a significant difference could be noticed and the minimum dose appears to be 0.42 g/100 g. With tryptophan there were no differences between the different concentrations. At 0.06 g/100 g the maximum growth was already reached. Ultimately, a report will be drawn up with the formulation of standards and guidelines, with special attention to the translation of the research into practice with attention to sustainability.

Session 78 Theatre 3

Evaluation of the suitability of hemp production side-streams for the rearing of edible insects

A. Kolorizos[1], G. Baliota[1], C. Adamaki-Sotiraki[1], I. Malikentzos[2], C.I. Rumbos[1] and C.G. Athanassiou[1]
[1]University of Thessaly, Lab of Entomology and Agrucultural Zoology, Phytokou str., 38446, Volos, Greece, [2]Kannabio – Hemp Hellas, 110 Skoufa Str., 38334, Volos, Greece; crumbos@uth.gr

In recent years, hemp cultivation has been considerably increased across EU, as hemp can be exploited for a number of applications of the textile and the pharmaceutical industry, as well as of the food and feed sector. During hemp production, several side-streams are produced, but these materials remain largely untapped. Insects can serve as efficient bioconverters of agricultural side-streams contributing to the upcycling of these residual resources to high-value end-products (i.e. insect meal and oil). However, studies on the suitability of hemp side-streams for insect rearing are limited. Therefore, the aim of the present study was to evaluate the suitability of four by-products of hemp production as insect feed for the larvae of the yellow mealworm, *Tenebrio molitor*, the lesser mealworm, *Alphitobius diaperinus*, the superworm, *Zophobas morio*, and the black soldier fly, *Hermetia illucens*. In this regard, we studied three side-streams of the hemp seed production *(Cannabis sativa* L. var. Fedora19), i.e. cold-pressed hempseed press cake, hemp plant biomass leftovers on the field (hemp stalks, leaves and stems) and the by-product of hemp seed cleaning process (small and broken hempseeds), and one side-stream of the hemp flower bud production *(C. sativa* var. Futura75; grinded buds and leave remains, secondary stems, hempseeds). In a series of laboratory trials, early-instar larvae were fed on each of the by-products tested and larval growth parameters were recorded. Based on the results, larval growth and performance varied depending on the insect species and the by-product tested. In general, the hempseed press cake could efficiently support the larval growth and development of all species tested, as in most cases a similar to the control growth rate was recorded when larvae were fed on this by-product. These results aim to integrate two relatively new and innovative agricultural activities, i.e. insect farming and hemp production, into a sustainable, based on circular economy, farming system. This research is supported by the EU-PRIMA program project ADVAGROMED (Prima 2021 – Section 2).

Session 78 Theatre 4

Dietary fat sources impact black soldier fly larvae performance

R. Zheng, S. Karanjit and A. Hosseini
Protix, Van Konijnenburgweg 86, 4612 PL Bergen op Zoom, the Netherlands; ruilongzheng@126.com

The use of Black soldier fly larvae (BSFL, *Hermetia illucens*) as a sustainable protein and lipid source for animal feed has gained significant momentum in recent years. Nutrition is a critical component of BSFL production, with dietary fat (F) and carbohydrates (C) serving as essential energy sources. However, the optimal balance of F and C in BSFL diets is not well understood. This study aimed to address this knowledge gap by investigating the influence of different dietary F sources on BSFL rearing. We used chemically semi-defined diets containing three different F sources: coconut oil, milk-based butter, and avocado oil. We took an integrative experimental approach to evaluate the impact of these F sources on key performance indicators in BSFL production, including bioconversion efficiency (BE), yield, and survival. Our findings revealed that the source of F influenced BSF larvae performance. Avocado oil enhanced dry yield, while coconut oil and butter were not effective when combined with high C levels. However, all F sources showed potential as replacements for C in BSFL diets without negatively affecting yield or BE when used in moderation. Our study also found that low C levels negatively impacted larval survival. Furthermore, increased levels of both C and F resulted in lower BE. Overall, this study provides new insights into the impact of different F sources on BSFL rearing and highlights potential strategies for optimizing BSFL diets commercial production.

Session 78 Theatre 5

Hermetia illucens **production parameters and the effect of dietary energy source**

E.Y. Gleeson and E. Pieterse
Stellenbosch University, Animal Sciences, Mike de Vries Building, Merriman ave, Stellenbosch Central, 7600, South Africa; 19539037@sun.ac.za

Hermetia illucens is arguably the most extensively researched insect species in agriculture and food sciences. The larva of this species is a great candidate for the valorisation of pre-consumer food waste. The larval nutritional composition is significantly affected by a number of factors. The first factor that was known to have an influence is age. However, recent studies have identified other factors such as rearing substrate composition and environmental conditions. Based on previous work on the *H. Illucens* larval nutritional requirements, the goal of formulating nutritionally balanced larval diets aimed at tailoring the larval composition to its intended use is more attainable than ever before. Although larval composition is of great importance in production, certain production parameters also have a significant influence on the success of a production system. Therefore, it is crucial to understand the effect of rearing substrate composition on production parameters of *H. illucens* larvae. This study aimed to determine the influence of dietary energy source on larval production parameters. The energy sources that were evaluated included oils i.e. sunflower and fish, and carbohydrates i.e. sucrose, fructose and corn starch. These food grade energy sources were chosen as they represent the most common energy sources found in pre-consumer waste products. The production parameters that were determined included feed conversion ratio, growth rate, and substrate reduction rate. Growth rate was unaffected by treatment diets ($P>0.05$). This indicates that the energy source does not have an influence if the diets are formulated to supply the nutrient requirements of the animals. There were significant difference between the treatments in terms of substrate reduction rate and feed conversion ratio. The carbohydrates exhibited higher substrate reduction rates, fructose having the highest reduction rate at 72.83% ($P<0.001$). The oils exhibited better feed conversion, fish oil treatment having the best feed conversion at 2.47 ($P<0.001$). These results could provide parameters for the determination of optimum inclusion of a number of energy sources in larval diets, thereby optimizing production based on the chosen energy sources.

Starch digestion in *H. illucens* conversion: exploring the role of amylases from larvae and substrate

J.B. Guillaume[1,2,3], S. Mezdour[4], F. Marion-Poll[2,5], C. Terrol[1], C. Brouzes[1] and P. Schmidely[3]
[1]Agronutris, R&D Department, 35 blv du Libre Echange, 31650 Saint-Orens de Gameville, France, [2]CNRS, IRD, Université Paris-Saclay, Laboratoire EGCE, IDEEV, 12 route 128, 91190 Gif-sur-Yvette, France, [3]Université Paris-Saclay, INRAE, AgroParisTech, UMR Modélisation Systémique Appliquée aux Ruminants, 22 place de l'Agronomie, 91120 Palaiseau, France, [4]Université Paris-Saclay, INRAE, AgroParisTech, Sayfood, Campus Palaiseau, 22 place de l'Agronomie, 91120 Palaiseau, France, [5]Université Paris-Saclay, AgroParisTech, 22 place de l'Agronomie, 91120 Palaiseau, France; jeremy.guillaume@agroparistech.fr

The capacity of black soldier fly larvae (BSFL; *Hermetia illucens*) to transform organic substrates into body proteins and lipids suitable for animal nutrition is receiving growing attention. Carbohydrate content of BSFL diet is positively related to larval growth, fat body content and adult egg production, but underlying mechanisms of feed conversion remain to be explained. Previous studies reported amylase activity in larval midgut, and it has recently been shown that BSFL were highly efficient at digesting starch. This study focused on the effect of starch content and type on BSFL amylase activity and starch Estimated Digestibility (ED). BSFL were fed on five plant-based diets with different starch contents and types, and larvae and substrates were sampled after 4, 7 and 11 days of feeding, along with initial BSFL and diets. Each sample was ground in phosphate-buffered saline with protease inhibitor and centrifuged to collect water phase. Amylase activity was assessed using the Bernfeld technique and reported to total soluble protein measured according to Bradford. For substrate samples that could contain enzymes from the plant material, microbes or larvae, BSFL amylases were detected by Western-Blot using antibodies specific to insect amylases. This approach offers insight into larval amylase regulation mechanisms and the role of extra-oral digestion of starch in BSFL conversion systems.

Investigating the nutritional requirements of black soldier fly larvae using artificial substrates

L. Broeckx, L. Frooninckx, A. Wuyts, S. Berrens, M. Van Peer and S. Van Miert
Thomas More University of Applied Sciences, RADIUS, Kleinhoefstraat, 4 2440 Geel, Belgium; laurens.broeckx@thomasmore.be

Due to the rising need for alternative protein, the black soldier fly gained interest of researchers worldwide. One reason for this is the insect's remarkable efficiency at converting feed into larval biomass, rich in high-quality protein. Moreover, they are able to grow on a high variety of organic side- and waste streams, allowing local production of high-quality protein, while also valorising organic waste. Even though their remarkable plasticity towards growing on a wide range of organic substrates has been proven in many studies, optimization is still necessary to increase the sector's viability. In order to optimize the sector, one important approach is to tailor the nutritional requirements of black soldier fly larvae. By doing so, models can be developed that predict larval growth based on substrate nutrient compositions, which can help combine different organic side-streams to create a more nutritionally complete substrate for black soldier fly rearing. Therefore in this study, multiple designs were developed and tested to study the effects of macronutrient contents on the growth of black soldier fly larvae through the use of artificial substrates. Substrates were composed of casein, sunflower oil, potato starch and cellulose, which allowed precise formulation of substrates, and could reduce the background noise such as the presence of complex sugars and unbalances in amino acid profiles. In this study, a model was built, showing significant main, interaction and quadratic effects of substrate protein and fat contents. A maximal larval growth was acquired at a substrate protein content of 30.12% and a fat content of 8.75%. The effect of carbohydrates on larval growth was not significant in this study, however this could be due to poor digestibility of raw potato starch by the larvae. While larvae tend to grow best on high-protein substrates, research shows that the protein conversion efficiency is inversely proportional to substrate protein content, meaning we may need to reconsider our approach to substrate optimization if we want to balance between maximizing larval growth and improving protein conversion efficiency.

Session 78 Theatre 8

Km0 diets for black soldier fly larvae: the link between insect rearing and biogas systems

S. Bellezza Oddon, I. Biasato, Z. Loiotine, A. Resconi, C. Caimi and L. Gasco
University of Turin, Department of Agriculture, Forest and Food Sciences, Largo Paolo Braccini, 2, 10095 Grugliasco, Italy; sara.bellezzaoddon@unito.it

Energy consumption is one of the aspects that need to be considered to improve insect farming sustainability. In biogas systems part of the thermal energy is dissipated. Since black soldier fly (BSF) is reared at 28-30 °C and is able to bio-convert organic materials, the link between its farming and biogas plants could valorise the use of thermal energy, surplus of by-products (BP) for the anaerobic digestion and surrounding BP from food production. In the frame of the POWERFOOD project where a climatic room is warmed by biogas heat, this study aimed to identify the optimal BSF rearing substrates formulated using BP. Substrates were: biogas BP [BioS] (triticale, ryegrass silage, sorghum, corn mash and chopped), sugar beet BP [SB] (filtering waste, dry and liquid vinasse), beer BP [B] (brewer's spent grain and spent yeast) and milk BP [M] (milk whey). Since the BioS were km0, it was tested alone and included in all diet formulations as follows: (D1) BioS by-products, (D2) BioS + SB – except filtering waste, (D3) BioS + B, (D4) BioS + SB + B, (D5) BioS + B + M, (D6) BioS + SB + M and (D7) BioS + SB + B + M. 6-day-old larvae were reared (100 individuals/box; 6 replicates/treatment) until 5% reached the prepupae instar. At the trial end, survival rate (SR), growth rate (GR), reduction rate (RR), waste reduction index (WRI) and bioconversion efficiency corrected for residue (BER) were calculated. Data were analysed by one-way ANOVA (post-hoc: Tukey). The SR was high in all dietary treatments (>97%). Among treatments, D3 and D5 led to the best results in terms of weight (P<0.001). D1 larvae were lighter when compared to the other groups (P<0.001), with the exception of D2 and D6. D3 and D5 displayed the greatest WRI, and higher GR and BER than D1, D6 and D7 (P<0.001). The RR of the D1 was equal to D3 and D5, and higher when compared to the other groups (P<0.001). In short, diets with the B ingredients showed the best growth performance and bioconversion efficiency, with the exception of the D7 that was formulated with the highest inclusion of triticale, which is rich in fibre and may have caused reduced metabolization of nutrients.

Session 78 Theatre 9

Sugar processing by-products for black soldier fly farming: does the rearing scale matter?

I. Biasato, S. Bellezza Oddon, A. Resconi, Z. Loiotine and L. Gasco
University of Torino, Largo Paolo Braccini 2, 10095 Grugliasco (TO), Italy; ilaria.biasato@unito.it

Rearing scale may influence black soldier fly larvae (BSFL) traits, but different feeding substrates have not been tested yet. This study evaluated the effects of sugar processing by-products-based diets on performance and nutritional profile of BSFL reared in different scales. Four diets (D1-D2 [isonitrogenous, isolipidic and isoenergetic]; D3-D4 [1:1 and 1:2 as protein to carbohydrate ratios]) were tested in 3 rearing scales (4 replicate boxes/diet, constant volume [0.84 cm^3]/larva and feed [0.7 g]/larva): (1) small (S; 12×12 cm, substrate height [H]: 4 cm, 686 6-day-old larvae [6-DOL]/box); (2) medium (M, 32×21 cm, substrate H: 7 cm, 5,600 6-DOL/box); and (3) large (L, 60×40 cm, substrate H: 7 cm, 20,000 6-DOL/box). Larval weight was recorded at the beginning of trial and every 4 days, and growth rate (GR), specific growth rate (SGR), feed conversion ratio (FCR), survival, bioconversion efficiency corrected for residue, reduction rate (RR), and waste reduction index (WRI) calculated at the end of larval growth (frass dry matter [DM]≥55%). Substrate pH, T and H were measured at the beginning, every 4 days, and end of trial. Larval proximate composition was analysed at the end of trial. Data were analysed by GLMM (SPSS software, P<0.05). D1 larvae showed higher weight, GR, SGR and WRI (along with higher substrate T) than D2 in M scale, while increased SGR and FCR – as well as decreased survival, RR and WRI – were observed in D2 larvae in S scale (P<0.05). Larval crude protein (CP) and ether extract (EE) contents were influenced by M and L scales only, being higher in D2 group than D1 (P<0.001). Differently, decreased ash was recorded in D2 larvae when reared in S and M scales, while L scale revealed higher ash in D2 group than D1 (P>0.001). D3 larvae displayed greater weight, SGR, survival, RR and WRI (along with greater substrate T) than D4 in M scale, with increased survival and substrate T being also highlighted in L scale (P<0.05). D3 larvae also showed lower DM and EE – as well as higher CP – than D4 in all the rearing scales (P<0.001). In conclusion, D1 and D3 led to better BSF larval performance and nutritional profile in M and L scales, attributable to the initial higher substrate H and higher substrate T achieved.

Session 78 Theatre 10

Optimization of a hatchery residues fermentation process to feed black soldier fly larvae

M. Dallaire-Lamontagne[1,2], G.W. Vandenberg[2], L. Saucier[1,2] and M.H. Deschamps[1,2,3]
[1]Institute of Nutrition and Functional Foods, 2440 Bd Hochelaga, Québec, QC, G1V0A6, Canada, [2]Dép. des sciences animales, U. Laval, 2425 rue de l'Agriculture, Québec, QC, G1V0A6, Canada, [3]CLEIC, https://cleic.fsaa.ulaval.ca/, G1V0A6, Canada; marieve.dallaire-lamontagne.1@ulaval.ca

The conventional management of hatchery residues (HR; unmarketable chicks and eggs) is associated with environmental issues and health risks. Fermenting HR could reduce the loads of pathogenic microorganisms and odours while minimizing energy costs compared to thermal rendering. Fermented HR could then be processed into animal feed through bioconversion of black soldier fly larvae (BSFL; *Hermetia illucens*), in a circular manner. Given the low sugar content of the HR, a low-cost co-product must be incorporated to optimize the fermentation. HR were thus fermented in semi-anaerobic conditions with or without the addition of a starter culture (Lacult-SAX-01; 0.3%, HR) and with supplementation of whey permeate (dry or at 55% moisture) for lactose inclusion rates of 0, 5, 15, 25 and 35% (dry matter) for 2 weeks. The pH, microbiological quality (total mesophilic aerobes, lactic acid bacteria (LAB), coliforms, *Escherichia coli*), and volatile fatty acids were measured at 0, 3, 7 and 14 days to evaluate the efficiency of the process. Fermentation of HR resulted in a maximum pH reduction (<5) after 7 days for treatments with lactose inclusion of 15-35%. The reduction in pH, the formation of lactic acid (50,6 mg/g HR) and acetic acid (7,4 mg/g HR) as well as the maintenance of LAB loads in the HR contributed to the average 5 log cfu/g reduction in coliform and *E. coli* loads for treatment at 25-35% of lactose inclusion. However, the time to reach the detection limit for these microorganisms (<1,70 log cfu/g) varied from 7 to 14 days depending on the initial contamination of the HR and the addition of a starter culture. The initial loads of LAB (>7 log cfu/g) were sufficient to initiate spontaneous fermentation of the HR. However, addition of a starter culture could still be relevant since its efficiency was demonstrated in heat pre-treated HR. This project will improve the sustainability of livestock waste management systems and provide data to support the accreditation of BSFL ingredients reared on animal co-products.

Session 78 Theatre 11

Fish side residues as a substrate for *Hermetia illucens*

M.G. Zhelezarova
NASEKOMO, Novoseltsi 174, 2100, Bulgaria; mirena.zhelezarova@nasekomo.life

Black soldier fly (BSF) is becoming the preferred insect solution to valorise non-utilized biomass from agro-industries. For fish industries, these side streams present ecological and economic benefits that the insect industry may process. However, previous studies show that high inclusion levels of aquaculture residues can have negative impacts on BSF larval development. Thus, a solution on how to efficiently include fish biomass into BSF feed to achieve significant protein yields without triggering high mortality rates and poor Feed Conversion Ratio (FCR) is needed. Understanding interactions with other raw ingredients in the recipe is also necessary. 'EcoeFISHent' is a project funded by the European Union H2020 program (GA 101036428), to develop an industrial symbiosis system able to efficiently exploit fish-processing side streams and create a variety of valuable products. It involves a consortium of thirty-four parties and representatives of different sectors. NASEKOMO`s mission in this project is to establish a methodology to convert such side streams into added value resources through insect bioconversion. Four diets with different saltwater fish residues inclusions were formulated in mixture with regular local raw materials and tested against benchmark recipes containing the same ingredients and proportions except for the fish ingredient being replaced by non-digestible pure cellulose. Survival rates (6-12 days old) were above 95% in all fish residues feed groups. Fish side streams diets resulted in higher weight per individual and insect biomass total yield, but lower FCR compared to their benchmarks. BSF larvae were able to feed and survive on the set of raw materials provided locally by 'Nasekomo' and mixed with aquatic animal production side-streams. High lipid concentrations in aquaculture side residues are challenging for BSF bioconversion but can be diluted in the feed with other ingredients. A variety of vitamins and sterols are also needed to satisfy nutritional requirements of the animals and need to be provided by other feed components. Substrate texture is also showed to be crucial but ultimately depends on the fish residues composition, nutritional values, and water salinity along with other ingredients characteristics and inclusions in the feed recipe.

Session 78 Poster 12

Black soldier fly larvae production is optimized by the presence of HMTBa

K. Luyt[1], G. Crielaard[1], M. Briens[2], J.A. Conde-Aguilera[2] and M. Ceccantini[2]
[1]Entobel Ltd Company, 531a Upper Cross Street, #04-95 Hong Lim Complex, Singapore, [2]Adisseo France S.A.S., 10 place du Général de Gaulle, 92160 Antony, France; mickael.briens@adisseo.com

Insects are becoming a promising source of proteins and fats for aquaculture, pet and poultry feed, and can partially replace vegetal proteins. It represents an important way to upcycle food and feed waste and decrease the carbon footprint of meat production. Still, the development of insect meal production has to be improved to have efficient and economical ways to produce insect larvae. In this study, black soldier fly (*Hermetia Illucens*) raised on food waste were tested for their efficiency on performance and cost, when a source of methionine (Hydroxy-methylthiobutanoic acid [HMTBa]) was added or not in the larvae feed. A 7-day trial starting with 1st instar black soldier fly larvae (BSFL) was run with 4 dietary treatments × 6 crate replicates: control, and control plus HMTBa at doses 0.08, 0.16 or 0.32%. The average initial weight was 2.41 mg/larvae and the larvae were equally inoculated in 12 kg of feed commonly used in Entobel production system. The HMTBa additive was added during the fermentation of feed for 3 days prior to the trial. Results: HMTBa used at 0.16 and 0.32% presented 11.9 and 10.4% significatively (P<0.05%) higher fresh larvae biomass production than control on whole period. Considering larvae biomass and feed residue on the whole period, the feed conversion of these treatments was statistically different (3.49^a; 3.14^b; 2.74^c; 2.75^d, respectively). The crude fat content of the fresh larvae was not modified by treatments containing HMTBa when compared to control (27.6; 28.9; 30.3; 29.0%, respectively). The crude protein content was not affected as well (52.7; 50.9; 48.8; 50.6%, respectively). Besides that, the cost of insect meal per treatment demonstrated to be lower by 20.3% when HMTBa was added in the larvae feed. Conclusion: the use of HMTBa brings high benefits in terms of performance and cost to produce BSFL insect proteins and fats. Further studies are necessary to confirm those effects and demonstrate the combined effect of different feed additives.

Session 78 Poster 13

Black soldier fly larvae production is optimized by the presence of a multi-carbohydrase

K. Luyt[1], G. Crielaard[1], M. Briens[2], J.A. Conde-Aguilera[2] and M. Ceccantini[2]
[1]Entobel Ltd Company, Hong Lim Complex, #04-95 Hong Lim Complex, Singapore, [2]Adisseo France S.A.S., 10 place du Général de Gaulle, 92160 Antony, France; mickael.briens@adisseo.com

Insects are becoming a promising source of proteins and fats for aquaculture, pet and poultry feed, and can partially replace vegetal proteins. It represents an important way to upcycle food and feed waste and decrease the carbon footprint of meat production. Still, the development of insect meal production must be improved to have efficient and economical ways to produce insect larvae. In this study, black soldier fly (*Hermetia Illucens*) raised on food waste were tested for their efficiency on performance and cost, when a multi-carbohydrase additive (containing xylanase [12,500 U/ml], glucanase [8,600 U/ml] and arabinofuranosidase) was added or not in the larvae feed. A 7-day trial starting with 1st instar black soldier fly larvae (BSFL) was run with 4 dietary treatments × 6 crate replicates: control, and control plus a multi-carbohydrase dose of 100, 200 or 400 ml/ton larvae feed. The average initial weight was 2.41 mg/larvae and those larvae were equally inoculated in 12 kg of feed commonly used in Entobel production system. The additive was added during the fermentation of feed for 3 days prior to the trial. Results: the treatments with multi-carbohydrase at 100 and 200 ml/ton presented 8.5 and 12.7% significantly (P<0.05) better fresh larvae biomass production than control on the whole period, respectively. Considering larvae biomass and feed residue on the whole period, the feed conversion of these treatments was statistically different (3.49^a; 2.55^b; 2.26^c; 2.15^c; respectively). The fresh larvae presented just numerical increase in crude fat content for all treatments when compared to control (27.6; 33.0; 34.1; 37.0%, respectively), and no differences on the crude protein content (52.7; 47.7; 47.8; 46.7%(. Besides that, the cost of insect meal per treatment demonstrated to be lower by 32% when the additive was added in the larvae feed. Conclusion: the use of multi-carbohydrases brings high benefits in terms of performance and cost to produce BSFL by the better feed conversion and fat production. Further studies are necessary to confirm those results.

Session 78 — Poster 14

Effect of water-soluble complementary feed on performance in nursery of black soldier fly

L. Schneider[1], M. Brake[2], A. Heseker[2], W. Westermeier[3] and G. Dusel[1]
[1]University of Applied Science Bingen, Department of Animal Nutrition, Berlinstraße 109, 55411 Bingen am Rhein, Germany, [2]MIAVIT GmbH, Robert-Bosch-Str. 3, 49632 Essen, Germany, [3]FarmInsect GmbH, Münchner Str. 10, 85232 Bergkirchen, Germany; l.schneider@th-bingen.de

The starter feed is one of the most important diets of livestock insects and especially for newly hatched black soldier fly (BSF) neonates. Proper nutrition during the starter phase is essential for adequate growth and development, which will ultimately affect the overall performance of the flock. The aim of the study was to evaluate the performance and development of BSF neonates reared in a controlled environment on an iso-nitrogenous and iso-energetic standard diet (wheat bran, chicken feed, water) with either water-soluble mineral complementary feed (*WSCF1*, 1%, *WSCF2*, 2%, *WSCF3* 3%) or a non-water-soluble mineral feed addition for poultry (*MFP*, 3%) or a control without mineral feed addition (*CON*). A total of 62 g BSF eggs (1 g eggs = 0.45 g neonates) were placed randomly per unit (6 replicates/treatment, 60×40×15 cm). At the end of the nursery phase the mean bodyweight of 100 larvae (BW), dry matter (DM) and total weight of 5-day-old larvae output per unit as well as total weight of larval frass was observed. The results showed that the mean BW of *WSCF3*- and *WSCF2*-fed (BW, 0.12 g/100 BSFL) larvae were nearly similar and increased by 33% ($P<0.05$) compared to *CON*-fed larvae (BW, 0.09 g/100 BSFL). Additionally, the DM (%) of *WSCF3*-fed larvae showed an 8% increase compared to *MFP*-fed larvae. *WSCF2*-fed larvae showed a 13% higher ($P<0.05$) total weight of young larvae compared to *CON*-fed larvae (*WSCF2*, 678.5 g; *CON*, 588.4 g). The total weight of the larval frass at the end of the nursery phase of group *WSCF2* was 31% lower ($P<0.05$) than that of the *CON*-group (*WSCF2*, 1,797.3 g; *CON* 2,616.6 g). In conclusion, this first study demonstrates the potential of a water-soluble complementary feed on the performance and development of young BSF in the crucial pre-starter phase. Further investigations are necessary to recommend a possible efficient and sustainable preconditioning effect in BSF nursery.

Session 78 — Poster 15

Digestibility in *H. illucens* larvae: resolving faeces collection and ingesta quantification issues

J.B. Guillaume[1,2,3], S. Mezdour[4], F. Marion-Poll[2,5], C. Terrol[1] and P. Schmidely[3]
[1]Agronutris, R&D Department, 35 bld du Libre Echange, 31650 Saint-Orens de Gameville, France, [2]CNRS, IRD, Université Paris-Saclay, Laboratoire EGCE, IDEEV, 12 route 128, 91190 Gif-sur-Yvette, France, [3]Université Paris-Saclay, INRAE, AgroParisTech, UMR Modélisation Systémique Appliquée aux Ruminants, 22 place de l'Agronomie, 91120 Palaiseau, France, [4]Université Paris-Saclay, INRAE, AgroParisTech, Sayfood, 22 place de l'Agronomie, 91120 Palaiseau, France, [5]Université Paris-Saclay, AgroParisTech, 22 place de l'Agronomie, 91120 Palaiseau, France; jeremy.guillaume@agroparistech.fr

Black soldier fly larvae (BSFL; *Hermetia illucens*) are increasingly studied for their ability to convert organic substrates into body proteins and lipids that can be used for animal nutrition. Although many studies have used BSFL high weight gain to highlight their strong feed conversion efficiency, little is known about the inherent efficiency of each of the four feed conversion stages: ingestion, digestion, absorption, and metabolic utilisation. Assessing digestibility requires quantifying the amount of feed ingested and the associated faeces produced. However, this is challenging in BSFL because they feed and release excreta in the same substrate, which also hosts complex microbiota participating in digestion. This study introduced a new indicator called Estimated Digestibility (ED), defined as the difference between distributed feed and frass macronutrient weight, divided by macronutrient weight in distributed feed. The evolution of ED was assessed with increasing larval density in order to ensure complete feed ingestion and frass free from refused feed. ED was measured on a standard diet with densities from 0 to 29 larvae/cm^2 for dry matter (DM), starch, nitrogen, ether extract (EE), neutral detergent fibre, acid detergent fibre, acid detergent lignin, ash and energy. The results showed a sigmoidal pattern for ED of all fractions except fibres. Asymptotic ED was 80.3±1.3% (mean ± standard error) for DM, 99.0±2.3% for starch, 78.6±1.1% for nitrogen, 95.3±1.5% for EE, 58.4±1.0% for ash and 80.6±1.2% for energy. Asymptotic ED is the closest estimation of digestibility as defined in other species. It offers perspective on the understanding of BSFL digestive efficiency and could be used for diet formulation.

Session 78 Poster 16

Black soldier fly larvae as tools for the bioconversion of sludge from wastewater treatments

C. Ligeiro[1], I. Lopes[1,2], T. Ribeiro[1,2], I. Rehan[3], K. Silvério[4], M. De Fátima[4] and D. Murta[1,2]
[1]Ingredient Odyssey SA, EntoGreen, Portugal, [2]CiiEM – Centro de Investigação Interdisciplinar Egas Moniz, Campus Universitário, Portugal, [3]Polo de Inovação da Fonte Boa, Instituto Nacional de Investigação Agrária e Veterinária, Portugal, [4]Instituto Politécnico de Beja, Departamento de Tecnologias e Ciências Aplicadas, Portugal; carolina.ligeiro@entogreen.com

The project NETA (New Strategies in Wastewater Treatment) proposes a solution for the problems related to wastewater treatments. After the wastewater treatment, a sludge is obtained and is a secondary product that must be further processed. In this study, we evaluated the bioconversion of the-derived sludge with black soldier fly larvae (BSFL). For such, 7-day-old BSFL were inoculated in plastic boxes containing a diet formulated with sludge from the treatment of 2 wastewaters: water from a dairy industry and from a general wastewater collection tank, which receives waste from toilets, slaughterhouses and animal farms. The larvae diets using sludge contained sludge, wheat bran and water (T1, with sludge from the dairy industry and T2, with sludge from the collection tank) were formulated to be isoproteic, and a control diet (CT) was formulated using wheat bran and water. All diets were balanced to a 30% DM content. The temperature of the diets and the weight of the larvae were evaluated periodically throughout the experiment. In the last day, it was possible to sift the box content and the larvae were harvested and the total larval biomass and fertilizer (frass) produced were assessed. Samples from both larvae and frass were analysed for bromatological and nutrient composition. The larvae performed well in T1 and T2, being heavier in T1 (0.893±0.06 g per 10 larvae) in comparison to the CT (0.642±0.04 g per 10 larvae). In addition, the larval biomass was higher in T1 as well (2.32 kg per experimental unit in comparison to 1.96 kg in CT). Production of frass was statistically higher in the treatments containing sludge. It was concluded that BSFL performed well in sludge-based diet and can be considered a viable biological tool for the bioconversion and treatment of this wastewater sludge. This study was conducted under the scope of the NETA project: POCI-01-0247- FEDER-046959 funded by PORTUGAL2020.

Session 78 Poster 17

Mediterranean agricultural by-products as insect diet ingredients: the ADVAGROMED perspective

C.G. Athanassiou[1], S. Bellezza Oddon[2], V. Zambotto[3], T. Ribeiro[4], R. Rosa García[5], A. El Yaacoubi[6], C. Adamaki-Sotiraki[1], I. Biasato[2], A. Resconi[2], D. Murta[4], C.I. Rumbos[1] and L. Gasco[2]
[1]University of Thessaly, Phytokou Str., 38446, Volos, Greece, [2]University of Turin, largo P. Braccini 2, 10095, Italy, [3]Italian National Research Council, Via Amendola, 122/O, 70126 Bari, Italy, [4]Ingredient Odyssey Lda, R. Eng. Albertino Filipe Pisca Eugénio 140, 2005-079, Portugal, [5]Servicio Regional de Investigación y Desarrollo Agroalimentario, Ctra. AS-267, PK 19, 33300, Villaviciosa, Spain, [6]University of Sultan Moulay Slimane, Avenue Mohamed V., B 591, Beni Mellal, Morocco; laura.gasco@unito.it

The upcycling of agricultural by-products through insect bioconversion can help the Mediterranean countries valorise these locally available resources for the production of animal feeds and subsequently increase the resilience of Mediterranean farming systems. In this framework, within the ADVAGROMED project, a broad spectrum of agricultural by-products that are generated in Greece, Italy, Portugal, Spain and Morocco were identified, collected and chemically characterized, as a first step towards their valorisation as diet ingredients for the two most commonly reared edible insect species, i.e. *Hermetia illucens* and *Tenebrio molitor*. Specifically, by-products of the seed cleaning process of cereals and legumes (lupin, triticale, oats, barley, peas), by-products of the cotton, rice, grape, olive and hemp production, food waste (e.g. cookies, crackers, wafer, bread), vegetable and fruit by-products (carrot, pepper, green pea, tomato, zucchini, lettuce, apple, kiwi, blueberries, melon, watermelon), as well as algae were collected and analysed. Based on the results, there was a high variability in the nutritional composition of the tested by-products. Indicatively, their protein content ranged from 0.7 to 48.3%, whereas their ether extract content varied between 0.1 and 35.0%. Similarly, variable results were obtained for their dry matter (3.6-97.9%), ash (0.3-58.8%) and fibre content (1-48.4%). As a next step, compound diets will be formulated and evaluated for the rearing of *H. illucens* and *T. molitor* based on the composition of the selected by-products and the insect nutrient requirements. This research is supported by the EU-PRIMA program project ADVAGROMED (Prima 2021–Section 2).

Session 78 Poster 18

Multigenerational and nutritional traits of BSF reared on seaweed or selenium enriched substrates

M. Ottoboni[1], L. Ferrari[1], A. Moradei[1], F. Defilippo[2], P. Bonilauri[2] and L. Pinotti[1]
[1]University of Milan, Via dell'Università, 6, 26900, Lodi, Italy, [2]Istituto Zooprofilattico Sperimentale della Lombardia e dell'Emilia Romagna Bruno Ubertini, Via Pitagora, 2, 42124 Reggio Emilia, Italy; matteo.ottoboni@unimi.it

This study evaluated the effect of inclusion of brown algae (*Ascophyllum nodosum*) or sodium selenite in BSF growing substrate on selected insect life cycle traits and the nutritional composition of the deriving prepupae. The larvae were reared on three different substrates: (1) Gainesville diet, (CTR); (2) A. nodosum diet (AN30%), with 30% substitution of the alfalfa meal with brown algae; and (3) selenium diet (Se), based on Gainesville diet fortified with 0.3 mg/kg of sodium selenite. Eggs collected from a stable colony of BSF were placed for hatching on three experimental substrates CTR, AN30%, and Se. All experiments were carried out under dark condition, at 25 °C with 70% relative humidity, and repeated for three consecutive generations. Recorded data were: live performance (i.e. percentage of mortality in each stage, development period, larvae weight, percentage of adult emergence), and reproductive performance (i.e. number of eggs laid and percentage of hatched eggs). Rearing substrate and deriving prepupae were analysed for proximate and fatty acid composition. Mortality and larval weight did not show any difference among groups during the three-generation study. In contrast, BSF development period was lower ($P<0.05$) in CTR than in AN30% and Se group. Further differences were observed in term of percentage of adult emergence that was higher in CTR (90.4%) and Se (90.4%) compared to AN30% (78.7%). All prepupae contained a high level of saturated fatty acids. The inclusion of seaweed in the rearing substrate did not affect the fatty acid profile in the AN30% prepupae compared to CTR and Se dietary groups. Concluding, best values for number of eggs laid and percentage of hatched eggs were observed in the case of CTR. Reproductive performance dramatically decreased in AN30%, and Se groups. Combining obtained results, BSF larvae can grow on media containing up to 30% *A. nodosum* or selenium with some implication for both live and reproductive performance. Further investigations are required for optimising BSF meal production for specific feed or food purposes.

Session 78 Poster 19

Bacterial biomass improves performance and antibacterial activity of black soldier fly larvae

N. N. Moghadam[1], K. Dam Nielsen[1], A. Simongiovanni[2] and T. Chalvon-Demersay[2]
[1]Danish Technological Institute, Kongsvang Allé 29, 8000 Aarhus C, Denmark, [2]METEX Animal Nutrition, 32 rue Guersant, 75017, France; aude.simongiovanni@metex-noovistago.com

Black soldier fly larvae (BSFL) has a great capacity to identify, inactivate, and eliminate microbial load in different sources of raw materials and to valorise them into high quality protein and oil. In this experiment, we assessed the effects of a bacterial biomass as a feed additive on the performance and antibacterial activity of BSFL. The experiment consisted of 4 diets (a control and 3 treatment diets) and performed in triplicate. The treatment diets had 2, 4 and 6% inclusion rates of the bacterial biomass which was derived from an amino acid producing bacteria. The inclusion of the bacterial biomass was done at the expense of maize gluten to reach similar protein level in all diets. The diets were used at the production stage and effects of the diets were assessed on the larval production (weight and biomass), development time, survival rate, feed conversion ratio (FCR), and activity against three different bacterial strains (*Escherichia coli*, *Staphylococcus aureus* and *Pseudomonas aeruginosa*). Overall, inclusion of the bacterial biomass had a positive effect on the BSFL weight and biomass and reduced the FCR without affecting the survival rate. Interestingly, the presence of the bacterial biomass in the BSFL diet induced the antibacterial activity of the BSFL against *E. coli* and *P. aeruginosa*. This response was more pronounced in the BSFL grown on the diet containing 4% bacterial biomass. In conclusion, inclusion of the bacterial biomass in the BSFL diet improved the larval performance in general and induced the larval activity against the two tested gram-negative bacterial strains (*E. coli* and *P. aeruginosa*). The latter finding needs to be evaluated in future studies.

Session 78 Poster 20

Impact of zinc supply on black soldier fly larvae growth, bioconversion and microbiota

L. Frooninckx[1], L. Broeckx[1], D. Vandeweyer[2], C. Keil[3], M. Maares[3] and S. Van Miert[1]
[1]Thomas More University of Applied Sciences, RADIUS, Kleinhoefstraat 4, 2440 Geel, Belgium, [2]KU Leuven, Department of Microbial and Molecular Systems, Research Group for Insect Production, Kleinhoefstraat 4, 2440 Geel, Belgium, [3]Technical University of Berlin, Institute of Food Technology and Food Chemistry, Department of Food Chemistry and Toxicology, Gustav-Meyer-Allee 25, Berlin, Germany; laurens.broeckx@thomasmore.be

The larvae of the black soldier fly (BSFL), *Hermetia illucens*, can be efficiently grown on a wide range of organic residues. Moreover, BSFL are a nutritious source of protein which can be used as a feed ingredient for poultry, fish, and other livestock. Therefore it has potential to play an important role in sustainable waste management, food security, and bio-based economy. Little is known about the specific metabolic and nutritional response to alterations in trace element supply for BSFL. Zinc salts are known to improve livestock animals' immune systems and growth when fed in appropriate doses. Zinc supplementation has, however, not been well studied when it comes to food/feed-relevant insects such as BSFL. In this study, BSFL were grown on Gainesville diet and food waste spiked with three different zinc concentrations. The influence of zinc on growth, survival, bioconversion, zinc-bioaccumulation and the bacterial biota of BSFL was investigated. High zinc concentrations negatively affected growth, survival and bioconversion of BSFL. There is a clear correlation between the zinc concentration in the diets and the levels in the BSFL. Although bioconversion differs between the two diets, no difference in zinc concentration in the BSFL between the two diets was observed.

Session 79 Theatre 1

Review: recent outcomes associating time to PSPB increase with pregnancy loss in dairy cows

T. Minela, A. Santos and J.R. Pursley
Michigan State University, 474 S Shaw Ln, 48824, East Lansing, USA; thainaminela@gmail.com

In the last 4 years, our laboratory has focused on understanding the role of time to increase in pregnancy specific protein B (PSPB) during early pregnancy establishment in lactating dairy cows. We utilize a daily, within-cow sampling regimen that allows us to capture PSPB increases on an individual basis. This regimen covers three important periods of embryonic development: pre-, peri- and post-attachment of the conceptus to the uterus. We utilize PSPB as a marker of conceptus viability. Characteristically, most cows have a significant PSPB increase, around d 20 (25.3%), and 21 (25%). Yet, some cows may experience a delayed increase in PSPB, on d 22 (8.7%) or later (up to d 26; 3.2%). Inarguably, an extended period to onset of increase in PSPB was associated with greater early pregnancy losses (occurring before ~d 35). Pregnancy losses totalled 76% when increase of PSPB occurred ≥23 d post-ovulation, 30% for cows with increase in PSPB at d 22, 12% for d 21 and 10% for ≤20 d post-ovulation. This phenotype is also associated with parity. Parous cows are more prone to have a later increase in PSPB than their nulliparous counterparts (nulliparous 21.1 d, primiparous 23.6 d and multiparous 24.3 d). There are three distinct patterns of PSPB profiles identified across studies. Cows with an initial PSPB increase that subsequently maintain pregnancy, cows with an initial PSPB increase that lose pregnancy by the first pregnancy diagnosis, and cows with no marked increases, thus non-pregnant cows. Cows that maintain pregnancy have greater PSPB concentrations between d 23 and 28 in comparison with cows that lose pregnancies. A more pronounced and premature increase in PSPB may be associated with greater potential for embryonic development capacity and pregnancy success. This novel way to utilize daily PSPB concentrations may provide an accurate indication of time to conceptus attachment and may be an excellent model for gaining a greater understanding of pregnancy loss in dairy cows.

Session 79 Theatre 2

Improving cow fertility by immunizing against inhibin and P4 supplementation

Z.D. Shi, F. Chen and R.H. Guo
Jiangsu Academy of Agricultural Sciences, Institute of Animal Science, 50 Zhongling Street, Xuanwu District, 210014, China, P.R.; zdshi@jaas.ac.cn

Pursuing higher and higher milk production has led to severe decreases in dairy cow fertility that is especially prominent under hot climatic conditions. Tremendous effort has been invested to solve this problem worldwide with little success. We utilized the method of immunoneutralization of inhibin bioactivity in order to enhance ovarian development capacity by stimulating FSH secretion by the pituitary gland and enhancing follicle response to FSH. This approach has been previously shown to enhance the ovarian follicle development, oestradiol secretion and conception rate (CR) in cows subjected to Ovsynch protocol treatment. In addition, antibodies to inhibin alpha subunit treatment of the *in vitro* matured oocytes also improved oocyte maturation and the resultant pathogenic embryo development. In the latest experiments, we incorporated immunization against inhibin into the Ovsynch protocol together with supplementation of progesterone to treat the post-partum cows. In experiment 1 that was conducted under severe heat stress of July in the tropical region, CR increased from 8.1% (5/57) in the control cows to 43.9% (25/62) in the immunized animals. In experiment 2 conducted under mild heat stress of August in the temperate region, CR was increased from 26.7% (4/15) in the control cows to 64.7% (11/17) in the immunized ones. The 3rd experiment was conducted on the same farm of experiment 2 but the comfortable April, and CR was increased from 44.2% (19/43) to 71.1% (32/45) by immunization against inhibin. In experiment 2, immunization against inhibin increased blood concentration of FSH, activin A during both follicular and luteal phases, and of E2 in the follicular phase, but decreased P4 concentrations during luteal phase. However, interferon-tau concentrations in blood around the time of pregnancy recognition were doubled in the inhibin immunized cows. In conclusion, immunization against inhibin plus P4 treatment enhances ovarian follicle and the subsequent early embryo developments that help to greatly improve the conception rate to the range of 60 to 70% in Holstein dairy cows.

Session 79 Theatre 3

Characterization of sex chromosomes-linked lncRNAs in Holstein spermatozoa under stress conditions

A. Yousif, G. Thaller and M. Saeed-Zidane
Institute of Animal Breeding and Husbandry, Christian-Albrechts-University of Kiel, Animal Breeding and Genetics Group, Olshausenstraße 40, 24098 Kiel, Germany; ayousif@tierzucht.uni-kiel.de

Long non-coding RNAs (lncRNA) are known to be involved in regulation of numerous biological processes in most types of cells studied so far including sperms. LncRNAs play a vital role during the spermatogenesis and post-fertilization stages. However, very little is known about the potential role of lncRNAs located on sex chromosomes for sperm quality and functionality under stress conditions. Therefore, the aim of this study is to investigate the genetic characteristics of X and Y chromosome-linked lncRNAs and the association of sperm-borne-lncRNAs in spermatogenesis and post-fertilization. To achieve our goals, two different groups of qualified (n=6) and non-qualified (n=6) Holstein bulls were subjected to semen collection at four different seasons. Thereafter, sperm-extracted DNA was used for genotype analysis of thymosin beta 15b lncRNA (TMSB15B) located on the Y chromosome using Sanger sequencing. The RNA level of sperm-borne lncRNAs located on sex chromosomes was analysed in all experimental semen samples collected across seasons (winter, spring, summer, and autumn). The preliminary results showed that one SNP (A/G) was found in the 350 bases upstream region of TMSB15B gene of two qualified bulls compared to bovine reference genome (NCBI). On the transcriptional level, TMSB15B lncRNA was significantly higher in the semen of qualified bulls compared with non-qualified ones. Furthermore, X inactive-specific transcript (XIST) lncRNA located on the X chromosome showed significantly higher RNA levels in non-qualified bull's semen collected in autumn compared to winter and spring. Additionally, transcript X2 was the most abundant one among the three transcripts of XIST, particularly in qualified bull's semen collected in winter and non-qualified bull's semen collected in spring. On the other hand, sperm-born lncRNA KANTR was significantly higher only in winter semen of qualified bulls. In conclusion, the sperm-borne-lncRNAs located on sex chromosomes level can have a profound prediction for pre-and post-fertilization success. Therefore, further investigations are ongoing on analysing further sperm-borne lncRNAs with the application of *in vitro* fertilization.

Session 79 Theatre 4

Inbreeding affects the freezing ability in sperm samples of Pura Raza Español stallions

Z. Peña[1], M. Valera[2], A. Molina[1], N. Laseca[1] and S. Demyda-Peyrás[3]
[1]Universidad de Córdoba, Campus Rabanales, 14071, Córdoba, Spain, [2]Universidad de Sevilla, Utrera 2, Sevilla, Spain, [3]FCV – UNLP, Calle 60 y 118, La Plata, Argentina; zahiramaria@hotmail.com

Inbreeding is a genetic condition produced by mating relatives, which affects negatively the fertility of individuals. This condition is particularly important in horses, where close-related matings are common. In this study, we analysed the effect of inbreeding on sperm quality and freezing ability (endogamic depression) of Pura Raza Español Stallions. We evaluated 720 ejaculates from 108 PRE individuals collected by natural mating on the artificial vagina. Samples were centrifugated and diluted and equilibrated for 5 min in BOTUCRIO™ extender to 50×10^6 spz/ml. In addition, a sperm aliquot was frozen following a controlled-ramp protocol. The assessment of kinetic parameters in fresh and frozen-thawed samples was performed at 5 and 60 m of incubation using a CASA analyser. The effect of inbreeding was assessed using a multivariate mixed model with REML methodology (total (MOT) and progressive (PROG) motility, VSL, VAP, VCL, ALH, and BCF), including as fixed effects age at jump, and year of birth. The inbreeding value (F) was included as a linear covariate. The variance components were estimated using the BLUPF90 software package. Results showed that the increase of inbreeding produced higher EBVs in MOT and PROG, as well as in VSL, VAP, and BCF on fresh samples. However, estimates for both motilities and BCF after 60 min of incubation were significantly lower. No effect on kinetic parameters was detected. These results agree with the hypothesis that inbreeding increases the percentage of hyperactivated spz at early stages, thus decreasing the longevity of the sperm. On the contrary, the increase in inbreeding produced a negative effect on MOT and PROG in frozen-thawed sperm which estimates were significantly lower in individuals with higher inbreeding rates. To our knowledge, this is the first attempt to model the kinetic parameters of stallion sperm from a genetic point of view. With its method, we detected and quantified a negative effect of inbreeding in kinetic parameters and freezing ability in stallions. Interestingly, similar patterns were previously reported in captive populations of wild animals, but not in livestock.

Session 79 Theatre 5

Sperm transcripts and seminal plasma metabolome reveal the regulatory mechanism of bull semen quality

W.L. Li and Y. Yu
China Agricultural University, Animal Breeding and Genetics, no.2 Yuanmingyuan WRD, China Agricultural University, 100193, Beijing, China, P.R.; yuying@cau.edu.cn

Semen quality of dairy bulls is very important for selecting excellent stud bulls. Recent studies have found that some transcripts are not only the residue of sperm maturation, but also related to semen quality. Moreover, seminal plasma can affect sperm transcriptome and further influence semen quality. However, no research has integrated sperm transcriptome and seminal plasma metabolome to analyse the molecular regulatory mechanisms of bull semen quality traits. Number of motile sperm per ejaculate (NMSPE) is an integrated trait to evaluate semen quality of stud bulls. In the present study, we selected 7 bulls with higher NMSPE (5,698.55±945.40 million) as group H and 7 bulls with lower NMSPE (2,279.76±1,305.69 million) as group L from 53 stud bulls. The differentially expressed genes (DEGs) in sperm were evaluated between the two groups (H vs L). Gene co-expression network analysis of groups H and L bulls was conducted by weighted gene co-expression network analysis (WGCNA) to screen candidate genes for NMSPE. A total of 1,099 DEGs were identified in the sperm of H and L groups. The DEGs were primarily concentrated in energy metabolism and sperm transcription. The significantly enriched KEGG pathways of the 57 differential metabolites were the aminoacyl-tRNA biosynthesis pathway and vitamin B6 metabolism pathway. The expression level of candidate gene *FBXO39* could be regarded as a reliable biomarker of bull NMSPE. We also observed that sperm cells expressed genes correlated with seminal plasma metabolites were not only located near the QTLs of reproduction-related traits, but also were enriched in the GWAS signals of sire conception rate traits. Moreover, three metabolites (mesaconic acid, 2-coumaric acid and 4-formylaminoantipyrine) might regulate *FBXO39* expression through potential pathways. These data provide new insights into the improvement of bovine semen quality traits.
Funding: 2021YFD1200903, NK20221201.

Application of an oligo-based FISH method in fertility assessment of boars by chromosomal analysis

W. Poisson[1,2,3], J. Prunier[3], A. Bastien[3], A. Carrier[1,2,3], I. Gilbert[1,2,3] and C. Robert[1,2,3]
[1]Réseau Québécois en reproduction, 3200 Rue Sicotte, Saint-Hyacinthe, QC, J2S 2M2, Canada, [2]Centre de Recherche en Reproduction, Développement et Santé Intergénérationnelle, 2440, boul Hochelaga, Québec, QC, G1V 0A6, Canada, [3]Université Laval, 2325 Rue de l'Université, Québec, QC, G1V 0A6, Canada; william.poisson.1@ulaval.ca

Many genetic fields have benefited from the use of Fluorescence *in situ* hybridization (FISH) since its first implementation over 35 years ago. The use of vectors or flow-sorted and microdissected chromosomes are frequent means of producing FISH probes. With the improvement of oligonucleotide (oligo) production using massively parallel synthesis, these short DNA fragments become more affordable and available for use as probes for FISH experiments. Recently, we implement a new method using 207,847 oligo to evaluate chromosome integrity in pigs. Karyotyping boars before their use in artificial insemination centres is crucial since chromosomal rearrangements are recognized as a potential cause of hypoprolificity. Traditionally, karyotyping is performed using a bichrome banding pattern and requires experienced analysts for proper analysis. Alternatively, the new method called Oligo-banding produces 96 chromosomal fluorescent bands that can be analysed without prior expertise. It has been successfully used to detect and confirm four translocations and an unbalanced material addition. We also tested its versatility on other Suidae and on spermatozoa DNA. The probes generated clear banding patterns when applied to wild boar and warthog chromosomes. Hybridization on spermatozoa has shown clear signals, but analysis of translocation inheritance remains complex. Overall, the oligo-based technology generated defined patterns on chromosomes useful to evaluate genomic constitution in somatic and sperm cells with applications in fertility assessment. This project is supported by a grant from the Ministère de l'agriculture et des pêcheries du Québec, the Natural Sciences and Engineering Research Council of Canada and the Fonds de recherche du Québec – Nature et technologies.

Artificial insemination success: a new trait for French dairy goats

V. Clément, A. Piacère and M. Chassier
Institut de l'Elevage, Chemin de Borde Rouge, 31320 Castanet-Tolosan, France; virginie.clement@idele.fr

Breeding scheme of French dairy goats relies on creation of genetic progress and its dissemination by frozen artificial insemination (AI). So, AI success is a major concern to increase genetic level of commercial flocks. The selection criterion is a total merit index which combines production traits (protein and fat yields and protein and fat contents), udder morphology and somatic cell counts as an indicator of udder health. Our objective was to study AI success as a new trait to select. The first step was to define the phenotype. The result of an AI is considered a success if, for a given kidding, there is a unique breeding event that coincides with the date of the AI. All AI that are not followed by a kidding are considered failures. The heritability of AI success was estimated to be 0.052 in the Saanen breed and 0.045 in the Alpine breed. Despite these low values, the genetic standard deviation is significant (9 AI points in Alpine and 11 AI points in Saanen breeds). Genome wide association studies have been conducted with Blupf90 software using genomic information from genotyped animals with the Illumina GoatSNP50 Bead Chip (800 females and 2,342 males in Saanen breed, 1,123 females and 2,937 males in Alpine breed). The two breeds were analysed separately. Analysis showed, in the Saanen breed, a zone on CHI 19 (25.6-28.4 Mb) significantly associated with AI success, with a percentage of variance explained by 20 adjacent SNP window higher than 5%. This result is consistent with previous studies that have shown that this genomic region is associated with a large number of traits as udder morphology, protein yield, buck semen production, suggesting pleiotropic effects. Finally, in order to select goats for AI success, a genomic evaluation has been implemented on this trait which has been introduced in the total merit index in January 2023. The associated weights (15% in Saanen breed and 16% in Alpine breed) make it possible to improve this new trait while continuing to progress on the other traits under selection.

Session 79 Theatre 8

The male effect as an alternative to eCG in oestrus induction and synchronization treatment in ewes

N. Debus[1], G. Besche[1], S. Fréret[2], A. Hardy[3], M.T. Pellicer-Rubio[2], A. Tesniere[1] and J.-B. Menassol[1]
[1]UMR SELMET, Place Viala, 34000 Montpellier, France, [2]UMR PRC, INRAE, 37380 Nouzilly, France, [3]EPL La Cazotte, Route de Bournac, 12400 Saint-Affrique, France; nathalie.debus@inrae.fr

We tested if the male effect (ME) can be an alternative to eCG in oestrus induction treatment used prior to artificial insemination (AI) in ewes. Two groups of 50 Arles Merino (M) (2021) or Lacaune dairy (L) (2022) ewes were monitored during anoestrus season: (1) FGA group: ewes were treated with intra-vaginal sponges impregnated with progestagen for 14 days prior to ME; (2) C group: ewes received no hormonal treatment before ME. First, ME was performed for 14 days with 4 vasectomized rams equipped with the electronic oestrus detectors to monitor oestrus kinetics and introduced with the ewes on the day of sponge removal (D0). Then, the ewes were mated with 4 entire rams for 17 days. Progesterone was measured before FGA treatment to determine the cyclic status of the ewes prior to ME and for 11 days from D0 to characterise the ovulatory response to ME. We observed a high percentage of cyclic ewes before ME (from 46 to 74%). Most cyclic and non-cyclic ewes ovulated after ME in FGA and C groups (respectively 98 and 68% for M ewes; 98 and 63% for L ewes). The FGA treatment suppressed short cycles in response to ME (0 vs 12% for M ewes; 0 vs 6,1% for L ewes, $P<0.01$). The percentage of ewes in oestrus from D0 to D14 was higher in the FGA group than in the C group (96 vs 44% for M ewes; 96 vs 68% for L ewes; $P<0.001$). The onset of oestrus after D0 was earlier and better grouped in the FGA group (Median=34.34 h [Q1=31.03 h, Q3=43.25 h] for M ewes; 42.16 h [33.73 h, 56.85 h] for L ewes) compared to the C group (137.83 h [71.61 h, 246.74] for M ewes; 147.14 h [51.02 h, 232.35 h] for L ewes) ($P<0.001$). Our previous results showed a higher fertility rate for ewes inseminated between 0 and 35 h after the onset of oestrus . In the present study, 90% of M ewes and 68% of L ewes in the FGA group came into oestrus within a time window compatible with AI performed 53 h (M ewes) or 60 h (L ewes) after sponge removal, compared to 4% (M) and 10% (L) in the C group ($P<0.001$). The combination of ME and FGA pre-treatment is an alternative method to the use of eCG in AI protocols with a unique and fixed timing of insemination.

Session 79 Theatre 9

Circulating anti-Müllerian hormone from 5-month old Merino ewe lambs predicts first birthing rates

J. Daly, J. Kelly, K. Kind and W. Van Wettere
University of Adelaide, School of Animal and Veterinary Sciences, Roseworthy, 5371, Australia; william.vanwettere@adelaide.edu.au

Anti-Müllerian hormone (AMH) is an accurate phenotypic marker of antral follicle numbers and responsiveness to stimulation protocols in cattle and sheep. However, the relationship between pre-pubertal AMH and fertility of mature breeding females is poorly understood. Fertility at first mating and AMH has been correlated in Rasa Aragonesa and Sarda ewe lambs. The relationship between pre-pubertal AMH and the fertility of Australian Merino ewes has not yet been investigated. This study determined whether pregnancy and birthing rates of Australian Merinos following their first mating was related to plasma AMH concentration at 5 months old. At 5 months of age, a single blood sample was collected from 86 Merino ewe lambs managed under routine husbandry at a Research Centre in South Australia. At 18 months of age, ewes were housed with harnessed Merino rams for 35 days (24 hours / day). Ultrasound was used to determine pregnancy status, at an average of day 60 pregnancy. AMH levels in plasma were measured using an ovine specific AMH ELISA kit (Ansh laboratories, Texas, USA). Differences in AMH between groups of ewes that became pregnant and lambed and those that did not were determined using an ANOVA, unbalanced design (Genstat 19th Edition; VSC International). Data presented as mean ± SEM. Ewes scanned pregnant (n=70) had higher AMH at 5 months of age than those scanned not-pregnant (n=16) (2.74±0.22 vs 1.64±0.48 ng/ml; $P<0.05$). AMH was higher ($P<0.05$) for ewes which birthed (n=71) a lamb compared with those which did not (n=15) (2.83±0.21 vs 1.26±0.48 ng/ml). AMH concentrations at 5 months of age were higher ($P<0.05$) for singleton and twin-bearing ewes compared with non-lambing ewes (2.74±0.23 and 3.34±0.81 vs 1.26±0.48 ng/ml, respectively), but similar for singleton and twin-bearing ewes. These are the first data in the Merino to demonstrate that circulating AMH concentrations at 5 months of age are higher for hoggets which produce a lamb following their first mating. This finding is consistent with previous evidence that AMH can predict fertility in Rasa Aragonesa and Sarda ewes. Ongoing studies by our group will determine the relationship between AMH at weaning and lifetime fertility and productivity of Merino ewes.

Session 79 — Poster 10

Impact of pinecone oil on reproductive performance, milk composition and serum parameters in sows

Q.Q. Zhang[1], J.H. Song[1], C.B. Lim[1], S. Biswas[1], J.S. Yoo[2] and I.H. Kim[1]
[1]Dankook University, Department of Animal Resource and Science, Cheonan, 31116, Korea, South, [2]Daehan feed company, 13 Bukseongpo-gil, Jung-gu, Incheon, 22300, Korea, South; 1844700948@qq.com

Pinecone oil (PO) of *Pinus koraiensis* mainly contains alpha-pinene, beta-pinene, and limonene that may ameliorate animal well-being and growth performance. This study evaluated its effects on feed intake, milk composition and yield, serum parameters, and litter growth of sows. Twenty-seven pregnant sows (parity 2 to 4) were distributed to 3 dietary treatments. The trial started on day 107 of gestation and ended on day 21 of lactation. Sows were given either a basal diet or the basal diet + 200 or 400 mg/kg PO. Each treatment contained 9 sows and each sow was considered an experimental unit. Results showed that the average daily gain and weaned weight of piglets from the sows fed 400 mg/kg PO supplements were higher ($P<0.05$) than the piglets from the control sows. Lactose content in colostrum samples and fat content in milk samples were higher ($P<0.05$) in 200 and 400 mg/kg PO-treated sows, respectively than those from the sows fed basal diet. Additionally, cortisol concentration for assessing stress in sow serum was lowered ($P<0.05$) by dietary 200 and 400 mg/kg PO on d 21 of lactation. Aspartate aminotransferase concentration in sow serum was lowered ($P<0.05$) by supplementing 400 mg/kg PO. In conclusion, supplementation of 400 mg/kg PO during late gestation and lactation contributed to greater offspring growth performance, possibly by enhanced milk quality and alleviated maternal stress.

Session 79 — Poster 11

Maternal *Forsythia suspensa* extract alleviated oxidative stress and improved gut health in sows

S. Long, Z. Chen and T. He
China Agricultural University, No.2 Yuanmingyuan West Road, 100193, Beijing, China, P.R.; longshenfei@cau.edu.cn

This experiment aimed to investigate the effects of maternal *Forsythia suspensa* extract (FSE) supplementation from d 85 of gestation to lactation on alleviating oxidative stress and improving intestinal health of sows and suckling piglets. Fifty gestating sows (Landance × Yorkshire, average parity 1.61±0.68, initial weight 199.28±37.23 kg) were selected and divided into 2 treatments, 25 sows per treatment. The trial was started from d 85 of gestation and ended on d 21 of lactation (weaning day). The treatments included the control group (corn-soybean meal basal diet; CON) and FSE group (basal diet + 100 mg/kg FSE). The results showed that compared with the CON, the litter weight gain of piglets and the apparent total tract digestibility of dry matter and organic matter in sows during lactation were increased ($P<0.05$) in FSE group. Moreover, the colostrum fat content, serum IgG and IL-10 levels of sow were increased ($P<0.05$) at weaning, while the serum MDA and weaning serum hydroxyl free radical levels of sows were reduced ($P<0.05$) at farrowing in FSE group. The serum T-AOC and GSH-Px activities of piglets were increased ($P<0.05$), while the serum hydroxyl free radicals, serum and colostrum IL-1β level of piglets were reduced ($P<0.05$) in FSE group. Maternal dietary FSE supplementation increased the levels of SOD and CAT in duodenum ($p\leq0.05$) and the level of GSH-Px in ileum ($P<0.05$) of piglets. Moreover, dietary FSE supplementation increased ($P<0.05$) the relative abundance of Bacilli at class level, Lactobacillus and Prevotellaceae_UCG-004 at genus level in caecal digesta of suckling piglets. In conclusion, maternal FSE supplementation from d 85 of gestation and lactation period improved the reproductive performance, milk composition, antioxidant status, immune response and gut microbial community of sow and their offsprings.

Session 79 Poster 12

Effects of PGF2α on luteal tissue morphology, expression of relative genes in Hu sheep

Y.Q. Liu, Y. Li, C.H. Duan and Z.P. Song
Hebei Agricultural University, Lekai Street, Baoding, 071000, China, P.R.; liuyueqin66@126.com

The purpose of this experiment was to study the effects of prostaglandin F2α(PGF2α) on luteal tissue morphology, expression of PGF2α receptors and programmed necrosis-associated genes at different luteal stages in Hu sheep. Forty eight healthy ewes with similar body weight were randomly divided into 6 groups, namely the early-luteal phase experiment group and control group, the mid-luteal phase experiment group and control group, and the late-luteal phase experiment group and control group. The three experimental groups were injected with 1ml PGF2α (0.1 mg) on day 6 (early-luteal phase), day 11 (mid-luteal phase) and day 16 (late-luteal phase), respectively. The corpus luteum tissue was collected at 3 h after each injection for gene expression detection and the morphological changes of corpus luteum. The results showed that the morphology of corpus luteum was changed after injection of PGF2α,with inflammatory cell infiltration and apoptosis, especially at the mid-luteal phase with a higher degree of degeneration. In the early-luteal phase, injection of PGF2α for ewes significantly reduced the expression of PGF2α receptor A (PA) ($P<0.05$), and significantly increased the expression of PGF2α receptor B (FPB), tumour necrosis factor receptor one(TNFR1), mixed lineage kinase domain-like protein (MLKL) ($P<0.05$), and had no effect on the expression of tumour necrosis factor α (TNF-α), tumour necrosis factor receptor one(TNFR1), and receptor interacting protein kinases three(RIPK3) ($P>0.05$). In the mid-luteal phase, injection of PGF2α for ewes significantly up-regulated the expression of FPB, TNF-α, RIPK1 and MLKL ($P<0.05$), but has no effect on the expression of FPA, TNFR1 and RIPK3 ($P>0.05$); In the late-luteal phase, injection of PGF2α for ewes significantly up-regulated the expression of RIPK1 ($P<0.05$), has no effect on the expression of FPA, FPB, TNF-α, TNFR1, RIPK3 and MLKL ($P>0.05$). The mRNA expression level of FPA was positively correlated with RIPK3 and TNFR1 ($P<0.05$), and the mRNA expression level of FPB was positively correlated with MLKL and TNF-α ($P<0.05$). Our findings demonstrate that PGF2α may initiate necroptosis through the TNF-α/TNFR1 pathway by binding to its receptors FPA and FPB, and promote the degeneration of the corpus luteum.

Session 79 Poster 13

Relationship between some micronutrients in serum and their effect on the seminal quality of bulls

A. Benito-Diaz[1], A.S. Sarmiento-García[2], M. Montañes-Foz[1], R. Bodas-Rodríguez[1] and J.J. García-García[1]
[1]Instituto Tecnológico Agrario de Castilla y León, Subdirección de Investigación, Autovía de Castilla, 119, 47009 Valladolid, Spain, [2]Universidad de Salamanca, Área de Producción Animal, Departamento de Construcción y Agronomía, C/Filiberto Fillalobos1 119, 37007 Salamanca, Spain; asarmg00@usal.es

Many factors influence reproductive efficiency in beef cattle, among them, the bioavailability of micronutrients (such as vitamins and mineral traces) has been described as having an essential role in the basic motility characteristics of semen. The current research aimed to demonstrate the influence of some micronutrient serum concentrations (vitamin E, copper, and zinc) on the seminal quality of beef bulls (*Bous Taurus*). 44 seminal and serum samples from bulls of 2 through 8 years of international (Limousine, Charolais) and national (Avileña Negra Ibérica, Morucha, and Lidia) breeds which were allocated in Castilla y León and Extremadura (Spain) over 2020 were collected. The semen was collected from the electroejaculation procedure, while blood samples were drawn from the coccygeal vein. Blood samples were centrifuged, split into two aliquots, and frozen (-20 °C) until they were sent to the laboratory to determine the levels of micronutrients. Vitamin E was determined by ultra-performance liquid chromatography with a photodiode detector, while, copper and zinc concentrations were quantified by inductively coupled plasma mass spectrometry. Semen samples were analysed within two hours of the collection with CASA instruments. RStudio was used for data management and statistical analysis. Progressive motility was the most critical semen quality variable: Concentrations of copper, zinc, and vitamin E were: 67.55±16.03 μg/dl, 116.91±46.02 μg/dl, and 3.47±1.73 mg/l respectively. Concerning progressive motility, the average percentage was 49,57%. A positive effect was observed ($P<0.05$) between vitamin E concentration and the sperm motility of the bull. These findings suggest that enhancing vitamin E bioavailability would be desirable to improve fertility rates in bulls.

Session 79 Poster 14

Genetic correlations on kinetic sperm parameters in a closed population of PRE stallions

Z. Peña[1], A. Molina[1], C. Medina[1], M. Valera[2] and S. Demyda-Peyrás[3]
[1]Universidad de Córdoba, Campus Rabanales, 14014, Cordoba, Spain, [2]Universidad de Sevilla, Utrera 1, 41005 Sevilla, Spain, [3]FCV-UNLP, Calle 60 y 118 SN, 1900 La Plata, Argentina; zahiramaria@hotmail.com

Sperm quality is an important factor in horse production systems. Nowadays, computer-assisted sperm analysis (CASA)it is possible to analyse kinetic parameters which characterize the motility pattern of a given stallion. Hereby, we modelled, the genetic relationships among 5 kinetic sperm parameters and sperm motility in a closed population of Pura Raza Español stallions. The dataset included 720 ejaculates from 108 PRE individuals collected by the artificial vagina during 10 reproductive seasons in Spain. Raw samples were collected, centrifuged, and diluted to 50×106 spz/ml, and equilibrated by 5 min. Thereafter, total (MOT) and progressive (PROG) motility, and 5 kinetic parameters (VSL, VAP, VCL, ALH, and BCF) were assessed using an Androvision™ CASA analyser. EBV's for the 7 sperm traits were obtained using a multivariate REML mixed model which included age at the jump, and year of birth as fixed effects, the permanent environmental effect (PEA) and additive genetic effect (2,005 pedigree records). Inbreeding value (F) was included as a linear covariate using the BLUPF90 software package. Results showed low-to-moderate h^2 values for MOT and PROG (0.08 and 0.28), but moderate-higher estimates on kinetic traits (ranging 0.25 to 0.46). Interestingly PEA was much higher in motilities than in kinetic traits. MOT was positively correlated with velocities (~0.4) and head movement (~0.3), whereas PROG was only correlated positively with VSL (0.47) and negatively with ALH (-0.54). The genetic correlations among velocities (VAP, VSL, and VCL) were very high (ranging from 0.84 to 0.96), as well as with head movements but to a lesser extent (around 0.6). However, VSL and ALH; and VCL and BCF (0.47) correlations were lower. Finally, the correlation between both head displacement parameters was also high and positive (0.57). To our knowledge, this is the first attempt to model the genetic relationships among kinetic traits in horse sperm. However, the particularities (high relatedness and inbreeding values) and limited size of the population analysed suggest caution in extrapolate the results to further equine populations.

Session 79 Poster 15

Gestation length in Braunvieh cows

M. Fanger, A. Burren and H. Jörg
Bern University of Applied Sciences, School of Agricultural, Forest and Food Sciences HAFL, Länggasse 85, 3052, Switzerland; melanie.fanger@students.bfh.ch

In the present study, the influence of sire breed on pregnancy duration in Braunvieh (BV) and Original Braunvieh (OB) cows was investigated. Braunvieh Schweiz provided data from 2,096,865 BV and 132,731 OB cows from 2000-2021 with 14 and 11 different sire breeds, respectively. The different sire breeds were represented as follows: Angus (AN) 20'119/780, Blue Belgian (BB) 22'630/316, Blonde d'Aquitaine (BD) 28'191/316, Brown Swiss (BS) 288'845/395, Braunvieh (BV) 1'177'947/3'159, Charolais (CH) 24'828/262, Holstein (HO) 11'718/0, Eringer (HR) 2'679/101, Jersey (JE) 2'793/0, Limousin (LM) 368'229/11'451, Original Braunvieh (OB) 96'716/114'809, Pinzgauer (PI) 5'469/168, Swiss Fleckvieh (SF) 1'938/0 and Simmentaler (SI60) 44'763/974. The relationship between gestation length and sire breed was examined using a mixed linear model. The model included fixed effects calving year and month, milk linear and quadratic, calving age linear and quadratic, number of calves born by sex and sire breed, and random effects mother, sire, and farm. The longest gestation period resulted in BV and OB cows with sire breed BD (289.7 days) and (289.9 days), respectively. This was followed by PI (289.2), OB (289.2), LM (288.4) and SI (287.8) breeds in BV cows and PI (289.3), OB (289.2), LM (288.6) and SI (288.3) in OB cows. In BV cows, the sire breed HO (282.4) and in OB cows, the sire breed AN (283.2) achieved the shortest pregnancy durations. Further, both sex and the number of calves born had an influence on pregnancy duration. For a male calf, the gestation period is longer than for a female (BV: f=288.6, m=290.4; OB: f=289.8, m=291.5). Twin births result in a slightly shorter gestation period (BV: ff=283.6, mf=283.9, mm=285.3; OB: ff=284.3, mf=284.8, mm=286.0). The results show that breeders can influence the gestation length by their selection of sire breed.

Session 79 Poster 16

Distribution of adipokines in reproductive tract and embryonic annexes in hen

O.B. Bernardi[1,2], A.E. Estienne[1], M.R. Reverchon[2], A.B. Brossaud[1], C.R. Rame[1] and J.L. Dupont[1]
[1]INRAE, UMR PRC, 37380 Nouzilly, France, [2]SYSAAF, UMR BOA, 37380 Nouzilly, France; ophelie.bernardi@inrae.fr

Nowadays, it is well known that adipokines, molecules secreted predominantly by white adipose tissue, have endocrine and paracrine functions involved in the regulation of many biological processes including energy metabolism, inflammation and reproduction. In mammals, it has been shown that adipokines are synthetized locally in the reproductive tract and could influence reproductive functions such as steroidogenesis. In avian species, we previously showed that chemerin is highly abundant in the albumen, compared to the yolk. Moreover, chemerin and its receptors are more expressed in the part of the oviduct called the magnum, where the albumen is formed. In addition, we observed that chemerin is expressed by allantoic and amniotic membranes and fluids during the embryo development. So in chicken, chemerin is present in the albumen, the magnum of the oviduct and expressed in embryo annexes. Here we investigated other adipokines such as adiponectin, visfatin, apelin and adipolin in the egg, the reproductive tract and embryo annexes in order to better understand their distribution and role during the embryo development. By using Western blot, RT-qPCR analysis and immunohistochemistry, we demonstrated the expression of different adipokines in the egg albumen (visfatin) and the reproductive tract (adiponectin, visfatin, apelin, adipolin). However, in case of regressive tract no adipokines were expressed suggesting a potential inhibition by steroids. We also measured the expression of adipokines in the allantoic and amniotic membranes at different embryonic days ED: 7, 9, 11, 14, 16 and 18. We showed that the amount of adipokines was dependent of the embryo development. Taken together, adipokines and their receptors are expressed in the egg, the reproductive tract and the embryonic annexes in chickens. Datas suggest an important function of theses metabolic hormones during the early embryo development in *Gallus gallus domesticus*. Thus, adipokines could be used as predictive biomarkers for embryo development.

Session 79 Poster 17

The effects of PROK 1, alone or in combination with IFN gamma on endometrial immune response in pig

S.E. Song and J. Kim
Dankook University, Animal Resources Science, 119, Dandae-ro, Dongnam-gu, Cheonan-si, Chungcheongnam-do, Republic of Korea, 31116, Korea, South; thdtjdms@naver.com

Porcine endometrium undergoes fine-regulated inflammatory response during implantation as pig conceptuses secrete the pro-inflammatory cytokine interferon gamma (IFNG) and prokineticin1 (PROK1). IFNG promotes inflammatory responses in endometrium by increasing endometrial chemokines recruiting T cells and mast cells to maternal-foetal interface of pig. PROK1 acts pleiotropically as an embryonic signal mediator that regulates endometrial receptivity by increasing the expression of the genes and proteins involved in implantation. however, the underlying molecular regulatory mechanisms are not fully elucidated *in vitro*. This study investigated loading the secretory protein PROK1 into porcine endometrial cells and measured the migration of porcine peripheral blood mononuclear cells (PBMCs) toward the cytokine and PROK1 loaded endometrial cells. We also analysed the expression of T cell receptors. The results of this study demonstrate that PROK1 loaded endometrial cells increase migration of CD4+, CD8+, and CD4+CD8+ T cells in PBMCs compared to non-loaded cells. Long term exposure (24 h) to a combination of PROK1 and IFNG increased the expression of T cell co-signalling receptors including programmed cell death 1 (PDCD1), CD28, as well as chemokines CXCL9, CXCL10 and CXCL11 compared to IFNG treated alone. These results suggest that PROK1 may be critical for the establishment of pregnancy by regulating inflammatory environment and various subpopulations of T cell recruitment into the endometrium during the implantation period in pigs.

Session 80 Theatre 1

Creating a culture of openness for the responsible use of video observation in animal sciences

M.F. Giersberg and F.L.B. Meijboom
Utrecht University, Veterinary Medicine, Yalelaan 2, 3584 CM Utrecht, the Netherlands; m.f.giersberg@uu.nl

Video observation is a popular and frequently used tool in animal behaviour and welfare research. However, in addition to the object of research, video recordings often provide unintended and unexpected information about the progress of the study, the animals or the people involved. This information can cause uncertainty in the observer when conflicting interests between animal welfare, social expectations and research integrity are weighed against each other. How should they deal with what they saw? Does for instance the welfare of the animals outweigh the privacy of the caretaker? In this study, we propose a framework which enables a culture of openness in dealing with unexpected and unintended events observed during video analysis in animal sciences. To deal with these events, the general principles of privacy and research integrity are an essential starting point. People involved in a video-based animal study need to be aware of their institutions' policies or the absence of such, for instance on consent or data confidentiality. However, in animal studies, a single focus on the prevention of privacy issues is not sufficient. That is because study animals are in a vulnerable position and depend on human care. The assumptions and responsibilities underlying such a duty of care or the expected human behaviour towards animals should therefore be made explicit in a similar way as privacy or integrity issues. Informing, educating or reminding people of these responsibilities during the planning of the study provides them with so called 'good conditions' under which to choose their behaviour. During running and reporting of the study, the risk of animal welfare and research integrity issues can be mitigated by making conflicts discussible and offering realistic opportunities on how to deal with them. A practice which is outlined and guided by conversation will prevent a mere compliance-based approach centred on checklists and decision trees. Based on this framework, we are currently developing a workshop for PhD candidates, which includes serious gaming approaches and offers room to share own experiences. With this we aim to foster reflection, co-creation and application of ethical practice.

Session 80 Theatre 2

Salivary oxytocin and lachrymal caruncle temperature as indicators of anticipation in growing pigs

G.A. Franchi[1], L.R. Moscovice[2], H. Telkänranta[3] and L.J. Pedersen[1]
[1]Aarhus University, Department of Animal and Veterinary Sciences, Blichers Allé 20, 8830 Tjele, Denmark, [2]Research institute for Farm Animal Biology (FBN), Institute of Behavioural Physiology, Wilhelm-Stahl-Allee 2, 18196 Dummerstorf, Germany, [3]Arador Innovations, Kamreerintie 10, 02770 Espoo, Finland; amorimfranchi@anivet.au.dk

Conventional pigs are typically raised under intensive conditions that do not meet pigs' basic needs of behavioural expression and cognitive challenge, likely threatening their welfare. Thus, providing pigs an environment that promotes positive affective states and, consequently, positive welfare should be encouraged. Affective states and valuing of rewards can be examined through anticipation, i.e. a response to a stimulus or context based on expectations about the future. Hence, we investigated, as part of the H2020 project PIGWEB, the potential of salivary oxytocin (sOXT) and lachrymal caruncle temperature (LCT), measured by infrared thermography, as indicators of anticipatory responses in 32 conventional Yorkshire × Landrace pigs (mean ± SD body weight: 81±7 kg/pig) housed in 8 identical pens (4 pigs/pen; 3.28 m^2/pig) during anticipation of either a positive (delivery of 200 g/pig of fresh straw; POS) or a negative context (straw delivery omission; NEG). Individual saliva samples were collected at -10 and +20 min and thermograms were recorded at -10, -5 and +10 min relative to the moment of conditioned stimulus presentation. Additionally, based on footage, play behaviour (locomotor-rotational, LOC; social, SOC; object, OBJ) was sampled continuously and tail posture (high; not high) was instantaneously sampled at 1-min interval, both at the individual level, from -30 to +30 min relative to the conditioned stimulus presentation moment. We found higher LCT (37.9±0.47 vs 37.3±0.47 °C; P=0.013) and sOXT concentration (110±7.64 vs 80±6.70 pg/ml; P=0.002) in POS than in NEG. No high tail was labelled in NEG, whereas 66% of pigs displayed high tail (P<0.001). The odds of pigs displaying OBJ were higher in POS than in NEG (OR, 95%CI:6.7, 1.39-32.47; P=0.015). Additionally, pigs tended to engage more in SOC in POS than in NEG (P=0.077). Our results suggest thermographic LCT, sOXT, tail posture, and play behaviour have potential as non-invasive measures of anticipation and affective processing in pigs.

Session 80 Theatre 3

Pig oxidative stress model: behaviour as a potential indicator of oxidative stress in pigs

R.D. Guevara[1,2], J.J. Pastor[3], S. López-Vergé[3], X. Manteca[2], G. Tedo[3] and P. Llonch[2]
[1]AWEC Advisors S.L., Farm Animal Welfare Education Centre (FAWEC), Research Park UAB, Campus UAB, 08193, Spain, [2]Universitat Autònoma de Barcelona, Department of Animal and Food Science, Edifici V, Travessera dels Turons, 08193, Spain, [3]Lucta, Animal Science Innovation Division, Research Park UAB, Campus UAB, 08193, Spain; raul.guevara@awec.es

The goal of the current study was to develop a pig experimental model to investigate oxidative stress with fewer negative impacts on piglet welfare. Four independent trials (A, B, C, and D) were performed using a single intraperitoneal shot of lipopolysaccharide (immune stimulator, LPS) as an immune challenge, aiming to assess the minimal LPS dose and the most appropriate pigs' age at the oxidative challenge to trigger a measurable acute response. The challenge time and dose for each trial were the following, Trial A and B: 21 days post-weaning (p.w.) piglets (A: n=24, 6.98±1.97 kg; B: n=24, 9.20±1.09 kg) with 25 µg/kg BW of LPS; Trial C: 28 days p.w. piglets (n=24, 9.08±3.25 kg), with 25 µg/kg BW of LPS; and Trial D: 41 days p.w. piglets (n=20, 11.86±5.95 kg), with 25 µg/kg BW. The pigs were randomly allocated either to T1) Control diet + saline solution (CON), or T2) CON + LPS challenge (LPS). The oxidative response was measured through plasma glutathione peroxidase (GPx), glutathione-S-transferase (GST), superoxide dismutase (SOD), and catalase (CAT). Intestinal relative gene expression of oxidative and inflammatory markers was assessed. Faecal myeloperoxidase (MPO) activity was measured to evaluate intestinal inflammation. Illness-related behaviours (panting, prostration, trembling, and vomits) were also recorded. Plasmatic oxidative response was not consistent across the four trials even when the dose and pig age were similar (studies A & B). Pigs' individual variability may be an explanation for this outcome. Relative gene expression of inflammatory (IL10) and antioxidant (iNOS, GPx4, MnSOD, and CAT) activity detected differences between CON and LPS treatment (P<0.05). Behavioural observations were sensitive to the LPS dose relative to CON (P<0.05). These results suggested that behavioural observations may be used as a non-invasive methodology to measure the impact of oxidative stress in pigs.

Session 80 Theatre 4

Comparison of aversiveness of 8 different inert gas (mixtures) to CO_2 for stunning pigs at slaughter

I. Wilk[1], J. Gelhausen[2], T. Friehs[2], T. Krebs[2], D. Mörlein[2], J. Tetens[2] and J. Knöll[1]
[1]Friedrich-Loeffler-Institut, Institute for Animal Welfare and Animal Husbandry, Dörnbergstraße 25/27, 29223 Celle, Germany, [2]Georg-August-Universität Göttingen, Department of Animal Sciences, Burckhardtweg 2, 37077 Göttingen, Germany; jonas.knoell@fli.de

Despite concerns over its aversiveness, stunning of pigs at the time of slaughter using a high concentration of CO_2 is the most common method in Europe. Inert gases and mixtures of inert gases with low concentration of CO_2 have been proposed as an alternative, but have so far not been considered market-ready due to concerns of gas stability, meat quality, costs and stunning effectiveness. As part of the project for Testing Inert Gases in order to Establish Replacements for high concentration CO_2 stunning for pigs at the time of slaughter (TIGER), experiments were conducted in a commercial Dip-Lift system using a new gassing system with residual oxygen concentrations <1%. Pigs were stunned to screen 8 different inert gas mixtures for their suitability in terms of animal welfare and meat quality. In two separate series for argon-based and nitrogen-based gas mixtures, pigs were exposed to one of four inert gas mixtures with 0, 10, 20 or 30% CO_2, or high concentration CO_2 control atmospheres. In nitrogen-based gas mixtures, argon was added with up to 30% to aid gas stability. Here we report results of the aversiveness to the gases in the induction phase. For each of the four measurement days, video sequences of the stunning procedures were randomized and cut to a new video. Aversive and other events were captured by two observers according to an ethogram. Most pigs showed an initial retreat attempt (62.5%), shortly after start of the descend of the gondola. Other aversive reactions before loss of balance were statistically significantly reduced (P<0.05) for all argon-based gas mixtures (7 to 21%) and for nitrogen without CO_2 (21%) compared to high concentration CO_2 (59%). Of those aversive events, hyperventilation was only observed for high concentration CO_2 (25%) and nitrogen with 30% CO_2 (15%). We conclude that all tested inert gas mixtures were beneficial with respect to reduced aversiveness in the induction phase compared to high concentration CO_2, with gas mixtures with less CO_2 showing less or shorter aversions.

Session 80 Theatre 5

Detecting onset of farrowing using CUSUM-charts based on sows' activity

T. Wilder, B. Baude and J. Krieter
Christian-Albrechts-University, Institute of Animal Breeding and Husbandry, Olshausenstr. 40, 24098 Kiel, Germany; twilder@tierzucht.uni-kiel.de

With the latest changes in legislation in Germany, sows are only allowed to be confined in farrowing crates for five days to prevent piglet crushing. Thus, it is important to estimate the onset of farrowing correctly to close the crate just before the farrowing and open it again after the first days to reduce negative effects on sow welfare. Monitoring the activity level of the sow, which increases about 16 to 20 h ante partum due to nest-building activities and restlessness, it is possible to estimate the onset of farrowing. However, doing this with human observation is not feasible under commercial conditions. Therefore, the activity of sows (n=125) was estimated using optical flow (OF, changing of pixels from one frame to the other) of video footage and cost efficient passive infrared motion sensors (PIR) during 96 h ante partum. CUSUM-charts were used to detect a change in activity separately for OF and PIR. To minimise the influence of farm staff moving in the pen, the mean activity was calculated over different time intervals (15, 30, 60 min). Afterwards, the individual activity of every sow was standardised using z-transformation ($(x-\mu)/\sigma$). Thus, every sow had a mean activity of 0 and a standard deviation of 1. This data was used in univariate one-sided CUSUM-charts. The best results of the OF were achieved with mean activity of the 60 min time interval and the combination of h=9 and k=0.25. Here, the increase in activity was correctly detected in 97% of times at a mean of 9 h 7 min (±5 h 31 min standard deviation) before the actual farrowing. The number of false positive alarms (earlier than 24 h before farrowing) was 0.23 per sow. For the PIR, the best results were achieved with 15 min time interval and the combination of h=1.5 and k=0.5. In 91% of times increase of activity was detected at a mean of 16 h 46 min (±7 h 11 min) before the farrowing. Here, the number of false positive alarms was 4.84 per sow. Therefore, using activity monitoring, it is possible to time the closing of the crate more precisely. However, the number of false positive alarms of the PIR motion sensor has to be improved to be of practical interest.

Session 80 Theatre 6

Impact of environmental enrichment on the behaviour and immune cell transcriptome of pregnant sows

M.M. Lopes[1], C. Clouard[1], J. Chambeaud[1], M. Brien[1], N. Villain[2], C. Gerard[2], F. Hérault[1], A. Vincent[1], I. Louveau[1], R. Resmond[3], H. Jammes[4] and E. Merlot[1]
[1]INRAE, PEGASE, 16 Le Clos, 35590 Saint-Gilles, France, [2]Chambre Régionale d'Agriculture de Bretagne, Maurice le Lannou, 35042 Rennes, France, [3]INRAE, IGEPP, La Motte au Vicomte, 35653 Le Rheu, France, [4]INRAE, BREED, All. de Vilvert, 78352 Jouy-en-Josas, France; mariana.mescouto-lopes@inrae.fr

The ability to assess farm animals' mood is important to evaluate their welfare, but practical assessment tools are still lacking. Human research has demonstrated a link between psychological states and the transcriptome of blood immune cells. Therefore, this study aimed to investigate whether blood immune cell transcriptome can be used to assess the animals' mood using environmental enrichment as a method to generate contrasted welfare states. Pregnant sows of mixed parities were housed in two contrasting conditions throughout gestation (0 to 105 days): a conventional system on a slatted floor (C, n=36) or an enriched system on accumulated straw with additional space per sow (E, n=35). The behaviour of multiparous sows of low (2nd and 3rd gestation; n=29) and high (4th gestation or higher; n=31) parity was observed from G99 to G104 and 14 sows per system were selected for biological sampling. Cortisol concentrations in saliva (G35 and G98), and in the hair (G98), were lower in E sows ($P<0.04$). E sows spent more time exploring the pen ($P<0.001$), less time chewing enrichment material ($P<0.001$) or exhibiting stereotypic behaviours ($P=0.04$), and had lower frequencies of agonistic behaviours ($P=0.04$). High-throughput sequencing of the blood mononuclear cell transcriptome (G98) identified only 24 differentially expressed genes (DEGs) between C and E sows (adjusted P-value<0.1, FC<0.8 or >1.2). However, parity (894 DEGs) and social dominance (437 DEGs) had a greater effect, and most of the DEGs were related to innate and adaptive immunity pathways. Therefore, these results confirm that long-term environmental enrichment decreases cortisol concentration and positively influences sow behaviour. The blood transcriptome did not allow discrimination between housing conditions but it was influenced by another important factor for welfare, such as social dominance in the group.

The knowns and unknowns about feather pecking in laying hens

A. Harlander and N. Van Staaveren
University of Guelph, Animal Biosciences, Campbell Centre for the Study of Animal Welfare, Stone Road East, N1G2W1, Canada; aharland@uoguelph.ca

Feather pecking (FP), which leads to feather damage, is a behaviour commonly performed by millions of birds kept for egg-laying worldwide. It is a unique behaviour as FP is both an indicator and a cause of reduced welfare. Given its prevalence and impact to hen health, there is an urgent need to prevent FP behaviour. Ethologists have long considered FP to be reflective of frustration, arising from a lack of foraging opportunities within commercial settings. However, FP also occurs when birds are provided outdoor access, allowing them to express normal foraging behaviour, indicating additional contributing factors. This talk will summarize recent research linking FP in laying hens to the gut-microbiome-brain axis. In humans, there is an overlap between behavioural disorders, such as autism spectrum disorder, the gut microbiota and 5-HT neurotransmission between the brain and gut. Similar associations have been recently established in laying hens. Indeed, the avian serotonergic system and the amino acid precursor TRP are intimately linked to FP. Moreover, a strong desire to eat feathers, altered intestinal motility, lower gut bacterial diversity and a decreased presence of *Lactobacillaceae* has been reported in feather peckers vs non-peckers. Interestingly, the administration of *Lactobacillaceae* as a probiotic to laying hens impacted the immune system via tryptophan metabolism by reducing plasma indoleamine-pyrrole 2,3-dioxygenase activity and kynurenine concentration. More importantly, the probiotic *Lactobacillaceae* treatment prevented pecking. To conclude, the field of research to better understand FP through a neurobiological gut-brain axis lens is rich with opportunities and the potential for preventive treatment.

Lower redness of the facial skin is a marker of a positive human-hen relationship

D. Soulet[1], A. Jahoui[1], M.-C. Blache[1], B. Piégu[1], G. Lefort[1], L. Lansade[1], K. Germain[2], F. Lévy[1], S. Love[1], A. Bertin[1] and C. Arnould[1]
[1]CNRS, IFCE, INRAE, Université de Tours, PRC, 37380, Nouzilly, France, [2]INRAE, UE EASM, Le Magneraud, CS 40052, 17700, Surgères, France; delphine.soulet@inrae.fr

In certain mammals, facial expressions are reliable markers of the emotional state of an individual. In humans and parrots, emotional stimuli can induce a rapid change of face colour that is linked to blood flow. The aim of this study was to test whether the redness of the face in domestic hens could be used as an emotional marker in the context of the human-hen relationship. Two groups of hens (Sussex) were studied: a group habituated to a human with daily positive interaction (n=13) and a non-habituated group (n=12). Behaviour and skin colour of wattles, cheeks, ear lobes and comb were analysed from two video-recorded tests (novel environment and reactivity to human tests) conducted after 5 and 6 weeks of habituation. In the novel environment test, usually used for testing general underlying fearfulness, the hens were tested alone in an unknown environment. In the reactivity to human test, the hens were tested again in this environment but with the familiar human sitting inside. The human presence should only be stressful for the non-habituated group. As expected, the behaviour of the two groups did not differ during the novel environment test. During the reactivity to human test, the habituated hens were more relaxed than the non-habituated hens: they took their first step faster (P=0.03), came faster into contact with the human (P=0.03), explored him longer (P=0.0003) and displayed more comfort behaviours (P=0.01). Face redness differed between the two groups during only the reactivity to human test: lower redness was observed in habituated hens for 3 out of the 4 regions (wattles P=0.01, cheeks P<0.0001, ear lobes P=0.0003). These results show that more relaxed hens are less red, thus face redness could be a reliable marker of the quality of the human-hen relationship. As such, it could be used as a tool for human to infer the emotional state of hens and more broadly, to evaluate their welfare.

Do sheep differentiate emotional cues conveyed in human body odour?

I. Larrigaldie[1], F. Damon[1], S. Mousqué[1,2], B. Patris[1], L. Lansade[3], B. Schaal[1] and A. Destrez[1,2]
[1]UMR Centre des Sciences du Goût, CNRS, Institut Agro, INRAE, Univ. Bourgogne, 9E boulevard Jeanne d'Arc, 21000 Dijon, France, [2]Institut Agro, Dijon, France, 26 Bd Dr Petitjean, 21079 Dijon, France, [3]UMR Physiologie de la Reproduction et des Comportements, CNRS, IFCE, INRAE, Univ. Tours, INRAE, 37380 Nouzilly, France; izia.larrigaldie@u-bourgogne.fr

Humans are significant sources of stimuli for domestic animals, especially by visual/vocal cues. Evidence that horses, dogs, and cattle do react to cues carried in human body odours is recent. Sheep perception of human visual/vocal cues has been previously reported, their reactivity to human body odor is yet to be demonstrated. We assessed whether sheep detect human odours and differentiate stressed vs non-stressed individuals. Axillary secretions were sampled from 34 students (age: 23 y, 31 ♀) before an oral examination vs a standard class, conveying stress (SO) and non-stress (nSO) odours, respectively. These samples were frozen until testing that consisted in a habituation-dishabituation procedure (n=29 sheep, age: 7-8 m, 14 ♀). An odour n°1 (habituation stimulus) was presented one minute four times at animal, then odour n°2 (dishabituation stimulus) was presented one minute once time. Sheep were either habituated with nSO and dishabituted with SO or the reverse (ability to detect and discriminate human nSO/SO). Tests were video recorded to blindly code behaviours indicating attraction, aversion or indifference (i.e. approach/withdrawal, locomotor spatiality, sniffing, ingestion, vocalization). Ewes (8.4; P=0.038) and rams (8.6; P=0.037) displayed more avoidance/stress (frequency of ears pointing backwards) when facing the dishabituation odour, but regardless of its assumed cueing of human stress. While sheep were expected to express specific different behaviours when exposed to nSO and SO, these results ultimately suggest that sheep avoid the odours of unfamiliar humans (although the discrimination of familiar/unfamiliar humans remains to be tested). Overall, further studies are required to unveil ovine olfactory performance toward humans along dimensions of familiarity and emotionality, and to provide a better understanding of subtle aspects of the human-animal relationship.

Evaluation of inter-observer reliability of dichotomous and four-level animal-based indicators

B. Torsiello[1], M. Giammarino[2], L. Battaglini[1], M. Battini[3], S. Mattiello[3], P. Quatto[4] and M. Renna[5]
[1]University of Turin, Dept. of Agricultural, Forest and Food Sciences, Largo P. Braccini 2, 10095 Grugliasco, Italy, [2]Asl TO3, Veterinary Service, Dept. of Prevention, Via Trento 1, 10045 Piossasco, Italy, [3]University of Milan, Dept. of Agricultural and Environmental Sciences – Production, Landscape, Agroenergy, Via Celoria 2, 20133 Milano, Italy, [4]University of Milan-Bicocca, Dept. of Economics, Management and Statistics, Piazza dell'Ateneo Nuovo, 1, 20126 Milano, Italy, [5]University of Turin, Dept. of Veterinary Sciences, Largo P. Braccini 2, 10095 Grugliasco, Italy; benedetta.torsiello@unito.it

This study focuses on the problem of assessing inter-observer reliability (IOR) in the case of dichotomous and four-level animal-based welfare indicators, using and comparing the performance of the agreement indexes available in literature. Udder asymmetry was evaluated as a dichotomous variable in 160 dairy goats by three observers (A, B and C). As four-level variables, Earposture (EP) and Eyewhite (EW) were scored by two observers from 436 photos of dairy cows during the lactating (L) and dry (D) periods. Krippendorff's α, Fleiss' K and Quatto's S were implemented to assess IOR for the dichotomous variable with three observers, while Cohen's K, K_C, K_{PABAK}, Krippendorff's α and Quatto's S were implemented to assess IOR for the four-level variables with two observers. In the case of the dichotomous variable, α and Fleiss' K were affected by the paradox effect: these indexes gave low agreement values despite a high concordance rate (P_0) (P_{0ABC}=91%; α, K=0.44). Similarly, regarding the four-level indicators, in some cases K_{PABAK} showed the paradoxical behaviour (P_{0EWL}=64%; K_{PABAK}=0.29; P_{0EWD}=57%; K_{PABAK}=0.14). S index is suggested to evaluated IOR in the case of dichotomous indicators (P_{0ABC}=91%; S=0.83). Cohen's K, α and S index are suggested to evaluate IOR in the case of four-level variables (P_{0EPL}=79%; K, α=0.67; S=0.73; P_{0EPD}=83%; K, α=0.73; S=0.78; P_{0EWL}=64%; K, S=0.53; α=0.52; P_{0EWD}=57%; K, α=0.42; S=0.43).

Session 80 Theatre 11

Social network analysis of dairy cows' group structure at the feeding trough

T. Wilder, P. Hasenpusch, A. Seidel, G. Thaller and J. Krieter
Christian-Albrechts-University, Institute of Animal Breeding and Husbandry, Olshausenstr. 40, 24098 Kiel, Germany; twilder@tierzucht.uni-kiel.de

As individual behaviour of social animals is always affected by behaviour of other animals, it is important to consider the group structure as well as the animal's position within this structure when analysing animal behaviour. This applies to the behaviour of group housed animals like dairy cows as well. Social network analysis is a tool which provides standardised parameters to do this. However, data acquisition of animal behaviour by human observation is very laborious and needs time to produce a sufficient data basis. Therefore, feeding troughs (animal to feeding trough ratio of 2:1) which automatically identify the cows with every visit were used to acquire data on animal behaviour, group structure and management measures. Using information on time, trough number and animal number, two types of networks were built: an undirected network of adjacent eating (NAE), in which a contact was defined as eating at adjacent troughs for at least 90 sec, and a directed network of displacement behaviour (NDB), in which a contact was defined as an animal change at one trough within 30 sec, using the succeeding cow as origin of the contact and the displaced cow as target. In NAE, cows have a greater effective degree (number of relevant contacts), if they have a greater dominance index (correlation coefficient R=0.31, $P<0.05$). In NDB, cows with a greater effective outdegree have a greater dominance index (R=0.40, $P<0.05$), while cows with a greater lactation number have a lower effective outdegree and a lower effective indegree (both R=-0.27, $P<0.05$), indicating less participation in displacement behaviour. Furthermore, comparing NDB at a period after the troughs were refilled with fresh feed and a control period shows higher correlation coefficients for the outdegree (R=0.21, $P<0.05$) than the indegree (R=0.10, $P<0.05$), indicating cows performing displacement behaviour tend to do this in both periods, whereas cows receiving displacement behaviour are rather randomly targeted. Combining these networks with networks based on data of an indoor positioning system, which will be done in the next step, promise new insights into group structure and behaviour of dairy cows.

Session 80 Theatre 12

Roughage and type of dispensers: what consequences on horses' feeding behaviour?

M. Roig-Pons and S. Briefer
Agroscope, SNSF, Les Longs Prés 1, 1580 Avenches, Switzerland; marie.roig-pons@agroscope.admin.ch

A great amount of hay dispensers are currently available on the market, with some of them designed to decrease speed ingestion ('slowfeeders'). This preliminary study aimed at evaluating the effect of dispenser and roughage on horses' feeding behaviour and posture. The study was divided in two experiments. (1) In 2021, we compared 3 slowfeeders (SF): HayBag (HB), HeuToy (HT) and PortaGrazer (PG) to loose hay (LOO) on the floor, the latter being traditionally considered as the gold standard. 4 Franches-Montagnes stallions housed at the SNSF were habituated to each SF during 10 days and then fed hay 3 times a day for 4 days, following a Latin square design. Horses were filmed during the first hour of their meals and 'explore', 'gather', 'bite' and 'chew' were labelled. The posture of the neck was also noted. The first video analysis revealed a difference of feeding behaviour between LOO and SF, especially in terms of bite rate ($mean(nBites)_{LOO}/min = 0.8\pm1.1$ vs $mean(nBites)_{SF}/min = 9.4\pm4.6$, Student-test, $P=2.2e-16$). (2) We then investigated if the feeding behaviour displayed under LOO treatment, often considered as reference, was similar to the one expressed under 'natural' conditions, namely horses grazing in pasture (PAS). To this aim, in 2022, 4 additional geldings were filmed when feeding under LOO and PAS condition. In total, 44 hours of video were analysed: (1): 32 h; (2) 12 h. The results highlighted that HT and HB decreased the ingestion speed (0.85 and 0.87 vs 1.61 kg/h for LOO), but could lead to a twisted neck position during up to 58% of the meal. In addition, horses displayed more bites resp. less chews per min under PAS (44.3±10.9, resp. 32.9±6.6) compared to LOO (0.2±0.3, resp. 69.2±4.8, Wilcoxon-test, both $P=0.001$). To conclude, our experimental plan did not allow for direct comparison between PAS and SF, but results tended to show that horses express a more similar feeding behaviour between PAS and SF compared to PAS and LOO, especially for the time spent exploring and the chewing rate. The findings of this preliminary study questioned the use of loose hay as reference in many studies. An analytical study is needed to assess the influence of roughage and dispenser on horses' feeding behaviour and to discuss the use of LOO treatment as reference.

Session 80 Poster 13

Caudectomy effect on the severity of tail lesions and abscesses' frequency in pig carcasses

P. Trevisi[1], D. Luise[1], S. Virdis[1], S. Dalcanale[1] and U. Rolla[2]
[1]University of Bologna, DISTAL, Viale Fanin 46, 40126 Bologna, Italy, [2]Gruppo Martini, Via Emilia 2614, 47020 Budrio di Longiano, Italy; sara.virdis3@unibo.it

The aim of this study was to evaluate the implication of the presence of the undocked tail on the prevalence and severity of tail lesions, abscess frequency and thigh defects in heavy pig carcasses at Italian slaughter. For this purpose, a total of 5,216 carcasses belonging to 45 batches were inspected in the slaughterhouse between April and September 2022. Tail lesions severity was scored from 0 to 2 according to the Welfare Quality® (2009) protocol and Lesion Score Index was calculated (LSI). Abscesses located on the thigh, spine, throat (lymph nodes), thoracic lymph node, abdomen skin, sternum/ribs, lower abdomen, navel and neck were recorded and reported as frequencies per batch. Thigh defects such as hematomas, petechiae and veins, PSE, arthritis and underweight were assessed by qualified slaughterhouse staff and reported as frequencies per batch. The LSI was higher in undocked tail pigs (P<0.0001). Furthermore, a higher frequency of both thigh lesions (hematomas and petechiae) and arthritis was found in the carcasses of undocked pigs compared with docked pigs (P<0.0001). No difference in any of the abscess frequencies was observed between docked or undocked carcasses. A positive correlation was observed between the increase of LSI and thigh defects such as hematomas (r=0.51; P<0.0001) and petechiae (r=0.64; P<0.0001) and the frequency of arthritis (r=0.56; P<0.0001). In conclusion, the absence of caudectomy appears a risk factor for the prevalence and severity of tail lesions. Although there is a correlation between tail length, thigh defects and arthritis, these two problems may be related to other risk factors. Tail length, on the other hand, did not appear to be associated with the presence of abscesses.

Session 80 Poster 14

Characterising cattle daily activity patterns using accelerometer data

S. Hu[1], A. Reverter[1], R. Arablouei[2], G. Bishop-Hurley[1] and A. Ingham[1]
[1]CSIRO, Agriculture and Food, 306 Carmody Rd, St Lucia QLD 4067, Australia, [2]CSIRO, Data61, 1 Technology Ct, Pullenvale QLD 4069, Australia; shuwen.hu@csiro.au

Livestock production systems have recently benefited from the adoption of automated monitoring technologies (e.g. Global Navigation Satellite Systems, inertial sensors, cameras, and microphones) facilitating the continuous surveillance of animals. In this study, cattle were fitted with a smart ear tag containing a triaxial accelerometer from which a simple measure of activity was computed. To better recognise the differences between day and night, each 24 h period was divided into 6 intervals of 4 hours. There are three-day intervals (D1, D2 and D3) and three-night intervals (N1, N2 and N3) for each day. The level of activity for each 4-hour interval was computed from the standard deviation of the accelerometer vector norm averaged over 5-minute windows. Averaged across all animals, the daily activity is 3.01×10^6 for cattle yarded overnight (environment 1, ENV1), while for free-roaming cattle (ENV2) the total daily activity is 6.15×10^6. To characterise patterns of activity, a metric named daily differential activity (DDA) was calculated from the difference between each two intervals (e.g. D1-N1, 30 possible pairs in total). For each animal, the DDA has three components: (1) DDA_v, the maximum difference in activity for all pairs; (2) DDA_p, the pair generating DDA_v; (3) DDA_e, the information entropy (Does one pair frequently provide the highest DDA_v or is this equally spread across all pairs?). In ENV1, the mean of DDA_v is 3,086.4 and ranges from 2,378.4 to 4,598.4; while in ENV2, the mean of DDA_v is 1,087.9 and ranges from 837.5 to 1,814.1. Across all animals in both environments, we observe a strong negative correlation between DDA_v and DDA_e (r=-0.896) which indicates that cattle with regular behaviour (a low entropy) attain a high DDA_v. We concluded that the assessment of daily activity using the DDA metrics allows a better understanding of animal responses to different environments and, if related to performance (e.g. growth), potentially provides an indicator of animal welfare. Further, our results suggest that any approach to detect anomalies in animal behaviour based on accelerometer activity needs to be able to handle the variation observed here.

Effects of olfactory exposure to twelve essential oils on behaviour and health of cows with mastitis

R. Nehme[1,2], C. Michelet[2], E. Vanbergue[2], O. Rampin[3], S. Bouhallab[1], A. Aupiais[2] and L. Andennebi-Najar[2,4]
[1]INRAE, Institut Agro, STLO, Rennes, 35042, France, [2]IDELE Institute, Quality and Health Department, Paris, 75012, France, [3]Université Paris Saclay, INRAE, AgroParisTech, PNCA, Jouy-en-Josas, 78350, France, [4]CRSA, Sorbonne University, INSERM UMRS, Paris, 75012, France; ralph.nehme@idele.fr

Essential oils (EO) are known for their anti-bacterial, anti-inflammatory and relaxing properties. They are used nowadays by a large number of farmers and considered as promising compounds against mastitis. We have recently found that applying *Thymus capitatus* EO on the mammary gland of dairy cows through massage didn't have any effects on subclinical mastitis (SM). The aim of this study was to evaluate the effects of a scent of 12 EO belonging to different chemical families on cow's EO preference and the follow up of SM in dairy cows. 29 Holstein cows were involved in this experiment among them 19 with signs of SM. After 2 days adaptation, animals had free access (2 days) to 12 EO distributed randomly in a fenced pilot. We combined both video recording and biological sampling to study: (1) animal behaviour and preferences towards each of the 12 EO; and (2) follow the evolution of SM. Blood cortisol concentrations were also evaluated to estimate the stress level after EO's exposure. Nasal microbiota was studied from the nasal mucosa samples. For each of the healthy and SM cows, we did not observe any significant preference towards a particular EO during the whole experimentation. However, SM cows were significantly more attracted to monoterpene (MN) and monoterpenol (ML) EO's families compared to healthy ones ($P<0.05$). Interestingly, EO with predominant ketones family were exclusively scented by the healthy cows ($P<0.05$). For health evaluation, no statistical differences on SCC and bacteriological status in milk were highlighted at day 21 days after EO's exposure, but we did observe a significant increase in the cortisol rate in blood compared to healthy ones ($P<0.05$). Our findings suggest that SM cows attraction to MN and ML families did not alleviate the signs of SM at day 21. Additional evaluation of the specific effects of MN and ML families on SM is needed to test different doses and other experimental conditions before any wide recommendation to farmers.

Assessment of stress in dairy cattle industry via analysis of cortisol residue in commercial milk

M. Ataallahi, G.W. Park, E. Nugrahaeningtyas, M. Dehghani, J.S. Lee and K.H. Park
Kangwon National University, College of Animal Life Sciences, Chuncheon, Gangwon, Republic of Korea, 24341, Korea, South; ataallahim@kangwon.ac.kr

Chronic stress has been a concern in dairy cattle industry with significant impact on animal health, productivity, and welfare. Cortisol is released in response to stress and there is a positive correlation between stress level and blood cortisol concentration. However, blood collection is an invasive procedure with additional effects on increasing cortisol secretion. Thus, it is essential to find less stressful procedures to measure cortisol concentration. It has been confirmed that cortisol transfers to milk and resists the high temperature during milk processing. Therefore, analysis of cortisol residue in commercial milk may be a suitable procedure to estimate the degree of stress in lactating cattle. This study was conducted to evaluate the relationship between commercial milk cortisol concentration (MCC) and temperature-humidity index (THI) at milk production date. A total of 11 commercial pasteurized and sterilized milk samples were purchased in Chuncheon, Korea with production dates from July to October in 2021. The MCC was extracted using diethyl ether and measured using enzyme immunoassay. The THI was calculated based on microclimate data from the Korea Meteorological Administration. The average of THI was 77±0.8, 75±1.4, 69±1.4, and 58±1.8, in July, August, September, and October, respectively. The average of MCC was 211.9±95.1, 173.5±63.8, 109.6±53.2, and 106.7±33.7 pg/ml in July, August, September, and October, respectively. The MCC in July was higher than August, September, and October ($P<0.05$), whereas it was lower in September and October compared with August ($P<0.05$). Overall, monitoring cortisol residue in commercial milk can be an indicator of stress in lactating cattle in farms.

Session 80 Poster 17

The physiological and behavioural effects of COVID-19 restrictions on lactating dairy cows

S.J. Morgan and D. Barrett
Dalhousie University, Faculty of Agriculture, Department of Animal Science and Aquaculture, 62 Cumming Drive, B2N 4H5, Bible Hill, Nova Scotia, Canada; david.barrett@dal.ca

During the height of the global pandemic, COVID-19 restrictions resulted in altered human-animal interactions with many species worldwide, which may have negatively impacted their welfare. However, few studies have assessed the impact of these restrictions on domestic animals. Thus, this study aimed to identify the impact of reduced handling and foot traffic due to the COVID-19 pandemic restrictions on the physiology and behaviour of dairy cows. Fifteen lactating dairy cows located at a research/teaching barn that saw significant changes in staff/personnel during the pandemic were used for this study. Subjects were followed from January 2021 – September 2022, through Seven COVID-19 restriction phases. As the phases progressed, cows were exposed to increasing human interactions and handling. During each phase cows were assessed for potential physiological or behavioural signs of stress. Physiological measurements included heart rate, respiration rate, and rectal temperature. Cow behaviour was assessed using cameras and a behavioural ethogram to analyse cow time budgets. Statistical analysis was carried out using a mixed model for physiological data and non-parametric Kruskal-Wallis tests for behavioural data. From Phase 1 to 7 significant decreases ($P<0.05$) in heart rate (82.6±1.6 vs 62.9±4.59 beats per minute) and respiration rate (38.2±1.5 vs 29.1±2.1 breathes per minute) were noted in the dairy cows. No significant differences in rectal temperature or behaviours were noted among any of the restriction phases. Thus, our findings suggest that the dairy cows experienced a state of physiological stress at the onset of the pandemic, with a reduced stress response as they were re-introduced to pre-COVID-19 levels of handling and foot traffic. We therefore conclude that sudden alterations in human handling/ interaction induce a state of stress in dairy cows, whereas prior experiences and gradual changes in human interaction do not invoke strong stress responses in cattle. Further studies are needed to determine what may be helpful in offsetting the stress response of dairy cows to a sudden unavoidable environmental change, or how to mitigate these changes should these situations arise again in the future.

Session 80 Poster 18

Effects of the living environment on the behaviour of rabbits

M. Fetiveau, M. Besson, V. Fillon, M. Gunia and L. Fortun-Lamothe
GenPhySe, INRAE, 24 Chemin de Borde Rouge, 31326 Castanet-Tolosan, France; laurence.lamothe@inrae.fr

Today, rabbits are mainly reared (>90%) in wired unenriched cages, which limits the behavioural repertoire of the animals. To contribute to the design of welfare-friendly systems, we compared the influence of the living environment on the behaviour of growing rabbits. Rabbits of two genetic types (control or selected for disease resistance) were distributed at weaning (35 d of age) in two contrasting rearing systems: (1) in a confined building equipped with wire cages housing 5 rabbits (0.9×0.9 m, modality 1); or (2) in a mobile house equipped with 8 roofless pens (2×1 m, modality 2) giving free access to an enriched (hay rack, straw bale shelters) grassy paddock of 1,940 m^2. The behaviour of the animals was assessed using scan sampling by direct observation twice a day (morning and afternoon) at 45, 56 and 66 days of age. We distinguished between locomotion (walking, pouncing, running), maintenance (browsing, eating hay or pellets, resting, gnawing), exploration (scratching, observing, training) and social interaction behaviours (side by side, nose to nose). The genetic type of the animals only influenced the very weakly expressed nose-to-nose behaviour ($P<0.05$). Indoors, the dominant behaviour was resting (Resting: 67.8 and 56.6% vs 31.2% in pens, cages and paddock, respectively; $P<0.001$) followed by watching (11.4 vs 3.7% and 4.4%, respectively; $P<0.001$) and pellet ingestion (8.6 and 6.7% in cages and pens). On the contrary, the rabbits were more active on the paddock (pouncing: 8.2% on the run vs 2.1 and 0% in pens or cages; $P<0.001$), the only place where they ran (2.2% of observations). On the course, 36% of the rabbits were observed grazing. Social interactions were frequent (18% of behaviours, of which side by side 16.7% (NS). Other behaviours (Walking, Hay eating, Gnawing or Training) were rarely observed (<5%). In conclusion, these results show that the living environment has a very strong influence on the animals' behaviour. Providing access to a grassy paddock allows rabbits to satisfy their need to graze and considerably stimulates their locomotion. It also allows a spatialization of activities, with resting inside and locomotion outside.

Session 80 Poster 19

Behaviour test and study of sheepdogs' abilities

B. Lasserre[1], B. Ducreux[1], M. Chassier[1], L. Joly[1], T. Le Morzadec[1], P. Cacheux[1] and C. Gilbert[2]
[1]Institut de l'Elevage, 23 Rue Jean Baldassini, 69007, France, [2]Ecole Nationale Vétérinaire, 7 avenue du Général de Gaulle, 94704 Maisons-Alfort, France; boris.lasserre@idele.fr

Currently, sheepdogs' selection only relies on trial performance, in which the trainer's expertise is dominant. The objective of this study was to set up a behaviour test with a grid composed of unbiased criteria to assess the natural abilities of a sheepdog from an early age. If these criteria are heritable, a selection could be put in place. The test is composed of 2 steps: a 1-minute pre-test to let the dog familiarise with the new environment, and a 2-minute test with a fetch, the driver surrounding the flock twice and moving away from it. Each step is separated from the other by a 15-second unmoving phase. The test is conducted by a driver unknown of the dog, around 10-15 ewes, without any command. After the test, a blood sample was taken on each dog, to collect genetic information. The test was deployed in 15 French regions where 460 Border collies between 8 and 24 months old have been tested since 2019. At the end of the test, each dog is evaluated according to the phenotype considered: Not Use Value (Not VU) or Use Value (VU). A VU dog is defined as able to avoid flights, to maintain animals in a group close to human, to confront animals in a respectful way, if need be, and to provoke a movement without preying (chasing, disturbing, and biting). 23 indicators were also assessed during the test. Some focus on the dog's behaviour: fetch, movement when driver surrounding, reference to human, movement when driver walks, movement to make ewes move, biting, focusing and motivation. The other indicators focus on the ewes' reactions to the dog: mobility when fetch starts, reaction to fetch, position after 15sec unmoving, reaction when surrounded, movement when driver walks, gregarious behaviour, distance to driver walking, attempt to confront dog. The 10 first dimensions of the MCA explain 48% of VU variance. The grid correctly detects our phenotype (VU or not) with a 96% prediction rate, and ranks dogs in 10 different profiles, with a 92,5% prediction rate. 17% of tested dogs are VU, and the link between this phenotype and environmental variables was also studied. For instance, dogs under 11 months old are significantly more aggressive than older dogs (P<0.001).

Session 80 Poster 20

Innovative evaluation method of behavioural reactivity for 'foie gras' ducks

C.M. Monestier[1], A.W. Warin[1], S.L. Lombard[1], S.L.M. Laban-Mele[2] and W.M. Massimino[3]
[1]Bureau Bankiva, department of R&D, 4 impasse de la fin de chêne, 21410 Gergueil, France, [2]Labeyrie Finde Foods SAS, Department of Corporate Social Responsibility, 39 route de Bayonne, 40230 Saint-Geours de Maremne, France, [3]Lur Berri, Department of Innovation, R&D, Route de Sauveterre, 64120 Aïciris, France; chloe.monestier@bankiva.fr

Achieving significant advancements in farm animal welfare requires assessments that are scientifically validated. To date, the scientific literature still lacks a behavioural assessment methodology covering all animals at all rearing stages such as the overfeeding stage in 'foie gras' farms. The purpose of our study was to implement an objective measurement of behavioural responses in ducks during overfeeding. In 2021, trials were conducted in an experimental station at two separate sites and repeated four times. Behavioural and health data were recorded for 307 male mule ducks 2 h before and during overfeeding, three times throughout the duration of overfeeding. We developed a new metric conducted during overfeeding that allowed us to discriminate animal reactivity creating two behavioural indicators (struggling and overall excitement profiles) associated with the wings flapping behaviour. Ducks housed in collective cages were more excited and struggled during overfeeding more than ducks housed in pens altogether interpreted as a higher reaction to a stressful situation than ducks housed in pens (P<0.05). We also found that ducks housed in collective cages showed more discomfort behaviours while ducks housed in pens expressed more natural behaviours before overfeeding (P<0.05). The results obtained before and during overfeeding are complementary and consistent. In addition, the before overfeeding results reinforce even more those obtained with the creation of the new behavioural metrics. Moreover, as ducks express these behaviours while being handled, it is relevant to also take them into account when considering the day-to-day work conditions of farmers. Our preliminary data are encouraging in terms of reproducibility and repeatability of measurements, and we have implemented them in current trials to evaluate the individual behaviour and rearing conditions of ducks throughout overfeeding. The measured reactivity addresses the welfare of both animals and farmers, aligning with the 'one welfare' approach.

Session 80 Poster 21

Use of scaring devices to avoid roe deer fawns getting injured or killed during mowing

J. Mačuhová, T. Wiesel and S. Thurner
Bavarian State Research Center for Agriculture, Institute for Agricultural Engineering and Animal Husbandry, Vöttinger Str. 36, 85354 Freising, Germany; juliana.macuhova@lfl.bayern.de

Every year by forage cuts in May and June, roe deer fawns can get injured or killed during mowing. The aim of this study was to evaluate the use of scaring devices set in fields before mowing to persuade the animals to leave or not enter the fields. A query was performed using a data collection sheet (available for download from one web page for everybody applying this measure and interested in the study) during the season 2020. The goal was to obtain information such as type of used scaring devices, time of setting up and removal of scaring devices, the number of fields with scaring devices, total number of set up scaring devices on fields with scaring devices, use of other measures on fields with scaring devices, and number of fawns seen, rescued, and injured or killed during mowing. One data collection sheet should be filled in for all fields with scaring devices mowed on one day by one farmer or in one hunting ground. 171 out of 183 received data collection sheets could be evaluated. The number of fields per sheet varied between 1 to 31 (2.89±3.67; n=160 (n=number of data collection sheets)). Scaring devices with light and sound signals were mainly used alone (n=82) or in combination with plastic bags (n=57). Moreover, one (n=52) or more (n=26) additional measures were applied at least at one field before mowing. 79 fawns (n=35) were seen running away during moving or were found during or after moving (27 of them injured or killed (n=15; thereof n=6 applying at least one and n=8 applying even two or more additional measures)). Unfortunately, it cannot be excluded that data regarding the number of seen and found fawns were recorded for all fields mowed at a certain term and not only for fields with scaring devices in some data sheets. Nevertheless, it can be concluded that the use of scaring devices did not prevent completely mowing death. This could be caused not only due to a possible missing effect (or incorrect effect) but also due to an improper application (not setting up the scaring devices on all mowed fields; the fawns could be already on fields without scaring devices or they moved there from fields with scaring devices).

Session 81 Theatre 1

Precision feeding, recent advances for gestating sows and dairy cows

C. Gaillard, C. Ribas and M. Durand
Institut Agro, PEGASE, INRAE, 16 Le Clos, 35590, Saint Gilles, France; charlotte.gaillard@inrae.fr

Precision feeding aims to define the right feeding strategy according to individual's nutrient requirements. Few years ago, an individual feeding strategy for gestating sows based on energy and lysine adjustment of daily supplies has been shown to reduce feed costs and environmental loads. The nutritional requirements were calculated thanks to an updated version of the InraPorc model requiring input data (i.e. sows' characteristics). A third variable to take into account during the adjustment (i.e. phosphorus requirement) is currently being added into the model and simulations are being run. More recently, on one hand the integration of the individual and daily physical activity measurements in the calculation of the energy requirements is becoming possible thanks to the development of a software able to analyse automatically and continuously the sow's postures. On the other hand, machine learning algorithms allows to predict daily nutritional requirements using only data measured by sensors or automatons. For example, energy and lysine requirements can be predicted thanks to physical activity and feeder data, respectively. Combining these data with the sows' characteristics improved the prediction accuracy. This result indicate a possible way to simplify the application of precision feeding on farm. Finally, the long term effect of a precision feeding strategy on the sows' performances is being evaluated in two different experimental farms. Concerning dairy cows, individualized feeding strategies are more complicated to set up and models or criteria to calculate precisely the individual daily requirements are not available. Preliminary experiments are taking place to determine criteria to adjust the energy of each cow's ration, at the scale of the week, by varying the amount of concentrates distributed in the automatic feeder. Furthermore, biomarkers and production variables are being screened to determine if they can be used in early lactation to predict the lactation persistency, and then adapt the feeding strategy and the reproduction timing based on the objective of the farmer and the potential of each cow. Precision feeding is therefore a burning and growing topic for farm animals, with promising results.

Session 81 Theatre 2

Individual ingestion time prediction with RGB-D cameras in beef cattle: a machine learning approach

P. Guarnido-Lopez[1], F. Ramirez-Agudelo[2], S. Tomozyk[1], A. Kjorvel[1], A. Carvalho[1], F. Jezegou-Bernard[1], P. Gauthier[1] and M. Benaouda[1]

[1]Institut AgroDijon, Université de Bourgogne, Animal production, 26 Bd Dr Petitjean, 21000, Dijon, France, [2]Université Paris-Saclay, INRAE, AgroParisTech, UMR Modélisation Systémique Appliquée aux Ruminants, 22 place de l'Agronomie – CS 80022, 91120, Palaiseau, France; pabloguarnido@hotmail.com

New machine learning approaches have been developed in recent years to help cattle farmers more easily and affordably measure individual dry matter intake, a key factor in overall farm efficiency and animal performance. The objective of the present work was to assess the performance of a computer vision model for classifying activities in steers and predict ingestion time (IT). For this, a total of 6 young Charolais bulls (581±37kg BW) were recorded at the same moment (meal's distribution) for 7 days. Videos (15-20 mins each) were recorded through an RGB-D camera (Intel® RealSense TMD455) at 15 frames per second. The model used for images analysis was the YOLO (V5) and the whole process entailed two steps: (1) Characterization of individual activity, where we identified animal's behaviour among 5 different activities (biting, ruminating, visiting chewing and others). A total of 530 images were randomly extracted from the videos, and the activities were manually identified using a labelling software. The images were divided into three sets for training, validation, and testing, (60, 20, and 20%, respectively. (2) The IT was determined per animal per video by summing the predicted biting and chewing times, and these predictions were compared to the actual individual ingestion time measured manually using a chronometer. Results showed a high precision of biting activity over the test images (P=0.73, R=0.98) while a lower precision in rumination (P=0.59, R=0.47). In addition, the CM showed a high (r=0.78) confusion between 'visiting' and 'rumination'. Then, we observed a high correlation (r=0.92, P=0.001) between predicted and measured IT. This work, shows the potential of ML models to predict accurately the IT, which may be used to help decision-making of beef cattle producers both in farm efficiency (detection of efficient individuals) and in animal's welfare (earlier disease detection).

Session 81 Theatre 3

Using image classification to estimate feed intake in weaned piglets under commercial conditions

T. Van De Putte, J. Degroote and J. Michiels

Ghent University, Department of Animal Production and Aquatic Ecology, Coupure Links 653, 9000 Ghent, Belgium; thomvdpu.vandeputte@ugent.be

Assessing feed intake in weaned piglets is critical for ensuring optimal nutrition without overconsumption, particularly during the first few days post-weaning when digestive capacity is limited. However, commercial conditions often hinder accurate assessment via computer vision due to limitations in visualizing the entire pen and identifying individual piglets. To address this challenge, we tested a simple image classifier in an experiment involving 288 pigs housed in 24 pens, monitored via camera surveillance for the first three days post-weaning. Of these pens, 12 had access to two distinct feeders: creep feed (familiar to the piglets) and the unfamiliar weaner diet, while the remaining 12 were exclusively fed the weaner diet. To create a deep learning dataset, 17,000 images were cropped around the vicinity of both feeders and classified by a human expert based on the number of eating animals, with non-nutritional feeder visits and head-lifting events excluded. Using the ResNet50 architecture as a base model, with two dense layers and a prediction layer added, the model was trained using the Tensorflow Library in Python, achieving a precision and recall of 93% each on the test dataset. Additionally, feed intake for each diet was measured daily by weighing excess feed from both feeders and compared to a daily feed index, determined by subjecting all video data to the image classification algorithm. The correlation yielded an R-squared of 0.90 (weaner diet) and 0.77 (creep feeder), indicating the algorithm provides a reliable estimate of actual daily feed intake in group-housed piglets. Further analysis revealed that eating speed increased daily, with piglets consuming at an average speed of 31 mg/s during the first 72 hours post-weaning, and spending 7% of their time eating near the end of this period. Preference for creep feed was observed when it was available, although there was no significant difference in total feed intake or total feeding time between the animals exclusively fed the weaner diet and those with access to both diets.

Session 81 Theatre 4

The value of precision feeding technologies: economic, productivity and environmental aspect

J.K. Niemi
Natural Resources Institute Finland (Luke), Kampusranta 9 C, 60320 Seinäjoki, Finland; jarkko.niemi@luke.fi

Precision livestock farming (PLF) can play an important role in making animal production economically, environmentally and socially more sustainable. In the context of feeding, PLF allows feeding animals according to their potential to utilize nutrients, thus reducing economic and environmental losses caused by insufficient or excessive feeding. While precision feeding can provide important economic and environmental benefit and enhance farm productivity, investing in precision feeding is not always economically viable. This presentation summarizes the current knowledge concerning economic, productivity and environmental aspects of precision feeding technologies in pig, poultry and cattle production. It provides examples on quantified economic costs and benefits of precision feeding especially in pig farming. Finally, it examines factors that contribute to the farmer's decision to adopt or not to adopt new precision feeding technology solutions. Besides costs and benefits themselves, these include factors such as the heterogeneity of herd and the accuracy and uncertainty of information on animal's current state of nature that is needed to determine the most preferred diet.

Session 81 Theatre 5

Precision feeding in dairy cows: what do professionals in ruminant nutrition think about it?

A. Igier[1], A. Petillon-Pronk[1], J. Martin[1], E. Hertault[1], N. Gaudillière[2], J. Jurquet[3], A. Fischer[3] and Y. Le Cozler[1]
[1]Institut Agro Rennes-Angers, 65 rue de St-Brieuc, 35000 Rennes, France, [2]Eliance, 149 rue de Bercy, 75595 Paris cedex 12, France, [3]IDELE, 149 rue de Bercy, 75012 Paris, France; yannick.lecozler@agrocampus-ouest.fr

Modern dairy farming aims at improving feed efficiency and decreasing negative environmental impacts. This can be achieved through precision feeding strategies, which consists in offering the right diet (in quantity and quality) at the right time to the right animal. However, this concept of feeding is not well-known and can have diverse definitions. A survey was performed on 50 French professionals involved in dairy cattle nutrition to identify their needs and to analyse their expectations about precision feeding. Their answers highlighted that precision feeding combined the adequate diet delivering to the requirements of the animal (8/47), the individualization of feeding (6/47), but also the need of automatic concentrate dispensers (ACD; 3/47) or robots (5/47). According to participants, the main advantages of this feeding method are to improve herd productivity (12/37), and to enhance the potential and individual performance of each cow (6/37). Among the surveyed persons, 14 claimed that precision feeding was not cost-effective whereas 10 said that it was cost effective. Complexity (6/35) and time-consumption (8/35) were two additional dis-advantages that were cited. For most professionals, adopting precision feeding should result in maximizing individual performance (12/38), improving cows' health (11/38) and improving economic profitability at the herd level (10/38). To a lower extend, the surveyed persons also expected a better feed efficiency (3/38) and reduced environmental impacts (2/38). But for most of them, adopting precision feeding must be considered farm by farm. It is concluded that different, and sometimes, opposite opinions exist between professionals working in the field of dairy cows nutrition regarding precision feeding, resulting in different expectations that need to be considered for potential development of practical tools or consultancy services in precision feeding.

Session 81 Theatre 6

Characterizing dairy cow's individual reaction to a decrease in production concentrate distribution

A. Fischer[1], R. Lehuraux[2] and J. Jurquet[1]
[1]Idele, 149 rue de Bercy, 75012 Paris, France, [2]Institut Agro, 26 Bd Dr Petitjean, 21000 Dijon, France; julien.jurquet@idele.fr

The individual allocation of concentrates in lactating dairy cows is often based on each cow's milk production and lactation stage only, without considering the other requirements. The objective was to identify clusters of individual patterns of cow's reaction subsequently to a decrease in production concentrate distribution. The hypothesis were that the cows would offset this challenge by a combination of a decrease in net energy exported in milk (NEMilk), increase of 'basis' diet intake and body reserves mobilization. Sixty Holstein dairy cows were fed *ad libitum* with a unique basis and balanced diet distributed in individual weighing bins and supplemented individually at an electronic feeder with 4.0 kg/d/cow for 3 weeks (P4Kg) and 1.0 kg/d/cow of production concentrate the following 3 weeks (P1Kg). The first half of cows (group AdLib) had *ad libitum* access to the basis diet, whereas the second half (group Limited) had *ad libitum* access to the basis diet during P4Kg and limited access during P1Kg. This limit was set for each cow as its average intake of basis diet during P4Kg. The results highlighted that AdLib group had an average intake of 22.4 kg DM/d/cow in period P4Kg (181 MJ/d/cow) which increased by 1.4 kgDM/d/cow in period P1Kg (-8 MJ/d/cow), with a steady NEMilk for both periods of 110 MJ/d/cow. The Limited group had an average intake of 22.7 kgDM/d/cow of basis diet during both periods (173 MJ/d/cow) and produced on average 110 MJ/d/cow NEMilk during P4Kg which decreased by 7 MJ/d/cow during P1Kg. Those results show that only cows that were able to offset the decrease in concentrate distribution by an increased intake of basis diet, i.e. that had *ad libitum* access to the basis diet maintained their NEMilk. Three clusters of individual patterns were identified: cows that compensated mostly by higher intake of basis diet without decreasing their NEMilk; cows with a low or no increase in the basis diet and a decrease in NEMilk; and cows, mostly with a lower milk production, that had small changes in their performance. To have a full overview and validate those results, the next step will have to include the change in body reserves as a third lever available to each cow to adapt to the concentrate reduction.

Session 81 Theatre 7

Longitudinal patterns for feeding traits in the two parental populations of the mule duck

H. Chapuis[1], C. Ribas[1], H. Gilbert[1], M. Lagüe[2] and I. David[1]
[1]INRAE, GenPhySE, 31326 Castanet Tolosan, France, [2]INRAE, UEPFG, 40280 Benquet, France; herve.chapuis@inrae.fr

In France, the mule duck contributes to more than 90% of the production of foie gras. The mule duck is the sterile hybrid obtained by crossing a Muscovy drake with a Pekin dam. Our objective was to study the dynamics of feeding behaviour traits during the growth of mule duck parental populations and ultimately evaluate if they can be used to improve their offspring foie gras production. Data were recorded on 453 Muscovy males and 740 Pekin ducks of both sexes using electronic feeders. Records were available for 5 weeks, starting from d15 in Pekin and d22 in Muscovy. Traits were computed on a daily basis as meal and feed intake (MFI and DFI), feeding time (DFT), meal duration (MD), feeding rate (FR), number of meals (NM) and body weight (BW). Traits were analysed using random regression models. The combined effects of age, batch and sex were used as fixed effects, while additive and environmental effects were fitted using Legendre polynomials of order up to 3. Daily heterogeneous residual variances were accounted for, through a structured antedependence approach. Average performances largely differed between populations: for instance, Muscovy ducks spent daily almost twice as much time feeding as Pekin ducks, while they ingested 12% less feed. However, despite these differences feeding traits exhibited similar longitudinal patterns between populations: BW, FR, DFI and MFI showed a positive trend throughout the trial in both Muscovy and Pekin populations, contrary to DFT and NM which decreased over time. In both populations, heritability estimates were low (0.10 for Pekin DFI) to moderate (0.35 for Pekin MD) at start, and increased to reach a maximum around 0.5, six to ten days later. Afterwards, in Pekin, heritabilities of feeding traits remained moderately high over time, while in Muscovy they showed a slight decreased over time (DFI, DFT, FR) or a sharp drop towards the end of the period (MFI, NM). This analysis would profitably be extended to estimate the genetic correlations of FR, DFT and MD, recently evidenced as potential selection criteria for an improved foie gras production, with the foie gras weight of mule offspring.

Models predicting methane emissions and methane conversion factor of Finnish Nordic Red dairy cows

J.S. Adjassin[1,2], A. Guinguina[1], M. Eugène[3] and A.R. Bayat[1]
[1]Natural Resources Institute Finland (LUKE), Animal Nutrition, Tietotie 2C, 31600 Jokioinen, Finland, [2]Laboratory of Ecology, Health, and Animal Production (LESPA), Arafat, P.O Box 123 Parakou, Benin, [3]INRAE, Université Clermont Auvergne, VetAgro Sup, UMR 1213 Herbivores, 63122, Saint-Genès-Champanelle, France; josias-steve.adjassin@inrae.fr

Accurate estimates of methane (CH_4) emission factor (Y_m) are essential for greenhouse gas inventories, which can be obtained by direct measurements or model predictions. We aimed at quantifying the Y_m from Finnish Nordic Red dairy cows based on data obtained from respiration chambers and ruminal SF_6 gas tracer technique (2011-2022). Overall, 391 individual records from 13 experiments were compiled and divided randomly into a training data set (9 experiments, n=319 records) for model development and a validation data set (4 experiments, n=72 records) for model evaluation. Data were analysed using the mixed model procedure of SAS with experiment as a random effect. Root MSPE (RMSPE) and the ratio of RMSPE to standard deviation of observed values (RSR) were used to compare the models. A total of 10 and 14 linear models were developed for predicting Y_m (% of energy intake) and CH_4 emission (g/d), respectively. The predictor variables were DM intake (DMI), body weight (BW), diet composition (EE, NDF, CP, Ash, starch), energy corrected milk (ECM), milk fat and protein, organic matter digestibility and rumen acetate to propionate ratio. To avoid multicollinearity, only significant ($P<0.05$) predictors with variance inflation factor <10 were kept in the models. Methods of measurement were not different ($P>0.05$) in terms of Y_m despite lower daily CH_4 emission and higher variability from SF_6 tracer technique compared to respiration chambers (446±105 vs 515±71.7 g/d). The average Y_m was 6.42±0.88%. Prediction equations based on DMI, BW, dietary EE, milk fat and AP for daily CH_4 emission performed the best (RSR=0.54 and RMSPE=8.52%). Using only BW and DMI for predicting Y_m had a low performance (RSR=0.92 and RMSPE=12.6%) and adding dietary composition variables improved the model significantly (RSR=0.71 and RMSPE=9.70%). In conclusion, accurate prediction of Y_m requires information on DMI and dietary factors such as crude fat, Ash and NDF concentrations.

Use of automated head-chamber systems as exclusive concentrate dispensers during bulls fattening

N. Lorant[1], E. Henrotte[2], Y. Beckers[3], F. Forton[4], A. Vanlierde[1], M. Mathot[1] and A. Mertens[1]
[1]Walloon Agricultural Research Centre, Rue du Serpont 100, 6800 Libramont, Belgium, [2]Inovéo, Chemin du Tersoit 32, 5590 Ciney, Belgium, [3]Gembloux Agro-Bio Tech, Passage des Déportés 2, 5030 Gembloux, Belgium, [4]Dumoulin SRL, Rue Bourrie 18, 5300 Seilles, Belgium; n.lorant@cra.wallonie.be

To help the beef sector to reach European climate objectives, the 'Blanc Bleu vert' project (SPW Research funding) aims at mitigating greenhouse gases emission and optimizing feed efficiency of Belgian blue bulls (BBB) through innovative feed solutions and genetic selection. In order to optimize the methane emission measurement, a trial is performed to test an exclusive use of an automated head-chamber system (AHCS) for the fattening of BBB bulls. The methodology explores the capacity to distribute the whole concentrated feed during the fattening period of BBB with AHCS and the interconnected variability and uncertainties associated to the methane measurement when the frequency of visits is increased. This research was carried out between May 2022 and August 2022 on 4 groups of 7-8 bulls, with the same concentrate/straw finishing diet. During 19 days, methane emission and ingestion of BBB, with an average live weight of 546±46 kg, were monitored with AHCS. Depending on the fattening stage and weight, each bull received up to 18 feeding periods per day (840 g/visit: 15 drops/visit with 15 seconds between drop), with free access to straw. For a good measure of methane emission, only visits with a duration of more than 3 minutes were taken into account in the calculation. As expected, it is possible to feed more than 90% of a complete finishing bulls diet through the AHCS. The AHCS provided an average of 9.4 kg diet/bull/d. This daily consumption is not different from average consumption at the automatic concentrate feeding system in the previous 3 weeks (9.1 kg diet/bull/d). The dispersion coefficient related to the average daily methane emission of an animal is 3.7±1.1%, which corresponds to the accuracy of the device itself. To reach this precision 9.4±1.6 visits/d/bull with good duration (321±26 sec on average) are considered among the 9.7±1.8 daily visits. With this methodology, a three week measurement allows to reach the desired accuracy to carry out the project.

Session 81 Poster 10

On-farm nutritional supplementation to improve sows and piglets' performances: field report

P. Engler[1], G. Gemo[2], S. Cissé[1] and J.M. Garcia[1]
[1]Nor-Feed SAS, 3 Rue Amedeo Avogadro, 49070 Beaucouzé, France, [2]G&G Consultenza e Sviluppo, Località Brodi nr. 6, 29010 Castelvetro Piacentino (PC), Italy; sekhou.cisse@norfeed.net

Lactation is a critical aspect of swine production for both the sow, which mobilises different body reserves to produce milk, and for the piglets, which depend on efficient transmission of essential nutrients and biological substances through the colostrum and milk. In the recent years, a particular focus has been put on the importance of colostrum in early-life management for piglets and its impact on pre-weaning mortality. Nutritional studies have showed that the use of nutritional solutions around farrowing could improve colostrum quality and beneficial nutrient transfer from the sow to the piglets. The aim of this experiment was to study the effects of supplementation with a commercial combination of standardised grape extract (SGE, Nor-Grape®, Nor-Feed, France), calcium and vitamin D on sows and piglets' performances. 80 sows of similar parity rank were randomly divided into 2 groups at their entry in nursery, a control group (CTL, n=45) and a supplemented group (SGE, n=35). The supplementation of the commercial combination was administered to the SGE group through the drinking water at a rate of 50 g/sow/day, from their entry in nursery (5-6 days pre-farrowing), until 5 days post farrowing. Litter size parameters at farrowing, after cross-fostering reorganization and at weaning (28 days) were recorded for each sow. Whilst no significant difference was observed at birth in terms of total litter size, and after cross-fostering reorganization of the litters, the supplementation tended to result in a higher level of weaned piglets in the SGE group with, on average +1.2 weaned piglet/sow (10.6±0.5 vs 11.8±0.5 weaned piglet/sow for the CTL and SGE group respectively, P=0.10). These findings are in accordance with previous ones showing that a supplementation with a SGE significantly improves colostrum quality in sows and healthier piglets. This phenomenon could thus explain the higher level of weaned piglets in sows supplemented a few days around farrowing. These results underline the importance of peri-partum management in sows and how nutritional interventions around this time can result in beneficial effects on sows and piglets performances.

Session 81 Poster 11

Effects of monosaccharides on *in vitro* NDF and starch digestibility, pH and volatile fatty acids

E. Raffrenato[1], G. Esposito[2] and L. Bailoni[1]
[1]University of Padova, Department of Comparative Biomedicine and Food Science (BCA), Viale dell'Universitá 16, 35020 Legnaro PD, Italy, [2]University of Parma, Department of Veterinary Science, Via Taglio 8, 43126 Parma, Italy; giulia.esposito@unipr.it

Sugars are known to improve NDF digestibility (NDFd) in lactating dairy cows. However, not much information is known about specific monosaccharides and inclusion levels. A study was conducted to investigate the effects of glucose, fructose, xylose and rhamnose (selected from a preliminary study) on *in vitro* NDFd, pH and VFA concentration. Oat hay (OH), alfalfa hay (AH) or a TMR were fermented with each sugar for 12 and 24 h using inclusion levels of 0 (control), 10, 15 or 20% of the NDF content of the substrates used. Fermentations were run in triplicates, across 3 runs. Bacterial communities were profiled in the *in vitro* fluids using ARISA technique. Data were analysed according to a randomized complete block design with a factorial arrangement of treatments. Each sugar was analysed separately with level, substrate and their interaction as fixed effects and run as random factor. The addition of fructose and rhamnose increased NDFd at the 15 and 20% inclusion levels (P<0.05), respectively, while glucose and xylose were significant at 10% of the NDF (P=0.035 and P=0.022, respectively). Overall, the sugars were more effective in increasing NDFd when added to the TMR (P<0.05), followed by LH and OH. Independently of the sugar and sample used, the *in vitro* system was able to maintain the pH above 6.8 (P>0.05). While acetic acid was variable, both propionate and butyrate were highest with 20% inclusion level for all sugars (P<0.05). Differences were observed in bacterial communities between the control and most of the sugar treatments. TMR showed no similarities in the bacterial communities between inclusion levels while LH had the most similarities in bacterial communities across sugars levels. Both bacterial richness and diversity were higher when glucose was added compared to the control (P=0.0311). This preliminary work demonstrates the potential of specific monosaccharides and the need to explore their possible effects *in vivo* on diets of dairy cows as a way of improving the rumen microenvironment.

Session 81 Poster 12

Benefit of full matrix application for a novel phytase using a phased-dosing or fixed dosing strateg

B.C. Hillen[1], S. Gilani[1], R.D. Gimenez-Rico[2], K.M. Venter[3], P. Plumstead[3] and Y. Dersjant-Li[1]
[1]Danisco Animal Nutrition & Health, IFF, 4 Willem Einthovenstraat, 2342 BH, Oegstgeest, the Netherlands, [2]Danisco Animal Nutrition & Health, IFF, Paseo de la Castellana, 149, 28046, Madrid, Spain, [3]Neuro Livestock Research, Kameeldrift, Brits, South Africa; barthold-christian.hillen@iff.com

Application of full matrix for phytase including digestible phosphorus (P), calcium (Ca), sodium (Na), digestible amino acids (AA) and metabolizable energy (ME) is an important strategy to reduce feed cost and improve production benefit. This study validated the application of the full matrix for a novel consensus bacterial 6-phytase variant (PhyG) supplemented at a phased-dose or fixed dose in broilers. Ross 308 broilers were randomly assigned to 5 dietary treatments (12 replicates of 55 birds/pen). The diets were based on corn, wheat, SBM, sunflower meal and rapeseed meal, in pelleted form in 3 phases (starter, 1-10 d; grower, 10-21 d and finisher, 21-32 d). The treatments included: (1) a nutrient and energy adequate positive control (PC); (2) PC reduced in Ca, dig P, Na, dig AA and ME based on the contribution of PhyG at the specific phased-dose (NC1); (3) NC1 supplemented with PhyG at 2,000, 1,500, and 1,000 FTU/kg (NC1+PhyG) in starter, grower and finisher phase, respectively; (4) PC reduced in Ca, dig P, Na, dig AA and ME based on the contribution of PhyG at the fixed dose of 1000 FTU/kg (NC2); (5) NC2 supplemented with PhyG at 1,000 FTU/kg in all phases (NC2+PhyG). Data were analysed using JMP 16.1 and means separated by Tukey test. NC1 and NC2 decreased ($P<0.05$) 32 d BW, carcass weights, bone ash contents at d 10 and 32 and increased 1-32 d FCR vs PC. The PhyG supplementation maintained all these parameters vs PC. The estimated overall feed cost (€/kg BWG) was reduced by 7.2 and 6.8% and carbon footprint, g CO_2 eq/kg BWG was 12 and 11% lower with phased-dose and fixed dose of PhyG, respectively vs PC ($P<0.05$). In conclusion, supplementation of the novel phytase with both dosing strategies, with full matrix application maintained broiler growth performance, bone ash, carcass characteristics and resulting production benefit which could contribute to sustainable broiler production, with the phased dosing strategy being more cost effective.

Session 81 Poster 13

Use of fodder beets in French dairy diets: farmers points of views and experimental results

V. Brocard[1], A. Marsault[1], L. Vivenot[2], E. Tranvoiz[3] and J. Jurquet[1]
[1]Institut de l'Elevage, 149 rue de Bercy, 75012 Paris, France, [2]Union Laitière de la Meuse, CS 20149, 55104 Verdun, France, [3]Chambre d'agriculture de Bretagne, 24 route de Cuzon, 29322 Quimper, France; valerie.brocard@idele.fr

In France the use of fodder beets in diets for dairy cows is regularly questioned: new interest of producers followed by periods of oblivion. Currently 2.5% of the dairy farms are using fodder beets in average during 3 to 4 months per year, either on maize or grass silage or hay-based systems. The experience developed in 3 regions has been gathered to analyse the strengths and weaknesses of this crop. In Southwest of France, the network of farmers involved in the AccelAir program has shown the difficulties related to the crop itself and the variable yields resulting from its monitoring (seedbed, density, pesticides program, irrigation). In Eastern France, a network of dairy farmers of the ULM dairy cooperative related a positive effect of the incorporation of 2 kg DM of fodder beets on grass silage-based diets: higher productions and milk solids. To validate this impression, a trial was led during two winters in Trevarez experimental farm, Brittany (Chamber of Agriculture, Idele). 4 kg DM of fodder beets were added to a diet made of 5 kg DM of grass silage, maize silage *ad libitum* and 3.6 kg of rapeseed cakes. The cows with fodder beets decreased their intake of maize silage with a 1/1 substitution rate. No effect on fat and protein produced was noted, but the diet with the fodder beets led to a significant decrease by 1 kg of milk/cow/d. The conclusion of these 3 regional experiences does not question the quality of fodder beets to produce milk. But the efficiency seems higher on grass-based systems than on maize-based systems, notably when maize is of good quality and fed *ad libitum*. Whatever the regional background, their main asset is their capacity to secure the forage yields as their growing period takes place after the summer: they can contribute to face global warming by producing a regular, high-energy forage for lactating cows. Though, in conventional farming, their high index of pesticides use is a weakness and accounts for a high cost per hectare. In organic production, the workload related to the weed control is limiting their adoption.

Effect of wilting grass silage on milk production of dairy cows fed two concentrate protein levels

M. Grøseth[1,2], L. Karlsson[1], H. Steinshamn[3], M. Johansen[4], A. Kidane[2] and E. Prestløkken[2]
[1]Felleskjøpet Fôrutvikling, Nedre Ila 20, 7018 Trondheim, Norway, [2]Norwegian University of Life Sciences, P.O. Box 5003, 1432 Ås, Norway, [3]Norwegian Institute of Bioeconomy Research (NIBIO), Gunnars veg 6, 6630 Tingvoll, Norway, [4]Aarhus University, AU Viborg – Research Centre Foulum, Blichers Allé 20, 8830 Tjele, Denmark; martha.groseth@fkf.no

Increasing the protein value in grass silages for dairy cows is of interest to increase use of homegrown protein sources and reduce nitrogen (N) losses to the environment. Studies have shown that wilting of grass silage can improve the metabolizable protein (MP) value by increasing the rumen microbial protein yield (MCP) and rumen escaped feed protein. We hypothesised that feeding wilted grass silage can improve milk and milk protein production in dairy cows and reduce the need for MP, estimated as amino acids absorbed in the small intestine (AAT), in concentrate. To test this, a continuous feeding experiment with 48 early to mid-lactation Norwegian Red dairy cows, kept in a loose housing system was conducted. Treatments were first cut grass silages from round bales, harvested at early booting from a sward of timothy (*Phleum pratense*), perennial rye grass (*Lolium perenne*) and meadow fescue (*Festuca pratensis*), wilted to 260 and 417 g dry matter (DM)/kg fresh matter. The grass silage was fed *ad libitum* and supplied with 8.3 kg/d of concentrate, either low (108 g AAT/kg DM) or high (125 g AAT/kg DM) in MP concentration, in a 2×2 factorial arrangement. The experiment lasted for 11 weeks, with the 2 first weeks, where cows received same feeding, used as covariate, and the last 4 weeks were used as data collection period. Wilting reduced fermentation products, ammonia and soluble N in the grass silage, while increased residual water-soluble carbohydrates, like expected. However, there was no difference between treatments in daily silage DM intake (13.1 kg) and milk yield (30.2 kg) or milk content, but feeding high MP concentrate increased urea and uric acid in urine. No major differences were found for rumen pH, amino acids in blood plasma or purine derivatives over creatinine index, as indirect estimate for MCP. In conclusion, high silage DM and high MP in concentrate did not increase the milk production in this study.

Zootechnical performance of dairy cattle fed ensiled Italian ryegrass or winter rye

L. Vandaele, J.L. De Boever, T. Van Den Nest and K. Goossens
ILVO, Scheldeweg 68, 9090 Melle, Belgium; leen.vandaele@ilvo.vlaanderen.be

After harvest of silage maize both, winter rye (*Secale cereale* L.) (WR) and Italian ryegrass (*L. multiflorum* Lam.) (IR) are potential forage intercrops (harvest in spring) before planting new maize. WR normally results in higher dry matter (DM) and crude protein (CP) production in comparison with IR. Furthermore, WR needs less N fertilization, so that the following maize crop can receive sufficient N. However, in practice many farmers have doubts about the production capacity, feed quality and feed intake of WR harvested as forage. In this trial a parcel of 5,8 ha was split in two plots. Both IR and WR were sown on Sept 20th 2020 and mowed on April 25th, prewilted, cut and ensiled in silage bags on April 28th. The production of IR and WR was 3,924 and 2,165 kg DM per ha, respectively. The DM content of IR was higher than for WR (513 vs 368 g/kg), with 125 and 138 g CP/kg DM, 84 and 74 g DPI/kg DM and 6.73 and 5.86 MJ NE_L/kg DM, respectively. In the next year a trial was set up with 24 high producing dairy cows. Maize silage, IR or WR and pressed beet pulp (50/43/7 on DM) mixed with soybean meal (10.5% for IR, 9.2% for WR), corn meal (2% for WR) and feed urea (0.5% for IR and 0.3% for WR) were fed *ad libitum*. Both diets were supplemented with (rumen protected) soybean meal and balanced concentrates to meet 105% of the energy and protein requirements of each individual cow and to attain a rumen degraded protein balance of 180 g/d. DM intake (DMI) and milk production were registered daily, whereas milk composition was determined in the last week of each period. Intake and performance data were compared using ANOVA ($P<0.05$). Cows achieved a higher total DMI and roughage DMI with IR (24.4±0.5 and 18.3±0.5 kg) than with WRF (23.1±0.5 and 16.9±0.5 kg) ($P<0.05$), which resulted in a trend for higher milk production for IR as compared with WR (33.2±1.0 vs 32.3±1.0 kg). Milk fat and protein content were not different. On the other hand, WR showed a higher nitrogen (32.4±0.7% for WR and 32.3±0.7 for IR) and feed efficiency (1.55±0.04 for WRF and 1.48±0.04 for IR). In conclusion, unexpectedly winter rye showed a lower DM yield, a lower feed intake and trend for lower production, but showed potential in increasing feed and nitrogen efficiency in dairy cattle.

Performance of NIR spectrometry to predict amino acids content in forages

L. Bahloul[1], V. Larat[1], P. Riche[2] and B. Sloan[3]
[1]Adisseo, CERN, Commentry, 03600, France, [2]Adisseo, Carat, Commentry, 03600, France, [3]Adisseo, NA, Alpharetta, 30022, USA; lahlou.bahloul@adisseo.com

Near Infrared Spectroscopy (NIRS) has been used as a precise and cost-effective tool to predict chemical composition and nutritive values of ingredients. Nutrient-based feeding systems are required in ruminants to formulate more efficient, lower cost and sustainable diets. Progress towards this goal is possible with better characterization of ingredients for amino acids (AA) especially forages that are usually not analysed in feed analysis. The objective of this work was to develop NIRS calibrations to demonstrate the interest to predict individual amino acids profile in grass silages. 135 grass silages samples were collected between November 2021 and July 2022 from west and centre regions of France. The samples were dried at 70 °C for 48 h before being ground with a 1 mm sieve and further analysed for proximate parameters as well as AA. Spectra of these samples were collected from both undried and dried forms, using a DS 2500 NIR instrument from Foss, over the range 400-2,500 nm. The set of spectra was split into a calibration set of 119 samples and a test set of 16 samples. Lysine (Lys) and methionine (Met) are of utmost interest as they are usually considered the first limiting AA for milk production health and reproduction. The database represents a large variation range (CP: 13.7±3.19, Met: 0.145±0.05, Lys 0.39±0.15 as g/100 g).The NIR calibrations performed on the 1 mm ground samples using the Partial Least Squares (PLS) regression reached good predictive precision for both Lys (R^2=0.93, Standard Error of Prediction (SEP)=0.047 g/100 g, Ratio of percent deviation (RPD)=2.7) and Met (R^2=0.91, SEP=0.013 g/100 g, RPD=2.8), as well as crude protein (CP) (R^2=0.95, SEP=0.74 g/100 g, RPD=4.5). Interestingly, the SEP obtained with NIR calibrations surpass in precision the standard error obtained if the AA concentration is estimated by applying a fixed coefficient to the CP wet chemical analyses or NIR predictions (respectively 0.098 and 0.099 g/100 g for Lys; 0.021 and 0.023 g/100 g for Met). In conclusion, the results show the ability of NIR spectroscopy to accurately estimate AA in grass forages and its interest to move forward precise formulation of dairy rations.

Estimation of feed weight of dairy cows using computer vision

M. Taghavi, T. Izquierdo and I. Fodor
Wageningen University & Research, Wageningen, 6700 AH, the Netherlands; marjaneh.taghavirazavizadeh@wur.nl

Monitoring the feed intake of individual dairy cows is important for assessing not only their feed efficiency but also to detect health events. Roughage intake control bins are commonly used to measure the feed intake of individual cows in research but are not practical for commercial farms due to high costs and frequent cleaning requirements. Computer vision approaches based on deep learning algorithms are increasingly being studied as a potential solution. Our aim is to develop and validate a deep learning-based method that is scalable and affordable for commercial farms. An Intel® RealSenseTM D455 depth camera was installed at the Dairy Campus research facility of Wageningen University and Research (Leeuwarden, the Netherlands). The camera was placed on a frame above a pile of feed (Total Mixed Ration), and the feed weight was measured using a digital scale with 0.1 kg accuracy. RGB-depth images of the feed were captured within a 0-20 kg range (with 0.5 kg increments) at two different vertical distances from the camera (1 m and 2.6 m), under two light conditions (morning and afternoon light) in the barn. Five different shapes were captured for each feed weight in each setup. In total, 800 RGB-depth images were collected. The images of 4 different shapes from each setup (1 m and 2.6 m in the morning and afternoon) for each weight were randomly selected for training, and the image of 1 shape per setup and weight was selected for validation. The multi-channel input models were trained using both RGB and depth images. The models consist of well-known backbones from image classification for image feature extraction, and the classification head was removed in favour of a regression head. The training and validation process was performed on the dataset for 1 m only, 2.6 m only, and a combination of both datasets. The model trained on the 1 m dataset was more accurate (expressed in mean squared error) than the one trained on 2.6 m, as expected, because the depth information is more accurate at the smaller distance of 1 meter. Prediction accuracy is continuously being improved. Further results will be presented at the conference.

Session 82 — Theatre 1

The performance of cross fostered lambs from birth to weaning on commercial sheep flocks in Ireland

F.P. Campion, J. Molloy and M.G. Diskin
Teagasc, Animal & Bioscience Department, Mellows Campus, Athenry, Co. Galway, H65 R718, Ireland; francis.campion@teagasc.ie

The management and care of lambs born in large litters (>2) is often discussed and debated. Allowing ewes to rear large litters to weaning and the use of artificial milk feeders have been shown to be effective. The success of both of these systems is largely depending on flock management and can be expensive. Cross fostering is also employed by some producers whereby excess lambs are fostered to single bearing ewes with the process usually occurring once the fostering ewe is lambing. This study aimed to investigate how the performance of the fostered lamb (F) compared to the ewes own lamb (O) from birth until weaning time (14 weeks post-partum). Over three production years' ewes of mixed age and breed were performance recorded on five commercial mid-season lambing, grass based sheep flocks in Ireland. All of these flocks were enrolled in the Teagasc BETTER farm sheep programme and performance recording was done in conjunction with trained research staff. At parturition each lamb was linked to its dam and its foster dam where necessary. Lambs had their live weight recorded at birth, seven weeks and 14 weeks post-partum. The weights were used to calculate average daily (ADG). Only ewes that gave birth to a live single lamb and that had a lamb from either a triplet or quadruplet litter fostered to them were considered for the final analysis. Approximately 7% of the lambs recorded as F or O at parturition in this data set did not present for weighing at either seven or 14 weeks post-partum. Lamb ADG from birth to seven weeks post-partum differed between F lambs and O lambs (274 vs 300 g/day, ±12.9 g/day; $P<0.05$). However, there was no difference in ADG between F and O lambs from birth to weaning (245 vs 251 g/day, ±10.5 g/day; $P>0.05$) or from seven weeks to 14 weeks post-partum (229 vs 226 g/day, ±15.1 g/day; $P>0.05$). Live weight between F and O lambs differed at birth, seven and 14 weeks post-partum ($P<0.05$) with F lambs being lighter at all three time points. While the ADG of the F lambs was behind the O lambs for the first seven weeks the performance was still satisfactory, particularly when the increased efficiency of a single bearing ewe rearing an additional lamb are considered.

Session 82 — Theatre 2

Latent factors analysis of protein profile, composition, and cheese-making traits of goat milk

N. Amalfitano[1], G. Secchi[1], M. Pazzola[2], G.M. Vacca[2], M.L. Dettori[2], F. Tagliapietra[1], S. Schiavon[1] and G. Bittante[1]
[1]University of Padova, DAFNAE department, Viale dell'Università 16, 35020 Legnaro, Italy, [2]University of Sassari, DVM department, Via Vienna 2, 07100 Sassari, Italy; nicolo.amalfitano@unipd.it

The present study aimed to analyse the relationships between the detailed protein profile, the composition, and the cheese-making ability traits of goat milk identifying the latent factors controlling them. This study is part of the GOOD-MILK project (D.M. 9367185 – 09/12/2020) which involved the collection of at least 800 goats belonging to 6 different breeds (Camosciata delle Alpi, Saanen, Murciano-Granadina, Maltese, Sarda, and Sarda Primitiva). The detailed protein profile was obtained using the reverse-phase high-performance liquid chromatography technique (RP-HPLC). Milk composition was estimated using Fourier-transform infrared spectroscopy (FTIR) and cheese-making ability traits were obtained through a laboratory cheese-making procedure. The complete dataset was analysed for latent factors in the R environment. All the traits together yield 8 latent explanatory factors. With protein profile expressed in g/l, the first factor is the 'cheese-yield' affecting positively almost all the cheese-making traits, the fat content, and the total caseins and β-CN, and negatively the pH. The second is the 'curd firmness' factor affecting, beyond the curd firmness traits, the total casein, β-CN, and κ-CN and the recoveries of milk fat and energy in cheese. The third is the 'coagulation time' factor. The fourth is the 'udder health' factor influencing positively the lactose, the β-LG, and α-LA and negatively the somatic cells score and the NaCl. The fifth is the 'curd dynamics' factor affecting positively the curd-firming and syneresis rates and negatively the curd firmness after 30 and 45 min from rennet addition. The sixth is the 'cheese protein' factor whereas the seventh and eighth are, respectively, the 'α_{S2}-CN + κ-CN' and the 'α_{S1}-CN' factors. With protein profile expressed in % of total protein, the latent factors present some differences. This information may help to better understand the link between composition, protein profile, and cheese-making ability of goat milk.

Session 82 Theatre 3

Pyrenean wools: how improving the organisation of the upstream and downstream sectors in the massif

C. Viguié[1], S. Fichot[1] and G. Brunschwig[2]

[1]Association des Chambres d'Agriculture des Pyrénées, 32 Av. du Général de Gaulle, 09000 Foix, France, [2]VetAgro Sup – INRAE, UMR Herbivores, 89 avenue de l'Europe, BP 35, 63370 Lempdes, France; gilles.brunschwig@vetagro-sup.fr

For decades, the price of sheep's wool has been falling: a few years ago it was paid at around €1/kg, but now it is around €0.10/kg, a derisory sum that no longer covers shearing costs. With the closure of historic markets such as China and North Africa, the situation is deteriorating further, and in some departments the wool is not even collected on the farms and has to be stored. In several Pyrenean departments, breeders deplore a 'dramatic situation, with no solution'. In 2021, wool sector contracts were signed by several French regions and define an action plan for the next 5 years. The study presents the situation in Pyrenean chain. A cartographic inventory was carried out using regional data extractions. It provided a visual representation of the main sheep basins, and therefore of the available wool. This territorial approach was then completed by the analysis of a questionnaire sent to 2,600 farms. Focusing on breeding and shearing practices, it provides an insight into the diversity of breeding systems and practices, as well as the types of wool encountered in the Pyrenean massif. Three types of breeders can be distinguished: (1) breeders with little training in wool and who pay little attention to its cleanliness; (2) breeders who produce clean wool but do not sort it and who store it because of a lack of outlets; (3) breeders who are trained in wool sorting and who, by sorting it, value it at a higher price and are more satisfied with their situation. Furthermore, no link appears between the size and location of the farms and the outlets for their wool, especially as the strategies are varied (local craftsmen, cooperatives or private traders). This study provides an initial representation of wool production in the Pyrenees and its challenges. It contributes to the structuring of the sector, coupled with other projects carried out in parallel on the identification of artisanal initiatives or the prospecting and solicitation of French companies ready to engage in the transformation of this resource.

Session 82 Theatre 4

The Gentile di Puglia merino sheep breeds preliminary wool quality assessment

V. Landi[1], E. Ciani[2], G. Molina[1], R. Topputi[2], F.M. Sarti[3], A. D'Onghia[4], G. Mangini[4], F. Pilla[5], S. Grande[6], A. Maggiolino[1] and P. De Palo[1]

[1]University of Bari Aldo Moro, Veterinary Medicine, SP. 62 km. 3, Valenzano (BA), 70010, Italy, [2]University of Bari Aldo Moro, Department of Biosciences, Biotechnologies and Biopharmaceutics, Via Amendola 165 Bari, 70126, Italy, [3]Univerity of Perugia, Borgo XX Giugno, Perugia, 06128, Italy, [4]ARA Puglia, Strada S. Nicola, 2, Putignano BA, 70017, Italy, [5]University of Molise, Department of Agriculture, Environment, and Food, Via De Sanctis, Campobasso, 86100, Italy, [6]ASSONAPA, Via XXIV Maggio, 44, Roma, 00187, Italy; vincenzo.landi@uniba.it

Wool quality assessment was carried out in the Gentile di Puglia sheep breed after more than three decades. The study was conducted on three farms in Apulia, Italy, where individual wool samples were collected from registered herds. Individual samples (100 for each herd) were collected, and the total fleece weight measured (TW). Diameter (FD) was assessed using the Fibrelux instrument both on greasy and scoured wool. The results showed that the TW ranged from the mean value of 1.54±0.53 kg to 3.44±0.79 for animals of 2 and 3 year more respectively. the average diameter is concentrated around 20 microns, with values between 11.3 and 28.1 (standard deviation 3.35). In greasy wool, the diameter range is between 14.82 and 31.48 (standard deviation 3.18). The average FD of the greasy wool was 23.99±3.184 microns against the 20.78±3.37 of the washed wool. The average washing weight yield was 0.56%, which, considering a range between 63 and 67% that is relatively low showing that there is ample room for improvement in animal management. The average FD for greasy wool for animals of 2, 3 and 4 years was 22.38±1.71, 23.35±2.75 and 23.18±2.61 respectively while variation coefficient was 5.87±6.79, 8.69±9.88 and 8.383±9.03. These results will help to restart a selection plan for wool in this breed. Founded by the Agritech National Research Center and received funding from the European Union Next-Generation EU (PNRR) – Missione 4-2, Investimento 1.4 – D.D. 1032 17/06/2022, CN00000022). This manuscript reflects only the authors' views and opinions, neither the UE can be considered responsible for them. We thank to C. Carrino, G. Bramante and F. D'innocenzio farms for making the animals available for the collection of samples.

Session 82 Theatre 5

Using indoor and outdoor finishing systems to finish hill lambs to carcass weights between 12-16 kg

M. Dolan[1], T. Boland[2], N. Claffey[1] and F. Campion[1]
[1]Teagasc, Animal & Bioscience Department, Animal & Grassland Research & Innovation Centre, Mellows Campus, Athenry, Galway, H65R718, Ireland, [2]University College Dublin, School of Agriculture and Food Science, Belfield, Dublin 4, D04 V1W8, Ireland; mark.dolan@teagasc.ie

Fattening lambs indoors to 12-16 kg carcass weights is widely adopted by primary hill and mountain producers in parts of Europe. Research has shown that 'light' lambs finished indoors on *ad libitum* concentrates can meet the market specifications. However, there is a paucity of information on fattening these type of lambs to 12-16 kg carcass weights on autumn pasture. The objective of the study was to examine the performance of Scottish Blackface (SBF) ram and castrate lambs when offered two dietary treatments from September to January; *ad libitum* concentrates indoors (ALC) and permanent pasture (predominantly *Lolium perenne*) outdoors supplemented with 500 g of concentrate (PP). The study was a randomised complete block design with lambs blocked by weight and balanced for sex. The experiment was carried out on two separate years with 50 ram and 50 castrate lambs allocated to each treatment each year. Lamb live weight (LW) was recorded at the start of the experiment and every fourteen days up until the point of slaughter. Ram and castrate lambs started the experiment at 23 kg LW. Ram and castrate lambs were drafted for slaughter as they reached at least 34 kg and 33 kg LW respectively with adequate fat cover. Lambs on ALC treatment had a higher average daily gain than PP lambs (189 and 124±4.1 g/day; $P<0.05$). There was no difference in carcass weights (14.9 and 15.0±0.12 kg; $P>0.05$), carcass conformation (2.8 and 2.8±0.03; $P>0.05$) or fat scores (2.8 and 2.7±0.03; $P>0.05$) while diet type also had no effect on kill out percentage (42.6 and 42.4% ±0.25; $P>0.05$). The average days to slaughter was significantly lower for ALC lambs when compared to PP lambs (65 and 101±1.9 days; $P<0.05$). Results from this study show that hill bred lambs can be satisfactory slaughtered to carcass weights of between 12-16 kg from both dietary treatments. However, ALC had higher growth rates and reached the target slaughter weight significantly quicker than PP lambs which is an important consideration for these types of systems.

Session 82 Theatre 6

Influence of climatic disturbances on the lactation curve in dairy goats

A. Harnois Gremmo[1,2], N. Gafsi[1,2], F. Bidan[1], O. Martin[2] and L. Puillet[2]
[1]IDELE, MNE, 75595 Paris, France, [2]Université Paris-Saclay, INRAE, AgroParisTech, UMR Modélisation Systémique Appliquée aux Ruminants, 91120 Palaiseau, France; aurore.harnoisgremmo@agroparistech.fr

In the context of climate change, intensification of temperature and humidity fluctuations are inducing unquantified impacts on animal performance. The goat is described as adapted to harsh environments, however the limits of its adaptability to climatic stress are not clearly defined. The aim of this study was to investigate the relationship between lactation dynamics and climatic disturbances. This study used the daily milk production data related to 1,174 lactations from 585 goats of the experimental goat farm of INRAE MoSAR, between 2006 and 2021. The data were fitted with the Perturbed Lactation Model, which provides a theoretical-unperturbed lactation curve and characterizes the detected perturbations in the milk production time-series. Daily temperature and hygrometry data were collected at the nearest meteorological station. To explore the impact of climatic conditions, each perturbation detected was confronted with the climate conditions at the date of the perturbation and seven days prior to it. Climatic stress, cold or hot, was estimated using three indicators: a temperature accumulation and two equations of temperature humidity index. Statistical analyses were performed to evaluate if goat lactation curve perturbations were related to climatic stress indicators or to major herd management events. The results showed that for 1,174 lactations 9,010 perturbations were detected. 5.7% of these perturbations in milk production occurred on the period of reproduction (synchronization followed by artificial insemination). The climatic environment study indicated that about 25% of the perturbations occurred during cold stress periods. These results highlight the benefit of combining mathematical models with time-series data to explore animal responses to the environment, especially climatic stress. They would help to precise the thermal tolerance threshold of goats in temperate zones. Furthermore, they allow to unravel the relative effects of thermal fluctuations and other environmental factors that may alter goat metabolism, and therefore to better understand the conditions necessary to rear goats in an environment more respectful of their well-being.

Session 82 Theatre 7

Genetic parameters for fleece uniformity in Alpacas

J.P. Gutiérrez[1], A. Cruz[2], R. Morante[2], A. Burgos[2], N. Formoso-Rafferty[3] and I. Cervantes[1]
[1]Universidad Complutense de Madrid, Departamento de Producción Animal, 28040 Madrid, Spain, [2]INCA TOPS S.A., Miguel Forga, 348, Arequipa, Peru, [3]Universidad Politécnica de Madrid, Departamento de Producción A, 28040 Madrid, Spain; gutgar@ucm.es

Fiber diameter is the main selection objective and criterion in alpaca breeding programs, but it can differ across anatomic regions of the animal. Fleece uniformity is one of the requests of the textile industry. Fiber diameter is usually registered from a unique sample from the mid side of the body, and therefore fibre diameter uniformity within fleece is never taken into account, when phenotypic and genetic differences may exist for fleece uniformity in alpaca populations. The objective of this work was to estimate genetic parameters for fleece uniformity in alpacas. Fiber diameters were measured in three different locations and used as repeated records of the same animal. The dataset counted with 1,641 records of white females of Huacaya ecotype, and 1,085 animals in the pedigree file, being studied fitting a model that considers heterogeneous the residual variance of the model. The logarithm of the standard deviation of the three measures was also used as a measure of the fleece variability. Model included same effects for mean trait and its variability: the month-year of shearing, the body location and the age as covariate as systematic effects. The environmental permanent effect was included as random effect besides the residual and genetic effects. GSEVM program was used to solve the model. Shoulder location showed to be more consistent than mid side and thigh location. Age importantly affected to the residual variance leading to less uniform fleece in old animals. The additive genetic variance of the environmental variability estimate (standard error) was moderate to high, 0.43 (0.14), suggesting a considerable existence of genetic differences between individuals to select to increase fleece uniformity. Genetic correlation (standard error) of the environmental variability of the fibre diameter within fleece with the fibre diameter itself was 0.76 (0.13) indicating that fleece uniformity would be indirectly selected when reducing the fibre diameter. Therefore, and due to the cost of registering, there seems to be no need to consider uniformity as a selection criterion in alpacas breeding programs.

Session 82 Theatre 8

Farmers' competitions: stimulating new management practices by Peruvian alpaca and llama farmers

E. Quina[1], T. Felix[2], M. Aguilar[1], G. Gutierrez[2], J. Candio[2], J. Gamarra[2], M. Mamani[1,2], A. Mejía[1] and M. Wurzinger[3]
[1]DESCOSUR-Centro de Estudios y Promoción del Desarrollo del Sur, Calle Málaga Grenet 678, Arequipa, Peru, [2]Universidad Nacional Agraria La Molina, Av. La Molina s/n, Lima, Peru, [3]BOKU-University of Natural Resources and Life Sciences, Vienna, Gregor-Mendel-Strasse 33, 1180 Vienna, Austria; maria.wurzinger@boku.ac.at

The Puna ecoregion of the Andes is located above 3,800 masl and is a key ecoregion for the national and local economies, providing various products and ecosystem services. Most of Peru's extensive livestock farming is concentrated here. In these low-input, pasture-based systems, 4,4 million alpacas and 1,0 million llamas – autochthonous species of the Andes – are kept by about 83,000 farmers. The production level of these farms is low due to many reasons. The NGO Descosur has been co-developing mechanisms to stimulate the implementation of different management strategies with the aim of improving the management of alpacas and llamas. Descosur's area of operation is the national park 'Reserva Salinas y Aguada Blanca' in the province of Arequipa in Southern Peru. In a participatory approach, farmers' competitions were designed with local municipalities, farmers' representatives and extension workers. Since 2018 the competitions have been advertised publicly in three different districts, and interested parties could register. A jury of experts and farmers judges whether the specified measures have been implemented and to what extent. Measures that were implemented by participants were: water harvesting techniques (construction of water channels, ponds), fencing of pasture areas, sowing of improved pastures, and data collection on productive and reproductive data of animals. The five best farmers in the competition received tools and materials and were officially recognized by the Ministry of Agriculture. All other participants also received small incentives. This participatory approach was very successful as farmers showed much interest. The common design ensured that only management practices that farmers could implement were included in the competition. The winning farmers can be seen as pioneers, and their farms are models for neighbouring farmers.

Session 82 Theatre 9

YoGArt project: milk quality and oxidative status of ewes fed microalgae blend

A. Mavrommatis, P. Kyriakaki, F. Satolias and E. Tsiplakou
Agricultural University of Athens, Animal Science, Iera Odos 75, Athens, 11855, Greece; eltsiplakou@aua.gr

Enriching dairy ruminant diets with polyunsaturated fatty acids (PUFA) appears to be a sustainable strategy for improving milk's nutritional value. However, long-chain PUFAs have been recognized for their susceptibility to oxidation increasing the risk of oxidative stress emergence which in turn can activate a pro-inflammatory response. Hence, this study aims to examine the effect of microalgae blends dietary supplementation on ewes' milk quality and immune-oxidative status, assessing oxidative biomarkers and antioxidant enzymes activity along with the transcriptional profiling of immune genes from ewes' monocytes and neutrophils. Thirty-two dairy ewes at early lactation were divided into four homogeneous groups; the control group (CON) had no microalgae, the SC30 group was supplemented with 30 g *Schizochytrium* spp./ewe/day while MB30 and MB40 groups were supplemented with 30 and 40 g microalgae blend /ewe/day, respectively. Microalgae supplementation did not significantly affect milk yield while fat content has significantly reduced in MB30 ewes. In milk though, the concentration of Malondialdehyde (MDA) was decreased in the MB30 group while Protein Carbonyls (PC) were increased in MB40 compared to the MB30 group. In blood plasma, Catalase and Glutathione transferase activities were significantly increased in the MB40 group while PC was significantly reduced in the MB30 group. Regarding the immune system, the mRNA levels of TLR4 in monocytes were significantly downregulated in SC30 compared to the CON group. Additionally, IFNG, NFKB, TNFA, and CCL5 relative transcript levels tended to decrease in monocytes and neutrophils of the treated ewes. The combination of *Schizochytrium* spp. with other autotrophic algae species rich in antioxidant compounds alleviated oxidative stress induced by PUFA overload in ewes' diet.

Session 82 Theatre 10

Goat AI programs based on the male effect using less or no hormones progress with suitable results

A. Fatet[1], L. Johnson[2], L. Jourdain[2], F. Bidan[3] and P. Martin[4]
[1]INRAE UE1373 FERLus, Les Verrines, 86600 Lusignan, France, [2]Innoval, Rue E. Tabarly, 35530 Noyal-Sur-Vilaine, France, [3]Institut de l'Elevage, 42 Rue G. Morel, 49070 Beaucouzé, France, [4]Capgènes, 2135 Rte de Chauvigny, 86550 Mignaloux-Beauvoir, France; alice.fatet@inrae.fr

In 2021 in France, 70,648 caprine inseminations were performed (Capgènes). In the national database, the type of preparation of the goats before artificial insemination (AI) is individually recorded for 92% of performed AI. Goat insemination in France is mainly practiced after hormonal synchronisation. However, preparation techniques with less or no hormones have been made available and advertised by the goat sector industries over the past years. This led to a sensible reduction in hormonal synchronization and a noticeable progression of the use of alternative methods based on the buck effect. Across the north-west quarter of France, where insemination cooperative Innoval performs goat AI, the proportion of AI performed after detected natural heats, buck effect (BE) or vaginal sponge and buck effect (VS-BE) preparation has grown from 6% in 2019 (n=33,870 total AI) to 15% in 2021 (n=36,076 total AI) showing hormonal synchronization is slowly decreasing. Regional practice is very diverse with a rather high adoption of less or non-hormonal techniques in Bretagne (35% of total AI), medium in Pays de la Loire (17%) and Poitou-Charente (15%) and rather low in Centre (9%; Innoval AI database). Innoval has been following closely 23 pilot breeders using BE or VS-BE preparation techniques for the past 3 years. Fertility results were quite acceptable in comparison with national fertility results after AI (58.2% in 2021 including all breeds, seasons, preparation techniques, source: Capgènes). BE resulted in 57% fertility (n=891 AI). VS-BE preparation resulted in 59% fertility (n=1,650 AI). This study allowed identification of key steps ensuring good fertility results when using these alternative techniques: female:male ratio, well prepared males (photoperiodic treatment, sexual awakening), free interactions between males and female in the same pen, quality of heat detection, heat detection to AI interval.

Session 82 Theatre 11

Fiber characteristics of cashmere goat in Mongolia

S.B. Baldan[1,2], M.P. Purevdorj[1], G.M. Mészáros[2] and J.S. Sölkner[2]
[1]Research Institute of Animal Husbandry, Animal Science Sector, Zaisan street, Khan Uul, 17024 Ulaanbaatar, Mongolia, [2]University of Natural Resources and Life Sciences, Institute for Livestock Sciences, Gregor-Mendel-Straße 33, 1180 Wien, Austria; johann.soelkner@boku.ac.at

Producing up to 40% of world's raw cashmere annually, Mongolian herders largely depend on cashmere goats for their livelihoods. Increased economic value led to escalated growth in the goat population posing environmental challenges in a vast pastureland. On the other hand, the intention to improve production volume brought a decline in fibre quality due to mismanaged breeding practices. This situation requires many actions to tackle environmental and productivity issues. One of them is to evaluate current production and biological resources in order to improve animal production efficiency. This paper aims to define the fibre characteristics of cashmere goats in Mongolia for their cashmere yield, fibre length, fibre diameter and cashmere content. A total of 621 goats from 14 populations were sampled for their cashmere and genetic material. Genotype was obtained by 50K goat SNP array. Sampled populations were distributed in different economic regions in the country. We intend to define the genetic structure of sampled goat populations and to study whether there is a regional difference among populations for their cashmere quality traits. Variations in cashmere production traits if observed could help to define breeding objectives for different goat populations depending on their specific quality traits. Additionally, we would examine whether there are genome-wide association signals for fibre characteristics.

Session 82 Poster 12

Anco fit improves feed efficiency, milk production and milk solids in lactating ewes

C. Panzuti[1], R. Breitsma[2] and C. Gerard[1]
[1]ADM, Talhouet, 56250 SAINT NOLFF, France, [2]ADM, Linzer Straße, 3100 St. Poelten, Austria; clemence.panzuti@adm.com

Sheep milk is an excellent source of nutrients and is mainly used for cheese production due to its high total solids content. Dietary administration of some bioactive compounds have demonstrated a positive effect on ewe milk production or composition. The aim of this study was to evaluate the dietary supplementation of Anco FIT (AF), containing bioactive substances derived from ginger, lemon balm and cinnamon, yeast components and clays, on feed efficiency, milk production and composition in dairy ewes. Eighty multiparous Lacaune ewes were allocated into two homogenous groups based on milk yield, milk protein and fat contents, lactation number, body weight and body condition score. The animals were housed in two separate pens (n=40 ewes/pen). Each animal received a standard diet consisting of forage (1/3 grass, 1/3 corn silages, 1/3 alfalfa hay) *ad libitum* (~2.3 kg dry matter (DM)/day (d)), and 1.55 kg/d as-fed concentrate composed of 0.45 kg barley, 0.60 kg dehydrated alfalfa, and 0.5 kg of an experimental concentrate without or with supplementation of AF providing 3.8 g AF/head/d. The trial lasted from 1 to 4.5 months post-partum. Milk yield, milk fat and protein were measured at the beginning of the trial and then individually every 2 weeks. Yields of milk fat, protein and solids (sum of protein and fat) were calculated. Within 2 weeks interval forage intake was measured per pen and adjusted for feed refusals. The amount of the distributed concentrate feed was recorded daily per pen. Statistical analyses were performed using Anova mixed model with R software. In comparison to the control group, the supplementation of AF ($P \leq 0.05$) increased milk production by 10% (2.1 vs 2.3 kg/d) and the yields of milk protein (135 vs 151 g/d), fat (125 vs 139 g/d), and total solids (260 vs 290 g/d) by 11, 12 and 12%, respectively. In AF group, the animals had higher ($P \leq 0.05$) forage intake + 0.2 kg DM/d and the feed efficiency was enhanced ($P \leq 0.05$) by gaining +0.03 kg milk/kg DM intake (I) and +5.3 g solids/kg DMI. The results of this study showed that AF can be used as an efficient dietary strategy to increase feed efficiency, milk production and solids in dairy ewes.

Session 82 Poster 13

No effect of meslin flattening on dairy ewes

B. Fança[1], A. Hardy[2], M. Rinn[2] and L. Buisson[2]
[1]Institut de l'Elevage, 149 rue de Bercy, 75012 Paris, France, [2]EPLEFPA La Cazotte, Route de Bournac, 12400 Saint-Affrique, France; barbara.fanca@idele.fr

The protein self-sufficiency of farms can be improved by producing and consuming their own concentrates. Beyond the agronomic aspect, which was not covered in this study, the objective was to evaluate the use of meslin (a grain mixture of cereals and protein crops) in the diet of dairy ewes and, in particular, the interest of distributing it after flattening. Three homogeneous groups of 40 ewes were formed. After an adaptation period, 9 milk control points were recorded for a total trial duration of 3 months. The ewes were fed with the same fodder feed ration and their supplementation varied. To accompany the decline in production, the quantities of concentrates were adjusted. At the beginning of the trial, at the start of the milking period, the Control group received the standard daily supplementation of the farm, i.e. 0.45 kg of barley, 0.5 kg of a nitrogen corrector and 0.6 kg of dehydrated alfalfa. The two Experimental groups received 0.25 kg of the commercial feed, 0.8 kg of dehydrated alfalfa and meslin equivalent to 0.3 kg of barley and 0.3 kg of peas, presented flattened or not. No significant difference in milk production or milk quality was demonstrated between the three groups. Replacing half of the corrector with the mix of grain had no effect on the quantity and quality of milk produced by the dairy ewes. Moreover, the flattening of the barley-pea mixture, representing an additional cost of about 50 euros per ton, and requiring handling and therefore time, had no positive effect on the ewes' zootechnical performance. Finally, if cereals and protein crops are produced on the farm, the part of purchased concentrates in the ration at the beginning of milking drops from 53% (Control) to 30% (Experimental). Similar results were obtained for goats on the Pradel farm (Ardèche, France) with a mixture of barley and faba beans.

Session 82 Poster 14

Quality of cheeses made from the milk of native breeds of goats

A. Kawęcka, M. Pasternak, J. Sikora and M. Puchała
National Research Institute of Animal Production, Department of Sheep and Goat Breeding, 2 Sarego Street, 31-047 Krakow, Poland; jacek.sikora@iz.edu.pl

Goat milk and products made from it are very popular among consumers around the world and are perceived as ecological, natural, with a high content of bioactive ingredients. Many of them are made from the milk of local goat breeds, the importance of which has been growing recently. In Poland, such a breed is The Carpathian goat, typical for mountain and foothill regions, whose population is dynamically developing and is an extremely valuable element of biodiversity and potential for the development of the local market of goat milk products. The aim of this work was to study the chemical characteristics of experimental cheeses from milk of native Carpathian goat (C) in comparison to Anglonubian (A) and crossbred goats (Saannese and Alpine breed, SA). Cheese milk was obtained three times during the grazing season: in May, July, and September. In the farm's cheese factory, the milk was processed into rennet cheese according to a traditional recipe. Cheese samples were analysed for the content of dry matter, protein and fat, vitamin A and E and determination of fatty acids in fat. The obtained results indicate that the goat breed had an impact on the chemical composition of the cheeses. Cheeses from milk of SA contained the most dry matter (20.5 vs 17.5%) and vitamin A. Native breed milk (C) contained the most ash (4.6 vs 3.3-3.8%), while Anglo-Nubian (A) more total protein (20.3 vs 16.7-18.40%) than the other groups. Significant differences were found in the fatty acid profile of the cheese fat. Carpathian goat milk cheeses contained more unsaturated (UFA) than saturated (SFA) acids compared to the other groups less short-chain C8 and C10 acids than the other two groups, twice as many monounsaturated C16-1 and C18-1 acids and the CLA c9-t11 isomer. The PUFA-6/3 ratio was lower in group C, 1.4 vs 1.93-2.16 in the other groups. From the point of view of the nutritional value, the composition of the tested cheeses varied depending on the breed of goat. The cheeses from the milk of the native Carpathian breed were characterized by a more favourable profile of fatty acids, important from the point of view of human health.

Session 82 Poster 15

The impact of live-weight and body condition score on reproductive success in ewes

P. McCarron, N. McHugh, N. Fetherstone, H. Walsh and F.M. McGovern
Teagasc, Animal and Bioscience, Animal & Grassland Research & Innovation Centre, Mellows Campus, Athenry, Co. Galway, H65 R718, Ireland; patrick.mccarron@teagasc.ie

Reproductive efficiency is deemed one of the most important traits governing overall farm productivity. In sheep production systems body condition score (BCS), which is interrelated with live-weight, is regularly monitored as a key performance indicator for benchmarking flock performance and efficiency. The objective of this study was to investigate the relationship between ewe live-weight and BCS pre-mating on a plethora of key traits associated with mating and lambing performance. A total of 2,046 data records collected from ewes (females, >18 months of age), representing two main breeds (Suffolk and Texel) over eight breeding seasons were available. Prior to mating ewe live-weight was recorded and BCS assessed, by a trained observer, using a 1 (emaciation) to 5 (over-fatness) scale, in increments of 0.5. Ewe overall conception rate and pregnancy scanned litter size were subsequently recorded within each season. At lambing the total number of lambs born, litter mortality and litter birth and pre-weaning weight were recorded. Data were analysed using linear mixed models with ewe breed, experimental group, year and ewe age included as fixed effects; animal was included as the random effect. Ewe BCS or live-weight had no effect on conception rate ($P>0.05$). Mating BCS influenced pregnancy scan rate, a 1 unit increase in BCS increasing pregnancy scan rate by 0.2 lambs ($P<0.01$). Ewes with a mating BCS of 3.5 or greater gave birth to litters 2.5% heavier when compared to ewes with a BCS of 3.0 or less ($P<0.01$). This difference increased to 5.3% in litter pre weaning live-weight. Similarly every 1 kg increase in ewe live-weight resulted in a higher pregnancy scan rate, number of lambs born and heavier litter birth weights ($P<0.05$). Additionally, ewe age at lambing had a significant effect whereby ewes aged 3 and 4 gave birth to a larger number of lambs with heavier combined litter birthweights than ewes aged 2 years ($P<0.01$). Results from this study show that both BCS and live-weight, recorded at mating, influence reproductive and lambing traits in ewes thus emphasising the importance of having ewes at BCS 3.5 or greater at mating.

Session 82 Poster 16

Genetic parameters of medullation types in alpaca fibre

A. Cruz[1,2], Y. Murillo[1], A. Burgos[2], A. Yucra[2], M. Quispe[3], E. Quispe[1] and J.P. Gutiérrez[4]
[1]Universidad Nacional Agraria de la Molina, La Molina, 15024 Lima, Peru, [2]INCA TOPS S.A., Miguel Forga 348, Arequipa, Peru, [3]Maxcorp, La Molina, Lima, Peru, [4]Universidad Complutense de Madrid, Avda Puerta de hierro s/n, 28040, Spain; gutgar@ucm.es

Alpaca fibre industry competes with other noble fibres in an international market. Improving the fibre quality is mandatory with this aim, referring to reduce the incidence of objectionable fibres. Alpaca breeding programs have been successful reducing fibre diameter (FD), but medullation seems to be greatly involved in prickling, and its reduction has been recently introduced as selection criterion in breeding programs. It has been measured by opacity using OFDA100 (optical-based fibre diameter analyser) or using projection microscopy. Fiber Med device has been recently developed, providing medullation measures, and differentiating medullation degrees. The objective of this work was to estimate the genetic parameters of medullation types and their relationship with their respective diameters in alpaca fibre. A total of 3,149 fibre samples collected between 2020 and 2022 from 1,626 alpacas aged from 1 to 16 years old were used, the total pedigree consisting of 14,457 records. Percentages of medullation types were transformed to their centre log-ratio as indicated for compositional data. The types were Strongly Medullated (SM), Continuously Medullated (CM), Uncontinuously Medullated (UM), Fragmented Medullated (FM) and No Medullated fibre (NM), with their respective diameters. A multitrait animal model was used to estimate the variance components using VCE 6.0 program. The heritabilities ± standard errors were 0.11±0.01, 0.20±0.01, 0.10±0.01, 0.18±0.01 and 0.25±0.01 for SM, CM, UM, FM and NM respectively and 0.10±0.02, 0.25±0.02, 0.30±0.02, 0.35±0.02, 0.27±0.02 and 0.29±0.03 for respective FD_SM, FD_CM, FD_UM, FD_FM and FD_NM, and global FD. Genetic correlations of medullation types with FD ranged from -0.75 (with NM) to 0.77 (with CM), obtaining moderate to low genetic correlations between medullation types and their respective diameters (from 0.01 to -0.42). The findings suggest that the reduction of FD and total percentage of medullation would perform optimal results in removing the coarsest fibers and consequently the itching.

Session 82 Poster 17

Blood and gastrointestinal nematode profile of sheep-fed diets supplemented with fossil shell flour

O.O. Ikusika and C.T. Mpendulo
University of Fort Hare, Livestock and Pasture Science, Kings Williams Road, Ailce, 5700, South Africa; oikusika@ufh.ac.za

Fossil shell flour (FSF) is a naturally occurring fossilized atom and has recently been used as a feed additive in livestock production. Therefore, this study investigated the effects of feeding different amounts of FSF on hematobiochemical profiles and the gastrointestinal parasitic loads of Dohne Merino wethers. Twenty-four Dohne Merino wethers were randomly assigned in a completely randomized design with six wethers per treatment. The wethers were fed a basal diet, 0% FSF, basal diet +2%FSF, +4% FSF or a basal diet + 6% FSF of diet. Blood and faecal samples were collected on days 0, 25, 50, 75, and 100 of the feeding trials. The red and white blood counts increased in wethers-fed FSF-supplemented diets from day 25 to 100 compared to the control ($P<0.05$). Blood urea and serum creatinine were significantly lower in wethers with a 4% FSF diet compared to the control ($P<0.05$). Total protein concentration, bilirubin, Na, K, and glucose were within the normal range. Haemanthus and Coccidian egg count reduced in FSF-supplemented diets ($P<0.01$). Providing a diet supplemented with FSF to sheep improves hematobiochemical parameters and remarkably reduces parasite loads, especially at a 4% inclusion level.

Session 82 Poster 18

Synchrotron FTIR microspectroscopy investigation on the sperm cryopreservation of Sanan goat

S. Ponchunchoovong[1], F. Suwor[1], K. Tammanu[2] and S. Siriwong[2]
[1]Suranaree University of Technology, School of Animal Technology and Innovation, 111 University Avenue, Muang District, Nakhon Ratchasima 30000, Thailand, [2]Synchrotron Light Research Institute, 111 University Avenue, Muang District, Nakhon Ratchasima 30000, Thailand; samorn@sut.ac.th

Cryopreservation of semen and artificial insemination (AI) represent a powerful tool for livestock breeding. Several efforts have been to improve the efficiency of sperm cryopreservation in different ruminant species. However, amount of sperm still suffers considerable cryodamage, which may affect sperm quality and fertility. The success of cryopreservation depends on several factors like cooling-thawing rates, the types of extenders or cryoprotectants used to assist with the stabilization of cells during the freezing and thawing process. Although being a crucial step for Assisted Reproduction Technologies (ART) success, to date sperm selection is based only on morphology, motility, and concentration characteristics. Consideration the many possible alterations, there is a great need for analytical approaches allowing more effective sperm selections. The use of Synchrotron Fourier transform infrared (FTIR) spectroscopy is an advanced and powerful optical technique for obtaining absorption spectra in the infrared region may represent an interesting possibility, being able to reveal many macromolecular changes in a single measurement in a non-destructive way. This study found that Tris base extenders with the combination of egg yolk (10%) and glycerol (5%) yielded the higher motility and viability rates compared to the other treatments, but it was lower than the commercial used (Andromed) and fresh sperm ($P<0.05$). This result was related to synchrotron FTIR study. We obtained important clues on the macromolecular changes, mainly lipid, protein modifications and nucleic acid. In our study claimed that synchrotron FTIR spectroscopy can be proposed as a new smart diagnostic tool for semen quality assessment on sperm cryopreservation of Saanen goat.

Session 82 Poster 19

A need for knowledge on the utilisation of forage by goats according to the distribution methods

B. Fança and B. Bluet
Institut de l'Elevage, 149 rue de Bercy, 75012 Paris, France; barbara.fanca@idele.fr

Goat farming is facing multiple constraints: economic, environmental, and societal. Feed management is among the first lever of action available for breeders to adapt. The management of feed distribution is supposed to have a strong impact on feed intake. However, there are few references as to the real impact of distribution methods on the use of forage, even though the consequences on technical and economic results and on the workload and organisation of work can be significant. The aim of this preliminary study was to measure the diversity of distribution practices and to identify the issues that farmers and their advisors are facing today. Two online surveys were therefore constructed in parallel, for livestock advisors and for farmers. 41 advisors and 119 farmers representing all the French goat sector regions and feeding systems responded and took part in the survey. A first observation is the great diversity of practices, particularly concerning the number of forage distributions ranging from 1 to 8 per day. Farmers practicing a single distribution per day or less are in the minority: 9%. However, 46% of the farmers consider the work involved in distribution to be painful, and 25% of them consider the time allocated to distribution to be not acceptable. The majority opinion of the advisors also tends to prefer 2 or 3 distributions per day, to foster the milk production. 69% of farmers distribute at least two different forages per day. However, no majority opinion emerges concerning the order of distribution of forages during the day (when they are note mixed) according to their respective qualities (feed values, fibrosity, humidity, etc.), both among farmers and advisors. The surveys distinguished the different food systems according to the nature or method of harvesting the main forages. The diversity of practices and opinions can be observed in all systems. However, systems based on hay are considered easier to manage than those on fresh grass. These results show the importance of objectivizing the consequences of different practices according to the different systems so that breeders can choose a distribution method that reconciles technical results and work organization.

Session 82 Poster 20

Changes in fattening and slaughter traits of old-type Polish merino between 2010 and 2020

M. Puchała, A. Kawęcka, J. Sikora and M. Pasternak
National Research Institute of Animal Production, Department of Sheep and Goat Breeding, 2 Sarego Street, 31-047 Kraków, Poland; michal.puchala@iz.edu.pl

Stationary evaluation of rams based on progeny is an important element of breeding value assessment, especially in paternal (meat) herds. In addition to growth rate and feed consumption, slaughter traits are determined and analysed in stationary evaluation. The information obtained allows for accurate and objective evaluation and ongoing monitoring of traits. In further mating, it is recommended to prefer rams characterized by high values of individual traits (indices). The inspection station evaluates animals from breeding herds with a minimum of two herd rams, belonging to different breeds. At collection from the breeder, the age of individual rams should be between 56 and 75 days, and the average weight for the group should be about 18-19 kg. They are fed at will with a complete pelleted mix. A total of 36 old-type Polish Merino rams were evaluated at 209 progeny from 2010-2020 ram control stations. During the period, changes in the values of selected indicators of fattening traits were recorded. There was a marked daily gain during the fattening period from 282.84 g in 2010 to 349.18 in 2020, an increase of 18.99%. During the same period, there was a decrease in feed consumption per kg of weight gain during the fattening period by 18.18% from 4.73 kg in 2010 to 3.87 kg in 2020. The average age of fattening onset also decreased during the analysed period from 89.4 to 73.0 days in 2020, a difference of 18.38%. There was a reduction in fattening time from 83.06 to 67.39 days (a decrease of 18.86%). There was no change in the parameters of selected indicators of slaughter traits, such as cold carcass weight, slaughter yield and share of carcass cuttings in half-carcasses, while a slight increase in the proportion of meat in half-carcasses was observed from 63.22 to 64.79% (an increase of 2.42%) and a decrease in the proportion of fat in half-carcasses by 9.18%, from 15.33% in 2010 to 14.04% in 2020. During the period under review, improvements were observed in the most important fattening traits and selected slaughter traits, from the economic point of view of the highly important feed consumption and shortened fattening period.

Session 82 Poster 21

Milk fat nutritional indices – a comparative study between retail goat and cow milk

G. Papatzimos[1], P. Mitlianga[2], M.A. Karatzia[3], Z. Basdagianni[4] and E. Kasapidou[1]
[1]University of Western Macedonia, Department of Agriculture, Florina, 53100, Greece, [2]University of Western Macedonia, Department of Chemical Engineering, Kozani, 50100, Greece, [3]Research Institute of Animal Science, HAO-Demeter, Paralimni, 58100, Greece, [4]Aristotle University of Thessaloniki, School of Agriculture, Department of Animal production, Thessaloniki, 54124, Greece; papatzimos@hotmail.com

Milk is a nutritive food essential to people's diets throughout their lives. Lately, there is increased consumer interest in goat milk due to its health-related benefits. However, information on milk nutrient profile at retail levels is limited, whereas few studies contemporaneously compare the nutritional value of goat and cow milk. This study investigated differences in the nutritional indices related to healthy fat consumption between goat and cow retail milk produced in Greece. Samples from conventionally produced retail cow (n=48) and goat (n=48) milk were collected monthly for one year from four major food retailers. The same manufacturers produced both goat and cow milk samples. Milk fatty acid profiles were analysed by GC flame ionisation detection, and the following indices were determined: Hypocholesterolemic: hypercholesterolemic fatty acid ratio (h/H), Health Promoting Index (HPI) and Desirable Fatty acids (DFA). The h/H is a new index assessing the effect of FA composition on cholesterol. HPI is mainly used in research on dairy products such as milk and cheese. Finally, DFA represent the sum of stearic acid and unsaturated fatty acids, which are all anti-atherogenic. The latter indices are less frequently applied in evaluating the lipid nutritional quality of milk and dairy products. Differences between goat and cow milk were determined using independent t-tests. The h/H ratio was almost 0.62 in both types of milk. HPI was 0.43 and 0.82 for goat and cow milk, and the percent of DFA was almost 43 in milk from both species. There were no statistically significant differences (P>0.05) in all examined parameters in both types of milk. However, further research is needed to identify differences in milk fatty acid composition between seasons and between production systems. Support: 'AGROTOUR' Project (MIS 5047196), 'Reinforcement of the Research and Innovation Infrastructure', NSRF 2014-2020.

Session 82 Poster 22

Mountain caprine milk – the higher the better?

M.A. Karatzia[1], M. Amanatidis[2], G. Papatzimos[2], E. Kasapidou[2], P. Mitliagka[2] and Z. Basdagianni[3]
[1]Hellenic Agricultural Organization-DEMETER, Paralimni, 58100, Greece, [2]University of Western Macedonia, Kozani, 50100, Greece, [3]Aristotle University of Thessaloniki, Thessaloniki, 54124, Greece; karatzia@elgo.gr

Mountain dairy products are widely recognized as of higher quality and taste in comparison to the ones produced at the plains. The advanced quality characteristics of mountain milk originate from animals grazing in mountain pastures and meadows. As nomadic or transhumant livestock farming has been practiced for centuries in Greece, it is mainly based on highland pasture utilisation through low input goat and sheep management systems. Within this scope, a preliminary study was undertaken, to compare the chemical composition of mountain goat milk and milk produced at the plains of Western Macedonia. A total of 32 bulk goat milk samples were collected at a bi-weekly rate from two flocks during mountain transhumance from June until September (M-16 samples at 1,300 m altitude) and their stay at the plains during March-May and upon their return from the highlands in October (P-16 samples at 300 m altitude). Both farms reared Greek breed goats (*Capra prisca*). Chemical composition and total bacterial count were determined in all milk samples and results were statistically analysed using IBM SPSS v.27.0 (mean ± SD). Total solids (M-14.14±1.07%; P-13.81±0.48%), water (M-85.86±1.07%; P-86.19±0.48%), protein (M-3.91±0.48%; P-3.66±0.21%) and non-fat milk constituents (M-8.89±0.47%; P-9.09±0.19%) percentages did not differ significantly between M and P groups. Milk fat and lactose levels did not follow the aforementioned trend. The mean fat percentage in mountain milk samples was significantly higher than in plain milk samples (M-5.25±0.67%; P-4.77±0.38%, $P \leq 0.05$). Additionally, mean lactose level in mountain milk samples was significantly lower than in plain milk samples (M-4.15±0.13%; P-4.60±0.08%, $P \leq 0.05$). Total bacterial count was within the normal range in all samplings. Significantly elevated milk fat levels of mountain goat milk support a quality distinction between it and milk produced in the plains. Funding received by 'AGROTOUR' Project (MIS 5047196), 'Reinforcement of the Research and Innovation Infrastructure', NSRF 2014-2020.

Session 82 Poster 23

Results of milk performance of Carpathian goats

J. Sikora, A. Kawęcka, M. Puchała and M. Pasternak
National Research Institute of Animal Production, Department of Sheep and Goat Breeding, 2 Sarego Street, 31-047 Krakow, Poland; jacek.sikora@iz.edu.pl

Milk is one of the basic raw materials used by humans in their diet. It is a natural food produced by mothers for their offspring. At a later age, children, adolescents and adults use the milk of other mammals to meet the body's need for protein and essential amino acids. There are many breeds of goats used for dairy purposes in Poland. Among other things, imported breeds, such as: Saanen or Alpine and domestic breeds, which are currently represented by the Carpathian breed. Carpathian goats represent a primitive breed of goats that have been present in Poland for centuries, recognized as an extinct breed at the end of the 20th century. In the years 2005-2010, at the National Research Institute of Animal Production in Balice, an attempt was made to restitute this breed, and in 2014 Carpathian goats were included in the genetic resources protection program. Currently, there are about 400 mother goats in the country. Since 2007, goats of this breed have been subject to milk performance control. The goat population built from scratch consisted mainly of young animals, whose milk yields were obtained in the first or second lactation – about 80%. Older goats, milked in the third and fourth lactation, they constituted a minority in the first years of the assessment, respectively 15 and 5%. In the following years, the group of older animals, assessed in higher lactations, slightly increased, however, the constant influx of young animals shaped the average level of lactation of the breed. In the first year of milk performance control, the average yield of the Carpathian goat population was 450 kg of milk, 2.8% protein, 2.9% fat. In the following years, the average milk yield was at different levels. In 2012, the average yield was 322 kg, 2.87% protein, 3.67% fat. In 2020, the average yield was 255 kg, 3.2% protein, 3.25% fat. The decrease in the level of average lactation can be explained by the introduction of young goats to evaluate milk yield and a strong emphasis on features related to the exterior of selected animals for breeding, and other features, including the level of milk production, which are omitted at this stage of breed reproduction.

Session 82 Poster 24

Barley in concentrates for dairy goats – effects of alkaline and mechanical treatments

A. Martinsen[1], D. Galméus[1], H. Volden[1], K. Hove[1], M. Silberberg[2] and M. Eknæs[1]
[1]Faculty of Biosciences, Norwegian University of Life Sciences, Oluf Thesens vei 6, 1433 Ås, Norway, [2]Université Clermont Auvergne, INRAE, Unité Mixte de Recherche sur les Herbivores (UMRH), Site de Theix, 63122 Saint-Genès-Champanelle, France; anita.martinsen@nmbu.no

Goat milk production in Norway is partly dependent on imported feed raw materials. Increasing the use of Norwegian cereals in commercial concentrate for goats is interesting when attempting an effective and sustainable goat milk production. We proposed to increase the use of domestic barley in pelleted concentrate feeds to goats. However, barley contains a large proportion of rapidly digestible starch. Fed in high amounts, barley will reduce ruminal pH and cause subclinical ruminal acidosis (SARA). We used processing methods like alkaline and mechanical treatments before pelleting to modify changes in ruminal pH. Nine rumen cannulated lactating goats were fed three different types of concentrates containing: (1) rolled barley treated with alkali (Alka); (2) rolled barley (Rolled); and (3) finely ground barley (Ground). The experiment was conducted as a 3×3 Latin square, with three replicates and was implemented as a dose-response experiment with three periods. Silage was fed *ad libitum* and daily concentrate rations were divided into six equal meals fed every fourth hour. Daily concentrate allowances were increased by 150 g DM every 4th day until the goats stopped eating or developed symptoms of SARA. Neither Alka barley nor mechanical treatment affected silage intake, whereas Alka barley led to lower concentrate consumption per 100 kg ECM (P<0.01) compared to Rolled barley. There was a higher milk yield (P<0.01) from the Alka barley diet compared to Rolled barley and the difference increased with increasing level of concentrate. Milk yield was not affected by mechanical treatment. Fat and protein content in milk were equal among all treatments. Ruminal pH tended to be lower (P=0.069) for Alka barley compared to Rolled barley, but the ratio between concentration of acetic and propionic acid was unaffected. Alka barley had a positive effect by increasing milk production on less feed. The effects of added enzymes in Alka barley and increased level of non-protein nitrogen probably resulted in improved fibre digestion.

Session 83 Theatre 1

The effect of population history, mutation and recombination on genomic selection

D. Adepoju[1], T. Klingström[1], A.M. Johansson[1], E. Rius-Vilarrasa[2] and M. Johnsson[1]
[1]Swedish University of Agricultural Sciences, Department of Animal Breeding and Genetics, Box 7023, 750 07 Uppsala, Sweden, [2]Växa Sverige, P.O. Box 30204, 104 25 Stockholm, Sweden; martin.johnsson@slu.se

Population genetic features such as population history, mutation rate and recombination rate affect the distribution of genetic variants and their association, and thus the performance of genomic prediction. In our ongoing research on the genome dynamics of livestock breeding, we explore the effects of population history, recombination and mutation in farm animals. We have estimated the recent population history of Swedish cattle breeds (Swedish Red cattle, Swedish Mountain cattle, Fjällnära cattle, Ringamåla cattle, Bohus Polled, Väne cattle, and Swedish Red Polled), using linkage disequilibrium-based methods on both SNP chip genotypes and whole-genome sequence data. While the estimates are affected by method and data type, they consistently suggest that recent population size has been relatively large (on the order of thousands) up until the onset of systematic breeding, when it shrunk rapidly. It also suggests that efforts by breed organisations to maintain the effective population size of breeds like the Swedish Mountain cattle during the 20th century have been successful. In the case of recombination, sequence data shows the long-term effects of recombination rate on genetic diversity, in the form of a correlation between local recombination rate and the number of variants detected in whole-genome sequence data. For the effect of recombination rate on breeding, we have built a quantitative genetic simulation that includes variable recombination rate, based on real linkage maps. This model confirms that high recombination rate contributes to somewhat higher additive genetic variance in the long run, but that a higher recombination rate lowers the accuracy of genomic selection. The benefit to variance happens in the long run, but the decrease in accuracy is immediate. This suggests that, on balance, high recombination rate is not beneficial for breeding.

Session 83 Theatre 2

The detection of putative recessive lethal haplotypes in Irish sheep populations

R. McAuley[1], N. McHugh[1], T. Pabiou[2] and D.C. Purfield[3]
[1]Teagasc, Animal and Grassland Research Centre, Teagasc, Animal & Grassland Research Centre, Moorepark, Fermoy, Co. Cork, P61 C996, Ireland, [2]Sheep Ireland, Sheep Ireland, Link Road, Ballincollig, Co. Cork, P31 D452, Ireland, [3]Munster Technological Univeristy, Biological Sciences, Munster Technological University – Cork Campus, Bishopstown, Cork, T12 P928, Ireland; rory.mcauley@teagasc.ie

In livestock populations, recessive lethal alleles are a known contributor to poor reproductive performance due to embryonic death in homozygous individuals. Despite their lethal effect in the recessive form, these alleles may be maintained at high frequencies among carrier animals because of their positive pleiotropic effects on economically important traits. Although several such recessive alleles have been identified in cattle and pig populations, limited studies have been completed in sheep, and none within Irish sheep populations. Genotype data for 69,034 animals from 5 major Irish sheep breeds genotyped on a variety of panels was available for this study. Only animals and single nucleotide polymorphisms (SNPs) with a call rate >90%, and a minor allele frequency >0.01 were retained. Non-autosomal SNPs and SNPs that did not adhere to Mendelian inheritance patterns were discarded. Following imputation to medium density, 43,951 SNPs remained across 66,996 animals, which included 32,256 verified progeny-sire-dam and 2,089 verified progeny-sire-maternal grandsire trios. To identify haplotypes with homozygous deficiency, a sliding haplotype window of 10 SNPs was used to scan the genome of all genotyped animals. Haplotype lethality was determined using a chi-square test between the number of observed and expected haplotypes. Two haplotypes showed a significant deficiency of homozygote animals with expected frequency observations ranging from 80 to 88% and a carrier frequency ranging from 15 to 16%. These haplotypes were located on chromosomes 8 and 19 and found solely in the Vendeen population. Comparison of at risk-matings (between carriers) and safe matings, demonstrated that these haplotypes were associated with an increased still-birth rate ($P<0.05$). This finding can be incorporated into Irish sheep breeding programs to avoid carrier-by-carrier matings to reduce lamb mortality.

Session 83 Theatre 3

Assessing the impact of inbreeding on mastitis and digital dermatitis in German dairy cattle

J. Mugambe
Christian-Albrechts University of Kiel, Dithmarscher Straße 9, 24113, Germany; jmugambe@tierzucht.uni-kiel.de

Understanding the impact of inbreeding on health traits is of economic importance for dairy cattle farmers. However, there is currently no study on the impact of inbreeding on mastitis (MAS) and digital dermatitis (DD) in German Holstein dairy cattle. Here we investigated the impact of inbreeding on MAS and DD using both pedigree-based and genomic-based inbreeding estimators. Data included records of 24,489 cows with both phenotypes and genotypes. The pedigree used in the analysis comprised 819,021 animals born after 2012 with a pedigree completeness index (PCI) greater than 85% tracing 5 generations back. To investigate the effects of inbreeding depression, inbreeding coefficients from various estimators were included separately in single-trait threshold models to account for the binary nature of the health traits. For easier interpretation of results, solutions from the analyses were transformed from a liability scale to a probability scale with respect to the phenotypic incidence of 17 and 27% for MAS and DD, respectively. The mean inbreeding coefficients from all estimators ranged from -0.003 to 0.110, with negative values observed for most genomic-based methods. The probability distributions resulting from a 1% increase in inbreeding for all estimators ranged from 0.38 to 0.76 for MAS and 0.46 to 0.87 for DD. Despite the significant effect of inbreeding on both traits, inbreeding depression was larger for DD compared to MAS in German Holstein dairy cattle. Given the high economic costs associated with MAS and DD in dairy production, inbreeding can negatively impact the financial returns for dairy farmers. Thus, there is a strong motivation to regulate and manage the rate of inbreeding in dairy populations in order to minimize the incidence and spread of these diseases.

Session 83 Theatre 4

Genomic inbreeding in the Austrian Turopolje pig population

G. Mészáros[1], B. Berger[2], C. Draxl[3] and J. Sölkner[1]
[1]University of Natural Resources and Life Sciences, Vienna, Gregor-Mendel-Str. 33, 1180 Vienna, Austria, [2]HBLFA Raumberg-Gumpenstein, Austraße 10, 4600 Thalheim bei Wels, Austria, [3]Österreichische Schweineprüfanstalt Gesellschaft mbH, Unter den Linden 1a, 2004 Streitdorf, Austria; gabor.meszaros@boku.ac.at

The Turopolje pig is an endangered, local breed from Austria. The population was set up based on a small number of animals recovered from Croatia during the war in the early 1990ies. Due to the small number of founders and lack of further imports, the population is highly inbred. With the start of routine SNP genotyping, it was possible to set up approaches to manage the diversity of the breed. In this study we aim to assess the genomic inbreeding within the breed. The data set of Turopolje consisted of 184 animals, genotyped with custom chip with 77,122 SNPs. For comparison, pigs from all commercial breeds in Austria were used, including 82 Large White, 76 Landrace, 74 Duroc and 70 Piétrain pigs, genotyped with a Porcine 60K SNP chip. Only the common SNPs between chips were considered. The data were controlled for quality, removing SNP and individuals with more than 10% missingness. After quality control a total of 485 pigs and 54,075 autosomal SNPs remained. The genomic inbreeding was assessed based on runs of homozygosity (ROH) with the cgaTOH software, considering common ancestors up to 6 generations in the past (minimal length of ROH was 8 Mb). The average inbreeding coefficient in Turopolje was 0.18 (sd 0.06). This was similar to Duroc and Piétrain breeds (both 0.17, sd 0.04), but higher than in Landrace (0.10, sd 0.03) and Large White (0.11, sd 0.02). The variance of inbreeding levels was the highest in the Turopolje breed, ranging from 0.06 to 0.41. The genetic diversity measures have to be prioritized in the Turopolje population, with an aim to decrease overall inbreeding levels in order to maintain the viability of the breed.

A genetic diversity study of a local Belgian chicken breed using the new IMAGE SNP array

R. Meyermans[1], W. Gorssen[1], J. Bouhuijzen Wenger[1], O. Heylen[2], J. Martens[2], S. Janssens[1] and N. Buys[1]
[1]Center for Animal Breeding and Genetics, KU Leuven, Department of Biosystems, Kasteelpark Arenberg 30, Box 2472, 3001 Leuven, Belgium, [2]Steunpunt Levend Erfgoed vzw, Spiegel 1, 9860 Oosterzele, Belgium; roel.meyermans@kuleuven.be

Screening for genetic diversity in livestock species breeds is important, especially for local, small populations at the risk of extinction. Moreover, recent developments have made genotyping also affordable for smaller breeds and populations. One of these new developments is the IMAGE_01 SNP genotyping array, that includes SNPs for six different species (cow, pig, sheep, chicken, horse and goat, 10K SNPs per species). This multiple species array can be a valuable addition to screen multiple breeds at once. Moreover, the array allows genotyping a small number of samples per species and combining samples where possible. In this study, we used the IMAGE genotyping array to genotype a local Belgian chicken breed, the Turkey-headed Malines (*Mechelse Kalkoenkop / Coucou de Malines à tête de dindon*). 110 DNA samples were collected for this breed, together with 29 samples from four possibly related local breeds. Genetic diversity of our local breed was examined via a principal component analysis and a F_{is} study. This revealed the breed to be closest related to the Malines chicken, and less related to the Coucou des Flandres than previously assumed. Next, an inbreeding analysis based on runs of homozygosity (F_{ROH}) and effective population size estimation was tested. Average F_{ROH} was estimated as rather high (±20%) and the effective population size was found to be very low (<35). Finally, our data were matched to internationally available open access chicken genotypes, although the SNP overlap of the IMAGE array with other commercial arrays was rather low. Valuable insights on the breed composition were generated and shared with the breeders to optimize their breeding practices. Moreover, this study is the first to evaluate the genetic diversity of a Belgian chicken breed and can serve as pilot for future genetic research on local Belgian chicken breeds.

Genetic diversity in Dutch sheep breeds shaped by geography, history, use and genetic management

J.N. Hoorneman, W.J. Windig and M.A. Schoon
Centre for Genetic Resources, the Netherlands, & Livestock Research, Wageningen University & Researc, P.O. Box 338, 6700 AH Wageningen, the Netherlands; noelle.hoorneman@wur.nl

The Netherlands has a rich variety of native sheep breeds from different geographic regions with different historic and present uses. Based on the number of breeding females, ten of these breeds are currently classified at risk with associated risks for high inbreeding levels. To gain more insight in the genetic structure of the Dutch sheep population, we studied the distinctiveness of the native breeds, their genetic diversity both within and between breeds, and how this diversity is shaped by geography, purpose, history and genetic management practices. Semen samples of 171 rams of 11 native Dutch sheep breeds were genotyped with the IMAGE001 10K multispecies SNP chip. Principal component analysis (PCA), population structure and phylogenetic analyses all clearly differentiated the breeds according to their historical origins and geographical distribution, except for the Texel and two Texel related breeds. Breeds bred for milk or meat production on rich grasslands in the north and west of the Netherlands, and breeds bred for grazing of nature areas and manure collection on poorer grounds in the east and south were separated in different clusters. The graphical distribution of the PCA mirrored the geographical distribution of the origin of the breeds. Within breed diversity was larger for the two breeds with largest population sizes plus the one breed, Veluwe Heathsheep, that consistently has used a breeding circle to mitigate inbreeding rates. Largest unique diversity was found for the Flevolander a breed more recently developed with the use of foreign breeds. All other breeds shared large parts of their genetic diversity with either the production breeds from rich soils or the grazing breeds from poorer soils. This study not only provides more insight in the structure of the Dutch native sheep population, but is valuable as well for assessing the effects of selection and genetic management measures. Additionally, insight in genetic diversity of the Dutch genebank collection, in contrast to the *in situ* population, can support identification of unique males in the live population to be added to the genebank collection.

Session 83 Theatre 7

Genetic diversity and population structure of the Cyprus Chios sheep and Damascus goat breeds

A.C. Dimitriou, G. Maimaris, A.N. Georgiou and G. Hadjipavlou
Agricultural Research Institute, Animal Production, P.O. Box 22016, 1516, Lefkosia, Cyprus; adimitriou@ari.moa.gov.cy

Cyprus lies within a global biodiversity hotspot and it is directly threatened by climate change and hence by desertification given its geographic location at the eastern Mediterranean Sea basin. At the same time, Cyprus has the second largest population growth rate among European countries, fact that highlights the need to optimize applied primary production strategies aiming to secure food production and minimize the impact on local natural environment in a cost and time effective way. Considering the population growth and the recent registration of locally produced cheese (Halloumi) as Protected Designation of Origin, the demand of goat and sheep milk is anticipated to increase dramatically. According to constructed scenarios taking into account the current milk production on the island, goat and sheep livestock should be doubled (from ~500k to 1000k) in order to meet the expected needs. Alternatively, this goal could be fulfilled by improving the existing livestock using genetic data and modelling. For this purpose, the first high-throughput genomic data for local Cyprus sheep and goat breeds were generated by the Agricultural Research Institute (ARI), employing available SNP genotyping microarray assays. The datasets analysed included genomic data for more than 1000 animals per species, representing six farms. Estimated population metrics indicated the existence of at least four genetically distinct groups in both sheep and goat breeds, as well as an ongoing gene flow between specific farms included in our dataset. These findings are also corroborated by calculated population fixation indices in both cases. Produced data call for a more thorough, large-scale analysis including individuals from more farms aiming to describe the local livestock diversity in a comprehensive way. Towards this direction, within the framework of an EU funded project led by ARI, a significant number of farms are expected to participate in the national genetic and genomic improvement program regarding Cyprus Chios sheep and Damascus goats.

Session 83 Theatre 8

Genome-wide analysis of Greek goats with global breeds revealing population structure and diversity

E. Tosiou[1], V. Tsartsianidou[1,2], S. Vouraki[3], V. Papanikolopoulou[3], L.V. Ekateriniadou[4], E. Boukouvala[4], I.G. Bouzalas[4], I. Sakaridis[4], G. Arsenos[3] and A. Triantafyllidis[1,2]
[1]Aristotle University of Thessaloniki, School of Biology, University Campus, 54124 Thessaloniki, Greece, [2]Center for Interdisciplinary Research and Innovation, Genomics and Epigenomics Translational Research, Balkan Center, 57001 Thessaloniki, Greece, [3]Aristotle University of Thessaloniki, School of Veterinary Medicine, University Campus, 54124 Thessaloniki, Greece, [4]Hellenic Agricultural Organization-DIMITRA, Veterinary Research Institute, Campus of Thermi, 57001 Thessaloniki, Greece; tsarvale@bio.auth.gr

Goat farming is of major importance for the Greek livestock sector, while the national herd is one of the largest in Europe, mainly consisting of Eghoria and Skopelos breeds. However, little genetic information is available, for the Skopelos breed only. We studied the genome-wide diversity and population structure of Greek goats, collectively analysed with 49 global breeds from AdaptMap project. A total of 287 goats sampled from seven indigenous Greek populations (Aridaia, Chalkidiki, Eghoria, Serres, Paggaio, Skopelos, Crossbreed) and three introduced breeds widely reared in Greece (Anglo-Nubian, Damascus and Murciano-Granadina) were genotyped using the Axiom Caprine Genotyping v2 Array. High average genetic diversity levels were estimated for Greek goats including observed and expected heterozygosity (Ho=0.36-0.41, He=0.37-0.41), similar to values estimated for other Mediterranean breeds. According to global-scale principal component analyses (PCA) results, most Greek populations were closely clustered and adequately differentiated from introduced and global goat breeds. Greek goats were placed closer to Turkish, Romanian and Mediterranean (Maltese sarda, Maltese, Garganica, Jonica, Rossa Mediterranea) breeds. These results contribute to the genome-wide understanding of Greek goats' population stratification, diversity and demographic history, while they provide genetic information on the downstream development of a cost-effective panel of markers for population assignment aiming to differentiate Greek from other goat breeds. A representative set of Greek goat genomes will be fully sequenced within the framework of this study, funded by the GRAEGA CHEESE project (KMP6-0083632).

Session 83 — Theatre 9

Genome-wide scan for runs of homozygosity in South Americam camelids

S. Pallotti[1], M. Picciolini[2], M. Antonini[3], C. Renieri[4] and V. Napolioni[1]
[1]School of Biosciences and Veterinary Medicine, University of Camerino, Via Gentile III Da Varano s/n, 62032 Camerino (MC), Italy, [2]SYNBIOTEC Laboratori s.r.l., Località Torre del Parco, 62032 Camerino (MC), Italy, [3]Italian National Agency for New Technologies, Energy and Sustainable Development (ENEA), Centro Ricerche Casaccia, Via Anguillarese, 301, 00123 S.M. di Galeria (RM), Italy, [4]School of Pharmacy, University of Camerino, Via Gentile III Da Varano s/n, 62032 Camerino (MC), Italy; stefano.pallotti@unicam.it

Four species of South American camelids (SAC) are distributed over the Andean high-altitude grasslands. These include 2 wild species guanacos (*Lama guanicoe*) and vicugna (*Vicugna vicugna*), and 2 domestic species, alpaca (*Vicugna pacos*) and llama (*Lama glama*) bred for meat and fibre production. The genome-wide scan for runs of homozygosity (Roh), which are contiguous lengths of homozygous segments of the genome where the two haplotypes inherited from the parents are identical, offers the possibility to unveil and compare the signatures of selection along the genome. Here, we investigated the occurrence and distribution of ROH in the four SAC species starting from whole-genome sequencing (WGS) data retrieved from public available repository and new WGS of alpacas. Sample consisted of 12 WGS from alpacas, 7 WGS from llama, 6 WGS from guanaco and 6 WGS from vicugna retrieved from NCBI. Moreover we added to our sample 7 new WGS of alpacas. The genomic variants calling was performed, then data were converted to plink files and used to perform the genome-wide scan for Roh longer than 100 kb and 500 kb. Vicugna and guanaco showed a slightly higher number of Roh longer than 100 kb than llama and alpaca. In particular, llama was the species with the lower number of Roh found. Roh longer than 500 kb were found only in alpacas genome. The inbreeding coefficient based on Roh (Froh) was low for all four species ranging from 0.04 to 0.02 and ANOVA showed no significant differences between the four species. The analysis of the overlapping Roh within the species found genomic regions which harbour candidate genes selected for fibre and body growth traits in domestic species and for local adaptation to environmental stress for both domestic and wild species.

Session 83 — Theatre 10

Exploring global cattle genealogy with tree sequence

G. Mafra Fortuna, J. Obsteter, A. Kranis and G. Gorjanc
The Roslin Institute, The University of Edinburgh, Easter Bush Campus, Midlothian, EH25 9RG, United Kingdom; g.mafra-fortuna@sms.ed.ac.uk

The two cattle subspecies showcase distinct aptitudes. Acknowledging and exploring these distinctions at genomic level might be critical to improving their crosses. Two evolutionary events separate *Bos taurus* and *Bos indicus*: an ancestor split and domestication. The ancestor split occurred over 100,000 years ago. Domestication happened 10,000 and 8,500 years ago, respectively. Another 100+ years of contrasting breeding practices and environmental challenges added to the substantial differences in economic traits between subspecies. Crossing *B. taurus* and *B. indicus* is necessary for many dairy cattle systems to increase production while securing environmental resilience. Despite some boost in yield, genetic composition within admixed herds fluctuates largely. Expectedly, the genetic variation translates into noteworthy phenotypic variation. These characteristics hamper the genetic improvement of admixed populations. To overcome this issue, we require genetic evaluation methods to work across breeds but especially across subspecies. For example, methods must capture differences in genetic architecture and gene frameworks between subspecies. We propose using ancestral recombination graphs. Here we use the novel tree sequence methodology to describe cattle's genome evolution from breed differentiation to family segregation. The tree sequence succinctly encodes genealogy for each genomic region by tracking haplotype lineages in the ancestral recombination graph accounting for recombination and mutation. We inferred global cattle genealogy from 1,181 whole-genome sequences from the 1000 Bull Genome Project public data. The dataset comprised 116M SNP on 29 autosomal chromosomes, 21 *Bos taurus*, 6 *Bos indicus* breeds, and 26 admixed individuals. We inferred genealogy using the python libraries tsinfer and tsdate and analysed with tskit. The resulting tree sequences spanned 11,997,863 genome regions (local trees), 257,997,301 haplotypes (edges), and 269,727,991 mutations, all efficiently stored in 49 GB. Our results draw a detailed profile of cattle genomic evolution and the differentiation of its subspecies and breeds. The next step is applying the knowledge to improve breeding methods for admixed populations.

Session 83 Theatre 11

Which diversity measures are important for small breeds like German Black Pied Cattle (DSN)?

G.A. Brockmann[1], G.B. Neumann[1], P. Korkuc[1], M.J. Wolf[2], K. May[2] and S. König[2]
[1]Humboldt-Universität zu Berlin, Albrecht Daniel Thaer-Institute, Unter den Linden 6, 10099 Berlin, Germany, [2]Justus-Liebig-Universität, Institute of Animal Breeding and Genetics, Ludwigstrasse 21, 35390 Giessen, Germany; gudrun.brockmann@hu-berlin.de

DSN, a dual-purpose breed, is considered an ancestral population of Holstein, the high-yielding and most-widely used dairy breed in the world. DSN has about 2,500 herdbook animals. The breed has been replaced almost entirely because of its low milk yield, which is 2,500 kg less than in Holstein cattle. Therefore, breeding goals are improving milk production while maintaining the dual-purpose characteristics, robustness and genetic diversity. For accessing genomic peculiarities of DSN, we sequenced 304 DSN animals and developed the DSN-specific bovine DSN200k SNP chip representing 182,154 sequence variants of DSN. Using this chip and whole genome sequencing information, we investigated the diversity of DSN and performed genome-wide association analyses. DSN showed a close genetic relationship with breeds from the North Sea region. The nucleotide diversity in DSN (0.151%) was higher than in Holstein (0.147%) and other breeds. The F_{Hom} and F_{RoH} values in DSN were among the lowest. Regions with high F_{ST} between DSN and Holstein, significant XP-EHH regions, and RoH islands detected in both breeds harbour candidate genes that had been reported for milk, meat, fertility, production, and health traits, including one locus detected in DSN for endoparasite infection resistance. Using whole genome sequencing data in genome-wide association studies, we identified 13 significant genomic loci for improving milk production. The highest significance was found for MGST1 affecting milk fat content (log10(p)=11.93, MAF=0.23, (βMA)=0.151%). Key loci for protein content were regions around CSN1S1 (log10(p)=8.47, MAF=049, βMA=0.055%) and GNG2 (log10(p)=10.48, MAF=0.34, βMA=0.054%). Loci containing HTR3C, TLE4, and TNKS were suggestive for milk yield. Different from Holstein, DGAT1 was fixed (0.97) for the alanine protein variant for high milk and protein yield.

Session 83 Theatre 12

Characterisation of free-living and exotic animal species at species and individual level

M. Zorc[1], M. Cotman[2], A. Dovč[2], J. Zabavnik-Piano[2] and P. Dovc[1]
[1]University of Ljubljana, Biotechnical Faculty, Department of Animal Science, Jamnikarjeva 101, 1000 Ljubljana, Slovenia, [2]University of Ljubljana, Veterinary Faculty, Gerbičeva 60, 1000 Ljubljana, Slovenia; peter.dovc@bf.uni-lj.si

As more and more species of wild animals are kept in captivity, including dangerous and poisonous species, there is an urgent need to regulate throughout the EU which species may be kept as pets in the future. With the amendment of the Animal Protection Act, Slovenia introduced a ban on keeping dangerous, aggressive, poisonous and difficult-to-care-for wild animal species, including endangered species. The already established methodology for identification of species (speciation), individual (individualization) and sex in various native mammal, bird and fish species represents a starting point for the development of a similar methodology for exotic animal species. Important information on DNA barcoding in a variety of animal species is available in the database BOLD (https://www.boldsystems.org/). For species identification purposes, mtDNA is widely used as a genetic marker (cytochrome oxidase I and D-loop region), whereas the use of STR markers, which are often conserved within taxonomic families, provides a reliable and cost-effective method for individualization. In cases where genomic data are not available, RAPD markers can be used. Recently, we have developed a methodology based on mtDNA and STR for species identification and individualization of the Hermann's tortoise (*Testudo hermani*), which is an example of an endangered species often used for commercial purposes, and a molecular test for sex determination of the Linne's two toed sloth (*Choloepus didactylus*). Molecular genetic tests allow identification of the specimen or species and can be used to determine parentage, sex, or for forensic purposes in case of animals from the prohibited species list.

Session 83 Theatre 13

Comparative analysis of heterozygosity-enriched regions in two autochthonous Italian cattle breeds

G. Schiavo[1], S. Bovo[1], F. Bertolini[1], A. Ribani[1], V. Taurisano[1], S. Dall'olio[1], M. Bonacini[2] and L. Fontanesi[1]
[1]University of Bologna, Department of Agricultural and Food Sciences, Division of Animal Sciences, Viale Giuseppe Fanin 46, 40127 Bologna, Italy, [2]Associazione Nazionale Allevatori Bovini di Razza Reggiana (ANABORARE), Via Masaccio 11, 42124 Reggio Emilia, Italy; giuseppina.schiavo2@unibo.it

Reggiana and Modenese are autochthonous cattle breeds mainly reared in Emilia Romagna region, in the North of Italy. The production of two breed-branded Parmigiano-Reggiano cheeses assures the sustainable conservation of these cattle genetic resources. Both breeds experienced a reduction in population size in the 1980', followed by a slow recovery. Therefore, the monitoring of inbreeding and the definition of proper conservation programs are fundamental. Information derived from pedigree and from genomic regions based on runs of homozygosity had previously been evaluated for both breeds. Here, we explored Runs of Heterozygosity (ROHet), defined as regions of continuous single nucleotide polymorphisms (SNPs) with heterozygous genotype. In this work, almost two thirds of the actual population for both breeds have been genotyped with the GGP Bovine 150k SNP chip panel. ROHet were identified with the R package 'detectRuns' and ROHet islands were defined by analysing the proportion of SNPs occurring in ROHet over all genotyped cattle. The average number of ROHet in Reggiana and Modenese cattle was 14.24±3.8 and 12.91±3.3, respectively. The average length of the genome covered by ROHet was 2,532.38±858 kb in Reggiana cattle and 2,291.26±781 kb in Modenese cattle. According to the functional annotation of the ROHet islands, these regions might have an impact on reproduction and other fitness-related traits. Results will be useful to dissect genetic mechanisms affecting breed-specific traits and define breeding and conservation programmes to further exploit these animal genetic resources. *Acknowledgements:* The research was funded by the PSRN (Programma di Sviluppo Rurale Nazionale) Dual Breeding 2 (co-funded by the European Agricultural Fund for Rural Development of the European Union and by the MASAF).

Session 83 Theatre 14

Recovering latent population stratification using ADMIXTURE and metafounders in Brown Swiss

C. Anglhuber[1], C. Edel[1], E.C.G. Pimentel[1], R. Emmerling[1], K.-U. Götz[1] and G. Thaller[2]
[1]Bavarian State Research Center for Agriculture, Institute for Animal Breeding, Prof. Duerrwaechter Platz 1, 85586 Grub, Germany, [2]Christian-Albrechts-Universität, Institute for Animal Breeding and Husbandry, Olshausenstraße 40, 24098 Kiel, Germany; christine.anglhuber@lfl.bayern.de

Unknown pedigree information or hidden stratification within a breeding population may lead to an incomplete formulation of the conventional numerator relationship matrix (A). When such an A is combined with a genomic relationship matrix (G) in a single-step approach for genetic evaluation the inconsistency may cause bias in the resulting estimates. The objective of this study was to identify relevant stratifications within the Brown Swiss population using genomic data. Genotypes and pedigree information of European and North American Brown Swiss animals (85.249) were available for the analysis. Using ADMIXTURE, a software designed to find ancestral origins in humans, animals, and plants, we developed an iterative approach by stepwise increasing the number of potential stratifications (K) in an unsupervised manner and introducing this information into matrix A, using the Metafounder methodology. Improvements in consistency between matrix G and the resulting matrices A^{Γ} were evaluated by visual inspection of graphs from principal component analyses, by regression analysis and through the comparison of the mean and mean diagonal values of both matrices. Analyses with ADMIXTURE were initially performed on the full set of genotypes (S1). In addition, we sampled another dataset where we avoided selecting animals with close relationships (S2). Results of the regression analyses of standard A on G were -0.489, 0.780 and 0.647 for intercept, slope and fit. In the analysis of S2 our approach reached an optimum fit of 0.818 for A^{Γ} on G considering K=23, with intercept and slope being close to 0 and 1. With this optimum K practically no differences in mean and mean diagonal values between both matrices were observed. Similar improvements were found in S1 for K>7 although there was no clear optimum K. The family structure in S1 affected the ability of ADMIXTURE to detect additional stratification information beyond the information already present in A.

Session 83 Poster 15

Genomic diversity analysis of the Swedish Landrace goat

B. Hegedűs[1,2], A.M. Johansson[1] and P. Bijma[2]
[1]Swedish University of Life Sciences, Animal Breeding and Genetics, Box 7023, 750 07 Uppsala, Sweden, [2]Wageningen University and Research, Animal Breeding and Genomics, P.O. Box 338, 6700 AH Wageningen, the Netherlands; bernadett.hegedus@wur.nl

There are four native goat breeds present in Sweden. However, genetic diversity in these breeds has not been investigated. This study therefore aimed to describe the population structure and the level of inbreeding in the largest Swedish goat breed, the Swedish Landrace goat. Forty-eight (48) samples from eight farms were genotyped with the Goat SNP50 Bead Chip. To study the population structure, a principal coordinate analysis (PCoA) and a Structure analysis were conducted. The level of inbreeding was investigated with three measures; observed heterozygosity, inbreeding coefficient based on runs of homozygosity (F_{ROH}) and approximated coancestry. The results from the PCoA and the Structure analysis show that there is some structuring in the population and this structure is not solely due to the geographic location of the farms. The different inbreeding measures give a similar ranking of the farms. The inbreeding level of the Swedish Landrace goat based on observed heterozygosity is comparable to that of non-island goat populations from Norway, Spain and Italy. Furthermore, a potential pattern of selection was identified on Chromosome 6, with ROH in the region of the Casein genes. Further research is needed to calculate the effective population size and the rate of inbreeding in this breed. To calculate the rate of inbreeding and the effective population size pedigree data of several generations needs to be accessible.

Session 83 Poster 16

Effect of interpopulation distance on average heterosis in crosses

A. Legarra[1,2], D. Gonzalez-Dieguez[3], A. Charcosset[4] and Z.G. Vitezica[2]
[1]Council on Dairy Cattle Breeding, Bowie, 20716 MD, USA, [2]INRAE, UMR 1388 GenPhySE, 31326 Castanet-Tolosan, France, [3]International Maize and Wheat Improvement Center, Texcoco, 56237 Texcoco, Mexico, [4]GQE-Le Moulon, INRAE, Univ. Paris-Sud, CNRS, AgroParisTech, Université Paris-Saclay, Gif-sur-Yvette, France; zulma.vitezica@inrae.fr

Crosses between parental populations are widely used in animal and plant breeding for exploiting heterosis and complementarity. In livestock, the parental populations are not too-distant and often co-selected for traits expressed in the crosses. Assuming that the allelic frequencies differ between populations, the variance of dominance deviations (var.D) is proportional to the product of the heterozygosities of both parental populations. The distance between two populations can be expressed as Nei's distance or allele frequency correlations. We show analytically that var.D decreases as a function of distance of parental populations as far as allele frequencies are positively correlated. Heterosis (H) can be modelled as the sum of dominance deviations with non-zero expectation. Thus, we provide a general expression for the expected H as a function of genetic distance between populations. the relationship between H and genetic distance is linear. We illustrate the relationships between distance, var.D and H across two pig lines (1 and 2). Using genomic data (40,634 SNP markers) provided by Genus plc., two lines of pig showed low Nei's minimum genetic distance between them (0.07) and an allele frequency correlation of 0.55. The var.D in F1 is lower than into the parental lines. The ratio of expected H / sqrt(var.D) was 186, 185 and 241 in lines 1,2 and the F1, respectively. These results show that the ratio is larger in crosses than in parental populations. In practice, var.D is a measure of the importance of selection within purebreds as opposed to assortative matings in crosses.

Session 83 — Poster 17

Genome of the extinct Gotland cattle breed

M. Johnsson and A.M. Johansson
Swedish University of Agricultural Sciences, Department of Animal Breeding and Genetics, Box 7023, 75007 Uppsala, Sweden; anna.johansson@slu.se

The extinct cattle breed Gotland cattle (Gotlandsko in Swedish) lived on the island of Gotland in the Baltic Sea until the beginning of the 1950s. Gotland cattle were of small size and often had yellow coat colour and big horns. We sequenced the genomes of two Gotland cattle, based on bones samples from skulls. The skulls originate from a local museum in Viklau in the middle of Gotland. DNA isolation and library preparation was performed under ancient DNA conditions at the Ancient DNA Unit at SciLifeLab, Uppsala. The depth of coverage was 2.7X and 3.3X, respectively, with a breadth of coverage of 85 and 89%. Based on coverage of the sex chromosomes, both samples appeared to be female. We detected 19 million SNPs and 2.8 million indels in the joint dataset of Gotland cattle jointly called with 47 modern Swedish cattle. In the Gotland cattle samples, 15 and 9.3% of the SNPs had missing values, which can be compared to an average missingness of 1.3% in the modern samples. In a principal component analysis, the two Gotland cattle placed the closest to Swedish Red Cattle, rather than among the southern or northern traditional breeds. In terms of the first principal component, the second closest breed was Ringamåla cattle. In terms of the second principal component, Gotland cattle samples also had similar values as the northern breed Fjällnära cattle. When performing the principal components analysis with subsets of the data as a robustness check, these qualitative patterns were preserved. In terms of mitochondrial haplotypes, the two Gotland cattle differed by one substitution. We compared them to the whole mitochondrial genomes of 30 Swedish cattle and a sample of D-loop sequences from Nordic and Baltic cattle. The Gotland cattle haplotypes were not identical to any of the other mitochondrial haplotypes, neither among the local Swedish cattle or a sample of D-loop sequences. However, they were similar to clusters of related haplotypes involving multiple other breeds, including Swedish Mountain Cattle and Swedish Red Polled. In summary, our results suggest that Gotland cattle were genetically dissimilar to the extant traditional Swedish breeds, and places it closer to the ancestors of Swedish Red cattle.

Session 83 — Poster 18

Genetic variability in Italian Mediterranean buffalo: implications for conservation and breeding

M.M. Gómez[1], R. Cimmino[1], D. Rossi[1], G. Zullo[1], G. Campanile[2], G. Neglia[2] and S. Biffani[3]
[1]Italian National Association of Buffalo Breeders, V. Petrarca, 42-44, 81100 Caserta, Italy, [2]Department of Veterinary Medicine and Animal Production Federico II University Naples, Via Federico Delpino, 1, 80137 Naples, Italy, [3]National Research Council (CNR), Institute of Agricultural Biology and Biotechnology, V. Edoardo Bassini, 15, 20133 Milano, Italy; m.gomezcarpio@anasb.it

The Italian Mediterranean Buffalo (IMB) is part of the historical and cultural heritage of Italy and its preservation is an indisputable priority. Knowing the structure of a population as well as its variability and gene flow are necessary information for setting up any selection program and for the sustainability of the population. In this study, we sought to investigate the genetic diversity and the structure of the IMB in 8 Italian geographic areas. Information from 2,441 animals genotyped with the Axiom Buffalo Genotyping Array 90K was used. The total mean values of observed and expected heterozygosity were 0.404 and 0.399, respectively. These values suggest the presence of an acceptable degree of genetic variability. The mean inbreeding coefficient (FIS) was positive in two southern areas, namely Caserta and Salerno. Incidentally, both areas have the highest buffalo population density, with 70.8 and 22.5 head per square km, respectively. Moreover, a neighbour network analysis grouped animals from those areas in the same branch, a predictable result since these areas are in geographically close relationships. On average the genetic differentiation between areas, based on the fixation index (Fst), was 2.06%. Finally, a multidimensional scaling plot analysis was performed to investigate the population structure between the 8 areas. Again, individuals from Caserta's area formed a well-defined and homogenous cluster. Results from the present study depict the current genetic structure of the IMB. Overall, no concerns can be raised as regards genetic diversity, however positive FIS values observed in two geographic areas suggest the need for correct selection and mating strategies. Acknowledgements: This research was funded by Italian Ministry of Agriculture. Project: 'BIG' Prot. N. 0215513 11/05/2021 and by Programma di ricerca per la Biosicurezza delle Aziende Bufaline- articolo 4 bis dell'OPCM n. 3634 del 21/12/2007.

Session 83 — Poster 19

A comparative genome analysis across pig breeds can help to identify putative deleterious alleles

M. Ballan[1], S. Bovo[1], G. Schiavo[1], F. Bertolini[1], M. Bolner[1], M. Cappelloni[2], S. Tinarelli[2], M. Gallo[2] and L. Fontanesi[1]
[1]University of Bologna, Department of Agricultural and Food Sciences, Viale G. Fanin 46, 40127 Bologna, Italy, [2]Associazione Nazionale Allevatori Suini, Via Nizza 53, 00198 Roma, Italy; luca.fontanesi@unibo.it

The small effective population size of many highly selected livestock breeds or nuclei can lead to an increased frequency of unfavourable and usually deleterious alleles, in particular if they are in linkage disequilibrium with other favourable QTL alleles. Animals that are homozygous for these negative alleles are, however, not observed in the adult population as the genotype is not usually compatible with a productive life. In this study, we used high density single nucleotide polymorphisms (SNPs) analysed in more than 10,000 heavy pigs belonging to three breeds (Italian Large White, Italian Landrace and Italian Duroc) and whole genome sequencing data obtained in a subset of these animals to define a comparative-across breed strategy that can identify putative deleterious alleles segregating in one or more breeds. The across-breed strategy made it possible to filter out genotyping errors and other genotyping biases that would create statistical biases in the identification of candidate genomic regions harbouring deleterious alleles. Combining this information with whole genome sequencing data, these results provided genomic markers that can be used to design targeted breeding and selection programs aimed to reduce the frequency of deleterious alleles in the Italian pig population.

Session 83 — Poster 20

The effect of recessive genetic defects on pregnancy loss in Swedish dairy cattle

P. Ask-Gullstrand[1], E. Strandberg[1], R. Båge[2], E. Rius-Vilarrasa[3] and B. Berglund[1]
[1]Swedish University of Agricultural Sciences, Animal Breeding and Genetics, Ulls väg 26, 750 07, Uppsala, Sweden, [2]Swedish University of Agricultural Sciences, Clinical Sciences, Ulls väg 26, 750 07, Uppsala, Sweden, [3]Växa Sverige, Ulls väg 29A, 756 51 Uppsala, Sweden; patricia.gullstrand@slu.se

The effect of carrier status of ten lethal recessive genetic defects on pregnancy maintenance was examined in Swedish dairy cattle. The genetic defects were Ayrshire haplotype 1, Ayrshire haplotype 2, *Bos taurus* autosome 12 (BTA12), *Bos taurus* autosome 23, and Brown Swiss Haplotype 2 for Red Dairy Cattle (RDC), and Holstein Haplotypes 1, 3, 4, 6, and 7 (HH1 – HH7) for Swedish Holstein (SH). The data were derived from 1,432 milk recording herds, and carrier status of genetic defects were obtained from the Nordic Cattle Genetic Evaluation. In total, 165,472 inseminations were available from 99,039 lactations from 28,473 RDC and 22,093 SH cows. Pregnancy status was defined based on pregnancy diagnosis, insemination events, calving and culling data. Pregnancy loss traits were defined as embryonic loss (1 d to 41 d after AI), foetal loss (from 42 d after AI until calving), and total pregnancy loss. Least squares means (LSM±SE, %) of pregnancy loss were estimated using linear mixed models. Effect of cow and service sire haplotype, parity (0, 1, 2, ≥3) and insemination number were considered as fixed effects, and herd by year and season of insemination, cow, service sire, and permanent environmental effect as random effects. We observed a negative effect of the ten genetic defects on pregnancy maintenance. The data permitted separate analyses of BTA12 and HH3, however, the remaining defects had too low carrier frequencies to enable further analysis. Carrier frequencies of BTA12 and HH3 were 14.1 and 4.2% in sires, and 5.6 and 4.5% in cows, respectively. The largest effect of BTA12 was in total pregnancy loss with estimates of 68.9±2.8% in at-risk matings compared with 54.8±0.7% in noncarrier matings. For HH3, the largest impact of the genetic defect was in embryonic loss with estimates of 57.5±4.7% in at-risk matings compared with 42.6±1.0% in noncarrier matings. Pregnancy loss may be reduced by avoiding at-risk matings.

Session 84 Theatre 1

EuroFAANG – an infrastructure for farmed animal genotype to phenotype research in Europe and beyond

E.L. Clark[1], E. Giuffra[2], A. Granados-Chapette[3], M. Groenen[4], P.W. Harrison[5], C. Kaya[3], S. Lien[6], M. Tixier-Boichard[2] and C. Kuehn[7]

[1]University of Edinburgh, The Roslin Institute, Edinburgh, EH25 9RG, United Kingdom, [2]INRAE, AgroParisTech, Université Paris Saclay, 78352 Jouy en Josas, France, [3]European Forum of Farm Animal Breeders, 1040, Brussels, Belgium, [4]Wageningen University, 6708 PB, Wageningen, the Netherlands, [5]EMBL European Bioinformatics Institute, Hinxton, CB10 1SD, United Kingdom, [6]Norwegian University of Life Sciences, 1432, Ås, Norway, [7]Research Institute for Farm Animal Biology (FBN), 18196, Dummerstorf, Germany; emily.clark@roslin.ed.ac.uk

The aim of the EuroFAANG infrastructure, which is currently in the concept development phase, is to realise the full potential of genotype to phenotype (G2P) research across species, breeds and populations of farmed animals in Europe and globally. This goal will be achieved through the following four objectives: (1) Creation of a common data structure and data access service. Creating this service for the large variety of datasets generated for farmed animals will consolidate research efforts and minimise redundancy across Europe. (2) Development, curation and biobanking of *in vitro* cellular models, for farmed animal species. A durable framework of connected biobanks, that is accessible to stakeholders across Europe, will enable scientists to use and access *in vitro* cellular models, and popularise the use of such models for G2P research. (3) Sharing and expanding capabilities in new breeding, phenotyping, and genomic technologies. Including, for example, genome editing, and artificial intelligence tools and capabilities in linking *in vitro* and *in vivo* phenotyping and exploitation of genomic data, to provide key routes to application of FAANG data. (4) Connecting with existing projects and infrastructures to consolidate G2P research in farmed animals across Europe. Links include, for example, AQUAEXCEL, PigWeb and Elixir and initiatives for linking G2P in farmed animals in the US including AG2PI and AgBioData. The EuroFAANG infrastructure concept, builds on the foundation provided by the six current H2020 EuroFAANG projects and will lead to a better alignment of the research infrastructure landscape for the advancement of excellent farmed animal science and frontier G2P research in Europe.

Session 84 Theatre 2

GENE-SWitCH: improving the functional annotation of pig and chicken genomes for precision breeding

E. Giuffra[1], H. Acloque[1], A.L. Archibald[2], M.C.A.M. Bink[3], M.P.L. Calus[4], P.W. Harrison[5], C. Kaya[6], W. Lackal[7], F. Martin[5], A. Rosati[8], M. Watson[2] and J.M. Wells[4]

[1]Université Paris-Saclay, INRAE, AgroParisTech, Animal Genetics, UMR GABI, Domaine de Vilvert, 78350 Jouy-en-Josas, France, [2]The Roslin Institute and R(D)SVS, University of Edinburgh, Easter Bush, EH25 9RG, Edinburgh, United Kingdom, [3]Hendrix Genetics BV, Research and Technology Center (RTC), Villa 'de Körver', Spoorstraat 69, 5831 CK Boxmeer, the Netherlands, [4]Wageningen University & Research, Animal Breeding and Genomics, P.O. Box 338, 6700 AH Wageningen, the Netherlands, [5]EMBL-European Bioinformatics Institute, Wellcome Genome Campus, Hinxton, CB10 1SD, United Kingdom, [6]EFFAB, Rue de Trèves 61, 1040 Brussels, Belgium, [7]Epigenetics R&D, Diagenode S.A., Liège Science Park, Ru du Bois Saint-Jean 3, 4102 Liège, Belgium, [8]EAAP, Via G. Tomassetti 3/A, 00161 Roma, Italy; elisabetta.giuffra@inrae.fr

The H2020 GENE-SWitCH (the regulatory GENomE of SWine and CHicken: functional annotation during development) project has delivered extensive functional genomics information at three developmental stages (early and late organogenesis, newborn) for a panel of seven selected tissues. Data are publicly available (https://projects.ensembl.org/gene-switch); they are being processed to characterize biological switches during development and to generate the first Ensembl Regulatory Builds for the chicken (both layer and broiler) and the pig. To assess the interest of using genome annotations for precision animal breeding, two approaches were taken, coupled with dissemination and outreach activities aimed to transfer the outcomes of the project to stakeholder communities. A 'diet × epigenetic' experiment in pigs has investigated how the fibre content of the maternal diet can affect the epigenetic profiles of the liver and muscle of the offspring at the foetal and postweaning stages. In the second approach, genomic prediction models have been extended and tested to evaluate how prediction accuracy can be improved by incorporating different functional and overlapping annotations. The two next presentations will highlight the potential of this approach for swine and poultry breeding. GENE-SWitCH has received funding from the European Union's Horizon 2020 Research and Innovation Programme under the grant agreement n° 817998.

Session 84 Theatre 3

Genome-wide association studies for body weight in broilers using sequencing and SNP chip data

E. Tarsani[1], V. Riggio[1] and A. Kranis[1,2]
[1]The Roslin Institute and R(D)SVS, University of Edinburgh, Easter Bush Campus, EH25 9RG, United Kingdom, [2]Aviagen Ltd., Newbridge, Midlothian, EH28 8SZ, United Kingdom; etarsani@exseed.ed.ac.uk

Body weight (BW) is an economically important trait for the chicken industry. Detecting which variants affect chicken growth is of great benefit to the genetic improvement of this important agricultural species. A powerful resource for identifying trait-associated variants in genome-wide association studies (GWAS) is using whole genome sequencing (WGS) data. In this study, we used a dataset comprising of 60,558 chickens genotyped with a 50k SNP chip. These SNP genotypes were imputed to sequence level for a total of 12.8 variants with high levels of accuracy (>0.95). To evaluate the power of sequencing data, we conducted two GWAS for BW using the chip and WGS genotypes, respectively. When using the 50k SNP chip data, we obtained high genomic heritability (0.40) and identified 18 SNPs across eight autosomes (2-6, 8, 20 and 23) reaching genome-wide significance (FDR *P*-value<0.10) for BW. The majority of those SNPs were introns. With the use of sequencing data, results were similar to those obtained with SNP chip data. However, additional GWAS signals were found, including variants in regions harbouring functional elements for BW. The results suggest an improvement in power and precision when using WGS data. A significant increase in computational time to conduct such a large-scale WGS GWAS was noted, which can limit the number of animals to include in the study. Taken together, the present study enriched our knowledge of the genetic mechanisms regulating chicken growth traits. Next steps will focus on improving the computational tools to enable larger scale analyses to further increase the statistical power.

Session 84 Theatre 4

Using methylation annotation to improve genomic prediction of gene expression in pigs

B.C. Perez[1], J. De Vos[2], D. Crespo-Piazuello[3], M.C.A.M. Bink[1], M. Ballester[3], M.J. Mercat[4], O. Madsen[2] and M.P.L. Calus[2]
[1]Hendrix Genetics BV, Research and Technology Center (RTC), Villa 'de Körver', Spoorstraat 69, 5831 CK, the Netherlands, [2]Wageningen University & Research, Animal Breeding and Genomics, P.O. Box 338, 6700 AH Wageningen, the Netherlands, [3]IRTA, Animal Breeding and Genetics Program, Torre Marimon, 08140 Caldes de Montbui, Spain, [4]IFIP, Institut du porc and Alliance R&D, B.P. 104, 35651 Le Rheu, France; bruno.perez@hendrix-genetics.com

Methylation of DNA has been linked to variation in gene expression and could be used to improve prediction of gene transcripts. Within the GENE-SWitCH project, we aimed to incorporate genome-wide methylation annotation in genomic prediction to improve predictive accuracy for gene expressions. Data comprised of whole genome sequence from 300 pigs (100 Duroc, 100 Landrace and 100 Large White). We considered 2 Kb upstream and 0.2 Kb within genes as regulatory regions and selected across all known genes the SNP placed within those regions, resulting in a panel of 65,486 SNP. Gene transcripts from 10 genes (in 3 tissues, totalizing 30 traits) measured in the same individuals were used as phenotypes. Methylation from the same tissues and two time points (30 d. post-fert. and newborn) were used to weight SNP in the genomic relationship matrix (GRM). These GRM were used in GBLUP with either including (mGBLUP) or not (GBLUP) methylation annotation, and their ability to predict gene-expression were compared using a 5-fold cross validation scheme. Fully methylated sites received a weight of 0, reflecting the assumption that those sites are not expressed. The performance of mGBLUP was quite variable across genes and tissues analysed, together with the methylation state considered as SNP weights. In general, the best mGBLUP results were obtained when methylation information came from the same tissue for which gene-expressions were predicted. Across traits, average difference in predictive accuracy between mGBLUP and GBLUP was +2.6±3.9%. Although mGBLUP showed competitive predictive performance compared to GBLUP when predicting gene-expression for some genes, this pattern was not consistent across all gene-tissue combinations. GENE-SWitCH has received funding from the European Union's Horizon 2020 Research and Innovation Programme under the grant agreement n° 817998.

The path from functional annotation towards improved genomic prediction in cattle

C. Kühn
Research Institute for Farm Animal Biology (FBN), Wilhelm-Stahl-Allee, 18196 Dummerstorf, Germany; kuehn@fbn-dummerstorf.de

In farmed animal breeding there is still a considerable gap in understanding how the (epig)genome translates into the phenotypes of interest, particularly for complex traits. However, this knowledge is a prerequisite for precision breeding and tailored management in livestock. The EU Horizon 2020 funded BovReg project provides multiple layers of annotated functionally active genomic features in the bovine genome. Besides information obtained from FAANG core assays, also positions with significant effect on zootechnical and molecular phenotypes contributed to the improved functional annotation map. Together with a new custom transcriptome annotation, the eQTL analysis conducted with a new Nextflow-based bioinformatic pipeline highlighted previously unknown regions with potential impact on expression regulation. Finally, the functional annotation map is also complemented with regions sensitive to epigenetic modifications after environmental challenges of whole animals or embryos. All functional annotation information feeds into new models for improved genotype-to-phenotype prediction in cattle. The clustering of respective activities together with five other EU H2020-funded projects was the seed for EuroFAANG, an new EU Horizon Europe project aiming to build an infrastructure to consolidate genotype-phenotype research in farm animals in Europe. This project has received funding from the European Union's Horizon 2020 research and innovation programme under grant agreement no. 815668.

Session 84 Theatre 6

Biology-driven genomic predictions for dry matter intake within and across breeds using WGS data

R. Bonifazi[1], G. Plastow[2], M. Heidaritabar[2], A.C. Bouwman[1], L. Chen[2], P. Stothard[2], J. Basarb[2], C. Li[2,3], Bovreg Consortium[4] and B. Gredler-Grandl[1]
[1]Wageningen University & Research, Animal Breeding and Genomics, Droevendaalsesteeg 1, 6700 AH Wageningen, the Netherlands, [2]Livestock Gentec, Department of Agricultural, Food and Nutritional Science, University of Alberta, Livestock Gentec, AB T6G 2HI Edmonton, Canada, [3]Lacombe Research and Development Centre, Agriculture and Agri-Food Canada, AB T4L 1W1 Lacombe, Canada, [4]https://www.bovreg.eu/project/consortium/, the Netherlands; renzo.bonifazi@wur.nl

Biology-driven genomic predictions incorporating functional genomic features (GF) such as expression quantitative trait loci (eQTL) have the potential to increase the accuracy of estimated genomic breeding values (GEBVs) and predicted phenotypes by accounting for casual variants underlying phenotypic variation. To this end, functional GF can be used in genomic prediction models to pre-select and weight single-nucleotide-polymorphisms (SNPs) from whole-genome-sequence data (WGS). Such prioritization can improve genomic predictions in a single and multi-breed context with lowly related animals, especially for scarcely recorded traits such as feed intake and associated traits where large datasets are not yet available. However, validation studies using real data are lacking. Hence, our aim is to validate the usage of biology-driven genomic predictions for feed efficiency within and across breeds. From the Netherlands, 4,996 Holstein-Frisian animals with dry matter intake (DMI) phenotypes and WGS imputed DNA variants are available. From Canada, more than 10,000 animals with phenotypes for DMI and WGS imputed DNA variants are available: ~3,000 for Kinsella composite and ~7,000 for commercial crossbred cattle. Significant variants for several GF (using meta-GWAS QTL, eQTL, mQTL, ChiPseq, and mobile genetic elements identified within the European Union's Horizon 2020 project BovReg) will be used to prioritize SNPs for genomic predictions. We will use a novel Bayesian approach that allows modelling multiple overlapping GF. Different scenarios in which each GF are considered separately along with SNPs from a commercial 50K panel, and where all GF are combined within and across breeds will be tested. Scenarios including GF are expected to improve the accuracy of predicted phenotypes.

Session 84 Theatre 7

Newly annotated genomic features for biology-driven genome selection: the BovReg contribution

G.C.M. Moreira[1], L. Tang[1], S. Dupont[1], M. Bhati[2], H. Pausch[2], D. Becker[3], M. Salavati[4], R. Clark[5], E.L. Clark[4], G. Plastow[6], C. Kühn[3], C. Charlier[1] and The Bovreg Consortium[7]

[1]GIGA, Liège, 4000, Belgium, [2]ETH, Zürich, 8092, Switzerland, [3]Research Institute for Farm Animal Biology (FBN), Dummerstorf, 18196, Germany, [4]The Roslin Institute, Edinburgh, EH25 9RG, United Kingdom, [5]Genetics Core, Edinburgh Clinical Research Facility, The University of Edinburgh, Edinburgh, EH16 4SA, United Kingdom, [6]University of Alberta, Edmonton, T6G 2R3, Canada, [7]the BovReg consortium, https://www.bovreg.eu/project/consortium/, 18196, Germany; gcosta@uliege.be

Transcriptome (mRNA, totalRNA, small-RNA and CAGE), ATAC- and ChIP-Seq (H3K4me3, H3K4me1, H3K27me3, H3K27ac and CTCF) assays were compiled in a catalogue of 129 tissue samples collected from six individuals of both sexes, different ages, kept in different environments and from three divergent dairy and beef cattle breeds/crosses. From the de novo transcriptome assembly, 43,117k gene models including ≥15k potentially novel transcripts were assembled. 1,265 (638 known and 627 novel) miRNAs were detected, and for the majority, potential primary transcripts were identified. We fine mapped 51,295 transcription start sites (TSS) and 2,328 TSS-Enhancer regions shared across the three populations. On average, we identified 105,245, 28,187, 152,646, 127,855, 77,967 and 71,868 peaks (q-value≥0.05) per sample for ATAC, H3K4me3, H3K4me1, H3K27me3, H3K27ac and CTCF, respectively, covering 2.89, 1.52, 7.96, 5.16, 3.95 and 2.49% of the bovine genome on average. Regulatory regions are being characterized by virtue of epigenetic features including open chromatin regions, histone marks and transcription factor as well as, expression data from the different RNA-Seq assays compiled in the diverse catalogue of tissue samples. Moreover, using WGS data from different breeds, a catalogue of unfixed mobile genetics elements was built and a panel of tests for 727 selected insertion sites have been deployed at the public part of EuroGenomics array, widely used in European cattle breeding. The newly annotated genomic features will be used on biology-driven genomic predictions within and across different cattle breeds. The BovReg project has received funding from the European Union's Horizon 2020 research and innovation programme under grant agreement No 815668.

Session 84 Theatre 8

AQUA-FAANG: decoding genome function to enhance genotype-to-phenotype prediction in farmed finfish

D. MacQueen[1] and S. Lien[2]

[1]University of Edinburgh, The Roslin Institute, Easter Bush Campus, EH25 9RG, United Kingdom, [2]Norwegian University of Life Sciences, Department of Animal and Aquacultural Sciences, Centre for Integrative Genetics (CIGENE), 1432, Ås, Norway; daniel.macqueen@roslin.ed.ac.uk

AQUA-FAANG is a European Commission funded research consortium and part of the EuroFAANG initiative. The key project objective is to deliver the first comprehensive genome functional annotations for six commercially important finfish species, namely Atlantic salmon, rainbow trout, common carp, European seabass, Gilthead seabream and turbot. To achieve this goal in support of genotype-to-phenotype prediction, extensive functional genomics datasets were generated for standardized sample types in all six species, inclusive of embryonic development, a panel of tissues from sexually immature and mature adult fish, and immune tissues/derived cells stimulated immunologically. The functional annotation data generated are thus comprehensive in terms ontogeny and of great relevance to production and welfare traits. Datasets include hundreds of RNA-Seq (mRNA-Seq and small RNA-Seq), ChIP-Seq (H3K27ac, H3K4me1, H3K4me3, H3K27me3) and ATAC-Seq libraries per species. This vast catalogue of data is currently being presented and shared on the latest genome assemblies for each species via the Ensembl Genome Browser, including annotations of regulatory elements derived from ATAC-Seq and ChIP-Seq. As of Ensembl version 109, ATAC-Seq data has been shared for several species. AQUA-FAANG data has vast applications spanning fundamental questions about fish genome biology through to practical uses in support of precision breeding. This includes fast-tracking identification and prioritization of candidate causative genetic variants overlapping regulatory elements controlling gene expression, in addition to target genes and variants for future genome editing studies. AQUA-FAANG has received funding from the European Union's Horizon 2020 research and innovation programme under grant agreement No 817923.

Session 84 Theatre 9

Integrating functional annotation data in genomic prediction of VNN resistance in European sea bass

S. Faggion[1], R. Mukiibi[2], L. Peruzza[1], M. Babbucci[1], R. Franch[1], G. Dalla Rovere[1], S. Ferraresso[1], D. Robledo[2] and L. Bargelloni[1]

[1]University of Padova, Department of Comparative Biomedicine and Food Science, viale dell'Università, 16, 35020, Legnaro, Italy, [2]The University of Edinburgh, The Roslin Institute and Royal (Dick) School of Veterinary Studies, Easter Bush Campus, Midlothian, EH25 9RG, United Kingdom; sara.faggion@unipd.it

For the European sea bass (*Dicentrarchus labrax* L.) industry, viral nervous necrosis (VNN) is a major threat. Selective breeding to enhance disease resistance is considered a feasible approach to prevent and control mortality from VNN outbreaks; hence, one of the major goals of AQUA-FAANG is the integration of functional data to improve genomic prediction accuracy of breeding values for VNN resistance in European sea bass. 1,016 juvenile fish (body weight: 6 to 20 gr) produced in a full-factorial mating (25 sires × 25 dams) were subjected to a 29-days VNN challenge test. VNN resistance was recorded both as a binary trait and as time to death. The experimental fish were genotyped using a high-density SNP panel (~27,740 SNPs after quality control), while their parents were whole-genome sequenced and used to impute the offspring to whole-genome genotypes. Genome-wide association analyses revealed a major QTL associated with VNN resistance phenotypes on linkage group 12. A total of 528 and 578 SNP markers were identified as significantly (FDR<0.05) associated with VNN resistance as a binary trait and time to death, respectively. Using evidence from eQTLs and chromatin accessibility data, a putative causal variant was identified. SNP genotypes were used as predictors of resistance and to estimate breeding values (EBV). Genomic predictions were performed integrating prior categorizations of SNPs derived from functional information. The accuracy of the model was assessed in a cross-validation approach generating training and test sets consisting of animals of varying genetic relatedness according to the genomic relationship matrix. *Acknowledgements*. Funding: Horizon 2020 Grant n. 817923 (AQUA-FAANG). Animals used in the experiment: Valle Cà Zuliani Società Agricola srl (Italy).

Session 84 Theatre 10

GEroNIMO (Genome and Epigenome eNabled breedIng in MOnogastrics)

S. Lagarrigue[1], F. Pitel[2] and T. Zerjal[3]

[1]Institut Agro, Rennes, 35042, France, [2]INRAE, Castanet Tolosan, 31320, France, [3]INRAE, Jouy-en-Josas, 78350, France; sandrine.lagarrigue@institut-agro.fr

Pig and chicken are the two main animal protein sources for human consumption in the world. To face today's challenges (climate change, socio-cultural evolution, human population growth, etc.), the GEroNIMO project is rethinking pig and chicken breeding to build more sustainable production models promoting efficient resource use, animal resilience, health, welfare, and preserving genetic diversity. Applying omics technologies, GEroNIMO contributes to improve knowledge on genome-to-phenome relationships accounting for both genetic and non-genetic mechanisms controlling traits related to production (quantity and quality), efficiency, persistency, fertility, resilience and welfare. GEroNIMO is a 5-year program that started in June 2021, involving 21 partners from 11 countries and organized in 8 work packages (WP). Four of them have objectives that contribute to the EuroFAANG mission by: (1) identifying the underlying genetics and epigenetics biological mechanisms affecting trait variation (WP1); (2) determining how changes in environment, generation, and age affect epigenetic molecular traits (WP2); (3) characterizing genetic and epigenetic diversity and propose strategies to optimize local breed conservation (WP3); 4) developing methods to improve selection strategies integrating genetic and non-genetic factors (WP4). GEroNIMO is currently generating molecular data in both species for large numbers of animals/samples including about 6,000 genotypes, 10,000 methylations (by GBS-MeDIP), and 2,000 RNAseq (mainly polyA RNA, but also small RNA and ribo depleted RNA). We already proposed a new annotation for the chicken genome by combining multiple reference annotations and providing a functional gene annotation across 47 tissues using 1,400 samples. This new atlas considers the latest release of the chicken genome assembly (GRCg7b) and gathers both associated reference annotations, 'Ensembl' integrating GENE-SWitCH data and 'NCBI-RefSeq', as well as additional FAANG and NONCODE resources. The Refseq and Ensembl gene atlases respectively increased from 18,010 and 17,007 to 24,102 protein-coding genes and from 5,789 and 11,946 to 44,428 lncRNAs. Project funded by the European GEroNIMO programme under grant agreement N°101000236.

Session 84 — Theatre 11

RUMIGEN: new breeding tools in a context of climate change

S. Mattalia[1], A. Vinet[2], M.P.L. Calus[3], H.A. Mulder[3], M.J. Carabaño[4], C. Diaz[4], M. Ramon[5], S. Aguerre[1], J. Promp[1], R. Vallée[1], B.C.D. Cuyabano[2], D. Boichard[2], E. Pailhoux[6] and J. Vandenplas[3]
[1]Idele, Domaine de Vilvert, 78350 Jouy en Josas, France, [2]INRAE, GABI, Domaine de Vilvert, 78350 Jouy en Josas, France, [3]Wageningen University & Research, Animal Breeding and Genomics, P.O. Box 338, 6700 AH Wageningen, the Netherlands, [4]INIA, Departamento de Mejora Genética Animal, Ctra. de La Coruña km 7.5, 28040 Madrid, Spain, [5]IRIAF, 13300, Valdepeñas, Spain, [6]INRAE, BREED, Domaine de Vilvert, 78350 Jouy en Josas, France; sophie.mattalia@idele.fr

RUMIGEN is a project financially supported by the EU that aims to develop breeding programs able of managing the trade-offs between efficient production and resilience to extreme climate conditions. RUMIGEN is designed under a multi-disciplinary approach that mixes competencies in both genetics and social sciences. The genetic approach aims to enhance genomic selection using three levers: quantitative genetics, genome editing, and epigenetics. One of the objectives of RUMIGEN is to enlarge selection criteria and to provide genomic tools to select heat tolerant dairy cows. Studies are dedicated to the definition of heat-tolerance traits based on production, reproduction and health records, as well as to the study of the trade-offs between these traits, and with those already included in selection indexes. In France, Spain and the Netherlands, performances recorded in commercial herds (i.e. milk production traits, somatic cell scores and conception rate after first AI) were combined with meteorological data obtained from the nearest weather stations, and analysed in order to measure the impact of heat stress. First results obtained for different breeds and in a large range of farming and climatic scenarios showed that the combination of both types of information was relevant at the population level. Some differences between thresholds for optimal THI were observed between countries for some traits, which could be explained by different factors (farm management, exposure to outside temperatures, mitigation practices, etc.). However a decrease was observed on all performance traits with increasing temperatures, with consistent patterns of slopes between breeds and countries. Therefore this approach appears relevant to define novel traits related to heat tolerance for different breeds and different climates.

Session 84 — Theatre 12

The genetic basis of ruminant microbiomes – contribution of the HoloRuminant project

Y. Ramayo-Caldas[1], I. Mizrahi[2], P. Pope[3], C. Creevey[4], J.P. Sanchez[1], R. Quintanilla[1] and D. Morgavi[5]
[1]IRTA, IRTA Torre Marimon, 08026, Spain, [2]Ben-Gurion University of the Negev, Department of Life Sciences, Negev, Be'er-Sheva, 8443944, Israel, [3]Norwegian University of Life Sciences, Faculty of Biosciences, Elizabeth Stephansens v. 15, 1430, Norway, [4]Queen's University Belfast, School of Biological Sciences, Institute for Global Food Security, Belfast, BT9 5DL, United Kingdom, [5]Université Clermont Auvergne, INRAE, VetAgro Sup, UMR Herbivores, Saint-Genès-Champanelle, 63122, France; yuliaxis.ramayo@irta.cat

Microbiome-host interactions have profound effects on metabolic functions and physiological processes that influence growth, production efficiency, animal welfare, and the robustness of the ruminant holobiont. Interactions between the host and its microbiome are influenced by both external and host-specific factors. While the influence of the external environment has long been recognised, it has also been reported that some rumen microbes in the core microbiome are transferred from dam to offspring, with the heritability of the rumen microbiome being low to moderate. In addition, genetic variants and candidate genes associated with individual variation in microbial traits have been reported in the host ruminant genome. However, existing information is scarce, based on reduced numbers of animals and mainly focused on the rumen microbiome. In HoloRuminant, we plan to assess the host genetic influence on microbiomes by using a holistic multi-omics approach to characterise the establishment and dynamics of microbiomes from different body sites during key life events. Planned activities also include the development of an open access database (HoloR) of existing and novel data, together with a repository (HoloR tools) of standardised bioinformatics and analytical pipelines developed in the project. Hologenomic data collected at multiple meta-omics levels will then be used to explore associations between microbial functions and host phenotypes, as well as links between host genetics, microbial, and host traits. The results of the project will improve our understanding of the mechanisms driving host-microbiome interactions in ruminants and may enable the implementation of novel conservation and breeding programs from a holobiont perspective.

Session 84 Poster 13

The effect of production systems on adipose tissue gene expression in Krškopolje pig

K. Poklukar[1], M. Čandek-Potokar[1], N. Batorek-Lukač[1], M. Vrecl[2], G. Fazarinc[2] and M. Škrlep[1]
[1]Agricultural Institute of Slovenia, Hacquetova 17, 1000 Ljubljana, Slovenia, [2]Veterinary faculty, Gerbičeva 60, 1000 Ljubljana, Slovenia; meta.candek-potokar@kis.si

Krškopolje pig is a Slovenian autochthonous breed, which is characterized by lower muscularity and higher fat deposition. The breed is raised in various, mainly low-input production systems. The present study aimed to characterize the transcriptome of adipose tissue of Krškopolje pigs raised indoors (IND) and outdoors (OUT) (n=2×24). At 330 days of age, the animals were slaughtered and samples of back fat were collected. After RNA extraction, the samples were sequenced using Illumina NovaSeq. In total, sequencing yielded 66.8 million paired-end reads, with more than 92.2% of the reads uniquely assigned to the Sscrofa11.1 genome. Differential expression analysis revealed 798 genes, of which 304 were upregulated and 494 were downregulated in the OUT group compared with IND. The upregulated genes in the OUT group were involved in lipid metabolism (FASN, ME1, and SCD genes with log2FC=1.4, 1.2, and 1.1, respectively). The most interesting downregulated genes in the OUT group were involved in collagen synthesis (COL1A; log2FC=4.6), energy homeostasis (LEPR; log2FC=1.9), or triglyceride metabolism (MOGAT; log2FC=2.5). Functional enrichment analysis of downregulated genes in the OUT group revealed biological processes such as negative regulation of angiogenesis GO:0016525), collagen fibril organization (GO:0030199), and endothelial cell morphogenesis (GO:0001886). Up-regulated genes in OUT were connected with the immune response (GO:0006955). The results of the present study provide the first insights into the genetic regulation of Krškopolje pigs kept in different production systems. Acknowledgement: Slovenian Agency of Research (P4-0133, P4-0053, J4-3094, V4-2201), GEroNIMO (EU H2020 GA no. 101000236).

Session 84 Poster 14

A lncRNA gene-enriched atlas for GRCg7b chicken genome using Ensembl, RefSeq and two FAANG databases

F. Degalez[1,2], M. Charles[1], S. Foissac[1], H. Zhou[3], D. Guan[3], L. Fang[4], C. Klopp[1], F. Lecerf[1,2], T. Zerjal[1], F. Pitel[1] and S. Lagarrigue[1,2]
[1]INRAE, 16 Le clos, 35590 Saint-Gilles, France, [2]Institut Agro, 65 rue de Saint Brieuc, 35000 Rennes, France, [3]University of California Davis, Meyer Hall, 1 Shields Ave, CA95616 Davis, USA, [4]Aarhus University, C.F. Møllers Allé 3, 8000 Aarhus C, Denmark; fabien.degale@inrae.fr

With the release of new genome sequences, gene atlases in livestock species are steadily improving. Furthermore, genome annotations greatly vary across databases depending on data resources and bioinformatics pipelines used. These differences are particularly important for long non coding RNAs (lncRNA) compared to protein coding genes (PCG) due to their higher tissue- and stage- specificity. As previously done in 2020 for the galgal5 and GRCg6a chicken assemblies, we provide a new lncRNA-enriched atlas by considering the latest GRCg7b genome assembly. This new chicken gene atlas gathers: (1) both EMBL-EBI Ensembl/GENCODE databases which integrate GENE-SWitCH data and NCBI-RefSeq, considered as references; (2) 3 databases from independent projects in particular from the University of California Davis and Fr-AgENCODE based on FAANG multi-tissues resources; and (3) NONCODE, a lncRNA dedicated database. We characterized the overlap rate of gene models (max of 90% for PCGs and 39% for lncRNAs) between databases two by two and calculated concordance between gene TSS and CAGE peaks from FANTOM (max of 60% for PCGs and 40% for lncRNAs) for each database. Based on these characteristics and prioritizing the reference databases, we determined the order of gathering to maximize the quality of the final annotation. In total, the Ensembl and Refseq catalogues respectively grew from 17,007 and 18,010 to 24,102 PCGs and from 11,946 and 5,789 to 44,428 lncRNAs for a total of 78,323 genes. In addition, we provide a 'functional' gene annotation with 1,400 transcriptomes across 47 tissues and found 35,257 (79.4%) lncRNAs and 22,468 (93.2%) PCGs with an expression of TPM≥0.1. Each gene of this atlas is provided with 'genomic' and 'functional' information and the corresponding RefSeq and Ensembl gene identifiers (http://www.fragencode.org). Project funded by the European Union's Horizon 2020 research and innovation program under grant agreement N°101000236 and by ANR CE20 under EFFICACE program.

Session 85 Theatre 1

Effect of the Rumitech in a high-forage diet on methane production and performance of dairy cows

A. Cieslak[1], H. Huang[1], B. Nowak[1], M. Kozłowska[1], D. Lechniak[2], P. Pawlak[2], M. Szumacher-Strabel[1] and J. Dijkstra[3]
[1]Poznań University of Life Sciences, Department of Animal Nutrition, Wołyńska 33, 60-637 Poznań, Poland, [2]Poznań University of Life Sciences, Department of Genetics and Animal Breeding, Wołyńska 33, 60-637 Poznan, Poland, [3]Wageningen University and Research, Animal Nutrition Group, De Elst 1, 6708 WD Wageningen, the Netherlands; adam.cieslak@up.poznan.pl

The essential oils blend Rumitech was evaluated in three experiments (*in vitro*, cannulated cows, and lactating dairy cows) to investigate its effects on methane emission, milk production, and milk composition of dairy cows. Initially, an *in vitro* study (Exp. 1; Hohenheim test) was performed; 0.35 mg of Rumitech (0.1% on diet DM basis) was supplemented to a commercial TMR (rich in brewer's grain and beet pulp; 20% of total diet DM) which corresponded to 20 g/cow per day fed to dairy cows under commercial conditions. The addition of Rumitech decreased methane yield by 6%. Based on the results of the *in vitro* experiment, two consecutive long-term (39 days) *in vivo* experiments were conducted using 4 Polish Holstein-Friesian cows fitted with rumen cannulas (Exp. 2) and 22 lactating dairy cows (Exp. 3) fed TMR (76% forage on DM basis) in a replicated 2 (diet) × 2 (period) crossover design. Methane production decreased by 10 and 8% in the groups fed the Rumitech diet compared to the control in Exp. 2 and 3, respectively. Moreover, in Exp. 3 milk yield and milk protein and lactose yield increased (by 5, 8 and 8%, respectively) in the groups fed the Rumitech diet. Project CCCfarming National Centre for Research and Development (SUSAN/II/CCCFARMING/03/2021).

Session 85 Theatre 2

Sorghum hybrids as viable forage alternative to corn silage when water availability is limited

D. Duhatschek[1], J. Bell[2], D. Druetto[3], L.F. Ferraretto[4], K. Raver[5], J. Goeser[5], J. Smith[1], S. Paudyal[1], G.M. Schuenemann[6] and J.M. Piñeiro[1]
[1]Department of Animal Science, Texas A&M Univ., College Station, TX, USA, [2]Department of Soil and Crop Sciences, Texas A&M Univ., College Station, TX, USA, [3]Richardson Seeds Ltd., Vega, TX, USA, [4]Department of Animal & Dairy Sciences, Univ. of Wisconsin, Madison, WI, USA, [5]Rock River Laboratory, Watertown, WI, USA, [6]Department of Veterinary Preventive Medicine, The Ohio State Univ., Columbus, OH, USA; schuenemann.5@osu.edu

Over a third of milk produced in U.S. takes place in regions facing water scarcity challenges. Sorghum could be a viable forage alternative to corn silage when water availability is limited. Objective was to assess the effect male-sterile vs non-sterile sorghum hybrids on carbohydrates and CP. Two male-sterile (Non-BMR, F465; and BMR12, F430) and three non-sterile sorghum hybrids (non-BMR F10; BMR12, F382, and F431; Richardson Seeds) were used in a split-plot design to reduce cross-pollination. Plots were harvested 148 days after seeding with non-sterile hybrids after the hard dough stage and male-sterile hybrids 7 weeks after boot stage and harvested with a theoretical length of cut of 12 mm and roll gap setting at 2 mm. Analyses were performed using mixed linear regression models. Values are reported as percent (%) dry matter basis. Compared to non-sterile hybrids, male sterile hybrids had lower DM (28.4 and 40.8%; $P<0.0001$). WSC was 2.5-fold higher (18.0±0.73%) for the sterile hybrids compared to non-sterile (7.21±0.63%), whereas starch was 2.2-fold greater for non-sterile (28.6±0.95%) compared to sterile hybrids (13.0±1.1%; $P<0.0001$). Non-fibre carbohydrates were higher for non-sterile over sterile sorghum hybrids (46.2±0.68 vs 39.6±0.75%; $P<0.003$). Male sterile hybrids had increased aNDF (42.3±0.59%) compared to non-sterile sorghum (35.0±0.51%; $P<0.0001$). CP was significantly lower for sterile compared with non-sterile hybrids (8.29 and 10.2%; $P<0.0001$). The trade-off between WSC and starch in male-sterile sorghum hybrids is because sugars are not translocated to the grain where starch is developed as in non-sterile hybrids. These findings show that low sorghum starch digestibility can be overcome by using male-sterile sorghum hybrids that will, at least partially, compensate for the reduced starch yield by increasing water-soluble carbohydrates storage.

Session 85 | Theatre 3

Feed intake and milk production of dairy cows fed with a ration with ensiled tall fescue

M. Cromheeke[1], L. Vandaele[1], D. Van Wesemael[1], J. Baert[1], M. Cougnon[1], D. Reheul[2] and N. Peiren[1]
[1]Flanders Research Institute for Agriculture, Fisheries and Food – ILVO, Scheldeweg 68, 9090 Melle, Belgium, [2]Ghent university, Faculty of Bioscience Engineering, Department of Plants and Crops, Coupure Links 653, 9000 Gent, Belgium; maarten.cromheeke@ilvo.vlaanderen.be

Voluntary intake and digestibility of ensiled tall fescue (*Festuca arundinacea* Schreb.) is proved to be lower compared to perennial ryegrass (*Lolium perenne* L.). However, new varieties of tall fescue have been bred to improve palatability and digestibility, but there is a paucity of results on intake and animal performance trials with these new varieties. This knowledge gap inhibits adoption in practice. At the same time, there is an increased interest in tall fescue as an alternative for perennial ryegrass in North-West European dairy production because of increased frequency of periods of summer drought that jeopardizes the production of high quality forage grass. In this feeding trial we investigated differences in feed intake and milk production (MY) of dairy cows fed with a ration with ensiled tall fescue compared to (diploid and tetraploid) perennial ryegrass. We measured intake and milk production of Holstein Friesian dairy cows fed with a forage ration that was based on 1/3 of maize silage and 2/3 of ensiled (1) tall fescue (Fa), (2) perennial ryegrass diploid (Lp2) or (3) perennial ryegrass tetraploid (Lp4). Cows were randomized in 6 groups of 5 cows which subsequently received each of the three treatments using a balanced Latin-square design. Results showed a significantly lower total dry matter intake (DMI) of Fa (21.9 kg) compared to the two Lp treatments (22.8 kg Lp2 and 22.9 kg Lp4) ($P<0.05$). This difference was related to lower roughage DMI. Fat- and protein- corrected milk (FPCM) was significantly lower for Fa in comparison with Lp2 and Lp4 (respectively 32.8, 34.6 and 35.5 kg). There were no significant differences in milk composition (fat, protein and lactose) except for milk urea. These findings confirm lower DMI and FPCM of dairy cows fed with a ration with ensiled tall fescue compared to ensiled perennial ryegrass. Similar trials will be set-up in the VLAIO KlimGras project.

Session 85 | Theatre 4

Effect of supplementing live bacteria on methane production in lactating dairy cows

O. Ramirez-Garzon[1], D.G. Barber[2], J. Alawneh[3], L. Huanle[1] and M. Soust[1]
[1]Terragen Biotech Pty Ltd, Unit 6/39-41 Access Cres, Coolum Beach QLD, 4573, Australia, [2]DairyNEXT, Marburg, QLD, 4346, Australia, [3]GCP Veterinary Consulting Ltd Pty., Brisbane, QLD, 4069, Australia; martins@terragen.com.au

Various feed supplements have been investigated to determine their association with methane (CH_4) production in dairy cattle. This longitudinal study assessed the effect of supplementing lactating dairy cows with direct-fed microbes (DFM) on CH_4 column density (ppm-m; used as a proxy for CH_4 emissions) in a commercial Australian dairy farm. A total of 150 multiparous Holstein cows were randomly selected from the main herd (n=360) based on parity (2±1 lactations) and days in milk (DIM; 127±55 DIM). The animals were randomly allocated to Group A (control; n=75) and Group B (DFM; n=75) and managed as two separate groups, housed and fed on a feed pad separately. The diet consisted of a silage-based partial mixed ration (PMR) with concentrate fed in the dairy twice daily and pasture to meet the milk production target relative to the stage of lactation. Both groups were fed the same amount of PMR once daily at 06:00 and pasture *ad libitum*. Group A (control) received the PMR ration whilst Group B (DFM) received the PMR supplemented with DFM, top-dressed using a manual sprayer. The DFM consisted of three strains of Lactobacillus sourced from a commercial product (Mylo, Terragen Biotech Australia). From each group, a subset of 25 Holstein cows (25 control, 25 DFM) were randomly selected and followed longitudinally throughout the study duration to measure CH_4 emissions. The subset of animals were assessed every 2-3 months (from September 2021 to January 2023) at different times of the day (00.00, 05.00, 09.00, 13.00 and 19.00) using a laser methane detector (LMD, Tokyo Gas Engineering solutions, Japan) for 5 minutes whilst restrained in a crush. There was an upward trend in CH_4 emissions across the year in both groups, which was correlated with the PMR:forage:concentrate ratio, and a significant transient reduction in CH_4 emissions ($P<0.05$) after supplementation with the DFM that persisted for approximately 4 hours after supplementation. These findings suggest that DFM could play a role in reducing methane emissions from dairy cows.

Session 85 Theatre 5

Greenhouse gas emissions from dairy cows fed best practice diets

M. Managos[1], C. Lindahl[2], S. Agenäs[1], U. Sonesson[3] and M. Lindberg[1]
[1]Swedish University of Agricultural Sciences, Department of Animal Nutrition and Management, P.O.Box, 75007 Uppsala, Sweden, [2]Lantmännen Lantbruk foder, S:t Göransgatan 160A, 112 17 Stockholm, Sweden, [3]Research Institute of Sweden, Sven Hultins 5, 412 58 Göteborg, Sweden; markos.managos@slu.se

During the last years, the global COVID-19 pandemic and armed conflicts have impacted, among other things, the agricultural supply chain creating uncertainty, resulting in price volatility, and affecting the availability of resources. This has highlighted that, in addition to sustainably increasing food production without exceeding the Earth's biophysical limit, it is important to consider aspects of resilience to climate change and unexpected events. This study investigated milk production and methane emissions (CH_4) from high-yielding dairy cows fed with three different concentrate mixtures with varying degrees of climate impact (CI) in the production stage. Forty-eight Swedish Holstein cows with a milk yield of 45±6.4 kg/d (mean ±SD) at the start of the experiment, blocked by parity in 2 groups, were used in a randomized complete block design with a 2-week adaptation and 7 weeks of data collection. Individual daily feed intake and milk yield were recorded, and the GreenFeed system was used for continuous measurements of CH_4. All cows received the same high-quality grass/clover silage, while three different concentrates were used in the experiment. A commercial concentrate was selected as control, and of the experimental concentrates, the first was based on by-products (BYP) where the ingredients were chosen to reduce feed-food competition, and the second was based on ingredients that can be produced domestically in Sweden. The experimental diets were formulated to have similar theoretical nutritional value as the control and the ingredients' CI was used as the optimization parameter. The sum of CH_4 emissions calculated as carbon dioxide equivalents and emissions from feed production did not differ between diets, while milk yield decreased in the group fed BYP. No difference was observed between treatment groups in the CI per kg milk or kg energy corrected milk. In conclusion, our results highlight the importance of accounting for all sources of emissions in relation to production when the CI of dairy production is evaluated.

Session 85 Theatre 6

Effects of cashew nutshell liquid on milk production and methane emission of dairy cows

R.J.F. Gaspe[1,2], T. Obitsu[2], T. Sugino[2], Y. Kurokawa[2] and Y. Kuroki[2]
[1]Capiz State University, Agriculture, Natividad Pilar Capiz, 5804, Philippines, [2]Hiroshima University, Graduate School of Integrated Sciences for Life, 1-4-4 Kagamiyama Higashi Hiroshima, 739-8528, Japan; rjunfredie@gmail.com

Cashew nutshell liquid (CNSL) contains phenolic compounds and was reported to inhibit gram-positive bacteria and increase rumen propionate concentration, thereby reducing enteric methane emissions from dairy cows. This study aimed to clarify the efficacy of CNSL in methane emissions, milk production, and rumen fermentation from lactating cows in practical condition. Ten Holstein lactating cows (BW 660±51 kg) were used in a completely randomized design with repeated measurement for a 28-day experiment in a free-stall barn with an automatic milking system. Two treatments were arranged as control (no CNSL additive, n=5) or top dressing of the additive to provide 10 g/d of CNSL (n=5) for 21 days, after 7 days of preliminary period (no addition). The average ratio of CH_4/CO_2 in respiration air during milking every 7 days was applied to predict each cow's daily methane production and methane conversion factor (MCF). Milk component and rumen VFA concentration were measured in the fourth week. *In vitro* rumen gas production was also tested using rumen fluid collected from individual cows in the fourth week. Daily dry matter intake, milk production, and methane production were not affected by the CNSL addition. However, methane production per dry matter intake and MCF tended to be lower ($P<0.15$) for the CNSL cows than those for the control cows. Especially, MCF tended to reduce by 12% in the third week compared to those in the first week in CNSL cows, but these were relatively stable in Control cows over the experimental period. Milk composition was not affected by CNSL, but ruminal total VFA concentration and acetate proportion tended to be lower ($P<0.15$) for CNSL cows. A tendency to decrease ($P<0.10$) in total gas, methane, and total VFA production was observed in the *in vitro* incubation with the rumen fluid obtained from the CNSL cows, compared to those from Control cows. These results suggest that microbial activity was slightly inhibited by top-dressing of the CNSL, thereby reducing methane production of cows in practical conditions.

In vivo testing of a methane-suppressing feed additive that acts by altering rumen redox potential

C. O'Donnell[1], C. Thorn[2], A.C.V. Montoya[1], S. Nolan[2], M. McDonagh[1,2], E. Dunne[3], F. McGovern[3], R. Friel[2], S. Waters[4] and V. O'Flaherty[1,2]
[1]Microbial Ecology Laboratory, School of Natural Sciences, University of Galway, Galway, H91 TK33, Ireland, [2]GlasPort Bio Ltd, Business Innovation Centre, University of Galway, Galway, H91 TK33, Ireland, [3]Teagasc Animal and Grassland Research and Innovation Centre, Mellows Campus, Athenry, Co. Galway, H65 R718, Ireland, [4]Animal and Bioscience Research Department, Teagasc, Grange, Dunsany, Co. Meath, C15 PW93, Ireland; C.odonnell64@nuigalway.ie

Enteric methane (CH_4) emissions from livestock is a leading cause of anthropogenic greenhouse gases. A promising CH_4 mitigation strategy is anti-methanogenic feed additives. A novel mode of action for feed additives involves modulation of the rumen oxidation reduction potential (ORP) to favourably alter the rumen fermentation pathway away from methanogenesis. Prior research, using the rumen simulation technique (Rusitec) system, successfully demonstrated the potential of oxygen based additives to alter ORP and suppress methane production. The objective of this study was to evaluate the efficacy of the most promising additive tested via Rusitec (Calcium Peroxide (CaO_2)) *in vivo* using a six-week sheep trial. Sixty ewes were individually penned, housed indoors and fed a basal diet of *ad libitum* grass silage (~1.4 kg DM) and 0.4 kg barley-based concentrates. Animals were randomly assigned to one of three treatments (n=20); no additive (control), and CaO_2 supplemented at either 1.47% or 0.73% of daily DMI. Methane was measured using portable accumulation chambers (PAC) during week 6. Rumen fluid samples were extracted transoesophageally for ORP measurement, dry matter intake (DMI) was measured daily and live weight measured weekly. CaO_2 (1.47%) reduced CH_4 by 14% in terms of g/DMI (P=0.04) and elevated ORP by 22% (P=0.04). The lower CaO_2 dose had no effect on CH_4 emissions or ORP levels (p>0.05). Both CaO_2 treatments had no effects on DMI or live weight gain (p>0.05). This study represents the first *in vivo* trial of a novel ORP modulating additive demonstrating the safety and efficacy of such additives. Further optimisation on delivery mechanisms and inclusion rates to improve efficacy will be necessary for future studies.

Diurnal variation and repeatability of CO_2 production in GreenFeed studies with lactating cows

A. Guinguina[1,2] and P. Huhtanen[1,2]
[1]Swedish University of Agricultural Sciences, Animal Nutrition and Management, Skogsmarksgränd, 901 83 Umeå, Sweden, [2]Natural Resources Institute Finland, Production Systems, Tietotie 2 C, 31600 Jokioinen, Finland; abdulai.guinguina@luke.fi

Not long ago, the concept of residual CO_2 (RCO_2) was proposed as a proxy of feed efficiency using respiration chamber (RC) data from lactating dairy cows. However, the RC is costly and can reduce animal dry matter intake (DMI). The GreenFeed (GF-C-Lock, Inc., Rapid City, SD) system, an affordable and easy to use substitute to the RC can also estimate CO_2 production (g/d), which can be used to calculate RCO_2. However, due its spot-sampling nature, the reliability of the GF has been debated given the diurnal pattern and variability of animal gas production. Thus, a meta-analysis of an individual-cow data set was conducted to examine the diurnal pattern, between-cow variation, and repeatability of CO_2 production from lactating dairy cows. Data were taken from 12 (11 change-over and one continuous) studies consisting of a total of 829 cow/period observations. The experimental diets were based on grass silage with cereal grains or by-products as energy supplements, and rapeseed meal as protein supplement. Mean forage: concentrate ratio across all diets on a dry matter basis was 56:44. The diurnal pattern was determined using CO_2 production data from the continuous study. Variance components and repeatability estimates of CO_2 production were determined using data from the change-over studies with diet, period and cow within experiment as random effects in PROC MIXED of SAS. The average diurnal pattern for CO_2 production did not significantly change between feeding periods during the day. The between cow variation in CO_2 production was 9%. The repeatability estimate was also high (0.78) and practically the same as the repeatability of DMI from this dataset. Production of CO_2 was positively associated (r=0.51, P<0.001) with DMI. It is concluded that if the management of feed allotment remains constant then the diurnal pattern of CO_2 production from dairy cows will not necessarily alter with time. The high repeatability and close association between CO_2 production and DMI suggests that measurements of CO_2 made by the GF could be used to estimate feed efficiency in lactating dairy cows on-farm.

Session 85 Theatre 9

Repeatability and correlations of residual carbon dioxide and feed efficiency in Nordic Red cattle

A. Chegini, M.H. Lidauer, T. Stefanski, A.R. Bayat and E. Negussie
Natural Resources Institute Finland (Luke), Tietotie 4, 31600 Jokioinen, Finland; arash.chegini@luke.fi

Feed utilization efficiency (FE) is an important trait in dairy production and plays a significant role in reducing feed costs and lowering methane emission. One of the metrics used to measure FE in dairy cows so far is residual feed intake (RFI). This metric requires routine measurement of feed intake which is difficult and expensive. Recently, studies have found high correlation between heat and carbon dioxide (CO_2) production of cows and hence residual CO_2 could be a suitable proxy for FE in dairy cattle. Data was collected from 2 to 305 days in milk (DIM) from 46 Nordic Red dairy cows recorded for CO_2 output using two GreenFeed units and FE traits. RCO_2 and RFI were calculated by subtracting expected CO_2 production and DMI from actual CO_2 production and DMI, respectively. A bivariate model was fitted to estimate repeatability and correlations between RCO_2 and RFI traits. The models fitted included fixed effects of year-month of recording and lactation month, fixed regression (4th order Legendre plus Wilmink term) as well as random regression (1st order Legendre plus Wilmink term) for animal effect. Heterogenous residual variance considered. Repeatabilities and animal and phenotypic correlations between RCO_2 and RFI between selected DIM (DIM; 6, 36, …, 276 and 305) were calculated. Repeatability of RCO_2 was high at the beginning of lactation (0.72 at DIM 6) and decreased around peak yield (0.27 at DIM 96) and again increased gradually towards the end of lactation. Similarly, RFI had high repeatability at the beginning (0.86 at DIM 6), however, it decreased in mid-lactation and increased towards the end of lactation. Animal correlations between RCO_2 and RFI were moderate to high on same DIM and ranged from 0.37 for DIM 96 to 0.88 at DIM 6. However, when the selected DIMs were far apart, correlations between RCO_2 and RFI were low, indicating different mechanisms controlling DMI and CO_2 exhalation at different stages of lactation. This was an initial study with 5,995 daily average records and analyses of a larger data set is needed to verify the results with higher accuracy.

Session 85 Theatre 10

Methane emission from purebred Holstein, Nordic Red and F1 crossbred cows of the two breeds

A.L.F. Hellwing[1], M. Vestergaard[1] and M. Kargo[2]
[1]Aarhus University, Campus Viborg, AU Foulum, Department of Animal and Veterinary Sciences, Blichers Allé 20, 8830 Tjele, Denmark, [2]Aarhus University, Center for quantitative genetics and genomics, C. F. Møllers Allé 3, 8000 Aarhus, Denmark; annelouise.hellwing@anivet.dk

Results in the literature indicate that crossbred cows eat less per kg milk and are healthier than purebred cows. Would that also lead to lower methane emission per kg milk? The aim of the current experiment was to investigate methane emission of purebred Holstein, Nordic Red and F1 crossbred cows fed the same feed ration. Twenty-four cows, 8 Danish Holstein (HOL), 8 Nordic Red (RED) and 8 crossbred cows (CROSS) were used. The feed ration was made up of 34.5% maize silage, 17.4% grass-clover silage, 9.8% barley, 15.5% rapeseed cake, 10.2% rapeseed meal, 11.4% dried sugar beet pulp and 1.1% minerals and vitamins based on dry matter (DM), which is a typical ration fed to Danish dairy cows. The methane emissions were measured over three days by means of indirect calorimetry. DM intake and milk yield and composition was measured during chamber stay. The first day in the chamber was considered as an adaptation and only data from the last two days were used. Data were analysed with PROC GLM in SAS (version 9.4) with breed as the main effect. All cows were in second lactation and were 167±31 (mean ±s td) days in milk. The average weight of the cows was 670±66 kg. The DM intake (DMI) was 22.2, 19.4 and 21.3 kg for HOL, RED and CROSS, respectively (P=0.40). The daily methane emission did not differ between breeds and was on average 388 g/day. The methane emission was 17.9, 19.2 and 18.4 g/kg DMI for HOL, RED and CROSS, respectively (P=0.19). The production of energy corrected milk (ECM) differed between the breeds (P=0.01) where HOL yielded 35.1, RED 27.0, and CROSS 30.1 kg ECM/day during the chamber period. The CH_4 production per kg ECM was 11.5, 13.9, and 13.0 for HOL, RED and CROSS, respectively (P=0.14). In conclusion, it was not possible to show any breed differences in total methane production or methane per kg DMI or ECM yield when cows were fed on typical Danish feed ration.

How does a beef × dairy calving affect the dairy cow's following lactation?

R.E. Espinola Alfonso[1], W.F. Fikse[2], M.P.L. Calus[3] and E. Strandberg[1]
[1]Swedish University of Agricultural Sciences, Department of Animal Breeding and Genetics, P.O. Box 7023, 750 07 Uppsala, Sweden, [2]Växa Sverige, Ulls väg 29A, 75651 Uppsala, Sweden, [3]Wageningen University Research, Department of Animal Sciences, P.O. Box 338, 6700 AH Wageningen, the Netherlands; erling.strandberg@slu.se

For beef semen usage on dairy cows, much of the research has focused on the performance of the crossbred calves, yet little focus has been given to the subsequent performance of the cow itself. This study aimed to evaluate the performance of dairy cows for milk yield, fertility, and survival after giving birth to crossbred calves and compare this to the performance after giving birth to purebred dairy calves. Further, we aimed to study if the effect of a difficult calving was the same regardless of whether the calf was purebred or crossbred. Phenotypic records from 4,980,886 calving events distributed in 4,509 herds from 1997 to 2020 were collected from the Swedish milk recording system from cows of the dairy breeds Swedish Red (SR) and Swedish Holstein (SH). The data were analysed for all traits for first and later parities separately using mixed linear models, with a focus on the estimates of dam breed by sire breed combinations. Overall, milk yield was lower after mating beef × dairy compared with the purebred matings. The largest effects were found on total lactation yields and in later parities, with lower effects for earlier yields and yields in first parity. The largest decrease was about 17 kg for cumulative fat yield (corresponding to 14% of a phenotypic SD) when breeding heavy beef breed sires with purebred SR dams. For fertility traits, for most breed combinations, the effects were not large enough to be significant. Conversely, all dam-sire crossbred combinations showed significantly lower survival to the next lactation, and mostly also for last day in milk (lactation length). There was generally an unfavourable effect of a difficult calving on all traits, however, there were only some significant interactions between calving performance and dam by sire breed combination, and never in first parity. Beef × dairy difficult calvings were generally affecting performance less than corresponding difficult purebred calvings.

Extended lactation and milk yield – a randomized controlled trial in high yielding older cows

A. Hansson[1], C. Kronqvist[2], R. Båge[3] and K. Holtenius[2]
[1]Växa, P.O. Box 30204, 10435 Stockholm, Sweden, [2]Swedish University of Agricultural Sciences, Department of Animal Nutrition and Management, P.O. Box 7024, 750 07, Sweden, [3]Swedish University of Agricultural Sciences, Department of Clinical Sciences, P.O. Box 7024, 750 07, Sweden; annica.hansson@slu.se

The current recommendation is that Swedish dairy herds should aim for 12.5 months calving interval. In practice, farmers are advised to strive for a voluntary waiting period (VWP) of 50 days. However recent studies have shown that primiparous cows with prolonged VWP showed improved fertility results, without negative effects on milk production. It appears to be a general opinion that older cows are less suitable than primiparous for a longer VWP due to reduced milk production and too long dry periods. However published randomized controlled studies presenting effects of delayed VWP in older cows are scares. The aim of this study was to evaluate effects of an extended VWP on milk production in second lactation cows in herds with high milk production. Fifteen herds connected to the Swedish milk recording scheme volunteered for a randomized controlled study. Average herd milk production exceeded 10,800 kg ECM, compared with the 2018/19 national average of 10,400 kg ECM. Cows having their second calf from Dec. 2019 until Dec. 2020 were randomized to 50 days VWP (n=504) or 140 days VWP (n=481). In total 371 cows in VWP50 and 324 cows in VWP140 completed the second lactation with a third calf. Preliminary results show that the average calving interval in cows with VWP50 was 12.6 months and in VWP140 13.9 months (P<0.001). Dry period was 65 days in both groups. In VWP50 average 305 days lactation yield was 12,190 kg ECM and in VWP140 12,868 (P<0.001). Average daily milk yield over the lactation including the dry period was in VWP50 33.9 kg ECM and in VWP140 34.7 kg ECM (P=0.05). In conclusion, these preliminary results show that second lactation cows in high yielding herds with extended lactation had higher 305 days lactation yield and tended to have higher daily milk yield over the lactation than cows with traditional calving interval. The length of the dry period was not affected by VWP.

Effect of usage BWB sire on Holstein cow performances

K. Elzinga[1], C. Schrooten[1], G. De Jong[1] and P. Duenk[2]
[1]Cooperative CRV, Animal Evaluation Unit, Rijksweg-West 2, 6842 BD Arnhem, the Netherlands, [2]Wageningen University & Research, Animal Breeding & Genomics Group, Droevendaalsesteeg 1 Building 107, 6708 PB Wageningen, the Netherlands; kirsten.elzinga@crv4all.com

Mating a Holstein Friesian (HF) dairy cow with a Belgian White Blue (BWB) sire is done as the crossbred calves are more profitable compared with calves when using a HF sire. To investigate to what extent the use of a BWB sire on a HF dam, compared with a HF sire on a HF dam, can affect the milk production and fertility of the dam in the following lactation, data was used from cows (n=268,475) in lactation 1 to 3 in the period 2016 to 2022. Cows were mated with a HF service sire as well as, in another year, a BWB service sire. Regarding production, the 305-day production of milk, fat and protein was included. Regarding fertility, number of inseminations, conception rate, non-return 56 days, calving interval, interval calving – first insemination, and interval first – last insemination was investigated. First, the effect of breed of service sire on production and fertility was studied, second the effect of the sire estimated breeding value (EBV) for calving ease was included in the model, and lastly the effect of calf characteristics (calf birth weight, calf sex, and calving ease) was studied. The use of a BWB service sire compared with a HF service sire decreased milk-, fat-, and protein production in the following lactation of the dam. HF cows mated with a HF service sire compared with a BWB service sire had better fertility in the subsequent lactation. More inseminations were needed for HF dams mated with a BWB sire and intervals between calvings and inseminations were longer. Also, conception rate for cows mated to a BWB sire was lower compared with cows mated to a HF sire. An increase in mating sire EBV for calving ease resulted in lower production and worse fertility in the following lactation. However, effects of the EBV for calving ease were low. A normal calving (class 2) resulted in the highest production and best fertility, whereas after a caesarean section (class 4) cows had the lowest production and worst fertility.

Combined analysis: genetic and phenotypic trends in German Holstein dairy cattle

L. Hüneke[1], J. Heise[1], D. Segelke[1], S. Rensing[1] and G. Thaller[2]
[1]IT Solutions for Animal Production (vit), Heinrich-Schröder-Weg 1, 27283 Verden (Aller), Germany, [2]Christian-Albrechts-University Kiel, Institute of Animal Breeding and Husbandry, Hermann-Rodewald-Straße 6, 24118 Kiel, Germany; laura.hueneke@vit.de

Dairy cattle breeding programs have experienced drastic changes during the last 20 years, especially due to the implementation of genomic selection, which accelerated the breeding progress. In the context of intensifying discussions about animal welfare and public reports of the detrimental impact of breeding, scientifically objective information are needed. This study investigates the phenotypic and genetic trends of milk yield, longevity, calving ease, and stillbirth in the German Holstein population from the year 2000 on. Phenotypic data and breeding values of about 12 million cows were available. The means of breeding values and phenotypic values per year were calculated to analyse genetic and phenotypic trends. A parallel increase in the average breeding value and the phenotypic milk yield was observed for the milk production traits. On the other hand, the trait longevity showed clear differences in the development of phenotype and breeding value, as the strongly increasing genetic trend did not become fully visible in the phenotypic progress. This could be due to management decisions based on high replacement rates and the resulting early culling of cows due to limited space in the barn. The rates of stillbirth and calving difficulty indicated higher risk in first calving and for the birth of male calves. Nevertheless, both traits showed a declining trend over the last 20 years and incidence rates of stillbirth and calving difficulty were reduced by approximately 50%. However, the stillbirth rate was almost twice the rate of calving difficulty, which may indicate that many calves were born easily but then died for unknown reasons. The genetic trend of the calving traits only started to rise in the last 10 years, which could be caused by low weights in the selection indexes combined with low heritabilities. Moreover, it is likely that the calving traits have benefited from genomic selection. The results from this study will contribute immensely to the discussion of animal welfare in Germany.

Session 85 — Poster 15

Whole-crop maize forage preservation improvement with silage additives

I. Nikodinoska[1], E. Wambacq[2], G. Haesaert[2] and C.A. Moran[3]

[1] Alltech European Headquarters, Summerhill Road, A86 X006, Dunboyne, Co. Meath, Ireland, [2]University of Applied Sciences and Arts, Research Centre AgroFoodNature, School of Bioscience and Industrial Technology, V. Vaerwyckweg, 9000 Ghent, Belgium, [3]Alltech SARL, Rue Charles Amand, Vire, 14500, France; ivana.nikodinoska@alltech.com

The efficacy of 6 silage additives on whole-crop maize preservation was examined in an ensiling trial using 2.75 l PVC microsilos. Maize was harvested at the Research Farm Bottelare (Belgium) in November 2021, at a dry matter (DM) content of 28% and chopped to 6-8 mm. The starting material contained 2.19% of water-soluble carbohydrates in the fresh material (FM) and classified as moderately difficult to ensile. The crop was inoculated with one out of 6 preparations: *Lactiplantibacillus plantarum* (LP) IMI 507026:*Pediococcus pentosaceus* (PP) IMI 507024 (50:50) at 5×10^5 cfu/g FM, LP IMI 507026:PP IMI 507024 (50:50) at 1×10^5 cfu/g FM, LP IMI 507027:PP IMI 507025 (75:25) at 5×10^5 cfu/g FM, LP IMI 507026:PP IMI 507025 (75:25) at 5×10^5 cfu/g FM + 25 mg/kg $MnSO_4$, LP IMI 507026:PP IMI 507025 (75:25) at 1×10^5 cfu/g FM + 25 mg/kg $MnSO_4$, or treated with an equal amount of water (Control-T1). Five replicates per treatment were prepared at a density of 150 kg DM/m^3. Lactic acid (LA), acetic acid (AA), were determined by HPLC; ethanol was determined on an aqueous extract by NIR absorption. DM content at silo opening was determined by air drying at 60 °C and corrected for loss of volatiles during drying. The DM loss during the 100-days ensiling period at 20 °C was calculated. Crude protein and ammonia were analysed and the NH_4-N,%TN calculated as the ratio of ammonia nitrogen to total nitrogen. The pH was determined on an aqueous extract. Data were analysed according to a Mixed Effect model (Minitab v21.1) and Dunnett's post-hoc test, to compare the difference between T1 and the six inoculated treatments (T2-T7) ($P<0.05$). All treatments significantly reduced the ethanol content compared to T1. Treatments T2, T3, T4, T6 and T7 significantly reduced the pH, whereas T3-T6 significantly increased the LA in treated forages compared to T1. Acetic acid, NH4-N,%TN and DM losses did not significantly differ between T1 and T2-T7. These findings suggest that the homofermentative pathway was dominant, efficiently improving the silage fermentation process.

Session 85 — Poster 16

Gallic acid alleviates the negative effect of *Asparagopsis armata* on milk yield in dairy cows

R. Huang[1], P. Romero[2], C. Martin[1], E.M. Ungerfeld[3], A. Demeter[4], A. Belanche[5], D.R. Yáñez-Ruiz[2], M. Popova[1] and D. Morgavi[1]

[1]Université Clermont Auvergne, INRAE, VetAgro Sup, UMR Herbivores, Saint-Genès Champanelle, 63122, France, [2]Estación Experimental del Zaidín (CSIC), Profesor Albareda 1, Granada, 18008, Spain, [3]Instituto de Investigaciones Agropecuarias INIA, Centro Regional de Investigación Carillanca, Vilcún, La Araucanía, 4880000, Chile, [4]Volta Greentech, AB Nanna Svartz väg 2-4, Solna, 17165, Sweden, [5]Universidad de Universidad de Zaragoza-CITA, Departamento de Producción Animal y Ciencia de los Alimentos, Miguel Servet 177, Zaragoza, 50013, Spain; rongcai.huang@inrae.fr

We have shown that when ruminal methanogenesis is inhibited *in vitro*, gallic act as a dihydrogen acceptor and improve ruminal fermentation. In this work, we tested the effect of gallic acid and the methanogenesis inhibitor *Asparagopsis armata* on dihydrogen emissions and milk production in dairy cows. Lactating Holstein cows (n=28) were separated into four treatment groups as follow: CON, basal hay:concentrate diet; AS, basal diet with 0.25% *A. armata*; GA, basal diet with 0.8% gallic acid; and MIX, basal diet with 0.25% *A. armata* and 0.8% gallic acid. This study used a randomized complete block design with 6 wk of covariate and 5 wk of experimental measurements. The AS treatment decreased methane yield by 25%, increased dihydrogen emissions threefold, and reduced the acetate to propionate ratio in the rumen. In contrast, gallic acid did not affect methane or dihydrogen emissions, nor acetate or propionate molar proportions. Feed intake decreased by 10% in the AS group with a consequent decrease in milk yield of 18%. In the MIX group, feed intake was also decreased (8%), but the drop in milk yield was less severe (11%). This study confirms the antimethanogenic effect of *A.armata* but also shows a negative effect on feed intake and milk yield. While gallic acid did not decrease dihydrogen loss, it partially alleviated the negative effect of *A. armata* on milk yield.

Session 85 Poster 17

Milk production in primiparous cows with customized voluntary waiting period

A. Edvardsson Rasmussen[1], E. Strandberg[2], R. Båge[3], M. Åkerlind[4], K. Holtenius[1] and C. Kronqvist[1]
[1]Swedish University of Agricultural Sciences, Department of Animal Nutrition and Management, P.O. Box 7024, 750 07 Uppsala, Sweden, [2]Swedish University of Agricultural Sciences, Department of Animal Breeding and Genetics, P.O. Box 7023, 750 07 Uppsala, Sweden, [3]Swedish University of Agricultural Sciences, Department of Clinical Sciences, P.O. Box 7054, 750 07 Uppsala, Sweden, [4]Växa Sverige, P.O. Box 288, 751 05 Uppsala, Sweden; anna.edvardsson.rasmussen@slu.se

Randomized trials of extended voluntary waiting period (VWP) in primiparous cows have shown no difference in milk yield per day in the calving interval between cows with extended compared to conventional VWP. We hypothesized that individually customized VWP would improve daily milk yield. A randomized field trial in 18 commercial dairy herds was conducted. For primiparous cows calving during six months, the farmers regularly reported calving history, diseases and daily milk yields. Three criteria were used to select cows assumed to benefit from extended VWP: High genetic lactation persistency index before the start of the study (top 10%, n=41), a difficult calving and or disease during the first month of lactation (n=59) or above average daily milk yield during day 4-33 in lactation (n=247). Cows were randomized within each criterion to an extended VWP (from 185 DIM) resulting in a 16.3 mo calving interval (ExtExt, n=174) or a conventional VWP (up to 90 DIM) resulting in a 12.4 mo calving interval (ExtConv, n=173). Cows not fulfilling any of the three criteria (ConvConv, n=183) were appointed a conventional VWP resulting in a 12.0 month calving interval. Results, including cows receiving the intended VWP and having a second calf show that 305-d yield was highest in the ExtExt group (10,383±394 kg) second highest in the ExtConv group (9,817±391 kg) and lowest in the ConvConv group (8,115±391 kg) ($P<0.01$). Milk yield per day in the calving interval was lower ($P<0.01$) in the ConvConv (22.6±1.1 kg/day) compared to the ExtConv (27.3±1.1 kg/day) and ExtExt group (27.7±1.1 kg/day), and there was no difference between the ExtConv and ExtExt groups ($P=0.63$). Cows selected for conventional or extended VWP based on calving history, diseases and daily yields produced equal average daily yield independent on VWP group.

Session 85 Poster 18

Association of nisin and *Cymbopogon citratus* against *S. aureus* isolated from bovine mastitis

L. Castelani, T.M. Mitsunaga, L.C. Roma Jr. and L.E.F. Zadra
Institute of Animal Science and Pastures (IZ), Heitor Penteado, 56, 13380-011, Brazil; liviacastelani@gmail.com

Bovine mastitis caused by *Staphylococcus aureus* represent losses to the dairy industry and risks to consumer health. The replacement of conventional therapies by alternative biocompounds is fundamental for the sustainability of milk production. The Nisin (NS) is a bacteriocin produced by of *Lactococcus lactis* strains, and has action against several of Gram-positive bacteria species. The *Cymbopogon citratus* (CC) essential oil has anti-inflammatory and antimicrobial properties mainly due to the presence of tannins and flavonoids in its composition. The aim of this study was to evaluate the *in vitro* antibacterial activity of nisin, *C. citratus* and the NS/CC complexes against multidrug-resistant *S. aureus* (MDR) strains, isolated from bovine mastitis and belonging to an International Collection of Culture Strains (i.e. ATCC). *C. citratus* was solubilized in Muëller Hinton broth supplemented with 0.5% Tween 80, and nisin was solubilized in 0.02N HCl, decontaminated by filtration (0.45 μm). NS/CC complexes were obtained after 1 hour of interaction at 25 °C. The *in vitro* activity of the nisin, *C. citratus* and NS/CC complexes was evaluated by minimum bactericidal concentration (MBC), checkerboard method and time-kill against 27 strains of *S. aureus* exhibiting an MDR profile (aminoglycoside, B-lactam, cephalosporin, folate inhibitor, lincosamide, macrolide, quinolone, tetracycline), isolated from bovine mastitis; and *S. aureus* BAA-976™, BAA-1026™, MU-50. The MBC_{50} value of nisin was 400 μg/ml, whereas the MBC_{50} of *C. citratus* was 625 μg/ml. The MBC of the NS/CC complex was 100/156 μg/ml, and this combination was interpreted as a synergistic effect ($\Sigma FBC \leq 0.5$). According to time kill assay, for the NS/CC complexes, a ≥3 log10 reduction (= 99.9% reduction of cfu/ml) of the ATCC BAA-976™ strain was achieved after 1 hour of interaction. The association *in vitro* of *C. citratus* and the nisin showed potential for the control of MDR *S. aureus* strains isolated from bovine mastitis. The combination of biomolecules can be useful for preventing bacterial resistance, in addition to reducing the administered dose and side effects. (Financial Support FAPESP 2018/16531-9; FAPESP 2020/09249-5).

Session 86 Theatre 1

Quick sustainability scan calculator for intensive and extensive pig farms

S. Sanz-Fernández[1], C. Reyes- Palomo[1], P. Meatquality Consortium[2] and V. Rodríguez-Estévez[1]
[1]Universidad de Córdoba, Cátedra de Producción Ecológica Ecovalia-Clemente Mata, Ctra. Madrid-Cádiz km 396, 14071, Cordoba, Spain, [2]Project mEATquality, Consortium, https://meatquality.eu/, Denmark; v22repac@uco.es

One of the main objectives of the H2020 mEATquality project is to support the development of techniques to assess the pig production sustainability, in order to improve it through the best practices and to assess the environmental, social and economic impacts of pig production. This project will assess different pig husbandry systems around Europe (80 fattening pig farms from Denmark, Italy, Poland and Spain). Thus, two quick sustainability scan calculators have been developed: one for intensive pig farms and another for extensive ones. These calculators propose each evaluation schemes, integrating the score assigned individually to a farm trough the evaluation of 10 aspects related to livestock management: (1) certifications; (2) water management; (3) feed; (4) waste and residues management; (5) energy efficiency; (6) socio-economic contribution to the farm territory; (7) farm associated businesses; (8) animal management; (9) management of pastures and soil, and biodiversity; (10) stocking rate and pigs density. The last two aspects are only evaluated for extensive farms. Each of these aspects receives a score according to its importance and contributes in a different proportion to a final score, which is organized into 3 categories: environmental, social, and economic impact The maximum achievable score in each category is 100 points. These quick scans are designed to be easily answered (yes/no), and the margin of error for over or underestimation the contribution of some aspect is small, since the number of questions (n=62 for extensive farms; n=64 for intensive farms) is distributed among the different aspects and none force excessively the final score. With these calculators it is intended to obtain some benefits, since it is useful as a self-assessment tool for the farmers themselves, allows comparison (benchmarking) and identifies the weak points of each farm, enables participatory certification, and helps to provide confidence to consumers. This project has received funding from the European Union's Horizon 2020 research and innovation program under Grant Agreement No 101000344.

Session 86 Theatre 2

Bulk milk and farms characterization in the Parmigiano Reggiano Consortium area: the INTAQT project

M. Berton, M.A. Ramirez Mauricio, N. Amalfitano, L. Gallo, A. Cecchinato and E. Sturaro
DAFNAE-University of Padova, Viale dell'Università 16, 35020 Legnaro, Italy; marco.berton.1@unipd.it

Parmigiano Reggiano (PR) is probably the most valuable Italian cheese, produced under the regulations of a specific Consortium that supervises one of the main dairy chain in Italy. The study aimed to characterize structure, management and milk traits (milk yield, kg/d per cow –MY, protein content, % -CP, and fat content, % -FAT) of the PR-producing farms enrolled in the INTAQT project. Data originated from farm inspections (altitude zone -AZ, herd size, housing type, genetic type of cows -GT, use of total mixed rations -TMR, proportion of concentrate inclusion in the lactating cow diets -CONC) and official milk recording system (test-day MY, CP, FAT). All farms were scored for animal welfare related features (management -A, structure -B, and animal-based measure -C; 0-100 score range per category) according to the official national method for animal welfare assessment. A total of 4,558 milk records from 804 farms were considered. Altitude zone was classified as mountain and plain, GT as mostly Brown Swiss, mostly Holstein Friesian (HF+), only Holstein Friesian (HF) and local breeds, CONC into 3 classes (<25, 25-55 and >55% of the average dry matter) and welfare scores into 3 classes (mean±0.5 SDs). Test-day were grouped as seasons (SE; winter, spring, summer). Milk yield, CP and FAT were analysed with a mixed model with AZ, GT, CONC, welfare A, B, C scores and SE classes as fixed effects and farm as random one. All the fixed effects except SE were nested into farm. Farms were mostly located in the plain (77% of farms), with free-stall housing (68%), rearing HF cows (41% HF; 45% HF+) and not using TMR (44%). Lactating cows averaged 108±106 and were fed diets that included 42±8% CONC. Welfare scores A, B and C averaged 78±9, 71±10 and 79±6, respectively. Most of the fixed effects affected MY, with greater values for plain-located farms, rearing HF cows, with TMRs, high levels of CONC, scores A and C. Milk protein was affected by GT, TMR, score C and SE, whereas FAT by AZ and SE. These results could contribute to drawing interventions aiming to improve both intrinsic quality of milk destined for cheese production and extrinsic one related to environmental sustainability and animal welfare.

Session 86 Theatre 3

Bridging environmental sustainability and intrinsic quality traits of pork

M. Gagaoua, F. Gondret, F. Garcia-Launay and B. Lebret
PEGASE, INRAE, Institut Agro, 35590, Saint-Gilles, France; mohammed.gagaoua@inrae.fr

The perceptions by consumers of fresh pork quality determine their re-purchase behaviour, which is key to the success of the pork industry. Consumers rely on intrinsic (product-related) and extrinsic (production-related) cues to form their opinions about pork. The intrinsic cues (organoleptic, technological and nutritional properties) are usually used to assess pork quality. Extrinsic cues refer to the ethical, cultural and environmental dimensions of pork production, and are becoming increasingly important to consumers who are notably seeking pork produced in sustainable and environmentally friendly practices. Notwithstanding, studies integrating environmental impacts into pork quality assessment are constrained. This work intends to address this drawback through two approaches. First, we provide an overview of the current knowledge on the expectations of consumers towards environmental sustainability of pork production with an emphasize on the gap between attitudes (expectations) and behaviours (willingness-to-purchase). Second, a proof of concept integrating environmental footprints with intrinsic qualities of meat was provided. For that, we used an experiment from farm-to-fork designed to improve intrinsic and extrinsic dimensions of pork quality. The trial included two pig genotypes (Duroc and Pietrain NN crossbreeds) and two feeding regimens: an experimental diet with extruded faba bean (national origin) as protein and extruded linseed as omega-3 fatty acid sources, while the control diet was based on oilseed meal (imported soybean, rapeseed and shelled sunflower). Raising Duroc crossbred pigs with the diet containing extruded faba bean and linseed was found as a favourable strategy to jointly improve the sensory, technological and nutritional properties of pork, while meeting the challenge of relocating protein feed resources. We further calculated and compared the environmental impacts of the four combinations of pig genotype and feeding strategy using life cycle assessment, and addressed their relationships with intrinsic meat quality traits using multivariate and clustering analyses. Incorporating environmental impacts into pork quality assessment can help promoting sustainable pork production practices and guide consumers in their purchase decisions.

Session 86 Theatre 4

Improving milk intrinsic quality: considering synergies and antagonisms of farming practices

L. Rey-Cadilhac[1], A. Ferlay[1], M. Gelé[2], S. Léger[3] and C. Laurent[1]
[1]UCA, INRAE, VetAgro Sup, UMR12133 Herbivores, 63122 St-Genès-Champanelle, France, [2]Institut de l'Elevage, 148, rue de Bercy, 75595 Paris Cedex12, France, [3]UCA, Laboratoire de Mathématiques Blaise Pascal, UMR6620-CNRS, 63178 Aubière Cedex, France; lucille.rey-cadilhac@inrae.fr

While over the past 30 years many studies have focused on the effects of farming practices on a few milk compounds, few have considered combinations of farming practices on all compounds characterizing the intrinsic quality of milk. However, these practices could interact with each other and have antagonistic or synergistic effects on different compounds. Therefore, this study aimed to investigate the effect of combinations of farming practices on overall intrinsic quality of milk using data obtained from private farms. Ninety-nine dairy farms were visited to collect a sample of bulk tank milk and to conduct a survey on the farming practices applied to produce this milk (herd characteristics, feeding management, housing conditions, milking and milk storage conditions). The milk sample was analysed according to different compounds involved in intrinsic quality. A multi-criteria assessment was then conducted: it allows to implement about 30 indicators from the analyses carried out on milk, which, when aggregated, give scores (from 0 to 10) of the 4 dimensions (sensory, technological, nutritional and health) of the intrinsic quality and of the overall intrinsic quality. This assessment is implemented for 2 targeted products: raw milk cheese and semi-skimmed ultra-high-temperature milk. These indicator values, dimension and overall intrinsic quality scores were then predicted from combinations of farming practices using the regression tree method. Regression trees allowed to identify the combinations of practices that lead to the best indicator values, dimension and overall intrinsic quality scores, and to prioritize the importance of these practices. By comparing the trees, it was also possible to determine synergistic and antagonistic effects of farming practices among the different indicators, dimensions and targeted products. This method could be adapted to other topics such as the 'One quality' by considering both extrinsic aspects of quality in order to determine the trade-offs to be made among different quality aspects.

Imagining futures of husbandry farming: new business models for the sustainable transition?

B. Smulders[1], R. Valkenburg[1] and M. Anastasi[2]
[1]Eindhoven University of Technology, Den Dolech 1, 5612 AE Eindhoven, the Netherlands, [2]Cyprus Research & Innovation Center Ltd, 72, 28th Octovriou Avenue, 2414 Nicosia, Cyprus; b.smulders@tue.nl

Husbandry farming is under pressure and faces challenges, including discussions on emissions, intensive vs extensive agriculture, energy use, and health-related issues. Society at large is pressing increasingly towards sustainability, but what are potential sustainable future perspectives for husbandry farming? And what are potential business models? In this study, we support the thinking of opportunities by reimagining the future to think beyond current problems. The future is not necessarily an extrapolation of the past and present, but rather a *phenomenological* quality that allows proposing a *frame* that is discontinuous with what preceded. Future scenarios were crafted through several steps. First, a broad range of experts was interviewed, using a selection of general trends from horizon scanning reports as input. These trends were thematically coded and organized into seven dimensions – technology, users and markets, societal and environmental effects, industry and production, infrastructure, policy and regulations, and socio-cultural traditions. In the analysis of the interviews, 3-6 drivers for change per dimension were identified as signified by fault lines or missing elements from the past, discontinuities, or desires for change. Future scenarios were then constructed by creatively synthesizing the input across dimensions. This effort resulted in six images and vivid, but intentionally vague narratives that describe alternative environments for husbandry farming. The scenarios of alternative environments form the basis for reimagining value propositions. The *what-if reality* implied by the alternative value proposition allows the reconstruction of farming system elements by reinterpreting the activities, actors, artefacts, and their relations. In an interactive session, the six future scenarios will be presented during which people will be challenged to rethink their own practices, resulting in opportunities for change toward more sustainable and animal-friendly farming systems with business potential. The work presented is supported by the European Union's Horizon 2020 program under grant agreement No. 101000216 – Project Code Refarm.

Current prediction and authentication tools used and needed in EU livestock product chains

S. McLaughlin[1], F. Albert[2], F. Bedoin[2], C. Couzy[2], I. Legrand[2], A. Nicolazo De Barmon[2], C. Laithier[2], C. Manuelian[3], F. Klevenhusen[4], S. De Smet[5], E. Sturaro[6] and N. Scollan[1]
[1]Queen's University Belfast, School of Biological Sciences, Institute for Global Food Security, 19 Chlorine Gardens, Stranmillis, BT9 5DL, Belfast, United Kingdom, [2]IDELE, 149, rue de Bercy, 75595 Paris cedex 12, France, [3]University of Barcelona, Plaça Cívica, 08193 Bellaterra, Barcelona, Spain, [4]BfR, Max-Dohrn-Straße 8-10, 10589 Berlin, Germany, [5]Ghent University, Coupure Links 653, 9000 Gent, Belgium, [6]University of Padova, DAFNAE, Viale dell'Università 16, 35020 Legnaro, Italy; s.mclaughlin@qub.ac.uk

Introduction: There is a need to develop innovative and cost-effective analytical tools to enable: (1) the authentication of husbandry systems with different levels of intensification and of animal breeds and strains; and (2) the prediction of intrinsic quality traits of milk and dairy products, beef, and chicken meat. Aim: To obtain an overview of current prediction and authentication tools used and needed in European livestock product chains. Methods: Quantitative and qualitative survey administered to laboratories, processors, advisors and retailers in France, Germany, Italy, and the UK. 93 responses with 42 completed reports for all survey models. Results: Pathogen and microbiological quality analyses are widely used by the industry. Authentication methods were used most commonly to assess specific processes (e.g. meat cuts, ageing of meat, heat treatment used for milk). 56% of the sample are currently using NIR and/or MIR devices. There is great interest to develop tools and methods that rapidly assess quality using pathogens, antibiotic resistance, antibiotic residues, and shelf life as the parameters of interest. Developed tools must have the correct software and calibration to enable laboratories and companies to use such devices. The cost of equipment in terms of accessibility to end user should be considered. Conclusions: There is a growing need for novel, rapid methods to assess and authenticate product quality.

Session 86 | Theatre 7

Milk quality assessment for intensive and extensive goat farming of the Skopelos breed in Greece

Z. Basdagianni[1], I. Stavropoulos[1], G. Manessis[1], C.G. Biliaderis[2] and I. Bossis[1]
[1]Aristotle University of Thessaloniki, Department of Animal Production, School of Agriculture, Thessaloniki, 54124, Greece, [2]Aristotle University of Thessaloniki, Department of Food Science and Technology, School of Agriculture, Thessaloniki, 54124, Greece; basdagianni@agro.auth.gr

The objective of the study was to evaluate the quality of goat milk produced in an intensive (I) and an extensive (E) farming system throughout the lactation period in the Greek Skopelos dairy goats. 938 individual milk samples from 235 goats (105 and 135, for the intensive and the extensive system respectively) were collected at 4 different lactation stages starting immediately after weaning. Milk composition (fat, protein, lactose, and total solids content), physicochemical characteristics (pH, electrical conductivity, brix value, and refractive index), fatty acid composition, somatic cells count (SCC) and total bacterial count (TBC) were measured. A mixed linear model was used to determine the effects ($P \leq 0.01$) of farming system, and lactation stage on the milk quality assessment parameters. The farming system had no significant effect on the chemical composition and physicochemical properties of milk. Farming system however, significantly affected ($P \leq 0.01$) SCC (509.02 vs 841.29 cell/ml × 1000, for the I and E, respectively) and TBC (21.49 vs E:326.45 cfu × 1000 for the I and E, respectively). In addition, farming system affected certain parameters of FA composition. Total unsaturated FAs were higher ($P \leq 0.01$) in the E system (27.39%) compared to the I system (21.99%). More specifically, polyunsaturated fatty acids were 3.58 and 4.24% and monounsaturated FAs were 18.41 and 23.15% for the I and the E systems, respectively. Differences in SCC and TBC between the two farming systems are probably due to milking practices and overall hygiene status. In contrast, the superior lipid profile in the extensive farming system could probably be attributed to grazing for the majority of the lactation period. This work was funded by EU (HORIZON H2020) under the 'Code Re-farm: Consumer-driven demands to reframe farming systems' (Grant No: 101000216).

Session 86 | Theatre 8

Rheological evaluation of rennet-induced curdling of goat milks from different farming systems

K. Kotsiou[1], M. Andreadis[1], A. Lazaridou[1], C.G. Biliaderis[1], Z. Basdagianni[2], I. Bossis[2] and T. Moschakis[1]
[1]Aristotle University of Thessaloniki, Department of Food Science and Technology, Thessaloniki, 541 24, Greece, [2]Aristotle University of Thessaloniki, Department of Animal Production, School of Agriculture, Thessaloniki, 541 24, Greece; basdagianni@agro.auth.gr

The present study focuses on the rheological assessment of rennet-induced coagulation of milk from dairy goats of the Greek Skopelos breed. The goats were raised by either an extensive or an intensive farming system at two separate farms in Greece. The milk samples were collected after weaning. Dynamic rheological measurements were used to assess the kinetics of milk coagulation at 35 °C following the addition of rennet as well as the mechanical properties of the formed coagulums (cheese curds). The rennet coagulation time (defined as the time of the abrupt increase of storage modulus, G') of the milk samples from the extensive goat farming system was shorter than that for samples of the intensive farming system, without affecting the curd firming rate, determined by the increment of elasticity (IE = (dlogG'/dt)max), a parameter used as a measure of the gelation rate. Curds obtained from goat milks raised by the extensive farming system exhibited higher G'35 °C and complex viscosity (η*) values, compared to those of the intensive system; i.e. stronger gel network structures were noted for the extensive farming system samples. The observed differences in the rheological behaviour may be ascribed mainly to differences in the compositional and/or structural characteristics of the protein components (e.g. casein fractions) content between the two groups of samples. The rheological findings make an interesting contribution to the field of animal science and cheesemaking, by demonstrating how the goat feeding system may affect the mechanical properties of rennet-coagulated cheese products. This work was funded by EU (HORIZON H2020) under the 'Code Re-farm: Consumer-driven demands to reframe farming systems' (Grant No: 101000216).

Session 86 Theatre 9

Technological tools for authentication assessment of animal products

C.L. Manuelian
Universitat Autònoma de Barcelona (UAB), G2R, Dept. Ciència Animal i dels Aliments, Facultat de Veterinària, 08193 Bellaterra (Barcelona), Spain; carmen.manuelian@uab.cat

Consumers are demanding more information on the products' origin (e.g. country, mountain areas), method of production (e.g. organic, free-range) or health benefits (e.g. low fat, rich in Calcium). However, authentication methods are needed to protect consumers' interests as labelling claims are link with an increase in price. Traditional methods are expensive, time consuming, need skilled personal, and require toxic solvents and reagents. Therefore, non-invasive, cost-effective, rapid, easy-to-use and environmentally friendly methods for authentication are needed. Infrared (IR) and energy-dispersive X-ray fluorescence (EDXRF) spectroscopy accomplishes all these criteria. The IR produces a 'fingerprint' involving C-H, O-H and C-H chemical bonds. Benchtop devices interest is moving from gross to detailed composition and authentication. In general, individual or groups of FA present at medium-to-high concentration are more easily predicted, as well as using ground samples. Ability to predict technological, sensorial quality traits, and minerals is still limited. The ability of IR to predict minerals has been link to their bond with organic molecules. Economic portable miniaturized IR devices have recently arrived to the market and have been tested in slaughterhouses and processed dairy products showing promising results. The main advantages are their low cost making them affordable for small-sized producers or processors. The EDXRF technique is based on an X-ray impulse which passes through the sample and to which sample minerals emit a fluorescent radiation in response. The energy of this emitted response is indicative of the mineral, and the intensity depends on the amount of each mineral in the matrix. It does not need a prior sample digestion and simultaneously quantify minerals and trace element. This technique application in animal-derived food is quite new, and some studies have been conducted in milk powder, fluid milk, and cheese with promising results and suggesting water interference in fluid milk. Researcher funded with a María Zambrano grant of the Spanish Ministry of Universities and the European Union-Next Generation EU.

Session 86 Theatre 10

Potential of milk infrared spectroscopy to discriminate farm characteristics: the INTAQT project

M.A. Ramirez Mauricio, D. Giannuzzi, L. Gallo, M. Berton, A. Cecchinato and E. Sturaro
University of Padova, DAFNAE, Viale dell'Università 16, 35020, Legnaro, Padova, Italy; marcoaurelio.ramirezmauricio@unipd.it

The Fourier-transform infrared spectroscopy (FTIR) is one of the most developed and implemented tool for the analysis of milk chemical compounds. Besides this, FTIR can also be used to determine the fingerprint of milk for authentication purposes to certify the area of origin or the farming system in which the milk is produced. The present study, carried out within the INTAQT EU project, aimed at assessing the effectiveness of FTIR applied to bulk milk samples in discriminating dairy herds of Parmigiano Reggiano Consortium (PRC) for their structural and management characteristics. Dairy farm information included altitude zone (AZ), herd size (HS), housing type (HT), dairy cows genetic type (GT), use of total mixed rations (MR), and proportion of concentrate inclusion in the cow diets (CONC). This database was merged with milk data obtained from the official milk recording system along with FTIR spectral data of bulk milk, stored by the Breeders Association of Emilia Romagna Region lab (ARAER, Reggio Emilia, Italy) from January to August 2022. Overall data set comprised 4,610 bulk milk FTIR spectra from 940 farms, with a mean of 4.9 (±1.1) observations per farm respectively. Each spectrum contained absorbance values at 1,060 different wavenumbers (5,000 to 930 × cm^{-1}). Quality control of spectral data involved centering and scaling, removing spectral samples using a Mahalanobis distance greater than 3 SD, and removing the water region. A Partial Least Squares Discriminant Analysis (PLS-DA) and Linear Discriminant Analysis (LDA) were fitted to estimate the probability of each observation (i.e. farm) belonging to a specific group (e.g. AZ, HS, etc.). The 60% of the data was used as training set and 40% as testing set. Finally, a prediction of the testing set was performed with both methods. PLS-DA gained an accuracy of 85, 74, 86, 82 and 95% for classifying the proportion of CONC, AZ, HT, MR and GT respectively. In the case of LDA the accuracy values obtained were 63, 73, 86, 82 and 97% respectively. These results suggest the potential of FTIR to determine the fingerprint of milk for authentication purposes in the PRC area.

Session 86 Theatre 11

A longitudinal cohort study of health and welfare status in extensively and intensively reared goats

V. Korelidou, A.I. Kalogianni and A.I. Gelasakis
Agricultural University of Athens, Department of Animal Science, Iera Odos 75, 11855 Athens, Greece; vkorelidou@aua.gr

Goats are challenged by various health and welfare issues depending on the production system and management. The objective of this study was to prospectively evaluate health and welfare status of dairy goats reared under intensive and extensive farming systems. A population of 133 and 105 purebred adult Skopelos goats, were randomly selected from a typical extensive (Farm A) and intensive farm (Farm B), respectively and were monitored during one milking period. For each individual goat, a detailed physical examination of the head, body, limbs, udder, and lymph nodes was performed by the same veterinarian every 50 days. Various health and welfare indicators were recorded using a modified protocol of AWIN (Animal Welfare Indicators). Data were analysed to estimate period prevalence, incidence rate (new cases per 1000 goat-months), and cumulative incidence. Higher prevalence, incidence rate, and cumulative incidence were observed in goats of Farm B compared to those of Farm A, with the respective values for specific conditions being as follows: (1) lameness 3.8%, 6.0, and 2.9% over 1.5%, 3.2, and 1.5%; (2) overgrown hooves 29.5%, 68.5, and 26.7% over 3.8%, 8.1, and 3.8%; (3) teeth problems 10.5%, 20.6, and 9.6% over 2.3%, 4.8, and 2.3%; and (4) swollen submandibular lymph nodes 21.9%, 56.5, and 17.2% over 5.3%, 10.1, and 3.1%. On the contrary, higher prevalence, incidence rate, and cumulative incidence were found in Farm A compared to Farm B for: (1) anaemia 48.9%, 210.9, and 46.0%, over 15.2%, 13.3, and 4.3%; (2) faecal soiling 8.3%, 16.7, and 7.6% over 3.8%, 4.0, and 1.9%; (3) poor hair quality 60.9%, 64.3, and 25.7% over 49.5%, 60.5, and 26.4%; (4) swollen prefemoral lymph nodes 28.6%, 53.5, and 15.2% over 1.9%, 5.8, and 1.9%; and (5) swollen prescapular lymph nodes 25.6%, 47.2, and 13.2% over 15.2%, 29.6, and 9.2%, respectively. The results showed that health and welfare issues occur with different frequencies in intensively and extensively reared goats. Hence, it is important to further study the underlying causative agents and relevant risk factors on a case-specific basis. This project has received funding from the European Union's Horizon 2020 research and innovation programme under grant agreement No 101000216.

Session 86 Theatre 12

New tool for accurate analysis of gas concentrations in barns

O. Bonilla-Manrique[1], A. Moreno-Oyervides[1], H. Moser[2], J.P. Waclawek[2], B. Lendl[2] and P. Martín-Mateos[1]
[1]Universidad Carlos III de Madrid, Av. Universidad 30, 28911 Leganés, Spain, [2]Technische Universität Wien, Getreidemarkt 9, 1060 Vienna, Austria; obonilla@ing.uc3m.es

Controlling the concentration of gases inside livestock buildings is essential to ensure both animal health and product quality. Compounds such as carbon dioxide and ammonia are of particular relevance, as they are produced in significant quantities by animals. The accumulation of these gases in the ambient air, which is of particular concern in the winter months when ventilation is kept to a minimum, causes significant stress on the respiratory system of the animals and has been demonstrated to affect the quality of products. There are other compounds that are also of great interest, such as, especially for ruminants, methane. Besides affecting the respiratory tract, it also provides an indication of digestibility and feed intake, and several studies indicate that methane concentration could be used to improve animal productivity and product quality. However, accurate, reliable, and real-time determination of the above-mentioned gas concentrations is currently a challenge. Several portable or fixed-location gas analyser technologies can be used today to estimate the concentration of gases inside barns. However, measuring at a single, or a few, sampling points has significant weaknesses, as it is easily affected by changes in airflow patterns and provides results that may not be representative of actual values. The main objective of the work presented here is the experimental validation of a new gas measurement tool, specifically designed for use in livestock barns. An optical system, based on molecular dispersion spectroscopy, has been designed so that an array of laser beams sweeps the barn at a height similar to that of the animals, thus providing a highly reliable and representative measurement of gases concentration to which the animals are actually exposed. This tool will make available to farmers and scientists unprecedented data that could be of great use in the challenging task of optimizing the balance of factors that ensures the maximum return from a livestock farm. The work presented was supported by the European Union's Horizon 2020 programme under grant agreement No. 101000216 – Project Code Refarm.

Session 86 Poster 13

Implementation of husbandry practices improving quality and sustainability: a living lab approach

E. Sturaro, C. Berri, D. Berry, R. Eppenstein, C. Laithier, A. Cartoni Mancinelli, B. Martin and F. Leiber
INTAQT Consortium, Rte de Theix, 63122 Saint-Genès-Champanelle, France; enrico.sturaro@unipd.it

The living lab approach to innovation is receiving increasing attention also in the agricultural sector in view of the current environmental, economic, and social challenges. This contribution presents some preliminary results of INTAQT project (EU Horizon 2020), which aims to perform an in- depth multi-criteria assessment of the relationships between animal husbandry and qualities of products. In specific, this research aims to identify and implement on-farm changes in the production processes (e.g. feeding regimes, outdoor access, herd management), which are expected to improve intrinsic quality traits of the products and/or sustainability traits of the farms. A participatory approach was used to establish farmer field-groups (living labs) representative of the different geographic regions and of the main production systems involved in the project. Each farmer field group involves from 5 to 8 farms. The groups are established considering different husbandry systems according to a gradient of intensification (extensive vs intensive systems): 3 groups for dairy farms (Ireland, northern Italy and France); two groups for beef farms (Switzerland and northern Italy); two groups for poultry (France and Italy). The methodological approach is based on 5 steps: (1) tarting analytical phase: a critical analysis of trade-offs / synergies between sustainability and quality traits for each farm-field group; (2) decision phase: development of practices to improve the identified synergies / mitigate trade-offs; (3) implementation phase: implementation of practices for at least one year. During this time, 2-3 meetings of the whole farmers group on farms allow farmers discussions about their experiences, successes and drawbacks; (4) concluding analytical phase: the aim is to analyse the effects of the implementation of the practices during a last meeting in the farmer's groups and presentation of the analysis results; (5) scientific data analysis and interpretation. The first results of this approach will be presented and discussed. The ambition is to establish a network of living labs usable as pilot and demonstration enterprises regarding practice improvements for better food quality and sustainability.

Session 86 Poster 14

Muscle proteomics towards molecular understanding of colour and water-holding biochemistry in meat

S. Yigitturk
Wageningen University & Research, Department of Agrotechnology and Food Sciences, Food Quality and Design Group, Bornse Meadows 9, 6708 WG Wageningen, P.O. Box 17, 6700 AA Wageningen, the Netherlands; seren.yigitturk@wur.nl

Poultry meat is projected to constitute 47% of the protein consumed globally from meat sources and is the primary driver of growth in meat production. The physiological and metabolic functions of the animals are influenced by many factors such as extensive husbandry practices, with consequent impact on the animal welfare and intrinsic quality of the broiler meat. However, to what extent the extensiveness of animal husbandry affects intrinsic meat quality and animal welfare still needs to be confirmed. This study, which is part of the EU-funded H2020-FNR-05 *mEATquality* project, aims to elucidate colour and water-holding biochemistry in chicken meat and clarify the function of proteomics biomarkers in the complex molecular pathways of muscle-to-meat conversion. We hypothesize that animal welfare linked to extensification factors such as genetics, diet, space availability and environmental enrichment changes the proteome of pectoralis major, leading to differences in broiler meat quality. In this project, variations in levels of structural proteins, mitochondrial enzymes, oxidative enzymes, chaperones and heat shock proteins in response to stressors would be expected to be associated with variations in colour and water-holding capacity in meat. We propose that animal welfare would change the dynamics of the aforementioned protein interactions, leading to differences in light scattering and water-holding capacity on broiler breast muscle. Final description to follow later.

Session 86 — Poster 15

Exploring YOLO deep learning model for goat detection in animal welfare and behavioural monitoring

A. Temenos, A. Voulodimos, D. Kalogeras and A. Doulamis
Institute of Communications and Computer Systems, National Technical University of Athens, Patision 42, 157 80 Athens, Greece; tasostemenos@mail.ntua.gr

In this paper, a fully automated monitoring system that detects welfare and behavioural traits in intensive farming systems is introduced. Animal health and welfare is strongly correlated with milk yield and composition, affecting dairy products' quality and safety, livestock farming trends and the dairy supply chain. Traditionally, animal monitoring is done through regular visits by veterinarians, who identify health and welfare issues based on the physical examination of the animals. However, the regular monitoring of large herds can be time-consuming and expensive for farmers. Automated procedures can detect early-stage behavioural changes and health issues (e.g. lameness) on a cost-efficient basis, without the continuous demand of an expert. You Only Look Once (YOLO) is a deep neural network capable of detecting objects in video scenes, utilizing results from a single convolutional network that predicts multiple bounding boxes and class probabilities. Its strengths are speed, accuracy, and well-generalization in real-world applications. YOLOv5 is a variation of the initial architecture that: (1) reduces number of layers; (2) reduces number of parameters; and (3) increases speed without any impact of performance. We train a YOLOv5 model to detect goats walking inside the barn, with annotated data from videos. The final dataset was split into 510 images for training, 25 for validation, and 48 for testing, after applying image processing algorithms for data augmentation (e.g. affine transformation). The proposed model achieved a 99.5% mean Average Precision. The YOLOv5 model presented in this work can detect goats inside the barn. Such detectors can be utilized for the automatic monitoring of behaviour (e.g. drinking and eating activity), and the early-stage detection of alterations in health and welfare status (e.g. lameness notification). In future work, we intend to apply tacking algorithms reading RFID tags, produce speed diagrams, and classify videos from healthy and non-healthy groups of animals. This work was funded by Code Re-farm Horizon 2020 project: 'Consumer-driven demands to reframe farming systems' (Grant No: 101000216).

Session 86 — Poster 16

The association between body condition score and the udder skin surface temperature in goats

V. Korelidou, A.I. Kalogianni and A.I. Gelasakis
Agricultural University of Athens, Department of Animal Science, Iera Odos 75, 11855 Athens, Greece; vkorelidou@aua.gr

The objective of the study was to assess the association between the body condition score (BCS) and the udder skin surface temperature (USST) in goats. For this reason, a prospective (S1) and a cross-sectional (S2) study were conducted. For the S1, 104 Skopelos goats from an intensive farm (A) were enrolled and recorded across one milking period. For the S2, 132 Skopelos goats from an extensive farm (B) and the 104 goats from farm A were studied at mid-lactation. Udder skin surface images were captured using a thermal camera (FLIR E8-XT, FLIR Systems Inc.), and the appropriate software was used to estimate the maximum and mean temperatures of teats (T1, T2), udder cleft (T3, T4), and udder halves (T5, T6), respectively. In addition, for each animal, the BCS was assessed using a 5-degree scale. Data were analysed using SPSS v26, and two linear regression models were built, with age, udder fibrosis occurrence (S1 and S2), lactation stage (S1), and farm (S2) being forced into the models as fixed effects, and the animal as random effect (S1), to estimate the effect of BCS on the USST (T1-T6). In S1, the range of means of T1, T2, T3, T4, T5, T6, and BCS were 32.9-37.3, 31.9-36.9, 36.3-38.0, 35.1-37.4, 37.1-38.4, 35.2-37.5, and 2.9-3.2 °C, respectively, while, a one-degree decrease in BCS was associated with 0.7, 0.9, and 0.6 °C increase in T3, T4, T6 ($P<0.001$), respectively, and 0.3 °C in T5 ($P<0.01$). In S2, mean values (±SD) of T1, T2, T3, T4, T5, T6, and BCS were 34.2±0.99, 33.0±0.95, 36.8±0.88, 35.2±1.28, 37.6±0.68, 35.6±0.87, and 3.2±0.33 °C, respectively, for farm A, and 35.1±1.10, 34.2±1.16, 37.3±1.12, 35.8±1.34, 37.9±0.86, 36.2±0.98, and 2.9±0.22 °C, respectively, for farm B. In S2, a one-degree decrease in BCS was associated with 1.3 °C increase in T4 ($P<0.001$), 0.8, 0.5, and 0.7 °C in T3, T5, and T6 ($P<0.01$), respectively. Metabolic rhythms, and udder fat deposition are likely to be associated with the USST. However, further research to validate these findings and elucidate underlying mechanisms is warranted. This project has received funding from the European Union's Horizon 2020 research and innovation programme under grant agreement No 101000216.

Session 87 Theatre 1

Leveraging the measuring and managing of livestock farms on the journey towards Net Zero

J. Gilliland
Queen's University Belfast, School of Biological Sciences, BT9 5DL Belfast, United Kingdom; john.gilliland@brookhall.org

John Gilliland was recently appointed as a Professor of Practice at Queen's University Belfast, for his work on climate-smart farming. John started his journey on climate-smart farming on his farm back in 1988. Since then, he has participated in many different roles which have given him a breadth of knowledge which he has put to good use in accelerating livestock farmers' journey to Net Zero. Between his previous roles as a practitioner; farm leader; policy advocate; regulator; researcher; and now chair of the EIP-Agri Operational Group, ARC Zero, his hands-on experience has delivered unprecedented engagement of the farming and the agriculture policy communities. John will share some of this journey which has secured the forensic measuring of on-farm emissions and carbon stocks, resulting in the empowerment of actual farmers with their net-carbon position, allowing them to make better quality and more informed decisions, thus accelerating their journey to Net Zero. In addition, this project has delivered dynamic peer-to-peer learning and secured public policy buy-in across Northern Ireland, not just on delivering Net Zero, but also in delivering other public goods, such as food production and improved water quality.

Session 87 Theatre 2

The role of carbon sequestration in organic dehesas ruminant farms

R. Casado[1], M. Escribano[1], P. Gaspar[2] and A. Horrillo[1]
[1]Faculty of Veterinary Science, Department of Animal Production and Food Science, Avda. de la Universidad, s/n, 10003 Caceres, Spain, [2]Faculty of Agriculture, Department of Animal Production and Food Science, Avda. Adolfo Suarez, s/n, 06007 Badajoz, Spain; andreshg@unex.es

Agri-food production must not only guarantee food security for the growing world population, but also ensure the conservation of ecosystems by avoiding overexploitation of natural resources. Therefore, one of the main objectives for the agricultural sector is to adapt its production systems to new challenges, including its contribution to the fight against climate change and the reduction of greenhouse gas (GHG) emissions. The main objective of the regional project 'MitigaDehex' (IB20070) is to analyse extensive livestock systems in Extremadura as sustainable and active production models against climate change. In this sense, one of the systems most in line with these principles are the organic ruminant farms in dehesas in the southwest of the Iberian Peninsula (Spain). The aim of this study was to calculate the balance of GHG emissions. For this purpose, life cycle assessment was used, including the calculation of the carbon footprint and carbon sequestration in two different time horizons, one at 20 years and the other at 100 years. The Carbon Footprint was performed following the IPCC 2019 guidelines and the proposed national GHG emissions inventories adapted to the characteristics of extensive livestock systems. For carbon sequestration, the livestock-manure-grassland system proposed by Petersen in 2013 was used. The functional unit used was kg of CO_2 equivalent per kilogram of live weight (FU) and also per hectare (ha). Organic meat sheep and beef cattle farms were analysed, which were differentiated according to their farming system. The carbon footprint of the sheep meat farms was 12.27 kg CO_2 eq/FU in the group 'sale lambs 18.5 to 20 kg' and 11.27 kg CO_2 eq/FU in the group 'sale lambs 20 to 23 kg'. In beef cattle, the carbon footprint was 14.92 kg CO_2 eq/ FU in calf farms and 7.04 kg CO_2 eq/ FU in yearling farms. Regarding carbon sequestration, the results show that organic farms in dehesa areas can function as true carbon sinks, since carbon sequestration through pasture and manure can offset a large part of the GHG emissions produced by livestock farming.

Session 87 Theatre 3

Estimating soil carbon sequestration gaps for ruminant systems across the globe

Y. Wang[1], I. Luotto[2], Y. Yigini[2], R. Vargas[2], D. Wisser[3], T.P. Robinson[3], U.M. Persson[4], C. Cederberg[4], R. Ripoll-Bosch[1], I.J.M. De Boer[1] and C.E. Van Middelaar[1]

[1]Wageningen University & Research, Animal Production Systems group, De Elst 1, 6700 AH Wageningen, the Netherlands, [2]Food and Agriculture Organization of the United Nations, Global Soil Partnership, Vialle de Terme di Caracalla, Rome, Italy, [3]Food and Agriculture Organization of the United Nations, Animal Production and Health Division, Vialle de Terme di Caracalla, Rome, Italy, [4]Chalmers University of Technology, Physical Resource Theory, Department of Space, Earth & Environment, Chalmersplatsen 4, 412 96 Göteborg, Sweden; corina.vanmiddelaar@wur.nl

Soil carbon sequestration is a potential way of greenhouse gas (GHG) mitigation in livestock systems, especially for ruminants, a major contributor to methane (CH_4) emissions. The sequestration, however, is time-limited, and there are intrinsic differences between long- and short-lived GHGs, which cannot be reflected by the current widely used GHG metrics. We use GHG metric parametrization to equate the climate impact of carbon sequestration with a continues flow of emissions, showing that one ton of carbon can offset 0.99 kg CH_4 or 35 kg nitrous oxide (N_2O) per year over 100 years. The conversion factor was applied to estimate the 'carbon sequestration gap', being defined as the difference between current and required carbon sequestration levels to offset emissions, for ruminant systems across the globe, using regional estimates of ruminant emissions and soil carbon sequestration values for grasslands. Results show that at global level, current sequestration levels are far less than the 135 gigatonnes of carbon required to offset annual emissions of CH_4 and N_2O from the ruminant sector. To effectively contribute to climate change mitigation, a combination of strategies will be required, including reduction in animal numbers and improved land use management. This study calls for critical thought regarding the approach and implications of valuing soil carbon sequestration as a means to offset the climate impact of GHG emissions by ruminant systems.

Session 87 Theatre 4

Agronomic and environmental impacts of sheep integration in cover crop management in Wallonia

N. Lorant, B. Huyghebaert and D. Stilmant

Walloon Agricultural Research Centre, Rue du Serpont 100, 6800 Libramont, Belgium; nicolas.lorant@cra.wallonie.be

Cover crops valorisation by sheep offer a new partnership between breeder and cash crop producer to support the development of the sheep sector in Wallonia, which covers only 16% of national sheep meat needs. During 2 years, in the 'SERVEAU' project (SPGE funding), 10 trials including three levels of grazing intensity (non-grazing, partial grazing and total grazing) were performed. Between 15 July and 10 September, a cover crop was sown. The grazing period extends from late October to mid-January. During the time spent on the plot, sheep destroyed 80 to 93% of the available biomass reaching an ADG of 70 g/d. To characterize the risks of nitrogen leaching, soil samples (0-30, 30-60, 60-90 cm) have been taken at different periods: cover crop sowing, after grazing, in the middle of winter and at the end of winter. According to the trials, spring crops were sown post-grazing (sugar beets, chicory, peas, beans, potatoes or corn) and their developments were followed. Soil analyses carried out at the implantation of cover crops show great variability in nitrogen residues. However, at each site, the amount of mineral nitrogen present in the soil has decreased thanks to cover crops. Soil analyses indicate a significant increase (+11 kg N/ha) in nitrogen content in the first 30 centimetres of soil following grazing, regardless of the intensity applied. This difference was no longer noticeable when the entire profile was considered (0-90 cm). Similarly, at the end of winter, there was no significant differences between the treatments. For the next spring crop, no significant differences were observed in either the emergences or the yields. Cover crops grazing could also lead to a reduction in cropping interventions for their destruction. Destruction by sheep allows the farmer to avoid a mechanical destruction, generating a gain of 384€/ha. Moreover, a breeding system combining rearing in sheepfolds and cover crop grazing makes it possible to reduce the environmental impact by 33% in kilograms of CO_2eq/kg of meat produced. This renewed interest in this practice offers several diversification opportunities within the farm but also at territorial scale.

Carbon footprint assessment of Korean native beef cattle and options to achieve net-zero emissions

R. Ibidhi[1], T. Kim[2], J. Byun[2], R. Bharanidharan[3], Y. Lee[4], S. Kang[2,3] and K. Kim[2,3]
[1]Agroécologie, INRAE, Institut Agro, Univ. Bourgogne, Univ. Bourgogne Franche-Comté, 17 rue Sully, Dijon Cédex, 21065, France, [2]Graduate School of International Agricultural Technology, Seoul National University, Pyeongchang, 25354, Korea, South, [3]Institutes of Green-Bio Science and Technology, Seoul National University, Pyeongchang, 25354, Korea, South, [4] National Institute of Animal Sciences, Rural Development Administration, Jeonju-si, 54875, Korea, South; ibidhi_ridha@hotmail.fr

In the context of global climate change, many countries are committed to achieve carbon neutrality by 2050. The carbon footprint (CF) indicator was proposed as an metric to assess greenhouse gas emissions (GHG) from beef production systems. The aims of the study were (1) to determine the CF of Korean native beef cattle (Hanwoo), and (2) to identify the major contributors to GHG emissions using life cycle assessment (LCA) method at farm-gate, expressed as a carbon dioxide equivalent (CO_2-eq) per kg live body weight (LBW). This study, based on data from a sample of 107 beef farms from nine Korean provinces. This analysis was used to identify options for mitigating GHG emissions using artificial neural networks (ANN). We generated 35 randomized scenarios of activity combinations at farm scale and evaluate associated GHG flux. The CF averaged to 8.2 CO_2-eq/kg of LBW and ranged from 3.9 to 13.8 CO_2-eq/kg of LBW. Results indicated that the largest source of GHG comes mostly from enteric fermentation (73%), followed by manure management (16%), manure and fertilizer land application (8%) and energy consumption (3%). Methane accounted for 86% of total emissions, originating from enteric fermentation and manure management. Nitrous oxide and CO_2 accounted for 11.6 and 2.8% of total GHG emissions, respectively. The modelled combination scenarios using ANN indicated that shortening the fattening period, limit the dry matter intake, coupling fattening with crop production and using organic fertilizer instead of synthetic fertilizer) seems to be the most effective strategy that can reduce the CF of Hanwoo meat by 56%. This study, suggest as well to consider the importance of soil carbon sequestration in developing GHG mitigation strategies.

Carbon Footprint of organic beef from dairy male calves

L. Mogensen[1], T. Kristensen[1], C. Kramer[2], A. Munk[3], P. Spleth[3] and M. Vestergaard[4]
[1]Aarhus University, Department of Agroecology, Blichers Alle 20, 8830, Denmark, [2]Center for Frilandsdyr, Randers, DK 8940, Denmark, [3]SEGES, Aarhus N, 8200, Denmark, [4]Aarhus University, Department of Animal and Veterinary Sciences, Blichers Alle 20, 8830, Denmark; lisbeth.mogensen@agro.au.dk

Organic beef production has some challenges compared to conventional production, among other things because organic male calves in Denmark typically are raised as steers, which gives a high feed consumption per kg live weight gain and thus a high carbon footprint per kg carcass. Instead, raising organic young bulls will meet the desire, from both consumers and the political system, for organic foods with a lower carbon footprint per kg carcass. The objective was to develop new strategies for rearing organic young bulls, which consider genetic type, feeding intensity, time of year of birth, and age at slaughter with a focus on achieving the lowest possible climate impact. Ten strategies to produce organic young bulls – either pure Holstein breed or crossbred bull calves – were set up, as a combination of the time of year the calves are born, slaughter age, and feeding intensity. The effect of different production strategies was estimated by modelling. The carbon footprint per kg carcass including contribution from soil carbon changes was 32% lower for raising organic young bulls compared with that of organic steers. A strategy with high feeding intensity and slaughter at 13 months had the lowest carbon footprint per kg carcass (CF) due to the lowest feed consumption per kg LW gain and use of a high proportion (35% of DM) of concentrate. A strategy with slaughter at 17 months and high feeding intensity increased CF around 6% due to an increased feed consumption and reduction in feed conversion due to higher age at slaughter. A strategy with low feeding intensity and a slaughter at 17 months gave the highest CF. Using crossbred bull calves compared to purebred dairy bulls gave a 10-15% lower CF due to higher dressing percentage, higher carcass weight and higher growth rates.

Session 87 Theatre 7

Model to calculate the impact of interventions on the carbon footprint of dairy

A. Esselink
Trouw Nutrition, Stationsstraat 77, 3811 MH Amersfoort, the Netherlands; allard.esselink@trouwnutrition.com

An LCA model for production of milk at farm level was developed with Pré consultants and an external expert. Goal is to quantify and compare the potential environmental impact of theoretical and actual milk production with different feeding and farming practices, such as implementation of nutritional programs, use of feed additives and manure management strategies for the respective geographical scope of the tool. The tool was developed conform ISO 14040:2006 and ISO 14044:2006 standards but also more specific guidelines, i.e. IDF (2022), PEF Guide, PEFCR of dairy products, PEFCR of animal feed. A herd dynamics model based on McLeod *et al.* is implemented to assure the number of animals born, calves for finishing and cull cows leaving the farm are in balance. Input parameters can be modified in such a way that the almost every farm situation across the globe can be described. The model can differentiate between the direct effect of an additive and the indirect effect of an intervention, like the reduction of age at first calving, which will reduce the total carbon footprint of milk. For instance, land use change, acidification, water use and eutrophication are calculated. In the application of the model, there are differences between direct effects of an additive that reduces enteric methane production and indirect effects of, for instance, a difference in milk production or animal numbers. In addition to the different global warming potential (GWP) figures, other environmental impact factors are also calculated. Using the tool, Trouw Nutrition can do an evaluation on the environmental effect of the different interventions in different situations. As an example, we took an average Dutch dairy farm, where in the model we applied the LifeStart concept. As input figures, age at first calving was reduced from 25.5 to 22.0 months, and the replacement rate from 30 to 25%. This intervention reduced the carbon footprint of the milk produced by 5%. The tool and the report written about the effect of the Trouw Nutrition solutions is externally reviewed to assure conformity with the mentioned ISO standards.

Session 87 Theatre 8

Tools to optimise the carbon footprint of milk production

D. Schwarz
FOSS Analytical A/S, Nils Foss Alle 1, 3400 Hilleroed, Denmark; das@foss.dk

It is estimated that about 14% of the total anthropogenic greenhouse gas (GHG) originates from global livestock production. Dairy farming is responsible for approximately 30% of these emissions. Dairy herd improvement (DHI) testing, meaning monthly collection and analysis of milk samples from individual cows, is broadly used for herd health, feeding, and management purposes around the globe. The objective of this study is to investigate the effect of DHI on GHG emission of dairy farms. Regular DHI results from Denmark (n=193,321) and Thuringia, Germany, (n=399,428) were available for data analysis. Mastitis is known to impair the performance of dairy cows and milk somatic cell count is used as a proxy for detection of (subclinical) mastitis. We found that the corrected (a generalized linear mixed models was applied) daily milk production of cows with 250,000 to 1,000,000 cells/ml (15% of test day results) was 3 kg and of those with >1,000,000 cells/ml (5% of test day results) was 6 kg below the production of cows with <250,000 cells/ml. These losses, in turn, translate into considerable GHG emissions, which were estimated to be 75 t of CO_2 and 78 t of CH_4 per day in Denmark alone assuming that the CO_2eq per kg milk is 1.28 kg. Such milk losses also lead to an increased GHG emission per kg of milk produced. The risk for having ketosis, a metabolic disorder in high-yielding dairy cows, can be estimated by determining milk beta-hydroxybutyrat and evident differences in milk yield between low (70% of test days) and high risk (30% of test days) cows were seen. Moreover, milk urea results provide highly valuable information on the protein content in feed and about 20% of test day results had elevated urea results indicating an oversupply of dietary protein. Milk fatty acid profiles can be used as another indicator to evaluate and optimise dairy cow feeding. In conclusion, our findings revealed that there is still potential to optimise productivity of cows and GHG emission per kg milk produced with respect to mastitis, ketosis and feeding. In this context, DHI testing programmes represent a practical and inexpensive tool for dairy farmers to manage and optimise herd health and feeding and thus productivity of their cows. This, in turn, is an essential component to achieve low GHG emission per kg of milk produced.

Session 87 Theatre 9

Evaluation of feed additives under varying diets of dairy cows on performances and enteric emissions

M.E. Uddin[1], M.R.A. Redoy[1], S. Ahmed[1], M. Bulness[1], D.H. Kleinschmit[2], J. Lefler[3] and C. Marotz[3]
[1]South Dakota State University, SD, 57007, USA, [2]Zinpro Corporation, MN, 55344, USA, [3]Native Microbials Inc., CA, 92101, USA; mdelias.uddin@sdstate.edu

This abstract summarizes the effects of supplementing two feed additives, isoacids [IsoFerm (ISO); Zinpro Inc.] and a microbial product [Galaxis Frontier (GF); Native Microbials Inc.] on performances and enteric emissions. For ISO study, 64 Holsteins were used in a randomized complete block design trial (10-wk). In ISO study, cows were assigned to 1 of 4 diets (n=16; 16.5% CP, 1.29 Mcal/kg DM, and 28% NDF) with a 2×2 factorial arrangement, 2 forage level (FL) comprising 18 (LF) and 23% forage NDF (HF), without or with ISO supplementation (40 g/d). For GF study, 56 Holsteins were assigned to 1 of 2 diets in a randomized complete block design from -22 to 100 days in milk (DIM). A control diet (CON; n=29) and GF diet (CON + 5 g/d of GF; n=29), consisting of a similar basal close-up (1.29 Mcal/kg DM and 10.8% CP) and lactation diets (1.67 Mcal/kg DM and 15.3% CP) were used. Enteric methane (CH_4) was measured using GreenFeed (C-Lock Inc.). Supplementation of ISO did not affect dry matter intake (DMI) but increased milk yield (MY, 35.0 vs 30.8 kg/d; P<0.01) and fat-and-protein corrected milk (FPCM; 36.8 vs 33.4 kg/d; P=0.01) only in the HF diet. Similarly, ISO increased feed efficiency (FPCM/DMI) in the HF diet but decreased in the LF diet (P=0.03). Importantly, ISO decreased CH_4 intensity by 9% (9.7 vs 10.7 g/kg MY; P=0.01) and tended to reduce CH_4 production (332 vs 361 g/d; P=0.07) regardless of FL. Supplementation of GF tended to increase MY (39.7 vs 37.1 kg/d, P=0.08) and feed efficiency (+0.11, MY/DMI) during early (0 to 30 DIM) and mid-lactations (31 to 100 DIM), respectively. However, GF reduced plasma glucose and paraoxonase but increased BHB (0.49 vs 0.37 mmol/l; P=0.03), with a tendency to surge the abundance of lactate-utilizing bacteria (*Megasphaera elsdenii*) and drop the cellulolytic bacteria (*Fibrobacter succinogens*; P=0.07). Overall, ISO and GF enhanced the performances and efficiency of cows, while GF altered rumen bacterial abundance and ISO decreased enteric CH_4 emissions. Rumen microbiota modulation by the ISO and the potential of GF to reduce CH_4 under varying diets need to be tested.

Session 87 Theatre 10

Supplementation with a calcium peroxide additive mitigates enteric methane emissions in beef cattle

E. Roskam[1,2], D.A. Kenny[2,3], V. O'Flaherty[1], A.K. Kelly[3] and S.M. Waters[1,2]
[1]University of Galway, School of Natural Sciences and Ryan Institute, Galway, H91TK33, Ireland, [2]Teagasc Grange, Animal and Bioscience Research Department, Dunsany, Co. Meath, Ireland, [3]University College Dublin, School of Agricultural and Food Science, Dublin, D04 V1W8, Ireland; emily.roskam@teagasc.ie

In Ireland, the agricultural sector is responsible for 38% of GHGs, 60% of which are methane (CH_4). Globally, GHGs need to be reduced rapidly to adhere to legally binding targets. While research into dietary supplementation with anti-methanogenic feed additives has proliferated, there are still limited products showing consistent CH_4 reductions, many of which cannot be pelleted due to sensitivity to temperature and pressure. Calcium peroxide (CaO_2) is a widely available, non-toxic, inexpensive compound, which has been shown to have anti-methanogenic effectiveness based on modulating the oxidation-reduction potential of the rumen. In this study, the effect of twice daily supplementation of CaO_2 to beef bulls on CH_4 and H_2 emissions and animal productivity was assessed. Seventy two beef bulls offered a 60:40 forage:concentrate diet were allocated silage each morning (09:00 h) and concentrates at 08:00 and 15:00 h. Following a 7 d covariate period, animals were offered 1 of 4 treatments; Control (unsupplemented), Low (1.35% CaO_2), High (2.25% CaO_2) and Pellet (High, pelleted) (n=18). All concentrates were formulated in a coarse ration, except for Pellet. CH_4 and H_2 were measured using GreenFeed systems. Data were analysed using ProcMixed, SAS 9.4. No effect on animal ADG or FCR was observed (P>0.1), however High reduced DMI compared to Control (P<0.05). All treatments reduced CH_4 emissions when expressed per day, per kg ADG and per kg DMI (P<0.0001). Low reduced CH_4 parameters by 17-21%. High and Pellet reduced CH_4 g/d, g/kg ADG and g/kg DMI by 28 and 27, 29 and 28, and 20 and 27% (P<0.0001) respectively. A 36, 35 and 32% reduction in H_2 (P<0.0001) was observed for Low, High and Pellet. CaO_2 can successfully endure the pelleting process and reduce CH_4 and H_2 emissions with no negative effects on animal production, making it a viable CH_4 mitigation option for pasture based systems with concentrate supplementation.

Methane emission from rosé veal calves feed a corn cob silage-based or grass silage-based ration

A.L.F. Hellwing and M. Vestergaard
Aarhus University, Campus Viborg, AU Foulum, Department of Animal and Veterinary Sciences, Blichers Allé 20, 8830 Tjele, Denmark; mogens.vestergaard@anivet.au.dk

The rosé veal calf production in Denmark is characterized by high daily growth rates and energy rich diets. Farmers feeding total mixed rations normally use whole-crop or corn-cob silage. However, use of grass silage-based rations has never been used in specialized rosé veal calf operations. Corn cob silage and grass silage differ considerable in starch and NDF, and this might affect the methane emission from rations based on either corn cob or grass silage. The aim of the experiment was to investigate the methane emission from rosé veal calves fed energy rich rations based on either grass silage and grains or corn cob silage and grains. Thirty-two Holstein calves were selected from a production experiment in which the calves were fed these rations from 3½ month of age. The grass silage-based ration (GB) included rolled barley, 1. cut grass silage (25-35% of dry matter) and untoasted milled fava beans, and the corn cob silage-based ration (CB) included corn-cob silage (40% of dry matter), rolled barley and rapeseed meal. Crude protein, NDF, starch, and NE were similar for the two rations. The enteric methane was measured when the calves were 8-month old by means of indirect calorimetry over three days. The data was analysed with PROC MIXED in SAS with ration as a fixed effect. The results showed that the dry matter intake did not differ between the two rations (P=0.88) and was on average 7.1±1.1 kg (mean ± std). The average weight was 337±25 kg with no difference between the two treatments. The methane emission tended to differ (P=0.07) with 128 g/day for GB and 112 g/day for CB. The methane emission per kg dry matter was 18.1 and 16.1 g/day for GB and CB, respectively (P=0.14). The methane emission per kg metabolic body size tended to differ (P=0.07) and was 1.63 and 1.42 g/kg $BW^{0.75}$ for GB and CB, respectively. In conclusion, the results did not show any difference in DMI, methane emission per day or per kg dry matter intake in rosé veal calf feed either a grass silage-based or a corn cob silage-based ration with similar chemical composition.

Feeding a starch and protein binding agent for mitigating enteric methane emissions in sheep

P. Prathap[1], S.S. Chauhan[1], J.J. Cottrell[1], B.J. Leury[1] and F.R. Dunshea[1,2]
[1]The University of Melbourne, School of Agriculture and Food, Faculty of Science, Melbourne, VIC 3010, Australia, [2]The University of Leeds, Faculty of Biological Sciences, LS2 9JT Leeds, United Kingdom; pprathap@student.unimelb.edu.au

This study investigated the effect of heat stress (HS) and wheat treated with a starch and protein binding agent (Bioprotect®) on the enteric methane (CH_4) emissions in sheep. All experimental procedures were approved (Ethics ID:1914950.1) by University of Melbourne animal ethics committee. Twenty-four one-year-old merino lambs (42.6±3.6 kg) were randomly allocated to 3 dietary treatment groups that included a wheat based diet (WD), a 2% Bioprotect treated wheat-based diet (BD) and a maize based diet (MD). All diets contained 50% grain, 25% oaten chaff and 25% lucerne chaff. Animals were subjected to 3 consecutive experimental periods of 1 week each. During period 1 (P1), animals were fed 1.7×Maintenance energy (ME) level and kept under a thermoneutral environment (18-21 °C and 40-50% relative humidity; RH). During period 2 (P2) sheep fed at 1.7×ME level were exposed to HS (28-40 °C and 30-50% RH). During period 3 (P3) sheep were fed at 2×ME level during HS. Sheep fed BD had lower CH_4 production than WD fed sheep with MD being intermediate (25.0, 19.3, and 32.4 g/day for WD, BD, and MD respectively; P<0.001). Sheep showed the highest CH_4 production under thermoneutral conditions compared to during HS (28.5, 23.0 and 25.2 g/day for P1, P2, and P3 respectively; P<0.04). There was no significant interaction between diet and period, indicating that BD reduced CH_4 emission in sheep irrespective of controlled environmental conditions. Further, whole tract dry matter digestibility was greater for BD fed sheep followed by WD and MD fed sheep (83.0, 84.2 and 81.4% for WD, BD, and MD respectively; P<0.001) and was higher during P2 (82.7, 83.4 and 82.6% for P1, P2, and P3 respectively; P<0.001) than P1 and P3. In conclusion, sheep consuming the WD had lower CH_4 emissions than those fed MD and this was further decreased by treating the wheat component Bioprotect®. Further, the results indicate that HS exposure decreases CH_4 production in Merino sheep.

Session 87 Poster 13

Neutralization of bovine enteric methane emission by the presence of trees in a silvopastoral system

J.R.M. Pezzopane, H.B. Brunetti, P.P.A. Oliveira, A.C.C. Bernardi, A.R. Garcia, A. Berndt, A.F. Pedroso, A.L.J. Lelis and S.R. Medeiros
Embrapa Pecuaria Sudeste, Rod. Washington Luis, Km 334, 13563-970, São Carlos, Brazil; jose.pezzopane@embrapa.br

Trees in silvopastoral systems (SPS) serve as carbon sink, especially when used for lumber. Our aim was to evaluate whether carbon fixation by trees could neutralize the bovine enteric methane emission in a SPS following the Neutral Carbon Brazilian Beef (NCBB) protocol. The study was conducted at Embrapa Pecuária Sudeste, São Carlos, SP, Brazil from Dec-2017 to Feb-2019 in a SPS comprised of *Urochloa brizantha* (syn. *Brachiaria brizantha)* (Hochst ex A. Rich.) Stapf cv. BRS Piatã and *Eucaliptus urograndis* (GG100 clone). The SPS was established in 2011 spaced 15×2 m (oriented east-west) and thinned to 15×4 m in 2016. In four 3 ha modules with six paddocks each, Nellore and Canchim (5/8 Charolais + 3/8 Zebu) finishing bulls were managed under rotational stocking. The enteric methane emission was estimated by the IPCC Tier 2 equation using average live weight (ALW), average daily gain (ADG), forage digestibility and intake of each grazing cycle as inputs. Allometric equations were used to estimate tree stem biomass, using the diameter at breast height (DBH) and height of fifteen trees per paddock. The results were multiplied by 0.4549 to adjust for carbon content. The mean ALW was of 473.16 kg with ADGs ranging from 0.48 in the winter to 0.94 kg/day in the summer, resulting in an emission rate of 2.2 Mg CO_2 eq./AU/yr. The DBH and tree height at the end of the experiment were of 29.9 cm and 30.5 m, respectively, resulting in 12.7 Mg stem biomass/ha and a fixation rate of 20.9 Mg CO_2 eq./ha/yr. When considering three scenarios: all the trees (those thinned in 2016 and the remaining ones); only the trees remaining after the thinning; and only 40% of the stem volume in the second scenario as suitable to be used as lumber, the trees in the SPS neutralized the emission of enteric methane relative to 9.4, 6.7, and 2.3 AU/ha, respectively. Even in the worst scenario, the trees neutralized the emission of bovine enteric methane relative to a stocking rate above the Brazilian average. Acknowledgment: Fapesp 2019/04528-6 and IABS/Rede ILPF.

Session 87 Poster 14

Dairy goat farms in Extremadura: carbon sources and sinks in several management systems

L. Madrid[1], M. Escribano[1], P. Gaspar[2] and A. Horrillo[1]
[1]Faculty of Veterinary Science, Department of Animal Production and Food Science, Avda. de la Universidad, s/n, 10003 Caceres, Spain, [2]Faculty of Agriculture, Department of Animal Production and Food Science, Avda. Adolfo Suarez, s/n, 06007 Badajoz, Spain; andreshg@unex.es

The fight against climate change has become a major challenge for animal food production systems due to the emission of greenhouse gases (GHG). Nowadays, society links the consumption of food of animal origin to a deterioration of the environment. This is certainly not only detrimental to food production, but all animal production systems are considered to have the same impact. This study analysed the Carbon Footprint (CF) of several management systems of goat farms in northern Extremadura. A total of seven cases were selected, grouped into three production models: intensive, semi-intensive and semi-extensive. The study is part of a regional project (MitigaDehex; IB20070) whose main objective is to analyse extensive dehesa livestock systems in Extremadura (SW Spain) as sustainable and active production models against climate change, and its specific objective is to calculate the GHG balance of these livestock production systems. The analysis was carried out using Life Cycle Assessment (LCA) and the HC was calculated following the IPCC 2019 methodology. In addition, carbon sequestration was included in this calculation, with the aim of assessing the contribution of goat farming to climate change depending on its production and feed management. The results show that enteric fermentation is the main source of on-farm emissions (50-58.9%), followed by emissions from off-farm feeding (15.9-30.71%), whose variability depends on the type of production system. In addition, there are differences among the three groups in the study, with an increase in GHG emissions per litre of milk produced (FU) linked to the intensification of livestock systems. Thus, the role of soils and their capacity to act as carbon sinks in grazing systems is relevant, showing mitigation potential with a carbon sequestration of 0.29 kg CO_2 eq /FU up to 1.9 kg CO_2 eq /FU depending on the type of farm. At present, studies that analyse the real impact of GHG emissions from livestock farming under different production models are necessary and, in particular, goat milk production is one of the most neglected.

Session 87 Poster 15

Life cycle assessment of IntelliBond on the carbon footprint of a dairy farm

D. Brito De Araujo, K. Perryman and J.G.H.E. Bergman
Nutreco, Selko, Stationsstraat 77, 3811 MH Amersfoort, the Netherlands; davi.araujo@selko.com

According to FAO, 14.5% of global greenhouse gas emissions are related to farming. Out of the total emissions, 58% originate directly from cows, with 42% coming from farm operations. Pressure from milk processors on farmers to reduce their carbon footprint continues to increase. Thus, a life cycle assessment accessing the impact of Selko IntelliBond trace minerals was conducted to determine its impact on the carbon footprint of dairy cattle. A meta-analysis of 11 peer reviewed studies has demonstrated that by replacing sulphate-based zinc, copper, and manganese with IntelliBond, an improvement in NDFd of 1.7% can be achieved. University research has demonstrated that each one point of difference in NDF can result in a predicted increase of Energy Corrected Milk (ECM) of 0.25-0.3 kg. These studies were then evaluated by an independent, third-party expert utilizing an ISO compliant model (ISO 14040, 14044) to predict the potential impact of IntelliBond vs sulphate trace minerals on the carbon footprint per kg of ECM. Data from this exercise indicates that a reduction of the carbon footprint of 1.5-2.0% per kg of ECM can be expected when sulphates are replaced by IntelliBond.

Session 87 Poster 16

Effects of supplementation of *Asparagopsis taxiformis* as a means to mitigate methane emission

M. Angellotti, M. Lindberg, M. Ramin, S.J. Krizsan and R. Danielsson
Swedish University of Agricultural Sciences, Depart. Animal Nutrition & Management, Ulls väg 26, 753 23 Uppsala, Sweden; melania.angellotti@slu.se

Methane emissions from feed digestion in ruminants contribute significantly to anthropogenic greenhouse gas emissions globally. The red algae *Asparagopsis taxiformis* (AT) has been reported to have high potential to reduce methane emissions when fed to ruminants. AT is rich in halogenated methane analogues, such as bromoform, which is able to block the methanogenesis in the rumen, but all mechanisms and effects on animal physiology are not well understood. Hence, the aim of this study was to investigate addition of AT in dairy cow diets and evaluate the effects on enteric methane and hydrogen production, feed intake and milk production in dairy cows. Thirty cows fed a total mixed ration were blocked according to parity and days in milk and randomly assigned to one of three dietary treatment groups with different inclusion levels of AT in % of organic matter; a control group with no AT (CON), a group with 0.15% AT (LOW) and a group with 0.3% AT (HIGH) for 12 experimental weeks. Individual feed intake and milk yield were recorded automatically throughout the trial. Milk composition was determined by collecting milk samples fortnightly, enteric methane and hydrogen levels were measured continuously by the GreenFeed system. Statistical analyses were performed by using the MIXED procedure of Rstudio, treatment and experimental week was included as fixed effects and the random effect of block was used. Week was considered as repeated measurement. The results showed a significant reduction in methane production in the HIGH group by 30%, with a concomitant increase in hydrogen from the breath. However, the level of decrease in methane was not stable over time. Feed intake and energy corrected milk yield were lower in the HIGH group compared to the other two groups, which decreased by 6 and 2%, respectively. In conclusion, using AT as additive in dairy cows' feed rations allows for mitigation of enteric methane emissions, however, a reduction in feed intake with a concomitant decrease in milk production was also observed. Additional research on AT as enteric methane inhibitor is necessary since the effects appear to be dosage-dependent and change over time.

Session 88 Theatre 1

Trace element concentrations in feed ingredients: previously overlooked, now under the spotlight

M. López-Alonso
University of Santiago de Compostela, Animal Pathology, Veterinary Faculty, 27002 Lugo, Spain; marta.lopez.alonso@usc.es

Trace minerals are part of numerous enzymes and coordinate a great number of biological processes, being essential to maintain animal health and productivity. Trace element deficiencies are very frequent worldwide. They are linked to an insufficient intake in trace elements (primary deficiencies) or to the interaction with other elements (by competitive and non-competitive mechanisms) that inhibits its absorption (secondary deficiencies). To avoid trace element deficiencies and maintain a high standard of production, diets of animals have been traditionally supplemented well above the physiological requirements, greatly overlooking the background trace element concentration of the feed materials. This was possible because rations can be formulated with large 'safety margins' so that trace element intakes generously exceed requirements without important risk for the animal health, assuming that timely supplementation of complete trace minerals was an inexpensive insurance that well worth the cost. These large safety margins even allowed to use some minerals as growth promoters: the classical example being Cu and Zn supplementation in pigs and poultry. The extensive use of these trace elements in intensive farming has led to serious environmental problems that need an urgent solution. Even representing less than 1% of total feed ingredients consumed in Europe they have a great environmental footprint, showing the highest contribution to freshwater and marine ecotoxicity and to a metal depletion due to the fact they are non-renewable resources. In times of higher production costs and changing consumer and environmental demands a tailor-made mineral supplementation is needed, what has been called precision mineral feeding, to really provide the trace elements each animal demand. Not all animals need the same trace element intakes and aspects as growth rates, stress factors (in example heat) or antibiotic-free diets should be addressed. Having in mind that quality and not only quantity of trace elements should be considered, and background concentrations of feed materials be carefully analysed, both quantitative and qualitatively, since they will be in most cases the main sources of trace elements in the animal diet.

Session 88 Theatre 2

Review on native Se concentrations in feed ingredients

D. Cardoso and A. Hachemi
Adisseo, 10 Pl. du Général de Gaulle, 92160, France; denise.cardoso@adisseo.com

This review aims to assess the amount of Se in feed ingredients and its species, that can influence the bioavailability and the metabolism in the animal. Selenium (Se) is an essential nutrient that plays a vital role in the antioxidant system of animals. In feed, the total Se concentration is recommended to be of 0.5 ppm to meet animal requirements. However, Se content of feed ingredients greatly varies depending on many different factors. Se concentration in corn and rice grown in normal and high Se areas can vary 100-500-fold. Animal by product such as fish meal can represent up 5 ppm of Se, but it known to have a low bioavailability. Indeed, average data on Se content in feedstuffs presented in various tables are not suitable for diet balancing and Se supplementation is a routing practice in commercial animal production, which is why a Se supplementation is a common practice worldwide. Selenium in feed is mainly derived from the soil, with its concentration varying significantly between 0.1 and 2 ppm. However, the bioavailability of Se for plants is largely dependent on its form rather than its total concentration. The forms of Se in soil include selenides, elemental Se, selenites, selenates, and organic Se compounds. The transfer of the selenium from soil to plant and distribution into various parts also depends on plant species; that differ markedly in their ability to incorporate Se into tissues. Se in plants exists in different forms, being the majority in the form of seleno aminoacids, such as selenomethionine (SeMet), that makes up more than half of the Se in cereal grains, grassland legumes, and soybeans. SeMet is primarily stored in the grain and the root, while lower concentrations are found in the stems and leaves. Interestingly, plants supplemented with different forms of Se metabolize Se differently. For example, selenite-supplied plants accumulate organic Se, likely SeMet, while selenate-supplied plants accumulate selenate. In summary, the Se concentration in feed will mainly depend on the Se soil concentration and its specie.

Session 88 Theatre 3

Does USEtox predicts adequately Cu and Zn ecotoxicity in soils amended with animal effluents?

E.P. Clement[1,2], M.N. Bravin[2], A. Avadi[2,3] and E. Doelsch[2]

[1]Ademe, 20 avenue du Grésillé BP-90406, 90406 Angers, France, [2]CIRAD, Avenue Agropolis TA B-78/01, 34398 Montpellier cedex 5, France, [3]ESA/INP-HB, Yamoussoukro, Cote d Ivoire; emma.clement@cirad.fr

Animal effluents (AE) are of interest to substitute mineral fertilizers in agriculture. They are, however, a major contributor to soil chronic copper (Cu) and zinc (Zn) contamination that are provided in animal feeds and may exert toxic effects on soil organisms and reduce soil fertility. These negative impacts question the long-term sustainability of agricultural AE recycling. To estimate the terrestrial ecotoxicity of Cu and Zn, a novel method was proposed involving the USEtox model for life cycle impact assessment. It computes the comparative toxicity potential (CTP) of trace elements via pedo-transfer functions and by decomposing it in four characterization factors: fate, availability, bioavailability and toxicity towards soil organisms. Beyond the interest of this model, its relevance in the context of AE recycling in agriculture is unknown. We empirically assessed how this model is adapted to estimate ecotoxicological impact of Cu and Zn in an AE-amended agricultural soil. A 26 day-long incubation of the soil with or without 31 AE was carried out in controlled lab conditions that mimic a field application. The AE were pig or broiler faeces, showing distinct Cu and Zn concentrations due to different feed. Initial soil properties (clay, soil organic matter, pH and Al/Fe oxides) were analysed to inform the model. Soil properties (pH, dissolved organic matter – DOM) and Cu and Zn availability in soil and soil solution, matching the CTP characterization factors, were measured at the end of the incubation in each soil-AE mixture. Empirically-determined CTPs were different by up to 1.5 log unit from the model estimations for Cu, 0.3 for Zn. The CTP variability measured after AE application in our experiment was comparable to the variability observed for world-wide soils. One explanation is the overestimation of DOM concentration and the lack of consideration of DOM binding properties in soil. Our results overall support the need to improve the model for its application to agricultural AE recycling. This will fuel our ability to assess the environmental impacts of feed trace minerals in animal productions.

Session 88 Theatre 4

Challenges in establishing the mineral composition of feed materials

G. Tran and V. Heuzé

Association française de zootechnie, 22 place de l'Agronomie CS 20040, 91123 Palaiseau Cedex, France; gilles.tran@zootechnie.fr

Knowledge of the mineral composition of feeds is critical for livestock nutrition. Inadequate mineral supply results in performance and health problems while excess dietary minerals is detrimental to the environment. However, minerals remain secondary in feed analysis. To examine this issue, we will use the French Feed Database (FFD), which has collected since 1989 about 3 million analytical values on feed materials from laboratories. The first problem is that of data availability. Mineral analyses account for 12% of FFD data. There are fewer values for the 33 mineral elements in the FFD than there are for protein. These data are skewed toward macro minerals. Ca, P, K, Mg, Na and Cl represent 81% of the mineral data, with Ca and P each accounting for 29%. The main trace elements (Cu, Mn, Zn, Fe, S) represent 12% of mineral data, while the 22 other elements represent less than 5%. There are about 200,000 values for Ca and P while iodine is represented by only 550 values. Data are sufficient for macro minerals in major feed materials such as maize grain, but, for lesser feeds, it is common that few values are available for macro minerals, and less than 5 or none for trace elements. The second issue is that of variability. The mineral composition of feeds depends on factors such as geography, soil type, climate, and farming practices, but the range of mineral values is much greater than that of constituents like protein, with factors of ten or more between the minimum and maximum values: in maize, Ca varies from 0.1 to 2.2 g/kg DM while Mn varies from 3 to 56 mg/kg. This large variability, which can be sometimes explained by soil contamination at harvest or during storage, makes the notion of representative data much harder to assess and correlate with other factors. This problem is compounded by the lack of relations of mineral elements with other constituents, including crude ash. In conclusion, while significant numbers of data are available to establish the mineral composition of feeds, there are still gaps in knowledge that need to be addressed. The development of quicker analytical methods and the collection of more comprehensive data on the mineral composition of both traditional and novel feed materials will be crucial in improving the nutritional management of livestock.

Session 88 — Theatre 5

Swiss feed database: mineral and trace element composition of feedstuffs

M. Lautrou[1], E. Manzocchi[2] and P. Schlegel[1,2]
[1]Agroscope, Swine Research Group, Rte de la Tioleyre 4, 1725 Posieux, Switzerland, [2]Agroscope, Nutrition and Emissions Research Group, Rte de la Tioleyre 4, 1725 Posieux, Switzerland; marion.lautrou@agroscope.admin.ch

Feeding animals as close as possible to their needs has become a priority, with the idea of limiting excretions of certain elements such as copper or zinc and reducing the use of limited resources such as mineral phosphate. For this purpose, it is necessary to precisely know the animal requirements but also to properly characterize the native mineral and trace element composition of feedstuffs. It is for this second objective that the Swiss feed database is a useful tool. The Swiss Feed database gathers the chemical composition of feedstuffs used in Switzerland since the late 1980s until today. This online database presents the mineral (Ca, P, Mg, K, Na, S and Cl) and trace element (Fe, Cu, Zn, Mn, Co and Se) composition of any feedstuff analysed by Agroscope and by two private laboratories. The list of feedstuffs is large: cereal grains (barley, maize, rye, wheat, and others), cereal co-products from milling, starch industry or distillery, oil seeds and co-products, tuber and roots, milk co-products, mineral feeds and so on. In addition to the nature of the raw material, the year of production, the canton of origin and the altitude are available. These data can be used individually but are also exploited to define the Swiss reference values of each feedstuff. Then, in contrast to reference tables available in other countries, the values available for trace element are all from feedstuffs analysis and not from the literature. This Swiss feed database provides the opportunity to regularly create reference tables, according to origin or years, for the actors of the animal production field. Based on this database, we will present the evolution of the mineral and trace element composition of selected feedstuffs through years, and study the effect of their origin. Then, we will also compare these values to others reference values (from France and North America). We will also present the contribution of minerals and trace elements supplied by feedstuffs relative to the requirements of pigs.

Session 88 — Theatre 6

Overview of international aquaculture feed formulation database with reference to mineral nutrition

D.P. Bureau[1], L. Manomaitis[2] and F.M. Damasceno[3]
[1]University of Guelph, Dept. of Animal Biosciences, 50 Stone Rd East, N1G2W1, Guelph, Ontario, Canada, [2]USSEC, SEA Aquaculture, 541 Orchard Rd, Singapore 238881, 238881, Singapore, [3]Wittaya Aqua International, 1 University Avenue, Toronto, Ontario, Canada; dbureau@uoguelph.ca

The great diversity in species cultivated and culture conditions encountered in aquaculture result in a need for a great diversity of feeds of different composition. These feeds are formulated with an ever-increasing portfolio of ingredients. Cost-effective feed formulation requires accurate information on composition of ingredients and requirements of animals across their life cycle. The International Aquaculture Feed Formulation Database (IAFFD.com) is a free online open-access resource supported by USSEC (ussec.org) and Wittaya Aqua International (wittaya-aqua.ca) that helps meet the need of the industry. It is comprised of two main sub-databases: (1) Aquaculture Species Nutritional Specifications (ASNS); and (2) Feed Ingredient Composition Database (FICD). The databases are designed to be easily imported in least-cost feed formulation software. The ASNC contains detailed nutritional specifications for over 35 species at different weight ranges. Most of these nutritional specifications are generated using a nutritional modelling approach. The models and specifications are constantly updated. The FICD contains detailed information on the chemical composition for more than 700 ingredients. The information is compiled from different sources, reconciled, and regularly updated. The ASNS provides specifications for macro and micro minerals. The large number of species cultivated makes it a challenging task to develop accurate recommendations. Macro and micro-mineral composition of feed ingredients are reported in the FICD. However, information on the concentration in minerals of ingredients is often limited. There is also limited information on the digestibility and bioavailability of minerals of ingredient. This is complicated by the fact that digestibility of minerals is largely influenced by the GI physiology of animals and interactions with dietary components. Many stakeholders are involved in R&D efforts on minerals in animal nutrition and their contribution to improving the IAFFD would be highly valuable.

Session 88 Theatre 7

Mineral analysis: opportunities and limitations

P. Berzaghi[1], R. Fornaciari[2], M. Dorigo[2] and G. Cozzi[1]
[1]University of Padua, Animal Medicine Production and Health, Viale Dell'Università 16, 35020, Italy, [2]Nutristar S.p.A, Via del Paracadutista 9, 42122 Reggio Emilia, Italy; paolo.berzaghi@unipd.it

Feed and forage analysis has been traditionally limited to feed industries and large analytical laboratories. Methods of minerals analysis have been laborious and expensive and for these reasons used only when strictly necessary. The replacement of the traditional Atomic Absorption with Inductively Coupled Plasma (ICP) techniques have reduced costs and facilitated a greater use of these analysis. Yet costs and accessibility of mineral determination in feed are still limiting its wider use, particularly for forages that may have large variation in their composition. Near Infrared Spectroscopy (NIR) has gained popularity for mineral determination, because of its rapidity and very low costs and this despite the knowledge that mineral have little or not signal in the NIR spectral region. In recent years X-Ray fluorescence (XRF) has been able to greatly simplify mineral determination in feed, contributing significantly to a wider use of mineral analysis even by small feed plants and farms. As a rapid (few minutes) and non-destructive methods, XRF easily complements NIR in feed laboratories as it requires the same sample preprocessing (drying and fine grinding). XRF technique requires a calibration with a few (<20) samples for best performances and it spans from lower atomic weight of Na till heavier element like Pb. Elements under 10 ppm may not be accurately predicted by XRF and for this reason it is normally adopted for the determination of all major macro and some micro minerals (Na, Mg, Al, Si, P, S, Cl, K, Ca, Mn, Fe, Cu and Zn). Calibration for these elements in feed have great performances with R-square normally greater than 0.95 and prediction errors generally, below 5% in relative units. Low concentration is a limitation, yet mineral supplements because of their greater concentration of micro mineral can be easily monitored improving consistency during their production also for element like Se, I, or Mo that are at very low concentration in feed.

Session 88 Theatre 8

Determination of forage mineral analysis with portable X-ray Fluorescence device

R. Balegi[1], F. Penen[1], M. Lemarchand[2] and A. Boudon[2]
[1]ANIMINE, 10 rue Léon Rey Grange, 74960, France, [2] INRAE-Institut Agro Rennes Angers, UMR 1348 PEGASE, Saint-Gilles, France, 35650, France; rbalegi@animine.eu

Mineral supplementation of dairy herds is necessary to compensate imbalances in mineral content of many basal diets. Supplementation must be assessed as accurately as possible to avoid excessive environmental and economic losses and to allow good coverage of cow nutritional requirement. However, mineral content of forages is variable and seldom known because of high cost of reference analytical methods used in laboratories. Developing an innovative handheld analytical tool, based on X-Ray Fluorescence (XRF) technique, is an opportunity to perform more accurately mineral supplementation in a cheap and immediate way at the farm. This work aims to validate the analysis of selected minerals (Ca, P, K, Mg, Na, S, Zn, Cu, Mn, Fe and Mo) in 89 forages with a portable XRF device in comparison to inductively coupled plasma – optical emission spectrometry (ICP-OES). Twenty samples of grass, 21 of hay, 11 of haylage, 21 of maize silage and 16 samples of grass silage were collected all over Auvergne-Rhônes-Alpes area in France, then dried and ground on a 0.5 mm grid. Relationship between XRF and ICP-OES methods was assessed thanks to linear regression with ICP-OES results as predicted variable. Accuracy of the relationship was given by the coefficient of determination (R^2) and mean absolute error (MAE). The correlation between the two methods results depended on the chemical element. Copper, iron and sulphur concentrations obtained by XRF and ICP-OES were correlated, with R^2 of respectively 0.70, 0.78 and 0.94 and MAE of respectively 13, 24 and 8% of the average concentrations obtained by ICP-OES. For the other minerals (Ca, P, K, Zn and Mn), R^2 ranged between 0.89 and 0.96 and MAE between 6 and 18%. For Mg, the correlation was weak (R^2=0.02) mainly because of XRF limit. Sodium was not detected by XRF. The XRF performance were also limited for Mo (R^2=0.49, MAE=61%) likely due to its very low concentration (between 0.4 and 3.7 mg/kg DM) compared with the limit of quantification of ICP-OES. Although, the portable XRF analysis showed globally promising results to quantify mineral content in forage, further analytical development is needed for magnesium, sodium and molybdenum.

A new phosphorus feeding system for the sustainability of swine and poultry production

M.P. Létourneau Montminy[1], M. Lautrou[2], M. Reis[3], C. Couture[4], N. Sakomura[3], B. Meda[5], C.R. Angel[6] and A. Narcy[5]
[1]Laval University, 2425 rue de l'Agriculture, G1W3C1 Quebec, Canada, [2]Agroscope, Posieux, 1725 Posieux, Switzerland, [3]UNESP, 14884-900, Jaboticabal, Brazil, [4]Trouw Nutrition, Saint-Hyacinthe, J2R 1S5, Canada, [5]INRAE, UMR BOA, Centre de Nouzilly, 37380, France, [6]University Maryland, College Park, 20742-2311, USA; marie-pierre.letourneau@fsaa.ulaval.ca

Phosphorus (P) is essential for all types of life but the natural P cycle is disrupted. A parsimonious use of phosphate is therefore essential. This requires a P precision feeding system for livestock. A precise prediction of P availability of raw materials as well as of animal P requirements is needed. To answer this research priority, a new P feeding system combining mechanistic and empirical modelling is proposed based on apparent digestible P. This robust multicriteria approach has been developed for growing pigs and broilers. First, a mechanistic model that simulates the fate of dietary P through 3 sub-modules, digestion, soft tissue, and ash was developed. The soft tissue module simulates the growth of the protein and lipid. The ash module simulates the partitioning of absorbed calcium (Ca) and P into the bone, protein, and lipid compartments as well as urinary excretion. Then, using inversion principles, these models were adapted to predict the amount of digestible P needed to maximize P retention in soft and bone tissues. In a second step, these needs must be met by diet formulation which requires knowledge of the digestible P value of diets. In pigs, digestible P value of feedstuff are available which is not the case in broilers. Using a single P value for each feedstuff does not consider interactions with other components of the diet. To overcome this drawback, an empirical equation has been developed in pigs and broilers to quantify the impact of the different forms of dietary P, calcium (Ca) and exogenous phytases, and their interactions on digestible P. These equations predict the P digestibility based on the chemical analysis and appears more robust than a fixed value per feedstuff. Equations has been evaluated and the accuracy has been confirmed in swine with similar evaluation ongoing in broilers. This new and generic system for both species opens relevant perspectives in terms of precision feeding for P.

Phosphorus digestibility of porcine processed animal proteins (PAPs) in broiler diets

J. Van Harn and P. Bikker
Wageningen University and Research, Wageningen Livestock Research, P.O. Box 338, 6700 AH, the Netherlands; paul.bikker@wur.nl

Since 2021, EU legislation allows the use of porcine processed animal proteins (PAPs) in poultry diets. These by-products of the slaughter of pigs are a valuable source of protein, energy and minerals and their use can improve resource utilisation and circularity of poultry production. Their use requires adequate insight in digestible nutrient content since feed tables were largely based on meat and bone meal available before the feed ban, and not representative for present species-specific PAPs. Our recent studies provided insight in the digestibility of energy and amino acids but information on digestibility of phosphorus (P) and calcium (Ca) for use in poultry diets was lacking. This study was conducted to determine the pre-caecal (ileal) digestibility of P in six porcine PAPs varying in ash, Ca and P content and produced with different methods. The P content of the PAPs varied between 16 and 64 g/kg, depending on the proportion of bone in the product. At 14 days of age, 640 healthy broilers were allocated to 64 floor pens (1.2 m^2) with a flexible slatted floor, and ten birds per pen. The broilers received one of ten experimental diets: a low P basal diet, a diet with mono calcium phosphate (MCP) as reference or a diet with 1.5 g P from one of six different PAPs at a constant Ca:P ratio of 1.25:1, added to the basal diet. These products reflected processing methods 1, 4 and 7 according to commission regulation (EU) 142/2011, with two method 7-products coarse and finely ground to determine the effect of particle size. At 22 days of age chyme of the terminal ileum was collected and analysed to determine pre-caecal digestibility of P and Ca. The left tibia bone of three birds per pen was collected to determine the ash, Ca and P content. Analysis of variance with pen as experimental unit was used to determine differences between treatments. The pre-caecal P-digestibility of MCP was 91%. The P-digestibility of the PAPs varied between 80 and 95%, while processing method of the PAPs and particle size did not significantly influence the pre-caecal P-digestibility and the tibia mineral content. Hence, porcine PAPs are a valuable source of nutrients with high P digestibility in poultry diets and can support circularity of food production.

Session 88 Theatre 11

Prediction of magnesium absorption in dairy cows: An update

R. Khiaosa-Ard and Q. Zebeli
University of Veterinary Medicine Vienna, Veterinaeplatz 1, 1210 Vienna, Austria; ratchaneewan.khiaosa-ard@vetmeduni.ac.at

Earlier research (e.g. Schonewille *et al.*) proposed the quantitative prediction of magnesium (Mg) absorption based on Mg intake and dietary K level. Different findings emerged in newer studies, which might be related to cow factor and feeding. We updated the database used by Schonewille *et al.* and reevaluated their predictions. In total, 21 published studies with 94 dietary treatments were included in the present analysis. Of these, 41 treatments originated from non-lactation cows and 53 from lactation cows. The dietary Mg content ranged from 0.45-17.4 g/kg DM for non-lactation cows and 1.1-6.25 g/kg DM for lactation cows and the Mg intake was 2-124 g/d and 12-120 g/d, respectively. The dietary K content ranged from 11.2-75.6 g/kg DM for non-lactation cows and 6.9-41.1 g/kg DM for lactation cows. True Mg absorption (g/d) was Mg intake – faecal Mg output + endogenous Mg secretion (which was 0.004 × BW). The study was included as a random factor. The evaluation showed that lactation and non-lactation cows absorb similar amounts of Mg at a given Mg intake and the average Mg absorption was 20% Mg at Mg intake ≥20 g/d (range 10-40%). Of the factors tested, Mg intake (g/d) and dietary K (g/kg DM) were the significant predictors for true Mg absorption (g/d) similar to the recommendation of Schonewille *et al.* However, the new data revealed that instead of treating dietary K as a quantitative predictor, categorizing it increased the accuracy of the prediction. The data suggested two different equations as follows: True Mg absorption (g/d) = -1.927 (±1.16, P=0.11) + 0.34 (±0.025, P<0.001) × Mg intake (g/d), when dietary K≤20 g/kg DM, and = 0.154 (±1.06, P=0.05) + 0.209 (±0.026, P<0.001) × Mg intake (g/d), when dietary K>20 g/kg DM. The root mean square error was 2.19 and the concordance correlation coefficient (ccc) was 0.923. Estimation of the new data when using the model from Schonewille *et al.*: true Mg absorption (g/d) 3.6 + 0.2 × Mg intake (g/d) – 0.08 × dietary K (g/kg DM) resulted in a ccc of 0.854. This study refined the prediction of Mg absorption in dairy cows related to the influence of dietary K.

Session 88 Theatre 12

Se and Co balance in dairy cows: longitudinal study from late lactation to subsequent mid-lactation

J.B. Daniel and J. Martín-Tereso
Trouw Nutrition R&D, P.O. Box 299, 3800 AG, Amersfoort, the Netherlands; jean-baptiste.daniel@trouwnutrition.com

We quantitatively evaluated Se and Co balance in dairy cows from late lactation through the subsequent mid-lactation. Twelve Holstein dairy cows were housed in a tie-stall from 10 weeks before to 16 weeks after parturition. When lactating, cows were fed a total mixed ration (TMR) with 0.44 ppm Se and 0.52 ppm Co, and when dry, a TMR with 0.16 ppm Se and 0.23 ppm Co. After 2 weeks of adaptation to the facility and diet, Co and Se balances were determined at weekly intervals, by calculating the difference between total intakes and complete faecal, urinary and milk outputs, with the latter three fluxes quantified over a 48-h period. Weekly serum samples were also collected. Repeated measures mixed models were used to evaluate the effects on trace mineral balances over time. Between the dry period and week 11-16 of lactation, intakes of Se and Co increased from 2.3 to 10.7 mg/d, and from 3.1 to 12.6 mg/d, respectively. As a result, faecal Se and Co output increased from 1.7 to 6.6 mg/d, and from 3.1 to 11.3 mg/d, respectively. Apparent absorption of Se and Co were lowest during the dry period (24 and 2% respectively) as compared to lactation (32-38%, 8-12%, respectively). Milk Se concentration was highest at the end of the previous lactation (19.3 µg/l), and lowest between week 6 and 10 (15.2 µg/l). Milk Co concentration was not influenced by stage of lactation and averaged 0.36±0.06 µg/l. Although urinary Co excretion was correlated to Co intake, amount excreted were negligeable (from 14 to 33 µg/d). In contrast, urinary Se excretions were larger, and increased with Se intake from 0.4 to 2.9 mg/d. From late lactation to week 5, Se balance was not different from 0, but increased linearly from calving to week 16, up to 0.5 mg/d. During the dry period, Co balance did not differ from 0, but increased to 0.9-1.3 mg/d during lactation, with no effect of stage of lactation. Serum Se and Co were lowest during the dry period (0.7 and 3.9 µmol/l, respectively) and highest between week 6 and 16 (1.1 and 6.0 µmol/l, respectively). Further research is needed to elucidate if increasing Se and Co balance during lactation reflect actual body retention of Se and Co. Evaluating potential consequences on health would also be valuable.

Session 88 Poster 13

Mineral analysis in feed using a new automated analytical method

D. Schwarz and D. Aden
FOSS Analytical A/S, Nils Foss Alle 1, 3400 Hilleroed, Denmark; das@foss.dk

Wet-chemistry methods for mineral analysis are cumbersome and time-consuming as they are heavy on sample preparation. Hence, a high-throughput and chemical-free instrument based on Laser Induced Breakdown Spectroscopy (LIBS) has been developed. With this method, sample preparation is reduced to a minimum and total turn-around time (i.e. sample preparation and analysis) is as fast as two minutes. It is possible to analyse up to 60 samples per hour for their mineral content. Feed materials tested so far were the following: grass silage, maize silage, alfalfa/legume hay, hay, fresh grass, haylage, mixed hay, sorghum, straw, grass and clover mixture, and barley silage (whole crop silage). Macro and micro minerals in above listed feed materials were determined using classical wet-chemistry methods and the new LIBS-based method. High correlations between the two methods were found for the macro minerals calcium (Ca), magnesium (Mg), phosphorus (P), potassium (K), sodium (Na), sulphur (S) and the micro minerals aluminium (Al), boron (B), copper (Cu), iron (Fe), Manganese (Mn), Zinc (Zn). The newly developed LIBS-based method may be used complementary to existing wet-chemistry methods and open up the opportunity of cost-efficient mineral analysis of feed material. This, in turn, could help to test larger amounts of samples, thus help to better understand varying mineral contents in, e.g. grass silage, and eventually contribute to precision mineral feeding of livestock.

Session 88 Poster 14

Trace mineral trends in British and Irish forages

A.H. Clarkson and N.R. Kendall
University of Nottingham, Sutton Bonington Campus, Leicestershire, LE12 5RD, United Kingdom; andrea.clarkson@nottingham.ac.uk

Forage is the primary diet for grazing ruminants in the UK & Ireland. Problems with availability are easy to detect but trace mineral issues are harder to determine. The mineral content of forages can vary with, rainfall, soil, sward, maturity, season, and location, as well as product and fertiliser use. It is important to understand the supply of trace minerals, and their antagonists, to better refine and optimise mineral supply for animal growth, health, and production, without the risks of oversupply. This work looks at trends in trace mineral supply from forage data submitted for commercial analysis, compiling data from farms across the UK and the Republic of Ireland. UK and Irish forage sample data (n=1,453) submitted for commercial analysis (2020-2023) was used to calculate means, standard error and plot graphed data (Microsoft Excel v2301). The pooled averages for Mn, Co, Se, and Fe were typically all within or above sheep, beef, and dairy requirements (based on NRC data) across all years. However, the average Cu concentration in the UK was below the lowest requirements (<8 mg/kg DM) every year. In Ireland Cu was below this threshold in 2021 & 2022 and was below cattle requirements (<10 mg/kg DM) in 2020. Furthermore, S was above the threshold for potential copper antagonism (>2.5 g/kg DM) in 2020 & 2021 (Ireland), remaining ~2 g/kg DM the other years. Fe was consistently above the threshold (>250 mg/kg DM) for copper problems in all years except 2021 (Ireland). Mo was lowest in 2022 (Ireland) at 1.32 mg/kg DM, remained around 2 mg/kg DM (UK & Ireland) the rest of the time, but had risen to 4.31 mg/kg DM in 2023 (UK). All above the threshold for potential copper problems (>0.5 mg/kg DM). Zn was below the lowest requirements (<30 mg/kg DM) in 2021 (UK & Ireland), and 2023, remaining below dairy requirements (<40 mg/kg DM) in 2020 (UK), and 2022 (UK & Ireland). Iodine was below the lowest requirements (<0.45 mg/kg DM) in 2020 (UK), 2021 (Ireland), and 2023 (UK). Whilst there is substantial individual variation, these data suggest that grazing ruminants in the UK and Ireland would benefit from further trace mineral investigation and supplementation. The authors thank Bimeda UK & ROI for the use of their anonymised data.

Does size matter? Comparing liver sample sizes for trace element status

A.H. Clarkson[1], J. Angel[2] and N.R. Kendall[1]
[1]University of Nottingham, Sutton Bonington Campus, Leicestershire, LE12 5RD, United Kingdom, [2]Wern Vets, Department of Research and Innovation, Unit 11, Lon Parcwr Industrial Estate, Ruthin, Denbighshire, LL15 1NJ, United Kingdom; andrea.clarkson@nottingham.ac.uk

Liver tissue samples are valuable for defining trace mineral status in ruminants to estimate supplemental need. Liver can be used for essential trace elements such as copper, cobalt, selenium, manganese, and zinc. These minerals play important roles in reproduction and production and their deficiency or over-supply can result in compromising health conditions or death. The use of liver tissue samples provides different information to blood, providing a longer-term picture. For these elements the hepatic concentration adjusts slowly over several months compared to blood, and therefore liver concentrations allow a better understanding of changing trends and directions for informing supplemental decisions across a herd, flock, or management group. Due to the refinement of biopsy technique, liver recovery sizes are much lower (~40% less) than when the technique was first developed. The advantages of this are considerable, but it carries the risk that the smaller sample size may not provide sufficient tissue for accurate decision making. The minimum sample size of liver required for accurate representation is unclear. It is necessary to determine if there is a minimum liver tissue recovery size (from slaughter or biopsy) which can be used as a reliable indicator of trace mineral status. Three different livers >300 g were prepared into duplicate samples within four wet weight ranges; 0.04-0.14 g, 0.15-0.30 g, 0.31-0.65 g and 0.66-0.80 g. Samples were frozen then freeze-dried prior to acid-digestion and ICP-MS analysis. Data was analysed for descriptive statistics and statistical significance using GLM ANOVA with Tukey comparisons using Minitab (v.19) As expected, mineral concentration varied substantially between livers. However, there was no significant difference in trace mineral concentration between the weight ranges for Zn, Cu, Mn, Se, and Co in any of the individual livers. Indicating that the use of small size liver samples and biopsies as a measurement for hepatic concentration of trace minerals is effective in comparison to the use of larger cut liver samples.

Investigation on the variation in phosphorus of wheat bran: effect in diets of fattening pigs

R. Puntigam, P. Riesinger, A. Honig, M. Schaeffler, W. Windisch and H. Spiekers
Bavarian State Research Center for Agriculture, Prof.-Dürrwaechter-Platz 3, 85586, Germany; wilhelm.windisch@tum.de

Wheat (W) is the third most produced crop worldwide and the production volume increased markedly. Hence, the amount of wheat bran (WB) increased accordingly as it is a by-product of flour production. Although WB is locally processed, the high amount of phytate-bound phosphorus (P), the risk of mycotoxin contamination and its bulkiness limit the use. Especially the reduction of P in diets is an important strategy to minimise its excretion and thereof environmental pollution. Contrary it gains more interest in regard of animal welfare and reducing the competition for food between humans and pig production due to very low human-edible fraction. Since fibre and P are concentrated in the outer layers, the efficiency of the milling process is crucial for the composition of the fractions afterwards. Our hypotheses were that: (1) cultivar contribute to variation in the nutritive value of WB; (2) there is a relationship between the nutrient content in W and WB; and (3) WB is a suitable feedstuff in diets of fattening pigs. The material comprised dry milled wheat samples of eleven cultivars which were grown on three locations in Germany (3 replicates per cultivar per location; 99 W samples and 99 WB samples). Samples were analysed using the official methods of the VDLUFA. Data were analysed using the MIXED procedure and pairwise t tests of SAS (SAS Institute Inc.). The amount of WB was 220±19 g (min.: 162; max. 255 g) per kg W. Cultivar influenced CP and P content in WB ($P<0.05$). Mean content of CP in W was 125±14 g (min.: 89; max. 153 g) and P 3.4±0.7 g (min.: 2.2; max. 4.7 g) while in WB P content was 14.1±1.5 g (min.: 11.0; max. 17.7 g) per kg DM. The P in WB was increased by 322±63% (min.: 454; max. 187%). There was no correlation between the content of CP or P in W and the content of P in WB ($r^2<0.01$). Based on P content (min.: 9.8; mean: 12.6; max. 15.8 g/kg) diets for fattening pigs were calculated. 5% in the starter-, 10% in grower- and 15% in finisher-period of WB with increasing amount of P resulted in mean content of P in the diets of 4.2, 4.5 and 4.8 g/kg, respectively. Diet calculated using these levels of WB (approx. 26 kg) reduced 20 kg wheat per fattening pig.

Session 88 Poster 17

Swiss Feed Database: a closer look at minerals and trace elements in 10 years of roughage surveys

E. Manzocchi[1], *M. Lautrou*[2] *and P. Schlegel*[1,2]
[1]*Agroscope, Ruminant Nutrition and Emissions, Tioleyre 4, 1725 Posieux, Switzerland,* [2]*Agroscope, Swine Research Group, Tioleyre 4, 1725 Posieux, Switzerland; elisa.manzocchi@agroscope.admin.ch*

The native mineral and trace elements supply from roughages in the basal diet of ruminants, requires full attention if a sustainable and economical supply of these nutrients is to be achieved. However, mineral and trace element concentrations in roughages need to be analysed regularly or can be retrieved from feed tables with reference values. In Switzerland, an annual survey of the quality of conserved roughages, including barn-dried and field-dried hay, as well as grass and whole-plant maize silage, has been carried out since 2012. A large proportion of all roughage samples collected on Swiss farms are analysed in two commercial laboratories for major organic nutrients and estimation of energy and protein value. At the explicit request of the farmers or extension services supplying the samples, concentrations of macro minerals (Ca, P, Mg, Na, K, S), trace minerals (Cu, Fe, Mn, Zn) and trace elements (Mo, Co, I, Se) are analysed by wet chemistry (ICP-AES/MS). All data are compiled annually in the Swiss Feed Database. Over the last 10 years the database has been expanded each year with 698±77 (mean ± standard deviation), 406±170, and 108±51 samples for macro minerals, trace minerals, and trace elements in hay, respectively; with 121±39, 84±38, and 10±8 samples analysed for macro minerals, trace minerals and trace elements in grass silage, respectively. The database on whole-plant maize silage was increased annually by 48±25, 23±16 and 6±3 samples for macro minerals, trace minerals, and trace elements, respectively. Samples are geolocated to regions and altitudes, and herbages are attributed to one of seven main grassland types as declared by farmers or advisors. The data collected from this nationwide survey deliver unique, annually updated information on mineral and trace element concentrations in a large number of roughages produced on Swiss farms. Based on these data, we present trends in mineral and trace element concentrations in roughages, compare them with existing reference values, and finally show the importance of the native mineral and trace element supply from roughages in typical roughage-based dairy diets.

Session 88 Poster 18

Mineral forage value in the INRAE feeding system for ruminants

A. Boudon[1] *and G. Maxin*[2]
[1]*PEGASE, INRAE, Institut Agro, 35590 St-Gilles, France,* [2]*UMR Herbivores, INRAE-VetagroSup, Univ. Clermont Auvergne, 63122 St-Genès-Champanelle, France; anne.boudon@inrae.fr*

Mineral supplementation is necessary for ensuring optimal growth, health and productivity of productive ruminants for almost all diets. However, it also represents an economic cost and can induce supplementary losses of mineral elements in the environment. Thus, mineral supplementation must be adjusted as accurately as possible to insure that the total mineral supplies for animals are equivalent to their requirements. For ruminants, a significant part of the nutrient, and more specifically mineral, supplies are provided by the forages. The mineral contents of forages is highly variable, even within a farm, if we considered the evolution of plant mineral content according to the species present in the field, the meteorological conditions and the plant physiological stage. Thus, the provision of tables of forage mineral composition seems to be an interesting tool to allow a first assessment of mineral supplies allowed by the basal diet before choosing adequate supplementation. The aim of the presentation is be to make a synthesis on the table of forage mineral contents backed in the INRAE feeding system and to share some perspective of evolution. The INRAE feed tables for forage is based on 740 mineral analyses in fresh forages (mainly herbages). Correlations with forage crude protein contents were used to adjust the effect of the plant physiological stage of mineral contents. For macro-element (Ca, P, Mg, Na, K, Cl and S), the effect of conservation was assessed from comparisons between fresh forages and hays and fresh forages and silages. Given that P, Ca and Mg requirements are predicted thanks to a factorial approach in absorbable amounts, true absorption coefficient are determined according to the forage family and the conservation methods according to syntheses from published data. New analytic methods could now allow cheaper and quicker mineral analyses of forage. Thus, emerging questions about these tables are to evaluate their representativeness considering the diversity of pedoclimatic context and species grown in France and to determine if their use can allow a reasonable assessment of mineral supplies allowed by the basal diets, especially for trace elements.

Session 88 Poster 19

Minerals in cattle nutrition – meet the needs!

A.C. Honig[1], V. Inhuber[2], H. Spiekers[1], W. Windisch[2], K.-U. Götz[1], G. Strauß[1] and T. Ettle[1]
[1]Bavarian State Research Center for Agriculture, Prof.-Duerrwaechter-Platz 3, 85586 Poing, Germany, [2]Technical University of Munich, Liesel-Beckmann-Strasse 2, 85354 Freising, Germany; wilhelm.windisch@tum.de

Adequate macro mineral supply in cattle feed ensures optimal body functions and animal growth. Feeding cattle according to their nutrient requirements is a crucial factor in reducing environmental impact. Particularly phosphorus (P) is an important issue in livestock nutrition because high P excretions can cause environmental pollution due to excessive phosphate leaching into surface water. Furthermore, P is a finite resource and its use in feed should be limited to the animals' requirements. Mineral requirements depend on the animals' physiological needs for maintenance and performance, which change during animal growth. To assess these requirements, a feeding experiment ending with a serial slaughter trial was conducted with 72 growing Fleckvieh (German Simmental) bulls. For the fattening period, bulls were randomly allocated to normal energy and high energy treatment groups fed 11.6 and 12.4 MJ ME/kg DM, respectively. Differences in the TMRs' energy concentrations were reached by varying the percentage of maize silage and concentrates. The bulls were slaughtered in five final live weight groups of 120, 200, 400, 600, and 780 kg and their entire empty bodies were analysed for their macro mineral (Ca, P, Na, K, S, Mg) concentration and accretion. Increasing the amount of concentrate and mineral supplement in the diet increased the animals' daily macro mineral intake but did not affect body mineral composition or accretion. Experimental data can be used to adjust the feeding recommendations for mineral requirements of growing Fleckvieh bulls. Comparing calculated daily P requirements and daily P intake reveals a P surplus in higher weight groups. The excess of P arose because mineral intake increased during growth, but mineral accretion declined. This was especially true for bulls in the high energy group, which showed higher P intake due to high concentrate feeding. Hence, phase feeding should be used to feed growing cattle according to their mineral requirements and to reduce P excretion and the resulting environmental impact.

Session 88 Poster 20

Ruminant feed rations rich in clay minerals may induce Zn deficiency

M. Schlattl, M. Buffler and W. Windisch
Technical University of Munich, Chair of Animal Nutrition, Liesel-Beckmann-Str. 2, 85354 Freising, Germany; wilhelm.windisch@tum.de

In ruminants, zinc (Zn) requirement is low compared to monogastric livestock and native dietary Zn contents may be sufficient at low levels of performance. But ruminants may also ingest high amounts of clay from soils via roughages, which may depress Zn bioavailability. The present study tested the potential of a total mixed ration (TMR) rich in clay to induce Zn deficiency in animals living at maintenance level. The TMR consisted of hay, alfalfa, concentrate and clay (45.9, 28.2, 14.3 and 11.2% in dry matter (DM)). Crude ash (CA) in DM was 20.5%, which is within the range of CA in grass silage. Nutritional properties of the TMR met recommendations except for Zn (27.4 ppm). The TMR was either fed as is (ZN[-]) or was added with Zn (83.1 ppm) to ensure sufficient Zn supply (ZN[+]). TMR variants were fed to eight fistulated, non-lactating dairy cows at restricted amounts (7.91 kg DM/d) according to a balanced Latin Square Design involving two time periods of each 6 weeks adaptation and 2 weeks quantitative collection of feed intake, faeces and urine, and sampling of duodenal chyme and blood. Samples were analysed for Ca, Mg, Cu, Fe, Mn, Zn, and duodenal flux was estimated via TiO_2 (0.5% in TMR). Statistics included ANOVA with animal as fixed block and dietary treatment (ZN[-], ZN[+]). Data on Ca, Mg, Cu, Fe, and Mn were not affected by Zn treatment. Compared to ZN[+], feeding ZN[-] depressed Zn intake (659 vs 216 mg/d; $P<0.05$)), and faecal Zn excretions (505 vs 254 mg/day; $P<0.05$). Apparent digestibility dropped from 22.2 to -14.4% ($P<0.05$) indicating a net loss of Zn from body stores when fed ZN[-]. Concomitant duodenal Zn fluxes of 324 mg/d at Zn[-] exceeded Zn intake, and indicated an Zn influx from body stores into the digestive tract. In ZN[+] duodenal Zn flux of 588 mg/d ($P<0.05$) indicated Zn disappearance from feed towards inside the body. Plasma Zn did not markedly differ (708 vs 685 ppm, $P>0.05$). In conclusion, the TMR rich in clay minerals seemed to bind feed zinc along the entire digestive tract and hence to negatively affected Zn bioavailability. Native dietary Zn contents were not sufficient to meet the animals' requirements of Zn. In total, dietary contaminations with clay seem to be a risk factor for Zn deficiency in ruminants.

Session 88 Poster 21

Zn, Cu, Mn and Fe balance in dairy calves fed milk replacer or whole milk at two feeding allowances

T. Chapelain, J.B. Daniel, J.N. Wilms, L.N. Leal and J. Martín-Tereso
Trouw Nutrition R&D, P.O. Box 299, 3800 AG, Amersfoort, the Netherlands; tchapela@uoguelph.ca

Trace minerals inclusion in milk replacer (MR) largely exceed natural mineral occurrence in whole milk (WM). Our objective was to quantify trace mineral balance in calves fed either WM or MR at two feeding allowances (L: 4.5 l/d and H: 9.0 l/d) during their first 13 weeks of life. Forty-eight newborn Holstein-Friesian calves (2±1.0 d of age; 45.0±4.37 kg body weight) were enrolled after receiving a standardized colostrum administration. Upon arrival at the facility, calves were blocked by age and arrival date and randomly assigned to one of the four treatments. The MR was fed at 169 g/l of final product to provide the same amount of metabolizable energy as WM (3.4 MJ/l). The concentrations of zinc (Zn), copper (Cu), manganese (Mn) and iron (Fe; mg/l) in MR and WM were: 10.14 vs 4.64, 1.82 vs 0.07, 4.95 vs 0.03, and 12.19 vs 0.25, respectively. Starter feed, and straw were introduced at week 6, and calves were gradually weaned from week 6 to 10 and studied up to week 13. Complete 24 h collection of urine and faeces were performed on weeks 2, 4, 5, 7, 9, 11 and 13. All data were analysed using PROC MIXED (SAS 9.4). Preweaning, Cu, Mn and Fe, balances were higher (P<0.01) for MR fed calves than WM fed calves. Increasing milk allowance increased Cu and Fe balances for calves fed MR but not for calves fed WM (P<0.01). For Zn, increasing milk allowance increased Zn balance for WM calves, but decreased Zn balance for MR calves. During weaning, feeding higher milk allowance resulted in lower balances of Zn, Cu and Mn (P<0.01), and Fe (P=0.08). During that time, Cu balance was further reduced for WM fed calves (P=0.04), whereas Zn balance was increased by feeding WM (P<0.01) as compared to feeding MR. Postweaning, greater Cu (P<0.01) and Fe (P=0.04) balances were observed for WM vs MR-fed calves. Although trace mineral balance was lower in WM-fed calves during preweaning, no negative impacts on health and performance were observed. These results highlight the opportunity for further research to review mineral inclusion in MR.

Session 88 Poster 22

Impact of drinking water salinity on lactating cows

A. Iritz[1,2] and Y. Ben Meir[2]
[1]The Robert H. Smith Faculty of Agriculture, Food, and Environment, The Hebrew University of Jerusal, Animal Science, Herzl Street 229, 7610001 Rehovot, Israel, [2]Agricultural Research Organization, Ruminant Science, 68 HaMaccabim Road, 7505101 Rishon LeZion, Israel; adi.iritz@mai.huji.ac.il

Increase use of desalinated sea water in Israel for urban, industrial and agricultural uses led to drinking water with low minerals and salts content as reflected by its electrical conductivity (EC). We measured the EC through the year and observed changes from 0.3 mS/cm to 0.9 mS/cm, depends on the precipitation. Daily free water intake (FWI) of lactating cows in the Volcani Center is circa 140 l/d. The large amount of consumed water, combined with the recent changes in the water composition, encouraged us to explore the possible effects of this on various parameters related to cow's physiology and performance. To enable the recording of the cows individual FWI and to control its minerals composition, we designed and build a systems composed of 4 individual troughs with gates allowing only designated cow to enter. Each trough were fed by 500 l water tank that was placed 4 meter higher. The tanks were filled using an irrigation robot with water and salt solution at the predetermined EC level. When the cow drank from the trough, a tap opened and water flowed from the tanks. A hydrometer was installed at the exit of each tank to record the amount and time of the water flow. Four multiparous lactating Israeli Holstein cows were offered, in a Latin square design trial, drinking water at four different EC levels: 0.4, 0.6, 0.8 and 1.0 mS/cm. Each period of the trial lasted 18 days, including 5 days of adaptation and 13 days for measurements of feed and water intake, milk yield and rumen pH. Faeces, urine and rumen fluid sampled at each period to determine digestibility, VFA content and profile, and minerals balance. Drinking water salinity affect feed intake and milk yield that was highest when cows offered water with higher EC (intake of 29.7 kg/d, 29.6 kg/d, 28.3 kg/d and 27.5 kg/d for EC of 1.0 mS/cm and 0.8 mS/cm, 0.6 mS/cm and 0.4 mS/cm respectively, P=0.01 and milk yield of 47 kg/d, 45.4 kg/d, 45.4 kg/d and 44.6 kg/d for EC of 1.0 mS/cm and 0.8 mS/cm, 0.6 mS/cm and 0.4 mS/cm respectively, P=0.01). Water salinity did not affect FWI that was 142 l/day.

Session 88 Poster 23

Seasonal variation of trace essential and toxic minerals in milk and blood of dairy ewes

A. Nudda[1], M.F. Guiso[1], G. Sanna[2], A. Cesarani[1], M. Deroma[1], G. Pulina[1] and G. Battacone[1]
[1]University of Sassari, Department of Agriculture, Viale Italia 39A, 07100, Italy, [2]University of Sassari, Department of Chemical, Physical, Mathematical and Natural Sciences, Via Vienna 2, 07100, Italy; battacon@uniss.it

A survey was carried out to characterize the trace essential and toxic minerals in milk and blood of Sarda dairy ewes during whole lactation. Milk samples were collected from sheep dairy farms in Sardinia (Italy) from the beginning to the end of lactation every 1.5 months. Ten animals per farms were selected. Microminerals were measured in milk, blood, and feeds samples. Minerals were determined on samples using the Inductively Coupled Plasma Optical Emission Spectrometry preceded by mineralization. Data were analysed with a linear mixed model with diet, sampling, and their interaction as fixed effects and animal as random effect. The potential of MIRS to predict the essential and toxic minerals content of sheep milk has been evaluated. Days in milk significantly affected all mineral contents. Selenium (Se), zinc (Zn) and copper (Cu) showed a peak at the beginning of lactation and then decreased during lactation. The manganese content had a peak in the spring. The lead (Pb) had a peak at the beginning of lactation and then declined, while cadmium (Cd) had the opposite pattern; at all stages of lactation the values were below the EU upper limit in milk products. The nickel (Ni) content in blood exhibited no clear pattern during lactation. The mineral contents of feeds partly explain the mineral concentrations found in milk and blood. This study provided new information on the mineral quality of milk from Sarda dairy sheep and the relationship with minerals in feeds. Moreover, this research could help to identify faster and innovative methods such as mid-infrared spectroscopy for the determination of minerals in sheep milk. Acknowledgement: This research was funded by the Autonomous Region of Sardinia SELOVIN project (CUP: J86C17000190002).

Session 89 Theatre 1

Measuring the behaviour of lambs in an isolated environment with artificial intelligence methods

B. Benet and R. Lardy
University of Clermont Auvergne, INRAE, VetAgro Sup, UMR Herbivores, 63122 Saint-Genest Champanelle, France; bernard.benet@inrae.fr

An experiment was conducted in an enclosed 16 m^2 test pen, gridded into 1 m^2 zones, to study the behaviour of a group of 39 lambs placed individually in the pen, for periods of 5 minutes. The aim was to study the behaviour of two groups, focusing on positioning, orientation and mobility, in order to characterise their emotional reactivity. The two groups corresponded to lambs that had previously been in an enriched environment (platforms, balls, brushes, etc.). For this study, videos were recorded for 5 days (6 hours/day). Artificial intelligence methods combined with image processing techniques were developed to automatically measure lambs' behaviour. From each 5-minute video recorded for a given lamb, the developed software first recognised the lamb and determined its position in the pen, using a GoogleNet-type neural network, with 90% accuracy, taking into account the different orientations and positions of the animal in the scene. It then automatically identified the two parts of the lamb's body (head and hindquarters). For this second step, a database of 1,500 images divided into two parts (80% for learning and 20% for testing) and annotated by experts, was used to develop an Inception-type neural network (using the Tensorflow tool in Python) that identified the front and back parts of the lamb. The success rate of this body part identification was 90%. Finally, an image processing was used to identify the orientation of the animal, in order to determine the areas in its field of vision. These operations made it possible to measure the time spent by the lambs in the different areas of the pen, to measure their mobility by looking at the speed of their movements, the evolution of their movements and their orientation over time, to identify the areas of interest of the lambs in the pen, and to make comparisons between the lambs but also between the different days for each lamb. This lamb behaviour study showed the potential of artificial intelligence to automatically analyse certain animal behaviours from video, thus facilitating behavioural analysis and extending the measurement ranges over long experimental periods.

Session 89 Theatre 2

Estimating sow posture from computer vision: influence of the sampling rate

M. Bonneau[1], J.A. Vayssade[1] and L. Canario[2]
[1]INRAE, ASSET, Domaine Duclos, 97170 Petit-Bourg, Guadeloupe, [2]INRAE, UMR1388 GenPhySE, 31326 Castanet-Tolosan, Guadeloupe; mathieu.bonneau@inrae.fr

Advances in computer vision (CV) offer a valuable tool to monitor animal behaviour over the long term. CV tools are particularly interesting as they are non invasive and could be used to derive fine behaviour traits. On the other hands, CV quickly produces a large amount of data. This could be problematic in practice, as it implies an important organization in terms of data storage and transfer, but also for data analysis. One way to reduce the amount of generated data is the sampling rate. In this article, we were interested in estimating the daily sow activity time budget. More precisely to estimate the proportion of time the sow spent in one of these 6 postures: (1) knee; (2) sitting; (3) standing; (4) sternal; (5) udder left; and (6) udder right. The analysis was based on 23 different samples. One sample is the monitoring of a sow posture during one day. 12 different animals were considered, some of them for several days. Monitoring was done after parturition on sows kept in crate. For each sample, a reference postural time budget (PTB) was estimated from the raw camera data, at 12 frames per second. For each frame, the posture was estimated using a convolutional neural networks. Then, the PTB was also estimated using different frame rate, from one frame per second to one frame per hour. For each frame rate, the difference with the reference PTB was computed. We found that the average difference with the reference PTB was not significantly different using the lower sampling rates. On the contrary, the variance of the difference was different and increases with a decreasing sampling rate. Our data showed that using a sampling rate of 1 image every 30 seconds was sufficient to obtain less than 2% of difference with the reference PTB, with probability 95%.

Session 89 Theatre 3

Estimating pig mass in a high-density pig group during transportation

V. Bloch[1], A. Valros[2], C. Munsterhjelm[2], M. Heinonen[2], M. Tuominen-Brinkas[2], H. Koskikallio[2] and M. Pastell[1]
[1]Natural Resources Institute Luke (Finland), Latokartanonkaari 9, 00790 Helsinki, Finland, [2]University of Helsinki, Research Centre for Animal Welfare, Faculty of Veterinary Medicine, Department of Production Animal, P.O. Box 57, 00014, Helsinki, Finland; victor.bloch@luke.fi

Measuring mass of pigs would allow improved assessment of pig quality when pigs are sold as finishers. Many farms are not equipped with individual scales, hence a convenient possibility for estimation of the pig mass and it's diversity can be in connection to transportation of the pigs. Usually, the pig mass is estimated with the help of 3D cameras when different factors influencing the estimation accuracy are eliminated, particularly, when a pig is separated from other pigs. However, during the transportation, the pig density is high, which causes multiple contacts and occlusions. For these conditions, a method accurately separating individual pigs and detecting their body parts must be used. In this study, a mask R-CNN model was used to extract the pig silhouette in RGBD images. The silhouette was used to define a point cloud related to the pig's back. The pig mass was estimated based on the volume calculated under the surface of the pig's back. In an experiment conducted on a commercial farm, 100 pigs were transported in four groups of 22-26 pigs. The entire group was recorded by a 3D camera installed above a door of a transportation truck in a bounded elevating platform 2×4 m^2. The pigs were marked and weighed before recording. The mask R-CNN was trained on 20 images including more than 400 annotated pigs. A pretrained model was trained on available datasets and used for transfer learning. Preliminary result indicated that the mass estimation based on a 3D point cloud and extraction of a pig point cloud from a dense group can be used for practical purpose on commercial farms.

Session 89 Theatre 4

Assessment of piglet maturity at birth using computer vision

L. Maignel[1], R. Mailhot[2], A. Carrier[2] and P. Gagnon[2]
[1]Canadian Centre for Swine Improvement, 960 Carling avenue Building 75, K1A 0C6 Ottawa, Ontario, Canada, [2]Centre de développement du porc du Québec, Tour Belle Cour, Place de la Cité, Tour Belle Cour 450, 2590, Boulevard Laurier, G1V 4M6 Quebec City, Quebec, Canada; laurence@ccsi.ca

Piglet maturity at birth is strongly linked to early survival. It can be assessed using specific morphologic characteristics, mainly based on skull dimensions and body/head proportions. Mature piglets have well developed organs at birth, especially the liver and small intestine, and more body fat, both of which are important characteristics for postnatal survival and growth. Immature piglets, also called IUGR (intra uterine growth retardation) piglets can be recognized by a specific head shape, as well as characteristic eyes and ears. Maturity scoring can be done visually after some basic training and can be assessed when piglets are weighed at birth. Developments in artificial intelligence have led to new tools such as the Pic'Let system (Neotec Vision, France) developed to classify piglets into normal/mild/severe maturity level at birth using machine learning for the classification of digital pictures of piglets' heads collected shortly after birth. A group of 692 commercial piglets born from 47 Yorkshire × Landrace sows were tracked at the Sow Research and Training Barn of the *Centre de développement du porc du Québec* (CDPQ) in Armagh, Québec. Each piglet was weighed and classified with the Pic'Let prototype within 24 hours after birth. Each animal was also assigned a subjective maturity score (0=normal/1=mildly immature/2=severely immature) at the barn, as well as a visual maturity score based on piglet digital pictures. Various body measurements were also collected including body, head and femur lengths, chest circumference and distance between the eyes. Preliminary results showed moderate concordance levels between the in-barn visual scores and Pic'Let classification (66, 59 and 34% for scores 0, 1 and 2, respectively). The accuracy increases when the comparison is made on piglets where both subjective evaluations agree (67, 80 and 83% for scores 0, 1 and 2, respectively). Overall, the Pic'Let system provided valuable information on crossbred piglets in this study, but its accuracy could be improved by expanding its reference photo bank.

Session 89 Theatre 5

Validation of new software (r-Algo) for predicting meat chemical composition from ultrasound images

B. Ahmadi[1], T. Schwarz[2] and P.M. Bartlewski[1]
[1]University of Guelph, Department of Biomedical Sciences, 50 Stone Road, Guelph N1G 2W1, Canada, [2]University of Agriculture in Krakow, Department of Animal Genetics, Breeding and Ethology, 24/28 Mickiewicza Avenue, 30-059 Cracow, Poland; ahmadib@uoguelph.ca

There is no non-invasive method to accurately and repeatedly predict the proximate chemical composition and fatty acid profile of meat from observations in live animals. The objective of this study was to validate a novel computerized method of ultrasound image analysis to determine chemical composition of pectoralis major muscles in broiler chickens. Ultrasonograms of the pectoral muscles in the longitudinal and transverse plane were obtained from forty birds on the day of slaughter. All chemical constituents of muscle samples were determined with the validated and standardized laboratory techniques, and the results served as a benchmark for developing the present algorithmic estimates of chicken meat composition. An in-house developed algorithm (r-Algo) was used to normalize the ultrasonograms and to identify pixel intensity ranges for which linear correlations between mean numerical pixel values (NPV) and the content of various chemical constituents were the strongest (based on the values of correlation coefficients), using a stepwise sequestration of ultrasound bitmaps. Percentages of chemical constituents were the dependent (accepted) variables and the results of echotextural analyses (luminance or pixel intensity), carried out with a commercially available image analysis software (ImageProPlus), were the explanatory variables. The predictive regression equations were determined in thirty randomly selected algorithm-training experimental units, and their accuracy was tested in a subset of ten birds allocated to the algorithm-validation group. Significant determination coefficients were found for all chemical constituents studied, with the accuracy ranging from 62.70% (linoleic acid, transverse plane, pixel range of 141-142) to 96.65% (total hypocholesterolemic acids, longitudinal plane, pixel range of 136-150). The results of the present validation study indicate that accurate prediction of muscle chemical composition using echotextural image analyses is feasible after identifying specific pixel intensity ranges.

Session 89 Theatre 6

Coupling a sow herd model with a bioclimatic model of gestation rooms: development and evaluation

E. Dubois[1], F. Garcia-Launay[1], N. Quiniou[2], M. Marcon[2], J.Y. Dourmad[1], D. Renaudeau[1] and L. Brossard[1]
[1]PEGASE, INRAE, Institut-Agro, 35590 Saint-Gilles, France, [2]IFIP, Institut du Porc, 35651 Le Rheu, France; dubois@itavi.asso.fr

Global warming and the increased frequency of heat waves intensify the risk for reproductive sows to be exposed to temperatures above their thermoneutrality zone. This affects the overall sow herd performance through feeding behaviour perturbation, reduced prolificacy, unsuccessful insemination, or increased mortality. A model was developed on Python to simulate dynamic interactions between ambient temperature in gestation rooms and individual performances of sows in the herd. Dynamics of ambient temperature result from thermal balance between heat losses (walls, air renewal) and heat sources (animals, heaters) in the room, depending on feeding management, the fan and heating systems management, walls' thickness and conductivity, and outside temperature. The model runs on an hourly time step and couples an individual-based module of herd management and feeding practices with a model of thermal balance in the room. The herd module applies discrete events associated to sow reproduction and farmer's practices. It also describes rooms with reproduction stage, number of animals and occupation time. The heat production of each sow is modelled as a function of its gestation stage, live body weight, production objectives (backfat thickness and litter size at farrowing) and diet. Data from three batches of sows were used to evaluate the model's ability to predict sows' performance and the effective ambient temperature within a gestation room. The sows' ingestion and live weight evolution were similar between observed and simulated data. Simulated temperatures were similar to the observed temperatures (mean error=0.3, 0.2 and 0.8 °C; RMSEP=0.8, 0.9 and 1.2 °C, respectively for the three batches). However, the simulated variation amplitude is larger than the observed one, especially during heat peaks. A better description of the hourly distribution of feed intake, individual activity level, and regulation rules of fan and heating systems should improve the accuracy of model predictions. The model will be integrated into a decision support tool to assess the vulnerability of swine systems to climate change.

Session 89 Theatre 7

Supervised machine learning as a tool to improve farrowing monitoring and stillborn rate in sows

C. Teixeira Costa[1], G. Boulbria[1], C. Dutertre[2], C. Chevance[1], T. Nicolazo[1], V. Normand[1], J. Jeusselin[1] and A. Lebret[1]
[1]Rezoolution, ZA de Gohélève, 56920 Noyal-Pontivy, France, [2]HumanoIA, 338 route de Philondex, 64410 Cabidos, France; c.teixeira-costa@rezoolution.fr

On average, more than 60% of sows give birth to stillborn in French farms. It is an important cause of piglet mortality which continue to increase with the constant improvement of sow's prolificacy. The objective of this study was to build a predictive model of stillborn rate. This study was performed in two farrow-to-finish farms and one farrowing farm located in Brittany. In each farm, number of total born (TB), born alive (BA), stillborn piglets (S), same data at previous farrowing (TBn-1, BAn-1 and Sn-1), backfat thickness (BFT) just before farrowing and at weaning, and parity rank were added to our dataset. In total, 3,686 farrowing data were recorded. Bayesian networks as an integrated modelling approach were used for analysing stillborn rate at farrowing using BayesiaLab® software. Our results propose a hybrid model to predict the stillborn percentage during farrowing. Three significant main risk factors were retained by the model, namely, the parity rank (percentage of total mutual information: MI=64%), Sn-1 (MI=25%) and TBn-1 (MI=11%). Additionally, a fourth factor (BFT just before farrowing) was also retained for sows of parity five or more (MI=0.4%). Practically, for example, in the best conditions (i.e. low litter rank, less than 8% of stillborn and a prolificacy lower than 14 piglets at the previous farrowing), our model predicted a stillborn rate almost divided by two from 6.5% (mean risk of our dataset) to 3.5% for a sow at the next farrowing. On the opposite (i.e. older sows with a BFT<15 mm, more than 15% of stillborn and a prolificacy higher than 18 piglets at the previous farrowing), the risk would be multiplied by 2.5 from 6.5 to 15.7%. In our knowledge, it is the first study updating the Blackwell grid published in 1987. Our results highlight the impact of previous prolificacy and stillborn rate on the probability of stillborn. Moreover, the importance of backfat thickness, especially on old sows, is to be considered. These hopeful results would allow the farmer to classify sows according to their risk of give birth to stillborn and to manage them accordingly.

Session 89 Theatre 8

Use of accelerometers to predict the behaviour of growing rabbits

M. Piles[1], J.P. Sánchez[1], L. Riaboff[2,3], I. David[4] and M. Mora[1]
[1]IRTA, Torre Marimon s/n, Caldes de Montbui (Barcelona), 08140, Spain, [2]VistaMilk SFI Research, Teagasc Moorepark, Fermoy, Co. Cork, P61 C996, Ireland, [3]School of Computer Science, University College Dublin, Belfield, Dublin 4, D04 V1W8, Ireland, [4]GenPhySe, Université de Toulouse, INRAE, Castanet Tolosan, France, 31320, France; miriam.piles@irta.cat

One of the challenges of precision livestock is to find tools to quantify individually the time that animals spend on each of the different activities. In this study, we propose to use accelerometers to predict the type of activity that growing rabbits are engaging at each moment. Eight rabbits (4 fed under restriction and 4 fed *ad libitum*) were equipped with an accelerometer and filmed simultaneously. A total of 4.5 hours of video were annotated, identifying 7 types of activity: eating (E), drinking (D), moving in-situ (M), walking (W), grooming (G), lying (L) and sitting (S). These types were grouped into 4 classes of activities: eating (E), drinking (D), resting (R, L+S) and general movement (GM, M+W+G). Accelerometer signals and video annotations were manually synchronized. The signal was segmented into windows of 2 seconds, in which a total of 47 features were extracted from the three axis in time and frequency domain. They included position and dispersion parameters. A total of 6,658 windows were retained. Using the training set (80%), a 3-fold cross validation was conducted to optimize the hyperparameters and to select the 20 most informative features with random forest. This algorithm was implemented to construct the prediction model with all training data given the hyperparameters previously obtained. The overall classification accuracy was 0.74. The best predicted class was E with an accuracy and a recall of 0.83 and 0.75, respectively. The worst predicted class was GM with an accuracy and a recall of 0.66 and 0.50, respectively, primarily confused with class R. These results are expected to improve by increasing data from less frequent classes synthetically and/or by annotating more images in other batches of animals. These results show that the accelerometer seems to be a suitable tool to quantify the time that rabbits devote to different daily activities. However, more data should be annotated and other algorithms are worth testing to improve predictive models.

Session 89 Theatre 9

Use of accelerometry data to detect kidding in goats

P. Gonçalves[1], M.R. Marques[2], A.T. Belo[2], A. Monteiro[3] and F. Braz[4]
[1]Universidade de Aveiro, Campus Universitário de Santiago, 3810-193 AVEIRO, Portugal, [2]Instituto Nacional de Investigação Agrária e Veterinária, Av. Professor Vaz Portugal, 2005-424 Vale de Santarém, Portugal, [3] Escola Superior Agrária de Viseu, Quinta da Alagoa – Av. Dr. António Almeida Henriques, 3500-606 Viseu, Portugal, [4]Instituto Federal Catarinense, Campus Araquari, Araquari 89245-000, Brazil; pasg@ua.pt

Kidding assistance for goats is a critical factor in reducing problems such as mother and kid mortality arising from birth complications. Creating an autonomous mechanism that allows continuous monitoring of the process would thus allow for timely human intervention. The application of Internet of Things technologies to monitor animal behaviour during kidding events, specifically through the use of wearable inertial sensors, allows for the analysis of the dynamic animal behaviour, the identification of behaviour patterns, as well as the detection of their changes. Present work consisted in the analysis of a public goat kidding dataset generated from the sensorization of 16 pregnant and two non-pregnant Charnequeira goats. The data, obtained by continuous sampling every 10 s during the experiment (day and night), include ultrasound measurements from neck to floor height, and accelerometry data measured by an accelerometer existing at the monitoring collar. It also includes the time the kiddings occurred, how many offspring they produced, and additional remarks were used to annotate the accelerometry data. A deep learning mechanism solution to identify behaviour pattern changes was created, using several concept drift mechanisms to detect and adapt the learning model to the changes in the data stream. Hourly intervals of 5, 4, 3, 2, 1, before, during, and one hour after the kidding interval, were analysed and the accuracy of kidding prediction was measured for each of the algorithms in the referred intervals. Despite the small sample size of animals and kidding events, the results obtained are very promising, especially when using the XGBoost and Random Forest algorithms, with accuracies close to 60%, revealing that the test should be continued so that new kidding events can be observed and the learning model can be improved. Future work plans include integrating the detection mechanism into a gateway to create a trial enabled with real-time kidding detector.

Session 89 Theatre 10

FTIR milk fatty acids quantification for non-invasive monitoring of rumen health in dairy cows

F. Huot[1], S. Claveau[2], A. Bunel[2], D. Warner[3], D.E. Santschi[3], R. Gervais[1] and E.R. Paquet[1]
[1]Université Laval, 2325 rue de l'université, Québec, Qc, G1V0A6, Canada, [2]Agrinova, 640, rue Côté Ouest, Alma, Qc, G8B7S8, Canada, [3]Lactanet, 555, boul. des Anciens-Combattants, Ste-Anne-de-Bellevue, Qc, H9X3R4, Canada; felix.huot.1@ulaval.ca

Our objective was to validate the possibility to detect subacute ruminal acidosis (SARA) from milk FTIR fatty acids (FA) and machine learning. For the prediction of SARA, a total of 562 milk samples (67 Holstein cows, 7 Canadian farms) were used. Every milk sample was associated with its corresponding SARA classification (SARA-positive if the cow spent 300 minutes or more with a reticuloruminal pH under 6.0 or SARA-negative otherwise). Data were split into a training and a validation set consisting of 70 and 30% of animals, respectively. Partial least square models were validated using 3 cross-validation (CV) scenarios: (1) completely random; (2) individual cow allocated to either training or test sets; (3) individual farm allocated to either training or test sets. We also tested 3 sets of predictors: (1) Milk major components, namely fat, protein, lactose, MUN, and SCC (MMC); (2) Milk FA composition (MFA); and (3) MMC coupled with MFA (MCFA). Area under the ROC curve (AUC) was used to evaluate model performance. The effect of the set of predictors was assessed using a mixed model and higher AUC were obtained for both FA and MCFA models (*P*<0.01), suggesting that milk FA provided essential information to the models. We then evaluated models on the validation set and compared the performance of the CV scenarios. Training models on the completely random CV led to the lowest prediction performance on new data (*P*<0.01). Thus, using different animals/farms in the CV is of importance for a better performance evaluation. Overall, the best predictions were obtained when test and training sets were selected based on individual cows and with MCFA predictors (CV-AUC=0.75; Validation-AUC=0.65). Milk FTIR FA quantification enables monitoring of SARA on commercial farms with moderate performance.

Session 89 Theatre 11

Optimizing breeding performance through algorithmic approaches to maximize meat quality in livestock

J. Albechaalany[1,2], M.P. Ellies-Oury[1,2], J.F. Hocquette[2], C. Berri[3] and J. Saracco[4]
[1]Bordeaux Science Agro, 1, cours du Général de Gaulle, CS 40201, 33170 Gradignan Cedex, France, [2]INRAE, UMR1213, Theix, 63122, France, [3]INRAE, UMR BOA, 37380 Nouzilly, France, [4]INIRIA, IMB, 33400 Talence, France; john.albechaalany@agro-bordeaux.fr

Consumers are now increasingly aware of the impact of meat production on animal welfare and the environment. Simultaneously, there has been a decline in meat consumption and a demand for high-quality meat (in terms of sensory as well as nutritional quality). This study aims to propose a methodological approach that uses breeding practices to estimate meat quality, aiming to achieve optimal quality and meet consumer demand. To achieve this goal, we have developed an updated version of NSGA-II (Non-dominated Sorting Genetic Algorithm II). This algorithm generates a set of candidate solutions, selects the best individuals based on their fitness, and applies genetic operators such as crossover and mutation to generate new offspring. The decision space is defined by the variables X related to the management of breeding practices, while the objective space Y represents the variables related to the sensory and/or nutritional quality of the meat to optimize. To ensure accuracy and precision, the fitness value of each objective is assessed using a multiple linear regression model. An AIC (Akaike Information Criterion) approach is then mobilized to select the most relevant model for each objective. Once a new population is evaluated using the selected models, the Pareto front approach is utilized to identify the non-dominant variables in the multi-objective space. In order to prevent the algorithm from getting trapped in local maximum scenarios, a crowding distance method is employed to maintain population variability and to ultimately reach the global maximum. With this approach, we can generate the best breeding practices for each breed/type of animal and optimize quality. Using the hypervolume approach, we can compare the different optimum front scenarios and recommend, for example, the best breed according to the objectives. In conclusion, this study presents an updated methodological approach for estimating meat quality using breeding practices, which has the potential to improve meat quality and meet consumer demands.

Session 89 Theatre 12

Detection of multiple feeding behaviours in calves using noseband and accelerometer sensors

S. Addo, K.A. Zipp, M. Safari, F. Freytag and U. Knierim
University of Kassel, Faculty of Organic Agricultural Sciences, Nordbahnhofstr. 1a, 37213 Witzenhausen, Germany; sowah.addo@uni-kassel.de

Behavioural activities provide a valuable source of information for accessing animal health and welfare. However, gathering such information through direct personal observation is laborious, which makes the use of digital tools a viable alternative. The current study aimed at investigating suckling, feeding and ruminating behaviours of calves using a combination of noseband and accelerometer sensor variables. Ten calves were fitted, each with a ruminating halter containing an accelerometer at the cheek, and a noseband to measure acceleration and pressure, respectively, over a three-day period. Additionally, direct observational data on the various behaviours lasting about ten hours were concurrently recorded. Eighteen features were generated from the raw sensor data for five different epoch length categories (1 s, 5 s, 10 s, 30 s and 60 s). A sixth epoch category was defined by combining the afore-mentioned epoch lengths. Finally, different machine learning algorithms including random forest (RF), classification and regression tree (CART) and conditional inference tree (ctree) were applied to each epoch category to simultaneously predict suckling, feeding and ruminating events. Using 70% of each dataset for training and the remaining as test set, preliminary results on two animals showed very high accuracies for the mixed epoch dataset: 99.7% (RF), 97.8% (ctree) and 85% (CART), while the 60 s epoch dataset had the lowest, being 72% (CART). From the mixed epochs, the estimates of sensitivity, specificity and precision for all behaviours were above 95% considering RF and ctree. Features including the variance in pressure and movement intensity at 60 s intervals proved to be very important in our prediction models. Our findings have a potential for studying changes in feeding behaviours and welfare of calves undergoing natural weaning.

Session 89 Theatre 13

Numerical detection of productive anomalies induced by heat stress in dairy cows

M. Bovo, M. Ceccarelli, C. Giannone, S. Benni, P. Tassinari and D. Torreggiani
University of Bologna, Department of Agricultural and Food Sciences, Viale Giuseppe Fanin 48, 40127, Italy; marco.bovo@unibo.it

Since dairy cattle are in an intensive housing system for most of their lives, facilities have a significant impact on the animals' welfare. Despite the growing interest in finding new animal housing and equipment management strategies for reducing impacts, and the interesting results mostly concerning the daily production data, there are lack of studies investigating the factors that can lead to productive anomalies. On the other hand, the use of automatic milking robots, milking parlours, collars and pedometers allows the precise monitoring of dairy cows, providing farmers with real time information. In this context, the early detection of production anomalies is fundamental for animal health and safety. In this work, two numerical methods for detecting daily milk production anomalies are presented and applied to three different farms selected as case studies. The methods described in this paper provide a numerical procedure having the scope of detecting milk yield anomalies. Both the algorithms presented hereinafter are based on statistical calculations and take as input daily resting time and daily milk yield recorded respectively by the pedometers worn by the cows and by the automatic milking system of the barns. The first method take into consideration two indicators, namely the Difference in Relative Production (DRP) and the Daily Rest time (DR). DRP is defined as the relative difference in daily milk yield between real-time data of a single animal and a baseline curve considered as an ideal trend. An anomaly (i.e. a deviation from the normal value) is determined, for a single cow, for a specific day, if two conditions on DRP and DR are contemporary verified. In the second method, starting from the Wood function, maybe the most famous model to fit the production of the cow in dependence of day in milk, the concept of reliability of robust statistics has been introduced in order to obtain, for each animal, a more solid and realistic lactation curve since not affected by outlier values.

Session 89 Poster 14

Using existing slaughterhouse data to assist detection of boar taint

B. Callens[1], M. Aluwé[1], R. Klont[2], M. Bouwknegt[2] and J. Maselyne[1]
[1]ILVO, Burg Van Gansberghelaan 92 bus 1, 9820 Merelbeke, Belgium, [2]Vion Food Group, Boseind 15, 5281 RM Boxtel, the Netherlands; jarissa.maselyne@ilvo.vlaanderen.be

This study aimed to investigate whether existing slaughter data could be used to identify pig carcasses or batches with (a high risk of) boar taint presence. The human nose method is currently used for this purpose as an instrumental method is lacking, but it is labour-intensive and requires close follow-up. The dataset analysed consisted of more than one million records of entire male pigs collected by Vion over a period of 28 months. The data contains carcass quality statistics like muscle- and fat thickness and lean meat percentage determined by a fibre optic probe, as well as the estimated weight and yield of the different meat cuts determined by the Frontmatec Autofom III. Furthermore, waiting time of the pigs in the slaughterhouse, month of slaughter and the cold carcass weight were included. Principle Component Analysis was used to exclude some outliers. Ten-fold cross-validation was used for the results, and z-score normalisation was applied. A Random Under Sampling Boosting Tree Ensemble model was used to handle the imbalanced data (only around 3.1% boar taint), resulting in an accuracy of 59.5%. Other machine learning methods (including XGB, various SVM, Trees, KNN, ANN, Naïve Bayes, Discriminants) achieved higher accuracy on the majority class (no boar taint), but failed to classify the minority class of boar taint present. A balanced dataset was created by randomly sampling the majority class, but the accuracy remained low even for the best performing balanced models (58-59%). When these balanced models were applied on the full dataset, accuracy was similar (57-59%). Variable importance was determined using local model-specific and model-agnostic methods but the results were not consistent. Overall, it does not seem possible to create a practical working model to identify pig carcasses or batches with a higher risk of boar taint using existing slaughter data alone. Additional data from other sources such as the farm might be needed to create a better performing risk assessment model, as boar taint is a multi-factorial problem. A dedicated boar taint sensor would also be a huge step forward for automation.

Session 89 Poster 15

Evaluation of SARA risk prediction models based on non-invasive measurements in dairy cows

V. Leroux, C. Jaffres, A. Budan and N. Rollet
Neolait, Cargill Animal Nutrition, Voie de la ville Louze, 22120 Yffiniac, France; vincent_leroux@cargill.com

The subacute ruminal acidosis (SARA) is an energy metabolism disorder affecting frequently dairy cows. The measurement of this disorder is difficult to evaluate in farms. The objective of this study is to define the prediction performance of different multiparametric models based on non-invasive measurements to predict SARA in dairy cows. The accuracy of 12 multiparametric prediction models of SARA was evaluated using data coming from 65 articles published in peer review journals, where SARA was evaluated according to specific pH thresholds. The efficiency of the prediction of SARA (EF) ranged from 50 to 81%. The sensitivity and the specificity of the best model were respectively 83 and 80%. The EF reached 85% when three prediction models were combined, with a sensitivity of 78% and a specificity of 89%. One out of 12 models and a combination of 3 of them, all based on faecal pH, milk urea, DMI and fat milk/protein milk were able to predict in a simple and non-invasive way SARA in dairy farms with an efficiency higher than 80%.

Session 90 Theatre 1

Use of linear data for characterization and selection of sport horses with highest genetic potential

K.F. Stock[1], A. Hahn[2], I. Workel[2] and W. Schulze-Schleppinghoff[2]
[1]IT Solutions for Animal Production (vit), Heinrich-Schroeder-Weg 1, 27283 Verden, Germany, [2]Oldenburger Pferdezuchtverband e.V., Grafenhorststrasse 5, 49377 Vechta, Germany; friederike.katharina.stock@vit.de

Routine linear profiling and genetic evaluation for linear traits is giving access to detailed information on the individual horse and its genetics regarding conformation, gaits and jumping. Previous studies have indicated the suitability of genetic linear profiles for selecting horses for performance in riding sport, with generally minor role of conformation and particular value of distinct aspects of gaits for dressage and several linear jumping traits for show jumping. The aim of this study was to quantify the possible impact of strong focus on top sport performance on development towards extremes and risk of overemphasis of certain traits. Analyses were performed for the mare population of the Oldenburg studbooks OL and OS and based on phenotypic data and estimated breeding values (EBVs) for 46 linear traits included in the routine genetic evaluation and on EBVs for sport traits, using rank-based (R) and summarizing (highest level achieved, L) trait definitions for reflecting dressage (D) and show jumping (J) competition performance. Mares were categorized depending on whether or not they belonged to the 10% (25%) with the highest sport EBVs, followed by statistical analyses of these categories vs linear trait phenotypes and EBVs using SAS software. Only mares with sport EBV reliability ≥30% were considered (n=2,873 for DL to n=5,187 for JR), and inclusion criterion of own phenotypic data and/or at least two progeny with phenotypic data for the linear traits reduced the studied mare sample to n=1,084 (JL) to 2,737 (DR). Analyses of variance revealed that ranking among the mares with highest sport EBV significantly increased the probability of extreme linear trait EBV with discipline-specific pattern (all gait and few conformation aspects for D, jumping aspects for J). On the phenotypic level, more extreme values were rarely found in the mares and their progeny, and none of the category differences reaching significance indicated strong shifts towards extreme trait expressions which may harm long-term functional integrity. The use of linear data for continued monitoring is recommended.

Session 90 Theatre 2

Effect of mitochondrial genetic variability on performance of endurance horses

A. Ricard[1,2], S. Dhorne-Pollet[2], C. Morgenthaler[2], J. Speke Katende[2], C. Robert[2,3] and E. Barrey[2]
[1]Institut Français du Cheval et de L'Equitation, Pole développement, innovation et recherche, 61310 Exmes, France, [2]Université Parus-Saclay, AgroParisTech, INRAE, GABI, 78350 Jouy-enJosas, France, [3]Ecole Nationale Vétérinaire d'Alfort, 7 Avenue du Général de Gaulle, 94700 Maisons-Alfort, France; anne.ricard@inra.fr

Endurance races in horse require a high level of aerobic energy production to fulfil the demand of the cardio-respiratory system and skeletal muscles. At the cellular level, the mitochondrial respiratory chain (RC) performs ATP synthesis via the phosphorylation oxidative pathway. We performed mtDNA sequencing followed by SNP calling and GWAS analysis on 434 horses (Arabian 83.4% and Anglo-Arabian 9.6%) descended from 232 sires with an average family size of 1.8 descendants per sire. The endurance performances were measured by two traits: average speed and status at the arrival (finishing or not). Among 1,268 SNP detected with some heteroplasmy, 458 were finally used after selection with minimum calling frequency of 85% and minimum minor allele frequency of 2.3%. The GWAS analysis was performed using a mixed animal model to estimate the SNP allele effects (up to 4 copies) on the two performances corrected for fixed effects (age, sex and race). Random polygenic effect was included in the model using pedigree information (5,382 ancestors). ASREML software was used. The model could detect a total of 15 SNP significantly associated (raw *P*-value<0.01) to the performance traits. Seven SNP were significantly associated with both average speed and status at the arrival. In these cases, the reference allele was frequent (≥89%) and its effect was favourable on both traits (from +0.30 to +0.64 in phenotypic standard deviation unit). Five SNP were significantly associated with average speed with the same configuration: reference allele frequent (>92%) and favourable (+0.40 to +0.49). Three SNP were significantly associated with status at the arrival, with high reference allele frequency (≥93%) and favourable effect of reference allele for one SNP (+0.45) and unfavourable for two SNPs (-0.39 and -0.45).

Early life jumping traits' potential as proxy for jumping performance in Belgian Warmblood horses

L. Chapard, R. Meyermans, W. Gorssen, N. Buys and S. Janssens
Center for Animal Breeding and Genetics, KU Leuven, Department of Biosystems, Kasteelpark Arenberg 30, Box 2472, 3001 Leuven, Belgium; lea.chapard@kuleuven.be

The breeding goal of most Warmblood studbooks is to breed successful show jumpers. However, competition results are only available rather late in life (horses start competing at 6 years old on average). Moreover, performance in competitions was shown to be lowly heritable. To improve the breeding practice, having good proxy phenotypes that are available from an early age on and can be used to indirectly select on the trait of interest would be of interest for studbooks. In this study, we investigated the potential use of early life jumping traits as proxy for later success in competitions in Belgian Warmblood horses. Therefore, we studied 2,280 free jumping and 1,768 jumping under saddle records on 2,170 and 1,588 horses and eleven early life jumping traits assessed during young horses contests: scope, take-off (power/quickness), technique of forelegs, technique of back, technique of haunches, attitude (willingness), care, stride length, impulsion, elasticity and balance. These traits were scored by trained assessors on a 9-point scale. Additionally, almost 675,000 national show jumping competition records on 26,351 horses and more than 80,000 horses in the pedigree were used in this study. Later success in competitions was expressed as 'adjusted fence height' which combines fence height and ranking in competitions. Genetic correlations between early life jumping traits and adjusted fence height were estimated using bivariate animal models. They were moderate to high (from 0.40 for attitude (willingness) assessed freely and under saddle to 0.65 for scope assessed freely). These results show that early life jumping traits provide valuable information for the target trait (success in competitions) and could help achieving more genetic progress in the Belgian Warmblood populations by including early life jumping traits in their breeding program.

Genetic covariance components of conformation, movement and athleticism traits in Irish Sport Horses

J.L. Doyle[1], S. Egan[1] and A.G. Fahey[2]
[1]Horse Sport Ireland, Naas, Co. Kildare, W91 TK7N, Ireland, [2]University College Dublin, School of Agriculture and Food Science, Dublin 4, D04 V1W8, Ireland; jdoyle@horsesportireland.ie

Conformation, movement, and athleticism traits have been routinely captured on Irish Sport Horses (ISH) via linear profiling at studbook inspections of mares and stallions since 2010. The scoring system used in the ISH studbook is derived from the linear descriptive system that was first introduced by the Dutch Warmblood studbook in 1989 and has since been widely used as inspiration for other warmblood studbooks. Genetic parameters of linear descriptive traits have been extensively researched in continental warmbloods but, no such research has been conducted in the ISH population. Therefore, the objective of the present study was to estimate the genetic parameters of the linear scored conformation, movement, and athleticism (jumping) traits in the ISH and to determine their suitability for inclusion in a national genetic evaluation for sport horses. A total of 37 linear scored traits on 2,129 ISH were included in the analysis. Data were analysed using an animal linear mixed model that included the fixed effects of sex of the horse, age at scoring, the year of scoring, and the chairperson of the inspection panel that scored the horse. The heritability estimates for the conformation traits ranged from 0.03 (standard error (SE)=0.04) for length of croup to 0.23 (SE=0.06) for head-neck connection and shape of feet. The lowest heritability estimate within the movement traits was for walk correctness (0.09; SE=0.04) while heritability estimates for the rest of the traits within this group ranged from 0.23 (SE=0.06) for canter balance to 0.40 (SE=0.07) for trot length of stride. Scope (0.31; SE=0.07) had the highest heritability of the athleticism traits while attitude (0.02; SE=0.04) had the lowest heritability of these traits. Overall, the genetic correlations among the traits were in the same direction as the phenotypic correlations among the same traits but were generally stronger in magnitude. In general, the strongest genetic correlations were found among the movement traits and the athleticism traits. Results of the present study may be used for future genetic evaluations in the ISH population.

Session 90 Theatre 5

Start status in Swedish Warmblood horses

Y. Blom, S. Eriksson and Å. Gelinder Viklund
Swedish University of Agricultural Sciences, Dept. of Animal Breeding and Genetics, P.O. Box 7023, 75007, Sweden; asa.gelinder.viklund@slu.se

The breeding goal for the Swedish Warmblood horse (SWB) is to produce internationally competitive horses in dressage and show jumping. In the current genetic evaluation, lifetime accumulated points is used as measure of competition performance. Points are given to the 25% best horses in each competition. There is a preselection of horses that enter competition, and some competing horses never receive points. The aim of this study was to analyse the all-or-none trait start status in competition for possible use in genetic evaluation. Since 2007, all started horses in competitions are recorded. The studied population was restricted to SWB horses born between 2003 and 2018 that had the possibility to compete during the period from 2007 until 2022. Horses were categorized into disciplines according to their sire's and grandsire's discipline category. In total, 23,125 jumping (J) horses and 14,470 dressage (D) horses were studied separately. Information on start status in show jumping or dressage, lifetime accumulated points, assessed gaits and jumping traits from young horse test (YHT) and riding horse test (RHT) were available. The genetic analyses were performed with BLUP animal models. For J horses, 12,837 (55.5%) had competed in show jumping, whereof 9,413 (73.3%) had received points. Out of the competing J horses, 43.1% had also assessment from YHT (4,208) and/or RHT (1,905). For D horses, 4,964 (34.3%) had competed in dressage, whereof 2,722 (54.8%) had received point. About 55% of the competing D horses also had assessment from YHT (2,408) and/or RHT (997). The heritability for start status in show jumping was estimated to 0.30 on the observable 0/1-scale and 0.47 when transformed to the underlying continuous scale. For start status in dressage the corresponding heritability estimates were 0.20 and 0.34. Genetic correlations were strong between start status in show jumping and jumping traits at YHT and RHT (0.78-0.93) and moderate to strong between start status in dressage and gait traits at YHT and RHT (0.46-0.88). The genetic correlations between start status and accumulated lifetime point were strong, 0.93 for show jumping and 0.86 for dressage. We conclude that start status has potential to be used in the genetic evaluation.

Session 90 Theatre 6

Genetic analysis of the precocity potential in trotting races of Spanish Trotter Horses

M. Ripollés-Lobo, D.I. Perdomo-González, M.D. Gómez, M. Ligero and M. Valera
Universidad Sevilla, Agronomy, Ctra.Utrera km 1, 41005 Sevilla, Spain; marriplob@alum.us.es

Precocity is a quality that is sought after in all racehorses in order to obtain better results in the short term, as there is a perceived better opportunity for return on training costs in prize money. The Spanish Trotter Horse (STH) can be considered as a composite of the main World trotter populations. In the STH, a strong selection for sport performance in trotting races is underway. So far, the potential of horses for training and early participation in trotting races has not been evaluated from a genetic point of view. The aim of this paper is to study the precocity in trotting races of STH, estimating the genetic parameters of different variables: (1) age at first placed race (position 1, 2 or 3); (2) percentage of races placed in five years of life; (3) number of races needed for the animal to reach the first placed race. For this purpose, we used 176,137 racing records from 4,947 horses during the five-year period from the birth of each animal. All the known ancestors of the recorded animals were included in the pedigree file (11,488 horses), making sure that at least four generations were included. The genetic parameters were estimated using Gibbsf90+ modules of BLUPF90 package software. The genetic model of variable 1, included race time as covariate and gender (2 classes), type of race (2 classes), distance meters (3 classes), racecourse (4 classes), and race earnings (3 classes) as fixed effects. The model of variable 2, included number of races in 5 years and age at first race as covariate, and gender (2 classes), total earnings in 5 years/number of races placed, year of birth of the horse (7 classes) as fixed effects. The model of variable 3 included age at first placed race minus age at first race participated as covariable and gender (2 classes) and age at the first race placed (3 classes) as fixed effects. The estimates of heritability were 0.14±0.030 for variable 1, 0.32±0.037 for variable 2 and 0.08±0.023 for variable 3. The precocity evaluated through these new criteria, is a predictive tool for selection of horses for early racing performance and for informing training decisions.

Is the ability to race barefoot a heritable trait in Standardbred trotters?

P. Berglund[1], S. Andonov[1], A. Jansson[2], T. Lundqvist[3], C. Olsson[3], E. Strandberg[1] and S. Eriksson[1]
[1]Swedish University of Agricultural Sciences, Dept. of Animal Breeding and Genetics, P.O. Box 7023, 75007 Uppsala, Sweden, [2]Swedish University of Agricultural Sciences, Dept. of Anatomy, Physiology and Biochemistry, P.O. Box 7011, 75007 Uppsala, Sweden, [3]Swedish Trotting Association, P.O. Box 20151, 16102 Bromma, Sweden; paulina.berglund@slu.se

Racing barefoot is a common practice to increase speed of the horse in trotting races. By removing all four shoes or only the hind shoes at races, the horse does not only run faster, but the risk of gallop and disqualification also increases. Not all horses have sufficient hoof quality to race barefoot, and the hind hooves are often the limiting factor. Therefore, the aim of this study was to investigate the possibilities to select for the ability to race barefoot, as an indirect measurement of hoof quality in Swedish Standardbred trotters by estimating heritability. The data contained recordings of shoeing information from Swedish trotting races from the years 2005-2022. Only races in the period from March 1 to November 30 were included, because barefoot racing is not allowed during winter. Only horses that had started at least 10 races in the studied period and with at least one race with barefoot hind hooves were included in the study. In total 546,492 observations of 19,261 horses born 2002-2018 and raced at 3-10 years of age were included. Ability to race barefoot was defined as relative frequency of races with unshod hind hooves. This figure increased by birth years from 0.27 in 2002 to 0.31 in 2018. A univariate mixed linear animal model with sex and year of birth as fixed effects was analysed with the BLUPF90 suites of programmes using REML to estimate variance components. Heritability of 0.21±0.02 was estimated. Further studies using repeated observations are needed to verify the genetic contribution to the ability to race barefoot and genetic correlations to the traits used in today's genetic evaluation. In conclusion, these preliminary results show that there exists a genetic variation in ability to race with barefoot hind hooves.

Session 90 Theatre 8

Searching for genomic regions associated with conformation traits in the Pura Raza Español horse

N. Laseca[1], C. Ziadi[1], D.I. Perdomo-González[2], M. Valera[2], P. Azor[3], S. Demyda-Peyrás[1] and A. Molina[1]
[1]University of Cordoba, Department of Genetics, Campus de Rabanales. N-IV, km 396, 14014 Córdoba, Spain, [2]University of Seville, Department of Agronomie, ETSIA, 41013 Seville, Spain, [3]Asociación Nacional de Criadores de Caballo de Pura Raza Española, ANCCE, Edificio Indotorre, 41012 Seville, Spain; ge2lagan@uco.es

Conformation is of great importance in equine breeds like Pura Raza Español (PRE) horse, not only because it reflects the horse's appearance, but also because it is related to it functionality, especially classical dressage in this breed. The aim of the study was to identify markers associated with 5 zoometric measurements (scapular-ischial length (SiL), length of back (LB), dorsal-sternal diameter (DsD), thoracic perimeter (TP), perimeter of anterior cannon bone (PACB)), of the 50 zoometric measurements analysed in the PRE horse. For this, a mixed model was implemented using the ssREML animal methodology with the BLUPF90+ package. In addition to the animal effect, sex, age, coat colour, and country of origin of the animal were included as fixed effects. A total of 7,152 horses from 1,615 studs from 6 countries were included. From them, 2,916 horses were genotyped with 61,271 SNPs. The relationship matrix included ancestors up to the 5th generation with 41,889 animals. The average molecular relatedness between the genotyped animals was 0.073 and the genealogical relatedness between the animals with records and the genotyped animals was 0.029. Our results showed 562, 27, 49, 168, and 70 significant SNPs associated with SiL, LB, DsD, TP, and PACB, respectively ($P<10^{-6}$). Furthermore, 17 common significant markers were found for all traits, of which 14 SNPs were located on chromosome 3 and 3 SNPs on chromosome 17. In conclusion, although more studies are required to increase the number of animals and the number of zoometric measurements analysed, our findings indicate that genomic markers associated with conformational traits are located on regions of chromosomes 3 and 17.

Successful genotype imputation from medium to high density in Belgian Warmblood horses

L. Chapard[1], R. Meyermans[1], W. Gorssen[1], B. Van Mol[2], F. Pille[2], N. Buys[1] and S. Janssens[1]
[1]Center for Animal Breeding and Genetics, KU Leuven, Department of Biosystems, Kasteelpark Arenberg 30, Box 2472, 3001 Leuven, Belgium, [2]Ghent University, Faculty of Veterinary Medicine, Deartment of Large Animal Surgery, Anaesthesia and Orthopaedics, Salisburylaan 133, 9820 Merelbeke, Belgium; lea.chapard@kuleuven.be

Genotype imputation is a method that infers unobserved genotypes in a sample of individuals using a reference population. Genotype imputation permits increasing the density of available genotypes within a population and combining genotypes from different density SNP arrays and can therefore improve the power of genomic studies. Here, we investigate the feasibility of genotype imputation from medium to high density SNP arrays in Belgian Warmblood horses. 628 horses were genotyped during the course of 3 projects using the Affymetrix Axiom Equine 670K. SNP positions were updated from EquCab 2.0 to EquCab 3.0 and common variants between the three projects (74%) were kept for further analysis. Quality control (QC) was performed on the 31 autosomes and the 628 horses with PLINK v1.9 and consisted of removing individuals with a call-rate≤0.95, outlying heterozygosity and duplicated individuals and SNPs with a call-rate≤0.95, minor allele frequency≤0.02 and Hardy-Weinberg equilibrium≤0.00001. 616 horses and 416,143 SNPs were retained after QC. Haplotype phasing and genotype imputation were performed using Beagle 5.1. A cross-validation scheme was used to assess imputation accuracy for the GGP 70K array. The dataset was randomly divided five times into two sets: a reference set (75% of the horses) and a validation set (25% of the horses). Complete genotypes were retained for horses in the reference set whereas only 52,454 common SNPs between the 670K array and the GGP Equine 70K were kept for horses in the validation set. Imputation accuracy (correlation between true and inferred genotype) was calculated for each horse. Imputation accuracies were high and ranged from 0.93 to 0.98 with a mean value of 0.97. These results show that a reference population of 616 horses can be used for genotype imputation in Belgian Warmblood horses. This opens perspectives to routinely genotype at 70K and impute with this reference population in order to aid in the ongoing and future genomic research in Warmbloods.

Inbreeding in the Belgian equine warmblood population: current degree and evolution

B. Van Mol[1,2], H. Hubrechts[2], R. Meyermans[2], L. Chapard[2], W. Gorssens[2], M. Oosterlinck[1], N. Buys[2], F. Pille[1] and S. Janssens[2]
[1]Ghent University, Department of Large Animal Surgery, Anaesthesia and Orthopaedics, Faculty of Veterinary Medicine, Salisburylaan 133, 9820 Merelbeke, Belgium, [2]Center for Animal Breeding and Genetics, KU Leuven, Department of Biosystems, Kasteelpark Arenberg 30, Box 2472, 3001 Leuven, Belgium; bram.vanmol@ugent.be

A tremendous evolution has taken place in the field of equine reproduction during the past decade. Advanced reproduction techniques, such as ovum pick up combined with intracytoplasmic sperm injection and embryo transfer, have become more and more routine procedures in warmblood breeding. These techniques allow 'popular' mares to produce multiple foals per year via surrogate mares. The latter evolution and the impression that only a limited number of popular stallions are preferred by the equine breeders due to strong selection, could generate a fast-paced increase in autozygosity. Consequently, research is needed to assess the current levels of inbreeding in the equine warmblood population and to map the evolution of inbreeding. In this study we estimated the level of inbreeding of the Belgian warmblood population by analysing the HD genotypes (670K, Affymetrix) of over 700 horses by runs of homozygosity (ROH). Horses of the study sample were born in the period 2001-2020 and were divided in 2 time-windows to compute the evolution in the genomic inbreeding coefficient (F_{ROH}). We assessed the F_{ROH} at both the population and the individual level using PLINK. We examined the number and length of ROH to determine the evolution of inbreeding and studied ROH islands to achieve insights in signatures of positive selection. Furthermore, we quantified the genetic diversity by calculating the effective population size (N_e) of the Belgian equine warmblood population. The analyses so far showed restricted recent inbreeding (F_{ROH}±2%) and quite extensive old inbreeding (F_{ROH}±18%). N_e is ±90, which means the population is at risk (N_e<100). The final goal of this study is to give equine breeders insights on the inbreeding that occurs in this population and as such help them re-think and adjust their mating choices.

Population structure assessment using genome wide molecular information in Martina Franca donkey

V. Landi[1], E. Ciani[2] and P. De Palo[1]
[1]University of Bari Aldo Moro, Veterinary Medicine, SP. 62 per Casamassima km. 3, Valenzano (BA), 70010, Italy, [2]University of Bari Aldo Moro, Department of Biosciences, Biotechnologies and Biopharmaceutics, Via Amendola 165/a Bari, 70126, Italy; vincenzo.landi@uniba.it

The Martina Franca donkey is indigenous to the region of Puglia in southern Italy. It is a medium-sized animal that is renowned for its strength, stamina, and tame temperament. These donkeys are primarily used for agricultural purposes, as they are well adapted to the stony land of the region. The history of the Martina Franca donkey can be traced back to the Roman Empire, where they were used for transportation and agriculture. The breed continued to thrive throughout the Middle Ages, and by the 17th century, they had become a prized possession of the local nobility. However, with the advent of modern mechanization, the demand for donkeys declined, and the breed was almost lost during the decade of '70- '80. Recently, a renewed interest in both meat-derived products and their use in tourism and onotherapy practices, but above all the use of milk as a nutraceutical food for infants and to produce cosmetics has made possible a new increase in the number of animals raised but in the absence of a structured selective plan. To analyse the population structure and compare the pedigree inbreeding data with those obtained from molecular data we analysed 100 samples from 10 farms in the Puglia region using the ddRAD Seq technique. 41,796 genotyped loci were obtained, and finally 28,347 passing the frequency and missingness filters. The call rate for SNPs shows more than 90% of the markers have values greater than 95%. The observed heterozygosity for SNPs is 0.26±0.16 and that for individual is 0.23±0.018. The results relating to the genetic distances and to the assignment with the Admixture software show how the Farm of the Apulia region (a conservation nucleus) is confirmed as a point of dispersion of breeding animals. The analysis of linkage disequilibrium decay shows some herds, with very low levels in the range 0.4 and 0.6. The divergence of some groups of animals could however be due to exotic genetic introgression phenomena (with other Italian or Spanish breeds) but also the presence of very different genetic lines interesting from a conservation point of view.

Unravelling genomic regions with transmission ratio distortion in horse

N. Laseca[1], A. Cánovas[2], M. Valera[3], S. Id-Lahoucine[4], D.I. Perdomo-González[3], P.A.S. Fonseca[5], S. Demyda-Peyrás[1] and A. Molina[1]
[1]University of Cordoba, Department of Genetics, N-IV, km 396, 14014, Córdoba, Spain, [2]Center of Genetic Improvement of Livestock, Department of Animal Biosciences, University of Guelph, Guelph, ON, Canada, [3]University of Sevilla, Department of Agronomie, ETSIA, Sevilla, Spain, [4]Scotland's Rural College, Department of Animal and Veterinary Science, Easter Bush, United Kingdom, [5]University of Leon, Department of Animal Production, Leon, Spain; ge2lagan@uco.es

Transmission ratio distortion (TRD, or the deviation from the expected Mendelian inheritance) can be due to multiple biological mechanisms affecting gametogenesis, embryo development and postnatal viability. The TRD analysis estimates the frequency with which heterozygous parents transmit each of their two different alleles at a locus. The TRD was characterized across the Pura Raza Español horse genome with a SNP-by-SNP approach using 277 genotyped horses in trios (offspring-stallion-mare) including 554,634 SNPs, by a genotypic and allelic parameterization using TRDscan v.2.0 software. Evidence of TRD was found for 166 SNPs. Among them, 38, 76 and 63 SNPs exhibited overall, stallion- and mare-TRD, respectively. Moreover, among the SNPs identified by the genotypic TRD model, 5 presented additive TRD effect, 3 dominance TRD effect and 3 SNPs showed both additive and dominance TRD. A total of 790 functional candidate genes were annotated in allelic- and genotypic-TRD regions. Functional analyses revealed significant functional categories related to spermatogenesis, gametogenesis, oocyte division, embryonic development, and hormonal activity. Among them, 26 candidate genes were identified playing an important role in fertility and reproductive processes. Twenty-one were associated with allelic TRD (*FGF8, CXCL12, EPHA2, PRLH, STAT6, SPIRE1, NOS2, RELA, PDPK1, GCG, CDC25B, PDGFB, PPARA, NTF3, CATSPER1, PRSS21, BAG6, KCNU1, DDX17, MSH5* and *PRNP)* and 6 with genotypic TRD *(HORMAD1, YTHDC2, HSPA1L, EHMT2, BAG6* and *ASTL*). Interestingly, the *BAG6* gene encompassed a region with allelic-and genotypic-TRD. To our knowledge, this is the most extensive study performed to evaluate from a genomic perspective the presence of alleles and functional candidate genes with transmission ratio distortion in horse.

Session 90 Theatre 13

Microsatellite-based detection of transmission ratio distortion in the Pura Raza Española horse

D.I. Perdomo-González[1], S. Id-Lahoucine[2], A. Molina[3], A. Cánovas[4], N. Laseca[3], P.J. Azor[5] and M. Valera[1]
[1]Universidad de Sevilla, Ctra Utrera km1, 41005 Sevilla, Spain, [2]Scotland's Rural College, Easter Bush, EH25 9RG Midlothian, United Kingdom, [3]Universidad de Córdoba, Ctra.Madrid-Cadiz, Km.396, 14071 Córdoba, Spain, [4]Center of Genetic Improvement of Livestock, University of Guelph, N1G 2W1 Guelph, Canada, [5]Real Asociación Nacional de Criadores de Caballos de Pura Raza Española, Av. del Reino Unido, 41012 Sevilla, Spain; dperdomo@us.es

Transmission ratio distortion (TRD) is a genetic phenomenon widely demonstrated in most livestock species, which occurs when certain alleles/genotypes are over- or under-represented in the offspring. It can be caused by a variety of factors: embryonic lethality, gamete viability, genomic imprinting, or inbreeding depression. Although there is ample evidence that the expected Mendelian inheritance is altered in certain molecular markers in horses, the TRD has not yet been investigated. A total of 126,394 trios (offspring-stallion-mare) of Pura Raza Española horses genotyped by a set of 17 neutral microsatellite markers have been analysed. The number of alleles available for each marker ranged from 13 to 18, thus a total number of 268 alleles were investigated. The TRDscan software was used with the biallelic procedure to investigate the inheritance of each allele separately. After completing the analysis, a total of 12 alleles (out of 11 microsatellites) were identified with decisive evidence for genotypic TRD; 3 with additive and 9 with heterosis patterns. In addition, 19 alleles (out of 10 microsatellites) were identified displaying allelic TRD. Among them, 14 were parent-nonspecific and 5 were stallion-mare specific TRD. A total of 24 positional candidate genes were annotated within those genomic regions with signals of TRD. Functional analysis using the annotated genes identified molecular functions and biological processes related to cholesterol metabolism and homeostasis which results in unspecific symptoms of reduced growth, fertility, and health. TRD results show the significant impact on the inheritance of certain genetic traits in horses. Further analysis and validation are needed to better understand the TRD impact before the potential implementation in the horse breeding program strategies.

Session 90 Poster 14

Estimation of the genetic parameters for temperament in Haflinger horses

T.H. Zanon[1], B. Fürst Waltl[2], S. Gruber[1] and M. Gauly[1]
[1]Free University of Bolzano, Piazza Universitá 5, 39100, Italy, [2]University of Natural Resources and Life Sciences (BOKU), Gregor-Mendel-Straße 33/II, 1180, Austria; thomas.zanon@unibz.it

From the literature, there are indications that the personality and temperament of a horse are sufficiently genetically determined. This background motivated continued collecting of phenotypic information on the temperament in the South Tyrolean Haflinger horses to have it for marketing purpose and to possibly estimate genetic parameters on a long-term basis. The latter may open the possibility to include temperament traits in the breeding program. Therefore, the aim of the present study was to consider for the very first-time phenotypic information from multi-year character testing for estimating genetic parameters for temperament in Haflinger horses. For the present study, data from character testing of 210 South Tyrolean Haflinger mares between 2019 and 2021 were considered. A multivariate estimation of heritabilities (h2) and genetic correlations (r) was carried out for the criteria 'Interest', 'Activity' and 'Excitement'. In addition, univariate and bivariate estimation runs were also performed. Heritability for the behavioural trait Interest was lowest with 0.07 and a standard error of ±0.04. For the other two criteria, Activity and Excitement, h^2 was 0.19±0.05 and 0.21±0.04, respectively. The genetic correlation between Interest and Activity and between Interest and Excitement was in both cases negative with -0.52 and -0.61, respectively, while the genetic correlation between Excitement and Activity was positive (0.99). Although the quality of the phenotypic performance data collected during the character testing was good, the restricted number of animals limited the accuracy of the estimated genetic parameters. As next step, a continuation of the collection of phenotypic temperament performance on mares and possibly also on stallions is crucial for increasing the robustness of the data and therefore, improve the accuracy in heritability estimation.

Session 90 Poster 15

Estimation of the genetic propensity to suffer hock osteochondrosis in Pura Raza Española horses

M. Ripollés[1], A. Molina[2], M. Novales[3], C. Ziadi[2], E. Hernández[3] and M. Valera[1]
[1]Universidad Sevilla, Ctra Utrera, 41005 Sevilla, Spain, [2]Universidad Córdoba, Genética, Rabanales, 14014 Córdoba, Spain, [3]Universidad Córdoba, Medicina y Cirugía Animal, Rabanales, 14014 Córdoba, Spain; marriplob@alum.us.es

Osteochondrosis (OC) is a Developmental Orthopaedic Disease caused by an imbalance in bone metabolism that weakens the bone structure. It can cause pain, swelling or lameness. It has importance in most sporting breeds, including the Pura Raza Española horse (PRE). The hock OC in PRE horses has been evaluated at the level of 2 zones in both limbs, left and right (Eict: Intermediate eminence of tibial cochlea, Llta: Lateral trochlea of talus) in 3,053 horses. The clinical veterinary hospital of University of Cordoba performed the clinical diagnosis of the digital radiographs. Prevalence of 13.33% was determined for hock OC, being the highest prevalence in Eict_Right (6.58%) and the lowest in Llta_Right (1.97%). Two genetic approaches have been used for the estimation of the genetic propensity to suffer OC of the hock: (1) by a threshold model (absence/presence of OC); (2) by Bayesian lineal model (using OC grades: a linear scale from 0 (absence) to 3 (maximum grade)). Both at the general level and at the level of each of the zones and extremities (right and left) using different genetic models: (1) threshold model (absence/presence of hock OC); (2) Bayesian lineal model (using OC grades: 0,1,2,3); with a pedigree of 12,032 animals. The model to estimate the heritability (h^2) of hock OC included as fixed effects: age at radiological diagnosis, gender, coat colour, *and* size *of the* breeder's stud farm. Genetic parameters were estimated using different modules of BLUPF90 package software. The h^2 for presence/absence of hock OC, independent of location, was 0.28±0.112 in model 1 and 0.03±0.018 in model 2. Estimates for different locations, ranged from 0.19±0.131 (Llta_Right) to 0.21±0.124 (Eict_Left) for model 1 *and* ranged from 0.01±0.007 (Llta_Left) to 0.05±0.025 (Eict_Right) for model 2. The best results were obtained by Threshold Model. It would be necessary to increase the animals examined, but both prevalence and h^2 (depending on the methodology used) are similar to those found in other equine breed. These results imply that selection against OC is possible and genetic risk of offspring could be reduced.

Session 90 Poster 16

Genetic characterization of white facial marking in Pura Raza Español horses depending on coat colour

A.E. Martínez, M.J. Sánchez Guerrero, A. López, M. Ligero and M. Valera Córdoba
Universidad de Sevilla, Ciencias Agroforestales, Carretera de Utrera km 1, 41031, Sevilla, España., Spain; anaencmar1@alum.us.es

White facial markings are a relatively common horse coat characteristic of the Pura Raza Español horses (PRE) population. These white facial markings in horses are the result of a local lack of melanocytes in the skin. The aim of this study was to calculate the within-breed prevalence of white facial markings in a representative population sample of PRE and determine its prevalence depending on the coat colour and the other environmental effects (year of birth, sex, birth stud geographical area and inbreeding coefficient). The appearance of white facial markings has been classified into 5 score: 0 no white facial markings, 1 small white marking on the forehead, 2 big white markings in the forehead, 3 white markings on front-nose and 4 white markings all over the head. A total of 27,374 Pura Raza Español horses were analysed. Of this PRE population, a total of 64.91% had no white facial markings and 35.09% had white facial markings. Genetic parameters were estimated using a Bayesian procedure with the BLUPF90 software. Systematic effects were included in the model according to significant results in analyses of variance (Generalized Non-Linear Model (GLZ)): year of birth, sex, birth stud geographical area and inbreeding coefficient. The pedigree information included a minimum of 4 generations (74,675 horses). Heritability of white facial marking ranged from 0.687 (chestnut, bay, and grey in the dichotomic heterogeneity model) to 0.348 (chestnut population in the linear heterogeneity model). It has been demonstrated that factors (such as the coat colour but also year of birth, sex, birth stud geographical area and inbreeding coefficient) were significant. Our results showed that white facial markings were more prevalent in inbred chestnut males born in Europe (including Spain). The additive genetic base of white facial markings in PRE, which presents moderate to high heritability, shows that the prevalence of this horse coat characteristic could be effectively managed by genetic selection.

Session 90 Poster 17

Climate effects on foaling event: genotype-by-environment interactions in horse fertility

C. Sartori[1], E. Mancin[1], G. Gomez-Proto[1], B. Tuliozi[2] and R. Mantovani[1]
[1]University of Padova, Dept. of Agronomy Food Natural resources Animals and Environment, Viale dell'Universita 16, 35020 Legnaro (PD), Italy, [2]Duke University, 080 Duke University Road, Durham, NC 27708, USA; roberto.mantovani@unipd.it

Horses are a species with a low reproductive capacity, in which fertility is influenced by many environmental and management factors, often difficult to disentangle from individual effects like inbreeding (F) and individual additive genetic value. Few estimations of genetic parameters have been performed, and literature on the possible impact of genotype-by-environment interactions (G×E) is lacking. This study aimed to estimate genetic parameters for horse fertility and investigate the G×E on the trait using climate data. The study subject was the local Italian Heavy Draught Horse breed, and target trait was the individual outcome of the mating (or foaling event): 1 was successful foaling event and 0 no successful foaling (e.g. absence of pregnancy, abortion). The time lag between two parities was also considered. Data over 30 years of more than 22,600 foaling events were considered for about 3,350 mares and joined with individual F computed from the stud book. Multiple combinations of environmental units (EU) were realized for studs in the same geographical area and rearing system (stable, feral or semi-feral), and included in a traditional animal model (M1) with age at foaling, F, permanent environment and additive genetic effects. Then, climate information of the studs' municipalities were obtained from the NASA's Climate Data Services and referred to the time in which mares conceived (about 1 month around the date of conceiving). Temperature-humidity index (THI) was calculated and included as environmental covariate in a single-step reaction norm model (M2) updating M1. About 84% of data were successful foaling events. Heritability (h2) estimates varied from 0.11 to 0.14 on liability scale (0.05 to 0.06 on observed scale). Results from M2 showed that part of the phenotypic covariance was due to G×E, and h2 changed consequently. Some re-ranking occurred between individual breeding values under M1 and M2. Results suggest the importance of accounting for climate information and G×E in genetic evaluation of horse breeds, typically reared under a variety of environments and management conditions.

Session 91 Theatre 1

Impact of raw and digested pig manure on ammonia volatilization after land application

F.M.W. Hickmann[1,2,3,4], R. Rajagopal[4], N. Bertrand[3], I. Andretta[2], M.-P. Létourneau-Montminy[1] and D. Pelster[3]
[1]Université Laval, 2325 Rue de l'Université, G1V 0A6, Québec, Canada, [2]Universidade Federal do Rio Grande do Sul, Av. Bento Gonçalves, 91540-000, Porto Alegre, Brazil, [3]Quebec Research and Development Center, Agriculture and Agri-Food Canada, 2560 Bd Hochelaga, G1V 2J3, Quebec, Canada, [4]Sherbrooke Research and Development Center, Agriculture and Agri-Food Canada, 2000 College Street, J1M 0C8, Sherbrooke, Canada; felipe-mathias.weber-hickmann.1@ulaval.ca

Improved pig manure management practices can mitigate the environmental impacts of pig production by reducing nitrogen (N) losses such as ammonia (NH_3) volatilization. Anaerobic digestion (AD) is a promising technology for transforming pig manure into bio-based fertilizers (i.e. digestate). However, little is known about the impact of such technology on NH_3 volatilization after land application with manure that differs in terms of N content. Thus, this study aimed to evaluate the NH_3 volatilization after land application of raw and digested pig manure from animals fed different low-protein diets. Different N concentrations of raw pig manure (6,600 and 6,100 mg/kg) and digested manure (4,000 and 3,200 mg/kg) were tested. In a temperature-controlled room (25±1 °C), 16 cylinders (4 treatments × 4 replicates) with sandy loam soil were assigned to a 2×2 factorial design: factor 1 (raw or digestate) and factor 2 (control or low N). In each cylinder, manure was applied on soil surface to reach the 160 kg N/ha recommendation for corn. Data were analysed with PROC mixed and comparison of means through the Tukey test (SAS software). Differences in raw and digested pig manure began to be observed 4 h after land application (29.05 vs 102.25 mg N-NH_3/h). Overall, applying digested manure has three times more NH_3 volatilization than the raw manure application throughout all periods analysed (4 h, 24 h, 1 week, 2 weeks, and 3 weeks after land application; $P<0.001$). On the other hand, there was no effect of N level in any period ($P=0.624$). These results show some trade-offs of AD that increases NH_3 volatilization given the N organic form after digestion. Moreover, differences in N concentrations were not enough to affect NH_3 volatilization as manure was applied to reach the same recommended application rate.

Session 91 Theatre 2

Influence of housing manure management on ammonia emissions from broiler and laying hen productions

E. Caron[1], P. Le Bras[2] and M. Hassouna[3]
[1]ITAVI – Institut Technique de l'Aviculture, Rue Maurice le Lannou, 35000 Rennes, France, [2]IFIP – Institut du Porc, La Motte au Vicomte, 35650 Le Rheu, France, [3]INRAe UMR SAS, 65 rue de Saint Brieuc, 35000 Rennes, France; caron@itavi.asso.fr

In France, poultry productions contribute for about 0.06% of national greenhouse gases emissions and 6.3% of national ammonia (NH_3) emissions (CITEPA, 2022). The emissions come from the different steps of manure management on farm and vary in function of the breeding practices, the manure management inside the building and climate conditions. Nowadays there is a need of more accurate emissions factors (EFs) for different kind of purposes. This study aims to refine national EFs for broiler and laying hen productions according to different housing manure managements by using all data collected in the ELFE (ELevage et Facteurs d'Emissions, i.e. livestock and emission factors) database. ELFE structure was built in 2017 and permits an exhaustive description of the publications through 979 variables for the housing. Emission values have been converted into references units (kg NH_3-N/place/year) by using data given in the corresponding papers. In January 2022, the database contains 1,027 emissions values from 106 publications for the housing. About 50% of the values relate to the NH_3 compound (551), and 97% of the values refer to broilers and laying hens. After the cleaning process and the conversion to reference units, 25% of NH_3 values was deleted. Within the analysis, the type of floor has been selected as an influential variable. Several EFs are given, one for each technical itinerary. The number of emission values decreases according to the complexity of the technical itinerary. Standard deviation is computed for each value, to assess the accuracy of the result. These values take into account publications from the USA for more than a half of the database and experimental results, from experimental facilities with specific breeding conditions. This has to be interpreted within the results. A comparison to international references like the EMEP guideline has also been done. The limits of the study were the small number of literature references for the poultry sector and the lack of information describing the experiments.

Session 91 Theatre 3

NH_3, N_2O and CH_4 emission from piggeries: comparison of several manure management techniques

N. Guingand and P. Le Bras
IFIP Institut du Porc, La motte au vicomte, 35651, France; nadine.guingand@ifip.asso.fr

In France, the contribution of the pig farming to national emissions is 8% for ammonia (NH_3), 7% for nitrous oxide (N_2O) and 9% for methane (CH_4). Although low, this contribution will have to be drastically reduced in the foreseeable future to limit pig production impact on climate change and answer to social expectations. Housing is the first step in the emission process. Building design and practices such as manure management can affect emissions from building but also from storage and land application. The aim of this study is to compare different manure management techniques within the building on NH_3, N_2O and CH_4 emissions. The comparison is based on the use of the database called ELFE described by Vigan *et al.* This database gathers emission factors available in the literature but also metadata concerning breeding conditions and measurement methodology. In December 2022, 2,116 NH_3, N_2O and CH_4 emission factors of pig buildings published from 22 countries between 1964 and 2022 are collected in the database. Emission factors of sows, post-weaning piglets and growing-finishing pigs are established for four different manure management techniques: storage in the pit under animal, frequent gravity evacuation, manure collection in water, flushing and V-scraper. The comparison of emissions factors is discussed and argued in relation with the advantages and limits of each technique both in terms of implementation and relative effectiveness on each gas. Because French piggeries are mostly old, technical constraints can become a real obstacle to the deployment of a national policy to reduce gas emission from piggeries.

Feeding pigs with low-protein diets: impact of pig manure nitrogen content on biogas production

F.M.W. Hickmann[1,2,3], I. Andretta[3], L. Cappelaere[2], B. Goyette[1], M.-P. Létourneau-Montminy[2] and R. Rajagopal[1]
[1]Agriculture and Agri-Food Canada, Sherbrooke, J1M 1Z7, Quebec, Canada, [2]Université Laval, Département des Sciences Animales, 2325 Rue de l'Université, G1V 0A6, Québec, Canada, [3]Universidade Federal do Rio Grande do Sul, Department of Animal Science, Av. Bento Gonçalves, 91540-000, Porto Alegre, Brazil; felipe-mathias.weber-hickmann.1@ulaval.ca

Lowering dietary crude protein levels is a nutritional strategy recognized to decrease both nitrogen (N) excretion and the use of feed ingredients with high environmental impacts. Improved pig manure management practices can further mitigate the environmental impacts associated with pig production towards carbon neutrality and net-zero emissions. Anaerobic digestion (AD) is a promising technology for transforming pig manure into energy as biogas and into bio-based fertilizers (i.e. digestate from AD). However, little is known about the effects of pig manure N content on AD. Thus, this study aimed to evaluate the impact of pig manure N content on biogas production through the AD of manure from pigs fed low-protein diets. Three pig manure N concentrations were tested: T_1= 5,873, T_2= 5,421, and T_3= 5,149 total Kjeldahl nitrogen (TKN, mg/l). Throughout 5 fed-batch cycles (25±4 days/cycle), biogas production and its composition (CH_4, CO_2, and H_2S) were measured. In a temperature-controlled room (20±1 °C), 6 digesters (3 treatments × 2 replicates) were operated as single-stage reactors to digest pig slurry (mixture of urine and faeces, TS: 5.6%) inoculated with a liquid inoculum (TS: 2.3%) to improve manure-microbe interactions. Data were analysed as repeated measures with PROC mixed and comparison of means through the Tukey test (SAS software). Decreasing pig manure N content showed a tendency to reduce biogas (-20% in T_3 vs T_1; *P*-value=0.078) and methane (-22% in T_3 vs T_1; *P*-value=0.0578) production per cycle. Regarding biogas composition, CH_4/biogas and CH_4/CO_2 decreased with N content (-3 and -4% in T_3 vs T_1; $P<0.01$). These results suggest that a reduction in pig manure N content reduces biogas production and its quality (ratio of CH_4 to CO_2). This latter parameter is important for biogas efficiency; thus, reducing crude protein levels in pig diets may affect biogas production in AD.

A tool to reduce worker exposure from ammonia and particles in swine and poultry housing

N. Guingand[1], S. Lagadec[2], K. Amin[2], D. Bellanger[3], A.L. Boulestreau-Boulay[3], C. Delaqueze[4], C. Depoudent[2], L. Gabriel[3], E. Koulete[3], V. Le Gall[4], P. Lecorguille[5], L. Leroux[3], G. Manac'h[6], S. Roffi[2] and M. Ruch[2]
[1]IFIP institut du Porc, La motte au vicomte, 35651 Le Rheu, France, [2]Chambre d'Agriculture de Bretagne, Rue de St Brieuc, 35000 Rennes, France, [3]Chambre d'Agriculture des Pays de Loire, Rue A. Brouard, 49105 Angers, France, [4]MSA 49, Rue C. Lacretelle, 49070 Beaucouzé, France, [5]MSA Armorique, Rue de Paimpont, 22025 Saint Brieuc, France, [6]Porc Armor Evolution, Impasse Monge, 22600 Loudéac, France; nadine.guingand@ifip.asso.fr

Air quality inside pig and poultry housing is already known to be concentrated in ammonia and particles. Regular exposure is associated with a high prevalence of respiratory dysfunctions and pathologies such as chronic bronchitis or asthma. The aim of the project is to develop an assessment tool to (1) allow workers to evaluate the air quality inside buildings, (2) promote the implantation of techniques to prevent the risk of exposure and (3) advise workers on the existence/use of personal protective equipment (PPE). For the first step, a questionnaire grouping technical aspects, working organisation and health history of workers has been elaborated by the partners. A group of 24 farmers (pig and poultry) has been formed to be directly implicated in the test of the self-diagnostic tool. Visits and collection of data will be completed in summer 2023. The next steps are the elaboration of fact sheets highlighting the different ways of reduction of gas and particles inside building and the different type of PPE with a focus on strengths and weaknesses. After having collected data on PPE available in the market, a study will be conducted on their assessment on the ground directly by farmers and people working in the experimental farms of partners. Results of this step will be the base of exchange between partners and manufacturers for the improvement of one or two PPE. In relation with a previous work done on best practices to reduce ammonia and particles, a multi-criteria evaluation grid on mitigation techniques to reduce gas and particles has been built. The writing of fact sheets in progress and contacts have been already taken with manufacturers of PPE. First results are available on www.qualiair.chambres-agriculture.fr.

Session 91 — Theatre 6

Methodological redesign of the poultry mass-balance excretion and variability of nitrogen excretion

V. Blazy, E. Caron, D.D. Djenontin-Agossou and Y. Guyot
ITAVI, The French Technical Insitute for Poultry, environment division, rue Maurice le Lannou, 35000 Rennes, France; blazy@itavi.asso.fr

Nowadays the excretion levels are an essential indicator for the administration, the farmers and their advisors. Excretion values are used in environmental assessment and to dimension land spreading plans. The industrial emission directive defines best available technics associated total nitrogen (N) excreted and annual excretion monitoring for different poultry productions. This monitoring can be carried out by using manure analysis or mass balance calculation based on feed intakes and animal performance. Based on daily mass balance computation, a new excretion calculation methodology is developed in this study. It considers the feeding program (with the different nutrients levels for each diet) during the rearing period. Calculation is based on the generation of a posteriori growth, feed intake and mortality dynamics from users zootechnical data and references curves for a given poultry production. Thus, with help of deposition values, excretion dynamics can be simulated for 6 elements (including N), and so a more accurate balance-sheet can be computed on the rearing span. The methodological redesign tends to make the excretion computation tool more readable and conceptually more accurate. Computing the mass-balance at daily scale allows the user a better understanding of how each input data affects the final excretion results. In parallel, a quantification of N excretion variability is carried out for laying hens production (as example). For that purpose, a sensitivity analysis has been performed for each data input on the mass balance results. The data input having the most influence on the excretion value is the flow of ingested feed. Then, this is the mass of egg production with the slaughter weight. Finally the growing span and the mortality have a very poor influence. Thus, mass-balances simulations including interquartile and extreme ranges of feed consumption and egg production can roughly allow assessing the variability of the laying hens N excretion. Results indicated that based on average performances, laying hens excreted 912 gN/ animal. Interquartile and extreme ranges respectively corresponded to excretion results of 814-1057 and 786-1153 gN/ animal.

Session 91 — Theatre 7

Access to bedding and an outdoor run for growing-finishing pigs and their impact on the environment

A.K. Ruckli[1,2], S. Hörtenhuber[1], S. Dippel[3], P. Ferrari[4], M. Gebska[5], J. Guy[6], M. Heinonen[7], J. Helmerichs[3], C. Hubbard[6], H. Spoolder[8], A. Valros[7], C. Winckler[1] and C. Leeb[1]
[1]Univ. of Natural Resources and Life Sciences, Gregor-Mendel-Str. 33, 1180 Vienna, Austria, [2]Bern Univ. of Applied Sciences, Länggasse 85, 3052 Zollikofen, Switzerland, [3]Friedrich-Loeffler-Inst., Dörnbergstr. 25/27, 29223 Celle, Germany, [4]Fondazione CRPA Studi e Ricerche, Viale Timavo 43/2, 42121 Reggio Emilia, Italy, [5]Warsaw Univ. of Life Sciences, Nowoursynowska 166, 02-787 Warszawa, Poland, [6]Newcastle University, Kings Road, NE1 7RU Newcastle upon Tyne, United Kingdom, [7]Univ. of Helsinki, P.O. Box 57, 00014 University of Helsinki, Finland, [8]Wageningen University & Research, De Elst 1, 6708 WD Wageningen, the Netherlands; antonia.ruckli@bfh.ch

Husbandry aspects intended to improve animal welfare (AW) such as bedding or access to an outdoor run are currently financially supported through AW labels in several countries. At the same time, they may impact the environment through possible changes in feed efficiency and manure management. Therefore, we compared farms differing in AW relevant husbandry aspects regarding pig welfare, global warming (GWP), acidification potential (AP), fresh water (FEP) and marine (MEP) eutrophication potential. We collected data on 50 farms with growing-finishing pigs in seven European countries: NOBED (31 farms without bedding and outdoor run), BED (11 farms with bedding only) and BEDOUT (8 farms with bedding and outdoor run). Pigs in NOBED farms manipulated enrichment (e.g. straw, objects) less, pen fixtures more and showed more oral stereotypies than on BED and BEDOUT farms. AP of BEDOUT and BED farms was significantly higher (67.1±7.5, 73.5±6.3 g SO_2-eq/kg BMNG) compared to NOBED farms (44.3±3.8, P=0.002) due to higher ammonia emissions allocated to solid manure. BEDOUT farms had higher MEP than NOBED farms (BEDOUT: 36.6±5.0, BED: 25.1±4.3, NOBED: 19.0±2.6 g N-eq/kg BMNG, P=0.035). GWP and FEP did not differ, probably due to large variability within farm systems regarding feed composition and -conversion. Nevertheless, the large variation within farm systems suggests that the trade-off may be minimised, e.g. through manure management (frequent cleaning) and feed composition (protein adjustment).

Session 91 Theatre 8

Reduction of protein and potassium to improve welfare and environmental footprint in broilers

T. De Rauglaudre[1,2,3], B. Méda[2], W. Lambert[4], S. Fournel[1] and M.P. Létourneau-Montminy[3]
[1]Université Laval, Département des sols et de génie agroalimentaire, Québec, G1V 0A6, Canada, [2]INRAE, BOA, Nouzilly, 37380, France, [3]Université Laval, Département des sciences animales, Québec, G1V 0A6, Canada, [4]METEX Animal Nutrition, Paris, 75017, France; theophane.de-rauglaudre.1@ulaval.ca

A proper management of litter moisture management is important to limit nitrogen (N) volatilization and footpad lesions. In broilers, it is known that both dietary crude protein (CP) and electrolyte balance (DEB) influence water intake, and consequently litter moisture. Since soybean meal (SBM) is rich in potassium (K), reducing CP in feed also reduces K intake, and thus DEB. The objective of this study was therefore to dissociate the effects of reducing CP and K on water intake, total water excretion, and N balance. For this purpose, three experimental treatments were tested, a control diet (C; CP=20.5%; DEB=224 mE/kg), a low CP diet (LCP; CP=17.5%; DEB=145 mE/kg), and a low CP diet supplemented with K carbonate (LCP+K; CP=17.7%; DEB=224 mE/kg) to bring back DEB to the level of C. The three diets were iso energy (AMEn=3,050 kcal) and SID lysine (9.5 g/kg), and feed-grade amino acids (AA) have been used to reach indispensable AA requirements. At 20 days of age, 72 Ross 308 male were randomly allocated to 18 cages of 4 birds each. After 7 days of adaptation to the experimental diets, total excreta, feed and water consumption were measured daily per cage for 3 days and a representative sample of excreta was collected. In comparison to C, only LCP treatment have an effect on water intake (-31%; $P<0.01$), water excretion (-39%; $P<0.001$), humidity of excreta (-9%; $P<0.001$), and water:feed ratio (-21%; $P<0.01$). Comparatively to C treatment, N excretion was reduced by 34% ($P<0.001$) and 30% ($P<0.001$) for LCP and LCP+K treatments respectively, while the N retention efficiency was increased by 6% ($P<0.05$) and 11% ($P<0.001$). Only simultaneous reduction of DEB and CP affected water intake and water excreted, while the effect of CP alone was not significant. As expected, N balance was only impacted by the CP. Thus, decreasing K and CP together is an effective strategy to reduce N excreted, water intake and improve litter DM which will lead to a reduction in both N volatility and pododermatitis.

Session 91 Theatre 9

Effects of reducing protein content in broiler diets on environmental impacts

J. Gickel[1], V. Wilke[1,2], C. Ullrich[2] and C. Visscher[2]
[1]University of Veterinary Medicine Hannover, Foundation, Science and Innovation for Sustainable Poultry Production (WING), Heinestraße 1, 49377 Vechta, Germany, [2]University of Veterinary Medicine Hannover, Foundation, Institute of Animal Nutrition, Bischofsholer Damm 15, 30173, Germany; julia.gickel@tiho-hannover.de

The use of extracted soybean meal (SBM) is resulting in a higher carbon footprint of poultry production in Europe. Therefore, efforts are being made to reduce the amounts in the diets. Moreover, generally lower contents of crude protein (CP) and the use of certain amino acids in the diets might reduce negative environmental impacts, too. In this study, 360 day-old broilers, where divided in four treatment groups, kept for 35 days and fed diets (starter/grower/finisher) containing different amounts of CP/added amino acids (CPC [control]: 22.2%/20.5%/19.9%; CP-1: 21.5%/19.4%/19.2%; CP-2: 20.2%/ 18.6%/ 18.1%; CP-3: 19.2%/17.7%/17.0%). Starter and grower phase lasted 7 days each, followed by 21 days finisher phase; performance parameters were measured weekly. The assessment of the environmental impact per t carcass weight (according to ISO 14040) was performed using the online software application Opteinics® (BASF Lampertsheim GmbH, Germany), based on the Global Feed LCA Institute (GFLI) database. In week 1 and 2 body weight (BW) was similar between CPC, CP-1 and CP-2, while the BW of CP-3 was lower. At the end of the trial CP-2 showed the highest BW followed by CP-1, CPC and CP-3. No significant differences were found regarding the feed conversion ratio (FCR). Highest values occurred for CP-3 (week 1-4) and CPC (week 5). The impact on climate change throughout the production (related to carcass weight) was 4.4 t per t for CPC and could be reduced by 6% (CP-1), 8% (CP-3) and 12% (CP-2). For CP-2, the lowest environmental impact of the production was also found in the following categories: eutrophication (marine/freshwater), land use, particulate matter and use of resources and water. The impact on acidification and eutrophication (terrestrial) showed the highest reduction for CP-3. The impact on ozone depletion was reduced for CP-1 but higher for CP-2 and CP-3 (compared to CPC). The results showed that the highest CP-reduction did not result in the lowest environmental impact for all categories mainly due to a lower efficiency of the production.

Session 91 Theatre 10

Effect of reduced-crude protein diets on performance, meat yield, and nitrogen production in broiler

C. Gayrard[1], T. Wise[2], J.D. Davis[2], J.C. De Paula Dorigam[1], V.D. Naranjo[1] and W.A. Dozier, Iii[2]
[1]Evonik Nutrition & Care GmbH, Animal nutrition, Hanau, Germany, [2]Auburn University, 36849, Alabama, USA; cecile.gayrard@evonik.com

The impact of feeding reduced crude protein (CP) diets to 1,500 Ross × Ross 708 male broilers while providing adequate essential amino acid concentrations, on growth performance, nitrogen concentrations, mortality rate and carcass characteristics was examined from 1 to 33 d of age. Broilers were randomly distributed into 60 floor pens (25 birds/pen) and fed 1 of 6 dietary treatments (10 pens/treatment) varying in CP content. Diet 1 (control) was supplemented with DL-Met, L-Lys, and L-Thr and contained 23.0, 19.8, 18.9% CP in the starter (1-14 d of age), grower (15-25 d of age), and finisher (26-33 d of age) periods respectively. Additional L-Val, Gly, L-Ile, L-Arg, and L-Trp were sequentially supplemented in the order of limitation respective of diets 2 through 6. Dietary CP was reduced from diets 1 to 6 to 20.0, 17.4, and 16.7% in the starter, grower and finisher periods, respectively. Broilers fed diets 2 to 6 had similar average daily gain (65 g/d, *P*-values ranging 0.10 to 0.99) and feed intake (86.4 g/d, *P*=0.22) compared with broilers fed the control diet. There was no treatment effect on feed conversion ratio (FCR) except for the lowest CP-diet which was significantly higher than all other treatments (1.45 vs 1.41 to 1.43; *P*<0.05). Dietary treatments did not affect carcass traits (*P*-values ranging 0.11 to 0.60). The mortality rate was low (1.5%) and not impacted by the treatments (*P*=0.31). Nitrogen intake (*P*<0.001) was decreased in the low CP-diet 6 vs the control diet. The N excretion values were compared using a direct and an indirect methods, and decreased from diet 1 to 6 by 32% (*P*<0.001) when using the indirect method. With both methods, N excretion per kg of meat was reduced (*P*=0.046 vs *P*<0.001) from diet 1 to 6. Nitrogen utilization was improved from 65.5 to 73.5% (*P*<0.001) from diet 1 to 6 (indirect method) while it was not affected (76.5%) with the direct method (*P*=0.89). Results from this study indicated that the supplementation of amino acids from DL-Met up to L-Trp allowed reduction of dietary CP content without depressing growth rate and meat yield of broilers from 1 to 33 d of age while reducing Nitrogen meat excretions emission values.

Session 91 Theatre 11

Effect of reducing dietary protein content on performance and environmental impact of pig production

E. Gonzalo[1], A. Simongiovanni[1], C. De Cuyper[2], B. Ampe[2], M. Aluwe[2], W. Lambert[1] and S. Millet[2]
[1]METEX Animal Nutrition, 32 rue Guersant, 75017 Paris, France, [2]ILVO, Animal Sciences, Scheldeweg 68, 9090 Melle, Belgium; enrique.gonzalo@metex-noovistago.com

The objective of the present trial was to study the effect of a decrease in dietary crude protein (CP) content on the growth performance, carcass quality, nitrogen efficiency and global warming potential (GWP) in fattening pigs. Thus, 288 pigs were distributed in 48 pens of 6 animals (male-female parity) during three feeding phases: 10-15 (phase 1), 15-20 (phase 2) and 20 weeks of age – slaughter (phase 3). Six dietary treatments tested a reduction in CP (RCP). The SID Lys/CP levels were kept stable between the feeding phases at 5.4, 5.7, 6.0, 6.3, 6.7 and 7.1%, respectively for T1 to T6 treatments: dietary CP varied from 17.4 to 13.2% in phase 1, 15.9 to 12.1% in phase 2, and 14.2 to 10.9% in phase 3, with SID Lys levels of 9.4, 8.6 and 7.7%, for phase 1, 2 and 3, respectively. Essential amino acid (AA) levels in relation to lysine were maintained. Pen was the experimental unit. Over the entire period, RCP had no effect on average daily feed intake and carcass yield. In contrast, RCP with SID Lys/CP≥6.3% degraded average daily gain (825±36 g/d in T1 vs 730±36 g/d in T6) and feed conversion ratio (2.33±0.06 in T1 vs 2.60±0.15 in T6) (*P*<0.001). Carcass characteristics were maintained up to SID Lys/CP≤6.7%. Nitrogen efficiency increased linearly from 43.6±1.1% (T1) to 51.8±2.9% (T6). The effect on GWP (kg CO_2eq/kg BW gain with AA from Europe or China) showed a significant interaction between treatment and AA origin (*P*<0.001). The GWP calculated with European AA decreased from T1 to T6 (1.45±0.04 vs 1.29±0.07 kg CO_2eq/kg BW gain), while results with Chinese AA increased from T1 to T6 (1.51±0.04 vs 1.82±0.11 kg CO_2eq/kg BW gain). This was due to high values for GWP for Chinese compared to European AA in the dataset used (Agribalyse 3.1), as a result of differences in the production process. It can be concluded that reducing the dietary CP level while maintaining pig performance is possible up to a SID Lys/CP level of 6%. This RCP is accompanied by a beneficial effect on GWP when using AA of European origin.

Session 91 Theatre 12

Impact of genotype×feed-interactions in nitrogen- and phosphorus-reduced ration on of fattening pigs

C. Große-Brinkhaus[1], B. Bonhoff[1], I. Brinke[1], E. Jonas[2], S. Kehraus[1], K.H. Südekum[1] and E. Tholen[1]
[1]University of Bonn, Institute of Animal Science, Endenicher Allee 15, 53115 Bonn, Germany, [2]Association for Bioeconomy Research (FBF), Adenauerallee 174, 53113 Bonn, Germany; cgro@itw.uni-bonn.de

Strict governmental regulations for the distribution of manure coupled with high production costs lead lowered nitrogen (N) and phosphorus (P) reduced concentrations in diets for growing pigs. This feeding strategy requires high availability of nutrients in diets ingredients, and efficient nutrient utilization by pigs. It is assumed that the genotype of an animal is the reason for variation in adaptability to dietary changes, potentially leading to a re-ranking of genotypes: a genotype×feed-interaction (G×F). Two trials were conducted to assess the effects of lowered dietary N and P concentrations on N and P excretion and to investigate the existence of G×F. In total, 900 pigs (♂ and ♀) of Pietrain × crossbred sows, and 1000 Landrace boars were divided into one of two feeding groups, the control- or low protein/P-group before fattening phase. Animals were tested for performance in a 3- or a 2-phase *ad libitum* growth period until 115kg. A subset of 200 pigs of each line were kept individually in order to collect faeces samples after diet change and to determine the digestibility and excretion in feaces (ex_{fe}) of N and P. Blood samples were collected from 100 pigs to estimate urinary N excretion (ex_{ur}). First results revealed a h^2 ranging from 0 for P ex_{fe} to 0.56 for N ex_{ur}. For all traits the effect of sire×feeding group was below 0.15. Although the influence of the sire and the interaction group was not significant, sire rank correlations of 0.27 for N- and 0.07 for P ex_{fe} indicate the existence of a G×F. The results of the present study provide further evidence for the existence of G×F in diets with lowered N and P concentrations for growing pigs. In addition, possible consequences on bone composition and density due to lowered nutrient concentrations and the observed interindividual variability of digestibility values will be investigated.
These studies are funded by the Ministry of Environment, Agriculture, Nature and Consumer Protection of North Rhine-Westphalia and by the German Government's Special Purpose Fund held at Landwirtschaftliche Rentenbank (FKZ 28-RZ-3.104).

Session 91 Theatre 13

Pig performance, carcass composition and meat quality can be maintained with high CP reduction

L. Cappelaere[1,2], F. Garcia-Launay[1], W. Lambert[3], A. Simongiovanni[3] and M.-P. Létourneau-Montminy[2]
[1]PEGASE, INRAE, Institut Agro, 16 le Clos, 35590 Saint Gilles, France, [2]Université Laval, Département des sciences animales, rue de l'agriculture, G1V 0A6 Québec, Canada, [3]METEX Animal Nutrition, 32 rue Guersant, 75017 Paris, France; lea.cappelaere.1@ulaval.ca

Low crude protein (CP) diets reduce nitrogen (N) born environmental impacts in pig production. Newly available feed-grade amino acids allow to reduce dietary CP to seldom tested levels, with balanced diets. The objective of this trial was to test the effect on growth performance, N balance, carcass composition and meat quality of CP reduction permitted by L-Ile and L-His supplementation. Twenty-four pens of 3 barrows and 3 gilts were fed *ad libitum* one of three soybean meal-corn based dietary treatments. Control CP levels in the four feeding phases (25-50 kg, 50-80 kg, 80-100 kg and 100-135 kg) were 180, 161, 143 and 126 g/kg. Treatments were 12 and 24 g/kg CP reductions, in all phases. Bodyweight (BW) gain and feed intake (FI) were recorded for each phase. Body composition was analysed by two-photon X-ray absorptiometry (DXA) in one pig per pen at the beginning and end of the trial. Carcass parameters were measured at the slaughterhouse for all pigs. Longissimus dorsi samples were taken for meat quality analysis 24 h after slaughter for two pigs per pen. A general linear model was used to test the effect of CP reduction. Dietary CP did not affect FI (2.86 vs 2.93 vs 2.93±0.12 kg/d, P=0.34), BW gain (1.18 vs 1.19 vs 1.16±0.04 kg/d, P=0.42), or feed conversion ratio (2.73 vs 2.81 vs 2.76±0.10, P=0.11). N intake was linearly decreased by 6.7% between treatments and N excretion by 11% (P<0.001). N efficiency was linearly improved by 2.8 percentage points between treatments (P<0.001). Carcass composition was not affected: DXA fat (20 kg; P=0.91) and lean (56 kg; P=0.44) deposition, carcass weight (109 kg, P=0.39), muscle depth (67 mm, P=0.13), backfat thickness (21 mm, P=0.93) and meat percentage (59.9%, P=0.87). Meat colour, intramuscular fat and pH at 24 h were not affected (P>0.1) but drip loss tended to increase (P=0.054) with CP reduction. In this trial, a 24 g/kg CP reduction efficiently decreased N excretion while maintaining performance and carcass and meat quality.

Alpha mannan polysaccharide supplement increase growth performance and reduce gas emission in pig

J.H. Song[1], C.B. Lim[1], M.D.M. Hossain[1], S. Biswas[1], J.S. Yoo[2], I.H. Kim[1] and Q.Q. Zhang[1]
[1]Dankook University, Department of Animal Resource and Science, Cheonan, 31116, Korea, South, [2]Daehan feed company, 13 Bukseongpo-gil, Jung-gu, Incheon, 22300, Korea, South; thdwnsh@naver.com

The objective of this study was to evaluate the effects of supplementing α-mannan polysaccharides to the diet of weaning pigs on their growth performance, apparent nutrient digestibility, faecal microbiota, faecal score, and faecal gas emission. A total of 160 crossbred weaning pigs [(Yorkshire × Landrace) × Duroc, 28 days old] were randomly assigned to four groups based on their initial body weight (7.70±1.27). The experimental period was 42 days (Phase 1: week 1; Phase 2: weeks 2-3; Phase 3: weeks 4-6). Dietary treatments consisted of pigs receiving basal diet without α-mannan polysaccharides as the control group and treatment groups consisted of pigs fed basal diet supplemented with 0.1, 0.2, 0.3% α-mannan polysaccharides respectively. The addition of α-mannan polysaccharide resulted in a linear increase ($P<0.05$) in daily gain over Weeks 1, 3, and throughout the trial. Supplementation of α-mannan polysaccharide to the diet of weaned pigs led to a linear increase in the gain to feed ratio. There was a linear spike in dry matter digestibility ($P<0.05$) in piglets that had been fed a diet with α-mannan polysaccharide. The bacteria count in the faeces of weaned pigs was not affected by the addition of α-mannan polysaccharide to their diet. The faecal score was not significantly affected. H_2S and total mercaptan levels have been demonstrated to linearly decrease ($P<0.10$) with increasing amounts of mannan in the diet of weaned pigs. The addition of α-mannan polysaccharide in the diet of weaning pig could be beneficial to enhance the daily gain and gain to feed ratio and, improved the dry matter digestibility and decreased faecal H_2S and total mercaptan gas emission.

Characteristic of odour substances from broiler farms

S.Y. Seo, J.S. Park, S.Y. Park and M.W. Jung
National Institute of Animal Science, Animal Envionment Division, #1500, Kongjwipatjiwi-ro, Ison-myeon, Wanju-gun, Jeollabuk-do, 55365, Korea, South; seosi@korea.kr

An increase in meat consumption led to the growth of the livestock industry. However, with growth of the livestock industry, it brings increase of livestock manure and complaints of odour. Even this situation, so far odour research mostly has been conducted on swine farm. For study how to reduce livestock odour, we should evaluate odour of other species like dairy cow, beef cattle and poultry. The main aim of this study was to evaluate odour from broiler farms and find main odour substances of broiler farms. The measurements were performed by investigating complex odour and the concentration of ammonia, acetealdehyde, volatile fatty acids(acetic acid, propionic acid, n-butyric acid, i-valeric acid, n-valeric acid) and evaluating odour activity value(OAV). Sampling point was in front of ventilation fan of the broiler farm and sampling was started at one week after all in. Four times, once a week, of sampling was conducted and samples were brought to lab to measure concentration by Gas Chromatography. OAV was calculated with minimum detection concentration. The result showed that ammonia and other VFAs were increased with time. Ammonia was 0.15 ppm at 1st week, 1.10 ppm, 1.37 ppm, 4.00 ppm. However, complex odour dilution factor and concentration of acetealdehydes were decreased by time. Acetealdehydes was 0.33 ppm, 0.12 ppm, 0.07 ppm and last week was 0.01 ppm. The highest OAV was acetealdehydes(66.26), and propionaldehydes(32.45), ammonia(16.55), trimethylamine(7.16), n-butyric acid(1.29) follow. It seems acetealdehydes is main odour substance which affect complex odour most and as broiler grows, the ammonia and VFAs are getting more emission.

Session 91 — Poster 16

Environmental assessment of layer-type male chicks breeding

E. Dubois and M. Quentin
ITAVI, environment, 7 rue du Faubourg Poissonnière, 75009 Paris, France; dubois@itavi.asso.fr

Since 2022, due to ethical concerns, French legislation prohibited the culling of layer-type male chicks after hatching. In order to deal with this prohibition, three methods can be applied. The first one is *in ovo* sexing, resulting in the culling of male eggs before hatching. A second one is the selective breeding of layer-type male chicks. The third method is the use of dual-purpose breeds, selected for balanced performances in meat and egg productions. Since the genetic selection of current poultry breeds has led to a negative correlation between meat and egg production performances, layer-type male or male from dual-purpose breeds have variable growth performances and a different environmental impact. Thus, this study aimed to assess the environmental impact of male chicks breeding, using life cycle assessment (LCA) method from cradle to slaughterhouse. A trial was set up to evaluate the meat production performances of two different layer genetics (Lohmann Brown, H&N SuperNick White) and one dual-purpose breed (Hendrix ISA-DUAL) compared to a meat-type slow growing breed (Hubbard JA S757N) used in French 'label-rouge' production. Each genetic type was raised at two different market targets: cockerel or broiler, with an objective live-weight at slaughter of respectively 850 to 1000 g or 1,800 to 2,200 g. The results of this study were carried out through eight LCA scenarios, one for each genetic at each growth stage. The functional unit is kg of carcass weight. The indicators assessed will be climate change, land use, acidification, water use and eutrophication. Indicators will be compared for each scenario in order to quantify the environmental burden of alternatives to the culling of layer-type male chicks. The environmental impact is expected to be higher for broilers than for cockerels, with longer growth duration and higher feed conversion ratio. Breeding male chicks of layer-type has a higher environmental impact than conventional broilers. These results could help the poultry sector to find environmental, economic and technical compromises to face an ethical challenge related to the prohibition of layer-type male chicks' culling. Further environmental assessment of the couple male chicks and laying hens will be led for the layer-type and dual purpose breeds, in order to assess their global impact.

Session 92 — Theatre 1

Milk metabolites differ with feed efficiency during early lactation but not during feed restriction

J. Pires[1], T. Larsen[2], S. Bes[1], I. Constant[1], D. Roux[3], M. Tourret[1], A. De La Torre[1], I. Ortigues-Marty[1], F. Blanc[1] and I. Cassar-Malek[1]
[1]INRAE, UCA, VetAgro Sup, UMRH, Theix, 63122, France, [2]Aarhus University, D. Animal Sci., Tjele, 8830, Denmark, [3]INRAE, UE1414 Herbipôle, Theix, 63122, France; jose.pires@inrae.fr

The objective was to study associations among indicators of feed efficiency and selected milk metabolites during early lactation, and in response to experimental feed restriction (FR) for the discovery of non-invasive proxies. 28 cows divergent in phenotypic residual feed intake (RFI; positive or negative) were selected, each RFI group composed of 14 Holstein and 14 Montbéliarde cows. Energy balance (EB) and conversion efficiency (ECE; NE_L secreted/intake) were calculated. Early lactation RFI (RFIearlylact) was measured during the first 10 wks of lactation. Starting at 87±9 DIM, cows underwent four 4-day periods of FR to meet 50% of individual energy requirements (FR1-FR4). Mid-lactation RFI (RFImidlact) was measured during 5 wks following FR4, as described (doi.org/10.1016/j.anscip.2022.07.181). Milk isocitrate, glucose, glucose-6-phosphate, malate, glutamate and free amino groups (NH_2-groups) were measured on wk 1, 2, 3, 4, 6 and 8, and daily during FR1 only. Data was analysed using mixed models with repeated measures, and spearman correlations. RFIearlylact ranged from -0.44 to -2.39 for negative RFIearlylact (i.e. more efficient), and from 0.42 to 2.34 kg DMI/d for positive RFIearlylact groups. Early lactation ECE was higher, whereas EB, plasma glucose, milk glucose, NH_2-groups and glutamate were lower for the negative RFIearlylact group. Average early lactation milk glutamate and NH_2-groups from wk1 to 8 were correlated with RFIearlylact (r= 0.44 and 0.41), but not with RFImidlact. No milk metabolite RFI group differences were observed during FR1. Weekly EB was positively correlated with milk glucose and malate (r=0.30 and 0.32), and negatively correlated with milk isocitrate and NH_2-groups (r=-0.25 and -0.17), and the inverse correlations were observed for ECE. RFIearlylact associations with plasma glucose, milk glucose, NH_2-groups and glutamate during early lactation may reflect in prioritization of nutrients towards milk secretion in more efficient cows. Conversely, these effects may be due to the lower EB experienced by cows with negative RFIearlylact.

Session 92 Theatre 2

Dairy cow inflammatory status is modulated by physiological stage and feed restriction

C. Delavaud[1], A. De La Torre[1], D. Durand[1], S. Bes[1], D. Roux[2], A. Thomas[1], M. Tourret[1], I. Ortigues-Marty[1], I. Cassar-Malek[1], M. Bonnet[1] and J. Pires[1]
[1]INRAE, UCA, VetAgro Sup, UMR Herbivores, Theix, 63122 Saint-Genès-Champanelle, France, [2]INRAE, U1414 Herbipole, Theix, 63122 Saint-Genès-Champanelle, France; carole.delavaud@inrae.fr

In dairy cows, the transition period is characterized by an altered inflammatory status (IS) whose physiological origin has not been clarified. The objectives were to describe the IS around parturition, and to evaluate the effects of feed restriction (FR) after peak lactation, in order to evaluate effects of DMI on IS separately from physiological changes of transition period. Fourteen Holstein and 14 Montbéliarde multiparous cows were studied from weeks -4 to + 23 of lactation. Starting at 87±9 DIM, cows underwent four 4-day periods of FR to meet 50% of individual net energy requirements (FR1 wk12, FR2 wk14, FR3 wk15, FR4 wk 16). Dry matter intake was measured, and blood samples were collected from coccygeal vessels before am feeding on weeks -3 (wk-3), 2 (wk+2), 12 (wk+12) and 22 (wk+22) of lactation. Two samples were collected on days 0 (i.e. before) and 4 of FR1 wk 12. Plasma was stored at -80 °C and the IS was assessed by analysis of a panel of chemokines, pro and anti-inflammatory cytokines (IL-1α, IL-6, IL-8, IL-10, MCP-1, MIP-1β, TNFα and VEGFα; Milliplex^R Bovine; BCYT1-33K-10, Merck, DE). Antioxidant capacity was assessed by FRAP assay and lipomobilization by plasma NEFA. Data were analysed using mixed models and repeated measures with cow as random effect. For Holstein and Montbéliarde cows, IL-1α, IL-10, MCP-1, IL-6 and MIP-1β were significantly lower on wk+2 compared to wk-3 and remained steady over the course of lactation. IL-8 concentration was higher on wk+2 compared with wk-3, whereas VEGFα declined on d0-wk+12 and wk+22 compared with wk-3 and +2. IL-8, TNFα and VEGFα were higher in Holstein than Montbéliarde cows. The 4-day FR1 significantly decreased IL-10, MCP-1 and MIP-1β concentrations. Plasma FRAP was negatively correlated with NEFA (r=-0.40), and decreased with FR. The cytokines measured were not correlated with either plasma FRAP or NEFA. To conclude, the lactation stage modulates IS in dairy cows, and short FR after lactation peak does not seem to cause inflammation despite lipomobilization and altered redox state.

Session 92 Theatre 3

Energy balance clusters in relation to metabolic and inflammatory status and disease in dairy cows

J. Ma[1], A. Kok[1], R. Bruckmaier[2], E. Burgers[1], R. Goselink[1], J. Gross[2], T. Lam[3], A. Minuti[4], E. Saccenti[1], E. Trevisi[4] and A. Van Knegsel[1]
[1]Wageningen University & Research, De Elst 1, Wageningen, the Netherlands, [2]University of Bern, Bremgartenstrasse 109a, Bern, Switzerland, [3]Utrecht University, Yalelaan 7, Utrecht, the Netherlands, [4]Università Cattolica del Sacro Cuore, Via Emilia Parmense 84, Piacenza, Italy; ariette.vanknegsel@wur.nl

The early lactation period in dairy cows is characterized by complex interactions among energy balance (EB), disease and alterations in metabolic and inflammatory status. The objective of this study was to evaluate relationships between EB, disease, metabolic and inflammatory status in dairy cows in early lactation. Holstein-Friesian dairy cows (n=154) were selected and monitored for disease treatments during week 1 to 6 in lactation. Weekly EB was calculated and plasma samples were analysed for metabolic and inflammatory variables. First, cows were clustered based on time profiles of EB (SP: stable positive; MN: mild negative; IN: intermediate negative; SN: severe negative). Cluster of EB was related to plasma non-esterified fatty acids and β-hydroxybutyrate concentration. Cows in SN cluster had a higher milk yield, lower dry matter intake, and lower insulin concentration compared with cows in SP cluster, and lower glucose and IGF-1 concentration compared with cows in SP and MN cluster. Cows in SN cluster tended to have a higher concentration of ceruloplasmin compared with cows in SP cluster. Second, cows were grouped based on disease treatments (CHP: cows with treatment for clinical health problem including endometritis, fever, clinical mastitis or retained placenta; OHP: cows with no CHP but treatment for other health problem; NHP: cows with no treatments). Energy balance was not different among disease groups. The CHP cows had lower albumin and paraoxonase concentration and a higher haptoglobin concentration compared with OHP and NHP cows. In week 1 after calving, NHP cows had a higher concentration of glucose, insulin and IGF-1 concentration compared with OHP and CHP cows. Overall, EB time profiles were associated with metabolic status of dairy cows in early lactation, but limitedly related with inflammatory status. Inflammatory status was related to disease events in early lactation.

Session 92 Theatre 4

Effects of parity on metabolism, redox status and cytokines in early lactating dairy cows

A. Corset[1,2], A. Boudon[2], A. Remot[3], S. Philau[2], P. Poton[2], O. Dhumez[2], B. Graulet[4], P. Germon[3] and M. Boutinaud[2]
[1]Biodevas Laboratoires, ZA de L'Épine, 72460 Savigné-l'Évêque, France, [2]INRAE-Institut Agro Rennes Angers, UMR 1348 PEGASE, 16 Le Clos, 35590 Saint Gilles, France, [3]INRAE-Université de Tours, UMR 1282 ISP, Centre de recherche Val de Loire, 37380 Nouzilly, France, [4]INRAE-VetAgro Sup, UMR Herbivores, Theix, 63122 Saint-Gènes-Champanelle, France; marion.boutinaud@inrae.fr

In dairy cows during early lactation, inflammation and oxidative stress can occur and lead to a risk of cell damages and may be related to the metabolic status. Our objective was to investigate the influence of parity on the metabolic, redox status and immunity of dairy cows. Fifteen Holstein cows were classified into four groups according to calving date with 7 primiparous and 8 multiparous cows. Blood samples were collected 3 weeks before calving and 4, 8 and 12 weeks postpartum and after calving milk was sampled at the same times. Plasma metabolic and redox status markers were analysed by spectrophotometry. In leucocytes collected from milk, the expression of genes coding for antioxidant enzymes were analysed by RTqPCR. Plasma cytokines were analysed after an *ex vivo* challenge of whole-blood cells with or without heat-killed *Escherichia coli*. Data were analysed with a mixed model including parity, week, and their interaction as fixed effects, and cow as a random effect. As expected, milk production was higher in multiparous than in primiparous cows. Plasma non-esterified fatty acids and β-hydroxybutyrate were higher in multiparous than in primiparous respectively at the 2nd and the 4th weeks of lactation, whereas glucose was lower. Multiparous cows have higher d-ROM at 8 weeks in plasma, higher GPX activity in erythrocytes at 4 weeks, and higher Sod1 expression levels at 4 weeks in milk leucocytes. Associated with these results, multiparous cows had a higher level of vitamin E but lower plasma concentrations of cytokines CXCL10, CCL2, IL1Rα and IFNγ. After *E. coli* challenge, at the 3rd week before and the 8th week after calving, the immunity of multiparous cows would be less effective with lower IL1α and TNFα. The increased energy metabolism is accompanied by an increase in the antioxidant response to resolve redox balance, and by a probably less effective immunity in multiparous compared with primiparous cows.

Session 92 Theatre 5

Tissue distribution and pharmacological characterization of bovine free fatty acids-sensing GPCRs

T.C. Michelotti[1], M. Bonnet[1], V. Lamothe[2], S. Bes[1] and G. Durand[1,2]
[1]INRAE, INRAE, Université Clermont Auvergne, Vetagro Sup, UMRH, 63122 Saint-Genès-Champanelle, France, [2]Bordeaux Sciences Agro, 1, cours du général de Gaulle, 33175 Gradignan Cedex, France; tainara.michelotti@inrae.fr

More than just an energy source, free fatty acids (FFAs) are signalling molecules known to affect metabolic functions, such as regulation of systemic energy homeostasis and immune response. FFAs mediate their action through the interaction with different receptors: PPAR nuclear receptor, glucocorticoid receptor, and FFA receptors (FFARs). Recently discovered, FFARs have been well studied in humans and mice, although their characterization in bovine species is currently scarce. Therefore, our objective is to characterize 5 bovine FFARs (FFAR1 to 4, and GPR84) in regards of tissue distribution and pharmacological properties. Samples from 6 tissues [liver, ileum, rectum, spleen, longissimus thoracis (LT), and perirenal adipose tissue (PRAT)] were collected from 16 Charolais bulls (16-18 months old) at slaughter. Total RNA was extracted and *FFARs*' gene expression was assessed by RT-qPCR. Results show that *FFAR1, FFAR2, FFAR3,* and *GPR84* were expressed in all studied tissues, while *FFAR4* expression was restricted to ileum, rectum, and PRAT. Furthermore, in order to unravel the mode of action of FFAs at the level of the different bovine FFARs, we are conducting *in vitro* experiments using HEK293A cells. Cells are transfected to individually express the 5 bovine FFARs. Through bioluminescence resonance energy transfer (BRET) assay, pharmacological properties of the FFARs (e.g. G protein selectivity, G protein dependent and β-arrestin-dependent signalling) are assessed in response to a broad range of FFAs. Preliminary results show that FFAR1 and FFAR2 are coupled to G-protein subunits $G\alpha_{q1}$ and $G\alpha_{i/o}$, and present β-arrestin2 recruitment. While FFAR1 is activated by medium and long chain FFAs, FFAR2 is activated exclusively by short chain FFAs. Overall, we believe that results from the *in vitro* experiments, aligned with tissue expression data, could revel novel information regarding FFA effects in bovine metabolism, which might affect animal production, adaptation to physiological conditions, or development of metabolic disorders.

Session 92 — Theatre 6

Maternal nutrition carry-over effects on beef cow colostrum but not on milk fatty acid composition

N. Escalera-Moreno[1], B. Serrano-Pérez[1], E. Molina[1], L. López De Armentia[2], A. Sanz[2] and J. Álvarez-Rodríguez[1]
[1]University of Lleida, Lleida, 25198, Spain, [2]CITA de Aragón-IA2, Zaragoza, 50059, Spain; javier.alvarez@udl.cat

Prepartum nutrition (60 vs 100% their energy requirements during 3 months before calving, LOW vs HIGH) effects on beef cow colostrum and milk fatty acids (FA) profile were evaluated (n=80 cows, half from Parda de Montaña and Pirenaica beef breeds). The postpartum cows were fed 100% their energy requirements (0.7[illegible]% of total FA, 37.1% C18:2 n-6, 29.8% C16:0, 12.9% c9-C18:1, 9.7% 18:3 n-3). Colostrum was manually milked <12 h post-partum and milk was machine milked by oxytocin technique at week 3 post-partum. Samples were freeze-dried, and FA were analysed by gas chromatography using C11:0 as internal standard and reference standards to compare identified peaks. In colostrum, de novo synthetized saturated FA (SFA) (C4:0 to C15:0) content were lower in LOW than in HIGH fed cows (17.73 vs 21.31±0.41%, $P<0.001$) whereas odd-chain SFA were higher in LOW than in HIGH fed cows (1.98 vs 1.77±0.04%, $P<0.001$). Desaturase index was higher in LOW than in HIGH fed cows ($P<0.001$), leading to greater total trans- and cis-monounsaturated FA (MUFA) in LOW than in HIGH cows (2.05 vs 1.51±0.07%, 26.32 vs 21.35±0.52%, $P<0.001$). Polyunsaturated FA (PUFA) n-3 were higher in LOW than in HIGH cows (2.26 vs 1.96±0.08%, $P<0.001$), mainly as a result of greater C18:3 n-3, whereas PUFA n-6 did not differ across groups ($P>0.05$). Likewise, colostrum rumenic acid content was higher in LOW than in HIGH cows (0.97 vs 0.71±0.031%, $P<0.001$). The c9-C18:1/C15:0 ratio, which is a proxy of negative energy balance, was higher in LOW than in HIGH cows (27.8 vs 20.4±0.99, $P<0.001$). However, total FA in colostrum did not differ across pre-partum feeding levels (4.64±0.52%, $P>0.05$). Subsequently, there were no differences between treatments in any milk FA group, except in total PUFA n-3, that were lower in LOW than in HIGH cows (0.85 vs 0.93±0.03%, $P<0.05$), although C18:3 n-3 did not differ anymore. The total FA in milk did not differ across pre-partum feeding levels either (3.71±0.10, $P>0.05$). In conclusion, when the cows were underfed during the last third of pregnancy, less SFA but more MUFA and essential PUFA were found in colostrum. Later, the milk FA differences were nearly vanished.

Session 92 — Theatre 7

Different types and doses of colostrum to optimize the passive immune transfer and health in lambs

A. Belanche[1], F. Canto[1] and O. Calisici[2]
[1]University of Zaragoza, Animal Production and Food Science, Miguel Servet 177, Zaragoza, 50013 Spain, [2]Phytobiotics Futterzusatzstoffe GmbH, Wallufer Str. 10a, Eltville am Rhein, 65353, Germany; belanche@unizar.es

Optimizing the artificial rearing of young ruminants represents a challenge in modern dairy production systems. However, access to sufficient quantity and quality of colostrum represents a limitation in many farms, which can lead to failure in the passive immune transfer. The objective of this experiment was to investigate the effects of a vacuum-dried bovine colostrum powder (BCP) on the health and of lambs. At birth, fifty-five newborn lambs were separated from their dams and randomly divided into five experimental groups (n=11). Animals received equal volumes pooled ovine colostrum (OC), pooled bovine colostrum (BC), reconstituted BCP at medium (BCPM) or high dose (BCPH). All these lambs received artificial milk feeding whereas a fifth group of lambs sucked maternal colostrum (in unknown quantities) followed by a natural lactation on de dam as control (CTL). Colostrum was provided by oro-gastric intubation at 2 h (6% of BW) and 6 h (6% of BW) whereas at 12 h after birth lambs received milk a similar dose of milk replacer (expect for the BCPH which received BCP). Blood samples were collected at 0, 1, 3, 14, 45 and 52 days of age to monitor plasma metabolites. Results showed a greater IgG concentration in OC than in BC or BCP resulting on a higher IgG intake. All artificially reared lambs (OC, BC, BCPM and BCPH) showed a similar concentrations of IgG at 24 h of age leading to a lower efficiency of IgG absorption for the OC and BCPH treatments ($P=0.09$). No differences were noted among these treatments in terms of plasma metabolites, health and productivity. The presence of anaemia due to the use of heterologous colostrum was discarded. However, CTL lambs had 1.8 to 2.0 times higher serum IgG concentration than artificially reared lambs when measured by ELISA or refractometer, respectively. Control lambs also had lower incidence of diarrhoea and higher BW gain during the post-weaning period. In conclusion, the use of pooled OC, pooled BC or bovine colostrum powder can be considered as successful strategies to artificially rear ruminants, despite maternal rearing provides extra health and productive benefits.

Oxidative status in female Holstein calves fed with or without transition milk

C.S. Ostendorf[1], M. Hosseini Ghaffari[1], B. Heitkönig[1], C. Koch[2] and H. Sauerwein[1]
[1]University of Bonn, Bonn, 53111, Germany, [2]Hofgut Neumühle, Münchweiler, 67728, Germany; costendo@uni-bonn.de

This study examined the effects of feeding transition milk to calves on their blood oxidative status, comparing two groups: transition milk (TRANS, n=14) and milk replacer (MR, n=15). All calves received 3.5-4 l of colostrum milked from their dams within 2 h after birth, and 1-1.5 l of colostrum 11.5 h later. Calves were fed according to their feeding group with 12 l/d maximal feed allowance until d 6 of life. All calves received MR (12 l/d; 140 g/l) and were weaned at d 98 of life. Samples from colostrum and from transition milk collected before every feeding were pooled per calf and day. Blood was sampled before colostrum feeding, 12 h after birth, daily during the first 5 d, weekly until weaning (wk 14), and at wk 16. Antioxidant capacity was assessed as the ferric-reducing ability of plasma (FRAP), and reactive oxygen metabolites by the dROM test. An oxidative stress index (OSi) was calculated: dROM/FRAP × 100, and total protein (TP) was also quantified via the Bradford assay. Data were analysed using mixed models with group, time, and the interaction therefrom as fixed effects. Time was significant for all variables assessed ($P<0.05$). Except for TP, neither group nor the group by time interaction affected dROM, FRAP, and OSi. The observed interaction for TP was due to initially greater TP concentrations in the TRANS which levelled off later. The dROM values increased with colostrum intake and remained relatively constant during the remaining wks. The FRAP concentrations were highest after birth and then showed a continuous decrease with the lowest value observed at wk 16. The OSi values increased after colostrum feeding and were then fluctuating including an increase around weaning (wk 13-16), and reaching the greatest values in wk 16. TP and FRAP levels in milk were highest in colostrum and decreased thereafter but did not affect serum levels in calves. Overall, these results provide insight into the oxidative status of dairy calves in the preweaning period but suggest that the oxidative status as characterized by the variables investigated herein is neither affected during feeding transition milk nor thereafter.

Effects of concentrate feeding and hay quality on adipogenesis and inflammation in dairy calves

R. Khiaosa-Ard, A. Sener-Aydemir, S. Kreuzer-Redmer and Q. Zebeli
University of Veterinary Medicine Vienna, Veterinaeplatz 1, 1210 Vienna, Austria; ratchaneewan.khiaosa-ard@vetmeduni.ac.at

We previously showed that concentrate feeding increased the intake of n-6 polyunsaturated fatty acids and the contents of their metabolites, which are lipid mediators, in the adipose tissue of young calves as compared to hay-only feeding. Here, we investigated whether concentrate feeding also affects the cellularity and gene expression related to metabolism and inflammation of the adipose tissue. From the first week to 15 weeks of life, 20 Holstein-Friesian calves (17 males and 3 females) were adapted to one of the four diets differing in the concentrate amount: 0 or 70% on a DM basis, milk excluded) and hay quality: medium-quality hay (MQH) or high-quality hay (HQH). The daily intake of hay, concentrate and milk was recorded. Two regions of kidney fat: proximal and distal to the kidneys were taken for analysis. For qPCR results, data were normalized by the housekeeping genes ACTB and RPL19 and reported as fold changes relative to the average of MQH-only samples. Data were analysed as a 2×2×2 factorial design with respect to concentrate feeding, hay quality, and adipose region. The proximal region showed larger adipocyte size and higher expression of several adipose genes including LEP (leptin), GLUT4 (glucose transporter type4), PPARG (peroxisome proliferator-activated receptor gamma), FASN (fatty acid synthase), and ADIPOR1 and ADIPOR2 (adiponectin receptor 1 and 2) than the distal region ($P<0.01$). An effect of concentrate feeding on adipose cellularity and adipose gene expression was region dependent. Both concentrate feeding groups and HQH-only enlarged the adipocyte size. However, very large adipocytes (>5,000 μm^2) were more frequent with concentrate feeding. Concentrate feeding increased the expression of TNFα, LEP, FASN, and PPARG in the proximal region and decreased the expression of NF-kB, IL-6, IL-10, and ADIPOR2 in the distal region. Using ΔCt values, ADIPOR2 positively correlated with TNFα, NF-kB, and IL-1β, and IL-10 and GLUT4 with NF-kB, IL-1β, and IL-6, while LEP negatively correlated with IL-10 ($r=0.45-0.56$, $P<0.01$). This study marked the role of concentrate feeding in adipogenesis and inflammatory regulation.

Session 92 Theatre 10

Synergistic benefits of marine derived bioactives to maintain the gut barrier in in an *ex vivo* model

M.A. Bouwhuis[1], I. Nooijen[2], E. Van Der Steeg[2] and S. O'Connell[1,3]
[1]Celtic Sea Minerals, Strand Farm, Carrigaline, Co. Cork, Ireland, [2]TNO, Sylviusweg 71, 2333 BE Leiden, the Netherlands, [3]Munster Technological University, Plant Biostimulant Group, Shannon ABC, Clash Road, Tralee, co. Kerry, Ireland; m.bouwhuis@celticseaminerals.com

A healthy gut barrier restricts entry of pathogens, bacteria and other toxins in the diet/environment to submucosal tissue and bloodstream of pigs. Two marine derived bioactives from red seaweed *Lithothamnion sp.* (LG) and brown seaweed *Ascophyllum nodosum* (ANE) have shown individual potential *in vitro* to strengthen the gut barrier. Hence, the objective of the current experiment was to investigate the effects of LG and ANE and their combination on gut permeability in the *ex-vivo* InTESTine™ system. Ileal tissue from one pig was mounted into the InTESTine™ system and incubated for 5 hours. Eleven test conditions were used with 3 control conditions. Due to the calcium (Ca) level of LG (30%w/w), 3 control conditions were used (2.0, 3.8 and 5.3 mM Ca) to match Ca levels of the test conditions. The eleven test conditions consisted of solubilised LG at 63.2 µg/well, ANE (liquid form) at 2.3 µg/well and the remaining 9 conditions were set up as a 3×3 factorial design with LG at 37.8, 63.2 and 88.7 µg/well and ANE at 2.3, 4.5 and 9.1 µg/well. All 14 test conditions were incubated with and without *Salmonella enterica enteritidis* (SEE) (n=4 for each condition). Tissue permeability (P_{app}; 10^{-6} cm/s) was measured by paracellular transfer of mannitol from 0-120, 120-240 and 240-300 minutes. Data was analysed by ExcelStat (16.7, 2022.4.1) using Tukey's adjustment. The results showed an interaction between time, SEE challenge and test condition ($P<0.001$).The SEE challenge resulted in an 6.73 fold increase in gut permeability ($P<0.001$). Between 120-240 minutes, no effect of different Ca levels was observed ($P>0.05$). However, inclusion of both LG+ANE (2.3 µg/well ANE with any level of LG, as well as 4.5 µg/well ANE with either 37.8 or 63.2 µg/well LG) reduced P_{app}, showing less damage to the gut barrier compared to the control treatments or LG or ANE individually ($P<0.05$). Hence, it can be concluded that an SEE challenge disrupts the gut barrier and that LG and ANE work synergistically together to maintain a healthy gut barrier.

Session 92 Theatre 11

Effects of milk replacer diets on the intestinal development of sucked piglets

G.Y. Duan[1,2], C.B. Zheng[3], J. Zheng[1,2], P.W. Zhang[3], M.L. Wang[3], J.Y. Yu[1,2], B. Cao[4], M.M. Li[4], F. Cong[4], Y.L. Yin[1,2,3] and Y.H. Duan[1,2]
[1]University of Chinese Academy of Sciences, Beijing, 100049, China, P.R., [2]Institute of Subtropical Agriculture, Chinese Academy of Sciences, Hunan Provincial Key Laboratory of Animal Nutritional Physiology and Metabolic Process, Changsha, 410125, China, P.R., [3]Hunan Agricultural University, College of Animal Science and Technology, Changsha, 410128,China, P.R., [4]Fengyi (Shanghai) Biotechnology R&D Center Co., Ltd, Shanghai, 200120, China, P.R.; duanyehui@isa.ac.cn

This study aimed to investigate the effect of milk replacer diets on the intestinal development of sucking piglets. Sixteen healthy seven-day-old crossbred (Duroc × Landrance × Yorkshire) male piglets with similar body weight (3.98±0.14 kg) were selected and randomly divided into two groups (n=8): (1) suckling group (continued to be suckled by the sow); (2) milk replacer-feeding group (separated from the sow at 7 days of age and bottle-fed by a milk replacer). The trial lasted for 21 days. Results showed that the final body weight has no significant difference between the two groups ($P>0.05$). However, compared to the suckling group, piglets in the milk replacer-feeding group had higher serum concentrations of proinflammatory cytokines (TNF-α, IL-1β, and IL-6) and lower protein expression levels of TLR4 and NF-κB in the small intestine ($P<0.05$). In addition, compared to the suckling group, piglets fed with milk replacer diet exhibited lower villi folds in the small intestine, accompanied by reduced protein expression of Ki67, mTOR, lipogenesis-related proteins (PPARγ and SREBP1), and tight junction proteins (Occludin, Claudin and ZO-1) as well as downregulated mRNA expression levels of MUC2 ($P<0.05$). Otherwise, compared to the suckling group, piglets in the milk replacer-feeding group illustrated higher protein expression levels of ULK1, TFEB, Parkin, and PNK1 ($P<0.05$). Moreover, milk replacer feeding significantly increased the activity of duodenal alkaline phosphatase and sucrose ($P<0.05$). In summary, piglets given a milk replacer displayed damaged intestinal morphology and function and elevated intestinal autophagy without effects on body weight within the 3-week feeding period.

Session 92 Theatre 12

The impact of early thermal manipulation on the hepatic energy metabolism of mule duck

C. Andrieux[1], M. Morisson[2], V. Coustham[1], S. Panserat[1] and M. Houssier[1]
[1]UPPA, INRAE UMR1419 NuMéA, 173 Route de Saint-Jean-de-Luz, 64310 Saint Pée sur Nivelle, France, [2]ENVT, INRAE, GenPhySE, 24 Chemin de Borde Rouge, 31320 Castanet-Tolosane, France; charlotte.andrieux@univ-pau.fr

Increasing egg incubation temperature has been shown to have a significant impact on energy metabolism in poultry, and more precisely on hepatic metabolism in mule duck. We therefore investigated the effects of thermal manipulation (TM) on the hepatic metabolism and *foie gras* production performance. The impacts of the TM were measured throughout the life of the mule duck and during two different metabolic challenges, from embryogenesis to slaughter day. One hour after the rise in temperature, the relative expression of around 10% of the genes studied in the liver (8/78 genes), involved in lipid and carbohydrate metabolisms, was significantly modulated showing a direct effect of temperature on metabolism (Anova tests, 8 ducks). Thereafter, hatchability sometimes decreased depending on the time and intensity of the TM (Chi tests, 265 to 279 ducks), while under most conditions, hatching body weight and body temperature were lower (Student tests, 100 ducks). Several months after the thermal stimulus, refeeding after a 23 h-fast induced a change in energy metabolism in TM duck livers compared to control ducks, with an increase in hepatic cell size (+1 µm on average), a change in liver lipid composition (+4% of saturated fatty acids 4 h after the meal) and changes in the relative expressions of metabolic genes (14 genes, Anova tests, 8 ducks). On the other hand, overfeeding (OF) induced an 8% increase in *foie gras* weight in the TM group compared to the control group without any negative impact on the quality of the *foie gras* (Student tests, 36 to 58 ducks), but again, differences in lipid composition were noticed (student test, 8 livers) associated with modulations in the expression of hepatic genes involved in energy metabolism. Our results confirm that embryonic TM can program liver metabolism of mule duck, and that this programming can be revealed by different feeding challenges. TM could be a lever to reduce the number of OF meals, thereby improving animal welfare and reducing production costs. Finally, this new technique could be combined with optimized breeding methods to produced fattened livers without force-feeding.

Session 92 Poster 13

Feed efficiency and responses of plasma and milk isotopic signatures in Charolais beef cows

A. De La Torre[1], J. Pires[1], I. Cassar-Malek[1], I. Ortigues-Marty[1], F. Blanc[1], L. Barreto-Mendes[1], G. Cantalapiedra-Hijar[1] and C. Loncke[2]
[1]Univ. Clermont Auvergne, INRAE, VetAgro Sup, UMR Herbivores, Theix, 63122 Saint-Genès-Champanelle, France, [2]Univ. Paris-Saclay, INRAE, AgroParisTech, UMR Modélisation systémique appliquée aux ruminants, 91123 Palaiseau, France; anne.de-la-torre-capitan@inrae.fr

Productive and metabolic responses are supported by body reserves changes in beef cows when facing feed restriction (FR). Because dynamics of natural abundance of ^{13}C ($\delta^{13}C$) and ^{15}N ($\delta^{15}N$) in milk may reflect body changes in dairy cows, the objectives were: (1) to test if plasma and milk isotope signatures change in response to FR in beef cows; and (2) to study their associations with phenotypic feed efficiency. Twenty-two Charolais cows were studied from calving until wk 21 of lactation. Residual feed intake (RFI) was calculated on wks 1 to 8 of lactation when cows were allowed *ad libitum* intake of the same diet. On wk 9, cows underwent a 4-day FR to meet 50% of net energy requirements, followed by *ad libitum* intake. Individual production (MY) and metabolic (NEFA) responses were monitored from d-10 to d19 relative to FR. The $\delta^{13}C$ and $\delta^{15}N$ were analysed in plasma [before (d-10, d-2, d1); during (d2-5) and after (d6-d8, d11 and d19) FR] and in milk [before (d-10, d-2, d1); during (d4, d5) and after (d11, d19) FR]. The effects of FR and RFI on plasma and milk $\delta^{13}C$ and $\delta^{15}N$ were tested using mixed model and Spearman correlations. RFI ranged from -0.01 to -0.69 for negative RFI and from 0.04 to 1.15 kg DMI/d for positive RFI and had no effect on plasma and milk isotopic signatures. Plasma NEFA and $\delta^{15}N$ increased, and $\delta^{13}C$ decreased during FR ($P<0.0001$). Accordingly, milk $\delta^{15}N$ increased and $\delta^{13}C$ decreased during FR ($P<0.0001$). Plasma NEFA was correlated across the experiment with plasma $\delta^{15}N$ (r=0.41, $P<0.0001$) and $\delta^{13}C$ (r=0.40, $P<0.0001$), whereas no correlation was shown with milk isotopes signatures. Plasma $\delta^{15}N$ was correlated with milk $\delta^{15}N$ (r=0.66, $P<0.001$), but no relation was observed between plasma and milk $\delta^{13}C$. These results suggest that only plasma isotope signatures respond to body reserves changes in Charolais cows but correlations with plasma NEFA were moderate. To date, no clear relations with RFI were found. Further work is ongoing to model kinetics of $\delta^{13}C$ and $\delta^{15}N$ to explore links with feed efficiency.

Session 92 — Poster 14

How the composition of Holstein cow colostrum differs according to the immunoglobulins G level

A. Goi[1], A. Costa[2] and M. De Marchi[1]
[1]University of Padova, viale università, Legnaro 35020, Italy, [2]University of Bologna, via Tolara di Sopra, Ozzano dell'Emilia 40064, Italy; angela.costa2@unibo.it

In cattle, the narrow-sense quality of colostrum depends on the immunoglobulins G (IgG) concentration. Conventionally, IgG must be >50 g/l to consider colostrum acceptable for feeding newborn calves and administration must take place as soon as possible after birth to exploit the maximum gut permeability. High- and low-quality colostrum are expected to differ in terms of composition traits different from IgG, such as amino acids. By using 521 colostrum samples collected within 6 h from calving, we compared the composition of samples belonging to different quality levels: <50 (I), 50-69.99 (II), and ≥70 g/l (III) of IgG. Samples belonged to 521 Holstein cows reared in 9 farms that calved between 2019 and 2020. The traits analysed comprise content of fat, total protein, ash, somatic cell score (SCS) and concentration of essential AA, namely Arg, His, Ile, Leu, Lys, Met, Phe, Thr, Trp, and Val. All traits were determined with their respective gold standard and analysis of variance accounted for the fixed effect of farm (n=9), calving season (Jan-Feb, Mar-Apr, May-Jun, Jul-Aug, Sep-Oct, Nov-Dic), parity (3 classes: 1, 2, and 3-8), quality level, and the interactions of quality level with parity and calving season. Except for fat, all the traits were the greatest when IgG was ≥70 g/l and the lowest when <50 g/l. Somatic cell count was not different in samples belonging to the II and III quality class and the estimates of fat content were 5.57±0.41 (I), 4.82±0.25 (II), and 4.36±0.26% (III), with class II being similar (P>0.05) to both I and III. The interaction with parity was significant for fat and SCS. Samples in class III presented greater SCS compared to I and II. Regardless of the quality level, primiparous cows' colostrum had in general greater fat content compared to the others. Such findings suggest that colostrum composition vary according to the IgG level, other than to parity and season. Although colostrum yield (kg) was not available to us to evaluate the presence of a dilution effect, we can conclude that the narrow-sense quality of colostrum (IgG-based) goes in parallel with the broad-sense quality.

Session 92 — Poster 15

An educational kit for effective prevention and management of energy deficit in dairy cows

M. Gelé[1], M. Boutinaud[2], A. Bouqueau[1], M. Marguerit[1], J. Jurquet[1], Y. Le Cozler[2] and J. Guinard-Flament[2]
[1]IDELE, 149 rue de Bercy, 75595 Paris, France, [2]PEGASE, INRAE, Institut Agro, 35590 Saint-Gilles, France; yannick.lecozler@agrocampus-ouest.fr

Energy deficit (ED) occurs when energy intake is lower than the animal's needs at early lactation or later when cow is underfed. Severe ED can alter milk production, reproductive and health performances, and thus the economy of the dairy farm. Despite this, a recent survey showed there is an inappropriate knowledge on ED in the dairy sector (dairy farmers, farmer advisers, etc.). That is why the partners of the Biomarq'lait project, financed by the French Ministry of Agriculture and Food, decided to synthesise current knowledge on the ED to provide an educational kit for French agricultural and agronomy students. A survey showed a strong interest of teachers for this topic and enabled the kit to be structured into three materials: a video, three summary sheets, and an evaluation test. The video presents the key points of prevention and management of energy deficit in dairy cows, while the three sheets give more details about the knowledge on ED. The evaluation test is in a fun format and allows students to assess their level of knowledge before and/or after the training session. This kit will be offered to all agricultural and agronomic schools at the start of the 2023-2024 school year.

Session 92 Poster 16

Inflammation status weakly modulated by short-term feed restriction in lactating beef cows

C. Delavaud[1], J. Pires[1], M. Barbet[2], I. Ortigues-Marty[1], I. Cassar-Malek[1], M. Bonnet[1] and A. De La Torre[1]
[1]INRAE, UCA, VetAgroSup, UMR Herbivores, Theix, 63122 Saint-Genès-Champanelle, France, [2]INRAE, U1414 Herbipole, Theix, 63122 Saint-Genès-Champanelle, France; carole.delavaud@inrae.fr

Variations in feed intake modify metabolism and may have putative effects on inflammation status. There is scarce data on suckling cows experiencing acute shifts in feed allowance. The objective was to study inflammation and metabolic responses to a short nutritional stress induced by experimental feed restriction (FR) after 9 weeks of lactation in lactating beef cows. Twenty-two primiparous Charolais cows underwent a 4-day FR to meet 50% of net energy requirements (FR1 wk9, FR2 wk12, FR3 wk13, FR4 wk14). Dry matter intake was measured and residual feed intake (RFI) was calculated on wk 1 to 8 of lactation when cows were allowed *ad libitum* intake of the same diet. RFI ranged from -0.01 to -0.69 for negative RFI (RFI-) and from 0.04 to 1.15 kg DMI/d for positive RFI (RFI+). Blood samples were collected from coccygeal vessels before am feeding on days -13, -6, -1, +4, and +91 relative to FR1. Plasma was stored at -80 °C until analysed for IL-1α, IL-8, IL-10, IL-17α, MCP-1, MIP-1β, TNFα and VEGFα using Milliplex® Bovine Cytokine/Chemokine Panel (BCYT-33K-10, Merck, DE). Plasma stored at -20 °C was analysed for NEFA, glucose, β-hydroxybutyrate (βOH) and urea. Data were analysed using mixed models and repeated measures with cow as random effect. In both RFI group, FR1 increased NEFA, βOH and urea concentrations, whereas glucose decreased between d-16 to d+4. IL-1α, IL-10, MCP-1 and TNFα were closely correlated (Spearman: r>+0.8, P<0.0001). Whatever the RFI group, IL-8 and VEGFα were increased by FR1. A significant interaction between RFI and day was observed for IL-1α and TNFα with constant increase between d-6 to d+4 for RFI- cows, whereas no change was observed for RFI+. The relation between the effect of FR on cytokines and chemokines and RFI was observed despite large inter-individual variations. Although FR appears to increase the chemokine IL-8 and proinflammatory VEGFα, the lack of effects of FR on the other cytokines suggests that, in the presence of a weak modulation of the metabolic status, it has no marked effect on the inflammation status.

Session 92 Poster 17

Metabolic profiles of grazing dairy cows with or without feed restriction during early lactation

A.L. Astessiano, M. Garcia-Roche, D. Custodio, G. Ortega, P. Chilibroste and M. Carriquirry
Facultad de Agronomia, Garzon 780, 11600, Uruguay; lauaste@gmail.com

Twenty-four multiparous Holstein cows were selected to assess the effect of feed restriction on energy metabolism (468±73 kg body weight and 2.75±0.16 body condition score (BCS) units). Cows were randomly assigned at 31±14 days in milk (DIM) to ADLIB or RESTR treatments (36.5 vs 16.5 kg DM/cow/d pasture allowance and 5.3 vs 2.6 kg DM/cow/d energy concentrate). From 31 to 40 DIM cows went through an adaptation period (fed ADLIB), from 41 to 46 DIM cows were fed according to their treatments and from 46 to 57 DIM cows went through a flushing period (fed ADLIB). Milk production was recorded daily, BCS and plasma samples were recorded or collected twice during every period. Plasma metabolites and insulin were measured with colorimetric and radioimmunoassay commercial kits, respectively. Data were analysed using a mixed model that included period, treatment, and their interaction as fixed effects. Milk yield was lower for the RESTR group during feed restriction (25.0 vs 29.8±0.7, P<0.01) and BCS was greater during adaptation (2.77 vs 2.60 and 2.62±0.05 for 38, 48 and 56 DIM, P<0.01). Concentration of insulin in plasma was lower at 46 DIM (12.8, 9.7, 9.5, 6.4, 9.0 and 10.9±1.1 μU/ml for 31, 38, 44, 46, 51 and 56 DIM, P<0.01), while glucose concentration was lower for the RESTR than the ADLIB group (3.54 vs 3.78±0.09 mmol/l, P<0.05), non-esterified fatty acid concentration was greater for the RESTR than the ADLIB group at 44, 46 and 51 DIM (0.46 vs 0.27, 0.61 vs 0.38, 0.36 vs 0.24±0.04 mmol/l, P<0.01) and beta-hydroxybutyrate was greater for the RESTR group than the ADLIB group at 46 DIM (0.66 vs 0.39±0.05 mmol/l, P<0.01). Finally, urea concentration decreased progressively with lactation (6.5, 7.4, 7.1, 5.7, 5.5 and 4.9±0.27 mmol/l for 31, 38, 44, 46, 51 and 56 DIM, P<0.0001), and no differences were found for cholesterol or gamma-glutamyl transferase. Our results show that feed restriction in early lactation decreased glucose concentration in plasma and increased lipid mobilization.

Hepatic mitochondrial function in grazing dairy cows with or without feed restriction

M. García-Roche, D. Custodio, A. Astessiano, G. Ortega, P. Chilibroste and M. Carriquiry
Universidad de la República, 809 Garzón, 11200, Uruguay; mercedesg@fagro.edu.uy

To assess the effect of feed restriction on hepatic mitochondrial function during early and mid-lactation, multiparous Holstein cows (n=24, 468±73 kg body weight and 2.75±0.16 body condition score units) were randomly assigned to: *ad libitum* (ADLIB, n=12) or restriction (RESTR: 50% of ADLIB offered intake, n=12) treatments in three period trials at 31 and 146±17 days in milk (DIM). All cows were fed ADLIB in the adaptation period (31 to 40 and 146 to 157 DIM), then fed according to their treatments (41 to 46 and 157 to 162 DIM) and finally went through a flushing period (fed ADLIB, 46 to 57 and 162 to 170 DIM). Pasture height and milk production were recorded daily, milk composition was analysed during every period and liver biopsies were collected and cryopreserved at the end of every period. Mitochondrial function was assessed measuring oxygen consumption rates in liver biopsies. Pasture dry matter intake (DMI) was estimated from pre and post grazing pasture height measured with a plate meter. Energy corrected milk yield (ECM) and mitochondrial function were analysed using a repeated model that included treatment, experimental period, stage of lactation as fixed effects and their interactions. Pasture DMI was 9.4±1.1 and 17.0±2.4 kg DM/d and 11.5±1.4 and 9.0±4.4 kg DM/d for RESTR vs ADLIB cows during early and mid-lactation, respectively, in the latter ADLIB cows were supplemented with 9.5±1.7 kg DM/d ryegrass haylage due to low pasture cover. Supplementation with an energy concentrate represented 20% of total DMI. The interaction between treatment, period and stage of lactation was not significant for the variables analysed. However, the interaction between period in treatment and between period and stage affected ECM as it was lower for the RESTR group during feed restriction (24 vs 28±1, $P<0.05$, RESTR vs ADLIB respectively) only during early lactation. Respiration linked to ATP synthesis was greater during mid than early lactation (10.0 vs 6.9±1.7 pmolO_2/min/mg, $P<0.01$) and greater for the ADLIB than RESTR group (11.0 vs 6.0±1.4 pmolO_2/min/mg, $P<0.01$). Our results show that mitochondrial function is decreased during early lactation and during feed restriction.

On-farm emergency slaughtered dairy cows: causes and haematological biomarkers

F. Fusi[1,2], I.L. Archetti[2], S.M. Chisari[2], V. Lorenzi[2], C. Montagnin[2], R. Salonia[3], L. Bertocchi[2] and G. Cascone[3]
[1]University of Parma, Food & Drug, Parco Area delle Scienze 27a, PR, 43100, Italy, [2]IZS Lombardia ed Emilia Romagna, Via Bianchi 9, BS, 25124, Italy, [3]IZS Sicilia, Via Marinuzzi 3, PA, 90129, Italy; francesca.fusi@izsler.it

The management of end-of-career cows and their on-farm killing are major animal welfare (AW) issues to be addressed. Since Reg.(EC) No 853/2004 allows 'an otherwise healthy animal which suffered an *accident* that prevented its transport' to be emergency slaughtered on-farm (OFES), blood samples were collected from 33 OFES cows during *ante-mortem* inspections run by official veterinarians on 25 cattle farms in the province of Ragusa (Sicily, Italy) from May 2021 to April 2022. Any time an OFES cow was sampled, another healthy cow with similar characteristics (farm, age, production stage) was sampled too, as a negative control. The following parameters were analysed: total proteins, globulins, albumin/globulin ratio (A/G), urea, glucose, non-esterified fatty acids (NEFA), beta-hydroxybutyrate, total cholesterol, triglycerides, total bilirubin, aspartate aminotransferase (AST), gamma-glutamyl transferase, creatinine, magnesium, calcium, phosphorus, iron, potassium, chlorine, sodium, creatine kinase, oxidative stress (ROMs), positive (haptoglobin) and negative (albumin) acute phase proteins. Average herd size was 147 (min=21, max=437) and 18 out of 25 farms were assessed for their AW level through the ClassyFarm system, with an average total AW score of 73 (min=62; max=88). Average age of OFES cows was 4.7 yrs (min=1.8; max=9.35); out of 33, 30 were lactating cows and 3 dry cows. The causes of OFES were: traumatic leg injuries associated with slips and falls in 69.6% (n=23) cows; traumatic injuries associated with calving in 27.2% (n=9) cows; and mastitis in 3% (n=1) cows (according to Italian rules in force until early 2022). Comparing OFES and healthy cows, significant differences were found for the levels of globulins, A/G ratio ($P<0.05$) and of albumin, urea, haptoglobin, glucose, NEFA, total bilirubin, AST, calcium and iron ($P<0.01$), confirming some findings of previous studies. These are preliminary results of a larger project on metabolomics and osteocalcin in OFES cows, funded by the Italian Ministry of Health, grant IZSSI 09/20 RC.

Session 92 Poster 20

The effect of early feed restriction of ewe lambs on milk miRNAome of the filial generation

A. Martín[1], F. Ceciliani[2], F.J. Giráldez[1], R. Calogero[3], C. Lecchi[2] and S. Andrés[1]
[1]Instituto de Ganadería de Montaña (CSIC-Universidad de León), Finca Marzanas s/n, 24346, Grulleros (León), Spain, [2]Università degli Studi di Milano, Department of Veterinary Medicine and Animal Sciences, Via dell'Università, 6, 26900, Lodi, Italy, [3]Università di Torino, Department of Biotechnology and Health Sciences, Molecular Biotechnology Center, 10126, Turin, Italy; alba.martin@csic.es

Feed restriction during the early life of ewe lambs (F0) triggers the transfer of epigenetic marks to the next generation (F1). However, the effects of this factor on milk production and composition, including its abundance in regulatory miRNA (with a role in the modulation of immune response of the offspring), has not been tested so far. Therefore, in this study, the replacement ewe lambs (F0) obtained in a previous project (a group of ewes fed milk replacer *ad libitum* (ADL) vs a group of ewes restricted (RES) to 62.5% the intake level of milk replacer during the suckling period) were raised under similar post-weaning conditions and mated to obtain the progeny (F1). The F1 dairy ewes were also mated, and six of these F1 dairy ewes were selected and divided into two groups: the F1-ADL group (n=3) and the F1-RES group (n=3). Milk production was controlled during the peak lactation period, and milk samples were obtained for each gland separately to measure chemical composition, somatic cell counts (SCC), bacteriology and miRNA. The miRNAome was determined following a Next Generation Sequencing approach. The first preliminary results indicate significant differences in the abundance of five miRNAs; thus, oar-miR-150, oar-miR-221, oar-miR-23a, oar-miR-27a, oar-miR-376c were all down modulated in F1-RES when compared to F1-ADL. Most of these miRNAs play a role in development, apoptosis, muscle differentiation, reproduction, or milk production. Moreover, SCC were significantly reduced in milk samples of F1-RES dairy sheep. No significant differences were found in milk production, the chemical composition of milk (fat, protein, lactase), nor in bacteriology (colony forming units, cfu). These results provide some evidence of the effects of nutritional programming events on the milk's bioactive components.

Session 92 Poster 21

Quercetin supplementation in perinatal sows diet influencing suckling piglets and its mechanism

Y. Li, Q.L. Yang, Y.X. Fu, S.S. Zhou, J.Y. Liu and J.H. Liu
Northeast Agricutural University, Animal Nutrition and FeedScience, #600 Changjiang Road, Xiangfang District, Harbin, Heilongjiang Province, 150030, China, P.R.; liyaolzw@163.com

Similar 24 Landrace × Large white sows were randomly divided into control and three experimental groups, fed by basal diet supplemented with 0.000, 0.025, 0.050 and 0.075% quercetin, respectively, with 6 replicates in each group. The experiment started from the 100th day of gestation to the end of the next oestrus mating. Compared with control, 0.075% quercetin significantly increased weaning weight at day 28 (P<0.05). Quercetin significantly reduced serum MDA content, 0.050 and 0.075% quercetin significantly increased content of serum and liver GSH, reduced serum ROS content, and 0.075% quercetin significantly reduced content of liver MDA and NO (P<0.05). 0.050%, 0.075% quercetin significantly reduced content of serum IL-6, TNF-α, MCP-1 and IL-1β (P<0.05). Quercetin significantly increased content of IGF-1 and DA, decreased content of NE and EPI in serum (P<0.05). At the genus level, 0.050 and 0.075% quercetin significantly increased the relative abundance of Lactobacillus, Dorea and Peptococcus, respectively; reduced the relatively abundance of Escherichia-Shigella; Quercetin significantly increased the relative abundances of Phascolarctobacterium, Prevotellaceae_NK3B31_group and unclassified_f_Lachnospiracea; 0.075% quercetin significantly reduced the relative abundance of norank_f_Muribaculaceae (P<0.01). Quercetin significantly affected metabolites of sphingolipid metabolism, histidine metabolism and arginine biosynthesis. 0.050% quercetin significantly increased L-ornithine, N-acetyl-L-glutamic acid and N-acetyl-citrulline; 0.050 and 0.075% quercetin significantly increased L-aspartic acid and L-histidine; quercetin significantly reduced 1-methylhistidine, histamine, SM (d18:0/16:1(9Z)), ceramide and phytosphingosine, and significantly increased sphingosine content (P<0.01). In conclusion, optimal quercetin supplementation in diet of perinatal sows increased weaning weight in piglets via relieving oxidative stress and inflammatory response, regulating colonic microflora and metabolic pathways of sphingolipid metabolism, histidine metabolism, arginine synthesis.

Session 92 — Poster 22

Development of muscle injury in weanling piglets under chronic immune stress

Y. Duan

Institute of Subtropical Agriculture, Chinese Academy of Sciences, Changsha, Hunan, 410125, China, P.R.; duanyehui@isa.ac.cn

This study aimed to explore the effects of lipopolysaccharide (LPS)-induced chronic immune stress on muscle growth, inflammatory state, mitochondrial morphology and function, and oxidative status in piglets. Eighty healthy Duroc × Landrace × Yorkshire piglets (21±2 days old, barrow, 6.98±0.14 kg body weight) were selected and randomly allotted to two groups: control group (pigs were injected with 0.90% sterile saline) and LPS-challenged group (pigs were challenged with LPS on days 1, 3, 5, 7, 9, 11, 13, 15 of the trail at an initial dose of 80 μg per kg body weight, which was increased by 30% at each following injection). The pigs were slaughtered at 1, 5, 9, and 15 d of LPS injection. Results showed that, compared with the CON group, the chronic immune stress weakened the muscle growth manifested as decreased muscle mass and myofiber area from 1 d to the last day, accompanied by increased serum pro-inflammatory cytokines (IL-1β, IL-6, and TNF-α) and reduced protein expression of CD163 from 1 d to 15 d ($P<0.05$). The impaired mitochondrial morphology and function were correlated with excessive mitochondrion-derived ROS, reduced membrane potential, depletion of mitochondrial ATP, exceed mitophagy, and unbalanced fusion and division of mitochondria ($P<0.05$). In addition, AMPK signalling pathway was activated whereas the mTOR signalling was inactivated in *longissimus dorsi muscle* (LDM) from 5 d to 15 d ($P<0.05$). In terms of redox system, chronic immune stress markedly increased the accumulation of MDA and MPO in LDM, and the protein expression of 8-OHDG in LDM and *soleus muscle* (SM) ($P<0.05$). Collectively, the chronic immune stress was extremely detrimental to the muscle growth and development accompanied by alteration of mitochondrial function and crumble redox system, which then activated signal transduction, such as the autophagy system and AMPK.

Session 93 — Theatre 1

The great power of being tiny: *in vitro* gut systems in livestock research

S.K. Kar

Wageningen Livestock Research, Animal Nutrition, De Elst1, 6708 WD, the Netherlands; soumya.kar@wur.nl

A better understanding of digestive physiology requires an interdisciplinary approach because of the complexity of the subject. As alternatives to animal experimentation, researchers break down the processes involved in digestion to simple *in vitro* models that provide insight and clarity into the digestive physiology. These *in vitro* models cannot fully or accurately mimic the phenomenon of digestive physiology under *in vivo* conditions. Undoubtedly, such *in vitro* gut systems have their own advantages over *in vivo* models in conducting research, such as no ethical constraints, speed, high throughput, and the ability to provide more mechanistic insights. The available *in vitro* gut systems for (livestock) research can be broadly divided into two categories: *in vitro* digestion models and cell culture models. *In vitro* digestion models offer specific features such as (multiple) compartments and static or (semi)dynamic systems that allow researchers to study either (bio) availability/digestibility or digestion kinetics. Cell culture models are usually based on primary cell cultures or immortalized cell lines from the gut. In addition, researchers have developed intestinal organoids, often referred to as 'mini-guts', an enhanced, near-physiological *in vitro* system. Researchers use cell culture-based *in vitro* systems to study gut health and functionality, including nutrient uptake and barrier function. In addition, researchers use such cell culture models to generate robust host (gut) tissue data that enable them to test/screen food or feed ingredients. Over time, *in vitro* gut systems are becoming more sophisticated as researchers add additional biological layers such as the microbiome and immune cells, or technologies such as 'gut-on-a-chip', a biomimetic system for a physiological organ on a microfluidic chip. Certainly, the advances made by multiple (scientific) disciplines to understand the complexities in digestive physiology can be applied by livestock researchers to enhance the development of *in vitro* gut systems and their use in livestock research.

Session 93 Theatre 2

Comparison of adult intestinal stem cell derived organoids between different ages and sex in pigs

R.S.C. Rikkers[1], O. Madsen[2], S.K. Kar[1], L. Kruijt[1], A.A.C. De Wit[1], E. Van Der Valk[1], L.M.G. Verschuren[3], S. Verstringe[4] and E.D. Ellen[1]

[1]Wageningen Livestock Research, Droevendaalsesteeg 1, 6700 AH Wageningen, the Netherlands, [2]Wageningen University & Research, Droevendaalsesteeg 1, 6700 AH Wageningen, the Netherlands, [3]Topigs Norsvin Research Center B.V., P.O. Box 43, 6640 AA Beuningen, the Netherlands, [4]Nutrition Sciences NV, Booiebos 5, 9031 Drongen, Belgium; lisanne.verschuren@topigsnorsvin.com

Adult intestinal stem cell derived organoids are an *in vitro* research tool that is used in livestock animal research. Organoids are self-renewing and self-organizing 'mini-organs' and can be used to study gut health and function, and to unravel complex phenotypes. The transcriptomic resemblance of intestinal tissue and tissue-derived organoids of pigs has been studied before, but a comparison between ages and sex is missing. The aim of this study was to investigate the transcriptomic similarities in cultured organoids between: (1) four different ages in pigs to determine a potential age effect; (2) males and females to determine a potential sex effect. Ileal tissue was collected from 40 male pigs at four different ages (1, 4, 11 and 26 weeks) and 9 female pigs (26 weeks). These tissue samples were used to culture 3D organoids. Transcriptomic RNA profiles of the organoids were compared between the four ages and between sex. Average Pearson correlations of transcriptomic profiles within the 1, 4, 11 and 26 weeks were respectively, 0.96±0.04, 0.93±0.06, 0.98±0.03 and 0.90±0.09, suggesting low variability within age groups. Furthermore, the correlation between all 40 male samples was 0.93±0.05, suggesting a high similarity between ages. Within males and females (26 weeks), we observed correlations of 0.90±0.09 and 0.91±0.15, respectively, and overall of 0.90±0.12. These first results indicate that the resemblance of intestinal derived organoids between the ages is high and the age at which the intestinal stem cells are derived has limited influence. Moreover, the resemblance between males and females is high. Although additional in depth analysis is required, we suggest that organoids could be a promising tool to study complex phenotypes for gut health and function slaughter animals (26 weeks), reducing the number of experimental animals.

Session 93 Theatre 3

Butyric glycerides directly and indirectly enhance chicken enterocyte resistance to pathogens

A. Mellouk, N. Vieco-Saiz, V. Michel, O. Lemâle, T. Goossens and J. Consuegra

Adisseo SAS, Research in Nutrition, 8 route noire, 03600 Malicorne, France

Butyric acid supplementation to the feed is mainly used for its benefits on chickens' gut health. As a source of butyrate, mono-, di-, and triglycerides of butyric acid provide a better distribution of butyrate in the entire gastrointestinal tract. Here, we aim to study direct and indirect effects of these butyric glycerides' mixture on avian enterocyte's resistance to pathogen colonization. The Chic clone-8E11 immortalized chicken enterocytes (8E11) was used to conduct this study *in vitro*. First, we measured, by CellELISA, their resistance to *C. jejuni* (C.j.) and *S. Typhimurium* (S.T.) adhesion after a preincubation with sodium butyrate (SB) or butyric glycerides enriched with monobutyrin (MB). At 2 mM, MB significantly reduced the adhesion of C.j. and S.T. to 8E11 cells by 84 and 60%, respectively. SB reduced only S.T adhesion by 20% (n=3, $P<0.001$). MB also protected 8E11 cells from the toxicity of *C. perfringens* α toxin measured by Resazurin cell survival test. MB maintained 8E11 survival levels at 80 vs 15 and 20% in the control and SB groups, respectively (n=3, $P<0.001$). Next, we measured, by Seahorse®, the effects of the digested and non-digested butyric glyceride treatments on 8E11 cell mitochondrial activity. Only SB and butyrate released by lipase cleavage significantly increased, up to twice, the initial and maximal oxygen consumption rate (n=8, $P<0.001$). In mammals, this metabolic response is known to induce an intestinal ecology shift to an anoxic profile favouring anaerobic microbiota and disfavouring aerobic pathogens. In summary, we show *in vitro* how butyric glyceride mix confers a triple antimicrobial effect to enhance animal resilience against pathogenic challenges. Besides the described effect of α-monoglycerides, we demonstrate that butyric glycerides act directly on enterocytes making them more resistant to pathogens adhesion and toxicity. In addition, butyric acid, when released by lipase, become a source of energy not only for mammalian but also avian enterocytes and can therefore indirectly drive a microbial shift towards a resistant profile to pathogen invasions.

Session 93 Theatre 4

Spheroid and organoid models to study animal reproduction

K. Reynaud, M. Ta, C. Pucéat, M. Billet, C. Mahé, L. Schmaltz, L. Laffont, P. Mermillod and M. Saint-Dizier
INRAE, Reproductive Physiology and Behaviour, INRAE, UMR PRC, 37380, France; karine.reynaud@inrae.fr

For decades, primary cell or cell line culture models have been used to answer a multitude of scientific questions. In reproductive biology, the presence of cells in coculture makes it possible, for example, to improve oocyte maturation and then the rate of embryo development to the blastocyst stage and the survival after freezing. However, these cell cultures are performed in monolayers and the cells rapidly dedifferentiate. The 3D culture models, spheroids and organoids, could allow the preservation of cellular characteristics and thus better mimic the organ microenvironment. Spheroids are groups of primary cells that form spontaneously *in vitro* and do not adhere to the substrate. It is possible to create spheroids of multiple organs and their name will depend on the organ of origin (nerve tissue/neurospheres, colon/colospheres, mammary gland / mammospheres, etc.). An example of oviduct spheroids produced in bovine and porcine species will be described as well as potential applications to better understand the physiology of the organ, to evaluate the fertility of the males (tests of binding with spermatozoids) or to support the development of embryos *in vitro*. Organoids, on the other hand, are self-organising biological systems, derived from stem cells and reproducing the structure and cell types of organs as well as some of their functions. The cells used to generate organoids can be adult stem cells/aSCs or induced pluripotent stem cells/iPSCs. Adult stem cells do not require any treatment before culture, unlike induced stem cells, for which the differentiation protocols can be complex, with purification (cell sorting with antibodies) and cell expansion steps. Cells are grown in matrigel domes with culture medium and growth factors specific to the organ of origin. Over the last 15 years, intensive research has made it possible to create organoids from a very large number of tissues: nerve tissue, lung, stomach, colon, pancreas, liver, kidney and bladder. In reproduction, organoids of cervix, uterus, oviduct and ovary have been described. Examples of oviduct organoids from different domestic mammals and their applications in research and development will be described.

Session 93 Theatre 5

3D-cell culture systems; a breakthrough to improve the human and animal disease research

E.S. Smirnova[1], K.A. Adhimoolam[2] and T.M. Min[3]
[1]Jeju National University, Department of Animal Biotechnology, Jeju International Animal Research Center (JIA) and Sustainable Agriculture Research Institute, Jeju, 63243, Korea, South, [2]Jeju National University, Subtropical Horticulture Research Institute, Jeju, 63243, Korea, South, [3]Jeju National University, Department of Animal Biotechnology, Bio-Resources Computing Research Center, Sustainable Agriculture Research Institute (SARI), Jeju, 63243, Korea, South; tsmin@jejunu.ac.kr

Many disease models are available for studying the disease mechanisms and screening therapies in humans and animals. These models help us to understand the disease biology; however, they have considerable issues. In particular, a dearth of good *in vitro* and *in vivo* models has impeded progress in understanding the fundamental mechanisms underlying diseases. In recent times, innovative techniques and the modification of current techniques in preclinical models have ascended. In particular, 3D-cell culture systems (based on the spheroids, organoids, and tissue heterogeneity) are promising since they are more realistic and can imitate the biological and physiological properties and functionalities of grown cells and tissues. Here, we demonstrated the reliability and efficiency of 3D-cell culture of Caov-3 from a human ovarian adenocarcinoma over typical monolayer culture on a flat surface in cancer research. It created a more natural complex environment and architecture of tumours. Besides, the use of curcumin, which has shown anti-cancer effects, against 3D-generated Caov-3 tumour spheroids has shown improved drug penetration, promising more predictive drug responses for preclinical studies, and also 3D system resulted in higher tumour take rates than single cell suspension retrieval after injection, thus reducing overall animal experiment time. Collectively, if the potential advantages of 3D culture systems are given a chance to become a standard for *in vitro* research, it will be helpful to improve human and animal disease research.

Session 93 Theatre 6

Slurry chemical characteristics and ammonia emissions assessed by different methods

L. Sarri[1], E. Fuertes[1], E. Pérez-Calvo[2], A.R. Seradj[1], R. Carnicero[1], J. Balcells[1] and G. De La Fuente[1]
[1]Universitat de Lleida – Agrotecnio Center, Ciència Animal, Av. Alcalde Rovira Roure, 191, 25198 Lleida, Spain, [2]DSM Nutritional Products, Wurmisweg 576, 4303 Kaiseraugst, Switzerland; laura.sarri@udl.cat

Reduction of crude protein (CP) and addition of feed additives are practical and cost-effective strategies to reduce ammonia (NH_3) emissions from slurry in order to achieve the Paris Agreement targets. Moreover, animal welfare limitations in *in vivo* (VV) trials encourage the development of *in vitro* (VT) techniques. The aim of this study was to compare the chemical characteristics and NH_3 emissions of slurry from pigs fed different dietary treatments in a pig farm (VV) and under VT conditions. A 2×2 factorial design was conducted with 80 growing pigs fed 2 CP levels [standard (16%) vs low (14.5%)], combined with the addition or absence of a mixture of additives (MA; carbohydrases, organic acids, and essential oils). Pigs were distributed in 4 separate modules, dividing the dietary treatments, which drained into 4 slurry pits. Portable flux chambers were used to measure NH_3 at the pit level. Four pigs per treatment were allocated in metabolic cages for individual urine and faeces collection, which were mixed as slurry per pig and placed in artificial slurry pit simulators. Subsamples of the same mixture were added daily. Chemical characteristics and NH_3 emissions were determined at 2 time points between 4 weeks. Statistics were performed using JMP®, with GLM procedures and Tukey's test. Chemical characteristics differed due to the presence of feed in VV slurries. Ammonia emission was higher in VT, as it is a non-leaking system, and decreased considerably over the 4 weeks. An interaction between CP and MA was identified in VV on pH, where MA only reduced pH in low CP diet, whereas in standard CP it may promote microbial growth. Higher N content was observed in high CP diets and MA supplementation. Reduction NH_3 emissions with the inclusion of MA was detected in both methods, but especially in VT method, which was accompanied by an acidic pH. Overall, NH_3 was reduced around 40-50%, or 17% for low CP diet in VV method. In conclusion, VT method was capable to detect changes on NH_3 emissions when MA was included, but VT was not sensitive enough to detect changes due to protein level.

Session 93 Theatre 7

***In vitro* protein fractionation methods for ruminant feeds**

B.Z. Tunkala[1], K. Digiacomo[1], P.S. Alvarez Hess[2], F.R. Dunshea[1,3] and B.J. Leury[1]
[1]The University of Melbourne, Agricultural Sciences, Parkville campus, 3010, Parkville, Australia, [2]Agriculture Victoria Research, 1301 Hazeldean Road, 3821, Ellinbank VIC, Australia, [3]Faculty of Biological Sciences, The University of Leeds, Leeds, LS2 9JT, United Kingdom; btunkala@student.unimelb.edu.au

Estimating protein fractions of feeds is vital to ensure optimum protein supply and degradation in the digestive system of ruminants. This study investigated the possibility of using the ANKOM gas production system and preserved rumen fluid to estimate the protein fractions and *in vitro* degradability of protein-rich feeds. Three *in vitro* methods: (1) gas production method (2) Cornell Net Carbohydrate and Protein System (CNCPS), and (3) the unavailable nitrogen (uN) assay of Ross (uN_{Ross}) were used to quantify the protein fractions of four feeds (lupin meal, vetch grain, Desmanthus hay, and soybean meal). Rumen fluid mixed with 5% dimethyl sulfoxide (DMSO) and frozen at -20 °C (D-20°C) was also compared against fresh rumen fluid in the gas production and uNRoss methods. The *in vitro* degradable protein (IVDP) was higher for vetch grain (46 and 70%) at the 4th and 8th hours of incubation, whereas soybean meal (85%) exceeded the other feeds after the 16th hour of incubation (P<0.001). The greatest ammonia-N concentration was from soybean meal (1.27 mg/g) and lupin meal (0.87 mg/g) fermented for four hours using fresh rumen fluid. Protein fraction a is degradable in rumen. Fraction b is degradable in intestine. Fraction c is undegradable fraction. The proportion of fraction 'b' for soybean (82.1% CP) and lupin meals (39.4% CP) from the CNCPS method were not different (P>0.5) from the fraction 'b' estimation of the gas production method for the same feeds (r=0.99). Regardless of the methods, a greater fraction a was found from vetch grain (39.6-46.6% CP), and the proportion of fraction c in Desmanthus hay (39.1-41.5% CP) exceeded other substrates (P<0.001). All three methods ranked the feeds identically in the proportions of available or ('a+b') protein fractions as vetch grain, soybean meal, lupin meal, and Desmanthus hay in decreasing order. The positive correlation between fractions across different methods and identical ranking of feeds suggests the possibility of using ANKOM gas production system for protein fractionation.

Session 93 Theatre 8

Storage conditions of rumen inoculum: impact on gas productions in mini dual flow fermenters

V. Berthelot[1], M. Charef-Mansouri[1], A.-M. Davila[2] and L.P. Broudiscou[3]
[1]Université Paris-Saclay, INRAE, AgroParisTech, UMR Modélisation Systémique Appliquée aux Ruminants, 22 place de l'agronomie, 91120 Palaiseau, France, [2]Université Paris-Saclay, INRAE, AgroParisTech, UMR Physiologie de la Nutrition et Alimentation, 22 place de l'agronomie, 91120 Palaiseau, France, [3]INRAE, UPPA, UMR Nutrition, Métabolisme, Aquaculture, 173 Route Saint Jean de Luz, 64310 Saint-Pée-sur-Nivelle, France; valerie.berthelot@agroparistech.fr

A mini dual flow fermenter, with a working volume of 80 ml compared to 1 l in previous devices, has been designed to study rumen microbial metabolism, potentially without the need of fresh inoculum. The study aimed to optimise a method of rumen content preservation to restore normal *in vitro* fermentations and methanogenesis after a 5-day adaptation period in the device. Different conditions (cryoprotectant, temperature of storage) and times of storage were tested. Two cannulated dry goats were used as donors of rumen inoculum. Each condition for each goat inoculum was tested in triplicate for 7 days with a diet (hay 0.6; barley 0.2, soybean meal 0.2) given twice a day at 08:30 h and 15:30 h. The first experiment tested the effect of freezing storage temperature (-20 vs -80 °C) with DMSO (5%) used as a cryoprotectant at 3 times of storage (fresh as control, 1 and 4 months). Daily gas volume and methane productions after 5 days of adaptation were similar in devices with fresh inoculum and with rumen inoculum stored at -20 and -80 °C for 1 month. After 4 months of storage, gas volume and methane production of inoculum stored at -80 °C were closed to control ones but at -20 °C methanogenesis was greatly impaired with a significant interaction between temperature and animal donor inoculum. The second experiment tested the use of 2 cryoprotectants (glycerol or DMSO) at 5%, at 3 times of storage at -80 °C (fresh, 2.5 and 12 months). Gas volume and methane productions were closed to productions of fresh inoculum up to 12 months of storage. Even though gas volumes were higher with glycerol compared to DMSO, methanogenesis was similar with the 2 cryoprotectants. Based on these data, rumen inoculum can be kept for at least one year at -80 °C without any gas production and methanogenesis impairment in our mini dual flow fermenters.

Session 93 Theatre 9

Dynamics of rumen microbiome and methane emission during *in vitro* rumen fermentation

R. Dhakal[1], R. Sapkota[2], P. Khanal[3], A. Winding[2] and H.H. Hansen[1]
[1]University of Copenhagen, Department of Veterinary and Animal Sciences, Grønnegårdsvej 3, Frederiksberg C, 1870, Denmark, [2]Aarhus University, Department of Environmental Science, Roskilde, Denmark, 4000, Denmark, [3]Nord University, Faculty of Biosciences and Aquaculture, Skolegata 22, Steinkjer, 7713, Norway; dbk638@ku.dk

Dynamics of the rumen microbial communities during *in vitro* rumen fermentation of maize silage was studied to understand rumen fermentation characteristics. Maize silage was used as a basal feed in three fermentations, and the process was stopped after 6, 12, 24, 36, and 48 hours. Total gas production, dry matter degradability, and methane concentration were measured at these time points, and rumen fluid samples were collected for microbiome analysis. Amplicon sequencing of 16sRNA genes (V4 region) was used to profile the microbial community structure in the rumen during the fermentation process. The following linear relationship between methane concentration and methane yield with total gas production and degraded dry matter was obtained for: Methane concentration (%/gm DM) = 0.0503×Total gas – 0.3247 (R^2=0.9082); Methane yield (ml/g DM) = 0.1301 × (Total gas) – 5.689 (R^2=0.9337); Methane yield (%/gm DM) = 15.688 × (degraded dry matter %) + 0.279 (R^2=0.8553); Methane yield (ml/g DM) = 40.258 × (degraded dry matter %) – 3.8595 (R^2=0.8584). Microbiome analysis revealed the dominance of Bacteroidota, Cyanobacteria, Desulfobacterota, Euryarchaeorta, Fibrobacterota, Firmicutes, Patescibacteria, Protobacteria, Spirochaeota, and Verrucomicrobiota. Significant temporal variation was observed in Bacterodiata, Campilobacterota, Firmicutes, Proteobacteria, and Spirochaeota. Alpha diversity estimates using Observed and Shannon index analyses found hardly any difference among the different time points during fermentations. In conclusion, the *in vitro* fermentation characteristics can be predicted based on a few measured *in vitro* parameters for a given feed type in rumen fluid from similar cows. These results indicate that the dynamics of the rumen microbes change with the advancement of time, resulting in the changes seen in the parameters of fermentation kinetics.

Session 93 Theatre 10

Effects of *Asparagopsis armata* inclusion on methane production in semi-continuous fermenters

P. Romero[1], S.M. Waters[2], D.R. Yañez-Ruiz[1], A. Belanche[3] and S.F. Kirwan[2]
[1]Estación Experimental del Zaidín (CSIC), Profesor Albareda, 1, 18008, Granada, Spain, [2]Teagasc, Animal and Bioscience Research Department, Grange, Dunsany, Co. Meath, C15PW93, Ireland, [3]Universidad de Zaragoza-CITA, Departamento de Producción Animal y Ciencia de los Alimentos, Miguel Servet 177, 50013 Zaragoza, Spain; stuart.kirwan@teagasc.ie

Recent studies have highlighted the potential of *Asparagopsis taxiformis* to reduce CH_4 production from ruminants, despite its availability being scarce in temperate regions. Alternatively, *Asparagopsis armata* can be grown in temperate regions and it also contains bromoform as the main bioactive compound to reduce rumen methanogenesis. The objective of this study was to evaluate the effects of *A. armata* at different inclusion levels on CH_4 production *in vitro* using the artificial Rumen Simulation Technique. The experiment was run over 21 days using 16 rusitec vessels (n=4) which were randomly allocated to 4 dietary treatments: Control, without seaweed (CTL); *A. taxiformis* supplemented at 25 g/kg DM (AT) or *A. armata* supplemented at 30 (AA1) or 40 g/kg DM (AA2). The *in vitro* basal diet consisted of 50:50 grass silage: concentrate on a dry matter (DM) basis. Rumen inoculum was sourced from four rumen-cannulated beef cattle on grass silage *ad libitum* plus 2 kg concentrates. Seaweed supplementation had no effect on pH in the fermentation or in the overflow effluent vessels (P>0.05). There was no difference observed in total gas production between treatments (P>0.05). However, vessels that received AA1 reduced CH_4 in terms of gas percentage, CH_4 production by 59.76 and 66.66%, respectively, compared to the control treatment (P>0.01). Under *in vitro* conditions, the inclusion of *A. armata* supplemented at 30 g/kg demonstrated the greatest CH_4 mitigation potential.

Session 93 Theatre 11

Calcium propionate mitigated adverse effects of incubation temperature shift on *in vitro* fermentation

T. He, S. Long and Z. Chen
China Agricultural University, College of Animal Science and Technology, No. 2, Yuanmingyuan West Road, Haidian District, Beijing, 100194, Beijing, China, P.R.; hetengfei@cau.edu.cn

This study aimed to investigate the comprehensive effects of calcium propionate (CaP) and different incubation temperature modes on rumen methane production, fermentation characteristics, and microbial communities *in vitro*. A 2×2 factorial experiment with four treatments was conducted over a 72-hour period, including with or without 2.5% CaP (dry matter basis) in the substrate under the mode of constant incubation temperature of 39 °C or incubation temperature shift (39 °C for 4 h followed by 30 °C for 2 h, 12 cycles, ITS). The results showed that ITS inhibited (P<0.01) total gas production, methane production, and methane concentration at 12 h and 72 h and reduced the concentration of ammonium nitrogen, propionate, butyrate, and total volatile fatty acids in fermentation liquid. The addition of 2.5% CaP significantly increased (P<0.05) the gas production of 72 h, theoretical maximum gas production, concentrations of propionate and valerate, while it had no observable effect on the production or concentration of methane. Furthermore, 2.5% CaP demonstrated a significant (P<0.05) increase in the relative abundance of phylum Bacteroides. The relative abundance of *Methanomicrobiales* at genus level decreased under the ITS condition and was positively correlated (r=0.52, P<0.05) with methane production at 72 h. In addition, CaP increased the relative abundance of *Ruminococcus* at genus level, which was positively correlated (r=0.63, P<0.05) with the maximum theoretical gas production. Overall, these findings suggest that including 2.5% CaP might alleviate the negative impact of ITS on *in vitro* fermentation parameters by regulating the microbial composition and could sustain the reduction in methane production.

In vitro generation of ovine monocyte-derived macrophages for SRLV infection studies

E. Grochowska[1], M. Ibrahim[1], M. Wu[2] and K. Stadnicka[3]
[1]Bydgoszcz University of Science and Technology, Department of Animal Biotechnology and Genetics, Mazowiecka 28, 85084, Poland, [2]University of Molise, Department of Agricultural, Environmental and Food Sciences, Via Francesco De Sanctis 1, 86100, Italy, [3]Collegium Medicum, Nicolaus Copernicus University, Faculty of Health Sciences, Łukasiewicza 1, 85821, Poland; grochowska@pbs.edu.pl

Animal health has a great impact on animal productivity, quality, and quantity of given products, and consequently, on the economic outcome of animal production. Among animal pathogens, viruses are on top as the biggest threats to animal health and safety. The infection of small ruminant lentivirus (SRLV) in sheep causes pain, lower productivity, inevitable death, and consequently significant economic losses. There is no effective cure or vaccination. Monocytes and macrophages play crucial roles in SRLV infection. However, the transcriptomes of infected monocytes/macrophages are not very well studied. The study aimed to select the most suitable conditions of cell culture to differentiate sheep blood monocytes into macrophages and to establish an effective RNA extraction protocol to obtain RNA of sufficient quantity for transcriptome sequencing. Peripheral blood mononuclear cells (PBMC) were separated from the blood of Polish Merino male lambs by Histopaque density gradient centrifugation. Monocytes were purified from PBMCs by two methods: adherence and magnetic-activated cellular sorting (MACS). Monocytes or PBMCs were suspended in RPMI 1640 medium with glutamine, 10% FBS, 1% Penicillin-Streptomycin, 1% Amphotericin B, macrophage colony-stimulating factor (m-CSF), and interleukin 4 (IL-4). Five conditions of monocyte culture in a medium with different concentrations of m-CSF and IL-4 were tested. The use of the MACS cell sorting technique resulted in a higher purity of the monocyte population than adherence method. The medium with the addition of 100 ng/ml m-CSF + 20 ng/ml IL-4 enabled the most effective generation of monocyte-derived macrophages (MDM). The established protocol with MACS monocyte sorting step and addition of 100 ng/ml m-CSF + 20 ng/ml IL-4 enables generation of MDMs for transcriptome studies in SRLV infected sheep. This research was supported by National Science Centre, Poland (grant no. 2020/39/I/NZ9/01304).

Resistance of probiotics to antibiotics and antagonism to poultry pathogens

N. Akhavan[1,2], K. Stadnicka[1], D. Thiem[2] and K. Hrynkiewicz[2]
[1]Collegium Medicum, Nicolaus Copernicus University, Health Science, Jagiellońska 13/15, 85-067 Bydgoszcz, Poland, [2]Nicolaus Copernicus University, Microbiology, Faculty of Biological and Veterinary Sciences, Lwowska 1, 87-100 Torun, Poland; niloo@doktorant.umk.pl

The performance of broilers is dependent on the function and growth of the digestive system, which is stimulated by early colonization by beneficial bacteria. During the last few years, there were a lot of studies showing that *in ovo* injection with probiotics can improve performance efficiency, stimulate the immune system and decrease poultry pathogens. For this reason, in our studies, the candidate probiotics (three *Lactobacillus* strains, *Bifidobacterium lactis* and *Carnobacterium divergens*) were characterized in terms of their resistance to common antibiotics with the use of an E-test and determination of MIC (Minimum inhibitory concentration) values. In addition, we analysed the potentially antagonistic properties of probiotics against the major poultry pathogens (*Listeria monocytogenes, Salmonella Typhimurium, Salmonella Enteritidis and Escherichia fergusonii*) with the agar well diffusion assay and cell-free supernatant assay (CFS). According to the results of the ANOVA test, all analysed probiotics were resistant towards ciprofloxacin in the range of 0.002 to 32 μg/ml, whereas *L. rhamnosus H25* and *L. plantarum* showed resistance against amoxicillin and tetracycline too. Determined MIC values depend on the used antibiotic. In general growth of *Lactobacillus sp.* and *B. lactis* was inhibited in higher antibiotic concentrations (100 mg/l) than *C. divergens* (25 mg/l). In our antagonistic study, we observed that *L. rhamnosus H25* and *L. plantarum* revealed an inhibition against *L. monocytogenes*. Potential resistance to common antibiotics and antipathogenic activities are addressing the broader ecosystem and is not limited to the host performance only. Therefore, the obtained results provide a piece of relevant information about the probiotic's bioactivity, before their application on farms. National Science Centre OVOBIOM_UMO-2019/35/B/NZ9/03186, IDUB program NCU Poland.

Session 93 Poster 14

Metabolic footprint of prebiotics and probiotics in Chick8E11 and Caco-2 intestinal cell lines

S. Zuo[1], W. Studziński[2], K. Stadnicka[1] and P. Kosobucki[2]
[1]Ludwik Rydygier Collegium Medicum Nicolaus Copernicus University in Torun, Faculty of Health Sciences, 85-821, Bydgoszcz, Poland, [2]Bydgoszcz University of Science and Technology, Department of Food Analysis and Environmental Protection, 85-326, Bydgoszcz, Poland; sanling.zuo@doktorant.umk.pl

Probiotics and prebiotics have been widely studied in poultry as an early strategy to program host health and performance by manipulating gut microbiota towards optimal metabolism. This study aimed to identify metabolic footprints of candidate probiotics in intestinal cells *in vitro* to explain the molecular mechanism of their effects in chicken intestinal cells. Prebiotic candidates, special seaweed extract and protein hydrolysate, were tested to stimulate the probiotic growth. Probiotic was cultivated in medium with addition of 2%[v/v] prebiotic solution. The new chicken cell line Chic8E11 and Caco-2 (as reference) were inoculated with probiotic *Bifidobacterium lactis*, respectively, at a ratio of 30:1(bacteria:cells) for 48 h(replicates=3). The supernatants were subjected to the targeted analysis with gas chromatography-mass spectrometry to reveal changes in organic acids including short chain fatty acids (SCFAs). It was found that *B. lactis* produced different metabolites depending on the type of prebiotics stimulating its growth. Its lactic and valeric acid levels were changed (P<0.05) by protein hydrolysate. In the co-culture of cells and probiotic, the change in acetic acid was found in the Caco-2 system, whilst the acetate and propionate levels were changed in the Chick8E11 system. Also, significant changes were found in organic acid levels (succinic acid and L-alanine), and for essential and non-essential amino acids (leucine and serine)(P<0.05). SCFAs play an important role in maintaining the gut normal function. Other organic acids are closely related to metabolic pathways such as cellular glucose metabolism, TCA cycle, and G protein-coupled receptors. The results obtained from *in vitro* studies may serve as a basis to predict the potential of bioactive candidates' function in the host. The results of this study will be compared with metabolic profile of *in ovo* stimulated chicken gut. The research was supported by grant 2019/35/B/NZ9/03186(OVOBIOM) funded by the National Science Centre and IDUB program at NCU Poland.

Session 93 Poster 15

Cellular imagery evaluation of butyric acid biological activities in Caco-2 cell line

D. Gardan-Salmon[1], B. Saldaña[2], J.I. Ferrero[2] and M. Arturo-Schaan[1]
[1]DELTAVIT – CCPA GROUP, ZA du Bois de Teillay, 35150 Janzé, France, [2]NOVATION – CCPA GROUP, C/ Marconi 9, 28823 Coslada (Madrid), Spain; dgardan-salmon@ccpa.com

In the context of reducing antibiotic use in animal production, greater attention is given to natural alternatives to improve intestinal health and welfare. Relevant *in vitro* models may be employed to evaluate the biological properties and to better understand the mode of action of these natural alternatives. Butyric acid (BA) is a short chain fatty acid naturally formed in the colon of monogastric animals by microbial fermentation. Recent studies show that BA is a regulator of inflammatory processes in the small intestine and can enhance intestinal barrier function and mucosal immunity. Thus, BA is an adequate candidate for dietary intervention for livestock animals facing stress and diseases impairing gut health and homeostasis. Our study aimed at investigating the ability of BA to modulate inflammation and oxidative stress *in vitro* in an intestinal cell line. IL-1β-induced Caco-2 cell line was used to study inflammation through fluorescent immunostaining imagery of NF-κB translocation into the nuclei. Oxidative stress mitigation was measured using menadione as reactive oxygen species inducer in Caco-2 cells and Deep Red CellROX fluorescent probe for imagery. Briefly, Caco-2 cells were cultured for 4 days (37 °C, 5% CO_2); cells were then pre-treated with BA (8 to 100 μg/ml) for 1 h; and then co-incubated with IL-1β (100 ng/ml) or menadione (50 μM) for 1 h prior to cellular imagery. BA alleviated inflammation induced by IL-1β by significantly reducing the % of NF-κB positive nuclei (-36% with 8 μg/ml of BA; P<0.01). BA reduced oxidative stress induced by menadione by reducing oxidized probe quantity (-19% to -23% with 8 μg/ml and 20 μg/ml of BA respectively; P<0.05). Using cellular imagery on Caco-2 cells, our results showed that BA can exert anti-inflammatory effect through NF-κB modulation as well as antioxidant effects in enterocytes. This *in vitro* approach is a promising sustainable and original imagery tool to better evaluate the biological properties of feed ingredients prior to *in vivo* evaluation.

Effect of silage sample preparation on rumen fermentation (*in vitro*)

J.A. Huntington[1], T. Snelling[1], L.A. Sinclair[1], H. Warren[2], D.A. Galway[1] and D.R. Davies[3]
[1]Harper Adams University, Agriculture and Environment, Edgmond, Newport, Shropshire, TF10 8NB, United Kingdom, [2]Alltech, European Biosciences Centre, Dunboyne, Co. Meath, A86X006, Ireland, [3]Silage Solutions Ltd, Bwlch y Blean, Ceredigion, SY23 2HN, United Kingdom; jhuntington@harper-adams.ac.uk

Sample preparation for in *in vitro* rumen fermentation kinetics model can affect resultant fermentation product profiles. Second cut ryegrass ensiled in varying conditions of compaction density, hygiene and sealing were used to determine the effect of sample preparation for rumen fermentation kinetics (*in vitro*). Samples of the ensiled ryegrass with a dry matter ranging between 397 and 408 g/kg FM were either oven dried (60 °C) then milled (2 mm screen), freeze dried or frozen fresh (freeze dried & frozen fresh silages were not milled) prior rumen fermentation (*in vitro*). Samples (1.0 g DM based on oven DM, 60 °C) were incubated in 200 ml buffered (Mould's buffer) rumen fluid (20%) at 39 °C for 48 h. Cumulative increase in headspace pressure was used to determine rumen fermentation profile. Following rumen fermentation (*in vitro*) final culture headspace gas was collected and methane concentration measured. Final rumen cultures were then filtered (P1 glass sinters) to determine 48 h DM degradability. Method of sample preparation was found to affect (P<0.001) DM degradability, fermentation profile and methane evolution. Drying the sample reduced DM degradability with mean values of 0.772^a, 0.692^b and 0.571^c SED=1.949; P<0.001 for fresh frozen, freeze dried and oven dried +milled silages, respectively. Rumen fermentation profiles were different for the different sample preparations and mean gas pressure was also found to be reduced by drying, with mean PSI of 13.04^a, 9.82^c and 10.60^b for frozen fresh, freeze dried and oven dried +milled, respectively (SED=0.304; P<0.001). Freeze dried samples generated the highest concentration of methane whilst fresh frozen the lowest; mean methane concentrations were 71.92^a, 57.16^b and 52.90^c % of headspace gas for freeze dried, oven dried +milled and fresh frozen, respectively (SED=2.632; P<0.001). It was concluded that method of sample preparation had a substantial effect on rumen fermentation parameters with consequences for data interpretation and method standardisation.

Repeatability of fermentation kinetics using *in vitro* gas fermentation

B. Jantzen and H.H. Hansen
University of Copenhagen, Department of Veterinary and Animal Sciences, Groennegaardsvej 3, 1870, Denmark; htz757@sund.ku.dk

In vitro gas production (IVGP) fermentation kinetics are widely used to screen feeds and feed additives to reduce the number of animals needed, which in turn, reduces costs. However, information about repeatability is scarce. The objective of this study was to evaluate the variation from IVGP fermentations in the same laboratory using the same feed sample. Rumen fluid from two different farms with either producing dairy cows or dry fasting heifers were used as donor animals. Seventeen 24-hour fermentations, undertaken during a year, were used to evaluate the variation between the following parameters: best fit-gas curve, baseline corrected gas production (TGP: ml at Standard Temperature and Pressure/gram dry matter), methane concentration and yield, pH and apparent dry matter degradability (dDM). Significant differences between donor animal types were found for methane concentration and yield, pH of individual animals within type and pH of the fermented fluid. Observed variation was non-significant within donor types for methane concentration and yield and pH after fermentation. The rate of early gas production was significantly different by donor types, but TGP was not significantly different at 24 hours when corrected for baseline gas production. No dDM differences after 24 hours of fermentation between and within donor types were detected. Best fit curves were different by donor type, being either a monophasic version of the sigmoidal model or an exponential curve for the heifers or the production animals respectively. However, no differences were observed within type. A high repeatability was seen, when using the same donor types.

Evaluation of rumen bypass protein in processed soybean meal products

A. Chariopolitou, A. Plomaritou, A. Tzamourani, K. Droumtsekas and A. Foskolos
University of Thessaly, Department of Animal Science, Campus Gaiopolis, Larissa, 41222, Greece; anplomaritou@uth.gr

To meet increased protein requirements due to increased milk production, various feedstuffs have been developed to escape rumen degradation providing directly metabolizable protein to the ruminant. However, important differences among commercially available products do exist. Thus, the objective of this study was to evaluate commercially available bypass soybean meals with a fast and reliable *in vitro* method. Four different soybean meal products were selected: unprocessed soybean meal (SBM) and three commercially available processed soybean meal products (PROT1, PROT2 and PROT3). Eight samples of each product were collected and assessed for their protein digestibility using the Daisy^{II} incubator. Briefly, 1 g of sample was weighted into fibre bags (Ankom F57) in duplicate. Rumen fluid from five cows was collected before feed delivery using rumen fluid extractor. Bags were incubated in Daisy incubation bottles containing 1:1 rumen fluid and buffer solution. After 24 h of ruminal incubation bags were removed and washed in a washing machine. All the original samples and the residue in bags were analysed for Dry Matter (DM; 24 h in a 55 °C oven) and N content (Kjeldahl). Dry matter 24 h digestibility (DMD) ranged from 39.3 to 69.4% and crude protein 24 h digestibility (CPD) ranged from 12.5 to 53.2%, with significant differences between the SBM and the bypass products. Furthermore, significant differences were observed among bypass products, where PROT3 had the lowest DMD (39.3 vs 56.6 and 55.5% for PROT3 vs PROT1 and PROT2, respectively; $P<0.0001$) and the lowest CPD (12.5 vs 31.1 and 29.9% for PROT3 vs PROT1 and PROT2, respectively; $P<0.0001$). When CPD of bypass products was corrected for SBM rumen digestibility, the % of ruminal protection was estimated. In this case, even though the PROT1 and PROT2 did not show statistically significant differences between them (41.4 and 43.7%, respectively), the PROT3 provided greater ruminal protection than the other two (76.5%; $P<0.0001$). In conclusion, significant differences were found among the available processed soybean meal products and frequent assessment should take place in order to best describe thein nutritional value.

Effect of rumen inoculum on predicted *in vivo* methane production from barley and oats

P. Fant, M. Ramin and P. Huhtanen
Swedish University of Agricultural Sciences, Department of Animal Nutrition and Management, Skogsmarksgränd, 907 36 Umeå, Sweden; petra.fant@slu.se

The first objective was to examine the effect of rumen inoculum (RI) on predicted *in vivo* CH_4 production in an *in vitro* gas production experiment. The second objective was to compare predicted CH_4 values with observed CH_4 values from a feeding trial conducted simultaneously using the same diets. The feeding trial was a 4×4 Latin square design with 16 Nordic Red dairy cows. Diets consisted of 60% grass silage and 40% grain: barley, oats with hulls (hulled oats), dehulled oats, or a 50:50 mix of hulled and dehulled oats on dry matter (DM) basis. The *in vitro* experiment was a 2×4 factorial design replicated in 4 runs with 2 types of RI and the same 4 diets as in the feeding trial. The RI was obtained by stomach tubing from cows in the feeding trial fed either hulled oats (2 cows) or barley (2 cows) as grain component. Predicted *in vivo* CH_4 production was estimated from the *in vitro* according to Ramin and Huhtanen. Data were analysed in SAS by the MIXED procedure. Orthogonal contrasts were specified for barley vs all oat diets, and linear and quadratic effects of replacing hulled oats with dehulled oats. There were no interactions between diet and RI ($P=0.40$). *In vitro* organic matter (OM) digestibility was not affected by RI ($P=0.28$), but predicted CH_4 tended to be higher ($P=0.07$) in the RI from oat fed cows than from barley fed cows (25.5 and 24.6 g/kg DM, respectively). *In vitro* OM digestibility increased linearly ($P<0.01$) from 884 to 919 g/kg OM when hulled oats was replaced by dehulled oats, but predicted total gas ($P=0.21$) and CH_4 ($P=0.23$) were not affected. However, predicted CH_4 tended to be lower ($P=0.06$) in the oat diets than in the barley diet (24.8 and 25.8 g/kg DM, respectively). The CH_4 ranking of the oat diets was consistent with the results in feeding trial. Due to low number of observations (n=4), the relationship between predicted and observed CH_4 (g/kg DM intake) was not significant ($P=0.25$, R2=0.56, RMSE 2.2% of observed CH_4 mean). In conclusion, rumen inoculum did not affect the diet comparison in this study indicating a minimum effect of rumen microbiome on observed differences in *in vivo* CH_4 emission between barley and oats.

Session 93 Poster 20

Comparison of *in vitro* and *in vivo* methane production in dairy cows

D.W. Olijhoek, É. Chassé, M. Battelli, M.V. Curtasu, M. Thorsteinsson, M.H. Kjeldsen, W.J. Wang, G. Giagnoni, M. Maigaard, C.F. Børsting, M.R. Weisbjerg, P. Lund and M.O. Nielsen
AU Viborg – Research Center Foulum, Aarhus University, Dept. of Animal and Veterinary Sciences, Blichers Allé 20, 8830 Tjele, Denmark; dana.olijhoek@anivet.au.dk

Ideally, *in vivo* methane production from individual cows fed different diets could be predicted by a low-cost *in vitro* method. Therefore, this study aimed to assess the correlation between *in vivo* methane emission from dairy cows measured using respiration chambers or GreenFeed (C-Lock Inc, Rapid City, USA) and *in vitro* methane production when identical diets were incubated in a system simulating rumen fermentation (Ankom RF, Ankom Technology, Macedon, USA). Two experiments were conducted. In *in vitro* Exp. 1, total gas and methane production were determined when dried samples (0.5 g) of 43 different TMRs (from 11 feeding trials) and silages (maize and grass) as control feeds were incubated in 90 ml buffered rumen fluid obtained from 3 heifers (pooled). In *in vitro* Exp. 2, TMR samples (0.5 g) with high (63% of DM) and low content (35% of DM) of forage were incubated in 90 ml buffered rumen liquid obtained and separately tested for 16 Danish Holstein cows fed each of these diets in 2 different experimental periods. All *in vitro* incubations lasted 48 h to record total gas production and gas samples were collected at 24 h to determine methane production. Data were analysed as Pearson correlations. Preliminary results of Exp. 1 showed a weak correlation (r=0.35; P=0.02) between methane production from the diets determined *in vitro* (ml CH_4/g incubated DM) and *in vivo* (L CH_4/kg DMI). Graphical visualization of *in vitro* and *in vivo* methane production of Exp. 2 showed clustering by the diet fed to the cow, but not by the diet incubated *in vitro*. The correlation between *in vitro* and *in vivo* methane production was stronger when cows were fed a diet low in forage (r=0.56; P=0.001) than high in forage (r=0.17; P=0.34). Further statistical analyses will be performed. These first results highlight the impact of the diet fed to cows that deliver rumen liquid for *in vitro* trials and the lack of considerable consistency in methane production assessed *in vitro* compared to *in vivo*, and this was particularly pronounced when cows were fed diets high in forage.

Session 94 Theatre 1

What impairs laying hens' foot health? A retrospective German study

N. Volkmann[1,2], A. Riedel[1,2], N. Kemper[1] and B. Spindler[1]
[1]University of Veterinary Medicine Hannover, Foundation, Institute for Animal Hygiene, Animal Welfare and Farm Animal Behaviour (ITTN), Bischofsholer Damm 15, 30173 Hannover, Germany, [2]University of Veterinary Medicine Hannover, Foundation, Science and Innovation for Sustainable Poultry Production (WING), Heinestraße 1, 49377 Vechta, Germany; nina.volkmann@tiho-hannover.de

One of the most common foot health problems and welfare issues in laying hens housed in non-cage systems is food pad dermatitis (FPD) with an inflammation of the subcutaneous tissue of food pad. To identify potential risk factors for this disorder, the present retrospective study investigated 22 German laying hen herds during production, while housing system, herd size, observation time, season, litter type and quality as well as foot health were recorded. Both feet of randomly selected animals (n=4,534) were evaluated with regard to the occurrence and extent of footpad changes using a modified 4-point scoring system (FPD Score); the highest degree of FPD per animal was noted, as well as whether the animal was affected on one or both sides of its feet. Statistical analyses were carried out using SAS 9.4 software with a significance levels of P less than 0.05. Effects on the FPD Score were analysed using a generalized linear mixed model; pairwise comparisons were carried out using Tukey Kramer t-tests. In total, 73.4% of the hens had healthy feet with no alterations, 26.1% showed moderate FPD in at least one foot, 0.4% were classified as those with a moderate FPD2 and one hen showed severe FPD. Of the affected animals (n=1,205), 45.7% had FPD on one foot and 54.3% on both sides. While only few animals showed FPD at the first two visits, the ratio of affected animals increased with the laying peak. At the end of the laying period 37.5% of the hens showed moderate alterations. FPD Score was statistically affected by the time of visit (P<0.0001), the season (P<0.0001) and the type of litter (P=0.0157). This retrospective study showed that more than a third of the animals had alterations in their footpads at the end of laying period. In addition to the age of the hens, effects of litter type and the season as risk factors could be determined.

Session 94 Theatre 2

Pododermatitis in broilers: prevalence and risk factors present in reused beddings

J. Montalvo, M. Cevallos and M. Cisneros
Central University of Ecuador, Faculty of Veterinary Medicine and Zootechnics, Jeronimo Leiton y gato sobral, 170129, Ecuador; mcisneros@uce.edu.ec

The purpose of this research was to study the prevalence and possible risk factors that predispose broiler chickens raised in recycled beddings to develop contact pododermatitis. The information was collected from eight sheds, at 11 times between day 1 to day 40 in two different types of beds (bed reused one time or treatment 1 (T1) and bed reused two times or treatment 2 (T2)). 23 birds were sampled in each shed, by giving a total of 2,024 individuals at the end of the investigation. To analyse the data obtained, a nested mixed model was necessary to use, and it analysed fixed and covariates effects, as well as random ones. For the analysis, the R software, version 4.1.3, and the Ime4 package were used. As important results, it can be mentioned that on day 40, 320 birds (51.9%) with pododermatitis in T1 beds and 669 birds (47.5%) with pododermatitis in T2 beds were quantified. It is important to clear up that, although the prevalence among treatments is almost similar, birds exposed to T2 beds presented more severe pododermatitis lesions. The mixed model analysis showed that the severity of footpad dermatitis increased by 0.228 in birds exposed to T2 litters; likewise, the birds presented a daily increase in the severity of footpad dermatitis of 0.046 or 0.46 per 10 days in the breeding bed. In conclusion, the overuse of recycling of breeding beds and the age of the birds (variable related to weight and time in breeding) were the main risk factors that could predispose meat-producing birds to develop pododermatitis by contact.

Session 94 Theatre 3

Digital dermatitis in French young bulls fattening farms

A. Waché[1], M. Petitprez[2], E. Dod Ioan[3], M. Delacroix[4] and C. Guibier[5]
[1]Livestock Institute, Morel street, 49071 Beaucouzé, France, [2]Animal Health Protection Group of Picardie, De Gennes street, 02007 Barenton-Bugny, France, [3]Eurolia Vet clinic, St Quentin street, 80400 Ham, France, [4]Veterinarian, Souleyant, 42130 Marcoux, France, [5]Chamber of Agriculture of Aisne, Blondelle street, 02007 Laon, France; aurore.wache@idele.fr

Digital dermatitis (DD) has been discovered in some young bulls fattening farms of north of France suffering economic losses. Because little is known on DD in this kind of farming, a study was set up to describe this disease and make hypotheses on the risk factors favouring its development. This is the 1st study carried out in France on this issue in young bull fattening.14 pens from 8 fattening farms were followed for several months. In each pen, the 4 feet of bulls were observed by a professional trimmer at the beginning, the middle and the end of fattening period. All visible podal lesions without trimming were recorded and DD lesions were described. At each visit, litter's temperature of pens was recorded and data on farming conditions and practices were collected. Descriptive statistical analyses, multiple correspondence analysis and hierarchical bottom-up classification were then performed. By the end of the study, animals affected by DD were observed in 7 out of the 8 farms. A total of 242 bulls were included in the analysis, fattened in lots from 11 to 32 animals. Results showed a significant difference in intra-pen prevalence between farms (e.g. 4% in farm E2 against 85% in farm E1 at mid-fattening) and also between pens of some same farms. A total of 257 DD lesions were described, 49% of which were located on front feet. Several risk factors have been identified including: overcrowding, a bedding temperature >40 °C, moving animals of pen without changing the litter, composition of batches with young cattle from more than 5 different suckler-cows farms, scraping area that communicate, presence of the infirmary-pen next to the fattening pens. A larger scale study would be needed to validate these hypotheses regarding risk factors. However, this 1st study allowed to describe and to better understand this disease in French fattening farms and to identify practices on which it could be possible to act to limit the development of the disease, and therefore economic losses.

Session 94 Theatre 4

Prevalence and factors associated with teat-end hyperkeratosis in dairy cows

F.G. Silva[1,2], C. Antas[1], A.M.F. Pereira[1] and C. Conceição[1]
[1]Mediterranean Institute for Agriculture Environment and Development & Change, University of Évora, Department of Zootechnic, Largo dos Colegiais 2, 7004-516 Évora, Portugal, [2]Veterinary and Animal Research Centre & Al4AnimalS, University of Trás-os-Montes e Alto Douro, Department of Animal Science, Quinta de Prados, 5000-801 Vila Real, Portugal; fsilva@uevora.pt

Teat-end hyperkeratosis is a welfare problem associated with dairy cows' production system, that causes pain and discomfort to the animal. Although a lower degree of teat-end hyperkeratosis can be considered normal, more severe cases highly impact animal welfare. The identification of risk factors allows the dairy farmer to take preventive measures to reduce severe teat-end hyperkeratosis. This study aimed to identify the risk factors associated with teat-end hyperkeratosis in dairy cows. From a Holstein-Friesian dairy farm, 492 cows (mean of 3.3 in the number of lactations) were observed during the milking time, and the degree of hyperkeratosis of all teats (n=1,968) were evaluated in a 4 points scale (no lesion (1), smooth ring (2), rough (3) and very rough skin (4)). Parity, days in milking, teat length and teat-end shape, udder health, milking procedures, and behaviour parameters were all evaluated to identify risk factors for teat-end hyperkeratosis. From all the teats evaluated, 41.3% had no lesion (1), 43.1% had a score of 2, 14.2% had a score of 3 and 1.4% showed severe hyperkeratosis signs (4). On average, cows had a degree of hyperkeratosis of 1.8±0.7. The sum of the worst cases of hyperkeratosis (scores 3 and 4) was 15.6%, a value reported in the literature within the recommended limits. Cows with more lactations, longer and rounded shape teats, with mastitis at least once, and with a higher milking time had a superior degree of teat-end hyperkeratosis ($P<0.05$). Although the prevalence of severe teat-end hyperkeratosis (scores 3 and 4) was not very high in the present study, there were several animal and management-based factors, that increased the degree of teat-end hyperkeratosis. Furthermore, the factors influencing the degree of teat-end hyperkeratosis may also be related to culling on a commercial dairy farm, which may conceal the actual impact on animal welfare.

Session 94 Theatre 5

Milk calcium content to detect hypocalcaemia in dairy cows at the onset of lactation

T. Aubineau[1], R. Guatteo[2] and A. Boudon[3]
[1]Innoval, Tabarly, 35530 Noyal-Sur-Vilaine, France, [2]INRAE, BIOEPAR, Chantrerie, 44300 Nantes Cedex, France, [3]INRAE, PEGASE, Prise, 35590 Saint-Gilles, France; thomas.aubineau@innoval.com

Subclinical and clinical hypocalcaemia are major health issues in dairy cattle. The main objective of this study was to describe calcium (Ca) dynamics in both milk and blood at the onset of lactation in dairy cows to assess the possibility of identifying cows with hypocalcaemia from their milk Ca content. We performed a longitudinal analysis of 32 multiparous and 18 primiparous cows starting two weeks before the expected date of calving. Blood samples were collected 15 days before expected calving. Then, blood and milk samples were collected on a daily basis from day 1 (day of calving) to day 7, and again on day 15. These data were used to describe the dynamics of plasma and milk Ca, using generalized linear mixed models with repeated values. We identified the sampling days that demonstrated the strongest correlations between plasma Ca and milk Ca. Then, we assessed the ability of milk Ca to distinguish between two different patterns of hypocalcaemia (delayed or persistent) via receiver-operating characteristic analysis. Blood Ca dynamics were largely similar in both multiparous and primiparous cows, with a strong decline from prepartum sampling (98.75±1.63 mg/l and 98.13±2.54 mg/l, respectively) to day 1 (76.87±2.35 mg/l and 88.21±1.03 mg/l), followed by a progressive increase that reached initial values on day 5 (multiparous, 91.58±1.54 mg/l) or 6 (primiparous, 98.13±2.54 mg/l). In both multiparous and primiparous cows, a decline in milk Ca levels was observed, with a sharp decrease from day 1 (1,917±67 mg/kg and 1,823±94 mg/kg, respectively) to day 3 (1,342±34 mg/kg and 1,341±34 mg/kg), then relative stability until day 15 (1,282±39 mg/kg and 1,226±25 mg/kg). We observed moderate correlations between blood Ca and milk Ca. Nevertheless, we could identify cut-off values of milk Ca allowing to discriminate between two patterns of persistent or delayed hypocalcaemia, with good to fair sensitivity (90 and 84.6%, respectively) and specificity (79.5 and 80.6%, respectively). The next step will be to confirm these results in broader field conditions in order to assess their accuracy and estimate predictive values in a field context.

Session 94 Theatre 6

PRRSV stabilisation programs in French farrow-to-finish farms: a way to reduce antibiotic use

C. Teixeira Costa, G. Boulbria, V. Normand, C. Chevance, J. Jeusselin, T. Nicolazo and A. Lebret
Rezoolution, ZA de Gohélève, 56920 Noyal-Pontivy, France; c.teixeira-costa@rezoolution.fr

PRRSV-1 affects more than 60% of farrow-to-finish farms in Brittany, with serious economic consequences. Moreover, agricultural advisors describe an increased use of antibiotics when PRRSV-1 is circulating in pig farms. The objective of this study was to assess the impact of PRRSV-1 stabilization programs on reducing antibiotic usages. The study was carried out on 19 farrow-to-finish farms that successfully implemented a PRRSV-1 stabilization protocol between 2007 and 2019. For each, antibiotic usages (expressed in mg/PCU and ALEA) were compared one year before implementation of control measures (P1) and one year following the end of PRRSV-1 monitoring signing the success of the plan (P2). Then, the evolutions of antibiotic usages (differences between P1 and P2, expressed in percentages) were calculated and analysed taking into account the level of consumption in P1. For this, according to our dataset, three categories of level's consumption in P1 were created for ALEA (high, medium and low consumers) and two for mg/PCU (high and low consumers). All comparisons were realised using non-parametric tests with R Studio software. Concerning the overall antibiotic usages, our results showed a significant decrease in the antibiotic consumption between P1 and P2 for mg/PCU ($P=0.049$) and a tendency for ALEA ($P=0.061$). The evolution of antibiotic usages, expressed in ALEA, differed significantly between high and low levels of usage in P1 ($P<0.01$) and a tendency existed between high and medium levels of consumption ($P=0.08$). Likewise, a tendency was also observed between high and low consumers for mg/PCU ($P=0.07$). In our knowledge, it is the first study showing the impact of PRRSV-1 stabilization programs on antibiotic usages. Indeed, our results highlight the impact of a stabilization protocol against PRRSV-1 on antibiotic consumption, especially on farms that use high levels. These hopeful results show the interest of further investigations about the relationship between PRRSV-1 and antibiotic usages.

Session 94 Theatre 7

Monitoring of antimicrobial use on French pig farms from 2010 to 2019 using INAPORC panels

A. Poissonnet[1], I. Correge[1], C. Chauvin[2] and A. Hemonic[1]
[1]IFIP Institut du Porc, Ille et Vilaine, La Motte au Vicomte, 35650 Le Rheu, France, [2]Anses, 22, 31 rue des fusillés, 22440 Ploufragan, France; anne.hemonic@ifip.asso.fr

There has been a strong implication of both the French swine industry and the national authorities on reducing the use of antimicrobials in pig production since 2010. The annual monitoring of antimicrobial sales by the French Veterinary Medicines Agency (Anses-ANMV) provides overall estimations but not detailed results on actual use of antimicrobials in pig farms. This study's objective was to understand the main areas of reduction in antibiotic use based on INAPORC panels, which were surveys of representative samples of pig farms performed in 2010, 2013, 2016 and 2019. The INAPORC panels included between 119 and 171 farms, randomly selected from the BDPORC national database. The same method was applied for each panel. For each antimicrobial they bought, farmers described, during a phone call, their antimicrobial usage pattern, like category of animals treated and indications of treatment. The amounts of antibiotics used in the corresponding age group was expressed in number of Animal Daily Dose / animal (ADD/a). Over this nine-year period, the mean ADD/a for sows decreased by 52%, mainly in the last three years (-48%). Since 2016, premixes and oral medicines (powders and solutions) decreased by 65 and 42%, respectively. The main reason for treatment remains urogenital problems (64%), mainly treated with tetracyclines. A significant decrease of 66% on antibiotic use for suckling piglets was observed, which was partly due to the reduction in oral medicines and the stop of antibiotics in premix. Exposure of weaned piglets to antibiotics decreased by 78% between 2010 and 2019. Since 2016, this reduction began to slow down (-28%). Colistin was no more used as a premix in 2019, contrary to the previous periods, and was no more the most used antibiotic during the post-weaning period, even if digestive problems were always the first reasons of treatment. In fattening pigs, antibiotic use decreased by 81% between 2010 and 2019 and by 36% between 2016 and 2019. In conclusion, the INAPORC panels provide detailed references on antibiotic use in the French pig production and demonstrate the continued commitment to improving current practices.

Session 94 — Theatre 8

Digital drug registration and its relation to veterinary slaughter findings

H. Görge[1], I. Dittrich[1], N. Kemper[2] and J. Krieter[1]
[1]Kiel University, Institute of Animal Breeding and Husbandry, Olshausenstr. 40, 24098 Kiel, Germany, [2]University of Veterinary Medicine Hannover, Institute of Animal Hygiene, Animal Welfare and Farm Animal Behavior, Bischofsholer Damm 15, 30173 Hannover, Germany; hgoerge@tierzucht.uni-kiel.de

Smart solutions to support farmers in their daily routine have to meet a growing demand for documentation, transparency and animal welfare standards of the public and legislature. For this purpose, a digital tool to collect treatment data at animal level at the time of application was implemented at a pig farm that met the documentation requirements of the European Union. The digitally collected data was used for descriptive analyses to determine the status quo of the farm's health status and put into relation with veterinary slaughter findings. Data was collected between August 2020 and September 2022 on a combined farm housing 147 sows per production day with 2.3 litters and 31 piglets weaned per sow per year. The treatments related to the locomotor system stand out in all the production stages until slaughter. Out of 8,713 suckling piglets 1.8% were treated, 4% were treated as rearing piglets (n=7,803) and 4.7% were treated in the fattening stable (n=6,989). The other frequent treatment reasons vary in the production stages: gastrointestinal treatments (1.5%) for the suckling piglets, respiratory treatments (4.5%) for the rearing piglets and treatments of tail lesions (2.3%) for the fattening pigs. To examine the relation to slaughter findings linear logistic regression models were used. Treatments as suckling and rearing piglets (Odds ratio (OR) 1.6, P=0.0077) as well as treatments as fattening pigs (OR 1.9, P<0.0001) increase the chance of slaughter findings compared to chances of pigs having received no treatments. Interestingly, treatments in early and late stages of the pigs' lives had no effect on the chance of slaughter findings. Batches were investigated in the model with varying chances for slaughter findings, which could be traced back to accumulated treatment events either in the early or late stages of the pigs' lives. Overall, the tool is useful for digitalising treatments and provides a data basis of good quality for further inquiries.

Session 94 — Theatre 9

A critical lens to regulations around veterinary antimicrobial use across countries and species

G. Olmos Antillón[1] and I. Blanco-Penedo[1,2]
[1]SLU, Veterinary Epidemiology Unit, Department of Clinical Sciences, Uppsala, 750 07, Sweden, [2]UdL, Department of Animal Science, Lleida, 25198, Spain; gabriela.olmos.antillon@slu.se

Central to addressing the antimicrobial resistance (AMR) crisis has been curving antimicrobial use (AMU) in all sectors, emphasising the livestock sector. The primary approach has been surveillance and antibiotic stewardship overseen by evolving national AMR programs (NPs). However, these strategies' meaning and practical significance for veterinarians and animal caretakers are contested. We seek to situate veterinary-AMU practices within a paradigm framed by social practice theory. The relevance of such an approach focuses on the life course of AMU across four clinical conditions in two contrasting species (dairy cattle and dogs) in countries with contrasting veterinary-AMU (Brazil, Spain and Sweden). For that, NPs, legislation and guidelines framing AMU were critically appraised and coupled with a thematic analysis of in-depth interviews (n=20) with national experts who described current national best practices. As a result, the data provide a situated understanding of how the veterinary profession at different geographies and species reconciles practices in entirely different scenarios. Our analysis indicates that NPs expect results in human behaviour change by focusing on increasing awareness and top-down regulatory and policy imperatives. Guidelines are often based on assumptions unaligned with the prevalent practices or heterogeneity of conditions within their respective countries. A cross-country theme is a call to follow the basic principles of AMU, which are based on the assumption that there is a valid veterinary client-patient relationship. However, failures are found on at least one of the many steps for such a relationship to be successfully achieved (e.g. recording of outcome follow-up). It is also clear that NPs' social accountability is not equally shared among stakeholders. A more significant burden is placed on veterinarians working in the livestock sector and farmers, which may further undermine the moral license of these stakeholders currently challenged by other world problems. Our finding supports the need for NPs to be revised based on contextual knowledge of specific settings to be meaningful and practical.

Session 94 Theatre 10

An approach of the place of animal health and role of veterinarian in sustainable farming systems

A. Scholly-Schoeller[1] and G. Brunschwig[2]
[1]VetAgro Sup – INRAE, UMR Herbivores, 1 avenue Bourgelat, 69280 Marcy l'Etoile, France, [2]VetAgro Sup – INRAE, UMR Herbivores, 89 avenue de l'Europe, BP 35, 63370 Lempdes, France; gilles.brunschwig@vetagro-sup.fr

Research has majorly ignored herd health in agroecological systems. Therefore, this study intends to bolster the value of livestock health, and develop a better understanding of health management in agroecosystems. We first explore the needs and expectations of farmers regarding the support of health currently provided. A sociological study was conducted, based on observations and semi-structured interviews in sixteen farms. The choice of the farms was based on the selection of mixed farming systems with a large animal species diversity in various regions of France and Switzerland, in order to explore a diversity of situations, without claiming to be representative. These interviews dealt with four main subjects: (1) farm presentation; (2) sustainability in the agrosystems; (3) health practices used by the farmer; and (4) expectations for veterinary medicine. The data gathered allows the establishment of profiles of sustainable farms. This also enabled us to identify the existing practices of health management which are involved in sustainable farms. The healthcare practices used by the farmers interviewed were classified into eight categories: (1) choose animals adapted to the production system: breeds resilient in harsh environments, ability to maximize the forage resource; (2) develop herd immunity and balance: resistance to parasites and pathologies; (3) feeding practices: systems based on local and natural feeding resources, as pastoralism; (4) agronomic practices: soil and forage management involved in feed quality; (5) husbandry practices: reproduction and milk practices, rearing of the young; (6) medical practices: minimize chemical drugs, using of alternatives medicines; (7) managing parasitism: a result of a plurality of practices; (8) take time, observe, and learn: farmers' knowledge is based on observation of animals in their context and way of life, to prevent and understand their needs. Finally, all the interviewed farmers believe that the current veterinary approach, mainly based on curative treatments, is not sufficient to support an agroecological evolution and that a holistic perspective has to be developed.

Session 94 Theatre 11

Developing an animal welfare benchmarking framework for Australian lot-fed cattle

T. Collins[1], E. Taylor[1], A. Barnes[1], E. Dunston-Clarke[1], D. Miller[1], D. Brookes[2], E. Jongman[2] and A. Fisher[2]
[1]Murdoch University, School of Veterinary Medicine, South Street, Murdoch, WA 6150, Australia, [2]University of Melbourne, Faculty of Science, Parkville, VIC 3010 Australia, Australia; t.collins@murdoch.edu.au

Societal concerns about sustainability and animal welfare in commercial livestock production systems are increasing. Yet increased intensification of beef cattle production to meet rising global demand for animal protein is evident with over one million head of cattle on-feed in Australian feedlots. Despite the feedlot context presenting numerous welfare challenges, including animal, resource, and management factors, no recognised welfare assessment protocol for the Australian industry exists. This paper describes a framework for benchmarking animal welfare in lot-fed cattle that aims to: (1) measure welfare performance over time and enable validation of care; and (2) drive animal welfare improvement. Following consultation with the industry, a refined pilot assessment framework was trialled over seven months in eight feedlots in Queensland, Western Australia, and South Australia. A pen-side assessment technique was developed based on the four principles of the Welfare Quality® (WQ) framework, including forty-eight measures deemed applicable and feasible in the feedlot context. Measures that captured the health, behaviour and demeanour of livestock were collected in home pens monthly. Principal Components Analysis and Generalised Linear Mixed Modelling compared the effect of environmental and management measures on cattle behaviour within home pens. Recommendations for the timing of data collection (early morning and midafternoon) and sample size requirements (2 pens and ideally 3 replicate pens, per feed program and breed) were made with significant variability in some cattle outcomes (e.g. cattle activity and rest) found between visits, feedlot, observation time, and feeding program. The implementation of an evidence-based animal welfare benchmarking framework will enable feedlot producers to track their performance in animal welfare over time, and for the feedlot industry to further demonstrate its commitment to animal welfare.

Session 94 Theatre 12

H2020 mEATquality: on-farm animal welfare assessment in slaughter pigs

T. Rousing, L.D. Jensen, M.L.V. Larsen and L.J. Pedersen
Aarhus University, Animal and Veterinary Science, Blichers Allé 20, P.O. Box 50, 8830 Tjele, Denmark; lene.juulpedersen@anivet.au.dk

In the Horizon mEATquality project (Project id: 101000344-mEAT quality-H2020-FNR-2020-2, Project home page: https://meatquality.eu/), one of the tasks is to improve the quality of pork from conventional and organic farming systems by understanding how it is affected by extensive vs intensive production factors on-farm. One of the quality parameters that is in focus is animal welfare. In the project, an animal welfare assessment protocol has been developed including a farmer interview regarding management and caretaking procedures (e.g. medicine usage, mortality rates, invasive procedures, grouping strategies and feeding management), as well as a protocol for direct registrations of slaughter pig health (e.g. Body condition score, fouling, lesions, lameness and symptoms of respiratory disease) and behaviour (e.g. avoidance, aggressive behaviour, mounting and play) as well as resource measurements (space allowance, water and feed quality and quantity, etc.) available for the animals. Before applying the protocol in practice, two training sessions were arranged in the countries of project partners doing the assessments. In the autumn of 2022 and in the spring 2023, on-farm data has been collected to assess animal welfare in a total of 80 slaughter pig herds in Spain, Italy, Poland and Denmark with 20 herds per country. The herds were selected to reflect the variety of herds found in the respective countries and are thus including both conventional and organic farms, full-line farms (including sows, weaners as well as slaughter pig groups), different breeds, smaller and larger herds, etc. The EAAP presentation will include a presentation of the welfare assessment protocol, the procedures for calibrating observers across countries, as well as results of the animal welfare assessments for different types of farms.

Session 94 Theatre 13

Variability in pig hair cortisol concentrations at the end of the fattening period

P. Levallois[1], M. Leblanc-Maridor[1], S. Gavaud[2], B. Lieubeau[2], G. Morgant[1], C. Fourichon[1], J. Hervé[2] and C. Belloc[1]
[1]Oniris, INRAE, BIOEPAR, 101, route de Gachet, 44300, France, [2]Oniris, IECM, 101, route de Gachet, 44300, France; pierre.levallois@oniris-nantes.fr

Hair cortisol is an indicator of chronic stress and could be used to assess the exposure of pigs to stressors in the weeks/months prior to a non-invasive hair sampling. In particular, hair cortisol could be an indicator to identify stressful management practices for pigs. The objective of this study was to describe hair cortisol concentration (HCC) variability between farms, between batches of a same farm and within a batch. We assumed that stressors could differ between farms and/or between batches and would lead to a variability of HCC. Moreover, within a batch, individual behaviour and individual response to stressors may also induce variability between pigs. Twenty farrow-to-finish pig farms were recruited in western France considering the diversity of their management practices and of their health statuses. Hair was sampled in two distinct batches, eight months apart. The necks of 24 fattening pigs were clipped during each sampling session the week before slaughtering. The management practices and health statuses of the farms were described. To describe the variability in HCC, a linear model was used with three explanatory variables: batch, farm and the interaction between both. A hierarchical clustering on principal components was also performed, with four active variables: the means and the standard-deviations of the two batches per farm. HCC ranged from 0.4 to 121.6 pg/mg, with a mean of 25.9±16.2 pg/mg. Within a batch, the coefficient of variation of HCC ranged from 15 to 71%. On six farms, HCC differed significantly between the two batches. HCC varied significantly among farms and three clusters of farms were identified: low and homogeneous concentrations (n=3 farms), heterogeneous concentrations with either between-batch significant increase (n=8) or decrease (n=9). The diversity of management practices and health statuses allowed us to discuss explanatory hypotheses for the observed variations in cortisol concentration. The variations of HCC described herein confirm the interest in considering this biomarker in an animal welfare perspective.

Session 94 Poster 14

Identification and characterization of IUGR piglets

C. Ollagnier, R. Ruggeri, J. Bellon and G. Bee
Agroscope, Rte de la Tioleyre 4, 1725, Switzerland; catherine.ollagnier@agroscope.admin.ch

Intra Uterine Growth Restriction (IUGR) has economical and animal welfare impact. Research to resume or prevent IUGR conditions are impaired by a poor diagnosis. Three methods to diagnose IUGR were compared in one study involving 268 piglets. These methods were based on the head shape, birth weight (BtW) and the brain-to-liver weight ratio (BrW/LW). In addition, muscle histology and cardiac anatomy were analysed. The brain and liver volume of all piglets born alive and stillborn were determined using the computerized tomography scan (CT). Two days after birth, piglets with typical IUGR dolphin head shape, bulging eyes, and wrinkles perpendicular to the mouth were scored 3. If none of the IUGR characteristics applied, the piglet was defined as normal and scored 1. The rest of the piglets were scored 2. On the same day, 30 piglets were euthanized to sample the heart, the semitendinosus muscle (STM), the brain and the liver. The volume and weight of the liver and the brain were measured to calculate their densities. The brain and liver volumes measured by CT were then converted to BrW/LW to assess the IUGR status. Piglets with BrW/LW>0.78 were considered IUGR and normal, otherwise. The heart and its left ventricle were weighted, the diameter of the aorta, and the wall thickness of the left ventricle and of the aorta were measured. The mRNA expression levels of myosin heavy chain (MHC) I, IIa, IIb, and IIx were evaluated in the STM. Our results showed a discrepancy between the three methods of IUGR identification. Indeed, 68% of the piglets with a score 3 based on the head shape and 25% of the piglets with a birth weight <1 kg, were considered as normal according to their BrW/LW. The IUGR pigs had a lighter heart than the normal piglets. For a same heart weight, IUGR piglets had a larger relative aortic diameter (*P*<0.001), relatively thicker aorta (*P*<0.001) and larger left ventricular posterior wall (*P*<0.001). In the STM, IUGR pigs had a lower relative mRNA expression level of MHC IIa (*P*<0.01), IIb (*P*<0.01) and IIx (*P*=0.01). The identification of IUGR piglets based only on the head morphology and/or BtW does not always correspond with an increased BrW/LW. This project has received funding from the European Union's Horizon 2020 research and innovation program under grant agreement N° 955374.

Session 94 Poster 15

Investigating the association between sole ulcers and sole temperature in dairy cattle using IRT

A. Russon, A. Anagnostopoulos, M. Barden, B. Griffiths, C. Bedford and G. Oikonomou
University of Liverpool, UK, Department of Livestock and One Health, Leahurst Chester High Road Neston Merseyside, CH64 7TE, United Kingdom; amyrusson16@icloud.com

Our aim was to explore the association between the presence of sole ulcers and the sole temperature using infrared thermography (IRT). We enrolled 2,334 cows from four farms. Each cow was assessed at four different time points: pre-calving, just after calving, during early lactation and late lactation. The variables tested included: sole ulcers and their corresponding severity grade (0-3), ambient temperature, days in milk, height, body condition score, mobility, parity, type of farm, and the foot imaged (left or right). At each assessment, a thermal image of the sole of each hindfoot was taken using a Flir E8 Wi-Fi Thermal Camera, collating in 12,698 images. Statistical analysis was conducted using R studio. A multiple variable linear regression model was calculated to determine the influence of various explanatory variables on the sole temperature. Across the four farms, the sole temperature significantly increased with the presence of a sole ulcer. The most remarkable temperature difference was between sole ulcers of grades 0 and 3, with a difference of 5.13 °C (SE=0.76, p≤0.001), indicating that a greater amount of radiation was emitted when more damage to the sole occurred. Lactation stage and the ambient temperature were critical variables in the IRT measurements, influencing the sole temperature significantly. At the early lactation assessment, the mean sole temperature was decreased by 1.0 °C (SE=0.067, p≤2e16), whereas at the fresh stage, the mean sole temperature was increased by 0.16 °C (SE=0.060, *P*=0.0082). These results could be explained by various external factors. As through multivariable regression analysis, 66% of the sole temperature results were influenced by the variables tested. Other factors not assessed could include the diverse management practices of each farm. Further studies investigating how thermal images combined with relevant information (e.g. ambient temperature) could be utilised as a monitoring tool for sole ulcers on farm. These results, along with published literature, suggest that infrared thermography can be a reliable, early detection tool, for the presence of sole ulcers.

Session 94 Poster 16

Osteopathic manipulative treatment: a complementary approach to promote milk production in cows

C. Omphalius[1], A. Hardouin[2], M. Launay Ventelon[2] and S. Julliand[1]
[1]Lab To Field, 26 bd Dr Petitjean, 21000 Dijon, France, [2]Ecole Française d'Ostéopathie Animale, 6 pl Boston, 14200 Hérouville Saint-Clair, France; cleo.omphalius@lab-to-field.com

The use of osteopathy is increasingly popular in the dairy industry with the ambition to treat certain health issues and improve milk production. However, data on the impact of osteopathic manipulative treatment (OMT) is scarce. Our objective was to evaluate the effect of an OMT on milk production during one month through a monocentric longitudinal study with 100 Holstein cows (32 primiparous, 68 multiparous) managed similarly. Milk yield (MY) was individually monitored (automatic milking system) for 2 weeks before OMT and 2 homogeneous groups (Control [C, n=50] and Treated [T, n=50]) were formed at d0 according to the cows' basal MY (daily average between d-14 and d0), parity, and stage of lactation. At d0, cows in group T received an OMT by an animal practitioner (n=4, trained in the same school) using similar practices, while group C did not (sham treatment). Individual MY, fat and protein contents were daily recorded and averaged over 2 weeks after d0 (d0-d14) or d15 (d15-d28). Data analysis included the effects of OMT, practitioner, and their interaction using a MIXED procedure (SAS) with basal yield as covariable. No interaction or practitioner effect was observed. Cows in group T (30.6 kg/d) tended to produce more milk than those in group C (29.8 kg/d; P=0.054) for d0-d14, but no difference was observed during d15-d28 (P=0.767). The higher MY in group T was exacerbated when only multiparous cows were considered (C=30.0 kg/d and T=31.3 kg/d for d0-d14; P=0.036). A trend of increasing protein yield was also observed in multiparous cows (+37 g/d for d0-d14; P=0.068) but fat yield was not altered. The increased MY may be related to a higher mobility of treated cows. Indeed, the number of milkings per day was numerically higher in group T than C for multiparous for d0-d14 (2.87 vs 2.76 milkings/d; P=0.153). The insignificant effect in primiparous cows could be due to their naturally higher activity. This first study describing the OMT effect on dairy cows highlights possible benefits for short-term production. Additional observations would be valuable to assess when OMT is the most beneficial for performance, health, and welfare.

Session 94 Poster 17

The relationship between stress levels and skin allergy or atopy problem by analysing hair cortisol

G.W. Park[1,2], M. Ataallahi[2] and K.H. Park[2]
[1]CO-ANI, Kangwondo, Chuncheonsi, Kangwondaehakgil 1, Dept of animal life sciences 1st building Room#420, 24341, Korea, South, [2]Kangwon national university, Animal life sciences, Kangwondo, Chuncheonsi, Kangwondaehakgil 1, Dept of animal life sciences 1st building Room#420, Kang, 24341, Korea, South; kpark74@kangwon.ac.kr

The cortisol as stress indicator for dogs can be affected by various factors such as feed, health, animal skin conditions, especially allergy or atopy, and walk frequency. Using dog hair to measure the cortisol is one of methods to measure chronic stress. It is less stressful method than another invasive method like blood sampling. The objective of this study was to assess that allergy or atopy of dogs can affect the chronic stress level. For this reason, we have used the most raised dog's hair in Korea to evaluate the cortisol concentration. For five species of dogs, Bichon fries (n=49), Maltese (n=27), Poodle (n=69), Pomeranian (n=22), and mixed breeds (n=27) were selected. The cortisol hormone was extracted by using methanol and analysed via enzyme-linked immunosorbent assay (ELISA). For statistical treatment unpaired t-test was used because the subjects which had allergies or not were randomly assigned. The result showed that the cortisol concentration of dogs suffering from allergies or atopy was 9% higher in the Bichon frise, 14% higher in the Maltese, 18% higher in the Poodle, 23% higher in the Pomeranian, and 21% higher in mixed breeds. Even though the dogs which have atopy or allergy problem showed higher cortisol level, there was no significant difference in whether dogs had skin problems or not. Although the results didn't show a significant difference, dogs are needed to feed high quality protein sources to prevent allergy or atopy. It can make the dog's stress level lessened.

Session 94 Poster 18

SECURIVO: self-assessment tools for biosecurity in veal calves' farms

M. Chanteperdrix[1], M. Drouet[1], M. Mounaix[1], M. Tourtier[1], D. Le Goic[2], C. Jaureguy[2], P. Briand[3], M. Coupin[4], A. Hemonic[5] and N. Rousset[6]

[1]Institut de l'Elevage, 8 route de Monvoisin, 35650 Le Rheu, France, [2]SNGTV, 5 rue Moufle, 75011 Paris, France, [3]CRAB, Rue Maurice Le Lannou – CS 74 223, 35042 Rennes Cedex, France, [4]CRAPL, BP 70510, 9 Rue André Brouard, 49105 Angers, France, [5]IFIP, 5 rue Lespagnol, 75020 Paris, France, [6]ITAVI, 7 Rue du Faubourg Poissonnière, 75009 Paris, France; manuel.tourtier@idele.fr

In France, controlling health in veal calf farming is one of the major and priority challenges of the sector. Biosecurity, whether external or internal, has an impact on health as a preventive factor against the diseases' introduction or spread in farms. Carrying out biosecurity self-assessments is one of the ways of raising farmers' awareness about the need to comply with recommendations and thus improve compliance. In 2015, the interprofession INTERBEV launched a major awareness-raising campaign by creating an interprofessional charter for good health control and proper use of medicinal treatments. In addition, to support farmers in an approach to reduce the use of antibiotics, the French Livestock Institute in partnership with SNGTV and CRAB has developed a specific training course thanks to which, since September 2018, more than 200 farmers have been able to improve their skills on the reasoned use of antibiotics and on the levers to reduce the risks of emergence of infectious diseases during a dozen training sessions. These training sessions highlighted areas of progress that vary from farm to farm but are undeniable in the sector despite the awareness-raising actions carried out in recent years. In this context, the SECURIVO project has created: (1) 12 technical brochures on biosecurity to help veal calf farmers to improve the biosecurity level of their farms, (2) 2 self-assessments to enable veal calf farmers to self-assess the biosecurity level of their farms: a 'detailed' self-assessment for a complete and precise assessment of each biosecurity topic in 123 questions and a 'quick' self-assessment for a quick and general assessment in 30 basic questions, (3) A dedicated website to enable veal calf farmers to carry out the self-assessments, to access personalized assessments and to consult the technical brochures.

Session 94 Poster 19

Temporally synchronized ruminal vs reticular fluid pH and short chain fatty acids in dairy cows

A. Kidane, K.S. Eikanger and M. Eknæs

Faculty of Biosciences, Norwegian University of Life Sciences, Norway, Oluf Thesens vei 6, 1430 Ås, Norway; alemayes@nmbu.no

Sub-acute ruminal acidosis (SARA) is a metabolic disorder, typically of dairy cows fed diets high in fermentable carbohydrates and low in fibre content. The fluctuation in rumen pH, and the extent and duration of its depression determine the severity of SARA. Timely detection of SARA is challenging due to lack of overt clinical symptoms. Indwelling electronic sensors used to monitor behaviour, health and fluctuations in the rumen pH often reside in the reticulum. This technique may have limitations, especially for detection of SARA, due to spatial variation in pH in the reticulo-rumen. Here, we assessed the correlations between ruminal fluid (RumF) vs reticular fluid (RetF) pH and short chain fatty acids (VFA) in lactating dairy cows getting *ad libitum* access to grass silage supplemented with concentrate feed formulated to meet requirements according to the Nordic feed evaluation system (NorFor). Time synchronized RumF (n=72) and RetF (n=72) samples were collected from 8 rumen cannulated Norwegian red dairy cows at 3 time points in each of the 3 sampling days separated by a 35-day interval. The pH of the samples was measured immediately, and then aliquots of both RumF and RetF samples were analysed for VFA. Correlation analysis indicated statistically significant relationship between RumF vs RetF pH (r= 0.716; P<0.0001). However, mean RetF pH was higher than RumF pH by 0.358 (SEM= 0.0189 units; Paired t-test, P<0.0001), and this difference was larger at the lower RumF pH range (i.e. increasing mean difference with decreasing RumF pH). Furthermore, the RumF pH showed larger fluctuation (SD 0.230 units) than RetF pH (SD 0.171 units). The higher pH in the RetF was accompanied by lower total VFA concentration in the location (98.1 mM; SEM 1.06) compared to that of RumF (108.4 mM; SEM 1.65). The observed differences in both pH and VFA with the temporally synchronized samples from the two locations could be explained by differences in location specific rates of substrate fermentation, VFA absorption and digesta passage. These results suggest that using reticular pH as a proxy for rumen pH may undermine any potential SARA cases in dairy cows due to inflated reticular pH, especially at the lower physiological range.

Session 94 Poster 20

Antimicrobial use and main causes of treatment in Italian buffalo farms

G. Di Vuolo[1], F. Scali[2], C. Romeo[2], V. Lorenzi[2], C.D. Ambra[1], M. Serrapica[1], G. Cappelli[1], F. Fusi[2], E. De Carlo[1], G.L. Alborali[2], L. Bertocchi[2] and D. Vecchio[1]

[1]Istituto Zooprofilattico Sperimentale del Mezzogiorno, via della salute, 2 Portici, 80055, Italy, [2]Istituto Zooprofilattico Sperimentale della Lombardia e dell'Emilia Romagna, via Bianchi, 9 Brescia, 25124, Italy; domenico.vecchio@izsmportici.it

Antimicrobial resistance (AMR) is a major global health emergency; 25,000 people die each year in Europe from infections sustained by resistant bacteria. Inappropriate antimicrobial use (AMU) in humans and animals is reinforcing AMR, threatening public health. For this reason, there is growing institutional attention to implementing actions for the optimization of AMU. In particular, the World Health Organization's Global Plan of Action on Antimicrobial Resistance recommends a close intersectoral collaboration between public health and animal health (One Health). The purpose of this study was to report the first data on AMU in Italian dairy buffalo farms and the principal causes of treatment. Data from 102 farms were collected over a three-year period (2015-2017); AMU was estimated separately by age group (adults, heifers and calves) using the Defined Daily Dose for Italy (DDDAit). The causes of treatment were clustered in urogenital, gastrointestinal, respiratory, locomotor, osteoarticular, mastitis, dry therapy, septicaemia, cutaneous, nervous, and unknown. The farms involved in the study had an average size of 228 adults, 116 heifers, and 144 calves. Antimicrobials were used almost exclusively on adults, where overall AMU was relatively low, averaging 1.56 DDDAit/head. The three most common causes of AMU in adult buffaloes were: urogenital (37.3%), dry cow therapy (26.9%), and mastitis (22.3%). Although direct comparisons may have some limitations, AMU seems to be lower in buffalo than in other species reared in Italy. The identification of the principals causes of AMU in buffalo represents an important step towards a more efficient AMU analysis of structural and management critical aspects connected to the buffalo breeding system finalized to reduce in sustainable mode the AMU and their connected AMR.

Session 94 Poster 21

Herd factors associated with levels of parasitism in alternative pig farms

M. Delsart[1], N. Rose[2], B. Dufour[1], J.M. Répérant[3], R. Blaga[4], F. Pol[5] and C. Fablet[2]

[1]EnvA, Laboratoire de Santé Animale USC EPIMAI, 7 avenue du Général de Gaulle, 94704, France, [2]Anses Ploufragan-Plouzané-Niort, Unité Épidémiologie, Santé et Bien-Etre, 41 rue de Beaucemaine, 22440 Ploufragan, France, [3]Anses Ploufragan-Plouzané-Niort, Unité Virologie, Immunologie, Parasitologie Aviaires et Cunicoles, 31 rue des fusillés, 22440 Ploufragan, France, [4]Anses, INRAE, École Nationale Vétérinaire d'Alfort, Laboratoire de Santé Animale, BIPAR, 7 avenue du Général de Gaulle, 94700 Maisons-Alfort, France, [5]Oniris, 101 Rte de Gachet, 44300 Nantes, France; christelle.fablet@anses.fr

Parasitism may constitute a critical point in pig farms on litter or with outdoor access. This study aimed to identify the herd factors associated with the level of parasite infestation. A study was carried out in 112 alternative French pig farms where faecal samples were taken from 10 sows, 10 pigs aged 10-12 weeks and/or 10 pigs at the end of the fattening period for coprological analysis, as well as blood samples from 10 sows and/or 10 pigs at the end of fattening period for serological analysis for *Ascaris suum* and *Toxoplasma gondii*. Data on the facilities and management of the farm were collected during the on-farm visit and were analysed using principal component analysis to determine farm profiles regarding infestation and farm characteristics. Coccidia oocysts were observed in pig faeces in the largest proportion of farms (84%) followed by eggs of strongyles (55%), *Trichuris suis* (32%) and *A. suum* (16%). The rates of *A. suum* and *T. gondii* seropositive farms were 89 and 71% respectively. Hygiene and especially decontamination of facilities were factors associated with low levels of parasitism in farms. Conversely, keeping animals outdoors or on litter, poor building maintenance, farms with small numbers of animals and summer were parameters associated with high levels of parasitism. The use of multiple anthelmintic treatments on growing pigs was associated with low levels of *T. suis* egg infestation and high levels of *A. suum* seroprevalence. This study highlighted the internal parasite burden in alternative farms. Even if some factors are not under farmers' control (season, outdoor rearing), there is still some room for improvement via hygiene and appropriate use of treatments.

Session 94 Poster 22

Biosecurity and Animal Welfare relationship on buffalo farms through ClassyFarm assessments

D. Vecchio[1], G. Santucci[2], V. Lorenzi[2], C. Caruso[3], C.D. Ambra[1], M. Serrapica[1], G. Di Vuolo[1], G. Cappelli[1], F. Fusi[2], E. De Carlo[1], C. Romeo[2], F. Scali[2], G.L. Alborali[2], A.M. Maisano[2] and L. Bertocchi[2]

[1]Istituto Zooprofilattico Sperimentale del Mezzogiorno, via della salute, 2 Portici, 80055, Italy, [2]Istituto Zooprofilattico sperimentale della Lombardia e dell'Emilia Romagna, Via Bianchi, 9 Brescia, 25124, Italy, [3]ASL Cuneo 1, P.zza Gallo Racconigi, 12035, Italy; domenico.vecchio@izsmportici.it

Biosecurity in ruminant production is often an overlooked area, instead it is key aspect of the interactions among the animal production systems, the environment, and human health in a One-Health approach. ClassyFarm is an integrated system for categorising of farms according to the risk assessment methodology. ClassyFarm gathers and processes data referred to biosecurity, animal welfare (AW), health and antimicrobial usage. It can be applied to several livestock species, including water buffalo, for which biosecurity is evaluated on 15 items. The assessment of AW includes 63 non-animal-based measures (N-ABMs) and 17 animal-based measures (ABMs). The N-ABMs are divided into 'Management' (32 items) and 'Housing' (31 items). Each item can be scored as 'insufficient', 'improvable' or 'optimal'. For each item, a weight was determined using an expert opinion elicitation process. Total and partial scores are expressed in percentage on a scale from 0 to 100%. The Biosecurity & AW were assessed on 102 buffalo farms, with an average size of 470 heads. The mean overall AW score was 61.5±13.0% and the Biosecurity was 43.3±18.2%. Statistical analysis was performed by Spearman's rank correlation using GraphPad Prism 8.0.1 (GraphPad Software, San Diego, CA, USA). Biosecurity and AW were positively correlated (ρ=0.501; P<0.001). A positive correlation was observed between Biosecurity and both 'management'(ρ=0.759; P<0.001) and 'housing' (ρ=0.610; P<0.001). No correlation was found between Biosecurity and ABM area. Knowing the Biosecurity and AW levels, their relationship and critical point in animal farm, represent a tool to plan the strategic priorities of human resources and investments, to increase the Farmers' awareness and the health, ethical sustainability of the farm.

Session 94 Poster 23

Immune function of pre-weaned calves fed a fortified milk replacer under heat stress conditions

A.A.K. Salama[1], S. Serhan[1], L. Ducrocq[2], M. Biesse[2], A. Joubert[2] and G. Caja[1]

[1]Universitat Autonoma de Barcelona, Ruminant Research Group (G2R), Campus UAB, 08193 Bellaterra, Spain, [2]Bonilait protéines, 5 Route de Saint-Georges, 86361 Chasseneuil-du-Poitou, France; ahmed.salama@uab.cat

Heat stress (HS) negatively affects the immunity and health of pre-weaned calves, which needs to be alleviated. Angus × Holstein heifer calves (n=32) were used until weaning (d 56) to evaluate the effects of fortifying milk replacer with a combination of antioxidants (vit E, Se, flavonoids, terpenes) under HS conditions. From d 14 to 27 all calves were kept in thermal-neutral (TN) conditions (THI=62) and fed a control milk replacer (CON; n=20) or the fortified replacer (FRT; n=12). From d 28 to 56, 12 CON and the 12 FRT calves were exposed to HS (day, THI=83; night, THI=77), whereas the remaining 8 CON animals continued in TN. This resulted in 3 treatments (ambient-diet): TN-CON, HS-CON, and HS-FRT. Both milk replacers were produced by Bonilait-Protéines. Calves were milk-fed twice daily (125 g/l) with free access to starter, straw, and water. Rectal temperature (RT, °C) and respiratory rate (RR, bpm) were daily recorded. At d 42, blood samples were obtained and *ex vivo* challenged with *E. coli* lipopolysaccharide (LPS) at low (0.01 µg/ml) and high (5 µg/ml) doses. Fifteen cytokines produced after the LPS challenge were measured by MILLIPLEX Bovine Cytokine Magnetic Bead Panel 1 (EMD Millipore Corp., Burlington, MA). Data were analysed by PROC MIXED v.9 of SAS for repeated measures. HS markedly increased RT and RR (+0.52 °C and +51 bpm; P<0.01). At the high LPS dose, both IL-1α (innate immunity response) and MIP-1α were lower (P<0.05) in HS-CON (2,334 and 9,244 pg/ml) than in HS-FRT (3,264 and 16,483 pg/nl). IL-6 (anti-inflammatory functions) was higher in TN-CON (4,434 pg/ml) than HS-CON (3,137 pg/ml), but similar to HS-FRT (4,176 pg/ml). Regardless of the LPS dose, levels of IL-17α (sensing 'danger' role) were greater (P<0.05) in TN-CON and HS-FRT (30.1 and 26.3 pg/ml, on average) than HS-CON (18.6 pg/ml, on average). In conclusion, calves fed with the fortified milk replacer under HS conditions exhibited an enhanced ability to face infections by producing more anti-inflammatory cytokines, better sensing of danger, and improved immune cell trafficking.

The effect of udder health on milk composition

Z. Nogalski and M. Sobczuk-Szul
University of Warmia and Mazury in Olsztyn, Department of Animal Feeding, Feed Sciences and Cattle Breeding, Olsztyn; ul. Oczapowskiego 5, 10-719 Olsztyn, Poland; zena@uwm.edu.pl

Mastitis is classified as clinical if symptoms of mammary dysfunction and defective milk are evident, or as subclinical if no clinical signs are visible. Subclinical mastitis is most challenging because it is asymptomatic, and the quality of milk may deteriorate when milk with high somatic cell counts (SCC) is included in the bulk tank. The aim of this study was to determine the effect of udder health on milk composition and in particular on the fatty acid profile. A total of 576 milk samples were collected in six herds of Holstein-Friesian cows. Similar feeding was used in all herds. None of the cows showed clear symptoms of clinical mastitis or was treated for mastitis. Samples were collected four times, in winter. Fresh milk samples were analysed to determine their chemical composition by infrared spectrophotometry, and SCC by flow cytometry. Milk fat was extracted and the proportions of 43 fatty acids were determined by gas chromatography. Udder health was evaluated based on SCC classes, in thousands of cells per ml of milk: ≤200 – healthy udder, 201-400 – risk of mastitis; 401-1000 – subclinical mastitis, >1000 – severe subclinical mastitis. One-way analysis of variance SCC class, was performed. An increase in SCC was accompanied by a significant decrease in daily milk yield and lactose concentration. The difference in daily milk yield between groups ≤200 SCC and >1000 SCC reached 8.6 kg ($P<0.01$). Milk with higher SCC contained more ($P<0.05$) polyunsaturated fatty acids and less ($P<0.05$) monounsaturated fatty acids. The proportion of medium-chain fatty acids (MCFAs) in milk fat increased ($P<0.05$), whereas the proportion of long-chain fatty acids (LCFAs) decreased ($P<0.05$) with increasing SCC. The proportions of butyric acid and oleic acid decreased ($P<0.05$) with increasing SCC. Conclusions: The lack of clinical mastitis symptoms did not mean low somatic cell content and stable milk composition. Project financially supported by the Minister of Education and Science under the program entitled 'Regional Initiative of Excellence' for the years 2019-2023, Project No. 010/RID/2018/19, amount of funding 12.000.000 PLN.

An attempt to predict dairy cows chronic stress biomarkers using milk MIR spectra

C. Grelet[1], H. Simon[1], J. Leblois[2], M. Jattiot[3], C. Lecomte[4], R. Reding[5], J. Wavreille[1], E.J.P. Strang[6], F.J. Auer[7], Happymoo Consortium[2] and F. Dehareng[1]
[1]Walloon Agricultural Research Center, 5030, Gembloux, Belgium, [2]Walloon Breeders Association Group, 5590, Ciney, Belgium, [3]Innoval, 35530, Noyal sur Vilaine, France, [4]Eliance, 75795, Paris, France, [5]Convis, 9042, Ettelbruck, Luxembourg, [6]LKV Baden Württemberg, 70190, Stuttgart, Germany, [7]LKV Austria, 1200, Vienna, Austria; c.grelet@cra.wallonie.be

Having a possibility to regularly monitor chronic stress of dairy cows would be beneficial regarding several aspects. Indeed, when stress becomes chronic (long duration and inability of animals to cope), it is likely to affect emotional state, health, immunity, fertility and milk production of cows, hence impacting welfare, economics and social acceptability of dairy farms. In a previous step of the HappyMoo project, two molecules were highlighted as chronic stress biomarkers: hair cortisol and blood fructosamine. The aim of this study was therefore to attempt predicting these two chronic stress biomarkers using milk mid-infrared (MIR) spectra, to enable routine monitoring. For this purpose, approximately 1,400 individual dairy cows were sampled for hair, blood and milk in Belgium, Luxembourg, Germany, France and Austria in 72 commercials farms. Herds were selected locally with the objective of gathering cows with various levels of stress. Hair samples were collected at the tail switch and analysed by ELISA for cortisol concentration, and blood samples analysed by ELISA for serum fructosamine concentration. Milk was analysed with MIR and all generated spectra were standardized. Quantitative approach using PLS and qualitative approach using PLS-DA were used. Preliminary results show that quantitative models were not able to accurately predict the chronic stress biomarkers ($R^2cv<0.3$). However, a qualitative approach might be a possibility to discriminate stressed from non-stressed cows with a cross-validation correct percentage of good classification (accuracy) of 75%. Further work is needed to better exploit this dataset, however this first study suggests a possibility of using MIR milk spectra to highlight stressed cows. This might enable improvement of welfare, health and production of animals.

Session 95 Theatre 2

Prediction of body condition score for the entire lactation in Walloon Holstein cows

H. Atashi[1,2], J. Chelotti[2,3] and N. Gengler[2]
[1]Shiraz University, Department of Animal Science, Shiraz, 7144113131, Iran, [2]University of Liège, GxABT, Pass. des Déportés 2, 5030 Gembloux, Belgium, [3]Instituto de Investigación en Señales, Sistemas e Inteligencia Computacional, FICH-UNL/CONICET, 3000 Santa Fe, Argentina; hadi.atashi@uliege.be

Body condition score (BCS) is a subjective measure of stored energy reserves of dairy cows. Monitoring changes in body condition throughout the lactation is more valuable than identifying absolute measures of body condition. However, recording BCS values is a time-consuming process and usually there is one BCS record per animal per lactation. The aim was to investigate the ability of a random regression test-day model (RR-TDM) to predict BCS values for the entire lactation in Walloon Holstein cows. Milk yield, fat percentage, protein percentage and BCS records collected on the first-parity cows calved between 2012 and 2022 in 785 herds were used. Records from days in milk (DIM) between 5 d and 365 d were used. Number of test-day records on milk yield traits were 801,095 on 96,770 animals, while the number of records on BCS was 98,390 on 87,231 animals. The (co)variance components and breeding values for the considered traits were estimated based on the integration of RR-TDM using the a multiple-trait (four traits), single-lactation model. Gibbs sampling was used to obtain marginal posterior distributions for the various parameters using a single chain of 200,000 iterates with a burn-in period of 50,000 iterates. Predicted test-day BCS were compared with observed values. The Pearson and Spearman correlation estimated between the predicted and observed BCS values was 0.99 and 0.93, respectively. The root mean squared error (RMSE) and the mean absolute error (MAE) of the prediction were 0.17 and 0.14 BCS unit, respectively. In the next step, BCS records of 1,500 animals were replaced by 0, and were predicted and compared to the observed values. The Pearson and Spearman correlation estimated between the predicted and observed BCS values was 0.57 and 0.56, respectively. The RMSE and the MAE of the prediction were 0.69 and 0.51 BCS unit, respectively. This study showed that MT-RR-TDM has the potential to interpolate the trend of BCS of animals with a single BCS record across the lactation.

Session 95 Theatre 3

Conditions to develop successful mid-infrared (MIR) based BCS and BCS change predictions

J. Chelotti[1,2], H. Atashi[2,3], C. Grelet[4], M. Calmels[5], J. Leblois[6] and N. Gengler[2]
[1]Instituto de Investigación en Señales, Sistemas e Inteligencia Computacional, FICH-UNL/CONICET, 3000 Santa Fe, Argentina, [2]ULiège – GxABT, 5030, Gembloux, Belgium, [3]Shiraz University, Department of Animal Science, 71441-13131 Shiraz, Iran, [4]Walloon Agricultural Research Center, 5030, Gembloux, Belgium, [5]Seenovia, Le Mans, 72000, France, [6]Awé groupe, Ciney, 5590, Belgium; nicolas.gengler@uliege.be

Usefulness of Body condition score (BCS) and BCS changes have been validated in many studies and the development of mid-infrared (MIR) based BCS and BCS change predictions is an ongoing effort. Literature shows the importance of several key factors. First type of reference data, quantity and quality, and measurement methods are important. Reference BCS needs also to be synchronized with MIR spectral measurements. Moreover, especially at the beginning of the lactation, closer repetitions yielded the best literature values but results obtained were only valid in the first 120 days. In the HappyMoo project two efforts were made to use existing data: (1) from an experience using the BodyMat system; and (2) from Walloon conformation recording (1 record / lactation). In both cases, reference values were not synchronized and therefore BCS values had to be imputed across the lactation using appropriate modelling strategies. Also 75 vs 25% calibration vs. validation data was used during the MIR calibration process. For 1) 26,207 BCS records on 3,038 cows (mainly Holstein and Montbéliarde breeds) were used. Partial least squares (PLS) were used as the reference method for prediction, achieving a performance of r=0.58. Machine learning (ML) methods were also assessed. Among them, Support Vector Machine (SVM) achieved best results (r=0.59). For Walloon data 129,870 MIR test-day records on 17,292 animals were available. PLS method achieved the best performance over tested ML methods with r=0.42. Studies showed the difficulties to get enough and synchronized BCS records. Imputations are always second-best options. Also observed variability was suboptimal, with little extreme animals. Therefore, conditions for successful future research will be to optimize selection of herds (i.e. most variable for BCS), of BCS recording (e.g. automatic device-based) and methods.

Session 95 Theatre 4

Large-scale analysis of chronic stress in dairy cows using hair cortisol and blood fructosamine

H. Simon[1], C. Grelet[1], S. Franceschini[2], H. Soyeurt[2], J. Leblois[3], M. Jattiot[4], C. Lecomte[5], R. Reding[6], J. Wavreille[1], E.J.P. Strang[7], F.J. Auer[8] and F. Dehareng[1]

[1]Walloon Agricultural Research Center, Gembloux, 5030, Belgium, [2]Gembloux Agro-Bio Tech, Gembloux, 5030, Belgium, [3]Walloon Breeders Association Group, Ciney, 5590, Belgium, [4] Innoval, Noyal-sur-Vilaine, 35530, France, [5]Eliance, Pôle données et Elevage, Paris, 75595, France, [6]Convis, Ettelbruck, 9085, Luxembourg, [7]LKV Baden Württemberg, Stuttgart, 70190, Germany, [8]LKV-Austria, Vienna, 1200, Austria; h.simon@cra.wallonie.be

Stress in dairy herds can come from many sources. When stress becomes chronic due to long duration and inability of animals to adapt, it can affect the emotional state, health, immunity, fertility and milk production of cows. It therefore has a negative impact on the welfare, economy and social acceptability of dairy farms. Knowing the practices that influence chronic stress would allow proposing strategies to improve animal welfare on farms. The study aims to better understand the relationship between two chronic stress biomarkers and potential explanatory variables related to production, management, health and housing. From a past research, two molecules were identified as biomarkers of chronic stress: capillary cortisol and blood fructosamine. Based on 1,372 ELISA measurements of hair cortisol and blood fructosamine, we studied their variability from 23 qualitative and quantitative features related to the herd management practices collected in 77 farms from 5 different countries (Austria, Belgium, France, Germany, and Luxembourg). The qualitative variables included information on cow diet type, management and housing system. The quantitative features included information on cow characteristics, milk composition, housing conditions, udder health, body condition score, and lameness. Principal components analysis and multivariate regressions highlighted the differences in cortisol and fructosamine results between farms, and relate them to the different herd management practices implemented on each farm. Preliminary results show that there may be multiple sources of stress, but that cortisol and fructosamine levels appear to be strongly influenced by overcrowding. These first and subsequent results could help to identify sources of chronic stress and provide useful guidance to dairy farmers to improve animal welfare.

Session 95 Theatre 5

Calculation of dry matter intake and energy balance based on MIR spectral data at European level

V. Wolf[1], L. Dale[2], M. Gelé[3], U. Schuler[4], U. Müller[5], N. Gengler[6], J. Leblois[7] and M. Calmels[8]

[1]CEL 25-90, Roulans, 25640, France, [2]LKV Baden-Wuerttemberg, Stuttgart, 70190, Germany, [3]IDELE, Paris, 75595, France, [4]Qualitas AG, Zug, 6300, Switzerland, [5]University of Bonn, Bonn, 53115, Germany, [6]Gembloux Agro-Bio Tech, University of Liège, Gembloux, 5030, Belgium, [7]Awé group, Ciney, 5590, Belgium, [8]Seenovia, Le Mans, 72000, France; valerie.wolf@cel2590.fr

Calculated energy balance (EB) status of cows at the beginning of lactation is a challenge for farmers. Failure to cover energy needs can lead to metabolic disorders, stress of the immune system, breeding problems and reduced milk production. Working on EB calculation models at European level is a challenge because the evaluation of feeding systems differs between countries. In this study, combination of four datasets coming from France (25,876 daily cow data), Germany (2,801 daily cow data,) and Switzerland (38,000 daily cow data) was made possible thanks to common descriptors: the net energy of lactation (NEL) and the dry matter intake (DMI) of each feed components. For each cow with individual intake, EB, DMI and feed efficiency (as a function of milk yield, fat, and protein content and DMI) were calculated. From the datasets, equations with the MIR spectral data were developed. A country effect was checked on residuals and it appeared that it was not significant. To develop the models, all spectra were standardized. The best models were obtained based on corrected MIR spectral data for days in milk (DIM) with Legendre polynomial correction and by using canonical powered partial least square (CPPLS) and Lasso and Elastic-Net Regularized Generalized Linear Model and by adding fix effects (parity, breed, milking moment, milk yield). The validation R^2 were of 0.70 for DMI, 0.76 for EB, and 0.75 for feed efficiency. This study shows that it is possible to concatenate and align European databases on EB calculation and that EB can be for dairy cows estimated from MIR spectra with a good accuracy to be used on the field.

Validation of HappyMoo MIR energy balance models on external datasets with feeding restriction

M. Calmels[1], J. Pires[2], M. Boutinaud[3], A. Leduc[3], C. Leroux[2], J. Chelotti[4], L. Dale[5], C. Grelet[6], A. Tedde[4], J. Leblois[7] and M. Gelé[8]
[1]Seenovia, Le Mans, 72000, France, [2]INRAE, UCA, VetAgro Sup, UMRH, Saint-Genès-Champanelle, 63122, France, [3]INRAE, Institut Agro Rennes Angers, UMR 1348 PEGASE, Saint-Gilles, 35590, France, [4]Gembloux Agro-Bio Tech, University of Liège, Gembloux, 5030, Belgium, [5]LKV Baden-Wuerttemberg, Stuttgart, 70190, Germany, [6]Walloon Agricultural Research Center, Gembloux, 5030, Belgium, [7]Awé groupe, Ciney, 5590, Belgium, [8]IDELE, Paris, 75595, France; marion.calmels@seenovia.fr

The Interreg HappyMoo project developed three sets of equations related to energy balance (EB). The first set concerns EB, dry matter intake (DMI) and feed efficiency. The second predicts fine milk components (oleic acid, beta-hydroxybutyrate, acetone, citrate), and the third concerns bodyweight and BCS. All models were developed based on European databases, including different countries, breeds and systems. An external validation was performed using feed restriction (FR) trials to ensure the ability of these models to detect energy deficit in case of FR. In the first external dataset, 30 Holstein dairy cows were studied at 21 days in milk and in mid-lactation, after which all cows were submitted to FR. The FR aimed to reduce the intake by 20% by adding straw to dilute energy and protein content in the ration. The second external dataset included 8 Holstein and 10 Montbéliarde mid-lactation cows that underwent 6 days FR with feed allowance reduced to meet 50% of individual energy requirements. The milk samples from these two datasets were analysed for fatty acids and selected metabolites and mid-infrared spectra were recorded and standardized before applying HappyMoo equations. The best performances were obtained for oleic acid (R^2=0.94 and 0.91 for the first and second dataset, respectively), then EB and DMI ($R^2 \geq 0.55$) but with a strong correlation with oleic acid. For BCS and bodyweight, the shape of the predicted curves coincided with observed values. For BHB, the lack of variability in the validation dataset did not allow to test the equations. In this study, oleic acid seems to be the best indicator to determine the nutritional status of the cow, presenting a good response to variations of nutrient allowance, even in mid-lactation.

Genetic analyses of principal components of milk mid-infrared spectra from Holstein cows

Y. Chen[1], P. Delhez[1], H. Atashi[1,2], H. Soyeurt[1] and N. Gengler[1]
[1]ULiège-GxABT, Passage des Déportés 2, 5030 Gembloux, Belgium, [2]Shiraz University, Ghasro, 71441-13131 Shiraz, Iran; yansen.chen@uliege.be

Milk mid-infrared (MIR) spectra are widely used to predict milk composition and phenotypes linked to animal health, efficiency, emissions, resilience, and even milk processability. However, the milk phenome represented by the large number of individual MIR wavenumbers (hereafter called MIR spectral traits) could allow strategies to develop innovative and holistic monitoring and decision-making PLF tools by avoiding phenotypic calibrations. The objectives of this study were to start exploring the complete genetic architecture of MIR spectral traits by studying principal components of milk MIR traits from Holstein cows and their genetic correlations with milk fat percent. In total, 27,855 records for 311 MIR spectral traits and fat percent from 3,303 first-parity Holstein cows were used. All principal components (PCs) of 311 wavenumbers were extracted and single-traits and bivariate (PC and fat percentage) repeatability models were used. The first 75 PCs explained already 99.99% of the phenotypic variance across the 311 wavenumbers. However, the genetic variances of these 75 PCs explained only 85.54% of the total genetic variances of the 311 PCs. Their range of heritability was from 0.00 to 0.47. The mean and standard deviation of genetic correlations (absolute value) between 311 PCs and fat percent were 0.25 and 0.28, respectively. After first 75 PCs, some PCs (e.g. PC 86) still had genetic correlations (here 0.37) with fat percent. This preliminary study showed that phenotypic rank reduction using first 75 PCs might be dangerous, also as even some PCs (e.g. PC 145) with very low phenotypic variances (<0.001%) showed heritability (here 0.05) over 0.01. By estimating heritability and in future studies genetic correlations among MIR spectral traits associated with genomics, understanding the genetic architecture of MIR spectral traits will not only help us optimize their use for breeding, but also manage by avoiding phenotypic calibration steps.

MastiMIR – MIR prediction for udder health

L.M. Dale[1], A. Werner[1], E.P.J. Strang[1], Happymoo Consortium[2] and J. Bieger[1]
[1]Regional association for performance testing in livestock breeding of Baden-Wuerttemberg (LKVBW), Heinrich Baumann Str. 1-3, 70190, Germany, [2]http://www.happymoo.eu, 4 Rue des Champs Elysées, 5590, Ciney, Belgium; ldale@lkvbw.de

Udder health plays a major role in many dairy farms. In addition to the great economic impact, the well-being of the animal is also of primary importance. Especially subclinical inflammations, which are not visible at first sight, are problematic and costly. An important tool to monitor udder health and identify problem areas are udder health metrics. These key figures are an integral part of the milk performance test for every LKV member and can be viewed in the LKV Herd Manager. Another tool that has been evaluated as part of the HappyMoo project is MastiMIR. MastiMIR is an early warning system for udder disease that evaluates monthly milk spectral data. As milk compositions are related to a cow's health and metabolism, milk MIR spectra can be used to detect mastitis, a process already established with KetoMIR for early detection of ketosis. The goal of procedures such as KetoMIR and MastiMIR is to detect animal health problems early enough to take appropriate countermeasures quickly. Accompanied with bacterial examinations MastiMIR was tested in 25 pilot farms over two years during the HappyMoo project. In consultations with farmers, vets and other experts the focus was on the joint assessment of the animals that had been classified as risk animals on the basis of the spectral data. In addition, the corresponding bacteriological examination results of these animals were viewed. The comparison showed that most of the animals that appeared red on the report (high risk) were also infected with a pathogen. The new tool MastiMIR offers a lot of potential to become another building block for farmers, veterinarians and advisors to improve udder health in Baden-Württemberg farms.

Prediction of lameness and hoof lesions using MIR spectral data

M. Jattiot[1], L.M. Dale[2], M. El Jabri[3], J. Leblois[4], M.-N. Tran[4] and HappyMoo Consortium[5]
[1]Innoval, rue Eric Tabarly, CS 80038, 35538 Noyal-sur-Vilaine, France, [2]LKV Baden-Württemberg, Heinrich-Baumann-Str. 1-3, 70190 Stuttgart, Germany, [3]IDELE, 149 rue de Bercy, 75012 Paris, France, [4]Elevéo, Association Wallonne des Eleveurs, 4 Rue des Champs Elysées, 5590 Ciney, Belgium, [5]http://www.happymoo.eu, 4 rue des Champs Elysées, 5590 Ciney, Belgium; manon.jattiot@innoval.com

Lameness is currently one of the most severe health issues in dairy cows' farming systems. Thus, the project HappyMoo, aiming at predicting welfare related problems using the milk mid-infrared (MIR) spectroscopy has chosen to work on lameness prediction. Two types of data were studied: lameness score (using ICAR scale 1-5) and hoof lesions (using ICAR codes), associated with milk analysis, animal information, and MIR spectral data. Concerning lameness score data, 22,576 observations was used. Pre-processing of the MIR spectral data has been done before comparing different machine learning models to find the best one, based on sensitivity. The variable to predict was binary: 'no lame' vs 'lame'. We found a variability on the performances, accuracy varying from 0.61 (DECISION TREE) to 0.68 (GBM or NEURAL NETWORK), but sensitivity from 0.12 (GBM) to 0.65 (LOGISTIC REGRESSION). Another tested approach was CCPLS – GLMNET model, which had satisfying results in calibration model, but the performances on external validation were not good enough to be used in routine (best sensitivity=0.29). Regarding hoof lesions, the models were first tested by discriminating animals with at least one lesion on one of the 4 limbs vs animals with no lesion. Like for lameness, different models have been applied. The results showed a good prediction of healthy animals (best specificity=0.94 with CNND-N) but sensitivity was bad (0.12). Other groups were constructed according to the type of lesions encountered. The best results in calibration (spe=0.72, sen=0.64) was obtained with CPPLS-GLMNET model for the following discrimination: healthy animals vs ill, considering only the infectious disorders. In conclusion, several models were tested on lameness score and hoof lesions data but the results do not yet seem to be able to predict lameness in dairy cows through milk spectra. More research needs to be done on this topic.

MIR spectral prediction based on heat stress in dairy cattle

L.M. Dale[1], M. Jattiot[2], A. Werner[1], E.J.P. Strang[1], P. Lemal[3], Happymoo Consortium[4], Klimaco Consortium[5] and N. Gengler[3]
[1]Regional association for performance testing in livestock breeding of Baden-Wuerttemberg (LKVBW), Heinrich Baumann Str. 1-3, 70190, Germany, [2]INNOVAL, 35538, Noyal-sur-Vilaine, France, [3]Gembloux Agro-Bio Tech, University of Liège, Passage des Déportés 2, 5030 Gembloux, Belgium, [4]http://www.happymoo.eu, 4 Rue des Champs Elysées, 5590, Ciney, Belgium, [5]https://agroecologie-rhin.eu/de/klimaco-2/, 2 rue de Rome, 67013, Schiltigheim, France; ldale@lkvbw.de

It is well known that heat stress (HS) negatively affects the profitability of dairy farms. However, the effects of HS on detailed milk composition have been much less studied. Until now, early results in different environments (e.g. Belgium, Tunisia, Germany) were reported based on MIR spectrum responses to HS in dairy cows. The effects of HS on milk production traits and the MIR spectrum are reflecting the detailed information on milk samples in Baden-Württemberg. The model was built in the period 2012 to 2019 and the external validation was made between 2020 and 2022. Data from public weather station, temperature and humidity data, was merged with milk recording data. Values for the daily average temperature-humidity index (THI) were calculated. All statistical analysis was performed in R using 'glmnet', and spectral data were first standardized and then pre-processed by 1st derivative. Results showed that there were differences in MIR spectra recorded under HS and thermoneutral conditions; certain wavenumbers of the MIR spectrum responded differently. The THI index was build based on the report between individual cow THI and the mean THI value of the farm. Pearson correlations based on the THI index and milk parameters were calculated with 'corrplot' library in R. The negative correlation with THI index were on milk traits, such as milk yield: 0.15, and lactose: 0.12, acetate: 0.33, blood NEFA: 0.2 and positive correlations with fat: 0.59, protein content: 0.40, blood BHB: 0.25, blood glucose: 0.30, blood calcium: 0.21 and fatty acids more as 0.35 Further analysis is needed to identify potential milk-based MIR phenotypes that could be used in this context for herd management and breeding for HS resistance.

Use of MIR spectra-based indicator for genetic evaluation of heat stress in dairy cattle

P. Lemal[1], L.M. Dale[2], M. Jattiot[3], J. Leblois[4], M. Schroyen[1] and N. Gengler[1]
[1]ULiège – GxABT, Passage des Déportés, 2, 5030 Gembloux, Belgium, [2]LKVBW, Heinrich Baumann Str. 1-3, 70190 Stuttgart, Germany, [3]INNOVAL, Rue Eric Tabarly, 35538 Noyal-sur-Vilaine, France, [4]Elevéo, Rue des Champs Elysées 4, 5590 Ciney, Belgium; pauline.lemal@uliege.be

Heat stress (HS) detection and its genetic evaluation is still a challenge. Indeed, milk yield and milk content represent a small fraction of variability induced by HS. A solution could be to use the information of whole mid-infrared (MIR) spectra that are already routinely measured to evaluate fat and protein content. Recently, a MIR spectra-based HS indicator (called hereafter MIR indicator) has been developed to evaluate HS. The objective of this study was to evaluate the interest of using this new MIR indicator for genetic evaluation of HS by comparing it with milk yield. Milk yield data and the MIR indicator for 53,842 Holstein cows were obtained from 2015 to 2022 for a total of 837,492 records. The threshold at which milk yield starts to be affected by HS was estimated at a temperature-humidity index (THI) of 61. The MIR indicator follows a linear relation with the THI and thus no clear threshold is observable. However, a higher relative variation along the THI scale is observed for the MIR indicator compared to milk yield (1.2 and 0.15 respectively). A two-trait random regression reaction norm model was fitted with a threshold of 61 for milk yield and 50 for the MIR indicator. Heritability was 0.32 for the MIR indicator at the thresholds and this value was similar at high THI. Regarding reaction norm effects, the MIR indicator showed a negative genetic correlation with milk yield. Because a low value of the MIR indicator is expected at high THI for heat-tolerant cows, a negative genetic correlation between the regression on the THI for milk yield and the MIR indicator was considered favourable. If confirmed, this new MIR indicator could be a valuable tool for genetic evaluation of HS because of its variation along the THI scale and the high heritability values, including at high THI.

Session 95 — Theatre 12

A unique overview of enteric methane emissions by dairy cows in Bretagne, France

S. Mendowski[1], O. Garcia[1], C. Bruand[2], M. Tournat[1], T. Viot[2] and L. Meriaux[2]
[1]Eco-Sens, La Messayais, 35210 Combourtillé, France, [2]Eilyps Group, 17 boulevard Nominoë, 35740 Pacé, France; s.mendowski@eco-sens.com

As enteric methane (CH_4) represents 53% of the GHG emissions in dairy farms, its decrease is an effective way to reduce their carbon footprint. The CH_4 emissions occurred during ruminal fermentations, so adapting dairy cow diets is a good way to reduce them. What are the real levels of CH_4 emissions in French dairy farms? By combining the equation used by Eco-Sens to predict CH_4 from milk fatty acids and milk yield and the data collected by Eilyps, a unique overview of dairy cows CH_4 emissions in Bretagne, west of France, was performed. The database contains 59.561 CH_4 intensity (iCH_4) data from 2.481 farms, predicted between 2019 and 2021, linked to other data that allows to study the effect of years, months, forage (pastured grass, stored grass, maize) and breeds on iCH_4. Results show iCH_4 variations between and within the different systems. The iCH_4 vary from 11 to 20 g/l of milk (mean of 16.2 g/l). The iCH_4 are not constant during a year: April and December are the months with the lowest and highest iCH_4 means (15.3 and 16.9 g/l of milk, respectively), mainly due to the large use of pasture in this area in spring. Prim'Holstein is the breed for which iCH_4 are the lowest in average (15.8 g/l of milk), because their slightly higher level of milk yield. In average, farms with more than 75% of maize silage in diet emit less CH_4 than farms with more than 75% of grass (fresh or stored) in diet. However, pasture-based diets allow the lowest CH_4 emissions, due to the improvement of milk fatty acid profile reflecting fermentations in rumen that emit less CH_4. Individual farm results show that some farms using more than 75% of grass have lower CH_4 emissions than some farms using more than 75% of maize silage. This study confirms that diet has an impact on reducing CH_4 emissions: from a technical point of view, reducing CH_4 emissions is concomitant with feed efficiency, with high quality and well valuated forages. From an economic point of view, reducing CH_4 emissions is linked to increasing food cost margin. This study will make it possible to position each farm in relation to its CH_4 emissions and to implement a concrete reduction action plan.

Session 95 — Theatre 13

Genetic parameters analysis of milk citrate for Holstein cows in early lactation

H. Hu[1], Y. Chen[1], C. Grelet[2] and N. Gengler[1]
[1]ULiège-GxABT, Passage des Déportés 2, 5030 Gembloux, Belgium, [2]Walloon Agricultural Research Center (CRA-W), Chaussée de Namur 24, 5030 Gembloux, Belgium; hongqing.hu@doct.uliege.be

Delivering innovative and holistic monitoring and decision-making PLF tools relies on the availability of critical biomarkers. Negative energy balance is a difficult trait complex as there is a difference between perceived imbalance and physiological imbalance. Milk citrate is considered to be an early biomarker of negative energy balance for dairy cows in early lactation, but its genetic analysis is lacking. The objectives of this study were to (1) show the distribution of milk citrate content in early lactation; (2) analyse the genetic parameters of milk citrate. The coefficient of determination (R^2) and root mean square error (RMSE) of the predicted milk citrate model by milk mid-infrared (MIR) spectra in external validation were 0.86 and 0.76 mmol/l, and available from DIM 5 to 50 d. Records were divided into three traits according to the first (citrate1), second (citrate2), and from third to fifth party (citrate3+). After editing, the data included 134,517 records, from 52,198 cows, and 4,479 animals in the pedigree with 566,170 SNPs. A multiple-trait repeatability model was used in this study. The citrate is decreasing in early lactation, on average from 10.04 to 8.58 mmol/l from DIM 5 to 50 d. When cows start to be in energy balance (DIM $\approx$ 40 d), milk citrate was 8.82 mmol/l. The average of citrate1 was 8.93 mmol/l; citrate2 was 8.93 mmol/l; citrate3+ was 9.17 mmol/l. The heritability for citrate1 was 0.40; for citrate2, 0.37 and for citrate3+, 0.35. The ranges of genetic correlations between the three traits were from 0.98 to 0.99, and of phenotypic correlations, from 0.41 to 0.42. This study showed that considering MIR-based milk citrate as an indicator to identify negative energy balance should be possible in early lactation, and this indicator could help select for animals less affected by negative energy balance.

Session 95 Poster 14

Milk fluctuations in daily milk yield associated with diseases in Chinese Holstein cattle

A. Wang[1], D.K. Liu[2], C. Mei[3] and Y.C. Wang[1]
[1]China Agricultural University, No.2 Yuanmingyuan West Road, Haidian District, Beijing, 100193, China, P.R., [2]Hebei Sunlon Modern Agricultural Technology Co. Ltd., Dingzhou, 073000, China, P.R., [3]Dongying Austasia Modern Dairy Farm Co. Ltd., Dongying, 257000, China, P.R.; wangyachun@cau.edu.cn

Disease-associated milk loss is one of the key constraints on herd profitability and sustainable productivity. In this context, our study aimed to quantify phenotypic variability of milk fluctuations caused by diseases and to explore fluctuation traits that can contribute to disease resilience breeding from yield perspective. By combining high-frequency daily milk yield with disease records, we calculated milk variability trends in a fixed window around the treatment day of each case of five diseases: udder health, reproductive disorders, metabolic disorders, digestive disorders, and hoof health. The average milk yield for all diseases decreased rapidly from 6 to 8 days before treatment day, with the largest decreases observed on treatment day. Furthermore, significant fluctuations caused by diseases were determined. Milk fluctuation was defined as a period of at least 10 successive days of negative deviations in which yield dropped at least once below 90% of the expected values. We defined development and recovery phases using 3,847 fluctuations caused by 4,185 disease cases, and estimated genetic parameters of fluctuation traits such as milk loss, duration and variation rate for each phase. In general, each disease-associated milk fluctuation lasted for 21.19±10.36 days with a milk loss of 115.54±92.49 kg. Milk fluctuations in recovery phase were 3.3 days longer in duration, 11.04 kg higher in loss, and 0.35 kg/d slower in variation rate compared to development phase. Fluctuation traits are inheritable with heritability estimates ranging from 0.02 to 0.19, and favourable moderate to high genetic correlations with milk yield, milk loss during entire lactation, and resilience indicators, which may support breeding for high disease resilience cows through fluctuation traits. Overall, this study confirms the high impact of diseases on milk yield and provides important insights into the interrelationship of relevant traits in Holstein cattle breeding programs, and shows the potential of using automatic monitoring of milk yield to assist disease management in dairy cows.

Session 96 Theatre 1

Impact of genomic selection for methane emissions in a high performance sheep flock

S.J. Rowe[1], T. Bilton[1], P. Johnson[1], S. Hickey[2], A. Jonker[3], N. Amyes[2], K. McRae[1], S. Clarke[1] and J. McEwan[1]
[1]AgResearch Ltd, Invermay Agricultural Centre, Puddle Alley, Mosgiel 9092, New Zealand, [2]AgResearch Ltd, Ruakura Research Centre, 10 Bisley Road, Hamilton, 3214, New Zealand, [3]AgResearch Ltd, Grasslands campus, Tennent Drive, 11 Dairy Farm Road, Palmerston North 4442, New Zealand; suzanne.rowe@agresearch.co.nz

The use of breeding for mitigation of methane emissions in livestock has been demonstrated under single trait selection. The objective of this longitudinal study is to demonstrate the inclusion of methane emissions into a selection index. Genomic breeding values for methane were included in the selection index of a high performance maternal sheep flock of 750 composite ewes for a full generation. During that period. growth, meat quality, health, production and methane emissions were closely monitored. We present genetic trends for all traits and demonstrate increases in productivity with a sustained lowering of methane emissions. Economic weighting relies on government policy, pricing mechanisms and the expression of the trait in a format that is auditable. Here we discuss the technical, political and economic challenges associated with ruminant breeding schemes for mitigation in methane in New Zealand. We show physical impact of incorporating methane emissions into the breeding scheme and the economic gains under two different accounting strategies. We review the measurement technologies available together with the potential impact of incorporating genomic selection for methane emissions into the national breeding goal.

Session 96 Theatre 2

Improving feed efficiency in meat sheep increases CH_4 emissions measured indoor or on pasture

F. Tortereau[1], J.-L. Weisbecker[1], C. Coffre-Thomain[2], Y. Legoff[2], D. François[1], Q. Le Graverand[1] and C. Marie-Etancelin[1]
[1]GenPhySE, Université de Toulouse, INRAE, ENVT, Chemin de Borde Rouge, 31326 Castanet-Tolosan Cedex, France, [2]INRAE Experimental unit P3R, La Sapinière, 18390 Osmoy, France; flavie.tortereau@inrae.fr

Ruminants are often criticized when it comes to environmental impacts, due to CH_4 production and to the part of concentrates in the diets. A major objective in breeding ruminants is to limit both feed-food competition and greenhouse gas (GHG) emissions. We divergently selected, over 4 generations, Romane meat sheep on their Residual Feed Intake (RFI) from 3 to 5 months old under a 100% concentrate diet. We present here two results of GHG measurements (performed with sheep GreenFeed, C-Lock) on males and females belonging to these divergent lines (animals were from the RFI- (efficient) or RFI+ (inefficient) line). A total of 124 males belonging to the 3rd and 4th generations of selection, reared indoor, had both GHG and forage intake measurements. On average, males weighed 64.6 kg and emitted 1,345 g/d of CO_2 and 39.26 g/d of CH_4. Positive correlations were estimated between CO_2 and CH_4 (0.59) and between gases and forage intake (0.24 for CH_4 and 0.54 for CO_2) and body weight (0.38 for CH_4 and 0.78 for CO_2). No significant difference in body weight was observed between RFI- and RFI+ males. RFI- males ate less forage (1.13 kg/d for RFI- and 1.22 kg/d for RFI+) but they emitted significantly more CH_4 (38.86 g/d for RFI- and 37.50 g/d for RFI+), and less CO_2 (1,326 g/d for RFI- and 1,335 g/d for RFI+). We confirmed this result with 85 ewes from the 4th generation of selection, first fed indoor with dry forage and then fed on pasture: RFI- ewes emitted significantly more CH_4 than RFI+ ewes, whatever the diet. GHG emissions were positively correlated between both diets (0.48 for CH_4 and 0.56 for CO_2). Animal effects for GHG emissions, obtained from a repeatability model were significantly correlated under the two diets (from 0.66 for CH_4 to 0.78 for CO_2 intensity). Both studies highlight unfavourable relationship between feed efficiency and gas emissions: the most efficient animals ate less, but emitted more CH_4 than the least efficient ones. Moreover, phenotyping GHG indoor under a forage-based diet is of high interest when phenotyping GHG on pasture is not feasible.

Session 96 Theatre 3

Effects of sire and diet on rumen volume and relationships with feed efficiency

N.R. Lambe, A. McLaren, K.A. McLean, J. Gordon and J. Conington
SRUC, SRUC Hill and Mountain Research Centre, FK20 8RU, United Kingdom; nicola.lambe@sruc.ac.uk

The rumen plays an important role in the digestion of food by ruminants and previous research has found that larger reticulo-rumen volumes, as measured by CT scanning (RRvol), are associated with increased methane emissions. Across two years, twin-born Texel × Scotch Mule lambs from 10 sires were split across two finishing systems (diets) at weaning (~12 weeks old) for a total of eight weeks: outdoors (n=242, grazed on grass) and indoors (n=237, fed pelleted grass nuts and recorded through individual feed intake recording equipment). At the end of the study, lambs finished on grass nuts indoors were significantly heavier, grew quicker, and had more fat and muscle (as measured by ultrasound and CT scanning) at the same live weight, than lambs grazing lowland pastures ($P<0.001$ for all). Lambs grazed outdoors had significantly larger liveweight-adjusted RRvol post-trial ($P<0.001$). Sire had a significant effect on all measured traits. However, diet interactions with sire were non-significant for growth, carcass traits or RRvol, suggesting that lambs from the same sires ranked similarly in each system. Residual feed intake (RFI) was calculated for each lamb finished indoors, by adjusting average daily dry matter intake for live weights, average daily liveweight gain, fat and muscle weights, as well as sex and year. RRvol had no significant effect on RFI ($P>0.05$), regardless of whether RRvol was measured before or after the feeding trial, or whether or not it was adjusted for lamb live weight. These results add to our understanding of traits associated with feed efficiency and methane emissions from sheep systems, and the relationships among them.

Session 96 Theatre 4

Evaluating the effect of herbage composition on methane output in sheep

F.M. McGovern, P. Creighton, E. Dunne and S. Woodmartin
Teagasc, Animal and Bioscience, Animal & Grassland Research & Innovation Centre, Mellows Campus, Athenry, Co. Galway, Ireland, H65 R718, Ireland; fiona.mcgovern@teagasc.ie

Understanding the influence of diet composition on methane (CH_4) output in ruminant animal production systems is of critical importance to the development of mitigation strategies. In Ireland enteric CH_4 emissions account for 60.7% of the national greenhouse gas emissions from the agricultural sector. The aim of this study was to investigate the possible relationship(s) between CH_4 output, diet quality and dry matter intake (DMI) in sheep. A 5×5 Latin square design experiment was undertaken to investigate five dietary treatments: Perennial ryegrass (*Lolium perenne* L.; PRG) only or PRG plus white clover (*Trifoluim repens* L.), red clover (*Trifoluim pratense* L.), chicory (*Chicorium intybus* L.) or plantain (*Plantago lanceolate* L.) at a ratio of 75% PRG and 25% of the respective forage on a dry matter (DM) basis. Twenty wether sheep were housed in metabolism crates across five feeding periods. Daily offered herbage samples were collected from each animal prior to freeze drying and chemical analysis. Methane measurements were collected from each animal on the final day of each feeding period using portable accumulation chambers. The final dataset contained 600 records of individual DMI and 100 records of methane output. Data were analysed using linear mixed models with relationships between variables of interest examined using Pearson rank correlations. Dry matter intake ranged from 1.55±0.038 (PRG) to 1.76±0.038 kg DM/animal/day (PRG + Chicory). Methane output ranked highest for animals offered the PRG or the PRG + Chicory diets ($P<0.05$). Results showed a Pearson rank correlation between DMI and methane output of 0.46 while this ranged from 0.38 to 0.76 depending on the diet type offered. The overall percentage of leaf and stem offered to animals influenced methane output with correlations of -0.42 and 0.57, respectively. In addition, the regression coefficient between methane output and DMI had an R^2=062. Data from this study highlights the impact of diet quality on methane output in small ruminants and shows the importance of optimising sward management in pasture based production systems.

Session 96 Theatre 5

Dry matter intake across life stages in sheep

E. O'Connor[1,2], N. McHugh[1], E. Dunne[1], T.M. Boland[2] and F.M. McGovern[1]
[1]Teagasc, Animal and Grassland Research and Innovation Centre, Athenry, Co. Galway, H65 R718, Ireland, [2]University College Dublin, School of Agriculture and Food Science, Belfield, Dublin 4, D04 V1W8, Ireland; edel.oconnor@teagasc.ie

Dry matter intake (DMI) is a key livestock production metric as it determines not only the feed budget required for the flock but also the animals realised growth potential. The objective of this study was to investigate the relationship between estimated DMI across life stages in sheep. Dry matter intake was estimated over a seven-year period from 2016 to 2022 using the n-alkane technique on animals grazing perennial ryegrass swards or based on a silage diet during the winter indoor housing period. For DMI estimated during the grazing period, animals were orally dosed with a n-alkane bolus for 11 consecutive days, with faecal samples collected from day 7 to 12. To determine DMI on a silage based diet lambs and pregnant ewes were individually penned indoors and the quantity of feed offered and refused was measured daily. Dry matter intake on a total of 621 female animals was collected across various life stages including: as lambs (<12 months), nulliparous hoggets (12 to 24 months) and ewes (primiparous or greater; 24 to 72 months). Ewes were further classified as pregnant, lactating or dry (non-pregnant and non-lactating). Animal live-weight and methane output, coinciding with DMI measurements were also available over a three year period from 2020 to 2022. Factors affecting DMI were modelled using a linear mixed model with life stage, breed and live-weight of the animal included as fixed effects and animal included as a random effect. The average DMI (expressed in kg DM per day) measured at each life-stage was 0.91 kg DM/day (lambs), 1.17 kg DM/day (hoggets), 1.19 kg DM/day (pregnant ewes), 2.30 kg DM/day (lactating ewes) and 1.31 kg DM/day (dry ewes). Lactating ewes had a significantly higher DMI compared to all other life stages ($P<0.01$); while DMI did not differ amongst lambs, hoggets and pregnant ewes ($P>0.05$). Moderate positive correlations were estimated between DMI and methane output (0.38) and animal live-weight (0.64). Results from this study show that DMI differs across life stage and is a contributing factor to the amount of methane the animal produces.

Session 96 Theatre 6

Selecting feed-efficient sheep with concentrates alters their efficiency with forages and behaviour

C. Marie-Etancelin[1], J.L. Weisbecker[1], D. Marcon[2], L. Estivalet[2], Q. Le Graverand[1] and F. Tortereau[1]
[1]INRAE- UMR GENPHYSE, chemin de Borde Rouge, 31326 Castanet Tolosan Cedex, France, [2]INRAE- UE P3R, Domaine de la Spinière, 18390 Osmoy, France; christel.marie-etancelin@inrae.fr

Selecting to improve feed efficiency is an objective shared by all animal productions, with the aim of reducing feed consumption and rejects while maintaining production abilities. To make such a selection, the Residual Feed Intake (RFI) is the most convenient trait. We divergently selected Romane meat sheep on their RFI under a 100% concentrate diet. In the 3rd and 4th generations of selection, a total of 332 Romane males, belonging either to the RFI- (efficient) or RFI+ (inefficient) lines with a difference of 2 genetic standard deviations between lines, were phenotyped. Feed efficiency and feeding behaviour traits were recorded during 2 phases: a first phase under a 100% concentrate diet from 3 to 5 months of age and then a second phase under a mixed diet (forage *ad libitum* + 700 g of concentrate) from 6 to 8 months of age. The significant difference in RFI between lines under a concentrate diet (131 g/d) was still significant under the mixed diet although reduced (51 g/d), with RFI- being more efficient than RFI+. In terms of feed intake (FI), the RFI+ males ingested more feed than the RFI- ones whatever the diet: +123 g/d of concentrate during the first phase and +80 g/d of forage with no difference in restricted concentrate during the second phase. Between the 2 diets, correlations were of +0.27 for RFI and +0.22 for FI. During the concentrate diet phase, RFI- males had a lower daily duration of intake and fewer daily visits to the feeders. During the mixed diet phase, there was only a trend toward fewer daily visits (P=6%) at forage feeder for RFI- sheep, no difference in restricted concentrate intake behaviour, but a significantly slower water drinking rate for the efficient animals. Analysis of feeding times per hour highlighted that efficient animals ate more often after peak feeder times, about one hour later. Could efficient sheep be dominated while accessing feeders?

Session 96 Theatre 7

Genome-wise association study of footrot and mastitis in UK Texel sheep

K. Kaseja[1], S. Mucha[1], J. Yates[2], E. Smith[2], G. Banos[1] and J. Conington[1]
[1]Scotland's Rural College (SRUC), Roslin Institute Building, EH25 9RG, Edinburgh, United Kingdom, [2]The British Texel Sheep Society, Stoneleigh Park, CV8 2LG, Warwickshire, United Kingdom; karolina.kaseja@sruc.ac.uk

The aim of this study was to investigate the genetic background of footrot and mastitis using available phenotypes and genomic data provided by UK Texel Sheep Association. Initially, 10,193 genotypes underwent quality control leaving 9,505 genotypes for further analysis. Selected genotypes were imputed to a subset of 45,686 markers from 50k array, distributed on 27 chromosomes. Phenotypes were collected on 32 farms across the UK and included 9,123 records for footrot (FRT) and 4,787 records for mastitis – recorded as California Mastitis Test (CMT), which is highly correlated with the somatic cell count. Both traits were recorded on five points scale, ranging from zero indicating no infection to four indicating severe infection. De-regressed estimated breeding values (EBV) were used as pseudo-phenotypes for GWAS. No genome-wise significant SNPs were found for either trait. Three SNPs were found to be significant on the chromosome-wise level for FRT on chromosomes 19, 23 and 26 and one SNP on chromosome 17 for CMT. SNPs on chromosomes 23 and 26 were very close of the genome-wise significance level of 5.96, reaching 5.50 and 5.23, respectively. SNP reported on chromosome 26 was within the previously reported QTLs associated with health traits such as FEC worm count and FEC eggs per worm. SNP reported on chromosome 17 was within the 14 Mbp from somatic cell score QTL. These results indicate that both traits are highly polygenic. Further research which includes more genotyped and phenotyped animals with perhaps denser DNA array is required in order to increase the chance to discover potentially important SNPs affecting FRT and CMT in the UK Texel breed. This work was supported by SMARTER H2020 project no. 772787.

Session 96 Theatre 8

Genetic parameters of nematode resistance in dairy sheep

B. Bapst[1], K. Schwarz[2], S. Thüer[2] and S. Werne[2]
[1]Qualitas AG, Genetic/Genomic evaluation, Chamerstrasse 56, 6300 Zug, Switzerland, [2]Research Institute for Organic Agriculture (FiBL), Ackerstrasse 113 / Postfach 219, 5070 Frick, Switzerland; beat.bapst@qualitasag.ch

Resistance of gastrointestinal nematodes (GIN) in sheep to anthelmintics is rapidly increasing. The results of several studies indicate that selection for increased nematode resistance in sheep is possible, and the trait of interest is faecal egg count (FEC). Phenotyping of FEC is rather time-consuming and often costly, auxiliary traits that are more effective and easier would be greatly appreciated. The extent of *Haemonchus contortus* infestation is often measured by the coloration of the conjunctiva and the so-called FAMACHA© test. In our study, we aimed to test whether FAMACHA could also be used as auxiliary trait for FEC. Additionally we aimed to provide genetic parameters for FEC, milk yield and packed cell volume (PCV). We phenotyped 1,150 naturally infected Lacaune ewes on 15 commercial Swiss farms. Phenotypic correlation between FEC and FAMACHA as well as FAMACHA and PCV resulted in 0.25 (SE 0.03) and -0.35 (SE 0.08), respectively. Subsequent genetic analysis was carried out with a multi-trait animal model in order to estimate the genetic parameters of the traits mentioned above. We found moderate heritabilities of FEC, FAMACHA, PCV and milk yield in the range of 0.30 to 0.36 (SE 0.08 for all traits). The particular focus was on the genetic correlation of FEC and FAMACHA and was estimated to be 0.03 (SE 0.22). The distribution of the FAMACHA score indicated a medium to high worm infestation, but FEC was relatively low compared to other studies. Eventually the infection pressure was not high enough to yield a good genetic correlation of FEC and FAMACHA. In general, it can be concluded that the heritability for FEC is very appealing and that a selection program could be based on them. FAMACHA as an auxiliary trait could not be confirmed in this study. Either further studies need to be designed differently to better explore the relationship between FAMACHA and FEC or other possible auxiliary traits need to be sought and developed. This study was funded by SMARTER H2020 project no. 772787.

Session 96 Theatre 9

Contrasting genetic resistance to GIN on growth performance and feed efficiency of Corriedale lambs

E.A. Navajas[1], G. Ciappesoni[1] and I. De Barbieri[2]
[1]Instituto Nacional de Investigación Agropecuaria, Las Brujas, 90100, Uruguay, [2]Instituto Nacional de Investigación Agropecuaria, Tacuarembó, 45000, Uruguay; enavajas@inia.org.uy

Genetic improvement of resistance to gastrointestinal nematodes (GIN) in sheep is a well-known strategy to reduce the negative consequences of GIN, whose control is limited by resistance to anthelmintics, on sheep health and performance. Genetic selection using faecal egg count (FEC) as selection criteria is in place in sheep breeding programmes of different countries, including Uruguay. The assessment of ocular mucous membrane colour using a chart to determine anaemia (FAMACHA©), which is easier to record than FEC, is also used as indicator of GIN infection. The aim of this study was to explore the association between FEC EBV and phenotypic FEC and FAMACHA, as well as body weights (BW), body condition score (BCS) and residual feed intake (RFI) by comparing resistant (R) and susceptible (S) lambs, according to their FEC EBVs. Lambs with the lowest (R, n=83) or highest (S, n=78) 25% FEC EBV (estimated in 2023) were identified in a dataset of 317 Corriedale female lambs of four cohorts (2018-2022). Uruguayan genetic evaluation protocol was used to measure FEC. When faecal samples were taken for FEC, FAMACHA score (five classes) was assessed, as well as BW and BCS. RFI is the difference between actual and predicted feed intake based on average daily gain (ADG) and metabolic body weight (MW), using data recorded in 46-day tests. Lineal models that included year and R/S group as class fixed effects were used. Least-square mean comparison indicated that R lambs had significantly lower FEC values than S lambs($P<0.05$) as expected, as well as lower FAMACHA scores ($P<0.05$), which agrees with the moderate genetic and phenotypic correlations estimated in this breed. In contrast, non-significant differences between R and S ($P>0.05$) were found for BW, BCS and RFI. Our results are aligned with the non-significant correlations of FEC EBV with feed efficiency traits (feed intake, MW and ADG) reported in commercial lambs and in a comparison of Corriedale divergent FEC selection lines under natural and artificial GIN. Overall, improving genetic resistance to GIN, reduces FEC and FAMACHA, without unfavourable effects on production or feed efficiency.

Session 96 Poster 10

Australian Merino: animal welfare and resilience in extensive systems

M. Del Campo[1], J.L. De Araújo Pimenta[2], I. De Barbieri[1], P. Lorenze[1], F. Rovira[1] and J.M. Soares De Lima[1]
[1]INIA – National Institute of Agricultural Research, Ruta 5 km 386, 45000 Tacuarembó, Uruguay, [2]UNESP, Jaboticabal, SP, CEP 14884-900, Brazil; mdelcampo@inia.org.uy

In extensive systems, perinatal lamb's losses could be enhanced by the sum of factors such as lamb weight at birth (BW) and low resistance to climatic adversities, live weight (LWC) or body condition of the mother at calving (BCC), calving difficulty (CD), type of delivery (DT), maternal ability (MA). The objective of this project (financed by Smarter H2020 and Rumiar) was to characterize these variables in a flock under genetic evaluation, made up of 400 dams and their progeny for 4 years: 2018 to 2021. The variables were: (1) LWC; (2) BCC (scale from 0 to 5); (3) CD, scale from: 0 (without assistance and of short duration) to 4 (with veterinary assistance); (4) MA at the time of identifying the lamb (scale from 1 to 5) where 1 is when she abandons the lamb and 5 is when she is in permanent contact with the lamb; (5) TP (single, twin, triplet); (6) BW and g) survival of lambs at 72 hours after birth (S72). Ordinal and binary logistic regressions (SAS, v9.3) were performed. Main variables that affected CD were BCC, DT and BW (P<0.05). Calving without difficulty (CD=0) was associated with BCC<3.2 for single, twin or triplet births, regardless of the mother's category. Twin and triplet births presented higher CD compared to single births (P<0.05) also regardless of the mother category. Mortality average at 72 hours for the 4 years was 8.6%, mainly affected by CD, DT and BW. In relation to CD, for values of 2 (calving that needs assistance although low.) S72 was decreased to 50%. S72 was lower in multiple births (93.7% single, 91.2% twins, 68% triplets). Lambs from multiple births had a lower BW than single lambs (single 4.8 kg, twins 4.1 kg, triplets 3.4 kg, P<0.05. In the present experiment, MA becomes very important in multiple births, where S72 increases from 87.5 to 94.6% when passing from MA3 to MA4. BW has a high incidence in S72 for both sheep and ewes. In the case of ewes, this is compromised with BW values <3.8 kg. In summary, a high CD compromises the sheep welfare and would affect lambs survival mainly in multiple births, where S72 is also affected by low BW.

Session 96 Poster 11

Prediction of feed efficiency related traits from plasma NMR spectra

A. Marquisseau, F. Tortereau, N. Marty-Gasset, C. Marie-Etancelin and Q. Le Graverand
GenPhySE, Université de Toulouse, INRAE, ENVT, Chemin de Borde Rouge, 31326 Castanet-Tolosan Cedex, France; flavie.tortereau@inrae.fr

Feed efficiency is a key trait to integrate in breeding programs, particularly in order to limit the feed-food competition and the environmental impact of livestock production. The calculation of feed efficiency criteria requires individual feed intakes to be recorded, which is too expensive in small ruminants to be reasonably proposed. An option to make this trait more affordable is to predict feed intake or efficiency from a variety of predictors that can be easily recorded. As it has been evidenced that the animal metabolism is one of the main biological function underlying feed efficiency, we propose to examine the predictability of feed efficiency related traits from plasma metabolome. Plasma samples from 265 Romane male lambs fed a 100% concentrate diet were analysed with NMR. NMR spectra were divided into 877 buckets of 0.01 ppm, and the considered values were the area under the curve of each bucket. These variables were CLR transformed before multivariate analyses (sparse Partial least squares, sPLS) used for prediction. Prediction performances of feed intake and RFI were assessed through 5-fold nested cross-validation repeated 50 times, i.e. over 250 models. Accuracies of prediction from NMR buckets were compared to the accuracy obtained from body weights, growth, body composition (called zootechnical traits hereafter). As a result, we highlighted that buckets did not improve the prediction of feed intake from zootechnical traits: an average R^2 of 0.7 was obtained from zootechnical traits with or without buckets, against 0.2 from buckets only. For RFI, R^2 were below 0.1 whatever the set of predictors. Considering whole spectra did not help predict feed efficiency nor feed intake. However, the main buckets involved in RFI prediction were consistent with metabolites previously associated to feed efficiency: such as beta-hydroxyisovaleric acid or L-tyrosine.

Session 96 Poster 12

Metabolism in lambs from two feed-efficiency genetic lines subjected to different early rearing practices

G. Cantalapiedra-Hijar[1], M.M. Milaon[1], S. Parisot[2], C. Durand[2], M. Vauris[1], F. Tortereau[3] and C. Ginane[1]
[1]INRAE, UMR Herbivores, Clermont-Ferrand-Theix, 63000, France, [2]INRAE, UE0321 La Fage, Roquefort-sur-Soulzon, France, [3]INRAE, UMR13888 GenPhySE, Auzeville Tolosane, France; gonzalo.cantalapiedra@inrae.fr

The genetic selection for feed efficiency may have negative impacts on animal resilience, as both traits rely on energy-demanding mechanisms that may conflict with each other. This study aimed to investigate whether metabolism differed in lambs from two genetic lines with divergent feed efficiency when subjected to two contrasting early rearing practices. Eighty Romane ewe lambs from two divergent genetic lines for residual feed intake (RFI), were reared either indoors with artificial rearing (AR) or in the rangeland with maternal rearing (MR) from birth to 3 months. After this initial period, they were allocated by 20 lambs into four experimental rangeland plots, 2 AR and 2 MR, balanced for the RFI line effect. After 3 weeks in the plots, blood samples from 8 animals per condition (line × rearing) were randomly selected for plasma analysis (n=32). Plasma samples were analysed by LC-MS to quantify around 270 targeted metabolites. Results were analysed through Metaboanalyst software using both the statistical (two-way ANOVA) and enrichment analysis options. No interaction between the RFI line and early rearing practice was found (FDR>0.05) on plasma metabolome. Nine plasma metabolites were affected by the RFI line (P<0.05), including two belonging to the urea cycle (ornithine and citrulline) and having lower plasma concentration in efficient RFI lambs. However, none reached significance after FDR corrections (FDR>0.05). Accordingly, enrichment analysis found no significant metabolic pathway associated with the RFI line (FDR<0.05). On the other hand, several plasma metabolites belonging to the families of phosphatidylcholines, acyl-carnitines, sphingomyelins, and ceramides showed higher concentrations in AR vs MR lambs (FDR<0.05). Additionally, twelve metabolic pathways, most of them related to lipid metabolism, were found to be enriched in the AR vs MR lambs (FDR<0.05). Overall, our results do not support different metabolic regulation in lambs genetically divergent for RFI when subjected to contrasting early rearing practices and then exposed to challenging environment.

Session 96 Poster 13

Genetic link between fertility and resilience in sheep and goat divergent selection experiments

R. Rupp[1], C. Oget-Ebrad[1], S. Parisot[2], T. Fassier[3], G. Tosser Klopp[1] and S. Freret[4]
[1]INRAE, Université de Toulouse, ENVT, GenPhySE, Castanet Tolosan, France, [2]INRAE, UE0321, Domaine de La Fage, Roquefort Sur Soulzon, France, [3] INRAE, P3R, Small Ruminants Phenotyping Facility, Osmoy, France, [4]INRAE, IFCE, CNRS, Université de Tours, PRC, Nouzilly, France; gwenola.tosser@inrae.fr

The challenge for breeding is to improve resilience traits simultaneous with efficiency and reproduction. However very little information exits about possible trade-off between resilience and reproduction in small ruminants. The aim of this study was to investigate such possible trade-off in four divergent selection experiments in dairy sheep and goat. In each experiment we analysed the success to artificial insemination (AI) using dates of AI and natural mating for return to oestrus and dates of lambing/kidding. The first experiment was a divergent selection in Lacaune sheep on mastitis (SCS). The second was an experiment in Lacaune sheep selected for the SOCS2 gene associated to susceptibility to mastitis, with favourable (CC) and unfavourable (CT and TT) genotypes. The third experiment was a divergent selection in Alpine goats on mastitis (SCS). The fourth experiment was a divergent selection in Alpine goats on longevity. The data included 1,254, 5,375, 481, and 305 females and 3,671, 5,375, 788, and 617 AI events, respectively – and the fertility rate in these four experiments was 68.8, 70, 51.9, and 59.4%. We further estimated the odds ratio (OR) for success to AI using a logistic regression according to the genetic line, with year and lactation number as fixed effects. No effect on AI fertility was associated with selection on longevity in goats or mastitis (SCS) in sheep. However, an important favourable effect on AI fertility was associated: (1) with divergent selection on mastitis (SCS) in goat: $OR_{SCS-\ vs.\ SCS+}$ =1.65 (1.22-2.23); and (2) SOCS2 genotype in sheep: $OR_{CC\ vs.\ Tx}$=1.36 (1.08-1.70). The results suggest that selection for resilience in dairy sheep and goats is not associated with an adverse effect on reproduction in these species. On the contrary, selection for some resilience traits seems to drive significant beneficial effects in terms of fertility after AI. This study has received funding from the European Union's Horizon 2020 research and innovation program under grant agreement No 772787 (SMARTER).

Phenotypic and genetic variability of health and welfare traits in French dairy goats

I. Palhiere[1], A. Bailly-Salins[2], A. Gourdon[2], M. Chassier[3], R. De Cremoux[3], M. Berthelot[4] and R. Rupp[1]
[1]GenPhySE, Université de Toulouse, INRAE, INPT, ENVT, Chemin de Borde Rouge, 31326 Castanet-Tolosan, France, [2]Capgenes, 2135 Route de Chauvigny, 86550 Mignaloux-Beauvoir, France, [3]French Livestock Institute, 149 Rue de Bercy, 75595 Paris Cedex 12, France, [4]French Agency for Food, Environmental and Occupational Health and Safety (Anses), 60 rue de Pied de Fond, 79024 Niort, France; rachel.rupp@inrae.fr

The objective of the study was to investigate the phenotypic and genetic variability of health and welfare traits in French dairy goats. A total of 1,977 primiparous goats of Alpine and Saanen breeds, from 14 farms, were involved in the study during the years 2020-2021. Eleven indicators were assessed individually and once for each goat: abscess, arthritis, nasal discharge, ocular discharge, dirty and light soiling hindquarters, lameness, body condition, bag-shaped udder, dehorning issues, hair coat condition, claw issues. For the study, all the indicators were considered as binary traits (0: absence of disorder; 1: presence of disorder). The total number of disorders per animal was defined as the sum of the disorders for the eleven indicators. The disorder frequency ranged between 0.5 and 23% depending on the 11 indicators. Two groups of indicators were observed: those with a frequency lower than 5% (n=7) and those with a frequency about 20% (n=4: dehorning issues, bag-shaped udder, abscess, claw issues). The total number of disorders per goat was 0.94, on average in the total dataset, and ranged between 0 and 5. 40% of goats showed no disorder. Heritabilites were estimated using linear models for 4 traits suspected to be under genetic control and with sufficient frequency (>5%). They ranged between 4% (abscess) and 26% (bag-shaped udder). Arthritis and the total number of disorders had intermediate values (11 and 15%, respectively). These first results suggest that selection may be a potential strategy to improve health and welfare traits in dairy goats. This study has received funding from the European Union's Horizon 2020 research and innovation program under grant agreement No 772787 (SMARTER).

Assessment of phenotypic and genetic variability of rumen temperatures in goats

I. Palhiere[1], C. Huau[1], T. Fassier[2], R. Rupp[1] and L. Bodin[1]
[1]GenPhySE, Université de Toulouse, INRAE, INPT, ENVT, Chemin de Borde Rouge, 31326 Castanet-Tolosan, France, [2]P3R, INRAE, Small Ruminants Phenotyping Facility, Osmoy, 31326, France; rachel.rupp@inrae.fr

The aim of this study was to identify the factors of variation of rumen temperatures in goats and to estimate the genetic parameters of that trait. 97 Alpine goats from an INRAE experimental farm were monitored continuously during their first lactation (267 days on average, from kidding to dry off) using Medria rumen temperature boluses (ThermoBolus San'Phone®). Temperatures were collected every 5 min (288 records per day) and have been corrected to account for drinking events. The average rumen temperature was 39.68±0.51 °C on the whole data set (n=7,273,000). The average temperature during the day increased from about 5.00am to about 5.00pm and decreased afterwards: the minimum temperatures (39.28 °C) were observed in early morning (between 5.00am and 5.30am) whereas the maximum temperatures were observed in late afternoon (39.94 °C between 4.30pm and 5.00pm). In addition to the time, the phenotypic variability was also explained by the date, the birth year, the stage of lactation and the milk yield of the goat. Genetic parameters were estimated using a linear repeatability model. Data were spread into 5 periods according to their raw mean temperature and standard deviation (4.30am-6.00am, 6.40am-1.20pm, 3.30pm-8.00pm, 9.00pm-3.00am) and each one was analysed separately. Repeatability varied from 0.20 (4.30am-6.00am) to 0.32 (3.30pm-8.00pm), and heritability from 0.03 (4.30am-6.00am) to 0.12 (9.00pm-3.00am). For this last period, breeding values ranged from -0.10 to +0.14 °C. These results suggest that body temperature measured by internal sensors has a genetic control, which is potentially linked to basal metabolism or response to various stressors (heat stress, infection). This study has received funding from the European Union's Horizon 2020 research and innovation program under grant agreement No 772787 (SMARTER).

Session 96 Poster 16

Heritability of novel metabolite-based resilience biomarkers in dairy goat

M. Ithurbide[1], T. Fassier[2], M. Tourret[3], J. Pires[3], T. Larsen[4], N.C. Friggens[5] and R. Rupp[1]
[1]INRAE, Genphyse, Castanet Tolosane, France, [2]INRAE, P3R, Small Ruminants Phenotyping Facility, Osmoy, France, [3]INRAE, Université Clermont Auvergne, Vetagro Sup, 3UMR Herbivores, Saint-Genès-Champanelle, France, [4]Aarhus University, Faculty of Agricultural Sciences, Research Centre Foulum, 8830 Tjele, Denmark, [5]INRAE, AgroParisTech, Université Paris-Saclay, UMR 0791 Modélisation Systémique Appliquée aux Ruminants, Paris, France; marie.ithurbide@inrae.fr

The aim of this study was to estimate the heritability (h^2) of metabolite-based resilience biomarkers in dairy goat. Metabolites were measured repeatedly during two periods of stress: around parturition and during a feeding challenge (48 h with straw only) in early first lactation in two INRAE facilities (P3R Bourges and Paris). The 4 blood metabolites were: glucose (Glu), beta-hydroxy-butyrate (BOHB), urea and non-esterified fatty acids. The 14 milk metabolites were: BOHB, Glu, urea, glucose-6-phosphate (Glu6P), galactose, isocitrate, glutamate, NH_2-groups, lactate dehydrogenase (LDH), choline, malate, urate, triacylglycerol, and cholesterol. The metabolite trajectories were described both by simple mean concentration per challenge period and by a functional PCA method. Variance components were estimated using an animal model (wombat®) for blood metabolites around kidding, and for trajectories of blood and milk metabolites upon the feeding challenge on 201, 228 and 138 goats respectively. The model included the fixed effects of facility and year. The total pedigree included 1,148 animals. We found 25 blood and milk metabolites parameters (out of 159) that were significantly heritable (h^2-2SE >0) with h^2 estimates ranging from 0.30 (±0.03) to 1.00(±0.25) with SE from 0.03 to 0.33. Among them: milk urate, BOHB, LDH, and Glu6P during feeding challenge and blood Glu around kidding. The heritability estimation of 250-d milk yield, fat and protein content were respectively 0.26 (±0.16), 0.53 (±0.19) and 0.68 (±0.19), showing the consistency of our dataset despite the small number of individuals. These results show the potential of metabolite-based biomarkers for genetic selection of resilience. This study has received funding from the European Union's Horizon 2020 research and innovation program under grant agreement No 772787 (SMARTER).

Session 96 Poster 17

Rumen size of sheep: difference between a modern and a native Norwegian sheep breed

B.A. Åby[1], M.A. Bhatti[2] and G. Steinheim[1]
[1]NMBU, Department of Animal and Aquacultural Sciences, Box 5003, 1432 Ås, Norway, [2]NMBU, Department of International Environment and Development Studies, Box 5003, 1432 Ås, Norway; bente.aby@nmbu.no

Globally, sheep account for approx. 4% of the GHG emissions from the livestock sector, mainly through enteric methane (CH_4) emissions. Rumen size is shown to be relevant for enteric CH_4 emissions, with lager rumen volumes being associated with higher emissions. Breed variation in rumen size, relative to body size, may thus be one factor contributing to between-breed differences in CH_4 emissions. Two of the most contrasting sheep breeds in terms of selection history in Norway are the Norwegian white sheep (NWS; large, long tailed composite breed) and the Old Norwegian Spæl (ONS; small Nordic Short-tailed landrace). These breeds were previously compared in two experiments at NMBU, where adult ONS ewes had lower enteric CH_4, corrected for dry matter intake, when fed harvested grass silage and fresh cut grass. In this pilot study, we therefore hypothesized that there would be significant differences in rumen size between the breeds. The test animals (23 adult ewes, 40 lambs) came from two farms in Viken county rearing both breeds. On October 18, 2022, the animals were slaughtered at a commercial abattoir (Nortura Gol) shortly after the summer grazing period. Rumens were collected and ruminal contents were removed before the organs were cleaned and visceral fat removed. The rumens were then weighed after being squeezed dry. The data for lambs and adult ewes were analysed separately using two general linear models. Both models included the effects of carcass weight, breed, and farm. Additionally, the effect of birth year was included for adult ewes, while age at slaughter was included for lambs. For adult ewes, rumen weight corrected for carcass weight, was lower ($P<0.05$) for ONS compared to NWS (LSMEANS 1,589 vs 1,832 grams), while there were no breed differences for lambs. Our results suggest that previously observed breed differences between ONS and NWS in enteric CH_4 emissions (i.e. lower enteric methane emissions, corrected for dry matter intake, for ONS) may be linked to differentiation in rumen size. However, this needs to be confirmed in a controlled study also including enteric CH_4 measurements on the test animals.

Diurnal feed intake pattern of two sheep breeds fed different silage qualities

I. Dønnem, B.A. Åby and G. Steinheim
Norwegian University of Life Sciences, Faculty of Bioscience, Oluf Thesens vei 6, 1433 Ås, Norway; ingjerd.donnem@nmbu.no

Sheep farming in Norway is mainly based on using grass silage as feed in the indoor period during winter. When formulating a feed ration based on silage for sheep, knowledge about feed intake is important. Feed intake is affected by factors such as size and anatomical differences between animals, silage characteristics, and feeding management. The aim of this study was to investigate effects of grass silage quality and sheep breed on feed intake and the diurnal intake pattern. Grass silage qualities investigated were: (1) early; cut in the boot stage; and (2) average; cut in the heading stage. The two breeds were the modern Norwegian White Sheep (NWS), selected for production and maternal traits and the Old Norwegian Spæl (ONS), a lighter, rural breed. Mean weights of NWS and ONS were 91 and 60 kg. Mature ewes in early gestation, 20 of each breed, were individually fed the two different silage qualities *ad libitum* in three weeks periods in a 2×2 crossover design. Feed was offered morning and afternoon, and intake was continuously logged throughout the experiment, by an automatic system from BioControl. The results showed significant differences in feed intake between silage qualities. Average daily dry matter intake of early and average silage was 2.48 and 1.84 kg for NWS, and 1.34 and 1.14 kg for ONS. The diurnal feed intake pattern was different between the two silage qualities. The average harvested silage had two distinct peaks during the day, coinciding with feeding hours. Early harvested silage had a more even curve during the day and had higher frequency of eating bouts. The ewes also consumed more of this quality at night. Rumination behaviour was not measured, but the results indicate that silage intake is limited as a result of increased rumination time of low-quality silage. The feed intake pattern was similar for the two breeds, however the ONS had higher frequency of eating bouts, but lower intake per bout. This reflects the anatomically and behaviourally differences between the two breeds, including the ONS having a smaller rumen relative to size. In conclusion, improved silage quality increased eating frequency during both day and night, and breed differences should be accounted for when predicting silage intake.

Author index

A

Aaskov, M. 800
Aass, L. 753
Abanikannda, O.T.F. 195, 198, 199, 336, 342
Abbate, S. 658
Abbott, D.W. 560
Abdedaim, N. 192
Abdel-Shafy, H. 386
Abd El-Wahab, A. 845
Abdennebi-Najar, L. 589
Abe, H. 805
Abe, T. 233
Abomselem, S. 801
Abraham, R.K. 315
Abreu, M.J.I. 713
Abreu, M.L.C. 259
Abuoul Naga, A. 722
Åby, B.A. 753, 1039, 1040
Accatino, F. 557, 847
Acciaro, M. 249, 502, 653, 654, 656, 657
Achkakanova, E. 180
Achour, I. 419
Acloque, H. 482, 924
Adachi, K. 519
Adamaki-Sotiraki, C. 865, 872
Addes, M. 450
Addo, S. 973
Adebambo, S.M. 198
Adebambo, S.O. 195
Adekale, D. 786
Aden, D. 962
Adepoju, D. 914
Adhimoolam, A.K. 286
Adhimoolam, K.A. 1004
Adjassin, J.S. 897
Adler, R.A. 617
Adriaens, I. 494, 495, 773
Aerts, C. 348
Aerts, M. 478
Afonso, A. 780
Agabriel, J. 331
Agazzi, A. 285, 354
Agboola, L. 200
Agenäs, S. 154, 365, 933
Agouros, A. 207
Aguerre, S. 665, 809, 929
Aguilar, I. 787
Aguilar, M. 905
Aguirre-Lavin, T. 779
Aguirre-Saavedra, O. 425
Agyemang, K. 590
Agy Loureiro, B. 216
Ahmadi, B. 969
Ahmad, S.M. 480
Ahmad, Z.A. 342
Ahmed, R.H. 528
Ahmed, S. 952
Ahmed Waqas, M. 265
Ahn, J.W. 227
Ahvenjärvi, S. 323, 333
Aihara, M. 283
Ait-Sidhoum, A. 628
Ajasa, A. 793
Ajmone Marsan, P. 230
Akachi, K. 661, 667
Åkerlind, M. 939
Akhavan, N. 1008
Akraim, F. 604
Akram, M.Z. 610
Alawneh, J. 565, 932
Albanell, E. 506
Alban, L. 206
Albechaalany, J. 595, 972
Albert, F. 825, 826, 827, 942
Alberto Stanislao Atzori, A.A.S. 273
Alborali, G.L. 1022, 1023
Albouy-Kissi, B. 316
Albuquerque, L.G. 282
Aldai, N. 454
Alderkamp, L.M. 173
Aldridge, M.N. 148, 277
Aleksovski, G. 622
Alexandre, P.A. 487, 632
Alexandri, R. 313
Alfaia, C. 186
Alfonso, L. 429
Alhamada, M. 250
Aliakbari, A. 200
Alkhoder, H. 786
Allain, C. 376, 486, 520, 626, 704
Allais, S. 794
Allart, L. 440, 442
Allen, K. 860
Almadani, M. 731
Al-Marashdeh, O. 438
Almeida, A.M. 186, 493, 518, 734, 833, 839
Almeida, J.C. 685
Almeida, J.M. 295, 297, 426
Almeida, K.V. 842
Almeida, M. 185, 780
Al-Soufi, S. 425, 426, 716
Aluwé, M. 355, 360, 417, 471, 974, 988
Alvanitakis, M. 848
Alvarez Hess, P.S. 294, 1005
Álvarez, I. 386, 388
Alvarez Munera, A. 787
Álvarez-Rodríguez, J. 994
Álvarez, S. 505
Alves, M. 294
Alves, S.P. 484, 685
Alves, T.C. 772

Alyagor, I.A. 617
Amalfitano, N. 902, 940
Amalraj, A. 597, 600
Amanatidis, M. 912
Amâncio, B.R. 574, 575, 583, 584
Amaral, A.J. 824
Amaral, I. 480
Ambra, C.D. 1022, 1023
Ambrosino, S. 425
Amelchanka, S.L. 569, 570, 840
Amer, P.R. 801
Amin, K. 985
Ammer, S. 415, 645
Ammon, C. 772, 775
Amon, T. 756, 769, 772, 775
Ampe, B. 988
Amposta, N. 442, 704
Ampuero Kragten, S. 375, 797
Amstutz, F. 834
Amyes, N. 278, 1031
Anagnostopoulos, A. 767, 1019
Anagnostopoulos, E.C. 286, 836, 855
Anastasi, M. 942
And Almeida, A.M. 596, 604
Andennebi-Najar, L. 890
Andersen, I.L. 238
Andersen, J. 288
Anderson, F. 204, 312, 313, 314, 316, 428
Andersson, B. 798
Andonov, S. 622, 978
Andreadis, M. 943
André, C. 398
Andreou, S. 253, 255, 804
Andrés, S. 235, 1001
Andretta, I. 983, 985
Andrews, L. 264
Andriamandroso, A.L.H. 245
Andrieu, B. 569
Andrieux, C. 997
Andueza, D. 179, 752
Angel, C.R. 960
Angeles-Hernandez, J.C. 259, 260
Angel, J. 963
Angellotti, M. 955
Ângelo, M. 480
Anger, B. 214
Anger, J.C. 332, 835
Angevin, F. 156
Anglhuber, C. 920
Angón, E. 270
Anita, A. 343
An, J.W. 214
An, N.R. 525
Anselmet, S. 728
Anselmo, A. 348
Antas, C. 1014
Antia, R. 756
Antoni-Gautier, A. 553
Antonini, M. 675, 918
Antonios, S. 382
Antonis, A.F.G. 346
Antonopoulou, E. 218, 228, 350
Antúnez, L. 423
Anwar, K. 232
Anzai, H. 248
Aparicio, P. 446
Appel, A.K. 415, 812
Appel, M. 346
Apps, R. 310
Aquilani, C. 678
Arablouei, R. 889
Arango, S. 574
Arantes, A. 185
Araujo, M. 622
Araújo, R.D. 490
Arbizu, C. 393
Arce, N. 358
Archetti, I.L. 1000
Archibald, A.L. 924
Archimede, H. 289
Arends, D. 386
Ares, G. 423
Arévalo Sureda, E. 736
Argenti, G. 678
Argüello, A. 505
Argyriadou, A. 555, 671
Arias Escobar, M.A. 823
Arias, K.D. 386, 388
Arles, S. 704
Armone, R. 284
Arnal, M. 811
Arnaud, E.A. 337, 612
Arndt, C. 148, 263, 273, 729
Arnould, C. 886
Arquet, R. 289
Arranz, J.J. 629
Arsenos, G. 355, 369, 528, 529, 531, 555, 671, 672, 917
Arshad, M.A. 842
Arturo-Schaan, M. 573, 1009
Artus, J. 482
Asadollahpour Nanaei, H. 169
Asakuma, S. 686
Asenov, A. 196
Ask-Gullstrand, P. 923
Aspeholen Åby, B. 274
Assouma, M.H. 699, 700, 702, 706
Astaptsev, A. 281
Astessiano, A.L. 640, 999, 1000
Astruc, J.M. 382, 510, 802, 852
Ataallahi, M. 359, 855, 890, 1020
Atamer Balkan, B. 327
Atanassova, S. 179
Atashi, H. 803, 1025, 1027
Athanasiadis, S. 636
Athanassiou, C.G. 218, 227, 228, 350, 865, 872
Attig, I. 214
Atxaerandio, R. 298, 299

Atzori, A.S. 274, 299, 323, 327, 633, 653, 654, 656, 657, 658, 659
Aubert, H. 756
Aubert, P.-M. 823
Aubineau, T. 1014
Aubin, J. 169
Aubron, C. 850, 851
Aubry, A. 269, 563
Auclair-Ronzaud, J. 453, 458, 861
Audebert, C. 516
Auer, F.J. 413, 1024, 1026
Auffret, M.D. 625
Aupiais, A. 496, 890
Auppakhun, W. 197
Auvray, A.A. 361
Avadi, A. 957
Avril, C. 510
Ayala, M.C. 268
Ayanfe, N. 333
Ayres, L. 785, 791
Azad, M. 385
Azevedo, G.N. 772
Azevedo, I. 529
Azevedo, J. 185
Azhir, D. 568
Aznar, M.J. 225
Azor, P.J. 978, 981
Azzena, M.G.D. 299

B

Babbucci, M. 928
Bachmann, I. 650
Bacou, E. 834
Baczkiewicz, K. 812
Badino, P. 217
Badr, A. 801
Baert, J. 932
Baes, C.F. 159, 275, 406, 527
Baeten, V. 348, 478
Båge, R. 923, 936, 939
Bagnato, A. 168
Bahloul, L. 901
Bahrndorff, S. 618
Bai, B.Q. 433
Baik, M. 187, 489, 583
Bailly-Caumette, E. 822
Bailly, J. 243, 379, 756
Bailly-Salins, A. 1038
Bailly, Y. 482
Bailoni, L. 574, 898
Bai, M.M. 580, 765
Bainville, S. 850
Bain, W. 281
Bakare, K.O. 199
Bakke, K.A. 277, 374
Bakker, W. 567
Bakuła, T. 226
Balan, W. 755
Balcells, J. 305, 306, 1005
Balcha, E. 273
Baldan, S.B. 907
Baldassini, W.A. 236, 522, 837
Baldi, A. 285, 695
Baldinger, L. 730, 731
Baldwin, R. 256
Balegi, R. 959
Balikowa, D. 590
Baliota, G. 865
Ballan, M. 923
Ballard, V. 579
Ballester, M. 925
Balzar, L. 670
Banchero, G. 726
Banhazi, T. 644
Bannister, I. 715
Banos, G. 355, 528, 531, 671, 1034
Bapst, B. 1035
Barahona, M. 422, 427
Barasc, H. 781
Barbari, M. 303
Barbat, A. 664
Barbé, F. 745
Barber, D.G. 565, 932
Barbet, M. 167, 999
Barbey, S. 413
Barbier, E. 749
Barbieri, S. 499, 535
Barbier, L. 717
Barchilon, N. 595
Barclay, D. 317
Barde, D. 289
Barden, M. 1019
Bargelloni, L. 928
Barioni Junior, W. 705
Barker, J. 860
Barlier, E. 465
Barnes, A. 497, 1017
Barnes, K.D. 842
Baro, A. 700
Barragán-Fonseca, K.B. 342
Barreto, A.N. 705, 772
Barreto-Mendes, L. 334, 689, 997
Barrett, D. 891
Barrey, E. 975
Barrieu, J. 232
Barro, A. 309, 318, 433
Barron, L.J.R. 454
Barroso, H. 480
Bar-Shamai, A. 501, 503, 776, 777, 778
Bartlewski, P.M. 969
Barzola Iza, C. 557
Basarb, J. 926
Basdagianni, Z. 509, 703, 912, 943
Bassignana, M. 675
Basso, B. 619
Basterra-García, A. 803
Bastianelli, D. 697, 699, 702, 706
Bastien, A. 877

Bastien, D. 365, 373, 551, 554, 562
Bastin, C. 803
Batorek-Lukač, N. 930
Battacone, G. 967
Battaglia, A. 603
Battaglini, L. 887
Batta, U. 343
Battelli, M. 1012
Battheu-Noirfalise, C. 815
Battini, M. 887
Baude, B. 885
Baude, R. 479
Bauer, J. 787, 808
Baumgartner, M. 650
Baumgärtner, W. 335
Baumont, R. 156, 331, 586
Baumung, R. 264, 381, 718
Bausson, C. 545
Bayat, A.R. 323, 333, 897, 935
Bazan, S. 697
Beal, T. 153
Beatson, P. 275
Beauchemin, K.A. 560
Beauclercq, S. 590, 862
Beaumatin, F. 193
Beaumont, M. 230
Becciolini, V. 301, 303
Becker, D. 927
Becker, E. 514
Becker, J. 533, 630
Beckers, Y. 815, 897
Bédère, N. 416, 782, 810
Bedford, C. 1019
Bednarczyk, M. 234, 411
Bédoin, F. 628, 823, 825, 826, 827, 829, 942
Beechener, S. 497, 503
Bee, G. 212, 338, 339, 461, 463, 472, 494, 495, 608, 611, 612, 1019
Beglinger, C. 249
Béguin, E. 545
Belaid, M.A. 262
Belanche, A. 563, 938, 994, 1007
Belay, T.K. 785
Beline, F. 709, 854
Bellagi, R. 328
Bellamy, L. 543
Bellanger, D. 985
Bell, D.J. 532, 538
Bellezza Oddon, S. 217, 219, 225, 475, 868, 872
Bell, J. 175, 656, 931
Belloc, C. 464, 1018
Belloir, P. 745, 762
Belloli, A.G. 530
Bellon, J. 1019
Belloumi, D. 226, 760
Belo, A.T. 971
Beltramo, M. 459
Beltran De Heredia, I. 776, 777
Belz, E. 213, 468
Belz, J. 581
Benaissa, S. 771
Benaouda, M. 894
Benarbia, A. 213
Benarbia, M.E.L.A. 468, 735
Benavides, C. 798
Ben Braiek, M. 398
Benchaar, C. 846
Bendtsen, S.B. 800
Benedeti, P.D.B. 575
Benedetti Del Rio, E. 720
Benet, B. 967
Bénézet, C. 457
Benhamou-Prat, A. 720
Benito-Diaz, A.B.-D. 462, 880
Ben Meir, Y. 966
Bennewitz, J. 372, 408, 417, 514, 517, 526, 664
Benni, S. 973
Benoist, A. 848, 851
Benoit, M. 184, 846
Benoit, O. 247
Ben Sassi, M. 229
Ben Zaabza, H. 804
Berard, J. 565
Bérard, J. 244, 249
Berchoux, A. 327, 570, 579, 585
Berger, B. 915
Berger, M. 586
Berger, Q. 416, 810
Bergh, A. 201
Berglund, B. 923
Berglund, P. 978
Bergman, J.G.H.E. 334, 560, 576, 577, 955
Bergmann, T. 645
Bergoug, H.B. 194, 218
Berg, P. 384, 391
Béri, B. 816
Bermann, M. 782, 787, 790
Bermejo-Poza, R. 370, 660
Bernadet, M.-D. 232
Bernalier-Donadille, A. 834
Bernardeau, M. 737
Bernardi, A.C.C. 705, 954
Bernardi, C.E.M. 338
Bernardi, O.B. 882
Bernard, L. 255, 261, 485
Bernard, M. 263, 738, 778, 779
Berndt, A. 561, 954
Bernier Gosselin, V. 630
Bernstein, R. 619
Béroulle, V. 728
Berrens, S. 346, 867
Berri, C. 595, 605, 730, 823, 825, 826, 827, 946, 972
Berrocal, R. 290, 422
Berry, D. 817, 946
Bertelsen, M. 818
Berteselli, G.V. 498, 499, 503, 535
Berthelot, M. 1038
Berthelot, V. 325, 332, 1006

Bertin, A. 886
Bertocchi, L. 367, 1000, 1022, 1023
Bertolini, F. 513, 920, 923
Berton, M. 192, 825, 826, 827, 940, 944
Bertram, H.C. 611
Bertrand, C. 664
Bertrand, E. 271
Bertrand, N. 983
Bertron, J.J. 180, 677
Beruete, M. 424
Berzaghi, P. 959
Bes, A. 561
Besbes, B. 718
Besche, G. 247, 878
Besharati, M. 568
Besnard, F. 161
Bes, S. 485, 991, 992, 993
Bessa, R.J.B. 295, 297, 734, 833, 839
Besson, M. 891
Bethard, M. 532
Bharanidharan, R. 583, 950
Bhatia, A. 343
Bhati, M. 927
Bhatti, M.A. 1039
Bhérer-Breton, P. 274
Biada, I. 515
Bianchini, M. 244
Bian, P.P. 523
Biard, K. 330
Biasato, I. 217, 219, 868, 872
Biau, S. 457
Bickhart, D. 615, 627
Bidanel, J. 200, 201, 208
Bidan, F. 904, 906
Bidaux, R. 835
Bieber, A. 556, 810, 820
Bièche-Terrier, C. 548
Bieger, J. 771, 1028
Biesse, M. 1023
Biffani, S. 922
Bigot, G. 452
Bijma, P. 159, 171, 379, 406, 783, 921
Bikker, P. 462, 844, 960
Biliaderis, C.G. 943
Billaudet, L. 556
Billet, M. 1004
Billon, Y. 200, 230, 243, 379, 483, 491, 492
Bilton, T. 278, 1031
Bindelle, J. 245
Bink, M.C.A.M. 924, 925
Binnendijk, G. 844
Birkner, J. 728
Birnie, J. 269
Birolo, M. 740
Birwal, P. 385
Biscarini, F. 710
Bishop-Hurley, G. 889
Bisutti, V. 190, 230, 634
Biswas, S. 600, 742, 743, 879, 990
Bittante, G. 902
Bittar, C.M. 712
Bizjak, M. 654
Bjerring, M. 278, 539
Blache, M.-C. 886
Blackburn, H. 804
Blaga, R. 1022
Blair, A. 172, 755
Blanc, F. 334, 647, 689, 991, 997
Blanchard, M. 698, 700
Blanchard, T. 626, 717
Blanchet, B. 208
Blanc, L. 745
Blanco-Alibes, M. 650
Blanco-Doval, A. 454
Blanco-Penedo, I. 270, 1016
Blanfort, V. 697, 726
Blanvillain, V. 577
Blasco, A. 483, 515
Blas, E. 216
Błażejak-Grabowska, J. 226
Blazy, V. 986
Blichfeldt, T. 415, 655
Bloch,.V. 968
Blom, Y. 977
Bloor, J. 174, 680
Blouin, M. 587
Bluet, B. 581, 911
Bluy, L.E. 232
Boby, C. 482, 485
Boccardo, A. 354, 530
Bochu, J.L. 726
Bocquier, F. 250
Bodas-Rodríguez, R. 880
Bodin, L. 232, 1038
Boerboom, G.M. 708
Boettcher, P. 264, 381
Bohnert, D.W. 246
Bohuon, E. 751
Boichard, D. 161, 229, 231, 276, 407, 409, 413, 519, 664, 781, 790, 929
Bois, B. 699, 706
Boison, S. 793
Boissard, K. 652
Boivent, C. 719
Boivin, X. 768
Bokkers, E.A.M. 497, 498, 501, 533
Boklund, A. 206
Boland, T. 437, 564, 657, 676, 688, 904, 1033
Bolduc, N. 823
Boll, E.J. 592
Bolner, M. 923
Bompa, J.F. 756
Bonacini, M. 920
Bonani, W. 727
Bonardi, C. 824
Bonato, M. 405
Bonekamp, G. 674
Bonelli, F. 652

Bonfatti, V. 209
Bongiorno, V. 219
Bonhoff, B. 989
Bonifazi, R. 158, 785, 926
Bonilauri, P. 873
Bonilha, S.F.M. 282, 583, 584, 704, 795
Bonilla-Manrique, O. 945
Bonin, C. 508
Bonnafe, G. 779
Bonnal, L. 699, 706
Bonnard, L. 304, 438
Bonneau, M. 379, 968
Bonnefont, C.M.D. 232, 483, 491, 492
Bonnefous, C. 605, 729, 730
Bonnet, A. 232, 483, 491, 492, 789
Bonnet, M. 189, 255, 316, 482, 485, 488, 683, 992, 993, 999
Bonnin, C. 457
Bonos, E. 218, 228, 350, 764
Bontempo, L. 440
Bontempo, V. 710
Boonanuntanasarn, S. 197
Boonen, J. 540
Bord, C. 680
Bordignon, F. 192, 369, 740
Borek, R. 265
Boré, R. 271, 272, 276, 579, 841
Borges, M.S. 282
Borghuis, A. 346, 347
Børsting, C.F. 177, 1012
Borsuk-Stanulewicz, M. 571, 582
Bos, A.P. 267
Bosco, D. 623
Bosi, C. 727
Bosi, P. 339, 715
Bosoni, G. 367
Bossis, I. 943
Boucherot, J. 779
Bouchet, M. 453
Bouchez, O. 230
Bouchon, M. 328, 547, 680
Boudesocque-Delaye, L. 861
Boudes, P.B. 453
Boudinot, P. 192
Boudon, A. 330, 709, 717, 959, 964, 993, 1014
Boudrez, M. 865
Boudry, P. 669
Bouglé, L. 371
Bouhallab, S. 890
Bouhuijzen Wenger, J. 916
Boukouvala, E. 917
Boukrouh, S. 510
Boulbria, G. 206, 647, 970, 1015
Boulestreau-Boulay, A.L. 985
Boulet, L. 545
Boulling, A. 407, 522
Boumans, I.J.M.M. 497, 498, 501
Bouqueau, A. 998
Bouquet, A. 407
Bourassa, R. 582
Bourasseau, M. 658
Bourgeois-Brunel, L. 522
Bourin, M. 595, 825, 826, 827
Boussaha, M. 161, 229, 231, 407, 519, 522
Boutinaud, M. 791, 993, 998, 1027
Boutou, O. 558
Bouwhuis, M.A. 716, 996
Bouwknegt, M. 974
Bouwman, A.C. 380, 614, 926
Bouy, M. 506
Bouzalas, I.G. 917
Boval, M. 324
Bovo, M. 973
Bovo, S. 513, 623, 920, 923
Bovreg Consortium 926
Boyd, C. 246
Boyer, C. 258, 659, 728
Boyer, S. 451
Boyle, L. 239, 240, 646, 732
Boyle, L.A. 215, 238
Bozakova, N. 182
Bozakova, N.A. 182
Bozzi, R. 824
Braamhaar, D.J.M. 567
Bradley, D. 165
Bragina, L. 654, 655, 656
Brajon, S.B. 557, 558
Brake, M. 871
Brambilla, F. 192
Brameld, J. 476, 748
Branco, R.H. 561, 574, 575, 583, 584, 795
Brand, T. 188, 243, 368, 651
Brard-Fudulea, S. 385
Brasseur, C. 272
Brasseur, M.A. 677
Braun, E. 527
Bravin, M.N. 957
Bravo, C. 491, 492
Braz, F. 971
Breceda, A. 721
Breen, J. 269, 681
Breen, V. 158
Breitsma, R. 907
Brenaut, P. 789
Brenig, B. 830
Brennan, J. 172
Brennan, J.R. 326, 679, 755
Bretaudeau, A. 762
Breton, S. 779
Briand, P. 540, 1021
Bric, M. 178
Bridi, A.M. 309
Bridier, A. 214
Briefer, S. 368, 888
Brien, M. 885
Briens, M. 870
Brinke, I. 989
Brisseau, N. 371, 536

Brisson, G. 587
Brito, A.F. 842
Brito De Araujo, D. 713, 714, 955
Brito, G. 423
Brito, J.A.A. 480
Brocard, V. 540, 545, 815, 821, 899
Brocart, M. 728
Brocas, C. 666
Brockmann, G.A. 341, 386, 781, 919
Broeckx, L. 867, 874
Broekema, R. 265
Bronzo, V. 530
Brookes, D. 1017
Brossard, L. 970
Brossaud, A.B. 882
Brosse, C. 828
Brouard, S. 677
Broudiscou, L.P. 332, 1006
Brouklogiannis, I.P. 286, 836, 855
Brouzes, C. 867
Brown, D.J. 170, 313, 380, 401, 673
Browne, N. 742, 746, 815
Brscic, M. 364
Bruand, C. 1030
Bruckmaier, R. 488, 678, 992
Bruder, T. 481
Bruins, M.E. 346, 347
Brulin, L. 516
Brun, A. 296, 319
Bruneau, N. 738
Bruneteau, E. 156
Brunetti, H.B. 727, 954
Brunet, V. 183
Bruni, M.A. 573
Brunschwig, G. 549, 903, 1017
Bruyas, M. 376
Bry-Chevalier, T. 691
Brzoza, A. 199
Bubnič, J. 476, 620
Buccioni, A. 565
Buchet, A. 424
Buchli, C. 535
Buckley, C. 653, 654, 655, 656
Buckley, F. 628
Buczinski, S. 582
Budan, A. 643, 974
Bu, D.P. 573
Buffler, M. 965
Bühl, V. 825, 826, 827
Bui, H. 468
Buisson, D. 398
Buisson, L. 908
Bulness, M. 952
Bunel, A. 972
Bungenstab, E. 577
Bureau, D.P. 191, 958
Burger, P. 381
Burgers, E. 534, 992
Burgos, A. 905, 909
Burke, C.R. 513, 801
Burlot, T. 416, 647, 810
Burmańczuk, A. 537
Burren, A. 318, 806, 881
Bus, J.D. 497, 501
Büşra, A.T. 323
Bussiere, F.I. 223
Bussiman, F. 784
Butenholz, K. 205
Butler, F. 732
Büttner, K. 335
Butty, A.M. 406
Buys, N. 335, 337, 340, 402, 404, 417, 916, 976, 979
Byrne, N. 167, 676
Byun, J. 950

C

Cabaraux, J.-F. 510
Cabau, C. 486
Cabello, T. 763
Cabezas, A. 370, 660
Cabrita, A.R.J. 734, 833, 839
Cacheux, P. 892
Cachucho, L. 292, 294
Cadogan, D.J. 341
Cado, O. 782
Caetano, P. 529
Cafiso, A. 861
Cagliari, A.R. 575
Caillat, H. 156, 652, 779
Caimi, C. 217, 225, 868
Caja, G. 249, 251, 484, 499, 500, 503, 504, 506, 1023
Calandreau, L. 647, 730, 856
Calenge, F. 647, 738
Caligiani, A. 639
Calisici, O. 994
Callaway, T.R. 843
Callens, B. 974
Calmels, M. 1025, 1026, 1027
Calnan, H. 310, 312
Calogero, R. 1001
Calsamiglia, S. 636
Calus, M.P.L. 159, 171, 382, 388, 394, 398, 785, 787, 791, 924, 925, 929, 936
Calvet, S. 760
Camargo, G.S. 575
Camarinha-Silva, A. 517, 526, 739
Cambra-López, M. 216
Camin, F. 440
Cammack, K. 172, 755
Campanile, D. 443
Campanile, G. 922
Campion, F. 657, 904
Campion, F.P. 902
Campo, M.M. 290, 422, 427, 593
Canale, C. 845
Canali, E. 498, 499, 503, 535
Canario, L. 243, 379, 483, 491, 733, 756, 968
Čandek-Potokar, M. 824, 930

Candela, I. 252
Candio, J. 905
Candrák, J. 390
Canesin, R.C. 583, 584, 702, 795
Canibe, N. 610
Canlet, C. 483, 492
Cannas, A. 511, 659
Cannas, S. 503
Cano, C. 226
Cánovas, A. 980, 981
Cantalapiedra-Hijar, G. 181, 328, 330, 488, 561, 626, 683, 997, 1037
Canto, F. 994
Cao, B. 996
Cao, L. 384
Cao, Y. 222, 638, 832
Capitan, A. 161, 519, 522, 781
Cappelaere, L. 985, 989
Cappelli, G. 1022, 1023
Cappelloni, M. 923
Cappellozza, B. 537, 571, 574, 575, 592, 631, 632
Cappone, E.E. 219
Capra, E. 230
Caputi Jambrenghi, A. 443
Carabaño, M.J. 663, 929
Cardenas, G. 491
Cardenas, L.M. 641
Cardona, E. 193
Cardoso, D. 956
Cargo-Froom, C. 329
Carillier-Jacquin, C. 208
Carloto, D. 185
Carlu, C. 466
Carnicero, R. 1005
Carnier, P. 209, 336
Caro, I. 300
Caron, E. 984, 986
Carozzi, M. 557
Carrier, A. 877, 969
Carrión, D. 296
Carriquirry, M. 999
Carriquiry, M. 640, 1000
Carta, A. 776, 777
Carta, S. 633
Cartoni Mancinelli, A. 825, 826, 827, 946
Caruso, C. 1023
Caruso, F. 367
Carvalho, A. 894
Carvalho, D.F. 493, 518, 596, 734, 833, 839
Carvalho, I.P. 490
Carvalho, J. 668
Casadio, R. 513
Casado, R. 948
Casagrande, A.C. 575
Casal, A. 573
Casasús, I. 722
Cascone, G. 1000
Casellas, J. 203
Casey, A. 564
Casey, N.H. 357
Cason, E.D. 288, 674, 701, 807
Cassandro, M. 408
Cassar-Malek, I. 334, 482, 488, 991, 992, 997, 999
Casserly, R. 716
Castanheira, R.C. 371
Castelani, L. 939
Castellan, E. 545, 718
Castellini, C. 730
Castiglioni, B. 710
Castillo-Lopez, E. 835
Castonguay, F. 582
Casto-Rebollo, C. 757
Castro, N. 505
Catalan, O. 320
Catalán, O. 563
Catellani, A. 634
Caulfield, M. 729
Cavallini, D. 531
Cavillot, E. 852
Cebo, C. 255, 261
Cebron, N. 486
Ceccantini, M. 870
Ceccarelli, M. 973
Cecchinato, A. 190, 230, 633, 705, 775, 940, 944
Ceccinato, A. 634
Ceciliani, F. 354, 1001
Cederberg, C. 949
Cegarra, E. 425, 426, 716
Ceppatelli, A. 722
Cerisuelo, A. 226, 760
Cerjak, M. 427
Cerón, J. 240
Cerpa Aguila, F. 761, 766
Cerqueira, J.O.L. 529
Cervantes, I. 202, 208, 389, 404, 414, 629, 798, 800, 809, 813, 905
Cervantes, M. 358
Cesarani, A. 157, 168, 967
Cestonaro, G. 463
Cevallos, M. 1013
Chaalia, B. 251
Chabrillat, T. 466
Chadwick, D. 729
Chadyiwa, M.C. 671
Chagunda, M. 664
Chagunda, M.G.G. 372, 548
Chai, H.H. 525
Chaiyabutr, N. 511
Cha, J.H. 525
Chakkingal Bhaskaran, B. 340, 417
Chakraborty, D. 385
Chakroun, S. 761, 766
Chalvon-Demersay, T. 470, 473, 838, 873
Chambeaud, J. 553, 885
Chambellon, E. 223
Chamberlain, A.J. 513
Chamberland, J. 587
Chang, K.C. 298

Chang, S.Y. 214, 227
Chanteperdrix, M. 540, 551, 1021
Chantzi, P. 350
Chapard, L. 335, 340, 402, 404, 976, 979
Chapelain, T. 536, 966
Chapoutot, P. 327, 331
Chapuis, H. 232, 896
Chára, J. 150
Charcosset, A. 921
Chardulo, L.A. 236, 522, 837
Charef-Mansouri, M. 1006
Charef, S. 298, 299
Charfeddine, N. 414, 813
Chariopolitou, A. 1011
Charles, M. 930
Charlier, C. 927
Charroin, T. 853
Charton, S. 518
Chassaing, C. 717
Chassé, É. 1012
Chassier, M. 811, 877, 892, 1038
Chaudhry, A.S. 580
Chauhan, S.S. 953
Chaulot-Talmon, A. 231
Chauveau, A. 474
Chauvin, C. 1015
Chaves, A.A.M. 734
Checa, M. 798
Chegini, A. 935
Cheikh, F. 377
Cheli, F. 695
Chelotti, J. 1025, 1027
Chen, F. 875
Cheng, C. 784
Cheng, Y.H. 298, 743, 744
Chen, J.X. 507
Chen, L. 926
Chentouf, M. 510
Chen, W.J. 298, 743, 744
Chen, X. 586, 840, 842
Chen, Y. 803, 1027, 1030
Chen, Y.X. 434
Chen, Z. 352, 362, 879, 1007
Chereau, A. 861
Chesneau, G. 588, 758
Chessa, F. 502
Chevance, A. 540
Chevance, C. 647, 970, 1015
Chevaux, E. 569, 745
Chevillon, P. 203
Chew, L. 340
Cheype, A. 768
Chikwanha, O.C. 428
Chilibroste, P. 573, 640, 999, 1000
Chillemi, G. 721
Chincarini, M. 658
Chinthakayala, L. 756
Chiofalo, B. 284
Chiron, G. 596
Chiron, P. 552
Chisari, S.M. 1000
Chizzotti, M.L. 317
Cho, H. 629
Cho, H.A. 227
Choi, H.J. 525
Choijilsuren, B. 660
Choi, J.Y. 459
Choisis, J.-P. 852
Choi, S.Y. 525
Choi, Y.H. 741
Cho, J.-H. 214, 227, 521
Chombart, M. 612
Chouinard, P.Y. 587
Chow, K. 498
Chriki, S. 320, 429, 690, 696, 828
Christensen, M. 311, 318
Christensen, O.F. 512, 790
Christofi, T. 253, 254, 255, 416
Christophe, O. 306, 797
Christou, I.C.C. 262
Chud, T.C.S. 406
Chung, W. 194
Church, J. 755
Chu, T.T. 405
Ciani, E. 903, 980
Ciappesoni, G. 1035
Ciarelli, C. 369
Cidrini, I.A. 713, 714
Cieślak, A. 174, 175, 301, 931
Cieslar, S. 329
Ciganda, V. 726
Cilloni, A. 805
Cimmino, R. 922
Cinardi, G. 147
Cisneros, M. 1013
Cissé, S. 735, 898
Claeys, J. 344, 346
Claffey, N. 657, 904
Clariget, J. 726
Clark, E. 486, 924, 927
Clarke, J. 590
Clarke, S. 419, 1031
Clark, R. 927
Clark, S.A. 170
Clarkson, A.H. 962, 963
Clasen, J.B. 403, 405
Claveau, S. 972
Clement, E.P. 957
Clément, V. 811, 877
Cleveland, M.A. 155
Cloete, S.W.P. 243, 368, 405, 651, 672
Cloez, B. 502
Clouard, C. 206, 885
Coates, T.W. 560
Cobo, E. 232
Cocco, E. 518
Cochrane, T. 543
Cockburn, M. 650

Coelho, M.S. 537
Coello, C. 377
Coeugnet, P. 549
Coffey, M. 486
Coffre-Thomain, C. 1032
Cohen-Zinder, M. 168, 422
Cohrs, I. 488
Cole, J.B. 401
Coletti, G. 303
Coli, L. 381
Colin, F. 203
Collas, C. 679
Colleluori, R. 531
Collet, J. 605
Collin, A. 232, 605, 729, 730, 731
Collin-Chenot, A. 730
Collins, T. 1017
Colnago, L.A. 521
Colombié, S. 719
Colonna, M.A. 443
Comer, L. 610, 736
Conceição, C. 529, 1014
Conceição, L. 227
Conde-Aguilera, J.A. 870
Confessore, A. 678
Cong, F. 996
Conington, J. 181, 386, 1032, 1034
Connaughton, S. 310, 312, 313
Conneely, M. 533
Conroy, S. 280
Constancis, C. 506, 552
Constant, I. 991
Constantin, S. 207
Consuegra, J. 602, 746, 1003
Conte, G. 338, 565, 652, 721, 826
Cooke, A.S. 265
Cooke, R. 236, 713
Copani, G. 571, 592, 631, 632
Coppa, M. 219, 561
Coquereau, G. 257, 508
Corbiere, F. 510
Cordero, J. 713
Cordonnier, A. 348, 478
Corlett, M. 204
Corlett, M.T. 314
Cornelison, A. 712, 831
Corniaux, C. 704
Cornuez, A. 232
Corrêa, D.C.C. 726
Correa, F. 338, 463, 612, 715, 759, 760
Corre, C. 271, 272
Correddu, F. 168, 633
Corredor, F.-A. 393
Correge, I. 1015
Correia, B.S.B. 611
Correia Da Silva, P. 480
Correia-Gomes, C. 237
Corset, A. 993
Cosby, S.L. 485, 624
Costa, A. 377, 998
Costa, C. 292, 294
Costa, J.M.S. 292, 294
Costa, L.N. 705
Costa, M. 186
Costanzo, E. 463
Costa, R. 499, 500
Costa, S. 452
Costa, T.C. 490
Costes, V. 229
Cotman, M. 919
Cottrell, J.J. 953
Coudron, C.L. 344, 346, 475
Cougnon, M. 932
Couldrey, C. 162
Coulmier, D. 728
Coupin, M. 1021
Coustham, V. 997
Coutinho, M. 309
Couture, C. 960
Couvreur, S. 546
Couzy, C. 548, 823, 825, 826, 827, 829, 942
Coville, J.L. 738
Cozzi, G. 366, 639, 959
Craig, A. 716
Crane, T.A. 164
Creevey, C. 841, 929
Creighton, P. 437, 676, 688, 1033
Crémilleux, M. 722, 725
Cremonesi, P. 710
Cresci, R. 327
Crespo-Piazuello, D. 925
Cretenet, M. 737
Crielaard, G. 870
Cristobal-Carballo, O. 562, 563, 586, 840, 841
Crofton, E. 688
Croiseau, P. 158, 516, 786, 790
Cromheeke, M. 932
Crompton, L.A. 572
Crosson, P. 181, 269, 681
Cruz, A. 905, 909
Cui, Y.M. 187, 517
Cullen, S.A. 703
Cupido, C.F. 445
Cupido, M. 238
Curik, I. 381, 393
Curi, R. 236, 522, 837
Curnel, Y.C. 682
Currie, D. 716
Curry, D. 174
Curtasu, M.V. 1012
Curtil-Dit-Galin, M. 650
Custodio, D. 640, 999, 1000
Cuthbertson, H. 307, 308, 310, 320, 692
Cuyabano, B. 664, 665, 666, 786, 790, 929
C., Y.D. 291
Cyrillo, J.N.S.G. 282, 583, 584, 702, 704, 795
Czeglédi, L. 816
Cziszter, L.T. 249, 503

Czycholl, I. 335, 643, 859

D

D4dairy Consortium 375
Daddam, J.R. 353
Dagel, A.K. 326
Dahle, B. 622
Dai, X.L. 523
Dalaka, E. 190
Dalby, F.R. 177
Dalcanale, S. 759, 760, 889
Dale, L. 374, 771, 1026, 1027, 1028, 1029
D'Alessio, R.M. 237
Dalim, M. 216
Dalla Costa, E. 498, 499, 503, 535, 656, 861
Dallaire-Lamontagne, M. 869
Dalla Rovere, G. 928
Dall'olio, S. 513, 920
Dall-Orsoletta, A.C. 573
Daly, J. 878
Damasceno, F.M. 958
Dam Nielsen, K. 873
Damon, F. 887
Danchin, C. 381, 389, 670
Danese, T. 641, 642
Daniel, D. 353
Daniel, J.B. 324, 536, 564, 961, 966
Danielsson, R. 955
Dankowiakowska, A. 599
Daoulas, G. 751
Daraï, L. 481
Darani, P. 329
Darbot, V. 520
Dardevet, D. 834
Da Ros, L. 440
Daş, G. 212, 611
Da Silva Siqueira, T.T. 852
Daskalova, A. 179
Daudet, A. 573
David, I. 165, 379, 474, 896, 971
David, J. 328
Davies, D.R. 1010
Davila, A.-M. 1006
Davis, J.D. 988
Davis, S.R. 661
Davis, T.A. 494
Davoli, R. 805
Dawid Słomian, D.S. 792
Day, L. 420
De Almeida, A.M. 484, 524, 685
De Araújo Pimenta, J.L. 1036
Dearden, P.K. 622, 754
Debaeke, P. 395
De Barbieri, I. 1035, 1036
Debevere, S. 545
Debode, F. 750
De Boer, H. 175
De Boer, I.J.M. 268, 442, 547, 849, 949
De Boever, J.L. 900
De Boyer Des Roches, A. 183
Debrez, F. 552
Deb, S. 705
Debus, N. 250, 878
Decandia, M. 249, 502, 657
De Carlo, E. 1022, 1023
Dechaux, T. 677
De Cremoux, R. 1038
Decruyenaere, V. 467
De Cuyper, C. 460, 471, 758, 988
Dedieu, B. 550
Deermann, A. 415
De Faria Lainetti, P. 285
De Fátima, M. 872
Defilippo, F. 873
De Gaiffier, W. 207, 237
Degalez, F. 794, 930
Degano, L. 157
De Goede, M.L. 478
De Greef, K.H. 469, 739
Dégremont, L. 669
Degroote, J. 894
De Gussem, M. 600
De Haas, Y. 148, 277
Dehareng, F. 306, 797, 1024, 1026
Dehghani, M. 359, 555, 890
De Iorio, M.G. 623
De Jong, G. 937
Dejonghe, L. 545
De Jong, I.C. 220, 287, 737
De Jong, M. 206
De Jong, M.C.M. 406
Dekkers, M.H. 703
De Klerk, B. 380
De Koning, D.J. 154, 798
De Laat, J. 844
Delaby, L. 304, 331, 413, 438, 635
Delacroix, M. 1013
Delafosse, A. 407
De La Fuente, G. 305, 306, 1005
De La Fuente, J. 370, 660
Delagarde, R. 331, 444, 581, 659
Delahaye, Q. 189
De La Llave-Propín, A. 370
De Lange, E. 352
Delaqueze, C. 985
De La Torre, A. 167, 331, 334, 689, 991, 992, 997, 999
Delattre, L. 257, 376
Delavaud, A. 255, 485, 488, 683
Delavaud, C. 992, 999
Delaveau, J. 356
Delbès, C. 680, 728
Del Campo, M. 423, 1036
Delcenserie, V. 735
De Leonardis, D. 607
Delepouve, N. 457
Delhez, P. 1027
Del Hierro, O. 653, 654, 656

Deloison, H. 385
Delord, B. 586
Delosière, M. 255
Delouard, J.M. 257
Del Pino, L. 796
Delsart, M. 1022
Demarbaix, A. 246
De Marchi, M. 179, 377, 998
Demarquet, F. 246
Demars, J. 230, 231
De Matos, L.G. 354
De-Meo-Filho, P. 641
Demeter, A. 938
Demyda-Peyrás, S. 876, 881, 978, 980
Deneux - Le Barh, V. 857, 858
De Neve, N. 630
Dengler, F. 835
Denis, M. 214, 519
Denman, S. 632
De Noni, M. 705
Denoyelle, C. 551
Dentinho, M.T.P. 292, 294
De Olde, E. 154
De Olde, E.M. 547, 849
Depalo, P. 354
De Palo, P. 903, 980
De Paula Dorigam, J.C. 988
Depoudent, C. 985
De Prekel, L. 355, 360
Depuille, L. 246, 776, 777, 778
Depuydt, J. 404
De Quelen, F. 709, 854
De Rauglaudre, T. 597, 605, 987
Derbez, F. 853
Dernat, S. 723
Deroma, M. 967
Dersjant-Li, Y. 899
Déru, V. 200, 201, 208
Deruytter, D. 346, 475
Desaint, B. 730
Deschamps, C. 602
Deschamps, M.H. 869
De Smet, J. 477
De Smet, S. 314, 417, 837, 942
Desnica, N. 590
De-Sousa, K.T. 371
Despinasse, M. 508
Desrousseaux, G. 591
Destrez, A. 887
Dettori, M.L. 902
Deurenberg, R. 346
Devailly, G. 230
Devaux, M.F. 586
Devillard, E. 602
Devincenzi, T. 796
De Vos, J. 235, 353, 925
De Vries, A. 302, 818
De Vries, M. 175, 267
De Vries, T.J. 541
Dewhurst, R.J. 625
De Wit, A.A.C. 616, 1003
De With, P.H.N. 379
Dewulf, J. 597, 600
Dhakal, R. 1006
Dhorne-Pollet, S. 975
Dhumez, O. 993
Diao, Q.Y. 566
Dias, J. 195
Dias, S. 668
Díaz, C. 663, 929
Díaz, M.T. 370, 660
Di Battista, D. 658
Dichou, K.D. 682
Dicke, M. 220
Dickhoefer, U. 456
Dieudonné, A. 677, 681
Difford, G. 277
Digiacomo, K. 294, 1005
Di Giuseppe, P. 247
Dijkstra, J. 931
Dillon, E. 153
Dimauro, C. 168
Dimauro, S. 354
Dimitriou, A. 253, 383, 804, 917
Dimov, D. 188, 254
Ding, X. 789
Dinh Khanh, T. 698
Dinis, M.T. 195
Diot, V. 548
Di Paolo, L. 658
Dippel, S. 986
Diskin, M.G. 902
Dißmann, L. 756
Dittmann, A. 484, 493
Dittrich, I. 1016
Di Vuolo, G. 1022, 1023
Djenontin-Agossou, D.D. 986
Djikeng, A. 670
Dodds, K. 655
Dod loan, E. 1013
Doekes, H.P. 379
Doelman, J. 564
Doelsch, E. 710, 957
Doerper, A. 220
Doeschl-Wilson, A. 317, 414
Dohlman, T.M. 712
Dohme-Meier, F. 293, 565, 681
Dolan, M. 657, 904
Dolinar, A. 178
Dollé, J.B. 654, 656
Domínguez, S. 809
Dominique, S. 538
D'Onghia, A. 903
Dong, L.F. 566
Do Nguyen, D.K. 511
Donkersloot, E.G. 661
Donkpegan, A. 618
Donnadieu, C. 519, 781

Donnellan, T. 153
Dønnem, I. 1040
Doornewaard, G.J. 267
Doornweerd, J.E. 380
Dorbe, A. 174
Dorenlor, V. 214
Dorigo, M. 959
D'Orleans, L. 345
Dortmans, B. 747
Dossa, L.H. 699, 706
Dos Santos, T.C. 537
D'Ottavio, P. 244
Doublet, M. 794
Douet, C. 862
Douhard, F. 166, 650
Douhay, D. 443
Douhay, J. 246
Douine, C. 496, 778, 779
Doulamis, A. 947
Dounies, B. 215
Dourmad, J.Y. 709, 970
Doussal, P. 424
Doutart, E. 496
Dovč, A. 919
Dovč, P. 392, 919
Do, Y. 376
Doyle, J.L. 976
Dozier, W.A. 988
Drachmann, F.F. 319
Dragomir, C. 654, 656
Drašler, D. 178
Draxl, C. 915
Drique, C. 470
Drouet, M. 258, 1021
Droumtsekas, K. 262, 636, 1011
Druet, T. 383, 390, 394
Druetto, D. 931
Drusch, S. 853
Druyan, S. 595, 606
D'Souza, D. 209, 465
Duan, C.H. 880
Duan, G.Y. 996
Duan, X.H. 508
Duan, Y. 996, 1002
Duarte, D. 410
Duarte, M.S. 490
Dubarry, K. 486
Dubois, B. 750
Dubois, E. 970, 991
Dubois, P. 758
Dubois, S. 683
Dubreuil, D. 779
Duclos, D. 389
Duclos, M.J. 711
Ducreux, B. 892
Ducro, B. 280, 376
Ducrocq, L. 1023
Ducrocq, S. 516
Ducrocq, V. 784, 807
Dudek, K. 224, 349, 621
Dudek, M. 224, 621
Duenk, P. 783, 787, 806, 937
Dufils, A. 156, 853
Dufour, A. 482
Dufour, B. 1022
Dufrasne, I. 687
Dufreneix, F. 841
Dugué, C. 768
Duhatschek, D. 931
Dulaurent, A.-M. 751
Du, M. 619
Dumas, C. 619
Dumas, G. 378
Dumesny, M. 759
Dumm, R. 532
Dumont, B. 440, 442, 683, 861
Duncan, A. 780
Duncan, J. 264, 497, 780
Dunière, L. 569
Dunne, E. 248, 934, 1033
Dunne, R. 681
Dunn, I. 798
Dunshea, F. 209, 294, 953, 1005
Dunston-Clarke, E. 1017
Dupain, R. 619
Dupire, O. 853
Düpjan, S. 471
Dupont, J. 356, 882
Dupont, S. 927
Duprat, N. 492
Dupuis, J. 240
Durand, C. 559, 779, 1037
Durand, D. 992
Durand, G. 993
Durand, J.-L. 395
Durand, M. 893
Durosoy, S. 710
Duroy, S. 496
Dürr, J. 163
Dusel, G. 871
Dutertre, C. 970
Duthie, C.-A. 317, 532, 538
Dutot, S. 331
Duval, J. 547, 549, 550
Duval, M. 570, 585
Duvnjak, M. 569
Du, W. 542
Du, X. 268, 305
Dwyer, C. 249, 496, 497
Dycus, M.M. 843
Dzama, K. 243, 368, 672

E

Earley, B. 485, 624, 627
Eché, C. 519
Eckhardt, M.E. 188
Edel, C. 783, 920
Edouard, N. 173, 301, 444, 841

Edvardsson Rasmussen, A. 939
Eekelder, J.T. 302
Egan, M. 439
Egan, S. 976
Egger-Danner, C. 375, 413, 770
Egger, J. 747
Eggerschwiler, L. 293, 565
Ehlert, K. 172, 679
Eiberger, C. 454
Eichinger, J. 293
Eikanger, K.S. 296, 1021
Eikje, L.S. 785
Einkamerer, O.B. 261
Eiperle, A. 751
Eisele, J. 266
Ekateriniadou, L.V. 369, 917
Eknæs, M. 296, 913, 1021
Eleftheriou, E. 749
El Faro, L. 371
El-Haddad, S. 572
Elhadi, A. 249, 499, 500, 504
Elissen, H.J.H. 347
El Jabri, M. 510, 1028
Ellen, E.D. 380, 614, 616, 1003
Ellies-Oury, M.-P. 320, 424, 429, 433, 558, 595, 690, 696, 829, 972
Ellis, J.L. 329
Elluin, G. 545
El-Ouazizi El-Kahia, L. 629, 800
El Yaacoubi, A. 872
Elzinga, K. 937
Emery, S. 255, 485
Emissioncow Consortium 771
Emmerling, R. 783, 920
Emond, P. 834
Encina, A. 860
Enez, F. 669
Engelbrecht, A. 405, 672
Engler, P. 898
Englmaierová, M. 285
Enriquez-Hidalgo, D. 641
Eono, F. 214
Eppenstein, R. 556, 820, 825, 826, 827, 946
Eriksson, S. 799, 977, 978
Escalera-Moreno, N. 994
Escouflaire, C. 519
Escribano, D. 296
Escribano, M. 948, 954
Eska, N. 855
Eslan, C. 452
Espinola Alfonso, R.E. 936
Espinoza, E.M. 617
Esposito, G. 367, 898
Esselink, A. 578, 951
Estany, J. 525
Esteban Blanco, C. 629
Estellé, J. 516
Estellés, F. 772
Estienne, A. 356, 882
Estivalet, L. 1034
Estrada, R. 393
Etienne, R. 723
Ettle, T. 965
Eugene, M. 680
Eugène, M. 897
Eugenio, F.A. 471
Evans, R. 264, 280, 857
Even, G. 516
Eveno, E. 214
Everaert, N. 467, 610, 735, 736
Evrard, D. 348
Ewing, D. 264
Eymard, A. 779

F

Fabbri, G. 366
Fablet, C. 214, 1022
Fabre, S. 398
Facchini, E. 614, 615
Faggion, S. 209, 928
Fahey, A.G. 976
Fahy, D. 167
Falco, M. 354
Falcou, T. 553
Falcucci, A. 147
Falker-Gieske, C. 408, 417, 514
Fança, B. 510, 776, 777, 908, 911
Fanelli, F. 513
Fanger, M. 881
Fang, L. 930
Fanizza, C. 192
Fannes, F. 735
Fant, P. 1011
Fanzo, J. 395
Faraut, T. 781
Farias, I.M.S.C. 837
Farizon, Y. 626
Farkas, J. 393
Farouk, M. 420, 423
Farr, R. 632
Farruggia, A. 289
Fassier, T. 261, 507, 779. 1037, 1038, 1039
Fatemeh, H. 265
Fatet, A. 652, 779, 906
Faucon, M.-P. 751
Faulhaber, M.S. 307
Faure, P. 179, 308
Fauvel, Y. 173
Favale, C. 697
Faverdin, P. 321, 331
Fayet, T. 694
Fazarinc, G. 930
Fazilleau, J. 754
Fehmer, L. 301
Feidt, C. 679, 699
Feldmann, K.P. 843
Felix, T. 905
Fennessy, P.F. 622, 754

Fenske, K. 838
Ferchaud, S. 482, 559, 733
Ferdouse, J. 693
Ferlay, A. 157, 287, 941
Fernandes, A.M. 259, 260
Fernandes, E.A. 596, 604
Fernandes Lazaro, S. 808
Fernandes, M.H.M.R. 504
Fernandes, P. 824
Fernández, C. 226
Fernández, I. 386, 388
Fernández, J. 397, 533
Fernandez-Turren, G. 636
Fernandez, X. 394
Ferneborg, S. 365, 366
Ferrand, N. 589
Ferraresso, S. 928
Ferraretto, L.F. 931
Ferrari, L. 873
Ferrari, P. 464, 822, 986
Ferreira De Oliveira, T. 320
Ferreira, I.M. 714
Ferreira, L. 185, 685
Ferreira, M. 246
Ferreira, V. 647
Ferrero, J.I. 1009
Ferriz, M. 730
Ferrulli, V. 530
Fetherstone, N. 909
Fetiveau, M. 853, 891
Fève, K. 200, 230
Fichot, S. 903
Fidelle, F. 398
Field, N.L. 533
Fievez, V. 630
Figueiroa, F.J.F. 712
Figueroa, D. 393
Figueroa, J. 763
Figueroa, V. 688
Fikse, F. 769, 799, 936
Filipe, J.F.S. 354
Filippone Pavesi, L. 530
Fillon, V. 853, 891
Finocchiaro, R. 408
Finocchi, M. 721, 826
Fiorbelli, E. 297, 323
Fiorilla, E. 219
Firman, C.A. 341
Fischer, A. 895, 896
Fisher, A. 1017
Fissore, P. 488
Fito, L. 506
Flanigan, D. 315
Flatrès-Grall, L. 165, 201
Flechard, C. 173
Fleming, A. 275
Fleurance, G. 861
Fleury, J. 663
Flynn, A. 534
Fodor, I. 220, 221, 223, 347, 520, 767, 773, 901
Foggi, G. 565
Fois, G. 168
Foissac, S. 482, 930
Foldager, L. 242
Folorunsho, O.A. 336
Fonseca, A.J.M. 734, 833, 839
Fonseca, P.A.S. 980
Fontaine, S. 597
Fontanesi, L. 513, 623, 920, 923
Fontez, B. 502
Font-I-Furnols, M. 319, 471
Forano, E. 571
Forder, R.E.A. 341
Formigoni, A. 531
Formoso-Rafferty, N. 202, 208, 389, 404, 629, 798, 800, 905
Fornaciari, R. 959
Forneris, N.S. 383, 390
Forouzandeh, A. 710
Forslund, A. 395
Forster, P. 174
Förster, T. 773
Fortomaris, P. 555
Forton, F. 897
Fortun-Lamothe, L. 546, 551, 891
Foskolos, A. 262, 635, 636, 637, 638, 1011
Fossaert, C. 677
Fossey, M. 265
Fotiadou, V. 555, 671
Fotou, K. 764
Foucras, G. 229, 486
Fouéré, C. 231
Fourcot, A. 699
Fourdin, S. 545
Fourichon, C. 407, 464, 1018
Fournel, S. 987
Fournier, A. 699
Fournier, D. 690
Fournier, F. 485
Foury, A. 200
Foy, D. 766
Foyer, C. 174
Frambourg, A. 231
Franceschini, S. 1026
Franchi, G. 497, 883
Franch, R. 928
Francino, P. 760
Francioni, M. 244
Franciosi, E. 288
François, D. 1032
Franco, M. 295
Freire, J.P.B. 493, 518, 734, 833, 839
French, P. 681
Fréret, S. 878, 1037
Fresco, S. 276
Fressinaud, S. 659, 728
Frétaud, M. 198
Fréville, M. 356

Freytag, F. 973
Friehs, T. 204, 242, 884
Friel, R. 934
Friggens, N. 165, 331, 507, 753, 782, 1039
Fritz, S. 276, 407, 519
Fröhlich, T. 482
Froidmont, E. 306, 815
Froment, P. 356
Fromm, K. 769
Frooninckx, L. 867, 874
Frossard, E. 747
Fuchs, P. 678
Fuentes Pardo, P. 500
Fuentes-Pila, J.F. 694
Fuerst, C. 413
Fuerst-Waltl, B. 375, 413
Fuertes, E. 305, 306, 1005
Fukuzawa, Y. 661, 667
Fulghesu, F. 659
Fulínová, D. 808
Fumagalli, F. 634
Fumagalli, T. 727
Fumière, O. 478
Fumo, V. 357
Fu, Q.Y. 233
Furbeyre, H. 758
Furgał-Dierzuk, I. 537
Fürst Waltl, B. 981
Furtado, A.J. 727
Fusaro, I. 658
Fusi, F. 367, 1000, 1022, 1023
Fu, Y.X. 1001

G

Gabarrou, J.-F. 591
Gabarrou, J.G. 361
Gaborit, M. 276, 413
Gabriel, L. 985
Gachora, J. 164
Gafsi, M.G. 453
Gafsi, N. 904
Gagaoua, M. 941
Gagnepain-Germain, F. 474, 479, 618
Gagnon, P. 378, 969
Gaiani, N. 522
Gai, F. 217, 225, 227, 344
Gaillard, C. 791, 893
Galama, P. 301, 303
Galama, P.J. 175, 816
Galaverna, G. 639
Galimberti, G. 513
Galli, M. 240
Gallimore, S.E. 711
Gall, M. 536
Gallo, A. 297, 323, 449, 634
Gallo, L. 336, 633, 775, 940, 944
Gallo, M. 513, 923
Galluzzo, F. 408
Galma, P. 175
Galméus, D. 913
Galoro Leite, N. 782
Galschioet, C. 632
Galvagnon, C. 553
Galway, D.A. 1010
Gamarra, J. 905
Gambier, E. 762
Gambino, M. 410
Gameiro, A.H. 702
Gancárová, B. 257, 259, 260, 262
Gangloff, H. 786
Gangnat, I.D.M. 747
Ganier, P. 208
Gan, S.Q. 523
Gao, H. 788
Gao, Y. 206
Garcia, A. 563, 705, 772, 954
Garcia Baccino, C. 424
García-Casco, J.M. 404
García-García, J.J. 880
García García, M.J. 648
García-Gudiño, J. 270
Garcia, I.M.C. 282
García, J. 425, 426, 716
Garcia, J.M. 898
Garcia-Launay, F. 844, 850, 941, 970, 989
Garcia, O. 1030
García-Rebollar, P. 760
García-Roche, M. 640, 999, 1000
García-Rodríguez, A. 298, 299
Garcia-Santos, S. 185
Garcia-Solares, F. 168, 422
Garcia, T. 573
Garcia-Vazquez, C. 328
García Viñado, I. 472, 612
Garçon, C.J.J. 472
Gardan-Salmon, D. 1009
Gardiner, G.E. 337, 608, 612
Gardner, C. 153
Gardner, G. 204, 309, 310, 311, 312, 313, 314, 315, 316, 380, 428, 431
Gariglio, M. 219
Garnier, M.-G. 658
Garrabos, P. 618
Garrick, D. 275
Garrick, D.J. 275, 801
Gasco, L. 217, 219, 225, 227, 475, 752, 868, 872
Gaspa, G. 168
Gaspar-García, P. 720, 723, 813
Gaspar, P. 948, 954
Gaspe, R.J.F. 933
Gastli, M. 227
Gatphayak, K. 830
Gatsas, K. 636
Gaudillière, N. 895
Gauly, M. 251, 613, 981
Gauthier, P. 894
Gauthier, V. 768
Gautier, D. 263, 496, 778, 779

Gautier, J.-M. 496, 497, 776, 777, 778
Gaüzere, Y. 728
Gavaud, S. 1018
Gayrard, C. 762, 988
Gazzonis, A. 530, 861
Gbenou, G.X. 699, 700, 706
Gebbie, S. 278
Gebska, M. 986
Geddes, E. 264, 780
Gedgaudas, E. 304
Geerse, H. 849
Geibel, J. 387, 399
Geisler, S. 241
Gelasakis, A.I. 945, 947
Gelé, M. 941, 998, 1026, 1027
Gelhausen, J. 204, 242, 884
Gelinder Viklund, Å. 977
Gemo, G. 898
Generalovic, T. 613, 621
Gengler, N. 276, 382, 388, 394, 410, 662, 803, 1025, 1026, 1027, 1029, 1030
Georgieva, S. 178, 179, 180, 182, 184
Georgiou, A.N. 253, 383, 416, 804, 917
Gerard, C. 231, 714, 885, 907
Gérard, M. 550
Gerdes, K. 387
Gerevini, M. 824
Gerfault, V. 749
Gergovska, Z. 185
Germain, K. 605, 730, 886
Germon, P. 993
Gervais, R. 582, 587, 972
Gesek, M. 226
Ghaderi Zefreh, M. 414
Ghaffari, M.H. 488, 773
Ghallabi, N. 563
Ghilardelli, F. 297, 634
Giagnoni, G. 1012
Giagnoni, L. 705
Giammarco, M. 658
Giammarino, M. 887
Gianelle, D. 440
Gianesella, M. 190
Giannenas, I. 218, 228, 350, 764, 765
Giannico, F. 443
Giannone, C. 973
Giannuzzi, D. 190, 230, 288, 336, 633, 775, 944
Gianotten, N. 479
Gian Simone Sechi, G.S.S. 273
Gianvecchio, S.B. 282
Giavasis, I. 218, 228
Gibbons, J. 729
Gichuki, L. 164
Gickel, J. 696, 845, 987
Gidenne, T. 762
Giersberg, M.F. 645, 734, 883
Giger-Reverdin, S. 325, 331, 333
Giglio, L. 822
Gilani, S. 899
Gilbert, C. 892
Gilbert, H. 200, 201, 208, 230, 844, 896
Gilbert, I. 877
Gillier, M. 841
Gilliland, J. 948
Gimbert, F. 497
Giménez, A. 721
Gimenez-Rico, R.D. 899
Ginane, C. 1037
Gindri, M. 753
Gionbelli, M.P. 490
Giovanetti, V. 249, 502, 503, 657, 776, 777, 778
Giráldez, F.J. 235, 256, 300, 1001
Girard, C. 466
Girardie, O. 379
Girard, M. 339, 463, 471
Girard, P. 550
Giraud, A. 673
Giromini, C. 285, 695
Gispert, M. 319
Gitau, J.W. 164
Giua, E. 652
Giuffra, E. 924
Gjøen, H. 793
Gjuvsland, A.B. 785
Gleeson, E.Y. 345, 866
Gligorescu, A. 616
Gloria, L.S. 259, 260
Gobbo Oliveira Erünlü, N. 244, 249, 650
Gobeti Barro, A. 424
Godber, O.F. 848, 854
Goddard, M.E. 513
Göderz, H. 670
Godo, A. 501
Godoc, B. 718
Godoy-Santos, F. 563
Goenaga, I. 424
Goers, S. 607
Goeser, J. 931
Goethals, S. 337, 340
Gofflot, S. 348, 750
Goglio, P. 265
Goi, A. 179, 377, 998
Goiri, I. 298, 299
Gold, M. 747
Golomazou, E. 197
Gol, S. 378, 757
Gombault, P. 762
Gomes, G. 577
Gomes, M.J. 685
Gomes, V.M. 727
Gómez, E.A. 226
Gómez Herrera, Y. 499
Gómez, M.D. 977
Gómez, M.M. 922
Gomez-Proto, G. 384, 983
Gómez, Y. 498
Gonçalves, A. 195
Gonçalves, D. 294

Gonçalves, L.L. 480
Gonçalves, P. 971
Gondret, F. 200, 471, 482, 514, 941
Gondro, C. 666
Gong, M. 523
Gonzales, J. 393
González-Cabrera, M. 505
González De Chavarri, E. 370
Gonzalez-Dieguez, D. 921
González, E. 249
González-García, E. 247, 250, 442, 496, 502, 559, 704, 753, 778, 779
González Garoz, R. 370
González-González, R. 504
Gonzalez, L. 573
González-Luna, S. 251, 506
Gonzalez Recio, O. 229
Gonzalo, E. 988
Gonzalo, G. 563
Goodell, G.M. 532
Goossens, K. 843, 900
Goossens, T. 746, 1003
Gordon, C. 578, 639
Gordon, J. 181, 1032
Goren, A.G. 617
Göres, N. 488
Görge, H. 1016
Goris, K. 844
Gorjanc, G. 160, 170, 402, 620, 670, 918
Gorrens, E. 477
Gorssens, W. 979
Gorssen, W. 335, 340, 402, 404, 417, 916, 976, 979
Gort, G. 220
Goselink, R. 534, 992
Gotoh, T. 233, 592
Gottardo, F. 366, 639
Götz, K.-U. 783, 920, 965
Goudet, G. 459, 862
Goujon, M. 421
Gould, L.M. 242
Gourdine, J.L. 663
Gourdon, A. 1038
Gourichon, D. 647
Gourlez, E. 709
Gouva, E. 765
Gouzenes, A. 398
Goyache, F. 386, 388
Goyenetche, M. 778, 779
Goyette, B. 985
Graat, E.A.M. 734
Grace, C. 268
Graczyk-Bogdanowicz, M. 812
Grafl, B. 611
Graham, M. 263, 729
Granados, A. 155, 526
Granados-Chapette, A. 924
Granado-Tajada, I. 803
Grande, S. 903
Gras, M.A. 654, 656
Graulet, B. 680, 717, 993
Graviou, D. 633
Gredler-Grandl, B. 926
Green, A. 343
Grelet, C. 1024, 1025, 1026, 1027, 1030
Gresse, R. 571
Gress, L. 232, 483, 491, 492, 789
Gresta, F. 284
Greyling, J.P.C. 725
Griela, E. 286, 836
Griffith, B.A. 641
Griffiths, B. 767, 1019
Griffon, L. 809
Grigoriadou, A. 350
Grillot, M. 441, 853
Grimm, P. 859
Grindflek, E. 377
Grinnell, N.A. 245, 441
Grisot, P.-G. 246, 776, 777
Grobler, S.M. 725
Grochowska, E. 234, 411, 599, 1008
Grodzki, B. 481
Groenen, M. 924
Groenewoud, A. 849
Grohs, C. 519, 522, 781
Gröndahl, G. 858
Groot, J.C.J. 268
Groot Koerkamp, P. 303
Grøseth, M. 900
Große-Brinkhaus, C. 812, 989
Grossiord, B. 558, 829
Gross, J. 992
Gross, N. 440
Grøva, L. 249, 503, 776, 777, 778
Gruber, M. 251
Gruber, S. 981
Grudzinski, C.E. 772
Grundy, L. 281
Gruninger, R.J. 560
Gual-Crau, A. 834
Gualdron Duarte, J.L. 390
Guan, D. 930
Guan, L.L. 625
Guan, X. 410
Guarnido-Lopez, P. 894
Guatteo, R. 371, 407, 536, 1014
Guedes, C. 185, 780
Guégnard, F. 861
Guerreiro, D. 480
Guesdon, V. 647, 730
Guevara, L. 259, 260
Guevara, R.D. 884
Guibert, J. 483, 520
Guibier, C. 1013
Guidou, C. 223, 762
Guignard, C. 797
Guillaume, J.B. 867, 871
Guillet, S. 750
Guillevic, M. 758

Guillon, F. 586
Guilloteau, L.A. 730
Guimbert, F. 496
Guinan, F. 163
Guinard-Flament, J. 332, 791, 998
Guingand, N. 984, 985
Guinguina, A. 897, 934
Guiocheau, S. 754
Guiso, M.F. 967
Guldbrandtsen, B. 793
Gunia, M. 891
Gunnarsdottir, H. 281
Guo, R.H. 875
Gupta, D. 754
Gurgul, A. 210
Gurman, P.M. 401
Gutierrez, G. 905
Gutiérrez, J.P. 202, 208, 386, 388, 389, 404, 414, 629, 798, 800, 809, 813, 905, 909
Gutierrez Vallejos, J. 824
Guy, F. 677
Guy, J. 986
Guyomard, H. 395
Guyonneau, J.D. 247
Guyot, Y. 595, 986
Guz, P. 411
Guzzo, N. 574

H

Haak, T. 375, 797
Haas, V. 526
Habeanu, M. 653
Habermann, B. 164
Hachemi, A. 956
Haddache, N. 214
Haddad, L. 862
Hadjigeorgiou, I. 855
Hadjipavlou, G. 165, 253, 254, 255, 383, 416, 804, 917
Haenen, O.L.M. 346
Haesaert, G. 938
Haffray, P. 669
Haga, S. 489
Hager-Theodorides, A.L. 291, 509
Hahn, A. 975
Haile, A. 152
Haj Chahine, G. 548
Hajduk, P. 794
Hakansson, F. 649
Halachmi, I. 249, 501, 595, 606, 776, 777, 778
Halas, V. 336
Halil, M. 182
Hall, E.J. 711
Hallez, L. 828
Hall, M. 334, 560, 576, 578
Hall, N. 307
Halmemies-Beauchet-Filleau, A. 588
Ha, M. 209
Hamidi, D. 245, 441, 684
Hamidi, M. 245, 441
Hamon, A. 332
Hamon, B. 543
Han Anh, T. 698
Han, D. 213
Handcock, R. 275
Hanlon, A. 237
Hanlon, M.E. 635, 636, 637
Hannas, M.I. 317
Han, R. 821
Hansen, H.H. 1006, 1010
Hansen, L.S. 614, 615, 618
Hansen, S.V. 709
Hansson, A. 936
Hansson, I. 769, 770
Han, Y. 634
Hao, L.Z. 433
HappyMoo Consortium 1024, 1028, 1029
Haque, M.N. 332
Hardouin, A. 1020
Hardy, A. 496, 878, 908
Hargreaves, P. 174, 175, 301
Harlander, A. 886
Harlizius, B. 397
Harnois Gremmo, A. 904
Harris, B.L. 162
Harrison, P.W. 924
Härter, C.J. 504
Harvatine, K.J. 450
Harvey, K. 713
Ha, S. 349, 350, 363, 364
Hasenpusch, P. 768, 888
Hashiba, K. 279
Haskell, M.J. 183, 532, 538
Hassanat, F. 846
Hassan, M. 563
Hassanpour, A. 399
Hassenfratz, C. 208
Hassouna, M. 984
Hassoun, P. 331, 510
Hayes, M. 839, 842
Hayoz, B. 681
Hazard, D. 559
Hearn, C. 439
Hébrard, W. 243, 756
Heddi, I. 400
Hediger, F. 810
Hedonukun, M.S. 195, 198
Heetkamp, M. 471
Hegedűs, B. 171, 921
Heidaritabar, M. 926
Heijmans, N. 479
Heikkilä, U. 244, 249, 650
Heinicke, J. 769, 775
Heinonen, M. 968, 986
Heinzl, G.C. 354
Heirbaut, S. 630
Heise, J. 408, 417, 514, 937
Heitkönig, B. 995

Heldt, J.S. 708
Heller, R. 523
Helloin, E. 223
Hellstrand, S. 266
Hellwing, A.L.F. 473, 539, 935, 953
Helmerichs, J. 986
Hely, F.S. 622
Hemingway, C. 851
Hemmert, K.J. 773
Hemonic, A. 1015, 1021
Henchion, M. 423
Henderson, K.C. 538
Hennart, S. 527
Henne, H. 415, 812
Hennessy, D. 438, 439
Henrotte, E. 897
Henry, D. 756
Henry, M. 197
Henryon, M. 615
Herath, H.M.G.P. 438
Hérault, F. 885
Hercule, J. 553
Heringstad, B. 277, 374
Herlin, A.H. 266
Hermant, E. 570, 585
Hermisdorff, I.C. 406
Hernandez, A. 166
Hernández-Castellano, L.E. 505
Hernández, E. 982
Hernandez, G.M.P. 246
Hernández, P. 203, 400, 483
Herrero, M. 153
Herring, W. 158
Herron, J. 547
Herskin, M.S. 242
Hertault, E. 895
Hervé, J. 1018
Herve, L. 498
Hervo, F. 322, 711, 762
Heseker, A. 871
Heseker, P. 645, 649
Hesse, D. 341
He, T. 352, 362, 879, 1007
Hetta, M. 154, 446
Heuel, M. 747
Heuer, C. 374
Heurtault, J. 322
Heuze, V. 327
Heuzé, V. 344, 957
Hewitt, R. 209
Hewitt, R.J.E. 465
He, X.H. 234
Heylen, O. 916
Hickey, R. 609
Hickey, S. 278, 1031
Hickmann, F.M.W. 983, 985
Hidalgo, J. 158
Hiemstra, S. 670, 674
Hietala, S. 281, 682
Hillen, B.C. 899
Hill, V. 476
Hiltpold, M. 166
Hinrichs, D. 731
Hitchman, S. 281
Hlohlongoane, M.N. 428
Hockenhull, J. 860
Hocquette, É. 429, 690
Hocquette, J.-F. 179, 189, 308, 309, 318, 320, 394, 429, 433, 435, 595, 690, 696, 828, 972
Hoek - Van Den Hil, E.F. 346, 347, 348, 747
Hoelzle, L.E. 454
Hoffman, L.C. 345
Hoffmann, G. 756, 769, 772, 775
Hoffmans, Y. 346, 347
Hofmanova, B. 393
Hogeveen, H. 821
Højberg, O. 610
Holgersson, A.-L. 858
Holinger, M. 732
Holland, J. 497
Holl, J. 784
Holshof, G. 173, 680
Holtenius, K. 936, 939
Holzhauer, A. 644
Home, R. 556
Honig, A. 963
Honig, A.C. 965
Hood, J. 641
Hooft, J.M. 191
Hoogstra, A.G. 849
Hoorneman, J.N. 916
Hoorweg, F. 530
Hooyberghs, K. 335, 337, 340, 404
Hoppe, A. 619
Horan, B. 817
Horgan, K. 742, 746
Hornick, J.-L. 510, 687
Horn, J. 245, 441
Horrillo, A. 720, 948, 954
Hörtenhuber, S. 986
Hossain, M.D.M. 990
Hosseindoust, A. 349, 350, 363, 364
Hosseini, A. 866
Hosseini Ghaffari, M. 490, 995
Hosseini, S. 830
Hoste, H. 263, 762
Hostiou, N. 550
Houaga, I. 670
Houard, E. 214
Houben, D. 751
Houdayer, C. 214
Houée, P. 214
Houndafoche, G. 425
Houry, B. 214
Houssier, M. 997
Hou, Y. 280
Hove, K. 913
Hoving, A.H. 674

Hozé, C. 229, 231, 407
Hristakieva, P. 225
Hrynkiewicz, K. 1008
Hsiao, S.H. 744
Hsu, K. 751
Hua, K.F. 743, 744
Huang, H. 931
Huang, L. 235
Huang, R. 938
Huang, X. 290
Huang, Y.Y. 433
Huang, Y.Z. 523
Huanle, L. 565, 932
Huau, C. 261, 507, 1038
Huau, G. 663
Hubbard, C. 986
Hubin, X. 803
Hubrechts, H. 979
Hu, D.X. 523
Huenul, E. 763
Hüe, T. 409
Huet, A.C. 410
Huet, B. 856
Hugo, A. 261
Hug, S. 840
Hu, H. 1030
Huhtanen, P. 934, 1011
Huisman, A. 159, 335, 403
Hulsegge, I. 394
Hulst, A.D. 406
Humphries, D. 588
Hüneke, L. 937
Huntington, J.A. 715, 1010
Huot, F. 587, 972
Hurtaud, C. 255, 261
Hu, S. 889
Huson, H.J. 804
Husson, C. 215
Hustinová, M. 392
Hütten, A.-L. 293
Hu, W.F. 222
Huws, S. 174, 562, 563, 841, 842
Huyghebaert, B. 949
Hu, Y.M. 444

I

Iampietro, C. 519
Iannotti, L. 153
Ibáñez-Escriche, N. 203, 378, 515, 598, 757
Ibáñez, N. 400
Ibarruri, J. 298, 299
Ibeagha-Awemu, E.M. 229
Ibidhi, R. 425, 583, 950
Ibrahim, M. 234, 1008
Id-Lahoucine, S. 980, 981
Idrees, M.F. 604
Igier, A. 552, 895
Iii 988
Ijiri, D. 606
Ikusika, O.O. 910
Ikuta, K. 489
Iliadis, I.V. 509
Imbert, A. 483, 485, 488, 492
Iñarra, B. 298, 299
Ingham, A. 889
Ingrand, S. 394, 689, 723
Inhuber, V. 341, 965
Inoe, F.N. 537
Insausti, K. 424, 433
Invernizzi, G. 357
Iritz, A. 966
Irshad, N. 285
Isele, L. 569, 840
Ishak, S. 536
Ishii, K. 661, 667
Isselstein, J. 245, 441, 455, 684
Ithurbide, M. 1039
Ivanković, A. 391, 427
Ivanova, I. 225
Ivkić, Z. 391
Iwamoto, E. 489
Izquierdo, T. 901

J

Jaafar, M. 804
Jacintho, M.A.C. 705
Jacobs, A.-K. 696
Jacotot, A. 173
Jacques, L. 192, 287
Jacquiet, P. 683, 852
Jacquot, A.-L. 550, 653
Jaffres, C. 643, 974
Jagusiak, W. 418
Jahagirdar, S. 567
Jahoui, A. 886
Jakimowicz, M. 788
Jakobsen, J. 415, 655
Jammes, H. 229, 231, 885
Jamrozik, J. 275
Jandl, J. 164
Janelle, J. 673
Jang, M.J. 525
Jang, S.S. 525
Janković, D. 392
Jankowska-Makosa, A. 644
Janodet, E. 819, 844
Jansen, C.A. 221
Jansen, L. 577, 639, 708, 727
Janssen, E. 348
Janssen, P. 278
Janssens, S. 335, 337, 340, 402, 404, 417, 916, 976, 979
Janss, L. 512
Jansson, A. 978
Jantasaeng, O. 668
Jantzen, B. 1010
Januarie, D.A. 701
Janvier, E. 207, 237

Jardat, P. 856
Jarde, E. 854
Jardet, D. 738
Jarzaguet, M. 834
Jasieniak, A. 696
Jaster-Keller, J. 747
Jattiot, M. 1024, 1026, 1028, 1029
Jaureguy, C. 1021
Jean-Louis, U. 496
Jeanneaux, P. 153
Jenko, J. 785
Jenn, P. 834
Jensen, D.B. 649
Jensen, E.H. 818
Jensen, J. 405, 800
Jensen, L.D. 242, 1018
Jensen, M. 610
Jensen, M.B. 497, 539, 818
Jensen, R.B. 473
Jensen, R.H. 473
Jensen, S.K. 449
Jeong, S. 629
Jeong, S.W. 293
Jeong, Y.D. 741
Jeon, K.H. 214, 227
Jeon, S. 325, 629
Jep, L. 192
Jerez-Bogota, K. 610
Jerónimo, E. 292, 294
Jesús, J.C. 500
Jeuffroy, M.H. 551
Jeusselin, J. 647, 970, 1015
Jeyanathan, J. 630
Jezegou-Bernard, F. 894
Jiang, J. 401
Jiang, L. 234
Jiang, X.Z. 187, 517
Jiang, Y. 523
Jiang, Z. 593
Jiggins, C. 613, 621
Ji, J. 435
Jiménez-Belenguer, A.I. 760
Jiménez Caparros, F. 500
Jiménez-Montenegro, L. 429
Jiménez-Moreno, E. 296
Jimeno, V. 660
Jimoh, A.A. 199
Jing, X.P. 630
Jin, L.Z. 764
Joanna Szyda, J.S. 792
Johansen, M. 471, 900
Johansson, A.M. 914, 921, 922
Johnsen, J.F. 366
Johnson, D. 246
Johnson, K. 147, 623
Johnson, L. 906
Johnson, P. 419, 420, 1031
Johnson, T. 281
Johnsson, M. 160, 798, 914, 922
Johnston, D. 624
Johnston, J. 708
Joly, F. 264, 440, 683
Joly, L. 892
Jonas, E. 989
Jongeneel, R. 153
Jongman, E. 1017
Jonker, A. 278, 1031
Jorgensen, G. 496
Jorgensen, G.H.M. 249
Jorge-Smeding, E. 328, 358
Jörg, H. 881
Jornet, B. 419
Joseph, W. 771
Jost, J. 658, 659
Joubert, A. 1023
Jouffroy, M. 570, 579, 585
Jourdain, J. 161, 781
Jourdain, L. 906
Jourdan, P. 619
Jouven, M. 250, 252
Jouy, M. 751
Jo, Y.H. 359
Joy, M. 722
Józefiak, D. 224, 621
Juch, A. 684
Juigné, C. 514
Julliand, S. 218, 591, 859, 1020
Junes, P. 614
Jung, L. 731
Jung, M.W. 990
Jung, S.A. 525
Juniper, D. 588
Juodka, R. 177
Jurjanz, S. 679
Jurquet, J. 635, 841, 895, 896, 899, 998
Juska, R. 177
Juškienė, V. 175, 177, 301

K

Kaakoosh, S. 422
Kacper Żukowski, K.Ż. 792
Kaczmarek, S. 349
Kadziene, G. 177
Kadžiene, G. 301
Kaewsatuan, P. 487
Kagai, J. 729
Kagawa, R. 795
Kalenga Tshingomba, U. 250
Kallas, Z. 420
Kalogeras, D. 947
Kalogianni, A.I. 945, 947
Kamegawa, M. 606
Kamer, H. 353, 566
Kaminska-Gibas, T. 199
Kang, H. 629
Kang, K. 629
Kang, S. 360, 950
Kantas, D. 635, 636, 637, 638

Karagiannis, A. 634
Karaiskou, N. 555
Karaman, E. 283, 512
Karamanlis, X. 355
Karam, C. 718
Karanjit, S. 866
Karaouglanis, D. 555
Karatzia, M.A. 509, 703, 912
Karatzinos, T. 197
Kareem-Ibrahim, K.O. 195, 198, 199, 336, 342
Kargo, M. 391, 407, 615, 935
Karisch, B. 713
Karlengen, I.J. 296
Karlsson, L. 900
Karolyi, D. 824
Karpiesiuk, K. 741
Kar, S. 221, 353, 520, 737, 1002, 1003
Kasaiyan, S. 300
Kasapidou, E. 509, 703, 912
Kasarda, R. 390, 392
Kaseja, K. 1034
Kašná, E. 412, 799
Kasper, C. 202, 208, 212, 271
Katafuchi, A. 606
Kater, J. 287
Katsumata, S. 606
Kavlak, A.T. 468
Kawęcka, A. 447, 908, 911, 913
Kaya, C. 924
Kazana, P. 529
Keady, T. 503, 653, 655
Keady, T.W.G. 777
Keady, T.W.J. 249, 654, 656, 776, 777, 778
Kearney, M. 167, 269, 681
Kechovska, S. 344
Kefalas, G. 836
Kehraus, S. 989
Keil, C. 874
Kellali, N. 292
Kell, S. 693
Kelly, A.K. 437, 726, 839, 952
Kelly, J. 878
Kelly, T. 532
Kelton, D. 527
Kemp, B. 213, 534, 567
Kemper, N. 205, 241, 370, 598, 601, 603, 645, 649, 1012, 1016
Kendall, N.R. 711, 962, 963
Kenéz, Á. 358
Kennedy, E. 534, 689
Kennedy, J. 268
Kenny, D. 628
Kenny, D.A. 487, 726, 825, 826, 827, 839, 952
Kenyon, F. 264, 497, 776, 777, 778, 780
Keogh, K. 487, 726
Kerhoas, N. 587, 588, 589
Kerouanton, A. 214
Kerr, B. 831
Kerros, S. 466
Keßler, F. 664
Ketterings, Q.M. 848, 854
Kettrukat, T. 599, 608
Kettunen, H. 410
Khajehmiri, Z. 350
Khanal, P. 1006
Khattab, A. 801
Khazzar, S. 639
Khelef, Y. 186, 524
Khempaka, S. 668, 744
Khiaosa-Ard, R. 961, 995
Khoshnam, N. 568
Khounsaknalath, S. 233
Kidane, A. 296, 900, 1021
Kiefer, H. 229, 231
Kiema, A. 697, 698
Kiendrebeogo, T. 699, 706
Kierończyk, B. 224, 621
Kihanguila, W. 748
Kilcline, K. 268
Kimata, M. 667
Kim, C.H. 741
Kim, D.H. 525
Kim, E.J. 359
Kim, H. 227, 521, 629, 686, 833
Kim, H.J. 684
Kim, I.H. 600, 742, 743, 879, 990
Kim, J. 349, 350, 363, 364, 686, 882
Kim, J.E. 741
Kim, J.G. 684
Kim, J.W. 485
Kim, K. 950
Kim, K.H. 583
Kim, M. 489, 724
Kim, S.Y. 187, 489
Kim, T. 950
Kindermann, M. 634
Kind, K. 878
Kindmark, A. 798
Kinghorn, M.G. 807
King, L.T. 204
King, M.T.M. 541
Kinugasa, T. 660
Kinukawa, M. 519
Kirwan, S.F. 634, 1007
Kischel, S.G. 366
Kistemaker, G. 275
Kistler, T. 619
Kjeldsen, M.H. 276, 1012
Kjorvel, A. 894
Kleinpeter, V. 848
Klein, R. 776, 777, 778
Kleinschmit, D.H. 952
Klevenhusen, F. 942
Kliem, K.E. 572
Klimaco Consortium 1029
Klimek, P.K. 770
Klingler, M. 548
Klingström, T. 914

Klont, R. 974
Klootwijk, C.W. 173, 680
Klop, A. 680
Klopčič, J. 178
Klopčič, M. 178, 540, 819, 821
Klopp, C. 930
Klumpp, K. 174
Klünemann, M. 193
Kluss, C. 176
Knap, P.W. 155, 403, 862
Knecht, D. 644
Knierim, U. 973
Knol, E.F. 397, 403
Knöll, J. 204, 242, 884
Knudsen, M.T. 847
Kobayashi, E. 233
Kobek-Kjeldager, C. 242
Koch, C. 488, 490, 773, 995
Köchle, B. 630
Köck, A. 375, 413
Koerkamp, P.W.G. 302
Kofler, J. 413
Kogiannou, D. 197
Ko, H.-L. 497, 500
Kojima, T. 279
Kok, A. 534, 821, 992
Kokemohr, L. 527
Kokkonen, T. 682
Kolmogorov, M. 627
Kolorizos, A. 865
Kołoszycz, E. 544
Komainda, M. 245, 441, 455, 684
Kombolo, M. 179, 308, 429, 433, 690
Kombolo-Ngah, M. 318
Komlósi, I. 816
Kondo, M. 831
Kong, X.F. 765
Koniali, L. 253, 254, 255
König, S. 544, 781, 919
Koning, L. 326, 680
Konjačić, M. 391, 427
Konkol, D. 644
Kononoff, P.J. 494
Konopoka, A.L. 842
Kootstra, G. 380
Korczynski, M. 644
Korelidou, V. 945, 947
Korir, D. 263, 567
Korkuc, P. 919
Korkuć, P. 781
Koseniuk, A. 210
Koskikallio, H. 968
Kosobucki, P. 1009
Kotlarz, K. 793, 794
Kotsiou, K. 943
Kouchner, C. 619
Kougioumtzis, A. 528, 531
Koulete, E. 985
Kour, K. 385
Kousoulaki, K.K. 194
Kovačić, M. 622
Kövér, G. 393
Kowalski, E. 417, 825, 826, 827
Kozera, W. 741
Kozłowska, M. 931
Kra, G. 353
Kramer, C. 950
Kramer, M. 370, 603
Kramer, T. 178
Kranenbarg, R. 311
Kranis, A. 598, 605, 918, 925
Krebs, T. 204, 242, 884
Kreismane, D. 174, 176
Kreuzer, M. 747
Kreuzer-Redmer, S. 341, 835, 995
Krieger, M. 731
Krieter, J. 335, 643, 768, 859, 885, 888, 1016
Kring, R.D. 403
Kristensen, N.B. 539
Kristensen, T. 618, 950
Kriszt, T. 247
Krizanac, A.-M. 408, 417, 514
Krizsan, S.J. 955
Krogh, R. 278
Kronqvist, C. 936, 939
Kruijt, L. 616, 1003
Krupa, E. 210, 212, 392, 808
Krupová, Z. 210, 212, 392, 412
Krzyścin, P. 796
Kubota, S. 487, 668
Kuehn, C. 924
Kühn, C. 926, 927
Kuipers, A. 175, 303, 814, 816, 821
Kumaishi, M. 248
Kumar, D. 385
Kuntzer, T. 249
Kunz, L. 484, 493
Kuoppala, K. 682, 815
Kupczynski, R. 644
Kurilo, C. 482
Kurogi, K. 519
Kurokawa, Y. 933
Kuroki, Y. 933
Kwok, M.Y.W. 743
K., X.F. 291
Kyriakaki, P. 906
Kyriazakis, I. 594, 646

L

Labadie, K. 618, 749
Labanca, H. 728
Laban-Mele, S.L.M. 892
Labatut, J. 549
Labrune, Y. 200
Labrunne, Y. 852
Labussière, E. 207, 208, 471, 562, 758
Lachemot, L. 484
Lackal, W. 924

Ladeira, M.M. 317
Laffont, L. 1004
Laflotte, A. 579, 679
Lagadec, S. 470, 985
Lagarrigue, S. 416, 738, 794, 810, 928, 930
Lagerlund, K. 858
Lagkouvardos, I. 764
Lagneaux, S.L. 682
Lagoda, M. 240
Lagoutte, L. 794
Lagriffoul, G. 510
Lagüe, M. 896
Laithier, C. 728, 823, 825, 826, 827, 829, 942, 946
Lakhdara, N. 292
Laloë, D. 379, 824
Lamadon, A. 331
Lamarque, M. 442, 704
Lambe, N. 181, 317, 497, 1032
Lambert, M. 652
Lambert, W. 597, 987, 988, 989
Lambolez, E. 453
Lamichhane, U. 843
Lamont, K. 780
Lamothe, L. 853
Lamothe, V. 993
Lamraoui, M. 186, 524
Lam, T. 992
Lamy, A. 458
Lamy, E. 529
Landi, V. 903, 980
Landry, M. 587
Landwehr, N. 756
Lange, A. 415
Langenhuizen, P. 379
Langevin, C. 192, 198
Lansade, L. 856, 886, 887
Lanzoni, D. 285, 695
Lanzoni, L. 653, 656, 658
Laplaize, A.C. 750
Laporta, J. 354
Laramee, A. 809
Larat, V. 901
Lardic, L. 762
Lardner, H.H. 149
Lardy, Q. 446
Lardy, R. 967
Larrigaldie, I. 887
Larroque, H. 255, 261, 802
Larsberg, F. 341
Larsen, M. 497, 1018
Larsen, R. 761
Larsen, T. 991, 1039
Larzul, C. 733
Laschon, L. 453
Laseca, N. 876, 978, 980, 981
Lashkari, S. 177, 449
Lassen, J. 276, 278, 373
Lasserre, B. 892
Latruffe, L. 153
Latvala, T. 628
Launay, F. 413, 635
Launay Ventelon, M. 1020
Laurain, J. 213, 735
Laurent, C. 941
Lauridsen, T. 311
Laursen, S.F. 618
Lautrou, M. 958, 960, 964
Lauvie, A. 673
Lavon, Y. 353
Lavrenčič, E. 670
Lawlor, P.G. 337, 608, 612
Law, R. 320
Lawrence, T.E. 188
Lazar, B. 234
Lazaridou, A. 943
Leal, L.N. 536, 966
Leandro, M.A. 372
Lebas, S. 237
Lebeuf, Y. 587
Le Bihan-Duval, E. 605, 730
Leblanc, M. 856
Leblanc-Maridor, M. 464, 1018
Leblois, J. 797, 1024, 1025, 1026, 1027, 1028, 1029
Le Bot, M. 213
Le Boucher, R. 194
Le Bouquin-Leneveu, S. 603
Le Bourgeois, T. 697
Le Bourhis, M.-C. 232
Le Bras, P. 984
Lebret, A. 206, 647, 970, 1015
Lebret, B. 559, 733, 941
Lebreton, A. 246, 376
Lê Cao, K.A. 518
Lecchi, C. 285, 354, 1001
Lecerf, F. 794, 930
Le Chenadec, H. 659, 728
Lechevestrier, Y. 207
Lechniak, D. 931
Leclerc, H. 161
Leclercq, C.C. 518
Lecoeur, A. 647, 738
Lecomte, C. 1024, 1026
Lecorguille, P. 985
Le Cozler, Y. 189, 562, 635, 653, 895, 998
Lecrenier, M.-C. 348, 478
Le Danvic, C. 229
Ledda, A. 511, 633, 659
Ledoux, D. 183
Le Dreau, A. 424
Leduc, A. 1027
Le Du, L. 551
Leeb, C. 986
Lee, C.S. 194
Lee, H. 363
Lee, H.-G. 359, 521
Lee, J. 583
Lee, J.A. 459
Lee, J.S. 359, 466, 890

Lee, M. 152, 270, 420, 629
Lee, S. 629
Lee, S.H. 187
Lee, Y. 950
Le Faouder, P. 255
Lefebvre, R. 276, 413
Lefebvre, T. 345, 351, 474, 479, 618, 749
Lefèvre, A. 834
Lefler, J. 952
Le Floc'h, N. 472
Lefort, A.-C. 603
Lefort, G. 886
Lefoul, V. 365, 373, 551
Lefranc, M.L. 194
Legako, J.F. 188
Le Gall, M. 207
Le Gall, V. 985
Legarra, A. 401, 784, 790, 921
Legein, L. 527
Léger, S. 941
Legoff, Y. 1032
Le Goic, D. 1021
Legoueix, S. 200, 492
Legrand, I. 180, 308, 432, 825, 826, 827, 829, 942
Legrand, T. 632
Le Graverand, Q. 518, 1032, 1034, 1036
Legris, M. 570, 585, 728
Legros, S. 710
Le, H. 191
Lehébel, A. 371, 536
Leheup, M. 345, 351
Lehuraux, R. 896
Leiber, F. 810, 946
Leishman, E.M. 329
Leitner, S. 263
Lelis, A.L.J. 954
Lemâle, O. 746, 1003
Lemal, P. 662, 1029
Lema, M. 796
Lemarchand, M. 959
Le Marechal, C. 854
Lemée, E. 728
Leme, P.R. 561, 584
Lemon, K. 485
Lemonnier, N. 553
Le Morvan, A. 219, 586
Le Morzadec, T. 892
Lemosquet, S. 325, 332, 835
Lendl, B. 945
Lendormi, T. 854
Lenerts, A. 175, 176
Lenoir, G. 165, 201
Lenoir, H. 824
Lenoir, M. 726
Lensches, C. 644, 649
Leonard, F. 239
León-Ecay, S. 424
Lepar, J. 501
Lepeltier, F. 640
Lepers, A. 809
Lepori, A. 202, 208
Lerch, S. 189, 330, 677, 681, 683
Leroux, C. 485, 1027
Leroux, L. 985
Leroux, V. 643, 974
Le Roux, Y. 699
Leroy, G. 264, 381, 718
Le-Roy, P. 416, 782, 810
Lescane, A. 409
Lescoat, P. 726
Lesne, R. 658
Lesoudard, C. 458
Lessire, F. 687
Lessire, M. 232
Leterrier, C. 412, 729
Le Thi Thanh, H. 698, 700
Létourneau-Montminy, M.-P. 322, 324, 597, 711, 761, 766, 960, 983, 985, 987, 989
Le Trouher, A. 698, 700
Leung, Y.H. 358
Leury, B.J. 294, 953, 1005
Le, V. 474
Levallois, P. 464, 1018
Levesque, J. 761
Levrad, O. 371
Levrault, C.M. 302
Lev-Ron, T. 606
Lévy, F. 886
Lewis, E. 563
L., H.G. 291
Lhoste, E. 853
Liaubet, L. 483, 491, 492, 663, 789
Li, B. 290
Li, C. 222, 926
Li, C.J. 222
Lidauer, M.H. 281, 788, 935
Lidon, F. 292
Liebhart, D. 611
Lieboldt, M. 644, 649
Liénart, J. 735
Lien, S. 924, 927
Liere, P. 862
Lieubeau, B. 1018
Lifshitz, L. 566
Ligda, C. 776, 777
Ligeiro, C. 872
Ligero, M. 977, 982
Ligonesche, B. 201, 424
Ligonniere, J. 468
Li, H. 646, 832
Lillehammer, M. 793
Lima, J. 625
Lima, M.L.P. 372
Lim, C.B. 600, 742, 743, 879, 990
Lim, D.J. 525
Limier, J.-Y. 465
Lim, J.A. 525
Li, M.M. 996

Lim, Y.J. 525
Lindahl, C. 933
Lindberg, M. 933, 955
Lind, V. 446
Linke, K. 413
Lin, P. 338, 339
Lipkens, Z. 843
Lipkin, E. 168
Li, Q. 474, 479, 602, 618
Li, R. 256, 523
Li, S. 542
Lisiecka, J. 621
Lissy, A.-S. 301
List, D.J. 567
Liu, D.K. 1031
Liu, G.E. 515
Liu, H.N. 580, 765
Liu, J. 179, 308, 690
Liu, J.H. 1001
Liu, J.J. 429
Liu, J.Y. 1001
Liu, K. 794
Liu, M. 434
Liu, S. 593
Liu, S.J. 433
Liu, T. 352
Liu, X.M. 235
Liu, Y. 524
Liu, Y.Q. 880
Liu, Z. 417, 514, 786
Liu, Z.Y. 523
Li, W.L. 876
Li, X. 209
Li, Y. 432, 686, 847, 880, 1001
Li, Y.F. 684
Li, Z. 436
Lizarralde, J. 720, 723, 813
Lizzi, L. 244
Llach, I. 502, 777, 779
Llach-Martinez, I. 776
Llonch, P. 497, 498, 499, 500, 501, 884
Lluch, J. 230
L. Manuelian, C. 506
Lobón, S. 650, 722
Loeffen, M.P.F. 420
Loges, R. 176, 814, 817
Loh, G. 341
Loi, A. 651
Loick, N. 641
Loiotine, Z. 868
Lollivier, V.L. 557, 558
Lombard, S. 205, 730, 759, 892
Loncke, C. 324, 332, 835, 997
Long, S. 352, 362, 879, 1007
Loof, J. 376
Looft, C. 264, 381
Lopes, I. 228, 872
Lopes, M.M. 231, 885
Lopes, M.S. 397
López, A. 982
López-Alonso, M. 425, 426, 707, 716, 956
López-Carbonell, D. 203, 402
López-Catalina, A. 229
López De Armentia, L. 994
López, M. 593
López-Maestresalas, A. 424
López-Paredes, J. 414
López-Vergé, S. 884
Lopez-Villalobos, N. 252
Lorant, N. 897, 949
Lordelo, M. 596, 604
Loregian, K.E. 575
Lorenze, P. 1036
Lorenzi, V. 367, 1000, 1022, 1023
Lorenzo, J. 185, 426
Lorrette, B. 481, 750
Louadj, L. 589
Lounglawan, P. 445
Lourd, C. 857
Lourenço, A.L. 574
Lourenco, D. 158, 163, 782, 784, 787
Lourenco, J.M. 843
Lourenço, M. 460
Louro-Lopez, A. 641
Louveau, I. 482, 885
Lou, W. 376
Louwagie, I. 545
Løvendahl, P. 278, 279
Love, S. 886
Loy, D.D. 712
Lozada-Soto, E.A. 401
Lozano-Jaramillo, M. 820
Luan, T. 384
Luan, Y.Y. 234
Lucau-Danila, C.L. 682
Lucherk, L.W. 188
Lucht, H.L. 357
Luciano, A. 299
Lu, D. 213
Luimes, P. 582
Luise, D. 338, 339, 463, 715, 759, 760, 889
Luisier-Sutter, H. 569, 840
Luković, Z. 824
Lumpkins, B.S. 340
Luna, D. 763
Lund, M.S. 283
Lund, P. 561, 1012
Lundqvist, T. 978
Lundy-Woolfolk, E.L. 712
Luo, F.N. 523
Luo, H.L. 187, 430, 434, 435, 436, 517
Luotto, I. 949
Lurette, A. 250, 650, 700, 722
Lussiana, C. 219
Lute, C. 567
Lutz, T.A. 511
Luyt, K. 870
Luzi, F. 705

Lv, W.L. 434
L., W. 291
L., Y. 291
Lymbery, A. 428
Lynch, C. 527
Lynch, M.B. 439
Ly, P. 852
Lyubov, B. 653

M

Maares, M. 874
Maasdam, R. 303
Macciotta, N. 157, 168, 633
MacDougall, H. 497
Macedo Mota, L.F. 775
Machado Neto, O. 236
Macheboeuf, D. 571
Machefert, C. 510, 802
Macken-Walsh, Á. 628
Mackie, R.I. 625
MacLeod, I.M. 513
MacNeil, M.D. 671
MacQueen, D. 927
Mačuhová, J. 258, 893
Mačuhová, L. 257, 258, 259, 260, 262, 757
Madoui, M.A. 618, 749
Madrid, A. 650, 718
Madrid, L. 954
Madsen, J.G. 461
Madsen, O. 235, 353, 925, 1003
Maes, D. 355, 360
Ma, F. 356
Mafra Fortuna, G. 160, 918
Magalhaes, J. 537
Maggiolino, A. 354, 903
Magklaras, G. 218, 228, 764, 765
Magnani, E. 561, 574, 575, 584
Magrin, L. 366
Maguire, T. 320
Mahé, C. 1004
Maia, M.R.G. 734
Maicelo, J. 393
Maigaard, M. 1012
Maigné, E. 789
Maignel, L. 378, 969
Mailhot, R. 969
Maillet, G. 156
Maimaris, G. 253, 383, 804, 917
Maiorano, G. 467
Mairesse, G. 588
Maisano, A.M. 1023
Ma, J. 992
Makanjuola, B. 159, 666
Makgahlela, M.L. 671
Makkar, H. 590
Makkar, H.P.S. 344
Makridis, L. 638
Ma, L. 573
Malchiodi, F. 275, 406
Malecki, I.A. 405
Malgwi, I.H. 336
Malheiros, J.M. 521
Malikentzos, I. 865
Maljean, J. 348
Malsa, J. 861
Maltecca, C. 159, 401, 516
Mamani, M. 905
Maman, S. 491, 492
Mammi, L. 531
Manac'h, G. 985
Managos, M. 933
Manceau, J. 768
Manceau, P. 482
Mancilla-Leyton, J.M. 720, 723, 813
Mancin, E. 384, 983
Mandaluniz, N. 720, 723, 813
Manessis, G. 943
Mangan, M. 599, 609
Mangini, G. 903
Mangin, T. 479
Maniaki, M. 197
Maniatis, G. 598
Manoli, C. 717
Manomaitis, L. 958
Manoni, M. 570
Manse, H. 232
Manteca, X. 497, 499, 500, 884
Mante, J. 409
Mantino, A. 652, 721, 826
Mantovani, G. 641, 642
Mantovani, R. 384, 983
Mäntysaari, E.A. 788
Manuelian, C. 942
Manuelian, C.L. 944
Manzanilla-Pech, C.I.V. 276, 278, 373
Manzano, J.A. 560
Manzocchi, E. 958, 964
Mapiye, C. 428
Marandel, L. 193
Marcato, F. 530, 737
Marcatto, J.O.S. 282
Marchetti, L. 710
Marchewka, J. 240
March, M. 183
Marcon, D. 559, 779, 1034
Marcondes, M.I. 575
Marcon, M. 970
Maresca, A. 265
Maretto, L. 440
Marfaing, H. 841
Marguerie, J. 603
Marguerit, M. 998
Marichatou, H. 700
Marie-Etancelin, C. 510, 518, 520, 626, 1032, 1034, 1036
Marie, F. 421, 759
Marien, A. 348, 750
Marimuthu, J. 315

Marina, H. 769, 770
Marin, C. 717
Marinič, A. 622
Marinov, I. 185, 188
Marinov, M. 186
Marion-Poll, F. 867, 871
Markey, A. 410
Markland, L. 239
Markou, P. 253
Marnet, P.-G. 257, 653
Maroto Molina, F. 648
Marotz, C. 952
Marounek, M. 285
Marques, M.R. 292, 426, 971
Marquisseau, A. 1036
Marsault, A. 899
Martel, G. 719, 850
Martelli, P.L. 513
Martel, S. 516
Martens, J. 916
Martens, L. 771
Martens, S. 613
Martikainen, K. 162
Martín, A. 235, 300, 1001
Martin, B. 680, 725, 823, 825, 826, 827, 946
Martin, C. 328, 330, 561, 699, 706, 938
Martín-Collado, D. 650, 720, 722, 723, 813
Martineau, C. 539, 551, 562
Martínez, A.E. 982
Martínez-Álvaro, M. 400, 625
Martinez Del Olmo, D. 262, 660
Martínez-Fernández, J. 721
Martínez, M. 226
Martínez-Paredes, E. 216
Martínez-Talaván, A. 226
Martínez Villalba, A. 370
Martin, F. 924
Martín-García, A.I. 563
Martinić, O. 366
Martinidou, E. 613
Martini, V. 354
Martin, J. 895
Martín-Mateos, P. 945
Martin, O. 904
Martin, P. 257, 276, 409, 413, 809, 906
Martins, C.F. 596, 734
Martinsen, A. 913
Martinsen, K.H. 377
Martins, F. 482
Martins, L. 529, 596, 604
Martins, M. 668
Martín-Tereso, J. 490, 536, 564, 961, 966
Martin Tome, N. 216
Martin, X. 232
Marty-Gasset, N. 483, 492, 789, 1036
Marume, U. 428, 736
Marusi, M. 408
Mary-Huard, T. 786
Marzano, A. 511
Masagounder, K. 193
Masaki, T. 489
Maselyne, J. 974
Mason, B. 513
Mason, C.S. 532, 538
Massaro, S. 288, 574
Masselin-Sylvin, S. 548
Masseron, A. 474, 479, 618
Massimino, W.M. 892
Mastroeni, C. 297
Masuda, M. 279
Masuda, Y. 802, 805
Masuero, D. 613
Mata, K. 310, 316
Matamoros, C. 450
Matamura, M. 831
Mata-Nicolás, E. 256, 292, 300
Mateescu, R.G. 166
Mateo, J. 300
Mateos, I. 256, 292, 300
Mateos, J. 262, 660
Mateus-Vargas, R.H. 205
Mathieu, G. 451
Mathieux, S. 545
Mathot, M. 527, 897
Mathy, D.M. 682
Mathys, A. 343, 747
Matilla, J. 660
Matoni, L. 731
Mattalia, S. 664, 665, 784, 929
Mattia, A. 303
Mattiauda, D. 573
Mattiello, S. 887
Mattioli, S. 730
Matton, B. 837
Matzhold, C.M. 770
Maugan, L.H. 784
Mavrommatis, A. 300, 906
Maxa, J. 774
Ma, X.H. 436
Maxin, G. 219, 331, 717, 964
Mayeres, P. 410
May, F. 598, 601
Ma, Y.H. 234, 524
May, K. 544, 781, 919
Mazur-Kuśnirek, M. 571
Mazza, A. 511
Mazzoleni, S. 338, 461
Mazzoni, M. 338, 760
Mbuthia, J.M. 212
McAllister, T.A. 625
McAloon, C. 237, 534, 627
McAuley, R. 914
McCabe, M.S. 485, 624
McCarron, P. 909
McCarthy, B. 439
McCarthy, K. 689
McCarthy, M.M. 708
McCaughern, J.H. 715

McClearn, B. 776, 777, 778
McCormack, A. 746
McCormack, U.M. 834
McDermott, A. 420
McDonagh, M. 934
McDougall, H. 264
McEwan, J. 278, 281, 655, 1031
Mc Fadden, M. 534
McGarr-O'Brien, K. 547
McGee, M. 181, 487
McGilchrist, P. 310, 313
McGovern, F. 248, 437, 657, 688, 909, 934, 1033
McGrane, L. 676
McGuire, R. 174, 269
McHugh, N. 167, 248, 676, 909, 914, 1033
McKay, S. 804
McKay, Z. 564
McKenzie, A.M. 715
McLaren, A. 181, 249, 497, 776, 777, 778, 1032
McLaughlin, S. 825, 826, 827, 829, 942
McLean, K.A. 181, 1032
McNaughton, L.R. 275, 661
McNeilly, T.N. 625
McNicol, L. 729
Mc Pherson, S. 534
McRae, K. 419, 1031
Meale, S.J. 703
Meatquality Consortium, P. 827, 940
Méda, B. 322, 465, 546, 597, 599, 605, 711, 762, 960, 987
Medale, F. 193
Medeiros, S.R. 954
Medina, C. 881
Medina-Fernandez, S. 737
Medina, P. 393
Medjadbi, M. 298, 299
Meehan, D. 315
Mehtiö, T. 281
Mei, C. 1031
Meier, S. 513, 801
Meijboom, F.L.B. 645, 883
Meijer, N. 347, 747
Meikle, A. 573
Mei, Q. 790
Meister, M. 318
Mejía, A. 905
Melbaum, H. 387
Mele, M. 338, 565, 652, 721, 826
Meli, G. 357
Mellouk, A. 746, 1003
Melville, L. 264, 780
Même, N. 647, 762
Mena-Guerrero, Y. 720, 723, 813
Menant, O. 732
Menasseri, S. 854
Menassol, J.-B. 247, 249, 250, 553, 878
Mendes, N.S.R. 318
Mendizabal, J.A. 429
Mendowski, S. 588, 1030
Menendez, H. 172, 679
Menendez Iii, H.M. 326, 755
Menghi, A.M. 814
Meng, Q. 430, 432, 433
Menino, R. 228
Menoury, V. 287
Mens, A. 541
Mentschel, J. 488
Merbold, L. 153, 263
Mercadante, M.E.Z. 282, 521, 583, 584, 702, 704, 795
Mercandalli, S. 550
Mercat, M.J. 201, 824, 925
Mercerand, F. 232
Mercier, Y. 472
Merhaz, M. 198
Meriaux, L. 1030
Merino, N. 292, 300
Merle, L.-A. 768
Merlini, M. 303
Merlino, V.M. 828
Merlin, S. 516
Merlot, E. 206, 231, 557, 558, 885
Mermillod, P. 668, 1004
Merritt, T. 543
Mertens, A. 527, 897
Mes, J.J. 221
Messman, M. 845
Mészáros, G. 381, 408, 907, 915
Metges, C. 212, 607, 609, 611
Mettauer, R. 441
Meunier, B. 189, 316
Meurisse, S. 258, 261, 548
Meuwissen, T. 158, 384, 785, 787, 791
Mevel, M. 666
Meyer, M. 454, 455
Meyermans, R. 335, 340, 402, 404, 417, 916, 976, 979
Meylan, M. 533
Meynadier, A. 510, 518, 520, 626, 717
Mezdour, S. 867, 871
Mhlongo, N.L. 774
Miara, M.M. 453
Michalski, M.C. 680
Michaud, A. 720, 722, 725
Michelet, C. 298, 299, 890
Michelotti, T.C. 993
Michel, V. 602, 1003
Michenet, A. 615
Michez, D. 348
Michiels, J. 894
Michot, P. 522
Michou, T. 637
Middelkoop, A. 410, 740
Mielczarek, M. 199, 793, 794
Miglior, F. 159, 275, 406, 527
Mignon-Grasteau, S. 465, 605, 730
Mikołajczak, Z. 224, 621
Milanesi, M. 721

Milaon, M.M. 1037
Miles, A. 804
Milgen, J.V. 863
Milkevych, V. 278, 279, 407, 512
Miller, D. 1017
Miller, G. 317
Miller, S. 692
Miller, S.M. 315
Millet, S. 337, 340, 417, 460, 758, 864, 988
Milochau, G. 708
Milojevic, V. 859
Milora, N. 631
Milotic, D. 428
Miltiadou, D. 291, 509
Minatchy, N. 289
Mincheva, N. 225
Mindus, C. 768
Minela, T. 874
Minero, M. 861
Minoudi, S. 555
Minozzi, G. 623
Min, T.M. 286, 1004
Minuti, A. 992
Minviel, J.J. 556
Min, Y.J. 741
Miquel, M. 778, 779
Miralles-Bruneau, M. 851
Miranda, C.O. 372
Miranda, M. 425, 426, 707
Misztal, I. 158, 160, 163, 782, 784, 787
Mitchell, G. 264
Miteva, T. 188
Mitev, J. 188
Mitliagka, P. 912
Mitlianga, P. 509, 703, 912
Mitsunaga, T.M. 939
Mizrahi, I. 929
Moakes, S. 265, 822
Moallem, U. 353, 566
Modzelewska-Kapituła, M. 582
Mogensen, L. 539, 847, 950
Mohamed-Brahmi, A. 722
Moisan, M.P. 200
Molee, A. 487
Molee, W. 487
Molina, A. 876, 881, 978, 980, 981, 982
Molina, E. 994
Molina, G. 903
Molinatto, G. 623
Molinero, E. 290, 525
Molist, F. 410, 740
Molle, G. 502
Møller, H. 847
Molloy, J. 902
Moll, X. 251
Momot, M. 546
Monaghan, A. 437, 688
Mondet, F. 619
Mondry, R. 590
Monestier, C.M. 554, 892
Monfoulet, L.E. 485
Montagne, L. 207, 421, 559, 562
Montagnin, C. 367, 1000
Montaholi, Y.R. 755
Montalvo, J. 1013
Montanari, C. 822
Montanari, M. 574
Montañes-Foz, M. 880
Monteiro, A. 709, 710, 828, 971
Montossi, F. 423
Montoya, A.C.V. 934
Moon, J. 833
Mora-Cuadrado, V. 798
Moradei, A. 865, 873
Moradi, M. 665
Moraes, M.J. 772
Moraine, M. 441
Morales, A. 358
Mora, M. 648, 971
Moran, B. 268
Moran, C.A. 295, 938
Morante, R. 905
Morardet, N. 728
Moravčíková, N. 210, 390, 392
Moreau, S. 733
Moreira, G.C.M. 927
Moreira, O. 228, 295, 297
Morel, I. 189, 677, 681, 683
Moreno, E. 809
Moreno-Oyervides, A. 945
Moretti, M. 826
Moretti, P. 530
Morey, L. 725
Morgan-Davies, C. 249, 496, 497, 503, 776, 777, 778, 779
Morgan, S.J. 891
Morgant, G. 1018
Morgavi, D. 149, 626, 633, 929, 938
Morgenthaler, C. 975
Morge, S. 728
Morin, J.-F. 750
Morin, L. 640
Morisset, F. 576
Morisson, M. 232, 997
Mörlein, D. 204, 242, 884
Morris, D.W. 485, 624
Morrison, S. 269, 563
Morvezen, R. 669
Moschakis, T. 943
Moscovice, L.R. 883
Moser, H. 945
Moskala, P. 687
Moškrič, A. 622
Mosnier, C. 440, 527
Motiang, M.D. 820
Mottet, A. 150, 153, 263, 718, 846
Mouhrim, N. 343
Moula, N. 510

Moulin, C.-H. 698
Mounaix, M. 1021
Mountzouris, K.C. 286, 738, 836, 855
Mourits, M. 821
Mousavi, M. 568
Mousqué, S. 887
Moya, D. 755
Moyano Lopez, F. 225
Moyes, S.M. 309
Mpendulo, C.T. 910
Mrode, R. 670
Mucha, A. 469
Mucha, S. 1034
Muenger, A. 565
Mugambe, J. 528, 915
Mugnier, S. 215
Muhlig, B. 539
Muigai, A.W.T. 164
Muíños, A. 425, 426, 716
Mukiibi, R. 928
Mulder, H. 161, 335, 376, 404, 662, 929
Mullaart, E. 409
Mullan, S. 238, 732, 860
Müller, F.L. 445
Müller, K. 840
Müller, U. 1026
Munezero, O. 742
Münger, A. 271, 293
Mun, J. 349, 350, 363, 364
Munk, A. 950
Muñoz, J.A. 583, 584, 795
Muñoz, M. 404
Muñoz-Tamayo, R. 165, 323
Muns, R. 646
Munsterhjelm, C. 968
Murawska, D. 226, 361
Murillo, Y. 909
Mur, L. 290, 422
Muroya, S. 592
Murphy, J.P. 534, 689
Murphy, M. 304
Murray, Á. 439
Murta, D. 228, 351, 480, 872
Musa, A.A. 792
Müsse, J. 598, 601
Muvhali, P.T. 405
Mwacharo, J.M. 152

N

Nacarati Da Silva, I. 260
Nadal-Desbarats, L. 862
Nade, S. 830
Nagai, R. 519
Naglis-Liepa, K. 175, 176
Nagy, I. 393
Naji, M. 390
Nakagawa, S. 805
Nakahori, Y. 805
Nalovic, A.N. 361
Nanchen, C. 535
Napolioni, V. 675, 918
Napora-Rutkowski, L. 199
Naranjo, V.D. 988
Narayana, S. 275, 406
Narcy, A. 322, 709, 711, 960
Naser El Deen, S. 220, 221, 223, 347, 520
Nasri, W. 442, 704
Nassar, M.K. 386
Nassy, G. 265
Nataloni, L. 463
Natonek-Wiśniewska, M. 796
Natterer, C. 771
Nauta, S. 734
Navajas, E.A. 796, 1035
Nava, V. 284
Naves, M. 409, 673
Nazari Ghadikolaei, A. 799
Ndiwa, N. 164
Ndung'u, P. 263
Neary, J.N. 767
Neave, H.W. 818
Nedelkov, K. 181
Nedeva, I. 184
Neglia, G. 922
Negrão, J.A. 372
Negrini, C. 338, 463, 715, 759, 760
Negro, S. 203
Negussie, E. 281, 935
Nehme, R. 890
Neira, L.M. 772
Nejad, J.G. 359
Nelli, A. 764, 765
Nemutandani, K.R. 672
Nengas, I. 197
Neofytou, M.C. 291, 509
Nery, J. 828
Neser, F.W. 261, 671, 674, 701, 725, 807
Neto, O.R.M. 522, 837
Neuenschwander, S. 339
Neumann, G.B. 386, 781, 919
Neupane, M. 804
Neveu, A. 179, 307, 308, 310, 318, 320, 692
Neveux, C. 860
Nevo Yassaf, I.N.Y. 617
Newberry, R.C. 238
Newman, S.-A. 419
Newton, E.E. 590
Ng Kai Lin, J. 194
Nguluma, A. 670
Nguyen-Ba, H. 157
Nguyen, N.V. 191
Nguyen, T.V. 513
Nhara, R.B. 736
Nicholas-Davies, P.K. 308
Nicklas, D. 774
Nicolas, E. 246
Nicolás-Jorrillo, C. 296

Nicolazo De Barmon, A. 180, 432, 825, 826, 827, 829, 942
Nicolazo, T. 206, 647, 970, 1015
Niderkorn, V. 571
Ní Dhufaigh, K. 624
Niehoff, T. 171, 388, 398
Nielsen, H.M. 407, 614, 615, 618
Nielsen, L. 206
Nielsen, M.O. 473, 1012
Nielsen, P.P. 769
Nielsen, R.K. 279
Nielsen, T.S. 709
Niemi, J. 501, 628, 729, 731, 895
Niermans, K. 348, 747
Niewind, P. 370, 603
Nijland, H.J. 442
Nikodinoska, I. 295, 938
Nikolaou, K. 765
Nikoloudaki, C. 197
Nilforooshan, M.A. 162
Nilsson, K. 201, 202, 864
Ninane, V. 348
Ning, J.X. 233
Nisbet, H. 317
Nishikawa, Y. 279
Nishino, D. 233
Nishio, M. 667
Nishio, N. 661
Nishiura, A. 283, 411, 805, 807
Niu, J.Z. 433
N. Moghadam, N. 873
Nobrega Cardoso, R.K. 216
Noël, F. 308, 659
Nogalski, Z. 546, 1024
Nolan, S. 934
Nonaka, I. 283
Nooijen, I. 996
Nordbø, Ø. 377
Nordhagen, S. 153
Nørgaard, J.V. 461, 709
Normand, J. 180, 258, 316, 432, 828
Normand, V. 206, 647, 970, 1015
Notenbaert, A.M.O. 151
Note, P. 184
Noutfia, A. 510
Novales, M. 982
Nowak, B. 931
Noziere, P. 331
Nozière, P. 287, 325, 328, 561, 586, 683
Nudda, A. 967
Nugrahaeningtyas, E. 359, 466, 890
Nunes, C.L.N. 317
Nunez Andrade, A.N. 166
Núñez, P. 378, 757
Nwosu, E.U. 417
Nyamiel, A. 559
Nziku, Z. 670

O

Obach, A. 191
Obari, C. 159
Obinata, R. 795
Obitsu, T. 933
Obšteter, J. 160, 620, 918
Ocak Yetisign, S. 776, 777
Ocepek, M. 238
O'Connell, S. 716, 996
O'Connor, E. 1033
O'Connor, R.C. 246
Oczkowicz, M. 210, 493
Ødegård, J. 791
Odintsov Vaintrub, M. 244, 247
Odo, A. 646
O'Doherty, J.V. 337, 612
O'Donnell, C. 934
O'Donoghue, S. 485, 624
O'Donovan, M. 304, 438, 439, 689
O' Driscoll, J. 676
O'Driscoll, K. 236, 237, 239, 240, 608, 732
O'Driscoll, K.M. 238
O'Flaherty, V. 839, 934, 952
Ogawa, S. 661, 667, 795
Oget-Ebrad, C. 486, 1037
Ogink, N.W.M. 302
Ogino, A. 519
O'Hara, E. 628
Oh, J. 583
Oh, S. 363
Ohtsuka, A. 606
Oikonomou, G. 767, 1019
Ojango, J. 164, 670
Ojeda-Marín, C. 389
Okamura, T. 661, 667
Okeyo, M. 670
Okorski, A. 741
Okrathok, S. 668, 744
Oladosu, O.J. 611
Olejnik, K. 644
Olgun, O.O. 462
Olijhoek, D.W. 1012
Oliveira, C. 426
Oliveira, H.R. 275, 406
Oliveira Júnior, G.A. 406
Oliveira, P.P.A. 727, 954
Ollagnier, C. 471, 472, 611, 612, 1019
Olleta, J.L. 422, 427, 593
Olmos Antillón, G. 1016
Olsen, H.B. 415
Olsen, H.F. 274, 847
Olson, K. 155, 326
Olsson, C. 978
Oltenacu, P.A. 166
Olukoya, F.O. 195, 198, 199
O'Mara, F. 152, 270
Omphalius, C. 218, 591, 859, 1020
Omri, H. 422
Ondé, D. 723

O'Neill, H.A. 261
Oono, Y. 795
Oosterlinck, M. 979
Oosting, S.J. 280, 376, 567
Oostvogels, V. 440, 442
Opalinski, S. 644
Oravcová, M. 257, 258, 259, 260, 757
O'Reilly, R. 311, 431
O'Riordan, E. 269, 681
Orlianges, M. 540, 551
Ormston, S. 588, 840
Ortega, G. 640, 999, 1000
Ortigues-Marty, I. 334, 494, 991, 992, 997, 999
Ortín-Bustillo, A. 240
Ortiz-Chura, A. 626, 633
Ortiz, J. 170
Ortuño, J. 477
Orvalho, T. 294
Oscarsson, H. 154
Osório, H. 524
Ostendorf, C.S. 995
Østergaard, S. 403, 814
Oster, M. 471
Oteri, M. 284
Ottati, S. 623
Ottoboni, M. 873
Oudshoorn, F.W. 823
Ouedraogo, L. 697, 698
Ouedraogo, S. 697, 698
Oundjian, C. 345, 351
Oyieng, E. 164
Ozarak, E. 673

P

Pabiou, T. 280, 914
Pacífico, C. 835
Paciocco, F. 244
Paes, C. 745
Paganoni, B. 380
Pagotto, U. 513
Pailhoux, E. 929
Pain, B. 482
Paiva, M.I. 371
Palamidi, I. 190, 286, 738, 855
Palhière, I. 381, 397, 398, 811, 1038
Pallotti, S. 675, 918
Palmans, S. 460
Palmieri, L. 613
Palomo, R. 763
Palumbo, F. 463
Panagiotaki, P. 197
Panayidou, S. 253, 254, 416
Paniagua, M. 837
Pannebakker, B.A. 616
Pannier, L. 309, 311, 312, 431
Panousis, N. 528, 529, 531
Panserat, S. 997
Panunzi, L. 749
Panzuti, C. 714, 907
Papadomichelakis, G. 286
Papadopoulos, V. 703
Papadouli, C. 197
Papanikolopoulou, P. 672
Papanikolopoulou, V. 369, 555, 671, 917
Paparamborda, I. 446, 688
Papatzimos, G. 509, 703, 912
Pappas, A. 286
Paquet, É. 587
Paquet, E.R. 972
Paraskeuas, V.V. 286, 738, 836, 855
Parias, C. 856
Parisot, S. 442, 704, 779, 1037
Parker, J.K. 588
Park, G.W. 359, 855, 890, 1020
Park, H.J. 741
Park, J. 363
Park, J.-E. 521
Park, J.S. 990
Park, K.H. 359, 466, 555, 855, 890, 1020
Park, K.K. 359
Park, S. 349, 350, 363, 364
Park, S.H. 214, 227
Park, S.Y. 990
Park, W. 521
Park, W.C. 525
Parra, M.C. 703
Parreiras, M. 295
Parr, T. 476, 748
Partridge, G. 428
Pascal, G. 738
Pascual, J.J. 216
Pascual, M. 648, 811
Pasomboon, P. 197
Pasquereau, T. 756
Pasquiet, B. 457
Pasquini Neto, R. 727
Pasri, P. 668
Pastell, M. 501, 968
Pasternak, M. 447, 908, 911, 913
Pastor, J.J. 884
Patris, B. 887
Patsios, S.I. 249, 503
Paudyal, S. 931
Paul, A. 216
Paula, E.M. 561, 574, 575, 584
Pauler, C.M. 678
Paul, K. 474, 618
Paulos, K. 292, 294
Pausch, H. 927
Pavlík, I. 390
Pavlov, B. 622
Pawlak, P. 931
Payet, A.-L. 852
Payet, V. 828
Payne, C. 312, 380, 651
Pazzola, M. 902
P., C. 291
Peana, I. 168

Péchernart, E. 465
Pećina, M. 391
Pečnik, Ž. 654
Pedeches, R. 850
Pedersen, L.J. 497, 501, 883, 1018
Pediconi, D. 675
Pedro, M.F. 833, 839
Pedro, S. 529
Pedroso, A.F. 727, 954
Peeters, C.F.W. 302
Peeters, K. 159, 335, 614, 615
Peetz Nielsen, P. 154
Pegolo, S. 190, 230, 633, 775
Peguero, D. 343
Peiren, N. 932
Peláez Acero, A. 260
Pelech, I. 353
Pellicer-Rubio, M.T. 878
Pellikaan, W.F. 567
Pelster, D. 983
Pena, R.N. 515, 525
Peña, Z. 876, 881
Penchev, I. 179, 180
Penen, F. 710, 959
Penev, T. 188
Peng, C.F. 580
Peng, O. 222
Peng, P. 222
Penndu, D. 424
Perčič, T. 178
Perdomo-González, D.I. 977, 978, 980, 981
Perea, J. 270
Pereira, A.M.F. 529, 705, 1014
Pereira, G.L. 522, 837
Pereira, V. 716
Perez, B.C. 925
Pérez, C. 370
Pérez-Cabal, M.A. 414, 813
Pérez-Calvo, E. 470, 834, 1005
Pérez-Ibarra, I. 721
Pérez, J.F. 607
Pérez Marín, C.C. 648
Pérez Marín, D.C. 648
Perreten, V. 533
Perricone, V. 285
Perryman, K. 955
Persson, U.M. 949
Perucho, L. 777
Peruzza, L. 928
Pesenti Rossi, G. 503, 530, 535
Pestana, J. 186
Petersen, G. 620
Petersen, G.E.L. 622, 754
Peterson, C.B. 708
Pethick, D. 308
Pethick, D.W. 309, 313, 431
Petillon-Pronk, A. 895
Petiot, V. 697
Petitprez, M. 1013
Petit, T. 546
Petre, R. 266
Petrova, A. 225
Petrova, T. 186
Petrov, Z. 186
Petrusán, J.-I. 227
Pétursdóttir, Á. 588, 590
Pevzner, P. 627
Peyraud, J.L. 156
Peyrichou, F. 345, 351, 749
Pezzopane, J.R.M. 705, 727, 954
Pfeifer, C. 265, 810, 822
Pfeiffer, M. 456
Pflanzer, S. 429
Pflanzer, S.B. 690
Pham, D.H. 191
Philau, S. 993
Philibert, A. 809
Philippeau, C. 215, 425
Philippon, O. 850
Phocas, F. 169, 619
Phomvisith, O. 592
Phyn, C.V.C. 513, 801
Piacère, A. 877
Piano, A. 862
Piantoni, P. 845
Picard, D. 201
Picciolini, M. 675, 918
Picconi, S. 657
Piedrafita, J. 484
Piégu, B. 886
Pierce, K.M. 817
Pieterse, E. 345, 866
Pignagnoli, A. 827
Piirsalu, P. 776, 777, 778
Piles, M. 648, 811, 971
Pilla, F. 903
Pillan, G. 369
Pille, F. 979
Pimentel, E.C.G. 783, 920
Pimpão, M. 426
Pinard Van-Der-Laan, M.H. 647
Pineda-Quiroga, C. 803
Piñeiro, J.M. 542, 931
Pinel, K. 193
Pinho, L.F. 772
Pinloche, E. 746
Pinotti, L. 293, 299, 338, 461, 570, 873
Pinto De Andrade, L. 668
Pinton, A. 781
Pinto, S. 772, 775
Piqué, J. 237
Piquer, L. 226, 760
Piquer, O. 760
Pirault, J. 759
Pires, J. 334, 485, 507, 991, 992, 997, 999, 1027, 1039
Pirlo, G. 527
Pishgarkomileh, H. 301

Pissard, A. 750
Pitchford, W. 313, 315, 692
Pitel, F. 928, 930
Pitino, R. 642
Pitino, R.G. 641, 642
Pittois, D. 797
Pi, Y. 213
Planchenault, D. 470
Plante-Dubé, M. 582
Plasman, L. 348, 478
Plastow, G. 926, 927
Płatosz, N. 588
Pleasants, A.B. 431
Pleissner, D. 227
Plesch, G. 825, 826, 827
Plets, D. 771
Plieninger, T. 396
Ploegaert, J.P.M. 302
Plomaritou, A. 262, 635, 636, 637, 1011
Pluk, P. 844
Plumstead, P. 899
Pluschke, H. 730
Poccard-Chapuis, R.J.M. 726
Pocrnic, I. 160, 170, 402, 670
Pogorzelska-Nowicka, E. 696
Pogorzelska-Przybyłek, P. 546, 582
Pogorzelski, G. 696
Pohlmann, W. 684
Point, S. 841
Poissonnet, A. 733, 1015
Poisson, W. 877
Poklukar, K. 930
Polasik, D. 211
Pol, F. 1022
Politis, I. 190, 286, 738, 855
Polkinghorne, R. 307, 308, 310, 320, 692
Pomiès, D. 547
Pommaret, A. 258, 659
Pomport, P.H. 579
Ponchunchoovong, S. 910
Pong-Wong, R. 414
Ponieważ, A. 361
Pook, T. 171, 387, 398, 399, 785
Poompramun, C. 487
Poonyachoti, S. 511
Pope, P. 929
Popiela, E. 644
Popova, M. 626, 633, 938
Portes, D. 704, 779
Portnick, Y. 566
Portugal, A.P. 297
Portugal, P.V. 292
Poton, P. 993
Potot, S. 470
Pouil, S. 169
Poulain, S. 745
Poulet, J.L. 508, 754
Poulopoulou, I. 613
Poupin, M. 243
PPILOW Consortium 729
Prache, S. 184, 683, 752
Prat-Benhamou, A. 723, 813
Prates, J.A.M. 186, 493, 518
Prathap, P. 953
Prat, N. 631
Pravettoni, D. 530
Prešern, J. 476, 620, 622
Prestløkken, E. 900
Preston, F.L. 315
Prevedello, P. 366
Prévéraud, D. 746
Pringle, T.D. 843
Prišťák, J. 390
Priymenko, N. 717
Probo, M. 678
Probst, J. 645, 649
Probst, S. 806
Prodhomme, O. 658
Prokešová, M. 192
Pröll-Cornelissen, M.J. 812
Promp, J. 809, 929
Proust, M. 658
Prunier, A. 205, 559, 733, 752
Prunier, J. 877
Pryce, J. 417, 513, 514, 801
Pszczola, M. 494, 495
Ptáček, M. 257
Pucéat, C. 1004
Puchała, M. 447, 908, 911, 913
Puech, T. 289
Puff, C. 335
Pugh, G. 841
Pugliese, C. 678
Puillet, L. 321, 331, 753, 904
Pulina, G. 967
Pulkoski, M. 475
Pulley, S. 641
Puntigam, R. 963
Purevdorj, M.P. 907
Purfield, D. 280, 676, 914
Pursley, J.R. 874
Purslow, P. 435
Purwin, C. 571, 582
Puškadija, Z. 622
Puterflam, J. 603
Pyoos, G.M. 671

Q

Qi, F. 646
Qin, N. 588
Qiu, Y. 593
Quatto, P. 887
Queiroz Silva, B. 693
Quéméneur, K. 207
Quenol, H. 543
Quentin, M. 731, 991
Quevedo Cascante, M. 265
Quilcate, C. 393

Quilichini, N. 358
Quina, E. 905
Quiniou, N. 203, 207, 240, 970
Quintanilla, R. 929
Quintas, A. 480
Quispe, E. 909
Quispe, M. 909

R

Radev, V. 184
Radko, A. 210
Radomski, P. 687
Raemy, M. 797
Raffrenato, E. 367, 898
Rafiq, R. 558
Ragionieri, L. 561
Rahman, M.A. 842
Rahmatalla, S.A. 386
Rahmel, L. 713
Raineri, C. 704
Rajagopal, R. 983, 985
Rakngam, S. 668
Ramalho, J. 529
Ramalho-Ribeiro, A. 195
Ramanzin, M. 440
Ramayo-Caldas, Y. 929
Ramé, C. 356
Rame, C.R. 882
Ramin, M. 446, 955, 1011
Ramirez-Agudelo, F. 894
Ramirez-Agudelo, J.F. 331
Ramirez-Garzon, O. 565, 932
Ramirez Mauricio, M.A. 940, 944
Ramljak, J. 391, 427
Ramón, M. 663, 929
Ramonteu, S. 156
Ramos, G.G. 772
Rampado, N. 179
Rampin, O. 890
Ramsbottom, G. 817
Ramsey, R. 656
Ranches, J. 246
Ranger, B. 156
Ranilla, M.J. 256, 292, 300
Raniolo, S. 440
Raoul, J. 397, 398, 401
Rapey, H. 215
Rapinel, V. 287
Rastello, L. 219
Rat, C. 356
Ravanetti, F. 561
Raver, K. 931
Ravon, L. 230, 605
Rawski, M. 224
Raymundo, A. 596, 604
Raynaud, S. 728
Ray, P.P. 572
Real, D. 651
Realini, C.E. 420, 423
Rebel, J.M.J. 287, 353, 541, 737
Rebucci, R. 695, 710
Recio, A. 499, 500
Recoules, E. 762
Recous, S. 728
Reding, E.R. 682
Reding, R. 797, 1024, 1026
Redoy, M.R.A. 952
Rees, R. 174, 175, 656
Reeves, M. 264, 497
Regnier, E. 823
Regrain, N. 602
Rehan, I. 228, 872
Reheul, D. 932
Reiche, A.-M. 293
Reigner, F. 459, 856, 861, 862
Reignier, S. 243
Reijs, J.W. 267
Reimer, C. 387, 417, 514
Reinikainen, A. 682
Reinsch, N. 792
Reinsch, T. 176
Reisdorffer, L. 707
Reisinger, N. 835
Reis, M. 281, 960
Reißmann, M. 386, 781
Reixach, J. 378, 757
Rell, J. 535
Relun, A. 371, 536
Rema, P. 195
Remissiondairy Consortium 771
Rémond, D. 834
Remot, A. 993
Renaudeau, D. 240, 663, 759, 970
Renaut, J. 518
Ren, G.D. 434
Ren, I. 769
Renieri, C. 675, 918
Renna, M. 217, 219, 828, 887
Rensing, S. 937
Reolon, H. 575
Répérant, J.M. 1022
Resconi, A. 868, 872
Resconi, V. 422, 423, 427, 593
Resende, F.D. 713, 714
Resende, V. 362
Resmond, R. 885
Restoux, G. 158, 381, 385, 824
Revell, C. 651
Reverchon-Billot, L.R.B. 554
Reverchon, M. 605, 730, 882
Reverter, A. 487, 632, 889
Rey-Cadilhac, L. 752, 941
Reyes, D.C. 842
Reyes-Palomo, C. 827, 940
Reynaud, K. 1004
Reynolds, E.G.M. 162
Reynolds, J.P. 766
Rezaei Far, A. 220, 221, 223, 347, 520

Rezende-De-Souza, J.H. 690
Rezende, F.M. 166
Rezende, V.T. 702, 704
Riaboff, L. 971
Ribani, A. 513, 623, 920
Ribas, C. 893, 896
Ribeiro, D.M. 484, 493, 518, 524
Ribeiro, E.G. 372
Ribeiro, R.V. 522, 837
Ribeiro, T. 480, 872
Ribeiro, V. 824
Ricard, A. 975
Ricard, E. 756
Ricci, A. 828
Ricciardi, C. 345, 351, 749
Ricci, S. 835
Richard, J. 479, 481
Riche, P. 901
Rico, D.E. 229, 358, 587, 761, 766
Rico, J.E. 761, 766
Riddersholm, K.V. 800
Riedel, A. 601, 1012
Rieder, S. 244, 249, 650
Riesch, F. 245, 441
Riesinger, P. 963
Rigaudeau, D. 192, 198
Riggio, V. 605, 925
Righi, F. 367, 636, 641, 642
Rigolot, C. 723
Rigon, F. 575
Rigos, G. 197
Rikkers, R.S.C. 469, 616, 773, 1003
Rinne, M. 295, 815
Rinn, M. 496, 908
Rios, H.N. 712
Ripamonti, A. 652, 826
Ripoll-Bosch, R. 268, 442, 949
Ripollés-Lobo, M. 977
Ripollés, M. 860, 982
Riquet, J. 200, 663
Rischkowsky, B. 152
Rispal, E. 328
Ristic, D. 344
Rius-Vilarrasa, E. 914, 923
Riuzzi, G. 639
Riva, M.G. 499, 861
Rivas, I. 707
Rivero, M.J. 265
Rivet, P.F. 320
Riviere, J. 522
Robert, C. 877, 975
Robert-Granié, C. 261, 802
Robert, J. 774
Robin, L. 658
Robin, P. 301
Robinson, F. 227
Robinson, T.P. 949
Robledo, D. 928
Rocha, B.M. 575
Rochat, T. 192
Roche, J.R. 817
Rochette, Y. 328
Rochus, C. 159, 406
Rodeghiero, M. 440
Rodehutscord, M. 517, 526, 739
Ródenas, L. 216
Rodenburg, T.B. 731, 734
Rödiger, M. 556
Rodrigues, G.R.D. 702, 704
Rodrigues, L. 236, 831
Rodrigues Silva, R. 320
Rodriguez, A. 798
Rodriguez, E.E. 166
Rodríguez-Estévez, V. 827, 940
Rodríguez-Prado, M. 636
Rodríguez-Ramilo, S.T. 382, 397
Rodríguez Silva, T. 824
Roehe, R. 625
Roehrig, N. 823
Roffi, S. 985
Rogina, S. 178
Rohmer, T. 474
Röhrig, N. 265
Roh, S. 489
Roig-Pons, M. 368, 650, 888
Roinsard, A. 205, 730
Rojas De Oliveira, H. 808
Rojo, S. 226
Rolla, U. 889
Rollet, N. 643, 974
Roma Jr., L.C. 939
Roman-Garcia, Y. 572, 845
Rombouts, T. 314
Romeo, A. 710
Roméo, A. 715
Romeo, C. 1022, 1023
Romero-Huelva, M. 265
Romero, J. 422, 427, 593
Romero, P. 563, 938, 1007
Romero, S. 427
Romijn, H. 825
Rondia, P. 467, 545
Rönnegård, L. 769, 770
Röös, E. 267, 847
Ropka-Molik, K. 211
Rosa García, R. 872
Rosati, A. 924
Rose, N. 1022
Rosen, B. 381
Rose, V. 214
Ros-Freixedes, R. 160, 525
Roskam, E. 839, 952
Rossano, A. 533
Rossi, D. 922
Rossi, G. 303
Rossi, U. 303
Rostellato, R. 784, 852
Rothacher, M. 681

Roth, C. 739
Roth, K. 812
Rothmann, C. 288
Rouger, R. 385, 824
Rouillé, B. 640, 841
Rousing, T. 1018
Rousset, N. 553, 603, 1021
Roux, D. 485, 991, 992
Rovadoscki, G. 309
Rovere, G. 666
Rovira, F. 1036
Rowe, S. 278, 281, 419, 1031
Rowe, T. 308
Royer, E. 844
Roy, J. 193
Rozier, A. 673
Ruch, M. 985
Ruckli, A.K. 986
Rueda, S. 290
Ruelle, E. 304, 438
Ruet, A. 457
Rufener, C. 535
Ruggeri, R. 611, 1019
Ruggia, A. 796
Ruíz-González, A. 358
Ruiz, R. 653, 654, 656, 720, 776, 777
Rumbos, C.I. 227, 865, 872
Rupp, R. 261, 486, 507, 1037, 1038, 1039
Rusev, R. 196
Ruska, D. 175, 176, 301
Russell, J. 712, 713
Russon, A. 1019
Ryan, C.V. 280
Rychen, G. 699
Rydal, M.P. 410
Rydhmer, L. 267
Rymer, C. 572
Ryschawy, J. 441
Rzewuska, K. 812

S

Sabbah, M. 589
Sabetti, M.C. 641, 642
Sabri, A. 735
Sabrià, D. 631
Sacarrão-Birrento, L. 484, 685
Saccenti, E. 567, 992
Saci, S. 230
Sacy, A. 745
Sadaillan, J.-M. 851
Sader, G. 614
Sadoud, M. 189
Sadri, H. 490
Saedi, N. 283
Saeed-Zidane, M. 232, 875
Safari, M. 973
Sagevik, R. 377
Sagot, L. 263, 778, 779
Sahana, G. 278, 279, 614, 618
Sahar, M. 329
Sahraoui, N. 186, 524
Saint-Dizier, M. 1004
Saint-Hilaire, M. 699
Saintilan, R. 786, 790
Sairanen, A. 333
Saito, A. 233
Saito, K. 233
Saito, Y. 411, 807
Sakaridis, I. 369, 917
Sake, B. 601
Sakkas, P. 358
Sakomura, N. 960
Sala, G. 354, 530
Salah, E. 722
Salah, N. 328
Salama, A.A.K. 251, 484, 499, 500, 1023
Salami, S.A. 188, 340
Salari, S.P. 478, 747
Salaris, S. 776, 777
Salavati, M. 927
Salazar, L.C. 215
Salazar, W. 393
Saldaña, B. 1009
Sales, J.R. 604
Sales, P. 652
Salgado, P. 442, 702, 704
Salimiyekta, Y. 800
Salis, L. 328
Sallam, M. 798
Sallé, G. 861
Salles, M.S.V. 371, 372, 712
Salomone-Caballero, M. 505
Salonia, R. 1000
Salter, A. 476, 748
Sami, D. 623
Samson, A. 207, 237
Samsonstuen, S. 753, 847
Samuels, M.I. 445
Sanchez, E. 798
Sánchez-Esquiliche, F. 404
Sánchez-Guerrero, M.J. 860, 982
Sanchez, I. 502
Sánchez, J.P. 648, 811, 929, 971
Sanchez, L. 217, 481
Sanchez, M.P. 231, 407, 516
Sandrini, S. 285
Sandrock, C. 747
San Martin, D. 298, 299
Sanna, G. 967
Sanogo, S. 699, 700, 706
Sans, A. 823
Santacreu, M.A. 515
Santander, D. 726
Sant'anna, A.C. 755
Santiago, B.M. 522
Santinello, M. 179, 429
Santos, A. 560, 874
Santos, A.R. 246

Santos, A.S. 155, 828
Santos, J.E.P. 448
Santos, M.M. 490
Santos, M.V. 351, 480
Santos-Silva, J. 292, 294
Santos, V. 185
Santschi, D.E. 587, 972
Santucci, G. 1023
Sanz, A. 994
Sanz-Fernandez, M.V. 448, 564
Sanz-Fernández, S. 827, 940
Sapkota, R. 1006
Saracco, J. 595, 972
Šaran, M. 392
Saran Netto, A. 712
Saremi, B. 837
Sari, N.F. 572
Sarlo Davila, K.M. 166
Sarmiento-García, A.S.-G. 462, 880
Saro, C. 256, 292, 300
Sarpong, N. 517
Sarri, L. 1005
Sarrou, E. 613
Sarry, J. 230, 486
Sarti, F.M. 903
Sartin, J. 494
Sartori, C. 384, 983
Sarviaho, K. 162
Sarzeaud, P. 853
Sasaki, O. 283, 411, 807
Sa, S.J. 741
Satoh, M. 211, 411, 519, 661, 667
Satolias, F. 906
Sattin, E. 463
Saucier, L. 869
Sauerwein, H. 485, 488, 490, 773, 995
Saunders, C. 423
Saunders, S. 420
Saury, J. 856
Sautier, M. 552, 819, 852
Sautot, L. 250, 252
Sauvant, D. 324, 325, 331, 333
Savary-Auzeloux, I. 834
Savary, P. 535
Savietto, D. 853
Savio, R.L. 575
Savoini, G. 285, 357
Savvidou, A. 190
Sayers, G.P. 533
Scali, F. 1022, 1023
Scarafoni, A. 354
Scarlato, S. 688
Schaafstra, F. 346, 347
Schaal, B. 887
Schaeffler, M. 963
Schei, I. 296
Schellander, K. 812
Schenkel, F. 159, 275, 406, 527
Schep, C. 175, 267
Scheumann, M. 645
Schiavo, G. 513, 623, 920, 923
Schiavone, A. 219, 344
Schiavon, S. 288, 336, 574, 633, 775, 902
Schick, M. 569, 840
Schilling, T. 454
Schivazappa, C. 805
Schjelde, M. 631
Schlattl, M. 965
Schlebusch, S. 400
Schlegel, P. 322, 709, 958, 964
Schlotterbeck, E. 460
Schmaltz, L. 1004
Schmeisser, J. 834
Schmidely, P. 324, 867, 871
Schmid, M. 517
Schmidtmann, C. 528
Schmitt, B. 395, 587, 589
Schmitt, E. 614
Schmitz, A. 455
Schmutz, M. 798
Schneider, H. 408
Schneider, L. 871
Schneider, M. 678
Schodl, K. 375, 413, 770
Schokker, D. 353, 737
Scholly-Schoeller, A. 1017
Scholtz, M.M. 671, 725
Schönfeldt, L. 548
Schönleben, M. 488
Schoon, M. 670, 674, 916
Schori, F. 271, 375, 797
Schrauf, M. 161, 398, 785
Schröder, K. 643
Schrøder-Petersen, D.L. 242
Schroeder, G. 572, 845
Schrooten, C. 409, 937
Schroyen, M. 662, 735, 1029
Schuchardt, S. 488
Schuenemann, G.M. 542, 931
Schuler, U. 1026
Schultz, E.B. 317
Schulze-Schleppinghoff, W. 975
Schulz, J. 205, 601
Schut, A.G.T. 849
Schwartz-Zimmermann, H. 835
Schwarz, D. 951, 962
Schwarzenbacher, H. 375, 789
Schwarz, K. 1035
Schwarz, T. 969
Schweer, W. 831
Schweer, W.P. 712
Schweingruber, K. 840
Sciascia, Q.L. 607
Scobie, D. 419
Scollan, N. 174, 269, 829, 942
Scollo, A. 464
Scordia, D. 284
Scully, S. 627

Searle, A. 278
Sebbane, M. 458
Šebek, L.B. 326
Sebsibe, A. 590
Secchi, G. 288, 902
Sechi, G.S. 274
Sécula, A. 232
Sedano, L. 223
Seedum, S. 197
Seefried, F.R. 789
Seegers, J. 853
Seelig, S. 730
Segarra, S. 740
Segato, S. 639
Segelke, D. 786, 937
Seguin, N. 457
Segura, A. 631
Segura, C. 641
Sehested, J. 791
Seidel, A. 374, 768, 888
Seifert, J. 517, 624, 739
Seignon, M. 425
Seiliez, I. 193
Sellem, E. 474, 479, 618
Sell-Kubiak, E. 202, 208
Semsirmboon, S. 511
Sener, A. 835
Sener-Aydemir, A. 995
Šen, G. 178
Senga Kiesse, T. 325
Seo, J. 629, 833
Seo, J.K. 293
Seo, S. 629
Seo, S.Y. 990
Sepchat, B. 167, 184, 331
Sepulveda, J. 484
Sequeira, A. 295
Seradj, A.R. 1005
Seradj, R. 305, 306
Serenius, T. 468
Sergeant, K. 518
Serhan, S. 251, 1023
Serra, M.G. 653, 657
Serrano-Pérez, B. 994
Serrapica, M. 1022, 1023
Serreau, D. 861
Serva, L. 639
Servin, B. 381
Seshoka, M.M. 671
Sevier, S. 278
Sevillano, C.A. 397
Seymour, D.J. 564
Seynaeve, M. 314
Sfakianaki, E. 291, 509
Sghieyer, M.M. 604
Shabtay, A. 168, 422
Shadpour, S. 275
Shakya, Y. 387
Shalloo, L. 547
Shamai, A.B. 249
Shan, Q. 356
Shao, Y.R. 580
Shaqura, I. 441
Sheng, Z.Y. 434
Shewbridge-Carter, L. 183
Shewmaker, D. 806
Shihabi, M. 393
Shimshoni, I. 501
Shin, D. 363, 521
Shin, M.C. 459
Shinoda, M. 660
Shinoda, Y. 686
Shin, S.M. 459
Shi, R. 280, 376
Shitta, S.A. 199
Shi, Z. 268, 305
Shi, Z.D. 875
Shi, Z.X. 646
Shokor, F. 786
Shopeyin, Z.F. 199
Shor-Shimoni, E. 168, 422
Shpigelman, A. 227
Shpirer, J. 566
Shrestha, K. 614, 615
Siachos, N. 529, 767
Siamito, R. 164
Sib, O. 699, 700, 706
Sibra, C. 717
Siebler, D. 859
Siede, C. 455, 684
Siegenthaler, R. 677, 681, 683
Siegmann, S. 533
Sigolo, S. 297
Sikora, J. 447, 908, 911, 913
Silacci, P. 338, 461, 463, 570, 683
Silberberg, M. 328, 330, 913
Silva Carvalho, R. 216
Silva, E.A. 371
Silva, F.G. 529, 1014
Silva, J.A. 185, 282, 685
Silva, J.A.I.I.V. 521
Silva, L.F.P. 665
Silva, P.F. 521
Silva Rodrigues Mendes, N. 320
Silva, S. 185, 484, 780
Silva, S.R. 529
Silva, T.H. 561, 574, 575, 584
Silvera, A. 863
Silvério, K. 872
Silvestre, A. 223
Silvi, A. 652
Simataa, M. 825
Simčič, M. 654
Simili, F.F. 372, 712
Simitzis, P. 286
Simoncini, N. 805
Simongiovanni, A. 838, 873, 988, 989
Simon, H. 1024, 1026

Simoni, M. 636, 641, 642
Simon, S. 853
Sinclair, L.A. 715, 1010
Sini, M. 511, 659
Sioutas, D. 511
Siqueira, G.R. 713, 714
Siqueira, T.T.S. 673, 851
Sirakov, I. 196
Sirisopapong, M. 744
Siriwong, S. 910
Siwek, M. 609
Skele, N. 261
Skiba-Cassy, S. 192, 193
Skiba, K. 370, 603
Skibba, B. 576
Škorput, D. 824
Skorupka, M. 175
Skoufos, I. 218, 228, 350, 764, 765
Skoufos, S. 765
Skřivan, M. 285
Skřivanová, E. 285
Škrlep, M. 824, 930
Slagboom, M. 407, 615
Slavov, T. 184
Śliwiński, B. 537
Sloan, B. 901
Słomian, D. 788
Smetana, S. 227, 343, 344, 693
Smirnova, E.S. 1004
Smith, C. 315
Smith, E. 1034
Smith, E.G. 170
Smith, J. 931
Smith, L.G. 265, 823
Smith, P. 628
Smith, P.E. 627
Smith, R.F. 767
Smith, T. 627
Smith, T.R. 766
Smołucha, G. 210
Smulders, B. 942
Snelling, T. 1010
Soares De Lima, J.M. 1036
Soares, V. 537
Sobczuk-Szul, M. 546, 1024
Sobrero, L. 861
Soca, P. 446, 688
Socha, M. 831
Soede, N. 213
Soelkner, H. 396
Soelkner, J. 381
Soffiantini, C.S.S. 814
Sölkner, J. 168, 526, 907, 915
Soller, M. 168
Somera, A. 852
Sommer, K. 840
Sommerseth, J.K. 296
Sonck, B. 771
Sonesson, U. 933
Song, D.C. 214, 227
Song, J.H. 600, 742, 743, 879, 990
Song, S.E. 882
Song, Z.P. 880
Son, J.W. 525
Son, S. 363
Sørby, J. 366
Sørensen, C. 403
Sørensen, J.G. 616, 618
Sørensen, L.P. 391
Soriano, B. 720, 723, 813
Sorin, V. 407
Sort, M. 499
Sosa-Madrid, B.S. 598
Sosin, E. 537
Sossidou, E.M. 249
Sossidou, E.N. 503
Soto, M. 358
Soucat, E. 474
Souchaud, F. 214
Souchère, V. 850
Soufleri, A. 355, 528, 531
Souillard, R. 597, 603
Soulard, T. 658
Soulet, D. 886
Soullier, G. 550
Soumet, C. 214
Soust, M. 565, 932
Souvignet, P. 331
Souza, A.P. 504
Sow, F. 701
Soyeurt, H. 388, 1026, 1027
Spadavecchia, L. 174
Spanu, C. 705
Sparaggis, D. 291, 509
Speidel, L.T. 456, 460
Speiser, L. 318
Speke Katende, J. 975
Spelman, R.J. 275, 661
Spiekers, H. 374, 963, 965
Spindler, B. 370, 598, 601, 603, 1012
Spindola, A. 475
Spleth, P. 950
Šplíchal, J. 787, 808
Spoelstra, M. 670
Spoelstra, S. 303
Spoolder, H.A.M. 863, 986
Spranghers, T. 865
Sprechert, M. 341
Springer, G. 717
Spyrou, A. 165
Squair, W. 755
Squartini, A. 440, 705
Squillace, F. 303
Srihi, H. 203
Srikanth, K. 804
S. Schenkel, F. 808
Stachowicz, J. 678
Stadnicka, K. 234, 411, 599, 1008, 1009

Stafford, M. 628
Stange, L.M. 859
Stankeviciene, D. 177
Stark, F. 441, 442, 650, 704, 722
Starling, D. 289
Stavropoulos, I. 943
Steeghs, N. 479
Steele, C. 310
Steele, N. 801
Steele, P. 625
Steenfeldt, S. 731
Stefanski, T. 935
Stefos, G.C. 190
Steg, A. 493
Stege, P.B. 353
Stein, H. 284
Steinheim, G. 1039, 1040
Steininger, F. 375, 770
Steinshamn, H. 900
Stejskal, V. 192
Stella, A. 381
Stepancheva, T. 185
Stephansen, R.B. 373
Stephen, M.A. 801
Stepura, L. 543
Stergiadis, S. 562, 572, 588, 590, 840
Stergioudi, R.A. 703
Steuer, S. 774
Stevanato, P. 705, 740
Stewart, S. 310, 311, 313
Steyn, S. 651
Stibal, J. 212
Stilmant, D. 527, 815, 949
Stocchetti, A. 659, 666
Stöcker-Gamigliano, C. 488
Stock, J. 372
Stock, K.F. 975
Stöckl, J. 482
Stomp, M.S. 554
Stothard, P. 406, 926
Stoyanchev, T. 179
Stoynov, M. 188
Strachan, L. 620
Stracke, J. 598, 601
Strandberg, E. 923, 936, 939, 978
Strandén, I. 788, 804
Strang, E.J.P. 771, 1024, 1026, 1029
Strang, E.P.J. 1028
Strauß, G. 965
Štrbac, L.J. 392
Strillacci, M.G. 168
Ströbele-Benschop, N. 548
Strydom, P.E. 307
Studziński, W. 1009
Stuper-Szablewska, K. 224
Sturaro, E. 192, 440, 720, 722, 940, 942, 944, 946
Stygar, A. 628
Stygar, A.H. 498, 501
Šubara, G. 391, 427
Subileau, O. 658
Suchocki, T. 199, 418, 788
Such, X. 251, 484
Südekum, K.-H. 375, 989
Sudo, K. 686
Sugino, T. 933
Sugrue, K. 533
Suin, A. 491, 492
Sullivan, I. 183
Sullivan, P. 275
Sundaram, T.S. 695
Sung, K. 724
Sun, P. 356
Suntinger, M. 375, 413
Sun, X. 542
Sun, Z. 847
Šuran, E. 427
Sureda, E.A. 610, 735
Sureshbabu, A.S. 286
Sürie, C. 205
Suwor, F. 910
Suzuki, K. 279
Suzuki, T. 279, 283
Svanes, E. 274
Svartedal, N. 163
Sverdrup, H.U. 266
Svitojus, A. 304
Swamy, H. 634
Swan, A.A. 401
Sweeney, T. 337, 612
Sweett, H. 275
Świątkiewicz, M. 493
Sylvestre, C. 582
Symeou, S. 291, 509
Syp, A. 265
Szejner, A. 174
Szmatoła, T. 210
Sztuka, M. 794
Szumacher, M. 174, 301
Szumacher-Strabel, M. 931
Szyda, J. 199, 788, 793, 794
Szymkowiak, P. 224
Szyndler-Nędza, M. 469

T

Tada, O. 672
Tada, S. 686
Taghavi, M. 773, 901
Taghouti, M. 838
Tagliapietra, F. 288, 574, 902
Taillandier, M. 255
Tajudeen, H. 349, 350, 363, 364
Takahashi, H. 661, 667
Tallet, C.T. 557, 558
Ta, M. 1004
Tamassia, L.F.M. 634
Tamim, B. 748
Tammanu, K. 910
Tan, B. 761, 832

Tančin, V. 257, 258, 259, 260, 262, 757
Tanga, C.M. 567
Tang, L. 927
Tang, M.X. 291
Tanguy-Roump, Y. 425
Tanrattana, N. 749
Tan Shun En, L. 194
Tapingkae, W. 830
Tapio, I. 323
Tarantola, M. 828
Tardif, V. 658
Tarricone, S. 443
Tarsani, E. 605, 925
Tas, B. 188, 340
Tashima, K. 661, 667
Taskinen, M. 788
Tassinari, P. 973
Tassoni, A. 227
Tatebayashi, R. 283, 807
Taube, F. 173, 176, 817
Taurisano, V. 513, 623, 920
Taussat, S. 809
Tavaniello, S. 467
Tavares, L.B.B. 575
Tavares, O. 730
Taverne, M. 547
Taylor, E. 1017
Taylor, J.B. 151
Taylor, J.F. 485
Tedde, A. 1027
Tedeschi, L. 323, 324, 327, 755
Tedo, G. 884
Tedone, L. 443
Teillard, F. 263
Teisseire, M. 250
Teissier, M. 200, 385
Teixeira Costa, C. 206, 647, 970, 1015
Teixeira, D.L. 215
Teixeira, I.A.M.A. 504
Tejeda, M. 481
Telkänranta, H. 883
Tella, M. 710
Temenos, A. 947
Tempio, G. 147
Ten Berge, A. 287
Tenhunen, S. 391
Ten Napel, J. 171, 541, 785
Tenza-Peral, A. 721, 722
Terada, F. 279, 283, 489
Terman, A. 211
Termeer, C.J.A.M. 849
Terranova, M. 569, 570, 747, 840
Terré, M. 491, 631
Terrier, F. 192
Terrol, C. 867, 871
Terry, S.A. 560
Tesnière, A. 559, 878
Tesnière, G. 249, 496, 497, 503
Tetens, J. 204, 242, 408, 417, 514, 786, 884
Thadee, A.-L.J. 549
Thaller, G. 232, 374, 408, 417, 514, 528, 768, 875, 888, 920, 937
Thammacharoen, S. 511
The Bovreg Consortium 927
Thenard, V. 546
Theodoridou, K. 477, 562, 586, 840, 842
Theodorou, G. 190, 286
Therkildsen, M. 319, 599, 608
Theron, P.G. 243, 368
Thiebeau, P. 728
Thiem, D. 1008
Thimm, G. 649
Thobe, P. 594, 731
Thodberg, K. 242
Tholen, E. 812, 989
Thomas, A. 992
Thomasen, J.R. 391, 403, 407
Thompson, J.J. 562
Thomson, A. 497
Thonart, P. 735
Thorey, P. 728
Thorn, C. 934
Thorne, F. 153
Thorsteinsson, M. 1012
Thoumy, L. 179, 308
Throude, S. 654, 656, 657, 658
Thudor, A.S. 778, 779
Thüer, S. 1035
Thurner, S. 774, 893
Tian, R. 169
Tian, Y.J. 585
Tibi, A. 395
Tichelaar, R. 410
Tiezzi, F. 401, 516
Tikasz, I. 644
Tilkens, N. 245
Tillard, E. 673
Timm, T.G. 575
Timonen, K. 682
Tinarelli, S. 923
Tiplady, K.M. 162
Titton, G. 463
Tixier-Boichard, M. 924
Tkacz, K. 582
Toft, H. 311
Toghiani, S. 401
Toledo Alvarado, H. 775
Tomaru, T. 279
Tomberlin, J.K. 222
Tomiyama, M. 667
Tomozyk, S. 894
Tomple, B.M. 583
Tonda, A. 343
Tondello, A. 705
Tonn, B. 455
Topolski, P. 418
Topputi, R. 903
Török, E. 816

Toro-Mujica, P.M. 685
Torreggiani, D. 973
Torrent, A. 571
Torres, A. 505
Torres, R.N.S. 837
Torsiello, B. 887
Tortadès, M. 491
Tortereau, F. 518, 559, 1032, 1034, 1036, 1037
Toscano, A. 190, 230, 336
Tosetti, L.B. 561, 584
Tosiou, E. 917
Tos, P. 700
Tosser Klopp, G. 1037
Tosser-Klopp, G. 486
Touitou, F. 717
Tourillon, M. 371
Tournat, M. 1030
Tournayre, J. 482
Tourret, M. 485, 507, 991, 992, 1039
Tourtier, M. 539, 540, 554, 1021
Trakooljul, N. 490
Tran, G. 327, 344, 957
Tran, H.Q. 192
Tran, M.-N. 662, 1028
Tran, T.L.T. 191
Tran, V.K. 191
Tranvoiz, E. 899
Traore, E. 289
Traore, E.H. 701
Traulsen, I. 241, 245, 415, 441, 644, 645, 649
Tremblais, D. 635
Trespeuch, C. 762
Trespuech, C. 223
Tretola, M. 293, 338, 461, 472, 570
Trevisi, E. 634, 775, 992
Trevisi, P. 338, 339, 463, 611, 612, 715, 759, 760, 889
Triantafyllidis, A. 555, 671, 672, 917
Tribout, T. 784
Trier-Kreutzfeldt, K.-E. 823
Trivunović, S. 392
Trocino, A. 192, 369, 740
Trossat, P. 261
Troude, S. 653
Trydeman Knudsen, M. 265
Trytsman, M. 445
Tsartsianidou, V. 671, 672, 917
Tseng, E. 627
Tsiamadis, V. 355, 528, 531
Tsigkas, A. 262
Tsinas, A. 218, 228, 765
Tsiokos, D. 776
Tsiplakou, E. 300, 637, 642, 906
Tsukahara, H. 795
Tsuruta, S. 158, 163, 782
Tuliozi, B. 384, 983
Tunkala, B.Z. 294, 1005
Tuominen-Brinkas, M. 968
Turgeon, J.-G. 378
Turini, L. 565, 652, 721
Turner, S.A. 252
Tuyttens, F. 731, 771
Tuyttens, F.A.M. 734
Tvarožková, K. 257, 259, 260, 262
Tyra, M. 211, 469
Tzamaloukas, O. 291, 509
Tzamourani, A. 1011
Tzora, A. 218, 228, 350, 764, 765

U

Uchisawa, K. 279
Uddin, M.E. 952
Ueda, K. 244, 249, 650
Ueda, Y. 686
Uemoto, Y. 279, 283, 489, 519
Uerlings, J. 735
Ugarte, E. 803
Uhrinčať, M. 257, 258, 259, 260, 262, 757
Uimari, P. 162, 468, 487
Ullah, M. 377
Ullrich, C. 987
Umstätter, C. 678
Ungerfeld, E.M. 938
Unseld, H. 455
Urakawa, M. 795
Urban, D. 540
Urrutia, O. 429
Usai, D. 657
Utsunomiya, Y.T. 381
Utzeri, V.J. 623
Uzawa, S. 831
Uzunov, A. 622

V

Vacca, G.M. 902
Vacherie, B. 618, 749
Vaggeli, A. 190
Vakondios, I. 637
Valenchon, M. 860
Valente, J.P.S. 282
Valentin, C. 214
Valera Córdoba, M. 982
Valera, M. 860, 876, 881, 977, 978, 980, 981, 982
Valergakis, G.E. 355, 528, 529, 531
Valero, L. 215
Valière, S. 230
Välimaa, A.-L. 682
Valkenburg, R. 825, 942
Vall, E. 700
Vallée, R. 664, 929
Vallet, L. 828
Valliere, C. 470
Valros, A. 968, 986
Van Amburgh, M.E. 641
Van Asseldonk, L. 352
Van Asten, A. 241
Van Baelen, C. 559
Van Barneveld, R.J. 465
Vanbergue, E. 628, 890

Vanblaere, T. 843
Van Breukelen, A.E. 148, 277
Vandaele, L. 630, 771, 843, 900, 932
Vandenberg, G.W. 869
Van Den Berg, I. 513
Van Den Boer, E. 614
Van Den Brand, H. 530
Van Den Broeke, A. 355, 360
Van Den Heuvel, J. 616
Vanden Hole, C. 734
Van Den Nest, T. 900
Vandenplas, J. 161, 662, 785, 929
Van De Putte, T. 894
Van Der Borght, M. 477
Van Der Fels-Klerx, H.J. 348, 352, 747
Van Der Heide, M. 461
Vanderick, S. 803
Van Der Linden, A. 173, 280, 376
Van Der Sluis, M. 380, 469, 739
Van Der Steeg, E. 996
Van Der Steen, F.T.H.J. 311
Van Der Valk, E. 616, 1003
Van Der Veeken, B. 557
Van Der Weide, R.Y. 347
Van Der Werf, J.H.J. 673
Van De Weijer, T.M. 311
Vandeweyer, D. 874
Van De Weyer, D. 477
Vandicke, J. 843
Van Dijk, L.L. 533
Van Dixhoorn, I.D.E. 541, 767
Van Dongen, K.C.W. 348, 352
Van Dooren, H.J. 175
Van Doormaal, B. 275
Van Duinkerken, G. 171
Van Eck, L. 221
Van Eekeren, N. 173
Van Emous, R. 221
Van Erp-Van Der Kooij, E. 767
Vangen, O. 163
Van Groenestijn, J.W. 346, 347
Van Hal, O. 849
Van Harn, J. 220, 221, 960
Vanhatalo, A. 682
Van Helvoort, M. 462
Van Horne, P. 594
Van Ittersum, M.K. 849
Van Kaam, J.B.C.H.M. 408
Van Knegsel, A. 992
Van Knegsel, A.T.M. 534, 567
Van Kuijk, S. 297, 634
Van Kuijk, S.J.A. 577
Vanlierde, A. 276, 306, 897
Van Loon, J.J.A. 348, 747
Van Looveren, N. 477
Van Meirhaeghe, H. 597, 600
Van Middelaar, C.E. 173, 280, 949
Van Mierlo, K. 265
Van Miert, S. 867, 874
Van Milgen, J. 321, 470, 472, 473, 494
Van Mol, B. 979
Vannier, C. 543
Van Peer, M. 346, 867
Van Pelt, M.L. 662
Vanraden, P.M. 401
Van Reenen, C.G. 533
Van Reenen, K. 530
Van Rozen, K. 347
Van Son, M. 397
Van Staaveren, N. 527, 886
Van Tassell, C.P. 804
Van Wesemael, D. 932
Van Wettere, W.H.E.J. 341, 878
Van Wikselaar, P. 220, 221, 223, 346, 347, 520
Van Wyk, A.E. 445
Van Wyk, J.B. 674
Van Zadelhoff, K. 346
Van Zijderveld, S. 572
Vanzin, A. 190, 230, 633
Van Zyl, J. 651
Vardali, S. 197
Vařeka, J. 412, 799
Väre, M. 731
Vargas, J.A.C. 504
Vargas, R. 949
Varino, R. 295
Varkonyi, E. 234
Varlyakov, I. 184
Varona, L. 203, 382, 402
Várzea Rodrigues, J. 668
Vasa, S.R. 337, 608
Vasconcelos, J.L.M. 537
Vasquez, H. 393
Vauris, M. 1037
Vayssade, J.A. 968
Vayssières, J. 848
Vazeille, K. 184, 633
Vecchio, D. 1022, 1023
Veerkamp, R.F. 148, 277, 380
Vegni, J. 805
Veissier, I. 183, 556
Velasco Gil, G. 718
Veldkamp, F. 287
Veldkamp, T. 220, 221, 223, 346, 347, 520, 752
Velichkova, K. 196
Velikov, K. 225
Veloso, C.M. 317
Venâncio, C.A. 484, 685
Venardou, B. 765
Venter, K.M. 899
Verbeeck, M. 641
Vercesi Filho, A.E. 372
Verdier-Metz, I. 680
Vereijken, A. 615
Vergé, X. 173, 301
Vermeer, H.M. 287
Vermeulen, P.D. 288, 674
Vernunft, A. 607

Verschuren, L. 862, 1003
Verstringe, S. 1003
Veslot, J. 452
Vestergaard, M. 539, 935, 950, 953
Veys, P. 348
Veysset, P. 157, 184, 527, 556, 719
Veyssières, J. 851
Vezzaro, G. 535
Vialaneix, N. 491, 492
Vial, C. 452, 458
Vial, R. 653, 666
Vian, M. 223
Vicario, D. 157
Vicente, A. 824
Vicentini, R.R. 755
Vieco-Saiz, N. 1003
Vieira-Aller, C.V.-A. 462
Vieira, I. 351, 480
Vigh, A. 714
Vignal, A. 619
Vignali, G. 721
Vigne, M. 848, 851
Vignola, G. 656, 658
Vigors, S. 564, 573, 863
Viguié, C. 903
Vilar, C.S.M.M. 521
Vilcinskas, A. 864
Vilela, R.S.R. 317
Villagra, A. 772
Villain, N. 885
Villalon, J. 427
Villarroel, M. 370
Villette, A. 658
Villot, C. 330, 569
Villumsen, T. 276
Villumsen, T.M. 278, 279, 405
Vilotte, M. 522
Vincent, A. 885
Vinet, A. 664, 809, 929
Violleau, F. 328
Viot, T. 1030
Virdis, S. 338, 339, 463, 759, 760, 889
Virgili, R. 805
Visentin, G. 408
Visscher, C. 845, 987
Visser, B. 379
Vitezica, Z.G. 382, 921
Vivenot, L. 899
Vogeler, I. 176
Voidarou, C. 764
Voland, L. 633
Volden, H. 913
Volkmann, N. 205, 649, 1012
Volmerange, L. 328
Von Kerssenbrock, F. 215
Von Rüden, F. 603
Vostra-Vydrova, H. 393
Vostry, L. 393
Voulodimos, A. 947
Vouraki, S. 369, 555, 671, 672, 917
Vourc'h, G. 549
Vouzaras, D. 638
Vrecl, M. 930
Vršková, M. 259
Vrtková, I. 210, 212
Vuattoux, J. 673
Vuorenmaa, J. 410
Vuylsteke, I. 545

W

Waang, L. 686
Waché, A. 536, 1013
Wachekowski, G.M. 575
Waclawek, J.P. 945
Wahmhoff, J. 241
Walachowski, S. 486
Walkenhorst, M. 556, 810, 820
Walker, A. 497
Walker, N.D. 634
Walkom, S.F. 170, 313, 380, 673
Wallenbeck, A. 154, 201, 202, 471
Wallgren, T. 494, 495
Wall, H. 798
Walraven, M. 218
Walsh, H. 909
Wambacq, E. 938
Wan, F.C. 765
Wang, A. 1031
Wang, B. 187, 517, 588
Wang, F. 523
Wang, G. 315
Wang, J. 213, 761, 764
Wang, L.L. 684
Wang, L.M. 638, 832
Wang, L.Y. 585
Wang, M. 229
Wang, M.L. 996
Wang, Q. 268, 305
Wang, S.S. 580
Wang, T.X. 235
Wang, W. 172, 542
Wang, W.J. 1012
Wang, X. 352, 567
Wang, Y. 162, 280, 290, 376, 534, 949
Wang, Y.C. 1031
Wang, Z. 379
Warburton, C. 665
Warin, A. 554, 892
Warin, L. 647
Warner, D. 972
Warner, R. 209
Warren, H. 1010
Watanabe, T. 519
Waterhouse, A. 497
Waterhouse, T. 249
Waters, S. 485, 624, 627, 628, 634, 839, 934, 952, 1007
Watson, M. 625, 924

Watson, R. 307
Wavreille, J. 410, 467, 735, 1024, 1026
Waxenberg, K. 656
Weaver, C.M. 421
Webb, E.C. 820
Webb, L. 530
Weens, M. 585
Weigel, S. 747
Weigend, A. 387
Weigend, S. 387
Weill, P. 587, 589
Weisbecker, J.-L. 518, 1032, 1034
Weisbjerg, M.R. 1012
Weiss, W.P. 706
Welch, C.B. 843
Welham, M. 174
Wellmann, R. 399, 664
Wellnitz, O. 244, 249
Wells, J.M. 924
Werner, A. 374, 771, 1028, 1029
Werne, S. 1035
Westendarp, H. 838
Westermeier, W. 871
Westin, R. 471
Whatford, L. 656
Wicki, M. 401
Wider, J. 670
Wientjes, Y. 398
Wientjes, Y.C.J. 159, 161, 171, 787
Wierzbicka, A. 493, 696
Wierzbicki, J. 307, 692, 696
Wiesel, T. 893
Wiggans, G. 163
Wijnen, H.J. 353
Wilczyński, A. 544
Wilder, T. 768, 859, 885, 888
Wilke, V. 845, 987
Wilk, I. 204, 242, 884
Williams, A. 204, 312, 313
Williams, A.P. 729
Wilmer, H.N. 151
Wilmot, H. 382, 388, 394
Wilms, J.N. 536, 966
Wilms, L. 441
Wilson, A. 632
Wilson, D.J. 532
Wilson, P. 798
Wimel, L. 453, 458, 861
Winckler, C. 986
Windhorst, H.-W. 696
Windig, J.J. 381, 674
Windig, W.J. 916
Winding, A. 1006
Winding, J.J. 158
Windisch, W. 963, 965
Winter, D. 454, 455, 456, 460
Winters, C. 335, 404
Wise, T. 988
Wishna-Kadawarage, R.N. 609
Wisser, D. 147, 949
Witkowska, D. 226, 361
Witte, F. 838
Wittenburg, D. 789
Witt, J. 643
Wolf, M.J. 781, 919
Wolf, V. 1026
Wolgust, V. 522
Wolter, S.C.M. 455
Wolthuis-Fillerup, M. 530
Woodhouse, A. 274
Woodmartin, S. 437, 688, 1033
Workel, I. 975
Workman, K. 848, 854
Worku, S. 278
Worth, G. 275
Worth, G.M. 661
Wo, Y. 356
Woyengo, T.A. 709
Woźniak, K. 211
Woźniakowska, A. 741
Wu, C. 194
Wu, H. 432
Wu, K.C. 743
Wulster-Radcliffe, M. 494
Wu, M. 1008
Wurzinger, M. 164, 669, 905
Wu, S.J. 234
Wutke, M. 415, 644
Wuyts, A. 867
Wu, Z. 384
Wyburn, G.L. 464
W., Y.L. 291

X

Xavier, C. 189, 677
Xaysana, P. 583
Xia, J. 542
Xiang, R. 513
Xiang, Y. 433
Xiccato, G. 192, 369, 740
Xie, Q. 761
Xiong, X. 580
Xue, C.Y. 222
Xu, J.C. 573
Xu, M. 447
Xu, Z.H. 222
X., X. 291

Y

Yagoubi, Y. 722
Yahav, M.O.R. 691, 695
Yamaguchi, S. 411, 807
Yamazaki, T. 411, 805, 807
Yáñez-Ruiz, D.R. 265, 563, 822, 938, 1007
Yang, C.Y. 222
Yang, H. 435, 436, 630
Yang, J. 384, 787
Yang, Q.L. 1001

Yang, R.C. 508
Yang, X. 290, 593
Yan, T. 562, 563, 586, 840, 841, 842
Yao, J.H. 638, 832
Yao, K. 542
Yao, W.L. 233
Yarga, H.P. 697
Yarga H Paul, H.P. 698
Yasuo, S. 233
Yates, J. 1034
Y., C. 291
Ye, H. 213
Yekoye, Y.D. 332
Yigini, Y. 949
Yigitturk, S. 946
Yi, H.B. 591
Yilmaz, S. 595
Yin, Y.L. 996
Yi, R. 573
Yitzhaky, Y. 606
Yngvesson, J. 265
Yoo, D.K. 293
Yoo, J.S. 600, 742, 743, 879, 990
Yoon, S. 351, 364
Yoshihara, Y. 660
Yousif, A. 875
Youssouf, R. 852
Ytournel, F. 201
Yucra, A. 909
Yu, J.Y. 996
Yu, Y. 187, 517, 686, 876
Yu, Y.H. 298, 743, 744
Yu, Y.S. 684

Z

Zaalberg, R.M. 405
Zabala, S. 427
Zabavnik-Piano, J. 919
Zaccaria, E. 520
Zacharis, C. 218, 228, 764
Zachut, M. 353
Zadra, L.E.F. 372, 939
Żak, G. 211, 741
Žáková, E. 210, 212, 392
Zambonelli, P. 805
Zamboni, C. 354
Zambotto, V. 225, 872
Zamora Restan, W.A. 216
Zanchetta, R. 499, 503
Zanetti, L.K. 772
Zanon, T. 251
Zanon, T.H. 981
Zaoui, M. 589
Zappaterra, M. 805
Zardinoni, G. 633, 740
Zare, M. 192
Żarnecki, A. 418
Zavadilová, L. 412, 799
Zayas, G.A. 166
Zbikowski, A. 597
Zebeli, Q. 835, 961, 995
Zemb, O. 200
Zenk, M. 212
Zerjal, T. 416, 738, 810, 928, 930
Zetouni, L. 398
Zhang, B.Y. 187, 517
Zhang, C. 434
Zhang, G. 602
Zhang, H. 434
Zhang, L. 523
Zhang, M. 580
Zhang, P.W. 996
Zhang, Q.Q. 600, 742, 743, 879, 990
Zhang, R. 420
Zhang, S.S. 436
Zhang, S.W. 508
Zhang, W. 435
Zhang, X.M. 523
Zhang, X.Y. 508
Zhang, Y. 222, 447
Zhang, Y.H. 580, 765
Zhang, Y.J. 508
Zhao, G. 602
Zhao, Q.J. 524
Zhao, X.D. 646
Zhao, X.G. 434, 436
Zhao, Y. 222
Zhelezarova, M.G. 481, 869
Zhelyazkova, P. 254
Zheng, C.B. 996
Zheng, J. 602, 996
Zheng, R. 866
Zhou, H. 930
Zhou, H.L. 434
Zhou, S.S. 1001
Zhou, Z.M. 431, 432
Zhuang, Z.J. 298
Ziadi, C. 978, 982
Zijlstra, J. 814, 816
Zipp, K.A. 973
Zoda, A. 795
Zorc, M. 392, 919
Zubiri-Gaitán, A. 400, 483
Zullo, G. 922
Zuo, S. 1009
Zurbrügg, C. 747
Zylowsky, T. 265